S0-ASF-902

THE Biomedical Engineering HANDBOOK

SECOND EDITION

The Electrical Engineering Handbook Series

Series Editor
Richard C. Dorf
University of California, Davis

Titles Included in the Series

The Avionics Handbook, Cary R. Spitzer
The Biomedical Engineering Handbook, 2nd Edition, Joseph D. Bronzino
The Circuits and Filters Handbook, Wai-Kai Chen
The Communications Handbook, Jerry D. Gibson
The Control Handbook, William S. Levine
The Digital Signal Processing Handbook, Vijay K. Madisetti & Douglas Williams
The Electrical Engineering Handbook, Richard C. Dorf
The Electric Power Engineering Handbook, L.L. Grigsby
The Electronics Handbook, Jerry C. Whitaker
The Engineering Handbook, Richard C. Dorf
The Handbook of Formulas and Tables for Signal Processing, Alexander D. Poularikas
The Industrial Electronics Handbook, J. David Irwin
Measurements, Instrumentation, and Sensors Handbook, John Webster
The Mechanical Systems Design Handbook, Osita D.I. Nwokah
The Microwave Engineering Handbook, J. Michael Golio
The Mobile Communications Handbook, 2nd Edition, Jerry D. Gibson
The Ocean Engineering Handbook, Ferial El-Hawary
The Technology Management Handbook, Richard C. Dorf
The Transforms and Applications Handbook, Alexander D. Poularikas
The VLSI Handbook, Wai-Kai Chen
The Electromagnetics Handbook, Aziz Inan and Umran Inan
The Mechatronics Handbook, Robert Bishop

THE Biomedical Engineering HANDBOOK

SECOND EDITION

VOLUME II

Editor-in-Chief

JOSEPH D. BRONZINO

The Vernon Roosa Professor of Applied Science
Trinity College
and
Director
Biomedical Engineering Alliance for Connecticut
(BEACON)

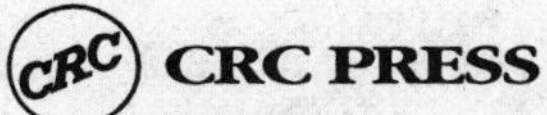

A CRC Handbook Published in Cooperation with IEEE Press

Library of Congress Cataloging-in-Publication Data

Catalog record is available from the Library of Congress.

International Standard Book Number 0-8493-0462-8
Printed in the United States of America 1 2 3 4 5 6 7 8 9 0
Printed on acid-free paper

Introduction and Preface

As we enter the new millennium, the prospects for the field of Biomedical Engineering are bright. Individuals interested in pursuing careers in this field continue to increase and the fruits of medical innovation continue to yield both monetary rewards and patient well being. These trends are reflected in this second edition of the *Biomedical Engineering Handbook*. When compared to the first edition published in 1995, this new two-volume set includes new sections on "Transport Phenomena and Biomimetic Systems" and "Ethical Issues Associated with Medical Technology". In addition, over 60% of the chapters has been completely revised, incorporating the latest developments in the field. therefore, this second edition is truly an updated version of the "state-of-the-field of biomedical engineering". As such, it can serve as an excellent reference for individuals interested not only in a review of fundamental physiology, but also in quickly being brought up to speed in certain areas of biomedical engineering research. It can serve as an excellent textbook for students in areas where traditional textbooks have not yet been developed, and serve as an excellent review of the major areas of activity in each biomedical engineering subdiscipline, such as biomechanics biomaterials, clinical engineering, artificial intelligence, etc., and finally it can serve as the "bible" for practicing biomedical engineering professionals by covering such topics as a "Historical Perspective of Medical Technology, the Role of Professional Societies and the Ethical Issues Associated with Medical Technology".

Biomedical Engineering is no longer an emerging discipline; it has become an important vital interdisciplinary field. Biomedical engineers are involved in many medical ventures. They are involved in the design, development and utilization of materials, devices (such as pacemakers, lithotripsy, etc.) and techniques (such as signal processing, artificial intelligence, etc.) for clinical research and use; and serve as members of the health care delivery team (clinical engineering, medical informatics, rehabilitation engineering, etc.) seeking new solutions for difficult heath care problems confronting our society. To meet the needs of this diverse body of biomedical engineers, this handbook provides a central core of knowledge in those fields encompassed by the discipline of biomedical engineering as we enter the 21st century. Before presenting this detailed information, however, it is important to provide a sense of the evolution of the modern health care system and identify the diverse activities biomedical engineers perform to assist in the diagnosis and treatment of patients.

Evolution of the Modern Health Care System

Before 1900, medicine had little to offer the average citizen, since its resources consisted mainly of the physician, his education, and his "little black bag." In general, physicians seemed to be in short supply, but the shortage had rather different causes than the current crisis in the availability of health care professionals. Although the costs of obtaining medical training were relatively low, the demand for doctors' services also was very small, since many of the services provided by the physician also could be obtained from experienced amateurs in the community. The home was typically the site for treatment and recuperation, and relatives and neighbors constituted an able and willing nursing staff. Babies were delivered by midwives, and those illnesses not cured by home remedies were left to run their natural, albeit frequently fatal, course. The contrast with contemporary health care practices, in which specialized physicians and nurses located within the hospital provide critical diagnostic and treatment services, is dramatic.

The changes that have occurred within medical science originated in the rapid developments that took place in the applied sciences (chemistry, physics, engineering, microbiology, physiology, pharmacology, etc.) at the turn of the century. This process of development was characterized by intense interdisciplinary cross-fertilization, which provided an environment in which medical research was able to take giant strides in developing techniques for the diagnosis and treatment of disease. For example, in 1903, Willem Einthoven, the Dutch physiologist, devised the first electrocardiograph to measure the electrical activity of the heart. In applying discoveries in the physical sciences to the analysis of biologic process, he initiated a new age in both cardiovascular medicine and electrical measurement techniques.

New discoveries in medical sciences followed one another like intermediates in a chain reaction. However, the most significant innovation for clinical medicine was the development of x-rays. These "new kinds of rays," as their discoverer W. K. Roentgen described them in 1895, opened the "inner man" to medical inspection. Initially, x-rays were used to diagnose bone fractures and dislocations, and in the process, x-ray machines became commonplace in most urban hospitals. Separate departments of radiology were established, and their influence spread to other departments throughout the hospital. By the 1930s, x-ray visualization of practically all organ systems of the body had been made possible through the use of barium salts and a wide variety of radiopaque materials.

X-ray technology gave physicians a powerful tool that, for the first time, permitted accurate diagnosis of a wide variety of diseases and injuries. Moreover, since x-ray machines were too cumbersome and expensive for local doctors and clinics, they had to be placed in health care centers or hospitals. Once there, x-ray technology essentially triggered the transformation of the hospital from a passive receptacle for the sick to an active curative institution for all members of society.

For economic reasons, the centralization of health care services became essential because of many other important technological innovations appearing on the medical scene. However, hospitals remained institutions to dread, and it was not until the introduction of sulfanilamide in the mid-1930s and penicillin in the early 1940s that the main danger of hospitalization, i.e., cross-infection among patients, was significantly reduced. With these new drugs in their arsenals, surgeons were able to perform their operations without prohibitive morbidity and mortality due to infection. Furthermore, even though the different blood groups and their incompatibility were discovered in 1900 and sodium citrate was used in 1913 to prevent clotting, full development of blood banks was not practical until the 1930s, when technology provided adequate refrigeration. Until that time, "fresh" donors were bled and the blood transfused while it was still warm.

Once these surgical suites were established, the employment of specifically designed pieces of medical technology assisted in further advancing the development of complex surgical procedures. For example, the Drinker respirator was introduced in 1927 and the first heart-lung bypass in 1939. By the 1940s, medical procedures heavily dependent on medical technology, such as cardiac catheterization and angiography (the use of a cannula threaded through an arm vein and into the heart with the injection of radiopaque dye for the x-ray visualization of lung and heart vessels and valves), were developed. As a result, accurate diagnosis of congenital and acquired heart disease (mainly valve disorders due to rheumatic fever) became possible, and a new era of cardiac and vascular surgery was established.

Following World War II, technological advances were spurred on by efforts to develop superior weapon systems and establish habitats in space and on the ocean floor. As a by-product of these efforts, the development of medical devices accelerated and the medical profession benefited greatly from this rapid surge of "technological finds." Consider the following examples:

1. Advances in solid-state electronics made it possible to map the subtle behavior of the fundamental unit of the central nervous system — the neuron — as well as to monitor various physiologic parameters, such as the electrocardiogram, of patients in intensive care units.
2. New prosthetic devices became a goal of engineers involved in providing the disabled with tools to improve their quality of life.

3. Nuclear medicine — an outgrowth of the atomic age — emerged as a powerful and effective approach in detecting and treating specific physiologic abnormalities.
4. Diagnostic ultrasound based on sonar technology became so widely accepted that ultrasonic studies are now part of the routine diagnostic workup in many medical specialties.
5. "Spare parts" surgery also became commonplace. Technologists were encouraged to provide cardiac assist devices, such as artificial heart valves and artificial blood vessels, and the artificial heart program was launched to develop a replacement for a defective or diseased human heart.
6. Advances in materials have made the development of disposable medical devices, such as needles and thermometers, as well as implantable drug delivery systems, a reality.
7. Computers similar to those developed to control the flight plans of the *Apollo* capsule were used to store, process, and cross-check medical records, to monitor patient status in intensive care units, and to provide sophisticated statistical diagnoses of potential diseases correlated with specific sets of patient symptoms.
8. Development of the first computer-based medical instrument, the computerized axial tomography scanner, revolutionized clinical approaches to noninvasive diagnostic imaging procedures, which now include magnetic resonance imaging and positron emission tomography as well.

The impact of these discoveries and many others has been profound. The health care system consisting primarily of the "horse and buggy" physician is gone forever, replaced by a technologically sophisticated clinical staff operating primarily in "modern" hospitals designed to accommodate the new medical technology. This evolutionary process continues, with advances in biotechnology and tissue engineering altering the very nature of the health care delivery system itself.

The Field of Biomedical Engineering

Today, many of the problems confronting health professionals are of extreme interest to engineers because they involve the design and practical application of medical devices and systems — processes that are fundamental to engineering practice. These medically related design problems can range from very complex large-scale constructs, such as the design and implementation of automated clinical laboratories, multiphasic screening facilities (i.e., centers that permit many clinical tests to be conducted), and hospital information systems, to the creation of relatively small and "simple" devices, such as recording electrodes and biosensors, that may be used to monitor the activity of specific physiologic processes in either a research or clinical setting. They encompass the many complexities of remote monitoring and telemetry, including the requirements of emergency vehicles, operating rooms, and intensive care units. The American health care system, therefore, encompasses many problems that represent challenges to certain members of the engineering profession called *biomedical engineers.*

Biomedical Engineering: A Definition

Although what is included in the field of biomedical engineering is considered by many to be quite clear, there are some disagreements about its definition. For example, consider the terms *biomedical engineering, bioengineering,* and *clinical* (or *medical*) *engineering* which have been defined in Pacela's *Bioengineering Education Directory* [Quest Publishing Co., 1990]. While Pacela defines *bioengineering* as the broad umbrella term used to describe this entire field, *bioengineering* is usually defined as a basic research–oriented activity closely related to biotechnology and genetic engineering, i.e., the modification of animal or plant cells, or parts of cells, to improve plants or animals or to develop new microorganisms for beneficial ends. In the food industry, for example, this has meant the improvement of strains of yeast for fermentation. In agriculture, bioengineers may be concerned with the improvement of crop yields by treatment of plants with organisms to reduce frost damage. It is clear that bioengineers of the future will have a tremendous impact on the quality of human life. The potential of this specialty is difficult to imagine. Consider the following activities of bioengineers:

- Development of improved species of plants and animals for food production
- Invention of new medical diagnostic tests for diseases
- Production of synthetic vaccines from clone cells
- Bioenvironmental engineering to protect human, animal, and plant life from toxicants and pollutants
- Study of protein-surface interactions
- Modeling of the growth kinetics of yeast and hybridoma cells
- Research in immobilized enzyme technology
- Development of therapeutic proteins and monoclonal antibodies

In reviewing the above-mentioned terms, however, *biomedical engineering* appears to have the most comprehensive meaning. Biomedical engineers apply electrical, mechanical, chemical, optical, and other engineering principles to understand, modify, or control biologic (i.e., human and animal) systems, as well as design and manufacture products that can monitor physiologic functions and assist in the diagnosis and treatment of patients. When biomedical engineers work within a hospital or clinic, they are more properly called *clinical engineers*.

Activities of Biomedical Engineers

The breadth of activity of biomedical engineers is significant. The field has moved significantly from being concerned primarily with the development of medical devices in the 1950s and 1960s to include a more wide-ranging set of activities. As illustrated below, the field of biomedical engineering now includes many new career areas, each of which is presented in this Handbook. These areas include:

- Application of engineering system analysis (physiologic modeling, simulation, and control) to biologic problems
- Detection, measurement, and monitoring of physiologic signals (i.e., *biosensors* and *biomedical instrumentation*)
- Diagnostic interpretation via signal-processing techniques of bioelectric data
- Therapeutic and rehabilitation procedures and devices (rehabilitation engineering)
- Devices for replacement or augmentation of bodily functions (*artificial organs*)
- Computer analysis of patient-related data and clinical decision making (i.e., medical informatics and artificial intelligence)
- Medical imaging, i.e., the graphic display of anatomic detail or physiologic function
- The creation of new biologic products (i.e., *biotechnology* and *tissue engineering*)

Typical pursuits of biomedical engineers, therefore, include:

- Research in new materials for implanted artificial organs
- Development of new diagnostic instruments for blood analysis
- Computer modeling of the function of the human heart
- Writing software for analysis of medical research data
- Analysis of medical device hazards for safety and efficacy
- Development of new diagnostic imaging systems
- Design of telemetry systems for patient monitoring
- Design of biomedical sensors for measurement of human physiologic systems variables

- Development of expert systems for diagnosis of disease
- Design of closed-loop control systems for drug administration
- Modeling of the physiologic systems of the human body
- Design of instrumentation for sports medicine
- Development of new dental materials
- Design of communication aids for the handicapped
- Study of pulmonary fluid dynamics
- Study of the biomechanics of the human body
- Development of material to be used as replacement for human skin

Biomedical engineering, then, is an interdisciplinary branch of engineering that ranges from theoretical, nonexperimental undertakings to state-of-the-art applications. It can encompass research, development, implementation, and operation. Accordingly, like medical practice itself, it is unlikely that any single person can acquire expertise that encompasses the entire field. Yet, because of the interdisciplinary nature of this activity, there is considerable interplay and overlapping of interest and effort between them. For example, biomedical engineers engaged in the development of biosensors may interact with those interested in prosthetic devices to develop a means to detect and use the same bioelectric signal to power a prosthetic device. Those engaged in automating the clinical chemistry laboratory may collaborate with those developing expert systems to assist clinicians in making decisions based on specific laboratory data. The possibilities are endless.

Perhaps a greater potential benefit occurring from the use of biomedical engineering is identification of the problems and needs of our present health care system that can be solved using existing engineering technology and systems methodology. Consequently, the field of biomedical engineering offers hope in the continuing battle to provide high-quality health care at a reasonable cost; if properly directed toward solving problems related to preventive medical approaches, ambulatory care services, and the like, biomedical engineers can provide the tools and techniques to make our health care system more effective and efficient.

Joseph D. Bronzino
Editor-in-Chief

The Discipline of Biomedical Engineering

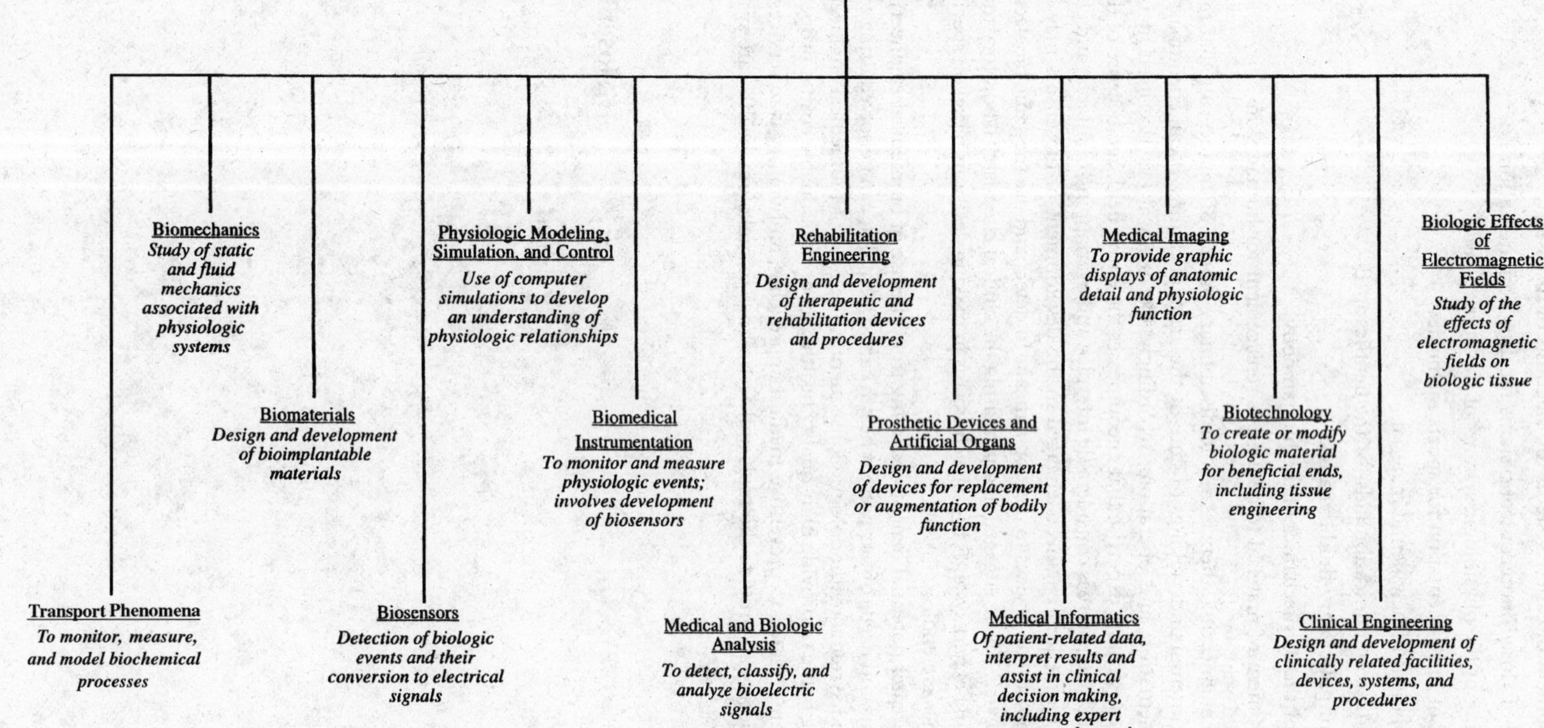

Transport Phenomena
To monitor, measure, and model biochemical processes
Biomechanics
Study of static and fluid mechanics associated with physiologic systems
Biomaterials
Design and development of bioimplantable materials
Biosensors
Detection of biologic events and their conversion to electrical signals
Physiologic Modeling, Simulation, and Control
Use of computer simulations to develop an understanding of physiologic relationships
Biomedical Instrumentation
To monitor and measure physiologic events; involves development of biosensors
Medical and Biologic Analysis
To detect, classify, and analyze bioelectric signals
Rehabilitation Engineering
Design and development of therapeutic and rehabilitation devices and procedures
Prosthetic Devices and Artificial Organs
Design and development of devices for replacement or augmentation of bodily function
Medical Informatics
Of patient-related data, interpret results and assist in clinical decision making, including expert systems and neural networks
Medical Imaging
To provide graphic displays of anatomic detail and physiologic function
Biotechnology
To create or modify biologic material for beneficial ends, including tissue engineering
Clinical Engineering
Design and development of clinically related facilities, devices, systems, and procedures
Biologic Effects of Electromagnetic Fields
Study of the effects of electromagnetic fields on biologic tissue

Editor-in-Chief

Joseph D. Bronzino, Ph.D., P.E., Vernon Roosa Professor of Applied Science at Trinity College, Hartford, Connecticut, and director of the Biomedical Engineering Alliance for Connecticut (BEACON), teaches graduate and undergraduate courses in biomedical engineering in the fields of clinical engineering, electrophysiology, signal analysis, and computer applications in medicine. He earned his B.S. in electrical engineering from Worcester Polytechnic Institute, M.S. in electrical engineering from the Naval Postgraduate School, and Ph.D. in electrical engineering also from Worcester Polytechnic Institute. Deeply concerned with the discipline of biomedical engineering, as well as with ethical and economic issues related to the application of technology to the delivery of health care, Dr. Bronzino has written and lectured internationally. He is the author of over 200 articles and 8 books: *Technology for Patient Care* (C.V. Mosby, 1977), *Computer Applications for Patient Care* (Addison-Wesley, 1982), *Biomedical Engineering: Basic Concepts and Instrumentation* (PWS Publishing Co., 1986), *Expert Systems: Basic Concepts and Applications* (Research Foundation of the State University of New York, 1989), *Medical Technology and Society: An Interdisciplinary Perspective* (MIT Press, 1990), *Management of Medical Technology* (Butterworth/Heinemann, 1992), *The Biomedical Engineering Handbook* (CRC Press, 1995), *The Introduction to Biomedical Engineering* (Academic Press, 1999), and *The Biomedical Engineering Handbook, 2nd Edition* (CRC Press, 2000). Dr. Bronzino is a fellow of both the Institute of Electrical and Electronic Engineers (IEEE) and the American Institute of Medical and Biological Engineering (AIMBE), a past president of the IEEE Engineering in Medicine and Biology Society (EMBS), a past chairman of the Biomedical Engineering Division of the American Society for Engineering Education, and a charter member of the American College of Clinical Engineering (ACCE) and the Connecticut Academy of Science and Engineering (CASE). Dr. Bronzino has extensive experience in the formulation of public policy regarding the utilization and regulation of medical technology. He has served as a past chairman of both the IEEE Health Care Engineering Policy Committee (HCEPC) and the IEEE Technical Policy Council in Washington, D.C.

Advisory Board

Contributors

James J. Abbas
University of Kentucky
Lexington, Kentucky

Joseph Adam
Premise Development Corporation
Hartford, Connecticut

Patrick Aebischer
Lausanne University Medical School
Lausanne, Switzerland

Robert C. Allen
Emory University
Atlanta, Georgia

John G. Aunins
Merck Research Laboratories
Rahway, New Jersey

Dennis D. Autio
Dybonics, Inc.
Portland, Oregon

James W. Baish
Bucknell University
Lewisburg, Pennsylvania

Pamela J. Hoyes Beehler
University of Texas at Arlington
Arlington, Texas

Ravi Bellamkonda
Lausanne University Medical School
Lausanne, Switzerland

Jan E. W. Beneken
Eindhoven University of Technology
Eindhoven, the Netherlands

François Berthiaume
Surgical Services, Massachusetts General Hospital, Harvard Medical School, and the Shriners Hospital for Children
Cambridge, Massachusetts

Jeffrey S. Blair
IBM Health Care Solutions
Atlanta, Georgia

Joseph D. Bronzino
Trinity College/Biomedical Engineering Alliance for Connecticut (BEACON)
Hartford, Connecticut

Susan V. Brooks
University of Michigan
Ann Arbor, Michigan

Mark E. Bruley
ECRI
Plymouth Meeting, Pennsylvania

Ewart R. Carson
City University
London, United Kingdom

Andrea Caumo
San Raffaele Scientific Institute
Milan, Italy

Joseph A. Chinn
Sulzer Carbomedics
Austin, Texas

Vivian H. Coates
ECRI
Plymouth Meeting, Pennsylvania

Claudio Cobelli
University of Padova
Padua, Italy

Clark K. Colton
Massachusetts Institute of Technology
Cambridge, Massachusetts

Rory A. Cooper
University of Pittsburgh, VA Pittsburgh Health Care System
Pittsburgh, Pennsylvania

Derek G. Cramp
City University
London, United Kingdom

Yadin David
Texas Children's Hospital
Houston, Texas

Benoit M. Dawant
Vanderbilt University
Nashville, Tennessee

Paul D. Drumheller
Gore Hybrid Technologies, Inc.
Flagstaff, Arizona

G. M. Drzewiecki
Rutgers University
Piscataway, New Jersey

Karen A. Duca
University of Wisconsin
Madison, Wisconsin

Graham A. Dunn
King's College
London, United Kingdom

John Denis Enderle
University of Connecticut
Storrs, Connecticut

John A. Faulkner
University of Michigan
Ann Arbor, Michigan

Stanley M. Finkelstein
University of Minnesota
Minneapolis, Minnesota

Robert J. Fisher
University of Connecticut
Storrs, Connecticut

J. Michael Fitzmaurice
Department of Health and Human Services
Rockville, Maryland

L.E. Freed
Harvard University
Cambridge, Massachusetts

Catherine Garbay
Laboratoire TIMC/IMAG
Grenoble, France

Leslie A. Geddes
Purdue University
West Lafayette, Indiana

John W. Goethe
The Institute of Living
Hartford, Connecticut

Michael L. Gullikson
Texas Children's Hospital
Houston, Texas

Katya Hill
Edinboro University of Pennsylvania
and University of Pittsburgh
Edinboro, Pennsylvania

Douglas Hobson
University of Pittsburgh
Pittsburgh, Pennsylvania

Jeffrey A. Hubbell
California Institute of Technology
Pasadena, California

H. David Humes
University of Michigan
Ann Arbor, Michigan

Sheik N. Imrhan
University of Texas at Arlington
Arlington, Texas

Marcos Intaglietta
University of California
La Jolla, California

Michel Jaffrin
Université de Technologie de
Compiègne
Compiègne, France

Hugo O. Jauregui
Rhode Island Hospital
Providence, Rhode Island

Stephen B. Johnson
Columbia University
New York, New York

Richard D. Jones
Christchurch Hospital and
University of Otago
Christchurch, New Zealand

Craig T. Jordan
Somatix Therapy Corp.
Alameda, California

Thomas M. Judd
Kaiser Permanente
Atlanta, Georgia

Kurt A. Kaczmarek
University of Wisconsin at Madison
Madison, Wisconsin

Robert Kaiser
University of Washington
Seattle, Washington

Peter L.M. Kerkhof
Medwise Working Group
Maarssen, the Netherlands

Tao Ho Kim
Harvard University and Boston
Children's Hospital
Boston, Massachusetts

Manfred R. Koller
Oncosis
San Diego, California

George V. Kondraske
University of Texas at Arlington,
Human Performance Institute
Arlington, Texas

David N. Ku
Georgia Institute of Technology
Atlanta, Georgia

Casimir Kulikowski
Rutgers University
Piscataway, New Jersey

Luis G. Kun
CIMIC/Rutgers University
New Brunswick, New Jersey

Robert S. Langer
Massachusetts Institute of
Technology
Cambridge, Massachusetts

Douglas A. Lauffenburger
Massachusetts Institute of
Technology
Cambridge, Massachusetts

Swamy Laxminarayan
New Jersey Institute of Technology
Newark, New Jersey

Joseph M. Le Doux
Center for Engineering in Medicine,
and Surgical Services,
Massachusetts General Hospital,
Harvard Medical School, and the
Shriners Burns Hospital
Cambridge, Massachusetts

Ann L. Lee
Merck Research Laboratories
Rahway, New Jersey

John K-J. Li
Rutgers University
Piscataway, New Jersey

E.N. Lightfoot
University of Wisconsin
Madison, Wisconsin

Michael W. Long
University of Michigan
Ann Arbor, Michigan

Marilyn Lord
King's College
London, United Kingdom

Michael J. Lysaght
Brown University
Providence, Rhode Island

Vasilis Z. Marmarelis
University of Southern California
Los Angeles, California

Kenneth J. Maxwell
Departure Technology
Fort Worth, Texas

Joseph P. McClain
Walter Reed Army Medical Center
Washington, D.C.

Larry V. McIntire
Rice University
Houston, Texas

Evangelia Micheli-Tzanakou
Rutgers University
Piscataway, New Jersey

David J. Mooney
Massachusetts Institute of
Technology
Cambridge, Massachusetts

John Moran
Vasca, Inc.
Tewksbury, Massachusetts

Jeffrey R. Morgan
Center for Engineering in Medicine, and Surgical Services, Massachusetts General Hospital, Harvard Medical School, and the Shriners Burns Hospital
Cambridge, Massachusetts

Robert L. Morris
Dybonics, Inc.
Portland, Oregon

Tatsuo Nakamura
Kyoto University
Kyoto, Japan

Brian A. Naughton
Advanced Tissue Sciences, Inc.
La Jolla, California

Robert M. Nerem
Georgia Institute of Technology
Atlanta, Georgia

A. Noordergraaf
University of Pennsylvania
Philadelphia, Pennsylvania

Patrick R. Norris
Vanderbilt University
Nashville, Tennessee

Pirkko Nykänen
VIT Information Technology
Tampere, Finland

Joseph L. Palladino
Trinity College
Hartford, Connecticut

Bernhard Ø. Palsson
University of Michigan
Ann Arbor, Michigan

Mohamad Parnianpour
Ohio State University
Columbus, Ohio

Charles W. Patrick, Jr.
Rice University
Houston, Texas

Chi-Sang Poon
Massachusetts Institute of Technology
Cambridge, Massachusetts

Dejan B. Popović
University of Belgrade
Belgrade, Yugoslavia

T. Allan Pryor
University of Utah
Salt Lake City, Utah

Yuchen Qiu
The Pennsylvania State University
University Park, Pennsylvania

Gerard Reach
Hôpital Hotel-Dieu Paris
Paris, France

Charles J. Robinson
Louisiana Tech University
Ruston, Louisiana
The Overton Brooks VA Medical Center
Shreveport, Louisiana

Barry Romich
Prentke Romich Company
Wooster, Ohio

Gerson Rosenberg
Pennsylvania State University
Hershey, Pennsylvania

Eric Rosow
Hartford Hospital
Hartford, Connecticut

Charles M. Roth
Center for Engineering in Medicine, Massachusetts General Hospital, Harvard Medical School, and the Shriners Burns Hospital
Cambridge, Massachusetts

Alan J. Russell
University of Pittsburgh
Pittsburgh, Pennsylvania

Maria Pia Saccomani
University of Padova
Padua, Italy

Pamela S. Saha
Clemson University
Clemson, South Carolina

Subrata Saha
Clemson University
Clemson, South Carolina

W. Mark Saltzman
Cornell University
Ithaca, New York

Rangarajan Sampath
Rice University
Houston, Texas

Niilo Saranummi
VIT Information Technology
Tampere, Finland

Soumitra Sengupta
Columbia University
New York, New York

Yasuhiko Shimizu
Kyoto University
Kyoto, Japan

Michael L. Shuler
Cornell University
Ithaca, New York

Srolamtj Simdara
Rutgers University
Piscataway, New Jersey

Steven M. Slack
The University of Memphis
Memphis, Tennessee

Susan S. Smith
Texas Woman's University
Dallas, Texas

Giovanni Sparacino
University of Padova
Padua, Italy

Ron Summers
Loughborough University
Leicestershire, United Kingdom

Srikanth Sundaram
Rutgers University
Piscataway, New Jersey

Karl Syndulko
UCLA School of Medicine
Los Angeles, California

John M. Tarbell
The Pennsylvania State University
University Park, Pennsylvania

William D. Timmons
University of Akron
Akron, Ohio

Gianna Maria Toffolo
University of Padova
Padua, Italy

Mehmet Toner
Massachusetts General Hospital, Harvard Medical School, and the Shriners Burns Hospital
Cambridge, Massachusetts

Elaine Trefler
University of Pittsburgh
Pittsburgh, Pennsylvania

Alan Turner-Smith
King's College
London, United Kingdom

Shiro Usai
Toyohashi University of Technology
Toyohashi, Japan

Joseph P. Vacanti
Harvard University and Boston Children's Hospital
Boston, Massachusetts

Robert F. Valentini
Brown University
Providence, Rhode Island

Gary Van Zant
University of Kentucky Medical Center
Lexington, Kentucky

Gregg Vanderheiden
University of Wisconsin
Madison, Wisconsin

Paul J. Vasta
University of Texas at Arlington
Arlington, Texas

Chenzhao Vierheller
University of Pittsburgh
Pittsburgh, Pennsylvania

David B. Volkin
Merck Research Laboratories
Rahway, New Jersey

G. Vunjak-Novakovic
Massachusetts Institute of Technology
Cambridge, Massachusetts

Alvin Wald
Columbia University
New York, New York

S. Patrick Walton
Center for Engineering in Medicine, Massachusetts General Hospital, Harvard Medical School, and the Shriners Burns Hospital
Cambridge, Massachusetts

Robert M. Winslow
SANGART, Inc.
La Jolla, California

Daniel E. Wueste
Clemson University
Clemson, South Carolina

Ioannis V. Yannas
Massachusetts Institute of Technology
Cambridge, Massachusetts

David M. Yarmush
Rutgers University
Piscataway, New Jersey

Martin L. Yarmush
Massachusetts General Hospital, Harvard Medical School, and the Shriners Burns Hospital
Cambridge, Massachusetts

Ajit P. Yoganathan
Georgia Institute of Technology
Atlanta, Georgia

Daniel A. Zahner
Rutgers University
Piscataway, New Jersey

Craig Zupke
Massachusetts General Hospital and the Shriners Burns Institute
Cambridge, Massachusetts

Andrew L. Zydney
University of Delaware
Newark, Delaware

Contents

SECTION XII Tissue Engineering

SECTION XIII Prosheses and Artificial Organs

SECTION XIV Rehabiliation Engineering

SECTION XV Human Performance Engineering

SECTION XVIII Medical Informatics

SECTION XIX Artificial Intelligence

SECTION XX Ethical Issues Associated with the Use of Medical Technology

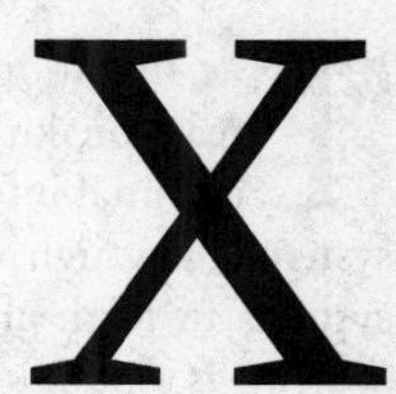

Transport Phenomena and Biomimetic Systems

Robert J. Fisher
University of Connecticut

The intention of this section is to couple the concepts of transport phenomena with chemical reaction kinetics and thermodynamics to introduce the field of reaction engineering. This is essential information needed to design and control engineering devices, particularly flow reactor systems. Through extension of these concepts, combined with materials design, to mimic biological systems in form and function, the field of biomimicry has evolved. The development of Biomimetic Systems is a rapidly emerging technology with expanding applications. Specialized journals, e.g., Biomimetics, devoted to exploring both the analysis of existing biological materials and processes, and the design and production of synthetic analogs that mimic biological properties are now emerging. These journals blend a biological approach with a materials/engineering science viewpoint covering topics that include: analysis of the design criteria used by organisms in the selection of specific biosynthetic materials and structures; analysis of the optimization criteria used in natural systems; development of systems modeled on biological analogs; and applications of "intelligent" or "smart" materials in areas such as biosensors, robotics, and aerospace. This biomimicry theme is prevalent throughout all the chapters in this section, including one specifically devoted to these concepts with explicit examples of applications to reacting and transport processes.

The field of transport phenomena traditionally encompasses the subjects of momentum transport (viscous flow), energy transport (heat conduction, convection, and radiation), and mass transport (diffusion). In this section the media in which the transport occurs is regarded as continua; however, some molecular explanations are discussed. The continuum approach is of more immediate interest to engineers, but both approaches are required to thoroughly master the subject. The current emphasis in engineering education is on understanding basic physical principles vs. "blind" use of empiricism. Consequently, it is imperative that the reader seek further edification in classical transport phenomena texts; general [Bird et al., 1960], with chemically reactive systems [Rosner, 1986], and more specifically, with a biologically oriented approach [Lightfoot, 1974]. The laws (conservation principles, etc.) governing such transport will be seen to influence: (1) the local rates at which reactants encounter one another; (2) the temperature field within body regions (or compartments in pharmacokinetics modeling); (3) the volume (or area) needed to accomplish the desired turnover or transport rates; and (4) the amount and fate of species involved in mass transport and metabolic rates. For biomedical systems such as dialysis, in which only physical changes occur, the same general principles, usually with more simplifications, are used to design and analyze these devices. The transport laws governing non-reactive systems can often be used to make rational predictions of the behavior of "analogous" reacting systems. The importance of relative time scales will be discussed throughout this section by all authors. It is particularly important to establish orders-of-magnitude and to make realistic limiting calculations. Dimensional analysis and pharmacokinetic modeling techniques (see other sections of the handbook and [Enderle et al., 1999; Fisher, 1999] are especially attractive for these purposes; in fact, they may permit unifying the whole of biological transport. As the reader progresses, the importance of transport phenomena in applied biology becomes steadily more apparent. "In all living organisms, but most especially the higher animals, diffusional and flow limitations are of critical importance; moreover, we live in a very delicate state of balance with respect to these two processes" [Lightfoot, 1974].

Each chapter in this section is somewhat self-contained. Similar concepts are brought forth and reinforced through applications and discussions. However, to further enhance the benefits obtained by the reader, it is prudent to first discuss some elementary concepts; the most relevant being control volume selection and flow reactors.

When applying the conservation laws to fluid matter treated as a continuum, the question arises as to the amount of matter to be considered. Typically this decision is based on convenience and/or level of detail required. There is no single choice. Many possibilities exist that can lead to the same useful predictions. The conservation laws of continuum dynamics can be applied to the fluid contained in a volume of arbitrary size, shape, and state of motion. The volume selected is termed a "control volume." The simplest is one where every point on its surface moves with the local fluid velocity. It is called a "material" control volume since, in the absence of diffusion across its interface, it retains the material originally present within its control surface. Although conceptually simple, they are not readily used since they move through space, change their volume, and deform. An analysis of the motion of material control volumes is usually termed "Legrangian" and time derivatives are termed "material" or "substantial" derivatives.

Another simple class of control volumes is defined by surfaces fixed in physical space, through which the fluid flows. These "fixed" control volumes are termed "Eularian" and may be either macroscopic or differential in any or all directions. The fluid contained within a Eularian control volume is said to be, in thermodynamic terms, an "open" (flow) system.

The most general type of control volume is defined by surfaces that move "arbitrarily", i.e., not related to the local fluid velocity. Such control volumes are used to analyze the behavior of non-material "waves" in fluids, as well as moving phase boundaries in the presence of mass transfer across the interface [Crank, 1956; Fisher, 1989].

Characterization of the mass transfer processes in bioreactors, as used in cell or tissue culture systems, is essential when designing and evaluating their performance. Flow reactor systems bring together various reactants in a continuous fashion while simultaneously withdrawing products and excess reactants. These reactors generally provide optimum productivity and performance [Levenspiel, 1989; Shuler and Kargi, 1992]. They are classified as either tank or tube type reactors. Each represents extremes in the behavior of the gross fluid motion. Tank type reactors are characterized by instant and complete mixing of the contents and are therefore called perfectly mixed, or backmixed, reactors. Tube type reactors are characterized by the lack of mixing in the flow direction and are called plug flow, or tubular, reactors. The performance of actual reactors, though not fully represented by these idealized flow patterns, may match them so closely that they can be modeled as such with negligible error. Others can be modeled as combinations of tank and tube type over various regions.

An idealized backmixed reactor is analyzed as follows. Consider a stirred vessel containing a known fluid volume, into which multiple streams may be flowing that contain reactants, enzymes (bio-catalysts), nutrients, reaction medium, and so on. When all these components are brought together in the vessel, under properly controlled conditions such as temperature, pressure, and concentrations of each component, the desired reactions occur. The vessel is well mixed to promote good contacting of all components and hence an efficient reaction scheme can be maintained. The well-mixed state is achieved when samples withdrawn from different locations, including the exit, at the same instance in time are indistinguishable. The system is termed "lumped" vs. "distributed", as in a plug flow system where location matters. The response characteristics of a backmixed reactor are significantly different from those of a plug flow reactor. How reaction time variations affect performance and how quickly each system responds to upsets in the process conditions are key factors. The backmixed system is far more sluggish than the plug flow system. To evaluate the role of reaction time variations, the concept of residence time and how it is determined must be discussed. A brief analysis of a batch reactor will be useful in understanding the basic principles involved.

In batch systems, there is no flow in or out. The feed (initial charge) is placed in the reactor at the start of the process and the products are withdrawn all at once at some later time. The time-for-reaction is thus readily determined. This is significant since the conversion of reactants to products is a function of time, and can be obtained from knowledge of the reaction rate and its dependence upon composition and process variables. Since these other factors are typically controlled at a constant value, the time for reaction is the key parameter in the reactor design process. In a batch reactor, the concentration of reactants change with time and, therefore, the rate of reaction as well. Conversion is now related to reaction time through the use of calculus. For this system, residence time, equal to the processing time, is the time for reaction.

In a backmixed reactor, the conversion of reactants is controlled by the average length of time fluid elements remain in the reactor (their residence time). The ratio of the volume of fluid in the tank to the volumetric flow rate of the exit stream determines the residence time. Recall that in an ideal system of this type, operating at steady state, concentrations in the vessel are uniform and equal to those in the exit stream and thus the rate is also constant. Conversion is determined simply by multiplying this rate by the residence time. With imperfect mixing, fluid elements have a distribution of residence times and the performance of the reactor is clearly altered.

The plug flow reactor, which is based on the premise that there is no mixing in the flow direction, implies no interaction between neighboring fluid elements as they traverse the length of the reactor. This idealization permits the use of the prior results from batch reactor analysis. Each fluid element in the tubular reactor functions as a small batch reactor, undisturbed by its neighboring elements, and its reaction time is well defined as the ratio of the length of the tube to the fluid velocity. Concentration varies along the length of the reactor (also, rate), and can be simply related to reaction time. The mathematical analysis and prediction of performance is similar to that for a batch reactor. This analysis shows that the plug flow configuration yields higher conversions, given equal residence times, for the same processing conditions. However, the plug flow reactor responds more quickly to system upsets and is more difficult to control.

Most actual reactors deviate from these idealized systems primarily because of non-uniform velocity profiles, channeling and bypassing of fluids, and the presence of stagnant regions caused by reactor shape and internal components such as baffles, heat transfer coils, and measurement probes. Disruptions to the flow path are common when dealing with heterogeneous systems, particularly when solids are present. To model these actual reactors, various regions are compartmentalized and represented as combinations of plug flow and backmixed elements. For illustration, a brief discussion of recycle, packed bed, and fluidized bed reactor systems follows.

Recycle reactors are basically plug flow reactors with a portion of the exit stream recycled to the inlet, which provides multiple passes through the reactor to increase conversion. It is particularly useful in bio-catalytic reactor designs, in which the use of a packed bed is desired because of physical problems associated with contacting and subsequent separation of the phases. The recycle reactor provides an excellent means to obtain backmixed behavior from a physically configured tubular reactor. Multiple reactors of various types, combined in series and/or parallel arrangements, can improve performance and meet other physical requirements. Contact patterns using multiple entries into a single reactor can emulate these situations. A system demonstrating this characteristic is a plug flow reactor with uniform side entry. If these entry points are distributed along the length with no front entry, the system will perform as a single backmixed system. The significance is that backmixed behavior is obtained without continual stirring or the use of recycle pumps. Furthermore, if these side entry points are limited to only a portion of the reactor length, then the system functions as backmixed reactors in series with plug flow reactors.

Special consideration needs to be given to heterogeneous reactors, in which interaction of the phases is required for the reactions to proceed. In these situations, the rate of reaction may not be the deciding factor in the reactor design. The rate of transport of reactants and/or products from one phase to another can limit the rate at which products are obtained. For example, if reactants cannot get to the surface of a solid catalyst faster than they would react at the surface, then the overall (observed) rate of the process is controlled by this mass transfer step. To improve the rate, the mass transfer must be increased. It would be useless to make changes that would affect only the surface reaction rate. Furthermore, if products do not leave the surface rapidly, they may block reaction sites and thus limit the overall rate. Efficient contacting patterns need to be utilized. Hence, fluidized bed reactors (two-phase backmixed emulator), trickle bed systems (three-phase packed bed emulator), and slurry reactors (three-phase backmixed emulator) have evolved as important bioreactors. They are readily simulated, designed, scaled up, and modified to meet specific contacting demands in cell and tissue culture systems.

Flow reactors of all shapes, sizes, and uses are encountered in all phases of life. Examples of interest to bioengineers include: the pharmaceutical industry, to produce aspirin, penicillin, and other drugs; the biomass processing industry, to produce alcohol, enzymes and other specialty proteins, and value added products; and the biotechnologically important tissue and cell culture systems. The type of reactor used depends on the specific application and on the scale desired. The choice is based on a number of factors.

The primary ones are the reaction rate (or other rate-limiting process), the product distribution specifications, and the catalyst or other material characteristics, such as chemical and physical stability.

Using living systems require more thorough discussions because, in these systems, the solid phase may change its dimensions as the reaction proceeds. Particles that increase in size can fall out of the reaction zone and change flow patterns at the inlet. This occurs when gravity overcomes fluid buoyancy and drag forces. In some instances, this biomass growth is the desired product; however usually, the substances produced by reactions catalyzed by the cellular enzymes that are the desired products. By reproducing themselves, more enzymes are generated and productivity increases. Note, however, that there are both advantages and disadvantages to using these whole cell systems vs. the enzymes directly. The cell membranes can be resistances to transport; the consumption of reactant as a nutrient reduces the efficiency of raw material usage; and special precautions must be taken to maintain a healthy environment and thus productivity. The payoff, however, is that the enzymes within their natural environment are generally more active and safer from poisons or other factors that could reduce their effectiveness. Interesting examples are microbial fermentation processes, as discussed earlier. Backmixed flow reactors are used in these applications. They are best suited to maintain the cell line at a particular stage in its life cycle to obtain the desired results. The uniform and constant environment provided for the cells minimizes the adjustments that they need to make concerning nutrient changes, metabolic wastes, and so forth, so that production can proceed at a constant rate. The term "chemostat" is used when referring to backmixed reactors used in biotechnology applications. These are typically tank type systems with mechanical agitation. All the reactor types discussed earlier, however, are applicable. Recall that mixed flow characteristics can be obtained in tubular reactors if recycle and/or side entry is employed. Thus, air lift systems using a vertical column with recycle and side entry ports is a popular design.

The design, analysis, and simulation of reactors thus becomes an integral part of the bioengineering profession. The study of chemical kinetics, particularly when coupled with complex physical phenomena, such as the transport of heat, mass, and momentum, is required to determine or predict reactor performance. It thus becomes imperative to uncouple and unmask the fundamental phenomenological events in reactors and to subsequently incorporate them in a concerted manner to meet the objectives of specific applications. This need further emphasizes the role played by the physical aspects of reactor behavior in the stability and controllability of the entire process. The following chapters in this section demonstrate the importance of all the concepts presented in this introduction.

References

Srinivasan, A.V. and Vincent, J.F.V., Eds., (1996), *Biomimetics,* Plenum Publishing, New York.

Bird, R.B., Stewart, W.E., and Lightfoot, E.N. (1960), *Transport Phenomena,* John Wiley and Sons, New York.

Crank, J. (1956), *The Mathematics of Diffusion,* Oxford University Press, Oxford.

Enderle, J., Blanchard, S., and Bronzino, J.D., Eds. (1999), *Introduction to Biomedical Engineering,* Academic Press, New York.

Fisher, R.J. (1989), Diffusion with immobilization in membranes: Part II, in *Biological and Synthetic Membranes,* Butterfield, A., Ed., Alan R. Liss, Inc., New York, 138-151.

Fisher, R.J. (1999), Compartmental analysis, in *Introduction to Biomedical Engineering,* Academic Press, New York, Chap. 8.

Levenspiel, O. (1989), *The Chemical Reactor Omnibook,* Oregon State University Press, Corvallis, OR.

Lightfoot, E.N. (1974), *Transport Phenomena and Living Systems,* John Wiley and Sons, New York.

Rosner, D.E. (1986), *Transport Processes in Chemically Reacting Flow Systems,* Butterworth Publishers, Boston, MA.

Shuler, M.L. and Kargi F. (1992), *Bioprocess Engineering: Basic Concepts,* Prentice-Hall, Englewood Cliffs, NJ.

The [illegible] ones are those concerned with [illegible] to produce [illegible] and the analysis of other [illegible].

Thus living systems [illegible] because [illegible] the [illegible]. [illegible]

[illegible]

References

[illegible]

[illegible] (Eds.) [illegible]

[illegible] Oxford University Press [illegible]

[illegible] New York.

[illegible] Academic Press [illegible]

[illegible] Wiley and Sons, New York.

[illegible] Publishers [illegible]

[illegible] Prentice Hall [illegible]

95

Biomimetic Systems

Robert J. Fisher
University of Connecticut

Humankind has always been fascinated by the phenomenological events, biological and physical in nature, revealed to us by our environment. Curiosity drives us to study these observations and understand the fundamental basis of the mechanisms involved. The goals have been to control them for our safety and comfort and/or design processes to mimic the beneficial aspects. Experience has demonstrated the complexity and durability of these natural processes and that adaptability with multi-functionality is a must for biological systems to survive. Evolution of these living systems to adapt to new environmental challenges is at the molecular scale. Our need to be molecular scientists and engineers is thus apparent. The knowledge of the molecular building blocks used in the architectural configurations of both living and non-living systems, along with the understanding of the design and processes used for implementation, is but one step. To control and utilize these resources is another. The ability to mimic demonstrates a sufficient knowledge base to design systems for most conceivable applications requiring controlled functionality. To perfect this approach, a series of sensor/reporter systems must be available. A particularly attractive feature of living systems is their unique ability to diagnose and repair localized damages through a continuously distributed sensor network. Mimicry of these networks is an integral component of this emerging research area. The primary approach is to couple transformation and separation technology with detection and control systems. A major research emphasis has been toward the development of intelligent membranes, specifically sensor/reporter technology.

The concept of intelligent barriers and substrates, such as membranes, arises from the coupling of this sensor/reporter technology with controlled chemistry and reaction engineering, selective transport phenomena, and innovative systems design. Engineered membrane mimetics are required to respond and adapt to environmental stresses, whether intentionally imposed or stochastic in nature. These intelligent

0-8493-0462-8/00/$0.00+$.50

membranes may take the form of polymeric films, composite materials, ceramics, supported liquid membranes, or as laminates. Their important feature is specific chemical functionality, engineered to provide selective transport, structural integrity, controlled stability and release, and sensor/reporter capabilities. Applications as active transport mimics and their use in studying the consequences of environmental stresses, such as ultrasonic irradiation and electromagnetic fields on enzymatic functions, will create valuable insight into cellular mechanisms.

95.1 Concepts of Biomimicry

Discoveries that have emerged from a wide spectrum of disciplines, from biotechnology and genetics to polymer and molecular engineering, are extending the range of design and manufacturing possibilities for mimetic systems that was once incomprehensible. Understanding of the fundamental concepts inherent in natural processes has led to a broad spectrum of new processes and materials modeled on these systems [Srinivasan et al., 1991]. Natural processes, such as active transport systems functioning in living systems, have been successfully mimicked and useful applications in other fields, such as pollution prevention, have been demonstrated [Thoresen and Fisher, 1995]. It is only through understanding the mechanisms on the molecular scale which living systems utilize to develop their processes and implement designs, that success at the macro-scale can be assured. Multi-functionality, hierarchical organization, adaptability, reliability, self-regulation, and reparability are the key elements that living systems rely upon and that we must mimic to develop "intelligent systems". Successes to date have been based on the use of molecular engineering of thin films, neural networks, reporter/sensor technology, morphology and properties developments in polymer blends, transport phenomena in stationary and reacting flow systems, cell culture and immobilization techniques, controlled release and stability mechanisms, and environmental stress analyses. A few examples in the bioengineering field are molecular design of supported liquid membranes to mimic active transport of ions; non-invasive sensors to monitor *in vivo* glucose concentrations; detection of microbial contamination by bioluminescence; immuno-magnetic capture of pathogens; improved encapsulation systems and carrier molecules for controlled release of pharmaceutics; *in situ* regeneration of co-enzymes electro-chemically; and measurement of transport and failure mechanisms in "smart" composites. All these successes were accomplished through interdisciplinary approaches, essentially using three major impact themes, as follows: Morphology and Properties Development; Molecular Engineering of Thin Films; and Biotechnology, Bioreaction Engineering, and Systems Development.

Morphology and Properties Development

Polymer blends are the current focus of this research area since they represent the most rapidly growing area of polymer technology [Weiss et al., 1995]. Technology, however, has outpaced a detailed understanding of many facets of the science, which impedes the development and application of new materials. The purpose of this research is to develop the fundamental science that influences the phase behavior, phase architecture, and morphology and interfacial properties of polymer blends. The main medical areas where polymers and composites (polymeric) have found wide use are artificial organs, the cardiovascular systems, orthopedics, dentistry, ophthalmology, drugs, and drug delivery systems. Success has been related to the wide range of mechanical properties, transformation processes (shape possibilities), and low production costs. The limitation has been their interaction with living tissue. To overcome the biological deficiencies of synthetic polymers and enhance mechanical characteristics, a class of bioartificial polymeric materials has been introduced based on blends, composites, and interpenetrating polymer networks of both synthetic and biological polymers. Preparations from biopolymers such as fibrin, collagen, and hyaluronic acid with synthetic polymers such as polyurethane, poly(acrylic acid), and poly(vinyl alcohol) are available [Giusti et al., 1993].

Molecular Engineering of Thin Films

The focus in this research impact area is to develop a fundamental understanding of how morphology can be controlled in (1) organic thin film composites prepared by Langmuir-Blodgett (LB) monolayer and multilayer techniques and (2) molecular design of membrane systems using ionomers and selected supported liquids. The ability to form controlled structures of this nature will find immediate application in several aspects of smart materials development, particularly in micro-sensors.

Surfaces, interfaces, and microstructures play an important role in many research frontiers. Exploration of structural property relationships at the atomic and molecular level, investigating elementary chemical and physical transformations occurring at phase boundaries, applying modern theoretical methods for predicting chemical dynamics at surfaces, and integration of this knowledge into models that can be used in process design and evaluation are within the realm of surface and interfacial engineering. The control of surface functionality by proper selection of the composition of the LB films can mimic many functions of a biologically active membrane. An informative comparison is that between inverted erythrocyte ghosts [Dinno et al., 1991] and their synthetic mimics when environmental stresses are imposed on both systems. These model systems can assist in mechanistic studies to understand the functional alterations that result from ultrasound, EM fields, and UV radiation. The behavior of carrier molecules and receptor site functionality must be mimicked properly along with simulating disturbances in the Proton Motive Force of viable cells. Use of ion/electron transport ionomers in membrane-catalyst preparations is beneficial for programs such as electro-enzymatic synthesis [Zawalich and Fisher, 1998]. Inventing new membranes for artificial organs and advances in micelle reaction systems has resulted from these efforts.

Biotechnology, Bioreaction Engineering, and Systems Development

This research impact area is focused on: (1) sensor/receptor reporter systems and detection methods; (2) transport processes in biological and synthetic membranes; (3) biomedical and bio-conversion process development; and (4) smart film development and applications for intelligent barrier systems. These topics require coupling with the previously discussed areas and the use of biochemical reaction engineering techniques. Included in all of these areas is the concept of metabolic engineering, the modification of the metabolism of organisms to produce useful products. Extensive research in bio-conversion processes is currently being directed to producing important pharmaceutics. Future efforts also need to be directed toward impacting the emerging field of cell and tissue engineering, i.e., the manipulation or reconstruction of cell and tissue function using molecular approaches.

95.2 Biomimetic Membranes for Ion Transport

A cell must take nutrients from its extra-cellular environment if it is to grow or maintain its metabolic activity. The selectivity and rate that these molecular species enter can be important in regulatory processes. The mechanisms involved depend upon the size of the molecules to be transported across the cell membrane. These biological membranes consist of a continuous double layer of lipid molecules in which various membrane proteins are imbedded. Individual lipid molecules are able to diffuse rapidly within their own monolayer; however, they rarely "flip-flop" spontaneously between the two monolayers. These molecules are amphoteric and assemble spontaneously into bilayers when placed in water. Sealed compartments are thus formed, which reseal if torn.

The topic of membrane transport is discussed in detail in many texts (e.g., [Alberts et al., 1989; Lehninger, 1988]). The following discussion is limited to membrane transport of small molecules, hence excluding macromolecules such as polypeptides, polysaccharides, and polynucleotides. The lipid bylayer is a highly impermeable barrier to most polar molecules and thus prevents the loss of the water soluble contents of the cell interior. Consequently, cells have developed special means to transport these species across their membranes. Specialized transmembrane proteins accomplish this, each responsible for the transfer of a specific molecule or group of closely related molecules. The mechanism can be either energy independent, as in passive and facilitated diffusion, or energy dependent, as in active transport and group translocation.

In passive diffusion, molecules are transported with (or "down") a concentration gradient that is thermodynamically favorable and can occur spontaneously. Facilitated diffusion utilizes a carrier molecule, imbedded in the membrane, that can combine specifically and reversibly with the molecule to be transported. This carrier protein undergoes conformational changes when the target molecule binds and again when it releases that molecule on the transverse side of the membrane. This binding is dependent on favorable thermodynamics related to the concentration of free vs. bound species. An equilibrium is established, as in a Langmuir isotherm, on both sides of the membrane. Thus, the rate of transport is proportional to concentration differences maintained on each side of the membrane and the direction of flow is down this gradient. Active transport is similar to facilitated transport in that a carrier protein is necessary; however, it occurs against (up) a concentration gradient, which is thermodynamically unfavorable and thus requires energy. Group translocation requires chemical modification of the substance during the process of transport. This conversion process traps the molecule on a specific side of the membrane due to its asymmetric nature and the essential irreversibility of the transformation. These complexities lead to difficulties in mimicry; thus, research in this area is slow in developing.

Several energy sources are possible for active transport, including electrostatic or pH gradients of the proton motive force (PMF), secondary gradients derived from the PMF by other active transport systems, and by the hydrolysis of ATP. The development of these ion gradients enables the cell to store potential energy in the form of these gradients.

It is essential to realize that simple synthetic lipid bylayers, i.e., protein-free, can mimic only passive diffusion processes since they are impermeable to ions but freely permeable to water. Thermodynamically, virtually any molecule should diffuse across a protein-free, synthetic lipid bylayer down its concentration gradient. However, it is the rate of diffusion that is of concern, which is highly dependent upon the size of the molecule and its relative solubility in oil (i.e., the hydrophobic interior of the bylayer). Consequently, small non-polar molecules such as O_2 readily diffuse. If small enough, uncharged polar molecules such as CO_2, ethanol, and urea can diffuse rapidly, whereas glycerol is more difficult and glucose is essentially excluded. Water, because it has such a small volume and is uncharged, diffuses rapidly even though it is polar and relatively insoluble in the hydrophobic phase of the bylayer. Charged particles, on the other hand, no matter how small, such as Na^+ and K^+, are essentially excluded. This is due to the charge and the high degree of hydration preventing them from entering the hydrocarbon phase. Quantitatively, water permeates the bylayer at a rate 10^3 faster than urea, 10^6 faster than glucose, and 10^9 faster than small ions such as K^+. Thus, only non-polar molecules and small uncharged polar molecules can cross the cellular lipid membrane directly by simple (passive) diffusion; others require specific membrane transport proteins, as either carriers or channels. Synthetic membranes can be designed for specific biomedical applications that can mimic the above discussed transport processes [Michaels, 1990]. Membrane selectivity and transport are enhanced with the aid of highly selective complexing agents, impregnated as either fixed site or mobile carriers. To use these membranes to their full potential, the mechanism of this diffusion needs to be thoroughly understood.

Active Transport Biomimetics

Extensive theoretical and experimental work has previously been reported for supported liquid membrane systems (SLMS) as effective mimics of active transport of ions [Thoresen and Fisher, 1995; Stockton and Fisher, 1998; Cussler et al., 1989; Kalachev et al., 1992]. This was successfully demonstrated using di-(2 ethyl hexyl)-phosphoric acid as the mobile carrier dissolved in n-dodecane, supported in various inert hydrophobic micro-porous matrices (e.g., polypropylene), with copper and nickel ions as the transported species. The results showed that a pH differential between the aqueous feed and strip streams, separated by the SLMS, mimics the PMF required for the active transport process that occurred. The model for transport in an SLMS is represented by a five-step resistance-in-series approach, as follows: (1) diffusion of the ion through a hydrodynamic boundary layer; (2) desolvation of the ion, where it expels the water molecules in its coordination sphere and enters the organic phase via ion exchange with the mobile carrier at the feed/membrane interface; (3) diffusion of the ion-carrier complex across the SLMS to the strip/membrane interface; (4) solvation of the ion as it enters the aqueous strip solution via ion exchange; and (5) transport of the ion through the hydrodynamic boundary layer to the bulk stripping solution. A local Peclet number is used to characterize the hydrodynamics and the mass transfer occurring at the fluid/SLMS interface. The SLMS itself is modeled as a heterogeneous surface with mass transfer and reaction occurring only at active sites; in this case, the transverse pores. Long-term stability and toxicity problems limit their application, as configured above, in the biomedical arena. Use in combination with fixed site carrier membranes as entrapping barriers has great potential and is an active research area. Some success has been obtained using (1) reticulated vitreous carbon as the support matrix and Nafion® for the thin film "active barrier"; and (2) an ethylene-acrylic acid ionomer, utilizing the carboxylic acid groups as the fixed site carriers. The most probable design for biomedical applications appears to be a laminate composite system that incorporates less toxic SLMSs and highly selective molecularly engineered thin film entrapping membranes. Use of fixed site carrier membranes in these innovative designs requires knowledge of transport characteristics. Cussler et al. [1989] have theoretically predicted a jumping mechanism for these systems. Kalachev et al. [1992] have shown that this mechanism can also occur in an SLMS at certain carrier concentrations. This mechanism allows for more efficient transport than common facilitated diffusion. Stability over time and a larger range of carrier concentrations where jumping occurs make fixed carrier membranes attractive for biomedical applications. A brief discussion of these jumping mechanisms follows.

Mechanism for Facilitated Diffusion in Fixed Carrier Membranes

A theory for the mechanism of diffusion through a membrane using a fixed carrier covalently bound to the solid matrix, was developed previously [Cussler et al., 1989]. The concept is that the solute molecule jumps from one carrier to the next in sequence. Facilitated diffusion can occur only if these "chained" carriers are reasonably close to each other and have some limited mobility. The advantages of using a chained carrier in a solid matrix vs. a mobile carrier in a liquid membrane are that the stability is improved; there is no potential for solvent loss from the system, and the transport may actually be enhanced. Their theory is compared to that for the mobile carriers in the SLMS. For the fixed carrier (chained) system, the assumptions of fast reactions and that they take place only at the interface are also used. The major difference is that the complex formed cannot diffuse across the membrane since the carrier is covalently bound to the polymer chain in the membrane. Although the complex does not diffuse in the classical random walk concept, it can "jiggle" around its equilibrium position. This movement can bring it into contact range with an uncomplexed carrier also "jiggling," and result in a reversible interaction typical to normal receptor/ligand surface motion. It is assumed that no uncomplexed solute can pass through the membrane; it would be immobilized and taken from the diffusion process. The diffusion process that is operable is best explained by viewing the chained carrier membrane as a lamella structure where each layer is of thickness L. Every carrier can move a distance X around its neutral position and is a length L away from its neighbors. Diffusion can occur only over the distance X. Therefore, there is a specific concentration where a solute flux is first detected, termed percolation threshold, occurring when L = X. This threshold concentration is estimated as $C = 1/(L^3 N_a)$, where C is the average concentration, L is the distance between carrier molecules, and N_a is Avogadro's number. In summary, the mechanism is that of intramolecular diffusion; each chained carrier having limited mobility within the membrane. A carrier at the fluid membrane interface reacts with the species to be transported and subsequently comes in contact with an uncomplexed carrier and reacts with it, repeating this transfer process across the entire width of the membrane.

Jumping Mechanism in Immobilized Liquid Membranes

Facilitated diffusion was studied in immobilized liquid membranes using a system composed of a microporous nitrocellulose film impregnated with tri-n-octylamine (TOA) in n-decane [Kalachev et al., 1992]. Experiments were monitored by measuring the conductivity of the feed and strip streams. The transport of ions (cobalt and iron) from an acidic feed (HCl) to a basic strip solution (NH_4OH) was accomplished. Their results suggest that there are three distinct transport regimes operable in the membrane. The first occurs at short times and exhibits very little ion transport. This initial time is termed the ion penetration time and is simply the transport time across the membrane. At long times, a rapid increase in indiscriminate transport is observed. At this critical time and beyond, there are stability problems; that is, loss of solvent from the pores leading to the degradation of the membrane and the formation of channels that compromise the ion selective nature of the system and its barrier properties.

Recall that their experiments were for a selective transport with (not against) the ion gradient. It is only in the intermediate time regime that actual facilitated transport occurs. In this second region, experiments were conducted using a cobalt feed solution for various times and carrier concentrations; all experiments showed a peak in flux. The velocity of the transported species can be obtained from these results and the penetration time vs. carrier concentration is available. At the threshold carrier concentration these researchers claim that the mechanism of transport is by jumping as proposed earlier for fixed site carriers. The carrier molecules are now close enough to participate in a "bucket brigade" transport mechanism. The carrier molecules use local mobility, made possible by a low viscous solution of n-decane, to oscillate, passing the transported species from one to another. This motion results in faster transport than common facilitated transport, which relies on the random walk concept and occurs at lower TOA concentrations. It is in this low concentration region that the carrier molecules are too far apart to participate in the jumping scheme. At higher concentrations, well above the threshold value, the increased viscosity interferes with carrier mobility; the jumping is less direct or does not occur because of the increased bonding sites and hence removal of the species from the transport process.

95.3 Assessing Mass Transfer Resistances in Biomimetic Reactors

Uncoupling Resistances

Characterization of mass transfer limitations in biomimetic reactors is essential when designing and evaluating their performance. When used in Cell Culture Analog (CCA) systems, the proper mimicry of the role of intrinsic kinetics and transport phenomena cannot be over-emphasized. Lack of the desired similitude will negate the credibility of the phenomenological observations as pertaining to toxicity and/or pharmaceutical efficacy. The systems must be designed to allow manipulation, and thus control, of all interfacial events. The majority of material transfer studies for gaseous substrates are based on the assumption that the primary resistance is at the gas/liquid interface. Studies examining the use of hollow fiber membranes to enhance gas/liquid transport have been successfully conducted [Grasso et al., 1995]. The liquid/cell interfacial resistance is thus uncoupled from that of the gas/liquid interface and they can now be examined separately to evaluate their potential impacts. A reduction in the mean velocity gradient, while maintaining a constant substrate flux into the liquid, resulted in a shift in the limiting resistance from the gas/liquid to the liquid/cell interface. This shift manifested itself as an increase in the Monad apparent half-saturation constant for the chemo-autotrophic methanogenic microbial system selected as a convenient analog. The result of these studies significantly influences the design and/or evaluation of reactors used in the BME research area, especially for the animal surrogate or CCA systems. Although a reactor can be considered as well mixed based on spatial invariance in cell density, it was demonstrated that significant mass transfer resistance may remain at the liquid/cellular boundary layer.

There are three major points to be stressed. First, the liquid/cellular interface may contribute significantly to mass transfer limitations. Second, when mass transfer limitations exist the intrinsic biokinetics parameters cannot be determined. In biochemical reactor design, intrinsic parameters are essential to model adequately the system performance. Furthermore, without an understanding of the intrinsic biokinetics, one cannot accurately study transport mechanisms across biological membranes. The determination of passive or active transport across membranes is strongly affected by the extent of the liquid/cellular interfacial resistance.

Use in Physiologically Based Pharmacokinetics Models and Cell Culture Analog Systems

The potential toxicity of, and/or the action of, a pharmaceutical is tested primarily using animal studies. Since this technique can be problematic from both a scientific and ethical basis [Gura, 1997], alternatives have been sought. *In vitro* methods using isolated cells [Del Raso, 1993] are inexpensive, quick, and generally present no ethical issues. However, the use of isolated cell cultures does not fully represent the full range of biochemical activity as in the whole organism. Tissue slices and engineered tissues have also been studied but not without their inherent problems, such as the lack of interchange of metabolites among organs and the time dependent exposure within the animal. An alternative to both *in vitro* and animal studies is the use of computer models based on physiologically based pharmacokinetics (PBPK) models [Connolly and Anderson, 1991]. These models mimic the integrated, multi-compartment nature of animals and thus can predict the time dependent changes in blood and tissue concentrations of the parent chemical and its metabolites. The obvious limitations lie in that a response is based on assumed mechanisms; therefore, secondary and "unexpected" effects are not included, and parameter estimation is difficult. Consequently, the need for an animal surrogates or CCA systems is created. The pioneering work of M.L. Shuler's group at Cornell University [Sweeney et al., 1995; Shuler et al., 1996; Mufti and Shuler, 1998] and many others has led to the following approach for CCA systems.

These CCA systems are physical representations of the PBPK structure where cells or engineered tissues are used in organ compartments. The fluid medium that circulates between compartments acts as a "blood surrogate". Small scale bioreactors housing the appropriate cell types are the physical compartments that represent organs or tissues. This concept combines attributes of PBPK and *in vitro* systems. Furthermore, it is an integrated system that can mimic dose release kinetics, conversion into specific metabolites from each organ, and the interchange of these metabolites between compartments. Since the CCA system permits dose exposure scenarios that can replicate those of animal studies, it works in conjunction with a PBPK as a tool to evaluate and modify proposed mechanisms. Thus, bioreactor design and performance evaluation testing is crucial to the success of this animal surrogate concept.

Efficient transfer of substrates, nutrients, stimulants, etc. from the gas phase across all interfaces may be critical for the efficacy of certain biotransformation processes and in improving blood compatibility of biosensors monitoring the compartments. Gas/liquid mass transfer theories are well established for microbial processes [Cussler, 1984], however, biotransformation processes also involve liquid/cellular interfacial transport. In these bioreactor systems, a gaseous species is transported across two interfaces. Each could be a rate determining step and can mask intrinsic kinetics modeling studies associated with cellular growth and/or substrate conversion and product formation.

A methanogenic chemo-autotrophic process was selected for study because of its relative simplicity and strong dependence on gaseous nutrient transport, thus establishing a firm quantitative base case [Grasso et al., 1995]. The primary objective was to compare the effect of fluid hydrodynamics on mass transfer across the liquid/cellular interface of planktonic cells and the subsequent impact upon growth kinetics. Standard experimental protocol to measure the gas/liquid resistance was employed [Cussler, 1984; Grasso et al., 1995]. The determination of the liquid/cellular resistance is more complex. The thickness of the boundary layer was calculated under various hydrodynamic conditions and combined with molecular diffusion and mass action kinetics to obtain the transfer resistance. Microbial growth kinetics associated with these hydrodynamic conditions can also be examined. Since Monad models are commonly applied to describe chemo-autotrophic growth kinetics [Ferry, 1993], the half-saturation constant can be an indicator of mass transfer limitations. The measured (apparent) value will be greater than the intrinsic value, as demonstrated in these earlier studies.

95.4 Electro-Enzymatic Membrane Reactors as Electron Transfer Chain Biomimetics

Mimicry of *In Vivo* Co-Enzyme Regeneration Processes

In many biosynthesis processes, a co-enzyme is required in combination with the base enzymes to function as high efficiency catalysts. A regeneration system is needed to repeatedly recycle the co-enzyme to reduce operating costs in continuous *in vitro* synthesis processes, mimicking the *in vivo* regenerative process involving an electron transfer chain system. Multiple reaction sequences are initiated as in metabolic cycles. NAD(H) is one such co-enzyme. Because of its high cost, much effort has focused on improving the NAD(H) regeneration process [Chenault and Whitesides, 1987], with electrochemical methods receiving increased attention. The direct regeneration on an electrode has proven to be extremely difficult [Paxinos et al., 1991]. Either acceleration of protonation or inhibition of intermolecular coupling of NAD^+ is required. Redox mediators have permitted the coupling of enzymatic and electrochemical reactions; the mediator accepts the electrons from the electrode and transfers them to the co-enzyme via an enzymatic reaction, and thus regeneration/recycling of the co-enzyme during a biosynthesis reaction can be accomplished [Hoogvliet et al., 1988]. The immobilization of mediator and enzyme on electrodes can reduce the separation procedure, increase the selectivity, and stabilize the enzyme activity [Fry et al., 1994]. Various viologen mediators and electrodes have been investigated for the NADH system in batch configurations [Kunugi et al., 1990]. The mechanism and kinetics were investigated by cyclic voltammetry, rotating disk electrode, and impedance measurement techniques. The performance of electrochemical regeneration of NADH on an enzyme immobilized electrode for the biosynthesis of lactate in a packed bed flow reactor [Zawalich and Fisher, 1998] is selected as a model system to illustrate an electron transfer chain biomimetic.

Electro-Enzymatic Production of Lactate from Pyruvate

The reaction scheme is composed of a three-reaction sequence: (1) the NADH-dependent enzymatic (lactate dehydrogenase: LDH) synthesis of lactate from pyruvate; (2) the regeneration of NADH from NAD^+ and enzymatic (lipoamide dehydrogenase: LipDH) reaction with the mediator (methyl viologen); and (3) the electrochemical (electrode) reaction. The methyl viologen (MV^{2+}) accepts electrons from the cathode and donates them to the NAD^+ via the LipDH reaction. The regenerated NADH in solution is converted to NAD^+ in the enzymatic (LDH) conversion of pyruvate to lactate. A key feature of this system is the *in situ* regeneration of the co-enzyme NADH. A flow-by porous reactor utilizes the immobilized enzyme system (LipDH and methyl viologen as a mediator) within the porous graphite cathodes, encapsulated by a cation exchange membrane (Nafion® 124). The free-flowing fluid contains the pyruvate/lactate reaction mixture, the LDH, and the NADH/NAD^+ system. Lactate yields up to 70% have been obtained when the reactor system was operated in a semi-batch (i.e., recirculation) mode for 24 h, as compared to only 50% when operated in a simple batch mode for 200 h. The multi-pass, dynamic input operating scheme permitted optimization studies to be conducted on system parameters. This includes concentrations of all components in the free solution (initial and dynamic input values could be readily adjusted through recycle conditioning), flow rates, and electrode composition and their transport characteristics. By varying the flow rates through this membrane reactor system, operating regimes can be identified that determine the controlling mechanism for process synthesis (i.e., mass transfer vs. kinetics limitations). Procedures for operational map development are thus established.

References

Alberts, B., D. Bray, J. Lewis, M. Raff, K. Roberts and J. D. Watson. (1989). *Molecular Biology of the Cell,* 2nd ed., Garland, New York.

Chenault, H.K. and G. H. Whitesides. (1987). *Applied Biochemistry and Biotechnology,* 14, 147.

Connolly, R. B. and M.E. Anderson. (1991). Biologically Based Phamacodynamic Models: Tool for Toxicological Research and Risk Assessment. *Annu. Rev. Pharmacol. Toxicol.* 31:503.

Cussler, E. L. (1984). *Diffusion: Mass Transfer in Fluid Systems,* Cambridge University Press, New York.

Cussler, E., R. Aris and A. Bhown. (1989). On the Limits of Facilitated Diffusion, *J. Membr. Sci.*, 43, 149-164.

Del Raso, N.J. (1993). *In Vitro* Methodologies for Enhanced Toxicity Testing. *Toxicol. Lett.* 68:91.

Dinno, M. A., R. J. Fisher, J. C. Matthews, L. A. Crum and W. Kennedy. (1991). Effects of Ultrasound on Membrane Bound ATPase Activity, *J. Acous. Soc. AM.* 90(4).

Ferry, J. G. (1993). *Methanogenesis,* Chapman and Hall, New York.

Fry, A.J., S.B. Sobolov, M.D. Leonida and K.I. Viovodov. (1994). *Denki Kagaku,* 62, 1260

Giusti, P., L. Lazzeri, and L. Lelli. (1993). Bioartificial Polymeric Materials, *TRIP,* 1(9).

Grasso, K., K. Strevett and R. Fisher. (1995). Uncoupling Mass Transfer Limitations of Gaseous Substrates in Microbial Systems. *Chem. Eng. J.* 59.

Gura, T. (1997). Systems for Identifying New Drugs are Faulty, *Science,* 273, 1041.

Hoogvliet, J.C., L.C. Lievense, C. V. Kijk and C. Veeger. (1998). *Eur. J. Biochem.*, 174, 273.

Kalachev, A. A., L. M. Kardivarenko, N. A. Plate and V. V. Bargreev. (1992). Facilitated Diffusion in Immobilized Liquid Membranes: Experimental Verification of the Jumping Mechanism and Percolation Threshold in Membrane Transport, *J. Membr. Sci.*, 75, 1-5.

Kunugi, S., K. Ikeda, T. Nakashima, and H. Yamada. (1990). *Polymer Bulletin,* 24, 247.

Michaels, A. L. (1988). Membranes, Membrane Processes and Their Applications: Needs, Unsolved Problems, and Challenges of the 1990s. *Desalination,* 77, 5-34.

Matthews, J. C., W. L. Harder, W. K. Richardson, R. J. Fisher, A. M. Al-Karmi, L. A. Crum, and M. A. Dinno. (1993). Inactivation of Firefly Luciferase and Rat Erythrocyte ATPase by Ultrasound, *Membr. Biochem.*, 10, 213-220.

Mufti, N. A. and M. L. Shuler. (1998). Different *In Vitro* Systems Affect CYPIA1 Activity in Response to 2,3,78-tetrachlorodibenzo-p-dioxin. *Toxicol. In Vitro.* 12:259.

Paxinos, A.S., H. Gunther, D.J.M. Schmedding and H. Simon. (1995). *Bioelectro. Bioenerg.*, 25, 425.

Shuler, M. L., A. Ghanem, D. Quick, M. C. Wang and P. Miller. (1996). A Self-Regulating Cell Culture Analog Device to Mimic Animal and Human Toxicological Responses. *Biotechol. Bioeng.* 52:45.

Srinivasan, A. V., G. K. Haritos and F. L. Hedberg. (1991). Biomimetics: Advancing Man-Made Materials Through Gruidance From Nature, *Appl. Mechanics Reviews,* 44, 463-482.

Stockton, E. and R. J. Fisher. (1998). Designing Biomimetic Membranes for Ion Transport, *Proc. NEBC/IEEE Trans.*, 24(4), 49-51.

Sweeney, L.M., M. L. Shuler, J. G. Babish and A. Ghanem. (1995). A Cell Culture Analog of Rodent Physiology: Application to Naphthalene Toxicology. *Toxicol. In Vitro,* 9, 307.

Thoresen, K. and R. J. Fisher. (1995). Use of Supported Liquid Membranes as Biomimetics of Active Transport Processes, *BIOMIMETICS,* 3(1), 31-66.

Weiss, R. A., C. Beretta, S. Sasonga, and A. Garton. (1995). *Appl. Polym. Sci.* 41:491,

Zawalich, M. E. and R. J. Fisher. (1998). *In Situ* Regeneration of Co-Enzyme NADH with Electro-Enzymatic Biomimetic Membranes, *Proc. NEBC/IEEE Trans.*, 24(4), 85-88.

96

Diffusional Processes and Engineering Design

E. N. Lightfoot
University of Wisconsin

Mass transport and diffusional processes play key roles in biomedical engineering at several levels, both within the body and in *extracorporeal* circuits. The roles of mass transfer in tissue function are discussed in Chapter 115 (Lightfoot and Duca), and here we will concentrate on external processes and design of therapeutic regimens. Moreover, we will depend heavily upon major references [Bird et al., 1960; Schmidt-Nielsen,, 1997; Bassingthwaighte et al., 1998; Lightfoot and Lightfoot, 1997; Welling, 1997; Noble and Stern, 1995; Ho and Sirkar, 1992] and seek understanding rather than detailed descriptions.

All mass transport processes, which can be defined as the technology for moving one species in a mixture relative to another, depend ultimately upon diffusion as the basis for the desired selective motion. Diffusion takes many forms, and a general description is provided in Table 115.7 of Chapter 115. However, a great deal of information can often be obtained by carefully written statements of simple constraints, and that of conservation of mass is the most useful for our purposes. We shall begin with examples where this suffices and show how one can determine the validity of such a simple approach. We then proceed to situations where more detailed analysis is needed.

96.1 Applications of Allometry

Much of biomedical engineering requires transferring information obtained from animal experiments to humans, and here *interspecies* similarities have proven quite useful. Many properties of living systems scale simply with body mass according to the simple *allometric* equation [Schmidt-Nielsen, 1997]:

$$P = aM^b \tag{96.1}$$

Here P is any property, M is average species mass, and a and b are species independent constants. Among the most important of such relations is that of basal metabolic rate of warm blooded animals, for which the total rate of oxygen consumption is given by

$$R_{O_2,tot} \approx 3.5\,M^{3/4} \tag{96.2}$$

0-8493-0462-8/00/$0.00+$.50

Here $R_{O_2,tot}$ is ml O_2 (STP) consumed per hour, and M is body mass in grams. Other important properties, for example blood volume per unit mass and decrease of total oxygen content per unit volume of blood on passing through the arterial system, are *invariant*: b is zero. It follows that average blood circulation time

$$T_{circ} \propto M^{1/4} \tag{96.3}$$

This is an important result because it means that *flow limited* body processes scale with the 1/4-power of body mass. Moreover, T_{circ} is the first of the *time constants* which we shall find to govern most diffusional processes.

An important application is the prediction of human drug elimination kinetics from animal experiments. For many it is found that

$$cV/m_0 = f\left(t/T_{circ}\right) \tag{96.4}$$

where c is blood concentration of the drug, V is body volume, and m_0 is the initial drug dose. The function f can be quite complex but, to a first order, it is independent of species (see for example [Dedrick et al., 1970; also Lightfoot, 1974, p. 353]). Moreover, V can be calculated from mass assuming a species independent density close to 1g/ml. Success of Eq. (96.4) requires that drug distribution between blood and body tissues be independent of drug concentration and that elimination be assumed flow limited. We now ask when this last assumption can be justified.

96.2 Flow Limited Processes

Processes in which diffusion and reaction rates are fast relative to mean solute residence times are said to be flow limited because change in diffusion or kinetic rates has little effect on the process under investigation. Although very few biomedically interesting processes are flow limited in this sense *at a detailed level of description*, it is often found that the global behavior of a complex system can be so described with little error. To understand this we look briefly at Fig. 96.1a: a closed system of constant volume V with a single inlet and single outlet. Blood or other fluid is flowing through this system at a constant flow rate Q and with a decaying inlet solute concentration

$$c_{in}\left(t\right) = c_0 e^{-t/T_{BC}} \tag{96.5}$$

We now calculate the difference between this inlet concentration and the outlet concentration $c_{out}(t)$, for two quite different situations: *plug flow in our system* (PF), and *perfect mixing* (*CSTR* for continuous stirred tank reactor). In both cases the mean residence time is $T_m \equiv V/Q$. The results are:

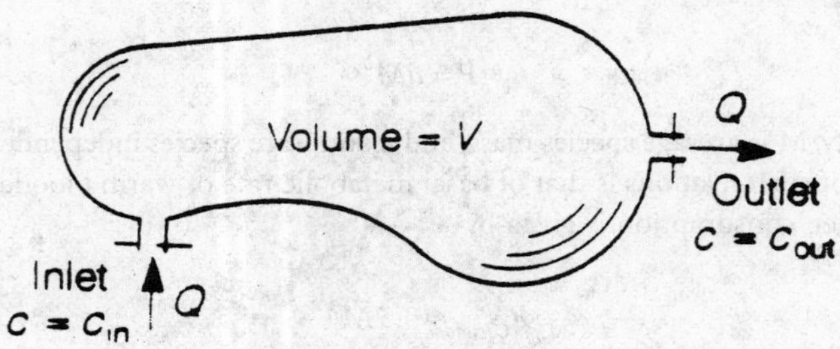

FIGURE 96.1a A simple flow system.

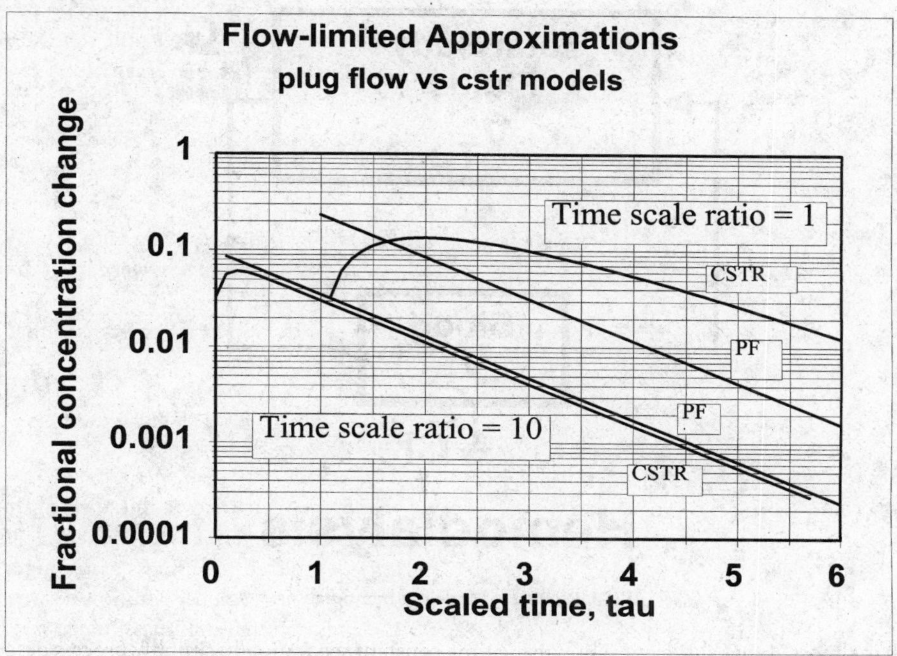

FIGURE 96.1b Flow limited approximations.

PF: Here, there is just a time delay without other change, so that

$$t < T_m: \quad c_{out} - c_{in} = -c_0 e^{-t/T_{BC}} \tag{96.6a}$$

$$t > T_m: \quad c_{out} - c_{in} = c_0 e^{-t/T_{BC}} \left[e^{T_M/T_{BC}} - 1 \right] \tag{96.6b}$$

CSTR: Here the internal concentration is uniform at the outlet level, and

$$c_{out} - c_{in} = \left(\frac{R}{R-1} \right) \left[e^{-\tau} - e^{-R\tau} \right] - e^{-\tau t}; \; R = T_{BC} / T_M \tag{96.7}$$

These results are plotted in Fig. 96.1b for two *time-constant ratios,* T_{BC}/T_M, = 1 and T_{BC}/T_M = 10. The two flow conditions are seen to produce very different behavior for equal time constants, but these differences quickly become minor when $T_m \ll T_{BC}$. More specifically, the effect of our system flow conditions on the inlet concentration becomes insensitive to these conditions when two criteria are met:

$$T_m \ll T_{BC} \tag{96.8a}$$

$$t = T_{OBS} \geq T_m \tag{96.8b}$$

Here T_{OBS} is the *observer time,* i.e., the time before which there is no interest in the system behavior. These are the conditions of *time constant separation,* of great practical importance in all engineering design calculations. Usually "much less than" can be taken to be less than a third, and vice versa, and

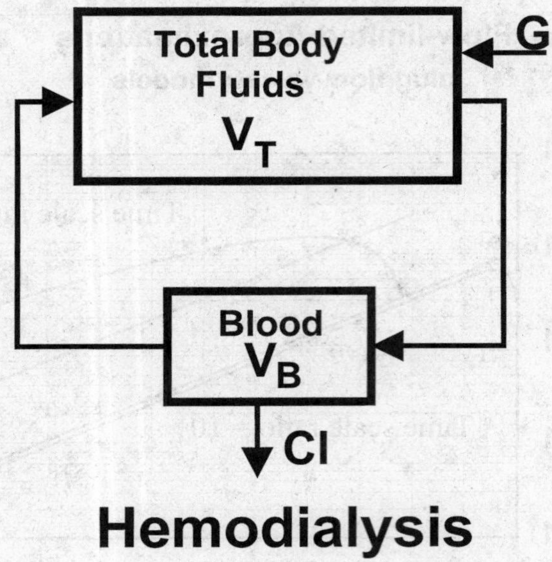

FIGURE 96.2 Modeling hemodialysis.

one tenth is almost always sufficient. Thus, using time constant separation to simplify process descriptions is usually referred to as an order-of-magnitude approximation. Returning to Fig. 96.1a we may now write a macroscopic mass balance [Bird et al., 1960, Ch. 22] of the form

$$V dc/dt \approx Q\left(c_{in} - c\right) + V\langle R\rangle \tag{96.9}$$

where c is both average internal and exit concentration, and ⟨R⟩ is the average rate of solute formation by chemical reaction. Here we have used the CSTR approximation as the simplest to handle mathematically. This expression is valid only under the constraints of Eq. (96.8), but these are commonly met in conditions of medical interest.

The major utility of Eq. (96.9) is in the description of networks such as the organ network of the human body, and applications include transient drug distribution, anaesthesia, and the study of metabolic processes [Welling, 1997; Bassingthwaighte et al., 1998]. Here an equation similar to Eq. (96.8) must be written for each element of the network, but individual organs can be combined, so long as the resulting element conforms to Eq. (96.8). These processes are often lumped under the heading of *pharmacokinetics*, and a large literature has developed [Welling, 1997]. An example of great economic importance is hemodialysis for the treatment of end stage kidney disease. Here the body can be approximated by the simple diagram of Fig. 96.2, and the defining equations reduce to

$$V_T dc_T/dt = Q_B\left(c_B - c_T\right) + G \tag{96.9a}$$

$$V_B dc_B/dt = -Q_B\left(c_B - c_T\right) - Cl \cdot c_B \tag{96.9b}$$

Here the subscript B and T refer to the blood and (whole body) tissue respectively; G is the rate of toxin generation, and Cl is the *clearance* of the dialyzer. Comparison of prediction with observation is shown for creatinine in Fig. 96.3. Here the parameters of Eq. (96.9) were determined for the first day's treatment. It may be seen that the pharmacokinetic approximation permits accurate extrapolation for four additional days.

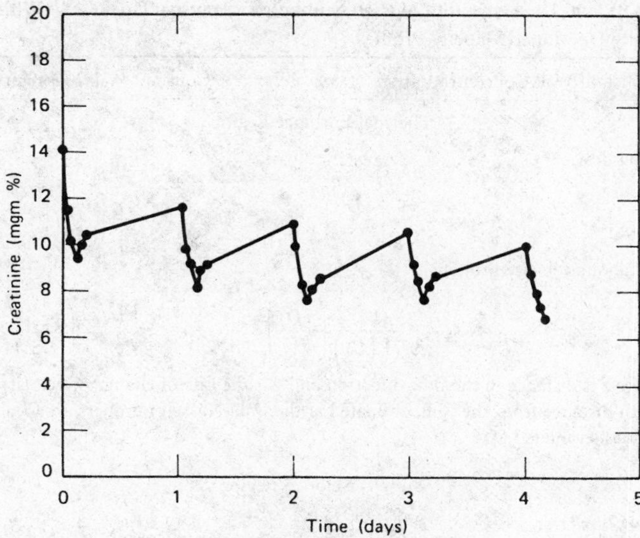

FIGURE 96.3 Actual creatinine dynamics compared to prediction.

96.3 Extracorporeal Systems

Next we look at the problem of designing extracorporeal systems, and these can normally be classified into a small number of categories. We shall consider steady-state membrane separators, chromatographic devices, and flow reactors.

Membrane Separators

The purpose of these devices is to transfer solute from one flowing steam to another, and there are two subcategories distinguished by the ratio of transmembrane flow induced (convective) and diffusional solute. This ratio in turn is defined by a Péclet number,

$$Pe \equiv \langle v \rangle / P \tag{96.10}$$

$$P \equiv N_i / \Delta c_i \tag{96.11}$$

Here ⟨v⟩ is the observable transmembrane solvent velocity, and P is the membrane solute diffusional permeability. The permeability in turn is defined as the ratio of the molar flux of solute transport, moles/area-time, to the solute concentration difference causing this transport. The most familiar examples of low-Pe devices are blood oxygenators and hemodialyzers. High-Pe systems include micro-, ultra-, and nano-filtration and reverse osmosis. The design and operation of membrane separators is discussed in some detail in standard references [Noble and Stern, 1995; Ho and Sirkar, 1992], and a summary of useful predictions is provided in Section 96.4.

Low-Pe devices are by far the simpler. Solute transport in dialyzers is essentially unaffected by the small amount of transmembrane water flow, and one may therefore use standard design techniques based on membrane permeabilities, usually supplied by the vendor, and mass transfer coefficients in the adjacent fluids. Local fluxes can be described by the simple expression

$$N_i = K_c \left(c_{ib} - c_{ie} \right) \tag{96.12}$$

TABLE 96.1 Asymptotic Nusselt Numbers for Laminar Duct Flow (With Fully Developed Velocity Profiles)

	Constant Wall Concentration	Constant Wall Mass Flux
	Thermal Entrance Region	
Plug flow:	$Nu_{loc} = \frac{1}{\sqrt{\pi}}\left(\frac{vD^2}{Đz}\right)^{1/2}$	$Nu_{loc} = \frac{\sqrt{\pi}}{2}\left(\frac{vD^2}{Đz}\right)^{1/2}$
Parabolic velocity profile:	$Nu_{loc} = \frac{1}{9^{1/3}\Gamma(4/3)}\left(\frac{gD^3}{Đz}\right)^{1/3}$	$Nu_{loc} = \frac{\Gamma(2/3)}{9^{1/3}}\left(\frac{gD^3}{Đz}\right)^{1/3}$
Here z is distance in the flow direction and g is the rate of change of velocity with distance from the wall, evaluated at the wall. Nusselt numbers are local values, evaluated at z.		
	Fully Developed Region	
Plug flow:	$Nu_{loc} = 5.783$	$Nu_{loc} = 8$
Parabolic Flow:	$Nu = 3.656$	$Nu = 48/11$

where N_i is the molar flux of the solute "i" across the membrane, moles/area-time, K_c is the overall mass transfer coefficient, c_i is molar solute concentration while the subscripts "b" and "e" refer to blood and the external fluid, respectively. The overall mass transfer coefficient must be calculated from the two fluid phase coefficients and membrane permeability. Hemodialysis solutes tend to distribute uniformly between blood (on a cell-free basis), and one may use the simple approximation

$$1/K_c = (1/k_b) + (1/P) + (1/k_e) \tag{96.13}$$

Here k_b and k_e are the mass transfer coefficients for the blood and external fluid, here dialysate, respectively. These phase mass transfer coefficients can usually be estimated to an acceptable accuracy from simple asymptotic formulas [Lightfoot and Lightfoot, 1997]. Examples of the latter are given in Table 96.1. For unequally distributed solutes, Eq. (96.3) must be appropriately modified [Lightfoot and Lightfoot, 1997], and, for blood this can be something of a problem [Popel, 1989].

Equipment performance for dialyzers is normally expressed in terms of *clearance*

$$Cl = fc_{i,in}Q \tag{96.14}$$

where f is the fraction of incoming solute "i" removed from the blood, $c_{i,in}$ is concentration of solute "i" in the entering blood, and Q is volumetric blood flow rate. Clearance is a convenient measure as it is independent of solute concentration, and it is easily determined experimentally. Prediction is useful for device design, but in operation, clearance is usually determined along with effective blood flow rates and tissue water volumes from numerical analysis of a test dialysis procedure. The efficiency of blood oxygenators can be dominated by either membrane permeability or mass transfer in the flowing blood, and it is complicated by the kinetics and thermodynamics of the oxygen/hemoglobin system. These aspects are discussed in detail by Popel [1989].

High-Pe devices are dominated by transmembrane water transport, and detailed discussion must be left to the above cited references. However, it is important to recognize their primary function is to remove water and undesired solutes while retaining one solute which is desired. Rejection of the desired product increases toward an asymptote as water flux increases, and one should operate near this asymptote if at all possible. The relation between water flux and rejection is perhaps best determined experimentally.

However, as water flux increases the rejected solute concentration at the interface between the feed stream and the membrane also increases. This process, usually known as *concentration polarization*, typically produces a significant increase in osmotic pressure which acts to reduce the flow. Polarization is a complex process, but to a good approximation the trans-membrane water velocity is given by

$$\langle v \rangle \approx 1.4 \ln\left(c_{s\delta}/c_{s0}\right) \cdot \left(\rho_\delta/\rho_0\right) \cdot \Theta \cdot k_c \tag{96.15}$$

Here c_s is concentration of the rejected solute, ρ is solution density, k_c is the concentration-based mass transfer coefficient *in the absence of water flux*, and the subscripts δ and 0 refer to conditions at the membrane surface and bulk of the fed solution, respectively. The factor Θ is a correction for variable viscosity and diffusivity, approximated at least for some systems by

$$\Theta \equiv \left\langle Đ^* \right\rangle^{2/3} \left\langle 1/\mu^* \right\rangle^{1/3} \tag{96.16}$$

and $\langle Đ^* \rangle$ and $\langle 1/\mu^* \rangle$ are the averages of solute diffusivity and reciprocal solution viscosity at the membrane surface and bulk solution divided by these quantities in the bulk solution. These equations are reasonable once appreciable polarization has occurred, if rejection is high, and they are a modification of earlier boundary-layer analyses of Kozinski and Lightfoot [1972]. They are more accurate than the more recent results made using simple film theory discussed in [Noble and Stern, 1995; Ho and Sirkar, 1992], both in incorporating the coefficient 1.4 and in the corrections for variable diffusivity and viscosity. The coefficient of 1.4 is a correction to account for boundary-layer compression accompanying transmembrane flow, not allowed for in film theory. Typical geometry insensitive boundary layer behavior is assumed, and these equations are not restricted to any given rejected solute or equipment configuration. However, in order to calculate the pressure drop required to obtain this flow, one must use the expression

$$\Delta p = \pi + v/k_h \tag{96.17}$$

where Δp is transmembrane pressure drop, π is solute osmotic pressure at the membrane surface, and k_h is the hydraulic permeability of the membrane. The osmotic pressure in turn is a function of solute concentration at the membrane surface, and thus is different for each rejected solute.

Chromatographic Columns

Chromatography is very widely used in biomedical analyses and to a significant extent for extracorporeal processing of blood and other body fluids. Recovery of proteins from blood is of particular importance, and these applications can be expected to grow. Good basic texts are available for underlying dynamic theory [Guiochon et al., 1994] and chemistry [Snyder et al., 1997], and a series of design papers is recommended [Lightfoot et al., 1997; Lode et al., 1998; Yuan et al., in press; Athalye et al., 1992]. Differential chromatography, in which small mixed-solute pulses are separated by selective migration along a packed column, is the simplest, and much of the chromatographic literature is based on concepts developed for this basic process.

In differential chromatography, the individual solutes do not interact, and the effluent curve for each is usually close to a Gaussian distribution:

$$c_f\left(L,t\right) = c_0 \exp\left[-\left(\frac{t}{\bar{t}} - 1\right)^2\right] \Big/ \left(2\pi\right)^{1/2} \left(\sigma/\bar{t}\right) \tag{96.18}$$

where $c_f(L,t)$ is the fluid phase concentration leaving a column of length L at time t, $\bar{t}$ is the mean solute residence time, and σ is the standard deviation of the distribution; c_0 is the maximum effluent concentration.

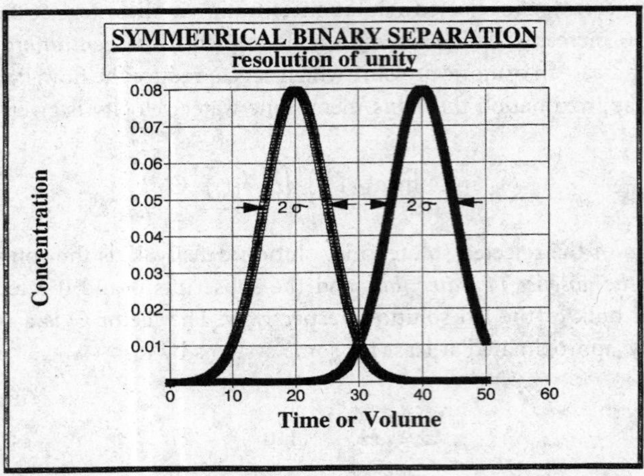

$$R_{12} = (1/2)|T_1 - T_2|/(\sigma_1 + \sigma_2)$$

FIGURE 96.4 Resolution in differential chromatography.

Degree of separation is defined in terms of the *resolution*, R_{12} defined as in Fig. 96.4 where separation with a resolution of unity is shown. Here the distance between the two mean residence times ($\bar{t}_1$-$\bar{t}_2$), 40 and 20 in Fig. 96.4, increase in direct proportion to column length, while the standard deviations increase with its square root. Thus resolution is proportional to the square root of column length. Column performance is normally rated in terms of the *number of equivalent theoretical plates*, N defined by

$$N \equiv \left(\bar{t}/\sigma\right)^2 . \tag{96.19}$$

Methods of predicting N from column design and operating conditions are described in the above mentioned references (see for example [Athalye et al., 1992]).

Flow Reactors

Development of small flow reactors for diagnostic purposes is a fast growing field. Patient monitoring of blood glucose is probably the largest current example, but new applications for a wide variety of purposes are rapidly being developed. These are essentially miniaturized variants of industrial reactors and normally consist of channels and reservoirs with overall dimensions of centimeters and channel diameters of millimeters or even tenths of millimeters. The key elements in their description are convective mass transfer, dispersion, and reaction, and they differ from larger systems only in parameter magnitudes. In particular, flow is almost invariably laminar and Schmidt numbers

$$Sc \equiv \mu/\rho Đ \tag{96.20}$$

are always high, never less than 10^5. Here μ is solvent viscosity, ρ is its density, and Đ is effective solute diffusivity through the solution. These aspects are briefly described in the next section.

96.4 Useful Correlations

Convective Mass Transfer

Mass transfer coefficients k_c are best correlated in the dimensionless form of mass transfer *Nusselt numbers* defined as

$$Nu \equiv k_c D / Ð \tag{96.21}$$

where D is any convenient characteristic length, typically diameter for tubes. For the high-Sc systems considered above, these Nusselt numbers are functions only of a dimensionless ratio L*, system geometry, and boundary conditions [Lightfoot and Lightfoot, 1997]. The scaled length L* in turn is just the ratio of mean solvent residence time to lateral diffusion time:

$$L^* \equiv \left(L/\langle v \rangle\right) / \left(D^2/24Ð\right) \tag{96.22}$$

For all geometries these functions have the same three characteristics: an entrance region solution for L* "much" less than unity, a constant asymptotic value for L* "much" greater than unity (the "fully developed" region), and a relatively unimportant transition region for L* close to unity. Normally the entrance region is of most practical importance and can be used without great error up to L* of unity. Typical results are shown in Table 96.1. The entrance region results are valid for both tubes (circular cross-section) and between parallel plates; it may be noted that the reference length D cancels. It appears to the same power on both sides of the equations. The correlations for laminar flow are also valid for non-Newtonian fluids if the appropriate wall velocity gradients are used. The solutions for fully developed flow are, however, only useful for round tubes and either plug or Poiseuille flow as indicated. These are much less important as they result in very low transfer rates. Plug and Poiseuille (parabolic) flows are limiting situations, and most real situations will be between them. Plug flow is approximated in ultrafiltration where the high solute concentrations produce a locally high viscosity. Poiseuille flow is a good approximation for dialysis.

Convective Dispersion and the One-Dimensional Convective Diffusion Equation

Non-uniform velocity profiles tend to reduce axial concentration gradients, a result which is qualitatively similar to a high axial diffusivity. As a result, it is common practice to approximate the diffusion equation for duct flow as

$$\partial \bar{c}/\partial t + \langle v \rangle \partial \bar{c}/\partial z \equiv Ð_{eff}\, \partial^2 \bar{c}/\partial z^2 + \langle R_{i,eff} \rangle \tag{96.23}$$

Here the overline indicates an average over the flow cross-section, usually further approximated as the cup-mixing or bulk concentration [Bird et al., 1960, p. 297), $\langle v \rangle$ is the flow average velocity, $Ð_{eff}$ is an effective diffusivity and $\langle R_{i,eff} \rangle$ is the rate of solute addition per unit volume by reaction plus mass transfer across the wall:

$$\langle R_{i,eff} \rangle = \langle R_{i,chem} \rangle + (C/S) K_c (c_e - \bar{c}) \tag{96.24}$$

Here $\langle R_{i,chem} \rangle$ is the average volumetric rate of formation of species "*i*", moles/volume-time, *C* is the duct circumference, *S* is its cross-sectional area, and c_e is the solute concentration outside the duct wall, i.e., in the surrounding fluid.

For Newtonian tube flow at L* greater than unity, it is reasonable to write

$$Ð_{eff} / D \approx 1 + (1/192)\left(D\langle v \rangle / Ð\right)^2 \tag{96.25}$$

The restrictions on this result and some additional useful approximations are widely discussed in the mass transfer literature (see for example [Ananthkrishnan et al., 1965]). For flow in packed beds, one often sees

$$Ð_{eff}/D \approx 0.4\left(Dv_2/Ð\right) \qquad (96.26)$$

but more accurate expressions are available [Athalye et al., 1992]. Convective dispersion is a very complex subject with a large literature, and successful application requires careful study [Brenner, 1962; Ananthkrishnan et al., 1965; Edwards et al., 1991].

References

Anthakrishnan, V, WN Gill, and AJ Barduhn, 1965, Laminar Dispersion in Capillaries, *AIChEJ,* **11**(6), 1063-1072.

Athalye, AM, SJ Gibbs, and EN Lightfoot, 1992, Predictability of Chromatographic Separations: Study of Size Exclusion Media with Narrow Particle Size Distributions, *J. Chromatogr A,* **589**, 71-85.

Bassingthwaighte, JB, CA Goresky and JH Linehan, 1998, *Whole Organ Approaches to Cellular Metabolism,* Springer, New York.

Bird, RB, WE Stewart and EN Lightfoot, 1960, *Transport Phenomena,* Wiley, New York.

Brenner, H, 1962, The diffusion model of longitudinal mixing in beds of finite length, *CES,* **17**, 229-243.

Dedrick, RL, KB Bischoff, and DS Zaharko, 1970, Interspecies Correlation of Plasma Concentration History of Methotrexate (NSC-740), *Cancer Ther Rep,* Part I, **54**, 95.

Edwards, DA, M. Shapiro, H. Brenner, and M. Shapira, 1991, Transport in Porous Media.

Guiochon, G, SG Shirazi, and AM Katti, 1994, *Fundamentals of Preparative and Non-Linear Chromatography,* Academic Press, New York.

Ho, WSW, and KK Sirkar, 1992, *Membrane Handbook,* Van Nostrand Reinhold, New York.

Kozinski, AA, and EN Lightfoot, 1972, Protein Ultrafiltration, *AIChEJ,* **18**, 1030-1040.

Lightfoot, EN, 1974, *Transport Phenomena and Living Systems,* Wiley-Interscience, New York.

Lightfoot, EN and Lightfoot, EJ, 1997, Mass Transfer in *Kirk-Othmer Encyclopedia of Separation Technology.*

Lightfoot, EN and KA Duca, 2000, The Roles of Mass Transfer in Tissue Function, in *The Biomedical Engineering Handbook,* 2nd ed., Bronzino, J., Ed., CRC Press, Boca Raton, FL, chap. 115.

Lightfoot, EN, JL Coffman, F Lode, TW Perkins, and TW Root, 1997, Refining the Description of Protein Chromatography, *J. Chromatgr A,* **760, 130.**

Lode, F, A Rosenfeld, QS Yuan, TW Root and EN Lightfoot, 1998, Refining the Scale-up of Chromatographic Separations, *J. Chromatogr A,* **796**, 3-14.

Noble, RD, and SA Stern, 1995, *Membrane Separations Technology,* Elsevier, London.

Popel, AS, 1989, Theory of oxygen transport to tissue, *Clin Rev Biomed Eng* **17**(3), 257.

Schmidt-Nielsen, 1997, *Animal Physiology,* 5th ed., Cambridge University Press, Cambridge.

Snyder, LR, JJ Kirkland, and JL Glajch, 1997, Wiley, New York.

Welling, PG, 1997, *Pharmacokinetics,* American Chemical Society.

Yuan, QS, A Rosenfeld, TW Root, DJ Klingenberg, and EN Lightfoot, in press, Flow Distribution in Chromatographic Columns, *J. Chromatogr A.*

97

Animal Surrogate Systems

Michael L. Shuler
Cornell University

97.1 Background

Animal surrogate or cell culture analog (CCA) systems mimic the biochemical response of an animal or human when challenged with a chemical or drug. A true animal surrogate is a device that replicates the circulation, metabolism, or adsorption of a chemical and its metabolites using interconnected multiple compartments to represent key organs. These compartments make use of engineered tissues or cell cultures. Physiologically based pharmacokinetic models (PBPK) guide the design of the device. The animal surrogate, particularly a human surrogate, can provide important insights into toxicity and efficacy of a drug or chemical when it is impractical or imprudent to use living animals (or humans) for testing. The combination of a CCA and PBPK provides a rational basis to relate molecular mechanisms to whole animal response.

Limitations of Animal Studies

The primary method used to test the potential toxicity of a chemical or action of a pharmaceutical is to use animal studies, predominantly with rodents. Animal studies are problematic. The primary difficulties are that the results may not be meaningful to assessment of human response [Gura, 1997]. Because of the intrinsic complexity of a living organism and the inherent variability within a species, animal studies are difficult to use to identify unambiguously the underlying molecular mechanism for action of a chemical. The lack of a clear relationship among all of the molecular mechanisms to whole animal response makes extrapolation across species difficult. This factor is particularly crucial when extrapolation of rodent data to humans is an objective. Further, without a good mechanistic model it is difficult to rationally extrapolate from high doses to low doses. However, this disadvantage due to complexity can be an advantage; the animal is a “black box” and provides response data even when the mechanism of action is unknown. Further disadvantages reside in the high cost of animal studies, the long period of time often necessary to secure results, and the potential ethical problems in animal studies.

0-8493-0462-8/00/$0.00+$.50

Alternatives to Animal Studies

In vitro methods using isolated cells (e.g., [Del Raso, 1993]) are inexpensive, quick, and have almost no ethical constraints. Because the culture environment can be specified and controlled, the use of isolated cells facilitates interpretation in terms of a biochemical mechanism. Since human cells can be used as well as animal cells, cross-species extrapolation is facilitated.

However, these techniques are not fully representative of human or animal response. Typical *in vitro* experiments expose isolated cells to a static dose of a chemical or drug. It is difficult to relate this static exposure to specific doses in a whole animal. The time-dependent change in the concentration of a chemical in an animal's organ cannot be replicated. If one organ modifies a chemical or prodrug which acts elsewhere, these situations would not be revealed by the normal *in vitro* test.

A major limitation on the use of cell cultures is that isolated cells do not fully represent the full range of biochemical activity of the corresponding cell type when in a whole animal. Engineered tissues, especially co-cultures [Bhatia et al., 1998], can provide a more "natural" environment which can improve (i.e., make normal) cell function. Another alternative is use of tissue slices, typically from the liver [Olinga et al., 1997]. Tissue slices require the sacrifice of the animal, there is intrinsic variability, and biochemical activities can decay rapidly after harvest. The use of isolated tissue slices also does not reproduce interchange of metabolites among organs and the time-dependent exposure that occurs within an animal.

An alternative to both animal and *in vitro* studies is the use of computer models based on PBPK models [Connolly and Anderson, 1991]. PBPK models can be applied to both humans and animals. Because PBPK models mimic the integrated, multicompartment nature of animals, they can predict the time-dependent changes in blood and tissue concentrations of a parent chemical or its metabolites. Although construction of a robust, comprehensive PBPK is time-consuming, once the PBPK is in place many scenarios concerning exposure to a chemical or treatment strategies with a drug can be run quickly and inexpensively. Since PBPKs can be constructed for both animals and humans, cross-species extrapolation is facilitated. There are, however, significant limitations in relying solely on PBPK models. The primary limitation is that a PBPK can only provide a response based on assumed mechanisms. Secondary and unexpected effects are not included. A further limitation is the difficulty in estimating parameters, particularly kinetic parameters.

None of these alternatives satisfactorily predict human response to chemicals or drugs.

97.2 The Cell Culture Analog Concept

A CCA is a physical replica of the structure of a PBPK where cells or engineered tissues are used in organ compartments to achieve the metabolic and biochemical characteristics of the animal. Cell culture medium circulates between compartments and acts as a "blood surrogate". Small-scale bioreactors with the appropriate cell types in the physical device represent organs or tissues.

The CCA concept combines attributes of a PBPK and other *in vitro* systems. Unlike other *in vitro* systems, the CCA is an integrated system that can mimic dose dynamics and allows for conversion of a parent compound into metabolites and the interchange of metabolites between compartments. A CCA system allows dose exposure scenarios that can replicate the exposure scenarios of animal studies.

A CCA is intended to work in conjunction with a PBPK as a tool to test and refine mechanistic hypotheses. A molecular model can be embedded in a tissue model which is embedded in the PBPK. Thus, the molecular model is related to overall metabolic response. The PBPK can be made an exact replica of the CCA; the predicted response and measured CCA response should exactly match if the PBPK contains a complete and accurate description of the molecular mechanisms. In the CCA all flow rates, the number of cells in each compartment, and the levels of each enzyme can be measured independently, so no adjustable parameters are required. If the PBPK predictions and CCA results disagree, then the description of the molecular mechanisms is incomplete. The CCA and PBPK can be used in an iterative manner to test modifications in the proposed mechanism. When the PBPK is extended to describe the whole animal, failure to predict animal response would be due to inaccurate description of transport (particularly within an organ), inability to accurately measure kinetic parameters (e.g., *in vivo* enzyme levels or activities), or the presence *in vivo* or metabolic activities not present in the cultured cells or tissues. Advances in tissue engineering will provide tissue constructs to use in a CCA that will display more aúthentic metabolism than isolated cell cultures.

The goal is predicting *human* pharmacological response to drugs or assessing risk due to chemical exposure. A PBPK that can make an accurate prediction of both animal CCA and animal experiments would be "validated". If we use the same approach to constructing a human PBPK and CCA for the same compound, then we would have a rational basis to extrapolate animal response to predict human response when human experiments would be inappropriate. Also, since the PBPK is mechanistically based, it would provide a basis for extrapolation to low doses. The CCA/PBPK approach complements animal studies by potentially providing an improved basis for extrapolation to humans.

Further, PBPKs validated as described above provide a basis for prediction of human response to mixtures of drugs or chemicals. Drug and chemical interactions may be synergistic or antagonistic. If a PBPK for compound A and a PBPK for compound B are combined, then the response to any mixture of A and B should be predictable since the mechanisms for response to both A and B are included.

97.3 Prototype CCA

A simple three-component CCA mimicking rodent response to a challenge by naphthalene has been tested by Sweeney et al. [1995]. While this prototype system did not fulfill the criteria for a CCA of physically realistic organ residence times or ratio of cell numbers in each organ, it did represent a multi-compartment system with fluid recirculation. The three components were liver, lung, and other perfused tissues. These experiments used cultured rat hepatoma (H4IIE) cells for the liver and lung (L2) cells for the lung compartment. No cells were required in "other tissues" in this model since no metabolic reactions were postulated to occur elsewhere for naphthalene or its metabolites. The H4IIE cells contained enzyme systems for activation of naphthalene (cytochrome P450IA1) to the epoxide form and conversion to dihydriol (epoxide hydrolase) and conjugation with glutathione (glutathione-S-transferase). The L2 cells had no enzymes for naphthalene activation. Cells were cultured in glass vessels as monolayers. Experiments with this system using lactate dihydrogenase release (LDH) and glutathione levels as dependent parameters supported a hypothesis where naphthalene is activated in the "liver" and reactive metabolites circulate to the "lung" causing glutathione depletion and cell death as measured by LDH release. Increasing the level of cytochrome p450 activity in the "liver" by increasing cell numbers or by preinducing H4IIE cells led to increased death of L2 cells. Experiments with "liver"-blank; "lung"-"lung", and "lung"-blank combinations all supported the hypothesis of a circulating reactive metabolite as the cause of L2 cell death.

This prototype system [Sweeney et al., 1995] was difficult to operate, very non-physiologic, and made time course experiments very difficult. An alternative system using packed bed reactors for the "liver" and "lung" compartments has been tested [Shuler et al., 1996; Ghanem, 1998]. This system successfully allowed time course studies, was more compact and simpler to operate, and was physiological with respect to the ratio of "liver" to "lung" cells. While liquid residence times improved in this system, they still were not physiologic (i.e., 114s vs. an *in vivo* value of 21s in the liver and 6.6s vs. *in vivo* lung value of about 1.5s) due to physical limitations on flow through the packed beds. Unlike the prototype system, no response to naphthalene was observed.

This difference in response of the two CCA designs was explained through the use of PBPK models of each CCA. In the prototype system, the large liquid residence times in the liver (6 min) and the lung (2.1 min) allowed formation of large amounts of naphthol from naphthalene oxide and presumably the conversion of napthol into quinones that were toxic. In the packed bed system, liquid residence times were sufficiently small so that the predicted naphthol level was negligible. Thus, the PBPK provided a mechanistic basis to explain the differences in response of the two experimental configurations.

Using a very simple CCA, Mufti and Shuler [1998] demonstrated that response of human hepatoma (HepG2) to exposure to dioxin (2,3,7,8-tetrachlorodibenzo-p-dioxin) is dependent on how the dose is delivered. The induction of cytochrome p450IA1 activity was used as a model response for exposure to dioxin. Data were evaluated to estimate dioxin levels giving cytochrome P450IA1 activity 0.01% of maximal induced activity. Such an analysis mimics the type of analysis used to estimate risk due to chemical exposure. The "allowable" dioxin concentration was 4×10^{-3} nM using a batch spinner flask, 4×10^{-4} nM using a one-compartment system with continuous feed, and 1×10^{-5} nM using a simple two-compartment CCA. Further, response could be correlated to an estimate of the amount of dioxin bound to the cytosolic Ah receptor with a simple model for two different human hepatoma cell lines. This work illustrates the potential usefulness of a CCA approach in risk assessment.

Ma et al. [1997] have discussed an *in vitro* human placenta model for drug testing. This was a two-compartment perfusion system using human trophoblast cells attached to a chemically modified polyethylene terephatholate fibrous matrix as a cell culture scaffold. This system is a CCA in the same sense as the two-compartment system used to estimate response to dioxin.

The above examples are the first that attempt to mimic circulation and metabolic response to model an animal as an integrated system. However, others have used engineered tissues as a basis for testing the efficacy of drugs or toxicity of chemicals. These tissues are important in themselves and could become elements in an integrated CCA.

97.4 Use of Engineered Tissues or Cells for Toxicity/Pharmacology

The primary use of engineered tissues for toxicity testing has been with epithelial cells that mimic the barrier properties of the skin, gut, or blood brain barrier. The use of isolated cell cultures has been of modest utility due to artifacts introduced by dissolving test agents in medium and due to the extreme sensitivity of isolated cells to these agents compared to *in vivo* tissue. One of the first reports on the use of engineered cells is that by Gay et al. [1992] reporting on the use of a living skin equivalent as an *in vitro* dermatotoxicity model. The living skin equivalent consists of a co-culture of human dermal fibroblasts in a collagen-containing matrix overlaid with human keratinocytes that have formed a stratified epidermis. This *in vitro* model used a measurement of mitochondrial function (i.e., the colormetric thiazolyl blue assay) to determine toxicity. Eighteen chemicals were tested. Eleven compounds classified as non-irritating had minimal or no effect on mitochondrial activity. For seven known human skin irritants, the concentration that inhibited mitochondrial activity by 50% corresponded to the threshold value for each of these compounds to cause irritation on human skin. However, living skin equivalents did not fully mimic the barrier properties of human skin; water permeability was 30-fold greater in the living skin equivalent than in human skin. Kriwet and Parenteau [1996] report on permeabilities of 20 different compounds in *in vitro* skin models. Skin cultures are slightly more permeable (two- or threefold) for highly lipophilic substances and considerably more permeable (about tenfold) for polar substances than human-cadaver or freshly excised human skin.

The above tests are static. Pasternak and Miller [1996] have tested a system combining perfusion and a tissue model consisting of MDCK (Madin-Darby canine kidney) epithelial cells cultured on a semi-porous cellulose ester membrane filter. The system could be fully automated using measurement of trans-epithelial electrical resistance (TER) as an end point. A decrease in TER is an indicator of cell damage. The system was tested using nonionic surfactants and predicted the relative occular toxicity of these compounds. The perfusion system mimics some dose scenarios (e.g., tearing) more easily than a static system and provides a more consistent environment for the cultured cells. The authors cite as a major advantage that the TER can be measured throughout the entire exposure protocol without physically disturbing the tissue model and introducing artifacts in the response.

Probably the most used cell-based assay is the Caco-2 model of the intestine to determine oral availability of a drug or chemical. The Caco-2 cell cultures are derived from a human colon adenocarcinoma cell line. The cell line, C2Bbel, is a clonal isolate of Caco-2 cells that is more homogeneous in apical brush border expression than the Caco-2 cell line. These cells form a polarized monolayer with an apical brush border morphologically comparable to the human colon. Tight junctions around the cells act to restrict passive diffusion by the paracellular route mimicking the transport resistance in the intestine. Hydrophobic solutes pass primarily by the transcellular route and hydrophilic compounds by the paracellular route. Yu and Sinko [1997] have demonstrated that the substratum (e.g., membrane) properties upon which the monolayer forms can become important in estimating the barrier properties of such *in vitro* systems. The barrier effects of the substratum need to be separated from the intrinsic property of the monolayers. Further, Anderle et al. [1998] have shown that the chemical nature of substratum and other culture conditions can alter transport properties. Sattler et al. [1997] provide one example (with hypericin) of how this model system can be used to evaluate effects of formulation (e.g., use of cyclodextrin or liposomes) on oral bioavailability. Another example is the application of the Caco-2 system to transport of paclitaxel across the intestine [Walle and Walle, 1998]. Rapid passive transport was partially counter-balanced by an efflux pump (probably P-glycoprotein) limiting oral bioavailability.

Another barrier of intense interest for drug delivery is the blood-brain barrier. The blood-brain barrier is formed by the endothelial cells of the brain capillaries. A primary characteristic is the high resistance of the capillary due to the presence of complex tight junctions inhibiting paracellular transport and the low endocytic activity of this tissue. Acceptable *in vitro* models have been more difficult to formulate but one commercial system marketed by Cellco, Inc. is available that uses a co-culture of endothelial cells and astrocytes to form a barrier.

A recent example of a different *in vitro* system is described by Glynn and Yazdanian [1998] who used bovine brain microvessel endothelial cells grown on porous polycarbonate filters to compare the transport of nevirapine, a reverse transcriptase inhibitor to other HIV antiretroviral agents. Nevirapine was the most permeable antiretroviral agent and hence may have value in HIV treatment in reducing levels of HIV in the brain.

These isolated cultures mimic an important aspect of cell physiology (oral uptake or transport into the brain). In principle, they could be combined with other tissue mimics to form a CCA that would be especially useful in testing pharmaceuticals.

97.5 Future Prospects

Significant progress is being made in the construction of tissue engineered constructs [Baldwin and Saltzman, 1996]. These efforts include highly vascularized tissues, which are especially important for toxicological or pharmacological studies in the liver. Liver constructs, often based on co-cultures, have become increasingly more normal in behavior (see [Griffith et al., 1997]). For toxicological studies multiple test systems operated in parallel are desirable which suggests the need for miniaturization to conserve cells, reagents, and time. Bhatia et al. [1998] have described a microfabricated device for hepatocyte-fibroblast co-cultures.

A CCA based on the concepts described here and incorporating these advanced-engineered tissues could become a powerful tool for preclinical testing of pharmaceuticals. While drug leads are expanding rapidly in number, the capacity to increase animal and human clinical studies is limited. It is imperative that preclinical testing and predictions for human response become more accurate. A CCA should become an important tool in preclinical testing.

Defining Terms

Animal surrogate: A physiologically based cell or tissue multi-compartmented device with fluid circulation to mimic metabolism and fate of a drug or chemical.

Engineered tissues: Cell culture mimic of a tissue or organ; often combines a polymer scaffold and one or more cell types.

Physiologically based pharmacokinetic model (PBPK): A computer model that replicates animal physiology by subdividing the body into a number of anatomical compartments, each compartment interconnected through the body fluid systems; used to describe the time-dependent distribution and disposition of a substance.

Tissue slice: A living organ is sliced into thin sections for use in toxicity studies; one primary organ can provide material for many tests.

References

Anderle P, Niederer E., Werner R., Hilgendorf C, Spahn-Langguth H, Wunderu-Allenspach H, Merkle HP, Langguth P. 1998. P-Glycoprotein (P-gp) mediated efflux in Caco-2 cell monolayers: the influence of culturing conditions and drug exposure on P-gp expression levels. J. Pharm. Sci. 87:757

Bhatia SN, Balis UJ, Yarmush ML, Toner M. 1998. Microfabrication of hepatocyte/fibroblast co-cultures: role of homotypic cell interactions. Biotechnol. Prog. 14:378

Baldwin SP, Saltzman WM. 1996. Polymers for tissue engineering. Trends Polym. Sci. 4:177

Connolly RB, Andersen ME. 1991. Biologically based pharmacodynamic models: tool for toxicological research and risk assessment. Annu. Rev. Pharmacol. Toxicol. 31:503

Del Raso NJ. 1993. *In vitro* methodologies for enhanced toxicity testing. Toxicol. Lett. 68:91

Gay R, Swiderek M, Nelson D, Ernesti A. 1992. The living skin equivalent as a model *in vitro* for ranking the toxic potential of dermal irritants. Toxic. *In Vitro* 6:303.

Ghanem A. 1998. Application of a novel packed bed cell culture analog bioreactor and corresponding pharmacokinetic model to naphthalene toxicology. Ph.D. Thesis. Cornell University, Ithaca, New York.

Glynn SL, Mehran Y. 1998. *In vitro* blood-brain barrier permeability of nevirapine compared to other HIV antiretroviral agents. J. Pharm. Sci. 87:306.

Griffith LG, Wu B, Cima MJ, Powers M, Chaignaud B, Vacanti JP. 1997. *In vitro* orgarogensis of vacularized liver tissue. Ann. N.Y. Acad. Sci. 831:382.

Gura T. 1997. Systems for identifying new drugs are often faulty. Science 273:1041

Kirwet K, Parenteau NL. 1996. *In vitro* skin models. Cosmetics & Toiletries 111 (Feb):93

Ma T, Yang S-T, Kniss DA. 1997. Development of an *in vitro* human placenta model by the cultivation of human trophoblasts in a fiber-based bioreactor system. Am. Inst. Chem. Eng. Ann. Mtg., Los Angeles, Ca, Nov. 16-21.

Mufti NA, Shuler ML. 1998. Different *in vitro* systems affect CYPIA1 activity in response to 2,3,7,8-tetrachlorodibenzo-p-dioxin. Toxicol. in Vitro 12:259.

Olinga, P, Meijer DKF, Slooff, MJH, Groothuis, GMM. 1997. Liver slices in *in vitro* pharmacotoxicology with special reference to the use of human liver tissue. Toxicol. *In Vitro* 12:77.

Pasternak AS, Miller WM. 1996. Measurement of trans-epitheial electrical resistance in perfusion: potential application for *in vitro* ocular toxicity testing. Biotechnol. Bioeng. 50:568.

Sattler S, Schaefer U, Schneider W, Hoelzl J, Lehr C-M. 1997. Binding, uptake, and transport of hypericin by Caco-2 cell monolayers. J. Pharm. Sci. 86:1120.

Shuler ML, Ghanem A, Quick D, Wong MC, Miller P. 1996. A self-regulating cell culture analog device to mimic animal and human toxicological responses. Biotechnol. Bioeng. 52:45.

Sweeney LM, Shuler ML, Babish JG, Ghanem A. 1995. A cell culture analog of rodent physiology: application to naphthalene toxicology. Toxicol. *In Vitro* 9:307.

Walle UK, Walle T. Taxol transport by human intestinal epithelial Caco-2 cells. Drug Metabol. Disposit. 26:343.

Yu H, Sinko PJ. 1997. Influence of the microporous substratum and hydrodynamics on resistances to drug transport in cell culture systems: calculation of intrinsic transport parameters. J. Pharm. Sci. 86:1448.

98

Microvascular Heat Transfer

James W. Baish
Bucknell University

98.1 Introduction and Conceptual Challenges

Models of microvascular heat transfer are useful for optimizing thermal therapies such as hyperthermia treatment, for modeling thermoregulatory response at the tissue level, for assessing environmental hazards that involve tissue heating, for using thermal means of diagnosing vascular pathologies, and for relating blood flow to heat clearance in thermal methods of blood perfusion measurement. For example, the effect of local hyperthermia treatment is determined by the length of time that the tissue is held at an elevated temperature, nominally 43°C or higher. Since the tissue temperature depends on the balance between the heat added by artificial means and the tissue's ability to clear that heat, an understanding of the means by which the blood transports heat is essential. This chapter outlines the general problems associated with such processes, while more extensive reviews and tutorials on microvascular heat transfer may be found elsewhere [1-4].

The temperature range of interest for all of these applications is intermediate between freezing and boiling, making only sensible heat exchange by conduction and convection as important mechanisms of heat transfer. At high and low temperatures, such as those present during laser ablation or electrocautery and cryopreservation or cryosurgery, the change of phase and accompanying mass transport present problems beyond the scope of this section. See, for example, [5].

While the equations that govern heat transport are formally similar to those that govern passive mass transport, the processes involve the microvasculature in fundamentally different ways because the thermal diffusivity of most tissues is roughly two orders of magnitude greater than the diffusivity for mass transport of most mobile species (1.5×10^{-7} m^2/s for heat vs. 1.5×10^{-9} m^2/s for O_2). Mass transport is largely restricted to the smallest blood vessels, the capillaries, arterioles, and venules, whereas heat transport occurs in somewhat larger, so-called thermally significant blood vessels with diameters in the range from 80 μm to 1 mm. The modes of heat transport differ from those of mass transport, not simply

0-8493-0462-8/00/$0.00+$.50

because these vessels are larger, but because they have a different geometrical arrangement than the vessels primarily responsible for mass transport. Many capillary beds approximate a uniformly spaced array of parallel vessels that can be well modeled by the Krogh cylinder model. In contrast, the thermally significant vessels are in a tree-like arrangement that typically undergoes several generations of branching within the size range of interest and are often found as countercurrent pairs in which the artery and vein may be separated by one vessel diameter or less. Moreover, the vascular architecture of the thermally significant vessels is less well characterized than that of either the primary mass exchange vessels or the larger, less numerous supply vessels that carry blood over large distances in the body. There are too few supply vessels to contribute much to the overall energy balance in the tissue, but they are often far from thermal equilibrium with the surrounding tissue producing large local perturbations in the tissue temperature. Much of the microvascular heat exchange occurs as the blood branches from the larger supply vessels into the more numerous and densely spaced thermally significant blood vessels.

While the details of the vascular architecture for particular organs have been well characterized in individual cases, variability among individuals makes the use of such published data valid only in a statistical sense. Imaging technology to map the thermally significant blood vessels in any given individual is not readily available.

An additional challenge arises from the spatial and temporal variability of the blood flow in tissue. The thermoregulatory system and the metabolic needs of tissues can change the blood perfusion rates in some tissues by a factor as great as 15 to 25.

98.2 Basic Concepts

For purposes of thermal analysis, vascular tissues are generally assumed to consist of two interacting subvolumes, a solid tissue subvolume and a blood subvolume which contains flowing blood. These subvolumes thermally interact through the walls of the blood vessels where heat, but little mass, is exchanged. Because the tissue subvolume can transport heat by conduction alone, it may be modeled by the standard heat diffusion equation [6]

$$\nabla \cdot k_t \nabla T_t(\vec{r},t) + \dot{q}_t'''(\vec{r},t) = \rho_t c_t \frac{\partial T_t(\vec{r},t)}{\partial t} \qquad (98.1)$$

where T_t is the local tissue temperature, k_t is the thermal conductivity of the tissue, $\dot{q}_t'''$ is the rate of volumetric heat generation from metabolism or external source, ρ_t is the tissue density, and c_t is the tissue specific heat. The properties used in Eq. (98.1) may be assumed to be bulk properties that average over the details of the interstitial fluid, extracellular matrix, and cellular content of the tissue. In the blood subvolume, heat may also be transported by advection which adds a blood velocity dependent term as given by [6]

$$\nabla \cdot k_b \nabla T_b(\vec{r},t) - \rho_b c_b \vec{u}_b(\vec{r},t) \cdot \nabla T_b(\vec{r},t) + \dot{q}_b'''(\vec{r},t) = \rho_b c_b \frac{\partial T_b(\vec{r},t)}{\partial t} \qquad (98.2)$$

where $\vec{u}_b$ is the local blood velocity and all other parameters pertain to the local properties of the blood. Potential energy, kinetic energy, pulsatility, and viscous dissipation effects are normally taken to be negligible.

At the internal boundary on the vessel walls we expect a continuity of heat flux $k_b \nabla T_b(\vec{r}_w, t) = k_t \nabla T_t(\vec{r}_w, t)$ and temperature $T_b(\vec{r}_w, t) = T_t(\vec{r}_w, t)$ where $\vec{r}_w$ represents points on the vessel wall. Few attempts have been made to solve Eqs. (98.1) and (98.2) exactly, primarily due to the complexity of the vascular architecture and the paucity of data on the blood velocity field in any particular instance. The sections that follow present approaches to the problem of microvascular heat transport that fall broadly into the categories of vascular models that consider the response of one or a few blood vessels to their immediate surroundings and continuum models that seek to average the effects of many blood vessels to obtain a single field equation that may be solved for a local average of the tissue temperature.

98.3 Vascular Models

Most vascular models are based on the assumption that the behavior of blood flowing in a blood vessel is formally similar to that of a fluid flowing steadily in a roughly circular tube (See Fig. 98.1), that is [7]

$$\pi r_a^2 \rho_b c_b \bar{u} \frac{d\bar{T}_a(s)}{ds} = q'(s) \tag{98.3}$$

where $\bar{T}_a$ is the mixed mean temperature of the blood for a given vessel cross section, r_a is the vessel radius, $\bar{u}$ is mean blood speed in the vessel, $q'(s)$ is the rate at which heat conducts into the vessel per unit length, and s is the spatial coordinate along the vessel axis. For a vessel that interacts only with a cylinder of adjacent tissue, we have

$$q'(s) = U' 2\pi r_a \left(\bar{T}_t(s) - \bar{T}_a(s)\right) \tag{98.4}$$

where U' is the overall heat transfer coefficient between the tissue and the blood. Typically, the thermal resistance inside the blood vessel is much smaller than that in the tissue cylinder so we may approximate $U' 2\pi r_a \approx k_t \sigma$ where the conduction shape factor σ relating local tissue temperature $\bar{T}_t(s)$ to the blood temperature is given by

$$\sigma \approx \frac{2\pi}{\ln\left(\frac{r_t}{r_a}\right)} \tag{98.5}$$

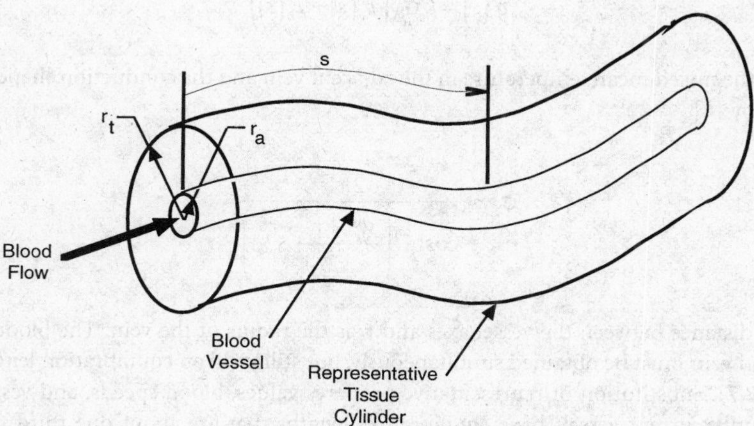

FIGURE 98.1 Representative tissue cylinder surrounding a blood vessel showing the radial and axial position coordinates.

Equilibration Lengths

One of the most useful concepts that arises from the simple vascular model presented above is the equilibration length L_e, which may be defined as the characteristic length over which the blood changes temperature from an inlet temperature $\overline{T}_{a_o}$ to eventually equilibrate with tissue at a constant temperature $\overline{T}_t$. The solution for Eqs. (98.3) and (98.4) under these conditions is given by

$$\frac{\overline{T}_a(s)-\overline{T}_t}{\overline{T}_{a_o}-\overline{T}_t}=\exp\left(-\frac{s}{L_e}\right) \tag{98.6}$$

where the equilibration length is given by

$$L_e=\frac{\pi r_a^2 \rho_b c_b \overline{u}}{k_t \sigma} \tag{98.7}$$

Chen and Holmes [8] found that vessels with diameters of about 175 μm have an anatomical length comparable to their thermal equilibration length, thus making vessels of this approximate size the dominant site of tissue-blood heat exchange. Accordingly, these vessels are known as the *thermally significant blood vessels.* Much smaller vessels, while more numerous, carry blood that has already equilibrated with the surrounding tissue. Much larger vessels, while not in equilibrium with the surrounding tissue, are too sparsely spaced to contribute significantly to the overall energy balance [7]. Even though the larger vessels do not exchange large quantities of heat with the tissue subvolume, they cannot be ignored because these vessels produce large local perturbations to the tissue temperature and form a source of blood for tissues that is at a much different temperature than the local tissue temperature.

Countercurrent Heat Exchange

Thermally significant blood vessels are frequently found in closely spaced countercurrent pairs. Only a slight modification to the preceding formulas is needed for heat exchange between adjacent arteries and veins with countercurrent flow [9]

$$q'(s)=k_t\sigma_\Delta\left(\overline{T}_v(s)-\overline{T}_a(s)\right) \tag{98.8}$$

where $\overline{T}_v(s)$ is the mixed mean temperature in the adjacent vein and the conduction shape factor is given approximately by [6]

$$\sigma_\Delta \approx \frac{2\pi}{\cosh^{-1}\left(\dfrac{w^2-r_a^2-r_v^2}{2r_a r_v}\right)} \tag{98.9}$$

where w is the distance between the vessel axes and r_v is the radius of the vein. The blood temperatures in the artery and vein must be obtained simultaneously, but still yield an equilibration length of the form given in Eq. (98.7). Substitution of representative property values, blood speeds, and vessel dimensions reveals that countercurrent vessels have equilibration lengths that are about one third that of isolated vessels of similar size [9]. Based on this observation, the only vessels that participate significantly in the overall energy balance in the tissue are those larger than about 50 μm in diameter.

TABLE 98.1 Shape Factors for Various Vascular Geometries

Geometry	Ref.
Single vessel to skin surface	[9]
Single vessel to tissue cylinder	[8]
Countercurrent vessel to vessel	[9, 40]
Countercurrent vessels to tissue cylinder	[24]
Countercurrent vessels with a thin tissue layer	[41]
Multiple parallel vessels	[42]
Vessels near a junction of vessels	[43]

Typical dimension of blood vessels are available in Chapter 1, Tables 1.3 and 1.4 of this handbook.

The shape factors given above are only rough analytical approximations that do not include the effects of finite thermal resistance within the blood vessels and other geometrical effects. The reader is referred to Table 98.1 for references that address these issues.

98.4 Heat Transfer Inside of a Blood Vessel

A detailed analysis of the heat transfer between the blood vessel wall and the mixed mean temperature of the blood can be done using standard heat transfer methods

$$q'(s) = h\pi d\left(T_w(s) - T_b\right) \tag{98.10}$$

where d is the vessel diameter, $T_w(s)$ is the vessel wall temperature, and the convective heat transfer coefficient h may be found from Victor and Shah's [10] recommendation that

$$\overline{Nu}_D = \frac{hd}{k_b} = 4 + 0.155\exp\left(1.58\log_{10} Gz\right) \quad Gz < 10^3 \tag{98.11}$$

where Gz is the Graetz number defined as

$$Gz = \frac{\rho_b c_b \bar{u} d^2}{k_b L} \tag{98.12}$$

where L is the vessel length. See also Barozzi and Dumas [11].

98.5 Models of Perfused Tissues

Continuum Models

Continuum models of microvascular heat transfer are intended to average over the effects of many vessels so that the blood velocity field need not be modeled in detail. Such models are usually in the form of a modified heat diffusion equation in which the effects of blood perfusion are accounted for by one or more additional terms. These equations then can be solved to yield a sort of moving average of the local temperature that does not give the details of the temperature field around every individual vessel, but provides information on the broad trends in the tissue temperature. The temperature they predict may be defined as

TABLE 98.2 Representative Thermal Property Values

Tissue	Thermal Conductivity (W/m-K)	Thermal Diffusivity (m^2/s)	Perfusion (m^3/m^3-sec)
Aorta	0.461 [16]	1.25×10^{-7} [16]	
Fat of spleen	0.3337 [44]	1.314×10^{-7} [44]	
Spleen	0.5394 [44]	1.444×10^{-7} [44]	0.023 [45]
Pancreas	0.5417 [44]	1.702×10^{-7} [44]	0.0091 [45]
Cerebral cortex	0.5153 [44]	1.468×10^{-7} [44]	0.0067 [46]
Renal cortex	0.5466 [44]	1.470×10^{-7} [44]	0.077 [47]
Myocardium	0.5367 [44]	1.474×10^{-7} [44]	0.0188 [48]
Liver	0.5122 [44]	1.412×10^{-7} [44]	0.0233 [49]
Lung	0.4506 [44]	1.307×10^{-7} [44]	
Adenocarcinoma of breast	0.5641 [44]	1.436×10^{-7} [44]	
Resting muscle bone	0.478 [50]	1.59×10^{-7} [50]	0.0007 [48]
Whole blood (21°C)	0.492 [50]	1.19×10^{-7} [50]	
Plasma (21°C)	0.570 [50]	1.21×10^{-7} [50]	
Water	0.628 [6]	1.5136×10^{-7} [6]	

All conductivities and diffusivities are from humans at 37°C except the value for skeletal muscle which is from sheep at 21°C. Perfusion values are from various mammals as noted in the references. Significant digits do not imply accuracy. The temperature coefficient for thermal conductivity ranges from –0.000254 to 0.0039 W/m-K-°C with 0.001265 W/m-K-°C typical of most tissues as compared to 0.001575 W/m-K-°C for water [44]. The temperature coefficient for thermal diffusivity ranges from -4.9×10^{-10} m^2/s-°C to 8.4×10^{-10} m^2/s-°C with 5.19×10^{-10} m^2/s-°C typical of most tissues as compared to 4.73×10^{-10} m^2/s-°C for water [44]. The values provided in this table are representative values presented for tutorial purposes. The reader is referred to the primary literature for values appropriate for specific design applications.

$$\overline{T}_t(\vec{r},t) = \frac{1}{\delta V}\int_{\delta V} T_t(\vec{r}',t)\,dV' \tag{98.13}$$

where δV is a volume that is assumed to be large enough to encompass a reasonable number of thermally significant blood vessels, but much smaller than the scale of the tissue as a whole. Much of the confusion concerning the proper form of the bioheat equation stems from the difficulty in precisely defining such a length scale. Unlike a typical porous medium, such as water percolating through sand where the grains of sand fall into a relatively narrow range of length scales, blood vessels form a branching structure with length scales spaning many orders of magnitude.

Formulations

Pennes Heat Sink Model

In 1948, physiologist Harry Pennes modeled the temperature profile in the human forearm by introducing the assumptions that the primary site of equilibration was the capillary bed and that each volume of tissue has a supply of arterial blood that is at the core temperature of the body. The Pennes' Bioheat equation has the form [12]

$$\nabla \cdot k\nabla \overline{T}_t(\vec{r},t) + \dot{\omega}_b(\vec{r},t)\rho_b c_b\left(T_a - \overline{T}_t(\vec{r},t)\right) + \dot{q}'''(\vec{r},t) = \rho c\frac{\partial \overline{T}_t(\vec{r},t)}{\partial t} \tag{98.14}$$

where $\dot{\omega}_b$ is taken to be the blood perfusion rate in volume of blood per unit volume of tissue per unit time and T_a is an arterial supply temperature which is generally assumed to remain constant and equal to the core temperature of the body, nominally 37°C. The other thermal parameters are taken to be effective values that average over the blood and tissue subvolumes. Major advantages of this formulation are that it is readily solvable for constant parameter values, requires no anatomical data, and in the absence of independent measurement of the actual blood rate and heat generation rate gives two adjustable parameters ($\dot{\omega}_b(\vec{r},t)$ and T_a) that can be used to fit the majority of the experimental results available. On the downside, the model gives no prediction of the actual details of the vascular temperatures, the actual blood perfusion rate is usually unknown and not exactly equal to $\dot{\omega}_b$ that best fits the thermal data, the assumption of constant arterial temperature is not generally valid, and, based on the equilibration length studies presented in the previous section, thermal equilibration occurs prior to the capillary bed. Despite these weaknesses, the Pennes formulation is the primary choice of modelers. Equilibration prior to the capillary bed does not invalidate the model as long as the averaging volume is large enough to encompass many vessels of the size in which equilibration does occur and as long as the venous return does not exchange significant quantities of heat after leaving the equilibrating vessels. As long as $\dot{\omega}_b$ and T_a are taken as adjustable, curve-fitting parameters rather than literally as the perfusion rate and arterial blood temperature, the model may be used fruitfully, provided that the results are interpreted accordingly.

Directed Perfusion

Some of the shortcomings of the Pennes model were addressed by Wulff [13] in a formulation that is essentially the same as that used for common porous media

$$\nabla \cdot k\nabla \overline{T}_t(\vec{r},t) - \rho c\vec{u}(\vec{r},t)\cdot\nabla\overline{T}_t(\vec{r},t) + \dot{q}'''(\vec{r},t) = \rho c\frac{\partial \overline{T}_t(\vec{r},t)}{\partial t} \tag{98.15}$$

where $\vec{u}$ is a velocity averaged over both the tissue and blood subvolumes. Among the difficulties with this model are that it is valid only when the tissue and blood are in near-thermal equilibrium and when the averaging volume is small enough to prevent adjacent arteries and veins from canceling out their contributions to the average velocity, thus erroneously suggesting that the blood perfusion has no net effect on the tissue heat transfer. Eq. (98.15) is rarely applied in practical situations, but served as an important conceptual challenge to the Pennes formulation in the 1970s and 1980s.

Effective Conductivity Model

The oldest continuum formulation is the effective conductivity model

$$\nabla \cdot k_{eff}\nabla\overline{T}_t(\vec{r},t) + \dot{q}'''(\vec{r},t) = \rho_t c_t\frac{\partial \overline{T}_t(\vec{r},t)}{\partial t} \tag{98.16}$$

where the effective conductivity is comprised of the intrinsic thermal conductivity of the tissue and a perfusion dependent increment. In principle, an effective conductivity can be defined from any known heat flow and temperature difference, that is

$$k_{eff} = \frac{q}{\Delta T} f\left(\frac{L}{A}\right) \tag{98.17}$$

where $f\left(\frac{L}{A}\right)$ is a function of geometry with dimensions length^{-1} (for example, $\Delta x/A$ in a slab geometry). Originally introduced as an empirical quantity [14], the effective conductivity has been linked to the Pennes formulation in the measurement for blood perfusion rates via small, heated, implanted probes [15-17]. In 1985, Weinbaum and Jiji [18] theoretically related the effective conductivity to the blood flow and anatomy for a restricted class of tissues and heating conditions which are dominated by a closely spaced artery-vein architecture and which can satisfy the constraint [19]

$$\frac{d\overline{T}_t}{ds} \approx \frac{1}{2}\frac{d\left(\overline{T}_a + \overline{T}_v\right)}{ds} \tag{98.18}$$

Here the effective conductivity is a tensor quantity related to the flow and anatomy according to [18]

$$k_{eff} = k_t\left(1 + \frac{\pi^2 \rho_b^2 c_b^2 n r_a^4 \overline{u}^2 \cos^2\phi}{k_t^2 \sigma_\Delta}\right) \tag{98.19}$$

where the enhancement is in the direction of the vessel axes and where n is the number of artery-vein pairs per unit area and ϕ is the angle of the vessel axes relative to the temperature gradient. The near equilibrium between the tissue and the blood vessels needed for this model to be valid is likely to occur only in the outer few millimeters near the skin and when the volumetric heat source is neither intense nor localized. Artery-vein pairs have been shown to act like a single highly conductive fiber even when the near equilibrium condition in Eq. (98.18) is violated [20]. The radius of the thermally equivalent fiber is given by

$$r_{fiber} = \left(w r_a\right)^{1/2} \tag{98.20}$$

and its conductivity is given by

$$k_{fiber} = \frac{\left(\rho_b c_b \overline{u}\right)^2 r_a^3 \cosh^{-1}\left(w/r_a\right)}{w k_t} \tag{98.21}$$

Under these non-equilibrium conditions, the tissue-blood system acts like a fiber-composite material, but cannot be well modeled as a single homogeneous material with effective properties.

Combination

Recognizing that several mechanisms of heat transport may be at play in tissue, Chen and Holmes [8] suggested the following formulation which incorporates the effects discussed above

$$\nabla \cdot k_{eff}\left(\vec{r},t\right)\nabla\overline{T}_t\left(\vec{r},t\right) + \dot{\omega}_b\left(\vec{r},t\right)\rho_b c_b\left(T_a^* - \overline{T}_t\left(\vec{r},t\right)\right) - \rho_b c_b \vec{u}_b\left(\vec{r},t\right)\cdot\nabla\overline{T}_t\left(\vec{r},t\right) + \dot{q}'''\left(\vec{r},t\right) = \rho_t c_t \frac{\partial\overline{T}_t\left(\vec{r},t\right)}{\partial t} \tag{98.22}$$

where T_a^* is the temperature exiting the last artery that is individually modeled. The primary value of this formulation is its conceptual generality. In practice, this formulation is difficult to apply because it requires knowledge of a great many adjustable parameters, most of which have not been independently measured to date.

Heat Sink Model with Effectiveness

Using somewhat different approaches, Brinck and Werner [21] and Weinbaum et al. [22] have proposed that the shortcomings of the Pennes model can be overcome by introducing a heat transfer effectiveness factor ε to modify the heat sink term as follows:

$$\nabla \cdot k_t \nabla\overline{T}_t\left(\vec{r},t\right) + \varepsilon\left(\vec{r},t\right)\dot{\omega}_b\left(\vec{r},t\right)\rho_b c_b\left(T\left(\vec{r},t\right)_t - T_a\right) + \dot{q}'''\left(\vec{r},t\right) = \rho_t c_t \frac{\partial\overline{T}_t\left(\vec{r},t\right)}{\partial t} \tag{98.23}$$

where $0 \le \varepsilon \le 1$. In the Brinck and Werner formulation, ε is a curve-fitting parameter that allows the actual (rather than the thermally equivalent) perfusion rate to be used [21]. Weinbaum et al. provide an analytical result for ε that is valid for blood vessels smaller than 300 μm diameter in skeletal muscle [22]. In both formulations, $\varepsilon < 1$ arises from the countercurrent heat exchange mechanism that shunts heat directly between the artery and vein without requiring the heat-carrying blood to first pass through the smaller connecting vessels. Typical values for the effectiveness are in the range $0.6 \le \varepsilon \le 0.8$ [22].

Multi-Equation Models

The value of the continuum models is that they do not require a separate solution for the blood subvolume. In each continuum formulation, the behavior of the blood vessels is modeled by introducing assumptions that allow solution of only a single differential equation. But by solving only one equation, all detailed information on the temperature of the blood in individual blood vessels is lost. Several investigators have introduced multi-equation models that typically model the tissue, arteries, and veins as three separate, but interacting, subvolumes [9, 23-27]. As with the other non-Pennes formulations, these methods are difficult to apply to particular clinical applications, but provide theoretical insights into microvascular heat transfer.

Vascular Reconstruction Models

As an alternative to the three equation models, a more complete reconstruction of the vasculature may be used along with a scheme for solving the resulting flow, conduction, and advection equations [28-33]. Since the reconstructed vasculature is similar to the actual vasculature only in a statistical sense, these models provide the mean temperature as predicted by the continuum models, as well as insight into the mechanisms of heat transport, the sites of thermal interaction, and the degree of thermal perturbations produced by vessels of a given size, but they cannot provide the actual details of the temperature field in a given living tissue. These models tend to be computationally intensive due to the high spatial resolution needed to account for all of the thermally significant blood vessels.

98.6 Parameter Values

Thermal Properties

The intrinsic thermal properties of tissues depend strongly on their composition. Cooper and Trezek [34] recommend the following correlations for thermal conductivity:

$$k = \rho x 10^{-3}\left(0.628\, f_{water} + 0.117\, f_{proteins} + 0.231\, f_{fats}\right) W/m-K; \tag{98.24}$$

specific heat:

$$c_p = 4{,}200\, f_{water} + 1{,}090\, f_{proteins} + 2{,}300\, f_{fats}\ J/kg-K; \tag{98.25}$$

and density:

$$\rho = \frac{1}{f_{water}/1{,}000 + f_{proteins}/1{,}540 + f_{fats}/815}\ kg/m^3 \tag{98.26}$$

where f_{water}, $f_{proteins}$, and f_{fats} are the mass fractions of water, proteins, and fats, respectively.

Thermoregulation

Humans maintain a nearly constant core temperature through a combination of physiological and behavioral responses to the environment. For example, heat loss or gain at the skin surface may be modified by changes in the skin blood flow, the rate of sweating, or clothing. In deeper tissues, the dependence of the blood perfusion rate, the metabolic heat generation rate, and vessel diameters depend on the environmental and physiological conditions in a complex, organ-specific manner. The blood perfusion varies widely among tissue types and for some tissues can change dramatically depending on the metabolic or thermoregulatory needs of the tissue. The situation is further complicated by the feedback control aspects of the thermoregulatory systems that utilize a combination of central and peripheral temperature sensors as well as local and more distributed actuators.

The following examples are provided to illustrate some of the considerations, not to exhaustively explore this complicated issue. A model of the whole body is typically needed even for a relatively local stimulus, especially when the heat input represents a significant fraction of the whole body heat load. The reader is referred to extensive handbook entries on environmental response for more information [35, 36]. Whole body models of the thermoregulatory system are discussed in Wissler [37].

Chato [1] suggests that the temperature dependence of the blood perfusion effect can be approximated by a scalar effective conductivity

$$k_{eff} = 4.82 - 4.44833\left[1.00075^{-1.575^{(T_t-25)}}\right] W/m-K \tag{98.27}$$

which is intended for use in Eq. (98.16).

Under conditions of local hyperthermia, where the heated volume is small compared to the body as a whole, the blood perfusion rate in skin and muscle may increase by roughly nine- or tenfold as the tissue warms to about 44°C and then drop again for higher temperatures. In contrast, tumors, the typical target of local hyperthermia, thermoregulate erratically and undergo a lesser change in blood flow up to 42°C and then may drop to near zero for higher temperatures [38].

The metabolic rate may also undergo thermoregulatory changes. For example, the temperature dependence of the metabolism in the leg muscle and skin may be modeled with [39]

$$\dot{q}_m''' = 170(2)^{\left[(T_o - T_t)/10\right]} W/m^3 \tag{98.28}$$

The metabolic rate and blood flow may also be linked through processes that reflect the fact that sustained increased metabolic activity generally requires increased blood flow.

Clinical Heat Generation

Thermal therapies such as hyperthermia treatment rely on local heat generation rates several orders of magnitude greater than that produced by metabolism. Under these circumstances, the metabolic heat generation is often neglected with little error.

98.7 Solutions of Models

The steady-state solution with spatially and temporally constant parameter values including the rate of heat generation for a tissue half-space with a fixed temperature on the skin T_{skin} is given by

$$\overline{T}_t(x) = T_{skin}\exp\left[-\left(\frac{\dot{\omega}_b\rho_b c_b}{k_t}\right)^{1/2} x\right] + \left(T_a + \frac{\dot{q}'''}{\dot{\omega}_b\rho_b c_b}\right)\left\{1 - \exp\left[-\left(\frac{\dot{\omega}_b\rho_b c_b}{k_t}\right)^{1/2} x\right]\right\} \tag{98.29}$$

This solution reveals that perturbations to the tissue temperature decay exponentially with a characteristic length of

$$L_c = \left(\frac{k_t}{\dot{\omega}_b \rho_b c_b} \right)^{1/2} \tag{98.30}$$

which for typical values of the perfusion rate $\dot{\omega}_b = 0.1 \times 10^{-3}$ to 3.0×10^{-3} m³/m³-sec yields $L_c = 6.5 \times 10^{-3}$ to 36×10^{-3} m (Fig. 98.2).

The transient solution of Pennes' bioheat equation with constant perfusion rate for an initial uniform temperature of T_o, in the absence of any spatial dependence, is

$$\overline{T}_t(t) = T_o \exp\left[-\left(\frac{\dot{\omega}_b \rho_b c_b}{\rho_t c_t} \right) t \right] + \left(T_a + \frac{\dot{q}'''}{\dot{\omega}_b \rho_b c_b} \right) \left\{ 1 - \exp\left[-\left(\frac{\dot{\omega}_b \rho_b c_b}{\rho_t c_t} \right) t \right] \right\} \tag{98.31}$$

Here the solution reveals a characteristic time scale

$$t_c = \frac{\rho_t c_t}{\dot{\omega}_b \rho_b c_b} \tag{98.32}$$

that has typical values in the range of $t_c = 300$ to 10,000 sec (Fig. 98.3). This solution is valid only when thermoregulatory changes in the perfusion rate are small or occur over a much longer time than the characteristic time scale t_c.

Numerical solution of the heat sink model is readily obtained by standard methods such as finite differences, finite element, boundary element, and Green's functions provided that the parameter values and appropriate boundary conditions are known.

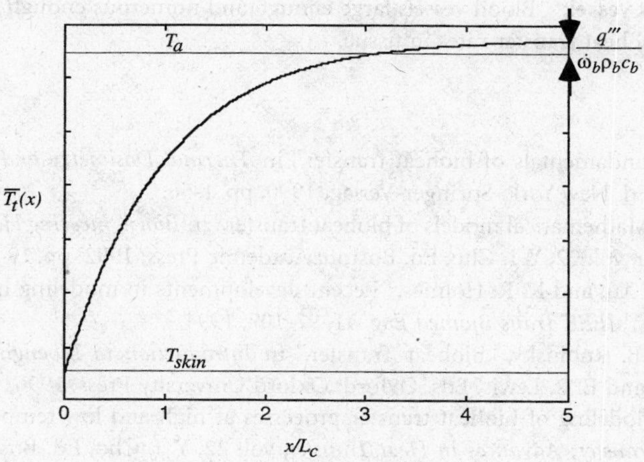

FIGURE 98.2 One-dimensional steady-state solution of Pennes bioheat equation for constant parameter values.

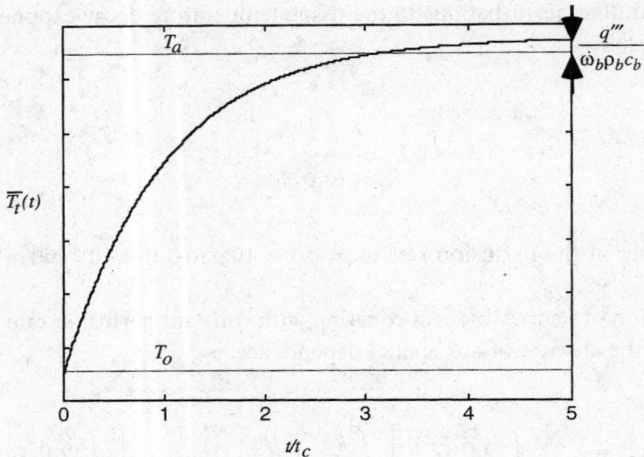

FIGURE 98.3 Transient solution of Pennes bioheat heat equation for constant parameter values in the absence of spatial effects.

Defining Terms

Conduction shape factor: Dimensionless factor used to account for the geometrical effects in steady-state heat conduction between surfaces at different temperatures.

Effective conductivity: A modified thermal conductivity that includes the intrinsic thermal conductivity of the tissue as well as a contribution from blood perfusion effects.

Equilibration length: Characteristic length scale over which blood in a blood vessel will change temperature in response to surrounding tissue at a different temperature.

Perfusion rate: Quantity of blood provided to a unit of tissue per unit time.

Specific heat: Quantity of energy needed for a unit temperature increase for a unit of mass.

Thermal conductivity: Rate of energy transfer by thermal conduction for a unit temperature gradient per unit of cross-sectional area.

Thermally significant vessel: Blood vessels large enough and numerous enough to contribute significantly to overall heat transfer rates in tissue.

References

1. J. C. Chato, "Fundamentals of bioheat transfer," in *Thermal Dosimetry and Treatment Planning*, M. Gautherie, Ed. New York: Springer-Verlag, 1990, pp. 1-56.
2. C. K. Charny, "Mathematical models of bioheat transfer," in *Bioengineering Heat Transfer: Advances in Heat Transfer*, vol. 22, Y. I. Cho, Ed. Boston: Academic Press, 1992, pp. 19-155.
3. M. Arkin, L. X. Xu, and K. R. Holmes, "Recent developments in modeling heat transfer in blood perfused tissues," *IEEE Trans Biomed Eng*, 41, 97-107, 1994.
4. T. K. Eto and B. Rubinsky, "Bioheat Transfer," in *Introduction to Bioengineering*, S. A. Berger, W. Goldsmith, and E. R. Lewis, Eds. Oxford: Oxford University Press, 1996, pp. 203-227.
5. K. R. Diller, "Modeling of bioheat transfer processes at high and low temperatures," in *Bioengineering Heat Transfer: Advances in Heat Transfer*, vol. 22, Y. I. Cho, Ed. Boston: Academic Press, 1992, pp. 157-357.
6. F. P. Incropera and D. P. DeWitt, *Fundamentals of Heat and Mass Transfer*, 4th ed. New York: John Wiley & Sons, 1996.
7. J. C. Chato, "Heat transfer to blood vessels," *J Biomechan Eng*, 102, 110-118, 1980.

8. M. M. Chen and K. R. Holmes, "Microvascular contributions in tissue heat transfer," *Ann NY Acad Sci*, 325, 137-50, 1980.
9. S. Weinbaum, L. M. Jiji, and D. E. Lemons, "Theory and experiment for the effect of vascular microstructure on surface tissue heat transfer-part I: anatomical foundation and model conceptualization," *J Biomech Eng*, 106, 321-330, 1984.
10. S. A. Victor and V. L. Shah, "Steady state heat transfer to blood flowing in the entrance region of a tube," *Int J Heat Mass Trans*, 19, 777-783, 1976.
11. G. S. Barozzi and A. Dumas, "Convective heat transfer coefficients in the circulation," *J Biomech Eng*, 113, 308-313, 1991.
12. H. H. Pennes, "Analysis of tissue and arterial blood temperatures in the resting forearm," *J Appl Phys*, 1, 93-122, 1948.
13. W. Wulff, "The energy conservation equation for living tissue," *IEEE Trans Biomed Eng*, 21, 494-495, 1974.
14. H. C. Bazett and B. McGlone, "Temperature gradients in tissues in man," *Am J Physiol*, 82, 415-428, 1927.
15. W. Perl, "Heat and matter distribution in body tissues and the determination of tissue bloodflow by local clearance methods," *J Theoret Biol*, 2, 201-235, 1962.
16. J. W. Valvano, L. J. Hayes, A. J. Welch, and S. Bajekal, "Thermal conductivity and diffusivity of arterial walls," 1984.
17. H. Arkin, K. R. Holmes, and M. M. Chen, "A technique for measuring the thermal conductivity and evaluating the "apparent conductivity" concept in biomaterials," *J Biomech Eng*, 111, 276-282, 1989.
18. S. Weinbaum and L. M. Jiji, "A new simplified bioheat equation for the effect of blood flow on local average tissue temperature," *J Biomech Eng*, 107, 131-139, 1985.
19. S. Weinbaum and L. M. Jiji, "The matching of thermal fields surrounding countercurrent microvessels and the closure approximation in the Weinbaum-Jiji equation," *J Biomech Eng*, 111, 271-275, 1989.
20. J. W. Baish, "Heat Transport by countercurrent blood vessels in the presence of an arbitrary temperature gradient," *J Biomech Eng*, 112, 207-211, 1990.
21. H. Brinck and J. Werner, "Efficiency function: improvement of classical bioheat approach," *J Appl Phys*, 77, 1617-1622, 1994.
22. S. Weinbaum, L. X. Xu, L. Zhu, and A. Ekpene, "A new fundamental bioheat equation for muscle tissue: part I blood perfusion term," *J Biomech Eng*, 119, 278-288, 1997.
23. L. M. Jiji, S. Weinbaum, and D. E. Lemons, "Theory and experiment for the effect of vascular microstructure on surface tissue heat transfer-part II: model formulation and solution," *J Biomech Eng*, 106, 331-341, 1984.
24. J. W. Baish, P. S. Ayyaswamy, and K. R. Foster, "Small-scale termperature fluctuations in perfused tissue during local hyperthermia," *J Biomech Eng*, 108, 246-250, 1986.
25. J. W. Baish, P. S. Ayyaswamy, and K. R. Foster, "Heat transport mechanisms in vascular tissues: a model comparison," *J Biomech Eng*, 108, 324-331, 1986.
26. C. K. Charny and R. L. Levin, "Heat transfer normal to paired arterioles and venules embedded in perfused tissue during hyperthermia," *J Biomech Eng*, 110, 277-282, 1988.
27. C. K. Charny and R. L. Levin, "Bioheat transfer in a branching countercurrent network during hyperthermia," *J Biomech Eng*, 111, 263-270, 1989.
28. J. W. Baish, "Formulation of a statistical model of heat transfer in perfused tissue," *J Biomech Eng*, 116, 521-527, 1994.
29. H. W. Huang, Z. P. Chen, and R. B. Roemer, "A counter current vascular network model of heat transfer in tissues," *J Biomech Eng*, 118, 120-129, 1996.
30. J. F. Van der Koijk, J. J. W. Lagendijk, J. Crezee, J. De Bree, A. N. T. J. Kotte, G. M. J. Van Leeuwen, and J. J. Battermann, "The influence of vasculature on temperature distributions in MECS interstitial hyperthermia: importance of longitudinal control," *Int J Hyperthermia*, 13, 365-386, 1997.

31. G. M. J. van Leeuwen, A. N. T. J. Kotte, J. Crezee, and J. J. W. Lagendijk, "Tests of the geometrical description of blood vessels in a thermal model using counter-current geometries," *Phys Med Biol*, 42, 1515-1532, 1997.
32. G. M. J. van Leeuwen, A. N. T. J. Kotte, J. de Bree, J. F. van der Koijk, J. Crezee, and J. J. W. Lagendijk, "Accuracy of geometrical modeling of heat transfer from tissue to blood vessels," *Phys Med Biol*, 42, 1451-1460, 1997.
33. G. M. J. Van Leeuwen, A. N. T. J. Kotte, and J. J. W. Lagendijk, "A flexible algorithm for construction of 3-D vessel networks for use in thermal modeling," *IEEE Trans Biomed Eng*, 45, 596-605, 1998.
34. T. E. Cooper and G. J. Trezek, "Correlation of thermal properties of some human tissue with water content," *Aerospace Med*, 42, 24-27, 1971.
35. ASHRAE, "Physiological Principles and Thermal Comfort," in *1993 ASHRAE Handbook: Fundamentals.* Atlanta, Georgia: ASHRAE Inc., 1993, 8.1-8.29.
36. M. J. Fregly and C. M. Blatteis, "Section 4: Environmental Physiology," in *Handbook of Physiology*, vol. I. New York: American Physiological Society, 1996.
37. E. H. Wissler, "Mathematical simulation of human thermal behavior using whole body models," in *Heat Transfer in Medicine and Biology: Analysis and Applications*, vol. 1, A. Shitzer and R. C. Eberhart, Eds. New York: Plenum Press, 1985, 325-373.
38. G. M. Hahn, "Blood Flow," in *Physics and Technology of Hyperthermia*, S. B. Field and C. Franconi, Eds. Dordrecht: Martinus Nijhoff Publishers, 1987, 441-447.
39. J. W. Mitchell, T. L. Galvez, J. Hangle, G. E. Myers, and K. L. Siebecker, "Thermal response of human legs during cooling," *J Appl Phys*, 29, 859-856, 1970.
40. E. H. Wissler, "An analytical solution of countercurrent heat transfer between parallel vessels with a linear axial temperature gradient," *J Biomech Eng*, 110, 254-256, 1988.
41. L. Zhu and S. Weinbaum, "A model for heat transfer from embedded blood vessels in two-dimensional tissue preparations," *J Biomech Eng*, 117, 64-73, 1995.
42. Cousins, "On the Nusselt number in heat transfer between multiple parallel blood vessels," *J Biomech Eng*, 119, 127-129, 1997.
43. J. W. Baish, J. K. Miller, and M. J. Zivitz, "Heat transfer in the vicinity of the junction of two blood vessels," in *Advances in Bioheat and Mass Transfer: Microscale Analysis of Thermal Injury Processes, Instrumentation, Modeling and Clinical Applications*, vol. HTD-Vol. 268, R. B. Roemer, Ed. New York: ASME, 1993, 95-100.
44. J. W. Valvano, J. R. Cochran, and K. R. Diller, "Thermal conductivity and diffusivity of biomaterials measured with self-heated thermistors," *Int J Thermophys*, 6, 301-311, 1985.
45. M. A. Kapin and J. L. Ferguson, "Hemodynamic and regional circulatory alterations in dog during anaphylactic challenge," *Am J Physiol*, 249, H430-H437, 1985.
46. C. W. Haws and D. D. Heistad, "Effects of nimodipine on cerebral vasoconstrictor responses," *Am J Physiol*, 247, H170-H176, 1984.
47. J. C. Passmore, R. E. Neiberger, and S. W. Eden, "Measurement of intrarenal anatomic distribution of krypton-85 in endotoxic shock in dogs," *Am J Physiol*, 232, H54-H58, 1977.
48. R. C. Koehler, R. J. Traystman, and J. Jones, M. D., "Regional blood flow and O_2 transport during hypoxic and CO hypoxia in neonatal and sheep," *Am J Physiol*, 248, H118-H124, 1985.
49. W. C. Seyde, L. McGowan, N. Lund, B. Duling, and D. E. Longnecker, "Effects of anesthetics on regional hemodynamics in normovolemic and hemorrhaged rats," *Am J Physiol*, 249, H164-H173, 1985.
50. T. A. Balasubramaniam and H. F. Bowman, "Thermal conductivity and thermal diffusivity of biomaterials: A simultaneous measurement technique," *Trans ASME, J Biomech Eng*, 99, 148-154, 1977.

99

Interstitial Transport in the Brain: Principles for Local Drug Delivery

W. Mark Saltzman
Cornell University

99.1 Introduction

Traditional methods for delivering drugs to the brain are inadequate. Many drugs, particularly water-soluble or high molecular weight compounds, do not enter the brain following systemic administration because they permeate through blood capillaries very slowly. This blood–brain barrier (BBB) severely limits the number of drugs that are candidates for treating brain disease.

Several strategies for increasing the permeability of brain capillaries to drugs have been tested. Since the BBB is generally permeable to lipid soluble compounds which can dissolve and diffuse through endothelial cell membranes [1, 2], a common approach for enhancing brain delivery of compounds is chemical modification to enhance lipid solubility [3]. Unfortunately, lipidization approaches are not useful for drugs with molecular weight larger than 1,000. Another approach for increasing permeability is the entrapment of drugs in liposomes [4], but delivery may be limited by liposome stability in the plasma and uptake at other tissue sites.

Specific nutrient transport systems in brain capillaries can be used to facilitate drug entry into the brain. L-dopa (L-3,4-dihydroxyphenylalanine), a metabolic precursor of dopamine, is transported across endothelial cells by the neutral amino acid transport system [5]. L-dopa permeates through capillaries into the striatal tissue, where it is decarboxylated to form dopamine. Therefore, systemic administration of L-dopa is often beneficial to patients with Parkinson's disease. Certain protein modifications, such as cationization [6] and anionization [7], produce enhanced uptake in the brain. Modification of drugs [8, 9] by linkage to an anti-transferrin receptor antibody also appears to enhance transport into the brain.

0-8493-0462-8/00/$0.00+$.50

This approach depends on receptor-mediated transcytosis of transferrin-receptor complexes by brain endothelial cells; substantial uptake also occurs in the liver.

The permeability of brain capillaries can be transiently increased by intra-arterial injection of the hyperosmolar solutions, which disrupt interendothelial tight junctions [10]. But BBB disruption affects capillary permeability throughout the brain, enhancing permeability to all compounds in the blood, not just the agent of interest. Intraventricular therapy, where agents are administered directly into the CSF of the ventricles, results in high concentrations within the brain tissue, but only in regions immediately surrounding the ventricles [11, 12]. Because the agent must diffuse into the brain parenchyma from the ventricles, and because of the high rate of clearance of agents in the CNS into the peripheral circulation, this strategy cannot be used to deliver agents deep into the brain.

Because of the difficulty in achieving therapeutic drug levels by systemic administration, methods for direct administration of drugs into the brain parenchyma have been tested. Drugs can be delivered directly into the brain tissue by infusion, implantation of a drug-releasing matrix, or transplantation of drug-secreting cells [13]. These approaches provide sustained drug delivery that can be limited to specific sites, localizing therapy to a brain region. Because these methods provide a localized and continuous source of active drug molecules, the total drug dose can be less than needed with systemic administration. With polymeric controlled release, the implants can also be designed to protect unreleased drug from degradation in the body and to permit localization of extremely high doses (up to the solubility of the drug) at precisely defined locations in the brain. Infusion systems require periodic refilling; the drug is usually stored in a liquid reservoir at body temperature and many drugs are not stable under these conditions.

This chapter describes the transport of drug molecules that are directly delivered into the brain. For purposes of clarity, a specific example is considered: polymeric implants that provide controlled release of chemotherapy. The results can be extended to other modes of administration [13, 14] and other types of drug agents [15].

99.2 Implantable Controlled Delivery Systems for Chemotherapy

The kinetics of drug release from a controlled release system are usually characterized *in vitro*, by measuring the amount of drug released from the matrix into a well-stirred reservoir of phosphate buffered water or saline at 37°C. Controlled release profiles for some representative anticancer agents are shown in Fig. 99.1; all of the agents selected for these studies—1,3-bis(2-chloroethyl)-1-nitrosourea (BCNU), 4-HC, cisplatin, and taxol—are used clinically for chemotherapy of brain tumors. The controlled release period can vary from several days to many months, depending on properties of the drug, the polymer, and the method of formulation. Therefore, the delivery system can be tailored to the therapeutic situation by manipulation of implant properties.

The release of drug molecules from polymer matrices can be regulated by diffusion of drug through the polymer matrix or degradation of the polymer matrix. In many cases (including the release of BCNU, cisplatin, and 4HC from the degradable matrices shown in Fig. 99.1), drug release from biodegradable polymers appears to be diffusion-mediated, probably because the time for polymer degradation is longer than the time required for drug diffusion through the polymer. In certain cases linear release, which appears to correlate with the polymer degradation rate, can be achieved; this might be the case for taxol release from the biodegradable matrix (Fig. 99.1), although the exceedingly low solubility of taxol in water may also contribute substantially to the slowness of release.

For diffusion-mediated release, the amount of drug released from the polymer is proportional to the concentration gradient of the drug in the polymer. By performing a mass balance on drug molecules within a differential volume element in the polymer matrix, a conservation equation for drug within the matrix is obtained:

$$\frac{\partial C_p}{\partial t} = D_p \nabla^2 C_p \tag{99.1}$$

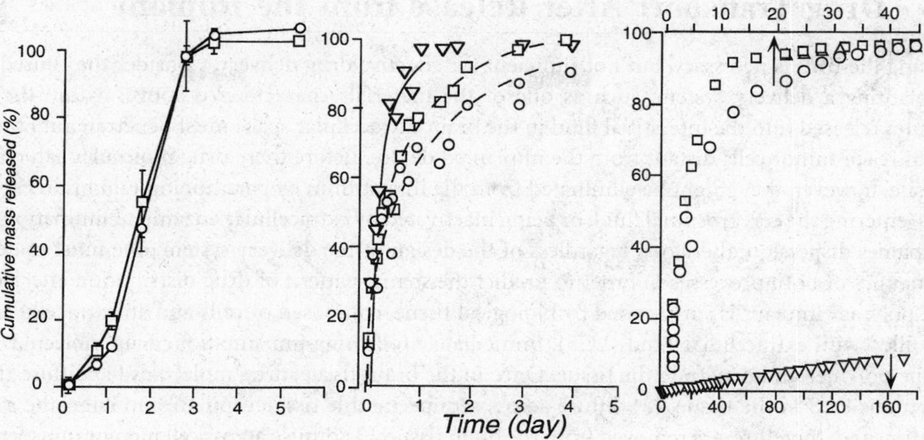

FIGURE 99.1 Controlled release of anticancer compounds from polymeric matrices. (a) Release of cisplatin (circles) from a biodegradable copolymer of fatty acid dimers and sebacic acid, p(FAD:SA), initially containing 10% drug (see [32, 33] for details). (b) Release of BCNU from EVAc matrices (circles), polyanhydride matrices p(CPP:SA) (squares), and p(FAD:SA) (triangles) matrices initially containing 20% drug. (c) Release of BCNU (squares), 4HC (circles), and taxol (triangles) from p(CPP:SA) matrices initially containing 20% drug. Note that panel (c) has two time axes: the lower axis applies to the release of taxol and the upper axis applies to the release of BCNU and 4HC.

where C_p is the local concentration of drug in the polymer and D_p is the diffusion coefficient of the drug in the polymer matrix. This equation can be solved, with appropriate boundary and initial conditions, to obtain the cumulative mass of drug released as a function of time; the details of this procedure are described elsewhere [16]. A useful approximate solution, which is valid for the initial 60% of release, is:

$$M_t = 4M_o\sqrt{\frac{D_{i:p}\,t}{\pi L^2}} \tag{99.2}$$

where M_t is the cumulative mass of drug released from the matrix, M_o is the initial mass of drug in the matrix, and L is the thickness of the implant. By comparing Eq. (99.2) to the experimentally determined profiles, the rate of diffusion of the agent in the polymer matrix can be estimated (Table 99.1).

TABLE 99.1 Diffusion Coefficients for Chemotherapy Drug Release from Biocompatible Polymer Matrices[a]

Drug	Polymer	Initial loading (%)	Dp (cm2/sec)
Cisplatin	P(FAD:SA)	10	6.8×10^{-9}
BCNU	EVAc	20	1.6×10^{-8}
BCNU	P(FAD:SA)	20	6.9×10^{-8}
BCNU	P(CPP:SA)	20	2.3×10^{-8} (panel b)
			2.0×10^{-8} (panel c)
4HC	P(CPP:SA)	20	3.1×10^{-10}
Taxol	P(CPP:SA)	20	n.a.

[a] Diffusion coefficients were obtained by comparing the experimental data show in Fig. 99.1 to Eq. (99.2) and determining the best value of the diffusion coefficient to represent the data. Abbreviations: not applicable (n.a.).

99.3 Drug Transport After Release from the Implant

Bypassing the BBB is necessary, but not sufficient for effective drug delivery. Consider the consequences of implanting a delivery system, such as one of the materials characterized above, within the brain. Molecules released into the interstitial fluid in the brain extracellular space must penetrate into the brain tissue to reach tumor cells distant from the implanted device. Before these drug molecules can reach the target site, however, they might be eliminated from the interstitium by partitioning into brain capillaries or cells, entering the cerebrospinal fluid, or being inactivated by extracellular enzymes. Elimination always accompanies dispersion; therefore, regardless of the design of the delivery system, one must understand the dynamics of both processes in order to predict the spatial pattern of drug distribution after delivery.

The polymer implant is surrounded by biological tissue, composed of cells and an extracellular space (ECS) filled with extracellular fluid (ECF). Immediately following implantation, drug molecules escape from the polymer and penetrate the tissue. Once in the brain tissue, drug molecules (1) diffuse through the tortuous ECS in the tissue, (2) diffuse across semipermeable tissue capillaries to enter the systemic circulation and, therefore, are removed from the brain tissue, (3) diffuse across cell membranes by passive, active, or facilitated transport paths, to enter the intracellular space, (4) transform, spontaneously or by an enzyme-mediated pathway, into other compounds, and (5) bind to fixed elements in the tissue. Each of these events influence drug therapy: diffusion through the ECS is the primary mechanism of drug distribution in brain tissue; elimination of the drug occurs when it is removed from the ECF or transformed; and binding or internalization may slow the progress of the drug through the tissue.

A mass balance on a differential volume element in the tissue [17] gives a general equation describing drug transport in the region near the polymer [18]:

$$\frac{\partial C_t}{\partial t} + \bar{v} \bullet \nabla C_t = D_b \nabla^2 C_t + R_e\left(C_t\right) - \frac{\partial B}{\partial t} \tag{99.3}$$

where C is the concentration of the diffusible drug in the tissue surrounding the implant (g/cm³ tissue), v is the fluid velocity (cm/sec), D_b is the diffusion coefficient of the drug in the tissue (cm²/sec), $R_e(C)$ is the rate of drug elimination from the ECF (g/sec-cm³ tissue), B is the concentration of drug bound or internalized in cells (g/cm³ tissue), and t is the time following implantation.[1] In deriving this equation, the conventions developed by Nicholson [20], based on volume-averaging in a complex medium, and Patlak and Fenstermacher [18] were combined. In this version of the equation, the concentrations C and B and the elimination rate $R_e(C)$ are defined per unit volume of tissue. D_b is an effective diffusion coefficient, which must be corrected from the diffusion coefficient for the drug in water to account for the tortuosity of the ECS.

When the binding reactions are rapid, the amount of intracellular or bound drug can be assumed to be directly proportional, with an equilibrium coefficient K_{bind}, to the amount of drug available for internalization or binding:

$$B = K_{bind}\, C \tag{99.4}$$

Substitution of Eq. (99.4) into Eq. (99.3) yields, with some simplification:

$$\frac{\partial C_t}{\partial t} = \frac{1}{1 + K_{bind}}\left(D_b \nabla^2 C_t + R_e\left(C_t\right) - \bar{v} \bullet \nabla C_t\right) \tag{99.5}$$

[1]An alternate form of this equation, which accounts more rigorously for transfer of drug between different phases in the tissue, is also available [19].,

The drug elimination rate, $R_e(C)$, can be expanded into the following terms:

$$R_e\left(C_t\right) = k_{bbb}\left(\frac{C_t}{\varepsilon_{ecs}} - C_{plasma}\right) + \frac{V_{\max} C_t}{K_m + C_t} + k_{ne} C_t \tag{99.6}$$

where k_{bbb} is the permeability of the BBB (defined based on concentration in the ECS), C_{plasma} is the concentration of drug in the blood plasma, V_{max} and K_m are Michaelis-Menton constants, and k_{ne} is a first order rate constant for drug elimination due to non-enzymatic reactions. For any particular drug, some of the rate constants may be very small, reflecting the relative importance of each mechanism of drug elimination. If it is assumed that the permeability of the BBB is low ($C_{pl} \ll C$) and the concentration of drug in the brain is sufficiently low so that any enzymatic reactions are in the first order regime ($C \ll K_m$), Eq. (99.6) can be reduced to:

$$-R_e\left(C_t\right) = \frac{k_{bbb}}{\varepsilon_{ecs}} C_t + \frac{V_{\max}}{K_m} C_t + k_{ne} C_t = k_{app} C_t \tag{99.7}$$

where k_{app} is a lumped first order rate constant. With these assumptions, Eq. (99.4) can be simplified by definition of an apparent diffusion coefficient, D^*, and an apparent first order elimination constant, k^*:

$$\frac{\partial C_t}{\partial t} = D^* \nabla^2 C_t + k^* C_t - \frac{\bar{v} \bullet \nabla C_t}{1 + K_{bind}} \tag{99.8}$$

where $k^* = \frac{k_{app}}{1 + K_{bind}}$ and $D = \frac{D_b}{1 + K_{bind}}$.

Boundary and initial conditions are required for solution of differential Eq. (99.8). If a spherical implant of radius R is implanted into a homogeneous region of the brain, at a site sufficiently far from anatomical boundaries, the following assumptions are reasonable:

$$C_t = 0 \text{ for } t = 0;\ r > R \tag{99.9}$$

$$C_t = C_i \text{ for t} > 0;\ r = R \tag{99.10}$$

$$C_t = 0 \text{ for t} > 0;\ r \to \infty \tag{99.11}$$

In many situations, drug transport due to bulk flow can be neglected. This assumption (v is zero) is common in previous studies of drug distribution in brain tissue [18]. For example, in a previous study of cisplatin distribution following continuous infusion into the brain, the effects of bulk flow were found to be small, except within 0.5 mm of the site of infusion [21]. In the cases considered here, since drug molecules enter the tissue by diffusion from the polymer implant, not by pressure-driven flow of a fluid, no flow should be introduced by the presence of the polymer. With fluid convection assumed negligible, the general governing equation in the tissue, Eq. (99.8), reduces to:

$$\frac{\partial C_t}{\partial t} = D^* \nabla^2 C_t + k^* C_t \tag{99.12}$$

The no-flow assumption may be inappropriate in certain situations. In brain tumors, edema and fluid movement are significant components of the disease. In addition, some drugs can elicit cytotoxic edema.

Certain drug/polymer combinations can also release drugs in sufficient quantity to create density-induced fluid convection.

Equation (99.12), with conditions (99.9) through (99.11), can be solved by Laplace transform techniques [13] to yield:

$$\frac{C_t}{C_i} = \frac{1}{2\zeta}\left\{\exp\left[-\phi(\zeta-1)\right]\operatorname{erfc}\left[\frac{\zeta-1}{2\sqrt{\tau}} - \phi\sqrt{\tau}\right] + \exp\left[\phi(\zeta-1)\right]\operatorname{erfc}\left[\frac{\zeta-1}{2\sqrt{\tau}} + \phi\sqrt{\tau}\right]\right\} \tag{99.13}$$

where the dimensionless variables are defined as follows:

$$\zeta = \frac{r}{R};\ \tau = \frac{D^* t}{R^2};\ \phi = R\sqrt{\frac{k^*}{D^*}} \tag{99.14}$$

The differential equation also has a steady-state solution, which is obtained by solving Eq. (99.12) with the time derivative set equal to zero and subject to the boundary conditions (99.9) and (99.10):

$$\frac{C_t}{C_i} = \frac{1}{\zeta}\exp\left[-\phi(\zeta-1)\right] \tag{99.15}$$

Figure 99.2 shows concentration profiles calculated using Eqs. (99.13) and (99.15). In this situation, which was obtained using reasonable values for all of the parameters, steady-state is reached approximately 1 h after implantation of the delivery device. The time required to achieve steady-state depends on the rate of diffusion and elimination, as previously described [22], but will be significantly less than 24 h for most drug molecules.

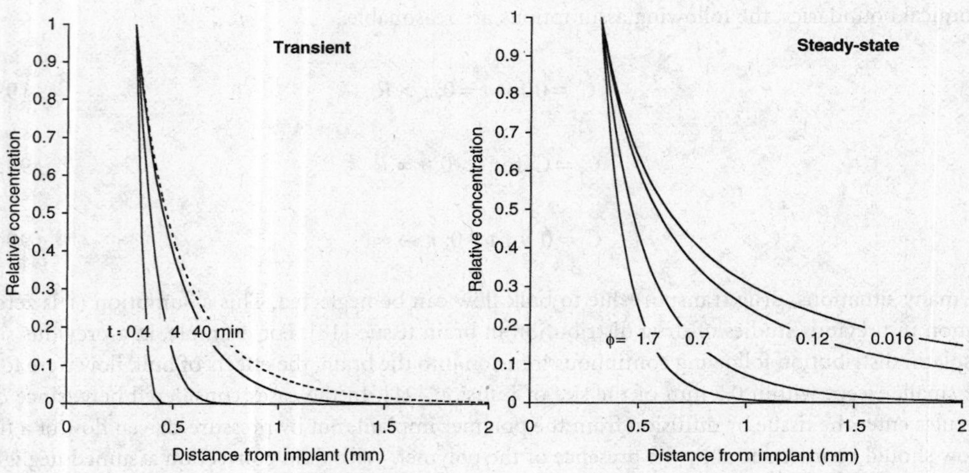

FIGURE 99.2 Concentration profiles after implantation of a spherical drug-releasing implant. (Panel a, Transient) Solid lines represent the transient solution to Eq. (99.12) [i.e., Eq. (99.13)] with the following parameter values: $D^* = 4 \times 10^{-7}$ cm²/s; R = 0.032 cm; $k^* = 1.9 \times 10^{-4}$ s^{-1} ($t_{1/2}$ = 1 h). The dashed line represents the steady-state solution [i.e., Eq. (99.15)] for the same parameters. (Panel b, Steady-state) Solid lines in this plot represent Eq. (99.15) with the following parameters: $D^* = 4 \times 10^{-7}$ cm²/s; R = 0.032 cm. Each curve represents the steady-state concentration profile for drugs with different elimination half-lives in the brain, corresponding to different dimensionless moduli, ϕ: $t_{1/2}$ = 10 min (ϕ = 1.7); 1 h (0.7); 34 h (0.12); and 190 h (0.016).

99.4 Application of Diffusion-Elimination Models to Intracranial BCNU Delivery Systems

The preceding mathematical analysis, which assumes diffusion and first-order elimination in the tissue, agrees well with experimental concentration profiles obtained after implantation of controlled release polymers (Fig. 99.3). At 3, 7, and 14 d after implantation of a BCNU-releasing implant, the concentration profile at the site of the implant was very similar. The parameter values (obtained by fitting Eq. (99.15) to the experimental data) were consistent with parameters obtained using other methods [19], suggesting that diffusion and first-order elimination were sufficient to account for the pattern of drug concentration observed during this period. Parameter values were similar at 3, 7, and 14 d, indicating that the rates of drug release, drug dispersion, and drug elimination did not change during this period. This equation has been compared to concentration profiles measured for a variety of molecules delivered by polymer implants to the brain—dexamethasone [22], molecular weight fractions of dextran [23], nerve growth factor in rats [24, 25], BCNU in rats [19], rabbits [26], and monkeys [27]. In each of these cases, the steady-state diffusion-elimination model appears to capture most of the important features of drug transport.

This model can be used to develop guidelines for the design of intracranial delivery systems. Table 99.2 lists some of the important physical and biological characteristics of a few compounds that have been considered for interstitial delivery to treat brain tumors. When the implant is surrounded by tissue, the maximum rate of drug release is determined by the solubility of the drug, C_s, and the rate of diffusive transport through the tissue:

$$\left(\frac{dM_t}{dt}\right)_{\max} = \left(Maximum\ flux\right) \times \left(Surface\ area\right) = -D^* \left.\frac{\partial C_t}{\partial r}\right|_R 4\pi R^2 \qquad (99.16)$$

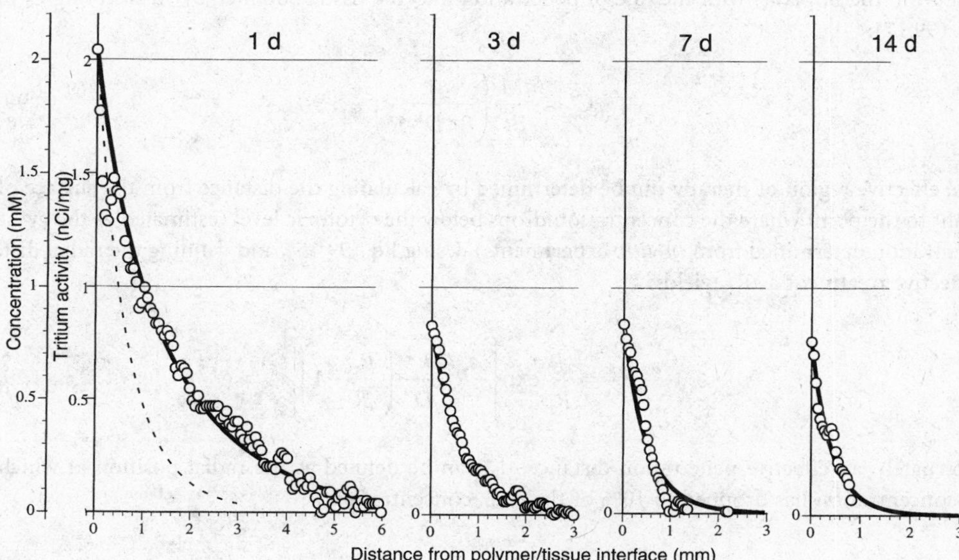

FIGURE 99.3 Concentration profiles after implantation of a BCNU-releasing implant. Solutions to Eq. (99.15) were compared to experimental data obtained by quantitative autoradiographic techniques. The solid lines in the three panels labeled 3, 7, and 14 d were all obtained using the following parameters: R = 0.15 cm; ϕ = 2.1; and C_I = 0.81 mM. The solid line in the panel labeled 1 d was obtained using the following parameters: R = 0.15 cm; ϕ = 0.7; and C_I = 1.9 mM. Modified from [19].

TABLE 99.2 Implant Design Applied to Three Chemotherapy Compounds[a]

	BCNU	4HC	Methotrexate
Molecular weight (daltons)	214	293	454
Solubility (mM)	12	100	100
$\log_{10}K$	1.53	0.6	−1.85
k* (h)	70	70	1
D* (10^{-7} cm²/sec)	14	14	5
Toxic concentration in culture (μM)	25	10	0.04
Maximum release rate (mg/d)	1.2	14	17
Implant lifetime at max rate (d)	0.85	0.07	0.06
Maximum concentration in tissue for 1-week-releasing implant (mM)	1.5	1.1	1.8
RT (mm)	1.3	2.5	5

[a] K is the octanol:water partition coefficient, k is the rate of elimination due to permeation through capillaries, D_b is the diffusion coefficient of the drug in the brain. The following values are assumed, consistent with our results from polymer delivery to rats and rabbits: radius of spherical implant, R = 1.5 mm; mass of implant, M = 10 mg; drug loading in implant, Load = 10%.

Evaluating the derivative in Eq. (99.16) from the steady-state concentration profile [Eq. (99.15)] yields:

$$\left(\frac{dM_t}{dt}\right)_{\max} = 8\pi D^* C_s R \tag{99.17}$$

Regardless of the properties of the implant, it is not possible to release drug into the tissue at a rate faster than determined by Eq. (99.17). If the release rate from the implant is less than this maximum rate, C_i (the concentration in the tissue immediately outside the implant) is less than the saturation concentration, C_s. The actual concentration C_i can be determined by balancing the release rate from the implant (dM_t/dt, which can be determined from Eq. (99.2) provided that diffusion is the mechanism of release from the implant) with the rate of penetration into the tissue obtained by substituting C_I for C_s in Eq. (99.17):

$$C^* = \frac{dM_t}{dt}\left(\frac{1}{8\pi D^* R}\right) \tag{99.18}$$

The effective region of therapy can be determined by calculating the distance from the surface of the implant to the point where the concentration drops below the cytotoxic level (estimated as the cytotoxic concentration determined from *in vitro* experiments). Using Eq. (99.15), and defining the radial distance for effective treatment as R_T, yields:

$$\frac{C_{cytotoxic}}{C_i} = \frac{R}{R_T}\exp\left\{-R\sqrt{\frac{k^*}{D^*}}\left(\frac{R_T}{R}-1\right)\right\} \tag{99.19}$$

Alternately, an effective penetration distance, d_P, can be defined as the radial position at which the drug concentration has dropped to 10% of the peak concentration:

$$0.10 = \frac{R}{d_P}\exp\left\{-R\sqrt{\frac{k^*}{D^*}}\left(\frac{d_P}{R}-1\right)\right\} \tag{99.20}$$

These equations provide quantitative criteria for evaluating the suitability of chemotherapy agents for direct intracranial delivery (Table 99.2).

99.5 Limitations and Extensions of the Diffusion-Elimination Model

Failure of the Model in Certain Situations

The previous section outlined one method for analysis of drug transport after implantation of a drug-releasing device. A simple pseudo-steady-state equation [Eq. (99.15)] yielded simple guidelines [Eq. (99.16 through 99.19] for device design. Because the assumptions of the model were satisfied over a substantial fraction of the release period (days 3 to 14, based on the data shown in Fig. 99.3), this analysis may be useful for predicting the effects of BCNU release from biodegradable implants. Pseudo-steady-state assumptions are reasonable during this period of drug release, presumably because the time required to achieve steady-state (which is on the order of minutes) is much less than the characteristic time associated with changes in the rate of BCNU release from the implant (which is on the order of days).

But experimental concentration profiles measured 1 d after implantation were noticeably different: the peak concentration was substantially higher and the drug penetration into the surrounding tissue was deeper (see the left-hand panel of Fig. 99.3). This behavior cannot be easily explained by the pseudo-steady-state models described above. For example, if the difference observed at 1 d represents transient behavior, the concentration observed at a fixed radial position should increase with time (Fig. 99.2); in contrast, the experimental concentration at any radial position on Day 1 is higher than the concentration measured at that same position on subsequent days.

Effect of Drug Release Rate

Alternately, the observed difference at 1 d might be due to variability in the rate of BCNU release from the polymer implant over this period, with transport characteristics in the tissue remaining constant. When similar BCNU-releasing implants are tested *in vitro*, the rate of drug release did decrease over time (Fig. 99.1). Equation (99.18) predicts the variation in peak concentration with release rate; the twofold higher concentration observed at the interface on Day 1 (as compared to Days 3 through 14) could be explained by a twofold higher release rate on Day 1. But the effective penetration distance, d_p, does not depend on release rate. Experimentally measured penetration distances are ~1.4 mm on Days 3, 7, and 14 and ~5 mm on Day 1. This difference in penetration is shown more clearly in the Day 1 panel of Fig. 99.3: the dashed line shows the predicted concentration profile if k^* and D^* were assumed equal to the values obtained for Days 3, 7, and 14. Changes in the rate of BCNU release are insufficient to explain the differences observed experimentally.

Determinants of Tissue Penetration

Penetration of BCNU is enhanced at Day 1 relative to penetration at Days 3, 7, and 14. For an implant of fixed size, penetration depends only on the ratio of elimination rate to diffusion rate: k^*/D^*. Increased penetration results from a decrease in this ratio (Fig. 99.2), which could occur because of either a decreased rate of elimination (smaller k^*) or an increased rate of diffusion (larger D^*). But there are no good reasons to believe that BCNU diffusion or elimination are different on Day 1 than on Days 3 through 14. With its high lipid solubility, BCNU can diffuse readily through brain tissue. In addition, elimination of BCNU from the brain occurs predominantly by partitioning into the circulation; since BCNU can permeate the capillary wall by diffusion, elimination is not a saturable process. Perhaps the enhanced penetration of BCNU is due to the presence of another process for drug dispersion, such as bulk fluid flow, which was neglected in the previous analysis.

The diffusion/elimination model compares favorably with available experimental data, but the assumptions used in predicting concentration profiles in the brain may not be appropriate in all cases. Deviations from the predicted concentration profiles may occur due to extracellular fluid flows in the brain, complicated patterns of drug binding, or multistep elimination pathways. The motion of interstitial fluid in the vicinity of the polymer and the tumor periphery may not always be negligible, as mentioned above. The interstitial fluid velocity is proportional to the pressure gradient in the interstitium; higher interstitial pressure in tumors—due to tumor cell proliferation, high vascular permeability, and the absence of functioning lymphatic vessels—may lead to steep interstitial pressure gradients at the periphery of the tumor [28]. As a result, interstitial fluid flows within the tumor may influence drug transport. A drug at the periphery of the tumor must overcome outward convection to diffuse into the tumor. Similarly, local edema after surgical implantation of the polymer may cause significant fluid movement in the vicinity of the polymer. More complete mathematical models that include the convective contribution to drug transport are required.

Effect of Fluid Convection

When bulk fluid flow is present ($v \neq 0$), concentration profiles can be predicted from Eq. (99.8), subject to the same boundary and initial conditions [Eqs. (99.9) through (99.11)]. In addition to Eq. (99.8), continuity equations for water are needed to determine the variation of fluid velocity in the radial direction. This set of equations has been used to describe concentration profiles during microinfusion of drugs into the brain [14]. Relative concentrations were predicted by assuming that the brain behaves as a porous medium (i.e., velocity is related to pressure gradient by Darcy's law). Water introduced into the brain can expand the interstitial space; this effect is balanced by the flow of water in the radial direction away from the infusion source and, to a lesser extent, by the movement of water across the capillary wall.

In the presence of fluid flow, penetration of drug away from the source is enhanced (Fig. 99.4). The extent of penetration depends on the velocity of the flow and the rate of elimination of the drug. These calculations were performed for macromolecular drugs, which have limited permeability across the brain capillary wall. The curves indicate steady-state concentration profiles for three different proteins with

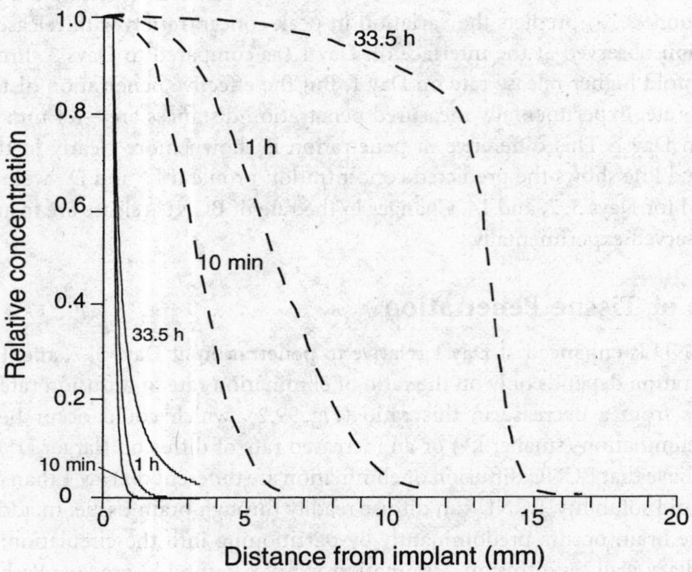

FIGURE 99.4 Concentration profiles predicted in the absence (solid lines) and presence (dashed lines) of interstitial fluid flow. Solid lines were obtained from Eq. (99.15) with the following parameter values: $R = 0.032$ cm; $D = 4 \times 10^{-7}$ cm^2/sec; and $k^* = \ln(2)/t_{1/2}$ where $t_{1/2}$ is either 10 min, 1 h, or 33.5 h as indicated on the graph. Dashed lines were obtained from [14] using the same parameter values and an infusion rate of 3 μL/min. The dashed line indicating the interstitial flow calculation for the long-lived drug ($t_{1/2} = 33.5$ h) was not at steady-state, but at 12 h after initiation of the flow.

TABLE 99.3 Interstitial Fluid Velocity as a Function of Radial Position During Microinfusion[a]

Radial Position (mm)	Interstitial Velocity (μm/sec)
2	5.0
5	0.8
10	0.2
20	0.05

[a] Calculated by method reported in [14].

metabolic half-lives of 10 min, 1 h, or 33.5 h. In the absence of fluid flow, drugs with longer half-lives penetrate deeper into the tissue [solid lines in Fig. 99.4 were obtained from Eq. (99.15)]. This effect is amplified by the presence of flow (dashed lines in Fig. 99.4).

During microinfusion, drug is introduced by pressure-driven fluid flow from a small catheter. Therefore, pressure gradients are produced in the brain interstitial space, which lead to fluid flow through the porous brain microenvironment. Volumetric infusion rates of 3 μL/min were assumed in the calculations reproduced in Fig. 99.4. Since loss of water through the brain vasculature is small, the velocity can be determined as a function of radial position:

$$v_r = \frac{q}{4\pi r^2 \varepsilon} \tag{99.21}$$

where q is the volumetric infusion rate and ε is the volume fraction of the interstitial space in the brain (approximately 0.20). Fluid velocity decreases with radial distance from the implant (Table 99.3); but at all locations within the first 20 mm of the implant site, predicted velocity was much greater than the velocities reported previously during edema or tumor growth in the brain.

The profiles predicted in Fig. 99.4 were associated with the introduction of substantial volumes of fluid at the delivery site. Flow-related phenomena are probably less important in drug delivery by polymer implants. Still, this model provides useful guidelines for predicting the influence of fluid flow on local rates of drug movement. Clearly, the effect of flow velocity on drug distribution is substantial (Fig. 99.4). Even relatively low flows, perhaps as small as 0.03 μm/sec, are large enough to account for the enhancement in BCNU penetration observed at Day 1 in Fig. 99.3.

Effect of Metabolism

The metabolism, elimination, and binding of drug are assumed to be first order processes in our simple analysis. This assumption may not be realistic in all cases, especially for complex agents such as proteins. The metabolism of drugs in normal and tumor tissues is incompletely understood. Other cellular factors (e.g., the heterogeneity of tumor-associated antigen expression and multidrug resistance) that influence the uptake of therapeutic agents may not be accounted for by our simple first order elimination.

Finally, changes in the brain that occur during the course of therapy are not properly considered in this model. Irradiation can be safely administered when a BCNU-loaded polymer has been implanted in monkey brains, suggesting the feasibility of adjuvant radiotherapy. However, irradiation also causes necrosis in the brain. The necrotic region has a lower perfusion rate and interstitial pressure than tumor tissue; thus, the convective interstitial flow due to fluid leakage is expected to be smaller. Interstitial diffusion of macromolecules is higher in tumor tissue than in normal tissue, as the tumor tissue has larger interstitial spaces [29]. The progressive changes in tissue properties—due to changes in tumor size, irradiation, and activity of chemotherapy agent—may be an important determinant of drug transport and effectiveness of therapy in the clinical situation.

99.6 New Approaches to Drug Delivery Suggested by Modeling

Mathematical models, which describe the transport of drug following controlled delivery, can predict the penetration distance of drug and the local concentration of drug as a function of time and location. The calculations indicate that drugs with slow elimination will penetrate deeper into the tissue. The modulus ϕ, which represents the ratio of elimination to diffusion rates in the tissue, provides a quantitative criterion for selecting agents for interstitial delivery. For example, high molecular weight dextrans were retained longer in the brain space, and penetrated a larger region of the brain than low molecular weight molecules following release from an intracranial implant [23]. This suggests a strategy for modifying molecules to improve their tissue penetration by conjugating active drug molecules to inert polymeric carriers. For conjugated drugs, the extent of penetration should depend on the modulus ϕ for the conjugated compound as well as the degree of stability of the drug-carrier linkage.

The effects of conjugation and stability of the linkage between drug and carrier on enhancing tissue penetration in the brain have been studied in a model system [30]. Methotrexate (MTX)-dextran conjugates with different dissociation rates were produced by linking MTX to dextran (molecular weight 70,000) through a short-lived ester bond (half-life $\approx$ 3 d) and a longer-lived amide bond (half-life > 20 d). The extent of penetration for MTX-dextran conjugates was studied in three-dimensional human brain tumor cell cultures; penetration was significantly enhanced for MTX-dextran conjugates and the increased penetration was correlated with the stability of the linkage. These results suggest that modification of existing drugs may increase their efficacy against brain tumors when delivered directly to the brain interstitium.

99.7 Conclusion

Controlled release polymer implants are a useful new technology for delivering drugs directly to the brain interstitium. This approach is already in clinical use for treatment of tumors [31], and could soon impact treatment of other diseases. The mathematical models described in this paper provide a rational framework for analyzing drug distribution after delivery. These models describe the behavior of chemotherapy compounds very well and allow prediction of the effect of changing properties of the implant or the drug. More complex models are needed to describe the behavior of macromolecules, which encounter multiple modes of elimination and metabolism and are subject to the effects of fluid flow. In addition, variations on this approach may be useful for analyzing drug delivery in other situations.

References

1. Lieb, W. and W. Stein, Biological Membranes behave as Non-Porous Polymeric Sheets with Respect to the Diffusion of Non-Electrolytes. *Nature*, 1969. **224**: 240-249.
2. Stein, W.D., The Movement of Molecules across Cell Membranes. 1967, New York: Academic Press.
3. Simpkins, J., N. Bodor, and A. Enz, Direct Evidence for Brain-Specific Release of Dopamine from a Redox Delivery System. *Journal of Pharmaceutical Sciences*, 1985. **74**: 1033-1036.
4. Gregoriadis, G., The carrier potential of liposomes in biology and medicine. *The New England Journal of Medicine*, 1976. **295**: 704-710.
5. Cotzias, C.G., M.H. Van Woert, and L.M. Schiffer, Aromatic amino acids and modificatino of parkinsonism. *The New England Journal of Medicine*, 1967. **276**: 374-379.
6. Triguero, D., J.B. Buciak, J. Yang, and W.M. Pardridge, Blood-brain barrier transport of cationized immunoglobulin G: enhanced delivery compared to native protein. *Proceedings of the National Academy of Sciences USA*, 1989. **86**: 4761-4765.
7. Tokuda, H., Y. Takakura, and M. Hashida, Targeted delivery of polyanions to the brain. *Proceed. Intern. Symp. Control. Rel. Bioact. Mat.*, 1993. **20**: 270-271.

8. Friden, P., L. Walus, G. Musso, M. Taylor, B. Malfroy, and R. Starzyk, Anti-transferrin receptor antibody and antibody-drug conjugates cross the blood-brain barrier. *Proceedings of the National Academy of Sciences USA*, 1991. **88**: 4771-4775.
9. Friden, P.M., L.R. Walus, P. Watson, S.R. Doctrow, J.W. Kozarich, C. Backman, H. Bergman, B. Hoffer, F. Bloom, and A.-C. Granholm, Blood-brain barrier penetration and *in vivo* activity of an NGF conjugate. *Science*, 1993. **259**: 373-377.
10. Neuwelt, E., P. Barnett, I. Hellstrom, K. Hellstrom, P. Beaumier, C. McCormick, and R. Weigel, Delivery of melanoma-associated immunoglobulin monoclonal antibody and Fab fragments to normal brain utilizing osmotic blood-brain barrier disruption. *Cancer Research*, 1988. **48**: 4725-4729.
11. Blasberg, R., C. Patlak, and J. Fenstermacher, Intrathecal chemotherapy: brain tissue profiles after ventriculocisternal perfusion. *The Journal of Pharmacology and Experimental Therapeutics*, 1975. **195**: 73-83.
12. Yan, Q., C. Matheson, J. Sun, M.J. Radeke, S.C. Feinstein, and J.A. Miller, Distribution of intracerebral ventricularly administered neurotrophins in rat brain and its correlation with Trk receptor expression. *Experimental Neurology*, 1994. **127**: 23-36.
13. Mahoney, M.J. and W.M. Saltzman, Controlled release of proteins to tissue transplants for the treatment of neurodegenerative disorders. *Journal of Pharmaceutical Sciences*, 1996. **85**(12): 1276-1281.
14. Morrison, P.F., D.W. Laske, H. Bobo, E.H. Oldfield, and R.L. Dedrick, High-flow microinfusion: tissue penetration and pharmacodynamics. *American Journal of Physiology*, 1994. **266**: R292-R305.
15. Haller, M.F. and W.M. Saltzman, Localized delivery of proteins in the brain: Can transport be customized? *Pharmaceutical Research*, 1998. **15**: 377-385.
16. Wyatt, T.L. and W.M. Saltzman, Protein delivery from non-degradable polymer matrices, in *Protein Delivery-Physical Systems,* L. Saunders and W. Hendren, Editors. 1997, Plenum Press: New York, NY. p. 119-137.
17. Bird, R.B., W.E. Stewart, and E.N. Lightfoot, *Transport Phenomena*. 1960, New York: John Wiley & Sons. 780.
18. Patlak, C. and J. Fenstermacher, Measurements of dog blood-brain transfer constants by ventriculocisternal perfusion. *American Journal of Physiology*, 1975. **229**: 877-884.
19. Fung, L., M. Shin, B. Tyler, H. Brem, and W.M. Saltzman, Chemotherapeutic drugs released from polymers: distribution of 1,3-bis(2-chloroethyl)-1-nitrosourea in the rat brain. *Pharmaceutical Research*, 1996. **13**: 671-682.
20. Nicholson, C., Diffusion from an injected volume of a substance in brain tissue with arbitrary volume fraction and tortuosity. *Brain Research*, 1985. **333**: 325-329.
21. Morrison, P. and R.L. Dedrick, Transport of cisplatin in rat brain following microinfusion: an analysis. *Journal of Pharmaceutical Sciences*, 1986. **75**: 120-128.
22. Saltzman, W.M. and M.L. Radomsky, Drugs released from polymers: diffusion and elimination in brain tissue. *Chemical Engineering Science*, 1991. **46**: 2429-2444.
23. Dang, W. and W.M. Saltzman, Dextran retention in the rat brain following controlled release from a polymer. *Biotechnology Progress*, 1992. **8**: 527-532.
24. Krewson, C.E., M. Klarman, and W.M. Saltzman, Distribution of nerve growth factor following direct delivery to brain interstitium. *Brain Research*, 1995. **680**: 196-206.
25. Krewson, C.E. and W.M. Saltzman, Transport and elimination of recombinant human NGF during long-term delivery to the brain. *Brain Research*, 1996. **727**: 169-181.
26. Strasser, J.F., L.K. Fung, S. Eller, S.A. Grossman, and W.M. Saltzman, Distribution of 1,3-bis(2-chloroethyl)-1-nitrosourea (BCNU) and tracers in the rabbit brain following interstitial delivery by biodegradable polymer implants. *The Journal of Pharmacology and Experimental Therapeutics*, 1995. **275**(3): 1647-1655.
27. Fung, L.K., M.G. Ewend, A. Sills, E.P. Sipos, R. Thompson, M. Watts, O.M. Colvin, H. Brem, and W.M. Saltzman, Pharmacokinetics of interstitial delivery of carmustine, 4-hydroperoxycyclophosphamide, and paclitaxel from a biodegradable polymer implant in the monkey brain. *Cancer Research*, 1998. **58**: 672-684.

28. Jain, R.K., Barriers to drug delivery in solid tumors. *Scientific American*, 1994. **271**(1): 58-65.
29. Clauss, M.A. and R.K. Jain, Interstitial transport of rabbit and sheep antibodies in normal and neoplastic tissues. *Cancer Research*, 1990. **50**: 3487-3492.
30. Dang, W.B., O.M. Colvin, H. Brem, and W.M. Saltzman, Covalent coupling of methotrexate to dextran enhances the penetration of cytotoxicity into a tissue-like matrix. *Cancer Research*, 1994. **54**: 1729-1735.
31. Brem, H., S. Piantadosi, P.C. Burger, M. Walker, R. Selker, N.A. Vick, K. Black, M. Sisti, S. Brem, G. Mohr, P. Muller, R. Morawetz, S.C. Schold, and P.-B.T.T. Group, Placebo-controlled trial of safety and efficacy of intraoperative controlled delivery by biodegradable polymers of chemotherapy for recurrent gliomas. *Lancet*, 1995. **345**: 1008-1012.
32. Dang, W. and W.M. Saltzman, Controlled release of macromolecules from a biodegradable polyanhydride matrix. *Journal of Biomaterials Science, Polymer Edition*, 1994. **6**(3): 291-311.
33. Dang, W., *Engineering Drugs and Delivery Systems for Brain Tumor Therapy.* 1993, The Johns Hopkins University:

100

Arterial Wall Mass Transport: The Possible Role of Blood Phase Resistance in the Localization of Arterial Disease

John M. Tarbell
The Pennsylvania State University

Yuchen Qiu
The Pennsylvania State University

Atherosclerosis is a disease of the large arteries which involves a characteristic accumulation of high molecular weight lipoprotein in the arterial wall [1]. The disease tends to be localized in regions of curvature and branching in arteries where fluid shear stress (shear rate) is altered from its normal patterns in straight vessels [2]. The possible role of fluid mechanics in the localization of atherosclerosis has been debated for many years [3,4]. One possibility considered early on was that the blood phase resistance to lipid transport, which could be affected by local fluid mechanics, played a role in the focal accumulation of lipid in arteries. Studies by Caro and Nerem [5], however, showed that the uptake of lipid in arteries could not be correlated with fluid phase mass transport, leading to the conclusion that the wall (endothelium) and not the blood, was the limiting resistance to lipid transport. This suggested that fluid mechanical effects on macromolecular transport were the result of direct mechanical influences on the transport characteristics of the endothelium.

0-8493-0462-8/00/$0.00+$.50

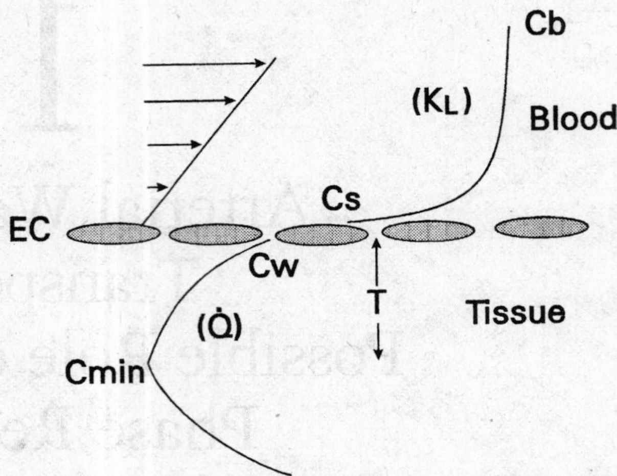

FIGURE 100.1 Schematic diagram of arterial wall transport processes showing the concentration profile of a solute which is being transported from the blood, where its bulk concentration is C_b, to the surface of the endothelium, where its concentration is C_s, then across the endothelium, where the subendothelial concentration is C_w, and finally to a minimum value within the tissue, C_{min}. Transport of the solute in the blood phase is characterized by the mass transport coefficient, k_L; consumption of the solute at the endothelial surface is described by a first-order reaction with rate constant, k_r; movement of the solute across the endothelium depends on the permeability coefficient, Pe; and reaction of the solute within the tissue volume is quantified by a zeroeth order consumption rate, $\mathring{Q}$.

While the transport of large molecules such as low density lipoprotein (LDL) and other high molecular weight materials, which are highly impeded by the endothelium, may be limited by the wall and not the fluid (blood), other low molecular weight species which undergo rapid reaction on the endothelial surface (e.g., adenosine triphosphate—ATP) or which are consumed rapidly by the underlying tissue (e.g, oxygen) may be limited by the fluid phase. With these possibilities in mind, the purpose of this short review is to compare the rates of transport in the blood phase to the rates of reaction on the endothelial surface, the rates of transport across the endothelium, and the rates of consumption within the wall of several important biomolecules. It will then be possible to assess quantitatively the importance of fluid phase transport; to determine which molecules are likely to be affected by local fluid mechanics; to determine where in blood vessels these influences are most likely to be manifest; and finally, to speculate about the role of fluid phase mass transport in the localization of atherosclerosis.

100.1 Steady-State Transport Modeling

Reactive Surface

Referring to Fig. 100.1, we will assume that the species of interest is transported from the blood vessel lumen, where its bulk concentration is C_b, to the blood vessel surface, where its concentration is C_s, by a convective-diffusive mechanism which depends on the local fluid mechanics and can be characterized by a fluid-phase mass transfer coefficient k_L (see [6] for further background). The species flux in the blood phase is given by

$$J_s = k_L \left(C_b = C_s \right) \tag{100.1}$$

At the endothelial surface, the species may undergo an enzyme-catalyzed surface reaction (e.g., the hydrolysis of ATP to ADP) which can be modeled using classical Michaelis-Menten kinetics with a rate given by

$$V = \frac{V_{max} C_s}{k_m + C_s} \tag{100.2}$$

where V_{max} is the maximum rate (high C_s) and k_m is the Michaelis constant. When $C_s \ll k_m$, as is often the case, then the reaction rate is pseudo-first order

$$V = k_r C_s \tag{100.3}$$

with the rate constant for the surface reaction given by $k_r = V_{max}/k_m$.

At steady state, the transport to the surface is balanced by the consumption at the surface so that

$$k_L \left(C_b - C_s \right) = k_r C_s \tag{100.4}$$

It will be convenient to cast this equation into a dimensionless form by multiplying it by d/D, where d is the vessel diameter and D is the diffusion coefficient of the transported species in blood, or the media of interest. Equation (100.4) then becomes

$$Sh\left(C_b - C_s \right) = Da_r C_s \tag{100.5}$$

where

$$Sh \equiv \frac{k_L d}{D} \tag{100.6}$$

is the Sherwood number (dimensionless mass transfer coefficient), and

$$Da_r \equiv \frac{k_r d}{D} \tag{100.7}$$

is the Damkhöler number (dimensionless reaction rate coefficient). Solving Eq. (100.5) for the surface concentration one finds

$$C_s/C_b = \frac{1}{1 + Da_r/Sh} \tag{100.8}$$

When $Da_r \ll Sh$,

$$C_s = C_b \tag{100.9}$$

and the process is termed "wall-limited" or "reaction-limited". On the other hand, when $Da_r \gg Sh$,

$$C_s = \left(\frac{Sh}{Da_r} \right) C_b \tag{100.10}$$

and the process is termed "transport-limited"or "fluid phase-limited". It is in this transport-limited case that the surface concentration, and in turn the surface reaction rate, depend on the fluid mechanics which determines the Sherwood number. It will therefore be useful to compare the magnitudes of Da_r and Sh to determine whether fluid mechanics plays a role in the overall transport process of a surface reactive species.

Permeable Surface

Many species will permeate the endothelium without reacting at the luminal surface (e.g., albumin, LDL) and their rate of transport (flux) across the surface layer can be described by

$$J_s = Pe\left(C_s - C_w\right) \tag{100.11}$$

where Pe is the endothelial permeability coefficient and C_w is the wall concentration beneath the endothelium. If the resistance to transport offered by the endothelium is significant, then it will be reasonable to assume

$$C_w \ll C_s \tag{100.12}$$

so that at steady state when the fluid and surface fluxes balance,

$$k_L\left(C_b - C_s\right) = PeC_s \tag{100.13}$$

Multiplying Eq. (100.13) by d/D to introduce dimensionless parameters and then solving for the surface concentration leads to

$$C_s/C_b = \frac{1}{1 + Da_e/Sh} \tag{100.14}$$

where Sh was defined in Eq. (100.6) and

$$Da_e \equiv \frac{Ped}{D} \tag{100.15}$$

is a Damkhöler number based on endothelial permeability. Equation (100.14) shows that when $Da_e \ll Sh$, the transport process is again "wall-limited". When $Da_e \gg Sh$, fluid mechanics again becomes important through the Sherwood number.

Reactive Wall

Oxygen is transported readily across the endothelium (Hellums), but unlike most proteins, is rapidly consumed by the underlying tissue. In this case it is fair to neglect the endothelial transport resistance (assume $C_w = C_s$), and then by equating the rate of transport to the wall with the (zeroeth order) consumption rate within the wall we obtain

$$k_L\left(C_b - C_s\right) = \mathring{Q}\,T \tag{100.16}$$

where $\mathring{Q}$ is the tissue consumption rate and T is the tissue thickness (distance from the surface to the minimum tissue concentration—see Fig. 100.1). For the specific case of O_2 transport, it is conventional to replace concentration (C) with partial pressure (P) through the Henry's law relationship C = KP, where

K is the Henry's law constant. Invoking this relationship and rearranging Eq. (100.16) into a convenient dimensionless form, we obtain

$$\frac{P_s}{P_b} = 1 - \frac{Da_w}{Sh} \tag{100.17}$$

where Sh was defined in Eq. (100.6), and Da_w is another Damkhöler number based on the wall consumption rate

$$Da_w = \frac{\mathring{Q}Td}{KD\,P_b} \tag{100.18}$$

Clearly when $Da_w \ll Sh$, the process is wall limited. But, as $Da_w \rightarrow Sh$, the process becomes limited by transport in the fluid phase ($P_s \rightarrow 0$), and fluid mechanics plays a role. Because we are treating the tissue consumption rate as a zeroeth order reaction, the case $Da_w > Sh$ is not meaningful ($P_s < O$). In reality, as Sh is reduced, the tissue consumption rate must be reduced due to the lack of oxygen supply from the blood.

100.2 Damkhöler Numbers for Important Solutes

A wide range of Damkhöler numbers characterize the transport of biomolecular solutes in vessel walls of the cardiovascular system, and in this section we focus on four important species as examples of typical biotransport processes: adenosine triphosphate (ATP), a species that reacts vigorously on the endothelial surface, albumin, and low density lipoprotein (LDL), species which are transported across a permeable endothelial surface; and oxygen, which is rapidly consumed within the vessel wall. Since most vascular disease (atherosclerosis) occurs in vessels between 3 and 10 mm in diameter, we use a vessel of 5 mm diameter to provide estimates of typical Damkhöler numbers.

Adenosine Triphosphate (ATP)

ATP is degraded at the endothelial surface by enzymes (ectonucleotidases) to form adenosine diphosphate (ADP). The Michaelis-Menten kinetics for this reaction have been determined by Gordon et al. [7] using cultured porcine aortic endothelial cells: $k_m = 249$ μM, $V_{max} = 22$ nmol/min/10^6 cells. V_{max} can be converted to a molar flux by using a typical endothelial cell surface density of 1.2×10^5 cells/cm^2, with the result that the pseudo-first order rate constant (Eq. 100.3) is $k_r = 1.77 \times 10^{-4}$ cm/sec. Assuming a diffusivity of 5.0×10^{-6} cm^2/s for ATP [8], and a vessel diameter of 5 mm, we find

$$Da_r = 17.7$$

Albumin and LDL

These macromolecules are transported across the endothelium by a variety of mechanisms including non-specific and receptor-mediated trancytosis, and paracellular transport through normal or "leaky" inter-endothelial junctions [9, 10]. In rabbit aortas, Truskey et al. [11] measured endothelial permeability to LDL and observed values on the order of $Pe = 1.0 \times 10^{-8}$ cm/sec in uniformly permeable regions, but found that permeability increased significantly in punctate regions associated with cells in mitosis to a level of $Pe = 5 \times 10^{-7}$ cm/sec. Using this range of values for Pe, assuming a diffusivity of 2.5×10^{-7} cm^2/sec for LDL, and a vessel diameter of 5 mm, we find

$$Da_e = 0.02 - 1.0 \qquad \text{(LDL)}$$

For albumin, Truskey et al. [12] reported values of the order $Pe = 4.0 \times 10^{-8}$ cm/sec in the rabbit aorta. This presumably corresponded to regions of uniform permeability. They did not report values in punctate regions of elevated permeability. More recently, Lever et al. [13] reported Pe values of similar magnitude in the thoracic and abdominal aorta as well as the carotid and renal arteries of rabbits. In the ascending aorta and pulmonary artery, however, they observed elevated permeability to albumin on the order of $Pe = 1.5 \times 10^{-7}$ cm/sec. Assuming a diffusivity of 7.3×10^{-7} cm²/sec for albumin, a vessel diameter of 5mm, and the range of Pe values described above, we obtain

$$Da_e = 0.027 - 0.10 \qquad \text{(albumin)}$$

Oxygen

The first barrier encountered by oxygen after being transported from the blood is the endothelial layer. Although arterial endothelial cells consume oxygen [14], the pseudo-first order rate constant for this consumption is estimated to be an order of magnitude lower than that of ATP, and it is therefore reasonable to neglect the endothelial cell consumption relative to the much more significant consumption by the underlying tissue. Liu et al. [15] measured the oxygen permeability of cultured bovine aortic and human umbilical vein endothelial cells and obtained values of 1.42×10^{-2} cm/sec for bovine cell monolayers and 1.96×10^{-2} cm/sec for human cell monolayers. Because the endothelial permeability to oxygen is so high, it is fair to neglect the transport resistance of the endothelium and to direct attention to the oxygen consumption rate within the tissue.

To evaluate the Damkhöler number based on the tissue consumption rate [Eq. (100.17)], we turn to data of Buerk and Goldstick [16] for $\mathring{Q}/(KD)$ measured both *in vivo* and *in vitro* in dog, rabbit, and pig blood vessels. The values of $\mathring{Q}/(KD)$ reported by Buerk and Goldstick are based on tissue properties for KD. To translate these tissue values into blood values, as required in our estimates [Eq. (100.17)], we use the relationship $(KD)_{tissue} = ⅓\,(KD)_{water}$ suggested by Paul et al. [17] and assume $(KD)_{blood} = (KD)_{water}$. In the thoracic aorta of dogs, $\mathring{Q}/(KD)$ ranged from 1.29×10^5 torr/cm² to 5.88×10^5 torr/cm² in the tissue. The thickness (distance to the minimum tissue O_2 concentration) of the thoracic aorta was 250 μm and the diameter is estimated to be 0.9 cm [18]. P_{O_2} measured in the blood (P_b) was 90 torr. Introducing these values into Eq. (100.17) we find:

$$Da_w = 10.8 - 49.0 \qquad \text{(thoracic aorta)}$$

In the femoral artery of dogs, $\mathring{Q}/(KD)$ ranged from 35.2×10^5 torr/cm² to 46.9×10^5 torr/cm² in the tissue. The thickness of the femoral artery was 50 μm and the estimated diameter is 0.4 cm [18]. P_{O_2} measured in the blood was about 80 torr. These values lead to the following estimates:

$$Da_w = 29.3 - 39.1 \qquad \text{(femoral artery)}$$

100.3 Sherwood Numbers in the Circulation

Straight Vessels

For smooth, cylindrical tubes (a model of straight blood vessels) with well-mixed entry flow, one can invoke the thin concentration boundary layer theory of Lévêque [6] to estimate the Sherwood number in the entry region of the vessel where the concentration boundary is developing. This leads to

$$Sh = 1.08 x^{*-1/3} \quad \text{(constant wall concentration)} \qquad (100.19a)$$

$$1.30\, x^{*-1/3} \quad \text{(constant wall flux)} \qquad (100.19b)$$

TABLE 100.1 Transport Characteristics in a Straight Aorta

Species	Sc	x*	Sh	Da
O_2	2,900	4.1×10^{-5}	31.1	10.8–49.0
ATP	7,000	1.7×10^{-5}	41.8	17.7
Albumin	48,000	2.5×10^{-6}	79.2	0.027–0.100
LDL	140,000	8.6×10^{-7}	114	0.02–1.00

Note: d = 1 cm, x = 60 cm, Re = 500, ν = .035 cm²/sec.

where

$$x^* = \frac{x/d}{\text{Re} \cdot \text{Sc}} \tag{100.20}$$

is a dimensionless axial distance which accounts for differing rates of concentration boundary layer growth due to convection and diffusion. In Eq. (100.20), Re = vd/ν is the Reynolds number, Sc = ν/D is the Schmidt number, and their product is the Péclet number. Equation (100.19) is quite accurate for distances from the entrance satisfying $x^* < .001$. Sh continues to drop with increasing axial distance as the concentration boundary layer grows, as described by the classical Graetz solution of the analogous heat transfer problem [19]. When the concentration boundary layer becomes fully developed, Sh approaches its asymptotic minimum value,

$$\text{Sh} = 3.66 \quad \left(\text{constant wall concentration}\right) \tag{100.21a}$$

$$\text{Sh} = 4.36 \quad \left(\text{constant wall flux}\right) \tag{100.21b}$$

For a straight vessel, Sh cannot drop below these asymptotic values. Equations (100.19) and (100.21) also indicate that the wall boundary condition has little effect on the Sherwood number.

It is instructive to estimate Sh at the end of a straight tube having dimensions and flow rate characteristics of the human aorta (actually a tapered tube). Table 100.1 compares Sh and Da (for O_2, ATP, albumin, and LDL) at the end of a 60-cm long model aorta having a diameter of 1 cm and a flow characterized by Re = 500.

Table 100.1 clearly reveals that for a straight aorta, transport is in the entry or Lévêque regime ($x^* < 10^{-3}$). For albumin and LDL, Da ≪ Sh, and transport is expected to be "wall-limited". For O_2 and ATP, Da ~ Sh, and the possibility of "fluid phase-limited" transport exists. At the lowest possible rates of wall mass transport in a straight vessel (Sh = 3.66 – 4.36), transport is still expected to be "wall-limited" for albumin and LDL, whereas it would be "fluid-phase limited" for oxygen and ATP.

100.4 Non-Uniform Geometries Associated with Atherogenesis

Sudden Expansion

Flow through a sudden expansion (Fig. 100.2) at sufficiently high Reynolds number induces flow separation at the expansion point followed by reattachment downstream. This is a simple model of physiological flow separation. The separation zone is associated with low wall shear stress since this quantity is identically zero at the separation and reattachment points.

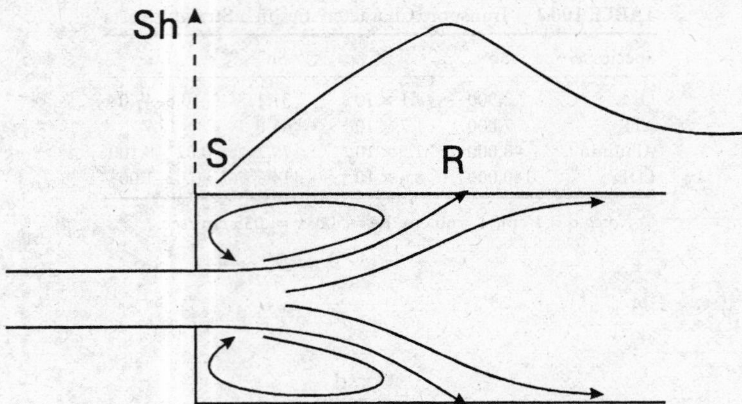

FIGURE 100.2 Schematic diagram showing the spatial distribution of the Sherwood number downstream of a sudden expansion. The flow separates (S) from the wall at the expansion point and reattaches (R) downstream. The Sherwood number is reduced near the separation point (radial velocity away from the wall) and elevated near the reattachment point (radial velocity toward the wall).

An experimental study of oxygen transport in saline and blood for an area expansion ratio of 6.7 and Reynolds numbers in the range 160 to 850 [20] displayed the general spatial distribution of Sh displayed in Fig. 100.2. The minimum value of Sh was observed near the separation point and the maximum value appeared near the reattachment point. The maximum Sh ranged between 500 and several thousand depending on the conditions and the minimum value was approximately 50. A numerical study of the analogous heat transfer problem by Ma et al. [21] at a lower area expansion ratio (4.42) and Schmidt number (Sc = 105) showed the same qualitative trends indicated in Fig. 100.2, but the Sherwood numbers were considerably lower (between 4 and 40) due to the lower expansion ratio, lower Schmidt number, and different entrance conditions. In both of these studies, Sh did not drop below its fully developed tube flow value at any axial location.

The sudden expansion flow field provides insight into the mechanism controlling the spatial distribution of the Sherwood number in separated flows. Near the reattachment point, the radial velocity component convects solute toward the wall (enhancing transport), whereas near the separation point, the radial velocity component convects solute away from the wall (diminishing transport). The net result of this radial convective transport superimposed on diffusive transport (toward the wall) is a maximum in Sh near the reattachment point and a minimum in Sh near the separation point.

Stenosis

Flow through a symmetric stenosis at sufficiently high Reynolds number (Fig. 100.3) will lead to flow separation at a point downstream of the throat and reattachment further downstream. Again, because the wall shear stress is identically zero at the separation and reattachment points, the separation zone is a region of low wall shear stress. Conversely, converging flow upstream of the stenosis induces elevated wall shear stress.

Schneiderman et al. [22] performed numerical simulations of steady flow and transport in an axisymmetric, 89% area restriction stenosis with a sinusoidal axial wall contour at various Reynolds numbers for a Schmidt number typical of oxygen transport. Relative to a uniform tube, Sh was always elevated in the converging flow region upstream of the stenosis, and when Re was low enough to suppress flow separation, transport remained elevated downstream of the stenosis as well. At high Re (Fig. 100.3), flow separation produced a region of diminished transport (relative to a uniform tube) while reattachment induced elevated transport. The lowest value of Sh in the diminished transport regime was approximately 10. This occurred at Re = 16, a flow state for which separation was incipient. At high Re, the minimum Sh was elevated above 10, and the region of diminished transport was reduced in axial extent.

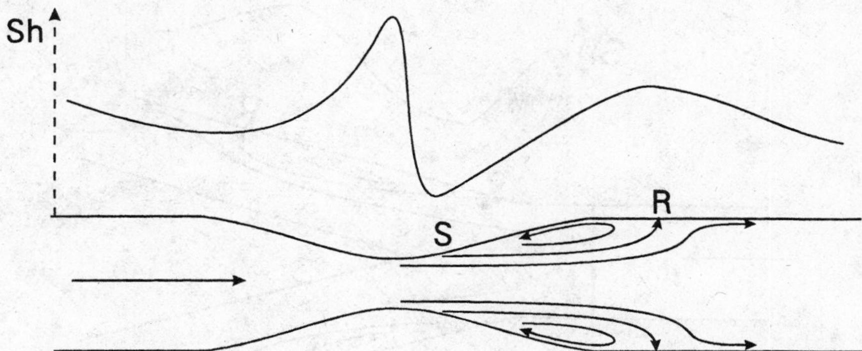

FIGURE 100.3 Schematic diagram showing the spatial distribution of the Sherwood number around a symmetric stenosis. The flow converges upstream of the stenosis where the Sherwood number is elevated (radial velocity toward the wall). The flow separates (S) from the wall just downstream of the throat, if the Reynolds number is high enough, and reattaches (R) downstream. The Sherwood number is reduced near the separation point (radial velocity away from the wall) and elevated near the reattachment point (radial velocity toward the wall).

More recent numerical simulations by Rappitsch and Perktold [23, 24] for a 75% stenosis in steady flow (Re = 448, Sc = 2,000 (oxygen)) and sinusoidal flow ($\bar{\text{Re}}$ = 300, Sc = 46,000 (albumin)) show the same basic trends depicted in Fig. 100.3. Again, Sh was reduced in a narrow region around the separation point, but did not drop below ~10.

Moore and Ethier [25] also simulated oxygen transport in a symmetric stenosis, but they accounted for the binding of oxygen to hemoglobin and solved for both the oxygen and oxyhemoglobin concentrations. They determined axial Sh profiles throughout the stenosis, which showed the same basic tendencies indicated in Fig. 100.3. However, because hemoglobin has a much higher molecular weight and Schmidt number than oxygen (about 100 times higher Sc), the local values of Sh were higher than computed on the basis of free oxygen transport alone. This is expected since simple transport considerations (Lévêque solution) suggest Sh α $Sc^{1/3}$.

The Sh profile for the stenosis (Fig. 100.3) reflects the same underlying mechanisms that were operative in the sudden expansion (Fig. 100.2): radial flow directed toward the wall enhances transport near the reattachment point, while radial flow directed away from the wall diminishes transport near the separation point.

Bifurcation

A few numerical studies of flow and transport in the carotid artery bifurcation have been reported recently as summarized in Fig. 100.4. The carotid artery bifurcation is a major site for the localization of atherosclerosis, predominantly on the outer wall (away from the flow divider) in the flow separation zone which is a region of low and oscillating wall shear stress [26]. Perktold et al. [27] simulated O_2 transport in a realistic pulsatile flow through an anatomically realistic three-dimensional carotid bifurcation geometry using a constant wall concentration boundary condition. Ma et al. [28] simulated oxygen transport in steady flow through a realistic, three-dimensional, carotid bifurcation with a constant wall concentration boundary condition.

As in the sudden expansion and stenosis geometries, the bifurcation geometry can induce flow separation on the outer wall with reattachment downstream. Again there is a region of attenuated transport near the separation point and amplified transport near the reattachment point. Perktold et al. [27] predicted minimum Sherwood numbers close to zero in the flow separation zone. Ma et al. [28] predicted the same general spatial distribution, but the minimum Sherwood number was ~25. Differences in the minimum Sherwood number may be due to differing entry lengths upstream of the bifurcation as well as differences in flow pulsatility.

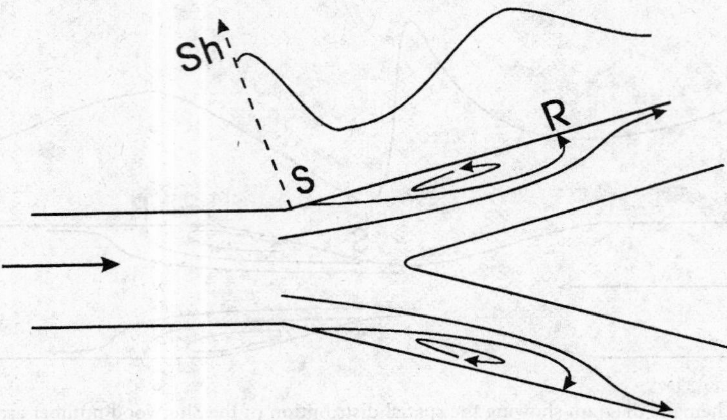

FIGURE 100.4 Schematic diagram showing the spatial distribution of the Sherwood number along the outer wall of a bifurcation. The flow separates (S) if the Reynolds number is high enough and there is an increase in cross-sectional area through the bifurcation, and reattaches (R) downstream. The Sherwood number is reduced near the separation point (radial velocity away from the wall) and elevated near the reattachment point (radial velocity toward the wall).

Curvature

Localization of atherosclerosis has also been associated with arterial curvature [2]. For example, the inner curvature of proximal coronary arteries as they bend over the curved surface of the heart has been associated with plaque localization [29]. Qiu [30] carried out three-dimensional, unsteady flow computations in an elastic (moving wall) model of a curved coronary artery for O_2 transport with a constant wall concentration boundary condition. He obtained the results shown in Fig. 100.5. Because Qiu assumed uniform axial velocity and concentration profiles (with no secondary flow) at the curved tube entrance (0 degrees), there is an entrance region (~25 degrees) where Sh essentially follows a Lévêque boundary layer development. Eventually (~50 degrees), the secondary flow effects become manifest, and marked differences in transport rates between the inside wall (low transport: Sh ~2) and the outside wall (high transport: Sh ~55) develop. The regions of high and low transport in the curved vessel geometry cannot be associated with axial flow separation (as in the expansion, stenosis, and bifurcation) because flow separation does not occur at the modest curvature levels in the coronary artery simulation.

The secondary flow (in the plane perpendicular to the axial flow—Fig. 100.5) determines, in large measure, the differences in transport rates between the inside and outside walls. The radial velocity is directed toward the wall at the outside, leading to enhancement of transport by a convective mechanism. At the inside of the curvature, the radial velocity is directed away from the wall, and transport is impeded by the convective mechanism. This secondary flow mechanism produces transport rates that are 25 times higher on the outside wall than the inside wall. This is to be contrasted with wall shear stress values which, for the coronary artery condition, are less than 2 times higher on the outside wall than on the inside wall [31]. Earlier studies of fully developed, steady flow and transport in curved tubes [32] are consistent with the above observations.

100.5 Discussion

The considerations of mass transport in the fluid (blood) phase and consumption by the vessel wall described in the preceding sections indicate that only highly reactive species such as O_2 and ATP can be "transport-limited". Larger molecules such as albumin and LDL, which are not rapidly transformed within the vessel wall, are not likely to be transport-limited.

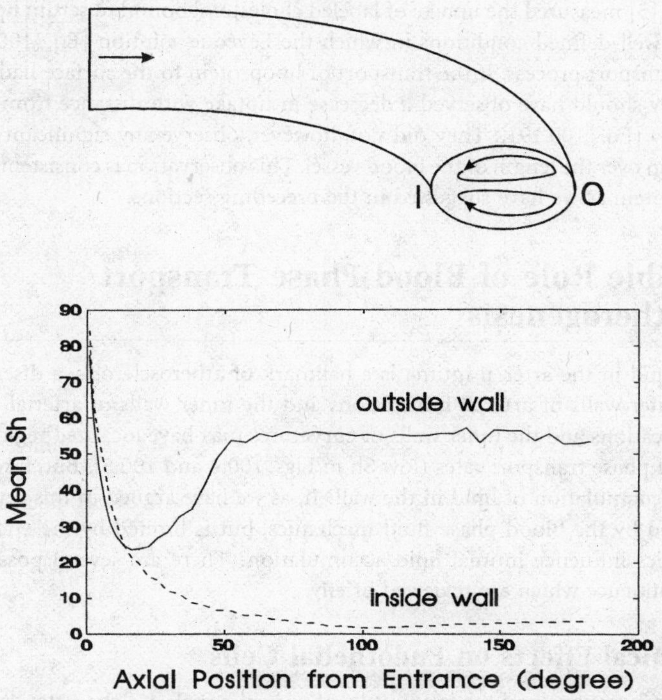

FIGURE 100.5 Schematic diagram showing the spatial distribution of the Sherwood number along the inner (I—toward the center of curvature) and outer (O—away from the center of curvature) walls of a curved vessel. In the entry region, before the secondary flow has developed, the Sherwood number follows a Lévêque distribution. As the secondary flow evolves, the Sherwood number becomes elevated on the outer wall where the radial velocity of the secondary flow is toward the wall, and diminished on the inner wall where radial velocity of the secondary flow is away from the wall.

For O_2, ATP, and other molecules characterized by large values of the Damkhöler number, transport limitation will occur in regions of the circulation where the Sherwood number is low (Sh < Da). Localized regions of low Sh arise in non-uniform geometries around flow separation points (not reattachment points) where radial flow velocities are directed away from the wall, and in secondary flow regions where the secondary velocity is directed away from the wall (inside wall of a curved vessel).

Considerable supporting experimental evidence for the above conclusions is available, and a few representative studies will be mentioned here. Santilli et al. [33] measured transarterial wall oxygen tension gradients at the dog carotid bifurcation. Oxygen tensions at the carotid sinus (outer wall in Fig. 100.4) were significantly decreased in the inner 40% of the artery wall compared to an upstream control location. Oxygen tensions at the flow divider (inner wall in Fig. 100.4) were increased significantly throughout the artery wall compared with control locations. These observations are consistent with fluid phase-limited oxygen transport at the outer wall of the carotid bifurcation.

Dull et al. [8] measured the response of intracellular free calcium in cultured bovine aortic endothelial cells (BAECs) after step changes in flow rate, using media containing either ATP or a non-reactive analog of ATP. In the presence of ATP, which is rapidly degraded by enzymes on the endothelial cell surface, step changes in flow rate induced rapid changes in intracellular calcium which were not apparent when the inactive ATP analog was used in place of ATP. The interpretation of these experiments was that increases in flow increased the rate of mass transport of ATP to the cell surface and exceeded the capacity of the surface enzymes to degrade ATP. This allowed ATP to reach surface receptors and stimulate intracellular calcium. This study thus provided evidence that ATP transport to the endothelium could be fluid phase (transport)-limited.

Caro and Nerem [5] measured the uptake of labeled cholesterol bound to serum lipoprotein in excised dog arteries under well-defined conditions in which the Lévêque solution [Eq. (100.19)] described the fluid phase mass transport process. If the transport of lipoprotein to the surface had been controlled by the fluid phase, they should have observed a decrease in uptake with distance from the vessel entrance following a $x^{-1/3}$ law [Eq. (100.19)]. They did not, however, observe any significant spatial variation of uptake of lipoprotein over the length of the blood vessel. This observation is consistent with "wall-limited" transport of lipoprotein as we have suggested in the preceding sections.

100.6 Possible Role of Blood Phase Transport in Atherogenesis

Accumulation of lipid in the arterial intima is a hallmark of atherosclerosis, a disease that tends to be localized on the outer walls of arterial bifurcations and the inner walls of arterial curvatures [2]. The outer walls of bifurcations and the inner walls of curvatures may have localized regions characterized by relatively low blood phase transport rates (low Sh in Figs. 100.4 and 100.5). But, how can low transport rates lead to high accumulation of lipid in the wall? If, as we have argued in this review, LDL transport is really not affected by the blood phase fluid mechanics, but is limited by the endothelium, how can local fluid mechanics influence intimal lipid accumulation? There are several possible scenarios for a fluid mechanical influence which are reviewed briefly.

Direct Mechanical Effects on Endothelial Cells

The outerwalls of bifurcations and the inner walls of curved vessels are characterized by low mean wall shear stress, and significant temporal oscillations in wall shear stress direction (oscillatory shear stress) [2]. Endothelial cells in this low, oscillatory shear environment tend to assume a polyhedral, cobblestone enface morphology, whereas endothelial cells in high shear regions tend to be elongated in the direction of flow. It has been suggested that these altered morphologies, which represent chronic adaptive responses to altered fluid mechanical environments, are characterized by distinct macromolecular permeability characteristics [2, 10] as direct responses of the endothelial layer to altered mechanical environments. In addition, a number of studies have shown that fluid shear stress on the endothelial surface can have an acute influence on endothelial transport properties both *in vitro* [34, 35] and *in vivo* [36, 37].

Hypoxic Effect on Endothelial Cells

Hypoxia (low oxygen tension), which can be induced by a blood phase transport limitation, can lead to a breakdown of the endothelial transport barrier either by a direct effect on the endothelial layer or by an indirect mechanism in which hypoxia up-regulates the production of hyperpermeabilizing cytokines from other cells in the arterial wall. A number of recent studies have shown that hypoxia increases macromolecular transport across endothelial monolayers in culture due to metabolic stress [38-40]. These studies describe direct effects on the endothelial layer since other cells present in the vessel wall were not present in the cell culture systems.

Hypoxia Induces VEGF

Many cell lines express increased amounts of vascular endothelial growth factor (VEGF) when subjected to hypoxic conditions as do normal tissues exposed to hypoxia, functional anemia, or localized ischemia [41]. VEGF is a multifunctional cytokine that acts as an important regulator of angiogenesis and as a potent vascular permeabilizing agent [41, 42]. VEGF is believed to play an important role in the hyperpermeability of microvessels in tumors, the leakage of proteins in diabetic retinopathy, and other vascular pathologies [43]. Thus, a plausible scenario for the increase in lipid uptake in regions of poor blood phase mass transport is the following: Hypoxia up-regulates the production of VEGF by cells within the vascular wall and the VEGF in turn permeabilizes the endothelium, allowing increased transport of lipid into the wall. This mechanism can be depicted schematically as follows:

$$O_2 \downarrow \rightarrow \text{VEGF} \uparrow \rightarrow \text{Pe} \uparrow$$

In support of this view, several recent studies have shown that VEGF is enriched in human atherosclerotic lesions [44, 45]. Smooth muscle cells and macrophages appear to be the predominant sources of VEGF in such lesions. Thus, a mechanism in which hypoxia induces VEGF and hyperpermeability is plausible, but at the present time it must only be considered a hypothesis relating fluid phase transport limitation and enhanced macromolecular permeability.

References

1. Ross, R., Atherosclerosis: A defense mechanism gone awry, *Am. J. Pathol.*, 143, 987, 1993.
2. Nerem, R., Atherosclerosis and the role of wall shear stress, in: *Flow-Dependent Regulation of Vascular Function*, Bevan, J. A., Kaley, G., and Rubany, G. M., Eds., Oxford University Press, New York, 1995.
3. Caro, L. G., Fitz-Gerald, J. M., and R. C. Schroter, Atheroma and arterial wall shear: observation, correlation and proposal of a shear dependent mass transfer mechanism for atherogenesis, *Proc. R. Soc. London* (*Biol.*), 177, 109, 1971.
4. Fry, D. L., Acute vascular endothelial changes associated with increased blood velocity gradients, *Circ. Res.*, 22, 165, 1968.
5. Caro, C. G. and Nerem, R. M., Transport of ^{14}C-4-cholesterol between serum and wall in the perfused dog common carotid artery, *Circ. Res.*, 32, 187, 1973.
6. Basmadjian, D., The effect of flow and mass transport in thrombogenesis, *Ann. Biomed. Eng.*, 18, 685, 1990.
7. Gordon, E. L., Pearson, J. D., and Slakey, L. L., The hydrolysis of extra cellular adenine nucleotides by cultured endothelial cells from pig aorta, *J. Biol. Chem.*, 261, 15496, 1986.
8. Dull, R. O., Tarbell, J. M., and Davies, P. F., Mechanisms of flow-mediated signal transduction in endothelial cells: kinetics of ATP surface concentrations, *J. Vasc. Res.*, 29, 410, 1992.
9. Lin, S. J., Jan, K. M., Schuessler, G., Weinbaum, S., and Chien, S., Enhanced macromolecular permeability of aortic endothelial cells in association with mitosis, *Atherosclerosis*, 17, 71, 1988.
10. Weinbaum, S. and Chien, S., Lipid transport aspects of atherogensis, *J. Biomech. Eng.*, 115, 602, 1993.
11. Truskey, G. A., Roberts, W. L., Herrmann, R. A., and Malinauskas, R. A., Measurement of endothelial permeability to ^{125}I-low density lipoproteins in rabbit arteries by use of en face preparations, *Circulation Research*, 71, 883, 1992.
12. Truskey, G. A., Colton, C. K., and Smith, K. A., Quantitative analysis of protein transport in the arterial wall, in *Structure and Function of the Circulation*, 3, Schwarts, C. J., Werthessen, N. T., and Wolf, S., Eds., Plenum Publishing Corp., 1981, 287.
13. Lever, M. J., Jay, M. T., and Coleman, P. J., Plasma protein entry and retention in the vascular wall: possible factors in atherogenesis, *Can. J. Physiol. Pharmacol.*, 74, 818, 1996.
14. Motterlini, R., Kerger, H., Green, C. J., Winslow, R. M., and Intaglietta, M., Depression of endothelial and smooth muscle cell oxygen consumption by endotoxin. *Am. J. Physiol.*, 275, H776, 1998.
15. Liu, C. Y., Eskin, S. G., and Hellums, J. D., The oxygen permeability of cultured endothelial cell monolayers, paper presented at the 20th International Society of Oxygen Transport to Tissue Conference, August 26-30, Mianz, Germany, 1992.
16. Buerk, D. G. and Goldstick, T. K., Arterial wall oxygen consumption rate varies spatially, *Am. J. Physiol.*, 243, H948, 1982.
17. Paul, R. J., Chemical energetics of vascular smooth muscle, in *Handbook of Physiology. The Cardiovascular System. Vascular Smooth Muscle*, 2, Am. Physiol. Soc., Bethesda, MD, 1980, 201.

18. Caro, C. G., Pedley, T. J., Schroter, R. C., and Seed, W. A., *The Mechanics of the Circulation*, Oxford University Press, Oxford, 1978.
19. Bennett, C. O. and Meyers, J. E., *Momentum, Heat and Mass Transfer*, McGraw-Hill, New York, 1962.
20. Thum, T. F. and Diller, T. E., Mass transfer in recirculating blood flow, *Chem. Eng. Commun.*, 47, 93, 1986.
21. Ma, P., Li, X., and Ku, D. N., Heat and mass transfer in a separated flow region of high Prandtl and Schmidt numbers under pulsatile conditions, *Int. J. Heat Mass Transfer*, 1994.
22. Schneiderman, G., Ellis, C. G., and Goldstick, T. K., Mass transport to walls of stenosed arteries: Variation with Reynolds number and blood flow separation, *J. Biomechanics*, 12, 869, 1979.
23. Rappitsch, G. and Perktold, K., Computer simulation of convective diffusion processes in large arteries, *J. Biomechanics*, 29, 207, 1996.
24. Rappitsch, G. and Perktold, K., Pulsatile albumin transport in large arteries: A numerical simulation study, *J. Biomech. Eng.*, 118, 511, 1996.
25. Moore, J. A. and Ethier, C. R., Oxygen mass transfer calculations in large arteries, *J. Biomech. Eng.*, 119, 469, 1997.
26. Ku, D. N., Giddens, D. P., Zarins, C. K., and Glagov, S., Pulsatile flow and atherosclerosis in the human carotid bifurcation: positive correlation between plaque location and low and oscillating shear stress, *Arteriosclerosis*, 5, 293, 1985.
27. Perktold, K., Rappitsch, G., Hofer, M., and Karner, G., Numerical simulation of mass transfer in a realistic carotid artery bifurcation model, Proceedings of the 1997 Bioengineering Conference (ASME), *BED*, 35, 85, 1997.
28. Ma, P., Li, X., and Ku, D. N., Convective mass transfer at the carotid bifurcation, *J. Biomechanics*, 30, 565, 1997.
29. Chang, L. J. and Tarbell, J. M., A numerical study of flow in curved tubes simulating coronary arteries, *J. Biomechanics*, 21, 927, 1988.
30. Qiu, Y., Numerical simulation of oxygen transport in a compliant curved tube model of a coronary artery, M.S. Thesis, The Pennsylvania State University, University Park, PA, 1999.
31. Qiu, Y. and Tarbell, J., Numerical simulation of pulsatile flow in a compliant curved tube model of a coronary artery, *J. Biomech. Eng.*, in review, 1999.
32. Kalb, C. E. and Seader, J. D., Heat and mass transfer phenomena for viscous flow in curved circular tubes, *Int. J. Heat Mass Transfer*, 15, 801, 1972.
33. Santilli, S. M., Stevens, R. B., Anderson, J. G., Payne, W. D., and Caldwell, M. D-F., Transarterial wall oxygen gradients at the dog carotid bifurcation, *Am. J. Physiol.*, 268, H155, 1995.
34. Jo, H., Dull, R. O., Hollis, T. M., and Tarbell, J. M., Endothelial permeability is shear-dependent, time-dependent, and reversible, *Am. J. Physiol.*, 260, H1992, 1991.
35. Sill, H. W., Chang, Y. S., Artman, J. R., Frangos, J. A., Hollis, T. M., and Tarbell, J. M., Shear stress increases hydraulic conductivity of cultured endothelial monolayers, *Am. J. Physiol.*, 268, H535, 1995.
36. Williams, D. A., Thipakorn, B., and Huxley, V. H., In situ shear stress related to capillary function, *FASEB J.*, 8, M17, 1994.
37. Yuan, Y., Granger, H. J., Zawieja, D. C., and Chilian, W. M., Flow modulates coronary venular permeability by a nitric oxide-related mechanism, *Am. J. Physiol.*, 263, H641, 1992.
38. Fischer, S., Renz, D., Schaper, W., and Karliczek, G. F., Effects of barbiturates on hypoxic cultures of brain derived microvascular endothelial cells, *Brain Res.*, 707, 47, 1996.
39. Kondo, T., Kinouchi, H., Kawase, M., and Yoshimoto, T., Astroglial cells inhibit the increasing permeability of brain endothelial cell monolayer following hypoxia/reoxygeneration, *Neurosci. Lett.*, 208, 101, 1996.
40. Plateel, M., Teissier, E., and Cecchelli, R., Hypoxia dramatically increases the nonspecific transport of blood-borne proteins to the brain, *J. Neurochemistry*, 68, 874, 1997.

41. Brown, L., Detmer, M., Claffey, K., Nagy, J., Peng, D., Dvorak, A., and Duorak, H., Vascular permeability factor/vascular endothelial growth factor: a multifunctional angiogenic cytokine, in *Regulation of Angiogenesis*, Goldberg, I. D. and Rosen, E. M., Eds., Birkhäuser Verlag, Basel, Switzerland, 1997.
42. Ferrara, N. and Davis-Smyth, T., The biology of vascular endothelial growth factor, *Endocrine Rev.*, 18, 4, 1997.
43. Chang, Y. S., Munn, L. L., Hillsley, M. V., Dull, R. O., Gardner, T. W., Jain, R. K., and Tarbell, J. M., Effect of vascular endothelial growth factor on cultured endothelial monolayer transport properties, *Microvascular Res.*, in review, 1998.
44. Couffinhal, T., Kearney, M., Witxzenbichler, B., Chen, D., Murohara, T., Losordo, D. W., Symes, J., and Isner, J. M., Vascular endothelial growth factor/vascular permeability factor (VEGF/VPF) in normal and atherosclerotic human arteries, *Am. J. Pathol.*, 150, 1673, 1997.
45. Ramos, M. A., Kuzuya, M., Esaki, T., Miura, S., Satake, S., Asai, T., Kanda, S. Hayashi, T., and Iguchi, A., Induction of macrophage VEGF in response to oxidized LDL and VEGF accumulation in human atherosclerotic lesions, *Arterioscler. Thromb. Vasc. Biol.*, 18, 1188, 1998.

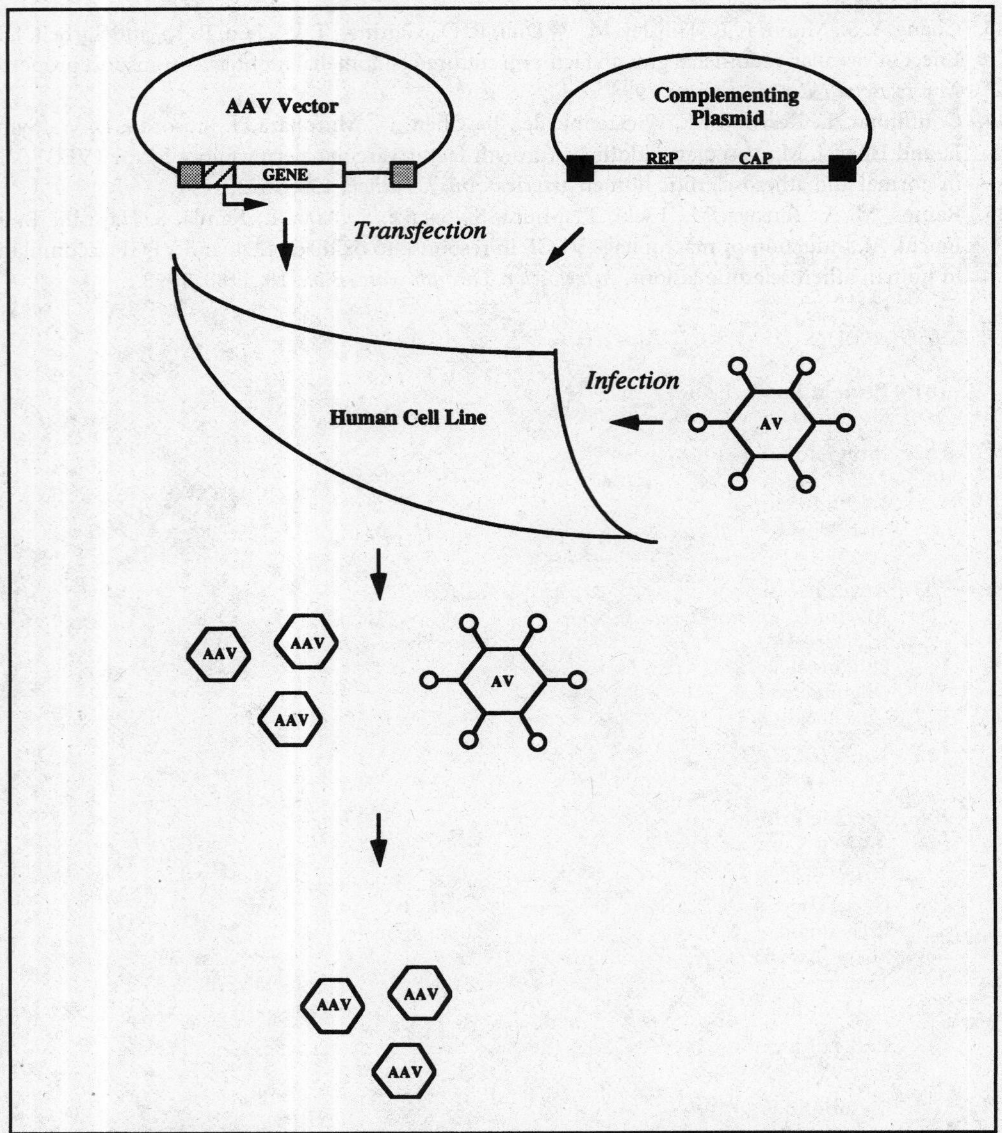

Production of recombinant adeno-associated virus (AAV), which is used in treating cystic fibrosis patients.

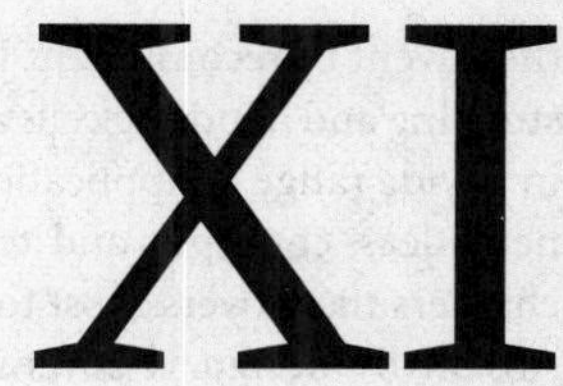

Biotechnology

Martin L. Yarmush and Mehmet Toner
Massachusetts General Hospital, Harvard Medical School, and the Shriners Burns Hospital

THE TERM *BIOTECHNOLOGY* HAS UNDERGONE significant change over the past 50 years or so. During the period prior to the eighties, biotechnology referred primarily to the use of microorganisms for large-scale industrial processes such as antibiotic production. Since the 1980s, with the advent of recombinant DNA technology, monoclonal antibody technology, and new technologies for studying and handling cells and tissues, the field of biotechnology has undergone a tremendous resurgence in a wide range of applications pertinent to industry, medicine, and science in general. It is some of these new ideas, concepts, and technologies that will be covered in this section. We have assembled a set of chapters that covers most topics in biotechnology that might interest the practicing biomedical engineer. Absent by design is coverage of agricultural, bioprocess, and environmental biotechnology, which is beyond the scope of this *Handbook*.

Chapter 101 deals with our present ability to manipulate genetic material. This capability, which provides the practitioner with the potential to generate new proteins with improved biochemical and physicochemical properties, has led to the formation of the field of protein engineering. Chapter 102 discusses the field of monoclonal antibody production in terms of its basic technology, diverse applications, and ways that the field of recombinant DNA technology is currently "reshaping" some of the earlier constructs. Chapters 103 and 104 describe applications of nucleic acid chemistry. The burgeoning field of antisense technology is introduced with emphasis on basic techniques and potential applications to AIDS and cancer, and Chapter 104 is dedicated toward identifying the computational, chemical, and machine tools which are being developed and refined for genome analysis. Applied virology is the implied heading for Chapters 105 and 106, in which viral vaccines and viral-mediated gene therapy are the main foci.

Finally, Chapters 107 and 108 focus on important aspects of cell structure and function. These topics share a common approach toward quantitative analysis of cell behavior in order to develop the principles for cell growth and function. By viewing the world of biomedical biotechnology through our paradigm of proteins and nucleic acids to viruses to cells, today's biomedical engineer will hopefully be prepared to meet the challenge of participating in the greater field of biotechnology as an educated observer at the very least.

101

Protein Engineering

Alan J. Russell
University of Pittsburgh

Chenzhao Vierheller
University of Pittsburgh

Enzymes have numerous applications in both research and industry. The conformation of proteins must be maintained in order for them to function at optimal activity. Protein stability is dependent on maintaining a balance of forces that include hydrophobic interactions, hydrogen bonding, and electrostatic interactions. Most proteins are denatured, i.e., lose their active conformation in high-temperature environments (exceptions to this are found among the enzymes from thermophilic microorganisms). Therefore, understanding and maintaining enzyme stability are critical if enzymes are to be widely used in medicine and industry. Protein engineering [1] is used to construct and analyze modified proteins using molecular biologic, genetic engineering, biochemical, and traditional chemical methods (Table 101.1). The generation of proteins with improved activity and stability is now feasible. Recent developments in molecular biology also have enabled a rapid development of the technologies associated with protein engineering (Table 101.2).

101.1 Site-Direction Mutagenesis

Since site-directed mutagenesis was described by Hutchinson et al. [2] in 1978, it has become a powerful tool to study the molecular structure and function of proteins. The purpose of site-directed mutagenesis is to alter a recombinant protein by introducing, replacing, or deleting a specific amino acid. The technique enables a desired modification to be achieved with exquisite precision [3]. Site-directed mutagenesis has been used to change the activity and stability of enzymes, as well as substrate specificity and affinity. Indeed, much of the biologic detergent enzyme sold in the United States (~$200 million per year) is a protein-engineered variant of the native enzyme.

The basic idea behind site-directed mutagenesis is illustrated in Fig. 101.1. The first step of site-directed mutagenesis involves cloning the gene for the protein to be studied into a vector. This nontrivial exercise is discussed elsewhere in this *Handbook*. Next, an oligonucleotide is designed and synthesized. The oligonucleotide will have a centrally located desired mutation (usually a mismatch) that is flanked by sequences of DNA which are complementary to a specific region of interest. Thus the oligonucleotide is designed to bind to a single region of the target gene. The mutation can then be introduced into the gene by hydridizing the oligonucleotide to the single-stranded template. Single-stranded templates of a target gene may be obtained by either cloning the gene to a single-stranded vector, such as bacteriophage M13 [4], or by using phagemids (a chimeric plasmid containing a filamentous bacteriophage replication origin that directs synthesis of single-stranded DNA with a helper bacteriophage) [5]. Alternatively, single-stranded template may be generated by digesting double-stranded DNA with exonuclease III following nicking of the target DNA with DNase or a restriction endonuclease [6].

0-8493-0462-8/00/$0.00+$.50

TABLE 101.1 Selected Methods for Modification of Proteins

Chemical/biochemical methods	Side chain/amino acid residue modification
	Immobilization
Biologic methods	Site-directed mutagenesis
	Random mutagenesis

The second strand is synthesized with DNA polymerase using the oligonucleotide as a primer, and the DNA can be circularized with DNA ligase. The vector that carries the newly synthesized DNA (in which one strand is mutated) can be introduced into a bacterial host, where, because DNA duplicates in a semiconservative mode, half the newly synthesized cells containing the DNA will theoretically contain the mutation. In reality, premature DNA polymerization, DNA mismatch repair, and strand displacement synthesis result in much lower yield of mutants. Different strategies have been developed to obtain a higher level of mutation efficiency in order to minimize the screening of mutant. A second mutagenic oligonucleotide containing an active antibiotic resistance gene can be introduced at the same time as the site-directed oligonucleotide. Therefore, the wild-type plasmid will be eliminated in an antibiotic-selective medium [7] (Fig. 101.2). Alternatively, introducing a second mutagenic oligonucleotide can result in the elimination of a unique restriction endonuclease in the mutant strand [8]. Therefore, wild-type plasmid can be linearized with this unique restriction endonuclease, while the circular mutant plasmid can be transformed into bacteria host. Another alternative method is to perform second-strand extension in the presence of 5-methyl-dCTP, resulting in resistance of number of restriction enzymes, including HpaII, MspI, and Sau3A I, in the mutant strand only [9]. The template DNA can then be nicked with these enzymes, followed by digestion with exonuclease III, to increase mutagenesis efficiency.

The recent development of the polymerase chain reaction provides a new approach to site-direct mutagenesis. In 1989, Ho and colleagues [10] developed a method named *overlap extension* (Fig. 101.3), in which four oligonucleotides are used as primers. Two of the primers containing the mutant are complementary to each other. The two other primers are complementary to the opposite strand of the ends of the cloned genes. Polymerase chain reactions are performed three times. The first two polymerase chain reactions are carried out using one end primer and one mutant primer in each reaction. The products contain one double-stranded DNA from one end to the mutation point and the other double-stranded DNA from another end to the mutation point. In other words, the two DNA products are overlapped in the mutated region. The polymerase chain reaction products are purified and used as templates for the third polymerase chain reaction with two end primers. This generates the whole length of DNA with the desired mutation. The advantage of this method is that it can be done quickly with nearly 100% efficiency.

After screening and selection, a mutation is generally confirmed by DNA sequencing. The mutant can then be subcloned into an expression vector to test the effect of mutation on the activity of the enzyme.

TABLE 101.2 Potentially Useful Modifications to Proteins

Stability	Increased thermostability
	Increased stability at extremes of pH
	Increased stability in organic solvents
	Resistance to oxidative inactivation
	Resistance to proteolysis
Kinetics	Increased maximum velocity
	Altered affinity for substrate
	Altered substrate specificity
	Resistance to substrate/product inhibition
Biology	Altered spectrum of activity
	Altered substrate specificity

Source: Modified from Primrose SB. 1991. In Molecular Biotechnology, 2d ed. New York, Blackwell Scientific.

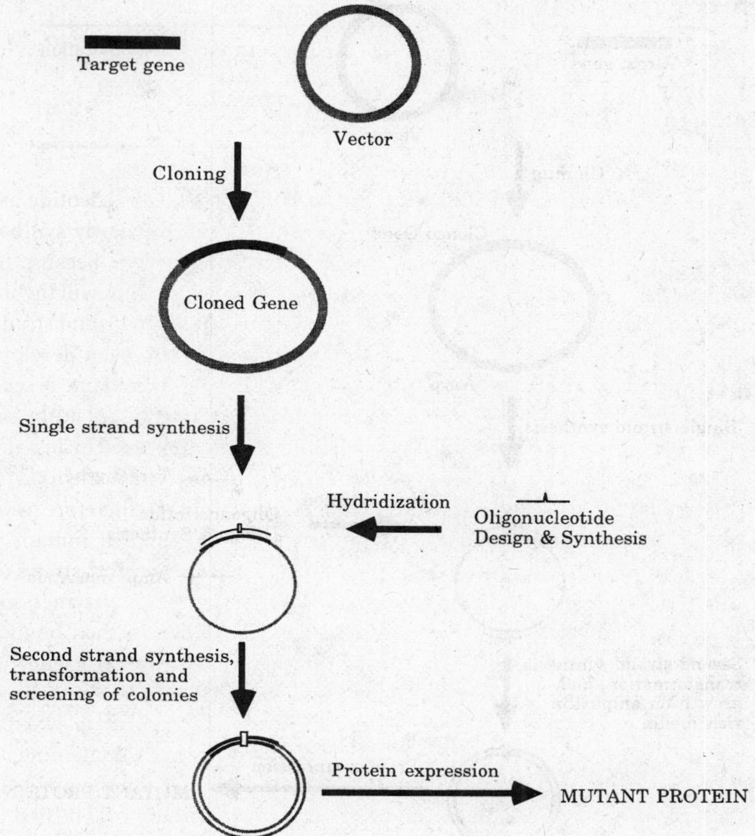

FIGURE 101.1 Overview of site-directed mutagenesis. See text for details.

Over the last decade, site-directed mutagenesis has become a somewhat trivial technical process, and the most challenging segment of a protein engineering research project is unquestionably the process of determining why a mutant protein has altered properties. The discussion of this particular enterprise lies beyond the scope of this brief review.

101.2 Solvent Engineering

An enzyme-catalyzed reaction can be simplified to its most basic components by considering it as the transfer of a substrate molecule from solvent to the surface of an enzyme molecule. The exchange of substrate-solvent and enzyme-solvent interactions for enzyme-substrate interactions then enables the chemistry of catalysis to take place. *Protein engineering,* as described above, is the process of changing the enzyme in a predictable and precise manner to effect a change on the catalytic process. Since the enzyme is only one side of the balance, however, any changes in the rest of the equation also will alter the catalytic process. Until the late 1980s, for example, substrate specificity could be altered by either protein engineering of the enzyme or by changing the substrate. *Solvent engineering* is now also emerging as a powerful tool in rational control of enzyme activity. Using solvents other than water has successfully led to enzymes with increased thermostability, activity against some substrates, pH dependence, and substrate and enantiospecificity.

The simplicity of the solvent-engineering approach is clear. If a protein is not inactivated by a solvent change, then its activity will be dependent on the solvent in which the enzyme and substrate are placed.

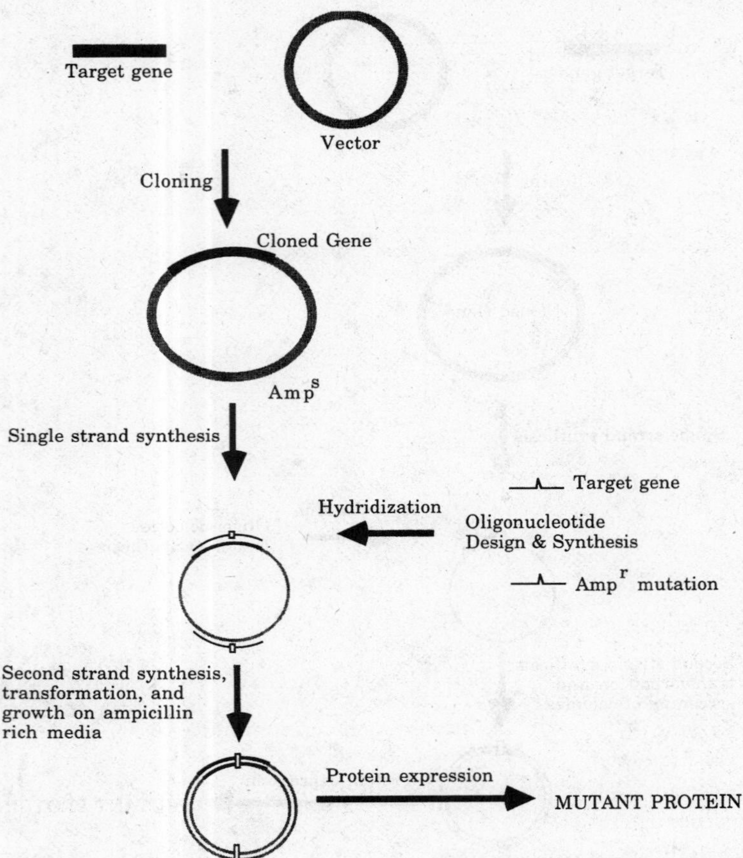

FIGURE 101.2 Selective antibiotic genes as strategy for improvement of site-directed mutagenesis. See text for details.

This strategy has been discussed in detail in recent reviews [11], and we will only summarize the most pertinent information here.

Enzymes that have been freeze-dried and then suspended in anhydrous organic solvents, in which proteins are not soluble, retain their activity and specificity [12]. The enzyme powders retain approximately a monolayer of water per molecule of enzyme during the freeze-drying process. As long as this monolayer of water remains associated with the enzyme in an organic environment, the structure of the enzyme is not disrupted [13], and hence enzyme activity in essentially anhydrous organic solvents can be observed.

The physical properties of a solvent in which an enzyme is placed can influence the level of activity and specificity in a given reaction. For instance, alcohol dehydrogenase can catalyze oxidation-reduction reactions equally well in buffer and heptane, but the level of activity in heptane is sharply dependent on the thermodynamic activity of water in the system. In general, activity can be increased by increasing the hydrophobicity of the solvent used. Interestingly, however, the specificity (both stereo- and substrate) of an enzyme or antibody dispersed in an organic solvent is generally greater in more hydrophilic solvents [14]. Water is involved as a substrate in many of the chemical modification processes that lead to irreversible inaction of proteins. Not surprisingly, therefore, proteins in anhydrous environments are stable at temperatures exceeding 100°C [15].

Considerable attention is now being given to developing the predictability of the solvent-engineering approach. Elucidating the structure-function-environment relationships that govern the activity and

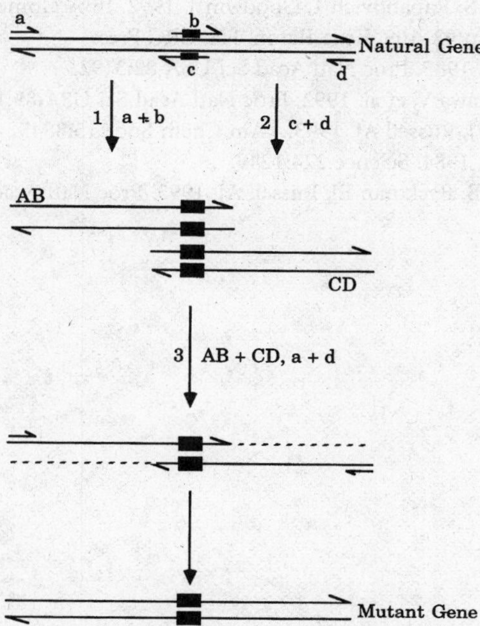

FIGURE 101.3 Site-directed mutagenesis by overlap extension. See text for details.

specificity of an enzyme is a crucial step for learning how to apply the power of solvent engineering. In recent years, supercritical fluids (materials at temperatures and pressures above their critical point) have been used as a dispersant for enzyme-catalyzed reactions. In addition to being excellent process solvents, the physical properties of supercritical fluids are pressure-dependent. Thus it may be possible to detect which physical properties of a nonaqueous medium have a role in determining a given enzyme function [16].

101.3 Conclusions

The goals of protein and solvent engineering are similar. Ideally, one should be able to predictably alter a given property of an enzyme by either changing the enzyme or its environment. In reality, both approaches are still somewhat unpredictable. Both protein engineering and solvent engineering have been used successfully to alter the protein properties that appear in Table 101.1. Further research is needed, however, before the results of any given attempt at such biologic engineering can be predicted.

References

1. Alvaro G, Russell AJ. 1991. Methods Enzymol 202:620.
2. Hutchinson C, Phillips S, Edgell M. 1978. J Biol Chem 253:6551.
3. Moody P, Wilkinson A. 1990. Protein Engineering, pp 1–3. IRL Press.
4. Zoller M, Smith M. 1983. Methods Enzymol 100:468.
5. Dente L, Cesareni G, Cortese R. 1983. Nucl Acid Res 11:1645.
6. Rossi J, Zoller M. 1987. In D Oxender, C Fox (eds), Protein Engineering, pp 51–63. New York, Alan R Liss.
7. Promega Protocols and Application Guide, 2d ed, pp 98–105. 1991.
8. Deng W, Nickoloff J. 1992. Anal Biochem 200:81.
9. Vandeyar M, Weiner M, Hutton C, Batt C. 1988. Gene 65:129.
10. Ho N, Hunt H, Horton R, et al. 1989. Gene 77:51.

11. Russell AJ, Chatterjee S, Rapanovich I, Goodwin J. 1992. In A Gomez-Puyon (ed), Biomolecules in Organic Solvents, pp 92–109. Boca Raton, Fla, CRC Press.
12. Zaks A, Klibanov AM. 1985. Proc Natl Acad Sci USA 82:3192.
13. Affleck R, Xu Z-F, Suzawa V, et al. 1992. Proc Natl Acad Sci USA 89:1100.
14. Kamat SV, Beckman EJ, Russell AJ. 1993. J Am Chem Soc 115:8845.
15. Zaks A, Klibanov AM. 1984. Science 224:1249.
16. Kamat S, Iwaskewycz B, Beckman EJ, Russell AJ. 1993. Proc Natl Acad Sci USA 90:2940.

102

Monoclonal Antibodies and Their Engineered Fragments

Srikanth Sundaram
Rutgers University

David M. Yarmush
Rutgers University

Antibodies are a class of topographically homologous multidomain glycoproteins produced by the immune system that display a remarkably diverse range of binding specificities. The most important aspects of the immune system are that it is diverse and driven to produce antibodies of the highest possible antigen affinity. The primary repertoire of antibodies consists of about 10^9 different specificities, each of which can be produced by an encounter with the appropriate antigen. This diversity is known to be produced by a series of genetic events each of which can play a role in determining the final function of the antibody molecule. After the initial exposure to the antigen, additional diversity occurs by a process of somatic mutation so that, for any selected antigen, about 10^4 new binding specificities are generated. Thus the immunologic repertoire is the most diverse system of binding proteins in biology. Antibodies also display remarkable binding specificity. For example, it has been shown that antibodies are able to distinguish between *ortho-*, *meta-*, and *para-*forms of the same haptenic group [Landsteiner, 1945]. This exquisite specificity and diversity make antibodies ideal candidates for diagnostic and therapeutic agents.

Originally, the source of antibodies was antisera, which by their nature are limited in quantity and heterogeneous in quality. Antibodies derived from such sera are termed *conventional antibodies* (*polyclonal antibodies*). Polyclonal antibody production requires methods for the introduction of immunogen into animals, withdrawal of blood for testing the antibody levels, and finally exsanguination for collection of immune sera. These apparently simple technical requirements are complicated by the necessity of choosing a suitable species and immunization protocol that will produce a highly immune animal in a short time. Choice of animal is determined by animal house facilities available, amount of antiserum required (a mouse will afford only 1.0 to 1.5 ml of blood; a goat can provide several liters), and amount of

0-8493-0462-8/00/$0.00+$.50

immunogen available (mice will usually respond very well to 50 μg or less of antigen; goats may require several milligrams). Another consideration is the phylogenic relationship between the animal from which the immunogen is derived and that used for antibody production. In most cases, it is advisable to immunize a species phylogenetically unrelated to the immunogen donor, and for highly conserved mammalian proteins, nonmammals (e.g., chickens) should be used for antibody production. The polyclonal antibody elicited by an antigen facilitates the localization, phagocytosis, and complement-mediated lysis of that antigen; thus the usual polyclonal immune response has clear advantages in vivo. Unfortunately, the antibody heterogeneity that increases immune protection *in vivo* often reduces the efficacy of an antiserum for various *in vitro* uses. Conventional heterogeneous antisera vary from animal to animal and contain undesirable nonspecific or cross-reacting antibodies. Removal of unwanted specificities from a polyclonal antibody preparation is a time-consuming task, involving repeated adsorption techniques, which often results in the loss of much of the desired antibody and seldom is very effective in reducing the heterogeneity of an antiserum.

After the development of hybridoma technology [Köhler and Milstein, 1975], a potentially unlimited quantity of homogeneous antibodies with precisely defined specificities and affinities (*monoclonal antibodies*) became available, and this resulted in a step change in the utility of antibodies. Monoclonal antibodies (mAbs) have gained increasing importance as reagents in diagnostic and therapeutic medicine, in the identification and determination of antigen molecules, in biocatalysis (catalytic antibodies), and in affinity purification and labeling of antigens and cells.

102.1 Structure and Function of Antibodies

Antibody molecules are essentially required to carry out two principal roles in immune defense. First, they recognize and bind nonself or foreign material (antigen binding). In molecular terms, this generally means binding to structures on the surface of the foreign material (antigenic determinants) that differ from those on the host. Second, they trigger the elimination of foreign material (biologic effector functions). In molecular terms, this involves binding of effector molecules (such as complement) to the antibody-antigen complex to trigger elimination mechanisms such as the complement system and phagocytosis by macrophages and neutrophils, etc.

In humans and other animals, five major immunoglobulin classes have been identified. The five classes include immunoglobulins G (IgG), A (IgA), M (IgM), D (IgD), and E (IgE). With the exception of IgA (dimer) and IgM (pentamer), all other antibody classes are monomeric. The monomeric antibody molecule consists of a basic four-chain structure, as shown in Fig. 102.1. There are two distinct types of chains: the light (L) and the heavy chains (H). The chains are held together by disulfide bonds. The light and heavy chains are held together by interchain disulfides, and the two heavy chains are held together by numerous disulfides in the hinge region of the heavy chain. The light chains have a molecular weight of about 25,000 Da, while the heavy chains have a molecular weight of 50,000 to 77,000 Da depending on the isotope. The L chains can be divided into two subclasses, kappa (κ) and lambda (λ), on the basis of their structures and amino acid sequences. In humans, about 65% of the antibody molecules have κ chains, whereas in rodents, they constitute over 95% of all antibody molecules. The light chain consists of two structural domains: The carboxy-terminal half of the chain is constant except for certain allotypic and

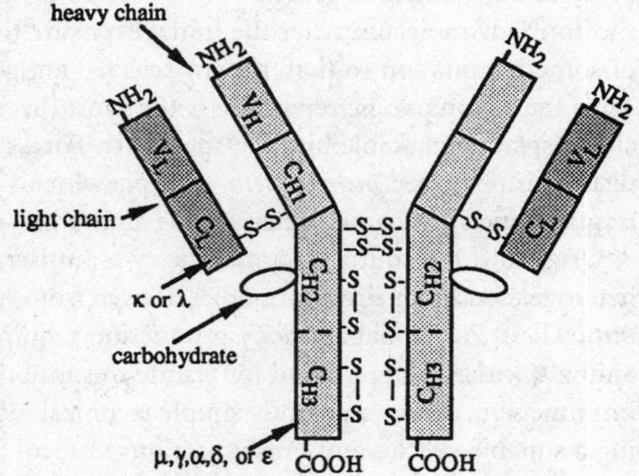

FIGURE 102.1 The structure of a monomeric antibody molecule.

isotype variations and is called the C_L (constant: light chain) region, whereas the amino-terminal half of the chain shows much sequence variability and is known as the V_L (variable: light chain) region. The H chains are unique for each immunoglobulin class and are designated by the Greek letter corresponding to the capital letter designation of the immunoglobulin class (α for the H chains of IgA, γ for the H chains of IgG). IgM and IgA have a third chain component, the J chain, which joins the monomeric units. The heavy chain usually consists of four domains: The amino-terminal region (approximately 110 amino acid residues) shows high sequence variability and is called the V_H (variable: heavy chain) region, whereas there are three domains called C_{H1}, C_{H2}, and C_{H3} in the constant part of the chain. The C_{H2} domain is glycosylated, and it has been shown that glycosylation is important in some of the effector functions of antibody molecules. The extent of glycosylation varies with the antibody class and, to some extent, the method of its production. In some antibodies (IgE and IgM), there is an additional domain in the constant part of the heavy chain called the C_{H4} region. The hinge region of the heavy chain (found between the C_{H1} and C_{H2} domains) contains a large number of hydrophilic and proline residues and is responsible for the segmental flexibility of the antibody molecule.

All binding interactions to the antigen occur within the variable domains (V_H and V_L). Each variable domain consists of three regions of relatively greater variability, which have been termed *hypervariable* (HV) or *complementarity-determining regions* (CDR) because they are the regions that determine complementarity to the particular antigen. Each CDR is relatively short, consisting of from 6 to 15 residues. In general, the topography of the six combined CDRs (three from each variable domain) produces a structure that is designed to accommodate the particular antigen. The CDRs on the light chain include residues 24 to 34 (L1), 50 to 56 (L2), and 89 to 73 (L3), whereas those on the heavy chain include residues 31 to 35 (H1), 50 to 65 (H2), and 95 to 102 (H3) in the Kabat numbering system. The regions between the hypervariable regions are called the *framework regions* (FR) because they are more conserved and play an important structural role. The effector functions are, on the other hand, mediated via the two C-terminal constant domains, namely, the C_{H2} and C_{H3}.

Each variable region is composed of several different genetic elements that are separate on the germ line chromosome but rearrange in the B cell to produce a variable-region exon. The light chain is composed of two genetic elements, a variable (V) gene and a joining (J) gene. The V_κ gene generally encodes residues 1 to 95 in a kappa light chain, and the J_κ gene encodes residues 96 to 107, which is the carboxyl end of the variable region. Thus the kappa V gene encodes all the first and second hypervariable regions (L1 and L2) and a major portion of the third hypervariable region (L3). In the human system, there are estimated to be about 100 different V_κ genes and about 5 functional J_κ genes, and the potential diversity is increased by the imprecision of the VJ joining. The heavy-chain variable region is similarly produced by the splicing together of genetic elements that are distant from each other in the germ line. The mature V_H region is produced by the splicing of a variable gene segment (V) to a diversity segment (D) and to a joining (J) segment. There are about 300 V_H genes, about 5 D_H genes, and 9 functional J_H genes [Tizard, 1992]. With respect to the CDRs of the heavy chain, H1 and H2 are encoded by the V_H gene, while the H3 region is formed by the joining of the V, D, and J genes. Additional diversity is generated by recombinational inaccuracies, light- and heavy-chain recombination, and N-region additions.

The three-dimensional structure of several monoclonal antibodies has been determined via x-ray crystallography and has been reviewed extensively [Davies et al., 1990]. Each domain consists of a sandwich of β sheets with a layer of four and three strands making up the constant domains and a layer of four and five strands making up the variable domains. The CDRs are highly solvent exposed and are the loops at the ends of the variable region β strands that interact directly with the antigen. The superimposition of several V_L and V_H structures from available x-ray crystallographic coordinates using the conserved framework residues shows that the framework residues of the antibody variable domains are spatially closely superimposable. The CDRs varied in shape and length, as well as in sequence. Similarities in the structures of the CDRs from one antibody to another suggest that they are held in a fixed relationship in space, at least within a domain. Also, the N and C terminal of the CDRs were rigidly conserved by the framework residues. Among the CDRs, L1, L2, H2, and H3 show the most conformational variability.

102.2 Monoclonal Antibody Cloning Techniques

The original, and still dominant method for cloning murine monoclonal antibodies is through the fusion of antibody-producing spleen B cells and a myeloma line [Köhler and Milstein, 1975]. More recently, "second generation" monoclonal antibodies have been cloned without resorting to hybridoma technology via the expression of combinatorial libraries of antibody genes isolated directly from either immunized or naive murine spleens or human peripheral blood lymphocytes. In addition, display of functional antigen-binding antibody fragments (Fab, sFv, Fv) on the surface of filamentous bacteriophage has further facilitated the screening and isolation of engineered antibody fragments directly from the immunologic repertoire. Repertoire cloning and phage display technology are now readily available in the form of commercial kits [e.g., ImmunoZap and SurfZap Systems from Stratacyte, Recombinant Phage Antibody System (RPAS) from Pharmacia] with complete reagents and protocols.

Hybridoma Technology

In 1975, Kohler and Milstein [1975] showed that mouse myeloma cells could be fused to B lymphocytes from immunized mice resulting in continuously growing, specific monoclonal antibody-secreting somatic cell hybrids, or *hybridoma* cells. In the fused hybridoma, the B cell contributes the capacity to produce specific antibody, and the myeloma cell confers longevity in culture and ability to form tumors in animals.

The number of fusions obtained via hybridoma technology is small, and hence there is a need to select for fused cells. The selection protocol takes advantage of the fact that in normal cells there are two biosynthetic pathways used by cells to produce nucleotides and nucleic acids. The de novo pathway can be blocked using aminopterin, and the salvage pathway requires the presence of functional hypoxanthine guanine phosphoribosyl transferase (HGPRT). Thus, by choosing a myeloma cell line that is deficient in HGPRT as the "fusion partner," Kohler and Milstein devised an appropriate selection protocol. The hybrid cells are selected by using the HAT (hypoxanthine, aminopterin, and thymidine) medium in which only myeloma cells that have fused to spleen cells are able to survive (since unfused myeloma cells are HGPRT–). There is no need to select for unfused spleen cells because these die out in culture.

Briefly, experimental animals (mice or rats) are immunized by injecting them with soluble antigen or cells and an immunoadjuvant. Three of four weeks later, the animals receive a booster injection of the antigen without the adjuvant. Three to five days after this second injection, the spleens of those animals which produce the highest antibody titer to the antigen are excised. The spleen cells are then mixed with an appropriate "fusion partner," which is usually a nonsecreting hybridoma or myeloma cell (e.g., SP2/0 with 8-azaguanine selection) that has lost its ability to produce hypoxanthine guanine phosphoribosyl transferase (HGPRT–). The cell suspension is then briefly exposed to a solution of polyethylene glycol (PEG) to induce fusion by reversibly disrupting the cell membranes, and the hybrid cells are selected for in HAT media. In general, mouse myelomas should yield hybridization frequencies of about 1 to 100 clones per 10^7 lymphocytes. After between 4 days and 3 weeks, hybridoma cells are visible, and culture supernatants are screened for specific antibody secretion by a number of techniques (e.g., radioimmunoassay, ELISA, immunoblotting). Those cultures wells which are positive for antibody production are then expanded and subcloned by limiting dilution or in soft agar to ensure that they are derived from a single progenitor cell and to ensure stability of antibody production. Reversion of hybridomas to nonsecreting forms can occur due to loss or rearrangement of chromosomes. The subclones are, in turn, screened for antibody secretion, and selected clones are expanded for antibody production *in vitro* using culture flasks or bioreactors (yield about 10 to 100 μm/liter) or *in vivo* as ascites in mice (1 to 10 mg/ml).

The most commonly used protocol is the polyethylene glycol (PEG) fusion technique, which, even under the most efficient conditions, results in the fusion of less than 0.5% of the spleen cells, with only about 1 in 10^5 cells forming viable hybrids. Several methods have been developed to enhance conventional hybrid formation that have led to only incremental improvements in the efficiency of the fusion process (see Neil and Urnovitz [1988] for review). Methods that enhance conventional hybridoma formation include pretreatment of myeloma cells with colcemid and/or *in vitro* antigen stimulation of the spleen cells prior to fusion, addition of DMSO or phytohemagglutinin (PHA) to PEG during fusion, and addition of insulin, growth factors, human endothelial culture supernatants (HECS), etc. to the growth medium after fusion. In addition, improvements in immunization protocols such as suppression of dominant responses, *in vitro* or intrasplenic immunization, and antigen targeting also have been developed. Suppression of dominant immune responses is used to permit the expression of antibody-producing cells with specificity for poor immunogens and is achieved using the selective ability of cyclophosphoamide to dampen the immune response to a particular antigen, followed by subsequent immunization with a similar antigen. In *in vitro* immunization, spleen cells from nonimmunized animals are incubated with small quantities of antigen in a growth medium that has been "conditioned" by the growth of thymocytes. This technique is used most commonly for the production of human hybridomas, where *in vivo* immunization is not feasible. Intrasplenic immunization involves direct injection of immunogen into the spleen and is typically used when only very small quantities of antigen are available. Other advantages include a shortened immunization time, ability to generate high-affinity monoclonals, and improved diversity in the classes of antibodies generated. Finally, in antigen targeting, the immune response is enhanced by targeting the antigen to specific cells of the immune system, for example, by coupling to anti-class II monoclonal antibodies.

There also have been several advances in fusion techniques such as electrofusion, antigen bridging, and laser fusion. In electrofusion, cells in suspension are aligned using an ac current resulting in cell-cell contact. A brief dc voltage is applied which induces fusion and has resulted in a 30- to 100-fold increase in fusion frequencies in some selected cases. Further improvement in electrofusion yields have been obtained by an antigen bridging, wherein avidin is conjugated with the antigen and the myeloma cell membranes are treated with biotin. Spleen cells expressing immunoglobulins of correct specificity bind to the antigen-avidin complex and are in turn bound by the biotinylated cell membranes of the myeloma cells. Finally, laser-induced cell fusion in combination with antigen bridging has been used to eliminate the tedious and time-consuming screening process associated with traditional hybridoma techniques. Here, rather than carry out the fusion process in "bulk," preselected B cells producing antibody of desired specificity and affinity are fused to myeloma cells by irradiating each target cell pair (viewed under a microscope) with laser pulses. Each resulting hybridoma cell is then identified, subcloned, and subsequently expanded.

Despite these improvements, many of the limitations of the hybridoma technique persist. First, it is slow and tedious and labor- and cost-intensive. Second, only a few antibody-producing hybridoma lines are created per fusion, which does not provide for an adequate survey of the immunologic repertoire. Third, as is the case with most mammalian cell lines, the actual antibody production rate is low. Fourth, it is not easy to control the class or subclass of the resulting antibodies, a characteristic that often determines their biologic activity and therefore their usefulness in therapeutic applications. Finally, the production of human monoclonal antibodies by conventional hybridoma techniques has not been very successful due to a lack of suitable human fusion partners, problems related to immunization, difficulty in obtaining human lymphocytes, etc.

Repertoire Cloning Technology

The shortcomings of the hybridoma technology and the potential for improvement of molecular properties of antibody molecules by a screening approach have lead to the expression of the immunologic repertoire in *Escherichia coli* (repertoire cloning). Two developments were critical to the development of this technology. First, the identification of conserved regions at each end of the nucleotide sequences encoding the variable domains enabled the use of the polymerase chain reactions (PCR) to clone antibody Fv and Fab genes from both spleen genomic DNA and spleen messenger RNA [Orlandi et al., 1989; Sastry et al., 1989]. The amplification of a target sequence via PCR requires two primers, each annealing to one end of the target gene. In the case of immunoglobulin variable-region genes, the J segments are sufficiently conserved to enable the design of universal "downstream" primers. In addition, by comparing the aligned sequences of many variable-region genes, it was found that 5′ ends of the V_H and V_L genes are also relatively conserved so as to enable the design of universal "upstream" primers as well. These primers were then used to establish a repertoire of antibody variable-region genes. Second, the successful expression of functional antigen-binding fragments in bacteria using a periplasmic secretion strategy enabled the direct screening of libraries of cloned antibody genes for antigen binding [Better et al., 1988; Skerra and Pluckthun, 1988].

The first attempt at repertoire cloning resulted in the establishment of diverse libraries of V_H genes from spleen genomic DNA of mice immunized with either lysozyme or keyhole-limpet hemocyanin (KLH). From these libraries, V_H domains were expressed and secreted in *E. coli* [Ward et al., 1989]. Binding activities were detected against both antigens in both libraries; the first library, immunized against lysozyme, yielded 21 clones with lysozyme activity and 2 with KLH activity, while the second library, immunized against KLH, yielded 14 clones with KLH activity and 2 with lysozyme activity. Two V_H domains were characterized with affinities for lysozyme in the 20 n*M* range. The complete sequences of 48 V_H gene clones were determined and shown to be unique. The problems associated with this single-domain approach are (1) isolated V_H domains suffer from several drawbacks such as lower selectivity and poor solubility and (2) an important source of diversity arising from the combination of the heavy and light chains is lost.

In the so-called combinatorial approach, Huse et al. [1989] used a novel bacteriophage λ vector system (λ-ZAP technology) to express in *E. coli* a combinatorial library of Fab fragments of the murine antibody repertoire. The combinatorial expression library was constructed from spleen mRNA isolated from a mouse immunized with KLH-coupled *p*-nitrophenyl phosphonamidate (NPN) antigen in two steps. In the first step, separate heavy-chain (Fd) and light-chain (κ) libraries were constructed. These two libraries were then combined randomly, resulting in the combinatorial library in which each clone potentially coexpresses a heavy and a light chain. In this case, they obtained 25 million clones in the library, approximately 60% of which coexpressed both chains. One million of these were subsequently screened for antigen binding, resulting in approximately 100 antibody-producing, antigen-binding clones. The light- and heavy-chain libraries, when expressed individually, did not show any binding activity. In addition, the vector systems used also permitted the excision of a phagemid containing the Fab genes; when grown in *E. coli*, these permitted the production of functional Fab fragments in the periplasmic supernatants. While the study did not address the overall diversity of the library, it did establish repertoire cloning as a potential alternative to conventional hybridoma technology.

Repertoire cloning via the λ-ZAP technology (now commercially available as the ImmunoZap kit from Stratacyte) has been used to generate antibodies to influenza virus hemagglutinin (HA) starting with mRNA from immunized mice [Caton and Kaprowski, 1990]. A total of 10 antigen-binding clones was obtained by screening 125,000 clones from the combinatorial library consisting of 25 million members. Partial sequence analysis of the V_H and V_κ regions of five of the HA-positive recombinants revealed that all the HA-specific antibodies generated by repertoire cloning utilized a V_H region derived from members of a single B-cell clone in conjunction with one of two light-chain variable regions. A majority of the HA-specific antibodies exhibited a common heavy–light-chain combination which was very similar to one previously identified among HA-specific hybridoma monoclonal antibodies. The relative representation of these sequences and the overall diversity of the library also were studied via hybridization studies and sequence analysis of randomly selected clones. It was determined that a single functional V_H sequence was present at a frequency of 1 in 50, while the more commonly occurring light-chain sequence was present at a frequency of 1 in 275. This indicates that the overall diversity of the gene family representation in the library is fairly limited.

The λ-ZAP technology also has been used to produce high-affinity human monoclonal antibodies specific for tetanus toxoid [Mullinax et al., 1990; Persson et al., 1991]. The source of the mRNA in these studies was peripheral blood lymphocytes (PBLs) from human donors previously immunized with tetanus toxoid and boosted with the antigen. Mullinax et al. [1990] estimated that the frequency of positive clones in their library was about 1 in 500, and their affinity constants ranged from 10^6 to 10^9 *M* (Molar)$^{-1}$. However, the presence of a naturally occurring SacI site (one of the restriction enzymes used to force clone PCR-amplified light-chain genes) in the gene for human C_κ may have resulted in a reduction in the frequency of positive clones. Persson et al. [1991] constructed three different combinatorial libraries using untreated PBLs, antigen-stimulated PBLs (cells cultured in the presence of the antigen), and antigen-panned PBLs (cells that were selected for binding to the antigen). Positive clones were obtained from all three libraries with frequencies if 1 in 6000, 1 in 5000, and 1 in 4000, respectively. Apparently binding constants were estimated to be in the range of 10^7 to 10^9 M^{-1}. Sequence analysis of a limited number of clones isolated from the antigen-stimulated cell library indicated a greater diversity than that described for HA or NPN. For example, of the eight heavy-chain CDR3 sequences examined, only two pairs appeared to be clonally related. The λ-ZAP technology also has been used to rescue a functional human antirhesus D Fab from an EBV-transformed cell line [Burton, 1991].

In principle, repertoire cloning would allow for the rapid and easy identification of monoclonal antibody fragments in a form suitable for genetic manipulation. It also provides for a much better survey of the immunologic repertoire than conventional hybridoma technology. However, repertoire cloning is not without its disadvantages. First, it allows for the production of only antibody fragments. This limitation can be overcome by mounting the repertoire cloned variable domains onto constant domains that possess the desired effector functions and using transfectoma technology to express the intact immunoglobulin genes in a variety of host systems. This has been demonstrated for the case of a human Fab fragment to tetanus toxoid, where the Fab gene fragment obtained via repertoire cloning was linked to an Fc fragment gene and successfully expressed in a CHO cell line [Bender et al., 1993]. The second limitation to the use of "immunized" repertoires, which has serious implications in the applicability of this technology for the production of human monoclonal antibodies. The studies reviewed above have all used spleen cells or PBLs from immunized donors. This has resulted in relatively high frequency of positive clones, eliminating the need for extensive screening. Generating monoclonal antibodies from "naive" donors (who have not had any exposure to the antigen) would require the screening of very large libraries. Third, the actual diversity of these libraries is still unclear. The studies reported above show a wide spectrum ranging from very limited (the HA studies) to moderate (NPN) to fairly marked diversity (tetanus toxoid). Finally, the combinatorial approach is disadvantageous in that it destroys the original pairing of the heavy and light chains selected for by immunization. Strategies for overcoming some of these limitations have already been developed and are reviewed below.

Phage Display Technology

A critical aspect of the repertoire cloning approach is the ability to screen large libraries rapidly for clones that possess desired binding properties, e.g., binding affinity or specificity, catalysis, etc. This is especially the case for "naive" human repertoires, wherein the host has not been immunized against the antigen of interest for ethical and/or safety reasons. In order to facilitate screening of large libraries of antibody genes, *phage display* of functional antibody fragments has been developed, which has resulted in an enormous increase in the utility of repertoire cloning technology. In phage display technology, functional antibody fragments (such as the sFv and Fab) are expressed on the surface of filamentous bacteriophages, which facilitates the selection of specific binders (or any other property such as catalysis, etc.) from a large pool of irrelevant antibody fragments. Typically, several hundreds of millions of phage particles (in a small volume of 50 to 100 ml) can be tested for specific binders by allowing them to bind to the antigen of interest immobilized to a solid matrix, washing away the nonbinders, and eluting the binders using a suitable elution protocol.

Phage display of antibody fragments is accomplished by coupling of the antibody fragment to a coat protein of the bacteriophage. Two different coat proteins have been used for this purpose, namely, the major coat protein encoded for by gene VIII and the adsorption protein encoded for by gene III. The system based on gene VIII displays several copies of the antibody fragment (theoretically there are 2000 copies of gene VIII product per phage) and is used for the selection of low-affinity binders. The gene III product is, on the other hand, present at approximately four copies per phage particle and leads to the selection of high-affinity binders. However, since the native gene III product is required for infectivity, at least one copy on the phage has to be a native one.

The feasibility of phage display of active antibody fragments was first demonstrated by McCafferty et al. [1990] when the single-chain Fv fragment (single-chain antibody) of the anti-hen egg white lysozyme (HEL) antibody was cloned into an fd phage vector at the N-terminal region of the gene III protein. This study showed that complete active sFv domains could be displayed on the surface of bacteriophage fd and that rare phage displaying functional sFv (1 in 10^6) can be isolated. Phage that bound HEL were unable to bind to turkey egg white lysozyme, which differs from HEL by only seven residues. Similarly, active Fab fragments also have been displayed on phage surfaces using gene VIII [Kang et al., 1991]. In this method, assembly of the antibody Fab molecules in the periplasm occurs in concert with phage morphogenesis. The Fd chain of the antibody fused to the major coat protein VIII of phage M13 was coexpressed with κ chains, with both chains being delivered to the periplasmic space by the pelB leader sequence. Since the Fd chain is anchored in the membrane, the concomitant secretion of the κ chains results in the assembly of the two chains and hence the display of functional Fab on the membrane surface. Subsequent infection with helper phage resulted in phage particles that had incorporated functional Fab along their entire surface. Functionality of the incorporated Fab was confirmed by antigen-specific precipitation of phage, enzyme-linked immunoassays, and electron microscopy. The production of soluble antibody fragments from selected phages can now be accomplished without subcloning [Hoogenboom et al., 1991]. The switch from surface display to soluble antibody is mediated via the use of an amber stop codon between the antibody gene and phage gene protein. In a *supE* suppresser strain of *E. coli*, the amber codon is read as Glu and the resulting fusion protein is displayed on the surface of the phage. In nonsuppresser strains, however, the amber codon is read as a stop codon, resulting in the production of soluble antibody.

The combination of repertoire cloning technology and phage display technology was initially used to screen antibody fragments from repertoires produced from immunized animals, namely, the production of Fv fragments specific for the hapten phenyloxazolone using immunized mice [Clackson et al., 1991], human Fab fragments to tetanus toxoid [Barbas et al., 1991], human Fab fragments to gp120 using lymphocytes isolated from HIV-positive individuals [Burton et al., 1991], and human Fab fragments to hepatitis B surface antigen from vaccinated individuals [Zebedee et al., 1992]. These studies established the utility of phage display as a powerful screening system for functional antibody fragments. For example, attempts to generate human Fab fragments against gp120 using the λ-ZAP technology failed to produce any binders. Phage display, on the other hand, resulted in 33 of 40 clones selected via antigen panning possessing clear reactivity with affinity constants of the order of 10^{-8} M^{-1}. In the case of the tetanus toxoid studies, phage display was used to isolate specific clones from a library that included known tetanus toxoid clones at a frequency of 1 in 170,000.

Bypassing Immunization

The next step was the application of phage display technology to generate antibodies from unimmunized donors (naive repertoires). Marks et al. [1991] constructed two combinatorial libraries starting from peripheral blood lymphocytes of unimmunized human donors, namely, an IgM library using μ-specific PCR primers and an IgG library using γ-specific primers. The libraries were then screened using phage display and sFv fragments specific for nonself antigens such as turkey egg white lysozyme, bovine serum albumin, phenyloxazolone, and bovine thyroglobulin, as well as for self antigens such as human thyroglobulin, human tumor necrosis factor α, cell surface markers, carcinoembryonic antigen, and mucin, and human blood group antigens were isolated [Hoogenboom et al., 1992]. The binders were all isolated from the IgM library (with the exception of six clones for turkey egg white lysozyme isolated from the IgG library), and the affinities of the soluble antibody fragments were low to moderate (2×10^6 to 10^7 M^{-1}). Both these results are typical of the antibodies produced during a primary response.

The second stage of an immune response *in vivo* involves affinity maturation, in which the affinities of antibodies of the selected specificities are increased by a process of somatic mutation. Thus one method by which the affinities of antibodies generated from naive repertoires may be increased is by mimicking this process. Random mutagenesis of the clones selected from naive repertoires has been accomplished by error-prone PCR [Gram et al., 1992]. In this study, low-affinity Fab fragments (10^4 to 10^5 M^{-1}) to progesterone were initially isolated from the library, and their affinities increased 13- to 30-fold via random mutagenesis. An alternative approach to improving the affinities of antibodies obtained from naive repertoires involves the use of chain shuffling [Marks et al., 1992]. This study describes the affinity maturation of a low-affinity antiphenyloxazolone antibody (3×10^7 M^{-1}). First, the existing light chain was replaced with a repertoire of *in vivo* somatically mutated light chains from unimmunized donors resulting in the isolation of a clone with a 20-fold increase in affinity. Next, shuffling of the heavy chain in a similar manner with the exception of retaining the original H3 resulted in a further increase in affinity of about 16-fold. The net increase in affinity (320-fold) resulted in a dissociation constant of 1 n*M*, which is comparable with the affinities obtained through hybridoma technology.

Other approaches to bypassing human immunization involve the use of semisynthetic and synthetic combinatorial libraries [Barbas et al., 1992] and the immunization of SCID mice that have been populated with human peripheral blood lymphocytes [Duchosal et al., 1992].

102.3 Monoclonal Antibody Expression Systems

Several expression systems are currently available for *in vitro* production of antibodies such as bacteria, yeast, plants, baculovirus, and mammalian cells. In each of these systems, cloned antibody genes are the starting point for production. These are obtained either by traditional cloning techniques starting with preexisting hybridomas or by the more recent repertoire cloning techniques. Each of the aforementioned systems has its own advantages and drawbacks. For example, bacterial expression systems suffer from the following limitations: They cannot be used for producing intact antibodies, nor can they glycosylate antibodies. Unglycosylated antibodies cannot perform many of the effector functions associated with normal antibody molecules. Proper folding may sometimes be a problem due to difficulty in forming disulfide bonds, and often, the expressed antibody may be toxic to the host cells. On the other hand, bacterial expression has the advantage that it is cheap, can potentially produce large amounts of the desired product, and can be scaled up easily. In addition, for therapeutic products, bacterial sources are to be preferred over mammalian sources due to the potential for the contamination of mammalian cell lines with harmful viruses.

Bacterial Expression

Early attempts to express intact antibody molecules in bacteria were fairly unsuccessful. Expression of intact light and heavy chains in the cytoplasm resulted in the accumulation of the proteins as nonfunctional inclusion bodies. *In vitro* reassembly was very inefficient [Boss et al., 1984; Cabilly et al., 1984]. These results could be explained on the basis of the fact that the *E. coli* biosynthetic environment does not support protein folding that requires specific disulfide bond formation, post-translational modifications such as glycosylation, and polymeric polypeptide chain assembly.

There has been much more success obtained with antibody fragments. Bacterial expression of IgE "Fc-like" fragments has been reported [Kenten et al., 1984; Ishizaka et al., 1986]. These IgE fragments exhibited some of the biologic properties characteristic of intact IgE molecules. The fragments constituted 18% of the total bacterial protein content but were insoluble and associated with large inclusion bodies. Following reduction and reoxidation, greater than 80% of the chains formed dimers. The fragment binds to the IgE receptor on basophils and mast cells and, when cross-linked, elicits the expected mediator (histamine) release.

Cytoplasmic expression of a Fab fragment directed against muscle-type creatinine kinase, followed by *in vitro* folding, resulted in renaturation of about 40% of the misfolded protein, with a total active protein yield of 80 μg/ml at 10°C [Buchner and Rudolph, 1991]. Direct cytoplasmic expression of the so-called single-chain antibodies, which are novel recombinant polypeptides composed of an antibody V_L tethered to a V_H by a designed "linker" peptide that links the carboxyl terminus of the V_L to the amino terminus of the V_H or vice versa, was the next important step. Various linkers have been used to join the two variable domains. Bird et al. [1988] used linkers of varying lengths (14 to 18 amino acids) to join the two variable domains. Huston et al. [1988] used a 15 amino acid "universal" linker with the sequence $(GGGGS)_3$. The single-chain protein was found to accumulate in the cell as insoluble inclusion bodies and needed to be refolded *in vitro.* However, these proteins retained both the affinity and specificity of the native Fabs.

In 1988, two groups reported, for the first time, the expression of functional antibody fragments (Fv and Fab, respectively) in *E. coli* [Better et al., 1988; Skerra and Pluckthun, 1988]. In both cases, the authors attempted to mimic in bacteria the natural assembly and folding pathway of antibody molecules. In eukaryotic cells, the two chains are expressed separately with individual leader sequences that direct their transport to the endoplasmic reticulum (ER), where the signal sequences are removed and correct folding, disulfide formation, and assembly of the two chains occur. By expressing the two chain fragments (V_L and V_H in the case of Fv expression and the Fd and **K** chains in the case of Fab expression) separately with bacterial signal sequences, these workers were successful in directing the two precursor chains to the periplasmic space, where correct folding, assembly, and disulfide formation occur along with the removal of the signal sequences, resulting in fully functional antibody fragments. Skerra and Pluckthun [1988] report the synthesis of the Fv fragment of MOPC 603 which has an affinity constant identical to that of the intact Ab. Better et al. [1988] report the synthesis of the Fab fragment of an Ig that binds to a ganglioside antigen. While Skerra and Pluckthun obtained a yield of 0.2 mg/liter after a periplasmic wash, Better et al. found that the Fab fragment was secreted into the culture medium with a yield of 2 mg/liter. However, previous attempts to synthesize an active, full-size Ig in *E. coli* by coexpression and secretion mediated by procaryotic signal sequences resulted in poor synthesis and/or secretion of the heavy chain.

Since these early reports, several additional reports have been published (for reviews, see Pluckthun [1992] and Skerra [1993]) which have established the two aforementioned strategies (namely, direct cytoplasmic expression of antibody fragments followed by *in vitro* refolding and periplasmic expression of functional fragments) as the two standard procedures for the bacterial expression of antibody fragments. Expression of the protein in the periplasmic space has advantages: (1) the expressed protein is recovered in a fully functional form, thereby eliminating the need for *in vitro* refolding (as is required in the case of cytoplasmically expressed fragments), and (2) it greatly simplifies purification. On the other hand, direct cytoplasmic expression may, in some cases, reduce problems arising from toxicity of the expressed protein and also may increase the total yield of the protein.

Several improvements have been made in the past few years so as to simplify the expression and purification of antibody fragments in bacteria. These include the development of improved vectors with strong promoters for the high-level expression of antibody fragments, the incorporation of many different signal sequences, and the incorporation of cleavable "affinity" handles that simplify purification. Many expression vector systems are now commercially available that enable the rapid cloning, sequencing, and expression of immunoglobulin genes in bacteria within a matter of 2 to 3 weeks.

Expression in Lymphoid and Nonlymphoid Systems (Transfectoma Technology)

Expression of immunoglobulin genes by transfection into eukaryotic cells (*transfectoma technology*) such as myelomas and hybridomas is an alternative approach for producing monoclonal antibodies [Wright et al., 1992; Morrison and Oi, 1989]. Myelomas and hybridomas are known to be capable of high-level expression of endogenous heavy- and light-chain genes and can glycosylate, assemble, and secrete functional antibody molecules and therefore are the most appropriate mammalian cells for immunoglobulin gene transfection. Nonlymphoid expression in CHO and COS cells also has been examined as a potential improvement over expression in lymphoid cells. The biologic properties and effector functions, which are very important considerations for applications involving human therapy and diagnostics, are completely preserved in this mode of expression. However, transfectoma technology still involves working with eukaryotic cell lines with low antibody production rates and poor scale-up characteristics.

Transfectoma technology provides us with the ability to genetically manipulate immunoglobulin genes to produce antibody molecules with novel and/or improved properties. For example, production of "chimeric" and "reshaped" antibodies, wherein the murine variable domains or CDRs are mounted onto a human antibody framework in an attempt to reduce the problem of immunogenicity in administering murine antibodies for *in vivo* diagnostic and/or therapeutic purposes, would not be possible without the techniques of transfectoma technology. It is also possible to change the isotype of the transfectoma antibodies in order to change their biologic activity. It also has enabled the fusion of antibodies with nonimmunoglobulin proteins such as enzymes or toxins, resulting in novel antibody reagents used in industrial and medicinal applications.

The most commonly used vectors are the pSV2 vectors that have several essential features. First, they contain a plasmid origin of replication and a marker selectable for procaryotes. This makes it relatively easy to obtain large quantities of DNA and facilitates genetic manipulation. Second, they contain a marker expressible and selectable in eukaryotes. This consists of a eukaryotic transcription unit with an SV40 promoter, splice, and poly A addition site. Into this eukaryotic transcription unit is placed a dominant selectable marker derived from procaryotes (either the *neo* gene or the *gpt* gene).

In order to create the immunoglobulin molecules, the two genes encoding the heavy and light chains must be transfected and both polypeptides must be synthesized and assembled. Several methods have been used to achieve this objective. Both the heavy- and light-chain genes have been inserted into a single vector and then transfected. This approach generates large, cumbersome expression vectors, and further manipulation of the vector is difficult. A second approach is to transfect sequentially the heavy- and light-chain genes. To facilitate this, one gene is inserted into a vector with the *neo* gene, permitting selection with antibiotic G418. The other gene is placed in an expression vector containing the *gpt* gene, which confers mycophenolic acid resistance to the transfected cells. Alternatively, both genes may be introduced simultaneously into lymphoid cells using protoplast fusion.

Heavy- and light-chain genes, when transfected together, produced complete, glycosylated, assembled tetrameric antigen-binding antibody molecules with appropriate disulfide bonds. Under laboratory conditions, these transfected cells yield about 1 to 20 mg/liter of secreted antibody. A persisting problem has been the expression level of the transfected immunoglobulin gene. The expression of transfected heavy-chain genes is frequently seen to approach the level seen in myeloma cells; however, efficient expression of light-chain genes is more difficult to achieve.

Expression in Yeast

Yeast is the simplest eukaryote capable of glycosylation and secretion and has the advantages of rapid growth rate and ease of large-scale fermentation. It also retains the advantage that unlike other mammalian systems, it does not harbor potentially harmful viruses.

Initial attempts to express λ and μ immunoglobulin chains specific for the hapten NP in the yeast *Sacchromyces cerevisiae* under the control of the yeast 3-phosphoglycerate kinase (PGK) promoter resulted in the secretion into culture medium at moderate efficiency (5% to 40%), but secreted antibodies had no antigen-binding activity [Wood et al., 1985]. Subsequent attempts to coexpress heavy and light chains with the yeast invertase signal sequence under the control of the yeast phosphoglycerate kinase (PGK) promoter were more successful, presumably due to differences in the efficiency of different yeast signal sequences in directing the secretion of mammalian proteins from yeast. Culture supernatants contained significant quantities of both light and heavy chains (100 and 50 to 80 mg/liter, respectively) with about 50% to 70% of the heavy chain associated with the light chain. The yeast-derived mouse-human chimeric antibody L6 was indistinguishable from the native antibody in its antigen-binding properties [Horowitz et al., 1988; Better and Horowitz, 1989]. Furthermore, it was superior to the native antibody in mediating antibody-dependent cellular cytotoxicity (ADCC) but was incapable of eliciting complement-dependent cytolysis (CDC). Yeast-derived L6 Fab also was indistinguishable from proteolytically generated Fab as well as recombinant Fab generated from *E. coli.*

Expression in Baculovirus

The baculovirus expression system is potentially a very useful system for the production of large amounts of intact, fully functional antibodies for diagnostic and even therapeutic applications. Foreign genes expressed in insect cell cultures infected with baculovirus can constitute as much as 50% to 75% of the total cellular protein late in viral replication. Immunoglobulin gene expression is achieved by commercially available vectors that place the gene to be expressed under the control of the efficient promoter of the gene encoding the viral polyhedrin protein. The levels of expression seen in this system can be as much as 50- to 100-fold greater per cell than in procaryotes while retaining many of the advantages of an eukaryotic expression system such as glycosylation and extracellular secretion. However, scale-up is not straightforward, since viral infection eventually results in cell death.

There have been at least two reports of antibody secretion in a baculovirus system [Hasemann and Capra, 1990; Putlitz, 1990]. In both cases, the secreted antibody was correctly processed, glycosylated (albeit differently than hybridoma-derived antibody), and assembled into a normal functional heterodimer capable of both antigen binding and complement binding. The secreted antibodies were obtained at a yield of about 5 mg/liter.

Expression in Plants

The development of techniques for plant transformation has led to the expression of a number of foreign genes, including immunoglobulin genes, in transgenic plants [Hiatt and Ma, 1992]. The most commonly used plant cell transformation protocol employs the ability of plasmid Ti of *Agrobacterium tumefaciens* to mediate gene transfer into the plant genome.

The expression of a murine anti-phosphonate ester catalytic antibody in transgenic plants was accomplished by first transforming the heavy-chain and light-chain genes individually into different tobacco plants [Hiatt et al., 1989]. These were then sexually crossed to obtain progeny that expressed both chains simultaneously. It was shown that leader sequences were necessary for the proper expression and assembly of the antibody molecules. The level of antibody expression was determined to be about 3 ng/mg of total protein, and the plant-derived antibody was comparable with ascites-derived antibody with respect to binding as well as catalysis.

102.4 Genetically Engineered Antibodies and Their Fragments

The domain structure of the antibody molecule allows the reshuffling of domains and the construction of functional antibody fragments. A schematic representation of such genetically engineered antibodies and their fragments is shown in Figs. 102.2 and 102.3.

Among intact engineered constructs are chimeric, humanized, and bifunctional antibodies (see Wright et al. [1992] and Sandhu [1992] for reviews). A major issue in the long-term use of murine monoclonal antibodies for clinical applications is the immunogenicity of these molecules or the so-called HAMA response (human antimouse antibody response). Simple chimeric antibodies were constructed by linking murine variable domains to human constant domains in order to reduce immunogenicity of therapeutically administered murine monoclonals. The approach has been validated by several clinical trials that show that chimeric antibodies are much less likely to induce a HAMA response compared with their murine counterparts. In a more sophisticated approach, *CDR grafting* has been used to "humanize" murine monoclonal antibodies for human therapy by transplanting the CDRs of a murine monoclonal antibody of appropriate antigenic specificity onto a human framework. Humanized antibodies are, in some cases, even better than their chimeric counterparts in terms of the HAMA response.

Bifunctional antibodies that contain antigen-specific binding sites with two different specificities have been produced via genetic engineering as well as chemical techniques [Fanger et al., 1992]. The dual specificity of bispecific antibodies can be used to bring together the molecules or cells that mediate the

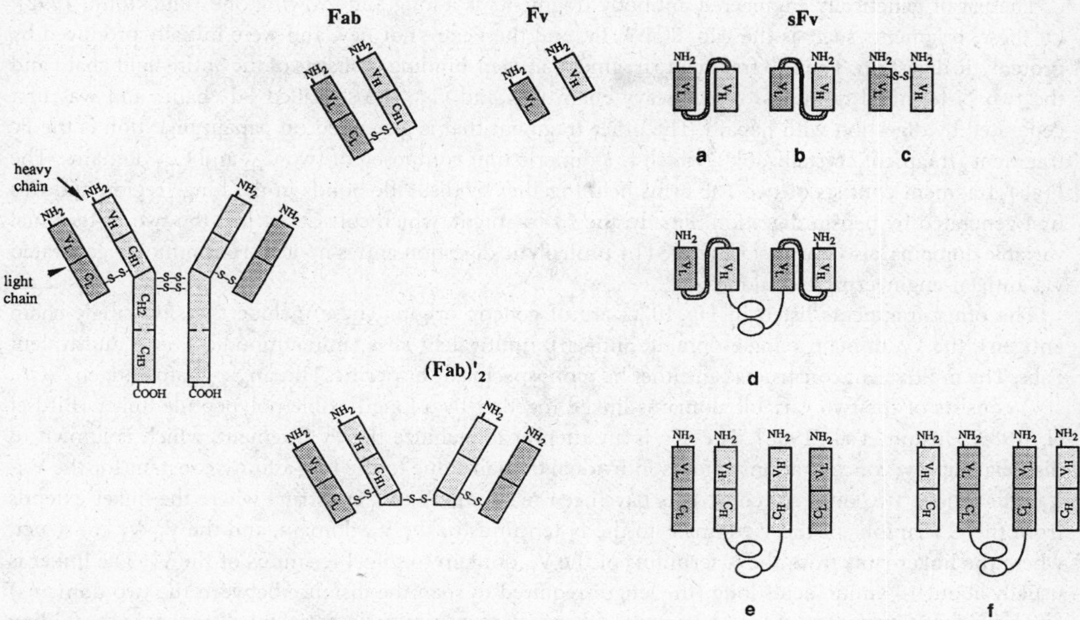

FIGURE 102.2 Schematic of an antibody molecule and its antigen binding fragments. (*a–e*) Examples of the fragments expressed in *Escherichia coli:* (*a*) single-chain Fv fragment in which the linking peptide joins the light and heavy Fv segments in the following manner: V_L (carboxyl terminus)–peptide linker–V_H (amino terminus); (*b*) single-chain Fv fragment with the following connection: V_L (amino terminus)–peptide linker–V_H (carboxyl terminus); (*c*) disulfide-linked Fv fragment; (*d*) "miniantibody" comprised of two helix-forming peptides each fused to a single-chain Fv; (*e*) Fab fragments linked by helix-forming peptide fused to heavy chain; (*f*) Fab fragments linked by helix-forming peptide to light chain.

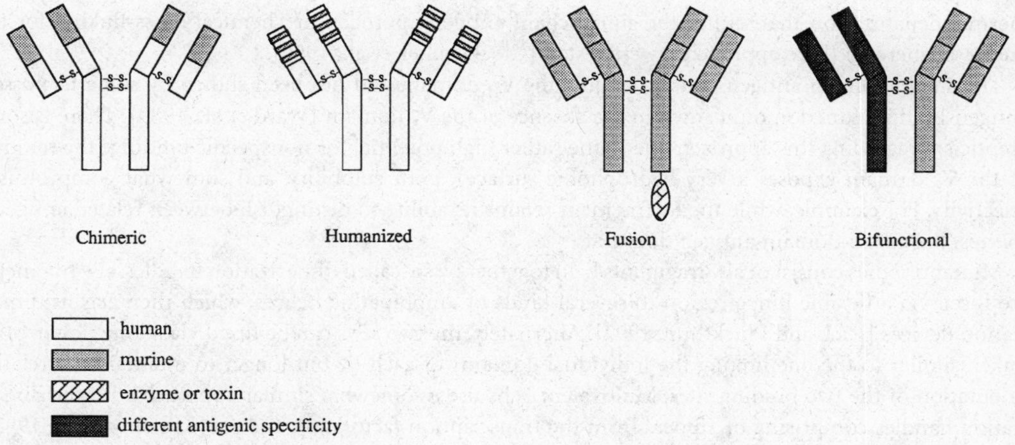

FIGURE 102.3 Schematic of genetically engineered intact monoclonal antibodies prepared with antibody-engineering techniques.

desired effect. For example, a bispecific antibody that binds to target cells such as a tumor cell and to cytotoxic trigger molecules on host killer cells such as T cells has been used to redirect the normal immune system response to the tumor cells in question. Bispecific antibodies also have been used to target toxins to tumor cells.

The list of genetically engineered antibody fragments is a long and growing one [Pluckthun, 1992]. Of these, fragments such as the Fab, $F(ab)'_2$, Fv, and the Fc are not new and were initially produced by proteolytic digestion. The Fab fragment (fragment antigen-binding) consists of the entire light chain and the two N-terminal domains of the heavy chain (V_H and C_{H1}, the so-called Fd chain) and was first generated by digestion with papain. The other fragment that is generated on papain digestion is the Fc fragment (fragment crystallizable), which is a dimeric unit composed of two C_{H2} and C_{H3} domains. The $F(ab)'_2$ fragment consists of two Fab arms held together by disulfide bonds in the hinge region and was first generated by pepsin digestion. Finally, the Fv fragment, which consists of just the two N-terminal variable domains, also was first generated by proteolytic digestion and is now more commonly generated via antigen-engineering techniques.

The other fragments listed in Fig. 102.2 are of genetic origin. These include the sFv (single-chain antigen), the V_H domain (single-domain antigen), multivalent sFvs (miniantibodies), and multivalent Fabs. The multivalent constructs can either be monospecific or bispecific. The single-chain antigen (SCA, sFv) consists of the two variable domains linked together by a long flexible polypeptide linker [Bird et al., 1988; Huston et al., 1988]. The sFv is an attempt to stabilize the Fv fragment, which is known to dissociate at low concentrations into its individual domains due to the low-affinity constant for the V_H-V_L interaction. Two different constructs have been made: the V_L-V_H construct where the linker extends from the C terminus of the V_L domain to the N terminus of the V_H domain, and the V_H-V_L construct, where the linker runs from the C terminus of the V_H domain to the N terminus of the V_L. The linker is usually about 15 amino acids long (the length required to span the distance between the two domains) and has no particular sequence requirements other than to minimize potential interferences in the folding of the individual domains. The so-called universal linker used by many workers in the area is $(GGGGS)_3$. Other strategies to stabilize the Fv fragment include chemical cross-linking of the two domains and disulfide-linked domains [Glockshuber et al., 1990]. Chemical cross-linking via glutaraldehyde has been demonstrated to be effective in stabilizing the Fv fragment in one instance; here, the cross-linking was carried out in the presence of the hapten (phosphorylcholine) to avoid modification of the binding site, an approach that may not be feasible with protein antigens. In the disulfide-linked sFv, Cys residues are introduced at suitable locations in the framework region of the Fv so as to form a natural interdomain disulfide bond. This strategy was shown to be much more effective in stabilizing the Fv fragment against thermal denaturation than either the single-chain antigen approach or chemical cross-linking for the one case where all three approaches were tested [Glockshuber et al., 1990].

The single-domain antigen consists of just the V_H domain and has been shown by some to possess antigen-binding function on its own in the absence of the V_L domain [Ward et al., 1989]. There is some skepticism regarding this approach due to the rather high potential for nonspecific binding (the removal of the V_L domain exposes a very hydrophobic surface), poor solubility, and somewhat compromised selectivity. For example, while the Fv fragment retains its ability to distinguish between related antigenic species, the single-domain antigen does not.

Miniantibodies consist of sFv fragments held together by so-called dimerization handles. sFv fragments are fused via a flexible hinge region to several kinds of amphipathic helices, which then acts as dimerization devices [Pack and Pluckthun, 1992]. Alternately, the two sFvs can be fused via a long polypeptide linker, similar to the one linking the individual domains of each Fv but longer to maintain the relative orientation of the two binding sites. Multivalent Fabs use a somewhat similar approach with the dimerization handles comprising of zippers from the transcription factors *jun* and *fos* [Kostelny et al., 1992].

In addition to the constructs described above, a whole new set of genetically engineered fusion proteins with antibodies has been described [Wright et al., 1992]. Antibody fusion proteins are made by replacing a part of the antibody molecule with a fusion partner such as an enzyme or a toxin that confers a novel function to the antibody. In some cases, such as immunotoxins, the variable regions of the antibody are retained in order to retain antigen binding and specificity, while the constant domains are deleted and replaced by a toxin such as ricin or *Pseudomonas* exotoxin. Alternately, the constant regions are retained (thereby retaining the effector functions) and the variable regions replaced with other targeting proteins (such as CD4 for AIDS therapy and IL-2 for cancer therapy).

102.5 Applications of Monoclonal Antibodies and Fragments

The majority of applications for which monoclonal antibodies have been used can be divided into three general categories: (1) purification, (2) diagnostic functions (whether for detecting cancer, analyzing for toxins in food, or monitoring substance abuse by athletes), and (3) therapeutic functions. From the time that monoclonal antibody technology was introduced almost 20 years ago, application methodologies using whole antibody have gradually been transformed into methodologies using antibody fragments such as the Fab_2, Fab′, and Fv fragments and even synthetic peptides of a CDR region. Antibody conjugates have come to include bound drugs, toxins, radioisotopes, lymphokines, and enzymes [Pietersz and McKenzie, 1992] and are largely used in the treatment of cancer. Tables 102.1 and 102.2 list some typical examples of monoclonal antibodies and fragments used for diagnostic and therapeutic applications.

Thousands of murine monoclonal antibodies have been made to human carcinomas since the introduction of antibody technology, but very few, if any, of these monoclonal antibodies are entirely specific for malignant cells. In the vast majority of cases, these monoclonal antibodies define tumor-associated differentiation antigens (TADAs), which are either proteins, mucins, proteoglycans, or glycolipids (gangliosides). Examples of TADA proteins are carcinoembryonic antigen (CEA) and α-fetoprotein (AFP), both well-known diagnostic markers. An anti-CEA antibody has been conjugated with the enzyme carboxypeptidase, which, in turn, activates a prodrug at the site of the tumor [Bagshawe et al., 1992]. This strategy overcomes the inability of monoclonal antibodies conjugated with drugs to deliver a therapeutic dose. Examples of TADA gangliosides are those referred to as GD2 and GD3, for which the respective unmodified monoclonal antibodies have shown to be effective therapeutics, particularly when intralesionally administered [Irie et a., 1989]. An example of a TADA mucin is the antigens found in colorectal and ovarian carcinoma that react with the antibody 72.3. Chimeric monoclonal antibodies as well as fragments of antibody 72.3 have been constructed and tested [Khazaeli et al., 1991; King et al., 1992].

As mentioned above, monoclonal antibodies defining different TADAs have been used passively (unmodified) and as carriers of, for example, radioisotopes and enzymes. From the results brought forth to date, the passive mode of antibody therapy has produced relatively few remissions in patients, and in those cases where it has shown effects, it is likely that the ability of the antibody to mediate ADCC (antibody-dependent cellular cytotoxicity) and CDC (complement-dependent cytotoxicity) has contributed to the remission. For the case of modified monoclonal antibodies (e.g., toxin and radioisotope conjugates), success of treatment is varied depending on the type of neoplasm. Antibody conjugate

TABLE 102.1 Uses of Whole MAb Derived from Hybridoma, Other Cell Fusions, and Genetically Engineered Cell Lines

MAb	Antigen	Ab Source	Actual or Potential Uses
C23	Cytomegalovirus glycoprotein of 130 and 55 kDa	Humab—fusion of human lymphocyte and mouse myeloma (p3 × 63Ag8ul)	Prophylactic agent for viral infection
EV2-7	Cytomegalovirus protein of 82 kDa	Trioma: human × (human × mouse heteromyeloma)	Prophylactic agent for CMV infection
OKT3	CD3 antigen	Murine MAb	Eradication of T lymphocytes involved in graft rejection
Campath 1H	Lymphocyte antigen	"Humanized" chimeric rat/human MAb; murine CDRs	Prevention of bone marrow and organ rejection
R24	GD2 ganglioside TADA tumor-associated differentiation antigen	Murine MAb	Treatment of melanoma
L72, L55	GD3 ganglioside TADA	Human MAb	Treatment of melanoma
	Digoxin	Murine MAb	Immunodiagnostic for cardiac glycoside digoxin
6H4	*Salmonella* flagella	Murine MAb	Detection of *Salmonella* bacteria in food

TABLE 102.2 Immunoconjugates Having Potential for Cancer Therapy

MAb	Conjugate	Antibody Source	Use
A7	Neocarzinostatin	Murine monoclonal (from fusion with murine myeloma P3, X63.Ag8.653)	Eradication of colon cancer
30.6, I-1	*N*-acetylmelphalan	Murine monoclonal	Eradication of colon cancer
	Adriamycin, Mitomycin C	Murine monoclonal	Various cancers
RFB4	RFB4(Fab′)-Ricin A (αCD22 (Fab′)-Ricin A)	Murine Fab′	B-cell lymphoma
	Anti-CD19-blocked ricin	Murine monoclonal	B-cell lymphoma
	Xomazyme-Mel	Murine monoclonal	Metastatic melanoma
Anti-CEA (carcinoembryonic antigen)	Carboxypeptidase G2	Murine MAb $F(ab')_2$ fragment chemically bound to enzyme	Colon cancer
Recombinant anti-CEA MAb BW431	(DNA coded) human β-glucuronidase	Transfectoma	
B72.3	^{131}I	Murine monoclonal	Ovarian cancer
B72.3	^{131}I	Mouse-human chimeric Ab	Colon cancer
B72.3		Mouse-human chimeric Fab′ fragment	Colon cancer
B72.3 (Oncoscint)	Chelated ^{111}In	Murine MAb	Diagnostic imaging agent for colorectal and ovarian cancers

treatment of leukemia and lymphoma results in a relatively greater remission rate than that found in treatments of malignancies having solid tumors (i.e., carcinoma of the ovary, colon, and lung).

The rare case of complete remission for solid tumors is probably due to the inaccessibility of antibody to that tumor. Several barriers impeding access of antibody to cancer cells have been pointed out [Jain, 1988]. A few of these barriers include (1) the high interstitial fluid pressure in tumor nodules, (2) heterogeneous or poor vascularization of tumors, and (3) the long distances extravasated monoclonal antibodies must travel in the interstial mesh of proteoglycans in the tumor. There also exists the possibility that tumor antigen shed from the surface is limiting antibody buildup. In the case of bound toxins or drugs, there is the added concern that organs such as the kidney and liver are quickly processing and eliminating the antibody conjugates. In this respect, immunoconjugates based on antibody fragments (such as the sFv) can be very advantageous. For example, it has been shown that the sFv exhibits rapid diffusion into the extravascular space, increased volume of distribution, enhance tumor penetration, and improved renal elimination. An assessment of solid tumor therapy with modified antibody has led Riethmüller et al. [1993] to recommend that current cancer therapy be directed toward minimal residual disease, the condition in which micrometastatic cells exist after curatively resecting solid tumors.

With regard to purification, the research literature is replete with examples of immunoaffinity purification of enzymes, receptors, peptides, and small organic molecules. In contrast, commercial applications of immunoaffinity chromatography, even on industrially or clinically relevant molecules, are far less widespread (Table 102.3). Despite its potential utility, immunoadsorption is an expensive process. A significant portion of the high cost is the adsorbent itself, which is related to the cost of materials, preparation, and most important, the antibody. In addition, the binding capacity of the immunoadsorbent declines with repeated use, and a systemic study has shown that significant deactivation can typically take place over 40 to 100 cycles [Antonsen et al., 1991]. A number of factors can contribute to this degradation, including loss of antibody, structural change of the support matrix, nonspecific absorption of contaminating proteins, incomplete antigen elution, and loss of antibody function. In most cases, this degradation is associated with repeated exposure to harsh elution conditions. Noteworthy commercial applications of immunoaffinity chromatography on useful molecules include the separation of factor VIII used to treat hemophilia A and factor IX, another coagulation factor in the blood-clotting cascade

TABLE 102.3 Clinically or Industrially Relevant Proteins Purified by Immunoaffinity Chromatography

Protein	Use (Actual and Potential)
Factor VIII	Treatment of hemophilia A
Factor IX	Blood coagulant
α-Galactosidase	Improve the food stabilizing properties of guar gum
Alkaline phosphatase	Purify enzyme (a particular glycoform) used as a tumor marker for diagnostic tests
Interferon (recombinant)	Immunotherapeutic
Interleukin 2	Immunotherapeutic

[Tharakan et al., 1992]. The immunoaffinity purification step for factor VIII was one of several additional steps added to the conventional preparation methodology in which plasma cryoprecipitates were heat-treated. The new method contains a virus-inactivation procedure that precedes the immunoaffinity column, followed by an additional chromatographic step (ion exchange). The latter step serves to eliminate the eluting solvent and further reduce virus-inactivating compounds. The often mentioned concern of antibody leakage from the column matrix did not appear to be a problem. Furthermore, with the relatively mild elution conditions used (40% ethylene glycol), one would expect little change in the antibody-binding capacity over many elution cycles.

Typical immunoaffinity matrices contain whole antibody as opposed to antibody fragments. Fragmentation of antibody by enzymatic means contributes additional steps to immunoadsorbent preparation and adds to the overall cost of the separation and is thus avoided. However, fragmentation can lead to a more efficient separation by enabling the orientation of antibody-binding sites on the surface of the immunomatrix [Prisyazhnoy et al., 1988; Yarmush et al., 1992]. Intact antibodies are bound in a random fashion, resulting in a loss of binding capacity upon immobilization. Recombinant antibody fragments could prove to be more useful for immunoaffinity applications due to the potential for production of large quantities of the protein at low cost and improved immobilization characteristics and stability [Spitznagel and Clark, 1993]. In what one could consider the ultimate fragment of a antibody, some investigators have utilized a peptide based on the CDR region of one of the chains (termed *minimal recognition units*) to isolate the antigen. Welling et al. [1990] have synthesized and tested a 13-residue synthetic peptide having a sequence similar to one hypervariable region of an antilysozyme antibody.

Important diagnostic uses of antibody include the monitoring in clinical laboratories of the cardiac glycoside digoxin and the detection of the *Salmonella* bacteria in foods (Table 102.1). These two examples highlight the fact that despite the exquisite specificity offered by monoclonal antibodies, detection is not failure-proof. Within digoxin immunoassays there are two possible interfering groups: endogenous digoxin-like substances and digoxin metabolites; moreover, several monoclonal antibodies many be necessary to avoid under- or overestimating digoxin concentrations. In the case of bacteria detection in food, at least two antibodies (MOPC 467 myeloma protein and 6H4 antibody) are needed to detect all strains of *Salmonella.*

102.6 Summary

The domain structure of antibodies, both at the protein and genetic levels, facilitates the manipulation of antibody properties via genetic engineering (antibody engineering). Antibody engineering has shown tremendous potential for basic studies and industrial and medical applications. It has been used to explore fundamental questions about the effect of structure on antigen binding and on the biologic effector functions of the antibody molecules. A knowledge of the rules by which the particular sequences of amino acids involved in the binding surface are chosen in response to a particular antigenic determinant would enable the production of antibodies with altered affinities and specificities. Understanding the structures and mechanisms involved in the effector function of antibodies is already starting to result in the production of antibodies with novel biologic effector functions for use as diagnostic and therapeutic reagents. In addition, the production of antibodies via immunoglobulin gene expression has enabled the engineering of novel hybrid, chimeric, and mosaic genes using recombinant DNA techniques and the transfection and expression of these genetically engineered genes in a number of different systems such as bacteria and yeast, plant cells, myeloma or hybridoma cells, and nonlymphoid mammalian cells.

References

Antonsen KP, Colton CK, Yarmush ML. 1991. Elution conditions and degradation mechanisms in long-term immunoadsorbent use. Biotechnol Prog 7:159.

Bagshawe KD, Sharma SK, Springer CJ, et al. 1992. Antibody directed enzyme prodrug therapy (ADEPT). Antibody Immunoconj Radiopharm 54:133.

Barbas CF, Kang AS, Lerner RA, Benkovic SJ. 1991. Assembly of combinatorial antibody libraries on phase surfaces: The gene III site. Proc Natl Acad Sci USA 88:7978.

Barbas CF, Bain JD, Hoekstra DM, Lerner RA. 1992. Semisynthetic combinatorial antibody libraries: a chemical solution to the diversity problem. Proc Natl Acad Sci USA 89:4457.

Bender E, Woof JM, Atkin JD, et al. 1993. Recombinant human antibodies: Linkage of an Fab fragment from a combinatorial library to an Fc fragment for expression in mammalian cell culture. Hum Antibod Hybridomas 4:74.

Better M, Chang CP, Robinson RR, Horwitz AH. 1988. Escherichia coli secretion of an active chimeric antibody fragment. Science 240:1041.

Better M, Horowitz AH. 1989. Expression of engineered antibodies and antibody fragments in microorganisms. Methods Enzymol 178:476.

Bird RE, Hardman KD, Jacobson JW, et al. 1988. Single-chain antibody-binding proteins. Science 242:423.

Boss MA, Kenten JH, Wood CR, Emtage JS. 1984. Assembly of functional antibodies from immunoglobulin heavy and light chains synthesized in E. coli. Nucleic Acids Res 12:3791.

Buchner J, Rudolph R. 1991. Renaturation, purification, and characterization of recombinant Fab fragments produced in Escherichia coli. Biotechnology 9:157.

Burton DR. 1991. Human and mouse monoclonal antibodies by repertoire cloning. Trends Biotechnol 9:169.

Burton DR, Barbas CF, Persson MAA, et al. 1991. A large array of human monoclonal antibodies to type 1 human immunodeficiency virus from combinatorial libraries of asymptomatic seropositive individuals. Proc Natl Acad Sci USA 88:10134.

Cabilly S, Riggs AD, Pande H, et al. 1984. Generation of antibody activity from immunoglobulin polypeptide chains produced in Escherichia coli. Proc Natl Acad Sci USA 81:3273.

Caton AJ, Koprowski H. 1990. Influenza virus hemagglutinin-specific antibodies isolated from a combinatorial expression library are closely related to the immune response of the donor. Proc Natl Acad Sci USA 87:6450.

Clackson T, Hoogenboom HR, Griffiths AD, Winter G. 1991. Making antibody fragments using phage display libraries. Nature 352:624.

Davies DR, Padlan EA, Sheriff S. 1990. Antigen-antibody complexes. Annu Rev Biochem 59:439.

Duchosal MA, Eming SA, Fischer P, et al. 1992. Immunization of hu-PBL-SCID mice and the rescue of human monoclonal Fab fragments through combinatorial libraries. Nature 355:258.

Fanger MW, Morganelli PM, Guyre PM. 1992. Bispecific antibodies. Crit Rev Immunol 12:101.

Glockshuber R, Malia M, Pfitzinger I, Pluckthun A. 1990. A comparison of strategies to stabilize immunoglobulin fragments. Biochemistry 29:1362.

Gram H, Lore-Anne M, Barbas CF, et al. 1992. In vitro selection and affinity maturation of antibodies from a naive combinatorial immunoglobulin library. Proc Natl Acad Sci USA 89:3576.

Hassemann CA, Capra JD. 1990. High-level production of a functional immunoglobulin heterodimer in a baculovirus expression system. Proc Natl Acad Sci USA 87:3942.

Hiatt A, Cafferkey R, Bowdish K. 1989. Production of antibodies in transgenic plants. Nature 342:76.

Hiatt A, Ma JK-C. 1992. Monoclonal antibody engineering in plants. FEBS Lett 307:71.

Hoogenboom HR, Griffiths AD, Johnson KS, et al. 1991. Multi-subunit proteins on the surface of filamentous phage: Methodologies for displaying antibody (Fab) heavy and light chains. Nucleic Acids Res 19:4133.

Hoogenboom HR, Marks JD, Griffiths AD, Winter G. 1992. Building antibodies from their genes. Immunol Rev 130:41.

Horowitz AH, Chang PC, Better M, et al. 1988. Secretion of functional antibody and Fab fragment from yeast cells. Proc Natl Acad Sci USA 85:8678.

Huse WD, Sastry L, Iverson SA, et al. 1989. Generation of a large combinatorial library of the immunoglobulin repertoire in phage lambda. Science 246:1275.

Huston JS, Levinson D, Mudgett-Hunter M, et al. 1988. Protein engineering of antibody binding sites: Recovery of specific activity in an anti-digoxin single-chain Fv analogue produced in Escherichia coli. Proc Natl Acad Sci USA 85:5879.

Ishizaka T, Helm B, Hakimi J, et al. 1986. Biological properties of a recombinant human immunoglobulin ϵ-chain fragment. Proc Natl Acad Sci USA 83:8323.

Jain RK. 1988. Determinants of tumor blood flow: A review. Cancer Res 48:2641.

Kang AS, Barbas CF, Janda KD, et al. 1991. Linkage of recognition and replication functions by assembling combinatorial antibody Fab libraries along phage surfaces. Proc Natl Acad Sci USA 88:4363.

Keneten J, Helm B, Ishizaka T, et al. 1984. Properties of a human immunoglobulin ϵ-chain fragment synthesized in Escherichia coli. Proc Natl Acad Sci USA 81:2955.

Khazaeli MB, Saleh MN, Liu TP, Meredith RF. 1991. Pharmacokinetics and immune response of 131I-chimeric mouse/human B72.3 (human γ4) monoclonal antibody in humans. Cancer Res 51:5461.

King DJ, Adair JR, Angal S, et al. 1992. Expression, purification, and characterization of a mouse-human chimeric antibody and chimeric Fab′ fragment. Biochem J 281:317.

Kostelny SA, Cole MS, Tso JY. 1992. Formation of bispecific antibody by the use of leucine zippers. J Immunol 148:1547.

Köhler G, Milstein C. 1975. Continuous cultures of fused cells secreting antibody of predefined specificity. Nature 256:495.

Landsteiner K. 1945. The Specificity of Serological Reactions. Cambridge, Mass, Harvard University Press.

Marks JD, Hoogenboom HR, Bonnert TP, et al. 1991. Bypassing immunization: Human antibodies from V-gene libraries displayed on phase. J Mol Biol 222:581.

Marks JD, Griffiths AD, Malmqvist M, et al. 1992. Bypassing immunization: Building high affinity human antibodies by chain shuffling. Biotechnology 10:779.

McCafferty J, Griffiths AD, Winter G, Chriswell DJ. 1990. Phage antibodies: Filamentous phage displaying antibody variable domains. Nature 348:552.

Morrison SL, Oi VT. 1989. Genetically engineered antibody molecules. Adv Immunol 41:65.

Mullinax RL, Gross EA, Amberg JR, et al. 1990. Identification of human antibody fragment clones specific for tetanus toxoid in a bacteriophage λ immunoexpression library. Proc Natl Acad Sci USA 87:8095.

Neil GA, Urnovitz HB. 1988. Recent improvements in the production of antibody-secreting hybridoma cells. Trends Biotechnol 6:209.

Orlandi R, Gussow DH, Jones PT, Winter G. 1989. Cloning immunoglobulin variable domains for expression by the polymerase chain reaction. Proc Natl Acad Sci USA 86:3833.

Pack P, Pluckthun P. 1992. Miniantibodies: Use of amphipathic helices to produce functional flexibility linked dimeric Fv fragments with high avidity in Escherichia coli. Biochemistry 31:1579.

Perrson MAA, Caothien RH, Burton DR. 1991. Generation of diverse high-affinity human monoclonal antibodies by repertoire cloning. Proc Natl Acad Sci USA 88:2432.

Pietersz GA, McKenzie IFC. 1992. Antibody conjugates for the treatment of cancer. Immunol Rev 129:57.

Plückthun A. 1992. Mono- and bivalent antibody fragments produced in Escherichia coli: Engineering, folding and antigen-binding. Immunol Rev 130:150.

Prisyazhnoy VS, Fusek M, Alakhov YB. 1988. Synthesis of high-capacity immunoaffinity sorbents with oriented immobilized immunoglobulins or their Fab′ fragments for isolation of proteins. J Chromatogr 424:243.

Putlitz JZ, Kubasek WL, Duchene M, et al. 1990. Antibody production in baculovirus-infected insect cells. Biotechnology 8:651.

Riethmüller G, Schneider-Gädicke E, Johnson JP. 1993. Monoclonal antibodies in cancer therapy. Curr Opin Immunol 5:732.

Sandhu JS. 1992. Protein engineering of antibodies. Crit Rev Biotechnol 12:437.

Sastry L, Alting-Mees M, Huse WD, et al. 1989. Cloning of the immunoglobulin repertoire in Escherichia coli for generation of monoclonal catalytic antibodies: Construction of a heavy chain variable region-specific cDNA library. Proc Natl Acad Sci USA 86:5728.

Skerra A, Plückthun A. 1988. Assembly of a functional immunoglobulin Fv fragment in Escherichia coli. Science 240:1038.

Skerra A. 1993. Bacterial expression of immunoglobulin fragments. Curr Opin Biotechnol 5:255.

Spitznagel TM, Clark DS. 1992. Surface density and orientation effects on immobilized antibodies and antibody fragments. Biotechnology 11:825.

Tizard IR. 1992. The genetic basis of antigen recognition. In Immunology: An Introduction, 3d ed. Orlando, Fla, Saunders Coolege Publishing.

Ward ES, Gussow D, Griffiths AD, et al. 1989. Binding activities of a repertoire of single immunoglobulin variable domains secreted from Escherichia coli. Nature 341:544.

Wood CR, Boss MA, Kenten JH, et al. 1985. The synthesis and in vivo assembly of functional antibodies in yeast. Nature 314:446.

Welling GW, Guerts T, Van Gorkum J, et al. 1990. Synthetic antibody fragment as ligand in immunoaffinity chromatography. J. Chromatogr 512:337.

Wright A, Shin S-U, Morrison SL. 1992. Genetically engineered antibodies: progress and prospects. Crit Rev Immunol 12:125.

Yarmush ML, Lu X, Yarmush D. 1992. Coupling of antibody-binding fragments to solid-phase supports: Site-directed binding of F(ab′)2 fragments. J Biochem Biophys Methods 25:285.

Zebedee SL, Barbas CF, Yao-Ling H, et al. 1992. Human combinatorial libraries to hepatitis B surface antigen. Proc Natl Acad Sci USA 89:3175.

103

Antisense Technology

S. Patrick Walton
Center for Engineering in Medicine, Massachusetts General Hospital, Harvard Medical School, and Shriners Burns Hospital, Boston

Charles M. Roth
Center for Engineering in Medicine, Massachusetts General Hospital, Harvard Medical School, and Shriners Burns Hospital, Boston

Martin L. Yarmush
Center for Engineering in Medicine, Massachusetts General Hospital, Harvard Medical School, and Shriners Burns Hospital, Boston

Antisense molecules can selectively inhibit the expression of one gene among the 100,000 present in a typical human cell. This inhibition is likely based on simple Watson-Crick base-pairing interactions between nucleic acids and makes possible, in principle, the rational design of therapeutic drugs that can specifically inhibit any gene with a known sequence. The intervention into disease states at the level of gene expression may potentially make drugs based on antisense techniques significantly more efficient and specific than other standard therapies. Indeed, antisense technology is already an indispensable research tool and may one day be an integral part of future antiviral and anticancer therapies.

Zamecnik and Stephenson were the first to use antisense DNA to modulate gene expression. They constructed antisense oligonucleotides complementary to the 3′ and 5′ ends of Rous sarcoma virus (RSV) 35S RNA and added them directly to a culture of chick embryo fibroblasts infected with RSV [1]. Remarkably, viral replication was inhibited, indicating that the cells had somehow internalized the antisense DNA and that the DNA was somehow interrupting the viral life cycle.

Natural antisense inhibition was first observed in bacteria as a means of regulating the replication of plasmid DNA [2, 3]. RNA primers required for the initiation of replication were bound by (i.e., formed duplexes with) antisense RNA. The concentration of these RNA primers, and therefore the initiation of replication, was controlled by the formation of these duplexes. Shortly after this discovery, investigators developed antisense RNA constructs to control gene expression in mammalian cells. Antisense RNA, encoded on expression plasmids that were transfected into mouse cells, successfully blocked expression of target genes [4]. These early successes launched what is now a significant effort to expand the use of antisense molecules for research and therapeutic purposes.

103.1 Background

Antisense oligonucleotides are DNA or RNA molecules whose sequences are complementary to RNA transcribed from the target gene. These molecules block the expression of a gene by interacting with its

0-8493-0462-8/00/$0.00+$.50

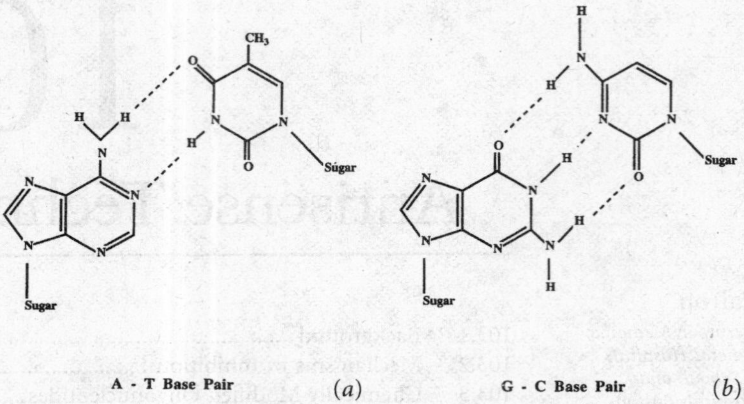

A - T Base Pair (*a*) G - C Base Pair (*b*)

FIGURE 103.1 Watson-Crick base pairing interactions between adenosine and thymidine (A-T) and between guanosine and cytidine (G-C). The sugar is 2′-deoxyribose.

RNA transcript. Antisense molecules are typically short, single stranded oligonucleotides with a sequence complementary to a sequence within the target RNA transcript. The oligonucleotides bind to this sequence via Watson-Crick base pairing (adenosine binds to thymidine (DNA) or uracil (RNA), and guanosine binds to cytidine), form a DNA:RNA duplex, and block translation of the RNA (see Fig. 103.1).

Nucleic acids have been used in a variety of strategies to modulate gene expression. For example, antisense RNA, encoded by a plasmid transfected into the target cells, has been used to inhibit gene expression. Cells are transfected with a plasmid encoding the antisense RNA, the plasmid is transcribed, and the resulting antisense RNA transcript inhibits gene expression. Unfortunately, difficulties in controlling the expression of the transfected plasmid hinder the effective use of antisense RNA for therapeutic purposes. More recently, antisense RNA work has focused on the use of hammerhead ribozymes. These synthetic RNA oligonucleotides have the ability to cleave other RNA strands and they appear more therapeutically viable. Other methods for using nucleic acids to block gene expression are being explored, such as:

1. nucleic acids that competitively bind to proteins via their 3-dimensional structure (aptamers),
2. nucleic acids designed to prevent transcription by forming a DNA triplex with the target gene (antigene), and
3. nucleic acids that mimic binding sites for transcription and translation complexes (decoys).

Although promising, these approaches are preliminary and are beyond the scope of this article. The interested reader is directed to reviews concerning these topics [5-7]. Only antisense DNA molecules, which mimic the antisense strand of the target gene and thus hybridize to RNA transcribed from the gene, will be discussed in detail in this article. For a recent review on antisense RNA, see Rossi [8].

Many technical issues limit the therapeutic usefulness of current antisense oligonucleotides; for instance, they are highly susceptible to degradation by nucleases. Also, our understanding of the mechanism of antisense inhibition must improve before the widespread development of therapeutically useful antisense molecules is possible. Advances in oligonucleotide chemistry have begun to resolve the issue of their intracellular stability; however, the reaction of natural systems to these unnatural species has not been fully explored. Many oligonucleotides still lack adequate specificity for *in vivo* use, and their delivery to cells is often non-specific and inefficient. The impact of these problems and approaches to solving them will be discussed. In addition, potential applications of antisense techniques to antiviral and anticancer therapies and their evaluation in animal models and clinical trials will also be highlighted.

103.2 Mechanisms of Inhibition

The inhibition of gene expression by antisense molecules is believed to occur by a combination of two mechanisms: ribonuclease H (RNase H) degradation of the RNA and steric hindrance of the processing

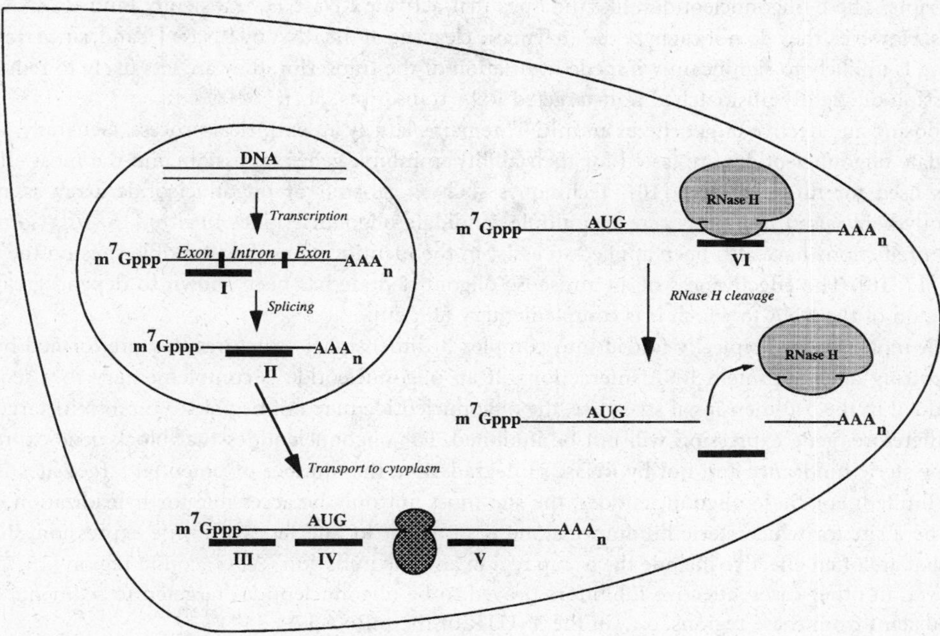

FIGURE 103.2 Several possible sequence specific sites of antisense inhibition. Antisense oligodeoxynucleotides are represented by black bars. Antisense oligodeoxynucleotides can interfere with (I) splicing, (II) transport of the nascent mRNA from the nucleus to the cytoplasm, (III) binding of initiation factors, (IV) assembly of ribosomal subunits at the start codon or (V) elongation of translation. Inhibition of capping and polyadenylation is also possible. (VI) Antisense oligodeoxynucleotides that activate RNase H (e.g., oligonucleotides with phosphodiester and phosphorothioate backbones) can also inhibit gene expression by binding to their target mRNA, catalyzing the RNase H cleavage of the mRNA into segments that are rapidly degraded by exonucleases.

of the RNA [9]. Some oligonucleotides, after hybridizing to the target RNA, activate RNase H, which specifically recognizes an RNA:DNA duplex and cleaves the RNA strand. The antisense oligonucleotide is not degraded and is free to bind to and catalyze the destruction of other target RNA transcripts [10]. Oligonucleotides that activate RNase H, therefore, are potentially capable of destroying many RNA transcripts in their lifetime, which suggests that lower concentrations of these oligonucleotides may be sufficient to significantly inhibit gene expression [9].

However, not all oligonucleotides activate RNase H enzymes. Some oligonucleotides inhibit gene expression by interfering with the RNA's normal life cycle. Newly transcribed RNA must be spliced, polyadenylated, and transported to the cytoplasm before being translated into protein. In theory, virtually any step in an RNA's life cycle can be blocked by an oligonucleotide (see Fig. 103.2). Oligonucleotides that do not activate RNase H can block RNA maturation by binding to sites on the RNA important for post-transcriptional processing or for ribosomal assembly and translation initiation [11]. Oligonucleotides designed to block the elongation of translation by binding to coding sequences downstream from the initiation codon region have rarely succeeded, presumably because ribosomes destabilize and read through DNA:RNA duplexes [11]. Inhibition by both RNase H degradation and steric hindrance was demonstrated in a study in which the expression of an intracellular adhesion molecule (ICAM-1) was blocked using two different oligonucleotides [12].

The mechanism of inhibition influences the specificity of antisense molecules. Oligonucleotides that activate RNase H can inactivate slightly mismatched RNA transcripts by transiently binding to them and catalyzing their cleavage and destruction by RNase H [13]. This non-specific inactivation of RNA could lead to unwanted side effects. In contrast, oligonucleotides that do not activate RNase H, and therefore can only inhibit gene expression by steric hindrance, are less likely to inactivate slightly mismatched RNA transcripts. These oligonucleotides, like the ones that activate RNase H, transiently bind to RNA transcripts. However, they do not catalyze the enzymatic cleavage of the RNA by RNase H, and, since transient binding is unlikely to significantly impede translation of the transcript, they are less likely to reduce the expression of slightly mismatched non-targeted RNA transcripts. [14].

Choosing an effective target site in an mRNA remains largely an empirical process. Generally, several candidate oligonucleotides are tested for their ability to inhibit gene expression, and the most effective one is used for future studies [10]. Techniques such as gel-shift or oligonucleotide array assays are currently being used to rapidly screen multiple candidate oligonucleotides *in vitro* [15, 16]. Computational predictions have also been applied to assist in the identification of susceptible sites on the target RNA [17, 18]. The effectiveness of an antisense oligonucleotide has been shown to depend greatly on the region of the RNA to which it is complementary [19, 20].

RNA molecules are typically folded into complex 3-dimensional structures that are formed by self-base pairing and by protein-RNA interactions. If an oligonucleotide is complementary to a sequence embedded in the 3-dimensional structure, the oligonucleotide may not be able to bind to its target site, and, therefore, gene expression will not be inhibited. For oligonucleotides that block gene expression only by steric hindrance and not by RNase H degradation, the number of potential target sites is even more limited. For these oligonucleotides, the site must not only be accessible for hybridization, but it must be a site for which steric hindrance alone is sufficient to interfere with gene expression. Binding sites that are often effective include the 5′ cap region and the initiation (AUG) codon region [15, 21-23]. However, in other cases, effective inhibitors proved to be oligonucleotides targeted to segments of the RNA distant from these regions, e.g., in the 3′-UTR of the mRNA [20, 23].

103.3 Chemically Modified Oligonucleotides

Natural oligonucleotides with a phosphodiester backbone (PO oligonucleotides) are very susceptible to nuclease degradation, having half-lives in the presence of purified exonucleases as short as fifteen minutes [11, 24, 25]. Almost no full sequence PO oligonucleotide remained following a 1h incubation in human serum or cell extracts [26]. The need for oligonucleotides that are stable *in vivo* prompted the development of chemical derivatives of the phosphodiester backbone. The first derivatives were methylphosphonate and phosphorothioate backbone oligonucleotides. Methylphosphonate oligonucleotides have a methyl group in place of one of the non-bridge backbone oxygens. They are highly lipophilic with a neutral backbone. Recently, however, little attention has been afforded to methylphosphonates as they appear less therapeutically viable due to their poor cellular uptake and the inability of their duplexes to be recognized by RNase H [10, 11].

Phosphorothioate oligonucleotides, on the other hand, remain prominent. Phosphorothioate (PS) DNA derivatives have a sulfur atom substituted for a non-bridge backbone oxygen atom. PS oligonucleotides are significantly more stable than PO oligonucleotides with a half-life of > 5 h in the presence of purified nucleases [25]. Like PO oligonucleotides, PS oligonucleotides are believed to be internalized by receptor mediated endocytosis [27]. Their affinity for complementary RNA is not as high as that observed with PO oligonucleotides, and they are more likely to bind non-specifically to proteins. Nevertheless, PS oligonucleotides have been shown to efficiently inhibit gene expression [28]. One concern with PS oligonucleotides (and all chemically modified oligonucleotides) is their potential for toxicity. The metabolic byproducts of PS oligonucleotides are not native to the cell. If these were incorporated into cellular DNA, they could cause mutations [29]. Studies have also shown that PS oligonucleotides activate complement, resulting in immunological complications for *in vivo* applications [30]. These complications and issues of affinity for the RNA, RNase H activity, stability, and cellular uptake have led to investigation of other types of modifications (e.g., N3′-P5′ phosphoramidates) and chimeric oligonucleotides (see Fig. 103.3 and Table 103.1).

One particularly promising modification focuses on the 2′ position of the sugar ring. RNA:RNA hybrids are more stable than RNA:DNA hybrids, presumably due to the formation of A-form helices rather than

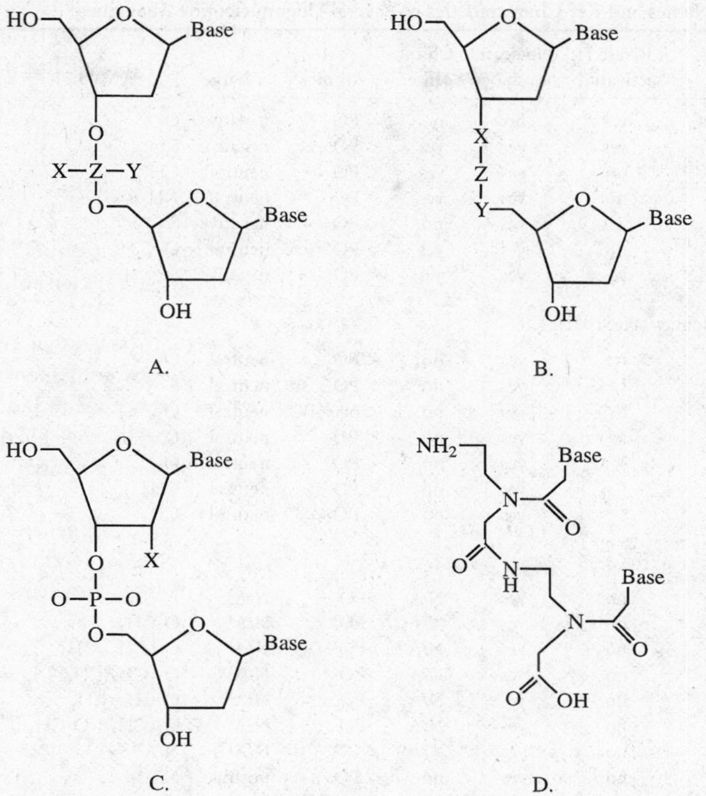

FIGURE 103.3 Analogues of natural oligonucleotides. (A) Backbone modifications in which the phosphorus bridge atom is retained. (B) Backbone modifications in which the phosphorus bridge atom is replaced. (C) 2′ ribose modifications. (D) Peptide nucleic acids—the entire backbone is replaced with amino acids. See Table 103.1 for legend.

the A:B-form of the heteroduplex. However, with a hydroxyl at the 2′ position, RNA is exceptionally labile to nucleases. Thus, to maintain the A-form geometry (the preferred geometry for RNA helices) and prevent degradation, replacement of the 2′-hydroxyl is being investigated. A study by Monia, et al. investigated both 2′-halide and 2′-*O*-alkyl substitutions, finding the affinity for the target ranked 2′-fluoro > 2′-*O*-methyl > 2′-*O*-propyl > 2′-*O*-pentyl > 2′-*O*-deoxy [31]. However, fully modified oligonucleotides were ineffective in inhibiting Ha-ras gene expression in HeLa cells unless the interior of the molecule contained at least 5 2′-deoxy nucleotides to initiate RNase H cleavage. Hence, chimeric oligonucleotides have begun to receive more attention as the next generation of antisense therapeutics.

Chimeric oligonucleotides take advantage of the properties of multiple modifications in the same molecule. Earlier chimeric oligonucleotides used end caps of nuclease resistant bases with unmodified central bases [32]. More recently, "mixed-backbone" oligonucleotides have been synthesized with both backbone end modifications (primarily PS) and central sugar modifications (primarily 2′-*O*-methoxy) [33]. These molecules have been shown to have improved target affinity over PS oligonucleotides while maintaining the ability to initiate RNase H cleavage of the RNA.

Investigators have also replaced the phosphodiester backbone entirely, as with peptide nucleic acids (PNAs) in which the entire ribose-phosphodiester backbone is replaced with a polyamide (see Fig. 103.3C). PNAs, with a covalently bound terminal lysine residue to prevent self-aggregation, have been shown to preferentially bind to complementary sequences and inhibit gene expression *in vitro* [34, 35]. PNAs have an unusually high affinity for their complementary target sequences, possibly because they can form a triple helix with homopurine stretches of RNA (PNA:RNA-PNA) [36]. Duplex forming

TABLE 103.1 The Names and Key Characteristics of Several Oligonucleotide Analogues

Phosphorus Analogues (Fig. 103.3a)	RNase H activation	Nuclease resistance	Chiral center	Affinity	Charge	X (Fig. 103.3)	Y (Fig. 103.3)	Z (Fig. 103.3)
phosphodiester (PO)	yes	no	no	= PO	negative	O^-	O	P
phosphorothioate (PS)	yes	yes	yes	< PO	negative	S^-	O	P
methylphosphonate (MP)	no	yes	yes	< PO	neutral	CH_3	O	P
phosphoramidate	no	yes	yes	< PO	neutral	NH-R	O	P
phosphorodithioate	yes	yes	no	< PO	negative	S^-	S	P
phosphoethyltriester	no	yes	yes	> PO	neutral	$O\text{-}C_2H_5$	O	P
phosphoroselenoate	yes	yes	yes	< PO	negative	Se^-	O	P
Non-Phosphorus Analogues (Fig. 103.3b)								
formacetal	?	yes	no	< PO	neutral	O	O	CH_2
3′ thioformacetal	?	yes	no	> PO	neutral	S	O	CH_2
5′-N-carbamate	?	yes	no	< or > PO	neutral	O	NH	C=O
sulfonate	?	yes	no	< PO	neutral	O	CH_2	SO_2
sulfamate	?	yes	no	< PO	neutral	O	NH	SO_2
sulfoxide	?	yes	no	< PO	neutral	CH_2	CH_2	SO
sulfide	?	yes	no	< PO	neutral	CH_2	CH_2	S
2′ Modified Analogues (Fig. 103.3c)								
fluoro	no	yes	N/A	>PO	N/A[1]	F		
methoxy	no	yes	N/A	>PO	N/A[1]	$O\text{-}CH_3$		
propoxy	no	yes	N/A	>PO	N/A[1]	$O\text{-}(CH_2)_2CH_3$		
pentoxy	no	yes	N/A	>PO	N/A[1]	$O\text{-}(CH)_4CH_3$		
O-allyl	no	yes	N/A	>PO	N/A[1]	$O\text{-}CH{=}CH_2$		
methoxyethoxy	no	yes	N/A	>PO	N/A[1]	$O\text{-}(CH_2)_2\text{-}O\text{-}CH_3$		
α-analogues	no	yes	N/A[1]	< PO	N/A[1]	N/A[2]	N/A[2]	N/A[2]
Peptide Nucleic Acids (Fig. 103.3d)	no	yes	no	> PO	positive[3]	N/A[4]	N/A[4]	N/A[4]

? = unknown.

[1] Chirality and charge depends on backbone structure used.

[2] Structure not drawn; The bond between the sugar and base (an *N*-glycosidic bond) of α-analogues have the reverse orientation (α-configuration) from natural (β-configuration) oligonucleotides.

[3] Typically, the C-terminus is covalently linked to a postively charged lysine residue, giving the PNA a positive charge.

[4] See Fig. 103.3D for chemical structure.

X, Y, Z: these columns reference Fig. 103.3; Replace the designated letter in Fig. 103.3 with the molecule indicated in the Table I to determine the chemical structure of the oligo.

PNAs have shown inhibition of translation when targeted to the mRNA start codon, while triple helix formation is required to inhibit protein elongation [37]. The ability of PNAs to invade the duplex of DNA and form a stable triple helix structure remains an issue for their *in vivo* application.

Other modifications to the ribose sugars (e.g., α-anomeric) and modifications to the nucleoside bases (e.g., 5-methyl or 5-bromo-2′-deoxycytidine and 7′-deazaguanosine and 7′-deazaadenosine oligonucleotides) have also been examined [10]. Oligonucleotides covalently linked to active groups (e.g., intercalators such as acridine, photoactivated crosslinking agents such as psoralens and chelating agents such as EDTA-Fe) are also being actively investigated as potential antisense molecules [9]. Whether for clinical application or for research/diagnostic tools, studies with modified oligonucleotides will provide valuable information regarding the mechanistic steps of antisense oligonucleotide activity. These future generations of antisense “oligonucleotides” may bear little structural resemblance to natural oligonucleotides, but their inhibition of gene expression will still rely on sequence specific base pairing. The reader is directed to several reviews for a comprehensive treatment of all classes of chemically modified oligonucleotides [9, 38-40].

103.4 Oligonucleotide Synthesis and Purification

The growth of the field of antisense therapeutics has also resulted in a need for increased production of oligonucleotides for *in vitro* and *in vivo* applications. As a result, new synthesis procedures are being explored to increase batch size and synthesis rate. Large-scale automated synthesizers have been developed, with which batches of 100 mmol have been produced [http://www.hybridon.com]. This is a significant improvement given that a large batch on a standard synthesizer is 1 μmol.

Standard procedures begin with the 3′ base of the oligonucleotide attached to a solid controlled pore glass (CPG) support, although other supports have also been proposed [41]. The process then continues with deprotection, monomer introduction, activated coupling, oxidation, and capping in a cycle for the addition of each oligonucleotide. Depending on the type of monomer being used (e.g., β-cyanoethyl phosphoramidite or hydrogen phosphonate), sulfurization to generate PS oligonucleotides occurs either during the synthesis cycling or following the termination of synthesis. Given the imperfections in the synthesis procedures, the resulting oligonucleotide product must then be purified to remove failure products.

Oligonucleotides are typically purified using reverse-phase high performance liquid chromatography (RP-HPLC). During synthesis, a hydrophobic dimethoxy-trityl (DMT) group is removed from the phosphoramidite to allow the next nucleoside to attach. Leaving this group on the final nucleoside of the synthesis increases the hydrophobicity of the full, correct oligo, increasing its retention time on an RP-HPLC column. The eluent is then fractionated to collect only the full sequence product.

However, PS oligonucleotides, unlike PO oligonucleotides, are chiral molecules, i.e., molecules that are not identical to their mirror image. Current synthesis techniques do not control the orientation of each internucleotide bond, and the resultant oligonucleotides are therefore a mixture of $2^{(n-1)}$ diastereomers (where $n =$ the number of bases in the complete oligo). The resulting sequences elute with similar retention times on RP-HPLC as all of the complete sequence diastereomers should have the hydrophobic DMT group attached. However, analysis following removal of the DMT group can be used to confirm or deny the presence of multiple diastereomers under higher resolution conditions. Capillary gel electrophoresis and anion-exchange HPLC are being examined as ways to enhance the selectivity of purification of PS oligonucleotides. Additional details on the synthesis and purification of oligonucleotides are available for the interested reader [42-44].

103.5 Specificity of Oligonucleotides

The ability to block the expression of a single gene without undesired side effects (specificity) is the major advantage, in principle, of antisense based strategies. This specificity is primarily determined by the length (i.e., the number of bases) of the oligonucleotide. Experimental and theoretical data suggest that there is an optimum length at which specific inhibition of gene expression is maximized and non-specific effects are minimized [13, 14].

Affinity and specificity limit the minimum effective length of oligonucleotides. Oligonucleotides that are too short do not inhibit gene expression, because they do not bind with sufficient affinity to their substrates. The shortest oligonucleotide reported to affect gene expression in mammalian cells was 7 bases in length (7-mer) [45]. This oligonucleotide was highly modified, containing 5-(1-propynyl) cytosines and uracils as well as a phosphorothioate backbone. An oligonucleotide that is too short is less likely to represent a unique sequence in a given cell's genome and thus more likely to bind to a non-targeted RNA and inhibit its expression. Woolf et al. estimated the minimum sequence length that will be statistically unique in a given pool of mRNAs [13]. Since each position in a given sequence can be occupied by any of four nucleotides (A, C, G or U), the total number of different possible sequences of length N bases is 4^N. Letting R equal the total number of bases in a given mRNA pool and assuming that it is a random and equal mixture of the four nucleotides, then the frequency (F) of occurrence in that pool of a sequence of length N is given by:

$$F = \frac{R}{4^N} \tag{103.1}$$

For a typical human cell, which contains approximately 10^4 unique mRNA species whose average length is 2000 bases, R is approximately equal to 2×10^7. Therefore, for a sequence to be unique ($F < 1$), N must be greater than or equal to 13 bases [13]. The minimum oligonucleotide length will also be constrained by the minimum binding affinity to form a stable complex.

However, oligonucleotides cannot be made arbitrarily long, because longer oligonucleotides are more likely to contain internal sequences complementary to non-targeted RNA molecules. This has also been expressed mathematically [13]. The expected number of complementary sites (S) of length L for an oligonucleotide of length N in an mRNA pool with R bases is given by:

$$S = \frac{\left[\left(N - L + 1\right) \times R\right]}{4^L} \tag{103.2}$$

For example, an 18-mer ($N = 18$) has 6 internal 13-mers. Since a 13-mer is expected to occur 0.3 times in a mRNA pool containing 2×10^7 bases, the 18-mer is expected to match 1.8 (i.e., 6×0.3) 13-mers in the mRNA pool. Equation (103.2) also gives this result ($N = 18$, $L = 13$, and $R = 2 \times 10^7$; therefore, $S = 1.8$).

Woolf et al. have demonstrated that significant degradation of non-targeted mRNAs can occur [13]. They compared the effectiveness of three different 25-mers in suppressing the expression of fibronectin mRNA in *Xenopus* oocytes. Nearly 80% of the fibronectin mRNA was degraded after the oocytes were microinjected with a 25-mer in which all 25 of its bases were complementary to the mRNA. However, when the oocytes were microinjected with 25-mers that had only 17 or 14 complementary bases flanked by random sequences, greater than 30% of their fibronectin mRNA was still degraded. They also showed that a single mismatch in an oligonucleotide did not completely eliminate its antisense effect. Over 40% of the target mRNA was degraded when oocytes were treated with a 13-mer with one internal mismatch, though the mismatch left a 9 base consecutive complementary sequence that showed nearly the same activity as a 13-mer with 10 complementary bases in succession.

Although these studies were conducted at lower temperatures and therefore under less stringent hybridization conditions than those found in mammalian cells, they clearly showed that complementary oligonucleotides flanked by unrelated sequences and even oligonucleotides with mismatched sequences can lead to significant degradation of RNA. The possibility of undesired inhibition of partially complementary sequences must be considered when designing and testing any antisense oligonucleotide. Hence, a useful heuristic is that an oligonucleotide should be long enough to be unique and possess high affinity for its target but short enough to minimize side effects due to degradation of non-targeted mRNAs.

The efforts to sequence the human genome have begun to provide more information about the actual specificity of oligonucleotides. It is possible to scan for sequences against all of the genomic information available using resources such as BLAST at the National Center for Biotechnology Information [46; http://www.ncbi.nlm.nih.gov]. The results will determine the uniqueness of the oligonucleotide within the known database sequences. As the completeness of the databases grows, computational comparisons of target RNA sequences against the database will provide more reliable assessments of the uniqueness of these target sequences within the genome.

103.6 Oligonucleotide Delivery

In many cases, the concentrations of antisense oligonucleotides required to achieve a biological effect are currently too large to be of therapeutic value. Achieving the necessary concentration in the vicinity of the target cells is thought to be limited at least in part by charge repulsion between the negatively-charged oligonucleotide and the negatively-charged cell surface. Four major methods are currently being investigated to overcome this and other extracellular transport barriers: (1) chemical modification of the oligonucleotide to increase its hydrophobicity; (2) conjugation of the oligonucleotide to a polycation; (3) conjugation of the oligonucleotide to a ligand specific for a cellular receptor; and (4) encapsulation of the oligonucleotide in a liposome.

Increasing the hydrophobicity of oligonucleotides was first accomplished by synthesis of methylphosphonate backbones. Despite their neutral backbone charge, cellular uptake of MP oligonucleotides was shown to be very inefficient, perhaps due in part to their inability to escape lysosomal degradation and gain entry into the cytoplasm. The hydrophobicity of an oligonucleotide can also be increased by conjugation with hydrophobic moieties such as cholesterol derivatives (chol-oligonucleotide). The increased hydrophobicity of chol-oligonucleotides reportedly improves their association with cell membranes and their internalization by cells [47]. Cellular accumulation of chol-oligonucleotides after 2 hours has been shown to occur when no accumulation was seen with unconjugated oligonucleotides [48]. The mechanism by which chol-oligonucleotides are more easily and rapidly taken up by cells has not been elucidated. Krieg [49] showed that chol-oligonucleotides with cholesterol at the 5′ end could be bound by low density lipoprotein (LDL) and that this markedly increased the association of the oligonucleotide with the cell membrane and its internalization *in vitro.* LDL-associated chol-oligonucleotides were effective at 8-fold lower concentration than PO oligonucleotide controls.

Oligonucleotides conjugated with polycations such as poly-L-lysine (PLL) also have improved cellular uptake. Covalent conjugates can be constructed by coupling the 3′ end of the oligonucleotide to the epsilon amino groups of the lysine residues [50]. Oligo-PLL conjugates complementary to the translation initiation site of the *tat* gene (a key HIV regulatory protein) protected cells from HIV-1 infection with concentrations 100-fold lower than non-conjugated oligonucleotides [51]. Low concentrations of oligo-PLL conjugates (100 nM), in which the 15-mer oligonucleotide was complementary to the initiation region of the viral N protein mRNA, also inhibited vesicular stomatitis virus (VSV) infection [50].

Internalization of oligonucleotides has also been improved by conjugation to ligands. One such ligand, transferrin, is internalized by receptor mediated endocytosis. Mammalian cells acquire iron carried by transferrin. Oligo-PLL complexes have been conjugated to transferrin to take advantage of this pathway. For example, an 18-mer complementary to *c-myb* mRNA (an oncogene that is responsible for the hyperproliferation of some leukemia cells) was complexed with a transferrin PLL conjugate and rapidly internalized by human leukemia (HL-60) cells [52]. The expression of *c-myb* was greatly reduced and the uncontrolled proliferation of these cells inhibited. Because an oligo-PLL conjugate (without transferrin) was not tested, it is not clear whether the improved antisense effects were due to the PLL moiety, the transferrin moiety, or a combination thereof.

Oligo-ligand conjugates have also been used to target oligonucleotides to cells with specific cellular receptors, such as the asialoglycoprotein (ASGP) receptor, which is expressed uniquely by hepatocytes. These receptors bind and internalize ASGPs, serum glycoproteins that have exposed galactose residues at the termini of their glycosylation chains. Oligonucleotides complexed to ASGP are rapidly internalized by hepatocytes. A number of researchers have used ASGP conjugated to cationic PLL. When the ASGP:PLL is conjugated to oligonucleotides, their cellular uptake and antisense effectiveness *in vitro* is increased [53-55]. However, toxicity of the PLL moiety and instability of the noncovalent complexes between ASGP-PLL and oligonucleotides have limited their *in vivo* applicability [56]. One possible solution is the direct conjugation of the oligonucleotides to ASGP via a cleavable disulfide linkage [57].

Another promising means to deliver oligonucleotides is lipofection. Oligonucleotides are mixed with cationic lipids that condense around the negatively charged oligonucleotide forming a lipid vesicle (liposome). The positively charged lipids reduce the electrostatic repulsion between the negatively-charged oligonucleotide and the similarly charged cell surface. Bennett et al. made liposomes using DOTMA {*N*-{1-(2,3-dioleyloxy)propyl)}-*N*,*N*,*N*-trimethylammonium chloride} as the cationic lipid and an oligonucleotide complementary to the translation initiation codon of human intracellular adhesion molecule 1 (ICAM-1), increasing the potency of the oligonucleotide by >1000-fold [12]. A recent study has shown that liposomally encapsulated oligonucleotides at 0.01 nM were as active as free oligonucleotides at 1.5 μM at inhibiting HIV-1 replication in chronically infected cells [58]. It was found that the oligonucleotides separate from cationic liposomes following cellular internalization [59]. Cationic lipids subsequently accumulate in cytoplasmic structures and the plasma membrane, suggesting that they have fused with the endosomal vesicles and are being recirculated throughout the secretory pathways [59-61]. The increase in oligonucleotide activity suggests that cellular uptake may be a significant limitation in the effectiveness of antisense oligonucleotides. *In vitro* studies using electroporation, streptolysin O or α-toxin permeabilization, and particle bombardment to enhance cellular uptake of nucleic acids support this hypothesis [62-65].

Efforts have been made to use antibodies to target liposomes containing oligonucleotides to specific cell types. Protein A, which binds to many IgG antibodies, is covalently bound to the liposome to form a liposome-oligo-protein A complex. An antibody specific for the target cell's surface is bound to the protein A, and these complexes are incubated with the target cells. Leonetti [66] synthesized such a complex to target the mouse major histocompatibility complex H-2K molecule on mouse L929 cells. The complex contained a 15-mer complementary to the 5′ end of mRNA encoding the N protein of VSV. The complexes inhibited viral replication by more than 95%. Unconjugated liposomes, and liposomes conjugated to antibodies specific for a nonexpressed antigen, had no effect.

103.7 Potential Applications of Antisense Oligonucleotides

The potential applications for antisense oligonucleotides are limited only by the genetic information available. Antisense oligonucleotides can be developed against any target in which the inhibition of protein production or the inhibition of RNA processing yields the therapeutic result. Currently, clinical trials are underway using antisense oligonucleotides to treat rheumatoid arthritis, psoriasis, renal transplant rejection, and inflammatory bowel disease (Crohn's disease) [http://www.phrma.org]. However, the primary targets remain refractory viral diseases and cancers for which the necessary target genetic information is typically available.

Viral Diseases

One of the many potential therapeutic applications of antisense technology is the treatment of infectious viral diseases. The use of antisense oligonucleotides as anti-viral agents is particularly promising because viral nucleic acid sequences are unique to the infected cell and are not found in normal healthy cells. The goal of antiviral therapy is to block the expression of key viral proteins that are vital to the life cycle of the virus. This has been achieved *in vitro* with several viruses including HIV, HSV, influenza and human papilloma virus [10]. The *in vitro* work with HIV is representative of other viruses and will be highlighted here.

Human Immunodeficiency Virus

Retroviruses, and HIV in particular, have high rates of mutation and genetic recombination. The effectiveness of many anti-HIV drugs has been severely reduced because drug resistant viral strains often arise after prolonged drug treatment. This is especially relevant to strategies that use an antisense approach that relies on a specific nucleotide sequence. One strategy to inhibit HIV replication is to target conserved sequences in key regulatory proteins such as *tat* and *rev*. Part of the *rev* sequence is highly conserved, with the known 16 isolates of HIV differing by at most one base pair in this conserved region [67]. The importance of targeting a highly conserved region as is found in *rev* was demonstrated by Lisziewicz, who investigated the efficacy of antisense oligonucleotides directed against various viral proteins [68]. Although several 28-mers initially inhibited viral replication, resistant mutant viruses developed after 25 days to all of the oligonucleotides with the exception of those directed at the highly conserved *rev* regions.

A second concern in treating HIV and other viruses is that viral replication can restart after antisense treatment is stopped [67]. Can oligonucleotides be continuously administered to prevent viral replication without unwanted side effects? These issues and others common to all antisense based therapies (oligonucleotide stability, specificity, affinity, and delivery) must be addressed prior to the successful implementation of antisense-based HIV therapies.

Cancer

In principle, antisense technology can be used against cancer, but the target is more challenging. Oncogenes are typically genes that play a vital regulatory role in the growth of a cell, the mutation or inappropriate expression of which can result in the development of cancer. In the case of mutation, it is often difficult to distinguish an oncogene from its normal counterpart, because they may differ by as little as one base. Thus, attempts to inhibit oncogene expression might block the expression of the normal gene in non-cancerous cells and cause cytotoxic effects. Despite these challenges, steady progress has been made in the development of effective antisense oligonucleotides that have inhibited many types of oncogenes including *bcr-abl*, *c-myc*, *c-myb*, *c-ras*, Ha-*ras*, *neu/erbB2* and *bFGF* [11, 25, 61, 69, 70].

Studies targeting *ras* oncogenes are encouraging and are representative of progress against other classes of oncogenes. Chang [71] targeted a *ras p21* gene point mutation with an antisense oligonucleotide. Only the expression of the mutated genes was inhibited, suggesting that it is possible to selectively inhibit an oncogene that differs by only a single base from the normal gene. Using a 17-mer PS oligo, it was shown that ~98% inhibition of Ha-*ras* expression could be achieved at 1 μM concentration [25].

Successes in inhibiting oncogene expression *in vitro* are encouraging, but many problems remain to be solved before antisense oligonucleotides are therapeutically useful for cancer patients. For example, finding a suitable oncogene to target is difficult. Even if a genetic defect is common to all cells of a given cancer, there is no guarantee that inhibition of that oncogene will halt cancer cell proliferation [70]. Even if cell growth is inhibited, tumor growth may restart after antisense treatment is stopped. Oligonucleotides are needed that induce cancer cell terminal differentiation or death [70]. In order to avoid toxic side effects by inhibiting gene expression of regulated oncogenes in normal tissues, oligonucleotides that specifically target cancer cells are required. Even if appropriate antisense oligonucleotides are developed, they will still be ineffective against solid tumors unless they can reach the interior of the tumor at biologically effective concentrations.

103.8 *In Vivo* Pharmacology

The first *in vivo* antisense studies were designed to test the biodistribution and toxicity of antisense oligonucleotides. These studies demonstrated that PO oligonucleotides are rapidly excreted from the body with PS oligonucleotides being retained longer. One study in mice noted that completely modified PS oligonucleotides were retained significantly longer than a chimeric 20-mer that contained 15 phosphodiester bonds and only 4 phosphorothioate bonds [72]. Only 30% of the PS oligonucleotide was excreted in the urine after 24 hours, whereas 75% of the chimeric oligonucleotide was excreted after only 12 hours. A similar study demonstrated that PS oligonucleotides were retained in body tissues of adult male rats with a half-life of 20 to 40 hours [73].

These studies suggested that antisense oligonucleotides may need to be administered repeatedly to maintain a therapeutic effect for chronic disorders. At least with PS oligonucleotides, this has proven to be problematic. Continuous or repeated administration of PS oligonucleotides results in immunologic complement activation. It has been shown that infusion of PS oligonucleotides to rhesus monkeys results in almost immediate (within 10 min.) decreases in white blood cell count and the fraction of neutrophils among the white cells [30]. At 40 min. post-infusion, the neutrophil counts had increased to higher than baseline levels. A recent study indicated that a plasma concentration of >50 μg/ml results in complement activation [74]. These difficulties have further hastened the development of novel oligonucleotide chemistries and chimeric oligonucleotides for which the immunologic properties are expected to be improved due to the possible reduction in the use of PS backbone linkages [33, 75].

The targeted delivery of oligonucleotides has been proposed as a way to improve the effectiveness of systemically administered antisense oligonucleotides by oligonucleotide concentration in the desired cells and minimizing non-specific side effects. Targeting oligonucleotides to a specific organ has been demonstrated. As previously discussed, a soluble DNA carrier system has been developed to specifically target hepatocytes [76]. This system was used to target a 67-mer PO oligonucleotide (complementary to the

5′ end of rat serum albumin mRNA) to the liver [56]. Following tail vein injection into rats, the complex rapidly and preferentially accumulated in the liver, but the efficiency of this targeting method was limited by the rapid dissociation of the oligonucleotide from the ASGP:PLL complex (30% dissociated within 7 minutes).

103.9 Animal Models

The effectiveness of antisense oligonucleotides at inhibiting gene expression *in vivo* has been demonstrated in several animal models. Immunodeficient mice, bearing a human myeloid leukemia cell line, were continuously infused with 100 μg/day of a 24-mer PS oligonucleotide complementary to *c-myb* mRNA [77]. Mice that received the antisense oligonucleotide survived on average 8.5 times longer than control mice that received no treatment and 5.7 times longer than sense-treated controls. The treated animals also had a significantly lower tumor burden in the ovaries and brain.

Antisense oligonucleotides have also inhibited the growth of solid tumors *in vivo*. Immuno-deficient mice bearing a fibrosarcoma or melanoma were injected subcutaneously twice weekly with 1.4 mg of PS oligonucleotides complementary to the 5′ end of the p65 subunit of the mRNA for the NF-κB transcription factor (NF-κB activates a wide variety of genes and is believed to be important in cell adhesion and tumor cell metastasis) [78]. Greater than 70% of antisense treated mice exhibited a marked reduction in their tumor size. Administration of control oligonucleotides complementary to GAPDH and *jun-D* had no effect on tumor size.

The capability of antisense oligonucleotides to inhibit viral replication *in vivo* has also been studied. Fourteen one day old ducklings were infected with duck hepatitis B virus (DHBV), which is closely related to the strain of hepatitis B virus that is a major cause of chronic liver disease and cancer in humans [79]. Two weeks later, the ducks were injected daily, for 10 consecutive days, with 20 μg/gm body weight of an 18-mer PS oligonucleotide complementary to the start site of the Pre-S region of the DHBV genome. The antisense oligonucleotides blocked viral gene expression and eliminated the appearance of viral antigens in the serum and liver of all treated ducks. No toxic effects due to the oligonucleotides were noted. Unfortunately, residual amounts of DNA precursors of viral transcripts were detected in the nuclei of liver cells, which resulted in a slow restart of viral replication after antisense treatment was stopped. Further studies are needed to determine if prolonged treatment with antisense will eliminate this residual viral DNA.

Though many of the oligonucleotides under investigation target viruses or cancer, other refractory targets are also being investigated. Antisense oligonucleotides have been evaluated in the treatment of restenosis, a narrowing of an artery following corrective surgery caused by mitogen-induced proliferation of vascular smooth muscle cells (SMC). Several studies have tested the ability of antisense oligonucleotides to inhibit genes whose expression is important in SMC proliferation, including c-*myc*, c-*myb*, *cdc2* and *cdk2* [80-82]. In one study with an antisense therapeutic complementary to c-*myc* mRNA, the formation of a thick intimal layer (neointima) in treated areas was significantly reduced in antisense treated rats when compared to control rats with maximum c-*myc* expression reduced by 75% relative to controls [82]. Dextran sodium sulfate-induced colitis in Swiss-Webster mice (as measured by disease activity index, DAI) was reduced by ~64% following 5 days of daily subcutaneous administration of an antisense oligonucleotide complementary to ICAM-1 at 0.3 mg/kg/day [83]. Interestingly, in this study, higher oligonucleotide concentrations (> 5 mg/kg/day) were ineffective in reducing inflammation.

103.10 Clinical Trials

The FDA recently approved the first antisense therapeutic, fomivirsen, a treatment for cytomegalovirus (CMV) retinitis in AIDS patients [http://www.isip.com]. CMV infection is fairly common (~50% of people over 35 are seropositive); however, it results in complications only in immunocompromised individuals. The oligonucleotide is complementary to the immediate-early viral RNA, a region encoding

TABLE 103.2 A Sample of Current Clinical Trials of Antisense Oligodeoxynucleotide Therapeutics

Compound	Type of Oligo	Disease Target	Gene Target	Phase
ISIS13312	2′-modified—2nd Gen.	CMV Retinitis	N/A	I/II
ISIS2302	PS	Crohn's Disease	ICAM-1	II/Pivotal
		Psoriasis	ICAM-1	II
		Renal Transplant Rejection	ICAM-1	II
		Rheumatoid Arthritis	ICAM-1	II
		Ulcerative Colitis	ICAM-1	II
G3139	N/A	Cancer	N/A	I
ISIS2503	PS	Cancer	Ha-ras	I
ISIS3521	PS	Cancer	PKC-α	II
ISIS5132	PS	Cancer	c-raf	II
LR-3001	PS	Chromic Myelogenous Leukemia	bcr-abl	I
LR-3280	PS	Restenosis	myc	II

N/A—Not available

Potential applications of antisense technology include the treatment of cancer, infectious diseases, and inflammatory diseases.

proteins that regulate virus gene expression [84]. The activity of the oligonucleotide therapeutic is potentiated by traditional drugs such as ganciclovir. Additional studies suggested, however, that the oligonucleotide acts not only via an antisense mechanism, though complementarity to the target was required for maximal inhibitory capacity [85].

Clinical trials on many other antisense oligonucleotides have advanced to Phase II. However, complications associated with antisense oligonucleotides have also been found. An antisense compound targeted to the gag region of HIV was pulled from clinical trials due to the resulting decreases in platelet counts in 30% of the patients after 10 days of daily administration [http://www.hybridon.com]. The immunologic complications are likely related to the activation of complement discussed earlier. Perhaps, fomivirsen avoids these complications due to its local administration by intravitreal injection rather than systemic infusion. Table 103.2 lists clinical trials that are currently ongoing and the disease targets. Published results from these trials are not yet available.

103.11 Summary

Antisense oligonucleotides have the potential to selectively inhibit the expression of any gene with a known sequence. Although there have been some remarkable successes, realizing this potential is proving difficult because of problems with oligonucleotide stability, specificity, affinity, and delivery. Natural oligonucleotides contain a phosphodiester backbone that is highly susceptible to nuclease degradation. Chemically modified oligonucleotides (PS and chimeric oligonucleotides) have been synthesized that are more stable. However, the increased stability of PS oligonucleotides comes at a cost, in particular the activation of complement. Another issue for chemically modified oligonucleotides is their large scale production and purification. Alternative chemistries are needed that are more biologically active and more amenable to large scale production. Targeted delivery of antisense oligonucleotides is needed to minimize side effects and maximize oligonucleotide concentration in target cells. Attempts to improve delivery include chemical modification of the oligonucleotides to increase their cellular permeability, the conjugation of oligonucleotides to specific ligands to utilize more efficient receptor-mediated internalization pathways, and the use of antibody conjugated liposomes to deliver oligonucleotides to specific cells.

The potential applications of antisense oligonucleotides to the treatment of disease are vast. Antisense-based therapies are under development for the treatment of infectious diseases such as CMV, HIV, herpes simplex virus, influenza, hepatitis and human papilloma virus as well as the treatment of complex genetic disorders like cancer. Animal models are being used to determine if (1) antisense oligonucleotides can be delivered to target cells at high enough concentrations to be effective, (2) repeated treatments with oligonucleotides are toxic or elicit an immune response, and (3) antisense oligonucleotides directed against a single gene can be effective against complex genetic diseases such as cancer. Multiple clinical trials against viral, cancer, and other targets are now in progress with one drug having been approved for the clinic. Improvements in our understanding of the mechanisms of antisense inhibition, the pharmacology of antisense oligonucleotides *in vivo* and the development of chemically modified oligonucleotides with high affinity, specificity and stability are needed to realize the clinical potential of antisense-based strategies for the treatment of a wide variety of diseases.

Defining Terms

Antisense: Any DNA or RNA molecule whose sequence is complementary to the RNA transcribed from a target gene.

Chiral: A molecule whose configuration is not identical with its mirror image.

Complementary: A nucleic acid sequence is complementary to another if it is able to form a perfectly hydrogen-bonded duplex with it, according to the Watson-Crick rules of base pairing (A opposite U or T, G opposite C).

Diastereomer: Optically active isomers that are not enantiomorphs (mirror images).

Exonuclease: An enzyme that catalyzes the release of one nucleotide at a time, serially, from one end of a polynucleotide.

***In vitro*:** In an artificial environment, referring to a process or reaction occurring therein, as in a test tube or culture dish.

***In vivo*:** In the living body, referring to a process or reaction occurring therein.

Lipofection: Delivery of therapeutic drugs (antisense oligonucleotides) to cells using cationic liposomes.

Lipophilic: Capable of being dissolved in lipids (organic molecules that are the major structural elements of biomembranes).

Liposome: A spherical particle of lipid substance suspended in an aqueous medium.

Plasmid: A small, circular extrachromosomal DNA molecule capable of independent replication in a host cell.

Receptor mediated endocytosis: The selective uptake of extracellular proteins, oligonucleotides and small particles, usually into clathrin coated pits, following their binding to cell surface receptor proteins.

Restenosis: A narrowing of an artery or heart valve following corrective surgery on it.

RNase H: An enzyme that specifically recognizes RNA:DNA duplexes and cleaves the RNA portion, leaving the DNA portion intact.

References

1. Zamecnik, P. C. and Stephenson, M. L. 1978. Inhibition of Rous sarcoma virus replication and cell transformation by a specific oligodeoxynucleotide, *Proc. Natl. Acad. Sci. USA*, 75, 280.
2. Tomizawa, J. and Itoh, T. 1981. Plasmid ColE1 incompatibility determined by interaction of RNA I with primer transcript, *Proc. Natl. Acad. Sci. USA*, 78, 6096.
3. Tomizawa, J., Itoh, T., Selzer, G., and Som, T. 1981. Inhibition of ColE1 RNA primer formation by a plasmid-specified small RNA, *Proc. Natl. Acad. Sci. USA*, 78, 1421.
4. Izant, J. G. and Weintraub, H. 1985. Constitutive and conditional suppression of exogenous and endogenous genes by anti-sense RNA, *Science*, 229, 345.
5. Stull, R. A. and Szoka, F. C., Jr. 1995. Antigene, ribozyme and aptamer nucleic acid drugs: progress and prospects, *Pharm Res*, 12, 465.
6. Hélène, C., Giovannangeli, C., Guieysse-Peugeot, A. L., and Praseuth, D. 1997. Sequence-specific control of gene expression by antigene and clamp oligonucleotides, *Ciba Found. Symp.*, 209, 94.

7. Gewirtz, A. M., Sokol, D. L., and Ratajczak, M. Z. 1998. Nucleic acid therapeutics: state of the art and future prospects, *Blood*, 92, 712.
8. Rossi, J. J. 1997. Therapeutic applications of catalytic antisense RNAs (ribozymes), *Ciba Found. Symp.*, 209, 195.
9. Hélène, C. and Toulme, J. J. 1990. Specific regulation of gene expression by antisense, sense and antigene nucleic acids, *Biochem. Biophys. Acta.*, 1049, 99.
10. Milligan, J. F., Matteucci, M. D., and Martin, J. C. 1993. Current concepts in antisense drug design, *J. Med. Chem.*, 36, 1923.
11. Nagel, K. M., Holstad, S. G., and Isenberg, K. E. 1993. Oligonucleotide pharmacotherapy: an antigene strategy, *Pharmacotherapy*, 13, 177.
12. Bennett, C. F., Condon, T. P., Grimm, S., Chan, H., and Chiang, M. Y. 1994. Inhibition of endothelial cell adhesion molecule expression with antisense oligonucleotides, *J. Immunol.*, 152, 3530.
13. Woolf, T. M., Melton, D. A., and Jennings, C. G., 1992. Specificity of antisense oligonucleotides in vivo, *Proc. Natl. Acad. Sci. USA*, 89, 7305.
14. Herschlag, D. 1991. Implications of ribozyme kinetics for targeting the cleavage of specific RNA molecules *in vivo*: more isn't always better, *Proc. Natl. Acad. Sci. USA*, 88, 6921.
15. Stull, R. A., Zon, G., and Szoka, F. C., Jr. 1996. An *in vitro* messenger RNA binding assay as a tool for identifying hybridization-competent antisense oligonucleotides, *Antisense Nucleic Acid Drug Dev.*, 6, 221.
16. Milner, N., Mir, K. U., and Southern, E. M. 1997. Selecting effective antisense reagents on combinatorial oligonucleotide arrays, *Nat. Biotechnol.*, 15, 537.
17. Jaroszewski, J. W., Syi, J. L., Ghosh, M., Ghosh, K., and Cohen, J. S. 1993. Targeting of antisense DNA: comparison of activity of anti-rabbit beta- globin oligodeoxyribonucleoside phosphorothioates with computer predictions of mRNA folding, *Antisense Res. Dev.*, 3, 339.
18. Sczakiel, G., Homann, M., and Rittner, K. 1993. Computer-aided search for effective antisense RNA target sequences of the human immunodeficiency virus type 1, *Antisense Res. Dev.*, 3, 45.
19. Lima, W. F., Monia, B. P., Ecker, D. J., and Freier, S. M. 1992. Implication of RNA structure on antisense oligonucleotide hybridization kinetics, *Biochemistry*, 31, 12055.
20. Monia, B. P., Johnston, J. F., Geiger, T., Muller, M., and Fabbro, D. 1996. Antitumor activity of a phosphorothioate antisense oligodeoxynucleotide targeted against C-raf kinase, *Nat. Med.*, 2, 668.
21. Goodchild, J., Carroll, E. D., and Greenberg, J. R. 1988. Inhibition of rabbit beta-globin synthesis by complementary oligonucleotides: identification of mRNA sites sensitive to inhibition, *Arch. Biochem. Biophys.*, 263, 401.
22. Wakita, T. and Wands, J. R. 1994. Specific inhibition of hepatitis C virus expression by antisense oligodeoxynucleotides. *In vitro* model for selection of target sequence, *J. Biol. Chem.*, 269, 14205.
23. Peyman, A., Helsberg, M., Kretzschmar, G., Mag, M., Grabley, S., and Uhlmann, E. 1995. Inhibition of viral growth by antisense oligonucleotides directed against the IE110 and the UL30 mRNA of herpes simplex virus type-1, *Biol. Chem. Hoppe Seyler*, 376, 195.
24. McKay, R. A., Cummins, L. L., Graham, M. J., Lesnik, E. A., Owens, S. R., Winniman, M., and Dean, N. M. 1996. Enhanced activity of an antisense oligonucleotide targeting murine protein kinase C-alpha by the incorporation of 2′-O-propyl modifications, *Nucleic Acids Res.*, 24, 411.
25. Monia, B. P., Johnston, J. F., Sasmor, H., and Cummins, L. L. 1996. Nuclease resistance and antisense activity of modified oligonucleotides targeted to Ha-ras, *J. Biol. Chem.*, 271, 14533.
26. Iribarren, A. M., Cicero, D. O., and Neuner, P. J. 1994. Resistance to degradation by nucleases of (2'S)-2′-deoxy-2′-C-methyloligonucleotides, novel potential antisense probes, *Antisense Res. Dev.*, 4, 95.
27. Loke, S. L., Stein, C. A., Zhang, X. H., Mori, K., Nakanishi, M., Subasinghe, C., Cohen, J. S., and Neckers, L. M. 1989. Characterization of oligonucleotide transport into living cells, *Proc. Natl. Acad. Sci. USA*, 86, 3474.
28. Stein, C. A., Subasinghe, C., Shinozuka, K., and Cohen, J. S. 1988. Physicochemical properties of phosphorothioate oligodeoxynucleotides, *Nucleic Acids Res.*, 16, 3209.

29. Neckers, L. and Whitesell, L. 1993. Antisense technology: biological utility and practical considerations, *Am. J. Physiol.*, 265, L1.
30. Galbraith, W. M., Hobson, W. C., Giclas, P. C., Schechter, P. J., and Agrawal, S. 1994. Complement activation and hemodynamic changes following intravenous administration of phosphorothioate oligonucleotides in the monkey, *Antisense Res. Dev.*, 4, 201.
31. Monia, B. P., Lesnik, E. A., Gonzalez, C., Lima, W. F., McGee, D., Guinosso, C. J., Kawasaki, A. M., Cook, P. D., and Freier, S. M. 1993. Evaluation of 2′-modified oligonucleotides containing 2′-deoxy gaps as antisense inhibitors of gene expression, *J. Biol. Chem.*, 268, 14514.
32. Pickering, J. G., Isner, J. M., Ford, C. M., Weir, L., Lazarovits, A., Rocnik, E. F., and Chow, L. H. 1996. Processing of chimeric antisense oligonucleotides by human vascular smooth muscle cells and human atherosclerotic plaque. Implications for antisense therapy of restenosis after angioplasty, *Circulation*, 93, 772.
33. Agrawal, S., Jiang, Z., Zhao, Q., Shaw, D., Cai, Q., Roskey, A., Channavajjala, L., Saxinger, C., and Zhang, R. 1997. Mixed-backbone oligonucleotides as second generation antisense oligonucleotides: *in vitro* and *in vivo* studies, *Proc. Natl. Acad. Sci. USA*, 94, 2620.
34. Hanvey, J. C., et al. 1992. Antisense and antigene properties of peptide nucleic acids, *Science*, 258, 1481.
35. Nielsen, P. E., Egholm, M., Berg, R. H., and Buchardt, O. 1993. Peptide nucleic acids (PNAs): potential antisense and anti-gene agents, *Anticancer Drug Des.*, 8, 53.
36. Egholm, M., Nielsen, P. E., Buchardt, O., and Berg, R. H. 1992. Recognition of guanine and adenine in DNA by cytosine and thymine containing peptide nucleic acids (PNA), *J. Am. Chem. Soc.*, 114, 9677.
37. Good, L. and Nielsen, P. E. 1997. Progress in developing PNA as a gene-targeted drug, *Antisense Nucleic Acid Drug Dev.*, 7, 431.
38. Uhlmann, E. and Peyman, A. 1990. Antisense oligonucleotides: A new therapeutic principle, *Chem. Rev.*, 90, 544.
39. Cook, P. D. 1991. Medicinal chemistry of antisense oligonucleotides—future opportunities, *Anticancer Drug Des.*, 6, 585.
40. Matteucci, M. 1997. Oligonucleotide analogues: an overview, *Ciba Found. Symp.*, 209, 5.
41. Fearon, K. L., Hirschbein, B. L., Chiu, C. Y., Quijano, M. R., and Zon, G. 1997. Phosphorothioate oligodeoxynucleotides: large-scale synthesis and analysis, impurity characterization, and the effects of phosphorus stereochemistry, *Ciba Found. Symp.*, 209, 19.
42. Righetti, P. G. and Gelfi, C. 1997. Recent advances in capillary electrophoresis of DNA fragments and PCR products, *Biochem. Soc. Trans.*, 25, 267.
43. Righetti, P. G. and Gelfi, C. 1997. Recent advances in capillary electrophoresis of DNA fragments and PCR products in poly(n-substituted acrylamides), *Anal. Biochem.*, 244, 195.
44. Schlingensiepen, R., Brysch, W., and Schlingensiepen, K.-H. 1997. *Antisense—from technology to therapy*, Blackwell Wissenschaft Berlin, Vienna, Austria.
45. Wagner, R. W., Matteucci, M. D., Grant, D., Huang, T., and Froehler, B. C. 1996. Potent and selective inhibition of gene expression by an antisense heptanucleotide, *Nat. Biotechnol.*, 14, 840.
46. Altschul, S. F., Madden, T. L., Schaffer, A. A., Zhang, J., Zhang, Z., Miller, W., and Lipman, D. J. 1997. Gapped BLAST and PSI-BLAST: a new generation of protein database search programs, *Nucleic Acids Res.*, 25, 3389.
47. Boutorine, A. S. and Kostina, E. V. 1993. Reversible covalent attachment of cholesterol to oligodeoxyribonucleotides for studies of the mechanisms of their penetration into eucaryotic cells, *Biochimie*, 75, 35.
48. Alahari, S. K., Dean, N. M., Fisher, M. H., Delong, R., Manoharan, M., Tivel, K. L., and Juliano, R. L. 1996. Inhibition of expression of the multidrug resistance-associated P- glycoprotein of by phosphorothioate and 5′ cholesterol-conjugated phosphorothioate antisense oligonucleotides, *Mol. Pharmacol.*, 50, 808.

49. Krieg, A. M., Tonkinson, J., Matson, S., Zhao, Q., Saxon, M., Zhang, L. M., Bhanja, U., Yakubov, L., and Stein, C. A. 1993. Modification of antisense phosphodiester oligodeoxynucleotides by a 5′ cholesteryl moiety increases cellular association and improves efficacy, *Proc. Natl. Acad. Sci. USA*, 90, 1048, 1993.
50. Leonetti, J. P., Degols, G., Clarenc, J. P., Mechti, N., and Lebleu, B. 1993. Cell delivery and mechanisms of action of antisense oligonucleotides, *Prog. Nucleic Acid Res. Mol. Biol.*, 44, 143.
51. Degols, G., Leonetti, J. P., Benkirane, M., Devaux, C., and Lebleu, B. 1992. Poly(L-lysine)-conjugated oligonucleotides promote sequence-specific inhibition of acute HIV-1 infection, *Antisense Res. Dev.*, 2, 293.
52. Citro, G., Perrotti, D., Cucco, C., I, D. A., Sacchi, A., Zupi, G., and Calabretta, B. 1992. Inhibition of leukemia cell proliferation by receptor-mediated uptake of *c-myb* antisense oligodeoxynucleotides, *Proc. Natl. Acad. Sci. USA*, 89, 7031.
53. Bonfils, E., Depierreux, C., Midoux, P., Thuong, N. T., Monsigny, M., and Roche, A. C. 1992. Drug targeting: synthesis and endocytosis of oligonucleotide- neoglycoprotein conjugates, *Nucleic Acids Res.*, 20, 4621.
54. Wu, G. Y. and Wu, C. H. 1992. Specific inhibition of hepatitis B viral gene expression *in vitro* by targeted antisense oligonucleotides, *J. Biol. Chem.*, 267, 12436.
55. Roth, C. M., Reiken, S. R., Le Doux, J. M., Rajur, S. B., Lu, X.-M., Morgan, J. R., and Yarmush, M. L. 1997. Targeted antisense modulation of inflammatory cytokine receptors, *Biotechnol. Bioeng.*, 55, 72.
56. Lu, X. M., Fischman, A. J., Jyawook, S. L., Hendricks, K., Tompkins, R. G., and Yarmush, M. L. 1994. Antisense DNA delivery in vivo: liver targeting by receptor-mediated uptake, *J. Nucl. Med.*, 35, 269.
57. Rajur, S. B., Roth, C. M., Morgan, J. R., and Yarmush, M. L. 1997. Covalent protein-oligonucleotide conjugates for efficient delivery of antisense molecules, *Bioconjugate Chem.*, 8, 935.
58. Lavigne, C. and Thierry, A. R. 1997. Enhanced antisense inhibition of human immunodeficiency virus type 1 in cell cultures by DLS delivery system, *Biochem. Biophys. Res. Commun.*, 237, 566.
59. Zelphati, O. and Szoka, F. C., Jr. 1996. Mechanism of oligonucleotide release from cationic liposomes, *Proc. Natl. Acad. Sci. USA*, 93, 11493.
60. Marcusson, E. G., Bhat, B., Manoharan, M., Bennett, C. F., and Dean, N. M. 1998. Phosphorothioate oligodeoxyribonucleotides dissociate from cationic lipids before entering the nucleus, *Nucleic Acids Res.*, 26, 2016.
61. Bhatia, R. and Verfaillie, C. M. 1998. Inhibition of BCR-ABL expression with antisense oligodeoxynucleotides restores beta1 integrin-mediated adhesion and proliferation inhibition in chronic myelogenous leukemia hematopoietic progenitors, *Blood*, 91, 3414.
62. Bergan, R., Connell, Y., Fahmy, B., and Neckers, L. 1993. Electroporation enhances c-myc antisense oligodeoxynucleotide efficacy, *Nucleic Acids Res.*, 21, 3567.
63. Schiedlmeier, B., Schmitt, R., Muller, W., Kirk, M. M., Gruber, H., Mages, W., and Kirk, D. L. 1994. Nuclear transformation of Volvox carteri, *Proc. Natl. Acad. Sci. USA*, 91, 5080.
64. Lesh, R. E., Somlyo, A. P., Owens, G. K., and Somlyo, A. V. 1995. Reversible permeabilization. A novel technique for the intracellular introduction of antisense oligodeoxynucleotides into intact smooth muscle, *Circ Res*, 77, 220.
65. Spiller, D. G. and Tidd, D. M. 1995. Nuclear delivery of antisense oligodeoxynucleotides through reversible permeabilization of human leukemia cells with streptolysin O, *Antisense Res. Dev.*, 5, 13.
66. Leonetti, J. P., Machy, P., Degols, G., Lebleu, B., and Leserman, L. 1990. Antibody-targeted liposomes containing oligodeoxyribonucleotides complementary to viral RNA selectively inhibit viral replication, *Proc. Natl. Acad. Sci. USA*, 87, 2448.
67. Stein, C. A. and Cheng, Y. C. 1993. Antisense oligonucleotides as therapeutic agents—is the bullet really magical?, *Science*, 261, 1004.

68. Lisziewicz, J., Sun, D., Klotman, M., Agrawal, S., Zamecnik, P., and Gallo, R. 1992. Specific inhibition of human immunodeficiency virus type 1 replication by antisense oligonucleotides: an *in vitro* model for treatment, *Proc. Natl. Acad. Sci. USA*, 89, 11209.
69. Monia, B. P., Johnston, J. F., Ecker, D. J., Zounes, M. A., Lima, W. F., and Freier, S. M. 1992. Selective inhibition of mutant Ha-ras mRNA expression by antisense oligonucleotides, *J. Biol. Chem.*, 267, 19954.
70. Carter, G. and Lemoine, N. R. 1993. Antisense technology for cancer therapy: does it make sense?, *Br. J. Cancer*, 67, 869.
71. Chang, E. H., Miller, P. S., Cushman, C., Devadas, K., Pirollo, K. F., Ts'o, P. O., and Yu, Z. P. 1991. Antisense inhibition of *ras* p21 expression that is sensitive to a point mutation, *Biochemistry*, 30, 8283.
72. Agrawal, S., Temsamani, J., and Tang, J. Y. 1991. Pharmacokinetics, biodistribution, and stability of oligodeoxynucleotide phosphorothioates in mice, *Proc. Natl. Acad. Sci. USA*, 88, 7595.
73. Iversen, P. 1991. *In vivo* studies with phosphorothioate oligonucleotides: pharmacokinetics prologue, *Anticancer Drug Des.*, 6, 531.
74. Henry, S. P., Giclas, P. C., Leeds, J., Pangburn, M., Auletta, C., Levin, A. A., and Kornbrust, D. J. 1997. Activation of the alternative pathway of complement by a phosphorothioate oligonucleotide: potential mechanism of action, *J. Pharmacol. Exp. Ther.*, 281, 810.
75. Yu, D., Iyer, R. P., Shaw, D. R., Lisziewicz, J., Li, Y., Jiang, Z., Roskey, A., and Agrawal, S. 1996. Hybrid oligonucleotides: synthesis, biophysical properties, stability studies, and biological activity, *Bioorg. Med. Chem.*, 4, 1685.
76. Wu, C. H., Wilson, J. M., and Wu, G. Y. 1989. Targeting genes: delivery and persistent expression of a foreign gene driven by mammalian regulatory elements in vivo, *J. Biol. Chem.*, 264, 16985.
77. Ratajczak, M. Z., Kant, J. A., Luger, S. M., Hijiya, N., Zhang, J., Zon, G., and Gewirtz, A. M. 1992. *In vivo* treatment of human leukemia in a scid mouse model with c-myb antisense oligodeoxynucleotides, *Proc. Natl. Acad. Sci. USA*, 89, 11823.
78. Higgins, K. A., Perez, J. R., Coleman, T. A., Dorshkind, K., McComas, W. A., Sarmiento, U. M., Rosen, C. A., and Narayanan, R. 1993. Antisense inhibition of the p65 subunit of NF-kappa B blocks tumorigenicity and causes tumor regression, *Proc. Natl. Acad. Sci. USA*, 90, 9901.
79. Offensperger, W. B., Offensperger, S., Walter, E., Teubner, K., Igloi, G., Blum, H. E., and Gerok, W. 1993. *In vivo* inhibition of duck hepatitis B virus replication and gene expression by phosphorothioate modified antisense oligodeoxynucleotides, *Embo. J.*, 12, 1257.
80. Simons, M., Edelman, E. R., DeKeyser, J. L., Langer, R., and Rosenberg, R. D. 1992. Antisense *c-myb* oligonucleotides inhibit intimal arterial smooth muscle cell accumulation in vivo, *Nature*, 359, 67.
81. Abe, J., Zhou, W., Taguchi, J., Takuwa, N., Miki, K., Okazaki, H., Kurokawa, K., Kumada, M., and Takuwa, Y. 1994. Suppression of neointimal smooth muscle cell accumulation *in vivo* by antisense cdc2 and cdk2 oligonucleotides in rat carotid artery, *Biochem. Biophys. Res. Commun.*, 198, 16.
82. Bennett, M. R., Anglin, S., McEwan, J. R., Jagoe, R., Newby, A. C., and Evan, G. I. 1994. Inhibition of vascular smooth muscle cell proliferation *in vitro* and *in vivo* by *c-myc* antisense oligodeoxynucleotides, *J. Clin. Invest.*, 93, 820.
83. Bennett, C. F., Kornbrust, D., Henry, S., Stecker, K., Howard, R., Cooper, S., Dutson, S., Hall, W., and Jacoby, H. I. 1997. An ICAM-1 antisense oligonucleotide prevents and reverses dextran sulfate sodium-induced colitis in mice, *J. Pharmacol. Exp. Ther.*, 280, 988.
84. Azad, R. F., Brown-Driver, V., Buckheit, R. W., Jr., and Anderson, K. P. 1995. Antiviral activity of a phosphorothioate oligonucleotide complementary to human cytomegalovirus RNA when used in combination with antiviral nucleoside analogs, *Antiviral Res.*, 28, 101.
85. Anderson, K. P., Fox, M. C., Brown-Driver, V., Martin, M. J., and Azad, R. F. 1996. Inhibition of human cytomegalovirus immediate-early gene expression by an antisense oligonucleotide complementary to immediate-early RNA, *Antimicrob. Agents Chemother*, 40, 2004.

Further Information

The book *Antisense Nucleic Acids and Proteins: Fundamentals and Applications*, edited by Joseph N. M. Mol and Alexander R. van der Krol, is a collection of reviews in the use of antisense nucleic acids to modulate or downregulate gene expression. The book *Antisense RNA and DNA*, edited by James A. H. Murray, explores the use of antisense and catalytic nucleic acids for regulating gene expression.

The journal *Antisense and Nucleic Acid Drug Development*, published by Mary Ann Liebert, Inc., presents original research on antisense technology. For subscription information, contact Antisense and Nucleic Acid Drug Development, Mary Ann Liebert, Inc., 2 Madison Avenue, Larchmont, NY 10538; (914) 834-3100, Fax (914) 834-3688, email: info@liebertpub.com. The biweekly journal *Nucleic Acids Research* publishes papers on physical, chemical, and biologic aspects of nucleic acids, their constituents, and proteins with which they interact. For subscription information, contact Journals Marketing; Oxford University Press, Inc., 2001 Evans Road, Cary, NC 27513; 1-800-852-7323, Fax, (919) 677-1714, email: jnlorders@oup-usa.org.

104

Tools for Genome Analysis

Robert Kaiser
University of Washington

The development of sophisticated and powerful recombinant techniques for manipulating and analyzing genetic material has led to the emergence of a new biologic discipline, often termed *molecular biology.* The tools of molecular biology have enabled scientists to begin to understand many of the fundamental processes of life, generally through the identification, isolation, and structural and functional analysis of individual or, at best, limited numbers of genes. Biology is now at a point where it is feasible to begin a more ambitious endeavor—the complete genetic analysis of entire genomes. Genome analysis aims not only to identify and molecularly characterize all the genes that orchestrate the development of an organism but also to understand the complex and interactive regulatory mechanisms of these genes, their organization in the genome, and the role of genetic variation in disease, adaptability, and individuality. Additionally, the study of homologous genetic regions across species can provide important insight into their evolutionary history.

As can be seen in Table 104.1, the genome of even a small organism consists of a very large amount of information. Thus the analysis of a complete genome is not simply a matter of using conventional techniques that work well with individual genes (comprised of perhaps 1000 to 10,000 base pairs) a sufficient (very large) number of times to cover the genome. Such a brute-force approach would be too slow and too expensive, and conventional data-handling techniques would be inadequate for the task of cataloging, storing, retrieving, and analyzing such a large amount of information. The amount of manual labor and scientific expertise required would be prohibitive. New technology is needed to provide high-throughput, low-cost automation and reduced reliance on expert intervention at intermediate levels in the processes required for large-scale genetic analysis. Novel computational tools are required to deal with the large volumes of genetic information produced. Individual tools must be integrated smoothly to produce an analytical system in which samples are tracked through the entire analytical process, intermediate decisions and branch points are few, a stable, reliable and routine protocol or set of protocols is employed, and the resulting information is presented to the biologic scientist in a useful and meaningful format. It is important to realize that the development of these tools requires the interdisciplinary efforts of biologists, chemists, physicists, engineers, mathematicians, and computer scientists.

Genome analysis is a complex and extended series of interrelated processes. The basic processes involved are diagrammed in Fig. 104.1. At each stage, new biologic, chemical, physical (mechanical, optical), and computational tools have been developed within the last 10 years that have begun to enable large-scale (megabase) genetic analysis. These developments have largely been spurred by the goals of

0-8493-0462-8/00/$0.00+$.50

TABLE 104.1 DNA Content of Various Genomes in Monomer Units (Base Pairs)

Organism	Type	Size
Phage T4	Bacteriophage (virus)	160,000
Escherichia coli	Bacterium	4,000,000
Saccharomyces	Yeast	14,000,000
Arabidopsis thaliana	Plant	100,000,000
Caenorhabditis elegans	Nematode	100,000,000
Drosophila melanogaster	Insect (fruit fly)	165,000,000
Mouse	Mammal	3,000,000,000
Human	Mammal	3,500,000,000

Source: Adapted from [1] and [2].

the Human Genome Project, a worldwide effort to decipher the entirety of human genetics. However, biologists are still a significant ways away from having a true genome analysis capability [3], and as such, new technologies are still emerging.

This chapter cannot hope to describe in depth the entire suite of tools currently in use in genome analysis. Instead, it will attempt to present the basic principles involved and to highlight some of the recent enabling technological developments that are likely to remain in use in genome analysis for the foreseeable future. Some fundamental knowledge of biology is assumed; in this regard, an excellent beginning text for individuals with a minimal background in molecular biology is that by Watson et al. [1].

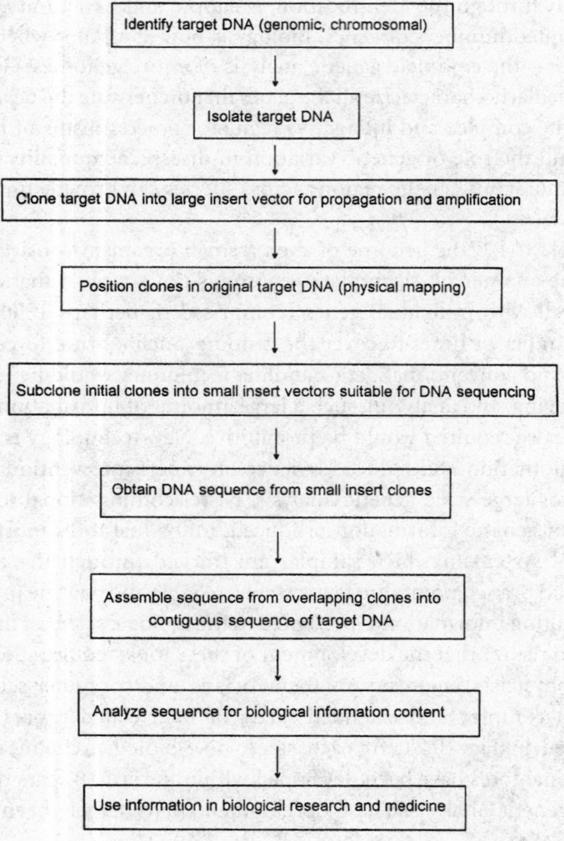

FIGURE 104.1 Basic steps in genome analysis.

TABLE 104.2 Enzymes Commonly Used in Genome Analysis

Enzyme	Function	Common Use
Restriction endonuclease	Cleave double-stranded DNA at specific sites	Mapping, cloning
DNA polymerase	Synthesize complementary DNA strand	DNA sequencing, amplification
Polynucleotide kinase	Adds phosphate to 5′ end of single-stranded DNA	Radiolabeling, cloning
Terminal transferase	Adds nucleotides to the 3′ end of single-stranded DNA	Labeling
Reverse transcriptase	Makes DNA copy from RNA	RNA sequencing, cDNA cloning
DNA ligase	Covalently joins two DNA fragments	Cloning

104.1 General Principles

The fundamental blueprint for any cell or organism is encoded in its genetic material, its deoxyribonucleic acid (DNA). DNA is a linear polymer derived from a four-letter biochemical alphabet—A, C, G, and T. These four letters are often referred to as *nucleotides* or *bases*. The linear order of bases in a segment of DNA is termed its *DNA sequence* and determines its function. A gene is a segment of DNA whose sequence directly determines its translated protein product. Other DNA sequences are recognized by the cellular machinery as start and stop sites for protein synthesis, regulate the temporal or spatial expression of genes, or play a role in the organization of higher-order DNA structures such as chromosomes. Thus a thorough understanding of the DNA sequence of a cell or organism is fundamental to an understanding of its biology.

Recombinant DNA technology affords biologists the capability to manipulate and analyze DNA sequences. Many of the techniques employed take advantage of a basic property of DNA, the molecular complementarity of the two strands of the double helix. This complementarity arises from the specific hydrogen-bonding interactions between pairs of DNA bases, A with T and C with G. Paired double strands of DNA can be denatured, or rendered into the component single strands, by any process that disrupts these hydrogens bonds—high temperature, chaotropic agents, or pH extremes. Complementary single strands also can be renatured into the duplex structure by reversing the disruptive element; this process is sometimes referred to as *hybridization* or *annealing*, particularly when one of the strands has been supplied from some exogenous source.

Molecular biology makes extensive use of the DNA-modifying enzymes employed by cells during replication, translation, repair, and protection from foreign DNA. A list of commonly used enzymes, their functions, and some of the experimental techniques in which they are utilized is provided in Table 104.2.

104.2 Enabling Technologies

The following are broadly applicable tools that have been developed in the context of molecular biology and are commonly used in genome analysis.

Cloning. Cloning is a recombinant procedure that has two main purposes: First, it allows one to select a single DNA fragment from a complex mixture, and second, it provides a means to store, manipulate, propagate, and produce large numbers of identical molecules having this single ancestor. A cloning vector is a DNA fragment derived from a microorganism, such as a bacteriophage or yeast, into which a foreign DNA fragment may be inserted to produce a chimeric DNA species. The vector contains all the genetic information necessary to allow for the replication of the chimera in an appropriate host organism. A variety of cloning vectors have been developed which allow for the insertion and stable propagation of foreign DNA segments of various sizes; these are indicated in Table 104.3.

Electrophoresis. Electrophoresis is a process whereby nucleic acids are separated by size in a sieving matrix under the influence of an electric field. In free solution, DNA, being highly negatively charged by virtue of its phosphodiester backbone, migrates rapidly toward the positive pole of an electric field. If

TABLE 104.3 Common Cloning Vectors

Vector	Approximate Insert Size Range (Base Pairs)
Bacteriophage M13	100–5000
Plasmid	100–10,000
Bacteriophage lambda	10,000–15,000
Cosmid	25,000–50,000
Yeast artificial chromosome (YAC)	100,000–1,000,000

the DNA is forced instead to travel through a molecularly porous substance, such as a gel, the smaller (shorter) fragments of DNA will travel through the pores more rapidly than the larger (longer) fragments, thus effecting separation. Agarose, a highly purified derivative of agar, is commonly used to separate relatively large fragments of DNA (100 to 50,000 base pairs) with modest resolution (50 to 100 base pairs), while cross-linked polyacrylamide is used to separate smaller fragments (10 to 1,000 base pairs) with single base-pair resolution. Fragment sizes are generally estimated by comparison with standards run in another lane of the same gel. Electrophoresis is used extensively as both an analytical and a preparative tool in molecular biology.

Enzymatic DNA Sequencing. In the late 1970s, Sanger and coworkers [4] reported a procedure employing DNA polymerase to obtain DNA sequence information from unknown cloned fragments. While significant improvements and modifications have been made since that time, the basic technique remains the same: DNA polymerase is used to synthesize a complementary copy of an unknown single-stranded DNA (the template) in the presence of the four DNA monomers (deoxynucleotide triphosphates, or dNTPs). DNA polymerase requires a double-stranded starting point, so a single-stranded DNA (the primer) is hybridized at a unique site on the template (usually in the vector), and it is at this point that DNA synthesis is initiated. Key to the sequencing process is the use of a modified monomer, a dideoxynucleotide triphosphate (ddNTP), in each reaction. The ddNTP lacks the 3′-hydroxyl functionality (it has been replaced by a hydrogen) necessary for phosphodiester bond formation, and its incorporation thus blocks further elongation of the growing chain by polymerase. Four reactions are carried out, each containing all four dNTPs and one of the four ddNTPs. By using the proper ratios of dNTPs to ddNTP, each reaction generates a nested set of fragments, each fragment beginning at exactly the same point (the primer) and terminating with a particular ddNTP at each base complementary to that ddNTP in the template sequence. The products of the reactions are then separated by electrophoresis in four lanes of a polyacrylamide slab gel. Since conventional sequencing procedures utilize radiolabeling (incorporation of a small amount of ^{32}P- or ^{35}S-labeled dNTP by the polymerase), visualization of the gel is achieved by exposing it to film. The sequence can be obtained from the resulting autoradiogram, which appears as a series of bands (often termed a *ladder*) in each of the four lanes. Each band is composed of fragments of a single size, the shortest fragments being at the bottom of the gel and the longest at the top. Adjacent bands represent a single base pair difference, so the sequence is determined by reading up the ladders in the four lanes and noting which lane contains the band with the next largest sized fragments. The enzymatic sequencing process is diagrammed in Fig. 104.2. It should be noted that although other methods exist, the enzymatic sequencing technique is currently the most commonly used DNA sequencing procedure due to its simplicity and reliability.

Polymerase Chain Reaction (PCR). PCR [5] is an in vitro procedure for amplifying particular DNA sequences up to 10^8-fold that is utilized in an ever-increasing variety of ways in genome analysis. The sequence to be amplified is defined by a pair of single-stranded primers designed to hybridize to unique sites flanking the target sequence on opposite strands. DNA polymerase in the presence of the four dNTPs is used to synthesize a complementary DNA copy across the target sequence starting at the two primer sites. The amplification procedure is performed by repeating the following cycle 25 to 50 times (see Fig. 104.3). First, the double-stranded target DNA is denatured at high temperature (94 to 96°C). Second, the mixture is cooled, allowing the primers to anneal to their complementary sites on the target single

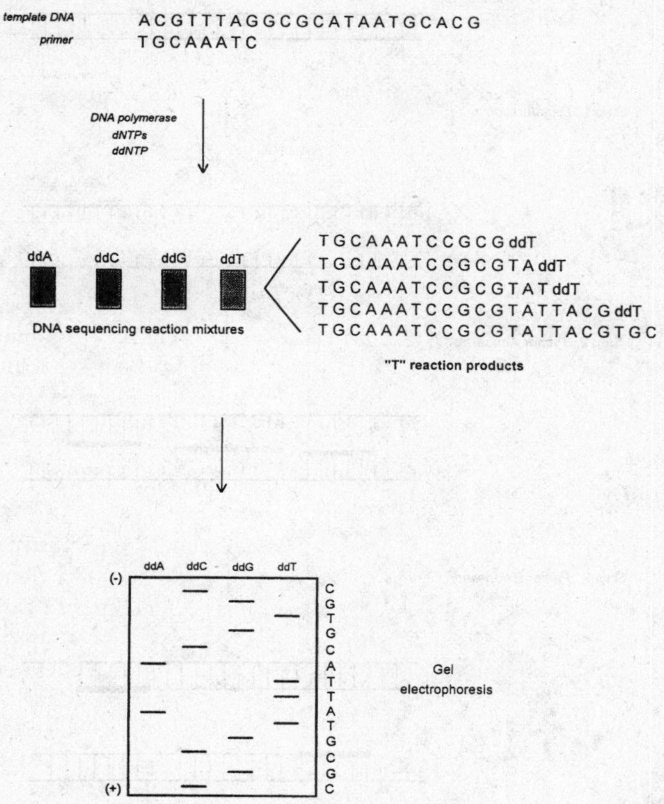

FIGURE 104.2 Enzymatic DNA sequencing. A synthetic oligonucleotide primer is hybridized to its complementary site on the template DNA. DNA polymerase and dNTPs are then used to synthesize a complementary copy of the unknown portion of the template in the presence of a chain-terminating ddNTP (see text). A nested set of fragments beginning with the primer sequence and ending at every ddNTP position is produced in each reaction (the ddTTP reaction products are shown). Four reactions are carried out, one for each ddNTP. The products of each reaction are then separated by gel electrophoresis in individual lanes, and the resulting ladders are visualized. The DNA sequence is obtained by reading up the set of four ladders, one base at a time, from smallest to largest fragment.

strands. Third, the temperature is adjusted for optimal DNA polymerase activity, initiating synthesis. Since the primers are complementary to the newly synthesized strands as well as the original target, each cycle of denaturation/annealing/synthesis effectively doubles the amount of target sequence present in the reaction, resulting in a 2^n amplification (n = number of cycles). The initial implementation of PCR utilized a polymerase that was unstable at the high temperatures required for denaturation, thus requiring manual addition of polymerase prior to the synthesis step of every cycle. An important technological development was the isolation of DNA polymerase from a thermophilic bacterium, *Thermus aquaticus* (*Taq*), which can withstand the high denaturation temperatures [6]. Additionally, the high optimal synthesis temperature (70 to 72°C) of *Taq* polymerase improves the specificity of the amplification process by reducing spurious priming from annealing of the primers to nonspecific secondary sites in the target.

While PCR can be performed successfully manually, it is a tedious process, and numerous thermal cycling instruments have become commercially available. Modern thermal cyclers are programmable and capable of processing many samples at once, using either small plastic tubes or microtiter plates, and are characterized by accurate and consistent temperature control at all sample positions, rapid temperature ramping, and minimal temperature over/undershoot. Temperature control is provided by a variety of means (Peltier elements, forced air, water) using metal blocks or water or air baths. Speed, precise temperature control, and high sample throughput are the watchwords of current thermal cycler design.

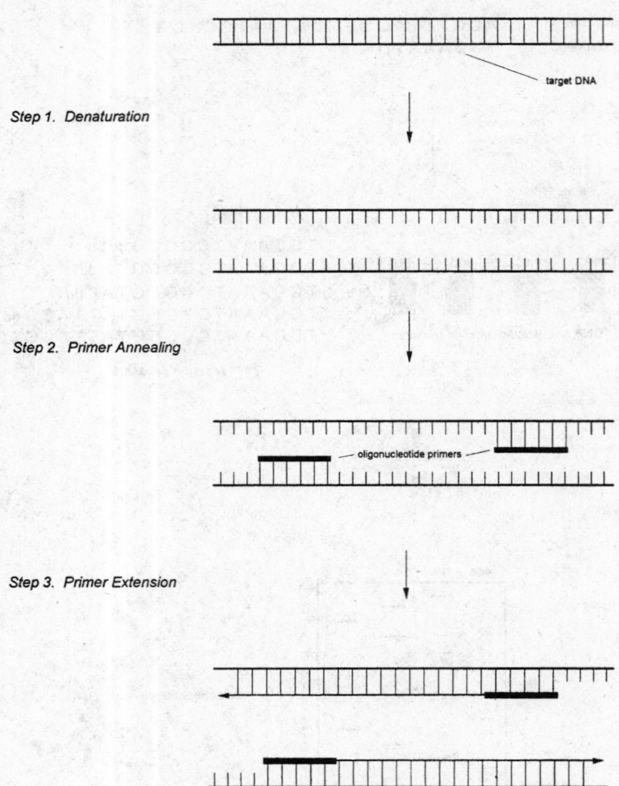

FIGURE 104.3 The first cycle in the polymerase chain reaction. In step 1, the double-stranded target DNA is thermally denatured to produce single-stranded species. A pair of synthetic primers, flanking the specific region of interest, are annealed to the single strands to form initiation sites for DNA synthesis by polymerase (step 2). Finally, complementary copies of each target single strand are synthesized by polymerase in the presence of dNTPs, thus doubling the amount of target DNA initially present (step 3). Repetition of this cycle effectively doubles the target population, affording one million-fold or greater amplification of the initial target sequence.

PCR technology is commonly used to provide sufficient material for cloning from genomic DNA sources, to identify and characterize particular DNA sequences in an unknown mixture, to rapidly produce templates for DNA sequencing from very small amounts of target DNA, and in cycle sequencing, a modification of the enzymatic sequencing procedure that utilizes *Taq* polymerase and thermal cycling to amplify the products of the sequencing reactions.

Chemical Synthesis of Oligodeoxynucleotides. The widespread use of techniques based on DNA polymerase, such as enzymatic DNA sequencing and the PCR, as well as of numerous other techniques utilizing short, defined-sequence, single-stranded DNAs in genome analysis, is largely due to the ease with which small oligodeoxynucleotides can be obtained. The chemical synthesis of oligonucleotides has become a routine feature of both individual biology laboratories and core facilities. The most widely used chemistry for assembling short (<100 base pair) oligonucleotides is the phosphoramidite approach [7], which has developed over the past 15 years or so. This approach is characterized by rapid, high-yield reactions and stable reagents. Like modern peptide synthesis chemistry, the approach relies on the tethering of the growing DNA chain to a solid support (classically glass or silica beads, more recently cross-linked polystyrene) and is cyclic in nature. At the end of the assembly, the desired oligonucleotide is chemically cleaved from the support, generally in a form that is sufficiently pure for its immediate use in a number of applications. The solid phase provides two significant advantages: It allows for the use of large reagent excesses, driving the reactions to near completion in accord with the laws of mass action while reducing the removal of these excesses following the reactions to a simple matter of thorough washing, and it enables the reactions to be performed in simple, flow-through cartridges, making the entire synthesis procedure easily automatable. Indeed, a number of chemical DNA synthesis instruments ("gene machines") are commercially available, capable of synthesizing one to several oligonucleotides at once. Desired sequences are programmed through a keyboard or touchpad, reagents are installed, and DNA is obtained a few hours later. Improvements in both chemistry and instrument design have been aimed at increasing synthesis throughput (reduced cycle times, increased number of simultaneous sequence assemblies), decreasing scale (most applications in genome analysis require subnanomole quantities of any particular oligonucleotide), and concomitant with these two, reducing cost per oligonucleotide.

104.3 Tools for Genome Analysis

Physical Mapping. In the analysis of genomes, it is often useful to begin with a less complex mixture than an entire genome DNA sample. Individual chromosomes can be obtained in high purity using a technology known as *chromosome sorting* [8], a form of flow cytometry. A suspension of chromosomes stained with fluorescent dye is flowed past a laser beam. As a chromosome enters the beam, appropriate optics detect the scattered and emitted light. Past the beam, the stream is acoustically broken into small droplets. The optical signals are used to electronically trigger the collection of droplets containing chromosomes by electrostatically charging these droplets and deflecting them into a collection medium using a strong electric field. Chromosomes can be differentiated by staining the suspension with two different dyes that bind in differing amounts to the various chromosomes and looking at the ratio of the emission intensity of each dye as it passes the laser/detector. Current commercial chromosome sorting instrumentation is relatively slow, requiring several days to collect sufficient material for subsequent analysis.

As mentioned previously, whole genomes or even chromosomes cannot yet be analyzed as intact entities. As such, fractionation of large nucleic acids into smaller fragments is necessary to obtain the physical material on which to perform genetic analysis. Fractionation can be achieved using a variety of techniques: limited or complete digestion by restriction enzymes, sonication, or physical shearing through a small orifice. These fragments are then cloned into an appropriate vector, the choice of which depends on the size range of fragments involved (see Table 104.3). The composite set of clones derived from a large nucleic acid is termed a *library.* In general, it is necessary to produce several libraries in different cloning vectors containing different-sized inserts. This is necessary because the mapping of clones is facilitated by larger inserts, while the sequencing of clones requires shorter inserts.

The library-generating process yields a very large number of clones having an almost random distribution of insert endpoints in the original fragment. It would be very costly to analyze all clones in a library, and unnecessary as well. Instead, a subset of overlapping clones is selected whose inserts span the entire starting fragment. These clones must be arrayed in the linear order in which they are found in the starting fragment; the process for doing this is called *physical mapping.* The conventional method for physically mapping clones uses restriction enzymes to cleave each clone at enzyme-specific sites, separating the products of the digestion by electrophoresis, and comparing the resulting patterns of restriction fragment sizes for different clones to find similarities. Clones exhibiting a number of the same-sized fragments likely possess the same subsequence and thus overlap. Clearly, the longer the inserts contained in the library, the faster a large genetic region can be covered by this process, since fewer clones are required to span the distance. Physical mapping also provides landmarks, the enzyme cleavage sites in the sequence, that can be used to provide reference points for the mapping of genes and other functional sequences. Mapping by restriction enzyme digestion is simple and reliable to perform; however, manual map assembly from the digest data is laborious, and significant effort is currently being expended in the development of robust and accurate map assembly software.

Normal agarose gel electrophoresis can effectively separate DNA fragments less than 10,000 base pairs and fragments between 10,000 and 50,000 base pairs less effectively under special conditions. However, the development of very large insert cloning vectors, such as the yeast artificial chromosome [9], necessitated the separation of fragments significantly larger than 10,000 base pairs to allow for use in physical mapping. In order to address this issue, a technology called *pulsed-field gel electrophoresis* (PFGE) was developed. Unlike conventional electrophoresis, in which the electric field remains essentially constant, homogeneous, and unidirectional during a separation, PFGE utilizes an electric field that periodically changes its orientation. The principle of PFGE is thought to be as follows: When DNA molecules are placed in an electric field, the molecules elongate in the direction of the field and then begin to migrate through the gel pores. When the field is removed, the molecules relax to a more random coiled state and stop moving. Reapplication of the field in another orientation causes the DNA to change its conformation in order to align in that direction prior to migration. The time required for this conformational change to occur has been found to be very dependent on the size of the molecules, with larger molecules reorienting more slowly than small ones. Thus longer DNAs move more slowly under the influence of the constantly switching electric field than shorter ones, and size-based separation occurs. PFGE separations of molecules as large as 10 million base pairs have been demonstrated. Numerous instruments for PFGE have been constructed, differing largely in the strategy employed to provide electric field switching [10].

Physical maps based on restriction sites are of limited long-term utility, since they require the provision of physical material from the specific library from which the map was derived in order to be utilized experimentally. A more robust landmarking approach based on the PCR has been developed recently [11], termed *sequence-tagged site* (STS) *content mapping.* An STS is a short, unique sequence in a genome that can be amplified by the PCR. Clones in a library are screened for the presence of a particular STS using PCR; if the STS is indeed unique in the genome, then clones possessing that STS are reliably expected to overlap. Physical mapping is thus reduced to choosing and synthesizing pairs of PCR primers that define unique sequences in the genome. Additionally, since STSs are defined by pairs of primer sequences, they can be stored in a database and are thus universally accessible.

DNA Sequencing. Early in the development of tools for large-scale DNA analysis, it was recognized that one of the most costly and time-consuming processes was the accumulation of DNA sequence information. Two factors, the use of radioisotopic labels and the manual reading and recording of DNA sequence films, made it impossible to consider genome-scale (10^6 to 10^9 base pairs) sequence analysis using the conventional techniques. To address this, several groups embarked on the development of automated DNA sequencing instruments [12–14]. Today, automated DNA sequencing is one of the most highly advanced of the technologies for genome analysis, largely due to the extensive effort expended in instrument design, biochemical organization, and software development.

Key to the development of these instruments was the demonstration that fluorescence could be employed in the place of autoradiography for detection of fragments in DNA sequencing gels and that the use of fluorescent labels enabled the acquisition of DNA sequence data in an automated fashion in real time. Two approaches have been demonstrated: the "single-color, four-lane" approach and the "four-color, single-lane" approach. The former simply replaces the radioisotopic label used in conventional enzymatic sequencing with a fluorescent label, and the sequence is determined by the order of fluorescent bands in the four lanes of the gel. The latter utilizes a different-colored label for each of the four sequencing reactions (thus A might be "blue," C, "green," G, "yellow," and T, "red"). The four base-specific reactions are performed separately and upon completion are combined and electrophoresed in a single lane of the gel, and the DNA sequence is determined from the temporal pattern of fluorescent colors passing the detector. For a fixed number of gel lanes (current commercial automated DNA sequencers have 24 to 36), the four-color approach provides greater sample throughput than the single-color approach. Instruments employing the four-color technology are more widely used for genome analysis at present, and as such, this strategy will be discussed more fully.

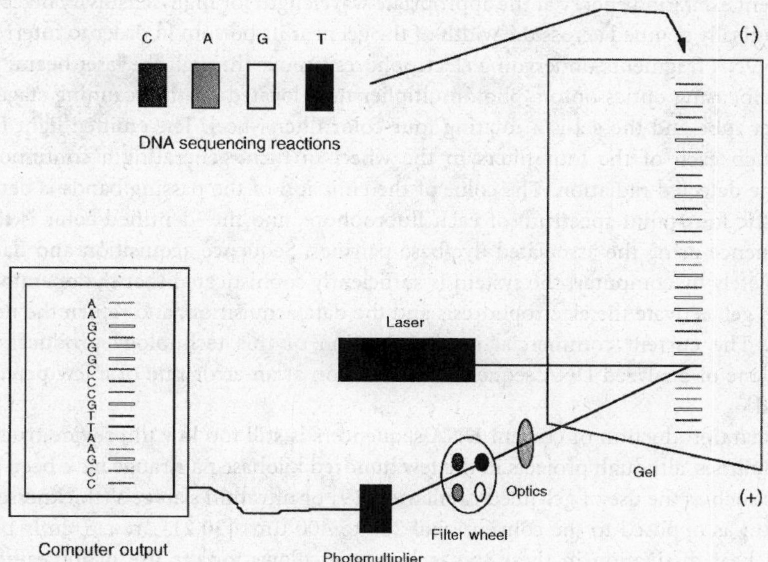

FIGURE 104.4 Schematic illustration of automated fluorescence-based DNA sequencing using the "four-color, single-lane" approach. The products of each of the four enzymatic sequencing reactions are "color-coded" with a different fluorescent dye, either through the use of dye-labeled primers or dye-terminators. The four reaction mixtures are then combined and the mixture separated by gel electrophoresis. The beam of an argon ion laser is mechanically scanned across the width of the gel near its bottom to excite the labeled fragments to fluorescence. The emitted light is collected through a four-color filter wheel onto a photomultiplier tube. The color of each fluorescing band is determined automatically by a computer from the characteristic four-point spectrum of each dye, and the order of colors passing the detector is subsequently translated into DNA sequence information.

In order to utilize fluorescence as a detection strategy for DNA sequencing, a chemistry had to be developed for the specific incorporation of fluorophores into the nested set of fragments produced in the enzymatic sequencing reactions. The flexibility of chemical DNA synthesis provided a solution to this problem. A chemistry was developed for the incorporation of an aliphatic primary amine in the last cycle of primer synthesis (i.e., at the 5′ terminus) using standard DNA synthesis protocols [15,16]. This amine was then conjugated with any of several of readily available amine-reactive fluorochromes that had been developed previously for the labeling of proteins to produce the desired labeled sequencing primers. The purified dye-primer was demonstrated to perform well in DNA sequencing, exhibiting both efficient extension by the polymerase and the necessary single-base resolution in the electrophoretic separation [12]. A set of four spectrally discriminable, amine-reactive fluorophores has been developed [16] for DNA sequencing.

While dye-primers are relatively easy to obtain by this method, they are costly to prepare in small quantities and require sophisticated chromatographic instrumentation to obtain products pure enough for sequencing use. Thus they are generally prepared in large amounts and employed as vector-specific "universal" primers, as in situations in which a very large number of inserts cloned in a given vector need to be sequenced [17]. For occasions where small amounts of sample-specific primers are needed, as in the sequencing of products from the PCR, a simpler and more economical alternative is the use of dideoxynucleotides covalently coupled to fluorescent dyes (so-called dye-terminators), since these reagents allow the use of conventional unlabeled primers [18].

Special instrumentation (Fig. 104.4) has been developed for the fluorescence-based detection of nucleic acids in DNA sequencing gels. An argon ion laser is used to excite the fluorescent labels in order to provide sufficient excitation energy at the appropriate wavelength for high-sensitivity detection. The laser beam is mechanically scanned across the width of the gel near its bottom in order to interrogate all lanes. As the labeled DNA fragments undergoing electrophoresis move through the laser beam, their emission is collected by focusing optics onto a photomultiplier tube located on the scanning stage. Between the photomultiplier tube and the gel is a rotating four-color filter wheel. The emitted light from the gel is collected through each of the four filters in the wheel in turn, generating a continuous four-point spectrum of the detected radiation. The color of the emission of the passing bands is determined from the characteristic four-point spectrum of each fluorophore, and the identified color is then translated into DNA sequence using the associated dye/base pairings. Sequence acquisition and data analysis are handled completely by computer; the system is sufficiently sophisticated that the operator can load the samples on the gel, activate the electrophoresis and the data acquisition, and return the next day for the analyzed data. The current commercial implementation of this technology produces about 450 to 500 bases per lane of analyzed DNA sequence information at an error rate of a few percent in a 12- to 14-hour period.

The rate of data production of current DNA sequencers is still too low to provide true genome-scale analytical capabilities, although projects in the few hundred kilobase pair range have been accomplished. Improvements such as the use of gel-filled capillaries [19] or ultrathin slab gels (thicknesses on the order of 50 to 100 μm as opposed to the conventional 200 to 400 μm) [20,21] are currently being explored. The improved heat dissipation in these thin-gel systems allows for the use of increased electric field strengths during electrophoresis, with a concomitant reduction in run time of fivefold or so.

The greatly increased throughput of the automated instruments over manual techniques has resulted in the generation of a new bottleneck in the DNA sequencing process that will only be exacerbated by higher-throughput systems: the preparation of sufficient sequencing reaction products for analysis. This encompasses two preparative processes, the preparation of sequencing templates and the performance of sequencing reactions. Automation of the latter process has been approached initially through the development of programmable pipetting robots that operate in the 96-well microtiter plate format commonly used for immunoassays, since a 96-well plate will accommodate sequencing reactions for 24 templates. *Sequencing robots* of this sort have become important tools for large-scale sequencing projects. Template preparation has proven more difficult to automate. No reliable system for selecting clones, infecting and culturing bacteria, and isolating DNA has been produced, although several attempts are in progress. It is clear that unlike in the case of the sequencing robots, where instrumentation has mimicked manual manipulations with programmable mechanics, successful automation of template preparation will require a rethinking of current techniques with an eye toward process automation. Furthermore, in order to obtain true genome-scale automation, the entire sequencing procedure from template preparation through data acquisition will need to be reengineered to minimize, if not eliminate, operator intervention using the principles of systems integration, process management, and feedback control.

Genetic Mapping. Simply stated, genetic mapping is concerned with identifying the location of genes on chromosomes. Classically, this is accomplished using a combination of mendelian and molecular genetics called *linkage analysis*, a complete description of which is too complex to be fully described here (but see Watson et al. [1]). However, an interesting approach to physically locating clones (and the genes contained within them) on chromosomes is afforded by a technique termed *fluorescence in situ hybridization* (FISH) [22]. Fluorescently labeled DNA probes derived from cosmid clones can be hybridized to chromosome spreads, and the location of the probe-chromosome hybrid can be observed using fluorescence microscopy. Not only can clones be mapped to particular chromosomes in this way, but positions and distances relative to chromosomal landmarks (such as cytogenetic bands, telomere, or centromere) can be estimated to as little as 50,000 base pairs in some cases, although 1 million base pairs or larger is more usual. The technique is particularly useful when two or more probes of different colors are used to order sequences relative to one another.

Computation. Computation plays a central role in genome analysis at a variety of levels, and significant efforts has been expanded on the development of software and hardware tools for biologic applications. A large effort has been expended in the development of software that will rapidly assemble a large contiguous DNA sequence from the many smaller sequences obtained from automated instruments. This assembly process is computationally demanding, and only recently have good software tools for this purpose become readily available. Automated sequencers produce 400 to 500 base pairs of sequence per template per run. However, in order to completely determine the linear sequence of a 50,000-base-pair cosmid insert (which can be conceptually represented as a linear array of 100 adjacent 500-base-pair templates), it is necessary to assemble sequence from some 300 to 1000 clones to obtain the redundancy of data needed for a high-accuracy finished sequence, depending on the degree to which the clones can be preselected for sequencing based on a previously determined physical map. Currently, the tools for acquiring and assembling sequence information are significantly better than those for physical mapping; as such, most large-scale projects employ strategies that emphasize sequencing at the expense of mapping [23]. Improved tools for acquiring and assembling mapping data are under development, however, and it remains to be seen what effect they will have on the speed and cost of obtaining finished sequence on a genome scale relative to the current situation.

Many software tools have been developed in the context of the local needs of large-scale projects. These include software for instrument control (data acquisition and signal processing), laboratory information-management systems, and local data-handling schemes. The development of process approaches to the automation of genome analysis will necessitate the continued development of tools of these types.

The final outcome of the analysis of any genome will be a tremendous amount of sequence information that must be accessible to researchers interested in the biology of the organism from which it was derived. Frequently, the finished genomic sequence will be the aggregate result of the efforts of many laboratories. National and international information resources (databases) are currently being established worldwide to address the issues of collecting, storing, correlating, annotating, standardizing, and distributing this information. Significant effort is also being expended to develop tools for the rapid analysis of genome sequence data that will enable biologists to find new genes and other functional genetic regions, compare very large DNA sequences for similarity, and study the role of genetic variation in biology. Eventually, as the robust tools for predicting protein tertiary structure and function from primary acid sequence data are developed, genome analysis will extend to the protein domain through the translation of new DNA sequences into their functional protein products.

104.4 Conclusions

Genome analysis is a large-scale endeavor whose goal is the complete understanding of the basic blueprint of life The scale of even the smallest genomes of biologic interest is too large to be effectively analyzed using the traditional tools of molecular biology and genetics. Over the past 10 years, a suite of biochemical techniques and bioanalytical instrumentation has been developed that has allowed biologists to begin to probe large genetic regions, although true genome-scale technology is still in its infancy. It is anticipated that the next 10 years will see developments in the technology for physical mapping, DNA sequencing, and genetic mapping that will allow for a 10- to 100-fold increase in our ability to analyze genomes, with a concomitant decrease in cost, through the application of process-based principles and assembly-line approaches. The successful realization of a true genome analysis capability will require the close collaborative efforts of individuals from numerous disciplines in both science and engineering.

Acknowledgments

I would like to thank Dr. Leroy Hood, Dr. Maynard Olson, Dr. Barbara Trask, Dr. Tim Hunkapiller, Dr. Deborah Nickerson, and Dr. Lee Rowen for the useful information, both written and verbal, that they provided me during the preparation of this chapter.

References

1. Watson JD, Gilman M, Witkowski J, Zoller M (eds). 1992. Recombinant DNA, 2d ed. New York, Scientific American Books, WH Freeman.
2. Lewin B. 1987. Genes III, 3d ed. New York, Wiley.
3. Olson MV. 1993. The human genome project. Proc Natl Acad Sci USA 90:4338.
4. Sanger F, Nicklen S, Coulson AR. 1977. DNA sequencing with chain-terminating inhibitors. Proc Natl Acad Sci USA 74:5463.
5. Saiki RK, Scharf SJ, Faloona F, et al. 1985. Enzymatic amplification of betaglobin sequences and restriction site analysis for diagnosis of sickle cell anemia. Science 230:1350.
6. Saiki RK, Gelfand DH, Stoffel S, et al. 1988. Primer-directed enzymatic amplification of DNA with a thermostable DNA polymerase. Science 239:487.
7. Gait MJ (ed). 1984. Oligonucleotide Synthesis: A Practical Approach. Oxford, England, IRL Press.
8. Engh Gvd. 1993. New applications of flow cytometry. Curr Opin Biotechnol 4:63.
9. Burke DT, Carle GF, Olson MV. 1987. Cloning of large segments of exogenous DNA into yeast by means of artificial chromosome vectors. Science 236:806.
10. Lai E, Birren BW, Clark SM, Hood L. 1989. Pulsed field gel electrophoresis. Biotechniques 7:34.
11. Olson M, Hood L, Cantor C, Botstein D. 1989. A common language for physical mapping of the human genome. Science 245:1434.
12. Smith LM, Sanders JZ, Kaiser RJ, et al. 1986. Fluorescence detection in automated DNA sequence analysis. Nature 321:674.
13. Prober JM, Trainor GL, Dam RJ, et al. 1987. A system for rapid DNA sequencing with fluorescent chain terminating dideoxynucleotides. Science 238:336.
14. Ansorge W, Sproat B, Stegemann J, et al. 1987. Automated DNA sequencing: Ultrasensitive detection of fluorescent bands during electrophoresis. Nucleic Acid Res 15:4593.
15. Smith LM, Fung S, Hunkapiller MW, et al. 1985. The synthesis of oligonucleotides containing an aliphtic amino group at the 5′ terminus: Synthesis of fluorescent DNA primers for use in DNA sequencing. Nucleic Acids Res 15:2399.
16. Connell C, Fung S, Heiner C, et al. 1987. Biotechniques 5:342.
17. Kaiser R, Hunkapiller T, Heiner C, Hood L. 1993. Specific primer-directed DNA sequence analysis using automated fluorescence detection and labeled primers. Methods Enzymol 218:122.
18. Lee LG, Connell CR, Woo SL, et al. 1992. DNA sequencing with dye-labeled terminators and T7 DNA polymerase: Effect of dyes and dNTPs on incorporation of dye-terminators and probability analysis of termination fragments. Nucleic Acids Res 20:2471.
19. Mathies RA, Huang XC. 1992. Capillary array electrophoresis: an approach to high-speed, high-throughput DNA sequencing. Nature 359:167.
20. Brumley RL Jr, Smith LM. 1991. Rapid DNA sequencing by horizontal ultrathin gel electrophoresis. Nucleic Acid Res 19:4121.
21. Stegemann J, Schwager C, Erfle H, et al. 1991. High speed on-line DNA sequencing on ultrathin slab gels. Nucleic Acid Res 19:675.
22. Trask BJ. 1991. Gene mapping by in situ hybridization. Curr Opin Genet Dev 1:82.
23. Hunkapiller T, Kaiser RJ, Koop BF, Hood L. 1991. Large-scale and automated DNA sequencing. Science 254:59.

105

Vaccine Production

John G. Aunins
Merck Research Laboratories

Ann L. Lee
Merck Research Laboratories

David B. Volkin
Merck Research Laboratories

Vaccines are biologic preparations that elicit immune system responses that protect an animal against pathogenic organisms. The primary component of the vaccine is an antigen, which can be a weakened (attenuated) version of an infectious pathogen or a purified molecule derived from the pathogen. Upon oral administration or injection of a vaccine, the immune system generates humoral (antibody) and cellular (cytotoxic, or killer T cell) responses that destroy the antigen or antigen-infected cells. When properly administered, the immune response to a vaccine has a long-term memory component, which protects the host against future infections. Vaccines often contain adjuvants to enhance immune response, as well as formulation agents to preserve the antigen during storage or upon administration, to provide proper delivery of antigen, and to minimize side reactions.

Table 105.1 presents a simple classification scheme for vaccines according to the type of organism and antigen, *Live, attenuated whole-organism vaccines* have been favored for simplicity of manufacture and for the strong immune response generated when the organism creates a subclinical infection before being overwhelmed. These are useful when the organism can be reliably attenuated in pathogenicity (while maintaining immunogenicity) or when the organism is difficult to cultivate ex vivo in large quantities and hence large amounts of antigen cannot be prepared. Conversely, *subunit antigen vaccines* are used when it is easy to generate large amounts of the antigen or when the whole organisms is not reliably attenuated. Since there is no replication in vivo, subunit vaccines rely on administrating relatively large amounts of antigen mass and are almost always adjuvanted to try to minimize the antigen needed. Subunit preparations have steadily gained favor, since biologic, engineering, and analytical improvements make them easily manufactured and characterized to a consistent standard. *Passive vaccines* are antibody preparations from human blood serum. These substitute for the patient's humoral response for immune-suppressed persons, for postexposure prophylaxis of disease, and for high infection risk situations where immediate protection is required, such as for travelers or medical personnel. Even more so than subunit vaccines, large amounts of antibodies are required. These vaccines do not provide long-term immune memory.

The organism and the nature of the antigen combine to determine the technologies of manufacture for the vaccine. Vaccine production generally involves growing the organism or its antigenic parts

0-8493-0462-8/00/$0.00+$.50

TABLE 105.1 Examples of Vaccines Against Human Pathogens, and Their Classification

Type of Organism	Live, Attenuated Cells/Particles	Killed, Inactivated Cells/Particles	Subcellular/Particle Vaccines	Passive Immunization
Virus	Measles, mumps, rubella, polio, yellow fever, varicella*, rotavirus*	Polio, hepatitis B, rabies, yellow fever, Japanese encephalitis, influenza, hepatitis A*	*Virus-like particles* Hepatitis B, HIV gp120-vaccinia* *Characterized single/combined antigens* Influenza HA + NA, herpes simplex gD*	*Immune serum globulin* Rabies, hepatitis B, cytomegalovirus *Monoclonal antibody* Anti-HIV gp120*
Bacterium	Tuberculosis (BCG), typhoid	Pertussis, cholera, plague bacillus	Toxoids Tetanus, diphtheria *Characterized single/combined antigens* Pertussis PT + FHA + LPF + PSA + 69kD, *Pneumococcus* polysaccharides (23), *Meningococcus* polysaccharides *Polysaccharide-protein conjugates* *Haemophilus* b, *Pneumococcus**	Immune serum globulin Tetanus, Nonspecific Ig

* Denotes vaccines in development.

(cultivation), treating it to purify and/or detoxify the organism and antigen (downstream processing), and in many cases further combining the antigen with adjuvants to increase its antigenicity and improve storage stability (formulation). These three aspects of production will be discussed, in addition to future trends in vaccine technology. This chapter does not include vaccines for parasitic disease [see Barbet, 1989], such as malaria, since such vaccines are not yet commercially available.

105.1 Antigen Cultivation

Microbial Cultivation

Bacterial Growth

As far as cultivation is concerned, the fundamental principles of cell growth are identical for the various types of bacterial vaccines. Detailed aspects of bacterial growth and metabolism can be found in Ingraham et al. [1983]. In the simplest whole-cell vaccines, obtaining cell mass is the objective of cultivation. Growth is an autocatalytic process where the cell duplicates by fission; growth is described by the differential equation

$$\frac{dX}{dt} = \mu X_v \tag{105.1}$$

where X is the total mass concentration of cells, X_v is the mass concentration of living, or viable cells, and μ is the specific growth rate in units of cells per cell time. In this equation, the growth rate of the cell is a function of the cells' physical and chemical environment:

$$\mu = f\left(\text{temperature, pH, dissolved oxygen}, C_1, C_2, C_3, \ldots, C_n\right) \tag{105.2}$$

All bacteria have a narrow range of temperatures permissive to growth; for human pathogens, the optimal temperature is usually 37°C; however, for attenuated strains, the optimal temperature may be purposefully lowered by mutation. At the end of culture, heat in excess of 50°C is sometimes used to kill pathogenic bacteria. Dissolved oxygen concentration is usually critical, since the organism will either require it or be inhibited by it. Cultivation of *Clostridium tetani*, for example, is conducted completely anaerobically.

In Eq. (105.2), C_i is the concentration of nutrient i presented to the cells. Nutrient conditions profoundly affect cell growth rate and the ultimate cell mass concentration achievable. At a minimum, the cells must be presented with a carbon source, a nitrogen source, and various elements in trace quantities. The carbon source, usually a carbohydrate such as glucose, is used for energy metabolism and as a precursor for various anabolites, small chemicals used to build the cell. The nitrogen source, which can be ammonia, or a complex amino acid mix such as a protein hydrolysate, also can be used to produce energy but is mainly present as a precursor for amino acid anabolites for protein production. The elements K, Mg, Ca, Fe, Mn, Mo, Co, Cu, and Zn are cofactors required in trace quantities for enzymatic reactions in the cell. Inorganic phosphorus and sulfur are incorporated into proteins, polysaccharides, and polynucleotides.

Historically, bacterial vaccines have been grown in a batchwise fashion on complex and ill-defined nutrient mixtures, which usually include animal and/or vegetable protein digests. An example is *Bordetella pertussis* cultivation on Wheeler and Cohen medium [World Health Organization, 1977c], which contains corn starch, an acid digest of bovine milk, casein, and a dialyzed lysate of brewer's yeast cells. The starch serves as the carbon source. The casein digest provides amino acids as a nitrogen source. The yeast extract provides amino acids and vitamins, as well as some trace elements, nucleotides, and carbohydrates. Such complex media components make it difficult to predict the fermentation performance, since their atomic and molecular compositions are not easily analyzed, and since they may contain hidden inhibitors of cell growth or antigen production. More recently, defined media have been used to better reproduce cell growth and antigen yield. In a defined medium, pure sugars, amino acids, vitamins, and other antibodies are used. When a single nutrient concentration limits cell growth, the growth rate can be described by Monod kinetics

$$\mu = \frac{\mu_{max} C_i}{C_i + K} \tag{105.3}$$

Here μ_{max} is the maximum specific rate, and K is the Monod constant for growth on the nutrient, an empirically determined number. It is seen from this equation that at high nutrient concentrations, the cell grows at maximum rate, as if the substrate were not limiting. At low concentration as the nutrient becomes limiting, the growth rate drops to zero, and the culture ends.

Antigen Production

Cultivation of microbes for subunit vaccines is similar to that of whole-cell vaccines, but here one is concerned with the production of the antigenic protein, polysaccharide, or antigen-encoding polynucleotide as opposed to the whole cell; thus cell growth is not necessarily the main objective. The goal is to maintain the proper environmental and nutritional factors for production of the desired product. Similar to Eqs. (105.1) and (105.2) above, one can describe antigen production by

$$\begin{gathered} \frac{dP}{dt} = q_p X_v - k_d P \text{ where } q_p \text{ and} \\ k_d = f\left(\text{temperature pH, dissolved } O_2, C_1, \ldots, C_n\right) \end{gathered} \tag{105.4}$$

where P is the product antigen concentration, q_p is the cell-specific productivity, and k_d is a degradation rate constant. For example, the production of pertussis toxin occurs best at slightly alkaline pH, about 8.0 [Jagicza et al., 1986]. *Corynebacterium diphtheriae* toxin production is affected by the concentration of iron [Reylveld, 1980]. Nutrient effects on degradation can be found for hepatitis B virus vaccine, whose production is actually a microbial cultivation. Here, recombinant DNA inserted into *Saccharomyces cerevisiae* yeast cells produces the hepatitis B surface antigen protein (HBsAg), which spontaneously assembles into virus-like particles (about 100 protein monomers per particle) within the cell. Prolonged starvation of the yeast cells can cause production of cellular proteases that degrade the antigen protein to provide nutrition. Consequently, maximum production of antigen is accomplished by not allowing the yeast culture to attain maximum cell mass but by harvesting the culture prior to nutrient depletion.

Cultivation Technology

Cultivation vessels for bacterial vaccines were initially bottles containing stagnant liquid or agar-gelled medium, where the bacteria grew at the liquid or agar surface. These typically resulted in low concentrations of the bacteria due to the limited penetration depth of diffusion-supplied oxygen and/or due to the limited diffusibility of nutrients through the stagnant liquid or agar. Despite the low mass, growth of the bacteria at the gas-liquid interface can result in cell differentiation (pellicle formation) and improved production of antigen (increased q_p). For the past several decades, however, glass and stainless steel fermentors have been used to increase production scale and productivity. Fermentors mix a liquid culture medium via an impeller, thus achieving quicker growth and higher cell mass concentrations than would be achievable by diffusive supply of nutrients. For aerobic microbe cultivations, fermenters oxygenate the culture by bubbling air directly into the medium. This increases the oxygen supply rate dramatically over diffusion supply. Due to the low solubility of oxygen in water, oxygen must be continuously supplied to avoid limitation of this nutrient.

Virus Cultivation

Virus cultivation is more complex than bacterial cultivation because viruses are, by themselves, nonreplicating. Virus must be grown on a host cell substrate, which can be animal tissue, embryo, or ex vivo cells; the host substrate determines the cultivation technology. In the United States, only Japanese encephalitis virus vaccine is still produced from infected mature animals. Worldwide, many vaccines are produced in chicken embryos, an inexpensive substrate. The remainder of vaccines are produced from ex vivo cultivated animal cells. Some virus-like particle vaccines are made by recombinant DNA techniques in either microbial or animal cells, e.g., hepatitis B virus vaccine, which is made in yeast as mentioned above, or in Chinese hamster ovary cells. An interesting synopsis of the development of rabies vaccine technology from Pasteur's use of animal tissues to modern use of ex vivo cells can be found in Sureau [1987].

In Vivo Virus Cultivation

Virus cultivation in vivo is straightforward, since relatively little control can be exercised on the host tissue. Virus is simply inoculated into the organ, and after incubation, the infected organ is harvested. Influenza virus is the prototypical in vivo vaccine; the virus is inoculated into the allantoic sac of 9- to 11-day-old fertilized chicken eggs. The eggs are incubated at about 33°C for 2 to 3 days, candled for viability and lack of contamination from the inoculation, and then the allantoic fluid is harvested. The process of inoculating, incubating, candling, and harvesting hundreds of thousands of eggs can be highly automated [Metzgar and Newhart, 1977].

Ex Vivo Virus Cultivation

Use of ex vivo cell substrates is the most recent technique in vaccine cultivation. In the case of measles and mumps vaccines, the cell substrate is chicken embryo cells that have been generated by trypsin enzyme treatment that dissociates the embryonic cells. Similarly, some rabies vaccines use cells derived from fetal rhesus monkey kidneys. Since the 1960s, cell lines have been generated that can be characterized, banked, and cryopreserved. Cryopreserved cells have been adopted to ensure reproducibility and freedom from contaminating viruses and microorganisms and to bypass the ethical problem of extensive use of animal tissues. Examples of commonly used cell lines are WI-38 and MRC-5, both human embryonic lung cells, and Vero, an African green monkey kidney cell line.

Ex vivo cells must be cultivated in a bioreactor, in liquid nutrient medium, with aeration. The principles of cell growth are the same as for bacterial cells described above, with some important additions. First, ex vivo animal cells cannot synthesize the range of metabolites and hormones necessary for survival and growth and must be provided with these compounds. Second, virus production is usually fatal to the host cells. Cells also become more fragile during infection, a process that is to an extent decoupled from cell growth. Virus growth and degradation and cell death kinetics can influence the choice of process or bioreactor. Third, all ex vivo cells used for human vaccine manufacture require a surface to adhere to in order to grow and function properly. Finally, animal cells lack the cross-linked, rigid polysaccharide cell wall that gives physical protection to microorganisms. These last three factors combine to necessitate specialized bioreactors.

Cell Growth. Supplying cells with nutrients and growth factors is accomplished by growing the cells in a complex yet defined medium, which is supplemented with 2% to 10% (v/v) animal blood serum. The complex medium will typically contain glucose and L-glutamine as energy sources, some or all of the 20 predominant L-amino acids, vitamins, nucleotides, salts, and trace elements. For polio virus, productivity is a function of energy source availability [Eagle and Habel, 1956] and may depend on other nutrients. For polio and other viruses, the presence of the divalent cations Ca^{2+} and Mg^{2+} promotes viral attachment and entry into cells and stabilizes the virus particles. Serum provides growth-promoting hormones, lipids and cholesterol, surface attachment-promoting proteins such as fibronectin, and a host of other functions. The serum used is unusually of fetal bovine origin, since this source is particularly rich in growth-promoting hormones and contains low levels of antibodies that could neutralize virus. After cell growth, and before infection, the medium is usually changed to serum-free or low-serum medium. This is done both to avoid virus neutralization and to reduce the bovine protein impurities in the harvested virus. These proteins are immunogenic and can be difficult to purify away, especially for live-virus vaccines.

Virus Production Kinetics. An intriguing aspect of ex vivo virus cultivation is that each process depends on whether the virus remains cell-associated, is secreted, or lyses the host cells and whether the virus is stable in culture. With few exceptions, a cell produces a finite amount of virus before dying, as opposed to producing the virus persistently. This is because the host cell protein and DNA/RNA synthesis organelles are usually commandeered by virus synthesis, and the host cannot produce its own proteins, DNA, or RNA. The goal then becomes to maximize cell concentration, as outlined above for bacterial vaccines, while maintaining the cells in a competent state to produce virus. Since infection is transient and often rapid, cell-specific virus productivity is not constant, and productivity is usually correlated with the cell state at inoculation. Specific productivity can be a function of the cell growth rate, since this determines the available protein and nucleic acid synthetic capacity. Although virus production can be nutrient-limited, good nutrition is usually ensured by the medium exchange at inoculation mentioned above. For viruses that infect cells slowly, nutrition is supplied by exchanging the medium several times, batchwise or by continuous perfusion.

For many viruses, the degradation term in Eq. (105.4) can be significant. This can be due to inherent thermal instability of the virus, oxidation, or proteases released from lysed cells. For an unstable virus, obtaining a synchronized infection can be key to maximizing titers. Here, the multiplicity of infection (MOI), the number of virus inoculated per cell, is an important parameter. An MOI greater than unity results in most cells being infected at once, giving the maximum net virus production rate.

Cultivation Technology. The fragility of animal cells during infection and the surface attachment requirement create a requirement for special reactors [Prokop and Rosenberg, 1989]. To an extent, reactor choice and productivity are determined by the reactor surface area. Small, simple, and uncontrolled vessels, such as a flat-sided T-flask or Roux bottle, made of polystyrene or glass, can be used for small-scale culture. With these flasks the medium is stagnant, so fragile infected cells are not exposed to fluid motion forces during culture. Like bacterial cultures, productivity can be limited due to the slow diffusion of nutrients and oxygen. To obtain larger surface areas, roller bottles or Cell Factories are used, but like egg embryo culture, robotic automation is required to substantially increase the scale of production. Even larger culture scales (≥50 liters) are accommodated in glass or stainless steel bioreactors, which are actively supplied with oxygen and pH controlled. The growth surface is typically supplied as a fixed bed of plates, spheres, or fibers or, alternately, as a dispersion of about 200-μm spherical particles known as *microcarriers* [Reuveny, 1990]. Microcarrier bioreactors are used for polio and rabies production [Montagnon et al., 1984]. The stirred-tank microcarrier reactors are similar to bacterial fermentors but are operated at much lower stirring speeds and gas sparging rates so as to minimize damage to the fragile cells. Packed-bed reactors are used for hepatitis A virus vaccine production [Aboud et al., 1994]. These types of reactors can give superior performance for highly lytic viruses because they subject the cells to much lower fluid mechanical forces. Design considerations for animal cell reactors may be found in Aunins and Henzler [1993].

105.2 Downstream Processing

Following cultivation, the antigen is recovered, isolated in crude form, further purified, and/or inactivated to give the unformulated product; these steps are referred to collectively as the *downstream process*. The complexity of a downstream process varies greatly depending on whether the antigen is the whole organism, a semipurified subunit, or a highly purified subunit. Although the sequence and combination of steps are unique for each vaccine, there is a general method to purification. The first steps reduce the volume of working material and provide crude separation from contaminants. Later steps typically resolve molecules more powerfully but are limited to relatively clean feed streams. Some manufacturing steps are classic small laboratory techniques, because many vaccine antigens are quite potent, requiring only micrograms of material; historically, manufacturing scale-up has not been a critical issue.

Purification Principles

Recovery

Recovery steps achieve concentration and liberation of the antigen. The first recovery step consists of separating the cells and/or virus from the fermentation broth or culture medium. The objective is to capture and concentrate the cells for cell-associated antigens or to clarify the medium of particulates for extracellular antigens. In the first case, the particle separation simultaneously concentrates the antigen. The two methods used are centrifugation and filtration. Batch volume and feed solids content determine whether centrifugation or filtration is appropriate; guidelines for centrifugation are given by Datar and Rosen [1993] and Atkinson and Mavituna [1991]. Filtration is used increasingly as filter materials science improves to give filters that do not bind antigen. Filtration can either be dead-end or cross-flow. Dead-end filters are usually fibrous depth filters and are used when the particulate load is low and the antigen is extracellular. In cross-flow filtration, the particles are retained by a microporous (≥0.1 μm) or ultrafiltration (≤0.1 μm) membrane. The feed stream is circulated tangential to the membrane surface at high velocity to keep the cells and other particulates from forming a filter cake on the membrane [Hanisch, 1986].

If the desired antigen is subcellular and cell-associated, the next recovery step is likely to be cell lysis. This can be accomplished by high-pressure valve homogenization, bead mills, or chemical lysis with detergent or chaotropic agents, to name only a few techniques. The homogenate or cell lysate may subsequently be clarified, again using either centrifugation or membrane filtration.

Isolation

Isolation is conducted to achieve crude fractionation from contaminants that are unlike the antigen; precipitation and extraction are often used here. These techniques rely on large differences in charge or solubility between antigen and contaminants. Prior to isolation, the recovered process stream may be treated enzymatically with nucleases, proteases, or lipases to remove DNA/RNA, protein, and lipids, the major macromolecular contaminants. Nonionic detergents such as Tween and Triton can serve a similar function to separate components, provided they do not denature the antigen. Subsequent purification steps must be designed to remove the enzyme(s) or detergent, however.

Ammonium sulfate salt precipitation is a classic method used to concentrate and partially purify various proteins, e.g., diphtheria and tetanus toxins. Alcohol precipitation is effective for separating polysaccharides from proteins. Cohn cold alcohol precipitation is a classic technique used to fractionate blood serum for antibody isolation. Both techniques concentrate antigen for further treatment.

Liquid-liquid extraction, either aqueous-organic or aqueous-aqueous, is another isolation technique suitable for vaccine purification. In a two-phase aqueous polymer system, the separation is based on the selective partitioning of the product from an aqueous liquid phase containing, e.g., polyethylene glycol (PEG), into a second, immiscible aqueous liquid phase that contains another polymer, e.g., dextran, or containing a salt [see Kelley and Hatton, 1992].

Final Purification

Further purification of the vaccine product is to remove contaminants that have properties closely resembling the antigen. These sophisticated techniques resolve molecules with small differences in charge, hydrophobicity, density, or size.

Density-gradient centrifugation, although not readily scalable, is a popular technique for final purification of viruses [Polson, 1993]. Either rate-zonal centrifugation, where the separation is based on differences in sedimentation rate, or isopycnic equilibrium centrifugation, where the separation is based solely on density differences, is used. Further details on these techniques can be found in Dobrota and Hinton [1992].

Finally, different types of chromatography are employed to manufacture highly pure vaccines; principles can be found in Janson and Ryden [1989]. Ion-exchange chromatography (IEC) is based on differences in overall charge and distribution of charge on the components. Hydrophobic-interaction chromatography (HIC) exploits differences in hydrophobicity. Affinity chromatography is based on specific stereochemical interactions common only to the antigen and the ligand. Size-exclusion chromatography (SEC) separates on the basis of size and shape. Often, multiple chromatographic steps are used, since the separation mechanisms are complementary. IEC and HIC can sometimes be used early in the process to gain substantial purification; SEC is typically used as a final polishing step. This technique, along with ultrafiltration, may be used to exchange buffers for formulation at the end of purification.

Inactivation

For nonattenuated whole organisms or for toxin antigens, the preparation must be inactivated to eliminate pathogenicity. This is accomplished by heat pasteurization, by cross-linking using formaldehyde or glutaraldehyde, or by alkylating using agents such as β-propiolactone. The agent is chosen for effectiveness without destruction of antigenicity. For whole organisms, the inactivation abolishes infectivity. For antigens such as the diphtheria and tetanus toxins, formaldehyde treatment removes the toxicity of the antigen itself as well as killing the organism. These detoxified antigens are known as *toxoids.*

The placement of inactivation in the process depends largely on safety issues. For pathogen cultures, inactivation traditionally has been immediately after cultivation to eliminate danger to manufacturing personnel. For inactivation with cross-linking agents, however, the step may be placed later in the process in order to minimize interference by contaminants that either foil inactivation of the organism or cause carryover of antigen-contaminant cross-linked entities that could cause safety problems, i.e., side-reactions in patients.

Purification Examples

Examples of purification processes are presented below, illustrating how the individual techniques are combined to create purification processes.

Bacterial Vaccines

Salmonella typhi Ty21, a vaccine for typhoid fever, and BCG (bacille Calmette-Guérin, a strain of *Mycobacterium bovis*) vaccine against *Myobacterium tuberculosis*, are the only vaccines licensed for human use based on live, attenuated bacteria. Downstream processing consists of collecting the cells by continuous-flow centrifugation or using cross-flow membrane filtration. For *M. bovis* that are grown in a liquid submerged fermenter culture, Tween 20 is added to keep the cells from aggregating, and the cultures are collected as above. Tween 20, however, has been found to decrease virulence. In contrast, if the BCG is grown in stagnant bottles as a surface pellicle, the downstream process consists of collecting the pellicle sheet, which is a moist cake, and then homogenizing the cake using a ball mill. Milling time is critical, since prolonged milling kills the cells and too little milling leaves clumps of bacteria in suspension.

Most current whooping cough vaccines are inactivated whole *B. pertussis.* The cells are harvested by centrifugation and then resuspended in buffer, which is the supernate in some cases. This is done because some of the filamentous hemagglutinin (FHA) and pertussis toxin (PT) antigens are released into the supernate. The cell concentrate is inactivated by mild heat and stored with thimerosal and/or formaldehyde. The inactivation process serves the dual purpose of killing the cells and inactivating the toxins.

C. diphtheria vaccine is typical of a crude protein toxoid vaccine. Here the 58 kDa toxin is the antigen, and it is converted to a toxoid with formaldehyde and crudely purified. The cells are first separated from the toxin by centrifugation. Sometimes the pathogen culture is inactivated with formaldehyde before centrifugation. The supernate is treated with formaldehyde to 0.75%, and it is stored for 4 to 6 weeks at 37°C to allow complete detoxification [Pappenheimer, 1984]. The toxoid is then concentrated by ultrafiltration and fractionated from contaminants by ammonium sulfate precipitation. During detoxification of crude material, reactions with formaldehyde lead to a variety of products. The toxin is internally cross-linked and also cross-linked to other toxins, beef peptones from the medium, and other medium proteins. Because detoxification creates a population of molecules containing antigen, the purity of this product is only about 60% to70%.

Due to the cross-linking of impurities, improved processes have been developed to purify toxins before formaldehyde treatment. Purification by ammonium sulfate fractionation, followed by ion-exchange and/or size-exclusion chromatography, is capable of yielding diphtheria or tetanus toxins with purities ranging from 85% to 95%. The purified toxin is then treated with formaldehyde or glutaraldehyde [Relyveld and Ben-Efraim, 1983] to form the toxoid. Likewise, whole-cell pertussis vaccine is being replaced by subunit vaccines. Here, the pertussis toxin (PT) and the filamentous hemagglutinin (FHA) are purified from the supernate. These two antigens are isolated by ammonium sulfate precipitation, followed by sucrose density-gradient centrifugation to remove impurities such as endotoxin. The FHA and PT are then detoxified with formaldehyde [Sato et al., 1984].

Bacterial components other than proteins also have been developed for use as subunit vaccines. One class of bacterial vaccines is based on capsular polysaccharides. Polysaccharides from *Meningococcus*, *Pneumococcus*, and *Haemophilus influenzae* type b are used for vaccines against these organisms. After separating the cells, the polysaccharides are typically purified using a series of alcohol precipitations. As described below, these polysaccharides are not antigenic in infants. As a consequence, the polysaccharides are chemically cross-linked, or conjugated, to a highly purified antigenic protein carrier. After conjugation, there are purification steps to remove unreacted polysaccharide protein, and small-molecular-weight cross-linking reagents. Several manufacturers have introduced pediatric conjugate vaccines against *H. influenzae* type b (Hib-conjugate).

Viral Vaccines

Live viral vaccines have limited downstream processes. For secreted viruses such as measles, mumps, and rubella, cell debris is simply removed by filtration and the supernate frozen. In many cases it is necessary to process quickly and at refrigerated temperatures, since live virus can be unstable [Cryz and Gluck, 1990].

For cell-associated virus such as herpesviruses, e.g., varicella zoster (chickenpox virus), the cells are washed with a physiologic saline buffer to remove medium contaminants and are harvested. This can be done by placing them into a stabilizer formulation (see below) and then mechanically scraping the cells from the growth surface. Alternately, the cells can be harvested from the growth surface chemically or enzymatically. For the latter, the cells are centrifuged and resuspended in stabilizer medium. The virus is then liberated by disrupting the cells, usually by sonication, and the virus-containing supernate is clarified by dead-end filtration and frozen.

Early inactivated influenza virus vaccines contained relatively crude virus. The allantoic fluid was harvested, followed by formaldehyde inactivation of the whole virus, and adsorption to aluminum phosphate (see below), which may have provided some purification as well. The early vaccines, however, were associated with reactogenicity. For some current processes, the virus is purified by rate-zonal centrifugation, which effectively eliminates the contaminants from the allantoic fluid. The virus is then inactivated with formaldehyde or β-propiolactone, which preserves the antigenicity of both the hemagglutinin (HA) and neuraminidase (NA) antigens. Undesirable side reactions have been even further reduced with the introduction of *split vaccines*, where the virus particle is disrupted by an organic solvent or detergent and then inactivated. By further purifying the split vaccine using a method such as zonal centrifugation to separate the other virion components from the HA and NA antigens, an even more highly purified HA + NA vaccine is available. These antigens are considered to elicit protective antibodies against influenza [Tyrrell, 1976].

Recently, an extensive purification process was developed for hepatitis A vaccine, yielding a >90% pure product [Aboud et al., 1994]. The intracellular virus is released from the cells by Triton detergent lysis, followed by nuclease enzyme treatment for removal of RNA and DNA. The virus is concentrated and detergent removed by ion-exchange chromatography. PEG precipitation and chloroform solvent extraction purify away most of the cellular proteins, and final purification and polishing are achieved by ion-exchange and size-exclusion chromatography. The virus particle is then inactivated with formaldehyde. In this case, inactivation comes last for two reasons. First, the virus is attenuated, so there is no risk to process personnel. Second, placing the inactivation after the size-exclusion step ensures that there are no contaminants or virus aggregates that may cause incomplete inactivation.

The first hepatitis B virus vaccines were derived from human plasma [Hilleman, 1993]. The virus is a 22-nm-diameter particle, much larger than most biologic molecules. Isolation was achieved by ammonium sulfate or PEG precipitation, followed by rate zonal centrifugation and isopycnic banding to take advantage of the large particle size. The preparation was then treated with pepsin protease, urea, and formaldehyde or heat. The latter steps ensure inactivation of possible contaminant viruses from the blood serum. More recently, recombinant DNA–derived hepatitis B vaccines are expressed as an intracellular noninfectious particle in yeast and use a completely different purification process. Here, the emphasis is to remove the yeast host contaminants, particularly high levels of nucleic acids and polysaccharides. Details on the various manufacturing processes have been described by Sitrin et al. [1993].

Antibody Preparations

Antibody preparation starts from the plasma pool prepared by removing the cellular components of blood. Cold ethanol is added in increments to precipitate fractions of the blood proteins, and the precipitate containing IgG antibodies is collected. This is further redissolved and purified by ultrafiltration, which also exchanges the buffer to the stabilizer formulation. Sometimes ion-exchange chromatography is used for further purification. Although the plasma is screened for viral contamination prior to pooling, all three purification techniques remove some virus.

105.3 Formulation and Delivery

Successful vaccination requires both the development of a stable dosage form for in vitro storage and the proper delivery and presentation of the antigen to elicit a vigorous immune response in vivo. This is done by adjuvanting the vaccine and/or by formulating the adjuvanted antigen. An adjuvant is defined as an agent that enhances the immune response against an antigen. A formulation contains an antigen in a delivery vehicle designed to preserve the (adjuvenated) antigen and to deliver it to a specific target organ or over a desired time period. Despite adjuvanting and formulation efforts, most current vaccines require multiple doses to create immune memory.

Live Organisms

Live viruses and bacteria die relatively quickly in liquid solution (without an optimal environment) and are therefore usually stored in the frozen state. Preserving the infectivity of frozen live-organism vaccines is typically accomplished by lyophilization or freeze-drying. The freeze-drying process involves freezing the organism in the presence of stabilizers, followed by sublimation of both bulk water (primary drying) and more tightly bound water (secondary drying). The dehydration process reduces the conformational flexibility of the macromolecules, providing protection against thermal denaturation. Stabilizers also provide conformational stability and protect against other inactivating mechanisms such as amino acid deamidation, oxidation, and light-catalyzed reaction.

Final water content of the freeze-dried product is the most important parameter for the drying process. Although low water content enhances storage stability, overdrying inactivates biologic molecules, since removal of tightly bound water disrupts antigen conformation. Influenza virus suspensions have been shown to be more stable at 1.7% (w/w) water than either 0.4% to 1% or 2.1% to 3.2% [Greiff and Rightsel, 1968]. Other lyophilization parameters that must be optimized pertain to heat and mass transfer, including (1) the rate of freezing and sublimation, (2) vial location in the freeze-drier, and (3) the type of vial and stopper used to cap the vial. Rates of freezing and drying affect phase transitions and compositions, changing the viable organism yield on lyophilization and the degradation rate of the remaining viable organisms on storage.

Stabilizers are identified by trial-and-error screening and by examining the mechanisms of inactivation. They can be classified into four categories depending on their purpose: specific, nonspecific, competitive, and pharmaceutical. *Specific stabilizers* are ligands that naturally bind biologic macromolecules. For example, enzyme antigens are often stabilized by their natural substrates or closely related compounds. Antigen stabilizers for the liquid state also stabilize during freezing. *Nonspecific stabilizers* such as sugars, amino acids, and neutral salts stabilize proteins and virus structures via a variety of mechanisms. Sugars and polyols act as bound water substitutes, preserving conformational integrity without possessing the chemical reactivity of water. Buffer salts preserve optimal pH. *Competitive inhibitors* outcompete the organism or antigen for inactivating conditions, such as gas-liquid interfaces, oxygen, or trace-metal ions [Volkin and Klibanov, 1989]. Finally, *pharmaceutical stabilizers* may be added to preserve pharmaceutical elegance, i.e., to prevent collapse of the lyophilized powder during the drying cycle, which creates difficult redissolution. Large-molecular-weight polymers such as carbohydrates (dextrans or starch) or proteins such as albumin or gelatin are used for this purpose. For example, a buffered sorbitol-gelatin medium has been used successfully to preserve the infectivity of measles virus vaccine during lyophilized storage for several years at 2 to 8°C [Hilleman, 1989]. An example of live bacterium formulation to preserve activity on administration is typhoid fever vaccine, administered orally, *S. typhi* bacteria are lyophilized to a powder to preserve viability on the shelf, and the powder is encapsulated in gelatin to preserve bacterial viability when passing through the low-pH stomach. The gelatin capsule dissolves in the intestine to deliver the live bacteria.

Oral polio vaccine is an exception to the general rule of lyophilization, since polio virus is inherently quite stable relative to other viruses. It is formulated as a frozen liquid and can be used for a limited time after thawing [Melnick, 1984]. In the presence of specific stabilizers such as $MgCl_2$, extended 4°C stability can be obtained.

Subunit Antigens

Inactivated and/or purified viral and bacterial antigens inherently offer enhanced stability because whole-organism infectivity does not need to be preserved. However, these antigens are not as immunogenic as live organisms and thus are administered with an adjuvant. They are usually formulated as an aqueous liquid suspension or solution, although they can be lyophilized under the same principles as above. The major adjuvant recognized as safe for human use is alum. *Alum* is a general term referring to various hydrated aluminum salts; a discussion of the different alums can be found in Shirodkar et al. [1990]. Vaccines can be formulated with alum adjuvants by two distinct methods: adsorption to performed aluminum precipitates or precipitation of aluminum salts in the presence of the antigen, thus adsorbing and entrapping the antigen. Alum's adjuvant activity is classically believed to be a "depot" effect, slowly delivering antigen over time in vivo. In addition, alum particles are believed to be phagocytized by macrophages.

Alum properties vary depending on the salt used. Adjuvants labeled aluminum hydroxide are actually aluminum oxyhydroxide, AlO(OH). This material is crystalline, has a fibrous morphology, and has a positive surface charge at neutral pH. In contrast, aluminum phosphate adjuvants are networks of platelike particles of amorphous aluminum hydroxyphosphate and possess a negative surface charge at neutral pH. Finally, alum coprecipitate vaccines are prepared by mixing an acidic alum solution of $KAl(SO_4)_2 \cdot 12H_2O$ with an antigen solution buffered at neutral pH, sometimes actively pH-controlled with base. At neutral pH, the aluminum forms a precipitate, entrapping and adsorbing the antigen. The composition and physical properties of this alum vary with processing conditions and the buffer anions. In general, an amorphous aluminum hydroxy(buffer anion)sulfate material is formed.

Process parameters must be optimized for each antigen to ensure proper adsorption and storage stability. First, since antigen adsorption isotherms are a function of the antigen's isoelectric point and the type of alum used [Seeber et al., 1991], the proper alum and adsorption pH must be chosen. Second, the buffer ions in solution can affect the physical properties of alum over time, resulting in changes in solution pH and antigen adsorption. Finally, heat sterilization of alum solutions and precipitates prior to antigen adsorption can alter their properties. Alum is used to adjuvant virtually all the existing inactivated or formaldehyde-treated vaccines, as well as purified subunit vaccines such as HBsAg and Hib-conjugate vaccines. The exception is for some bacterial polysaccharide vaccines and for new vaccines under development (see below).

An interesting vaccine development challenge was encountered with Hib-conjugate pediatric vaccines, which consist of purified capsular polysaccharides. Although purified, unadjuvanted polysaccharide is used in adults, it is not sufficiently immunogenic in children under age 2, the population is greatest risk [Ellis, 1992; Howard, 1992]. Chemical conjugation, or cross-linking, of the PRP polysaccharide to an antigenic protein adjuvant elicits T-helper cell activation, resulting in higher antibody production. Variations in conjugation chemistry and protein carriers have been developed; example proteins are the diphtheria toxoid (CRM 197), tetanus toxoid, and the outer membrane protein complex of *N. meningitidis* [Ellis, 1992; Howard, 1992]. The conjugated polysaccharide is sometimes adsorbed to alum for further adjuvant action.

105.4 Future Trends

The reader will have noted that many production aspects for existing vaccines are quite archaic. This is so because most vaccines were developed before the biotechnology revolution, which is creating a generation of highly purified and better-characterized subunit vaccines. As such, for older vaccines "the process defines the product," and process improvements cannot readily be incorporated into these poorly characterized vaccines without extensive new clinical trials. With improved scientific capabilities, we can understand the effects of process changes on the physicochemical properties of new vaccines and on their behavior in vivo.

Vaccine Cultivation

Future cultivation methods will resemble existing methods of microbial and virus culture. Ill-defined medium components and cells will be replaced to enhance reproducibility in production. For bacterial and ex vivo cultivated virus, analytical advances will make monitoring the environment and nutritional status of the culture more ubiquitous. However, the major changes will be in novel product types—single-molecule subunit antigens, virus-like particles, monoclonal antibodies, and gene-therapy vaccines, each of which will incorporate novel processes.

Newer subunit vaccine antigens will be cultivated via recombinant DNA in microbial or animal cells. Several virus-like particle vaccines are under development using recombinant baculovirus (nuclear polyhedrosis virus) to infect insect cells (spodoptera frugipeeda or trichoplusia ni). Like the hepatitis B vaccine, the viral antigens spontaneously assemble into a noninfectious capsid within the cell. Although the metabolic pathways of insect cells differ from vertebrates, cultivation principles are similar. Insect cells do not require surface attachment and are grown much like bacteria. However, they also lack a cell wall and are larger and hence more fragile than vertebrate cells.

Passive antibody vaccines have been prepared up to now from human blood serum. Consequently, there has been no need for cultivation methods beyond vaccination and conventional harvest of antibody-containing blood from donors. Due to safety concerns over using human blood, passive vaccines will likely be monoclonal antibodies or cocktails thereof prepared in vitro by the cultivation of hybridoma or myeloma cell lines. This approach is under investigation for anti-HIV-1 antibodies [Emini et al., 1992]. Cultivation of these cell lines involves the same principles of animal cell cultivation as described above, with the exception that hybridomas can be less fastidious in nutritional requirements, and they do not require surface attachment for growth. These features will allow for defined serum-free media and simpler cultivation vessels and procedures.

For the gene-therapy approach, the patient actually produces the antigen. A DNA polynucleotide encoding protein antigen(s) is injected intramuscularly into the patient. The muscle absorbs the DNA and produces the antigen, thereby eliciting an immune response [Ulmer et al., 1993]. For cultivation, production of the DNA plasmid is the objective, which can be done efficiently by bacteria such as *Escherichia coli*. Such vaccines are not sufficiently far along in development to generalize the factors that influence their production; however, it is expected that producer cells and process conditions that favor high cell mass, DNA replication, and DNA stability will be important. A potential beauty of this vaccination approach is that for cultivation, purification, and formulation, many vaccines can conceivably be made by identical processes, since the plasmids are inactive within the bacterium and possess roughly the same nucleotide composition.

Downstream Processing

Future vaccines will be more highly purified in order to minimize side effects, and future improvements will be to assist this goal. The use of chemically defined culture media will impact favorably on downstream processing by providing a cleaner feedstock. Advances in filtration membranes and in chromatographic support binding capacity and throughput will improve ease of purification. Affinity purification methods that rely on specific "lock and key" interactions between a chromatographic support and the antigen will see greater use as well. Techniques amenable to larger scales will be more important to meet increased market demands and to reduce manufacturing costs. HPLC and other analytical techniques will provide greater process monitoring and control throughout purification.

As seen during the evolution of diphtheria and tetanus toxoid vaccines, the trend will be to purify toxins prior to inactivation to reduce their cross-linking with other impurities. New inactivating agents such as hydrogen peroxide and ethyl dimethylaminopropyl carbodiimide have been investigated for pertussis toxin, which do not have problems of cross-linking or reversion of the toxoid to toxin status.

Molecular biology is likely to have an even greater impact on purification. Molecular cloning of proteins allows the addition of amino acid sequences that can facilitate purification, e.g., polyhistidine or polyalanine tails for metal ion, or ion-exchange chromatography. Recent efforts also have employed genetic manipulation to inactivate toxins, eliminating the need for the chemical treatment step.

Vaccine Adjuvants and Formulation

Many new subunit antigens lack the inherent immunogenicity found in the natural organism, thereby creating the need for better adjuvants. Concomitantly, the practical problem of enhancing worldwide immunization coverage has stimulated development of single-shot vaccine formulations in which booster doses are unnecessary. Thus future vaccine delivery systems will aim at reducing the number of doses via controlled antigen release and will increase vaccine efficacy by improving the mechanism of antigen presentation (i.e., controlled release of antigen over time or directing of antigen to specific antigen-presenting cells). Major efforts are also being made to combine antigens into single-shot vaccines to improve immunization rates for infants, who currently receive up to 15 injections during the first 2 years of life. Coadministration of antigens presents unique challenges to formulation as well.

Recent advances in the understanding of in vivo antigen presentation to the immune system has generated considerable interest in developing novel vaccine adjuvants. The efficacy of an adjuvant is judged by its ability to stimulate specific antibody production and killer cell proliferation. Developments in biology now allow analysis of activity by the particular immune cells that are responsible for these processes. Examples of adjuvants currently under development include saponin detergents, muramyl dipeptides, and lipopolysaccharides (endotoxin), including lipid A derivatives. As well, cytokine growth factors that stimulate immune cells directly are under investigation.

Emulsion and liposome delivery vehicles are also being examined to enhance the presentation of antigen and adjuvant to the immune system [Edelman, 1992; Allison and Byars, 1992]. Controlled-release delivery systems are also being developed that encapsulate antigen inside a polymer-based solid microsphere. The size of the particles typically varies between 1 and 300 μm depending on the manufacturing process. Microspheres are prepared by first dissolving the biodegradable polymer in an organic solvent. The adjuvanted antigen, in aqueous solution or lyophilized powder form, is then emulsified into the solvent-polymer continuous phase. Microspheres are then formed by either solvent evaporation, phase-separation, or spray-drying, resulting in entrapment of antigen [Morris et al., 1994]. The most frequently employed biodegradable controlled-released delivery systems use FDA-approved poly(lactide-co-glycolide) copolymers (PLGA), which hydrolyze in vivo to nontoxic lactic and glycolic acid monomers. Degradation rate can be optimized by varying the microsphere size and the monomer ratio. Antigen stability during encapsulation and during in vivo release from the microspheres remains a challenge. Other challenges to manufacturing include encapsulation process reproducibility, minimizing antigen exposure to denaturing organic solvents, and ensuring sterility. Methods are being developed to address these issues, including the addition of stabilizers for processing purposes only. It should be noted that microsphere technology may permit vaccines to be targeted to specific cells; they can potentially be delivered orally or nasally to produce a mucosal immune response.

Other potential delivery technologies include liposomes and alginate polysaccharide and poly(dicarboxylatophenoxy)phosphazene polymers. The latter two form aqueous hydrogels in the presence of divalent cations [Khan et al., 1994]. Antigens can thus be entrapped under aqueous conditions with minimal processing by simply mixing antigen and soluble aqueous polymer and dripping the mixture into a solution of $CaCl_2$. The particles erode by Ca^{2+} loss, mechanical and chemical degradation, and macrophage attack. For alginate polymers, monomer composition also determines the polymer's immunogenicity, and thus the material can serve as both adjuvant and release vehicle.

For combination vaccines, storage and administration compatibility of the different antigens must be demonstrated. Live-organism vaccines are probably not compatible with purified antigens, since the former usually require lyophilization and the latter are liquid formulas. Within a class of vaccines, formulation is challenging. Whereas it is relatively straightforward to adjuvant and formulate a single antigen, combining antigens is more difficult because each has its own unique alum species, pH, buffer ion, and preservative optimum. Nevertheless, several combination vaccines have reached the market, and others are undergoing clinical trials.

105.5 Conclusions

Although vaccinology and manufacturing methods have come a considerable distance over the past 40 years, much more development will occur. There will be challenges for biotechnologists to arrive at safer, more effective vaccines for an ever-increasing number of antigen targets. If government interference and legal liability questions do not hamper innovation, vaccines will remain one of the most cost-effective and logical biomedical technologies of the next century, as disease is prevented rather than treated.

Challenges are also posed in bringing existing vaccines to technologically undeveloped nations, where they are needed most. This problem is almost exclusively dominated by the cost of vaccine manufacture and the reliability of distribution. Hence it is fertile ground for engineering improvements in vaccine production.

Defining Terms

Adjuvant: A chemical or biologic substance that enhances immune response against an antigen. Used here as a verb, the action of combining an antigen and an adjuvant.

Antigen: A macromolecule or assembly of macromolecules from a pathogenic organism that the immune system recognizes as foreign.

Attenuation: The process of mutating an organism so that it no longer causes disease.

Immunogen: A molecule or assembly of molecules with the ability to invoke an immune system response.

Pathogen: A disease-causing organism, either a virus, mycobacterium, or bacterium.

References

Aboud RA, Aunins JG, Buckland BC, et al. 1994. Hepatitis A Virus Vaccine. International patent application, publication number WO 94/03589, Feb. 17, 1994.

Allison AC, Byars NE. 1992. Immunologic adjuvants and their mode of action. In RW Ellis (ed), Vaccines: New Approaches to Immunological Problems, p 431. Reading, Mass, Butterworth-Heinemann.

Atkinson B, Mavituna F. 1991. Biochemical Engineering and Biotechnology Handbook, 2d ed. London, Macmillan.

Aunins JG, Henzler H-J. 1993. Aeration in cell culture bioreactors. In H-J Rehm et al. (eds), Biotechnology, 2d ed., vol 3, p 219. Weinheim, Germany, VCH Verlag.

Bachmayer H. 1976. Split and subunit vaccines. In P. Selby (ed), Influenza Virus, Vaccines, and Strategy, p 149. New York, Academic Press.

Barbet AF. 1989. Vaccines for parasitic infections. Adv Vet Sci Comp Med 33:345.

Cryz SJ, Reinhard G. 1990. Large-scale production of attenuated bacterial and viral vaccines. In GC Woodrow, MM Levine (eds), New Generation Vaccines, p 921. New York, Marcel Dekker.

Datar RV, Rosen C-G. 1993. Cell and cell debris removal: Centrifugation and crossflow filtration. In H-J Rehm et al. (eds), Biotechnology, 2d ed., vol 3, p 469. Weinheim, Germany, VCH Verlag.

Dobrota M, Hinton R. 1992. Conditions for density gradient separations. In D Rickwood (ed), Preparative Centrifugation: A Practical Approach, p 77. New York, Oxford U Press.

Eagle H, Habel K. 1956. The nutritional requirements for the propagation of poliomyelitis virus by the HeLa cell. J Exp Med 104:271.

Edelman R. 1992. An update on vaccine adjuvants in clinical trial. AIDS Res Hum Retrovir 8(8):1409.

Ellis RW. 1992. Vaccine development: Progression from target antigen to product. In JE Ciardi et al. (eds), Genetically Engineered Vaccines, p 263. New York, Plenum Press.

Emini EA, Schleif WA, Nunberg JH, et al. 1992. Prevention of HIV-1 infection in chimpanzees by gp120 V3 domain-specific monoclonal antibodies. Nature 355:728.

Greiff D, Rightsel WA. 1968. Stability of suspensions of influenza virus dried to different contents of residual moisture by sublimation in vacuo. Appl Microbiol 16(6):835.

Hanisch W. 1986. Cell harvesting. In WC McGregor (ed), Membrane Separations in Biotechnology, p 66. New York, Marcel Dekker.

Hewlett EL, Cherry JD. 1990. New and improved vaccines against pertussis. In GC Woodrow, MM Levine (eds), New Generation Vaccines, p 231. New York, Marcel Dekker.

Hilleman MR. 1989. Improving the heat stability of vaccines: Problems, needs and approaches. Rev Infect Dis 11(suppl 3):S613.

Hilleman MR. 1993. Plasma-derived hepatitus B vaccine: A breakthrough in preventive medicine. In R Ellis (ed), Hepatitus B Vaccines in Clinical Practice, p 17. New York, Marcel Dekker.

Howard AJ. 1992. Haemophilus influenzae type-b vaccines. Br J Hosp Med 48(1):44.

Ingraham JL, Maaløe O, Neidhardt FC. 1983. Growth of the Bacterial Cell. Sunderland, Mass, Sinauer.

Jagicza A, Balla P, Lendvai N, et al. 1986. Additional information for the continuous cultivation of Bordetella pertussis for the vaccine production in bioreactor. Ann Immunol Hung 26:89.

Janson J-C, Ryden L (eds). 1989. Protein Purification Principles, High Resolution Methods, and Applications. Weinheim, Germany, VCH Verlag.

Kelley BD, Hatton TA. 1993. Protein purification by liquid-liquid extraction. In H-J Rehm et al. (eds), Biotechnology, 2d ed, vol 3, p 594. Weinheim, Germany, VCH Verlag.

Khan MZI, Opdebeeck JP, Tucker IG. 1994. Immunopotentiation and delivery systems for antigens for single-step immunization: Recent trends and progress. Pharmacol Res 11(1):2.

Melnick JL. 1984. Live attenuated oral poliovirus vaccine. Rev Infect Dis 6(suppl 2):S323.

Metzgar DP, Newhart RH. 1977. U.S. patent no. 4,057,626, Nov. 78, 1977.

Montagnon B, Vincent-Falquet JC, Fanget B. 1984. Thousand litre scale microcarrier culture of vero cells for killed polio virus vaccine: Promising results. Dev Biol Stand 55:37.

Morris W, Steinhoff MC, Russell PK. 1994. Potential of polymer microencapsulation technology for vaccine innovation. Vaccine 12(1):5.

Pappenheimer AM. 1984. Diphtheria. In R Germanier (ed), Bacterial Vaccines, p 1. New York, Academic Press.

Polson A. 1993. Virus Separation and Preparation. New York, Marcel Dekker.

Prokop A, Rosenberg MZ. 1989. Bioreactor for mammalian cell culture. In A Fiechter (ed), Advances in Biochemical Engineering, vol 39: Vertebrate Cell Culture II and Enzyme Technology, p 29. Berlin, Springer-Verlag.

Rappuoli R. 1990. New and improved vaccines against diphtheria and tetanus. In GC Woodrow, MM Levine (eds), New Generation Vaccines, p 251. New York, Marcel Dekker.

Relyveld EH. 1980. Current developments in production and testing of tetanus and diphtheria vaccines. In A Mizrahi et al. (eds), Progress in Clinical and Biological Research, vol 47: New Developments with Human and Veterinary Vaccines, p 51. New York, Alan R Liss.

Relyveld EH, Ben-Efraim S. 1983. Preparation of vaccines by the action of glutaraldehyde on toxins, bacteria, viruses, allergens and cells. In SP Colowic, NO Kaplan (eds), Methods in Enzymology, vol 93, p 24. New York, Academic Press.

Reuveny S. 1990. Microcarrier culture systems. In AS Lubiniecki (ed), In Large-Scale Mammalian Cell Culture Technology, p 271. New York, Marcel Dekker.

Sato Y, Kimura M, Fukumi H. 1984. Development of a pertussis component vaccine in Japan. Lancet 1(8369):122.

Seeber SJ, White JL, Helm SL. 1991. Predicting the adsorption of proteins by aluminum-containing adjuvants. Vaccine 9:201.

Shirodkar S, Hutchinson RL, Perry DL, et al. 1990. Aluminum compounds used as adjuvants in vaccines. Pharmacol Res 7(12):1282.

Sitrin RD, Wampler DE, Ellis R. 1993. Survey of licensed hepatitis B vaccines and their product processes. In R Ellis (ed), Hepatitus B Vaccines in Clinical Practice, p 83. New York, Marcel Dekker.

Sureau P. 1987. Rabies vaccine production in animal cell cultures. In A Fiechter (ed), Advances in Biochemical Engineering and Biotechnology, vol 34, p 111. Berlin, Springer-Verlag.

Tyrrell DAJ. 1976. Inactivated whole virus vaccine. In P Selby (ed), Influenza, Virus, Vaccines and Strategy, p 137. New York, Academic Press.

Ulmer JB, Donnelly JJ, Parker SE, et al. 1993. Heterologous protection against influenza by injection of DNA encoding a viral protein. Science 259(5102):1745.

Volkin DB, Klibanov AM. 1989. Minimizing protein inactivation. In TE Creighton (ed), Protein Function: A Practical Approach, pp 1–12. Oxford, IRL Press.

Further Information

A detailed description of all the aspects of traditional bacterial vaccine manufacture may be found in the World Health Organization technical report series for the production of whole-cell pertussis, diphtheria, and tetanus toxoid vaccines:

World Health Organization. 1997a. BLG/UNDP/77.1 Rev. 1. Manual for the Production and Control of Vaccines: Diphtheria Toxoid.

World Health Organization. 1997b. BLG/UNDP/77.2 Rev. 1. Manual for the Production and Control of Vaccines: Tetanus Toxoid.

World Health Organization. 1997c. BLG/UNDP/77.3 Rev. 1. Manual for the Production and Control of Vaccines: Pertussis Vaccine.

A description of all the aspects of cell culture and viral vaccine manufacture may be found in Spier RE, Griffiths JB. 1985. Animal Cell Biotechnology, vols 1 to 3. London, Academic Press.

For a review of virology and virus characteristics, the reader is referred to Fields BN, Knipe DM (eds). 1990. Virology, 2d ed, vols 1 and 2. New York, Raven Press.

106
Gene Therapy

Joseph M. Le Doux
The Center for Engineering in Medicine, and Surgical Services, Massachusetts General Hospital, Harvard Medical School, and the Shriners Burns Hospital

Jeffrey R. Morgan
The Center for Engineering in Medicine, and Surgical Services, Massachusetts General Hospital, Harvard Medical School, and the Shriners Burns Hospital

Martin L. Yarmush
The Center for Engineering in Medicine, and Surgical Services, Massachusetts General Hospital, Harvard Medical School, and the Shriners Burns Hospital

Gene therapy, the transfer of genes into cells for a therapeutic effect, is a new approach to the treatment of disease. The first clinically applicable system for efficiently delivering genes into mammalian cells was developed in the early 1980s and was based on a genetically engineered **retrovirus**, which, as part of its lifecycle, stably integrates its genome into the target cell's chromosomal DNA. Using **recombinant** DNA technology perfected in the mid-1970s, investigators replaced the viral genes with therapeutic genes and the resulting recombinant retrovirus shuttled these genes into the target cells. The potential applications of gene therapy are far reaching (Table 106.1) since there are over 4000 known human genetic diseases (many of which have no viable treatment) and virtually every human disease is profoundly influenced by genetic factors [Anderson, 1992]. In one of the first clinical applications, gene therapy was used to treat patients with ADA (adenosine deaminase) deficiency, a genetic defect which causes severe combined immune deficiency (SCID) and death at an early age [Anderson, 1992]. The patient's lymphocytes were isolated and **transduced** (insertion of a foreign gene into the genome of a cell) by a recombinant retrovirus encoding a functional ADA gene. These transduced cells were expanded in culture then reinfused back to the patient. These protocols were conducted after an exhaustive peer review process that laid the groundwork for future gene therapy protocols. Successful treatment of other genetic diseases is likely to be achieved in the future [Levine and Friedmann, 1993; Roemer and Friedmann, 1992]. In addition to inherited diseases, other viable targets for gene therapy include more prevalent disorders that show a complex genetic dependence (i.e., cancer and heart disease) as well as infectious diseases (i.e., human immunodeficiency virus (HIV) and applications in tissue engineering [Anderson, 1992; Morgan and Yarmush, 1998].

106.1 Background

Gene therapy protocols conduct gene transfer in one of two settings; either ***ex vivo*** or ***in vivo*** [Mulligan, 1993]. For *ex vivo* gene therapy, target cells or tissue are removed from the patient, grown in culture, genetically modified and then reinfused or retransplanted into the patient [Ledley, 1993]. *Ex-vivo* gene therapy is limited to those tissues which can be removed, cultured ***in vitro*** and returned to the patient

0-8493-0462-8/00/$0.00+$.50

TABLE 106.1 Target Diseases for Gene Therapy

Target Disease	Target Tissues	Corrective Gene	Gene Delivery Systems Used
Inherited			
ADA deficiency	Hematopoietic cells	ADA	RV, RM
Alpha-1 antitrypsin deficiency	Fibroblasts	Alpha-1 antitrypsin	RV
	Hepatocytes		RV
	Lung epithelia cells		AV
	Peritoneal mesothelial cells		AV
Alzheimer's Disease	Nervous system	Nerve growth factor	RV, AV, HSV
Cystic fibrosis	Lung epithelia cells	CFTR	AV,AAV, RM, L
Diabetes	Fibroblasts	Human insulin	RV
	Hepatocytes		L
Duchenne muscular dystrophy	Muscle cells	Dystrophin	AV, DI
Familial hypercholesterolemia	Hepatocytes	LDL receptor	RV, AV, RM
Gaucher disease	Hematopoietic cells	Glucocerebrosidase	RV
	Fibroblasts		RV
Growth hormone deficiency	Endothelial cells	Human growth hormone	RV
	Fibroblasts		TR
	Keratinocytes		RV
	Muscle cells		RV
Hemoglobinopathies	Hematopoietic cells	α or β-globin	RV, AAV
Hemophilia	Fibroblasts	Factor VIII, IX	RV, DI, L
	Keratinocytes		RV
	Hepatocytes		RV, AV, RM, L
	Muscle cells		RV
Leukocyte adhesion deficiency	Hematopoietic cells	CD-18	RV
Parkinson's disease	Nervous system	Tyrosine hydroxylase	RV, AV, HSV
Phenylketonuria	Hepatocytes	Phenylalanine hydroxylase	RV
Purine nucleoside phosphorylase deficiency	Fibroblasts	Purine nucleoside phosphorylase	RV
Urea cycle disorders	Hepatocytes	Ornithine transcarbamylase or arginosuccinate synthetase	AV
Acquired			
Cancer	Acute lymphoblastic leukemia	p53	RV
	Brain tumors	HSV thymidine kinase	RV
	Carcinoma	γ-interferon	TR
	Melanoma	Tumor necrosis factor	RV
	Retinoblastoma	Retinoblastoma gene	RV
Infectious diseases	HIV	Dominant negative Rev	RV
		TAR decoy	RV
		RRE decoy	TR
		Diptheria toxin A	RV
Cardiomyopathy	Muscle cells	(Used reporter gene)	DI
Emphysema	Lung epithelial cells	Alpha-1 antitrypsin	AV
Local thrombosis	Endothelial cells	Anti-clotting factors	RV
Vaccines	Muscle cells	Various	DI

Note: AAV = recombinant adeno-associated viruses
AV = recombinant adenoviruses
DI = direct injection
HSV = recombinant herpes simplex virus
L = lipofection
RM = receptor mediated
RV = recombinant retroviruses
TR = transfection

TABLE 106.2 Physical Characteristics of Recombinant Virions

Characteristic	Units	Recombinant Retroviruses	Recombinant Adenoviruses	Recombinant AAV
Genome type		ss RNA (2 per virion)	ds DNA	ss DNA
Genome size	Bases	8300	36000	4700
Genome MW	Daltons	3×10^6	$20–25 \times 10^6$	$1.2–1.8 \times 10^6$
Particle diameter	nm	90–147	65–80	20–24
Particle mass	Grams	3.6×10^{-16}	2.9×10^{-16}	1.0×10^{-17}
Composition				
DNA/RNA	%	2	13	26
Protein	%	62	87	74
Lipid	%	36	0	0
Density	g/cm³ CsCl	1.15–1.16	1.33–1.35	1.39–1.42
Enveloped?	Yes/no	Yes	No	No
Shape		Spherical	Icosahedral	Icosahedral
Surface projections	Yes/no	Yes	Yes	No
Number		~60–200	12	
Length	nm	5	25–30	
Max diameter	nm	8	4	
Virus titer	pfu/ml	$10^6–10^7$	$10^{10}–10^{12}$	$10^5–10^7$
Integration?	—	Yes, random	No, episomal	Yes, chromosome 19

and cannot be applied to many important target tissues and organs such as the lungs, brain and heart. For *in vivo* gene therapy, the gene transfer agent is delivered directly to the target tissue/organ, and gene transfer occurs in the patient rather than in the tissue culture dish [Mulligan, 1993]. Both strategies have inherent advantages and disadvantages and current research is evaluating which approach can best meet the needs of a particular disease.

Gene delivery systems can be classified as either viral or non-viral [Friedmann, 1989]. For viral gene transfer, one of several different types of viruses is engineered to deliver genes. Typically, viral genes are removed to prevent self-replication of the virus and to provide room for the insertion of one or more therapeutic genes that the recombinant virus will carry. To further ensure the safety of the recombinant viruses, specialized **packaging cell lines** have been developed to produce the recombinant viruses and minimize the production of infectious wild-type viruses. Some viruses are able to integrate the therapeutic genes into the target cell's nuclear DNA (retroviruses, adeno associated viruses), whereas others are not (adeno viruses) (Table 106.2).

Non-viral gene transfer systems are based on a variety of technologies that employ physical/chemical means to deliver genes [Felgner and Rhodes, 1991]. These technologies include direct **plasmid** injection, bombardment with DNA coated microprojectiles, and DNA complexed with **liposomes**. Some non-viral transfection techniques are too inefficient (e.g., coprecipitation of DNA with calcium phosphate [Chen and Okayama, 1987], DNA complexed with DEAE-dextran [Pagano et al., 1967], electroporation [Neumann et al., 1982]) or laborious (e.g., microinjection of DNA [Capecchi, 1980]) for clinical use. Only those gene delivery systems (viral and non-viral) with potential for clinical application will be discussed in this article. The main features of these technologies (Table 106.3) will be described and specific examples of their applications highlighted.

106.2 Recombinant Retroviruses

Many of the approved clinical trials have utilized recombinant retroviruses for gene delivery. Retroviral particles contain two copies of identical RNA genomes that are wrapped in a protein coat and further encapsidated by a lipid bilayer membrane. The virus attaches to specific cell surface receptors via surface proteins that protrude from the viral membrane. The particle is then internalized and its genome is released into the cytoplasm, reverse transcribed from RNA to DNA, transported into the nucleus and

TABLE 106.3 Features of the Various Gene Transfer Systems

Features	Retrovirus	AAV	Adenovirus	Non-viral
Maximum transgene size	8 kb	4.7 kb	36 kb	$\gg$36 kb
Maximum concentration (vectors/ml)	$\sim 10^7$	$\sim 10^{12}$	$\sim 10^{12}$	$\gg$36 kb
Transfers genes to quiescent cells	No/Yes*	Yes	Yes	Yes
Integrates transgene into target cell genome	Yes	Yes	No	No
Persistence of gene expression	wks–yrs	yrs	wks–mos	days–wks
Immunological problems	Few	None known	Extensive	None
Pre-existing host immunity	No	Yes	Yes	No
Stability	Poor	Good	Good	Good
Ease of large scale production	Difficult	Difficult	Easy	Easy
Safety concerns	Insertional mutagenesis	Inflammation toxicity	Inflammation toxicity	Toxicity

* Recombinant lentiviruses, such as human immunodeficiency virus, are capable of transducing quiescent cells.

then integrated into the cell's chromosomal DNA. The integrated viral genome has LTRs (long terminal repeats) at both ends which encode the regulatory sequences that drive the expression of the viral genome [Weiss et al., 1982].

Retroviruses used for gene transfer are derived from wild-type murine retroviruses. The recombinant viral particles are structurally identical to the wild type virus but carry a genetically engineered genome (retroviral vector) which encodes the therapeutic gene of interest. These recombinant viruses are incapable of self-replication but can infect and insert their genomes into a target cell's genome [Morgan et al., 1993].

Recombinant retroviruses, like all other recombinant viruses, are produced by a two part system composed of a packaging cell line and a recombinant vector (Fig. 106.1) [Anderson, 1992; Levine and Friedmann, 1993]. The packaging cell line has been engineered to express all the structural viral genes (*gag, pol* and *env*) necessary for the formation of an infectious virus particle. *gag* encodes the capsid proteins and is necessary for encapsidation of the vector. *pol* encodes the enzymatic activities of the virus including reverse transcriptase and integrase. *env* encodes the surface proteins on the virus particle which are necessary for attachment to the target cell's receptors.

The retroviral vector is essentially the wild-type genome with all the viral genes removed. This vector encodes the transgene(s) and the regulatory sequences necessary for their expression as well as a special packaging sequence, (ψ), that is required for encapsidation of the genome into an infectious viral particle [Morgan et al., 1993]. To produce recombinant retrovirus particles, the retroviral vector is transfected into the packaging cell line. The structural proteins expressed by the packaging cell line recognize the packaging sequence on RNAs transcribed from the transfected vector and encapsidate them into an infectious virus particle that is subsequently exocytosed by the cell and released into the culture medium. This medium containing infectious recombinant viruses is harvested and used to transduce target cells.

Many different retroviral vector designs have been used (Fig. 106.2). A commonly used vector encodes two genes, one expressed from the LTR and the other from an internal promoter (Fig. 106.2c) [Miller, 1992]. Often, one gene expresses a therapeutic protein and the other a selectable marker that makes the modified cells resistant to selective media or a drug. This allows the investigator to establish a culture composed solely of transduced cells by growing them under the selective conditions. Several configurations are possible, but the optimum design is often dictated by the transgene(s) being expressed and the cell type to be transduced. Vector configuration is crucial for maximizing viral titer and transgene expression [Roemer and Friedmann, 1992].

As with all gene transfer technologies, there are advantages and disadvantages to the use of recombinant retroviruses. Retroviruses can only transduce dividing cells since integration requires passage of the target cells through mitosis [Roe et al., 1993]. This limits the use of recombinant retroviruses for *in vivo* gene therapy, since few normal cells are actively dividing. Recently, however, new retroviral vectors based on lentiviruses have been developed that are capable of transducing non-dividing cells [Naldini et al., 1996].

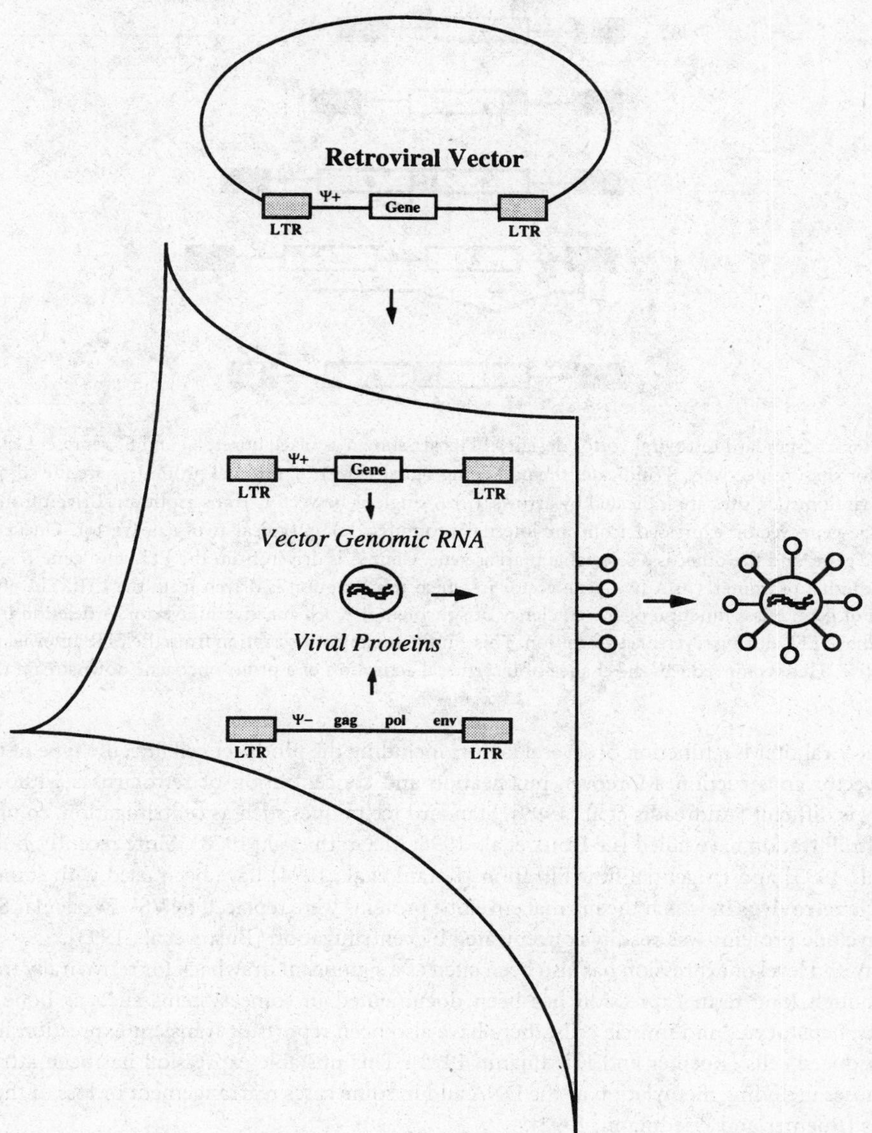

FIGURE 106.1 Packaging cell line for retrovirus. A simple retroviral vector composed of 2 LTR regions which flank sequences encoding the packaging sequence (Ψ) and a therapeutic gene. A packaging cell line is transfected with this vector. The packaging cell line expresses the three structural proteins necessary for formation of a virus particle (*gag*, *pol*, and *env*). These proteins recognize the packaging sequence on the vector and form an infectious virion around it. Infectious virions bud from the cell surface into the culture medium. The virus-laden culture medium is filtered to remove cell debris and then either immediately used to transduce target cells or the virions are purified and/or concentrated and frozen for later use.

These HIV based vectors were able to achieve stable gene transfer after injection into a non-dividing tissue (brain) *in vivo*.

Other drawbacks of retroviral vectors include: (a) a limitation to the size of inserted genes (<8 kilobases) [Roemer and Friedmann, 1992]; (b) the particles are unstable and lose activity at 37°C with a half-life of 5-7 hours [Kotani et al., 1994; Le Doux et al., 1999]; and (c) virus producer cell lines typically produce retrovirus in relatively low titers (10^5 to 10^7 infectious particles per ml) [Paul et al.,

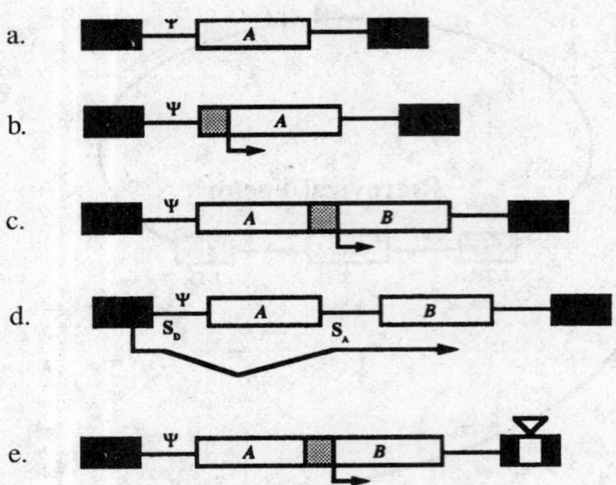

FIGURE 106.2 Standard retroviral vector designs. LTRs are shown as black boxes. S_D and S_A represent splice donor and acceptor sites, respectively. Ψ indicates the packaging signal sequence. Internal promoters are the stippled boxes and transcription start sites are indicated by arrows. (*a*) A single-gene vector. Transcription is driven from the LTR. (*b*) A single-gene vector expressed from an internal promoter. (*c*) A typical two-gene vector. One gene is the therapeutic gene, and the other is a selectable marker gene. Gene *A* is driven from the LTR, and gene *B* is expressed from the internal promoter. (*d*) A two-gene vector in which transcription is driven from the LTR. The efficiency of expression of gene *B* is a function of the efficiency of splicing. (*e*) A self-inactivating vector. A deletion in 3′ LTR is copied to the 5′ LTR during reverse transcription. This eliminates any transcription from the LTR. Internal promoters are still active. This vector reduces the chance of insertional activation of a proto-oncogene downstream from the 3′ LTR.

1993]. The viral titer is a function of several factors including the producer cell line, the type of transgene and the vector construction. Moreover, purification and concentration of retroviruses without loss of infectivity is difficult [Andreadis et al., 1999]. Standard techniques such as centrifugation, column filtration or ultrafiltration have failed [Le Doux et al., 1998; McGrath et al., 1978]. More recently, hollow fiber [Paul et al., 1993] and tangential flow filtration [Kotani et al., 1994] have been used with some success. A **chimeric retrovirus** in which the normal envelope proteins were replaced by VSV (Vesicular Stomatitis Virus) envelope proteins was readily concentrated by centrifugation [Burns et al., 1993].

Stability and level of expression has also been cited as a significant drawback for retrovirally transduced cells. Although long term expression has been documented in some systems such as bone marrow, fibroblasts, hepatocytes and muscle cells, there have also been reports of transient expression in retrovirally transduced cells [Roemer and Friedmann, 1992]. This unstable expression has been attributed to several causes including methylation of the DNA and in some cases rearrangement or loss of the proviral sequences [Roemer and Friedmann, 1992].

The use of recombinant retroviruses has raised two major safety concerns [Roemer and Friedmann, 1992; Temin, 1990]. Older packaging cell lines occasionally produced replication competent virus by homologous recombination between the retroviral vector and the packaging cell line's retroviral sequences. New packaging cell lines have made the production of replication competent viruses essentially impossible [Danos and Mulligan, 1988]. The other safety concern was the possibility that the integration of the recombinant retrovirus would activate a proto-oncogene. The probability of this event is very low, and since several mutations are required for a cell to become cancerous, the risk of cellular transformation is extremely low and is typically outweighed by the potential therapeutic benefits [Temin, 1990].

106.3 Recombinant Adenoviruses

Recombinant adenoviruses have a number of properties that make them a useful alternative to recombinant retroviruses for human gene transfer. Recombinant adenoviruses are well-characterized, relatively easy to manipulate, can be grown and concentrated to very high titers (up to 10^{13} infectious particles/ml), are stable particles, and can transduce a wide variety of cell types [Crystal, 1995; Hitt et al., 1997].

Recombinant adenoviruses are also a relatively safe gene transfer system. Replication competent viruses, should they contaminate virus stocks or form by recombination with wild-type adenoviruses from latent infections, do not pose any significant risk since they elicit only a mild, self-limiting infection [Parks et al., 1996]. Furthermore, genes transferred by recombinant adenoviruses do not integrate, eliminating the risk of insertional mutagenesis of the chromosomal DNA of the target cell [Crystal, 1995]. Most important, recombinant adenoviruses efficiently transfer genes to non-dividing, as well as dividing, cells which makes possible the *in vivo* transduction of tissues composed of fully differentiated or slowly dividing cells such as the liver and lung [Mulligan, 1993].

Adenoviruses consist of a large double stranded DNA genome (about 36 kilobase pairs long) packaged within a non-enveloped icosahedral capsid that is primarily composed of three virus-encoded proteins (hexon, penton base, and fiber proteins) [Horwitz, 1990]. The fiber proteins protrude from the surface of the virus and mediate its attachment to target cells via a high affinity interaction with cell surface receptors. The virus is then internalized into endosomal vesicles via specific interactions between the penton base proteins and α_v integrins [Wickham et al., 1993]. Adenoviruses escape these vesicles by an acid-induced endosomolytic activity and are transported to the nucleus, into which they enter via pores in the nuclear membrane [Greber et al., 1993].

The wild type adenovirus genome consists primarily of five early genes (E1-E5) each of which is expressed from their own promoters [Horwitz, 1990]. There are also five late genes (L1-L5) which are expressed from the major late promoter (MLP). The first generation of recombinant adenoviruses were based on a mutant adenovirus in which the E1 region (and in some cases the E3 region) was deleted. The E1 region is required for replication. Nevertheless, E1 minus mutants can be grown on specialized packaging cell lines (293 cells) which express the E1 gene and therefore provide the necessary functions for virus production [Levine and Friedmann, 1993].

To generate recombinant adenoviruses, a plasmid that contains the gene of interest, flanked by adenovirus sequences, is transfected into 293 cells in which adenoviruses that lack the E1 region are actively replicating (Fig. 106.3). The virus stock is screened for the rare recombinants in which the gene of interest has correctly recombined with the E1 minus mutant by homologous recombination. These recombinant virions are purified and grown to high titer on 293 cells [Morgan et al., 1994].

Recombinant adenoviruses have been successfully used for *in vivo* gene transfer in a number of animal models and in several clinical protocols, including those which delivered a functional copy of the cystic fibrosis transmembrane conductance regulator (CFTR) gene into the nasal epithelium and lungs of cystic fibrosis patients [Boucher and Knowles, 1993; Welsh, 1993; Wilmott and Whitsett, 1993]. Unfortunately, transgene expression was short-lived (5 to 10 days) [Verma and Somia, 1997]. Short-lived expression is a disadvantage for the treatment of genetic or chronic disorders because, in order to maintain the therapeutic effect of the transgene, recombinant adenoviruses would have to be administered to the patient repeatedly. Repeat administration often fails due to the presence of neutralizing antibodies that form in response to the first administration of the recombinant adenoviruses [Hitt et al., 1997]. Alternatively, short lived high expression from adenovirus vectors might be therapeutically useful for applications where transient expression is preferred. Such applications include the stimulation of angiogenesis by adenoviruses expressing vascular endothelial growth factor (VEGF) [Magovern et al., 1997; Morgan and Yarmush, 1998].

Short-lived gene expression is currently one of the most significant limitations of recombinant adenoviruses for use in human gene therapy. Although short-lived gene expression could be the result of vector cytotoxicity, promoter shutoff, or loss of the transgene DNA, most investigators believe it is primarily due to the elimination of transduced cells by transgene or adenovirus-specific cytotoxic T-lymphocytes [Dai et al., 1995; Engelhardt et al., 1994; Scaria et al., 1998; Zsengeller et al., 1995]. As a result, most efforts to increase the longevity of gene expression have focused on eliminating the immune rejection of cells transduced by recombinant adenoviruses. Two major strategies have been pursued. One strategy has been to construct recombinant adenoviruses that encode fewer immunogenic adenoviral proteins, such as by deleting portions of the E2 region or the E4 region. Such vectors have the added benefit of being able to accommodate the cloning of much larger inserts of therapeutic DNA (to as much

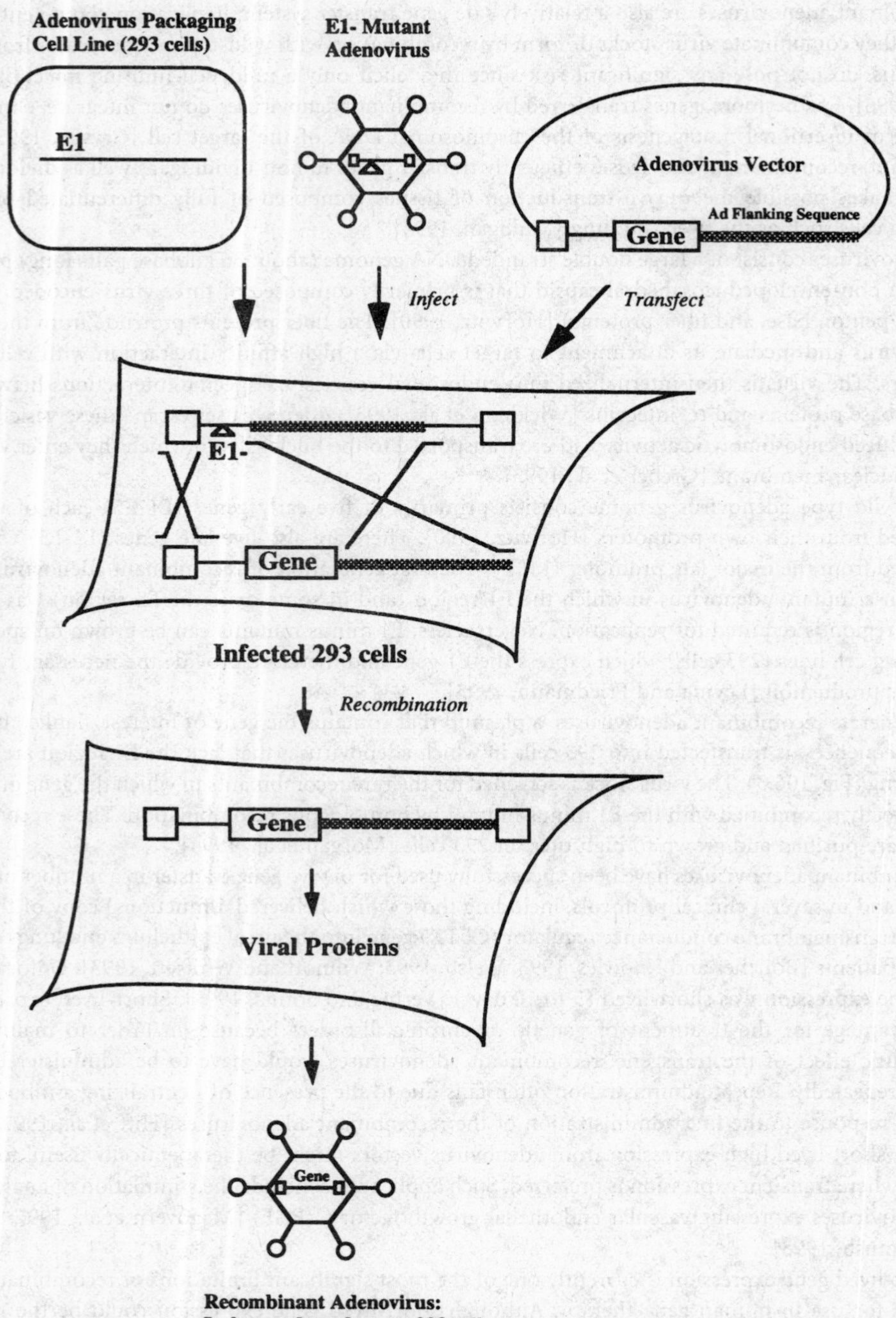

FIGURE 106.3 Isolation of recombinant adenovirus. A packaging cell line (293 cells) which expresses the E1 gene is infected with an E1-minus mutant adenovirus. The adenovirus is derived from a plasmid encoding the wild-type adenovirus genome. The E1 region of the adenovirus genome is replaced with the therapeutic gene, and the resultant plasmid (the adenovirus vector) is transfected into the 293 cell line which is infected by the mutant adenovirus. Since the therapeutic gene is flanked by adenoviral sequences, the mutant adenovirus genome and the adenovirus vector will occasionally undergo homologous recombination and form an infectious recombinant adenovirus whose genome encodes the therapeutic gene. These rare recombinants are isolated then grown on another 293 cell line.

as 36 kilobases) than was possible with the first generation (E1 and E3 minus) recombinant adenoviruses which could accommodate only 7 kilobases of foreign DNA [Morsy et al., 1998; Parks et al., 1996].

The second strategy is to block the host immune response. For example, investigators have co-injected recombinant adenoviruses with immunomodulatory compounds (e.g., deoxyspergualin, IL-12, IFN-gamma) that transiently suppressed the immune system, eliminated the humoral response, and made possible repeat administration of the vector [Hitt et al., 1997]. Others have co-injected vectors that encode for immunosuppressive compounds that induce tolerance to recombinant adenoviruses (e.g., CTLA4Ig), or are developing recombinant adenoviruses whose backbones contain genes that encode for immunosuppressive proteins (e.g., Ad5 E3 19-kDa protein or HSV ICP47) [Nakagawa et al., 1998; Wold and Gooding, 1989; York et al., 1994].

As a result of these and similar efforts, the persistance of gene expression *in vivo* has been extended to as long as 5 months [Kay et al., 1995; Scaria et al., 1998]. It remains to be determined, however, if these methods can eliminate the immune response against recombinant adenoviruses and their products, whether or not elimination of the immune response results in long-term gene expression, and if these methods will be effective in human gene therapy protocols.

106.4 Recombinant Adeno-Associated Viruses

Adeno-associated viruses (AAV) are another virus-based gene transfer system that has significant potential for use in human gene therapy. AAV are small, non-enveloped human parvoviruses that contain a single-stranded DNA genome (4.7 kilobases) and encode two genes required for replication, *rep* and *cap* [McCarty and Samulski, 1997]. AAV are stable, have a broad host range, can transduce dividing and non-dividing cells, and do not cause any known human disease [Verma and Somia, 1997]. AAV require the presence of a helper virus (typically adenovirus) to replicate [Muzyczka, 1992]. When no helper virus is present, AAV do not replicate but instead tend to establish a latent infection in which they permanently integrate into the chromosomal DNA of the target cells [Inoue and Russell, 1998]. Wild-type AAV preferentially integrates into a specific site in chromosome 19 due to the interaction of the virus-encoded protein Rep with the host cell DNA [McCarty and Samulski, 1997]. Recombinant AAV, however, are Rep negative and as a result integrate randomly into the chromosomal DNA of the target cell.

To generate recombinant AAV, human 293 cells are transfected with the AAV vector, a plasmid that contains the therapeutic gene of interest flanked by the 145-bp AAV inverted terminal repeats (ITR) that are necessary for its encapsidation into a virus particle [Ferrari et al., 1997] (Fig. 106.4). The cells are also transfected with a helper plasmid, a plasmid that encodes for the virus proteins necessary for particle formation and replication (i.e., AAV *rep* and *cap* genes). The transfected cells are infected with wild-type adenovirus which supplies helper functions required for amplification of the AAV vector [Morgan and Yarmush, 1998]. Contaminating adenovirus is removed or inactivated by density gradient centrifugation and heat inactivation. Recombinant AAV generated by this and similar methods have been successfully used to achieve long term expression of therapeutic proteins in a number of cell types and tissues, including in the lung, muscle, liver, central nervous system, retina, and cardiac myocytes [Bartlett et al., 1998; Flannery et al., 1997; Flotte et al., 1993; Maeda et al., 1998; Monahan et al., 1998; Xiao et al., 1998]. Therapeutic effects have been achieved in a number of model systems, including in a dog model of hemophilia and a mouse model for obesity and diabetes [Monahan et al., 1998; Murphy et al., 1997].

Despite these early successes, several technical issues must be addressed before recombinant AAV can be used for human gene therapy on a routine basis. For example, little is known about the conditions or factors which enhance recombinant AAV transduction efficiency *in vitro* or *in vivo* [McCarty and Samulski, 1997]. Nor is it well-understood what controls the efficiency with which recombinant AAV integrate into the chromosomal DNA of target cells or why the efficiency of integration is so low (< 1%). Perhaps the most significant technical issue, however, is that the current methods for producing recombinant AAV (see above) are tedious, labor intensive, and not well-suited for producing clinical grade virus.

Current methods are not adequate for producing a clinical grade virus for several reasons. First, current methods do not produce enough virus particles to be useful for many gene therapies. For example, it

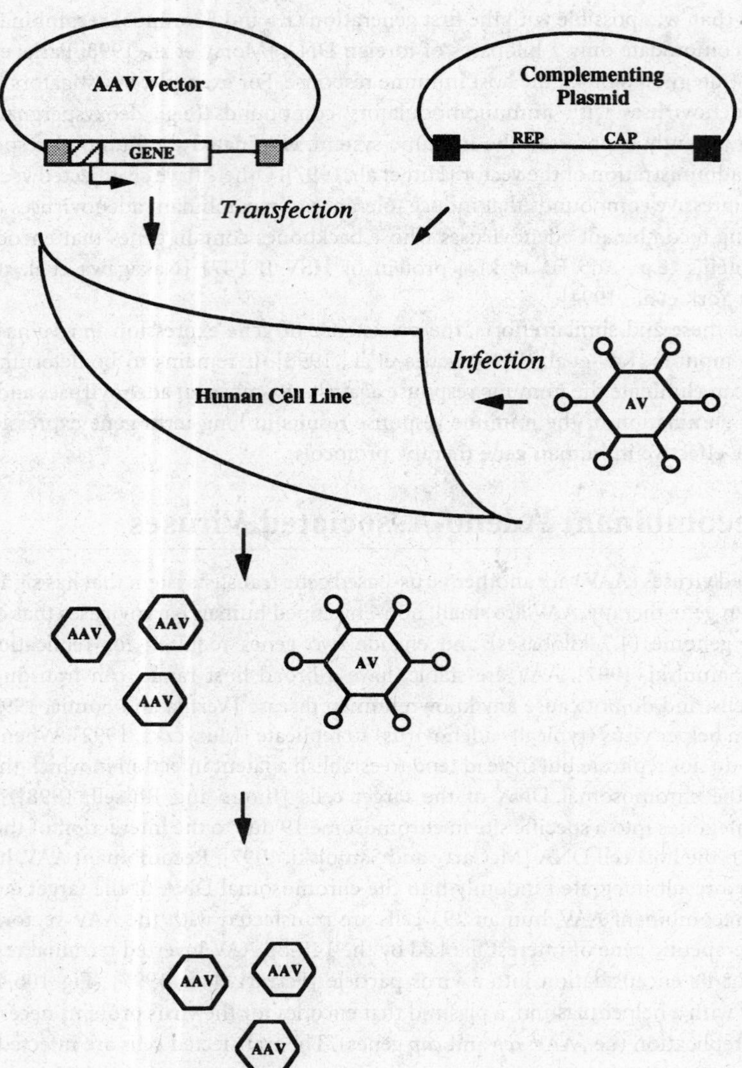

FIGURE 106.4 Production of recombinant AAV. Human cells are transfected with a plasmid encoding the therapeutic gene which is driven from a heterologous promoter (*cross-hatched box*) and flanked with AAV terminal repeats (*stippled boxes*). The cells are also transfected with a complementing plasmid encoding the AAV *rep* and *cap* genes, which cannot be packaged because they are not flanked by AAV terminal repeats. The *rep* and *cap* genes, whose products are required for particle formation, are flanked by adenovirus 5 terminal fragments (*black boxes*) which enhance their expression. The transfected cells are infected with wild-type adenovirus which supplies helper functions required for amplification of the AAV vector. The virus stock contains both AAV recombinant virus and adenovirus. The adenovirus is either separated by density gradient centrifugation or heat inactivated.

has been estimated that 10^{14} recombinant viruses will be needed to achieve systemic production of therapeutic levels of proteins such as factor IX or erythropoietin [Inoue and Russell, 1998]. With current production methods, which yield 10^8 to 10^9 viruses per ml, about 100 to 1000 liters of virus would have to be produced for each treatment. Second, current methods require the use of helper adenovirus which is a significant disadvantage because their removal is tedious and labor intensive. In addition, there is the risk that residual contaminating adenovirus proteins will be present and stimulate the immune rejection of transduced cells [Xiao et al., 1998]. Finally, the use of transient transfection is not amenable to scale up and significantly increases the likelihood that replication competent AAV will be generated by recombinogenic events [Inoue and Russell, 1998].

To overcome these problems, investigators are working towards the development of stable packaging cell lines that produce recombinant AAV. Substantial increases in the yield of recombinant AAV have been achieved by increasing in the virus producer cell lines the number of copies of the AAV vector or of the helper plasmid. For example, one group developed an AAV packaging cell line which contained integrated helper and vector constructs that were linked to the simian virus 40 replication origin [Inoue and Russell, 1998]. These packaging cells could be induced to express SV40 T antigen, which then amplified the helper and vector constructs by 4 to 10 fold, resulting in 5 to 20 fold increases in recombinant AAV production. Progress has also been made towards the elimination of the need for helper adenovirus. For example, adenovirus miniplasmids have been constructed that provide all of the adenovirus helper functions but do not support the production of adenovirus proteins or viruses [Xiao et al., 1998]. Use of adenovirus miniplasmids, in conjunction with an improved helper plasmid, have improved the yield of recombinant AAV by up to 40-fold. Most likely a combination of strategies such as these will be necessary to develop a stable producer cell line that will be free of helper adenovirus and that will generate recombinant AAV at levels that are adequate for use in human gene therapies.

106.5 Direct Injection of Naked DNA

Direct injection of plasmid DNA intramuscularly, intraepidermally, and intravenously is a simple and direct technique for modifying cells *in vivo*. DNA injected intramuscularly or intraepidermally is internalized by cells proximal to the injection site [Hengge et al., 1995; Wolff et al., 1990]. DNA injected intravenously is rapidly degraded in the blood ($t_{1/2}$ < 10 min.) or retained in various organs in the body, preferentially in the non-parenchymal cells of the liver [Kawabata et al., 1995; Lew et al., 1995]. Gene expression after direct injection has been demonstrated in skeletal muscle cells of rodents and non-human primates, cardiac muscle cells of rodents, livers of cats and rats, and in thyroid follicular cells of rabbits [Hengge et al., 1995; Hickman et al., 1994; Levine and Friedmann, 1993; Sikes et al., 1994].

The efficiency of gene transfer by direct injection is somewhat inefficient [Doh et al., 1997]. As little as 60-100 myocardial cells have been reported to be modified per injection [Acsadi et al., 1991]. Higher efficiencies were observed when plasmids were injected into regenerating muscle cells [Wells, 1993] or co-injected with recombinant adenoviruses [Yoshimura et al., 1993]. Injected DNA does not integrate, yet gene expression can persist for as long as two months [Wolff et al., 1990]. Levels of gene expression are often low but can be increased by improvements in vector design [Hartikka et al., 1996].

Despite the low efficiency of gene transfer and expression, direct injection of plasmid DNA has many promising applications in gene therapy, particularly when low levels of expression are sufficient to achieve the desired biological effect. For example, several patients suffering from critical limb ischemia were successfully treated by injection of DNA encoding human vascular endothelial growth factor (VEGF), a potent angiogenic factor [Isner et al., 1996; Isner et al., 1996; Melillo et al., 1997; Tsurumi et al., 1996]. Following injection of DNA into the ischemic limbs of ten patients, VEGF expression was detected in their serum, new blood vessels were formed in 7 of the 10 patients, and 3 patients were able to avoid scheduled below-the-knee amputations [Baumgartner et al., 1998].

Direct injection of DNA could also be used to vaccinate patients against pathogens as evidenced by the effects of direct injection of plasmid DNA encoding pathogen proteins or immunomodulatory cytokines [Donnelly et al., 1997; Sato et al., 1996]. Direct injection may also be an effective way to systemically deliver therapeutic proteins [Tripathy et al., 1996]. For example, the incidence of autoimmune diabetes in a mouse model of the disease was significantly reduced in mice that were injected intramuscularly with DNA encoding interleukin 10, an immunosuppressive cytokine [Nitta et al., 1998].

In contrast to viral-based delivery systems, there is little restriction on the size of the transgene that can be delivered by direct DNA injection. As a result, direct DNA injection is particularly well-suited for treating disorders that require the delivery of a large transgene. For example, Duchenne's muscular dystrophy, a genetic disease of the muscle caused by a defect in the gene for dystrophin (12 kilobases) can potentially be treated by direct DNA injection [Acsadi et al., 1991].

106.6 Particle-Mediated Gene Transfer

Particle-mediated gene transfer is an alternative method used to deliver plasmid DNA to cells. DNA coated gold particles are loaded onto a macro-projectile which is then accelerated through a vacuum chamber to high velocity by a burst of helium or a voltage discharge until it hits a stopping plate [Yang et al., 1996]. Upon impact, the DNA-coated microprojectiles are released through a hole in the stopping plate, penetrate the target tissue, and the transferred gene is expressed [Morgan and Yarmush, 1998; Yang et al., 1996]. Genes have been introduced and expressed in a number of cell types and tissues, including skin, liver, spleen, muscle, intestine, hematopoietic cells, brain, oral mucosa and epidermis, tumor explants, and cells of developing mouse embryos [Jiao et al., 1993; Keller et al., 1996; Mahvi et al., 1996; Morgan and Yarmush, 1998; Verma et al., 1998; Zelenin et al., 1993]. Similar to direct DNA injection, there are few constraints on the size of the DNA that can be delivered.

The efficiency of particle-mediated gene transfer varies with tissue type, but in general it is most efficient in the liver, pancreas, and epidermis of the skin, and least efficient in muscle, vascular, and cardiac tissues [Rakhmilevich and Yang, 1997]. For example, 20 percent of bombarded epidermal cells of the skin, but only 1 to 3 percent of bombarded muscle cells, expressed the transferred gene [Williams et al., 1991; Yang et al., 1990].

106.7 Liposome-Mediated Gene Delivery

Liposomes made from a mixture of neutral and cationic lipids have also been used to deliver plasmid DNA to cells. Liposomes are relatively easy to make, can be made with well-defined biophysical properties, and can accommodate virtually any size transgene [Kay et al., 1997; Tomlinson and Rolland, 1996]. Small unilamellar liposomes ranging from 20 to 100 nm in diameter are prepared by sonication of a mixture of cationic (e.g., DOTMA, *N*-{1-(2,3-dioleyloxy)propyl}-*N,N,N*-triethylammonium) and neutral (e.g., DOPE, dioleoyl-phosphatidylethanolamine) lipids, followed by extrusion through a porous polycarbonate

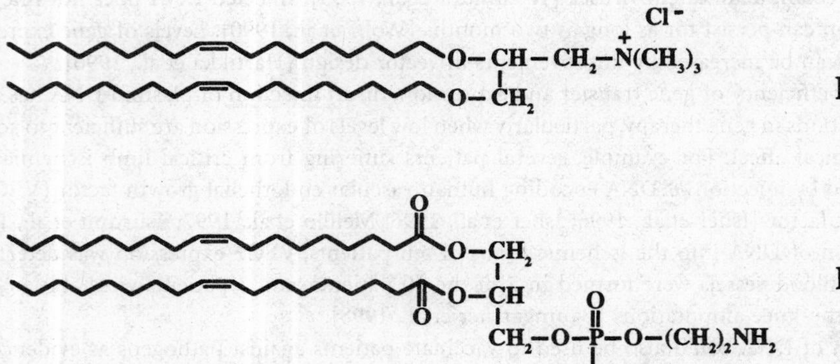

FIGURE 106.5 Common constituents of cationic liposomes. A typical cationic liposome is composed of a mixture of cationic lipids and neutral phospholipids. DOTMA {*N*-[1-(2,3,-dioleyloxy]-*N,N,N*-trimethylammonium chloride} is a synthetic cationic lipid which attaches to DNA using its positively charged head group. DOPE {1,2-Di[(*cis*)-9-octadecenoyl]-sn-glycero-3-phosphoethanolamine} is a neutral phospholipid which enhances the activity of these cationic liposomes. The resultant DNA-liposome complex is positively charged and reduces or eliminates the repulsive electrostatic forces between the negatively charged DNA and the negatively charged cell surface.

filter (e.g., 100 nm pore size) (Fig. 106.5) [Morgan and Yarmush, 1998; Radler et al., 1997]. DNA is added to the cationic liposomes, binds non-covalently to the positively charged cationic lipids, and induces a topological transition from liposomes to multilamellar structures composed of lipid bilayers alternating with DNA monolayers [Felgner and Ringold, 1989; Radler et al., 1997]. The size of the structures depends on their overall charge, with charged structures (negative or positive) being the smallest (about 100 nm in diameter) due to stabilization by electrostatic repulsion and neutral structures being the largest (greater than 3000 nm in diameter) due to aggregation as a result of van der Waals attractive forces [Radler et al., 1997]. The relationship between structure and transfection efficiency is not well-understood, although in general a slight excess of cationic lipid is needed for optimal gene transfer [Kay et al., 1997].

Gene transfer is accomplished by simply mixing or applying the lipid-DNA complexes to the target cells or tissue. Cationic liposome-DNA complexes have been used in a number of applications including transfer of genes to the arterial wall, lung, skin, and systemically by intravenous injection [Morgan and Yarmush, 1998; Zhu et al., 1993]. Genes delivered by cationic liposomes do not integrate so gene expression is transient and there is minimal risk of insertional mutagenesis. Liposomes do not carry any viral sequences or proteins, and are relatively non-toxic and non-immunogenic [Nabel et al., 1993; Zhu et al., 1993].

Liposome-mediated gene transfer is somewhat inefficient, however, in part due to the failure of a large fraction of lipid-complexed DNA to escape degradation in cellular endosomes [Xu and Szoka, 1996; Zabner et al., 1995]. Several strategies have been taken to overcome this limitation, including use of acidotropic bases to reduce the rate of degradation by raising the pH of the endosomes, coupling of liposomes to endosomolytic or fusogenic virus proteins, and developing new liposome formulations that use pH sensitive cationic lipids that become fusogenic in the acidic cellular endosomes [Budker et al., 1996; Legendre and Szoka, 1992; Yonemitsu et al., 1997]. Substantial improvements in the efficiency of liposome-mediated gene transfer will only be achieved when the fundamental mechanisms that govern the interactions of cationic lipid-DNA complexes with cells are better understood.

106.8 Other Gene Transfer Methods

Several other gene transfer technologies have been tested in clinical trials or are in various stages of development. These include recombinant viruses such as vaccinia virus [Qin and Chatterjee, 1996], herpes simplex virus [Glorioso et al., 1997], canarypox virus [Kawakita et al., 1997], and fowlpox [Wang et al., 1995], and non-viral vectors that mimic some of the properties of viruses, such as DNA conjugated to proteins that promote binding to specific cell-surface receptors, fusion, or localization to the nucleus [Wu et al., 1994]. These methods are not discussed here in greater detail because even though they are capable of transferring genes to cells they have not yet been well-developed or extensively tested in the clinic.

106.9 Summary and Conclusion

Several gene transfer systems have been developed that have successfully transferred genes to cells, and have elicited various biological effects. To date, nearly 300 clinical trials have been approved to test their safety and efficacy (Table 106.4). Each system has unique features, advantages, and disadvantages that determine if its use in a particular application is appropriate. One principal consideration is whether or not permanent or temporary genetic modification is desired. Other important considerations include what the setting of gene transfer will be (*ex vivo* or *in vivo*), what level of gene expression is needed, and whether or not the host has pre-existing immunity against the vector. No single gene transfer system is ideal for any particular application and it is unlikely that such a universal gene transfer system will ever be developed. More likely, the current gene transfer systems will be further improved and modified, and new systems developed, that are optimal for the treatment of specific diseases.

TABLE 106.4 Current Clinical Experience Using Gene Transfer (as of December 1, 1998)

Human gene transfer protocols	Numbe r	Percentage
Gene marking studies	30	11
Non-therapeutic protocols	2	1
Therapeutic protocols	238	88
Infectious diseases	25	9
Inherited diseases	35	13
Cancer	167	62
Other diseases	11	4
Strategies for gene delivery		
Ex vivo protocols	134	50
In vivo protocols	136	50
Gene transfer technologies		
Permanent genetic modification		
Recombinant retroviruses	128	47
Recombinant adeno associated virus	6	2
Temporary genetic modification		
Recombinant adenovirus	60	22
Cationic liposomes	37	14
Plasmid DNA	16	6
Particle mediated	2	1
Other		
Viral vectors	19	7
Non-viral vector	2	1

*'Other' viral vectors refers to recombinant canarypox virus, vaccinia virus, fowlpox, and herpes simplex virus. 'Other' non-viral vector refers to RNA transfer and antisense delivery.

Defining Terms

Chimeric retrovirus: A recombinant retrovirus whose structural proteins are derived from two or more different viruses.

***Ex-vivo*:** Outside the living body, referring to a process or reaction occurring therein.

***In-vitro*:** In an artificial environment, referring to a process or reaction occurring therein, as in a test tube or culture dish.

***In-vivo*:** In the living body, referring to a process or reaction occurring therein.

Liposome: A spherical particle of lipid substance suspended in an aqueous medium.

Packaging cell line: Cells that express all the structural proteins required to form an infectious viral particle.

Plasmid: A small, circular extrachromosomal DNA molecule capable of independent replication in a host cell, typically a bacterial cell.

Recombinant: A virus or vector that has DNA sequences not originally (naturally) present in their DNA.

Retrovirus: A virus that possesses RNA-dependent DNA polymerase (reverse transcriptase) which reverse transcribes the virus' RNA genome into DNA then integrates that DNA into the host cell's genome .

Transduce: To effect transfer and integration of genetic material to a cell by infection with a recombinant retrovirus.

Vector: A plasmid or viral DNA molecule into which a DNA sequence (typically encoding a therapeutic protein) is inserted.

References

Acsadi G, Dickson G, Love DR et al., 1991. Human dystrophin expression in mdx mice after intramuscular injection of DNA constructs [see comments]. *Nature.* 352:815-8.

Acsadi G, Jiao SS, Jani A et al., 1991. Direct gene transfer and expression into rat heart *in vivo. New Biol.* 3:71-81.

Anderson WF. 1992. Human gene therapy. *Science.* 256:808-13.

Andreadis ST, Roth CM, Le Doux JM et al., 1999. Large scale processing of recombinant retroviruses for gene therapy. *Biotechnol. Prog.* In press.

Bartlett JS, Samulski RJ, and McCown TJ. 1998. Selective and rapid uptake of adeno-associated virus type 2 in brain. *Hum. Gene Ther.* 9:1181-6.

Baumgartner I, Pieczek A, Manor O et al., 1998. Constitutive expression of phVEGF165 after intramuscular gene transfer promotes collateral vessel development in patients with critical limb ischemia [see comments]. *Circulation.* 97:1114-23.

Boucher RC, and Knowles MR. 1993. Gene therapy for cystic fibrosis using E1 deleted adenovirus: a phase I trial in the nasal cavity. Office of Recombinant DNA Activity, NIH.

Budker V, Gurevich V, Hagstrom JE et al., 1996. pH-sensitive, cationic liposomes: a new synthetic virus-like vector. *Nat. Biotechnol.* 14:760-4.

Burns JC, Friedmann T, Driever W et al., 1993. Vesicular stomatitis virus G glycoprotein pseudotyped retroviral vectors: concentration to very high titer and efficient gene transfer into mammalian and nonmammalian cells [see comments]. *Proc. Natl. Acad. Sci. USA.* 90:8033-7.

Capecchi MR. 1980. High efficiency transformation by direct microinjection of DNA into cultured mammalian cells. *Cell.* 22:479-88.

Chen C, and Okayama H. 1987. High-efficiency transformation of mammalian cells by plasmid DNA. *Mol. Cell Biol.* 7:2745-52.

Crystal RG. 1995. Transfer of genes to humans: early lessons and obstacles to success. *Science.* 270:404-10.

Dai Y, Schwarz EM, Gu D et al., 1995. Cellular and humoral immune responses to adenoviral vectors containing factor IX gene: tolerization of factor IX and vector antigens allows for long-term expression. *Proc. Natl. Acad. Sci. USA.* 92:1401-5.

Danos O, and Mulligan RC. 1988. Safe and efficient generation of recombinant retroviruses with amphotropic and ecotropic host ranges. *Proc. Natl. Acad. Sci. USA.* 85:6460-4.

Doh SG, Vahlsing HL, Hartikka J et al., 1997. Spatial-temporal patterns of gene expression in mouse skeletal muscle after injection of lacZ plasmid DNA. *Gene Ther.* 4:648-63.

Donnelly JJ, Ulmer JB, Shiver JW et al., 1997. DNA vaccines. *Ann. Rev. Immunol.* 15:617-48.

Engelhardt JF, Ye X, Doranz B et al., 1994. Ablation of E2A in recombinant adenoviruses improves transgene persistence and decreases inflammatory response in mouse liver. *Proc. Natl. Acad. Sci. USA.* 91:6196-200.

Felgner PL, and Rhodes G. 1991. Gene therapeutics. *Nature.* 349:351-2.

Felgner PL, and Ringold GM. 1989. Cationic liposome-mediated transfection. *Nature.* 337:387-8.

Ferrari FK, Xiao X, McCarty D et al., 1997. New developments in the generation of Ad-free, high-titer rAAV gene therapy vectors. *Nat. Med.* 3:1295-7.

Flannery JG, Zolotukhin S, Vaquero MI et al., 1997. Efficient photoreceptor-targeted gene expression *in vivo* by recombinant adeno-associated virus. *Proc. Natl. Acad. Sci. USA.* 94:6916-21.

Flotte TR, Afione SA, Conrad C et al., 1993. Stable *in vivo* expression of the cystic fibrosis transmembrane conductance regulator with an adeno-associated virus vector. *Proc. Natl. Acad. Sci. USA.* 90:10613-7.

Friedmann T. 1989. Progress toward human gene therapy. *Science.* 244:1275-81.

Glorioso JC, Goins WF, Schmidt MC et al., 1997. Engineering herpes simplex virus vectors for human gene therapy. *Adv. Pharmacol.* 40:103-36.

Greber UF, Willetts M, Webster P et al., 1993. Stepwise dismantling of adenovirus 2 during entry into cells. *Cell.* 75:477-86.

Hartikka J, Sawdey M, Cornefert-Jensen F et al., 1996. An improved plasmid DNA expression vector for direct injection into skeletal muscle. *Hum. Gene. Ther.* 7:1205-17.

Hengge UR, Chan EF, Foster RA et al., 1995. Cytokine gene expression in epidermis with biological effects following injection of naked DNA. *Nat. Genet.* 10:161-6.

Hickman MA, Malone RW, Lehmann-Bruinsma K et al., 1994. Gene expression following direct injection of DNA into liver. *Hum. Gene. Ther.* 5:1477-83.

Hitt MM, Addison CL, and Graham FL. 1997. Human adenovirus vectors for gene transfer into mammalian cells. *Adv. Pharmacol.* 40:137-206.

Horwitz MS. 1990. Adenoviridae and their replication, p. 1679. In BN Fields, et al. (ed.), *Fields Virology,* 2nd ed, vol. 2. Raven Press, Ltd., New York.

Inoue N, and Russell DW. 1998. Packaging cells based on inducible gene amplification for the production of adeno-associated virus vectors. *J. Virol.* 72:7024-7031.

Isner JM, Pieczek A, Schainfeld R et al., 1996. Clinical evidence of angiogenesis after arterial gene transfer of phVEGF165 in patient with ischaemic limb [see comments]. *Lancet.* 348:370-4.

Isner JM, Walsh K, Symes J et al., 1996. Arterial gene transfer for therapeutic angiogenesis in patients with peripheral artery disease. *Hum. Gene. Ther.* 7:959-88.

Jiao S, Cheng L, Wolff JA et al., 1993. Particle bombardment-mediated gene transfer and expression in rat brain tissues. *Biotechnology* (NY). 11:497-502.

Kawabata K, Takakura Y, and Hashida M. 1995. The fate of plasmid DNA after intravenous injection in mice: involvement of scavenger receptors in its hepatic uptake. *Pharm. Res.* 12:825-30.

Kawakita M, Rao GS, Ritchey JK et al., 1997. Effect of canarypox virus (ALVAC)-mediated cytokine expression on murine prostate tumor growth [see comments]. *J. Natl. Cancer Inst.* 89:428-36.

Kay MA, Holterman AX, Meuse L et al., 1995. Long-term hepatic adenovirus-mediated gene expression in mice following CTLA4Ig administration. *Nat. Genet.* 11:191-7.

Kay MA, Liu D, and Hoogerbrugge PM. 1997. Gene therapy. *Proc. Natl. Acad. Sci. USA.* 94:12744-6.

Keller ET, Burkholder JK, Shi F et al., 1996. *In vivo* particle-mediated cytokine gene transfer into canine oral mucosa and epidermis. *Cancer Gene Ther.* 3:186-91.

Kotani H, Newton PB, 3rd, Zhang S et al., 1994. Improved methods of retroviral vector transduction and production for gene therapy. *Hum. Gene. Ther.* 5:19-28.

Le Doux JM, Davis HE, Morgan JR et al., 1999. Kinetics of retrovirus production and decay. *Biotech. Bioeng. Prog.* In Press.

Le Doux JM, Morgan JR, and Yarmush ML. 1998. Removal of proteoglycans increases efficiency of retroviral gene transfer. *Biotechnol. Bioeng.* 58:23-34.

Ledley FD. 1993. Hepatic gene therapy: present and future. *Hepatology.* 18:1263-73.

Legendre JY, and Szoka FC, Jr. 1992. Delivery of plasmid DNA into mammalian cell lines using pH-sensitive liposomes: comparison with cationic liposomes. *Pharm. Res.* 9:1235-42.

Levine F, and Friedmann T. 1993. Gene therapy. *Am. J. Dis. Child.* 147:1167-74.

Lew D, Parker SE, Latimer T et al., 1995. Cancer gene therapy using plasmid DNA: pharmacokinetic study of DNA following injection in mice [see comments]. *Hum. Gene. Ther.* 6:553-64.

Maeda Y, Ikeda U, Shimpo M et al., 1998. Efficient gene transfer into cardiac myocytes using adeno-associated virus (AAV) vectors [In Process Citation]. *J. Mol. Cell Cardiol.* 30:1341-8.

Magovern CJ, Mack CA, Zhang J et al., 1997. Regional angiogenesis induced in nonischemic tissue by an adenoviral vector expressing vascular endothelial growth factor. *Hum. Gene. Ther.* 8:215-27.

Mahvi DM, Burkholder JK, Turner J et al., 1996. Particle-mediated gene transfer of granulocyte-macrophage colony- stimulating factor cDNA to tumor cells: implications for a clinically relevant tumor vaccine. *Hum. Gene. Ther.* 7:1535-43.

McCarty DM, and Samulski RJ. 1997. Adeno-associated viral vectors. In M Strauss, et al. (eds.), *Concepts in Gene Therapy,* p. 61-78. Berlin, Walter de Gruyter & Co.

McGrath M, Witte O, Pincus T et al., 1978. Retrovirus purification: method that conserves envelope glycoprotein and maximizes infectivity. *J. Virol.* 25:923-7.

Melillo G, Scoccianti M, Kovesdi I et al., 1997. Gene therapy for collateral vessel development. *Cardiovasc. Res.* 35:480-9.

Miller AD. 1992. Retroviral vectors. *Curr. Top. Microbiol. Immunol.* 158:1-24.

Monahan PE, Samulski RJ, Tazelaar J et al., 1998. Direct intramuscular injection with recombinant AAV vectors results in sustained expression in a dog model of hemophilia. *Gene Ther.* 5:40-9.

Morgan JR, Tompkins RG, and Yarmush ML. 1993. Advances in recombinant retroviruses for gene delivery. *Adv. Drug Delivery Rev.* 12:143-158.

Morgan JR, Tompkins RG, and Yarmush ML. 1994. Genetic engineering and therapeutics. In RS Greco (ed.), *Implantation Biology: The Host Responses and Biomedical Devices,* p. 387. CRC Press, Boca Raton, FL.

Morgan JR, and Yarmush ML. 1998. Gene therapy in tissue engineering, In CW Patrick Jr., et al. (eds.), *Frontiers in Tissue Engineering.* p. 278-310. Elsevier Science Ltd, Oxford.

Morsy MA, Gu M, Motzel S et al., 1998. An adenoviral vector deleted for all viral coding sequences results in enhanced safety and extended expression of a leptin transgene. *Proc. Natl. Acad. Sci. USA.* 95:7866-71.

Mulligan RC. 1993. The basic science of gene therapy. *Science.* 260:926-32.

Murphy JE, Zhou S, Giese K et al., 1997. Long-term correction of obesity and diabetes in genetically obese mice by a single intramuscular injection of recombinant adeno-associated virus encoding mouse leptin. *Proc. Natl. Acad. Sci. USA.* 94:13921-6.

Muzyczka N. 1992. Use of adeno-associated virus as a general transduction vector for mammalian cells. *Curr. Top. Microbiol. Immunol.* 158:97-129.

Nabel GJ, Nabel EG, Yang ZY et al., 1993. Direct gene transfer with DNA-liposome complexes in melanoma: expression, biologic activity, and lack of toxicity in humans. *Proc. Natl. Acad. Sci. USA.* 90:11307-11.

Nakagawa I, Murakami M, Ijima K et al., 1998. Persistent and secondary adenovirus-mediated hepatic gene expression using adenovirus vector containing CTLA4IgG [In Process Citation]. *Hum. Gene. Ther.* 9:1739-45.

Naldini L, Blomer U, Gallay P et al., 1996. *In vivo* gene delivery and stable transduction of nondividing cells by a lentiviral vector [see comments]. *Science.* 272:263-7.

Neumann E, Schaefer-Ridder M, Wang Y et al., 1982. Gene transfer into mouse lyoma cells by electroporation in high electric fields. *Embo J.* 1:841-5.

Nitta Y, Tashiro F, Tokui M et al., 1998. Systemic delivery of interleukin 10 by intramuscular injection of expression plasmid DNA prevents autoimmune diabetes in nonobese diabetic mice. *Hum. Gene. Ther.* 9:1701-7.

Pagano JS, McCutchan JH, and Vaheri A. 1967. Factors influencing the enhancement of the infectivity of poliovirus ribonucleic acid by diethylaminoethyl-dextran. *J. Virol.* 1:891-7.

Parks RJ, Chen L, Anton M et al., 1996. A helper-dependent adenovirus vector system: removal of helper virus by Cre-mediated excision of the viral packaging signal. *Proc. Natl. Acad. Sci. USA.* 93:13565-70.

Paul RW, Morris D, Hess BW et al., 1993. Increased viral titer through concentration of viral harvests from retroviral packaging lines. *Hum. Gene. Ther.* 4:609-15.

Qin H, and Chatterjee SK. 1996. Cancer gene therapy using tumor cells infected with recombinant vaccinia virus expressing GM-CSF. *Hum. Gene. Ther.* 7:1853-60.

Radler JO, Koltover I, Salditt T et al., 1997. Structure of DNA-cationic liposome complexes: DNA intercalation in multilamellar membranes in distinct interhelical packing regimes [see comments]. *Science.* 275:810-4.

Rakhmilevich AL, and Yang N-S. 1997. Particle-mediated gene delivery system for cancer research. In M Strauss, et al. (eds.), *Concepts in Gene Therapy,* p. 109-120, Berlin, Walter de Gruyter.

Roe T, Reynolds TC, Yu G et al., 1993. Integration of murine leukemia virus DNA depends on mitosis. *Embo. J.* 12:2099-108.

Roemer K, and Friedmann T. 1992. Concepts and strategies for human gene therapy. *Eur. J. Biochem.* 208:211-25.

Sato Y, Roman M, Tighe H et al., 1996. Immunostimulatory DNA sequences necessary for effective intradermal gene immunization. *Science.* 273:352-4.

Scaria A, St George JA, Jiang C et al., 1998. Adenovirus-mediated persistent cystic fibrosis transmembrane conductance regulator expression in mouse airway epithelium. *J. Virol.* 72:7302-9.

Sikes ML, O. Malley BW J, Finegold MJ et al., 1994. *In vivo* gene transfer into rabbit thyroid follicular cells by direct DNA injection. *Hum. Gene. Ther.* 5:837-44.

Temin HM. 1990. Safety considerations in somatic gene therapy of human disease with retrovirus vectors. *Hum. Gene. Ther.* 1:111-23.

Tomlinson E, and Rolland AP. 1996. Controllable gene therapy pharmaceutics of non-viral gene delivery systems. *J. Contr. Rel.* 39:357-372.

Tripathy SK, Svensson EC, Black HB et al., 1996. Long-term expression of erythropoietin in the systemic circulation of mice after intramuscular injection of a plasmid DNA vector. *Proc. Natl. Acad. Sci. USA.* 93:10876-80.

Tsurumi Y, Takeshita S, Chen D et al., 1996. Direct intramuscular gene transfer of naked DNA encoding vascular endothelial growth factor augments collateral development and tissue perfusion [see comments]. *Circulation.* 94:3281-90.

Verma IM, and Somia N. 1997. Gene therapy—promises, problems and prospects [news]. *Nature.* 389:239-42.

Verma S, Woffendin C, Bahner I et al., 1998. Gene transfer into human umbilical cord blood-derived CD34+ cells by particle-mediated gene transfer [In Process Citation]. *Gene Ther.* 5:692-9.

Wang M, Bronte V, Chen PW et al., 1995. Active immunotherapy of cancer with a nonreplicating recombinant fowlpox virus encoding a model tumor-associated antigen. *J. Immunol.* 154:4685-92.

Weiss R, Teich N, Varmus H et al., 1982. Molecular biology of tumor viruses. Cold Spring Harbor Laboratory, Cold Spring Harbor.

Wells DJ. 1993. Improved gene transfer by direct plasmid injection associated with regeneration in mouse skeletal muscle. *FEBS Lett.* 332:179-82.

Welsh MJ. 1993. Adenovirus-mediated gene transfer of CFTR to the nasal epithelium and maxillary sinus of patients with cystic fibrosis. Office of Recombinant DNA Activity, NIH.

Wickham TJ, Mathias P, Cheresh DA et al., 1993. Integrins alpha v beta 3 and alpha v beta 5 promote adenovirus internalization but not virus attachment. *Cell.* 73:309-19.

Williams RS, Johnston SA, Riedy M et al., 1991. Introduction of foreign genes into tissues of living mice by DNA-coated microprojectiles. *Proc. Natl. Acad. Sci. USA.* 88:2726-30.

Wilmott RW, and Whitsett J. 1993. A phase I study of gene therapy of cystic fibrosis utilizing a replication deficient recombinant adenovirus vector to deliver the human cystic fibrosis transmembrane conductance regulator cDNA to the airways. Office of Recombinant DNA Activity, NIH.

Wold WS, and Gooding LR. 1989. Adenovirus region E3 proteins that prevent cytolysis by cytotoxic T cells and tumor necrosis factor. *Mol. Biol. Med.* 6:433-52.

Wolff JA, Malone RW, Williams P et al., 1990. Direct gene transfer into mouse muscle *in vivo. Science.* 247:1465-8.

Wu GY, Zhan P, Sze LL et al., 1994. Incorporation of adenovirus into a ligand-based DNA carrier system results in retention of original receptor specificity and enhances targeted gene expression. *J. Biol. Chem.* 269:11542-6.

Xiao W, Berta SC, Lu MM et al., 1998. Adeno-associated virus as a vector for liver-directed gene therapy [In Process Citation]. *J. Virol.* 72:10222-6.

Xiao X, Li J, and Samulski RJ. 1998. Production of high-titer recombinant adeno-associated virus vectors in the absence of helper adenovirus. *J. Virol.* 72:2224-32.

Xu Y, and Szoka FC, Jr. 1996. Mechanism of DNA release from cationic liposome/DNA complexes used in cell transfection. *Biochem.* 35:5616-23.

Yang NS, Burkholder J, Roberts B et al., 1990. *In vivo* and *in vitro* gene transfer to mammalian somatic cells by particle bombardment. *Proc. Natl. Acad. Sci. USA.* 87:9568-72.

Yang NS, Sun WH, and McCabe D. 1996. Developing particle-mediated gene-transfer technology for research into gene therapy of cancer. *Mol. Med. Today.* 2:476-81.

Yonemitsu Y, Kaneda Y, Muraishi A et al., 1997. HVJ (Sendai virus)-cationic liposomes: a novel and potentially effective liposome-mediated technique for gene transfer to the airway epithelium. *Gene Ther.* 4:631-8.

York IA, Roop C, Andrews DW et al., 1994. A cytosolic herpes simplex virus protein inhibits antigen presentation to CD8+ T lymphocytes. *Cell.* 77:525-35.

Yoshimura K, Rosenfeld MA, Seth P et al., 1993. Adenovirus-mediated augmentation of cell transfection with unmodified plasmid vectors. *J. Biol. Chem.* 268:2300-3.

Zabner J, Fasbender AJ, Moninger T et al., 1995. Cellular and molecular barriers to gene transfer by a cationic lipid. *J. Biol. Chem.* 270:18997-9007.

Zelenin AV, Alimov AA, Zelenina IA et al., 1993. Transfer of foreign DNA into the cells of developing mouse embryos by microprojectile bombardment. *FEBS Lett.* 315:29-32.

Zhu N, Liggitt D, Liu Y et al., 1993. Systemic gene expression after intravenous DNA delivery into adult mice. *Science.* 261:209-11.

Zsengeller ZK, Wert SE, Hull WM et al., 1995. Persistence of replication-deficient adenovirus-mediated gene transfer in lungs of immune-deficient (nu/nu) mice. *Hum. Gene. Ther.* 6:457-67.

107

Cell Engineering

Douglas A. Lauffenburger
Massachusetts Institute of Technology

Cell engineering can be defined either academically, as "application of the principles and methods of engineering to the problems of cell and molecular biology of both a basic and applied nature" [Nerem, 1991], or functionally, as "manipulation of cell function by molecular approaches" [Lauffenburger and Aebischer, 1993]. However, defined, there is no question that cell engineering is becoming a central area of biomedical engineering. Applications to health care technology include design of molecular therapies for wound healing, cancer, and inflammatory disease, development of biomaterials and devices for tissue regeneration and reconstruction, introduction of gene therapies to remedy a variety of disorders, and utilization of mammalian cell culture technology for production of therapeutic proteins.

Cell engineering has its roots in revolutions that have occurred in the field of molecular cell biology over the past few decades (see Darnell et al., [1986, Chap. 1] for an excellent historical summary). Key advances upon which this field was built include electron microscopy in the 1940s, permitting intracellular structural features to be seen; the identification of DNA as the biochemical basis of genetics, also in the 1940s; the elucidation of the structure and function of DNA in the 1950s; the discovery of gene regulation in the 1960s; and progress in areas such as molecular genetics, light microscopy, and protein biochemistry in the 1970s, 1980s, and 1990s. Combining the powerful set of information and techniques from molecular cell biology with the analytical and synthetic approaches of engineering for purposeful exploitation and manipulation of cell function has given rise to the wide range of applications listed above. Engineering contributions include elucidation of kinetic, transport, and mechanical effects; identification and measurement of system parameters; and production of novel materials and devices required for control of cell and tissue functions by molecular mechanisms. All these aspects of engineering are essential to development of reliable technological systems requiring reproducible performance of cells and tissues.

The central paradigm of cell engineering is employment of an engineering perspective of understanding system function in terms of underlying component properties, with the cells and tissues being the system and the molecular mechanisms being the components. Thus cell and tissue functions must be quantified and related to molecular properties that serve as the key design parameters. This paradigm is made especially powerful by the capability derived from molecular cell biology to alter molecular properties intentionally, permitting rational system design based on the identified parameters.

107.1 Basic Principles

The fundamental aspects of cell engineering are twofold: (1) quantitative understanding of cell function in molecular terms and (2) ability to manipulate cell function through molecular mechanisms, whether

0-8493-0462-8/00/$0.00+$.50

by pharmacologic or genetic intervention or by introduction of materials or devices. Elucidation of basic technical principles that will stand a relatively long-term test of time is difficult because the field of molecular cell biology continues to experience incredibly rapid and unpredictable advances. However, an attempt can be made to at least outline central principles along the two fundamental lines cited above. First, one can consider the basic categories of cell function that may be encountered in cell engineering applications. These include proliferation, adhesion, migration, uptake and secretion, and differentiation. Second, one can consider basic categories of regulation of these functions. Two major loci of regulation are gene expression and structural/enzymatic protein activity; these processes act over relatively long-term (hours) and short-term (seconds to minutes) time spans, respectively. Thus a cell can change the identities and quantities of the molecular components it makes as well as their functional productivity. A third locus of regulation, which essentially encompasses the others, is receptor-mediated signaling. Molecular ligands interacting with cell receptors generate intracellular chemical and mechanical signals that govern both gene expression and protein activity and thus essentially control cell function.

At the present time, bioengineering efforts to manipulate cell behavior in health care technologies are primarily centered on development of materials, devices, and molecular or cellular therapies based on receptor-ligand interactions. These interactions are amenable to engineering approaches when relevant biochemical and biophysical properties of receptors and their ligands are identified [Lauffenburger and Linderman, 1993]. Examples of such efforts and the basic principles known at the present time to be involved are described below in three categories of current major activity: cell proliferation, cell adhesion, and cell migration.

107.2 Cell Proliferation

Proliferation of mammalian tissue and blood cells is regulated by signaling of growth factor receptors following ligand binding. In some cases, the resulting regulation is that of an on/off switch, in which the fraction of cells in a population that are stimulated to move into a proliferative state is related to the level of receptor occupancy; an example of this is the control of blood cell proliferation in the bone marrow [Kelley et al., 1993]. In other cases, the rate of progression through the cell cycle, governed by the rate of movement into the DNA synthesis phase, is related to the level of receptor occupancy. For two examples at least, epidermal growth factor stimulation of fibroblast proliferation [Knauer et al., 1984] and interleukin 2 stimulation of T-lymphocyte proliferation [Robb, 1982], DNA synthesis is linearly proportional to the number of growth factor-receptor complexes, perhaps with a minimum threshold level.

In engineering terms, two key principles emerge from this relationship. One is that the details of the biochemical signal transduction cascade are reflected in the coefficient relating DNA synthesis to the number of complexes, which represents an intrinsic mitogenic sensitivity. Thus, for a specified number of complexes, the cell proliferation rate can be altered by modifying aspects of the signal transduction cascade. The second principle is that mechanisms governing the number, and perhaps location, of growth factor-receptor complexes can influence cell proliferation with just as great an effect as signal transduction mechanisms. At the present time, it is more straightforward for cell engineering purposes to manipulate cell proliferation rates by controlling the number of signaling complexes using a variety of methods. Attempts to quantitatively analyze and manipulate components of receptor signal transduction cascades are also beginning [Mahama and Linderman, 1994; Renner et al., 1994].

Engineering methods helpful for providing desired growth factor concentrations include design of controlled-release devices [Powell et al., 1990] and biomaterials [Cima, 1994]. Endogenous alteration of ligand concentration and receptor number can be accomplished using gene transfer methods to introduce DNA that can increase ligand production or receptor expression or antisense therapy to introduce RNA that can decrease ligand production or reduce receptor expression. Levels of growth factor-receptor binding can be tuned by rational design of ligands; these might be antagonists that block binding of the normal growth factor and hence decrease complex levels or super-agonist growth factor mimics that provide for increased levels of complex formation over that of the normal growth factor. Rates or extents of receptor and ligand trafficking processes can be altered by modification of the receptor, ligand, or accessory components. Such alterations can affect the number of signaling complexes by increasing or decreasing the amount of receptor down-regulation, ligand depletion, or compartmentation of complexes from the cell surface to intracellular locations.

Regulation of fibroblast proliferation by epidermal growth factor (EGF) offers an instructive example of some of these approaches. Under normal circumstances, fibroblasts internalize EGF-receptor complexes efficiently via ligand-induced endocytosis, leading to substantial down-regulation of the EGF-receptor and depletion of EGF from the extracellular medium. The latter phenomenon helps account for the requirement for periodic replenishment of serum in cell culture; although growth factors present in serum could in principle operate as catalysts generating mitogenic signals through receptor binding, they generally behave as pseudonutrients because of endocytic internalization and subsequent degradation. Use of a controlled-release device for EGF delivery thus improved wound-healing responses compared with topical administration by helping compensate for growth factor depletion due to cell degradation as well as proteolysis in the tissue and loss into the bloodstream [Buckley et al., 1985]. Transfecting the gene for EGF into fibroblasts can allow them to become a self-stimulating autocrine system not requiring exogenous supplementation [Will et al., 1995]. Altering EGF/EGF-receptor trafficking can be similarly useful. Fibroblasts possessing variant EGF receptors for which ligand-induced endocytic internalization is abrogated exhibit dramatically increased proliferation rates at low EGF concentrations, primarily due to a diminished depletion of EGF from the medium [Reddy et al., 1994]. Increasing the proportion of ligands or receptors sorted in the endosome to recycling instead of degradation also reduces ligand depletion and receptor down-regulation, thus increasing the number of growth factor-receptor complexes. An instance of this is the difference in response of cells to transforming growth factor alpha compared with EGF [French et al., 1994]; the receptor-binding affinity of TGFα is diminished at endosomal pH, permitting most of the TGFα-receptor complexes to dissociate with consequently increased receptor recycling, in contrast to EGF, which remains predominantly bound to the receptor in the endosome, leading to receptor degradation. TGFα has been found to be a more potent wound-healing agent, likely due to this alteration in trafficking [Schultz et al., 1987]. Hence ligands can be designed to optimize trafficking processes for most effective use in wound-healing, cell culture, and tissue engineering applications.

Proliferation of many tissue cell types is additionally regulated by attachment to extracellular matrices via interactions between adhesion receptors and their ligands. The quantitative dependence of proliferation on adhesion receptor-ligand interactions is not understood nearly as well as that for growth factors, though a relationship between the extent of cell spreading and DNA synthesis has been elucidated [Ingber et al., 1987] and has been applied in qualitative fashion for engineering purposes [Singhvi et al., 1994]. It is likely that a quantitative dependence of proliferation on adhesion receptor-ligand binding, or a related quantity, analogous to that found for growth factors will soon be developed and exploited in similar fashion by design of attachment substrata with appropriate composition of attachment factors. Properties of the cell environment affecting molecular transport [Yarmush et al., 1992] and mechanical stresses [Buschmann et al., 1992] are also important in controlling cell proliferation responses.

107.3 Cell Adhesion

Attachment of mammalian tissue and blood cells to other cells, extracellular matrices, and biomaterials is controlled by members of the various families of adhesion receptors, including the integrins, selectins, cadherins, and immunoglobulins. Interaction of these receptors with their ligands—which are typically accessed in an insoluble context instead of free in solution—is responsible for both physical attachment and chemical signaling events. (For clarity of discussion, the term *receptor* will refer to the molecules on "free" cells, while the term *ligand* will refer to the complementary molecules present on a more structured substratum, even if that substratum is a cell layer such as blood vessel endothelium.) It is essential to note at the outset that biologic adhesion is a dissipative phenomenon, one which in general is nonreversible; the energy of adhesion is typically less than the energy of deadhesion. Thus cell adhesion cannot be analyzed solely in terms of a colloidal adhesion framework. Cell spreading following adhesion to a substratum is an active process requiring expenditure of energy for membrane protrusion and intracellular (cytoskeletal) motile force generation rather than being a purely passive, thermodynamic phenomenon. Similarly, migration of cells over two-dimensional and through three-dimensional environments requires involvement of adhesion receptors not only for physical transmission of motile force via traction to the substratum but also perhaps for active signaling leading to generation of the cytoskeletal motile force.

An engineering approach can be applied to design of cells, ligands, and ligand-bearing biomaterials to optimize attachment and subsequent signaling. For instance, material bearing synthetic peptides to which only a specific, desired cell type will adhere can be developed; an example is the immobilization of a unique amino acid sequence on a vascular-graft polymer to promote attachment and spreading of endothelial cells but not fibroblasts, smooth muscle cells, or platelets [Hubbell et al., 1991]. Moreover, the substratum density of a particular adhesive ligand required for proper cell spreading and leading to desired behavior can be quantitatively determined [Massia and Hubbell, 1991]. Design of substratum matrices with appropriate ligands for attachment, spreading, and proliferation in wound-healing and tissue-generation applications can be based on such information. These matrices also must possess suitable biodegradation properties along with appropriate mechanical properties for coordination with forces generated by the cells themselves as well as external loads [Cima et al., 1991b, Tranquillo et al., 1992].

Analyses of cell-substratum adhesion phenomena have elucidated key design parameters of receptor-ligand interactions as well as cell and substratum mechanics. An important principle is that the strength of adhesion is proportional to the receptor-ligand equilibrium binding affinity in a logarithmic manner [Kuo and Lauffenburger, 1993], so to alter adhesion strength by an order of magnitude requires a change in binding affinity by a factor of roughly 10^3. Cells possess the capability to accomplish this via covalent modification of proteins, leading to changes in effective affinity (i.e., avidity) either through direct effects on receptor-ligand binding or through indirect effects in which the state of receptor aggregation is modified. Aggregation of adhesion receptors into organized structures termed *focal contacts* leads to dramatic increases in adhesion strength because distractive forces are distributed over many receptor-ligand bonds together, as demonstrated experimentally [Lotz et al., 1989] and explained theoretically [Ward and Hammer, 1993]. Hence biochemical processes can lead to changes in biophysical properties. The fact that physical forces can affect distribution of receptor-ligand bonds and probably other protein-protein interactions suggests that biophysical processes can similarly alter cellular biochemistry [Ingber, 1993].

Cells are able to regulate the mode of detachment from a substratum, with a variety of alternative sites at which the linkage can be disrupted. Between the intracellular cytoskeleton and the extracellular matrix are generally a number of intermolecular connections involving the adhesion receptor. The receptor can be simply associated with membrane lipids or can be anchored by linkages to one or more types of intracellular components, some of which are also linked to actin filaments. Thus cell-substratum detachment may take place by simple ligand-receptor dissociation or extraction of the intact ligand-receptor bond from the membrane or extraction of the bond along with some intracellularly linked components. Biochemical regulation by the cell itself, likely involving covalent modification of the receptor and associated components, may serve as a means to control the mode of detachment when that is an important issue for engineering purposes. It is likely that the different superfamilies of adhesion receptors possess distinct characteristics of their micromechanical and biochemical regulation properties leading to different responses to imposed stresses generated either intracellularly or extracellularly.

A different situation involving the cell adhesion is that of cell capture by a surface from a flowing fluid. Instances of this include homing of white blood cells to appropriate tissues in the inflammatory and immune responses and metastatic spread of tumor cells, both by adhesion to microvascular endothelial cells, as well as isolation of particular cell subpopulations by differential cell adhesiveness to ligand-coated materials to facilitate cell-specific therapies. In such cases, it may be desired to express appropriate attachment ligands at required densities on either the target cells or capturing surfaces or to use soluble receptor-binding competitors to block attachment in the circulation. This is a complicated problem involving application of quantitative engineering analysis of receptor-ligand binding kinetics as well as cell and molecular mechanics. Indeed, properties of adhesion receptors crucial to facilitating attachment from flowing fluid are still uncertain; they appear to include the bond-formation rate constant but also bond mechanical parameters such as strength (i.e., ability to resist dissociation or distraction from the cell membrane) and compliance (i.e., ability to transfer stress into bond-dissociating energy).

Cell-substratum attachment during encounter in fluid flow conditions can be considered under equilibrium control if the strength or affinity of bonds is the controlling parameter, whereas it is under kinetic control when the rate of bond formation is controlling [Hammer and Lauffenburger, 1987]. Most physiologic situations of cell capture from the flowing bloodstream by blood vessel endothelium appear to be under kinetic control. That is, the crucial factor is the rate at which the first receptor-ligand bond can be formed; apparently the bond strength is sufficiently great to hold the cell at the vessel wall despite the distracting force due to fluid flow. Rolling of neutrophils along the endothelium is mediated by members of the selectin family, slowing the cells down for more stable adhesion and extravasation mediated by members of the integrin family. It is not yet clear what special property of the selectins accounts for their ability to yield rolling behavior; one candidate is a combination of both fast association and dissociation rate constants [Lawrence and Springer, 1991], whereas another is a low compliance permitting the dissociation rate constant to be relatively unaffected by stress [Hammer and Apte, 1992]. It will be important to determine what the key properties are for proper design of capture-enhancing or -inhibiting materials or regimens.

107.4 Cell Migration

Movement of cells across or through two- and three-dimensional substrata, respectively, is governed by receptor-ligand interactions in two distinct ways. First, active migration arising from intracellular motile forces requires that these forces can be effectively transmitted to the substratum as differential traction. Adhesion receptors, mainly of the integrin superfamily, are involved in this process. Integrin interactions with their extracellular matrix ligands can influence the linear cell translocation speed. Second, the intracellular motile forces must be stimulated under the specific conditions for which movement is desired. Often, the ligands that stimulate cytoskeletally generated motile forces can do so in a spatially dependent fashion; i.e., cells can be induced to extend membrane lamellipodia preferentially in the direction of higher ligand concentrations, leading to net movement up a ligand concentration gradient. These ligands are termed *chemotactic attractants*; many such attractants can additionally modulate movement speed and are thus also termed *chemokinetic agents*. Chemotactic and chemokinetic ligands include some growth factors as well as other factors seemingly more dedicated to effects on migration.

Two different types of intracellular motile forces are involved, one that generates membrane lamellipod extension and one that generates cell body contraction; it is the latter force that must be transmitted to the substratum as traction in order for locomotion to proceed. Indeed, it must be transmitted in a spatially asymmetrical manner so that attachments formed at the cell front can remain adhered, while attachments at the cell rear are disrupted. Since the contraction force is likely to be isotropic in nature, this asymmetry must arise at the site of the force-transmission linkage. Migration speed of a particular cell type on a given ligand-coated surface thus depends on (1) generation of intracellular motile forces leading to lamellipod extension and to subsequent cell body contraction, stimulated by chemokinetic and/or chemotactic ligand binding to corresponding receptors or by adhesion receptor binding to appropriate soluble or substratum-bound ligands, (2) an appropriate balance between intracellular contractile force and overall level of cell-substratum traction, and (3) a significant front versus rear difference in cell-substratum traction. Speed should be roughly proportional to the first and last quantities but dependent on the middle quantity in biphasic fashion—maximal at an intermediate level of motile force-to-traction-force ratio [DiMilla et al., 1991].

Since many adhesion ligands stimulate intracellular motile force generation, an important parameter for design purposes is thus the effective density at which they are present on the substratum. Migration should be maximized at an intermediate density at which the cell contraction force is roughly in balance with the cell-substratum adhesiveness, consistent with observations for both two-dimensional [DiMilla et al., 1993] and three-dimensional [Parkhurst and Saltzman, 1992] movement environments. Since adhesive force is related to the ligand-receptor binding affinity, use of ligands with different affinities provides a means for migration under conditions of different ligand densities. Recalling also that the degree of receptor organization into aggregates can dramatically increase cell-substratum adhesiveness, use of ligands yielding different degrees of receptor aggregation can similarly influence migration speed. Engineering attempts to develop useful substrata for cell migration, e.g., for wound-healing scaffolding or tissue-regeneration biomaterials also need to consider possible effects of microarchitecture, i.e., the physical geometry of three-dimensional matrices through which the cells may crawl and ligands or nutrients may diffuse [Cima and Langer, 1993].

Addition of soluble receptor-binding ligand competitors into the extracellular medium is another approach to affecting cell migration speed by means of altered cell-substratum adhesiveness. As the concentration of a competitor is increased, adhesiveness should decrease. However, since migration speed varies with adhesiveness in biphasic manner, the result of adding the competitor can be either a decrease or an increase in motility, depending on whether the adhesiveness in the absence of the competitor is below or above optimal [Wu et al., 1994]. For example, the rate of blood capillary formation, which is related to the speed of microvessel endothelial cell migration, can be either diminished or enhanced by addition of soluble integrin-binding competitors [Nicosia and Bonnano, 1991; Gamble et al., 1993].

Finally, migration speed may be altered genetically, e.g., by changing the expression level of an adhesion receptor [Schreiner et al., 1989]. An increase in the number of receptors may lead to either an increase or a decrease in motility, depending on whether the change in adhesiveness is helpful or detrimental. At the same time, signaling for motile force generation, if it occurs through the same receptor, will provide an accompanying positive influence. Receptors may be altered in their ligand-binding or cytoskeleton-coupling capabilities as well, providing alternative approaches for affecting force transmission.

Attempts to manipulate the direction of cell migration fall into three categories: soluble chemotactic ligands, immobilized haptotactic ligands, and contact guidance. *Haptotaxis* is commonly referred to as directed migration on a density gradient of an immobilized ligand, regardless of whether it is due to differential adhesiveness or to spatially dependent membrane extension. If the latter effect is causative, the phenomenon should be more precisely termed *chemotaxis* but merely with an immobilized ligand; the classic definition of haptotaxis requires an adhesion effect. Examples in which efforts have been made to manipulate the direction of cell migration including homing of cytotoxic lymphocytes to tumor sites, neovascularization of healing wound tissue, and neuron outgrowth.

Contact guidance can be produced using either patterned substrata, in which pathways for cell migration are delineated by boundaries of adhesiveness [Kleinfeld et al., 1988], or by a alignment of matrix fibers for directional orientation of cell movement [Dickinson et al., 1994]. A useful technology for generation of ligand concentration gradients is controlled release of chemotactic ligands from polymer matrices [Edelman et al., 1991]. Gradients required for effective cell guidance have been found to be approximately a few percent in concentration across a cell length, and they must persist for many multiples of the cell mean path length; this is the product of the persistence time (i.e., the mean time between significant direction changes) and the linear speed. Analysis has predicted that the affinity of a chemotactic ligand or, more precisely, the dissociation rate constant may be a key parameter in governing the sensitivity of a cell to a chemotactic ligand concentration gradient [Tranquillo et al., 1988]. Data on directional responses of neutrophil leukocytes to gradients of interleukin 8 show that these cells respond more sensitively to variants of IL-8 possessing lower binding affinities [Clark-Lewis et al., 1991]. Although this finding is consistent with theory, involvement of multiple forms of the IL-8 receptor with possibly varying signaling capabilities also must be considered as an alternative cause. Work with platelet-derived growth factor as a chemotactic attractant has elucidated features of the PDGF receptor involved in sensing ligand gradients, permitting approaches for pharmacologic or genetic manipulation of chemotactic receptors to modify directional migration responses [Kundra et al., 1994].

References

Buckley A, Davidson JM, Kamerath CD, et al. 1985. Sustained release of epidermal growth factor accelerates wound repair. Proc Natl Acad Sci USA 82:7340.

Buschmann MD, Gluzband YA, Grodzinsky AJ, et al. 1992. Chondrocytes in agarose culture synthesize a mechanically functional extracellular matrix. J Orthop Res 10:745.

Cima LG. 1994. Polymer substratas for controlled biological interactions. J Coll Biochem 56:155.

Cima LG, Vacanti JP, Vacanti C, et al. 1991. Tissue engineering by cell transplantation using degradable polymer substrates. ASME Trans J Biomec Eng 113:143.

Cima LG, Langer R. 1993. Engineering human tissue. Chem Eng Progr June:46.

Clark-Lewis I, Schumacher C, Baggiolini M, Moser B. 1991. Structure-activity relationships of IL-8 determined using chemically synthesized analogs. J Biol Chem 266:23128.

Darnell J, Lodish H, Baltimore D. 1986. Molecular Cell Biology. New York, Scientific American Books.

Dickinson RB, Guido S, Tranquillo RT. 1994. Correlation of biased cell migration and cell orientation for fibroblasts exhibiting contact guidance in oriented collagen gels. Ann Biomed Eng 22:342.

DiMilla PA, Barbee K, Lauffenburger DA. 1991. Mathematical model for the effects of adhesion and mechanics on cell migration speed. Biophys J 60:15.

DiMilla PA, Stone JA, Quinn JA, et al. 1993. An optimal adhesiveness exists for human smooth muscle cell migration on type-IV collagen and fibronectin. J Cell Biol 122:729.

Edelman ER, Mathiowitz E, Langer R, Klagsbrun M. 1991. Controlled and modulated release of basic fibroblast growth factor. Biomaterials 12:619.

French AR, Sudlow GP, Wiley HS, Lauffenburger DA. 1993. Postendocytic trafficking of EGF-receptor complexes is mediated through saturable and specific endosomal interactions. J Biol Chem 269:15749.

Gamble JR, Mathias LJ, Meyer G, et al. 1993. Regulation of in vitro capillary tube formation by anti-integrin antibodies. J Cell Biol 121:931.

Hammer DA, Apte SA. 1992. Simulation of cell rolling and adhesion on surfaces in shear flow: General results and analysis of selectin-mediated neutrophil adhesion. Biophys J 63:35.

Hammer DA, Lauffenburger DA. 1987. A dynamical model for receptor-mediated cell adhesion to surfaces. Biophys J 52:475.

Hubbell JA, Massia SP, Desai NP, Drumheller PD. 1991. Endothelial cell-selective materials for tissue engineering in the vascular graft via a new receptor. Biotechnology 9:568.

Ingber DE. 1993. The riddle of morphogenesis: a question of solution chemistry or molecular cell engineering? Cell 75:1249.

Ingber DE, Madri JA, Folkman J. 1987. Endothelial growth factors and extracellular matrix regulate DNA synthesis through modulation of cell and nuclear expansion. In Vitro Cell Dev Biol 23:387.

Kelly JJ, Koury MJ, Bondurant MC, et al. 1993. Survival or death of individual proerythroblasts results from differing erythropoietin sensitivities: A mechanism for controlled rate of erythrocyte production. Blood 82:2340.

Kleinfeld D, Kahler KH, Hockberger PE. 1988. Controlled outgrowth of dissociated neurons on patterned substrates. J Neurosci 8:4098.

Knauer DJ, Wiley HS, Cunningham DD. 1984. Relationship between epidermal growth factor receptor occupancy and mitogenic response: Quantitative analysis using a steady-state model system. J Biol Chem 259:5623.

Kundra V, Escobedo JA, Kazlauskas A, et al. 1994. Regulation of chemotaxis by the PDGF-receptor β. Nature 367:474.

Kuo SC, Lauffenburger DA. 1993. Relationship between receptor/ligand binding affinity and adhesion strength. Biophys J 65:2191.

Lauffenburger DA, Aebischer P. 1993. Cell and tissue engineering: overview. In Research Opportunities in Biomolecular Engineering: The Interface Between Chemical Engineering and Biology. US Dept of Health and Human Services Administrative Document, pp 109–113.

Lauffenburger DA, Lindermann JJ. 1993. Receptors: Models for Binding, Trafficking, and Signaling. New York, Oxford University Press.

Lotz MM, Burdsal CA, Erickson HP, McClay DR. 1989. Cell adhesion to fibronectin and tenascin: Quantitative measurements of initial binding and subsequent strengthening response. J Cell Biol 109:1795.

Mahama P, Lindermann JJ. 1994. Monte Carlo study on the dynamics of G-protein activation. Biophys J 67:1345.

Massia SP, Hubbell JA. 1991. An RGD spacing of 440 nm is sufficient for integrin $\alpha_v\beta_3$-mediated fibroblast spreading and 140 nm for focal contact and stress fiber formation. J Cell Biol 114:1089.

Nerem RM. 1991. Cellular engineering. Ann Biomed Eng 19:529.

Nicosia RF, Bonanno E. 1991. Inhibition of angiogenesis in vitro by RGD-containing synthetic peptide. Am J Pathol 138:829.

Parkhurst M. Saltzman WM. 1992. Quantification of human neutrophil motility in three-dimensional collagen gels: Effect of collagen concentrations. Biophys J 61:306.

Powell EM, Sobarzo MR, Saltzman WM. 1990. Controlled release of nerve growth factor from a polymeric implant. Brain Res 515:309.

Reddy CC, Wells A, Lauffenburger DA. 1993. Proliferative response of fibroblasts expressing internalization-deficient EGF receptors is altered via differential EGF depletion effects. Biotech Progr 10:377.

Renner WA, Hatzimanikatis V, Eppenberger HM, Bailey JE. 1993. Recombinant cyclin E overexpression enables proliferation of CHO K1 cells in serum- and protein-free medium. Preprint from Institute for Biotechnology, ETH-Hoenggerberg, Zurich, Switzerland.

Robb RJ. 1982. Human T-cell growth factor: Purification, biochemical characterization, and interaction with a cellular receptor. Immunobiology 161:21.

Schreiner CL, Bauer JS, Danilov YN, et al. 1989. Isolation and characterization of chinese hamster ovary cell variants deficient in the expression of fibronectin receptor. J Cell Biol 109:3157.

Schultz GS, White M, Mitchell R, et al. 1987. Epithelial wound healing enhanced by TGFα and VGF. Science 235:350.

Singhvi R, Kumar A, Lopez GP, et al. 1994. Engineering cell shape and function. Science 264:696.

Tranquillo RT, Lauffenburger DA, Zigmond SH. 1988. A stochastic model for leukocyte random motility and chemotaxis based on receptor binding fluctuations. J Cell Biol 106:303.

Tranquillo RT, Durrani MA, Moon AG. 1992. Tissue engineering science: Consequences of cell traction force. Cytotechnology 10:225.

Ward MD, Hammer DA. 1993. A theoretical analysis for the effect of focal contact formation on cell-substrate attachment strength. Biophys J 64:936.

108

Metabolic Engineering

Craig Zupke
Massachusetts General Hospital and the Shriners Burns Institute

Metabolic engineering can be defined as the modification of cellular metabolism to achieve a specific goal. Metabolic engineering can be applied to both prokaryotic and eukaryotic cells, although applications to prokaryotic organisms are much more common. The goals of metabolic engineering include improved production of chemicals endogenous to the host cell, alterations of the substrate required for growth and product synthesis, synthesis of new products foreign to the host cell, and addition of new detoxification activities. As metabolic engineering has evolved, some general principles and techniques have emerged. In particular, developments in genetic engineering enable highly specific modification of cellular biochemical pathways. However, the complexity of biologic systems and their control severely limits the ability to implement a rational program of metabolic engineering. Early attempts at metabolic engineering were based on a simplistic analysis of cellular biochemistry and did not live up to expectations [Stephanopoulos and Sinskey, 1993]. An important realization is that metabolism must be considered as a network of interrelated biochemical reactions. The concept of a single rate-limiting step usually does not apply—instead, the control of a metabolic pathway is frequently distributed over several steps. The choice of specific modifications to achieve a desired goal is difficult because of this distributed nature of control. These issues are especially relevant in medical applications of metabolic engineering, where there is the added complication of interacting organ systems.

The field of metabolic engineering is still young, and advancements are continuing, especially in the analysis of metabolic networks and their control. Recent advances include techniques for the experimental determination of kinetic or control parameters necessary to apply established mathematical formalisms to metabolic networks. Coupled with computer optimization and simulation, these techniques should lead to a greater success rate for the directed modification of cellular metabolism. In addition, the knowledge and experience gained from engineering cellular metabolism can be applied to individual organs or to the whole body to help guide the development and evaluation of new drugs and therapies.

0-8493-0462-8/00/$0.00+$.50

108.1 Basic Principles

Metabolic engineering is inherently multidisciplinary, combining knowledge and techniques from cellular and molecular biology, biochemistry, chemical engineering, mathematics, and computer science. The basic cycle of metabolic engineering is summarized in the flow sheet in Fig. 108.1. The first step in the process is to define the problem or goal to which metabolic engineering will be applied. The next step is to analyze the relevant biochemistry to determine the specific modifications that will be attempted. Recombinant DNA techniques are then used to implement the modifications, and the analysis of their impact completes one pass through the metabolic engineering cycle. This process is continued in an iterative fashion until the desired result is achieved. In the discussion that follows, each step of the cycle will be addressed individually.

108.2 Problem Definition

As mentioned above, the typical goals of metabolic engineering include improving the production of endogenous chemicals, altering the organism substrate requirements, synthesizing new products, and adding new detoxification activities. There are a variety of ways that these goals can be achieved through genetic engineering and mutation and selection. Through genetic engineering, heterologous enzymes can be added to a cell, either singly or in groups. The addition of new enzymatic activities can be used to meet any of the metabolic engineering goals. Mutation and selection can be used to affect the regulation of particular enzymes and thus are most useful for altering metabolite flow.

Addition of New Activities

The distinction between product formation, substrate utilization, and detoxification is simply a matter of emphasizing different aspects of a metabolic pathway. In product formation, the end of a metabolic pathway is the focal point. In substrate utilization and detoxification, it is the input to the pathway that is important. With detoxification, there is the additional requirement that the toxin is converted to less harmful compounds, but what they are is not important. In the discussion that follows, the addition of new activity for the synthesis of new products will be emphasized. However, since they all involve adding new activity to a metabolic pathway, the same techniques used for the synthesis of new products also can be used to alter substrate requirements or add new detoxification activity.

There are several possible strategies for the synthesis of new products. The choice of strategy depends on the biochemistry present in the host organism and the specific product to be synthesized. In many instances, pathway completion can be used, which only involves the addition of a few enzymes that catalyze the steps missing from the existing metabolic pathway. Entire biochemical pathways also can be transferred to a heterologous host. The motivation for this type of transfer is that the host organism may be more robust, have a more desirable substrate requirement, or have inherently higher productivities than the organism that naturally possesses the biochemical pathway of interest. The transfer of multistep pathways to heterologous hosts has been used frequently for the production of antibiotics because of the clustering of the genes involved [Malpartida, 1990]. When the genes are clustered, they can be cloned and inserted into the host organism as a single unit.

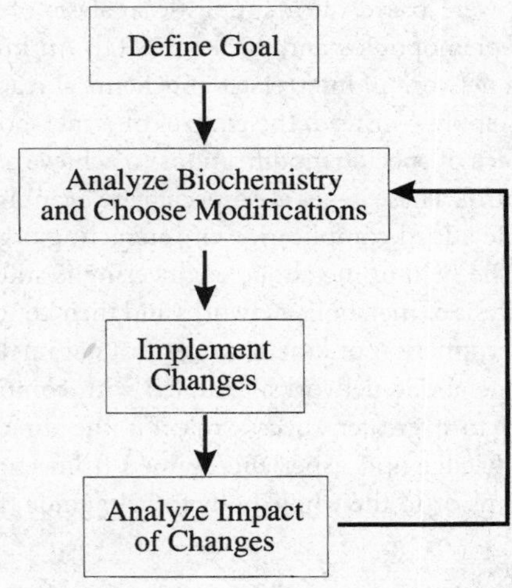

FIGURE 108.1 The metabolic engineering cycle.

The procedures described above involve the transfer of a known metabolic activity from one organism to another. It is also possible to create completely new products, because many enzymes can function with multiple substrates. If an enzyme is introduced into a cell and is thus presented with a metabolite similar to its normal substrate but which was not present in the donor cell, a completely new reaction may be catalyzed. Similarly, if a foreign enzymatic activity produces an intermediate foreign to the host cell, that intermediate may undergo further reactions by endogenous enzymes and produce a novel product. This phenomenon has been used to synthesize novel antibiotics in recombinant *Streptomyces* [McAlpine, 1987; Epp, 1989] and to engineer an *Escherichia coli* for the degradation of trichlorethylene [Winter, 1989]. Of course, the production of new products also may be an undesirable side effect of the introduction of a heterologous enzyme and could interfere with the production of the desired product. The complexity of enzyme activities makes the prediction of negative side effects extremely difficult.

Improving Existing Metabolism

Although a cell or organism may possess a particular metabolic pathway, the flow of material through that pathway (metabolic flux) may be suboptimal. Both recombinant DNA techniques and classic mutation-selection techniques can help to redirect metabolic fluxes. The principal ways to redirect metabolic fluxes are by increasing or decreasing the activity of specific enzymes and by changing their regulation. The choice of which enzymes and how to alter them can be very complicated because of the distributed nature of metabolic control. However, some basic concepts can be illustrated by the analysis of an idealized metabolic network for threonine synthesis, as shown in Fig. 108.2. Aspartate is the biosynthetic precursor for threonine, as well as for lysine, methionine, and isoleucine. If our goal is to maximize the production of threonine, then we may want to decrease the activity of the enzymes that catalyze the production of the other amino acids derived from aspartate. This would result in less of the carbon flow from aspartate being drawn off by the unwanted amino acids, and thus a larger flux to threonine might be achieved. An alternative way of increasing the flux to threonine is to amplify the activity of the enzymes along the pathway from aspartate to threonine. Finally, accumulation of the desired product, threonine, will lead to the reduction of flux from aspartate to ASA and from ASA to homoserine because of feedback inhibition. This would result in a reduction in threonine synthesis as well. If the feedback inhibition were removed, threonine could then accumulate without affecting its rate of synthesis. In practice, multiple modifications would probably be required to achieve the desired goal.

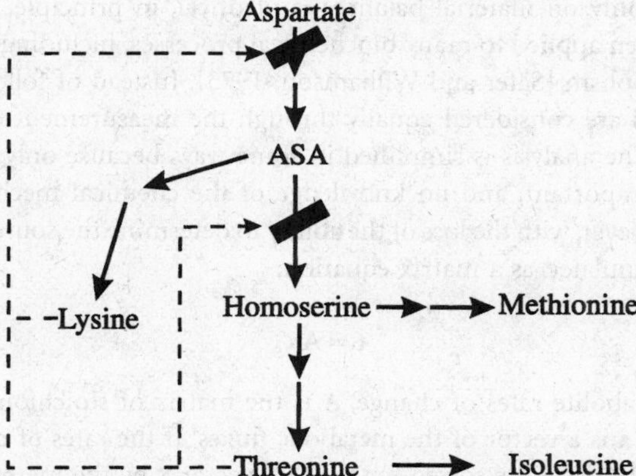

FIGURE 108.2 A simple metabolic network for threonine synthesis. The solid arrows represent enzyme catalyzed biochemical reactions. The dotted arrows indicate allosteric inhibition (ASA = aspartate semialdehyde).

108.3 Analysis of Metabolic Networks

Metabolic Flux Analysis

The measurement or estimation of metabolic fluxes is useful for the evaluation of metabolic networks and their control. Direct measure of intracellular fluxes is possible with in vivo nuclear magnetic resonance (NMR) spectroscopy. However, the inherent insensitivity of NMR limits its applicability. A more general strategy combines material balances with carefully chosen measurements to indirectly estimate metabolic fluxes. In isotopic tracer methods, a material balance is performed on the labeled isotope, and the measured isotope distribution is used to estimate the fluxes of interest. Alternatively, a total mass or carbon balance can be combined with extensive measurements of the rates of change of extracellular metabolic concentrations to calculate intracellular fluxes. Used individually or together, these methods can provide a great deal of information about the state of a biochemical system.

In isotope tracer methods, the cells to be studied are provided with a substrate specifically labeled with a detectable isotope (usually ^{14}C or ^{13}C). The incorporation of label into cellular material and by-products is governed by the fluxes through the biochemical pathways. The quantity and distribution of label are measured and combined with knowledge of the biochemistry to determine the intracellular fluxes. The choices of substrate labeling patterns, as well as which by-products to measure, are guided by careful analysis of the assumed biochemical network. These experiments are usually performed at isotopic steady state so that the flow of isotope into each atom of a metabolite equals the flux out. For the nth atom of the kth metabolite, the flux balance is [Blum, 1982]:

$$\sum_i V_{i,k} S_i(m) = S_k(n) \sum_o V_{o,k} \tag{108.1}$$

where $S_i(m)$ is the specific activity of the mth atom of the ith metabolite that contributes to the labeling of the nth atom of the kth metabolite, $S_k(n)$, and $V_{i,k}$ and $V_{o,k}$ are the input and output fluxes of the kth metabolite, respectively. The system of equations represented by Eq. (108.1) can be solved for the fluxes ($V_{i,k}$, $V_{o,k}$) from measurements of the specific activities. A similar analysis can be applied to the isotope isomers, or isotopomers, yielding more information from a single experiment. Isotopomer analysis is not just concerned with the average enrichment of individual atoms. Instead, it involves quantifying the amounts of the different isotopomers that occur at a specific metabolic state. The isotopomer distribution contains more information than that obtained from positional enrichments, making it generally a more powerful technique. Both mass spectrometry and NMR can be used to analyze isotopomer distributions.

Flux estimation based only on material balances is identical, in principle, to flux estimation using isotopic tracers and has been applied to many biochemical processes, including lysine synthesis [Vallino, 1989] and rat heart metabolism [Safer and Williamson, 1973]. Instead of following the fate of specific atoms (i.e., ^{13}C or ^{14}C), all are considered equally through the measurement of the rates of change of substrates and products. The analysis is simplified in some ways because only the stoichiometry of the biochemical reactions is important, and no knowledge of the chemical mechanisms is needed. Some resolving power is lost, however, with the loss of the ability to determine the source of individual reactions. The analysis is usually formulated as a matrix equation:

$$\mathbf{r} = \mathbf{A}\mathbf{x} \tag{108.2}$$

where $\mathbf{r}$ is a vector of metabolite rates of change, $\mathbf{A}$ is the matrix of stoichiometric coefficients for the biochemical network, and $\mathbf{x}$ is a vector of the metabolic fluxes. If the rates of change of the metabolites are measured, then Eq. (108.2) can be solved for the flux vector $\mathbf{x}$ by the method of least squares:

$$\mathbf{x} = \left(\mathbf{A}^T\mathbf{A}\right)^{-1}\mathbf{A}^T\mathbf{r} \tag{108.3}$$

Typically, extracellular metabolites are measured, while intracellular metabolites are assumed to be at pseudo-steady state and their rates of change are taken to be zero.

Metabolic Control Analysis

A *biochemical network* is a system of enzyme-catalyzed reactions that are interconnected through shared metabolites. At steady state, the fluxes through each pathway will be a function of the individual enzyme kinetic properties, as well as the network architecture. The activity of a particular enzyme will affect the concentration of its reactants and products and thus will influence the flux through pathways upstream and downstream. Metabolic control analysis (MCA) grew from the work originally presented by Kacser and Burns [1973] and Heinrich and Rapoport [1977]. MCA provides a framework for analyzing and quantifying the distributed control that enzymes exert in biochemical networks.

In the discussion of that follows, a metabolic network consists of enzymes *e*, metabolites *X*, substrates *S*, and products *P*. For simplicity, we assume that the reaction rate v_i is proportional to the enzyme concentration e_i. If this is not true, then some modifications of the analysis are required [Liao and Delgado, 1993]. The concentrations of the products and substrates are fixed, but the metabolite concentrations are free to change in order to achieve a steady-state flux *J*. An example of a simple, unbranched network is

$$S \xrightarrow{e_1} X_1 \xrightarrow{e_2} X_2 \xrightarrow{e_3} X_3 \xrightarrow{e_4} P$$

MCA is essentially a sensitivity analysis that determines how perturbations in a particular parameter (usually enzyme concentration) affect a variable (such as steady-state flux). The measures of the sensitivities are control coefficients defined as follows:

$$C_{e_i}^J \equiv \frac{e_i}{J}\frac{\partial J}{\partial e_i} = \frac{\partial \ln|J|}{\partial \ln e_i} \tag{108.4}$$

The flux control coefficient C_{ei}^j is a measure of how the flux *J* changes in response to small perturbations in the concentration or activity of enzyme *i*. The magnitude of the control coefficient is a measure of how important a particular enzyme is in the determination of the steady-state flux. The summation theorem relates the individual flux control coefficients:

$$\sum_i C_{e_i}^J = 1 \tag{108.5}$$

A large value for C_{ei}^j indicates that an increase in the activity of enzyme $i(e_i)$ should result in a large change in the metabolic flux. Thus enzymes with large flux control coefficients may be good targets for metabolic engineering. Usually, there are several enzymes in a network with comparably large *C* values, indicating that there is no single rate-limiting enzyme.

The challenge in analyzing a metabolic network is determination of the flux control coefficients. It is possible to determine them directly by "enzyme titration" combined with the measurement of the new steady-state flux. For in vivo systems, this would require alteration of enzyme expression through an inducible promoter and is not very practical for a moderately sized network. A more common method of altering enzyme activities involves titration with a specific inhibitor. However, this technique can be complicated by nonspecific effects of the added inhibitor and unknown inhibitor kinetics [Liao and Delgado, 1993].

Although the control coefficients are properties of the network, they can be related to individual enzyme kinetics through elasticity coefficients. If a metabolic concentration *X* is altered, there will be an effect on the reaction rates in which X is involved. The elasticity coefficient $\varepsilon_X^{v_i}$ is a measure of the effect changes in X have on v_i, the rate of reaction catalyzed by enzyme e_i:

$$\epsilon_X^{v_i} \equiv \frac{X}{v_i}\frac{\partial v_i}{\partial X} = \frac{\partial \ln v_i}{\partial \ln X} \tag{108.6}$$

The flux control connectivity theorem relates the flux control coefficients to the elasticity coefficients:

$$\sum_i C_{e_i}^J \epsilon_{X_k}^{v_i} = 0 \qquad \text{for all } k \tag{108.7}$$

In principle, if the enzyme kinetics and steady-state metabolite concentrations are known, then it is possible to calculate the elasticities and through Eq. (108.7) determine the flux control coefficients.

Flux Control Coefficients from Transient Metabolite Concentrations

Recent efforts to address the difficulty of determining MCA parameters include the use of transient measurements of metabolite concentrations to give good estimates of the flux control coefficients [Delgado and Liao, 1992]. The key assumption in the dynamic approach is that the reaction rates are reasonably linear around the steady state. When this is true and a transient condition is induced (by changing substrate or enzyme concentration), the transient fluxes $v_i(t)$ are constrained by the following equation:

$$\sum_{i=1}^{r} C_{e_i}^j v_i(t) = J \tag{108.8}$$

where r equals the number of reactions in the network. Measurements of transient metabolite concentration allow the determination of regression coefficients α_i from the following equation:

$$\sum_{i=1}^{n} \alpha_i \left[X_i(t) - X_i(0)\right] = 0 \tag{108.9}$$

where n equals the number of metabolites measured. The flux control coefficients are then determined from

$$\begin{bmatrix} C_{e_1}^J & C_{e_2}^J & \cdots & C_{e_r}^J \end{bmatrix} = J\begin{bmatrix} \alpha_1 & \alpha_2 & \cdots & \alpha_n \end{bmatrix}\mathbf{A} \tag{108.10}$$

where $\mathbf{A}$ is the matrix of stoichiometric coefficients. Using this method requires the measurements of transient metabolite concentrations, which are used to calculate the α_i values, and the accumulation or depletion of external metabolites to determine the steady-state flux J. The measurement of transient metabolite concentrations can be difficult but is possible with in vivo NMR or from cell extracts taken at different time points.

"Top Down" MCA

The traditional approach of MCA can be considered to be "bottom up," since all the individual enzyme flux control coefficients are determined in order to describe the control structure of a large network. The "top down" approach makes extensive use of lumping of reactions together to determine group flux control coefficients [Brown, 1990]. These can give some information about the overall control of a metabolic network without its complete characterization.

Consider a simple, multireaction pathway:

$$S \to \to \to X \to \to \to P$$

$$\underset{J_1}{\text{produces } X} \qquad \underset{J_2}{\text{consumes } X}$$

The reactions of a metabolic network are divided into two groups, those which produce a particular metabolite X and those which consume it. By manipulating the concentration X and measuring the resulting fluxes J_1 and J_2 "group" or "overall" elasticities $^*\varepsilon$ of the X producers and X consumers can be determined. Application of the connectivity theorem (Eq. 108.7) then permits the calculation of the group control coefficient for both groups of reactions. Each pathway can subsequently be divided into smaller groups centered around different metabolites, and the process repeated. The advantage of the top-down approach is that useful information about the control architecture of a metabolic network can be obtained more quickly. This approach is particularly appropriate for highly complex systems such as organ or whole-body metabolism.

Large Deviations

One of the limitations of MCA is that it only applies when the perturbations from the steady state are small. Experimentally, it is much easier to induce large changes in enzyme or metabolite concentrations, and in terms of metabolic engineering, the desired perturbations are also likely to be large. Small and Kacser [1993a, 1993b] have developed an analysis based on large deviations and related it to MCA.

In this discussion, e_i^0 is the original concentration of enzyme i, $e_i^r = re_i^0$, where r is noninfinitesimal. Thus e_i^r represents a large perturbation to the system. J^0 is the flux at the original steady state, and J^r is the flux after the large perturbation. A deviation index D is used to characterize the change from the original steady state:

$$D_{e_i}^{J^r} = \left(\frac{\Delta J}{\Delta e_i}\right)\frac{e_i^r}{J^r} \qquad (108.11)$$

where $\Delta J = J^r - J^0$ and $\Delta e_i = e_i^r - e_i^0$. By assuming that each individual enzymatic reaction rate is a linear function of the participating metabolites, it can be shown that the deviation index and control coefficients are equivalent:

$$D_{e_i^r}^{J^r} = C_{e_i}^{J_0^0} \qquad (108.12)$$

Similarly, an alternate deviation index $^*D$ can be defined as

$$^*D_{e_i^r}^{J^r} = \left(\frac{\Delta J}{\Delta e_i}\right)\frac{e_i^0}{J^0} \qquad (108.13)$$

and it is equivalent to the control coefficient at the new steady state:

$$D_{e_i^r}^{J^r} = C_{e_i^r}^{J^r} \qquad (108.14)$$

Thus, with a single large perturbation, the control coefficients at the original and new steady states can be estimated. This analysis has been extended to branched pathways, but relationships between the deviation indices and the flux control coefficients are more complicated and depend on the magnitude of the deviation (r). If the subscript a designates one branch and b another, the following relationship holds:

$$D^{J_b^r}_{e_a^r} = C^{J_b^0}_{e_a^0} \frac{1}{1-\left(C^{J_a^0}_{e_a^0} - C^{J_b^0}_{e_a^0}\right)\frac{r-1}{r}} \tag{108.15}$$

Metabolic control analysis, especially with the recent innovations for determining flux control coefficients, is a powerful tool for the analysis of metabolic networks. It can describe the control architecture of a biochemical reaction network and identify which steps are the most promising targets for efforts at metabolic engineering.

Pathway Synthesis

The diversity of biochemical reactions found in nature is quite extensive, with many enzymes being unique to a particular organism. Through metabolic engineering, it is possible to construct a metabolic network that performs a specific substrate-to-product transformation not found in nature. When exploring the possibility of synthesizing new biochemical pathways, there are several key issues that must be addressed. Given a database of possible enzymatic activities and a choice of substrate and product, one must first generate a complete set of possible biochemical reactions that can perform the desired conversion. Once they are generated, the set of possible biochemical pathways must be checked for thermodynamic feasibility and evaluated in terms of yields, cofactor requirements, and other constraints that might be present. In addition, the impact of an engineered metabolic pathway on the growth and maintenance of the host cell is also important to evaluate.

The problem of synthesizing a complete set of possible biochemical pathways subject to constraints on allowable substrates, intermediates, and by-products has been solved [Mavrovouniotis, 1990]. A computer algorithm allows the efficient determination of a complete and correct set of biochemical pathways that connect a substrate to a product. In addition, a complementary computer algorithm evaluates metabolic pathways for thermodynamic feasibility [Mavrovouniotis, 1993]. The key concept in the thermodynamic analysis is that evaluation of the feasibility of biologic reactions requires the specification of the concentrations of the products and reactants. The standard free-energy change ΔG^0 is not sufficient because physiologic conditions are significantly different from standard conditions. Both local and distributed thermodynamic bottlenecks can be determined by incorporating knowledge of the metabolite concentration ranges. This thermodynamic analysis of a biochemical pathway can pinpoint specific reactions or groups of reactions that should be modified or bypassed in order to better favor product formation.

The addition of new biochemical pathways or the modification of existing pathways is likely to affect the rest of the cellular metabolism. The new or altered pathways may compete with other reactions for intermediates or cofactors. To precisely predict the impact of a manipulation of a metabolic network is virtually impossible, since it would require a perfect model of all enzyme kinetics and of the control of gene expression. However, with relatively simple linear optimization techniques, it is possible to predict some of the behavior of a metabolic network. The procedure involves the solution of Eq. (108.2) for the metabolic fluxes for networks that have more unknowns $\mathbf{x}$ than knowns $\mathbf{r}$. An undetermined system of linear equations can be solved uniquely using linear optimization techniques if a "cellular objective function" is postulated. Examples of objective functions used in the literature include minimizing ATP or NADH production [Savinell, 1992] and maximizing growth or product formation [Varma, 1993]. Examination of the fluxes gotten from the linear optimization can indicate potential effects of a proposed metabolic change. For example, the maximum growth of *E. coli* was found to decrease in a piecewise linear fashion as leucine production increased [Varma, 1993]. In principle, linear optimization also could be applied to whole-body metabolism to evaluate the effects of metabolically active drug or genetic therapies.

108.4 Implementing Changes

The techniques available to implement specific changes in both eukaryotic and prokaryotic cells are quite powerful. Through classic mutation-selection and modern genetic engineering, it is possible to amplify or attenuate existing enzyme activity, add completely new activities, and modify the regulation of existing pathways. A detailed description of all the genetic engineering techniques available is beyond the scope of this discussion, and only a general overview will be presented.

Mutation-Selection

Mutation and selection constitute a method of manipulating the phenotype of a cell. Cells are exposed to a mutagenic environment and then placed in a selective medium in which only those cells which have a desired mutation can grow. Alternatively, cells can be screened by placing them in a medium in which it is possible to visibly detect those colonies which possess the desired phenotype. Although it is a random process, careful design of selection or screening media can result in alterations in specific enzymes. For example, growth in a medium that contains an allosteric inhibitor may be the result of the loss of the allosteric inhibition. By altering metabolite concentrations or adding substrate analogues, it is possible to select for increased or decreased activity or for changes in regulation. The major drawback of mutation and selection for metabolic engineering is the lack of specificity. There may be multiple ways that a particular phenotype can be generated, so there is no guarantee that the enzyme that was targeted was affected at all.

Recombinant DNA

Recombinant DNA techniques are very flexible and powerful tools for implementing specific metabolic changes. The cloning of genes encoding a specific enzyme is relatively routine, and the subsequent insertion and expression are also straightforward for many host cells. With the choice of a very active promoter, the activity of a cloned enzyme can be greatly amplified. In addition, insertion of a heterologous activity into a host cell can serve to alter the regulation of a metabolic pathway, assuming the foreign enzyme has different kinetics than the original. Enzyme activity also can be attenuated by antisense sequences that produce mRNA complementary to the endogenous mRNA and thus form double-stranded RNA complexes that cannot be translated. Finally, heterologous recombination can be used to completely remove a particular gene from the host genome.

108.5 Analysis of Changes

The techniques used to analyze the biochemistry at the beginning of the metabolic engineering cycle are also applicable for assessing the effects of any attempted changes. The effect of the specific change can be evaluated, as well as its impact on the activity of the whole metabolic network. It is quite possible for a desired change to be made successfully at the enzyme level but not have the desired effect on the metabolic network. If a change was partially successful, then the resulting cells can be put through another iteration of metabolic engineering in order to make further improvements. If, on the other hand, a particular change was not successful, then any information which that failure gives about the regulation and control of the metabolic network can be used to make another attempt at implementing the desired change.

108.6 Summary

Metabolic engineering is an evolving discipline that tries to take advantage of the advances which have been made in the genetic manipulation of cells. Through genetic engineering we have the capability of making profound changes in the metabolic activity of both prokaryotic or eukaryotic cells. However, the complexity of the regulation of biochemical networks makes the choice of modifications difficult. Recent advances in the analysis of metabolic networks via MCA or other techniques have provided some of the tools needed to implement a rational metabolic engineering program. Finally, metabolic engineering is best viewed as an iterative process, where attempted modifications are evaluated and the successful cell lines improved further.

Defining Terms

Flux: The flow of mass through a biochemical pathway. Frequently expressed in terms of moles of metabolite or carbon per unit time.

Heterologous enzyme: An enzyme from a foreign cell that has been expressed in a host cell. Frequently used to provide activity normally not present or to alter the control structure of a metabolic network.

Host: The cell that has had foreign DNA inserted into it.

Isotopomers: Isomers of a metabolite that contain different patterns of isotopes (for metabolic studies, the carbon isotopes ^{12}C, ^{13}C, and ^{14}C are the most commonly analyzed).

Metabolic engineering: The modification of cellular metabolism to achieve a specific goal. Usually performed with recombinant DNA techniques.

Metabolic network: A system of biochemical reactions that interact through shared substrates and allosteric effectors.

Pathway completion: The addition of a small number of heterologous enzymes to complete a biochemical pathway.

References

Blum JJ, Stein RB. 1982. On the analysis of metabolic networks. In RF Goldberger, KR Yamamoto (eds), Biological Regulation and Development, pp 99–125. New York, Plenum Press.

Brown GC, Hafner RP, Brand MD. 1990. A "top-down" approach to the determination of control coefficients in metabolic control theory. Eur J Biochem 188:321.

Delgado J, Liao JC. 1992. Determination of flux control coefficients from transient metabolite concentrations. Biochem J 282:919.

Epp JK, Huber MLB, Turner JR, et al. 1989. Production of a hybrid macrolide antibiotic in Streptomyces ambofaciens and Streptomyces lividans by introduction of a cloned carbomycin biosynthetic gene from Streptomyces thermotolerans. Gene 85:293.

Heinrich R, Rapoport TA. 1974. Linear steady-state treatment of enzymatic chains: General properties, control, and effector strength. Eur J Biochem 42:89.

Kascer H, Burns JA. 1973. Control of [enzyme] flux. Symp Soc Exp Biol 27:65.

Liao JC, Delgado J. 1993. Advances in metabolic control analysis. Biotechnol Prog 9:221.

Malpartida F, Niemi J, Navarrete R, Hopwood DA. 1990. Cloning and expression in a heterologous host of the complete set of genes for biosynthesis of the Streptomyces coelicolor antibiotic undercylprodigiosin. Gene 93:91.

Mavrovouniotis ML, Stephanopoulos G, Stephanopoulos G. 1990. Computer-aided synthesis of biochemical pathways. Biotechnol Bioeng 36:1119.

Mavrovouniotis ML. 1993. Identification of localized and distributed bottlenecks in metabolic pathways. In Proceedings of the International Conference on Intelligent Systems for Molecular Biology, Washington.

McAlpine JB, Tuan JS, Brown DP, et al. 1987. New antibiotics from genetically engineered Actinomycetes: I. 2-Norerythromycins, isolation and structural determination. J Antibiot 40:1115.

Safer B, Williamson JR. 1973. Mitochondrial-cytosolic interactions in perfused rat heart. J Biol Chem 248:2570.

Savinell JM, Palsson BO. 1992. Network analysis of intermediary metabolism using linear optimization: I. Development of mathematical formalism. J Theor Biol 154:421.

Small JR, Kacser H. 1993a. Responses of metabolic systems to large changes in enzyme activities and effectors: I. The liner treatment of unbranched chains. Eur J Biochem 213:613.

Small JR, Kacser H. 1993b. Responses of metabolic systems to large changes in enzyme activities and effectors: II. The linear treatment of branched pathways and metabolite concentrations. Assessment of the general non-linear case. Eur J Biochem 213:625.

Stephanopoulos G, Sinskey AJ. 1993. Metabolic engineering—Methodologies and future prospects. TIBTECH 11:392.

Vallino JJ, Stephanopoulos G. 1989. Flux determination in cellular bioreaction networks: Applications to lysine fermentations. In SK Sikdar et al (eds), Frontiers in Bioprocessing, pp 205–219. Boca Raton, Fla, CRC Press.

Varma A, Boesch BW, Palsson BO. 1993. Biochemical production capabilities of Escherichia coli. Biotechnol Bioeng 42:59.

Winter RB, Yen K-M, Ensley BD. 1989. Efficient degradation of trichloroethylene by recombinant Escherichia coli. Biotechnology 7:282.

Further Information

Good reviews of recent applications of metabolic engineering, as well as advances in metabolic engineering tools can be found in Cameron DC, Tong IT. 1993. Cellular and metabolic engineering: An overview. *Appl Biochem Biotechnol* 38:105, and in Cameron DC, Chaplen FWR. 1997, Developments in metabolic engineering. *Curr Opin Biotechnol* 8/2:175. A pioneering discussion of the emerging field of metabolic engineering can be found in Bailey JE. 1991. Toward a science of metabolic engineering. *Science* 252:1668. A good source of information about recombinant DNA techniques is Gilman M, Watson JD, Witkowski J, et al. 1992. *Recombinant DNA*, 2d ed. San Francisco, WH Freeman.

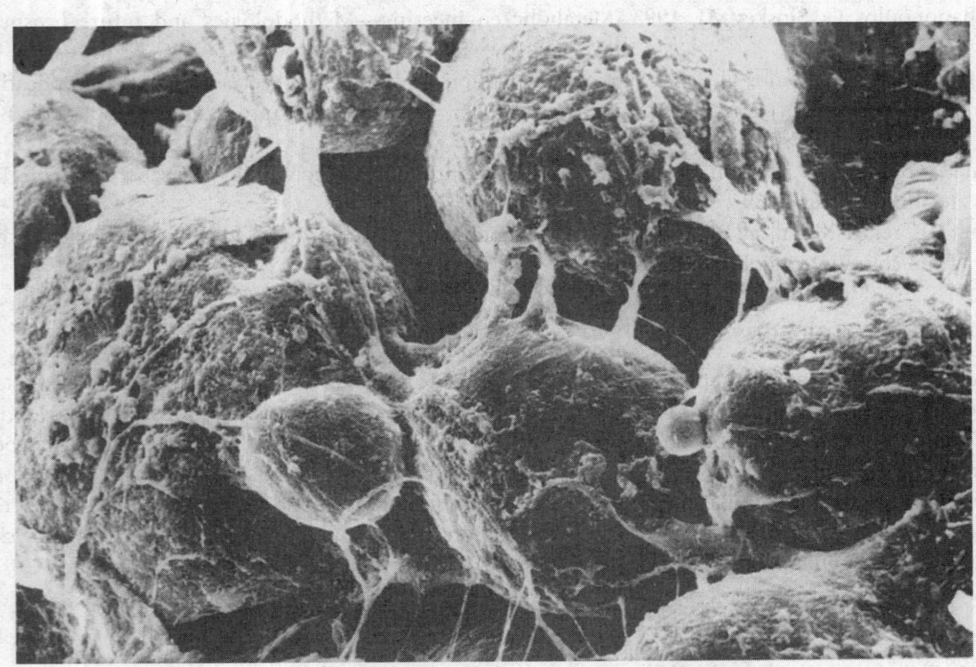

Scanning electron micrographs of bone marrow cultures on a nylon screen template showing myeloid cells of a hematopoietic colony.

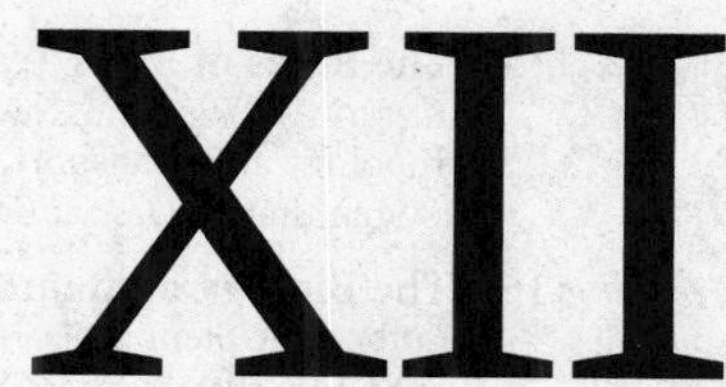

Tissue Engineering

Bernhard Ø. Palsson
University of Michigan

Jeffrey A. Hubbell
California Institute of Technology

TISSUE ENGINEERING IS A NEW field that is rapidly growing in both scope and importance within biomedical engineering. It represents a marriage of the rapid developments in cellular and molecular biology on the one hand and materials, chemical, and mechanical engineering on the other. The ability to manipulate and reconstitute tissue function has tremendous clinical implications and is likely to play a major role in cell and gene therapies during the next few years in addition to expanding the tissue supply for transplantation therapies.

Tissue function is complex and involves an intricate interplay of biologic and physicochemical rate processes. A major difficulty with selecting topics for this section was to define its scope. After careful consideration we focused on three goals: first, to cover some of the generic engineering issues; second, to cover generic cell biologic issues; and third, to review the status of tissue engineering of specific organs.

In the first category we have two common engineering themes of material properties and development and the analysis of physical rate processes. The role of materials is treated in terms of two length scales—the molecular and the cellular size scales. On the smaller size scale, Chapter 110 deals with engineering the biomolecular properties of a surface, and Chapter 111 deals with protein adsorption onto surfaces to which cells will be exposed. Chapters 112 and 113 deal with the engineering of scaffolds and templates on which cells are grown for transplantation purposes. The effects of physical rate processes are treated in Chapter 114 and 115. Both fluid mechanical forces and mass transfer rates influence important attributes of tissue function and relate to the physical constraints under which tissues operate. The issues associated with materials and the physical rate process represent important engineering challenges that need to be met to further the development of tissue engineering.

In the second category we have common cell biologic themes: stem cell biology, cell motion, the tissue microenvironment, and the role of stroma in tissue function. It is now believed that many tissues contain stem cells from which the tissue is generated. Chapter 116 describes the basic concepts of stem cell biology. Cell motion is an important process in tissue function and tissue development. Basic concepts of cell motility are covered in Chapter 117, and the manipulation of this process is likely to become an essential part of tissue engineering. It is well established that the cellular microenvironment in vivo is critical to tissue function. This complex set of tissues is treated in Chapter 118. The specific interaction between cells performing a tissue-specific function and the accessory cells (or stroma) found in tissues has proved to be a key to reconstituting tissue function ex vivo. The important role of the stroma in tissue engineering is described in Chapter 119. The understanding and mastery of these cell biologic issues are essential for the future tissue engineer.

Chapters in the third and last category describe progress with the engineering of specific tissues. Six tissues where much progress has been made were selected: bone marrow, liver, the nervous system, skeletal muscle, cartilage, and kidney. These chapters provide a concise review of the accomplishments to date and the likely clinical implications of these developments.

By focusing on these three categories we hope that the reader will acquire a good understanding of the engineering and cell biological fundamentals of tissue engineering and of the progress that has been made to date and will develop ideas for further development of this emerging and important field.

109

Tissue Engineering

François Berthiaume
Surgical Services, Massachusetts General Hospital, Department of Surgery, Harvard Medical School, and the Shriners Hospital for Children

Martin L. Yarmush
Surgical Services, Massachusetts General Hospital, Department of Surgery, Harvard Medical School, and the Shriners Hospital for Children

109.1 Introduction

Tissue engineering can be defined as the application of scientific principles to the design, construction, modification, growth, and maintenance of living tissues. Tissue engineering can be divided into two broad categories: (1) *in vitro* construction of bioartificial tissues from cells isolated by enzymatic dissociation of donor tissue, and (2) *in vivo* alteration of cell growth and function. The first category of applications includes bioartificial tissues (i.e., tissues which are composed of natural and synthetic substances) to be used as an alternative to organ transplantation. Besides their potential clinical use, reconstructed organs may also be used as tools to study complex tissue functions and morphogenesis *in vitro*. For tissue engineering *in vivo*, the objective is to alter the growth and function of cells *in situ*, an example being the use of implanted polymeric tubes to promote the growth and reconnection of damaged nerves. Some representative examples of applications of tissue engineering currently being pursued are listed in Table 109.1.

Conceptually, bioartificial tissues involve three-dimensional structures with cell masses that are orders of magnitude greater than that used in traditional two-dimensional cell culture techniques. In addition, bioartificial organ technology often involves highly differentiated somatic and parenchymal cells isolated from normal tissues. In this chapter, we will provide an overview on tissue reconstruction *in vitro* with particular emphasis on the techniques used to control cell function and organization and to scale up bioartificial tissues. Other equally important issues in tissue engineering, such as cell isolation and biomaterial fabrication will not be presented here. A chapter on cell preservation is presented later within this section.

109.2 Basic Principles and Considerations

Cell Type and Source

In tissue engineering, differentiated cells offer some advantages over tumor cell lines: (1) tumor cells often do not express the full spectrum of functions at the same level as the somatic cell lines they originated

0-8493-0462-8/00/$0.00+$.50

TABLE 109.1 Representative Applications of Tissue Engineering

Application	Examples
Implantable device	Endothelialized vascular grafts
Extracorporeal device	Bone and cartilage implants
Cell production	Bioartificial skin
In situ tissue growth and repair	Bioartificial pancreatic islets
	Neurotransmitter-secreting cells
	Bioartificial liver
	Hematopoiesis *in vitro*
	Nerve regeneration
	Artificial skin

from, (2) tumor cell growth can be very rapid and uncontrollable which can cause design problems (e.g., clogging of microchannels), (3) there is a potential risk of seeding tumor cells in the patient. Notwithstanding these limitations, appropraite selection procedures can be used to derive tumor cell lines which are easily propagated, are contact inhibited, and maintain most of the parent somatic cell features.

Certain human cell types can be easily propagated using standard culture techniques, such as human keratinocytes [Parenteau, 1991]. Conversely, other cell types, such as adult hepatocytes or pancreatic islet cells, do not replicate to any appreciable extent *in vitro.* Because there is currently a shortage of human organs, animal sources are also considered. Success in using xenogeneic cells will largely depend on our ability to control the immunological response of the host to these cells as well as the proteins that they produce. Genetic engineering offers the possibility of creating new cells (i.e., cells which perform a new function) which are easy to grow (e.g., fibroblasts), however, care must be taken to verify that these cells also exhibit the necessary control mechanisms for responding appropriately to the host's metabolic changes.

Control of Cell Function by the Extracellular Matrix

Since the vast majority of mammalian cells are anchorage-dependent, they must attach and spread on a substrate to proliferate and function normally. Cell adhesion is generally mediated by certain extracellular matrix proteins such as fibronectin, vitronectin, laminin, and collagen as well as various glycosaminoglycans. Adsorption of these proteins to surfaces and the conformation of the adsorbed proteins appear to be important factors influencing cell attachment and growth on synthetic substrates [Grinnell, 1981]. Small sequences of the cell binding region of these proteins (RGDS in fibronectin and vitronectin, YIGSR in laminin) can also be covalently attached to surfaces to promote cell-substrate adhesion [Massia, 1990].

In general, seeding density is important for normal cell function, especially if cell-cell communications must be established, either by direct cell-cell contacts or via the secretion of trophic factors by the cells. Efficient cell seeding will mainly depend on: (1) the affinity of certain cell surface proteins for extracellular matrix components, (2) the density of cell-substrate binding sites on the material surface, (3) the presence or absence of certain nutritional factors. While the first two factors can be controlled independently of the number of cells placed on the surface, the last issue can be problematic because attempts at seeding large numbers of cells can significantly deplete nutrients, with the result that less cells than expected will attach. This point is discussed below in the section on "metabolic requirements of cells".

After seeding, the cells spread on the surface and reach a stable shape. The final morphology of the cells depends on 3 factors: (1) adherence of the substrate for the cells, which is a function of the affinity and number of the adhesion sites, (2) rigidity of the substrate (i.e., ability to resist cell-generated tractional forces, (3) cell-cell adherence. The effect of substrate adherence and compliance on cell shape and function has been studied on hepatocytes. Increased adherence by increasing the amount of fibronectin adsorbed on a surface leads to increased cell spreading, increased DNA synthesis and reduced expression of liver-specific functions [Mooney, 1992]. Conversely, reduced adherence or the use of a compliant substrate where little spreading is observed helps maintain liver-specific function but reduces DNA synthesis

TABLE 109.2 Effect of Extracellular Matrix (ECM) on Cell Shape and Function

ECM Characteristics		Hepatocytes[a]		Capillary Endothelial
Surface Chemistry	Geometry	Cell shape[b]	Liver-Specific Function	Cell Shape[c]
1 ng/cm² Fibronectin on polystyrene	Single surface	Round (1.2)	+	
1000 ng/cm² Fibronectin on polystyrene	Single surface	Spread, flat (5)	—	
0.1% w/v Collagen gel	Single surface	Spread, flat (3.5)	—	Cobblestone, flat
	sandwich	Cuboidal, flat (1.5)	+	capillary network
Heat denatured collagen	Single surface	Round, cell aggregates	+	
Basement membrane extract	Single surface	Round, cell aggregates	+	Capillary network

[a] Data from [Mooney, 1992; Lindblad, 1990; Ezzell, 1993].
[b] The number between brackets indicates the projected surface area relative to that of isolated hepatocytes (~1200 μm²).
[c] Data from [Montesano, 1983; Madri, 1983].

[Lindblad, 1991]. The quantitative difference between cell-substrate and cell-cell adhesion strength on a rigid substrate is also a potential factor which may dramatically affect the organization of cells on the substrate. A thermodynamic view of the problem suggests that the overall system (consisting of the cells and the extracellular support) ultimately reaches an equilibrium state when the surface free energy is minimized [Martz, 1974]. According to this hypothesis, the existence of large cell-substrate adhesion forces relative to cell-cell adhesion forces may be a sufficient condition to prevent cell-cell overlapping. In contrast, the opposite situation would lead to cell clumping or multilayered growth on the substrate. This prediction is in agreement with the observation of cellular aggregate formation when hepatocytes are plated on a nonadherent surface as opposed to a highly adherent surface such as type I collagen [Koide, 1990].

One of the most striking features of endothelial and epithelial cells (as opposed to connective tissue cells such as fibroblasts) is the organization of the membrane, which exhibits distinct basal and apical domains, where different proteins and receptors are found. Most epithelial cells can be cultured on a single surface of plastic or extracellular matrix materials which allows the basal surface to be in contact with the substrate and the apical surface to be exposed to the liquid medium. Changing the extracellular matrix configuration can induce these cells to adopt different morphologies. For example, capillary endothelial cells grown on a single collagen gel reorganize into branching capillary-like structures when overlaid with a second layer of collagen [Montesano, 1983]. Unlike most epithelial cells, which exhibit only two distinct membrane domains (one apical and one basal), hepatocytes possess two basolateral surfaces separated by a belt of apical membrane. To express and maintain this particular phenotype, these cells have been cultured between two layers of extracellular matrix, thereby creating a "sandwich" configuration which closely mimics the *in vivo* geometry. Examples of the effect of the extracellular matrix on cell shape and function are summarized in Table 109.2.

Control of Tissue Organization

New ways of controlling cell distribution and shape on surfaces are being developed. These methods provide finer control of cell distribution and creation of microstructures (e.g., microchannels) at the a scale that rivals that of natural organs.

The current approach to control cell distribution at size scales down to cell size (10 μm) involves micropatterning techniques. A micropatterning technique based on the utilization of photoreactive cross-linking of hydrophobic or hydrophilic compounds to a polymeric surface has recently been described. Cells attach and grow along the less hydrophilic micropattened domains. Micropatterned surfaces produced with this technique have been used to produce two-dimensional neuronal networks with neuroblastoma cells [Matsuda, 1992]. Small micropatterned adherent squares of different sizes have been used to control the extent of spreading of hepatocytes [Bhatia, 1994; Singhvi, 1994].

The control of cell orientation may be important when nonisotropic connective tissues are desired. For example, cells can be mixed with a chilled solution of collagen in physiological buffer, followed by exposure to 37°C to induce the gelation of the collagen. The result is a loose gel comprised of small collagen fibrils where cells and fluid are entrapped. Fibroblasts embedded in a collagen gel cause the contraction of the gel [Bell, 1979]. The contraction process can be controlled to a certain extent by mechanically restricting the motion along certain directions; this also induces a preferential alignment of the collagen fibers as well as the cells within it. A mathematical description of the effect of cell tractional forces on the deformation of collagen-cell lattices and the resulting alignment of the collagen fibers has been presented by Tranquillo [1994]. Other strategies to align cells have been reported: (1) alignment of fibroblasts in collagen gels where collagen fibrils were aligned by a magnetic field during gelation [Guido, 1993], (2) repetitive stretching of endothelial cells cultured on an extensible surface [Banes, 1993], and (3) flow-induced alignment of endothelial cells [Girard, 1993].

Heterotypic cell systems or "co-culture" systems have been used for the production of skin grafts, in long-term cultures of hepatocytes, and in long-term cultures of mixed bone marrow cells. These systems take advantage of the trophic factors (for the most part unknown) secreted by "feeder" cells. Greater use of different cell types used in co-culture will enable engineered cell systems to closely mimic *in vivo* organization, with potential benefits including increased cell function and viability, and greater range of functions expressed by the bioartificial tissue. The organization of multicellular three-dimensional structures may not be obvious. Provided that the adherence of homotypic and heterotypic interactions is known, a thermodynamic analysis similar to that used to describe the morphology of a pure cell culture on a surface can be used to predict how cells will organize in these systems [Wiseman, 1972]. The process of cell-cell sorting in multicellular systems may be altered by changing the composition of the medium or altering the expression of proteins mediating cell-cell adhesion via genetic engineering [Kuhlenschmidt, 1982; Nose, 1988]. Furthermore, micropatterned surfaces can be used to control the initial interface between two cell types in co-culture systems [Bhatia et al., 1998].

Metabolic Requirements of Cells

Oxygen is very often the limiting nutrient in reconstructed tissues. The oxygen requirements vary largely with cell type. Among cells commonly used in bioartificial systems, hepatocytes and pancreatic islet cells are particularly sensitive to the availability of oxygen. Oxygen is required for efficient cell attachment and spreading to planar surfaces as well as microcarriers [Foy, 1993; Rotem, 1994].

Based on a simple mathematical model, we can estimate the critical distance at which the oxygen concentration at the cell surface becomes limiting (this concentration is arbitrarily set equal to K_m) if an unstirred aqueous layer is placed between the gas phase (air) and a confluent monolayer of cells [Yarmush, 1992]. In the case of a confluent monolayer of hepatocytes (2×10^5 cells/cm^2), this distance is 0.95 mm. Thus, in a first approximation, a successful bioartificial liver system containing hepatocytes will have to keep the diffusional distance between the oxygen-carrying medium and the cells to below 1 mm (assuming a confluent cell monolayer on the surface).

Figure 109.1 can be used to estimate the maximum half thickness of a cell mass surrounded by a membrane or external diffusion barrier before the nutrient concentration in the center falls below the Michaelis-Menten constant for nutrient uptake, a sign of nutrient limitation at the cellular level. Oxygen uptake rate parameters for different cell types are given by Fleischaker [1981]. Values for hepatocytes and pancreatic islets can be found in Yarmush [1992], Rotem [1992], and Dionne [1989]. For illustrative purposes, we will use cylindrical hepatocyte aggregates, as an example. We assumed the following: no external diffusion barrier (R_1/R_0) = 1; medium saturated with air at 37°C at the aggregate surface (160 mmHg = 190 nmol/cm^3); diffusivity of oxygen in aggregates (D_0) similar to that in water (2×10^{-5} cm^2/s); a packed cell mass (given a cell diameter of approximately 20 μm, this corresponds to 1.25×10^8 cells/cm^3). The oxygen uptake parameters for hepatocytes were μ_{max} = 0.4 nmol/10^{6}cells/s (thus 50 nmol/cm^3/s for the above cell concentration) and K_m = 0.5 mmHg (e.g., 0.6 nmol/cm^3). We obtain $C_1/K_m = C_0/K_m = 320$, $(\mu_{max} R^2)/(D_0 K_m) = 724$, 1370, and 2010 for the slab, cylindrical, and spherical

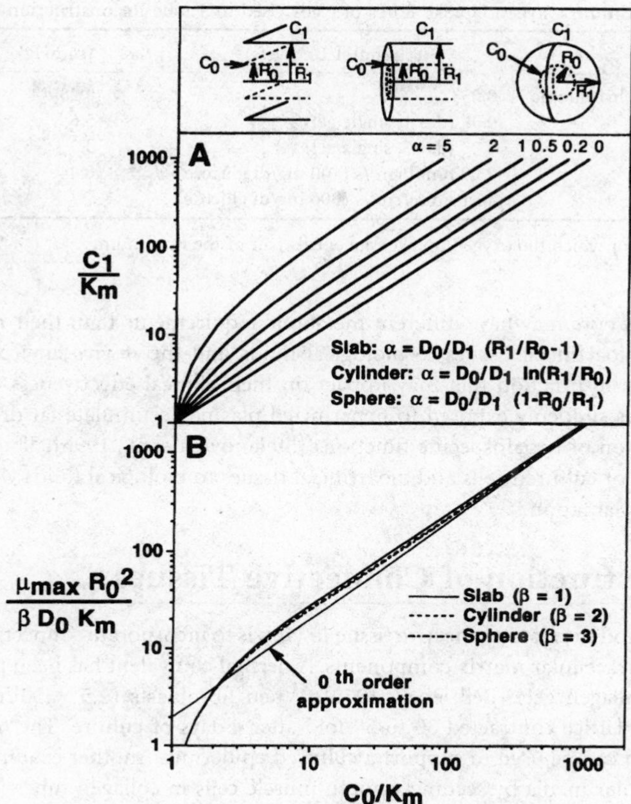

FIGURE 109.1 Correlation to predict the maximum half-thickness R of a cell mass surrounded by a shell of thickness R_1-R_0 without nutrient limitation, assuming that diffusion is the only transport mechanism involved. A: The nutrient concentration at the surface of the cell mass normalized to the Michaelis-Menten constant for the nutrient uptake by the cells (C_0/K_m) is obtained from the normalized bulk nutrient concentration (C_1/K_m), and the aspect ratio of the system (R_1/R_0). D_0, D_1 are the diffusivities for the nutrient within the cell mass and the external diffusion barrier, respectively. The partition coefficient between the cell mass and the surrounding shell is assumed to be equal to 1. B: The half thickness R for which $C/K_m = 1$ in the center ($R = 0$) is obtained from the value of the y-axis corresponding to C_0/K_m, knowing, in addition to the parameters listed above, the maximum nutrient uptake rate by the cells (μ_{max}). If K_m is unknown, a zero order approximation may be used, in which case K_m is set arbitrarily so that C_1/K_m falls in the linear portion of the curve in A. In B, R_0 is obtained using the line labeled "0th order approximation", and corresponds to the half thickness R for which $C = 0$ at the center ($R = 0$).

geometries, respectively, and thus the corresponding maximum half thicknesses obtained are R = 132, 181, and 220 μm, respectively. We now consider the case where there is a 100 μm thick membrane around a cylindrical cell aggregate assuming that the diffusivity of oxygen in the membrane (D_1) is the same as in the cell mass. An aspect ratio (α) must be assumed and the values of C_0/K_m, $(\mu_{max} R^2)/(D_0K_m)$, and R_o are calculated. R_1 is then obtained from the assumed aspect ratio. Calculations must be performed with several aspect ratios until the difference R_1-R_0 equates the membrane thickness. Here, we found that $\alpha = 0.6$ generates $C_0/K_m = 141$, $(\mu_{max} R^2)/(D_0K_m) = 622$, $R_0 = 122$ μm. For a cylinder, $\alpha = D_0/D_1 \ln(R_1/R_0)$, and thus in this case $R_1 = 222$ μm. Thus, the maximum half-thickness of the cell mass is 122 μm, as compared to 181 μm in the absence of the membrane. These estimates can be used as first guidelines to design a bioartificial liver, and they clearly suggest that the thickness of the cell mass must be limited to a few hundred microns to prevent the formation of an anoxic core.

TABLE 109.3 Maximum Oxygen Uptake Rates of Cells Used in Tissue Reconstruction

Cell type	μ_{max} (nmol/10^6cells/s)	$K_{0.5}$[a] (mmHg)	Reference
S_p2/0-derived mouse hybridoma	0.053	0.28	[Miller, 1987]
Hepatocytes	0.38 (day 1, single gel)	5.6	Oxyvice paper
	0.25 (day 3, single gel)	3.3	
Pancreatic islet cells	25.9 nmol/cm³/s (100 mg/dl glucose)	0.44	[Dionne, 1989]
	46.1 nmol/cm³/s (300 mg/dl glucose)		

[a] Oxygen tension for which the oxygen uptake rate equals half of the maximum.

Tissue assembled *in vitro* may have different metabolic requirements than their *in vivo* counterparts, so that exposure of bioartificial tissues to biological fluids and the *in vivo* environment may lead to unexpected alterations in function that may impact on their clinical effectiveness. For example, stable cultures of hepatocytes suddenly exposed to heparinized plasma accumulate fat droplets and exhibit a progressive deterioration of hepatospecific functions [Stefanovich et al., 1996]. Thus, it is important to consider the response of cultured cells and bioartificial tissues to biological fluids when evaluating their performance after implantation.

109.3 Reconstruction of Connective Tissues

The most simple method to create connective tissue *in vitro* is to incorporate connective tissue cells within a loose network of extracellular matrix components. A dermal equivalent has been produced by embedding fibroblasts in collagen gels [Bell, et al., 1979]. When fibroblasts (2.5×10^4/mL) were seeded in collagen, the collagen lattice contracted 30 to 50 fold after 4 days of culture. The result is a dense cell-collagen matrix which can be used to support a cultured epidermis. Another example is the production of a bioartificial vascular media by seeding smooth muscle cells in collagen tubes [Weinberg, 1986]. In the latter case, orientation of the cells along the circumference of the tube may be necessary in order to eventually obtain a tissue with contractile properties similar to that exhibited by native blood vessels. More recently, L'Heureux et al. [1998] showed that smooth muscle cells and fibroblasts cultured in the presence of ascorbic acid produce large amounts of extracellular matrix and generate cell-extracellular matrix composite membranes which can be harvested from the substrate. Vascular grafts generated by wrapping around these membranes had a higher burst strength than human saphenous veins commonly used for arterial reconstructions.

When reconstructing connective tissue, it may be advantageous to use a relatively low seeding density with the expectation that cell replication together with the migration of blood vessels from the host's surrounding tissue would occur after implantation. This approach may not be appropriate for tissues that grow slowly and that are subjected to high stresses after implantation. This is the case of bioartificial cartilage, which would be best implanted once its mechanical properties are similar to that of the authentic tissue. Chondrocytes seeded at high density (10^7 cells/cm³) in agarose gels and in serum-free medium retain their phenotype and remain viable for up to 6 months [Bruckner, 1989]. In this system, the deposition of type II collagen and highly charged proteoglycans can be observed over a period of 43 days [Buschmann, 1992]. The cartilage obtained in the latter studies had a stiffness of approximately 1/3 that of the cartilage explants used to isolate the cells. It should be recognized that in these experiments, the cell density was considerably lower than that found in the native cartilage explants (7.5×10^7 cells/cm³). Higher chondrocyte seeding density in the initial gel could potentially yield a material with mechanical properties closer to that of native cartilage.

Meshes made of slowly biodegradable polymers may be particularly suitable for dense connective tissue synthesis, such as bone and cartilage. Loose meshes made of lactic and glycolic acid copolymer seeded with calf chondrocytes (5×10^7/mL) have been shown to harbor the production of cartilage *in vitro*

[Freed, 1993]. In the same studies, meshes seeded with cells were implanted in nude mice and recovered at different time following implantation. Implants excised at the 7 week time point consisted almost entirely of cartilage-like tissue with an overall shape similar to that of the original synthetic matrix. Histochemical analysis of the specimens revealed the presence of type II collagen and sulfated glycosaminoglycans at week 7 and later.

Angiogenesis may be necessary if there is cell growth in the reconstructed tissue after implantation. Because open matrices are mainly used to reconstruct these tissues, the host's capillaries can migrate into the implant. The presence of angiogenic factors (e.g., heparin-binding growth factor-1) may be used to improve the kinetics of implant vascularization [Thompson, 1989]. Genetically modified cells have been used to provide stable long-term release of angiogenic factors by the implant [Eming et al., 1995].

109.4 Reconstruction of Epithelial or Endothelial Surfaces

Cells Embedded in Extracellular Matrix Materials

A simple way to maintain certain epithelial cells in a three-dimensional matrix is to seed them in gels or meshes in a similar manner as connective tissues are constructed. For example, hepatocytes can be maintained in the same mesh-type matrices made of biodegradable materials used to create bioartificial cartilage [Cima, 1991]. In this configuration, hepatocytes have a tendency to aggregate; such aggregates (sometimes called organoids) are known to contain cells which have maintained their phenotypic stability [Koide, 1990]. However, this process may be somewhat limited unless the aggregate size can be controlled to prevent the formation of large aggregates with anoxic cores. Also, this method may not be used to produce continuous epithelial cell sheets.

Culture on a Single Surface

Bioartificial vascular grafts have been produced by seeding endothelial cells on the luminal surface of small diameter (6 mm or less) synthetic vascular grafts. These prostheses have given poor results because of inflammatory responses at the anastomotic sites and the poor retention of the seeded endothelium under *in vivo* conditions. A more intrinsically biocompatible approach to vascular graft production is to reproduce more closely the organization of actual blood vessels. For instance, a bioartificial media has been produced by embedding vascular smooth muscle cells in a collagen gel annulus, and the intima re-established by seeding endothelial cells on the inside surface of the gel [Weinberg, 1986].

A dermal equivalent consisting of fibroblasts embedded in collagen has been used to support a stratified layer of human keratinocytes [Parenteau, 1991]. Exposure to air induces the terminal differentiation of the keratinocytes near the air-liquid interface, which is characterized by the formation of tight cell-cell contacts and cornified envelopes. The resulting bioartificial skin has a morphological and biochemical organization very similar to that of real skin including a stratum corneum which exhibits a high resistance to chemical damage. This type of bioartificial skin has been tested successfully in animals, but acceptance of allogeneic skin grafts remains problematic in humans, which means that the production of grafts requires that both fibroblasts and keratinocytes be obtained from the patient and propagated *in vitro.*

Culture in a Sandwich Configuration

This culture technique has been successfully used to maintain hepatocyte polarity and function *in vitro.* Hepatocytes are first plated on a single gel of collagen and then a top layer of type I collagen is placed on the cells after one day in culture. The resulting extracellular geometry closely mimics that found *in vivo* and maintains a wide spectrum of liver-specific functions, including protein secretion (e.g., albumin, transferrin, and fibrinogen), detoxification (e.g., cytochrome P450-dependent pathways) for up to 6 to 8 weeks, [Dunn, 1991].

109.5 Bioreactor Design in Tissue Engineering

A bioreactor is a system containing a large number of cells that transform an input of reactants into an output of products. Bioreactors have been designed primarily for use as bioartificial liver or pancreas, and more recently for the production of blood cells from hematopoietic tissue. These systems require the maintenance of the function of a large number of cells in a small volume. For example, a hypothetical bioartificial liver device possessing 10% of the detoxification and protein synthesis capacity of the normal human liver (a rough estimate of the minimum processing and secretory capacities that can meet a human body's demands) would contain a total of 10^{10} adult hepatocytes. Thus, to keep the total bioreactor volume within reasonable limits (1 L or less), 10^7 cells/mL or more are required. For comparison, the normal human liver contains approximately 10^8 hepatocytes/mL. Two main types of bioreactor design have been considered in tissue engineering: (1) hollow fiber systems, and (2) microcarrier systems.

Hollow Fiber Systems

Hollow fiber systems consist of a shell traversed by a large number of small diameter tubes. The cells may be placed within the fibers in the intracapillary space or on the shell side in the extracapillary space. The compartment which does not contain the cells is generally perfused with culture medium or the patient's plasma or blood. The fiber walls may provide the attaching surface for the cells and/or act as barrier against the immune system of the host. Some of these systems are essentially designed to be implanted as vascular shunts, but may also be perfused with the patient's blood or plasma extracorporeally.

For the selection of fiber dimensions, spacing, and reactor length, the reader is referred to previously published experimental and theoretical studies [Chresand, 1988; Piret, 1991]. If the pressure gradient across the fiber length is sufficiently high and relatively permeable fibers are used, the pressure difference between the inlet and outlet induces convective flow (called Starling flow) across the fiber wall and through the shell compartement [Kelsey, 1990; Pillarella, 1990]. The maintenance of Starling flow in implanted hollow fiber devices is contingent upon the prevention of a gradual decrease in fiber permeability due to protein deposition over time on the fiber walls, which may difficult to achieve with fluids containing high levels of proteins such as plasma.

There are several reports in the literature describing the use of hollow fiber systems in the development of a bioartificial pancreas [Colton, 1991; Ramírez, 1992]. Most designs place the islets on the shell side, while perfusing the fibers with the animal's plasma or blood. Recent studies using implantable devices connected to the vascular system of the patient have given some encouraging results [Sullivan, 1991], but it appears that none of the pancreatic islet devices tested to date have shown long term function in a reproducible manner. It is interesting to note that none of the studies involving pancreatic islets encapsulated in hollow fiber systems have recognized that oxygen transport can be severely compromised within the isolated islets alone, resulting in a substantial reduction in the insulin secretion capacity of the islets in response to glucose changes [Dionne, 1993].

It may be advantageous to place cells in the lumen of small fibers because the diffusional distance between the shell (where the nutrient supply would be) and the cells is essentially equal to the fiber diameter, which is easier to control than the inter-fiber distance. In one configuration, cells have been suspended in a collagen solution and injected into the lumen of fibers where the collagen is allowed to gel. Contraction of the collagen lattice by the cells creates a void in the intraluminal space [Scholz, 1991]. Such a configuration has been described for the construction of a bioartificial liver using adult hepatocytes [Nyberg, 1993]. It has been further proposed that the lumen be perfused with culture medium containing the appropriate hormones for cell maintenance while the patient's plasma would flow on the shell side.

Microcarrier-based Systems

Microcarriers are small beads (usually less than 500 μm in diameter) with surfaces treated to support cell attachment. The surface area available per microcarrier can be increased by using porous microcarriers, where cells can migrate and proliferate within the porous matrix as well as on the microcarrier surface. Porous ceramic beads have been used to support a rat cell line transfected with the human proinsulin gene [Park, 1993]; under certain conditions intraparticle flow significantly enhances mass transport (especially that of oxygen) to the cells in the beads [Stephanopoulos, 1989].

Maintaining a constant number of microcarriers while increasing the number of cells increases the surface coverage of the microcarriers as long as the concentration of nutrients (especially oxygen) is not limiting. With the adult rat hepatocyte-Cytodex 3 microcarrier system, we observed that sufficient supply of oxygen to the hepatocytes is required for the efficient attachment (approximately 90%) of cells to microcarriers. A simple oxygen diffusion-reaction model indicated that a cell surface oxygen partial pressure greater than 0.1 mm Hg was needed [Foy, 1993].

Two bioreactor configurations using microcarriers may find potential use in tissue engineering: (1) packed or fluidized bed, and (2) microcarriers incorporated in hollow fiber cartridges. A packed bed of microcarriers consists of a column filled with microcarriers with porous plates at the inlet and outlet of the column to allow perfusion while preventing microcarrier entrainment by the flow. Total flow rate is mainly dependent on cell number and the nutrient uptake rate of the cells. Reactor volume is proportional to the microcarrier diameter. From that point of view, it is therefore advantageous to reduce the microcarrier size as much as possible. However, packed beds with small beads may be potentially more prone to clogging and the cells may have tendency to accumulate in the channels between the microcarrier surfaces. The aspect ratio of the bed (height/diameter) is adjusted so that the magnitude of fluid mechanical forces (proportional to the aspect ratio) within the bed is below damaging levels. Given a fixed flow rate and microcarrier diameter, decreasing column diameter increases fluid velocity, and therefore increases the shear stress at the surface of the microcarriers and cells, with potential mechanical damage to cells. On the other hand, low aspect ratios may be difficult to perfuse evenly. Fluidized beds differ from packed beds in that the perfusing fluid motion maintains the microcarriers in suspension [Runstadler, 1988].

Packed bed systems have been used to support high densities (5×10^8 cells/mL) of anchorage-dependent cell lines seeded on microporous microcarriers (589 to 850 μm in diameter) [Park, 1993]. In addition, packed beads (1.5 mm diameter) have been used to entrap aggregates of hepatocytes. This system was shown to maintain a relatively stable level of albumin secretion (a liver-specific product) for up to 3 wk [Li, 1993].

Microcarriers have also been used as a way to provide an attachment surface for anchorage-dependent cells introduced in the shell side of hollow fiber devices, as in the case of a hollow fiber bioartificial liver device [Demetriou, 1986]. A flat bed device, which is in principle similar to a hollow fiber system (cells are separated from the circulating medium by a membrane), for the maintenance of cultured human bone marrow at high densities on porous microcarriers has been recently described [Palsson, 1993].

References

Banes AJ. 1993. Mechanical strain and the mammalian cell. In JA Frangos (ed), *Physical Forces and the Mammalian Cell*, pp. 81-123. San Diego, Academic Press.

Bell E, Ivarsson B, Merill C. 1979. Production of a tissue-like structure by contraction of collagen lattices by human fibroblasts of different proliferative potential *in vitro*. *Proc. Natl. Acad. Sci. USA.* 76: 1274.

Bhatia SN, Toner M, Tompkins RG, Yarmush ML. 1994. Selective adhesion of hepatocytes on patterned surfaces *in vitro*. *Ann. N.Y. Acad. Sci.* 745: 187.

Bhatia SN, Balis UJ, Yarmush ML, Toner M. 1998. Microfabrication of hepatocyte/fibroblast co-cultures: Role of homotypic cell interactions. *Biotechnol. Prog.* 14: 378.

Bruckner P, Hoerler I, Mendler M, Houze Y, Winterhalter KA, Eich-Bender SG, Spycher MA. 1989. Induction and prevention of chondrocyte hypertrophy in culture. *J. Cell Biol.* 109: 2537.

Buschmann MD, Gluzband YA, Grodzinsky AJ, Kimura JH, Hunziker EB. 1992. Chondrocytes in agarose culture synthesize a mechanically functional extracellular matrix. *J. Orthop. Res.* 10: 745.

Chresand TJ, Gillies RJ, Dale BE. 1988. Optimum fiber spacing in a hollow fiber bioreactor. *Biotechnol. Bioeng.* 32: 983.

Cima LG, Vacanti JP, Vacanti C, Ingber D, Mooney D, Langer R. 1991. Tissue engineering by cell transplantation using degradable polymer substrates. *J. Biomech. Eng.* 113: 143.

Colton CK, Avgoustiniatos ES. 1991. Bioengineering in development of the hybrid artificial pancreas. *J. Biomech. Eng.* 113: 152.

Demetriou A, Chowdhury NR, Michalski S, Whiting J, Schechner R, Feldman D, Levenson SM, Chowdhury JR. 1986. New method of hepatocyte transplantation and extracorporeal liver support. *Ann. Surg.* 204: 259.

Dionne KE, Colton CK, Yarmush ML. 1993. Effect of hypoxia on insulin secretion by isolated rat and canine islets of Langherans. *Diabetes.* 42: 12.

Dunn JCY, Tompkins RG, Yarmush ML. 1991. Long-term *in vitro* function of adult hepatocytes in a collagen sandwich configuration. *Biotechnol. Prog.* 7: 237.

Eming SA, Lee J, Snow RG, Tompkins RG, Yarmush ML, Morgan JR. 1995. Genetically modified human epidermis overexpressing PDGF-A directs the development of a cellular and vascular connective tissue stroma when transplanted to athymic mice. Implications for the use of genetically modified keratinocytes to modulate dermal generation. *J. Invest. Dermatol.* 105: 756.

Ezzell RM, Toner M, Hendricks K, Dunn JCY, Tompkins RG, Yarmush ML. 1993. Effect of collagen configuration on the cytoskeleton in cultured rat hepatocytes. *Exp. Cell Res.* 208: 442.

Fleischaker RJ Jr, Sinskey AJ. 1981. Oxygen demand and supply in culture. *European J. Appl. Microbiol. Biotechnol.* 12: 193.

Foy BD, Lee J, Morgan J, Toner M, Tompkins RG, Yarmush ML. 1993. Optimization of hepatocyte attachment to microcarriers: importance of oxygen. *Biotechnol. Bioeng.* 42: 579.

Freed LE, Marquis JC, Nohria A, Emmanual J, Mikos AG, Langer R. 1993. Neocartilage formation *in vitro* and *in vivo* using cells cultured on synthetic biodegradable polymers. *J. Biomed. Mat. Res.* 27: 11.

Girard PR, Helmlinger G, Nerem RM. 1993. Shear stress effects on the morphology and cytomatrix of cultured vascular endothelial cells. In JA Frangos (ed), *Physical Forces and the Mammalian Cell,* pp. 193-222. San Diego, Academic Press.

Grinnell F, Feld MK. 1981. Adsorption characteristics of plasma fibronectin in relationship to biological activity. *J. Biomed. Mater. Res.* 15: 363.

Guido S, Tranquillo RT. 1993. A methodology for the systematic and quantitative study of cell contact guidance in oriented collagen gels. *J. Cell Sci.* 105: 317.

Kelsey LJ, Pillarella MR, Zydney AL. 1990. Theoretical analysis of convective flow profiles in a hollow-fiber membrane bioreactor. *Chem. Eng. Sci.* 45: 3211.

Koide N, Sakaguchi K, Koide Y, Asano K, Kawaguchi M, Matsushima H, Takenami T, Shinji T, Mori M, Tsuji T. 1990. Formation of multicellular spheroids composed of adult rat hepatocytes in dishes with positively charged surfaces and under other nonadherent environments. *Exp. Cell Res.* 186: 227.

Kuhlenschmidt MS, Schmell E, Slife CF, Kuhlenschmidt TB, Sieber F, Lee YC, Roseman S. 1982. Studies on the intercellular adhesion of rat and chicken hepatocytes. Conditions affecting cell-cell specificity. *J. Biol. Chem.* 257: 3157.

L'Heureux N, Pâquet S, Labbé R, Germain L, Auger FA. 1998. A completely biological tissue-engineered human blood vessel. *FASEB J.* 12: 47.

Li AP, Barker G, Beck D, Colburn S, Monsell R, Pellegrin C. 1993. Culturing of primary hepatocytes as entrapped aggregates in a packed bed bioreactor: a potential bioartificial liver. *In Vitro Cell. Dev. Biol.* 29A: 249.

Lindblad WJ, Schuetz EG, Redford KS, Guzelian PS. 1991. Hepatocellular phenotype *in vitro* is influenced by biphysical features of the collagenous substratum. *Hepatology.* 13: 282.

Madri JA, Williams SK. 1983. Capillary endothelial cell cultures: phenotypic modulation by matrix components. *J. Cell Biol.* 97: 153.

Martz E, Phillips HM, Steinberg MS. 1974. Contact inhibition of overlapping and differential cell adhesion: a sufficient model for the control of certain cell culture morphologies. *J. Cell Sci.* 16: 401.

Massia SP, Hubbell JA. 1990. Covalently attached GRGD on polymer surfaces promotes biospecific adhesion of mammalian cells. *Ann. N. Y. Acad. Sci.* 589: 261.

Matsuda T, Sugawara T, Inoue K. 1992. Two-dimensional cell manipulation technology. An artificial neural circuit based on surface microphotoprocessing. *ASAIO J.* 38: M243.

Montesano R, Orci L, Vassalli P. 1983. *In vitro* rapid organization of endothelial cells into capillary-like networks is promoted by collagen matrices. *J. Cell Biol.* 97: 1648.

Mooney D, Hansen L, Vacanti J, Langer R, Farmer S, Ingber D. 1992. Switching from differentiation to growth in hepatocytes: control by extracellular matrix. *J. Cell. Physiol.* 151: 497.

Nose A, Nagafuchi A, Takeichi M. 1988. Expressed recombinant cadherins mediate cell sorting in model systems. *Cell.* 54: 993.

Nyberg SL, Shatford RA, Peshwa MV, White JG, Cerra FB, Hu W-S. 1993. Evaluation of a hepatocyte-entrapment hollow fiber bioreactor: a potential bioartificial liver. *Biotechnol. Bioeng.* 41: 194.

Palsson BO, Paek S-H, Schwartz RM, Palsson M, Lee G-M, Silver S, Emerson SG. 1993. Expansion of human bone marrow progenitor cells in a high cell density continuous perfusion system. *Bio/Technology.* 11: 368.

Parenteau NL, Nolte CM, Bilbo P, Rosenberg M, Wilkins LM, Johnson EW, Watson S, Mason VS, Bell E. 1991. Epidermis generated *in vitro*: practical considerations and applications. *J. Cell. Biochem.* 45: 245.

Park S, Stephanopoulos G. 1993. Packed bed bioreactor with porous ceramic beads for animal cell culture. *Biotechnol. Bioeng.* 41: 25.

Pillarella MR, Zydney AL. 1990. Theoretical analysis of the effect of convective flow on solute transport and insulin release in a hollow fiber bioartificial pancreas. *J. Biomech. Eng.* 112: 220.

Piret JM, Cooney CL. 1991. Model of oxygen transport limitations in hollow fiber bioreactors. *Biotechnol. Bioeng.* 37: 80-

Ramírez CA, López M, Stephens CL. 1992. *In vitro* perfusion of hybrid artificial pancreas devices at low flow rates. *ASAIO J.* 38: M443.

Rotem A, Toner M, Bhatia S, Foy BD, Tompkins RG, Yarmush ML. 1994. Oxygen is a factor determining *in vitro* tissue assembly: effects on attachment and spreading of hepatocytes. *Biotechnol. Bioeng.* 43: 654.

Runstadler PW, Cerneck SR. 1988. Large-scale fluidized-bed, immobilized cultivation of animal cells at high densities. In *Animal Cell Biotechnology,* (ed.) pp. 306-320. London, Academic Press.

Scholz M, Hu WS. 1991. A two-compartment cell entrapment bioreactor with three different holding times for cells, high and low molecular weight compounds. *Cytotechnology.* 4: 127.

Singhvi R, Kumar A, Lopez GP, Stephanopoulos GN, Wang DIC, Whitesides GM, Ingber DE. 1994. Engineering cell shape and function. *Science.* 264: 696.

Stephanopoulos GN, Tsiveriotis K. 1989. The effect of intraparticle convection on nutrient transport in porous biological pellets. *Chem. Eng. Sci.* 44: 2031.

Stefanovich P, Matthew HWT, Toner M, Tompkins RG, Yarmush ML. 1996. Extracorporeal perfusion of cultured hepatocytes: Effect of intermittent perfusion on hepatocyte function and morphology. *J. Surg. Res.* 66: 57.

Sullivan SJ, Maki T, Borland KM, Mahoney MD, Solomon BA, Muller TE, Monaco AP, Chick WL. 1991. Biohybrid artificial pancreas: long-term implantation studies in diabetic, pancreatomized dogs. *Science.* 252: 718.

Thompson JA, Haudenschild CC, Anderson KD, DiPietro JM, Anderson WF, Maciag T. 1989. Heparin-binding growth factor 1 induces the formation of organoid neovascular structures in vivo. *Proc. Natl. Acad. Sci. USA.* 86: 7928.

Tranquillo RT, Durrani MA, Moon AG. 1992. Tissue engineering science: consequences of cell traction force. *Cytotechnology.* 10: 225.

Weinberg CB, Bell E. 1986. A blood vessel model constructed from collagen and cultured vascular cells. *Science.* 230: 669.

Wiseman LL, Steinberg MS, Phillips HM. 1972. Experimental modulation of intercellular cohesiveness: reversal of tissue assembly patterns. *Dev. Biol.* 28: 498.

Yarmush ML, Toner M, Dunn JCY, Rotem A, Hubel A, Lee J, Tompkins RG. 1992. Hepatic tissue engineering: development of critical technologies. *Ann. N.Y. Acad. Sci.* 21: 472.

Further Information

Books, reviews and special issues of scientific and engineering journals on tissue engineering which may be of interest to the reader are listed below.

Morgan JR, Yarmush ML (eds). 1999. Tissue Engineering Methods and Protocols. Totowa, NJ, Humana Press.

Lanza RP, Langer RS. 1997. Principles of Tissue Engineering. San Diego, Academic Press.

Special issue on tissue engineering. 1996. *Biomaterials,* 17(2-3).

Vacanti CA, Mikos AG (eds). 1995- Tissue Engineering. The official journal of the Tissue Engineering Society. Larchmont, NY, Mary Ann Liebert, Inc.

Silver FH. 1995. Biomaterials, Medical Devices, and Tissue Engineering: An Integrated Approach. London, Chapman & Hall.

Special issue on tissue engineering. 1995. *Ann. Biomed. Eng.,* 43(7 and 8).

Special issue on tissue engineering. 1994. *Biotechnol. Bioeng.,* 43(7 and 8).

Berthiaume F, Toner M, Tompkins RG, Yarmush ML. 1993. Tissue engineering. In *Implantation Bi*ology: the Response of the Host, Boca Raton, FL, CRC Press, 363-386.

Special issue on tissue engineering. 1992. *Cytotechnology,* 10(3).

Special issue on tissue engineering. 1991. *J. Biomech. Eng.,* 113(2).

Special issue on tissue engineering. 1991. *J. Cell Biochem.,* 45(4-5).

110

Surface Immobilization of Adhesion Ligands for Investigations of Cell-Substrate Interactions

Paul D. Drumheller
Gore Hybrid Technologies, Inc.

Jeffrey A. Hubbell
California Institute of Technology

The interaction between cells and surfaces plays a major biologic role in cellular behavior. Cell membrane receptors responsible for adhesion may influence cell physiology in numerous ways. Examples of these interactions are the rolling and activation of leukocytes on vascular endothelium, the spatial differentiation of embryonic basement membranes in development, the extension of neurites on adhesion proteins, the proliferation of cells induced by mitogenic factors, and the spreading and recruitment of platelets onto the vascular subendothelium. These and other cell-surface interactions influence or control many aspects of cell physiology, such as adhesion, spreading, activation, recruitment, migration, proliferation, and differentiation.

There is great interest in understanding cell-surface interactions. This understanding is paramount to the development of pharmaceutical compounds to enhance or inhibit cell-substrate interactions, such as agents to enhance cell adhesion for tissue regeneration or biomaterials integration or to reduce cell adhesion in metastasis or fibrosis. In the emerging area of tissue engineering, the adhesion of cells to a culture support is essential for many products such as biohybrid dermal dressings, synthetic articular cartilage, and hepatocyte scaffolds. Investigations of cell-surface interactions are central to many areas of biomedicine and bioengineering, including tissue engineering, biomaterials design, immunology, oncology, haematology, and developmental biology.

To examine the interactions of cells with substrates and how they may influence cell behavior, simplified models that simulate basement membranes, cell surfaces, or extracellular matrices have been developed. These models involve the immobilization of biologically active ligands of natural and synthetic origin

0-8493-0462-8/00/$0.00+$.50

onto various substrates to produce chemically defined bioactive surfaces. Ligands that have been immobilized include cell-membrane receptor fragments, antibiotics, adhesion peptides, enzymes, adhesive carbohydrates, lectins, membrane lipids, and glycosaminoglycan matrix components. Functionalized two-dimensional surfaces are important tools to elucidate the molecular mechanisms of cell-mediated bioadhesion.

Techniques for preparing these ligand-functionalized two-dimensional substrates for investigation of cell adhesion are similar to methods for labeling ligands, preparing immunomatrices, or immobilizing enzymes. This brief tutorial review will focus primarily on schemes for immobilizing bioactive ligands specifically for purposes of investigating cell-surface interactions. Topics that will be addressed include: (1) general considerations of bioconjugation; (2) examples of surface bioconjugation; (3) preparation of surfaces for ligand immobilization; (4) examples of ligand-containing copolymers; (5) techniques for characterizing these surfaces; and (6) examples of applications of surface immobilized ligands in cell-substrate investigations.

110.1 General Considerations

Numerous reviews are available that describe the chemistry of ligand conjugation and immobilization, and the reader is encouraged to consult them for more detailed descriptions [1–12]. When immobilizing ligands for the preparation of cell-adhesive substrates, several factors must be considered to retain maximum bioactivity. Ligand surface immobilization may proceed in low yields due to sterically inaccessible reactive sites on the ligand molecule; the immobilized ligand may not have optimum cell-receptor interactions due to physical constraints imposed by the surface or by the active site being buried; or the immobilized ligand may be partially denatured upon immobilization. The inclusion of spacer arms on the surface may help relieve these limitations [13]. If the ligand has a directionality, then the cell adhesive response may vary depending upon the orientation of the immobilized ligand [14, 15] or upon the length of the spacer arm [16].

Ligands have been physicochemically absorbed onto polymeric substrates to investigate cell-surface interactions. The advantage of covalently immobilizing a ligand is that a chemical bond is present to prevent desorption. The general chemical scheme for grafting a ligand onto a surface is shown in Fig. 110.1. Activation usually is a necessary step to produce highly reactive species to enable surface immobilization under mild conditions. Typically the surface is activated, since the chemical steps are somewhat simpler than for ligand activation, but each strategy may have its own advantages and limitations.

The activation of ligands may require suitable protective schemes to prevent homopolymerization or cross-linking if these competing reactions are not desired. Ligands may also be activated at more than one site, some of which may remain unconsumed after coupling is complete and have biological activity. In contrast, the activation of surfaces usually does not require protective schemes; nonetheless, activated surfaces may contain unconsumed sites after immobilization that may require deactivation so as not to influence cell-mediated adhesive responses.

Because most ligands immobilized for cell-surface studies are biologic in origin, water is often the only choice of solvent as the coupling medium. Water is nucleophilic, and hydrolysis can be a competing reaction during activation and coupling, especially at higher pH (>9–10). To reduce hydrolysis, surfaces are commonly activated in nonaqueous solvents; ligands are then coupled using concentrated aqueous solutions (>1 mg/ml) for extended periods (>12 hr). Some ligands are soluble in polar solvents such as dimethylsulfoxide (DMSO), dimethylformamide (DMF), acetone, dioxane, or ethanol, or their aqueous cosolvents, and these combinations may be used to reduce hydrolysis during coupling.

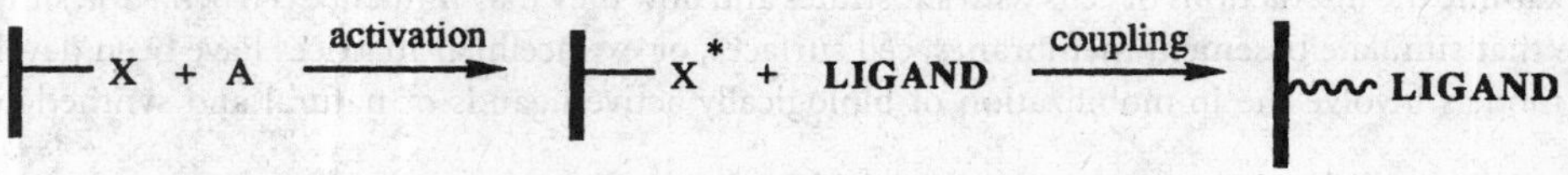

$$|\!-X + A \xrightarrow{\text{activation}} |\!-X^* + \text{LIGAND} \xrightarrow{\text{coupling}} |\!\sim \text{LIGAND}$$

FIGURE 110.1 General chemical scheme for grafting a ligand onto a surface.

110.2 Surface Bioconjugation

The typical chemical groups involved in immobilizing ligands for cell-surface studies are hydroxyls, amines, carboxylic acids, and thiols. Other groups, such as amides, disulfides, phenols, guanidines, thioethers, indoles, and imidazoles, have also been modified, but these will not be described.

Hydroxyls and carboxylic acids are commonly activated to produce more reactive agents for ligand acylation. Elimination is a possible competing reaction for secondary and aryl alcohols. Primary amines, present on many biologic ligands from N-amino acid termini and lysine, are good nucleophiles when unprotonated ($pK_a \sim 9$); moderately basic pH (8–10) ensures their reactivity. Thiols, present on cysteine ($pK_a \sim 8.5$), are stronger nucleophiles than amines, and as such they can be selectively coupled in the presence of amines at lower pH (5–7). Sulhydryl groups often exist as their disulfide form and may be reduced to free thiols using mild agents such as dithiothreitol.

The following paragraphs review activation schemes for particular chemical groups present on the surface, giving brief coupling schemes. Due to the advantages of surface versus ligand activation, only general methods involving surface activation will be discussed.

Immobilization to Surface Alcohols

A hydroxyl-bearing surface may be activated with numerous reagents to produce more reactive species for substitution, the most common example being the cyanogen bromide activation of cellulose and agarose derivatives [2–5, 11, 17]. Due to problems with high volatility of the cyanogen bromide, sensitivity of the activated species to hydrolysis, competing reactions during activation and coupling, and the desire to use culture substrates other than polysaccharides, other activation schemes have been developed for more defined activation and coupling. These schemes have been used to modify functionalized glasses and polymers for cell-mediated adhesion studies.

Alcohols react with sulfonyl halides [18–25], carbonyldiimidazole [26, 27], succinimidyl chloroformate [28], epoxides [29, 30], isocyanates [31–33] and heterocyclic [34, 35] and alkyl halides [21]. Activation of surface alcohols, with these agents can be performed in organic solvents such as acetonitrile, methylene chloride, acetone, benzene, dioxane, diethyl ether, toluene, or DMF.

Reactive Esters

Alcohols can be activated to reactive sulfonic ester leaving groups by reaction with sulfonyl halides [18–25] (Fig. 110.2) such as p-toluenesulfonyl chloride (tosyl chloride), trifluoroethanesulfonyl chloride (tresyl chloride), methanesulfonyl chloride (mesyl chloride), or fluorobenzenesulfonyl chloride (fosyl chloride). The resulting sulfonic esters are readily displaced in mild aqueous conditions with amines or thiols to produce amino- or thioether-bound ligands. Aryl alcohols should not be activated in this manner, since the sulfonate group may irreversibly transfer to the aromatic nucleus [36]. Sulfonic esters differ in their ease of nucleophilic substitution and resistance to hydrolysis, tosyl esters being low in coupling potential [23] and fosyl esters being high in hydrolysis resistance [24]. Activation can be performed in many organic solvents that are properly dry: Trace water and other species such as ammonia will react. DMSO should not be used as a solvent, as it will react with sulfonyl halides. A tert-amino base such as pyridine, dimethylaminopyridine, triethylamine, diisopropylethyl amine, or ethylmorpholine can be added in equimolar amounts to serve as a nucleophilic catalyst and to combine with the liberated HCl. It has been suggested that hydroxyls be converted to alkoxides prior to sulfonic ester activation in the presence of ethers as they may be sensitive to the HCl generated during activation [25].

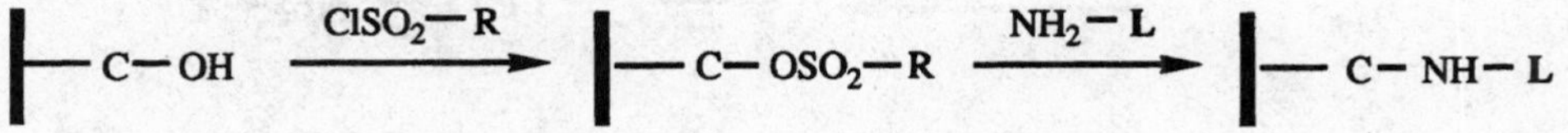

$$\vdash C-OH \xrightarrow{ClSO_2-R} \vdash C-OSO_2-R \xrightarrow{NH_2-L} \vdash C-NH-L$$

FIGURE 110.2 Alcohols can be activated with sulfonyl halides, which are readily displaced with amine- (shown) or thiol-containing (not shown) ligands.

An alcohol-containing surface is typically incubated with the sulfonyl halide for 0.5–6.0 hr. Any precipitated salts can be rinsed away with 1 mM HCl. The ligand is coupled to the surface for 12–24 hr in borate or carbonate buffer (pH 9–10) at a concentration of 1 mg/ml. Coupling can proceed at more mild pH (~7) with more concentrated solutions (>10 mg ligand/ml) or with thiol-bearing ligands. Excess sulfonic esters can be displaced with aqueous solutions (10–50 mM, pH = 8–9) of tris(hydroxymethyl)aminomethane, aminoethanol, glycine, or mercaptoethanol, or by hydrolysis.

Other Acylating Agents

Surface hydroxyls may be activated to groups other than sulfonic esters. Alcohols react readily with carbonyldiimidazole [26, 27] (Fig. 110.3) and succinimidyl chloroformate [28] (Fig. 110.4) to produce reactive imidazole-N-carboxylates and succinimidyl carbonates, respectively. These species acylate amines to urethane linkages. Activation typically proceeds in organic solvent for 2–6 hr. Dimethylaminopyridine catalyzes formation of the succinimidyl carbonate. Amine-bearing ligands (>10 mg/ml) are coupled to the surface in borate or phosphate buffer (pH 8–9) for 12–24 hr, or 4°C, 2–3 days. Thiol-bearing groups may also be immobilized onto these activated groups; however the thiocarbamate linkage can be sensitive to hydrolysis and may not be generally applicable to preparing well-defined substrates for long-term biologic investigations.

Bifunctional Bridges

In lieu of alcohol conversion to activated reaction groups, the hydroxyl may be added to homo- or heterobifunctional bridges, wherein alcholysis consumes one terminus to produce newly functionalized surfaces (Figure 110.5). Epoxide bridges can be added to surface-bound alcohols in aqueous base (to reduce hydrolysis or polymerization of the epoxide, 10–100 mM NaOH) or in organic solvent, 12 hr. Isocyanate bridges are added to surface alcohols in organic solvents with organotin catalysts (such as dibutyltin dilaurate), 12 hr. Halo alkylation of surface-bound alcohols or alkoxides proceeds in organic solvents, 2–3 days, or with heat (~80°C), 12 hr. The unconsumed free group (isocyanate, epoxide, or alkyl halide) is substituted with amine- or thiol-bearing ligands in buffered aqueous conditions, pH 8–10, 1–10 mg/ml, 12–24 hr. Hydroxyl-bearing ligands, such as carbohydrates, can be coupled in aqueous DMSO, dioxane, or DMF, 12–24 hr, 60–80°C. Hydroxyl coupling may also be performed in aqueous base (10–100 mM NaOH) for ligands that are resistant to ester- or amide hydrolysis, such as mono- or polysaccharides.

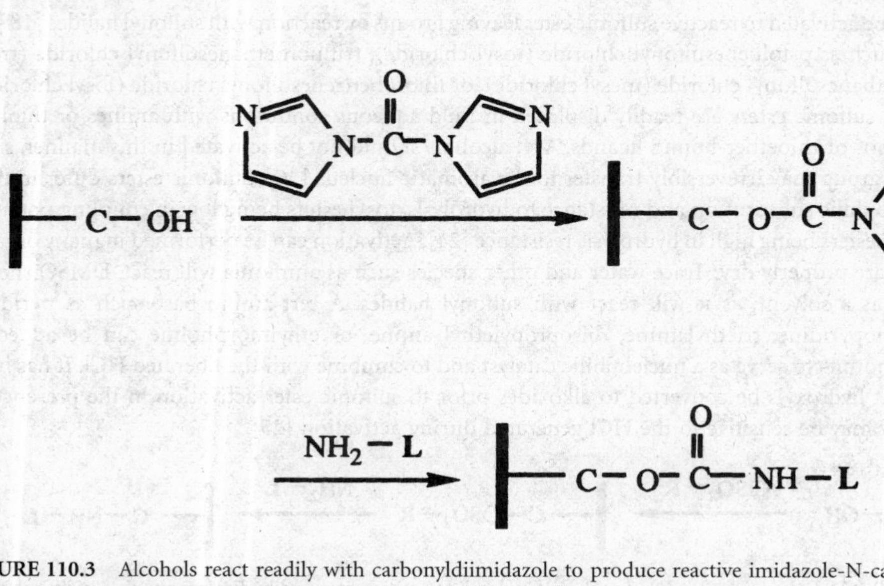

C–OH → C–O–C(=O)–N (imidazole) ; NH_2–L → C–O–C(=O)–NH–L

FIGURE 110.3 Alcohols react readily with carbonyldiimidazole to produce reactive imidazole-N-carboxylates for coupling to amine-containing ligands.

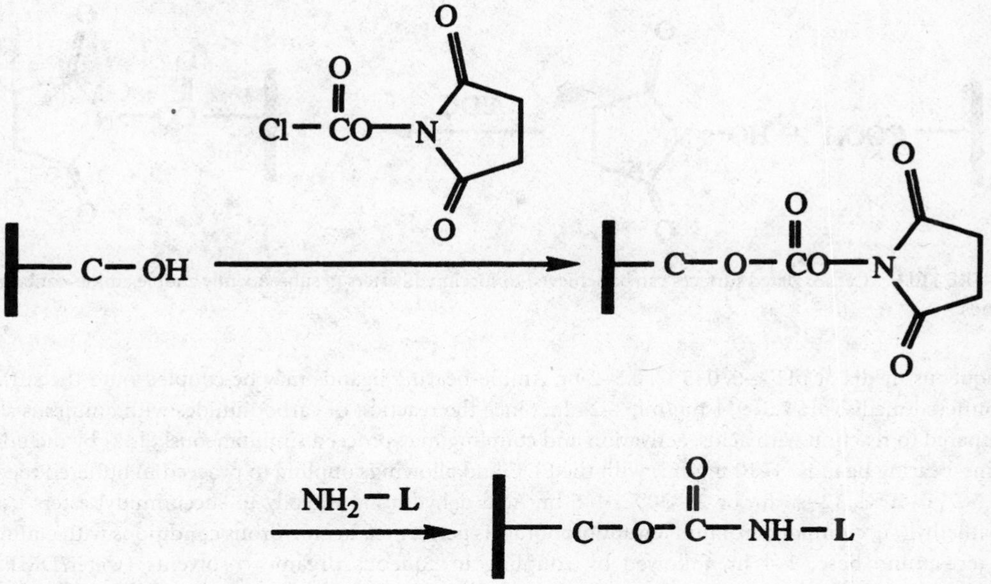

FIGURE 110.4 Alcohols react readily with succinimidyl chloroformate to produce a succinimidyl carbonate for coupling to amine-containing ligands.

Heterocyclic aryl halides, such as cyanuric chloride [34, 35], react with free hydroxyl groups in polar solvents containing sodium carbonate, 40–50°C, 30 min. Amine-bearing ligands are immobilized to the surface in borate buffer, pH 9, 1–10 mg/ml, 5–10°C, 12–24 hr.

Immobilization to Surface Carboxylic Acids

Acids may be converted to activated leaving groups by reaction with carbodiimides [36]. The generated O-acylureas react with amines to produce amide linkages and can react with alcohols to produce ester linkages with acid or base catalysts. An undesirable competing reaction is urea rearrangement to nonreactive N-acylurea; this effect may be accelerated in aprotic polar solvents such as DMF but is reduced at low temperatures (0–10°C) or by the addition of agents such as hydroxybenztriazole to convert the O-acylurea to benztriazole derivatives. Acids may also be activated with carbonyldiimidizole to produce easily amino- and alcoholyzed imidazolide intermediates; however, imidazolides are highly susceptible to hydrolysis and necessitate anhydrous conditions for activation and coupling [37].

Water-soluble carbodiimides such as ethyl(dimethylaminopropyl)-carbodiimide (EDC) can be used in either aqueous or organic media for the immobilization of ligands to produce bioactive substrates [20, 38, 39]. To reduce hydrolysis of the O-acylurea, conversion to more resistant succinimidyl esters can be performed using EDC and N-hydroxysuccinimide or its water-soluble sulfonate derivative, and base catalysts (Fig. 110.6). Alternatively, the acid may be converted directly to a succinimidyl ester via reaction with tetramethyl(succinimido)uronium tetrafluoroborate [40, 41]. The activation of surface-bound carboxylic acids with EDC in organic media (ethanol, DMF, dioxane) is complete with 1–2 hr at 0–10°C;

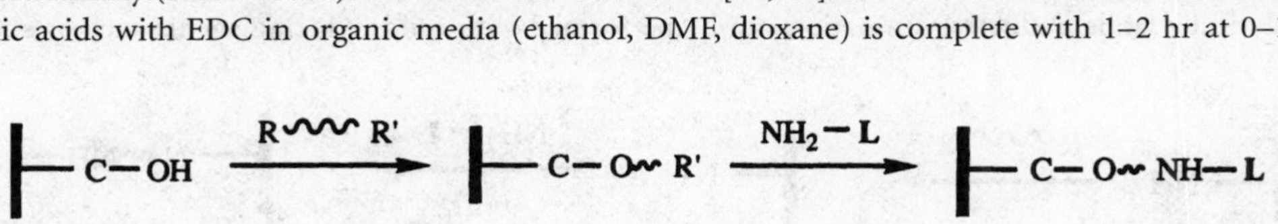

FIGURE 110.5 Bifunctional coupling agents may be used, e.g., to couple an amine-containing ligand to a hydroxyl-containing surface via a spacer.

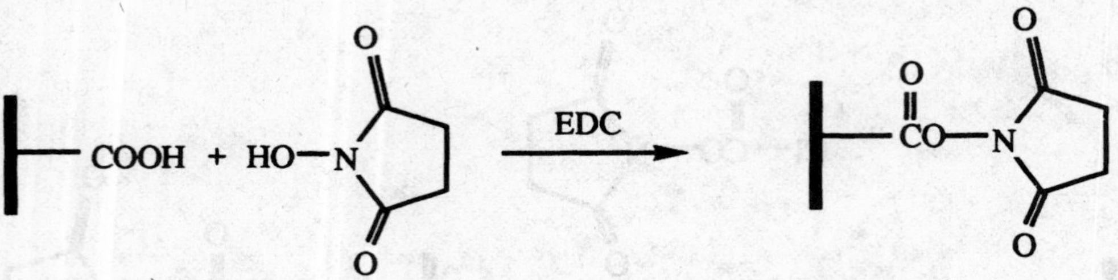

FIGURE 110.6 Carboxylated surfaces can be achieved to succimydil esters to subsequently couple amine-containing ligands.

in aqueous media at pH 4–6, 0–5°C, 0.5–2 hr. Amine-bearing ligands may be coupled onto the surface in buffered media, pH 7.5–9, 1 mg/ml, 2–24 hr. Since the reaction of carbodiimides with amines is slow compared to reaction with acids, activation and coupling may proceed simultaneously [42] by including amine-bearing ligands (1–10 mg/ml) with the EDC and allowing coupling to proceed in buffered media, pH 5–7, 0–10°C, 12–24 hr, or 25–30°C, 4–6 hr. Acid dehydration directly to succinimidyl esters using tetramethyl(succinimido)uronium tetrafluoroborate is performed in anhydrous conditions with equimolar tert-amino base, 2–4 hr, followed by coupling in aqueous organic cosolvents (water/DMF or water/dioxane), 1–10 mg/ml ligand, pH 8–9, 12–24 hr.

Immobilization to Surface Amines

Primary and secondary amine-containing surfaces may be reacted with homo- or heterobifunctional bridges (Fig. 110.7). Amines are more nucleophilic than alcohols, they generally do not require the addition of catalysts, and their addition is faster. These bridges may contain isocyanates [31, 33], isothiocyanates [43], cyclic anhydrides [44], succinimidyl esters [45–47], or epoxides [48, 49].

Isocyanates add to amines with good efficiency but are susceptible to hydrolysis; epoxides and cyclic anhydrides are somewhat less reactive yet are still sensitive to hydrolysis. Hydrolysis-resistant diisothiocyanates have been used for many years to label ligands with reporter molecules; however, the thiourea linkage may be hydrolytically labile (especially at lower pH) and may be unsuitable for investigations of cell-surface interactions. Succinimidyl esters, although not as resistant to hydrolysis as isothiocyanates, have very good reactivity to amines and form stable amide linkages. Hydrolysis is all these reagents is accelerated at higher pH (≥9–10).

Coupling to surface-bound amines is performed in organic conditions (DMF, DMSO, acetone) for 1–3 hr. Coupling of amine-bearing ligands onto immobilized bridges is performed in buffered media, pH 8–10.5, 2–12, 1–10 mg/ml. Excess reagent can be displaced with buffered solutions of tris(hydroxymethyl)aminomethane, aminoethanol, glycine, or mercaptoethanol, or hydrolysis.

Bifunctional aldehydes, such as glutaraldehyde and formaldehyde, have been used classically as crosslinkers for purposes of immunohistochemistry and ultrastructural investigations. They have been used also to couple ligands onto amine-bearing substrates [50–52]. Hydrolysis of aldehydes is usually not a concern, since the hydrolysis product, alkyl hydrate, is reversible back to the carbonyl. Amines add to aldehydes to produce imine linkages over a wide range of pH (6–10). These Schiff bases are potentially hydrolytically labile; reductive amination can be performed with mild reducing agents such as sodium

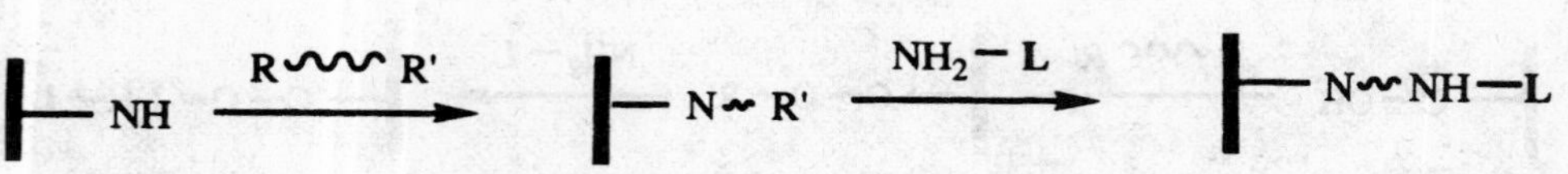

FIGURE 110.7 Primary and secondary amine-containing surfaces may be reacted with homo- or heterobifunctional bridges.

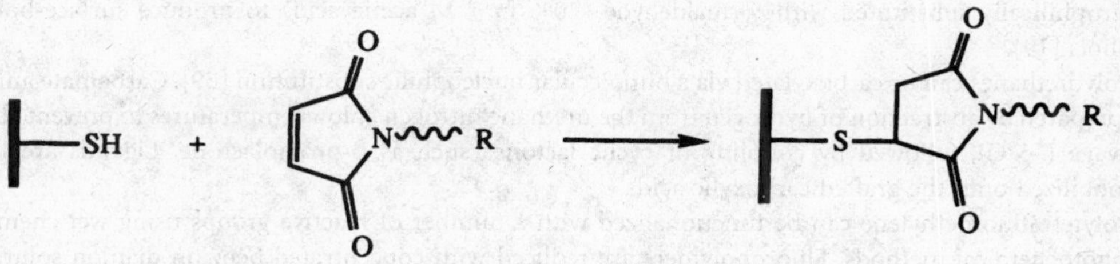

FIGURE 110.8 Thiol-containing surfaces may be coupled to malcimide activated species, either heterobifunctional linkers or ligands.

cyanoborohydride, pH 8–9, without substantial losses in ligand bioactivity [53]. Acetalization of polyhydric alcohols may commence in the presence of Lewis acid catalysts followed by dehydrating the hemiacetal linkage, an acetal [54, 55]. The dehydration conditions (air-drying followed by 70–90°C, 2 hr) may damage many biologic ligands.

Alkyl halide-bearing surfaces can be coupled to amine-bearing ligands [49, 56, 57]; their reaction is slower, but they are resistant to hydrolysis. Ligands can be immobilized in buffered medium (pH 9–10), 12–24 hr, 1–10 mg/ml, or in organic medium. Heat may be used to increase yields (60–80°C).

Immobilization to Surface Thiols

Thiols are more nucleophilic than amines or alcohols and may be selectively coupled in their presence at more neutral pH. Thiols can be reacted with reagents that are not reactive toward amines, such as homo- or heterobifunctional maleimide bridges [32, 58–60]. Surface-bound thiols react with maleimides in acetone, methanol, DMF, etc., for 1–2 hr, to produce a thioether linkage (Fig. 110.8). Thiol-bearing ligands can be immobilized to surface maleimides by incubating in mild buffered media, pH 5–7, 12–24 hr, 1–10 mg/ml. More basic pH should be avoided as hydrolysis of the maleimide to nonreactive maleamic acid may occur.

Photoimmobilization

Surfaces can be modified using photoactivated heterobifunctional bridges in one of three schemes: (1) the bridge may be immobilized onto a functionalized surface, followed by photocoupling of the ligand onto the photoactivated group [58, 61]; (2) the reagent may be photocoupled onto the surface, followed by immobilization onto the free terminus [58, 62]; or (3) the ligand may be coupled to the reagent, followed by photoimmobilization to the surface [63, 64]. The photoactivatible group may be light-sensitive azides, benzophenones, diazinines, or acrylates; polymerization initiator-transfer agent-terminators [65] and plasma-deposited free radicals [64] have also been used. Photoactivation produces highly reactive radical intermediates that immobilize onto the surface via a nonspecific insertion of the radical into a carbon-carbon bond. Photolabile reagents may be coupled to amine- or thiol-bearing ligands using succinimidyl esters or maleimides. One utility of photoimmobilization is the ability to produce patterns of ligands on the surface by lithography.

110.3 Surface Preparation

Polymeric materials can be functionalized using plasma deposition or wet chemical methods to produce surfaces containing reactive groups for ligand bioconjugation. Plasma polymerization has been widely used to deposit numerous functional groups and alter the surface chemistry of many medically relevant polymers [66, 67]. Polyesters such as polyethyleneterephthalate can be partially saponified with aqueous base (5% NaOH, 100°C, 30 min) to produce surface-bound carboxylic acids [39], partially aminolyzed with diamines such as ethylenediamine (50% aqueous, 12 hr) to produce surface-bound amines [68] or electrophilically substituted with formaldehyde (20% in 1 M acetic acid) to produce surface-bound alcohols [19].

Polyurethanes can be carboxylated via a bimolecular nucleophilic substitution [69]. Carbamate anions are prepared by abstraction of hydrogen from the urethane nitrogen at low temperatures to prevent chain cleavage (–5°C), followed by coupling of cyclic lactones such as β-propiolactone. Ligands are then immobilized onto the grafted carboxylic acids.

Polytetrafluoroethylene can be functionalized with a number of reactive groups using wet chemical or photochemical methods. Fluoropolymers are reduced with concentrated benzion dianion solutions to form carbonaceous surface layers which can then be halogenated, hydroxylated, carboxylated, or aminated [70]. Photochemical modification of fluoropolymers is possible by incubating the polymer in solutions of alkoxides or thiolate anions and exposing to UV light [71]. The patterning of surfaces is possible using this technique in conjunction with photolithography.

Glasses and oxidized polymers can be functionalized with bifunctional silanating reagents to generate surfaces containing alkyl halides, epoxides, amines, thiols, or carboxylic acids [19, 49, 59, 72]. The substrate must be thoroughly clean and free of any contaminating agents. Glass is soaked in strong acid or base for 30–60 min. Polymers are cleaned with plasma etching or with strong oxidizers such as chromic acid. Clean substrates are immersed in silane solutions (5–10% in acetone, toluene, or ethanol/water 95%/5%) for 1 hr and cured 50–100°C, 2–4 hr, or at 25°C, 24 hr, for oxidation-sensitive silanes such as mercaptosilanes. Prepared surfaces stored under Ar are stable for several weeks.

Metals such as gold can be functionalized via the chemisorption of self-assembled monolayers of alkanethiols [73–75]. Gold substrates, prepared by evaporating gold on chromium-primed silicon substrates, are immersed in organic solutions of alkanethiol (1–10 mM) for 5–60 min. The monolayer is adsorbed via a gold-sulfur bond; competitive displacement of the alkanethiol may occur [75], and it is unclear if these surfaces are applicable for use in reducing media. Similar substrates can be prepared from the chemisorption of carboxylic acids onto alumina [76]. Prepared substrates stored under Ar are stable for several months.

Amine-bearing surfaces have been produced by the adsorption of biologically inert proteins such as albumin. Cells do not have adhesion receptors for albumin, and for this reason albumin has been used to passivate surfaces against cell adhesion. Bioactive ligands can be amino-immobilized onto the albumin-coated surfaces [45, 46, 48].

Functional polymers and copolymers containing alcohols, amines, alkyl halides, carboxylic acids, or other groups, can be synthesized, coated onto a surface, and used as the substrate for immobilization of bioactive ligands [77, 78]. Examples include poly(vinyl alcohol) [31, 33], chloromethyl polystyrene [56], aminopolystyrene [79, 80], poly(acrylic acid) [20, 38], polyallylamine [80], poly(maleic acid anhydride) [12], poly(carbodiimide) [81], and poly(succinimide) [82].

110.4 Ligand/Polymer Hybrids

Hybrid copolymers may be synthesized in which one of the components is the biologically active ligand. These copolymers may then be coated onto a substrate or crosslinked into a three-dimensional network. Since the ligand is a component of the copolymer, no additional ligand immobilization may be necessary to produce bioactive substrates.

Examples are available for particular cases of hybrid copolymers (54, 55, 82–87], including gamma-irradiated crosslinked poly(peptide) [86, 88], dialdehyde crosslinked poly(vinyl alcohol)-glycosaminoglycan [54], poly(amino acid-etherurethane) [47, 87], poly(amino acid-lactic acid) [89], poly(amino acid-carbonate) [90], poly(peptide-styrene) [84], and linear [83] or crosslinked [82] poly(glycoside-acrylamide).

110.5 Determining Ligand Surface Densities

The concentration of ligands immobilized upon a surface can be measured using radiolabeling, photochrome labeling, surface analysis, or gravimetry. Since most materials can support the nonspecific adsorption of bioactive ligands, especially proteins, controls must be utilized to differentiate between covalent immobilization and physicochemical adsorption occurring during coupling. For example, ligands immobilized onto unactivated versus activated substrates give relative differences between non-specific adsorption and specific bioconjugation.

The surface immobilization of radiolabeled ligands with markers such as ^{3}H, ^{35}S, or ^{125}I, can be followed to give information on the kinetics of coupling, the coupling capacity of the substrate, and the surface density of ligands. Densities on the order of pmol-fmol/cm^2 are detectable using radiolabeled molecules.

The coupling capacity and density of immobilized acids, amines, and thiols can be evaluated using colorimetric procedures. Substrates can be incubated in solutions of Ellman's reagent [91, 92]; the absorbances of the reaction products give the surface density. Antigens which have been immobilized can be exposed to photochrome-labeled antibodies and surface concentrations calculated using standard enzyme immunosorbent assays [49, 93].

Verification of ligand immobilization may be performed using a number of surface analysis techniques. Mass spectroscopy, x-ray photoelectron spectroscopy, and dynamic contact angle analysis can give information on the chemical composition of the substance's outmost layers. Changes in composition are indicative of modification. Ellipsometry can be used to gauge the thickness of overlapping surface layers; increases imply the presence of additional layers. Highly sensitive gravimetric balances, such as quartz crystal microbalances, can detect *in situ* changes in the mass of immobilized ligand in the nanogram range [65].

110.6 Applications of Immobilized Ligands

Extracellular matrix proteins such as fibronectin, laminin, vitronectin and collagen, or adhesion molecules such as ICAM-1, VCAM-1, PCAM-1, and sialyl Lewis X, interact with cell surface receptors and mediate cell adhesion. The tripeptide adhesion sequence Arg-Gly-Asp (RGD) is a ubiquitous signal present in many cell adhesion proteins. It interacts with the integrin family of cell surface adhesion receptors, and comprises the best studied ligand-receptor pair [94–96]. In lieu of immobilizing complex multifunctional proteins for purposes of cell adhesion studies, synthetic RGD sequences have instead been immobilized onto many substrates as simplified models to understand various molecular aspects of cell adhesion phenomena. The following paragraphs cite examples of RGD-grafted substrates that have been used in biomedicine and bioengineering.

The density of RGD necessary to mediate cell adhesion has been determined in a number of fashions. RGD-containing peptides and protein fragments have been physicochemically adsorbed onto tissue culture substrates [97, 98] or covalently bound to albumin-coated substrates [45, 46] to titrate the dependency of cell adhesion function upon RGD surface densities. To remove potential complications due to desorption of ligands or albumin, RGD has been covalently bound onto functionalized substrates. Immobilization also restricts the number of conformations the peptide may assume, helping to ensure that all the peptide is accessible to the cells. RGD has been immobilized onto silanated glasses by its amino [19, 99] and carboxyl [15] termini. The effects of RGD density on cell adhesion, spreading, and cytoskeletal organization was examined [99] using this well-defined system. Other peptides have been immobilized in identical fashion to determine if they influence cell physiology [100].

RGD peptides have been immobilized onto highly cell-resistant materials to ensure that the peptide is the only cell adhesion signal responsible for cell adhesion to diminish signals borne of nonspecifically adsorbed serum proteins. Hydrogels of polyacrylamide [101], poly(vinyl alcohol) [31] and poly(ethylene glycol) [85] and nonhydrogel networks of polyacrylate/poly(ethylene glycol) [102] have been grafted with RGD; these background materials were highly resistant to the adhesion of cells even in the presence of serum proteins, demonstrating that the RGD sequence was solely responsible for mediating cell adhesion.

RGD-containing peptides have been immobilized onto medically relevant polymers in an effort to enhance their biocompatibilities by containing an adhered layer of viable cells. RGD-grafted surfaces can be more efficient in supporting the number and strength of cell adhesion by the peptide facilitating cell adhesion additionally to adsorbed adhesion proteins from the biological milieu. RGD has been conjugated onto surfaces by means of photoimmobilization [103] and plasma glow discharge [64]. In an effort to promote cell adhesion onto biodegradable implants, RGD peptides have been covalently grafted onto poly(amino acid-lactic acid) copolymers [89]. In this manner, cells adherent on the degradable material can eventually obtain a completely natural environment. Self-assembled monolayers of biologic ligands have been immobilized onto gold substrates by adsorbing functionalized thiol-containing bridges followed by covalent grafting of the ligand [93] or by adsorbing alkanethiol-containing ligands [104]. It has been suggested [105] that RGD-containing peptides could be immobilized onto gold substrates in these manners to engineer highly defined surfaces for cell culture systems.

These examples only partially illustrate the utility of ligand-grafted substrates for bioengineering and biomedicine. These substrates offer simplified models of basement membranes to elucidate mechanisms and requirements of cell adhesion. They have applications in biomedicine as biocompatible, cell-adhesive biomaterials for tissue engineering or for clinical implantation.

References

1. Brinkley M. 1992. A brief survey of methods for preparing protein conjugate with dyes, haptens, and cross-linking reagents. Bioconjugate Chem 3:2.
2. Means GE, Feeney RE. 1990. Chemical modification of proteins: History and applications. Bioconjugate Chem 1:2.
3. Wong SS. 1991. Chemistry of Protein Conjugation and Cross-Linking, Boca Raton, Fla, CRC Press.
4. Pharmacia Inc. 1988. Affinity chromatography principles and methods, Tech Bull, Uppsala, Sweden.
5. Trevan MD. 1980. Immobilized Enzymes: An Introduction and Applications in Biotechnology, New York, Wiley.
6. Wimalasena RL, Wilson GS. 1991. Factors affecting the specific activity of immobilized antibodies and their biologically active fragments. J Chromatogr 572:85.
7. Mason RS, Little MC. 1988. Strategy for the immobilization of monoclonal antibodies on solid-phase supports. J Chromatogr 458:67.
8. Wingard LB Jr, Katchalski-Katzir E, Goldstein L. 1976. Applied Biochemistry and Bioengineering: Immobilized Enzyme Principles, vol 1, New York, Academic.
9. Smalla K, Turkova J, Coupek J, et al. 1988. Influence on the covalent immobilization of proteins to modified copolymers of 2-hydroxyethyl methacrylate with ethylene dimethacrylate. Biotech Appl Biochem 10:21.
10. Schneider C, Newmanm RA, Sutherland DR, et al. 1982. A one-step purification of membrane proteins using a high efficiency immunomatrix. J Biol Chem 257:10766.
11. Scouten W. 1987. A survey of enzyme coupling techniques. Methods Enzymol 135:30.
12. Maeda H, Seymour LW, Miyamoto Y. Conjugates of anticancer agents and polymers: advantages of macromolecular therapeutics in vivo. Bioconjugate Chem 3:351.
13. Nojiri C, Okano T, Park KD, et al. 1988. Suppression mechanisms for thrombus formation on heparin-immobilized segmented polyurethane-ureas. Trans ASAIO, 34:386.
14. Fassina G. 1992. Oriented immobilization of peptide ligands on solid supports. J Chromatogr 591:99.
15. Hubbell JA, Massia SP, Drumheller PD. 1992. Surface-grafted cell-binding peptides in tissue engineering of the vascular graft. Ann NY Acad Sci 665:253.
16. Beer JH, Coller BS. 1989. Immobilized Arg-Gly-Asp (RGD) peptides of varying lengths as structural probes of the platelet glycoprotein IIb/IIIa receptor. Blood 79:117.
17. Axen R, Porath J, Ernback S. Chemical coupling of peptides and proteins to polysaccharides by means of cyanogen halides. Nature 214:1302.
18. Delgado C, Patel JN, Francis GE, et al. 1990. Coupling of poly(ethylene glycol) to albumin under very mild conditions by activation with tresyl chloride: Characterization of the conjugate by partitioning in aqueous two-phase systems. Biotech Appl Biochem 12:119.

19. Massia SP, Hubbell JA. 1991. Human endothelial cell interactions with surface-coupled adhesion peptides on a nonadhesive glass substrate and two polymeric biomaterials. J Biomed Mater Res 25:223.
20. Nakajima K, Hirano Y, Iida T, et al. 1990. Adsorption of plasma proteins on Agr-Gly-Asp-Ser peptide-immobilized poly(vinyl alcohol) and ethylene-acrylic acid copolymer films. Polym J 22:985.
21. Testoff MA, Rudolph AS. 1992. Modification of dry 1,2-dipalmitoylphosatidylcholine phase behavior with synthetic membrane-bound stabilizing carbohydrates. Bioconjugate Chem 3:203.
22. Fontanel M-L, Bazin H, Teoule R. 1993. End attachment of phenololigonucleotide conjugates to diazotized cellulose. Bioconjugate Chem 4:380.
23. Nilsson K, Mosbach K. 1984. Immobilization of ligands with organic sulfonyl chlorides. Methods Enzymol 104:56.
24. Chang Y-A, Gee A, Smith A, et al. Activating hydroxyl groups of polymeric carriers using 4-fluorobenzenesulfonyl chloride. Bioconjugate Chem 3:200.
25. Harris JM, Struck EC, Case MG. 1984. Synthesis and characterization of poly(ethylene glycol) derivatives. J Polym Sci Polym Chem Edn 22:341.
26. Sawhney AS, Hubbell JA. 1992. Poly(ethylene oxide)-graft-poly(L-lysine) copolymers to enhance the biocompatibility of poly(L-lysine)-alginate microcapsule membranes. Biomaterials 13:863.
27. Hearn MTW. 1987. 1,1′-Carbonyldiimidazole-mediated immobilization of enzymes and affinity ligands. Methods Enzymol 135:102.
28. Miron T, Wilchek M. 1993. A simplified method for the preparation of succinimidyl carbonate polyethylene glycol for coupling to proteins. Bioconjugate Chem 4:568.
29. Uy R, Wold F. 1977. 1,4-Butanediol diglycidyl ether coupling of carbohydrates to sepharose: affinity adsorbents for lectins and glycosidases. Anal Biochem 81:98.
30. Sundberg L, Porath J. 1974. Preparation of adsorbents for biospecific affinity chromatography: I. Attachment of group-containing ligands to insoluble polymers by means of bifunctional oxiranes. J Chromatogr 90:87.
31. Kondoh A, Makino K, Matsuda T. 1993. Two-dimensional artificial extracellular matrix: bioadhesive peptide-immobilized surface design. J Appl Polym Sci 47:1983.
32. Annunziato ME, Patel US, Ranade M, et al. 1993. P-Maleimidophenyl isocyanate: A novel heterobifunctional linker for hydroxyl to thiol coupling. Bioconjugate Chem 4:212.
33. Kobayashi H, Ikada Y. 1991. Covalent immobilization of proteins onto the surface of poly(vinyl alcohol) hydrogel. Biomaterials 12:747.
34. Shafer SG, Harris JM. 1986. Preparation of cyanuric-chloride activated poly(ethylene glycol). J Polym Sci Polym Chem Edn 24:375.
35. Kay G, Cook EM. 1967. Coupling of enzymes to cellulose using chloro-s-triazine. Nature (London) 216:514.
36. Bodanszky M. 1988. Peptide Chemistry, Berlin, Springer-Verlag.
37. Staab HA. 1962. Syntheses using heterocyclic amides (azolides). Agnew Chem Internat Edn 7:351.
38. Hirano Y, Okuno M, Hayashi T, et al. 1993. Cell-attachment activities of surface immobilized oligopeptides RGD, RGDS, RGDT, and YIGSR toward five cell lines. J Biomater Sci Polym Edn 4:235.
39. Ozaki CK, Phaneuf MD, Hong SL, et al. 1993. Glycoconjugate mediated endothelial cell adhesion to Dacron polyester film. J Vasc Surg 18:486.
40. Barnwarth W, Schmidt D, Stallard RL, et al. 1988. Bathophenanthroline-ruthenium(II) complexes as non-radioactive labels for oligonucleotides which can be measured by time-resolved fluorescence techniques. Helv Chim Acta 71:2085.
41. Drumheller PD, Elbert DL, Hubbell JA. 1994. Multifunctional poly(ethylene glycol) semi-interpenetrating polymer networks as highly selective adhesive substrates for bioadhesive peptide grafting. Biotech Bioeng 43:772.

42. Liu SQ, Ito Y, Imanishi Y. 1993. Cell growth on immobilized cell growth factor: 9. Covalent immobilization of insulin, transferrin, and collagen to enhance growth of bovine endothelial cells. J Biomed Mater Res 27:909.
43. Wachter E, Machleidt W, Hofner H, Otto J. 1973. Aminopropyl glass and its p-phenylene diisothiocyanate derivative, a new support in solid-phase Edman degradation of peptides and proteins. FEBS Lett 35:97.
44. Maisano F, Gozzini L, de Haen C. 1992. Coupling of DTPA to proteins: A critical analysis of the cyclic dianhydride method in the case of insulin modification. Bioconjugate Chem 3:212.
45. Streeter HB, Rees DA. 1987. Fibroblast adhesion to RGDS shows novel features compared with fibronectin. J Cell Biol 105:507.
46. Singer II, Kawka DW, Scott S, et al. 1987. The fibronectin cell attachment Arg-Gly-Asp-Ser promotes focal contact formation during early fibroblast attachment and spreading. J Cell Biol 104:573.
47. Nathan A, Bolikal D, Vyavahare N, et al. 1992. Hydrogels based on water-soluble poly(ether urethanes) derived from L-lysine and poly(ethylene glycol). Macromolecules 25:4476.
48. Elling L, Kula M-R. 1991. Immunoaffinity partitioning: Synthesis and use of polyethylene glycol-oxirane for coupling to bovine serum albumin and monoclonal antibodies. Biotech Appl Biochem 13:354.
49. Pope NM, Kulcinski DL, Hardwick A, et al. 1993. New applications of silane coupling agents for covalently binding antibodies to glass and cellulose solid supports. Bioconjugate Chem 4:166.
50. Werb Z, Tremble PM, Behrendtsen O, et al. 1989. Signal transduction through the fibronectin receptor induces collagenase and stromelysin gene expression. J Cell Biol 109:877.
51. Yamagata M, Suzuki S, Akiyama SK, et al. 1989. Regulation of cell-substrate adhesion by proteoglycans immobilized on extracellular substrates. J Biol Chem 264:8012.
52. Robinson PJ, Dunnill P, Lilly MD. 1971. Porous glass as a solid support for immobilization or affinity chromatography of enzymes. Biochim Biophys Acta 242:659.
53. Harris JM, Dust JM, McGill RA, et al. 1991. New polyethylene glycols for biomedical applications. ACS Symp Ser 467:418.
54. Cholakis CH, Zingg W, Sefton MV. 1989. Effect of heparin-PVA hydrogel on platelets in a chronic arterio-venous shunt. J Biomed Mater Res 23:417.
55. Cholakis CH, Sefton MV. 1984. Chemical characterization of an immobilized heparin: heparin-PVA. In SW Shalaby, AS Hoffman, BD Ratner, et al. (eds), Polymers as Biomaterials, New York, Plenum.
56. Gutsche AT, Parsons-Wingerter P, Chand D, et al. 1994. N-Acetylglucosamine and adenosine derivatized surfaces for cell culture: 3T3 fibroblast and chicken hepatocyte response. Biotech Bioeng 43:801.
57. Jagendorf AT, Patchornik A, Sela M. 1963. Use of antibody bound to modified cellulose as an immunospecific adsorbent of antigen. Biochim Biophys Acta 78:516.
58. Collioud A, Clemence J-F, Sanger M, et al. 1993. Oriented and covalent immobilization of target molecules to solid supports: Synthesis and application of a light-activatible and thiol-reactive crosslinking reagent. Bioconjugate Chem 4:528.
59. Bhatia SK, Shriver-Lake LC, Prior KJ, et al. Use of thiol-terminal silanes and heterobifunctional crosslinkers for immobilization of antibodies on silica surfaces. Anal Biochem 178:408.
60. Moeschler HJ, Vaughan M. 1983. Affinity chromatography of brain cyclic nucleotide phosphodiesterase using 3-(2-pyridyldithio)proprionyl-substituted calmodulin linked to thiol-Sepharose. Biochemistry 22:826.
61. Tseng Y-C, Park K. 1992. Synthesis of photoreactive poly(ethylene glycol) and its application to the prevention of surface-induced platelet activation. J Biomed Mater Res 26:373.
62. Yan M, Cai SX, Wybourne MN, et al. 1993. Photochemical functionalization of polymer surfaces and the production of biomolecule-carrying micrometer-scale structures by deep-UV lithography using 4-substituted perfluorophenyl azides. J Am Chem Soc 115:814.
63. Guire PE. 1993. Biocompatible device with covalently bonded biocompatible agent, U.S. Patent 5,263,992.

64. Ito Y, Suzuki K, Imanishi Y. 1994. Surface biolization by grafting polymerizable bioactive chemicals. ACS Symp Ser 540:66.
65. Nakayama Y, Matsuda T, Irie M. 1993. A novel surface photo-graft polymerization method for fabricated devices. ASAIO J 39:M542.
66. Ratner BD, Chilkoti A, Lopez GP. 1990. Plasma deposition and treatment for biomaterial applications. In R d'Agostino (ed), Plasma Deposition, Treatment, and Etching of Polymers, p 463, New York, Academic Press.
67. Ratner BD. 1992. Plasma deposition for biomedical applications: a brief review. J Biomater Sci Polym Edn 4:3.
68. Desai NP, Hubbell JA. 1991. Biological responses to polyethylene oxide modified polyethylene terephthalate surfaces. J Biomed Mater Res 25:829.
69. Lin H-B, Zhao Z-C, Garcia-Echeverria C, et al. 1992. Synthesis of a novel polyurethane co-polymer containing covalently attached RGD peptide. J Biomater Sci Polymer Edn 3:217.
70. Costello CA, McCarthy TJ. 1987. Surface-selective introduction of specific functionalities onto poly(tetrafluoroethylene). Macromolecules 20:2819.
71. Allmer K, Feiring AE. 1991. Photochemical modification of a fluoropolymer surface. Macromolecules 24:5487.
72. Ferguson GS, Chaudhury MK, Biebuyck HA, et al. 1993. Monolayers on disordered substrates: self-assembly of alkyltrichlorosilanes on surface-modified polyethylene and poly(dimethylsiloxane). Macromolecules 26:5870.
73. Plant AL. 1993. Self-assembled phospholipid/alkanethiol biomimetic bilayers on gold. Langmuir 9:2764.
74. Prime KL, Whitesides GM. 1993. Adsorption of proteins onto surfaces containing end-attached oligo(ethylene oxide): A model system using self-assembled monolayers. J Am Chem Soc 115:10714.
75. Biebuyck HA, Whitesides GM. 1993. Interchange between monolayers on gold formed from unsymmetrical disulfides and solutions of thiols: evidence for sulfur-sulfur bond cleavage by gold metal. Langmuir 9:1766.
76. Laibinis PE, Hickman JJ, Wrightson MS, et al. 1989. Orthogonal self-assembled monolayers: Alkanethiols on gold and alkane carboxylic acids on alumina. Science 245:845.
77. Veronese FM, Visco C, Massarotto S, et al. 1987. New acrylic polymers for surface modification of enzymes of therapeutic interest and for enzyme immobilization. Ann NY Acad Sci 501:444.
78. Scouten WH. 1987. A survey of enzyme coupling techniques. Methods Enzymol 135:30.
79. Mech C, Jeschkeit H, Schellenberger A. 1976. Investigation of the covalent bond structure of peptide-matrix systems by Edman degradation of support-fixed peptides. Eur J Biochem 66:133.
80. Iio K, Minoura N, Aiba S, et al. 1994. Cell growth on poly(vinyl alcohol) hydrogel membranes containing biguanido groups. J Biomed Mater Res 28:459.
81. Weinshenker NM, Shen C-M. 1972. Polymeric reagents: I. Synthesis of an insoluble polymeric carbodiimide. Tetrahedron Lett 32:3281.
82. Schnaar RL, Brandley BK, Needham LK, et al. 1989. Adhesion of eukaryotic cells to immobilized carbohydrates. Methods Enzymol 179:542.
83. Sparks MA, Williams KW, Whitesides GM. 1993. Neuraminidase-resistant hemagglutination inhibitors: acrylamide copolymers containing a C-glycoside of N-acetylneuramic acid. J Med Chem 36:778.
84. Ozeki E, Matsuda T. 1990. Development of an artificial extracellular matrix. Solution castable polymers with cell recognizable peptidyl side chains. ASAIO Trans 36:M294.
85. Drumheller PD. 1994. Polymer Networks of Poly(Ethylene Glycol) as Biospecific Cell Adhesive Substrates, PhD dissertation, University of Texas at Austin.
86. Nicol A, Gowda DC, Parker TM, et al. 1993. Elastomeric polytetrapeptide matrices: Hydrophobicity dependence of cell attachment from adhesive $(GGIP)_n$ to nonadhesive $(GGAP)_n$ even in serum. J Biomed Mater Res 27:801.
87. Kohn J, Gean KF, Nathan A, et al. 1993. New drug conjugates: attachment of small molecules to poly(PEG-Lys). Polym Mater Sci Eng 69:515.

88. Nicol A, Gowda DC, Urry DW. 1992. Cell adhesion and growth on synthetic elastomeric matrices containing ARG-GLY-ASP-SER. J Biomed Mater Res 26:393.
89. Barrera DA, Zylstra E, Lansbury PT, et al. 1993. Synthesis and RGD peptide modification of a new biodegradable copolymer: Poly(lactic acid-colysine). J Am Chem Soc 115:11010.
90. Pulapura S, Kohn J. 1993. Tyrosine-derived polycarbonate: Backbone-modified "pseudo"-poly(amino acids) designed for biomedical applications. Biopolymers 32:411.
91. Ngo TT. Coupling capacity of solid-phase carboxyl groups. Determination by a colorimetric procedure. Appl Biochem Biotech 13:207.
92. Ngo TT. 1986. Colorimetric determination of reactive amino groups of a solid support using Traut's and Ellman's reagents. Appl Biochem Biotech 13:213.
93. Duan C, Meyerhoff ME. 1994. Separation-free sandwich enzyme immunoassays using microporous gold electrodes and self-assembled monolayer/immobilized capture antibodies. Anal Chem 66:1369.
94. Albeda SM, Buck CA. 1990. Integrins and other cell adhesion molecules. FASEB J 4:2868.
95. Ruoslahti E. 1991. Integrins. J Clin Invest 87:1.
96. Humphries MJ. 1990. The molecular basis and specificity of integrin-ligand interactions. J Cell Sci 97:585.
97. Underwood PA, Bennett FA. 1989. A comparison of the biological activities of the cell-adhesive proteins vitronectin and fibronectin. J Cell Sci 93:641.
98. Yamada KM, Kennedy DW. 1985. Amino acid sequence specificities of an adhesive recognition signal. J Cell Biochem 28:99.
99. Massia SP, Hubbell JA. 1991. An RGD spacing of 44 nm is sufficient for integrin $\alpha_v\beta_3$-mediated fibroblast spreading and 140 nm for focal contact and stress fiber formation. J Cell Biol 114:1089.
100. Hubbell JA, Massia SP, Desai NP, et al. 1992. Endothelial cell-selective materials for tissue engineering in the vascular graft via a new receptor. Bio/Technology 9:568.
101. Brandley BK, Schnaar RL. 1989. Tumor cell haptotaxis on covalently immobilized linear and exponential gradients of a cell adhesion peptide. Dev Biol 135:74.
102. Drumheller PD, Elbert DL, Hubbell JA. 1994. Multifunctional poly(ethylene glycol) semi-interpenetrating polymer networks as highly selective adhesive substrates for bioadhesive peptide grafting. Biotech Bioeng 43:772.
103. Clapper DL, Daws KM, Guire PE. 1994. Photoimmobilized ECM peptides promote cell attachment and growth on biomaterials. Trans Soc Biomater 17:345.
104. Spinke J, Liley M, Guder H-J, et al. 1993. Molecular recognition at self-assembled monolayers: The construction of multicomponent multilayers. Langmuir 9:1821.
105. Singhvi R, Kumar A, Lopez GP, et al. 1994. Engineering cell shape and function. Science 264:696.

111

Biomaterials: Protein-Surface Interactions

Joseph A. Chinn
Sulzer Carbomedics

Steven M. Slack
The University of Memphis

111.1 Introduction

A common assumption in biomaterials research is that cellular interactions with natural and artificial surfaces are mediated through adsorbed proteins. Such diverse processes as **thrombosis** and **hemostasis**, hard and soft tissue healing, infection, and inflammation are each affected by protein adsorption to surfaces *in vivo.* Many *in vitro* diagnostic analyses, chromatographic separation techniques, and genetic engineering processes also involve protein adsorption at solid-liquid interfaces.

The adsorption of fibrinogen, a prevalent blood plasma protein, has been studied extensively because of its role in blood coagulation and thrombosis, as has the adsorption of albumin, because it is thought to inhibit the adhesion of blood platelets [Young et al., 1982]. The amount of protein adsorbed to a substrate is best measured directly using **radiolabel**ed proteins, whereas the thickness of an adsorbed protein film can be calculated from ellipsometric measurements. Further, the importance of the state of an adsorbed protein in mediating cellular interactions is now becoming evident. Molecularly sensitive indirect measurement techniques, e.g., circular dichroism (CD), differential scanning calorimetry (DSC), enzyme-linked immunosorbent assay (ELISA), Fourier transform infrared spectroscopy/attenuated total reflectance (FTIR/ATR), radio-immunoassay (RIA), or total internal reflection fluorescence (TIRF), can be used to characterize the **conformation** and **organization** of adsorbed proteins. (The amount of protein adsorbed to a substrate is best measured directly using **radiolabeled** proteins; the thickness of an adsorbed protein film can be calculated from ellipsometric measurements.) Highly specific monoclonal (MAb) (against specific protein **epitopes**) [Shiba et al., 1991] and polyclonal (PAb) (against multiple epitopes) [Lindon et al., 1986] antibodies provide direct probes of adsorbed protein conformation and organization. Thus, cellular responses can be compared not only with the amounts of proteins adsorbed but also the organization of the proteins on the surface of the substrate.

Whereas previous studies confirmed roles for adsorbed proteins in subsequent cell-surface interactions, much current research aims to better understand cell-protein-surface interactions on a molecular level. Recently, peptide sequences contained within the cell binding domains of adhesive proteins have been identified and characterized, synthesized, and demonstrated to bind cellular receptors known as **integrins** [Ruoslahti, 1991; Yamada, 1991]. Current and potential applications range from selective or enhanced *in vitro* cell culture to selective *in vivo* cellular responses such as endothelialization in the absence of inflammation, infection, or thrombosis [Hubbell et al., 1991].

0-8493-0462-8/00/$0.00+$.50

111.2 Fundamentals of Protein Adsorption

Detailed and comprehensive reviews of protein adsorption have been published [Andrade, 1985; Andrade & Hlady, 1986; Horbett, 1982; Norde & Lyklema, 1991]. A thorough understanding of key principles will prove helpful in critically evaluating reports in literature. Particularly important concepts are protein structure and heterogeneity, factors that dramatically affect the thermodynamics and kinetics of adsorption, reversibility of adsorption, and the dynamics of multi-component adsorption.

A protein is a complex molecule consisting of amino acid copolymer (polypeptide) chains that interact with each other to give the molecule a three-dimensional structure. Importantly, each amino acid in the polymer contributes to the chemical and physical properties of the protein. A dramatic example of this is the oxygen-carrying protein, hemoglobin, which consists of four polypeptide chains denoted $\alpha_2\beta_2$. A single amino acid substitution in the 146 amino acid β-chain results in the conversion of normal hemoglobin (HbA) to sickle-cell hemoglobin (HbS) and underlies the serious consequences of sickle-cell disease [Stryer, 1995]. Protein structure and function are relevant to protein adsorption and have been described on four different scales or orders [Andrade & Hlady, 1986]. Primary structure refers to the sequence and number of amino acids in a copolymer chain. The 20 amino acids that are polymerized to make proteins are termed **residues**. Of these, 8 have non-polar side chains, 7 have neutral polarity side chains, and 5 have charged polar side chains [Stryer, 1995]. Secondary structure results from hydrogen bonding associated with the amide linkages in the polymer chain backbone to form structures such as the α-helix and β-pleated sheet. Tertiary structure results from associations within chains, including hydrogen bonding, ionic and hydrophobic interactions, salt bridges, and disulfide bonds and dictates the three-dimensional structure adopted by protein molecules. Quaternary structure results from associations between chains. Many blood proteins contain polar, non-polar, and charged residues. In polar media such as buffered saline or blood plasma, hydrophilic residues tend to self-associate (often at the outer, water-contacting surface of the protein), as do hydrophobic residues (often "inside" the protein). This results in distinct domains (Fig. 111.1) that dictate higher order protein structure.

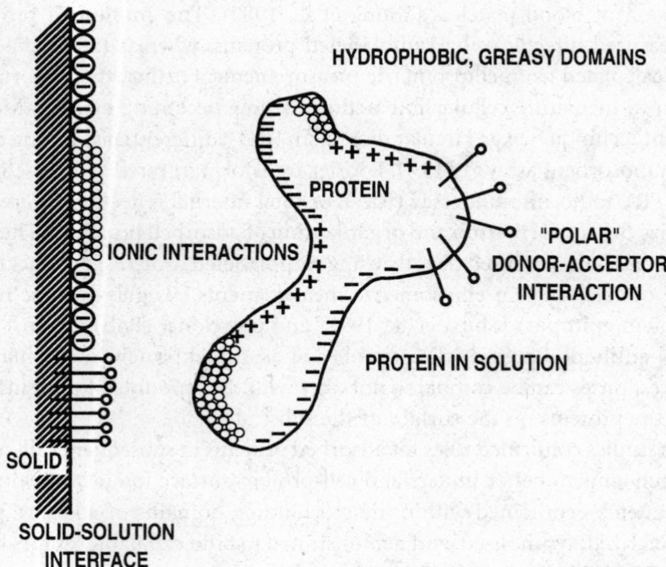

FIGURE 111.1 Schematic view of protein interacting with a well characterized surface. The protein has a number of surface domains with hydrophobic, charged, and polar character. The solid surface has a similar domain like character (Andrade & Hlady, 1986.)

When a single, static protein solution contacts a surface, the rate of adsorption depends upon transport of the protein from the bulk to the surface. Andrade and Hlady [1986] identified four primary transport mechanisms, namely diffusion, thermal convection, flow convection, and coupled convection-diffusion. In isothermal, parallel laminar flow or static systems, protein transport to the interface occurs exclusively by diffusion. In turbulent or stirred systems, each of the four transport modes can be significant.

When adsorption is reaction limited, the net rate of adsorption can sometimes be described by the classic Langmuir theory of gas adsorption [Smith, 1981]:

$$r_A = k_A * C_b * (1 - \Theta) - k_D * \Theta \qquad (111.1)$$

where r_A is the net adsorption rate, k_A is the adsorption rate constant, C_b is the bulk concentration of the protein in solution, Θ is fractional surface coverage, and k_D is the desorption rate constant. At equilibrium, the net rate of adsorption, r_A is zero, and Θ can be calculated from Eq. (111.1) as the Langmuir adsorption isotherm,

$$\Theta = \frac{K * C_b}{1 + K * C_b} \qquad (111.2)$$

where $K = k_A/k_D$. This model assumes reversible monolayer adsorption, no conformational changes upon adsorption, and no interactions between adsorbed molecules. It is most applicable to dilute solutions and non-hydrophobic substrates. When adsorption is diffusion limited, the initial rate of adsorption is equivalent to the rate of diffusion, described mathematically as [Andrade & Hlady, 1986]

$$r_A = r_{\mathcal{D}} = 2 * C_{\mathcal{D}} * \sqrt{\frac{\mathcal{D}}{\pi t}} \qquad (111.3)$$

where $\mathcal{D}$ is the diffusivity of the protein, and t is time.

Protein adsorption to hydrophobic substrates differs from that to hydrophilic materials. The primary driving force for adsorption to hydrophilic substrates is often enthalpic, whereas that to hydrophobic substrates is entropic [Norde, 1986]. Water near a hydrophobic surface tends to hydrogen bond to neighboring water molecules, resulting in a highly ordered water structure [Andrade & Hlady, 1986]. Disruption of this structure (dehydration) by adsorption of a protein to the surface increases the entropy of the system and is therefore thermodynamically favored. As a result, adsorption to hydrophilic substrates is generally reversible, whereas that to hydrophobic substrates is not. Denaturation of the adsorbed protein by hydrophobic interactions with the substrate can also contribute to irreversible adsorption [Feng & Andrade, 1994].

The amount of a specific protein adsorbed to a substrate can be measured directly if the protein is radiolabeled. Gamma emitting isotopes are preferred because their signal is directly proportional to the amount of protein present. Radioisotopes of iodine (^{125}I, ^{129}I, and ^{131}I) are commonly used because iodine readily attaches to tyrosine residues [Macfarlane, 1958]. An ^{125}I monochloride radiolabeling technique has been published [Horbett, 1986] but others, such as those using chloramine-T or lactoperoxidase, can also be used.

If neither the ^{125}I-protein nor the unlabeled protein preferentially adsorbs to the substrate at the expense of the other, then the amount of that protein adsorbed to a substrate from multi-component media such as plasma can be measured by adding a small amount of the ^{125}I-protein to the adsorption medium. Studies have shown that ^{125}I-fibrinogen generally behaves like its unlabeled analog [Horbett, 1981]. The total amount of the protein adsorbed (both unlabeled and ^{125}I-protein) is calculated by dividing the measured radioactivity by the **specific activity** of the protein in the medium. (Specific activity is determined by dividing the gamma activity in a measured aliquot of the adsorption medium by the total amount of that protein, both labeled and unlabeled, in the aliquot.) To verify that neither labeled nor unlabeled fibrinogen preferentially adsorbs to the substrates, the specific activity of the protein in a small aliquot of the plasma dilution from which adsorption was maximum should be increased 10x, and adsorption from that dilution again measured. Changes in calculated adsorption values should be attributable only to the variability in the data and differences in the signal to noise ratio, not the ratio of labeled to unlabeled fibrinogen in the plasma. Similarly, to verify that absorption into the sample materials of free ^{125}I in the buffer is not significant, adsorption from dilute plasma to which only 0.01M unlabeled free iodide is added should be measured. Sample calculations for measuring protein adsorption using this technique are illustrated in the next section.

The use of radioisotopes for the measurement of protein adsorption to surfaces offers a significant, though underused, advantage compared with other techniques. Because the radioisotopes, e.g., ^{125}I and ^{131}I, emit unique energy spectra, i.e., their peak radiation emissions occur at distinct energies, one can simultaneously measure the adsorption of two different proteins from a protein mixture. By labeling one protein with ^{125}I and the other with ^{131}I, the adsorption behavior of both can be determined in one experiment. Modern gamma counters come with preset energy windows specific to common isotopes, making such measurements routine. In the absence of such an instrument, one can still perform dual-labeling experiments by exploiting the fact that the half-lives of the isotopes differ. For example, the half-life of ^{125}I is sixty days whereas that of ^{131}I is only eight days. After measuring the ^{131}I emission immediately after the experiment, one can allow the radioactivity associated with it to decay to background levels (ten half-lives or 80 days for ^{131}I) and then measure the signal associated solely with ^{125}I [Dewanjee, 1992].

Indirect methods are also used to study proteins adsorbed to a substrate. ELISA and RIA analytical techniques exploit specific antibody-antigen interactions as follows. Antibodies against specific epitopes of an adsorbed protein are either conjugated to an enzyme (ELISA), or radiolabeled (RIA). Substrates are first incubated with the medium, then with a solution containing the antibody or antibody conjugate. In the case of ELISA, the substrates are subsequently incubated with substrate solution. As the substrate is converted to product, the color of the solution changes in proportion to the amount of bound antibody present and is measured spectrophotometrically. However, extensive calibration is required to quantify results. In the case of RIA, the radioactivity originating from the radiolabeled antibody (bound to the protein adsorbed to the substrate) is measured and the amount of antibody bound calculated. With both methods, the relative amount of the adsorbed protein to which the antibody has bound, rather than the total amount of protein adsorbed, is measured. Antibody binding is a function of not only the amount of protein adsorbed, but also the particular antibody used, total protein surface loading, and protein residence time. Thus, although antibody techniques provide direct probes of adsorbed protein conformation and organization, such measurements do not necessarily reflect the absolute amounts of protein adsorbed. Other indirect methods used to study adsorbed proteins include ellipsometry, electron microscopy, high performance liquid chromatography (HPLC), and staining techniques.

111.3 Example Calculations and Applications of Protein Adsorption

The following example illustrates how radiolabeled proteins are used to measure the amount of protein adsorbed to a substrate. The fibrinogen concentration in 10 ml plasma is determined to be 5.00 mg/ml by measuring the light absorbance at 280 nm of a redissolved, thrombin-induced clot [Ratnoff & Menzie, 1950]. The concentration and specific activity of a 10 μl aliquot of ^{125}I-fibrinogen are 1.00 mg/ml and 10^9 cpm/μg, respectively. Fibrinogen adsorption from dilute plasma to a series of polymer samples, each having 1.00 cm^2 total surface area (counting both sides of the sample) is to be measured. Based upon reports in literature, maximal adsorption of 250 ng/cm^2 is expected. The background signal in the gamma radiation counter is 25 cpm. To achieve a maximum signal/noise ratio of 10, the specific activity of fibrinogen in the plasma should be,

$$sp.ac. = \frac{(signal/noise) * noise}{mass\ adsorbed} = \frac{10 * 25\ cpm}{25\ ng/cm^2 * 1.00\ cm^2} * \frac{10^3\ ng}{\mu g} = 10^3\ cpm/\mu g \qquad (111.4)$$

The volume of ^{125}I-fibrinogen solution to be added to the plasma to obtain 10^3 cpm/g specific activity (neglecting the mass of ^{125}I-fibrinogen added) is calculated as,

$$\frac{10^3\ cpm/\mu g * 10^3\ \mu g/mg * 5.00\ mg/ml * 10\ ml}{10^9\ cpm/mg * 1.00\ mg/ml * ml/10^3\ \mu l} = 50\ \mu l \qquad (111.5)$$

Addition of 50 µl ^{125}I-fibrinogen solution should increase the total fibrinogen concentration in the plasma by only a small fraction,

$$\frac{50\ \mu l * ml/10^3\ \mu l * 1.00\ mg/ml}{10\ ml * 5.00\ mg/ml} = 10^{-3} \qquad (111.6)$$

To determine the amount of protein adsorbed to the substrate, the radioactivity of samples incubated with plasma is measured and compared with the specific activity of the protein in the plasma. In this example, if the amount of radioactivity retained by a sample is measured to be 137 cpm, then the mass of fibrinogen adsorbed is calculated as,

$$\frac{(137\ cpm/sample - 25\ cpm\ background) * 10^3\ ng/\mu g}{10^3\ cpm/\mu g\ fibrinogen * 1.00\ cm^2/sample} = 112\ ng/cm^2 \qquad (111.7)$$

Adsorption of proteins to polymeric substrates is measured because adsorbed proteins influence cellular processes. Adsorbed albumin is proposed to favor biocompatibility, whereas adsorbed fibrinogen is proposed to discourage biocompatibility because of its role in mediating initial adhesion of blood platelets [Young et al., 1982]. This simplified view inadequately describes biocompatibility *in vivo* for several reasons. First, the relationships between processes involved in thrombosis, hemostasis, inflammation, and healing (e.g., adhesion of platelets, fibroblasts, white blood cells, endothelial cells) and long term biocompatibility remain mostly unknown. For example, Sakariassen and colleagues [1979] proposed that exclusively adsorbed von Willebrand factor (vWF) mediates platelet adhesion to vascular subendothelial structures, yet although some people that lack serum vWF in their blood exhibit bleeding disorders, others remain asymptomatic. Second, biological processes that do not require fibrinogen mediated cell adhesion (e.g., contact activation [Kaplan, 1978], complement activation [Chenowith, 1988]) are also related to material biocompatibility. Third, biological factors and serum proteins other than fibrinogen and albumin (e.g., vWF, fibronectin, vitronectin, laminin) significantly affect cellular processes. Fourth, the reactivity of adsorbed proteins depends upon their organization upon the substrate.

Studies of fibrinogen adsorption *in vitro* illustrate the dynamic nature of protein adsorption from plasma. Vroman and Adams [1969] reported that oxidized silicon and anodized tantalum incubated 2s with plasma bind fibrinogen anti-serum, whereas the same materials incubated 25s with plasma do not. Ellipsometric measurements indicated that the observed decrease in antibody binding was not due to loss of protein. Brash and ten Hove [1984] and Horbett [1984] reported that maximal equilibrium fibrinogen adsorption to different materials occurred from intermediate dilutions of plasma. The adsorption maximum is sometimes called a **Vroman peak**, which describes the shape of the adsorption versus log (**plasma concentration**) curve referred to as an adsorption isotherm. Both the location and the magnitude of the peak depend upon the surface chemistry of the substrate. (For this reason, it is wise to fully characterize any substrate prior to measuring protein adsorption. Electron spectroscopy for chemical analysis, ESCA [Ratner & McElroy, 1986], and secondary ion mass spectroscopy, SIMS [Andrade, 1985], are often appropriate.)

Wojciechowski and colleagues [1986] reported that at short contact times, adsorption was greatest from undiluted plasma. As contact time increased, the plasma concentration at which adsorption was greatest decreased. The unusual observed adsorption behavior occurs because fibrinogen adsorption is driven initially by mass action, i.e., a gradient between surface and bulk concentration. However, as coverage of the surface increases, bulk proteins must compete for surface binding sites. The composition of the adsorbed layer continues to change as proteins of higher surface activity displace adsorbed proteins of lower surface activity [Horbett, 1984]. Vroman and colleagues [1980] called this process conversion of the fibrinogen layer, and proposed that fibrinogen adsorbed at early time is at later time displaced by other plasma proteins, possibly high molecular weight kininogen (HMWK). Because Vroman pioneered much of this work, these phenomena are collectively referred to as the **Vroman effect** [Slack & Horbett, 1995]. This principle is most applicable to hydrophilic substrates and is commonly used in biocompatibility studies to vary the composition of the protein layer adsorbed from plasma in a controlled manner.

Slack and Horbett [1989] postulated the existence of two distinct types of adsorbed fibrinogen molecules; displaceable and non-displaceable. Protein adsorption from plasma was modeled as competitive adsorption from a binary solution of fibrinogen and a hypothetical protein H (representing all other plasma components). In this model, protein H adsorbs to unoccupied surface sites in a non-displaceable state, and fibrinogen molecules first adsorb to unoccupied surface sites in a displaceable state, then are displaced by protein H, or spread to become resistant to displacement. The latter process is referred to as fibrinogen transition. Neglecting desorption and mass transfer limitations, rate equations for surface coverage by protein H, displaceable fibrinogen, and non-displaceable fibrinogen were solved simultaneously for a surface initially free of adsorbate. The analytical solution is given by Eq. (111.8) and Eq. (111.9).

$$\Theta_1(t) = \beta\left(e^{r_1^* t} - e^{r_2^* t}\right) \tag{111.8}$$

$$\Theta_2(t) = \left(\beta * k_2 / r_1 * r_2\right) * \left[r_1\left(1 - e^{r_2^* t}\right) - r_2\left(1 - e^{r_1^* t}\right)\right] \tag{111.9}$$

where Θ_1, Θ_2, and Θ_3 are fractional coverages by displaceable fibrinogen, non-displaceable fibrinogen, and the hypothetical protein, respectively, k_1 is the fibrinogen adsorption rate constant, k_2 is the fibrinogen transition rate constant, k_3 is the hypothetical protein adsorption rate constant, k_4 is the fibrinogen displacement rate constant, C_F is the bulk concentration of fibrinogen, C_H is the bulk concentration of the hypothetical protein in solution, and β, r_1, and r_2 are constants related to rate constants and bulk concentrations. This model predicts maximal fibrinogen coverage at intermediate adsorption time (Fig. 111.2), consistent with reported experimental results. Surface exclusion and molecular mobility arguments have also been proposed to explain the Vroman effect [Willems et al., 1991].

Changes in protein conformation upon adsorption have been inferred from a variety of indirect measurements. For example, Castillo and colleagues [1984] inferred substrate, adsorption time, and residence time dependent conformational changes in human serum albumin (HSA) adsorbed to different hydrogels based upon changes in FTIR/ATR and CD spectra. Specific conformational changes, i.e., decreased α-helix and increased random coil and β-pleated sheet content upon adsorption were proposed. Similarly, Norde and colleagues [1986] reported lower α-helix content in HSA first adsorbed to, then desorbed from different substrates compared with native HSA. Castillo and colleagues [1985] also reported that lysozyme became increasingly denatured with increased adsorption time and residence time, and that denatured lysozyme adsorbed irreversibly to contact lens materials. De Baillou and colleagues [1984] proposed conformational changes in fibrinogen upon adsorption to glass based upon DSC measurements. Similarly, Feng and Andrade [1994] proposed that low temperature isotropic (LTI)

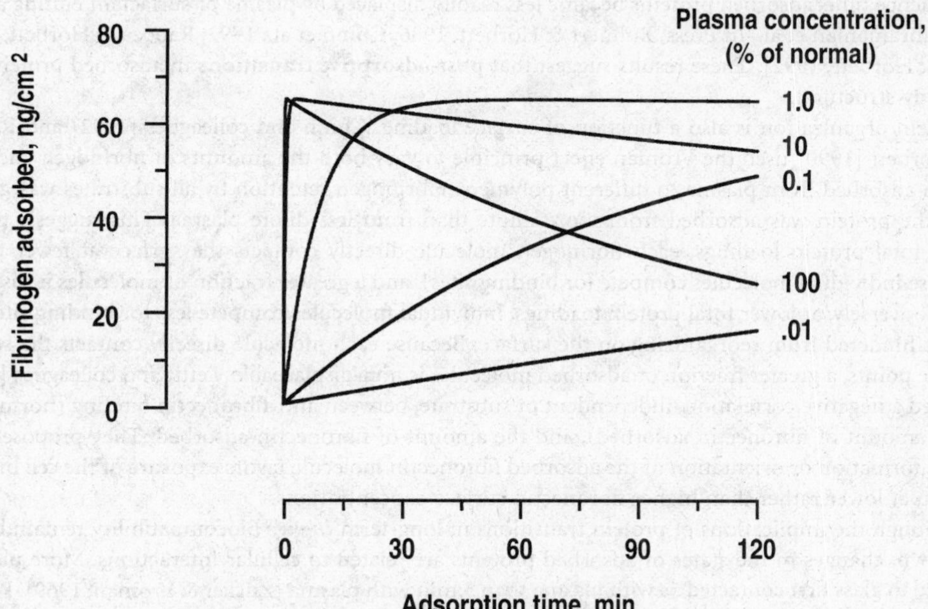

FIGURE 111.2 Time course of fibrinogen adsorption to a polymeric substrate as predicted by the model proposed by Slack and Horbett [1989].

carbon significantly denatures adsorbed proteins through hydrophobic interactions. Rainbow and colleagues [1987] proposed conformational changes in albumin upon adsorption to quartz, based upon changes in fluorescence lifetimes calculated from TIRF measurements. Finally, investigators noted that certain enzymes, following their adsorption to solid surfaces, lose a considerable amount of their enzymatic activity compared with enzymes in solution, suggesting that a surface-induced structural rearrangement had occurred [Sandwick & Schray, 1987; Sandwick & Schray, 1988].

Whereas indirect methods provide evidence for changes in protein organization upon adsorption, antibodies against specific protein epitopes provide direct evidence, as antibody-binding measurements directly reflect the availability of different protein epitopes. Proteins at interfaces can undergo both covalent (e.g., conversion of fibrinogen to fibrin) and non-covalent (e.g., denaturation, change in conformation, change in epitope accessibility) organizational changes. Therefore, epitopes inaccessible in solution might become available upon adsorption of the protein, and consequently, the biological activity of the adsorbed protein may differ from that of the same protein in solution. For example, receptor-induced binding site (RIBS) anti-fibrinogens bind to adsorbed but not free fibrinogen molecules [Zamarron et al., 1990]. Because an MAb binds to a single protein epitope, whereas a PAb binds to multiple epitopes on the protein, MAbs rather than PAbs should be more sensitive to such changes. Also, epitopes available at low surface loadings may become unavailable at higher loadings due to steric hindrance by proteins adsorbed to neighboring surface sites.

Horbett and Lew [1994] used the Vroman effect principle to maximize the amount of fibrinogen adsorbed from plasma to different polymers, and then measured the binding of different MAbs directed against various epitopes of fibrinogen. Binding was reported to be substrate-dependent and, with some MAbs, changed with protein residence time. Thus, different fibrinogen epitopes become more or less available as the adsorbed molecule reorganizes upon the surface. Soria and colleagues [1985] demonstrated that a MAb directed against fragment D of the fibrinogen molecule, which did not bind to the molecule in solution, did bind to it following its adsorption to a surface. Although this method cannot distinguish between changes in protein conformation (higher order structure) and changes in surface orientation (e.g., rotation, steric effects), several authors reported that with increased adsorption time or residence time, adsorbed proteins became less readily displaced by plasma or surfactant eluting agents [Balasubramanian et al., In Press; Bohnert & Horbett, 1986; Chinn et al., 1992; Rapoza & Horbett, 1990; Slack & Horbett, 1992]. These results suggest that **post-adsorptive transitions** in adsorbed proteins are primarily structural.

Protein organization is also a function of surface loading. Chinn and colleagues [1992] and Rapoza and Horbett [1990] used the Vroman effect principle to vary both the amounts of fibrinogen and total protein adsorbed from plasma to different polymers. Fibrinogen retention by all substrates was greater when the protein was adsorbed from more dilute than from less dilute plasma. This suggests that at higher total protein loadings, each fibrinogen molecule directly contacts the surface at fewer points (because individual molecules compete for binding sites), and a greater fraction of molecules is displaceable. Conversely, at lower total protein loadings, individual molecules compete less for binding sites and are not hindered from reorganizing on the surface. Because each molecule directly contacts the surface at more points, a greater fraction of adsorbed molecules is non-displaceable. Pettit and colleagues [1994] reported a negative correlation, independent of substrate, between anti-fibronectin binding (normalized to the amount of fibronectin adsorbed), and the amount of fibronectin adsorbed. They proposed that the conformation or orientation of the adsorbed fibronectin molecule favors exposure of the cell binding domain at lower rather than higher fibronectin surface concentrations.

Although the implications of protein transitions in long-term *in vivo* biocompatibility remain largely unknown, changes in the states of adsorbed proteins are related to cellular interactions. More platelets adhered to glass first contacted 5s with plasma than 3 min with plasma [Zucker & Vroman, 1969]. Platelet adhesion *in vitro* to polymers upon which the Vroman effect principle was used to vary the composition of adsorbed protein layer was reported related not to total fibrinogen binding, but to anti-fibrinogen binding [Lindon et al., 1986; Shiba et al., 1991], as well as the fraction of adsorbed protein that can be eluted by surfactant such as sodium dodecyl sulfate (SDS) [Chinn et al., 1991].

111.4 Summary, Conclusions, and Directions

Clearly, adsorbed proteins affect biocompatibility in ways that are not entirely understood. Fibrinogen has been extensively studied because blood platelets involved in thrombosis and hemostasis have a receptor for this protein [Phillips et al., 1988]. Adsorbed fibrinogen is often proposed to discourage biocompatibility, but this view does not consider that adsorbed proteins exist in different states, depending upon adsorption conditions, residence time, and substrate. Evidence suggests that fibrinogen adsorbed from blood to substrates upon which the protein readily denatures, e.g., LTI carbon, may in fact promote biocompatibility. Because it is the organization and not the amount of an adsorbed protein that determines its biological activity, what happens to proteins after they adsorb to the substrate must be determined to properly evaluate the biocompatibility of a material. Further, a material might be made blood compatible if protein organization can be controlled such that the cell binding epitopes become unavailable for cell binding [Horbett et al., 1994]. Much current research aims to understand the relationship between the states of adsorbed proteins and cell-protein-surface interactions.

Fundamental to better understanding of material biocompatibility is understanding the importance of protein and surface structure and heterogeneity in determining the organization of proteins at the solid-liquid interface, the dynamics of multi-component adsorption from complex media, e.g., the Vroman effect, and the significance of post-adsorptive events and subsequent cellular interactions. MAbs against specific protein epitopes provide direct evidence of changes in the states of adsorbed proteins, and indirect methods provide corroborative evidence. Molecular imaging techniques such as atomic (AFM), lateral force (LFM), and scanning tunneling (STM) microscopies have been adapted to aqueous systems to better define the states of adsorbed proteins with limited success [Sit et al., 1998], as have surface analysis techniques such as surface-matrix assisted laser desorption ionization (MALDI) [Kingshott et al., 1998].

Identification and characterization of the cell binding domains of adhesive proteins has led to better understanding of cell adhesion at the molecular level. Pierschbacher and colleagues[1981] used monoclonal antibodies and proteolytic fragments of fibronectin to identify the location of the cell attachment site of the molecule. Subsequently, residue sequences within the cell binding domains of other adhesive proteins were identified as summarized by Yamada [1991]. The Arg-Gly-Asp (RGD) amino acid sequence first isolated from fibronectin was later found present within vitronectin, osteopontin, collagens, thrombospondin, fibrinogen, and vWF [Ruoslahti & Pierschbacher, 1987]. Different adhesive peptides exhibit cell line dependent biological activity *in vitro*, but relatively few *in vivo* studies of adhesive peptides have been reported.

Haverstick and colleagues [1985] reported that addition of RGD-containing peptides to protein-free platelet suspension inhibited thrombin induced platelet aggregation, as well as platelet adhesion *in vitro* to fibronectin, fibrinogen, and vWF coated polystyrene. These results suggest that binding of the peptide to the platelet renders the platelet's receptor for the proteins unavailable for further binding. Similarly, Hanson and Harker [1988] used MAbs against the platelet glycoprotein IIb/IIIa (fibrinogen binding) complex to prevent thrombus formation upon Dacron vascular grafts placed within a chronic baboon AV shunt. Controlled release of either MAbs or adhesive peptides at the site of medical device implant might allow localized control of thrombosis. Locally administered adhesive peptides might also be used to selectively control cell behavior, e.g., Ruoslahti [1992] used RGD containing peptides to inhibit tumor invasion *in vitro* and dissemination *in vivo.* A recently developed drug delivery system offers great promise with respect to these therapeutic interventions [Markou et al., 1996; Markou et al., 1998].

Alternatively, adhesive peptides can be used to promote cell proliferation. Hubbell and colleagues [1991] reported that immobilization of different adhesive peptides resulted in selective cell response *in vitro.* Whereas the immobilized RGD and YIGSR peptides both enhanced spreading of human foreskin fibroblasts, human vascular smooth muscle cells, and human vascular endothelial cells, immobilization of REDV enhanced spreading of only endothelial cells. If this concept can be applied *in vivo,* then endothelialization in the absence of inflammation, infection, or thrombosis might be achieved.

Defining Terms

Conformation: Higher order protein structure that describes the spatial relationship between the amino acid chains that a protein comprises.

Epitope: Particular regions of a protein to which an antibody or cell can bind.

Hemostasis: Mechanism by which damaged blood vessels are repaired without compromising normal blood flow.

Integrin: Cellular transmembrane protein that acts as a receptor for adhesive extracellular matrix proteins such as fibronectin. The tripeptide RGD is the sequence recognized by many integrins.

Organization: The manner in which a protein resides upon a surface, in particular, the existence, arrangement, and availability of different protein epitopes.

Plasma concentration: Not the concentration of total protein in the plasma, but rather the volume fraction of plasma in the adsorption medium when protein adsorption from different dilutions of plasma is measured.

Post-adsorptive transitions: Changes in protein conformation and organization that occur when adsorbed proteins reside upon a surface.

Residue: The individual amino acids of a peptide or protein.

Specific activity: The amount of radioactivity detected per unit mass of a specific protein.

State: The reactive state of an adsorbed protein as determined by its conformation and organization.

Thrombosis: Formation of plug comprising blood platelets and fibrin that stops blood flow through damaged blood vessels. Embolized thrombus refers to a plug that has detached from the wound site and entered the circulation.

Vroman effect: Collective term describing (1) maximal adsorption of a specific protein from multi-component medium at early time, (2) maximal equilibrium adsorption from intermediate dilution, and (3) decrease with increased adsorption time in the plasma concentration at which adsorption is maximum.

Vroman peak: The adsorption maximum in the adsorption versus log (plasma concentration) curve when protein adsorption from different dilutions of plasma is measured.

References

Andrade JD. 1985. Principles of protein adsorption. In JD Andrade (ed), *Surface and Interfacial Aspects of Biomedical Polymers: Protein Adsorption,* vol 2, pp 1–80, Plenum, New York.

Andrade JD, Hlady V. 1986. Protein adsorption and materials biocompatibility: A tutorial review and suggested hypotheses. In *Advances in Polymer Science* 79. Biopolymers/Non-Exclusion HPLC, pp 1–63, Springer-Verlag, Berlin.

Andrade JD. 1985. Polymer surface analysis: Conclusions and expectations. In JD Andrade (ed), *Surface and Interfacial Aspects of Biomedical Polymers,* vol 1, Surface Chemistry and Physics, pp. 443-460, Plenum Press, Boca Raton.

Balasubramanian V, Grusin NK, Bucher RW, et al. In Press. Residence-time dependent changes in fibrinogen adsorbed to polymeric biomaterials. *J. Biomed. Mater. Res.*

Bohnert JL, Horbett TA. 1986. Changes in adsorbed fibrinogen and albumin interactions with polymers indicated by decreases in detergent elutability. *J. Coll. Interface Sci.* 111:363.

Brash JL, ten Hove P. 1984. Effect of plasma dilution on adsorption of fibrinogen to solid surfaces. *Thromb. Haemostas.* 51:326.

Castillo EJ, Koenig JL, Anderson JM et al., 1984. Characterization of protein adsorption on soft contact lenses. I. Conformational changes of adsorbed serum albumin. *Biomaterials* 5:319.

Castillo EJ, Koenig JL, Anderson JM et al., 1985. Characterization of protein adsorption on soft contact lenses. II. Reversible and irreversible interactions between lysozyme and soft contact lens surfaces. *Biomaterials* 6:338.

Chenowith DE. 1988. Complement activation produced by biomaterials. *Artif. Organs* 12:502.

Chinn JA, Posso SE, Horbett TA et al., 1991. Post-adsorptive transitions in fibrinogen adsorbed to Biomer: Changes in baboon platelet adhesion, antibody binding, and sodium dodecyl sulfate elutability. *J. Biomed. Mater. Res.* 25:535.

Chinn JA, Posso SE, Horbett TA et al., 1992. Post-adsorptive transitions in fibrinogen adsorbed to polyurethanes: Changes in antibody binding and sodium dodecyl sulfate elutability. *J. Biomed. Mater. Res.* 26:757.

De Baillou N, Dejardin P, Schmitt A et al., 1984. Fibrinogen dimensions at an interface: Variations with bulk concentration, temperature, and pH. *J. Coll. Interf. Sci.* 100:167.

Dewanjee MK. 1992. Radioiodination: Theory, Practice, and Biomedical Application, Kluwer Academic Press, Norwell.

Feng L, Andrade JD. 1994. Protein adsorption on low temperature isotropic carbon. I. Protein conformational change probed by differential scanning calorimetry. *J. Biomed. Mater. Res.* 28:735.

Hanson SR, Harker LA. 1988. Interruption of acute platelet-dependent thrombosis by the synthetic antithrombin D-phenylalanyl-L-prolyl-L-arginyl chloromethylketone. *Proc. Natl. Acad. Sci. USA* 85:3184.

Haverstick DM, Cowan JF, Yamada KM et al., 1985. Inhibition of platelet adhesion to fibronectin, fibrinogen, and von Willebrand factor substrates by a synthetic tetrapeptide derived from the cell-binding domain of fibronectin. *Blood* 66:946.

Horbett TA. 1981. Adsorption of proteins from plasma to a series of hydrophilic-hydrophobic copolymers. II. Compositional analysis with the prelabeled protein technique. *J. Biomed. Mater. Res.* 15:673.

Horbett TA. 1982. Protein adsorption on biomaterials. In SL Cooper, NA Peppas (eds), *Biomaterials: Interfacial Phenomena and Applications*, ACS Advances in Chemistry Series, vol 199, pp 233–244, Washington, D.C., American Chemical Society.

Horbett TA. 1984. Mass action effects on competitive adsorption of fibrinogen from hemoglobin solutions and from plasma. *Thromb. Haemostas.* 51:174.

Horbett TA. 1986. Techniques for protein adsorption studies. In DF Williams (ed), *Techniques of Biocompatibility Testing*, pp 183–214, CRC Press, Boca Raton.

Horbett TA, Grunkemeier JM, Lew KR. 1994. Fibrinogen orientation of a surface coated with a GP IIb/IIIa peptide detected with monoclonal antibodies. *Trans. Soc. Biomater.* 17:335.

Horbett TA, Lew KR. 1994. Residence time effects on monoclonal antibody binding to adsorbed fibrinogen. *J. Biomat. Sci. Polym. Edn.* 6:15.

Hubbell JA, Massia SP, Desai NP et al., 1991. Endothelial cell-selective materials for tissue engineering in the vascular graft via a new receptor. *Biotechnology (NY)* 9:568.

Kaplan AP. 1978. Initiation of the intrinsic coagulation and fibrinolytic pathway of man: The role of surfaces, Hageman factor, prekallikrein, high molecular weight kininogen, and factor XI. *Prog. Hemostas. Thromb.* 4:127.

Kingshott P, St. John HAW, Vaithianathan T et al., 1998. Study of protein adsorption at monolayer and sub-monolayer levels by surface-MALDI spectroscopy. *Trans. Soc. Biomater.* 21:253.

Lindon JN, McManama G, Kushner L et al., 1986. Does the conformation of adsorbed fibrinogen dictate platelet interactions with artificial surfaces? *Blood* 68:355.

Macfarlane AS. 1958. Efficient trace-labelling of proteins with iodine. *Nature* 182:53.

Markou CP, Chronos NF, Harker LA et al., 1996. Local endovascular drug delivery for inhibition of thrombosis. *Circulation* 94:1563.

Markou CP, Lutostansky EM, Ku DN et al., 1998. A novel method for efficient drug delivery. *Ann. Biomed. Engr.* 26:502.

Norde W, MacRitchie F, Nowicka, G et al. 1986. Protein adsorption at solid-liquid interfaces: reversibility and conformation aspects. *J. Colloid Interface Sci.* 112:447.

Norde W. 1986. Adsorption of proteins from solution at the solid-liquid interface. *Adv. Coll. Interf. Sci.* 25:267.

Norde W, Lyklema J. 1991. Why proteins prefer interfaces. *J. Biomater. Sci. Polym. Edn.* 2:183.

Pettit DK, Hoffman AS, Horbett TA. 1994. Correlation between corneal epithelial cell outgrowth and monoclonal antibody binding to the cell domain of fibronectin. *J. Biomed. Mater. Res.* 228:685.

Phillips DR, Charo IF, Parise LV et al., 1988. The platelet membrane glycoprotein IIb-IIIa complex. *Blood* 71:831.

Pierschbacher MD, Hayman EG, Ruoslahti E. 1981. Location of the cell-attachment site in fibronectin with monoclonal antibodies and proteolytic fragments of the molecule. *Cell* 26:259.

Rainbow MR, Atherton S, Eberhart RE. 1987. Fluorescence lifetime measurements using total internal reflection fluorimetry: Evidence for a conformational change in albumin adsorbed to quartz. *J. Biomed. Mater. Res.* 21:539.

Rapoza RJ, Horbett TA. 1990. Postadsorptive transitions in fibrinogen: Influence of polymer properties. *J. Biomed. Mater. Res.* 24:1263.

Ratner BD, McElroy BJ. 1986. Electron spectroscopy for chemical analysis: Applications in the biomedical sciences. In RM Gendreau RM (ed), *Spectroscopy in the Biomedical Sciences*, pp 107–140, CRC Press, Boca Raton.

Ratnoff OD, Menzie C. 1950. A new method for the determination of fibrinogen in small samples of plasma. *J. Lab. Clin. Med.* 37:316.

Ruoslahti E. 1991. Integrins. *J. Clin. Invest.* 87:1.

Ruoslahti E. 1992. The Walter Herbert Lecture: Control of cell motility and tumour invasion by extracellular matrix interactions. *Br. J. Cancer* 66:239.

Ruoslahti E, Pierschbacher MD. 1987. New perspectives in cell adhesion: RGD and integrins. *Science* 238:491.

Sakariassen KS, Bolhuis PA, Sixma JJ. 1979. Human blood platelet adhesion to artery subendothelium is mediated by factor VIII-Von Willebrand factor bound to the subendothelium. *Nature* 279:636.

Sandwick RK, Schray KJ. 1987. The inactivation of enzymes upon interaction with a hydrophobic latex surface. *J. Coll. Interf. Sci.* 115:130.

Sandwick RK, Schray KJ. 1988. Conformational states of enzymes bound to surfaces. *J. Coll. Interf. Sci.* 121:1.

Shiba E, Lindon JN, Kushner L et al., 1991. Antibody-detectable changes in fibrinogen adsorption affecting platelet activation on polymer surfaces. *Am. J. Physiol.* 260:C965.

Sit PS, Siedlecki CA, Shainoff JR et al. 1998. Substrate-dependent conformations of human fibrinogen visualized by atomic force microscopy under aqueous conditions. *Trans. Soc. Biomater.* 21:101.

Slack SM, Horbett TA. 1989. Changes in the state of fibrinogen adsorbed to solid surfaces: An explanation of the influence of surface chemistry on the Vroman effect. *J. Colloid Interface Sci.* 133:148.

Slack SM, Horbett TA. 1992. Changes in fibrinogen adsorbed to segmented polyurethanes and hydroxymethacrylate-ethylmethacrylate copolymers. *J. Biomed. Mater. Res.* 26:1633.

Slack SM, Horbett TA. 1995. The Vroman effect: A critical review. In TA Horbett, JL Brash (eds), *Proteins At Interfaces II,* pp 112–128, Washington, D.C., American Chemical Society.

Smith JM. 1981. *Chemical Engineering Kinetics,* 3rd ed, McGraw Hill, New York.

Soria J, Soria C, Mirshahi M et al., 1985. Conformational changes in fibrinogen induced by adsorption to a surface. *J. Coll. Interf. Sci.* 107:204.

Stryer L. 1995. *Biochemistry,* 4th ed, W.H. Freeman, New York.

Vroman L, Adams AL. 1969. Identification of rapid changes at plasma-solid interfaces. *J. Biomed. Mater. Res.* 3:43.

Vroman L, Adams AL, Fischer GL et al., 1980. Interaction of high molecular weight kininogen, factor XII and fibrinogen in plasma at interfaces. *Blood* 55:156.

Willems GM, Hermens WT, Hemker HC. 1991. Surface exclusion and molecular mobility may explain Vroman effects in protein adsorption. *J. Biomater. Sci. Polym. Edn.* 1:217.

Wojciechowski P, ten Hove P, Brash JL. 1986. Phenomenology and mechanism of the transient adsorption of fibrinogen from plasma (Vroman effect). *J. Colloid Interface Sci.* 111:455.

Yamada KM. 1991. Adhesive recognition sequences. *J. Biol. Chem.* 266:12809.

Young BR, Lambrecht LK, Cooper SL. 1982. Plasma proteins: Their role in initiating platelet and fibrin deposition on biomaterials. In SL Cooper, NA Peppas (eds), *Biomaterials: Interfacial Phenomena and Applications,* ACS Advances in Chemistry Series, vol 199, pp 317–350, Washington, D.C., American Chemical Society.

Zamarron C, Ginsberg MH, Plow EF. 1990. Monoclonal antibodies specific for a conformationally altered state of fibrinogen. *Thromb. Haemostas.* 64:41.

Zucker MB, Vroman L. 1969. Platelet adhesion induced by fibrinogen adsorbed on glass. *Proc. Soc. Exp. Med.* 131:318.

Further Information

The American Society for Artificial and Internal Organs (ASAIO) (P.O. Box C, Boca Raton, FL 33429-0468, Web address *http://www.asaio.com*) publishes original articles in ASAIO J through Lippincott-Raven Publishers (227 East Washington Square, Philadelphia, PA 19106-3780) and meeting transactions in *Trans. Am. Soc. Artif. Intern. Organs.*

The Society for Biomaterials (SFB) (6518 Walker St., Ste. 150, Minneapolis, MN 55426-4215, Web address *http://www.biomaterials.org*) publishes original articles in both *J. Biomed. Mater. Res.* and *J. Biomed. Mater. Res. Appl. Biomater.* through John Wiley and Sons, Inc. (605 Third Ave., New York, NY 10158) and meeting transactions in *Trans. Soc. Biomaterials.* The *J. Colloid Interface Sci.* often contains articles related to protein-surface interactions as well.

The American Association for the Advancement of Science (AAAS) (1200 New York Ave. NW, Washington, D.C. 20002, Web address *http://www.aaas.org*) publishes original articles in science, and often contains excellent review articles and very current developments in protein research. *Nature*, published through Macmillan Magazines (Porters South, 4 Crinan St., London N1 9XW), is similar in content but provides a decidedly European perspective.

Comprehensive references summarizing applications of protein adsorption and biocompatibility include: 1982. SL Cooper, NA Peppas (eds.), *Biomaterials: Interfacial Phenomena and Applications*, ACS Advances in Chemistry Series, vol 199, Washington, D.C., American Chemical Society; 1987. TA Horbett and JL Brash (eds.), *Proteins at Interfaces: Physicochemical and Biochemical Studies*, ACS Symposium Series, vol 343, Washington, D.C., American Chemical Society; and 1993. SM Factor (ed.), Cardiovascular Biomaterials And Biocompatibility. *Cardiovasc. Path.* 2(3) Suppl.

112

Engineering Biomaterials for Tissue Engineering: The 10–100 Micron Size Scale

David J. Mooney
Massachusetts Institute of Technology

Robert S. Langer
Massachusetts Institute of Technology

A significant challenge in tissue engineering is to take a biomaterial and process it into a useful form for a specific application. All devices for tissue engineering transplant cells and/or induce the ingrowth of desirable cell types from the host organism. The device must provide sufficient mechanical support to maintain a space for tissue to form or serve as a barrier to undesirable interactions. Also, the device can be designed to provide these functions for a defined period before biodegradation occurs or on a permanent basis.

Generally speaking, devices can be broken down into two types. Immunoprotective devices contain semipermeable membranes that prevent elements of the host immune system (e.g., IgG antibodies and lymphocytes) from entering the device. In contrast, open devices have large pores (>10 μm) and allow free transport of molecules and cells between the host tissue and transplanted cells. These latter devices are utilized to engineer a tissue that is completely integrated with the host tissue. Both types of devices can range in size from microns to centimeters or beyond, although the larger sizes are usually repetitions on the structure found at the scale of hundreds of microns.

A fundamental question in designing a device is whether to use synthetic or natural materials. Synthetic materials (e.g., organic polymers) can be easily processed into various structures and can be produced cheaply and reproducibly; it also is possible to tightly control various properties such as the mechanical strength, hydrophobicity, and degradation rate of synthetic materials. Whereas natural materials (e.g., collagen) sometimes exhibit a limited range of physical properties and can be difficult to isolate and process, they do have specific biologic activity. In addition, these molecules generally do not elicit unfavorable host tissue responses, a condition which is typically taken to indicate that a material is biocompatible. Some synthetic polymers, in contrast, can elicit a long-term inflammatory response from the host tissue [Bostman, 1991].

A significant challenge in fabricating devices is either to develop processing techniques for natural biomaterials that allow reproducible fabrication on a large-scale basis [Cavallaro et al., 1994] or to develop materials that combine the advantages of synthetic materials with the biologic activity of natural biomaterials [Barrera et al., 1993; Massia & Hubbell, 1991]. Computer-aided-design–computer-aided-manufacturing (CAD-CAM) technology may possibly be employed in the future to custom-fit devices with complex structures to patients.

0-8493-0462-8/00/$0.00+$.50

112.1 Fundamentals

The interaction of the host tissue with the device and transplanted cells can be controlled by both the geometry of the device and the internal structure. The number of inflammatory cells and cellular enzyme activity around implanted polymeric devices has been found to depend on the geometry of the device [Matlaga et al., 1976], with device geometries that contain sharp angles provoking the greatest response. The surface geometry, or microstructure, of implanted polymer devices also has been found to affect the types and activities of acute inflammatory cells recruited to the device as well as the formation of a fibrous capsule [Taylor & Gibbons, 1983].

The pore structure of a device dictates the interaction of the device and transplanted cells with the host tissue. The pore structure is determined by the size, size distribution, and continuity of the individual pores within the device. Porous materials are typically defined as microporous (pore diameter $d < 2$ nm), mesoporous (2 nm $< d <$ 50 nm), or macroporous ($d >$ 50 nm) [Schaeffer, 1994]. Only small molecules (e.g., gases) are capable of penetrating microporous materials. Mesoporous materials allow transport of larger molecules, such as small proteins, but transport of large proteins and cells is prevented. Macroporous materials allow free transport of large molecules, and, if the pores are large enough ($d >$ 10^4 nm), cells are capable of migrating through the pores of the device. The proper design of a device can allow desirable signals (e.g., a rise in serum sugar concentration) to be passed to transplanted cells while excluding molecular or cellular signals which would promote rejections of transplanted cells (e.g., IgG protein).

Fibrovascular tissue will invade a device if the pores are larger than approximately 10 μm, and the rate of invasion will increase with the pore size and total porosity of a device [Mikos et al., 1993*c*; Weslowski et al., 1961; White et al., 1981]. The degree of fibrosis and calcification of early fabric leaflet valves has been correlated to their porosity [Braunwald et al., 1965], as has the nonthrombogenicity of arterial prosthesis [DeBakey et al., 1964] and the rigidity of tooth implants and orthopedic prosthesis [Hamner et al., 1972; Hulbert et al., 1972].

It is important to realize that many materials do not have a unimodal pore size distribution or a continuous pore structure, and the ability of molecules or cells to be transported through such a device will often be limited by bottlenecks in the pore structure. In addition, the pore structure of a device may change over time in a biologic environment. For example, absorption of water into polymers of the lactic/glycolic acid family results in the formation first of micropores, and eventually of macropores as the polymer itself degrades [Cohen et al., 1991]. The porosity and pore-size distribution of a device can be determined utilizing a variety of techniques [Smith et al., 1994].

Specific properties (e.g., mechanical strength, degradability, hydrophobicity, biocompatibility) of a device are also often desirable. These properties can be controlled both by the biomaterial itself and by the processing technique utilized to fabricate the device. An advantage of fabricating devices from synthetic polymers is the variety of processing techniques available for these materials. Fibers, hollow fibers, and porous sponges can be readily formed from synthetic polymers. Natural biomaterials must be isolated from plant, animal, or human tissue and are typically expensive and suffer from large batch-to-batch variations. Although the wide range of processing techniques available for synthetic polymers is not available for these materials, cells specifically interact with certain types of natural biomaterials, such as extracellular matrix (ECM) molecules [Hynes, 1987]. The known ability of ECM molecules to mediate cell function in vitro and in vivo may allow precise control over the biologic response to devices fabricated from ECM molecules.

112.2 Applications

The applications of tissue engineering are very diverse, encompassing virtually every type of tissue in the human body. However, the devices utilized in these areas can be divided into two broad types. The first type, immunoprotective devices, utilizes a semipermeable membrane to limit communication between cells in the device and the host. The small pores in these devices ($d < 10$ nm) allow low-molecular-weight proteins and molecules to be transported between the implant and the host tissue, but they prevent large proteins (e.g., immunoglobulins) and host cells (e.g., lymphocytes) of the immune system from entering the device and mediating rejection of the transplanted cells. In contrast, open structures with large pores are typically utilized ($d > 10$ μm) if one desires that the new tissue be structurally integrated with the host tissue. Applications that utilize both types of devices are described below.

Immunoprotective Devices

Devices that protect transplanted cells from the immune system of the host can be broken down into two types, microencapsulation and macroencapsulation systems [Emerich et al., 1992]. Individual cells or small clusters of cells are surrounded by a semipermeable membrane and delivered as a suspension in microencapsulation systems (Fig. 112.1*a*). Macroencapsulation systems utilize hollow semipermeable membranes to deliver multiple cells or cell clumps (Fig. 112.1*b*). The small size, thin wall, and spherical shape of microcapsules all optimize diffusional transport to and from the microencapsulated cells. Macroencapsulation devices typically have greater mechanical integrity than microencapsule devices, and they can be easily retrieved after implantation. However, the structure of these devices is not optimal for diffusional transport. Nonbiodegradable materials are the preferred choice for fabricating both types of devices, as the barrier function is typically required over the lifetime of the implant.

A significant effort has been made to cure diabetes by transplanting microencapsulated pancreatic islet cells. Transplantation of nonimmunoprotected islets has led to short-term benefits [Lim & Sun, 1980], but the cells were ultimately rejected. To prevent this, islets have been immobilized in alginate (a naturally occurring polymer derived from seaweed) microbeads coated with a layer of poly(L-lysine) and a layer of polyethyleneimine [Lim & Sun, 1980]. Alginate is ionically crosslinked in the presence of calcium, and the permeability of alginate/poly(L-lysine) microbeads is determined by the formation of ionic or hydrogen bonds between the polyanion alginate and the polycation poly(L-lysine). This processing technique allows cells to be immobilized without exposure to organic solvents or high temperatures. The outer layer of polyethyleneimine was subsequently replaced by a layer of alginate to prevent the formation of fibrous capsules around the implanted microcapsules [O'Shea et al., 1984]. Smaller microbeads (250–400 μm) have been generated with an electrostatic pulse generator [Lum et al., 1992] to improve the in vivo survival and the response time of encapsulated cells [Chicheportiche & Reach, 1988]. These devices have been shown to be effective in a variety of animal models [Lim & Sun, 1980; Lum et al., 1992], and clinical trials of microencapsulated islets in diabetic patients are in progress [Soon-Shiong et al., 1994]. Synthetic analogs to alginate have also been developed [Cohen et al., 1990].

The superior mechanical stability of macroencapsulation devices, along with the possibility of retrieving the entire device, makes these types of devices especially attractive when the transplanted cells have limited lifetimes and/or when one needs to ensure that the transplanted cells are not migrating out of the device. Macroencapsulation devices have been utilized to transplant a variety of cell types, including pancreatic cells [Lacy et al., 1991], NGF-secreting cells [Winn et al., 1994], dopamine-secreting cells [Emerich et al., 1992], and Chromaffin cells [Sagen et al., 1993]. The nominal molecular mass cutoff of the devices was 50 kD, allowing immunoprotection without interfering with transport of therapeutic agents from the encapsulated cells. To prevent cell aggregation and subsequent large-scale cell death due to nutrient limitations, macroencapsulated islets have been immobilized in alginate [Lacy et al., 1991].

Devices with Open Structures

Devices with large, interconnected pores ($d > 10$ μm) are utilized in applications where one wishes the transplanted cells to interact directly with host tissue and form a structurally integrated tissue. The open structure of these devices provides little barrier to diffusional transport and often promotes the ingrowth of blood vessels from the host tissue. Degradable materials are often utilized in these applications, since once tissue structures are formed the device is not needed.

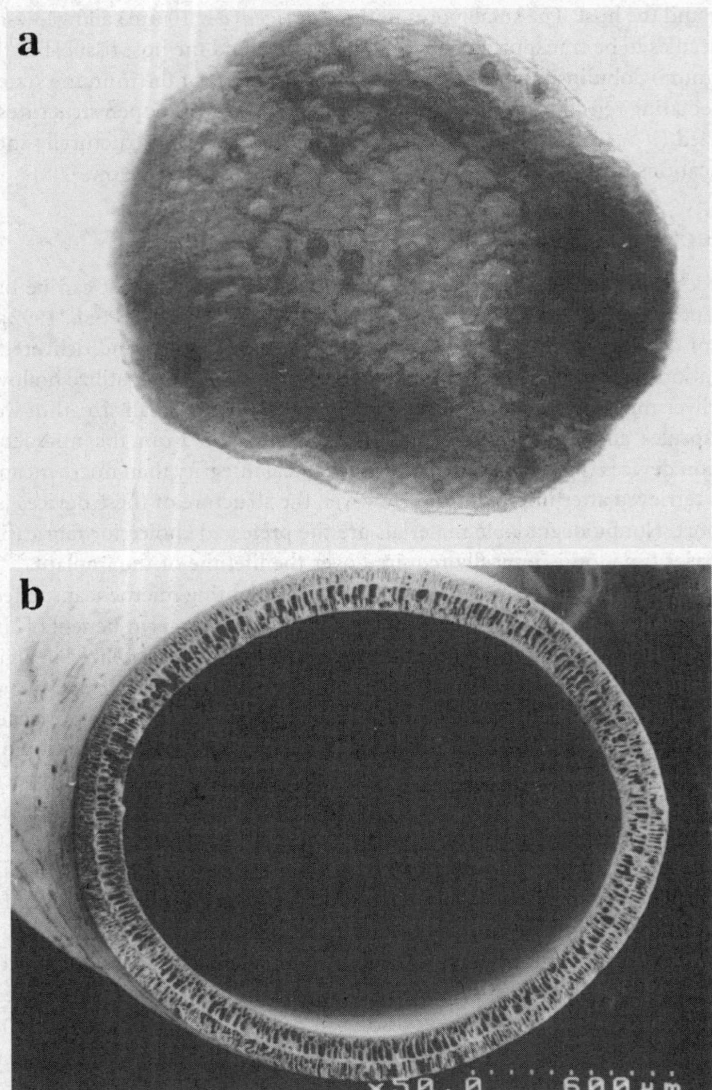

FIGURE 112.1 Examples of microencapsulated cells and a device used for macroencapsulation of cells. (*a*) Phase contrast photomicrograph of hybridoma cells encapsulated in a calcium crosslinked polyphosphazene gel [Cohen et al., 1993]. Original magnification was 100×. Used with permission of Editions de Sante. (*b*) A SEM photomicrograph of a poly(acrylonitrile-co-vinyl chloride) hollow fiber formed by phase inversion using a dry-jet wet-spinning technique [Schoichet et al., 1994]. Used with permission of John Wiley and Sons, Inc.

These types of devices typically fall into two categories. The first types is fabrics, either woven or nonwoven, of small-diameter (approximately 10–40 μm) fibers (Fig. 112.2*a*). High porosity (>95%) and large average pore size can be easily obtained with this type of material, and these materials can be readily shaped into different geometries. Fibers can be formed from synthetic, crystalline polymers such as polyglycolic acid by melt extrusion [Frazza & Schmitt, 1971]. Fibers and fabrics also can be formed from natural materials such as type I collagen, a type of ECM molecule, by extrusion of soluble collagen into a bath where gelling occurs followed by dehydration of the fiber [Cavallaro et al., 1994]. The tensile strength of fibers is dependent on the extent of collagen crosslinking, which can be controlled by the processing technique [Wang et al., 1994]. Processed collagen fibers can be subsequently spooled and

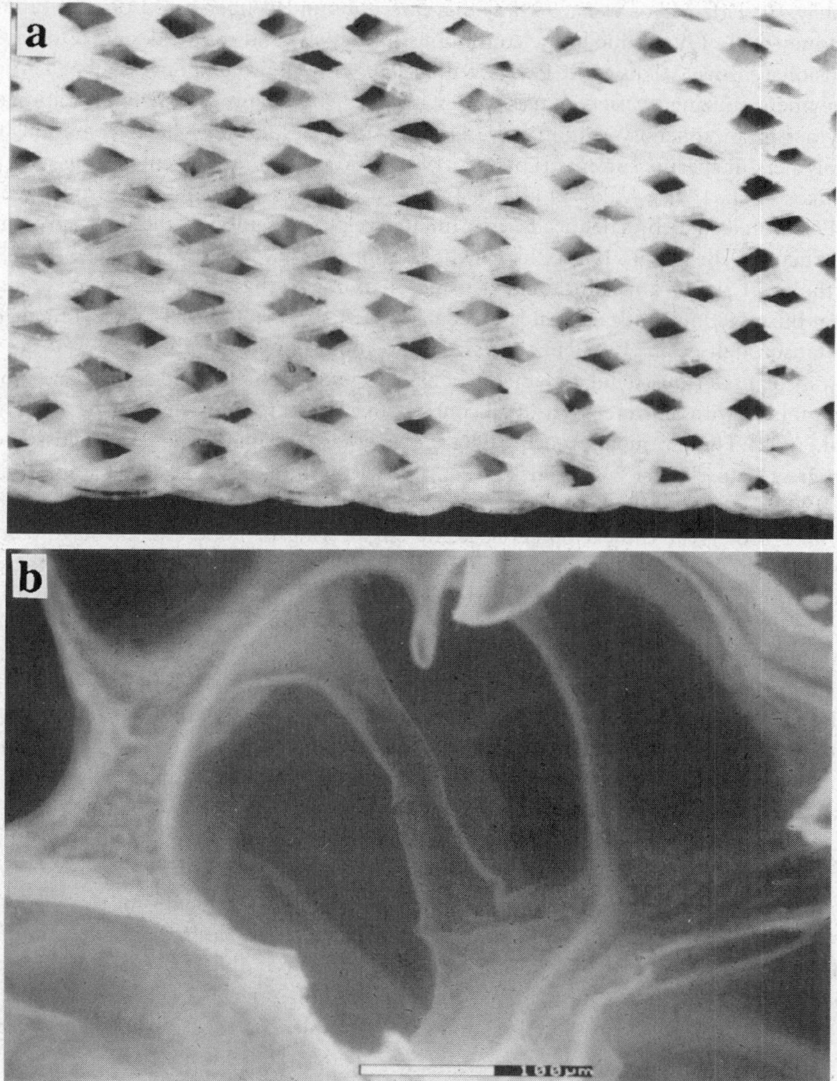

FIGURE 112.2 Examples of a fiber-based fabric and a porous sponge utilized for tissue engineering. (*a*) A photomicrograph of type I collagen fibers knitted into a fabric. Fiber diameters can be as small as 25 μm, and devices constructed from these fibers can be utilized for a variety of tissue engineering applications [Cavallaro et al., 1994]. Used with permission of John Wiley and Sons, Inc. (*b*) A SEM photomicrograph of a formaldehyde-crosslinked polyvinyl alcohol sponge. These devices have been utilized for a variety of applications, including hepatocyte transplantation [Uyama et al., 1993].

knitted to form fabrics [Cavallaro et al., 1994]. These devices are often ideal when engineering two-dimensional tissues. However, these fabrics typically are incapable of resisting large compressional forces, and three-dimensional devices are often crushed in vivo. Three-dimensional fiber-based structures have been stabilized by physically bonding adjacent fibers [Mikos et al., 1993*a*; Mooney et al., 1994*a*; Vacanti et al., 1992].

To engineer three-dimensional tissues, porous sponge devices (Fig. 112.2*b*) are utilized typically in place of fiber-based devices. These devices are better capable of resisting larger compressional forces (approximately 10^4 Pa) [Mikos et al., 1993*a*] than are unbounded fiber-based devices (approximately 10^2 Pa) [Mooney et al., 1994*a*] due to the continuous solid phase and can be designed to have complex, three-dimensional forms [Mikos et al., 1993*b*; White et al., 1972]. Porous sponges can be fabricated from synthetic polymers utilizing a variety of techniques, including performing the polymerization of a hydrophobic polymer in an aqueous solution [Chirila et al., 1993], exploiting phase separation behavior of dissolved polymers in specific solvents [Lo et al., 1994], and combining solvent casting with particulate leaching [Mikos et al., 1993*b*; Mikos et al., 1994]. Porous sponges can be formed from type I collagen and other ECM molecules by chemically crosslinking gels or assembling the collagen in nonnatural polymeric structures [Bell et al., 1981; Chvapil, 1979; Stenzel et al., 1974; Yannas et al., 1982].

Perhaps the most significant clinical effort using open devices has been expended to engineer skin tissue to treat burn victims. Both natural [Bell et al., 1981; Yannas et al., 1982] and synthetic degradable materials [Hansbrough et al., 1992] in the form of porous sponges or fiber-based fabrics have been utilized to transplant various cellular elements of skin. One device fabricated from ECM molecules has also been combined with an outer coat of silicone elastomer to prevent dehydration of the wound site [Yannas et al., 1982]. The various approaches have shown efficacy in animal models, and tissue-engineered skin has progressed to clinical trials [Burke et al., 1981; Compton et al., 1989; Heimbach et al., 1988; Stern et al., 1990].

Another area with great clinical potential is the engineering of bone and cartilage tissue. Various ceramics and biodegradable synthetic polymers have been utilized to fabricate devices for these purposes. Porous calcium phosphate devices loaded with mesenchymal stem cells have been shown to promote bone formation when implanted into soft tissue sites of animals [Goshima et al., 1991; Haynesworth et al., 1992]. Ceramics also have been coated onto prosthetic devices (e.g., hip replacements) to promote bone ingrowth and bonding between the prosthetic device and the host tissue [Furlong & Osborn, 1991]. The degradation rate [de Bruijn et al., 1994] and mechanical properties [Yoshinari et al., 1994] of these ceramics can be controlled by the deposition technique, which determines the crystallinity and chemical structure of the deposited ceramic. The brittleness of ceramic materials limits them in certain applications, and to bypass this problem composite ceramic/polymer devices have been developed [Stupp et al., 1993]. Fiber-based fabrics of biodegradable polymers have also been utilized to transplant cells derived from periosteal tissue and form new bone tissue [Vacanti et al., 1993]. To engineer cartilage tissue with specific structures such as an ear or nasal septum, devices have been fabricated from a nonwoven mesh of biodegradable synthetic polymers molded to the size and shape of the desired tissue. These devices, after seeding with chondrocytes and implantation, have been shown to induce the formation of new cartilage tissue with the same structure as the polymer device utilized as the template [Puelacher et al., 1993; Vacanti et al., 1991, 1992]. After tissue development is complete, the device itself degrades to leave a completely natural tissue.

Liver tissue [Uyama et al., 1993], ligaments [Cavallaro et al., 1994], and neural tissue [Guenard et al., 1992; Madison et al., 1985] also have been engineered with open devices. Tubular tissues, including blood vessels [Weinberg & Bell, 1986], intestine [Mooney et al., 1994*b*; Organ et al., 1993], and urothelial structures [Atala et al., 1992] have been engineered utilizing open devices fabricated into a tubular structure.

112.3 Conclusions

A variety of issues must be addressed to design and fabricate a device for tissue engineering. Do the transplanted cells need immunoprotection, or should they structurally integrate with the host tissue? If immunoprotection is desired, will a micro- or macroencapsulation device be preferred? If a structurally integrated new tissue is desired, will a fiber-based device or a porous sponge be more suitable? The specific roles that the device will play in a given application and the material itself will dictate the design of the device and the fabrication technique.

Defining Terms

Biocompatible: A material which does not elicit an unfavorable response from the host but instead performs with an appropriate host response in a specific application [Williams, 1987].

Biodegradation: The breakdown of a material mediated by a biologic system [Williams et al., 1992]. Biodegradation can occur by simple hydrolysis or via enzyme- or cell-mediated breakdown.

Extracellular matrix (ECM) molecules: Various substances present in the extracellular space of tissues that serve to mediate cell adhesion and organization.

Immunoprotective: Serving to protect from interacting with the immune system of the host tissue, including cellular elements (e.g., lymphocytes) and proteins (e.g., IgG).

Open devices: Devices with large ($d > 10$ μm) interconnected pores which allow unhindered transport of molecules and cells within the device and between the device and the surrounding tissue.

References

Barrera DA, Zylstra E, Lansbury PT, et al. 1993. Synthesis and RGD peptide modification of a new biodegradable copolymer: poly(lactic acid-co lysine). J Am Chem Soc 115:11010.

Bell E, Ehrlich HP, Buttle DJ, Nakatsuji T. 1981. Living tissue formed in vitro and accepted as skin-equivalent tissue of full thickness. Science 211:1052.

Bostman OM. 1991. Absorbable implants for the fixation of fractures. J Bone Joint Surg 73-A(1):148.

Braunwald NS, Reis RL, Pierce GE. 1965. Relation of pore size to tissue ingrowth in prosthetic heart valves: an experimental study. Surgery 57:741.

Burke JF, Yannas IV, Quinby WC, et al. 1981. Successful use of a physiological acceptable artificial skin in the treatment of extensive burn injury. Ann Surg 194:413.

Cavallaro JF, Kemp PD, Kraus KH. 1994. Collagen fabrics as biomaterials. Biotech Bioeng 43:781.

Chicheportiche D, Reach G. 1988. In vitro kinetics of insulin release by microencapsulated rat islets: effect of the size of the microcapsules. Diabetologia 31:54.

Chirila TV, Constable IJ, Crawford GJ, et al. 1993. Poly(2-hydroxyethyl methacrylate) sponges as implant materials: in vivo and in vitro evaluation of cellular invasion. Biomaterials 14(1):26.

Chvapil M. 1979. Industrial uses of collagen. In DAD Parry, LK Creamer (eds), Fibrous proteins: scientific, industrial, and medical aspects, London, L.K. Academic Press.

Cohen S, Allcock HR, Langer R. 1993. Cell and enzyme immobilization in ionotropic synthetic hydrogels. In AA Hincal, HS Kas (eds), Recent advances in pharmaceutical and industrial biotechnology, Editions de Sante, Paris.

Cohen S, Bano MC, Visscher KB, et al. 1990. Ionically cross-linkable phosphazene: a novel polymer for microencapsulation. J Am Chem Soc 112:7832.

Cohen S, Yoshioka T, Lucarelli M, et al. 1991. Controlled delivery systems for proteins based on poly(lactic/glycolic acid) microspheres. Pharm Res 87(6):713.

Compton C, Gill JM, Bradford DA. 1989. Skin regenerated from cultured epithelial autografts on full-thickness wounds from 6 days to 5 years after grafting: A light, electron microscopic, and immunohistochemical study. Lab Invest 60:600.

DeBakey ME, Jordan GL, Abbot JP, et al. 1964. The fate of dacron vascular grafts. Arch Surg 89:757.

De Bruijn JD, Bovell YP, van Blitterswijk CA. 1994. Structural arrangements at the interface between plasma sprayed calcium phosphates and bone. Biomaterials 15(7):543.

Emerich DF, Winn SR, Christenson L, et al. 1992. A novel approach to neural transplantation in Parkinson's disease: Use of polymer-encapsulated cell therapy. Neurosci Biobeh Rev 16:437.

Frazza EJ, Schmitt EE. 1971. A new absorbable suture. J Biomed Mater Res Symp 1:43.

Furlong RJ, Osborn JE. 1991. Fixation of hip prostheses by hydroxylapatite ceramic coatings. J Bone Jt Surg 73-B(5):741.

Goshima J, Goldberg VM, Caplan AI. 1991. The origin of bone formed in composite grafts of porous calcium phosphate ceramic loaded with marrow cells. Clin Orthop Rel Res 191:274.

Guenard V, Kleitman N, Morissey TK, Bunge RP, Aebischer P. 1992. Syngeneic schwann cells derived from adult nerves seeded in semipermeable guidance channels enhance peripheral nerve regeneration. J Neurosci 12:3310–3320.

Hamner JE, Reed OM, Greulich RC. 1972. Ceramic root implantation in baboons. J Biomed Mater Res Symp 6:1.

Hansbrough JF, Cooper ML, Cohen R, et al. 1992. Evaluation of a biodegradable matrix containing cultured human fibroblasts as a dermal replacement beneath meshed skin grafts on athymic mice. Surgery 111(4):438.

Haynesworth SE, Goshima J, Goldberg VM, et al. 1992. Characterization of cells with osteogenic potential from human marrow. Bone 13:81.

Heimbach D, Luterman A, Burke J, et al. 1988. Artificial dermis for major burns. Ann Surg 208(3):313.

Hulbert SF, Morrison SJ, Klawitter JJ. 1972. Tissue reaction to three ceramics of porous and nonporous structures. J Biomed Mater Res 6:347.

Hynes RO. 1987. Integrins: A family of cell surface receptors. Cell 48:549.

Lacy PE, Hegre OD, Gerasimidi-Vazeou A, et al. 1991. Maintenance of normoglycemia in diabetic mice by subcutaneous xenografts of encapsulated islets. Science 254:1782.

Lim F, Sun AM. 1980. Microencapsulated islets as bioartificial endocrine pancreas. Science 210:908.

Lo H, Kadiyala S, Guggino SE, et al. 1994. Biodegradable foams for cell transplantation. In R Murphy, A Mikos (eds), Biomaterials for drug and cell delivery, Materials Research Society Proceedings, vol 331.

Lum Z, Krestow M, Tai IT, et al. 1992. Xenografts of rat islets into diabetic mice. Transplantation 53(6):1180.

Madison R, Da Silva CR, Dikkes P, et al. 1985. Increased rate of peripheral nerve regeneration using bioresorbable nerve guides and a laminin-containing gel. Exp Neurol 88:767.

Massia SP, Hubbell JA. 1991. An RGD spacing of 440 nm is sufficient for integrin $\alpha_v\beta_3$-mediated fibroblast spreading and 140 nm for focal contact and stress fiber formation. J Cell Biol 115(5):1089.

Matlaga BF, Yasenchak LP, Salthouse TN. 1976. Tissue response to implanted polymers: the significance of sample shape. J Biomed Mater Res 10:391.

Mikos AG, Bao Y, Cima LG, et al. 1993*a*. Preparation of poly(glycolic acid) bonded fiber structures for cell attachment and transplantation. J Biomed Mater Res 27:183.

Mikos AG, Sarakinos G, Leite SM, et al. 1993*b*. Laminated three-dimensional biodegradable foams for use in tissue engineering. Biomaterials 14(5):323.

Mikos AG, Sarakinos G, Lyman MD, et al. 1993*c*. Prevascularization of porous biodegradable polymers. Biotech Bioeng 42:716.

Mikos AG, Thorsen AJ, Czerwonka LA, et al. 1994. Preparation and characterization of poly(L-lactic) foams. Polymer 35(5):1068.

Mooney DJ, Mazzoni CL, Organ GM, et al. 1994*a*. Stabilizing fiber-based cell delivery devices by physically bonding adjacent fibers. In R Murphy, A Mikos (eds), Biomaterials for Drug and Cell Delivery, Materials Research Society Proceedings, Pittsburgh, Pennsylvania, vol 331, 47–52.

Mooney DJ, Organ G, Vacanti JP. 1994*b*. Design and fabrication of biodegradable polymer devices to engineer tubular tissues. Cell Trans 3(2):203.

Organ GM, Mooney DJ, Hansen LK, et al. 1993. Enterocyte transplantation using cell-polymer devices causes intestinal epithelial-lined tube formation. Transplan Proc 25:998.

O'Shea GM, Goosen MFA, Sun AM. 1984. Prolonged survival of transplanted islets of Langerhans encapsulated in a biocompatible membrane. Biochim Biophys Acta 804:133.

Puelacher WC, Vacanti JP, Kim WS, et al. 1993. Fabrication of nasal implants using human shape specific polymer scaffolds seeded with chondrocytes. Surgical Forum 44:678–680.

Sagen J, Wang H, Tresco PA, et al. 1993. Transplants of immunologically isolated xenogeneic chromaffin cells provide a long-term source of pain-reducing neuroactive substances. J Neuroscience 13(6):2415.

Schaeffer DW. 1994. Engineered porous materials. MRS Bulletin April 1994:14.

Schoichet MS, Winn SR, Athavale S, et al. 1994. Poly(ethylene oxide)-grafted thermoplastic membranes for use as cellular hydrid bio-artificial organs in the central nervous system. Biotech Bioeng 43:563.

Smith DM, Hua D, Earl WL. 1994. Characterization of porous solids. MRS Bulletin April 1994:44.

Soon-Shiong P, Sandford PA, Heintz R, et al. 1994. First human clinical trial of immunoprotected islet allografts in alginate capsules. Society for Biomaterials Annual Meeting, Boston, Mass, abstract 356.

Stenzel KH, Miyata T, Rubin AL. 1974. Collagen as a biomaterial. Ann Rev Biophys Bioeng 3:231.

Stern R, McPherson M, Longaker MT. 1990. Histologic study of artificial skin used in the treatment of full-thickness thermal injury. J Burn Care Rehab 11:7.

Stupp SI, Hanson JA, Eurell JA, et al. 1993. Organoapatites: Materials for artificial bone: III. Biological testing. J Biomed Mater Res 27(3):301.

Taylor SR, Gibbons DF. 1983. Effect of surface texture on the soft tissue response to polymer implants. J Biomed Mat Res 17:205.

Uyama S, Takeda T, Vacanti JP. 1993. Delivery of whole liver equivalent hepatic mass using polymer devices and hepatotrophic stimulation. Transplantation 55(4):932.

Vacanti CA, Cima LG, Ratkowski D, et al. 1992. Tissue engineered growth of new cartilage in the shape of a human ear using synthetic polymers seeded with chondrocytes. In LG Cima, ES Ron (eds), Tissue Inducing Biomaterials, pp 367–374, Materials Research Society Proceedings, Pittsburgh, Pennsylvania, vol 252.

Vacanti CA, Kim W, Upton J, et al. 1993. Tissue engineered growth of bone and cartilage. Transplan Proc 25(1):1019.

Vacanti CA, Langer R, Schloo B, et al. 1991. Synthetic polymers seeded with chondrocytes provide a template for new cartilage formation. Plast Reconstr Surg 88(5):753.

Wang MC, Pins GD, Silver FH. 1994. Collagen fibers with improved strength for the repair of soft tissue injuries. Biomaterials 15:507.

Weinberg CB, Bell E. 1986. A blood vessel model constructed from collagen and cultured vascular cells. Science 231:397.

Weslowski SA, Fries CC, Karlson KE, et al. 1961. Porosity: Primary determinant of ultimate fate of synthetic vascular grafts. Surgery 50(1):91.

White RA, Hirose FM, Sproat RW, et al. 1981. Histopathologic observations after short-term implantation of two porous elastomers in dogs. Biomaterials 2:171.

White RA, Weber JN, White EW. 1972. Replamineform: A new process for preparing porous ceramic, metal, and polymer prosthetic materials. Science 176:922.

Williams DF. 1987. Definitions in Biomedicals. Progress in Biomedical Engineering, vol 4, New York, Elsevier.

Williams DF, Black J, Doherty PJ. 1992. Second consensus conference on definitions in biomaterials. In PJ Doherty, RL Williams, DF Williams, et al. (eds), Advances in Biomaterials, vol 10, Biomaterials-Tissue Interactions, pp 525–533, New York, Elsevier.

Winn SR, Hammang JP, Emerich DF, et al. 1994. Polymer-encapsulated cells genetically modified to secrete human nerve growth factor promote the survival of axotomized septal cholinergic neurons. Proc Natl Acad Sci 91:2324–2328.

Yannas IV, Burke JF, Orgill DP, et al. 1982. Wound tissue can utilize a polymeric template to synthesize a functional extension of skin. Science 215:174.

Yoshinari M, Ohtsuka Y, Derand T. 1994. Thin hydroxyapatite coating produced by the ion beam dynamic mixing method. Biomaterials 15:529.

Further Information

The Society for Biomaterials, American Society for Artificial Internal Organs, Cell Transplantation Society, and Materials Research Society all sponsor regular meetings and/or sponsor journals relevant to this topic. The following *Materials Research Society Symposium Proceedings* contain a collection of relevant articles: volume 252, *Tissue Inducing Biomaterials* (1992); volume 331, *Biomaterials for Drug and Cell Delivery* (1994). Another good source of relevant material is *Tissue Engineering*, edited by R Skalak and CF Fox, New York, Alan Riss (1988).

[illegible] Brown [illegible] Transplantation of [illegible]

[illegible] 1994. Collagen [illegible]

[illegible] 1983. [illegible]

[illegible] 1993. [illegible] Transplantation [illegible]

Vacanti [illegible] 1992. [illegible] In [illegible] Materials Research Society Proceedings [illegible]

[illegible] 1994. Collagen tubes [illegible]

White EA, [illegible] 1991. [illegible] J Biomed Mater Res [illegible]

[illegible] 1984. [illegible] Proc Natl Acad Sci USA [illegible]

113
Regeneration Templates

Ioannis V. Yannas
Massachusetts Institute of Technology

113.1 The Problem of the Missing Organ

Drugs typically replace or correct a missing function at the molecular scale; by contrast, regeneration templates replace the missing function at the scale of tissue or organ. An organ may be lost to injury or may fail in disease: The usual response of the organism is repair, which amounts to contraction and synthesis of scar tissue. Tissues and organs in the adult mammal typically do not regenerate. There are exceptions, such as epithelial tissues of the skin, gastrointestinal tract, genitals, and the cornea, all of which regenerate spontaneously; the liver also shows ability to synthesize substantial organ mass, though without recovery of the original organ shape. There are reports that bone and the elastic ligaments regenerate. These exceptions underscore the fact that the loss of an organ by the adult mammal almost invariably is an irreversible process, since the resulting scar tissue largely or totally lacks the structure and function of the missing organ. The most obvious examples involve losses due to injury such as the loss of a large area of skin following a burn accident or the loss of substantial nerve mass following an automobile accident. However, irreversible loss of function can also occur following disease, although over a lengthy time: Examples are the inability of a heart valve to prevent leakage during diastole as a result of valve tissue response to an inflammatory process (rheumatic fever), and the inability of liver tissue to synthesize enzymes due to its progressive replacement by fibrotic tissue (cirrhosis).

Five approaches have been used to solve the problem of the missing organ. In autografting, a mass of similar or identical tissue from the patient (autograft) is surgically removed and used to treat the area of loss. The approach can be considered to be spectacularly successful until one considers the long-term cost incurred by the patient. An example is the use of sheet autograft to treat extensive areas of full-thickness skin loss; although the patient incorporates the autograft fully with excellent recovery of function, the "donor" site used to harvest the autograft remains scarred. When the autograft is not

0-8493-0462-8/00/$0.00+$.50

available, as is common in cases of burns extending over more than 30% of body surface area autograft is meshed in order to make it extensible enough to cover the large wound areas. However, the meshed autograft provides cover only where the graft tissue provides direct cover; where there is no cover, scar forms, and the result is one of low cosmetic value. Similar problems of donor site unavailability and scarring must be dealt with in heart bypass surgery, another widespread example of autografting. In transplantation, the donor tissue is typically harvested from a cadaver, and the recipient has to cope with the problems of rejection and the risk of transmission of viruses from this allograft. Another approach has been based on efforts to synthesize tissues in vitro using autologous cells from the patient; this approach has yielded so far a cultured epidermis (a tissue which regenerates spontaneously provided there is a dermal substrate underneath) about 2–3 weeks after the time when the patient was injured. In vitro synthesis of the dermis, a tissue which does not regenerate, has not been accomplished so far. Perhaps the most successful approach from the commercial standpoint has been the one in which engineered biomaterials are used; these materials are typically required by their designers to remain intact themselves without interfering with the patient's physiologic functions during the entire lifetime; overwhelmingly, this requirement is observed in its breach. A fifth approach is based on the discovery that an analog of the extracellular matrix (ECM) induces partial regeneration of the dermis, rather than of scar, in full-thickness skin wounds in adult mammals (human, guinea pig, pig) where it is well known that no regeneration occurs spontaneously. This fifth approach of solving the problem of organ loss, in situ regeneration, will be described in this chapter.

Efforts to induce regeneration have been successful with only a handful of ECM analogs. Evidence of regeneration is sought after the ECM analog has been implanted in situ, i.e., at the lesion marking the site of the missing organ. When morphogenesis is clearly evident, based on tests of recovery both of the original tissue structure and function, the matrix which has induced these physiologic or nearly physiologic tissues is named a regeneration template. In the absence of evidence of such morphogenetic activity of the cell-free matrix, the latter is not referred to as a regeneration template.

113.2 Search Principles for Identification of Regeneration Templates

Several parameters have been incorporated in the search for organ regeneration templates. Briefly, these parameters account for the performance of the implant during the early or acute stage following implantation (physicochemical parameters) as well as for the long-term or chronic stage following implantation (biologic parameters).

Immediately upon making contact with the wound bed, the implant must achieve physicochemical nanoadhesion (i.e., adhesion at a scale of 1 nm) between itself and the lesion. Without contact of this type it is not possible to establish and maintain transport of molecules and cells between implant and host tissue. The presence of adequate adhesion can be studied by measurements of the force necessary to peel the implant from the wound bed immediately after grafting.

Empirical evidence has supported a requirement for an implant which is capable of isomorphous tissue replacement, i.e., the synthesis of new tissue at a rate which is of same order as the rate of degradation of the matrix.

$$\frac{t_b}{t_h} = O(1) \tag{113.1}$$

In Eq. (113.1), t_b denotes a characteristic time constant for biodegradation of the implant *at that tissue site*, and t_h denotes a time constant for healing, the latter occurring by synthesis of new tissue inside the implant.

A third requirement refers to the *critical cell path length* l_c, beyond which migration of a cell into the implant deprives it of an essential nutrient, assumed to be transported from the host tissue by diffusion alone. The emphasis is on the characteristic diffusion path for the nutrient during the early stages of wound healing, before significant angiogenesis occurs several days later. Calculation of the critical cell path can be done by use of the *cell lifeline number, S,* a dimensionless number expressing the relative importance of a chemical reaction, which leads to consumption of an essential nutrient by the cell, and of diffusion of the nutrient which alone makes the latter accessible to the cell. This number is defined as

$$S = \frac{rl^2}{Dc_o} \tag{113.2}$$

where r is the rate of consumption of the nutrient by the cell in mole/cm^3/s, l is the diffusion length, D is the diffusivity of the nutrient in the medium of the implant, and c_o is the nutrient concentration at or near the surface of the wound bed, in mole/cm^3. When $S = O(1)$ the value of l is the critical path length, l_c along which cells can migrate, away from host tissue, without requirement of nutrient in excess of that supplied by diffusion. Eq (113.2) can, therefore, be used to define the maximum implant thickness beyond which cells require the presence of capillaries.

The chemical composition of the implant which has induced regeneration was designed on the basis of studies of wound-healing kinetics. In most mammals, full-thickness skin wounds close partly by contraction of the wounds edges and partly by synthesis of scar tissue. Clearly, skin regeneration over the entire area of skin loss cannot occur unless the wound edges are kept apart and, in addition, the healing processes in the wound bed are modified drastically enough to yield a physiologic dermis rather than scar. Although several synthetic polymers, such as porous (poly)dimethyl siloxane, delay contraction to a small but significant extent, they do not degrade and therefore violate isomorphous tissue replacement, Eq. (113.1). Synthetic biodegradable polymers, such as (poly)lactic acid, can be modified, e.g., by copolymerization with glycolic acid, to yield polymers which satisfy Eq. (113.1); however, evidence is lacking that these synthetic polymers delay contraction and prevent synthesis of scar. By contrast, there is considerable evidence that a certain analog of the extracellular matrix (ECM analogs) not only delays contraction significantly but also leads to synthesis of partly physiologic skin. Systematic use of the delay in contraction as an essay has been made to identify the structural features of the dermis regeneration template (DRT), as shown schematically in Fig. 113.1.

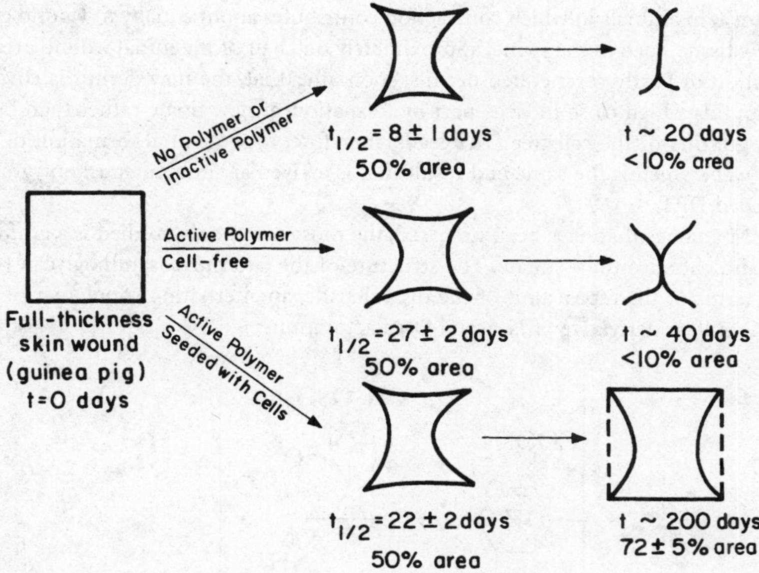

FIGURE 113.1 The kinetics of contraction of full-thickness guinea pig skin wounds can be used to separate collagen-*graft*-glycosaminoglycan copolymers into three classes, as shown. The wound half-life, $t_{1/2}$, is the number of days necessary to reduce the original wound area to 50%. (Courtesy of Massachusetts Institute of Technology.)

Summarized in Fig. 113.1 are three modes of wound-healing behavior, each elicited by an ECM analog of different design. Mode O of healing is described by a very short time for onset of contraction, followed by contraction and definitive closure of the wound with formation of a thin linear scar. Mode I is characterized by a significant delay in onset of contraction, following which contraction proceeds and eventually leads to a linear scar. Mode II is characterized by a significant delay in onset of contraction (somewhat smaller than in mode I), followed by contraction and then by reversal of contraction, with expansion of the original wound perimeter at a rate which exceeds significantly the growth rate of the entire animal. Mode II healing leads to synthesis of a partly physiologic dermis and a physiologic epidermis within the perimeter of the expanded wound bed. Mode O healing occurs when an ECM analog which lacks specificity is used to graft the wound. Mode O is also observed when the wound bed remains ungrafted. Mode I healing occurs when an ECM analog of highly specific structure (the DRT) is grafted on the wound bed. Mode II healing occurs when the DRT, previously identified as the ECM analog which leads to mode I healing, is seeded with autologous epidermal cells before being grafted. Although contraction is a convenient screening method for identification of structural features of ECM analogs which induce skin regeneration, a different procedure has been used to identify the features of an implant that induces regeneration of the peripheral nerve. The structural features of the nerve regeneration template (NRT) were identified using an essay focused on the long-term return of function of the nerve following treatment with candidate ECM analogs.

Recent studies have shown that it is possible to achieve regeneration both of a dermis and an epidermis in sequence, rather than simultaneously, provided that the animal model used is one in which wound closure does not take place almost overwhelmingly by wound contraction. The choice of animal model is, therefore, critical in this respect. Mode I behavior, illustrated in Fig. 113.1, is applicable to wound healing in rodents, such as guinea pigs and rats. In these animals, about 90% of wound closure is accounted for fully by wound contraction, the remainder is accounted for by formation of new tissue. Grafting of a full-thickness skin defect in a rodent with the ECM analog which possesses a highly specific structure (the dermis regeneration template or DRT; see below) leads to lengthy delay of the onset of contraction, eventually followed by contraction and formation of a linear scar-like tissue. Although the gross appearance of the tissue formed is that of a modified linear scar (see Fig. 113.1), there is some histological evidence that the small mass of connective tissue layer formed underneath the epidermis is not scar. In contrast, in animals in which contraction contributes about equally to wound closure as does formation of new tissue, such as the swine, approximately one half of the initial wound area is eventually closed by formation of partly regenerated dermis; soon after that, the new dermis is covered by a new epidermis. In wounds which close in large part by formation of new tissue rather than by contraction, as in the swine, grafting of the cell-free DRT leads, therefore, to sequential formation of a dermis and an epidermis as well, which is the same end result that is arrived at simultaneously by grafting with the keratinocyte-seeded DRT.

Of several ECM analogs that have been prepared, the most commonly studied is a graft copolymer of type I collagen and chondroitin 6-sulfate. The structure of the latter glycosaminoglycan (GAG) is illustrated below in terms of the repeat unit of the disaccharide, an alternating copolymer of D-glucuronic acid and of an O-sulfate derivative of N-acetyl D-galactosamine:

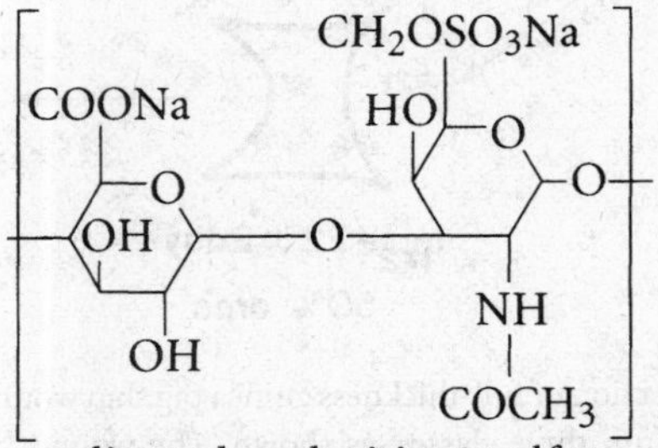

CH_2OSO_3Na, COONa, HO, O, OH, NH, $COCH_3$, n

sodium chondroitin 6-sulfate

The principle of isomorphous replacement, Eq. (113.1), cannot be satisfied unless the *biodegradation time constant* of the network, t_b, can be adjusted to an optimal level, about equal to the rate of synthesis of new tissue at that site. Reduction of the biodegradation rate of collagen can be achieved either by grafting GAG chains onto collagen chains or by crosslinking collagen chains to each other. The chemical grafting of GAG chains on polypeptide chains proceeds by previously coprecipitating the two polymers under conditions of acidic pH, followed by covalent crosslinking of the freeze-dried precipitate. A particularly useful procedure for crosslinking collagen chains to GAG, or collagen chains to each other, is a self-crosslinking reaction, requiring no use of crosslinking agent. This condensation reaction principally involves carboxylic groups from glutamyl/aspartyl residues on polypeptide chain P_1 and ϵ-amino groups of lysyl residues on an adjacent chain P_2 to yield covalently bonded collagen chains; as well as condensation of amine groups of collagen with carboxylic groups of glucuronic acid residues on GAG chains to yield *graft*-copolymers of collagen and GAG:

$$P_1{-}COOH + P_2{-}NH_2 \rightarrow P_2{-}NHCO{-}P_1 + H_2O \tag{113.3a}$$

$$GAG{-}COOH + P_2{-}NH_2 \rightarrow P_2{-}NHCO{-}GAG + H_2O \tag{113.3b}$$

In each case above the reaction proceeds to the right, with formation of a three-dimensional crosslinked network when the moisture content of the protein, or protein-GAG coprecipitate, drops below about 1 wt%. As illustrated by Eq. (113.3), removal of water, the volatile product of the condensation, is favored by conditions which drive the reaction towards the right, with formation of a crosslinked network. Thus, the reaction proceeds to the right when both high temperature and vacuum are used. Another crosslinking reaction, used extensively in preparing implants that have been employed in clinical studies as well as in animal studies, involves use of glutaraldehyde. Dehydration crosslinking, which amounts to self-crosslinking as described above, obviously does not lend toxicity to these implants. Glutaraldehyde, on the other hand, is a toxic substance, and devices treated with it require thorough rinsing before use until free glutaraldehyde cannot be detected in the rinse water. Network properties of crosslinked collagen-GAG copolymers can be analyzed structurally by studying the swelling behavior of small specimens. The method is based on the theory of Flory and Rehner, who showed that the volume fraction of a swollen polymer v_2 depends on the average molecular weight between crosslinks, M_c, through the following relationship:

$$\ln\left(1-v_2\right)+v_2+\chi v_2-\left(\rho V_1/M_c\right)\left(v_2^{1/3}-v_2/2\right)=0 \tag{113.4}$$

In Eq. (113.4), V_1 is the molar volume of the solvent, ρ is the density of the dry polymer, and χ is a constant characteristic of a specific polymer-solvent pair at a particular temperature.

Although the chemical identify of collagen-GAG copolymers is a necessary element of their biologic activity, it is not a sufficient one. In addition to chemical composition and the crosslink density, biologic activity also depends strongly on the *pore architecture* of these ECM analogs. Pores are incorporated first by freezing a very dilute suspension of the collagen-GAG coprecipitate and then by inducing sublimation of the ice crystals by exposing to vacuum at low temperatures. The resulting pore structure is, therefore, a negative replica of the network of ice crystals (dendrites). It follows that control of the conditions of ice crystal nucleation and growth can lead to a large variety of pore structures. In practice, the average pore diameter decreases with decreasing temperature of freezing while the orientation of pore channel axes also depends on the magnitude of the heat flux vector during freezing. The dependence of pore channel orientation on heat transfer parameters is illustrated by considering the dimensionless Mikic number *Mi*, a ratio of the characteristic freezing time of the aqueous medium of the collagen-GAG suspension, t_f, to the characteristic time for entry, t_e, of a container filled with the suspension which is lowered at constant velocity into a well-stirred cooling bath:

$$Mi = \frac{t_f}{t_e} = \frac{\rho_w h_{fg} r V}{10 k_j \Delta T} \tag{113.5}$$

In Eq. (113.5), ρ_w is the density of the suspension, h_{fg} is the heat of fusion of the suspension, r is an arbitrary length scale, V is the velocity with which the container is lowered into the bath, k_j is the thermal conductivity of the jacket, and ΔT is the difference between freezing temperature and bath temperature. The shape of the isotherms near the freezing front is highly dependent on the value of Mi. The dominant heat flux vector is normal to these isotherms, i.e., ice dendrites grow along this vector. It has been observed that, for $Mi < 1$ (slow cooling), the isotherms are shallow, flat-shaped parabolae, and the ice dendrites exhibit high axial orientation. For $Mi > 1$ (rapid cooling), the isotherms are steep parabolae, and ice dendrites are oriented along the radial direction.

The structure of the porous matrix is defined by quantities such as the volume fraction, specific surface, mean pore size, and orientation of pores in the matrix. Determination of these properties is based on principles of stereology, the discipline which relates the quantitative statistical properties of three-dimensional structures to those of their two-dimensional sections or projections. In reverse, stereologic procedures allow reconstruction of certain aspects of three-dimensional objects from a quantitative analysis of planar images. A plane which goes through the two-phase structure of pores and collage-GAG fibers may be sampled by random points, by a regular pattern of points, by a near-total sampling using a very dense array of points, or by arranging the sampling points to form a continuous line. The volume fraction of pores, V_V, is equal to the fraction of total test points which fall inside pore regions, P_P, also equal to the total area fraction of pores, A_A, and, finally, equal to the line fraction of pores, L_L, for a linear point array in the limit of infinitely close point spacing

$$V_V = P_P = A_A = L_L \tag{113.6}$$

Whether cells of a particular type should be part of a regeneration template depends on predictions derived from models of developmental biology as well as empirical findings obtained with well-defined wound-healing models. During morphogenesis of a large variety of organs, an interaction between epithelial and mesenchymal cells, mediated by the basal lamina which is interleaved between the two types of cells, is both necessary and sufficient for development of local physiologic structures involving two types of tissue in juxtaposition. In particular, skin morphogenesis in a full-thickness wound model requires the presence of this interaction between the two cell types and the basal lamina over a critical period. In skin wound–healing experiments with adult mammals, wound healing proceeds with formation of scar, rather than physiologic skin, if the wound bed contains epithelial cells and mesenchymal cells (fibroblasts) but no ECM structure which could act temporarily as a basal lamina. If, by contrast, an analog of the basal lamina is present, wound healing proceeds with nearly physiologic morphogenesis of skin. Furthermore, no epidermis is formed unless epithelial cells become involved in wound healing early during wound healing and continue being involved until they have achieved confluence. It is also known that no dermis forms if fibroblasts are not available early during wound healing. These observations suggest the requirements for a DRT which is an analog of the basal lamina and is designed to encourage the migration and interaction of both epithelial cells and fibroblasts within its volume.

Nerve regeneration following injury essentially amounts to elongation of a single nerve cell across a gap resulting from the injury. During nerve development many nerve cells are elongated by processes which eventually become axons. Interaction of elongating processes with basal lamina are credited as being essential in the formation of nerve during development, and it will be assumed here that such interactions are essential in regeneration following adult injury as well. It is also known that Schwann cells, which derive from neural crest cells during development, are essential contributors to regeneration following injury to the adult peripheral nerve. These considerations suggest a nerve regeneration template which is structured as an analog of the basal lamina, interacting with the elongating axons in the presence of Schwann cells.

113.3 Structural Specificity of Dermis Regeneration Template (DRT)

The major events accompanying skin wound healing can be summarized as contraction and scar synthesis. Conventional wisdom prescribes the need for a treatment that accelerates wound healing. A large number of devices that claim to speed up various aspects of the healing process have been described in the literature. The need to achieve healing within as short a time as possible is certainly well founded, especially in the clinical setting where the risk to patient's life as well as the morbidity increase with extension of time to heal. However, the discovery of partial skin regeneration by use of the skin regeneration template has introduced the option of a drastically improved healing result for the patient in exchange for a slightly extended hospital stay.

The DRT was optimized in studies with animals in which it was observed that skin regeneration did not occur unless the test ECM analog effectively delayed, rather than accelerated, wound contraction. The length of delay in onset of contraction eventually was used as a quantitative basis for preliminary optimization of the structural features of the DRT. Optimization studies are currently continuing on the basis of a new criterion, namely, the fidelity of regeneration achieved.

Systematic use of the criterion of contraction inhibition has been made to select the biodegradation rate and the average pore diameter of DRT. The kinetics of contraction of full-thickness skin wounds in the guinea pig have been studied for each of the three modes of healing schematically presented in Fig. 113.1. The results, presented in Fig. 113.2, show that mode O healing is characterized by early onset of contraction, whereas mode I and mode II healing show a significant delay in onset of contraction. A measure of the delay is the wound half-life $t_{1/2}$, the time required for the wound area to decrease to 50% of the original value. Use of this index of contraction rate has been made in Figs. 113.3 and 113.4, which present data on the variation of wound half-life with average pore diameter and degradation rate for a

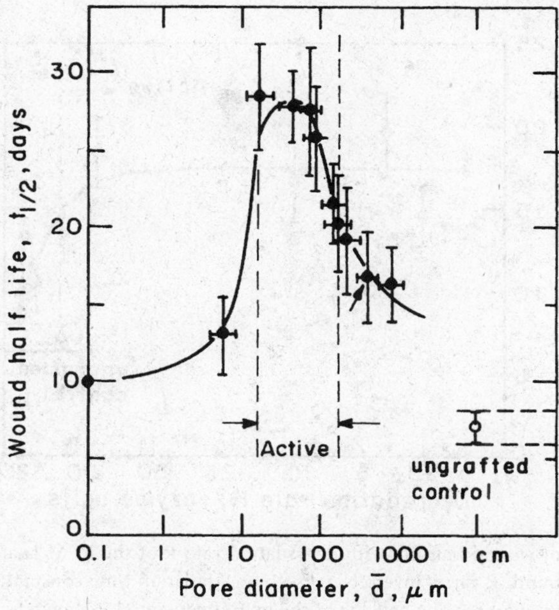

FIGURE 113.2 The kinetics of guinea pig skin wound contraction following grafting with three classes of ECM analogs. Inactive ECM analogs delay the onset of contraction only marginally over the ungrafted wound, whereas active cell-free ECM analogs delay the onset of contraction significantly. When seeded with epithelial cells, not only does an active ECM analog delay the onset of contraction significantly, but it also induces formation of a confluent epidermis and then arrests and reverses the direction of movement of wound edges, leading to expansion of the wound perimeter and to synthesis of partly physiologic skin. (Courtesy of Massachusetts Institute of Technology.)

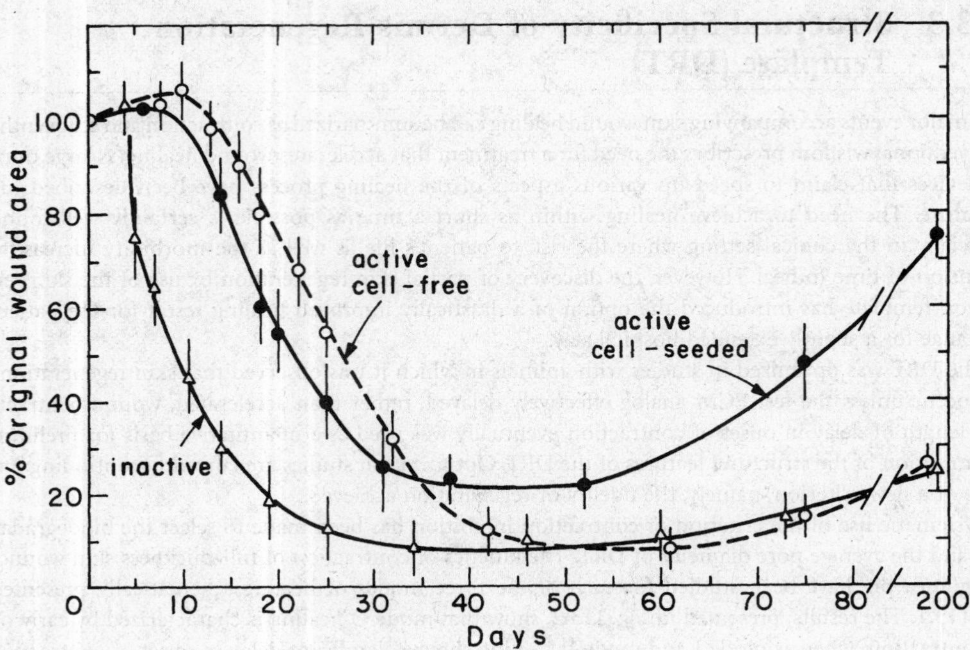

FIGURE 113.3 The half-life wounds, $t_{1/2}$, grafted with ECM analogs varies with the average pore diameter of the analog. The range of pore diameters where activity is maximal is shown by broken lines. The half-life of the ungrafted wound is shown for comparison. (Courtesy of Massachusetts Institute of Technology.)

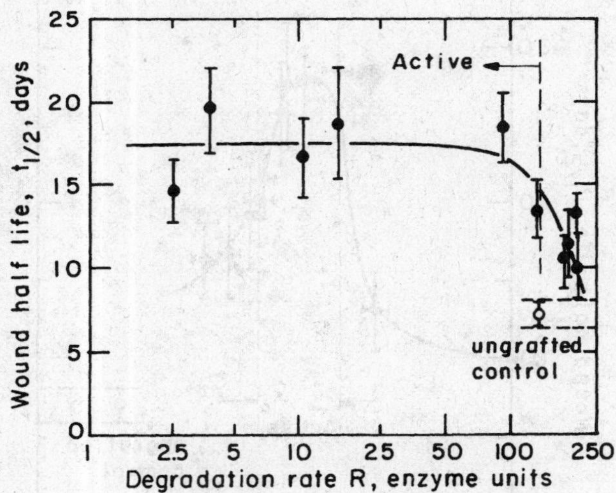

FIGURE 113.4 Variation of wound half-life with degradation rate *R* of the ECM analog used as graft for a full-thickness guinea pig skin wound. *R* varies inversely as the biodegradation time constant, tf. The region of maximal activity is indicated by the broken line. The half-life of the ungrafted wound is shown for comparison. (Courtesy of Massachusetts Institute of Technology.)

type I collagen-chondroitin 6-sulfate copolymer. In Fig. 113.2, mode I kinetics are observed using a cell-free copolymer with average pore diameter and biodegradation rate that correspond to the regions of Figs. 113.3 and 113.4 in which the values of half-life are maximal; these regions characterize the structure of SRT which possesses maximal activity. Maximum delay in wound half-life up to 27 ± 3 days is seen to have occurred when the average pore diameter has ranged from values as low as 20 ± 4 μm to an upper limit of 125 ± 35 μm. In addition, significant delay in wound healing has been observed when the degradation rate has become less than 115 ± 25 enzyme units. The latter index of degradation has been based on the in vitro degradation rate of the copolymer in a standardized solution of bacterial collagenase.

The upper limit in degradation rate, defined in Fig. 113.4, is consistent with the requirement of a lower limit in t_b, Eq. (113.1), below which the implant is biodegraded too rapidly to function as a scaffold over the period necessary for synthesis of new tissue; the latter is characterized by t_h. The lower limit in pore diameter, defined in Fig. 113.3, suggests a dimension which is on the order of two cell diameters; adequate space for migration of mesenchymal cells (fibroblasts) from the wound bed into the DRT is thereby guaranteed. At an estimated velocity of 0.2 mm/day for fibroblasts these cells would be expected to migrate across a 0.5-mm thickness of the DRT within very few days, provided that adequate supplies of critical nutrients could be made available to them from host tissue. Use of Eq. (113.2) leads to an estimated critical cell path length l_c of order 100 μm. These estimates suggest that fibroblasts can migrate across at least one-fifth the thickness of the porous implant without requiring the presence of capillaries. However, taking into account the observation that wound exudate (comprising primarily serum, with growth factors and nutrients) fills at least one-half the thickness of the implant within no more than a few hours following grafting, we conclude that the boundary of "host" tissue has moved clearly inside the implant.

Epithelial cells, as well as fibroblasts and a basal lamina analog, also are required for morphogenesis. They can be supplied in a variety of forms. They are always available as epithelial cell sheets, migrating from the wound edges toward the center of the wound bed. However, when the area of skin loss is several centimeters, as with a severely burned patient, these cell sheets, migrating with speeds of about 0.5 mm/day from opposite edges, would not be expected to cover one-half the characteristic wound dimension in less than the time constant t_h for synthesis of new tissue. In the absence of epithelial cells, therefore, at the center of the wound bed, Eq. (113.1) wound be violated. To overcome this limitation, which is imposed by the scale of the wound, it has been necessary to resort to a variety of procedures. In the first, uncultured autologous epidermal cells, extracted from a skin biopsy by controlled enzymatic degradation, have been seeded into ECM analogs by centrifugation into the porous matrix prior to grafting of the latter on the wound bed. Tested with animals, the procedure leads to formation of a confluent epidermis by about 2 weeks, provided that at least 5×10^4 epithelial cells per cm^2 of DRT area have been seeded. In another procedure, a very thin epidermal layer has been surgically removed from an intact area of the patient and has been grafted on the dermal layer which has been synthesized about 2 weeks after grafting with the DRT. The latter procedure has been tested clinically with reproducible success. A third procedure, studied with animals, has made use of cultured epithelia, prepared by a 2- to 3-week period of culture of autologous cells in vitro and grafted on the newly synthesized dermal bed. Approximately equivalent fidelity of skin regeneration has been obtained by each of these procedures for supplying epithelial cells to the DRT in the treatment of skin wounds of very large area.

113.4 In Situ Synthesis of Skin with DRT

The skin regeneration template DRT induces regeneration of skin to a high degree of fidelity. Fidelity of regeneration has been defined in terms of the degree of recovery of structural and functional features which are present in the intact organ.

The first test of fidelity of skin regeneration following use of the DRT was a study of treatment of full-thickness skin loss in guineas pigs. In this study the lesion was produced by surgery on healthy animals. The characteristic dimension of the wound was about 3 cm, and the desired period for cover by a confluent epidermis was 2 weeks. Covering a wound of such a scale within the prescribed time would have been out of reach of epithelial cells migrating from the wound edges. Accordingly, autologous epithelial cells were extracted from a skin biopsy and seeded into the DRT under conditions of carefully controlled centrifugation.

A clear and unmistakable difference between healing in the presence and absence of DRT was provided by observing the gross anatomy of the wound (Fig. 113.2). In the absence of the DRT, the wound contracted vigorously and closed up with formation of a linear scar by about day 30. In the presence of a cell-seeded DRT, the wound perimeter started contracting with a delay of about 10 days, and contraction was completely arrested and then reversed between days 30 and 40. The wound then continued to expand at a rate that was clearly higher than that expected from the rate of growth of the animal. The long-term appearance of the wound treated with the cell-seeded DRT was that of an organ that appeared grossly identical in color, texture, and touch to intact skin outside the wound perimeter. However, the newly synthesized skin was totally hairless, and the total area of new skin was smaller in area from the original wound area by about 30% (Fig. 113.2).

Morphologic studies of newly synthesized skin in the presence of cell-seeded DRT included comparison with intact skin and scar. Optical microscopy and electron microscopy were supplemented by laser light scattering, the latter used to provide a quantitative measure of collagen fiber orientation in the dermal layer. It was concluded that, in most respects, partly regenerated skin was remarkably similar to intact guinea pig skin. The epidermis in regenerated skin was often hyperplastic; however, the maturation sequence and relative proportion of all cell layers were normal. Keratohyaline granules of the neoepidermis were larger and more irregular in contour than those of the normal granular cell layer. The new skin was characterized by melanocytes and Langerhans cells, as well as a well-formed pattern of rete ridges and interdigitations with dermal papillae, all of which appear in normal skin. Newly synthesized skin was distinctly different morphologically from scar. Scar showed characteristic thinning (atrophy) of the epidermis, with absence of rete ridges and of associated dermal papillae. Elastic fibers in regenerated skin formed a delicate particulate structure, in contrast with scar, where elastic fibers were thin and fragmented. The dermal layer in regenerated skin comprised collagen fibers which were not oriented in the plane of the epidermis, as well as fibroblasts which were not elongated; in scar, collagen fibers were highly oriented in the plane and fibroblasts were elongated. Both normal and regenerated skin comprised unmyelinated nerve fibers within dermal papillae, closely approximated to the epidermis; scar had few, if any, nerves. There were no hair follicles or other skin appendages either in regenerated skin or in scar.

Laser light–scattering measurements of the orientation of collagen fibers in tissue sections of the dermal layer were based on use of the Hermans orientation function

$$f = 2\left\langle \cos^2 \alpha \right\rangle - 1 \qquad (113.7)$$

In Eq. (113.7), α is the angle between an individual fiber and the mean axis of the fibers, and $\langle \cos^2\alpha \rangle$ is the square cosine of α averaged over all the fibers in the sample. For a random arrangement of fibers, $\langle \cos^2\alpha \rangle$ equals 1/2, while for a perfectly aligned arrangement it is equal to 1. Accordingly, S varies from 0 (truly random) to 1 (perfect alignment). Measurements obtained by use of this procedure showed that S took the values 0.20 ± 0.11, 0.48 ± 0.05 and 0.75 ± 0.10 for normal dermis, regenerated dermis, and scar dermis, respectively. These results provided objective evidence that regenerated dermis had a morphology of collagen fibers that was clearly not scar ($n = 7$; $p < 0.001$), and was intermediate between that of scar and normal dermis.

Functional studies of regenerated skin showed that the moisture permeability of intact skin and regenerated skin had values of 4.5 ± 0.8 and 4.7 ± 1.9 g/cm/h, insignificantly different from each other ($n = 4$; $p < 0.8$). Mechanical behavior studies showed a positive curvature for the tensile stress-strain curve of regenerated skin as well as of normal skin. However, the tensile strength of regenerated skin was 14 ± 4 MPa, significantly lower than the strength of intact skin, 31 ± 4 MPa ($n = 4$; $p < 0.01$).

The second test of fidelity of skin regeneration was a study of treatment of 106 massively burned humans with the cell-free DRT. The characteristic dimension of the wound in this study was of order 15 cm, and the desired period for cover by a confluent epidermis was 2 weeks. The scale of the wound necessitated the introduction of epithelial cells, and this was accomplished by grafting the newly synthesized dermal layer, 2 weeks after grafting with DRT, with a very thin epidermal layer (epidermal autograft), which was removed from an intact area of the patient. The results of the histologic study of the patient population showed that physiologic dermis, rather than scar, had been synthesized in sites that had been treated with DRT and had later been covered with an epidermal graft.

Progress has been made in clarification of the mechanism by which DRT induces regeneration of the dermis. There is considerable evidence, some of which was presented above, supporting the hypothesis that inhibition of wound contraction is required for regeneration. The evidence also suggests strongly that DRT competitively inhibits formation of specific binding interactions between contractile cells and endogenous ECM.

113.5 Advantages and Disadvantages of Clinical Treatment of Skin Loss with DRT

When skin is the missing organ, the patient faces threats to life posed by severe dehydration and infection. These threats can be eliminated permanently only if the area of missing skin is covered with a device that controls moisture flux within physiologic limits and presents an effective barrier to airborne bacteria. Both of these functions can be returned over a period of about 2–4 weeks by use of temporary dressings. The latter include the allograft (skin from a cadaver) and a very large variety of membranes based on synthetic or natural polymers. None of these dressings solves the problem of missing skin over the lifetime of the patient: The allograft does not support synthesis of physiologic skin and must be removed to avoid rejection, and the devices based on the vast majority of engineered membranes do not make effective biologic contact with the patient's tissues, and all lead to synthesis of scar.

Three devices have been tested extensively for their ability to provide long-term physiologic cover to patients with massive skin loss: the patient's own skin (autograft) the dermis regeneration template (DRT), and cultured epithelia (CEA). All three of these treatments have been studied extensively with massively burned patients. Of these, the autograft and DRT are effective even when the loss of skin extends through the full thickness of the dermis (e.g., third-degree burns), whereas CEA is effective when the loss of skin is through part of its thickness only.

The basis for the differences among the three treatments lies in the intrinsic response of skin to injury. Skin comprises three layers: the epidermis, a 100-μm thick cellular layer; the dermis, a 1- to 5-mm layer of connective tissue with very few cells; and the subcutis, a 2- to 4-mm layer of primarily adipose tissue. In the adult mammal, an epidermis lost through injury regenerates spontaneously provided that a dermal substrate is present underneath. When the dermis is lost, whether through part thickness or full thickness, none of the injured mass regenerates; instead, a nonphysiologic tissue, scar, forms. Scar is epithelialized and can, therefore, control moisture flux within physiologic limits as well as provide a barrier to bacterial invasion. However, scar does not have the mechanical strength of physiologic skin. Also, scar synthesis frequently proceeds well beyond what is necessary to cover the wound, and the result of such proliferation is hypertrophic scarring, a cosmetically inferior integument which, when extending over a large area or over hands or face, reduces significantly the mobility of the patient's joints as well as the patient's cosmetic appearance. Autograft can, if used without meshing, provide an excellent permanent cover; if, however, as commonly practiced, autograft is meshed before grafting in order to extend the wound area that becomes covered, the result is new integument which comprises part physiologic skin and part scar and provides the patient with a solution of much lower quality than unmeshed (sheet) autograft. The clinical use of skin regeneration template has led to a new integument comprising a dermal layer, which has been synthesized by the patient, and an epidermal layer, which has been harvested as a very thin epidermal autograft and has been placed on top of the newly synthesized dermis.

The chief advantages DRT over the meshed autograft are the shorter time that it takes to heal the donor site, from which the epidermal graft was harvested, and a superior cosmetic result at the site of the wound. The main disadvantage in the clinical use of DRT is the subjection of the patient to two surgical treatments rather than one (first graft DRT, then graft the epidermal layer after about 2 weeks).

Although it has been shown that a dermis and an epidermis are regenerated sequentially in DRT-treated wounds like that of the swine, in which contraction plays only a modest role in wound closure, the process of sequential skin regeneration is slower than that of simultaneous skin regeneration. Accordingly, the kinetics of wound closure suggest an advantage in the use of keratinocyte-seeded DRT (a two-stage treatment) relative to the unseeded DRT (one-stage treatment). The clinical advantages of CEA are the ability to grow a very large area, about 10,000 times as large as the area of the skin biopsy, thereby providing cover over the entire injured skin area of the patient. The disadvantages of cultured epithelia are the lengthy period required to culture the epidermis from the patient's biopsy and the inability of the CEA to form a mechanically competent bond with the wound bed.

113.6 Modifications of DRT: Use of a Living Dermal Equivalent

The design concept of a skin regeneration template outlined above has been adopted and modified. One modification involves the replacement of the collagen-GAG matrix with a biodegradable mesh consisting of either (poly)glycolic acid (PGA) or polyglactin-910 (PGL) fibers. The latter is used as a matrix for the in vitro culture of human fibroblasts isolated from neonatal skin. Fibroblasts synthesize extracellular matrix inside the synthetic polymeric mesh, and this "living dermal equivalent" has been cryopreserved for a specified period prior to use.

The living dermal equivalent has been used to graft full-thickness skin wounds in athymic mice. Following grafting of wounds, these PGA/PGL fibroblast grafts were covered with meshed allograft. The latter is human cadaver skin that was meshed in this study to expand it and achieve coverage of maximum possible wound area. This composite graft became vascularized and that, additionally, epithelial cells from the cadaver graft had migrated to the matrix underneath. After a period of about 100 days following grafting, the reported result was an epithelialized layer covering a densely cellular substratum that resembled dermis. A variant of this design, in which the meshed allograft is not used, has been reported; epidermal cells are cultured with the fibroblast mesh before grafting. Studies of these designs are in progress.

113.7 The Bilayered Skin-Equivalent Graft

In one widely reported development, in vitro cell culture procedures have been used to synthesize a bilayered tissue which has been reported to be a useful model of human skin. Briefly, fibroblasts from humans or from rats have been placed inside a collagen gel. Under these conditions, fibroblasts exert contractile forces, trapping the cells inside the contracted collagen lattice. Human epithelial cells, which have been plated onto this contracted dermal equivalent, have been observed to attach to the collagen substrate, multiply, and spread to form a continuous sheet. Differentiation of this sheet has led to formation of specialized epidermal structures, such as a multilayered cell structure with desmosomes, tonofilaments, and keratohyalin granules. Further differentiation events have included the formation of a basement membrane (basal lamina) in vitro when a rat epidermis was formed on top of a dermal equivalent produced from rat fibroblasts.

Grafts prepared from the bilayered skin equivalent have been grafted on animals. When grafted on animals, these structures have been reported to become well vascularized with a network of capillaries within 7 to 9 days. It has been reported that the best grafts have blocked wound contraction, but no systematic data have been presented which could be used to compare the in vivo performance of these skin equivalents with grafts based on DRT (see above). Gross observations of the area grafted with skin equivalents have shown pink hairless areas which were not hypertrophically scarred. A systematic comparison between scar and the tissue synthesized in sites that have been grafted with skin equivalents has not yet been made.

113.8 Structural Specificity of Nerve Regeneration Template (NRT)

The design principles for regeneration templates presented above have been used to design implants for regeneration of peripheral nerves. The medical problem typically involves the loss of innervation in arms and legs, leading to loss of motor and sensory function (paralysis). The nerves involved are peripheral, and the design problem becomes the regeneration of injured peripheral nerves, with recovery of function.

A widely used animal model for peripheral nerve injury is a surgically generated gap in the sciatic nerve of the rat. Interruption of nerve function in this case is localized primarily in the region of plantar muscles of the foot involved. This relatively well defined area of loss of function can then be studied neurologically in relative isolation from other neurologic events. Furthermore, the rate of recovery of function can be studied by electrophysiologic methods, a procedure which provides continuous data over the entire period of healing. In the peripheral nerve the healing period extends to about 10 weeks, clearly longer than healing in skin, which occurs largely within a period of only 3 weeks.

When the sciatic nerve is cut and a gap forms between the two nerve ends, the distal part of the nerve is isolated from its cell body in the spinal cord. Communication between the central nervous system and the leg is no longer possible. The lack of muscle innervation leads to inactivity, which in turn leads to muscle atrophy. At the site of injury there is degeneration of the myelin sheath of axons, dissociation of Schwann cells and formation of scar. It has been hypothesized that the formation of scar impedes, more than any other single cause, the elongation of axons across the gap. Axonal elongation through a gap of substantial length becomes, therefore, a parameter of prime importance in the design of a nerve regeneration template (NRT).

Intubation of severed nerve ends is a widely used procedure for isolating axons from the tissue environment of the peripheral nerve. The lumen of the tube serves to isolate the process of nerve regeneration from wound-healing events involving connective tissues outside the nerve; the tube walls, for example, prevent proliferation of scar tissue inside the tube and the subsequent obstruction of the regenerating nerve. Silicone tubes are the most widely used method of intubation. These tubes are both nonbiodegradable and nonpermeable to large molecules. In this isolated environment it is possible to study the substrate preferences of elongating axons by incorporating well-defined ECM analogs and studying the kinetics of functional recovery continuously with electrophysiologic procedures for about 40 weeks following implantation. As in studies of dermal regeneration described above, the ECM analogs which were used in the study of substrate preferences of axons were graft copolymers of type I collagen and chondroitin 6-sulfate. Controls used included the autograft, empty silicone tubes, as well as tubes filled with saline. In these studies, it was necessary to work with a gap dimension large enough to preclude spontaneous regeneration, i.e., regeneration in the absence of an ECM analog. It was observed that a gap length of 10 mm was occasionally innervated spontaneously in the rat, whereas no instances of spontaneous regeneration were observed with a 15-mm gap length. Gap lengths of 10 and 15 mm were used in these studies with rats.

Three structural parameters of ECM analogs were varied systematically. The degradation rate had a significant effect on the fidelity of regeneration, and an abbreviated optimization procedure led to an ECM analog which degraded much faster than the DRT. In combination with Eq. (113.1), this empirical finding suggests that a healing nerve wound contains a much smaller concentration of the degrading enzyme, collagenase; this suggestion is qualitatively consistent with observations of collagenolytic activity in injured nerves. The average diameter also was found to have a significant effect on fidelity of regeneration, and the optimization procedure led to a value of 5 μm, significantly smaller than the average pore diameter in the DRT. Finally, use of Eq. (113.5) led to procedures of preparing ECM analogs, the pore channels of which were either highly aligned along the tube axis, randomly oriented, or radially oriented. ECM analogs with axially aligned pore channels were found to be superior to analogs with other types of alignment.

TABLE 113.1 Design Parameters for Two Regeneration Templates

Design Parameter of ECM Analog	DRT	NRT
Degradation rate, enzyme units	<120	>150
Average pore diameter, μm	20–125	5
Pore channel orientation	random	axial

These results have led to a design for an NRT consisting of a specified degradation rate, average pore diameter, and pore channel alignment as shown in Table 113.1 in which the structural parameters of NRT are contrasted to those of DRT.

Studies of Nerve Regeneration Using Degradable Tubes

Porous collagen tubes without matrix content have been extensively studied as guides for peripheral nerve regeneration in rodents and nonhuman primates. The walls of these collagen tubes had an average pore diameter which was considered sufficiently large for transport of molecules as large as bovine serum albumin (MW = 68 kDa). A 4-mm gap in the sciatic nerve of the rat was the standard injury studied. Other injury models that were studied included the 4-mm gap and the 15-mm gap in the adult monkey.

The use of an empty tube did not allow for any degree of optimization of tube parameters to be achieved in this study. Nevertheless, the long-term results showed almost complete recovery of motor and sensory responses, at rates that approximated the recovery obtained following use of the nerve autograft, currently the best conventional treatment in cases of massive loss of nerve mass. The success obtained suggests that collagen tubes can be used instead of autografts, the harvesting of which subject the patient to nerve-losing surgery.

Even more significant improvement in quality of nerve regeneration is obtained when collagen tubes are filled with NRT. Although lower than normal, the histomorphometric and electrophysiological properties of the regenerate resulting from use of an NRT-filled collagen tube have been statistically indistinguishable from those of the autograft control. Probably, however, the most important effect observed following implantation of the NRT-filled collagen tube has been the observation that, while the total number of myelinated axons in regenerated nerves had reached a plateau by 30 weeks after injury, the number of axons with diameter larger than 6 μm continues to increase at substantial rates through the completion of a recent study (60 weeks). Axons with a diameter larger than 6 μm have been uniquely associated with the A-fiber peak of the action potential, which determines the maximum conduction velocity of the regenerates. Thus, kinetic evidence has been presented which supports the view that the nerve trunk maturation process continues beyond 60 weeks after injury, resulting in a nerve trunk which increasingly approaches the structure of the normal control.

113.9 In Situ of Synthesis of Meniscus Using a Meniscus Regeneration Template (MRT)

The meniscus of the knee performs a variety of functions which amount to joint stabilization and lubrication, as well as shock absorption. Its structure is that of fibrocartilage, consisting primarily of type I collagen fibers populated with meniscal fibrochondrocytes. The architecture of collagen fibers is complex. In this tissue, which has a shape reminiscent of one-half a doughnut, the collagen fibers are arranged in a circumferential pattern which is additionally reinforced by radially placed fibers. The meniscus can be torn during use, an event which causes pain and disfunction. Currently, the accepted treatments are partial or complete excision of torn tissue. The result of such treatment is often unsatisfactory, since the treated meniscus has an altered shape which is incompatible with normal joint motion and stability. The long-term consequence of such incompatibility is joint degeneration, eventually leading to osteoarthritis.

An effort to induce regeneration of the surgically removed meniscus was based on the use of a type I collagen-chondroitin 6-sulfate copolymer. The precise structural parameters of this matrix have not been reported, so it is not possible to discuss the results of this study in terms of possible similarities and differences with DRT and NRT. The study focused on the canine knee joint, since the latter is particularly sensitive to biomechanical alterations; in this model, joint instabilities rapidly lead to osteoarthritic changes, which can be detected experimentally. Spontaneous regeneration of the canine meniscus following excision is partial and leads to a biomechanically inadequate tissue which does not protect the joint from osteoarthritic changes. The latter condition provides the essential negative control for the study. In this study, knee joints were subjected to 80% removal (resection) of the medial meniscus, and the lesion was treated either by an autograft or by an ECM analog or was not treated at all. Evaluation of joint function was performed by studying joint stability, gait, and treadmill performance and was extended up to 27 months. No evidence was presented to show that the structure of the ECM analog was optimized to deliver maximal regeneration.

The results of this study showed that two-thirds of the joints implanted with the ECM analog, two-thirds of the joints which were autografted, and only 25% of the joints which were resected without further treatment showed regeneration of meniscal tissue. These results were interpreted to suggest that the ECM analog, or meniscus regeneration template (MRT), supported significant meniscal regeneration and provided enough biomechanical stability to minimize degenerative osteoarthritis in the canine knee joint.

Recently, a clinical trial of the feasibility of MRT treatment was conducted on nine patients with either an irreparable tear of the meniscal cartilage or major loss of meniscal cartilage, and who remained in the study for at least 36 months. Following deletion of irreparably damaged meniscal tissue, MRT was implanted at the site of the missing meniscal tissue mass. The results showed that implantation of MRT induced regeneration of meniscal cartilage while the template was undergoing degradation. Compared to patients who were observed during the same period after meniscectomy (removal of damaged meniscal tissue without MRT treatment), patients who were implanted with MRT showed significant reduction in pain as well as greatly improved resumption of strenuous activity.

Defining Terms

Autograft: The patient's own tissue or organ, harvested from an intact area.

Cell lifeline number, S: A dimensionless number that expresses the relative magnitudes of chemical reaction and diffusion. This number, defined as S in Eq. (113.2), can be used to compute the maximum path length l_c over which a cell can migrate in a scaffold while depending on diffusion alone for transport of critical nutrients that it consumes. When the critical length is exceeded, the cell requires transport of nutrients by angiogenesis in order to survive.

Cultured epithelia: A mature, keratinizing epidermis synthesized in vitro by culturing epithelial cells removed from the patient by biopsy. A relatively small skin biopsy (1 cm^2) can be treated to yield an area larger by about 10,000 in 2–3 weeks and can then be grafted on patients.

Dermal equivalent: A term which has been loosely used in the literature to describe a device that replaces, usually temporarily, the functions of the dermis following injury.

Dermis: A 1–5 mm layer of connective tissue populated with quiescent fibroblasts which lies underneath the epidermis. It is separated from the former by a very thin basement membrane. The dermis of adult mammals does not regenerate spontaneously following injury.

Dermis regeneration template (DRT): A graft copolymer of type I collagen and chondroitin 6-sulfate, average pore diameter 20–125 μm, degrading in vivo to an extent of about 50% in 2 weeks, which induces partial regeneration of the dermis in wounds from which the dermis has been fully excised. When seeded with keratinocytes prior to grafting, this analog of extracellular matrix has induced simultaneous synthesis both of a dermis and an epidermis.

ECM analog: A model of extracellular matrix, consisting of a highly porous graft copolymer of collagen and a glycosaminoglycan.

Epidermis: The cellular outer layer of skin, about 0.1 mm thick, which protects against moisture loss and against infection. An epidermal graft, e.g., cultured epithelium or a thin graft removed surgically, requires a dermal substrate for adherence onto the wound bed. The epidermis regenerates spontaneously following injury, provided there is a dermal substrate underneath.

Isomorphous tissue replacement: A term used to describe the synthesis of new, physiologic tissue within a regeneration template at a rate of the same order as the degradation rate of the template. This relation, described by Eq. (113.1), is the defining equation for a biodegradable scaffold which couples with, or interacts in this unique manner with, the inflammatory response of the wound bed.

Meniscus regeneration template (MRT): A graft copolymer of type I collagen and an unspecified glycosaminoglycan, average pore diameter unspecified, which has induced partial regeneration of the knee meniscus in dogs following 80% excision of the meniscal tissue.

Mikic number, *Mi*: Ratio of the characteristic freezing time of the aqueous medium of the collagen-GAG suspension, t_f, to the characteristic time for entry, t_e, of a container filled with the suspension which is lowered at constant velocity into a well-stirred cooling bath. *Mi*, defined by Eq. (113.5), can be used to design implants which have high alignment of pore channels along a particular axis, or no preferred alignment.

Morphogenesis: The shaping of an organ during embryologic development or during wound healing, according to transcription of genetic information and local environmental conditions.

Nerve regeneration template (NRT): A graft copolymer of type I collagen and chondroitin 6-sulfate, average pore diameter 5 mm, degrading in vivo to an extent of about 50% in 6 weeks, which has induced partial regeneration of the sciatic nerve of the rat across a 15-mm gap.

Regeneration: The synthesis of new, physiologic tissue at the site of a tissue (one cell type) or organ (more than one cell type) which either has been lost due to injury or has failed due to a chronic condition.

Regeneration template: A biodegradable device which, when attached to a missing organ, induces its regeneration.

Scar: The end result of a repair process in skin and other organs. Scar is morphologically different from skin, in addition to being mechanically less extensible and weaker than skin. The skin regeneration template induces synthesis of nearly physiologic skin rather than scar. Scar is also formed at the site of severe nerve injury.

Self-crosslinking: A procedure for reducing the biodegradation rate of collagen, in which collagen chains are covalently bonded to each other by a condensation reaction which is driven by drastic dehydration of the protein. This reaction illustrated by Eqs. (113.3*a*) and (113.3*b*), is also used to graft glycosaminoglycan chains to collagen chains without use of an extraneous crosslinking agent.

Stereology: The discipline which relates the quantitative statistical properties of three-dimensional structures to those of their two-dimensional sections or projections. In reverse, stereologic procedures allow reconstruction of certain aspects of three-dimensional objects from a quantitative analysis of planar images. Its rules are used to determine features of the pore structure of ECM analogs.

References

Archibald SJ, Krarup C, Sheffner J, Li S-T, Madison RD. 1991. A collagen-based nerve guide conduit for peripheral nerve repair: An electrophysiological study of nerve regeneration in rodents and non-human primates. J Comp Neurol 306:685–696.

Archibald SJ, Sheffner J, Krarup C, Madison RD. 1995. Monkey median nerve repaired by nerve graft or collagen nerve guide tube. J Neurosci 15:4109–4123.

Butler CE, Orgill DP, Yannas IV, Compton CC. 1998. Effect of keratinocyte seeding of collagen-glycosaminoglycan membranes on the regeneration of skin in a porcine model. Plast Reconstr Surg 101:1572–1579.

Chamberlain LJ, Yannas IV, Hsu H-P, Strichartz G, Spector M. 1998. Peripheral nerve regeneration through a collagen device is comparable with an autograft across 10-mm gaps after 60 weeks. Exp Neurol, in press.

Chang AS, Yannas IV. 1992. Peripheral Nerve Regeneration. In B Smith, G Adelman (eds), Neuroscience Year (Supplement 2 to the Encyclopedia of Neuroscience), pp 125–126, Boston, Birkhauser.

Compton CC, Butler CE, Yannas IV, Warland G, and Orgill DP. 1998. Organized skin structure is regenerated *in vivo* from collagen-GAG matrices seeded with autologous keratinocytes. J Invest Dermatol 110:908–916.

Hansbrough JF, Boyce ST, Cooper ML, Foreman TJ. 1989. Burn wound closure with autologous keratinocytes and fibroblasts attached to a collagen-glycosaminoglycan substrate. JAMA, vol 262, pp 2125–2141.

Hull BE, Sher SS, Rosen S, Church D, Bell E. 1983. Structural integration of skin equivalents grafted to Lewis and Sprague-Dawley rats. J Invest Derm, vol 8, pp 429–436.

Orgill DP, Butler CE, Regan JF. 1996. Behavior of collagen-GAG matrices as dermal replacements in rodent and porcine models, Wounds 8:151–157.

Stone KR, Rodkey WK, Webber RJ, McKinney L, Steadman JR. 1990. Collagen Based Prostheses for Meniscal Regeneration, Clin Orth, vol 252, pp 129–135.

Stone KR, Steadman R, Rodkey WG, Li S-T. 1997. Regeneration of meniscal cartilage with use of a collagen scaffold. J Bone Joint Surg 79A:1770–1777.

Yannas IV, Burke JF. 1980. Design of an artificial skin. I. Basic design principles. J Biomed Mater Res, vol 14, pp 65–81.

Yannas IV, Burke JF, Orgill DP, Skrabut EM. 1982. Wound tissue can utilize a polymeric template to synthesize a functional extension of skin. Science 215:174–176.

Yannas IV, Lee E, Orgill DP, Skrabut EM, Murphy GF. 1989. Synthesis and characterization of a model extracellular matrix that induces partial regeneration of adult mammalian skin. Proc Natl Acad Sci USA, vol 86, pp 933–937.

Yannas IV. 1997. Models of organ regeneration processes induced by templates. In A Prokop, D Hunkeler, AD Cherrington (eds), Bioartificial Organs, Ann NY Acad Sci 831:280–293.

Yannas IV. 1998. Studies on the biological activity of the dermal regeneration template. Wound Rep Regen, in press.

Further Information

Chamberlain LJ, Yannas IV, Arrizabalaga A, Hsu H-P, Norregaard TV, Spector M. 1998. Early peripheral nerve healing in collagen and silicone tube implants. Myofibroblasts and the cellular response. Biomaterials, in press.

Chamberlain LJ, Yannas IV, Hsu H-P, Spector M. 1997. Histological response to a fully degradable collagen device implanted in a gap in the rat sciatic nerve. Tissue Eng 3:353–362.

Eldad A, Burt A, Clarke JA, Gusterson B. 1987. Cultured epithelium as a skin substitute. Burns 13:173–180.

FR Noyes (eds), Biology and Biomechanics of the Traumatized Synovial Joint: The Knee as a Model, pp 221–231, AAOS.

Hansbrough JF, Cooper ML, Cohen R, Spielvogel R, Greemnleaf G, Bartel RL, Naughton G. 1992. Evaluation of a biodegradable matrix containing cultured human fibroblasts as a dermal replacement beneath meshed skin grafts on athymic mice. Surg vol 111, pp 438–446.

Heimbach D, Luterman A, Burke J, Cram A, Herndon D, Hunt J, Jordan M, McManus W, Solem L, Warden G, Zawacki B. 1988. Artificial dermis for major burns. Ann Surg vol 208, pp 313–320.

Madison RD, Archibald SJ, Krarup C. 1992. Peripheral Nerve Injury. In IK Cohen, RF Diegelmann, WJ Lindblad (eds), Wound Healing, pp 450–487, Philadelphia, Saunders.

Orgill DP, Yannas IV. 1998. Design of an artificial skin. IV. Use of island graft to isolate organ regeneration from scar synthesis and other processes leading to skin wound closure. J Biomed Mater Res 36:531–535.

Rodkey WG, Stone KR, Steadman JR. 1992. Prosthetic Meniscal Replacement. In GAM Finerman, FR Noyes (eds), Biology and Biomechanics of the Traumatized Synovial Joint: The Knee as a Model, pp 221–231, Amer Acad Orth Surg, Chicago, IL.

Yannas IV, Burke JF, Gordon PL, Huang C. 1977. Multilayer membrane useful as synthetic skin. US Patent 4,060,081, Nov. 29.

Yannas IV, Burke JF, Orgill DP, Burke JF. 1982. Regeneration of skin following closure of deep wounds with a biodegradable template. Trans Soc Biomat, vol 5, pp 1–38.

Yannas IV. 1988. Regeneration of skin and nerves by use of collagen templates. In ME Nimni (ed), Collagen, vol 3, pp 87–115, Boca Raton, CRC Press.

Yannas IV. 1989. Skin. Regeneration Templates. Encyclopedia of Polymer Science and Engineering, vol. 15, pp. 317–334.

Yannas IV. 1990. Biologically active analogues of the extracellular matrix: Artificial skin and nerve. Angew Chemie Int Ed Engl, vol 29, pp 20–35.

114

Fluid Shear Stress Effects on Cellular Function

Charles W. Patrick, Jr.
Rice University

Rangarajan Sampath
Rice University

Larry V. McIntire
Rice University

Cells of the vascular system are constantly exposed to mechanical (hemodynamic) forces due to the flow of blood. The forces generated in the vasculature include the frictional force or a fluid shear stress caused by blood flowing tangentially across the endothelium, a tensile stress caused by circumferential vessel wall deformations, and a net normal stress caused by a hydrodynamic pressure differential across the vessel wall. We will restrict our discussion to examining fluid shear stress modulation of vascular cell function. The endothelium is a biologically active monolayer of cells providing an interface between the flowing blood and tissues of the body. It can synthesize and secrete a myriad of vasoconstrictors, vasodilators, growth factors, fibrinolytic factors, cytokines, adhesion molecules, matrix proteins, and mitogens that modulate many physiologic processes, including wound healing, hemostasis, vascular remodeling, vascular tone, and immune and inflammatory responses. In addition to humoral stimuli, it is now well accepted that endothelial cell synthesis and secretion of bioactive molecules can be regulated by the hemodynamic forces generated by the local blood flow. These forces have been hypothesized to regulate neovascularization and the structure of the blood vessel [Hudlicka, 1984]. Clinical findings further show that arterial walls undergo an endothelium-dependent adaptive response to changes in blood flow, with blood vessels in high flow regions tending to enlarge and vessels in the low flow region having reduced lumen diameter, thereby maintaining a nearly constant shear stress at the vessel wall [Zarins et al., 1987]. In addition to playing an active role in the normal vascular biology, hemodynamic

0-8493-8594-6/00/$0.00+$.50

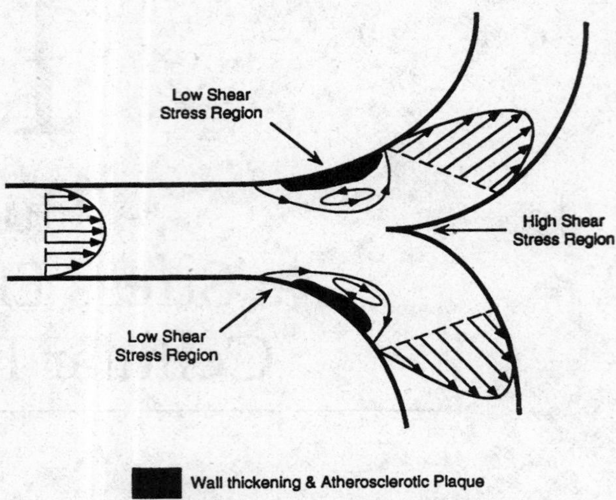

FIGURE 114.1 Atherosclerotic plaques develop in regions of arteries where the flow rate (and resultant wall shear stress) is relatively low, which is often downstream from a vessel bifurcation, where there can be flow separation from the outer walls and regions of recirculation. These regions of low shear stress are pathologically prone to vessel wall thickening and thrombosis.

forces have also been implicated in the pathogenesis of a variety of vascular diseases. Atherosclerotic lesion-prone regions, characterized by the incorporation of Evans blue dye, enhanced accumulation of albumin, fibrinogen, and LDL, increased recruitment of monocytes, and increased endothelial turnover rates exhibit polygonal endothelial cell morphology typically seen in a low-shear environment, as opposed to nonlesion regions that have elongated endothelial cells characteristic of high-shear regions [Nerem, 1993].

In vivo studies of the distribution of atherosclerosis and the degree of intimal thickening have shown preferential plaque localization in low-shear regions. Atherosclerotic lesions were not seen in random locations but were instead found to be localized to regions of arterial branching and sharp curvature, where complex flow patterns develop, as shown in Fig. 114.1 [Asakura & Karino, 1990; Friedman et al., 1981; Gibson et al., 1993; Glagov et al., 1988; Ku et al., 1985; Levesque et al., 1989; Zarins et al., 1983]. Morphologically, intact endothelium over plaque surfaces showed variation in shape and size and loss of normal orientation, characteristic of low-shear conditions [Davies et al., 1988]. Flow in the vascular system is by and large laminar, but extremely complex time dependent flow patterns can develop in localized regions of complex geometry. Zarins and coworkers [1983] have shown that in the human carotid bifurcation, regions of moderate-to-high shear stress where flow remains unidirectional and axially aligned, were relatively free of intimal thickening, whereas extensive plaque localization was seen in the regions where low shear stress, fluid separation from the wall, and complex flow patterns were predominant. Asakura and Karino [1990] presented a technique for flow visualization by using transparent arteries where they directly correlated flow patterns to regions of plaque formation in human coronary arteries. They observed that the flow divider at the bifurcation point, a region of high shear stress, was relatively devoid of plaque deposition, whereas the outer wall, a region of low shear stress, showed extensive plaque formation. Similar patterns were also seen in curved vessels, where the inner wall or the hip region of the curve, a region of low shear and flow separation, exhibited plaque formation. In addition to the direct shear-stress-mediated effects, the regions of complex flow patterns and recirculation also tend to increase the residence time of circulating blood cells in susceptible regions, whereas the blood cells are rapidly cleared away from regions of high wall shear and unidirectional flow [Glagov et al., 1988]. This increased transit time could influence plaque deposition by favoring margination of monocytes and platelets, release of vasoactive agents, or altered permeability at the intercellular junctions to extracellular lipid particles and possible concentration of procoagulant materials [Nollert et al., 1991].

The endothelium forming the interface between blood and the surrounding tissues is believed to act as a sensor of the local hemodynamic environment in mediating both the normal response and the pattern of vascular diseases. In vivo studies have shown changes in the actin microfilament localization in a shear-dependent manner. Actin stress fibres have been observed to be aligned with the direction of flow in high-shear regions, whereas they were mostly present in dense peripheral bands in the low-shear regions [Kim et al., 1989]. Langille and Adamson [1981] showed that the cells in large arteries, away from the branches, were aligned parallel to the long axes of the artery in rabbit and mouse. Similar results were shown in coronary arteries of patients undergoing cardiac transplantation [Davies et al., 1988]. Near the branch points of large arteries, however, cells aligned along the flow streamlines. In smaller blood vessels, where secondary flow patterns did not develop, the cell alignment followed the geometry of the blood vessel. In fact, endothelial cell morphology and orientation at the branch may be a natural marker of the detailed features of blood flow [Nerem et al., 1981].

Another cell type that is likely to be affected by hemodynamic forces is the vascular smooth muscle cell (SMC) present in the media. In the early stages of lesion development, SMCs proliferate and migrate into the intima [Ross, 1993; Schwartz et al., 1993]. Since the endothelium is intact in all but the final stages of atherosclerosis, SMCs are unlikely to be directly affected by shear stress. Most of the direct force they experience comes from the cyclical stretch forces experienced by the vessel itself due to pressure pulses [Grande et al., 1989]. Shear forces acting on the endothelium, however, can release compounds such as endothelin, nitric oxide, and platelet-derived growth factors that act as SMC mitogens and can modulate SMC tone. Ono and colleagues [1991] showed that homogenate of shear-stressed endothelial cells contained increased amounts of collagen and stimulated SMC migration in vitro, compared to endothelial cells grown under static conditions. The local shear environment could thus act directly or indirectly on the cells of the vascular wall in mediating a physiologic response.

This chapter presents what is currently known regarding how the mechanical agonist of shear stress modulates endothelial cell function in the context of vascular physiology and pathophysiology. In discussing this topic we have adopted an outside-in approach. That is, we first discuss how shear stress affects endothelial cell-blood cell interactions, then progress to how shear stress affects the endothelial cell cytoskeleton, signal transduction, and protein secretion, and finally end with how shear stress modulates endothelial cell gene expression. We close with gene therapy and tissue engineering considerations related to how endothelial cells respond to shear stress. Before proceeding, however, we discuss the devices and methodology used for studying the endothelial cell responses to shear stress in vitro.

114.1 Devices and Methodology Used for in Vitro Experiments

In vivo studies aimed at understanding cellular responses to shear forces have the inherent problem that they cannot quantitatively define the exact features of the hemodynamic environment. Moreover, it is very difficult to say if the resultant response is due to shear stress or some other feature associated with the hemodynamic environment. Cell culture studies and techniques for exposing cells to a controlled shear environment in vitro have been increasingly used to elucidate cellular responses to shear stress and flow. Mechanical-force-induced changes in cell function have been measured in vitro using mainly two cell types, cultured monolayers of bovine aortic endothelial cells (BAECs) and human umbilical vein endothelial cells (HUVECs). Shear stress is typically generated in vitro by flowing fluid across endothelial cell monolayers under controlled kinematic conditions, usually in the laminar flow regime. Parallel plate and cone-and-plate geometries have been the most common. Physiologic levels of venous and arterial shear stress range between 1–5 dynes/cm^2 and 6–40 dynes/cm^2, respectively.

The use of the parallel plate flow chamber allows one to have a controlled and well-defined flow environment based on the chamber geometry (fixed) and the flow rate through the chamber (variable).

In addition, individual cells can be visualized in real time using video microscopy. Assuming parallel plate geometry and Newtonian fluid behavior, the wall shear stress on the cell monolayer in the flow chamber is calculated as

$$\tau_w = \frac{6Q\mu}{\left(bh^2\right)} \tag{114.1}$$

where Q is the volumetric flow rate, μ is the viscosity of the flowing fluid, h is the channel height, b is the channel width, and τ_w is the wall shear stress. The flow chambers are designed such that the entrance length is very small compared to the effective length of the chamber [Frangos et al., 1985]. Therefore, entry effects can be neglected, and the flow is fully developed and parabolic over nearly the entire length of the flow chamber. Flow chambers usually consist of a machined block, a gasket whose thickness determines in part the channel depth, and a glass coverslip to which is attached a monolayer of endothelial cells (Fig. 114.2*a*). The individual components are held together either by a vacuum, or by evenly torqued screws, thereby ensuring a uniform channel depth. The flow chamber can have a myriad of entry ports, bubble ports, and exit ports. For short-term experiments, media are drawn through the chamber over the monolayer of cells using a syringe pump. For long-term experiments, the chamber is placed in a flow loop. In a flow loop (Fig. 114.2*b*), cells grown to confluence on glass slides can be exposed to a well-defined shear stress by recirculation of culture medium driven by gravity [Frangos et al., 1985]. Culture medium from the lower reservoir is pumped to the upper reservoir at a constant flow rate such that there is an overflow of excess medium back into the lower reservoir. This overflow serves two purposes: (1) It maintains a constant hydrostatic head between the two reservoirs, and (2) it prevents entry of air bubbles into the primary flow section upstream of the flow chamber that could be detrimental to the cells. The pH of the medium is maintained at physiologic levels by gassing with a humidified mixture of 95% air and 5% CO_2. The rate of flow in the line supplying the chamber is determined solely by gravity and can be altered by changing the vertical separation between the two reservoirs. A sample port in the bottom reservoir allows periodic sampling of the flowing medium for a time-dependent assay.

As mentioned above, cone-and-plate geometries can also be utilized. A schematic of a typical cone-and-plate viscometer is shown in Fig. 114.2*c*. Shear stress is produced in the fluid contained between the rotating cone and the stationary plate. Cells grown on coverslips can be placed in the shear field (up to 12 at a time). For relatively small cone angles and low rates of rotation, the shear stress throughout the system is independent of position. The cone angle compensates for radial effects seen in plate-and-plate rheometers. The wall shear stress (τ_w) on the cell monolayer in the cone-and-plate viscometer is calculated as

$$\tau_w = \frac{3T}{2\pi R^3} \tag{114.2}$$

where T is the applied torque and R is the cone radius. The flow becomes turbulent, however, at the plate's edge and at high rotational speeds. Modifications from the basic design have allowed use of an optical system with the rheometer, enabling direct microscopic examination and real-time analysis of the cultured cells during exposure to shear stress [Dewey et al., 1981; Schnittler et al., 1993]. For a more complete description of in vitro device design and applications, refer to the text edited by Frangos [Tran-son-Tay, 1993].

114.2 Shear Stress-Mediated Cell-Endothelium Interactions

Cell-cell and cell-substrate interactions in the vascular system are of importance in a number of physiologic and pathologic situations. Lymphocytes, platelets, or tumor cells in circulation may arrest at a

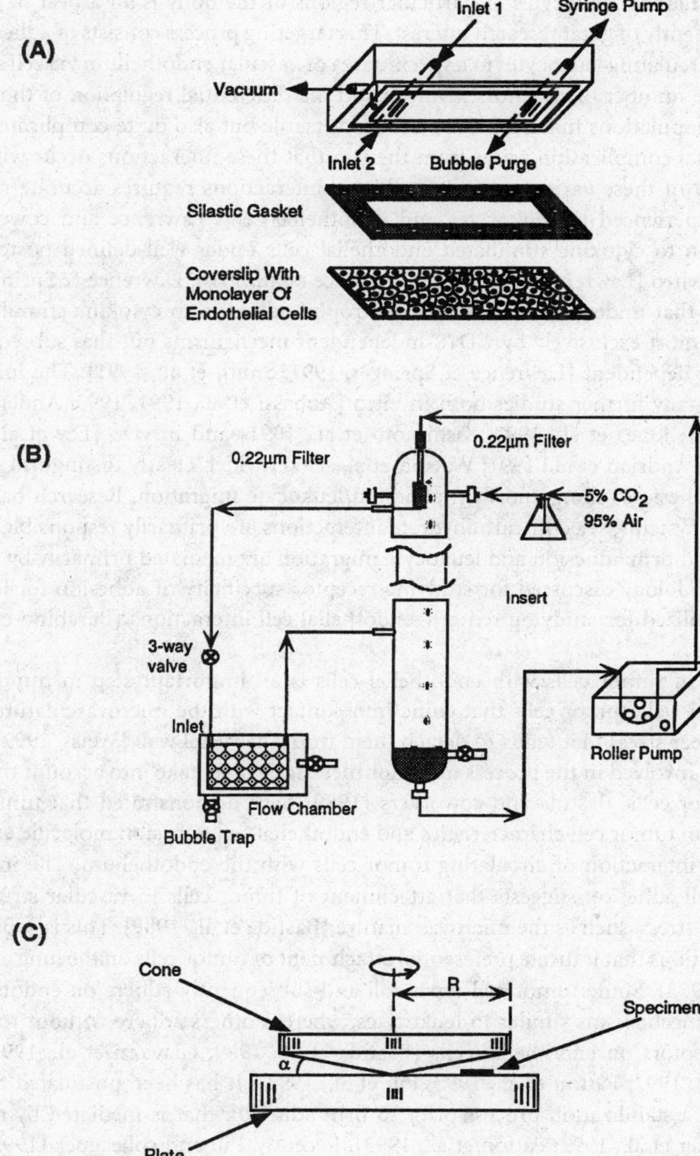

FIGURE 114.2 Devices used for in vitro study of shear stress effects on vascular endothelial cells. (*a*) Parallel plate flow chamber, (*b*) flow loop, (*c*) cone-and-plate viscometer.

particular site as a result of interaction with the endothelium or the subendothelial matrix. While the margination of leukocyte/lymphocyte to the vessel wall be a normal physiologic response to injury or inflammation, adhesion of blood platelets to the subendothelium and subsequent platelet aggression could result in a partial or complete occlusion of the blood vessel leading to thrombosis or stroke. In addition, the adhesion of tumor cells to the endothelium is often the initial step in the development of secondary metastases. The adhesion of leukocytes, platelets, and tumor cells is not only mediated by adhesion molecules on the endothelium but also mediated by the hemodynamic force environment present in the vasculature. In fact, the specific molecular mechanisms employed for adhesion often vary with the local wall shear stress [Alevriadou et al., 1993; Lawrence et al., 1990].

Targeting of circulating leukocytes to particular regions of the body is an aspect of immune system function that is currently of great research interest. This targeting process consists of adhesion of a specific subpopulation of circulating leukocytes to a specific area of vascular endothelium via cell surface adhesion receptors. The large number of receptors involved and the differential regulation of their expression on particular cell subpopulations make this process very versatile but also quite complicated. An extremely important additional complication arises from the fact that these interactions occur within the flowing bloodstream. Study of these various types of adhesive interactions requires accurate recreation of the flow conditions experienced by leukocytes and endothelial cells. Lawrence and coworkers examined neutrophil adhesion to cytokine-stimulated endothelial cells under well-defined postcapillary venular flow conditions in vitro [Lawrence et al., 1987; Lawrence et al., 1990; Lawrence & Springer, 1991]. They also demonstrated that under flow conditions neutrophil adhesion to cytokine-stimulated endothelial cells is mediated almost exclusively by CD18-independent mechanisms but that subsequent neutrophil migration is CD18-dependent [Lawrence & Springer, 1991; Smith et al., 1991]. The initial flow studies were followed by many further studies both in vitro [Abbassi et al., 1991, 1993; Anderson et al., 1991; Hakkert et al., 1991; Jones et al., 1993; Kishimoto et al., 1991] and in vivo [Ley et al., 1991; Perry & Granger, 1991; von Andrian et al., 1991; Watson et al., 1991] which clearly distinguish separate mechanisms for initial adhesion/rolling and firm adhesion/leukocyte migration. Research has further shown that in a variety of systems, selectin/carbohydrate interactions are primarily responsible for initial adhesion and rolling, and firm adhesion and leukocyte migration are mediated primarily by integrin/peptide interactions. Methodology discussed for studying receptor specificity of adhesion for leukocytes under flow can also be utilized for studying red cell–endothelial cell interactions [Barabino et al., 1987; Wick et al., 1987, 1993].

The interaction of tumor cells with endothelial cells is an important step in tumor metastasis. To adhere to the vessel wall, tumor cells that come into contact with the microvasculature must resist the tractive force of shear stress that tends to detach them from the vessel wall [Weiss, 1992]. Hence, studies of the mechanisms involved in the process of tumor metastasis must take into account the physical forces acting on the tumor cells. Bastida and coworkers [1989] have demonstrated that tumor cell adhesion depends not only on tumor cell characteristics and endothelial cell adhesion molecule expression but on shear stress in the interaction of circulating tumor cells with the endothelium. The influence of shear stress on tumor cell adhesion suggests that attachment of tumor cells to vascular structures occurs in areas of high shear stress, such as the microvasculature [Bastida et al., 1989]. This is supported by earlier pathologic observations that indicate preferential attachment of tumor cells on the lung and liver capillary system [Warren, 1973]. Some tumor cell types roll and subsequently adhere on endothelial cells using selectin-mediated mechanisms similar to leukocytes, whereas others adhere without rolling using integrin-mediated receptors on endothelial cells [Bastida et al., 1989; Giavazzi et al., 1993; Kojima et al., 1992; Menter et al., 1992; Patton et al., 1993; Pili et al., 1993]. It has been postulated that some tumor cell types undergo a stabilization process prior to firm adhesion that is mediated by transglutaminase crosslinking [Menter et al., 1992; Patton et al., 1993]. Recently, Pili and colleagues [1993] demonstrated that tumor cell contact with endothelial cells increases Ca^{2+} release from endothelial cell intracellular stores, which may have a fundamental role in enhancing cell-cell adhesion. The factors and molecular mechanisms underlying tumor cell and endothelial cell interactions remain largely undefined and are certain to be further explored in the future.

The endothelium provides a natural nonthrombogenic barrier between circulating platelets and the endothelial basement membrane. However, there are pathologic instances in which the endothelium integrity is compromised, exposing a thrombogenic surface. Arterial thrombosis is the leading cause of death in the United States. Among the most likely possibilities for the initiation of arterial platelet thrombi are: (1) adhesion of blood platelets onto the subendothelium of injured arteries and arterioles or on ruptured atherosclerotic plaque surfaces, containing collagen and other matrix proteins, with subsequent platelet aggregation, or (2) shear-induced aggregation of platelets in areas of the arterial circulation partially constricted by atherosclerosis or vasospasm. Various experimental models have been developed to investigate the molecular mechanisms of platelet attachment to surfaces (adhesion) and platelet cohesion to each other (aggregation) under shear stress conditions. Whole-blood perfusion studies using annular or parallel-plate perfusion chambers simulate the first proposed mechanism of arterial thrombus formation in vivo, that may occur as a result of adhesion on an injured, exposed subendothelial or atherosclerotic plaque surface [Hubbel & McIntire, 1986; Weiss et al., 1978, 1986]. Platelets arriving subsequently could, under the right conditions, adhere to each other, forming aggregates large enough to partially or completely occlude the blood vessel. In vitro studies have shown an increase in thrombus growth with local shear rate, an event that is believed to be the result of enhanced arrival and cohesion of platelets near the surface at higher wall shear stresses [Turitto et al., 1987]. Video microscopy provides information on the morphologic characteristics of thrombi and enables the reconstruction of three-dimensional models of thrombi formed on surfaces coated with biomaterials or endothelial cell basement membrane proteins. Macroscopic analysis of thrombi can provide information on platelet mass transport and reaction kinetics with the surface, and microscopic analysis allows dynamic real-time study of cell-surface and intercellular interactions. Such a technology enables the study of key proteins involved in mural thrombus formation, such as vWF [Alevriadou et al., 1993; Folie & McIntire, 1989]. In addition, tests of antithrombotic agents and investigation of the thrombogenicity of various purified components of the vessel wall or polymeric biomaterials can be performed.

114.3 Shear Stress Effects on Cell Morphology and Cytoskeletal Rearrangement

In addition to mediating cell–endothelial cell interactions, shear stress can act directly on the endothelium. It has been demonstrated for almost a decade that hemodynamic forces can modulate endothelial cell morphology and structure [Barbee et al., 1994; Coan et al., 1993; Eskin et al., 1984; Franke et al., 1984; Girard & Nerem, 1991, 1993; Ives et al., 1986; Langille et al., 1991; Levesque et al., 1989; Wechezak et al., 1985]. Conceivably, the cytoskeletal reorganization that occurs in endothelial cells several hours after exposure to flow may, in conjunction with shape change, transduce mechanical signals to cytosolic and nuclear signals (mechanotransduction), thereby playing a role in gene expression and signal transduction [Ingber, 1993; Resnick et al., 1993; Watson, 1991]. In fact, investigators have shown specific gene expression related to cytoskeletal changes [Botteri et al., 1990; Breathnach et al., 1987; Ferrua et al., 1990; Werb et al., 1986]. F-actin has been implicated as the principal transmission element in the cytoskeleton and appears to be required for signal transduction [Watson, 1991]. Actin filaments are anchored in the plasma membrane at several sites, including focal adhesions on the basal membrane, intercellular adhesion proteins, integral membrane proteins at the apical surface, and the nuclear membrane [Davies & Tripathi, 1993]. Substantiating this, Barbee and coworkers [1994] have recently shown, utilizing atomic force microscopy, that F-actin fiber stress bundles are formed in the presence of fluid flow and the fibers are coupled to the apical membrane. Moreover, endothelial cell shape change and realignment with flow can be inhibited by drugs that interfere with microfilament turnover [Davies & Tripathi, 1993]. Although all evidence leads one to believe that the endothelial cell cytoskeleton can respond to flow and that F-actin is involved, the exact mechanism involved remains to be elucidated.

114.4 Shear Stress Effects on Signal Transduction and Mass Transfer

Shear stress and resultant convective mass transfer are known to affect many important cytosolic second messengers in endothelial cells. For instance, shear stress is known to cause ATP-mediated increases in calcium ions (Ca^{2+}) [Ando et al., 1988; Dull & Davies, 1991; Mo et al., 1991; Nollert & McIntire, 1992]. Changes in flow influence the endothelial boundary layer concentration of ATP by altering the convective transport of exogenous ATP, thereby altering ATP's interaction with both the P_{2y}-purinoreceptor and ecto-ATPase. At low levels of shear stress, degradation of ATP by ecto-ATPase exceeds the rate of diffusion, and the steady state concentration of ATP remains low. In contrast, at high levels of shear stress, convection enhances the delivery of ATP from upstream to the P_{2y}-purinoreceptor, and diffusion exceeds the rate of degradation by the ecto-ATPase [Mo et al., 1991]. Whether physiologic levels of shear stress can directly increase intracellular calcium remains unclear. Preceding the calcium increases are increases in inositol-1,4,5 trisphosphate (IP_3) [Bhagyalakshmi et al., 1992; Bhagyalakshmi & Frangos, 1989*b*; Nollert et al., 1990; Prasad et al., 1993], which binds to specific sites on the endoplasmic reticulum and causes release of Ca^{2+} from intracellular stores, and diacylglycerol (DAG) [Bhagyalakshmi & Frangos, 1989*a*]. Elevated levels of both DAG and Ca^{2+} can activate several protein kinases, including protein kinase C (PKC). Changes in Ca^{2+}, IP_3, and DAG are evidence that fluid shear stress activated the PKC pathway. In fact, the PKC pathway has been demonstrated to be activated by shear stress [Bhagyalakshmi & Frangos, 1989*b*; Hsieh et al., 1992, 1993; Kuchan & Frangos, 1993; Milner et al., 1992]. In addition to the PKC pathway, pathways involving cyclic adenosine monophosphate (cAMP) and cyclic guanosine monophosphate (cGMP) may also be modulated by shear stress, as evidenced by increases in cAMP [Bhagyalakshmi & Frangos, 1989*b*] and cGMP [Kuchan & Frangos, 1993; Ohno et al., 1993] with shear stress. Moreover, it has been shown recently that shear stress causes acidification of cytoslic pH [Ziegelstein et al., 1992]. Intracellular acidification of vascular endothelial cells releases Ca^{2+} into the cytosol [Ziegelstein et al., 1993]. This Ca^{2+} mobilization may be linked to endothelial synthesis and release of vasodilatory substances during the pathological condition of acidosis.

114.5 Shear Stress Effects on Endothelial Cell Metabolite Secretion

The application of shear stress in vitro is accompanied by alterations in protein synthesis that are detectable within several hours after initiation of the mechanical agonist. Shear stress may regulate the expression of fibrinolytic proteins by endothelial cells; tPA is an antithrombotic glycoprotein and serine protease that is released from endothelial cells. Once released, it rapidly converts plasminogen to plasmin which, in turn, dissolves fibrin clots. Diamond and coworkers [1989] have shown that venous levels of shear stress do not affect tPA secretion, whereas arterial levels increase the secretion rate of tPA. Arterial flows lead to a profibrinolytic state that would be beneficial in maintaining a clot-free artery. In contrast to tPA, neither venous nor arterial levels of shear stress cause significant changes in plasminogen activator inhibitor-1 (PAI-1) secretion rates [Diamond et al., 1989; Kuchan & Frangos, 1993]. Increased proliferation of smooth muscle cells is one of the early events of arteriosclerosis. The secretion rate of endothelin-1 (ET-1), a potent smooth muscle cell mitogen, has been shown to increase in response to venous levels of shear stress and decrease in response to arterial levels of shear stress [Kuchan & Frangos, 1993; Milner et al., 1992; Nollert et al., 1991; Sharefkin et al., 1991; Yoshizumi et al., 1989]. Both the in vitro tPA and ET-1 results are consistent with in vivo observations that atherosclerotic plaque development occurs in low shear stress regions as opposed to high shear stress regions. The low shear stress regions usually occur downstream from vessel branches. In these regions we would expect low secretion of tPA and high secretion of ET-1, leading to locally increased smooth muscle cell proliferation (intimal thickening of the vessel) and periodic problems with clot formation (thrombosis). Both of these observations are observed pathologically and are important processes in atherogenesis [McIntire, 1994].

Important metabolites in the arachidonic acid cascade via the cyclooxygenase pathway; prostacylin (PGI_2) and prostaglandin (PGF_{2a}) are also known to increase their secretion rates in response to arterial levels of shear stress [Frangos et al., 1985; Grabowski et al., 1985; Nollert et al., 1989]. In addition, endothelial cells have been shown to increase their production of fibronectin (FN) in response to shear stress [Gupte & Frangos, 1990]. Endothelial cells may respond to shear stress by secreting FN in order to increase their attachment to the extracellular matrix, thereby resisting the applied fluid shear stress. Endothelial cells have been shown to modulate their receptor expression of intercellular adhesion molecule-1 (ICAM-1), vascular cell adhesion molecule-1 (VCAM-1), and monocyte chemotactic peptide-1

(MCP-1) in response to shear stress [Sampath et al., 1995; Shyy et al., 1994]. The adaptive expression of these adhesion molecules in response to hemodynamic shear stress may aid in modulating specific adhesion localities for neutrophils, leukocytes, and monocytes.

114.6 Shear Stress Effects on Gene Regulation

In many cases, the alterations in protein synthesis observed with shear stress are preceded by modulation of protein gene expression. The molecular mechanisms by which mechanical agonists alter the gene expression of proteins are under current investigation. Diamond and colleagues [1990] and Nollert and colleagues [1992] have demonstrated that arterial levels of shear stress upregulate the transcription of tPA mRNA, concomitant with tPA protein secretion. The gene expressions of ET-1 and PDGF, both of which are potent smooth muscle cell mitogens and vasoconstrictors, are also known to be modulated by hemodynamic forces. For instance, arterial levels of shear stress cause ET-1 mRNA to be downregulated [Malek et al., 1993; Malek & Izumo, 1992; Sharefkin et al., 1991]. ET-1 mRNA downregulation is sensitive to the magnitude of the shear stress in a dose-dependent fashion, reaching a saturation at 15 dynes/cm^2. Conversely, Yoshizumi and coworkers [1989] have shown that venous levels of shear stress cause a transient increase in ET-1 mRNA, peaking at 4 hours and returning to basal levels by 12–24 hours. However, they had previously reported downregulation of ET-1 mRNA expression [Yanagisawa et al., 1988]. PDGF is expressed as a dimer composed of PDGF-A and PDGF-B subunits. There are conflicting reports as to how arterial shear stresses affect PDGF-B mRNA expression. Malek and colleagues [1993] have shown that arterial levels of shear stress applied to BAECs cause a significant decrease in PDGF-B mRNA expression over a 9-hr period. In contrast, Mitsumata and coworkers [1993] and Resnick and coworkers [1993] have shown increases in BAEC mRNA expression of PDGF-B when arterial shear stresses were applied. The discrepancy in the results may be attributed to differences in BAEC cell line origins or differences in the passage number of cells used. In support of the latter two investigators, Hsieh and colleagues [1992] have reported upregulation of PDGF-B mRNA in HUVECs in the presence of arterial shear stress. As with the B chain of PDGF, there are conflicting reports as to the affect of arterial shear stress on PDGF-A. Mitsumata and coworkers [1993] have reported no change in PDGF-A mRNA expression, whereas Hsieh and coworkers [1991, 1992] have reported a shear-dependent increase in the mRNA expression from 0–6 dynes/cm^2 which then plateaus from 6–51 dynes/cm^2. Endothelial cell expression of ET-1 and PDGF may be important in blood vessel remodeling. In blood vessels exposed to increased flow, the chronic vasculature response is an increase in vessel diameter [Langille & O'Donnell, 1986]. Hence, it is tempting to postulate that hemodynamic-force-modulated alterations in ET-1 and PDGF mRNA expression may account for much of this adaptive change.

The gene expression of various adhesion molecules involved in leukocyte recruitment during inflammation and disease has also been investigated. ICAM-1 mRNA has been shown to be transiently upregulated in the presence of venous or arterial shear stresses [Nagel et al., 1994; Sampath et al., 1995]. Its time-dependent response peaked at 1–3 hr following exposure to shear, before declining below basal levels with prolonged exposure of 6–24 hr. In contrast, VCAM-1 mRNA level was downregulated almost immediately upon onset of flow and was found to drop significantly below basal levels within 6 hr of initiation of flow at all magnitudes of shear stresses. E-selectin mRNA expression appeared to be generally less responsive to shear stress, especially at the lower magnitudes. After an initial downward trend 1 hr following exposure to shear stress (2 dynes/cm^2), E-selectin mRNA remained at stable levels for up to 6 hr [Sampath et al., 1995]. Recent evidence shows that the expression of MCP-1, a monocyte-specific chemoattractant expressed on endothelial cells, also follows a similar biphasic response with shear stress like ICAM-1 [Shyy et al., 1994]. In addition to adhesion molecules, the gene expression of several growth factors has been investigated. The mRNA expression of heparin-binding epidermal growth factor like growth factor (HB-EGF), a smooth muscle-cell mitogen, transiently increases in the presence of minimal arterial shear stress [Morita et al., 1993]. Transforming growth factor-β1 (TGF-β1) mRNA has been reported to be upregulated in the presence of arterial shear stresses within 2 hr and remain elevated for

12 hr [Ohno et al., 1992]. In addition, Malek and coworkers [1993] have shown that basic fibroblast growth factor (bFGF) mRNA is upregulated in BAECs in the presence of arterial shear stresses. In HUVECs, however, no significant changes in bFGF message were observed [Diamond et al., 1990]. This difference is probably due to differences in cell source (human versus bovine).

Proto-oncogenes code for binding proteins that either enhance or repress transcription and, therefore, are ideal candidates to act as gene regulators [Cooper, 1990]. Komuro and colleagues [1990, 1991] were the first to demonstrate that mechanical loading causes upregulation of c-*fos* expression. Recently, Hsieh and coworkers [1993] investigated the role of arterial shear stress on the mRNA levels of nuclear proto-oncogenes c-*fos*, c-*jun*, and c-*myc*. Gene expression of c-*fos* was transiently upregulated, peaking at 0.5 hr and returning to basal levels within an hour. In contrast, both c-*jun* and c-*myc* mRNA were upregulated to sustainable levels within an hour. The transcribed protein products of c-*fos*, c-*jun*, and c-*myc* may act as nuclear-signaling molecules for mechanically induced gene modulation [Nollert et al., 1992; Ranjan & Diamond, 1993].

114.7 Mechanisms of Shear Stress-Induced Gene Regulation

Although no unified scheme to explain mechanical signal transduction and modulation of gene expression is yet possible, many studies provide insight in an attempt to elucidate which second messengers are involved in gene regulation mediated by hemodynamic forces. There is substantial evidence that the PKC pathway may be involved in the gene regulation. A model of the PKC transduction pathway is shown in Fig. 114.3. Mitsumata and colleagues [1993] have shown that PDGF mRNA modulation by shear stress could be partially attributed to a PKC pathway. Hsieh and coworkers [1992] have shown that shear-induced PDGF gene expression in HUVECs is mainly mediated by PKC activation and requires Ca^{2+} and the involvement of G proteins. In addition, they demonstrated that cAMP and cGMP dependent protein kinases are not involved in PDGF gene expression. Morita and colleagues [1993] have shown that shear-induced HB-EGF mRNA expression is mediated through a PKC pathway and that Ca^{2+} may be involved in the pathway. PKC was also found to be an important mediator in flow-induced c-*fos* expression, with the additional involvement of G proteins, phospholipase C (PLC), and Ca^{2+} [Hsieh et al., 1993]. Moreover, Levin and coworkers [1988] have shown that tPA gene expression is enhanced by PKC activation and Iba and coworkers [1992] have shown that cAMP is not involved in tPA gene expression. As depicted in Fig. 114.3, the PKC pathway may be involved in activating DNA-binding proteins via phosphorylation.

In addition to second messengers, it has been proposed that cytoskeletal reorganization may play a role in regulating gene expression [Ingber, 1993; Resnick et al., 1993]. Morita and colleagues [1993] have shown that shear stress-induced ET-1 gene expression is mediated by the disruption of the actin cytoskeleton and that microtubule integrity is also involved. Gene expression of other bioactive molecules may also be regulated by actin disruption. The actual molecular mechanisms involved in cytoskeletal-mediated gene expression remain unclear. However, the cytoskeleton may activate membrane ion channels and nuclear pores.

In addition to second messengers and the cytoskeletal architecture, it has been postulated by Nollert and others [1992] that transcriptional factors that bind to the DNA may play an active role mediating the signal transduction between the cytosol and nucleus (Fig. 114.4). It is known that nuclear translocation of transcriptional factors and DNA-binding activity of the factors can be mediated by phosphorylation [Bohmann, 1990; Hunter & Karin, 1992]. Many of the transcription factors are protein products of proto-oncogenes. It has previously been stated that shear stress increases expression of c-*fos* and c-*jun*. The gene products of c-*fos* and c-*jun* form protein dimers that bind to transcriptional sites on DNA promoters and act as either transcriptional activators or repressors. Two of the transcriptional sites to which the *fos* and *jun* family dimers bind are the TRE (tumor promoting agent response element) and CRE (cAMP response element). These have consensus sequences of TGACTCA and TGACGTCA, respectively. It is known that the promoter regions of tPA, PAI-1, ET, and TGF-β1 possess sequences of homology to the CRE and TRE [Nollert et al., 1992]. In addition to mediating known transcription factor binding

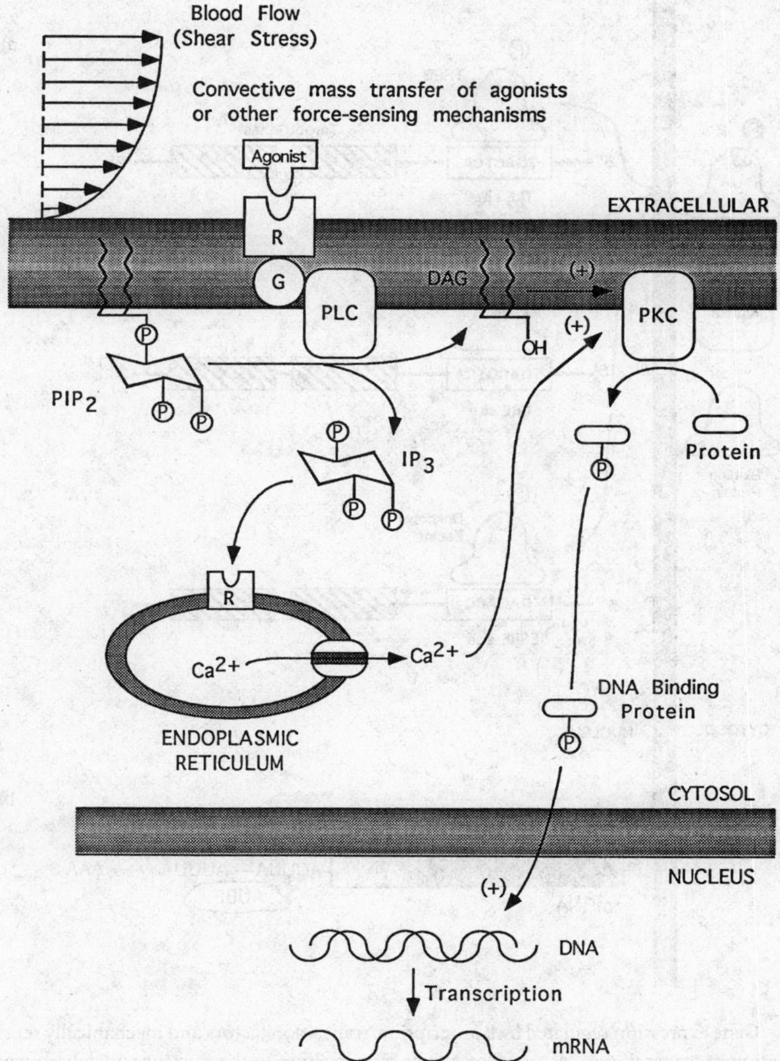

FIGURE 114.3 Model for shear stress induced protein kinase C (PKC) activation leading to modulation of gene expression. Mechanical agonists enhance membrane phosphoinositide turnover via phospholipase C (PLC), producing inositol-1,4,5 trisphosphate (IP_3), and diacylglycerol (DAG). DAG can then activate PKC. IP_3 may also activate PKC via Ca^{2+} release. PKC phosphorylates DNA-binding proteins, thereby making them active. The nuclear binding proteins then alter mRNA transcription. R = receptor, G = G protein, PIP_2 = phosphatidylinositol 4,5-bisphosphate, (+) = activation.

in perhaps novel ways, mechanical perturbations may initiate molecular signaling through specific stress- or strain-sensitive promoters that can activate gene expression. In fact, Resnick and colleagues [1993] have described a cis-acting shear stress response element (SSRE) in the PDGF-B promoter, and this sequence is also found in the promoters of tPA, ICAM-1, TGF-β1, *c-fos*, and *c-jun* but not in VCAM-1 or E-Selectin promoters. A core-binding sequence of GAGACC was identified that binds to transcriptional factors found in BAEC nuclear extracts. The identify of transcriptional factors that bind this sequence remains unknown. In addition, Malek and coworkers [1993] have shown that shear-stress mediated ET-1 mRNA expression is not dependent on either the PKC or cAMP pathways, but rather shear stress regulates the transcription of the ET-1 gene via an upstream cis element in its promoter. It remains to be seen if

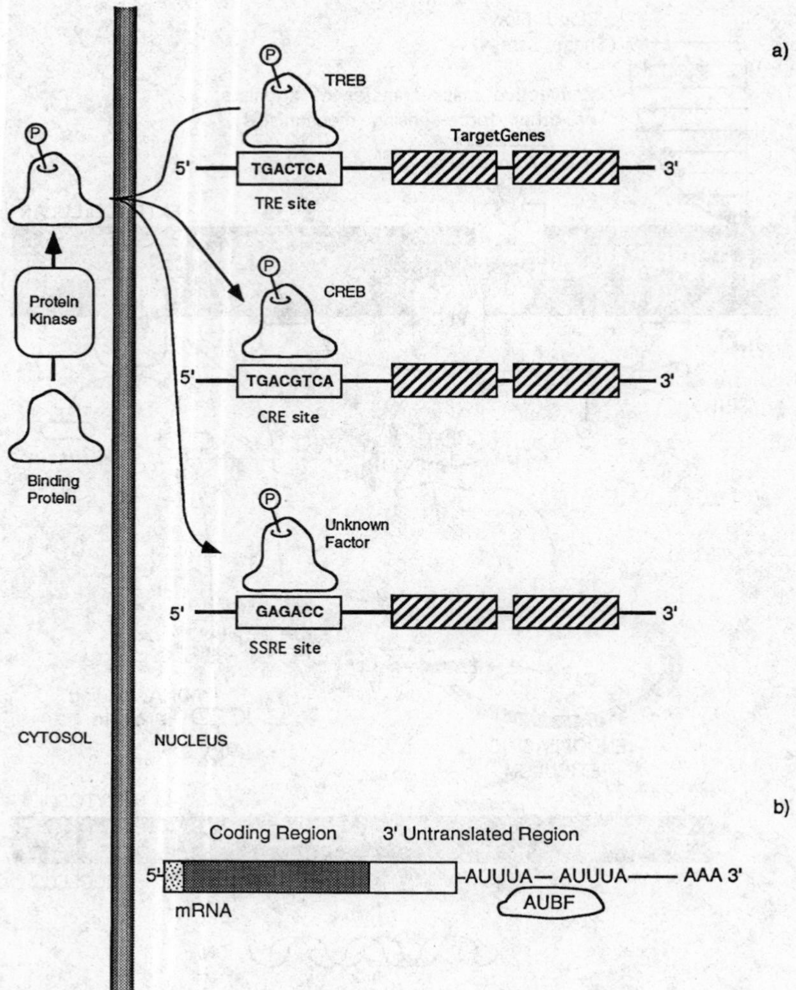

FIGURE 114.4 Gene expression regulated by transcription/translation factors and mechanically sensitive promoters. (*a*) Phosphorylation of transcription factors allows their translocation to the nucleus and subsequent DNA binding. The transcription factors can activate or inhibit transcription. TRE = tumor promoting agent response element, TREB = tumor promoting agent response element binding protein, CRE = cAMP response element, CREB = cAMP response element binding protein, SSRE = shear sensitive response element. (*b*) The AU-binding factor binds the AUUUA motif located in the 3′ untranslated regions of some mature mRNA and could play an important role in mediating stability and/or transport of these mRNAs; AUBF = AU binding factor.

ET-1 is shown to have the same response element as that proposed by Resnick and colleagues. As mentioned previously, Mitsumata and coworkers [1993] found that PDGF mRNA modulation was not solely dependent on a PKC pathway. The PKC independent mechanism may very well involve the cis-acting shear stress response element that Resnick and colleagues described, though other factors are probably important in controlling the complex temporal pattern of gene expression seen in response to mechanical stimuli. The biphasic response of MCP-1 was shown to be regulated at the transcriptional level, possibly involving activation of AP-1, sites and the subsequent binding to the TRE site. The putative SSRE binding site was also identified in the 5′ flanking region of the cloned MCP-1 gene suggesting an interactive mechanism, wherein these cis-acting elements collectively regulate the transcriptional activation of MCP-1 gene under shear stress [Shyy et al., 1994].

Another important determinant in the role played by shear stress on gene regulation could be at the level of mRNA stability. It is not known whether the shear-induced downregulation that has been reported, especially in the case of adhesion molecules, is due to a decrease in the transcriptional rate of the gene or due to a decrease in the stability of the transcript. The β-adrenergic receptors, which are part of a large class of G-protein linked cell surface receptors, are also susceptible to desensitization and downregulation [Hadcock & Malbon, 1988]. A study of the molecular basis for agonist-induced instability revealed the existence of an M_r 35,000 protein that specifically binds mRNA transcripts which contain the sequence AUUUA in their 3′ untranslated region (3′-UTR) [Port et al., 1992]. Agonists that decrease the β-adrenergic RNA transcript levels increased the level of this protein. Further, it has been shown that tandem repeats (three or more) of this sequence are needed for efficient binding of the destabilizer protein [Bohjanen et al., 1992]. Recent reports, however, suggest that AUUUA motif alone does not destabilize the mRNA transcripts and that the presence of additional elements in the AU-rich 3′-UTR may be necessary for the observed mRNA instability [Peppel et al., 1991]. Muller and others [1992] also report that binding of certain AU-binding factors (AUBF) to AUUUA motif accelerates the nuclear export of these mRNAs (Fig. 114.4). Further, their study indicates that this AUBF may protect AU-rich mRNAs against degradation by endoribonuclease V, by binding AUUUA motifs, thereby stabilizing the otherwise unstable mRNA.

Analysis of the 3′-UTR of some of the adhesion molecules studied revealed the presence of the AUUUA motif in all three cases. Human VCAM-1 [Cybulsky et al., 1991] has 6 dispersed AUUUA repeats in its 3′-UTR. Human ICAM-1 [Staunton et al., 1988] has 3 AUUUA repeats within 10 bases of each other, and 3-selectin [Hession et al., 1990] has 8 AUUUA repeats, which include 1 tandem repeat and 3 others that are within 10 bases of each other. The constitutively expressed gene human GAPDH [Tokunaga et al., 1987] does not have any AUUUA repeats in the 3′ region analyzed. Of the other shear responsive genes, human PDGF-B (c-sis) gene has 1 AUUUA repeat [Ratner et al., 1985]; human endothelin gene [Bloch et al., 1989] has 4, three within 10 bases of each other; human tissue plasminogen activator (tPA) [Reddy et al., 1987] has 2; and human thrombomodulin [Jackman et al., 1987] has 9 AUUUA boxes, including 1 tandem repeat. The steady state levels of the tPA and PDGF have been shown to be increased, and those of the adhesion molecules and endothelin have been shown to be decreased by arterial shear stress [Diamond et al., 1990; Hsieh et al., 1992; Sampath et al., 1995], suggesting a possible mechanism of shear-induced destabilization of adhesion molecule mRNA. However, the observed changes can be at the transcriptional level, with the AUUUA motifs playing a role only in the transport of these mRNAs to the cytoplasm. As shown by Muller and colleagues [1992], transport of low-abundance mRNAs like cytokine mRNAs, whose cellular concentrations are one-hundredth or less of that of the high abundance mRNAs, are the ones that are prone to modulation by transport-stimulatory proteins such as AUBF. This can explain the presence of AUUUA boxes in the 3′-UTR of adhesion molecules and other modulators of vascular tone and not in that of the constitutively expressed GADPH, which is an essential enzyme in the metabolic pathway and is expressed in high copy numbers. A more detailed study using transcription run-on assays is needed to address these specific issues.

114.8 Gene Therapy and Tissue Engineering in Vascular Biology

Endothelial cells, located adjacent to the flowing blood stream, are ideally located for use as vehicles for gene therapy, since natural or recombinant proteins of therapeutic value can be expressed directly into the blood to manage cardiovascular diseases or inhibit vascular pathology. One could drive gene expression only in regions of vasculature where desired by using novel cis-acting stress- or strain-sensitive promoter elements or stress-activated transcription factors. For instance, endothelial cells in regions of low shear stress could be modified to express tPA so as to inhibit atherosclerotic plaque formation. An appropriate vector driven by the ET-1 promoter (active at low stress but downregulated at high stress) attached to the tPA gene would be a first logical construct in this application. Likewise, endothelial cells could be modified to increase proliferation rates in order to endothelialize vascular grafts or other vascular prostheses, thereby inhibiting thrombosis. In addition, endothelial cells could be modified to decrease expression of smooth muscle cell mitogens (ET-1, PDGF) to prevent intimal thickening and restenosis. The endothelial cells could even be modified so as to secrete components that would inactivate toxic substances in the blood stream. Work has already begun to develop techniques to express foreign proteins in vivo [Nabel et al., 1989; Wilson et al., 1989; Zwiebel et al., 1989]. Vical and Gen Vec are currently examining ways to express growth factors in coronary arteries to stimulate angiogenesis following balloon angioplasty [Glaser, 1994]. In addition, Tularik Inc is modifying endothelial cells to over express low density lipoprotein (LDL) receptors to remove cholesterol from the blood stream so as to prevent atherosclerotic plaque formation. The application of gene therapy to cardiovascular diseases is in its infancy and will continue to grow as we learn more about the mechanisms governing endothelial cell gene expression. Since endothelial cells lie in a mechanically active environment, predicting local secretion rates of gene products from transfected endothelial cells will require a knowledge of how these mechanical signals modulate gene expression for each target gene and promoter [McIntire, 1994].

In addition to gene therapy applications, there are tissue engineering applications that can be realized once one gains a fundamental understanding of the function-structure relationship intrinsic to vascular biology. This includes understanding how hemodynamic forces modulate endothelial cell function and morphology. Of primary concern is the development of an artificial blood vessel for use in the bypass and replacement of diseased or occluded arteries [Jones, 1982; Nerem, 1991; Weinberg & Bell, 1986]. This is particularly important in the case of small-diameter vascular grafts (such as coronary bypass grafts), which are highly prone to reocclusion. The synthetic blood vessels must provide the structural support required in its mechanically active environment as well as provide endothelial-like functions, such as generating a nonthrombogenic surface.

114.9 Conclusions

This chapter has demonstrated the intricate interweaving of fluid mechanics and convective mass transfer with cell metabolism and the molecular mechanisms of cell adhesion that occur continuously in the vascular system. Our understanding of these mechanisms and how they are modulated by shear stress is really in the initial stages—but this knowledge is vital to our understanding of thrombosis, atherosclerosis, inflammation, and many other aspects of vascular physiology and pathophysiology. Knowledge of the fundamental cellular and molecular mechanisms involved in adhesion and mechanical force modulation of metabolism under conditions that mimic those seen in vivo is essential for real progress to be made in vascular biology and more generally in tissue engineering.

Defining Terms

CRE: *c*AMP *r*esponse *e*lement; its consensus sequence TGACGTCA is found on genes responsive to cAMP agonists such as forskolin.

Gene regulation: Transcriptional and posttranscriptional control of expression of genes in eukaryotes where regulatory proteins bind specific DNA sequences to turn a gene either on (positive control) or off (negative control).

Gene therapy: The modification or replacement of a defective or malfunctioning gene with one that functions adequately and properly, for instance, addition of gene regulatory elements such as specific stress- or strain-sensitive response elements to specifically drive gene expression only in regions of interest in the vasculature so as to control proliferation, fibrinolytic capacity, etc.

Shear: Shear refers to the relative parallel motion between adjacent fluid (blood) planes during flow. The difference in the velocity between adjacent layers of blood at various distances from the vessel wall determines the local shear rate, expressed in cm/s/cm, or s^{-1}.

Shear stress: Fluid shear stress, expressed in dynes/cm^2, is a measure of the force required to produce a certain rate of flow of a viscous liquid and is proportional to the product of shear rate and blood viscosity. Physiologic levels of venous and arterial shear stresses range between 1–5 dynes/cm^2 and 6–40 dynes/cm^2, respectively.

SSRE: *S*hear *s*tress *r*esponse *e*lement; its consensus GAGACC has been recently identified in genes responsive to shear stress.

Transcription factor: A protein that binds to a cis-regulatory element in the promoter region of a DNA and thereby directly or indirectly affects the initiation of its transcription to an RNA.

TRE: *T*umor promoting agent *r*esponse *e*lement: its consensus sequence TGACTCA is commonly located in genes sensitive to phorbol ester stimulation.

References

Abbassi O, Kishimoto TK, McIntire LV, et al. 1993. E-selectin supports neutrophil rolling in vitro under conditions of flow. J Clin Invest 92:2719.

Abbassi OA, Lane CL, Krater S, et al. 1991. Canine neutrophil margination mediated by lectin adhesion molecule-1 in vitro. J Immunol 147:2107.

Alevriadou BR, Moake JL, Turner NA, et al. 1993. Real-time analysis of shear-dependent thrombus formation and its blockade by inhibitors of von Willebrand factor binding to platelets. Blood 81:1263.

Anderson DC, Abbassi OA, McIntire LV, et al. 1991. Diminished LECAM-1 on neonatal neutrophils underlies their impaired CD18-independent adhesion to endothelial cells in vitro. J Immunol 146:3372.

Ando J, Komatsuda T, Kamiya A. 1988. Cytoplasmic calcium response to fluid shear stress in cultured vascular endothelial cells. In Vitro Cell Devel 24:871.

Asakura T, Karino T. 1990. Flow patterns and spatial distribution of atherosclerotic lesions in human coronary arteries. Circ Res 66:1045.

Barabino GA, McIntire LV, Eskin SG, et al. 1987. Endothelial cell interactions with sickle cell, sickle trait, mechanically injured, and normal erythrocytes under controlled flow. Blood 70:152.

Barbee KA, Davies PF, Lal R. 1994. Shear stress-induced reorganizations of the surface topography of living endothelial cells imaged by atomic force microscopy. Circ Res 74:163.

Bastida E, Almirall L, Bertomeu MC, et al. 1989. Influence of shear stress on tumor-cell adhesion to endothelial-cell extracellular matrix and its modulation by fibronectin. Int J Cancer 43:1174.

Bhagyalakshmi A, Berthiaume F, Reich KM, et al. 1992. Fluid shear stress stimulates membrane phospholipid metabolism in cultured human endothelial cells. J Vasc Res 29:443.

Bhagyalakshmi A, Frangos JA. 1989*a*. Mechanism of shear-induced prostacyclin production in endothelial cells. Biochem Biophys Res Comm 158:31.

Bhagyalakshmi A, Frangos JA. 1989*b*. Membrane phospholipid metabolism in sheared endothelial cells. Proc 2nd Int Symp Biofluid Mechanics and Biorheology, Munich, Germany 240.

Bloch KD, Friedrich SP, Lee ME, et al. 1989. Structural organization and chromosomal assignment of the gene encoding endothelin. J Biol Chem 264:10851.

Bohjanen PR, Petryniak B, June CH, et al. 1992. AU RNA-binding factors differ in their binding specificities and affinities. J Biol Chem 267:6302.

Bohmann D. 1990. Transcription factor phosphorylation: A link between signal transduction and the regulation of gene expression. Cancer Cells 2:337.

Botteri FM, Ballmer-Hofer K, Rajput B, et al. 1990. Disruption of cytoskeletal structures results in the induction of the urokinase-type plasminogen activator gene expression. J Biol Chem 265:13327.

Breathnach R, Matrisian LM, Gesnel MC, et al. 1987. Sequences coding part of oncogene-induced transin are highly conserved in a related rat gene. Nucleic Acid Res 15:1139.

Coan DE, Wechezak AR, Viggers RF, et al. 1993. Effect of shear stress upon localization of the Golgi apparatus and microtubule organizing center in isolated cultured endothelial cells. J Cell Sci 104:1145.

Cooper GM. 1990. Oncogenes, Boston, Jones and Bartlett.

Cybulsky MI, Fries JW, Williams AJ, et al. 1991. Gene structure, chromosomal location, and basis for alternative mRNA splicing of the human VCAM-1 gene. Proc Natl Acad Sci USA 88:7859.

Davies MJ, Woolf N, Rowles PM, et al. 1988. Morphology of the endothelium over atherosclerotic plaques in human coronary arteries. Br Heart J 60:459.

Davies PF, Tripathi SC. 1993. Mechanical stress mechanisms and the cell: An endothelial paradigm. Circ Res 72:239.

Dewey CF Jr, Bussolari SR, Gimbrone MA Jr, et al. 1981. The dynamic response of vascular endothelial cells to fluid shear stress. J Biomech Eng 103:177.

Diamond SL, Eskin SG, McIntire LV. 1989. Fluid flow stimulates tissue plasminogen activator secretion by cultured human endothelial cells. Science 243:1483.

Diamond SL, Sharefkin JB, Dieffenbach C, et al. 1990. Tissue plasminogen activator messenger RNA levels increase in cultured human endothelial cells exposed to laminar shear stress. J Cell Physiol 143:364.

Dull RO, Davies PF. 1991. Flow modulation of agonist (ATP)-response (Ca^{++}) coupling in vascular endothelial cells. Am J Physiol 261:H149.

Eskin SG, Ives CL, McIntire LV, et al. 1984. Response of cultured endothelial cells to steady flow. Microvasc Res 28:87.

Ferrua B, Manie S, Doglio A, et al. 1990. Stimulation of human interleukin-1 production and specific mRNA expression by microtubule-disrupting drugs. Cell Immunol 131:391.

Folie BJ, McIntire LV. 1989. Mathematical analysis of mural thrombogenesis. Biophys J 56:1121.

Frangos JA, Eskin SG, McIntire LV, et al. 1985. Flow effects on prostacyclin production by cultured human endothelial cells. Science 227:1477.

Franke RP, Grafe M, Schnittler H. 1984. Induction of human vascular endothelial stress fibres by shear stress. Nature 307:648.

Giavazzi R, Foppolo M, Dossi R, et al. 1993. Rolling and adhesion of human tumor cells on vascular endothelium under physiological flow conditions. J Clin Invest 92:3038.

Girard PR, Nerem RM. 1991. Fluid shear stress alters endothelial cell structure through the regulation of focal contact-associated proteins. Adv Bioeng 20:425.

Girard PR, Nerem RM. 1993. Endothelial cell signaling and cytoskeletal changes in response to shear stress. Front Med Biol Eng 5:31.

Glagov S, Zarins CK, Giddens DP, et al. 1988. Hemodynamics and atherosclerosis. Arch Pathol Lab Med 112:1018.

Glaser V. 1994. Targeted injectable vectors remain the ultimate goal in gene therapy. Genetic Eng News 14:8.

Grabowski EF, Jaffe EA, Weksler BB. 1985. Prostacyclin production by cultured endothelial cell monolayers exposed to step increases in shear stress. J Lab Clin Med 105:36.

Grande JP, Glagov S, Bates SR, et al. 1989. Effect of normolipemic and hyperlipemic serum on biosynthesis response to cyclic stretching of aortic smooth muscle cells. Arteriosclerosis 9:446.

Gupte A, Frangos JA. 1990. Effects of flow on the synthesis and release of fibronectin by endothelial cells. In Vitro Cell Devel Biol 26:57.

Hadcock JR, Malbon CC. 1988. Down-regulation of beta-adrenergic receptors: Agonist-induced reduction in receptor mRNA levels. Biochemistry 95:5021.

Hakkert BC, Kuijipers TW, Leeuwenberg JFM, et al. 1991. Neutrophil and monocyte adherence to and migration across monolayers of cytokine-activated endothelial cells: the contribution of CD18, ELAM-1, and VLA-4. Blood 78:2721.

Hession C, Osborn L, Goff D, et al. 1990. Endothelial leukocyte adhesion molecule-1: Direct expression cloning and functional interactions. Proc Natl Acad Sci USA 87:1673.

Hsieh HJ, Li NQ, Frangos JA. 1991. Shear stress increases endothelial platelet-derived growth factor mRNA levels. Am J Physiol H642.

Hsieh H, Li NQ, Frangos JA. 1992. Shear-induced platelet-derived growth factor gene expression in human endothelial cells is mediated by protein kinase C. J Cell Physiol 150:552.

Hsieh H, Li NQ, Frangos JA. 1993. Pulsatile and steady flow induces *c-fos* expression in human endothelial cells. J Cell Physiol 154:143.

Hubbel JA, McIntire LV. 1986. Technique for visualization and analysis of mural thrombogenesis. Rev Sci Instrum 57:892.

Hudlicka O. 1984. Growth of vessels—historical review. In F Hammerson, O Hudlicka (eds), Progress in Applied Microcirculation Angiogenesis, vol 4, 1–8, Basel, Karger.

Hunter T, Karin M. 1992. The regulation of transcription by phosphorylation. Cell 70:375.

Iba T, Mills I, Sumpio BE. 1992. Intracellular cyclic AMP levels in endothelial cells subjected to cyclic strain in vitro. J Surgical Res 52:625.

Ingber D. 1993. Integrins as mechanochemical transducers. Curr Opin Cell Biol 3:841.

Ives CL, Eskin SG, McIntire LV. 1986. Mechanical effects on endothelial cell morphology: In vitro assessment. In Vitro Cell Dev Biol 22:500.

Jackman RW, Beeler DL, Fritze L, et al. 1987. Human thrombomodulin gene is intron depleted: Nucleic acid sequences of the cDNA and gene predict protein structure and suggest sites of regulatory control. Proc Natl Acad Sci USA 84:6425.

Jones DA, Abbassi OA, McIntire LV, et al. 1993. P-selectin mediates neutrophil rolling on histamine-stimulated endothelial cells. Biophys J 65:1560.

Jones PA. 1982. Construction of an artificial blood vessel wall from cultured endothelial and smooth muscle cells. J Cell Biol 74:1882.

Kim DW, Gotlieb AI, Langille BL. 1989. In vivo modulation of endothelial F-actin microfilaments by experimental alterations in shear stress. Arteriosclerosis 9:439.

Kishimoto TK, Warnock RA, Jutila MA, et al. 1991. Antibodies against human neutrophil LECAM-1 and endothelial cell ELAM-1 inhibit a common CD18-independent adhesion pathway in vitro. Blood 78:805.

Kojima N, Shiota M, Sadahira Y, Handa K, Hakomori S. 1992. Cell adhesion in a dynamic flow system as compared to static system. J Biol Chem 267:17264–17270.

Komuro I, Kaida T, Shibazaki Y, et al. 1990. Stretching cardiac myocytes stimulates protooncogene expression. J Biol Chem 265:3595.

Komuro I, Katoh Y, Kaida T, et al. 1991. Mechanical loading stimulates cell hypertrophy and specific gene expression in cultured rat cardiac myocytes. J Biol Chem 266:1265.

Kuchan MJ, Frangos JA. 1993. Shear stress regulates endothelin-1 release via protein kinase C and cGMP in cultured endothelial cells. Am J Physiol 264:H150.

Langille BL, Adamson SL. 1981. Relationship between blood flow direction and endothelial cell orientation at arterial branch sites in rabbit and mice. Circ Res 48:481.

Langille BL, Graham JJ, Kim D, et al. 1991. Dynamics of shear-induced redistribution of F-actin in endothelial cells in vivo. Arterioscler Thromb 11:1814.

Langille BL, O'Donnell F. 1986. Reductions in arterial diameter produced by chronic diseases in blood flow are endothelium-dependent. Science 231:405.

Lawrence MB, McIntire LV, Eskin SG. 1987. Effect of flow on polymorphonuclear leukocyte/endothelial cell adhesion. Blood 70:1284.

Lawrence MB, Smith CW, Eskin SG, et al. 1990. Effect of venous shear stress on CD18-mediated neutrophil adhesion to cultured endothelium. Blood 75:227.

Lawrence MB, Springer TA. 1991. Leukocytes roll on a selectin at physiologic flow rates: distinction from and prerequisite for adhesion through integrins. Cell 65:859.

Levesque MJ, Sprague EA, Schwartz CJ, et al. 1989. The influence of shear stress on cultured vascular endothelial cells: The stress response of an anchorage-dependent mammalian cell. Biotech Prog 5:1.

Levin EG, Santell L. 1988. Stimulation and desensitization of tissue plasminogen activator release from human endothelial cells. J Biol Chem 263:9360.

Ley K, Gaehtgens P, Fennie C, et al. 1991. Lectin-like adhesion molecule 1 mediates leukocyte rolling in mesenteric venules in vivo. Blood 77:2553.

Malek AM, Gibbons GH, Dzau VJ, et al. 1993*a*. Fluid shear stress differentially modulates expression of genes encoding basic fibroblast growth factor and platelet-derived growth factor B chain in vascular endothelium. J Clin Invest 92:2013.

Malek AM, Greene AL, Izumo S. 1993*b*. Regulation of endothelin 1 gene by fluid shear stress is transcriptionally mediated and independent of protein kinase C and cAMP. Proc Natl Acad Sci 90:5999.

Malek A, Izumo S. 1992. Physiological fluid shear stress causes downregulation of endothelin-1 mRNA in bovine aortic endothelium. Am J Physiol 263:C389.

McIntire LV. 1994. Bioengineering and vascular biology. Ann Biomed Eng 22:2.

Menter DG, Patton JT, Updike TV, et al. 1992. Transglutaminase stabilizes melanoma adhesion under laminar flow. Cell Biophys 18:123.

Milner P, Bodin P, Loesch A, et al. 1992. Increased shear stress leads to differential release of endothelin and ATP from isolated endothelial cells from 4- and 12-month-old male rabbit aorta. J Vasc Res 29:420.

Mitsumata M, Fishel RS, Nerem RM, et al. 1993. Fluid shear stress stimulates platelet-derived growth factor expression in endothelial cells. Am J Physiol 263:H3.

Mo M, Eskin SG, Schilling WP. 1991. Flow-induced changes in Ca^{2+} signaling of vascular endothelial cells: Effect of shear stress and ATP. Am J Physiol 260:H1698.

Morita T, Yoshizumi M, Kurihara H, et al. 1993. Shear stress increases heparin-binding epidermal growth factor-like growth factor mRNA levels in human vascular endothelial cells. Biochem Biophys Res Comm 197:256.

Muller WEG, Slor H, Pfeifer K, et al. 1992. Association of AUUUA-binding Protein with A + U-rich mRNA during nucleo-cytoplasmic transport. J Mol Biol 226:721.

Nabel EG, Plautz G, Boyce FM, et al. 1989. Recombinant gene expression in vivo within endothelial cells of the arterial wall. Science 244:1342.

Nagel T, Resnick N, Atkinson W, et al. 1994. Shear stress selectivity upregulates intercellular adhesion molecule-1 expression in cultured vascular endothelial cells. J Clin Invest 94:885.

Nerem RM. 1991. Cellular engineering. Ann Biomed Eng 19:529.

Nerem RM. 1993. Hemodynamics and the vascular endothelium. J Biomech Eng 115:510.

Nerem RM, Levesque MJ, Cornhill JF. 1981. Vascular endothelial morphology as an indicator of the pattern of blood flow. J Biomech Eng 103:172.

Nollert MU, Diamond SL, McIntire LV. 1991. Hydrodynamic shear stress and mass transport modulation of endothelial cell metabolism. Biotech Bioeng 38:588.

Nollert MU, Eskin SG, McIntire LV. 1990. Shear stress increases inositol trisphosphate levels in human endothelial cells. Biochem Biophys Res Comm 170:281.

Nollert MU, Hall ER, Eskin SG, et al. 1989. The effect of shear stress on the uptake and metabolism of arachidonic acid by human endothelial cells. Biochim Biophys Acta 1005:72.

Nollert MU, McIntire LV. 1992. Convective mass transfer effects on the intracellular calcium response of endothelial cells. J Biomech Eng 114:321.

Nollert MU, Panaro NJ, McIntire LV. 1992. Regulation of genetic expression in shear stress stimulated endothelial cells. Ann NY Acad Sci 665:94.

Ohno M, Gibbons GH, Dzau VJ, et al. 1993. Shear stress elevates endothelial cGMP. Role of potassium channel and G protein coupling. Circulation 88:193.

Ohno M, Lopez F, Gibbons GH, et al. 1992. Shear stress induced TGFβ1 gene expression and generation of active TGFβ1 is mediated via a K^+ channel. Circulation 86:I-87.

Ono O, Ando J, Kamiya A, et al. 1991. Flow effects on cultured vascular endothelial and smooth muscle cell functions. Cell Struct Funct 16:365.

Patton JT, Menter DG, Benson DM, et al. 1993. Computerized analysis of tumor cells flowing in a parallel plate chamber to determine their adhesion stabilization lag time. Cell Motility and the Cytoskeleton 26:88.

Peppel K, Vinci JM, Baglioni C. 1991. The AU-rich sequences in the 3′ untranslated region mediate the increased turnover of interferon mRNA induced by glucocorticoids. J Exp Med 173:349.

Perry MA, Granger DN. 1991. Role of CD11/CD18 in shear rate-dependent leukocyte-endothelial cell interactions in cat mesenteric venules. J Clin Invest 87:1798.

Pili R, Corda S, Passaniti A, et al. 1993. Endothelial cell Ca^{2+} increases upon tumor cell contact and modulates cell-cell adhesion. J Clin Invest 92:3017.

Port JD, Huang LY, Malbon CC. 1992. β-adrenergic agonists that down-regulate receptor mRNA upregulate a M_r 35,000 protein(s) that selectively binds to β-adrenergic receptor mRNAs. J Biol Chem 267:24103.

Prasad AR, Logan SA, Nerem RM, et al. 1993. Flow-related responses of intracellular inositol phosphate levels in cultured aortic endothelial cells. Circ Res 72:827.

Ranjan V, Diamond SL. 1993. Fluid shear stress induces synthesis and nuclear localization of c-fos in cultured human endothelial cells. Biochem Biophys Res Comm 196:79.

Ratner L, Josephs SF, Jarrett R, et al. 1985. Nucleotide sequence of transforming human c-sis-cDNA clones with homology to platelet derived growth factor. Nucl Acids Res 13:5007.

Reddy VB, Garramone AJ, Sasak H, et al. 1987. Expression of human uterine tissue-type plasminogen activator using BPV vectors. DNA 6:461.

Resnick N, Collins T, Atkinson W, et al. 1993. Platelet-derived growth factor B chain promoter contains a cis-acting fluid shear-stress-responsive-element. Proc Natl Acad Sci 90:4591.

Ross R. 1993. The pathogenesis of atherosclerosis: a perspective for the 1990s. Nature 362:801.

Sampath R, Kukielka GL, Smith CW, et al. 1995. Shear stress mediated changes in the expression of leukocyte adhesion receptors on human umbilical vein endothelial cells in vitro. Ann Biomed Eng.

Schnittler HJ, Franke RP, Akbay U, et al. 1993. Improved in vitro rheological system for studying the effect of fluid shear stress on cultured cells. Am J Physiol 265:C289.

Schwartz CJ, Valente AJ, Spargue EA. 1993. A modern view of atherogenesis. Am J Cardiol 71:9B.

Sharefkin JB, Diamond SL, Eskin SG, et al. 1991. Fluid flow decreases preproendothelin mRNA levels and suppresses endothelin-1 peptide release in cultured human endothelial cells. J Vasc Surg 14:1.

Shyy YJ, Hsieh HJ, Usami S, et al. 1994. Fluid shear stress induces a biphasic response of human monocyte chemotactic protein 1 gene expression in vascular endothelium. Proc Natl Acad Sci (USA) 91:4678.

Smith CW, Kishimoto TK, Abbassi OA, et al. 1991. Chemotactic factors regulate LECAM-1 dependent neutrophil adhesion to cytokine-stimulated endothelial cells in vitro. J Clin Invest 87:609.

Staunton DE, Marlin SD, Stratowa C, et al. 1988. Primary structure of ICAM-1 demonstrates interaction between members of the immunoglobulin and integrin supergene families. Cell 52:925.

Tokunaga K, Nakamura Y, Sakata K, et al. 1987. Enhanced expression of a glycearldehyde-3-phosphate dehydrogenase. Cancer Res 47:5616.

Tran-son-Tay R. 1993. Techniques for studying the effects of physical forces on mammalian cells and measuring cell mechanical properties. In JA Frangos (ed), Physical Forces and the Mammalian Cell, p 1, New York, Academic Press.

Turitto VT, Weiss HJ, Baumgartner HR, et al. 1987. Cells and aggregates at surfaces. In EF Leonord, VT Turitto, L Vroman (eds), Blood in Contact with Natural and Artificial Surfaces, vol 516, pp 453–467, New York, Annals of the New York Academy of Sciences.

Von Andrian UH, Chambers JD, McEvoy LM, et al. 1991. Two-step model of leukocyte-endothelial cell interaction in inflammation. Proc Natl Acad Sci 88:7538.

Warren BA. 1973. Evidence of the blood-borne tumor embolus adherent to vessel wall. J Med 6:150.

Watson PA. 1991. Function follows form: Generation of intracellular signals by cell deformation. FASEB J 5:2013.

Watson SR, Fennie C, Lasky LA. 1991. Neutrophil influx into an inflammatory site inhibited by a soluble homing receptor-IgG chimaera. Nature 349:164.

Wechezak AR, Viggers RF, Sauvage LR. 1985. Fibronectin and F-actin redistribution in cultured endothelial cells exposed to shear stress. Lab Invest 53:639.

Weinberg CB, Bell E. 1986. A blood vessel model constructed from collagen and cultured vascular cells. Science 231:397.

Weiss HJ, Turitto VT, Baumgartner HR. 1978. Effect of shear rate on platelet interaction with subendothelium in citrated and native blood. J Lab Clin Med 92:750.

Weiss HJ, Turitto VT, Baumgartner HR. 1986. Platelet adhesion and thrombus formation on subendothelium in platelets deficient in GP IIb-IIIa, Ib, and storage granules. Blood 67:905.

Weiss L. 1992. Biomechanical interactions of cancer cells with the microvasculature during hematogenous metastasis. Cancer Metastasis Rev 11:227.

Werb Z, Hembry RM, Murphy G, et al. 1986. Commitment to expression of metalloendopeptidases, collagenase, and stromelysin: Relationship of inducing events to changes in cytoskeletal architecture. J Cell Biol 102:697.

Wick TM, Moake JL, Udden MM, et al. 1987. ULvWF multimers increase adhesion of sickle erythrocytes to human endothelial cells under controlled flow. J Clin Invest 80:905.

Wick TM, Moake JL, Udden MM, et al. 1993. ULvWF multimers preferentially promote young sickle and non-sickle erythrocyte adhesion to endothelial cells. Am J Hematol 42:284.

Wilson JM, Birinyi LK, Salomon RN, et al. 1989. Implantation of vascular grafts lined with genetically modified endothelial cells. Science 244:1344.

Yanagisawa M, Kurihara H, Kimura S, et al. 1988. A novel potent vasoconstrictor peptide produced by vascular endothelial cells. Nature 332:411.

Yoshizumi M, Kurihara H, Sugiyama T, et al. 1989. Hemodynamic shear stress stimulates endothelin production by cultured endothelial cells. Biochem Biophys Res Commun 161:859.

Zarins C, Zatina MA, Giddens DP. 1987. Shear stress regulation of artery lumen diameter in experimental atherogenesis. J Vasc Surg 5:413.

Zarins CK, Giddens DP, Bharadvaj BK, et al. 1983. Carotid bifurcation atherosclerosis. Circ Res 53:502.

Ziegelstein RC, Cheng L, Blank PS, et al. 1993. Modulation of calcium homeostasis in cultured rat aortic endothelial cells by intracellular acidification. Am J Physiol 265:H1424.

Ziegelstein RC, Cheng L, Capogrossi MC. 1992. Flow-dependent cytosolic acidification of vascular endothelial cells. Science 258:656.

Zwiebel JA, Freeman SM, Kantoff PW, et al. 1989. High-level recombinant gene expression in rabbit endothelial cells transduced by retroviral vectors. Science 243:220.

115
The Roles of Mass Transfer in Tissue Function

Edwin N. Lightfoot
University of Wisconsin

Karen A. Duca
University of Wisconsin

Mass transfer lies at the heart of physiology and provides major constraints on the metabolic rates and anatomy [Pries et al., 1996; Bunk, 1998] of living organisms, from the organization of organ networks to intracellular structures. Limitations on mass transport rates are major constraints for nutrient supply, waste elimination and information transmission at all of these levels. The primary functional units of metabolically active tissue, e.g., the Krogh tissue cylinders of muscle and brain, liver lobules and kidney nephrons, have evolved to just eliminate significant mass transfer limitations in the physiological state [Lightfoot, 1974]. Turnover rates of highly regulated enzymes are just on the slow side of diffusional limitations [Weisz, 1973]. Signal transport rates are frequently mass transport limited [Lauffenburger and Linderman, 1993], and very ingenious mechanisms have evolved to speed these processes [Berg and von Hippel, 1985; Bray, 1998; Francis and Palsson, 1997; Valee and Sheets, 1996]. In contrast elaborate membrane barriers organize and control intracellular reactions in even the simplest organisms.

Understanding tissue mass transport is important both for the engineer and scientist. Examples of engineering interest include the design of extracorporeal devices and biosensors [Fishman et al., 1998]. For the scientist it is important to understand the sometimes complex interactions of transport and reaction to interpret transport based experiments [Bassingthwaighte, Goresky and Linehan, 1998]. Here we shall concentrate on qualitative behavior and orders of magnitude, with particular emphasis on time constants. These are the normal starting points of any serious study. A general background in biology [Campbell, 1996] and physiology [Johnson, 1998] will be helpful.

0-8493-0462-8/00/$0.00+$.50

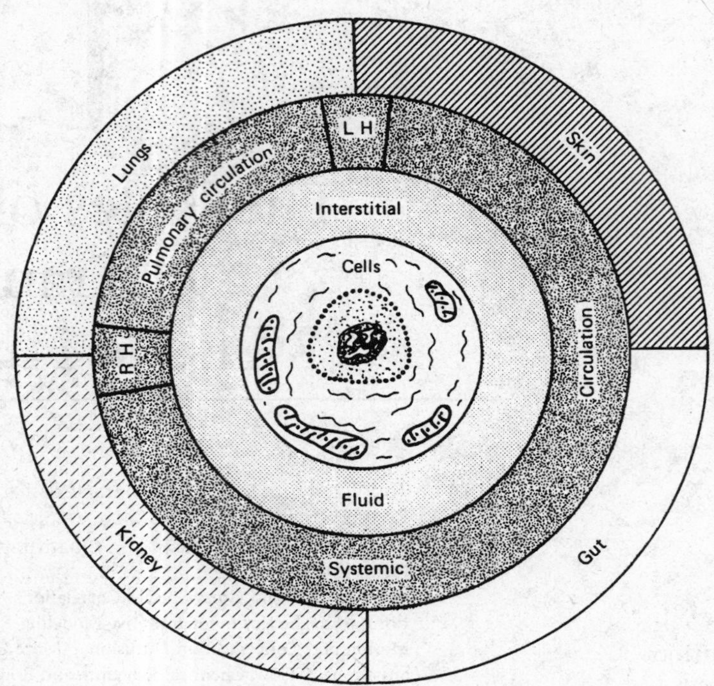

FIGURE 115.1 Diffusional topology of the mammalian body. Convective transport (bulk flow) dominates in the major blood vessels and airways but becomes comparable to diffusion and reaction in the functional units surrounding capillaries and sinusoids. At the cellular and subcellular levels concentration diffusion complicated by electrical effects and a wide variety of carrier transport processes, interacts in complex ways with enzymatic and genetic reactions. Inspired by Dr. Peter Abbrecht, University of Michigan. Reprinted with permission from Lightfoot (1974). Copyright 1974 by John Wiley and Sons, Inc.

115.1 Topology and Transport Characteristics of Living Organisms [Schmidt-Nielsen 1983, 1984, Calder 1984, Berg 1993 Bassingthwaighte et al. 1994]

If one includes viruses, the size scales of living systems range from nanometers to meters: spanning a linear ratio of 10^9 and a mass ratio of 10^{27}, which is greater than Avogadro's number! As indicated in Fig. 115.1, the higher animals are organized into spatial hierarchies of discrete structures, which span this whole size range (Table 115.1). At the largest scales animals may be considered as organ networks connected by major blood vessels, with each organ carrying out its own set of specialized tasks.

Organs are in turn composed essentially of large numbers of microcirculatory *functional units* organized in parallel and perfused by capillaries or other microscopic ducts which supply oxygen and other nutrients, carry away waste products, and interchange via a variety of chemical messengers [Lauffenburger and Linderman, 1993]. The Krogh tissue cylinder of Fig. 115.2 is representative and corresponds approximately to the functional units of the brain (see, however, Federspiel and Popel, 1986; Popel, 1989; Hellums et al., 1996; Vicini et al., 1998; Bassingthwaighte et al., 1998).

TABLE 115.1 The Spatial Hierarchy in Mammals, Characteristic Lengths

Entity	Length Scale, m
Whole body	10^{-1}–10^{0}
Organs	10^{-2}–10^{-1}
Microcirculatory units	10^{-4}
Cells	10^{-5} (eukaryiots) 10^{-6} (prokaryiots)
Intracellular organelles	10^{-6}
Molecular complexes	10^{-8}

Nutrients and metabolic end products are transported at the size scale of the tissue cylinder axially by convection.

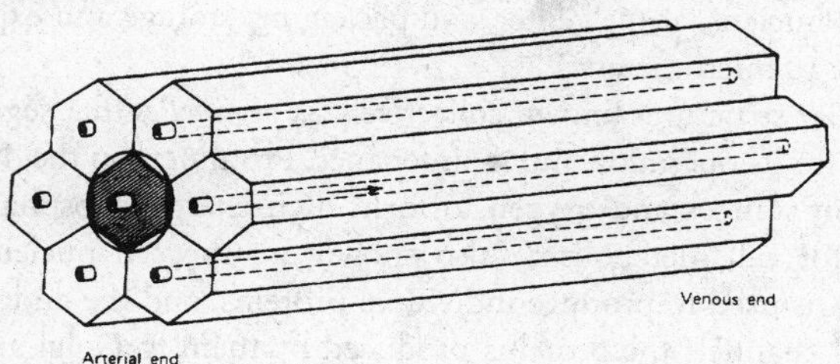

FIGURE 115.2 Krogh tissue cylinder. A circular cylinder is used as an approximation to close-packed hexagonal elements.

Radial transport is primarily by concentration diffusion or *Brownian motion* [Lightfoot and Lightfoot, 1997] and, to a lesser extent by *Starling flow*, slow seepage across the *proximal* region of the capillary driven by the relatively high pressure of entering capillary blood. Flow takes place through clefts between the *endothelial cells* forming the capillary wall and between the *parenchymal* cells forming the tissue cylinder. The clefts are narrow and *permselective*, rejecting large solutes. Starling flow was once thought to reverse direction in the *distal* (downstream) regions of the tissue cylinder as a result of lower hydrodynamic pressure and increased colloid osmotic pressure caused by the permselectivity of the clefts. However, it is now accepted that such reverse flow does not occur to a significant extent and that the seepage ends up primarily in the lymph ducts and on to the venous blood.

The microcirculatory units are in turn comprised of cells which are themselves complex structures. Sketched in Fig. 115.3 is a pancreatic beta cell, which produces, stores, and secretes insulin. Its organization is similar to that of a chemical plant, with raw materials and energy input at the bottom where mitochondria produce the chemical energy for cell metabolism, a synthesis and transport area, plus the

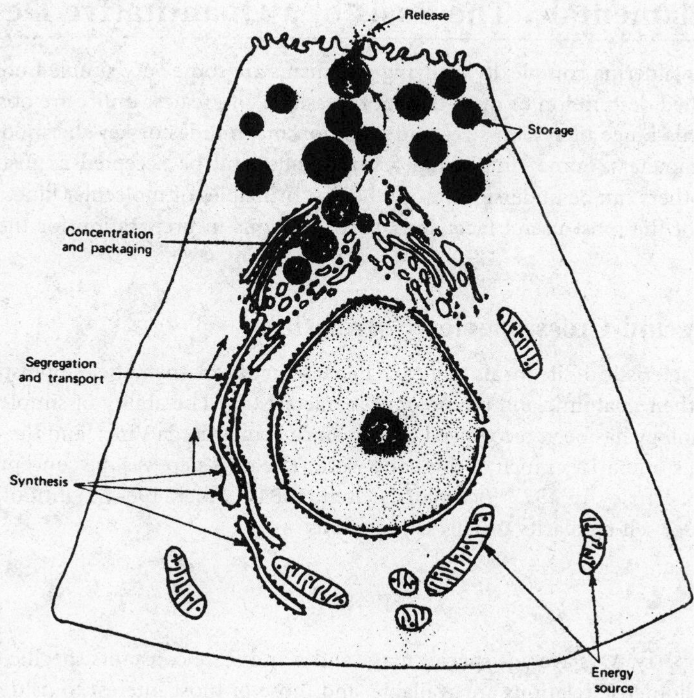

FIGURE 115.3 Structure and organization of a representative eukaryotic cell. Schematic cross-section of a pancreatic beta cell. Reprinted with permission from Lightfoot (1974). Copyright 1974 by John Wiley and Sons.

control region of the cell nucleus, at the center, and packaging, storage and export at the top. All of this is accomplished in about 10 micrometers.

The cell contains smaller structures known collectively as *organelles* that segregate and coordinate the many processes required for its operation and maintenance. Prominent in the diagram are mitochondria, which use various carbon sources and oxygen to form high-energy phosphate bonds as in ATP—the energy sources needed for cell metabolism. Also shown are the cell nucleus where DNA is stored, ribosomes in which RNA is used to produce individual proteins, and the endoplasmic reticulum which holds the ribosomes and channels the proteins produced by them to Golgi apparatus (not shown) for packaging. Also not shown are *microtubules* and other filaments comprising the *cytoskeleton* [Alberts et al., 1945; Campenot et al., 1996; Maniotis et al., 1997]. The latter provide for chromosome segregation, help to maintain cellular shape, and in some cases provide mobility [Vallee and Sheetz]. They also act as transport networks along with motor proteins, discussed below. Organization and structure of the cells is described in standard works [Alberts et al., 1945].

At the smallest level are enzyme clusters and substructures of cell membranes used for selective transport across the cell boundary [Lauffenberger and Linderman, 1993]. Underlying these structures are biochemical reaction networks that are largely shared by all species, and all are composed of the same basic elements, primarily proteins, carbohydrates, and lipids. As a result there are a great many interspecies similarities [Schmidt-Nielsen, 1983].

This elaborate organization just described is constrained to a very large extent by mass transfer considerations and in particular by the effect of characteristic time and distance scales of the effectiveness of different mass transport mechanisms. At the larger size scales, only flow or convection is fast enough to transport oxygen and major metabolites, and convective transport is a major function of the larger blood vessels. Diffusive transport begins to take precedence at the level of microcirculatory units. At the cellular level and below, diffusion may even be too fast and undirected, and selectively permeable membranes, have evolved to maintain spatial segregation against the randomizing forces of diffusion.

115.2 Fundamentals: The Basis of a Quantitative Description

Underneath the bewildering complexity of living organisms are some very simple underlying principles which make a unified description of mass transport feasible. Of greatest utility are observed similarities across the enormous range of system sizes and common magnitudes of key thermodynamic, reaction, and transport parameters. Some simplifying features must still be accepted as justified by repeated observation, and others can be understood from the first principles of molecular kinetic theory. Here we summarize some of the most useful facts and approximations in preparation for the examples of the next section.

Self Similarity and Cross-Species Correlations

It is a striking characteristic of life forms that each increase in our mathematical sophistication has shown new regularity in their anatomic and physiologic characteristics. The ability of simple geometric forms to describe morphology has been recognized at least since Leonardo da Vinci, and the definitive work of D'Arcy Thompson summarizes much of this early work. The next step was the concept of *allometry*, first introduced by J. S. Huxley in 1927 [Schmidt-Nielsen, 1984; Calder, 1984; Lightfoot, 1974]. This is a rudimentary form of self similarity usually expressed as

$$P = aM^b \tag{115.1}$$

where P is any property, M is average species mass, and a and b are constants specific to the property. A large number of allometric relations are available, and those of most interest to us deal with metabolic rate. For example, total basal rates of oxygen consumption for whole animals is given to a good approximation by:

TABLE 115.2 Characteristic Cerebral Tissue Cylinder

Item	Magnitude
Outer radius	30 μm
Capillary radius	3 μm
Length	180 μm
Blood velocity	400 μm/s
Arterial oxygen tension	0.125 atm
Arterial oxygen concentration (total)	8.6 mM
Arterial oxygen concentration (dissolved O_2)	0.12
Venous oxygen tension	0.053 atm
Venous oxygen concentration (total)	5.87 mM
Venous oxygen concentration (dissolved O_2)	0.05
Tissue oxygen diffusivity	1.5 E-5 cm^2/s (estd.)
Oxygen respiration rate (zero order)	0.0372 mmols O_2/liter-s

$$R_{O,tot} \times 3.5M^{3/4} \tag{115.2}$$

Here $R_{O,tot}$ is the oxygen consumption rate in ml O_2 (STP)/hr, and M is body mass in grams. Under basal conditions, fat is the primary fuel and heat generation is about 4.7 kcal per milliliter of oxygen (STP) consumed. Small animals have higher specific metabolic rates than large ones, but this is in part because a higher proportion of their body mass is made up of highly active body mass, and for the important case of brain tissue it is invariant at about:

$$R_{O_2}\left(\text{brain}\right) \approx 3.72 \cdot 10^{-5}\,\text{mmolsO}_2/\text{cm}^3,\text{s} \tag{115.3}$$

for all species. For the liver and kidneys, specific metabolic activity is somewhat lower and falls off slowly with an increase in animal size, but this may be due to an increasing proportion of supporting tissue such as blood vessels and connective tissue. Accurate data valid under physiologic conditions are difficult to find, and to a first approximation specific metabolic activity of parenchymal cells, those actually engaged in the primary activity of the organ, may be close to that of brain for both liver and kidneys.

The sizes of both microcirculatory units and cells in vertebrates are also very insensitive to animal size. Capillaries are typically about 3 to 4 μm in radius and about 50 to 60 μm apart. Typical mammalian cells are about 10 to 50 μm in diameter, and organelles such as mitochondria are about the size of prokaryotic cells, about 1 μm in diameter. Approximate characteristics of a cerebral tissue cylinder are given in Table 115.2. The oxygen-carrying capacity of blood, ionic makeup of body fluids, solubility of gases and oxygen diffusivities in body fluids are also largely invariant across species, and some representative data are provided in Tables 115.3, 115.4, and 115.5.

TABLE 115.3 Oxygen Solubilities

Solvent	Temperature, °C	O_2 Pressure, atm	Concentration, mM/atm
Water	25		1.26
	30		1.16
	35		1.09
	40		1.03
Plasma	37		1.19
Red cell interior (dissolved O_2 only)	37		1.18
Extracellular tissue (estd.)	37		1.1
Oxygen gas	37		39.3
Alveolar air	37	0.136	
Air (0.21 atm of oxygen)	37	0.21	

TABLE 115.4 Effective Oxygen Diffusivities

Solvent	Temperature, °C	Pressure	Diffusivity, cm²/s
Water	25		2.1 E-5
Water	37		3.0 E-5
Blood plasma	37		2.0 E-5
Normal blood	37		1.4 E-5
Red cell interior	37		0.95 E-5
Air	25	1 atm	0.20

TABLE 115.5 Intracellular Diffusion Coefficients

Compound	MW	Radius, Å	Diffusivity in Water $cm^2/s \times 10^7$	Intracell Diffusivity; $cm^2/s \times 10^7$	Diffusivity Ratio
Sorbitol	170	2.5	94	50	1.9
Methylene blue	320	3.7	40	15	2.6
Sucrose	324	4.4	52	20	2.6
Eosin	648	6.0	40	8	5.0
Dextran	3600	12.0	18	3.5	5.0
Inulin	5500	13.0	15	3.0	5.0
Dextran	10,000	23.3	9.2	2.5	3.7
Dextran	24,000	35.5	6.3	1.5	4.2
Actin	43,000	23.2	5.3	0.03	167
Bovine serum albumin	68,000	36.0	6.9	0.10	71

Allometric correlations are essentially empirical and cannot be predicted from any fundamental physical principles. Moreover, most measurements, except solubility and diffusivity data, are of doubtful accuracy.

Recently a more sophisticated form of self similarity, fractal geometry has been found useful for the description of a wide variety of biological applications of mass transfer interest [Bassingthwaighte et al., 1994], applications of nonlinear dynamics are fast increasing [Griffith, 1996].

Time Constants: The Key to Quantitative Modeling

The first step in system modeling is to establish orders of magnitudes of key parameters, and most particularly time constants: estimates of the time required for a given transient process to be "effectively complete." Time constant or order of magnitude analysis is useful because dynamic response times are insensitive to geometric detail and boundary conditions at the order of magnitude level of approximation, i.e., within a factor of ten. Time constants are, however, essentially heuristic quantities and can only be understood on the basis of experience [Lightfoot and Lightfoot, 1997].

Once characteristic system time scales are established, one can restrict attention to processes with response times of the same order: Those an order of magnitude faster can be treated as instantaneous and those ten times slower as not happening at all. Both fast and slow terms in system description, e.g., the diffusion equation of transport phenomena [Bird et al., 1960], can then be eliminated. Such simplification often provides valuable insights as well as simplifying integration. Quantitative descriptions are particularly valuable at the microcirculatory level, for example in diagnostic procedures, and they have been well studied [Bassingthwaighte and Goresky]. Here we shall stay with relatively simple examples to illustrate selected characteristics of tissue mass transfer, and we begin with diffusion. We then briefly introduce time constants characterizing flow, chemical reaction, and boundary conditions.

Brownian Motion and Concentration Diffusion

The basis of most species selective transport is the relatively slow observable motion of molecules or particles resulting from intermolecular collisions [Lightfoot and Lightfoot, 1997]. Such *Brownian motion* does not take any predictable direction, but the particles under observation do tend to move farther from their starting point with increasing time. The extent of diffusional motion can be described as the probability of finding any reference particle a distance r from its initial position. For an unbounded quiescent fluid this is [Einstein, 1905]:

$$P(r) = e^{-r^2/4D_{PF}t} \Big/ \left[8(\pi D_{PF}t)^{3/2}\right]; \quad 4\pi\int_0^\infty P(r) r^2 dr = 1. \tag{115.4a, 4b}$$

This equation defines the *Brownian* diffusivity, D_{PF}, of a particle or molecule relative to a surrounding fluid. Here P is a spherically symmetrical normalized probability density. Here r is distance (in spherical coordinates) from the initial position, and t is time. The mean net distance traveled from the initial point in unbounded quiescent three-dimensional space is easily determined from the above distribution as:

$$r_m^2 = 4\pi\int_0^\infty r^2 \cdot r^2 P(r) dr = 6D_{PF}t, \; r_m^2 = 4D_{PF}t, \; x_m^2 = 2D_{PF}t \tag{115.5a, 5b, 5c}$$

for distances from a point, line, and plane respectively. These results provide useful insight in suggesting characteristic diffusion times t as the time required for "most" of a transient diffusion process to be complete. Some commonly accepted values are shown in Table 115.6. The numbers in the last column are the fractional changes in solute inventory for a sudden change in surface concentration on a particle initially at a different uniform concentration. For the hollow cylinder, the outer surface of radius R_T is assumed impermeable to diffusion, and the inner surface of radius R_C is permeable. The length L is assumed large compared to either radius. Fractional completion depends in a complex way on the radius ratio, but the diffusion time given is a good approximation.

For large numbers of particles or molecules nonuniformly distributed in a moving fluid, following the Brownian motion of individual molecules becomes too cumbersome to use. We must then average behavior over large numbers of molecules to obtain a *continuum approximation* known as Fick's law [Bird et al., 1960], which describes the relative motion of solute and solvent and may be written in the form

$$\left(v_P - v_F\right)\left(\frac{x_P x_F}{D_{PF}}\right) = -\nabla x_P \tag{115.6}$$

Here v_p is observable velocity of species P in a mixture of P and F, x_p is mole fraction of species P in the mixture, and D_{PF} is binary mutual mass diffusivity of solute P relative to solvent fluid F. For situations of interest here, the Fick's law diffusivity may be considered equal to the Brownian diffusivity.

TABLE 115.6 Characteristic Diffusional Response Times

Shape	T_{dif}	L	Fractional Completion
Sphere	$L^2/6D_{PF}$	Radius	>0.99
Cylinder	$L^2/4D_{PF}$	Radius	>0.99
Slab	$L^2/2D_{PF}$	Half-thickness	>0.93
Hollow cylinder	$(L^2/2D_{PF})\ln(R_T/R_c)$	Outer radius, R_T	

Equation 115.6 is valid for binary or dilute solutions of liquids, gases, and homogeneous, and some typical magnitudes for biologic situations are shown in Tables 115.4 and 115.5. For dilute solutions, hydrodynamic diffusion theory [Bird et al., 1960] provides useful insight:

$$\pi R_P D_{PF} \mu_F / \kappa T = C \qquad (115.7)$$

where C is equal to (1/6) if the molecular radius of the solute P is much less than that of the solvent F and (1/4) if they are about equal. Here μ is solvent viscosity, R_P and R_F are effective spherical solute and solvent radii, κ is the Boltzmann constant, and T is absolute temperature. Hydrodynamic diffusion theory has been extended to solutes in small fluid-filled pores [Deen, 1987] for characterizing transport in microporous membranes.

More Complex Situations

Most biologic transport processes occur in multicomponent solutions. A generalized diffusion equation [Hirschfelder et al., 1954; Lightfoot and Lightfoot, 1997; Taylor and Krishna, 1993] is available, but lack of data usually forces neglect of multicomponent effects. For our purposes transport of both molecules and particles is adequately described by the simplified equations in Table 115.7, which simply state that relative velocity of a particle or molecule P through a fluid F is proportional to the sum of "driving forces" acting on it.

Flow, Chemical Reaction, and Boundary Conditions

We now introduce additional time constants in order to characterize the interactions of diffusion with flow and chemical reaction and to show that boundary conditions can have an important bearing on

TABLE 115.7 Particle-Molecular Analogs Dilute Binary Diffusion in a Quiescent Continuum

Particles

$$\left(v_p - v_F\right) = D_{PF}\left\{\nabla \ln n_p + \frac{1}{\kappa T}\left[V_P\left(1-\rho_p^o/\rho\right)\left(\nabla p\right)_\infty - F_{em} + \left(\rho_p^o + \tfrac{1}{2}\rho_f\right)V_P \frac{dv_P}{dt}\right.\right.$$

$$\left.\left. -6\left(\pi\mu\rho_F^o\right)^{1/2} R_P^2 I + \text{thermal diffusion}\right]\right\}$$

where:

$$I = \int_0^t \left[v_P'(\tau) / (t-\tau)^{1/2} \right] d\tau$$

Molecules (interdiffusion of species P and F)

$$\left(v_P - v_F\right) = -D_{PF}\left\{\nabla \ln x_p + \left(1/RT\right)\left[V_P\left(1-\rho_P^o/\rho\right)\nabla_p - F_{em}\right]\right\}$$

Both:

$$D_{PF} = \frac{\kappa T}{6\pi\mu R_P}$$

Here:
κ = the Boltzmann constant, or molecular gas constant
n_p = number concentration of particles
R_p = particle or molecular radius
V_p = particle volume or partial molal volume of species P
F_{em} = total electromagnetic force per particle
F_{em} = molar electromagnetic force on species P
ρ = density of fluid phase
ρ_P^o = density of particle or reciprocal of partial specific volume of species P in solution
v' = instantaneous acceleration of particle P
The subscript ∞ refers to conditions near the particle but outside its hydrodynamic boundary layer.

system behavior. We begin with flow where mean solute residence time, T_m, forms a convenient time scale. For flow through a cylindrical duct of length L with constant volumetric average velocity <v>, we may write:

$$T_m \equiv L/<v> \tag{115.8}$$

More generally, for constant volumetric flow at a rate Q through a system of volume V with a single inlet and outlet and no significant diffusion across either, T_m is equal to V/Q. For chemical reactions we choose as our system, a reactive solid open to diffusion over at least part of its surface where concentration of solute i under consideration is maintained at a uniform value c_{i0} and where the average volumetric rate of consumption of that solute is $<R_i>$. We then define a reaction time constant as:

$$T_{rxn} \equiv c_{i0}/<R_i> \tag{115.9}$$

Note that both c_{i0} and $<R_i>$ are measurable, at least in principle [Damköhler, 1937; Weisz, 1973]. Finally we consider as an illustrative situation decay of solute concentration c_i in the inlet stream to a flow system according to the expression

$$c_i(t, \text{inlet}) = c_i(0, \text{inlet}) \exp(-t/T_{BC}) \tag{115.10}$$

where t is time and T_{BC} is a constant. We now have enough response times for the examples we have selected below, and we turn to illustrating their utility.

115.3 Characteristic Behavior: Selected Examples

We now consider some representative examples to illustrate the mass transfer behavior of living tissues.

Alveolar Transients and Pulmonary Mass Transport: At the distal ends of the pulmonary network are the alveoli, irregular sacs of air surrounded by blood-filled capillaries, with effective diameters of about 75-300 μm. The blood residence time is of the order of a second, and it is desired to determine whether gas-phase mass transfer resistance is appreciable. To answer, we take the worst possible scenario: flat-plate geometry with a half-thickness of 150 μm or 0.0105 cm. From Table 115.4 we find the oxygen diffusivity to be about 0.2 cm²/s. It follows that alveolar response time is:

$$T_{dif} \approx (0.015)/0.4 = 0.56\,\text{ms}$$

This is extremely fast, with respect to both the 1s residence time assumed for alveolar blood and the one-twelfth of a minute between breaths: gas-phase mass transfer resistance is indeed negligible. This is typical for absorption of sparingly soluble gases, and a similar situation occurs in cell culture vessels.

Mass Transfer Between Blood and Tissue: We next identify blood vessels capable of transferring dissolved solute between blood and surrounding tissue. To be effective we assume that mean blood residence time should at least equal the radial diffusion time:

$$\frac{L}{<v>} \geq \frac{R^2}{4D_{PB}} = \frac{D^2}{16D_{PB}};\ \frac{LD_{PB}}{D^2<v>} \geq 1/16 = 0.0625$$

where D is vessel diameter. From the data for a 13-kg dog in Table 115.8, we find the great majority of vessels much too short and that only the three smallest classes—arterioles, capillaries, and venules—are at all capable of transferring appreciable mass. These, especially the capillaries, are quite effective. They

TABLE 115.8 Mass Transfer Effectiveness of Blood Vessels (13-kg Dog, #39)

Vessel	Radius, cm	Lenght, cm	$<v>$, cm/s	$LD^2/d_{PF}<v>$
Aorta	0.5	10	50	0.00003
Larger arteries	0.15	20	13.4	0.00003
Secondary arteries	0.05	10	8	0.005
Terminal arteries	0.03	1.0	6.0	0.002
Arterioles	0.001	0.2	0.032	6.25
Capillaries	0.0004	0.1	0.07	89.4
Venules	0.0015	0.2	0.07	12.7
Terminal veins	0.075	1.0	1.3	0.0014
Secondary veins	0.12	10	1.48	0.0047
Larger veins	0.3	20	3.6	0.0006
Venae cavae	0.625	40	33.4	0.00003

have long been classified as the microcirculation on the basis of being invisible to the unaided eye. This simple order of magnitude analysis provides a functional definition and a guide to the design of hollow fiber cell culture vessels. More refined analyses [Lightfoot, 1974] based on the parameter magnitudes of Table 115.2 shows that lateral diffusional resistance of capillaries is in fact rather small, but recent analyses [Popel, 1989] suggest a more complex picture.

Convection and diffusion in parallel [Lightfoot and Lightfoot, 1997] is also of interest, but it is more complicated because of convective dispersion [Lightfoot and Lightfoot, 1997]. However, the lung shows a sharp transition between convective ducts and those which can be assumed well-mixed by axial diffusion [Hobbs and Lightfoot, 1979; Hubal, 1996; Farhi,]. This has long been known to pulmonary physiologists who often model the adult human lung as a plug-flow channel ("dead space") of about 500 ml leading to a well-mixed volume of about 6,000 ml.

Intercapillary Spacing in the Microcirculation: It has been shown [Damköhler, 1937; Weisz, 1973] that optimized commercial catalysts normally exhibit ratios of diffusion to reaction times, as defined in Table 115.6 in the relatively narrow range:

$$1/3 < T_{dif}/T_{rxn} < 1$$

Moreover, Weisz has shown that this ratio holds for many biologic systems as well so that enzyme activities can often be inferred from their location and function. Here we compare this expectation for cerebral tissue cylinders using the data of Tables 115.2 and 115.6. Throughout the tissue cylinder:

$$T_{dif} = \frac{9 \cdot 10^{-6}\,\text{cm}^2}{3 \cdot 10^{-5}\,\text{cm}^2/\text{s}} \cdot \ln(10) = 0.69\text{s}$$

For venous conditions, which are the most severe,

$$T_{rxn} = \frac{0.0372}{0.0584} = 0.64\text{s} \text{ and } T_{dif}/T_{rxn} \approx 1.$$

These numbers are reasonable considering the uncertainty in the data and the approximation used for diffusion time. The brain, for example, is far from homogeneous. More elaborate calculations suggest somewhat more conservative design. However, these figures are correct in suggesting that the brain is "designed" for oxygen transport and that the safety factor is small [Neubauer and James, 1998]. Sections of the brain become anoxic at about 2/3 of normal blood flow. The body's control mechanisms will shut down other vital organs such as the gut, liver, and kidneys when blood cardiac output oxygen supply drops to keep the brain as well supplied as possible. All of these comments are for the physiological (normal) state. Lowering brain temperature can greatly decrease oxygen demand and permit survival under otherwise fatal conditions. This effect is important in cryosurgery and drowning: whereas about 6 minutes of anoxia damages the adult brain at normal temperature, about one hour is required at 62°F.

Intracellular Dispersion: We now ask how long it will take a protein initially concentrated in a very small region to disperse through the interior of a cell, and we shall assume the cell to be a sphere with a 1 μm radius. We begin by noting that the cell interior or cytoplasm is a concentrated solution of polymeric species and that the diffusivity of large molecules is considerably slowed relative to water. We therefore assume a diffusivity of 10^{-8} cm²/s as suggested by the last entry in Table 115.5. If the protein is originally at the center of the sphere, the dispersion time is:

$$T_{dis} \equiv T_{dif} = \frac{R_{cell}^2}{6Đ_{pc}}\left(1 \cdot 10^{-4}\right)^2 \Big/ \left(6 \cdot 10^{-8}\right) \approx \left(1/6\right) s$$

If the protein is initially near the cell periphery, diffusion will take about 4 times as long, or about 2/3 s. Reliable *intracellular* diffusion coefficients are important: the corresponding numbers for aqueous solution would have been about 2.4 ms and 10 ms, respectively! Cell interiors—the *cytoplasm*—are crowded.

Diffusion Controlled Reaction: We now calculate the rate at which a very dilute protein of 2.5 nm radius solution in a cell of 1 micron radius is adsorbed on a spherical adsorbent particle of radius 2.5 nanometers if the adsorption is *diffusion controlled*. That is, we assume that the free protein concentration immediately adjacent to the adsorbent surface is always effectively zero because of the speed and strength of the adsorption reaction. In diffusion only the region within a few diameters of a sphere's surface offers effective resistance to transport [Carslaw and Jaeger, 1959; see also Özisik, 1993]. We can thus assume rapid equilibration over the entire cell volume, except for a thin diffusional boundary layer adjacent to the target surface: the ratio of protein plus adsorbent to cell diameter is only about 0.05. The rate at which protein is adsorbed on the sphere surface per unit area is then [Lightfoot, 1974]:

$$N_P = \frac{D_{PF} c_{P\infty}}{\left(R_P + R_{ads}\right)} \cdot \left(1 + \frac{1}{\sqrt{\pi\tau}}\right)$$

where $\mathbf{C}_{P\infty}$ is the protein concentration far from the sphere, assumed uniform over the bulk of cell volume since we are about to see that it adjusts rapidly on the time scale of the adsorption process. The radius of the protein is R_P, that of the adsorbent R_{ads}, and the dimensionless time τ defined by:

$$\tau \equiv \frac{tD_{PF}}{\left(R_P + R_{ads}\right)^2} = \frac{t}{2.5 \cdot 10^{-5} s}$$

Adsorption rate is within 10% of its asymptotic value when $\pi\tau = 100$, and:

$$t_{1/10} = \left(2.5 \cdot 10^{-5} s\right)\left(100/\pi\right) \approx 0.8 \, ms$$

This is much smaller than distribution time, and transients can be neglected: the time scales of the transient and distribution are *well separated*. We now write a mass balance for the rate of change of bulk protein concentration:

$$-V_{cell}\frac{dc_{P\infty}}{dt} \approx -V_{cell}\frac{dc_P}{dt} = A_{ads}N_P; \, d\ln c_P/dt = 3\left[\left(R_P + R_{ads}\right)/R_{cell}^3\right]Đ_{PF} = 1/T_{rxn}$$

Here V_{cell} is the volume of the cell, and A_{ads} is the effective area of the adsorbent; T is the characteristic reaction time, that required to reduce cellular concentration by a factor e or about 2.7. For our conditions, reasonably representative of a prokaryotic cell,

$$T_{rxn} = \left[\frac{3 \cdot 5 \cdot 10^{-7}}{\left(10^{-4}\right)^{3}} \cdot 10^{-8}\right]^{-1} s = 67s$$

This is clearly a long time compared to the dispersion time of the last example, and the assumed time scale separations of all three adsorption and dispersion processes are amply justified. Using time scale separations has greatly simplified what was originally a major numerical task [Berg and von Hipple, 1985].

If the protein concentration is now interpreted as the probability of finding a protein in any position, and the target is a DNA site, T_{rxn} is the time required for gene expression, and it can be seen that it is very slow! Comparison of these predictions with experiment showed that both prokaryotic cells respond much faster and this led to a major discovery [Berg and von Hipple, 1985]: DNA exhibits a general binding affinity for promoter molecules which permits the adsorbed protein to undergo one-dimensional diffusion along the DNA chain—thus increasing effective binding site size. Eukaryotes, which are roughly ten times larger in diameter, have elaborate internal barriers [Holsstege et al., 1998] and are much more complex diffusionally.

Many other diffusion controlled reactions occur in living systems, for example, protein folding, and here brute force molecular dynamic calculations are difficult even with supercomputers. One can greatly speed calculation by taking judicious time scale separation [Rojnuckarin et al., 1998].

The Energy Cost of Immobility

Many cell constituents, from small metabolites to organelles, are free to move in the cytoplasm and yet must be limited in their spatial distribution. Many such entities are transported by mechanochemical enzymes known as *protein motors* which depend upon consumption of metabolic energy. Here we estimate the energy cost of such processes to maintain spatial segregation, without detailed knowledge of the mechanisms used. The basis of our analysis [Okamoto and Lightfoot, 1992] is the Maxwell-Stefan equation of Table 115.7 and more particularly the fact that migration velocities are related to the motive force of Brownian motion and any mechanically transmitted force through the diffusivity. The heart of our argument is that the mechanical force applied by the protein motors must produce a migration equal and opposite to that resulting from the dispersive force of Brownian motion:

$$\left(v_P - v_F\right)_{tot} = \left(v_P - v_F\right)_{dif} + \left(v_P - v_F\right)_{mech} = 0,$$

$$\left(v_P - v_F\right)_{dif} = D_{PF} \nabla \ln n_P \qquad \left(v_P - v_F\right)_{mech} = FD_{PF}/\kappa T$$

where n_P is the number concentration of proteins, and F is the mechanical force required to produce the motion. Now the power P_P required for transporting a particle back against diffusional motion is the product of the mechanical force and mechanical migration velocity:

$$P_P = F \cdot \left(v_P - v_F\right)_{mech} = \kappa T \left(\nabla \ln n_P\right)^2$$

For a particle of mass $\mathbf{m_P} = \mathbf{(4/3)\pi R_P^3 \rho_P}$ where ρ_P is particle density, the power requirement per unit mass is:

TABLE 115.9 Energetics of Forced Diffusion

		Dilute Solution		Cytoplasm	
					Estimated
Particle	Radius, Å	Rel. Diff.	$\hat{P}/\Delta G$	Rel. Diff.	$\hat{P}/\Delta G$
Small metabolite	3	1	7×10^6	1	7×10^6
Globular protein	30	1	7×10^2	0.01	7
Typical organelle	10^4	1	6×10^{-8}	$10^{-1/2}$	6×10^{-12}

$$P_P / m_P \equiv \hat{P} = \frac{\left(\kappa T \nabla \ln n_P\right)^2}{8\pi\mu_{eff}\rho_P R_P^4}.$$

Here μ_{eff} is the effective viscosity of the cytoplasm, ρ_p is particle density, and R_p is effective particle radius. This very strong effect of particle radius suggests that protein motors will be most effective for larger particles, and calculations of Okamoto and Lightfoot [1992] bear out this suggestion.

If the particle is to be held for the most part within one micrometer of the desired position and the free energy transduction of the cell as a whole corresponds to that of brain tissue, Table 115.9 shows the cost of mechanical motion is negligible for organelles, prohibitive for small metabolites and problematic for proteins. However, for small amounts of critically important metabolites and problematic for proteins. However, for small amounts of critically important metabolites, mechanical transport may still take place. The quantity, $\hat{\mathbf{P}}/\Delta\mathbf{G}$, is the ratio of power consumption to mean cellular energy transduction, both per unit mass.

It remains to be noted that estimating the cost of organelle transport accurately requires additional information, and some is available in [Okamoto and Lightfoot, 1992]. It is impossible for the cell to transport the large number of small metabolites, and a number of alternate means of segregation have developed. Among them are permselective membranes and compact enzyme clusters, making it difficult for intermediate metabolites to escape into the general cytoplasmic pool. These topics must be discussed elsewhere.

Defining Terms[1]

Allometry: A special form of similarity in which the property of interest scales across species with some constant power of species mass.

Alveolus: A pulmonary air sac at the distal end of an airway.

Arterioles: The smallest subdivision of the arterial tree proximal to the capillaries.

ATP: Abbreviation of adenosine triphosphate, the source of chemical energy for a wide variety of metabolic reactions.

Capillaries: The smallest class of blood vessel, between an arteriole and venule, whose walls consist of a single layer of cells.

Cell: The smallest unit of living matter capable of independent functioning, composed of protoplasm and surrounded by a semipermeable plasma membrane.

Convection: Mass transport resulting directly from fluid flow.

Cytoplasm: The protoplasm or substance of a cell surrounding the nucleus, carrying structures within which most of the life processes of the cell take place.

Cytoskeleton: An intracellular network of microtubules.

---

[1]For more complete definitions see the *Oxford Dictionary of Biochemistry and Molecular Biology*, 1997.

Distal: At the downstream end of a flow system.

Endothelium: A single layer of thin flattened cells lining blood vessels and some body cavities.

Microcirculation: The three smallest types of blood vessels-arterioles, capillaries, and venules.

Mitochondrion: Compartmentalized double-membrane self-reproducing organelle responsible for generating useable energy by formation of ATP In the average cell there are several hundred mitochondria each about 1.5 micrometers in length.

Microtubule: Long, generally straight elements of the cytoskeleton, formed of the protein tubulin.

Organelle: A specialized cytoplasmic structure of a cell performing a specific function.

Parenchyma: The characteristic tissue of an organ or a gland, as distinguished from connective tissue.

Proximal: At the upstream end of a flow system.

Venules: The smallest vessels of the venous tree, distal to the capillaries.

References

Adolph EF. 1949. Quantitative relations in the physiological constitutions of mammals, *Science* 109: 579.

Alberts, B et al., 1945. *The Molecular Biology of the Cell*, 3rd Ed., Garland, NY.

Bassingthwaighte JB, Goresky CA. Modeling in the Analysis of Solute and Water Exchange in the Microvasculature, Chap. 13, *Handbook of Physiology-The Cardiovascular System IV.*

Bassingthwaighte, JB, Liebovitch, LS, and West, BJ, 1994. Fractal Physiology, Oxford.

Bassingthwaighte, JB, Goresky, CA, and Linehan, JH. 1998. Whole Organ Approaches to Cellular Metabolism, Springer.

Berg, OG. 1993. Random Walks in Biology, expanded edition, Princeton.

Berg OG, von Hippel PH. 1985. Diffusion controlled macromolecular interactions. *Ann. Rev. Biophys. Chem.* 14:13 1.

Bird RB, Stewart WE, Lightfoot EN. 1960. *Transport Phenomena*, Wiley, New York.

Bray, D. Signaling Complexes: Biophysical Constraints on Intracellular Communication, *Ann. Rev. Biophys. Biomol. Struct.* 1998, 27, 59-75.

Bunk, S. Do Energy Transport Systems Shape Organisms?, 1998. *The Scientist*, Dec. 14-15.

Calder WA. 1984. Size, Function and Life History, Harvard University Press, Cambridge, Mass.

Campenot, RB, Lund, K, and Senger, DL. 1996. Delivery of Newly Synthesized Tubulin to Rapidly Growing Distal Axons of Rat Sympathetic Neurons in Compartmented Cultures, *J. Cell Biol.*, 135, 701-709.

Campbell, NA, 1996. *Biology*, 4th ed, Benjamin/Cummins.

Carslaw HS, Jaeger JC. 1959. Conduction of Heat in Solids, 2nd ed, Oxford.

Damköhler G. 1937. Einfluss von Diffusion, Strömung und Wärmetransport auf die Ausbeute bei chemische-technische Reaktionen, *Der Chemieingenieur*, Bd. 3, p 359.

Deen WM. 1987. Hindered Transport of Large Molecules in Liquid-Filled Pores, *A I Ch E 1*, 33 (9): 1409.

Einstein A. 1905. *Annalen der Physik.* 17:549.

Farhi LE. Ventilation-perfusion relationships, Chap. 11 In *Hdbk. of Physiol.—the Resp. System IV.*

Federspiel WJ and Popel AS. 1986. A Theoretical Analysis of the Effect of the Particulate Nature of Blood on Oxygen Release in Capillaries, *Microvasc. Res.*, 32:164-189

Fishman, HA, Greenwald DR, Zare RN. 1998. Biosensors in Chemical Separations, *Ann. Rev. Biophys. Biol. Mol. Struct.*, 27:165-198.

Francis K and Palsson BO. 1997. Effective intercellular communication distances, etc., *Proc. Natl. Acad. Sci.* 94:12258-12262.

Griffith, TM. 1996. Temporal chaos in the microcirculation, *Cardiovasc. Res.*, 31 (3), 342-358.

Gruenberg J and Maxfield FR. 1995. Membrane Transport in the Endocytic Pathway, Current Opinion ion Cell Biology, 7:552-563.

Hellums, JD et al., 1996. Simulation of Intraluminal Gas Transport Processes in the Microcirculation, *Ann. Biomed. Eng.*, 24:1-24

Hirschfelder JO, Curtiss CF, Bird RB. 1954. *Molecular Theories of Gases and Liquids*, Wiley, New York.

Hobbs SH, Lightfoot EN. 1979. A Monte-Carlo simulation of convective dispersion in the large airways. *Resp. Physiol.* 37:273.

Holsstege FCP et al., 1998. Dissecting the Regulatory Circuitry of of a Eukaryotic Genome, *Cell,* 95:717-728.

Hubal, EA et al., 1996, Mass transport models to predict toxicity of inhaled gases in the upper respiratory tract, *J. App. Physiol.,* 80(4):1415-1427.

Johnson, LR. 1998. Essential Medical Physiology, 2nd ed., Lippincott-Raven.

Lauffenberger DA, Linderman JJ. 1993. *Receptors,* Oxford.

Lightfoot EN. 1974. Transport Phenomena and Living Systems, Wiley-Interscience, New York.

Lightfoot EN, Lightfoot EJ. 1997. *Mass Transfer in Kirk-Othmer Encyclopedia of Separation.*

Maniotis, AJ, Chen CS, Ingber DE. 1997. Demonstration of mechanical connections between integrins, cytoskeletal filaments, and nucleoplasm that stabilize nuclear structure, *Proc. Natl. Acad. Sci.,* 94:849-854.

Neubauer RA and James P. 1998. Cerebral oxygenation and the recovereable brain, *Neurolog. Res.* 20, Supplement 1, S33-36.

Okamoto GH, Lightfoot EN. 1992. Energy cost of intracellular organization. *Ind. Eng. Chem. Res.* 31 (3):732.

Ozisik MN. 1993. Heat Conduction, 2nd ed, Wiley, New York.

Popel AS. 1989. Theory of oxygen transport to tissue. *Clin. Rev. Biomed. Eng.* 17 (3):257.

Pries, AR, TW Secomb and P Gaeghtgens, 1996, Biophysical aspects of blood flow in the microvasculature, *Cardiovasc. Res.,* 32, 654-667.

Rojnuckarin A, Kim S, Subramanian S. 1998. Brownian dynamic simulation of protein folding: Access to milliseconds time scale and beyond, *Proc. Natl. Acad. Sci.* 95:4288-4292.

Schmidt-Nielsen Knut. 1983. *Animal Physiology: Adaptation and Environment,* 3rd ed, Cambridge.

Schmidt-Nielsen Knut. 1984. *Scaling: Why Is Animal Size So Important?.* Cambridge.

Suominen, PK et al., 1997. Does water temperature affect outcome of nearly drowned children?, *Resuscitation,* 35(2), 111-115.

Taylor R, Krishna R. 1993. Multicomponent Mass Transfer, New York, Wiley.

Thompson DW. 1961. On Growth and Form (an abridged edition, JT Bonner, ed.) Cambridge.

Vallee RB, Sheetz MP. 1996.Targeting of Motor Proteins, *Science,* 271, 1539-1544.

Vicini, Paolo et al., 1998. Estimation of Blood Flow Heterogeneity in Skeletal Muscle et al., *Ann. Biomed. Eng.,* 26, 764-774.

Weisz PB. 1973. Diffusion and chemical transformation-an interdisciplinary excursion. *Science,* 179:433.

Welling PG. 1997. *Pharmacokinetics,* American Chemical Society.

116

The Biology of Stem Cells

Craig T. Jordan
Somatix Therapy Corp.

Gary Van Zant
University of Kentucky Medical Center

Life for most eukaryotes, and certainly all mammals, begins as a single totipotent stem cell, the zygote. This cell contains the same complement of genes—no more and no less—as does every adult cell that will make up the organism once development is complete. Nonetheless, this cell has the unique characteristic of being able to implement every possible program of gene expression and is thus totipotent. How is this possible? It is now known that the selective activation and repression of genes distinguishes cells with different developmental potentials. Unraveling this complex series of genetic changes accompanying the progressive restriction of developmental potential during ontogeny is the realm of modern developmental biology. In contrast to the zygote, which has unlimited developmental potential, an intestinal epithelial cell or a granulocyte, for example, is a highly developed cell type that is said to be differentiated. These cells are fixed with respect to their developmental potential and thus no longer posses the ability to contribute to other tissue types. Indeed, intestinal epithelial cells and granulocytes are incapable of undergoing further division and are said to be terminally differentiated. These mature cells have therefore undergone a process whereby they each have acquired a unique and complex repertoire of functions. These functions are usually associated with the cellular morphologic features and/or enzymatic profiles required to implement a specific developmental or functional program. We will come back to the tissue dynamics of these two cell types in a later section of this chapter.

Between the extremes of developmental potency represented by the zygote and terminally differentiated cells, there is obviously a tremendous number of cell divisions (roughly 2^{44}) and an accompanying restriction of this potential in zygotic progeny. The human body, for example, is composed of greater than 10^{13} cells—all ultimately derived from one, the zygote. Where during the developmental sequence, does restriction occur? Is it gradual or quantal? These questions are fundamental to an understanding of developmental biology in general and stem cell biology in particular.

116.1 Embryonic Stem Cells

Let us consider first the ultimate human stem cell, the zygote, in more detail. As cellular growth begins, the early embryonic cells start to make a series of irreversible decisions to differentiate along various developmental pathways. This process is referred to as developmental commitment. Importantly, such

0-8493-0462-8/00/$0.00+$.50

decisions do not occur immediately; rather, the zygote divides several times and proceeds to the early blastocyst stage of development while maintaining totipotency in all its daughter cells. This is evident most commonly in the phenomenon of identical twins, where two distinct yet genetically matched embryos arise from the same zygote. The ability of early embryonic cells to maintain totipotency has been utilized by developmental biologists as a means to experimentally manipulate these embryonic stem cells, or ES cells, as they are commonly known. In 1981, two scientists at Cambridge, Evans and Kaufman, were able to isolate ES cells from a blastocyst-stage mouse embryo and demonstrate that such cells could be cloned and grown for many generations in vitro [Kaufman et al., 1983]. Remarkably, under the appropriate culture conditions, such cells remained completely totipotent. That is to say, upon reimplantation into another embryo, the stem cells could grow and contribute to the formation of an adult mouse. Importantly, the ES cells retained the developmental potential to differentiate into all the possible adult phenotypes, thereby proving their totipotency.

Thus, in culture, the ES cells were said to self-renew without any subsequent loss of developmental potential. Upon reintroduction into the appropriate environment, the ES cells were able to differentiate into any of the various mature cell types. The basic decision of whether to self-renew or differentiate is a common theme found in stem cells of many developmental systems. Generally self-renewal and differentiation go hand in hand, i.e., the two events are usually inseparable. The important findings of Evans and Kaufman demonstrated that self-renewal and differentiation could be uncoupled and that an extended self-renewal phase of growth was attainable for mammalian embryonic cells.

The advent of ES cell technology has had an enormous impact on the field of mammalian molecular genetics. The ability to culture ES cells was quickly combined with molecular techniques which allow for the alteration of cellular DNA. For example, if an investigator were interested in the function of a gene, he or she might elect to mutate the gene in ES cells so that it was no longer functional. The genetically altered ES cells would then be used to generate a line of mice which carry the so-called gene knockout [Koller & Smithies, 1992; Robertson, 1991]. By examining the consequences of such a mutation, clues to the normal function of a gene may be deduced. Techniques such as these have been widely used over the past 10 years and continue to develop as even more powerful means of studying basic cellular function become available.

116.2 Control of Stem Cell Development

The concepts of self-renewal and differentiation are central to the description of a stem cell; indeed, the potential to manifest these two developmental options is the only rigorous criterion used in defining what constitutes a true stem cell. Consequently, in studying stem cells, one of the most important questions to consider is how the critical choice whether to self-renew or differentiate is made. As seen in the case of ES cells, the environment of the stem cell or extrinsic signals determine the outcome of the self-renewal decision. In other words, such cells are not intrinsically committed to a particular developmental fate; rather, their environment mediates the differentiation decision. Surprisingly, for ES cells this decision is dictated by a single essential protein, or growth factor, known as Leukemia inhibitory factor (LIF) [Hilton & Gough, 1991; Smith et al., 1992]. In the presence of sufficient concentrations of LIF, ES cells will self-renew indefinitely in culture. Although ES cells eventually lose their totipotency in vitro, this is thought to be due to technical limitations of ex vivo culture rather than developmental decisions by the cells themselves.

Interestingly, the default decision for ES cells appears to be differentiation, i.e., unless the cells are prevented from maturing by the presence of LIF, they will quickly lose their totipotent phenotype. Upon beginning the differentiation process, ES cells can be steered along a variety of developmental pathways simply by providing the appropriate extrinsic signal, usually in the form of a growth factor.

Unfortunately, the control of other types of stem cells has proven more difficult to elucidate. In particular, the control of blood-forming, or hematopoietic stem cells, has been extensively studied, but as yet the developmental control of these cells is poorly understood.

116.3 Adult Stem Cells

As the mammalian embryo develops, various organ systems are formed, and tissue-specific functions are elaborated. For the majority of mature tissues, the cells are terminally differentiated and will continue to function for extended periods of time. However, some tissues operate in a much more dynamic fashion, wherein cells are continuously dying and being replenished. Thee tissues require a population of stem cells in order to maintain a steady flow of fresh cells as older cells turn over. Although there are several examples of such tissue types, perhaps the best characterized are the hematopoietic system and the intestinal epithelia. These two cell types have population parameters that call for their continuous and rapid production: Both cell types occur in very large numbers (approximately 10^{11} to 10^{12} for the human hematopoietic system) and have relatively short life spans that can often be measured in days or sometimes even hours [Kroller & Pallson, 1993]. These two tissues in adults are therefore distinct (along with skin epithelium) in that they require tissue-specific stem cells in order to satisfy the inherent population dynamics of the system.

Stem cells of this nature represent a population arrested at an intermediate level of developmental potency that permits them to perform the two classic stem cell functions: They are able to replenish and maintain their own numbers through cell divisions that produce daughter cells of equivalent potency, that is, self-renew. And, they have the capacity, depending on need, to differentiate and give rise to some, if not all, of the various mature cell types of that tissue. Stem cells of the small intestine, to the best of our knowledge, give rise to at least four highly specialized lineages found in the epithelium: Paneth, goblet, enteroendocrine, and enterocytes; these stem cells are therefore pluripotent (i.e., they have the potential to give rise to many different, but not all, lineages) [Potten & Loeffler, 1990]. Similarly, pluripotent stem cells of the hematopoietic system give rise to an even wider variety of mature cells, including at least eight types of blood cells: the various lymphocytes, natural killer cells, megakaryocytes, erthroid cells, monocytes, and three types of granulocytes [Metcalf & Moore, 1971].

Proof of the existence of gut and hematopoietic stem cells came from studies of the effects of ionizing radiation on animals. This research, in the 1940s and 1950s, was spurred by concern over military and peaceful uses of atomic energy. It became recognized that the organ/tissue systems most susceptible to radiation damage were those that normally had a high turnover rate and were replenished by stem cell populations, i.e., the gut lining and the hematopoietic system. In particular, the latter was found to be the radiation dose-limiting system in the body. It was subsequently discovered that mice could be rescued from imminent death from radiation "poisoning" by the transfusion of bone marrow cells following exposure [Barnes et al., 1959]. Initially it was not clear whether the survival factor was humoral or cellular, but mounting evidence pointed to a cell-mediated effect, and, as the dose of bone marrow cells was titrated to determine the number required for survival, it was found that low numbers of cells resulted in the development of macroscopic nodules on the spleens of irradiated mice.[1] These nodules were composed of cells of some but not all lineages of blood formation—lymphopoiesis was notably missing [Till & McCulloch, 1961]. Low-level radiation was then used to induce unique chromosomal aberrations in bone marrow cells prior to transplantation into lethally irradiated recipients. In this way, unique microscopically identifiable translocations would be passed on to all progeny of an altered cell. It was found that the spleen nodules were, in fact, colonies of hematopoietic cells, all possessing an identical chromosomal marker [Becker al., 1963]. This observation strongly suggested that the nodule was a clonal population, derived from a single stem cell. Since several lineages were represented in the spleen colony, the stem cell responsible was pluripotent. Moreover, single spleen colonies could be isolated and injected into secondary irradiated recipients and give rise to additional spleen colonies. This suggested that the cell giving rise to a spleen colony was capable of some degree of self-renewal as well as multilineage

[1]Injected bone-marrow-derived stem cells lodge and develop in several types of hematopoietic tissue, including spleen. Apparently, the splenic microenvironment can at least transiently support stem cell growth and development.

differentiation [Till et al., 1964]. The cells which give rise to spleen colonies were termed CFU-S for colony-forming unit-spleen and have been studied extensively in the characterization of pluripotent hematopoietic stem cells.

More recent studies employing a similar strategy have used retroviruses to mark stem cells. The site of viral integration in an infected host cell genome is random and is passed with high fidelity to all progeny, thus by molecular means stem cell clones may be identified. Such an approach allows for the analysis not only of spleen colonies but of all anatomically dispersed lymphohematopoietic sites, including bone marrow, spleen, thymus, lymph nodes, and mature blood cells in the circulation [Dick et al,., 1985; Keller et al., 1985; Lemischka et al., 1986]. These analyses unequivocally showed that the same stem cell may give rise to all lineages, including lymphocytes. Repetitive blood cell sampling and analysis gave a temporal picture of the usage and fate of the stem cell population. In the first few weeks and months after transplant of nonlimiting numbers of stem cells, polyclonal hematopoiesis was the rule; however, after approximately 4-6 months, the number of stem cell clones was reduced. In fact, in some cases, a single stem cell clone was responsible for all hematopoiesis for over a year–about half the mouse's lifetime [Jorday & Lemischka, 1990]. These data were interpreted to mean that many stem cell clones were initially active in the irradiated recipient mice, but over time a subset of stem cells grew to dominate the hematopoietic profile. This implies either that not all the stem cells were equivalent in their developmental potential or that the stem cells were not all seeded in equivalent microenvironments and therefore manifest differing developmental potentials. Both these possibilities have been supported by a variety of subsequent experiments; however, the details of this observation remain cloudy.

One piece of evidence suggesting intrinsic differences at the stem cell level comes from studies of allophenic mice [Mintz, 1971]. These artificially generated strains of laboratory mice are created by aggregating the early embryos of two distinguishable mouse strains. As mentioned previously, early embryonic cells are totipotent; thus, upon combining such cells from two strains, a chimeric mouse will arise in which both cell sources contribute to all tissues, including the stem cell population. Patterns of stem cell contribution in allophenic mice show that one strain can cycle more rapidly and thus contribute to mature blood cells early in life, whereas the other slow-growing strain will arise to dominate at later times. Importantly, upon reimplantation of allophenic bone marrow into a secondary irradiated recipient, the two phases of stem cell activity are recapitulated. Thus, the stem cells which had become completely quiescent in the primary animal were reactivated initially, only to be followed by a later phase of activity from the second strain [Van Zant, 1992]. These data suggest that intrinsic differences at the stem cell level rather than the local microenvironment can mediate differences in stem cell activity.

Unlike most organ systems, mature cells of the lymphohematopoietic system are dispersed either in the circulation or in scattered anatomic sites such as the thymus, lymph, nodes, spleen, and, in the case of macrophages, in virtually all tissues of the body. The site of production of most of the mature cells, the bone marrow, is a complex tissue consisting of stromal elements, stem and progenitor cells, maturing cells of multiple lineages, and capillaries and sinusoids of the circulatory system into which mature cells are released. Spacial relationships between these different components are not well understood because of their apparently diffuse organization and because of the paucity of some of the critical elements, most notably the stem cells and early progenitors.

In contrast, the small intestinal epithelium has a much more straightforward organization that has expedited the understanding of some of the critical issues having to do with stem cell differentiation. Numerous fingerlike projections of the epithelium, called *villi*, extend into the intestinal lumen to effectively increase the surface area available for absorption. Each villus is covered with approximately 3500 epithelial cells, of which about 1400 are replaced on a daily basis [Potten & Loeffler, 1990]. Surrounding each villus are 6–10 crypts from which new epithelial cells are produced. They subsequently migrate to villi, and as senescent cells are shed from the villus tip, a steady progression of epithelial cells proceed unidirectionally to replace them. Crypts consist of only about 250 cells, including what is now estimated to be 16 stem cells. Since there are about 16 cells in the circumference of the crypt, stem cells occupy one circumferential ring of the crypt interior. This ring has been identified as the fourth from the bottom, directly above the Paneth cells. In addition, the fifth circumferential ring is occupied by direct progeny of stem cells which retain pluripotency and, in emergencies, may function as stem cells. Given the detailed quantitative information available regarding stem cell numbers and epithelial cell turnover rates, the number of stem cell doublings occurring during a human life span of 70 years has been estimated to be about 5000. Whether this demonstrates that tissue-specific stem cells are immortal is the topic of the following section.

116.4 Aging of Stem Cells

Given the zygote's enormous developmental potential and that ES cells represent cells of apparently equivalent potency that can be propagated as cell lines, it is reasonable to ask whether aging occurs at the cellular level. Put another way, can normal cells, other than ES cells, truly self-replicate without differentiation or senescence? Or do ES cells represent unique examples of cells capable of apparently indefinite self-renewal without differentiation? One of the definitions of hematopoietic stem cells alluded to above, is that they self-replicate. Without self-renewal, it might be argued, a stem cell population may be exhausted in a time-frame less than a lifetime of normal hematopoiesis or in far less time in the event of unusually high hematopoietic demands associated with disease or trauma. If, for example, hematopoietic stem cells can be propagated in vitro without differentiation, it could have tremendous impact on a number of clinically important procedures including bone marrow transplantation and gene therapy.

In classic experiments studying fibroblast growth in vitro, Hayflick [1965] observed that there were a finite number of divisions (about 50) that a cell was capable of before reaching senescence. It has been thought that totipotent and pluripotent stem cells may be exempt from this constraint. An analysis above of intestinal epithelial stem cells suggested that in a lifetime they undergo several thousand replications, apparently without any loss in developmental potential. However, several lines of evidence call into question the immortality of hematopoietic stem cells and point to at least a limited self-renewal capacity. For example, studies in which marrow was serially transplanted from primary recipients to secondary hosts, and so on, the number of effective passages is only about four to five [Siminovitch et al., 1964]. After the first transplant, normal numbers of relatively late progenitors are produced, but the number of repopulating stem cells is either diminished or the cells' developmental potential attenuated, or both, resulting in a requirement for successively larger numbers of transplanted cells to have achieve engraftment. Another interpretation of these results is that the transplantation procedure itself is responsible for the declining repopulating ability of marrow, rather than an intrinsic change in the self-renewal capacity of the stem cells. According to this argument, the repetitive dissociation of stem cells from their normal microenvironmental niches in the marrow, and the required reestablishment of those contacts during seeding and serial engraftment, irreversibly alter their self-renewal capacity [Harrison et al., 1978]. A mechanistic possibility for this scenario is that differentiation is favored when stem cells are not in contact with their stromal microenvironment. In this context, exposure to growth factors has an overwhelming differentiating influence on stem cells in suspension that is normally tempered by stromal associations.

Recently, an intriguing series of findings has emerged which may at least partially explain cellular aging. At the end of chromosomes there is a specialized region of DNA known as a telomere. This segment of DNA is comprised of hundreds of short six-nucleotide repeats of the sequence TTAGGG. It has been found that the length of telomeres varies over the life of a cell. Younger cells have longer telomeres, and as replication occurs the telomeres can be seen to shorten. It is thought that via the normal DNA replication process, the last 50-200 nucleotides of the chromosome fail to be synthesized, and thus telomeres are subject to loss with every cell division (reviewed in Blackburn [1992] and Greider [1991]). Consequently, telomeric shortening may act as a type of molecular clock. Once a cell has undergone a certain number of divisions, i.e., aged to a particular extent, it becomes susceptible to chromosome destabilization and subsequent cell death. Importantly, the rate at which telomeric sequence is lost may not be constant. Rather, some cells have the ability to regenerate their telomeres via an enzymatic activity known as telomerase. By controlling the loss of telomeric sequence, certain cell types may be able to extend their ability to replicate. Perhaps primitive tissue such as ES cells, when cultured with LIF, are able to express high levels of telomerase and thereby maintain their chromosomes indefinitely. Similarly, perhaps early hematopoietic stem cells express telomerase, and as differentiation occurs, the telomerase activity is downregulated. Although intriguing, these hypotheses are very preliminary, and much more basic research will be required to elucidate the mechanisms of the stem cell replication and aging.

In contrast to normal mechanisms of preserving replicative ability, a type of aberrant self-renewal is observed in the phenomenon of malignant transformation, or cancer. Some hematopoietic cancers are thought to originate with a defect at an early stem or progenitor cell level (e.g., chronic myelogenous leukemia). In this type of disease, normal differentiation is blocked, and a consequent buildup of immature, nonfunctional hematopoietic cells is observed. Malignant or neoplastic growth comes as a consequence of genetic damage or alteration. Such events range from single nucleotide changes to gross chromosomal deletions and translocations. Mechanistically, there appear to be two general types of mutation which cause cancer. One, activation of so-called oncogenes, is a dominant event and only needs to occur in one of a cell's two copies of the gene. Second, inactivation of tumor-suppressor genes removes the normal cellular control of growth and results in unchecked replication. These two categories of genetic alteration are analogous to stepping on a car's accelerator versus releasing the brake; both allow movement forward. Importantly, malignancy often comes as the result of a multistep process, whereby a series of genetic alterations occur. This process has been shown to involve different combinations of genes for different diseases [Vogelstein & Kinzler, 1993].

116.5 Other Types of Stem Cells

Other tissues may also have stem cell populations contributing to the replacement of effete mature cells. For example, in the liver only a very small faction ($2.5–5 \times 10^{-5}$) of hepatocytes is dividing at any given time, resulting in a complete turnover time of about 1 year [Sell, 1994]. This compares with the complete turnover of intestinal epithelia or granulocytes in a period of a few days. Nonetheless, growing evidence, some of which remains controversial, suggests that hepatic stem cells play a role in this tissue turnover. A moderate loss of liver tissue due to mild or moderate insult is probably replaced by the division of mature hepatocytes. However, a severe loss of hepatic tissue is thought to require the enlistment of the putative stem cell population, morphologically identified as oval cells.

Similarly, Noble's group in London has identified cells in the rat optic nerve that have the requisite functions of stem cells [Wren et al., 1992]. These cells, called *oligodendrocyte-type 2 astrocyte (O-2A) progenitors*, are capable of long-term self-renewal in vitro and give rise to oligodendrocytes through asymmetric divisions resulting in one new progenitor and one cell committed to differentiation. Conversion in vitro of O-2A progenitors into rapidly dividing and differentiating cells has been shown to be regulated extrinsically by platelet-derived growth factor (PDGF) and the basic form of fibroblast growth factor (bFGF) [Wolswijk & Noble, 1992]. Since these two growth factors are known to be produced in vivo after brain injury, a mechanism is suggested for generation of the large numbers of new oligodendrocytes required subsequent to trauma and demyelination.

116.6 Summary

- The zygote is the paradigm of a totipotent stem cell.
- ES cells derived from early embryos can be propagated indefinitely and maintain totipotency when cultured in the presence of LIF. Experimental control of differentiation and self-renewal in adult stem cells is being extensively investigated.
- Stem cells are defined by two characteristic traits: (1) Stem cells can self-renew, and (2) they can produce large numbers of differentiated progeny. Stem cells possess the intrinsic ability to manifest either trait; however, extrinsic factors mediate their developmental fate.
- Tissue-specific stem cells are pluripotent but not totipotent.
- Intestinal, epithelial, and hematopoietic tissues are classical self-renewing systems of the adult. In addition, recent studies have indicated the presence of stem cells in several other tissues (e.g., liver, nervous system).

- Although stem cells clearly have an extensive replication potential, it is not clear whether they are truly immortal. Stem cells may possess the ability to circumvent normal cellular processes that determine aging at the cellular level.
- Mutations at the DNA level can alter normal cellular control of stem cell replication and differentiation. This type of even can lead to aberrant development and subsequent malignancy.

Defining Terms

Commitment: The biologic process whereby a cell decides which of several possible developmental pathways to follow.

Differentiation: Expression of cell- or tissue-specific genes which results in the functional repertoire of a distinct cell type.

ES cells: Mouse stem cells originating from early embryonic tissue, capable of developing into any of the adult cell types.

Gene knockout: Deletion or alteration of a cellular gene using genetic engineering technology (generally performed on ES cells).

Hematopoietic: Blood forming.

Ontogeny: The process of development, generally referring to development from the zygote to adult stages.

Pluripotent: Capable of differentiation into multiple cell types.

Self-renew: Term describing cellular replication wherein no developmental commitment or differentiation takes place.

Terminally differentiated: The final stage of development in which all cell-specific features have been attained and cell division is no longer possible.

Totipotent: Capable of differentiation into all possible cell types.

References

Barnes DWH, Ford CE, Gray SM, et al. 1959. Progress in Nuclear Energy, series VI: Spontaneous and Induced Changes in Cell Populations in Heavily Irradiated Mice, London, Pergamon Press.

Becker AJ, McCulloch EA, Till JE. 1963. Cytological demonstration of the clonal nature of spleen colonies derived from transplanted mouse marrow cells. Nature 197:452.

Blackburn EH. 1992. Telomerases. Annu Rev Biochem 61:113.

Dick JE, Magil MC, Huszar D, et al. 1985. Introduction of a selectable gene into primitive stem cells capable of long-term reconstruction of the hemopoietic system of W/W^v mice. Cell 42:71.

Evans MJ, Kaufman MH. 1981. Establishment in culture of pluripotent cells from mouse embryos. Nature 292:154.

Greider CW. 1991. Telmeres. Curr Opin Cell Biol 3(3):444.

Harrison DE, Astle CM, Delaittre JA. 1978. Loss of proliferative capacity in immumohemopoietic stem cells caused by serial transplantation rather than aging. J Exp Med 147:1526.

Hayflick L. 1965. The limited in vitro lifetime of human diploid cell strains. Exp Cell Res 37:614.

Hilton DJ, Gough NM. 1991 Leukemia inhibitory factor: A biological perspective. J Cell Biochem 46(1):21.

Jordan CT, Lemischka IR. 1990. Clonal and systemic analysis of long-term hematopoiesis in the mouse. Genes Dev 4:220.

Kaufman MH, Robertson EJ, Handyside AH, et al. 1983. Establishment of pluripotential cell lines from haploid mouse embryos. J Embryol Exp Morphol 73:249.

Keller G, Paige C, Gilboa E, et al. 1985. Expression of a foreign gene in myeloid and lymphoid cells derived from multipotent hematapoietic precursors. Nature 318:149.

Koller BH, Smithies O. 1992. Altering genes in animals by gene targeting. Annu Rev Immunol 10:705.

Koller MR, Palsson BØ. 1993. Tissue engineering: Reconstitution of human hematopoiesis ex vivo. Biotechnol Bioeng 42:909.

Lemischka IR, Raulet DH, Mulligan RC. 1986. Developmental potential and dynamic behavior of hematopoietic stem cells. Cell 45:917.

Metcalf D, Moore MAS. 1971. Haemopoietic Cells, Amsterdam, Elsevier/North-Holland.

Mintz B. 1971. Methods in Mammalian Embryology, San Francisco, WH Freeman.

Potten CS, Loeffler M. 1990. Stem cells: Attributes, cycles, spirals, pitfalls and uncertainties. Lessons for and from the crypt. Develop 110:1001.

Robertson EJ. 1991. Using embryonic stem cells to introduce mutations into the mouse germ line. Biol Reprod 44(2):238.

Sell S. 1994. Liver stem cells. Mod Pathol 7(1):105.

Siminovitch L, Till JE, McCulloch EA. 1964. Decline in colony-forming ability of marrow cells subjected to serial transplantation into irradiated mice. J Cell Comp Physiol 64:23.

Smith AG, Nichols J. Robertson M, et al. 1992. Differentiation inhibiting activity (DIA/LIF) and mouse development. Dev Biol 151(2):339.

Till JE, McCulloch EA. 1961. A direct measurement of the radiation sensitivity of normal mouse bone marrow cells. Radiat Res 14:213.

Till JE, McCulloch EA, Siminovitch L. 1964. A stochastic model of stem cell proliferation based on the growth of spleen colony-forming cells. Proc Natl Acad Sci 51:29.

Van Zant G. Scott-Micus K, Thompson BP, et al. 1992. Stem cell quiescence/activation is reversible by serial transplantation and is independent of stromal cell genotype in mouse aggregation chimeras. Exp Heatol 20:470.

Vogelstein B, Kinzler KW. 1993. The multistep nature of cancer. Trends Genet 9(4):138.

Wolswijk G, Noble M. 1992. Cooperation between PDGF and FGF converts slowly dividing O-2A adult progenitors cells to rapidly dividing cells with characteristics of O-2A perinatal progenitor cells. J Cell Biol 118(4):889.

Wren D, Wolswijk G, Noble M. 1992. In vitro analysis of the origin and maintenance of O-2A adult progenitor cells. J Cell Biol 116(10):167.

117

Cell Motility and Tissue Architecture

Graham A. Dunn
King's College London

The characteristic architecture of a tissue results from an interplay of many cellular processes. In addition to the secretion of extracellular matrix, we may distinguish between processes related to the cell cycle—cell growth, division, differentiation, and death—and processes related to cell motility—cell translocation, directed motile responses, associated movements, and remodeling of the extracellular matrix. These processes are controlled and directed by cell-cell interactions, by cell-matrix interactions, and by cellular interactions with the fluid phase of the tissue. It is known that all three types of interactions can control both the speed and direction of cell translocation. This control results in the directed motile responses, which are the main subject of this chapter. Learning how to manipulate these motile responses experimentally will eventually become an essential aspect of tissue engineering.

Probably the greatest challenge to tissue engineering lies in understanding the complex dynamic systems that arise as a result of feedback loops in these motile interactions. Not only is cell translocation controlled by the fluid phase, by the matrix, and by other cells of the tissue, but cell motility can profoundly influence the fluid phase, remodel the matrix, and influence the position of other cells by active cell-cell responses or by associated movements. It is often forgotten that, especially in "undifferentiated" types of tissue cells such as fibroblasts, the function of the cell's motile apparatus is not only to haul the cell through the extracellular matrix of the tissue spaces but also to remodel this matrix and to change the positions of other cells mechanically by exerting tension on cell-matrix and cell-cell adhesions. These complex dynamic systems lie at the heart of pattern formation in the tissue, and indeed in the developing embryo, and understanding them will require not only a knowledge of the motile responses but also an understanding of the mechanism of cell motility itself.

In the study of cell motility, a great deal is now known about the relative dispositions of specific molecules that are thought to contribute to the motile process, and the dynamics of their interactions are beginning to be unraveled, yet there appears to have been comparatively little progress toward a satisfactory explanation of how a cell moves. There are some molecular biologists who still believe that it is just a question of time before the current molecular genetic thrust will alone come up with the answers. But there is a rapidly growing climate of opinion, already prevalent among physicists and

0-8493-0462-8/00/$0.00+$.50

engineers, that nonlinear dynamic processes such as cell motility have emergent properties that can never be completely understood solely from a knowledge of their molecular basis. Cell locomotion, like muscle contraction, is essentially a mechanical process, and a satisfactory explanation of how it works will inevitably require a study of its mechanical properties. Unlike muscle, the cellular motile apparatus is a continuously self-organizing system, and we also need to know the overall dynamics of its capacity for reorganization. These outstanding areas of ignorance are essentially problems in engineering. There is a nice analogy for this conceptual gap between the molecular and engineering aspects of biologic pattern formation in Harrison's new book on the kinetic theory of living pattern [Harrison, 1993]: "…one cannot supplant one end of a bridge by building the other. They are planted in different ground, and neither will ever occupy the place of the other. But ultimately, one has a bridge when they meet in the middle." This chapter is intended to encourage the building of that bridge.

117.1 Directed Motile Responses in Vivo

Cellular Interactions with the Fluid Phase

Cellular responses to the fluid phase of tissue spaces are thought to be mediated largely by specific diffusible molecules in the fluid phase. By far the most important directed response is chemotaxis: the unidirectional migration of cells in a concentration gradient of some chemoattractant or chemorepellent substance. Its study dates back to long before the advent of tissue culture, since it is a widespread response among the free-living unicellular organisms. The most widely studied system in vertebrates is the chemotaxis of neutrophil leukocytes in gradients of small peptides. In the case of tissue cells, it has long been conjectured, usually on the basis of surprisingly little evidence, that the direction of certain cell migrations also might be determined by chemotaxis, and it is recently becoming clear that chemotaxis can in fact occur in several vertebrate tissue cells in culture, particularly in gradients of growth factors and related substances. Yet whether it does occur naturally, and under what circumstances, is still an area of dispute. The concentration gradients themselves are usually speculated to arise by molecular diffusion from localized sources, although a nonuniform distribution of sinks could equally explain them, and they are only likely to arise naturally in conditions of very low convective flow. Besides controlling the migration of isolated cells, there is some evidence that gradients of chemoattractants also may control the direction of extension of organized groups of cells such as blood capillary sprouts, and of cellular processes, such as nerve axons. Although the mechanisms of these responses may be closely allied to those chemotaxis, they should more properly be classed as chemotropism, since the effect is to direct the extension of a process rather than the translocation of an isolated cell. The *influence on tissue architecture* of chemotaxis and chemotropism is possibly to determine the relative positions of different cell types and to determine the patterns of angiogenesis and innervation—though this is still largely conjectural. Chemotaxis is potentially a powerful pattern generator if the responding cells can modify and/or produce the gradient field.

Apart from chemotaxis, there also exists the possibility of mechanically mediated cellular responses to flow conditions in the fluid phase. The principal situation in vivo where this is likely to be important is in the blood vessels, and the mechanical effects of flow on the endothelial cells that line the vessel walls has been investigated.

Cellular Interactions with the Acellular Solid Phase

In a tissue, the cellular solid phase is the extracellular matrix, which is usually a fibrillar meshwork, though it may take the laminar form of a basement membrane. One directed response to the fibrillar extracellular matrix is the guidance of cell migrations, during embryonic development, for example, along oriented matrix fibrils. The discovery of this response followed soon after the dawn of tissue culture and is generally attributed to Paul Weiss, who named it contact guidance in the 1930s, though Loeb and Fleisher had already observed in 1917 that cells tend to follow the "path of least resistance" along oriented matrix fibrils (see [Dunn, 1982]). In culture, contact guidance constrains cell locomotion to be predominantly bidirectional, along the alignment axis of the matrix, whereas many embryonic migrations are predominantly unidirectional. This raises the question of whether contact guidance alone can account for these directed migrations in vivo or whether other responses are involved. The *influence on tissue architecture* of contact guidance is less conjectural than that of chemotaxis, since matrix alignment is more easily observed than chemical gradients, and the cells themselves are often coaligned with the matrix. Directed motion can be inferred since, in culture, this orientation of the cell shape on aligned surfaces is strongly correlated with an orientation of locomotion. In conjunction with cellular remodeling of the matrix, contact guidance becomes a potentially powerful generator of pattern. Since the cells, by exerting tension, can align the matrix, and the matrix alignment can guide the cells, a mutual interaction can arise whereby cells are guided into regions of higher cell density. Harris and colleagues [1984] have shown that such a feedback loop can result in the spontaneous generation of a regular array of cell clusters from a randomly distributed field of cells and matrix. The pattern of feather formation in birds' skin, for example, might arise by just such a mechanism.

Several other types of directed response may be mediated by the properties of the solid phase of the cell's environment. One is chemoaffinity, which Sperry [1963] proposed could account for the specific connections of the nervous system by guiding nerve fibres along tracks marked out by specific molecules adsorbed to the surface of substratum. A similar response has been proposed to account for the directional migration of primordial germ cells.

Cellular Interactions with Other Cells

Neighboring cells in the tissue environment may be considered as part of the solid phase. Thus it has been reported that when cells are used as a substratum for locomotion by other cells, directed responses such as contact guidance or chemoaffinity may occur, and these responses may persist even if the cells used as a substratum are killed by light fixation. However, the reason that cell-cell interactions are dealt with separately here is that cells show a directed response on colliding with other living cells that they do not show on colliding with cells that have been lightly fixed. Contact inhibition of locomotion, discovered by Abercrombie and Heaysman [1954], is a response whereby a normal tissue cell, on collision with another, is halted and eventually redirected in its locomotion. The effect of the response is to prevent cells from using other cells as a substratum, and a distinguishing feature of the response to living cells is a temporary local paralysis of the cell's active leading edge at the site of contact, which does not generally occur with a chemoaffinity type of response. The *influence on tissue architecture* of contact inhibition is probably profound but not easily determined. It is possibly the main response by which cells are kept more or less in place within a tissue, rather than milling around, and a failure of contact inhibition is thought to be responsible for the infiltration of a normal tissue by invasive malignant cells. Contact inhibition can cause cell locomotion to be directed away from centers of population and thus gives rise to the radial pattern of cell orientation that is commonly observed in the outgrowths from explant cultures. A major question is whether this motile response is related, mechanistically, to the so-called contact inhibition of growth, which is also known to fail in malignant cells but which is probably mediated by diffusion of some signal rather than by cell-cell contact.

117.2 Engineering Directed Motile Responses in Vitro

The investigation of the directed motile responses, using tissue culture, can reveal many aspects of the mechanisms that control and direct cell motility in vivo and also give a valuable insight into the mechanism of cell motility. In fact, most of what we know about these responses in vivo is deduced from experiments in culture. On the other hand, many of the responses discovered in culture may never occur naturally, and yet their study is equally important because it may yield further clues to the mechanism of cell motility and result in valuable techniques for tissue engineering. However, responses to properties of the culture environment, such as electric or magnetic fields, that are not yet known to have any counterparts in vivo will not be dealt with here.

A general experimental approach to engineering motile responses is first to try to reproduce in culture some response that appears to occur in vivo. The main reason is that the cell behavior can be observed in vitro, whereas this is possible in vivo only in very few instances. Another important reason is that once achieved in vitro, the response may be "dissected," by progressively simplifying the culture environment, until we isolate one or more well-defined properties that can each elicit a response. To be successful, this approach not only requires the design of culture environments that each isolate some specific environmental property and simultaneously allow the resulting cell behavior to be observed but also requires methods for adequately quantifying this resulting behavior.

When designing an artificial environment, it is important to consider its information content. Obviously, a uniform field of some scalar quantity is isotropic and cannot, therefore, elicit a directed motile response. A nondirected motile response, such as a change in speed caused by a change in some scalar property, is generally called a kinesis. Anisotropic uniform fields may be vector fields or may be able to distinguish opposite directions in the field and exhibit a unidirectional response, generally known as a taxis. Or cell movement perpendicular to the vector may predominate, which is sometimes known as a diataxis. In the case of a uniform field of some symmetric second-order tensor, such as strain or surface curvature, there is simply no information to distinguish opposite directions in the field, and yet orthogonal axes may be distinguished. This can give rise to a bidirectional response or, in three-dimensional fields, also to a response where translocation along one axis is suppressed. There is no generally agreed on name to cover all such possible responses, but the term guidance will be used here. Some examples of culture environments with specific physiocochemical properties are given below.

Environments with Specific Properties of the Fluid Phase

A Linear Concentration Gradient

The most common method of reproducing the chemotaxis response in vitro is to use a Boyden chamber in which a gradient of some specific chemical is formed by diffusion across a membrane filter. The resulting directed cell translocation is inferred from the relative number of cells that migrate through the pores of the filter from its upper to its lower surface. While this system is very useful for screening for potential chemoattractants, its usefulness for investigating the mechanism of the motile response is strictly limited, since it does not fulfill our two main criteria for an in vitro system. In the Boyden chamber, the properties of the environment are not well defined (since the gradient within the narrow, often tortuous pores of the filter is unpredictable), and the cell response cannot be observed directly. The Zigmond chamber was introduced to overcome these difficulties, by allowing the cell behavior to be observed directly, but the gradient is very unstable and cannot be maintained reliably for longer than an hour or two. Zicha and colleagues [1991] have recently developed a direct viewing chemotactic chamber with much greater stability and better optical properties. The chamber is constructed from glass and has an annular bridge separating concentric wells (Fig. 117.1). When covered by a coverslip carrying the cells,

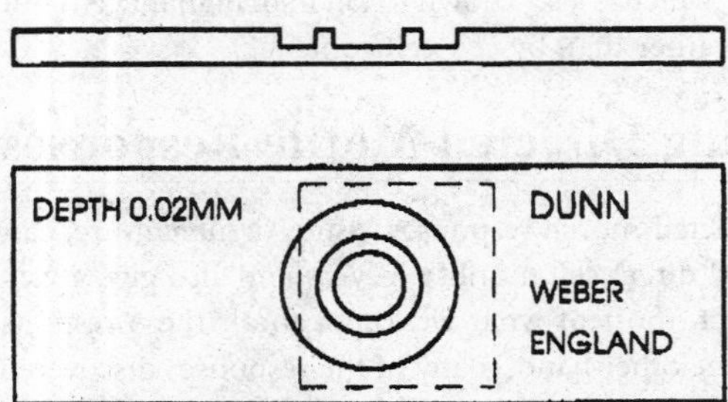

FIGURE 117.1 The Dunn Chemotaxis chamber.

a gap of precisely 20 μm is formed between coverslip and bridge in which the gradient develops. The blind central well confers the greater stability, which allows chemotactic gradients to be maintained for many hours and thus permits the chemotactic responses of slowly moving tissue cells and malignant cells to be studied for the first time. Weber Scientific International, Ltd., manufactures this as the Dunn chamber.

In use, both wells of the chamber are initially filled with control medium. The coverslip carrying the cells is then inverted over the wells, firmly seated, and sealed with wax in a slightly offset position (shown by the dashed lines) to allow access to the outer well. The outer well is then emptied using a syringe, refilled with medium containing the substance under test at known concentration, and the narrow opening is sealed with wax.

Assuming that diffusion is the only mechanism of mass transport in the 20-μm gap between coverslip and bridge, whereas convection currents keep the bulk contents of the two wells stirred, then the concentration in the 20-μm gap as a function of distance r from the center of the inner well is given by

$$C(r) = \frac{C_i \ln(b/r) + C_o \ln(r/a)}{\ln(b/a)} \tag{117.1}$$

where C_i and C_o are the concentrations in the inner and outer wells, respectively, and a and b are the inner and outer radii of the bridge. Because the bridge is annular, the gradient is slightly convex, but the deviation from linearity is very small.

Figure 117.2 shows the formation of the gradient during the first hour for a molecule with diffusion coefficient $D = 13.3 \times 10^{-5}$ mm²/s, such as a small globular protein of molecular weight 17 kD and chamber dimensions $a = 2.8$ mm, $b = 3.9$ mm. The equations describing gradient formation are given in Zicha et al. [1991], but it suffices here to show that the gradient is almost linear after 30 minutes. The flux from outer to inner well though the gap of height h (= 20 μm) is given by

$$\frac{dQ}{dt} = \frac{2\pi h D(C_o - C_i)}{\ln(b/a)} \tag{117.2}$$

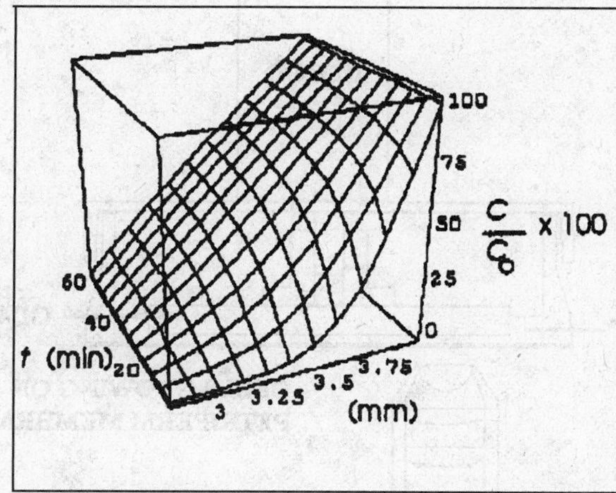

FIGURE 117.2 Formation of the gradient in the Dunn chemotaxis chamber (see text).

This flux tends to destroy the gradient, and the half-life of the gradient is equal to (ln 2)/*kD*, where *k* is a constant describing the geometric properties of the chamber. For a chamber with volumes v_o and v_i of the outer and inner wells, respectively, *k* is given by

$$k = \frac{2\pi h\left(v_i + v_o\right)}{\ln\left(b/a\right)v_i v_o} \tag{117.3}$$

Thus, for our small protein, a chamber with v_o = 30 μl, v_i = 14 μl, and other dimensions as before gives a gradient with a half-life of 33.6 hours. This is ample time to study the chemotaxis of slowly moving tissue cells, with typical speeds of around 1 μm per minute, as well as permitting the study of long-term chemotaxis in the more rapidly moving leukocytes with typical speeds around 10 μm per minute.

From the point of view of information content, a linear concentration gradient may be viewed as a nonuniform scalar field of concentration or as a uniform vector field of concentration gradient. Thus, if the cell can distinguish between absolute values of concentration, as well as being able to detect the gradient, then the chemotaxis response may be complicated by a superimposed **chemokinesis** response. A stable linear concentration gradient is a great advantage when trying to unravel such complex responses. Other sources of complexity are the possibilities that cells can modify the gradient by acting as significant sinks, can relay chemotactic signals by releasing pulses of chemoattractant, or can even generate chemotactic signals in response to nonchemotactic chemical stimulation. These possibilities offer endless opportunities for chaotic behavior and pattern generation.

A Gradient of Shear Flow

The flow conditions in the fluid phase of the environment can have at least two possible effects on cell motility. First, by affecting mass transport, they can change the distribution of molecules, and second, they can cause a mechanical shear stress to be exerted on the cells. Figure 117.3 shows a culture chamber designed by Dunn and Ireland [1985] that allows the behavior of cells to be observed under conditions of laminar shear flow. If necessary, the cells may be grown on the special membrane in Petriperm culture dishes (Heraeus), which allows gaseous exchange into the medium beneath the glass disk. Laminar shear

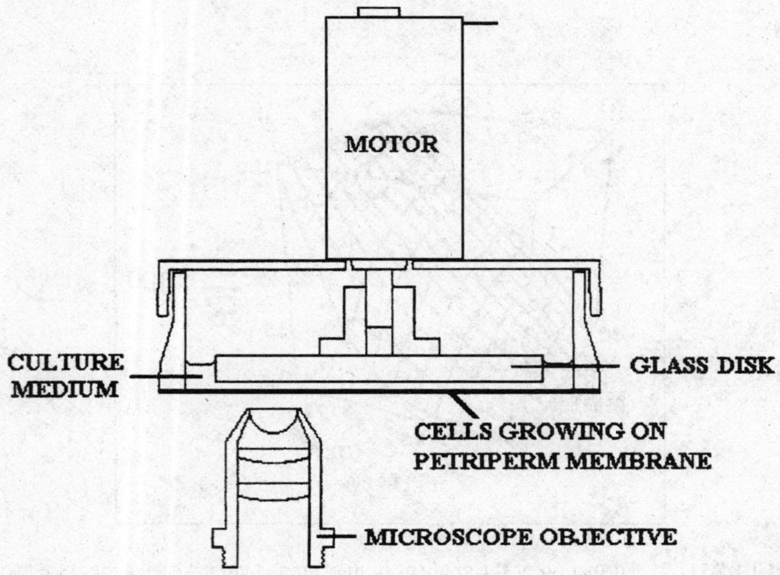

FIGURE 117.3 A culture chamber for laminar shear flow.

flow is probably the simplest and best defined flow regime, and the shear flow produced by an enclosed rotating disk has been described in detail [Daily and Nece, 1960]. Very low flow rates, caused by rotating the disk at around 1 rpm with a separation of about 1 mm between disk and cells, are useful for causing a defined distortion to diffusion gradients arising as a result of cell secretion or adsorption of specific molecules. Much information about, for example, chemoattractants produced by cells may be obtained in this way. Higher shear stresses, around 100 times greater, are known to affect the shape and locomotion of cells mechanically, and higher rotational speeds and/or smaller separations will be needed to achieve these.

Environments with Specific Properties of the Solid Phase

Aligned Fibrillar Matrices

The bidirectional guidance of cells by oriented fibrillar matrices is easily replicated in culture models. Plasma clots and hydrated collagen lattices are the most commonly used substrates, and alignment may be achieved by shear flow, mechanical stress, or strong magnetic fields during the gelling process. The magnitude and direction of orientation may be monitored, after suitable calibration, by measuring the birefringence in a polarizing microscope equipped with a Brace-Köhler compensator. These methods of alignment result in environments that are approximately described as uniform tensor fields, since mechanical strain is a familiar second-order tensor, and they generally give rise to bidirectional motile responses. In three dimensions, the strain tensor that results from applying tension along one axis can be described as a prolate ellipsoid, whereas applying compression along one axis results in an oblate ellipsoid. Thus we might distinguish between prolate guidance, in which bidirectional locomotion predominates along a single axis, and oblate guidance, in which locomotion is relatively suppressed along one of the three axes. But unidirectional information can be imposed on an oriented fibrillar matrix. Boocock [1989] achieved a unidirectional cellular response, called desmotaxis, by allowing the matrix fibrils to form attachments to the underlying solid substratum before aligning them using shear flow. This results in an asymmetrical linkage of the fibrils, and similar matrix configurations may well occur in vivo.

Fibrillar environments in culture are probably good models for the type of cell guidance that occurs in vivo, but they are so complex that it is difficult to determine which anisotropic physicochemical properties elicit the cellular response. Among the possibilities are morphologic properties (the anisotropic shape or texture of the matrix), chemical properties (an oriented pattern of adhesiveness or specific chemical affinity) and mechanical properties (an anisotropic viscoelasticity of the matrix). One approach to discovering which properties are dominant in determining the response is to try to modify each in turn while leaving the others unaffected. This is difficult, though some progress has been made. Another approach is to design simpler environments with better defined properties, as described in the sections that follow.

Specific Shapes and Textures

Ross Harrison, the inventor of tissue culture, placed spiders' webs in some cultures and reported in 1912 that "The behavior of cells … shows not only that the surface of a solid is a necessary condition but also that when the latter has a specific linear arrangement, as in the spider web, it has an action in influencing the direction of movement." It is hardly surprising to us today that a cell attached to a very fine fiber, in a fluid medium, is constrained to move bidirectionally. Nevertheless, such stereotropism or guidance by substratum availability is a guidance response and may have relevance to the problem of guidance by aligned fibrillar matrices. Moreover, in the hands of Paul Weiss, the guidance of cells on single cylindrical glass fibers was shown to have a more subtle aspect and it is now known that fibers up to 200 μm in diameter, which have a circumference approximately 10 times greater than the typical length of a fibroblast, will still constrain the locomotion to be parallel with the fiber axis. And so we may distinguish guidance by substratum shape, sometimes known as topographic or morphographic guidance, from guidance by substratum availability. The surface curvature of a cylinder is a uniform tensor field, and since opposite directions within the surface are always equivalent, the cellular response must be bidirectional.

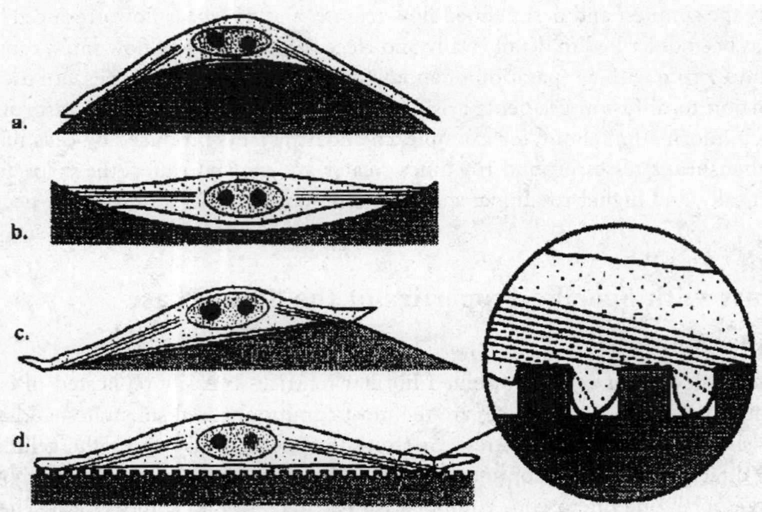

FIGURE 117.4 Diagrammatic cross-section of fibroblasts on substrata of various shapes.

Figure 117.4*a* is a diagrammatic cross section of a fibroblast attached to a convex cylindrical surface. These surfaces are easily made to any required radius of curvature by pulling glass rod in a flame. Dunn and Heath [1976] speculated that a cell must have some form of straightedge in order to detect slight curvatures of around 100 μm in radius. Obvious candidates were the actin cables that extend obliquely into the cytoplasm from sites of adhesion. These cables or stress fibers are continually formed as the fibroblast makes new adhesions to the substratum during locomotion and are known to contact and thereby to exert a tractive force on the substratum. The bundles of actin filaments are shown in the diagram as sets of parallel straight lines meeting the substratum at a tangent, and it is clear that the cables could not be much longer than this without being bent around the cylinder. Dunn and Heath proposed as an explanation of cell guidance along cylinders that the cables do not form in a bent condition and hence the traction exerted by the cell is reduced in directions of high convex curvature.

Further evidence for this hypothesis was found by observing cell behavior on substrata with other shapes. On concave cylindrical surfaces made by drawing glass tubing in a flame, the cells are not guided along the cylindrical axis but tend to become bipolar in shape and oriented perpendicular to the axis [Dunn, 1982]. This is to be expected, since, as shown in Fig. 117.4*b*, concave surfaces do not restrict formation of unbent cables but allow the cells to spread up the walls, which lifts the body of the cell clear of the substratum and thus prevents spreading along the cylinder axis. On substrata made with a sharp change in inclination like the pitched roof of a house, the hypothesis predicts that locomotion across the ridge is inhibited when the angle of inclination is greater than the angle at which the actin cables normally meet a plane substratum, as in Fig 117.4*c*. These substrata are more difficult to make than cylinders and require precision optical techniques for grinding and polishing a sharp and accurate ridge angle. Fibroblasts behaved as predicted on these substrata, the limiting angle being about 4 degrees, and high-voltage electron microscopy revealed the actin cables terminate precisely at the ridge on substrata with inclinations greater than this.

Substrata with fine parallel grooves have long been known to be very effective for eliciting morphographic guidance and are interesting because they can be a well-defined mimicry of some of the shape properties of an aligned fibrillar matrix while being mechanically rigid. Effective substrata can easily be made by simply scratching a glass surface with a very fine abrasive, but such substrata are not well defined and their lack of uniformity may give rise to variations in macroperties such as wettability, that cannot be ruled out as causing the guidance. Early attempts to make better-defined substrata used ruling engines such as those used to make diffraction gratings, but Dunn and Brown [1986] introduced electron beam lithography followed by ion milling to make grooves of rectangular cross section down to about 1 μm in width. Clark and colleagues [Clark et al., 1991] have now achieved rectangular grooves with spacings as low as 260 nm using the interference of two wavefronts, obtained by splitting an argon laser beam, to produce a pattern of parallel fringes on a quartz slide coated with photoresist. Groove depths a small as 100 nm can elicit a guidance response from certain cell types, and the main reason for pursuing this line of inquiry is now to discover the molecular mechanism responsible for this exquisite sensitivity.

Figure 117.4*c* shows a diagrammatic cross section of a fibroblast on a substratum consisting of a parallel array of rectangular grooves. One question that has been debated is whether the cells generally sink into the grooves, as shown here, or bridge across them. In the latter case, the wall and floor of the grooves are not an available substratum, and the cellular response might be a form of guidance by substratum availability. Ohara and Buck [1979] have suggested that this might occur, since the focal adhesions of fibroblasts are generally elongated in the direction of cell movement, and if they are forced to become oriented by being confined to the narrow spaces between grooves, this may force the locomotion into the same orientation. On the other hand, if the cells do generally sink into the grooves, then the Dunn and Heath hypothesis also could account for guidance even by very fine grooves, since individual actin filaments in the bundles, shown as dashed lines in the inset to the figure, would become bent and possibly disrupted if the cell made any attempt to pull on them other than in a direction parallel with the grooves. It is still therefore an unresolved issue whether different mechanisms operate in the cases of cylinders and grooves or whether a common mechanism is responsible for all cases of guidance by the shape of the substratum. Other groove profiles, particularly asymmetrical ones such as sawtooth profiles, will be needed for testing these and other rival hypotheses, and an intriguing possibility is that a unidirectional cell response might be achieved on microfabricated substrata with two orthogonal arrays of parallel sawtooth grooves, as shown in Fig. 117.5.

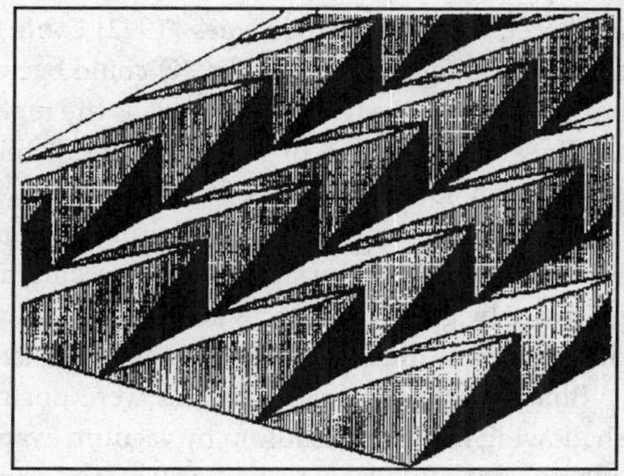

FIGURE 117.5 A proposed microfabricated substratum with two orthogonal arrays of parallel saw tooth grooves.

Specific Patterns of Adhesiveness

The equivalent in culture of the chemoaffinity response is **guidance by differential adhesiveness**, in which cell locomotion is confined to regions of higher adhesiveness patterned on the substrate. As with grooved surfaces, adhesive tracks that guide cells effectively are easily made by a variety of methods, including physically streaking nonadhesive viscous materials on an adhesive substratum or scratching through a nonadhesive film overlying an adhesive substratum. Again, however, these easily made surfaces are not well defined, and in particular, their anisotropic adhesiveness tends to be contaminated by anisotropic surface texture and sometimes by anisotropic mechanical properties. Carter [1965] was probably the first to describe a method of printing a well-defined pattern of adhesiveness onto a substratum; he used the vacuum evaporation of palladium, through a mask, onto a glass substratum made nonadhesive by first coating it with cellulose acetate. Clark and colleagues [1992] have now described a method for fabricating any required pattern of differential adhesiveness by using photolithography to obtain a hydrophobic pattern of methyl groups covalently coupled to a hydrophilic quartz substratum. The most recent developments in their laboratories are to use these patterns of hydrophobicity as templates for patterning specific proteins onto the substratum, and it seems that soon it will be possible to make almost any required pattern in any combination of proteins.

The explanation of the guidance of cells along tracks of higher adhesiveness appears to be obvious. In extreme cases, when the cells cannot adhere at all to the substrate outside the track, then the response is equivalent to guidance by substratum availability, and if the track happens to be sufficiently narrow, cell locomotion is restricted to the two directions along the track. But guidance along tracks of higher adhesiveness may still be very pronounced even when the cells can also adhere to and move on the regions of lower adhesiveness. The explanation in this case is that on encountering boundaries between regions of different adhesiveness, cells will cross them far more frequently in the direction from lower to higher adhesiveness. It is generally assumed that this results from a tug-of-war competition, since traction can be applied more effectively by the parts of the cell overlapping the region of higher adhesiveness. It is not known, however, how the traction fails in those parts of the cell which lose the competition, whether by breakage or slipping of the adhesions or by a relative failure of the contractile apparatus. Another possibility is simply that the cell spreads more easily over the more highly adhesive regions.

One reason for studying guidance by differential adhesiveness is to discover whether it can account for guidance by oriented extracellular matrices. An array of very narrow, parallel stripes of alternately high and low adhesiveness mimics the linear arrangement of substratum availability in an aligned matrix. Dunn [1982] found that if the repeat spacing is so small that a cell can span several stripes, there is no detectable cell orientation or directed locomotion even though an isolated adhesive stripe can strongly guide cells. Clark and colleagues [1992] confirmed this observation with one cell type (BHK) but found that cells of another type (MDCK) could become aligned even when spanning several stripes but would become progressively less elongated as the repeat spacing decreased. It is not yet clear, therefore, whether the linear arrangement of substratum availability in an aligned matrix might contribute to the guidance response in some cell types. It is clear from work of Clark and colleagues, however, that the adhesive stripes become less effective in eliciting guidance as their repeat spacing decreases, whereas the opposite is true for grooved surfaces. Thus it seems unlikely that substratum availability is the mechanism of guidance by grooved surfaces and, conversely, unlikely that adhesive stripes guide cells by influencing the orientation of the focal adhesions as suggested for grooved surfaces by Ohara and Buck [1979].

Binary patterns of adhesiveness were not the only ones studied by Carter [1965]. His technique of shadowing metallic palladium by vacuum evaporation onto cellulose acetate also could produce a graded adhesiveness. By placing a rod of 0.5 mm diameter on the substratum before shadowing, he found that the penumbral regions of the rod's shadow acted as steep gradients of adhesiveness that would cause cultured cells to move unidirectionally in the direction of increasing adhesiveness. This is therefore a taxis as distinct from a guidance response, and he named it haptotaxis. It is still not clear whether haptotaxis plays any role in vivo.

Specific Mechanical Properties

As yet, there has been no demonstration that anisotropic mechanical properties of the substratum can elicit directed motile responses. However, it is known that isotropic mechanical properties, such as the viscosity of the substratum, can influence cell locomotion[Harris, 1982], and it appears that changing the mechanical properties of aligned matrices can reduce cell guidance [Dunn, 1982; Haston et al., 1983], although it is probable that other properties are altered at the same time. Moreover, the phenomenon of desmotaxis suggest that it is the asymmetrical mechanical linkage of the fibrils that biases the locomotion. Guidance by anisotropic mechanical properties therefore remains a distinct possibility, but further progress is hampered by the difficulty of fabricating well-defined substrata. An ideal substratum would be a flat, featureless, and chemically uniform surface with anisotropic viscoelastic properties, and it is possible that liquid crystal surfaces will provide the answer.

Environments with Specific Arrangements of Neighboring Cells

Although contact inhibition of locomotion is of primary importance in determining patterns that develop in populations of cells, it is not easy to control the effects of contact inhibition in culture. If the cells are seeded on the substratum at nonuniform density, the response will generally cause cell locomotion to be biased in the direction of decreasing cell density. This can give a unidirectional bias if superimposed on a guidance response, and it has been conjectured that certain cell migrations in vivo are biased in this way. Cellular contact responses also can lead to the mutual orientation of confluent neighboring cells, and this can lead to wide regions of cooriented cells arising spontaneously in uniformly seeded cultures.

A typical culture arrangement from studying contact inhibition is to seed two dense populations of cells, often primary explants of tissue, about 1 mm apart on the substratum [Abercrombie and Heaysman, 1976]. Homologous contact inhibition causes the cells to migrate radially from these foci, usually as confluent sheets, until the two populations collide. With noninvasive cells, their locomotion is much reduced after the populations have met, and there is little intermixing of the population boundary. If one of the two populations is of an invasive type, however, failure of heterologous contact inhibition will cause it to infiltrate the other population, and their invasiveness can be measured by the depth of interpenetration.

Defining Terms

Associated movements: Occur when cells passively change position as a result of external forces, generated by cell motility elsewhere, that are transmitted either through the extracellular matrix or through cell-cell contacts.

Cell motility: A blanket term that covers all aspects of movement actively generated by a cell. It includes changes in cell shape, cell contraction, protrusion and retraction of processes, intracellular motility, and cell translocation.

Cell translocation, cell locomotion, or cell migration: All describe active changes in position of a cell in relation to its substratum. The translocation of tissue cells always requires a solid or semisolid substratum. In seeking a more rigorous definition, *positions* must first be defined for both the cell and its substratum. This is not always easy, since both may change continually in shape.

Chemoaffinity: The directional translocation of cells or extension of cellular protrusions along narrow tracks of specific molecules adsorbed to the substratum.

Chemokinesis: A *kinesis* (q.v.) in which the stimulating scalar property is the concentration of some chemical.

Chemotaxis: The directional translocation of cells in a concentration gradient of some chemoattractant or chemorepellent substance.

Chemotropism: The directional extension of a cellular protrusion or multicellular process in a concentration gradient of some chemoattractant or chemorepellent substance.

Contact guidance: The directional translocation of cells in response to some anisotropic property of the substratum.

Contact inhibition of locomotion: Occurs when a cell collides with another and is halted and/or redirected so that it does not use the other cell as a substratum.

Desmotaxis: Describes a unidirectional bias of cell translocation in a fibrillar matrix that is allowed to attach to a solid support and then oriented by shear flow.

Diataxis: A *taxis* (q.v.) in which translocation perpendicular to the field vector predominates. This leads to a bidirectional bias in two dimensions.

Directed motile responses: The responses of cells to specific properties of their environment that can control the direction of cell translocation or, in the case of nerve growth, for example, can control the direction of extension of a cellular protrusion.

Guidance: Used here to indicate a directed response to some high-order-tensor-like property of the environment in which opposite directions are equivalent. Translocation is biased bidirectionally in two dimensions.

Guidance by differential adhesiveness: A form of *guidance by substratum availability* (q.v.) in which cell locomotion is wholly or largely confined to narrow tracks of higher adhesiveness on the substratum. The response may be absent or much reduced if the cell spans several parallel tracks.

Guidance by substratum availability: Occurs when the translocation of a cell is confined to an isolated narrow track either because no alternative substratum exists or because the cell is unable to adhere to it.

Guidance by substratum shape, topographic guidance, or morphographic guidance: All refer to the guidance of cells by the shape or texture of the substratum. It is not known whether all types of morphographic guidance are due to a common mechanism.

Haptotaxis: The tendency of cells to translocate unidirectionally up a steep gradient of increasing adhesiveness of the substratum.

Heterologous contact inhibition: The *contact inhibition of locomotion* (q.v.) that may occur when a cell collides with another of different type. Contact inhibition is called nonreciprocal when the responses of the two participating cells are different. A cell type that is invasive with respect to another will generally fail to show contact inhibition in heterologous collisions.

Homologous contact inhibition: The *contact inhibition of locomotion* (q.v.) that may occur when a cell collides with another of the same type. It is not always appreciated that invasive cell types can show a high level of homologous contact inhibition.

Kinesis: The dependence of some parameter of locomotion, usually speed or rate of turning, on some scalar property of the environment. In an adaptive kinesis, the response is influenced by the rate of change of the scalar property, and if the environment is stable but spatially nonuniform, this can lead to behavior indistinguishable from a *taxis* (q.v.). Current nomenclature is inadequate to deal with such situations [Dunn, 1990].

Oblate guidance: A form of guidance in three dimensions in which translocation is suppressed along a single axis.

Prolate guidance: A form of guidance in three dimensions in which translocation along a single axis predominates.

Stereotropism: A form of *guidance by substratum availability* (q.v.) in which the only solid support available for locomotion consists of isolated narrow fibers.

Taxis: A directed response to some vectorlike property of the environment. Translocation is usually biased unidirectionally, either along the field vector or opposite to it, except in the case of *diataxis* (q.v.).

References

Abercrombie M, Heaysman JEM. 1954. Observations on the social behaviour of cells in tissue culture. Exp Cell Res 6:293.

Boocock CA. 1989. Unidirectional displacement of cells in fibrillar matrices. Development 107:881.

Carter SB. 1965. Principles of cell motility: The direction of cell movement and cancer invasion. Nature 208:1183.

Clark P, Connolly P, Curtis ASG, et al. 1991. Cell guidance by ultrafine topography in vitro. J Cell Sci 99:73.

Clark P, Connolly P, Moores GR. 1992. Cell guidance by micropatterned adhesiveness in vitro. J Cell Sci 103:287.

Daily JW, Nece RE. 1960. Chamber dimension effects on induced flow and frictional resistance of enclosed rotating disks. Trans AM Soc Mech Engrs D82:217.

Dunn GA. 1982. Contact guidance of cultured tissue cells: A survey of potentially relevant properties of the substratum. In R Bellairs, A Curtis, G Dun (eds), Cell Behaviour: A Tribute to Michael Abercrombie, pp 247–280. Cambridge, Cambridge University Press.

Dunn GA. 1990. Conceptual problems with kinesis and taxis. In JP Armitage, JM Lackie (eds), Biology of the Chemotactic Response, pp 1–13. Society for General Microbiology Symposium, 46. Cambridge, Cambridge University Press.

Dunn GA, Brown AF. 1986. Alignment of fibroblasts on grooved surfaces described by a simple geometric transformation. J Cell Sci 83:313.

Dunn GA, Heath JP. 1976. A new hypothesis of contact guidance in tissue cells. EXP Cell Res 101:1.

Dunn GA, Ireland GW. 1984. New evidence that growth in 3T3 cell cultures is a diffusion-limited process. Nature 312:63.

Harris AK. 1982. Traction and its relation to contraction in tissue cell locomotion. In R Bellairs, A Curtis, G Dunn (eds), Cell Behaviour: A Tribute to Michael Abercrombie, pp 109–134. Cambridge, Cambridge University Press.

Harris AK, Stopak D, Warner P. 1984. Generation of spatially periodic patterns by a mechanical instability: A mechanical alternative to the Turning model. J Emb Exp Morph 80:1.

Harrison LG. 1993. Kinetic Theory of Living Pattern, Cambridge, Cambridge University Press.

Haston WS, Shields JM, Wilkinson PC. 1983. The orientation of fibroblasts and neutrophils on elastic substrata. Exp Cell Res 146:117.

Ohara PT, Buck RC. 1979. Contact guidance in vitro: A light, transmission and scanning electron microscopic study. Exp Cell Res 121:235.

Sperry RW. 1963. Chemoaffinity in the orderly growth of nerve fiber patterns and connections. Proc Natl Acad Sci USA 50:703.

Zicha D, Dunn GA, Brown AF. 1991. A new direct-viewing chemotaxis chamber. J Cell Sci 99:769.

118

Tissue Microenvironments

Michael W. Long
University of Michigan

Tissue development is regulated by a complex set of events in which cells of the developing organ interact with each other, with general and specific growth factors, and with the surrounding extracellular matrix (ECM) [Long, 1992]. These interactions are important for a variety of reasons such as localizing cells within the microenvironment, directing cellular migration, and initiating growth-factor-mediated developmental programs. It should be realized, however, that simple interactions such as those between cells and growth factor are not the sole means by which developing cells are regulated. Further complexity occurs via interactions of cells and growth factors with extracellular matrix or via other interactions which generate specific developmental responses.

Developing tissue cells interact with a wide variety of regulators during their ontogeny. Each of these interactions is mediated by defined, specific receptor-ligand interactions necessary to stimulate both the cell proliferation and/or motility. For example, both chemical and/or extracellular matrix gradients exist which signal the cell to move along "tracks" of molecules into a defined tissue area. As well, high concentrations of the attractant, or other signals, next serve to "localize" the cell, thus stopping its nonrandom walk. These signals which stop and/or regionalize cells in appropriate microenvironments are seemingly complex. For example, in the hematopoietic system, complexes of cytokines and extracellular matrix molecules serve to localize progenitor cells [Long et al., 1992], and similar mechanisms of cell/matrix/cytokine interactions undoubtedly exist in other developing systems. Thus, the regulation of cell development, which ultimately leads to tissue formation is a complex process in which a number of elements work in cohort to bring about coordinated organogenesis: stromal and parenchymal cells, growth factors and extracellular matrix. Each of these is a key component of a localized and highly organized microenvironmental regulatory system.

Cellular interactions can be divided into three classes: cell-cell, cell-extracellular matrix, and cell-growth factor, each of which is functionally significant for both mature and developing cells. For example, in a number of instances blood cells interact with each other and/or with cells in other tissues. Immunologic cell-cell interactions occur when lymphocytes interact with antigen-presenting cells, whereas neutrophil or lymphocyte egress from the vasculature exemplifies blood cell-endothelial cell recognition. Interactions between cells and the extracellular matrix (the complex prontinaceous substance surrounding

0-8493-0462-8/00/$0.00+$.50

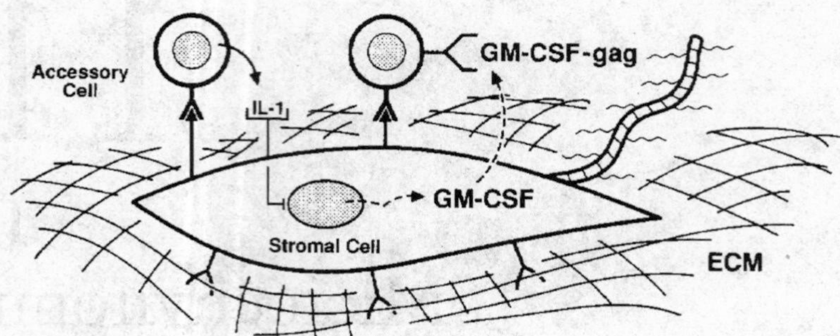

FIGURE 118.1 Hematopoietic cellular interactions. This figure illustrates the varying complexities of putative hematopoietic cell interactions. A conjectural complex is shown in which accessory cell–stromal cell, and stromal cell–PG-growth factor complexes localize developmental signals. ECM = extracellular matrix, gag = glycosaminoglycan side chain bound to proteoglycan (PG) core protein (indicated by cross-hatched curved molecule); IL-1 = interleukin-1; GM-CSF = granulocyte-macrophage colony-stimulating factor. Modified from Long [1992] and reprinted with permission.

cells) also play an important role. During embryogenesis matrix molecules are involved both in cell migration and in stimulating cell function. Matrix components are also important in the growth and development of precursor cells; they also serve either as cytoadhesion molecules for these cells or to compartmentalize growth factors within specific microenvironmental locales. For certain tissues such as bone marrow, a large amount of information exists concerning the various components of the nature of the microenvironment. For others such as bone, much remains to be learned of the functional microenvironment components. Many experimental designs have examined simple interactions (e.g., cell-cell, cell-matrix). However, the situation in vivo is undoubtedly much more complex. For example, growth factors are often bound to matrix molecules which, in turn, are expressed on the surface of underlying stromal cells. Thus, very complex interactions occur (e.g., accessory cell–stromal cell–growth factor–progenitor cell–matrix, see Fig. 118.1), and these can be further complicated by a developmental requirement for multiple growth factors.

The multiplicity of tissue-cell interactions requires highly specialized cell surface structures (i.e., receptors) to both mediate cell adhesion and transmit intracellular signals from other cells, growth factors, and/or the ECM. Basically, two types of receptor structures exist. Most cell surface receptors are proteins which consist of an extracellular ligand-binding domain, a hydrophobic membrane-spanning region, and a cytoplasmic region which usually functions in signal transduction. The amino acid sequence of these receptors often defines various families of receptors (e.g., immuno-globulin and integrin gene superfamilies). However, some receptors are not linked to the cell surface by protein, as certain receptors contain phosphotidylinositol-based membrane linkages. This type of receptor is usually associated with signal transduction events mediated by phospholipase C activation [Springer, 1990]. Other cell surface molecules important in receptor-ligand interactions are surface proteins which function as a coreceptors. Coreceptors function with a well-defined receptor, usually to amplify stimulus-response coupling.

The goal of this chapter is to examine the common features of each component of the microenvironment (cellular elements, soluble growth factors, and extracellular matrix). As each tissue or organ undergoes its own unique and complex developmental program, this review cannot cover these elements for all organs and tissue types. Rather, two types of microenvironments (blood and bone) will be compared in order to illustrate commonalties and distinctions.

118.1 Cellular Elements

Cells develop in a distinct hierarchical fashion. During the ontogeny of any organ, cells migrate to the appropriate region for the nascent tissue to form and there undergo a phase of rapid proliferation and differentiation. In tissues which retain their proliferative capacity (bone marrow, liver, skin, the gastrointestinal lining, and bone), the complex hierarchy of proliferation cells is retained throughout life. This is best illustrated in the blood-forming (hematopoiesis) system. Blood cells are constantly produced, such that approximately 70-times an adult human's body weight of blood cells is produced through the human life span. This implies the existence of a very primitive cell type that retains the capacity for self-renewal. This cell is called a *stem cell*, and it is the cell responsible for the engraphment of hematopoiesis in recipients of bone marrow transplantation. Besides a high proliferative potential, the stem cell also is characterized by its multipotentiality in that it can generate progeny (referred to as *progenitor cells)* which are committed to each of the eight blood cell lineages. As hematopoietic cells proliferate, they progressively lose their proliferative capacity and become increasingly restricted in lineage potential. As a result, the more primitive progenitor cells in each lineage produce higher colony numbers, and the earliest cells detectable in vitro produce progeny of 2–3 lineages (there is no in vitro assay for transplantable stem cell*s)*. Similar stem cell hierarchies exist for skin and other regenerating tissues, but fewer data exist concerning their hierarchical nature.

The regulation of bone cell development is induced during bone morphogenesis by an accumulation of extracellular and intracellular signals [Urist et al., 1983*a*]. Like other systems, extracellular signals are known to be transferred from both cytokines and extracellular matrix molecules [Urist et al., 1983*a*] to responding cell surface receptors, resulting in eventual bone formation. The formation of bone occurs by two mechanisms. Direct development of bone from mesenchymal cells (referred to as *intramembranous ossification*, as observed in skull formation) occurs when mesenchymal cells directly differentiate into bone tissue. The second type of bone formation (the endochondrial bone formation of skeletal bone) occurs via an intervening cartilage model. Thus, the familiar cell hierarchy exists in the development and growth of long bones, beginning with the proliferation of mesenchymal stem cells, their differentiation into ostogenic progenitor cells, and then into osteoblasts. The osteoblasts eventually calcify their surrounding cartilage and/or bone matrix to form bone. Interestingly, the number of osteoporgenitor cells in adult bone seems too small to replace all the large mass of bone normally remodeled in the process of aging of the skeleton [Urist et al., 1983*a*]. Observations from this laboratory confirm this concept by showing that one (unexpected) source of bone osteoprogenitor cells is the bone marrow [Long et al., 1990; Long & Mann, 1993]. This reduced progenitor cell number also implies that there is a disassociation of bone progenitor cell recruitment from subsequent osteogenic activation and bone deposition and further suggests multiple levels of regulation in this process (*vide infra*).

As mentioned, cell-cell interactions mediate both cellular development and stimulus-response coupling. When coupled with other interactions (e.g., cell-ECM), such systems represent a powerful mechanism for directing and/or localizing developmental regulation. Further, combinations of these interactions potentially can yield lineage-specific or organ-specific information. Much of out understanding of cell-cell interactions comes from the immune system and from the study of developing blood cells and their interactions with adjacent stromal cells [Dexter, 1982; Gallatin et al., 1986; Springer, 1990]. For example, the isolation and cloning of immune cell ligands and receptors resulted in the classification of gene families which mediate cell-cell interactions within the immune and hematopoietic systems, and similar systems undoubtedly play a role in the development of many tissues.

There are three families of molecules which mediate cell-cell interactions (Table 114.1). The immunoglobulin superfamily is expressed predominantly on cells mediating immune and inflammatory responses (and is discussed only briefly here). The integrin family is a large group of highly versatile proteins which is involved in cell-cell and cell-matrix attachment. Finally, the selectin family is comprised of molecules which are involved in lymphocyte, platelet, and leukocyte interactions with endothelial cells. Interestingly, this class of cell surface molecules utilizes specific glycoconjungates (encoded by glycoslytransferase genes) as their ligands on endothelial cell surfaces.

Immunoglobulin Gene Superfamily

These molecules function in both antigen recognition and cell-cell communication. The immunglobulin superfamily (Table 118.2) is defined by a 90–100 base pair immunoglobulinlike domain found within a

TABLE 118.1 Cell Adhesion Molecule Superfamilies

Immunoglobulin superfamily of adhesion receptors	
LFA 2 (CD2)	T-Cell receptor (CD3)
LFA 3 (CD58)	CD4 (TCR coreceptor)
ICAM 1 (CD54)	CD8 (TCR coreceptor)
ICAM 2	MHC class I
VCAM-1	MHC class II
Integrins	
β_1 integrins (VLA proteins)	
P150,95 ($CD11_c$/CD18)	
VLA 1–3,6	
VLA 4(LPAM 1, CO49d/CO29)	
Fibronectin receptor (VLA 5, CD-/CD29)	
LPAM 2	
β_2 Integrins	
LFA 1 ($CD11_a$/ CD18)	
Mac 1 or Mo 1 ($CD11_b$/CD18)	
β_3 Integrins	
Vitronectin receptor	
Platelet gp-IIb/IIa	
Selectin/LEC-CAMS	
Mel14 (LE-CAM-1, LHR, LAM-1, Leu 8, Ly 22, gp90 MEL)	
ELAM-1 (LE-CAM-2)	
GMP 140 (LE-CAM 3, PADGEM, CD 62)	

Source: Originally adapted from Springer [1990] and Brandley and coworkers [1990] and reprinted from Long [1992] with permission.

dimer of two antiparallel β strands [Sheetz et al., 1989; Williams & Barclay, 1988]; for a review, see Springer [1990]. Two members of this family, the T-cell receptor and immunoglobulin, function in antigen recognition. The T-cell receptor recognizes antigenic peptides in the context of two other molecules on the surface of antigen-presenting cells: major histocompatibility (MHC) class I and class II molecules [Bierer et al., 1989; Sheetz et al., 1989; Springer, 1990]. Whereas the binding of T-cell receptor to MHC/antigenic peptide complexes seems sufficient for cell-cell adhesion, cellular activation also requires the binding of either of two coreceptors, CD8 or CD4. Neither coreceptor can directly bind the MHC complex, but, rather, each seems to interact with the T-cell receptors to synergistically amplify intercellular signaling [Shaw et al., 1986; Spits et al., 1986; Springer, 1990].

Integrin Gene Superfamily

Integrin family members are involved in interactions between cells and extracellular matrix proteins [Giancotti & Ruoslahti, 1990]. Cell attachment to these molecules occurs rapidly (within minutes) and is a result of increased avidity rather than increased expression (see Lawrence and Springer [1991] and references therein). The binding sequence within the ligand for most, but not all, integrins is the tripeptide sequence arganine-glycine-asparagine (RGD) [Ruoslahti & Pierschbacher, 1987]. Structurally, integrins consist of two membrane-spanning alpha and beta chains. The alpha subunits contain three to four tandem repeats of a divalent-ion-binding motif and require magnesium or calcium to function. The alpha chains are (usually) distinct and bind with common or related β subunits to yield functional receptors [Giancotti & Ruoslahti, 1990]. The β subunits of integrins have functional significance, and integrins can be subclassified based on the presence of a given beta chain. Thus, integrins containing β1 and β3 chains are involved predominantly in cell-extracellular matrix interactions, whereas molecules containing the β2 subunits function in leukocyte-leukocyte adhesion (Tables 118.1 and 118.2). The cytoplasmic domain of many integrin receptors interacts with the cytoskeleton. For example, several

TABLE 118.2 Cell Surface Molecules Mediating Cell-Cell Interactions

Cell Receptor	Receptor: Cell Expression	Ligand, Co-, or Counter-Receptor	Ligand Co- or Counter-Receptor Cell Expression	References
Ig superfamily				
MHC I	Macroph, T cell	CD8,TCR	T cells	Springer 1990
MHC II	Macroph, T cell	CD4, TCR	T cells	Springer 1990
ICAM-1	Endo, neut. HPC, B cells, T cells, macroph*	LFA-1	Mono, T and B cells	Springer 1990
ICAM-2	Endo	LFA-1	Mono, T and B cells	Springer 1990
LFA-2	T cells	LFA-3	T cells, eryth	Springer 1990
Integrins				
Mac1	Macroph, neut	Fibrinogen, C3bi	Endo, plts	Springer 1990
LFA-1	Macroph, neut	(See above)	(See above)	
VCAM	Endo	VLA4	Lymphocytes, monocytes B cells	Brandley et al., 1990 Miyamake, 1990
gp150,95	Macro, neut	(See above)	(See above)	
FN-R	Eryth lineage	Fibronectin	N.A.	see Table 118.3
IIb/IIIa	Plts, mk	Fibrinogen, TSP, VN vWF	Endo	Springer, 1990
Selectins				
LEC-CAM-1 (Mel 14)	Endo	Addressins, neg. charged oligosaccrides	Lymphocytes	Brandley et al., 1990
ELAM-1 (LE-CAM-2)	Endo	sialyl-Lewis X†	Endo‡ Neut, tumor cells	Lowe et al., 1990
LEC-CAM-3 (GMP-140)	Plt gran Weible-Palade bodies, endo	Lewis X (CD15)	Endo Neut	Brandley et al., 1990

**Source*: Modified from Long [1992] and reprinted with permission. macroph = macrophage; Mono = monocyte; Endo = endothelial cell; Eryth = erythroid cells; Plts = platelets; Neut = neutrophil; Mk = megakarocyte.

†Constituatively expressed by few cells, upregulated by TNF- and IL-1.

‡Sialylated, fucosylated lactosaminoglycans. Modified from Long [1992] and reprinted with permission.

integrins are known to localize near focal cell contacts were actin bundles terminate [Giancotti & Ruoslahti, 1990; Springer, 1990]. As a result, changes in receptor binding offer an important mechanism for linking cell adhesion to cytoskeletal organization.

Selectins

The selectin family of cell-adhesion receptors contains a single N-terminus, calcium-dependent, lectin-binding domain, an EGF receptor (EFGR) domain, and a region of cysteine-rich tandem repeats (from two to seven) which are homologous to complement-binding proteins [Bevilacqua et al., 1989; Springer, 1990; Stoolman, 1989]. Selectins (e.g., MEL14, gp90MEL, ELAM-1, and GMP140/PADGEM, Table 118.2) are expressed on neutrophils and lymphocytes. They recognize specific glycoconjugate ligands on endothelial and other cell surfaces. Early studies demonstrated that fucose or mannose could block lymphocyte attachment to lymph node endothelial cells [Brandley et al., 1990]. Therefore, the observation that the selectin contain a lectin-binding domain [Bevilacqua et al., 1989] led to the identification of the ligands for two members of this family; for review see Brandley and coworkers [1990]. Lowe and coworkers first demonstrated that alpha (1,3/1,4) fucosyltransferase cDNA converted nonmyeloid COS or CHO cells to selectin (sialyl-Lewis X) positive cells which bound to both HL60 cells and neutrophils in an ELAM-1-dependent manner [Lowe et al., 1990]. Conversely, Goelz and coworkers screened an expression library using a monoclonal antibody which inhibited ELAM-mediated attachment which yielded a novel alpha (1,3) fucosyltransferase whose expression conferred ELAM binding activity on target cells [Goelz et al., 1990].

Unlike mature neutrophils and lymphocytes, information on cell-cell interactions among hematopoietic progenitor cells is less well developed (see Table 118.3). Data concerning the cytoadhesive capacities

TABLE 118.3 Hematopoietic Cell–Stromal Cell Interactions (Unknown Receptor-Ligand)

Cell Phenotype	Stromal Cell	References
B cells	Fibroblasts	Ryan et al., 1990; Witte et al., 1987
Pre-B cell	Heter stroma	Palacios et al., 1989
BFC-E	Fibroblasts	Tsai et al., 1986, Tsai et al., 1987
CFC-S	Fibroblasts (NIH 3T3)	Roberts et al., 1987
CFC-S, B1-CFC	Heter stroma*	Gordon et al., 1985, 1990*a*, 1990*b*
CFC-GM	Heter stroma	Tsai et al., 1987; Campbell et al., 1985
CFC-Mk	Heter stroma	Tsai et al., 1987; Campbell et al., 1985

*Methylprednisolone stimulated stromal cells, unstimulated fail to bind—see Gordon and coworkers [1985]. Heter = heterologous; BFC-E = burst forming cell-erythroid; CFC = colony-forming cell; S = spleen; B1 = blast; GM = granulocyte/macrophage; Mk = megakaryocyte. From Long [1992], reprinted with permission.

of hematopoietic progenitor cells deal with the interaction of these cells with underlying, preestablished stromal cell layers [Dexter, 1982]. Gordon and colleagues documented that primitive hematopoietic human blast-colony forming cells (Bl-CFC) adhere to performed stromal cell layers [Gordon et al., 1985, 1990*b*] and showed that the stromal cell ligand is not one of the known cell adhesion molecules [Gordon et al., 1987*a*]. Other investigators have shown that hematopoietic (CD34-selected) marrow cell populations attach to stromal cell layers and that the attached cells are enriched for granulocyte-macrophage progenitor cells [Liesveld et al., 1989]. Highly enriched murine spleen colony-forming cells (CFC-S) attach to stromal cell layers, proliferate, and differentiate into hematopoietic cells [Spooncer et al., 1985]. Interestingly, underlying bone marrow stromal cells can be substituted for by NIH 3T3 cells [Roberts et al., 1987], suggesting that these adherent cells supply the necessary attachment ligand for CFU-S attachment [Roberts et al., 1987; Yamazaki et al., 1989].

118.2 Soluble Growth Factors

Soluble specific growth factors are an obligate requirement for the proliferation and differentiation of developing cells. These growth factors differ in effects from the endocrine hormones such as anabolic steroids or growth hormone. Whereas the endocrine hormones affect general cell function and are required and/or important to tissue formation, their predominant role is one of homeostasis. Growth *factors*, however, specifically drive the developmental programs of differentiating cells. Whether these function in a permissive or an inductive capacity has been the subject of considerable past controversy, particularly with respect to blood cell development. However, the large amount of data demonstrating linkages between receptor-ligand interaction and gene activation argues persuasively for an inductive/direct action on gene expression and, hence, cell proliferation and differentiation.

Again, a large body of knowledge concerning growth factors comes from the field of hematopoiesis (blood cell development). Hematopoietic cell proliferation and differentiation is regulated by numerous growth factors; for reviews see Metcalf [1989] and Arai and colleagues [1990]. Within the last decade approximately 29 stimulatory cytokines (13 interleukins, M-CSF, erythroprotein, G-CFS, GM-CSF, c-kit ligand, gamma-interferon, and thrombopoietin) have been molecularly cloned and examined for their function in hematopoiesis. Clearly, this literature is beyond the scope of this review. However, the recent genetic cloning of eight receptors for these cytokines has led to the observation that a number of these receptors have amino acid homologies [Arai et al., 1990], showing that they are members of one or more gene families (Table 118.4). Hematopoietic growth factor receptors structurally contain a large extracellular domain, a transmembrane region, and a sequence-specific cytoplasmic domain [Arai et al., 1990]. The extracellular domains of interleukin-1, interleukin-6, and gamma-interferon are homologous with the immunoglobulin gene superfamily, and weak but significant amino-acid homologies exist among the interleukin-2 (beta chain), IL-6, IL-3, IL-4, erythropoietin, and GM-CSF receptors [Arai et al., 1990].

TABLE 118.4 Hematopoietic Growth Factor Receptor Families

Receptors with homology to the immunoglobulin gene family
Interleukin-1 receptor
Interleukin-6 receptor
Gamma-interferon receptor
Hematopoietic growth factor receptor family
Interleukin-2 receptor (β-chain)
Interleukin-3 receptor
Interleukin-4 receptor
Interleukin-6 receptor
Erythropoietin receptor
G/M-CSF receptor

Source: Modified from Long [1992]. Reprinted with permission.

Like other developing tissues, bone responds to bone-specific and other soluble growth factors. TGF-β is a member of a family of polypeptide growth regulators which affects cell growth and differentiation during developmental processes such as embryogenesis and tissue repair [Sporn & Roberts, 1985]. TGF-β strongly inhibits proliferation of normal and tumor-derived epithelial cells and blocks adipogenesis, myogenesis, and hematopoiesis [Sporn & Roberts, 1985]. However, in bone, TGF-β is a positive regulator. TGF-β is localized in active centers of bone differentiation (cartilage canals and osteocytes) [Massague, 1987], and TGF-β is found in high quantity in bone, suggesting that bone contains the greatest total amount of TGF-β [Gehron Robey et al., 1987; Massague, 1987]. During bone formation, TGF-β promotes chrondrogenesis [Massague, 1987]—an effect presumably related to its ability to stimulate the deposition of extracellular matrix (ECM) components [Ignotz & Massague, 1986]. Besides stimulating cartilage formation, TGF-β is synthesized and secreted in bone cell cultures and stimulates the growth of subconfluent layers of fetal bovine bone cells, thus showing it to be an autocrine regulator of bone cell development [Sporn & Roberts, 1985].

In addition to TGF-β, other growth factors or cytokines are implicated in bone development. Urist and coworkers have been able to isolate various regulatory proteins which function in both in vivo and in vitro models [Urist et al., 1983*b*]. Bone morphogenic protein (BMP), originally an extract of demineralized human bone matrix, has now been cloned [Wozney et al., 1988] and, when implanted in vivo, results in a sequence of events leading to functional bone formation [Muthukumaran & Reddi, 1985; Wozney et al., 1988]. The implanting of BMP is followed by mesenchymal cell migration to the area of the implant, differentiation into bone progenitor cells, deposition of new bone, and subsequent bone remodeling to allow the establishment of bone marrow [Muthukumaran & Reddi, 1985]. A number of additional growth factors which regulate bone development exist. In particular, bone-derived growth factors (BDGF) stimulate bone cells to proliferate in serum-free media. [Hanamura et al., 1980; Linkhart et al., 1986]. However, these factors seem to function at a different level from BMP [Urist et al., 1983*a*].

118.3 Extracellular Matrix

The extracellular matrix (ECM) varies in its tissue composition throughout the body and consists of various molecules such as laminin, collagens, proteoglycans, and other glycoproteins [Wicha et al., 1982]. Gospodarowicz and coworkers demonstrated that ECM components greatly affect corneal endothelial cell proliferation in vitro [Gospodarowicz et al., 1980; Gospodarowicz & Ill, 1980]. Studies by Reh and coworkers indicate that the ECM protein laminin is involved in inductive interactions which give rise to retinal-pigmented epithelium [Reh & Gretton, 1987]. Likewise, differentiation of mammary epithelial cells is profoundly influenced by ECM components; mammary cell growth in vivo and in vitro requires type IV collagen [Wicha et al., 1982]. A number of investigations elucidated a role for ECM and its components in hematopoietic cell function. These studies have identified the function of both previously known and newly identified ECM components in hematopoietic cell cytoadhesion (Table 118.5).

TABLE 118.5 Protiens and Glycoprotiens Mediating Hematopoietic Cell–Extracellular Matrix Interactions

Matrix Component	Cell Surface Receptor	Cellular Expression	References
Fibronectin	FnR	Erythroid; BFC-E; B cells; Lymphoid cells; HL60 cells	Patel, 1984, 1986, 1987, Patel et al., 1985; Ryan et al., 1990; Tsai et al., 1987; Van de Water et al., 1988
	IIb/IIIa	Platelets and megakaryocytes	Giancotti et al., 1987
Thrombospondin	TSP-R	Monocytes and platelets	Silverstein and Nachman, 1987; Leung, 1984
		Human CFC,	Long and Dixit, 1990
		CFC-GEMM	Long and Dixit, 1990
Hyaluronic acid	CD44	T and B cells	Aruffo et al., 1990, Dorshkind, 1989; Horst et al., 1990; Miyake et al., 1990
		Neutrophils Tumor cells	
Hemonectin	Unk	CFC-GM, BFC-E Immat. neutr. BFU-E	Campbell et al., 1985, 1987, 1990
Proteoglycans:			
Heparan sulfate	Unk	B1-CFC	Gordon, 1988; Gordon et al., 1988
Unfract ECM	Unk	B1-CFC, bm Stroma	Gordon et al., 1988; Campbell et al., 1985

R = receptor; BFC-E and erythroid progenitor cell = the burst-forming cell-erythrocyte; HL60 = a promyelocytic leukemia cell line; CFC = colony-forming cell; GEMM = granulocyte erythrocyte macrophage megakarocyte; GM = granulocyte/macrophage; unk = unknown; B1 = blast cell; bm = bone marrow. Modified and reprinted from Long [1992] with permission.

As mentioned, soluble factors, stromal cells, and extracellular matrix (the natural substrate surrounding cells in vivo) are critical elements of the hematopoietic microenvironment. Work by Wolf and Trenton on the hematopoietic microenvironment in vivo provided the first evidence that (still unknown) components of the microenvironment are responsible for the granulocytic predominance of bone marrow hematopoiesis and the erythrocytic predominance of spleen [Wolf & Trentin, 1968]. Dexter and coworkers observed that the in vitro development of adherent cell populations is essential for the continued proliferation and differentiation of blood cells in long-term bone marrow cell cultures [Dexter & Lajtha, 1974; Dexter et al., 1976]. These stromal cells elaborate specific ECM components such as laminin, fibronectin, and various collagens and proteoglycans, and the presence of these ECM proteins coincided with the onset of hematopoietic cell proliferation [Zuckerman & Wicha, 1983]. The actual roles for extracellular matrix versus stromal cells in supporting cell development remains somewhat obscure, as it often is difficult to disassociate stromal cell effects from those of the ECM, since stromal cells are universally observed to be enmeshed in the surrounding extracellular matrix.

Bone extracellular matrix contains both collagenous and noncollagenous proteins. A number of noncollagenous matrix proteins, isolated from demineralized bone, are involved in bone formation. Osteonectin is a 32 kDa protein which, binding to calcium, hydroxypatite, and collagen, is felt to initiate nucleation during the mineral phase of bone deposition [Termine et al., 1981]. In vivo analysis of osteonectin message reveals its presence in a variety of developing tissues [Holland et al., 1987; Nomura et al., 1988]. However, osteonectin is present in its highest levels in bones of the axial skeleton, skull, and the blood platelet (megakaryocyte) [Nomura et al., 1988]. Bone gla protein (BGP, osteocalcin) is a vitamin K-dependent, 5700 Da calcium-binding bone protein which is specific for bone and may regulate $Ca^2 +$ deposition [Price et al., 1976, 1981; Termine et al., 1981]. Other bone proteins seem to function as cytoadhesion molecules [Oldberg et al., 1986; Somerman et al., 1987] or have unresolved functions [Reddi, 1981]. Moreover, bone ECM also contains a number of the more common mesenchymal growth factors such as PDGF, basic, and acidic fibroblast growth factor [Canalis, 1985; Hauschka et al., 1986; Linkhart et al., 1986; Urist et al., 1983*a*]. These activities are capable of stimulating the proliferation of mesenchymal target cells (BALB/c 3T3 fibroblasts, capillary endothelial cells, and rat fetal osteoblasts). As well, bone-specific proliferating activities such as the BMP exist in bone ECM. Although these general and specific growth factors undoubtedly play a role in bone formation, little is understood concerning the direct inductive/permissive capacity of bone-ECM or bone proteins themselves on human bone cells or their progenitors. Nor is the role of bone matrix in presenting growth factors understood—such "matricrine" (factor-ECM) interactions may be of fundamental importance in bone cell development.

When bone precursor cells are cultured on certain noncollagenous proteins, they show an increase in proliferation and bone protein expression (MWL, unpublished observation). Moreover, we have shown, using the hematopoietic system as a model, that subpopulations of primitive progenitor cells require both a mitogenic cytokine and a specific extracellular matrix (ECM) molecule in order to proliferate [Long et al., 1992]. Indeed, without this obligate matrix-cytokine ("matricrine") signal, the most primitive of blood precursor cells fail to develop in vitro [Long et al., 1992]. Although poorly understood, a similar requirement exists for human bone precursor cells, and complete evaluation of osteogenic development (or that of other tissues) thus requires additional studies of ECM molecules. For example, we have demonstrated the importance of three bone ECM proteins in human bone cell growth: osteonectin, osteocalcin, and type I collagen [Long et al., 1990; Long et al., 1994]. Additional bone proteins such as bone sialoprotien and osteopontin are no doubt important to bone structure and function, but their role is unknown [Nomura et al., 1988; Oldberg et al., 1986].

The above observations on the general and specific effects of ECM on cell development have identified certain matrix components which seem to appear as a recurrent theme in tissue development. These are proteoglycans, thrombospondin, fibronectin, and the collagens.

Proteoglycans and Glycosaminoglycans

Studies on the role of proteoglycans in blood cell development indicate that both hematopoietic cells [Minguell & Tavassoli, 1989] (albeit as demonstrated by cell lines) and marrow stromal cells [Gallagher et al., 1983; Kirby & Bentley, 1987; Spooncer et al., 1983; Wight et al., 1986] produce various proteoglycans. Proteoglycans (PG) are polyanionic macromolecules located both on the stromal cell surface and within the extracellular matrix. They consist of a core protein containing a number of covalently linked glycosaminoglycan (GAG) side chains, as well as one or more O- or N-linked oligosaccharides. The GAGs consist of nonbranching chains of repeating N-acetylglucosamine or N-acetylglactosamine disaccharide units. With the exception of hyaluronic acid, all glycosaminoglycans are sulfated. Interesting, many extracellular matrix molecules (fibronectin, laminin, and collagen) contain glycosaminoglycan-binding sites, suggesting that complex interactions occur within the matrix itself.

Proteoglycans play a role in both cell proliferation and differentiation. Murine stromal cells produce hyaluronic acid, heparan sulfate, and chondroitin sulfate [Gallagher et al., 1983], and in vitro studies show PG to be differentially between stromal cell surfaces and the media, with heparan sulfate being the primary cell-surface molecule and chondroitin sulfate the major molecular species in the aqueous phase [Spooncer et al., 1983]. In contrast to murine cultures, the human hematopoietic stromal cells in vitro contain small amounts of heparan sulfate and large amounts of dermatin and chondroitin sulfate, which seem to be equally distributed between the aqueous phase and extracellular matrix [Wight et al., 1986]. The stimulation of proteoglycan/GAG synthesis is associated with an increased hematopoietic cell proliferation, as demonstrated by an increase in the percentage of cells in S-phase [Spooncer et al., 1983]. Given the general diversity of proteoglycans, it is reasonable to expect that they may encode both lineage-specific and organ-specific information. For example, organ-specific PGs stimulate differentiation, as marrow-derived ECM directly stimulates differentiation of human progranulocytic cells (HL60), whereas matrix derived from skin fibroblasts lacks this inductive capacity [Luikart et al., 1987]. Moreover, organ-specific effects are seen in studies of human blood precursor cell adhesion to marrow-derived heparan sulfate but not to heparan sulfates isolated from bovine kidney [Gordon et al., 1988].

Interestingly, cell-surface-associated PGs are involved in the compartmentalization or localization of growth factors within the microenvironment. Thus, the proliferation of hematopoietic cells in the presence of hematopoietic stroma is associated with a glycosaminoglycan-bound growth factor (GM-CSF) [Gordon et al., 1987*b*], and determination of the precise GAG molecules involved in this process (i.e., heparan sulfate) has showed that heparan sulfate side chains bind two blood cell growth factors: GM-CSF and interleukin-3 [Roberts et al., 1988]. These data imply that ECM components and growth factors combine to yield lineage-specific information and indicate that, when PG- or GAG-bound, growth factor is presented to the progenitor cells in a biologically active form.

Thrombospondin (TSP)

Thrombospondin is a large, trimeric disulfide-linked glycoprotein (molecular weight 450,000, subunit molecular weight 180,000) having a domainlike structure [Frazier, 1987]. Its protease-resistant domains are involved in mediating various TSP functions such as cell binding and binding of other extracellular matrix proteins [Frazier, 1987]. Thrombospondin is synthesized and secreted into extracellular matrix by most cells; for review see Lawler [1986] and Frazier [1987]. Matrix-bound TSP is necessary for cell adhesion [Varian et al., 1986], cell growth [Majack et al., 1986], and carcinoma invasiveness [Riser et al., 1988] and is differentially expressed during murine embryogenesis [O'Shea & Dixit, 1988].

Work from our laboratories shows that thrombospondin functions within the hematopoietic microenvironment as a cytoadhesion protein for a subpopulation of human hematopoietic progenitor cells [Long et al., 1992; Long & Dixit, 1990]. Interestingly, immunocytochemical metabolic labeling studies show that hematopoietic cells (both normal marrow cells and leukemic cell lines) synthesize TSP, deposit it within the ECM, and are attached to it. The attachment of human progenitor cells to thrombospondin is not mediated by its integrin-binding RGD sequence because this region of the TSP molecule is cryptic, residing within the globular carboxy-terminus of the molecule. Thus, excess concentrations of a tetrapeptide containing the RGD sequence did not inhibit attachment of human progenitor cells [Long & Dixit, 1990], and similar observations exist in other cell systems [Asch et al., 1987; Roberts et al, 1987; Varian et al., 1988]. Other studies from this author's laboratories show that bone marrow ECM also plays a major role in hematopoiesis in that complex ECM extracts greatly augment LTBMC cell proliferation [Campbell et al., 1985] and that marrow-derived ECM contains specific cytoadhesion molecules [Campbell et al., 1987, 1990; Long et al., 1990, 1992; Long & Dixit, 1990].

Fibronectin

Fibronectin is a ubiquitous extracellular matrix molecule that is known to be involved in the attachment of paryenchymal cells to stromal cells [Bentley & Tralka, 1983; Zuckerman & Wicha, 1983]. As with TSP, hematopoietic cells synthesize, deposit, and bind to fibronectin [Zuckerman & Wicha, 1983]. Extensive work by Patel and coworkers shows that erythroid progenitor cells attach to fibronectin in a developmentally regulated manner [Patel et al., 1985; Patel & Lodish, 1984, 1986, 1987]. In addition to cells of the erythroid lineage, fibronectin is capable of binding lymphoid precursor cells and other cell phenotypes [Bernardi et al., 1987; Giancotti et al., 1986]. Structurally, cells adhere to two distinct regions of the fibronectin molecule, one of which contains the RGD sequence; the other is within the carboxy terminal region and contains a high-affinity binding site for heparan [Bernardi et al., 1987].

Collagen

The role of various collagens in blood cell development remains uncertain. In vitro marrow cells produce types I, III, IV, and V collagen [Bentley, 1982; Bentley & Foidart, 1981; Castro-Malaspina et al., 1980; Zuckerman and Wicha, 1983], suggesting a role for these extracellular matrix components in the maintenance of hematopoiesis. Consistent with this, inhibition of collagen synthesis with 6-hydroxyprolene blocks or reduces hematopoiesis in vitro [Zuckerman et al., 1985]. Type I collagen is the major protein of bone, comprising approximately 90 percent of its protein.

118.4 Considerations for ex Vivo Tissue Generation

The microenvironmental complexities discussed above suggest that the ex vivo generation of human (replacement) tissue (e.g., marrow, liver) will be a difficult process. However, many of the needed tissues (liver, marrow, bone, and kidney) have a degree of regenerative or hyperplastic capacity which allows their in vitro cultivation. Thus, in vivo growth of many of these tissues types is routinely performed, albeit at varying degrees of success. The best example of this is bone marrow. If unfractionated human bone marrow is established in culture, the stromal and hematopoietic cells attempt to recapitulate in vivo hematopoiesis. Both soluble factors and ECM proteins are produced [Dexter & Spooncer, 1987; Long & Dixit, 1990; Zuckerman & Wicha, 1983], and relatively long-term hematopoiesis occurs. However, if long-term bone marrow cultures are examined closely, they turn out to not faithfully reproduce the in vivo microenvironment [Dexter & Spooncer, 1987; Spooncer et al., 1985; Schofield & Dexter, 1985]. Over a period of 8–12 weeks (for human cultures) or 3–6 months (for murine cultures), cell proliferation ceases, and the stromal/hematopoietic cells die. Moreover, the pluripotentiality of these cultures is rapidly lost. In vivo, bone marrow produces a wide variety of cells (erythroid, megakaryocyte/platelet, four types of myeloid cells, and B-lymphocytes). Human long-term marrow cultures produce granulocytes and megakaryocytes for 1–3 weeks, and erythropoiesis is only seen if exogenous growth factors are added. Thereafter, the cultures produce granulocytes and macrophages. These data show that current culture conditions are inadequate and further suggest that other factors such as rate of fluid exchange (perfusion) or that the three-dimensional structure of these cultures is limiting.

Recent work by Emerson, Palsson, and colleagues demonstrated the effectiveness of altered medium exchange rates in the expansion of blood cells in vitro [Caldwell et al., 1991; Schwartz et al., 1991*a*, 1991*b*]. These studies showed that a daily 50% medium exchange affected stromal cell metabolism and stimulated a transient increase in growth factor production [Caldwell et al., 1991]. As well, these cultures underwent a 10-fold expansion of cell numbers [Schwartz et al., 1991*b*]. While impressive, thee cultures nonetheless decayed after 10–12 weeks. Recently, this group utilized continuous-perfusion bioreactors to achieve a longer ex vivo expansion and showed a 10–20-fold expansion of specific progenitor cell types [Koller et al., 1993]. These studies demonstrate that bioreactor technology allows significant expansion of cells, presumably via better mimicry of in vivo conditions.

Another aspect of tissue formation is the physical structure of the developing organ. Tissues exist as three-dimensional structures. Thus, the usual growth in tissue culture flasks is far removed from the in vivo setting. Essentially, cells grown in vitro proliferate at a liquid/substratum interface. As a result, primary tissue cells grow until they reach confluence and then cease proliferating, a process known as *contact inhibition*. This, in turn, severely limits the degree of total cellularity of the system. For example, long-term marrow cultures (which do not undergo as precise a contact inhibition as do cells from solid tissues) reach a density of $1–2 \times 10^6$ per milliliter. This is three orders of magnitude less than the average bone marrow density in vivo. A number of technologies have been applied to this three-dimensional growth problem (e.g., hollow fibers). However, the growth of cells on or in a nonphysiologic matrix is less than optimal in terms of replacement tissue, since such implants trigger a type of immune reaction (foreign body reaction) or are thrombogenic.

Recently, another type of bioreactor has been used to increase the ex vivo expansion of cells. Rotating wall vessels are designed to result in constant, low-shear suspension of tissue cells during their development. These bioreactors thus simulate a microgravity environment. The studies of Goodwin and associates document a remarkable augmentation of mesenchymal cell proliferation in low-shear bioreactors. Their data show that mesenchymal cell types show an average three- to six-fold increase in cell density in these bioreactors, reaching a cellularity of approximately 10^7 cells/mL [Goodwin et al., 1993*a*, 1993*b*]. Importantly, this increase in cell density was associated with a 75% reduction in glucose utilization as well as an approximate 85% reduction in the enzymatic markers of anabolic cellular metabolism (SGOT and LDH) [Goodwin et al., 1993*a*]. Importantly, further work by Goodwin and colleagues show that the growth of mesenchymal cells (kidney and chondrocyte) under low-shear conditions leads to the formation of tissuelike cell aggregates which is enhanced by growing these cells on collagen-coated microcarriers [Duke et al., 1993; Goodwin et al., 1993*a*].

The physical requirements for optimal bone precursor cell (i.e., osteoprogenitor cells and preosteoblasts) proliferation both in vivo and ex vivo are poorly understood. In vivo, bone formation most often occurs within an intervening cartilage model (i.e., a three-dimensional framework referred to as endochondral ossification). This well-understood bone histogenesis is one of embryonic and postnatal chondrogenesis, which accounts for the shape of bone and the subsequent modification and calcification of bone cell ECM by osteoblasts. Recent work in this laboratory has examined the physical requirements for bone cell growth. When grown in suspension cultures (liquid/substratum interface), bone precursor cells develop distinct, clonal foci. These cells, however, express low amounts of bone-related proteins, and they rapidly expand as "sheets" of proliferating cells. As these are nontransformed (i.e., primary) cells, they grow until they reach confluence and then undergo contact inhibition and cease proliferating. We reasoned that growing these cells in a three-dimensional gel might augment their development. It is known from other systems that progenitor cell growth and development in many tissues requires the presence of at least one mitogenic growth factor, and that progenitor cell growth in a three-dimensional matrix results in the clonal formation of cell colonies by restricting the outgrowth of differentiated progeny [Metcalf, 1989]. Thus, bone precursor cells were overlayered with chemically defined, serum-free media containing a biopolymer, thus providing the cells with a three-dimensional scaffold in which to proliferate. In sharp contrast to bone precursor cell growth at a planar liquid/substratum interface, cells grown in a three-dimensional polymer gel show a marked increase in proliferative capacity and an increased per-cell production of the bone-specific proteins (MWL, unpublished observations).

118.5 Conclusions and Perspectives

As elegantly demonstrated by the composite data above, the molecular basis and function of the various components of tissue microenvironments are becoming well understood. However, much remains to be learned regarding the role of these and other, as yet unidentified molecules in tissue development. One of the intriguing questions to be asked is how each interacting molecule contributes to defining the molecular basis of a given microenvironment—for example, the distinct differences in hematopoiesis as it exists in the marrow versus the spleen (i.e., a predominance granulopoiesis and erythropoiesis, respectively) [Wolf & Trentin, 1968]. Another rapidly advancing area is the dissection of the molecular basis of cell trafficking into tissues. Again, the immune system offers for the assessment of this process. Thus, interaction of the lymphocyte receptors with specific glycoconjugates of vascular addressins [Goldstein et al., 1989; Idzerda et al., 1989; Lasky et al., 1989] suggests that similar recognition systems are involved in other tissues, particularly the bone marrow. Another interesting observation is that hematopoietic progenitor cells synthesize and bind to their own cytoadhesion molecules, independent of the matrix molecules contributed by the stromal cells. For example, developing cells in vitro synthesize and attach to fibronectin, thrombospondin, and hemonectin, suggesting that these molecules function solely in an autochthonous manner to localize or perhaps stimulate development. Such a phenomenon may be a generalized process, as we have noted similar patterns of ECM expression/attachment in osteopoietic cell cultures. Coupled with data from the bioreactor studies, this suggest that under appropriate biology/physical conditions tissue cells may spontaneously reestablish their structure. Finally, the elucidation of the various requirements for optimal progenitor cell growth (cell interactions, specific growth factors and matrix components, and/or accessory cells which supply them) should allow the improvement of ex vivo culture systems to yield an environment in which tissue reconstitution is possible. Such a system would have obvious significance in organ-replacement therapy.

References

Arai K, Lee F, Miyajima A, et al. 1990. Cytokines: Coordinators of immune and inflammatory responses. Annu Rev Biochem 59:783.

Aruffo A, Staminkovic I, Melnik M, et al. 1990. CD44 is the principal cell surface receptor for hyaluronate. Cell 61:1303.

Asch AS, Barnwell J, Silverstein RL, et al. 1987. Isolation of the thrombospondin membrane receptor. J Clin Invest 79:1054.

Bentley SA. 1982. Collagen synthesis by bone marrow stromal cells: a quantitative study. Br J Haematol 50:491.

Bentley SA, Froidart JM. 1981. Some properties of marrow derived adherent cells in tissue culture. Blood 56:1006.

Bentley SA, Tralka TS. 1983. Fibronectin-mediated attachment of hematopoietic cells to stromal elements in continuous bone marrow cultures. Exp Hematol 11:129.

Bernardi P, Patel VP, Lodish HF. 1987. Lymphoid precursor cells adhere to two different sites on fibronectin. J Cell Biol 105:489.

Bevilacqua MP, Stengelin S, Gimbrone MA, et al. 1989. Endothelial leukocyte adhesion molecule 1: An inducible receptor for neutrophils related to complement regulatory proteins and lectins. Science 243:1160.

Bierer BE, Sleckman BP, Ratnofsky SE, et al. 1989. The biologic roles of CD2, CD4, and CD8 in T-cell activation. Annu Rev Immunol 7:579.

Brandley BK, Sweidler SJ, Robbins PW. 1990. Charbohydrate ligands of the LEC cell adhesion molecules. Cell 63:861.

Caldwell J, Palsson PB, Locey B, et al. 1991. Culture perfusion schedules influence the metabolic activity and granulocyte-macrophage colony-stimulating factor production rates of human bone marrow stromal cells. J Cell Physiol 147:344.

Campbell A, Sullenberger B, Bahou W, et al. 1990. Hemonectin: A novel hematopoietic adhesion molecule. Prog Clin Biol Res 352:97.

Campbell A, Wicha MS, Long MW. 1985. Extracellular matrix promotes the growth and differentiation of murine hematopoietic cells in vitro. J Clin Invest 75:2085.

Campbell AD, Long MW, Wicha MS. 1987. Haemonectin, a bone marrow adhesion protein specific for cells of granuolocyte lineage. Nature 329:744.

Campbell AD, Long MW, Wicha MS. 1990. Developmental regulation of granulocytic cell binding to hemonectin. Blood 76:1758.

Canalis E. 1985. Effect of growth factors on bone cell replication and differentiation. Clin Orth Rel Res 193:246.

Castro-Malaspina H, Gay RE, Resnick G, et al. 1980. Characterization of human bone marrow fibroblast colony-forming cells and their progeny. Blood 56:289.

Coulombel L, Vuillet MH, Tchernia G. 1988. Lineage- and stage-specific adhesion of human hematopoietic progenitor cells to extracellular matrices from marrow fibroblasts. Blood 71:329.

Dexter TM. 1982. Stromal cell associated haemopoiesis. J Cell Physiol 1:87.

Dexter TM, Allen TD, Lajtha LG. 1976. Conditions controlling the proliferation of haemopoietic stem cells in vitro. J Cell Physiol 91:335.

Dexter TM, Lajtha LG. 1974. Proliferation of haemopoietic stem cells in vitro. Br J Haematol 28:525.

Dexter TM, Spooncer E. 1987. Growth and differentiation in the hemopoietic system. Annu Rev Cell Biol 3:423.

Dorshkind K. 1989. Hemopoietic stem cells and B-lymphocyte differentiation. Immunol Today 10:399.

Duke PJ, Danne EL, Montufar-Solis D. 1993. Studies of chondrogenesis in rotating systems. J Cell Biochem 51:274.

Frazier WA. 1987. Thrombospondin: A modular adhesive glycoprotein of platelets and nucleated cells. J Cell Biol 105:625.

Gallagher JT, Spooncer E, Dexter TM. 1983. Role of the cellular matrix in haemopoiesis: I. Synthesis of glycosaminoglycans by mouse bone marrow cell cultures. J Cell Sci 63:155.

Gallatin M, St John TP, Siegleman M, et al. 1986. Lymphocyte homing receptors. Cell 44:673.

Gehron Robey P, Young MF, Flanders KC, et al. 1987. Osteoblasts synthesize and respond to transforming growth factor-type beta (TGF-beta) in vitro. J Cell Biol 105:457.

Giancotti FG, Comoglio PM, Tarone G. 1986. Fibronectin-plasma membrane interaction in the adhesion of hemopoietic cells. J Cell Biol 103:429.

Giancotti FG, Languino LR, Zanetti A, et al. 1987. Platelets express a membrane protein complex immunologically related to the fibroblast fibronectin receptor and distinct from GPIIb/IIIa. Blood 69:1535.

Giancotti FG, Ruoslahti E. 1990. Elevated levels of the alpha 5 beta 1 fibronectin receptor suppress the transformed phenotype of Chinese hamster ovary cells. Cell 60:849.

Goelz SE, Hession C. Goff D, et al. 1990. ELFT: A gene that directs the expression of an ELAM-1 ligand. Cell 63:1349.

Goldstein LA, Zhou DF, Picker LJ, et al. 1989. A human lymphocyte homing receptor, the hermes antigen, is related to cartilage proteoglycan core and link proteins. Cell 56:1063.

Goodwin TJ, Prewett TI, Wolf DA, et al. 1993*a*. Reduced shear stress: A major component in the ability of mammalian tissues to form three-dimensional assemblies in simulated microgravity. J Cell Biochem 51:301.

Goodwin TJ, Schroeder WF, Wolf DA, et al. 1993*b*. Rotating vessel coculture of small intestine as a prelude to tissue modeling: Aspects of simulated microgravity. Proc Soc Exp Biol Med 202:181.

Gordon MY. 1988. The origin of stromal cells in patients treated by bone marrow transplantation. Bone Marrow Transplant 3:247.

Gordon MY, Bearpark AD, Clarke D, et al. 1990*a*. Haemopoietic stem cell subpopulations in mouse and man: discrimination by differential adherence and marrow repopulation ability. Bone Marrow Transplant 5:6.

Gordon MY, Clarke D, Atkinson J, et al. 1990*b*. Hemopoietic progenitor cell binding to the stromal microenvironment in vitro. Exp Hematol 18:837.

Gordon MY, Dowding CR, Riley GP, et al. 1987*a*. Characterization of stroma-dependent blast colony-forming cells in human marrow. J Cell Physiol 130:150.

Gordon MY, Hibbin JA, Dowding C, et al. 1985. Separation of human blast progenitors from granulocytic, erythroid, megakaryocytic, and mixed colony-forming cells by "panning" on cultured marrow-derived stromal layers. Exp Hematol 13:937.

Gordon MY, Riley GP, Clarke D. 1988. Heparan sulfate is necessary for adhesive interactions between human early hemopoietic progenitor cells and the extracellular matrix of the marrow microenvironment. Leukemia 2:804.

Gordon MY, Riley GP, Watt SM, et al. 1987*b*. Compartmentalization of a haematopoietic growth factor (GM-CSF) by glycosaminoglycans in the bone marrow microenvironment. Nature 326:403.

Gospodarowicz D, Delgado D, Vlodasvsky I. 1980. Permissive effect of the extracellular matrix on cell proliferation in vitro. Proc Natl Acad Sci USA 77:4094.

Gospodarowicz D, Ill C. 1980. Extracellular matrix and control of proliferation of vascular endothelial cells. J Clin Invest 65:1351.

Hanamura H, Higuchi Y, Nakagawa M, et al. 1980. Solubilization and purification of bone morphogenetic protein (BMP) from dunn osteosarcoma. Clin Orth Rel Res 153:232.

Hauschka PV, Marvrakos AE, Iafarati MD, et al. 1986. Growth factors in bone matrix: Isolation of multiple types by affinity chromatography on heparinsepharose. J Biol Chem 261:12665.

Holland PWH, Harmper SJ, McVey JH, et al, 1987. In vivo expression of mRNA for the Ca++-binding protein SPARC (osteonectin) revealed by in situ hybridization. J Cell Biol 105:473.

Horst E, Meijer CJML, Radaskiewicz T, Ossekoppele GP, VanKrieken JHJM, and Pals ST. 1990. Adhesion molecules in the prognosis of diffuse large-cell lymphoma: expression of a lymphocyte homing receptor (CD44), LFA-1 (CD11a/18), and ICAM-1 (CD54). Leukemia 4, 595-599.

Idzerda RL, Carter WG, Nottenburg C, et al. 1989. Isolation and DNA sequence of a cDNA clone encoding a lymphocyte adhesion receptor for high endothelium. Proc Natl Acad Sci USA 86:4659.

Ignotz RA, Massague J. 1986. Transforming growth factor-beta stimulates the expression of fibronectin and collagen and their incorporation into the extracellular matrix. J Biol Chem 261:4337.

Kirby SL, Bentley SA. 1987. Proteoglycan synthesis in two murine bone marrow stromal cell lines. Blood 70:1777.

Koller MR, Emerson SG, Palsson BO. 1993. Large-scale expansion of human stem and progenitor cells from bone marrow mononuclear cells in continuous perfusion cultures. Blood 82:378.

Lasky LA, Singer MS, Yednock TA, et al. 1989. Cloning of a lymphocyte homing receptor reveals a lectin domain. Cell 56:1045.

Lawler J. 1986. The structural and functional properties of thrombospondin. Blood 67, 1197-1209.

Lawrence MB, Springer TA. 1991. Leukocytes roll on a selectin at physiologic shear flow rates: Distinction from and prerequisite for adhesion through integrins. Cell 65:859.

Leung LLK. 1984. Role of thrombospondin in platelet aggregation. J Clin Invest 74:1764.

Liesveld JL, Abbound CN, Duerst RE, et al. 1989. Characterization of human marrow stromal cells: Role in progenitor cell binding and granulopoiesis. Blood 73:1794.

Linkhart TA, Jennings JC, Mohan S, et al. 1986. Characterization of mitogenic activities extracted from bovine bone matrix. Bone 7:479.

Long, MW. 1992. Blood cell cytoadhesion molecules. Exp Hematol 20:288.

Long MW, Ashcraft A, Mann KG. 1994. Regulation of human bone marrow-derived osteoprogenitor cells by osteogenic growth factors. Submitted.

Long MW, Briddell R, Walter AW, et al. 1992. Human hematopoietic stem cell adherence to cytokines and matrix molecules. J Clin Invest 90:251.

Long MW, Dixit VM. 1990. Thrombospondin functions as a cytoadhesion molecule for human hematopoietic progenitor cells. Blood 75:2311.

Long MW, Mann KG. 1993. Bone marrow as a source of osteoprogenitor cells. In MW Long, MS Wicha (eds), The Hematopoietic Microenvironment, pp 110–123, Baltimore, Johns Hopkins University.

Long MW, Williams JL, Mann KG. 1990. Expression of bone-related proteins in the human hematopoietic microenvironment. J Clin Invest 86:1387.

Lowe JB, Stoolman LM, Nair RP, et al. 1990. ELAM-1-dependent cell adhesion to vascular endothelium determined by a transfected human fucosyl transferase cDNA. Cell 63:475.

Luikart SD, Sackrison JL, Maniglia CA. 1987. Bone marrow matrix modulation of HL-60 phenotype. Blood 70:1119.

Majack RA, Cook SC, Bornstein P. 1986. Control of smooth muscle cell growth by components of the extracellular matrix: Autocrine role for the thrombospondin. Proc Natl Acad Sci USA 83:9050.

Massague J. 1987. The TGF-beta family of growth and differentiation factors. Cell 49:437.

Metcalf D. 1989. The molecular control of cell division, differentiation commitment and maturation in haematopoietic cells. Nature 339:27.

Minguell JJ, Tavassoli M. 1989. Proteoglycan synthesis by hematopoietic progenitor cells. Blood 73:1821.

Miyake K, Medina KL, Hayashi S, et al. 1990. Monoclonal antibodies to Pgp-1/CD44 block lymphhemopoiesis in long-term bone marrow cultures. J Exp Med 171:477.

Miyake K, Weissman IL, Greenberger JS, et al. 1991. Evidence for a role of the integrin VLA-4 in lymphohematopoiesis. J Exp Med 173:599.

Miyamake K, Medina K, Ishihara K, et al. 1991. A VCAM-like adhesion molecule on murine bone marrow stromal cells mediates binding of lymphocyte precursors in culture. J Cell Biol 114:557.

Muthukumaran N, Reddi AH. 1985. Bone matrix-induced local bone induction. Clin Orth Rel Res 200:159.

Nomura S, Wills AJ, Edwards DR, et al. 1988. Developmental expression of 2ar (osteopontin) and SPARC (osteonectin) RNA as revealed by in situ hybridization. J Cell Biol 106:441.

O'Shea KS, Dixit VM. 1988. Unique distribution of the extracellular matrix component thrombospondin in the developing mouse embryo. J Cell Biol 107:2737.

Oldberg A, Franzen A, Heinegard D. 1986. Cloning and sequence analysis of rat bone sialoprotein (osteopontin) cDNA reveals an Arg-Gly-Asp cell-binding sequence. Proc Natl Acad Sci USA 83:8819.

Palacios R, Stuber S, Rolink A. 1989. The epigenetic influences of bone marrow and fetal liver stroma cells on the developmental potential of pre-B lymphocyte clones. Eur J Immunol 19:347.

Patel VP, Ciechanover A, Platt O, et al. 1985. Mammalian reticulocytes lose adhesion to fibronectin during maturation to erythrocytes. Proc Natl Acad Sci USA 82:440.

Patel VP, Lodish HF. 1984. Loss of adhesion of murine erythroleukemia cells to fibronectin during erythroid differentiation. Science 224:996.

Patel VP, Lodish HF. 1986. The fibronectin receptor on mammalian erythroid precursor cells: Characterization and developmental regulation. J Cell Biol 102:449.

Patel VP, Lodish HF, 1987. A fibronectin matrix is required for differentiation of murine erythroleukemia cells into reticulocytes. J Cell Biol 105:3105.

Ploemacher RE, Brons NHC. 1988. Isolation of hemopoietic stem cell subsets from murine bone marrow: II. Evidence for an early precursor of day-12 CFU-S and cells associated with radioprotective ability. Exp Hematol 16:27.

Price PA, Ostuka AS, Poser JW, et al. 1976. Characterization of a gamma-carboxyglutamic acid-containing protein from bone, Proc Natl Acad Sci USA 73:1447.

Price PA, Lothringer JW, Baukol SA, et al. 1981. Developmental appearance of the vitamin K-dependent protein of one during calcification. Analysis of mineralizing tissues in human, calf, and rat. J Biol Chem 256:3781.

Reddi AH. 1981. Cell biology and biochemistry of endochondral bone development. Coll Res 1:209.

Reh TA, Gretton H. 1987. Retinal pigmented epithelial cells induced to transdifferentiate to neurons by laminin. Nature 330:68.

Riser BL, Varani J. Carey TE, et al. 1988. Thrombospondin binding and thrombospondin synthesis by human squamous carcinoma and melanoma cells: relationship to biological activity. Exp Cell Res 174:319.

Roberts DD, Sherwood JA, Ginsburg V. 1987*a*. Platelet thrombospondin mediates attachment and spreading of human melanoma cells. J Cell Biol 104:131.

Roberts RA, Spooncer E, Parkinson EK, et al. 1987*b*. Metabolically inactive 3T3 cells can substitute for marrow stromal cells to promote the proliferation and development of multipotent haemopoietic stem cells. J Cell Physiol 132:203.

Roberts R, Gallagher J, Spooncer E, et al. 1988. Heparan sulphate bound growth factors: A mechanism for stromal cell mediated haemopoiesis. Nature 332:376.

Ruoslahti E, Pierschbacher MD. 1987. New perspectives in cell adhesion: RGD and integrins. Science 238:491.

Ryan DH, Nuccie Bl, Abbound CN, et al. 1990. Maturation-dependent adhesion of human B cell precursors to the bone marrow microenvironment. J Immunol 145:477.

Schofield R, Dexter TM. 1985. Studies on the self-renewal ability of CFU-S which have been serially transferred in long-term culture or in vivo. Leuk Res 9:305.

Schwartz RM, Emerson SG, Clarke MF, et al. 1991*a*. In vitro myelopoiesis stimulated by rapid medium exchange and supplementation with hematopoietic growth factors. Blood 78:3155.

Schwartz RM, Palsson BO, Emerson SG. 1991*b*. Rapid medium perfusion rate significantly increases the productivity and longevity of human bone marrow cultures. Proc Natl Acad Sci USA 88:6760.

Shaw S, Luce GE, Quinones R, et al. 1986. Two antigen-independent adhesion pathways used by human cytotoxic T-cell clones. Nature 323:262.

Sheetz MP, Turney S, Qian H, et al. 1989. Nanometre-level analysis demonstrates that lipid flow does not drive membrane glycoprotein movements. Nature 340:248.

Silverstein RL, Nachman RL. 1987. Thrombospondin binds to monocytes-macrophages and mediates platelet-monocyte adhesion. J Clin Invest 79:867.

Somerman MJ, Prince CW, Sauk JJ, et al. 1987. Mechanism of fibroblast attachment to bone extracellular matrix: Role of a 44kilodalton bone phosphoprotein. J Bone Miner Res 2:259.

Sits H, van Schooten W, Keizer H, et al. 1986. Alloantigen recognition is preceded by nonspecific adhesion of cytotoxic T cells and target cells. Science 232:403.

Spooncer E, Gallagher JT, Krizsa F, et al. 1983. Regulation of haemopoiesis in long-term bone marrow cultures: IV. Glycosaminoglycan synthesis and the stimulation of haemopoiesis by beta-D-xylosides. J Cell Biol 96:510.

Spooncer E, Lord BI, Dexter TM. 1985. Defective ability to self-renew in vitro of highly purified primitive haematopoietic cells. Nature 316:62.

Sporn MB, Roberts AB. 1985. Autocrine growth factors and cancer. Nature 313:745.

Springer TA. 1990. Adhesion receptors of the immune system. Nature 346:425.

Stoolman LM. 1989. Adhesion molecules controlling lymphocyte migration. Cell 56:907.

Termine JD, Kleinman HK, Whison SW, et al. 1981. Osteonectin, a bone-specific protein linking mineral to collagen. Cell 26:99.

Tsai S, Sieff CA, Nathan DG. 1986. Stromal cell-associated erythropoiesis. Blood 67:1418.

Tsai S, Patel V, Beaumont E, et al. 1987. Differential binding of erythroid and myeloid progenitors to fibroblasts and fibronectin. Blood 69:1587.

Urist MR, DeLange RJ, Finerman GAM. 1983*a*. Bone cell differentiation and growth factors. Science 220:680.

Urist MR, Sato K, Brownell AG, et al. 1983*b*. Human bone morphogenic protein (hBMP). Proc Soc Exp Biol Med 173:194.

Van de Water L, Aronson D, Braman V. 1988. Alteration of fibronectin receptors (integrins) in phorbol ester-treated human promonocytic leukemia cells. Cancer Res 48:5730.

Varani J, Dixit VM, Fligiel SEG, et al. 1986. Thrombospondin-induced attachment and spreading of human squamous carcinoma cells. Exp Cell Res 156:1.

Varani J, Nickloff BJ, Risner BL, et al. 1988. Thrombospondin-induced adhesion of human kertinocytes. J Clin Invest 81:1537.

Wicha MS, Lowrie G, Kohn E, et al. 1982. Extracellular matrix promotes mammary epithelial growth and differentiation in vitro. Proc Natl Acad Sci USA 79:3213.

Wight TN, Kinsella MG, Keating A, et al. 1986. Proteoglycans in human long-term bone marrow cultures: Biochemical and ultrastructural analyses. Blood 67:1333.

Williams AF, Barclay AN. 1988. The immunoglobulin superfamily—domains for cell surface recognition. Annu Rev. Immunol 6:381.

Witte PL, Robbinson M, Henley A, et al. 1987. Relationships between B-lineage lymphocytes and stromal cells in long-term bone marrow cultures. Eur J Immunol 17:1473.

Wolf NS, Trentin JJ. 1968. Hemopoietic colony studies: V. Effect of hemopoietic organ stroma on differentiation of pluripotent stem cells. J Exp Med 127:205.

Wozney JM, Rosen V, Celeste AJ, et al. 1988. Novel regulators of bone formation: Molecular clones and activities. Science 242:1528.

Yamazaki K, Roberts RA, Spooncer E, et al. 1989. Cellular interactions between 3T3 cells and interleukin-3-dependent multipotent haemopoietic cells.: A model system for stromal-cell-mediated haemopoiesis. J Cell Physiol 139:301.

Zuckerman KS, Rhodes RK, Goodrum DD, et al. 1985. Inhibition of collagen deposition in the extracellular matrix prevents the establishment of a stroma supportive of hematopoiesis in long-term murine bone marrow cultures. J Clin Invest 75:970.

Zuckerman KS, Wicha MS. 1983. Extracellular matrix production by the adherent cells of long-term murine bone marrow cultures. Blood 61:540.

119

The Importance of Stromal Cells

Brian A. Naughton
Advanced Tissue Sciences, Inc.

All tissue is composed of *parenchymal* (from Greek, *that poured in beside*) and *stromal* (Greek, *framework or foundation*) cells. Parenchyma are the functional cells of a tissue (e.g., for liver, hepatic parenchymal cells or hepatocytes; for bone marrow, hematopoietic cells), where stroma comprises primarily connective tissue elements which, together with their products, form the structural framework of tissue. Parenchymal cells can be derivatives of any of the three germ layers, and during development they usually grow into areas populated by stromal cells or their progenitors. Under the strictest definition, stromal cells are derivatives of mesenchyme and include fibroblasts, osteogenic cells, myofibroblasts, and fat cells which appear to arise from a common stem/progenitor cell [Friedenstein et al., 1970; Owen, 1988] (Fig. 119.1). Some investigators apply the term *stromal cell* to all the nonparenchymal cells that contribute to the microenvironment of a tissue and include endothelial cells and macrophages (histiocytes) in this classification as well [Strobel et al., 1986]. However, the ontogency of both endothelial cells and macrophages is distinct from that of mesenchymal tissue-derived cells [Wilson, 1983]. In this chapter, the more expansive definition of stroma will be used. A partial listing of tissue cells that may influence the function of organ parenchyma is in Table 119.1. For the sake of brevity, migrating cells of bone marrow origin will not be discussed in the text (e.g., mast cells, B lymphocytes, natural killer cells), although these cells can influence parenchyma either directly or via cytokine-mediated modulation of stromal cell function.

Stromal and parenchymal cell components are integrated to form a multifunctional tissue in vivo. This chapter will focus on the contribution of stromal cells to the microenvironment and their use in culture to support parenchymal function.

119.1 Tissue Composition and Stromal Cells

Some similarities in the spatial organization of cells of different tissues are apparent. Epithelial cells are a protective and regulatory barrier not just for skin but for all surfaces exposed to blood or to the external environment (e.g., respiratory tract, tubular digestive tract). These cells rest atop a selectively permeable

0-8493-0462-8/00/$0.00+$.50

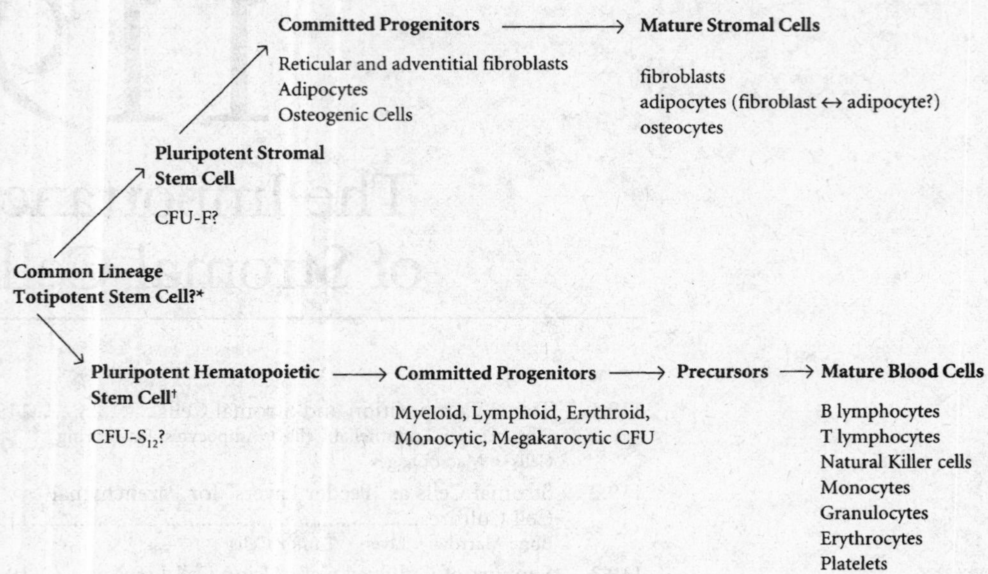

FIGURE 119.1 Hypothetical relationship between the ontogeny of stromal and parenchymal (hematopoietic) cells of the bone marrow. Note: that stromal cell is defined in this chart as only those cells that are derivatives of mesenchyme. Hematopoietic stem and progenitor cells express CD 34. This epitope also is expressed by stromal cell precursors (CFU-F) [Simmons & Torok-Storb, 1991] and adherent stromal cells develop from cell populations selected for the CD 34 antigen. In addition, fetal marrow elements with CD 34 +, CD 38–, HLA-DR-phenotypes were reported to develop not only the hematopoietic microenvironment but also the hematopoietic cells themselves [Huang & Terstappen, 1992].

*Indicates the possibility that, at least in the fetus, there may be a common stem cell for bone marrow stromal and hematopoietic elements.

†There are several schools of thought relating to single or multiple stem cell pools for the lymphoid versus the other lineages.

TABLE 119.1 Cells That Contribute to the Tissue Microenvironment

Stromal cells: derivatives of a common precursor cell
Mesenchyme
Fibroblasts
Myofibroblasts
Osteogenic/chondrogenic cells
Adipocytes
Stromal-associated cells: histogenically distinct from stromal cells, permanent residents of a tissue
Endothelial cells
Macrophages
Transient cells: cells that migrate into a tissue for host defense either prior to or following an inflammatory stimulus
B lymphocytes/plasma cells
Cytotoxic T cells and natural killer (NK) cells
Granulocytes
Parenchymal cells: cells that occupy most of the tissue volume, express functions that are definitive for the tissue, and interact with all other cell types to facilitate the expression of differentiated function

basement membrane. The composition of the underlying tissue varies, but it contains a connective tissue framework for parenchymal cells. If the tissue is an artery or a vein, the underlying connective tissue is composed of circularly arranged smooth-muscle cells which in turn are surrounded by an adventitia of loose connective tissue. Tissues within organs are generally organized into functional units around capillaries which facilitate metabolite and blood gas transport by virtue of their lack of a smooth muscle layer (*tunica media*) and the thinness of their adventitial covering. Connective tissue cells of the tissue underlying capillaries deposit extracellular matrix (ECM) which is a mixture of fibrous proteins (collagens) embedded in a hydrated gel of glycosaminoglycans (GAGs). The GAGs as distinguished by their sugar residues, are divided into five groups: hyaluronic acid, the chondiotin sulfates, dermatan sulfate, heparan and heparin sulfate, and keratan sulfate. All the GAGs except hyaluronic acid link with proteins to form proteoglycans. These proteoglycans bind to long-chained hyaluronic acid cores that interweave the crosslinked collagen fibers that form the framework of the tissue. The basic composition and density of deposition of stromal cell-derived ECM varies from tissue to tissue [Lin and Bissel, 1993]. Figure 119.2 is a scanning electron micrograph (SEM) depicting the intricacies of the ECM deposited by human bone marrow-derived stromal cells growing on a three-dimensional culture template. The ECM forms a sieve-like arrangement for diffusion in an aqueous environment and a number of bone marrow-derived migratory cells are present as well. Such an arrangement dramatically enhances the surface area for cell growth. Although these voids are primarily filled by parenchymal cells, normal tissue interstitium also contains migratory immunocompetent cells such as B lymphocyte/plasma cells, T cells, and natural killer cells as well as the ubiquitous macrophage (histiocyte). The numbers of these cells are enhanced during inflammatory episodes when neutrophils or other granulocytes and/or mononuclear leukocytes infiltrate the tissue. Cytokines and other proteins released by these leukocytes recruit stromal cells in the tissue repair process. This phenomenon can be exploited by tumor cells to enhance their invasiveness (see Section 119.2 Tumor Cells). Although tissue function in toto is measured by parenchymal cell output (e.g., for liver, protein synthesis, metabolism; for bone marrow, blood cell production), this is profoundly influenced by and in some instances orchestrated by the stromal cell microenvironment.

Stromal cells contribute to parenchymal cell function by synthesizing the unique mix of ECM proteins necessary for cell seeding/attachment of specific types of tissue cells, the modulation of gene expression in parenchymal cells by events triggered by cytoskeleton-mediated transduction [Ben-Ze'ev et al., 1988], the deposition of ECM proteins to sequester growth and regulatory factors for use by developing parenchyma [Roberts et al., 1988], deposition of ECM of the right density to permit diffusion of nutrients, metabolites, and oxygen (and egress of CO_2 and waste) to the extent necessary to maintain the functional state of the tissue, by forming the appropriate barriers to minimize cell migration or intrusion, and by synthesizing and/or presenting cytokins that regulate parenchymal cell function, either constituitively or following induction by other humoral agents (see Table 119.2).

Creating cultures that contain the multiplicity of cell types found *in vivo* presents a daunting task for several reasons: (1) culture media that are rich in nutrients select for the most actively mitotic cells at the expense of more mitotically quiescent cells and, (2) cell phenotypic expression and function is related to its location in tissue to some extent. It is difficult, especially using a flat (i.e., two-dimensional) culture template, to create a microenvironment that is permissive or inductive for the formation of tissue-like structures. (3) Localized microenvironmental niches regulate different parenchymal cell functions or, in the case of bone marrow, hematopoietic cell differentiation. These milieu are difficult to reproduce in culture, since all cells are exposed to essentially the same media components. However, a major goal of tissue culture is to permit normal cell-cell associations and the reestablishment of tissue polarity so that parenchymal cell function is optimized. The question of three-dimensionality will be addressed later in this chapter. A brief survey of the various types of stromal elements follows.

Fibroblasts

Fibroblasts are responsible for the synthesis of many GAGs and for the deposition and organization of collagens. Although present in most tissues, fibroblasts exhibit specialization with respect to the type of ECM that they secrete. For example, liver tissue contains type I and type IV collagen, whereas type II collagen is found in cartilaginous tissues. Bone marrow contains types I, III, IV, and V collagen [Zuckerman, 1984]. Heterogeneity of collagen deposition as well as GAG composition exists not only between different tissues but in developmentally different stages of the same tissue [Thonar & Kuettner, 1987].

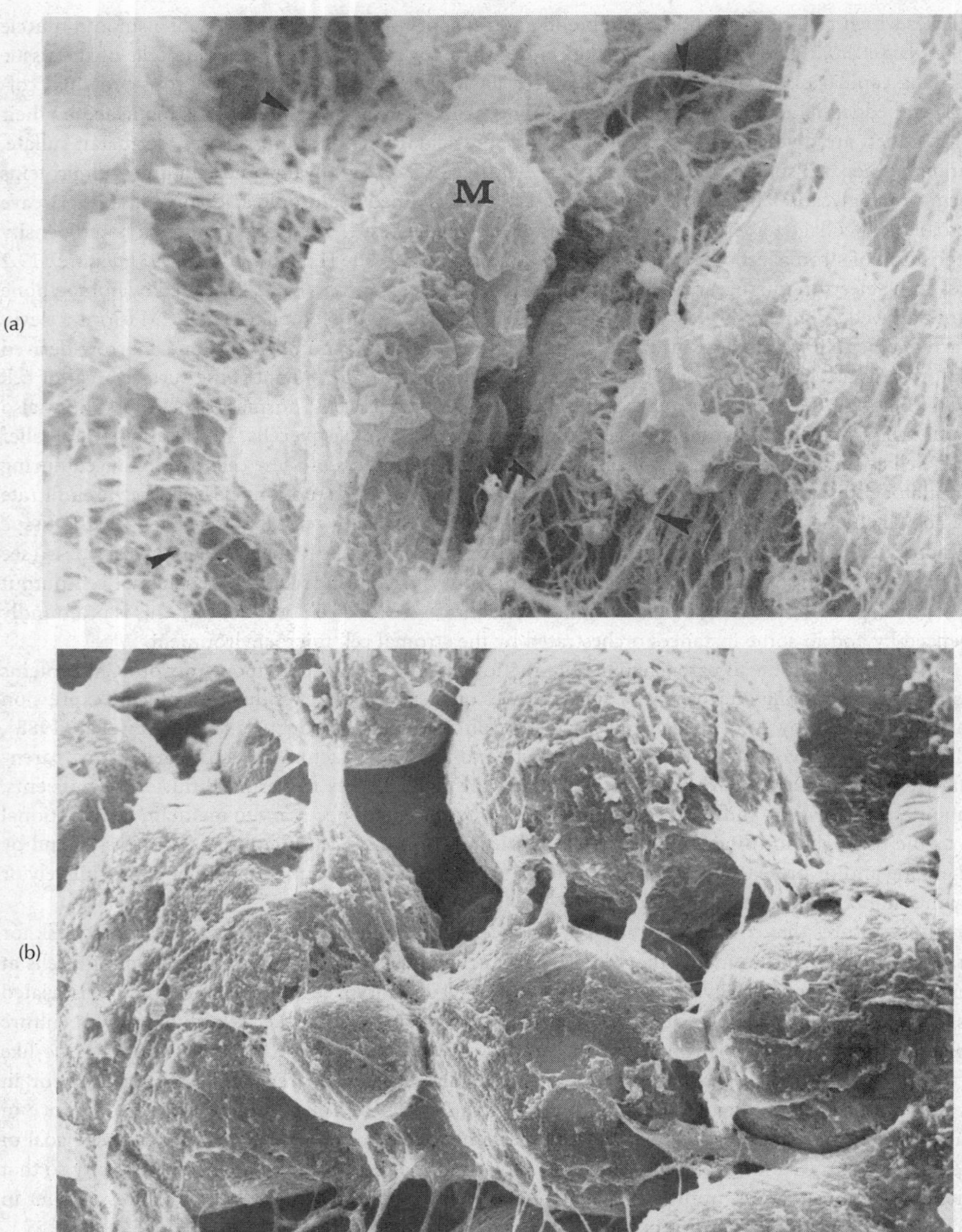

FIGURE 119.2 Scanning electron micrographs of bone marrow cultures on nylon screen templates (*a*) Photograph depicting a portion of a macrophage (M) associated with the numerous, delicate, interweaving strands of ECM (arrows) that are deposited between the openings of the nylon screen of a bone marrow stromal cell culture. (*b*) Myeloid cells of a hematopoietic colony growing in a coculture of human stromal cells and hematopoietic cells. Note the pattern of attachment of individual cells to matrix and the large open area between the cells for nutrient access. (*c*) The photograph depicts the intimate association of cells of a myeloid or mixed myeloid/monocytic colony (arrow) with enveloping fibroblastic cells (F) in a bone marrow coculture. A filament of the nylon (n) screen is also present in the field. (*d*) An erythroid colony (E) in a human bone marrow coculture on nylon screen. Note that the more mature erythroid cells are on the periphery of the colony and that it is in close apposition to a macrophage (M). A nylon filament is also present (n).

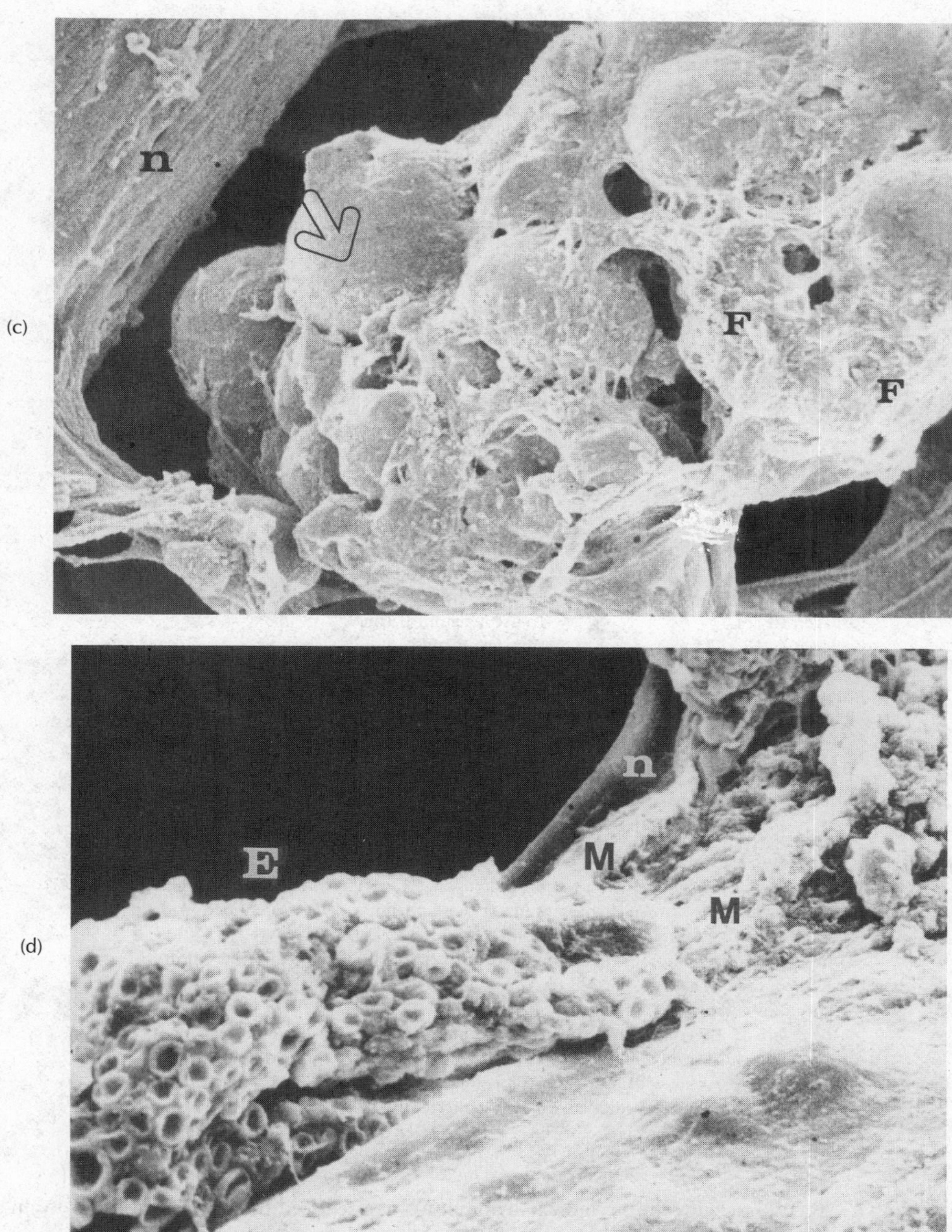

FIGURE 119.2 (continued)

In addition to matrix deposition, fibroblasts synthesize the cytokines of the fibroblast growth factor (FGF) family, a variety of interleukins, and GM-CSF as well as other regulatory cytokines (Table 119.2). Fibroblast activity can be modulated by circulating or locally diffusable factors including IL-1, TGF_{β}, TNF_{α}, and a host of other factors [Yang et al., 1988]. For example, IL-1 activated splenic stroma supported the growth of rat natural killer cells in long-term culture [Tjota et al., 1992]. In addition, treatment of bone

TABLE 119.2 Stromal Cell Phenotypes and Secretory Profiles

Phenotypes[a]	Cytokine/Protein Secretion
MHC I (all)	Il-1β (mφ, E, F, A)
MHC II (E, mφ)	M-CSF (mφ, E, F)
CD 10 (Neural endopeptidase)[b]	G-CSF (mφ, F)
CD 29 (β_2 integrin) (all)	GM-CSF (mφ, E)
CD 34 (CFU-F)	LIF (mφ, F)
CD 36 (IA 7 (mφ)	GM-colony enhancing factor (mφ)
CD 44 (H-CAM)[b]	TGFβ (mφ, F)
CD 49b (α_2 chain of VLA-2)[b]	TNFα (mφ, F)
CD 49d (α_4 chain of VLA-4)[b]	PAF (E)
CD 49e (α_5 chain of VLA-5)[b]	BPA (mφ, E)
CD 51 (vitronectin receptor)[b]	IFNα (mφ, F, E)
CD 54 (ICAM-1) (all)	IFNγ (F)
CD 58 (LFA-3) (all)	Il-6 (mφ)
CD 61 (GP 111a)	c-kit ligand (mφ, F)
VCAM-1 (E)	Acidic and basic FGF (F)
α SM actin (F)	Angiotensinogen (A)
Vimentin (F)	lipoprotein lipase (A)
Decorin (Mes)	Adipocyte P2 (A)
Fibronectin (F, mφ)	MIP-1α (mφ)
Laminin (E)	Complement proteins C1-C5 (mφ)
Collagen I (F)	Factor B, properdin (mφ)
Collagen III (F)	Transcobalamin II (mφ)
Collagen IV (F)	Transferrin (mφ)
von Willebrand factor (E)	Arachidonic acid metabolites (mφ, E)
Adipsin (A)	HGF/SF (F)
	flt3-ligand (S)
	Prolactin (F)
	LIF (S)[c]
	STR-3 (F)
	(IL-3, IL-7, IL-9, IL-11, IL-15, neuroleukin)[b]

[a] Phenotypic expression of stroma in bone marrow cultures. Parentheses indicate localization according to cell type (mφ = Macrophage F = Fibroblast, E = endothelial, A = adipocyte, Mes = mesenchyme). Many of these phenotypes were identified in the original work of Cicuttini et al., [1992] and Moreau et al., [1993].

[b] Indicates the presence of mRNA and /or protein expression in bone marrow cultures but not localized to a particular cell type.

[c] Found in fallopian tube stroma [Keltz et al., 1997].

Abbreviations: BPA—erythroid burst promoting activity, CD—cluster determinant, MHC—major histocompatibility complex, M—monocyte, G—granulocyte, CSF—colony stimulating factor, CFU-F—colony forming unit-fibroblast, MIP—macrophage inflammatory protein, IL—interleukin, IFN—interferon, TGF—transforming growth factor, TNF—tumor necrosis factor, PAF—platelet activating factor, FGF—fibroblast growth factor/scatter factor, LIF-leukemia inhibitory factor, STR-3—stromelysin 3.

marrow stromal cell cultures with the steroid methylprednisolone reduced the concentration of hyaluronic acid and heparan sulfate relative to other proteins [Siczkowski et al., 1993], perhaps making them more conducive to hematopoietic support. Horse serum, presumably because of its high content of hydrocortisone and other steroids, was an essential component of medium used to maintain hematopoiesis in early long term bone marrow cultures [Greenberger et al., 1979]. In our experience, medium supplementation with hydrocortisone or other corticosteroids enhances the ability of liver-derived stromal cells to support parenchymal hepatocyte function for longer-terms in vitro [Naughton et al., 1994], although the precise mechanism(s) has not been defined.

Fibroblastic cells in tissue appear to be heterogeneous with respect to support function. Early morphometric studies of bone marrow indicated that stroma supporting different types of hematopoiesis exhibited different staining patterns. The development of highly specific monoclonal antibodies in the intervening years made possible a much more detailed analysis of stromal cells of bone marrow and other tissues and provided a number of avenues to isolate these cells for study. Whereas bone marrow stromal cells can be separated from hematologic cells by virtue of their adherence to plastic, this technique was not optimal for all tissues. Monoclonal antibodies can now be used to "dissect" cells from the stromal microenvironment using a variety of techniques including flow cytometry, immune panning, immuno-magnetic microspheres, and affinity column methodologies. Monoclonal antibodies also make it possible to select for stromal cell progenitors like the fibroblast-colony forming unit (CFU-F) which is CD 34 positive.

Endothelial Cells

For nonglandular tissues such as bone marrow, the endothelial cells are vascular lining cells and can be distinguished by the expression of or the message for von Willebrand factor and surface major histo-compatibility complex-II (MHC-II) antigens. As cells of the vascular *tunica intima*, they, along with the basement membrane, form a selective barrier to regulate the transport of substances to and from the blood supply. In addition, their expression of integrins following stimulation with IL-1 or other mediators regulates the attachment of immunocompetent cells during acute (neutrophils) or chronic (T lymphocytes, monocytes) inflammation. Endothelial cells also regulate other vascular functions. They synthesize the vasodilatory effector nitric oxide after induction with acetylcholine [Furchgott & Zawadzki, 1980] and secrete regulatory peptides, the endothelins, which counteract this effect [Yanagisawa et al., 1988]. Specialized vascular endothelia are present in bone marrow, where they permit the egress of mature leukocytes from the marrow into the sinusoids, and in the liver, where the surfaces of the cells lining sinusoids are fenestrated to facilitate transport. In a more general context, endothelial cells also produce collagen IV for basement membranes and secrete several cytokines including IL-1, fibronectin, M-CSF, and GM-CSF and release platelet-activating factor (Table 119.2). Vascular endothelia are nonthrombogenic and contribute to angiogenesis by mitotically responding to locally secreted factors such as bFGF and aFGF. These endothelia possess LDL receptors and degrade this lipoprotein at substantially higher rates than other types of cells. These cells also have receptors for Fc, transferrin, mannose, galactose, and Apo-E, as well as for scavenger receptors [Van Eyken & Desmet, 1993]. Endothelia tend to assemble into tubular structures in culture, but their contribution to parenchymal cell growth in vitro is contingent on the generation of the proper tissue polarity.

In addition to vascular endothelial cells, some organs possess nonvascular endothelial cells such as the bile duct lining cells of the liver. These cells possess antigenic profiles and secretory potentials that are similar to vascular endothelia.

Adipocytes, Fat-Storing Cells

These cells are represented in varying concentrations in different tissues. They are related to fibroblasts (Fig. 119.2), and transformation of fibroblasts to adipocytes in vitro can be induced by supplementation of the medium with hydrocortisone or other steroids [Brockband & van Peer, 1983; Greenberger, 1979] which induce the expression of lipoproteinlipase and glycerolphosphate dehydrogenase as well as an increase in insulin receptors [Gimble et al., 1989]. Like fibroblasts, marrow adipocytes are heterogeneous and display different characteristics related to their distribution in red or yellow marrow [Lichtman, 1984]. These characteristics include insulin independence but glucocorticoid dependence in vitro, positive staining with perfomic acid-Schiff, and higher concentrations of neutral fats and unsaturated fatty acids in triglycerides. In bone marrow in vivo, there appears to be an inverse relationship between adipogenesis and erythropoiesis: phenylhydrazine-induced anemia causes a rapid conversion of yellow marrow (containing adipocytes) to red marrow due to compensatory erythroid heperplasia [Maniatis et al., 1971]. However, the role of the adipocyte in supporting hematopoiesis in culture is controversial. Although Dexter and coworkers [1977] associated declining myelopoiesis in long-term murine bone marrow cultures with the gradual disappearance of adipocytes from the stromal layer, these cells may not be necessary to support hematopoiesis in human bone marrow cultures [Touw & Lowenberg, 1983]. In this regard, IL-11 suppressed adipogenesis but simulated human CD34+ HL-DR + progenitor cells cocultured with human bone marrow stroma and enhanced the numbers of myeloid progenitor cells [Keller et al., 1993]. Hematopoiesis in vivo requires the proper ECM for cell attachment and differentiation as well as regulation by cytokines elaborated by stromal cells [Metcalf, 1993]. Although there is considerable redundancy in the synthesis of regulatory factors by different stromal cell populations, no single cell population has been identified that can provide an entire hematopoietic microenvironment. Further complicating the issue is a recent report indicating that CD 34+ hematopoietic progenitors produce soluble factors that control the production of some cytokines by stromal cells [Gupta et al., 1998]. If "cross-talk" between parenchyma and stroma is established as an important regulatory mechanism, the ratio between the relative numbers of stroma and parenchyma in a tissue culture will assume paramount importance. The use of cell lines to provide hematopoietic support will be discussed later in this chapter.

In the liver, purified fat-storing cells are capable of broad-scale synthetic activity that includes collagens I, III, IV, fibronectin, heparan sulfate, chondroitin sulfate, and dermatan sulfate [DeLeeuw et al., 1984]. Adipocytes also synthesize colony-stimulating factors and other regulatory cytokines (Table 119.2). In addition to the above characteristics, adipocytes are desmin positive and therefore are phenotypically related to myogenic cells or myofibroblasts. The function of adipocytes may vary depending on their location. Adipocytes in the bone marrow are in close apposition to the sinusoidal endothelial cells and in the liver are found in the space of Disse under the endothelial cells. Hepatic fat-storing cells are finely integrated into several contiguous parenchymal cells and contain fat droplets that are qualitatively different than those found in hepatic parenchymal cells in that they contain high levels of retinols. Liver adipocytes are responsible for the metabolism and storage of vitamin A [DeLeeuw et al., 1984], provide some of the raw materials for the synthesis of biologic membranes and also contribute to local energy metabolism needs. Adeipocytes of other tissues also act as a type of "progenitor" cell that can be converted to different phenotypes (e.g., osteoblasts or chondroblasts) under the appropriate conditions and are capable of stimulating osteogenesis via their secretion of cytokines [Benayahu et al., 1993].

Macrophages

Macrophages are derivatives of peripheral blood monocytes and are, therefore, bone-marrow-derived. They seed and remain on the surfaces of sinusoidal vessels in organs such as the liver and the spleen or migrate into the interstitial spaces of virtually all tissues. Macrophages are quintessential immunocompetent cells and are central components of many defense strategies, including randomized microbial phagocytosis and killing; antibody-dependent cellular cytotoxicity (ADCC), where they are directed against microbial or other cells that are opsonized with antibody; nonrandomized (specific) phagocytosis mediated by the association of immunoglobulins with a multiplicity of F_c receptors on their surfaces; the presentation of processed antigen to lymphocytes; secretion of and reaction to chemotaxins; and an enzymatic profile enabling them to move freely through tissue. The secretory capacity of macrophages is prodigious. In addition to plasma components such as complement proteins C1 through C5, they synthesize the ferric iron- and vitamin B_{12}–building proteins, transferrin and transcobalamin II, as well as a host of locally acting bioreactive metabolites of arachidonic acid. Macrophage secretory activity appears to influence two simultaneous events in vivo, inflammation and tissue repair. One monokine, IL-1, enhances the adhesion of neutrophils to vascular endothelial cells and activates B and T lymphocytes and other macrophages while stimulating the formation of acute phase proteins inducing collagen, ECM, and cytokine synthesis by fibroblasts and other stromal cells. IL-1 as well as a host of other humoral regulatory factors, originate in macrophages, including TNFα, IFNα, GM-CSF, and MIP-1α (Table 119.2). The ability of macrophages to secrete regulatory cytokines makes them an important contributor to the tissue microenvironment.

Macrophages have been intrinsically associated with erythropoiesis in the bone marrow [Bessis & Breton-Gorius, 1959] and [Naughton et al., 1979]. As such, these "nurse cells" destroy defective erythroblasts and the fetal and regenerating liver, and provide recycled iron stores for hemoglobin synthesis by developing red cells. They also synthesize and/or store erythropoietin.

119.2 Stromal Cells as "Feeder Layers" for Parenchymal Cell Culture

Irradiated stromal feeder layers were first used by Puck and Marcus [1955] to support the attachment and proliferation of HeLa cells in culture. Direct contact with feeder layers of cells also permits the growth of glioma cells and epithelial cells derived from breast tissue and colon [Freshney et al., 1982]. Enhanced attachment of parenchymal cells and production of factors to regulate growth and differentiation are two important benefits of coculturing parenchymal and stromal cells.

Bone Marrow

Bone marrow was the first tissue to be systematically investigated in regard to the influence of ECM and stromal cells on the production of blood cells. By the mid 1970s, the influence of microenvironmental conditions [Trentin, 1970; Wolf & Trentin, 1968] and ECM deposition [McCuskey et al., 1972; Schrock et al., 1973] upon hematopoiesis was well established in the hematology literature. Dexter and coworkers [1977] were the first to apply the feeder-layer-based coculture technology to hematopoietic cells by inoculating mouse bone marrow cells into preestablished, irradiated feeder layers of marrow-derived stromal (adherent) cells. These cultures remained hematopoietically active for several months. By comparison, bone marrow cells cultured in the absence of feeder layers or supplementary cytokines terminally differentiated within the first two weeks in culture; stromal cells were the only survivors [Brandt et al., 1990; Chang & Anderson, 1971]. If the cultures are not supplemented with exogenous growth factors, then myeloid cells, monocytic cells, and nucleated erythroblasts are present for the first 7–10 days of culture. This trilineage pattern is narrowed over successive weeks in culture so that from about 2–4 weeks myeloid and monocytic cells are produced, and, if cultured for longer terms, the products of the culture are almost entirely monocytic. However, the nature of hematopoietic support can be modulated by changing the environmental conditions. In this regard, cocultures established under identical conditions as the above produce B lymphocytes if the steroid supplementation of the medium is removed and the ambient temperature is increased [Whitlock & Witte, 1982]. Static bone marrow coculture systems are generally categorized as declining because of this and the finding that the hematopoietic progenitor cell concentrations decrease as a function of time in culture. However, this trend can be offset by coculturing bone marrow in bioreactors that provide constant media exchange and optimize oxygen delivery [Palsson et al., 1993] or by supplementing the cocultures with cocktails of the various growth factors elaborated by stromal cells [Keller et al., 1993]. Connective tissue feeder layers had originally been hypothesized to act by conditioning medium with soluble factors that stimulated growth and by providing a substratum for the selective attachment of certain types of cells [Puck & Marcus, 1955]. These are functions that have since been proven for bone marrow and other tissue culture systems.

It is very difficult to recreate the marrow microenvironment in vitro. One reason is that the marrow stromal cell populations proliferate at different rates; if expanded vigorously using nutrient-rich medium, the culture selects for fibroblastic cells at the expense of the more slowly dividing types of stroma. We established rat bone marrow cocultures using stromal cells that were passaged for 6 months, and these cultures produced lower numbers of progenitor cells (CFU-C) in the adherent zone as compared to cocultures established with stroma that was only passed three to four times (Fig. 119.3). Cells released from cultures using the older stroma were almost entirely monocytic, even at earlier terms of culture. In a related experiment, we suspended nylon screen cocultures of passage 3–4 stroma and bone marrow hematopoietic cells in flasks containing confluent, 6-month-old stromal cell monolayers and found that these monolayer cells inhibited hematopoiesis in the coculture [Naughton et al., 1989].

We use early passage stroma for our cocultures in order to retain a representation of all the stromal cell types. If hematopoietic cells from cocultures are removed after about a month in vitro and reinoculated onto a new template containing passage 3–4 stromal cells and cultured for an additional month, the progenitor cell concentrations of the adherent zones are considerably higher than in cocultures where no transfer took place [Naughton et al., 1994]. This experiment indicated that continued access to "fresh"

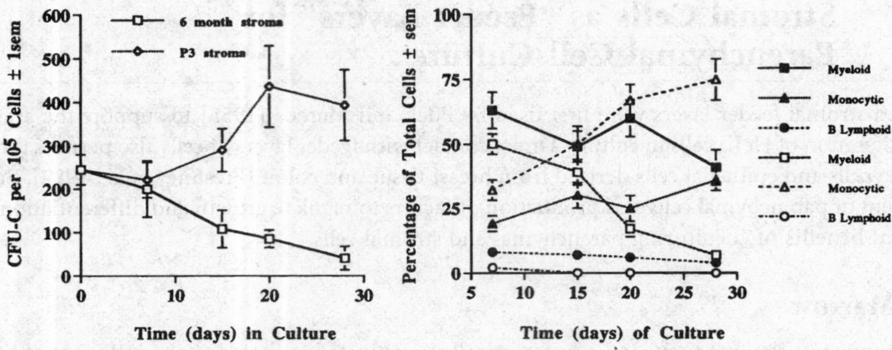

FIGURE 119.3 *Left.* Mean CFU-C progenitor concentration of the adherent zone of a three-dimensional coculture of rat bone hematopoietic cells and rat bone marrow stroma. Stock cultures to generate the stromal cells for the cocultures were seeded onto the three-dimensional template either after passage (P3) or following expansion in monolayer culture for 6 months. Stroma at early passage is substantially more supportive of CFU-C progenitor cells than "old" stroma that is primarily fibroblastic in nature. Vertical lines through the means = ±1 standard error of the mean (sem). *Right.* Analysis of the cellular content of the nonadherent zone of rat bone marrow hematopoietic cell: stromal cell cocultures on three-dimensional templates by flow cytometry. Cocultures were established either with P3 stroma (closed figures) or stroma that was grown in monolayer culture for 6 months (open figures). The mean percentages of cells recognized by the phenotyping antibodies MOM/3F12/F2 (myeloid), ED-1 (monocytic), and OX-33 (B lymphoid) (Serotec, UK) are depicted. Vertical lines through the means = ±1 sem. Whereas low passage stroma generates myeloid, B lymphoid, and monocytic cells in cocultures and releases them into the nonadherent zone, 6-month-old stroma supports mainly the production of macrophages.

stroma enhances hematopoiesis. The maintenance of mixed populations of stromal cells are desirable for a number of reasons. The same cytokine may be synthesized by different stromal cells, but its mode of action is usually synergistic with other cytokines for the differentiation of a specific blood cell lineage(s). Differentiation may not occur in the absence one of these cytokines, although some degree of redundancy in cytokine expression is intrinsic to the stromal system [Metcalf, 1993]. The fibroblastic cells that "grow out" of later passage human and rodent bone marrow stroma share some phenotypic properties with muscle cells (αSM actin+, CD34-, STRO-1-). These cells (endothelial cells and macrophages are absent or present in negligible quantities) supported hematopoietic progenitors for up to 7 weeks of culture but no longer [Moreau et al., 1993], indicating that other types of cells are necessary for long-term hematopoiesis in vitro.

Stromal cells derived from spleen or bone marrow were also shown to support the growth and maturation of rat natural killer cells in long-term three-dimensional culture [Tjota et al., 1992]. Natural killer cells produced by the cultures for more than 2 months in vitro continued to kill YAC-1 (NK sensitive) target cells when activated by IL-2. In more recent experiments, fetal thymus-derived stroma supported the production of CD45+/CD56+ human natural killer cells when cocultured with CD34+lin- progenitors [Vaz et al., 1998]. In related work, stromal cell-derived IL-15 was found to enhance the production of CD56+/CD3- natural killer cells from human CD34+ hematopoietic progenitors [Mrózek et al., 1996]. In addition, the process of natural killer cell maturation (but not early progenitor cell proliferation) appears to require direct contact with stromal cells [Tsuji and Pollack, 1995].

Cytoadhesive molecule expression and its modulation is also an important consideration for the engineering of bone marrow and other tissues. In the case of bone marrow, this regulates "homing" or seeding, restricting cells in specific areas within tissues both to expose them to regulatory factors and to the release of cells from the supportive stroma upon maturation. For example, the sequential expression of hemonectin, a cytoadhesion molecule of myeloid cells, in yolk sac blood islands, liver, and bone marrow during embryonic and fetal life parallels the granulopoiesis occurring in these respective organs [Peters et al., 1990]. Hematopoietic stem cells [Simmons et al., 1992] and B-cell progenitors [Ryan et al., 1991]

TABLE 119.3 Cells Contributing to the Hepatic Microenvironment

Cell Type	Size, μm	Relative Percent of Total Cells	Characteristics
Stroma			
Kupffer cells	12–16	8	MHC-1+, MHC-II+, Fcr, C3r, mannose and D-acetrylglucosamine receptors, acid phosphatase+, density = 1.076 gm/ml,* 1.036 gm/ml†
Vascular endothelia	11-12	9	MHC-1+, MHC II+, vWF+, F_cr, TF_r, mannose, apo-E, and scavenger receptors, density = 1.06–1.08 gm/ml,* 1.036 gm/ml†
Biliary endothelia	10–12	5	MHC-I+, Positive for cytokeratins 7, 8, 18, 19, β2 microglobulin+, positive for VLA-2,3,6 integrins, agglutinate with UEA, WGA, SBA, PNA, density = 1.075–1.1 gm/ml,* 1.0363gm/ml†
Fat-storing cells	14–18	3	MHC-I+, desmin+, retinol+, ECM expression, collagen I, IV expression., density =1.075–1.1 gm/ml*
Fibroblasts	11–14	7	MHC-I+, ECM expression, collagen I, IV expression, density = 1.025 gm/ml,* 1.063 gm/ml†
Pit cells	11-15	1-2(variable)	MHC-I,+, asialo-GM+, CD8+, CD-5-
Parenchymal cells			
Mononuclear (type 1)	17-22	35	MHC-I-, MHC-II-, blood group antigen-, density = 1.10–1.14 gm/ml,* 1.067 gm/ml†
Binuclear (type II)	20-27	27	MHC-I±, MHC-II-, blood group antigen-, density = 1.10-1.14 gm/ml,* 1.071 gm/ml†
Acidophilic (type III)	25-32	5	MHC-1±, MHC-II-, blood group antigen-, density = 1.038 gm/ml†

Abbreviations: ECM = extracellular matrix, MHC – major histocompatibility complex, PNA = peanut agglutinin, SBA = soybean agglutinin, WGA = wheat germ agglutinin, UEA = Ulex europaeus agglutinin.

bind to VCAM-1 on cultures stromal cells by expression of VLA-4. In addition, erythroid progenitor cells bind preferentially to the fibronectin component of the ECM and remain bound throughout their differentiation [Tsai et al., 1987]. A basic requirement of stromal feeder layers is the expression of the proper cytoadhesion molecule profile to permit attachment of parenchyma; in this instance these are hematopoietic stem and progenitor cells.

Liver

Liver is an hematopoietic organ during fetal life, and it contains many of the same stromal cell populations found in bone marrow (Table 119.3). Research trends in liver cell and bone marrow culture have followed a somewhat parallel course. Both tissues are difficult to maintain in vitro; the parenchymal cell numbers either declined (hematopoietic progenitors) or lost function and dedifferentiated (hepatocytes) over time in culture. When mixed suspensions of hepatic cells are inoculated into liquid medium, approximately 20% of the total cells attach. The nonadherent population remains viable for about 72 hours. The adherent cells, although they proliferate for only 24–48 hours after inoculation, can survive for substantially longer periods. However, many of these adherent parenchymal cells undergo drastic phenotypic alterations, especially if the medium is conditioned with serum. These changes include a flattening and spreading on plastic surfaces as well as a propensity to undergo nuclear division in the absence of concomitant cytoplasmic division. The appearance of these bizarre, multinucleated, giant cells herald the loss of liver-specific functions such as albumin synthesis and the metabolism of organic chemicals by cytochrome P450 enzymes. The percentage of hepaocytes attaching to the flask, the maintenance of rounded parenchymal cell phenotype, and the expression of specialized hepatic function in vitro can be enhanced by precoating the flasks with ECM components such as type I collagen [Michalopoulos & Pitot, 1975], fibronectin [Deschenes et al., 1980], homogenized liver tissue matrix [Reid et al., 1980], laminin, and type IV collagen [Bissell et al., 1987]. Hepatocyte survival and functional expression also improved when hepatocytes were cocultured with liver-derived [Fry & Bridges, 1980] and murine 3T3 [Kuri-Harcuch & Mendoza-Figueroa, 1989] fibroblasts as well as a preestablished layer of adherent liver epithelial cells [Guguen-Guillouzo et al., 1983]. As with bone marrow culture on feeder layers, cells derived from the liver itself usually provide the best support in culture. However, we have found that hepatic parenchyma will express a differentiated function in vitro if supported by bone marrow stroma. Conversely, liver-derived stromal cells support hematopoiesis in culture but his microenvironment favors erythropoisis (unpublished observations).

In contrast to the role of bone marrow stroma on hematopoiesis, the influence of hepatic stromal cells on the function and/or growth of parenchymal hepatocytes has not been exhaustively investigated. However, several studies indicate the Kupffer cells, fat-strong cells, and perhaps other stroma influence parenchymal cell cytochrome P450 enzyme expression [Peterson & Renton, 1984] and act in tandem with parenchymal cells to metabolize lipopolysaccharide [Treon et al., 1979]. In addition, adipocytes and hepatic fibroblasts synthesize collagen type I and the proteoglycans heparan sulfate, dermatan sulfate, and chondroitin sulfate as well as hyaluronic acid, which is unsulfated in the liver and occurs as a simple GAG. Stromal cells as well as parenchyma deposit fibronectin. Liver adipocytes, like their relatives in bone marrow, express phenotypes (vimentin+, actin+, tubulin+) linking them histogenically to fibroblasts and mogenic cells [DeLeeuw et al, 1984]. Fat-storing cells as well as vascular endothelia apparently synthesize the type IV collagen found in the space of Disse, and fibrin originating form parenchymal cell fibrinogen synthesis is a significant part of liver matrix, at least in culture. As with the hematopoietic cells in bone marrow cultures, hepatic parenchymal cell gene expression is modulated by attachment to ECM and influenced by factors released by stromal cells. For example, parenchymal cells attaching to laminin-coated surfaces express the differentiation-associated substance α-fetoprotein, whereas the synthesis of albumin synthesis is favored when parenchymal cells bind to type IV collagen [Michalopoulous & Pitot, 1975]. Hepatocytes bind to fibronectin using the $\alpha_5\beta$ integrin heterodimer and AGp110, a nonintegrin glycoprotein. Cytoadhesion molecule expression by heapatic cells and cells of other tissues changes with and perhaps controls development [Stamatoglou & Hughes, 1994]. Just as differential hemonectin expression occurred during the development of the hematopoietic system [Peters et al., 1990], a differential expression of liver proteins occurs during ontogeny. Although fibronectin and its receptor are strongly expressed in liver throughout life, AGp110 appears later in development and may guide the development of parenchyma into a polarized tissue. It will probably be necessary to incorporate the various stromal support cells found in liver into hepatocyte cultures.

Tumor Cells

Stromal cells contribute to the process of neoplastic invasion of tissue by responding in a paracrine fashion to signals released by tumor cells. This includes the tumor-stimulated release of angiogenic factors from fibroblasts such as acid and basic FGF and the release of vascular endothelia growth factor/vascular proliferation factor (VEGF/VPF) by tumor cells to recruit endothelial cells and stimulate their proliferation and development into blood vessels to feed the growing neoplasm (reviewed by Wernert [1997]). In addition to cooperating in the establishment of tumors, stromal cells also contribute to their invasiveness. Activation and/or release of stromal cell-derived Tissue Factor (TF) (whose release by damaged tissue initiates the extrinsic limb of the protease coagulation cascade) has been associated with the progression from early to invasive breast cancer [Vrana et al., 1996]. In this regard, upon direct contact with tumor cells, stromal cells produce stromelysin 3 (STR-3), a matrix metalloproteinase that accelerates the migration of metastatic cells through tissue by degrading tissue matrix [Mari et al., 1998]. STR-3 is overexpressed in a wide variety of tumor stroma in breast, lung, colon, and other cancers. It would be appropriate to include coculture studies (i.e., tumor and stroma) in the investigation of the biology of cancer in vitro as well as its reponsivity to potential treatments. We have previously demonstrated that the presence of stroma alters the hematologic responsiveness to chemotherapy agents such as cyclophosphamide [Naughton et al., 1992]. In related work, primary cultures of murine plasmacytomas requires

TABLE 119.4 Some Representative Stromal Cell Lines Used to Support Hematopoietic Cells or Hepatocytes

Cell line	Species	Phenotype*	Support Capability
Bone marrow			
AC-4	Mouse	A	Myeloid, monocytic, B lymphoid
ALC	Mouse	A	Myeloid, monocytic, B lymphoid
GM 1380	Human	F	Short-term myelo- + monopoiesis
GY-30	Mouse	A	Myeloid, monocytic, B lymphoid myelopoiesis
K-1	Mouse	A	
MBA-14	Mouse	F/M	Enhances CFU-C numbers myelopoiesis
MS-1	Mouse	F	
U2	Mouse	F†	Maintenance of CFU-S_{12}, myelopoiesis
3T3	Mouse	F	
10 T1/2	Mouse	F	Stimulates CFU-GEMM, -GM
10T1/2 clone D	Mouse	A	Stimulates CFU-GM only
266 AD	Mouse	A	Myeloid, monocytic, B lymphoid
Liver			
3T3	Mouse	F	Enhanced lipid metabolism, extends period of cytochrome P450 activity†.
Detroit 550	Human	F	Prolonged cytochrome P450 and NADPH-cytochrome C reductase activity‡

Source: Adapted from Anklesaria et al. [1987] and Gimble [1990].

*Major phenotype of the cell line (F = fibroblastic, M = monocytic, A = capable of supporting adipocyte phenotypes).

†A bone marrow-derived cell line that was transformed with the large T oncogene of simian virus 40.

‡Support functions are compared to liquid cultures of rat hepatocytes without stroma.

Abbreviations: CFU = colony-forming unit; G = granulocyte; M = macrophage; S = spleen; GEMM = granulocyte, erythrocyte, megakaryoctye, monocyte; NADPH = nicotinamide adenine dinucleotide phosphate.

a feeder layer. Alteration of ECM composition and other support functions of the stroma by anti-inflammatory drugs prevented plasmacytoma growth *in vivo* and *in vitro* [De Grassi et al., 1993].

119.3 Support of Cultured Cells Using Cell Lines

The precise roles of the various cellular constituents of the tissue microenvironment have not been fully defined. One approach to understanding these mechanisms is to develop stroma cell lines with homogeneous and well-defined characteristics and then ascertain the ability of these cells to support parenchyma in coculture. A brief survey of some of these lines is found in Table 119.4. Stromal cell lines derived from bone marrow are usually either fibroblastic (F) or a mixture of fibroblastic and adipocytic cells which contain a subpopulation of fibroblastic cells that can be induced to undergo lipogenesis with dexamethasone or hydrocortisone (A) (reviewed by Gimble, [1990]). In general, fibroblastic lines support myelopoiesis and monocytopoiesis and stroma with both fibroblastic and adipocytic cells supports myelopoiesis as well as B lymphopoiesis. The MBA-14 cell line, which consists of stroma bearing fibroblastic as well as monocytic phenotypes, stimulated the formation of CFU-C (bipotential myeloid/monocytic) progenitors in coculture. Different cell lines that were transformed using the large T oncogene of simian virus 40 (U2) were used as feeder layers for bone marrow hematopoietic cells. Cocultures established using bone-marrow-derived stroma exhibited considerably better maintenance of the primitive stem cell CFU-S_{12} in culture than similarly transfected skin, lung, or kidney tissue cells [Rios & Williams, 1990]. Although feeder layers share a number of characteristics [cytokine production, ECM deposition), it is probably best to derive your feeder cells from the tissue that you wish to coculture rather than use xenogeneic feeder cells or cells derived from a completely dissimilar tissue. In this respect, we found that total cell output and progenitor cell concentrations were substantially higher in rat bone marrow cocultures supported with rat bone marrow stroma as compared to those established using immortalized human skin fibroblasts or fetal lung cell lines [Naughton & Naughton, 1989].

119.4 Stereotypic (Three-Dimensional) Culture Versus Monolayer Culture

Tissue is a three-dimensional arrangement of various types of cells that are organized into a functional unit. These cells also are polarized with respect to their position within tissue and the microenvironment, and therefore the metabolic activity and requirements of the tissue are not uniform throughout. Three-dimensional culture was first performed successfully by Leighton [1951] using cellulose sponge as a template. Collagen gel frameworks [Douglas, 1980] also have been and are currently being employed to culture tissues such as skin, breast epithelium, and liver. In addition, tumor tissue cocultured with stroma in collagen gels respond to drugs in a manner similar to that observed in vivo [Rheinwald & Beckett, 1981].

We developed three-dimensional coculture templates using nylon filtration screens and felts made of polyester or bioresorbably polyglycolic acid polymers [Naughton et al., 1987, 1994]. Rodent or human bone marrow cocultures retained multilineage hematopoietic expression in these stereotypic cultures, a phenomenon that is possible in plastic flask or suspension cultures only if the medium is supplemented with cocktails of cytokines [Peters et al., 1993]. Cocultures of rat hepatic parenchymal and stromal cells on three-dimensional templates also displayed a number of liver-specific functions for at least 48 days in culture, including the active synthesis of albumin, fibrinogen, and other proteins and the expression of dioxin-inducible cytochrome P450 enzyme activity for up to 2 months in culture [Naughton et al., 1994]. Furthermore, hepatic parenchyma in these stereotypic cocultures proliferated in association with stromal elements and the ECM they deposited until all "open" areas within the template were utilized. Our method is different from others because we use stromal cells derived from the tissue we wish to culture to populate the three-dimensional template. These cells secrete tissue-specific ECM and other matrix components that are indigenous to the normal microenvironment of the tissue. Parenchymal cells associate freely within the template after their inoculation and bind to other cells and/or matrix based upon their natural cytoadhesion molecule profiles. We do not add exogenous proteins.

Three-dimensional scaffolds such as nylon screen or polyester felt provide large surface areas for cell attachment and growth. Although mass transfer limitations of diffusion dictate the maximum thickness (density) of a tissue culture, suspended three-dimensional cultures have the advantage of being completely surrounded by medium. This arrangement effectively doubles the maximum tissue thickness that is possible with a plastic flask-based culture. These stereotypic cultures also appear to form tissuelike structure in vitro and, when implanted after coculture on bioresorbable polymer templates, in vivo.

Defining Terms

When hematopoietic cells are inoculated into semisolid or liquid medium containing the appropriate growth factors, some of the cells are clonal and will form colonies after approximately 2 weeks in culture. These colonies arise from hematopoietic progenitor cells and are called **colony-forming units** or **CFU**. These colonies can consist of granulocytic cells (**G**), monocytic cells (**M**), a mixture of these two cell types (**GM**), or erythroid cells (**E**). Less mature progenitor cells have a greater potential for lineage expression. For example, a **CFU-GEMM** contains granulocytic, erythroid, megakaryocytic, and monocytic cells and is therefore the least mature progenitor of this group.

The text also mentions **CFU-S**. Whereas the other assays quantify colony formation in vitro, this is an in vivo assay. Briefly, irradiated mice are infused with meatopoietic cells. Some of these cells will colonize the spleen and will produce blood cells that "rescue" the animal. As with the in vitro assays, colonies that arise later in culture originate from less mature cells. The **CFU-S$_{12}$** therefore is a more primitive hematopoietic cell than the **CFU-S$_9$**.

References

Anklesaria P, Kase K, Glowacki J, et al. 1987. Engraftment of a clonal bone marrow stromal cell line *in vivo* stimulates hematopoietic recovery from total body irradiation. Proc Natl Acad Sci USA 84:7681.

Beneyahu D, Zipori D, Wientroub S. 1993. Marrow adipocytes regulate growth and differentiation of osteoblasts. Biochem Biophys Res Comm 197:1245.

Ben-Ze'ev A, Robinson GS, Bucher NLR, et al. 1988. Cell-cell and cell-matrix interactions differentially regulate the expression of hepatic and cytoskeletal genes in primary cultures of rat hepatocytes. Proc Nat Acad Sci (USA) 85:2161.

Bessis M, Breton-Gorius J. 1959. Nouvelles observations sur l'ilot erythroblastique et la rhopheocytose de la ferritin. Rev Hemato 14:165.

Bissell DM, Arenson DM, Maher JJ, et al. 1987. Support of cultured hepatocytes by a laminin-rich gel: Evidence for a functionally significant subendothelial matrix in normal rat liver. J Clin Invest 790:801.

Brandt J, Srour EF, Van Besien K, et al. 1990. Cytokine-dependent long term culture of highly enriched precursors of hematopoietic progenitor cells from human bone marrow. J Clin Invest 86:932.

Brockbank KGM, van Peer CMJ. 1983. Colony stimulating activity production by hemopoietic organ fibroblastoid cells in vitro. Acta Haematol 69:369.

Chang VT, Anderson RN. 1971. Cultivation of mouse bone marrow cells: I. Growth of granulocytes. J Reticuloendoth Soc 9:568.

Cicuttini FM, Martin M, Ashman L., et al. 1992. Support of human cord blood progenitor cells on human stromal cell lines transformed by SV_{40} large T antigen under the influence of an inducible (metallothionein) promoter. Blood 80:102.

Degrassi A, Hilbert DM, Rudikoff S, et al. 1993. In vitro culture of primary plasmacytomas requires stromal cell feeder layers. Proc Nat Acad Sci (USA) 90:2060.

DeLeeuw AM, McCarthy SP, Geerts A, et al. 1984. Purified rat liver fat-storing cells in culture divide and contain collagen. Hepatology 4:392.

Dexter TM, Allen TD, Lajtha LG. 1977. Conditions controlling the proliferation of haematopoietic stem cells in vitro. J Cell Physiol 91:335.

Douglas WHJ, Moorman GW, Teel RW. 1980. Visualization of cellular aggregates cultured on a three-dimensional collagen sponge matrix. In Vitro 16:306.

Freshney RI, Hart E, Russell JM. 1982. Isolation and purification of cell cultures from human tumours. In E Reid, GMW Cook, DJ Moore (eds), Cancer Cell Organelles. Methodological Surveys: Biochemistry, pp 97–110, Chichester, England, Horwood Press.

Friedenstein AJ, Chailakhyan RK, Gerasimov UV. 1970. Bone marrow osteogenic stem cells: In vitro cultivation and transplantation in diffusion chambers. Cell Tiss Kinet 20:263.

Furchgott RF, Zawadzki JV. 1980. The obligatory role of endothelial cells in the relaxation of arterial smooth muscle by actylcholine. Nature 286:373.

Gallagher JT, Spooncer E, Dexter TM. 1982. Role of extracellular matrix in haemopoiesis: I. Synthesis of glycosaminoglycans by mouse bone marrow cultures. J Cell Sci 63:155.

Gimble JM. 1990. The function of adipocytes in the bone marrow stroma. New Biol 2:304.

Gimble JM, Dorheim MA, Cheng Q, et al. 1989. Response of bone marrow stromal cells to adipogenetic antagonists. Mol Cell Biol 9:4587.

Greenberger JS. 1979. Corticosteroid-dependent differentiation of human marrow preadipocytes in vitro. In Vitro 15:823.

Guguen-Guilluozo C, Clement B, Baffet G, et al. 1983. Maintenance and reversibility of active albumin secretion by adult rat hepatocytes co-cultured with another cell type. Exp Cell Res 143:47.

Gupta P, Blazar BR, Gupta K, et al. 1998. Human CD34+ bone marrow cells regulate stromal production of interleukin-6 and granulocyte colony-stimulating factor and increase the colony-stimulating activity of stroma. Blood 91:3724.

Huang S, Terstappen LWMM. 1992. Formation of haematopoietic microenvironment and haematopoietic stem cells from single human bone marrow cells. Nature 360:745.

Keller D, Ou XX, Rour EF, et al. 1993. Interleukin-II inhibits adipogenesis and stimulates myelopoiesis in human long-term marrow cultures. Blood 82:1428.

Keltz MD, Atton E, Buradagunta S, et al. 1996. Modulation of leukemia inhibitory factor gene expression and protein biosynthesis in human fallopian tube. Am J Gyecol Obstet 175:1611.

Kuri-Harcuch W, Mendoza-Figueroa T. 1989. Cultivation of adult rat hepatocytes on 3T3 cells: Expression of various liver differentiated functions. Differentiation 41:148.

Leighton J. 1951. A sponge matrix method for tissue culture: Formation of organized aggregates of cells in vitro. J Natl Cancer Inst 12:545.

Lictman MA. 1984. The relationship of stromal cells to hemopoietic cells in marrow. In DG Wright, JS Greenberger (eds), Long-Term Bone Marrow Culture, pp 3–30, New York, A.R. Liss.

Lin CQ, Bissell MJ. 1993. Multi-faceted regulation of cell differentiation by extracellular matrix. FASEB J 7:737.

Maniatis A, Tavassoli M, Crosby WH. 1971. Factors affecting the conversion of yellow to red marrow. Blood 37:581.

Mari BP, Anderson IC, Mari SE, et al. 1998. Stromelysin-3 is induced in tumor/stroma cocultures and inactivated via a tumor-specific and basic fibroblast growth factor-dependent mechanism. J Biol Chem 273:618.

McCuskey RS, Meineke HA, Townsend SF. 1972. Studies of the hemopoietic microenvironment: I. Changes in the microvascular system and stroma during erythropoietic regeneration and suppression in the spleens of CF_1 mice. Blood 5:697.

Metcalf D. 1993. Hematopoietic growth factors. Redundancy or subtlety? Blood 82:3515.

Michalopoulos G, Pitot HC. 1975. Primary culture of parenchymal liver cells on collagen membranes. Exp Cell Res 94:70.

Moreau I, Duvert V, Caux C, et al. 1993. Myofibroblastic stromal cells isolated from human bone marrow induce the proliferation of both early myeloid and B-lymphoid cells. Blood 82:2396.

Mrózek E, Anderson P, and Aligiuri MA. 1996. Role of interleukin-15 in the development of human CD56+ natural killer cells from CD34+ human progenitor cells. Blood 87:2632.

Naughton BA, Kolks GA, Arce JM, et al. 1979. The regenerating liver: A site of erythropoiesis in the adult long-evans rat. Amer J Anat 156:159.

Naughton BA, Naughton GK. 1989. Hematopoiesis on nylon mesh templates. Ann NY Acad Sci 554:125.

Naughton BA, San Roman J, Sibanda B, et al. 1994. Stereotypic culture systems for liver and bone marrow: Evidence for the development of functional tissue in vitro and following implantation in vivo. Biotech Bioeng 43:810.

Naughton BA, Sibanda B, San Román J, et al. 1992. Differential effects of drugs upon hematopoiesis can be assessed in long-term bone marrow cultures established on nylon screens. Proc Soc Exp Biol Med 199:481.

Owen ME. 1988. Marrow stromal stem cells. J Cell Sci 10:63.

Palsson BO, Paek S-H, Schwartz RM, et al. 1993. Expansion of human bone marrow progenitor cells in a high cell density continuous perfusion system. Biotechnology 11:368.

Peters C, O'Shea KS, Campbell AD, et al. 1990. Fetal expression of hemonectin: An extracellular matrix hematopoietic cytoadhesion molecule. Blood 75:357.

Peterson TC, Renton KW. 1984. Depression of cytochrome P-450-dependent drug biotransformation in hepatocytes after the activation of the reticuloendothelial system by dextran sulfate. J Pharmacol Exp Ther 229:229.

Puck TT, Marcus PI. 1955. A rapid method for viable cell titration and clone production with HeLa cells in tissue culture: the use of X-irradiated cells to supply conditioning factors. Proc Natl Aca Sci (USA) 41:432.

Reid LM, Gaitmaitan Z, Arias I, et al. 1980. Long-term cultures of normal rat hepatocytes on liver biomatrix. Ann NY Acad Sci 349:70.

Rios M, Williams DA. 1990. Systematic analysis of the ability of stromal cell lines derived from different murine adult tissues to support maintenance of hematopoietic stem cells in vitro. J Cell Physiol 145:434.

Roberts R, Gallagher J, Spooncer E, et al. 1988. Heparan sulfate bound growth factors: A mechanism for stromal cell-mediated haemopoiesis. Nature 332:376.

Ryan DH, Nuccie BL, Abboud CN, et al. 1991. Vascular cell adhesion molecule-1 and the integrin VLA-4 mediate adhesion of human B cell precursors to cultured bone marrow adherent cells. J Clin Invest 88:995.

Schrock LM, Judd JT, Meineke HA, et al. 1973. Differences in concentration of acid mucopolysaccharides between spleens of normal and polycythemic CF1 mice. Proc Soc Exp Biol Med 144:593.

Siczkowski M, Amos As, Gordon MY. 1993. Hyaluronic acid regulates the function and distribution of sulfated glycosaminoglycans in bone marrow stromal cultures. Exp hematol 21:126.

Simmons PJ, Masinovsky B, Longenecker BM, et al. 1992. Vascular cell adhesion molecule-1 expressed by bone marrow stromal cells mediates the binding of hematopoietic progenitor cells. Blood 80:388.

Simmons PJ, Torok-Storb B. 1991. CD 34 expression by stromal precursors in normal human adult bone marrow. Blood 78:2848.

Strobel E-S, Gay RE, Greenberg PL. 1986. Characterization of the in vitro stromal microenvironment of human bone marrow. Int J Cell Cloning 4:341.

Stamatoglou SC, Hughes RC. 1994. Cell adhesion molecules in liver function and pattern formation. FASEB J 8:420.

Thonar EJ-MA, Kuttner KE. 1987. Biochemical basis of age-related changes in porteoglycans. In: TN Wight, RP Mecham (eds), Biology of Proteoglycans, pp 211–246, New York, Academic Press.

Tjota A, Rossi Tm, Naughton BA. 1992. Stromal cells derived from spleen or bone marrow support the proliferation of rat natural killer cells in long-term culture. Proc Soc Exp Biol Med 200:431.

Touw I, Lowenberg B. 1983. No simulative effect of adipocytes on hematopoiesis in long-term human bone marrow cultures. Blood 61:770.

Trentin JJ. 1970. Influence of hematopoietic organ stroma (Hematopoietic inductive microenvironment) on stem cell differentiation. In AS Gordon (ed), Regulation of Hematopoiesis, New York, Appple-ton-Century-Crofts.

Treon SP, Thomas P, Baron J. 1979. Lippopolysaccharide (LPS) processing by Kupffer cells releases a modified LPS with increased hepatocyte binding and decreased tumor necrosis-α stimulatory capacity. Proc Soc Exp Biol Med 202:153.

Tsai S, Patel V. Beaumont E, et al. 1987. Differential binding to erythroid and myeloid progenitors to fibroblasts and fibronectin. Blood 69:1587.

Tsuji JM and Pollack SB. 1995. Maturation of murine natural killer cells in the absence of exogenous cytokines requires contact with bone marrow stroma. Nat Immun 14:44.

Van Eyken P, Desmet VJ. 1993. Bile duct cells. In AV Le Bouton (ed), Molecular and Cell Biology of the Liver, pp 475-524, Boca Raton, Fla, CRC Press.

Vaz F, Srour EF, Almeida-Porada G, et al. 1998. Human thymic stroma supports natural killer (NK) cell development from immature progenitors. Cell Immunol 186:133.

Vrana JA, Stang MT, Grande JP, et al. 1996. Expression of tissue factor in tumor stroma correlates with progression to invasive human breast cancer: paracrine regulation by carcinoma cell-derived members of the transforming growth factor beta family. Cancer Res 56:5063.

Wernert N. 1997. The multiple roles of tumor stroma. Virchows Arch 430:433.

Whitlock CA, Witte ON. 1982. Long term culture of B lymphocytes and their precursors from murine bone marrow. Proc Nat Aca Sci (USA) 77:4756.

Wight TN, Kinsella MG, Keating A, et al. 1986. Proteoglycans in human long-term bone marrow cultures: Biochemical and ultrastructural analysis. Blood 67:1333.

Wilson D. 1983. The origin of the endothelium in the developing marginal vein of the chick wing-bud. Cell Differ 13:63.

Wolf NS, Trentin JJ. 1968. Hematopoietic colony studies: V. Effects of hemopoietic organ stroma on differentiation of pluripotent stem cells. J Exp Med 127:205.

Yanagisawa M, Jurihara HJ, Kimura S, et al. 1988. A novel potent vasoconstrictor peptide produced by vascular endothelial cells. Nature 332:411.

Yang Y-C, Tsai S, Wong GG, et al. 1988. Interleukin-1 regulation of hematopoietic growth factor production by human stromal fibroblasts. J Cell Physiol 134:292.

Zuckerman KS. 1984. Composition and function of the extracellular matrix in the stroma of long-term bone marrow cell cultures. In DG Wright, JS Greenberger (eds), Long-Term Bone Marrow Culture, pp 157–170, New York, A.R. Liss.

Zuckerman KS, Rhodes RK, Goodrum DD, et al. 1985. Inhibition of collagen deposition in the extracellular matrix prevents the establishment of stroma supportive of hematopoiesis in long term murine bone marrow cultures. J Clin Invest 75:970.

Further Information

There are a number of different methodologies for isolating cells including gradient density centrifugation, sedimentation at unit gravity, lectin agglutination, and reaction with specific antibodies followed by immunoselection via panning or immunomagnetic microspheres. These methods are described and illustrated well in volumes 1–5 of *Cell Separation: Methods and Selected Applications*, New York, Academic Press, 1987. For additional details concerning the relevance of ECM deposition to the development and functional expression of various types of tissue cells, the reader is referred to the serial reviews of this subject that appeared in *The FASEB Journal* from volume 7, number 9 (1993) to volume 8, number 4 (1994). Information about the various cell types of the liver, their interaction with matrix components, and their mechanisms of gene expression can be found in *Molecular and Cell Biology of the Liver*, Boca Raton, Florida, CRC Press (1993). Similarly, for more details concerning bone marrow cells and the mechanisms of hematopoiesis, consult *The Human Bone Marrow*, volume 1, Boca Raton, Florida, CRC Press (1992).

120

Tissue Engineering of Bone Marrow

Manfred R. Koller
Oncosis

Bernhard Ø. Palsson
University of California–San Diego

The human body consumes a staggering 400 billion mature blood cells every day, and this number increases dramatically under conditions of stress such as infection or bleeding. A complex scheme of multilineage proliferation and differentiation, termed *hematopoiesis* (Greek for blood forming), has evolved to meet this demand. This regulated production of mature blood cells from immature stem cells, which occurs mainly in the bone marrow (BM) of adult mammals, has been the focus of considerable research effort. Ex vivo models of human hematopoiesis now exist that have significant scientific value and promise to have an impact on clinical practice in the near future. This endeavor is spread across many fields, including cell biology, molecular biology, bioengineering, and medicine.

This chapter introduces the reader to the fundamental concepts of hematopoiesis, the clinical applications which drive much of the effort to reconstitute hematopoiesis ex vivo, and the progress made to date toward achieving this goal.

120.1 Biology of Hematopoiesis

The Hematopoietic System: Function and Organization

There are eight major types of mature blood cells which are found in the circulation (see Fig. 120.1). The blood cell population is divided into two major groups; the myeloid and lymphoid. The myeloid

0-8493-0462-8/00/$0.00+$.50

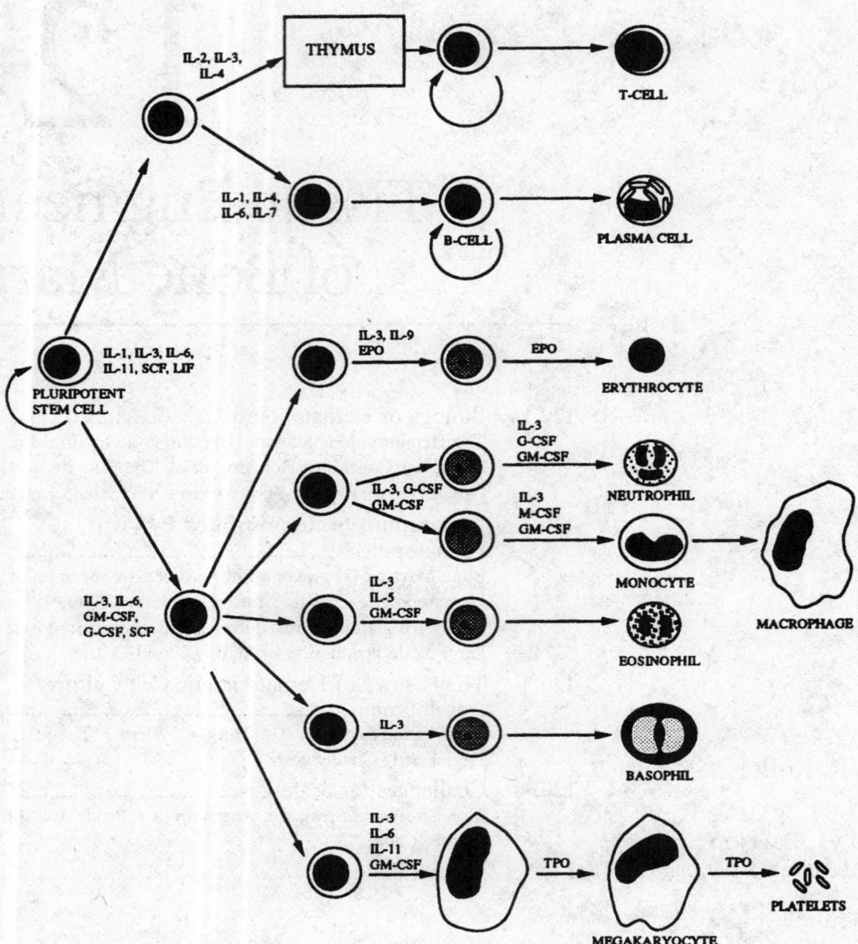

FIGURE 120.1 The hematopoietic system hierarchy. Dividing pluripotent stem cells may undergo self-renewal to form daughter stem cells without loss of potential or may experience a concomitant differentiation to form daughter cells with more restricted potential. Continuous proliferation and differentiation along each lineage results in the production of many mature cells. This process is under the control of many growth factors (GFs) (See Table 120.1). The site of action of some of the better-studied GFs are shown. The mechanisms that determine which lineage a cell will develop into are not understood, although many models have been proposed.

lineage includes erythrocytes (red blood cells), monocytes, the granulocytes (neutrophils, eosinophils, and basophils), and platelets (derived from noncirculating megakaryocytes). Thymus-derived (T) lymphocytes and BM-derived (B) lymphocytes constitute the lymphoid lineage. Most mature blood cells exhibit a limited lifespan in vivo. Although some lymphocytes are thought to survive for many years, it has been shown that erythrocytes and neutrophils have lifespans of 120 days and 8 hours, respectively [Cronkite, 1988]. As a result, hematopoiesis is a highly prolific process which occurs throughout our lives to fulfill this demand.

Mature cells are continuously produced from *progenitor cells*, which in turn are produced from earlier cells which originate from *stem cells*. There are many levels in this hierarchical system, which is usually diagrammed as shown in Fig. 120.1. At the left are the very primitive stem cells, the majority of which are in a nonproliferative state (G_o) [Lajtha, 1979]. These cells are very rare (1 in 100,000 BM cells) but collectively have enough proliferative capacity to last several lifetimes [Boggs et al., 1982; Spangrude et al., 1988]. Through some unknown mechanism, at any given time a small number of these cells are actively proliferating, differentiating, and self-renewing, thereby producing more mature progenitor cells while maintaining the size of the stem cell pool. Whereas stem cells (by definition) are not restricted to any lineage, their progenitor cell progeny do have a restricted potential and are far greater in number. The restricted nature of these progenitors has led to a nomenclature which describes their potential outcome. Those that develop into erythrocytes are called colony-forming unit-erythrocyte (CFU-E, the term colony-forming unit relates to the biological assay which is used to measure progenitor cells). Similarly, progenitors which form granulocytes and macrophages are called CFU-GM. Therefore, as the cells differentiate and travel from left to right in Fig. 120.1, they become more numerous, lose self-renewal ability, lose proliferative potential, become restricted to a single lineage, and finally become a mature cell of a particular type. The biology of stem cells is discussed elsewhere in this *Handbook* (Chapter 116). The need for identification of the many cell types present in the hematopoietic system has led to many types of assays. Many of these are biologic assays (such as *colony-forming assays*), which are performed by culturing the cells and examining their progeny, both in number and type [Sutherland et al., 1991*a*]. Another example is the *long-term culture-initiating cell* (LTC-IC) assay which measures a very early cell type through 5–16 weeks of in vitro maintenance [Koller et al., 1998a]. In contrast to these biologic assays, which are destructive to the cells being measured, is the real-time technique of *flow cytometry.* Flow cytometry has been used extensively in the study of the hematopoietic system hierarchy. Antibodies to antigens on many of the cell types shown in Fig. 120.1 have been developed (see [Brott et al., 1995]). Because of the close relation of many of the cell types, often combinations of antigens are required to definitively identify a particular cell. Recently, much effort has been focused on the identification of primitive stem cells, and this has been accomplished by analyzing increasingly smaller subsets of cells using increasingly complex antibody combinations. The first such antigen that was found in CD34, which identifies all cells from the stem through progenitor stage (typically about 2% of BM mononuclear cells (MNC), see Fig. 120.2) [Civin et al., 1984]. The CD34 antigen is stage-specific but not lineage-specific and therefore identifies cells that lead to repopulation of all cell lineages in transplant patients [Berenson et al., 1991]. However, the CD34 antigen is not restricted to hematopoietic cells because it is also found on certain stromal cells in the hematopoietic microenvironment (see Chapters 118 and 119). Although CD34 captures a small population which contains stem cells, this cell population is itself quite heterogeneous and can be fractionated by many other antigens. Over the past several years, many different combinations of antibodies have been used to fractionate the $CD34^+$ population. $CD34^+$ fractions which lack CD33, HLA-DR, CD38, or CD71 appear to be enriched in stem cells [Civin & Gore, 1993]. Conversely, $CD34^+$ populations which coexpress Thy-1 or *c-kit* appear to contain the primitive cells [Civin & Gore, 1993]. These studies have revealed the extreme rarity of stem cells within the heterogeneous BM population (see Fig. 120.2). Of the *mononuclear cell* (MNC) subset (~40% of whole BM), only ~2% are $CD34^+$, and of those, only ~5% may be $CD38^+$. Furthermore, this extremely rare population is still heterogeneous with respect to stem cell content. Consequently, stem cells as single cells have not yet been identified.

Molecular Control of Hematopoiesis: The Hematopoietic Growth Factors

A large number of hematopoietic growth factors (GFs) regulate both the production and functional activity of hematopoietic cells. The earliest to be discovered were the colony-stimulating factors (CSFs, because of their activity in the colony-forming assay), which include interleukin-3 (IL-3), granulocyte-macrophage (GM)-CSF, granulocyte (G)-CSF, and monocyte (M)-CSF. These GFs, along with erythropoeitin, have been relatively well characterized because of their obvious effects on mature cell production and/or activation. The target cells of some of the better-studied GFs are shown in Fig. 120.1. Subsequent intensive research continues to add to the growing list of GFs that affect hematopoietic cell proliferation, differentiation, and function (Table 120.1). However, new GFs have been more difficult to find and characterize because their effects are more subtle, often providing a synergistic effect which potentiates other known GFs. In addition, there appears to be a significant amount of redundancy and pleotropy in this GF network, which makes the discovery of new GFs difficult [Metcalf, 1993]. In fact, more recent

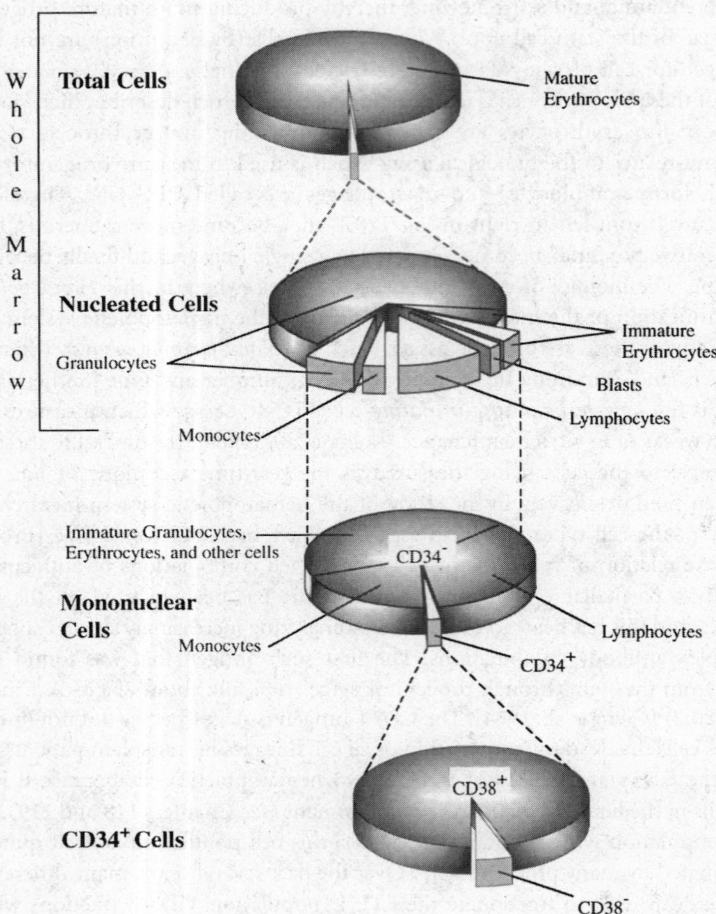

FIGURE 120.2 Relative frequency of the different cell subsets within the BM population. A BM aspirate typically contains 99% mature erythroid cells (mostly from blood contamination), and therefore usually only the nucleated cell fraction is studied. Simple density gradient centrifugation techniques, which remove most of the mature erythrocytes and granulocytes, yield what is known as the mononuclear cell (MNC) fraction (about 40% of the nucleated cells). $CD34^+$ cells, about 2% of MNC, can be isolated by a variety of methods to capture the primitive cells, although the population is quite heterogeneous. The most primitive cells are found in subsets of $CD34^+$ cells (e.g., $CD38^-$), which identify about 5% of the $CD34^+$ population. These rare subsets can be obtained by flow cytometry but are still somewhat heterogeneous with respect to stem cell content. Consequently, although methods are available to fractionate BM to a great extent, individual stem cells have not yet been identified. This diagram conveys the heterogeneous nature of different BM populations as well as the incredible rarity of the primitive cell subset, which is known to contain the stem cells.

discoveries have focused on potential receptor molecules on the target cell surface, which then have been used to isolate the appropriate ligand GF. Examples of such recently discovered GFs which exhibit synergistic interactions with other GFs to act on primitive cells include *c-kit* ligand and *flt-3* ligand. Another example is *thrombopoietin*, a stimulator of platelet production, a factor whose activity was described over 30 years ago but was cloned only recently [De Sauvage et al., 1994]. These and other GFs that act on primitive cells are the subject of intense study because of their potential scientific and commercial value. Already, several of the GFs have been developed into enormously successful pharmaceuticals, used in patients who have blood cell production deficiencies due to many different causes (see below).

TABLE 120.1 Hematopoietic Growth Factors

Growth Factor Name	Other Names	Abbreviations	Reference
Interleukin-1	Hemopoietin-1	IL-1	Dinarello et al., 1981
Interleukin-2		IL-2	Smith, 1988
Interleukin-3	Multicolony-stimulating factor	IL-3, Multi-CSF	Ihle et al., 1981
Interleukin-4	B-cell-stimulatory factor-1	IL-4, BSF-1	Yokota et al., 1988
Interleukin-5		IL-5	Yokota et al., 1988
Interleukin-6	B-cell-stimulatory factor-2	IL-6, BSF-2	Kishimoto, 1989
Interleukin-7		IL-7	Tushinski et al., 1991
Interleukin-8	Neutrophil activating peptide-1	IL-8, NAP-1	Herbert and Baker, 1993
Interleukin-9		IL-9	Donahue et al., 1990
Interleukin-10	Cytokine synthesis inhibitory factor	IL-10, CSIF	Zlotnik and Moore,1991
Interleukin-11		IL-11	Du and Williams, 1994
Interleukin-12	NK cell stimulatory factor	IL-12, NKSF	Wolf et al., 1991
Interleukin-13		IL-13	Minty et al., 1993
Interleukin-14	High molecular weight B cell growth factor	IL-14, HMW-BCGF	Ambrus Jr. et al., 1993
Interleukin-15		IL-15	Grabstein et al., 1994
Interleukin-16	Lymphocyte chemoattractant factor	IL-16; LCF	Center et al., 1997
Interleukein-17		IL-17	Yao et al., 1995
Interleukin-18	IFN-gamma-inducing factor	IL-18; IGIF	Ushio et al., 1996
Erythropoietin		Epo	Krantz, 1991
Monocyte-CSF	Colony-stimulating factor-1	M-CSF, CSF-1	Metcalf, 1985
Granulocyte-CSF		G-CSF	Metcalf, 1985
Granulocyte-macrophage-CF		GM-CSF	Metcalf, 1985
Stem cell factor	*c-kit* ligand, Mast cell growth factor	SCF, KL, MGF	Zsebo et al., 1990
Interferon-gamma	Macrophage activating factor	IFN-γ; MAF	Virelizier et al., 1985
Macrophage inflammatory protein-1	Stem cell inhibitor	MIP-1, SCI	Graham et al., 1990
Leukemia inhibitory factor		LIF	Metcalf, 1991
Transforming growth factor-beta		TGF-β	Sporn and Roberts, 1989
Tumor necrosis factor-alpha	Cachetin	TNF-α	Pennica et al., 1984
flk-2 ligand	*flk*-3 ligand	Fl	Lyman et al., 1994
Thrombopoietin	c-mpl ligand; Megakaryocyte growth and development factor	Tpo, ML; MGDF	de Sauvage et al., 1994

The Bone Marrow Microenvironment

Hematopoiesis occurs in the BM cavity in the presence of many accessory and support cells (or *stromal cells*). In addition, like all other cells in vivo, hematopoietic cells have considerable interaction with the extracellular matrix (ECM). These are the chief elements of what is known as the BM microenvironment. Further details on the function of stroma and the microenvironment, and their importance in tissue engineering, are found in Chapters 118 and 119.

Bone Marrow Stromal Cells

Due to the physiology of marrow, hematopoietic cells have a close structural and functional relationship with stromal cells. Marrow stroma includes fibroblasts, macrophages, endothelial cells, and adipocytes. The ratio of these different cell types varies at different places in the marrow and as the cells are cultured in vitro. The term stromal layer therefore refers to an undefined mixture of different adherent cell types which grow out from a culture of BM cells. In vitro, stem cells placed on a stromal cell layer will attach to and often migrate underneath the stromal layer [Yamakazi et al., 1989]. Under the stromal layer, some of the stem cells will proliferate, and the resulting progeny will be packed together, trapped under the stroma, forming a characteristic morphologic feature known as a cobblestone area (see below). It is widely believed that primitive cells must be in contact with stromal cells to maintain their primitive state. However, much of the effect of stromal cells has been attributed to the secretion of GFs. Consequently, there have been reports of successful hematopoietic cell growth with the addition of numerous soluble

GFs in the absence of stroma [Bodine et al., 1989; Brandt et al., 1990,1992; Haylock et al., 1992; Koller et al., 1992*a*; Verfaillie, 1992]. However, this issue is quite controversial, and stromal cells are still likely to be valuable because they synthesize membrane-bound GFs [Toksoz et al., 1992], ECM components [Long, 1992], and probably some as yet undiscovered cytokines. In addition, stromal cells can modulate the GF environment in a way that would be very difficult to duplicate by simply adding soluble GFs [Koller et al., 1995]. This modulation may be responsible for the observations that stroma can be both stimulatory and inhibitory [Zipori, 1989].

The Extracellular Matrix

The ECM of BM consists of collagens, laminin, fibronectin [Zuckerman & Wicha, 1983], victronectin [Coulombel et al., 1988], hemonectin [Campbell et al., 1987], and thrombospondin [Long & Dixit, 1990]. The heterogeneity of this system is further complicated by the presence of various proteoglycans, which are themselves complex molecules with numerous glycosaminoglycan chains linked to a protein core [Minguell & Tavassoli, 1989; Spooncer et al., 1983; Wight et al., 1986]. These glycosaminoglycans include chondroitin, heparan, dermatan, and keratan sulfates and hyaluronic acid. The ECM is secreted by stromal cells of the BM (particularly endothelial cells and fibroblasts) and provides support and cohesion for the marrow structure.

There is a growing body of evidence indicating that ECM is important for the regulation of hematopoiesis. Studies have shown that different glycosaminoglycans bind and present different GFs to hematopoietic cells in an active form [Gordon et al., 1987; Roberts et al., 1988]. This demonstrates that ECM can sequester and compartmentalize certain GFs in local areas and present them to hematopoietic cells, creating a number of different hematopoietically inductive microenvironments. Another important ECM function is to provide anchorage for immature hematopoietic cells. Erythroid precursors have receptors which allow them to attach to fibronectin. As cells mature through the BFU-E to the reticulocyte stage, adherence to fibronectin is gradually lost, and the cells are free to enter the circulation [Patel et al., 1985]. It has also been shown that binding to fibronectin renders these erythroid precursors more responsive to the effects of Epo [Weinstein et al., 1989]. Another adhesion protein termed hemonectin has been shown to selectively bind immature cells of the granulocyte lineage in an analogous fashion [Campbell et al., 1987]. This progenitor binding by ECM has led to the general concept for stem cell homing. When BM cells are injected into the circulation of an animal, a sufficient number are able to home to the marrow and reconstitute hematopoiesis [Tavassoli & Hardy, 1990]. It is therefore likely that homing molecules are present on the surface of the primitive cells. Studies suggest that lectins and CD44 on progenitor cells may interact with ECM and stromal elements of the marrow to mediate cellular homing [Aizawa & Tavassoli, 1987; Kansas et al., 1990; Lewinsohn et al., 1990; Tavassoli and Hardy, 1990]. These concepts have been reviewed in detail [Long, 1992].

120.2 Applications of Reconstituted Ex Vivo Hematopoiesis

The hematopoietic system, as described above, has many complex and interacting features. The reconstitution of functional hematopoiesis, which has long been desired, must address these features to achieve a truly representative ex vivo system. A functioning ex vivo human hematopoietic system would be a valuable analytic model to study the basic biology of hematopoiesis. The clinical applications of functional ex vivo models of human hematopoiesis are numerous and are just beginning to be realized. Most of these applications revolve around cancer therapies and, more recently, gene therapy. The large-scale production of mature cells for transplantation represents an important goal that may be realized in the more distant future.

Bone Marrow Transplantation

In 1980, when *bone marrow transplantation* (BMT) was still an experimental procedure, fewer than 200 BMTs were performed worldwide. Over the past decade, BMT has become an established therapy for many diseases. In 1996, over 40,000 BMTs were performed, primarily in the United States and Western

Europe, for more than a dozen different clinical indications [Horowitz and Rowling, 1997]. The number of BMTs performed annually is increasing at a rate of 20–30% per year and is expected to continue to rise in the foreseeable future.

BMT is required as a treatment in a number of clinical settings because the highly prolific cells of the hematopoietic system are sensitive to many of the agents used to treat cancer patients. Chemotherapy and radiation therapy usually target rapidly cycling cells, so hematopoietic cells are ablated along with the cancer cells. Consequently, patients undergoing these therapies experience neutropenia (low neutrophil numbers), thrombocytopenia (low platelet numbers), and anemia (low red blood cell numbers), rendering them susceptible to infections and bleeding. A BMT dramatically shortens the period of neutropenia and thromobocytopenia, but the patient may require repeated transfusions. The period during which the patient is neutropenic represents the greatest risk associated with BMT. In addition, some patients do not achieve *engraftment* (when cell numbers rise to safe levels). As a result, much effort is focused on reducing the severity and duration of the blood cell nadir period following chemotherapy and radiation therapy.

There are several sources of hematopoietic cells for transplantation. BMT may be performed with patient marrow (autologous) that has been removed and cryopreserved prior to administration of chemotherapy or with donor marrow (allogeneic). The numbers of autologous transplants outnumber allogenic transplants by a 2:1 ratio, but there are advantages and disadvantages with both techniques.

Autologous Bone Marrow Transplantation

Autologous BMTs have been used in the treatment of a variety of diseases including acute lymphoblastic leukemia (ALL), acute myelogenous leukemia (AML), chronic myelogenous leukemia (CML), various lymphomas, breast cancer, neuroblastoma, and multiple myeloma. Currently, autologous BM transplantation is hampered by the long hospital stay that is required until engraftment is achieved and the possibility of reintroducing tumor cells along with the cryopreserved marrow. In fact, retroviral marking studies have proven that tumor cells reinfused in the transplant can contribute to disease relapse in the patient [Rill et al., 1994; Deisseroth et al., 1994]. Long periods of neutropenia, anemia, and thrombocytopenia require parenteral antibiotic administration and repeated blood component transfusions.

Autologous transplantation could also be used in gene therapy procedures. The basic concept underlying gene therapy of the hematopoietic system is the insertion of a therapeutic gene into the hematopoietic stem cell, so that a stable transfection is obtained. Engineered retroviruses are the gene carriers currently used. However, mitosis of primitive cells is required for integration of foreign DNA [Bodine et al., 1989]. A culture system which contains dividing stem cells is therefore critical for the enablement of retroviral-based gene therapy of the hematopoietic system. This requirement holds true whether or not the target stem cell population has been purified prior to the transfection step. To date, only very limited success has been achieved with retroviral transfection of human BM cells, whereas murine cells are routinely transfected. A comprehensive and accessible accounting of the status of gene therapy has been presented elsewhere [Mulligan, 1993].

Allogeneic Bone Marrow Transplantation

In patients with certain hematologic malignancies, or with genetic defects in the hematopoietic population, allogeneic transplants are currently favored when suitable matched donors are available. With these leukemias, such as CML, it is likely that the patient's marrow is diseased and would not be suitable for autotransplant. A major obstacle in allogeneic transplantation, however, is the high incidence of *graft-versus-host disease* (GVHD), in which the transplanted immune cells attack the host's tissues as foreign.

Alternative Sources of Hematopoietic Cells for Transplantation

Although hematopoiesis occurs mainly in the BM of adult mammals, during embryonic development, pluripotent stem cells first arise in the yolk sac, are later found in the fetal liver, and at the time of delivery are found in high concentrations in umbilical cord blood. In adults, stem cells are found in peripheral blood only at very low concentrations, but the concentration increases dramatically after stem cell mobilization. Mobilization of stem cells into peripheral blood is a phenomenon that occurs in response to chemotherapy or GF administration. Therefore, hematopoietic stem cells can be collected from cord blood [Gluckman et al., 1989; Wagner et al., 1992] or from mobilized peripheral blood [Schneider et al., 1992] as well as from BM. Disadvantages of cord blood are the limited number of cells that can be obtained from the one individual and the question of whether this amount is sufficient to repopulate an adult patient. Mobilized peripheral blood results in more rapid patient engraftment than BM, and as a consequence, its use is becoming more prevalent, particularly in the autologous setting.

Tissue Engineering and Improved Transplantation Procedures

BMT would be greatly facilitated by reliable systems and procedures for ex vivo stem cell maintenance, expansion, and manipulation. For example, the harvest procedure, which collects 1–2 liters of marrow, is currently a painful and involved operating room procedure. The complications and discomfort of marrow donation are not trivial and can affect donors for a month or more [Stroncek et al., 1993]. Through cell expansion techniques, a small marrow specimen taken under local anesthesia in an outpatient setting could be expanded into the large number of cells required for transplant, thereby eliminating the large harvest procedure. Engraftment may be accelerated by increasing the numbers of progenitors and immature cells available for infusion. In addition, it may be possible to cryopreserve expanded cells to be infused at multiple time points, thereby allowing multiple cycles of chemotherapy (schedule intensification). Finally, the use of expanded cells may allow increasing doses of chemotherapy (dose intensification), facilitating tumor reduction while ameliorating myeloablative side effects.

The expansion of alternative hematopoietic cell sources would also facilitate transplant procedures. For example, multiple rounds of apheresis, each requiring ~4 hours, are required to collect enough mobilized peripheral blood cells for transplant. Expansion of a small amount of mobilized peripheral blood may reduce the number of aphereses required or eliminate them altogether by allowing the collection of enough cells from a volume of blood that has not been apheresed. With cord blood, there is a limit on the number of cells that can be collected from a single donor, and it is currently thought that this number is inadequate for an adult transplant. Consequently, cord blood transplants to date have been performed on children. Expansion of cord blood cells may therefore enable adult transplants from the limited number of cord blood cells available for collection.

Large-Scale Production of Mature Blood Cells

Beyond the ability to produce stem and progenitor cells for transplantation purposes lies the promise to produce large quantities of mature blood cells. Large-scale hematopoietic cultures could potentially provide several types of clinically important mature blood cells. These include red blood cells, platelets, and granulocytes. About 12 million units of red blood cells are transfused in the United States every year, the majority of them during elective surgery and the rest in acute situations. About 4 million units of platelets are transfused every year into patients who have difficulty exhibiting normal blood clotting. Mature granulocytes, which constitute a relatively low-usage market of only a few thousand units administered each year, are involved in combating infections. This need arises in situations when a patient's immune system has been compromised and requires assistance in combating opportunistic infections, such as during chemotherapy and the healing of burn wounds.

All in all, the market for these blood cells totals about $1–1.5 billion in the United States annually, with a worldwide market that is about three to four times larger. The ability to produce blood cells on demand ex vivo would alleviate several problems with the current blood cell supply. The first of these problems is the availability and stability of the blood cell supply. The availability of donors has traditionally been a problem, and, coupled with the short shelf-life of blood cells, the current supply is unstable and cannot meet major changes in demand. The second problem is the usual blood-type compatibility problem resulting in shortages of certain types at various times. The third problem is the safety of the blood supply. This last issue has received much attention recently due to the contamination of donated blood with the human immunodeficiency virus (HIV). However, the three forms of hepatitis currently pose an even more serious viral contamination threat to the blood supply.

Unlike ex vivo expansion of stem and progenitor cells for transplantation, the large-scale production of fully mature blood cells for routine clinical use is less developed and represents a more distant goal. The large market would require systems of immense size, unless major improvements in culture productivity are attained. For example, the recent discovery of thrombopoietin [De Sauvage et al., 1994] may make the large-scale production of platelets feasible. At present, there are ongoing attempts in several laboratories to produce large numbers of neutrophils from CD34-selected cell populations. Thus, large-scale production of mature cells may soon become technically feasible, although the economic considerations are still unknown.

120.3 The History of Hematopoietic Cell Culture Development

As outlined above, there are many compelling scientific and clinical reasons for undertaking the development of efficient ex vivo hematopoietic systems. Such achievement requires the use of in vivo mimicry, sophisticated cell culture technology, and the development of clinically acceptable cell cultivation devices. The foundations for these developments lie in the BM cell culture methods which have been developed over the past 20 years. The history of BM culture will therefore be described briefly, as it provides the backdrop for tissue engineering of the hematopoietic system. More complete reviews have been published previously [Dexter et al., 1984; Eaves et al., 1991; Greenberger, 1984].

The Murine System

In the mid 1970s, Dexter and coworkers were successful in developing a culture system in which murine hematopoiesis could be maintained for several months [Dexter et al., 1977]. The key feature of this system was the establishment of a BM-derived stromal layer during the first three weeks of culture which was then recharged with fresh BM cells. One to 2 weeks after the cultures were recharged, active sites of hematopoiesis appeared. These sites are often described as cobblestone regions, which are the result of primitive cell proliferation (and accumulation) underneath the stromal layer. Traditionally, the cultures are fed by replacement of one-half of the medium either once or twice weekly. In these so-called Dexter cultures, myelopoiesis proceeds to the exclusion of lymphopoiesis. The selection of a proper lot of serum for long-term BM cultures (LTBMCs) was found to be very important. In fact, when using select lots of serum, one-step LTBMCs were successfully performed without the recharging step at week 3 [Dexter et al., 1984]. It is thought that good serum allows rapid development of stroma before the primitive cells are depleted, and once the stroma is developed, the culture is maintained from the remaining original primitive cells without need for recharging. The importance of stroma has often been demonstrated in these Dexter cultures, because the culture outcome was often correlated with stromal development.

The Human System

The adaptation of one-step LTBMC for human cells was first reported in 1980 [Gartner & Kaplan, 1980]. A mixture of fetal bovine serum and horse serum was found to be required for human LTBMC, and a number of other medium additives such as sodium pyruvate, amino acids, vitamins, and antioxidants were found to be beneficial [Gartner & Kaplan, 1980; Greenberg et al., 1981; Meagher et al., 1988]. Otherwise, the culture protocol has remained essentially the same as that used for murine culture. Unfortunately, human LTBMCs have never attained the productivity or longevity which is observed in cultures of other species [Dexter et al., 1984; Greenberger et al., 1986]. The exponentially decreasing numbers of total and progenitor cells with time in human LTBMC [Eastment & Ruscetti, 1984; Eaves et al., 1991] renders the cultures unsuitable for cell expansion and indicates that primitive stem cells are lost over time. The discovery of hematopoietic GFs was an important development in human hematopoietic cell culture, because addition of GFs to human LTBMC greatly enhanced cell output. However, GFs did not prolong the longevity of the cultures, indicating that primitive cell maintenance was not improved [Coutinho et al., 1990; Lemoli et al., 1992]. Furthermore, although the total number of progenitors obtained was increased by GF's, it was still less than the number used to initiate the culture. Therefore, a net expansion in progenitor cell numbers was not obtained. The increased cell densities that were stimulated by GF addition were not well supported by the relatively static culture conditions.

The disappointing results from human LTBMCs led to the development of other culture strategies. The development of an increasing number of recombinant GFs was soon joined by the discovery of the CD34 antigen (see above). As protocols for the selection of $CD34^+$ cells became available, it was thought that the low cell numbers generated by enrichment could be expanded in GF-supplemented cultures without the impediment of numerous mature cells in the system. Because the enrichment procedure results in a cell population depleted of stromal cells, $CD34^+$ cell cultures are often called suspension cultures, due to the lack of an adherent stromal layer. A number of groups have reported experiments in which $CD34^+$ cells were incubated with high doses of up to seven recombinant GFs in suspension culture [Brandt et al., 1992; Haylock et al. 1992]. Although 500–1000-fold cell expansion numbers are often obtained, the magnitude of CFU-GM expansion is usually less than 10-fold, suggesting that differentiation, accompanied by depletion of primitive cells, is occurring in these systems. In fact, when LTC-IC have been measured, the numbers obtained after static culture of enriched cells have always been significantly below the input value [Sutherland et al., 1991*b*, 1993; Verfaillie, 1992]. A further consideration in $CD34^+$ cell culture is the loss of cells during the enrichment procedure. It is not uncommon to experience 70–80% loss of progenitors with most $CD34^+$ cell purification protocols [Traycoff et al., 1994], and this can be very significant when trying to maximize the final cell number obtained (such as in clinical applications, see below). Nevertheless, cultures of purified $CD34^+$ cells, and especially the smaller subsets (e.g., $CD33^-$, $CD38^-$), have yielded valuable information on the biology of hematopoietic stem cells.

An alternative approach has also been taken to improve human hematopoietic cell culture. Most of these advances have come from the realization that traditional culture protocols are highly nonphysiologic and that these deficiencies can be corrected by in vivo mimicry. Therefore, these techniques do not involve cell purification or high-dose cytokine stimulation. Because these cultures attain fairly high densities, it was thought that the tradition of changing one-half of the culture medium either once or twice weekly was inadequate. When Dexter-type cultures were performed with more frequent medium exchanges, progenitor cell production was supported for at least 20 weeks [Schwartz et al., 1991*b*]. This increase in culture longevity indicates that primitive cells were maintained for a longer period, and this was accompanied by an increase in progenitor cell yield. Although the precise mechanisms of increased medium exchange are unknown, the increased feeding rate enhances stromal cell secretion of GFs [Caldwell et al., 1991].

As previously noted, recombinant GFs can significantly improve culture productivity, but GF-stimulated cell proliferation exacerbates the problems of nutrient depletion because cell proliferation and the consumption of metabolites increases many-fold. Therefore, increased feeding protocols also benefit GF-stimulated cultures. Cultures supplemented with IL-3/GM-CSF/Epo and fed with 50% daily medium exchanges were found to result in significant cell and progenitor expansion while maintaining culture longevity [Schwartz et al., 1991*a*]. Optimization of these manually fed cultures has been published for both BM [Koller et al., 1996] and cord blood [Koller et al., 1998b].

Tissue Engineering Challenges

Although increased manual feeding significantly enhanced the productivity and longevity of hematopoietic cultures, the labor required to feed each culture is a daunting task. In addition, the cultures are subjected to physical disruption and large discontinuous changes in culture conditions and may be exposed to contamination at each feeding. Thee complications frustrate the optimization of the culture environment for production of hematopoietic cells and limit the clinical usefulness of the cultures. A perfusion system, if properly designed and constructed, would eliminate many of the problems currently associated with these cultures.

The success of this manual frequently fed culture approach led to development of continuously perfused bioreactors for human cord blood [Koller et al., 1993a; Koller et al., 1998b], BM [Palsson et al., 1993], and mobilized peripheral blood [Sandstrom et al., 1995] cell culture. Human BM MNC cultures have been performed in spinner flasks in a fed-batch mode as well [Zandstra et al., 1994]. Slow single-pass medium perfusion and internal oxygenation have given the best results to date, yielding cell densities in excess of 10^7 per ml accompanied by significant progenitor and primitive cell expansion [Koller et al., 1993b]. These systems have also been amenable to scale-up, first by a factor of ten, and then by a further factor of 7.5 [Koller et al., 1998b]. When an appropriate culture substrate is provided [Koller et al., 1998c], perfusion bioreactors support the development and maintenance of accessory cell populations, resulting in significant endogenous growth factor production which likely contributes to culture success [Koller et al., 1995a; Koller et al., 1997]. Importantly, stromal-containing cultures appear to generate greater numbers of primitive cells from a smaller initial cell sample, as measured by in vitro [Koller et al., 1995b; Sandstrom et al., 1995] and in vivo assays [Knobel et al., 1994], as compared with CD34-enriched cell cultures.

120.4 Challenges for Scale-Up

To gauge the scale-up challenges, one first needs to state the requirements for clinically useful systems. The need to accommodate stroma that supports active hematopoiesis represents perhaps the most important consideration for the selection of a bioreactor system.

Bioreactors and Stroma

If stroma is required, the choices are limited to systems that can support the growth of adherent cells. This requirement may be eliminated in future systems if the precise GF requirements become known and if the microenvironment that the stroma provides is not needed for hematopoietic stem and progenitor cell expansion.

Currently, there are at least three culture systems that may be used for adherent cell growth. Fluidized bed bioreactors with macroporous bead carriers provide one option. Undoubtedly, significant effort will be required to develop the suitable bead chemistry and geometry, since the currently available systems are designed for homogeneous cell cultures. Beads for hematopoietic culture probably should allow for the formation of functional colonies comprised of a mixed cell population within each bead. Cell sampling from fluidized beds would be relatively easy, but final cell harvesting may require stressful procedures, such as prolonged treatment with collagenase and/or trypsin.

Flatbed bioreactors are a second type of system which can support stromal development. Such units can be readily designed to carry the required cell number, and, further, they can allow for direct microscopic observation of the cell culture. Flatbed bioreactors provide perhaps the most straightforward scale-up and automation of LTBMC. In fact, such automated systems have been developed and used in human clinical trials for the treatment of cancer [Mandalam et al., 1998].

Finally, membrane-based systems, such as hollow fiber units, could be used to carry out hematopoietic cell cultures of moderate size. Special design of the hollow fiber bed geometry with respect to axial length and radial fiberspacing should eliminate all undesirable spatial gradients. Such units have been made already and have proved effective for their use for in vivo NMR analysis of metabolic behavior of homogenous cell cultures [Mancuso et al., 1990]. However, hematopoietic cell observation and harvesting may prove to be troublesome with this approach, as one report has suggested [Sardonini & Wu, 1993].

It is possible that the function of accessory cells may be obtained by the use of spheroids without classical adherent cell growth. This approach to the growth of liver cells in culture has meet with some success (see Chapter 121). In fact, there have been reports of functional heterogeneous cell aggregates within BM aspirates [Blazsek et al., 1990; Funk et al., 1994]. If successfully developed, suspension cultures containing these aggregates could be carried out in a variety of devices, including the rotating wall vessels that have been developed by NASA [Schwarz et al., 1992].

Alternatives

The precise arrangement of future large-scale systems will be significantly influenced by continuing advances in the understanding of the molecular and microenvironmental regulation of the hematopoietic process. Currently, the proximity of a supporting stromal layer is believed to be important. It is thought to function through the provision of both soluble and membrane-bound GFs and by providing a suitable microenvironment. The characterization of the microenvironment is uncertain at present, but its chemistry and local geometry are both thought to play a role. If the proximity of stroma is found to be unimportant, one could possibly design culture systems in which the stroma and BM cells are separated [Verfaillie, 1992], resulting in the ability to control each function separately. Finally, if the GF requirements can be defined and artificially supplied, and the stromal microenvironment is found to be unimportant, large-scale hematopoietic suspension cultures would become possible.

Production of Mature Cells

Culture systems for generic allogeneic BMT, or for the large-scale production of mature cells, will pose more serious scale-up challenges. The number of cells required for these applications, in particular the latter, may be significantly higher than that for autologous BMT. Of the three alternatives discussed above, the fluidized bed system is the most readily scalable. The flatbed systems can be scaled by a simple stacking approach, whereas hollow fiber units are known for their shortcomings with respect to large-scale use.

120.5 Recapitulation

Rapid advances in our understanding of hematopoietic cell biology and the molecular control of hematopoietic cell replication, differentiation, and apoptosis are providing some of the basic information that is needed to reconstitute human hematopoiesis ex vivo. Compelling clinical applications provide a significant impetus for developing systems that produce clinically useful cell populations in clinically meaningful numbers. Use of in vivo mimicry and the bioreactor technologies that were developed in the 1980s are leading to the development of perfusion-based bioreactor systems that will meet some of the clinical needs. Further tissue engineering of human hematopoiesis is likely to continue to grow in scope and sophistication and lead to definition of basic structure-function relationships and the enablement of many needed clinical procedures.

Defining Terms

Colony-forming assay: Assay carried out in semisolid medium under GF stimulation. Progenitor cells divide, and progeny are held in place so that a microscopically identifiable colony results after 2 weeks.

Differentiation: The irreversible progression of a cell or cell population to a more mature state.

Engraftment: The attainment of a safe number of circulating mature blood cells after a BMT.

Flow cytometry: Technique for cell analysis using fluorescently conjugated monoclonal antibodies which identify certain cell types. More sophisticated instruments are capable of sorting cells into different populations as they are analyzed.

Graft-versus-host disease: The immunologic response of transplanted cells against the tissue of their new host. This response is often a severe consequence of allogeneic BMT and can lead to death (acute GVHD) or long-term disability (chronic GVHD).

Hematopoiesis: The regulated production of mature blood cells through a scheme of multilineage proliferation and differentiation.

Lineage: Refers to cells at all stages of differentiation leading to a particular mature cell type, i.e., one branch on the lineage diagram shown in Fig. 120.1.

Long-term culture-initiating cell: Cell that is measured by a 7–18 week in vitro assay. LTC-IC are thought to be very primitive, and the population contains stem cells. However, the population is heterogeneous, so not every LTC-IC is a stem cell.

Miroenvironment: Refers to the environment surrounding a given cell in vivo.

Mononuclear cell: Refers to the cell population obtained after density centrifugation of whole BM. This population excludes cells without a nucleus (erythrocytes) and polymorphonuclear cells (granulocytes).

Progenitor cells: Cells that are intermediate in the development pathway, more mature than stem cells but not yet mature cells. This is the cell type measured in the colony-forming assay.

Self-renewal: Generation of a daughter cell with identical characteristics as the original cell. Most often used to refer to stem cell division, which results in the formation of new stem cells.

Stem cells: Cells with potentially unlimited proliferative and lineage potential.

Stromal cells: Heterogeneous mixture of support or accessory cells of the BM. Also refers to the adherent layer which forms in BM cultures.

References

Aizawa S, Tavassoli M. 1987. In vitro homing of hemopoietic stem cells is mediated by a recognition system with galactosyl and mannosyl specificities. Proc Nat Acad Sci 84:4485.

Ambrus Jr. JL, Pippin J, Joseph A, Xu C, Blumenthal D, Tamayo A, Claypool K, McCourt D, Srikiatchatochorn A, and Ford RJ. 1993. Identification of a cDNA for a human high-molecular weight B-cell growth factor. Proc Natl Acad Sci 90:6330–6334.

Armstrong RD, Koller MR, Paul LA, et al. 1993. Clinical scale production of stem and hematopoietic cells ex vivo. Blood 82:296a.

Berenson RJ, Bensinger WI, Hill RS, et al. 1991. Engraftment after infusion of CD34$^+$ marrow cells in patients with breast cancer or neuroblastoma. Blood 77:1717.

Blazsek I, Misset J-L, Benavides M, et al. 1990. Hematon, a multicellular functional unit in normal human bone marrow: Structural organization, hemopoietic activity, and its relationship to myelodysplasia and myeloid leukemias. Exp Hematol 18:259.

Bodine DM, Karlsson S, Nienhuis AW. 1989. Combination of interleukins 3 and 6 preserves stem cell function in culture and enhances retrovirus-mediated gene transfer into hematopoietic stem cells. Proc Natl Acad Sci 86:8897.

Boggs DR, Boggs SS, Saxe DF, et al. 1982. Hematopoietic stem cells with high proliferative potential. J Clin Invest 70:242.

Brandt JE, Briddell RA, Srour EF, Leemhuis TB, and Hoffman R. 1992. Role of *c-kit* ligand in the expansion of human hematopoietic progenitor cells. Blood 79:634–641.

Brandt JE, Srour EF, Van Besien K, et al. 1990. Cytokine-dependent long-term culture of highly enriched precursors of hematopoietic progenitor cells from human bone marrow. J Clin Invest 86:932.

Brott DA, Koller MR, Rummel SA, Palsson BO. 1995. Flow cytometric analysis of cells obtained from human bone marrow cultures. In M Al-Rubeai, AN Emery (eds) Flow cytometry applications in cell culture, pp 121–146, Marcel Dekker, New York.

Caldwell J, Palsson BØ, Locey B, et al. 1991. Culture perfusion schedules influence the metabolic activity and granulocyte-macrophage colony-stimulating factor production rates of human bone marrow stromal cells. J Cell Physiol 147:344.

Campbell AD, Long MW, Wicha MS. 1987. Haemonectin, a bone marrow adhesion protein specific for cells of granulocyte lineage. Nature 329:744.

Center DM, Kornfeld H, Cruikshand WW. 1997. Interleukin-16. Int J Biochem Cell Biol 29:1231–1234.

Civin CI, Gore SD. 1993. Antigenic analysis of hematopoiesis: A review. J Hematotherapy 2:137.

Civin CI, Strauss LC, Brovall C, et al. 1984. Antigenic analysis of hematopoiesis: III. A hematopoietic progenitor cell surface antigen defined by a monoclonal antibody raised against KG-Ia cells. J Immunol 133:157.

Coulombel L, Vuillet MH, Leroy C, et al. 1988. Lineage- and stage-specific adhesion of human hematopoietic progenitor cells to extracellular matrices from marrow fibroblasts. Blood 71:329.

Coutinho LH, Will A, Radford J, et al. 1990. Effects of recombinant human granulocyte colony-stimulating factor (CSF), human granulocyte macrophage-CSF, and gibbon interleukin-3 on hematopoiesis in human long-term bone marrow culture. Blood 75:2118.

Cronkite EP. 1988. Analytical review of structure and regulation of hemopoiesis. Blood Cells 14:313.

De Sauvage FJ, Hass PE, Spencer SD, et al. 1994. Stimulation of megakaryocytopoiesis and thrombopoiesis by the c-Mpl ligand. Nature 369:533.

Dexter TM, Allen TD, Lajtha LG. 1977. Conditions controlling the proliferation of haemopoietic stem cells in vitro. J Cell Physiol 91:335.

Dexter TM, Spooncer E, Simmons P, et al. 1984. Long-term marrow culture: An overview of techniques and experience 121–146. In DG Wright, JS Greenberger (eds), Long-Term Bone Marrow Culture, pp 57–96, New York, Alan R. Liss.

Dinarello CA, Rosenwasser LJ, Wolff SM. 1981. Demonstrating of a circulating suppressor factor of thymocyte proliferation during endotoxin fever in humans. J Immunol 127:2517.

Donahue RE, Yang Y-C, Clark SC. 1990. Human P40 T-cell growth factor (interleukin 9) supports erythroid colony formation. Blood 75:2271.

Du XX, Williams DA. 1994. Interleukin-11: A multifunctional growth factor derived from the hematopoietic microenvironment. Blood 83:2023.

Eastment CE, Ruscetti FW. 1984. Evaluation of hematopoiesis in long-term bone marrow culture: Comparison of species differences. In DG Wright, JS Greenberger (eds), Long-Term Bone Marrow Culture, pp 97–118, New York, Alan R. Liss.

Eaves CJ, Cashman JD, Eaves AC. 1991. Methodology of long-term culture of human hemopoietic cells. J Tiss Cult Meth 13:55.

Funk PE, Kincade PW, Witte PL. 1994. Native associations of early hematopoietic stem cells and stromal cells isolated in bone marrow cell aggregates. Blood 83:361.

Gartner S, Kaplan HS. 1980. Long-term culture of human bone marrow cells. Proc Natl Acad Sci 77:4756.

Gluckman E, Broxmeyer HE, Auerback AD, et al. 1989. Hematopoietic reconstitution in a patient with Fanconi's anemia by means of umbilical-cord blood from an HLA-identical sibling. NE J Med 321:1174.

Gordon MY, Riley GP, Watt SM, et al. 1987. Compartmentalization of a haematopoietic growth factor (GM-CSF) by glycosaminoglycans in the bone marrow microenvironment. Nature 326:403.

Grabstein KH, Eisenman J, Shanebeck K, Rauch C, Srinivasan S, Fung V, Beers C, Richardson J, Schoenborn MA, Ahdieh M. 1994. Cloning of a T cell growth factor that interacts with the beta chain of the interleukin-2 receptor. Science 264:965–968.

Graham GJ, Wright EG, Hewick R, et al. 1990. Identification and characterization of an inhibitor of haemopoietic stem cell proliferation. Nature 344:442.

Greenberg HM, Newburger PE, Parker LM, et al. 1981. Human granulocytes generated in continuous bone marrow culture are physiologically normal. Blood 58:724.

Greenberger JS. 1984. Long-term hematopoietic cultures. In DW Golde (ed), Hematopoiesis, pp 203–242, New York, Churchill Livingstone.

Greenberger JS, Fitzgerald TJ, Rothstein L, et al. 1986. Long-term culture of human granulocytes and granulocyte progenitor cells. In Transfusion Medicine: Recent Technological Advances, pp 159–185, New York, Alan R. Liss.

Haylock DN, To LB, Dowse TL, et al. 1992. Ex vivo expansion and maturation of peripheral blood CD34$^+$ cells into the myeloid lineage. Blood 80:1405.

Herbert CA, Baker JB. 1993. Interleukin-8: A review. Cancer Invest 11:743.

Hoffman R, Benz EJ Jr, Shattil SJ, et al. 1991. Hematology: Basic Principles and Practice, New York, Churchill Livingstone.

Horowitz MM, Rowlings PA. 1997. An update from the International Bone Marrow Transplant Registry and the Autologous Blood and Marrow Transplant Registry on current activity in hematopoietic stem cell transplantation. Curr Opin Hematol 4:359–400.

Ihle JN, Pepersack L, Rebar L. 1981. Regulation of T cell differentiation: In vitro induction of 20 alpha-hydroxysteroid dehydrogenase in splenic lymphocytes is mediated by a unique lymphokine. J Immunol 126:2184.

Kansas GS, Muirhead MJ, Dailey MO. 1990. Expression of the CD11/CD18, leukocyte adhesion molecule 1, and CD44 adhesion molecules during normal myeloid and erythroid differentiation in humans. Blood 76:2483.

Kishimoto T. 1989. The biology of interleukin-6. Blood 74:1.

Knobel KM, McNally MA, Berson AE, Rood D, Chen K, Kilinski L, Tran K, Okarma TB, Lebkowski JS. 1994. Long-term reconstitution of mice after ex vivo expansion of bone marrow cells: Differential activity of cultured bone marrow and enriched stem cell populations. Exp Hematol 22:1227–1235.

Koller MR, Bender JG, Miller WM, et al. 1993*a*. Expansion of human hematopoietic progenitors in a perfusion bioreactor system with IL-3, IL-6, and stem cell factor. Biotechnology 11:358.

Koller MR, Bender JG, Papoutsakis ET, et al. 1992*a*. Effects of synergistic cytokine combinations, low oxygen, and irradiated stroma on the expansion of human cord blood progenitors. Blood 80:403.

Koller MR, Bender JG, Papoutsakis ET, et al. 1992*b*. Beneficial effects of reduced oxygen tension and perfusion in long-term hematopoietic cultures. Ann NY Acad Sci 665:105.

Koller MR, Emerson SG, Palsson BØ. 1993*b*. Large-scale expansion of human stem and progenitor cells from bone marrow mononuclear cells in continuous perfusion culture. Blood 82:378.

Koller MR, Bradley MS, Palsson BØ. 1995. Growth factor consumption and production in perfusion cultures of human bone marrow correlates with specific cell production. Exp Hematol, 23:1275.

Koller MR, Manchel I, Palsson BO. 1997. Importance of parenchymal:stromal cell ratio for the ex vivo reconstitution of human hematopoiesis. Stem Cells 15:305–313.

Koller MR, Manchel I, Smith AK. 1998a. Quantitative long-term culture-initiating cell assays require accessory cell depletion that can be achieved by CD34-enrichment or 5-fluorouracil exposure. Blood 91:4056.

Koller MR, Manchel I, Maher RJ, Goltry KL, Armstrong RD, Smith AK. 1998b. Clinical-scale human umbilical cord blood cell expansion in a novel automated perfusion culture system. Bone Marrow Transplant 21:653.

Koller MR, Manchel I, Palsson MA, Maher RJ, Palsson, BØ. 1996. Different measures of human hematopoietic cell culture performance are optimized under vastly different conditions. Biotechnol. Bioeng. 50:505–513.

Koller MR, Palsson MA, Manchel I, Palsson, BØ. 1995b. LTC-IC expansion is dependent on frequent medium exchange combined with stromal and other accessory cell effects. Blood 86:1784–1793.

Koller MR, Palsson MA, Manchel I, Maher RJ, Palsson BØ. 1998c. Tissue culture surface characteristics influence the expansion of human bone marrow cells. Biomaterials (in press).

Krantz SB. 1991. Erythropoietin. Blood 77:419.

Lajtha LG. 1979. Stem cell concepts. Differentiation 14:23.

Lemoli RM, Tafuri A, Strife A, et al. 1992. Proliferation of human hematopoietic progenitors in long-term bone marrow cultures in gas permeable plastic bags is enhanced by colony-stimulating factors. Exp Hematol 20:569.

Lewinsohn DM, Nagler A, Ginzton N, et al. 1990. Hematopoietic progenitor cell expression of the H-CAM (CD44) homing-associated adhesion molecule. Blood 75:589.

Long MW. 1992. Blood cell cytoadhesion molecules. Exp Hematol 20:288

Long MW, Dixit VM. 1990. Thrombospondin functions as a cytoadhesion molecule for human hematopoietic progenitor cells. Blood 75:2311.

Lyman SD, James L, Johnson L, Brasel K, de Vries P, Escobar SS, Downey H, Splett RR, Beckmann MP, McKenna HJ. 1994. Cloning of human homologue of the murine flt3 ligand: A growth factor for early hematopoietic progenitor cells. Blood 83:2795–2801.

Mancuso A, Fernandez EJ, Blanch HW, et al. 1990. A nuclear magnetic resonance technique for determining hybridoma cell concentration in hollow fiber bioreactors. Biotechnology 8:1282.

Mandalam R, Koller MR, Smith AK. 1998. Ex vivo hematopoietic cell expansion for bone marrow transplantation. In R Nordon (ed), Ex vivo cell therapy. Landes Bioscience, Austin, TX. (in press)

Meagher RC, Salvado AJ, Wright DG. 1988. An analysis of the multilineage production of human hematopoietic progenitors in long-term bone marrow culture: Evidence that reactive oxygen intermediates derived from mature phagocytic cells have a role in limiting progenitor cell self-renewal. Blood 72:273.

Metcalf D. 1985. The granulocyte-macrophage colony-stimulating factors. Science 229:16.

Metcalf D. 1991. The leukemia inhibitory factor (LIF). Int J Cell Cloning 9:95.

Metcalf D. 1993. Hematopoietic regulators: Redundancy or subtlety? Blood 82:3515.

Minguell JJ, Tavassoli M. 1989. Proteoglycan synthesis by hematopoietic progenitor cells. Blood 73:1821.

Minty A, Chalon P, Derocq JM, et al. 1993. Interleukin-13 is a new human lymphokine regulating inflammatory and immune responses. Nature 362:248.

Mulligan RC. 1993. The basic science of gene therapy. Science 260:926–932.

Palsson BØ, Paek S-H, Schwartz RM, et al. 1993. Expansion of human bone marrow progenitor cells in a high cell density continuous perfusion system. Biotechnology 11:368.

Patel VP, Ciechanover A, Platt O, et al. 1985. Mammalian reticulocytes lose adhesion to fibronectin during maturation to erythrocytes. Proc Natl Acad Sci 82:440.

Pennica D, Nedwin GE, Hayflick JS, et al. 1984. Human tumor necrosis factor: Precursor structure, expression and homology to lymphotoxin. Nature 312:724.

Roberts R, Gallagher J, Spooncer E, et al. 1988. Heparan sulphate bound growth factors: A mechanism for stromal cell mediated haemopoiesis. Nature 332:376.

Sandstrom CE, Bender JG, Papoutsakis ET, Miller WM. 1995. Effects of $CD34^+$ cell selection and perfusion on ex vivo expansion of peripheral blood mononuclear cells. Blood 86:958–970.

Sardonini CA, Wu Y-J. 1993. Expansion and differentiation of human hematopoietic cells from static cultures through small scale bioreactors. Biotechnol Prog. 9:131.

Schneider JG, Crown J, Shapiro F, et al. 1992. Ex vivo cytokine expansion of CD34-positive hematopoietic progenitors in bone marrow, placental cord blood, and cyclophosphamide and G-CSF mobilized peripheral blood. Blood 80:268a.

Schwartz RM, Emerson SG, Clarke MF, et al. 1991*a*. In vitro myelopoiesis stimulated by rapid medium exchange and supplementation with hematopoietic growth factors. Blood 78:3155.

Schwartz RM, Palsson BØ, Emerson SG. 1991*b*. Rapid medium perfusion rate significantly increases the productivity and longevity of human bone marrow cultures. Proc Natl Acad Sci 88:6760.

Schwarz RP, Goodwin TJ, Wolf DA. 1992. Cell culture for three-dimensional modeling in rotating wall vessels: An application of simulated microgravity. J Tiss Cult Meth 14:51.

Smith KA. 1988. Interleukin-2: Inception, impact, and implications. Science 240:1169.

Spangrude GJ, Heimfeld S, Weissman IL. 1988. Purification and characterization of mouse hematopoietic stem cells. Science 241:58.

Spooncer E, Gallagher JT, Krizsa F, et al. 1983. Regulation of haemopoiesis in long-term bone marrow cultures: IV. Glycosaminoglycan synthesis and the stimulation of haemopoiesis by β-D-xylosides. J Cell Biol 96:510.

Sporn MB, Roberts AB. 1989. Transforming growth factor-β: Multiple actions and potential clinical applications. JAMA 262:938.

Stroncek DF, Holland PV, Bartch G, et al. 1993. Experiences of the first 493 unrelated marrow donors in the national marrow donor program. Blood 81:1940.

Sutherland HJ, Eaves AC, Eaves CJ. 1991*a*. Quantitative assays for human hemopoietic progenitor cells. In AP Gee (ed), Bone Marrow processing and Purging, pp 155–167, Boca Raton, Fla, CRC Press.

Sutherland HJ, Eaves CJ, Lansdorp PM, et al. 1991*b*. Differential regulation of primitive human hematopoietic stem cells in long-term cultures maintained on genetically engineered murine stromal cells. Blood 78:666.

Sutherland HJ, Hogge DE, Cook D, et al. 1993. Alternative mechanisms with and without steel factor support primitive human hematopoiesis. Blood 81:1465.

Tavassoli M, Hardy CL. 1990. Molecular basis of homing of intravenously transplanted stem cells to the marrow. Blood 76:1059.

Toksoz D, Zsebo KM, Smith KA, et al. 1992. Support of human hematopoiesis in long-term bone marrow cultures by murine stromal cells selectively expressing the membrane-bound and secreted forms of the human homolog of the steel gene product, stem cell factor. Proc Natl Acad Sci 89:7350.

Traycoff CM, Abboud CM, Abboud MR, Laver J, et al. 1994. Evaluation of the in vitro behavior of phenotyically defined populations of umbilical cord blood hematopoietic progenitor cells. Exp Hematol 22:215.

Tushinski RJ, McAlister IB, Williams DE, et al. 1991. The effects of interleukin 7 (IL-7) on human bone marrow in vitro. Exp Hematol 19:749.

Ushio S, Namba M, Okura T, Hattori K, Hukada Y, Akita K, Tanabe F, Konishi K, Micallef M, Fujii M, Torigoe K, Tanimoto T, Fukuda S, Ikeda M, Okamura H, Kurimoto M. 1996. Cloning of the cDNA for human IFN-gamma-inducing factor, espression in Escherichia coli, and studies on the biologic activities of the protein. J Immunol 156:4274–4279.

Verfaillie CM. 1992. Direct contact between human primitive hematopoietic progenitors and bone marrow stroma is not required for long-term in vitro hematopoiesis. Blood 79:2821.

Virelizier JL, Arenzana-Seisdedos F. 1985. Immunological functions of macrophages and their regulation by interferons. Med Biol 63:149–159.

Wagner JE, Broxmeyer HE, Byrd RL, et al. 1992. Transplantation of umbilical cord blood after myeloablative therapy: Analysis of engraftment. Blood 79:1874.

Weinstein R, Riordan MA, Wenc K, et al. 1989. Dual role of fibronectin in hematopoietic differentiation. Blood 73:111.

Wight TN, Kinsella MG, Keating A, et al. 1986. Proteoglycans in human long-term bone marrow cultures: Biochemical and ultrastructural analyses. Blood 67:1333.

Wolf SF, Temple PA, Kobayashi M, et al. 1991. Cloning of cDNA for natural killer cell stimulatory factor, a heterodimeric cytokine with multiple biologic effects on T and natural killer cells. J Immunol 146:3074.

Yamakazi K, Roberts RA, Spooncer E, et al. 1989. Cellular interactions between 3T3 cells and interleukin-3 dependent multipotent haemopoietic cells: A model system for stromal-cell-mediated haemopoiesis. J Cell Physiol 139:301.

Yao Z, Painter SL, Fanslow WC, Ulrich D, Macduff BM, Spriggs MK, and Armitage RJ. 1995. Human IL-17: A novel cytokine derived from T cells. J Immunol 155:5483-5486.

Yokota T, Arai N, de Vries JE, et al. 1988. Molecular biology of interleukin-4 and interleukin-5 genes and biology of their products that stimulate B cells, T cells and hemopoietic cells. Immunol Rev 102:137.

Zandstra PW, Eaves CJ, Piret JM. 1994. Expansion of hematopoietic progenitor cell populations in stirred suspension bioreactors of normal human bone marrow cells. Biotechnology 12:909–914.

Zipori D. 1989. Stromal cells from the bone marrow: Evidence for a restrictive role in regulation of hemopoiesis. Eur J Haematol 42:225.

Zlotnik A. Moore KW. 1991. Interleukin 10. Cytokine 3:366.

Zsebo KM, Wypych J, McNiece IK, et al. 1990. Identification, purification, and biological characterization of hematopoietic stem cell factor from buffalo rat liver-conditioned medium. Cell 63:195.

Zuckerman KS, Wicha MS. 1983. Extracellular matrix production by the adherent cells of long-term murine bone marrow cultures. Blood 61:540.

Further Information

The American Society of Hematology (ASH) is the premier organization dealing with both the experimental and clinical aspects of hematopoiesis. The society journal, *BLOOD*, is published twice per month, and can be obtained through ASH, 1200 19th Street NW, Suite 300, Washington, DC 20036 (phone: 202–857–1118). The International Society of Experimental Hematology (ISEH) publishes *Experimental Hematology* monthly.

121
Tissue Engineering of the Liver

Tao Ho Kim
Harvard University and Boston Children's Hospital

Joseph P. Vacanti
Harvard University and Boston Children's Hospital

Liver transplantation has been established as a curative treatment for end-stage adult and pediatric liver disease [Starlz et al., 1989], and over recent years, many innovative advances have been made in transplantation surgery. Unfortunately, a fundamental problem of liver transplantation has been severe donor shortage, and no clinical therapeutic bridge exists to abate the progression of liver failure (see Table 121.1). As the demand for liver transplantation surgery increases, still fewer than 3500 donors are available annually for the approximately 25,000 patients who die from chronic liver disease [National Vital Statistic System, 1991, 1992]. Currently, cadaveric and living-related donors are the only available sources. Xenograft [Starzl et al., 1993] and split liver transplantation [Merio & Campbell, 1991] are under experimental and clinical evaluation. The research effort to engineer a functional liver tissue has been vigorous, since tissue engineering of the liver offers, in theory, an efficient use of limited organ availability.

Whereas other experimental hepatic support systems such as extracorporeal bioreactors and hemoperfusion devices [Yarmusch et al., 1992] attempt to temporarily support the metabolic functions of liver, transplantation of hepatocyte systems is a possible temporary or permanent alternative therapy to liver transplantation for treatment of liver failure. An experimental model system of hepatocellular transplantation should provide optimal cell survival, proliferation, and maintenance of sufficient functional hepatocyte mass to replace liver function. Direct hepatocellular injection or infusion into various organs or tissues and a complex hepatocyte delivery system utilizing polymer matrices have been two major areas of research interest. Just as the liver is one of the most sophisticated organs in the human body, the science of hepatocyte or liver tissue construction has proved to be equally complex.

121.1 Background

The causes of end-stage liver disease are many, including alcoholic or viral cirrhosis, biliary atresia, inborn errors of metabolism, and sclerosing cholangitis. With chronic and progressive liver injury, hepatic necrosis occurs followed by fatty infiltration and inflammation. Scar tissue and nodular regeneration replace the normal liver architecture and increase the microcirculatory resistance, which results in portal hypertension; the liver is further atrophied as important factors in the portal blood which regulate liver growth and maintenance are shunted away from the liver. Currently, end-stage liver disease must be present to be considered for orthotopic liver transplantation therapy, but a significant difference exists

0-8493-0462-8/00/$0.00+$.50

between alcohol- or viral-induced liver injury and congenital liver diseases such as isolated gene defects and biliary atresia. With alcohol- or viral-induced chronic hepatic injury, the degree of liver injury is unknown until metabolic functions are severely impaired and signs of progressive irreversible hepatic failure including portal hypertension, coagulopathy, progressive jaundice, and hepatic encephalopathy have developed. In congenital liver diseases, normal hepatic metabolic functions exist until dangerous toxins build up and destroy the liver parenchyma. Hepatocellular transplantation could potentially *prevent* hepatic injury and preserve host hepatic function for congenital inborn errors of liver metabolism.

TABLE 121.1 UNOS Liver Transplantation Data Summary from 1989 to 1993 in the United States: Total Number of Liver Transplant Candidates and Deaths Reported per Year While on Transplant Waiting List

Year	No. of Patients	No. of Deaths Reported
1989	3096	39
1990	4008	45
1991	4866	67
1992	5785	104
1993	7040	141

Source: United Network for Organ Sharing and the Organ Procurement and Transplantation Network. Data as of January 12, 1994.

As liver transplantation emerged as an important therapeutic modality, research activity intensified to improve or understand many areas of liver transplantation such as immunological tolerance, preservation techniques, and the mechanism of healing after acute and chronic liver injury. Liver growth and regulation, in particular, have been better understood: For example, after partial hepatectomy, several *mitogens*—epidermal growth factor, alpha fibroblastic growth factor, hepatocyte growth factor, and transforming growth factor-alpha—are produced early after injury to stimulate liver regeneration. Comitogens—including insulin, glucagon, estrogen, norepinephrine, and vasopressin—also aid with liver regeneration [Michalopoulos, 1993]. These stimulation factors help govern the intricate regulatory process of liver growth and regeneration, but much about what controls these factors is still unknown. In vitro and in vivo experiments with mitogens such as hepatocyte growth factor, epidermal growth factor, and insulin have yielded only moderate improvement in hepatic proliferation.

The importance of hepatocyte proliferation and hepatic regeneration becomes evident when one considers the difficulty in delivering large numbers of hepatocytes in hepatocyte replacement systems. The potential advantage of hepatocyte cellular transplantation is to take a small number of hepatocytes and proliferate these cells, in vitro or in vivo, to create functional liver equivalents for replacement therapy. Without hepatocyte regeneration, delivery of a very large number of hepatocytes is required. Asonuma and coworkers [1992] have determined that approximately 12% of the liver by heterotopic liver transplantation can significantly correct hyperbilirubinemia in the Gunn rat, which is deficient in uridine diphosphate glucuronyl tranferase. Although long-term efficacy remains unclear, 10–12% of the liver is an approximate critical hepatocellular mass thought necessary to replace the metabolic functions of the liver. The inability to mimic normal liver growth and regeneration in *in vitro* and *in vivo* systems has been one significant obstacle for hepatocyte tissue construction thus far.

121.2 Hepatocyte Transplantation Systems

Hepatocyte transplant systems offer the possibilities of creating many functional liver equivalents, storing hepatocytes by cryopreservation for later application [Yarmush et al., 1992], and using autologous cells for gene therapy [Jauregui & Gann, 1991]. The two systems discussed below differ in the amount of hepatocytes delivered, use of implantation devices, and implantation sites and techniques. Yet, in both systems, one significant roadblock in proving the efficacy of hepatocyte transplantation has been in the lack of a definitive, reproducible isolated liver defect model. Previous studies using syngeneic rat models such as the jaundice Gunn rat, the analbuminemic Nagase rat, or acute and chronic liver injury rat models have attributed significant correction of their deficit from hepatocyte transplantation. However, either inconsistencies or variation of animal strains and lack of consistent reproducible results have made accurate scientific interpretations and conclusions difficult. For instance, a study assessing hepatocyte delivery with microcarrier beads reported significant decrease in bilirubin in the hyperbilirubinemic

Gunn rat and elevation of albumin in the Nagase analbuminemic rat after intraperitoneal implantation [Demetriou et al., 1986]; other studies have not demonstrated significant hepatocyte survival with intraperitoneal injection of hepatocytes with or without microcarriers; histology showed predominant cell necrosis and granuloma formation after 3 days [Henne-Bruns et al., 1991]. In *allogeneic* models, the possibility of immunological rejection further complicates the evaluation of the hepatocyte transplant system. Clearly, determining the efficacy of hepatocyte transplantation in liver metabolic-deficient models requires significant chemical results, and histologic correlation without confounding variables.

Hepatocellular Injection Model

Hepatocytes require an extracellular matrix for growth and differentiation, and the concept of utilizing existing **stromal** tissue as a vascular extracellular matrix is inviting. Isolated hepatocytes have been injected directly into the spleen or liver or in the portal or splenic vein. Several studies have reported significant but temporary correction of acute and metabolic liver defects in rat models as a result of intrahepatic, intraportal [Matas et al., 1976], or intrasplenic injections [Vroemen et al., 1985]. However, elucidating the efficacy of hepatocellular injection transplantation has been difficult for three significant reasons: (1) Differentiation of donor transplanted hepatocytes from host liver parenchyma has not been well established in an animal liver defect model, (2) how much hepatocyte mass needed to inject for partial or total liver function replacement has not been determined [Onodera et al., 1992], and (3) establishing a definitive animal liver defect model to prove efficacy of hepatocyte injection has been difficult.

Transgenic animal strains have been developed and offer a reproducible model in differentiating host from transplanted hepatocytes. Using transgenic mouse lines, donor hepatocytes injected into the spleen were histologically shown to migrate to the host liver, survive, and maintain function [Ponder et al., 1991]. Recently, a transgenic liver model in a mouse was used to evaluate the replicative potential of adult mouse hepatocytes. Normal adult mouse hepatocytes from two established transgenic lines were injected into the spleen of an Alb-uPA transgenic mouse that had an endogenous defect in hepatic growth potential and function (see Fig. 121.1). The adult mouse hepatocytes were shown to translocate to the liver and undergo up to 12 cell doublings; however, function of the transplanted hepatocytes was not fully reported [Rhim et al., 1994]. Rhim's study suggests that a hepatocyte has the potential to replicate many-fold so long as the structural and chemical milieu is optimal for survival and regeneration. If a small number of transplanted hepatocytes survive and proliferate many-fold in the native liver, sufficient hepatocyte mass to replace liver function would be accomplished, ameliorating the need to deliver a large quantity of donor hepatocytes.

Although many questions about hepatocyte injection therapy remain, the transgenic liver model has been used to test the safety and efficacy of ex vivo gene therapy for metabolic liver diseases. The Watanabe heritable hyperlipidemic rabbit, a strain deficient in low density lipoprotiens (LDL) receptors, has been used to evaluate the possible application of a gene therapy by hepatocyte injection. *Autologous* hepatocytes were obtained from a liver segment, genetically modified *ex vivo*, and infused into the inferior mesenteric vein through a catheter placed intraoperatively without postoperative sequelae to the rabbit. Decreased levels of LDL have been reported out to 6 months [Wilson et al., 1992]. Larger animal models have shown engraftment of the genetically altered hepatocytes for as long as 1.5 years. The first clinical application of hepatocellular injection has been performed on a patient diagnosed with homozygous familial hypercholesterolemia. The patient has had decreased levels of cholesterol and has maintained expression of the transfected gene after 18 months [Grossman et al., 1994]. This is an important step forward as we await long-term results.

Hepatocyte Transplantation on Polymer Matrices

Since the practical application of implanting few hepatocytes to proliferate and replace function is not yet possible, hepatocyte tissue construction, using polymer as a scaffold, relies on transplanting a large number of hepatocytes to allow survival of enough hepatocyte mass to replace function. The large surface

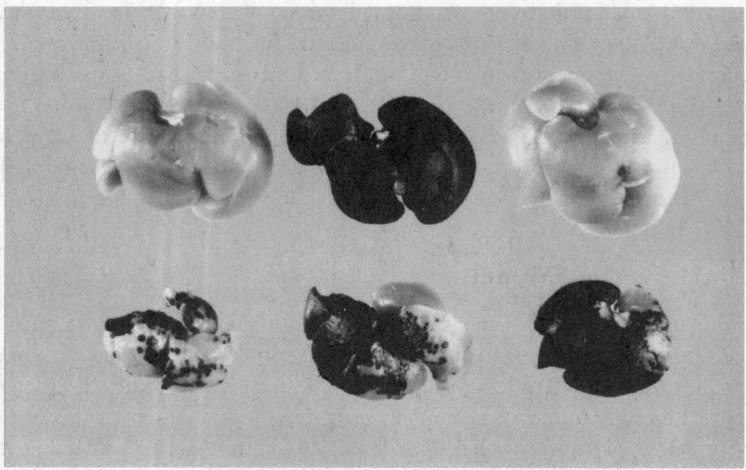

FIGURE 121.1 Control and transgenic liver specimens: a nontransgenic (normal mouse) control with normal liver color (top, left); a *transgenic* (*MT-lacZ*) liver with normal function which is stained in blue and served as a positive control (top, center); a *notransgenic* control transplanted with *transgenic MIT-lacZ* hepatocytes (top, right); and three livers with a different *transgene Alb-uPA* transplanted with *transgenic* (*MT-lacZ*) hepatocytes. A deficiency in hepatic growth potential and function is induced by the *Alb-uPA transgene*, resulting in a chronic stimulus for liver growth. The liver with *Alb-uPA transgene* has the same color as the *nontransgenic* control liver but is partially replaced by the blue-stained *transgencic* (*MT-lacZ*) normal functioning hepatocytes, showing the regenerative response of normal hepatocytes in a mouse liver with a chronic stimulus for live growth. (Rhim, Sandgren, and Brinster, University of Pennsylvania, reprinted with permission from American Association for the Advancement of Science.)

area of the polymer accommodates hepatocyte attachment in large numbers so that many cells may survive initially by diffusion of oxygen and other vital nutrients (see Fig. 121.2). The polymer scaffold is constructed with a high porosity to allow vascular ingrowth, and vascularization of surviving cells then can provide permanent nutritional access [Cima et al., 1991]. The cell-polymer system has been used in several other tissue-engineering applications such as cartilage, bone, intestine, and urologic tissue construction [Langer & Vacanti, 1993]. Hepatocytes adhere to the polymer matrix for growth and differentiation as well as locate into the interstices of the polymer. The polymer-hepatocyte interface can be manipulated with surface proteins such as laminin, fibronectin, and growth factors to improve adherence, viability, function, or growth [Mooney et al., 1992]. Hepatocyte proliferation by attaching mitogenic factors like hepatocyte growth factor, epidermal growth factor, or transforming growth factor-alpha is under current investigation.

Synthetic polymer matrices, both degradable and nondegradable, have been evaluated for hepatocyte-polymer construction. As degradable polymers were being studied and evaluated for tissue engineering, a nondegradable polymer, polyvinyl (PVA), was used for *in vitro* and *in vivo* systems. The polyvinyl alcohol sponge offered one significant advantage: a uniform, noncollapsible structure which allowed quantification of hepatocyte engraftment [Uyama et al., 1993]. *In vitro* and *in vivo* studies have demonstrated hepatocyte survival on the polyvinyl alcohol scaffold. However, the polyvinyl alcohol sponge will not degrade and could act as a nidus for infection and chronic inflammation. A degradable polymer conceptually serves as a better implantable scaffold for hepatocyte-polymer transplantation, since the polymer dissolves to leave only tissue. Polyglycolic acid (PGA), polyactic acid (PLA), and copolymer hybrids have been employed in several animal models of liver insufficiency (see Fig. 121.3). Histologic analyses have shown similar survival of hepatocytes when compared to studies with polyvinyl alcohol.

The experimental design for the hepatocyte-polymer model has been standardized as follows: (1) an end-to-side portacaval shunt is performed to provide hepatotrophic stimulation to the graft, (2) hepatocytes

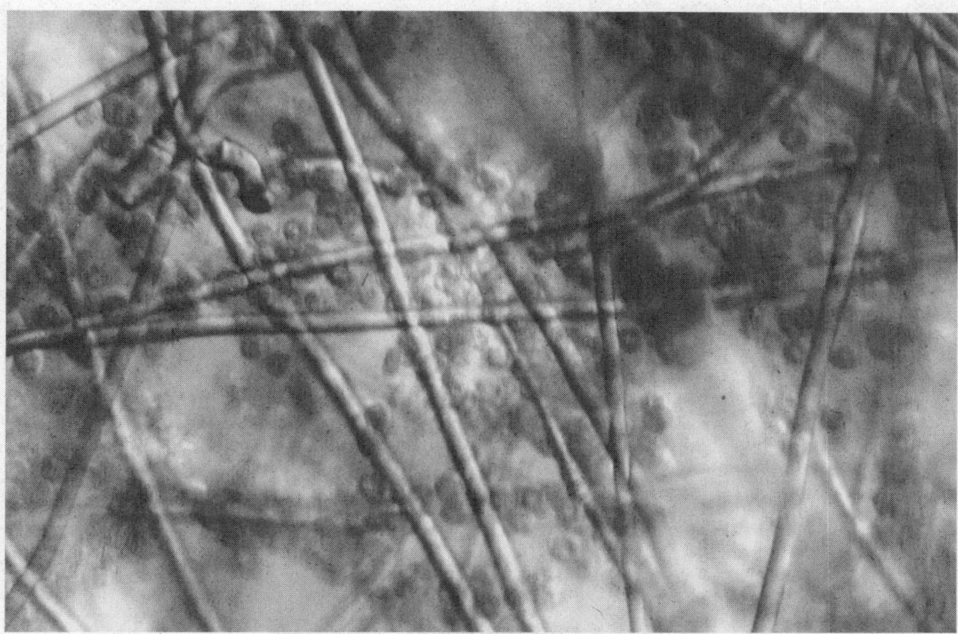

FIGURE 121.2 Hepatocytes are seen adhering to polyglycolic acid polymer in culture (Mooney, unpublished data).

are isolated and seeded onto degradable polyglycolic acid polymer, (3) the hepatocyte-polymer construct is implanted into the abdominal cavity on vascular beds of small intestinal mesentery and omentum, (4) pertinent chemical studies are obtained and analyzed at periodic intervals, and (5) histologic analysis in the specimens is performed at progressive time points. Early studies have shown that a large percentage of hepatocytes perish from hypoxia within 6–24 hours after implantation. To improve hepatocyte survival, implantation of hepatocytes onto large vascular surface areas for engraftment and exposing the hepatocytes to hepatotrophic factors were two important maneuvers. The small intestinal mesentery and omentum have offered large vascular surface areas. Hepatocyte survival has been reported at other sites such as the peritoneum [Demetriou et al., 1991], renal capsule [Ricordi et al., 1989], lung [Sandbichler et al., 1992], and pancreas [Vroemen et al., 1988], but these sites do not provide enough vascular surface area to allow survival of a large number of hepatocytes.

The concept of hepatotrophic factors originated when atrophy and liver insufficiency was observed with heterotopic liver transplantation. Later studies have confirmed that important factors regulating liver growth and maintenance existed in the portal blood [Jaffe et al., 1991]. Thus, when a portacaval shunt to redirect hepatotrophic factors from the host liver to the hepatocyte-polymer construct was performed, survival of heterotopically transplanted hepatocytes significantly improved [Uyama et al., 1993]; consequently, portacaval shunts were instituted in all experimental models. Studies thus far have shown survival of functional hepatocytes over 6 months in rat and dog models (unpublished data); replacement of liver function has been of shorter duration.

A study using the Dalmatian dog model of hyperuricosuria typifies the current status of the hepatocyte-polymer system. The hepatocyte membrane of the Dalmatian dog has a defect in the uptake of uric acid, which results in hyperuricemia and hyperuricosuria [Giesecke & Tiemeyer, 1984]. The study has shown a significant but temporary correction of the liver uric acid metabolic defect after implantation of normal beagle hepatocytes, which have normal uric acid metabolism, on degradable polyglycolic acid polymer [Takeda et al., 1994]. Cyclosporine was administered for immunosuppression. The results of the Dalmatian dog study suggest that (1) successful engraftment of the hepatocyte-polymer construct occurred,

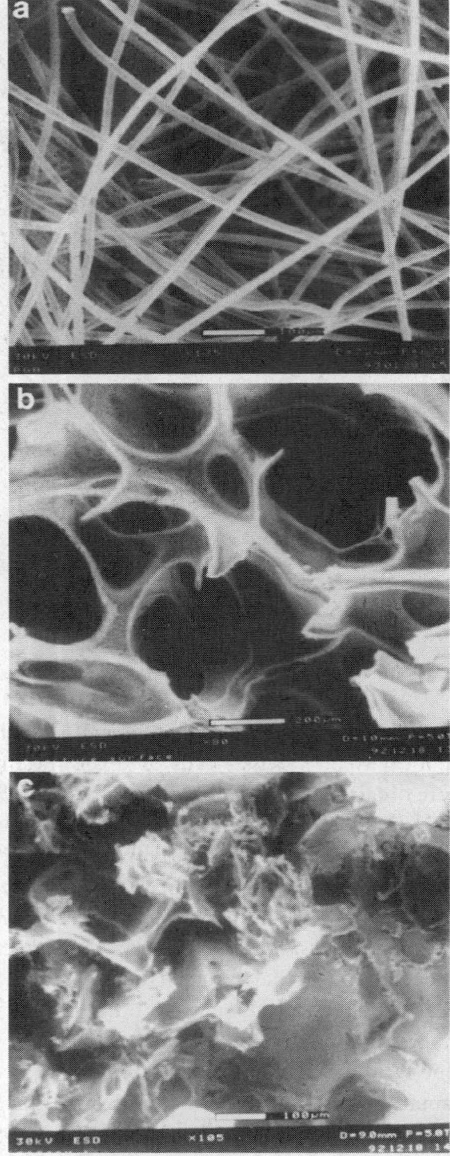

FIGURE 121.3 Scanning electron microscopic photographs of polyglyclic acid (top), polyvinyl alcohol (middle), and polylactic acid (bottom) demonstrate the high porosity of these polymers (Mooney, reprinted with permission from W.B. Saunders Company).

(2) maintenance of functional hepatocytes under current conditions was temporary, and (3) loss of critical hepatocyte mass to replace the liver uric acid metabolism defect occurred at 5–6 weeks after hepatocyte-polymer transplantation. The temporary effect could be attributed to suboptimal immunosuppression, suboptimal regulation of growth factors and hormones involved with hepatic growth, regeneration, and maintenance, or both.

Coculture of hepatocytes with pancreatic islet cells also has been under investigation to aid in hepatocyte survival, growth, and maintenance. Trophic factors from islet cells have been shown to improve hepatocyte survival [Ricordi et al., 1988], and cotransplantation of hepatocytes with islet cells on a polymer matrix has been shown to improve hepatocyte survival as well [Kaufmann et al., 1994]. Coculture with other cell types such as the biliary epithelial cell also may improve hepatocyte survival. Other studies with biliary epithelial cells have shown ductular formation in in vitro and in vivo models [Sirica et al., 1990], and vestiges of ductular formation in hepatocyte-polymer tissue have been histologically observed [Hansen & Vacanti, 1992]. A distinct advantage of the cell-polymer engineered construct is that one can manipulate the polymer to direct function. Thus far, diseases involving the biliary system, primarily biliary atresia, are not amenable to hepatocyte transplantation. An attempt to develop a biliary drainage system with biliary epithelial cells, hepatocytes, and polymer has been initiated. In the future, the potential construction of a branching polymer network could serve as the structural cues for the development of an interconnecting ductular system.

Ex vivo gene therapy with the hepatocyte-polymer system is also an exciting potential application as demonstrated by the recent clinical trial with hepatocyte injection therapy. Genetically altered hepatocytes transplanted on polymer constructs have been studied with encouraging results [Fontaine et al., 1993].

121.3 Conclusion

Studies in hepatocyte transplantation through tissue engineering methods have made important advances in recent years. The research in the hepatocellular injection and the research in the hepatocyte-polymer construct models have complemented each other in understanding the difficulties as well as the possibilities of liver replacement therapy. In order to make further advances with hepatocyte replacement systems, the process of liver development, growth, and maintenance need to be better understood. Currently, the amount of hepatocyte engraftment, proliferation, and the duration of hepatocyte survival remain undetermined in both systems. The amount of functional hepatocyte engraftment necessary may vary for different hepatic diseases. For instance, isolated gene defects of the liver may require a small number of functional transplanted hepatocytes to replace function, whereas end-stage liver disease may require a large amount of hepatocyte engraftment.

Current hepatic replacement models have both advantages and disadvantages. For hepatocellular injection, the application of ex vivo gene therapy for an isolated gene defect of liver metabolism is promising. However, the small amount of hepatocyte delivery and significant potential complications for patients with portal hypertension may preclude application of the hepatocellular injection method for end-stage liver disease. A significant amount of intrapulmonary shunting of hepatocytes was observed in rats with portal hypertension after intrasplenic injection of hepatocytes, which resulted in increased portal pressures, pulmonary hypertension, pulmonary infarction, and reduced pulmonary compliance [Gupta et al., 1993]. With the hepatocyte-polymer system, delivery of a large number of hepatocytes is possible. In patients with portal hypertension, portal blood-containing hepatotrophic factors are shunted away from the liver, obviating the need for a portacaval shunt. Thus, patients with end-stage liver disease and portal hypertension may need only transplantation of the hepatocyte-polymer construct. However, an end-to-side portacaval shunt operation is needed in congenital liver diseases with normal portal pressures to deliver hepatotrophic factors to the heterotopically place hepatocytes. In the future, each hepatocyte transplant system could have specific and different clinical applications. More important, both offer the hope of increasing therapeutic options for patients requiring liver replacement therapy. Approximately 260,000 patients out of 634,000 patients hospitalized for liver diseases have liver diseases which could have been considered for hepatic transplantation. The total acute care nonfederal hospital cost for liver diseases in 1992, which does not include equally substantial outpatient costs, exceeded $9.2 billion [HCIA Inc., 1992].

Defining Terms

Allogeneic: Pertaining to different genetic compositions within the same species.

Cadaveric: Related to a dead body. In transplantation, *cadaveric* is related to a person who has been declared brain dead; organs should be removed prior to cardiac arrest to prevent injury to the organs.

Heterotopic: Related to a region or place where an organ or tissue is not present in normal conditions.

Mitogens: Substances that stimulate mitosis or growth.

Orthotopic: Related to a region or place where an organ or tissue is present in normal conditions.

Portacaval shunt: A surgical procedure to partially or completely anastomose the portal vein to the inferior vena cava to divert portal blood flow from the liver to the systemic circulation.

Stroma: The structure or framework of an organ or gland usually composed of connective tissue.

Transgenic: Referred to introduction of a foreign gene into a recipient which can be used to identify genetic elements and examine gene expression.

Xengraft: A graft transferred from one animal species to another species.

References

Asonuma K, Gilber JC, Stein JE, et al. 1992. Quantitation of transplanted hepatic mass necessary to cure the Gunn rat model of hyperbilirubinemia. J Ped Surg 27(30):298.

Asonuma K, Vacanti JP. 1992. Cell transplantation as replacement therapy for the future. Pediatr Transplantation 4(2):249.

Cima L, Vacanti JP, Vacant C, et al. 1991. Tissue engineering by cell transplantation using degradable polymer substrates. J Biomech Eng 113:143.

Demetriou AA, Felcher A, Moscioni AD. 1991. Hepatocyte transplantation. A potential treatment for liver disease. Dig Dis Sci 12(9):1320.

Demetriou AA, Whiting JF, Feldman D, et al. 1986. Replacement of liver function in rats by transplantation of microcarrier-attached hepatocytes. Science 233:1190.

Fontaine MJ, Hansen, LK, Thompson S, et al. 1993. Transplantation of genetically altered hepatocytes using cell-polymer constructs leads to sustained human growth hormone secretion in vivo. Transplant Proc 25(1):1002–4.

Giesecke D, Tiemeyer W. 1984. Defect of uric acid in Dalmatian dog liver. Experientia 40:1415.

Grossman M, Roper SE, Kozarsky K, et al. 1994. Successful ex vivo gene therapy directed to liver in a patient with familial hypercholesterolemia. Nature Genetics 6:335.

Gupta S, Yereni PR, Vemuru RP, et al. 1993. Studies on the safety of intrasplenic hepatocyte transplantation: relevance to ex vivo gene therapy and liver repopulation in acute hepatic failure. Hum Gene Ther 4(3):249.

Hansen LK, Vacanti JP. 1992. Hepatocyte transplantation using artificial biodegradable polymers. 1992. In MA Hoffman (ed), Current Controversies in Biliary Atresia. The Medical Intelligence Unit Series pp 96–106, (CRC Press), Austin, Tex, R.G. Landes.

HCIA Inc. 1992. Survey of costs in non-Federal, acute care hospitals in the United States prepared for the American Liver Foundation, Baltimore.

Henne-Bruns D, Kruger U, Sumpelman D, et al. 1991. Intraperitoneal hepatocyte transplantation: morphological results. Virchows Arch A, Path Anat Histopathol 419(1):45.

Jaffe V, Darby H, Bishop A, et al. 1991. The growth of liver cells in the pancreas after intrasplenic implantation: the effects of portal perfusion. Int J Exp Pathol 72(3):289.

Jauregui HO, Gann KL. 1991. Mammalian hepatocytes as a foundation for treatment in human liver failure [Review]. J Cell Biochem 45(4):359.

Kaufmann P-M, Sano K, Uyama S, et al. 1994. Heterotopic hepatocyte transplantation using three dimensional polymers. Evaluation of the stimulatory effects by portacaval shunt or islet cell co-transplantation. Second International Congress of the Cell Transplant Society, Minneapolis, Minnesota.

Langer R, Vacanti J. 1993. Tissue Engineering Science 260:920.

Matas AJ, Sutherland DER, Steffes MW, et al. 1976. Hepatocellular transplantation for metabolic deficiencies: decrease of plasma bilirubin in Gunn rats. Science 192:892.

Merion RM, Campbell DA Jr. 1991. Split liver transplantation: One plus one doesn't always equal two. Hepatology 14(3):572.

Michalopoulos G. 1993. HGF and liver regeneration. Gasterologica Japonica 28(suppl 4):36.

Mooney DJ, Hansen LK, Vacanti JP, et al. 1992. Switching from differentiation to growth in hepatocytes: Control by extracellular matrix. J Cell Physiol 151:497.

National Vital Statistic System. 1991, 1992. Data derived from National Center for Health Statistics.

Onodera K, Ebata H, Sawa M, et al. 1992. Comparative effects of hepatocellular transplantation in the spleen, portal vein, or peritoneal cavity in congenitally ascorbic acid biosynthetic enzyme-deficient rats. Transplant Proc 24(6):3006.

Ponder KP, Gupta S, Leland F, et al. 1991. Mouse hepatocytes migrate to liver parenchyma and function indefinitely after intrasplenic transplantation. Proc Natl Acad Sci USA 88(4):1217.

Rhim JA, Sandgren EP, Degen JL, et al. 1994. Replacement of diseased mouse liver by hepatic cell transplantation. Science 263:1149.

Ricordi C, Lacy PE, Callery MP, et al. 1989. Trophic factors from pancreatic islets in combined hepatocyte-islet allografts enhance hepatocellular survival. Surgery 105:218.

Sandbichler P, Then P, Vogel W, et al. 1992. Hepatocellular transplantation into the lung for temporary support of acute liver failure in the rat. Gastroenterology 102(2):605.

Sirica AE, Mathis GA, Sano N, et al. 1990. Isolation, culture, and transplantation of intrahepatic biliary epithelial cells and oval cells. Pathobiology 58:44.

Starzl TE, Demetris AJ, Van Thiel D. 1989. Chronic liver failure: Orthotopic liver transplantation. N Eng J Med 321:1014.

Takeda T, Kim TH, Lee SK, et al. 1994. Hepatocyte transplantation in biodegradable polymer scaffolds using the Dalmatian dog model of hyperuricosuria. Fifteenth Congress of the Transplantation Society, Kyoto Japan. Submitted.

Uyama S, Takeda T, Vacanti JP. 1993. Delivery of whole liver equivalent hepatic mass using polymer devices and hetertrophic stimulation. Transplantation 55(4):932.

Uyama S, Takeda T, Vacanti JP. In press. Hepatocyte transplantation equivalent to whole liver mass using cell-polymer devices. Polymer Preprints.

Vacanti JP, Morse MA, Saltzman WM, et al. 1988. Selective cell transplantation using bioabsorbable artificial polymers as matrices. J Ped Surg 23(1):3.

Vroemen JPAM, Blanckaert N, Buurman WA, et al. 1985. Treatment of enzyme deficiency by hepatocyte transplantation in rats. J Surg Res 39:267.

Vroemen JPAM, Buurman WA, van der Linden CJ, et al. 1988. Transplantation of isolated hepatocytes into the pancreas. Eur Surg Res 20:1.

Wilson JM, Grossman M, Raper SE, et al. 1992. Ex vivo gene therapy of familial hypercholesterolemia. Human Gene Ther 3(2):179.

Yarmush ML, Toner M, Dunn JCY, et al. 1992. Hepatic tissue engineering: Development of critical technologies. Ann NY Acad Sci 665:238.

122

Tissue Engineering in the Nervous System

Ravi Bellamkonda
Lausanne University Medical School

Patrick Aebischer
Lausanne University Medical School

Tissue engineering in the nervous system facilitates the controlled application and/or organization of neural cells to perform appropriate diagnostic, palliative, or therapeutic tasks. As the word *tissue* implies, tissue engineering in general involves cellular components and their organization. Any cell, given its broad genetic program, expresses a particular phenotype in a manner that is dependent on its environment. The extracellular environment consists of cells, humoral factors, and the extracellular matrix. Research in genetic engineering and the intense focus on growth factors and extracellular matrix biology have made it possible to manipulate both the cell's genetic program and its phenotypic expression.

In the nervous system, degeneration or injury to neurons or glia and/or an aberrant extracellular environment can cause a wide variety of ailments. Diseases such as Parkinson's may required the replacement of diminished levels of a particular neurochemical, e.g., dopamine. Other pathologies such as injured nerves or reconnection of served neural pathways may require regeneration of nervous tissue.

Tissue engineering efforts in the nervous system have currently been addressing the following goals:

1. Functional replacement of a missing neuroactive component
2. Rescue or regeneration of damaged neural tissue
3. Human-machine interfaces: neural coupling elements

122.1 Delivery of Neuroactive Molecules to the Nervous System

Deficiency of specific neuroactive molecules has been implicated in several neurologic disorders. These factors may be neurotransmitters, neurotrophic agents, or enzymes. For example, part of the basal ganglia circuitry that plays a role in motor control consists of striatal neurons receiving dopaminergic input from

0-8493-0462-8/00/$0.00+$.50

Table 122.1 *Engineering* Solutions for Parkinson's Disease

Mode of Delivery	Rodents	Monkey	Human	Reference
Infusion	Rat striatum			Hargraves et al., 1987
		Cerebroventricular		De Yebenes et al., 1988
			Systemic	Hardie et al., 1984
Polymer slow release	Rat striatum (EVA rods)			Winn et al., 1989
	Rat subcutaneous (EVA rods)			Sabel et al., 1990
	Rat striatum (Si pellet)			Becker et al., 1990
	Rat striatum (liposomes)			During et al., 1992
Cell transplantation				
Fetal				
Substantia nigra	Rat striatum			Björklund et al., 1980
				Freund et al., 1985
Ventral mesencephalon	Rat striatum			Bolam et al., 1987
Human DA neurons			Putamen	Lindvall et al., 1990
Autologous primary				
Genetically altered skin fibroblasts	Rat striatum			Chen et al., 1991
Genetically altered skin fibroblasts with myoblasts	Rat striatum			Jiao et al., 1993
Encapsulated xenogeneic tissue				
Bovine adrenal chromaffin cells	Striatum (microcapsules)			Aebischer et al., 1991
PC12 cells	Rat striatum (microcapsules)			Winn et al., 1991
	Rat striatum (macrocapsule)			Aebischer et al., 1988
		Striatum (macrocapsule)		Aebischer et al., 1994
Mouse mesencephalon	Rat parietal cortex (macrocapsule)			Aebischer et al., 1988

the mesencephalic substantia nigra neurons. It has been shown that a lesioned nigrostriatal dopaminergic pathway is responsible for Parkinson's disease [Ehringer & Hornykiewicz, 1960]. In chronic cancer patients, the delivery of antinociceptive neurotransmitters such as enkephalins, endorphins, catecholamines, neuropeptide Y, neurotensin, and somatostatin to the cerebrospinal fluid may improve the treatment of severe pain [Akil et al., 1984; Joseph et al., in press]. Neurotrophic factors may play a role in the treatment of several neurodegenerative disorders. For example, local delivery of nerve growth factor (NGF) [Hefti et al., 1984; Williams et al., 1986] and/or brain-derived growth factor (BDNF) may be useful in the treatment of Alzheimer's disease [Anderson et al., 1990, Knüsel et al., 1991]. BDNF [Hyman et al., 1991; Knüsel et al., 1991] and glial cell line-derived nerve growth factor (GDNF) may be beneficial in Parkinson's' disease [Lin et al., 1993]. Other neurotrophins such as ciliary neurotrophic factor (CNTF) [Oppenheim et al., 1991]; Sendtner et al., 1990], BDNF [Yan et al., 1992], neurotrophin-3 (NT-3), neurotrophin-4 (NT-4/5) [Hughes et al., 19993] and GDNF [Zurn et al., 1994] also could have an impact on amyotrophic lateral sclerosis (ALS) or Lou Gehrigs's disease.

Therefore, augmentation or replacement of any of the above-mentioned factors in the nervous system would be a viable therapeutic strategy for the treatment of the pathologies listed above. There are several issues that ought to be considered in engineering a system to deliver these factors. The stability of the factors, the dosage required, the solubility, the target tissue, and possible side effects all are factors that influence the choice of the delivery mode. Pumps, slow-release polymer systems, and cells from various sources that secrete the compound of interest are the main modes by which these factors can be delivered. Table 122.1 lists all the means employed to deliver dopamine to alleviate the symptoms of Parkinson's disease.

Pumps

Pumps are used to deliver opiates epidurally to relieve severe pain [Ahlgren et al., 1987]. Pumps also have been used to deliver neurotrophic factors such as NGF intraventricularly as a prelude or supplement to transplantation of chromaffin cells for Parkinson's disease [Olson et al., 1991] or as a potential therapy for Alzheimer's disease [Olson et al., 1992]. Pumps also have been used to experimentally deliver dopamine or dopamine receptor agonists in Parkinson's disease models [DeYebenes et al., 1988; Hargraves et al., 1987]. While pumps have been employed successfully in these instances, they may need to be refilled every 4 weeks, and this may be a limitation. Other potential drawbacks include susceptibility to "dumping" of neuroactive substance due to presence of a large reservoir of neuroactive element, infections, and diffusion limitations. Also, some factors such as dopamine and ciliary neurotrophic factor (CNTF) are very unstable chemically and have short half-life periods, rendering their delivery by such devices difficult. Some of these problems may be eliminated by well designed slow-release polymer systems.

Slow-Release Polymer Systems

Slow-release polymer systems essentially trap the molecule of interest in a polymer matrix and release it slowly by diffusion over a period of time. Proper design of the shape and composition of the polymer matrix may achieve stabilization of the bioactive molecule and facilitate a steady, sustained release over a period of time. For instance, it has been shown that the dopamine precursor L- dopa may be effective in alleviating some motor symptoms of Parkinson's disease [Birkmayer, 1969], fluctuations in the plasma levels of L-dopa due to traditional, periodic oral administration and difficulties in converting L-dopa to dopamine may cause the clinical response to fluctuate as well [Albani et al., 1986; Hardie et al., 1984; Muenter et al., 1977; Shoulson et al., 1975]. It has been demonstrated that a slow-release ethylene vinyl acetate polymer system loaded with L-dopa can sustain elevated plasma levels of L-dopa for at least 225 days when implanted subcutaneously in rats [Sabel et al., 1990]. Dopamine can only be released directly in the CNS, since it does not pass the blood-brain barrier. It has been demonstrated that experimental parkinsonism in rats can be alleviated by intrastriatal implantation of dopamine-releasing ethylene vinyl acetate rods [Winn et al., 1989]. Silicone elastomer as well as resorbable polyester pellets loaded with dopamine also have been implanted intrastriatally and shown to induce a behavioral recovery in parkinsonian rats [Becker et al., 1990; McRae et al., 1991].

Slow-release systems also may be employed to deliver trophic factors to the brain either to avoid side affects that may come about due to systemic administration or to overcome the blood-brain barrier, getting the factor delivered to the brain directly [During et al., 1992; Hoffman et al, 1990]. Therefore, polymeric systems with the molecule of interest trapped inside may be able to achieve many of the goals of an ideal delivery system, including targeted local delivery and zero-order continuos release [Langer, 1981; Langer et al., 1976]. However, some of the disadvantages of this system are the finite amounts of loaded neuroactive molecules and difficulties in shutting off release or adjusting rate of release once the polymer is implanted. Also, for the long-term release in humans, the device size may become a limiting factor. Some of these limitations may be overcome by the transplantation of cells that release neuroactive factors to the target site.

Cell Transplantation

Advances in molecular biology and gene transfer techniques have given rise to a rich array of cellular sources, which have been engineered to secrete a wide range of neurologic compounds. These include cells that release neurotransmitters, neurotrophic factors, and enzymes. Transplantation of these cells leads to functional replacement or augmentation of the original source of these compounds in the host. They can deliver neuroactive molecules as long as they survive, provided they maintain their phenotype and/or transgene expression, in the case of gene therapy. Some of the disadvantages of the slow-release systems, such as the presence of a large reservoir for long-term release or reloading of an exhausted reservoir, can thus be overcome. Some of the tissues transplanted so far may be classified in the following manner.

Transplantation of Autologous Primary Cells

This technique involves the procurement of primary cells from the host, expanding them if necessary to generate requisite amounts of tissue, engineering them if so required by using gene transfer techniques, and then transplanting them back into the "donor" at the appropriate site. For instance, autologous Schwann cell shave been isolated and transplanted experimentally into the brain and have been shown to enhance retinal nerve regeneration, presumably by the release of factors that influence regeneration [Morrissey et al., 1991]. Schwann cells also express various neurologically relevant molecules [Bunge & Bunge, 1983; Muir et al., 1989]. Autologous Schwann cells also can work as nerve bridges to help reconstruction of rat sciatic nerve after axotomy [Guénard et al., 1992]. Primary skin fibroblasts also have been genetically engineered to secrete L-dopa and transplanted successfully into the autologous host's striatum in an experimental model of Parkinson's disease in rats [Chen et al., 1991]. The same group also has reported that nerve growth factor, tyrosine hydroxylase, glutamic acid decarboxylase, and choline acetyltransferase genes may be introduced successfully and expressed in primary fibroblasts [Gage et al., 1991]. More recently, muscle cells have been engineered to express tyrosine hydroxylase and transplanted successfully in a rat model of Parkinson's disease [Jiao et al., 1993].

Thus nonneural cells such as a fibroblast or muscle cells may be selected, in part, for the ease with which they can be engineered genetically and made neurologically relevant. It is therefore possible to engineer cells to suit particular pathologies as a step toward being able to engineer biomimetic tissues and then place them in appropriate locations and contexts in vivo. However, it may not always be technically possible to procure sufficient amounts of autologous tissue. Other sources of tissues have therefore been explored, and these include fetal tissue, used usually in conjunction with immunosuppression.

Fetal Tissue Transplantation

One of the important advantages of fetal tissue is its ability to survive and integrate into the host adult brain. Transplantation of fetal neural tissue allografts might be useful in treating Parkinson's disease [Lindvall et al., 1990]. While promising results have been reported using neural fetal tissue disease [Björklund, 1991; Björklund et al., 1980], availability of donor tissue and potential ethical issues involved with the technique may be potential shortcomings. One promising strategy to obtain allogeneic fetal tissue in large quantities is to isolate neural stem cells and have them proliferate in vitro. It has been demonstrated recently that CNS progenitor cells may be selected using epidermal growth factor (EGF). The cells thus selected have been shown to proliferate in vitro under appropriate culture conditions. With time and appropriate culture conditions, they differentiate into mature CNS neurons and glia [Reynolds et al., 1992; Vescovi et al., 1993]. When optimized, this technique could be useful to select and expand fetal neurons of interest in vitro and then transplant them. Transplantation of xenogeneic fetal tissue with immunosuppression using cyclosporin A is another alternative to using fetal tissue [Wictorin et al., 1992], and this approach might yield more abundant amounts of tissue for transplantation and overcome some of the possible ethical dilemmas in using human fetal tissue. However, immunosuppression may not be sufficient to prevent long-term rejection and may have other undesirable side-effects.

Transplantation of Encapsulated Xenogeneic Tissue

Polymeric encapsulation of xenogeneic cells might be a viable strategy to transplant cells across species in the absence of systemic immunosuppression [Aebischer et al., 1988; Tresco et al., 1992]. Typically, the capsules have pores large enough for nutrients to reach the transplanted tissue and let the neuroactive factors out, but the pores are too small to let the molecules and the cells of the immune system reach the transplanted tissue (Fig. 122.1). At the same time, this strategy retains all the advantages of using cells as controlled, local manufacturers of the neuroactive molecules. The use of encapsulation also eliminates the restriction of having to use postmitotic tissue for transplantation to avoid tumor formation. The physical restriction of the polymeric capsule prevents escape of the encapsulated tissue. Should the capsule break, the transplanted cells are rejected and eliminated by the host immune system. However, no integration of the transplanted tissue into the host is possible with this technique.

Some of the tissue engineering issues involved in optimizing the encapsulation technique are (1) the type and configuration of the encapsulating membrane, (2) the various cells to be used for encapsulation, and (3) the matrix in which the cells are immobilized.

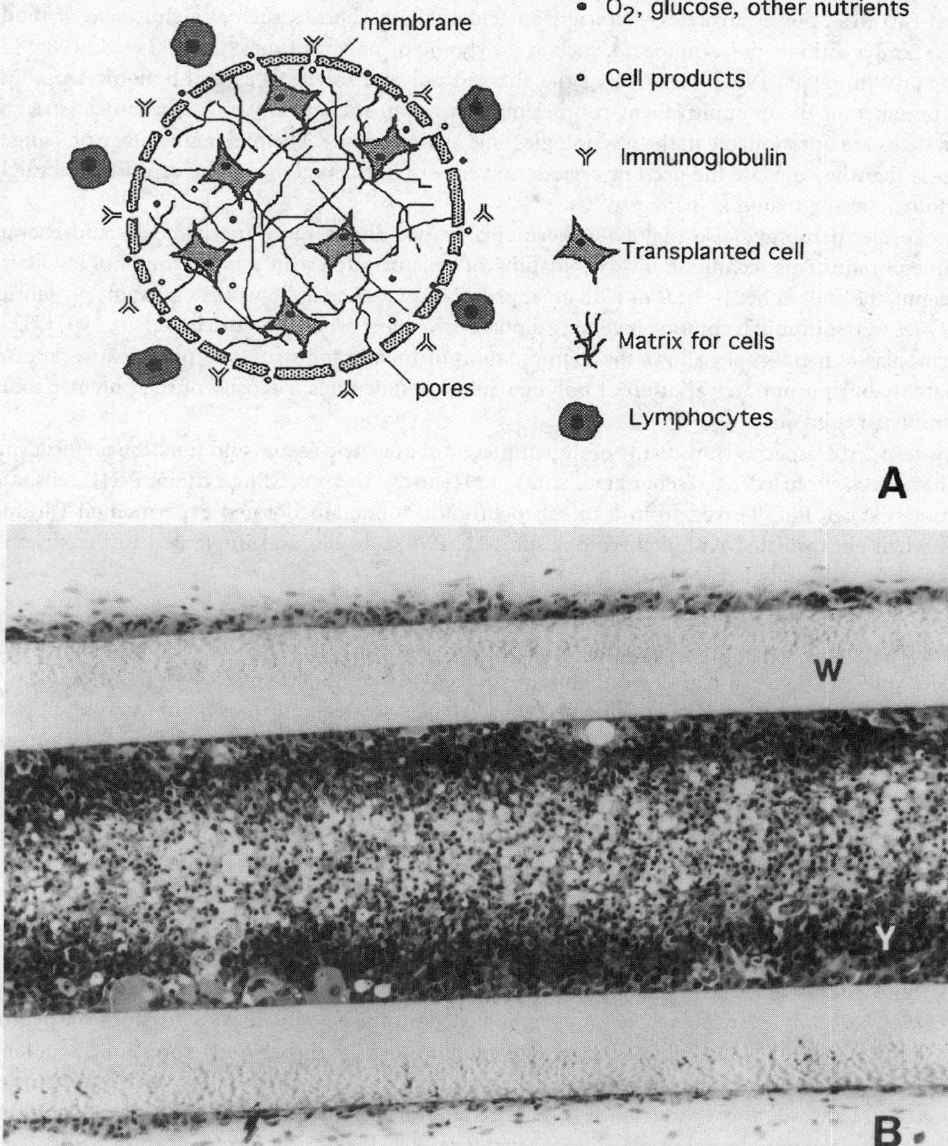

FIGURE 122.1 (a)Schematic illustration of the concept of immunoisolation involved in the transplantation of xenogenic tissue encapsulated in semipermeable polymer membranes. (b) Light micrograph of a longitudinal section shown baby hamster kidney cells encapsulated in a polyacryolonitrile-vinylchloride membrane and transplanted in an axotomized rat pup after 2 weeks in vivo. W denotes the capsule's polymeric wall, and Y shows the cells.

Type and Configuration of the Encapsulating Membrane. One important factor determining the size of the device is oxygen diffusion and availability to the encapsulated cells. This consideration influences the device design and encourages situations where the distances between the oxygen source, usually a capillary, and the inner core of transplanted tissue are kept as minimal as possible. The size and configuration of the device also influence the kinetics of release of the neuroactive molecules, the "response time" being slower in larger capsules with thicker membrane walls.

The capsule membrane may be a water-soluble system stabilized by ionic or hydrogen bonds formed between two weak polyelectrolytes—typically an acidic polysaccharide, such as alginic acid or modified cellulose, and a cationic polyaminoacid, such as polylysine or polyornithine [Goosen et al., 1985; Lim & Sun, 1980; Winn et al., 1991]. Gelation of the charged polyelectrolytes is caused by ionic cross-linking in the presence of di- or multivalient counterions. However, the stability and mechanical strength of these systems are questionable in the physiologic ionic environments. The major advantage of using these systems is that they obviate the need of organic solvent use in the making of the capsules and might be less cytotoxic in the manufacturing process.

The capsule membrane also may be a thermoplastic, yielding a more mechanically and chemically stable membrane. This technique involves loading of cells of interest in a preformed hollow fiber and then sealing the ends either by heat or with an appropriate glue. The hollow fibers are typically fabricated by a dry jet wet-spinning technique involving a phase-inversion process [Aebischer et al., 1991]. The use of thermoplastic membranes allows the manipulation of membrane structure, porosity, thickness, and permeability by appropriate variation of polymer solution flow rates, viscosity of the polymer solution, the nonsolvent used etc.

Long-term cross-species transplants of dopaminergic xenogeneic tissues and functional efficacy in the brain have been reported [Aebischer et al., 1991, 1994] using the preceding system. PC12 cells, a catecholaminergic cell line derived from a rat pheochromocytoma, ameliorated experimental Parkinson's disease when encapsulated within thermoplastic PAN/PVC capsules and implanted in the striatum of adult guinea pigs [Aebischer et al., 1991].

The Choice of Cells and Tissues for Encapsulation. Three main types of cells can be used for encapsulation. Cells may be postmitotic, cell lines that differentiate under specific conditions, or slow-dividing cells. The latter two types of cells lend themselves to genetic manipulation.

Postmitotic Cells. Primary xenogeneic tissue may be encapsulated and transplanted across species. For instance, chromaffin cells release various antinociceptive substances such as enkephalins, catecholamines, and somatostatin. Allografts of adrenal chromaffin cells have been shown to alleviate pain when transplanted in the subarachnoid space in rodent models and terminal cancer patients [Sagen et al., 1993]. Transplantation of encapsulated xenogeneic chromaffin cells may provide a long-term source of pain-reducing neuroactive substances [Sagen et al., 1993]. In our laboratory, clinical trials are currently underway using the transplantation of encapsulated bovine chromaffin tissue into human cerebrospinal fluid as a strategy to alleviate chronic pain in terminal cancer patients [Aebischer et al., 1994]. This may circumvent the problem of the limited availability of human adrenal tissue in grafting procedures of chromaffin tissue.

Cell Lines that Differentiate under Specific Conditions. Postmitotic cells are attractive for transplantation applications because the possibility of tumor formation, loss of phenotypic expression, is potentially lower. Also, when encapsulated, there is less debris accumulation inside the capsule due to turnover of dividing cells. However, the disadvantage is that the amount of postmitotic tissue is usually too limited in quantity for general clinical applications. Therefore, the use of cells that are mitotic and then are rendered postmitotic under specific conditions is attractive for transplantation applications.

Under appropriate culture conditions, primary myoblasts undergo cell division for at least 50 passages. Fusion into resting myoblasts can be obtained by controlling the culture conditions. Alternatively, a transformed myoblast cell line C_2C_{12} derived from a C_2H mouse thigh differentiates and forms myotubes when cultured under low serum conditions [Yaffe & Saxel, 1977]. Therefore, these cells could be genetically altered, expanded in vitro and then made postmitotic by varying the culture conditions. They can then be transplanted with the attendant advantages of using postmitotic tissue. Since myoblasts can be altered genetically, they have the potential to be a rich source for augmentation of tissue function via cell transplantation.

Slow-Dividing Cells. Slow-dividing cells that reach a steady state in a capsule either due to contact inhibition or due to the encapsulation matrix are an attractive source of cells for transplantation. Potentially, their division can result in a self-renewing supply of cells inside the capsule. Dividing cells are also easier to transfect reliably with retroviral methods and therefore lend themselves to genetic manipulation. It is therefore possible to envisage the transplantation of various genetically engineered cells for the treatment of several neurologic disorders. This technique allows access to an ever-expanding source of xenogeneic tissues that have been engineered to produce the required factor of interest. For instance, Horellou and Mallet [1989] have retrovirally transferred the human TH cDNA into mouse anterior pituitary AtT-20 cell line, potentially resulting in a plentiful supply of dopamine-secreting cells that can then be encapsulated and transplanted into the striatum.

Another promising area for the use of such techniques is in the treatment of some neurodegenerative disorders where a lack of neurotrophic factors is believed to be part of the pathophysiology. Neurotrophic factors are soluble proteins that are required for the survival of neurons. These factors often exert a trophic effect; i.e., they have the capability of attracting growing axons. The "target" hypothesis describes the dependence of connected neurons on a trophic factor that is retrogradely transported along the axons after release from the target neurons. In the absence of the trophic factors, the neurons shrink and die, presumably to avoid potential misconnections. Experimentally, fibroblast lines have been used in CNS transplantation studies because of their convenience for gene transfer techniques [Gage et al., 1991]. The transplantation of encapsulated genetically engineered fibroblasts to produce NGF has been shown to prevent lesion-induced loss of septal ChaT expression following a fimbria-fornix lesion [Hoffman et al., 1993]. The fimbria-fornix lesion is characterized by deficits in learning and memory resembling those of Alzheimer's disease.

Matrices for Encapsulation. The physical, chemical, and biologic properties of the matrix in which the cells have been immobilized may play an important role in determining the transplanted cell's state and function. Broadly, matrices can be classified into the following types: cross-linked polyelectrolytes, collagen in solution or as porous beads, naturally occurring extracellular matrix derivatives such as Matrigel, fibrin clots, and biosynthetic hydrogels with appropriate biologic cues bound to them to elicit a specific response from the cells of interest. The matrix has several functions: It can prevent the formation of large cell aggregates that lead to the development of central necrosis as a consequence of insufficient oxygen and nutrient access; it may allow anchorage-dependent cells to attach and spread on the matrix substrate; and it may induce differentiation of a cell line and therefore slow down or stop its division rate.

Negatively charged polyeletrolytes such as alginate have been used successfully for the immobilization of adrenal chromaffin cells [Aebischer et al., 1991]. Positively charged substrates, such as those provided by the amine groups of chitosan, allow attachment and spreading of fibroblasts [Zielinski et al., in press]. Biologically derived Matrigel induces differentiation of various cell lines such as Chinese hamster ovary (CHO) cells, astrocyte lines, or fibroblast lines (unpublished observations). Spongy collagen matrices, as well as fibrin matrices, seem to possess similar qualities. Our laboratory is currently evaluating the use of biosynthetic hydrogel matrices with biologically relevant peptides covalently bound to the polymer backbone. It is hypothesized that these matrices may elicit a specific designed response from the encapsulated cells.

122.2 Tissue Reconstruction: Nerve Regeneration

Most of the techniques described above were attempts at identifying the missing molecules of various neuropathologies, finding an appropriate source for these molecules and, if necessary, designing a cellular source via genetic engineering, and ultimately, choosing an optimal mode of delivery of the molecules, be it chemical or cellular. This approach, however, falls short of replacing the physical neuroanatomic synaptic circuits in the brain, which, in turn, may play an important role in the physiologic feedback regulating mechanisms of the system. Attempts to duplicate in vivo predisease neuronal structure have been made. For instance, the bridging of the nigrostriatal pathway, which when disrupted may cause Parkinson's disease, and the septohippocampal pathway, which may serve as a model for Alzheimer's disease, has attracted some attention. Wictorin and colleagues [1992] have reported long-distance directed axonal growth from human dopaminergic mesencephalic neuroblasts implanted along the nigrostriatal pathway in 6-hydroxydopamine-lesioned rats. In this section we shall examine the use of synthetic

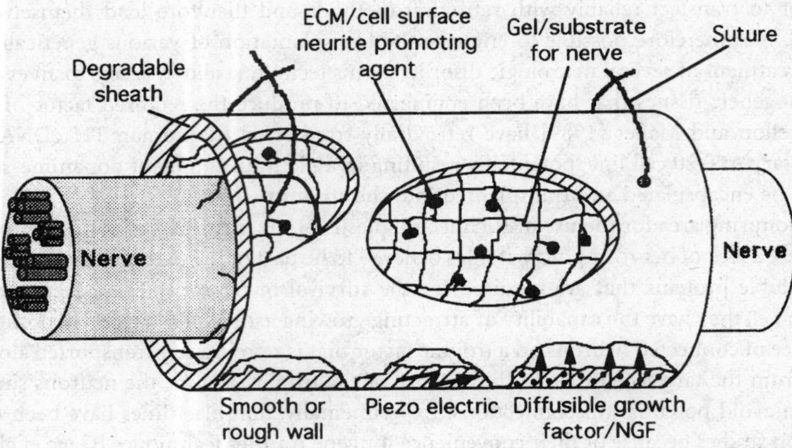

FIGURE 122.2 Schematic illustration of a nerve guidance channel and some of the possible strategies for influencing nerve regeneration.

guidance channels and extracellular matrix cues to guide axons to their appropriate targets. Thus a combination of all these techniques may render the complete physical and synaptic reconstruction of a degenerated pathway feasible.

The promotion of nerve regeneration is an important candidate task for tissue reconstruction in the nervous system. Synthetic nerve guidance channels (NGCs) have been used to study the underlying mechanisms of mammalian peripheral nerve regeneration after nerve injury and enhance the regeneration process. Guidance channels may simplify end-to-end repair and may be useful in repairing long nerve gaps. The guidance channel reduces tension at the suture line, protects the regenerating nerve from infiltrating scar tissue, and directs the sprouting axons toward their distal targets. The properties of the guidance channel can be modified to optimize the regeneration process. Nerve guidance channels also may be used to create a controlled environment in the regenerating site. In the peripheral nervous system, NGCs can influence the extent of nerve gap that can be bridged and the quality of regeneration. The channel properties, the matrix filling the NGC, the cells seeded within the channel lumen, and polymer-induced welding of axons all can be strategies used to optimize and enhance nerve regeneration and effect nervous tissue reconstruction (see Fig. 122.2 for a schematic). Table 122.2 lists some of the kinds of nerve guidance channels used so far.

The Active Use of Channel Properties

In the past, biocompatability of a biomaterial was evaluated by the degree of its passivity or lack of "reaction" when implanted into the body. However, the recognition that the response of the host tissue is related to the mechanical, chemical, and structural properties of the implanted biomaterial has led to the design of materials that promote a beneficial response from the host. In the context of a synthetic nerve guidance channel, this may translate to manipulation of its microstructural properties, permeability, electrical properties, and the loading of its channel wall with neuroactive components that might then be released locally into the regenerating environment. The strategy here is to engineer a tailored response from the host and take advantage of the natural repair processes.

Surface Microgeometry

The morphology of regenerating peripheral nerves is modulated by the surface microgeometry of polymeric guidance channels [Aebischer et al., 1990]. Channels with smooth inner walls give rise to organized, longitudinal fibrin matrices, resulting in discrete free-floating nerve cables with numerous myelinated axons. The rough inner surface channels, however, give rise to an unorganized fibrin matrix with nerve

TABLE 122.2 Nerve Guidance Channels

I. The Channel Wall	
1. *Passive polymeric channels*	
Silicone elastomer	Lundborg et al., 1982
Polyvinyl chloride	Scaravalli, 1984
Polyethylene	Madison et al., 1988
2. *Permeable polymer channels*	
Acrylonitrile vinychloride copolymer	Uzman and Villegas, 1983
Collagen	Archibald et al., 1991
Expanded polytetrafluroethylene	Young et al., 1984
3. *Resorbable polymer channels*	
Polyglycolic acid	Molander et al., 1989
Poly-L-lactic acid	Nyilas et al., 1983
Collagen	Archibald et al., 1991
4. *Electrically shaped polymer channels*	
Silicone channels with electrode cuffs	Kerns and Freeman, 1986; Kerns et al., 1991
Polyvinylidenefluoride (piezoelectric)	Aebischer et al., 1987
Polytetrofluoroethylene (electret)	Valentini et al., 1989
5. *Polymer channels releasing trophic factors*	
Ethylene vinylacetate copolymer	Aebischer et al., 1989
II. Intrachannel, luminal matrices	
1. Fibrin matrix	Williams et al., 1987
2. Collagen-glycosaminoglycan template	Yannas et al., 1985
3. Matrigel	Valentini et al., 1987
III. Cell seeded lumens for trophic support	
1. Schwann cell-seeded lumens (PNS)	Guénard et al., 1992
2. Schwann cell-seeded lumens (CNS)	Guénard et al., 1993; Kromer and Cornbrooks, 1985, 1987; Smith and Stevenson, 1988

fascicles scattered in a loose connective tissue filling the entire channel's lumen. Thus the physical textural properties and porosity of the channel can influence nervous tissue behavior and may be used to elicit a desirable reaction from the host tissue.

Molecular Weight Cutoff

The molecular weight cutoff of the NGCs influences peripheral nerve regeneration in rodent models [Aebischer et al., 1989]. The molecular weight cutoff may influence nerve regeneration possibly by controlling the exchange of molecules between the channel lumen and the external wound-healing environment. This may be important because the external environment consists of humoral factors that can play a role in augmenting regenerative processes in the absence of a distal stump.

Electrical Properties

In vivo regeneration following transection injury in the peripheral nervous system has been reported to be enhanced by galvantropic currents produced in silicone channels fitted with electrode cuffs [Kerns & Freeman, 1986; Kerns et al., 1991]. Polytetrafluoroethylene (PTFE) "electret" tubes show more myelinated axons compared with uncharged tubes in peripheral nerves [Valentini et al., 1989]. Dynamically active piezoelectric polymer channels also have been shown to enhance nerve regeneration in the sciatic nerves of adult mice and rats [Aebischer et al., 1987; Fine et al., 1991].

Release of Bioactive Factors from the Channel Wall

Polymer guidance channels can be loaded with various factors to study and enhance nerve regeneration. Basic fibroblast growth factor released from an ethylene-vinyl acetate copolymer guidance channel facilitates peripheral nerve regeneration across long nerve gaps after a rat sciatic nerve lesion [Aebischer et al., 1989]. The possible influence of interleukin-1 (IL-1) on nerve regeneration also was studied by the release of IL-1 receptor antagonist (IL-1ra) from the wall of an EVA copolymer channel [Guénard et al., 1991]. It is conceivable that the release of appropriate neurotrophic factors from the channel wall may enhance specifically subsets of axons, e.g., ciliary neurotrophic factor on motor neurons and nerve growth factor on sensory neurons.

Resorbable Channel Wall

Bioresorbable nerve guidance channels are attractive because once regeneration is completed, the channel disappears without further surgical intervention. Mice sciatic nerves have been bridged with poly-L-lactic acid channels [Nyilas et al., 1983] and polyester guidance channels [Henry et al., 1985]. Rabbit tibial nerves also have been bridged with guidance channels fabricated from polyglycolic acid [Molander et al., 1989]. Resorbable guidance channels need to retain their mechanical integrity over 4 to 12 weeks. At the same time, their degradation products should not interfere with the regenerative processes of the nerve. These issues remain the challenging aspects in the development of bioresorbable nerve guidance channels for extensive use in animals and humans.

Intraluminal Matrices for Optimal Organization of Regeneration Microenvironment

The physical support structure of the regenerating environment may play an important role in determining the extent of regeneration. An oriented fibrin matrix placed in the lumen of silicone guidance channels accelerates the early phases of peripheral nerve regeneration [Williams et al., 1987].

Silicone channels filled with a collagen-glycosaminoglycan template bridged a 15-mm nerve gap in rats, whereas no regeneration was observed in unfilled tubes [Yannas et al., 1985]. However, even matrices known to promote neuritic sprouting in vitro may impede peripheral nerve regeneration in semipermeable guidance channels if the optimal physical conditions are not ensured [Madison et al., 1988; Valentini et al., 1987]. Therefore, the structural, chemical, and biologic aspects of the matrix design may all play a role in determining the fate of the regenerating nerve. The importance of the effect of the physical environment on regeneration, mediated by its influence on fibroblast and Schwann cell behavior, has been demonstrated in several studies and has been reviewed by Schwartz [1987] and Fawcett & Keynes [1990]. Thus the choice of a hydrogel with physical, chemical, and biologic cues conducive to nerve regeneration may enhance nerve regeneration. This strategy is currently being explored [Belamkonda et al., in press]. Neurite-promoting oligopeptides from the basement membrane protein laminin (LN) were covalently coupled to agarose hydrogels. Agarose gels derivatized with LN oligopeptides specifically enhance neurite extension from cells that have receptors to the LN peptides in vitro [Bellamkonda et al., in press]. Preliminary results show that the presence of an agarose gel carrying the LN peptide CDPGYIGSR inside the lumen of a synthetic guidance channel enhances the regeneration of transected peripheral nerves (Fig. 122.3) in rats. Thus it is feasible to tailor the intraluminal matrices with more potent neurite-promoting molecules such as the cell adhesion molecules (CAMs) L1, N-CAM, or tenascin and "engineer" a desired response from the regenerating neural elements.

Cell-Seeded Lumens for Trophic Support

Cells secreting various growth factors may play an important role in organizing the regeneration environment, e.g., Schwann cells in the peripheral and central nervous system. It has been reported that regenerating axons do not elongate through acellular nerve grafts if Schwann cell migration was impeded [Hall et al., 1986]. Syngeneic Schwann cells derived from adult nerves and seeded in semipermeable guidance channels enhance peripheral nerve regeneration [Guénard et al., 1992]. Schwann cells in the preceding study orient themselves along the axis of the guidance channel, besides secreting various neurotrophic factors. Schwann cells could play a role in organizing the fibirin cable formed during the initial phases of nerve regeneration. Schwann cells may be effective in inducing regeneration in the CNS, too [Kromer & Cornbrook, 1985, 1987; Smith & Stevenson, 1988]. The use of tailored intraluminal matrices and presenting exogeneic Schwann cells to the regeneration environment in a controlled matter

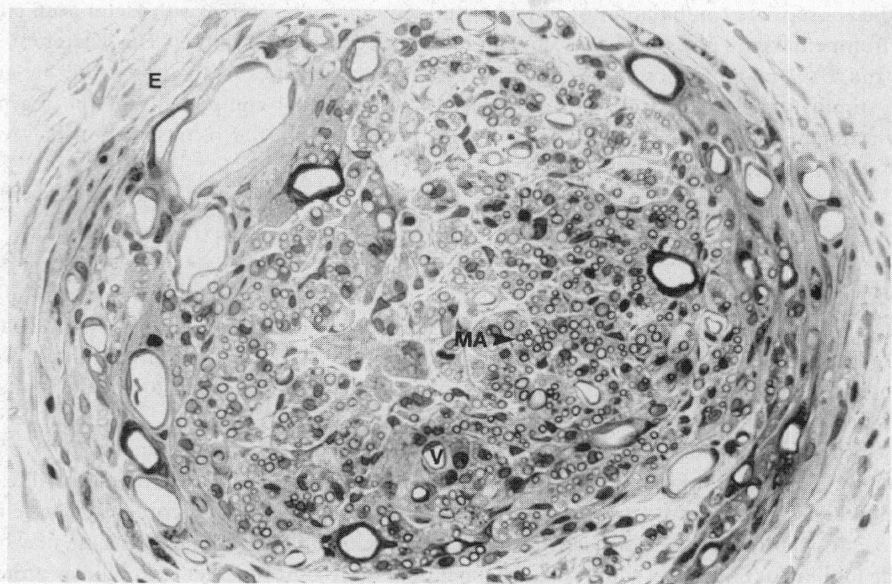

FIGURE 122.3 Light micrograph of a cross-sectional cut of a sural nerve regenerating through a polymer guidance channel 4 weeks after transection. The nerve guidance channel had been filled with a CDPGYIGSR derivatized agarose gel. E is the epineurium; V shows neovascularization; and MA is myelinated axon.

are strategies aimed at engineering the desired tissue response by creating the optimal substrate, trophic, and cellular environments around the regenerating nerves.

CNS glial cells have a secretory capacity that can modulate neuronal function. Astrocytes release proteins that enhance neuronal survival and induce neuronal growth and differentiation. When a silicone channel was seeded with astrocytes of different ages, ranging from P9 to P69 (postnatal), it was observed that while P9 astrocytes did not interfere with peripheral nerve regeneration, adult astrocytes downregulate axonal growth [Kalderon, 1988]. However, the presence of Schwann cells reverses the inhibition of PNS regeneration due to adult astrocytes [Guénard et al., 1994]. Thus the cellular environment in the site of injury may play an important role in determining the extent of regeneration. Knowledge of these factors also may be employed in designing optimal environments for nerve regeneration.

Polyethyleneglycol-Induced Axon Fusion

Rapid morphologic fusion of severed myelinated axons may be achieved by the application of polyethylene glycol (PEG) to the closely apposed ends of invertebrate-myelinated axons [Krause & Bittner, 1990]. Selection of appropriate PEG concentration and molecular mass, tight apposition, and careful alignment of the cut ends of the nerve may facilitate the direct fusion of axons. However, this technique is only applicable when the two ends of the severed nerve are closely apposed to each other, before the onset of wallerian degeneration.

CNS Nerve Regeneration

Most of the preceding studies have been conducted in the peripheral nervous system (PNS). In the CNS, however, endogenous components express poor support for axonal elongation. Significant regeneration may, however, occur with supporting substrates. Entubulation with a semipermeable acrylic copolymer tube allows bridging of a transected rabbit optic nerve with a cable containing myelinated axons [Aebischer et al., 1988]. Cholinergic nerve regeneration into basal lamina tubes containing Schwann cells has been reported in a transected septohippocampal model in rats [Kromer & Cornbrooks, 1985].

Thus the appropriate combination of physical guidance, matrices, and growth factors can create the right environmental cues and may be effective in inducing regeneration in the CNS. Therefore, both in the PNS and CNS, manipulation of the natural regenerative capacities of the host either by guidance factor or stimulation by electrical or trophic factors or structural components of the regenerating microenvironment can significantly enhance regeneration and help the reconstruction of severed or damaged neural tissue.

122.3 In Vitro Neural Circuits and Biosensors

The electrochemical and chemoelectrical transduction properties of neuronal cells can form the basis of a cell-based biosensing unit. The unique information-processing capabilities of neuronal cells through synaptic modulation may form the basis of designing simple neuronal circuits in vitro. Both the preceding applications necessitate controlled neuronal cell attachment, tightly coupled to the substrate and a sensitive substrate to monitor changes in the cell's electrical activity.

The use of bioactive material systems tailored to control neuronal cell attachment on the surface and still amenable to the incorporation of electrical sensing elements like a field effect transistor (FET) could be one feasible design. Therefore, composite material systems, which might incorporate covalently patterned bioactive peptides on their surface to control cell attachment and neurite extension, may be a step toward the fulfillment of the preceding goal. Oligopeptides derived from larger extracellular proteins like laminin have been shown to mediate specific cell attachment via cell surface receptors [Graf et al., 1987; Iwamoto et al., 1987; Kleinman et al., 1988]. Cell culture on polymeric membranes modified with the preceding bioactive peptidic components may give rise to a system where neuronal cell attachment and neuritic process outgrowth may be controlled. This control may help in designing microelectronic leads to complete the cell-electronic junction. Preliminary recordings from a FET-based neuron-silicon junction using leech Retzius cells [Fromherz et al., 1991] have been reported. Though there are many problems, such as attaining optimal coupling, this could form the basis of a "neural chip." A neural chip could potentially link neurons to external electronics for applications in neuronal cell-based biosensors, neural circuits, and limb prosthesis. Polymer surface modification and intelligent use of extracellular matrix components through selective binding could help attain this goal.

Studies in our laboratory have been trying to understand the underlying mechanisms involving protein adsorption onto polymeric substrates and their role in influencing and controlling nerve cell attachment [Ranieri et al., 1993]. Controlled neuronal cell attachment within a tolerance range of 20 μm may be achieved either nonspecifically monoamine surfaces or specifically via oligopeptides derived from ECM proteins like laminin and fibronectin [Fig. 122.4], mediated by integrin cell surface receptors [Ranieri et al., in press]. Molecular control of neuronal cell attachment and interfacing neuronal cells with electrodes may find applications in the design and fabrication of high-sensitivity neuron-based biosensors with applications in detection of low level neurotransmitters.

Studies are also currently in progress involving polymeric hydrogels and controlling neuronal cell behavior in a three-dimensional (3D) tissue culture environs as a step toward building 3D neuronal tissues [Bellamkonda et al., in press]. The choice of an appropriate hydrogel chemistry and structure, combined with the possibility of the gel serving as a carrier for ECM proteins or their peptidic analogues, can enable one to enhance regeneration when seeded in a nerve guidance channel. Also, the use of appropriate hydrogel chemistries in combination with the chemical modification of the polymer backbone by laser-directed photochemistry may be feasible in controlling the direction and differentiation of neuronal cells in three dimensions. Covalent binding of bioactive components like the laminin oligopeptides to the hydrogel backbone gives a specific character to the gel so that it elicits specific responses from anchorage-dependent neuronal cells [Bellamkonda et al., in press] (Fig. 122.5). Such a system could be useful in reorganizing nerves in 3D either for bridging different regions of the brain with nerve cables or for the 3D organization of nerves for optimal coupling with external electronics in the design of artificial limb prosthesis. In either case, development of such systems presents an interesting challenge for tissue engineering.

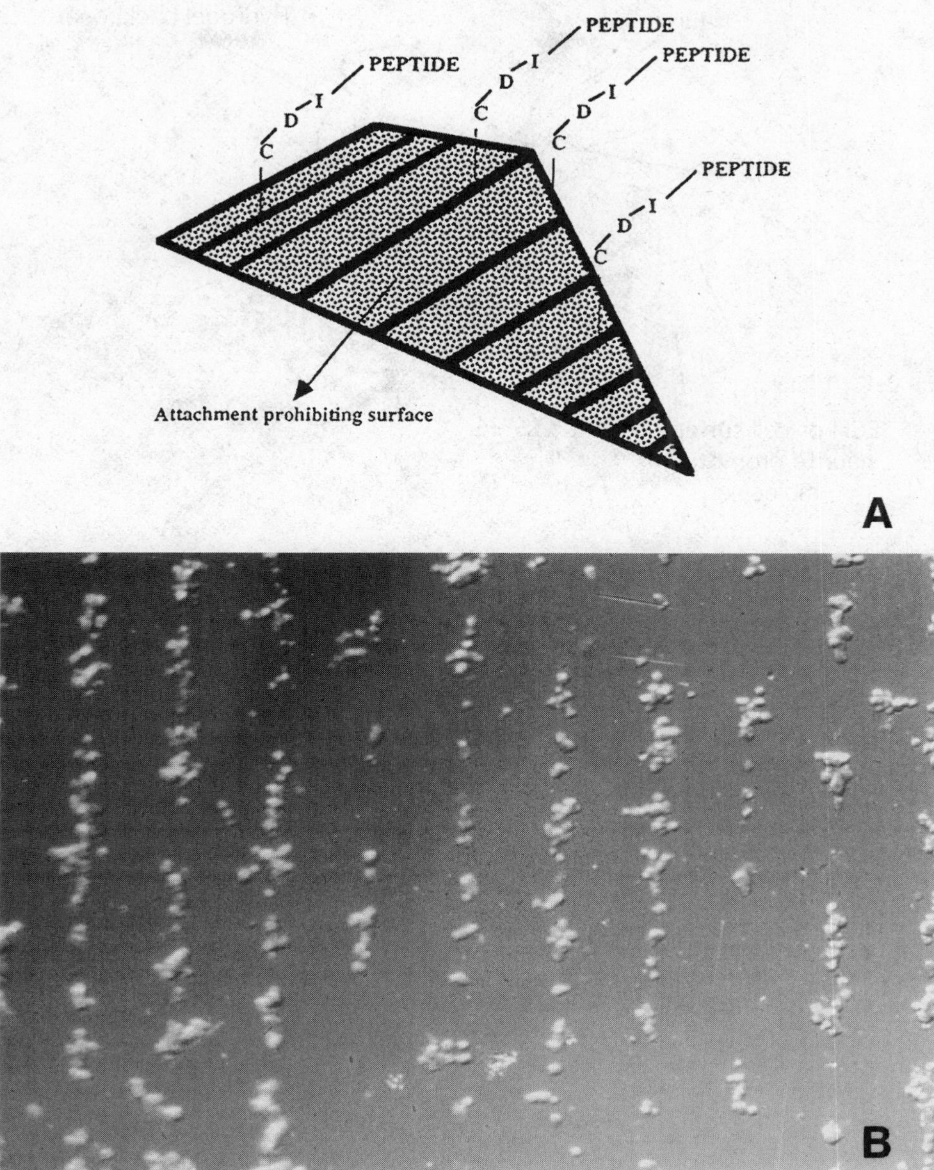

FIGURE 122.4 (*a*) Schematic of fluorinated ethylene propylene membrane modified in a "striped" fashion with bioactive oligopeptides using carbonyldiimidazole homobifunctional linking agent. (*b*) Light micrograph of Ng108-15 cells "striping" on FEP membrane surfaces selectively modified with CDPGYIGSR oligopeptide.

122.4 Conclusion

Advances in gene transfer techniques and molecular and cell biology offer potent tools in the functional replacement of various tissues of the nervous system. Each of these cells' functions can be optimized with the design and selection of its optimal extracellular environment. Substrates that support neuronal differentiation in two and three dimensions may play an important role in taking advantage of the advances in molecular and cell biology. Thus research aimed at tailoring extracellular matrices with the

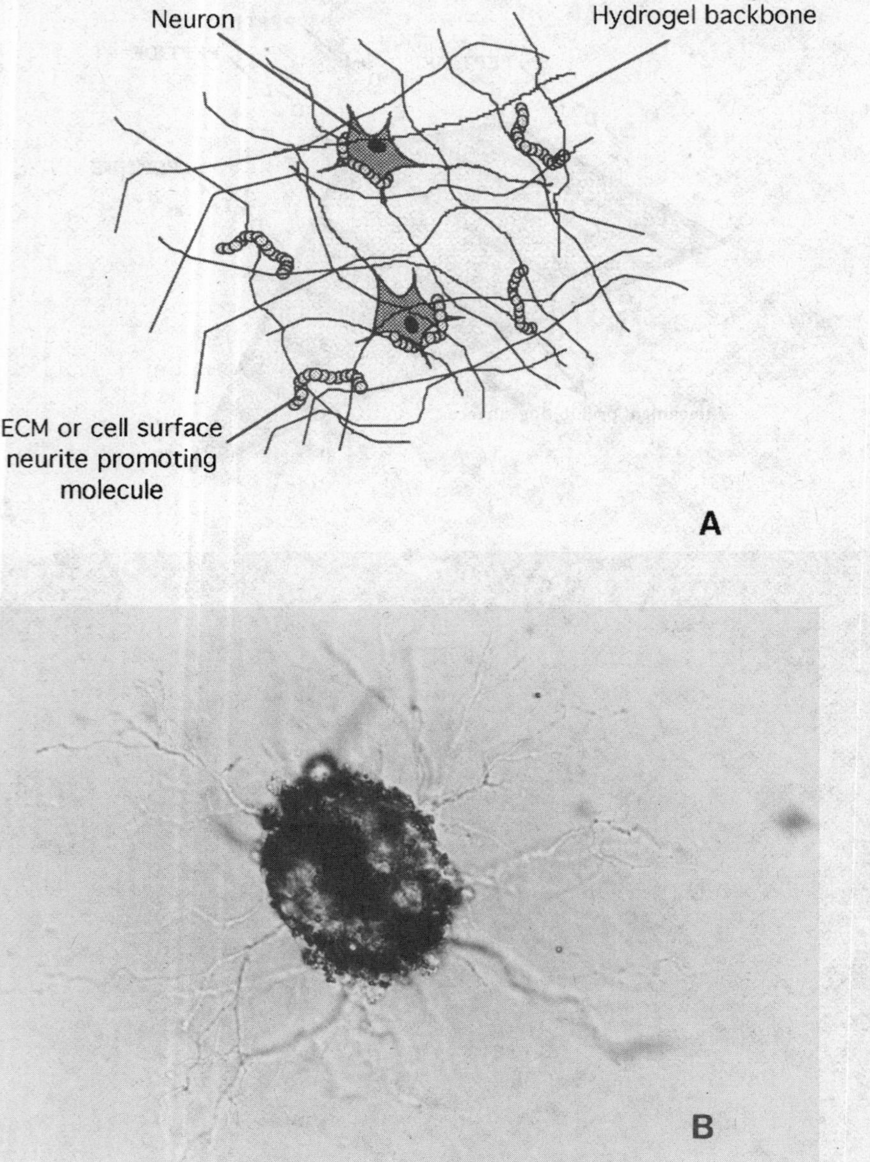

FIGURE 122.5 (*a*) Schematic of hydrogel derivatized with bioactive peptides with an anchorage-dependent neuron suspended in 3D. (*b*) Light micrograph of a E14 chick superior cervical ganglion suspended in 3D and extending neurites in an agarose gel derivatized with the laminin oligopeptide CDPGYIGSR.

appropriate physical, chemical, and biologic cues may be important in optimizing the function of transplanted cells, inducing nerve regeneration, or in the construction of neuronal tissues in two and three dimensions in a controlled fashion. Controlled design and fabrication of polymer hydrogels and polymer scaffolds on a scale that is relevant for single cells also may be important. This would presumably control the degree and the molecular location of permissive, attractive, and repulsive regions of the substrate and in turn control cellular and tissue response in vitro and in vivo.

Biologic molecules like laminin, collagen, fibronectin, and tenascin may provide attractive and permissive pathways for axons to grow. On the other hand, some sulfated proteoglycans have been shown to inhibit or repulse neurites [Snow & Letourneau, 1992]. The use of these molecules coupled with a clearer understanding of protein-mediated material-cell interaction may pave the way for neural tissue engineering, molecule by molecule, in three dimensions. Thus it is possible to tailor the genetic material of a cell to make it neurologically relevant and to control its expression by optimizing its extracellular environment. All the preceding factors make tissue engineering in the nervous system an exciting and challenging endeavor.

Acknowledgments

We wish to than Mr. Nicolas Boche for the illustrations.

References

Aebischer P, Goddard M, Signore A, Timpson R. 1994. Functional recovery in MPTP lesioned primates transplanted with polymer encapsulated PC12 cells. Exp. Neurol 26:1.

Aebischer P, Buchser E, Joseph JN. 1994. Transplantation in humans of encapsulated xenogeneic cells without immunosuppression. Transplantation 58:1275.

Aebischer P, Guénard V, Brace S. 1989a. Peripheral nerve regeneration through blind-ended semipermeable guidance channels: effect of the molecular weight cutoff. J Neurosci 9:3590.

Aebischer P, Guénard V, Valentini RF, 1990. The morphology of regenerating peripheral nerves is modulated by the surface microgeometry of polymeric guidance channels. Brain Res 531:211.

Aebischer P, Salessiotis AN, Winn SR. 1989b. Basic fibroblast growth factor released from synthetic guidance channels facilitates peripheral nerve regeneration across long nerve gaps. J Neurosci Res 28:282.

Aebischer P, Tresco PA, Winn SR, et al. 1991a. Long-term cross-species brain transplantation of a polymer-encapsulated dopamine-secreting cell line. Exp Neurol 111:269.

Aebischer P, Tresco PA, Sagen J, Winn SR. 1991b. Transplantation of microencapsulated bovine chromaffin cells reduces lesion-induced rotational asymmetry in rats. Brain Res 560:43.

Aebsicher P, Wahlberg L, Tresco PA, Winn SR. 1991c. Macroencapsulation of dopamine secreting cells by coextrusion with an organic polymer solution. Biomaterials 12:50.

Aebischer P, Winn SR, Galletti PM. 1998a. Transplantation of neural tissue in polymer capsules. Brain Res 448:364.

Aebischer P, Valentini RF, Dario P, et al. 1987. Piezoelectric guidance channels enhance regeneration in the mouse sciatic nerve after axotomy. Brain Res 436:165.

Aebischer P, Valentini RF, Winn SR, Galletti PM. 1988b. The use of a semipermeable tube as a guidance channel for a transected rabbit optic nerve. Brain Res 78:599.

Ahlgren FI, Ahlgren MB. 1987. Epidural administration of opiates by a new device. Pain 31:353.

Akil H, Watson SJ, Young E, et al. 1984. Endogenous opioids: biology and function. Annu Rev Neurosci 7:223.

Albani C, Asper R, Hacisalihzade SS, Baumgartner F. 1986. Individual levodopa therapy in Parkinson's disease. In Advances in Neurology: Parkinson's Disease, pp 497–501. New York, Raven Press.

Anderson RF, Alterman AL, Barde YA, Lindsay RM. 1990. Brain-derived neurotrophic factor increases survival and differentiation of septal cholinergic neurons in culture. Neuron 5:297.

Archibald SJ, Krarup C, Shefner J, et al. 1991. A collagen-based nerve guide conduit for peripheral nerve repair: an electrophysiological study of nerve regeneration in rodents and nonhuman primates. J Comp Neurol 306–685.

Becker JB, Robinson TE, Barton P, et al. 1990. Sustained behavioral recovery from unilateral nigrostriatal damage produced by the controlled release of dopamine from a silicone polymer pellet placed into the denervated striatum. Brain Res 508:60.

Bellamkonda R, Ranieri JP, Aebischer P. Laminin oligopeptide derivatized agarose gels allow three-dimensional neurite outgrowth in vitro. J Neurosci Res. In press.

Bellamkonda R, Ranieri JP, Bouche N, Aebischer P. In press. A hydrogel-based three-dimensional matrix for neural cells. J Biomed Nat Res. In press.

Birkmayer W. 1969. Experimentalle ergebnisse uber die kombinationsbehandlung des Parkinsonsyndroms mit 1-dopa und einem decarboxylasehemmer. Wiener Klin Wochenschr 81:677.

Björklund A. 1991. Neural transplantation—An experimental tool with clinical possibilities. TINS 14:319.

Björklund A, Dunnett SB, Stenevi U, et al. 1980. Reinnervation of the denervated striatum by substantia nigra transplants: Functional consequences as revealed by pharmacological and sensorimotor testing. Brain Res 199:307.

Bolam JP, Freund TF, Björklund A, et al. 1987. Synaptic input and local output of dopaminergic neurons in grafts that functionally reinnervate the host neostriatum. Exp Brain Res 68:131.

Bunge RP, Bunge MB. 1983. Interrelationship between Schwann cell function and extracellular matrix production. TINS 6:499.

Chen LS, Ray J, Fisher LJ, et al. 1991. Cellular replacement therapy for neurologic disorders: potential of genetically engineered cells. J Cell Biochem 45:252.

De Yebenes JG, Fahn S, Jackson-Lewis V, et al. 1988. Continuous intracerebroventricular infusion of dopamine and dopamine agonists through a totally implanted drug delivery system in animal models of Parkinson's disease. J Neural Transplant 27:141.

During MJ, Freese A, Deutch AY, et al. 1992. Biochemical and behavioral recovery in a rodent model of Parkinson's disease following stereotactic implantation of dopamine-containing liposomes. Exp Neurol 115:193.

Ehringer H, Hornykiewicz O. 1960. Vetreilung von Noradrenalin und Dopamin (3-Hydroxtyramin) im Gehirn des menschen und ihr Verhalten bei Erkrankungen des extrapyramidalen systems. Klin Ther Wochenschr 38:1236.

Fawcett JW, Keynes RJ. 1990. Peripheral nerve regeneration. Annu Rev Neurosci 13:43.

Fine EG, Valentini RF, Bellamkonda R, Aebischer P. 1991. Improved nerve regeneration through piezoelectric vinylidenefluoride-trifluoroethylene copolymer guidance channels. Biomaterials 12:775.

Fromherz P, Offenhausser A, Vetter T, Weis J. 1991. A neuron-silicon junction: A Retzius cell of the leech on an insulated-gate field-effect transistor. Science 252:1290.

Gage FH, Kawaja MD, Fisher LJ. 1991. Genetically modified cells: Applications for intracerebral grafting. TINS 14:328.

Goosen MFA, Shea GM, Gharapetian HM, et al. 1985. Optimization of microencapsualtion parameters: semipermeable microcapsules as a bioartificial pancreas. Biotech Bioeng 27:146.

Graf J, Ogle RC, Robey FA, et al. 1987. A pentrapeptide from the laminin B1 chain mediates cell adhesion and binds the 67,000 laminin receptor. Biochemistry 26:6896.

Guénard V, Aebischer P, Bunge R. 1994. The astrocyte inhibition of peripheral nerve regeneration is reversed by Schwann cells. Exp Neurol 126:44.

Guénard V, Dinarello CA, Weston PJ, Aebischer P. 1991. Peripheral nerve regeneration is impeded by interleukin 1 receptor antagonist released from a polymeric guidance channel. J Neurosci Res 29:396.

Guénard V, Kleitman N, Morrissey TK, et al. 1992. Syngeneic Schwann cells derived from adult nerves seeded in semipermeable guidance channels enhance peripheral nerve regeneration. J Neurosci 2:3310.

Guénard V, Xu XM, Bunge MB. 1993. The use of Schwann cell transplantation to foster central nervous system repair. Semin Neurosci 5:401.

Hall SM. 1986. The effect of inhibiting Schwann cells mitosis on the re-innervation of acellular autografts in the peripheral nervous system of the mouse. Neuropathol Appl Neurobiol 12:27.

Hardie RJ, Lees AJ, Stern GM. 1984. On-off fluctuations in Parkinson's disease: A clinical and neuropharmacological study. Brain 107:487.

Hargraves R, Freed WJ. 1987. Chronic intrastriatal dopamine infusions in rats with unilateral lesions of the substantia nigra. Life Sci 40:959.

Hefti F, Dravid A, Hartikka J. 1984. Chronic intraventricular injections of nerve growth factor elevate hippocampal choline acetyltransferase activity in adult rats with partial septo-hippocampal lesions. Brain Res 293:305.

Henry EW, Chiu TH, Nyilas E, et al. 1985. Nerve regeneration through biodegradable polyester tubes. Exp Neurol 90:652.

Hoffman D, Breakefield XO, Short MP, Aebischer P. 1993. Transplantation of a polymer-encapsulated cell line genetically engineered to release NGF. Exp Neurol 122:100.

Hoffman D, Wahlberg L, Aebischer P. 1990. NGF released from a polymer matrix prevents loss of ChaT expression in basal forebrain neurons following a fimbria-fornix lesion. Exp Neurol 110:39.

Horellou P, Guilbert B, Leviel V, Mallet J. 1989. Retroviral transfer of a human tyrosine hydroxylase cDNA in various cell lines: Regulated release of dopamine in mouse anterior pituitary AtT-20 cells. Proc Natl Acad Sci USA 86:7233.

Hughes RA, Sendtner M, Thoenen H. 1993. Members of several gene families influence survival of rat motoneurons in vitro and in vivo. J Neurosci Res 36:663.

Hyman C, Hofer M, Barde YA, et al. 1991. BDNF is a neurotrophic factor for dopaminergic neurons of the substantia nigra. Nature 350:230.

Iwamoto Y, Robey FA, Graf J, et al. 1987. YIGSR, a synthetic laminin pentapeptide, inhibits experimental metastasis formation. Science 238:1132.

Jiao S, Gurevich V, Wolff JA. 1993. Long-term correction of rat model of Parkinson's disease by gene therapy. Nature 362:450.

Joseph JM, Goddard MB, Mills J, et al. 1994. Transplantation of encapsulated bovine chromaffin cells in the sheep subarachnoid space: a preclinical study for the treatment of cancer pain. Cell Transplant 3:355.

Kalderon N. 1988. Differentiating astroglia in nervous tissue histogenesis regeneration: studies in a model system of regenerating peripheral nerve. J Neurosci Res 21:501.

Kerns JM, Fakhouri AJ, Weinrib HP, Freeman JA. 1991. Electrical stimulation of nerve regeneration in the rat: The early effects evaluated by a vibrating probe and electron microscopy. J Neurosci 40:93.

Kleinman H, Ogle RC, Cannon FB, et al. 1988. Laminin receptors for neurite formation. Proc Natl Acad Sci USA 85:1282.

Knüsel B, Winslow JW, Rosenthal A, et al. 1991. Promotion of central cholinergic and dopaminergic neuron differentiation by brain derived neurotrophic factor but not neurotrophin-3. Proc Natl Acad Sci USA 88:961.

Krause TL, Bittner GD. 1990. Rapid morphological fusion of severed myelinated axons by polyethylene glycol. Proc Natl Acad Sci USA 87:1471.

Kromer LF, Cornbrooks CJ. 1985. Transplants of Schwann cell culture cultures promote axonal regeneration in adult mammalian brain. Proc Natl Acad Sci USA 82:6330.

Kromer LF, Cornbrooks CJ. 1987. Identification of trophic factors and transplanted cellular environments that promote CNS axonal regeneration. Ann NY Acad Sci 495:207.

Langer R. 1981. Polymers for sustained release of macromolecules: Their use in a single-step method for immunization. IN JJ Langone, J Van Vunakis (eds), Methods of Enzymology pp 57–75. San Diego, Academic Press.

Langer R, Folkman J. 1976. Polymers for sustained release of proteins and other macromolecules. Nature 263:797.

Lim F, Sun AM. 1980. Microencapsulated islets as bioartificial endocrine pancreas. Science 210:908.

Lin HL-F, Doherty DH, Lile JD, et al. 1993. GDNF: A glial derived neurotrophic factor for midbrain dopaminergic neurons. Science 260:1130.

Lindvall O. 1991. Prospects of transplantation in human neurodegenerative diseases. TINS 14:376.

Lindvall O, Brundin P, Widner H, et al. 1990. Grafts of fetal dopamine neurons survive and improve motor function in Parkinson's disease. Science 247:574.

Lundborg G, Dahlin LB, Danielsen N, et al. 1982. Nerve regeneration in silicone chambers: Influence of gap length and of distal stump components. Exp Neurol 76:361.

Madison RD, Da Silva CF, Dikkes P. 1988. Entubulation repair with protein additives increases the maximum nerve gap distance successfully bridged with tubular prosthesis. Brain Res 447:325.

McRae A, Hjorth S, Mason DW, et al. 1991. Microencapsulated dopamine (DA)-induced restitution of function in 6-OHDA denervated rat striatum in vivo: Comparison between two microsphere excipients. J Neural Transplant Plast 2:165.

Molander H, Olsson Y, Engkvist O, et al. 1989. Regeneration of peripheral nerve through a polygalactin tube. Muscle Nerve 5:54.

Morrissey TK, Kleitman N, Bunge RP. 1991. Isolation and functional characterization of Schwann cells derived from adult nerve. J Neurosci 11:2433.

Muenter MD, Sharpless NS, Tyce SM, Darley FL. 1977. Patterns of dystonia (I-D-I) and (D-I-D) in response to 1-dopa therapy for Parkinson's disease. Mayo Clin Proc 52:163.

Muir D, Gennrich C, Varon S, Manthorpe M. 1989. Rat sciatic nerve Schwann cell microcultures: Responses to mitogens and production of trophic and neurite-promoting factors. Neurochem Res 14:1003.

Nyilas E, Chiu TH, Sidman RL, et al. 1983. Peripheral nerve repair with bioresorbable prosthesis. Trans Am Soc Artif Intern Organs 29:307.

Olson L, Backlund EO, Ebendal T, et al. 1991. Intraputaminal infusion of nerve growth factor to support adrenal medullary autografts in Parkinson's disease: One year follow-up of first clinical trial. Arch Neurol 48:373.

Olson L, Nordberg A, Von-Holst H, et al. 1992. Nerve growth factor affects ^{11}C-nicotine binding, blood flow, EEG, and verbal episodic memory in an Alzheimer patient (case report). J Neural Transm Park Dis Dement Sect 4:79.

Oppenheim RW, Prevette D, Yin QW, et al. 1991. Control of embryonic motoneuron survival in vivo by ciliary neurotrophic factor. Science 251:1616.

Ranieri JP, Bellamkonda R, Bekos E, et al. 1994. Spatial control of neural cell attachment via patterned laminin oligopeptide chemistries. Int J Dev Neurosci 12:725.

Ranieri JP, Bellamkonda R, Jacob J, et al. 1993. Selective neuronal cell attachment to a covalently patterned monoamine of fluorinated ethylene propylene films. J Biomed Mater Res 27:917.

Reynolds BA, Tetzlaff W, Weiss S. 1992. A multipotent EGF_responsive striatal embryonic progenitor cell produces neurons and astrocytes. J Neurosci 12:4565.

Sabel BA, Dominiak P, Hauser W, et al. 1990. Levodopa delivery from controlled release polymer matrix: Delivery of more than 600 days in vitro and 225 days elevated plasma levels after subcutaneous implantation in rats. J Pharmacol Exp Ther 255:914.

Sagen J, Pappas GD, Winnie AP. 1993a. Alleviation of pain in cancer patients by adrenal medullary transplants in the spinal subarachnoid space. Cell Transplant 2:259.

Sagen J, Wang H, Tresco PA, Aebischer P. 1993b. Transplants of immunologically isolated xenogeneic chromaffin cells provide a long-term source of pain-reducing neuroactive substances. J Neurosci 13:2415.

Scaravalli F. 1984. Regeneration of peineurium across a surgically induced gap in a nerve encased in a plastic tube. J Anat 139:411.

Schwartz M. 1987. Molecular and cellular aspects of nerve regeneration. CRC Crit Rev Biochem 22:89.

Sendtner M, Kreutzberg GW, Thoenen H. 1990. Ciliary neurotrophic factor prevents the degeneration of motor neurons after axotomy. Nature 345:440.

Shoulson I, Claubiger GA, Chase TN. 1975. On-off response. Neurology 25:144.

Smith GV, Stevenson JA. 1988. Peripheral nerve grafts lacking viable Schwann cells fail to support central nervous system axonal regeneration. Exp Brain Res 69:299.

Snow DM, Letourneau PC. 1992. Neurite outgrowth on a step gradient of chondroitin sulfate proteoglycan (CS-PG). J Neurobiol 23:322.

Tresco PA, Winn SR, Aebischer P. 1992. Polymer encapsulated neurotransmitter secreting cells; Potential treatment for Parkinson's disease. ASAIO J 38:17.

Uzman BG, Villegas GM. 1983. Mouse sciatic nerve regeneration through semipermeable tubes: A quantitative model. J Neurosci Res 9:325.

Valentini RF, Aebischer P, Winn SR, Galletti PM. 1987. Collagen- and laminin-containing gels impede peripheral nerve regeneration through semipermeable nerve guidance channels. Exp Neurol 98:350.

Valentini RF, Sabatini AM, Dario P, Aebischer P. 1989. Polymer electret guidance channels enhance peripheral nerve regeneration in mice. Brain Res 48:300.

Vescovi AL, Reynolds BA, Fraser DD, Weiss S. 1993. BFGF regulates the proliferative fate of unipotent (neuronal) and bipotent (neuronal/astroglial) EGF-generated CNS progenitor cells. Neuron 11:951.

Wictorin K, Brundin P, Sauer H, et al. 1992. Long distance directed axonal growth from human dopaminergic mesencephalic neuroblasts implanted along the nigrostriatal pathway in 6-hydroxydopamine lesioned rats. J Comp Neurol 323:475.

Williams LR, Danielsen N, Muller H, Varon S. 1987. Exogenous matrix precursors promote functional nerve regeneration across a 15-mm gap within a silicone chamber in the rat. J Comp Neurol 264:284.

Williams LR, Varon S, Peterson GM, et al. 1986. Continuos infusion of nerve growth factor prevents basal forebrain neuronal death after fimbria-fornix transection. Proc Natl Acad Sci USA 83:9231.

Winn SR, Tresco PA, Zielinski B, et al. 1991. Behavioral recovery following intrastriatal implantation of microencapsulated PC12 cells. Exp Neurol 113:322.

Winn SR, Wahlberg L, Tresco PA, Aebischer P. 1989. AN encapsulated dopamine-releasing polymer alleviates experimental parkinsonism in rats. Exp Neurol 105:244.

Yaffe D, Saxel O. 1977. Serial passaging and differentiation of myogenic cells isolated from dystrophic mouse muscle. Nature 270:725.

Yan Q, Elliott J, Snider WD. 1992. Brain-derived neurotrophic factor rescues spinal motor neurons from axotomy-induced cell death. Nature 360:753.

Yannas EV, Orgill DP, Silver J, et al. 1985. Polymeric template facilitates regeneration of sciatic nerve across 15 mm gap. Trans Soc Biomater 11:146.

Young BL, Begovac P, Stuart D, Glasgow GE. 1984. An effective sleeving technique for nerve repair. J Neurosci Methods 10:51.

Zielinski B, Aebischer P. 1994. Encapsulation of mammalian cells in chitosan-based microcapsules: effect of cell anchorage dependence. Biomaterials.

Zurn AD, Baetge EE, Hammang JP, et al. 1994. Glial cell line-derived neurotrophic factor (GDNF): A new neurotrophic factor for motoneurones. Neuroreport 6:113.

123

Tissue Engineering of Skeletal Muscle

Susan V. Brooks
University of Michigan

John A. Faulkner
University of Michigan

123.1 Introduction

Contractions of skeletal muscles generate the stability and power for all movement. Consequently, any impairment in skeletal muscle function results in at least some degree of instability or immobility. Muscle function can be impaired as a result of injury, disease, or old age. The goal of tissue engineering is to restore the structural and functional properties of muscles to permit the greatest recovery of normal movement. Impaired movement at all ages, but particularly in the elderly, increases the risk of severe injury, reduces participation in the activities of daily living, and impacts on the quality of life.

Contraction is defined as the activation of muscle fibers with a tendency of the fibers to shorten. Contraction occurs when an increase in the cytosolic calcium concentration triggers a series of molecular events that includes the binding of calcium to the muscle regulatory proteins, the formation of strong interactions between the myosin cross-bridges and the actin filaments, and the generation of the cross-bridge driving stroke. *In vivo*, muscles perform three types of contractions depending on the interaction between the magnitude of the force developed by the muscle and the external load placed on the muscle. When the force developed by the muscle is greater than the load on the muscle, the fibers shorten during the contraction. When the force developed by the muscle is equal to the load, or if the load is immovable, the overall length of the muscle remains the same. If the force developed by the muscle is less than the load placed on the muscle, the muscle is stretched during the contraction. The types of contractions are termed miometric, isometric, and pliometric, respectively. Most normal body movements require varying proportions of each type of contraction.

0-8493-0462-8/00/$0.00+$.50

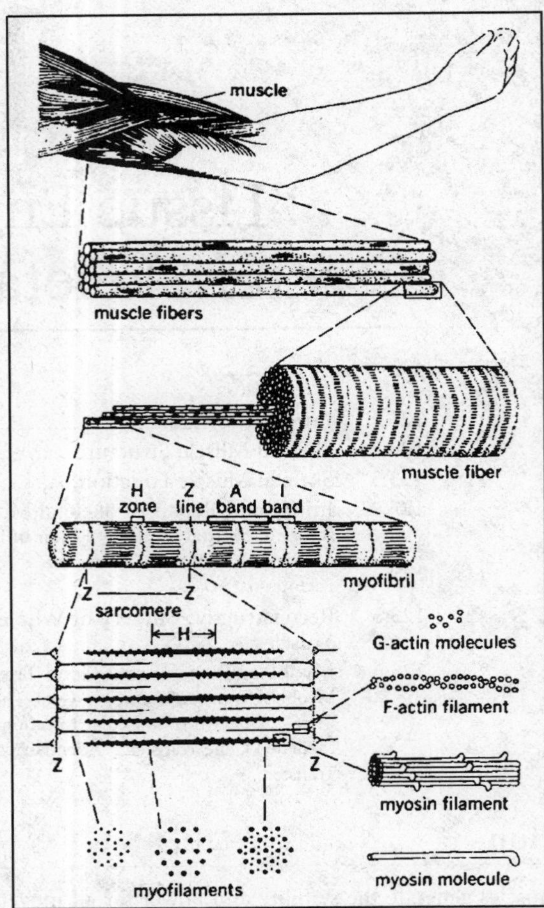

FIGURE 123.1 Levels of anatomical organization within a skeletal muscle. *Source:* Bloom, W. and Fawcett, D.W. 1968. *A Textbook of Histology,* 9th ed., Saunders, Philadelphia. With permission.

123.2 Skeletal Muscle Structure

Each of the 660 skeletal muscles in the human body is composed of hundreds to hundreds of thousands of single muscle fibers (Fig. 123.1). The plasma membrane of a muscle fiber is termed the sarcolemma. Contractile, structural, metabolic, regulatory, and cytosolic proteins, as well as many myonuclei and other cytosolic organelles are contained within the sarcolemma of each fiber (Fig. 123.2). The contractile proteins myosin and actin are incorporated into thick and thin myofilaments, respectively, which are arrayed in longitudinally repeated banding patterns termed sarcomeres (Fig. 123.1). Sarcomeres in series form myofibrils, and many parallel myofibrils exist within each fiber. The number of myofibrils arranged in parallel determines the cross-sectional area (CSA) of single fibers, and consequently, the force generating capability of the fiber. During a contraction, the change in the length of a sarcomere occurs as thick and thin filaments slide past each other, but the overall length of each actin and myosin filament does not change. An additional membrane, referred to as the basement membrane or the basal lamina, surrounds the sarcolemma of each fiber (Fig. 123.2).

In mammals, the number of fibers in a given muscle is determined at birth and changes little throughout the life span except in cases of injury or disease. In contrast, the number of myofibrils can change dramatically, increasing with normal growth or hypertrophy induced by strength training and decreasing

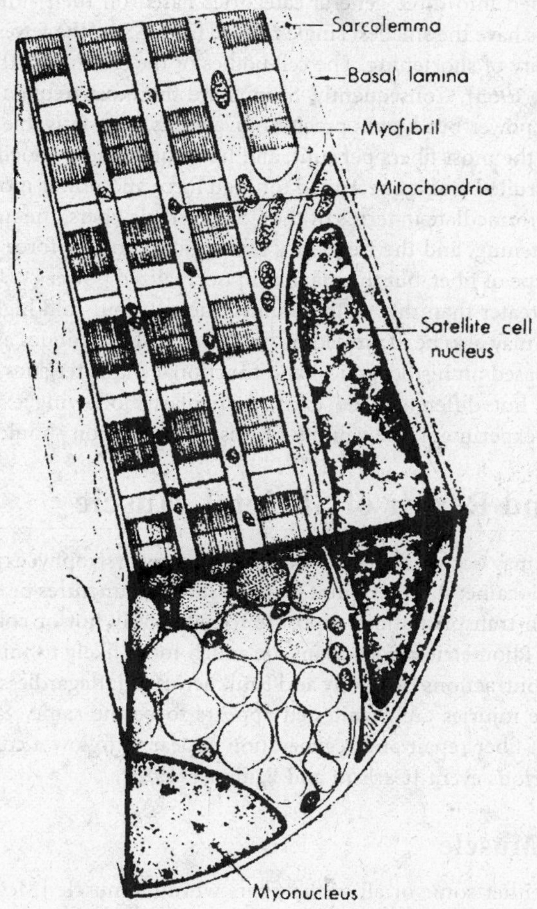

FIGURE 123.2 Drawing of a muscle fiber-satellite cell complex. Note that the satellite cell is located between the muscle fiber sarcolemma and the basal lamina. *Source:* Carlson and Faulkner, 1983. With permission.

with atrophy associated with immobilization, inactivity, injury, disease, or old age. A single muscle fiber is innervated by a single branch of a motor nerve. A motor unit is composed of a single motor nerve, its branches, and the muscle fibers innervated by the branches. The motor unit is the smallest group of fibers within a muscle that can be activated volitionally. Activation of a motor unit occurs when action potentials emanating from the motor cortex depolarize the cell bodies of motor nerves. The depolarization generates an action potential in the motor nerve that is transmitted to each muscle fiber in the motor unit, and each of the fibers contracts more or less simultaneously. Motor units range from small slow units to large fast units dependent on the CSA of the motor nerve.

123.3 Skeletal Muscle Function

Skeletal muscles may contract singly or in groups, working synergistically. On either side of limbs, muscles contract against one another or antagonistically. The force or power developed during a contraction depends on the frequency of stimulation of the motor units, the total number of motor units, and the size of the motor units recruited. The frequency of stimulation, particularly for the generation of power, is normally on the order of the frequency-power relationship. Consequently, the total number of motor units recruited is the major determinant of the force or power developed.

Motor units are classified into three general categories based on their functional properties [Burke et al., 1973]. Slow (S) units have the smallest single muscle fiber CSAs, the fewest muscle fibers per motor unit, and the lowest velocity of shortening. The cell bodies of the S units are the most easily depolarized to threshold [Henneman, 1965]. Consequently, S units are the most frequently recruited during tasks that require low force or power but highly precise movements. Fast-fatigable (FF) units are composed of the largest fibers, have the most fibers per unit, and have the highest velocities of shortening. The FF units are the last to be recruited and are recruited for high force and power movements. The fast fatigue-resistant (FR) units are intermediate in terms of the CSAs of their fibers, the number of fibers per motor unit, the velocity of shortening, and the frequency of recruitment. The force normalized per unit CSA is ~280 kN/m^2 for each type of fiber, but the maximum normalized power (W/kg) developed by FF units is as much as four-fold greater than that of the S units due to a four-fold higher velocity of shortening for FF units. Motor units may also be identified by histochemical techniques as Type I (S), IIA (FR), and IIB (FF). Classifications based on histochemical and functional characteristics are usually in good agreement with one another, but differences do exist, particularly following experimental interventions. Consequently, in a given experiment, the validity of this interpretation should be verified.

123.4 Injury and Repair of Skeletal Muscle

Injury to skeletal muscles may occur as a result of disease, such as dystrophy; exposure to myotoxic agents, such as bupivacaine or lidocaine; sharp or blunt trauma, such as punctures or contusions; ischemia, such as that which occurs with transplantation; exposure to excessively hot or cold temperatures; and contractions of the muscles. Pliometric contractions are much more likely to injure muscle fibers than are isometric or miometric contractions [McCully and Faulkner, 1985]. Regardless of the factors responsible, the manner in which the injuries are manifested appears to be the same, varying only in severity. In addition, the processes of fiber repair and regeneration appear to follow a common pathway regardless of the nature of the injurious event [Carlson and Faulkner, 1983].

Injury of Skeletal Muscle

The injury may involve either some or all of the fibers within a muscle [McCully and Faulkner, 1985]. In an individual fiber, focal injuries, localized to a few sarcomeres in series or in parallel (Fig. 123.3), as well as more widespread injuries, spreading across the entire cross section of the fiber, are observed using electron microscopic techniques [Macpherson et al., 1997]. Although the data are highly variable, many injuries also give rise to increases in serum levels of muscle enzymes, particularly creatine kinase, leading to the conclusion that sarcolemmal integrity is impaired [McNeil and Khakee, 1992; Newham et al., 1983]. This conclusion is further supported by an influx of circulating proteins, such as serum albumin [McNeil and Khakee, 1992], and of calcium [Jones et al., 1984]. An increase in intracellular calcium concentration may activate a variety of proteolytic enzymes leading to further degradation of sarcoplasmic proteins.

In cases when the damage involves a large proportion of the sarcomeres within a fiber, the fiber becomes necrotic. If blood flow is impaired, fibers remain as a necrotic mass of noncontractile tissue [Carlson and Faulkner, 1983]. In contrast, in the presence of an adequate blood supply, the injured fibers are infiltrated by monocytes and macrophages [McCully and Faulkner, 1985]. The phagocytic cells remove the disrupted myofilaments, other cytosolic structures, and the damaged sarcolemma (Fig. 123.4). The most severe injuries result in the complete degeneration of the muscle fiber, leaving only the empty basal lamina. The basal lamina appears to be highly resistant to any type of injury and generally remains intact [Carlson and Faulkner, 1983].

An additional indirect measure of injury is the subjective report by human beings of delayed onset muscle soreness, common following intense or novel exercise [Newham et al., 1983]. Because of the focal nature of the morphological damage, the variability of serum enzyme levels and the subjectivity of reports

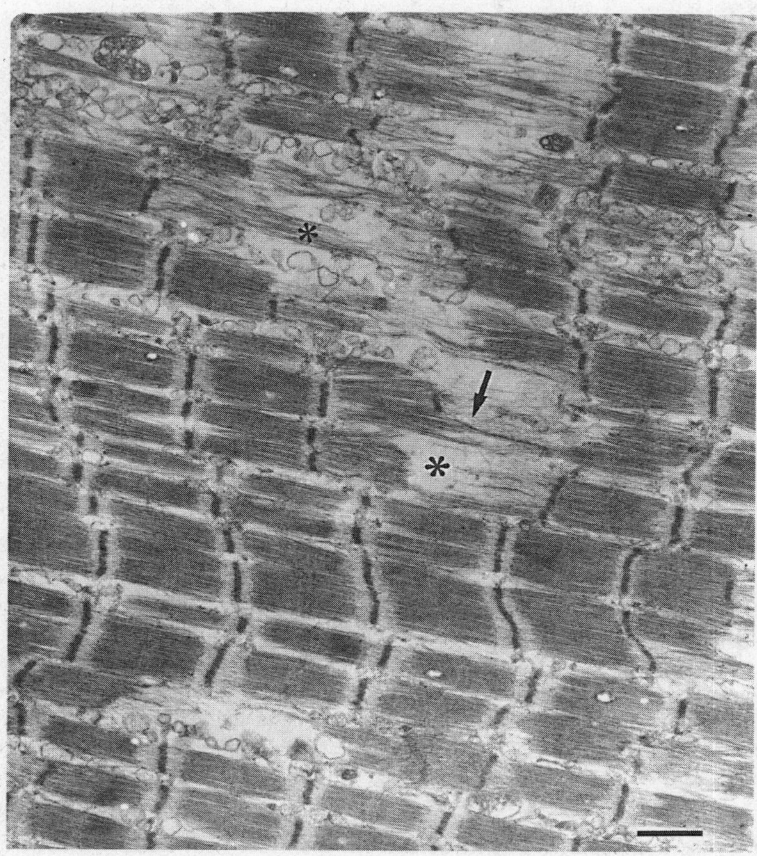

FIGURE 123.3 Focal areas of damage to single sarcomeres after a single 40% stretch of a single maximally activated rat skeletal muscle fiber (average sarcomere length 2.6 μm). Two types of damage are observed in this electron micrograph. The arrow indicates Z-line streaming and asterisks show disruption of the thick and thin filaments in the region of overlap, the A-band. A third type of damage, the displacement of thick filaments to one Z-line is not shown. Note that with the return of the relaxed fiber to an average sarcomere length of 2.6 μm, the sarcomeres indicated by asterisks are at 5.1 μm and 3.8 μm while sarcomeres in series are shortened to ~1.8 μm. Scale bar is 1.0 μm. *Source*: Modified from Macpherson et al., 1997. With permission.

of soreness, the most quantitative and reproducible measure of the totality of a muscle injury is the decrease in the ability of the muscle to develop force [McCully and Faulkner, 1985; Newham et al., 1983].

Repair of Injured Skeletal Muscles

Under circumstances when the injury involves only minor disruptions of the thick or thin filaments of single sarcomeres, the damaged molecules are likely replaced by newly synthesized molecules available in the cytoplasmic pool [Russell et al., 1992]. In addition, contraction-induced disruptions of the sarcolemma are often transient and repaired spontaneously, allowing survival of the fiber [McNeil and Khakee, 1992]. Following more severe injuries, complete regeneration of the entire muscle fiber will occur.

Satellite Cell Activation

A key element in the initiation of muscle fiber regeneration following a wide variety of injuries is the activation of satellite cells [Carlson and Faulkner, 1983]. Satellite cells are quiescent myogenic stem cells located between the basal lamina and the sarcolemma (Fig. 123.2). Upon activation, satellite cells divide

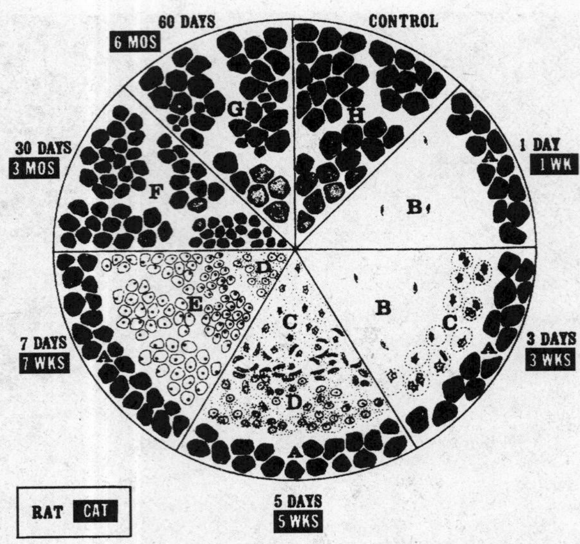

FIGURE 123.4 Schematic representation of the cellular responses during the processes of degeneration and regeneration following transplantation of extensor digitorum longus muscles in rats and cats. The diagram is divided into segments that represent the histological appearance of the muscle cross-section at various times after transplantation. The times given in days refer to rat muscles and those given in weeks refer to the larger cat muscles. The letters indicate groups of muscle fibers with similar histological appearances. A: surviving fibers, B: fibers in a state of ischemic necrosis, C: fibers invaded by phagocytic cells, D: myoblasts and early myotubes, E: early myofibers, F: immature regenerating fibers, G: mature regenerated fibers, H: normal control muscle fibers [Carlson and Faulkner, 1983]. *Source:* Mauro, A. 1979. *Muscle Regeneration*, p. 493-507, Raven Press, New York, NY. With permission.

mitotically to give rise to myoblasts. The myoblasts can then fuse with existing muscle fibers, acting as a source of new myonuclei. This is the process by which a muscle fiber increases the total number of myonuclei in fibers that are increasing in size [Moss and Leblond, 1971] and may be necessary to repair local injuries. Alternatively, the myoblasts can fuse with each other to form multinucleated myotubes inside the remaining basal lamina of the degenerated fibers [Carlson and Faulkner, 1983]. The myotubes then begin to produce muscle specific proteins and ultimately differentiate completely into adult fast or slow muscle fibers [Carlson and Faulkner, 1983]. Recent evidence also suggests that in addition to the activation of resident satellite cells, regenerating muscle may recruit undifferentiated myogenic precursor cells from other sources [Ferrari et al., 1998]

Following a closed contusion injury, mitotic activation of satellite cells has been observed within the first day after the injury and is correlated in time with the appearance of phagocytes and newly formed capillaries [Hurme and Kalimo, 1992]. Similarly, DNA synthesis by the satellite cells is observed within the first day following crush injuries [Bischoff, 1986] and exercise-induced injuries [Darr and Schultz, 1987] in rats. These observations are consistent with the hypothesis that the factors that activate satellite cells may be endogenous to the injured tissue itself or synthesized and secreted by platelets at the wound site, infiltrating neutrophils, and macrophages [reviewed in Husmann et al., 1996]. The primary candidates for the factors that activate and regulate satellite cell function include the fibroblast growth factors (FGFs), platelet-derived growth factor (PDGF), transforming growth factor beta (TGF-β), and insulin-like growth factors I and II (IGF-I, II).

The effects of these factors on muscle satellite cell proliferation and differentiation have been studied extensively in cell culture [reviewed in Florini and Magri, 1989]. FGF is a powerful mitogen for myogenic cells from adult rat skeletal muscle, but a potent inhibitor of terminal differentiation, i.e., myoblast fusion and expression of the skeletal muscle phenotype. PDGF also shows a strong stimulating effect on proliferation and inhibitory effect on differentiation of satellite cells [Jin et al., 1990]. Similarly, the presence of TGF-β prevents myotube formation as well as muscle-specific protein synthesis by rat embryo myoblasts and by adult rat satellite cells. In contrast to the previously mentioned growth factors, all of which inhibit myoblast differention, IGFs stimulate both proliferation and differentiation of myogenic cells [Ewton and Florini, 1980]. Despite our extensive knowledge of the actions of many individual growth factors *in vitro* and *in vivo*, the interactions between different growth factors have been less thoroughly investigated [Rosenthal et al., 1991]. A better understanding of the interactions between growth factors and the mechanisms that guide satellite cells through the regeneration process is necessary.

Myogenic Regulatory Factors

The conversion of pluripotent embryonic stem cells to differentiated muscle cells involves the commitment of these cells to the myogenic lineage and the subsequent proliferation, differentiation and fusion to form multinucleated myotubes and ultimately mature muscle cells. New muscle cell formation from muscle satellite cells resembles embryonic development in the sense that in regenerating muscle cells embryonic isoforms of the muscle proteins are expressed [Whalen et al., 1985]. The conversion of stem cells to mature fibers, in both developing muscle and regenerating muscle is directed by a group of related regulatory factors. These so-called muscle regulatory factors (MRFs) are part of a superfamily of basic helix-loop-helix DNA binding proteins that interact to regulate the transcription of skeletal muscle genes [reviewed in Weintraub et al., 1991]. MyoD was the first MRF identified followed by three other related genes including myogenin, MRF4 (also called herculin or Myf-6), and Myf-5 [Weintraub et al., 1991].

The observations that each of the MRFs has the ability to independently convert cultured fibroblasts into myogenic cells led to the original conclusion that functionally the MRFs were largely redundant. Subsequent gene targeting experiments, in which null mutations were introduced in each of the MRF genes, support separate and distinct roles in myogenesis for each MRFs [reviewed in Rudnicki and Jaenisch, 1995]. For example, mice that lack either Myf-5 or MyoD apparently have normal skeletal muscle but deletion of both genes results in the complete absence of skeletal myoblasts [Rudnicki et al., 1992; 1993; Braun et al., 1992]. While these observations suggest that Myf-5 and MyoD do have overlapping functions, characterization of the temporal-spatial patterns of myogenesis in Myf-5- and MyoD-deficient mouse embryos support the hypothesis that during normal development Myf-5 and MyoD primarily regulate epaxial (paraspinal and intercostal) and hypaxial (limb and abdominal wall) muscle development, respectively [Kablar et al., 1997]. In mice lacking myogenin, the number of myoblasts is not different from that of control mice, but skeletal muscles in myogenin-deficient mice display a marked reduction in the number of mature muscle fibers [Hasty et al., 1993; Nabeshima et al., 1993].

In summary, the MRF family can be divided into two functional groups. MyoD and Myf-5 are referred to as primary factors and are required for the determination of skeletal myoblasts whereas myogenin and MRF4 are secondary factors that act later and are necessary for differentiation of myoblasts into myotubes. How the MRFs control the series of events required for myoblast determination and differentiation are important questions for the future. In addition, the roles played by this family of proteins in regeneration, adaptation, and changes in skeletal muscle with aging are areas of active investigation [Jacobs-El et al., 1995; Marsh et al., 1997; Megeney et al., 1996].

123.5 Reconstructive Surgery of Whole Skeletal Muscles

When an injury or impairment is so severe that the total replacement of the muscle is required, a whole donor muscle must be transposed or transplanted into the recipient site [Faulkner et al., 1994a]. One of the most versatile muscles for transpositions is the latissimus dorsi (LTD) muscle. LTD transfers have been used in breast reconstruction [Moelleken et al., 1989], to restore elbow flexion, and to function as a heart assist pump [Carpentier and Chachques, 1985]. Thompson [1974] popularized the use of small free standard grafts to treat patients with partial facial paralysis. The transplantation of large skeletal muscles in dogs with immediate restoration of blood flow through the anastomosis of the artery and vein provided an operative technique with numerous applications [Tamai et al., 1970]. Coupled with cross-face nerve grafts, large skeletal muscles are transplanted with microneurovascular repair to correct deficits in the face [Harii et al., 1976] and adapted for reconstructive operations to treat impairments in function of the limbs, anal and urinary sphincters, and even the heart [Freilinger and Deutinger, 1992].

Transposition and transplantation of muscles invariably results in structural and functional deficits [Guelinckx et al., 1992]. The deficits are of the greatest magnitude during the first month and then a gradual recovery results in the stabilization of structural and functional variables between 90 and 120 days [Guelinckx et al., 1992]. In stabilized vascularized grafts ranging from 1 to 3 grams in rats to 90 grams in dogs, the major deficits are a ~30% decrease in muscle mass and in most grafts a ~40% decrease in maximum force [Faulkner et al., 1994a]. The decrease in power is more complex since it depends on both the average shortening force and the velocity of shortening. As a consequence, the deficit in maximum power may be either greater, or less than the deficit in maximum force [Kadhiresan et al., 1993]. Tenotomy and repair are major factors responsible for the deficits [Guelinckx et al., 1988]. When a muscle is transplanted to act synergistically with other muscles, the action of the synergistic muscles may contribute to the deficits observed [Miller et al., 1994]. Although the data are limited, skeletal muscle grafts appear to respond to training stimuli in a manner not different from that of control muscles [Faulkner et al., 1994a]. The training stimuli include traditional methods of endurance and strength training [Faulkner et al., 1994b], as well as chronic electrical stimulation [Pette and Vrbova, 1992]. In spite of the deficits, transposed and transplanted muscles develop sufficient force and power to function effectively to maintain posture and patent sphincters and to move limbs or drive assist devices in parallel or in series with the heart [Faulkner et al., 1994a].

123.6 Myoblast Transfer and Gene Therapy

Myoblast transfer and gene therapy are aimed at delivering exogenous genetic constructs to skeletal muscle cells. The implications of myoblast and gene therapy hold great promise for skeletal muscle research and for those afflicted with inherited myopathies such as Duchenne and Becker muscular dystrophy (DMD and BMD). Myoblast transfer is a cell-mediated technique designed to treat inherited myopathies by intramuscular injection of myoblasts containing a normal functional genome. The goal is to correct for a defective or missing gene in the myopathic tissue through the fusion of normal myoblasts with growing or regenerating diseased cells. Gene therapy presents a more complex and flexible approach, whereby genetically engineered DNA constructs are delivered to a host cell to specifically direct production of a desired protein. By re-engineering the coding sequence of the gene and its regulatory regions, the function, the quantity of expression, and the protein itself can be altered. For many years, cells have been genetically altered to induce the production of a variety of useful proteins, such as human growth hormone and interferon.

As a focus in skeletal muscle tissue engineering, DMD is an X-linked recessive disorder characterized by progressive muscle degeneration resulting in debilitating muscle weakness and death in the second or third decade of life as a result of respiratory failure [Emery, 1988]. DMD and the milder BMD are due to genetic defects that lead to the absence or marked deficiency in the expression or functional stability of the protein dystrophin [Bonilla et al., 1988; Hoffman et al., 1988]. Similarly, a mutation in the dystrophin gene leads to the complete absence of the protein in muscle and brain tissues of the *mdx* mouse [Bulfield et al., 1984; Sicinski et al., 1989]. The homology of the genetic defects in DMD patients and *mdx* mice support the use of *mdx* mice as a model of dystrophin deficiency to explore the processes of the dystrophic disease and test proposed therapies or cures.

Myoblast Transfer Therapy

The concept of myoblast transfer is based on the role satellite cells play in muscle fiber growth and repair [Carlson and Faulkner, 1983]. As a therapy for DMD, the idea is to obtain satellite cells containing a functional dystrophin gene from a healthy compatible donor, have the cells multiply in culture, and then inject the "normal" myoblasts into the muscles of the patient. The objective is for the injected myoblasts to fuse with growing or regenerating muscle fibers to form a mosaic fiber in which the cytoplasm will contain normal myoblast nuclei capable of producing a functional form of dystrophin.

Experiments involving *mdx* mice have had varying success. Several investigators have reported that implantation of healthy myoblasts into muscles of *mdx* mice led to the production of considerable quantities of dystrophin [Morgan et al., 1990; Partridge et al., 1989]. Others found that myoblasts injected into limb muscles of *mdx* and control host mice showed a "rapid and massive" die off shortly after injection [Fan et al. 1996] with large and permanent decreases in muscle mass and maximum force [Wernig et al., 1995]. The successful transfer and fusion of donor myoblasts may be enhanced by X-ray irradiation of *mdx* muscles prior to myoblast injection to prevent the proliferation of myoblasts endogenous to the host and encourage the growth of donor myoblasts [Morgan et al., 1990].

In contrast to the studies with mice, delivery of myoblasts to DMD patients has shown very low levels of fusion efficiency, transient expression of dystrophin, and immune rejection [Gussoni et al., 1992; Karpati et al., 1993; Mendell et al., 1995; Morgan, 1994]. The use of X-ray irradiation in an attempt to enhance transfer efficiency is not applicable to DMD boys due to substantial health risks. Furthermore, immunosuppression may be necessary to circumvent immune rejection, which carries risks of its own. A better understanding of the factors that govern the survival, fusion, and expression of donor myoblasts is required before the viability of myoblast transfer as a treatment of DMD can be evaluated.

Gene Therapy

The aim of gene therapy for DMD is to transfer a functional dystrophin gene directly into the skeletal muscle. The challenge behind gene therapy is not only obtaining a functional genetic construct of the dystrophin gene and regulatory region but the effective delivery of the gene to the cell's genetic machinery. Methods to transfer genetic material into a muscle cell include direct injection, and the use of retroviral and adenoviral vectors. Each of these strategies presents highly technical difficulties that to date remain unresolved.

Transgenic Mice

To explore the feasibility of gene therapy for DMD, Cox and colleagues [1993] examined the introduction of an exogenous dystrophin gene into the germ line of *mdx* mice to produce transgenic animals. The transgenic *mdx* mice expressed nearly 50 times the level of endogenous dystrophin found in muscles of control C57BL/10 mice and displayed the complete absence of any morphological, immunohistological, or functional symptoms of the murine muscular dystrophy with no deleterious side effects. Although transgenic technology does not provide an appropriate means for treating humans, these results demonstrated the efficacy of gene therapy to correct pathological genetic defects such as DMD.

Direct Intramuscular Injection

The straightforward gene delivery method of direct injection of plasmid DNA into skeletal and heart muscles [Lin et al., 1990; Wolff et al., 1990] has been proposed as a treatment for DMD and BMD. The idea is that dystrophic cells will incorporate the genetic constructs, whereby the genes will use the cell's internal machinery to produce the protein dystrophin. The advantages of direct injection of DNA as a gene delivery system are its simplicity, and it presents no chance of viral infection or the potential of cancer development that can occur with viral vectors [Morgan, 1994]. Although this approach is appealing in principal, direct intramuscular injection of human dystrophin plasmid DNA into the quadriceps muscles of *mdx* mice led to the expression of human dystrophin in only 1% to 3% of the muscle fibers [Acsadi et al., 1991]. In order for this method to be clinically effective, a much larger number of transfected myofibers must be achieved.

Retrovirus-Mediated Gene Transfer

Retroviruses reverse the normal process by which DNA is transcribed into RNA. A single-stranded viral RNA genome enters a host cell and a double helix comprised of two DNA copies of the viral RNA is created by the enzyme reverse transcriptase. Catalyzed by a viral enzyme, the DNA copy then integrates into a host cell chromosome where transcription, via the host cell RNA polymerase, produces large quantities of viral RNA molecules identical to the infecting genome [Alberts et al., 1989]. Eventually, new viruses emerge and bud from the plasma membrane ready to infect other cells. Consequently, retroviral vectors used for gene therapy are, by design, rendered replication defective. Once they infect the cell and integrate into the genome, they cannot make functional retroviruses to infect other cells. After the infective process is completed, cells are permanently altered with the presence of the viral DNA that causes the synthesis of proteins not originally endogenous to the host cell.

A primary obstacle to the efficiency of a retroviral gene delivery system is its dependence on host cell division. The requirement that a cell must be mitotically active for the virus to be incorporated into the host cell genetic machinery [Morgan, 1994] presents a problem for skeletal muscle tissue engineering since skeletal muscle is in a post-mitotic state. Nonetheless, as described above, recovery from injury in skeletal muscle involves the proliferation of myogenic precursor cells and either the incorporation of these cells into existing muscle fibers or the fusion of these cells with each other to form new fibers. Furthermore, since degeneration and regeneration of muscle cells are ongoing in DMD patients, viral transfection of myoblasts may be an effective route for gene delivery to skeletal muscle for treatment of DMD.

Another limitation of the retroviral gene delivery system is the small carrying capacity of the retroviruses of approximately 7 kilobases. This size limitation precludes the delivery of a full-length dystrophin construct of 12-14 kilobases [Dunckley et al., 1993; Morgan, 1994]. A 6.3 kilobase dystrophin construct, containing a large in-frame deletion resulting in the absence of ~40% of the central domain of the protein, has received a great deal of attention since its discovery in a BMD patient expressing a very mild phenotype. A single injection of retrovirus containing the Becker dystrophin minigene into the quadriceps or tibialis anterior muscle of *mdx* mice led to the sarcolemmal expression of dystrophin in an average of 6% of the myofibers [Dunckley et al., 1993]. Restoration of the 43-kDalton dystrophin-associated glycoprotein was also observed and expression of the recombinant dystrophin was maintained for up to nine months. Transduction of the minigene was significantly enhanced when muscles were pretreated with an intramuscular injection of the myotoxic agent bupivacaine to experimentally induce muscle regeneration.

Adenovirus-Mediated Gene Transfer

Adenoviruses have many characteristics that make the adenovirus-mediated gene transfer the most promising technology for gene therapy of skeletal muscle. The primary advantages are the stability of the viruses allowing them to be prepared in large amounts and the ability of adenoviral vectors to infect nondividing or slowly proliferating cells. In addition, adenoviral vectors have the potential to be used for systemic delivery of exogenous DNA. Through the use of tissue-specific promoters, specific tissues such as skeletal muscle may be targeted for transfection via intravenous injection.

Initial studies using adenoviral vectors containing the Becker dystrophin minigene driven by the Rous Sarcoma Virus promoter demonstrated that after a single intramuscular injection in newborn *mdx* mice, 50% of muscle fibers contained dystrophin [Ragot et al., 1993]. The truncated dystrophin was correctly localized to the sarcolemmal membrane and appeared to protect myofibers from the degeneration process characteristic of *mdx* muscles [Vincent et al., 1993]. Six months after a single injection, expression of the minigene was still observed in the treated muscle. More recently, these same investigators demonstrated that the injection of the adenoviral vector containing the dystrophin minigene into limb muscles of newborn *mdx* mice provided protection from the fiber damage and force deficit associated with a protocol of pliometric contractions that was administered at four months of age [Deconinck et al., 1996].

Despite the promise of adenovirus-mediated gene therapy, a number of limitations to its usefulness remain to be resolved. One major drawback of the system is the relatively brief duration of transgene expression observed following injection of adenoviral vectors into adult immunocompetent animals [Kass-Eisler et al., 1994]. The lack of long-term exogenous gene expression is likely the result of low level expression of endogenous viral proteins triggering an inflammatory response that attacks infected cells [Yang et al., 1994]. In addition to the difficulties resulting from the potent immune response triggered by adenovirus, direct cytotoxic effects of adenovirus injection on skeletal muscle have been reported [Petrof et al., 1995]. Current generation adenoviruses are also limited by their relatively small cloning capacity of ~8 kilobases. The development of adenoviral vectors with increased cloning capacity, the ability to evade host immune rejection, and no toxic effects are areas of active investigation [Kumar-Singh and Chamberlain, 1996; Petrof et al., 1996; Hauser et al., 1997].

Acknowledgments

The research in our laboratory and the preparation of this chapter was supported by grants from the United States Public Health Service, National Institute on Aging, AG-15434 (SVB) and AG-06157 (JAF) and the Nathan Shock Center for Basic Biology of Aging at the University of Michigan.

References

Acsadi, G., Dickson, G., Love, D.R., Jani, A., Walsh, F.S., Gurusinghe, A., Wolff, J.A., and Davies, K.E. 1991. Human dystrophin expression in *mdx* mice after intramuscular injection of DNA constructs. *Nature* 352:815-818.

Alberts, B., Bray, D., Lewis, J., Raff, M., Roberts, K., and Watson, J.D. 1989. *Molecular Biology of the Cell,* 2nd ed. Garland Publishing, Inc., New York, NY, p. 254.

Bischoff, R. 1986. A satellite cell mitogen from crushed adult muscle. *Dev. Biol.* 115:140-147.

Bonilla, E., Samitt, C.E., Miranda, A.F., Hays, A.P., Salviati, G., DiMauro, S., Kunkel, L.M., Hoffman, E.P., and Rowland, L.P. 1988. Duchenne muscular dystrophy: deficiency of dystrophin at the muscle cell surface. *Cell* 54:447-452.

Braun, T., Rudnicki, M.A., Arnold, H.H., and Jaenisch, R. 1992. Targeted inactivation of the muscle regulatory gene Myf-5 results in abnormal rib development and perinatal death. *Cell* 71:369-382.

Bulfield, G., Siller, W.G., Wight, P.A.L., and Moore, K.J. 1984. X chromosome-linked muscular dystrophy (*mdx*) in the mouse. *Proc. Natl. Acad. Sci. USA* 81:1189-1192.

Burke, R.E., Levin, D.N., Tsairis, P. and Zajac, F.E., III. 1973. Physiological types and histochemical profiles in motor units of the cat gastrocnemius muscle. *J. Physiol.* (*Lond.*) 234:723-748.

Carlson, B.M. and Faulkner, J.A. 1983. The regeneration of skeletal muscle fibers following injury: a review. *Med. Sci. Sports Exer.* 15:187-198.

Carpentier, A. and J.C. Chachques. 1985. Myocardial substitution with a stimulated skeletal muscle: first successful clinical case. *Lancet* 1(8440):1267.

Cox, G.A., Cole, N.M., Matsumura, K., Phelps, S.F., Hauschka, S.D., Campbell, K.P., Faulkner, J.A., and Chamberlain, J.S. 1993. Overexpression of dystrophin in transgenic *mdx* mice eliminates dystrophic symptoms without toxicity. *Nature* 364:725-729.

Darr, K.C. and Schultz, E. 1987. Exercise-induced satellite cell activation in growing and mature skeletal muscle. *J. Appl. Physiol.* 63:1816-1821.

Deconinck, N., Ragot, T., Maréchal, G., Perricaudet, M., and Gillis, J.M. 1996. Functional protection of dystrophic mouse (*mdx*) muscles after adenovirus-mediated transfer of a dystrophin minigene. *Proc. Natl. Acad. Sci. USA* 93:3570-3574.

Dunckley, M.G., Wells, D.J., Walsh, F.S., and Dickson, G. 1993. Direct retroviral-mediated transfer of a dystrophin minigene into *mdx* mouse muscle *in vivo. Hum. Mol. Genet.* 2:717-723.

Emery, A.E.H. 1988. *Duchenne Muscular Dystrophy,* 2nd ed. Oxford University Press, New York, NY.

Ewton, D.A., and Florini, J.R. 1980. Relative effects of the somatomedins, multiplication-stimulating activity, and growth hormone on myoblast and myotubes in culture. *Endocrinol.* 106:577-583.

Fan, Y., Maley, M., Beilharz, M., and Grounds, M. 1996. Rapid death of injected myoblasts in myoblast transfer therapy. *Muscle Nerve* 19:853-860.

Faulkner, J.A., Carlson, B.M., and Kadhiresan, V.A. 1994a. Whole muscle transplantation: Mechanisms responsible for functional deficits. *Biotech. and Bioeng.* 43:757-763.

Faulkner, J.A., Green, H.J., and White, T.P. 1994b. Response and adaptation of skeletal muscle to changes in physical activity. In *Physical Activity Fitness and Health*, ed. C. Bouchard, R.J. Shephard, and T. Stephens, p. 343-357. Human Kinetics Publishers, Champaign, IL.

Ferrari, G., Cusella-De Angelis, G., Coletta, M., Paolucci, E., Stornaiuolo, A., Cossu, G., and Mavilio, F. 1998. Muscle regeneration by bone marrow-derived myogenic progenitors. *Science* 279:1528-1530.

Florini, J.R. and Magri, K.A. 1989. Effects of growth factors on myogenic differentiation. *Am. J. Physiol.* 256 (*Cell Physiol.* 25):C701-C711.

Freilinger, G. and Deutinger, M. 1992. *Third Vienna Muscle Symposium*, Blackwell-MZV, Vienna, Austria.

Guelinckx, P.J, Faulkner, J.A., and Essig, D.A. 1988. Neurovascular-anastomosed muscle grafts in rabbits: Functional deficits result from tendon repair. *Muscle Nerve* 11:745-751.

Guelinckx, P.J., Carlson, B.M., and Faulkner, J.A. 1992. Morphologic characteristics of muscles grafted in rabbits with neurovascular repair. *J. Recon. Microsurg.* 8:481-489.

Gussoni, E., Pavlath, G.K., Lancot, A.M., Sharma, K.R., Miller, R.G., Steinman, L., and Blau, H.M. 1992. Normal dystrophin transcrips detected in Duchenne muscular dystrophy patients after myoblast transplantation. *Nature* 356:435-438.

Harii, K., Ohmori, K., Torii, S. 1976. Free gracilis muscle transplantation with microneurovascular anastomoses for the treatment of facial paralysis. *Plast. Recon. Surg.* 57:133-143.

Hasty, P., Bradley, A., Morris, J.H., Edmondson, J.M., Venuti, J.M., Olson, E.N., and Klein, W.H. 1993. Muscle deficiency and neonatal death in mice with a targeted mutation in the myogenin gene. *Nature* 364:501-506.

Hauser, M.A., Amalfitano, A., Kumar-Singh, R., Hauschka, S.D., and Chamberlain, J.S.1997. Improved adenoviral vectors for gene therapy of Duchenne muscular dystrophy. *Neuromusc. Dis.* 7:277-283.

Henneman, E., Somjen, G., and Carpenter, D. 1965. Functional significance of cell size in spinal motor neurons. *J. Neurophysiol.* 28:560-580.

Hoffman, E.P., Fischbeck, R.H., Brown, R.H., Johnson, M., Medori, R., Loike, J.D., Harris, J.B., Waterson, R., Brooke, M., Specht, L., Kupsky, W., Chamberlain, J., Caskey, C.T., Shapiro, F., and Kunkel, L.M. 1988. Characterization of dystrophin in muscle-biopsy specimens from patients with Duchenne's or Becker's muscular dystrophy. *N. Eng. J. Med.* 318:1363-1368.

Hurme, T. and Kalimo, H. 1992. Activation of myogenic precursor cells after muscle injury. *Med. Sci. Sports Exer.* 24:197-205.

Husmann, I., Soulet, L., Gautron, J., Martelly, I., and Barritault, D. 1996. Growth factors in skeletal muscle regeneration. *Cytokine Growth Factor Rev.* 7:249-258.

Jacobs-El, J., Zhou, M.Y., and Russell, B. 1995. MRF4, Myf-5, and myogenin mRNAs in the adaptive responses of mature rat muscle. *Am. J. Physiol.* 268:C1045-C1052.

Jin, P., Rahm, M., Claesson-Wesh, L., Heldin, C.-H., Sejerson, T. 1990. Expression of PDGF A chain and β-receptor genes during rat myoblast differentiation. *J. Cell Biol.* 110:1665-1672.

Jones, D.A., Jackson, M.J., McPhail, G., and Edwards, R.H.T. 1984. Experimental mouse muscle damage: the importance of external calcium. *Clin. Sci.* 66:317-322.

Kablar, B., Krastel, K., Ying, C., Asakura, A., Tapscott, S.J., and Rudnicki, M.A. 1997. MyoD and Myf-5 differentially regulate the development of limb versus trunk skeletal muscle. *Development* 124:4729-4738.

Kadhiresan, V.A., Guelinckx, P.J., and Faulkner, J.A. 1993. Tenotomy and repair of latissimus dorsi muscles in rats: implications for transposed muscle grafts. *J. Appl. Physiol.* 75:1294-1299.

Karpati, G., Ajdukovic, D., Arnold, D., Gledhill, R.B., Guttmann, R., Holland, P., Koch, P.A., Shoubridge, E., Spence, D., Vanasse, M., Watters, G.V., Abrahamowicz, M., Duff, C., and Worton, R.G. 1993. Myoblast transfer in Duchenne muscular dystrophy *Ann. Neurol.* 34:8-17.

Kass-Eisler, A., Falck-Pedersen, E., Elfenbein, D.H., Alvira, M., Buttrick, P.M., and Leinwand, L.A. 1994. The impact of developmental stage, route of administration and the immune system on adenovirus-mediated gene transfer. *Gene Therapy* 1:395-402.

Kumar-Singh, R., and Chamberlain, J.S. 1996. Encapsidated adenovirus minichromosomes allow delivery and expression of a 14 kb dystrophin cDNA to muscle cells. *Hum. Mol. Genet.* 5:913-921.

Lin, H., Parmacek, M.S., Morle, G., Bolling, S., and Leiden, J.M. 1990. Expression of recombinant genes in myocardium *in vivo* after direct injection of DNA. *Circulation* 82:2217-2221.

Macpherson, P.C.D., Dennis, R.G. and Faulkner, J.A. 1997. Sarcomere dynamics and contraction-induced injury to maximally activated single muscle fibres from soleus muscles of rats. *J. Physiol.* (*Lond.*), 500:523-533.

Marsh, D.R., Criswell, D.S., Carson, J.A., and Booth, F.W. 1997. Myogenic regulatory factors during regeneration of skeletal muscle in young, adult, and old rats. *J. Appl. Physiol.* 83:1270-1275.

McCully, K.K.and Faulkner, J.A. 1985. Injury to skeletal muscle fibers of mice following lengthening contractions. *J. Appl. Physiol.* 59:119-126.

McNeil, P.L. and Khakee, R. 1992. Disruptions of muscle fiber plasma membranes. Role in exercise-induced damage. *Am. J. Path.* 140:1097-1109.

Megeney, L.A., Kablar, B., Garrett, K., Anderson, J.E., and Rudnicki, M.A. 1996. MyoD is required for myogenic stem cell function in adult skeletal muscle. *Genes & Develop.* 10:1173-1183.

Mendell, J.R., Kissel, J.T., Amato, A.A., King, W., Signore, L, Prior, T.W., Sahenk, Z., Benson, S., McAndrew, P.E., Rice, R., Nagaraja, H., Stephens, R., Lantry L., Morris, G.E., and Burghes A.H.M. 1995. Myoblast transfer in the treatment of Duchenne's muscular dystrophy. *New Engl. J. Med.* 333:832-838.

Miller, S.W., Hassett, C.A., White T.P., and Faulkner, J.A. 1994. Recovery of medial gastrocnemius muscle grafts in rats: implications for the plantarflexor group. *J. Appl. Physiol.*, 77:2773-2777.

Moelleken, B.R.W., Mathes, S.A., and Chang, N. 1989. Latissimus dorsi muscle-musculocutaneous flap in chest-wall reconstruction. *Surgical Clinics of North America* 69(5):977-990.

Morgan, J.E. 1994. Cell and gene therapy in Duchenne muscular dystrophy. *Hum. Gene Ther.* 5:165-173.

Morgan, J.E., Hoffman, E.P., and Partridge, T.A. 1990. Normal myogenic cells from newborn mice restore normal histology to degenerating muscles of the *mdx* mouse. *J. Cell. Biol.* 111:2437-2449.

Moss, F.P. and Leblond, C.P. 1971. Satellite cells as the source of nuclei in muscles of growing rats. *Anat. Rec.* 170:421-436.

Nabeshima, Y., Hanaoka, K., Hayasaka, M., Esumi E., Li, S., Nonaka, I., and Nabeshima, Y. 1993. Myogenin gene disruption results in perinatal lethality because of a severe muscle defect. *Nature* 364:532-535.

Newham, D.J., McPhail, G., Jones, D.A., and Edwards, R.H.T. 1983. Large delayed plasma creatine kinase changes after stepping exercise. *Muscle Nerve* 6:380-385.

Partridge, T.A., Morgan. J.E., Coulton, G.R., Hoffman, E.P., and Kunkel, L.M. 1989. Conversion of *mdx* myofibers from dystrophin-negative to -positive by injection of normal myoblasts. *Nature* 337:176-179.

Petrof, B.J., Acsadi, G., Jani, A., Bourdon, J., Matusiewicz, N., Yang, L., Lochmüller, H., and Karpati, G. 1995. Efficiency and functional consequences of adenovirus-mediated *in vivo* gene transfer to normal and dystrophic (*mdx*) mouse diaphragm. *Am. J. Respir. Cell Mol. Biol.* 13:508-517.

Petrof, B.J., Lochmüller, H., Massie, B., Yang, L., Macmillan, C., Zhao, J.-E., Nalbantoglu, J., and Karpati, G. 1996. Impairment of force generation after adenovirus-mediated gene transfer to muscle is alleviated by adenoviral gene inactivation and host $CD8^+$ T cell deficiency. *Human Gene Ther.* 7:1813-1826.

Pette, D., and Vrbova, G. 1992. Adaptation of mammalian skeletal muscle fibers to chronic electrical stimulation. *Rev. Physiol. Biochem. Pharmcol.*, 120:115-202.

Ragot, T., Vincent, N., Chafey, P., Vigne, E., Gilgenkrantz, H., Couton, D., Cartaud, J., Briand, P., Kaplan, J., Perricaudet, M., and Kahn, A. 1993. Efficient adenovirus-mediated transfer of a human minidystrophin gene to skeletal muscle of *mdx* mice. *Nature* 361:647-650.

Rosenthal, S.M., Brown, E.J., Brunetti, A., and Goldfine, J.D.1991. Fibroblast growth factor inhibits insulin-like growth factor-II (IGF-II) gene expression and increases IGF-I receptor abundance in BC3H-1 muscle cells. *Mol. Endocrinol.* 5:678-684.

Rudnicki, M.A., Braun, T., Hinuma, S., and Jaenisch, R. 1992. Inactivation of MyoD in mice leads to up-regulation of the myogenic HLH gene Myf-5 and results in apparently normal muscle development. *Cell* 71:383-390.

Rudnicki, M.A., and Jaenisch, R. 1995. The MyoD family of transcription factors and skeletal myogenesis. *BioEssays* 17:203-209.

Rudnicki, M.A., Schnegelsberg, P.N.J., Stead, R.H., Braun, T., Arnold, H.H., and Jaenisch, R. 1993. MyoD or Myf-5 is required for the formation of skeletal muscle. *Cell* 75:1351-1359.

Russell, B., Dix, D.J., Haller, D.L., and Jacobs-El, J. 1992. Repair of injured skeletal muscle: a molecular approach. *Med. Sci. Sports Exer.* 24:189-196.

Sicinski, P., Geng, Y., Ryder-Cook, A.S., Barnard, E.A., Darlison, M.G., and Barnard, P.J. 1989. The molecular basis of muscular dystrophy in the *mdx* mouse: A point mutation. *Science* 244:1578-1580.

Tamai, S., Komatsu, S., Sakamoto, H., Sano, S., and Sasauchi, N. 1970. Free-muscle transplants in dogs with microsurgical neuro-vascular anastomoses. *Plast. Recon. Surg.* 46:219-225.

Thompson, N. 1974. A review of autogenous skeletal muscle grafts and their clinical applications. *Clin. Plast. Surg.* 1:349-403.

Vincent, N., Ragot, T, Gilgenkrantz, H., Couton, D., Chafey, P., Gregoire, A., Briand, P., Kaplan, J., Kahn, A., and Perricaudet, M. 1993. Long-term correction of mouse dystrophic degeneration by adenovirus-mediated transfer of a minidystrophin gene. *Nature Genetics* 5:130-134.

Weintraub, H., Davis, R., Tapscott, S., Thayer, M., Krause, M., Benezra, R., Blackwell, T.K., Turner, D., Rupp, R., Hollenberg, S., Zhuang, Y., and Lassar, A. 1991. The *myoD* gene family: nodal point during specification of the muscle cell lineage. *Science* 251:761-766.

Wernig, A., Irintchev, A., and Lange, G. 1995. Functional effects of myoblast implantation into histoincompatible mice with or without immunosuppression. *J. Physiol.* (*Lond.*) 484:493-504.

Whalen, R.G., Butler-Browne, G.S., Bugaisky, L.B., Harris, J.B., and Herliocoviez, D. 1985. Myosin isozyme transitions in developing and regenerating rat muscle. *Adv. Exp. Med. Biol.* 182:249-257.

Wolff, J.A., Malone, R.W., Williams, P., Chong, W., Acsadi, G., Jani, A., and Felgner, P.L. 1990. Direct gene transfer into mouse muscle *in vivo. Science* 247:1465-1468.

Yang, Y., Nunes, F.A., Berencsi, K., Gönczöl, E., Englelhardt, J.F., and Wilson, J.M. 1994. Inactivation of *E2a* in recombinant adenoviruses improves the prospect for gene therapy in cystic fibrosis. *Nature Genet.* 7:362-369.

124

Tissue Engineering of Cartilage

L.E. Freed
Harvard University

G. Vunjak-Novakovic
Massachusetts Institute of Technology

124.1 Scope

This chapter reviews the current state-of-the-art of articular cartilage tissue engineering, and focuses on a cell-polymer-bioreactor model system which can be used for controlled studies of chondrogenesis and the modulation of engineered cartilage by environmenal factors.

124.2 Cell-Based Approaches to Cartilage Tissue Engineering

Articular cartilage is an avascular tissue that contains only one cell type (the chondrocyte), and has a very limited capacity for self-repair. The chondrocytes are responsible for the synthesis and maintenance of their extracellular matrix (ECM), which is composed of a hydrated collagen network (~60% of the tissue dry weight, dw), a highly charged proteoglycan gel (PG, ~25% dw), and other proteins and glycoproteins (~15% dw).[1] Its high water content (70 to 80% of the tissue wet weight, ww) enables cartilage to withstand the compressive, tensile and shear forces associated with joint loading. None of the methods conventionally used for cartilage repair (e.g., tissue auto- or allografts) can predictably restore a durable articular surface to an osteoarthrotic joint.[2] Cell-based therapies, i.e., implantation of cells or engineered cartilage, represent an alternative approach to articular cartilage repair.

Figure 124.1 shows the cell-polymer-bioreactor system for cartilage tissue engineering. Constructs based on chondrogenic cells and polymer scaffolds are cultured in bioreactors to form 3-dimensional (3D) cartilaginous tissues. Engineered cartilage can be used either *in vivo,* to study articular cartilage repair or *in vitro,* for controlled studies of cell and tissue-level responses to molecular, mechanical, and genetic manipulations.

Table 124.1 lists selected, representative examples of recent articular cartilage tissue engineering studies in which the stated research goals were either *in vivo* cartilage repair or *in vitro* studies of chondrogenesis. Several key parameters varied from study to study: (i) cell source and expansion in culture, (ii) scaffold material and structure, (iii) *in vitro* cultivation (conditions and duration), (iv) additional components (e.g., perichondrium), (v) experimental animal (species and age), (vi) surgical model (defect location

0-8493-0462-8/00/$0.00+$.50

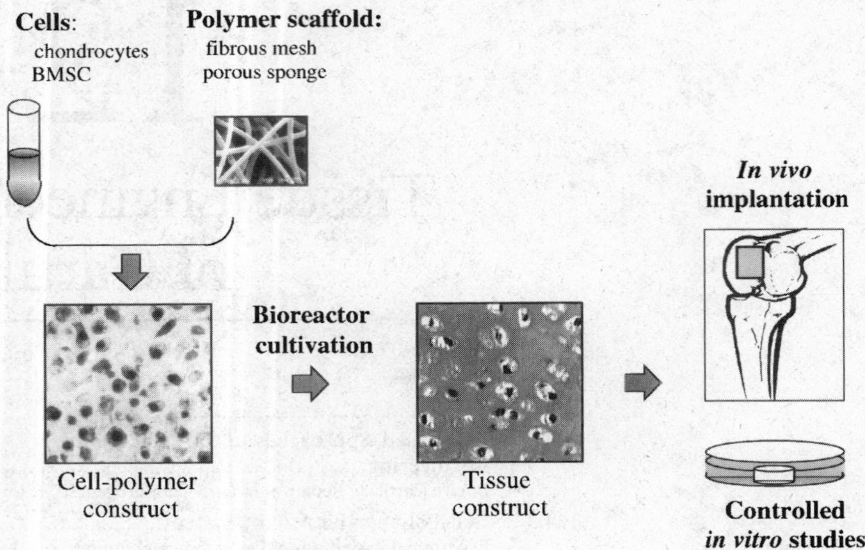

FIGURE 124.1 Cell-based approach to articular cartilage repair. Isolated cells, e.g., chondrocytes or bone marrow stromal cells (BMSC) are seeded onto 3-dimensional scaffolds, e.g., polymers formed as fibrous meshes or porous sponges, to form cell-polymer constructs. Constructs are first cultured *in vitro*, e.g., in bioreactors, and then used either as implants or for *in vitro* research.

and dimensions), (vii) graft type (autograft or allograft), (viii) duration of *in vivo* implantation, and (viii) methods used to assess the resulting tissue (e.g., histological, biochemical, and/or mechanical).

In Vivo Cartilage Repair

Brittberg et al.[3] reported a new cell-based procedure for the repair of human knee injuries that could potentially eliminate the need for more than half a million arthroplastic procedures and joint replacements performed annually in the United States. Autologous articular cartilage was harvested from a minor load-bearing area on the distal femur, and component cells were isolated, expanded *in vitro*, and reimplanted back into the patient. In particular, after excising the damaged cartilage down to the subchondral bone, defects (1.65 to 6.5 cm^2) were sutured closed with a periosteal flap creating a pocket into which the cultured cells were injected. Disabling symptoms (e.g., knee pain, swelling, locking) were markedly reduced over a follow-up period of up to 5 years. Femoral transplants were clinically graded as good or excellent in 88% of the cases, with 73% of the biopsy specimens resembling hyaline cartilage. In contrast, patellar transplants were graded as good or excellent in only 29% of the cases, with only 14% of the biopsies resembling hyaline cartilage. A therapy based on this technique was approved for clinical use by the Food and Drug Administration (FDA) in August, 1997 and is being marketed under the tradename Carticel®.[4] However, the original clinical study report has been criticized for not being a prospective controlled, randomized study and for its lack of a standard outcome analysis (i.e., for subjective evaluation with little quantitative biochemical or mechanical data).[5,6]

The Carticel® technique was also assessed in rabbits and dogs. In one study, cultured autologous chondrocytes were transplanted into patellar defects in conjunction with periosteal flaps and in some cases carbon fiber pads.[7] The repair tissue was characterized as hyaline-like and reported to develop *in vivo*, as assessed by an increasing degree of cellular columnarization. Histological scores were highest in defects treated with transplanted chondrocytes and periosteum, as compared to the combination of chondrocytes, carbon fibers and periosteum or periosteum only. The finding that the use of carbon fibers did not improve longterm repair was attributed to scaffold-induced diffusional limitations.

Table 124.1 Representative Studies of Articular Cartilage Tissue Engineering

Cells (source, expansion) and other components	Scaffold		In vitro culture		Animal model					
	material	dimensions	conditions	duration	species, age	implant site, defect size	graft type, duration	Assessment	Results	References (* = review)
1 *In vivo* cartilage repair										
Articular chondrocyte (human, expanded 2-3 wks)	None				Human 14–48 years old	Distal femur, patella 1.6-6.5 cm^2, full thickness	Autograft 16–66 months	Arthroscopy clinical signs	Symptomatic relief Femur repair better than patella.	Brittberg et al.[3]
Articular chondrocyte (rabbit, expanded 2 wks) and periosteum	None or carbon fibers		Static dish	Several hours	Rabbit 4 months old	Patella 3 mm dia, full thickness	Autograft 2–12 months	H (semiquantitative)	Cells plus periosteum better than cells on carbon meshes.	Brittberg et al.[7]
Articular chondrocyte (dog, expanded 3 wks) and periosteum					Dog 2yr–adult	Distal femur, 4 mm dia, full thickness	Autograft 1.5-18 months	H (semiquantitative) IH (collagen types I, II)	Fibroblastic and chondrocytic cells, collagen II early repair better than late repair.	Shortkroff et al.[8] Breinan et al.[9]
Precursor cells from bone marrow and periosteum; chondrocytes (rabbit, expanded); glue	None or collagen gel				Rabbit 3 mo.–adult	Distal femur up to 3×6× 3 mm ± protease treatment	Autograft precursor cells Allograft chondrocytes 1 week–1 year	H (semiquantitative); TEM mechanics (microindentation)	Hyaline cartilage (surface), new bone (deep); Precursor cells better than chondrocytes; Bone marrow better than periosteum; Integration improved by protease treatment.	*Caplan et al.[10] Wakitani et al.[11]
Precursor cells from bone marrow (goat, expanded)	Hyaluronic acid gel				Goat adult	Distal femur 4 mm dia, full thickness	Auto- and allografts 3 months	H (quantitative) IH (collagen type II, KS, PG)	Autograft better than allograft.	Butnariu-Ephrat et al.[13]
Precursor cells from bone marrow (embryonic and adult chick); chondrocytes (chick, expanded)	Hyaluronic acid gel				Chick 3 years old	Proximal tibia 3 mm dia × 2 mm deep	Auto- and allografts 3 months	H (semiquantitative) IH (collagen types I, II)	Autograft better than allograft; Embryonic cells better than adult.	Nevo et al.[14]
Perichondral chondrocyte (rabbit, expanded 3 weeks)	Poly(D,D,L,L) lactic acid	3.7 mm dia × 10 mm thick	Mixed vial	2 hours	Rabbit 9-12 months old	Distal femur 3.7 mm dia, 5 mm deep	Allograft 6 weeks– 12 months	H; amounts of GAG and collagen; IH (collagen types I, II); Confined compression modulus	Cell alignment determined by scaffold geometry Hyaline cartilage; collagen mainly type II GAG subnormal, mechanical properties normal.	Chu et al.[15,16]

Table 124.1 (continued) Representative Studies of Articular Cartilage Tissue Engineering

Cells (source, expansion) and other components	Scaffold		In vitro culture		Animal model			Assessment	Results	References (* = review)
	material	dimensions	conditions	duration	species, age	implant site, defect size	graft type, duration			
Articular chondrocyte (3 week old rats)	Collagen gel		Static dish	2 weeks	Rat 8-9 weeks old	distal femur 1.5 mm dia × 1.5 mm deep	auto- and allografts 2 weeks–12 months	H (semiquantitative)	Slight inflammation that resolved by 8 weeks 6-12 month autografts and allografts comparable.	Noguchi et al.[17]
Articular chondrocytes (2-8 month old rabbits)	Fibrous PGA	10 mm dia × 2 mm thick	Static dish	3-4 weeks	Rabbit 8 months old	distal femur 3 mm dia, full thickness	allograft 1–6 months	H (semiquantitative)	Fairly good repair at 6 months.	Freed et al.[18]
Muscle mesenchymal stem cell (expanded 3 wks)	Fibrous PGA	10 mm dia × 2 mm thick	Static dish	2.5 weeks	Rabbit 8 months old	distal femur 3 mm dia, full thickness	allograft 6 weeks–3 months	H (qualitative)	Fairly good repair at 3 months.	Grande et al.[19]
Articular chondrocyte (1 month old rabbits)	Collagen gel		Static dish	2 weeks	Rabbit 6 months old	distal femur 4 mm dia × 4 mm deep	allograft 1 day–6 months	H (semiquantitative)	Hyaline cartilage without endochondral ossification	Kawamura et al.[12]
2 *In vitro* chondrogenesis										
Articular chondrocyte (bovine)	Agarose gel	16 mm dia × 1 mm thick	Static dish; compression chamber	10 weeks 7 weeks				H; SEM; GAG and DNA content Incorporation of ^{35}S and ^{3}H Mechanical properties	Development of mechanically functional matrix (25% of normal) after 7 wks; Dynamic compression enhanced synthesis of GAG and collagen	Buschmann et al.[20,21]
Articular chondrocyte (human, 5-42 yr old)	Alginate beads	Not reported	Static dish	30 days				H; TEM; IH (keratan sulfate) Amounts of DNA, aggrecan, GAG, Incorporation of ^{35}S and ^{3}H	Formation of cartilaginous matrix composed of two compartments with different rates of proteoglycan turnover	Haeuselmann et al.[23]
Articular chondrocyte (bovine)	Fibrous PGA; porous PLLA fibrous PGA	10 mm × (5-10) mm × (2-3) mm 10 mm dia × (1-5) mm thick	Static and mixed dishes Mixed dish	6-8 weeks 6-8 weeks				H; IH (type II collagen) amounts of GAG, DNA, collagen and undegraded PGA	PGA better than PLLA 1.2-3.5 mm thick cartilaginous constructs; Mixing and high initial cellularity improved construct structures	Freed et al.[24,25,32,34]
Articular chondrocyte (bovine; expanded up to P2)	Fibrous PGA and nylon; PLGA mesh porous collagen	2 mm thick	Perfused teflon bag Static dish	5 weeks				H; GAG content incorporation of ^{35}S and ^{3}H	Scaffold materials affected GAG and collagen synthesis; Perfused bag better than static dish	Grande et al.[26]

Articular chondrocyte (human, 30-65 yr old)	PLGA and polydioxanon meshes coated with adhesion factors and agarose	Not reported	Perfused chamber	2 weeks	H; TEM; IH (collagen types I and II)	Evidence of collagen fibrils and proteoglycan	Sittinger et al.[27]
Articular chondrocyte (rabbit)	Fibrous PGA	10 mm dia × 2 mm thick	Perfused cartridge (1.2 mL) Static dish	4 weeks	H, IH (type II collagen, CS) amounts of GAG and collagen	2 mm thick constructs More tissue at edge than center	Dunkelman et al.[28]
Articular chondrocyte (equine; young and adult)	Fibrous PGA	10 mm × 10 mm × 1 mm	Compression chamber (3.4 or 6.9 MPa; 5s on, 30s off)	5 weeks	H; amounts of GAG and total collagen compressive stiffness	Cyclical loading promoted the production of GAG and collagen and increased compressive stiffness in constructs based on young cells	Carver and Heath[29,30]
Articular chondrocyte (human fetal)	None		Static dish	4 months	H, TEM; IH (collagen types I, II, IX; CS, KS) amounts of GAG and total collagen	Fetal chondrocytes cultured without serum formed 1.5-2 mm thick layer of hyaline cartilage.	Adkisson et al.[31]
Articular chondrocyte (bovine)	Fibrous PGA	10 mm dia × 5 mm thick	Static and mixed dishes Static and mixed flasks	8 weeks	H; amounts of GAG, DNA, and total collagen	2.7-4.8 mm thick constructs; Seeding and cultivation in mixed flasks yielded largest constructs and highest fractions of ECM	Vunjak-Novakovic et al.[33,35]
Articular chondrocyte (bovine)	Fibrous PGA	5 mm × 5 mm × 2 mm 5 mm dia × 2 mm thick	Rotating vessel	1 wk–7 months	H; TEM; IH (collagen types II, IX) amounts of DNA, GAG, and collagen Incorporation of ^{35}S and ^{3}H Mechanical properties	3-8 mm thick constructs. After 6 wks ECM was continuously cartilaginous. After 7 months, compressive stiffness became comparable to normal cartilage	Freed et al.[36,37]
Articular chondrocyte (bovine)	Fibrous PGA	5 mm dia × 2 mm thick	Rotating vessel	6 weeks	SEM; amounts of DNA, GAG and collagen (total and types II, IX and X); collagen pyridinium crosslinks	Structure and composition of collagen network in 6 wk constructs comparable to native cartilage; collagen content and crosslinking subnormal.	Riesle et al.[38]

Table 124.1 (continued) Representative Studies of Articular Cartilage Tissue Engineering

Cells (source, expansion) and other components	Scaffold		In vitro culture		Animal model			Assessment	Results	References (* = review)
	material	dimensions	conditions	duration	species, age	implant site, defect size	graft type, duration			
Articular chondrocyte (bovine)	Fibrous PGA	5 mm dia × 2 mm thick	Static and mixed flasks Rotating vessel	6 weeks				H; amounts of DNA, GAG and collagen (total and type II); mechanical properties	Construct structure and function could be modulated by the conditions of flow and mixing. Mechanical parameters correlated with wet weight fractions of GAG, collagen and water.	Vunjak-Novakovic et al.[39]
BMSC (chick embryo; bovine)	Fibrous PGA Porous PLGA	5 mm dia × 2 mm thick	Mixed dish	4 weeks				H; IH (collagen types II and X) amounts of DNA, GAG and collagen	Selective cell expansion and the cultivation on 3-dimensional scaffolds resulted in the formation of cartilaginous tissues	Martin et al.[42-44]

Note:

BMSC	bone marrow stromal cells	PG	proteoglycan
CS	chondroitin sulfate	PGA	polyglycolic acid
ECM	extracellular matrix	PLGA	polylactic-co-glycolic acid
GAG	glycosaminoglycan	PLLA	poly (L) lactic acid
H	histology	SEM	scanning electron microscopy
IH	immunohistochemistry	TEM	transmission electron microscopy
KS	keratan sulfate		

In other studies, cultured autologous chondrocytes in conjunction with periosteal flaps were transplanted into femoral defects in dogs and compared to periosteal flaps alone and to untreated defects.[8,9] Three phases of healing were demonstrated: formation of repair tissue (at 1.5 months), remodeling (at 3 and 6 months), and degradation (at 12 and 18 months). Neither periosteum nor transplanted chondrocytes enhanced healing after 1 year, at which time hyaline-like repair tissue appeared to be displaced by fibrocartilage. The differences between results obtained in rabbit and dog studies were attributed to several variables including species, subject age, defect location, surgical technique, and postoperative animal activity. Low retention of implanted cells in the lesion and failure of immature repair tissue subjected to high mechanical forces were listed as possible causes of degradation.

The group of Caplan et al. pioneered the use of autologous mesenchymal stem cells in the repair of osteochondral defects.[10] Osteoprogenitor cells were selected from whole bone marrow based on their ability to adhere to Petri dishes and expanded in monolayer culture prior to implantation. Autologous precursor cells were delivered in collagen gels into defects that were in some cases pretreated with proteolytic enzymes. Other implants were based on either autologous precursor cells derived from periosteum or allogeneic chondrocytes derived from articular cartilage. The dimensions of defects were up to 3 mm wide, 6 mm long, and 3 mm deep, among the largest reported for repair studies in rabbits. Repair was assessed histologically and in some cases mechanically, e.g., relative compliance using a microindentation probe.[11]

Autologous precursor cells in collagen gels formed a cartilaginous surface zone while tissue at the base of the defect hypertrophied, calcified, and was replaced by host-derived vasculature, marrow, and bone.[10] It was postulated that the different biological milieus of the synovial fluid at the joint surface and the osseus recepticle in the underlying bone played key roles in architecturally appropriate precursor cell differentiation. Improved integration of the graft with the surrounding host tissue was reported following partial digestion at the defect site (e.g., with trypsin) prior to implantation; this finding was attributed to enhanced interdigitation of newly synthesized ECM molecules. However, progressive thinning of the cartilaginous surface zone was observed over 6 months, which was more pronounced with periosteally-derived than with bone marrow-derived precursor cells. In contrast, allografted chondrocytes in collagen gels rapidly formed plugs of hyaline cartilage that filled the entire defect but failed to develop a region of subchondral bone at the base or to integrate with the surrounding host tissue even after 6 months. Similar results were obtained when chondrocytes were cultured *in vitro* in collagen gels prior to *in vivo* implantation[12] (as described below).

Nevo et al.[13,14] have explored the use of both chondrocytes and osteoprogenitor cells derived from bone marrow delivered in a hyaluronic acid gel. Autografts or allografts were used to repair tibial defects in goats and femoral defects in chicks. The autografts were superior to the allografts, which evoked a typical immune response and progressive arthrosis. In both animal models, autografted defects were repaired with a well-integrated tissue that resembled hyaline cartilage at the surface and bone at the base.

Amiel et al. qualitatively (i.e., histologically) and quantitatively (i.e., biochemically and mechanically) assessed osteochondral defects repaired with perichondral cells and porous polylactic acid in a rabbit allograft model.[15,16] After 6 weeks, the alignment of cells in the repair tissue followed the architecture of the scaffold. After 1 year, the repair tissue had variable histological appearance (only 50% of the subchondral bone reformed) and biochemical composition (dry weight fractions of GAG were 55% those of the host cartilage), but mechanical properties (i.e., modulus and permeability in radially confined compression) were comparable to those measured for the host cartilage. In this case, mechanical testing appeared to be less sensitive than histological and biochemical assessments; this finding was attributed to difficulties in mechanical measurements and the derivation of intrinsic parameters, due to the geometry and the non-homogeneous nature of the repair tissue, respectively.

Noguchi et al.[17] compared autologous and allogenic repair using chondrocytes cultured in collagen gels prior to implantation in inbred rats. After 2 to 4 weeks, the repair tissue consisted of hyaline articular cartilage with slight inflammatory cell infiltration in both groups; the immune response was somewhat more conspicuous in the allografts. However, inflammation resolved by 8 weeks and auto- and allografted repair tissues were almost identical after 6 months.

Freed et al.[18] cultured chondrocytes expanded by serial passage (P4) on polyglycolic acid (PGA) scaffolds for 3 to 4 weeks *in vitro* prior to implantation as allografts in rabbits. Compared to native rabbit articular cartilage, constructs contained 25% as much total collagen and 86% as much GAG per gram dry weight at the time of implantation. After 1 and 6 months, the histological scores of defects repaired with cell-PGA constructs did not differ significantly from those following implantation of PGA alone, except for qualitatively better surface smoothness, cell columnarization, and spatial uniformity of GAG in the defects grafted with cell-PGA constructs.

Grande et al.[19] cultured osteoprogenitor cells derived from skeletal muscle on PGA scaffolds for 2 to 3 weeks *in vitro* prior to implantation as allografts in rabbits. At the time of implantation, cells were attached to the scaffold but had not undergone chondrogenesis. After 3 months, defects repaired with cell-PGA constructs consisted of a cartilaginous surface region similar in thickness to the host cartilage and normal appearing subchondral bone, while implantation of PGA alone resulted in a patchy mixture of fibrous and hyaline cartilage.

Kawamura et al.[12] cultured chondrocytes in collagen gels for 2 weeks *in vitro* prior to implantation as allografts in rabbits. At the time of implantation, constructs appeared stiffer than uncultured cells in collagen gels and histologically resembled hyaline cartilage. After 6 months, the repair tissue still consisted of a thick plug of hyaline cartilage, i.e., the region at the base was neither vascularized nor replaced by bone. It was suggested that a mechanical mismatch between the thin layer of host cartilage and the thick plug of cartilaginous repair tissue, and the lack of regeneration of a proper subchondral bony base would contribute to long term implant failure.

In Vitro Chondrogenesis

Buschmann et al.[20,21] showed that high cell density cultures of calf articular chondrocytes in agarose gels formed a mechanically functional cartilaginous matrix. After 7 weeks *in vitro*, construct compositions and mechanical properties (i.e., fractions of DNA and GAG, compressive modulus, streaming potential) reached values of about 25% those of native calf cartilage. Constructs responded to mechanical forces in a manner similar to that of native cartilage: static compression supressed ECM synthesis by an amount that increased with increasing compression amplitude and culture time, while dynamic compression stimulated ECM synthesis by an amount that increased with ECM accumulation and culture time. The authors postulated that several mechanisms could be involved in mechanotransduction, including cell-ECM interactions and/or changes in interstitial fluid flow and streaming potential.

Hauselmann et al.[22] demonstrated the phenotypic stability of calf articular cartilage after prolonged (8 month) cultivation in alginate beads. The ECM formed by adult human chondrocytes cultured in gels for 30 days was similar to that of human articular cartilage and could be used to study PG turnover.[23] The ECM consisted of two compartments: a small amount of cell-associated matrix (corresponding to the pericellular and territorial ECM), and a larger amount of matrix further removed from the cells (corresponding to interterritorial ECM). Aggregated PGs in the cell-associated ECM turned over relatively quickly ($t_{1/2}$ of 29 days) as compared to those in the further-removed ECM ($t_{1/2}$ >100 days). These findings were attributed to the effects of proteolytic enzymes located at the cell membrane.

Freed et al.[24] characterized cartilaginous constructs based on calf articular or human rib chondrocytes and synthetic polymer scaffolds. Fibrous polyglycolic acid (PGA) meshes yielded constructs with higher cellularities and GAG production rates as compared to porous polylactic acid (PLLA) sponges; these findings were attributed to differences in polymer geometry (mesh vs. sponge, pore size distribution) and degradation (faster for PGA than PLLA). PGA scaffolds were subsequently characterized in detail and produced on a commercial scale.[25]

Grande et al.[26] showed that the choice of scaffold affected ECM synthesis rates, e.g., PGA and collagen scaffolds respectively enhanced GAG and protein synthesis rates of calf chondrocytes cultured in a perfused system. Sittinger et al.[27] observed PG and collagen deposition by human articular chondrocytes cultured on synthetic polyester meshes coated with poly-L-lysine or type II collagen and embedded in agarose gels. Dunkelman et al.[28] and Grande et al.[26] both reported that cell-polymer constructs were more likely to form cartilaginous ECM if cultured in perfused vessels rather than statically.

Carver and Heath[29,30] demonstrated that physiologicial levels of compression enhanced GAG deposition and improved the compressive moduli of engineered cartilage. In particular, constructs based on equine chondrocytes and PGA scaffolds were subjected to intermittent hydrostatic pressures of 500 and 1000 psi (3.4 and 6.9 MPa) in a semi-continuous perfusion system for 5 weeks. Structural and functional improvements were observed in constructs based on cells obtained from young but not adult horses. In a recent study, fetal human chondrocytes cultured at high density in serum-free medium formed 1.5 to 2.0 mm thick hyaline cartilage over 120 days whereas otherwise identical cultures containing serum could not be maintained for more than 30 days.[31]

Freed and Vunjak-Novakovic studied *in vitro* chondrogenesis using a variety of methods to seed and cultivate calf chondrocytes on polymer scaffolds as follows. Static seeding of scaffolds two or more mm thick resulted in bilaminar constructs with a fibrous upper region and a cartilaginous lower region.[32] In contrast, dynamic cell seeding in spinner flasks allowed relatively uniform cell seeding of scaffolds 2-5 mm thick at an essentially 100% yield.[33] An increase in the initial density of seeded cells resulted in comparable construct cellularities and collagen contents, but markedly higher GAG contents.[34]

Mixing during 3D culture markedly improved construct morphology and composition.[32,35] For example, 10 mm diameter × 5 mm thick scaffolds seeded and cultured in mixed flasks weighed twice as much as those seeded and cultured in mixed dishes, and contained about 2.5 times more of each GAG and total collagen.[35] However, constructs grown in mixed flasks formed outer capsules that were up to 300 μm thick and contained high concentrations of elongated cells and collagen and little GAG; this finding was attributed to the effects of turbulent flow conditions on cells at the construct surface.[35] In contrast, constructs cultured in rotating vessels had relatively uniform distributions of cells and ECM and were up to 5 mm thick after 6 weeks in culture.[36] With increasing culture time, construct ECM biosynthesis rates and the fractional loss of newly synthesized macromolecules into the culture medium both decreased[36] but construct GAG fractions and mechanical properties improved.[37] The collagen networks of constructs and native cartilage were similar with respect to fibril density and diameter and fractions of collagen types II, IX and X.[38]

Vunjak-Novakovic et al.[39] studied the relationships between construct compositions (fractions of water and ECM components) and mechanical properties in static and dynamic compression (equilibrium modulus, dynamic stiffness, hydraulic permeability, streaming potential) using three different culture environments: static flasks, mixed flasks, and rotating vessels. Constructs cultured in static and mixed flasks had lower concentrations of ECM and worse mechanical properties as compared to constructs cultured in rotating vessels. The structure-function relationships detected for chondrocyte-PGA constructs appeared consistent with those previously reported for chondrocytes cultured in agarose gels,[20] native calf cartilage,[40] and adult human cartilage.[41]

Martin et al.[42-44] demonstrated that bone marrow stromal cells (BMSC) expanded in monolayers, seeded onto scaffolds and cultured in mixed Petri dishes formed large, 3D cartilaginous tissues. The presence of a 3D scaffold was required, as demonstrated by the small size and noncartilaginous nature of control cell pellet cultures. The presence of fibroblast growth factor 2 (FGF) during 2D expansion promoted differentiation of BMSC during 3D cultivation. As compared to avian (embryonic chick) BMSC, mammalian (calf) BMSC required a more structurally stable 3D scaffold and the presence of additional biochemical signals.

124.3 Cell-Polymer-Bioreactor System

The literature reviewed in Section 124.2 above demonstrates the feasibility of using cell-based tissue engineering approaches to *in vivo* articular cartilage repair and *in vitro* studies of chondrogenesis. In the next section we will first describe the *in vitro* cell-polymer-bioreactor system and then use selected examples from our own work to illustrate its use in studies of the development and modulation of

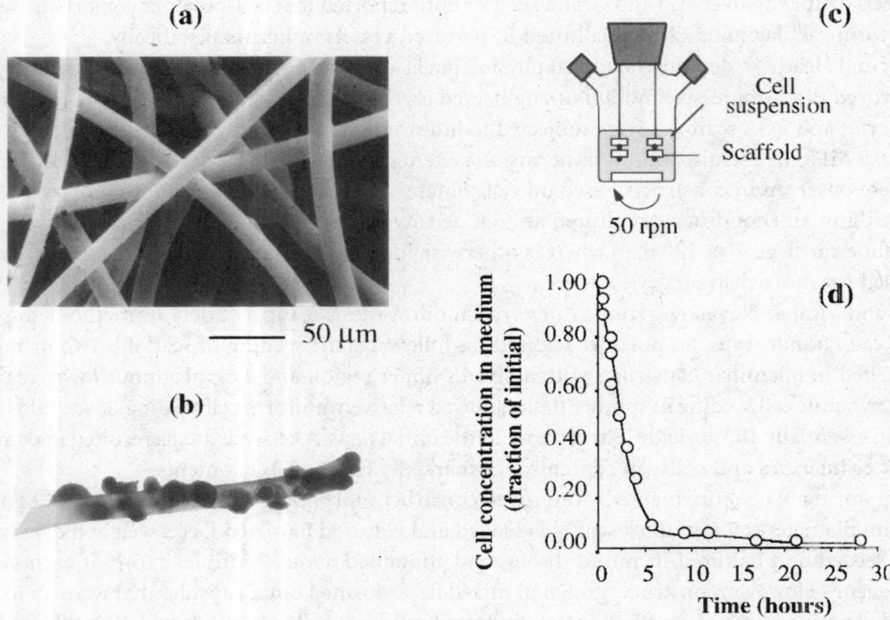

FIGURE 124.2 Cell seeding of polyglycolic acid (PGA) mesh. (a) PGA ultrastructure (scanning electron micrograph). (b) chondrocytes attach to PGA fibers and maintain their spherical morphology (H&E stained histological section). (c) dynamic cell seeding: magnetic stirring generates convective motion of cells into porous scaffolds that are fixed to needles embedded in the stopper of the flask. (d) chondrocytes attached to PGA scaffolds with an effective yield of 100% over 24 h.

construct structure and function. The unifying hypothesis of our work is that isolated chondrocytes or precursor cells can form functional cartilage tissue *in vitro,* if cultured on a 3D structural template in an environment that provides the necessary biochemical and physical signals. Ideally, engineered cartilage should display the key structural features of native cartilage, be able to withstand physiological loads, and be able to integrate with adjacent host tissues following *in vivo* implantation.

Experimental Methods

Cells: Cell types studied included articular chondrocytes and their precursors. *Chondrocytes* were obtained from the articular cartilage of 2 to 3 week old bovine calves[20,24] and used immediately after isolation. *Bone marrow stromal cells* (BMSC) were obtained from the marrow of 2 to 3 week old bovine calves[43] or 16 day embryonic chicks.[42]

Scaffolds: Our best characterized scaffolds are made of *polyglycolic acid* (PGA) in form of 97% porous nonwoven meshes of 13 μm diameter fibers (Fig. 124.2a).[25] Other scaffolds that have been studied include *polylactic acid* (PLLA)[24] and *polylactic-co-glycolic acid* (PLGA) sponges.[45]

Cell seeding: Polymer scaffolds (5 to 10 mm in diameter, 2 to 5 mm thick) were fixed in place and seeded with cells using well-mixed spinner flasks [33] (Fig. 124.2c).

Bioreactor cultivation: Tissue culture systems under investigation are schematically presented in Fig. 124.3. Static and mixed *dishes* contained one construct per well in 6 to 7 mL medium; mixed dishes were placed on an orbital shaker at 75 rpm.[32] In *spinner flasks,* scaffolds were fixed to needles (8 to 12 per flask, in 100 to 120 cm^3 medium), seeded with isolated cells under mixed conditions and cultured either statically or exposed to unidirectional turbulent flow using a non-suspended magnetic bar at 50-80 rpm.[35] *Rotating vessels* included the slow turning lateral vessel (STLV) and the high aspect ratio vessel (HARV), each of which was rotated around its central axis such that the constructs (8 to 12 per vessel, in 100 to

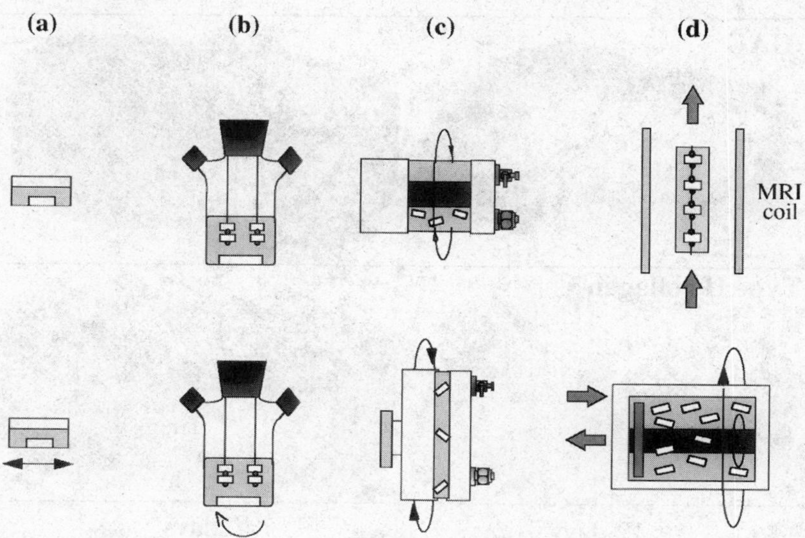

FIGURE 124.3 Tissue culture bioreactors. (a) static and orbitally mixed Petri dishes. (b) static and magnetically stirred flasks. (c) rotating vessels: the slow turning lateral vessel (above) and the high aspect ratio vessel (below). (d) perfused chamber in which cultured constructs are monitored using magnetic resonance imaging (MRI, above); perfused rotating vessel (below).

120 cm^3 medium) settled freely in a laminar flow field.[46,47] *Perfused vessels* included columns used to culture constructs and non-invasively monitor the progression of chondrogenesis using magnetic resonance imaging (MRI),[48] and a rotating perfused vessel developed by NASA for flight studies.[37] In all vessels, medium was exchanged batchwise (at a rate of ~3 cm^3 per construct and day), and gas was exchanged continuously, by surface aeration or by diffusion through a silicone membrane.

Analytical techniques: Construct size and distributions of cells and tissue components were assessed by image analysis.[33,36] Constructs for histological assessment were fixed in neutral buffered formalin, embedded in paraffin, sectioned (5-8 μm thick), and stained with hematoxylin and eosin (H&E for cells), safranin-O (for GAG), and monoclonal antibodies (for collagen types I, II, IX, and X).[36,38,42] Ultrastructural analyses included scanning and transmission electron microscopy (SEM, TEM).[38] Biochemical compositions were measured following papain or protease-K digestion of lyophilized constructs.[24,25,36] Cell number was assessed by measuring the amount of DNA using Hoechst 33258 dye.[49] Sulfated GAG content was determined by dimethylmethylene blue dye binding.[50] Total collagen content was determined from hydroxyproline content after acid hydrolysis and reaction with p-dimethylaminobenzaldehyde and chloramine-T.[51] Type II collagen content was determined by inhibition ELISA.[36,38,52] The presence of other collagen types (e.g., I, IX, X) was demonstrated and semiquantitatively measured using SDS-PAGE and Western blots.[38] GAG distribution was determined by MRI.[48,53] Synthesis rates of GAG, total protein and collagen were measured by incorporation and release of radiolabeled tracers.[36] Cell metabolism was assessed based on the ratio of lactate production and glucose consumption, and ammonia production rates.[54] Mechanical construct properties (e.g., compressive modulus, dynamic stiffness, hydraulic permeability, streaming potential) were measured in static and dynamic radially confined compression.[39,55]

Developmental Studies

As described above (Section 124.2, Table 124.1), articular cartilage repair has been enhanced following the transplantation of chondrocytes and osteochondral progenitor cells in conjunction with polymer scaffolds. However, identification of specific factors that influence *in vivo* regeneration and/or repair can be difficult due to host systemic responses (e.g., neuro-endocrinological, immunological). Controlled

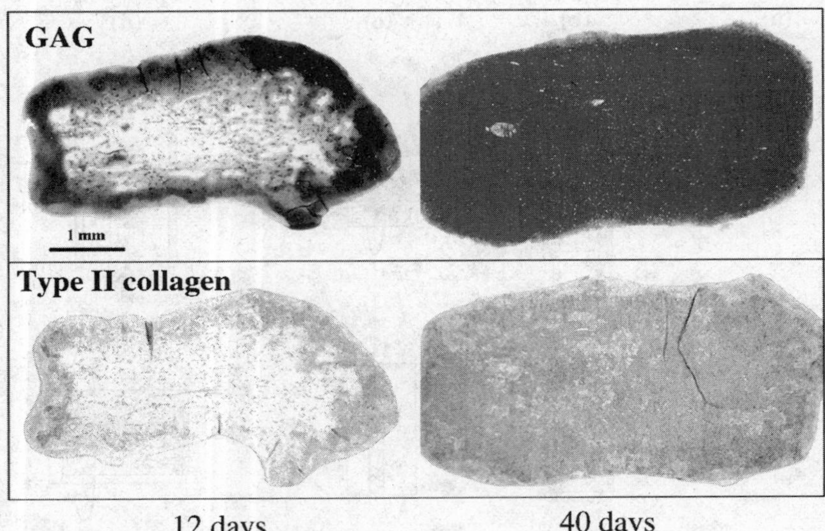

FIGURE 124.4 Temporo-spatial pattern of ECM deposition. Calf chondrocytes cultured on PGA scaffolds in rotating vessels. Chondrogenesis was initiated peripherally in 12 day constructs (left panels) and proceeded appositionally resulting in a continous cartilaginous ECM in 40 day constructs (right panels). Histological cross-sections were stained with safranin-O for GAG (top panels) or with a monoclonal antibody to collagen type II (bottom panels).

in vitro studies carried out in the cell-polymer-bioreactor system can thus provide useful, complementary information regarding the process of cartilage matrix regeneration starting from isolated cells (herein referred to as developmental studies).

Chondrogenesis *in vivo* and *in vitro* is thought to be initiated by precursor cell aggregation.[56] Cartilaginous ECM is deposited, with concurrent increases in the amounts of GAG and collagen types II and IX, both in mesenchymal cells obtained from embryonic chick limbs[57] and in human limbs at various developmental stages.[58] ECM deposition starts at the center of the developing limb bud and spreads peripherally,[57] in conjunction with cellular conditioning of their microenvironment.[59]

In calf chondrocyte-PGA constructs, amounts of GAG and collagen type II also increased concomittantly[36] (Fig. 124.4). By culture day 12, cartilaginous tissue had formed at the construct periphery; by day 40, constructs were continuously cartilaginous matrix over their entire cross sections (6.7 mm diameter × 5 mm thick). Chondrogenesis was initiated peripherally and progressed both inward towards the construct center and outward from its surface; these findings could be correlated with construct cell distributions, as described below. The temporo-spatial patterns of GAG distribution could also be monitored in living constructs using MRI.[48] Constructs cultured 6 weeks contained an interconnected network of collagen fibers that resembled that of calf articular collagen with respect to overall organization and fiber diameter (Fig. 124.5A). Type II collagen represented more than 90% of the total collagen, as quantitated by an ELISA[36,38] and demonstrated qualitatively in Fig. 124.5B. Construct collagen type II was not susceptible to extraction with a chaotropic agent (guanidine hydrochloride, GuHCl), suggesting some degree of cross-linkage. However, construct collagen had only 30% as many pyridinium crosslinks as native calf cartilage.[38]

Construct wet weight fractions of cartilaginous ECM increased, the amount of polymer decreased, and construct cellularity plateaued over 6 weeks of culture in rotating vessels[36] (Fig. 124.6). Over the same time interval, construct ECM synthesis rates decreased by approximately 40% and the fraction of newly synthesized macromolecules released into the culture medium decreased from about 25% of total at 4 days to less than 4%; the latter can be attributed in part to collagen network development.[36] The mass of the PGA scaffold decreased to about 40% of initial over 6 weeks.[25] The time constant for scaffold degradation

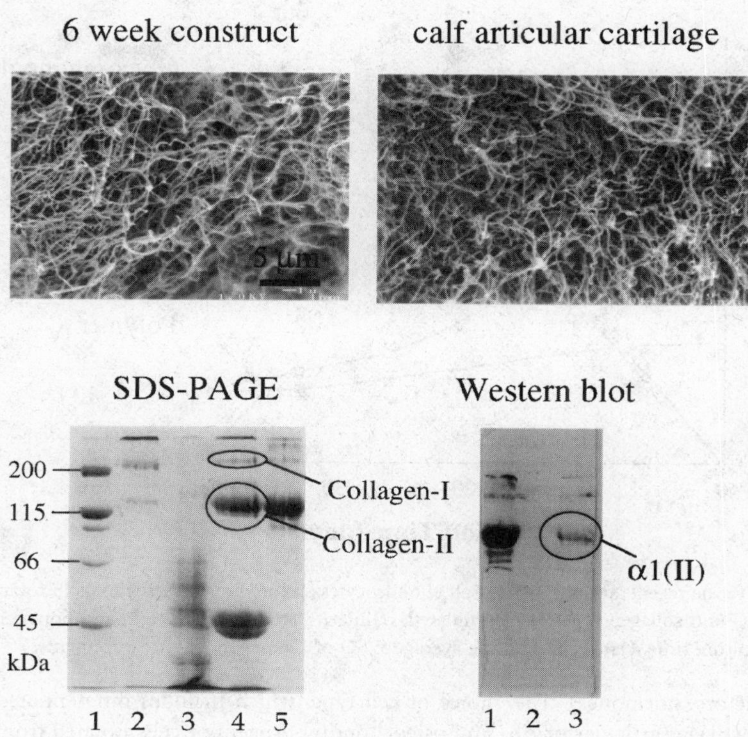

FIGURE 124.5 Collagen structure and type. Calf chondrocytes cultured on PGA scaffolds for 6 weeks in rotating vessels. (a) Scanning electron micrographs showed similar collagen network ultrastructures for constructs and native calf articular cartilage. (b) The presence of type II collagen in constructs was demonstrated using pepsin-digested constructs in conjunction with SDS-PAGE, and confirmed by Western blot (with a monoclonal antibody to αII). Lanes on SDS-PAGE: (1) molecular weight standards, (2) collagen type I, (3) GuHCl extract of construct, (4) pepsin extract of construct, (5) collagen type II; Lanes on Western blot: (1) collagen type II, (2) GuHCl extract of construct, (3) pepsin extract of construct.

was of the same order of magnitude as the time constant of ECM deposition, a situation associated with enhanced tissue regeneration according to the hypothesis of isomorphous tissue replacement.[60]

Modulation of Cartilaginous Structure and Function

The structure and function of articular cartilage are determined, at least in part, by environmental factors.[61] It is likely that the same factors that affect *in vivo* cartilage development, maintenance and remodeling also affect *in vitro* chondrogenesis. In the selected examples described below, we will describe the effects of: (1) cell, scaffold, and biochemical factors and (2) cultivation conditions and time on the structure and function of engineered cartilage. In principle, *in vitro* cultivation of a cell-polymer construct prior to *in vivo* implantation can help localize cell delivery and promote device fixation and survival, while maintaining an ability for integration at the graft-host interface. In contrast, cells transplanted in the absence of a carrier vehicle tended to leak away from the defect site,[9] while cells implanted immediately after loading in/on biomaterials were more vulnerable to mechanical forces and metabolic changes experienced during fixation or following implantation.[12]

Cells, the Scaffold and Biochemical Signals

Chondrogenesis depends on the cells themselves (e.g., type, *in vitro* expansion, density, spatial arrangement), the scaffold (e.g., material, structure, degradation rate), and the presence of biochemical factors

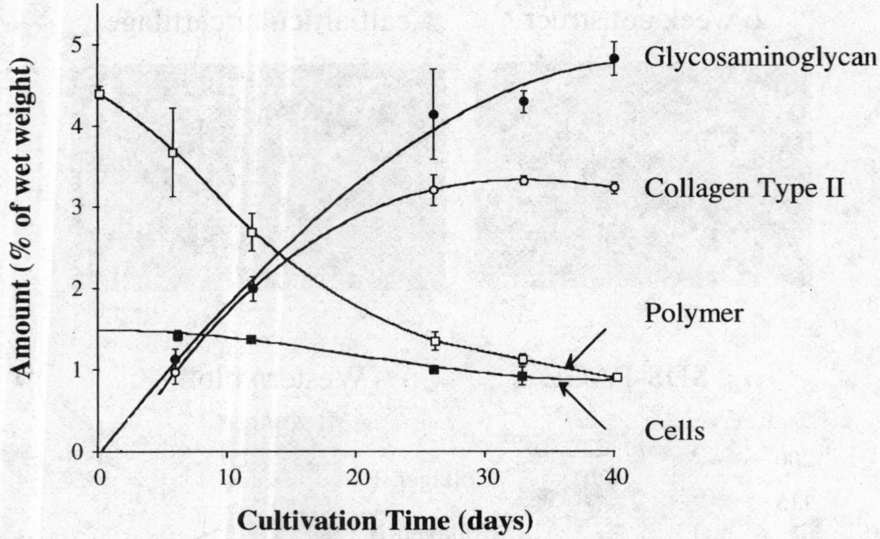

FIGURE 124.6 Tissue regeneration kinetics. Calf chondrocytes cultured on PGA scaffolds in rotating vessels. ECM components (GAG and collagen type II) accumulated, cellularity remained constant, and polymer mass decreased with increasing culture time. Data represent the average ± SD of 3 independent measurements.

(e.g., growth factors, hormones). The choice of cell type, which includes differentiated chondrocytes isolated from cartilage (articular or rib) and osteochondral progenitor cells isolated from bone marrow (herein referred to as bone marrow stromal cells, BMSC), can affect *in vitro* culture requirements (e.g., medium supplements) and *in vivo* construct function (e.g., integration potential). Polymer scaffolds vary with respect to surface properties (e.g., chemistry and wettability, which affect cell spreading and proliferation), geometry (e.g., dimensions, porosity, and pore size, which affect the spatial cell arrangement and the transmission of biochemical and mechanical signals), and physical properties (e.g., mechanical integrity and degradation rate, which determine whether the polymer can provide a structurally stable template for tissue regeneration).

Watt [61] suggested that a cell cultured *in vitro* will tend to retain its differentiated phenotype under conditions that resemble its natural *in vivo* environment. In the case of chondrocytes, phenotypic stability is enhanced by cultivation in alginate or agarose.[22,23,63] Moreover, chondrocytes dedifferentiated by serial passage in monolayers redifferentiated (i.e., reacquired a spherical shape, ceased dividing, and resumed the synthesis of GAG and collagen type II) when transferred into 3D cultures.[64,65] When cultured on 3D fibrous PGA scaffolds, chondrocytes retained their spherical shape and formed cartilaginous tissue.[25,66]

Chondrocytes cultured at high cell densities tended to express their differentiated phenotype.[63,67] It was hypothesized that the effective cross-talk between cells depends on the presence of homotypically differentiated cells in the immediate cell environment.[67] The term "community effect" was later coined by Gurdon[69] who suggested that the ability of a cell to respond to phenotypic induction is enhanced by, or even dependent on, other neighboring cells differentiating in the same way at the same time. A postulated underlying mechanism involves changes in gene transcription and translation caused by cell-cell/ECM interactions.[62,70,71]

In cell-polymer constructs, the density of cells initially seeded at the construct periphery was sufficient to initiate chondrogenesis in that region. In particular, corresponding tempero-spatial patterns of cartilaginous ECM deposition and cell distribution were observed (compare Figs. 124.4 and 124.7). The finding that cell densities were initially higher peripherally than centrally, which implies that cell seeding density and proliferation rate were relatively higher peripherally, can be attributed to enhanced rates of nutrient and gas transfer at the construct surface. Over the 6 week culture period, self-regulated cell

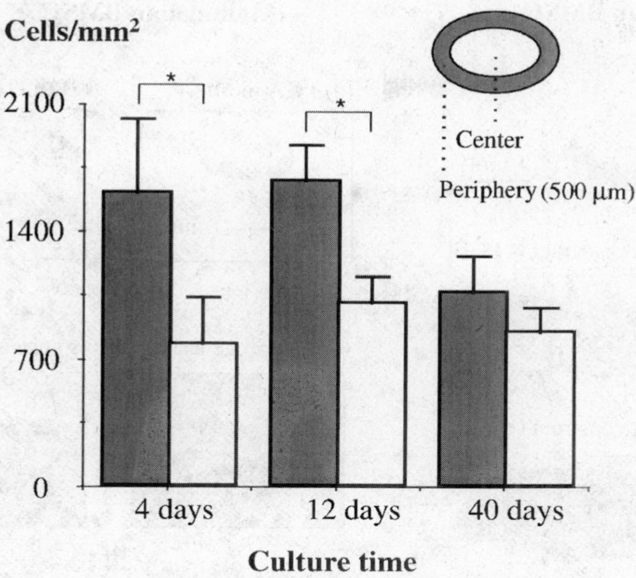

FIGURE 124.7 Temporo-spatial pattern of cell density. Calf chondrocytes cultured on PGA scaffolds in rotating vessels. Cell density was initially higher at the construct periphery (white bars) than at its center (shaded bars), while cell distribution in 40 day constructs was spatially uniform. Number of cells/mm^2 was assessed by image analysis of H&E stained histological sections (average ± SD of 24 independent measurements).

proliferation and ECM deposition resulted in constructs with physiological cellularities and spatially uniform distributions of components[36] (Figs. 124.4 and 124.7). Increases in the size, cellularity, and absolute amount of cartilaginous ECM were observed when chondrocyte-PGA constructs were cultured in medium supplemented with insulin-like growth factor I (IGF-I, 10-300 ng/mL) and serum (10%).[72]

In the case of osteochondral progenitor cells, Caplan et al.[10] suggested that principles of skeletal tissue engineering should be governed by the same motifs as embryonic development. BMSC differentiation is thought to be regulated by cell-to-cell contacts in an environment capable of activating the differentiation program.[73] *In vitro* (in BMSC aggregates), the induction of chondrogenesis depended on the presence of transforming growth factor-β1 (TGF β) and dexamethasone.[74] *In vivo* (in rabbits), osteochondral repair recapitulated embryonic events and depended on the spatial arrangement and density of precursor cells and the presence of specific bioactive factors.[10]

In cell-polymer constructs, chondrocytic differentiation of BMSC depended both on scaffold-related factors and on exogenous biochemical signals.[42-44] Avian BMSC cultured in the absence of polymer scaffolds formed small bilaminar tissues in which the lower region contained GAG and upper region appeared undifferentiated (Fig. 124.8A). The same BMSC formed constructs consisting of a single tissue phase when cultured on PGA scaffolds in mixed Petri dishes (Fig. 124.8B), while BMSC expansion in the presence of FGF prior to culture on PGA scaffolds resulted in the most cartilaginous ECM (Fig. 124.8C). When FGF-expanded mammalian BMSC were cultured on nonwoven PGA mesh (Fig. 124.2A), constructs first contracted and then collapsed (Fig. 124.8D), while the same cells cultured on a scaffold consisting of a continuous polymer phase (polylactic-co-glycolic acid and polyethylene glycol (PLGA/PEG) sponge)[45] maintained their original dimensions (Fig. 124.8E). When FGF-expanded mammalian BMSC were cultured on PLGA/PEG scaffolds in mixed dishes, chondrogenesis was observed in media supplemented with transforming growth factor beta 1 (TGF β), insulin, and dexamethasone (Fig. 124.8F) while osteogenesis was observed in media supplemented with betaglycerophosphate and dexamethasone (Fig. 124.8G). In the absence of these supplements, constructs consisted mainly of type I collagen and resembled loose connective tissue.

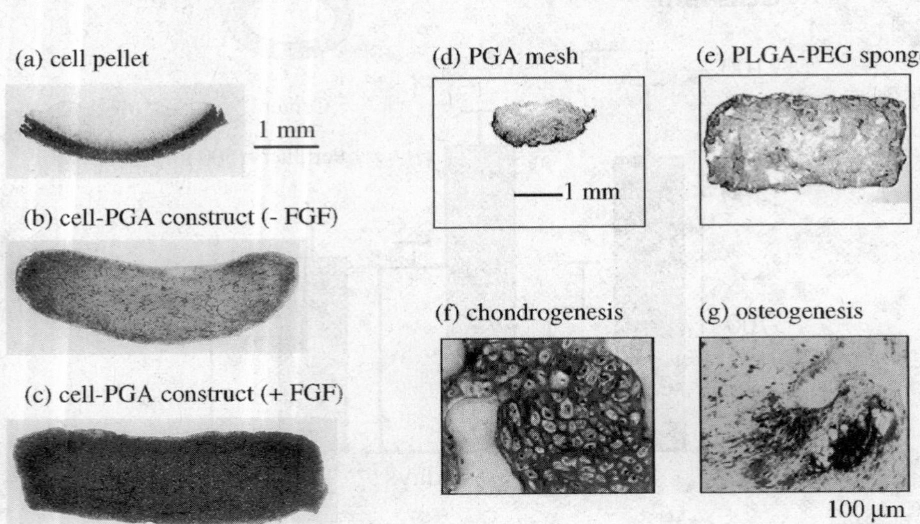

FIGURE 124.8 Chondrogenesis starting from bone marrow stromal cells (BMSC). (a-c) avian (chick embryo) BMSC cultured in media containing serum (10%) and ascorbic acid (50 μg/mL): (a) without polymer scaffolds (as cell pellets); (b) on PGA scaffolds, or (c) after expansion in the presence of FGF (1 ng/mL). (d-e) FGF-expanded mammalian (bovine calf) BMSC cultured on (d) PGA scaffolds (nonwoven meshes) or (e) PLGA/PEG scaffolds (sponges consisting of a continuous polymer phase). (f-g) FGF-expanded mammalian BMSC were cultured on PLGA/PEG sponges in media containing serum (10%), ascorbic acid (50 μg/mL) and either: (f) TGFβ (10 ng/mL), insulin (5 μg/mL), dexamethasone (100 mM), or (g) bGP (7 mM) and dexamethasone (10 mM).

Cultivation Conditions and Time

Tissue culture bioreactors permit the *in vitro* cultivation of larger, better organized engineered cartilage than can be grown in static Petri dishes.[47] Flow and mixing within bioreactors are expected to affect tissue formation in at least two ways: by enhancing *mass transfer* (e.g., of gases and nutrients) and by direct *physical stimulation* of the cells (e.g., by hydrodynamic forces).

The transport of chemical species lies at the heart of physiology and to a large extent determines tissue structure.[75,76] Cells communicate with each other by a combination of diffusion and convective flow, which are in turn driven by hydrodynamic, concentration and osmotic gradients. *In vivo*, mass transfer to chondrocytes involves diffusion and convective transport by the fluid flow that accompanies tissue loading[77]; the presence of blood vessels in immature cartilage can further enhance mass transfer. *In vitro*, mixing-induced convection can enhance mass transport at construct surfaces. In contrast, mass transfer within constructs, which occurs by diffusion only, can become the limiting factor in the cultivation of a large construct with a dense ECM.[78] As compared to constructs grown statically, constructs grown in orbitally mixed Petri dishes and in mixed spinner flasks were larger and contained higher amounts of tissue components.[32,35] Cell metabolism in constructs cultured in mixed and static flasks were respectively found to be aerobic and anaerobic, as assessed by lactate to glucose ratios and ammonia production rates.[54]

The form of a skeletal tissue represents a diagram of underlying forces transmitted across the ECM to the individual cells.[61] Mechanotransduction is thought to involve four steps: mechanocoupling, biochemical coupling, signal transmission, and effector cell response.[79] *In vivo*, load-bearing and immobilized articular cartilage respectively contained high and low GAG fractions.[80,81] *In vitro*, physiological levels of dynamic compression increased the rates of GAG and protein synthesis in cartilage explants,[82,83] while static loading supressed GAG synthesis.[82,84] Physiological levels of dynamic compression also increased the GAG content and improved the mechanical properties of engineered cartilage.[29,30] The motion of

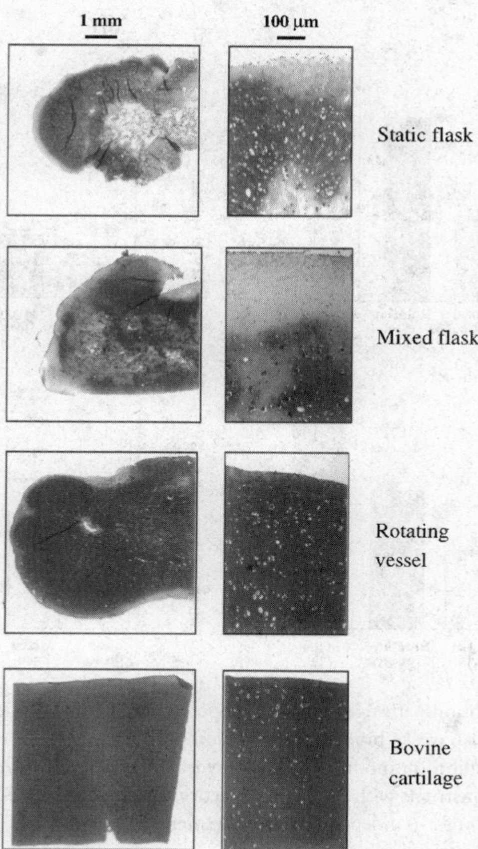

FIGURE 124.9 Culture conditions affect construct morphology. Representative cross sections of 6 week constructs (calf chondrocytes/PGA) cultured in static flasks, mixed flasks, and rotating vessels are compared to fresh calf cartilage (low and high power H&E stained histological sections).

medium in roller bottles stimulated chondrocytes to form cartilaginous nodules,[85,86] and fluid shear enhanced PG size and synthesis rate in chondrocyte monolayers.[87] Turbulent mixing in spinner flasks induced the formation of a fibrous outer capsule at the construct surface, the thickness of which increased with both the mixing intensity and the duration of tissue cultivation.[35] This finding was attributed to direct effects of mechanical forces, i.e., cells exposed to external forces tend to flatten and activate stress-protection mechanisms in order to remain firmly attached to their substrate[88] and increase their stiffnesses by cytoskeletal rearrangements.[89]

The effects of culture conditions on the morphology, composition, and mechanical properties of chondrocyte-PGA constructs were studied over 6 weeks using three different bioreactors: static flasks, mixed flasks, and rotating vessels.[39] In static cultures, GAG accumulated mostly at the periphery, presumably due to diffusional constraints of mass transfer (Fig. 124.9a). In mixed flasks, turbulent shear caused the formation of a thick outer capsule with little or no GAG (Fig. 124.9b). Only in rotating vessels, were GAG fractions high and spatially uniform (Fig. 124.9c). Construct fractions of GAG (Fig. 124.10a) and total collagen (Fig. 124.10b) increased in the following order: static flasks, mixed flasks, rotating vessels, native cartilage.[39] Construct equilibrium moduli and hydraulic permeabilities (Figs. 124.10c,d) varied in a manner consistent with sample composition.[39] As compared to native calf cartilage, 6-week constructs cultured in rotating bioreactors had similar cellularities, 75% as much GAG and 40% as much total collagen per unit wet weight, but only 20% the compressive modulus and 5-fold higher hydraulic

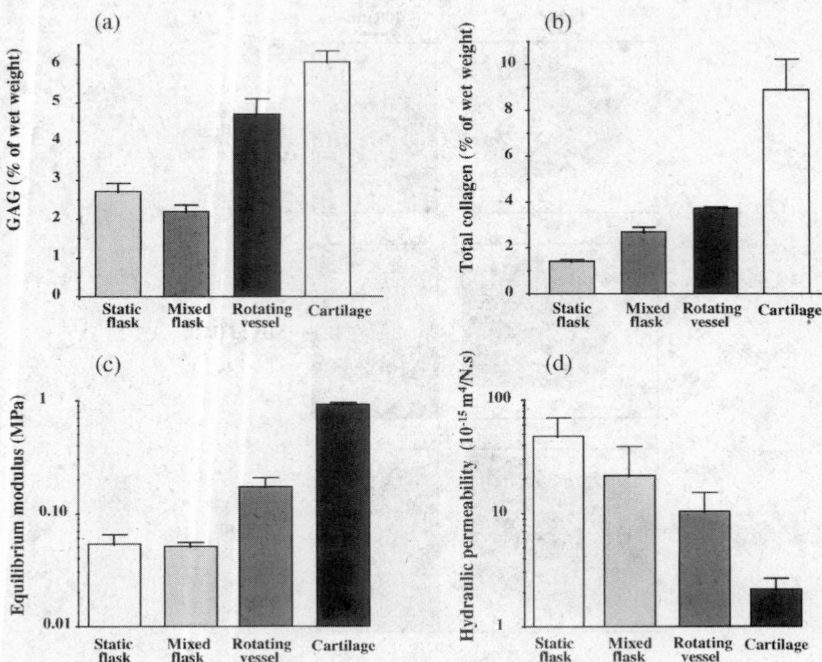

FIGURE 124.10 Culture conditions affect construct compositions and mechanical properties. Constructs (calf chondrocytes/PGA) cultured for 6 weeks bioreactors were compared with respect to: (a,b) wet weight fractions of GAG and total collagen, (c) equilibrium modulus, and (d) hydraulic permeability. As compared to static and mixed flasks, rotating vessels yielded constructs with the best properties, but these remained inferior to native calf cartilage. Data represent the average ± SD of 3-6 independent measurements.

permeability. The apparent lack of functional organization of ECM in constructs may be explained either by the use of immature cartilage (2 to 4 week old calves), or by the absence of specific factors in the *in vitro* culture environment which are normally present *in vivo.*

It is possible that the mechanisms by which the dynamic fluctuations in shear and pressure in rotating vessels enhanced chondrogenesis resembled those associated with dynamic loading *in vivo.* However, the acting hydrodynamic forces were different in nature and several orders of magnitude lower than those resulting from joint loading.[46,90,91] Studies of engineered cartilage subjected to physiological levels of dynamic compression[29,30] and shear should thus allow a more direct comparison of *in vitro* and *in vivo* tissue responses.

Chondrocytes can be phenotypically stable for prolongued periods of time in 3D cultures (e.g., for 8 months in alginate beads).[22] The effect of prolongued cultivation (7 months in rotating bioreactors) on the structure and function of chondrocyte-PGA constructs is shown in Fig. 124.11. As compared to native calf articular cartilage, 7 month constructs had comparable GAG fractions (Fig. 124.11a), 30% as much total collagen (Fig. 124.11b), comparable equilibrium moduli (Fig. 124.11c), and comparable hydraulic permeability (Fig. 124.11d). At 7 months, constructs were phenotypically stable (75% of the total construct collagen was type II) and consisted of metabolically active cells (component cells attached and spread in Petri dishes and were enzymatically active).[37]

A successful approach to cartilage tissue engineering must also provide the potential for constructs to integrate with the adjacent cartilage and subchondral bone. Most of the *in vivo* studies described in Section 2 addressed this issue. In general, the implantation of chondrocytes and BMSC without a cartilaginous ECM resulted in repair tissue that deteriorated with time *in vivo*,[9,11] while implantation of cartilaginous constructs integrated relatively poorly with adjacent host tissues.[10,12] *In vitro* systems can be used to study the effects of specific factors on construct integration in the absence of uncontrollable

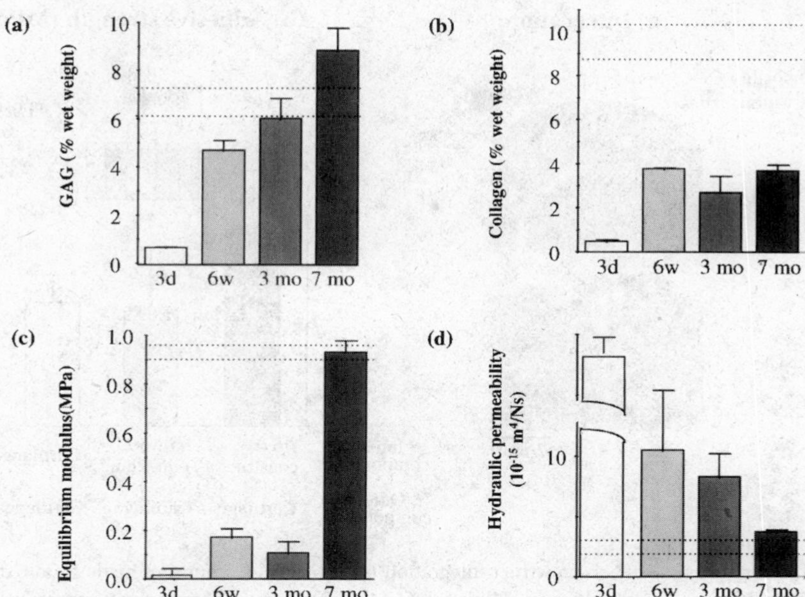

FIGURE 124.11 Culture time affects construct structure and function. Constructs (calf chondrocytes/PGA) cultured for 3 days (3 d), 6 weeks (6 w), 3 or 7 months (3 mo, 7 mo) in rotating vessels were compared with respect to: (a,b) wet weight fractions of GAG and total collagen, (c) equilibrium modulus, and (d) hydraulic permeability. All properties improved with culture time and approached values measured for native calf cartilage (normal ranges denoted by dotted lines). Data represent the average ± SD of 3 independent measurements.

variables intrinsic to *in vivo* studies. In one such study, constructs were cultured for various times, sutured into ring-shaped explants of native cartilage, and cultured for an additional period of time as composites.[92] The integration process involved cell migration into the construct-explant interface and the formation of a new tissue which was initially fibrous but became progressively cartilaginous with increasing culture time (Fig. 124.12a). Construct equilibrium modulus, which was negligible at the beginning of cultivation, increased to approximately 15% of native calf cartilage after 6 weeks (Fig. 124.11c). The adhesive strength at the construct-explant interface was approximately 65% higher for composites made with 6 day constructs, which consisted mainly of cells, as compared to composites made with 5 week constructs, which had a well-formed ECM (Fig. 124.12b).

124.4 Summary and Future Directions

Tissue engineering offers a cell-based approach to articular cartilage repair. In this chapter, we reviewed the state of the art of cartilage tissue engineering and focused on a cell-polymer-bioreactor system which can be used for controlled studies of chondrogenesis and the modulation of engineered cartilage by environmental factors.

A procedure in which autologous chondrocytes are obtained from an articular cartilage biopsy, expanded in culture, and transplanted in conjunction with a periosteal flap[3] is currently the only FDA-approved cell-based treatment for articular cartilage repair. However, *in vivo* studies in dogs had variable results and showed long-term degradation of the repair tissue.[8,9] The clinical study has been viewed with some skepticism[5,6] and long-term, prospective randomized clinical studies are needed to better evaluate the potential of this technique.

Alternatively, autologous BMSC have been isolated from bone marrow aspirates, expanded, and implantated in conjunction with various gels to repair osteochondral defects in experimental animals.[10,14] Following implanatation, the cells underwent a site-specific differentiation and formed a cartilaginous zone

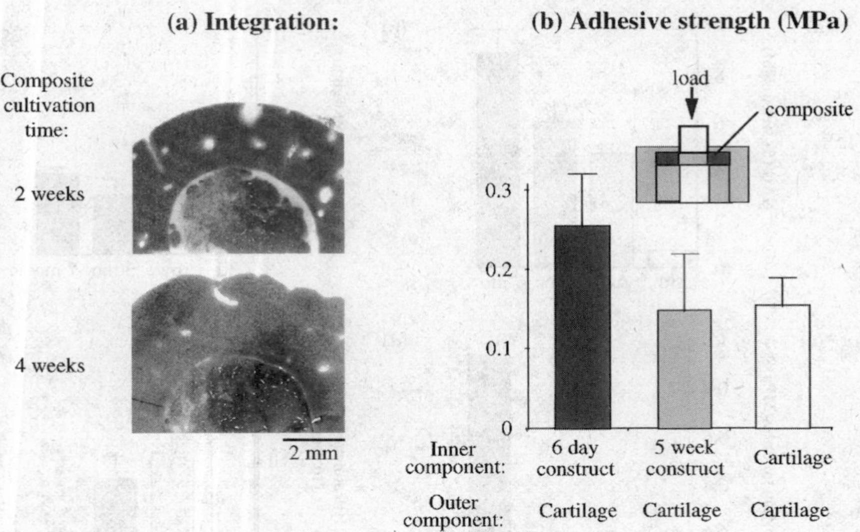

FIGURE 124.12 Culture time affects construct integration potential. (a) Composites made from 6 day constructs (calf chondrocytes/PGA) and cultured for an additional 2 or 4 weeks (upper or lower panel, respectively; safranin-O stained histological sections). (**b**) Adhesive strengths of composites cultured 4 weeks, including: construct-cartilage composites made from 6 day or 5 week constructs and cartilage-cartilage composites (adhesion was estimated by fixing the outer ring in place and uniaxially loading the newly integrated central portion to failure as shown in the inset).

at the surface of the defect and a vascularized bony tissue at its base. In contrast to grafts based on mature chondrocytes, which failed to develop subchondral bone and to fully integrate with the host tissue, grafts based on precursor cells suffered from the progressive thinning of the cartilaginous surface zone.

A 3D scaffold permits the *in vitro* cultivation of cell-polymer constructs that can be readily manipulated, shaped, and fixed to the defect site.[18] As compared to such a pre-formed cartilaginous construct, cells injected under a periosteal flap or immobilized in a hydrated gel are more vulnerable to environmental factors and mechanical forces.[12] The selection of an appropriate scaffold depends in part on the starting cell type. For example, nonwoven fibrous PGA meshes supported chondrogenesis starting from chondrocytes[25] and avian BMSC,[42] while a scaffold with more structural stability appeared to be required for chondrogenesis starting from mammalian BMSC (Fig. 124.8).

Tissue constructs resembling native cartilage were engineered *in vitro* using isolated cells, 3D polymer scaffolds, and bioreactors (Figs. 124.4 through 124.6). Construct structure (histological, biochemical) and function (mechanical, metabolic) depended on cultivation conditions and duration.[39] The cell-polymer-bioreactor system can thus provide a basis for controlled *in vitro* studies of the effects of time and specific biochemical and physical factors on chondrogenesis (Figs. 124.7 through 124.11). Moreover, *in vitro* studies can be used to assess the potential of a cell-polymer construct to integrate with adjacent tissues (Fig. 124.12). Some of the current research needs include:

1. *Development of design criteria.* Specific construct design criteria (e.g., required cell metabolic activity, ECM composition, and mechanical properties) need to be established based on the results of further *in vitro* and *in vivo* experimentation and phenomenological modeling of cell and tissue responses to environmental signals.
2. *Selection of a cell source.* Autologous BMSC and mature chondrocytes are the most likely immediate candidates; allogeneic fetal chondrocytes represent an option for the future. Both BMSC and chondrocytes maintain their chondrogenic potential when expanded in the presence of FGF and then cultured on 3D scaffolds.[42,43] BMSC-based grafts recapitulate embryonic events of endochondral bone formation in response to local environmental factors[10] resulting in repair tissue that

is relatively well integrated but can undergo ossification, leading to progressive thinning at the articular surface.[11] On the other hand, chondrocyte-based grafts do not integrate as well[12] but consist of phenotypically stable cartilage with a mechanically functional ECM[37] that may promote survival in the presence of mechanical loading. Human fetal articular chondrocytes were recently demonstrated to regenerate hyaline cartilage when cultured at high densities in serum-free medium.[31] However, this approach has unresolved ethical issues.

3. *Selection of an appropriate scaffold for human chondrocytes and BMSC.* Ideally, a tissue engineering scaffold should meet all of the following criteria: (1) reproducible processing into complex, 3D shapes, (2) highly porous structure that permits a spatially uniform cell distribution during cell seeding and minimizes diffusional constraints during *in vitro* cultivation, (3) controlled degradation at a rate matching that of cellular deposition of ECM (to provide a stable template during *in vitro* chondrogenesis) followed by complete elimination of the foreign material (to maximize long-term *in vivo* biocompatibility).
4. *Development of methods for in vitro seeding and cultivation of human cells.* Ideally, a tissue culture bioreactor should provide: (1) a means to achieve efficient, spatially uniform cell distribution throughout a scaffold (e.g., dynamic seeding), (2) uniform concentrations of biochemical species in the bulk phase and their efficient transport to the construct surface (e.g., well-mixed culture conditions), (3) steady state conditions (e.g., automated control based on biosensors triggering appropriate changes in medium and gas supply rates), to more closely mimic the *in vivo* cellular environment, and (4) applied physical forces (e.g., hydrostatic pressure and shear, dynamic compression), to more closely mimic the *in vivo* tissue environment.
5. *Development of methods to promote graft survival, integration, and maturation.* Articular cartilage repair refers to healing that restores a damaged articular surface without actually replicating the complete structure and function of the tissue while articular cartilage regeneration refers to the formation of a new tissue that is indistinguishable from normal articular cartilage, including the zonal organization, composition, and mechanical properties.[2] Ideally, engineered cartilage should meet the criteria for regeneration, since any other repair tissue represents a mechanical discontinuity likely to cause long-term device failure.[10] Local pre-treatment of the host tissues at the site of the defect (e.g., with proteolytic enzymes) may enhance graft integration.[10] Further *in vivo* studies are needed to evaluate whether implanted constructs develop characteristic architectural features of articular cartilage in conjunction with physiological loading.

This chapter describes technologies that can potentially lead to articular cartilage regeneration *in vitro* and *in vivo*. Representative studies are summarized in which cells (autologous or allogeneic chondrocytes or BMSC) were isolated, cultivated, and in some cases used to repair large full-thickness cartilage defects. Extension of these results to a human cell source and scale is expected to have a major clinical impact. At this time, tissue engineering studies are mainly observational. The increasing use of models to describe specific aspects of tissue formation (e.g., patterns of ECM deposition, structure-function correlations) is expected to help in the design of hypothesis-driven experiments and interpretation of their results.

Acknowledgments

This work was supported by the National Aeronautics and Space Association (grant NAG9-836). The authors would like to thank R. Langer for general advice and I. Martin for reviewing the manuscript.

Defining Terms

Bioreactors: Tissue culture vessels mixed by magnetic stirring or rotation.

Bone marrow stromal cell (BMSC): A bone marrow-derived precursor cell with the potential to differentiate into various tissues including cartilage.

Chondrocyte: A cartilage cell.

Chondrogenesis: The process of cartilage formation.

Extracellular matrix (ECM): The biochemical components present in the extracellular space of a tissue, e.g., collagen type II and glycosoaminoglycan (GAG) in articular cartilage.

Polymer scaffold: A synthetic material designed for cell cultivation, characterized by its specific chemical composition and 3D structure.

Tissue construct: The tissue engineered *in vitro* using isolated cells, polymer scaffolds, and bioreactors.

References

1. Buckwalter, J.A., Mankin, H.J.: Articular cartilage. Part I: tissue design and chondrocyte-matrix interactions. *J. Bone Joint Surg.* 79-A, 600, 1997a.
2. Buckwalter, J.A., Mankin, H.J. Articular cartilage repair and transplantation. *Arthrit. Rheum.* 41, 1331, 1998.
3. Brittberg, M., Lindahl, A., Nilsson, A., Ohlsson., C., Isaksson, O. Peterson, L. Treatment of deep cartilage defects in the knee with autologous chondrocyte transplantation. *NEJM* 331, 889, 1994.
4. Arnst, C., Carey, J. Biotech Bodies. *Business Week*, July 27, 56, 1998.
5. Messner, K., Gillquist, J. Cartilage repair: a critical review. *Acta Orthop.Scand.* 67, 523, 1996.
6. Newman, A.P. Articular cartilage repair. *Am. J. Sports Med.* 26, 309, 1998.
7. Brittberg, M., Nilsson, A., Lindahl, A., Ohlsson, C., Peterson, L. Rabbit articular cartilage defects treated with autologous cultured chondrocytes. *Clin. Orthop. Rel. Res.* 326, 270, 1996.
8. Shortkroff, S., Barone, L., Hsu, H.P., Wrenn, C., Gagne, T., Chi, T., Breinan, H., Minas, T., Sledge, C.B., Tubo, R., Spector, M. Healing of chondral and osteochondral defects in a canine model: the role of cultured chondrocytes in regeneration of articular cartilage. *Biomat.* 17, 147, 1996.
9. Breinan, H.A., Minas, T., Barone, L., Tubo, R., Hsu, H.P., Shortkroff, S., Nehrer, S., Sledge, C.B., Spector, M. Histological evaluation of the course of healing of canine articular cartilage defects treated with cultured autologous chondrocytes. *Tissue Eng.* 4, 101, 1998.
10. Caplan, A. I., Elyaderani, M., Mochizuki, Y., Wakitani, S., Goldberg, V. M. Principles of cartilage repair and regeneration. *Clin. Orthop. Rel. Res.* 342, 254, 1997.
11. Wakitani, S., Goto, T., Pineda, S.J., Young, R.G., Mansour, J.M., Caplan, A. I., Goldberg, V.M. Mesenchymal cell-based repair of large, full-thickness defects of articular cartilage. *J. Bone Joint Surg.*. 76A, 579, 1994.
12. Kawamura, S, Wakitani, S, Kimura, T, Maeda, A., Caplan, A.I., Shino, K, Ochi, T. Articular cartilage repair-rabbit experiments with a collagen gel-biomatrix and chondrocytes cultured in it. *Acta Orthop. Scand.* 69, 56, 1998.
13. Butnariu-Ephrat, M., Robinson, D., Mendes, D.G., Halperin, N., Nevo, Z. Resurfacing of goat articular cartilage from chondrocytes derived from bone marrow. *Clin. Orthop. Rel. Res.* 330, 234, 1996.
14. Nevo, Z., Robinson, D., Horowitz, S., Hashroni, A., Yayon, A. The manipulated mesenchymal stem cells in regenerated skeletal tissues. *Cell Transpl.* 7, 63, 1998.
15. Chu, C., Coutts, R.D., Yoshioka, M., Harwood, F.L., Monosov, A.Z., Amiel, D. Articular cartilage repair using allogeneic perichondrocyte-seeded biodegradable porous polylactic acid (PLA): A tissue-engineering study. *J. Biomed. Mat. Res.* 29, 1147, 1995.
16. Chu, C., Dounchis, J.S., Yoshioka, M., Sah, R.L., Coutts, R.D., Amiel, D. Osteochondral repair using perichondrial cells: a 1 year study in rabbits. *Clin. Orthop. Rel. Res.* 340, 220, 1997.
17. Noguchi, T., Oka,M., Fujino, M., Neo, M., Yamamuro, T. Repair of osteochondral defects with grafts of cultured chondrocytes: comparison of allografts and isografts. *Clin. Orthop. Rel. Res.* 302, 251, 1994.
18. Freed, L.E., Grande, D.A., Emmanual, J., Marquis, J.C., Lingbin, Z., Langer, R. Joint resurfacing using allograft chondrocytes and synthetic biodegradable polymer scaffolds, *J. Biomed. Mat. Res.* 28, 891, 1994.
19. Grande, D.A., Southerland, S.S., Manji, R., Pate, D.W., Schwartz, R.E., Lucas, P.A. Repair of articular cartilage defects using mesenchymal stem cells. *Tissue Eng.* 1, 345, 1995.

20. Buschmann, M.D., Gluzband, Y.A., Grodzinsky, A.J., Kimura, J.H., Hunziker, E.B. Chondrocytes in agarose culture synthesize a mechanically functional extracellular matrix. *J. Orthop. Res.* 10, 745, 1992.
21. Buschmann, M.D., Gluzband, Y.A., Grodzinsky, A.J., Hunziker, E.B. Mechanical compression modulates matrix biosynthesis in chondrocyte/agarose culture. *J. Cell Sci.* 108, 1497, 1995.
22. Hauselmann, H.J., Fernandes, R.J., Mok, S.S., Schmid, T.M., Block, J.A., Aydelotte, M.B., Kuettner, K.E., Thonar, E.J.-M. Phenotypic stability of bovine articular chondrocytes after long-term culture in alginate beads. *J. Cell Sci.* 107, 17, 1994.
23. Haeuselmann, H.J., Masuda, K., Hunziker, E.B., Neidha, M., Mok, S.S., Michel, B.A., Thonar E.J.-M. Adult human chondrocytes cultured in alginate form a matrix similar to native human articular cartilage. *Am. J. Physiol.* 271, C742, 1996.
24. Freed, L.E., Marquis, J.C., Nohria, A., Mikos, A.G., Emmanual, J., Langer, R. Neocartilage formation *in vitro* and *in vivo* using cells cultured on synthetic biodegradable polymers. *J. Biomed. Mat. Res.* 27, 11, 1993.
25. Freed, L.E., Vunjak-Novakovic, G., Biron, R., Eagles, D., Lesnoy, D., Barlow, S., Langer, R. Biodegradable polymer scaffolds for tissue engineering. *Bio/Technology* 12, 689, 1994.
26. Grande, D.A., Halberstadt, C., Naughton, G., Schwartz, R., Manji, R. Evaluation of matrix scaffolds for tissue engineering of articular cartilage grafts. *J. Biomed. Mat. Res.* 34, 211, 1997.
27. Sittinger, M., Bujia, J., Minuth, W.W., Hammer, C., Burmester, G.R. Engineering of cartilage tissue using bioresorbable polymer carriers in perfusion culture. *Biomaterials*, 15, 451, 1994.
28. Dunkelman, N.S., Zimber, M.P., LeBaron, R.G., Pavelec,R., Kwan, M., Purchio, A.F. Cartilage production by rabbit articular chondrocytes on polyglycolic acid scaffolds in a closed bioreactor system. *Biotech. Bioeng.* 46, 299, 1995.
29. Carver, S.E., Heath, C.A. A semi-continuous perfusion system for delivering intermittent physiological pressure to regenerating cartilage. *Tissue Eng.* (in press) 1998.
30. Carver, S.E., Heath, C.A. Increasing extracellular matrix production in regenerating cartilage with intermittent physiological pressure. *Biotech. Bioeng.* (in press) 1998.
31. Adkisson, H.D., Maloney, W.J., Zhang, J., Hruska, K.A. Scaffold-independent neocartilage formation: a novel approach to cartilage engineering. *Trans. Orth. Res. Soc.* 23, 803, 1998.
32. Freed, L.E., Marquis, J.C., Vunjak-Novakovic, G., Emmanual, J., Langer, R. Composition of cell-polymer cartilage implants, *Biotech. Bioeng.* 43, 605. 1994.
33. Vunjak-Novakovic, G., Obradovic, B., Bursac, P., Martin, I., Langer, R. Freed, L.E. Dynamic seeding of polymer scaffolds for cartilage tissue engineering. *Biotechnol. Progress* 14, 193, 1998.
34. Freed, L.E., Vunjak-Novakovic, G., Marquis, J.C., Langer, R. Kinetics of chondrocyte growth in cell-polymer implants, *Biotech. Bioeng.* 43, 597, 1994.
35. Vunjak-Novakovic, G., Freed, L.E., Biron, R.J., Langer, R. Effects of Mixing on the Composition and Morphology of Tissue Engineered Cartilage, *J. A.I.Ch.E.* 42, 850, 1996.
36. Freed, L.E., Hollander, A.P., Martin, I., Barry, J., Langer, R., Vunjak-Novakovic, G. Chondrogenesis in a cell-polymer-bioreactor system. *Exp. Cell Res.* 240, 58, 1998.
37. Freed, L.E., Langer, R., Martin, I., Pellis, N., Vunjak-Novakovic, G., Tissue engineering of cartilage in space, *PNAS* 94, 13885, 1997.
38. Riesle, J., Hollander, A.P., Langer, R., Freed, L.E., Vunjak-Novakovic, G. Collagen in tissue engineered cartilage: types, structure and crosslinks. *J. Cell. Biochem.* 71, 313, 1998.
39. Vunjak-Novakovic, G., Martin, I., Obradovic, B., Treppo, S, Grodzinsky, A.J., Langer, R., Freed, L. Bioreactor cultivation conditions modulate the composition and mechanical properties of tissue engineered cartilage. *J. Orthop. Res.* (in press) 1998.
40. Sah, R.L., Trippel, S.B., Grodzinsky, A.J. Differential effects of serum, insulin-like growth factor-I, and fibroblast growth factor-2 on the maintenance of cartilage physical properties during long-term culture. *J. Orthop. Res.* 14, 44, 1996.
41. Armstrong, C.G., Mow, V.C. Variations in the intrinsic mechanical properties of human articular cartilage with age, degeneration, and water content. *J. Bone Joint Surg.* 44A, 88, 1982.

42. Martin, I., Padera, R.F., Vunjak-Novakovic, G., Freed, L.E. *In vitro* differentiation of chick embryo bone marrow stromal cells into cartilaginous and bone-like tissues. *J. Orthop. Res.* 16, 181, 1998.
43. Martin, I., Shastri, V.P., Langer, R., Vunjak-Novakovic, G., Freed, L.E. Engineering autologous cartilaginous implants. BMES Annual Fall Meeting, *Ann. Biomed. Eng.* 26, S-139, 1998.
44. Martin, I., Shastri, V.P., Padera, R.F., Langer, R., Vunjak-Novakovic G., Freed, L.E. Bone marrow stromal cell differentiation on porous polymer scaffolds. *Trans. Orth. Res. Soc.* 24, 57, 1999.
45. Shastri, V.P., Martin, I., Langer, R. A versatile approach to produce 3-D polymeric cellular solids. *4th US-Japan Symposium on Drug Delivery Systems*, 1, 36, 1997.
46. Freed, L.E., Vunjak-Novakovic, G. Cultivation of cell-polymer constructs in simulated microgravity. *Biotechnol. Bioeng.* 46, 306, 1995.
47. Freed, L.E., Vunjak-Novakovic, G. Tissue Culture Bioreactors: Chondrogenesis as a Model System, In *Principles of Tissue Engineering*, R.P. Lanza, R. Langer and W.L. Chick, (eds.), Landes & Springer, 1997, chap. 11.
48. Williams, S.N.O., Burstein, D., Gray, M.L., Langer, R., Freed, L.E., Vunjak-Novakovic, G. MRI measurements of fixed charge density as a measure of glycosaminoglycan content and distribution in tissue engineered cartilage. *Trans. Orth. Res. Soc.* 23, 203, 1998.
49. Kim, YJ, Sah, RL, Doong, JYH et al. Fluorometric assay of DNA in cartilage explants using Hoechst 33258. *Anal. Biochem.* 174, 168, 1988.
50. Farndale, RW, Buttler, DJ, Barrett, AJ. Improved quantitation and discrimination of sulphated glycosaminoglycans by the use of dimethylmethylene blue. *Biochim. Biophys. Acta.* 883, 173, 1986.
51. Woessner, J.F. The determination of hydroxyproline in tissue and protein samples containing small proportions of this amino acid. *Arch. Biochem. Biophys.* 93, 440, 1961.
52. Hollander, AP, Heathfield, TF, Webber, C, Iwata, Y, Bourne, R, Rorabeck, C, Poole, RA: Increased damage to type II collagen in osteoarthritic articular cartilage detected by a new immunoassay. *J. Clin. Invest.* 93, 1722, 1994.
53. Bashir, A., Gray, M.L., Burstein, D. Gd-DTPA2 as a measure of cartilage degradation. *Magn. Res. Med.* 36, 665, 1996.
54. Obradovic, B., Freed, L.E., Langer, R., Vunjak-Novakovic, G. Bioreactor studies of natural and engineered cartilage metabolism. *Fall meeting of the AIChE*, Los Angeles, November, 1997.
55. Frank, E.H., Grodzinsky, A.J. Cartilage electromechanics: II. A continuum model of cartilage electrokinetics and correlation with experiments. *J. Biomech.* 20, 629, 1987.
56. Tachetti, C., Tavella, S., Dozin, B., Quarto, R., Robino, G. and Cancedda, R. Cell condensation in chondrogenic differentiation. *Exp. Cell Res.* 200, 26, 1992.
57. Kulyk, W.M., Coelho, C.N.D., Kosher, R.A. Type IX collagen gene expression during limb cartilage differentiation. *Matrix* 11, 282, 1991.
58. Treilleux, I., Mallein-Gerin, F., le Guellec, D., Herbage, D. Localization of the expression of type I, II, III collagen and aggrecan core protein genes in developing human articular cartilage. *Matrix* 12, 221, 1992.
59. Gerstenfeld, L.C., Landis, W.J. Gene expression and extracellular matrix ultrastructure of a mineralizing chondrocyte cell cul;ture system. *J. Cell Biol.* 112, 501, 1991.
60. Yannas, I.V. *In vivo* synthesis of tissues and organs, In *Principles of Tissue Engineering*, R.P. Lanza, R. Langer, and W. Chick (eds.), Academic Press & Landes, Austin, 1997, chap. 12.
61. Thompson, D.W. *On Growth and Form*. Cambridge University Press, New York, 1977.
62. Watt, F. The extracellular matrix and cell shape. *TIBS* 11, 482, 1986.
63. Bruckner, P., Hoerler, I., Mendler, M., Houze, Y., Winterhalter, K.H., Eiach-Bender, S.G., Spycher, M.A. Induction and prevention of chondrocyte hypertrophy in culture *J. Cell Biol.* 109, 2537, 1989.
64. Benya, P.D., Shaffer, J.D. Dedifferentiated chondrocytes reexpress the differentiated collagen phenotype when cultured in agarose gels. *Cell* 30, 215, 1982.
65. Bonaventure, J., Kadhom, N., Cohen-Solal, L., Ng, K.H., Bourguignon, J., Lasselin, C., Freisinger, P. Reexpression of cartilage-specific genes by dedifferentiated human articular chondrocytes cultured in alginate beads. *Exp. Cell Res.* 212, 97, 1994.

66. Vacanti, C., Langer, R., Schloo, B., Vacanti, J.P. Synthetic biodegradable polymers seeded with chondrocytes provide a template for new cartilage formation *in vivo. Plast. Reconstr. Surg.* 88, 753, 1991.
67. Watt, F. Effect of seeding density on stability of the differentiated phenotype of pig articular chondrocytes. *J. Cell Sci.* 89, 373, 1988.
68. Abbot, J., Holtzer, H. The loss of phenotypic traits by differentiated cells *J. Cell Biol.* 28, 473, 1966.
69. Gurdon, J.B. A community effect in animal development. *Nature* 336, 772, 1988.
70. Brockes, J.P. Amphibian limb regeneration: rebuilding a complex structure. *Science* 276, 81, 1997.
71. Zanetti, N.C., Solursh, M. Effect of cell shape on cartilage differentiation, In *Cell Shape: Determinants, Regulation and Regulatory Role.* Academic Press, New York, 1989, chap. 10.
72. Blunk, T., Sieminski, A.L., Nahir, M., Freed, L.E., Vunjak-Novakovic, G., Langer, R. Insulin-like growth factor-I (IGF-I) improves tissue engineering of cartilage *in vitro. Proc. Keystone Symp. on Bone and Collagen: Growth and Differentiation,* 1997. Paper #19
73. Osdoby, P., Caplan, A.I. Scanning electron microscopic investigation of *in vitro* osteogenesis. *Calcif. Tissue Int.* 30, 45, 1980.
74. Johnstone, B., Hering, T.M., Caplan, A.I., Goldberg, V.M., Yoo, J.U. *In vitro* chondrogenesis of bone marrow-derived mesenchymal progenitor cells. *Exp. Cell Res.* 238, 265, 1998.
75. Grodzinsky, A.J., Kamm, R.D., Lauffenburger, D.A. Quantitative aspects of tissue engineering: basic issues in kinetics, transport and mechanics, In *Principles of Tissue Engineering,* R.P.Lanza, R. Langer, and W. Chick (eds.), Academic Press & Landes, Austin, 1997, chap. 14.
76. Lightfoot, E.N. The roles of mass transfer in tissue function, In *The Biomedical Engineering Handbook.,* J.D. Bronzino (ed.), CRC Press, Boca Raton, 1995, chap. 111.
77. O'Hara, B.P., Urban, J.P.G., Maroudas, A. Influence of cyclic loading on the nutrition of articular cartilage. *Ann. Rheum. Dis.* 49, 536, 1990.
78. Bursac, P.M., Freed, L.E., Biron, R.J., Vunjak-Novakovic, G. Mass transfer studies of tissue engineered cartilage. *Tissue Eng.* 2, 141, 1996.
79. Dunkan, R.L., Turner, C.H. Mechanotransduction and the functional response of bone to mechanical strain. *Calc. Tissue Int.* 57, 344, 1995.
80. Slowman, S.D. and Brandt, K.D. Composition and glycosaminoglycan metabolism of articular cartilage from habitually loaded and habitually unloaded sites. *Arthrit. Rheum.* 29, 88, 1986.
81. Kiviranta, I., Jurvelin, J., Tammi, M., Saamanen, A.-M., Helminen, H.J. Weight-bearing controls glycosaminoglycan concentration and articular cartilage thickness in the knee joints of young beagle dogs. *Arthritis Rheum.* 3, 801, 1987.
82. Sah, R.L.-Y., Kim, Y.J., Doong, J.Y.H., Grodzinsky, A.J., Plaas, A.H., Sandy, J.D. Biosynthetic response of cartilage explants to dynamic compression. *J. Orthoped. Res.* 7, 619, 1989.
83. Parkinnen, J.J., Ikonen, J., Lammi, M.J., Laakkonen, J., Tammi, M., Helminen, H.J. Effect of cyclic hydrostatic pressure on proteoglycan synthesis in cultured chondrocytes and articular cartilage explants. *Arch. Biochem. Biophys.* 300, 458, 1993.
84. Schneidermann, R., Keret, D., Maroudas, A. Effects of mechanical and osmotic pressure on the rate of glycosaminoglycan synthesis in the human adult femoral head cartilage: an *in vitro* study. *J. Orthop. Res.* 4, 393, 1986.
85. Kuettner, K.E., Pauli, B.U., Gall, G., Memoli, V.A., Schenk, R. Synthesis of cartilage matrix by mammalian chondrocytes *in vitro.* Isolation, culture characteristics, and morphology. *J. Cell Biol.* 93, 743, 1982.
86. Kuettner, K.E., Memoli, V.A., Pauli, B.U., Wrobel, N.C., Thonar, E.J.-M.A., Daniel, J.C. Synthesis of cartilage matrix by mammalian chondrocytes *in vitro.* maintenance of collagen and proteoglycan. *J. Cell Biol.* 93, 751, 1982.
87. Smith, R.L., Donlon, B.S., Gupta, M.K., Mohtai, M., Das, P., Carter, D.R., Cooke, J., Gibbons, G., Hutchinson, N., Schurman, D.J. Effect of fluid-induced shear on articular chondrocyte morphology and metabolism *in vitro. J. Orthop. Res.* 13, 824, 1995.
88. Franke, R.P., Grafe, M., Schnittler, H., Seiffge, D., Mittermayer, C. Induction of human vascular endothelial stress fibers by fluid shear stress. *Nature* 307, 648, 1984.

89. Wang, N., Ingber, D.E. Control of cytoskeletal mechanics by extracellular matrix, cell shape and mechanical tension. *Biophys. J.* 66, 2181, 1994.
90. Berthiaume, F., Frangos, J. Effects of flow on anchorage-dependent mammalian cells-secreted products, In *Physical Forces and the Mammalian Cell*, J. Frangos (ed.), Academic Press, San Diego, 1993, chap. 5.
91. Buckwalter, J.A., Mankin, H.J. Articular cartilage, part II: degeneration and osteoarthrosis, repair, regeneration, and transplantation. *J. Bone Joint Surg.* 79A, 612, 1997.
92. Obradovic, B., Martin, I., Padera, R.F., Treppo, S., Freed, L. E., Vunjak-Novakovic, G. Integration of engineered cartilage into natural cartilage, *Annual Meeting of the AIChE*, Miami, Nov. 1998.

Further Information

1. Buckwalter, J.A., Ehrlich, M.G., Sandell, L.J., Trippel, S.B. (eds.): *Skeletal Growth and Development: Clinical Issues and Basic Science Advances*, American Academy of Orthopaedic Surgeons, 1998.
2. Buckwalter, J.A., Mankin, H.J. Articular cartilage repair and transplantation. *Arthrit. Rheum.* 41, 1331, 1998.
3. Caplan, A. I., Elyaderani, M., Mochizuki, Y., Wakitani, S., Goldberg, V. M. Principles of cartilage repair and regeneration. *Clin. Orthop. Rel. Res.* 342, 254, 1997.
4. Comper, W.D. (ed): *Extracellular Matrix: Molecular Components and Interactions*, Hardwood Academic Publishers, the Netherlands, 1996.
5. Newman, A.P. Articular cartilage repair. *Am. J. Sports Med.* 26, 309, 1998.

125

Tissue Engineering of the Kidney

H. David Humes
University of Michigan

Tissue engineering is one of he most intriguing and exciting areas in biotechnology due to its requirements for state-of-the-art techniques from both biologic and engineering disciplines. This field is on the threshold of the development of an array of products and devices comprised of cell components, biologic compounds, and synthetic materials to replace physiologic function of diseased tissues and organs.

Successful tissue and organ constructs depend on a thorough understanding and creative application of molecular, cellular, and organ biology and the principles of chemical, mechanical, and material engineering to produce appropriate structure-function relationships to restore, maintain, or improve tissue or organ function. This approach depends on the most advanced scientific methodologies, including stem cell culture, gene transfer, growth factors, and biomaterial technologies.

The kidney was the first solid organ whose function was approximated by a machine and a synthetic device. In fact, renal substitution therapy with hemodialysis or chronic ambulatory peritoneal dialysis (CAPD) has been the only successful long-term ex vivo organ substitution therapy to date [1]. The kidney was also the first organ to be successfully transplanted from a donor individual to an autologous recipient patient. However, the lack of widespread availability of suitable transplantable organs has kept kidney transplantation from becoming a practical solution to most cases of chronic renal failure.

Although long-term chronic renal replacement therapy with either hemodialysis or CAPD has dramatically changed the prognosis of renal failure, it is not complete replacement therapy, since it only provides filtration function (usually on an intermittent basis) and does not replace the homeostatic, regulatory, metabolic, and endocrine functions of the kidney. Because of the nonphysiologic manner in which dialysis performs or does not perform the most critical renal functions, patients with ESRD on dialysis continue to have major medical, social, and economic problems [2]. Accordingly, dialysis should be considered as renal substitution rather than renal replacement therapy.

Tissue engineering of a biohybrid kidney comprised of both biologic and synthetic components will most likely have substantial benefits for the patient by increasing life expectancy, increasing mobility and flexibility, increasing quality of life with large savings in time, less risk of infection, and reduced costs. This approach could also be considered a cure rather than a treatment for patients.

0-8493-0462-8/00/$0.00+$.50

125.1 Fundamentals of Kidney Function

The kidneys are solid organs located behind the peritoneum in the posterior abdomen and are critical to body homeostasis because of their excretory, regulatory, metabolic, and endocrinologic functions. The excretory function is initiated by filtration of blood at the glomerulus, which is an enlargement of the proximal end of the tubule incorporating a vascular tuft. The structure of the glomerulus is designed to provide efficient ultrafiltration of blood to remove toxic waste from the circulation yet retain important circulating components, such as albumin. The regulatory function of the kidney, especially with regard to fluid and electrolyte homeostasis, is provided by the tubular segments attached to the glomerulus.

The ultrafiltrate emanating from the glomerulus courses along the kidney tubule, which resorbs fluid and solutes to finely regulate their excretion in various amounts in the final urine. The kidney tubules are segmented, with each segment possessing differing transport characteristics for processing the glomerular ultrafiltrate efficiently and effectively to regulate urine formation. The segments of the tubule begin with the proximal convoluted and straight tubules, where most salt and water are resorbed. This segment leads into the thin and thick segments of Henle's loop, which are critical to the countercurrent system for urinary concentration and dilution of water. The distal tubule is next and is important for potassium excretion. The final segment is the collecting duct, which provides the final regulation of sodium, hydrogen, and water excretion. The functional unit of the kidney is therefore composed of the filtering unit (the glomerulus) and the regulatory unit (the tubule). Together they form the basic component of the kidney, the nephron. In addition to these excretory and regulatory functions, the kidney is an important endocrine organ. Erythropoietin, active forms of vitamin D, renin, angiotensin, prostaglandins, leukotrienes, and kallikrein-kinins are some of the endocrinologic compounds produced by the kidney.

In order to achieve its homeostatic function for salt and water balance in the individual, the kidney has a complex architectural pattern (Fig. 125.1). This complex organization is most evident in the exquisitely regulated structure-function interrelationships between the renal tubules and vascular structures to coordinate the countercurrent multiplication and exchange processes to control water balance with urinary concentration and dilution [3]. This complex structure is the result of a long evolutionary process in which animals adapted to many changing environmental conditions. Although well suited for maintaining volume homeostasis, the mammalian kidney is an inefficient organ for solute and water excretion. In humans, approximately 180 liters of fluid are filtered into the tubules daily, but approximately 178 liters must be reabsorbed into the systemic circulation. Because of the importance of maintaining volume homeostasis in an individual to prevent volume depletion, shock, and death, multiple redundant physiologic systems exist within the body to maintain volume homeostasis. No such redundancy, however, exists to replace renal excretory function of soluble metabolic toxic by products of metabolic activity. Accordingly, chronic renal disease becomes a clinical disorder due to loss of renal excretory function and buildup within the body of metabolic toxins which require elimination by the kidneys. Because of the efficiency inherent in the kidneys as an excretory organ, life can be sustained with only 5–10% of normal renal excretory function. With the recognition that the complexity of renal architectural organization is driven by homeostatic rather than excretory function and that renal excretory function is the key physiologic process which must be maintained or replaced to treat clinical renal failure, the approach to a tissue engineering construct becomes easier to entertain, especially since only a fraction of normal renal excretory function is required to maintain life.

For elimination of solutes and water from the body, the kidney utilizes simple fundamental physical principles which govern fluid movement (Fig. 125.2). The kidney's goal in excretory function is to transfer solutes and water from the systemic circulation into tubule conduits in order to eliminate toxic byproducts from the body in a volume of only several liters. Solute and fluid removal from the systemic circulation is the major task of the renal filtering apparatus, the glomerulus. The force responsible for this filtration process is the hydraulic pressure generated within the system circulation due to myocardial contraction and blood vessel contractile tone. Most of the filtered fluid and solutes must be selectively reabsorbed by the renal tubule as the initial filtrate courses along the renal tubules. The reabsorptive process depends upon osmotic forces generated by active solute transport by the renal epithelial cell and the colloid oncotic pressure within the peritubular capillary (Fig. 125.2). The approach of a tissue-engineering construct for renal replacement is to mimic these natural physical forces to duplicate filtration and reabsorption processes to attain adequate excretory function lost in renal disorders.

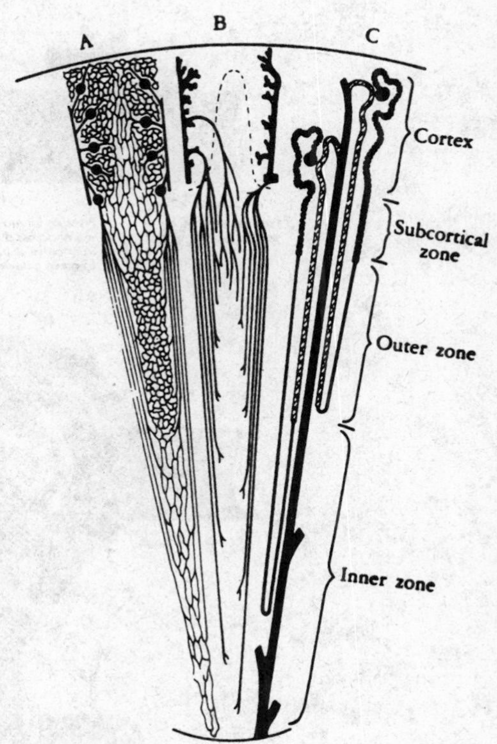

FIGURE 125.1 Representation of the complex morphologic architecture of a section of a mammalian kidney in which these components are schematized separately: (*a*) arterial and capillary blood vessels; (*b*) venous drainage; (*c*) two nephrons with their glomeruli and tubule segments (*Source*: From [3] with permission.)

Glomerular Ultrafiltration

The process of urine formation begins within the capillary bed of the glomerulus [4]. The glomerular capillary wall has evolved into a structure with the property to separate as much as one-third of the plasma entering the glomerulus into a solution of a nearly ideal ultrafitrate. The high rate of ultrafiltration across the glomerular capillary is a result of hydraulic pressure generated by the pumping action of the heart and the vascular tone of the preglomerular and postglomerular vessels as well as the high hydraulic permeability of the glomerular capillary walls. This hydraulic pressure as well as the hydraulic permeability of the glomerular capillary bed is at least two times and two orders of magnitude higher, respectively, than most other capillary networks within the body [5]. Despite this high rate of water and solute flux across the glomerular capillary wall, this same structure retards the filtration of important circulating macromolecules, especially albumin, so that all but the lower-molecular-weight plasma proteins are restricted in their passage across the filtration barrier.

A variety of experimental studies and model systems have been employed to characterize the sieving properties of the glomerulus. Hydrodynamic models of solute transport through pores have been successfully used to describe the size selective barrier function of this capillary network to macromolecules [6]. This pore model, in its simplest form, assumes the capillary wall to contain cylindrical pores of identical size and that macromolecules are spherical particles. Based upon the steric hindrances that macromolecules encounter in the passage through small pores, whether by diffusion or by convection (bulk flow of fluid), definition of the glomerular capillary barrier can be a fluid-filled cylindrical pore (Fig. 125.3). This modeling characterizes the glomerular capillary barrier as a membrane with uniform pores of 50Å radius [7]. This pore size predicts that molecules with radii smaller than 14 Å appear in the filtrate in the same concentration as in plasma water. Since there is no restriction to filtration, fractional clearance of this size molecule is equal to one. The filtration of molecules of increasing size decreases progressively, so that the fractional clearance of macromolecules the size of serum albumin (36 Å) is low.

The glomerular barrier, however, does not restrict molecular transfer across the capillary wall only on the basis of size (Fig. 125.4). This realization is based upon the observation that the filtration of the circulating protein, albumin, is restricted to a much greater extent than would be predicted from size alone. The realization that albumin is a polyanion at physiologic pH suggests that molecular charge, in addition to molecular size, is another important determination of filtration of macromolecules [8]. The

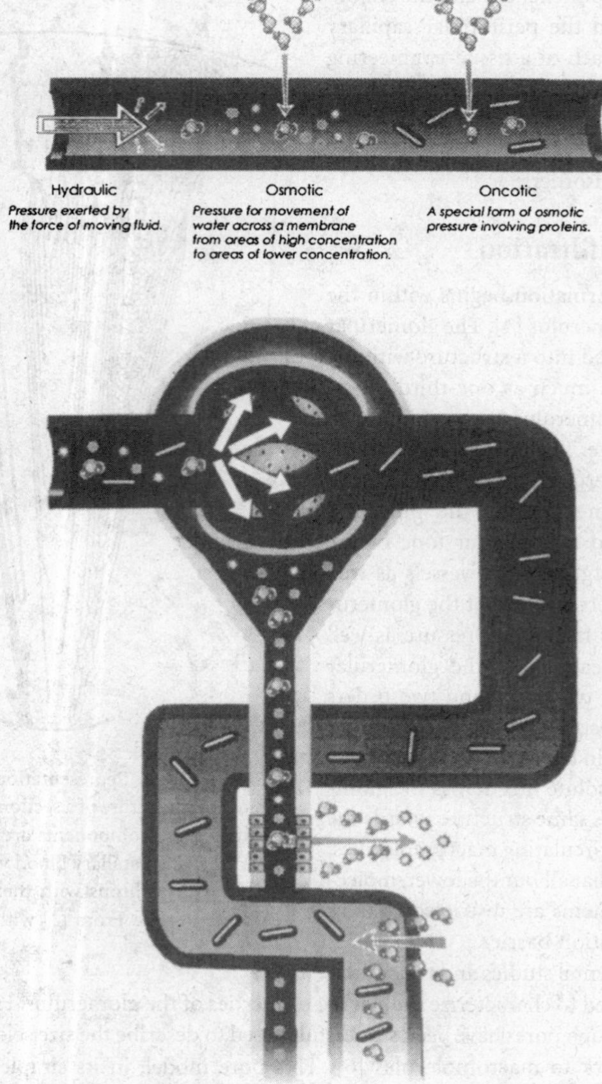

FIGURE 125.2 Physical forces which govern fluid transfer within the kidney.

greater restriction to the filtration of circulating polyanions, including albumin, is due to the electrostatic hindrance by fixed negatively charged components of the glomerular capillary barrier. These fixed negative charges, as might be expected, simultaneously enhance the filtration of circulating polycations.

Thus, the formation of glomerular ultrafiltrate, the initial step in urine formation, depends upon the pressure and flows within the glomerular capillary bed and the intrinsic permselectivity of the glomerular capillary wall. The permselective barrier excludes circulating macromolecules from filtration based upon size as well as net molecular charge, so that for any given size, negatively charged macromolecules are restricted from filtration to a greater extent than neutral molecules.

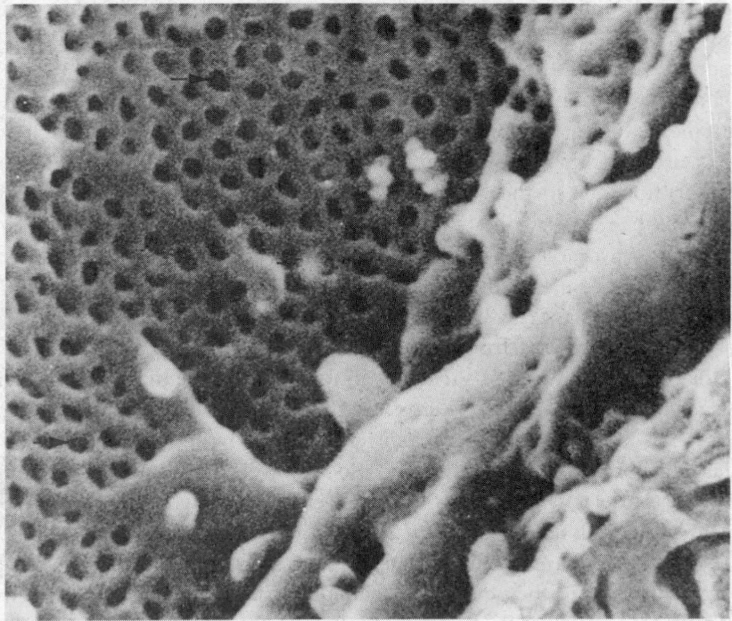

FIGURE 125.3 Scanning electron micrograph of the glomerular capillary wall demonstrating the fenestrae (pores) of the endothelium within the glomerulus (mag × 50,400). Reprinted with permission from Schrier RW, Gottschalk CW. 1988. Diseases of the Kidney, p 12, Boston, Little, Brown.

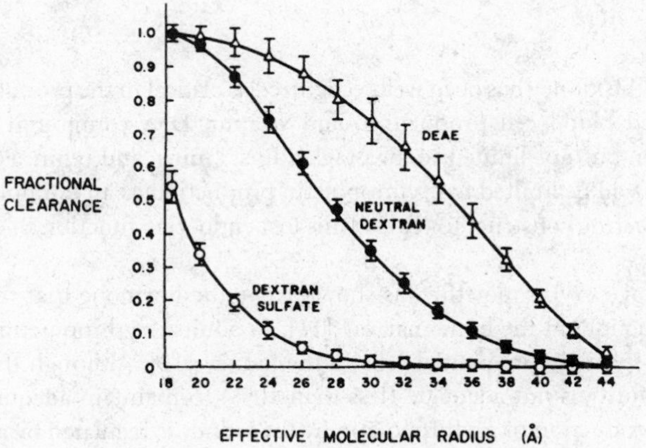

FIGURE 125.4 Fractional clearances of negatively charged (sulfate) dextrans, neutral dextrans, and positively charged (DEAE) dextrans of varying molecular size. These data demonstrate that the glomerular capillary wall behaves as both a size-selective and charge-selective barrier [8].

Tubule Reabsorption

Normal human kidneys form approximately 100 ml of filtrate every minute. Since daily urinary volume is roughly 2 L more than 98% of the glomerular ultrafiltrate must be reabsorbed by the renal tubule. The bulk of the reabsorption, 50–65%, occurs along the proximal tubule. Similar to glomerular filtration, fluid movement across the renal proximal tubule cell is governed by physical forces. Unlike the fluid transfer across the glomerular capillary wall, however, tubular fluid flux is principally driven by osmotic and oncotic pressures rather than hydraulic pressure (Fig. 125.2). Renal proximal tubule fluid reabsorption is based upon active Na^+ transport, requiring the energy-dependent Na^+K^+ ATPase located along the basolateral membrane of the renal tubule cell to promote a small degree of luminal hypotonicity [9]. This small degree of osmotic difference (2–3 mOsm/kgH_2O) across the renal tubule is sufficient to drive isotonic fluid reabsorption due to the very high diffusive water permeability of the renal tubule cell membrane. Once across the renal proximal tubule cell, the transported fluid is taken up by the peritubular capillary bed due to the favorable oncotic pressure gradient. This high oncotic pressure within the peritubular capillary is the result of the high rate of protein-free filtrate formed in the proximate glomerular capillary bed [10]. As can be appreciated, the kidney has evolved two separate capillary networks to control bulk fluid flow from various fluid compartments of the body. The glomerular capillary network has evolved an efficient structure to function as a highly efficient filter to allow water and small solutes such as urea and sodium, to cross the glomerular capillary wall while retaining necessary macromolecules, such as albumin. This fluid transfer is driven by high hydraulic pressures within the glomerular capillary network generated by the high blood flow rates to the kidney and a finely regulated vascular system. The permselectivity of the filter is governed by an effective pore size to discriminate macromolecular sieving based upon both size and net molecular charge. The high postglomerular vascular resistance and the protein-free glomerular filtrate results, respectively, in low hydraulic pressure and high oncotic pressure within the peritubular capillary system which follows directly in series from the glomerular capillary network. The balance of physical forces within the peritubular capillaries, therefore, greatly favors the uptake of fluid back into the systemic circulation. The addition of a renal epithelial monolayer with high rates of active Na^+ transport and high hydraulic permeability assists further in the high rate of salt and water reabsorption along the proximal tubule. Thus, an elegant system has evolved in the nephron to filter and reabsorb large amounts of fluid in bulk to attain high rates of metabolic product excretion while maintaining regulatory salt and water balance.

Endocrine

As an endocrine organ, the kidney has been well recognized as critical in the production of erythropoietin, a growth factor for red blood cell production, and vitamin D, a compound important in calcium metabolism, along with, but not limited to, prostaglandins, kinins, and renin. For the purposes of this chapter, this discussion will be limited to erythropoietin production as an example of a potential formulation of a tissue-engineering construct to replace this lost endocrine function in chronic end-stage renal disease.

More than 40 years ago erythropoietin was shown to be the hormone that regulates erythropoiesis, or red blood cell production, in the bone marrow [11]. In adults, erythropoietin is produced primarily (greater than 90%) by specialize interstitial cells in the kidney [12]. Although the liver also synthesizes erythropoietin, the quantity is not adequate (less than 10%) to maintain adequate red cell production in the body [13]. The production of erythropoietin by the kidney is regulated by a classic endocrinologic feedback loop system. As blood flows through the kidney, the erythropoietin-producing cells are in an ideal location to sense oxygen delivery to tissues by red cells in the bloodstream, since they are located adjacent to peritubular capillaries in the renal interstitium.

As demonstrated in Fig. 125.5, erythropoeitin production is inversely related to oxygen delivery to the renal interstitial cells. With hypoxemia or decline in red blood cell mass, a decline in oxygen delivery occurs to these specialized cells, and increased erythropoietin production develops. Upon return to normal oxygen delivery with normoxia and red blood cell mass, the enhanced production of erythropoeitin is suppressed, closing the classic feedback loop. Of importance, the regulation of erythropoietin in the kidney cells depends upon transcriptional control, not upon secretory control, as seen with insulin [14]. The precise molecular mechanism of the oxygen sensor for tissue oxygen availability has not been delineated but appears to depend on a heme protein.

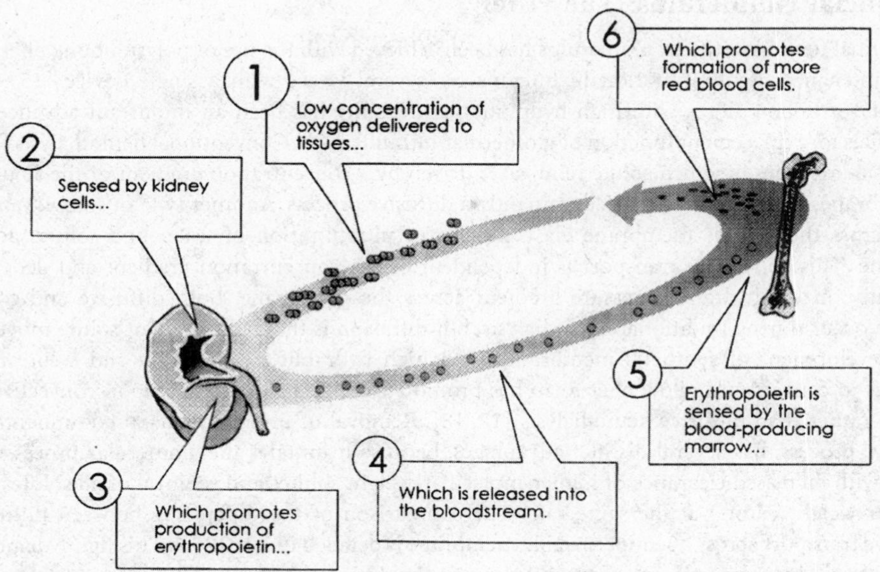

FIGURE 125.5 Endocrinologic back loop which regulates erythropoietin production by the kidney.

Once erythropoietin is released, it circulates in the bloodstream to the bone marrow, where it signals the marrow to produce red blood cells. In this regard, all blood cells, both white and red cells, originate from a subset of bone marrow cells, called *multipotent stem cells.* These stem cells develop during embryonic development and are maintained through adulthood via self-regulation. Under appropriate stimulation, stem cells proliferate and produce committed, more highly differentiated progenitor cells which are then destined to a specific differentiated line of blood cells, including neutrophils, lymphocytes, platelets, and red cells. Specifically, erythropoietin binds to receptors on the outer membrane of committed erythroid progenitor cells, stimulating the terminal steps in erythroid differentiation. Nearly 200 years ago, it was first recognized that anemia is a complication of chronic renal failure. Ordinarily an exponential increase in serum erythropoietin levels are observed when hemoglobin level in patients declines below 10 gm/dl. In the clinical state of renal disease, however, the normal increase in erythropoietin in response to anemia is impaired [14]. Although the oxygen delivery (or hemoglobin) to erythropoietin feedback loop is still intact, the response is dramatically diminished. In patients with end-stage renal disease (ESRD), the hematocrit levels are directly correlated to circulating erythropoietin levels. In fact, ERSD patients with bilateral nephrectomies have the lowest hematocrits and lowest rates of erythropoiesis, demonstrating that even the small amount of erythropoietin produced by the end-stage kidney is important. Thus, the loss of renal function due to chronic disease results in an endocrine deficiency of a hormone normally produced by the kidney and results in a clinical problem that complicates the loss of renal excretory function.

125.2 Tissue-Engineering Formulation Based upon Fundamentals

In designing an implantable bioartificial kidney for renal replacement function, essential functions of kidney tissue must be utilized to direct the design of the tissue-engineering project. The critical elements of renal function must be replaced, including the excretory, regulatory (reabsorptive), and endocrinologic functions. The functioning excretory unit of the kidney, as detailed previously, is composed of the filtering unit, the glomerulus, and the regulatory or reabsorptive unit, the tubule. Therefore, a bioartificial kidney requires two main units, the glomerulus and the tubule, to replace renal excretory function.

Bioartificial Golmerulus: The Filter

The potential for a bioartificial glomerulus has been achieved with the use of polysulphone fibers ex vivo with maintenance of ultrafiltration in humans for several weeks with a single device [15, 16]. The availability of hollow fibres with high hydraulic permeability has been an important advancement in biomaterials for replacement function of glomerular ultrafiltration. Conventional hemodialysis for ESRD has used membranes in which solute removal is driven by a concentration gradient of the solute across the membranes and is, therefore, predominantly a diffusive process. Another type of solute transfer also occurs across the dialysis membrane via a process of ultrafiltration of water and solutes across the membrane. This convective transport is independent of the concentration gradient and depends predominantly on the hydraulic pressure gradient across the membrane. Both diffusive and convective processes occur during traditional hemodialysis, but diffusion is the main route of solute movement.

The development of synthetic membranes with high hydraulic permeability and solute retention properties in convenient hollow fiber form has promoted ERSD therapy based upon convective homofiltration rather than diffusive hemodialysis [17, 18]. Removal of uremic toxins, predominantly by the convective process, has several distinct advantages, because it imitates the glomerular process of toxin removal with increased clearance of higher-molecular-weight solutes and removal of all solutes (up to a molecular weight cutoff) at the same rate. The comparison of the differences between diffusive and convective transport across a semipermeable membrane is detailed in Fig. 125.6. This figure demonstrates the relationship between molecular size and clearance by diffusion and convection. As seen, the clearance of a molecule by diffusion is negatively correlated with the size of the molecule. In contrast, clearance of a substance by convection is independent of size up to a certain molecular weight. The bulk movement of water carries passable solutes along with it in approximately the same concentration as in the fluid.

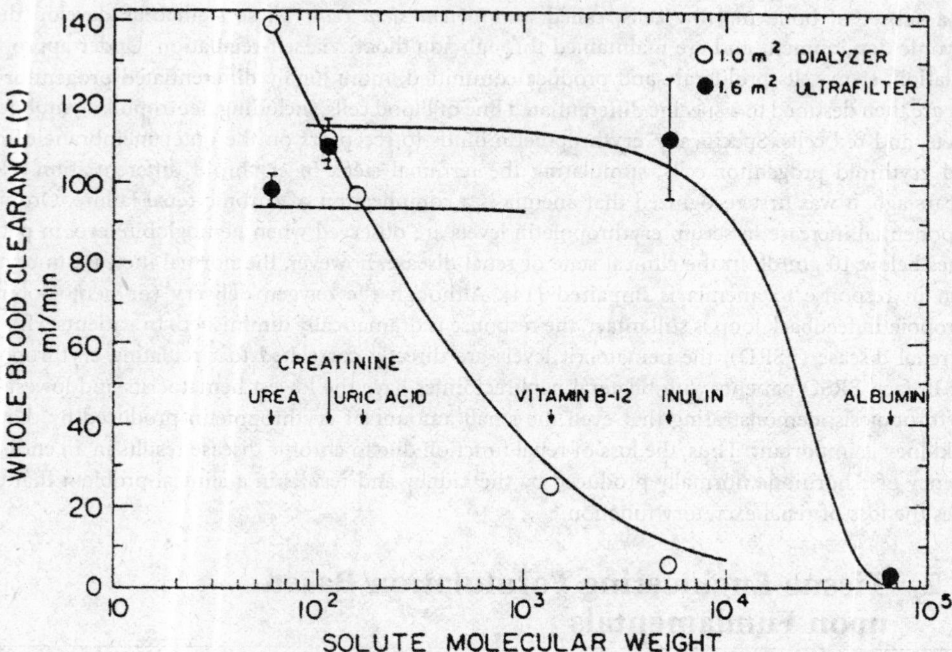

FIGURE 125.6 Relationship between solute molecular size and clearance by diffusion (open circles) and convection (closed circles). Left curve shows solute clearance for a 1.0 m² dialyzer with no ultrafiltration where solute clearance is diffusion. Right curve shows clearance for a 1.6 m² ultrafilter where solute clearance is by convection. Smaller molecules are better cleared by diffusion; larger molecules by convection. Normal kidneys clear solutes in a pattern similar to convective transport. (Figure adapted from [18].)

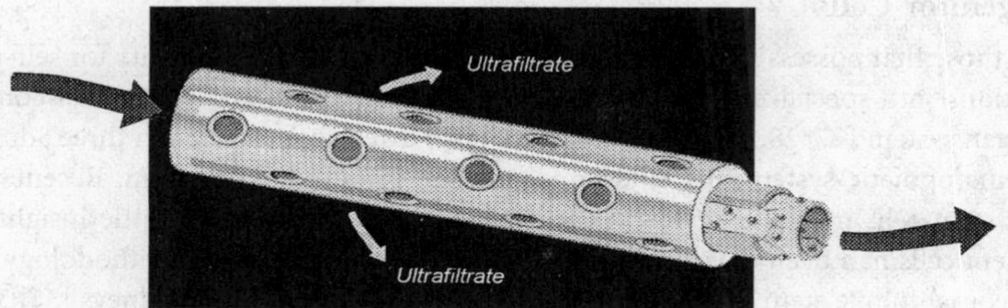

FIGURE 125.7 Conceptual schematization of bioartificial glomerulus.

Development of an implantable device which mimics glomerular filtration will thus depend upon convective transport. This physiologic function has been achieved clinically with the use of polymeric hollow fibers ex vivo. Major limitations to the currently available technology for long-term replacement of filtration function include bleeding associated with required anticoagulation, diminution of filtration rate due to protein deposition in the membrane over time or thrombotic occlusion, and large amounts of fluid replacement required to replace the ultrafiltrate formed from the filtering unit. The use of endothelial-cell-seeded conduits along filtration surfaces may provide improved long-term hemocompatibility and hemofiltaration in vivo [19, 20. 21], as schematized in Fig. 125.7.

In this regard, endothelial cell seeding of small-caliber vascular prosthesis has been shown experimentally to reduce long-term platelet deposition, thrombus formation, and loss of graft patency [21]. Recent results in humans have demonstrated success in autologous endothelial cell seeding in small caliber grafts after growth of these cells along the graft lumen ex vivo to achieve a confluent monolayer prior to implantation. Long-term persistent endothelialization and patency of the implanted graft has been reported [19]. A potential rate-limiting step in endothelial-cell-lined hollow fibers of small caliber is thrombotic occlusion, which limits the functional patency of this filtration unit. In this regard, gene transfer into seeded endothelial cells for constitutive expression of anticoagulant factors can be envisioned to minimize clot formation in these small-caliber hollow fibers. Since gene transfer for in vivo protein production has been clearly achieved with endothelial cells [22, 23], gene transfer into endothelial cells for the production of an anticoagulant protein is clearly conceivable.

For differentiated endothelial cell morphology and function, an important role for various components of the extracellular matrix (ECM) has been demonstrated [24, 25]. The ECM has been clearly shown to dictate phenotype and gene expression of endothelial cells, thereby modulating morphogenesis and growth. Various components of ECM, including collagen by type I, collagen type IV, laminin, and fibronectin, have been shown to affect endothelial cell adherence, growth, and differentiation. Of importance, ECM produced by MDCK cells, a permanent renal epithelial cell line, has the ability to induce capillary endothelial cells to produce fenestrations [25, 26]. Endothelial cell fenestrations are large openings which act as channels or pores for convective transport though the endothelial monolayer and are important in the high hydraulic permeability and sieving characteristics of glomerular capillaries. Thus, the ECM component on which the endothelial cells attach and grow may be critical in the functional characteristics of the lining monolayer.

Bioartificial Tubule: The Reabsorber

As detailed above, the efficiency of reabsorption, even though dependent upon natural physical forces governing fluid movement across biologic as well as synthetic membranes, requires specialized epithelial cells to perform vectorial solute transport. Critical to the advancement of this tubule construct, as well as for the tissue-engineering field in general, is the need for the isolation and growth in vitro of specific cells, referred to as *stem progenitor cells*, from adult tissues.

Stem or Progenitor Cells

These cells are those that possess stem-cell-like characteristics with a high capacity for self-renewal under defined conditions into specialized cells to develop correct structure and functional components of a physiologic organ system [27, 28, 29]. Stem cells have been extensively studied in three adult mammalian tissues: the hematopoietic system, the epidermis, and the intestinal epithelium. Recent work has also suggested that stem cells may also reside in the adult nervous system [30]. Little insight into possible renal tubule stem cells had been developed until recent data demonstrating methodology to isolate and grow renal proximal tubule stem or progenitor cells from adult mammalian kidneys [31, 32]. This series of studies was promoted by the clinical and experimental observations suggesting that renal proximal tubule progenitor cells must exist, because they have the ability to regenerate after severe neophrotoxic or ischemic injury to form a fully functional and differentiated epithelium [33, 34]. Whether proximal tubule progenitor cells are pluripotent, possessing the ability to differentiate into cells of other segments (such as loop of Henle, distal convoluted tubule) as in embryonic kidney development, is presently unclear; the clinical state of acute tubular necrosis certainly supports the idea that proximal tubule progenitor cells have the ability to replicate and differentiate into proximal tubule cells with functionally and morphologically differentiated phenotypes.

In this regard, recent data have demonstrated, using renal proximal tubule cells in primary culture, that the growth factors transforming growth factor-β1 (TGF-β1) and the epidermal growth factor (EGF), along with the retinoid, retinoic acid, promoted tubulogenesis in renal proximal tubule progenitor cells in tissue culture [31]. These observations defined a coordinated interplay between growth factors and retinoids to induce pattern formation and morphogenesis. This finding is one of the first definitions of inductive factors which may be important in the organogenesis of a mammalian organ. In addition, using immunofluorescence microscopy, retinoic acid induced laminin A- and B_1-chain production in these cells and purified soluble laminin completely substituted for retinoic acid in kidney tubulogenesis. These results clearly demonstrate the manner in which retinoic acid, as a morphogen, can promote pattern formation and differentiation by regulating the production of an extracellular matrix molecule.

Further work has demonstrated, in fact, that a population of cells resides in the adult mammalian kidney which have retained the capacity to proliferate and morphogenically differentiate into tubule structures in vitro [32]. These experiments have identified non-serum-containing growth conditions, which select for proximal tubule cells with a high capacity for self-renewal and an ability to differentiate phenotypically, collectively and individually, into proximal tubule structures in collagen gels. Regarding the high capacity for self-renewal, genetic marking of the cells with a recombinant retrovirus containing the *lacZ* gene and dilution analysis demonstrated that in vitro tubulogenesis often arose from clonal expansion of a single genetically tagged progenitor cell. These results suggest that a population of proximal tubule cells exist within the adult kidney in a relatively dormant, slowly replicative state, but with a rapid potential to proliferate, differentiate, and pattern-form to regenerate the lining proximal tubule epithelium of the kidney following severe ischemic or toxic injury.

Bioartificial Tubule Formulation

The bioartificial renal tubule is now clearly feasible when conceived as a combination of living cells supported on polymeric substrata [35]. A bioartificial tubule uses epithelial progenitor cells cultured on water and solute-permeable membranes seeded with various biomatrix materials so that expression of differentiated vectorial transport and metabolic and endocrine function is attained (Figs. 125.8 and 125.9). With appropriate membranes and biomatrices, immunoprotection of cultured progenitor cells can be achieved concurrent with long-term functional performance as long as conditions support tubule cell viability [35, 36]. The technical feasibility of an implantable epithelial cell system derived from cells grown as confluent monolayers along the luminal surface of polymeric hollow fibers has been achieved [35]. These previously constructed devices, however, have used permanent renal cell lines which do not have differentiated transport function. The ability to purify and grow renal proximal tubule progenitor cells with the ability to differentiate morphogenically may provide a capability for replacement renal tubule function.

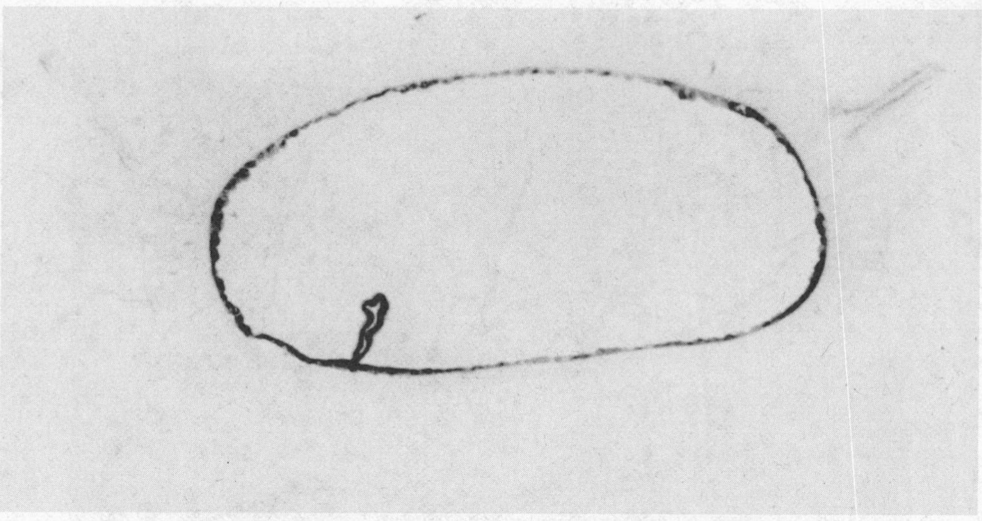

FIGURE 125.8 Ligh micrograph of an H&E section (100×) of hollow fiber lined with collagen type IV and confluent monolayer of human renal tubule epithelial cells along the inner component of the fiber. In this fixation process, the hollow fiber is clear with the outer contour of the hollow fiber identified by the irregular line (disregard artifact in lower left quadrant).

A bioartificial proximal tubule satisfies a major requirement of reabsorbing a large volume of filtrate to maintain salt and water balance within the body. The need for additional tubule equivalents to replace another nephronal segment function, such as the loop of Henle, to perform more refined homeostatic elements of the kidney, including urine concentration or dilution, may not be necessary. Patients with moderate renal insufficiency lose the ability to finely regulate salt and water homeostasis—because they are unable to concentrate or dilute, yet are able to maintain reasonable fluid and electrolyte homeostasis due to redundant physiologic compensation via other mechanisms. Thus, a bioartificial proximal tubule, which reabsorbs iso-osmotically the majority of the filtrate, may be sufficient to replace required tubular function to sustain fluid electrolyte balance in a patient with end-stage renal disease.

Bioartificial Kidney

The development of a bioartificial filtration device and a bioartificial tubule processing unit would lead to the possibility of an implantable bioartificial kidney, consisting of the filtration device followed in series by the tubule unit (Fig. 125.10). The filtrate formed by this device will flow directly into the tubule unit. The tubule unit should maintain viability, because metabolic substrates and low-molecular-weight growth factors are delivered to the tubule cells from the ultrafiltration unit. Furthermore, immunoprotection of the cells grown within the hollow fiber is achievable due to the impenetrance of immunologically competent cells through the hollow fiber. Rejection of transplanted cells will, therefore, not occur. This arrangement thereby allows the filtrate to enter the internal compartments of the hollow fiber network, which are lined with confluent monolayers of renal tubule cells for regulated transport function.

This device could be used either extracoporeally or implanted within a patient. In this regard, the specific implant site for a bioartificial kidney will depend upon the final configuration of both the bioartificial filtration and tubule device. As presently conceived, the endothelial-line bioartificial filtration hollow fibers can be placed into an arteriovenous circuit using the common iliac artery and vein, similar to the surgical connection for a renal transplant. The filtrate is connected in series to a bioartificial proximal tubule, which is embedded into the peritoneal membrane, so that reabsorbate will be transported into the peritoneal cavity and reabsorbed into the systemic circulation. The processed filtrate

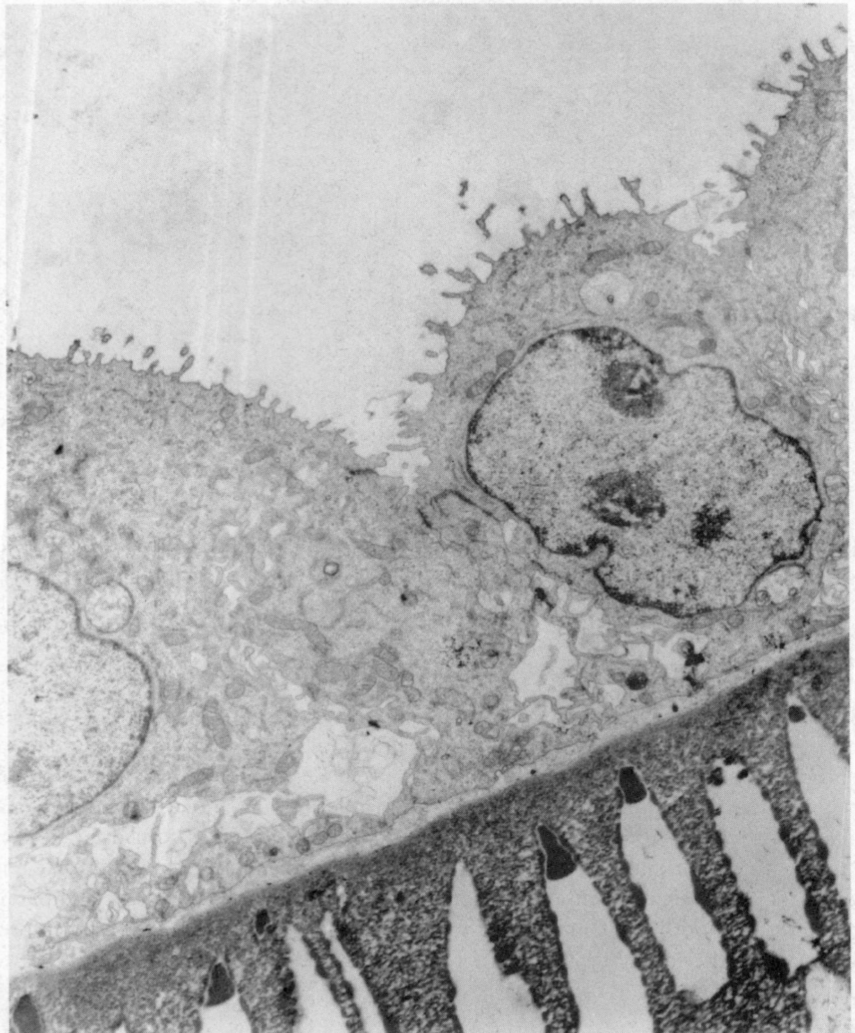

FIGURE 125.9 Electorn micrograph of a single hollow fiber lined with extracellular matrix and a confluent monolayer (see Fig. 125.8) of renal tubule cells along the inner component of the fiber. As displayed, the differentiated phenotype of renal tubule cells on the hollow fiber prelined with matrix is apparent. The well developed microvilli and apical tight junctions can be appreciated (14,000×).

exiting the tubule unit is then connected via tubing to the proximate ureter for drainage and urine excretion via the recipient's own urinary collecting system.

Although an implantable form of this therapy is in the conceptual phase of development, a functioning extracorporeal bioartificial renal tubule has been developed. The technical feasibility of an epithelial cell system derived from cells grown as confluent monolayers along the luminal surface of polymeric hollow fibers has been recently achieved [38]. Because of its anatomic and physiological similarities with humans and the relative simplicity with which it can be bred in large numbers, the pig is currently considered the best source of organs for both human xenotransplantation and immuno-isolated cell therapy devices [39–41]. Yorkshire breed pig renal tubule progenitor cells have been expanded and cultured on semi-permeable hollow fiber membranes seeded with various biomatrix materials, so that expression of differentiated vectorial transport, metabolic, and endocrine function has been attained [38, 42, 43]. This

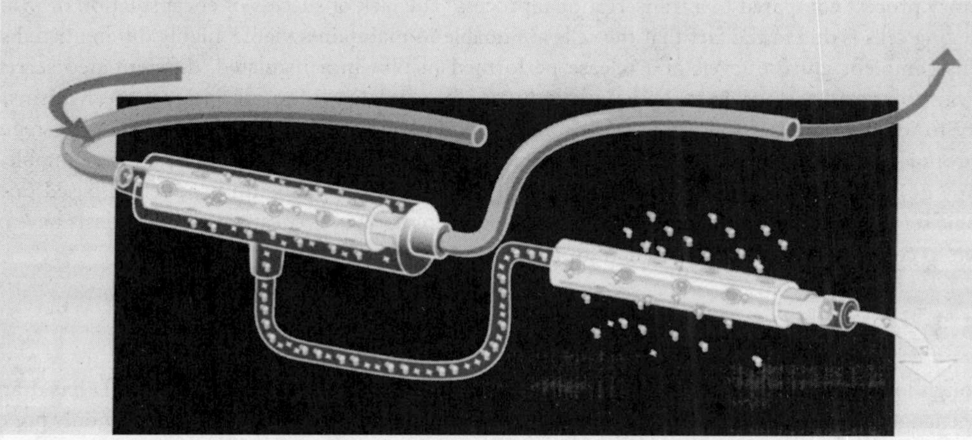

FIGURE 125.10 Conceptual schematization of an implantable tissue engineered bioartificial kidney with an endothelial-cell-lined hemofilter in series with a proximal tubule-cell-lined reabsorber.

bioartificial renal tubule has been evaluated in uremic and non-uremic large animals [44]. A customized extracorporeal circuit using standard arterial venous blood tubing with a dialysis machine has been developed which delivers the post-filtered blood through the extracapillary space and the ultrafiltrate through the liminal space of the bioartificial tubule. Experiments have confirmed the functional metabolic performance and active tubule transport properties of the bioartificial renal tubule using this extracorporeal circuit [44]. Proximal tubule cells derive embryonically from mesodermal progenitors closely related to bone marrow precursor cells and have retained many elements of immunologically competent cells, including the ability for antigen presentation and production of a variety of immunologically active cytokines [45]. Since the kidney participates in this complex and dynamic network of pro- and anti-inflammatory cytokines, cell replacement therapy with an extracorporeal bioartificial renal tubule assist device may play a critical role in the future treatment of renal failure.

Bioartificial Endocrine Gland

The Erythropoietin Generator

Because specialized cells are programmed to carry out specific biologic tasks, cell therapy may deliver several key proteins in a coordinated cascade to promote a biologic or physiologic process. Targeted delivery of a specific deficient protein, hormone, or neurotransmitter may be achieved with site-specific implantation of cells which can produce this deficient compound after being encapsulated in special polymeric membranes. The membranes allow cell nutrients into the encapsulated space to maintain cell viability and allow cellular metabolic wastes to exit along with the desired protein, hormone, or neurotransmitter while shielding the cells from the host's destructive immune response. This strategy is being employed, for example, to deliver dopamine produced by bovine adrenal cells to the substantia nigra where a deficiency of this neurotransmitter at this site leads to Parkinson's Disease [46].

Regulated and homeostatic drug dosing may also be achieved with cell therapy. For hormonal therapy, such as insulin for diabetes mellitus, appropriate insulin levels within the body are only crudely attained with a once-a-day or twice-a-day dosing. Any hormone-producing cell has a highly evolved biologic sensing system to monitor the ambient environment and respond with graded production and release of the hormone to regulate the sensed level of the moiety which is regulated. The circulating level of a protein or a hormone may be regulated at several levels: at the gene level by transcriptional mechanisms, at the protein level by translational processes, or at the secretory level by cellular processes. The complexity of regulation increases several-fold, as control progresses from transcriptional to translational to excretory processes. Accordingly, a more refined differentiated cell phenotype is required to maintain a regulated secretory process compared to a transcriptional process. The lack of success of encapsulation of insulin-producing cells is due to the fact that the cells are unable to maintain a viable, highly differentiated state to sense ambient glucose levels and release performed insulin in a regulated, differentiated secretory pathway. In contrast, since erythropoietin production is regulated by transcriptional mechanisms, the ability to identify and perhaps grow cells from adult mammalian kidneys with the ability to regulate erythropoietin production in response to oxygen delivery may allow the design of an implantable cell therapy device to sense circulating oxygen levels and regulate erythropoeitin production based upon a biologic sensing mechanism. Recently, genetically engineered polymer encapsulated myoblasts have been shown to continuously deliver human and mouse erythropoietin in mice [47].

125.3 Clinical and Economic Implications

Although long-term chronic renal replacement therapy with either hemodialysis or CAPD has dramatically changed the prognosis of renal failure, it is not complete replacement therapy, since it only provides filtration function (usually on an intermittent basis) and does not replace the excretory, regulatory, and endocrine functions of the kidney. Because of the nonphysiologic manner in which dialysis performs or does not perform the most critical renal functions, patients with ESRD on dialysis continue to have major medical, social, and economic problems. Renal transplant addresses some of these issues, but immunologic barriers and organ shortages keep this approach from being ideal for a large number of ESRD patients.

Although dialysis or transplantation therapies can prolong the life of a patient with ESRD disease, it is still a serious medical condition, with ESRD patients having only one-fifth the life expectancy of a normal age-matched control group. ESRD patients also experience significantly greater morbidity. Patients with ESRD have five times the hospitalization rate, nearly twice the disability rate, and five times the unemployment rate of age-matched non-ESRD individuals [2]. Accordingly, this new technology based upon the proposed bioengineering prototypes would most likely have substantial benefits to the patient by increasing life expectancy, increasing mobility and flexibility, increasing quality of life with large savings in time, less risk of infection, and reduced costs.

Besides the personal costs to the patient and family, care of chronic kidney failure is monetarily expensive on a per-capita basis in comparison to most forms of medical care [1, 2]. The 1989 estimated Medicare payment (federal only) per ESRD patient during the entire year averaged $30,900. The patient and private insurance obligations were an addition $6900 per patient. The total cost of a patient per year with end-stage renal disease, therefore, is approximately $40,000. In 1988 the expected life span after beginning dialysis for an ESRD patient was approximately four years; therefore, the total cost per patient of ESRD during his or her lifetime was approximately $160,000 in 1998.

The total cost in direct medical payments for ESRD, by both public and private payers, increased from $6 billion in 1989 to over $14 billion in 1996 [48]. These estimates do not include a number of indirect cost items, since they do not include patient travel costs and lost labor production. The number of patients receiving chronic dialytic therapy in the U.S. is presently over 250,000 with a current growth rate at 8 to 10% per year. It is conceivable that in the not too distant future, tissue engineering technology could supplant current treatments for ESRD. Although it is difficult to estimate the value of a technology, on a purely economic basis there may be an opportunity for major cost savings with this technology.

125.4 Summary

Three technologies will most likely dominate medical therapeutics in the next century. One is "cell therapy"—the implantation of living cells to produce a natural substance in short supply from the patient's own cells due to injury and destruction from various clinical disorders. Erythropoietin cell therapy is an example of this approach to replace a critical hormone deficiency in end-stage renal disease. A second therapy is tissue engineering, wherein cells are cultured to replace masses of cells that normally function in a coordinated manner. Growing a functional glomerular filter and tubule reabsorber from a combination of cells, biomaterials, and synthetic polymers to replace renal excretory and regulatory functions is an example of this formulation. Over the last few years, an extracorporeal bioartificial renal tubule has become a reality; demonstrating physiologic and biochemical regulatory function in large animal studies. Finally, a third technology that will dominate future therapeutics is gene therapy, in which genes are transferred into living cells either to deliver a gene product to a cell in which it is missing or to produce a foreign gene product by a cell to promote a new function. The use of genes which encode for anticoagulant proteins as a means to deliver in a targeted and local fashion an anticoagulant to maintain hemocompatibility of a tissue engineered hemofilter is an example of the application of this third technology.

The kidney was the first organ whose function was substituted by an artificial device. The kidney was also the first organ to be successfully transplanted. The ability to replace renal function with these revolutionary technologies in the past was due to the fact that renal excretory function is based upon natural physical forces which govern solute and fluid movement from the body compartment to the external environment. The need for coordinated mechanical or electrical activities for renal substitution was not required. Accordingly, the kidney may well be the first organ to be available as a tissue-engineered implantable device as a fully functional replacement part for the human body.

References

1. Iglehart JK. 1993. The American Health Care System: The End Stage Renal Disease Program. N Engl J Med 328:366.
2. Excerpts from United States Renal Data System 1991 Annual Data Report. Prevalence and cost of ESRD therapy. Am J Kidney Diseases 18(5)(supp)2:21.
3. Kriz W, Lever AF. 1969. Renal countercurrent mechanisms: Structure and function. Am Heart J 78:101.
4. Brenner BM, Humes HD. 1977. Mechanisms of glomerular ultra-filtration. N Engl J Med 297:148.
5. Landis EM, Pappenheimer JR. Exchange of substances through the capillary walls. In WF Hamilton, P Dow (eds), Handbook of Physiology: Circulation, sec 2, vol 2, p 961, Washington DC, American Physiological Society.
6. Anderson JL, Quinn JA. 1974. Restricted transport in small pores. A model for steric exclusion and hindered particle motion. Biophys J 14:130.
7. Chang RLS, Robertson CR, Deen WM, et al. 1975. Permselectivity of the glomerular capillary wall to macromolecules: I. Theoretical considerations. Biophys J 15:861.
8. Brenner BM, Hostetter TH, Humes HD. 1978. Molecular basis of proteinuria of glomerular origin. N Engl J Med 298:826.
9. Andreoli TE, Schafer JA. 1978. Volume absorption in the pars recta: III. Luminal hypotonic-sodium reabsorption. Circ Res 52:491.
10. Knox FG, Mertz JI, Burnett JC, et al. 1983. Role of hydrostatic and oncotic pressures in renal sodium reabsorption. Circ Res 52:491.
11. Jacobson LO, Goldwasser E, Fried W, et al. 1957. Role of kidney in erythropoiesis. Nature. 179:633.
12. Maxwell PH, Osmond MK, Pugh CW, et al. 1993. Identification of the renal erythropoietin-producing cells using transgenic mice. Kidney Int 441:1149.
13. Fried W. 1972. The liver as a source of extrarenal erythropoeitin production. Blood 49:671.
14. Jelkmann W. 1992. Erythropoietin: Structure, control of production, and function. Physiolog Rev 72(2):449.
15. Golper TA. 1986. Continuous arteriorvenous hemofiltration in acute renal failure. Am J Kidney Dis 6:373.
16. Kramer P, Wigger W, Rieger J, et al. 1977. Arterior-venous hemofiltration: A new and simple method for treatment of overhydrated patients resistant to diuretics. Klin Wochenschr 55:1121.
17. Colton CK, Henderson LW, Ford CA, et al. 1975. Kinetics of hemodiafiltration. In vitro transport characteristics of a hollow-fiber blood ultrafilter. J Lab Clin Med 85:355.

18. Henderson LW, Colton CK, Ford CA. 1975. Kinetics of hemodiafiltration: II. Clinical characterization of a new blood cleansing modality. J Lab Clin Med 85:372.
19. Kadletz M, Magometschnigg H, Minar E., et al. 1992. Implantation of in vitro endothelialized polytetrafluoroethylene grafts in human beings. J Thorac Cardiovasc Surg 104:736.
20. Schnider PA, Hanson SR, Price TM, et al. 1988. Durability of confluent endothelial cell monolayers of small-caliber vascular prostheses in vitro. Surgery 103:456.
21. Shepard AD, Eldrup-Jorgensen J, Keough EM, et al. 1986. Endothelial cell seeding of small caliber synthetic grafts in the baboon. Surgery 99:318.
22. Zweibel JA, Freeman SM, Kantoff PW, et al. 1989. High-level recombinant gene expression in rabbit endothelial cells transduced by retroviral vectors. Science 243:220.
23. Wilson JM, Birinyi LK, Salomon RN, et al. 1989. Implantation of vascular grafts lined with genetically modified endothelial cells. Science 244:1344.
24. Carey DJ. 1991. Control of growth and differentiation of vascular cells by extacellular matrix proteins. Annu Rev Physiol 53:161.
25. Carley WW, Milici AJ, Madri JA. 1988. Extracellular matrix specificity for the differentiation of capillary endothelial cells. Exp Cell Res 178:426.
26. Milici AJ, Furie MB, Carley WW. 1985. The formation of fenestrations and channels by capillary endothelium in vitro. Proc Natl Acad Sci 82:6181.
27. Garlick JA, Katz AB, Fenjves ES, et al. 1991. Retrovirus-mediated transduction of cultured epidermal keratinocytes. J Invest Dermatol 97:824.
28. Hall PA, Watt FM. 1989. Stem cells: The generation and maintenance of cellular diversity. Development 106:619.
29. Potten CS, Lieffler M. 1990. Stem cells; lessons for and from the crypt. Development 110:1001.
30. Reynolds BA, Weiss S. 1992. Generation of neurons and astrocytes from isolated cells of the adult mammalian central nervous system. Science 255:1707.
31. Humes HD, Cieslinski DA. 1992. Interaction between growth factors and retinoic acid in the induction of kidney tubulogenesis. Exp Cell Res 201:8.
32. Humes HD, Krauss JC, Cieslinski DA, et al. Tubulogenesis from isolated single cells of adult mammalian kidney: Clonal analysis with a recombinant retrovirus (submitted).
33. Coimbra T, Cieslinski DA, Humes HD. 1990. Exogenous epidermal growth factor enhances renal repair in mercuric chloride-induced acute renal failure. AM J Physiol 259:F 483.
34. Humes HD, Cieslinski DA, Coimbra T, et al. 1989. Epidermal growth factor enhances renal tubule cell regeneration and repair and accelerates the recovery of renal function in postischemic acute renal failure. J Clin Invest 84:1757.
35. Ip TK, Aebischer P. 1989. Renal epithelial-cell-controlled solute transport across permeable membranes as the foundation for a bioartificial kidney. Artif Organs 13:58.
36. Aebischer P, Wahlberg L, Tresco PA, et al. 1991. Macroencapsulation of dopamine-secreting cells by coextrusion with an organic polymer solution. Biomaterials 12:50.
37. Tai IT, Sun AM. 1993. Microencapsulation of recombinant cells: A new delivery system for gene therapy. FASEB J 7:1061.
38. McKay SM, Funke AJ, Buffington DA, Humes HD. Tissue engineering of a bioartificial renal tubule. *ASAIO Journal.* 44:179–183, 1998.
39. Cozzi E and White D. The generation of transgenic pigs as potential organ donors for humans. *Nature Medicine* 1:965–966, 1995.
40. Cooper DKC, Ye Y, Rolf JLL and Zuhdi N. The pig as potential organ donor for man. In: Cooper DKC, Kemp E, Reemtsma K, White DJG (eds): *Xeno-Transplantation.* Springer, Berlin. 1991. pp481–500.
41. Calne RY. Organ transplantation between widely disparate species. *Transplant Proc* 2:550–553, 1970.
42. Humes HD and Cieslinski DA. Interaction between growth factors and retinoic acid in the induction of kidney tubulogenesis. *Exp Cell Res* 201:8–15, 1992.

43. Humes HD, Krauss JC, Cieslinski DA and Funke AJ. Tubulogenesis from isolated single cells of adult mammalian kidney: clonal analysis with a recombinant retrovirus. *Am J Phsyiol: Renal* 271(40):F42–F49, 1996.
44. Weitzel WF, Browning A, Buffington DA, Funke AJ, Gupte A, MacKay S, Humes HD. Analysis of a Renal Proximal tubule Assist Device (RAD) during CVVH on uremic dogs, *J Am Soc of Nephrol* Abstract, 31st Annual Meeting, 1998 (In press).
45. Ong ACM and Fine LG. Tubular-derived growth factors and cytokines in the pathogenesis of tubulointerstitial fibrosis: implications for human renal disease progression. *Am J Kidney Dis* 23:205–209, 1994.
46. Aebischer P, Tresco PA, Winn SR, et al. Long-term cross-species brain transplantation of a polymer-encapsulated dopamine-secreting cell line. Exp Neurol 111:269.
47. Relulier E, Schneider BL, Deglon N, Beuzard Y, Aebischer P. Continuous delivery of human and mouse erythropoietin in mice by genetically engineered polymer encapsulated myoblasts. Gene Therapy 5:1014–1022, 1998.
48. Excerpts from the United States Renal Data System 1998 Annual Data Report. The economic cost of ESRD and Medicare spending for alternative modalities of treatment. J Kidney Diseases 32(2)(supp)1:S118.

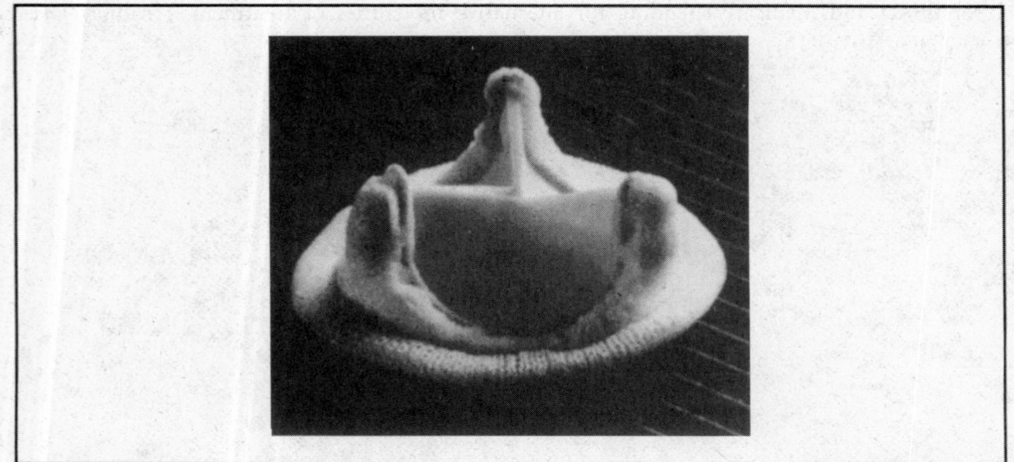

Photograph of Carpentier-Edwards pericardial valve.

XIII

Prostheses and Artificial Organs

Pierre M. Galletti (deceased) and Robert M. Nerem

XIII.1 Substitutive Medicine

Over the past 50 years, humanity has progressively discovered that an engineered device or the transplantation of organs, tissues, or cells can substitute for most, and perhaps all, organs and body functions. The devices are human-made, whereas the living replacement parts can be obtained from the patient, a relative, a human cadaver, or a live animal or can be prospectively developed through genetic engineering. The concept that a disease state may be addressed not only by returning the malfunctioning organ to health using chemical agents or physical means but also by replacing the missing function with a natural or an artificial counterpart has brought about a revolution in therapeutics. Currently in the United States alone, 2 to 3 million patients a year are treated with a human-designed spare part (assist device, prosthesis, or implant), with the result that over 20 million people enjoy a longer or better quality of life thanks to artificial organs. In comparison, a shortage of donor organs limits the number of transplantation procedures to about 20,000 a year, and the total population of transplant survivors is on the order of 200,000.

The fundamental tenet of substitutive medicine is that beyond a certain stage of failure, it is more effective to remove and replace a malfunctioning organ than to seek in vain to cure it. This ambitious proposition is not easy to accept. It goes against the grain of holistic views of the integrity of the person. It seems at odds with the main stream of twentieth-century scientific medicine, which strives to elucidate pathophysiologic mechanisms at the cellular and molecular level and then to correct them through a specific biochemical key. The technology of organ replacement rivals that of space travel in complexity and fanfare and strikes the popular imagination by its daring, its triumphs, and its excesses. Although the artificial organ approach does not reach the fundamental objective of medicine, which is to understand and correct the disease process, it is considerably more effective than drug therapy or corrective surgery in the treatment of many conditions, e.g., cardiac valve disease, heart block, malignant arrhythmia, arterial obstruction, cataract.

A priori, functional disabilities due to the destruction or wear of body parts can be addressed in two ways: the implantation of prosthetic devices or the transplantation of natural organs. For a natural organ transplant, we typically borrow a spare part from a living being or from an equally generous donor who before death volunteered to help those suffering from terminal organ failure. Transplanted organs benefit from refinements acquired over thousands of years of evolution. They are overdesigned, which means they will provide sufficient functional support even though the donated part may not be in perfect condition at the time of transfer to another person. They have the same shape and the same attachment needs as the body part they replace which means that surgical techniques are straightforward. The critical problem is the shortage of donors, and therefore only a small minority of patients currently benefit from this approach.

Artificial organs have different limitations. Seen on the scale of human evolution, they are still primitive devices, tested for 40 years at most. Yet they have transformed the prognosis of many heretofore fatal diseases, which are now allowed to evolve past what used to be their natural termination point. In order to design artificial organs, inventive engineers, physiologists, and surgeons think in terms of functional results, not anatomical structures. As a result, artificial organs have but a distant similarity to natural ones. They are mostly made of synthetic materials (often called biomaterials) which do not exist in nature. They use different mechanical, electrical, or chemical processes to achieve the same functional objectives as natural organs. They adapt but imperfectly to the changing demands of human activity. They cannot easily accommodate body growth and therefore are more beneficial to adults than to children. Most critically, artificial organs, as is the case for all machines, have a limited service expectancy because of friction, wear, or decay of construction materials in the warm, humid, and corrosive environment of the human body. Such considerations limit their use to patients whose life expectancy matches the expected service life of the replacement part or to clinical situations where repeated implantations are technically feasible. In spite of these obstacles, the astonishing reality is that millions of people are currently alive thanks to cardiac pacemakers, cardiac valves, artificial kidneys, or hydrocephalus drainage systems, all of which address life-threatening conditions. An even larger number of people enjoy the benefits of hip and knee prostheses, vascular grafts, intraocular lenses, and dental implants, which correct dysfunction, pain, inconvenience, or merely appearance. In short, the clinical demonstration of the central dogma of substitutive medicine over the span of two generations can be viewed demographically as the first step in a evolutionary jump which humans cannot yet fully appreciate.

Hybrid artificial organs, or bioartificial organs, are more recent systems which include living elements (organelles, cells, or tissues) as part of a device made of synthetic materials. They integrate the technology of natural organ transplantation and the refinements which living structures have gained through millions of years of evolution with the purposeful design approach of engineering science and the promises of newly developed synthetic materials. Table XIII.1 provides a current snapshot in the continuing evolution of substitutive medicine.

Depending upon medical needs and anticipated duration of use, artificial organs can be located outside of the body yet attached to it (paracorporeal prostheses or assist devices) or implanted inside the body in a appropriate location (internal artificial organs or implants). The application of artificial organs may be temporary, i.e., a bridge procedure to sustain life or a specific biologic activity while waiting for either recovery of natural function (e.g., the heart-lung machine), or permanent organ replacement (e.g., left ventricular assist devices). It can be intermittent and repeated at intervals over extended periods of time when there is no biologic necessity for continuous replacement of the missing body functions (e.g., artificial kidney). It can pretend to be permanent, at least within the limits of a finite life span.

Up to 1950, organ replacement technology was relatively crude and unimaginative. Wooden legs, corrective glasses, and dental prostheses formed the bulk of artificial organs. Blood transfusion was the only accepted form of transplantation of living tissue. Suddenly, within a decade, the artificial kidney,

TABLE XIII.1 Evolution of Organ Replacement Technology: A 1995 perspective

Current Status	Artificial Organs	Transplantation
Broadly accepted clinically	Heart-lung machine	Blood transfusion
	Large-joint prostheses	Corneal transplants
	Bone fixation systems	Banked bone
	Cardiac pacemakers	Bone marrow
	Implantable defibrillators	Kidney—living related donor
	Large vascular grafts	Kidney—cadaveric donor
	Prosthetic cardiac valves	Heart
	Intra-aortic balloon pump	Liver
	Intraocular lenses	
	Middle ear ossicle chain	
	Hydrocephalus shunts	
	Dental implants	
	Skin and tissue expanders	
	Maintenance hemodialysis	
	Chronic ambulatory peritoneal dialysis	
Accepted with reservations	Breast implants	Whole pancreas
	Sexual prostheses	Single and double lung
	Small joint prostheses	Combined heart-lung
	ECMO in children	
Limited clinical application	ECMO in adults	Cardiomyoplasty
	Ventricular assist devices	Pancreatic islets
	Cochlear prostheses	Liver lobe or segment
	Artificial tendons	Small Intestine
	Artificial skin	
	Artificial limbs	
Experimental stage	Artificial pancreas	Bioartificial pancreas
	Artificial blood	Bioartificial liver
	Intravenous oxygenation	CNS implants of secreting tissue
	Artificial esophagus	Gene therapy products
	Total artificial heart	
	Nerve guidance channels	
Conceptual stage	Artificial eye	Striated muscle implants
	Neurostimulator	Smooth muscle implants
	Blood pressure regulator	Cardiac muscle implants
	Implantable lung	Functional brain implants
	Artificial trachea	Bioartificial kidney
	Artificial gut	
	Artificial fallopian tube	

the heart-lung machine, the cardiac pacemaker, the arterial graft, the prosthetic cardiac valve, and the artificial hip joint provided the first sophisticated examples of engineering in medicine. More recently, the membrane lung, the implantable lens, finger and tendon prostheses, total knee replacements, and soft-tissue implants for maxillo-facial, ear, or mammary reconstruction have reached the stage of broad clinical application. Ventricular assist devices and the total artificial heart have been extensively tested in animals and validated for clinical evaluation. Artificial skin is increasingly used in the treatment of ulcers and burns. Soft- and hard-tissue substitutes function effectively for several years. Sexual and sensory prostheses offer promises for the replacement of complex human functions. Interfacing of devices with the peripheral and central nervous systems appears as promising today as cardiovascular devices were 30 years ago. Perhaps the brightest future belongs to "information prostheses" which bring to the human body, signals which the organism can no longer generate by itself (e.g., pacemaker functions), signals which need to be modulated differently to correct a disease state (e.g., electronic blood pressure regulators) or signals which cannot be perceived by the nervous system through its usual channels of information gathering (e.g., artificial eye or artificial ear).

XIII.2 Biomaterials

The materials of the first generation of artificial organs—those which are widely available at the moment—are for the most part standard commodity plastics and metals developed for industrial purposes. Engineers have long recognized the limitations of construction materials in the design and performance of machines. However, a new awareness arose when they started interacting with surgeons and biologic scientists in the emerging field of medical devices. In many cases the intrinsic and well established physical properties of synthetic materials such as mechanical strength, hardness, flexibility, or permeability to fluids and gases were not as immediately limiting as the detrimental effects deriving from the material's contact with living tissues. As a result, fewer than 20 chemical compounds among the 1.5 million candidates have been successfully incorporated into clinical devices. Yet some functional implants require material properties which exceed the limits of current polymer, ceramic, or metal alloy technology. This is an indirect tribute to the power of evolution, as well as a challenge to scientists to emulate natural materials with synthetic compounds, blends, or composites.

The progressive recognition of the dominant role of phenomena starting at the tissue-material interface has led to two generalizations:

1. All materials in contact with body fluids or living tissue undergo almost instantaneous and then continuing surface deposition of body components which alter their original properties.
2. All body fluids and tissues in contact with foreign material undergo a dynamic sequence of biologic reactions which evolve over weeks or months, and these reactions may remain active for as long as the contact persists and perhaps even beyond.

The recognition of biologic interactions between synthetic materials and body tissues has been translated into the twin operational concepts of biomaterials and compatibility.

Biomaterials is a term used to qualify materials which can be placed in intimate contact with living structures without harmful effects. Compatibility characterizes a set of material specifications and constraints which address the various aspects of material-tissue interactions. More specifically, hemocompatibility defines the ability of a biomaterial to stay in contact with blood for a clinically relevant period of time without causing alterations of the formed elements and plasma constituents of the blood or substantially altering the composition of the material itself. The term biocompatibility is often used to highlight the absence of untoward interactions with tissues other than blood (e.g., hard or soft tissues).

It is worth observing that hemocompatibility and biocompatibility are virtues demonstrated not by the presence of definable favorable properties but rather by the absence of adverse effects on blood or other tissues. Although these terms imply positive characteristics of the material, the presumption of compatibility is actually based on the accumulation of negative evidence over longer and longer periods of time, using an increasingly complex battery of tests, which must eventually be confirmed under the conditions of clinical use.

The clinical success of materials incorporated into actual devices is altogether remarkable, considering how limited our understanding is of the physical and biologic mechanisms underlying tissue-material interactions. Indeed the most substantial conclusion one can draw from a review of records of literally millions of implants is how few major accidents have been reported and how remarkably uncommon and benign have been the side effects of implanting substantial amounts of synthetic substances into the human body. Artificial organs are by no means perfect. Their performance must be appreciated within the same limits that the inexorable processes of disease and aging impose on natural organs.

XIII.3 Outlook for Organ Replacement

Now emerging is a second generation of implantable materials through the confluence of biomaterial science and cell biology (Fig. XIII.1). Cell culture technology, taking advantage of biotechnology products and progressing to tridimensional tissue engineering on performed matrices, now provides building

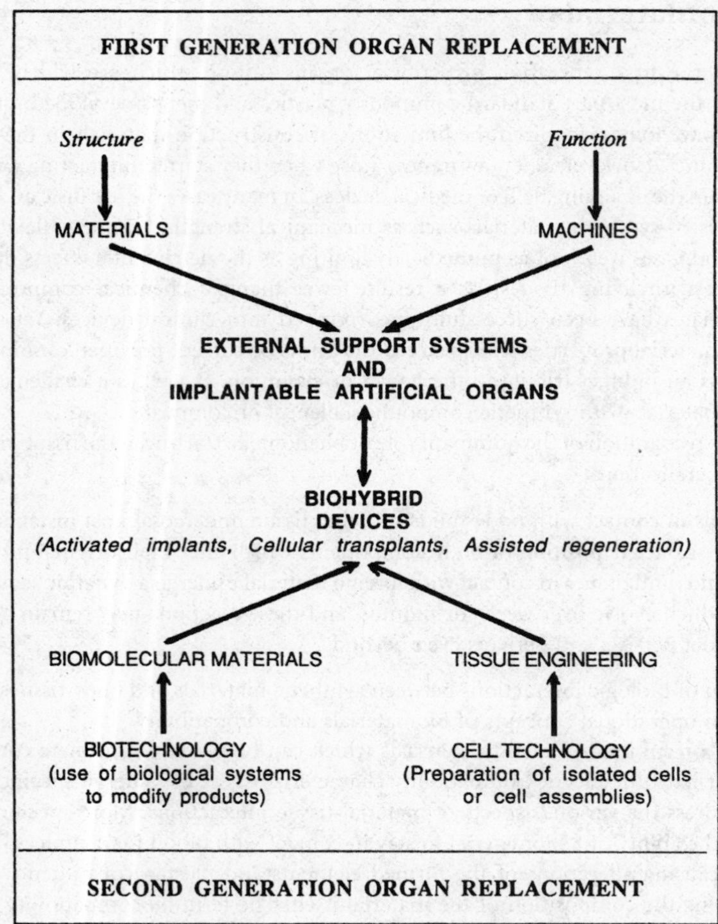

FIGURE XIII.1 Schematic description of the advances in engineering, biologic, and medical technology which led to the first generation of artificial organs (read from the top) and the newer developments in body replacement parts (read from the bottom).

blocks which incorporate the peptide or glycoprotein sequences responsible for cell-to-cell interactions. This combination leads to a new class of biohybrid devices which includes

1. Cellular transplants for continuing secretion of bioactive substances (e.g., transplants of insulin-producing xenograft tissue protected against immune rejection by permselective envelopes)
2. Composites of synthetic materials with living cells (often called organoids) to accelerate implant integration within the body (e.g. endothelial cell-lined polymer conduits designed for vascular grafts)
3. Replacement parts in which natural tissue regeneration is activated by the presence of supportive cells (e.g., Schwann cell-seeded nerve guidance channels)
4. Vehicles for gene therapy in which continued gene expression is enhanced by a synthetic polymer substrate with appropriate mechanical, chemical, or drug release properties (e.g., epicardial transplants of genetically modified skeletal or cardiac muscle grown on a distensible polymer matrix)

In many respects, the new wave of organ replacement exemplifies the synergy of the two original currents in substitutive medicine: prostheses and transplants. It expands the feasibility of cell and tissue transplantation beyond the boundaries of autografts and related donor allografts, opening the way to xenogeneic and engineered replacement parts. It also confronts the "foreign body" limitations of human-made synthetic implants by adding the molecular and cellular elements that favor permanent biointegration.

XIII.4 Design Considerations

Natural organ transplants, if ideally preserved, should be able to fulfill all functions of the original body part except for those mediated by the nervous system, since a transplanted organ is by definition a denervated structure. In actuality, transplants always present some degree of ischemic damage caused by interruption of the blood supply during transfer from donor to recipient. This may be reflected by a temporarily impaired function in the postoperative period or by permanent necrosis of the most delicate components of the transplant, resulting in some degree of functional limitation. In the long run, transplanted organs may also exhibit functional alterations because of cell or tissue damage associated with an underlying systemic disease. They may be damaged by the immunosuppression protocol, which at the current stage is needed for all organ replacements except for autografts, identical-twin homografts, and some types of fetal tissue transplants. The second-order limitations of transplanted organs are usually ignored, and the assumption is made that all original functions are restored in the recipient.

Artificial organs, however, can only replace those bodily functions which have been incorporated into their design because these functions were scientifically described and known to be important. Therefore, in the design of an artificial organ, the first task is to establish the specifications for the device, i.e., to describe in quantitative terms the function or functions which must be fulfilled by a human-made construct and the physical constraints that apply because the device must interface with the human body. Each human organ fulfills multiple functions of unequal importance in terms of survival. Consequently, it is critical to distinguish the essential functions which must be incorporated into an effective spare part from those which can be neglected.

Defining specifications and constraints is the first step in the conceptualization of an artificial organ. Only when this is done can one think realistically about design alternatives, the limitations of available materials, and the clinical constraints which will apply, of which the key ones are connections to the body and duration of expected service.

Once all these considerations have been integrated (modeling is often useful at that stage), the next step is typically the construction of a prototype. Ideally the device should achieve everything it was expected to do, but usually it exhibits some level of performance and durability which falls short of design specifications, either because of some misjudgment in terms of required function or because of some unanticipated problem arising at the interface between the device and the body.

The following step of development may be called *optimization*, if the specifications were well defined from the outset, or *reevaluation*, if they were not. More commonly it is the reconciliation of competition and at times contradictory design criteria which leads to a second prototype.

At this point, new experiments are needed to establish the reliability and effectiveness of the device in animal models of the target disease (if such exist) or at least in animals in which the natural organ can be removed or bypassed. This is the stage of *validation* of the device, which is first conducted in acute experiments and must later be extended to periods of observation approximating the duration of intended use in humans. These criteria, however, cannot always be met for long-term implants, since the life expectancy of most animals is shorter than that of humans. By this point, the diverse vantage points of the theoretician, the manufacturer, the performance evaluator, and the clinical user have been articulated for some specific devices and generalized in terms of quality control for classes of devices.

The final stage of design, for many artificial organs, is *individualization*, i.e., the ability to fit the needs of diverse individuals. Humans come in a wide range of body sizes. In some cases, the prostheses must fit very strict dimensional criteria, which implies that they must be fabricated over an extended range of sizes (e.g., cardiac valves). In other cases, there is enough reserve function in the device that one pediatric model and one adult size model may suffice (e.g., blood oxygenator for cardiac surgery).

XIII.5 Evaluation Process

The evaluation of an artificial organ typically is done in six phases:

1. *In vitro* bench testing
2. *Ex vivo* appraisal
3. *In vivo* studies with healthy experimental animals
4. *In vivo* studies with animal models of disease
5. Controlled clinical trials
6. General clinical use

In Vitro Bench Testing

In vitro bench testing of a completed prototype has three major purposes:

1. To observe the mode of operation of the device and assess its performance under tightly controlled circumstances
2. To define performance in quantitative terms over a wide range of environmental or input conditions
3. To assess the device's reliability and durability in a manner which can be extrapolated to the intended clinical use

For all its value, there are limitations to in vitro testing of devices. Devices are made to work while in contact with body fluids or body tissues. This complex environment modifies materials in ways which are not always predictable. To duplicate this effect as closely as possible a laboratory bench system can be made to match the body's environment in terms of temperature and humidity. Operating pressures and external forces can also be imitated but not perfectly reproduced (e.g., the complex pulsatile nature of cardiovascular events). Other fluid dynamic conditions such as viscosity, wall shear stress, and compliance of device-surrounding structures call for sophisticated laboratory systems and can only be approximated. The chemical environment is the most difficult to reproduce in view of the complexity of body fluids and tissue structures. Some in vitro testing systems make use of body fluids such as plasma or blood. This in turn brings in additional intricacies because these fluids are not stable outside of the body without preservatives and must be kept sterile if the experiment is to last more than a few hours.

Accelerated testing is a standard component in the evaluation of machines. It is critical for permanent implants with moving parts which are subject to the repeated action of external forces. Fatigue testing provides important information on progressive wear or catastrophic failure of device components. For example, the human heart beats about 40 million times per year. Manufacturers and regulatory agencies conduct testing of prosthetic cardiac valves over at least 400 million cycles. With a testing apparatus functioning at 1200 cycles per minute, this evaluation can be compressed by a factor of about 15, i.e., to about a year.

Ex Vivo Appraisal

Because of the difficulty of keeping blood in its physiologic state in a container, the evaluation of some blood processing or blood contacting devices is performed by connecting them through the skin to an artery or vein or both if the blood must be returned to the cardiovascular system to avoid excessive hemorrhage. Such experiments retain the advantage of keeping the device under direct observation while allowing longer experiments than are feasible in vitro, particularly if the animal does not require general anesthesia. It is also possible in some cases to evaluate several devices in parallel or sequentially under quite realistic conditions and therefore to conduct comparative experiments under reasonably standardized conditions.

In Vivo Evaluation with Healthy Experimental Animals

There comes a stage in the development of most devices where they must be assessed in their target location in a living body. The matching of device size and shape with available experimental sites in the appropriate animal species is a necessary condition. Such experiments typically last weeks, months, or years and provide information about body-device and tissue-material interactions either through non-invasive measurement techniques or through device retrieval at the end of the observation period. Rodents, felines, and dogs raised for research purposes are usually too small for the evaluation of human-sized devices. Farm animals such as sheep, goats, pigs, and calves are commonly used. Here again the limited life expectancy of experimental animals prevents studies for periods of service as long as can be expected with permanent implants in man.

In Vivo Evaluation with Animal Models of Disease

A first approximation of the effectiveness of a device in replacing a physiologic function can be obtained after removing the target organ in a normal animal. However, when the organ failure is only the cardinal sign of a complex systemic disease, the interactions between the device and the persisting manifestations of the disease occur spontaneously in some species and in other cases can be obtained by chemical, physical, or surgical intervention. Where such models of disease exist in animals which can be fitted with a device, useful information is obtained which helps to refine the final prototypes.

Controlled Clinical Trials

Although some devices can be evaluated with little risk in normal volunteers who derive no health benefit from the experiment, our culture frowns on this approach and legal considerations discourage it. Once reliability and effectiveness have been established through animal experiments and the device appears to meet a recognized clinical need, a study protocol is typically submitted to an appropriate ethics committee or institutional review board and, upon their approval, a series of clinical trials is undertaken. The first step often concentrates on the demonstration of safety of the device with a careful watch for side effects or complications. If the device passes this first hurdle, a controlled clinical trial will be carried out with patients to evaluate effectiveness as well as safety on a scale which allows statistical comparison with a control form of treatment. This protocol may extend from a few months to several years depending upon the expected benefits of the device and the natural history of the disease.

General Clinical Use

Once a device is deemed successful by a panel of experts, it may be approved by regulatory agencies for commercial distribution. Increasingly a third stage of clinical evaluation appears necessary, namely postmarket surveillance, i.e., a system of clinical outcomes analysis under conditions of general availability of the device to a wide range of doctors and patients.

Postmarket surveillance is a new concept which is not yet uniformly codified. It may take the form of a data collection and analysis network, a patient registry to allow continuing follow-up and statistical analysis, a device-tracking system aimed at early identification of unforseen types of failure, or ancillary controls such as inspection of facilities and review of patient histories in institutions where devices are used. Protocols of surveillance on a large scale are difficult and costly to implement and their cost-effectiveness is therefore open to question. They are also impaired by the shortage of broadly available and minimally invasive diagnostic methods for assessing the integrity or function of a device prior to catastrophic failure. Worthwhile postmarket surveillance requires a constructive collaboration between patients, doctors, device manufacturers, government regulatory agencies, and study groups assessing health care policy issues in the public and private sectors.

Acknowledgments

The chapters that follow are derived to a substantial extent from lecture notes used in graduate courses at Brown University, the Massachusetts Institute of Technology, and the Georgia Institute of Technology. Colleagues at other institutions have also contributed chapters in their own areas of specialization. The authors are also indebted to successive generations of students of biomedical engineering who through a fresh, unencumbered look at the challenges of organ replacement have demonstrated their curiosity, their creativity, and their analytical skills.

In Memoriam

This introduction was written by Pierre Galletti, Ph.D., who passed away in 1997. It is retained here as a memorial to his significant contributions to the field of Biomedical Engineering.

Defining Terms

Artificial organs: Human-made devices designed to replace, duplicate, or augment, functionally or cosmetically, a missing, diseased, or otherwise incompetent part of the body, either temporarily or permanently, and which require a nonbiologic material interface with living tissue.

Assist device: An apparatus used to support or partially replace the function of a failing organ.

Bioartificial organ: Device combining living elements (organelles, cells, or tissues) with synthetic materials in a therapeutic system.

Bicompatibility: The ability of a material to perform with an appropriate host tissue response when incorporated for a specific application in a device, prosthesis, or implant.

Biomaterial: Any material or substance (other than a drug) or combination of materials, synthetic or natural in origin, which can be used as a whole or as a part of a system which treats, augments, or replaces any tissue, organ, or function of the body.

Compatibility: A material property which encompasses a set of specifications and constraints relative to material-tissue interactions.

Device: Defined by Congress as "…an instrument, apparatus, implement, machine, contrivance, implant, in vitro reagent, or other similar or related article, including any component, part, or accessory, which is…intended for use in the diagnosis of disease or other conditions, or in the cure, mitigation, treatment, or prevention of disease, in man or other animals,…and which does not achieve any of its principal intended purposes through chemical action within or on the body of man or other animals and which is not dependent upon being metabolized for the achievement of any of its principal intended purposes."

Hemocompatibility: The ability of a biomaterial to stay in contact with blood for a clinically relevant period of time without causing alterations of the blood constituents.

Hybrid artificial organs: Synonym of *bioartificial organs*, stressing the combination of cell transplantation and artificial organ technology.

Implant: Any biomaterial or device which is actually embedded within the tissues of a living organism.

Organs: Differentiated structures or parts of a body adapted for the performance of a specific operation or function.

Organoid: An organlike aggregate of living cells and synthetic polymer scaffolds or envelopes, designed to provide replacement or support function.

Organ transplant: An isolated body part obtained from the patient, a living relative, a compatible cadaveric donor, or an animal and inserted in a recipient to replace a missing function.

Prosthesis: An artificial device to replace a missing part of the body.

Substitutive medicine: A form of medicine which relies on the replacement of failing organs or body parts by natural or human-made counterparts.

Tissue-material interface: The locus of contact and interactions between a biomaterial, implant, or device and the tissue or tissues immediately adjacent.

References

Cauwels JM. 1986. The Body Shop: Bionic Revolutions in Medicine, St. Louis, Mosby.

Galleti PM. 1991. Organ replacement: A dream come true. In R Johnson-Hegyeli, AM Marmong du Haut Champ (eds), Discovering New Worlds in Medicine, pp 262–277, Milan, Farmitalia Carlo Erba.

Galletti PM. 1992. Bioartificial organs. Artif Organs 16(1):55.

Galletti PM. 1993. Organ replacement by man-made devices. J Cardiothor Vasc Anesth 7:624.

Harker LA, Ratner BD, Didisheim P. 1993. Cardiovascular biomaterials and biocompatibility. A guide to the study of blood-tissue material interactions. Cardiovasc Pathol 2(3)(suppl):IS–2245.

Helmus MN. 1992. Designing critical medical devices without failure. Spectrum, Diagnostics, Medical Equipment Supplies Opthalmics, pp 37–1–37–17, Decision Resources, Inc.

Richardson PD. 1976. Oxygenator testing and evaluation: Meeting ground of theory, manufacture and clinical concerns. In WM Zapol, J Qvist (eds), Artificial Lungs for Acute Respiratory Failure. Theory and Practice. pp 87–102, New York, Academic Press.

Further Information

The articles and books listed above provide different overviews in the filed of organ replacement. Historically most contributions to the filed of artificial organs were described or chronicled in the 40 annual volumes of the *Transactions of the American Society of Artificial Internal Organs* (1955 to present). The principal journals in the field of artificial organs are *Artificial Organs* (currently published by Blackwell Scientific Publications, Inc.), the *ASAIO Journal* (published by J. B. Lippincott Company), and the *International Journal of Artificial Organs* (published by Wichtig Editore s.r.l. Milano, Italy). Publications of the Japanese Society for Artificial Organs (typically in Japanese with English abstracts, *Artificial Organs Today*, for example) contain substantial information.

126

Artificial Heart and Circulatory Assist Devices

Gerson Rosenberg
Pennsylvania State University

In 1812, LeGallois [1813] postulated the use of mechanical circulatory support. In 1934, DeBakey proposed a continuous flow blood transfusion instrument using a simple roller pump [DeBakey, 1934]. In 1961, Dennis et al., performed left heart bypass by inserting cannulae into the left atrium and returning blood through the femoral artery [Dennis, 1979]. In 1961, Kolff and Moulopoulos developed the first intra-aortic balloon pump [Moulopoulos et al., 1962]. In 1963, Liota performed the first clinical implantation of a pulsatile left ventricular assist device [Liotz et al., 1963]. In 1969, Dr. Denton Cooley performed the first total artificial heart implantation in a human [Cooley et al., 1969]. Since 1969, air driven artificial hearts and ventricular assist devices have been utilized in over 200 and 600 patients, respectively. The primary use of these devices has been as a bridge to transplant. Recently, electrically powered ventricular assist devices have been utilized in humans as a bridge to transplant. These electrically powered systems utilize implanted blood pumps and energy converters with percutaneous drive cables and vent tubes.

Completely implanted heart assist systems requiring no percutaneous leads have been implanted in calves with survivals over eight months. The electric motor-driven total artificial hearts being developed by several groups in the United States are completely implanted systems employing transcutaneous energy transmission and telemetry [Rosenberg, 1991. Appendix C, *The Artificial Heart, Prototypes, Policies, and Patients*]. One total artificial heart design has functioned in a calf for over five months [Snyder, 1993].

There has been steady progress in the development of artificial heart and circulatory assist devices. Animal survivals have been increasing, and patient morbidity and mortality with the devices has been decreasing. It does not appear that new technologies or materials will be required for the first successful clinical application of long-term devices. There is no doubt that further advances in energy systems, materials, and electronics will provide for smaller and more reliable devices, but for the present, sound engineering design using available materials and methods appear to be adequate to provide devices satisfactory for initial clinical trials.

0-8493-0462-8/00/$0.00+$.50

126.1 Engineering Design

A definition for design is given by Pahl and Beitz [1977]: "Designing is the intellectual attempt to meet certain demands in the best possible way. It is an engineering activity that impinges on nearly every sphere of human life, relies on the discoveries and laws of science, and creates the conditions for applying these laws to the manufacture of useful products." Designing is a creative activity that calls for sound grounding in mathematics, physics, chemistry, mechanics, thermodynamics, hydrodynamics, electrical engineering, production engineering, materials technology, and design theory together with practical knowledge and experience in specialist fields. Initiative, resolution, economic insight, tenacity, optimism, sociability, and teamwork are all qualities that will assist the designer. Engineering design has been broken into many steps or phases by various authors. In general, though, each of these definitions includes some common tasks. No matter what device is being designed, be it a complicated device such as a totally implantable artificial heart or a simpler device, such as a new fastener, sound engineering design principles utilizing a methodical approach will help ensure satisfactory outcome of the design process. The engineering design process can be broken down into at least four separate stages.

1. *Define the problem—clarification of the task.* At first it may appear that defining the problem or clarifying the task to be accomplished is an easy matter. In general, this is not the case. Great care should be taken in defining the problem, being very specific about the requirements and specifications of the system to be designed. Very often, a complex system can be reduced to the solution of one core problem. An excellent way to begin defining the problem or clarifying the task to be accomplished is to begin by writing a careful design specification or design requirement. This is a document that lists all the requirements for the device. Further, this design requirement may elucidate one or two specific problems which, when solved, will yield a satisfactory design.
2. *Conceptual design—plan treatment.* After the problem has been defined, the designer must plan the treatment of the problem and begin conceptual design. In this phase possible solutions to the problem at hand are examined. Various methods of determining possible solutions include brainstorming, focus groups, Delphi method. This is the phase of the design process where a thorough review of the literature and examination of similar problems is valuable. In this phase of design each of the proposed solutions to the problem should be examined in terms of a hazard analysis or failure modes and effects analysis to determine which solution appears the most feasible. Economic considerations should also be examined in this phase.
3. *Detailed design—Execute the plan.* In this phase of engineering design, a detailed design is formulated. Perhaps two designs may be evaluated in the initial detailed design phase. As the detailed design or designs near completion, they must be examined with reference to the design specifications. Here, each of the proposed designs may be evaluated with regard to its ability to perform. Such aspects as system performance, reliability, manufacturability, cost, and user acceptance are all issues which must be considered before a final design is chosen.
4. *Learn and generalize.* Finally, after the design is complete, the designer should be able to learn and generalize from the design. This educational process will include manufacturing of prototypes and testing. General concepts and principles may be gleaned from the design process that can be applied to further designs.

The remainder of this chapter will deal with the application of this engineering design method to the design of artificial heart and circulatory assist devices.

126.2 Engineering Design of Artificial Heart and Circulatory Assist Devices

Define the Problem—Clarification of the Task

In the broadest sense, this step can best be accomplished by writing the detailed design requirement or specification for the device. In defining the problem, it is often easiest to begin with the most obvious and imperative requirements and proceed to the subtler and less demanding. A general statement of the problem for a total artificial heart or assist device is "to develop a device (perhaps totally implantable) that when implanted in the human will provide a longer and better quality of life than conventional pharmacologic or transplant therapy." The devices considered will be assumed to be permanent implantable devices, not necessarily totally implantable; they may utilize percutaneous wires or tubes. In general, it will be assumed that they will be intended for long-term use (one year or longer) but may also be utilized for short-term applications.

Fit of the System

One must first decide who the device is intended for. Will it be used in men and women, and of what size? No matter how good the device is, it must first "fit" the patient. For our example, let us assume the device will be used in men and women in the size range of 70–100 kg. The device must then fit in these patients and cause minimal or no pathologic conditions. When considering the fit of the implanted device, one must consider the volume and mass of the device, as well as any critical dimension such as the length, width, or height and the location of any tubes, conduits, or connectors. Careful consideration must be given to the physical attributes of the system such as whether the system should be hard, soft, rough, or smooth and the actual shape of the system in terms of sharp corners or edges that may damage tissue or organs. The design specification must give the maximum length, height, and width of the device, these dimensions being dictated by the patient's anatomy. The designer must not be limited by existing anatomy; the opportunity to surgically alter the anatomy should be considered. Nontraditional locations for devices should be considered.

The device should not project heat in such a way that surfaces in contact with tissue or blood are subjected to a temperature rise 5°C above core temperature on a chronic basis. The use of heat spreaders or fins, along with insulation, may be required. Heat transfer analysis may be helpful.

The effect of device movement and vibration should be considered in the design specification. The acceptable sound levels at various frequencies must be specified. A device should meet existing standards for electromagnetic interference and susceptibility.

The use of any percutaneous tubes will require the choice of an exit site. This site must not be in a location of constant or excessive movement or tissue breakdown will be experienced at the interface.

Pump Performance

Pump performance must be specified in terms of cardiac output range. In a 70-kg person, a normal resting cardiac output is approximately 70ml/min/kg or 5 1/min. Choosing a maximum cardiac output of approximately 8 1/min will allow the patient the ability to perform light exercise. The cardiac output performance must be obtained at physiologic pressures, that is, the device, be it a heart assist or total artificial heart device, must be able to pump a cardiac output ranging up to 8 1/min with physiologic inlet and outlet pressures (central venous pressure ~ 5 mmHg mean, left atrial pressure ~ 7 mmHg mean, pulmonary artery pressure ~ 15 mmHg mean, aortic pressure ~ 100 mmHg mean).

Control of the device is critical and must be included in the design specification. For an assist pump, it may be as simple as stating that the pump will pump all the blood delivered to it by the native heart while operating under physiologic inlet and outlet pressures. Or, the design specification may include specific requirements for synchronization of the device with the natural heart. For the total artificial heart, the device must always maintain balance between the left and right pumps. It must not let left atrial pressure rise above a value that will cause pulmonary congestion (approximately 20 mmHg). The device must respond to the patient's cardiac output requirements. The device must either passively or though an active controller vary its cardiac output upon patient demand.

Bicompatibility

Bicompatibility has already been alluded to in the design requirements by saying that the device must not cause excessive damage to the biologic system. Specifically, the device must be minimally thrombogenic and minimally hemolytic. It should have a minimal effect on the immune system. It should not promote infection, calcification, or tissue necrosis. Meeting these design requirements will require careful design of the pumping chamber and controller and careful selection of materials.

Reliability

The design specification must assign a target reliability for the device. For total artificial hearts and circulatory assist devices, the NIH has proposed a reliability of 80% with an 80% confidence for a 2-year device life. This is the value that the NIH feels is a reliability goal to be achieved before devices can begin initial clinical trials, but the final design reliability may be much more stringent, perhaps 90% reliable with 95% confidence for a 5-year life. The design specification must state which components of the system could be changed if necessary. The design specification must deal with any service that the device may require. For instance, the overall design life of the device may be 5 years, but battery replacement at 2-year intervals may be allowed. The reliability issue is very complex and involves moral, ethical, legal, and scientific issues. A clear goal must be stated and is necessary before the detailed design can begin.

Quality of Life

The design specification must address the quality of life for the patient. The designers must specify what is a satisfactory quality of life. Again, this is not an easy task. Quantitative measures of the quality of life are difficult to achieve. One person's interpretation of a satisfactory quality of life may not be the same as another's. It must always be kept in mind that the quality of life for these patients without treatment would generally be considered unsatisfactory by the general public. The prognosis for patients before receiving artificial hearts and circulatory assist devices is very poor. The quality of life must thus be considered in relation to the patient's quality of life without the device without ignoring the quality of life of individuals unaffected by cardiac disease [Rosenberg, 1991].

The design specification must state the weight of any external power supplies. The designer must consider how much weight an older patient will be able to carry. How long should this patient be able to be free-roaming and untethered? How often will the energy source require a "recharge?" What sound level will be acceptable? These are all issues that must be addressed by the designer.

All the foregoing must be considered and included in the definition of the problem and clarification of the task. Each of these issues should be clearly described in the design specification or requirement. In many instances there are no right and wrong answers to these questions.

Conceptual Design—Plan Treatment

In the conceptual design phase, the designer must plan the treatment of the problem and consider various designs that meet the design specification. In the design specification, it must be stated whether the blood pump is to be pulsatile or nonpulsatile. If there is no requirement in the design specification, then in the conceptual design phase, the designer may consider various nonpulsatile and pulsatile flow devices. Nonpulsatile devices include centrifugal pumps, axial flow pumps, shear flow pumps, and peristaltic pumps. Pulsatile pumps have traditionally been sac- or diaphragm-type devices. At the present time there is no definitive work describing the absolute requirement for pulsatility in the cardiovascular system; thus, both types of devices can be considered for assist and total artificial hearts.

In the conceptual design phase, the designer should consider other nontraditional solutions to the problem such as devices that employ micro-machines or magneto-hydrodynamics. Careful consideration must be given to source of energy. Sources that have been considered include electrical energy stored in batteries or derived from piezoelectric crystals, fuel cells, and thermal energy created either thermonuclearly or through thermal storage. The performance of each of these energy sources must be considered in the conceptual design phase. Public considerations and cost for a thermonuclear source, at the present time, have essentially eliminated this as an implantable energy source. Thermal storage has been shown to be a feasible energy source, and adequate insulation has been developed and demonstrated to be reliable for short periods of time [Uringer, 1989].

Steady flow devices that employ seals within the blood stream deserve very careful consideration of the seal area. These seals have been prone to failure, causing embolization. Active magnetic levitation has been proposed for component suspension and may be considered as a possible design. A device that utilities a rotating member such as a rotating electric motor that would drive an impeller or mechanical mechanism will create forces on the tissue when the patient moves due to gyroscopic or coriolis accelerations. These forces must be considered.

Pump performance, in terms of purely hydraulic considerations, can be achieved with any of the pumping systems described. Control of these devices may be more difficult. Designs that include implanted sensors such as pressure or flow sensors must deal with drift of these signals. Very often, signal-to-noise ratios are poor in devices that must operate continuously for several years. Designs that employ few or no sensors are preferable.

Systems such as sac-type blood pumps have been described as having intrinsic automatic control. That is, these devices can run in a fill limited mode, and, as more blood is returned to the pump, it will increase its stroke volume. Intrinsic control is desirable, but, unfortunately, this generally provides for only a limited control range. Nearly all devices currently being developed employ some form of automatic electronic control system. These automatic control systems are utilized on both total artificial hearts and assist devices. In some designs system parameters such as blood pump fill time, electric current, voltage, or speed are utilized to infer the state of the circulatory system. These types of systems appear to demonstrate good long-term stability and eliminate the potential for device malfunction associated with transducer drift.

Consideration of pump performance and the interaction with the biologic system is important in the conceptual design phase. Two pump designs may be capable of pumping the same in a hydrodynamic sense in terms of pressures and flows, but one pump may have much higher shear stresses than the other and thus be more hemolytic. One device may have much more mechanical vibration or movement compromising surrounding tissue.

The subject of biocompatibility for circulatory assist and artificial hearts is a very complex one. No matter what type device is designed, its interaction with the environment is paramount. Blood-contacting materials must not cause thrombosis and should have a minimal effect on formed elements in the blood. In terms of tissue biocompatibility, the device should not have sharp corners or areas where pressure necrosis can occur. Both smooth and rough exterior surfaces of devices have been investigated. It appears that devices in contact with the pleura tend to form a thinner encapsulation when they are rough surfaced. Compliance chambers form much thinner capsules when they have a rough surface. In other areas, it is not entirely clear if a rough surface is advantageous. Tissue ingrowth into a rough surface makes removal of the device difficult.

The selection of materials in the design of these devices is limited. The designer's job would be made much easier if there were several completely biocompatible materials available. If the designer only had a perfectly nonthrombogenic material that had outstanding fatigue properties, the design of these devices would be greatly simplified! Unfortunately, at the present time the designer is limited to existing materials. Traditionally, metals that have been employed in blood pumps include various stainless steels, cobalt, cobalt chromium alloys, and titanium. Each of these materials has adequate performance when in contact with tissue and blood under certain circumstances. Ceramic materials such as pyrolytic carbon, alumina, and zirconia have been used in contact with both tissue and blood with varying degrees of success. The range of polymeric materials that have been utilized for these devices is much greater. These materials include various polyurenthanes, silicone rubber, Kel-F, Teflon, Delrin, butyl rubber, Hexsyn rubber, polysulfone, polycarbonate, and others [Williams, 1981]. The designer must carefully consider all the properties of these materials when employing them. The designer must look at the strength, durability, hardness, wear resistance, modulus of elasticity, surface energy, surface finish, and biocompatibility before choosing any of these materials.

The interaction between the biologic system and these materials is complex and may be strongly influenced by fluid mechanics if in contact with blood. The designer should carefully consider surface modification of any of these materials. Surface modification can be performed, including ion implantation or grafting of various substances to the surface to promote improved compatibility. Manufacturing and fabrication processes have a profound effect on the properties of these materials and must be carefully analyzed and controlled.

The designer must give a great deal of consideration to the fluid mechanics involved. It is well known that excessive shear stresses can promote hemolysis and activation of the clotting system, as well as damage to the other formed elements in the blood. Not only the actual design of the blood pump, but the operation of the blood pump can affect phenomena such as cavitation, which can be hemolytic and destructive to system components. Thrombosis is greatly affected by blood flow. Regions of stagnation, recirculation, and low wall shear stress should be avoided. The magnitude of "low" shear stress is reported in the literature with wide range [Folie & McIntire, 1989; Hashimoto et al., 1985; Hubbel & McIntire, 1986].

Many of the analytical tools available today in terms of computational fluid dynamics and finite element analysis are just beginning to be useful in the design of these devices. Most of these systems have complex flow (unsteady, turbulent, non-Newtonian with moving boundaries) and geometries and are not easily modeled.

Once a reliability goal has been established, the designer must ensure that this goal is met. The natural heart beats approximately 40 million times a year. This means that an artificial heart or assist device with a 5-year life may undergo as may as 200,000,000 cycles. The environment in which this device is to operate is extremely hostile. Blood and extracellular fluids are quite corrosive and can promote galvanic or crevice corrosion in most metallic materials. Devices that employ polymeric materials must deal with diffusion of mass across these materials. Water, oxygen, nitrogen, carbon dioxide, and so on may all diffuse across these polymeric materials. If temperature fluctuations occur or differences exist, liquid water may form due to condensation. With carbon dioxide present, a weak acid can be formed. Many polymeric materials are degraded in the biologic environment. A careful review of the literature is imperative. The use of any "new" materials must be based upon careful testing. The designer must be aware of the difficulty in providing a sealed system. The designer may need to utilize hermetic connectors and cables which can tolerate the moist environment. All these affect the reliability of the device.

External components of the system which may include battery packs or monitoring and perhaps control functions have reliability requirements that differ from the implanted components. System components which are redundant may not require the same level of reliability as do nonredundant implanted components. Externally placed components have the advantage of being amenable to preventative maintenance or replacement. Systems that utilize transcutaneous energy transmission through inductive coupling may have advantages over systems that utilize a percutaneous wire. Although the percutaneous wire can function for long periods of time, perhaps up to 1 year, localized infection is almost always present. This infection can affect the patient's quality of life and be a source of systemic infection.

In the conceptual design phase, one must carefully weigh quality of life issues when evaluating solutions to the problem. Careful consideration needs to be given to the traditional quality of life issues such as the patient's emotional, social, and physical well-being, as well as to the details of day-to-day use of the device. What are the frequency and sound level requirements for patient prompts or alarms, and are certain kinds of alarms useful at all? For visible information for the patient, from what angles can this information be viewed, how bright does the display need to be, should it be in different languages, or should universal symbols be used? A great deal of thought must be given to all aspects of device use, from charging of the batteries to dealing with unexpected events. All these issues must be resolved in the conceptual design phase so that the detailed phase will be successful.

Detailed Design—Execute Plan

This is the phase of engineering design where the designer and other members of the team must begin to do what is generally considered the designer's more traditional role, i.e., calculations and drawings. This phase of design may require some initial prototyping and testing before the detailed design can be complete. Akutsu and Koyangi [1993] provide several results of various groups' detailed designs for artificial hearts and circulatory assist devices.

Learn and Generalize

Substantial research and development of artificial hearts and circulatory assist devices has been ongoing for almost 30 years. During this period there have been literally thousands of publications related to this research, not only descriptions of device designs, but experimental results, detailed descriptions fluid dynamic phenomena, hemolysis, thrombosis, investigation of materials selection and processing, consideration of control issues, and so on. The artificial organs and biomaterials literature has numerous references to materials utilized in these devices. A thorough review of the literature is required to glean the general principles that can be applied to the design of these devices. Many of these principles or generalities apply only to specific designs. In some design circulatory assist devices, a smooth surface will function satisfactorily, whereas a rough surface may cause thrombosis or an uncontrolled growth of neointima. Yet, in other design devices, a textured or rough surface will have performance superior to that of an extremely smooth surface. Wide ranges of sheer stresses are quoted in the literature that can be hemolytic. Although considerable research has been performed examining the fluid mechanics of the artificial hearts and circulatory assist devices, there is really no current measure of what is considered a "good" blood flow within these devices. We know that the design must avoid regions of stasis, but how close to stasis one can come, or for how long, without thrombosis is unknown.

It is up to the designer to review current literature and determine what fundamental principles are applicable to his or her design. Then, when the design is complete, the designer can learn and generalize specific principles related to her or his device. Hopefully, general principles that can apply to other devices will be elucidated.

126.3 Conclusions

The design of artificial hearts and circulatory assist devices is a very complex process involving many engineering disciplines along with medicine and other life science areas. Social issues must enter into the design process. The design of such devices requires sound engineering design principles and an interdisciplinary design team dedicated to the development and ultimate clinical application of these devices.

References

Akutsu T, Koyanagi H. 1993. Heart Replacement. Artificial Heart 4, Tokyo, Springer-Verlag.

Cooley DA, Liota D, Hallman GL, et al. 1969. Orthotopic cardiac prosthesis for 2-stage cardiac replacement. AM J Card 24:723.

DeBakey ME. 1934. A simple continuous flow blood transfusion instrument. New Orleans Med Surg J 87:386.

Folie BJ, McIntire LV. 1989. Mathematical analysis of mural thrombogenesis, concentration profiles of platelet-activating agents and effects of viscous shear flow. Biophys J 56:1121.

Hashimoto S, Maeda H, Sasada T. 1985. Effect of shear rate on clot growth at foreign surfaces. Artif Organs 9:345.

Hubbell JA, McIntire LV. 1986. Visualization and analysis of mural thrombogenesis on collagen, polyurethane and nylon. Biomaterials 7:354.

LeGallois CJJ. 1813. Experience on the Principles of Life, Philadelphia, Thomas. (Translation of CJJ Le Gallois 1812. Experiences sur les Principles de Vie. Paris, France.)

Liota D, Hall CW, Walter SH, et al. 1963. Prolonged assisted circulation during and after cardiac and aortic surgery. Prolonged partial left ventricular bypass by means of an intra-corporeal circulation. AM J Card 12:399.

Moulopoulos D, Topaz SR, Kolff WJ. 1962. Extracorporeal assistance to the circulation at intra-aortic balloon pumping. Trans AM Soc Artif Intern Organs 8:86.

Pahl G, Beitz W. 1977. Engineering Design, Berlin, Heidelberg, Springer-Verlag (First English edition published 1984 by The Design Council, London.)

Rosenberg G. 1991. Technological opportunities and barriers in the development of mechanical circulatory support systems (Appendix C). In Institute of Medicine, The Artificial Heart, Prototypes, Policies, and Patients, National Academy Press.

Snyder AJ, Rosenberg G, Weiss WJ, et al. 1993. *In vivo* testing of a completely implanted total artificial heart system. ASAIO J 39(3):M415.

Unger Felix 1989. Assisted Circulation 3, Berlin, Heidelberg, Springer-Verlag.

Williams DF. 1981. Biocompatibility of Clinical Implant Materials, Boca Raton, Fla, CRC Press.

127

Cardiac Valve Prostheses

Ajit P. Yoganathan
Georgia Institute of Technology

The first clinical use of a cardiac valvular prosthesis took place in 1952, when Dr. Charles Hufnagel implanted the first artificial caged ball valve for aortic insufficiency. The Plexiglas cage contained a ball occluder and was inserted into the descending aorta without the need for cardiopulmonary bypass. It did not cure the underlying disease, but it did relieve regurgitation from the lower two-thirds of the body.

The first implant of a replacement valve in the anatomic position took place in 1960, with the advent of cardiopulmonary bypass. Since then, the achievements in valve design and the success of artificial heart valves as replacements have been remarkable [Roberts, 1976]. More than 50 different cardiac valves have been introduced over the past 35 years. Unfortunately, after many years of experience and success, problems associated with heart valve prostheses have not been eliminated. The most serious problems and complications [Bodnar, Frater, 1991; Butchart, Bodnar, 1992; Giddens et al. 1993; Roberts, 1976] are:

- Thrombosis and thromboembolism
- Anticoagulant-related hemorrhage
- Tissue overgrowth
- Infection
- Paravalvular leaks due to healing defects, and
- Valve failure due to material fatigue or chemical change.

New valve designs continue to be developed. Yet to understand the future of valve replacements, it is important to understand their history.

0-8493-0462-8/00/$0.00+$.50

127.1 A Brief History of Heart Valve Prostheses

This section on replacement valves highlights a relatively small number of the many various forms which have been made. However, those that have been included are either the most commonly used today or those which have made notable contributions to the advancement of replacement heart valves [Brewer, 1969; Yoganathan et al., 1992].

Mechanical Valves

The use of the caged-ball valve in the descending aorta became obsolete with the development in 1960 of what today is referred to as the *Starr-Edwards ball-and-cage valve.* Similar in concept to the original Hufnagel valve, it was designed to be inserted in place of the excised diseased natural valve. This form of intracardiac valve replacement was used in the mitral position and for aortic and multiple replacements. Since 1962 the Starr-Edwards valve has undergone many modifications to improve its performance in terms of reduced hemolysis and thromboembolic complications. However, the changes have involved materials and techniques of construction and have not altered the overall concept of the valve design in any way (Fig. 127.1*a*).

Other manufacturers have produced variations of the ball and cage valve, notably the Smeloff-Cutter valve and the Magovern Prosthesis. In the case of the former, the ball is slightly smaller than the orifice. A subcage on the proximal side of the valve retains the ball in the closed position with its equator in the plane of the sewing ring. A small clearance around the ball ensures easy passage of the ball into the orifice. This clearance also gave rise to a mild regurgitation which was felt, but not proven, to be beneficial in preventing thrombus formation. The Magovern valve is a standard ball-and-cage format which incorporates two rows of interlocking mechanical teeth around the orifice ring. These teeth are used for inserting the valve and are activated by removing a special valve holder once the valve has been correctly located in the prepared tissue annulus. The potential hazard of dislocation from a calcified annulus due to imperfect placement was soon observed. This valve is no longer in use.

Due to the high-profile design characteristics of the ball valves, especially in the mitral position, low-profile caged disc valves were developed in the mid-1960s. Examples of the caged disc designs are the Kay-Shiley and Beall prostheses, which were introduced in 1965 and 1967, respectively (Fig. 127.1*b*). These valves were used exclusively in the atrioventricular position. However, due to their inferior hemodynamic characteristics, caged disc valves are rarely used today.

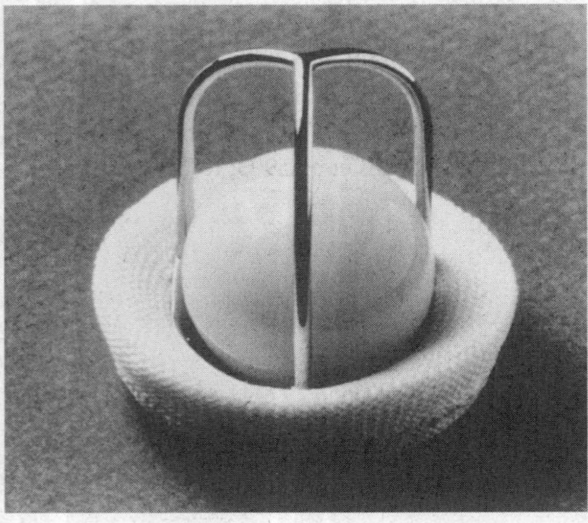

(a)

(b)

FIGURE 127.1 (*a*) Photograph of Starr-Edwards ball and cage valve; (*b*) photograph of Kay-Shiley disc valve; (*c*) photograph of Bjork-Shiley tilting disc valve; (*d*) photograph of Medtronic-Hall tilting disc valve; (*e*) photograph of St. Jude bileaflet valve; (*f*) photograph of CarboMedics bileaflet valve; (*g*) photograph of Parallel bileaflet valve.

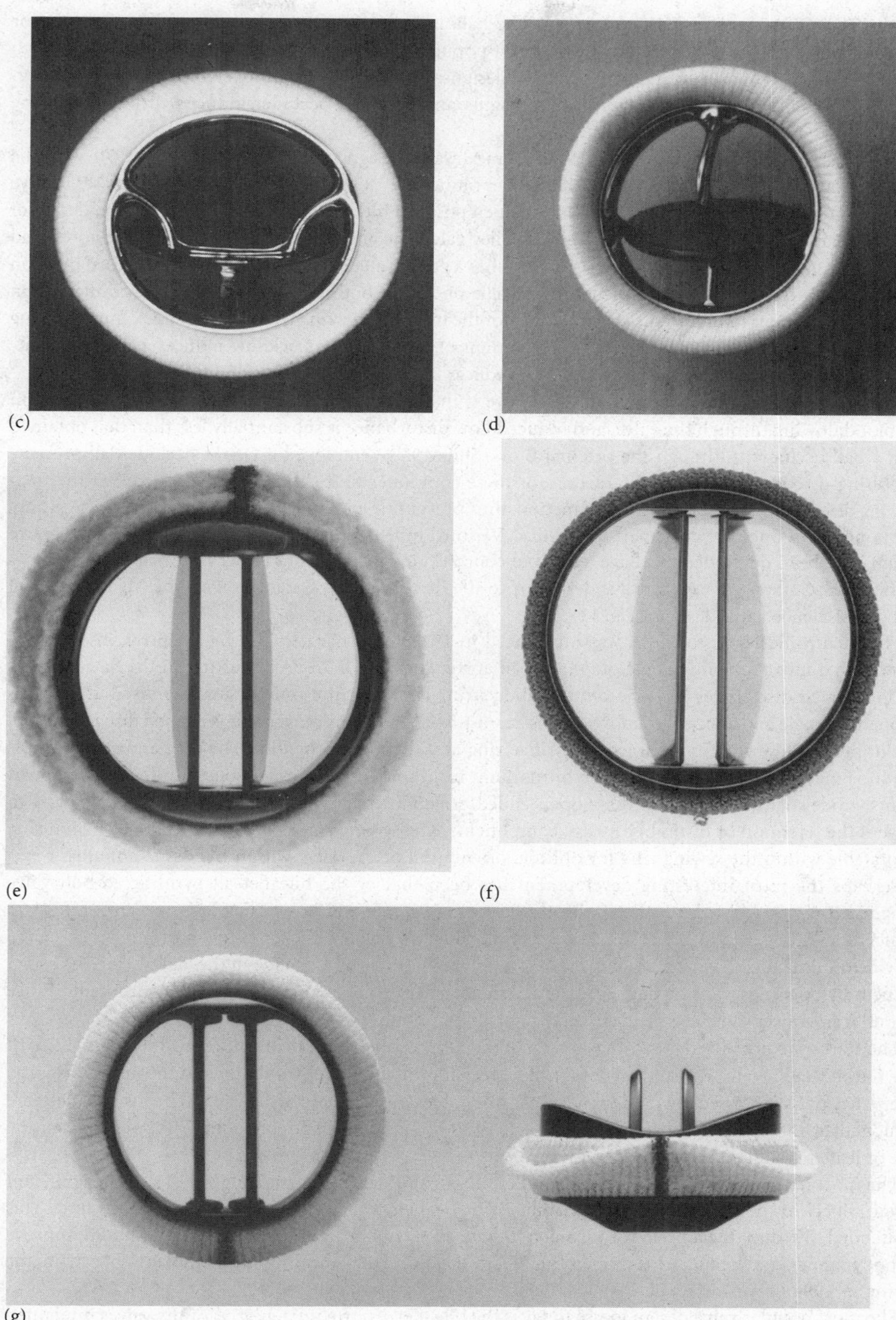

(c) (d) (e) (f) (g)

FIGURE 127.1 (continued)

Even after 35 years of valve development, the ball-and-cage format remains the valve of choice for some surgeons. However, it is no longer the most popular mechanical valve, having been superseded, to a large extent, by tilting-disc and bileaflet valve designs. These valve designs overcome two major drawbacks of the ball valve, namely, high profile heights and excessive occluder-induced turbulence in the flow through and distal to the valve.

The most significant developments in mechanical valve design occurred in 1969 and 1970 with the introduction of the Bjork-Shiley and Lillehei-Kaster tilting-disc valves (Fig. 127.1*c*). Both prostheses involve the concept of a free-floating disc which in the open position tilts to an angle depending on the design of the disc-retaining struts. In the original Bjork-Shiley valve, the angle of the tilt was 60° for the aortic and 50° for the mitral model. The Lillehei-Kaster valve has a greater angle of tilt of 80° but in the closed position is preinclined to the valve orifice plane by an angle of 18°. In both cases the closed valve configuration permits the occluder to fit into the circumference of the inflow ring with virtually no overlap, thus reducing mechanical damage to erythrocytes. A small amount of regurgitation backflow induces a "washing out" effect of "debris" and platelets and theoretically reduces the incidence of thromboemboli.

The obvious advantage of the tilting-disc valve is that in the open position it acts like an aerofoil in the blood flowing through the valve, and induced flow disturbance is substantially less than that obtained with a ball occluder. Although the original Bjork-Shiley valve employed a Delrin occluder, all present-day tilting-disc valves use pyrolitic carbon for these components. It should also be noted that the free-floating disc can rotate during normal function, thus preventing excessive contact wear from the retaining components on one particular part of the disc. Various improvements to this form of mechanical valve design have been developed but have tended to concentrate on alterations either to the disc geometry as in the Bjork-Shiley convexo-concave design or to the disc-retaining system as with the Medtronic-Hall and Omniscience valve designs (Fig. 127.1*d*).

The Medtronic-Hall prosthesis was introduced in 1977. It is characterized by a central, disc-control strut, with a mitral opening angle of 70° and an aortic opening of 75°. An aperture in the flat, pyrolitic carbon-coated disc affixes it to the central guide strut. This strut not only retains the disc but controls its opening angle and allows it to move downstream 1.5–2.0 mm; this movement is termed disc *translation* and improves flow velocity between the orifice ring and the rim of the disc. The ring and strut combination is machined from a single piece of titanium for durability. All projections into the orifice (pivot points, guide struts, and disc stop) are open-ended, streamlined, and in the region of highest velocity to prevent the retention of thrombi by valve components. The sewing ring is of knitted Teflon. The housing is rotatable within the sewing ring for optimal orientation of the valve within the tissue annulus.

Perhaps the most interesting development has been that of the bileaflet all-pyrolitic carbon valve designed by St. Jude Medical, Inc. and introduced in 1978 (Fig. 127.1*e*). This design incorporates two semicircular hinged pyrolitic carbon occluders (leaflets) which in the open position are intended to provide minimal disturbance to flow. The leaflets pivot within grooves made in the valve orifice housing. In the fully open position the flat leaflets are designed to open to an angle of 85°. The Duromedics valve is similar in concept to the St. Jude except that it incorporates curved leaflets.

The CarboMedics bileaflet prosthesis gained FDA approval for U.S. distribution in 1993 (Fig. 127.1*f*). The CarboMedics valve is also made of Pyrolite, which is intended for durability and thromboresistance. The valve has a recessed pivot design and is rotatable within the sewing ring. The two leaflets are semicircular, radiopaque, and open to an angle of 78°. A titanium stiffening ring is used to lessen the risk of leaflet dislodgment or impingement.

The most recent bileaflet design is the Parallel valve from Medtronic, Inc. (Fig. 127.1*g*). The significant design aspect of the Parallel valve is the ability of its leaflets to open to a position parallel to flow. This is intended to reduce the amount of turbulence that is created in the blood and therefore should improve hemodynamics and reduce thromboembolic complications. European clinical implants began in the Spring of 1994.

The most popular valve design in use today is the bileaflet. Approximately 75% of the valves implanted today are bileaflet prostheses.

Tissue Valves

Two major disadvantages with the use of mechanical valves is the need for life-long anticoagulation therapy and the accompanying problems of bleeding [Butchart & Bodnar, 1992]. Furthermore, the hemodynamic function of even the best designed valves differs significantly from that of the natural healthy heart valve. An obvious step in the development of heart valve substitutes was the use of naturally occurring heart valves. This was the basis of the approach to the use of antibiotic or cryotreated human aortic valves (homografts: from another member of the same species) removed from cadavers for implantation in place of a patient's own diseased valve.

The first of these homograft procedures was performed by Ross in 1962, and the overall results so far have been satisfactory. This is, perhaps, not surprising since the homograft replacement valve is optimum both from the point of view of structure and function. In the open position these valves provide unobstructed central orifice flow and have the ability to respond to deformations induced by the surrounding anatomical structure. As a result, such substitutes are less damaging to the blood when compared with the rigid mechanical valve. The main problem with these cadaveric allografts, as far as may be ascertained, is that they are no longer living tissue and therefore lack that unique quality of cellular regeneration typical of normal living systems. This makes them more vulnerable to long-term damage. Furthermore, they are only available in very limited quantities.

An alternative approach is to transplant the patient's own pulmonary valve into the aortic position. This operation was also the first carried out by Ross in 1967, and his study of 176 patients followed up over 13 years showed that such transplants continued to be viable in their new position with no apparent degeneration [Wain et al., 1980]. This transplantation technique is, however, limited in that is can only be applied to one site.

The next stage in development of tissue valve substitutes was the use of autologous fascia lata (a wide layer of membrane that encases the thigh muscles) either as free or frame-mounted leaflets. The former approach for aortic valve replacement was reported by Senning in 1966, and details of a frame-mounted technique were published by Ionescu and Ross in 1966 [Ionescu, 1969] The approach combined the more natural leaflet format with a readily available living autologous tissue. Although early results seemed encouraging, Senning expressed his own doubt on the value of this approach in 1971, and by 1978 fascia lata was no longer used in either of the above, or any other, form of valve replacement. The failure of this technique was due to the inadequate strength of this tissue when subjected to long-term cyclic stressing in the heart.

In parallel with the work on fascia lata valves, alternative forms of tissue leaflet valves were being developed. In these designs, however, more emphasis was placed on optimum performance characteristics than on the use of living tissue. In all cases the configuration involved a three-leaflet format which was maintained by the use of a suitably designed mounting frame. It was realized that naturally occurring animal tissues, if used in an untreated form, would be rejected by the host. Consequently, a method of chemical treatment had to be found which prevented this antigenic response but did not degrade the mechanical strength of the tissue.

Formaldehyde has been used by histologists for many years to arrest autolysis and "fix" tissue in the state in which it is removed. It had been used to preserve biologic tissues in cardiac surgery but, unfortunately, was found to produce shrinkage and increase the stiffness of the resulting material. For these reasons, formaldehyde was not considered ideal as a method of tissue treatment.

Glutaraldehyde is another histologic fixative which has been used especially for preserving fine detail for electron microscopy. It is also used as a tanning agent by the leather industry. In addition to arresting autolysis, glutaraldehyde produces a more flexible material due to increased collagen crosslinking. Glutaraldehyde has the additional ability of reducing the antigenicity of xenograft tissue to a level at which it can be implanted into the heart without significant immunologic reaction.

In 1969, Kaiser and coworkers described a valve substitute using an explanted glutaral dehyde-treated porcine aortic valve which was mounted on to a rigid support frame. Following modification in which the rigid frame was replaced by a frame having a rigid base ring with flexible posts, this valve became

(a)

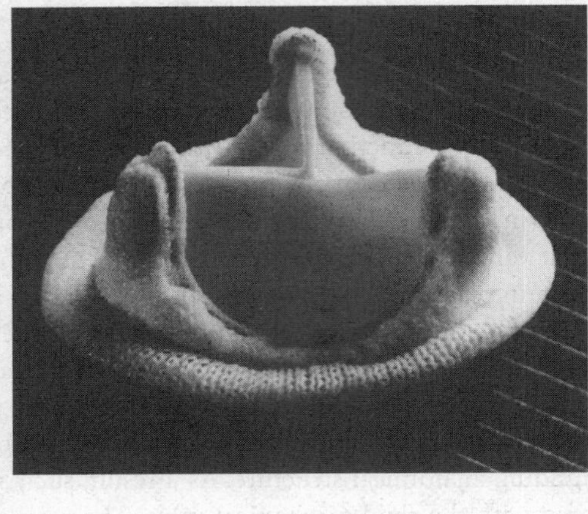
(b)

FIGURE 127.2 (*a*) Photograph of Hancock porcine valve; (*b*) photograph of Carpentier-Edwards pericardial valve.

commercially available as the Hancock Porcine Xenograft in 1970 (Fig. 127.2*a*). It remains one of the two most popular valve substitutes of this type, the other being the Carpentier-Edwards Bioprosthesis introduced commercially by Edwards Laboratories in 1976.This latter valve uses a totally flexible support frame.

In 1977 production began of the Hancock Modified Orifice (M.O.) valve, a refinement of the Hancock Standard valve. The Hancock M.O. is of a composite nature—the right coronary leaflet containing the muscle shelf is replaced by a noncoronary leaflet of the correct size from another porcine valve. This high-pressure fixed valve is mounted into a Dacron-covered polypropylene stent. The Hancock II and Carpentier-Edwards supra-annular porcine bioprostheses are second-generation bioprosthetic valve designs which were introduced in the early 1980s. The porcine tissue is initially fixed at 1.5 mmHg and then at high pressure. This fixation method is designed to ensure good tissue geometry. Both valves are treated with antimineralization treatments. Neither valve has been FDA approved for clinical use in the United States.

In porcine prostheses, the use of the intact biologically formed valve makes it unnecessary to manufacture individual valve cusps. Although this has the obvious advantage of reduced complexity of construction, it does require a facility for harvesting an adequate quantity of valves so that an appropriate range of valve sizes of suitable quality can be made available. This latter problem did not occur in the production of the three-leaflet calf pericardium valve developed by Ionescu and colleagues; the construction of this valve involved the molding of fresh tissue to a tricuspid configuration around a support frame. As the tissue is held in this position, it is treated with a glutaraldehyde solution. The valve, marketed in 1976 as the Ionescu-Shiley Pericardial Xenograft, was discontinued in the mid-1980s due to structural failure problems. Early clinical results obtained with tissue valves indicated their superiority to mechanical valves with respect to a lower incidence of thromboembolic complications [Bodnar & Yacoub 1991]. For this reason the use of tissue valves increased significantly during the late 1970s.

The Carpentier-Edwards pericardial valve consists of three pieces of pericardium mounted completely within the Elgiloy wire stent to reduce potential abrasion between the Dacron-covered frame and the leaflets. The pericardium is retained inside the stent by a Mylar button rather than by holding sutures. Its clinical implantation began in July 1980, and it is currently approved for clinical use in the United States (Fig. 127.2*b*).

Clinical experiences with different tissue valve designs have increasingly indicated time-dependent (5- to 7-year) structural changes such as calcification and leaflet wear, leading to valve failure and subsequent replacement [Ferrans et al., 1980; Oyer et al., 1979; Bodnar, Yacoub, 1986]. The problem of valve leaflet calcification is more prevalent in children and young adults. Therefore, tissue valves are rarely used in children and young adults at the present time. Such problems have not been eliminated by the glutaraldehyde tanning methods so far employed, and it is not easy to see how these drawbacks are to be overcome unless either living autologous tissue is used or the original structure of the collagen and elastin are chemically enhanced. On the latter point there is, as yet, much room for further work. For instance, the fixing of calf pericardium under tension during the molding of the valve cusps will inevitably produce "locked-in" stresses during fixation, thus changing the mechanical properties of the tissue.

127.2 Current Types of Prostheses

At present, over 175,000 prosthetic valves are implanted each year throughout the world. Currently, the four most commonly used basic types of prostheses are:

1. Caged ball
2. Tilting disc
3. Bileaflet
4. Bioprostheses (tissue valves)

Valve manufacturers continue to develop new designs of mechanical and tissue valves. The ideal heart valve prosthesis does not yet exist and may never be realized. However, the characteristics of the "perfect" prostheses should be noted. The ideal heart valve should:

- Be fully sterile at the time of implantation and be nontoxic
- Be surgically convenient to insert at or near the normal location of the heart
- Conform to the heart structure rather than the heart structure conforming to the valve (i.e., the size and shape of the prosthesis should not interfere with cardiac function)
- Show a minimum resistance to flow so as to prevent a significant pressure drop across the valve
- Have a minimal reverse flow necessary for valve closure, so as to keep the incompetence of the valve at a low level
- Show long resistance to mechanical and structural wear
- Be long-lasting (25 years) and maintain its normal functional performance (i.e., must not deteriorate over time)
- Cause minimal trauma to blood elements and the endothelial tissue of the cardiovascular structure surrounding the valve
- Show a low probability for thromboembolic complications without the use of anticoagulants
- Be sufficiently quiet so as not to disturb the patient
- Be radiographically visible
- Have an acceptable cost

127.3 Tissue Versus Mechanical

Tissue prostheses gained widespread use during the mid-1970s. The major advantage of tissue valves compared to mechanical valves is that tissue valves have a lower incidence of thromboembolic complications [Butchart & Bodnar, 1992]. Therefore, most patients receiving tissue valves do not have to take anticoagulants long-term. The major disadvantages to tissue valves are large pressure drops compared to some mechanical valves (particularly in the smaller valve sizes), jetlike flow through the valve leaflets, material fatigue and/or wear of valve leaflets, and calcification of value leaflets, especially in children and young adults. Valve deterioration, however, usually takes place slowly with tissue valves, and patients can be monitored by echocardiography and other noninvasive techniques.

The clear advantage of mechanical valves is their long-term durability. Current mechanical valves are manufactured from a variety of materials, such as pyrolitic carbon and titanium. Structural failure of mechanical valves is rare, but, when it occurs, is usually *catastrophic* [Giddens et al., 1993]. One major disadvantage of the use of mechanical valves is the need for continuous, life-long anticoagulation therapy to minimize the risk of thrombosis and thromboembolic complications. Unfortunately, the anticoagulation therapy may lead to bleeding problems; therefore, careful control of anticoagulation medication is essential for the patient's well-being and quality of life. Another concern is the hemodynamic performance of the prosthesis. The hemodynamic function of even the best designs of mechanical valves differs significantly from that of normal heart valves.

127.4 Engineering Concerns and Hemodynamic Assessment of Prosthetic Heart Valves

In terms of considerations related to heart valve design, the basic engineering concerns are

- Hydrodynamics/hemodynamics
- Durability (structural mechanics and materials)
- Biologic response to the prosthetic implant

The ideal heart valve design from the hemodynamic point of view should [Giddens et al., 1993]

- Produce minimal pressure gradient
- Yield relatively small regurgitation
- Minimize production of turbulence
- Not induce regions of high shear stress
- Contain no stagnation or separation regions in its flow field, especially adjacent to the valve superstructure

No valve as yet, other than normal native valves, satisfies all these criteria.

Pressure Gradient

The heart works to maintain adequate blood flow through a prosthetic valve; a well-designed valve will not significantly impede that blood flow and will therefore have as small a pressure gradient as possible across the valve.

Because of the larger separation region inherent in flow over bluff bodies, configurations such as the caged disc and caged ball have notably large pressure gradients. Porcine bioprostheses have relatively acceptable pressure gradients for larger diameter valves because they more closely mimic natural valve geometry and motion, but the small sizes (<23 mm) generally have higher pressure gradients than their mechanical valve counterparts, as shown in Fig. 127.3 [Yoganathan et al., 1984]. Tilting disc and bileaflet valve designs present a relatively streamlined configuration to the flow, and, although separation regions may exist in these designs, the pressure gradients are typically smaller than for the bluff shapes. The clinical importance of pressure gradients in predicting long-term performance is not clear. The fact that these gradients are a manifestation of energy losses resulting from viscous-related phenomena makes it intuitive that minimizing pressure gradients across an artificial valve is highly desirable in order to reduce the workload of the pump (i.e., left ventricle).

Effective Orifice Area (EOA)

The EOA is an index of how well a valve design utilizes its primary or internal stent orifice area. In other words, it is related to the degree to which the prosthesis itself obstructs the flow of blood. A larger EOA corresponds to a smaller pressure drop and therefore a smaller energy loss. It is desirable to have as large an EOA as possible. EOA is calculated from in vitro pressure drop measurements for a particular valve using the following formula [Yoganathan et al., 1984]:

$$\mathrm{EOA}\left(\mathrm{cm}^2\right) = \frac{Q_{\mathrm{rms}}}{51.6\sqrt{\Delta \bar{p}}}$$

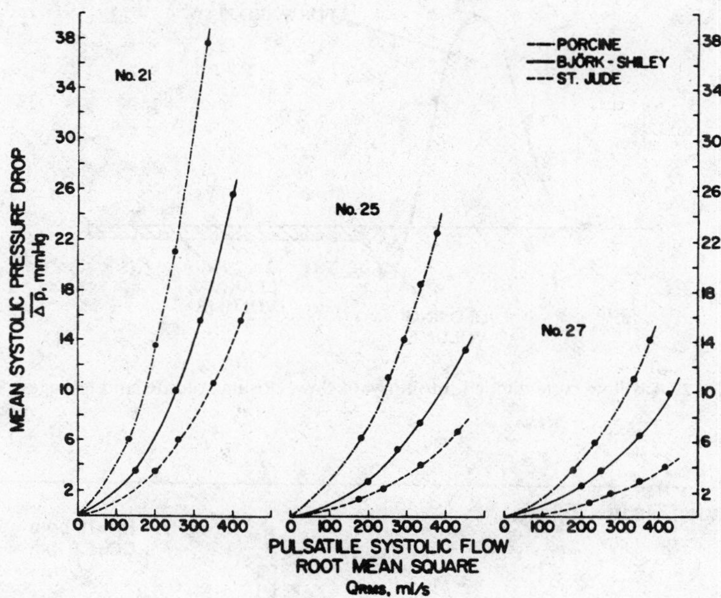

FIGURE 127.3 Examples of in vitro pulsatile flow pressure gradients across tilting disc (Bjork-Shiley convexo-concave), bileaflet (St. Jude Medical), and porcine aortic valves of three different sizes (27, 25, and 21 mm).

TABLE 127.1 Effecive Orifice Areas of Different Prosthetic Aortic Valve Designs

Valve Sewing Ring Diam, mm	Medronic-Hall Tilting Disc, cm^2	St. Jude Bileaflet cm^2	Carbomedics Bileaflet cm^2	Hancock I Porcine, cm^2	Hancock MO Porcine, cm^2	Carpentier-Edwards Pericardial, cm^2	Starr-Edwards 1260 Ball, cm^2
19/20	1.74	1.21	1.12	1.01	1.22	1.56	1.04
21	1.74	1.81	1.66	1.31	1.43	1.88	1.23
23	2.26	2.24	2.28	1.73	1.94	2.25	1.45
25	3.07	3.23	3.14	1.93	2.16	3.25	1.59
27	3.64	4.05	3.75	2.14	—	3.70	1.75

Q_{rms} is the root mean square systolic/diastolic flow rate (cm^3/s), $\Delta\bar{p}$ is the mean systolic/diastolic pressure drop (mmHg).

Table 127.1 lists EOAs obtained in vitro, for different size mechanical and tissue valve designs in clinical use today. These results illustrate the fact that, size for size, the newer mechanical valve designs have better pressure gradient characteristics than porcine bioprostheses in current clinical use.

Regurgitation

Regurgitation results from reverse flow created during valve closure and from backward leakage once closure is effected (see Fig. 127.4). Regurgitation reduces the net flow through the valve. Closing regurgitation is closely related to the valve shape and closing dynamics, and the percentage of stroke volume that succumbs to this effect ranges from 2.0–7.5% for mechanical valves. For tissue valves it is typically less: 0.1–1.5%. Leakage depends upon how well the orifices are "sealed" upon closure, and it has a reported incidence of 0–10% in mechanical valves and 0.2–3% in bioprosthetic valves. The overall tendency is for regurgitation to be less for the trileaflet bioprosthetic heart valves than for mechanical valve designs. Figure 127.5 illustrates in vitro regurgitant volumes (closing and leakage) measured on three commonly used mechanical valve designs in the aortic and mitral positions.

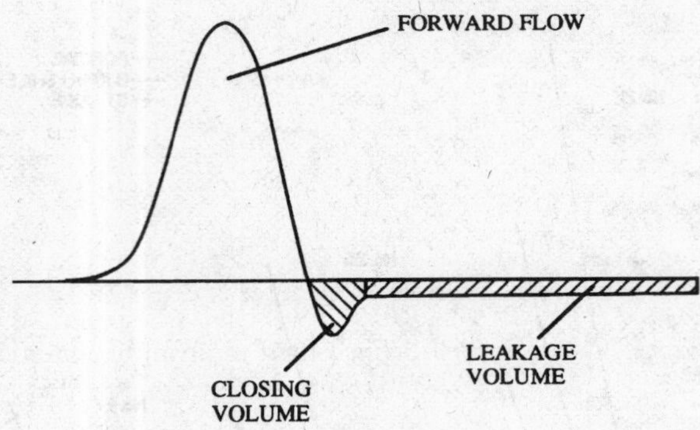

FIGURE 127.4 Flow cycle divided into forward flow, closing volume, and leakage volume.

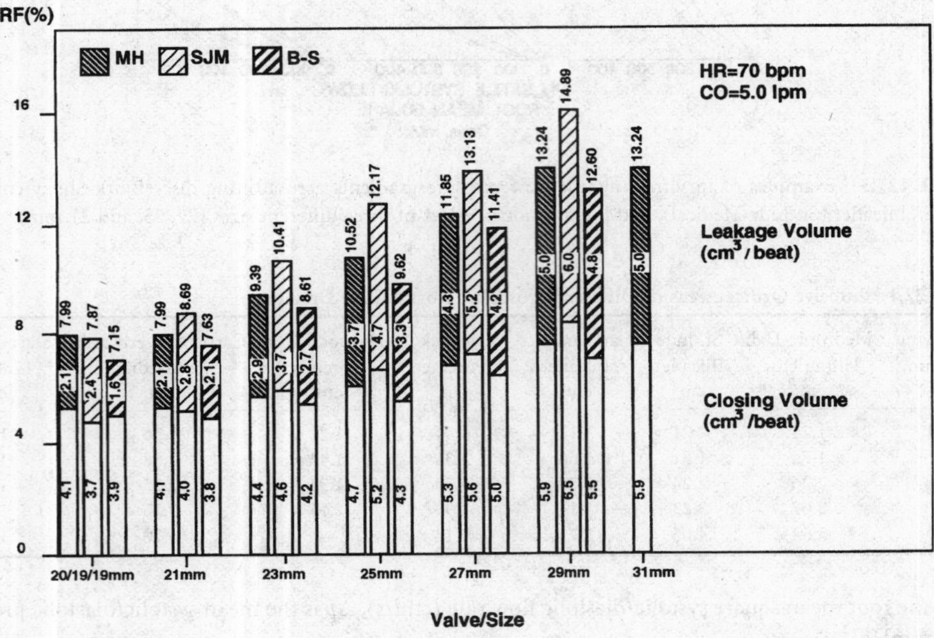

FIGURE 127.5 Examples of in vitro reguritant volumes (closing and leakage) with three mechanical valve designs (MH—Medtronic-Hall; SJM—St. Jude Medical; B—Bjork-Shiley mono-strut).

Regurgitation has implications other than simply for flow delivery. On the negative side, back flow through a narrow slit, such as can occur in leakage regurgitation through a bileaflet valve, can create relatively high laminar shear stresses, thus increasing the tendency toward blood cell damage [Baldwin, 1990; Cape et al., 1993]. However, regurgitation can have a beneficial effect in that the back flow over surfaces may serve to wash out zones that would otherwise have stagnant flow throughout the cycle. This is particularly true for the "hinge" region in some tilting disc and bileaflet designs.

Flow Patterns and Turbulent Shear Stresses

Thrombosis and embolism, tissue overgrowth, hemolysis, and damage to endothelium adjacent to the valve are directly related to the velocity and turbulence fields created by various valve designs and have been addressed in detail during the past decade by investigators studying cardiovascular fluid mechanics

[Chandran et al., 1983, 1984; Woo & Yoganathan 1985, 1986; Yoganathan et al., 1986*a*, 1986*b*; Yoganathan et al., 1988].

It has been established that shear stresses on the order of 1500–4000 dynes/cm^2 can cause lethal damage to red cells [Nevaril et al., 1969]. However, in the presence of foreign surfaces, red cells can be destroyed by shear stresses on the order of 10–100 dynes/cm^2 [Mohandas et al., 1974]. It has also been observed that sublethal damage to red cells can occur at turbulent shear stress levels of 500 dynes/cm^2 [Sutera & Merjhardi, 1975]. Platelets appear to be more sensitive to shear and can be damaged by shear stress on the order of 100–500 dynes/cm^2 [Ramstack, et al., 1979; Wurzinger et al., 1983]. Evidence that platelet activation, aggregation, and thrombosis is induced by fluid shear forces has been predominantly generated by viscometric studies performed under well-defined fluid mechanical conditions. In viscometers, the extent and reversibility of shear-induced platelet aggregation are a function of both the magnitude and duration of the applied shear stress. For example, at 150 dynes/cm^2, platelet aggregation is not observed until shear is applied for 300 s. But, as the intensity of the applied shear stress increases, platelet activation and aggregation occur more rapidly. For example, at a shear stress of 600 dynes/cm^2, platelet aggregation occurs within 30 s, and at 6500 dynes/cm^2, platelet activation occurs in fewer than 5 ms. As the magnitude of the applied shear increases, formed platelet aggregates tend not to separate when the shear forces are discontinued. Furthermore, platelet damage increases linearly with time of exposure to constant shear stress, indicating that shear-induced platelet damage is cumulative [Anderson & Hellums, 1978; Brown et al., 1977; Colantuoni et al., 1977].

Although the exact mechanism of turbulent stress damage to the cell is not precisely known, there is no disagreement that cell damage can be created by high turbulent stresses; minimizing these is conducive to better valve performance from the standpoints of thrombus formation, thromboembolic complications, and hemolysis and from energy loss considerations.

To illustrate the abnormal flow fields and elevated levels of turbulent shear stresses created by prosthetic valves, in vitro measurements conducted on 27-mm aortic valve designs in current clinical use in the United States (i.e., FDA approved) are presented below.

The figures of the in vitro flow field studies are presented as schematic diagrams and represent velocity and turbulence profiles obtained at peak systole, at a cardiac output of 6.01/min and a heart rate of 70 beats/min. All downstream distances are measured from the valve sewing ring. Tables 127.2 and 127.3 list the maximum and cross sectionally averaged mean turbulent shear stresses measured downstream of the valves at different times during systole and diastole.

Starr-Edwards Caged-Ball Valve (Model 1260)

The flow emerging from the valve formed a circumferential jet that separated from the ball, hit the wall of the flow chamber, and then flowed along the wall. The flow had very high velocities in the annular region. The maximum velocity, measured 12 mm downstream of the valve, was 220 cm/s at peak systole. The peak systolic velocity, measured 30 mm downstream of the valve, was 180 cm/s, as shown in Fig. 127.6. High-velocity gradients were observed at the edges of the jet. The maximum velocity gradient (1700 cm s^{-1}/cm) was observed in the annular region adjacent to the surface of the ball during peak systole. A large-velocity defect was observed in the central part of the flow chamber as a wake developed distal to the ball. A region of low velocity reverse flow was observed at peak systole and during the deceleration phase, with a diameter of about 8 mm immediately distal to the apex of the cage. The maximum reverse velocity measured was –25 cm/s; it occurred at peak systole 30 mm downstream of the valve (Fig. 127.6). The intensity of the reverse flow during the deceleration phase was not as high as that observed at peak systole. No reverse flow was observed during the acceleration phase. However, the velocity in the central part of the flow channel was low.

High turbulent shear stresses were observed at the edges of the jet. The maximum turbulent shear stress measured was 1850 dynes/cm^2, which occurred at the location of the highest-velocity gradient (Fig. 127.6). The intensity of turbulence during peak systole did not decay very rapidly downstream of the valve. Elevated turbulent shear stresses occurred during most of systole. Turbulent shear stresses as high as 3500 dynes/cm^2 were estimated in the annular region between the flow channel wall and the ball.

TABLE 127.2 Peak and Mean Turbulent Shear Stresses Measured Downstream of Different Aortic Valve Designs

		Acceleration Phase		Peak Systole		Deceleration Phase	
Valve	Location, mm	Peak, dynes/cm²	Mean, dynes/cm²	Peak, dynes/cm²	Mean, dynes/cm²	Peak, dynes/cm²	Mean, dynes/cm²
Starr-Edwards (1260)	26, centerline	750	450	1800	1000	1400	700
	30, centerline	600	200	1850	1100	1300	750
Medtronic-Hall	7, major orifice	450	250	1200	450	600	300
	7, minor orifice	950	400	1000	350	850	450
	13, centerline	1200	600	1000	550	800	400
	13, major orifice	400	320	1500	370	700	300
	13, minor orifice	1250	550	1450	700	700	450
	16, centerline	300	100	2000	1000	850	450
	15, 90 degree rotated (Fig.)	300	170	1450	700	900	450
St. Jude	8, centerline	820	450	1150	500	600	320
	8, 6.25mm lateral to centerline	1600	1050	2000	1000	1000	650
	13, centerline	950	470	1500	750	1400	900
	13, 6.25mm lateral to centerline	1400	800	2000	1050	1000	700
	11, 90-degree rotated (Fig.)	950	550	1700	1200	1000	700
Carpentier-Edwards	10, centerline	900	400	2750	1200	1750	1000
(2625 Porcine)	15, centerline	1400	700	4500	2000	1700	900
Hancock MO	10, centerline	1000	400	2900	1100	1150	550
(250 Porcine)	15, centerline	1750	950	2450	1900	2100	1400
Carpentier Edwards	17, centerline	500	100	850	200	1000	350
(Pericardial)	33, centerline	850	200	1130	450	900	370

Medtronic-Hall Tilting Disc Valve

High-velocity jetlike flows were observed from both the major and minor orifice outflow regions. The orientation of the jets with respect to the axial direction changed as the valve opened and closed. The major orifice jet was larger than the minor orifice jet and had a slightly higher velocity. The peak velocities measured 7 mm downstream of the valve were 210 cm/s and 200 cm/s in the major and the minor orifice regions, respectively (Fig. 127.7). A region of reverse flow was observed adjacent to the wall in the minor orifice region at peak systole, which extended 2 mm from the wall with a maximum reverse velocity of –25 cm/s. The size of this region increased during the deceleration phase to 8 mm from the wall. A small region of flow separation was observed adjacent to the wall in the major orifice region as illustrated by Fig. 127.7. In the minor orifice region, a profound velocity defect was observed 7 and 11 mm distal to the minor orifice strut (Fig. 127.7). Furthermore, the region adjacent to the wall immediately downstream from the minor orifice was stagnant during the acceleration and deceleration phases and had very low velocities (<15 cm/s) during peak systole.

In the major orifice region, high turbulent shear stresses were confined to narrow regions at the edges of the major orifice jet (Fig. 127.7). The peak turbulent shear stresses measured at peak systole were 1200 and 1500 dynes/cm², 7 and 13 mm downstream of the valve, respectively. During the acceleration and deceleration phases the turbulent shear stresses were relatively low. High turbulent shear stresses were more dispersed in the minor than those in the major orifice region as shown by Fig. 127.7. The turbulent shear stress profiles across the major and minor orifices 15mm downstream of the valve (Fig. 127.7) showed a maximum turbulent shear stress of 1450 dynes/cm² at the lower edge of the minor orifice jet.

St. Jude Medical Bileaflet Valve

The St. Jude Medical valve has two semicircular leaflets which divide the area available for forward flow into three regions: two lateral orifices and a central orifice. The major part of the forward flow emerged

TABLE 127.3 Peak and Mean Turbulent Shear Stresses Measured Downstream of Different Mitral Valve Designs

		Acceleration Phase		Peak Systole		Deceleration Phase	
Valve	Location, mm	Peak, dynes/cm^2	Mean, dynes/cm^2	Peak, dynes/cm^2	Mean, dynes/cm^2	Peak, dynes/cm^2	Mean, dynes/cm^2
Beall	11, centerline	500	150	1950	450	800	250
	11, 10 mm above centerline	360	150	1300	400	425	150
	17, centerline	375	125	700	225	400	175
Bjork-Shiley (cc)	8, 13 mm above centerline	320	70	235	60	240	60
	8, 10 mm below centerline	40	18	30	17	25	19
	12, centerline	100	40	330	75	150	55
	15, 90-degree rotated	310	85	375	100	280	60
	18, centerline	90	60	300	110	225	80
	18, 13 mm above centerline	155	80	350	240	210	140
	18, 10 mm below centerline	60	25	210	70	70	25
Medtronic-Hall	8, centerline	300	170	1800	400	650	180
	8, 12mm above line center	850	300	1150	450	500	200
	8, 9 mm below centerline	900	230	1200	350	225	130
	14, 90-degree rotated	620	230	1200	320	670	230
	18, centerline	340	100	500	230	300	150
	18, 12 mm above centerline	600	230	2600	900	950	500
	18, 9 mm below centerline	280	130	670	280	360	180
St. Jude	10, centerline	170	95	250	95	175	85
	10, 10 mm above centerline	310	120	440	150	185	100
	12, 90-degree rotated	725	270	770	300	670	250
	19, centerline	130	80	335	130	165	90
	19, 10 mm above centerline	115	50	575	170	250	110

from the two lateral orifices. The measurements along the centerline plane 8 mm downstream of the valve showed at peak systole a maximum velocity of 220 cm/s and 200 cm/s for the lateral and central orifice jets, respectively. The velocity of the jets remained about the same as the flow traveled from 8 to 13 mm downstream (Fig. 127.8). The velocity profiles showed two defects which corresponded to the location of the two leaflets. The velocity measurements indicated that the flow was more evenly distributed across the flow chamber during the deceleration than during the acceleration phase. Regions of flow separation were observed around the jets adjacent to the flow channel wall as the flow separated from the orifice ring. The measurements across the central orifice illustrated in Fig. 127.8 show that the maximum velocity in the central orifice was 220 cm/s. Small regions of low-velocity reverse flow were observed adjacent to the pivot/hinge mechanism of the valve (Fig. 127.8). More flow emerged from the central orifice during the deceleration phase than during the acceleration phase.

High turbulent shear stresses occurred at locations of high-velocity gradients and at locations immediately distal to the valve leaflets (Fig. 127.8). The flow along the centerline plane became more disturbed as the flow traveled from 8 to 13 mm downstream of the valve. The peak turbulent shear stresses measured along the centerline plane at peak systole were 1150 and 1500 dynes/cm^2 at 8 and 13 mm downstream of the valve, respectively. The profiles across the central orifice showed that the flow was very disturbed in this region. The maximum turbulent shear stress measured in the central orifice as shown in Fig. 127.8

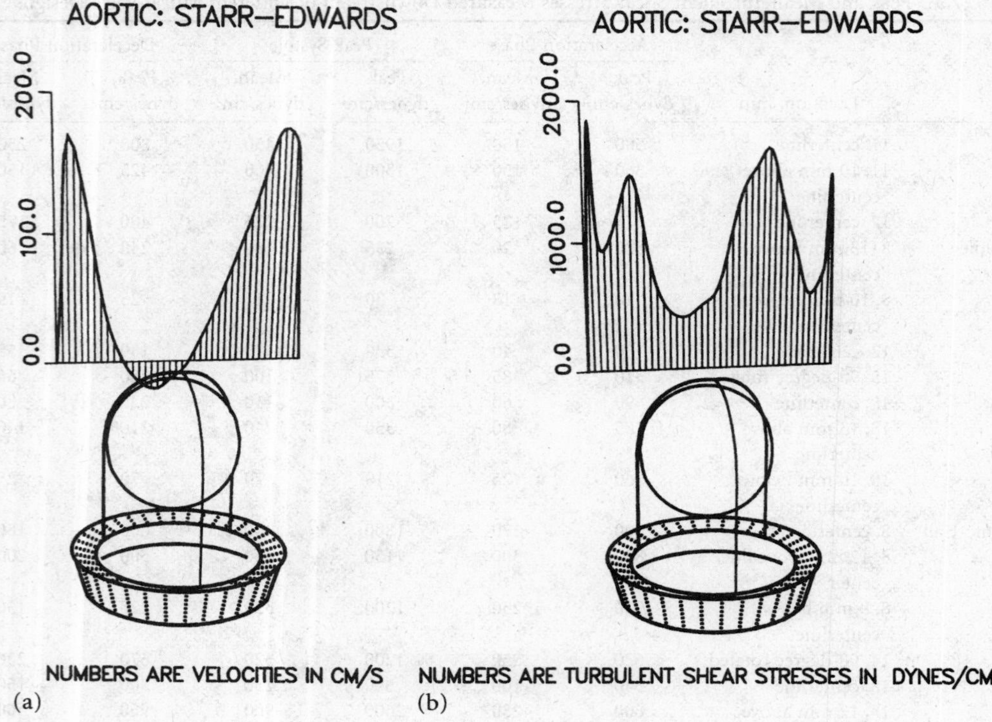

FIGURE 127.6 (*a*) Velocity profile 30 mm downstream on the centerline for the Starr-Edwards ball valve, at peak systole; (*b*) turbulent shear stress profile 30 mm downstream on the centerline for the Starr-Edwards ball valve, at peak systole.

(1700 dynes/cm^2) occurred peak systole. Since these high turbulent shear stresses across the central orifice were measured 11 mm downstream, it is probable that even higher turbulent shear stresses occurred closer to the valve.

Carpentier-Edwards Porcine Valve (Model 2625)

The velocity profiles taken 10 mm downstream of the valve, along the centerline plane, showed that the peak velocity of the jetlike flow emerging from the valve was as high as 220 cm/s at peak systole (Fig. 127.9). The peak velocities measured during the acceleration and deceleration phases were about the same, 175 and 170 cm/s, respectively. However, the flow was much more evenly distributed during the acceleration than during the deceleration phase. No regions of flow separation were observed throughout the systolic period in this plane of measurement. However, the annular region between the outflow surfaces of the leaflets and the flow chamber wall was relatively stagnant throughout systole. The velocity of the jet increased to about 370 cm/s at peak systole, as the flow traveled from 10 to 15 mm downstream of the valve. This indicated that the flow tended to accelerate toward the center of the flow channel.

High turbulent shear stresses occurred at the edge of the jet (Fig. 127.9). The maximum turbulent shear stress measured 10 mm downstream of the valve along the centerline plane at peak systole was 2750 dynes/cm^2. The turbulent shear stresses at the edge of the jet increased as the flow traveled from 10 to 15 mm downstream of the valve. The maximum and mean turbulent shear stress measured at peak systole increased to 4500 and 2000 dynes/cm^2, respectively (Table 127.2).

Hancock Modified Orifice Porcine Valve (Model 250)

In this design, a 25-mm valve was studied, since a 27-mm valve is not manufactured. The velocity measurements showed that this valve design also produced a high-velocity jetlike flow field with a

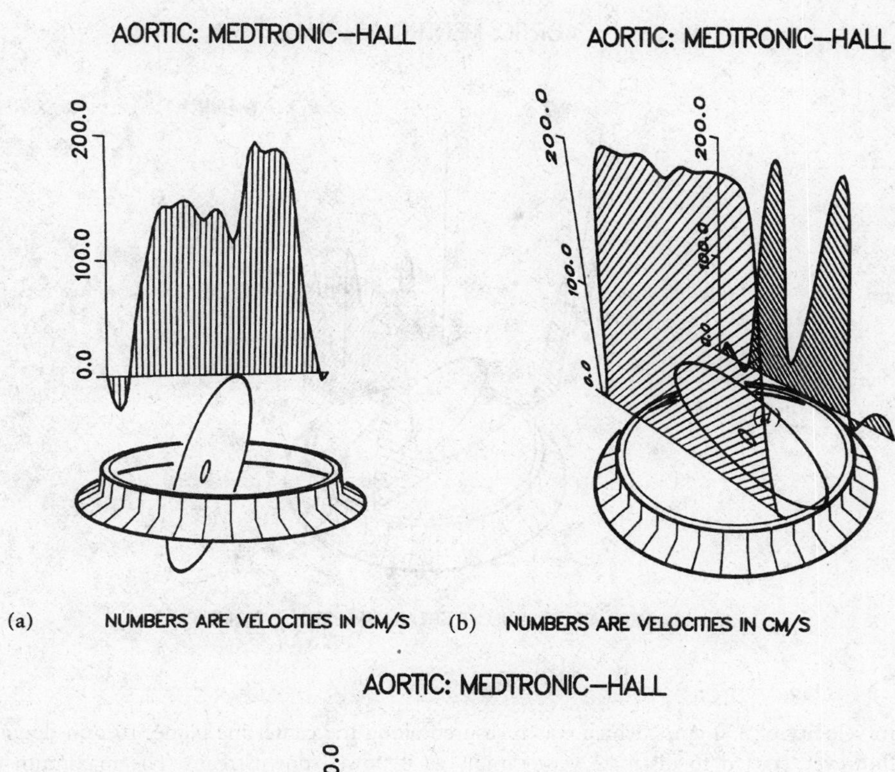

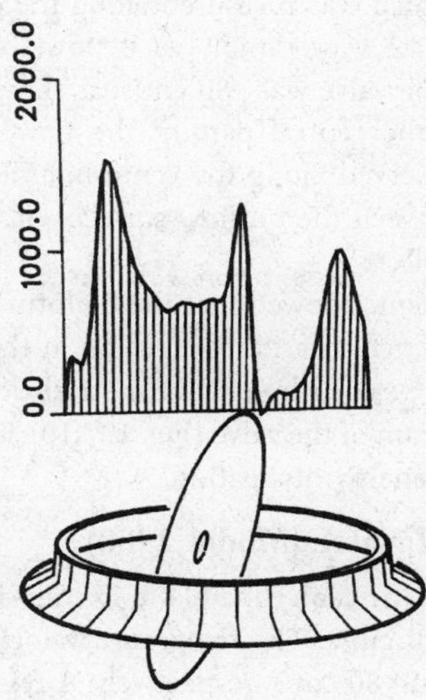

FIGURE 127.7 (*a*) Velocity profile 15 mm downstream on the centerline across the major and minor orifices of the Medtronic-Hall tilting disc valve (major orifice to the right), at peak systole; (*b*) velocity profiles 13 mm downstream in the major and minor orifices of the Medtronic-Hall tilting disc valve, at peak systole; (*c*) turbulent shear stress profile 15 mm downstream on the centerline across the major and minor orifices of the Medtronic-Hall tilting disc valve (major orifice to the right), at peak systole; (*d*) turbulent shear stress profiles 13 mm downstream in the major and minor orifices of the Medtronic-Hall tilting disc valve, at peak systole.

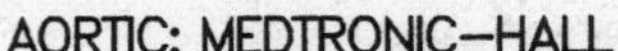

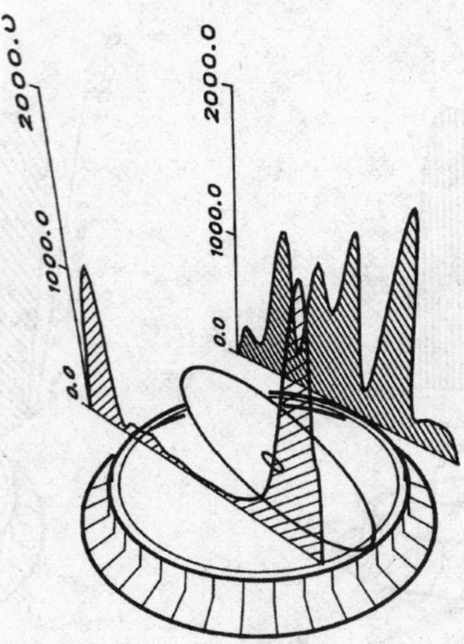

(d) NUMBERS ARE TURBULENT SHEAR STRESSES IN DYNES/CM^2

FIGURE 127.7 (continued)

maximum velocity of 330 cm/s, which was measured along the centerline plane, 10 mm downstream. The jet, however, started to dissipate very rapidly as it flowed downstream. The maximum velocity measured 15 mm downstream of the valve was 180 cm/s, as shown in Fig. 127.10. A velocity defect was observed 15 mm downstream in the central part of the flow channel peak systole and during the deceleration phase but was not observed along the centerline plane 10 mm downstream of the valve. Once again, the annular region between the outflow surfaces of the leaflets and the flow chamber wall was relatively stagnant during systole.

Turbulent shear stress measurements showed that the high turbulent shear stresses measured 10 mm downstream of the valve were confined to a narrow region at the edge of the jet, with a peak valve of 2900 dynes/cm^2 (Table 127.2). This peak turbulent shear stress decreased to 2400 dynes/cm^2 as the flow traveled from 10 to 15 mm downstream of the valve (Fig. 127.10). The region of high turbulence, however, became more diffuse as a result of energy dissipation.

Carpentier-Edwards Pericardial Valve (Model 2900)

The velocity profiles obtained along the centerline plane 17 mm downstream of the valve at peak systole showed a maximum velocity of 180 cm/s. The maximum velocities measured during the acceleration and deceleration phases were 120 and 80 cm/s, respectively. A region of flow separation which extended about 6 mm from the wall was observed at peak systole and during the deceleration phase. This region was relatively stagnant during the acceleration phase. The maximum velocity of the jet at peak systole did not change as the flow field traveled from 17 to 33 mm downstream of the valve (Fig. 127.11). However, the size of the region of flow separation decreased and extended only 1 mm from the wall.

Turbulent shear stress measurements taken along the centerline plane 17 mm downstream of the valve showed that, during the deceleration phase, elevated turbulent shear stresses were spread out over a wide region (with a maximum value of 100 dynes/cm^2). At peak systole, the high turbulent shear stresses were confined to a narrow region, with a maximum value of 850 dynes/cm^2 (Fig. 127.11). The intensity of turbulence at peak systole increased as the flow traveled from 17 to 33 mm downstream of the valve (Table 127.7).

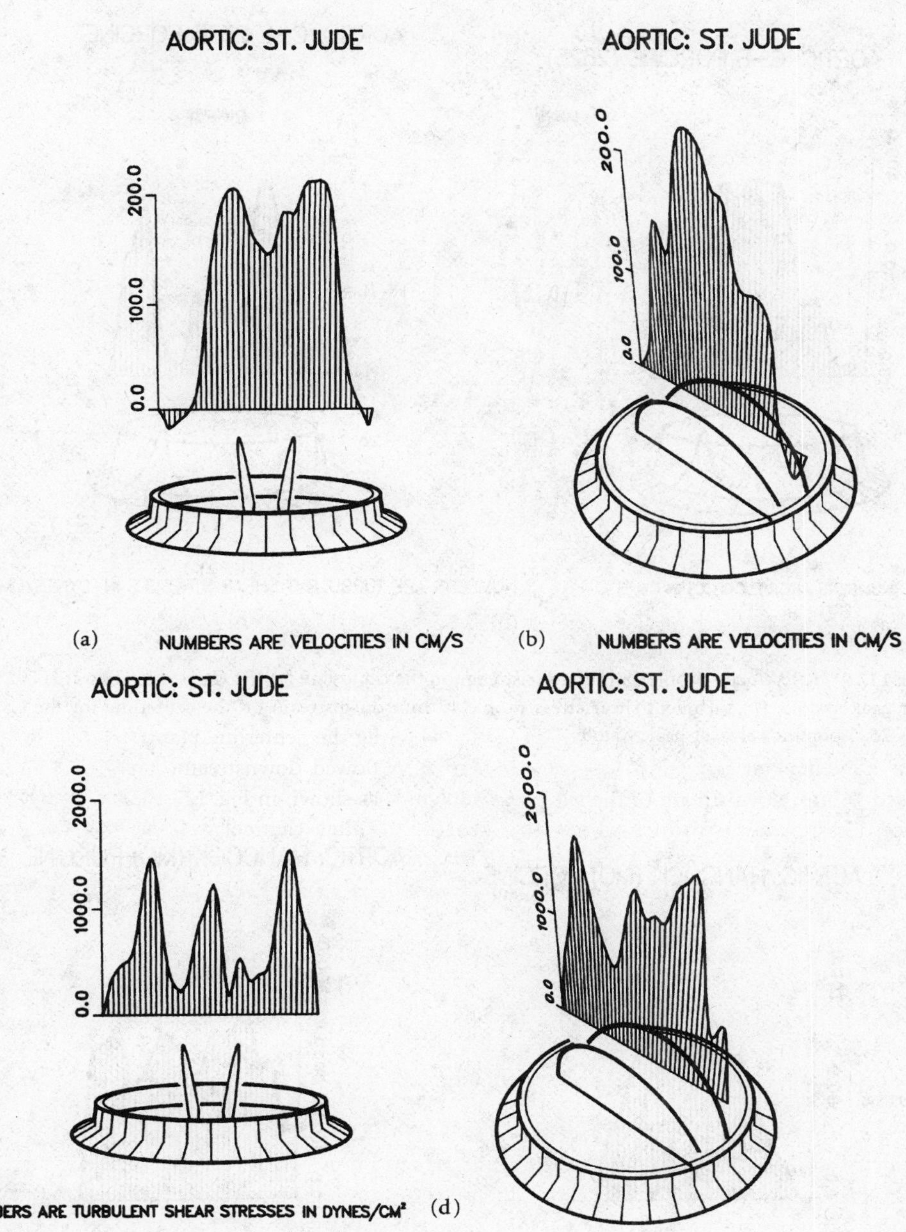

FIGURE 127.8 (*a*) Velocity profile 13 mm downstream on the centerline for the St. Jude Medical bileaflet valve, at peak systole; (*b*) velocity profile 13 mm downstream across the central orifice for the St. Jude Medical bileaflet valve, at peak systole; (*c*) turbulent shear stress profile 13 mm downstream of the centerline orifice for the St. Jude Medical bileaflet valve, at peak systole; (*d*) turbulent shear stress profile 13 mm downstream across the central orifice for the St. Jude Medical bileaflet valve, at peak systole.

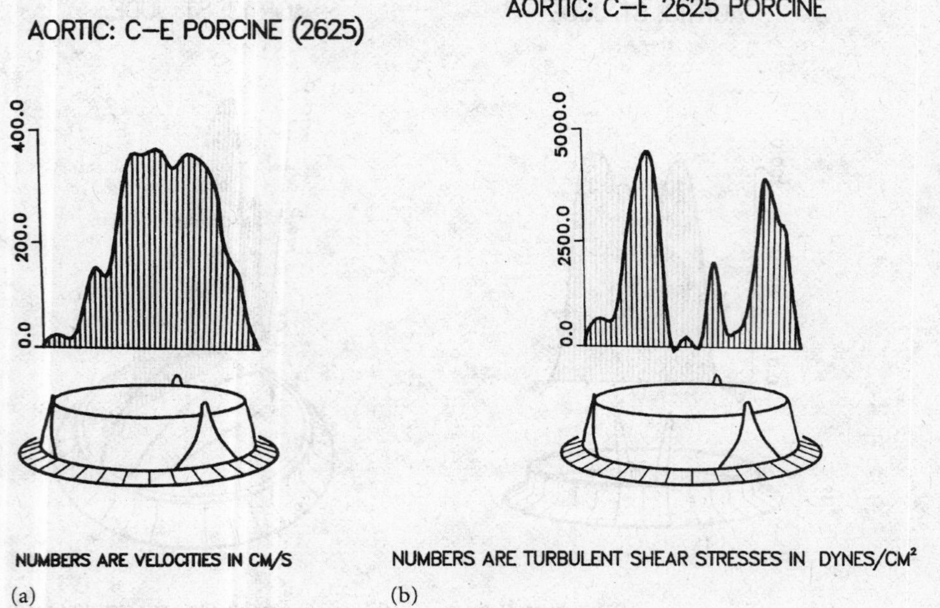

FIGURE 127.9 (*a*) Velocity profile 15 mm downstream on the centerline for the Carpentier-Edwards 2625 porcine valve, at peak systole; (*b*) turbulent shear stress profile 15 mm downstream on the centerline for the Carpentier-Edwards 2625 porcine valve, at peak systole.

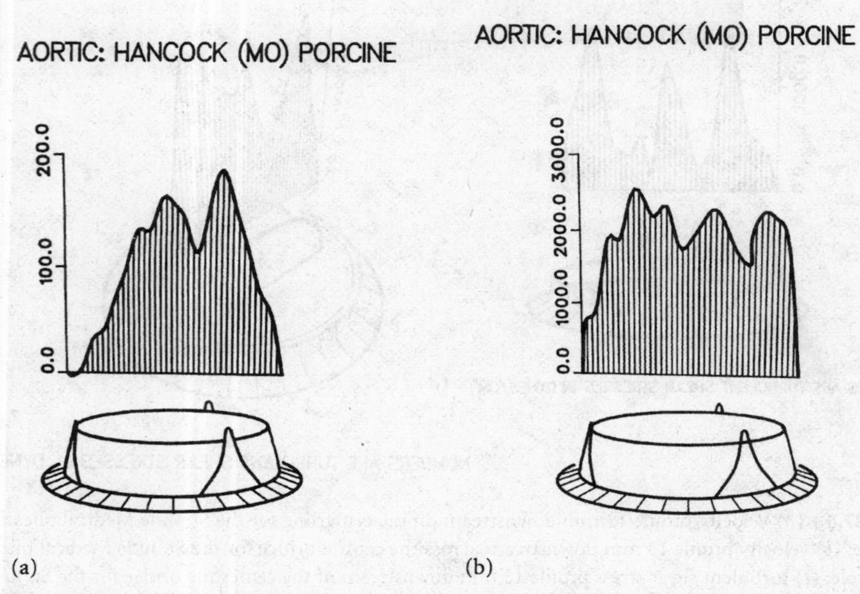

FIGURE 127.10 (*a*) Velocity profile 15 mm downstream on the center for the Hancock modified orifice porcine valve, at peak systole; (*b*) turbulent shear stress profile 15 mm downstream on the centerline for the Hancock modified orifice porcine valve, at peak systole.

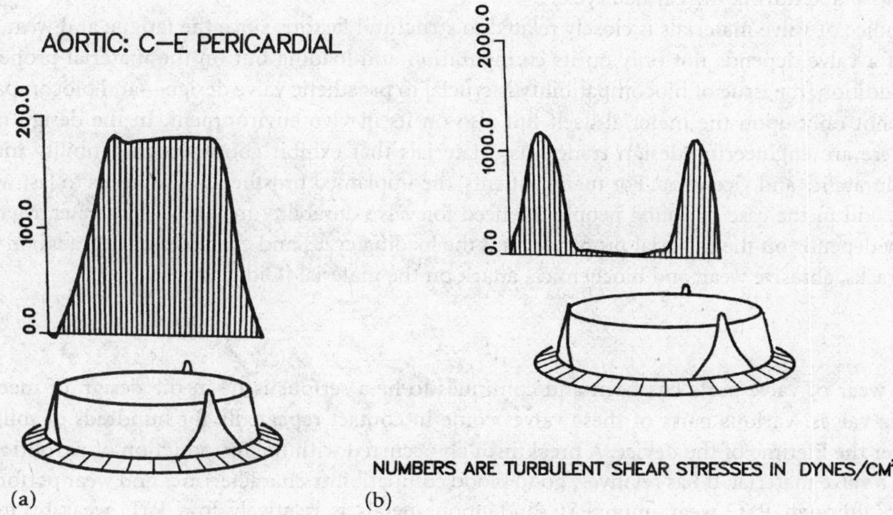

FIGURE 127.11 (*a*) Velocity profile 17 mm downstream on the centerline for the Carpentier-Edwards pericardial valve, at peak systole; (*b*) turbulent shear stress profile 17 mm downstream on the centerline for the Carpentier-Edwards pericardial valve, at peak systole.

127.5 Implications for Thrombus Deposition

In the vicinity of mechanical aortic heart valves, where peak shear stresses can easily exceed 1500 dynes/cm^2 and mean shear stresses are frequently in the range of 200–600 dynes/cm^2 (Table 127.2), platelet activation and aggregation can readily occur. Data indicating that shear-induced platelet damage is cumulative are particularly relevant to heart valves. During an individual excursion through the replacement valve, the combination of shear magnitude and exposure time may not induce platelet aggregation. However, as a result of multiple journeys through the artificial valve, shear-induced damage may accumulate to a degree sufficient to promote thrombosis and subsequent embolization.

All the aortic and mitral valve designs (mechanical and tissue) studied created mean turbulent shear stress in excess of 200 dynes/cm^2 during the major portion of systole and diastole (Tables 127.2 and 127.3), which could lead to damage to blood elements. In the case of mechanical prostheses, due to the presence of foreign surfaces, the chances for blood cell damage are increased. Furthermore, the regions of flow stagnation and/or flow separation that occur adjacent to the superstructures of these valve designs, could promote the deposition of damaged blood elements, leading to thrombus formation on the prosthesis.

127.6 Durability

The performance of prosthetic valves is in several ways related to structural mechanics. The design configuration affects the load distribution and dynamics of the valve components, which, in conjunction with the material properties, determine durability—notably wear and fatigue life. Valve configuration, in concert with the flow engendered by the geometry, also dictates the extent of low-wear (e.g., flow separation) and high-shear (e.g., gap leakage) regions. The hinges of bileaflet and tilting disc valves are vulnerable—their design can produce sites of stagnant flow, which may cause localized thrombosis, which may in turn restrict occluder motion. In addition, as discussed earlier, the rigid circular orifice ring is an unnatural configuration for a heart valve, since the elliptically shaped natural valve annulus changes in size and shape during the cardiac cycle.

The choice of valve materials is closely related to structural factors, since the fatigue and wear performance of a valve depends not only on its configuration and loading but on the material properties as well. In addition, the issue of biocompatibility is crucial to prosthetic valve design—and biocompatibility depends not only upon the material itself but also on its in vivo environment. In the design of heart valves there are engineering design trade-offs: Materials that exhibit good biocompatibility may have inferior durability and vice versa. For many patients, the implanted prosthetic valve needs to last well over a decade, and in the case of young people the need for valve durability may be even greater. Mechanical durability depends on the material properties and the loading cycle, and examples of degradation include fatigue cracks, abrasive wear, and biochemical attack on the material [Giddens et al., 1993].

Wear

Abrasive wear of valve parts has been and continues to be a serious issue in the design of mechanical prosthetic valves. Various parts of these valves come in contact repeatedly for hundreds of millions of cycles over the lifetime of the device. A breakthrough occurred with the introduction of pyrolitic carbon (PYC) as a valve material: It has relatively good blood compatibility characteristics and wear performance. However, although PYC wear upon PYC and upon metals is relatively low, PYC wear by metals is considerably greater [Shim & Schoen, 1974]. One example of this is a PYC disc mounted on a metallic orifice/hinge combination. The most durable wear couple is PYC-PYC, therefore PYC-coated components are very attractive. The first valve to employ a PYC-PYC couple was the St. Jude Medical valve, which has fixed pivots for the leaflets. Tests indicate that it would take 200 years to wear halfway through the PYC coating on a leaflet pivot [Gombrich et al., 1979]. By creating designs that allow wear surfaces to be distributed rather than focal (i.e., the Omnicarbon valve, which has a PYC-coated disc that is free to rotate), it is possible to reduce wear even further. Thus, materials technology continues to progress and in fact has reached the point where wear need not negatively impact the performance of prosthetic mechanical valves.

Fatigue

Metals are prone to fatigue failure. Their polycrystalline nature contains structural characteristics that may produce dislocations under mechanical loading. These dislocations can migrate when subjected to repeated loading cycles and can accumulate at intercrystalline boundaries, and the end result is tiny cracks. These tiny cracks are sites of stress concentration, and the fissures can worsen until fracture occurs. The Haynes 25 Stellite Bjork-Shiley valve, which used a chromium-cobalt alloy, experienced the most severe fatigue problem for a mechanical valve [Lindblom et al., 1986]. Previous investigations suggested that fatigue was not a problem for PYC; however, recent data contradict this, suggesting that cyclic fatigue-crack growth occurs in graphite/pyrolitic carbon composite material [Ritchie et al., 1990]. This work suggests a fatigue threshold that is as low as 50% of the fracture toughness, and those authors view cyclic fatigue as an essential consideration in the design and life prediction of heart valves constructed from PYC. The FDA now requires detailed characterization of PYC materials used in different valve designs (December 1993 FDA heart valve guidelines).

Mineralization

The major cause of both porcine aortic and pericardial bioprosthetic valve failure is calcification, which stiffens and frequently causes cuspal tears [Levy et al., 1991]. Calcific deposits occur most commonly at the commissures and basal attachments. Calcification is most extensive deep in (intrinsic to) the cusps in the spongiosa layer. Ultrastructurally, calcific deposits are associated with cuspal connective tissue cells and collagen. Degenerative cuspal calcific deposits are composed of calcium phosphates that are chemically and structurally related to physiologic bone mineral (hydroxyapatite). The flexing bladders in cardiac assist devices, flexing polymeric heart valves, and vascular grafts have also been found to be vulnerable to calcific deposits. In such cases, calcification is usually related to inflammatory cells adjacent to the blood-contacting surface, rather than to the implanted material itself.

The mechanisms of calcification, and the methods of preventing calcification are active areas of current research [Webb et al., 1991]. The most common methods of studying calcification involve valve tissue implanted either subcutaneously in 3-week old weanling rats or valves implanted as mitral replacements in young sheep or calves. Results of both types of studies show that bioprosthetic tissue calcifies in a fashion similar to clinical implants but at a greatly accelerated rate. The subcutaneous implantation mode is a well-accepted, technically convenient, economical, and quantifiable model for investigating mineralization issues. It is also very useful for determining the potential of new antimineralization treatments.

Host, implant, and biomechanical factors impact the calcification of tissue valves. Patients who are young or have renal failure are vulnerable to valve mineralization, but immunologic factors seem to be unimportant. Pretreatment of valve tissue with an aldehyde crosslinking agent has been found to cause calcification in rat subcutaneous implants; nonpreserved cusps do not mineralize. Calcification of bioprosthetic valves is greatest at the cuspal commissures and bases, where leaflet flexion is the greatest and deformations are maximal. Most data suggest that the basic mechanisms of tissue valve mineralization result from aldehyde pretreatment, which changes the tissue microstructure.

In both clinical and experimental bioprosthetic tissue, the earliest mineral deposits are observed to be localized to transplanted connective tissue cells. The collagen fibers are involved later. Mineralization of the connective tissue cells of bioprosthetic tissue is thought to result from glutaraldehyde-induced cellular "revitalization" and the resulting disruption of cellular calcium regulation. Normal animal cells have a low intracellular free calcium concentration (approximately 10^{-7} M), whereas extracellular free calcium is much higher (approximately 10^{-3} M), yielding a 10,000-fold gradient across the plasma membrane. In healthy cells, cellular calcium is maintained in low concentration by energy-requiring metabolic mechanisms. In addition, organellar and plasma membranes and cell nuclei, the observed sites of early nucleation of bioprosthetic tissue mineralization, contain considerable phosphorus, mainly in the form of phospholipids. In cells modified by aldehyde crosslinking, passive calcium entry occurs unimpeded, but the mechanisms for calcium removal are dysfunctional. This calcium influx reacts with the preexisting phosphorus and contributes to the mineralization [Schoen et al., 1986]. A very recent article discusses the approaches to preventing heart valve mineralization [Schoen, 1993].

127.7 Current Trends in Valve Design

The long-term clinical durability of bioprosthetic valves is the major impediment to their use. Before long-term durability data became available, improvement of hemodynamic performance was the focus of development. The clinical introduction of the bovine pericardial valve solved the hemodynamic problem, and such valves exhibit hemodynamics equal to or better than some mechanical prostheses.

The long-term durability of porcine and bovine bioprostheses can be improved through innovative stent designs that minimize stress concentrations and through improved processing fixation techniques that yield more pliable tissue. The most recent tissue valve design is the stentless bioprosthesis, used for aortic valve replacement. The aortic root bioprostheses are similar in concept to homografts. Absence of the stent is thought to improve hemodynamics, as there is less obstruction in the orifice. The absence of the stent is also thought to improve durability of the tissue, as there is less mechanical wear. Currently, three designs of stentless aortic valves (Medtronic's Freestyle, Baxter's Prima, and St. Jude's Toronto Non-Stented) are undergoing clinical evaluation in the United States and Europe. New antimineralization treatments are also being developed with the goal of increasing the durability of the tissue.

If the above-mentioned design challenges are met, so that bioprostheses can be produced that are durable and thromboresistant, and anticoagulant therapy is not required, there most likely will be another swing toward increased bioprosthesis use.

127.8 Conclusion

Direct comparison of the "total" performance of artificial heart valves is difficult, if not impossible. The precise definition of criteria used to benchmark valve performance varies from study to study. To study long-term performance, large numbers of patients and lengthy observation periods are required. During these periods there may be an evolution in valve materials, or design, and in the medical treatment of patients with prosthetic heart valves. The age of the patient at implant and the underlying valvular heart disease are extremely important factors in valve choice and longevity as well. A valve design suited for the aortic position may be inappropriate for the mitral position. Consequently, it is not possible to categorize a particular valve as *the best.* All valves currently in use, mechanical and bioprosthetic, produce relatively large turbulent stresses (that can cause lethal and or sublethal damage to red cells and platelets) and greater pressure gradients and regurgitant volumes than normal heart valves.

There are three promising directions for further advances in heart valves (and therefore three challenges for engineers designing new heart valves):

- Improved thromboresistance with new and better artificial materials
- Improved durability of new tissue valves, through the use of nonstented tissue valves, new anti-calcification treatments, and better fixation treatments
- Improved hemodynamics characteristics, especially reduction or elimination of low shear stress regions near valve and vessel surfaces and of high turbulent shear stresses along the edges of jets produced by valve outflow and/or leakage of flow.

Whereas the artificial valves certainly have room for further improvement, the superior prognosis for the patient with a replacement heart valve is dramatic and convincing.

Acknowledgments

The technical writing assistance of Ms. Julie Cerlson made the writing of this chapter an enjoyable experience.

References

Anderson GH, Hellums JD. 1978. Platelet lysis and aggregation in shear fields. Blood Cells 4:499.

Baldwin, T. 1990. An investigation of the mean fluid velocity and Reynolds stress fields within an artificial heart ventricle. Ph.D. thesis, Penn State University, PA.

Bodnar E, Frater R. 1991. Replacement Cardiac Valves, New York, Pergamon Press.

Bodnar E, Yacoub M. 1986. Biologic & Bioprosthetic Valves, New York, York Medical Books.

Brewer LA III. 1969. Prosthetic Heart Valves, Springfield, Ill, Charles C Thomas.

Brown CH III, Lemuth RF, Hellums JD, et al. 1977. Response of human platelets to shear stress. Trans ASAIO 21:35.

Butchart EG, Bodnar E. 1992. Thrombosis, Embolism and Bleeding, ICR Publishers.

Cape EG, Nanda NC, Yoganathan AP. 1993. Quantification of regurgitant flow through bileaflet heart valve prostheses: Theoretical and in vitro studies. Ultrasound Med Biol 19(6):461.

Chandran KG, Cabell GN, Khalighi B, et al. 1983. Laser anemometry measurements of pulsatile flow past aortic valve prostheses. J Biomech 16:865.

Chandran KG, Cabell GN, Khalighi B, et al. 1984. Pulsatile flow past aortic valve bioprosthesis in a model human aorta. J Biomech 17:609.

Colantuoni G, Hellums JD, Moake JL, et al. 1977. The response of human platelets to shear stress at short exposure times. Trans ASAIO 23:626.

Ferrans VJ, Boyce SW, Billingham ME, et al. 1980. Calcific deposits in porcine bioprostheses: Structure and pathogenesis. Am J Cardiol 46:721.

Giddens DP, Yoganathan Ap, Schoen FJ. 1993. Prosthetic cardiac valves. Cardiovasc Pathol 2(3)(suppl.):167S.

Gombrich PP, Villafana MA, Palmquist WE. 1979. From concept to clinical—the St. Jude Medical bileaflet pyrolytic carbon cardiac valve. Presented at Association for the Advancement of Medical Instrumentation, 14th Annual Meeting, Las Vegas, Nev.

Ionescu MF, Ross DN. 1969. Heart valve replacement with autologous fascia lata. Lancet. 2:335.

Lefrak EA, Starr A. 1979. Cardiac Valve Prostheses, New York, Appleton-Century-Crofts.

Levy RJ, Schoen FJ, Anderson HC, et al. 1991. Cardiovascular implant calcification: a survey and update. Biomaterials 12:707.

Lindblom D, Bjork VO, Semb KH. 1986. Mechanical failure of the Bjork-Shiley valve. J Thorac Cardiovasc Surg 92:894.

Mohandas H, Hochmuth RM, Spaeth EE. 1974. Adhesion of red cells to foreign surfaces in the presence of flow. J Biomed Mater Res 8:119.

Nevaril C, Hellums J, Alfrey C Jr, et al. 1969. Physical effects in red blood cell trauma. J Am Inst Chem Engr 15:707.

Oyer PE, Stinson EB, Reitz BA, et al. 1979. Long term evaluation of the porcine xenograft bioprosthesis. J Thorac Cardiovasc Surg 78:343.

Ramstack JM, Zuckerman L, Mockros, LF. 1979. Shear induced activation of platelets. J Biomech 12:113.

Ritchie RO, Dauskart RH, Yu W. 1990. Cyclic fatigue-crack propagation, stress-corrosion, and fracture-toughness in pyrolytic carbon-coated graphite for prosthetic heart valve applications. J Biomed Mater Res 24:189.

Roberts WC. 1976. Choosing a substitute cardiac valve: type, size, surgeon. Am J Cardiol 38:633.

Schoen FJ, Libby P, Diddersheim P. 1993. Future directions and therapeutic approaches. Cardiovasc. Pathol. 2(3) (Suppl.):2095.

Schoen FJ, Tsao JW, Levy RJ. 1986. Calcification of bovine pericardium used in cardiac valve bioprostheses: implications for the mechanisms of bioprosthetic tissue mineralization. Am J Pathol 123:134.

Shim HS, Schoen FJ. 1974. The wear resistance of pure and silicon alloyed isotropic carbons. Biomater Med Dev Artif Organs 2:103.

Sutera SP, Merjhardi MH. 1975. Deformations and fragmentation of human red cells in turbulent shear flow. Biophys J 15:1.

Wain EH, Greco R, Ignegen A, et al. 1980. Int J Artif Organs 3:169.

Webb CL, Schoen FJ, Alfrey AC, et al. 1991. Inhibition of mineralization of glutaraldehyde-pretreated bovine pericardium by $A1C1_3$ and other metallic salts in rat subdermal model studies. Am J Pathol 38:971.

Woo Y-R, Yoganathan AP. 1985. In vitro pulsatile flow velocity and turbulent shear stress measurements in the vicinity of mechanical aortic l heart valve prostheses. Life Support Syst 3:283.

Woo Y-R, Yoganathan AP. 1986. In vitro pulsatile flow velocity and shear stress measurements in the vicinity of mechanical mitral heart valve prostheses. J Biomech 19:39.

Wurzinger LJ, Opitz R, Blasberg P, et al. 1983. The role of hydrodynamic factors in platelet activation and thrombotic events. In G Schettler (ed), The Effects of Shear Stress of Short Duration, Fluid Dynamics as a Localizing Factor for Atherosclerosis, pp 91–102, Berlin, Springer.

Yoganathan AP, Chaux A, Gray R, et al. 1984. Bileaflet, tilting disc and porcine aortic valve substitutes: in vitro hydrodynamic characteristics. JACC. 3(2):313.

Yoganathan AP, Reul H, Black MM. 1992. Heart valve replacements: Problems and developments. In GW Hastings (ed), Cardiovascular Biomaterials, London, Springer-Verlag.

Yoganathan AP, Sung H-W, Woo Y-R, et al. 1988. In vitro velocity and turbulence measurements in the vicinity of three new mechanical aortic heart valve prostheses. J Thorac Cardiovasc Surg 95:929.

Yoganathan AP, Woo Y-R, Sung H-W, et al. 1986*a*. In vitro haemodynamic characteristics of tissue bioprostheses in the aortic position. J Thorac Cardiovasc Surg 92:198.

Yoganathan AP, Woo Y-R, Sung H-W. 1986*b*. Turbulent shear stress measurements in the vicinity of aortic heart valve prostheses. J Biomech 19:433.

[illegible]

[illegible] 1979 [illegible]

[illegible] 1998 [illegible]

[illegible] 1969 [illegible]

[illegible]

[illegible] 1993 [illegible]

[illegible]

[illegible]

[illegible]

[illegible]

[illegible]

[illegible]

[illegible]

[illegible]

[illegible]

[illegible]

[illegible]

[illegible]

[illegible]

[illegible]

[illegible]

[illegible]

[illegible]

[illegible]

128
Vascular Grafts

David N. Ku
Georgia Institute of Technology

Robert C. Allen
Emory University

The use of natural or synthetic replacement parts in vascular repair and vascular reconstruction is extensive, with a vascular graft market of approximately $200 million worldwide. A multitude of biologic grafts and synthetic prostheses are available, each with distinct qualities and potential applications [Rutherford, 1989; Veith et al., 1994]. The ideal vascular graft would (1) be biocompatible, (2) be nonthrombogenic, (3) have long-term potency (4) be durable yet compliant, (5) be infection resistant, and (6) be technically facile. There is currently no ideal conduit available, but overall, the autogenous saphenous vein is preferred for small-vessel reconstruction, and synthetic prostheses are best suited for large-vessel replacement. Large diameter (>10 mm) vascular grafts are predominantly used for aortic/iliac artery reconstruction with Dacron (80%) and PTFE (20%) being the standard construction materials. Synthetic grafts function well in these high-flow, low-resistance circuits with high long-term patencies. Small-caliber (<10 mm) vascular grafts are used for a variety of indications including lower-extremity bypass procedures, coronary artery bypass grafting (CABG), hemodialysis access, and extra-anatomic bypasses. Saphenous vein is the conduit of choice for lower-extremity revascularization and CABG procedures and shows superior patency rates compared to synthetic grafts. Internal mammary artery (IMA) grafts are used extensively for CABG with better long-term patency than saphenous vein grafts. Synthetic prostheses are used for hemodialysis access and extra-anatomic bypass grafting, especially PTFE due to its durability and resistance to external pressure. Bovine carotid heterografts are also used for hemodialysis access with fair success, but their value is questionable because of aneurysmal changes over time due to graft degeneration.

128.1 History of Vascular Grafts

Vascular surgery was first defined as a field with the initial publication of Alexis Carrel's work in 1902. Goyanes performed the first arterial graft in 1906 by using the popliteal vein *in situ* to replace an excised popliteal artery aneurysm. Lexer, in 1907, performed the first free autogenous vein graft by replacing an axillary artery defect with a segment of greater saphenous vein from the same patient. It was not until 1949, however, that Kunlin performed the first successful femoropopliteal bypass with reversed saphenous vein. The Korean War was the true advent of reconstructive surgery for arterial injuries as initiated by Shumacker and reported by Hughes and Bowers in 1952. The first report of a synthetic graft also appeared

0-8493-0462-8/00/$0.00+$.50

in 1952 when Voorhees described his initial work in dogs using tubes of Vinyon-N cloth to bridge arterial defects. In the late 1950s and 1960s DeBakey firmly established the clinical usefulness of Dacron grafts; Dacron arterial prostheses became commercially available in 1957. Microporous expanded polytetrafluoroethylene (ePTFE) grafts, introduced in the 1970s, were the next generation of successful synthetic grafts but bear no clear superiority to Dacron grafts.

128.2 Synthetic Grafts

Synthetic prostheses are the preferred conduit for large-vessel reconstruction and the primary alternative to saphenous vein in small-vessel repair. Thankfully, the saphenous vein is of adequate quality and length in 70–75% of patients requiring small-vessel grafting. the two major choices available for prosthetic reconstruction are Dacron (polyethylene terephthalate) and PTFE (polytetrafluoroethylene). The tensile strength of these grafts remains unchanged even after years of implantation, whereas Nylon (Polyamide), Ivalon, and Orlon lose significant tensile strength over a period of months.

Dacron grafts are constructed from multifilamentous yarn and fabricated by weaving or knitting [Rutherford, 1989; Sawyer, 1987]. Woven fabric is interlaced in an over-and-under pattern resulting in a nonporous graft with no stretch. Knitted fabrics have looped threads forming a continuous interconnecting chain with variable stretch and porosity. Knitted velour is a variant in which the loops of yarn extend upward at right angles to the fabric surface resulting in a velvety texture. The velour finish may be created on the internal, the external, or both surfaces to enhance preclotting and tissue incorporation. Porous knitted Dacron prostheses may be made impervious by impregnation or coating with albumin/collagen to obviate preclotting and minimize blood loss while maintaining the superior handling characteristics of knitted Dacron grafts. Dacron prostheses are often crimped to impart elasticity, maintain shape during bending, and facilitate vascular anastomosis formation. The use of external support for vascular grafts is an alternative to crimping which uses externally attached polyprophylene rings to avoid kinking with angulation and to create dimensional stability. Woven Dacron grafts possess poor handling characteristics and poor compliance and elicit a poor healing response and, therefore, are used predominantly in the repair of the thoracic aorta, ruptured abdominal aortic aneurysms, and patients with coagulation defects. Knitted Dacron prostheses are used in the remainder of cases involving the abdominal aorta, iliac, femoral, and popliteal vessels, and in extra-anatomic repair/bypass. Dacron grafts are not advocated in small-vessel reconstruction, such as the below-knee popliteal or tibial, because of poor patency rates. An active infective process or contaminated field contraindicate the use of Dacron prostheses. Vascular grafts are named after the vessel they replace or the arterial segments they connect.

Expanded Teflon (ePTFE) is a synthetic polymer of carbon and fluorine produced by mechanical stretching [Rutherford, 1989; Sawyer, 1987]. The result is a series of solid nodes of ePTFE with interconnecting small fibrils. The pore size is variable and averages 30 microns. The chief advantages of ePTFE in comparison to knitted Dacron include impermeability to blood, resistance to aneurysmal formation, and ease of declotting; ePTFE is used clinically mainly for femoropopliteal bypasses, hemodialysis angioaccess, and extra-anatomic reconstructions. The addition of an external support coil can be used to enhance graft stability against external pressure and to prevent kinking at points of angulation.

The primary area of failure in prosthetic grafting with Dacron and PTFE is the patency of small-caliber grafts, which remains poor. The inherent thrombogenicity of these prosthetic grafts in conjunction with neointimal hyperplasia, which is limited to the anastomotic area, are the prime factors. The seeding of prosthetic grafts with endothelial cells and subsequent proliferation of these cells into a confluent monolayer along the graft has been suggested as an attractive solution to this problem for a number of years. The technology and techniques devoted to develop this solution to small-caliber graft patency have been impressive, but the results have been variable and certainly not the panacea envisioned by early investigators. The clinical utility of endothelial cell seeding onto prosthetic grafts remains to be defined.

TABLE 128.1 Patency Data for Representative Vascular Grafts

Graft	Conduit	5-Year Patency, percent
Aortiobifemoral	Dacron	90
Aortiobifemoral	ePTFE	90
Femorofemoral	Dacron	80
Femorofemoral	ePTFE	80
Femoropopliteal	Saphenous vein	70
Femoropopliteal	ePTFE	50
Femoropopliteal	ePTFE	40
Dialysis access	ePTFE	60 (3 years)
Dialysis access	Bovine heterograft	25 (3 years)

128.3 Regional Patency

Patency data for any vascular graft depend on multiple factors including the conduit and vessel size. Large-diameter vessels have a high rate of patency, and therefore, synthetic grafts are the conduit of choice and display high short- and long-term patencies. Saphenous vein is the primary graft for small-caliber vessel reconstruction with good patency rates, and prosthetic grafts demonstrate acceptable short-term patencies but poor long-term results. Extra-anatomic bypasses utilize synthetic conduits with external support to reduce kinking. Dacron versus PTFE graft trials have shown no significant difference in patency rates [Rutherford, 1989]. Hemodialysis access grafts are constructed primarily of PTFE with a minority of bovine carotid heterografts still present. PTFE is preferred due to resistance to infection, decreased pseudoaneurysm formation, and improved patency rates. Patency data for representative vascular grafts is summarized in Table 128.1 by anatomical location and conduit [Imparato et al., 1972; Rutherford, 1989; Veith et al., 1994].

Vascular grafts have a finite life span which is usually reported in the form of a life-table of graft patency. The patency rate falls to about 50% in 4 years after implantation for the most common PTFE femoral-popliteal grafts. Graft failures are typically caused by: (1) thrombosis in the early postoperative period of 1 month, (2) tissue ingrowth or neointimal hyperplasia at the anastomoses after several years, and (3) graft infections. Thrombosis in the early postoperative period is usually caused by a technical error at time of implantation and is readily fixed. The longer-term growth of tissue at the anastomoses is more insidious. Late failure of these grafts as a result of neointimal hyperplasia is a major problem confronting the vascular surgeon. Increasing the longevity of these grafts would greatly benefit the patient and have an important economic impact, nationally.

128.4 Thrombosis

When an artificial vascular graft is first exposed to blood, a clot will typically form at the surface. The clot is formed initially of platelet aggregates, and then the coagulation cascade is activated to lay down fibrin and thrombin. The tissue which forms this inner lining of the graft is often called pseudointimal hyperplasia. Pseudointimal hyperplasia is devoid of cellular ingrowth and is composed mainly of fibrin, platelet debris, and trapped red blood cells [Greisler, 1991]. The blood clots immediately on contract and will occlude the vessel if the blood is stagnant or very sluggish. Platelet activity is generally most intense during the first 24 hours and subsidies to a very low level after 1 week.

Various artificial surfaces are more or less thrombogenic. Many approaches have been developed in an attempt to make the grafts less thrombogenic [Greisler, 1991]. Typically, the surface characteristics have been modified to prevent adherence or activation of platelets. The surfaces have been modified by seeding with endothelial cells and other exotic treatments such as photopolymerization, plasma-gas coatings, and antisense genetics. Although many of these theories have promise, few have been shown to yield long-term improvements over the existing materials of Dacron and PTFE.

128.5 Neointimal Hyperplasia

Tissue ingrowth onto the artificial surface of a PTFE graft occurs at the anastomoses and is called *neointimal hyperplasia.* Initially, platelets and thrombin/fibrin cover the entire graft surface within the first 24 hours. After this initial phase, smooth-muscle cells migrate from the native artery into the graft, variously referred to as pannus ingrowth, neointimal hyperplasia, and fibromuscular hyperplasia. The pannus advances from the edge of the graft at a rate of approximately 0.1 mm per week but does not cover the entire surface in humans. Typically, the maximum extent of ingrowth in humans is 1-2 cm. In the first few hours, a thrombotic stage of platelet, fibrin, and red blood cell accumulation occurs (stage I). Days 3-4 consist of cellular recruitment in which the thrombus is coated with an endothelial layer (stage II). Cell proliferation of actin-positive cells which resorb and replace the residual thrombus are the major events in stage III, at approximately 7–9 days. This description leads to a conclusion that thrombosis and cell proliferation are the two controlling events affecting graft patency. However, this process is one of normal healing which creates no significant stenosis in most cases. Recent findings indicate that a fourth stage should be added in which the tissue growths in volume primarily by synthesis and deposition of large amounts of extracellular substance over the following months, as shown in Fig. 128.1. This stage IV leads to a hemodynamically significant stenosis.

The ingrowth has been characterized by Glagov as having two different forms: intimal fibromuscular hypertrophy (IFMH) and intimal hyperplasia (IH) [Glagov et al., 1991]. The first form of IFMH is basically a normal healing process which consistently takes place on all grafts and does not become highly stenotic. Initial healing is well organized with smooth-muscle cells organized in lamellae and little extracellular proteins or mucopolysaccharide material. A second form of abnormal healing creates stenotic and occlusive thickening, called *intimal hyperplasia* (IH). Intimal hyperplasia is characterized by a chaotic appearance of randomly oriented cells embedded in a field of extracellular matrix material.

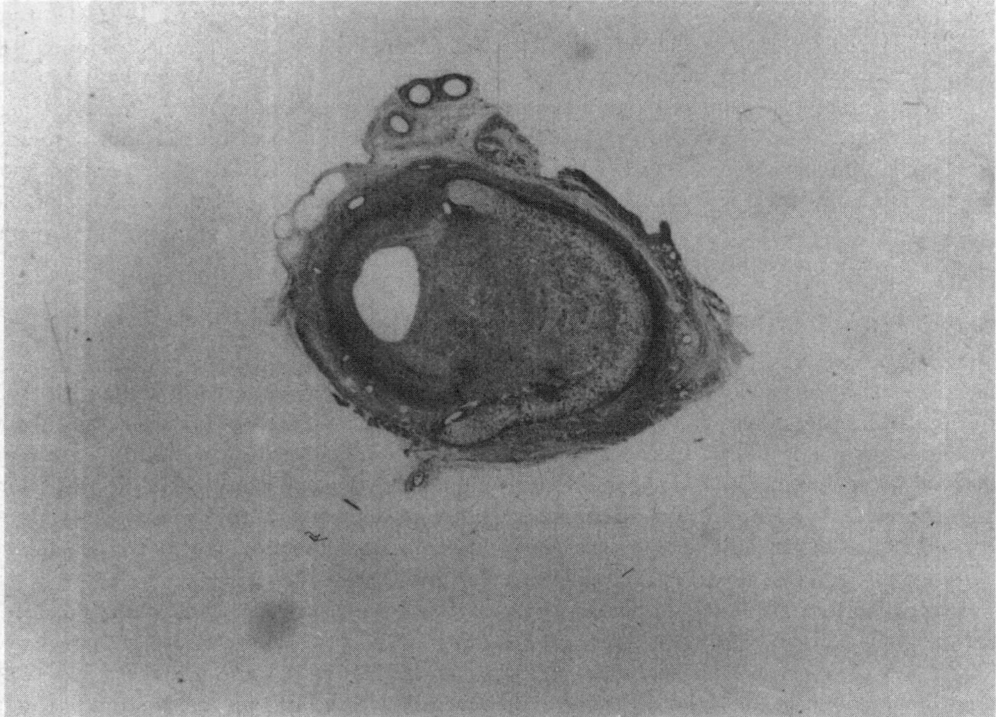

FIGURE 128.1 High-power sections of the anastomotic neointimal hyperplasia in a PTFE graft. This endothelium-lined layer is formed of highly disorganized spindle-shaped cells surrounded by a large amount of extracellular matrix material.

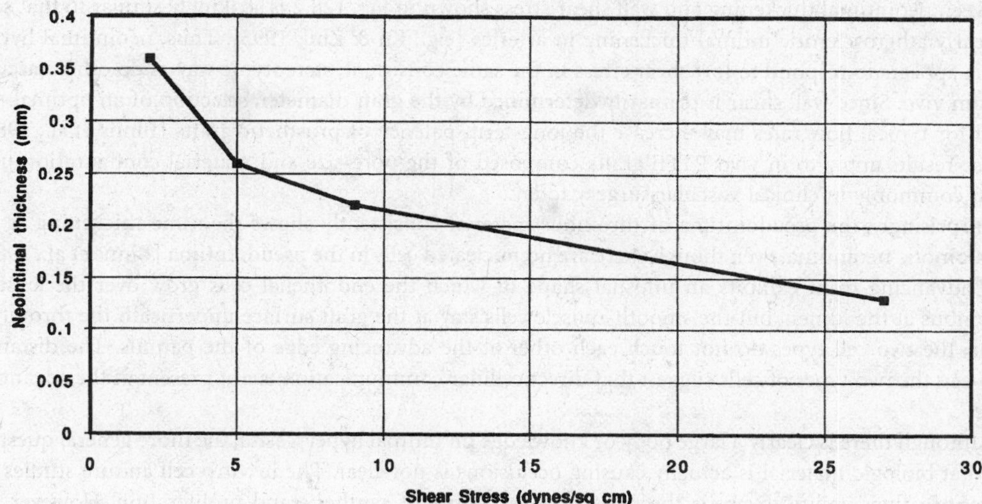

FIGURE 128.2 Graph of neointimal thickness versus mean shear stress for PTFE tapered grafts in a canine model of intimal hyperplasia.

The histologic appearance is one of hyperplasia or tumor-like behavior. Little is known on the exact stimulant which causes healing to proceed by IFMH or IH, although Glagov speculates that hemodynamic shear stress may play a role. The events in graft neointimal hyperplasia appear to be similar to that of restenotic hyperplasia.

Although the underlying etiology of neointimal hyperplasia has not been fully elucidated, a large number of theories have been proposed to explain this process. The theories generally fall into several broad categories: platelet deposition with local release of PDGF, other growth factor stimulation of SMC proliferation (e.g., TGF-, FGF), monocyte recruitment, complement activation, leukocyte deposition, chronic inflammation, and mechanical stimuli such as stress and shear abnormalities [Greisler, 1991].

Many studies have focused on preventing graft intimal thickening by attacking either thrombosis or cell proliferation. Cell culture studies clearly demonstrate that certain pharmacologic agents can profoundly affect the platelet adhesiveness and cell proliferation. However, the effectiveness of these agents in vivo is variable, depending on which animal species is used. A large number of human restenosis trials using a wide variety of antithrombotic agents, steroids, lipid-lowering therapies, and ACE inhibitors have largely been negative. Arteriovenous fistula graft studies in baboons suggest that modifying graft surfaces and infusing antithrombotic agents have little effect on the rate of obstruction of the prostheses.

Compliance mismatch, medial tension, and hemodynamic shear stress are additional putative agents which have been hypothesized to be important in arterial adaptation and neointimal hyperplasia. Arteries tend to adapt in size to maintain a specific wall shear stress under conditions of increased blood flow. Wall shear stress appears to play a major role in the pathogenesis of atherosclerosis and in arterially transplanted autogenous veins, and elevations in shear stress inhibits SMC proliferation and neointimal thickening in porous PTFE grafts in baboons.

We have recently characterized the quantitative relationship between the amount of shear stress and the associated neointimal thickness, shown graphically in Fig. 128.2.

This relationship can be written as the mathematical relationship

$$\text{Intimal thickness} = A \times \frac{1}{\text{wall shear rate}} + C \tag{128.1}$$

where C is a constant which would depend on the implantation age of the graft. This relationship appears to be dominant over proximal or distal positioning or absolute flow rate. This inverse, nonlinear relationship between neointimal thickening and wall shear stress shown in Fig. 128.2 is strikingly similar to that seen for early atherosclerotic intimal thickening in arteries [e.g., Ku & Zhu, 1993]. Thus, neointimal hyperplasia appears to respond to low shear stress in the same consistent, stereotypic way as do other vascular cells in vivo. Since wall shear is primarily determined by the graft diameter, selection of an optimal-size graft for typical flow rates may increase the long-term patency of prosthetic grafts [Binns et al., 1989]. These results apply to in vivo PTFE grafts composed of the pore-size and material configuration used most commonly in clinical vascular surgery today.

Surprisingly, the pseudointima of thrombus material consistently shows the same thickening as the anastomotic neointima, even though there are no nucleated cells in the pseudointima [Binns et al., 1989]. The advancing front exhibits an unusual shape in which the endothelial cells grow over the existing thrombus at the lumen, but the smooth-muscle cells stay at the graft surface underneath the thrombus. Thus, the two cell types do not touch each other at the advancing edge of the pannus. The distances between the two types of cells suggest that direct cellular communication is not present at the advancing edge.

Although there is clearly a large body of knowledge on intimal hyperplasisa, the more general question of what biologic material is actually causing occlusions is not clear. The in vitro cell culture studies are elegant in their ability to probe the molecular regulators of synthesis and proliferation. However, the response of cells grown as monolayers on plastic may be different from tissue growing in the complex milieu in vivo. Small animal studies are useful for initial testing of a wide variety of potential therapeutic agents. Unfortunately, many agents which are successful in rats and rabbits turn out to be unsuccessful in large animals or humans. Large animals such as dogs, pigs, and baboons appear to develop lesions similar to humans and respond to therapies in a similar manner. The major drawback, however, is that the lesions typically take approximately one year to develop into advanced stenoses which mimic the clinical human problem. An accurate large-animal model of high-grade neointimal hyperplastic lesions is needed in order to generate a comprehensive characterization of lesion development in vivo.

The cellular mechanism by which shear stress influences the development of neointimal hyperplasia is not well understood. The pannus ingrowth is likely to be composed predominantly of smooth-muscle cells. However, the fluid-wall shear stress interface is primarily experienced by endothelial cell. One mechanism for neointimal thickening may be that the endothelial cells sense the local wall shear and regulate the amount of underlying smooth muscle cell growth. Clearly, endothelial cells can change their physiologic behavior and morphologic appearance in response to wall shear stress [Nerem, 1992]. Shear-sensitive endothelial cells mechanoreceptors have been postulated as regulators of the adaptation process seen in arteries in response to increased shear stress. Such receptors, however, have not been identified. An "endothelial-derived relaxation factor" and an "endothelial-derived constriction factor" appear to play an important role in the acute response of the arterial wall to alterations in shear stress. A separate humoral factor acting on the media in response to alterations in shear stress has been suggested. The fact that most of these mediators are endothelium-derived or endothelium-dependent may explain the localization of neointimal hyperplasia in PTFE grafts to the juxta-anastomotic region. In contrast to native vessels, the commonly used human prosthetic grafts are rarely, if ever, fully lined with endothelium except for a short distance adjacent to the anastomosis.

The morphology of endothelial cells grown in low shear or static conditions are random in orientation without a preferred direction. As the intimal thickening progresses, one would expect the deeper layers to be more insulated from the shear effects of hemodynamics and to experience a more static environment. PTFE grafts are essentially nondistensible under physiologic pressure conditions. Thus, the perigraft cells would be expected to experience the least amount of stress from pressure pulses and circumferential stretch as well.

Recently, there has been greater recognition of the importance of extracellular matrix (ECM) in vascular lesions [Bell, 1992]. In general, we have observed that the deepest layers of cells in the perigraft region exhibited the most disorganization and extracellular matrix material. The appearance of the intimal hyperplasia in this region strongly suggests that growth of neointima in this region from the production of extracellular matrix is the dominant mechanism by which intimal hyperplasia becomes occlusive. ECM is dominant in advanced lesions, often to the virtual exclusion of cellular elements. The ECM is also directly responsible for a variety of tissue signals which can augment or reduce cellular proliferation and protein production. The ECM can act directly or act as a repository for other growth factors. Several investigators are attempting to recreate the detailed architecture of ECM in order to develop future biologic substrates for arteries.

Low compliance of PTFE grafts are associated with higher rates of intimal hyperplasia. Rodbard [1970] has hypothesized that the normal adaptive response of cells to static conditions is the production of large amounts of ECM. It is possible that the lack of stretch in the PTFE graft removes any mechanical stimulus to the smooth-muscle cells in this layer and transforms the cells into a synthetic mode with production of large amounts of ECM. The highly synthetic mode of these cells and the hypocellularity in this region would suggest that the major culprit of late graft occlusion is synthesis instead of proliferation. Studies of the response of endothelial cells to very low levels of wall shear stress and the response of underlying smooth muscle cells to endothelial cell products should provide important information on the cellular and molecular mechanisms producing the transformation of thin, organized healing into occlusive growth.

128.6 Graft Infections

Vascular graft infections are catastrophic and challenging problems that threaten life and limb. The incidence of synthetic graft infection is approximately 2% and has remained stable despite advances in aseptic surgical technique, vascular graft production, and immunology. Prosthetic grafts are involved in infections four times as often as autogenous vein. The infection rates for PTFE and Dacron prostheses are about the same. The mechanism of infectious complications may involve direct inoculation during surgery or hematogenous seeding, but most occur due to contamination at the time of surgery. Contact of the prosthesis with the skin of the patient is felt to be a key event. The highest incidence anatomically occurs in the inguinal areas due to perineal proximity, lymphatic disruption, and poor wound healing. Antibiotic prophylaxis, both systemically and locally, decreases the incidence of graft infection. The bonding of antibiotics to the prosthesis has been attempted in various forms to prevent prosthetic graft infection but to date is not clinically available. *Staphylococcus epidermidis* is the most common organism involved in graft infections, which *Staphylococcus aureus* also a prevalent organism. Early graft infections commonly are superficial infections (inguinal) with virulent organisms such as *Staphylococcus aureus* or *Pseudomonas aeruginosa.* Late graft infections usually involve contamination of an abdominal graft with less virulent species such as *Staphylococcus epidermidis.* The reported mortality rates vary with the anatomic location of the graft and the aggressiveness of the infecting organism. These may range from 5–10% for femoropopliteal grafts to 50–75% for aortobifemoral grafts. The associated amputation rate varies from 25–75%.

Vascular grafts are important tools in the clinical treatment of end-stage arteriosclerosis. Large-diameter grafts made of artificial material have high success and longevity. For small-diameter grafts, autogenous venous grafts are superior in longevity and patency than artificial grafts. The main problems for artificial vascular grafts are early thrombosis, late neointimal hyperplasia, and iatrogenic infection. Some active areas of research include the prevention of flowing blood thrombosis on surfaces, control of the pathogenic process causing neointimal hyperplasia occlusions, and facilitation of the immunogenic response to bacterial infection of a graft.

Defining Terms

Extra-anatomic bypass: Vascular graft used to bypass a blocked artery where the graft is placed in a different location than the original artery. One example is an axillo-femoral graft which brings blood from the arm to the leg though a tube placed under the skin along the side of the person.

Hemodialysis access: Arterial-venous connection surgically placed to provide a large amount of blood flow for hemodialysis. These are often loops of ePTFE placed in the arm.

Neointimal hyperplasia: Fibroblast and smooth-muscle cell growth covering a vascular graft on the inside surface.

Pannus: Neointimal hyperplasia tissue ingrowth at the anastomoses.

Pseudointimal hyperplasia: Fibrin/thrombin deposition on the inside surface of an arterial vascular graft. This accumulation of material is acellular.

Stenosis: Tissue ingrowth into vessel causing a narrow lumen and reduction of blood flow.

Vascular graft: Tube replacement of an artery or vein segment.

Vascular reconstruction: Reconstruction of an artery or vein after trauma, surgery, or blockage of blood flow from disease.

References

Bell E. 1992. Tissue Engineering, Current Perspectives, Boston, Birkhauser.

Binns RL, Du DN, Stewart MT, et al. 1989. Optimal graft diameter: Effect of wall shear stress on vascular healing. J Vasc Surg 10:326.

Glagov S, Giddens DP, Bassiouny H, et al. 1991. Hemodynamic effects and tissue reactions at grafts to vein anastomosis for vascular access. In BG Sommer, ML Henry (eds), Vascular Access for Hemodynamics—II, pp3–20, Precept Press.

Greisler HP. 1991. New Biologic and Synthetic Vascular Prosthesis, Austin, TX, RG Landes.

Imparato AM, Bracco A, Kim GE, et al. 1972. Intimal and neointimal fibrous proliferation causing failure of arterial reconstructions. Surgery 72:1007.

Ku DN, Zhu C. 1993. The mechanical environment of the artery. In B Sumpio (ed), Hemodynamic Forces and Vascular Cell Biology, pp 1–23, Austin, TX, RG Landes.

Nerem RM. 1992. Vascular fluid mechanics, the arterial wall, and atherosclerosis. J Biomech Eng 114:274.

Rodbard S. 1970. Negative feedback mechanisms in the architecture and function of the connective and cardiovascular tissues. Perspect Biol Med 13:507.

Rutherford RB. 1989. Vascular Surgery, pp 404–486, Philadelphia, Saunders.

Sawyer PN. 1987. Modern Vascular Grafts, New York, McGraw-Hill.

Veith FJ, Hobson RW, Williams RA, et al. 1994. Vascular Surgery, pp 523–535, New York, McGraw-Hill.

Further Information

Greisler HP. 1991. New Biologic and Synthetic Vascular Prosthesis, Austin, TX, RG Landes.

Ku DN, Salam TA, Chen C. 1994. The development of intimal hyperplasia in response to hemodynamic shear stress. Second World Congress of Biomechanics, 31b, Amsterdam.

Veith FJ, Hobson RW, Williams RA, et al. 1994. Vascular Surgery, pp 523–535, New York, McGraw-Hill.

129

Artificial Lungs and Blood-Gas Exchange Devices

Pierre M. Galletti
(deceased)

Clark K. Colton
Massachusetts Institute of Technology

The natural lung is the organ in which blood exchanges oxygen and carbon dioxide with the body environment. In turn, blood brings oxygen to all body tissues, so as to oxidize the nutrients needed to sustain life. The end products of the chemical reactions that take place in tissues (globally referred to as metabolism) include carbon dioxide, water, and heat, which must all be eliminated. In mammals, oxygen is obtained from the air we breathe through diffusion at the level of the pulmonary alveoli and then carried to the tissues by the hemoglobin in the red blood cells. The carbon dioxide produced by living cells is picked up by the circulating blood and brought to the pulmonary capillaries from where it diffuses into the alveoli and is conveyed out by ventilation through the airways. These processes can be slowed down to a fraction of resting levels by hypothermia or accelerated up to 20-fold when the demand for fuel increases, as for instance with hypothermia, fever, and muscular exercise.

Only a fraction of the oxygen in the air is actually transferred from the pulmonary alveoli to the pulmonary capillary blood, only a fraction of the oxygen carried by arterial blood is actually extracted by the tissues, and only a fraction of the oxygen present in tissues is actually replenished in a single blood pass. Similarly, only a fraction of the CO_2 present in tissues is conveyed to the circulating blood, only a fraction of the mixed venous blood CO_2 content is actually discharged in the alveoli, and only a fraction of the CO_2 in the aveolar gas is eliminated into each breath. Delicately poised physiologic mechanisms, further balanced by chemical buffer systems, maintain the gas exchange system in equilibrium.

The challenge of replacing the function of the natural lungs by an exchange device allowing continuous blood flow and continuous blood arterialization was first outlined by physiologists at the end of the 19th century but could not be met reliably until the 1950s. The large transfer areas needed for blood-gas exchange in an artificial lung were initially obtained by continuous foaming and defoaming in a circulating blood pool or by spreading a thin film of blood in an oxygen atmosphere. Because the direct blood-gas interface was found to be damaging to blood as well as difficult to stabilize over extended periods, membrane-mediated processes were introduced and are now almost universally preferred.

0-8493-0462-8/00/$0.00+$.50

129.1 Gas Exchange Systems

Artificial lungs are often called blood oxygenators because oxygen transfer has traditionally been seen as the most important function being replaced and, in most situations, has proved more limiting than CO_2 transfer. The change in blood color between inlet and outlet also encourages the term *oxygenator*, considering that changes in blood CO_2 content are not visible to the eye and are more difficult to measure than oxygen transfer.

As is the case for most artificial organs, artificial lungs may be called upon to *replace* entirely the pulmonary gas exchange function (when the natural organ is totally disabled or, while still sound, must be taken out of commission for a limited time to allow a surgical intervention) or to *assist* the deficient gas transfer capacity of the natural organ, either temporarily, with the hope that the healing process will eventually repair the diseased organ, or permanently, when irreversible lung damage leaves the patient permanently disabled.

Since most artificial lungs cannot be placed in the anatomical location of the natural lung, venous blood must be diverted from its normal path through the central veins, right heart, and pulmonary vascular bed and rerouted, via catheters and tubes, through an extracorporeal circuit which includes the artificial lung before being returned, by means of a pump, to the arterial system.

The procedure in which the pulmonary circulation is temporarily interrupted for surgical purposes and gas exchange is provided by an artificial lung is often referred to as extracorporeal circulation (ECC) because, for convenience sake in the operating room, the gas exchange device as well as the pumps which circulate the blood are located outside the body.

The vision of coupling extracorporeal blood pumping with extracorporeal gas exchange at a level of performance sufficient to permit unhurried surgical interventions in adult patients originated with Gibbon [1939] whose initial laboratory models relied on rotating cylinders to spread blood in a continuously renewed thin film, in the tradition of 19th-century physiologists. Gibbon's clinical model [1954], built with technical support from IBM, was a stationary film oxygenator, a bulky device in which venous blood was evenly smeared over a stack of vertical wire screen meshes in an oxygen atmosphere, flowing gently downward to accumulate in a reservoir from where blood could be returned to a systemic artery. The main problems with that design, besides its cumbersome dimensions, were to avoid blood streaming and to maintain a constant blood-gas exchange area. Other investigators tried to increase the flexibility of the system by replacing the stationary film support with rotating screens or rotating discs partly immersed in a pool of blood. This allowed some control of gas transfer performance by changing the rational speed, but minimizing the blood content of the device dictated a tight fit between the discs and the horizontal glass cylinder surrounding them. Foaming and hemolysis were encountered at high disc spinning velocity, and these designs were eventually abandoned.

The very first strategy of physiologists for exchanging oxygen and carbon dioxide in venous blood had been to bubble pure oxygen through a stationary blood pool. To turn this batch process into a continuous operation for totally body perfusion, blood was collected through cannulae from the central veins by a syphon or a pump, driven upward in a vertical chimney, mixed with a stream of oxygen gas bubbles, and finally passed through filters and defoaming sponges so as to collect bubble-free arterialized blood, which could then be used to perfuse the arterial tree. The efficiency of bubble oxygenators is extremely high because the smaller the bubbles, the larger the blood-oxygen exchange area developed by a steady current of gas. In the limiting case, it is even possible to saturate venous blood by introducing no more oxygen into the blood than is consumed by the tissues. This process, however, doe not remove any carbon dioxide. Since the partial pressure of CO_2 in the excess gas vented from the bubble oxygenator cannot exceed the CO_2 partial pressure in arterialized blood (and in actuality is much lower), it follows the carbon dioxide transfer rate of bubble oxygenators is a direct function of the volume inflow rate of oxygen, which must exceed oxygen uptake severalfold to transfer both O_2 and CO_2. Thus, the operating conditions for a bubble oxygenator are dictated to a major extent by the requirements for adequate Co_2 removal.

Three major advances propelled bubble oxygenators ahead of film oxygenators in the pioneer decades of cardiac surgery. The first was the identification of silicone-based defoaming compounds which could be smeared on top of the bubble chimney and proved much more effective in coalescing the blood foam than previously used chemicals. The second was the quantitative process analysis by Clark [1950] and Gollan [1956] which showed that, since small bubbles favor oxygen transfer and large bubbles are need for CO_2 removal, an optimum size could be found in between, or alternatively a mix of small and large bubbles should be used. The third and practically decisive advance was made simultaneously by Rygg [1956] and DeWall [1957], who replaced the assembly of glass, steel, and ceramic parts of early bubble oxygenators with inexpensive plastic components and thereby paved the way for the industrial manufacturing of disposable bubble oxygenators. As a result, reusable stationary film and disc oxygenators, which require careful cleaning and sterilization, slowly disappeared, and disposable bubble oxygenators dominated the field of extracorporeal gas exchange from 1960 to the early 1980s. The oxygenator and pumps designed during the pioneer phase of extracorporeal circulation are described in a book by Galletti and Brecher [1962].

In the 1970s,the bubble oxygenator was integrated with a heat exchanger (usually a stainless steel coil) and placed within a clean plastic container that also served as a venous reservoir with a capacity of several liters of blood. The level of blood-air interface allowed direct observation of changes of blood volume in the extracorporeal circuit. This simple design feature gave the equipment operator (who eventually became known as the perfusionist) plenty of time in which to react to a sudden blood loss and make compensating changes in operation.

Yet a number of problems occurred with bubble oxygenators because of their large blood-gas interface. If the blood foam did not coalesce completely, a gaseous microemboli could be carried into the arterialized bloodstream. Plasma proteins were denatured at the gas interface, leading to blood trauma associated with platelet activation and aggregation, complement activation, and hemolysis. These problems could be ameliorated by placement of a gas-permeable membrane between the blood and gas phases. The idea of using membranes permeable to respiratory gases in order to separate the blood phase from the gas phase in an artificial lung, and consequently avoid the risk of foaming or the formulation of thick blood rivulets inherent in bubble or film oxygenators, was stimulated after World War II by the growing availability of commercially produced thin plastic films for the packaging industry. The two major challenges for membrane oxygenators were, and indeed remain, that no synthetic membrane could be fabricated as thin as the pulmonary alveolar wall, and manifolding and blood distribution system could match the fluid dynamic efficiency of the pulmonary circulation, where a single feed vessel—the pulmonary artery—branches over a short distance and with minimal resistance to flow into millions of tiny gas exchange capillaries the size of an erythrocyte. Membrane oxygenators have progressively captured the largest share of the market for clinical gas exchange devices not only because their operation is less traumatic for blood but also because the blood content of the gas exchange unit is fixed, thereby limiting volume fluctuations to calibrated reservoirs and minimizing the risk of major shifts in intracorporeal blood volume during total body perfusion.

The interposition of a membrane between flowing gas and flowing blood reduces the gas transfer efficiency of the system as a consequence of the additional mass transfer resistances associated with the membrane itself and the geometry of the blood layer. The permeability of various polymeric materials to oxygen and carbon dioxide is summarized in Table 129.1. The very first plastic films used, such as thin sheets of polyethylene and polytetrafluoroethylene (PTFE), showed such a low diffusional permeability that 5–10 m^2 were needed to meet even the minimal oxygen needs of an anesthetized hypothermic adult [Clowes et al., 1956]. The advent of silicone elastomer films (either as solid sheets or cast over a textile support mesh) in the 1960s established the technical feasibility of membrane oxygenators [Bramson et al., 1965; Galletti et al., 1966; Peirce et al., 1967]. Silicone rubber is about 140 times more permeable to oxygen and 230 times more permeable to CO_2 than is PTFE for equivalent thickness. Even though silicone rubber and related elastomers cannot be cast as thin as PTFE, their permeability is so high that they became the standard material for early membrane oxygenator prototypes.

Extensive research was carried out in the 1960s and 1970s to develop improved membrane oxygenator designs with more efficient oxygen transport across the blood boundary layer. These designs are discussed later in this chapter. The clinical motivation was to provide continuous full or partial replacement of

TABLE 129.1 O_2 and CO_2 Permeability of Various Materials

Chemical Composition	Common Name	O_2 Permeability $\frac{\text{cm}^3\text{ (STP)}\cdot\text{mil}}{\text{min}\cdot\text{m}^2\cdot\text{atm}}$	$\frac{CO_2 \text{ Permeability}}{O_2 \text{ Permeability}}$
Air		5×10^7	0.8
Polydimethylsiloxane	Silicone rubber	1100	5
Water		150	18
Polystyrene		55	5
Polyisoprene	Natural rubber	50	6
Polybutadiene		40	7
Regenerated cellulose	Cellophane	25	18
Polyethylene		12	5
Polytetrafluoroethylene	Teflon	8	3
Polyamide	Nylon	0.1	4
Polyvinylidene chloride	Saran	0.01	6

Permeability is the product of the diffusion coefficient D and the Bunsen solubility coefficient α. The gas flux J [cm³(STP)/min•m²] across a membrane of thickness h with a partial pressure difference Δr is given by $J = P\Delta r/h$.

pulmonary gas exchanges for periods of weeks to patients with advanced respiratory failure, with the hope that in the interval the natural lung would recover. Since extensive blood trauma limited the use of blood oxygenators beyond 12–24 hours, membrane oxygenators became the key to this application. In the mid-1970s, an NIH-sponsored clinical study [Zapol, 1975] demonstrated that with the protocols used lung function was not regained after 1–3 weeks of extracorporeal circulatory with a membrane oxygenator. The major intended application of membrane oxygenators having vanished, and the high cost of silicone rubber making these devices much more expensive than bubble oxygenators, research and development almost came to a halt, leaving bubble oxygenators in control of the remaining field of use, namely cardiopulmonary bypass.

Two technical advances have over the following two decades reversed this trend: (1) the discovery that hydrophobic microporous membranes, through which gas can freely diffuse, have a high enough surface tension to prevent plasma filtration at the moderate pressures prevailing in the blood phase of an artificial lung; (2) the large-scale fabrication of defect-free hollow fibers of microporous polypropylene, which can be assembled in bundles, potted and manifolded at each extremity to form an artificial capillary bed of parallel blood pathways immersed in a cylindrical hard shell through which oxygen circulates.

Microporous hollow fiber membrane oxygenators now dominate the market. The most common embodiment of the hollow fiber membrane oxygenator features gas flow through the lumen of the fibers and blood flow in the interstices between fibers. This arrangement not only utilizes the larger outer surface area of the capillary tubes as gas transfer interface, instead of the luminal surface, it also promotes blood mixing in a manner which enhances oxygen transport, as will become apparent below.

129.2 Cardiopulmonary Bypass

Cardiopulmonary bypass (CPB), also called heart-lung bypass, allows the temporary replacement of the gas exchange function of the lungs and the blood-pumping function of the heart. As a result blood no longer flows through the heart and lungs, which then presents the surgeon with a bloodless operative field.

The terms pump-oxygenator and heart-lung machine graphically describe the equipment used. Cardiopulmonary bypass is the procedure. Open-heart surgery strictly speaking refers to interventions inside the heart cavities, which once provided the most frequent demand for cardiopulmonary bypass technology. By extension, the term is also applies to surgical procedures which take place primarily on the external aspects of the heart, such as creating new routes for blood to reach the distal coronary arteries from the aorta, e.g., coronary artery bypass grafting (CABG) or, popularly, *bypass surgery.* In these

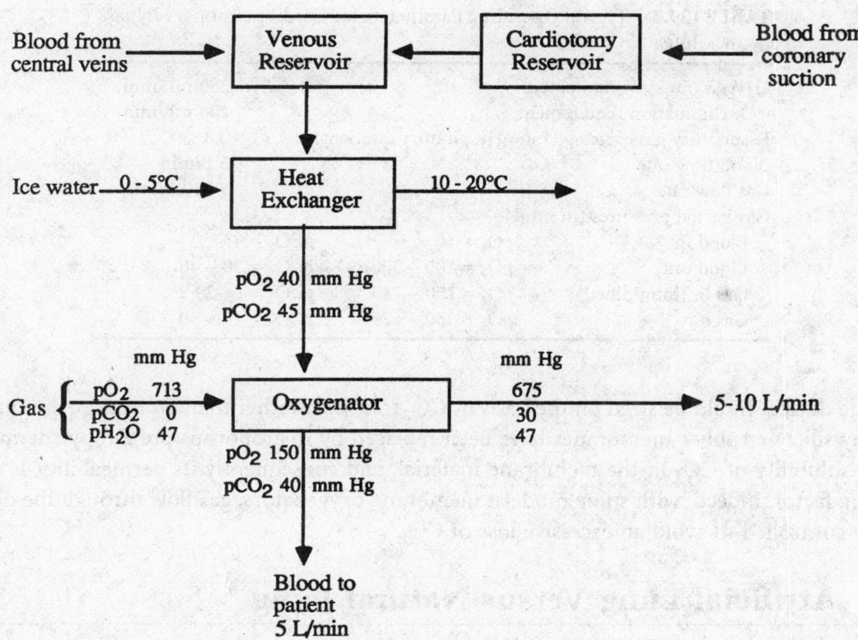

FIGURE 129.1 Scheme of standard operating conditions during cardiopulmonary bypass.

operations the cardiac cavities are temporarily vented (i.e., open to atmospheric pressure) to avoid the build-up of pressure which could damage the cardiac muscle.

As usually employed for cardiac surgery, the heart-lung machine is part of a total, venoarterial cardiopulmonary bypass circuit, meaning that all the venous blood returning to the right heart cavities is collected in the extracorporeal circuit and circulated through the gas exchange device, from where it is pumped back into the arterial tree, thereby "bypassing" the heart cavities and the pulmonary circulation. On the blood side, this procedure usually involves hemodilution, some degree of hypothermia, nonpulsatile arterial perfusion at a flow rate near the resting cardiac output, and continuous recirculation of blood in an extracorporeal circuit in series with the systemic circulation of the patient. On the gas side, oxygen or an oxygen-enriched gas mixture (with or without a low concentration of CO_2) flows from a moderately pressurized source in a continuous, nonrecycling manner and is vented to the room atmosphere.

CPB hinges on twin postulates: that blood circulation can be sustained by mechanical pumps while the heart is arrested and that venous blood can be artificially arterialized in an extracorporeal gas exchange device while blood flow is excluded from the lungs. Each of these claims was established separately through animal experimentation over the course of over 100 years (see Galletti and Brecher [1962]). Then surgeons and engineers combined the advances made by physiologists and pharmacologists and turned them, in the 1950s, into the basic tool of cardiac surgery. A recent update on the evolution of artificial membrane lungs for CPB surgery has been provided by Galletti [1993]. It is estimated that each year about 650,000 disposable membrane lungs are used in the operating room worldwide, with each gas exchange unit selling for a price in the range of \$250–400, i.e., a market close to \$2 billion.

Typical operating conditions for total cardiopulmonary bypass in an adult are summarized in Fig. 129.1 and Table 129.2. Most notable are the large differences in driving forces for 0_2 and CO_2. An approximate comparison can be made using the inlet conditions. Thus, under normothermia and with pure oxygen fed to the gas phase, initial driving force is approximately $45 - 0 = 45$ mmHg for CO_2 and $760 - 47 = 713$ mmHg for O_2 yielding a ratio of CO_2 to O_2 driving forces of about 0.06. The ratio of CO_2 to O_2 permeability for silicone rubber is roughly 5 (Table 129.2), and the corresponding flux ration is 0.35, less than half of the metabolically determined respiratory quotient. Under these conditions, CO_2 is the limiting

TABLE 129.2 Typical Operating Parameters for Cardiopulmonary Bypass in an Adult

Oxygen transfer requirement		250 mL/min
CO_2 elimination requirement		200 mL/min
Respiratory gas exchange ratio (respiratory quotient)		0.8
Blood flow rate		5 L/min
Gas flow rate		5 –10 L/min
Gas partial pressures (in mmHg)		
Blood in	$pO_2 = 40$	$pCO_2 = 45$
Blood out	$pO_2 = 100 - 300$	$pCO_2 = 30 - 40$
Gas in (humidified)	$pO_2 = 250 - 713$	$pCO_2 = 0 - 20$
Gas out	$pO_2 = 150 - 675$	$pCO_2 = 10 - 30$

gas, and the device should be sized on the basis of CO_2 transport rather than O_2 transport requirements. This is why silicone rubber membranes have been replaced by microporous polypropylene membrane, where the solubility of CO_2 in the membrane material, and consequently its permeability, is no longer the limiting factor. Indeed, with some modern membrane oxygenators, gas flow through the device may have to be controlled to avoid an excessive loss of CO_2.

129.3 Artificial Lung Versus Natural Lung

In the *natural lungs*, the factors underlying exchange across the alveolo-capillary barrier and transport by the blood can be grouped into four classes:

1. The ventilation of the lungs (the volume flow rate of gas) and the composition of the gas mixture to which mixed venous (pulmonary artery) blood will be exposed
2. The permeability of the materials which separate the gas phase from the blood phase in the pulmonary alveoli
3. The pattern of pressure and flow through the airways and through the pulmonary vascular bed and the distribution of inspired air and circulating blood among the various zones of the exchange system
4. The gas carrying capacity of the blood as regards oxygen and carbon dioxide (and secondarily nitrogen and anesthetic gases)

In an *artificial lung*, replacing the gas transfer function of the natural organ implies that blood circulation can be sustained by mechanical pumps for extended periods of time to achieve a continuous, rather than a batch process, and that venous blood can be arterialized in that device by exposure to a gas mixture of appropriate composition. The external gas supply to an artificial lung does not pose particular problems, since pressurized gas mixtures are readily available. Similarly the components of blood which provide its gas-carrying capacity are well identified and can be adapted to the task at hand. In clinical practice, it is important to minimize the amount of donor blood needed to fill the extracorporeal circuit, or priming volume. Therefore a heart-lung machine is generally filled with an electrolyte or plasma expander solution (with or without donor blood), resulting in hemodilution upon mixing of the contents of the extracorporeal and intracorporeal blood circuits. The critical aspects for the operation of an artificial lung are blood distribution to the exchanger, diffusion resistances in the blood mass transfer boundary layer, and stability of the gas exchange process.

Artificial lungs are expected to perform within acceptable limits of safety and effectiveness. The most common clinical situation in which an artificial lung is needed is typically of short duration, with resting or basal metabolism in anesthetized patients. Table 129.3 compares the structures and operating conditions of the natural lung and standard hollow fiber artificial membrane lungs with internal blood flow.

An artificial lung designed to replace the gas exchange function of the natural organ during cardiac surgery must meet specifications which are far less demanding than the range of capability of the

TABLE 129.3 Comparison of Nature and Artificial Membrane Lung

Natural Lung	Hollow Fiber Oxygenator
Performance	
Must meet demand at rest and during exercise, fever, etc.	Must meet demand at rest and under anesthesia
Constant temperature process around 37°C	Can be coupled with heat exchanger to lower body temperature.
O_2 and CO_2 transfer are matched to achieve the respiratory quotient imposed by foodstuff metabolism (around 0.8)	O_2 and CO_2 transfer are largely independent of each other and must be controlled by the operator
Continuous over a lifetime	Usually limited to few hours
Exchange surface area	
Wide transfer area (~ 70 m^2)	Limited transfer area (1 –3 m^2)
Highly permeable alveolar-capillary membrane	Diffusion barriers in synthetic membrane and blood oxygenation boundary layer
Short diffusion distances (1 –2 μm)	Relatively thick membranes (50 –100 μm)
Hydrophilic membrane	Hydrophobic polymers
No hemocompatibility problems	Hemocompatibility problems
Self-cleaning membrane	Protein build-up on membrane
Gas side	
To-and-fro ventilation	Steady cross flow gas supply
Operates with air	Operates with oxygen-rich mixture
Membrane structure sensitive to high oxygen partial pressure	Membrane insensitive to high oxygen partial pressure
Pressure below that in blood phase to avoid capillary collapse	Pressure below that on blood side to avoid bubble formation
Operates under water vapor saturation conditions	Can be clogged by water vapor condensation
Ventilation linked to perfusion by built-in control mechanisms	Ventilation dissociated from perfusion, with risks of hyper- or hypoventilation
Can be used for gaseous anesthesia	Can be used for gaseous anesthesia
Blood side	
Short capillaries (0.5 – 1 mm)	Long blood path (10 –15 cm)
Narrow diameter (3 – 7 μm)	Thick blood film (150 –250 μm)
Short exposure time (0.7s)	Long exposure time (5 –15s)
Low resistance to blood flow	Moderate to high resistance to blood flow
Sophisticated branching	Crude manifolding of entry and exit ports
Minimal priming volume	Moderate to large priming volume
Capillary recruitment capability	Fixed geometry of blood path
No recirculation	Possibility of recirculation
Limited venous admixture	Risk of uneven perfusion of parallel beds
No on-site blood mixing	Possibility of blood stirring and mixing
Operates with normal hemoglobin concentration	Hemodilution is common
Does not require anticoagulation	Requires anticoagulation

mammalian lung would suggest. Nonetheless, these specifications must embrace a range of performance to cover all metabolic situations which a patient undergoing cardiopulmonary bypass might present. These conditions range in terms of metabolic rate from the slightly depressed resting metabolism characteristic of an anesthetized patient, lightly clad in a cool operating room, to moderate (25–28°C) and occasionally deep (below 20°C) hypothermia. Hypothermia and high blood flow are occasionally encountered in patients with septic shock. In terms of body mass, patients range from 2–5-lb newborn with congenital cardiac malformations to the 250-lb obese, diabetic elderly patient suffering from coronary artery disease and scheduled for aortocoronary bypass surgery.

Whereas it is appropriate to match in advance the size and therefore the transfer capability of the gas exchange unit in the heart-lung machine to the size of the patient (largely out of concern for the volume of fluid needed to fill or "prime" the extracorporeal circuit), each gas exchange unit, once in use, must be capable of covering the patient's requirement under any circumstances. This is the responsibility of the perfusionist, who controls the system in the light of what is happening to the patient in the operative field. In fact, the perfusionist substitutes his or her own judgment for the natural feedback mechanisms which normally control ventilation and circulation to the natural lungs. The following analysis indicates how there requirements can be met in the light of the gas transport properties of blood and the characteristics of diffusional and convective transport in membrane lung devices.

129.4 Oxygen Transport

The starting point for analysis of O_2 transport is the conservation of mass relation, which is derived from a material balance in a differential element within the flowing blood. By way of example, the relation commonly employed for oxygen transport of blood flowing in a tube is

$$V\left(\frac{\partial[O_2]}{\partial x}+\frac{\partial[HbO_2]}{\partial x}\right)=DO_2\frac{1}{r}\frac{\partial}{\partial r}\left(r\frac{\partial[O_2]}{\partial x}\right) \quad (129.1)$$

where x is the axial coordinate, r is the radial coordinate, V is the velocity in the axial direction, DO_2 is the effective diffusion coefficient of O_2 in flowing blood, $[O_2]$ is the concentration of physically dissolved O_2, and $[HbO_2]$ is the concentration of O_2 bound to hemoglobin. The terms on the left side of Eq. (129.1) represent convection in the axial direction of O_2 in its two forms, dissolved in aqueous solution and bound to hemoglobin. The right-side term represents the diffusion of dissolved O_2 in the radial direction.

By making the usual assumption that the concentration of dissolved O_2 is linearly proportional to the O_2 partial pressure p, $[O_2] = \alpha p$, where α is the Bunsen solubility coefficient, and by using the definition of oxyhemoglobin saturation: $[HbO_2] = C_T S$ where C_T is the oxygen-carrying capacity of hemoglobin (per unit volume of blood) and S is its fractional saturation, Eq. (129. 1) can be rewritten in a modified form

$$V\frac{\partial p}{\partial x}\left(1+\frac{C_T}{a}\frac{\partial S}{\partial p}\right)=DO_2\frac{1}{r}\frac{\partial}{\partial r}\left(r\frac{\partial p}{\partial x}\right) \quad (129.2)$$

which shows that the convective contribution is proportional to the slope of the saturation curve. Equation (129.2) must be solved subject to the appropriate initial and boundary conditions, which for a tube are

$$x=0,\ p=p_i \quad (129.3)$$

$$r=0,\ \frac{\partial p}{\partial r}=0 \quad (129.4)$$

$$r=R,\ \alpha DO_2\frac{\partial p}{\partial r}=P_m\left(p_o-p\right) \quad (129.5)$$

where p_i and p_o are the oxygen partial pressures in the inlet blood and in the gas, respectively, and P_m is the membrane permeability for O_2 diffusion, defined so as to include the membrane thickness. These conditions represent, in order, a uniform inlet concentration, symmetry about the tube axis, and the requirement that the O_2 diffuison flux at the blood-membrane interface be equal to the O_2 diffusion flux through the membrane.

Implicit in the derivation of Eq. (129.2) are two important assumptions. The first is that, on a macroscopic scale, blood can be treated as a homogenous continuum, even though as a suspension of red blood cells in plasma, blood is microscopically heterogeneous. This simplification is acceptable as long as the volume element in which oxygen transport occurs is large compared to the size of a single red cell but small compared to the size of the overall diffusion path. This seems reasonable when applied to transport in membrane lungs, where the blood diffusion path thickness is usually equivalent to 20–30 red cell diameters or more.

The second important assumption is that the rate of reaction between O_2 and hemoglobin is sufficiently fast, when compared to the rate of diffusion of O_2 within the red cell, that the reaction can be considered at equilibrium, with the concentration of hemoglobin-bound O_2 in the red blood cells directly related to the concentration of dissolved O_2 in plasma via the oxyhemoglobin dissociation curve (ODC). Implicit in the use of this relationship is the assumption that the O_2 diffusion resistance of the red cell membrane is insignificant.

The human ODC for normal physiologic conditions is shifted to the right by increased temperature or decreased pH because of decreased hemoglobin affinity for O_2. The ODC is shifted to the right by an increase in the concentration of various organic phosphates, especially 2,3-diphosphoglycerate (DPG).

Under typical venous physiologic conditions (37°C, boundary O_2 partial pressures of 40 and 95 mmHg), the reaction between hemoglobin and oxygen reaches equilibrium during the time of contact between the blood and the gas phase. The ratio of the effective permeability of blood to that of plasma in a model system is 0.87 at the hematocrit of 45%; without any facilitation within the red cell, the ratio would be 0.75. Using the most reliable data for the O_2 solubility and diffusion coefficient in plasma leads to an estimate of 1.7×10^{-5} cm^2/s for the effective diffusion coefficient of O_2 in normal whole blood.

We can now return to the analysis of convective transport of oxygen in a membrane lung, specifically the solution of Eq. (129.2) or its equivalent for other geometrics, on the assumption of equilibrium for the hemoglobin oxygenation reaction in such devices. The various theoretical analyses that have been carried out differ primarily in the means by which the saturation curve is handled. The most common approach is to retain its intrinsic nonlinearity and to approximate it by a suitable analytical expression. The resulting nonlinear partial differential equation does not permit analytical solution, and it is therefore necessary to resort to numerical solution with a finite difference scheme on a digital computer. This approach yields numerical values for p and S as a function of x and r. To relate theoretical prediction with experimental measurement, one must calculate the velocity-weighted bulk average values of p and S [Colton, 1976]. The average O_2 partial pressure, p, and oxyhemoglobin saturation, S, that would be measured if the blood issuing from the tube were physically mixed are then calculated from

$$\frac{1}{\int_0^R Vr\,\mathrm{d}r}\left[\alpha\int_0^R Vpr\,\mathrm{d}r + C_T\int_0^R VSr\,\mathrm{d}r\right] = \alpha\bar{p} + C_T\bar{S} \tag{129.6}$$

Numerical solutions provide accurate predictions. However, they apply only to specific operating conditions and cannot be generalized for design purposes. Two methods have been used to simplify the ODC and provide approximate yet useful analytical solutions.

The first is known as the advancing front theory [Marx et al., 1960; Thews, 1957]. The oxyhemoglobin saturation curve is approximated as a step function between the saturation values corresponding to the O_2 partial pressure and that of the gas phase or the blood-membrane interface. The blood is treated as two regions of uniform oxyhemoglobin saturation separated by a front that moves rapidly inward. Outside the front, blood is saturated at a value corresponding to the blood-gas or blood-membrane interface, and the rest of the blood is relatively unsaturated. Oxygen diffuses through the saturated blood to the interface, where it reacts with unsaturated hemoglobin. The advancing front approximation reasonably represents the calculated saturation profiles and has proved useful in developing analytical design expressions for membrane lungs in terms of saturation changes effected. However, it can be in error by a very wide margin for the prediction of O_2 partial pressure changes. A modification of advancing front theory, involving approximation of the partial pressure changes. A modification of advancing front theory, involving approximation of the ODC by several straight line segments, retains the ability to provide an analytical solution and gives better accuracy than the conventional step function approximation of the ODC.

The second type of simplification is to approximate the ODC by a straight line drawn between the inlet and boundary O_2 partial pressures, which make $\partial S/dp$, the slope of the saturation curve constant and renders Eq. (129.2) linear. Use can then be made of existing solutions for analogous convective heat and mass transfer problems without chemical reaction.

For typical operating conditions in the clinic, the initial and boundary O_2 partial pressures lie on the upper portion of the ODC, and the advancing front solution provides an overestimation of the rate of O_2 transport, whereas the constant slope solution provides an underestimate. Conversely, on the lower portion of the ODC at very low O_2 partial pressures, the advancing front estimate of O_2 transport rate is too low and constant slope too high. The constant slope approximation is most accurate over the steep portion of the saturation cure, where it is nearly linear. Since the O_2 transport rate per unit membrane surface area is much more sensitive to blood inlet O_2 partial pressure than would be expected solely from the change in the overall driving force, comparative testing of membrane lungs must be carried out with identical inlet blood O_2 partial pressure and oxyhemoglobin saturation.

Theoretical prediction of membrane lung performance is useful for design purposes and for providing a guide to the effect of permissive design variables. However, theoretical prediction cannot substitute for experimental data under closely controlled conditions where control of pH, temperature, and CO_2 partial pressure in fresh blood allow the definition of the appropriate ODC.

129.5 CO_2 Transport

The CO_2 dissociation curve for normal human blood is far more linear in its normal operating range than the ODC. The fractional volume of CO_2 that is removed in the process of arterialization of venous blood is also considerably less than the corresponding fractional loss of oxygen (about 10 percent of blood Co_2 content, versus 25 percent for oxygen). As is the case for oxygen, the total CO_2 concentration is far larger than that of gas physically dissolved in the aqueous component of blood. Plasma accounts for about two-thirds of all the CO_2 carried in blood, whereas typically about 98% of O_2 is carried in the red cells.

The main vehicles for CO_2 transport in blood are bicarbonate, the primary carrier in both plasma and red cells, and carbamino hemoglobin, where CO_2 is combined with the amino groups of hemoglobin (Fig. 129.2*a*). Arrows on Fig. 129.2*b* indicate the direction and relative rate of each reaction whereby CO_2 is removed in a membrane lung. Carbonate and hydrogen ions form bicarbonate, which decomposes to CO_2 or combines with another hydrogen ion to form carbonic acid; the latter is dehydrated to liberate CO_2. Since the reactions that form CO_2 in plasma are very slow, biocarbonate is the predominant species. Biocarbonate can diffuse into the red cell, albeit slowly, in exchange for chloride, leading to the same chain of reactions. In the red cell, however, dehydration of carbonic acids is catalyzed by the enzyme carbonic anhydrase. This reaction liberates CO_2, which, in turn, diffuses out of the red cell into the plasma and then across the blood-membrane interface. Decomposition of carbamino hemoglobin is a significant additional source of CO_2. Carbamate compounds that arise from combination of CO_2 with plasma proteins have a much smaller effect because of the relatively unfavorable equilibria for their formation. Finally, various ionic species, such as organic and inorganic phosphates, amino acids, and proteins, behave as weak acids at pH 7.4. The buffering power of hemoglobin is particularly strong and has a marked effect in influencing the shape of the CO_2 dissociation curve.

Under clinical conditions of O_2 and CO_2 countertransport, two reciprocal phenomena occur which affect CO_2 and O_2 exchange. A decrease of CO_2 partial pressure causes a shift to the left of the oxyhemoglobin dissociation curve, leading to an increased affinity of hemoglobin when CO_2 is removed (Bohr effect). Meanwhile, because oxyhemoglobin is a stronger acid than hemoglobin, uptake of O_2 decreases the affinity of hemoglobin for CO_2, thereby releasing additional CO_2 from carbamino hemoglobin (Haldane effect). At the same time, increased acidity favors the conversion of more biocarbonate into carbonic acid, which then dissociates, releasing CO_2. The simultaneous occurrence of these two effects enhances transport rates of both gases.

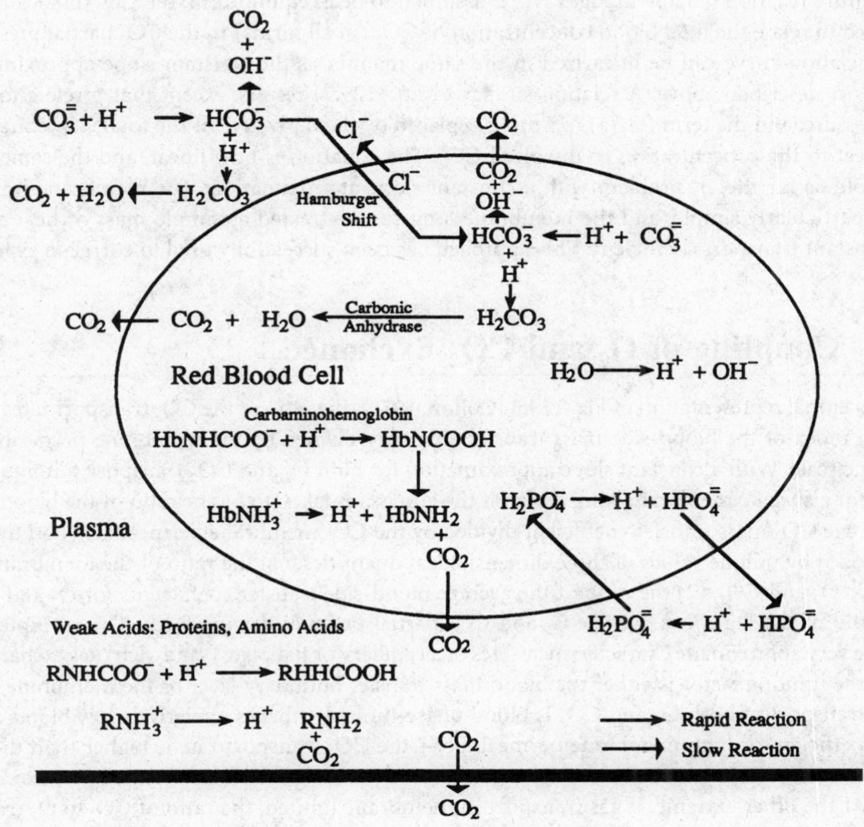

FIGURE 129.2a Schematic representation of CO_2 transfer from the red blood cell to plasma and into alveolar gas, emphasizing the various buffer systems involved.

CO_2 transport form		Mixed venous blood		Arterial blood		Veno-arterial difference	
		mM/L	%	mM/L	%	mM/L	%
Bicarbonate	plasma	14.41	61.8	13.42	62.4	.99	55.0
	RBC	5.92	25.4	5.88	27.3	.04	2.2
Dissolved CO_2	plasma	.76	3.3	.66	3.1	.10	5.6
	RBC	.51	2.2	.44	2.0	.07	3.9
Carbamino CO_2	plasma	-	-	-	-	-	-
	RBC	1.70	7.3	1.10	5.1	.60	33.3
Total in whole blood		23.30	100%	21.50	100%	1.80	100%

FIGURE 129.2b Blood CO_2 content. CO_2 transport forms in mixed venous and arterial blood, and as components of the veno-arterial CO_2 difference. Observe that the red blood cells (RBC) bicarbonate content, although it represents a large fraction of CO_2 transport, does not contribute significantly to the arteriovenous difference in CO_2 content. Conversely hemoglobin-bound CO_2, though less abundant than bicarbonate in red blood cells, constitutes a larger fraction of the veno-arterial difference.

If the entire reaction scheme in Fig. 129.2 is assumed to be at equilibrium, the CO_2 dissociation curve can be used to relate the total blood concentration of CO_2 (in all forms) to the CO_2 partial pressure. The CO_2 dissociation curve can be linearized in the same manner as the constant slope approximation for O_2 transport described above. A relation similar to Eq. (129.2) results, except that p refers to the CO_2 partial pressure, and the term (C_T/a) $(\partial S/\partial p)$ is replaced by the derivative of the total CO_2 concentration with respect to the concentration of dissolved CO_2. The equation is now linear, and the same simplifications hold as for the O_2 problem with a constant slope approximation. The membrane-limited case becomes particularly simple, and the membrane lung can be treated as simple mass or heat exchanger with a constant transport coefficient. This approach has been successfully used to correlate experimental data.

129.6 Coupling of O_2 and CO_2 Exchange

In the conceptual representation of Fig. 129.3 [Colton, 1976], the ratio of the CO_2 transport rates is plotted against an index of the blood-side mass transfer efficiency divided by the membrane permeability on a logarithmic scale. With a constant slope approximation for both O_2 and CO_2 transport, a unique curve is obtained for a specific membrane lung design if the abscissa is taken to be the ratio of the blood-side log-mean average CO_2 mass transfer coefficient divided by the CO_2 membrane permeability, and the curve is parameterized by unique values of three dimensionless quantities: (1) the ratio of the membrane permeabilities for O_2 and CO_2, (2) the ratio of the average blood-side transfer coefficients for O_2 and CO_2, and (3) the ratio of the log-mean average O_2 and CO_2 partial pressure driving force. The asymptotic limits plotted are very approximate characteristic values of a capillary or flat plate (sandwich) gas exchange device in which the limiting factor is either the blood mass transfer boundary layer or the membrane itself.

When transport of both O_2 and CO_2 is blood phase-limited—that is a relatively low blood-side mass transfer coefficient or high membrane permeability—the CO_2 transport rate is higher than the O_2 rate, and it is necessary to add CO_2 to the inlet gas or to decrease gas flow rate to prevent excessive CO_2 removal. At the other extreme, if gas transport in membrane limited, the ratio of CO_2 to O_2 transport is always lower than the physiological value (0.8), because no existing gas-permeable membrane has a sufficiently high CO_2/O_2 permeability ratio to overcome the unfavorable driving force ratio of the O_2 and CO_2 partial pressures (14:1) under the actual operating conditions of a membrane lung. Under either limiting condition, it is necessary to design the membrane lung on the basis of the gas that limits transport, O_2, for blood-limited conditions and CO_2 for membrane-limited conditions.

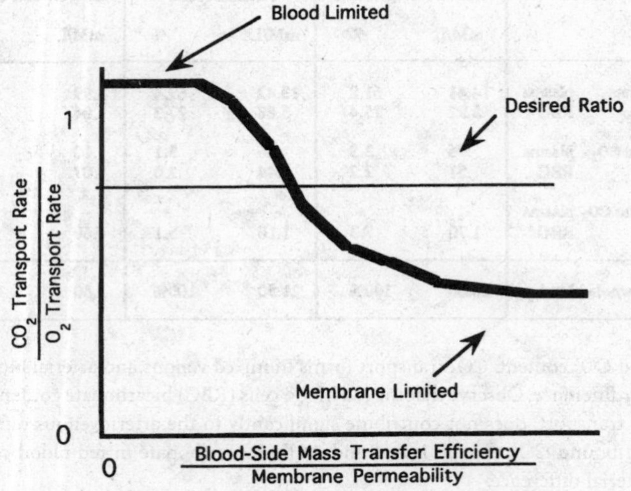

FIGURE 129.3 Relative CO_2 and O_2 transport in a membrane lung.

A priori, it seems undesirable to operate under blood-limited conditions because full advantage is not taken of the membrane permeability. Therefore, an increase in the blood-side mass transfer efficiency is valuable, but only to the point where the ratio of CO_2 to O_2 transport rates is equal to the respiratory quotient. Beyond that point, further improvements in design will not reduce the size of the device required unless membrane permeability to CO_2 is increased, thereby moving the operating point on the curve to the left and justifying the use of a more efficient exchange device. For example, conventional solid silicone rubber membranes cannot take advantage of high-efficiency designs because gas transport is membrane limited and has to be designed on the basis of CO_2 transport rate. To make effective use of the most effective devices, it is necessary to employ microporous membranes or ultrathin membranes on microporous substrates where the CO_2 permeability is no longer a limiting factor.

129.7 Shear-Induced Transport Augmentation and Devices for Improved Gas Transport

There is evidence that flow-dependent properties of blood can substantially influence the transport of O_2 and CO_2. The presence of velocity gradient in a stream can enhance mass transfer through blood either by shear-induced collision diffusion wherein interactions between red blood cells produce net lateral displacements and associated motions in the surrounding phase or by rotation of individual cells, which gives rise to local mechanical stirring of the adjacent fluid. Both mechanisms can lead to transport augmentation of species present in the dispersed or continuous phases. Only the first mechanism can cause dispersive migration of the particles themselves, and available evidence suggests that it is the dominant factor [Cha & Bessinger, 1994]. The shear-induced diffusion coefficient of particles in suspension increases linearly with the shear rate and can be orders-of-magnitude larger than the Brownian motion diffusivity. The effect of lateral cell movement in oxygen transport in capillary tubes depends on both the shear rate and the slope of the ODC, with a maximum at the steepest portion of the curve, i.e., below the operating range of a clinical oxygenator. The extent to which such phenomena occur in blood under the clinical operating conditions of membrane lungs in cardiopulmonary bypass has not been investigated, but the effect on oxygen transport is thought to be significant.

The earliest oxygenator configurations featuring rubber membranes made the blood flow in simple enclosed geometries, such as a flat plate or hollow fiber, that were inherently inefficient from a mass transfer standpoint. It was soon recognized that the gas transport rate was limited almost entirely by transport within the blood oxygenation boundary layer. There followed extensive efforts in the 1960s and early 1970s to investigate new approaches for improving gas transfer in membrane oxygenators, as summarized in Table 129.4. These approaches relied on one or more mechanisms: (1) increasing shear rate or producing turbulence; (2) keeping the oxygenation boundary layer very thin by using an appropriate contacting geometry or pulsatile blood flow; and (3) making use of secondary flow, which incorporates significant velocity components normal to the membrane surface. All these approaches are demonstrably effective under laboratory conditions, but few have achieved successful commercialization because of the constraints imposed by the geometry of available membrane materials (flat sheets and capillary tubes) and the clinical demand for simple, reliable, and inexpensive devices.

In Table 129.4, *passive designs* for inducing secondary flows are those for which no energy source is needed except that required for steady blood flow. A common technique has been to place obstacles, for

TABLE 129.4 Approaches for Improving Mass Transport in Membrane Oxygenators

Passive designs	Active designs
Obstacles in blood path	Oscillating toroidal chamber
Membrane undulations from external supports	Enclosed rotating disc
Membrane texturing or embossing	Pulsed flow vortex shedding
Helical flow systems	Couette flow
External blood flow over hollow fibers	Annulus, inner cylinder rotating axial flow, tangential flow.

example screens, in the blood path to induce secondary fluid movements and/or flow separation on a small scale and thereby increase blood mixing in flat plate devices. A similar result has been obtained by creating undulations in a flat membrane with grooved or multiple point supports or by using textured membranes to direct blood flow through the exchanger.

Active signs are those in which there is energy input to achieve high shear rates or create secondary flows. Highly efficient oxygenator prototypes have been described, and at least two, the enclosed rotating disc and a pulsed-flow vortex shedding device, have been commercialized, but neither has achieved widespread clinical acceptance, in part because of the complexity or cumbersomeness of their operating mechanisms in an operating room setting.

Another widely investigated approach has been to use flow geometries that naturally induce secondary flow, for example helical coils, where, superimposed on the primary flow, is a swirling motion in each half of the tube. The secondary flow trajectory which results from centrifugal effects takes particles from the periphery and carries them into the core, back to the periphery, back to the core in a continuously repeated fashion. Such secondary motion, by continuously sweeping oxygenated blood away from the membrane and replacing it with venous blood from the central core of the channel, can be extremely effective in increasing gas transport rates to the point where the dominant resistance to diffusion lies in the membrane. However, the practical difficulty of constructing such complex devices which are also disposable has thus far prevented industrial development.

When the potential market for continuous extracorporeal membrane oxygenation collapsed, so did the intensive research and development effort in developing new devices. However, three developments, two technical and one marketing-related, over the next decade eventually led to dominance of membrane oxygenators for cardiopulmonary bypass.

The first major technical advance was the fabrication of hydrophobic microporous capillary hollow fiber membranes at prices considerably lower than for silicon rubber membranes. The driving force for the initial development of these materials was their potential in another technology, membrane plasmapheresis for the separation of plasma from the cellular components of blood. Hollow fibers for the application had nominal pore sizes, around 0.5 μm, which were too large for membrane oxygenation because plasma could seep through the fiber wall under pressure leading to a catastrophic decrease in membrane permeability (129.1). In the early 1980s, microporous polypropylene hollow fibers with a nominal pore size around 0.1 μm became available and proved satisfactory for membrane oxygenation. In addition to reduced trauma and competitive pricing with bubble oxygenators, membrane lungs could be employed with a reduced, fixed blood-priming volume, thereby minimizing transfusion and hemodilution problems. Surprisingly, this advantage was initially viewed as a drawback by perfusionists who were used to having a large venous blood reservoir with a visible gas-blood interface.

The key marketing development of the mid-1980s was to make a membrane oxygenator by attaching it, along with a heat exchanger, to a clear plastic venous reservoir with a visible gas-blood interface. The astute move was breakthrough which put membrane oxygenators into the operating room. Their clear advantage in minimizing blood trauma and postoperative complications was so overwhelming that within a year membrane oxygenators captured the dominant market share. Shortly thereafter, bubble oxygenators were virtually eliminated from the U.S. marketplace.

The most recent technical development has been the inversion of the usual internal flow arrangement so that, now, gas flows through the lumen of the hollow fibers while blood is pumped over the external surface of the capillary membrane. This arrangement is most effective when the blood flow is at right angles to the hollow fiber. In that configuration the flowing blood successively encounters different fibers, and a new oxygenation boundary layer forms on each fiber. Because the boundary layer is thinnest where it begins (in this case, at the front of each fiber), and the mass transfer rate is correspondingly highest, the transport of oxygen averaged over the periphery of each fiber is much higher than with the conventional internal (luminal) flow of blood. Thus the transport of oxygen averaged over the periphery of each fiber is much higher than with the conventional internal (luminal) flow of blood. The performance of various membrane oxygenators is compared in Table 129.5. The data clearly demonstrate the superior performance attainable with microporous hollow fibers operated with external crossflow of blood.

TABLE 129.5 Comparison of Membrane Oxygenator Performance (Blood flow rate = gas flow rate = 51/min; AAMI conditions)

Model	Membrane Form	Blood Path Configuration	O_2, Flux cm³(STP)/min-m³	CO_2/O_2 Flux Ratio
Silicone Rubber				
Sci Med SM35	Sheet	Spiral coil embedded failure	90	0.6
Microporous polypropylene				
Cobe CMI	Sheet	Flat plate	130	1.4
Shiley M-2000	Sheet	Blood screens	120	1.0
Bentley Bos CM-40	Hollow fiber	Internal flow	70	0.7
Terumo Capiox II 43	Hollow fiber	Internal flow	60	0.9
Bard William Harvey 4000	Hollow fiber	External cross flow	140	0.9
Johnson & Johnson Maxima	Hollow fiber	External cross flow	150	0.9
Sarns	Hollow fiber	External cross flow	150	1.0
Bentely Univox	Hollow fiber	External cross flow	160	1.0

The state of the art is now fairly advanced for membrane blood oxygenators, but there is still room for improvement. Further increases in flux would further reduce cost and minimize priming volume and blood consumption. Now that membrane oxygenators are fully entrenched, it is more likely that the reduced priming volume and improved control of a closed system can be realized. Lastly, the residual gas-blood interface that still exists at the microporous membrane surface could be eliminated by coating with a very thin skin in a n asymmetric or composite structure.

Defining Terms

Advancing front theory: A type of exchanger theory addressing the limitation of oxygen transport in a blood film through a fully saturated boundary layer leading to a front where the hemoglobin in flowing blood reacts with the dissolved oxygen.

Arterialization: A gas exchange process whereby oxygen and carbon dioxide concentrations in venous blood are changed to levels characteristic of arterial blood.

Artificial lung: A device which allows for continuous exchange for oxygen and carbon dioxide between circulating blood and a controlled gas atmosphere.

Blood oxygenator: Synonymous with artificial lung, with the accent placed on oxygen transport, which is the most critical aspect of natural lung replacement, since the body oxygen reserves are very limited. Depending upon the physical process used for blood-gas transfer, artificial lungs are classified as bubble oxygenators, stationary or rotating film oxygenators, and membrane oxygenators.

Boundary layer: The film of blood adjacent to a permeable membrane, which, by reason of local fluid dynamics, is not renewed at the same rate as blood in the core of the flow path, thereby creating an additional diffusion barrier between the blood and the gas phase.

Bubble oxygenator: Blood-gas transfer device in which a large exchange surface is obtained by the dispersion of oxygen bubbles in a venous blood stream, followed by coalescence of the foam and venting of excess gas (cocurrent blood and gas flow) or by spreading of venous blood over a continuously renewed column of foam generated by bubbling oxygen at the bottom of a reservoir (countercurrent blood and gas flow).

Bypass: Derivation or rerouting of blood around an organ or body part, to diminish its blood supply, to abolish local circulation for the duration of a surgical intervention, or to increase blood flow permanently beyond an obstruction. The qualifier used with the word *bypass* designates the organ so isolated (e.g., *left ventricular bypass, coronary artery bypass*).

Cardiopulmonary bypass (CPB): A procedure whereby blood is prevented from circulating through the heart cavities and the lungs. Cardiopulmonary bypass (also known as *heart-lung bypass*) can be *partial* if no obstacle is placed on venous return to the right heart cavities and consequently some of the blood continues to flow through the pulmonary circulation. In that case the arterial system is fed in part by left ventricular output and in part by the arterialized blood perfusion from the hear-lung machine. The balance between the internal and extracorporeal blood circuits depends on the setting of the pumps and the relative resistance to flow in the two venous drainage pathways. During *total* cardiopulmonary bypass, the cardiac muscle must receive arterial blood from the extracorporeal circuit (intermittently or continuously) to prevent hypoxic myocardial damage. Since coronary venous blood drains into the cardiac cavities, this blood must be drained to the outside to prevent an intracavitary pressure increase which could be damaging to the heart.

Catheter: A long hollow cylinder designed to be introduced in a body canal to infuse or withdraw materials into or out of the body.

Coronary artery bypass graft (CABG): The construction of new blood conduits between the aorta (or other major arteries) and segments of coronary arteries beyond lesions which partially or totally obstruct the lumen of those vessels, for the purpose of providing an increased blood supply to regions of the myocardium made ischemic by those lesions.

Extracorporeal circulation: Artificial maintenance of blood circulation by means of pumps located outside of the body, with blood fed through catheters advanced in an appropriate blood vessel and returning the blood to another blood vessel.

Film oxygenator: Blood-gas transfer device in which a large exchange surface is obtained by spreading venous blood in a thin film over a stationary or moving physical support in an oxygen-rich atmosphere.

Heart-lung bypass: Synonymous with *cardiopulmonary bypass*.

Heart-lung machine: A mechanical system capable of pumping venous blood around the heart and the lungs and arterializing it in an appropriate gas exchange unit.

Hemodilution: Temporary reduction in blood erythrocyte concentration (and consequently hemoglobin content, hematocrit, oxygen-carrying capacity, and viscosity) resulting from mixing with the erythrocyte-free or erythrocyte-poor content of the liquid used to prime an extracorporeal circuit.

Hemolysis: The destruction of red blood cells with liberation of hemoglobin in surrounding plasma, caused by mechanical damage of the erythrocyte membrane, osmotic imbalance between intracorpuscular and extracorpuscular ion concentration, or uncontrolled freezing-thawing cycles.

Hollow fiber: A capillary tube of polymeric material produced by spinning a melted or dissolved polymer through an annular orifice.

Membrane: A solid or liquid phase which acts as a barrier to prevent coalescence of neighboring compartments while allowing restricted or regulated passage of one or more molecular species.

Membrane lung or **Membrane oxygenator:** Blood-gas transfer device in which the blood compartment is shielded from the gas phase by a porous or solid, hydrophobic polymer membrane permeable to gases but not to liquids (in particular, blood plasma).

Metabolism: The sum of the chemical reactions occurring within a living body including build up (anabolism) and break down (catabolism) of chemical substances.

Open heart surgery: Interventions taking place inside the cardiac cavities, such as for the replacement or reconstruction of cardiac valves, or the closure of abnormal communications between cardiac chambers, and which for reasons of convenience and safety, require the interruption of blood flow through the heart. By extension this stem is often used for cardiac interventions under total cardiopulmonary bypass, which address structures on the outside surface of the heart (such as coronary arteries) when drainage of the cardiac cavities through a vent is needed to avoid accumulation of coronary venous blood.

Oxygenation boundary layer: Stationary or slowly moving blood layer adjacent to a gas-permeable membrane, which progressively develops along the blood path and, once enriched with oxygen diffusion through the membrane, effectively becomes a barrier to oxygen transport perpendicular to the direction of flow.

Perfusion: A technique for keeping an organ or body part alive, though severed from its normal blood circulation, by introducing blood under pressure into the appropriate feeder artery.

Perfusionist: The operator of the heart-lung machine during cardiac surgery or respiratory assist procedures.

Priming volume: The volume of liquid (blood, plasma, synthetic plasma expanders, or electrolyte solutions) needed to fill all components of an extracoporeal circuit (oxygenator, heat exchanger, blood pumps, filter, tubing, and catheters) so as to avoid exsanguination once the intracorporeal and extracorporeal circulation systems are joined.

Pump-oxygenator: Equipment used to circulate blood through an extracorporeal circuit by means of mechanical blood pumps and to arterialize mixed venous blood by means of a gas exchange device. In most embodiments, the pump-oxygenator also serves to control blood temperature by means of a heat exchanger, typically incorporated in the gas exchange device. Synonymous with *heart-lung machine.*

Respiratory quotient: The ratio of carbon dioxide produced by tissues or eliminated by the lungs to oxygen consumed by tissues or taken in through the lungs.

Secondary flow: Any type of fluid motion, steady or periodic, in which the fluid is moving in a direction different from that of the primary flow. Secondary flow systems may be continuous in distribution, occupying the entire flow path, or comprise local elements that produce periodic remixing of the fluid.

Total body perfusion: Maintenance of blood circulation through the arterial and venous system by means of a positive displacement pump introducing blood into an artery under pressure and collecting it from a vein for continuous recirculation.

References

Bartlett Rh, Drinker PA, Galletti PM (eds). 1971. Mechanical Devices for Cardiopulmonary Assistance, Basel, Karger.

Bramson ML, Osborn JJ, Main FB, et al. 1995. A new disposable membrane oxygenator with integral heat exchange. J Thorac Cardiovasc Surg 50:391.

Cha W, Beissinger RL. In press. Augmented mass transport of macromolecules in sheared suspension to surfaces: Part B (Borine serum interface). J Colloid Interfac Sci.

Clark LC Jr., Gollan F, Gupta V. 1950. The oxygenation of blood by gas dispersion. Science 111:85.

Clowes GHA Jr, Hopkins AL, Neville WE. 1956. An artificial lung dependent upon diffusion of oxygen and carbon dioxide through plastic membranes. J Thor Surg 32:630.

Colton CK. 1976. Fundamentals of gas transport in blood. In WM Zapol, J Qvist (eds), Artificial Lungs for Acute Respiratory Failure. Theory and Practice, pp 3–41, New York, Academic Press.

Colton CK, Drake RF. 1971. Effect of boundary conditions on oxygen transport to flowing blood in a tube. Chem Eng Prog Symp 67(114):88.

Curtis RM, Eberhart RC. 1974. Normalization of oxygen transfer data in membrane oxygenators. Trans AM Soc Artif Intern Organs 20:210.

Dawids S, Engell HC (eds). 1976. Physiological and Clinical Aspects of Oxygenator Design, Amsterdam, Elsevier.

DeWall RA, Lillelei CW, Vareo RL, et al. 1957. The helix reservoir pump-oxygenator. Surg Gyn Obstet 104:699.

Dorson WJ Jr, Voorhees ME. 1974. Limiting models for the transfer of CO_2 and CO_2 in membrane oxygenators. Trans Am Soc Artif Intern Organs 20:219.

Dorson WJ Jr, Voorhees ME. 1976. Analysis of oxygen and carbon dioxide transfer in membrane lungs. In WM Zapol, J Qvist (eds), Artificial Lungs for Acute Respiratory Failure. Theory and Practice, pp 43–68.

Galletti PM. 1993. Cardiopulmonary bypass: A historical perspective. Artif Organs 17:675.

Galletti PM, Brecher GA. 1962. Heart-Lung Bypass, Principles and Techniques of Extracorporeal Circulation, New York , Grune & Stratton.

Galletti PM, Richardson PD, Snider MT, et al. 1972. A standardized method for defining the overall gas transfer performance of artificial lungs. Trans AM Soc Artif Internal Organs 18:359.

Galletti PM, Snider MT, Silbert-Aidan D. 1966. Gas permeability of plastic membrane for artificial lungs. Med Res Eng 20.

Gibbon JH Jr. 1939. An oxygenator with a large surface-volume ratio. J Lab Clin Med 24:1192.

Gibbon JH Jr. 1954. Application of a mechanical heart and lung apparatus to cardiac surgery. Minnesota Med 37:71.

Gollan, F. 1956. Oxygenation of circulating blood by dispersion, coalescence and surface tension separation. J Appl Physiol 8:571.

Hagl S, Klovekorn WP, Mayr N, et al. (eds). 1984. Thirty Years of Extracorporeal Circulation, Munich, Deutches Herzzentrum

Harris GW, Tompkins FC, de Filippi RP. 1970. Development of capillary membrane blood oxygenators. In D Hershey (ed), Blood Oxygenation, pp 333–354, New York, Plenum.

Marx TI, Snyder WE, St John AD, et al. 1960. Diffusion of oxygen into a film of whole blood. J Appl Physiol 15:1123.

Mockros LF, Gaylor JDS. 1975. Artificial lung design: Tubular membrane units. Med Biol Eng 13:171.

Peirce EC II. 1967. The membrane lung, its excuse, present status, and promise. J Mt Sinai Hosp 34:437.

Richardson PD. 1971. Effects of secondary flows in augmenting gas transfer in blood. Analytical considerations. In RH Bartlett, PA Drinker, PM Galletti (eds), Mechanical Devices for Cardiopulmonary Assistance, pp 2–16, Basel, Karger.

Richardson PD. 1976. Oxygenator testing and evaluation: Meeting ground of theory, manufacture and clinical concerns. In WM Zapol, J Qvist, (eds), Artificial Lungs for Acute Respiratory Failure. Theory and Practice, pp 87–102, New York, Academic Press.

Richardson PD, Galletti PM. 1976. Correlation of effects of blood flow rate, viscosity, and design features on artificial lung performance. In SG Davids, HC Engell (eds), Physiological and Clinical Aspects of Oxygenator Design, pp 29–44, Amsterdam, Elsevier.

Rygg IH, Kyvsgaard E. 1956. A disposable polyethylene oxygenator system applied in a heart-lung machine. Acta Chir Second 112:433.

Snider MT, Richardson PD, Friedman LI, et al. 1974. Carbon dioxide transfer rate in artificial lungs. J Appl Physiol 36:233.

Spaan JAE, Oomens JMM. 1976. Scaling rules for flat plate and hollow fiber membrane oxygenators. In SG Dawids, HC Engell (eds), Physiological and Clinical Aspects of Oxygenator Design, pp 13–28, Amsterdam, Elsevier.

Thews G. 1957. Verfahren zur Berechnung des O_2-Diffusionskoeffizienten aus Messungen der Sauerstoffdiffusion in Hemoglobin und Myoglobin Loesungen. Pfluegers Arch 2:138.

Villarroel F, Lanhyam CE, Bischoff K, et al. 1970. A mathematical model for the prediction of oxygen, carbon dioxide, and pH profiles with augmented diffusion in capillary blood oxygenators. In D Hershey (ed), Blood Oxygenation, pp 321–333, New York, Plenum.

Zapol WM. 1975. Membrane lung perfusion for acute respiratory failure. Surg Clin N Am 55:603.

Zapol WM, Qvist J (ed). 1967. Artificial Lungs for Acute Respiratory Failure. Theory and Practice, New York, Academic Press.

Further Information

Over the years, a number of books and monographs have reviewed the scientific and technical literature on gas exchange devices and their application. These include J.G. Allen, ed., *Extracorporeal Circulation*, Thomas, Springfield, 1958; P.M. Galletti and G.A. Brecher, *Heart-Lung Bypass, Principles and Techniques of Extracorporeal Circulation*, Grune and Stratton , New York, 1962; D. Hershey, ed., *Blood Oxygenation*, Plenum, New York, 1970; R.H. Bartlett, P.A. Drinker and P.M. Galletti, eds., *Mechanical Devices for Cardiopulmonary Assistance*, Karger, Basel, 1971; W.M. Zapol and J. Qvist, eds., *Artificial Lungs for Acute Respiratory Failure. Theory and Practice.* Academic Press, New York, 1976; G.G. Dawids and H.G. Engell, *Physiological and Clinical Aspects of Oxygenator Design,* Elsevier, Amsterdam, 1967; S. Hagl, W.P. Klövekorn, N. Mayr, and F. Sebening, *Thirty Years of Extracorporeal Circulation*, Deutsches Herzzentrum,

Munich, 1984; P.A. Casthely and D. Bregman, *Cardiopulmonary Bypass: Physiology, Related Complications and Pharmacology,* Futura, Mount Kisco, N.Y., 1991.

Topical advances in artificial lungs are typically published in the *Transactions of the American Society for Artificial Internal Organs* or in journals such s the ASAIO *Journal,* the *International Journal of Artificial Organs,* the *Japanese of Artificial Organs,* and the *Journal of Thoracic and Cardiovascular Surgery,* as well as the *Proceedings of the American Institute for Chemical Engineering* and, occasionally, *Chemically Engineering Symposium Series.*

130

Artificial Kidney

Pierre M. Galletti
(deceased)

Clark K. Colton
Massachusetts Institute of Technology

Michael J. Lysaght
Brown University

130.1 Structure of Function of the Kidney

The key separation functions of the kidney are:

1. To eliminate the water-soluble nitrogenous end-products of protein metabolism
2. To maintain electrolyte balance in body fluids and get rid of the excess electrolytes
3. To contribute to obligatory water loss and discharge excess water in the urine
4. To maintain acid-base balance in body fluids and tissues

To fulfill these functions, the kidney processes blood—or more accurately, plasma water—which in turn exchanges water and solutes with the extravascular water compartments: extracellular, intracellular, and transcellular. The solute concentrations in body fluids vary from site to site, yet all compartments are maintained remarkably constant in volume and composition despite internal and external stresses. The global outcome of normal renal function is a net removal of water, electrolyte, and soluble waste products from the blood stream. The kidney provides the major regulatory mechanisms for the control of volume, osmolality, and electrolyte and nonelectrolyte composition as well as pH of the body fluids and tissues.

Renal function is provided by paired, fist-sized organs, the kidneys, located behind the peritoneum against the posterior abdominal wall on both sides of the aorta. Each kidney is made up of over a million parallel mass transfer units which receive their common blood supply from the renal arteries, return the processed blood to the systemic circulation through the renal veins, and collect the waste fluids and solutes through the calyx of each kidney into the ureter and from there into the urinary bladder. These functional units are called *nephrons* and can be viewed as a sequential arrangement of mass transfer devices (glomerulus, proximal tubule, and distal tubule) for two fluid streams: blood and urine.

0-8493-0462-8/00/$0.00+$.50

Kidney function is served by two major mechanisms: *ultrafiltration*, which results in the separation of large amounts of extracellular fluid through plasma filtration in the glomeruli, and a combination of passive and active *tubular transport* of electrolytes and other solutes, together with the water in which they are dissolved, in the complex system provided in the rest of the nephron.

Glomerular Filtration

The volume of blood flowing into the natural kidneys far exceeds the amount needed to meet their requirements for oxygen and nutrients: The primary role of the kidneys is chemical processing. As blood flows through the glomerular capillaries, about one-fifth of the plasma water is forced through permeable membranes to enter the proximal portion of the renal tubule and form the primary urine, which henceforth becomes the second fluid phase of the renal mass exchanger. The concentrated blood remaining in the vascular system is collected in the efferent arterioles and goes on to perfuse the tubules via the peritubular capillaries of the "vasa recta" system, where it recovers some of the lost water and eventually coalesces with the other blood drainage channels to form the renal vein. The plasma water removed from the blood in the glomerulus is termed the *glomerular filtrate*, and the process of removal is called *glomerular filtration*. Glomerular filtrate normally contains no blood cells and very little protein. Glomerular filtration is a passive process driven by the differences in hydrostatic and oncotic pressures across the glomerular membrane. Solutes which are sufficiently small and not bound to large molecules pass quite freely through the glomerular membrane. All major ions, glucose, amino acids, urea, and creatinine appear in the glomerular filtrate at nearly the same concentrations as prevail in plasma.

The normal glomerular filtration rate (GFR) averages 120 ml/min. This value masks wide physiological fluctuations (e.g., up to 30% decrease during the night, and a marked increase in the postprandial period). Although the kidneys produce about 170 liters of glomerular filtrate per day, only 1–2 liters of urine is formed. A minimum volume of about 400 ml/day is needed to excrete the metabolic wastes produced under normal conditions (often called *obligatory water loss*).

Tubular Function

In the tubule, both solute and water transport take place. Some materials are transported from the lumen across the tubular epithelium into the interstitial fluid surrounding the tubule and thence into the blood of the peritubular capillaries. This process is called reabsorption and results in the return of initially filtered solutes to the blood stream. Other substances are transported from the peritubular blood to the interstitial fluid across the tubular epithelium and into the lumen. This process is called *secretion* and leads to elimination of those substances to a greater extent than would be possible solely through glomerular filtration. The return of filtered molecules from the kidney tubule to the blood is accompanied by the passive reabsorption of water through osmotic mechanisms.

The *proximal tubule* reabsorbs about two-thirds of the water and salt in the original glomerular filtrate. The epithelial cells extrude Na+ (and with it Cl–) from the glomerular filtrate into the interstitial fluid. Water follows passively and in proportionate amounts because of the osmotic pressure gradient (the proximal tubular membrane is freely permeable to water). In the loop of Henle, the glomerular filtrate, now reduced to one-third of its original volume, and still isoosmotic with blood, is processed to remove another 20% of its water content. The active element is the ascending limb of the loop of Henle, where cells pump out Na+, K+, and Cl– from the filtrate and move the Na+ and Cl– to the interstitial fluid. Because the ascending limb is not permeable to water, tubular fluid becomes increasingly more dilute along the ascending loop. The blood vessels around the loop do not carry back all the extruded salt to the general circulation, and therefore the Na+ concentration builds up down the descending limb of the loop of Henle, reaching a concentration 4–5 times higher than isoosmolar. As a result, the Na+ concentration in tubular fluid increases as its volume is decreased by passive transport into interstitial fluid and from there into the blood.

Countercurrent multiplication refers to the fact that the more Na+ the ascending limb extrudes the higher the concentration in the interstitial fluid, the more water removed from the descending limb by osmosis, and the higher the Na+ concentration presented to the ascending limb around the bend of the loop. Overall, the countercurrent multiplier traps salt in the medullary part of the nephron because it recirculates its locally. *Countercurrent exchange* refers to the interaction of the descending and ascending branches of the circulatory loops (vasa recta) with the loop of Henle which flows in the opposite direction. Substances which pass from the tubule into the blood accumulate in high concentrations in the medullary tissue fluid. Na+ and urea diffuse into the blood as it descends along the loop but then diffuse out of the ascending vessels and back into the descending vessels where the concentration is lower. Solutes are therefore recirculated and trapped (short-circuited) in the medulla, but water diffuses out of the descending vessel and into the ascending vessel to be transported out.

The distal tubule of the kidney, located in the cortex, is the site of fine adjustments of renal excretion. Here again the primary motor is the Na+/K+ pump in the baso-lateral membrane, which creates a Na+ concentration gradient. The walls of the collecting duct, which traverses through a progressively hypertonic renal medulla tissue, are permeable to water but not to Na+ and Cl–. As a result, water is drawn out and transported by capillaries to the general circulation. The osmotic gradient created by the countercurrent multiplier system provides the force for water reabsorption from the collecting duct. However, the permeability of the cell membrane in the collecting duct is modulated by the concentration of antidiuretic hormone (ADH). A decrease in ADH impairs the reabsorption of water and leads to the elimination of a larger volume of more dilute urine.

The terms *reabsorption* and *secretion* denote direction of transport rather than a difference in physiologic mechanism. In fact, a number of factors may impact on the net transport of any one particular solute. For example, endogenous creatinine (an end product of protein metabolism) is removed from plasma water through glomerular filtration in direct proportion to its concentration in plasma. Since it is neither synthesized nor destroyed anywhere in the kidney, and it is neither reabsorbed nor secreted in the tubule, its eventual elimination in the urine directly reflects glomerular filtration. Therefore, creatinine clearance can be used to measure glomerular filtration rate. However, glucose, which initially passes in the glomerular filtrate at the same concentration as in plasma, is completely reabsorbed from the tubular urine into peritubular capillaries as long as its plasma concentration does not exceed a threshold value somewhat above the level prevailing in normal subjects. As a result, there should be no glucose in the urine. When the threshold is exceeded, the amount of glucose excreted in the urine increases proportionately, producing glycosuria.

Several weak organic acids such as uric acid and oxalic acid, and some related but not naturally occurring substances such as p-aminohippuric acid (PAH), barbiturates, penicillates, and some x-ray contrast media, have the special property of being secreted in the proximal tubule. For example, PAH concentration in glomerular filtrate is the same as in plasma water. So avid is the tubular transport system for PAH that tubular cells remove essentially all the PAH from the blood perfusing them. Therefore, the removal of PAH is almost complete, and the rate of appearance of PAH in the urine mirrors the rate of presentation of PAH to the renal glomeruli, that is to say, renal plasma flow. Therefore, PAH clearance can be used, in association with the hematocrit, to estimate the rate of renal blood flow.

Urea appears in the glomerular filtrate at the same concentration as in plasma. However, one-third of urea diffuses back into the blood in the proximal tubule. In the distal nephron, urea (as an electrically neutral molecule without specific transport system) follows the fate of water (*solvent drag*). If large amounts of water are reabsorbed in the distal tubule and the collecting duct, then an additional third of the urea can be reabsorbed. However, if water diuresis is large, then correspondingly more urea is excreted.

130.2 Kidney Disease

The origin of kidney disease may be infectious, genetic, traumatic, vascular, immunologic, metabolic, or degenerative [Brenner & Rector, 1986]. The response of the kidneys to a pathologic agent may be rapid or slow, reversible or permanent, local or extensive. Under most circumstances, an abnormal body fluid composition is more likely to arise from the unavailability or excess of a raw material than from some intrinsic disturbance of renal function. This is why many clinical problems are corrected by fluid or electrolyte therapy and secondarily by dietary measures and pharmacologic agents which act on the kidney itself. Only as a treatment of last resort, where kidney disease progresses to renal failure, do clinicians use extracorporeal body fluid processing techniques that come under the generic concept of *dialysis.* These invasive procedures are intended to reestablish the body's fluid and electrolyte homeostasis and to eliminate toxic waste products. Processing can address the blood (e.g., *hemodialysis*) or a proxy fluid introduced in body cavities (e.g., *peritoneal dialysis*).

Even in healthy subjects, the GFR falls steadily from age 40 onward. Beyond age 80, it is only half of its adult value of 120 ml/min. However this physiologic deterioration is not extensive enough to cause symptoms. Since nature has provided kidneys with an abundance of overcapacity, patients do not become identifiably sick until close to 90% of original function has been lost. When kidneys keep deteriorating and functional loss exceeds 95%, survival is no longer possible without some form of replacement therapy.

Supplementation (as distinct from replacement) of renal function by artificial means is occasionally used in case of poisoning. Toxic substances are often excreted into the urine of glomerular filtration and active tubular secretion, but the body load at times exceeds the kidneys' clearing capacity. There are no methods known to accelerate the active transport of poisons into urine. Similarly, enhancement of passive glomerular filtration is not a practical means to facilitate elimination of toxic chemicals. Processing of blood in an extracorporeal circuit may be life-saving when the amount of poison in the blood is large compared to the total body burden and binding of the compound to plasma proteins is not extensive. In such cases (e.g., methanol, ethylene glycol, or salicylates poisoning) extracorporeal processing of blood for removing the toxic element from the body is indicated. If the poison is distributed in the entire extracellular space or tightly bound to plasma proteins, dialytic removal is unlikely to affect the clinical outcome because it can only eliminate a small fraction of the toxic solute.

Unfortunately in some situations either the glomerular or the tubular function of the kidneys, or both, fails and cannot be salvaged by drug and diet therapy. Failure can be temporary, self-limiting and potentially reversible, in which case only temporary substitution for renal function will be needed. Failure can also be the expression of progressive, intractable structural damage, in which case permanent replacement of renal function will eventually be needed for survival. However, the urgency of external intervention in end-stage renal disease (ESRD) is never as acute as is the case for the replacement of cardiac or respiratory function: The signs of renal dysfunction (water retention, electrolyte shifts, accumulation of metabolic end products normally eliminated by the kidneys) develop over days, weeks, or even months and are not immediately life threatening. Even in the end stage, renal failure can be addressed by intermittent rather than continuous treatment.

130.3 Renal Failure

There are two types of renal failure: acute (days or weeks) and chronic (months or years). *Acute renal failure* is typically associated with ischemia (reduction in blood flow), acute glomerulonephritis, tubular necrosis, or poisoning with "nephrotoxins" (e.g., heavy metals, some aminoglycosides, and excessive loads of free hemoglobin). *Chronic renal failure* is usually caused by chronic glomerulonephritis (of infectious or immune origin), pyelonephritis (ascending infection of the urinary tract), hypertension (leading to nephrosclerosis), or vascular disease (most commonly secondary to diabetes).

Renal insufficiency elicits the clinical picture of *uremia.* Although the word *uremia* means that there is too much urea in the blood, urea level in itself is not the cause of the problem. Uremia, often expressed in the United States as blood urea nitrogen concentration or BUN (which is actually half the urea concentration), serves as an indicator of the severity of renal disease. Urea is a metabolic end product in the catabolism of proteins that is hardly toxic even in high concentration. However, it mirrors the impaired renal elimination and the resulting accumulation in body fluids of other toxic substances, some of which have been identified (e.g., phenols, guanidine, diverse polypeptides); others remain unknown and are therefore referred to as *uremic toxins* or, for reasons to be discussed later, *middle molecules.* The attenuation of uremic symptoms by protein restriction in the diet and by various dialytic procedures underscores the combined roles of retention, removal, and metabolism in the constellation of signs of uremia. Toxicity may result from the synergism of an entire spectrum of accumulated molecules [Vanholder & Ringoir, 1992]. It may also reflect the imbalance that results from a specific removal through mechanisms which eliminate physiologic compounds together with potential toxins.

Not until the GFR (as estimated by its proxy, creatinine clearance) falls much below a third of normal do the first signs of renal insufficiency become manifest. At that point the plasma or extracellular concentration of substances eliminated exclusively through the glomeruli, such as creatinine or urea, increase measurably, and the possibility of progressive renal failure must be considered. In such cases, over a period of months to years, the kidneys lose their ability to excrete waste materials, to achieve osmoregulation, and to maintain water and electrolyte balance. The signs of ESRD become recognizable as creatinine clearance approaches 15 ml/min, eventually leading to uremic coma as water and solute retention depress the cognitive functions of the central nervous system. Empirically, it appears that the lowest level of creatinine clearance that is compatible with life is on the order of 8 ml/min, or 11.5 liters per day, or 80 liters per week. (These numbers have a bearing on the definition of adequate dialysis in ERSD patients, because they represent the time-averaged clearance which must be achieved by much more effective but intermittent blood processing). Human life cannot be sustained for more than 7–10 days in the total absence of kidney function. Clinical experience also shows that even a minimum of *residual renal clearance* (K_R) below the level necessary for survival can be an important factor of well-being in dialyzed patients, perhaps because the natural kidney, however sick, remains capable of eliminating middle molecular weight substances, whereas the artificial kidney is mostly effective in eliminating water and small molecules.

The incidence of ESRD (*incidence* is defined as the number of new patients entering treatment during a given year) has increased dramatically in the past 25 years in the United States and elsewhere. Whereas in the 1960s it was estimated at 700–1000 new cases a year in the United States (nearly three-quarters of them between the ages of 25 and 54), the number of new patients reached 16,000 per year at the end of the 1970s (still with the majority of cases under age 54) and 40,000 at the end of the 1980s, with the largest contingent between 65 and 74 years old. Serious kidney disease now strikes between in 1 in 5000 and 1 in 10,000 people per year in our progressively aging population. The fastest rate of growth is in the age group over 75, and the incidence of ESRD shows no signs of abating.

The prevalence of ESRD (*prevalence* is defined as the total number of patients present in the population at a specific time) has grown apace: In the United States, about 1000 people were kept alive by dialysis in 1969; 58,253 in 1979; 163,017 in 1989; close to 200,000 now. This is the result of a combination of factors which include longer survival of patients on hemodialysis and absolute growth of an elderly population suffering from an increasing incidence of diseases leading to ESRD such as diabetes. Worldwide, over 500,000 people are being kept alive by various modalities of "artificial kidney" treatment: about a third in the United States, a third in Europe, and a third in Japan and Pacific Rim countries. Another 500,000 or so have benefitted from dialytic treatment in the past but have since died or received transplants [Lysaght & Baurmeister, 1993]. Close to 85% of current patients are treated by maintenance hemodialysis, and 15% are on peritoneal dialysis. These numbers do not include about 100,000 people with a functional renal transplant, most of whom required hemodialysis support while waiting for a donor organ, and who may need it again, if only for a limited period, in case of graft rejection.

The mortality of ESRD patients in the United States has inched upward from 12%–16% per year in the 1970s and 1980s and has risen abruptly in recent years to levels in the order of 20%–25%. This has led to extensive controversy as to the origin of this deterioration, which has not been observed to the same extent in other regions with a similarly large population of ESRD patients, such as Western Europe and Japan, and may reflect for the United States insufficient dialysis as well as the burden of an increasingly older population.

130.4 Treatment of Renal Failure

Profound uremia, whether caused by an acute episode of renal failure or by the chronic progressive deterioration of renal function, used to be a fatal condition until the middle of the twentieth century.

The concept of clearing the blood of toxic substances while removing excess water by a membrane exchange process was first suggested by the experiments of Abel, Rowntree, and Turner at the Johns Hopkins Medical School. Back in 1913, these investigators demonstrated the feasibility of blood dialysis to balance plasma solute concentrations with those imposed by an appropriately formulated washing solution. However, their observation was not followed by clinical application, perhaps because experiments were limited by the difficulty of fabricating suitable exchange membranes, and blood anticoagulation was then extremely precarious. Collodion, a nitrocellulose film precipitated from an alcohol, ether, or acetone solution was the sole synthetic permeable membrane material available until the advent of cellophane in the 1930s. The unreliability of anticoagulants before the discovery of heparin also made continuous blood processing a hazardous process even in laboratory animals.

In 1944, Kolff in the Netherlands developed an artificial kidney of sufficient yet marginal capacity to treat acute renal failure in man. This device consisted of a long segment of cellophane sausage tubing coiled around a drum rotating in the thermostabilized bath filled with a hypertonic, buffered electrolyte solution, called the *dialysate*. Blood was allowed to flow from a vein into the coiled cellophane tube. Water and solute exchange occurred through the membrane with a warm dialysate pool, which had to be renewed every few hours because of the risk of bacterial growth. The cleared blood was returned to the circulatory system by means of a pump. After World War II, a somewhat similar system was developed independently by Alwall in Sweden. Because of the technical difficulty of providing repeated access to the patient's circulation, and the overall cumbersomeness of the extracorporeal clearing process, hemodialysis was limited to patients suffering from acute, and hopefully reversible, renal failure, with the hope that their kidneys would eventually recover. To simplify the equipment, Inouye and Engelberg [1953] devised a coiled cellophane tube arrangement that was stationary and disposable, and shortly thereafter Kolff and Watschinger (by then at the Cleveland Clinic) reported a variant of this design, the Twin Coil, that became the standard for clinical practice for a number of years.

Repeated treatment, as needed for chronic renal failure, was not possible until late 1959, when Scribner and Quinton introduced techniques for chronic access to the blood stream which, combined with improvements in the design and use of hemodialysis equipment, allowed the advent of chronic intermittent hemodialysis for long-term maintenance of ESRD patients. This was also the time when Kiil first reported results with a flat plate dialyzer design in which blood was made to flow between two sheets of cellophane supported by solid mats with grooves for the circulation of dialysate. This design—which had been pioneered by Skeggs and Leonard, McNeill, and Bluemle and Leonard—not only needed less blood volume to operate then the coiled tube devices, it also had the advantage of requiring a relatively low head of pressure to circulate the blood and the dialysate. This meant that the two fluids could circulate without high pressure differences across the membrane. Therefore, in contrast to coil dialyzers, where a long blood path necessitated a high blood pressure at the entrance of the exchanger, flat plate dialyzers could transfer metabolites through the membrane by diffusion alone, without coupling it with the obligatory water flux deriving from high transmembrane pressure. When ultrafiltration was needed, it could be achieved by circulating the dialysate at subatmospheric pressures.

Device development was also encouraged by the growing number of home dialysis patients. By 1965, the first home dialysate preparation and control units were produced industrially. Home dialysis programs based on the twin coil or flat plate dialyzers were soon underway. At that time the cost of home treatment was substantially lower than hospital care, and in the United States, Social Security was not yet underwriting the cost of treatment of ESRD.

In 1965 also, Bluemle and coworkers analyzed means to pack the maximum membrane area in the minimum volume, so as to reduce the bulkiness of the exchange device and diminish the blood loss associated with large dialyzers and long tubing. They concluded that a tightly packed bundle of parallel capillaries would best fit this design goal. Indeed by 1967, Lipps and colleagues reported the initial clinical experience with hollow fiber dialyzers, which have since become the mainstay of hemodialysis technology.

In parallel developments. Henderson and coworkers [1967] proposed an alternative solution to the problem of limited mass transfer achievable by diffusion alone with hemodialysis equipment. They projected that a purely convective transport (ultrafiltration) through membranes more permeable to water than the original cellulose would increase the effective clearance of metabolites larger than urea. The lost extracellular volume was to be replaced by infusing large volumes of fresh saline into the blood at the inlet or the outlet of the dialyzer to replace the lost water and electrolytes. The process was called *hemodiafiltration* or, sometimes, *diafiltration.* (The procedure in which solutes and water are removed by convective transport alone, using large pore membranes and without substantial replacement of the fluid, is now known as *hemofiltration* and is used primarily in patients presenting with massive fluid retention.)

As is intuitively apparent, the effectiveness of hemodialysis with a given devices is related to the duration of the procedure. In the pioneer years, dubbed "the age of innocence" by Colton [1987], patients were treated for as many as 30 hours a week. Economics and patient convenience promoted the development of more efficient transfer devices. Nowadays, intermittent maintenance dialysis can be offered with 10 hours (or even less) of treatment dividend in 3 sessions per week. Conversely, nephrologists have developed (mostly for use in the intensive care unit) the procedure known as *continuous arterio-venous hemodialysis* (with its variant continuous arterio-venous hemofiltration) in which blood pressure from an artery (aided or not by a pump) drives blood through the exchange device and back into a vein. Continuous operation compensates for the relatively low blood flow and achieves stable solute concentrations, as opposed to the seesaw pattern that prevails with periodic treatment.

The concept of using a biologic membrane and its blood capillary network to exchange water and solutes with a washing solution underlies the procedure known as *peritoneal dialysis,* which relies on the transfer capacity of the membranelike tissue lining the abdominal cavity and the organs it contains. In 1976, Popovich and Moncrief described continuous ambulatory peritoneal dialysis (CAPD), a procedure in which lavage of the peritoneal cavity is conducted as a continuous form of mass transfer through introduction, equilibration, and drainage of dialyzate on a repetitive basis 4–6 times a day. In CAPD, a sterile solution containing electrolytes and dextrose is fed by gravity into the peritoneal cavity through a permanently installed transcutaneous catheter. After equilibriation with capillary blood over several hours, this dialyzate is drained by gravity into the original container and the process is repeated with a fresh solution. During the dwell periods, toxins and other solutes are exchanged by diffusional processes. Water transfer is induced by the osmotic pressure difference due to the high dextrose concentration in the treatment fluid. This procedure is analyzed in detail in Chapter 131.

Plasmapheresis, i.e., the extraction of plasma from blood by separative procedures (see Chapter 132), has been used in the treatment of renal disease [Samtleben & Gurland, 1989]. However, the cost of providing fresh plasma to replace the discarded material renders plasmapheresis impractical for frequent, repeated procedures, and plasmapheresis is used mainly for other clinical indications.

Most hemodialysis is performed in free-standing treatment centers, although it may also be provided in a hospital or performed by the patient at home. The hemodialysis circuit consists of two fluid pathways. The blood circuitry is entirely disposable, though many centers reuse some or all circuit components in order to reduce costs. It comprises a 16-gauge needle for access to the circulation (usually through an arteriovenous fistula created in the patient's forearm), lengths of plasticized polyvinyl chloride tubing (including a special segment adapted to fit into a peristaltic blood pump), the hemodialyzer itself, a bubble trap and an open mesh screen filter, various ports for sampling or pressure measurements at the blood outlet, and a return cannula. Components of the blood side circuit are supplied in sterile and nonpyrogenic conditions. The dialysate side is essentially a machine capable of (1) proportioning out glucose and electrolyte concentrates with water to provide a dialysate of appropriate composition; (2) sucking dialysate past a restrictor valve and through the hemodialyzer at subatmospheric pressure; and (3) monitoring temperature, pressures, and flow rates. During treatment the patient's blood is anticoagulated with heparin. Typical blood flow rates are 200–350 ml/min; dialysate flow rates are usually set at 500 ml/min. Simple techniques have been developed to prime the blood side with sterile saline prior to use and to return to the patient nearly all the blood contained in the extracorporeal circuit after treatment. Whereas most mass transport occurs by diffusion, circuits are operated with a pressure on the blood side, which may be 100–500 mmHg higher than on the dialysate side. This provides an opportunity to remove 2–4 liters of fluid along with solutes. Higher rates of fluid removal are technically possible but physiologically unacceptable. Hemodialyzers must be designed with high enough hydraulic permeabilities to provide adequate fluid removal at low transmembrane pressure but not so high that excessive water removal will occur in the upper pressure range.

Although other geometries are still employed, the current preferred format is a "hollow fiber" hemodialyzer about 25 cm in length and 5 cm in diameter, resembling the design of a shell and tube heat exchanger. Blood enters at the inlet manifold, is distributed to a parallel bundle of capillary tubes (potted together with polyurethane), and exits at a collection manifold. Dialysate flows countercurrent in an external chamber. The shell is typically made of an acrylate or polycarbonate resin. Devices typically contain 6000–10,000 capillaries, each with an inner diameter of 200–250 microns and a dry wall thickness as low as 10 microns. The total membrane surface area in commercial dialyzers varies from 0.5 to 1.5 m^2, and units can be mass-produced at a relatively low cost (selling price around $10–$15, not including tubing and other disposable accessories). Several reference texts (see For Further Information) provide concise and comprehensive coverage of all aspects of hemodialysis.

130.5 Renal Transplantation

The uremic syndrome resembles complex forms of systemic poisoning and is characterized by multiple symptoms and side effects. Survival requires that the toxins be removed, and the resulting quality of life depends on the quantity of toxins which are actually eliminated. Ideally, one would like to clean blood and body fluids to the same extent as is achieved by normal renal function. This is only possible at the present time with an organ transplant.

The feasibility of renal transplantation as a therapeutic modality for ESRD was first demonstrated in 1954 by Murray and coworkers in Boston, and Hamburger and coworkers in Paris, in homozygous twins. Soon the discovery of the first immunosuppressive drugs led to the extension of transplantation practice to kidneys of live, related donors. Kidney donation is thought to be innocuous since removal of one kidney does not lead to renal failure. The remaining kidney is capable of hypertrophy, meaning that the glomeruli produce more filtrate, and the tubules become capable of increased reabsorption and secretion. A recent Canadian study indicates that the risk of ESRD is not higher among living kidney donors than in the general population, meaning that a single kidney has enough functional capacity for a lifetime. Nonetheless, cadaver donors now constitute the main organ source for the close to 10,000 renal transplants performed in the United States every year. Even though under ideal circumstances each cadaver donor allows two kidney transplants, the scarcity of donors is the major limitation to this form of treatment of ESRD. Most patients aspire to renal transplant because of the better quality of life it provides and the freedom from the time constraints of repeated procedures. However, the incidence of ESRD is such that only one patient in five can be kept alive by transplantation. Dialysis treatment remains a clinical necessity while waiting for a transplant, as a safety net in case of organ rejection, and for the many patients for whom transplantation is either contraindicated or simply not available.

130.6 Mass Transfer in Dialysis

In artificial kidneys, the removal of water and solutes from the blood stream is achieved by

1. Solute diffusion in response to concentration gradients
2. Water ultrafiltration and solute convection in response to hydrostatic and osmotic pressure gradients
3. Water migration in response to osmotic gradients

In most cases, these processes occur simultaneously and in the same exchange device, rather than sequentially as they do in the natural kidney with the cascade of glomerular filtration, tubular reabsorption, and final adjustments in the collecting tubule.

Mechanistically, the removal of water and solutes from blood is achieved by passive transport across thin, leaky, synthetic polymer sheets or tubes similar to those used in the chemical process call dialysis. Functionally, an artificial kidney (also called *hemodialyzer,* or *dialyzer* or *filter* for short) is a device in which water and solutes are transported from one moving fluid stream to another. One fluid stream is blood; the other is dialysate: a human-made solution of electrolytes, buffers, and nutrients. The solute concentration as well as the hydrostatic and osmotic pressures of the dialysate are adjusted to achieve transport in the desired direction (e.g., to remove urea and potassium ions while adding glucose or bicarbonate to the bloodstream).

Efficiency of mass transfer is governed by two and only two independent parameters. One, which derives from mass conservation requirements, is the ratio of the flow rates of blood and dialysate. The other is the rate constant for solute transport between the two fluid streams. This rate constant depends upon the overall surface area of membrane available for exchange, its leakiness or *permeability,* and such design characteristics as fluid channel geometry, local flow velocities, and boundary layer control, all of which affect the thickness of stationary fluid films, or diffusion barriers, on either side of the membrane.

130.7 Clearance

The overall mass transfer efficiency of a hemodialyzer is defined by the fractional depletion of a given solute in the blood as it passes through the unit. Complete removal of a solute from blood during a single pass defines the dialyzer clearance for that solute as equal to dialyzer blood flow. In other terms, dialyzer blood flow asymptotically limits the clearance of any substance in any device, however efficient.

Under conditions of steady-state dialysis, the mass conservation requirement is expressed as

$$N = Q_B\left(C_{Bi} - C_{Bo}\right) = Q_D\left(C_{Do} - C_{Di}\right) \tag{130.1}$$

where N is the overall solute transfer rate between blood and dialysate, Q_B and Q_D are blood flow and dialysate flow rates respectively, and C_{Bi}, C_{Bo}, C_{Di} and C_{Do}, are the solution concentrations C in blood, B, or dialysate, D, at the inlet, i, or the outlet, o of the machine.

Equation (130.1) about mass conservation leads to the first and oldest criterion for dialyzer effectiveness, namely clearance K, modeled after the concept of renal clearance. Dialyzer clearance is defined as the mass transfer rate N divided by the concentration gradient prevailing at the inlet of the artificial kidney.

$$K = \frac{N}{C_{Bi} - C_{Di}} \tag{130.2}$$

Since mass transfer rate also means the amount of solute removed from the blood per unit of time, which in turn is equal to the amount of solute accepted in the dialysate per unit of time, there are two expressions for dialysance

$$K_B = \frac{Q_B\left(C_{Bi} - C_{Bo}\right)}{C_{Bi} - C_{Di}} \tag{130.3}$$

$$= K_D = \frac{Q_D\left(C_{Do} - C_{Di}\right)}{\left(C_{Bi} - C_{Di}\right)} \tag{130.4}$$

which afford two methods for measuring it. Any discrepancy must remain within the error of measurements, which under the conditions of clinical hemodialysis easily approaches ± 10%. As in the natural kidney, the clearance of any solute is defined by the flow rate of blood which is completely freed of that solute while passing through the exchange device. The dimensions of clearance are those of flow (a *virtual* flow, one may say), which can vary only between zero and blood flow (or dialysate flow, whichever is smaller), much in the way the renal clearance of a substance can only vary between zero and effective renal plasma flow.

Since dialyzer clearance is a function of blood flow, a natural way to express the efficiency of a particular exchange device consists of "normalizing" clearance with respect to blood flow as a dimensionless ratio

$$\frac{K}{Q_B} = \frac{C_{Bi} - C_{Bo}}{C_{Bi} - C_{Di}} \tag{130.5}$$

or

$$\frac{K}{Q_B} = \frac{C_{Do} - C_{Di}}{C_{Bi} - C_{Di}} \left(\text{extraction fraction}\right) \tag{130.6}$$

K/Q_B can vary only between zero and one and represents the highest attainable solute depletion in the blood which is actually achieved in a particular device for a particular solute under a particular set of circumstances.

Another generalization of the dialysance concept may be useful in the case where the direction of blood flow relative to the direction of dialysate flow is either parallel, random, or undetermined, as occurs with the majority of clinical hemodialyzers. Under such circumstances, the best performance which can be achieved is expressed by the equality of solute concentration in outgoing blood and outgoing dialysate ($C_{Bo} = C_{Do} = C_e$ or equilibrium concentration). This limit defines, after algebraic rearrangement of Eqs. (130.3) and (130.4), the maximal achievable clearance at any combination of blood and dialysate flow rates without reference to solute concentrations.

$$K_{\max} = \frac{Q_B \times Q_D}{Q_B + Q_D} \tag{130.7}$$

Since blood and dialysate flows can usually be measured with a reasonable degree of accuracy, the concept of $K_{\max}$ provides a practical point of reference against which the effectiveness of an actual dialyzer can be estimated.

130.8 Filtration

So far we implicitly assumed that differences in concentration across the membrane provide the sole driving force for solute transfer. In clinical hemodialysis, however, the blood phase is usually subject to a higher hydrostatic pressure than the dialysate phase. As a result, water is removed from the plasma by ultrafiltration, dragging with it some of the solutes into the dialysate. Ultrafiltration capability is a necessary consequence of the transmural pressure required to keep the blood path open with flat sheet or wide tubular membranes. It is also clinically useful to remove the water accumulated in the patient's body in the interval of dialysis. Ultrafiltration can be enhanced by increasing the resistance to blood flow at the dialyzer outlet, and thereby raising blood compartment pressure, by subjecting the dialysate to a negative pressure or by utilizing membranes more permeable to water than the common cellophanes.

Whenever water is removed from the plasma by ultrafiltration, solutes are simultaneously removed in a concentration equal to or lower than that present in the plasma. For small, rapidly diffusible molecules such as urea, glucose, and the common electrolytes, the rate of solute removal almost keeps pace with that of water, and ultrafiltrate concentration is the same as that in plasma. With compounds characterized by a larger molecular size, the rate of solute removal lags behind that of water. Indeed with some of the largest molecules of biological interest, ultrafiltration leads to an actual increase in plasma concentration during passage through the artificial kidney.

Defining *ultrafiltration* as the difference between blood flow entering the dialyzer and blood flow leaving the dialyzer

$$F = Q_{Bi} - Q_{Bo}$$

one can rewrite the mass conservation requirement as

$$\begin{aligned} &Q_{Bi}C_{Bi}\left(\text{Amount of solute in the incoming blood}\right) \\ &= Q_{Bo}C_{Bo}\left(\text{amount of solute in the outgoing blood}\right) \\ &+ K_B\left(C_{Bi} - C_{Di}\right)\left(\text{amount cleared in the dialyzer}\right) \end{aligned}$$

The clearance equations can then be rewritten as

$$K_B = \frac{Q_{Bi}\left(C_{Bi} - \frac{Q_{Bo}}{Q_{Bi}} C_{Bo}\right)}{C_{Bi} - C_{Di}} \tag{130.8}$$

$$K_D = \frac{Q_{Di}\left(\frac{Q_{Do}}{Q_{Di}} C_{Do} - C_{Di}\right)}{C_{Bi} - C_{Di}} \tag{130.9}$$

The clearance is now defined as the amount of solute removed from the blood phase per unit of time, *regardless of the nature of the driving force,* divided by the concentration difference between incoming blood and incoming dialysate.

When $C_{Di} = 0$

$$K_B = Q_{Bi} - Q_{Bo}\left(\frac{Q_{Bo}}{Q_{Bi}}\right) \tag{130.10}$$

$$K_D = \frac{Q_{Do}C_{Do}}{C_{Bi}} \tag{130.11}$$

When $C_{Di} = 0$ and $C_{Bo} = C_{Bi}$

$$K_B = F \tag{130.12}$$

The practical value of these equations is somewhat limited, since their application requires a high degree of accuracy in the measurement of flows and solute concentrations. The special case where there is no solute in the incoming dialysate ($C_{Di} = 0$) is important for in vitro testing of artificial kidneys.

130.9 Permeability

The definition of clearance is purely operational. Based upon considerations of conservation of mass, it is focused primarily on the blood stream from which a solute must be removed, thus, in final analysis, on the patient herself or himself. Clearance describes the artificial kidney as part of the circulatory system and of the fluid compartments which must be cleared of a given solute. To relate the performance of a hemodialyzer to its design characteristics, clearance is of limited value.

To introduce into the picture the surface area of membrane and the continuously variable (but predictable) concentration difference between blood and dialysate within the artificial kidney, one must define the rate constant of solute transfer, or *permeability* P_Σ.

$$P_\Sigma = \frac{N}{A \times \Delta C} \tag{130.13}$$

where N is the overall solute transport rate between blood dialysate, A is the membrane area, and ΔC is the average solute concentration difference between the two moving fluids.

Permeability is defined by Eq. (130.13) as the amount of solute transferred per unit area and per unit of time, under the influence of a unit of concentration driving force. The proper average concentration, $\overline{\Delta C}$, driving force is the logarithmic mean of the concentration differences prevailing at the inlet and at the outlet

$$\overline{\Delta C} = \frac{\Delta C_i - \Delta C_o}{\ln \dfrac{\Delta C_i}{\Delta C_o}} \tag{130.14}$$

The boundary conditions on the concentration driving force (ΔC_i and ΔC_o) are uniquely determined by the geometry of the dialyzer. The three most common cases to consider are: (1) cocurrent flow of blood and dialysate; (2) laminar blood flow, with completely mixed dialysate flow; and (3) countercurrent blood and dialysate flow. The boundary conditions on concentration driving force follow.

Cocurrent flow is

$$C_{Bi} - C_{Di} \text{ and } C_{Bo} - C_{Do}$$

Mixed dialysate flow is

$$C_{Bi} - C_{Do} \text{ and } C_{DBo} - C_{Do}$$

Countercurrent flow is

$$C_{Bi} - C_{Do} \text{ and } C_{Bo} - C_{Di}$$

Thus permeability can be expressed as in the following equations. Cocurrent flow is

$$P_\Sigma = \frac{N \dfrac{\ln\left(C_{Bi} - C_{Di}\right)}{C_{Bo} - C_{Do}}}{A\left(C_{Bi} - C_{Di}\right) - \left(C_{Bo} - C_{Do}\right)} \tag{130.15}$$

Mixed dialysate flow is

$$P_\Sigma = \frac{N \dfrac{\ln\left(C_{Bi}-C_{Do}\right)}{C_{Bo}-C_{Do}}}{A\left(C_{Bi}-C_{Do}\right)-\left(C_{Bo}-C_{Do}\right)} \tag{130.16}$$

Countercurrent flow is

$$P_\Sigma = \frac{N \dfrac{\ln\left(C_{Bi}-C_{Do}\right)}{C_{Bo}-C_{Di}}}{A\left(C_{Bi}-C_{Do}\right)-\left(C_{Bo}-C_{Di}\right)} \tag{130.17}$$

By simultaneous solution of Eqs. (130.3), (130.4), and (130.12), and use of the formal definition of the logarithmic mean concentration driving force (130.14), the clearance ratio (K/Q_B) can be expressed as a function of two dimensionless parameters (Z and R), neither of which involves solute concentration terms [Leonard & Bluemle, 1959; Michaels, 1966]. Cocurrent flow is

$$\frac{K}{Q_B} = \frac{1}{1+Z}\left[1-\exp\left(-R\left(1+Z\right)\right)\right] \tag{130.18}$$

Mixed dialysate flow is

$$\frac{K}{Q_B} = \frac{1-\exp\left(-R\right)}{1+Z\left[1-\exp\left(-R\right)\right]} \tag{130.19}$$

Countercurrent flow is

$$\frac{K}{Q_B} = \frac{1-\exp\left[R\left(1-Z\right)\right]}{Z-\exp\left[R\left(1-Z\right)\right]} \tag{130.20}$$

where $Z = Q_B/Q_D$ and $R = P_\Sigma A/Q_B$

Michaels has expressed graphically Eqs. (130.8–130.20) as plots of clearance ratio (K/Q_B) versus flow ratio (Q_B/Q_D) with various solute transport ratios (P_Σ, A/Q_B) as parameters. These plots give an appreciation of the relative importance of the variables affecting dialyzer efficiency and permit one to recognize readily the factors which limit mass transfer under a particular set of conditions.

For the computation of actual permeability coefficients, from pooled data obtained at varying solute concentrations, Eqs. (130.15–130.17) can be rearranged, using definitions of N, C_{Bo}, and C_{Do} from Eqs. (130.2–130.4). Cocurrent flow is

$$P_\Sigma = \frac{Q_B Q_{D;}}{A\left(Q_B+Q_D\right)} \ln \frac{1}{1-K/Q_B-K/Q_D} \tag{130.21}$$

Mixed dialysate flow is

$$P_\Sigma = \frac{Q_B}{A} \ln \frac{1 - K/Q_D}{1 - K/Q_B - K/Q_D} \tag{130.22}$$

Countercurrent flow is

$$P_\Sigma = \frac{Q_B Q_D}{A\left(Q_D - Q_B\right)} \ln \frac{1 - K/Q_D}{1 - K/Q_B} \tag{130.23}$$

As remarked by Leonard and Bluemle [1959], when, and only when, Q_D is much greater than Q_B, the above equations (130.18–130.23) reduce to Renkin's [1956] formula

$$P_\Sigma = \frac{Q_B}{A} \ln \frac{1}{1 - K/Q_B} \tag{130.24}$$

or

$$\frac{K}{Q_B} = 1 - \exp\left(\frac{P_\Sigma A}{Q_B}\right) \tag{130.25}$$

Historically, Eq. (130.25) played an important role in pointing out to designers that the clearance ratio (K/Q_B) can be improved equally well by an increase in exchange area (A) or in permeability (P). However, caution is required in applying the equation to some of the efficient modern dialyzers. First, the assumption that dialysate flow is "infinitely" large with respect to blood flow is seldom verified. Furthermore, the functions relating permeability to dialysance, Eqs. (130.21–130.24), or to the individual solute concentrations and flows, Eqs. (130.15–130.17), have an exponential form. When the overall permeability approaches that of the membrane alone, when the outgoing solute concentrations approach the equilibrium conditions, or when clearance approaches blood flow, one deals with the steep part of that exponential function. Any slight error in the experimental measurements will lead to a disproportionately larger error in calculated permeability.

130.10 Overall Transport

In a dialyzer, separation occurs because small molecules diffuse more rapidly than larger ones and because the degree to which membranes restrict solute transport usually increases with permeant size (permselectivity). Fick's equation states that solute will move from a region of greater concentration to a region of lower concentration in a rate proportional to the difference in concentration on opposite sides of the membrane

$$\phi = -D\frac{\partial c}{\partial x} \tag{130.26}$$

where ϕ = unit solute flux in g/cm^2-s; D = diffusion coefficient cm^2/s; c = concentration in g/cm^3; and x = distance in cm. The minus sign accounts for the convention that flux is considered positive in the direction of decreasing concentration. Diffusion coefficient decreases roughly in proportion to the square root of molecular weight.

Ignoring boundary layer effects for the moment, and assuming that diffusion within the membrane is analogous to that in free solution, Eq. (130.26) can be integrated across a homogeneous membrane of thickness d to yield

$$\phi = \frac{S\,D_m\,\Delta c}{d} \tag{130.27}$$

where S represents the dimensionless solute partition coefficient, i.e., the ratio of solute concentration in external solution to that at the membrane surface, and D_m represents solute diffusion within the membrane and is assumed to be independent of solute concentration in the membrane. If two or more solutes are dialysing at the same time, the degree of separation or enrichment will be proportional to the ratio of their permeabilities. The closer the permeability of a membrane is to that of an equivalent thickness of free solution, the more rapid will be the resultant dialytic transport. Equation (130.27) is often further simplified to this expression for flux per unit of membrane area

$$\phi = P_\Sigma \overline{\Delta C} \tag{130.28}$$

where thickness is incorporated into an overall membrane mass transfer coefficient with units of cm/s, and $\overline{\Delta C}$ is the logarithmic mean concentration.

Chemical engineers provided a firm foundation for describing the overall performance of hemodialyzers recognizing the importance of understanding and describing mass transfer in each of the three phases of a hemodialyzer (blood, membrane, dialysate), the individual mass transfer resistances of which sum to the overall mass transfer resistance of the device [Colton, 1987]. Solutions adjacent to the membranes are rarely well mixed, and the resistance to transport resides not just in the membrane but also in the fluid regions termed *boundary layers,* on both the dialysate and blood side. Moreover, some dialyzers are designed to direct flow parallel to the surface of the membrane rather than expose it to a well-mixed bath. Boundary layer effects typically account for 25–75% of the overall resistance to solute transfer [Lysaght & Baurmeister, 1993]. In many exchanger designs, boundary layer effects can be minimized by rapid convective flow targeted to the surface of the membrane where fluid pathways are thin, flow near the membrane is laminar, and boundary layer resistance decreases with increasing wall shear rates. When geometry permits higher Reynolds numbers, flow becomes turbulent, and fluid resistance varies with net tangential velocity. Geometric obstacles (e.g., properly spaced obstacles) or fluid mechanical modulation (e.g., superimposed pulsation) are often-used tactics to minimize boundary layer effects, but all result in higher energy utilization. Quantitatively, the membrane resistance becomes part of an overall mass transfer parameter P_Σ which for conceptual purposes can be broken down into three independent and reciprocally additive components for the triple laminate: blood boundary layer (subscript B), membrane (subscript M), and dialysate boundary layer (subscript D), such that

$$\frac{1}{P_\Sigma} = \frac{1}{P_B} + \frac{1}{P_M} + \frac{1}{P_D} \tag{130.29}$$

or reciprocally

$$R_\Sigma = R_B + R_M + R_D \tag{130.30}$$

where P_Σ is the device-averaged mass transfer coefficient (or permeability) in cm/s and R_Σ is the device-averaged resistance in s/cm. D_B can be estimated for many relevant conditions of geometry and flow using mass transport analysis based upon wall Sherwood numbers [Colton et al., 1971]. P_M is best obtained by measurements employing special test fixtures in which boundary layer resistances are negligible or known [Klein et al., 1977]. P_D is more problematic and is usually obtained by extrapolations based upon Wilson plots [Leonard & Bluemle, 1960]. Boundary layer theory, as well as technique for correlation, estimation, and prediction of the constituent mass transfer coefficients, is reviewed in detail by Colton and coworkers [1971] and Klein and coworkers [1977]. Overall solute transport is obtained from local flux by mass balance and integration; for the most common case of countercurrent flow

$$\Phi = \left(C_{Bi} - C_{Di}\right)\frac{Q_B}{A}\,\frac{\exp\left[\dfrac{P_\Sigma A}{Q_B}\left(1-\dfrac{Q_B}{Q_D}\right)\right]-1}{\exp\left[\dfrac{P_\Sigma A}{Q_B}\left(1-\dfrac{Q_B}{Q_D}\right)\right]-\dfrac{Q_B}{Q_D}} \tag{130.31}$$

where C_{Bi} and C_{Di} represent inlet concentrations in the blood and dialysate streams in g/cm^3, A represents membrane surface area in cm^2, Q_B and Q_D are blood and dialysate flow rates in cm^3/min, and Φ and P_Σ are as defined in Eqs. (130.28) and (130.29). Derivations of this relationship and similar expressions for cocurrent or crossflow geometries can be found in reviews by Colton and Lowrie [1981] and Gotch and colleagues [1972].

As pointed out by Lysaght and Baurmeister [1993], hemodialysis is a highly constrained process. Molecular diffusion is slow, and the driving forces are set by the body itself, decreasing in the course of purification and not amenable to extrinsic augmentation. The permeant toxic species are not to be recovered, and their concentrations are necessarily more dilute in the dialysate than in the incoming blood. The flow and gentle nature of dialysis has a special appeal for biologic applications, particularly when partial purification of the feed stream, rather than recovery of a product, is intended.

130.11 Membranes

Hemodialysis membranes vary in chemical composition, transport properties, and, as we will see later, biocompatibility. Hemodialysis membranes are fabricated from these classes of materials: regenerated cellulose, modified cellulose, and synthetics [Lysaght & Baurmeister, 1993]. Regenerated cellulose is most commonly prepared by the cuproamonium process and are macroscopically homogenous. These extremely hydrophilic structures sorb water, bind it tightly, and form a true hydrogel. Solute diffusion occurs through highly water-swollen amorphous regions in which the cellulose polymer chains are in constant random motion and would actually dissolve if they were not tied down by the presence of crystalline regions. Their principles advantage is low unit cost, complemented by the strength of the highly crystalline cellulose, which allows polymer films to be made very thin. These membranes provided effective small-solute transport in relatively small exchange devices. The drawbacks of regenerated cellulose are their limited capacity to transport middle molecules and the presence of labile nucleophilic groups which trigger complement activation and transient leukopenia during the first hour of exposure to blood. The advantages appear to outweigh the disadvantages, since over 70% of all hemodialyzers are still prepared from cellulosics, the most common of which is supplied by Akso Faser AG under the trade name Cuprophan.

A variety of other hydrophilic polymers account for 20% of total hemodialyzer production, including derivatized cellulose, such as cellulose acetate, diacetate, triacetate, and synthetic materials such as polycarbonate (PC), ethylenevinylalcohol (EVAL), and polyacrylonitrile-sodium methallyl sulfonate copolymer (PAN-SO_3), which can all be fabricated into homogeneous films.

At the opposite end of the spectrum are membranes prepared from synthetic engineered thermoplastics, such as polysulfones, polyamides, and polyacylonitrile-polyvinylchloride copolymers. These hydrophobic materials, which account for about 10% of the hemodialyzer market, form asymmetric and anisotropic membranes with solid structures and open void spaces (unlike the highly mobile polymeric structure of regenerated cellulose). These membranes are characterized by a skin on one surface, typically a fraction of a micron thick, which contains very fine pores and constitutes the discriminating barrier determining the hydraulic permeability and solute retention properties of the membrane. The bulk of the membrane is composed of a spongy region, with interstices that cover a wide size range and with a structure ranging from open to closed cell foam. The primary purpose of the spongy region is to provide mechanical strength; the diffusive permeability of the membrane is usually determined by the properties of this matrix. As the convective and diffusive transport properties of these membranes are, to a large extent, associated independently with the properties of the skin and spongy matrix, respectively, it is possible to vary independently the convective and diffusive transport properties with these asymmetric structures. There is often a second skin on the other surface, usually much more open than the primary barrier. These materials are usually less activating to the complement cascade than are cellulosic membranes. The materials are also less restrictive to the transport of middle and large molecules. Drawbacks are increased cost and such high hydraulic permeability as to require special control mechanisms to avoid excess fluid loss and to raise concerns over the biologic quality of dialysate fluid because of the possibility of back filtration carrying pyrogenic substances to the blood stream.

The discovery of asymmetry membrane structures launched the modern era of membrane technology by motivating research on new membrane separation processes. Asymmetric membranes proved useful in ultrafiltration, and a variety of hydrophobic materials have been used including polysulfone (PS), polyacrylonitrile (PAN), its copolymer with polyvinylchloride (PVC), polyamide (PA), and polymethyl methacrylate (PMMA). PMMA does not form an obvious skin surface and should perhaps be placed in a class of its own.

130.12 Hemofiltration

Although low rates of ultrafiltration have been used routinely for water removal since the beginning of hemodialysis, the availability of membranes with very high hydraulic permeabilities led to radically new approaches to renal substitutive therapy. Such membranes allowed uniformly high clearance rates of solutes up to moderate molecular weights (several thousands) by the use of predominantly convective transport, thereby mimicking the separation capabilities of the natural kidney glomeruli. Progress in the development of this pressure-driven technique, which has come to be known as *hemofiltration,* has been reviewed by Henderson [1982], Lysaght [1986], and Ofsthum et al.]1986].

In ultrafiltration, the solute flux J_s (the rate of solute transport per unit membrane surface area) is equal to the product of the ultrafiltrate flux J_F (the ultrafiltrate flow rate per unit membrane surface area) and the solute concentration in the filtrate, c_P In turn c_F is related to the retentate concentration C_R in the bulk solution above the membrane by the observed rejection coefficient R:

$$J_S = J_F c_F = J_F\left(1-R\right) c_R \tag{130.32}$$

Thus, knowledge of the ultrafiltrate flux and observed rejection coefficient permits prediction of the rate of solute removal.

With increasing transmembrane pressure difference, the ultrafiltrate flux increases and then levels off to a pressure-independent value. This behavior arises from the phenomenon of *concentration polarization* [Colton 1987]. Macromolecules (e.g., proteins) that are too large to pass through the membrane build up in concentration in a region near the membrane surface. At steady state, the rate at which these rejected macromolecules are convected by the flow of fluid towards the membrane surface must be balanced by the rate of convective diffusion away from the surface. Estimation of the ultrafiltrate flux reduces largely to the problem of estimating the rate of back transport of macromolecules away from the membrane surface

$$J_F = k \ln \frac{c_{pw}}{c_{pb}} \tag{130.33}$$

where k is the mass transfer coefficient for back transport of the rejected species, and c_{pw} and c_{pb} are the plasma concentrations of rejected species at the membrane surface and in the bulk plasma, respectively. Attainment of an asymptotic, pressure-independent flux is consistent with the concentration at the wall c_{pw} reaching a constant value. As with diffusive membrane permeability, solute rejection coefficients must be measured experimentally, since the available theoretical models and details of membrane structure are inadequate for prediction.

In hemofiltration the magnitude of the maximum clearance is determined by the blood and ultrafiltrate flow rates and whether the substitution fluid is added before or after filtration. Solutes with molecular weights up to several thousand are cleared at essentially the same rate in hemofiltration, whereas there is a monotonic decrease with increasing molecular weight in hemodialysis. If a comparison is made with devices of equal membrane surface area, it is generally found that hemodialysis provides superior clearance for low-molecular-weight solutes such as urea. The superiority of hemofiltration becomes apparent at molecular weights of several hundred.

Hemodialysis and hemofiltration represent two extremes with membranes having relatively low and relatively high hydraulic permeabilities, respectively. As a variety of new membranes became available with hydraulic permeabilities greater than that of regenerated cellulose, various groups began to examine new treatment modalities in which hemodialysis was combined with controlled rates of ultrafiltration which were higher than those employed in conventional hemodialysis but smaller than those used in hemofiltration [Funck-Brenato et al., 1972; Lowrie et al., 1978; Ota et al., 1975]. The advantage of such an approach is that it retains the high clearance capabilities of hemodialysis for low-molecular-weight solutes while adding enhanced clearance rates for the high-molecular-weight solutes characteristic of hemofiltration. A variety of systems is now commercially available and in clinical use, mainly in Europe and Japan. The proliferation of mixed-mode therapies has led to a panopoly of acronyms: hemodialysis (HD), hemofiltration (HF), high-flux dialysis (HFD), hemodiafiltration (HDF), biofiltration (BF), continuous arteriovenous hemofiltration (CAVH), continuous arteriovenous hemodialysis (CAVHD), slow continuous ultrafiltration (SCUF), simultaneous dialysis and ultrafiltration (SDUF), and so on.

Rigorous description of simultaneous diffusion and convection in artificial kidneys has not yet been carried out. Available analyses span a wide range of complexity and involve, to varying degrees, simplifying assumptions. Their predictions have not been systematically compared with experimental data. In view of the growing interest in various “high-flux” membranes and their application for enhanced solute removal rates and/or shortened treatment times, further refinement may be helpful.

130.13 Pharmacokinetics

Whereas the above analysis is founded on understanding the solute-removal capabilities of hemodialyzers, clinical application must also consider the limitations imposed by the transport of solute between body fluid compartments. The earliest physiologic models were produced by chemical engineers [Bell et al., 1965; Dedrick & Bischoff, 1968] using techniques which had been developed to describe the flow of material in complex chemical processes and were applied to the distribution of drugs and metabolites in biologic systems. This approach has progressively found its way into the management of uremia by hemodialysis [e.g., Farrell, 1983; Gotch & Sargent, 1983; Lowrie et al., 1976; Sargent et al., 1978].

Pharmacokinetics summarizes the relationships between solute generation, solute removal, and concentration in the patient’s blood stream. It is most readily applied to urea as a surrogate for other uremic toxins in the quantitation of therapy and in attempts to define its adequacy. In the simplest case, the patient is assumed to have no residual renal function and to produce no urea during the relatively short periods of dialysis. Urea is generated in the body from the breakdown of dietary protein, which empirically has been found to approximate

$$G = 0.11\, I - 0.12 \tag{130.24}$$

where G is the urea generation rate and I the protein intake (both in mg/min). If reliable measurements of I are not available, one assumes an intake of 1 gram of protein per kg of body weight per day.

Urea accumulates in a single pool equivalent to the patient's total body water and is removed uniformly from that pool during hemodialysis. Mass balance yields the following differential equation:

$$\frac{d(cV)}{dt} = G - Kc \qquad (130.35)$$

where c is the blood urea concentration (equal to total body water urea concentration) in mg/ml; V is the urea distribution volume in the patient in ml; G is the urea generation rate in mg/min; t is the time from onset of hemodialysis in minutes; and K is the urea clearance in ml/min. V can be measured by tritiated water dilution studies but is usually 58% of body weight. Generation is calculated from actual measurement or estimate of the patient's protein intake (each gram of protein consumed produces about 250 mg of urea). Therefore, a 70-kg patient, consuming a typical 1.0 g of protein per kilogram of body weight per day, would produce 28 g of urea distributed over a fluid volume of 40.6 L. In the absence of any clearance, urea concentration would increase by 70 mg/100 ml every 24 hours. The reduction of urea concentration during hemodialysis is readily obtained from Eq. (130.32) by neglecting intradialytic generation and changes in volume:

$$c^t = c^i \exp\left(\frac{Kt}{V}\right) \qquad (130.36)$$

where c^i and c^t represent the urea concentrations in the blood at the beginning and during the course of treatment. A 3-1/2-hour treatment of a 70-kg patient ($V = 40.6$ L) with a urea clearance of 200 ml/min would lead to a 64% reduction in urea concentration or a value of 0.36 for the c^t/c^i ratio. (This parameter almost always falls between 0.30 and 0.45.)

The increase in urea concentration between hemodialysis treatments is obtained from Eq. (130.33), again assuming a constant V:

$$c^t = c^f + \frac{G}{V}t \qquad (130.37)$$

where c^f is the urea concentration in the patient's blood at the end of the hemodialysis and c^t the concentration at time t during the intradialytic interval. Urea concentration typically increases by about 50–100 mg/100 ml/24 hours. Even a small residual renal clearance will prove numerically significant. Therefore in oliguric patients who still exhibit a minimum of kidney function, one should use the slightly more complex equations given by Sargent and Gotch [1989] or Farrell [1988].

The exponential decay constant in Eq. (130.33), Kt/v, expresses the net normalized quantity of hemodialysis therapy received by a uremic patient. It is calculated simply by multiplying the urea clearance of the dialyzer (in ml/min) by the duration of hemodialysis (in min) and dividing by the distribution volume (in ml) which in the absence of a better estimate is taken as 0.58 × body weight. Gotch and Sargent [1983] first recognized that this parameter provides an index of the adequacy of hemodialysis. Based upon a retrospective analysis of various therapy formats, they suggested a value of 1.0 or greater as representing an adequate amount of hemodialysis for most patients. Although not immune to criticism, this approach has found widespread clinical acceptance and represents the current prescriptive norm in hemodialysis therapy.

TABLE 130.1 Uremic Syndrome under Dialysis

Adverse effects of uremia can be attributed to:

1. Retention of solutes normally degraded or excreted by the kidneys.
2. Overhydration associated with inadequate balance between fluid intake and water removal.
3. Absence of factors normally synthesized by the kidneys.
4. Pathophysiologic response to the decline in renal function on the part of other organ systems.
5. Pathologic response of the organism to repeated exposure to damaging procedures and foreign materials.

130.14 Adequacy of Dialysis

As outlined in Table 130.1, the uremic syndrome under dialysis is more complex than observed in ESRD before the institution of treatment. The pathology observed not only is related to insufficient removal of toxic solutes but also comprises some unavoidable adverse effects of extracorporeal blood processing, including the interactions of blood with foreign materials [Colton et al., 1994]. The attenuation of uremic syndrome symptoms by protein restriction in the patient's diet and by various dialytic procedures underscores the combined roles of retention, removal, and metabolism in the constellation of signs of the disease. Toxicity may result from the synergism of the entire spectrum of accumulated molecules, which is surprisingly large (see Table 130.2 and Vanholder and Ringoir [1992]. The uremic syndrome resembles complex forms of systemic poisoning and is characterized by multiple symptoms and side effects. Survival requires that the toxins be removed, and survival quality depends on the quantity of toxins that are actually eliminated. Ideally, one would like to clean blood and body fluids to the same extent as is achieved by normal renal function. This is possible with an organ transplant that works without interruption but is only asymptotically approached with intermittent dialysis.

There is a compelling need for objective definition of the adequacy of ESRD treatment: How much removal in how much time is necessary for each individual? The answer is indirect and approximate. Some define adequacy of dialysis by clinical assessment of patient well-being. More sophisticated procedures, such as electromyography, electroencephalography, and neuropsychologic tests, may refine the clinician's perception of inadequate dialysis. Yet inadequate therapy can remain unrecognized when therapeutic decisions are based exclusively on clinical parameters. The inverse is also true, and follow-up of dialysis adequacy should never be restricted to static markers of toxicity or dynamic biochemical parameters such as clearance, kinetic modeling, and the like.

Most patients undergoing dialysis do not work or function as healthy people do, and often their physical activity and employment status does not go beyond the level of taking care of themselves. In many centers, the best patients in a hemodialysis program are selectively removed for transplantation. Hospitalization rate is an approximate index of dialysis inadequacy. About 25 percent of all hospitalizations are due to vascular access problems. Comparison among centers may be difficult, however, because of differences in local conditions for hospital admission. Vanholder and Ringoir [1992] have attempted to relate the adequacy of dialysis to the relevant solute concentrations in blood and distinguish among solute-related factors, patient-related factors, and dialysis-related factors (Table 130.3). Their analysis constitutes a useful point of departure for adjusting the quantity of dialysis to the specific needs of an individual patient, which is a complex problem, since it requires not only an appreciation of what the removal process can do, but also of the generation rate of metabolic end products (related to nutrition, physical activity, fever, etc.) and the dietary load of water and electrolytes. Dialysis patients are partially rehabilitated, but their condition rarely compares to that of recipients of a successful renal transplant.

130.15 Outlook

The treatment of chronic renal failure by artificial kidney dialysis represents one of the most common, and certainly the most expensive, component of substitutive medicine. From an industrial viewpoint, 500,000 patients each "consuming" perhaps 100 hemodialysis filters per year (allowing for some reuse

TABLE 130.2 Uremic Solutes with Potential Toxicity

Urea	Middle molecules
Guanidines	Ammonia
Methylguanidine	Alkaloids
Guanidine	Trace metals (e.g., bromine)
β-guanidinipropionic acid	Uric acid
Guanidinosuccinic acid	Cyclic AMP
Gamma-guanidinobutyric acid	Amino acids
Taurocyanine	Myoinositol
Creatinine	Mannitol
Creatine	Oxalate
Arginic acid	Glucuronate
Homoarginine	Glycols
N-a-acetylarginine	Lysozyme
Phenols	Hormones
O-cresol	Parathormone
P-cresol	Natriuretic factor
Benzylalcohol	Glucagon
Phenol	Growth hormone
Tyrosine	Gastrin
Phenolic acids	Prolactin
P-hydroxyphenylacetic acid	Catecholamines
β-(m hydroxyphenyl)-hydracrylic acid	Xanthine
	Hypoxanthine
Hippurates	Furanpropionic acid
p-(OH)hippuric acid	Amines
o-(OH)hippuric acid	Putrescine
Hippuric acid	Spermine
Benzoates	Spermidine
Polypeptides	Dimethylamine
β_2-microglobulin	Polyamines
Indoles	Endorphins
Indol-3-acetic acid	Pseudouridine
Indoxyl sulfate	Potassium
5-hydroxyindol acetic acid	Phosphorus
Indo-3-acrylic acid	Calcium
5-hydroxytryptophol	Sodium
N-acetyltryptophan	Water
Tryptophan	Cyanides

from the 150 units per year that would be needed for 3 times per week treatment) means a production of 50 million filters. With each unit selling for an approximate price of 15 dollars, the world market is on the order of $750 million. From a public health viewpoint, if one is to take the U.S. figure of $30,000 for the world average annual cost of a single dialysis patient, the aggregate economic impact of the medical application of hemodialysis approaches $15 billion a year (of which less than 10 percent is spent on the purchase of technology; health care personnel costs are the most expensive component of the treatment).

Yet "maintenance dialysis on the whole is non-physiological and can be justified only because of the finiteness of its alternative" [Burton, 1976]. Dialytic removal remains nonspecific, with toxic as well as useful compounds eliminated indiscriminately. A better definition of disturbed metabolic pathways will be necessary to formulate treatment hypotheses and design adapted equipment. Sensors for on-line monitoring of appropriate markers may also help to evaluate the modeling of clearance processes. The confusing interference of interactions between the patient and the foreign materials in the dialysis circuit may be reduced as more compatible materials become available. A better clinical condition of the ESRD patient remains the ultimate goal of dialysis therapy because at the moment it seems unlikely that either preventative measures or organ transplantation will reduce the number of patients whose lives depend on the artificial kidney.

TABLE 130.3 Factors Influencing Solute Concentration in Dialyzed Patients

Solute-related factors	Patient-related factors
Compartmental distribution	Body weight
Intracellular concentration	Distribution volume
Resistance of cell membrane	Intake and generation of solutes metabolic precursors
Protein binding	Residual renal function
Electrostatic charge	Quality of vascular access
Steric configuration	Absorption from the intestine
Molecular weight	Hematocrit
Dialysis-related factors	Blood viscosity
Dialysis duration	Absorption of solutes on the membrane, on other parts of the circuit
Interdialytic intervals	Ultrafiltration rate
Blood flow	Intradialytic changes in efficiacy
Mean blood flow	Changes with indirect effect on solute-related factors
Blood flow pattern	Blood pH
Concentration gradients	Heparinization
Dialysate flow	Free fatty acid concentration
Dialyzer surface	
Dialyzer volume	
Dialyzer membrane resistance	
Dialyzer pore size	

Defining Terms

Arteriovenous fistula: A permanent communication between an artery and an adjacent vein, created surgically, leading to the formation of a dilated vein segment which can be punctured transcutaneously with large bore needles so as to allow connecting the circulatory system with an extracorporeal blood processing unit.

Artificial kidney: A blood purification device based on the removal of toxic substances through semipermeable membranes washed out by an acceptor solution which can safely be discarded.

Blood urea nitrogen (BUN): The concentration of urea in blood, expressed as the nitrogen content of the urea (BUN is actually 0.47 times, or approximately half, the urea concentration).

Boundary layer: The region of fluid adjacent to a permeable membrane, across which virtually all (99%) of the concentration change within the fluid occurs.

Catheter: A tube used to infuse a fluid in or out of the vascular system or a body cavity.

Clearance: A measure of the rate of mass removal expressed as the volume of blood which per unit of time is totally cleared of a substance through processing in a natural or artificial kidney. Clearance has the dimensions of a flow rate and can be defined only in relation to a specific solute. Clearance can also be viewed as the minimal volume flow rate of blood which would have to be presented to a processing device to provide the amount actually recovered in the urine or the dialysate if extraction of that material from blood were complete. Clearance is measured as the mass transfer rate of a substance divided by the blood concentration of that substance.

Continuous ambulatory peritoneal dialysis: A modality of peritoneal dialysis in which uninterrupted—although not evenly effective—treatment is provided by 4–6 daily cycles of filling and emptying the peritoneal cavity with a prepared dialysate solution. Solute removal relies on diffusive equalization with molecular species present in capillary blood. Water removal relies on the use of hyperosmotic dialysate.

Continuous arteriovenous hemodialysis: A dialytic procedure in which blood, propelled either by arterial pressure or by a pump, flows continuously at a low flow rate through a dialyzer, from where it returns to a vein, providing for uninterrupted solute and fluid removal and nearly constant equilibration of body fluids with the dialysate solution.

Dialysate: A buffered electrolyte solution, usually containing glucose at or above physiologic concentration, circulated through the water compartment of a hemodialyzer to control diffusional transport of small molecules across the membranes and achieve the blood concentrations desired.

Dialysis: A membrane separation process in which one or more dissolved molecular species diffuse across a selective barrier in response to a difference in concentration.

Dwell time: The duration of exposure of a solution used to draw waste products and excessive water out of the blood during peritoneal dialysis.

ESRD: End-stage renal disease.

Glomerular filtration rate: The volume of plasma water, or primary urine, filtered in the glomerulus per unit of time. Measured, for instance, by creatinine clearance, it expresses the level of remaining renal function in end-stage renal disease.

Hemodiafiltration: Removal of water and solutes by a combination of diffusive and convective transport (paired filtration-dialysis) across a dialysis membrane to achieve effective transport of small and middle molecules. To compensate for the water loss, a large volume of saline or balanced electrolyte solution must be infused in the blood circuit to prevent hemoconcentration.

Hemodialysis: A modality of extracorporeal blood purification in which blood is continuously circulated in contact with a permeable membrane, while a large volume of balanced electrolyte solution circulates on the other side of the membrane. Diffusion of dissolved substances from one stream to the other removes molecules that are in excess in the blood and replaces those for which there is a deficiency. Increased removal can be achieved by increasing the duration of the procedure, the overall membrane area, or the membrane permeability.

Hemofiltration: Removal of water and solutes by convective transport, controlled by a large hydrostatic pressure difference between blood and a liquid compartment across a large-pore, high-water-flux membrane.

Membrane: A thin film of natural or synthetic polymer which allows the passage of dissolved molecules and solvents in response to a concentration or pressure difference (diffusion or filtration) across the polymer.

Middle molecules: Molecules of intermediate molecular weight (roughly of 1000 to 30,000 daltons) which are presumed to be responsible for the toxic manifestations of end-stage renal disease and therefore should be eliminated by substitutive therapy.

Peritoneal dialysis: A process in which metabolic waste products, toxic substances, and excess body water are removed through a membranelike tissue that lines the internal abdominal wall and the organs in the abdominal cavity.

Permeability: The ability of a membrane to allow the passage of certain molecules while maintaining a physical separation between two adjacent phases.

Permselectivity: The property of a membrane whereby a differential rate of molecular transport between two phases is achieved based on characteristics such as molecular weight, molecular size, degree of hydration, affinity for membrane material, and electric charge. The most common feature leading to permselectivity is membrane pore size.

Residual renal clearance: The small level of renal function (measured as creatinine clearance by the diseased kidneys) remaining in some patients in end-stage renal disease, particularly in the early years of dialytic treatment.

Ultrafiltration: The process whereby plasma water flows through a membrane in response to a hydrostatic pressure gradient, dragging with it solute molecules at concentrations equal or lower to that prevailing in plasma.

Uremia: A condition in which the urea concentration in blood is chronically elevated, reflecting an inability to remove from the body the end products of protein metabolism.

Uremic toxins: Partly unidentified and presumably toxic substances appearing in the blood of patients in end-stage renal failure, which can be eliminated to a variable extent by chemical processing of body fluids.

References

Abel JJ, Rowntree LG, Turner BB. 1913. On the removal of diffusible substances from the circulating blood by means of dialysis. Trans Assoc Am Physicians 28:50.

Babb AL, Maurer CJ, Fry DL, et al. 1968. The determination of membrane permeabilities and solute diffusivities with application to hemodialysis. Chem Eng Prog Symp Ser 63:59.

Brunner H, Mann H. 1985. What remains of the "middle molecule" hypothesis today? Contr Nephrol 44:14.

Burton BJ. 1976. Overview of end stage renal disease. J Dialysis 1:1.

Chenoweth DE. 1984. Complement activation during hemodialysis: clinical observations, proposed mechanisms and theoretical implications. Artif Organs 8:281.

Colton CK. 1987*a*. Analysis of membrane processes for blood purification. Blood Purification 2:202.

Colton CK. 1987*b*. Technical foundations of renal prostheses. In HG Gurland (ed), Uremia Therapy, pp 187–217, Berlin, Heidelberg. Springer Verlag.

Colton CK, Smith KA, Merril EW, et al. 1971. Permeability studies with cellulosic membranes. J Biomed Mat Res 5:459.

Colton CK, Ward RA, Shaldon S. 1994. Scientific basis for assessment of biocompatibility in extracorporeal blood treatment. J Nephol Dial Transplant 9 (suppl 2):11.

Dedrick RA, Bischoff KB. 1968. Pharmacokinetics in applications of the artificial kidney. Chem Eng Prog Ser 64:32.

Gotch FA, Autian J, Colton CK, et al. 1972. The Evaluation of Hemodialyzers, Washington, DC, Department of Health, Education and Welfare Publ. No. NIH-72-103.

Gotch FA, Sargent JA, Keen ML, et al. 1974. Individualized, quantified dialysis therapy of uremia. Proc Clin Dial Transplant forum 4:27.

Jaffrin MY, Butruille Y, Granger A, et al. 1978. Factors governing hemofiltration (HF) in a parallel plate exchanger with highly permeable membranes. Trans Am Soc Artif Intern Organs 24:448.

Jaffrin MY, Gupta BB, Malbraneq JM. 1981. A one-dimensional model of simultaneous hemodialysis and ultrafiltration with highly permeable membranes. J Biomech Eng 103:261.

Kedem O, Katchalsky A. 1958. Thermodynamic analysis of the permeability of biological membranes to nonelectrolytes. Biochim Biophys Acta 27:229.

Kiil F. 1960. Development of a parallel flow artificial kidney in plastics. Acta Chir Scand [Suppl] 253:143.

Klein E, Holland FF, Donnaud A, et al. 1977. Diffusive and hydraulic permeabilities of commercially available cellulosic hemodialysis films and hollow fibers. J Membr Sci 2:349–364.

Leonard EF, Bluemle LW. 1960. The permeability concept as applied to dialysis. Trans Am Soc Artif Intern Organs 6:33.

Lysaght MJ, Colton CK, Ford CA, et al. 1978. Mass transfer in clinical blood ultrafiltration devices—a review. In TH Frost (ed), Technical Aspects of Renal Dialysis, pp 81–95, London, Pittman Medical.

Michaels AS. 1966. Operating parameters and performance criteria for hemodialyzers and other membrane separation devices. Trans Am Soc Artif Intern Organs 12:387.

Quinton W, Dillard D, Scribner BH. 1960. Cannulation of blood vessels for prolonged hemodialysis. Trans Am Soc Artif Intern Organs 6:104.

Renkin EM. 1956. The relationship between dialysance, membrane area, permeability, and blood flow in the artificial kidney. Trans Am Soc Artif Intern Organs 2:102.

Sargent JA, Gotch FA, Borah M, et al. 1978. Urea kinetics: A guide to nutritional management of renal failure. Am J Clin Nutr 31:1692.

Solomon BA, Castino F, Lysaght MJ, et al. 1978. Continuous flow membrane filtration of plasma from whole blood. Trans Am Soc Artif Intern Organs 24:21.

Vanholder R, Hsu C, Ringoir S. 1993. Biochemical definition of the uremic syndrome and possible therapeutic implications. Artif Organs 17:234.

Vanholder R, Ringoir S. 1992. Adequacy of dialysis, a critical analysis. Kidney Int 42:540.

Vanholder RC, De Smet RV, Ringoir SM. 1992. Validity of urea and other "uremic markers" for dialysis quantification. Clin Chem 38:1429.

Wilson EE. 1915. A basis for rational design of heat transfer apparatus. Trans Am Soc Mech Eng 37:47.

Further Information

An extensive review of renal pathophysiology is to be found in: B.M. Brenner and F.C. Rector, eds., *The Kidney,* 3d ed., Saunders Publishing Co., Philadelphia, 1986. Two volumes addressing the clinical aspects of dialysis are A.R. Nissanson, R. Fine, and D. Gentile, *Clinical Dialysis,* Appleton Lange Century Crofts, Norwalk, 1984; and H.J. Gurland, ed., *Uremia Therapy,* Springer Verlag, Berlin, 1987. The principles of designs and functions of dialysis therapy are outlined in P.C. Farrel, *Dialysis Kinetics, ASAIO Primers in Artificial Organs,* vol. 4, J.B. Lippincott, Philadelphia, 1988; and J.F. Maher, *Replacement of Renal Function of Dialysis,* 3d ed., Klumer, Boston, 1989. Recent reviews of specific aspects in the operation of artificial kidneys are C.K. Colton, "Analysis of Membrane Processes for Blood Purification," *Blood Purification* 5:202–251, 1987; C.K. Colton and E.G. Lowrie, "Hemodialysis Physical Principles and Technical Considerations," in B.M. Brenner and F.C. Rector, Jr., eds., *The Kidney,* 2d ed., vol 2, Saunders, Philadelphia; and C.K. Colton, R.A. Ward, and S. Shaldon, "Scientific Basis for Assessment of Biocompatibility in Extracorporeal Blood Treatment," *Nephrology Dialysis, Transplantation,* 9(Suppl. 2):11, 1994.

Ongoing contributions to the field of artificial kidney therapy are often found in biomaterials journals (e.g., the *Journal of Biomaterials Research*) and in artificial organ publications (e.g., the *Transactions of the American Society for Artificial Organs,* the *ASAIO Journal,* the *International Journal of Artificial Organs,* and *Artificial Organs*). Clinical contributions can be found in *Kidney International, Nephron,* and *Blood Purification.*

131

Peritoneal Dialysis Equipment

Michael J. Lysaght
Brown University

John Moran
Vasca, Inc.

Irreversible end-stage kidney disease occurs with an annual frequency of about 1 in 5000 to 10,000 in general population, and this rate is increasing. Until the 1960s, such disease was universally fatal. In the last four decades various interventions have been developed and implemented for preserving life after loss of all or most of a patient's own kidney function. Continuous ambulatory peritoneal dialysis (CAPD), the newest and most rapidly growing of renal replacement therapies, is one such process in which metabolic waste products, electrolytes, and water are removed through the peritoneum, an intricate membranelike tissue that lines the abdominal cavity and covers the liver, intestine, and other internal organs. This review begins with a brief summary of the development of CAPD and its role in the treatment of contemporary renal failure. The therapy format and its capacity for solute removal are then described in detail. Bioengineering studies of peritoneal transport, in which the peritoneum is described in terms analogous to the mass transfer properties of a planar membrane separating well-mixed pools of blood and dialysate, are then reviewed. The transport properties of the equivalent peritoneal membrane are summarized and compared to those of hemodialysis membranes. Models to describe and predict fluid and solute removal rates are examined. Finally, current developments and emerging trends are summarized.

The early history of peritoneal dialysis, as reviewed by Boen [1985], is contemporaneous with that of hemodialysis (HD). Small-animal experiments were reported in the 1920s and 1930s in the United States and Germany. Earliest clinical trials in acute reversible cases of kidney failure began in the late 1930s in the form of a continuous "lavage" in which dialysate was continuously infused and withdrawn from dual trochar access sites. Acute treatments were continued through the 1940s, and about 100 case reports appeared in the literature by 1950; sequential inflow, dwell, and withdrawal was increasingly favored over continuous flow. Chronic therapy was introduced in the early 1960s, followed shortly by indwelling peritoneal catheters. From 1960 onward, peritoneal dialysis clearly lagged behind HD, as the latter became more streamlined, efficient, and cost-effective. Although endorsed by a small group of enthusiasts and proponents, peritoneal dialysis had evolved into a specialized or niche therapy. This changed dramatically in 1976 when Popovich, a biomedical engineer, and Moncrief, a clinical nephrologist, announced the development of a new form of peritoneal dialysis in which ambulatory patients were continuously treated by two liters of dialysis dwelling in the peritoneal cavity and exchanged four times daily [Popovich et al., 1976]. Two years later Baxter began to offer CAPD fluid in flexible plastic containers, along with necessary ancillary equipment and supplies. The rapid subsequent growth of the process is tabulated in Fig. 131.1.

0-8493-0462-8/00/$0.00+$.50

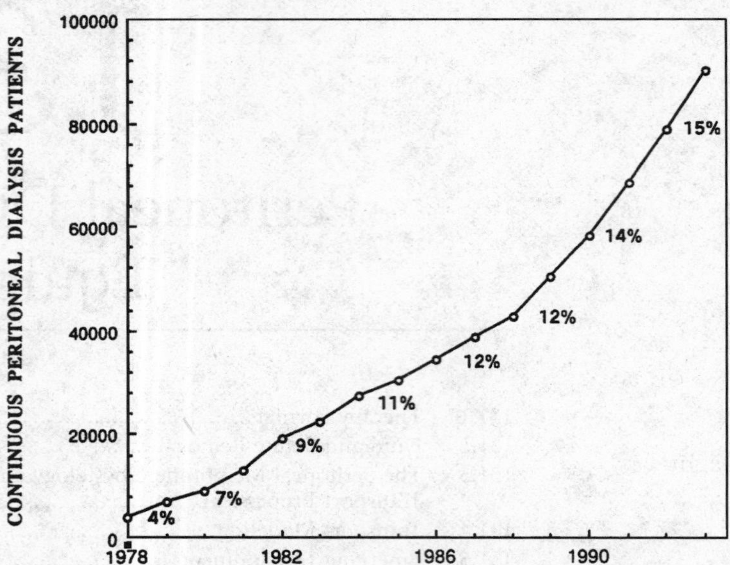

FIGURE 131.1 The growth of peritoneal dialysis. Line and points refer to the total estimated worldwide peritoneal dialysis population; the numbers adjacent to the points are PD patients as a percent of total dialysis population. At the end of 1993, 14,000 of the 90,000 peritoneal dialysis patients utilized some version of APD; the remainder were treated with CAPD. Data compiled taken from various patient registries and industrial sources.

At this writing, approximately 90,000 patients are treated by CAPD (versus 490,000 by HD and 130,000 with kidney transplants). A more recent development is the introduction of automated peritoneal dialysis (APD), in which all fluid exchanges are performed by a simple pump console, usually while the patient sleeps. About one peritoneal dialysis patient in six now receives some form of APD; this approach is discussed more fully in a later section on emerging developments.

Both CAPD and HD have advantages and disadvantages, and neither therapy is likely to prove better for all the patients all the time. The principal attraction of CAPD is that it frees the patient from the pervasive life-style invasions associated with thrice weekly in-center HD. CAPD is particularly popular with patients living in rural areas distant from a hemodialysis treatment center. The continuous nature of CAPD eliminates fluctuations in the concentrations of uremic metabolites and avoids the sawtooth pattern of hemodialysis peak toxin concentrations. Fluid and dietary constraints are less restrictive for patients on CAPD than those on HD. A major complication of CAPD is peritonitis. Th rate of peritonitis was initially around 2 episodes per patient year; this has fallen to fewer than 0.5 episodes per patient year due to advances in administration set design and use. The morbidity of peritonitis has also decreased with increased experience in its treatment. In most cases, detected early and treated promptly, peritonitis can be managed without requiring hospitalization. Peritonitis caused by certain organisms including Staphylococcus aureus, Pseudomonas, and fungi remains a clinical problem. Other drawbacks of CAPD include the daily transperitoneal administration of 100–150 g of glucose providing ~600 calories, and the tedium of the exchanges. APD and new solution formulations are being developed to address both issues. Little doubt now exists that risk-adjusted survival and morbidity for patients treated by CAPD is equivalent to that for patients treated with HD. On balance, the therapy seems well suited to many patients, and it continues to grow more rapidly than alternative treatment modalities.

131.1 Therapy Format

The process of CAPD is technically simple. Approximately 2 L of a sterile, nonpyrogenic, and hypertonic solution of glucose and electrolyte are instilled via gravity flow into the peritoneal cavity through an indwelling catheter 4 times per day. A single exchange is illustrated in Fig. 131.2. Intraperitoneal fluid partially equilibrates with solutes in the plasma, and plasma water is ultrafiltered due to osmotic gradients. After 4–5 hours, except at night where the exchange is lengthened to 9–11 hours to accommodate sleep, the peritoneal fluid is drained and the process repeated. Patients perform the exchanges themselves in 20–30 minutes, at home or in the work environment after a training cycle which lasts only 1–2 weeks. In APD, 10–15 L are automatically exchanged overnight; 2 L remain in the peritoneal cavity during the day for a "long dwell" exchange.

As will be discussed in more detail below, the drained fluid contains solute at concentrations around 90–100% of plasma for urea, 65–70% for creatinine, and 15–25% for inulin and β_2 microglobulin. Net fluid removal ranges up to 1000 ml per exchange. CAPD generally removes the same quantity of toxins and fluid as HD (a little thought will show that this is a requirement of steady state, provided that generation is unaltered between the two treatment formats); however, CAPD requires a higher plasma concentration as the driving force for this removal. Steady-state concentrations during CAPD are typically close to the peak, i.e., pretreatment, concentrations of small solutes during HD but much lower than the corresponding peaks for larger species.

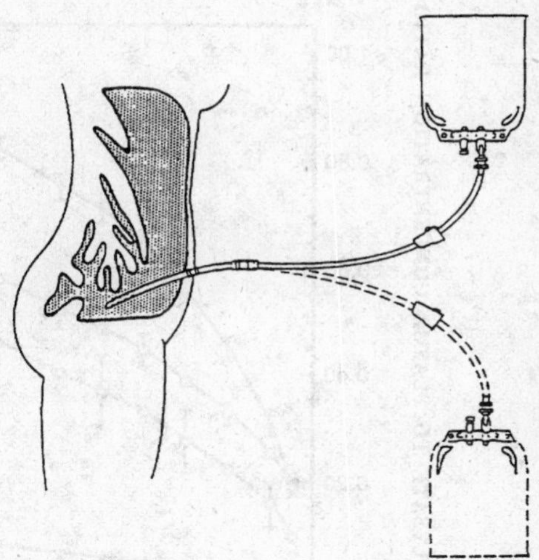

FIGURE 131.2 Illustration of the three steps involved in a single CAPD exchange: fluid infusion, dwell, and drain. Some administration sets require the bag to stay connected during dwell (it is rolled and fits in a girdle around the waist); others allow it to be disconnected. Drain and infusion take about 10 minutes each; three daytime dwells are 4–6 hours each; the overnight dwell lasts 8–10 hours.

Access to the peritoneum is usually via a double-cuff Tenchkhoff catheter, essentially a 50–100 cm length of silicone tubing with side holes at the internal end, a Dacron mesh flange at the skin line, and connector fittings at the end of the exposed end. Several variations have evolved, but little hard evidence supports the selection of one design format over another [Dratwa et al., 1986]. Most are implanted in a routine surgical procedure requiring about 1 hour and are allowed to heal for 1–2 weeks prior to routine clinical use. Sterile and nonpyrogenic fluid is supplied in 2-L containers fabricated from dioctyl phthalate plasticized polyvinyl chloride. The formulation is essentially potassium-free lactated Ringers to which has been added from 15-42.5 g/L of glucose (dextrose monohydrate). The solution is buffered to a pH of 5.1–5.5, since the glucose would caramelize during autoclaving at higher pH levels. Several different exchange protocols are in use. In the original design, the patient simply rolls up the empty bag after instillation and then drains into the same bag following exchange. The bag filled with drain fluid is disconnected and a fresh bag is reconnected. Patients are trained to use aseptic technique to perform the connect and disconnect. Many ingenious aids were developed to assist in minimizing breaches of sterility including enclosed ultraviolet-sterilized chambers and heat splicers. More recent approaches, known as the "O" set and "Y" set or more generically as "flush before fill" disconnect, invoke more complex tubing sets to allow the administration set to be flushed (often with antiseptic) prior to installation of dialysate and generally permit the patient to disconnect the empty bag during the dwell phase. Initial reports of the success of the protocols in reducing peritonitis were regarded with skepticism, but definite improvement over earlier systems has now been documented in a well-designed and carefully controlled clinical trials [Churchill et al., 1989].

131.2 Fluid and Solute Removal

The rate at which solutes are removed during peritoneal dialysis depends primarily upon the rate of equilibration between blood and instilled peritoneal fluid. This is usually quantified as the ratio of

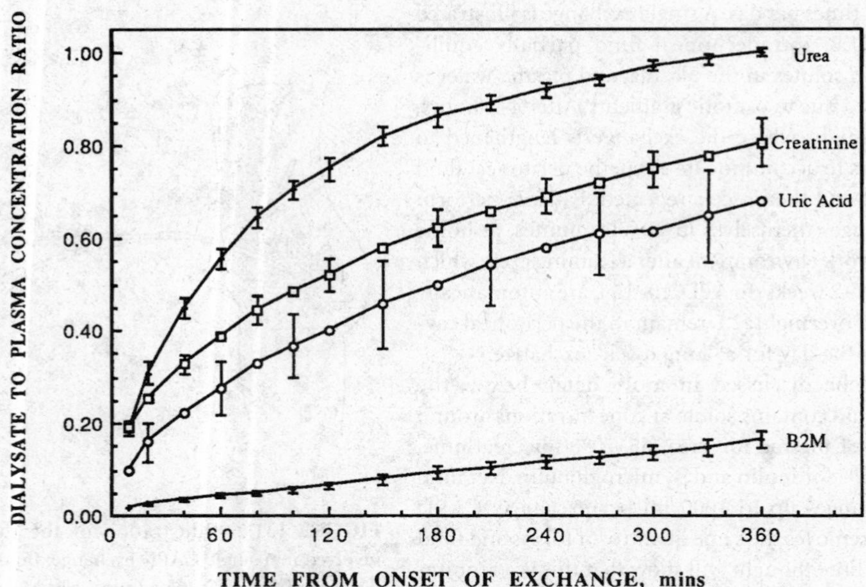

FIGURE 131.3 Ratio of plasma to dialysate concentration for urea (60 daltons), creatinine (113 daltons), uric acid (158 daltons), and β_2 microglobulin (~12,000 daltons). Data were obtained by withdrawing and analyzing a sample of dialysate at each time point and comparing it to plasma concentration. Each point is the average of two determinations on five patients. Error bars are standard error of the mean [Lysaght, 1989].

dialysate to plasma concentration as a function of dwell time, often in graphs called simply "D over P" (dialysate over plasma) curves. A typical plot of dialysate-to-plasma ratio for solutes of various molecular weight is given in Fig. 131.3. Smaller species equilibrate more rapidly than do larger ones, because diffusion coefficient varies in inverse proportion to the square root of a solute's molecular weight. Dialysate equilibrium rates vary considerably from patient to patient; error bars on the plot represent standard error of the mean for duplicate determinations with five patients.

The rate of mass removal during dialysis, ϕ, is simply the volume of fluid, V_D, removed from the peritoneal cavity at the end of a dwell period lasting time *t*, multiplied by the concentration C_D of the solute in the removed fluid

$$\phi = V_D C_D \tag{131.1}$$

The whole blood clearance, *Cl*, is the rate of mass removal divided by the solute concentration in blood C_B

$$Cl = \frac{\phi}{t\,C_B} = \frac{V_D C_D}{t\,C_B} \tag{131.2}$$

In Eqs. (131.1) and (131.2), time conventionally is reported in minutes, volume in milliliters, and concentration in any consistent units. Equations (131.1) and (131.2) are based on mass balances; they are thus general and unaffected by the complexity of underlying phenomena such as bidirectional selective connective transport and lymphatic uptake. Equation (131.2) requires that solute concentration in the denominator be reported as whole blood concentration, rather than as plasma concentration, which is often reported clinically. With many small solutes (urea, creatinine, and uric acid), only small error is introduced by considering blood and plasma concentration as interchangeable. With larger solutes,

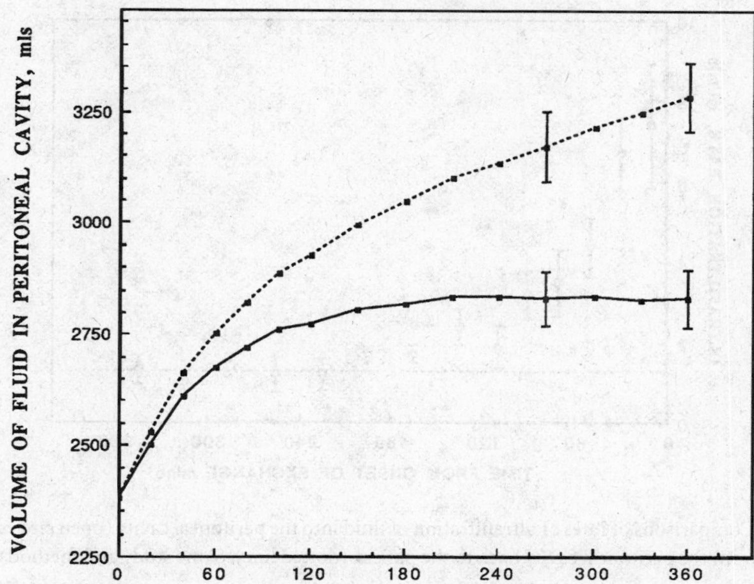

FIGURE 131.4 Volume of fluid in the peritoneal cavity versus time during an exchange with ~2.5% glucose dialysis fluid. Solid line is actual volume. Dotted line represents estimate of the volume in the absence of lymphatic flow. Results represent an average of duplicate determinations on five patients. Volume was estimated by dilution of radiolabeled tracers (too large to diffuse across the peritoneal membrane) added to dialysate prior to installation; lymphatic flow was calculated from a mass balance on net recovered marker. Each point is the average of two determinations on five patients. Error bars are standard error of the mean [Lysaght, 1989].

especially those excluded from the red blood cell, care must be taken to correct for differences in plasma and blood concentrations.

Since urea is nearly completely equilibrated during CAPD, i.e., $c_D/c_B = \sim 1.0$, urea clearance is commonly equated with total drainage volume. Four 2-L exchanges and 2 L of ultrafiltration would thus result in a continuous urea clearance of 10 L/day or ~7 ml/min. The situation is more complex with APD, which involves several (4–6) short exchanges at partial equilibrium and one very long exchange. In any case, no meaningful direct or *a priori* comparison of clearance with hemodialysis is possible because one therapy is intermittent and the other continuous.

The volume of fluid in the peritoneal cavity increases during an exchange but at a decreasing rate. The driving force for fluid transfer from the blood to the peritoneal cavity is the osmotic pressure of the glucose in the infused dialysate. Typical CAPD solutions contain ~1.5%, ~2.5%, or ~4.25% by weight of glucose monohydrate, leading to an initial maximum osmotic force (across an ideally semipermeable membrane) of approximately 1000–5000 mmHg. In the first few minutes of an exchange, the rate of ultrafiltration may be as high as 10–30 ml/min. The driving force rapidly dissipates as glucose diffuses from the peritoneal cavity into the bloodstream. After the first hour, rates of 1.0–2.0 ml/min are common. Throughout the exchange, the peritoneal lymphatics are draining fluid from the peritoneal cavity at a rate of 0.5–2.0 ml/min. Fluid balance is thus the difference between removal by a time-dependent rate of ultrafiltration and return via a more constant lymphatic drainage. Net fluid removal is very easily determined in the clinical setting simply by comparing the weight of fluid drained to that instilled. Instantaneous rates of ultrafiltration may be estimated in study protocols by a series of tedious mass balances around high-molecular-weight radiolabled markers added to the dialysate fluid. The results of a typical study are plotted in Figs. 131.4 and 131.5 showing both the instantaneous rate of ultrafiltration and the net intraperitoneal volume as a function of time. On average, these patients removed 500 ml of fluid in a single 6-hour exchange or roughly 2 L/day, which permits far more liberal fluid intake than

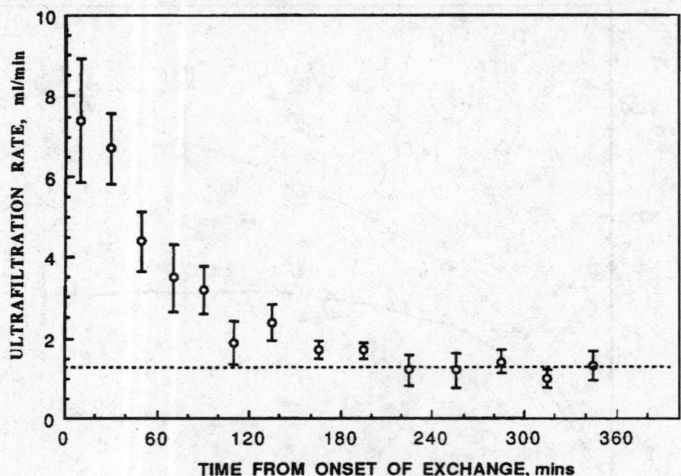

FIGURE 131.5 Comparisons of rates of ultrafiltration of fluid into the peritoneal cavity (open circles) and lymphatic drainage of fluid from the peritoneal cavity back to the patient (dotted line). Same study and methods as in Fig. 131.4.

is possible with patients on HD. But here again patient variation is high. Commercial CAPD fluid is available in a variety of solute concentrations; physicians base their prescription for a particular patient on his or her fluid intake and residual urine volume.

131.3 The Peritoneal Membrane: Physiology and Transport Properties

In contrast to synthetic membranes employed during HD, the peritoneum is not a simple selective barrier between two phases. As implied by its Latin root (*peritonere* = to stretch tightly around), the primary physiologic function of the peritoneum is to line the walls of the abdominal cavity and encapsulate its internal organs (stomach, liver, spleen, pancreas, and parts of the intestine). Most CAPD literature, including this review, uses the terms *peritoneum* and *peritoneal membrane* interchangeably and conveniently extends both expressions to include underlying and connective tissue. Overall adult peritoneal surface is approximately 1.75 ± 0.5 m^2, which generally is considered equal on an individual basis to skin surface area. The peritoneum is not physically homogenous. The visceral portion (~80%) covering the internal organs differs somewhat from the parietal portion overlaying the abdominal walls, which in turn is different from the folded or pleated mesentery connecting the two.

The physiology of the peritoneum, its normal ultastructure, and variations induced by CAPD have been increasingly elucidated over the past decade. Morphologically, the peritoneum is a smooth, tough, somewhat translucent sheath. Its thickness ranges from under 200 to over 1000 microns. The topmost layer, which presents to the dialysate during CAPD, is formed from a single layer of mesothelial cells, densely covered by microvilli (hairlike projections), although the latter tend to disappear gradually during the first few weeks of CAPD. Immediately underneath is the interstitium, a thick sheath of dense mucopolysaccharide hydrogel interlaced with collagenous fibers, microfibrillar structures, fibroblasts, adipocytes, and granular material. Most important for CAPD, the interstitium is perfused with a network of capillaries through which blood flows from the mesenteric arteries and the vasculature of the abdominal wall to the portal and systemic venous circulations. Blood-flow rate has been estimated to be in the range of 30–60 ml/min, but this is not well established. The interstitial layer is a hydrogel; its water content, and thus its transport properties, will vary in response to the osmolarity of the peritoneal dialysate.

Peritoneal mass transfer characteristics are most commonly obtained by back-calculating basic membrane properties from results in standard or modified peritoneal dialysis. Three membrane parameters will be described: *Lp*, the hydraulic permeability; *R*, the rejection coefficient; and K_oA, the mass transfer coefficient (= area A × diffusive permeability K_o). The formal definitions of these parameters are given in Eqs. (131.3) through (131.5), with R and K_oA defined for the limiting conditions of pure convection and pure diffusion.

$$LP = \frac{\text{Filtration rate}}{\text{Area} \cdot \text{pressure driving force}} = \frac{J_F}{A\left(\Delta P - \Sigma\sigma_i\pi_i\right)} \tag{131.3}$$

$$R = 1 - \frac{\text{Concentration in bulk filtrate}}{\text{Concentration in bulk retentate}} = \left[\frac{C_B - C_D}{C_B}\right]_{\phi d=0} \tag{131.4}$$

$$K_0A = \frac{\text{Solute transport}}{\text{Concentration driven force}} = \left[\frac{\phi_d}{C_B - C_D}\right]_{J_F R=0} \tag{131.5}$$

where J_F = filtration rate, A = area, σ = Staverman reflection coefficient, π = osmotic pressure, and other terms are as defined previously.

At the onset of a CAPD exchange using 4.25% dextrose, the ultrafiltration rate is 10–30 ml/min. Relative to a perfectly semipermeable membrane, the glucose osmotic pressure of the solution is 4400 mmHg. Overall membrane hydraulic permeability is the quotient of these terms and is thus of the order of 0.2 ml/hr-mmHg, in the units commonly employed for HD membranes. This estimate needs to be corrected for the osmotic back-pressure, which is primarily due to urea in the blood (conc ~ 1.3 g/L) as well as the fact that the membrane is only partially semipermeable. The best results are not obtained from a single point measurement but either from curve fitting to the ultrafiltration profile during the entire course of dialysis or from data taken at different osmotic gradients. A review [Lysaght & Farrell, 1989] of reports from several different investigators suggests an average value of ~ 2 ml/hr-mmHg or, roughly, 2 gal/ft^2/day (GSFD) at 100 PSI. This is higher than desalination membranes, just slightly lower than conventional regenerated cellulose hemodialysis membranes, and much lower than anisotropic ultrafiltration membranes.

Rejection coefficients, R, numerically equal to unity minus sieving coefficient are obtained either from kinetic modeling as described below or experimentally by infusing a hypertonic solution into the peritoneum with a permeant concentration equal to that in the plasma. After a suitable period of ultrafiltration, the ration of solute to water flux is calculated from the dilution of the recovered solution. Both methods are approximate and results from different investigators may vary substantially. Reported values are observed average rejection coefficients. These are often described as the Staverman rejection coefficient, σ, which is somewhat overreaching, since filtration velocity is not recorded and differences between bulk and wall concentrations are not known. Representative values, from a review of the literature [Lysaght & Farrell, 1989], are summarized in Table 131.1. Thus the membrane appears quite tight, possibly rejecting about 10–20% of urea and other small molecules, about 50% of intermediate-molecular-weight species, and over 99% of plasma proteins.

The diffusive permeability of the membrane is obtained by back calculation from measurements of blood and dialysate concentration versus time during an exchange, as will be further elaborated below. Values are given as the product of membrane permeability and estimated peritoneal area (K_oA), and the results of various investigators have been reasonably consistent. Critical values from a review of the literature are summarized in Table 131.1. A K_oA value of about 20 ml/min for urea is around one order of magnitude less than comparable values for contemporary hollow-fiber hemodialyzers. If the area of the peritoneum is taken as 1.75 m^2, then urea transfers through the peritoneum analogously to urea diffusing through a stagnant film of water roughly a centimeter thick. Alternatively, given a peritoneal

TABLE 131.1 Equivalent Transport Coefficients for the Peritoneal Membrane

Permeant Species, MW	Rejection Coefficient, dimensionless	K_oA cm³/min
Urea, 60	0.26 ± 0.08	21 ± 4
Creatinine, 113	0.35 ± 0.07	10 ± 2
Uric acid, 158	0.37	10
B-12, 1355		5
Inulin, 5200	0.5 ± 0.2	4 ± 1.5
$\beta_2$microglobulin, 12,000		0.8 ± 0.4
Albumin, 69,000	0.99	

Note: SD not given if $n < 3$. Equivalent ultrafiltration coefficient is ~2.0 ml/min-m²-mmHg. Data taken from a review by Lysaght and Farrell [1989].

thickness range of 200–2000 microns, the diffusion of urea inside the membrane is about 20% of what would be found in a film of stagnant water of the same thickness.

It should once more be noted that the physiologic peritoneum is a complex and heterogeneous barrier, and its transport properties would be expected to vary over different regions of its terrain. For example, studies in animal models have suggested that transport during peritoneal dialysis is little affected when large segments of the visceral membrane are surgically excised. It is also repeated for emphasis that the terms *Lp, R, and K_oA* do not describe this membrane itself but rather a hypothetical barrier that is functionally and operationally equivalent and thus capable of producing the same mass transfer characteristics in response to the same driving forces.

131.4 Transport Modeling

Several investigators have developed mathematic models to describe, correlate, and predict relationships among the time-course of solute removal, fluid transfer, treatment variables, and physiologic properties [Lysaght & Farrell, 1989; Vonesh et al., 1991; Waniewski et al., 1991].Virtually all kinetic studies start with the model illustrated in Fig. 131.6. The patient is considered to be a well-mixed compartment with a distribution volume V_B set equal to some fraction of total body weight. (For example, urea distributes over total body water, which is ~ 0.58 times body weight.) Dialysate occupies a second, much smaller compartment, V_D = 2–3 L, which is also considered well-mixed but which changes in size during the

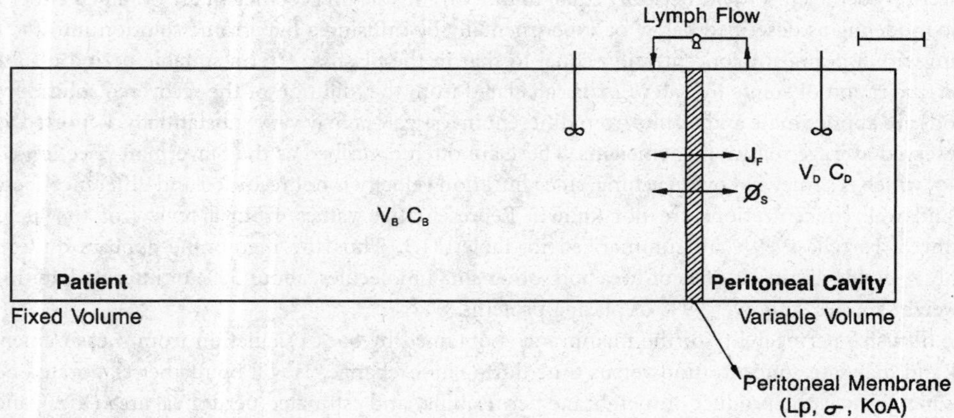

FIGURE 131.6 Single pool model for peritoneal dialysis. Solute diffuses across a planar selective membrane from a large well-mixed plasma space at constant volume and concentration to a smaller well-mixed space in which concentration and volume both increase with time. Fluid and solute are selectively ultrafiltered across the peritoneal membrane from plasma to dialysate; they are also nonselectively transported by the lymphatics from the dialysate to the body compartment.

course of exchange. These two compartments are separated by a planar membrane capable of supporting bidirectional transport and characterized by the terms *Lp, R,* and K_oA previously defined by Eqs. (131.3)–(131.5). Fluid drains from the peritoneum to the blood at a rate of Q_L. From this point forward, the complexity and appropriate utility of the models depend upon the investigators' choices of simplifying assumptions. The simplest model, proposed by Henderson and Nolph [1969], considers ultrafiltration rate and lymphatic flow to be negligible and treats all parameters except dialysate concentration as constant with time. The basic differential equation describing this model is

$$\frac{d\left(V_D C_D\right)}{dt} = V_D \frac{dC_D}{dt} = K_0 A\left(C_B - C_D\right) \tag{131.6}$$

Equation 131.6 may be readily solved, either to obtain K_oA from a knowledge of concentration versus time data Eq. (131.7), or to predict dialysate concentration from a knowledge of mass transfer coefficient, blood concentration, and initial dialysate concentration Eq. (131.8) where:

$$K_0 A = \frac{V_D}{t} \ln\left(\frac{C_B - C_D^0}{C_B - C_D^t}\right) \tag{131.7}$$

$$C_D^t = C_B - \left(C_B - C_D^0\right) e^{\frac{-K_0 A t}{V_D}} \tag{131.8}$$

In these equations, the superscript *t* represents the value at time *t*, and the superscript *0* designates the value at $t = 0$. This model provides a very easy way of measuring K_0A if it is applied during the isovolemic interval that often occurs ~ 30–90 min after the beginning of an exchange.

Several years later, investigators at the University of New South Wales [Garred et al., 1983] proposed a slightly more complex model that included ultrafiltration, subject to the assumptions that: (1) blood concentration was constant, (2) the membrane was nonselective (R = O), and (3) lymphatic involvement could be ignored. The appropriate differential equation is now:

$$\frac{d\left(V_D C_D\right)}{dt} = K_0 A\left(C_B - C_D\right) + C_B \frac{dV_D}{dt} \tag{131.9}$$

This equation can be solved in two ways. Over either relatively short time intervals or small differences in dialysate volume, an average volume V_D is obtained as the mean of initial and final volumes. In that case K_0A is given by

$$K_0 A = \frac{\overline{V}_D}{t} \ln\left[\frac{V_D^o\left(\overline{C}_B - C_D^o\right)}{V_D^t\left(\overline{C}_B - C_D^t\right)}\right] \tag{131.10}$$

where variables overlined with a solid diachrin are treated as constant during the integration of Eq. (131.9). The similarity of Eqs (131.9) and (131.10) to Eqs. (131.7) and (131.8) should be noted. Where a series of data points for blood and dialysate concentrations are available at various times during the treatment, Eq. (131.10) may be rewritten as

$$C_D^t = \overline{C}_B - \frac{V_D^o}{V_D^t}\left(\overline{C}_B - C_D^o\right) e^{\left(\frac{-K_0 A t}{V_D}\right)} \tag{131.11}$$

$$\ln\left[V_D^t\left(\overline{C}_B - C_D^t\right)\right] = \ln\left[V_D^o\left(\overline{C}_B - C_D^o\right)\right] - \frac{K_o A t}{V_D} \tag{131.12}$$

Data in the form of this equation may be readily regressed to obtain K_0A from a knowledge of V_D, C_B, and C_D at various times in an exchange. The values for peritoneal volume V_D may be obtained experimentally from tracer dilution studies, calculated from an algorithm, in which case it varies with time, or simply averaged between initial and final values, in which case it is assumed constant. Equations (131.11) and (131.12) are recommended for routine modeling of patient kinetics.

Several investigators, reviewed by Lysaght and Farrell [1989], have produced far more elaborate models which incorporate lymphatic drainage, deviations from ideal semipermeability of the peritoneal membrane, time-dependent ultrafiltration rates, and coupling between bidirectional diffusive and connective transport. Although potent in the hands of their developers, none of the numerical models has been widely adopted, and the current trend is toward simpler approaches. In comparative studies [Lysaght, 1989; Waniewski et al., 1991], only small differences were found between the numeric values of transport parameters calculated from simple analytic models [Eqs. (131.6)–(131.12)] and those we obtained by far more complex numerical methods. In peritoneal dialysis, solute is being exchanged through an inefficient membrane between a large body compartment through an inefficient membrane and a second compartment only 5% as large, and treatment times have been chosen so that the smaller compartment will reach saturation. These physical circumstances, and the very forgiving nature of exponential asymptotes, perhaps explain why simple analytic solutions perform nearly as well as their more complex numeric counterparts.

131.5 Emerging Developments

Modified therapy formats and new formulations for exchange solutions constantly are being proposed and evaluated. APD is the most successful of the new formats; at the end of 1993, about one in six peritoneal dialysis patients received some variant of automated overnight treatment. APD is carried out by a small console (Fig. 131.7) which automatically instills and drains dialysate at 1.5–3-hour intervals while the patient sleeps, typically over 8–10 hours each night. The peritoneum is left full during the day. Since the short exchanges do not permit complete equilibration even for urea, the process is somewhat wasteful of dialysate. However, reference to Fig. 131.3 will readily demonstrate that small-solute removal is most efficient in the early portion of an exchange; for example, two 2-hour exchanges will provide 75% more urea clearance than one 4-hour exchange. As currently prescribed, APD requires 84–105 L per week of dialysate (versus 56 for CAPD) and increases total small-solute clearance per 24 hours by up to 50% over that achieved by CAPD. The number of patients on APD is increasing by half every year, a phenomenon driven by two main factors. The first relates to quality of life; APD is far-and-away the least invasive of the maintenance dialysis protocols. The patient performs one connection at night and one disconnection in the morning and is thereby freed from the tedium and inconvenience of daily exchanges or the need to spend a significant portion of 3 days per week at an HD treatment facility. In addition, small-solute clearance is higher than in other continuous peritoneal therapies, which helps address increasing concern about the adequacy of the standard four 2-L CAPD exchanges per day, especially with large muscular patients and those with no residual renal function. A group of patients who may benefit from APD are those who have rapid transport of glucose across the peritoneal membrane; because of the consequent loss of the osmotic gradient, they have difficulty achieving adequate ultrafiltration. The short dwell times of ADP circumvent this problem. The counterbalancing disadvantage of APD is increased expense associated with the larger fluid consumption and the fluid cyclers.

Virtually all solution development comprises attempts to replace glucose with an alternative osmotic agent, preferably one which diffuses more slowly and thus provides a more stable osmotic gradient and one which obviates the obligatory load of about 600 calories of sugar. However glucose is cheap and safe, and it will be difficult to find a satisfactory alternative. A competing osmotic agent must be safe to

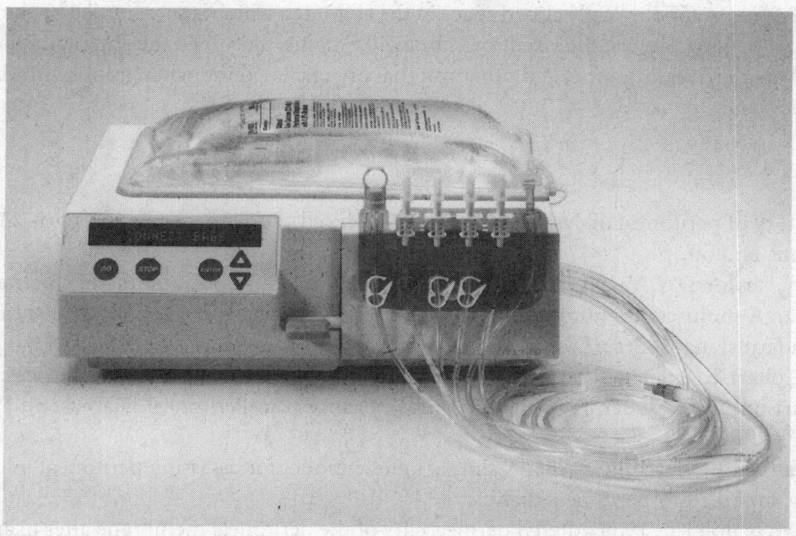

FIGURE 131.7 Contemporary equipment module for APD (Home Choice, Renal division, Baxter Healthcare) which automatically controls and monitors the delivery of 10–15 L of dialysate from 5-L bags via a multipronged disposable administration set. The console incorporates a diaphragm pump used to emulate gravity, and a derivative of the ideal gas law measures fluid volume, eliminating the need for scales. Setup and operation are designed to be straightforward and convenient.

administer in amounts of tens of grams per day over years to patients who have little or no ability to clear accumulated material via the kidney—but an osmotic agent which is readily metabolizable provides no caloric "advantage" over glucose. A glucose polymer, termed polyglucose, has been recently introduced in England [Mistry & Gokal, 1993]. This disperse oligodextrin has a weight-averaged MW of 18,700 daltons and number-averaged molecular weight of 7300 daltons. At a concentration of 7.5% (i.e., 30 g per 2-L exchange), it provides more stable ultrafiltration during long dwell exchanges; however, administration is limited to one exchange per day because of the accumulation of maltose and higher MW polysaccharides; an alternative approach, recently introduced in Europe and in clinical trials in the United States, is a solution in which glucose is replaced with 1.1% amino acids, enriched for essential amino acids [Jones et al., 1992]. This solution also improves nitrogen balance, a significant feature, since dialysate patients are frequently malnourished. Concern about excessive nitrogen intake, however, limits its use to one or two exchanges per day, and the amino acid solution is necessarily more expensive than glucose.

Defining Terms

Automated peritoneal dialysis (APD): A recent variant of CAPD in which fluid exchanges are performed by simple pumps, usually at night while the patient sleeps.

Clearance: The rate of mass removal divided by solute concentration in the body. Clearance represents the virtual volume of blood or plasma cleared of a particular solute per unit time.

Continuous ambulatory peritoneal dialysis (CAPD): A continuous process for the treatment of chronic renal failure in which metabolic waste products and excess body water are removed through the peritoneum with four exchanges of up to 3 L every 24 hours.

Diffusion: The molecular movement of matter from a region of greater concentration to lesser concentration at a rate proportional to the difference in concentration.

Hemodialysis (HD): Intermittent extracorporeal therapy for chronic renal failure. See Chapter 130.

Mass transfer coefficient: The proportionality constant between the rate of solute transport per unit area and the driving force.

Membrane: A thin barrier capable of providing directional selective transport between two phases.

Peritoneal cavity: A topologically closed space in the abdomen which is surrounded by the peritoneum.

Peritoneum: An intricate, vascularized, membranelike tissue that lines the internal abdominal walls and covers the liver, intestine, and other internal organs. Used interchangeably with the expression *peritoneal membrane.*

References

Boen ST. History of peritoneal dialysis.1985. In KD Nolph (ed), Peritoneal Dialysis, pp 1–22, The Hague, Martinus Nijhoff.

Churchill DN, Taylor DW, Vas SI, et al. 1989. Peritonitis in continuous ambulatory peritoneal dialysis (CAPD): A multi-centre randomized clinical trial comparing the Y connector disinfectant system to standard systems. Perit Dial Int 19:159.

Dratwa M, Collart F, Smet L. 1986. CAPD peritonitis and different connecting devices: A statistical comparison. In JF Maher, JF Winchester (eds), Frontiers in Peritoneal Dialysis, pp 190–197, New York, Field Rich.

Garred LJ, Canaud B, Farrell PC. 1983. A simple kinetic model for assessing peritoneal mass transfer in chronic ambulatory peritoneal dialysis. ASAIO J 6:131.

Henderson LW, Nolph KD. 1969. Altered permeability of the peritoneal membrane after using hypertonic peritoneal dialysis fluid. J Clin Invest 48:992.

Jones MR, Martis L, Algrim CE, et al. 1992. Amino acid solutions for CAPD: Rationale and clinical experience. Miner Electrolyte Metab 18:309.

Lysaght MJ. 1989. The Kinetics of Continuous Peritoneal Dialysis. PhD thesis, Center for Biomedical Engineering, University of New South Wales.

Lysaght MJ, Farrell PC. 1989. Membrane phenomena and mass transfer kinetics in peritoneal dialysis. J Mem Sci 44:5.

Mistry CD, Gokal R. 1993. Single daily overnight (12-h dwell) use of 7.5% glucose polymer (Mw 18700; Mn 7300) + 0.35% glucose solution: A 3-month study. Nephrol Dial Transplant 8:443.

Popovich RP, Moncrief JW, Decherd JF, et al. 1976. The definition of a novel portable/wearable equilibrium peritoneal technique. Abst AM Soc Artif Intern Organs 5:64.

Vonesh EF, Lysaght MJ, Moran J, et al. 1991. Kinetic modeling as a prescription aid in peritoneal dialysis. Blood Purif 9:246.

Waniewski J, Werynski A, Heimburger O, et al. 1991. A comparative analysis of mass transport models in peritoneal dialysis. ASAIO Trans 37:65.

Further Information

The literature on continuous peritoneal dialysis is abundant. Among several reference texts the most venerable and popular is *Peritoneal Dialysis* edited by K. Nolph and published by Kluwer; this is regularly updated. Also recommended is *Continuous Ambulatory Peritoneal Dialysis* edited by R Gokal and published by Churchill Livingston. The journal *Peritoneal Dialysis International* (published by MultiMed; Toronto) is published quarterly and is devoted exclusively to CAPD. The continuing education department of the University of Missouri-Columbia organizes a large annual conference on peritoneal dialysis with plenary lecture and submitted papers. The International Society of Peritoneal Dialysis holds its conference biannually and usually publishes proceedings. Peritoneal dialysis is also discussed in the meeting and journals of the other major artificial organ societies (American Society of Artificial Internal Organs; European Dialysis and Transplant Association; Japanese Society of Artificial Organs) and the American and International Societies of Nephrology. *Blood Purification* (published by Karger; Basel) attracts many outstanding papers dealing with engineering and transport issues in peritoneal dialysis. For the insatiable, Medline now contains over 10,000 citations to CAPD and peritoneal dialysis.

132

Therapeutic Apheresis and Blood Fractionation

Andrew L. Zydney
University of Delaware

Apheresis is the process in which a specific component of blood (either plasma, a plasma component, white cells, platelets, or red cells) is separated and removed with the remainder of the blood returned to the patient (often in combination with some type of replacement fluid). Donor apheresis is used for the collection of specific blood cells or plasma components from blood donors, resulting in a much more effective use of limited blood-based resources. Donor apheresis developed during World War II as a means for increasing the supply of critically needed plasma, and clinical trials in 1944 demonstrated that it was possible to safely collect donations of a unit of plasma (approximately 300 ml) on a weekly basis if the cellular components of blood were returned to the donor. Therapeutic apheresis is used for the treatment of a variety of diseases and disorders characterized by the presence of abnormal proteins or blood cells in the circulation which are believed to be involved in the progression of that particular condition. Therapeutic apheresis thus has its roots in the ancient practice of bloodletting, which was used extensively well into the 19th century to remove "bad humors" from the patient's body, thereby restoring the proper balance between the "blood, yellow bile, black bile, and phlegm."

The term *plasmapheresis* was first used by Abel, Rowntree, and Turner in 1914 in their discussion of a treatment for toxemia involving the repeated removal of a large quantity of plasma, with the cellular components of blood returned to the patient along with a replacement fluid [Kambic & Nosé, 1993]. The first successful therapeutic applications of plasmapheresis were reported in the late 1950s in the management of macroglobulinemia (a disorder characterized by a large increase in blood viscosity due to the accumulation of high-molecular-weight globulins in the blood) and in the treatment of multiple myeloma (a malignant tumor of the bone marrow characterized by the production of excessive amounts of immunoglobulins).

By 1990, there were well over 50 diseases treated by therapeutic apheresis [Sawada et al., 1990] with varying degrees of success. Plasmapheresis is used in the treatment of: (1) protein-related diseases involving excessive levels of specific proteins (e.g., macroglobulins in Waldenstrom's syndrome and lipoproteins in familial hypercholesterolemia) or excessive amounts of protein-bound substances (e.g., toxins in hepatic failure and thyroid hormone in thyrotoxicosis), (2) antibody-related or *autoimmune diseases* (e.g., glomerulonephritis and myasthenia gravis), and (3) immune-complex-related diseases (e.g., rheumatoid arthritis and systemic lupus erythematosus).

0-8493-0462-8/00/$0.00+$.50

Cytapheresis involves the selective removal of one (or more) of the cellular components of blood, and it has been used in the treatment of certain leukemias (for the removal of leukocytes) and in the treatment of polycythemia. Table 132.1 provides a more complete listing of some of the diseases and blood components that are removed during therapeutic apheresis. This list is not intended to be exhaustive, and there is still considerable debate over the actual clinical benefit of apheresis for a number of these diseases.

The required separation of blood into its basic components (red cells, white cells, platelets, and plasma) can be accomplished using centrifugation or membrane filtration; the more specific removal of one (or more) components from the separated plasma generally involves a second membrane filtration or use of an appropriate sorbent. The discussion that follows focuses primarily on the technical aspects of the different separation processes currently in use. Additional information on the clinical aspects of therapeutic apheresis is available in the references listed at the end of this chapter.

132.1 Plasmapheresis

The therapeutic application of plasmapheresis can take one of two forms: plasma exchange or plasma perfusion. In *plasma exchange* therapy, a relatively large volume of plasma, containing the toxic or immunogenic species, is separated from the cellular components of blood and replaced with an equivalent volume of a replacement fluid (either fresh frozen plasma obtained from donated blood or an appropriate plasma substitute). In *plasma perfusion*, the separated plasma is treated by an adsorptive column or second membrane filtration to remove a specific component (or components) from the plasma. This treated plasma is then returned to the patient along with the blood cells, thereby eliminating the need for exogenous replacement fluids. The different techniques that can be used for plasma perfusion are discussed subsequently.

The reduction in the concentration (C_i) of any plasma component during the course of a plasmapheresis treatment can be described using a single compartment pharmacokinetic model as

$$V_p \frac{dC_i}{dt} = -\alpha_i Q_p C_i + G_i \tag{132.1}$$

where V_p is the volume of the patient's plasma (which is assumed to remain constant over the course of therapy through the use of a replacement fluid or through the return of the bulk of the plasma after a plasma perfusion), Q_p is the volumetric rate of plasma collection, G_i is the rate of component generation, and α_i is a measure of the effectiveness of the removal process. In membrane plasmapheresis, α_i, is equal to the observed membrane sieving coefficient, which is defined as the ratio of the solute concentration in the filtrate collected through the membrane to that in the plasma entering the device; α_i is thus equal to 1 for a small protein that can pass unhindered through the membrane but can be less than 1 for large proteins and immune complexes. In plasma perfusion systems, α_i is equal to the fraction of the particular component removed from the collected plasma by the secondary (selective) processing step. The generation rate is typically negligible over the relatively short periods (fewer than 3 hours) involved in the actual plasmapheresis; thus the component concentration at the end of a single treatment is given as

$$\frac{C_i}{C_{io}} = \exp\left(-\frac{\alpha_i Q_p t}{V_p}\right) = \exp\left(-\frac{\alpha_i V_{exc}}{V_p}\right) \tag{132.2}$$

where $V_{exc} = Q_p t$ is the actual volume of plasma removed (or exchanged) during the process. Plasma exchange thus reduces the concentration of a given component by 63% after an exchange of one plasma volume (for $\alpha_i = 1$) and by 86% after two plasma volumes. This simple single compartment model has been verified for a large number of plasma components, although immunoglobulin G actually appears to have about 50% extravascular distribution with the re-equilibration between these compartments occurring within 24–28 hr following the plasmapheresis.

TABLE 132.1 Disease States Treated by Therapeutic Apheresis

Disease	Components Removed
Hematologic	
Hemophilia	AntiFactor VIII Ab
Idiopathic Thrombocytopenia Purpura	Antiplatelet Ab, immune complexes
Thrombotic Thrombocytopenia Purpura	Antiplatelet Ab, immune complexes
AIDS, HIV	Antilymphocyte Ab, immune complexes
Autoimmune Hemolytic Anemia	Anti-red cell Ab, red cells
Rh Incompatibility	Anti-Rh Ab
Cryoglobulinemia	Cryoglobulins
Hyperviscosity Syndrome	Macroglobulins, immunglobulin M
Waldenstrom's Syndrome	Immunoglobulin M
Paraproteinemia	Paraproteins
Sickle Cell Anemia	Red blood cells
Thrombocythemia	Platelets
Collagen/rheumatologic	
Systemic Lupus Erythematosus	Anti-DNA Ab, immune complexes
Progressive Systemic Sclerosis	Antinonhistone nuclear Ab
CREST Syndrome	Anticentromere Ab
Sjorgen Syndrome	Antimitochondrial Ab
Rheumatoid Arthritis	Rheumatoid factor, cryoglobulins, immunoglobulins
Periarteritis Nodosa	Cryoglobulins, immune complexes
Raynaud's Disease	Cryoglobulins, macroglobulins
Scleroderma	Immune complexes
Mixed Connective Tissue Diseases	Immunoglobulins
Neurologic	
Myasthenia Gravis	Antiacetylcholine receptor Ab, cryoglobulins
Multiple Sclerosis	Antimyelin Ab
Guillain-Barre Syndrome	Antimyelin Ab
Polyneuropathy	Cryoglobulins, macroglobulins
Polyradiculoneuropathy	Antibodies
Lambert-Eaton Syndrome	Antibodies
Hepatic	
Chronic Active Hepatitis	Antimitochondrial Ab
Hepatic Failure	Protein-bound toxins
Primary Biliary Cirrhosis	Protein-bound toxins, antimitochondrial Ab
Renal	
Goodpasture's Syndrome	Antiglomerular basement membrane Ab
Glomerulonephritis	Immune complexes
Lupus Nephritis	Immune complexes
Transplant Rejection	Immune complexes, anti-HLA Ab
Malignant diseases	
Cancer	Tumor-specific Ab, immune complexes
Multiple Myeloma	Immunoglobulins
Leukemia	Leukocytes
Miscellaneous	
Addison's Disease	Antiadrenal Ab
Autoimmune Thyroiditis	Antimicrosomal Ab
Chronic Ulcerative Colitis	Anticolonic epithelial cell Ab
Diabetes Mellitus	Antiinsulin receptor Ab
Hashimoto's Disease	Antithyroglobulin Ab
Insulin Autoimmune Syndrome	Antiinsulin Ab
Pemphigus	Antiepidermal cell membrane Ab
Ulcerative Colitis	Anticolonic lipopolysaccharide Ab
Asthma	Immunoglobulin E
Hypercholesterolemia	Cholesterol, lipoproteins
Hyperlipidemia	Low- and very low density lipoproteins
Thyrotoxicosis	Thyroid hormone

There is still considerable variability in the frequency and intensity of the plasmapheresis used in different therapeutic applications, and this is due in large part to uncertainties regarding the metabolism, pharmacokinetics, and pathogenicity of the different components that are removed during therapeutic apheresis. The typical plasma exchange therapy currently involves the removal of 2–3 L of plasma (approximately one plasma volume) at a frequency of 2–4 times per week, with the therapy continued for several weeks. There have also been a number of studies of the long-term treatment of several diseases via plasmapheresis, with the therapy performed on a periodic basis (ranging from once per week to once every few months) over as much as 5 years (e.g., for the removal of cholesterol and lipoproteins in the treatment of severe cases of hypercholesterolemia).

Centrifugal Devices

Initially, all plasmapheresis was performed using batch centrifuges. This involved the manual removal of approximately one unit (500 ml) of blood at a time, with the blood separated in a centrifuge so that the target components could be removed. The remaining blood was then returned to the patient before drawing another unit and repeating the entire process. This was enormously time-consuming and labor-intensive, requiring as much as 5 hours for the collection of only a single liter of plasma. Batch centrifugation is still the dominant method for off-line blood fractionation in most blood-banking applications, but almost all therapeutic plasmapheresis is performed using online (continuous) devices.

The first continuous flow centrifuge was developed in the late 1960s by IBM in conjunction with the National Cancer Institute, and this basic design was subsequently commercialized by the American Instrument Co. (now a division of Travenol) as the Aminco Celltrifuge [Nosé et al., 1983]. A schematic diagram showing the general configuration of this, and most other continuous flow centrifuges is shown in Fig. 132.1. The blood is input at the bottom of the rotating device and passes through a chamber in

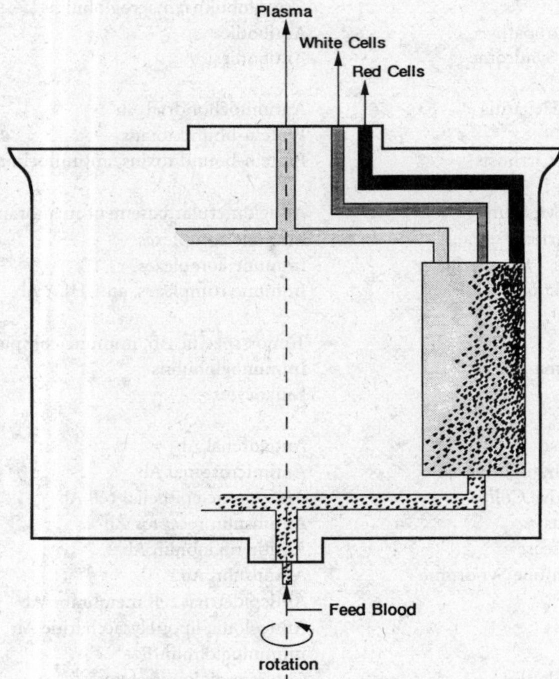

FIGURE 132.1 Schematic diagram of a generic continuous flow centrifuge for fractionation of blood into red cells, white cells/platelets, and plasma.

which the actual separation into plasma, white cells/platelets, and red cells occurs. Three separate exit ports are located at different radial positions to remove the separated components continuously from the top of the chamber using individual roller pumps. The position of the buffy coat layer (which consists of the white cells and platelets) is controlled by adjusting the centrifugal speed and the relative plasma and red cell flow rates to obtain the desired separation. Probably the most difficult engineering problem in the development of the continuous flow centrifuge was the design of the rotating seals through which the whole blood and separated components must pass without damage. The seal design in the original NCI/IBM device used saline lubrication to prevent intrusion of the cells between the contacting surfaces.

In order to obtain effective cell separation in the continuous flow centrifuge, the residence time in the separation chamber must be sufficiently large to allow the red cells to migrate to the outer region of the device. The degree of separation can thus be characterized by the packing factor

$$P = \frac{G V_{sed} t}{h} \tag{132.3}$$

where G is the g-force associated with the centrifugation, V_{sed} is the sedimentation velocity at 1 g, t is the residence time in the separation chamber, and h is the width of the separation chamber (i.e., the distance over which the sedimentation occurs). The packing factor thus provides a measure of the radial migration compared to the width of the centrifuge chamber, with adequate cell separation obtained when $P > 1$. The residence time in the separation chamber is inversely related to the blood flow rate (Q_B) as

$$t = \frac{AL}{Q_B} \tag{132.4}$$

where L and A are the length and cross-sectional area of the chamber, respectively.

Rotational speeds in most continuous flow centrifuges are maintained around 1500 rpm (about 100 g) to obtain a relatively clean separation between the red cells and buffy coat, to avoid the formation of a very highly packed (and therefore highly viscous) region of cells at the outer edge of the chamber, and to minimize excessive heating around the rotating seals [Rock, 1983]. The width of the separation chamber must be large enough to permit effective removal of the different blood components from the top of the device, while minimizing the overall extracorporeal blood volume. For example, a packing factor of 10 requires a residence time of about 20 s for a device operated at 100 g with a 0.1-cm-wide separation chamber. The blood flow rate for this device would need to be maintained below about 120 ml/min for a chamber volume of 40 ml. Most of the currently available devices operate at $Q_B \approx 50$ ml/min and can thus collect a liter of plasma in about 30–40 minutes.

More recent models for the continuous flow centrifuge have modified the actual geometry of the separation chamber to enhance the cell separation and reduce the overall cost of the device [Sawada et al., 1990]. Examples include tapering the centrifuge bowl to improve flow patterns and optimizing the geometry of the collection region to obtain purer products. The IBM 2997 (commercialized by Cobe Laboratories) uses a disposable semirigid plastic rectangular channel for the separation chamber, which eliminates some of the difficulties involved in both sterilizing and setting up the device. Fenwal Laboratories developed the CS-3000 Cell Separator (subsequently sold by Baxter Healthcare), which uses a continuous J-shaped mulitchannel tubing connected directly to the rotating element. This eliminates the need for a rotating seal, thereby minimizing the possibility of leaks. The tubing in this device actually rotates around the centrifuge bowl during operation, using a "jump rope" principle to prevent twisting of the flow lines during centrifugation.

In addition to the continuous flow centrifuge, Haemonetics has developed a series of intermittent flow centrifuges [Rock, 1983]. Blood flows into the bottom of a separation chamber similar to that shown in Fig. 132.1, but the red cells simply accumulate in this chamber while the plasma is drawn to the center of the rotating bowl and removed through an outlet port at the top of the device. When the process is complete, the pump action reverses, and the red cells are forced out of the bowl and reinfused into the patient (along with any replacement fluid). The entire process is then automatically repeated to obtain the desired level of plasma (or white cell) removal. This device was originally developed for the collection of leukocytes and platelets, but it is now used extensively for large-scale plasmapheresis as well. Maximum blood flow rates are typically about 70–80 ml/min, and the bowl rotates at about 5000 rpm. The device as a whole is more easily transported than most of the continuous flow centrifuges, but it requires almost 50% more time for the collection of an equivalent plasma volume due to the intermittent nature of the process. In addition, the total extracorporeal volume is quite high (about 500 ml compared to only 250 ml for most of the newer continuous flow devices) due to the larger chamber required for the red cell accumulation [Sawada et al., 1990].

All these centrifugal devices have the ability to carry out effective plasma exchange, although the rate of plasma collection tends to be somewhat slower than that for the membrane devices discussed in the next section. These devices can also be used for the collection of specific cell fractions, providing a degree of flexibility that is absent in the membrane systems. The primary disadvantage of the centrifugal units is the presence of a significant number of platelets in the collected plasma (typically about 10^5 platelets per μl). Not only can this lead to considerable platelet depletion during repeated applications of plasma exchange, it can also interfere with many of the secondary processing steps employed in plasma perfusion.

Membrane Plasmapheresis

The general concept of blood filtration using porous membranes is quite old, and membranes with pores suitably sized to retain the cellular components of blood and pass the plasma proteins have been available since the late 1940s. Early attempts at this type of blood filtration were largely unsuccessful due to severe problems with membrane plugging (often referred to as fouling) and red cell lysis. Blatt and coworkers at Amicon recognized that these problems could be overcome using a cross-flow configuration [Solomon et al., 1978] in which the blood flow was parallel to the membrane and thus perpendicular to the plasma (filtrate) flow as shown schematically in Fig. 132.2. This geometry minimizes the accumulation of retained cells at the membrane surface, leading to much higher filtration rates and much less cell damage than could be obtained in conventional dead-end filtration devices [Solomon et al., 1978]. This led to the development of a large number of membrane devices using either flat sheet or hollow-fiber membranes made from a variety of polymers including polypropylene (Travenol Laboratories, Gambro, Fresenius), cellulose diacetate (Asahi Medical), polyvinyl alcohol (Kuraray), polymethylmethacrylate (Toray), and polyvinyl chloride (Cobe Laboratories).

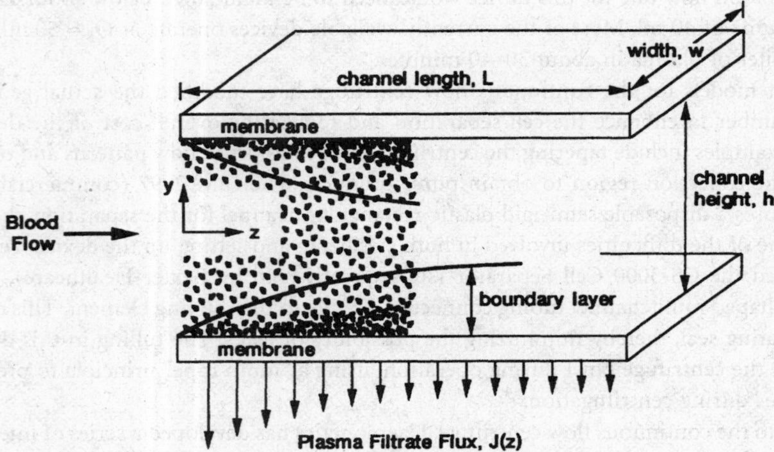

FIGURE 132.2 Schematic diagram of a parallel plate device for cross-flow membrane plasmapheresis.

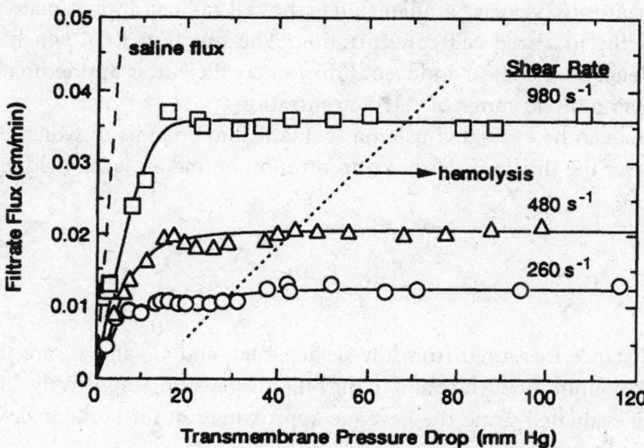

FIGURE 132.3 Experimental data for the filtrate flux as a function of the applied transmembrane pressure drop in a parallel plate membrane plasmaphersis device. Red cell hemolysis, defined as a filtrate hemoglobin concentration exceeding 20 mg/dl, occurs to the right of the dashed line. Data have been adapted from Zydney and Colton [1987].

These membrane devices all produce essentially cell-free plasma with minimal protein retention using membranes with pore sizes of 0.2–0.6 μm. In addition, these devices must be operated under conditions that cause minimal red cell lysis, while maintaining a sufficiently high plasma filtrate flux to reduce the cost of the microporous membranes. Typical experimental data for a parallel plate membrane device are shown in Fig. 132.3 [Zydney & Colton, 1987]. The results are plotted as a function of the mean transmembrane pressure drop (ΔP_{TM}) at several values of the wall shear rate (γ_w), where γ_w is directly proportional to the inlet blood flow rate (Q_B). The filtrate flux initially increases with increasing ΔP_{TM} reaching a maximum pressure independent value which increases with increasing shear rate. The flux under all conditions is substantially smaller than that obtained when filtering pure (cell-free) saline under identical conditions (dashed line in Fig. 132.3). No measurable hemolysis was observed at low ΔP_{TM} even when the flux was in the pressure-independent regime. Hemolysis does become significant at higher pressures, with the extent of hemolysis decreasing with increasing γ_w and with decreasing membrane pore size [Solomon et al., 1978]. The dashed diagonal line indicates the pressure at which the filtrate hemoglobin concentration exceeds 20 mg/dl.

The pressure-independent flux at high ΔP_{TM} generally has been attributed to the formation of a concentration polarization boundary layer consisting of a high concentration of the formed elements in blood, mainly the red blood cells, which are retained by the microporous membrane (Fig. 132.2). This dynamic layer of cells provides an additional hydraulic resistance to flow, causing the flux to be substantially smaller than that obtained during filtration of a cell-free solution. At steady-state, this boundary layer is in dynamic equilibrium, with the rate of convection of formed elements toward the membrane balanced by the rate of mass transport back into the bulk suspension. There has been some debate in the literature over the actual mechanism of cell transport in these devices, and the different models that have been developed for the plasma flux during membrane plasmapheresis are discussed elsewhere [Zydney & Colton, 1986].

Zydney and Colton [1982] proposed that cell transport occurs by a shear-induced diffusion mechanism in which cell-cell interactions and collisions give rise to random cell motion during the shear flow of a concentrated suspension. This random motion can be characterized by a shear-included diffusion coefficient, which was evaluated from independent experimental measurements as

$$D = a^2 \gamma f\left(C\right) \tag{132.5}$$

where γ is the local shear rate (velocity gradient), a is the cell radius (approximately 4.2 μm for the red blood cells), and C is the local red cell concentration. The function $f(C)$, which reflects the detailed concentration dependence of the shear-induced diffusion coefficient, is approximately equal to 0.03 for red cell suspensions over a broad range of cell concentrations.

The local filtrate flux can be evaluated using a stagnant film analysis in which the steady-state mass balance is integrated over the thickness of the concentration boundary layer yielding [Zydney & Colton, 1986]

$$J(z) = k_m \ln \frac{C_w}{C_b} \tag{132.6}$$

where z is the axial distance measured from the device inlet, and C_w and C_b are the concentrations of formed elements at the membrane surface and in the bulk suspension, respectively. The bulk mass transfer coefficient (k_m) can be evaluated using the Leveque approximation for laminar flow in either a parallel plate or hollow fiber device

$$k_m = 0.516\left(\frac{D^2\gamma_w}{z}\right)^{1/3} = 0.047\left(\frac{a^4}{z}\right)^{1/3}\gamma_w \tag{132.7}$$

where the second expression has been developed using the shear-induced diffusion coefficient given by Eq (132.5) with $f(C) = 0.03$. The wall shear rate is directly proportional to the blood flow rate with

$$\gamma_w = \frac{4Q_B}{N\pi R^3} \tag{132.8}$$

for a hollow-fiber device with N fibers of inner radius R and

$$\gamma_w = \frac{6Q_B}{wh^2} \tag{132.9}$$

for a parallel plate device with channel height h and total membrane width w.

At high pressures, C_w approaches its maximum value which is determined by the maximum packing density of the cells (about 95% under conditions typical of clinical plasmapheresis). The plasma flux under these conditions becomes independent of the transmembrane pressure drop, with this pressure-independent flux varying linearly with the wall shear rate and decreasing with increasing bulk cell concentration as described by Eqs. (132.6) and (132.7). A much more detailed numerical model for the flux [Zydney & Colton, 1987], which accounts for the concentration and shear rate dependence of both the blood viscosity and shear-induced cell diffusion coefficient as well as the compressibility of the blood cell layer that accumulates at the membrane, has confirmed the general behavior predicted by Eqs. (132.6) and (132.7).

The volumetric filtrate (plasma) flow rate (Q_p) in a hollow-fiber membrane filter can be evaluated by integrating Eqs. (132.6)–(132.8) along the length of the device accounting for the decrease in the blood flow rate (and thus γ_w) due to the plasma removal [Zydney & Colton, 1986]. The resulting expression for the fractional plasma yield is

$$\frac{Q_p}{Q_B} = 1 - \exp\left[-0.62\left(\frac{a^2 L}{R^3}\right)^{2/3} \ln \frac{C_w}{C_b}\right] \tag{132.10}$$

An analogous expression can be evaluated for a parallel plate device with the channel height h replacing R and the coefficient 0.62 becoming 0.84. Even though the development leading to Eq. (132.10) neglects the detailed variations in the bulk cell concentration and velocity profiles along the fiber length, the final expression has been shown to be in good agreement with experimental data for the plasma flow rate in actual clinical devices [Zydney & Colton, 1986]. Equation (132.10) predicts that the volumetric plasma flow rate is independent of the number of hollow fibers (or the membrane width for a parallel plate membrane device), a result which is consistent with a number of independent experimental investigations.

According to Eq. (132.10), the plasma flow rate increases significantly with decreasing fiber radius. There are, however, a number of constraints on the smallest fiber radius that can actually be employed in these hollow-fiber membrane devices. For example, blood clotting and fiber blockage can become unacceptable in very narrow bore fibers. The blood flow in such narrow fibers also causes a very high bulk shear stress, which potentially can lead to unacceptable levels of blood cell damage (particularly for white cells and platelets). Finally, the hollow-fiber device must be operated under conditions which avoid hemolysis.

Zydney and Colton [1982] developed a model for red cell lysis during membrane plasmapheresis in which the red cells are assumed to rupture following their deformation into the porous structure of the membranes. A given red cell is assumed to lyse if it remains in a pore for a sufficient time for the strain induced in the red cell membrane to exceed the critical strain for cell lysis. Since the red cell can be dislodged from the pore by collisions with other cells moving in the vicinity of the membrane or by the fluid shear stress, the residence time in the pore will be inversely related to the wall shear rate.

The tension (σ) in the red cell membrane caused by the deformation in the pore is evaluated using Laplace's law [Zydney & Colton, 1987]

$$\sigma = \frac{\Delta P_{TM} R_p}{2} \tag{132.11}$$

where R_p is the pore radius. Hemolysis is assumed to occur at a critical value of the strain in the red cell membrane (S); thus the time required for lysis is given implicitly by

$$S = \sigma\, g(t) = \sigma \left\{0.0010 + 0.0012\left[1 - \exp(-8t)\right] + 4.5\times 10^{-6}\, t\right\} \tag{132.12}$$

The function $g(t)$ represents the temporal dependence of the lytic phenomenon and has been evaluated from independent experimental measurements [Zydney & Colton, 1982]. Cell lysis occurs when $S \geq 0.03$ in Eq. (132.12) where σ is given in dyne/cm and t is in sec. This simple model has been shown to be in good agreement with experimental data for red cell lysis during cross-flow membrane plasmapheresis [Zydney & Colton, 1987].

This physical model for red cell lysis implies that hemolysis can be avoided by operating at sufficiently high shear rates to reduce the residence time in the membrane pores. However, operation at high shear rates also causes the inlet transmembrane pressure drop to increase due to the large axial pressure drop associated with the blood flow along the length of the device [Zydney & Colton, 1982]:

$$\Delta P_{TM}(0) = \Delta P_{TM}(L) + 2\mu\gamma_w \frac{L}{R} \tag{132.13}$$

where $\Delta P_{TM}(0)$ and $\Delta P_{TM}(L)$ are the inlet and exit transmembrane pressure drops, respectively, and μ is the average blood viscosity. $\Delta P_{TM}(L)$ is typically maintained at a small positive value (about 20 mmHg) to ensure that there is a positive transmembrane pressure drop along the entire length of the device. Since the increase in $\Delta P_{TM}(0)$ with increasing γ_w has a greater effect on hemolysis than the reduction in

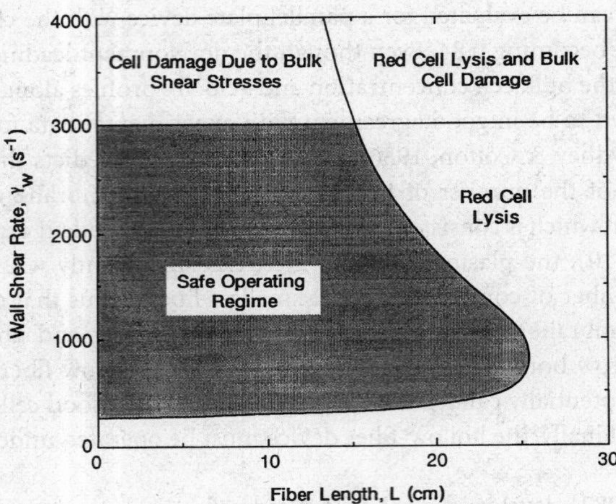

FIGURE 132.4 Schematic representation of the safe operating regime for a clinical membrane plasmapheresis device.

the residence time for the red cells in the membrane pores, there is also an upper bound on the shear rate for the safe operation of any given clinical device.

The predicted safe operating regime for a clinical membrane plasmapheresis device can be determined using Eq. (132.11), with the maximum transmembrane pressure drop occurring at the device inlet, Eq. (132.13). The results are shown schematically in Fig. 132.4. Hemolysis occurs at very low shear rates due to the long residence time in the membrane pores, whereas lysis at high shear rates is due to the large value of the inlet transmembrane pressure drop associated with the axial flow. Note that there is a critical fiber length [at fixed values of the fiber radius and $\Delta P_{TM}(L)$] above which there is no longer any safe operating condition.

To avoid some of the constraints associated with the design of both parallel plate and hollow-fiber membrane devices. Hemasciences developed a rotating membrane filter for use in both donor and therapeutic plasmapheresis. A nylon membrane is placed on an inner cylinder and rotated at about 3600 rpm inside a concentric outer cylindrical chamber using a magnetic coupling device. The rotating membrane causes a very high shear rate (on the order of 10,000 s^{-1}) in the narrow gap between the cylinders. However, these high shear rates do not result in a large axial pressure drop, as found in the parallel plate and hollow-fiber devices, due to the decoupling of the axial blood flow and the shear rate in this system (the shear is now due almost entirely to the membrane rotation). The fluid flow in this rotating cylinder system also leads to the development of fluid instabilities known as Taylor vortices, and these vortices dramatically increase the rate of cell mass transport away from the membrane and back into the bulk suspension. This leads to a dramatic increase in the plasma filtrate flux and a dramatic reduction in the required membrane area. The Autopheresis-C (the rotating filter currently sold by Baxter Healthcare) uses only 70 cm^2 of membrane, which is more than an order of magnitude less than that required in competitive hollow-fiber and parallel plate devices. The mathematical analysis of the plasma filtrate flux and the corresponding design equations for the rotating cylinder plasma filter are provided by Zeman and Zydney [1996].

132.2 Plasma Perfusion

In repeated applications of plasma exchange, it is necessary to use replacement fluids that contain proteins to avoid the risks associated with protein depletion. One approach to minimizing the cost of these protein-containing replacement fluids (either albumin solutions, fresh frozen plasma, or plasma protein fraction) is to use a saline or dextran solution during the initial stages of the process and to then switch to a protein-containing replacement fluid toward the end of the treatment. Alternatively, a number of techniques have been developed to selectively remove specific toxic or immunogenic components from the plasma, with this treated plasma returned to the patient along with the cellular components of blood. This effectively eliminates the need for any expensive protein-containing replacement fluids.

Plasma perfusion (also known as online plasma treatment) is typically performed using either membrane or sorbent-based systems. Membrane filtration separates proteins on the basis of size and is thus used to selectively remove the larger molecular weight proteins from albumin and the small plasma solutes (salts, sugars, amino acids, and so on). A variety of membranes have been employed for this type of plasma fractionation including cellulose acetate (Terumo), cellulose diacetate (Asahi Medical and Teijin), ethylene vinyl alcohol (Kuraray), and polymethylmetharcrylate (Toray). These membranes are generally hydrophilic to minimize the extent of irreversible protein adsorption, with pore sizes ranging from 100–600 Å depending on the specific objectives of the membrane fractionation.

The selectivity that can be obtained with this type of plasma filtration can be examined using available theoretical expressions for the actual sieving coefficient (S_a) for a spherical solute in a uniform cylindrical pore:

$$S_a = (1-\lambda)^2 \left[2-(1-\lambda)^2\right] \exp\left(-0.711\ \lambda^2\right) \tag{132.14}$$

where λ is the ratio of the solute to pore radius. Equation (132.14) is actually an approximate expression which has been shown to be in good agreement with more rigorous theoretical analyses. This expression for the actual sieving coefficient, where S_a is defined as the ratio of the protein concentration in the filtrate to that at the upstream surface of the membrane, is valid at high values of the plasma filtrate flux, since it implicitly assumes that the diffusive contribution to protein transport is negligible. To avoid excessive albumin loss, it is desirable to have $S_a > 0.8$, which can be achieved using a membrane with an effective pore size greater than about 160 Å (albumin has a molecular weight of 69,000 and a Stokes-Einstein radius of 36 Å). This membrane would be able to retain about 80% of the immunglobulin M (which has a molecular weight of about 900,000 and a Stokes-Einstein radius of 98 Å), but it would retain less than 40% of the immunglobulin G (with MW = 155,000 and a radius of 55 Å).

The protein retention obtained during an actual plasma filtration is substantially more complex than indicated by the above discussion. The polymeric membranes used in these devices actually have a broad distribution of irregularly shaped (noncylindrical) pores. Likewise, the proteins can have very different (nonspherical) conformations, and their transport characteristics also can be affected by electrostatic, hydrophobic, and van der Waals interactions between the proteins and the polymeric membrane, in addition to the steric interactions that are accounted for in the development leading to Eq. (132.14). Protein-protein interactions can also significantly alter the observed protein retention. Finally, the partially retained proteins will tend to accumulate at the upstream surface of the membrane during filtration (analogous to the concentration polarization effects described previously in the context of blood cell filtration).

This type of secondary plasma filtration, which is generally referred to in the literature as *cascade filtration*, is primarily effective at removing large immune complexes (molecular weight of approximately 700,000) and immunglobulin M (MW of 900,000) from smaller proteins such as albumin. Several studies have, however, found a higher degree of albumin-immunoglobulin G separation than would be expected based on purely steric considerations [Eq. (132.14)]. This enhanced selectivity is probably due to some type of long-range (e.g., electrostatic) interaction between the proteins and the membrane.

A number of different techniques have been developed to enhance the selectivity of these plasma filtration devices. For example, Malchesky and coworkers at the Cleveland Clinic [Malchesky et al., 1980] developed the process of cryofiltration in which the temperature of the plasma is lowered to about 10°C prior to filtration. A number of diseases are known to be associated with the presence of large amounts of cryo- (cold-) precipitable substances in the plasma, including a number of autoimmune diseases such as systemic lupus erythematosus and rheumatoid arthritis. Lowering the plasma temperature causes the aggregation and/or gelation of these cryoproteins, making it much easier for these components to be removed by the membrane filtration. About 10 g of cryogel can be removed in a single cryofiltration, along with significant amounts of the larger-molecular-weight immune complexes and IgM. The actual extent of protein removal during cyrofiltration depends on the specific composition of the plasma and thus on the nature as well as the severity of the particular disease state [Sawada et al., 1990]. There is thus considerable uncertainty over the actual components that are removed during cryofiltration under different clinical and/or experimental conditions. The cryogel layer that accumulates on the surface of the membrane also affects the retention of other plasma proteins, which potentially could lead to unacceptable losses even of small proteins such as albumin.

It is also possible to alter the selectivity of the secondary membrane filtration by heating the plasma up to or even above physiologic temperatures. This type of thermofiltration has been shown to increase the retention of low- (LDL) and very low (VLDL) density lipoproteins, and this technique has been used for the online removal of these plasma proteins in the treatment of hypercholesterolemia. LDL removal can also be enhanced by addition of a heparin/acetate buffer to the plasma, which causes precipitation of LDL and fibrinogen with the heparin [Sawada et al., 1990]. These protein precipitates can then be removed relatively easily from the plasma by membrane filtration. The excess heparin is subsequently removed from the solution by adsorption, with the acetate and excess fluid removed using bicarbonate dialysis.

An attractive alternative to secondary membrane filtration for the selective removal of plasma components is the use of sorbent columns such as: (1) activated charcoal or anion exchange resins for the removal of exogenous toxins, bile acids, and bilirubin; (2) dextran sulfate cellulose for the selective removal of cholesterol, LDL, and VLDL; (3) immobilized protein A for the removal of immunoglobulins (particularly IgG) and immune complexes; and (4) specific immobilized ligands like DNA (for the removal of anti-DNA Ab), tryptophan (for the removal of antiacetylcholine receptor antibodies), and insulin (for the removal of anti-insulin antibodies). These sorbents provide a much more selective separation than is possible with any of the membrane processes; thus they have the potential to significantly reduce the side effects associated with the depletion of needed plasma components. The sorbent columns generally are used in combination with membrane plasmaphersis, since the platelets that are present in the plasma collected from available centrifugal devices can clog the columns and interfere with the subsequent protein separation.

The development of effective sorbent technology for online plasma treatment has been hindered by the uncertainties regarding the actual nature of the plasma components that must be removed for the clinical efficacy of therapeutic apheresis in the treatment of different disease states. In addition, the use of biologic materials in these sorbent systems (e.g., protein A or immobilized DNA) presents particular challenges, since these materials may be strongly immunogenic if they desorb from the column and enter the circulation.

132.3 Cytapheresis

Cytapheresis is used to selectively remove one (or more) of the cellular components of blood, with the other components (including the plasma) returned to the patient. For example, leukocyte (white cell) removal has been used in the treatment of leukemia, autoimmune diseases with a suspected cellular immune mechanism (e.g., rheumatoid arthritis and myasthenia gravis), and renal allograft rejection. Erythrocyte (red cell) removal has been used to treat sickle cell anemia, severe autoimmune hemolytic anemia, and severe parasitemia. Plateletapheresis has been used to treat patients with thrombocythemia.

Most cytapheresis is performed using either continuous or intermittent flow centrifuges, with appropriate software and/or hardware modifications used to enhance the collection of the specific cell fraction. It is also possible to remove leukocytes from whole blood by depth filtration, which takes advantage of the strong adherence of leukocytes to a variety of polymeric materials (e.g., acrylic, cellulose acetate, polyester, or nylon fibers). Leukocyte adhesion to these fibers is strongly related to the configuration and the diameter of the fibers, with the most effective cell removal obtained with ultrafine fibers less than 3 μm in diameter. Available leukocyte filters (Sepacel, Cellsora, and Cytofrac from Asahi Medical Co.)

have packing densities of about 0.1–0.15 g fiber/cm^3 and operate at blood flow rates of 20–50 ml/min, making it possible to process about 2 L of blood in 1.5 hr.

Leukocyte filtration is used most extensively in blood-banking applications to remove leukocytes from the blood prior to transfusion, thereby reducing the likelihood of antigenic reactions induced by donor leukocytes and minimizing the possible transmission of white-cell associated viral diseases such as cytomegalovirus. The absorbed leukocytes can also be eluted from these filters by appropriate choice of buffer solution pH, making it possible to use this technique for the collection of leukocytes from donated blood for use in the subsequent treatment of leukopenic recipients. Depth filtration has also been considered for online leukocyte removal from the extracorporeal circuit of patients undergoing cardiopulmonary bypass as a means to reduce the likelihood of postoperative myocardial or pulmonary reperfusion injury which can be caused by activated leukocytes.

A new therapeutic technique that involves online cytapheresis is the use of extracorporeal photochemotherapy, which is also known in the literature as *photopheresis.* Photopheresis can be used to treat a variety of disorders caused by aberrant T-lymphocytes [Edelson, 1989], and it has become an established therapy for the treatment of advanced cutaneous T-cell lymphoma in the U.S. and several European countries. In this case, the therapy involves the use of photoactivated 8-methoxypsoralen, which blocks DNA replication causing the eventual destruction of the immunoactive T-cells. The psoralen compound is taken orally prior to the phototherapy. Blood is drawn from a vein and separated by centrifugation. The white cells and plasma are collected, diluted with a saline solution, and then pumped through a thin plastic chamber in which the cells are irradiated with a high-intensity UV light that activates the psoralen. The treated white cells are then recombined with the red cells and returned to the patient. Since the photoactivated psoralen has a half-life of only several microseconds, all its activity is lost prior to reinfusion of the cells, thereby minimizing possible side effects on other organs. The removal of the red cells (which have a very high adsorptivity to UV light) makes it possible to use a much lower energy UV light, thereby minimizing the possible damage to normal white cells and platelets.

Photopheresis has also been used in the treatment of scleroderma, systemic lupus erythematosus, and pemphigus vulgaris. The exact mechanism for the suppression effect induced by the photo-therapy in these diseases is uncertain, although the T-cell destruction seems to be highly specific for the immunoactive T-cells [Edelson, 1989]. The response is much more involved than simple direct photoinactivation of the white cells; instead, the photo-treated cells appear to undergo a delayed form of cell death which elicits an immunologic response possibly involving the production of anti-idiotypic antibodies or the generation of clone-specific suppressor T-cells. This allows for an effective "vaccination" against a particular T-cell activity without the need for isolating or even identifying the particular cells that are responsible for that activity [Edelson, 1989].

Phototherapy has also been used for virus inactivation, particularly in blood-banking applications prior to transfusion. This can be done using high-intensity UV light alone or in combination with specific photoactive chemicals to enhance the virus inactivation. For example, hematoporphyrin derivatives have been shown to selectively destroy hepatitis and herpes viruses in contaminated blood. This technique shows a high degree of specificity toward this type of enveloped virus, which is apparently due to the affinity of the photoactive molecules for the lipids and glycolipids that form an integral part of the viral envelope.

Another interesting therapeutic application involving cytapheresis is the ex vivo activation of immunologically active white cells (lymphokine-activated killer cells, tumor-infiltrating lymphocytes, or activated killer macrophages) for the treatment of cancer. The detailed protocols for this therapy are still being developed, and there is considerable disagreement regarding its actual clinical efficacy. A pool of activated cells is generated in vivo by several days of treatment with interleukin-2. These cells are then collected from the blood by centrifugal cytapheresis and further purified using density gradient centrifugation. The activated cells are cultured for several days in a growth media containing additional interleukin-2. These ex-vivo activated cells are then returned to the patient, where they have been shown to lyse existing tumor cells and cause regression of several different metastatic cancers.

132.4 Summary

Apheresis is unique in terms of the range of diseases and metabolic disorders which have been successfully treated by this therapeutic modality. This broad range of application is possible because apheresis directly alters the body's immunologic system though the removal or alteration of specific immunologically active cells and/or proteins.

Although there are a number of adverse reactions that can develop during apheresis (e.g., fluid imbalance, pyrogenic reactions, depletion of important coagulation factors, and thrombocytopenia), the therapy is generally well tolerated even by patients with severely compromised immune systems. This has, in at least some instances, led to the somewhat indiscriminate use of therapeutic apheresis for the treatment of diseases in which there was little physiologic rationale for the application of this therapy. This was particularly true in the 1980s, where dramatic advances in the available technology for both membrane and centrifugal blood fractionation allowed for the relatively easy use of apheresis in the clinical milieu. In some ways, apheresis in the 1980s was a medical treatment that was still looking for a disease. Although apheresis is still evolving as a therapeutic modality it is now a fairly well-established procedure for the treatment of a significant number of diseases (most of which are relatively rare) in which the removal of specific plasma proteins or cellular components can have a beneficial effect on the progression of that particular disease. Furthermore, continued advances in the equipment and procedures used for blood fractionation and component removal have, as discussed in this chapter, provided a safe and effective technology for the delivery of this therapy.

The recent advances in sorbent-based systems for the removal of specific immunologically active proteins and in the development of treatment for the activation or inactivation of specific cellular components of the immune system has provided exciting new opportunities for the alteration and even control of the body's immunologic response. This includes: (1) the direct removal of specific antibodies or immune complexes (using membrane plasmapheresis with appropriate immunosorbent columns), (2) the inactivation or removal of specific lymphocytes (using centrifugal cytapheresis in combination with appropriate extracorporeal phototherapy or chemotherapy), and/or (3) the activation of a disease-specific immunologic response (using cytapheresis and ex vivo cell culture with appropriate lymphokines and cell stimuli). New advances in our understanding of the immune system and in our ability to selectively manipulate and control the immunologic response should thus have a major impact on therapeutic apheresis and the future development of this important medical technology.

Defining Terms

Autoimmune diseases: A group of diseases in which pathological antibodies are produced that attack the body's own tissue. Examples include glomerulonephritis (characterized by inflammation of the capillary loops in the glomeruli of the kidney) and myasthenia gravis (characterized by an inflammation of the nerve/muscle junctions).

Cascade filtration: The combination of plasmapheresis with a second online membrane filtration of the collected plasma to selectively remove specific toxic or immunogenic components from blood based primarily on their size.

Cytapheresis: A type of therapeutic apheresis involving the specific removal of red blood cells, white cells (also referred to as *leukapheresis*), or platelets (also referred to as *plateletapheresis*).

Donor apheresis: The collection of a specific component of blood (either plasma or one of the cellular fractions), with the return of the remaining blood components to the donor. Donor apheresis is used to significantly increase the amount of plasma (or a particular cell type) that can be donated for subsequent use in blood banking and/or plasma fractionation.

Immune complexes: Antigen-antibody complexes that can be deposited in tissue. In rheumatoid arthritis this deposition occurs primarily in the joints, leading to severe inflammation and tissue damage.

Photopheresis: The extracorporeal treatment of diseases characterized by aberrant T-cell populations using visible or ultraviolet light therapy, possibly in combination with specific photoactive chemicals.

Plasma exchange: The therapeutic process in which a large volume of plasma (typically 3 L) is removed and replaced by an equivalent volume of a replacement fluid (typically fresh frozen plasma, a plasma substitute, or an albumin-containing saline solution).

Plasma perfusion: The therapeutic process in which a patient's plasma is first isolated from the cellular elements in the blood and then subsequently treated to remove specific plasma components. This secondary treatment usually involves a sorbent column designed to selectively remove a specific plasma component or a membrane filtration designed to remove a broad class of plasma proteins.

Plasmapheresis: The process in which plasma is separated from the cellular components of blood using either centrifugal or membrane-based devices. Plasmapheresis can be employed in donor applications for the collection of source plasma for subsequent processing into serum fractions or in therapeutic applications for the treatment of a variety of disorders involving the presence of abnormal circulating components in the plasma.

Therapeutic apheresis: A process involving the separation and removal of a specific component of the blood (either plasma, a plasma component, or one of the cellular fractions) for the treatment of a metabolic disorder or disease state.

References

Edelson RL. 1989. Photopheresis: A new therapeutic concept. Yale J Biol Med 62:565.

Kambic HE, Nosé Y. 1993. Plasmapheresis: Historical perspective, therapeutic applications, and new frontiers. Artif Organs 17(10):850.

Malchesky PS, Asanuma Y, Zawicki I, et al. 1980. On-line separation of macromolecules by membrane filtration with cryogelation. Artif Organs 400:205.

Nosé Y, Kambic HE, Matsubara S. 1983. Introduction to therapeutic apheresis. In Y Nosé, PS Malchesky, JW Smith, et al. (eds), Plasmapheresis: Therapeutic Applications and New Techniques, pp 1–22, New York, Raven Press.

Rock G. 1983. Centrifugal apheresis techniques. In Y Nosé, PS Malchesky, JW Smith, et al. (eds), Plasmapheresis: Therapeutic Applications and New Techniques, pp 75–80, New York, Raven Press.

Sawada K, Malchesky P, Nosé Y. 1990. Available removal systems: State of the art. IN UE Nydegger (ed), Therapeutic Hemapheresis in the 1990s, pp 51–113, New York, Karger.

Solomon BA, Castino F, Lysaght MJ, et al. 1978. Continuous flow membrane filtration of plasma from whole blood. Trans AM Soc Artif Intern Organs 24:21.

Zeman LJ, Zydney AL. 1986. Microfiltration and Ultrafiltration: Principles and Applications, pp. 471–489, New York, Marcel Dekker.

Zydney AL, Colton CK. 1982. Continuous flow membrane plasmaphersis: Theoretical models for flux and hemolysis prediction. Trans Am Soc Artif Intern Organs 28:408.

Zydney AL, Colton CK. 1986. A concentration polarization model for filtrate flux in cross-flow microfiltration of particulate suspensions. Chem Eng Commun 47:1.

Zydney AL, Colton CK. 1987. Fundamental studies and design analyses of cross-flow membrane plasmapheresis. In JD Andrade, JJ Brophy, DE Detmer (eds), Artificial Organs, pp 343-358, VCH Publishers.

Further Information

Several of the books listed above provide very effective overviews of both the technical and clinical aspects of therapeutic apheresis. In addition, the Office of Technology Assessment has published *Health Technology Case Study 23: The Safety, Efficacy, and Cost Effectiveness of Therapeutic Apheresis*, which has an excellent discussion of the early clinical development of apheresis. Several journals also provide more detailed discussions of current work in apheresis, including *Artificial Organs* and the *Journal of Clinical Apheresis*. The abstracts and proceedings from the meetings of the International Congress of the World Apheresis Association and the Japanese Society for Apheresis also provide useful sources for current research on both the technology and clinical applications of therapeutic apheresis.

133

Liver Support Systems

Pierre M. Galletti
(deceased)

Hugo O. Jauregui
Rhode Island Hospital

133.1 Morphology of the Liver

The liver is a complex organ that operates both in series and in parallel with the gastrointestinal tract. After entering the portal system, the products of digestion come in contact with the liver parenchymal cells, or hepatocytes, which remove most of the carbohydrates, amino acids, and fats from the feeder circulation, therefore preventing excessive increases throughout the body after a meal. In the liver, these products are then stored, modified, and slowly released to the better advantage of the whole organism.

The liver can be considered a complex large-scale biochemical reactor, since it occupies a central position in the *metabolism*, i.e., the sum of the physical and chemical processes by which living matter is produced, maintained, and destroyed, and whereby energy is made available for the functioning of liver cells as well as tissues from all other organs.

The adult human liver (weighing 1500 g) receives its extensive blood supply (on the order of 1 L/min or 20% of cardiac output) from two sources: the portal vein (over two-thirds) and the hepatic artery (about one-third). Blood from the liver drains through the hepatic veins into the inferior vena cava. Macroscopically, the liver is divided into 4 or 5 lobes with individual blood supply and bile drainage channels. Some of these lobes can be surgically separated, although not without difficulty.

Microscopically, human hepatocytes (250-500 $\times 10^9$ in each liver) are arranged in plates (Fig. 133.1) that are radially distributed around the central (drainage) vein [Jones & Spring-Mills, 1977] and form somewhat hexagonal structures, or liver lobules, which are much more clearly demarcated in porcine livers. Present in the periphery of these lobules are the so-called *portal triads*, in the ratio of three triads for each central vein. In each portal triad, there are tributaries of the portal vein, branches of the hepatic artery, and collector ducts for the bile (Fig. 133.1). Blood enters the liver lobule at the periphery from terminal branches of the portal vein and the hepatic arteries and is distributed into capillaries which separate the hepatocyte plates. These capillaries, called sinusoids, characteristically have walls lined by layers of endothelial cells that are not continuous but are perforated by small holes (fenestrae). Other cells are present in the sinusoid wall, e.g., phagocytic Kuppfer cells, fat-storing Ito cells, and probably a few yet undefined mesenchymal cells. It is important to emphasize that blood-borne products (with the exception of blood cells) have free access to the perisinusoidal space, called the space of Disse, which can be visualized by electron microscopy as a gap separating the sinusoidal wall from the hepatocyte plasma

0-8493-0462-8/00/$0.00+$.50

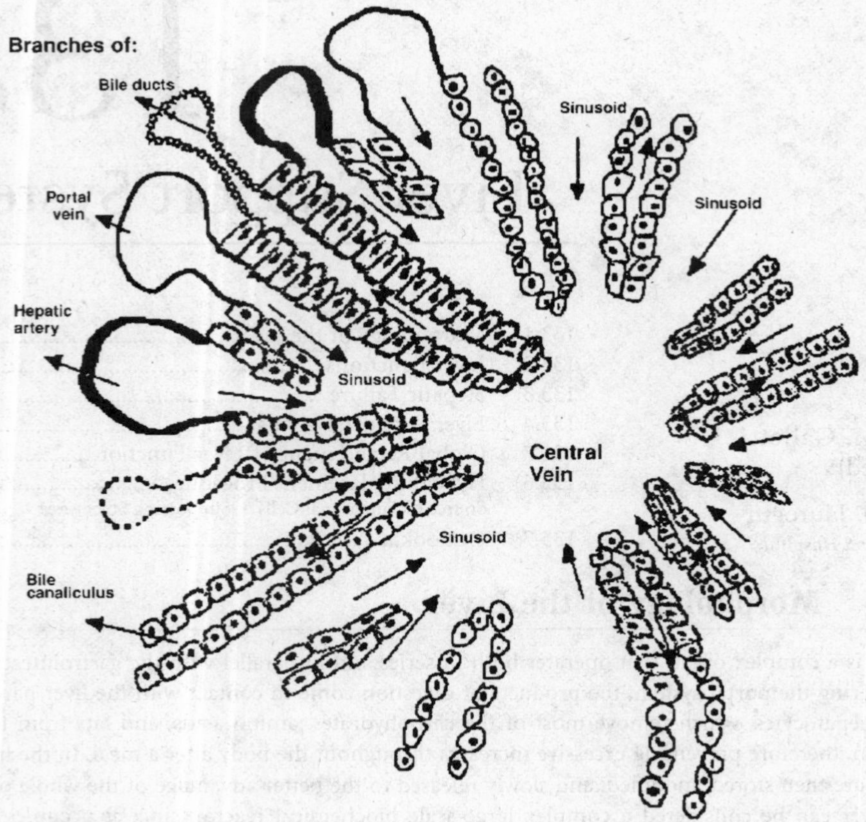

FIGURE 133.1 The liver lobule.

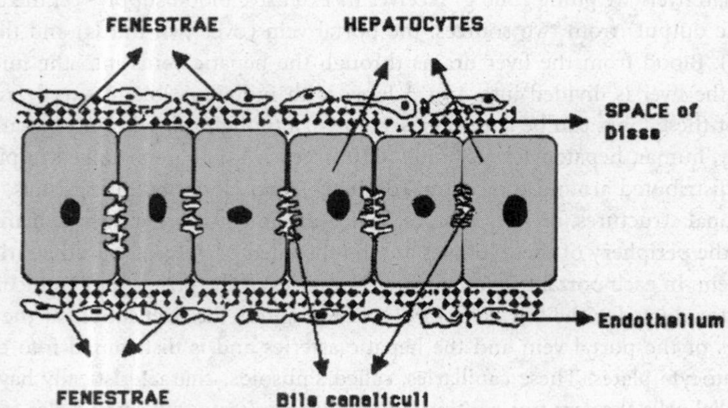

FIGURE 133.2 Hepatocyte relationships with the space of Disse and the sinusoid wall.

membrane (Fig. 133.2). In this space, modern immunomicroscopic studies have identified three types of collagens: Type IV (the most abundant), Type I, and Type III. Fibronectin and glycosaminoglycans are also found there, but laminin is only present in the early stages of liver development not in adult mammalian livers [Martinez-Hernandez, 1984].

The hepatocytes themselves are large (each side about 25 microns), multifaceted, polarized cells with an apical surface which constitutes the wall of the bile canaliculus (the channel for bile excretion) and basolateral surfaces which lie in close proximity to the blood supply. Hepatocytes constitute 80–90% of the liver cell mass. Kuppfer cells (about 2%) belong to the reticulo-endothelial system, a widespread class of cells which specialize in the removal of particulate bodies, old blood cells, and infectious agents from the blood stream.

The cytoplasm of hepatocytes contains an abundance of smooth and rough endoplasmic reticulum, ribosomes, lysosomes, and mitochondria. These organelles are involved in complex biochemical processes: fat and lipid metabolism, synthesis of lipoproteins and cholesterol, protein metabolism, and synthesis of complex proteins, e.g., serum albumin, transferrin, and clotting factors from amino acid building blocks. The major aspects of detoxification take place in the cisternae of the smooth endoplasmic reticulum, which are the site of complex oxidoreductase enzymes known collectively as the cytochrome P-450 system. In terms of excretion, hepatocytes produce bile, which contains bile salts and conjugated products. Hepatocytes also store large pools of essential nutrients such as folic acid, retinol, and cobalamin.

133.2 Liver Functions

The liver fulfills multiple and finely tuned functions that are critical for the homeostasis of the human body. Although individual pathways for synthesis and breakdown of carbohydrates, lipids, amino acids, proteins, and nucleic acids can be identified in other mammalian cells, only the liver performs all these biochemical transformations simultaneously and is able to combine them to accomplish its vital biologic task. The liver is also the principal site of biotransformation, activation or inactivation of drugs and synthetic chemicals. Therefore, this organ displays a unique biologic complexity. When it fails, functional replacement presents one of he most difficult challenges in substitutive medicine.

Under normal physiologic requirements, the liver modifies the composition and concentration of the incoming nutrients for its own usage and for the benefit of other tissues. Among the major liver functions, the detoxification of foreign toxic substances (*xenobiotics*), the regulation of essential nutrients, and the secretion of transport proteins and critical plasma components of the blood coagulation system are probably the main elements to evaluate in a successful organ replacement [Jauregui, 1991]. The liver also synthesizes several other critical proteins, excretes bile, and stores excess products for later usage, functions that can temporarily be dispensed with but must eventually be provided.

The principal functions of the liver are listed in Table 133.1. The challenge of liver support in case of organ failure is apparent from the complexity of functions served by liver cells and from our still imperfect ability to rank these functions in terms of urgency of replacement.

TABLE 133.1 Liver Functions

Carbohydrate metabolism: Glyconeogenesis and glycogenolysis
Fat and lipid metabolism: Synthesis of lipoproteins and cholesterol
Synthesis of plasma proteins, for example:
Albumin
Globulins
Fibrinogen
Coagulation factors
Transferrin
α-fetoprotein
Conjugation of bile acids; conversion of heme to bilirubin and biliverdin
Detoxification: Transformation of metabolites, toxins, and hormones into water-soluble compounds (e.g., cytochrome P-450 P-450 oxidation, glucuronyl transferase conjugation)
Biotransformation and detoxification of drugs
Metabolism and storage of vitamins
Storage of essential nutrients
Regeneration

133.3 Hepatic Failure

More than any other organ, the liver has the property of regeneration after tissue damage. Removal or destruction of a large mass of hepatic parenchyma stimulates controlled growth to replace the missing tissue. This can be induced experimentally, e.g., two thirds of a rat liver can be excised with no ill effects and will be replaced within 6 to 8 days. The same phenomenon can be observed in humans and is a factor in the attempted healing process characteristic of the condition called liver *cirrhosis.* Recent attempts at liver transplantation using a liver lobe from a living donor rely on the same expectation of recovery of lost liver mass. Liver regeneration is illustrated by the myth of Prometheus, a giant who survived in spite of continuous partial hepatectomy through the good auspices of a vulture (a surgical procedure inflicted on him as punishment for having stolen the secret fire from the gods and passing it on to humanity).

Hepatic failure may be acute or chronic according to the time span it takes for the condition to develop. Mechanisms and toxic by-products perpetuating these two conditions are not necessarily the same. Acute *fulminant hepatic failure* (FHF) is the result of massive necrosis of hepatocytes induced over a period of days or weeks by toxic substances or viral infection. It is characterized by jaundice and mental confusion which progresses rapidly to stupor or coma. The latter condition, *hepatic encephalopathy* (HE), is currently thought to be associated with diminished hepatic catabolism. Metabolites have been identified which impair synaptic contacts and inhibit neuromuscular and mental functions (Table 133.2). Although brain impairment is the rule in this condition, there is no anatomic damage to any of the brain structures, and therefore, the whole process is potentially reversible. The mortality rate of FHF is high (70–90%), and death is quite rapid (a week or two). Liver transplantation is currently the only effective form of treatment for FHF. Transplantation procedures carried out in life-threatening circumstances are much more risky than interventions in relatively better compensated patients. The earlier the transplantation procedure takes place, the greater is the chance for patient survival. However, 10–30% of FHF patients will regenerate their liver under proper medical management without any surgical intervention. Hence, liver transplantation presents the dilemma of choosing between an early intervention, which might be unnecessary in some cases, or proceeding to a late procedure with a statistically higher mortality [Jauregui et al., 1994].

Chronic hepatic failure, the more common and progressive form of the disease, is often associated with morphologic liver changes known as cirrhosis in which fibrotic tissue gradually replaces liver tissue as the result of long-standing toxic exposure (e.g., alcoholism) or secondary to viral hepatitis. More than 30,000 people died of liver failure in the United States in 1990.

In chronic hepatic failure, damaged hepatocytes are unable to detoxify toxic nitrogenous products that are absorbed by intestinal capillaries and carried to the liver by the portal system. Ammonia probably plays the major role in the deterioration of the patient's mental status, leading eventually to "hepatic coma." An imbalance of conventional amino acids (some abnormally high, some low) may also be involved in the pathogenesis of the central nervous system manifestation of hepatic failure, the most dramatic of which is cerebral edema. Impaired blood coagulation (due to decreased serum albumin and clotting factors), hemorrhage in the gastro-intestinal system (increased resistance to blood flow through the liver leads to portal hypertension, ascites formation, and bleeding from esophageal varices), and

TABLE 133.2 Metabolic Products with Potential Effects in Acute Liver Failure

Substance	Mode of action
Ammonia	Neurotoxic interaction with other neurotransmitters Contributes to brain edema
Benziodiazepinelike substances	Neural inhibition
GABA	Neural inhibition
Mercaptans	Inhibition of Na-K ATPase
Octopamine	Acts as a false neurotransmitter

hepatic encephalopathy with glial cell damage in the brain are the standard landmarks of chronic hepatic failure. In fact, HE in chronic liver failure is often precipitated by episodes of bleeding and infection, and progression to deep coma is an ominous sign of impending death.

Intensive management of chronic liver failure includes fluid and hemodynamic support, correction of electrolyte and acid-base abnormalities, respiratory assistance, and treatment of cerebral edema if present. Aggressive therapy can diminish the depth of the coma and improve the clinical signs, but the outcome remains grim. Eventually, 60–90% of the patients require transplantation. About 2500 liver transplants are performed every year in the United States, with a survival rate ranging from 68–92%. The most serious limitation to liver transplantation (besides associated interrelated diseases) remains donor scarcity. Even if segmented transplants and transplants from living related donors become acceptable practices, it is unlikely that the supply of organs will ever meet the demand. Further, the problem of keeping a patient alive with terminal hepatic failure, either chronic or acute, while waiting for an adequately matched transplant is much more difficult than the parallel problem in end-stage renal disease, where dialysis is a standardized and effective support modality.

An appreciation of the modalities of presentation of the two types of hepatic coma encountered in liver failure is needed for a definition of the requirements for the proper use of liver assist devices. In the case of FHF, the hepatologist wants an extracorporeal device that will circulate a large volume of blood through a detoxifying system [Jauregui & Muller, 1992] allowing either the regeneration of the patient's damaged liver (and the avoidance of a costly and risky liver transplantation procedure) or the metabolic support needed for keeping the patient alive while identifying a cadaveric donor organ. In the first option, the extracorporeal liver assist device functions as an organ substitute for the time it takes the liver to regenerate and recover its function; in the second, it serves as temporary bridge to transplantation.

In the case of chronic liver failure today, spontaneous recovery appears impossible. The damaged liver needs to be replaced by a donor organ, although not with the urgency of FHF. The extracorporeal liver assist device (LAD) is used as a bridge while waiting for the availability of a transplant. It follows that the two different types of liver failure may require different bioengineering designs.

133.4 Liver Support Systems

The concept of artificial liver support is predicated on the therapeutic benefit of removing toxic substances accumulating in the circulation of liver failure patients. These metabolites reflect the lack of detoxification by damaged hepatocytes, the lack of clearance of bacterial products from the gut by impaired Kupffer cells, and possibly the release of necrotic products from damaged cells which inhibit liver regeneration. Systemic endotoxemia as well as massive liver injury give rise to an inflammatory reaction with activation of monocytes and macrophages and release of cytokines which may be causally involved in the pathogenesis of multiorgan failure commonly encountered in liver failure.

Technologies for temporary liver support focus on the detoxifying function, since this appears to be the most urgent problem in liver failure. The procedures and devices which have been considered for this purpose include the following.

Hemodialysis

Hemodialysis with conventional cellulosic membranes (cut-off point around 2000 daltons) or more permeable polysulfone or polyacrylonitrile [de Groot et al., 1984] (cut-off 1500–5000 daltons) helps to restore electrolyte and acid-base balance and may decrease the blood ammonia levels but cannot remove large molecules and plasma protein-bound toxins. Improvement of the patient's clinical condition (e.g., amelioration of consciousness and cerebral edema) is temporary. The treatment appears to have no lasting value and no demonstrated effect on patient survival. In addition, hemodialysis may produce a respiratory distress syndrome caused by a complement-mediated poly-morphonuclear cell aggregation in the pulmonary circulatory bed. Because some of the clinical benefit seems related to the removal of toxic molecules, more aggressive approaches focused on detoxification have been attempted.

Hemofiltration

Hemofiltration with high cut-off point membranes (around 50,000 daltons with some polyacrylonitrile-polyvinyl chloride copolymers, modified celluloses, or polysulfones) clears natural or abnormal compounds within limits imposed by convective transport across the exchange membrane. These procedures again have a temporary favorable effect on hepatic encephalopathy (perhaps because of the correction of toxic levels of certain amino acids) with reversal of coma, but they do not clearly improve survival rates.

Hemoperfusion

Hemoperfusion, i.e., extracorporeal circulation of blood over nonspecific sorbents (e.g., activated charcoal) [Chang, 1975] or more complex biochemical reactors which allow the chemical processing of specific biologic products, such as ammonia, have not yet met clinical success in spite of encouraging experimental results, except in the case of hepatic necrosis induced by poisonous mushrooms such as Amanita phalloides. Anion exchange resins and affinity columns similar to those used in separative chromatography may help in removing protein-bound substances (e.g., bilirubin) which would not pass through dialysis or hemofiltration membranes, but nonspecific sorbents may also deplete the plasma of biologically important substances. Further, these techniques are complicated by problems of hemocompatibility, related in part to the entrainment of dust ("fines") associated with the sorbent material itself and in part to platelet activation in patients with an already compromised coagulation status. To minimize this problem direct blood or plasma contact with the sorbent material can be avoided by polymer coating of the sorbent particles using either albumin, cellulose nitrate, or similar thin films, but hemocompatibility remains a concern. Here again, there is anecdotal evidence of clinical improvement of hepatic failure with hemoperfusion, with some reports claiming a higher survival rate in hepatic encephalopathy, but these reports have not been supported by well-controlled studies. As is the case for hemodialysis and hemofiltration, the possible beneficial effect of hemoperfusion should be evaluated in the context of the clinical variability in the course of FHF.

Lipophilic Membrane Systems

Because lipophilic toxins dominate in fulminant hepatic failure, it is conceibable to eliminate such compounds with a hydrophobic (e.g., polysulfone) membrane featuring large voids filled with a nontoxic oil [Brunner & Tegtimeier, 1984]. After diffusion, the toxins can be made water-soluble through reaction with a NaOH-based acceptor solution, thereby preventing their return to the blood stream. A standard, high-flux dialyzer in series with the lipophilic membrane device allows the removal of hydrophilic solutes. Such a system has proved effective in removing toxins such as phenol and p-cresol as well as fatty acids without inducing detrimental side effects of its own.

Immobilized Enzyme Reactors

To address the problem of specificity in detoxification, enzymes such as urease, tyrosinase, L-asparaginase, glutaminase, and UDP-glucuronyl transferase have been attached to hollow fibers or circulated in the closed dialysate compartment of an artificial kidney or still incorporated into microcapsules or "artificial cells" exposed to blood. There is considerable in vitro evidence for the effectiveness of this approach, and some indication of therapeutic value from in vivo animal experiments [Brunner et al., 1979]. However, no clinical report has documented the superiority of enzyme reactors over the various modalitites of dialysis. Again there are clinical observations of clearing of the mental state of patients in hepatic coma, but no statistically demonstrated effect on survival. It is unclear whether the lack of success is due to the inability of specific enzymes to remove all offending toxins or is evidence of the need for more than detoxification for effective treatment.

Parabiotic Dialysis

Also referred to as *cross-dialysis,* parabolic dialysis is a variant of hemodialysis in which the dialysate compartment of a solute exchange device is perfused continuously with blood from a living donor. Because of membrane separation of the two blood streams, the procedure can be carried out even if the two subjects belong to different blood groups or different animal species. However, the risk of the procedure to a human donor (control of blood volume, transfer of toxic substances, mixing of blood streams in case of dialyzer leak) and the difficulty of introducing a live animal donor into the hospital environment have relegated this approach to the class of therapeutic curiosities.

Exchange Transfusion

Exchange transfusion, i.e., the quasi-total replacement of the blood volume of a patient in liver failure by alternating transfusion and bleeding, is occasionally used in severe hyperbilirubinemia of the newborn, which used to carry an ominous prognosis because of its association with cerebral edema. The rationale is that exchange transfusion will reduce the level of toxins and replenish the deficient factors in the blood stream while the underlying condition is corrected by natural processes or drugs [Trey et al., 1966]. With the advent of blood component therapy, specific plasma components can also be administered to treat identified deficiencies. Mortality rates of patients treated with exchange blood transfusions have been reported as greater than those observed with conventional therapies.

Plasmapheresis

Plasmapheresis, i.e., the combination of withdrawal of blood, centrifugation, or membrane processing to separate and discard the patient's plasma, and return of autologous cells diluted with donor plasma, was practiced initially as a batch process. Techniques now exist for a continuous exchange process, in which plasma and cells are separated by physical means outside of the body (membrane separation or centrifugation), and the patient's plasma replaced by banked plasma (up to 5000 ml per day) [Lepore et al., 1972]. There is evidence from controlled clinical trials for the effectiveness of this form of therapy, but the mortality rate remains high in patients with hepatic failure, whether from insufficient treatment or the risks of the procedure. It appears, however, that plasma exchange can be beneficial in the preoperative period prior to liver transplantation so as to correct severe coagulopathy. Plasmapheresis is used in conjunction with the placement of a hepatocyte-seeded extracorporeal hollow-fiber device to treat acute and chronic liver failure [Rozga et al., 1993].

Combined Therapy

Endotoxins and cytokines can be removed by hemoperfusion over activated charcoal and absorbent resins, but it may be more effective to process plasma than whole blood. This has led to the concept of combining plasmapheresis with continuous plasma treatment for removal of substances such as tumor necrosis factor (TNF), interleukin-6 (IL-6), and bile acids by a resin column, and then ultrafiltration or dialysis for fluid removal, since patients with liver failure often develop secondary renal failure.

133.5 Global Replacement of Liver Function

Because of the complexity and interplay of the various functions of the liver, more success can be expected from global approaches, which allow many or all hepatic functions to be resumed. These include the following.

Cross-Criculation

Cross-circulation of the patient in hepatic coma with a compatible, healthy donor is one approach. This procedure is more than a prolonged exchange transfusion since it allows the donor's liver to substitute for the patient's failing organ and to process chemicals from the patient's blood stream as long as the procedure lasts [Burnell, 1973]. It had been attempted in isolated cases, but reports of effectiveness are entirely anecdotal and the procedure has not been accepted clinically because of ambiguous results and the perceived risk for the donor.

Hemoperfusion over Liver Tissue Slices

The incorporation of active hepatocytes in a hemoperfusion circuit was suggested by the laboratory practice of biochemists who, since Warburg, have investigated metabolic pathways in tissue slices. For liver replacement, this technology has been pioneered primarily in Japan as a substitute for organ transplantation, which is culturally frowned upon in that country in spite of a major incidence of severe liver disease. The procedure may improve biochemical markets of liver failure but has no demonstrated clinical value [Koshino et al., 1975].

Ex Vivo Perfusion

Ex vivo perfusion uses an isolated animal liver (pig or baboon) connected to the patient's cardiovascular system [Saunders et al., 1968]. This is a cleaner and more acceptable form of treatment than cross-circulation of hemoperfusion over tissue pieces. Nevertheless, it is limited by the need for thorough washing of the animal's blood from the excised liver, the requirement for a virus-free donor organ source, the limited survival capacity of the excised, perfused organ, which must be replaced at intervals of approximately 24 hours, and the cost of the procedure. Success has recently been reported in isolated clinical trials [Chari, 1994].

Heterotopic Hepatocyte Transplantation

This procedure may someday offer an alternative to whole organ transplantation, especially in cases of chronic liver failure, if a sufficient number of cells can be grafted. Freshly isolated hepatocytes have damaged cell surfaces and must be cultured to regain their integrity and display the surface receptors needed for attachment or binding of xenobiotics or endogenous toxic products [McMillan et al., 1988]. At the clinical level, this procedure could rely in part on banking frozen hepatocytes from livers which are not usable for whole organ transplants but constitute a reliable source of cells [Bumgardener et al., 1988]. The procedure does not require removal of the recipients's liver and provides the followng advantages: (1) minimal surgery, (2) repeatability as needed, and (3) interim strategy to whole organ transplant. There is no agreement as to the best anatomic site for hepatocyte transplantation, the type of matrix needed for cell attachment and differentation, and the number of cells needed. It has been reported that hepatocyte culture supernatants were as effective as transplanted hepatocytes in treating rats with chemically induced liver failure [LaPlante-O'Neill et al., 1982]. The tridemensional reconstruction of high-density, functional liver tissue will be the greatest challenge in view of the cell mass required. Structural organization and differentiated functions may be achieved by using an asialoglycoprotein model polymer as the synthetic substrate for a primary culture of hepatocytes to develop functional modules for implantation in humans [Akaike et al., 1989].

Whole Organ in Situ Transplantation

This is currently the procedure of choice, most particularly in children. Although progress in clinical and surgical skills certainly accounts in large part for the growing success rate of this procedure over the past 20 years, it is worth noting that the introduction of extracorporeal ciruclation techniques in the surgical protocol to support the donor organ while waiting for completion of the anastomosis has paralleled the steepest increase in success rate of liver transplantation in the past few years. Whereas most liver transplants rely on the availability of cadaver organs, the recent advent of segmented transplants allows consideration of living related donors and possibly the sharing of a donor organ among two or more recipients.

133.6 Hybrid Replacement Procedures

The complexity of hepatic functions, coupled with shortage of human donor organs, has encouraged the development of procedures and devices which rely on xenogeneic living elements attached to synthetic structures and separated from the most host by a permselective membrane to replace temporarily, and perhaps someday permanently, the failing organ.

The *incorporation of liver microsomes* in microcapsules, hydrogels, or polymer sheets, with or without additonal enzymes or pharmacologic agents, goes one step beyond the immobilized enzyme reactor inasmuch as it calls on cell components endowed with a variety of enzymatic properties to process blood or other body fluids. The feasibility of this technique has been demonstrated in vitro [Denti et al., 1976] but has not been investigated extensively in animal experiments.

Cellular hybrid devices—the incorporation of functional cells in a device immersed in body fluids or connected to the vascular system (extracorporeally or somewhere inside the body of the recipient)—are a promising application of the concept of bioartificial organs. However, the problems faced by the "hepatocyte bioreactor" are formidable:

1. The mass of functional cells required is much larger than in the case of secretory or endocrine organs, since a normal liver weighs about 1.5 kg and since as much as 10–30% of that mass (i.e., several billion cells) may be required for life-sustaining replacement of function. Taking into account the need for supporting structures, the sheer size of the device will be an obstacle to implantation.
2. The liver features a double feeder circulation (portal and hepatic) and a complex secretory-excretory system which utilizes both blood and bile to dispose of its waste products. How to duplicate such a complex manifolding in an artificial organ and whether it is worth the resulting complexity are questions not yet resolved.
3. With membrane separation of recipient and donor cells, the size of some natural macromolecules to be exchanged (e.g., low-density lipoproteins) precludes the use of standard diffusion membranes. Allowing relatively free solute exchange between the compartments in a device without endangering the immune sequestration of the donor tissue remains a challenge for membrane technology. However, the low immunogenicity of hepatocytes (which lack type I HLA antigens) allows the consideration of relatively open membranes for the construction of extracorporeal reactors.

At the moment, most of the extracorporeal liver assist devices (ELAD) utilize xenogeneic mammalian hepatocytes seeded on solid, isotropic hollow fibers. Membrane selectivity limits the rate of diffusion of liposoluble toxins which are bound to plasma proteins for transport. Hence manipulation of membrane transport properties and concentrations of acceptor protiens can affect the clearance of polar materials.

Also, most devices focus on replacing the detoxification function of the natural liver and avoid the more complex "cascade" dialysis circuitry which could either (1) allow the macro-molecules synthesized by hepatocytes to return to the blood stream (or another body fluid compartment) though a high cut-off point membrane or (2) provide an excretory path to clear toxic products manufactured by the hepatoctyes, on the model of the bile excretory system. Nonetheless, such circuitry might be valuable to prolong the life of the seeded cells, since the combination of bile salts and bile acids has a damaging detergent effect on the lipid components of the hepatocyte membrane.

The development of bioreactors including cells capable of performing liver functions, and therefore capable of providing temporary liver support, finds applications in the treatment of acute, reversible liver failure or as a bridge for liver transplantation. Designs for a bioartificial liver can be classified according to (1) the type of cells selected to replace the hepatic functions or (2) the geometry and chemical nature of the polymer structure used organize the hepatocytes.

Source of Functional Cells

Two main methods of hepatocyte isolation (mechanical and enzymatic) can be used, separately or in combination. Mechanical methods (tissue dissociation) have largely failed in terms of long-term cell viability, although this is not always recognized by investigators developing liver assist systems. Collagenase perfusion [Seglen, 1976] is today the method of choice, yet there is evidence that hepatocytes lose some of their oligosaccharide-lectin binding capacity in the process and do not recover their glycocalyx until after a day or two in culture. Collagenase is thought to loosen cell junctions and secondarily to digest the connective tissue around the hepatocytes. Chemical methods using citrate, EDTA, and similar substances weaken cell junctions by depleting calicum ions without altering the cell membranes and presumably result in better preservation of natural enzymatic functions. Although the yield of chemical methods is lower than that of enzymatic dissociation methods, it may be of interest once a reliable technology for separating viable from nonviable cells on a large scale is perfected.

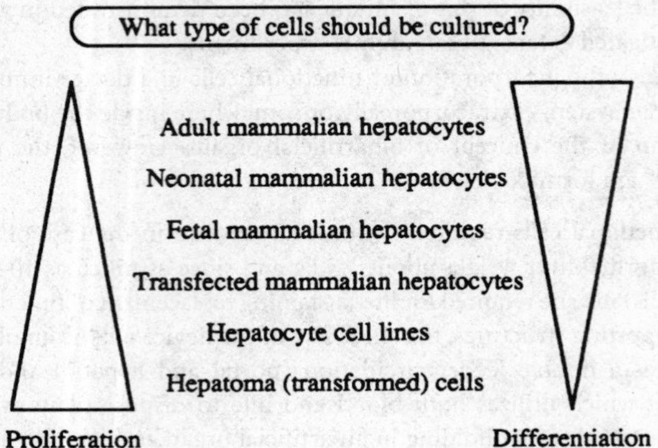

FIGURE 133.3 Cellular choices based on proliferation and differentiation.

A priori, an effective bioartificial liver would require either all the multifunctional characteristics of normal hepatocytes in vivo or only the specific functional activity that happens to be missing in the patient. Unfortunately, in most cases, clinical signs do not allow us to distinguish between these two extremes, justifying a preference for highly diffrentiated cells.

Differentiation and proliferation are usually at opposite ends of a biologic continuum in most cell types (Fig 133.3) [Jauregui, 1991]. Hence, there is a difficulty of obtaining large numbers of multifunctional hepatocytes. Several options are available:

1. The simplest approach is to use *adult mammalian liver cells* isolated in sufficient number from large animals. Porcine hepatocytes are preferred for clinical applications because of the availability of virus-free donors. Large-scale isolation of porcine hepatocytes is becoming a routine procedure, and Demetriou has been able to treat several patients with such a system. The expectation is that hepatocytes in suspension or attached to synthetic microcarriers will remain metabolically active even though separated from neighboring cells and normal supportive structures. This approach, which appeared almost beyond practicality a few years ago, seems now less formidable because of expanded knowledge of the molecular factors which favor both hepatocyte attachment to polymeric substrates and functional differentiation. The potential contribution of Kupffer cells to a bioartificial liver has not yet been extensively investigated.
2. The use of *liver tumor cells*—preferably nonmalignant hepatoma—has been pioneered by Wolfe and Munkelt [1975] because such cells can proliferate indefinitely and therefore require a minimal seeding dose. They are also less anchor-dependent than normal cells. The drawbacks are the loss of specialized hepatic functions often encountered in tumor cell lines, and the theroretical risk of escape of tumor cells in the recipient.
3. A modified, functionally differentiated, *human hepatoblastoma cell line* capale of growing to high densities, yet strongly contact-inhibited by containment with membranes, has been patented and used in animals and man by Sussman and colleagues [1992]. Evidence for metabolic effectiveness has been reported, and clinical trials are now in progress. The uncertainty associated with the use of a human tumor cell line remains a source of concern.
4. Potentially *replicating hepatocytes*, such as those obtained from embryonic liver, neonatal animals, or recently hepatectomized adults, have been proposed as a means to obtain a stem cell population with enhanced proliferation capacity. Neonatal hepatocyte-based bioartificial devices have been built on that principle [Hager et al., 1983] and have shown to produce albumin, to metabolize urea, to deaminate cytidine, and to detoxify drugs such as a diazepam. The reliance on "juvenile" cells has become less important with the identification of molecular mechanisms for growth control mitogens such as EGF, TGFa, hepatopoieten B, HSS (hepatic stimulatory substance) and HGF

TABLE 133.3 Present Choice of the Cellular Component for Extracorporeal Liver Assist Devices

Sussman & Kelly (1993)	Neuzil et al.	Jauregui & Muller (1992)
Hepatoblastoma-derived cell line	Porcine hepatocytes separated via portal vein perfusion	Porcine hepatocytes separated via hepatic vein perfusion
Cells divide indefinitely	Hepatocyte division has not been tested (5 to 8)	Limited number of hepatocyte doublings
Cells are cultured in the device	Hepatocytes are seeded on microcarriers and introduced in the device immediately before clinical use	Hepatocytes may be seeded or cultured in the device through a proprietary technology

(hepatoctye growth factor), co-mitogenic factors such as norepinephrine and growth inhibitors such as TGFβ, and interleukin-1β.

5. *Transfected or transgeic hepatocytes* may provide the ultimate solution to the cell supply problem. However, they are usually selected for the monoclonal expression of a single function and therefore may not be suitable except when a single cause of hepatic failure has been identified. Alternatively, a combination of different transfected hepatocytes may be considered.

Table 133.3 illustrates the choices made by three different groups of investigators for clinical liver assist devices.

Supporting Structures

Microencapsulation in a nutrient liquid or a polymer gel surrounded by a conformal membrane provides a suspension of metabolically active units which can either be placed in a container for extracorporeal hemoperfusion, introduced into the peritoneal cavity for implantation, or even infused into the portal vein for settling in the patient's liver. The blood or tissue reaction to multiple implants and the long-term in vivo stability of hydrogels based on polyelectrolytes remain unresolved problems.

A *flat sheet of membrane or a spongy matrix coated with attachment factors* can be used to anchor a suspension of functional cells. The limitation of this approach is the need for vascularization of the implant, since the cells are metabolically active and therefore are quite avid of oxygen and nutrients.

Microcarrier-attached hepatocytes are attractive because the technology of suspension cultures is amenable to the proliferation of a large number of cells. The bulk of the carrier beads probably limits this approach to the extracorporeal circuits similar to those used for hemoperfusion [Rozga et al., 1993].

Hollow-fiber ELADs are constructed by filling the interstices within a bundle of parallel hollow fibers with liver cells, using the lumen of the tubes to provide metabolic support and an excretion channel, typically by circulating blood from an extracorporeal circuit. Alternatively, a hybrid organ can be built by filling the lumen of the hollow fibers with functional cells and implanting the bundle in the body cavity to allow exchanges with the surrounding fluid. One could combine such a device with an oxygenation system, e.g., by filling the peritoneal cavity with a high-oxygen-capacity fluid such as a fluorocarbon and a gas exchange system for that fluid analogous to the intravenous oxygenator (IVOX).

133.7 Outlook

Medical trials with extracorporeal hollow-fiber systems are already in progress, although there are some unanswered questions in the proper design of a hepatocyte-seeded ELAD. A review of Table 133.4 will indicate that there is no consensus for implementing cellular choices in the construction of such systems.

Some researchers believe that the culture of hepatocytes on a synthetic matrix prior to chemical application is a complicated proposition that impairs the practical application of this technology. An alternative could be the isolation, freezing, and shipping of porcine hepatocytes to the medical centers treating patients with acute and chronic liver failure (usually transplantation centers). This approach may offer a direct and economically sound solution. Unfortunately, many FHF patients require emergency liver support and are managed in secondary care medical institutions which have neither a transplantation

TABLE 133.4 Consideration in the Choice of the Cellular Component of Extracorporeal Liver Assist Devices

Advantages	Disadvantages
Hepatoblastoma cell lines are easy to culture and are free of other cell contaminants	Hepatoblastoma cell lines are tumorigenic
Tumor cell lines are not anchorage dependent	Tumor cell lines may not respond to physiologic regulation
Porcine or primary hepatocytes respond to physiologic stimulation	Porcine hepatocytes have limited proliferation ability
Porcine hepatocytes express P450 (detoxification activity)	Porcine hepatocytes show limited life span

program nor a tissue culture facility for the seeding of hollow-fiber devices. Part of the argument for the porcine hepatocyte isolation, shipment, seeding, and immediate use in a bioartificial liver system relies on the earlier concept that primary mammalian hepatocytes do not grow in vitro or are very difficult to maintain in hollow-fiber devices. In fact, rodent hepatocytes have shown excellent detoxification activities when cultured on perfused hollow fiber cultures. Rabbit hepatocytes also survive in hollow-fiber bioreactors which have proved successful in treating one of the most representative animal modes of human FHF.

The technology for manufacturing ELADs may also need further attention. For instance, the efficiency of hollow fibers in maintaining hybridoma cultures and producing specific proteins is known to be inferior to that of cellular bioreactors based on microcarrier technology. The surface of hollow fibers has been optimized for blood compatibility but not for hepatocyte attachment, and therefore new materials or structures may have to be developed.

At the conceptual level, primary hepatocytes may need to be cultured either in combination with other cells that will provide parabiotic support or on substrates that imitate the composition of the extracellular matrix found in situ in the space of Disse. Our own experience [Naik et al., 1990] and that of others [Singhvi et al., 1994] raise some questions about the role of substrates in maintaining long-term hepatocyte viability in vitro. In fact, most of them operate in a rather indiscriminate fashion by providing an anchor for hepatocytes. Collagen, types I, III, IV, and fibonectin contribute to cytoplasmic spreading which in the long term does not favor maintenance of the pheontypic expression of hepatocytes. Polymer compositions expressing surface sugar residues responsible for hepatocyte attachment through the asailoglycoprotein receptor (a plasma membrane complex present mainly in the bile canalicular area which internalizes plasma asiaglycoprotein) appear able to maintain hepatocytes in culture with extended functional activities [Akaike et al., 1989]. Other investigators have shown the value of extracellular glycoproteins and glycosaminoglycans rich in laminin (e.g., Matrigel) [Caron & Bissell, 1989]. Hepatocytes immobilized on these substraes do not spread but maintain their tridimensional morphology, as well as their functional activity. Such observations suggest that long-term expression of hepaotcyte-specific functions depend on maintaining the in situ cell shape and their spacial interrelations [Koide et al., 1990]. Future ELAD designs should provide not only ideal polymer substrates for cell attachment but also a special configuration that will maintain the tridimensional structure of hepatic tissue.

Experience with artificial orans shows that the development of these devices tends to underestimate the effort and the technical advances needed to pass from a "proof of principle" prototype for animal evaluation to fabrictaion of a clinically acceptable system for human use. Full-scale design, cell procurement, cell survival, and device storage are all major bottlenecks on the way to a clinical product. One of the gray areas remains in the molecular weight cut-off of the hollow fiber wall. Which substances are responsible for the development of FHF and must therefore be cleared? Are they protein-bound, middle-molecular-, or low-molecular-size compounds? In the absence of clear answers to these questions, the designer of separation membranes is in an ambigous position; some investigators use low molecular cut-off membranes to guarantee immunoseparation of the xenograft, whereas others prefer high-molecular cut-off to enhance functionality. The use of microporous membranes in human subjects has not led, as of yet, to any hypersensitivity reactions in spite of potential passage of immunglobulins.

A reliable way to restore consciousness in animal models of FHF is to cross-circulate the blood with a normal animal. To the extent that ELADs may function in patients as in these animal models, they may prove successful only in relieving the symptoms of the disorder without increasing the survival rate. For instance, when 147 patients with advanced stages of HE were treated with charcoal hemoperfusion over a 10-year period, all showed symptomatic improvement, but the survival rate of 32% was the same as in five control groups. Therefore, enthusiasm generated by preliminary human trials with hollow-fiber devices should be tempered with a cautious approach. Without reliable control studies, we will remain in the position defined by Benhamou and coworkers [1972], "The best future one can wish for a sufferer from severe acute hepatic failure is to undergo a new treatment and have his case published—Be published or perish!"

Defining Terms

Bioartificial liver: A liver assist or liver replacement device incorporating living cells in physical or chemical processes normally performed by liver tissue.

Catabolism: The aspect of metabolism in which substances in living tissues are transformed into waste products or solutes of simpler chemical composition. (The opposite process is called *anabolism.*)

Cirrhosis: A degenerative process in the liver marked by excess formation of connective tissue destruction of functional cells, and, often, contraction of the organ.

Conjugated: The joining of two compounds to produce another compound, such as the combination of a toxic product with some substance in the body to form a detoxified product, which is then eliminated.

ELAD: Extracorporeal liver assist device.

Homeostasis: A tendency of stability in the normal body states (internal environment) of the organism. This is achieved by a system of control mechanisms activated by negative feedback, e.g., a high level of carbon dioxide in extracellular fluid triggers increased pulmonary ventilation, which in turn causes a decrease in carbon dioxide concentration.

Mesenchymal cells: The meshwork of embryonic connective tissue in the mesoderm from which are formed the connective tissues of the body and the blood vessels and lymphatic vessels.

Metabolism: The sum of all the physical and chemical processes by which living organized substance is produced and maintained (anabolism) and the transformation by which energy is made available for the uses of the organism (*catabolism*).

Parenchymal: The essential elements of an organ.

Phagoctyic: Pertaining to or produced by any cells that ingest microrganisms of other cells and foreign particles.

Portal triad: These are microscopic areas of collagen type I-III fibroblasts as well as other connective tissue elements. These triads have a branch of the portal vein, a branch of the hepatic artery, and intermediate caliber bile ducts.

Xenobiotic: A chemical foreign to the biologic system.

References

Akaike T, Kobayashi A, Kobayashi K, et al. 1989. Separation of parenchymal liver cells using a lactose-substituted styrene polymer substratum. J Bioact Compat Polym 4:51.

Benhamou JP, Rueff B, Sicot C. 1972. Severe hepatic failure: A critical study of current therapy. In F Orlandi, AM Jezequel (eds), Liver and Drugs, New York, Academic Press.

Brunner G, Holloway CJ, Lösgen H. 1979. The application of immobilized enzymes in an artificial liver support system. Artif Organs 3:27.

Brunner G, Tegtimeier F. 1984. Enzymatic detoxification using lipophilic hollow fiber membranes. Artif Organs 8:161.

Bumgardner GL, Fasola C, Sutherland DER. 1988. Prospects for hepatocyte transplantation. Hepatology 8:1158.

Burnell JM, Runge C, Saunders FC, et al. 1973. Acute hepatic failure treated by cross circulation. Arch Intern Med 132:493.

Caron JM, Bissel DM. 1989. Extracellular matrix induces albumin gene expression in cultured rat hepatocytes. Hepatology 10:636.

Chang TMS. 1975. Experience with the treatment of acute liver failure patients by hemoperfusion over biocompatible microencapsulated (coated) charcoal. In R Williams, IM Murray-Lyon (eds), Artificial Liver Support, Tunbridge Wells, England, Pitman Medical.

Chari RS, Collins BH, Magee JC, et al. 1994. Brief Report: Treatment of hepatic failure with ex vivo pig-liver perfusion followed by liver transplantation. N Engl J Med 331:134.

De Groot GH, Schalm SW, Schicht I, et al. 1984. Large-pore hemodialytic procedures in pigs with ischemic hepatic necrosis; a randomized study. Hepatogastroenterology 31:254.

Denti E, Freston JW, Marchisi M, Et al. 1976. Toward a bioartificial drug metabolizing system: Gel immobilized liver cell microsomes. Trans Am Soc Artif Intern Orans 22:693.

Hager JC, Carman R, Porter LE, et al. 1983. Neonatal hepatocyte culture on artificial capillaries: A model for drug metabolism and the artificial liver. ASAIO J 6:26–35.

Jauregui HO. 1991. Treatment of hepatic insufficiency based on cellular therapies. Int J Artif Organs 14:407.

Jauregui HO, Muller TE. 1992. Long-term cultures of adult mammalian hepatocytes in hollow fibers as the cellular component of extracorporeal (hybrid) liver assist devices. Artif Organs 16(2):209.

Jauregui HO, Naik S, Santangini H, et al. 1994. Primary cultures of rat hepatocytes in hollow fiber chambers. In Vitro Cell Dev BIol 30A:23.

Jones Al, Spring-Mills E. 1977. The liver and gallbladder. In Weiss, RO Greep, (eds), Histology, 4th ed, New York, McGraw-Hill.

Koide NH, Sakaguchi K, Koide Y, et al. 1990. Formation of multicellular spheroids composed of adult rat hepatocytes in dishes with positively charged surfaces and under other nonadherent environments. Exper Cell Res 186:227.

Koshino I, Castino F, Yoshida K, et al. 1975. A biological extracorporeal metabolic device for hepatic support. Trans Amer Soc Artif Intern Organs 21:492.

LaPlante-O'Neill P, Baumgarner D, Lewis WI, et al. 1982. Cell-free supernatant from hepatocyte cultures improves survival of rats with chemically induced acute liver failure. J Surg REs 32:347.

Lepore MJ, Stutman LJ, Bonnano CA, et al. 1972. Plasmapheresis with plasma exchange in hepatic coma. II. Fulminant viral hepatitis as a systemic disease. Arch Intern Med 129:900.

Martinez-Hernandez A. 1984. The hepatic extracellular matrix. I. Electron immunohistochemical studies in normal rat liver. Lab Invest 51:57.

McMillan PN, Hevey KA, Hixson DC, et al. 1988. Hepatocyte cell surface polarity as demonstrated by lectin binding to isolate and cultured hepatocytes. J Histochem Cytochem 36:1561.

Naik S, Santangini H, Jauregui HO. 1990. Culture of adults rabbit hepatoctyes in perfused hollow membranes. In Vitro Cell Dev Biol 26:107.

Neuzil DF, Rozga J, Moscioni AD, Ro MS, hakim R, Arnaout WS, Demetriou AA. 1993. Use of a novel bioartificial liver in a patient with acute liver insufficiency. Surgery 113:340.

Rozga J, Williams F, Ro M-S, et al. 1993. Development of a bioartificial liver: Properties and function of a hollow-fiber module inoculated with liver cells. Hepatology 17:258.

Saunders SJ, Bosman SCW, Terblanche J, et al. 1968. Acute hepatic coma treated by cross ciruclation with a baboon and by repeated exchange transfusions. Lancet 2:585.

Seglen PO. 1976. Preparation of isolated rat liver cells. Meth Cell Biol 13:29.

Singhvi R, Kumar A, Lopez GP, et al. 1994. Engineering cell shape and function. Science 264:696.

Sussman NL, Chong MG, Koussayer T, et al. 1992. Reversal of fulminant hepatic failure using an extracorporeal liver assist device. Hepatology 16:60.

Sussman NL, Kelly JH. 193. Extracorporeal liver assist in the treatment of fulminant hepatic failure. Blood Purif 11:170.

Trey C, Burns DG, Saunders SJ. 1966. Treatment of hepatic coma by exchange blood transfusion. N Eng J Med 274:473.

Wolf CFW, Munkelt BE. 1975. Bilirubin conjugation by an artificial liver composed of cultured cells and synthetic capillaries. Trans Amer Soc Artif Internal Organs 21:16.

Further Information

Two books of particular interest in hepatic encephalopathy are *Hepatic Encephalopathy: Pathophysiology and Treatment* edited by Roger F. Butterworth and Gilles Pomier Layrargues, published by The Humana Press (1989), and *Hepatology: A Textbook of Liver Disease* edited by David Zakim and Thomas D. Boyer, published by W. B. Saunders Company (1990).

Articles pertaining to extracorporeal liver assist devices appear periodically in the following journals: *ASAIO Journal*, the official journal for the American Society for Artificial Internal Organs (for subscription information, write to J.B. Lippincott, P.O. Box 1600, Hagerstown, MD 21741-9932); *Artificial Organs*, official journal of the International Society for Artificial Organs (for subscription information, write to Blackwell Scientific Publications, Inc., 328 Main Street, Cambridge, MA 02142); and *Cell Transplantation*, official journal of the Cell Transplantation Society (for subscription information, write to Elsevier Science, P.O. Box 64245, Baltimore, MD, 21264-4245).

134

Artificial Pancreas

Pierre M. Galletti
(deceased)

Clark K. Colton
Massachusetts Institute of Technology

Michel Jaffrin
Université de Technolgie de Compiègne

Gerard Reach
Hôpital Hotel-Dieu Paris

134.1 Structure and Function of the Pancreas

The pancreas is a slender, soft, lobulated gland (ca. 75 g in the adult human), located transversally in the upper abdomen in the space framed by the three portions of the duodenum and the spleen. Most of the pancreas is an *exocrine* gland which secretes proteolytic and lipolytic enzymes, conveying more than 1 L per day of digestive juices to the gastrointestinal tract. This liquid originates in cell clusters called *acini* and is collected by a system of microscopic ducts, which coalesce in a channel (the canal of Wirsung) which courses horizontally through the length of the organ and opens into the duodenum next to, or together with, the main hepatic duct (choledocous). Interspersed throughout the exocrine tissue are about 1 million separate, highly vascularized and individually innervated cell clusters known as islets of Langerhans, which together constitute the endocrine pancreas (1–2% of the total pancreatic mass). Blood is supplied to the islets by the pancreatic artery and drained into the portal vein. Therefore the entire output of pancreatic hormones is first delivered to the liver.

Human islets average 150 μm in diameter, each cluster including several endocrine cell types. Alpha cells (about 15% of the islet mass) are typically located at the periphery of the cluster. They are the source of glucagon, a fuel-mobilizing (catabolic) hormone which induces the liver to release into the circulation energy-rich solutes such as glucose and aceto-acetic and β-hydroxybutyric acids. Beta cells, which occupy the central portion of the islet, comprise about 80% of its mass. They secrete insulin, a fuel-storing (anabolic) hormone that promotes sequestration of carbohydrate, protein, and fat in storage depots in the liver, muscle, and adipose tissue. Delta cells, interposed between alpha and beta cells, produce somatostatin, one role of which seems to be to slow down the secretion of insulin and digestive juices, thereby prolonging the absorption of food. The PP cells secrete a "pancreatic polypeptide" of yet unknown significance. Other cells have the potential to produce gastrin, a hormone that stimulates the production of gastric juice.

0-8493-0462-8/00/$0.00+$.50

134.2 Endocrine Pancreas and Insulin Secretion

The term *artificial pancreas* is used exclusively for systems aimed at replacing the endocrine function of that organ. Although the total loss of exocrine function (following removal of the gland for polycystic disease, tumor, or trauma) can be quite debilitating, no device has yet been designed to replace the digestive component of the pancreas. Since insulin deficiency is the life-threatening consequence of the loss of endocrine function, the artificial pancreas focuses almost exclusively on insulin supply systems, even though some of the approaches used, such as islet transplantation, of necessity include an undefined element of delivery of other hormones.

Insulin is the most critical hormone produced by the pancreas because, in contradistinction to glucagon, it acts alone in producing its effects, and survival is not possible in its absence. Endogenous insulin secretion is greatest immediately after eating and is lowest in the interval between meals. Coordination of insulin secretion with fluctuating demands for energy production results from beta-cell stimulation by metabolites, other hormones, and neural signals. The beta cells monitor circulating solutes (in humans, primarily blood glucose) and release insulin in proportion to needs, with the result that in normal individuals, blood glucose levels fluctuate within quite narrow limits. The response time of pancreatic islets to an increase in blood glucose is remarkably fast: 10 to 20 sec for the initial burst of insulin release (primarily from intracellular stores), which is then followed, in a diphasic manner, by a gradual increase in secretion of newly synthesized hormone up to the level appropriate to the intensity and duration of the stimulus. Under fasting condition the pancreas secretes about 40μg (1 unit) of insulin per hour to achieve a concentration of insulin of 2–4 μg/ml (50 to 100μU/ml) in portal blood and of 0.5 μg/ml (12μU/ml) in the peripheral circulation.

134.3 Diabetes

Insulin deficiency leads to a disease called *diabetes mellitus*, which is the most common endocrine disease in advanced societies, affecting as many as 3–5% of the population in the United States, with an even higher incidence in specific ethnic groups.

Diabetes is a chronic systemic disease resulting from a disruption of fuel metabolism either because the body does not produce enough insulin or because the available insulin is not effective. In either case, the result is an accumulation of glucose in the blood or *hyperglycemia*, and, once the renal tubular threshold for glucose is exceeded, a spillover of glucose into the urine or "glycosuria." Hyperglycemia is thought to be the main determinant of microvascular alterations which affect several organs (renal glomerulus, retina, myocardium). It is considered as an important factor in large-blood-vessel pathology (aorta and peripheral arteries, such as carotid and lower limb vessels) often observed in diabetes. Neuropathies of the autonomic, peripheral, and central nervous systems are also common in diabetics, although the pathogenic mechanism is not known.

When blood glucose levels are abnormally high , an increasing fraction of hemoglobin in the red blood cells tends to conjugate with glucose forming a compound identifiable by chromatography and called HbA_{IC}. This reaction serves as a tool in diabetes management and control. The fraction of hemoglobin that is normally glycosylated (relative to total Hb) is between 6% and 9%. If elevated, HbA_{IC} reflects the time-averaged level of hyperglycemia over the lifetime of the red blood cells (3–4 months). It can be interpreted as an index of the severity of diabetes or the quality of control by dietary measures and insulin therapy.

There are two main forms of diabetes: *Type I*, juvenile onset diabetes (typically diagnosed before age 30), or insulin-dependent diabetes; and *type II*, maturity onset diabetes (typically observed after age 40), or non-insulin-dependent diabetes.

Type I appears abruptly (in a matter of days or weeks) in children and young people, although there is evidence that the islet destruction pattern may start much earlier but remain undiagnosed until close to 90% of the islets have been rendered ineffective. Type I represents only 10% of all cases of diabetes, yet it still affects close to 1 million patients in the United States alone. Endogenous insulin is almost totally absent, and an exogenous supply is therefore required immediately to avoid life-threatening metabolic accidents (ketoacidosis) as well as the more insidious degenerative processes which affect the cardiovascular system, the kidneys, the peripheral nervous system and the retina. For some unknown reason, 15% of juvenile diabetics never develop complications. Because the cellular and molecular mechanisms of hyperglycemic damage appear experi-mentally to be reversible by tight control of glucose levels, and the progress of vascular complications is said to be proportional to the square of the blood glucose concentration, early and rigorous control of insulin administration is believed to be essential to minimize the long term consequences of the disease.

TABLE 134.1 Pathogenesis of Diabetes

Type I Diabetes	Type II Diabetes
Autoimmune/genetic/ environmental factors	Unknown pathologic/genetic factors
|	|
Destruction of beta cells	Insulin resistance
|	|
Absolute insulin deficiency	|
|	|
Inability to maintain normoglycemia	Inability to maintain normoglycemia
|	|
Secondary lesions	Secondary lesions

In contrast, type II diabetes, which affects close to 90% of all patients, develops slowly and can often be controlled by diet alone or by a combination of weight loss and/or oral hypoglycemic drugs. Endogenous insulin is often present in normal and sometimes exaggerated amounts, reflecting an inability of cells to make use of available insulin rather than a true hormonal deficiency. Only 20–30% of type II diabetics benefit from insulin therapy. Keto-acidosis is rare, and the major problems are those associated with vascular wall lesions and arterial obstruction. Adult onset diabetics develop the same vascular, renal, ocular, and neural complications as type I diabetics, suggesting that even if the origin of the disease is not identical, its later evolution is fundamentally the same as in juvenile diabetics (Table 134.1).

Diabetes is the leading cause of blindness in the United States. It is also responsible for the largest fraction of the population of patients with end-stage renal disease who are being keep alive by maintenance dialysis. It is one of the most frequent causes of myocardial infraction and stroke, and the most common factor in arterial occlusion and gangrene of the lower extremities leading to limb amputation. With the diabetic population increasing by about 5% per year, diabetes has become one of the major contributors to health expenditures in advanced societies.

It is currently thought that destruction of the pancreatic islets in juvenile diabetes is the result of an autoimmune process, occurring in genetically predisposed individuals, perhaps in relation to an intercurrent infectious disease. This has led to clinical attempts to salvage the remaining intact islets at the earliest recognizable stage of the disease using standard immunosuppressive drugs such as cyclosporin A. The effectiveness of this approach has not been demonstrated.

Diabetologists have also recognized that, at the onset of juvenile diabetes, hyperglycemia exerts a deleterious effect on the function of the still surviving islets, and that in the short term, intensive exogenous insulin therapy may actually lead to an apparent cure of the disease. However, this phenomenon, occasionally referred to as the *honeymoon period* of insulin treatment is transitory. After a period of weeks to months, diabetes reappears, and an exogenous source of insulin administration becomes necessary for survival. The need for insulin to protect borderline functioning islets against hyperglycemic damage has also been recognized in the early stage of islet transplantation.

134.4 Insulin

Insulin is the mainstay for the treatment of diabetes type I and, to a lesser extent, type II. Since it is the key substance produced by beta cells, it is important to define its chemical nature and its mode of action.

Insulin is a 51 amino acid peptide with a molecular weight of about 5800 daltons, made up of two chains (A and B) connected by disulfide bridges. Insulin is formed in the beta cell as a cleavage product of a larger peptide, proinsulin, and is stored there as crystals in tiny intracellular vesicles until released into the blood. The other cleavage product, the "connecting" or C-peptide, though biologically inactive, passes into the blood stream together with insulin, and therefore its concentration reflects, in a patient receiving exogenous insulin, the amount of insulin secreted by the pancreas itself, if any.

Active extracts of islet tissue, obtained from animals following ligation of the pancreation duct to avoid autolysis, were first prepared by Banting and Best in 1922. Insulin was crystallized by Abel in 1926. In 1960, Sanger established the sequence of amino acids that make up the molecule shortly thereafter the hormone was synthesized.

The extract composition of insulin varies slightly among animal species. The presence of disulfide bonds between the A and the B chain is critical. Loss of the C-terminal alanine of the B chain by carboxypeptidase hydrolysis results in no loss of biologic activity. The octapeptide residue that remains after splitting off the last eight amino acids of the B chain has a biologic activity amounting to about 15% of the original insulin molecule. About one-half of the insulin disappears in its passage through the liver. Only a small fraction normally goes to peripheral tissues. Plasma half-life is on the order of 40 min, most of the degradation occurring in the liver and kidneys. No biologically active insulin is eliminated in the urine.

Since insulin is a polypeptide, it cannot be taken orally because it would be digested and inactivated in the gastrointestinal tract. In an emergency, insulin can be administered intravenously, but standard treatment relies on subcutaneous administration. This port of entry differs from physiologic secretion of insulin in two major ways: The adsorption is slow and does not mimic the rapid rise and decline of secretion in response to food ingestion, and the insulin diffuses into the systemic veins rather than into the portal circulation.

134.5 Insulin Therapy

Preparations of insulin are classified according to their species of origin and their duration of action. Human insulin (synthesized by recombinant DNA techniques) and porcine insulin (which is obtained by chemical extraction from slaughterhouse-retrieved organs and differs from human insulin by only one amino-acid at the carboxyl-terminus of the B chain) are in principle preferable to bovine insulin (which differs from the human form by three amino acid residues). In practice, all three are equipotent, and all three can, in a minority of patients, stimulate an immune response and cause hypersensitivity reactions. Because of insulin's relatively short duration of action, formulations delaying the absorption of subcutaneously injected hormone and hence prolonging its effectiveness have greatly facilitated the treatment of diabetes. The mode of administration of insulin influences its plasma concentration and bioavailability. The usual treatment schemes rely on the use of one or more types of insulin (Table 134.2).

Pharmacologists classify the available insulin formulations according to their latency and duration of action as fast acting, intermediate-acting, and slow acting using a terminology of *regular, semilente, lente,* and *ultralente* (Table 134.3). Crystalline insulin is prepared by precipitating the polypeptide in the presence of zinc in a suitable buffer solution. Insulin complexed with a strongly basic protein, protamine, and stabilized with zinc is relatively insoluble at physiological pH and is therefore released only slowly from the site of injection. However, one must keep in mind that onset and duration of action may be quite variable from one patient to another.

Whereas insulin was hailed as a life-saving drug in the years following its discovery (which it was, as evidenced by the much reduced incidence of hyperglycemic coma and the much longer life expectancy of juvenile diabetics, as compared to the period before 1930), it has not turned out to be quite the universal panacea it was expected to be. Some of the problems that have surfaced relate to the limitations associated with the mode of administration of the drug. Other problems are thought to derive from our inability to mimic the finely tuned feedback system which normally maintains the blood glucose levels within very narrow limits (Table 134.4).

TABLE 134.2 Types of Insulin

Fast acting	Multiple injections (basal and preprandial)
Intermediate	Two injections (morning and evening)
Slow acting	One injection (to prevent hyperglycemia at night)

TABLE 134.3 Properties of Insulin Preparations

Type	Appearance	Added Protein	Zinc Content mg/100 U	Time of Action (h) Onset	Peak	Duration
Rapid						
Crystalline (regular)	Clear	None	0.01–0.04	0.3–0.7	2–4	5–8
Insulin-zinc suspension (semilente)	Cloudy	None	0.2–0.25	0.5–1.0	2–8	12–16
Intermediate						
NPH (Isophane insulin suspension)	Cloudy	Protamine	0.016–0.04	1–2	6–12	18–24
Lente (insulin zinc suspension)	Cloudy	None	0.2–0.25	1–2	6–12	18–24
Slow						
Ultalente (extended insulin zinc suspension)	Cloudy	None	0.2–0.25	4–6	16–18	20–36
PZI (Protamine zinc insulin suspension)	Cloudy	Protamine	0.2–0.25	4–6	14–20	24–36

Source: Modified from Table 71-1 in *The Pharmacological Basis of Therapeutics*, 3rd ed., Goodman and Gillman

TABLE 134.4 Problems with Insulin Treatment

Problem	Answers
Poor compliance with multiple injections	Better syringes and needles
	Subcutaneous ports
	Intravenous ports
	Insulin pens
Inadequate pharmacokinetics of insulin preparations	Enhance immediate effect
	Prolong duration of effect
Hyperinsulinemia	Release insulin in portal circulation
Lack of feedback control	Servo-controlled administration
	Natural control by transplant
	Bioartificial pancreas

134.6 Therapeutic Options in Diabetes

Experimental studies of blood glucose regulation and empirical observations in diabetic patients have revealed three major characteristics of blood glucose regulation by the pancreas:

1. The natural system operates as a closed-loop regulatory mechanism within narrow limits.
2. Portal administration is more effective than systemic administration because insulin reaches the liver first.
3. Pulsatile administration of insulin is more effective than continuous administration.

There are several stages in the evolution of diabetes where a medical intervention might be helpful (Table 134.5). Many interventions have been attempted with various degrees of success. Since diabetes

TABLE 134.5 Potential Approaches to Diabetes Treatment

Prevent destruction of beta cells
Prevent beta cell exhaustion
Increase insulin output
Amplify glucose signal
Overcome insulin resistance
Replace beta cells by:
Insulin administration
Electromechanical delivery system
Whole organ transplantation
Islet transplantation
Biohybrid device

TABLE 134.6 Biologic and Engineered Insulin Delivery Systems

Insulin Administration Systems	
Standard Insulin	Routes of administration (subcutaneous, intravenous, intraperitoneal, nasal spray)
	Injection systems (syringes, pens)
Insulin release systems	Passive release from depot forms
	Bioresponsive insulin depot
	Implanted, permeable reservoirs
	Programmed release systems
Insulin delivery pumps	
Open loop	Osmotic pumps
	Piston and syringe pumps
	Peristaltic (roller) pumps
	Bellow frame pumps
	Pressurized reservoir pumps
Closed loop	Glucose sensors
	Electromechanical, servo-controlled delivery systems
Insulin Synthesis Systems	
Pancreas transplantation	Simultaneous with kidney transplant
	After kidney transplant
	Before kidney transplant
Islet transplantation	Autologous tissue
	Allogeneic tissue
	Xenogeneic tissue
Encapsulated islets or beta cell transplants	Microencapsulated cells
	Macroencapsulated cells
Genetically engineered cell transplants	Unprotected gene therapy
	Protected gene therapy

expresses a disturbance of biologic feedback mechanism between blood glucose levels and insulin secretion, there might be possibilities to influence the biologic regulation process through its natural sensing or amplifying mechanisms. However, no practical solution has yet emerged from this approach. As a result the accent has been placed primarily on pharmacologic forms of treatment (diverse insulin formulations and routes of administration), engineered delivery systems (extracorporeal or implanted pumps), and substitution by natural insulin production sources (organ, islet, or individual beta cell transplantation, with or without genetic manipulation and with or without membrane immunoprotection). An overview of the recently investigated approaches is given in Table 134.6.

134.7 Insulin Administration Systems

Syringes and Pens

Insulin traditionally has been administered subcutaneously by means of a syringe and needle, from a vial containing insulin at a concentration of 40 units per ml.

When insulin was first proposed to treat diabetic patients in 1922, glass syringes were used. The burden linked to repeated sterilization disappeared when disposable syringes became available in the early 1970s. Yet patients still had to perform the boring and, for some, difficult task of refilling the syringe from a vial. From a pragmatic viewpoint the development of pens in which insulin is stored in a cartridge represented a major advance in diabetes management. In France, for instance, a recent survey indicates that more than half of the patients use a pen for daily therapy. A needle is screwed on the cartridge and should ideally be replaced before each injection. Cartridges containing either 1.5 or 3 ml of 100U/ml regular or intermediate insulin are available. Unfortunately, long-acting insulin cannot be used in pens because it crystallizes at high concentrations. Attempts are being made toward developing soluble, slow-actin analogs of insulin to overcome this problem.

A common insulin regimen consists of injecting regular insulin before each meal with a pen and long-acting insulin at bedtime with a syringe. Typically half of the daily dosage is provided in the form of regular insulin. Since patients often need 50 units of insulin per day, a pen cartridge of 150 units of regular insulin must be replaced every 6 days or so. The setting of the appropriate dose is easy. However, most pens cannot deliver more than 36 units of insulin at a time, which in some patients may place a limit on their use. In those countries where insulin comes in vials of 40 U/ml for use with syringes, patients must be aware that insulin from cartridges is 2.5 times more concentrated than standard vial insulin. The major advantage of insulin pens is that they do not require refilling and therefore can be used discretely. Pens are particularly well accepted by teenagers.

Reservoirs, Depots, and Sprays

Attempts have been made to develop distensible, implantable reservoirs or bags made of silicone elastomers, fitted with a small delivery catheter, and refillable transcutaneously. In the preferred embodiment, insulin drains at a constant rate into the peritoneal cavity. Assuming uptake by the capillary beds in the serosal membranes which line the gastrointestinal tract, insulin reaches first the portal valve and hence the liver which enhances its effectiveness. This is an attractive concept but not without serious handicaps. The implanted reservoir may elicit an untoward tissue reaction or become a site of infection. The delivery catheter may be plugged by insulin crystals if a high insulin concentration is used, or the catheter may be obstructed by biologic deposits at the tip. The system only provides a baseline insulin delivery and needs to be supplemented by preprandial injections. Its overall capacity constitutes a potential risk should the reservoir accidentally rupture and flood the organism with an overdose of insulin.

Insulin depots, in the form of bioerodible polymer structures in which insulin—amorphous or in crystalline form—is entrapped and slowly released as hydrolytic decay of the carrier of the polymer progressively liberates it, present some of the same problems as reservoirs. In addition, the initial burst which often precedes zero-order release can prove unpredictable and dangerous. No reliable long-term delivery system has yet been developed on that principle.

New routes of introduction of insulin are being investigated, in particular, transmucosal adsorption. Nasal sprays of specially formulated insulin may become a practical modality of insulin therapy if means are found to control reliably the dose administered.

Insulin Pumps

Externally carried portable pumps, based on the motorized syringe or miniature roller pump designs, were evaluated in the 1970s with the anticipation that they could be both preprogrammed for baseline delivery and overridden for bolus injections at the time of meals without the need for repeated needle punctures. In actual practice, this system did not find wide acceptance on the part of patients, because it was just too cumbersome and socially unacceptable.

The first *implantable* insulin pump was evaluated clinically in 1980. Unlike heparin delivery or cancer chemotherapy, insulin administration requires programmable pumps with adjustable flow rates. Implantable pumps provide better comfort for the patient than portable pumps, since the former are relatively unobtrusive and involve no danger of infection at the skin catheter junction. They also permit intraperitoneal insulin delivery, which is more efficient than subcutaneous administration. However, these pumps operate without feedback control, since implantable glucose sensors are not yet available, and the patient must program the pump according to his/her needs.

The vapor pressure driven *Infusaid* pump relies on a remarkably astute mechanism for an implantable device. A rigid box is separated in two compartments by a metal bellow with a flat diaphragm and an accordion-pleated seal (bellow frame). One compartment is accessible from the outside through a subcutaneously buried filling port and contains a concentrated insulin solution. The other compartment (the bellows itself) is filled with a liquid freon which vaporizes at 37°C with a constant pressure of 0.6 bar. With no other source of energy but body heat (which is continuously replenished by metabolism and blood circulation), the freon slowly evaporates, and the pressure developed displaces the diaphragm and forces insulin at a constant flow rate through a narrow delivery catheter. The freon energy is restored each time the insulin reservoir is refilled, since the pressure needed to move the bellow liquefies the freon vapor within a smaller compartment.

The housing is a disc-shaped titanium box. The insulin reservoir contains 15 to 25 ml of insulin stabilized with polygenol at a concentration of either 100 or 400 U/ml, providing an autonomy of 1 to 3 months. The self-sealing septum on the filling port can be punctured through the skin up to 500 times. The pump is implanted under general or local anesthesia between skin and muscle in the lateral abdominal area. The tip of the catheter can be located subcutaneously or intraperitoneally.

The original device provided a constant flow of insulin which still needed to be supplemented by the patient at meal time. The most recent model (*Infusaid* 1000) weighs 272 g, contains 22 ml of insulin, and has a diameter of 9 cm and a height of 2.2 cm. It is designed for 100 U/ml insulin and is equipped with a side port which allows flushing of the catheter when needed. To dissipate the pressure generated by freon evaporation before it reaches the catheter, the insulin must cross a 0.22μm bacteriologic filter and pass through a steel capillary 3 cm long and 50μm in diameter. Basal flow rate can be adjusted from 0.001–0.5 ml/h, and a bolus can be superposed by releasing a precise amount of insulin stored in a pressurized accumulator through the opening of a valve leading to the catheter. These control features require an additional source of energy in the form of lithium thionile batteries with a service life of about 3 years. Control is achieved with a handheld electronic module or programs connected to the pump by telemetry.

The current *Minimed* pump (INIP 2001) has a dry weight of 162 g and a diameter of 8.1 cm and contains 14 ml of insulin at a concentration of 400 U/ml, allowing up to 3 months' autonomy between refills. This pump relies on a reservoir from which insulin can be delivered in a pulsatile form by a piston mechanism under the control of solenoid-driven valves. The basal rate can be adjusted 0.13–30 U/h (0.0003 to 0.07 ml/h), and boluses from 1 to 32 U over intervals 1–60 min can be programmed. The reservoir is permanently under negative pressure to facilitate filling.

The Siemens *Infusor* (1D3 model) replaced an earlier type based on a peristaltic miniature roller pump (discontinued in 1987). It is somewhat similar in design and dimensions to the minimed pumps, but it has now been withdrawn from the market.

The Medtronic *Synchromed* pump, which was originally designed for drug therapy and relied on a roller pump controlled by an external programmer, was evaluated for insulin therapy but is seldom used for that purpose because it lacks the required flexibility.

Between 1989 and 1992, 292 insulin pumps were implanted in 259 patients in France, where most of the clinical experience has been collected (205 Minimed, 47 Infusaid, and 7 Siemens). The treatment had to be permanently discontinued in 14 patients (10 because of poor tissue tolerance, 3 due to catheter obstruction, and 1 due to pump failure). The pump was replaced in 33 patients, with 28 cases of component failure; battery, microelectronic control, flow rate decline due to insulin precipitation in Infusaid pumps, insulin reflux in Minimed pumps, and perceived risk of leak from septum. The overall frequency of technical problems was 0.10 per patient-year. There were 46 surgical interventions without pump replacement because of catheter obstructions, and 24 related to poor tissue tolerance at the site of implantation. Glucose regulation was satisfactory in all cases, and the mean blood glucose concentration dropped appreciably after 6 and 18 months of treatment. A major gain with the intraperitoneal delivery was the reduction in the incidence of severe hypoglycemia associated with improved metabolic control. For the sake of comparison, other methods used for intensive insulin therapy showed a threefold increase in the frequency of serious hypoglycemic episodes.

Servo-Controlled Insulin Delivery Systems

Sensor-actuator couples which allow insulin delivery in the amount needed to maintain a nearly constant blood glucose level are sometimes designated as the *artificial beta cell* because of the analogy between

TABLE 134.7 Component Mechanisms of Insulin Delivery Systems

	Natural β cell	Artificial β cell
Sensor of glucose level	Intracellular glycolysis	Polarographic or optoelectronic sensor
Energy source	Mitochondria	Implantable battery
Logic	Nucleus	Minicomputer chip
Insulin reserve	Insulin granules	Insulin solution reservoir
Delivery system	Cell membrane	Insulin pump
Set point	Normally 80–120 mg%	Adaptable
Fail-safe	Glucagon and hunger	Glucose infusion

the components of the electromechanical system and the biologic mechanisms involved in sensing, controlling, and delivering insulin from the pancreas (Table 134.7).

From a technology viewpoint, the key aspects are the glucose sensor, the control systems applicable to a biologic situation where both hyperglycemia and hypoglycemia must be prevented, and the practicality of the overall system.

Glucose Sensors

A glucose sensor provides a continuous reading of glucose concentration. The device consists of a detection part, which determines for the specificity of the glucose measurement, and a transducing element, which transforms the chemical or physical signals associated with glucose recognition into an electric signal. The most advanced technology is based on enzymatic and amperometric detection of glucose. Glucose is recognized by a specific enzyme, usually glucose oxidase, layered on the anode of an electrode, generating hydrogen peroxide which is then oxidized and detected by the current generated in the presence of a fixed potential between the anode and the cathode. Alternatively, a chemical mediator may serve as a shuttle between the enzyme and the electrode to avoid the need for a potential difference at which other substances, such as ascorbate or acetaminophen, may be oxidized and generate an interfering current. Other approaches such as the direct electrocatalytic oxidation of glucose on the surface of the electrode or the detection of glucose by a combination with a lectin have not reached the stage of clinical evaluation.

Ideally a blood glucose sensor would be permanently installed in the bloodstream. Because of the difficulty of building a hemocompatible device and the inconvenience of a permanent blood access, several investigators have turned to the subcutaneous site for glucose sensing. Indeed, glucose concentration in the interstitial fluid is very close to that of blood and reflects quite directly changes associated with meals or physical activity as observed in diabetic patients. Two types of subcutaneous glucose sensors have been developed. Those with electrodes at the tip of a needle may be shielded from their environment by the inflammatory reaction in tissues contacting the electrode membrane and occasionally cause pain or discomfort. Nonetheless, reliable measurements have been obtained for up to 10 days, whereupon the sensor needed to be changed. Sensing can also be based on microdialysis through an implanted, permeable hollow fiber, providing fluid which is then circulated to a glucose electrode. This system has not yet been miniaturized and industrialized.

In either case, the concept is to develop an external monitor such as a wrist watch, which could receive its analyte from the implanted catheter and/or sensor and trigger a warning when glucose concentration is abnormally high or low. Such a monitor could eventually be incorporated in a closed-loop insulin delivery system, but this objective is not yet realizable.

Noninvasive technology has been suggested to monitor blood glucose, e.g., by near-infrared spectroscopy or optical rotation applied to transparent fluid media of the eye. No reliable technology has yet emerged.

The Artificial Beta Cell

The standard equipment for feedback-controlled insulin administration is the artificial beta cell, also referred to in the literature as the extracorporeal artificial pancreas. It consists of a system for continuous blood sampling, a blood glucose sensor, and minicomputer which, through preestablished algorithms, drives an insulin infusion pump according to the needs of the patient. Accessories provide a minute-by-minute recording of blood glucose levels, insulin delivery, and glucose delivery.

In the first clinically oriented artificial beta cell (Miles Laboratories' *Bioslator*), the glucose sensor was located in a flow-through chamber where blood from a patient's vein was sucked continuously, at the rate of 2 ml/h, through a double-lumen catheter. The double lumen allowed the infusion of a minute dose of heparin at the tip of the catheter to prevent thrombosis in the glucose analyzer circuit. Glucose diffused through a semipermeable membrane covering the electrode and was oxidized by the enzyme. The hydrogen peroxide generated by this reaction crossed a second membrane, the role of which was to screen off substances such as urate and ascorbate. The resulting electric current was fed to a computer which controlled the flow rate of the insulin pump when the readings were in the hyperglycemic range and administered a glucose solution in case of hypoglycemia.

Three forms of oversight of blood glucose levels and insulin needs can be achieved with the artificial beta cell:

Blood glucose monitoring is the technique employed to investigate the time course of blood glucose levels in a particular physiologic situation without any feedback control from the artificial beta cell, that is, without insulin administration to correct the fluctuations of blood glucose levels.

Blood glucose control is the standard feedback mode of application of the artificial beta cell, which entails an arbitrary choice of the level of blood glucose desired and provides a recording of the rate of administration and the cumulative dose of intravenous insulin needed to achieve a constant blood glucose level.

Blood glucose clamp involves intravenous insulin administration at a constant rate to obtain a stable, high blood insulin level. The desired value of blood glucose level is arbitrarily selected (typically in the normal range), and the feedback-control capability of the artificial beta cell is used to measure the amount of exogenous glucose needed to "clamp" blood glucose at the desired level in the presence of a slight excess of insulin. With this technique, a decrease in the rate of glucose administration signals a decrease in insulin biologic activity.

The artificial beta cell is primarily a clinical research tool, applicable also for therapy in high-risk conditions such as pregnancy in poor controlled diabetic mothers or cardiac surgery in brittle diabetics [Galletti et al., 1985], where hypothermia and a massive outpouring of adrenergic agents alter the response to insulin [Kuntschen et al., 1986]. The primary drawbacks of current servo-controlled insulin delivery systems are the instability of the glucose sensor (in terms of both risk of thrombosis and the need for intermittent recalibration), the complexity of the system (up to four pumps can be needed for continuous blood sampling, heparinization, insulin administration, and glucose administration), and the risks involved. The dose of intravenous insulin required for rapid correction of hyperglycemia may cause an overshoot, which in turn calls for rapid administration of glucose. Such fluctuations may also influence potassium uptake by cells and in extreme conditions lead to changes in potassium levels which must be recognized and corrected [Kuntschen et al., 1983]. Finally, the general cumbersomeness of the extracorporeal system, to which the patient must be tethered by sampling and infusion lines, limits its applicability. Modeling of blood glucose control by insulin has allowed the development of algorithms that are remarkably efficient in providing metabolic feedback in most clinical situations. The physiologic limits of closed-loop systems have been lucidly analyzed by Sorensen and coworkers [1982] and Kraegen [1983].

134.8 Insulin Production Systems

The exacting requirement placed on insulin dosage and timing of administration in diabetic patients, as well as the many years of safe and reliable treatments expected from the insulin delivery technology, have pointed to the advantages of implantable systems in which insulin would be synthesized as needed and made available to the organism on demand.

As already outlined in Table 134.6 four avenues have been considered and undergone chemical evaluation: whole organ transplantation, human islet and xenogeneic islet transplantation, immunisolation of normal or tumoral insulin-secreting tissue, and transplantation of cells genetically engineered to replace the functions of the beta cells.

Pancreas Transplantation

Human allograft transplantation, first attempted over 20 years ago, has been slow in reaching clinical acceptance because of the difficulty of identifying healthy cadaver organs (the pancreas is also a digestive gland which undergoes autolysis soon after death) and the need to deal with the gland exocrine secretion, which serves no useful purpose in diabetes but nonetheless must be disposed of. After many ingenious attempts to plug the secretory channels with room-temperature vulcanizing biopolymers, a preferred surgical technique has been developed whereby the pancreas is implanted in the iliac fossa, its arterial supply and venous drainage vessels anastomosed to the iliac artery and vein, and the Wirsung canal implanted in the urinary bladder. Surprisingly, the bladder mucosa is not substantially damaged by pancreatic enzymes, and the exocrine secretion dries up after a while.

The success rate of pancreatic tranplantation is now approaching that of heart and liver transplantation. Since it is mostly offered to immunosuppressed uremic diabetic patients who previously received or concurrently receive a kidney transplant, preventative therapy against organ rejection does not constitute an additional risk. The main limitation remains the supply of donor organs, which in the United States will probably not exceed 2000 per year, that is to say, very far from meeting the incidence of new cases of juvenile diabetes [Robertson & Sutherland, 1992]. Therefore, human organ transplantation is not a solution to the public health problems of diabetes and ensuing complications.

Human Islet Transplantation

Interest in the transplantation of islets of Langerhans, isolated from exocrine, vascular, and connective tissue by enzymatic digestion and cell separation techniques, has received a boost following the clinical demonstration that it can lead to insulin independence in type I diabetic patients [Scharp et al., 1990]. However, this "proof of principle" has not been followed by widespread application. Not only is the overall supply of cadaveric organs grossly insufficient in relating to the need (by as much as two orders of magnitude), but the islet separation techniques are complex and incompletely standardized. Therefore, it is difficult to justify at this point cutting into the limited supply of human pancreata for whole organ transplantation, the more so that whole organ replacement has become increasingly successful. Autologous islet transplantation can be successfully performed in relatively uncommon cases of pancreatic and polyscystic disease, where the risk of spilling pancreatic juice in the peritoneal cavity necessitates the removal of the entire organ. Allogeneic islet transplantation under the cover of pharmacologic immunosuppression is useful inasmuch as it provides a benchmark against which to evaluate islet transplantation from animal sources and the benefits of membrane immunoisolation.

Xenogeneic Islet Transplantation

The term bioartificial pancreas refers to any insulin-production-glycemia-regulation system combining living pancreatic beta cells or equivalents with a synthetic polymer membrane of gel to protect the transplant against immune rejection. A wide variety of device designs, cell sources and processing techniques, implant location, and biomaterial formulation and characterization have been investigated (Table 134.8). Common to all is the belief that if transplantation of insulin-producing tissue is to serve the largest number of insulin-dependent diabetic patients possible, xenogeneic tissue sources will have to be identified. In that context, protection by a semipermeable membrane may be the most effective way to dispense with drug immunosuppression therapy [Lysaght et al., 1994]. However, a number of issues must be addressed and resolved before the bioartificial pancreas becomes a clinically acceptable treatment modality [Colton, 1994].

TABLE 134.8 Membrane-Encapsulated Cell Transplants

Technology	Microencapsulation	
	Macroencapsulation	
	Coextrusion	
Location	Vascular release	Intravascular
		Perivascular
	Portal release	Intraperitoneal
		Intrahepatic
		Intrasplenic
	Systemic release	Subcutaneous
		Intramuscular
Cell source	Human islets or beta cells } fetal or adult	
	Animal islets or beta cells } fetal or adult	
	Functionally active tumor tissue	
	Immortalized cell lines	
	Genetically engineered cell lines	
Cell processing	Isolation and purification	
	Culture and banking	
	Cryopreservation	
	Implant manufacturing	
Membrane characterization	Envelope mechanical stability	
	Envelope chemical stability	
	Diffusion and filtration kinetics	
	Immunoportection	
	Bioacceptance	

Tissue Procurement

In the evolution of type I diabetes, the oral glucose tolerance test does not become abnormal until about 70% of the islets are destroyed. Therefore, roughly 250,000 human islets, or about 3500 islets per kilogram of body weight, should suffice to normalize blood glucose levels. In reality, clinical observations of transplanted patients indicate a need for a considerably larger number (3500–6000 islets/kg, and perhaps more). This suggests that many processed islets are either nonviable or not functional because they have been damaged during the isolation procedure or by the hyperglycemic environment of the transplanted patient [Eizirik et al., 1992].

To procure such large numbers of islets an animal source must be identified. Porcine islets are favored because of the size of the animal, the availability of virus-free herds, and the low antigenicity of porcine insulin. Pig islet separation procedures are not yet fully standardized, and there are still considerable variations in terms of yield, viability, and insulin secretory function. Quality control and sterility control present serious challenges to industrialization.

Alternative tissue sources are therefore being investigated. *Insulinomas* provided the first long-term demonstration of the concept of encapsulated endocrine tissue transplant [Altman et al., 1984]. Genetically engineered cell lines which can sense glucose concentration and regulate it with-in the physiologic range have been reported, but so far none has matched the secretory ability of normal pancreatic tissue or isolated islets. This approach remains nonetheless attractive for large-scale device production, even though no timetable can be formulated for successful development.

Device Design

Immunoisolation of allogeneic or xenogeneic islets can be achieved by two main classes of technology: microencapsulation and macroencapsulation. *Microencapsulation* refers to the formation of a spherical gel around each islet, cell cluster, or tissue fragment. Calcium alginate, usually surrounded with a polyanion such as poly-L-lysine to prevent biodegradation, and at times overcoated with an alginate layer to improve the biocompatibility, has been the most common approach, although other polymers may be substituted. Suspension of microcapsules are typically introduced in the peritoneal cavity to deliver insulin to the portal circulation.

Technical advances have permitted the fabrication of increasingly smaller microcapsules (order of 150–200 μm) which do not clump, degrade, or elicit too violent a tissue reaction. Control of diabetes has been achieved in mice, rats, dogs, and, on pilot basis, in humans [Soon-Siong et al., 1994]. An obstacle to human application has been the difficulty, if not the impossibility, of retrieving the very large number of miniature implants, many of which adhere to tissue. There is also a concern for the antigenic burden that may arise upon polymer degradation and liberation of islet tissue, whether living or necrotic.

Macroencapsulation refers to the reliance on larger, prefabricated envelopes in which a slurry of islets or cell clusters is slowly introduced and sealed prior to implantation. The device configuration can be tubular or planar. The implantation site can be perivascular, intravascular, intraperitoneal, subcutaneous, or intravascular.

Intravascular Devices

In the most successful intravascular device (which might be more accurately designated as *perivascular*), a semipermeable membrane is formed into a wide-bore tube connected between artery and vein or between the severed ends of an artery as if it were a vascular graft. The islets are contained in a gel matrix within a compartment surrounding the tube through which blood circulates. In other embodiments, the islets are contained in semipermeable parallel hollow fibers or coiled capillaries attached to the external wall of a macroporous vascular prosthesis. In all cases, a rising glucose concentration in the blood stream leads to glucose diffusion into the islet compartment, which stimulates insulin production, raises its concentration in the gel matrix, and promotes diffusion into the blood stream.

In an earlier type of perivascular device, blood is forced to circulate through the lumina of a tight bundle of hollow fibers, and the islet suspension is placed in the extracapillary space between fibers and circumscribed by a rigid shell, as was the case in the design which established the effectiveness of immunoprotected islets [Chick et al., 1987]. A truly intravascular device design has been proposed in which the islets are contained within the lumina of a bundle of hollow fibers which are plugged at both ends and placed surgically in the blood stream of a large vein or artery, in a mode reminiscent of the intravenous oxygenator (IVOX). Hemocompatibility has been the major challenge with that approach.

There have also been attempts to develop extracorporeal devices with a semipermeable tubular membrane connected transcutaneously, through flexible catheters, to an artery or vein. The islets are seeded in the narrow "extravascular" space separating the membrane from the device external wall and can therefore be inspected and, in case of need, replaced. This concept has evolved progressively to wider-bore (3–6 mm ID), spirally coiled tubes in a disc-shaped plastic housing. Such devices have remained patent for several months in dogs in the absence of anticoagulation and have shown the ability to correct hyperglycemia in spontaneous or experimentally induced diabetes.

In order to accelerate glucose and insulin transport across synthetic membranes and shorten the reactive time lag of immuno-protective islets, Reach and Jaffrin [1984] have proposed to take advantage of a Starling-type ultrafiltration cycle made possible, in blood-perfused devices, by the arteriovenous pressure difference between device inlet and outlet. An outward-directed ultrafiltration flux in the first half of the blood conduit is balanced by a reverse readsorption flux in the second half, since the islet compartment is fluid-filled, rigid, and incompressible. An acceleration of the response time to fluctuations in blood glucose levels has been demonstrated both by modeling and experimentally. This system may also enhance the transport of oxygen and nutrients and improve the metabolic support of transplanted tissue.

The primary obstacle to clinical acceptance of the intravascular bioartificial pancreas is the risk of thrombosis (including obstruction and embolism) of a device expected to function for several years. The smaller the diameter of the blood channel, and the greater the surface of polymeric material exposed to blood, the more likely are thrombotic complications. These devices also share with vascular grafts the risk of a small but definite incidence of infection of the implant and the potential sacrifice of a major blood vessel. Therefore, their clinical application is likely to remain quite limited.

Intratissular Devices

Tubular membranes with diameters on the order of 0.5–2 mm have also been evaluated as islet containers for subcutaneous, intramuscular, or intracavitary implantation: The islets are contained inside the membrane envelope at a low-volume fraction in a gel matrix. A problem is that small-diameter tubes require too much length to be practical. Large-diameter tubes are mechanically fragile and often display a core of necrotic tissue because diffusion distances are too long for adequate oxygen and nutrient transport to the islets. Both systems are subject to often unpredictable foreign body reactions and the development of scar tissue which further impairs metabolic support of the transplanted tissue. However materials that elicit a minimal tissue reaction have been identified [Galletti, 1992], and further device design and evaluation is proceeding at a brisk pace [Scharp et al., 1994].

Some microporous materials display the property of inducing neovascularization at the tissue-material interface and in some cases within the voids of a macroporous polymer. This phenomenon is thought to enhance mass transport for nutrient and secretory products by bringing capillaries in closer proximity to immuno-seperated cells. Pore size, geometry, and interconnections are critical factors in vascularization [Colton, 1994]. Some encapsulated cell types also stimulate vacularization beyond the level observed with empty devices. Chambers made by laminating a 5-μm porosity expanded polytetrafluorethylene (Gore PTFE) on a 9.45-μm Biopore (Millipore) membrane have shown the most favorable results [Brauker et al., 1992]. The laminated, vascualrized membrane structure has been fabricated in a sandwichlike structure that can be accessed through a port to inject an islet suspension once the empty device has been fully integrated in the soft tissues of the host. This organoid can nonetheless be separated from adjacent tissues, exteriorized and retrieved.

Polymers for Immunoisolation

The polymeric materials used in bioartificial endocrine devices serve two major purposes: (1) As a scaffold and an extracellular matrix, they favor the attachment and differentiation of functional cells or cell clusters, and keep them separate from one another, which in some cases has proven critical; (2) as permselective envelopes they provide immunoisolation of the transplant from the host, while inducing a surrounding tissue reaction which will maximize the diffusional exchange of solutes between the transplant and its environment. A number of materials can be used for these purposes (Table 134.9).

The matrix materials typically are gels made of natural or synthetic polyelectrolytes, with quite specific requirements in terms of viscosity, porosity, and electric charge. In some cases specific attachment or growth factors may be added.

For immunoisolation, the most commonly used envelopes are prepared from polyacrylonitrile-polyvinyl chloride copolymers. These membranes display the anistropic structure typical of most ultrafiltration membranes, in which a thin retentive skin is supported on a spongy matrix [Colton, 1994]. The

TABLE 134.9 Polymer Technology for Cell Transplantation

Component	Function	Polymer
Scaffold	Synthetic or semisynthetic extracellular matrix	Collagen Gelatin Alginate Agarose
	Physical separation of cells or cell aggregates	Chitosan Hyaluronan
Envelope		
Microencapsulation	Stabilization of cell suspension and immunoseperation	Alginate Poly-L-lysine
Macroencapsulation	Physical immunoseperation	Polyacrylonitrile Polyvinyl copolymers Polysulfones Modified celluloses
	Transport-promoting tissue-material interface	PTFE-Biopore laminate

shape and dimensions of the interconnecting voids and the microarchitecture of the inner and outer surfaces of the membrane are often critical characteristics for specific cell types and implant locations.

Protection from Immune Rejection

The central concept of immunoisolation is the placement of a semipermeable barrier between the host and transplanted tissue. It has been tacitly assumed that membranes with a nominal 50,000–100,000-daltons cutoff would provide adequate protection, because they would prevent the passage of cells from the host immune system and impair the diffusion of proteins involved in the humoral component of immune rejection.

This belief has been largely supported by empirical observations of membrane-encapsulated graft survival, with occasional failures rationalized as membrane defects.

However, the cut-off point of synthetic semipermeable membranes is not as sharp as one may believe, and there is often a small number of pores which theoretically at least allow the transport of much larger molecules than suggested by the nominal cut-off definition. Therefore, complex issues of rate of transport, threshold concentration of critical proteins, adsorption and denaturation in contact with polymeric materials, and interactions between proteins involved in immune reactions in an environment where their relative concentrations may be far from normal may all impact in immunoprotection. The possibility of antigen release from living or necrotic cells in the sequestered environment must also be considered. The cellular and luminal mechanisms of immunoseperation by semipermeable membranes and the duration of the protection they afford against both immune rejection of the graft and sensitizations of the host call for considerably more study.

134.9 Outlook

Replacement of the endocrine functions of the pancreas presents a special challenge in substitutive medicine. The major disease under consideration, diabetes, is quite common, but a reasonably effective therapy already exists with standard insulin administration. The disease is not immediately life-threatening, and therefore optimization of treatment is predicated on the potential for reducing the long-term complications of diabetes. Therefore, complete clinical validation will require decreases of observations, not merely short-term demonstration of effectiveness in controlling blood glucose levels.

Standard insulin treatment is also relatively inexpensive. Competitive therapeutic technologies will therefore be subject to a demanding cost-benefit analysis before they are widely recognized. Finally, the patient self-image will be a major factor in the acceptance of new diagnostic or treatment modalities. Already some demonstrably useful devices, such as subcutaneous glucose sensors, portable insulin pumps, and the extracorporeal artificial beta cell, have failed in the marketplace for reasons of excessive complexity, incompatibility with all activities of daily life, physicians' skepticism, or cost. Newer technologies involving the implantation of animal tissue or genetically engineered cells will bring about a new set of concerns, whether justified or imaginary. There is perhaps no application of the artificial organ concept where human factors are so closely intertwined with the potential of science and technology as is already the case with the artificial pancreas.

Defining Terms

Anabolism: The aspect of metabolism in which relatively simple building blocks are transformed into more complex substances for the purpose of storage or enhanced physiologic action.

Artificial beta cell: A system for the control of blood glucose levels based on servo-controlled administration of exogenous insulin based on continuous glucose level monitoring.

Artificial pancreas: A device or system designed to replace the natural organ. By convention, this term designates substitutes for the endocrine function of the pancreas and specifically glucose homeostasis though the secretion of insulin.

Autolysis: Destruction of the components of a tissue, following cell death, mediated by enzymes normally present in that tissue.

Bioartificial pancreas: A device or implant containing insulin-producing, glycemia-regulating cells in combination with polymeric structures for mechanical support and/or immune protection.

Catabolism: The aspect of metabolism in which substances in living tissues are transformed into waste products or solutes of simpler chemical composition.

Endogenous: Originating in body tissues.

Exogenous: Introduced in the body from external sources.

Extracorporeal artificial pancreas: An apparatus including a glucose sensor, a minicomputer with appropriate algorithms, an insulin infusion pump, and a glucose infusion pump, the output of which are controlled so as to maintain a constant blood glucose level. (Synonymous with *artificial beta cell.*)

Hyperglycemia: Abnormally high blood glucose level (in humans, above 140 mg/100 ml).

Hypoglycemia: Abnormally low blood glucose level (in humans, below 50 mg/100 ml).

Immunoisolation: Separation of transplanted tissue from its host by a membrane or film which prevents immune rejection of the transplant by forming a barrier against the passage of immunologically active solutes and cells.

Insulinoma: A generally benign tumor of the pancreas, originating in the beta cells and functionally characterized by a secretion of insulin and the occurrence of hypoglycemic coma. Insulinoma cells are thought to have lost the feedback function and blood-glucose-regulating capacity of normal cells, or to regulate blood glucose around an abnormally low set point.

Ketoacidosis: A form of metabolic acidosis encountered in diabetes mellitus, in which fat is used as a fuel instead of glucose (because of lack of insulin), leading to high concentration of metabolites such as acetoacetic acid, β hydroxybutyzic acid, and occasionally acetone in the blood and intestinal fluids.

Organoid: A device—typically an implant—in which cell attachment and growth in the scaffold provided by a synthetic polymer sponge or mesh leads to a structure resembling that of a natural organ including, in many cases, revascularization.

References

Altman JJ, Houlbert D, Callard P, et al. 1986. Long-term plasma glucose normalization in experimental diabetic rats using microencapsulated implants of benign human insulinomas. Diabetes 35:625.

Altman JJ, Houlbert D, Chollier A, et al. 1984. Encapsulated human islet transplantation in diabetic rats. Trans Am Soc Artif Intern Organs 30:382.

Brauker JH, Martinson LA, Young S, et al. 1992. Neovascularization at a membrane-tissue interface is dependent on microarchitecture. Abstracts, Fourth World Biomaterials Congress, Berlin FRG: 685.

Chick WL, Like AA, Lauris V, et al. 1975. Artificial pancreas using living beta cells: effects on glucose homeostasis in diabetic rats. Science 197:780.

Chick WL, Perla JJ, Lauris V, et al. 1977. Artificial pancreas using living beta cells: Effects on glucose homeostasis in diabetic rats. Science 197:780.

Colton CK.1992. The engineering of xenogeneic islet transplantation by immunoisolation. Diab Nutr Metab 5:145.

Colton CK. 1994. Engineering issues in islet immunoisolation. In R Lanza, W Chick (eds), Pancreatic Islet Transplantation, vol III: Immunoisolation of Pancreatic Islets, RG Landes.

Colton CK, Avgoustiniatos ES. 1991. Bioengineering in development of the hybrid artificial pancreas. J Biomech Eng 113:152.

Dionne KE, Colton CK, Yarmush ML. 1993. Effect of hypoxia on insulin secretion by isolated rat and canine islets of Langerhans. Diabetes 42:12.

Eizirik DL, Korbutt GS, Hellerstrom C. 1992. Prolonged exposure of human pancreatic islets to high glucose concentrations in vitro impairs the β-cell function. J Clin Invest 90:1263.

Galletti PM. 1992. Bioartificial organs. Artif Organs 16(1):55.

Galletti PM, Altman JJ. 1984. Extracorporeal treatment of diabetes in man. Trans AM Soc Artif Intern Organs 30:675.

Galletti PM, Kuntschen FR, Hahn C. 1985. Experimental and clinical studies with servo-controlled glucose and insulin administration during cardiopulmonary bypass. Mt Sinai J Med 52:500.

Jaffrin MY, Reach G, Notelet D. 1988. Analysis of ultrafiltration and mass transfer in a bioartificial pancreas. ASME J Biomech Eng 110:1.

Kraegen EW. 1983. Closed loop systems: Physiological and practical considerations. In P Brunetti, KGMM Albeti, AM Albisser, et al. (eds), Artificial Systems for Insulin Delivery, New York, Raven Press.

Kuntschen FR, Galletti PM, Hahn C. 1986. Glucose-insulin interactions during cardiopulmonary bypass. Hypothermia versus normothermia. J Thorac Cardiovas Surg 91:45.

Kuntschen FR, Taillens C, Hahn C, et al. 1983. Technical aspects of Biostator operation during coronary artery bypass surgery under moderate hypothermia. In Artificial Systems for Insulin Delivery, pp 555–559, New York, Raven Press.

Lanza RP, Butler DH, Borland KM, et al. 1991. Xenotransplantation of canine, bovine, and porcine islets in diabetic rats without immunosuppression. PNAS 88:11100.

Lanza RP, Sullivan SJ, Chick WL. 1992. Islet transplantation with immunoisolation. Diabetes 41:1503.

Lysaght MJ, Frydel B, Winn S, et al. 1994. Recent progress in immunoisolated cell therapy. J Cell Biochem 56:1.

Mikos AG, Papadaki MG, Kouvroukogiou S, et al. 1994. Mini-review: Islet transplantation to create a bioartificial pancreas. Biotechnol Bioeng 43:673.

Pfeiffer EF, Thum CI, Clemens AH. 1974. The artificial beta cell. A continuous control of blood sugar by external regulation of insulin infusion (glucose controlled insulin infusion system). Horm Metab Res 6:339.

Reach G, Jaffrin MY, Desjeux J-F. 1984. A U-shaped bioartificial pancreas with rapid glucose-insulin kinetics. In vitro evaluation and kinetic modeling. Diabetes 33:752.

Ricordi C. (ed). 1992. Pancreatic Islet Transplantation, Austin, Tex, RG Landes.

Robertson RP, Sutherland DE. 1992. Pancreas transplantation as therapy for diabetes mellitus. Annu Rev Med 43:395.

Scharp DW, Lacy PE, Santiago JV, et al. 1990. Insulin independence after islet transplantation into a Type I diabetes patient. Diabetes 39:515.

Scharp DW, Swanson CJ, Olack BJ, et al. 1994. Protection of encapsulated human islets implanted without immunosuppression in patients with Type I and II diabetes and in nondiabetic controls. Diabetes (accepted for publication in September 1994 issue).

Soon-Siong P, Heintz RE, Meredith N, et al. 1994. Insulin independence in a type 1 diabetic patient after encapsulated islet transplantation. Lancet 343:950.

Sorensen JT, Colton CK, Hillman RS, et al. 1982. Use of physiologic pharmacokinetic model of glucose homeostasis for assessment of performance requirements for improved requirements for improved insulin therapies. Diabetes Care 5:148.

Sullivan SJ, Maki T, Boreland KM, et al. 1991. Biohybrid artificial pancreas: Long-term implantation studies in diabetic, pancreatectomized dogs. Science 252:718.

Further Information

A useful earlier review of insulin delivery systems is to be found in: *Artificial Systems for Insulin Delivery*, edited by P. Brunetti, K.G.M.M. Alberti, A.M. Albisser, K.D. Hepp, and M. Massi Benedetti, Serono Symposia Publications from Raven Press, New York, 1983. An upcoming review, focused primarily on islet transplantation will be found in *Pancreatic Islet Transplantation*, vol. III: *Immunoisolation of Pancreatic Islets*, edited by R.P. Lanza and W.L. Chick, R.G. Landes Company, Austin, Texas, 1994. A volume entitled *Implantation Biology: The Host Response and Biomedical Devices*, edited by R.S. Greco, from CRC Press, 1994, provides a background in the multiple aspects of tissue response to biomaterials and implants.

Individual contributions to the science and technology of pancreas replacement are likely to be found in the following journals: *Artificial Organs*, the *Journal of Cell Transplantation*, and *Transplantation*. Reports on clinically promising devices are often published in *Diabetes* and *Lancet*.

[illegible] 1985. [illegible] clinical [illegible] and insulin [illegible] Diabetes [illegible]

[illegible] 1988. [illegible] artificial [illegible] 170.

[illegible] 1988. [illegible] practical [illegible] Artificial Systems for [illegible] New York [illegible] Press.

[illegible] 1986. Glucose-insulin [illegible] Diabetologia [illegible]

[illegible] New York, [illegible] Press.

[illegible]

[illegible] Pfeiffer [illegible] continuous [illegible] blood sugar [illegible]

[illegible]

Robertson [illegible]

Shapiro [illegible] patient [illegible]

[illegible] patients [illegible] Diabetes [illegible]

[illegible] transplantation [illegible]

[illegible] Diabetes Care [illegible]

Sullivan [illegible] pancreas [illegible]

Further Information

[illegible] Artificial Systems for [illegible] Diabetes [illegible] New York [illegible] the following [illegible] journal of [illegible] published in *Diabetes* [illegible]

135

Nerve Guidance Channels

Robert F. Valentini
Brown University

In adult mammals, including humans, the peripheral nervous system (PNS) is capable of regeneration following injury. The PNS consists of neural structures located outside the central nervous system (CNS), which is comprised of the brain and spinal cord. Unfortunately, CNS injuries rarely show a return of function, although recent studies suggest a limited capacity for recovery under optimal conditions. Neural regeneration is complicated by the fact that neurons, unlike other cell types, are not capable of proliferating. In successful regeneration, sprouting axons from the proximal nerve stump traverse the injury site, enter the distal nerve stump, and make new connections with target organs. Current surgical techniques allow surgeons to realign nerve ends precisely when the lesion does not require excision of a large nerve segment. Nerve realignment increases the probability that extending axons will encounter an appropriate distal neural pathway, yet the incidence of recovery in the PNS is highly variable, and the return of function is never complete. Surgical advances in the area of nerve repair seem to have reached an impasse, and biologic rather than technical factors now limit the quality of regeneration and functional recovery. The use of synthetic nerve guidance channels facilitates the study of nerve regeneration in experimental studies and shows promise in improving the repair of injured human nerves. Advances in the synthesis of biocompatible polymers have provided scientists with a variety of new biomaterials which may serve as nerve guidance channels, although the material of choice for clinical application has not yet been identified.

The purpose of this chapter is to review the biologic aspects of PNS regeneration (CNS regeneration will be discussed in a subsequent chapter) and the influence of nerve guidance channels on the regeneration process. The biologic mechanisms and the guidance channel characteristics regulating regeneration will be emphasized, since the rational design of guidance systems hinges on the integration of engineering (polymer chemistry, materials science, and so on) and biologic (cellular and molecular events) principles.

0-8493-0462-8/00/$0.00+$.50

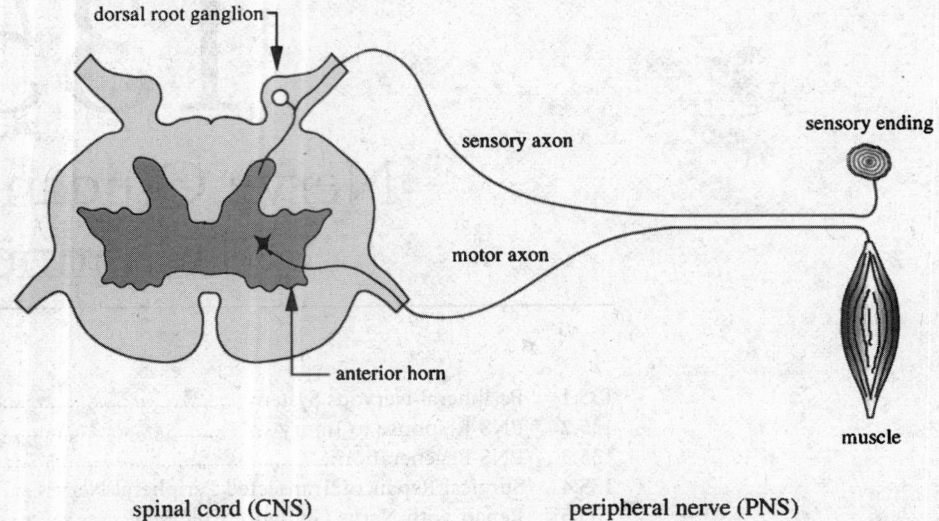

FIGURE 135.1 Spinal cord in cross-section. Relationship between motor and sensory cell bodies and their axons. Motor neurons (black) are located in the anterior horn of the gray matter within the spinal cord. Sensory neurons (white) are located in dorsal root ganglia just outside the spinal cord. Axons exit via dorsal (sensory) and ventral (motor) spinal roots and converge to form peripheral nerves, which connect to target structures (sensory endings and muscles). Note that the distance between axons and their corresponding cell bodies can be quite long.

135.1 Peripheral Nervous System

Peripheral nerve trunks are responsible for the innervation of the skin and skeletal muscles and contain electrically conductive fibers termed *axons*, whose cell bodies reside in or near the spinal cord (Fig. 135.1). Nerve cells are unique in that their cellular processes may extend for up to one meter or longer (e.g., for axons innervating the skin of the foot). Three types of neuronal fibers can be found in peripheral nerves; (1) motor fibers, whose axons originate in the anterior horn of the spinal cord and terminate in the neuromuscular ending of skeletal muscle; (2) sensory fibers, which are the peripheral projections of dorsal root ganglion neurons and terminate at the periphery either freely or in a variety of specialized receptors; and (3) sympathetic fibers, which are the postganglionic processes of neurons innervating blood vessels or glandular structures of the skin. These three fiber types usually travel within the same nerve trunk, and a single nerve can contain thousands of axons. All axons are wrapped by support cells named *Schwann cells.* Larger axons are surrounded by a multilamellar sheath of myelin, a phospholipid-containing substance which serves as an insulator and enhances nerve conduction. An individual Schwann cell may ensheath several unmyelinated axons but only one myelinated axon, within its cytoplasm. Schwann cells are delineated by a fine basal lamina and are in turn surrounded by a complex structure made of collagen fibrils interspersed with fibroblasts and small capillaries, forming a tissue termed the *endoneurium.* A layer of flattened cells with associated basement membrane and collagen constitute the *perineurium,* which envelops all endoneurial constituents. The presence of tight junctions between the perineurial cells creates a diffusion barrier for the endoeurium and thus functions as blood-nerve barrier. The perineurium and its endoneurial contents constitute a *fascicle,* which is the basic structural unit of a peripheral nerve. Most peripheral nerves contain several fascicles, each containing numerous myelinated and unmyelinated axons. Since the fascicular tracts branch frequently and follow a tortuous pathway, the cross-sectional fascicular pattern changes significantly along a nerve trunk. Outside the perineurium is the *epineurium,* a protective, structural connective tissue sheath made of several layers of flattened fibroblastic cells interspersed with collagen, blood vessels, and adipocytes (fat cells). Peripheral nerves are well vascularized, and their blood supply is derived either from capillaries located within the epineurium and endoneurium (i.e., vasa nervosum) or from peripheral vessels which penetrate into the nerve from surrounding arteries and veins.

135.2 PNS Response to Injury

Nerves subjected to mechanical, thermal, chemical, or ischemic insults exhibit a typical combination of degenerative and regenerative responses. The most severe from of injury results from complete transaction of the nerve, which interrupts communication between the nerve cell body and its target, disrupts the interrelations between neurons and their support cells, destroys the local blood-nerve barrier, and triggers a variety of cellular and humoral events. Injuries close to the nerve cell body are more detrimental than injuries occurring more peripherally.

The cell bodies and axons of motor and sensory nerves react in a characteristic fashion to transection injury. The central cell body (which is located within or just outside the spinal cord) and its nucleus swell. Neurotransmitter (e.g., the chemicals which control neuronal signaling) production is diminished drastically, and carbohydrate metabolism is shifted to the pentose-phosphate cycle. These changes indicate a metabolic shift toward production of the substrates necessary for reconstituting the cell membrane and other structural components. For example, the synthesis of the protein tubulin, which is the monomeric elements of the microtubule, the structure responsible for axoplasmic transport, is increased dramatically. Following injury, tubulin and other substrates produced in the cell soma are transported, via slow and fast axoplasmic transport, to the distal nerve fiber.

Immediately after injury, the tip of the proximal stump (i.e., the nerve segment closest to the spinal cord) swells to two or three times its original diameter, and the severed axons retract. After several days, the proximal axons begin to sprout vigorously and growth cones emerge. Growth cones are axoplasmic extrusions of the cut axons which flatten and spread when they encounter a solid, adhesive surface. They elicit numerous extensions (fillipodia) which extend outward in all directions until the first sprout reaches an appropriate target (i.e., Schwann cell basal lamina). The lead sprout is usually the only one to survive, and the others quickly die back.

In the distal nerve stump, the process of Wallerian degeneration begins within 1 to 2 hours after transection. The isolated axons and their myelin sheaths are degraded and phagocytized by Schwann cells and macrophages, leading to a complete dissociation of neurotubules and neurofilaments. This degradation process is accompanied by a proliferation of Schwann cells which retain the structure of the endoneurial tube by organizing themselves into ordered columns termed *bands of Bunger.* While the Schwann cells multiply, the other components of the distal nerve stump atrophy or are degraded, resulting in a reduction in the diameter of the overall structure. Severe retrograde degeneration can lead to cell atrophy and, eventually, cell death. Peripheral nerve transection also leads to marked atrophy of the corresponding target muscle fibers.

135.3 PNS Regeneration

In successful regeneration, axons sprouting from the proximal stump bridge the injury gap and encounter an appropriate Schwann cell column (band of Bunger). In humans, axons elongate at an average rate of 1 mm per day. The bridging axons are immediately engulfed in Schwann cell cytoplasm, and some of them become myelinated. Induction of myelogenesis by the Schwann cell is thought to depend on axonal contact. Some of the axons reaching the distal stump traverse the length of the nerve to form functional synapses, but changes in the pattern of muscular and sensory reinnervation invariably occur. The misrouting of growing fibers occurs primarily at the site of injury, since once they reach the distal stump their paths do not deviate further. The original Schwann cell basal lamina in the distal stump persists and may aid in the migration of axons and proliferating Schwann cells. Axons which fail to reach an appropriate end organ or fail to make a functional synapse eventually undergo Wallerian degeneration.

Abnormal regeneration may lead to the formation of a *neuroma*, a dense, irregular tangle of axons which can cause painful sensations. Some nerve injuries result from clean transections of the nerve which leave the fascicular pattern intact. More often, a segment of nerve is destroyed (e.g., during high-energy trauma or avulsion injuries in which tissue is torn), thus creating a nerve deficit. The resulting nerve gap is the separation between the proximal and distal stumps of a damaged nerve that is due to elastic retraction of the nerve stumps and tissue loss. The longer the nerve gap, the less likely regeneration will occur.

135.4 Surgical Repair of Transected Peripheral Nerves

In the absence of surgical reconnection, recovery of function following a transection or gap injury to a nerve is negligible, since (1) the cell bodies die due to severe retrograde effects, (2) the separation between nerve ends precludes sprouting axons from finding the distal stump, (3) connective tissue ingrowth at the injury site acts as physical barrier to neurite elongation, and (4) the proximal and distal fascicular patterns differ so much that extending axons cannot find an appropriate distal pathway.

Efforts to repair damaged peripheral nerves by surgical techniques date back many years although histologic evidence of regeneration was not reported by Ramon y Cajal until the first part of this century. Early attempts to reconnect severed peripheral nerves utilized a crude assortment of materials, and anatomic repair rarely led to an appreciable return of function. In the 1950s, the concept of end-to-end repair was refined by surgeons who directly reattached individual nerve fascicles or groups of fascicles. Further refinements occurred during the 1960s with the introduction of the surgical microscope and the availability of finer suture materials and instrumentation. Microsurgical nerve repair led to a significant improvement in the return of motor and sensory function in patients to the point that success rates of up to 70% can currently be achieved. In microsurgical nerve repair the ends of severed nerves are apposed and realigned by placing several strands of very fine suture material through the epineurial connective tissue without entering the underlying nerve fascicles. Nerve-grafting procedures are performed when nerve retraction or tissue loss prevents direct end-to-end repair. Nerve grafts are also employed when nerve stump reapproximation would create tension along the suture line, a situation which is known to hinder the regeneration process. In grafting surgery, an autologous nerve graft from the patient, such as the sural nerve (whose removal results in little functional deficit), is interposed between the ends of the damaged nerve.

135.5 Repair with Nerve Guidance Channels

In repair procedures using nerve guidance channels, the mobilized ends of a severed nerve are introduced in the lumen of a tube and anchored in place with sutures (Fig 135.2). Tubulation repair provides: (1) a direct, unbroken path between nerve stumps; (2) prevention of scar tissue invasion into the regenerating environment; (3) directional guidance for elongation neurites and migrating cells; (4) proximal-distal stump communication without suture-line tension in cases of extensive nerve deficit; (5) minimal number of epineurial stay sutures, which are known to stimulate connective tissue proliferation; and (6) preservation, within the guidance channel lumen, of endogenous trophic or growth factors released by the traumatized nerve ends. Guidance channels are also useful from an experimental perspective: (1) the gap distance between the nerve stumps can be precisely controlled; (2) the fluid and tissue entering the channel can be evaluated; (3) the properties of the channel can be varied; and (4) the channel can be filled with various drugs, gels, and the like. Nerves with various dimensions from several mammalian species, including mice, rats, rabbits, hamsters, and nonhuman primates have been tested. The regenerated tissue in the guidance channel is evaluated morphologically to quantify the outcome of regeneration. Parameters analyzed include the cross-sectional area of the regenerated nerve cable, the number of myelinated and unmyelinated axons, and the relative percentages of cellular constituents (i.e., epineurium, endoneurium, blood vessels, and so on). Electrophysiologic and functional evaluation can also be performed in studies conducted for long periods (e.g., several weeks or more).

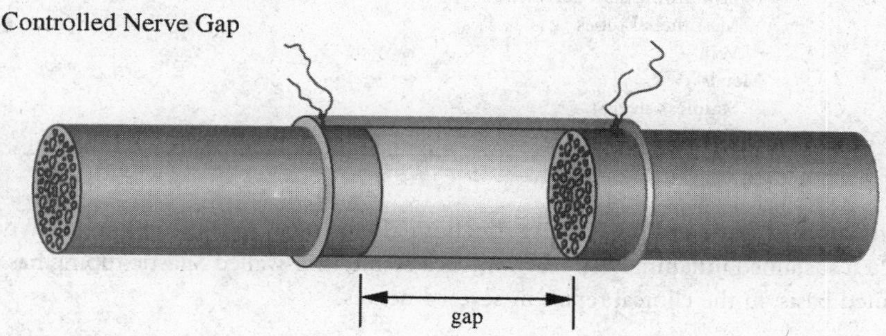

FIGURE 135.2 Tube to repair nerve. Surgical placement of nerve guidance channel.

The frequent occurrence of nerve injuries during the world wars of this century stimulated surgeons to seek simpler, more effective means of repairing damaged nerves. A variety of biologic and synthetic materials shaped into cylinders were investigated (Table 135.1). Bone, collagen membranes, arteries, and veins (either fresh or freeze-dried to reduce antigenicity) were used from the late 1800s through the 1950s to repair nerves. These materials did not enhance the rate of nerve regeneration when compared to regular suturing techniques, so clinical applications were infrequent. Magnesium, rubber, and gelatin tubes were evaluated during World War I, and cylinders of parchment paper and tantalum were used during World War II. The poor results achieved with these materials can be attributed to poor biocompatibility, since the channels elicited an intense tissue response which limited the ability of growing axons to reach the distal nerve stump. Following World War II, polymeric materials with more stable mechanical and chemical properties became available. Millipore (cellulose acetate) and Silastic (silicone elastomer) tubing received the greatest attention. Millipore, a filter material with a maximum pore size of 0.45μm, showed early favorable results. In human trials, however, Millipore induced calcification and eventually fragmented several months after implantation so that its use was discontinued. Silastic tubing, a biologically inert polymer with rubberlike properties, was first tested in the 1960s. Thin-walled Silastic channels were reported to support regeneration over large gaps in several mammalian species. Thick-walled tubing was

TABLE 135.1 Materials Used for Nerve Guidance Channels

Materials
Synthetic materials
Nonresorbable
Nonporous
Ethylene-Vinyl Acetate Copolymer (EVA)
Polytetrafluoroethylene (PTFE)
Polyethylene (PE)
Silicone elastomers (SE)
Polyvinyl chloride (PVC)
Polyvinylidene fluoride (PVDF)
Microporous
Gortex, expanded polytetrafluoroethylene (ePTFE)
Millipore (cellulose filter)
Semipermeable
Polyacrylonitrile (PAN)
Polyacrylonitrile/Polyvinyl chloride (PAN/PVC)
Polysulfone (PS)
Bioresorbable
Polyglycolide (PGA)
Polylactide (PLLA)
PGA/PLLA blends
Biologic materials
Artery
Collagen
Hyaluronic acid derivatives
Mesothelial tubes
Vein
Metals
Stainless steel
Tantalum

associated with nerve necrosis and neuroma production. The material showed no long-term degradation nor did it elicit a sustained inflammatory reaction. As a result, thin-walled Silastic tubing has been used, on a very limited basis, in the clinical repair of severed nerves.

135.6 Recent Studies with Nerve Guidance Channels

The availability of a variety of new biomaterials has led to a resurgence of tubulation studies designed to elucidate the mechanisms of nerve regeneration. The spatial-temporal progress of nerve regeneration across a 10-mm rat sciatic nerve gap repaired with a silicone elastomer tube has been analyzed in detail (Fig. 135.3). During the first hours following repair, the tube fills with a clear, protein-containing fluid exuded by the cut blood vessels in the nerve ends. The fluid contains the clot-forming protein, fibrin, as well as factors known to support nerve survival and outgrowth. By the end of the first week, the lumen is filled with a longitudinally oriented fibrin matrix which coalesces and undergoes syneresis to form a continuous bridge between the nerve ends. The fibrin matrix is soon invaded by cellular elements migrating from the proximal and distal nerve stumps, including fibroblasts (which first organize along the periphery of the fibrin matrix), Schwann cells, macrophages, and endothelial cells (which form capillaries). At 2 weeks, axons advancing from the proximal stump are engulfed in the cytoplasm of Schwann cells. After 4 weeks some axons have reached the distal nerve stump, and many have become myelinated. The number of axons reaching the distal stump is related to the distance the regenerating nerve has to traverse and the length of original nerve resected. Silicone guidance channels do not support regeneration if the nerve gap is greater than 10 mm and if the distal nerve stump is left out of the guidance channel. The morphology and structure of the regenerated nerve is far from normal. The size and number of axons and the thickness of myelin sheaths are less than normal. Electroymyographic evaluation of

Filling of Tube by Blood Derived Fluid and Proteins (day 1)

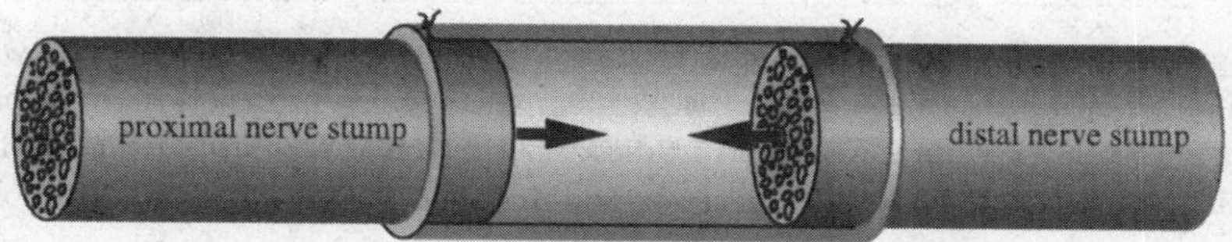

Formation and Coalescence of Fibrin Cable (days 2-6)

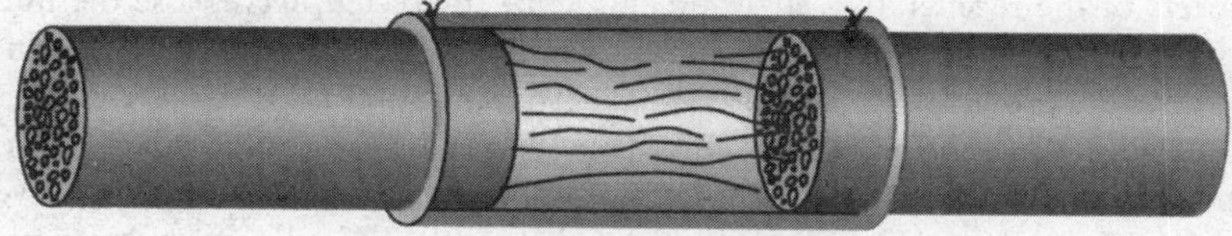

Invasion of Cable by Schwann Cells, Fibroblasts, and Endothelial Cells (days 7-14)

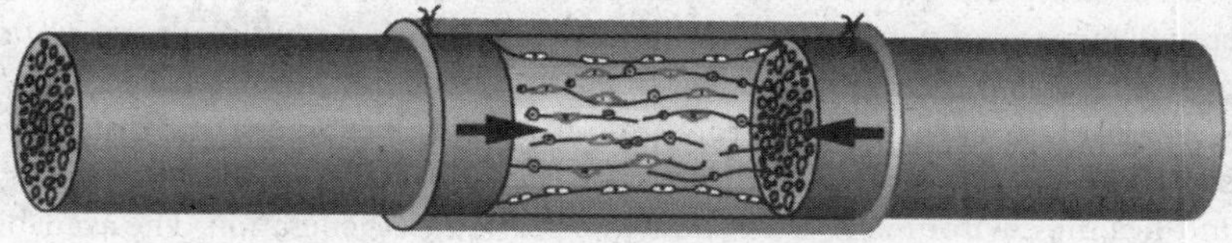

Axonal Elongation and Myelination (days 15-28)

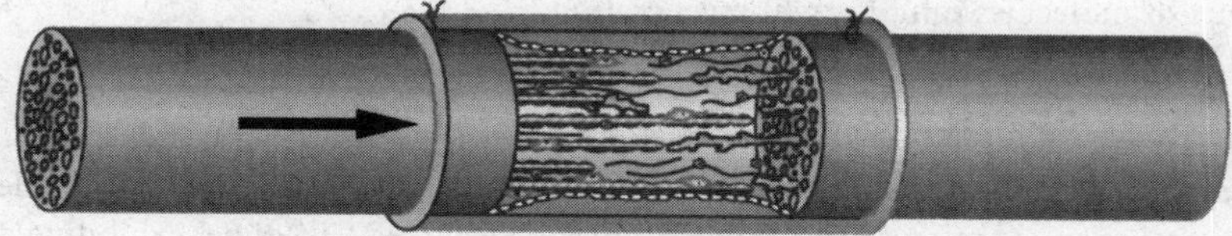

FIGURE 135.3 Regeneration process. Nerve regeneration through a guidance channel.

nerves regenerated through silicone tubes reveals that axons can make functional synapses with distal targets, although nerve conduction velocities and signal amplitudes are slower than normal, even after many months. Attempts to improve the success rate and quality of nerve regeneration have led to the use of other tubular biomaterials as guidance channels (Table 135.1). biodurable materials such as acrylic copolymers, polyethylene, and porous stainless steel and bioresorbable materials including polyglycolides, polylactides, and other polyesters have investigated. There is some concern that biodurable materials may cause long-term complications via compression injury to nerves or soft tissues. Biodegradable materials offer the advantage of disappearing once regeneration is complete, but success thus far has been limited by swelling, increased scar tissue induction, and difficulty in controlling degradation rates. In all cases, these materials have displayed variable degrees of success in bridging transected nerves, and the newly formed nerves are morphologically quite different from normal peripheral nerves. The general spatial-temporal progress of nerve regeneration, however, resembles that described for the silicone channel model.

135.7 Enhancing Regeneration by Optimizing Nerve Guidance Channel Properties

Manipulating the physical, chemical, and electrical properties of guidance channels allows control over the regenerating environment and optimization of the regeneration process. The following features of synthetic nerve guidance channels have been studied: (1) transmual permeability, (2) surface texture or microgeometry, (3) electric charges, (4) release of soluble factors, (5) inclusion of insoluble factors, and (6) seeding with neuronal support cells.

Transmural Permeability

The synthetic nerve guidance channel controls the regeneration process by influencing the cellular and metabolic aspects of the regenerating environment. Since transected nerves lose the integrity of their blood-nerve barrier (which controls oxygen and carbon dioxide tensions, pH, and the concentrations of nutrients and essential proteins), the guidance channel's transmural mass transfer characteristics modulate solute exchange between the regenerating tissue and the surrounding fluids Nerves regenerated through permselective tubes display more normal morphologic characteristics than nerves regenerated in impermeable silicone elastomer (SE) and polyethylene (PE) or freely permeable expanded polytetrafluoroethylene (ePTFE) tubes. Nerves found in semipermeable tubes feature more myelinated axons and less connective tissue. Nerve cables regenerated in semipermeable or impermeable tubes are both round-shaped and free from attachment to the inner wall of the guidance channel. Nerves regenerated in highly porous, open structures do not form a distinct cable but contain connective tissue and dispersed neural elements. The range of permeselectivity is very important, and optimal regeneration is observed with a molecular weight (MW) cut-off of 50,000–100,000 daltons (D). Permeselective PAN/PVC channels with an MW cut-off of 50,000 D support regeneration even in the absence of a distal nerve stump.

These observations suggest that controlled solute exchange between the internal regenerative and external wound-healing environments is essential in controlling regeneration. The availability of oxygen and other nutrients may minimize connective tissue formation in permeable PAN/PVC and PS tubes. Decreased oxygen levels and waste buildup may increase connective tissue formation in SE and PE tubes. Regeneration may also be modulated by excitatory and inhibitory factors released by the wound-healing environment. Semipermeable channels may sequester locally generated growth factors while preventing the inward flux of molecules inhibitory to regeneration.

Surface Texture or Microgeometry

The microgeometry of the luminal surface of the guidance channel plays an important role in regulating tissue outgrowth. Expanded microfibrilar polytetrafluorethylene (ePTFE) tubes exhibiting different internodal distances (1, 5, and 10 μm) were compared to smooth-walled impermeable PTFE tubes. Larger internodal distances result in greater surface irregularity and increased transmural porosity. Rough-walled tubes contained isolated fascicles of nerves disperses within a loose connective tissue stroma. The greater the surface roughness, the greater the spread of fascicles. In contrast, smooth-walled, impermeable PTFE tubes contained a discrete nerve cable delineated by an epineurium and located within the center of the guidance channel. Similar results were observed with semipermeable PAN/PVC tubes with the same chemistry and MW cut-off but with either smooth or rough surfaces. Nerves regenerated in tubes containing alternating sections of smooth and rough inner walls showed similar morphologies with an immediate change from single-cable to numerous fascicle morphology at the interface of the smooth and rough segments.

These studies suggest that the microgeometry of the guidance channel lumen also modulates the nerve regeneration process. Wall structure changes may alter the protein and cellular constituents of the regenerating tissue bridge. For example, the orientation of the fibrin matrix is altered in the presence of a rough inner wall. Instead of forming a single, longitudinally oriented bridge connecting the nerve ends, the fibrin molecules remain dispersed throughout the lumen. As a result, cells migrating in from the nerve stumps loosely fill the entire lumen rather than form a dense central structure.

Electric Charge Characteristics

Applied electric fields and direct dc stimulation are known to influence nerve regeneration in vitro and in vivo. Certain dielectric polymers may be used to study the effect of electric activity on nerve regeneration in vivo and in vitro. These materials are advantageous in that they provide electric charges without the need for an external power supply, can be localized anatomically, are biocompatible, and can be formed into a variety of shapes including tubes and sheets.

Electrets are a broad class of dielectric polymers which can be fabricated to display surface charges because of their unique molecular structure. True electrets, such as polytetrafluorethylene (PTFE), can be modified to exhibit a static surface charge due to the presence of stable, monopolar charges located in the bulk of the polymer. The sign of the charge depends on the poling conditions. Positive, negative, or combined charge patterns can be achieved. The magnitude of surface charge density is related to the number and stability of the monopolar charges. Charge stability is related to the temperature at which poling occurs. Crystalline piezoelectric materials such as polyvinylidene fluoride (PVDF) display transient surface charges related to dynamic spatial reorientation of molecular dipoles located in the polymer bulk. The amplitude of charge generation depends on the degree of physical deformation (i.e., dipole displacement) of the polymer structure. The sign of the charge is dependent on the direction of deformation, and the materials show no net charge at rest.

Negatively and positively poled PVDF and PTFE tubes have been implanted as nerve guidance channels. Poled PVDF and PTFE channels contain significantly more myelinated axons than unpoled, but otherwise identical, channels. In general, positively poled channels contained larger neural cables with more myelinated axons than negatively poled tubes. It is not clear how static or transient charge generation affects the regeneration process. The enhancement of regeneration may be due to electrical influences on protein synthesis, membrane receptor mobility, growth core mobility, cell migration, and other factors.

Release of Soluble Factors

The release of soluble agents, including growth factors and other bioactive substances from synthetic guidance channels, may improve the degree and specificity of neural outgrowth. Using single or multiple injections of growth factors has disadvantages including early burst release, poor control over local drug levels, and degradation in biologic environments. Guidance channels can be prefilled with drugs or growth factors, but the aforementioned limitations persist. Advantages of using a local, controlled delivery system are that the rate and amount of factor release can be controlled and that the delivery can be maintained for long periods (several weeks). Channels composed of an ethylene-vinyl acetate (EVA) copolymer have been fabricated and designed to release incorporated macromolecules in a predictable manner. The amount of drug loaded, its molecular weight, and geometry of the drug-releasing structure affect the drug release kinetics. It is also possible to restrict drug release to the luminal side of the guidance channel by coating its outer wall with a film of pure polymer.

Growth or neurontrophic factors that ensure the survival and general growth of neurons are produced by support cells (e.g., Schwann cells) or by target organs (e.g., muscle fibers). Some factors support neuronal survival, other support nerve outgrowth, and some do both. Numerous growth factors have been identified, purified, and synthesized through recombinant technologies (Table 135.2). In vivo, growth factors are found in solution in the serum or extracellular fluid or bound to extracellular matrix (ECM) molecules. Nerve guidance channels fabricated from EVA and designed to slowly release basic fibroblast growth factor (bFGF) or nerve growth factor (NGF) support regeneration over a 15-mm gap in a rat model. Control EVA tubes supported regeneration over a maximum gap of only 10 mm with no regeneration in 15-mm gaps. The concurrent release of growth factors which preferentially control the survival and outgrowth of motor and sensory neurons may further enhance regeneration, since the

TABLE 135.2 Growth Factors Involved in Peripheral Nerve Regeneration

Growth Factor	Possible function
NGF—nerve growth factor	Neuronal survival, axon-Schwann cell interaction
BDNF—brain-derived neurotophic factor	Neuronal survival
CNTF—ciliary neuronotrophic factor	Neuronal survival
NT-3—neuronotrophin 3	Neuronal survival
NT-4—neuronotrophin 4	Neuronal survival
IGF-1—insulinlike growth factor-1	Axonal growth, Schwann cell migration
IGF-2—Insulinlike growth factor-2	Motoneurite sprouting, muscle reinnervation
PDGF—platelet-derived growth factor	Cell proliferation, neuronal survival
aFGF—acidic fibroblast growth factor	Neurite regeneration, cell proliferation
bFGF—basic fibroblast growth factor	Neurite regeneration, neovascularization

TABLE 135.3 Neuronal Attachment and Neurite-Promoting Factors

Factor	Minimal Peptide Sequence
Collagen	RGD
Fibronectin	RGD
Laminim	RGD, SIKVAV, YIGSR
Neural cell adhesion molecule (N-CAM)	Unknown
N-caherin	Unknown

majority of peripheral nerves contain both populations. For example, NGF and b-FGF control sensory neuronal survival and outgrowth, whereas brain-derived growth factor (BDGF) and ciliary neurontrophic factor (CNTF) control motor neuronal survival and outgrowth. Growth factors released by guidance channels may also allow regeneration over large nerve deficits, and important consideration in nerve injuries with severe tissue loss. The local release of other pharmacologic agents (e.g., anti-inflammatory drugs) may also be useful in enhancing nerve growth.

Inclusion of Insoluble Factors

Several neural molecules found on cell membranes and in the extracellular matrix are potent modulators of neural attachment and outgrowth (Table 135.3). Proteins responsible for eliciting and stimulating axon elongation are termed *neurite promoting factors*. The glycoprotein laminin, an ECM component present in the balsa lamina of Schwann cells, has been reported to promote nerve elongation in vitro and in vivo. Other ECM products, including the glycoprotein fibronectin and the proteoglycan heparan sulphate, also have been reported to promote nerve extension in vitro. Some subtypes of the ubiquitous protein collagen also support neural attachment. Filling guidance channels with gels of laminin and collagen has been shown to improve and accelerate nerve repair. The addition of longitudinally oriented fibrin gels to SE tubes has also been shown to accelerate regeneration. Collagen, laminin, and glycosaminoglycans (another ECM component) introduced in the guidance channel lumen support some degree of regeneration over a 15–20 mm gap in adult rats. The concentration of ECM gel is important, as thicker gels impede regeneration in semipermeable tubes.

The activity of these large, insoluble ECM molecules (up to 10^6 D MW) can be mimicked by short sequences only 3–10 amino acids long (e.g., RGD, YIGSR) (Table 135.3). The availability of small, soluble, bioactive agents allows more precise control over the chemistry, conformation, and binding of neuron-specific substances. Additionally, their stability and linear structure facilitate their use instead of labile (and more expensive) proteins which require three-dimensional structure for activity. Two- and three-dimensional substrates containing peptide mimics have been shown to promote neural attachment and regeneration in vitro and in vivo.

Seeding Neuronal Support Cells

Adding neural support cells to the lumen of guidance channels is another strategy being used to improve regeneration or to make regeneration possible over otherwise irreparable gaps. For example, Schwann cells cultured in the lumen of semipermeable guidance channels have been shown to improve nerve repair in adult rodents. Cells harvested from inbred rats were first cultured in PAN/PVC tubes using various ECM gels as stabilizers. The Schwann cells and gel formed a cable at the center of the tube after several days in culture. Once implanted, the cells were in direct contact with the nerve stumps. A dose-dependent relationship between the density of seeded cells and the extent of regeneration was noted. Another approach toward nerve repair involves the use of Schwann cells, fibroblasts, and the like which are genetically engineered to secrete growth factors. The use of support cells which release neuronotrophic and neurite-promoting molecules may enable regeneration over large gaps. There is increasing evidence that PNS elements, especially Schwann cells, are capable of supporting CNS regeneration as well. For example, Schwann-cell-seeded semipermeable channels support regeneration at the level of the dorsal and ventral spinal cord roots and the optic nerve, which are CNS structures.

135.8 Summary

The permeability, textural, and electrical properties of nerve guidance channels can be optimized to impact favorably on regeneration. The release of growth factors, addition of growth substrates, and inclusion of neural support cells and genetically engineered cells also enhance regeneration through guidance channels. Current limitations in PNS repair, especially the problem of repairing long gaps, and in CNS repair, where brain and spinal cord trauma rarely result in appreciable functional return, may benefit from advances in engineering and biology. The ideal guidance system has not been identified but will most likely be a composite device that contains novel synthetic or bioderived materials and that incorporates genetically engineered cells and new products from biotechnology.

References

Aebischer P, Salessiotis AN, Winn SR. 1989. Basic fibroblast growth factor released from synthetic guidance channels facilitates peripheral nerve regeneration across nerve gaps. J Neurosci Res 23:282.

Dyck PJ, Thomas PK. 1993. Peripheral Neuropathy, 3rd ed, vol 1, Philadelphia, London, WB Saunders.

Guenard V, Kleitman N, Morrissey TK, et al. 1992. Syngeneic Schwann cells derived from adult nerves seeded in semi permeable guidance channels enhance peripheral nerve regeneration. J Neurosci 12:3310–3320.

LeBeau JM, Ellisman MH, Powell HC. 1988. Ultrastructural and morphometric analysis of long-term peripheral nerve regeneration through silicone tubes. J Neurocytol 17:161.

Longo FM, Hayman EG, Davis GE, et al. 1984. Neurite-promoting factors and extracellular matrix components accumulating in vivo within nerve regeneration chambers. Exp Neurol 81:756.

Lundborg G. 1987. Nerve regeneration and repair: A review. Acta Orthop Scand 58:145.

Lundborg G, Dahlin LB, Danielsen N, et al. 1982. Nerve regeneration in silicone chambers: Influence of gap length and of distal stump components. Exp Neurol 76:361.

Lundborg G, Longo FM, Varon S. 1982. Nerve regeneration model and trophic factors in vivo. Brain Res 232:157.

Raivich G, Kreutzberg GW. 1993. Peripheral nerve regeneration: role of growth factors and their receptors. Int J Dev Neurosci 11(3):311.

Sunderland S. 1991. Nerve Injuries and Their Repair, Edinburgh, London, Churchill Livingstone.

Valentini RF, Aebischer P. 1991. The role of materials in designing nerve guidance channels and chronic neural interfaces. In P Dario, G Sandini, P Aebischer (eds), Robots and Biological Systems: Towards a New Bionics?, pp 625–636, Berlin, Germany, Springer-Verlag.

Valentini RF, Aebischer P, Winn SR, et al. 1987. Collagen- and laminin-containing gels impede peripheral nerve regeneration through semi-permeable nerve guidance channels. Exp Neurol 98:350.

Valentini RF, Sabatini AM, Dario P, et al. 1989. Polymer electret guidance channels enhance peripheral nerve regeneration in mice. Brain Res 480:300.

Weiss P. 1944. The technology of nerve regeneration: A review. Sutureless tubulation and related methods of nerve repair. J Neurosurg 1:400.

Williams LR, Longo FM, Powell HC, et al. 1983. Spatial-temporal progress of peripheral nerve regeneration within a silicone chamber: Parameters for a bioassay. J Comp Neurol 218:460.

136

Tracheal, Laryngeal, and Esophageal Replacement Devices

Tatsuo Nakamura
Kyoto University

Yasuhiko Shimizu
Kyoto University

As the ability to reconstruct parts of the body has increased, so has the potential for complications associated with the replacement devices used to do so. Some of the most significant complications associated with replacement devices, such as vascular prostheses, cardiac valves, and artificial joints, are caused by infections at the implantation site. It is well known that the presence of a foreign material impairs host defenses and that prolonged infections cannot be cured unless the foreign material is removed from the site of infection. As the trachea, larynx, and esophagus are located at sites facing the "external environment," these prostheses are exposed to a high risk of infections and severe complications. The development of an artificial trachea, esophagus, and larynx is way behind that of artificial vascular grafts even though they are all tubular organs. In this chapter, we review conventional prostheses and their problems and limitations and discuss the current state of developments in this field.

136.1 Tracheal Replacement Devices

End-to-end anastomosis has been one of the standard operations for tracheal reconstruction. However, in patients in whom a large length of trachea has to be resected, this procedure is often difficult (the resectable limit is now considered to be approximately 6 cm). In such patients, alternative methods are required to reconstruct the airway. Such reconstructive methods can be classified into the following three categories: (1) reconstruction with autologous tissue, (2) reconstruction with nonautologous trachea (that is, transplantation), and (3) reconstruction with artificial material. The only one of these operative techniques which achieves good clinical results in the long term is the first (reconstruction with autologous tissue), particularly cervical tracheal reconstruction using an autologous skip flap conduit.

Designs of Artificial Tracheae

The artificial tracheae developed previously were designed according to one of two concepts. One is that the implanted prosthesis alone replaces the resected area of trachea, and the inner surface of the reconstructed

0-8493-0462-8/00/$0.00+$.50

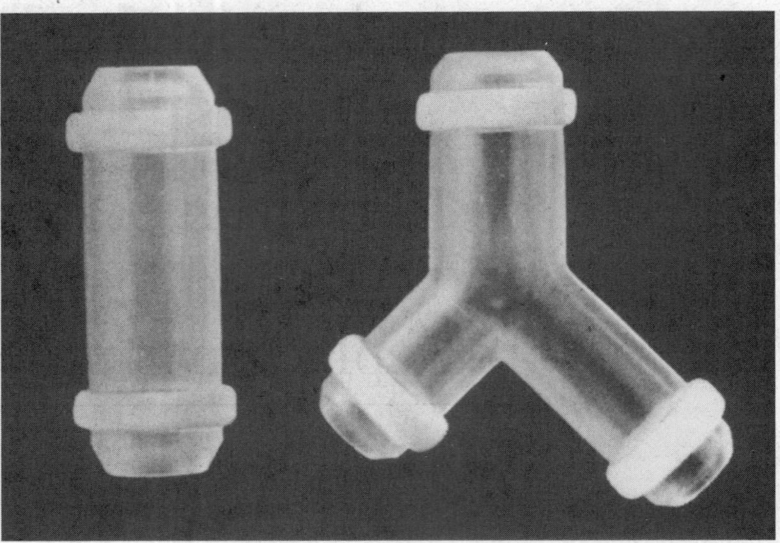

FIGURE 136.1 Neville artificial trachea constructed with a nonporous silicone tube with Dacron suture rings.

trachea is not endothelialized. The other is that the implanted prosthesis is incorporated by the host tissue and its inner surface is endothelialized. These two types of prosthesis are called *nonporous* and *mesh types*, respectively, which reflect the materials from which they are made.

Nonporous Tube—Type Artificial Tracheae

The study of nonporous tube-type trachea has a long history, and many materials have been tried repeatedly for artificial tracheae. However, the results have been unsatisfactory, and now only one prosthesis of this type remains on the market, the *Neville artificial trachea*, which comprises a silicone tube with two suture rings attached to each of its ends (Fig 136.1). Neville used this trachea in 62 patients, from 1970 to 1988, some of whom survived for a long time [1]. However, late complications, such as migration of the artificial trachea and granular tissue formation at the anastomosis, were inevitable in many cases and occurred within several months and even as late as 2 years after implantation. Therefore, the Neville prosthesis is not used for patients with benign tracheal disease, because clinically, even if a large portion of the trachea up to the bifurcation has to be resected, the airway can be reconstructed easily by tracheotomy using a skin flap with sternal resection. For the alleviation of tracheal stenosis, *silicone T-tubes* (Fig. 136.2) are widely indicated. The major advantage of such nonporous tubular prostheses is that airway patency can be ensured. Therefore, for patients for whom end-to-end anastomosis is impossible, nonporous prostheses may be used at last resort to avoid threatened suffocation (asphyxiation).

Tracheal replacement using the Neville artificial trachea requires the following operative procedure:

1. *Right-sided posterolateral skin incision and 4th or 5th intercoastal thoracotomy.* Up to the stage of tracheal resection, the operative procedure is similar to that for standard tracheal resection. However, because the tension at anastomosis is not as high as that with an end-to-end anastomosis, neither hilar nor laryngeal release is often necessary.
2. *Reconstruction with an artificial trachea.* After resection of the tracheal lesion, the oral intubation tube is drawn back, and the trachea is reintubated via the operative field (Fig. 136.3). In patients who require resection that reaches to near the bifurcation, the second intubation tube should be placed in the left main bronchus. An artificial trachea with a diameter similar to that of the tracheal end is used. Any small differences in their diameters can be compensated for by suturing. Anastomosis is carried out using 4–0 absorbable sutures, 2 mm apart, with interrupted suturing. The tracheal sutures are attached to the tracheal cartilage. On completion of the anastomosis on one

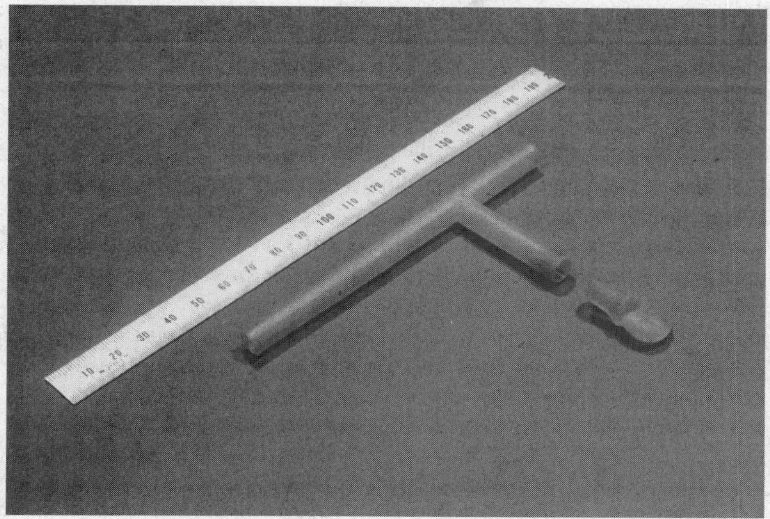

FIGURE 136.2 Silicone T-Tube.

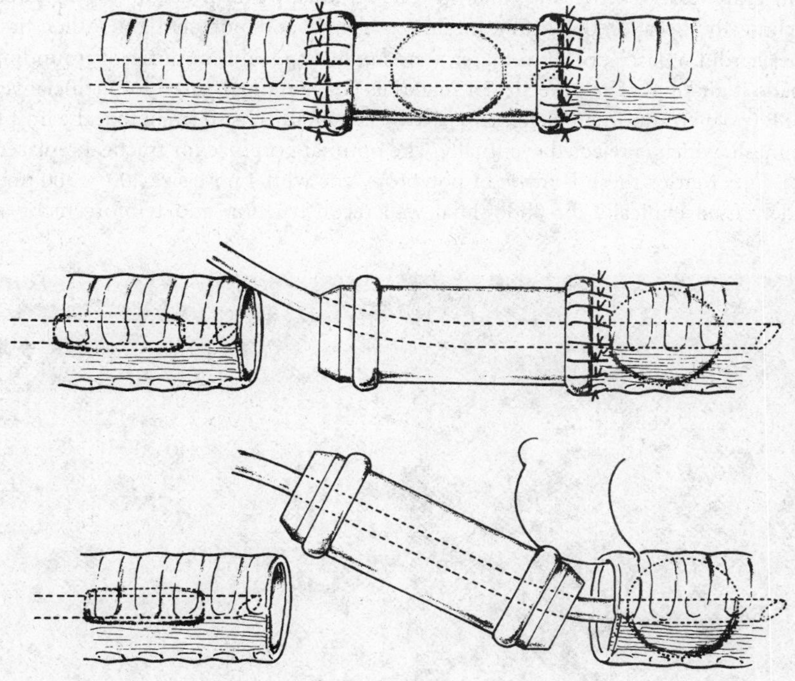

FIGURE 136.3 Operation proceeding of tracheal reconstruction using an artificial trachea.

side, the oral intubation tube is readvanced, and the other side is anastomosed under oral ventilation. After anastomosis, an air leakage test at pressure of 30 cmH_2O is carried out. In order to avoid rupture of the great vessels near the implanted artificial trachea, they should not be allowed to touch each other. In cases at risk of this occurring, wrapping of the artificial trachea with the grater omentum is also recommended.

3. *Postoperative care.* Frequent postoperative boncofiberscopic checks should be performed to ensure sputum does not come into contact with the anastomosis sites. The recurrent laryngeal nerves on both sides are often injured during the operation, so the movement of the vocal cord should be checked at extubation.

The major reason for the poor results with the Neville tracheal prosthesis during long-term follow-up was considered to be the structure of the prosthesis, that is, the nonelasticity of the tube and suture rings. A variety of improvements to this prosthesis have been tried, especially at the areas of interface with the host tissue by suture reinforcement with mesh skirts, increasing the flexibility of the tube, and application of hydroxyapatite to the suture rings. However, the problems at the anastomosis have not been conquered yet.

In attempt to overcome the problems described above, a new mesh-type artificial trachea has been designed that is intended to be incorporated into the host tissue so that eventually there is no foreign-body-external-environment interface.

Mesh-Type Artificial Tracheae

Porous artificial tracheae are called *mesh-type* because the prosthetic trunk is made of mesh. In the 1950s, several trials of tracheal reconstruction using metallic meshes made of tantalum and stainless steel were conducted. In the 1960s, heavy Marlex mesh was used clinically for tracheal reconstruction, and good short-term results were reported. However, long-term observations showed that this mesh caused rupture of the adjacent graft vessels, which was fatal, so it fell gradually out of use for tracheal reconstruction. When used clinically, because the heavy mesh was rough and not air-tight, other tissues, such as autografted pericardium, fascia, or dura mater, were applied to seal it until the surrounding tissue grew into it and made it air-tight. The pore size of materials conventionally used for artificial vessels, such as expanded PTFE (polytetrafluoroethylene, pore size of 15 ~ 30μm), is so small that the host tissue cannot penetrate the mesh, which is rejected eventually. The optimal pore size for tracheal replacement mesh is 200 ~ 300 μm. Fine Marlex mesh is made of polypropylene with a pore size 200 ~ 300 μm (Fig. 136.4). It is now widely used clinically for abdominal wall reconstruction and reinforcement after inguinal

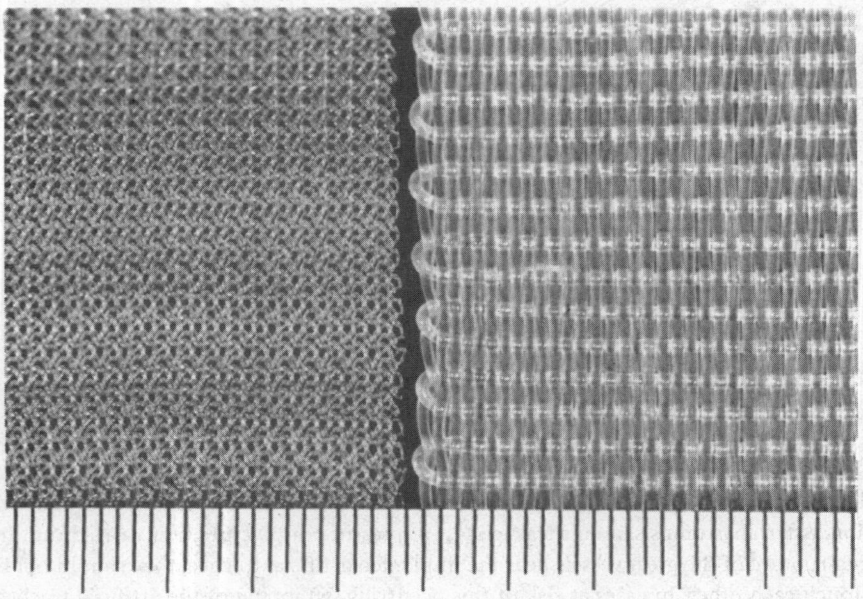

FIGURE 136.4 Marlex Mesh, fine (left) and heavy (right) (division of the scale; mm).

FIGURE 136.5 New artificial trachea made from collagen-conjugated fine Marlex mesh.

herniation. Collagen-grafted fine Marlex mesh is air-tight, and clinically, good tissue regeneration is achieved when it is used to patch-graft of the trachea. The grafted collagen has excellent biocompatibility and promotes connective tissue infiltration into the mesh. However, the fine mesh alone is too soft to keep the tube open, so a tracheal prosthesis was made of collagen-grafted fine Marlex mesh reinforced with a continuous polypropylene spiral (Fig. 136.5). In dogs, complete surgical resection of a 4-cm length of trachea, which was replaced with a 5-cm long segment of this type of artificial trachea, was performed, and the prostheses were incorporated completely by the host trachea and confluent formation of respiratory epithelium on each prosthetic lumen was observed (Fig. 136.6) [2]. These results indicate that this artificial trachea is highly biocompatible and promising for clinical application.

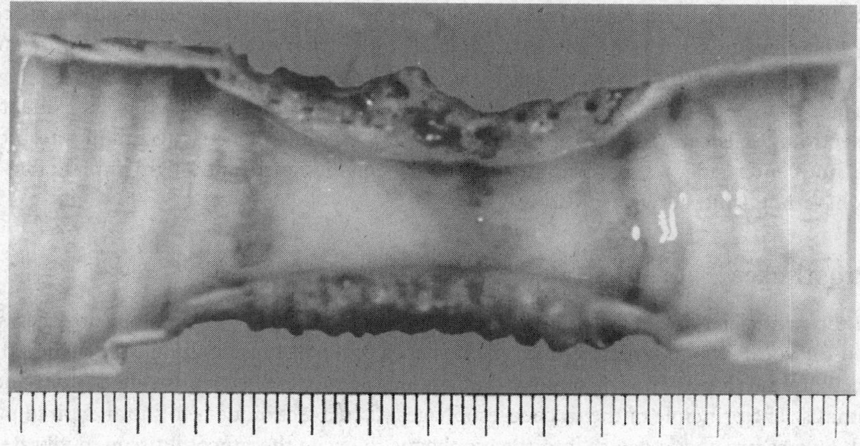

FIGURE 136.6 Macroscopic inner view of the reconstructed trachea 6 months after operation. Inner surface is covered with smooth and lustrous soft regenerated tissue.

136.2 Laryngeal Replacement Devices

Total laryngectomy is one of the standard operations for laryngeal carcinomas. As radiation therapy and surgery have progressed, the prognosis associated with laryngeal carcinoma has improved. The curability of total laryngeal carcinoma is now almost 70%, and therefore many patients survive for a long time after surgery. Individuals who have undergone laryngectomy are called *laryngectomees* or *laryngetomized patients*, and for them, laryngeal reconstruction is of the utmost importance. However, because the larynx is situated just beneath the oral cavity, where the danger of infection is high, successful reconstruction with foreign materials is very difficult. As yet, no total replacement device for the larynx has been developed, and laryngeal transplantation, although apparently feasible, is still at the animal experimental stage.

The larynx has three major functions: (1) phonation, (2) respiration, and (3) protection of the lower airway during swallowing. Of these, phonation is considered to be the most important. The conventional so-called *artificial larynx* can only substitute phonation. A variety of methods have been developed to recover phonation after total larygectomy, which is called *vocal rehabilitation.* Methods for vocal rehabilitation are classified as (1) esophageal speech, (2) artificial larynx, and (3) surgical laryngoplasty. Two-thirds of laryngectomees learn esophageal speech as their means of communication. For the rest, an artificial larynx and surgical laryngoplasty are indicated. The typical devices are the *pneumatic* and *electrical larynx* which are driven by the expiratory force and electric energy, respectively. Tracheo-esophageal (T-E) fistula with voice prosthesis is the most popular method in surgical laryngoplasty. Fig. 136.7 illustrates the mechanical structures of typical artificial larynxes.

Pneumatic Larynxes

The first pneumatic mechanical device was developed by Tapia in 1883. Several variations of pneumatic larynxes are now used. In Fig. 136.8, the pneumatic device uses expired air from the tracheo stoma to vibrate a rubber band or reed to produce a low frequency sound, which is transmitted to the mouth via a tube. Pneumatic transoral larynxes produce excellent natural speech, which is better than that with other artificial larynxes, but their disadvantages are that they are conspicuous and that regular cleaning and mopping of saliva leakage is necessary.

Electric Artificial Larynxes

The *transcervical electrolarynx* is an electric, handheld vibrator that is placed on the neck to produce sound (Fig. 136.9). The frequency used is 100 ~ 200 Hz. The vibrations of the electrolarynx are conducted to the neck tissue and create a low-frequency sound in the hypopharynx. This is the most popular artificial larynx. The *transoral artificial laryngeal device* is a handheld electric device that produces a low-pitched sound which is transmitted to the back of the mouth by a connecting tube placed in the patient's mouth. As microelectronic science progresses, great hopes of applying microelectric techniques to the artificial larynx are now entertained, and some devices have been designed, although implantable laryngeal prostheses have not achieved widespread use.

Voice Prostheses

As well as the artificial larynxes described above, tracheo-esophageal (T-E) fistula prostheses are now widely used for vocal rehabilitation, and excellent speech and voice results have been achieved. In 1980, Singer and Blom developed and introduced the first simple method, which is called Blom and Singer's voice prosthesis. The principle of the tracheo-esophageal fistula technique is to shunt expired pulmonary air through a voice prosthesis device into the esophagus to excite the mucosal tissue to vibrate. A fistula is made by puncturing the posterior wall of the trachea 5 mm below the upper margin of the tracheal stoma, and when a patient speaks, he/she manually occludes the stoma to control the expiratory flow

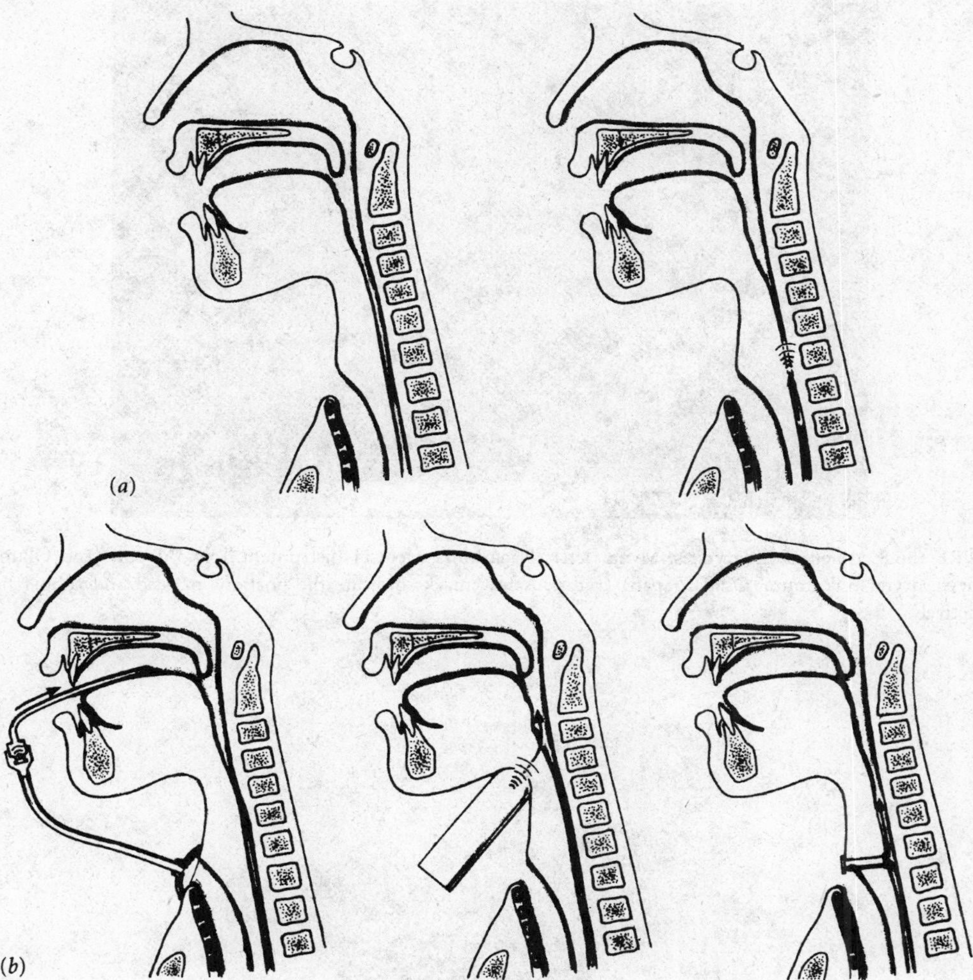

FIGURE 136.7 (*a*) Sagital views of the laryngectomee (left) and esophageal speech (right). Air flow from the esophagus makes the sound. (*b*) Pneumatic larynx (reed type) (left); electric artificial larynx (transcervical type), (center); voice prosthesis (T-E shut) of Blom & Singer method, right.

through the fistula to the oral cavity. The voice prosthesis has a one-way value to prevent saliva leakage (Fig. 136.10).

136.3 Artificial Esophagi (Esophageal Prostheses)

In patients with esophageal cancer, the esophagus is resected and reconstructed using a piece of pediculated alimentary tract, such as the gastric conduit, colon, or ileum. However, in some cases, it is impossible to use autologous alimentary tract, for example, in patients who have undergone gastrectomy. In such cases, an artificial esophagus is indicated. However, with the exception of extracorporeal-type esophagi and intraesophageal stent tubes, the esophageal replacement devices developed have remained far from useful in clinical reconstructive practice. The main reason is anastomosis dehiscence, which often leads to fatal infections, as well as prosthetic dislodgment, migration, and narrowing at a late stage.

The conventional artificial esophagi now used in the clinic can be broadly classified into two types: extracorporeal and intraesophageal stents. *Extracorporeal artificial esophagi* are used as bypasses from

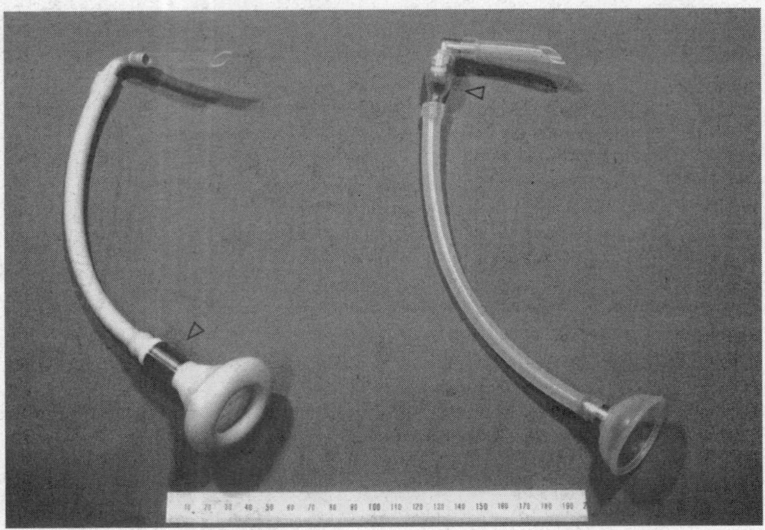

FIGURE 136.8 Pneumatic larynxes: Myna (left) (Nagashima Medical Instrument Tokyo, Japan) and Okumura Artificial larynx (Okumura, Osaka, Japan) (right). Arrow marks indicate the portions of reed and rubber band, respectively.

FIGURE 136.9 Electric artificial larynx (transcervical type): Servox (Dr Kuhn & Co. GmBH, Köln, Germany).

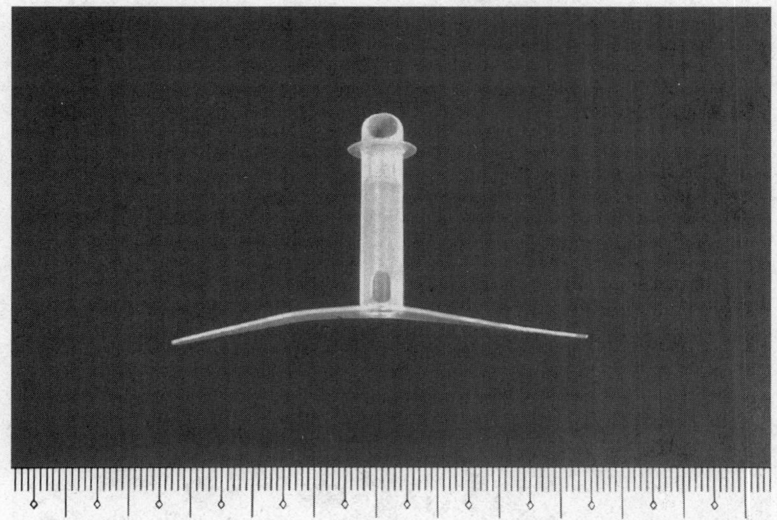

FIGURE 136.10 Voice prosthesis (Bivona, Indiana, USA). This valved tube is inserted into a surgically placed tracheo-esophageal fistula for voice restoration following laryngectomy.

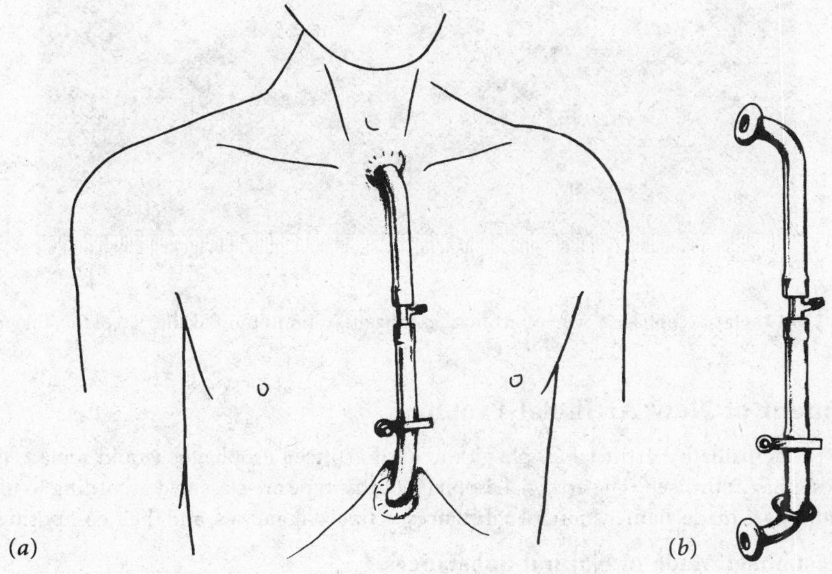

FIGURE 136.11 Extracorporeal artificial esophagus (Tokyo University type).

cervical esophageal to gastric fistulae (Fig. 136.11). They are used during the first stage of two-stage esophageal reconstruction. They are made of latex rubber or silicone, but since the development of IVH method, their use is now extremely rare.

Intraesophageal stent types of artificial esophagus are made from a rubber or plastic tube, which is inserted in the stenotic part of the esophagus (Fig. 136.12). They are used only for unresectable esophageal carcinoma, i.e., they are not indicated for resectable cases. Therefore, conventional artificial esophagi are only palliative devices for use as stop-gap measures.

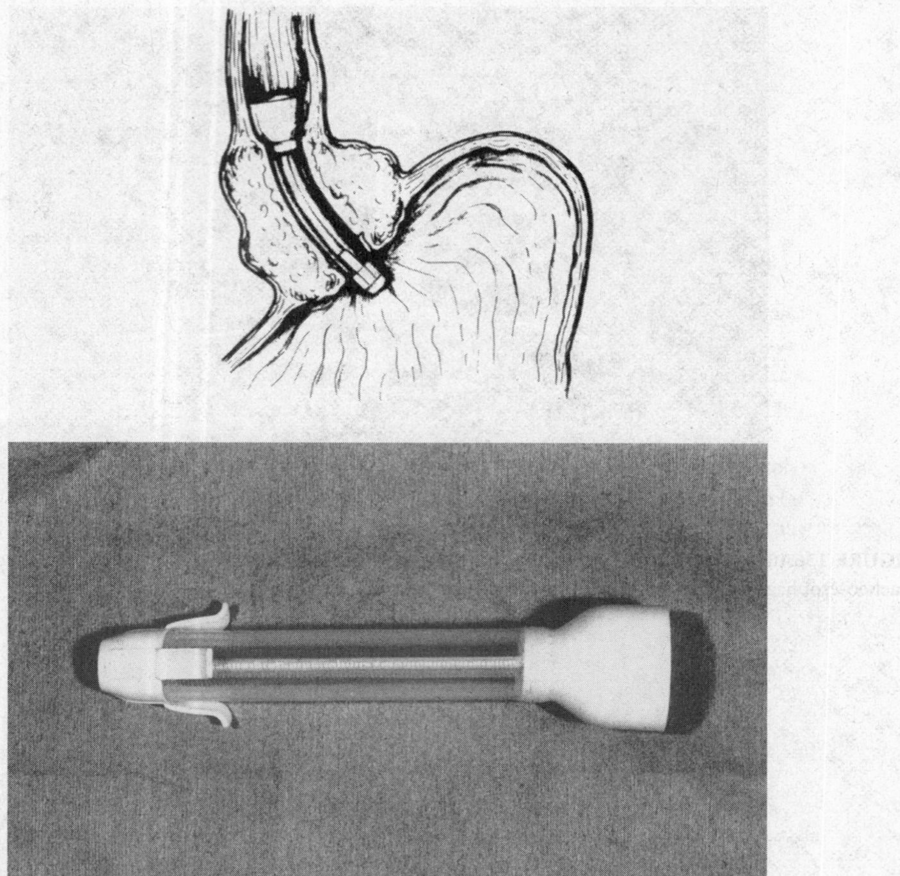

FIGURE 136.12 Intraesophageal type of artificial esophagus. (Sumitomo Bakelite Co., Ltd., Tokyo, Japan).

Development of New Artificial Esophagi

In contrast to the palliative artificial esophagi, the ideal artificial esophagus would replace the resected part of the esophagus by itself. The artificial esophagi of this type are classified according to the materials from which they are made, namely natural substances, artificial materials, and their composites (hybrids).

Artificial Esophagi Made of Natural Substances

The first report of an artificial esophagus made of a natural substance was the skin conduit developed by Bircher in 1907. Subsequently, esophageal reconstruction using a variety of natural substances—muscle fasciae, isolated jejunum, autologous aorta, aortic homograft, autologous esophagus, connective tissue conduit, trachea, and freeze-dried dura mater homograft—has been reported. However, non of these has overcome the problems of stenosis, which necessitates continuous bougieing, and other complications at the anastomosis sites.

Artificial Esophagi Made of Artificial Materials

The first trial of an esophagus made of an artificial material was carried out by Neuhof, who, in 1922, used a rubber tube to reconstruct the cervical esophagus. Subsequently, several artificial materials have been tried repeatedly, such as polyethylene, Dacron, stainless steel mesh, tantalum mesh, dimethylopolysiloxane, polyvinylformal sponge, Teflon, nylon, silicone rubber, collagen-silicone, Dacron-silastic, acryl-resin, and expanded PTFE tubes, but all the trials ended in failure. These materials were foreign bodies, after all, and the host tissues continuously rejected them, which caused anastomotic dehiscence followed by infections. Even rare cases, whose artificial esophagi escaped early rejection, did not avoid the late complications, such as prosthetic migration and esophageal stenosis.

Hybrid Artificial Esophagi

Cultured cells are now widely used for hybrid artificial organs, especially for metabolic organs, such as the liver and pancreas. A hybrid artificial esophagus comprising a latissimus dorsi muscle tube, the inner surface of which is epithelialized by human cultured esophageal epithelial cells, is being studied by the Keio University group [3]. Human esophageal epithelial cells cultured on collagen gel for 10 days were transplanted onto the surfaces of the lattisimus dorsi muscles of athymic mice, and new epithelial cells were observed, which indicates it may be possible to develop an artificial esophagus incorporating cultured epithelial cells.

New Concepts in Artificial Esophageal Design

The previous studies on artificial esophagi are a history of how to prevent the host tissue from recognizing the artificial esophagus as a foreign body, and all the trials, without exception, resulted in failure. Currently, an artificial esophagus made according to a completely new design concept is undergoing trials. The outer collagen sponge layer of the prosthesis is intended to be replaced with host tissue over a period of time, and its inner tube acts as a palliative (temporary) stent until the outer layer has been replaced by host tissue. This type of artificial esophagus comprises an inner silicone tube and an outer tube of collagen sponge (Fig. 136.13), which is nonantigenic and has excellent biocompatibility, as well as promoting tissue regeneration in the form of an extracellular matrix. In dogs implanted with a 5-cm length of this artificial esophagus, esophageal regeneration was accomplished in 3 weeks, and all the early

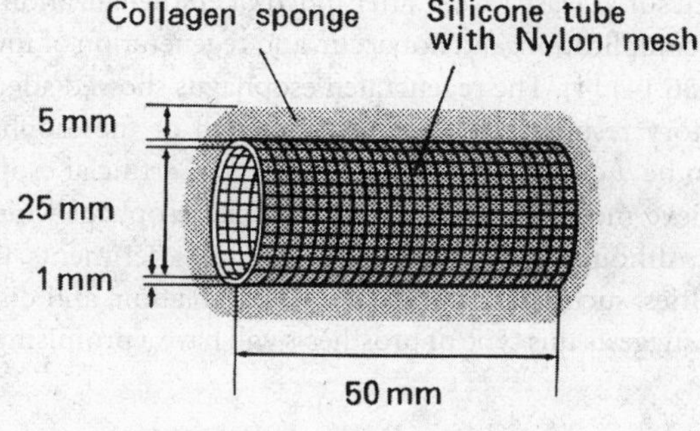

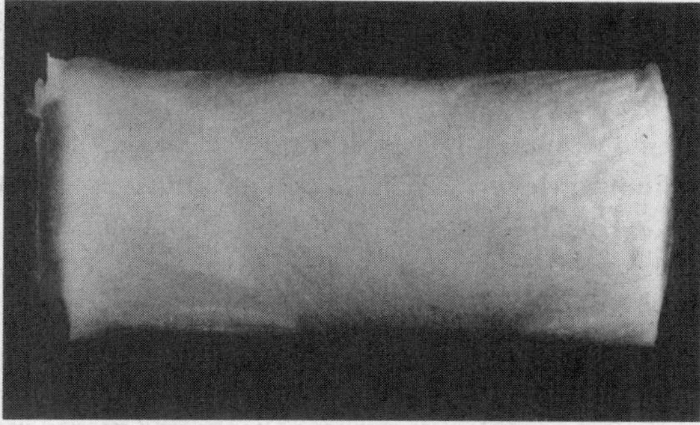

FIGURE 136.13 An artificial esophagus made of collagen sponge which is intended to be replaced with host tissue.

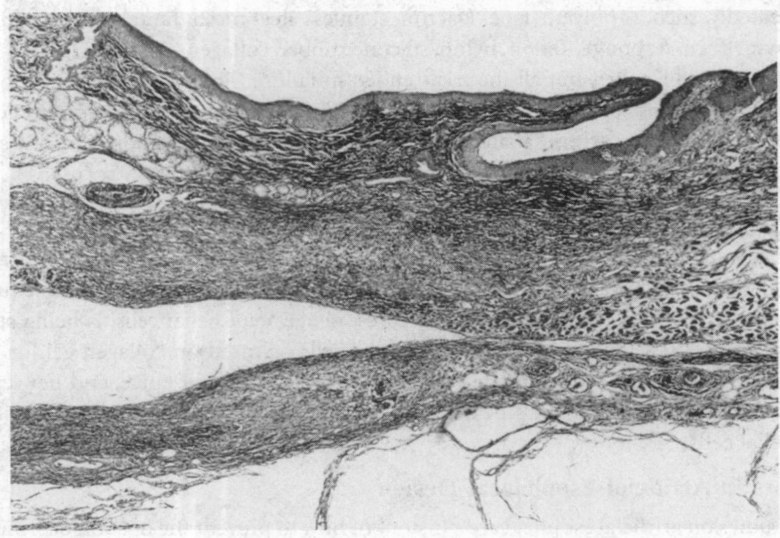

FIGURE 136.14 Pathologic finding of a regenerated esophageal tissue in a dog using the artificial esophagus (Fig. 136.13). Continuous epithelial layer and the muscle and esophageal gland were regenerated at the interposed area.

complications at the anastomosis were overcome. However, in dogs from which the inner stents were removed as soon as mucosal regeneration was accomplished, stenosis developed rapidly, and the regenerated esophagus began to constrict. Such stenosis and constriction could not be overcome, even when autologous buccal mucosal cells were seeded onto the collagen sponge. Accordingly, on the basis of the hypothesis that stenosis and constriction depend on the maturity of the regenerated submucosal tissue, stent removal was postponed for at least 1 week after mucosal epithelialization of the artificial esophagus was complete. Late stenotic complications did not occur, and regeneration of muscle tissue and esophageal glands was observed (Fig. 136.14) [4]. The regenerated esophagus showed adequate physiologic function and pathologically satisfactory results. The limit to the length of the esophagus that can be resected successfully was reported to be 9 cm, and the longest successful artificial esophagus developed hitherto was 5 cm. In order to achieve more widespread use, artificial esophagi which can be used for longer reconstructions are needed. Although still at the stage of animal experiments, the long-term observations indicate that typical difficulties, such as anastomotic leakage, ablation, and dislocation of the prosthesis, have been overcome, which suggests this type of prosthesis will have a promising future in clinical practice.

Defining Terms

Collagen: A main supportive protein of connective tissue, bone, skin, and cartilage. One-third of protein of vertebrate animal consists of collagen.

End-to-end anastomosis: An operative union after resecting the lesion, each end to be joined in a plane vertical to the ultimate flow through the structures.

Expanded PTFE (polytetrafluorethylene): PTFE is polymer made from tetrafluoroethylene ($CF_2 = CF_2$), with the structure of $–CF_2 –CF_2–$. It provides excellent stability chemically and thermally. Teflon is the trade mark of PTFE of Du Pont Co. Expanded PTFE has a microporous structure which has elasticity and antithrombogenisity in the body and is medically applied for vascular grafts or surgical seats.

Extracellular matrix (ECM): The substances which are produced by connective tissue cells. The two major components of the extracellular matrix are collagen and proteoglycans. There are a number of other macromolecules which provide important functions such as tissue growth, regeneration or aging.

Heavy Marlex mesh and fine Marlex mesh: Surgical mesh is used for reconstruction of the defect that results from massive resection. Most popular surgical meshes are Marlex meshes. Heavy mesh is made of high-density polyethylene and has been used for reconstruction of chest wall. Fine mesh is made of polypropylene and is now widely used for abdominal wall reconstruction or reinforcement of inguinal herniation operation.

Hilar release and laryngeal release: The pulmonis hilus is the depression of the mediastinal surface of the lung where the blood vessels and the bronchus enter. The hilus is fixed by the pulmonary ligament to downward. In order to reduce the tension at the tracheal anastomoses, the pulmonary ligament is released surgically. This method is called *hilar release.* At the resection and reconstruct of upper trachea, the larynx can be also released from its upper muscular attachment. This method is called *laryngeal release* and up to 5 cm of tracheal mobilization may be achieved.

Hydroxyapatite: An inorganic compound, $Ca_{10}(PO_4)_6(OH)_2$, found in the matrix of bone and the teeth which gives rigidity to these structures. The biocompatibility of hydroxyapatite has attracted special interest.

IVH (intravenous hyperalimentation): An alimentation method for patients who cannot eat. A catheter is placed in the great vessel, through which high concentration alimentation is given continuously.

Silicone T-tube: A self-retaining tube in the shape of a T which is made of silicone. Tracheal T-tube is used popularly for tracheal stenotic disease and serves both as a tracheal stent and tracheotomy tube. One side branchi of T-tube projects from the tracheotomy orifice.

References

1. Neville WE, Bolanowski PJP, Kotia GG. 1990. Clinical experience with the silicone tracheal prosthesis. J Thorac Cardiovasc Surg 99:604.
2. Okumura N, Nakamura T, Takimoto Y, et al. 1993. A new tracheal prosthesis made form collagen grafted mesh. ASAIO J 39:M475.
3. Sato M, Ando N, Ozawa S, et al. 1993. A hybrid artificial esophagus using cultured human esophageal cells. ASAIO J 39:M554.
4. Takimoto Y, Okumura N, Nakamura T, et al. 1993. Long-term follow up of the experimental replacement of the esophagus with a collagen-silicone composite tube. ASAIO J 39:M736.

Further Information

Proceedings of the American Society of Artificial Internal Organs Conference are published annually by the American Society of Artificial Internal Organs (ASAIO). These proceedings include the latest developments in the field of reconstructive devices each year. The monthly journal *Journal of Thoracic and Cardiovascular Surgery* reports advances in tracheal and esophageal prosthetic instruments. An additional reference is H.F. Mahieu, "Voice and speech rehabilitation following larygectomy, Groningen," the Netherlands, Rijksuniversiteit Groningen, 1988.

137

Artificial Blood

Marcos Intaglietta
University of California

Robert M. Winslow
SANGART Inc.

Blood is a key component of surgery and treatment of injury, and its availability is critical for survival in the presence of severe blood losses. Its present use under conditions of optimal medical care delivery is one unit (0.5 L) per 20 person-year. On a world-wide basis, however, its availability and use is limited to one unit per 100 person-year according to statistics from the World Health Organization. The gap between current optimal need and actual availability is further aggravated by the fact that the blood supply in many locations is not safe. Even in conditions of optimal testing and safeguards, blood per se has a number of inherent risks [Dodd, 1992], ranging from about 3% (3 adverse outcomes per 100 units transfused) for minor reactions, to a probability of 0.001% of undergoing a fatal hemolytic reaction. Superimposed on these risks is the possibility of transmission of infectious diseases such as hepatitis B (0.002%) and hepatitis non-A non-B (0.03%). The current risk of becoming infected with the human immunodeficiency virus (HIV) is about 1 chance in 400,000 under optimal screening conditions. These are risks related to the transfusion of one unit of blood, and become magnified in surgical interventions requiring multiple transfusions.

0-8493-0462-8/00/$0.00+$.50

The HIV epidemic caused many fatalities due to blood transfusion before stringent testing was introduced in 1984. The danger of contamination of the blood supply led the U.S. military to develop blood substitutes starting in 1985 with the objective of finding a product for battlefield conditions that was free of contamination, could be used immediately, and did not need special storage. The end of the Cold War, changes in the nature of military engagements, the slow progress in the development of a blood substitute, and the fact that blood testing caused the blood supply to be again safe and abundant in both the U.S. and Europe lowered the interest of the U.S. military in artificial blood, and the development passed almost exclusively into the hands of private industry.

While about a dozen products have been tested in humans, at present only three are in clinical trials for blood replacement. These products have side effects whose relevance is difficult to assess in the absence of massive clinical trials. A side effect common to hemoglobin-based products is hypertension, which was proposed to be of additional therapeutic value in the treatment of hypotensive conditions associated with severe blood losses and hemorrhagic shock. However, extensive clinical trials led by Baxter Healthcare, with their product HemAssist™ showed that their product caused twice the mortality found with the use of conventional therapies for volume resuscitation. This unfortunate outcome was fully predicted by results in academic research showing that hypertension per se causes a severe impairment of microvascular function.

The key transport parameters that determine the exchange of oxygen in the microcirculation were, until recently, incompletely understood. Most experimental studies underlying the development of artificial blood emphasized systemic, whole organ methods in order to assess efficacy and biocompatibility, without quantitative analysis of transport phenomena inherent to blood function, or data from the microcirculation. Advances in instrumentation and *in vivo* methods (Fig. 137.1) during the past few years have provided new understanding of how oxygen is distributed to the tissues by the microscopic blood vessels, and this information has proven to be crucial in identifying the transport characteristics needed for artificial blood to work effectively, and to counteract some of its inherent problems.

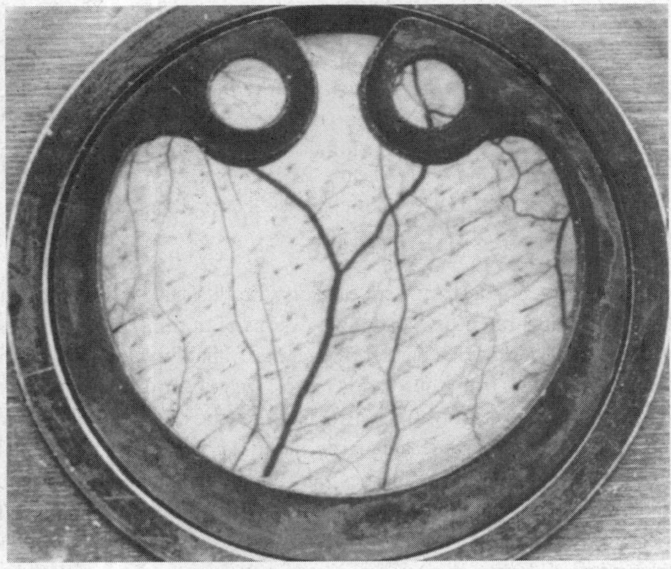

FIGURE 137.1 Experimental hamster skin fold model for the analysis of the microvasculature in the awake animal, without anesthesia and after the effects of surgery have subsided. This technology allows to quantify blood flow, blood and tissue pO_2, leukocyte activation, blood vessel permeability to macromolecules, and blood vessel tone throughout the arterioles, capillaries, and venules. The tissue under study consists of skeletal muscle and subcutaneous connective tissue. Arterial and venous catheters are used to monitor systemic data.

137.1 Oxygen Carrying Plasma Expanders and the Distribution of Transport Properties in the Circulation

To the present it has not been possible to obtain an artificial fluid with the same properties as blood, primarily due to its cellular nature. Consequently, artificial blood or blood substitutes are plasma expanding fluids that carry more oxygen than that dissolved in water or plasma. Their physical properties are significantly different from those of natural blood, and therefore it is necessary to determine the positive or negative effects on circulatory function and tissue oxygenation that may arise from these differences.

Under normal conditions, the transport properties of blood, such as hydraulic pressure, partial pressure of oxygen, oxygen content and viscosity, are distributed (Fig. 137.2). The distribution of transport properties of blood is matched to anatomical features of the macro- and microcirculation and set within narrow limits throughout the life of the organism. Hydraulic blood pressure, which changes continuously in the circulation due to viscous losses, is an example of this situation. Each branching order of the vasculature is adapted to recognize a narrow range of hydraulic blood pressures and react if pressure changes and the normal set point is not met. The distribution of blood pressure influences the cellular composition of blood vessels, and is regulated by the so-called myogenic response [Johnson, 1986], which causes vessels to constrict, thus increasing viscous losses, if pressure increases and vice versa.

The same type of matching is present for the distribution of oxygen tension and blood oxygen content in the vasculature. The shape of the oxygen dissociation curve for hemoglobin and the distribution of

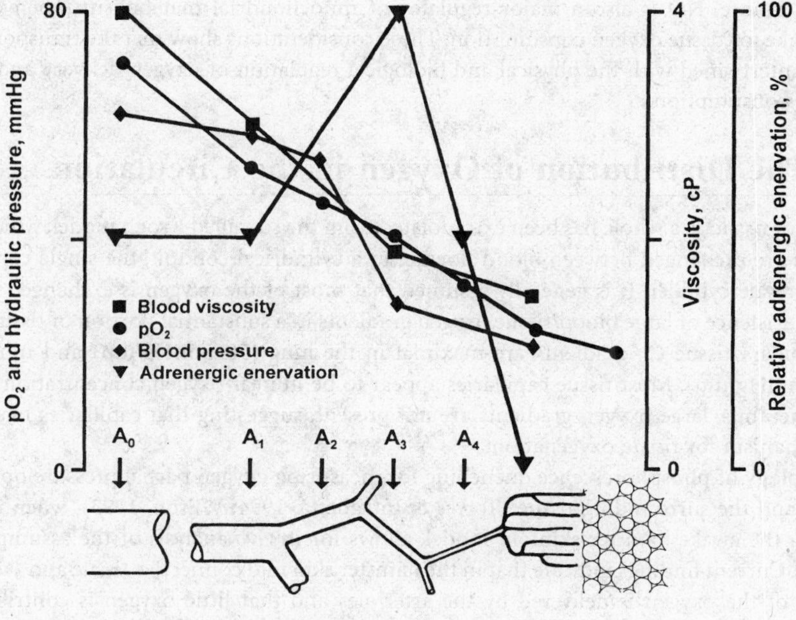

FIGURE 137.2 Distributed nature of transport properties in the hamster skinfold microcirculation. Arterial blood vessels are classified according to their hierarchical position starting with the A_0 small arteries (100 μm and greater diameter), and ending with A_4 terminal arterioles. Blood pressure, blood viscosity, and blood pO_2 fall continuously from the systemic value. Furthermore, the characteristics of the oxygen saturation of hemoglobin curve correspond to specific microvascular locations. This distribution is maintained through the life of an organism and is autoregulated by central and peripheral controls, including adrenergic nerve endings which have the highest density in the A_3 arterioles. Introduction of a molecular plasma expander fundamentally changes the distribution of viscosity, blood pressure, shear stress, and the availability of oxygen to the vessel wall. Autoregulatory processes sense these changes and cause the microcirculation to react. Data on viscosity is derived from [Lipowsky, 1987], data on blood pressure is from the hamster cheek pouch from [Joyner et al., 1972], other data is from our studies.

enervation of arterioles is such that the knee of the oxygen dissociation curve for hemoglobin is located in arterioles that present the highest density of adrenergic enervation [Saltzman et al., 1992].

Blood viscosity and blood flow velocity in the circulation are also distributed in such a fashion that together with the anatomical features of the local blood vessels, shear stress is caused to be virtually uniform throughout the circulation. This is also the consequence of the continuous variation of hematocrit from the systemic value in the larger blood vessels, to about half this value in the capillaries, due to the Faraheus-Lindquist effect (a property due to the cellular composition of blood) and the very strong dependance of blood viscosity on hematocrit. In larger arterioles, blood viscosity is about 3.5 to 4.0 centipoise (cP), while in the smaller arterioles viscosity falls to about that of plasma (1.1 to 1.2 cP). This has large effects on shear stress and shear stress dependant release of endothelial dependant relaxing factors (EDRF, NO, and prostaglandins), since the circulation is designed to maintain shear stress constant. In other words, shear stress is also present in a prescribed way in the microcirculation when the system is perfused by normal blood.

A molecular oxygen carrying plasma expander is not subjected to the Fahraeus-Lindquist effect (which determines the progressive decrease of hematocrit in the microcirculation), causing the whole microcirculation to be exposed to a uniform viscosity (see the mathematical model) and significantly changing shear stress distribution.

The significance of the mechanical traction at the interface between flowing blood and tissue is that shear stress modulates the production of vasoactive materials such as nitric oxide (NO) and prostacyclin [Frangos et al., 1985; Kuchan et al., 1994] and therefore is an active control of blood flow and the oxygen supply. Furthermore, NO is also a major regulator of mitochondrial metabolism [Shen et al., 1995], acting as a brake for tissue oxygen consumption. These considerations show that the transport properties of blood are intertwined with the physical and biological regulation of oxygen delivery and the level of tissue oxygen consumption.

137.2 The Distribution of Oxygen in the Circulation

Oxygenation of tissue as a whole has been extrapolated from the so-called Krogh model, which is focused on how gases are exchanged between blood flowing in a cylindrical conduit, the single capillary, and a surrounding tissue cylinder. It is generally assumed that most of the oxygen is exchanged at this level, implying the existence of large blood/tissue oxygen gradients in a substantial portion of tissue capillaries. However, capillary/tissue O_2 gradients are maximal in the lung (50 mmHg/μm) and minimal in the tissues (0.5 mmHg/μm). Most tissue capillaries appear to be in near oxygen concentration equilibrium with the tissue; thus, large oxygen gradients are not present, suggesting that capillaries may not be the primary mechanism for tissue oxygenation.

The technology of phosphorescence quenching for measuring oxygen partial pressure optically in the microvessels and the surrounding tissue [Torres & Intaglietta, 1994; Wilson, 1993), when used in conjunction with the awake hamster skinfold model, allows for *in vivo* analysis of the assumptions in the Krogh model. Current findings indicate that in the hamster skin fold connective tissue and skeletal muscle at rest, most of the oxygen is delivered by the arterioles and that little oxygen is contributed by the capillaries. In this tissue average capillary pO_2 is about 25 mmHg and venular pO_2 is higher. This indicates that at least half of the oxygen in blood exits the blood vessels prior to arrival in the tissue. In summary, these results and previous findings from other laboratories [Intaglietta et al., 1996] show that:

1. Capillary blood pO_2 is only slightly higher (about 5 mmHg) than tissue pO_2.
2. Arterio/venous capillary pO_2 differences are very small because tissue pO_2 is essentially uniform, and capillaries are close to pO_2 equilibrium with the tissue.
3. The only tissue domain where pO_2 exhibits large gradients is the immediate vicinity of the microvessels (vessels with diameter 80 μm and smaller), a tissue compartment whose main constituent is the microvascular wall.
4. A major portion of blood oxygen exits the circulation via the arterioles.

137.3 Oxygen Gradients in the Arteriolar Wall

The phosphorescence optical technology allows us to make an accurate mass balance between the decrease of oxygen content in the arterioles and the diffusion flux of oxygen out of the microvessels determined by the oxygen gradients in the surrounding tissue. These measurements made simultaneously inside and outside of the vessels show that oxygen exiting from arterioles is driven by steep oxygen gradients at the arteriolar wall, which is consistent with the hypothesis that the arteriolar wall is a high metabolism tissue and therefore a large oxygen sink [Tsai et al., 1998]. These steep gradients are present in arterioles, but not in capillaries and venules. The large oxygen consumption is due to biological activity in the endothelium and smooth muscle [Kjellstrom et al., 1987]. These findings lead to the following conceptualization for the design of artificial blood:

1. Endothelium and smooth muscle serve as a metabolic barrier to the passage of oxygen from blood to tissue, which in part protects the tissue from the high oxygen content (pO_2) of blood.
2. One of the goals of basal tissue perfusion is to supply oxygen to the endothelium and smooth muscle.
3. Oxygenation of working tissue (exercising skeletal muscle) results from three events:
   - Lowering of the vessel wall metabolic barrier.
   - Increased perfusion with oxygenated blood.
   - Deployment of a biochemical process that protects the tissue from high pO_2 levels.
4. Under basal conditions, tissue capillaries only partially serve to supply oxygen to the tissue. They may be a structure to expose the endothelium to blood in order to fulfill the large oxygen demand of these cells and provide for the extraction of CO_2.
5. The physical and biological properties of blood affect the oxygen consumption of the vessel walls.

137.4 Mathematical Modeling of Blood/Molecular Oxygen Carrying Plasma Expander

The previous discussion shows that tissue oxygenation is the result of the interplay of quantifiable physical events, and therefore it may be subjected to analytical modeling to identify key parameters and set the stage for rational design. The critical issue is to understand the changes due to the presence of molecular oxygen carriers. These changes are: (a) change in availability of oxygen to the vessel wall due to the molecular nature of the carrier; (b) distortion of the pattern of intraluminal oxygen distribution in the microcirculation; (c) lowered viscosity and decreased generation of EDRF due to lowered shear stress; (d) increased colloid osmotic pressure; and (e) increased vessel wall pO_2 gradient. The mathematical model is based on oxygen mass balance in subsequent vessel segments, starting with a major arterial vessel and terminating in the capillaries, with the objective of calculating the oxygen tension of blood in the capillaries, which should correspond to tissue pO_2.

The result of this analysis is the following equation, which expresses how the total loss of oxygen from the arterial and arteriolar network K_n is related to anatomical and transport features of the blood vessels and blood [Intaglietta, 1997]. Total oxygen exit is the summation of losses from n individual vascular segments i. This equation gives the functional relationship between transport parameters that determine capillary blood pO_2 for given changes in the physical properties of blood:

$$K_n = \sum k_i = \frac{128\mu}{F\left(Htc, C\right) m_t} \sqrt{\frac{g_o \alpha D}{2}} \sum_1^n \frac{L_i^2}{n_i d_i^3 \Delta P_i} \tag{137.1}$$

where μ is blood viscosity, $F(Htc,C)$ is the concentration of hemoglobin (red blood cells + molecular), m_t is the maximum amount of oxygen that can be dissolved in blood under normal conditions, D is the diffusion constant for oxygen in tissue, α is the solubility of oxygen in tissue, g_o is the oxygen consumption by the vessel wall, L_i is the length of each vessel segment, d_i is the diameter of each segment, and n_i is the slope of the oxygen dissociation curve for hemoglobin.

The summation shows two distinct groups of terms. One group is common to all vessel segments and includes blood viscosity, hematocrit or blood oxygen carrying capacity, and vessel wall metabolism. The second group is a summation where each term is specific to each vascular segment. This expression shows that, in principle, better capillary oxygenation results from lowering viscosity. However, increased capillary oxygenation is a signal that triggers auto regulatory mechanisms which strive to maintain capillary oxygen constant through vasoconstriction. Thus, lowering blood viscosity should be expected to lead to vasoconstriction.

The factor g_o represents vessel wall metabolism which is directly affected by the composition of blood and flow velocity. Furthermore g_o has a basal value representative of the baseline activity necessary for the living processes of the tissue in the microvessel wall and alteration of this activity, such as by an inflammatory process or increased tone, leads to an increase in tissue metabolism and therefore lowering of tissue oxygenation (since the increase in K_n lowers capillary pO_2).

137.5 Blood Substitutes and Hemodilution

In general, it is possible to survive very low hematocrits, corresponding to losses of the red blood cell mass of the order of 70%; however, our ability to compensate for comparatively smaller losses of blood volume is limited. A 30% deficit in blood volume can lead to irreversible shock if not rapidly corrected.

Maintenance of normovolemia is the objective of most forms of blood substitution or replacement, leading to the dilution of the original blood constituents. This hemodilution produces systemic and microvascular phenomena that underlie all forms of blood replacement and provides a physiological reference for comparison for blood substitutes. The fluids available to accomplish volume restitution can be broadly classified as crystalloid solutions, colloidal solutions, and oxygen carrying solutions. All of these materials significantly change the transport properties of blood, and therefore it is important to determine how these changes affect tissue oxygenation, a phenomenon that takes place in the microvasculature. Therefore, hemodilution must be analyzed in terms of systemic effects and how these, coupled with the altered composition of blood, influence the transport properties of the microcirculation.

137.6 Hematocrit and Blood Viscosity

Blood viscosity is primarily determined by the hematocrit in the larger vessels while it is a weaker function of the systemic hematocrit in the microcirculation. At a given shear rate, blood viscosity is approximately proportional to the hematocrit squared according to the relationship:

$$\mu = a_s + b_s H^2$$

while microvascular blood viscosity can be empirically described by a relation of the form:

$$\mu = a_m + b_m H$$

where μ is the blood viscosity in centipoise and a_i and the b_i are parameters that are shear rate and vessel size dependent [Dintenfass, 1979; Quemada, 1978]. In the microcirculation, blood viscosity is relatively insensitive to shear rate. These relationships show that when hematocrit is reduced, systemic viscous pressure losses decrease much more rapidly than those in the microvasculature, while in the microcirculation the A-V pressure drop is not very much affected. The net result is that if arterial pressure remains constant, hemodilution produces a significant pressure re-distribution in the circulation [Mirhashemi et al., 1987].

Hemodilution increases central venous pressure, which improves cardiac performance and increases cardiac output [Richardson and Guyton, 1959]. This effect causes increased blood flow velocity and therefore the maintenance of oxygen delivery capacity, since a lesser number of red blood cells arrive at a greater frequency. As a consequence of hemodilution, both systemic and capillary oxygen carrying capacities improve up to hematocrit 33% and are at the normal value at arterial hematocrits of the order of 27%. The maximum improvement in oxygen carrying capacity is about 10%.

This behavior of the heart and the circulation as determined by the viscosity of the circulating blood has been subjected to exhaustive experimental and clinical verification. The consequence of the adaptability of the circulation to changes in blood viscosity is that, in general, oxygen delivery capacity is not compromised up to red blood cell losses of about 50%. This fact determines the transfusion trigger, currently set at about 7 g Hb/dl. It should be noted that up to the transfusion trigger blood losses can be corrected with the use of a molecular plasma expander. Furthermore, if artificial blood is based on a molecular oxygen carrying material, its introduction in the circulation once the transfusion trigger is passed should be analyzed also as a phenomenon of extreme hemodilution. Conversely, introduction of an oxygen carrying plasma expander prior to reaching the transfusion trigger should show no improvement on oxygenation, since this point is the limit for compensatory adjustments in the circulation.

137.7 Regulation of Capillary Perfusion During Extreme Hemodilution in Hamster Skin Fold Microcirculation by High Viscosity Plasma Expander

Although capillaries do not appear to be the determinant structure for the supply of oxygen, our studies [Kerger et al., 1996] have shown that maintenance of functional capillary density (FCD, defined as the number of capillaries that possess red blood cell transit in a mass of tissue) in shock is a critical parameter in determining the outcome in terms of survival vs. non-survival, independent of tissue pO_2, suggesting that extraction of products of metabolism may be a more critical function of capillaries than oxygenation (Fig. 137.3).

When hemodilution is carried to extreme conditions, defined as the replacement of more than 80% of the red blood cell mass, blood viscosity falls to near plasma levels. A concomitant effect is the reduction of FCD to near pathological levels. The decrease in FCD observed in blood substitutions with plasma expanders is in part due to the lowered viscosity of the resulting blood in circulation. This has been demonstrated with conventional plasma expanders such as dextran 70 kDa, where reductions of systemic hematocrit were to 75% of control and final plasma viscosities of 1.38 cP lowered FCD to 53% of control, while hemodilution with a combination of dextrans of different molecular weights yielding a final plasma viscosity of 2.19 cP maintained FCD to near normal levels. In these experiments, vessel wall shear stress was not changed after low viscosity hemodilution while it increased by a factor of 1.3 in arterioles and 2.0 in venules relative to baseline with high viscosity [Tsai et al., 1998] exchanges. Therefore increased shear stress dependant release of endothelium derived relaxing factors are a possible mechanism that reverses the constrictor effects and decreased FCD due to low blood viscosity.

137.8 Crystalloid and Colloidal Solutions as Volume Expanders

Crystalloids are among the most widely used fluids for volume replacement. Ringer's lactate, for example, is administered in volumes that are as much as three times the blood loss, since the dilution of plasma proteins lowers the plasma oncotic pressure causing an imbalance of the fluid exchange favoring microvascular fluid extravasation and edema. The advantage of crystalloid solutions is that large volumes can be given over a short period of time with low danger of increasing pulmonary wedge pressure. Excess volumes are rapidly cleared from the circulation by diuresis, which in many instances is a beneficial side effect in the treatment of trauma. Blood volume replacement with Ringer's lactate lowers blood viscosity.

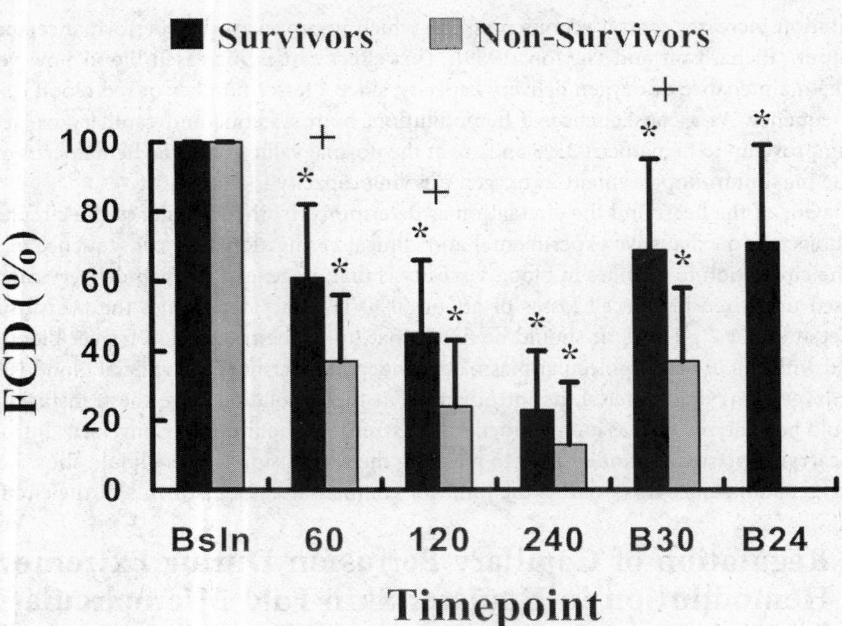

FIGURE 137.3 Changes of functional capillary density (FCD) in the hamster skin fold model during shock and resuscitation. FCD is the number of capillaries with red blood cell transit in a microscopic field of view. Hamsters were subjected to 4 h hemorrhagic shock to 40 mmHg and resuscitated with their own blood. FCD was the principal microvascular parameter that was the predictor of outcome, while microvascular and tissue pO_2 was the same for survivors and non-survivors. This result indicates that resuscitation solutions should ensure the restoration of FCD as well as re-establish tissue oxygenation [Kerger et al., 1996]. Bsln: Baseline. Timepoint: Time after induction of hemorrhage. B30: 30 min. after resuscitation. B24: 24 h after resuscitation. *: Significantly different from control, $p > 0.05$. +: Significantly different between survivors and non-survivors, $p > 0.05$.

The rapid clearance of crystalloid solutions from the intravascular compartment is at the expense of the expansion of the tissue compartment and edema. Peripheral edema has been speculated to impair oxygen delivery, wound healing, and the resistance to infection. Microvascular flow may be impaired by capillary compression from edema. However, given the wide clinical experience with the use of crystalloids, it appears that these are not major effects. A more important effect is the development of pulmonary edema, or adult respiratory distress syndrome (ARDS); however, the relationship between the use of large volumes of salt solutions and ARDS is not firm.

Larger colloidal molecular weight materials such as albumin have the advantage of longer retention time, they do not extravasate and therefore do not cause edema. Albumin is used as a plasma expander for emergency volume restitution, but due to cost considerations the synthetic colloids dextran and hydroxyethyl starch are used more frequently. These materials are free of viral contaminations, but may cause anaphylactic reactions and have a tendency to alter platelet function thus interfering with hemostasis. It is not well established whether these materials increase bleeding per se or the effect is due to improved perfusion.

The beneficial effect of this form of volume replacement resides in the high oncotic pressure which they generate, maintaining intravascular volume in the amount of about 20 ml of fluid per gram of circulating colloid. In this context, dextran 70 kDA and hydroxyethyl starch have the highest retention capacity, 22 and 17 ml/g, respectively, while albumin is 15 ml/g. Furthermore, this form of volume replacement lowers blood viscosity with all the associated positive transport effects described for hemodilution [Messmer et al., 1972].

137.9 Artificial Oxygen Carriers

The amount of oxygen carried in simple aqueous solutions is inconsequential in the overall oxygen requirements of the tissue, and therefore when intrinsic oxygen carrying capacity has to be restored, it is necessary to include an oxygen carrier. This can be presently accomplished through the use of modified hemoglobins which bind oxygen chemically and reversibly, and fluorocarbons which have a high capacity for dissolving oxygen. These compounds present important differences which will ultimately determine in what form of blood substitution and replacement they will be used.

Hemoglobin separated from the red cell membrane, i.e., stroma-free hemoglobin, carries oxygen with a high affinity. These materials have intrinsically high oxygen carrying capacity; however, they are based on a biologically active molecule, which in most instances requires specialized storage procedures. By contrast, fluorocarbons carry a limited amount of oxygen under normal atmospheric conditions, but are biologically inert, can be stored at room temperature, and are excreted as gas through the lungs.

137.10 Fluorocarbons

Fluorocarbons (perfluorochemicals, PFCs) are compounds that have a very high solubility of oxygen and their oxygen transport capacity is directly proportional to their concentration and pO_2. They are formulated in stable water soluble emulsions that are metabolically inert. These PFC emulsions consist in droplets of fluorochemicals in the range of 0.1 to 0.2 μm diameter coated by a thin film of egg yolk phospholipids. These products are heat sterilized and can be stored under ambient conditions ready for use. The technology for large scale manufacture is well established and the necessary materials for production are well characterized and commercially available. Their half life in the circulation is in the range of 6 to 9 h.

Pure PFCs have oxygen solubilities of the order of 50 ml O_2/100 ml at 37°C and 760 mmHg oxygen pressure. The solubility varies linearly as a function of gas pressure. These materials carry about 30% more oxygen than blood when diluted to 60% (the recommenced clinical form of administration) and with 100% inspired oxygen 100%. They have been given clinically at a maximal dosage of 2 g PFC/kg (vs. 15 g hemoglobin/kg for normal blood in humans) [Lamy et al., 1998]. At this dosage, in an isovolemic exchange they increase intrinsic oxygen carrying capacity of blood by about 20%, provided that 100% oxygen is inspired. Since normal tissues and the microcirculation operate in the range of 20 to 50 mmHg, the effect of this fundamentally altered blood pO_2 and oxygen content on the normal distribution of oxygen tension in the circulation is not known. However, since tissues regulate their blood oxygen supply and pO_2, it is unlikely that microvascular blood pO_2 would be much greater than 50 mmHg without eliciting autoregulatory adjustments; therefore, in the absence of hypertensive effects it may be assumed that microvascular blood pO_2 in the presence of PFCs does not exceed 50 mmHg and therefore in terms of tissue oxygen delivery the contribution is small relative to hemoglobin since at pO_2 50 mmHg they carry about 2 ml O_2/dl, while hemoglobin carries 15 ml O_2/dl.

The viscosity of PFCs is only slightly higher than water; therefore, when used in the amounts described they behave as low viscosity plasma expanders, and therefore should exhibit some of the beneficial effects to be derived from moderate hemodilution.

137.11 Hemoglobin for Oxygen Carrying Plasma Expanders

The hemoglobin molecule in solution presents unique characteristics as an oxygen carrying plasma expander, since it has a molecular weight similar to that of albumin and therefore comparable intravascular retention. Its ability to chemically bind oxygen determines that large amounts of oxygen can be associated chemically at relatively low pO_2, and its molecular weight is such that it does not extravasate. Consequently, it should provide an extended period of colloid osmotic effects that prevent the passage of water into the interstitium and therefore edema. However, when hemoglobin is diluted from the concentration present in red blood cells, the tetrameric molecule tends to spontaneously dissociate into smaller molecular weight dimers and monomers which rapidly extravasates from the vascular compartment. Until recently it was assumed that when hemoglobin is free in the circulation it has an oxygen affinity that is too high for tissue oxygenation. As a consequence, strategies for the modification of hemoglobin by chemical means at present are aimed at prolonging intravascular retention and reducing oxygen affinity [Winslow, 1997].

Hemoglobin can be prevented from dissociating by chemically crosslinking the tetramers such as the widely used agent glutaraldehyde. Direct crosslinking with this agent produces a spectrum of different high molecular weight hemoglobin derivatives since glutaraldehyde is non-specific and reacts with any amino group. A more homogenous product can be obtained by reacting human hemoglobin first with the 2,3-DPG analog PLP, which reduces oxygen affinity and then polymerizing the compound with glutaraldehyde. This product is called PLP-polyHB and has near normal oxygen carrying capacity and a plasma half-time retention (in rats) of about 20 h.

Diaspirins are another group of crosslinking agents. The reaction of native hemoglobin with bis(3,5-dibromosalicy)fumarate (DBBF) leads to the linking of the two α chains of the hemoglobin molecule, producing a compound called $\alpha\alpha$Hb which has an oxygen affinity (P_{50}) of 30 mm Hg and an intravascular retention time of the order of 12 h in rats. Phosphorylated compounds added to purified hemoglobin also change the P_{50} value of the resulting compound and have been studied extensively. There is an increasing variety of hemoglobin compounds that is being developed; however, at this time $\alpha\alpha$Hb is the most widely studied compound since it was produced in quantity by the U.S. Army Letterman Institute, San Francisco, and a similar compound was developed and used in extensive clinical trials by Baxter Healthcare Inc.

Hemoglobin can be produced as a recombinant protein leading to the construction of either the naturally occurring molecule or variants that have different crosslinking and oxygen affinity. It is presently possible to produce both alpha and beta chains of human hemoglobin in bacteria, yeast, and transgenic animals. While it is in principle possible to produce large amounts of a specific product by these means, there remain problems of purification, elimination of other bacterial products, and endotoxins that may be expressed in parallel.

Human hemoglobin genes have been induced into transgenic mammals, leading to the potential production of large amounts of human hemoglobin. However, the production is not 100% pure and some of the original animal material is also present, causing an important purification problem. This problem may be compounded by potential immunogenicity of animal hemoglobin and the potential transmission of animal diseases to humans.

A unique form of hemoglobin modification is the process of pegylation, consisting of attaching polyethylene glycol polymeric strands to the surface of the hemoglobin molecule. The resulting product termed PEG-hemoglobin interacts with the surrounding water, producing an aqueous barrier that surrounds the molecule. This phenomenon causes several beneficial effects, namely (1) renders the molecule invisible to the immune system; (2) draws water into the circulation increasing blood volume; (3) increases molecular size of the molecule, thus further limiting its capacity to extravasate; (4) decreases the diffusive oxygen delivery from HbO_2 by increasing path length; and (5) increases plasma viscosity. This type of molecule is presently manufactured by Sangart, Inc., La Jolla, and Apex Bioscience Inc., North Carolina, who uses human hemoglobin and Enzon Inc., New Jersey, who uses bovine hemoglobin.

137.12 Hemoglobin-Based Artificial Blood

Many hemoglobin-based artificial blood substitutes are being developed which may be classified relative to the technology used and whether the hemoglobin is either of human, animal, or bacterial origin. It is now established that hemoglobin containing red blood cell substitutes deliver oxygen to the issue and maintain tissue function. However, problems of toxicity and efficacy are not yet resolved, and are shared by many of the hemoglobin solutions presently available. Renal toxicity is one of the primary problems because hemoglobin toxicity to the kidney is a classic model for kidney failure. The mechanism of toxicity includes tubular obstruction, renal arterial vasospasm, and direct toxicity to tubular cells. It is not clear whether hemoglobin must be filtered by the kidney to be toxic, and the tolerance of the kidney to different types of hemoglobins and concentrations is not established [Paller, 1988].

There are many reports of anaphylaxis/anaphylactoid reactions to hemoglobin infusion. This may result from heme release from hemoglobin, the presence of endotoxins bound to hemoglobin or pre-existing in the circulation, or from oxygen-derived free radicals released after exposure to hemoglobin. Vasoconstriction of coronary, cerebral, and renal vessels has also been attributed to the presence of free hemoglobin in the circulation, a phenomenon that may be due to the intrinsic capacity of hemoglobin to scavenge nitric oxide or metabolic autoregulation.

Toxicity of hemoglobin may be due to special characteristics of the manufacturing procedure, the given molecule, or problems due to purification. It may also be inherent to the presence of large quantities of a molecular species for which the circulation has a very limited tolerance when present in solution. This problem may be circumvented by separating hemoglobin from the circulation by encapsulating this material in a synthetic membrane, in the same way that the red cell membrane encapsulates hemoglobin.

Encapsulation of hemoglobin has been accomplished by introducing the molecule into a liposome. To the present this technology presents several problems, namely (1) poor efficiency of hemoglobin incorporation into the liposome; (2) physical instability of liposomes in storage; and (3) chemical instability of the lipid-hemoglobin interface leading to lipid peroxidation and hemoglobin denaturation. These factors translate in decreased oxygen carrying capacity, increased viscosity of the circulating blood/liposome mixture, and reticuloendothelial blockade. This situation notwithstanding, lipid encapsulation may ultimately provide the artificial blood of choice once a quality hemoglobin becomes consistently available and liposome production can be scaled up to quantities commensurate with demand.

137.13 Results in the Microcirculation with Blood Substitution with $\alpha\alpha$-Hemoglobin

The effectiveness of $\alpha\alpha$-Hb (cell-free o-raffinose cross-linked) as a blood carrying plasma expander was tested in the microcirculation by implementing isovolemic hemodilution up to an exchange of 75% of the original cell mass. Exchanges with dextran 70 kDa molecular weight were used as control. Experiments were carried out in the microcirculation of the hamster window preparation, which was evaluated in terms of hemodynamic parameters and oxygen distribution at three successive levels of exchange. The oxygen delivery capacity of the microcirculation was determined in terms of the rate of arrival of hemoglobin (in red cell cells + molecular) to the capillaries. It was found that using molecular solutions of about 10 g/dl concentration, the oxygen delivery capacity of the microcirculation was identical to that of dextran 70 for substitutions up to 60% of the red blood cell mass. This was a consequence of the lack of increased cardiac output with the Hb solutions, which should be expected when viscosity is lowered, as is the case with colloidal solutions (starch and dextran).

This hemoglobin caused hypertension, and at the highest level of exchange exhibited abnormally low tissue pO_2, which fell from a normal value of 22.4 mmHg to about 5 mmHg. This lowered tissue pO_2 was accompanied by a significant decrease in functional capillary density, and increased vessel wall gradients. Consequently, this hemoglobin does not appear to function as a tissue oxygenator in the microcirculation, although the total amount of hemoglobin and its capacity to transport oxygen is similar to that of natural blood. The principal reasons for this outcome are hypertension elicited by lowered blood viscosity with consequent decrease of production of NO, increased scavenging of NO, and increased availability of oxygen to the vessel wall, which promotes autoregulatory responses aimed at limiting over oxygenation of the vessel wall. The linkage between blood supply and oxygen demand is due to the existence of chemical vasodilator signals transmitted from the tissue cells to the resistance vessels, which relax (increase their diameter) causing blood flow to increase when tissue oxygen tension falls below critical levels. Conversely, increasing blood oxygen tension over normal levels elicits vasoconstriction.

The most critical side effect in terms of microvascular function is the finding that molecular hemoglobin solutions significantly lower functional capillary density.

137.14 Hemoglobin and Nitric Oxide (NO) Binding

The principal adverse effect of hemoglobin in solution is to produce vasoconstriction. This is based on two fundamental observations: (1) Hb has a very high affinity to bind NO at heme sites, and (2) Hb has been shown to produce constriction of isolated aortic rings. However, there is no direct proof that these mechanisms exists because NO has a very short half-life in the circulation (1.8 ms) [Liu et al., 1998], making its detection difficult. Further complicating the story is the observation that NO can also bind to hemoglobin at its sulfhydryl residues [Stamler et al., 1997], and that this mechanism might be oxygen-linked. It should also be noted that NO binding as a cause for vasoconstriction has been primarily deduced from studies that have shown constriction of isolated aortic rings when exposed to hemoglobin solutions. However, *in vivo* the situation is more complicated, particularly because most of the important diameter reductions occur in specific microvessels. Alternative explanations are that this effect can be produced directly on the blood side of the endothelium, where the presence of molecular hemoglobin in the red cell free plasma layer could significantly distort the diffusion field from the endothelial cell, diverting NO from smooth muscle into blood. Another effect could arise if the protein leaves the vascular space and enters the interstitium, there to interfere with the diffusion of NO from endothelium to its target, vascular smooth muscle. Thus, Rohlfs et al. [1997], found that hemoglobins with differing effects on mean arterial blood pressure all had similar reaction kinetics with NO. These findings are consistent with the autoregulation hypothesis, and suggest that there is more to the story than NO scavenging.

A different explanation is that altered blood properties in extreme hemodilution affect the production of NO and prostacyclin (and other cytokines) in endothelial cells, which is dependent on the maintenance of some level of shear stress. Such shear stress, in a flowing system such as blood, is dependent on a number of factors, including the viscosity of the solution. Thus, it would be expected that hemoglobin solutions with low viscosity, such as $\alpha\alpha$-Hb, would decrease NO synthesis and induce vasospasm. In fact, vasoconstriction is probably a natural reaction of the organism to reduced blood viscosity.

The difficulties associated with the measurement of NO directly in biological systems has led to the formulation of theoretical models. This type of analysis shows that NO distribution between blood and smooth muscle is the consequence of its high diffusivity, the fact that its half-life is limited by its reaction with oxygen and hemoglobin, its diffusion into red cells, and the geometry of vessels in the microcirculation. A tentative conclusion, in part supported by direct studies in the microcirculation, indicates that effects should be maximal in arterioles of the order of 30 to 100 μm in diameter.

Thus far, the NO physiology has not been completely translated into the design of blood substitutes [Winslow, 1998]. The group at Somatogen have designed a series of hemoglobin mutants with a range of NO affinity and shown that there is a correlation between NO binding and the blood pressure response in rats [Doherty et al., 1998]. However, Rohlfs et al. [1998], have shown that a different series of chemically modified hemoglobins with different vasoactivities have essentially the same NO reactivity.

137.15 Rational Design of an Oxygen Carrying Molecular Plasma Expander

The products presently in clinical trials reflect the physiological know-how of the of 1970s, when their effects could not be analyzed at the level of microscopic blood vessels, blood flow regulation by NO was unknown, the theory of microvascular autoregulation had just been formulated, and the endothelium was viewed as a passive cellular lining which prevented the extravasation of plasma proteins. At that time, systemic experimental findings and the clinical experience with low viscosity plasma expanders used in hemodilution established that lowering blood viscosity was a safe and efficacious method for restoring blood volume, particularly if oncotic pressure was maintained within the physiological range through the use of colloids. These well-established tenets were incorporated into the design of artificial blood, or blood substitute, which was conceived as a fluid of low viscosity (i.e., lower than blood and closer to plasma in order to obtain the beneficial effects of hemodilution), moderate oncotic pressure (i.e., of the order of 25 mmHg), a right-shifted oxygen dissociation curve, and a concentration of hemoglobin, or equivalent oxygen carrying capacity, of 10 g Hb/dl.

This scenario was firmly established because there were no practical methods for assessing tissue pO_2 clinically, other than direct measurement of venous blood pO_2, while in experimental conditions tissue pO_2 could only be determined by extremely laborious and difficult to implement microelectrode methods. Thus, the distribution of oxygen in the tissue, even at the experimental level, was not known and related fundamental mechanisms directly involved with survival at the level of the microcirculation were equally unknown or misunderstood.

High resolution methods for measuring tissue pO_2 became available at the beginning of the 1990s, and when these were applied to the analysis of the effects of hemoglobin-based blood substitutes, they reversed several of the established principles. It was found that oxygen is delivered by the arterioles and not the capillaries, and that this rate of delivery was modulated by the consumption of oxygen in the arteriolar wall. Vasoconstrictor effects were found to increase the oxygen consumption of the arterioles to the detriment of tissue oxygenation. This was particularly true for hemoglobin solutions formulated with small molecules, such as the $\alpha\alpha$-crosslinked hemoglobin, which also presented another of the presumed required properties, namely low viscosity.

To the present, the vasoconstrictor effect has been attributed to NO scavenging by hemoglobin; however, other mechanisms may be involved and superimposed on the scavenging effect. The molecular nature of the presently developed hemoglobin solutions determines that the oxygen source in blood is very close to the endothelium, since the barrier determined by the plasma layer is no longer present. This configuration increases the oxygen availability to the microvascular wall, potentially eliciting autoregulatory vasoconstriction aimed at regulating oxygen delivery. It is now apparent that right-shifted materials further increase oxygen availability, thus enhancing the vasoconstrictor stimulus.

Concerning NO scavenging, it may well be that this shall remain an intrinsic property of the hemoglobin molecule dictated by the physical similarity of NO and O_2. In this context, two solutions to the problem may be implemented directly at the level of the oxygen carrier: One consists of producing an elevated (relative to normal) plasma viscosity, in such a fashion that shear stress dependant NO production remains constant or is elevated. Second, physically separating the hemoglobin molecule from the endothelial barrier as may be obtained with a pegylated material may retard hemoglobin–NO interaction.

There is increasing evidence that low viscosity is of no benefit once the transfusion trigger is passed, particularly if combined with vasoconstrictor effects that impede the increase of cardiac output. In fact, conditions of extreme hemodilution, as those that obtain purely on the basis of viscosity considerations when the red blood cell mass is reduced, can only survive when plasma viscosity is increased, so that blood viscosity is near normal, which has the effect of maintaining functional capillary density at basal levels. This beneficial effect is further augmented by added production of shear stress dependant NO, which induces vasodilation and adequate blood flow. The physiological consequences of introducing molecular solutions of hemoglobin in the circulation can be seen in full perspective when the products are ranked according to size (Fig. 137.4), where it becomes apparent that side effects such as hypertension and retention time are a direct result of molecular dimensions, which in turn determine viscosity and diffusivity.

In these conditions, tissue oxygenation is enhanced by the use of a left-shifted hemoglobin which effectively "hides" oxygen from the arterioles and facilitates its unloading in the regions of the microcirculation where oxygen consumption by the vessel wall is smaller.

Finally, a critical issue is how much hemoglobin is necessary to obtain the desired effects. Experimental evidence presently indicates that when the "counter-intuitive" formulation of product is implemented, the actual amount of hemoglobin needed may be as low as 3 g Hb/dl (Winslow et al., 1998). This

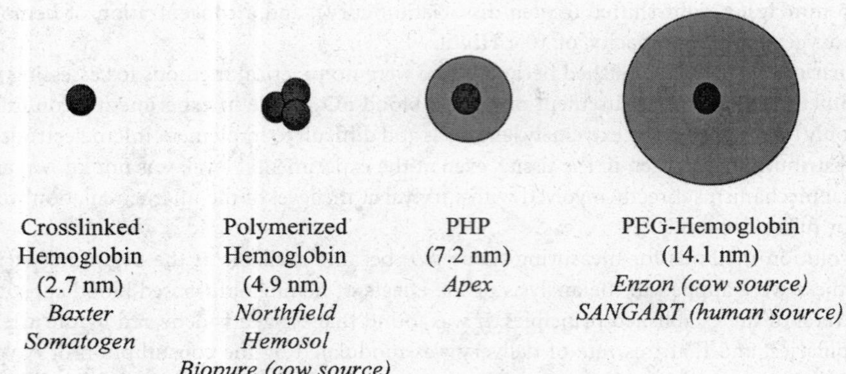

FIGURE 137.4 The physiological reaction to the presence of even small amounts of circulating hemoglobin can be predicted by grouping the products according to molecular dimension. Systemic blood pressure rise is maximal for the smallest molecules and virtually absent for the largest molecules. The effect on blood viscosity correlates with both the rise of blood pressure and the maintenance of functional capillary density, where small molecules, which lower blood viscosity, also significantly lower FCD, while the larger molecules, causing blood viscosity to remain similar to that of whole blood, have no effect on FCD. Intravascular retention time is also directly proportional to molecular size, varying from about 12 h for the smaller molecules to more than 48 h for PEG-hemoglobin. Products are identified according to their manufacturer, namely: Baxter Healthcare, Round Lake, IL; Somatogen, Inc., Boulder, CO; Northfield Laboratories Inc., Chicago, IL; Hemosol Ltd., Toronto, Canada; Biopure Pharmaceuticals Corp., Boston, MA; Apex Bioscience Inc., Research Triangle Park, NC.; Enzon Inc., Piscataway, NJ; SANGART Inc., La Jolla, CA. (Data from Vandegriff et al., 1997.)

consideration impacts on the cost of the product and the potential for production, since most hemoglobin modifications have at most a 50% yield. This is critical for products based on human blood, where the principal source is outdated material, which would impose an inherent limitation on the total production capacity if the product is formulated with 7 to 10 g Hb/dl concentration.

137.16 Conclusions

Lethality consequent to blood losses results from hypovolemia and not anemia, and a number of plasma expanders exist that restitute volume and ensure survival even though they carry oxygen only to the extent of its plasma solubility. This is a consequence of the "derived" increase of oxygen carrying capacity resulting from lowered viscosity and increased cardiac output. This approach serves as hemoglobin concentration reaches the transfusion trigger, beyond which restoration of oxygen carrying capacity is needed.

The use of oxygen carrying plasma expanders made with molecular solutions of hemoglobin, when the red blood cell mass is reduced below the transfusion trigger, causes extreme hemodilution associated with a significant reduction of blood viscosity and NO production, and reflex vasoconstriction, leading to decreased functional capillary density. This combination of events is difficult to survive because decreased NO availability increases the intrinsic oxygen consumption of the tissue. An oxygen carrying volume replacement with a capacity commensurate to blood is presently the perceived sine qua necessity for extreme blood losses. However, new findings indicate that this may not necessarily be the case.

Recent developments in the understanding of the physiology of extreme hemodilution, and the physical events associated with the substitution of red blood cells with molecular hemoglobin solutions have determined a shift in paradigm, indicating that a viable "artificial blood" will be obtained from a counterintuitive formulation of the product. In this formulation, viscosity is near normal for blood, the dissociation curve is left-shifted, oncotic pressure is high, and the concentration of hemoglobin is in the range of 3 to 5 g Hb/dl [Winslow and Intaglietta, 1998].

Acknowledgment

This work was supported in part by USPHS Grant HLBI 48018.

Defining Terms

Hemodilution: The replacement of natural blood with a compatible fluid that reduces the concentration of red blood cells

Stroma-free hemoglobin: Hemoglobin derived from red blood cells where all materials related to cell membrane and other components within the red blood cells have been removed.

Liposome: Microscopic phospholipid vesicle used to encapsulate materials for slow release. The use of liposomes to encapsulate hemoglobin is a departure from the conventional use of liposomes, exploiting encapsulation, and requires their modification to ensure sustained entrapment.

Oncotic: Refers to colligative properties due to the presence of macromolecules, for instance oncotic pressure, which is differentiated from osmotic pressure, which is that due to the presence of all molecular species in solution.

Oxygen carrying capacity: The total amount of oxygen that may be transported by blood or the fluid in the circulation. Differentiated from oxygen delivery capacity which involves considerations of flow rate.

Plasma viscosity: Viscosity of blood devoid of cellular elements, a parameter that is critically impacted by the introduction of molecular hemoglobin solutions.

Shear stress: Force per unit area parallel to the vessel wall or traction, experienced by the endothelium due to blood flow and blood viscosity.

Transfusion trigger: Concentration of blood hemoglobin beyond which blood is required to restore circulatory function.

References

Dintenfass, L. *Blood Microrheology: Viscosity Factors in Blood Flow, Ischemia and Thrombosis.* Appleton-Century-Crofts, New York, 1971.

Doherty, D. H., Doyle, M. P., Curry, S. R., Vali, R. J., Fattor, T. J., Olson, J. S., and Lemon, D. D. Rate of reaction with nitric oxide determines the hypertensive effect of cell-free hemoglobin. *Nature Biotechnology* 16:672, 1998.

Frangos, J. A., Eskin, S. G., McIntire, L. V., and Ives, C. L. Flow effects on prostacyclin production in cultured human endothelial cells. *Science* 227:1477, 1985.

Intaglietta, M., Johnson, P. C., and Winslow, R. M. Microvascular and tissue oxygen distribution. *Cardiovasc. Res.* 32:632, 1996.

Intaglietta, M. Whitaker Lecture 1996: Microcirculation, biomedical engineering and artificial blood. *Ann. Biomed. Eng.* 25:593, 1997.

Johnson, P. C. Brief Review: Autoregulation of blood flow. *Circ. Res.* 59:483, 1996.

Joyner, W. L., Davis, M. J., and Gilmore, J. P. Intravascular pressure distribution and dimensional analysis of microvessels in the hamsters with renovascular hypertension. *Microvasc. Res.* 22:1, 1974.

Kerger, H., Saltzman, D. J., Menger, M. D., Messmer, K., and Intaglietta, M. Systemic and subcutaneous microvascular pO_2 dissociation during 4-h hemorrhagic shock in conscious hamsters. *Am. J. Physiol.* 270:H827, 1996.

Kjellstrom, B. T., Ortenwall, P., and Risberg, R. Comparison of oxidative metabolism *in vitro* in endothelial cells from different species and vessels. *J. Cell. Physiol.* 132:578, 1987.

Kuchan, M. J., Jo, H., and Frangos, J. A. Role of G proteins in shear stress-mediated nitric oxide production by endothelial cells. *Am. J. Physiol.* 267:C753, 1994.

Lamy, M., Mathy-Hartert, M., and Deby-Dupont, G. Perfluorocarbons as oxygen carriers. In *Update in Intensive Care Medicine 33*, J.-L. Vincent, Ed., Springer-Verlag, Berlin, 1998, 332.

Lipowsky, H. H. Mechanics of blood flow in the microcirculation. In *Handbook of Bioengineering*, R. Skalak and S. Chien, Eds., McGraw-Hill, New York, 1987, ch. 8.

Liu, X., Miller, M. J., Joshi, M. S., Sadowska-Krowicka, H., Clark, D. A., and Lancaster, J. R. J. Diffusion-limited reaction of free nitric oxide with erythrocytes. *J. Biol. Chem.* 273:18709, 1998.

Messmer, K., Sunder-Plasman, L., Klövekorn, W. P., and Holper, K. Circulatory significance of hemodilution: Rheological changes and limitations. *Adv. Microcirculation*, 4:1, 1972.

Mirhashemi, S., Messmer, K., and Intaglietta, M. Tissue perfusion during normovolemic hemodilution investigated by a hydraulic model of the cardiovascular system. *Int. J. Microcirc.: Clin. Exp.* 6:123, 1987.

Paller, M. S. Hemoglobin and myoglobin-induced acute renal failure: Role of iron nephrotoxicity. *Am. J. Physiol.* 255:F539, 1988.

Quemada, D. Rheology of concentrated dispersed systems: III. General features of the proposed non-Newtonian model: Comparison with experimental data. *Rheol. Acta* 17:643, 1978.

Richardson, T. Q. and Guyton, A.C. Effects of polycythemia and anemia on cardiac output and other circulatory factors. *Am. J. Physiol.* 197:1167, 1959.

Rohlfs, R., Vandegriff, K., and Winslow, R. The reaction of nitric oxide with cell-free hemoglobin based oxygen carriers: Physiological implications. In: *Industrial Opportunities and Medical Challenges*, R. M. Winslow, K. D. Vandegriff, M. Intaglietta, Eds., Birkhäuser, Boston, 1997, 298.

Rohlfs, R. J., Bruner, E., Chiu, A., Gonzales, A., Gonzales, M. L., Magde, D., Magde, M. D. J., Vandegriff, K. D., and Winslow, R. M. Arterial blood pressure responses to cell-free hemoglobin solutions and the reaction with nitric oxide. *J. Biol. Chem.* 273:12128, 1998.

Saltzman, D., DeLano, F. A., and Schmid-Schönbein, G. W. The microvasculature in skeletal muscle. VI. Adrenergic innervation of arterioles in normotensive and spontaneously hypertensive rats. *Microvasc. Res.* 44:263, 1992.

Shen, W., Hintze, T. H., and Wolin, M. S. Nitric Oxide. An important signaling mechanism between vascular endothelium and parenchymal cells in the regulation of oxygen consumption. *Circulation* 92:3505, 1995.

Stamler, J. S., Jia, L., Eu, J. P., McMahon, T. J., Demchenko, I. T., Bonaventura, J., Gernert, K., and Piantadosi, C. A. Blood flow regulation by S-nitrosohemoglobin in the physiological oxygen gradient. *Science* 276:2034, 1997.

Torres Filho, I. P. and Intaglietta, M. Micro vessel pO_2 measurements by phosphorescence decay method. *Am. J. Physiol.* 265(34):H1434, 1994.

Tsai, A. G., Friesenecker, B., Mazzoni, M. C., Kerger, H., Buerk, D. G., Johnson, P. C., and Intaglietta, M. Microvascular and tissue oxygen gradients in the rat mesentery. *Proc. Natl. Acad. Sci. U.S.A.* 95:6590, 1998.

Tsai, A. G., Friesenecker, B., McCarthy, M., Sakai, H., Intaglietta, M. Plasma viscosity regulates capillary perfusion during extreme hemodilution in hamster skin fold mode. *Am. J. Physiol.*, 275:H2170, 1998.

Vandegriff, K. D., Rohlfs, R. J., and Winslow, R. M. Colloid osmotic effects of hemoglobin-based oxygen carriers. In *Advances in Blood Substitutes. Industrial Opportunities and Medical Challenges*, R. M. Winslow, K. D. Vandegriff, and M. Intaglietta, Eds., Birkhäuser, Boston, MA, 1997.

Wilson, D. F. Measuring oxygen using oxygen dependent quenching of phosphorescence: a status report. *Adv. Exp. Med. Biol.* 333:225, 1993.

Winslow, R. M. Blood substitutes. *Sci. Med.* 4:54, 1997.

Winslow, R. M. and Intaglietta, M. U.S. Patent 5,814,601. Methods and compositions for optimization of oxygen transport by cell free systems, 1998.

Winslow, R. M. Artificial blood: Ancient dream, modern enigma. *Nature Biotechnology.* 16:621, 1998.

Further Information

Blood Substitutes. Physiological Basis of Efficacy, R. M. Winslow, K. D. Vandegriff, and M. Intaglietta, Eds., Birkhäuser, Boston, 1995.

Blood Substitutes. New Challenges, R. M. Winslow, K. D. Vandegriff, and M. Intaglietta, Eds., Birkhäuser, Boston, 1996.

Advances in Blood Substitutes. Industrial Opportunities and Medical Challenges, R. M. Winslow, K. D. Vandegriff, and M. Intaglietta, Eds., Birkhäuser, Boston, 1997.

138

Artificial Skin and Dermal Equivalents

Ioannis V. Yannas
Massachusetts Institute of Technology

138.1 The Vital Functions of Skin

Skin is a vital organ, in the sense that loss of a substantial fraction of its mass immediately threatens the life of the individual. Such loss can result suddenly, either from fire or from a mechanical accident. Loss of skin can also occur in a chronic manner, as in skin ulcers. Irrespective of the time scale over which skin loss is incurred, the resulting deficit is considered life threatening primarily for two reasons: Skin is a barrier to loss of water and electrolytes from the body, and it is a barrier to infection from airborne organisms. A substantial deficit in the integrity of skin leaves the individual unprotected either from shock, the result of excessive loss of water and electrolytes, or from sepsis, the result of a massive systemic infection. It has been reported that burns alone account for 2,150,000 procedures every year in the United States. Of these, 150,000 refer to individuals who are hospitalized, and as many as 10,000 die.

Four types of tissue can be distinguished clearly in normal skin. The *epidermis*, outside, is a 0.1-mm-thick sheet, comprising about 10 layers of keratinocytes at levels of maturation which increase from the inside out. The *dermis*, inside, is a 2–5-mm-thick layer of vascularized and innervated connective tissue with very few cells, mostly quiescent fibroblasts. The dermis is a massive tissue, accounting for 15–20% of total body weight. Interleaved between the epidermis and the dermis is the *basement membrane*, an approximately 20–nm-thick multilayered membrane (Fig. 138.1). A fourth layer, the *subcutis*, underneath the dermis and 0.4–4-mm in thickness, comprises primarily fat tissue. In addition to these basic structural elements, skin contains several appendages (*adnexa*), including hair follicles, sweat glands, and sebaceous glands. The latter are mostly embedded in the dermis, although they are ensheathed in layers of epidermal tissue.

0-8493-0462-8/00/$0.00+$.50

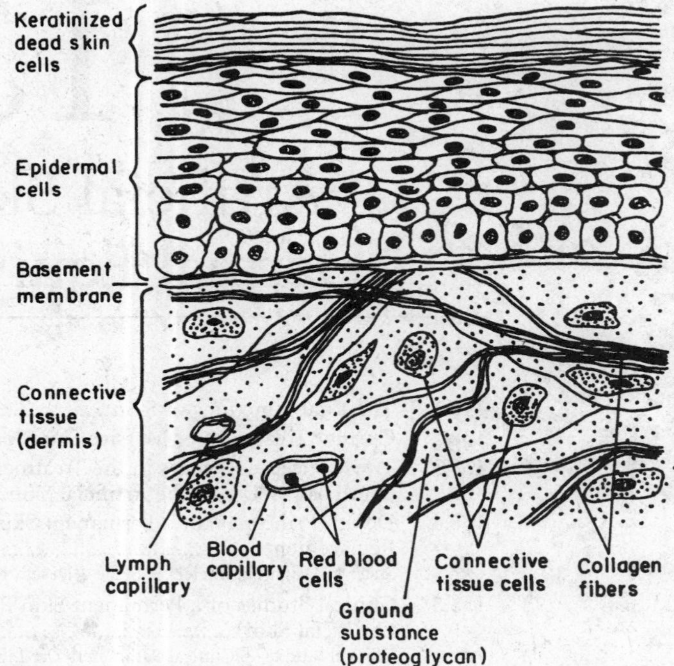

FIGURE 138.1 Schematic view of skin which highlights the epidermis, the basement membrane interleaved between the epidermis and the dermis, and the dermis underneath. Only a small fraction of the thickness of the dermis is shown. (Redrawn with permission from J. Darnell, J. Lodish, and D. Baltimore, *Molecular Cell Biology*, Scientific American Books, New York, Chapter 5, Fig. 552, 1986.)

The functions of skin are quite diverse, although it can be argued that the single function of skin is to provide a viable interface with the individual's environment. In addition to the specific vital functions mentioned above (protection from water and electrolyte loss, and from infection), skin also protects from heat and cold, mechanical friction, chemicals, and from UV radiation. Skin is responsible for a substantial part of the thermoregulatory and communication needs of the body, including the transduction of signals from the environment such as touch, pressure, and temperature. Further, skin transmits important emotional signals to the environment, such as paleness or blushing of the face and the emission of scents (pheromones). Far from being a passive membrane that keeps the internal organs in shape, skin is a complex organ.

138.2 Current Treatment of Massive Skin Loss

The treatment of skin loss has traditionally focused on the design of a temporary wound closure. Attempts to cover wounds and servere burns have been reported from historical sources at least as far back as 1500 B.C., and a very large number of temporary wound dressings have been designed. These include membranes or sheets fabricated from natural and synthetic polymers, skin grafts from human cadavers (homografts, or *allografts*), and skin grafts from animals (heterografts, or *xenografts*). Although a satisfactory temporary dressing helps to stem the tide, it does not provide a permanent cover. Polymeric membranes which lack specific biologic activity, such as synthetic polymeric hydrogels, have to be removed after several days due to incidence of infection and lack of formation of physiologic structures. Patients with cadaver allografts and xenografts are frequently immunosuppressed to avoid rejection; however, this is a stop-gap operation which is eventually terminated by removal of the graft after several

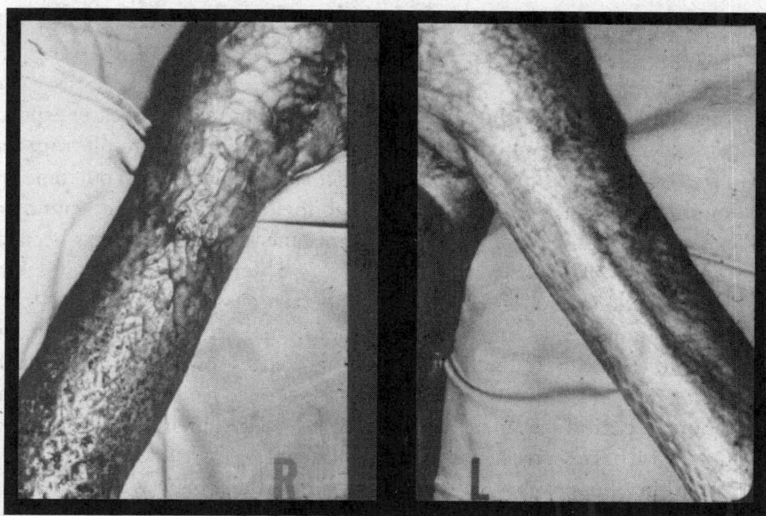

FIGURE 138.2 Comparision between treatment with the meshed autograft (R) and treatment with the artificial skin (L). Autograft is usually meshed before grafting; scar forms in areas coinciding with the open slits of the autograft. The artificial skin treatment consists of grafting the excised wound bed with a skin regeneration template, followed by grafting on about day 14 with a very thin epidermal autograft. (Photo courtesy of J.F. Burke.)

days. In all cases where temporary dressings have been used, the routine result has been an open wound. Temporary dressings are useful in delaying the time at which a permanent graft, such as an autograft, is necessary and are therefore invaluable aids in the management of the massively injured patient.

The use of an autograft has clearly shown the advantages of a permanent wound cover. This treatment addresses not only the urgent needs but also the long-term needs of the patient with massive skin loss. The result of treatment of a third-degree burn with a split-thickness autograft is an almost fully functional skin which has become incorporated into the patient's body and will remain functional over a lifetime. Autografts usually lack hair follicles and certain adnexa as well. However, the major price paid is the removal of the split thickness graft from an intact area of the patient's body: The remaining dermis eventually becomes epithelialized but not without synthesis of scar over the entire area of trauma (donor site). To alleviate the problem associated with the limited availability of autograft, surgeons have resorted to meshing, a procedure in which the sheet autograft is passed through an apparatus which cuts slits into the sheet autograft, allowing the expansion of the graft by several times and thereby extending greatly the area of use. An inevitable long-term result of use of these meshed autografts is scar synthesis in areas coinciding with the open slits of the meshed graft and a resulting pattern of scar which greatly reduces the value of the resulting new organ (Fig. 138.2). An important aspect of the use of the autograft is the requirement for early excision of dead tissue and the provision, thereby, of a viable wound bed to "take" the autograft.

The term "artificial skin" has been used to describe a cell-free membrane comprising a highly porous graft copolymer of type I collagen and chondroitin 6-sulfate which degrades at a specific rate in the wound and regenerates the dermis in dermis-free wounds in animal models and patients (see below: dermis regeneration template, DRT). "Skin equivalent" (SE) refers to a collagen lattice which has been prepared by contraction of a collagen gel by heterologous fibroblasts ("dermal equivalent" or DE) and has subsequently been overlayed with a keratinocyte culture to induce formation of a mature, cornified epidermis *in vitro* prior to grafting of skin wounds. Cultured epithelial autografts (CEA) consist of a mature, cornified epidermis which has been produced by culturing keratinocytes *in vitro*, prior to grafting on skin wounds. The major goal of these treatments has been to replace definitively the use of the autograft in the treatment of patients with massive skin loss.

138.3 Two Conceptual Stages in the Treatment of Massive Skin Loss by Use of the Artificial Skin

Loss of the epidermis alone can result from a relatively mild burn, such as an early exposure to the sun. Controlled loss of epidermis in a laboratory experiment with an animal can result from the repeated use of adhesive tape to peel off the keratinocyte layers. In either case, the long-term outcome is an apparently faithful regeneration of the epidermis by migration of epithelial cells from the wound edge, and from roots of hair follicles, over the underlying basement membrane and dermis. It has been shown that the epidermis can regenerate spontaneously provided there is a dermal substrate over which epithelial migration and eventual anchoring to the underlying connective tissue can occur.

Loss of the dermis has quite a different outcome. Once lost, the dermis does not regenerate spontaneously. Instead, the wound closes by contraction of the wound edges toward the center of the skin deficit, and by synthesis of *scar*. Scar is a distinctly different type of connective tissue than is dermis. The depth of skin loss is, therefore, a critical parameter in the design of a treatment for a patient who has a skin deficit. In the treatment of burns, physicians distinguish among a first-degree burn (loss of epidermis alone), a second-degree burn (loss of epidermis and a fraction of the thickness of the dermis), and a third-degree burn (loss of the epidermis and the entire dermis down to muscle tissue). A similar classification, based on depth of loss, is frequently applied to mechanical wounds, such as abrasion. The area of skin which has been destroyed needs also to be specified in order to assess the clinical status of the patient.

A massively injured patient, such as a patient with about 30% body surface area or more destroyed from fire through the full thickness of skin, presents an urgent problem to the clinician, since the open wound is an ongoing threat to survival. A large number of temporary wound coverings have been used to help the patient survive through this period while waiting for availability of autografts which provide permanent cover. If autograft is unavailable over a prolonged period while the patient has survived the severe trauma, contraction and scar synthesis occur over extensive areas. In the long run, the patient therefore has to cope with deep, disfiguring scars or with crippling contractures. Thus, even though the patient has been able to survive the massive trauma and has walked out of the clinic, the permanent loss of skin which has been sustained prevents, in many cases, resumption of an active, normal life.

138.4 Design Principles for a Permanent Skin Replacement

The analysis of the plight of the patient who has suffered extensive skin loss, presented above, leads logically to a wound cover which treats the problem in two stages. Stage 1 is the early phase of the clinical experience, one in which protection against severe fluid loss and against massive infection are defined as the major design objectives. Stage 2 is the ensuing phase, one in which the patient needs protection principally against disfiguring scars and crippling contractures. Even though the conceptual part of the design is separated in two stages for purposes of clarity, the actual treatment is to be delivered continuously, as will be become clear below. The sequential utilization of features inherent in stages 1 and 2 in a single device can be ensured by designing the graft as a bilayer membrane (Fig. 138.3). In this approach, the top layer incorporates the features of a stage 1 device, while the bottom layer delivers the performance expected from a stage 2 device. The top layer is subject to disposal after a period of about 10–15 days, during which time the bottom layer has already induced substantial synthesis of new dermis. Following removal of the top layer, the epidermal cover is provided either by covering with a thin epidermal graft or by modifying the device (cell seeding) so that an epidermis forms spontaneously by about 2 weeks after grafting.

Stage 1 Design Parameters

The overriding design requirement at this stage is based on the observation that air pockets ("dead space") at the graft-wound bed interface readily become sites of bacterial proliferation. Such sites can be prevented

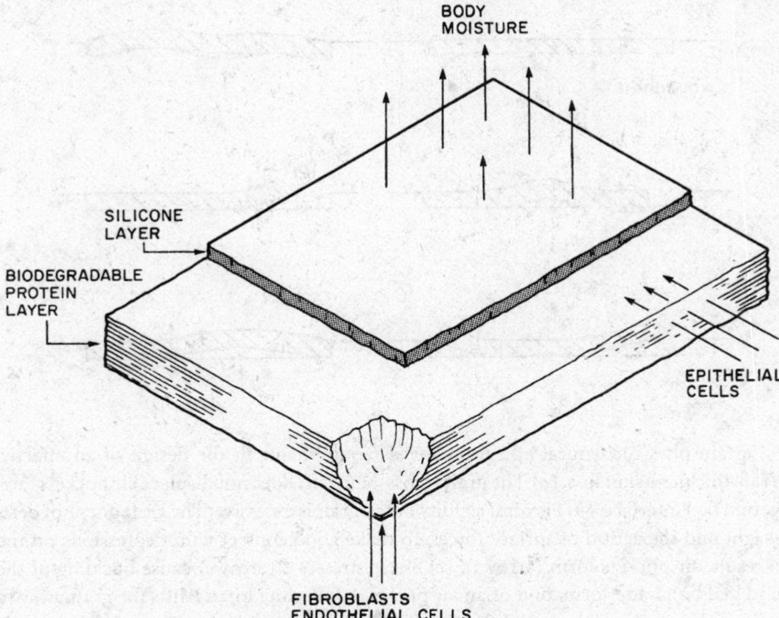

FIGURE 138.3 Schematic of the bilayer membrane which has become known as the artificial skin. The top layer is a silicone film which controls moisture flux through the wound bed to nearly physiologic levels, controls infection of the wound bed by airborne bacteria, and is strong enough to be sutured on the wound bed. The bottom layer is the skin regeneration template, which consists of a graft copolymer of type I collagen and chondroitin 6-sulfate, with critically controlled porosity and degradation rate. About 14 days after grafting, the silicone layer is removed and replaced with a thin epidermal autograft. The bottom layer induces synthesis of a nearly physiologic dermis and eventually is removed completely by biodegradation. (From Yannas IV, Burke JF, Orgill DP, et al. 1982. Science 215:74.)

from forming if the graft surface wets, in the physicochemical sense, the surface of the wound bed on contact and thereby displaces the air from the graft-tissue interface (Fig. 138.4). It follows that the physicochemical properties of the graft must be designed to ensure that this leading requirement is met, not only when the graft is placed on the wound bed but for several days thereafter, until the function of the graft has moved clearly into its stage 2, in which case the graft-wound bed interface has been synthesized de novo and the threat of dead space has been thereby eliminated indefinitely.

First, the flexural rigidity of the graft, i.e., the product of Young's modulus and moment of inertia of a model elastic beam, must be sufficiently low to provide for a flexible graft which drapes intimately over a geometrically nonuniform wound bed surface and thus ensures that the two surfaces will be closely apposed. In practice, these requirements can be met simply by adjusting both the stiffness in tension and the thickness of the graft to appropriately low values. Second, the graft will wet the wound bed if the surface energy of the graft-wound bed interface is lower than that of the air-wound bed surface, so that the graft can adequately displace air pockets from the air-wound bed surface. Although the measurement of a credible value of the surface energy is not a simple matter when the graft is based on certain natural polymers in the form of a hydrated gel, the requirement of adequate adhesion can be met empirically by chemical modification of the surface or by proper use of structural features such as porosity.

Third, the moisture flux through the graft must be maintained within bounds which are set by the following considerations. The upper bound to the moisture flux must be kept below the level where excessive dehydration of the graft occurs, thereby leading to alteration of the surface energy of the graft-wound bed interface and loss of the adhesive bond between graft and wound bed. Further, when the moisture flux exceeds the desired level, the graft is desiccated, and shrinkage stesses develop which pull

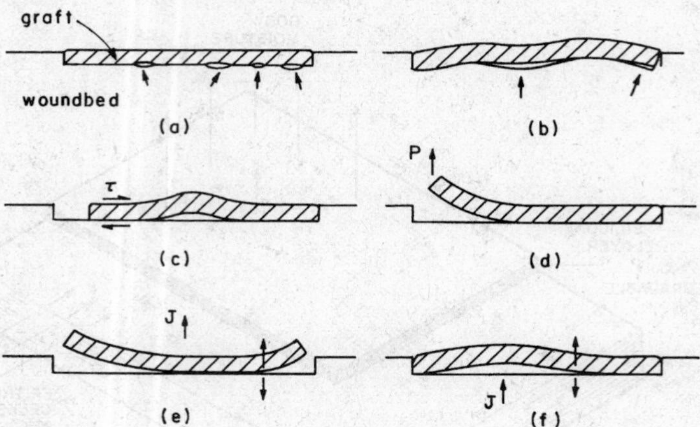

FIGURE138.4 Certain physicochemical and mechanical requirements in the design of an effective closure for a wound bed with full-thickness skin loss. (*a*) The graft (cross-hatched) does not displace air pockets (arrows) efficiently from the graft-wound bed interface. (*b*) Flexural rigidity of the graft is excessive. The graft does not deform sufficiently, under its own weight and the action of surface forces, to make good contact with depressions on the surface of the wound bed; as a result, air pockets form (arrows). (*c*) Shear stresses τ (arrows) cause buckling of the graft, rupture of the graft-wound bed bond and formation of an air pocket. (*d*) Peeling force *P* lifts the graft away from the wound bed. (*e*) Excessively high moisture flux rate *J* through the graft causes dehydration and development of shrinkage stresses at the edges (arrows), which cause lift-off away from the wound bed. (*f*) Very low moisture flux *J* causes fluid accumulation (edema) at the graft-wound bed interface and peeling off (arrows). (From Yannas IV, Burke JF. 1980. J Biomed Mater Res 14:65.)

the graft away from the wound bed. An estimate of the maximum normal stress σ_m can be obtained by modeling the desiccating graft in one dimension as a shrinking elastic beam bonded to a rigid surface

$$\sigma_m = 0.45\alpha\left(V_2 - V_1\right)E \tag{138.1}$$

In Eq. (138.1), α is the coefficient of expansion of a graft which swells in water, V_1 and V_2 are initial and final values of the volume fraction of moisture in the graft, and *E* is Young's modulus of the graft averaged over the range V_1 to V_2, the latter range being presumed to be narrow. If, by contrast, the moisture flux through the graft is lower than the desired low bound, water accumulates between the graft and the wound bed, and edema results with accompanying loss of the adhesive bond between the two surfaces.

Stage 2 Design Parameters

The leading design objectives in this stage are two: synthesis of new, physiologic skin and the eventual disposal of the graft.

The lifetime of the graft, expressed as the time constant of biodegradation t_b, was modeled in relation to the time constant for normal healing of a skin incision t_h. The latter is about 25 days. In preliminary studies with animals, it was observed that when matrices were synthesized to degrade at a very rapid rate, amounting to $t_b \ll t_h$, the initially insoluble matrix was reduced early to a liquidlike state, which was incompatible with an effective wound closure. At the other extreme, matrices were synthesized which degraded with exceptional difficulty within 3–4 weeks, compatible with $t_b \gg t_h$. In these preliminary studies it was observed that a highly intractable matrix, corresponding to the latter condition, led to formation of a dense fibrotic tissue underneath the graft which eventually led to loss of the bond between graft and wound bed. Accordingly, it was hypothesized that a rule of *isomorphous matrix replacement*, equivalent to assuming a graft degradation rate of order of magnitude similar to the synthesis rate for new tissue, and represented by the relation

$$\frac{t_b}{t_h} = 1 \tag{138.2}$$

would be optimal. Control of t_b is possible by adjustment of the crosslink density of the matrix. Equation (138.2) is the defining equation for a biodegradable scaffold which is coupled with, and therefore interacts with, the inflammatory process in a wound.

Migration of cells into the matrix is necessary for synthesis of new tissue. Such migration can proceed very slowly, defeating Eq. (138.2), when fibroblasts and other cells recruited below the wound surface are required to wait until degradation of a potentially solid-like matrix has progressed sufficiently. An easier pathway to migrating cells can be provided by modifying a solid-like matrix into one which has an abundance of pore channels, where the average pore is at least as large as one cell diameter (about 10 μm) for ready access. Although this rationale is supported by experiment, results with animal studies have shown that not only is there a lower limit to the average pore diameter, but there is also an upper limit (see below).

Migration of cells into the porous graft can proceed only if nutrients are available to these cells. Two general mechanisms are available for transport of nutrients to the migrating cells, namely, diffusion from the wound bed and transport along capillaries which may have sprouted within the matrix (angiogenesis). Since capillaries would not be expected to form for at least a few days, it is necessary to consider whether a purely diffusional mode of transport of nutrients from the wound bed surface into the graft could immediately supply the metabolic needs of the invading cells adequately. The cell has been modled as a reactor which consumes a critical nutrient with a rate r, in units of mole/cm^3/s; the nutrient is transported from the wound bed to the cell by diffusion over a distance l, the nutrient concentration at or near the surface of the wound bed is c_0, in units of mole/cm^3, and the diffusivity of the nutrient is D, in cm^2/s. The appropriate conditions were expressed in terms of a dimensionless number S, the *cell lifeline number*, which expresses the relative importance of reaction rate for consumption of the nutrient by the cell to rate of transport of the nutrient by diffusion alone:

$$S = \frac{rl^2}{Dc_0} \tag{138.3}$$

Eq. (138.3) suggests that when $S = 1$, the critical value of the path length, l_c, corresponds to the maximum distance along which cells can migrate inside the graft without requiring angiogenesis (vascularization) for nutrient transport. The value of l_c defines the maximum thickness of graft that can be populated with cells within a few hours after grafting, before angiogenesis has had time to occur.

These conceptual objectives have been partially met by designing the graft as an analog of *extracellular matrix* (ECM) which possesses morphogenetic activity since it leads to partial regeneration of dermis. The discovery of the specific ECM analog that possesses this activity has been based on the empirical observation that, whereas the vast majority of ECM analogs apparently do not inhibit wound contraction almost at all, one of the analogs does. The activity of this analog, for which the term *regeneration template* has been coined, is conveniently detected as a significant delay in the onset of wound contraction. When seeded with (uncultured) autologous keratinocytes, an active regeneration template is capable of inducing simultaneous synthesis both of a dermis and an epidermis in the guinea pig and in the swine (Yorkshire pig). The regeneration is almost complete; however, hair follicles and other skin adnexa are not formed. The resulting integument performs the two vital functions of skin, i.e., control of infection and moisture loss, while also providing physiologic mechanical protection to the internal organs and, additionally, providing a cosmetic effect almost identical to that of intact skin.

The morphogenetic specificity of the dermis regeneration template depends sensitively on retention of certain structural characteristics. The overall structure is that of an insoluble, three-dimensional covalently crosslinked network. The primary structure can be described as that of a *graft*-copolymer of type I collagen and a glycosaminoglycan (GAG) in the approximate ratio 98/2. The GAG can be either chondroitin 6-sulfate or dermatan sulfate; other GAGs appear capable of contributing approximately equal increments to morphogenetic specificity. The collagen fibers lack banding almost completely although the integrity of the triple helical structure is retained through the network. The resistance of the network to collagenase degradation is such that approximately two-thirds of the mass of the network becomes solubilized in vivo within about 2 weeks. The structure of the network is highly porous. The pore volume fraction exceeds 95% while the average pore diameter is maintained in the range 20–125 μm. The regeneration template loses its activity rapidly when these structural features are flawed deliberately in control studies.

The *dermis regeneration template*, a porous matrix unseeded with cells, induces synthesis of a new dermis and solves this old surgical problem. Simultaneous synthesis of a new, confluent epidermis occurs by migration of epithelial cell sheets from the wound edges, over the newly synthesized dermal bed. With wounds of relatively small characteristic dimension, e.g., 1 cm, epithelial cells migrating at speeds of about 0.5 mm/day from each wound edge can provide a confluent epidermis within 10 days. In such cases, the unseeded template fulfills all the design specifications set above. However, the wounds incurred by a massively burned patient are typically of characteristic dimension of several centimeters, often more than 20–30 cm. These wounds are large enough to preclude formation of a new epidermis by cell migration alone within a clinically acceptable timeframe, say 2 weeks. Wounds of that magnitude can be treated by seeding the porous collagen-GAG template, before grafting, with at lest 5×10^4 keratinocytes per cm^2 wound area. These uncultured, autologous cells are extracted by applying a cell separation procedure, based on controlled trypsinization, to a small epidermal biopsy.

Details of the synthesis of the dermis regeneration template, as well as of other templates which regenerate peripheral nerves and the knee meniscus, are presented elsewhere in this *Handbook* (see Chapter 13). The dermis regeneration template described in this section was first reported as a *synthetic skin* and as an *artificial skin*.

138.5 Clinical Studies of a Permanent Skin Replacement (Artificial Skin)

Clinical Studies

The skin regeneration template has been tested clinically on two occasions. In the first, conducted in the period 1979–1980, one clinical center was involved, and 10 severely burned patients were studied. In the second, conducted during 1986–1987, 11 clinical centers were involved, and 106 severely burned patients were treated in a prospective, randomized manner. In each case the results have been published in some detail. The second study led to a surgical report, a histologic report, and an immunologic report. There is now adequate information available to discuss the advantages and disadvantages of this prototype artificial skin in the treatment of the severely burned patient.

The artificial skin used in clinical studies so far consists of the bilayer device illustrated in Fig. 138.3. The outer layer is a silicone film, about 100 μm in thickness, which fulfills the requirements of stage 1 of the design (see above), and the inner layer is the skin regeneration template. In these clinical studies this device has not been seeded with keratinocytes. Closure of the relatively large wounds by formation of an epidermis has been achieved instead by use of a 100-μm-thin layer of the patient's epidermis (autoepidermal graft). The latter has been excised from an intact area of the patient's skin; the donor site can, however, be harvested repeatedly, since the excised epidermis regenerates spontaneously in a few days over the relatively intact dermal bed. Briefly, the entire procedure consists of preparation of the wound bed prior to grafting by excision of thermally injured tissue (*eschar*), followed by immediate grafting of the unseeded template on the freshly excised wound and ending, 3 weeks later, by replacing the outer, silicone layer of the device with a thin epidermal graft. The results of studies with a guinea pig model and a swine model have shown that seeding of the dermis regeneration template with fresh, uncultured autologous keratinocytes prior to grafting leads to simultaneous synthesis of an epidermis as well as a dermis in about 2 weeks. However, definitive clinical studies of the keratinocyte-seeded template have yet to be performed.

The discussion below focuses on the advantages and disadvantages of the (unseeded) artificial skin, as these emerge from clinical observations during the treatment as well as from a limited number of follow-up observations extending over several years after the treatment. The controls used in the clinical studies included meshed autograft, allograft, and xenografts. Comparative analysis of the clinical data will focus on each of the two stages of treatment for the massively burned patient, i.e., the early (acute) stage and the long term stage, the conceptual basis for which has been discussed above.

Clinical Parameters Used in the Evaluation

The clinical parameters during the early stage of treatment (about 3 weeks) include the *take* of the graft, expressed as a percentage of graft area which formed an adhesive bond of sufficient strength with the wound bed and became vascularized. In the case of the artificial skin treatment, two different measures of take are reported, namely, that of the bilayer membrane on the freshly excised wound bed and the take of the epidermal graft applied on the neodermal bed about 3 weeks later. Another parameter is the thickness of dermis that has been excised from the donor site in order to obtain the autograft that is used to close the wound definitively. An additional parameter which characterizes the cost of the donor site to the patient is the time to heal the donor site. The surgeon's overall qualitative evaluation of the treatment (relative to controls) during the early stage is also reported.

The long-term evaluation has extended at least 1 year in approximately one-quarter of the patients. The first long-term parameter is based on the patients' reports of the relative incidence of nonphysiologic sensations, including itching, dryness, scaliness, lack of elasticity (lack of deformability), sweating, sensation, and erythema. The second parameter is based on the physicians' report of the relative presence of hypertrophic scarring in the grafted area. A third parameter is the patient's evaluation of the physiologic feel and appearance of the donor sites. Finally, there is an overall evaluation and preference of the patients for a given grafted site as well as the physicians' evaluation of the same grafted site.

Short-Term Clinical Evaluation of Artificial Skin

The median percentage take of the artificial skin was 80%, compared with the median take of 95% for all controls. Use of the Wilcoxin Rank Sum Test for the bimodally distributed data led to the conclusion that the take of the artificial skin was lower than that of all controls with a p value of <0.0001. The reported difference reflected primarily the significantly lower take of the artificial skin relative to the meshed autograft. There was no significant difference in take when the artificial skin was compared with allograft (Wilcoxin Rank Sum $p > 0.10$). The take of the epidermal autograft was 86%.

Mean donor site thickness was 0.325 ± 0.045mm for control sites and only 0.15 ± 0.0625mm for epidermal grafts which were harvested for placement over the newly synthesized dermis; the difference was found to be significant by t test with a p value of < 0.001. The thinner donor sites used in the artificial skin procedures healed, as expected, significantly faster, requiring 10.6 ± 5.8 d compared to 14.3 ± 6.9 d for control sites, with a p value of < 0.001 by t test. It is worth noting that donor sites used in the artificial skin procedure were frequently reharvested sites from previous autografting; reharvested donor sites healed more slowly than primary sites.

The subjective evaluation of the operating surgeons at the conclusion of the acute stage of treatment was a response to the question, "Was artificial dermis (artificial skin) advantageous in the management of this particular patient?" Sixty-three percent of the comments were affirmative, whereas in 36% of the responses, the acute (early) results were believed to be no better than by use of routine methods. The physicians who responded positively to the use of artificial skin commented on the ability to use thin donor sites that healed quickly, relative to the thicker donor sites which were harvested in preparation for an autograft. Positive comments also cited the handling characteristics of the artificial skin relative to the allograft as well as the ability to close the wound without fear of rejection while awaiting healing of the donor site. Negative comments included a less-than-adequate drainage of serum and blood through the unperforated silicone sheet, the seemingly poor resistance of the artificial skin to infection, and the need for a second operation.

Long-Term Clinical Evaluation of Artificial Skin

One year after treatment, the allografted and xenografted sites had been long ago covered definitively with autograft; therefore, the experimental sites included, in the long term, either autografts (referred to occasionally as *controls* below) or the test sites, comprising the new integument induced as a result of treatment with the artificial skin (partially regenerated dermis closed with an epidermal graft).

The patients reported that itching was significantly less (Wilcoxin Rank Sum test $p < 0.02$) in the artificial skin site than in control sites. Dryness, scaliness, elasticity (deformability), sweating, sensation, and erythema were similar at both control and artificial skin sites. Hypertrophic scarring was reported to be less in artificial skin in 42% of sites and was reported to be equivalent on test and control sites 57% of the time. No patient reported that the artificial skin sites had more hypertrophic scar than the autografted sites. Even though donor sites that were used during treatment with the artificial skin were harvested repeatedly (recropping), 72% of patients reported that these artificial skin donor sites felt "more normal," 17% felt that there was no difference, and 11% felt that the control donor site was "more normal."

The results of the histologic study on this patient population showed that, in sites where the artificial skin was used, an intact dermis was synthesized as well as definitive closure of a complete epidermal layer with a minimum of scarring had occurred. The results of the immunologic study led to the conclusion that, in patients who had been treated with the artificial skin, there was increased antibody activity to bovine skin collagen, bovine skin collagen with chondroitin sulfate, and human collagen; however, it was concluded that these increased levels of antibodies were not immunologically significant.

The overall evaluation by the patients showed that 26% preferred the new integument generated by use of the artificial skin whereas 64% found that the sites were equivalent and 10% showed preference for the autografted site. Physician's overall evaluation showed that 39% preferred the artificial skin site, 45% found the sites to be equivalent, and 16% preferred the autografted site.

Clinical Results

The take of the artificial skin was comparable to all other grafts and was inferior only to the meshed autograft. The latter showed superior take in part because meshing reduces drastically the flexural rigidity of the graft (see above) leading thereby to greater conformity with the wound bed (see Fig. 138.4). The interstices in the meshed autograft also provided an outlet for drainage of serum and blood from the wound, thereby allowing removal of these fluids. By contrast, the continuity of the silicone sheet in the artificial skin accounted for the increased flexural rigidity of the graft and prevented drainage of wound fluids with a resulting increased incidence of fluid accumulation underneath the graft. Fluid accumulation was probably the cause of the reduced take of the artificial skin, since immediate formation of a physiochemical bond between the graft and the wound bed was thereby prevented (see Fig. 138.4). The development of infection underneath the artificial skin, noted by physicians in certain cases, can also be explained as originating in the layer of wound fluid which presumptively collected underneath the artificial skin. This analysis suggests that meshing of the silicone layer of the artificial skin, without affecting the continuity of the collagen-GAG layer, could lead to improved take and probably to reduced incidence of infection.

The healing time for donor sites associated with use of the artificial skin was shorter by about 4 days than for donor sites that were used to harvest autograft. An even shorter healing time for donor sites for artificial skin can be realized by reducing the thickness of the epidermal graft which is required to close the dermal bed. The average epidermal graft thickness reported in this study, 0.15mm, was significantly higher than thicknesses in the range 0.05–0.07mm, corresponding to an epidermal graft with adequate continuity but negligible amount of attached dermis. Increasing familiarity of surgeons with the procedure for harvesting these thin epidermal grafts is expected to lead to harvesting of thinner grafts in future studies. The importance of harvesting a thin graft cannot be overestimated, since the healing time of the donor site decreases rapidly with decreasing thickness of harvested graft. It has been reported that the mean healing time for donor sites for the artificial skin reported in this study, about 11 days, is reduced to 4 days provided that a pure epidermal graft, free of dermis, can be harvested.

Not only does the time to heal increase, but the incidence of hypertrophic scarring at a donor site also increases with the thickness of the harvested graft. This observation explains the higher incidence of hypertrophic scarring in donor sites associated with harvesting of autografts, since the latter were thicker by about 0.175mm than the epidermal grafts used with the artificial skin. An additional advantage associated with use of a thin epidermal graft is the opportunity to reharvest (recropping) within a few days; this reflects the ability of epithelial tissues to regenerate spontaneously provided there is an underlying dermal bed. When frequent recropping of donor graft is possible, the surface area of a patient that can be grafted within a clinically acceptable period increases rapidly. In the clinical study described here, a patient with deep burns over as much as 85% body surface area was covered with artificial skin grafts for 75 days while the few donor sites remaining were being harvested several times each.

In the long term rarely did a patient or a physician in this clinical study prefer the new skin provided by the autograft to that provided by the artificial skin treatment. This result is clearly related to the use of meshed autografts, a standard procedure in the treatment of massively burned patients. Meshing increases the wound area which can by autografted by between 1.5 and 6 times, thereby alleviating a serious resource problem. However, meshing destroys the dermis as well as the epidermis; although the epidermis regenerates spontaneously and fills in the defects, the dermis does not. The long-term result is a skin site with the meshed pattern permanently embossed on it. The artificial skin is a device that, in principle, is available in unlimited quantity; accordingly, it does not suffer from this problem (Fig. 138.2). The result is a smooth skin surface which was clearly preferred on average by patients and physicians alike.

It has been established that the artificial skin regenerates the dermis and, therefore, its use leads to complete inhibition of scar formation in full-thickness skin wounds. The regeneration is partial because skin adenexa (hair follicles, sweat glands) are not recovered. The results of studies of the mechanism by which the artificial skin regenerates the dermis in full-thickness skin wounds in animal models have been described elsewhere (see Chapter 113, Regeneration Templates).

Summary

The artificial skin leads to a new skin which appears closer to the patient's intact skin than does the meshed autograft. Take of the artificial skin is as good as all comparative materials except for the unmeshed autograft, which is superior in this respect. Donor sites associated with the artificial skin treatment heal faster, can be recropped much more frequently, and eventually heal to produce sites that look closer to the patient's intact skin than do donor sites harvested for the purpose of autografting. In comparison to the allograft, the artificial skin is easier to use, has the same take, does not get rejected, and is free of the risk of viral infection associated with use of allograft.

138.6 Alternative Approaches: Cultured Epithelial Autografts (CEA) and Skin Equivalents (SE)

The use of cultured epithelial autografts has been studied clinically. In this approach autologous epidermal cells are removed by biopsy and are then cultured in vitro for about 3 weeks until a mature keratinizing epidermis has formed; the epidermis is then grafted onto the patient. The epithelial cells spread and cover the dermal substrate, eventually covering the entire wound. Early reports on two pediatric patients were very encouraging, describing the life-saving coverage of one-half of body surface with cultured epithelial autografts and the technique was eventually used in a very large number of clinical centers in several countries. Later studies showed that the "take" of CEA was very good on partial thickness wounds but was questionable in full thickness wounds. In particular, blisters formed within 2 weeks in areas grafted by CEA, a problem which has recurred persistently in clinical studies. The mechanical fragility of the resulting integument resulting from use of CEA has been traced to lack of three structural features which are required for formation of a physiological dermal-epidermal junction at the grafted site, namely, the 7-S domain of type IV collagen, anchoring fibrils, and rete ridges. Early studies of the connective tissue underlying the CEA grafts have shown lack of a convincing dermal architecture as well as lack of elastin fibers.

In another development, a *skin equivalent* has been prepared by populating a collagen lattice with heterlologous fibroblasts, observing the contraction of the lattice by the cells and finally seeding the surface of the lattice with a suspension of epidermal cells from an autologous source or from cell banks. The latter attach, proliferate, and differentiate to form a multilayered epidermis in 7 to 10 days of exposure to the atmosphere and the resulting skin equivalent is then grafted on wounds. Clinical studies of the SE have been limited. In an early study (1988), the SE was used to cover partially full-thickness burn wounds covering over 15% of body surface area on eight patients. In every patient grafted with SE, an extensive lysis of the SE grafts was observed at the first dressing (48 h). In one patient only, a significant percentage of "take" (40%) was observed 14 days after grafting. It was concluded that the SE was not completely appropriate to serve routinely as a substitute for the autograft. In a later study (1995), the wounds treated were acute, mostly the result of excision of skin cancers. Twelve patients had clinical "takes" at the time of grafting and there was no evidence of rejection or toxicity following grafting with the SE. The wounds grafted with SE contracted by 10 to 15%, an extent larger than that observed following grafting with full-thickness skin. Biopsies of the grafted sites showed formation of scar tissue. The authors hypothesized that the SE was eventually replaced by host tissue. Recent studies of the SE have focused on patients with venous ulcers; these studies are in progress.

An attempt has been made to correct the erratic cover provided by split-thickness autografts; the latter are normally applied in a meshed form (meshed autograft) and, consequently, fail to cover the entire wound bed with a dermal layer. The attempted improvement consisted in grafting underneath the meshed autograft a living dermal tissue replacement, consisting of a synthetic polymeric mesh (polyglactin-910) which had been cultured in vitro over a period of 2 to 3 weeks with fibroblasts isolated from neonatal foreskin. Seventeen patients with full-thickness burn wounds were included in a preliminary clinical trial. Epithelialization of the interstices of the meshed autograft led to complete wound closure in 14 days in sites where the dermal living replacement had been grafted underneath the meshed autograft (experimental sites) and in those where it was omitted (control sites). Take of the meshed autograft was slightly reduced when the living dermal tissue replacement was underneath. Basement membrane structures developed both in control and experimental sites. Elastic fibers (elastin) were not observed in neodermal tissue either in control or experimental sites at periods up to one year after grafting. A subsequent clinical study explored the use of this device for the temporary closure of excised burn wounds. This 66-patient multicenter trial showed that the biosynthetic skin replacement was equivalent or superior to cadaver skin graft (frozen human cadaver allograft) with respect to its ability to prepare wounds for eventual closing with autograft.

Allograft is a human cadaver skin, which is frequently stored in frozen state in a skin bank. It is a temporary cover. If left on the wound longer than about 2 weeks, allograft is rejected by the severely burned patient, even though the latter is in an immunocompromised condition. When rejection is allowed to occur, the wound bed is temporarily ungraftable and is subject to infection. In a modification of this basic use of the allograft, the latter has been used as a dermal equivalent prior to grafting with cultured epithelia. Since the allograft is rejected if allowed to remain on the wound long enough for the epithelia to spread across the wound bed, the allograft has been treated in a variety of media in an effort to eliminate its immunogenicity.

Defining Terms

Adnexa: Accessory parts or appendages of an organ. Adnexa of skin include hair follicles and sweat glands.

Allograft: Human cadaver skin, usually maintained frozen in a skin bank and used to provide a temporary cover for deep wounds. About 2 weeks after grafting, the allograft is removed and replaced with autograft which has become available by that time. Previously referred to as *homograft.*

Artificial skin: A bilayer membrane consisting of an upper layer of silicone and a lower layer of dermis regeneration template. The template is a cell-free, highly porous analog of extracellular matrix.

Autograft: The patient's own skin, harvested from an intact area in the form of a membrane and used to graft an area of severe skin loss.

Basement membrane: An approximately 20-nm-thick multilayered membrane interleaved between the epidermis and the dermis.

Cell lifeline number: A dimensionless number which compares the relative magnitudes of chemical reaction and diffusion. This number, defined as *S* in Eq. (134.3) above, can be used to compute the maximum path length, l_c, over which a cell can migrate in a scaffold while depending on diffusion alone for transport of critical nutrients that it consumes. When the critical length is exceeded, the cell requires transport of nutrients by angiogenesis in order to survive.

Cultured epithelial autografts: A mature, keratinizing epidermis synthesized in vitro by culturing epithelial cells removed from the patient by biopsy. A relatively small skin biopsy (1 cm^2) can be treated to yield an area larger by about 10,000 in 2–3 weeks and is then grafted on patients with burns.

Dermal equivalent: A term which has been loosely used to describe a device that replaces, usually temporarily, the functions of the dermis following injury.

Dermis: A 2–5mm-thick layer of connective tissue populated with quiescent fibroblasts which lies underneath the epidermis. It is separated from the former by a very thin basement membrane. The dermis of adult mammals does not regenerate spontaneously following injury.

Dermis regeneration template: A graft copolymer of type *I* collagen and chondroitin 6-sulfate, average pore diameter 20–125 μm, degrading in vivo to an extent of about 50% in 2 weeks, which induces partial regeneration of the dermis in wounds from which the dermis has been fully excised. When seeded with keratinocytes prior to grafting, this analog of extracellular matrix has induced simultaneous synthesis both of a dermis and an epidermis.

Donor site: The skin site from which an autograft has been removed with a dermatome.

Epidermis: The cellular outer layer of skin, about 0.1-mm thick, which protects against moisture loss and against infection. An epidermal graft, e.g., cultured epithelium or a thin graft removed surgically, requires a dermal substrate for adherence onto the wound bed. The epidermis regenerates spontaneously following injury, provided there is a dermal substrate underneath.

Eschar: Dead tissue, typically the result of a thermal injury, which covers the underlying, potentially viable tissue.

Extracellular matrix: A largely insoluble, nondiffusible macromolecular network, consisting mostly of glycoproteins and proteoglycans.

Isomorphous matrix replacement: A term used to describe the synthesis of new, physiologic tissue within a skin regeneration template at a rate which is of the same order as the degradation of the template. This relation, Eq. (138.2) above, is the defining equation for a biodegradable scaffold which biologically interacts with the inflammatory process of the wound bed.

Living dermal replacement: A synthetic biodegradable polymeric mesh, previously cultured with fibroblasts, which is placed underneath a conventional meshed autograft.

Meshed autograft: A sheet autograft which has been meshed and then expanded by a factor of 1.5–6 to produce grafts with a characteristic pattern.

Regeneration template: A biodegradable scaffold which, when attached to a missing organ, induces its regeneration.

Scar: The result of a repair process in skin and other organs. Scar is morphologically different from skin, in addition to being mechanically less extensible and weaker than skin. The skin regeneration template induces synthesis of nearly physiologic skin rather than scar.

Sheet autograft: A layer of the patient's skin, comprising the epidermis and about one-third of the dermal thickness, which has not been meshed prior to grafting areas of severe skin loss.

Skin: A vital organ which indispensably protects the organism from infection and dehydration while also providing other functions essential to physiologic life, such as assisting in thermoregulation and providing a tactile sensor for the organism.

Skin equivalent: A collagen gel which has been contracted by fibroblasts cultured therein. Following culturing with keratinocyte, until a cornified epidermis is formed over the contracted collagen lattice, it is grafted on skin wounds.

Split-thickness autograft: An autograft which is about one-half or one-third as thick as the full thickness of skin.

Subcutis: A layer of fat tissue underneath the dermis.

Synthetic skin: A term used to describe the artificial skin in the early literature.

Take: The adhesion of a graft on the woundbed. Without adequate take, there is no physicochemical or biologic interaction between graft and wound bed.

Xenograft: Skin graft obtained from a different species: e.g., pig skin grafted on human. Synthetic polymeric membranes are often referred to as *xenografts*. Previously referred to as *heterograft*.

References

Boykin JV, Jr., Molnar JA. 1992. Burn scar and skin equivalents. In IK Cohen, RF Diegelmann and WJ Lindblad (eds). Wound Healing, pp523–540, Philadelphia, Saunders.

Burke JF, Yannas IV, Quinby WC, Jr., Bondoc CC, Jung WK. 1981. Successful use of a physiologically acceptable artificial skin in the treatment of extensive burn injury. Ann. Surg. 194:413–428.

Compton CC. 1992. Current concepts in pediatric burn care: the biology of cultured epithelial autografts: An eight-year study in pediatric burn patients. Eur. J. Pediatr. Surg. 2:216–222.

Compton CC, Gill JM, Bradford DA, Regauer S, Gallico GG, O'Connor NE. 1989. Skin regenerated from cultured epithelial autografts on full-thickness burn wounds from 6 days to 5 years after grafting. Lab. Invest. 60:600–612.

Eaglstein WH, Iriondo M, Laszlo K. 1995. A composite skin substitute (Graftskin) for surgical wounds. Dermatol. Surg. 21:839–843.

Gallico, GG, O'Connor NE, Compton CC, Kehinde O, Green H. 1984. Permanent coverage of large burn wounds with autologous cultured human epithelium, N. Engl. J. Med. 311:448–451.

Hansbrough, J.F., C. Dore and W.B. Hansbrough (1992b). Clinical trials of a living dermal tissue replacement placed beneath meshed, split-thickness skin grafts on excised burn wounds, J. Burn Care Rehab. 13:519–529.

Heimbach D, Luterman A, Burke J, Cram A, Herndon D, Hunt J, Jordan M, McManus W, Solem L, Warden G, Zawacki B. 1988. Artificial dermis for major burns. Ann. Surg. 208:313–320.

Michaeli D, McPherson M. 1990. Immunologic study of artificial skin used in the treatment of thermal injuries, J. Burn Care Rehab. 11:21–26.

Purdue GF, Hunt JL, Still JM Jr, Law EJ, Herndon DN, Goldfarb IW, Schiller WR, Hansbrough JF, Hickerson WL, Himel HN, Kealey P, Twomey J, Missavage AE, Solem LD, Davis M, Totoritis M, Gentzkow GD. 1997. A multicenter clinical trial of a biosynthetic skin replacement, Dermagraft-TC, compared with cryopreserved human cadaver skin for temporary coverage of excised burn wounds. J Burn Care Rehab. 18:52–57.

Sabolinski ML, Alvarez O, Auletta M, Mulder G, Parenteau NL. 1996. Cultured skin as a 'smart material' for healing wounds: experience in venous ulcers. Biomaterials 17:311–320.

Stern R, McPherson M, Longaker MT. 1990. Histologic study of artificial skin used in the treatment of full-thickness thermal injury, J. Burn Care Rehab. 11:7–13.

Tompkins RG, Burke JF. 1992. Artificial Skin. Surg. Rounds October: 881–890.

Waserman D, Sclotterer M, Toulon A, Cazalet C, Marien M, Cherruau B, Jaffray P. 1988. Preliminary clinical studies of a biological skin equivalent in burned patients. Burns 14:326–330.

Yannas IV, Burke JF. 1980. Design of an artificial skin I. Basic design principles, J. Biomed, Mater. Res. 14:65–81.

Yannas IV, Burke JF, Warpehoski M, Stasikelis P, Skrabut EM, Orgill D, Giard DJ. 1981. Prompt, long-term functional replacement of skin. Trans. Am. Soc. Artif. Intern. Organs 27:19–22.

Further Information

Bell E, Erlich HP, Buttle DJ, T. Nakatsuji. 1981. Living skin formed in vitro and accepted as skin-equivalent tissue of full thickness. Science 211:1052–1054.

Bell E, Ivarsson B, Merrill C. 1979. Production of a tissue-like structure by contraction of collagen lattices by human fibroblasts of different proliferative potential in vitro. Proc. Natl. Acad. Sci. USA 76:1274–1278.

Boyce ST, Hansbrough JF. 1988. Biologic attachment, growth, and differentiation of cultured human keratinocytes on a graftable collagen and chondroitin 6-sulfate substrate. Surg. 103:421–431.

Eldad A, Burt A, Clarke JA. 1987. Cultured epithelium as a skin substitute, Burns. 13:173–180.

Green H, Kehinde O, Thomas and J. 1979. Growth of cultured human epidermal cells into multiple epithelia suitable for grafting. Proc. Natl. Acad. Sci. USA 76:5665–5668.

Langer R, Vacanti JP. 1993. Tissue Engineering. Science. 260:920–926.

O'Connor, NE, Mulliken JB, Banks-Schlegel S, Kehinde O, Green H. 1981. Grafting of burns with cultured epithelium prepared from autologous epidermal cells. Lancet 1: 75–78.

Woodley, DT, Peterson HD, Herzog SR, Stricklin GP, Bergeson RE, Briggaman RA, Cronce DJ, O'Keefe EJ. 1988. Burn wounds resurfaced by cultured epidermal autografts show abnormal reconstitution of anchoring fibtils. JAMA 259:2566–2571.

Woodley, DT, Briggaman RA, Herzog SR, Meyers AA, Peterson HD, O'Keefe EJ. 1990. Characterization of "neo-dermis" formation beneath cultured human epidermal autografts transplanted on muscle fascia. J. Invest. Dermatol. 95:20–26.

Yannas, IV, Lee E, Orgill DP, Skrabut EM, Murphy GF. 1989. Synthesis and characterization of a model extracellular matrix that induces partial regeneration of adult mammalian skin. Proc. Natl. Acad. Sci. USA 86:933–937.

Yannas IV, Burke JF, Gordon PL, Huang C. 1977. Multilayer membrane useful as synthetic skin, US Patent 4,060,081, Nov. 29.

Yannas IV. 1982. Wound tissue can utilize a polymeric template to synthesize a functional extension of the skin. Science 215:174–176.

XIV

Rehabilitation Engineering

Charles J. Robinson
Louisiana Tech University and
the Overton Brooks VA Medical Center

Engineering advances have resulted in enormous strides in the field of rehabilitation. Individuals with reduced or no vision can be given "sight"; those with severe or complete hearing loss can "hear" by being provided with a sense of their surroundings; those unable to talk can be aided to "speak" again; and those without full control of a limb (or with the limb missing) can, by artificial means, "walk" or regain other movement functions. But the present level of available functional restoration for seeing, hearing, speaking, and moving still pales in comparison to the capabilities of individuals without disability. As is readily apparent from the content of many of the chapters in this Handbook, the human sensory and motor (movement) systems are marvelously engineered, both within a given system and integrated across systems. The rehabilitation engineer thus faces a daunting task in trying to design augmentative or replacement systems when one or more of these systems is impaired.

Rehabilitation engineering had its origins in the need to provide assistance to individuals who were injured in World War II. *Rehabilitation engineering* can be defined in a number of ways. Perhaps the most encompassing (and the one adopted here) is that proposed by Reswick [1982]—Rehabilitation engineering is the application of science and technology to ameliorate the handicaps of individuals with disabilities. With this definition, any device, technique, or concept used in rehabilitation that has a technological basis falls under the purview of rehabilitation engineering. This contrasts with the much narrower view that is held by some that rehabilitation engineering is only the design and production phase of a broader field called Assistive Technology. Lest one consider this distinction trivial, consider that the U.S. Congress has mandated that rehabilitation engineering and technology services be provided by all states; and an argument has ensued among various groups of practitioners about who can legally provide such services because of the various interpretations of what rehabilitation engineering is.

There is a core body of knowledge that defines each of the traditional engineering disciplines. Biomedical engineering is less precisely defined; but, in general, a biomedical engineer must be proficient in a traditional engineering discipline and have a working knowledge of things biological or medical. The rehabilitation engineer is a biomedical engineer who must not only be technically proficient as an engineer and know biology and medicine, but must also integrate artistic, social, financial, psychological, and physiological considerations to develop or analyze a device, technique, or concept that meets the needs of the population the engineer is serving. In general, rehabilitation engineers deal with musculoskeletal or sensory disabilities. They often have a strong background in biomechanics. Most work in a multidisciplinary team setting.

Rehabilitation engineering deals with many aspects of rehabilitation including applied, scientific, clinical, technical, and theoretical. Various topics include, but are not limited to, assistive devices and other aids for those with disability, sensory augmentation and substitution systems, functional electrical stimulation (for motor control and sensory-neural prostheses), orthotics and prosthetics, myoelectric devices and techniques, transducers (including electrodes), signal processing, hardware, software, robotics, systems approaches, technology assessment, postural stability, wheelchair seating systems, gait analysis, biomechanics, biomaterials, control systems (both biological and external), ergonomics, human performance, and functional assessment [Robinson, 1993].

In this section of the Handbook, we focus only on applications of rehabilitation engineering. The concepts of rehabilitation engineering, rehabilitation science, and rehabilitation technology are outlined in Chapter 139. Chapter 141 discusses the importance of personal mobility and various wheeled modes of transportation (wheelchairs, scooters, cars, vans, and public conveyances). Chapter 142 looks at other non-wheeled ways to enhance mobility and physical performance. Chapter 143 covers techniques available to augment sensory impairments or to provide a substitute way to input sensory information. Conversely, Chapter 144 looks at the output side.

For the purposes of this Handbook, many topics that partially fall under the rubric of rehabilitation engineering are covered elsewhere. These include chapters on Electrical Stimulation (Durand), Hard Tissue Replacement—Long Bone Repair and Joints (Goel), Biomechanics (Schneck), Musculoskeletal Soft Tissue Mechanics (Lieber), Analysis of Gait (Davis), Sports Biomechanics/Kinesiology (Johnson),

Biodynamics (Diggs), Cochlear Mechanics (Steele), Measurement of Neuromuscular Performance Capabilities (Smith), Human Factors Applications in Rehabilitation Engineering (Strauss and Gunderson), Electrical Stimulators (Peckham), Prostheses and Artificial Organs (Galletti), Nerve Guidance Channels (Valentini), and Tracheal, Laryngeal, and Esophageal Replacement Devices (Shimizu).

Rehabilitation engineering can be described as an engineering systems discipline. Imagine being the design engineer on a project that has an unknown, highly non-linear plant, with coefficients whose variations in time appear to follow no known or solvable model, where time (yours and your client's) and funding are severely limited, where no known solution has been developed (or if it has, will need modification for nearly every client so no economy of scale exists). Further, there will be severe impedance mismatches between available appliances and your client's needs. Or, the low residual channel capacity of one of your client's senses will require enormous signal compression to get a signal with any appreciable information content through it. Welcome to the world of the rehabilitation engineer!!

References

Reswick, J. 1982. What is a Rehabiliation Engineer? in *Annual Review of Rehabiltation*, Vol 2, E.L. Pan, T.E. Backer, and C.L. Vash, Eds., Springer-Verlag, New York.

Robinson, C.J., 1993. Rehabilitation engineering—An editorial, *IEEE Trans. Rehab. Eng.*, 1(1):1-2.

139

Rehabilitation Engineering, Science, and Technology

Charles J. Robinson
Louisiana Tech University and the Overton Brooks VA Medical Center

139.1 Introduction

Rehabilitation engineering requires a multidisciplinary effort. To put rehabilitation engineering into its proper context, we need to review some of the other disciplines with which rehabilitation engineers must be familiar. Robinson [93] has reviewed or put forth the following working definitions and discussions: *Rehabilitation is the (re)integration of an individual with a disability into society.* This can be done either by enhancing existing capabilities or by providing alternative means to perform various functions or to substitute for specific sensations.

Rehabilitation engineering is the "*application* of science and technology to ameliorate the handicaps of individuals with disabilities" [Reswick, 82]. In actual practice, many individuals who say that they practice "rehabilitation engineering" are not engineers by training. While this leads to controversies from practitioners with traditional engineering degrees, it also has the *de facto* benefit of greatly widening the scope of what is encompassed by the term "rehabilitation engineering."

Rehabilitation medicine is a clinical *practice* that focuses on the physical aspects of functional recovery, but that also considers medical, neurological and psychological factors. *Physical therapy, occupational therapy,* and *rehabilitation counseling* are professions in their own right. On the sensory-motor side, other medical and therapeutical specialties practice rehabilitation in vision, audition, and speech.

Rehabilitation technology (or *assistive technology*) narrowly defined is the selection, design, or manufacture of augmentative or assistive devices that are appropriate for the individual with a disability. Such devices are selected based on the specific disability, the function to be augmented or restored, the user's wishes, the clinician's preferences, cost, and the environment in which the device will be used.

Rehabilitation science is the *development* of a body of knowledge, gleaned from rigorous basic and clinical research, that describes how a disability alters specific physiological functions or anatomical structures, and that details the underlying principles by which residual function or capacity can be measured and used to restore function of individuals with disabilities.

0-8493-0462-8/00/$0.00+$.50

139.2 Rehabilitation Concepts

Effective rehabilitation engineers must be well versed in all of the areas described above since they generally work in a team setting, in collaboration with physical and occupational therapists, orthopedic surgeons, physical medicine specialists and/or neurologists. Some rehabilitation engineers are interested in certain activities that we do in the course of a normal day that could be summarized as *activities of daily living* (ADL). These include eating, toileting, combing hair, brushing teeth, reading, etc. Other engineers focus on *mobility* and the limitations to mobility. Mobility can be personal (e.g., within a home or office) or public (automobile, public transportation, accessibility questions in buildings). Mobility also includes the ability to move functionally through the environment. Thus, the question of mobility is not limited to that of getting from place to place, but also includes such questions as whether one can reach an object in a particular setting or whether a paralyzed urinary bladder can be made functional again. Barriers that limit mobility are also studied. For instance, an ill-fitted wheelchair cushion or support system will most assuredly limit mobility by reducing the time that an individual can spend in a wheelchair before he or she must vacate it to avoid serious and difficult-to-heal pressure sores. Other groups of rehabilitation engineers deal with *sensory disabilities*, such as sight or hearing, or with *communications disorders*, both in the production side (e.g., the non-vocal) or in the comprehension side. For any given client, a rehabilitation engineer might have all of these concerns to consider (i.e., ADLs, mobility, sensory and communication dysfunctions).

A key concept in physical or sensory rehabilitation is that of *residual function or residual capacity*. Such a concept implies that the function or sense can be quantified, that the performance range of that function or sense is known in a non-impaired population, and that the use of residual capacity by a disabled individual should be encouraged. These measures of human performance can be made subjectively by clinicians or objectively by some rather clever computerized test devices.

A rehabilitation engineer asks three key questions: Can a diminished function or sense be successfully augmented? Is there a substitute way to return the function or to restore a sense? And is the solution appropriate and cost-effective? These questions give rise to two important rehabilitation concepts: orthotics and prosthetics. An *orthosis* is an appliance that aids an existing function. A *prosthesis* provides a substitute.

An artificial limb is a *prosthesis*, as is a wheelchair. An ankle brace is an *orthosis*. So are eyeglasses. In fact, eyeglasses might well be the consumate rehabilitation device. They are inexpensive, have little social stigma, and are almost completely unobtrusive to the user. They have let many millions of individuals with correctable vision problems lead productive lives. But in essence, a pair of eyeglasses is an optical device, governed by traditional equations of physical optics. Eyeglasses can be made out of simple glass (from a raw material as abundant as the sands of the earth!) or complex plastics such as those that are ultraviolet sensitive. They can be ground by hand or by sophisticated computer-controlled optical grinders. Thus, crude technology can restore functional vision. Increasing the technical content of the eyeglasses (either by material or manufacturing method) in most cases will not increase the amount of function restored, but it might make the glasses cheaper, lighter and more prone to be used.

139.3 Engineering Concepts in Sensory Rehabilitation

Of the five traditional senses, vision and hearing most define the interactions that permit us to be human. These two senses are the main input channel through which data with high information content can flow. We read; we listen to speech or music; we view art. A loss of one or the other of these senses (or both) can have a devastating impact on the individual affected. Rehabilitation engineers attempt to restore the functions of these senses either through augmentation or via sensory substitution systems. Eyeglasses and hearing aids are examples of augmentative devices that can be used if some residual capacity remains. A major area of rehabilitation engineering research deals with *sensory substitution systems* [Kaczmarek, this Handbook].

The visual system has the capability to detect a single photon of light, yet also has a dynamic range that can respond to intensities many orders of magnitude greater. It can work with high contrast items and with those of almost no contrast, and across the visible spectrum of colors. Millions of parallel data channels form the optic nerve that comes from an eye; each channel transmits an asynchronous and quasi-random (in time) stream of binary pulses. While the temporal coding on any one of these channels is not fast (on the order of 200 bits per sec or less), the capacity of the human brain to parallel process the entire image is faster than any supercomputer yet built.

If sight is lost, how can it be replaced? A simple pair of eyeglasses will not work, since either the sensor (the retina), the communication channel (the optic nerve and all of its relays to the brain), or one or more essential central processors (the occipital part of the cerebral cortex for initial processing; the parietal and other cortical areas for information extraction) has been damaged. For replacement within the system, one must determine where the visual system has failed and whether a stage of the system can be artificially bypassed. If one uses another sensory modality (e.g., touch or hearing) as an alternate input channel, one must determine whether there is sufficient bandwidth in that channel and whether the higher-order processing hierarchy is plastic enough to process information coming via a different route.

While the above discussion might seem just philosophical, it is more than that. We normally read printed text with our eyes. We recognize words from their (visual) letter combinations. We comprehend what we read via a mysterious processing in the parietal and temporal parts of the cerebral cortex. Could we perhaps read and comprehend this text or other forms of writing through our fingertips with an appropriate interface? The answer surprisingly is yes! And, the adaptation actually goes back to one of the earliest applications of coding theory—that of the development of Braille. Braille condenses all text characters to a raised matrix of 2 by 3 dots (2^6 combinations), with certain combinations reserved as indicators for the next character (such as a number indicator) or for special contractions. Trained readers of Braille can read over 250 words per minute of grade 2 Braille (as fast as most sighted readers can read printed text!). Thus, the Braille code is in essence a rehabilitation engineering concept where an alternate sensory channel is used as a substitute and where a recoding scheme has been employed.

Rehabilitation engineers and their colleagues have designed other ways to read text. To replace the retina as a sensor element, a modern high resolution, high sensitivity, fast imaging sensor (CCD, etc.) is employed to capture a visual image of the text. One method, used by various page scanning devices, converts the scanned image to text by using optical character recognition schemes, and then outputs the text as speech via text-to-speech algorithms. This machine essentially recites the text, much as an sighted helper might do when reading aloud to the blind individual. The user of the device is thus freed of the absolute need for a helper. Such *independence* is often the goal of rehabilitation.

Perhaps the most interesting method presents an image of the scanned data directly to the visual cortex or retina via an array of implantable electrodes that are used to electrically activate nearby cortical or retinal structures. The visual cortex and retina are laid out in a topographic fashion such that there is an orderly mapping of the signal from different parts of the visual field to the retina, and from the retina to corresponding parts of the occipital cortex. The goal of stimulation is to mimic the neural activity that would have been evoked had the signal come through normal channels. And, such stimulation does produce the sensation of light. Since the "image" stays within the visual system, the rehabilitation solution is said to be *modality-specific.* However, substantial problems dealing with biocompatibility and image processing and reduction remain in the design of the electrode arrays and processors that serve to interface the electronics and neurological tissue.

Deafness is another manifestation of a loss of a communication channel, this time for the sense of hearing. Totally deaf individuals use vision as a substitute input channel when communicating via sign language (also a substitute code), and can sign at information rates that match or exceed that of verbal communication. Hearing aids are now commercially available that can adaptively filter out background noise (a predictable signal) while amplifying speech (unpredictable) using autoregressive, moving average (ARMA) signal processing. With the recent advent of powerful digital signal processing chips, true digital hearing aids are now available. Previous analog aids, or digitally programable analog aids, provided a set of tunable filters and amplifiers to cover the low, mid and high frequency ranges of the hearing spectrum. But the digital aids can be specifically and easily tailored (i.e., programmed) to compensate for the specific losses of each individual client across the frequency continuum of hearing, and still provide automatic gain control and one or more user-selectable settings that have been adjusted to perform optimally in differing noise environments.

An exciting development is occurring outside the field of rehabilitation that will have a profound impact on the ability of the deaf to comprehend speech. Electronics companies are now beginning to market universal translation aids for travellers, where a phrase spoken in one language is captured, parsed, translated, and restated (either spoken or displayed) in another language. The deaf would simply require that the visual display be in the language that they use for writing.

Deafness is often brought on (or occurs congenitally) by damage to the cochlea. The cochlea normally transduces variations in sound pressure intensity at a given frequency into patterns of neural discharge. This neural code is then carried by the auditory (eighth cranial) nerve to the brainstem where it is preprocessed and relayed to auditory cortex for initial processing and on to the parietal and other cortical areas for information extraction. Similar to the case for the visual system, the cochlea, auditory nerve, auditory cortex and all relays in between maintain a topological map, this time based on tone frequency (tonotopic). If deafness is solely due to cochlear damage (as is often the case) and if the auditory nerve is still intact, a cochlear implant can often be substituted for the regular transducer array (the cochlea) while still sending the signal through the normal auditory channel (to maintain modality-specificity).

At first glance, the design of a cochlear prosthesis to restore hearing appears daunting. The hearing range of a healthy young individual is 20 to 16,000 Hz. The transducing structure, the cochlea, has 3500 inner and 12000 outer hair cells, each best activated by a specific frequency that causes a localized mechanical resonance in the basilar membrane of the cochlea. Deflection of a hair cell causes the cell to fire an all-or-none (i.e., pulsatile) neuronal discharge, whose rate of repetition depends to a first approximation on the amplitude of the stimulus. The outputs of these hair cells have an orderly convergence on the 30,000 to 40,000 fibers that make up the auditory portion of the eighth cranial nerve. These afferent fibers in turn go to brainstem neurons that process and relay the signals on to higher brain centers [Klinke, 1983]. For many causes of deafness, the hair cells are destroyed, but the eighth nerve remains intact. Thus, if one could elicit activity in a specific output fiber by means other than the hair cell motion, perhaps some sense of hearing could be restored. The geometry of the cochlea helps in this regard as different portions of the nerve are closer to different parts of the cochlea.

Electrical stimulation is now used in the cochlear implant to bypass hair cell transduction mechanisms [Loeb, 1985; Clark et al., 1990]. These sophisticated devices have required that complex signal processing, electronic and packaging problems be solved. One current cochlear implant has 22 stimulus sites along the scala tympani of the cochlea. Those sites provide excitation to the peripheral processes of the cells of the eighth cranial nerve, which are splayed out along the length of the scala. The electrode assembly itself has 22 ring electrodes spaced along its length and some additional guard rings between the active electrodes and the receiver to aid in securing the very flexible electrode assembly after it is snaked into the cochlea's very small (a few mm) round window (a surgeon related to me that positioning the electrode was akin to pushing a piece of cooked spaghetti through a small hole at the end of a long tunnel). The electrode is attached to a receiver that is inlaid into a slot milled out of the temporal bone. The receiver contains circuitry that can select any electrode ring to be a source and any other electrode to be a sink for the stimulating current, and that can rapidly sequence between various pairs of electrodes. The receiver is powered and controlled by a radiofrequency link with an external transmitter, whose alignment is maintained by means of a permanent magnet imbedded in the receiver.

A digital signal processor stores information about a specific user and his or her optimal electrode locations for specific frequency bands. The object is to determine what pair of electrodes best produces the subjective perception of a certain pitch *in the implanted individual*, and then to associate a particular filter with that pair via the controller. An enormous amount of compression occurs in taking the frequency range necessary for speech comprehension and reducing it to a few discrete channels. At present, the optimum compression algorithm is unknown, and much fundamental research is being carried out in speech processing, compression and recognition. But, what is amazing is that a number of totally deaf individuals can relearn to comprehend speech exceptionally well without speech-reading through the use of these implants. Other individuals find that the implant aids in speech-reading. For some only an awareness of environmental sounds is apparent; and for another group, the implant appears to have had little effect. But, if you could (as I have been able to) finally converse in unaided speech with an individual who had been rendered totally blind and deaf by a traumatic brain injury, you begin to appreciate the power of rehabilitation engineering.

139.4 Engineering Concepts in Motor Rehabilitation

Limitations in mobility can severely restrict the quality of life of an individual so affected. A wheelchair is a prime example of a prosthesis that can restore personal mobility to those who cannot walk. Given the proper environment (fairly level floors, roads, etc.), modern wheelchairs can be highly efficient. In fact, the fastest times in one of man's greatest tests of endurance, the Boston Marathon, are achieved by the wheelchair racers. Although they do gain the advantage of being able to roll, they still must climb the same hills, and do so with only one-fifth of the muscle power available to an able-bodied marathoner.

While a wheelchair user could certainly go down a set of steps (not recommended), climbing steps in a normal manual or electric wheelchair is a virtual impossibility. Ramps or lifts are engineered to provide accessibility in these cases, or special climbing wheelchairs can be purchased. Wheelchairs also do not work well on surfaces with high rolling resistance or viscous coefficients (e.g., mud, rough terrain, etc.), so alternate mobility aids must be found if access to these areas is to be provided to the physically disabled. Hand-controlled cars, vans, tractors and even airplanes are now driven by wheelchair users. The design of appropriate control modifications falls to the rehabilitation engineer.

Loss of a limb can greatly impair functional activity. The engineering aspects of artificial limb design increase in complexity as the amount of residual limb decreases, especially if one or more joints are lost. As an example, a person with a mid-calf amputation could use a simple wooden stump to extend the leg, and could ambulate reasonably well. But such a leg is not cosmetically appealing and completely ignores any substitution for ankle function.

Immediately following World War II, the United States government began the first concerted effort to foster better engineering design for artificial limbs. Dynamically lockable knee joints were designed for artificial limbs for above-knee amputees. In the ensuing years, energy-storing artificial ankles have been designed, some with prosthetic feet so realistic that beach thongs could be worn with them! Artificial hands, wrists and elbows were designed for upper limb amputees. Careful design of the actuating cable system also provided for a sense of hand grip force, so that the user had some feedback and did not need to rely on vision alone for guidance.

Perhaps the most transparent (to the user) artificial arms are the ones that use electrical activity generated by the muscles remaining in the stump to control the actions of the elbow, wrist and hand [Stein et al., 1988]. This electrical activity is known as myoelectricity, and is produced as the muscle contraction spreads through the muscle. Note that these muscles, if intact, would have controlled at least one of these joints (e.g., the biceps and triceps for the elbow). Thus, a high level of modality-specificity is maintained since the functional element is substituted only at the last stage. All of the batteries, sensor electrodes, amplifiers, motor actuators and controllers (generally analog) reside entirely within these myoelectric arms. An individual trained in the use of a myoelectric arm can perform some impressive tasks with this arm. Current engineering research efforts involve the control of simultaneous multi-joint movements (rather than the single joint movement now available) and the provision for sensory feedback from the end effector of the artificial arm to the skin of the stump via electrical means.

139.5 Engineering Concepts in Communications Disorders

Speech is a uniquely human means of interpersonal communication. Problems that affect speech can occur at the initial transducer (the larynx) or at other areas of the vocal tract. They can be of neurological (due to cortical, brainstem or peripheral nerve damage), structural, and/or cognitive origin. A person might only be able to make a halting attempt at talking, or might not have sufficient control of other motor skills to type or write.

If only the larynx is involved, an externally applied artificial larynx can be used to generate a resonant column of air that can be modulated by other elements in the vocal tract. If other motor skills are intact, typing can be used to generate text, which in turn can be spoken via text-to-speech devices described above. And the rate of typing (either whole words or via coding) might be fast enough so that reasonable speech rates could be achieved.

The rehabilitation engineer often becomes involved in the design or specification of *augmentative communication aids* for individuals who do not have good muscle control, either for speech or for limb movement. A whole industry has developed around the design of symbol or letter boards, where the user can point out (often painstakingly) letters, words or concepts. Some of these boards now have speech output. Linguistics and information theory have been combined in the invention of acceleration techniques intended to speed up the communication process. These include alternative language representation systems based on semantic (iconic), alphanumeric, or other codes; and prediction systems, which provide choices based on previously selected letters or words. A general review of these aids can be found in Chapter 144, while Goodenough-Trepagnier [1994] edited a good publication dealing with human factors and cognative requirements.

Some individuals can produce speech, but it is dysarthric and very hard to understand. Yet the utterance does contain information. Can this limited information be used to figure out what the individual wanted to say, and then voice it by artificial means? Research labs are now employing neural network theory to determine which pauses in an utterance are due to content (i.e., between a word or sentence) and those due to unwanted halts in speech production.

139.6 Appropriate Technology

Rehabilitation engineering lies at the interface of a wide variety of technical, biological and other concerns. A user might (and often does) put aside a technically sophisticated rehabilitation device in favor of a simpler device that is cheaper, and easier to use and maintain. The cosmetic appearance of the device (or cosmesis) sometimes becomes the overriding factor in acceptance or rejection of a device. A key design factor often lies in the use of the *appropriate technology* to accomplish the task adequately given the extent of the resources available to solve the problem and the residual capacity of the client. Adequacy can be verified by determining that increasing the technical content of the solution results in disproportionately diminishing gains or escalating costs. Thus, a rehabilitation engineer must be able to distinguish applications where high technology is required from those where such technology results in an incremental gain in cost, durability, acceptance and other factors. Further, appropriateness will greatly depend on location. What is appropriate to a client near a major medical center in a highly developed country might not be appropriate to one in a rural setting or in a developing country.

This is not to say that rehabilitation engineers should shun advances in technology. In fact, a fair proportion of rehabilitation engineers work in a research setting where state-of-the-art technology is being applied to the needs of the disabled. However, it is often difficult to transfer complex technology from a laboratory to disabled consumers not directly associated with that laboratory. Such devices are often designed for use only in a structured environment, are difficult to repair properly in the field, and often require a high level of user interaction or sophistication.

Technology transfer in the rehabilitation arena is difficult, due to the limited and fragmented market. Advances in rehabilitation engineering are often piggybacked onto advances in commercial electronics. For instance, the exciting developments in text-to-speech and speech-to-text devices mentioned above are being driven by the commercial marketplace, and not by the rehabilitation arena. But such developments will be welcomed by rehabilitation engineers no less.

139.7 The Future of Engineering in Rehabilitation

The traditional engineering disciplines permeate many aspects of rehabilitation. Signal processing, control and information theory, materials design, computers are all in widespread use from an electrical engineering perspective. Neural networks, microfabrication, fuzzy logic, virtual reality, image processing and other emerging electrical and computer engineering tools are increasingly being applied. Mechanical engineering principles are used in biomechanical studies, gait and motion analysis, prosthetic fitting, seat cushion and back support design, and the design of artificial joints. Materials and metalurgical engineers provide input on newer biocompatable materials. Chemical engineers are developing implantable sensors. Industrial engineers are increasingly studying rehabilitative ergonomics.

The challenge to rehabilitation engineers is to find advances in *any* field, engineering or otherwise, that will aid their clients who have a disability.

Defining Terms

[n.b., The first five terms below have been proposed by the National Center for Medical Rehabilitation and Research (NCMRR) of the US National Institutes of Health (NIH)].

Disability: Inability or limitation in performing tasks, activities, and roles to levels expected within physical and social contexts

Functional Limitation: Restriction or lack of ability to perform an action in the manner or within the range consistent with the purpose of an organ or organ system

Impairment: Loss or abnormality of cognitive, emotional, physiological or anatomical structure or function, including all losses or abnormalities, not just those attributed to the initial pathophysiology

Pathophysiology: Interruption or interference with normal physiological and developmental processes or structures

Societal Limitation: Restriction, attributable to social policy or barriers (structural or attitudinal), which limits fulfillment of roles, or denies access to services or opportunities that are associated with full participation in society

Activities of daily living (ADL): Personal activities that are done by almost everyone in the course of a normal day including eating, toileting, combing hair, brushing teeth, reading, etc. ADLs are distinguished from hobbies and from work-related activities (e.g., typing).

Appropriate technology: the technology that will accomplish a task adequately given the resources available. Adequacy can be verified by determining that increasing the technological content of the solution results in diminishing gains or increasing costs.

Modality-specific: A task that is specific to a single sense or movement pattern.

Orthosis: A modality-specific appliance that aids the performance of a function or movement by augmenting or assisting the residual capabilities of that function or movement. An orthopaedic brace is an orthosis.

Prosthesis: an appliance that substitutes for the loss of a particular function, generally by involving a different modality as an input and/or output channel. An artificial limb, a sensory substitution system, or an augmentative communication aid are prosthetic devices.

Residual function or residual capacity: Residual function is a measure of the ability to to carry out one or more general tasks using the methods normally used. Residual capacity is a measure of the ability to to carry out these tasks using any means of performance. These residual measures are generally more subjective than other more quantifiable measures such as residual strength.

References

Clark, G.M., Y.C. Tong, and J.F. Patrick, 1990. *Cochlear Prostheses*, Edinburgh, Churchill Livingstone.

Goodenough-Trepagnier, C., 1994. Guest Editor of a special issue of *Assistive Technology* 6(1) dealing with mental loads in augmentative communication.

Kaczmarek, K.A., J.G. Webster, P. Bach-y-Rita and W.J. Tompkins, 1991. "Electrotactile and Vibrotactile Displays for Sensory Substitution", *IEEE Trans. Biomed. Engr.*, 38:1-16.

Klinke, R., 1983. "Physiology of the Sense of Equilibrium, Hearing and Speech." Chapter 12 in: *Human Physiology* (eds: R.F. Schmidt and G. Thews), Berlin, Springer-Verlag.

Loeb, G.E., 1985. "The Functional Replacement of the Ear," *Scientific American*, 252:104-111.

Reswick, J. 1982. "What is a Rehabiliation Engineer?" in *Annual Review of Rehabiltation*, Vol 2 (eds. E.L. Pan, T.E. Backer, C.L. Vash), New York, Springer-Verlag.

Robinson, C.J. 1993. "Rehabilitation Engineering—an Editorial," IEEE Transactions on Rehabilitation Engineering 1(1):1-2.

Stein, R.B., D. Charles, and K.B. James, 1988. Providing Motor Control for the Handicapped: A Fusion of Modern Neuroscience, Bioengineering, and Rehabilitation," *Advances in Neurology, Vol. 47: Functional Recovery in Neurological Disease*, (ed. S.G. Waxman), Raven Press, New York.

Further Information

Readers interested in rehabilitation engineering can contact RESNA—an interdisciplinary association for the advancement of rehabilitation and assistive technologies, 1101 Connecticut Ave., N.W., Suite 700, Washington, D.C. 20036. RESNA publishes a quarterly journal called *Assistive Technology.*

The United States Department of Veterans Affairs puts out a quarterly *Journal of Rehabilitation R&D.* The January issue of each year contains an overview of most of the rehabilitation engineering efforts occurring in the U.S. and Canada, with over 500 listings.

The IEEE Engineering in Medicine and Biology Society publishes the *IEEE Transactions on Rehabilitation Engineering,* a quarterly journal. The reader should contact the IEEE at PO Box 1331, 445 Hoes Lane, Piscataway, NJ 08855-1331 U.S.A. for further details.

140

Orthopedic Prosthetics and Orthotics in Rehabilitation

Marilyn Lord
King's College London

Alan Turner-Smith
King's College London

An *orthopedic prosthesis* is an internal or external device that *replaces* lost parts or functions of the neuroskeletomotor system. In contrast, a *orthopedic orthosis* is a device that *augments* a function of the skeletomotor system by controlling motion or altering the shape of body tissue. For example, an artificial leg or hand is a prosthesis, whereas a calliper (or brace) is an orthosis. This chapter addresses only orthoses and external orthopedic prostheses; internal orthopedic prostheses, such as artificial joints, are a subject on their own.

When a human limb is lost through disease or trauma, the integrity of the body is compromised in so many ways that an engineer may well feel daunted by the design requirements for a prosthetic replacement. Consider the losses from a lower limb amputation. Gone is the structural support for the upper body in standing, along with the complex joint articulations and muscular motor system involved in walking. Lost also is the multimode sensory feedback, from *inter alia* pressure sensors on the sole of the foot, length and force sensors in the muscles, and position sensors in the joints, which closed the control loop around the skeletomotor system. The body also has lost a significant percentage of its weight and is now asymmetrical and unbalanced.

We must first ask if it is desirable to attempt to replace all these losses with like-for-like components. If so, we need to strive to make a bionic limb of similar weight embodying anthropomorphic articulations with equally powerful motors and distributed sensors connected back into the wearer's residual neuromuscular system. Or, is it better to accept the losses and redefine the optimal functioning of the new unit of person-plus-technology? In many cases, it may be concluded that a wheelchair is the optimal solution for lower limb loss. Even if engineering could provide the bionic solution, which it certainly cannot at present despite huge inroads made into aspects of these demands, there remain additional problems inherent to prosthetic replacements to consider. Of these, the unnatural mechanical interface between the external environment and the human body is one of the most difficult. Notable, in place of weight bearing through the structures of the foot that are well adapted for this purpose, load must now be transferred to the skeletal structures via intimate contact between the surface of residual limb and prosthesis; the exact distribution of load becomes critical. To circumvent these problems, an alternative direct transcutaneous fixation to the bone has been attempted in limited experimental trials, but this

0-8493-0462-8/00/$0.00+$.50

brings its own problems of materials biocompatability and prevention of infection ingress around the opening through the skin. Orthotic devices are classified by acronyms that describe the joint which they cross. Thus an AFO is an ankle-foot orthosis, a CO is a cervical orthosis (neck brace or collar), and a TLSO is a thoracolumbosacral orthosis (spinal brace or jacket). The main categories are braces for the cervix (neck), upper limb, trunk, lower limb, and foot. Orthoses are generally simpler devices than prostheses, but because orthoses are constrained by the existing body shape and function, they can present an equally demanding design challenge. Certainly the interaction with body function is more critical, and successful application demands an in-depth appreciation of both residual function and the probable reaction to external interference. External orthotics are often classified as structural or functional, the former implying a static nature to hold an unstable joint and the latter a flexible or articulated system to promote the correct alignment of the joints during dynamic functioning. An alternative orthotic approach utilizes functional electrical stimulation (FES) of the patient's own muscles to generate appropriate forces for joint motion; this is dealt with in Chapter 142.

140.1 Fundamentals

Designers of orthotic and prosthetic devices are aware of the three cardinal considerations—function, structure, and cosmesis.

For requirements of function, we must be very clear about the objectives of treatment. This requires first an understanding of the clinical condition. Functional prescription is now a preferred route for the medical practitioner to specify the requirements, leaving the implementation of this instruction to the prosthetist, orthotist, or rehabilitation technologist. The benefits of this distinction between client specification and final hardware will be obvious to design engineers. Indeed, the influence of design procedures on the supply process is a contribution from engineering that is being appreciated more and more.

The second requirement for function is the knowledge of the biomechanics that underlies both the dysfunction in the patient and the function of proposed device to be coupled to the patient. Kinematics, dynamics, energy considerations, and control all enter into this understanding of function. Structure is the means of carrying the function, and finally both need to be embodied into a design that is cosmetically acceptable. Some of the fundamental issues in these concepts are discussed here.

To function well, the device needs an effective coupling to the human body. To this end, there is often some part that is molded to the contours of the wearer. Achieving a satisfactory mechanical interface of a molded component depends primarily on the shape. The internal dimensions of such components are not made an exact match to the external dimensions of the limb segment, but by a process of *rectification*, the shape is adjusted to relieve areas of skin with low load tolerance. The Shapes are also evolved to achieve appropriate load distribution for stability of coupling between prosthetic socket and limb or, in orthotic design, a system of usually three forces that generates a moment to stabilize a collapsing joint (Fig. 140.1). Alignment is a second factor influencing the interface loading. For lower limb prostheses particularly, the alignment of the molded socket to the remainder of the structural components also will be critical in determining the moments and forces transmitted to the interface when the foot is flat on the ground. The same is true for lower limb orthoses, where the net action of the ground reaction forces and consequent moments around the natural joints are highly dependent on the alignment taken up by the combination of orthosis and shoe. Adjustability may be important, particularly for children or progressive medical conditions. Functional components that enable desirable motions are largely straightforward engineering mechanisms such as hinges or dampers, although the specific design requirements for their dynamic performance may be quite complex because of the biomechanics of the body. An example of the design of knee joints is expanded below. These motions may be driven from external power sources but more often are passive or body-powered mechanisms. In orthoses where relatively small angular motions are needed, these may be provided by material flexibility rather than mechanisms.

The structural requirements for lower-limb prosthetics have been laid down at a consensus meeting (1978) bases on biomechanical measurement of forces in a gait laboratory, referred to as the *Philadelphia standards* and soon to be incorporated into an ISO standard (ISO 13404,5; ISO 10328).Not only are the

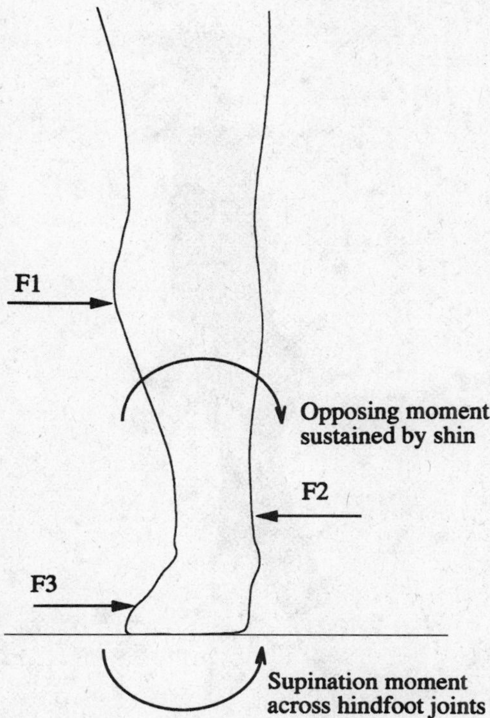

FIGURE 140.1 Three-force system required in an orthosis to control a valgus hindfoot due to weakness in the hindfoot supinators.

load level and life critical, but so is the mode of failure. Sudden failure of an ankle bolt resulting in disengagement of an artificial foot is not only potentially life-threatening to an elderly amputee who falls and breaks a hip but also can be quite traumatic to unsuspecting witnesses of the apparent event of autoamputation. Design and choice of materials should ensure a controlled slow yielding, not brittle fracture. A further consideration is the ability of the complete structure to absorb shock loading, either the repeated small shocks of walking at the heel strike or rather more major shocks during sports activities or falls. This minimizes the shock transmitted through the skin to the skeleton, known to cause both skin lesions and joint degeneration. Finally, the consideration of hygiene must not be overlooked; the user must be able to clean the orthosis or prosthesis adequately without compromising its structure or function.

Added to the two elements of structure and function, the third element of *cosmesis* completes the trilogy. Appearance can be of great psychological importance to the user, and technology has its contribution here, too. As examples, special effects familiar in science fiction films also can be harnessed to provide realistic cosmetic covers for hand or foot prostheses. Borrowing from advanced manufacturing technology, optical shape scanning linked to three-dimensional (3D) computer-aided design, and CNC machining can be pressed into service to generate customized shapes to match a contralateral remaining limb. Up-to-date materials and component design each contribute to minimize the "orthopedic appliance" image of the devices (Fig. 140.2). In providing cosmesis, the views of the user must remain paramount. The wearer will often choose an attractive functional design in preference to a life like design that is not felt to be part of his or her body.

Upper limb prostheses are often seen as a more interesting engineering challenge than lower limb, offering the possibilities for active motor/control systems and complex articulations. However, the market is an order of magnitude smaller and cost/benefit less easy to prove—after all, it is possible to function

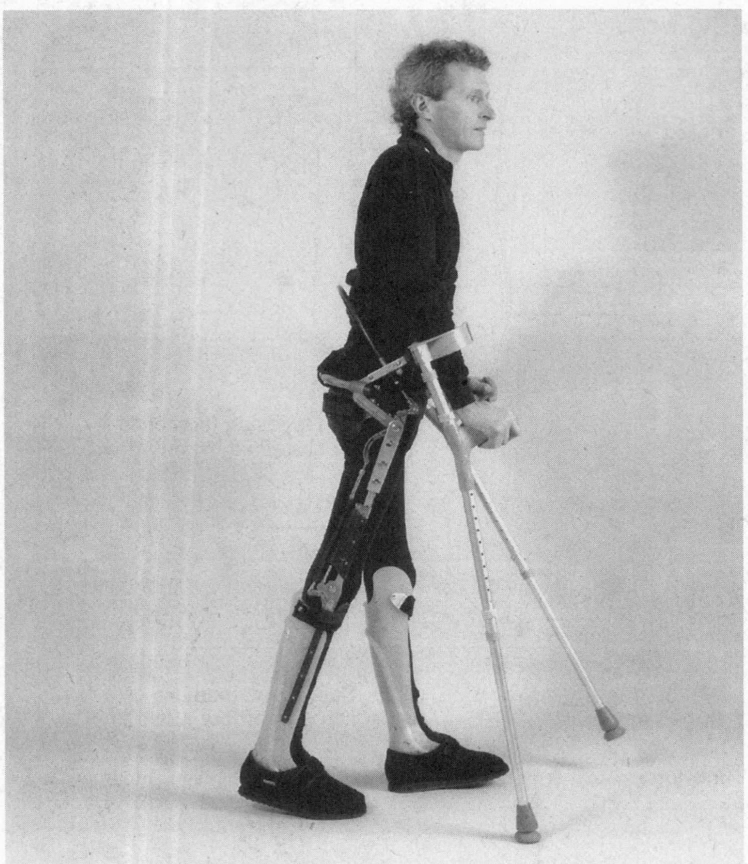

FIGURE 140.2 The ARGO reciprocating-gait orthosis, normally worn under the clothing, with structural components produced from 3D CAD. (Courtesy of Hugh Steeper, Ltd., U.K.)

fairly well with one arm, but try walking with one leg. At the simplest end, an arm for a below-elbow amputee might comprise a socket with a terminal device offering a pincer grip (hand or hook) that can be operated through a Bowden cable by shrugging the shoulders. Such body-powered prostheses may appear crude, but they are often favored by the wearer because of a sense of position and force feedback from the cable, and they do not need a power supply. Another, more elegant method of harnessing body power is to take a muscle made redundant by an amputation and tether its tendon through an artificially fashioned loop of skin: the cable can then be hooked through the loop [Childress, 1989].

Externally powered devices have been attempted using various power sources with degrees of success. Pneumatic power in the form of a gas cylinder is cheap and light, but recharging is a problem that exercised the ingenuity of early suppliers: where supplies were not readily available, even schemes to involve the local fire services with recharging were costed. Also, contemplate the prospect of bringing a loaded table fork toward your face carried on the end of a position-controlled arm powered with spongy, low-pressure pneumatic actuators, and you will appreciate another aspect of difficulties with this source. Nevertheless, gas-powered grip on a hand can be a good solution. Early skirmishes with stiffer hydraulic servos were largely unsuccessful because of power supply and actuator weight and oil leakage. Electric actuation, heavy and slow at first, has gradually improved to establish its premier position. Input control to these powered devices can be from surface electromyography or by mechanical movement of, for example, the shoulder or an ectromelic limb. Feedback can be presented as skin pressure, movement of a sensor over the skin, or electric stimulation. Control strategies range from position control around a

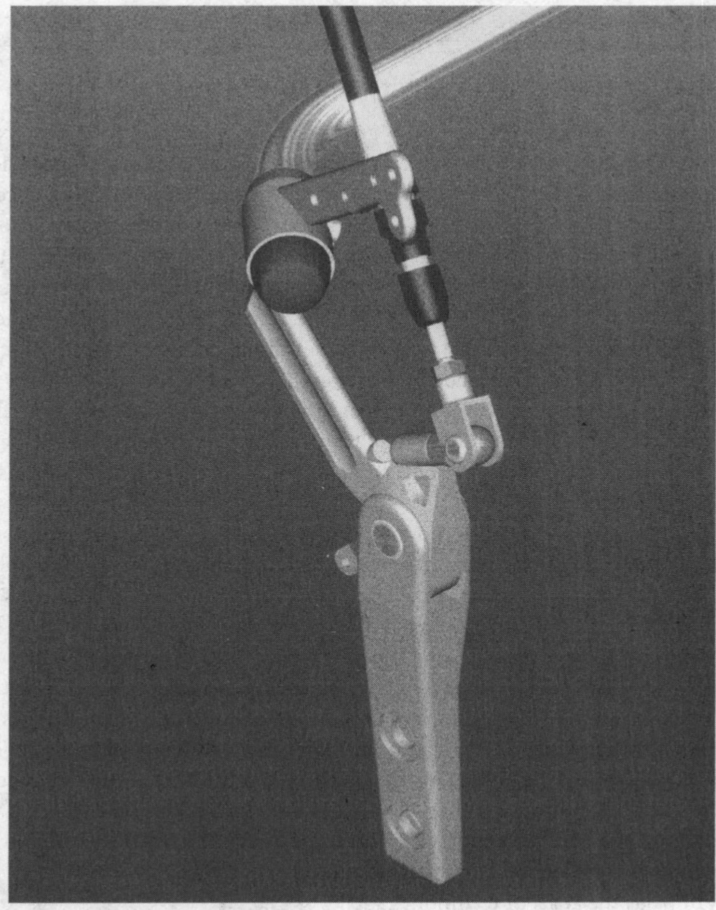

FIGURE 140.2 (continued)

single joint or group of related joints through combined position and force control for hand grip to computer-assisted coordination of entire activities such as feeding.

The physical designs in prosthetic and orthotic devices has changed substantially over the past decade. One could propose that this is solely the introduction of new materials. The sockets of artificial limbs have always been fashioned to suit the individual patient, historically by carving wood, shaping leather, or beating sheet metal. Following the introduction of thermosetting fiber-reinforced plastics hand-shaped over a plaster cast of the limb residuum, substitution of thermoforming plastics that could be automatically vacuum-formed made a leap forward to give light, rapidly made, and cosmetically improved solutions. Polypropylene is the favored material in this application. The same materials permitted the new concept of custom-molded orthoses. Carbon fiber composites substituted for metal have certainly improved the performance of structural components such as limb shanks. But some of the progress owes much to innovative thinking. The flex foot is a fine example, where a traditional anthropomorphic design with imitation ankle joint and metatarsal break is completely abandoned and a functional design adopted to optimize energy storage and return. This is based on two leaf springs made from Kevlar, joined together at the ankle with one splaying down toward the toes to form the forefoot spring and the other rearward to form the heel spring (Fig. 140.3). Apart from the gains for the disabled athletes for whom the foot was designed–and these are so remarkable that there is little point in competing now without this foot–clients across all age groups have benefited from the adaptability to rough ground and shock-absorption capability.

140.2 Applications

Computer-Aided Engineering in Customized Component Design

Computer-aided engineering has found a fertile ground for exploitation in the process of design of customized components to match to body shape. A good example is in sockets for artificial limbs. What prosthetists particularly seek is the ability to produce a well-fitting socket during the course of a single patient consultation. Traditional craft methods of casting the residual limb in plaster of paris, pouring a positive mold, manual rectification, and then socket fabrication over the rectified cast takes too long.

By using advanced technology, residual limb shapes can be captured in a computer, rectified by computer algorithms, and CNC machined to produce the rectified cast in under an hour so that with the addition of vacuum-formed machinery to pull a socket rapidly over the cast, the socket can be ready for trial fitting in one session. There are added advantages too, in that the shape is now stored in digital form in the computer and can be reproduced or adjusted whenever and wherever desired. Although such systems are still in an early stage of introduction, many practicing prosthetists in the United States have now had hands-on experience of this technology, and a major evaluation by the Veterans Administration has been undertaken [Houston et al., 1992].

Initially, much of the engineering development work went into the hardware components, a difficult brief in view of the low cost target for a custom product. Requirements are considerably different from those of standard engineering, e.g., relaxation in the accuracies required (millimeters, not microns);a need to measure limb or trunk parts that are encumbered by the attached body, which may resist being orientated conveniently in a machine and which will certainly distort with the lightest pressure; and a need to reproduce fairly bulky items with strength to be used as a sacrificial mold. Instrumentation for body shape scanning has been developed using methods of silhouettes, Moiré fringes, contact probes measuring contours of plaster casts, and light triangulation. Almost universally the molds are turned by "milling on a spit" [Duncan & Mair, 1983], using an adapted lathe with a milling head to spiral down a large cylindrical plug of material such as plaster of paris mix. Rehabilitation engineers watch with great interest, some with envy, the developments in rapid prototyping manufacture, which is so successful in reducing the cycle time for one-off developments elsewhere in industry, but alas the costs of techniques such as stereolithography are as yet beyond economic feasibility for our area.

Much emphasis also has been placed on the graphics and algorithms needed to achieve rectification. Opinions vary as to what extent the computer should simply provide a more elegant tool for the prosthetist to exercise his or her traditional skills using 3D modeling and on-screen sculpting as a direct replacement for manual plaster rectification or to what extent the computer system should take over the bulk of the process by an expert systems approach. Systems currently available tend to do a little of each. A series of rectification maps can be held as templates, each storing the appropriate relief or buildup to be applied over a particular anatomic area of the limb. Thus the map might provide for a ridge to be added down the front of the shin of a lower limb model so that the eventual socket will not press against the vulnerable bony prominence of the tibia (Fig. 140.4). Positioning of the discrete regions to match individual anatomy might typically be anchored to one or more anatomic features indicated by the prosthetist.

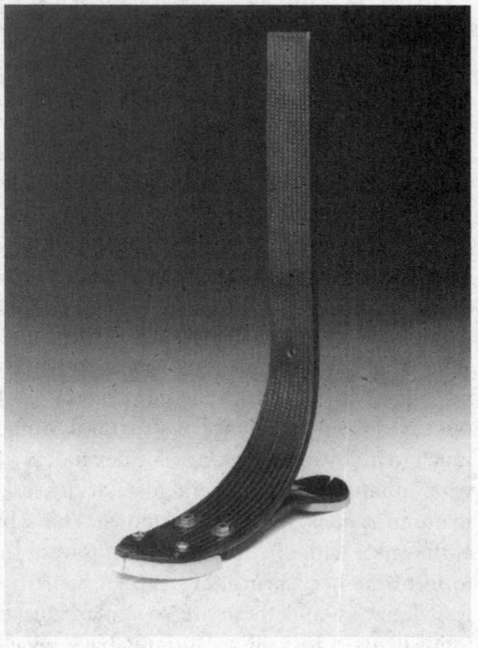

FIGURE 140.3 The Flex Foot.

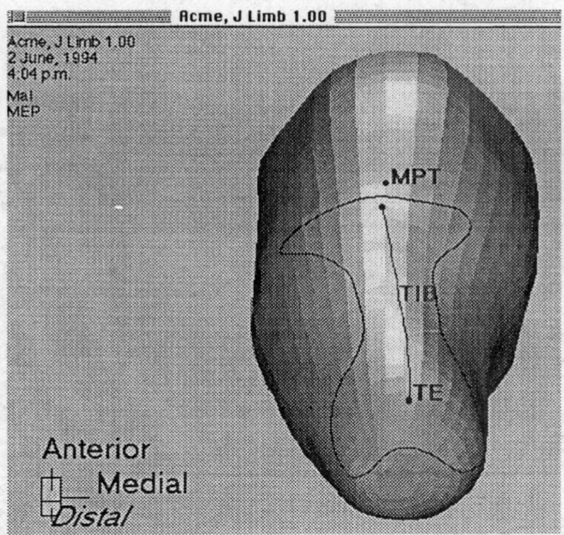

FIGURE 140.4 A rectification require defined over the tibia of lower limb stump using the Shapemaker application for computer-aided socket design.

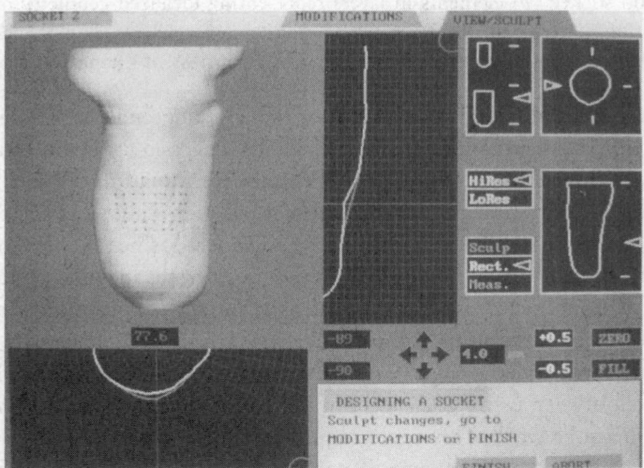

FIGURE 140.5 Adjusting a socket contour with reference to 3D graphics and cross-sectional profiles in the UCL CASD system. (Reproduced from Reynolds and Lord [1992] with permission.)

The prosthetist is also able to free-form sculpt a particular region by pulling the surface interactively with reference to graphic representation (Fig. 140.5); this is particularly useful where the patient has some unusual feature not provided for in the templates.

As part of this general development, finite-element analysis has been employed to model the soft tissue distortion occurring during limb loading and to look at the influence of severity of rectification in the resultant distribution of interface stress [e.g., Reynolds & Lord, 1992] (Fig. 140.6). In engineering terms, this modeling is somewhat unusual and decidedly nonlinear. For a start, the tissues are highly deformable but nearly incompressible, which raises problems of a suitable Poisson ratio to apply in the modeling. Values of $n = 0.3$ to $n = 0.49$ have been proposed, based on experimental matching of stress-strain curves from indentation of limb tissue in vivo. In reality, though, compression (defined as a loss of volume) may be noted in a limb segment under localized external pressure due to loss of mass as first the blood

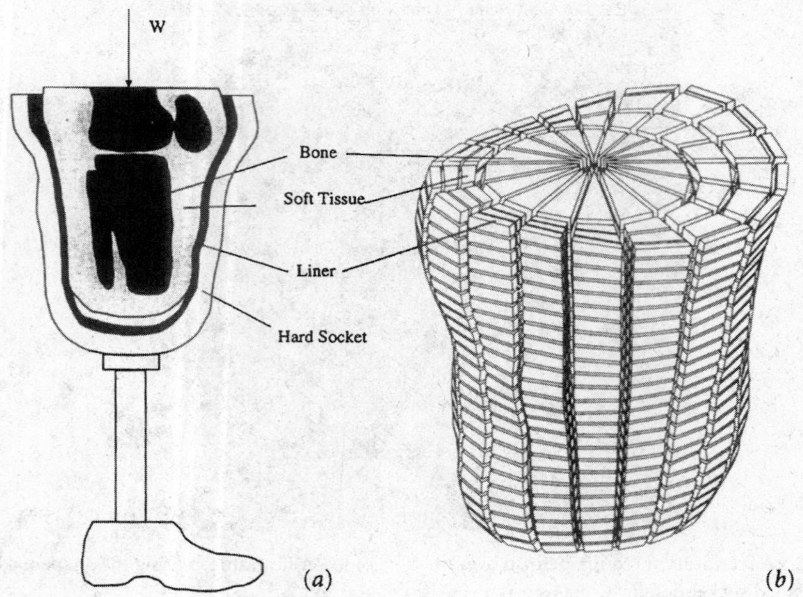

FIGURE 140.6 Finite-element analysis employed to determine the sensitivity of interface pressure to socket shape rectification: (*a*) limb and socket; (*b*) elements in layers representing idealized geometry of bone, soft tissue, and socket liner; (*c*) rectification map of radial differences between the external free shape of the limb and the internal dimensions of socket; and (*d*) FE predictions of direct pressure. (Courtesy of Zhang Ming, King's College London.)

is rapidly evacuated and then interstitial fluids are more slowly squeezed out. Also, it is difficult to define the boundaries of the limb segment at the proximal end, still attached to the body, where soft tissues can easily bulge up and out. This makes accurate experimental determination of the stress-strain curves for the tissue matrix difficult. A nonlinear model with interface elements allowing slip to occur between skin and socket at the limit of static friction may need to be considered, since the frictional conditions at the interface will determine the balance between shear and direct stresses in supporting body weight against the sloping sidewalls. Although excessive shear at the skin surface is considered particularly damaging, higher pressures would be required in its complete absence.

In a similar vein, computer-aided design (CAD) techniques are also finding application in the design of bespoke orthopedic footwear, using CAD techniques from the volume fashion trade modified to suit the one-off nature of bespoke work. This again requires the generation of a customized mold, or shoe last, for each foot, in addition to the design of patterns for the shoe uppers [Lord et al., 1991]. The philosophy of design of shoe lasts is quite different from that of sockets, because last shapes have considerable and fundamental differences from foot shapes. In this instance, a library of reference last shapes is held, and a suitable one is selected both to match the client's foot shape and to fulfill the shoemaking needs for the particular style and type of shoe. The schematic of the process followed in development of the Shoemaster system is shown in Fig. 140.7.

Design of shoe inserts is another related application, with systems to capture, manipulate, and reproduce underfoot contours now in commercial use. An example is the Ampfit system, where the foot is placed on a platform to which preshaped arch supports or other wedges or domes may first be attached. A matrix of round-ended cylinders is then forced up by gas pressure through both platform and supports, supporting the foot over most of the area with an even load distribution. The shape is captured from the cylinder locations and fed into a computer, where rectification can be made similar to that described for prosthetic sockets. A benchtop CNC machine then routs the shoe inserts from specially provided blanks while the client waits.

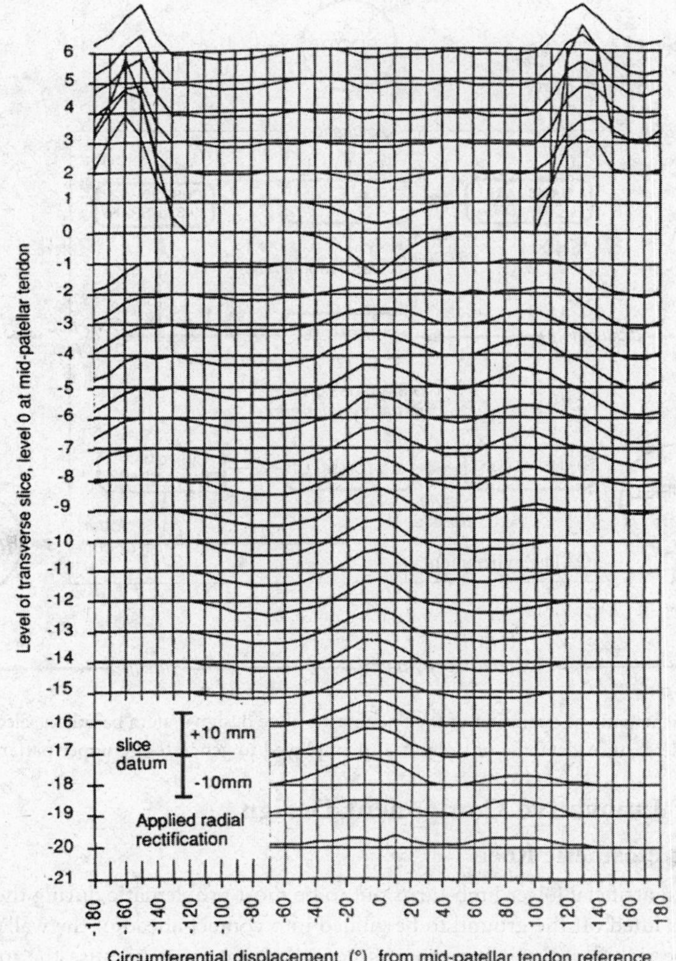

(*c*)

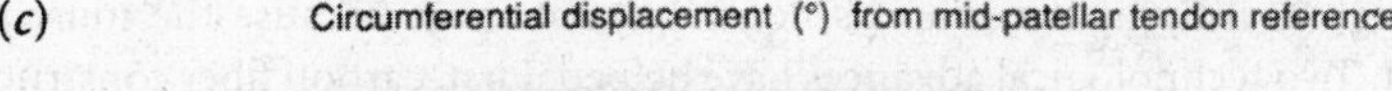

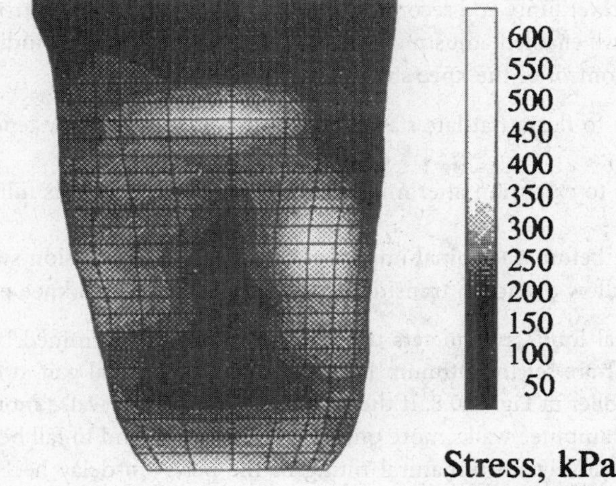

(*d*)

FIGURE 140.6 (continued)

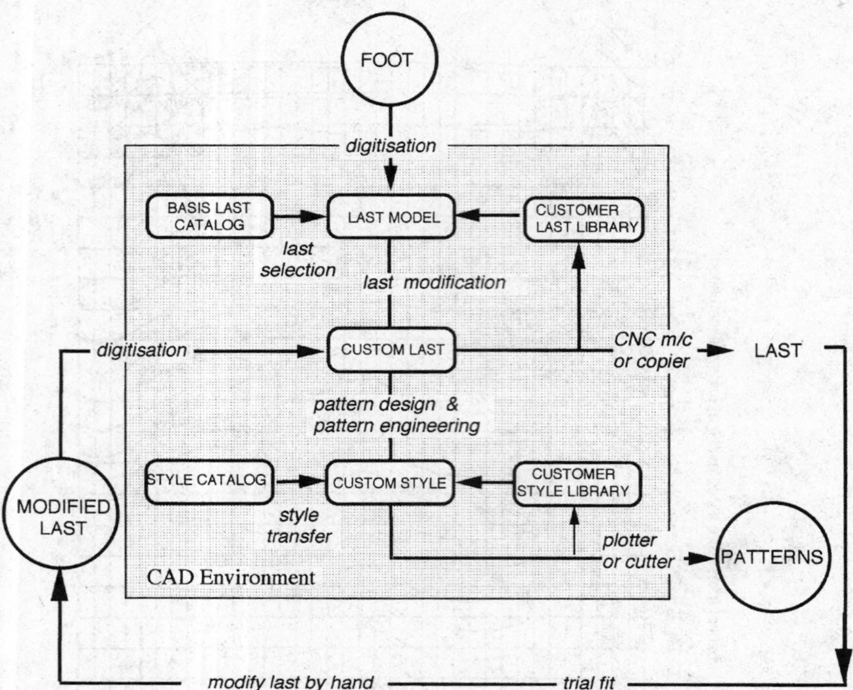

FIGURE 140.7 Schematic of operation of the Shoemaster shoe design system based on selection of a basis last from a database of model lasts. A database of styles is also employed to generate the upper patterns.

Examples of Innovative Component Design

An Intelligent Prosthetic Knee

The control of an artificial lower limb turns out to be most problematic during the swing phase, during which the foot is lifted off the ground to be guided into contact ahead of the walker. A prosthetic lower limb needs to be significantly lighter than its normal counterpart because the muscular power is not present to control it. Two technological advances have helped. First, carbon fiber construction has reduced the mass of the lower limb, and second, pneumatic or hydraulically controlled damping mechanisms for the knee joint have enabled adjustment of the swing phase to suit an individual's pattern of walking.

Swing-phase control of the knee should operate in three areas:

1. Resistance to flexion at late stance during toe-off controls any tendency to excessive heel rise at early swing.
2. Assistance to extension after midswing ensures that the limb is fully extended and ready for heel strike.
3. Resistance before a terminal impact at the end of the extension swing dampens out the inertial forces to allow a smooth transition from flexed to extended knee position.

In conventional limbs, parameters of these controls are determined by fixed components (springs, bleed valves) that are set to optimum for an individual's normal gait at one particular speed, e.g., the pneumatic controller in Fig. 140.8. If the amputee subsequently walks more slowly, the limb will tend to lead, while if the amputee walks more quickly, the limb will tend to fall behind; the usual compensatory actions are, respectively, an unnatural tilting of the pelvis to delay heel contact or abnormal kicking through of the leg.

In a recent advance, intelligence is built into the swing-phase controller to adjust automatically for cadence variations (Fig. 140.9). A 4-bit microprocessor is used to adjust a needle valve, via a linear stepper

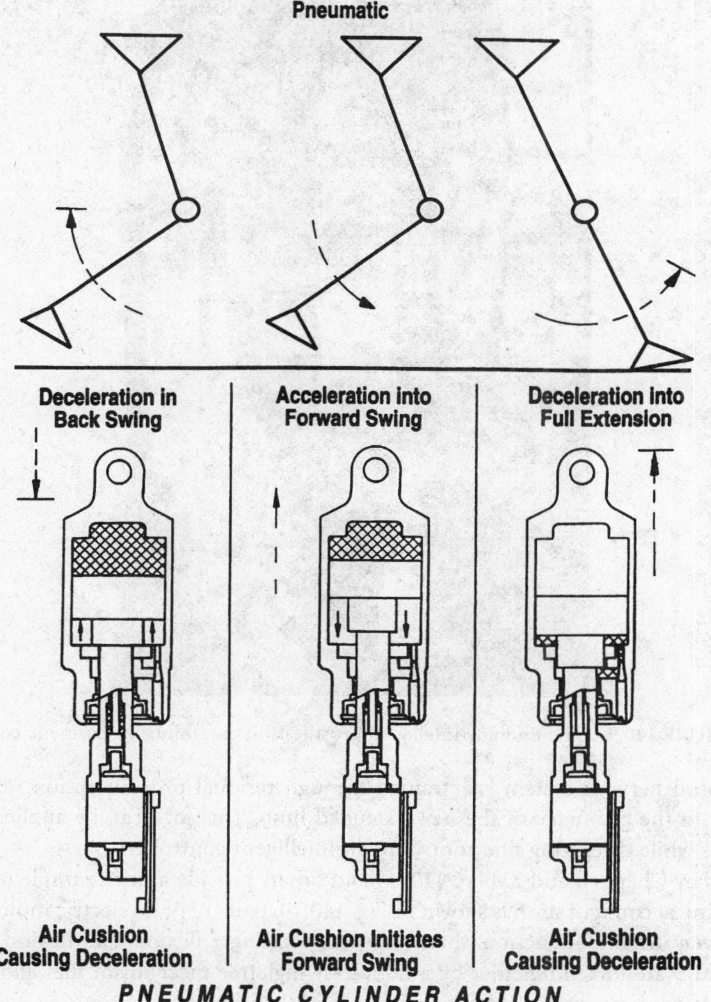

FIGURE 140.8 Pneumatic cylinder action in a swing phase controller. (Reprinted with permission from S. Zahedi, The Results of the Field Trial of the Endolite Intelligent Prosthesis, internal publication, Chas. A. Blatchford & Sons, U.K.)

motor, according to duration of the preceding swing phase [Zahedi, in press]. The unit is programmed by the prosthetist to provide optimal damping for the particular amputee's swing phase at slow, normal, and fast walking paces. Thereafter, the appropriate damping is automatically selected for any intermediate speed.

A Hierarchically Controlled Prosthetic Hand

Control of the intact hand is hierarchical. It starts with the owner's intention, and an action plan is formulated based on knowledge of the environment and the object to be manipulated. For gross movements, the numerous articulations rely on "preprogrammed" coordination from the central nervous system. Fine control leans heavily on local feedback from force and position sensors in the joints and tactile information about loading and slip at the skin. In contrast, conventional prostheses depend on the conscious command of all levels of control and so can be slow and tiring to use.

Current technology is able to provide both the computing power and transducers required to recreate some of a normal hand's sophisticated proprioceptive control. A concept of extended physiologic proprioception (EPP) was introduced for control of gross arm movement [Simpson & Kenworthy, 1973]

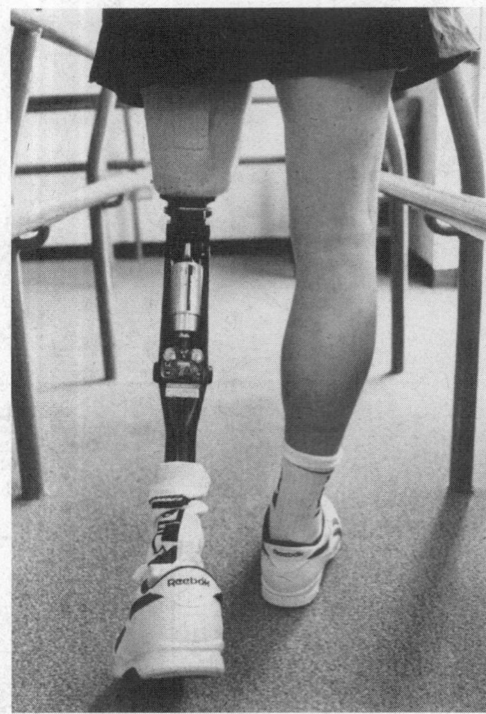

FIGURE 140.9 The Endolite intelligent prosthesis in use, minus its cosmetic covers.

whereby the central nervous system is retrained through residual proprioception to coordinate gross actions applying to the geometry of the new extended limb. This idea can be applied to initiate gross hand movements while delegating fine control to an intelligent controller.

Developments by Chappell and Kyberd [1991] and others provide a fine example of the possibilities. A suitable mechanical configuration is shown in Fig. 140.10. Four 12-V dc electric motors with gearboxes control, respectively, thumb adduction, thumb flexion, forefinger flexion, and flexion of digits 3,4, and 5. Digits 3, 4, and 5 are linked together by a double-swingletree mechanism that allows all three to be driven together. When one digit touches an object the other two can continue to close until they also touch or reach their limit of travel. The movement of the digits allows one of several basic postures:

- ***Three-point chuck:*** Precision grip with digits 1, 2, and 3 (thumb set to oppose the midline between digits 2 and 3); digits 4 and 5 give additional support.
- ***Two-point grip:*** Precision grip with digits 1 and 2 (thumb set to oppose forefinger); digits 3, 4, and 5 fully flexed and not used or fully extended.
- ***Fist:*** As two-point grip but with thumb fully extended to allow large objects to be grasped.
- ***Small Fist:*** As fist but with thumb flexed and abducted to oppose side of digit 2.
- ***Side, or key:*** Digits 2 to 5 half fully flexed with thumb opposing side of second digit.
- ***Flat hand:*** Digits 2 to 5 fully extended with thumb abducted and flexed, parked beside digit 2.

The controller coordinates the transition between these positions and ensures that trajectories do not tangle. Feedback to the controller is provided by several devices. Potentiometers detect the angles of flexion of the digits; touch sensors detect pressure on the palmer surfaces of the digits; and a combined contact force (Hall effect) and slip sensor (from acoustic frequency output of force sensor) is mounted at the fingertips. The latter detects movement of an object and so controls grip strength appropriate to the task–whether holding a hammer or an egg [Kyberd & Chappell, 1993].

The whole hand may be operated by electromyographic signals from two antagonistic muscles in the supporting forearm stump, picked up at the skin surface. In response to tension in one muscle, the hand opens progressively and then closes to grip with an automatic reflex. The second muscle controls the mode of operation as the hand moves between the states of touch, hold, squeeze, and release.

FIGURE 140.10 The Southampton hand prosthesis with four degrees of freedom in a power grip. An optical/acoustic sensor is mounted on the thumb. (Reprinted from Kyberd and Chappell [1993], Fig. 1.)

A Self-Aligning Orthotic Knee Joint

Knee orthosis are often supplied to resist knee flexion during standing and gait at an otherwise collapsing joint. The rigid locking mechanisms on these devices are manually released to allow knee flexion during sitting. Fitting is complicated by the difficulty of attaching the orthosis with its joint accurately aligned to that of the knee. The simple diagram in Fig. 140.11 shows how misplacement of a simple hinged orthosis with a notional fixed knee axis would cause the cuffs on the thigh and calf to press into the soft tissues of the limb (known as *pistoning*).

The human knee does not have a fixed axis, though, but is better represented as a polycentric joint. In a sagittal (side) view, it is easy to conceptualize the origin of these kinematics from the anatomy of the cruciate ligaments running crisscross across the joint, which together with the base of the femur and the head of the tibia form a classic four-bar linkage. The

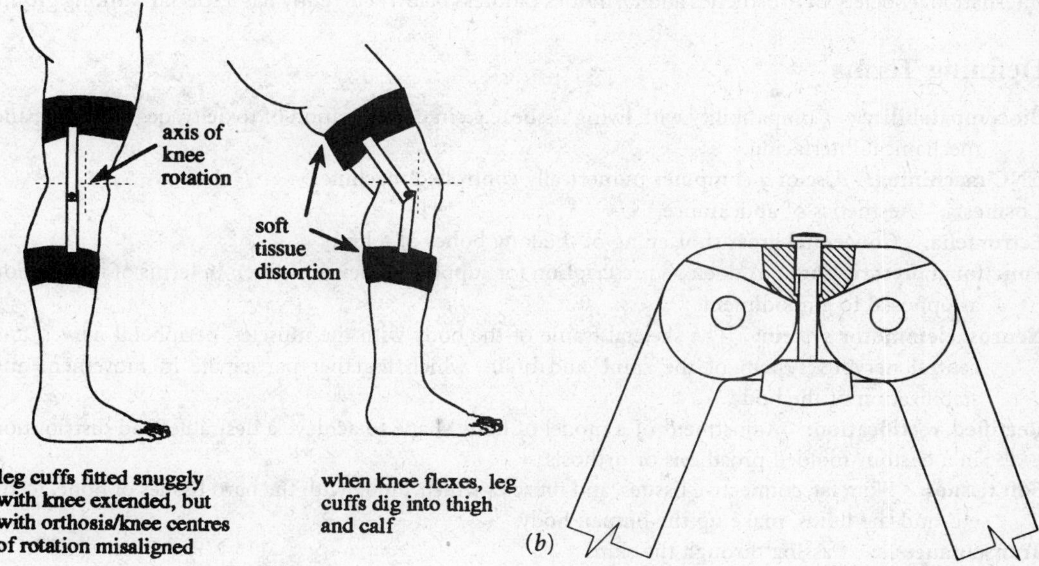

FIGURE 140.11 The problem caused by misplacement of a single-axis orthotic joint (*a*) is overcome by an orthosis (*b*) with a self-aligning axis. (The Laser system, courtesy of Hugh Steeper, Ltd., U.K.)

polycentric nature of the motion can therefore be mimicked by a similar geometry of linkage on the orthosis.

The problem of alignment still remains, however, and precision location of attachment points is not possible when gripping through soft tissues. In one attempt to overcome this specific problem, the knee mechanism has been designed with not one but two axes (Fig. 140.11). The center of rotation is then free to self-align. This complexity of the joint while still maintaining the ability to fixate the knee and meeting low weight requirements is only achieved by meticulous design in composite materials.

140.3 Summary

The field of prosthetics and orthotics is one where at present traditional craft methods sit alongside the application of high technology. Gradually, advanced technology is creeping into most areas, bringing vast improvements in hardware performance specifications and aesthetics. This is, however, an area where the clinical skills of the prosthetist and orthotist will always be required in specification and fitting and where many of the products have customized components. The successful applications of technology are those which assist the professional to exercise his or her judgment, providing him or her with good tools and means to realize a functional specification.

Since the demand for these devices is thankfully low, their design and manufacture are small scale in terms of volume. This taxes the skills of most engineers, both to design the product at reasonable up-front costs and to manufacture it economically in low volume. For bespoke components, we are moving from a base of craft manufacture through an era when modularization was exploited to allow small-batch production toward the use of CAD. In the latter, the engineering design effort is then embodied in the CAD system, leaving the prosthetist or orthotist to incorporate clinical design for each individual component.

Specific examples of current applications have been described. These can only represent a small part of the design effort that is put into prosthetics and orthotics on a continuing basis, making advances in materials and electronics in particular available.

We are also aware that in the space available, it has not been possible to include a discussion of the very innovative work that is being done in intermediate technology for the third world, for which the International Society of Prosthetics and Orthotics (address below) currently has a special working group.

Defining Terms

Biocompatability: Compatibility with living tissue, e.g., in consideration of toxicity, degradability, and mechanical interfacing.

CNC machining: Use of a computer numerically controlled machine.

Cosmesis: Aesthetics of appearance.

Ectromelia: Congenital gross shortening of the long bones of a limb.

Functional prescription: A doctor's prescription for supply of a device written in terms of its function as opposed to embodiment.

Neuroskeletomotor system: The skeletal frame of the body with the muscles, peripheral nerves, and central nervous system of the spine and brain, which together participate in movement and stabilization of the body.

Rectified, rectification: Adjustment of a model of body shape to achieve a desirable load distribution in a custom-molded prosthesis or orthosis.

Soft tissues: Skin, fat, connective tissues, and muscles which, along with the hard tissues of bone, teeth, etc. and the fluids, make up the human body.

Transcutaneous: Passing through the skin.

References

Chappell PH, Kyberd PJ. 1991. Prehensile control of a hand prosthesis by a microcontroller. J Biomed Eng 13:363.

Childress DS. 1989. Control philosophies for limb prostheses. In J Paul, et al. (eds), Progress in Bioengineering, pp 210–215. New York, Adam Hilger.

Duncan JP, Mair SG. 1983. Sculptured Surfaces in Engineering and Medicine. Cambridge, England, Cambridge Univ. Press.

Houston VL, Burgess EM, Childress DS, et al. 1992. Automated fabrication of mobility aids (AFMA): Below-knee CASD/CAM testing and evaluation. J Rehabil Res Dev 29:78.

Kyberd PJ, Chappell PH. 1993. A force sensor for automatic manipulation based on the Hall effect. Meas Sci Technol 4:281.

Lord M, Foulston J, Smith PJ. 1991. Technical evaluation of a CAD system for orthopaedic shoe-upper design. Eng Med Proc Instrum Mech Eng 205:109.

Reynolds DP, Lord M. 1992. Interface load analysis for computer-aided design of below-knee prosthetic sockets. Med Biol Eng Comput 30:419.

Simpson DC, Kenworthy G. 1973. The design of a complete arm prosthesis. Biomed Eng 8:56.

Zahedi S. In press, 1994. Evaluation and biomechanics of the intelligent prosthesis: A two-year study. Orthop Tech.

Further Information

Bowker P, Condie DN, Bader DL, Pratt DJ (eds). Biomechanical Basis of Orthotic Management. Oxford, Butterworth-Heinemann, 1993.

Murdoch G, Donovan RG (eds). 1988. Amputation Surgery and Lower Limb Prosthetics. Boston, Blackwell Scientific Publications.

Nordin M, Frankel V. 1980. Basic Mechanics of the Musculoskeletal System, 2d ed. Philadelphia, Lea & Febiger, 1980.

Smidt GL (ed). 1990. Gait in Rehabilitation. New York, Churchill-Livingstone.

Organizations

International Society of Prosthetics and Orthotics (ISPO), Borgervaenget 5,2100 Copenhagen Ø, Denmark [tel (31) 20 72 60].

Department of Veterans Affairs, VA Rehabilitation Research and Development Service, 103 Gay Street, Baltimore, MD 21202-4051.

Rehabilitation Engineering Society of North America (RESNA), Suite 1540, 1700 North Moore Street, Arlington, VA 22209-1903.

141

Wheeled Mobility: Wheelchairs and Personal Transportation

Rory A. Cooper
University of Pittsburgh
VA Pittsburgh Health Care System

141.1 Introduction

Centuries ago, people with disabilities who survived for an extended period of time were transported on hammocks slung between poles which were carried by others. This was the preferred means of transportation of the upper class and thus carried no stigma. Later the wheelbarrow was developed and soon became a common mode of transportation for people with disabilities. Because wheelbarrows were used to transport materials, during this period in history, people with disabilities were looked upon as outcasts from society. During the renaissance, the French court popularized the first wheelchairs. Wheelchairs were overstuffed arm chairs with wheels placed upon them. This enabled movement, with assistance, indoors. Later the wooden wheelchair with wicker matting was developed. This type of chair remained the standard until the 1930s. Franklin D. Roosevelt was not satisfied with the wooden wheelchair and had many common metal kitchen chairs modified with wheels. In the 1930s a young mining engineer, named Everest experienced an accident that left him mobility impaired. He worked with a fellow engineer Jennings to develop steel wheelchairs. Within a few years, they formed a company Everest & Jennings to manufacture wheelchairs. Following World War II, medical advances saved the lives of many veterans with spinal cord injuries or lower limb amputations, who would have otherwise died. Veterans medical centers issued these veterans steel framed wheelchairs with 18 inch seat widths. These wheelchairs were designed to provide the veteran some mobility within the hospital and home, and not to optimize ergonomic variables. Just as among the ambulatory population, mobility among people with disabilities varies. Mobility is more of a functional limitation than a disability related condition. Powered mobility can have tremendous positive psycho-social effects on an individual. Power wheelchairs provide greater independence to thousands of people with severe mobility impairments.

0-8493-0462-8/00/$0.00+$.50

Power wheelchairs began in the 1940's as standard cross-brace folding manual wheelchairs adapted with automobile starter motors and an automobile battery. The cross-braced wheelchair remained the standard for a number of years. When the rigid power wheelchair frame was developed, space became available under the seat for electronic controls, respirators, communication systems, and reclining devices. By the mid-1970s, wheelchairs had evolved to the point where people had acquired a significant level of mobility.

A personal automobile has a profound affect on a persons' mobility and ability to participate in society. A wheelchair is suitable for short distances, and for many situations where an unimpaired person would walk. Modifications to vehicles may be as simple as a lever attached to the brake and accelerator pedals or as complex as a complete joystick controlled fly-by-wire system. Modifications to other components of the vehicle may be required to provide wheelchair access. An automobile may not be appropriate for some people who travel distances too long to be convenient with a wheelchair, but not long enough to warrant an adapted automobile. Micro-cars, enlarged wheelchairs which travel at bicycle speeds, are convenient for many people who wish only to travel to the local grocery store or post office. Micro-cars are also useful for people who like to travel along bicycle paths or drive short distances off-road.

141.2 Categories of Wheelchairs

There are two basic classes of wheelchairs: manually powered, and externally powered. For practical purposes, externally powered wheelchairs are electrically powered wheelchairs. There are approximately 200,000 wheelchairs sold annually within the US of which about 20,000 are powered wheelchairs. Most wheelchairs are purchased by third-party-payers (e.g., insurance companies, government agencies). This requires the market to be responsive to wheelchairs user's needs, prescriber expertise and experience, third-party-payer purchase criteria, and competition from other manufacturers. Despite the complicated interaction between these components, and the regulation of products by several government agencies, a variety of wheelchairs and options are available.

Depot wheelchairs are intended for institutional use where several people may use the same wheelchair. Generally, these wheelchairs are inappropriate for active people who use wheelchairs for personal mobility. Depot wheelchairs are designed to be inexpensive, accommodate large variations in body size, to be low maintenance, and often to be attendant propelled. They are heavy and their performance is limited. A typical depot wheelchair will have swing away footrests, removable armrests, a single cross-brace frame and solid tires.

People who have impairment of an arm and one or both lower extremities may benefit from a one-arm drive wheelchair that uses a linkage connecting the rear wheels. This allows the user to push upon the pushrim of one wheel and to propel both wheels. To effectively turn the wheelchair, the user must have the ability to disengage the drive mechanism.

Some people have weakness of the upper and lower extremities and can gain maximal benefit from wheelchair propulsion by combining the use of their arms and legs or by using only their legs. The design and selection of a foot-drive wheelchair depends greatly upon how the user can take greatest advantage of their motor abilities.

Indoor spaces are more limited and one is often required to get close to furnishings and fixtures to use them properly. Indoor wheelchairs often use rear castors because of the manueverability of these designs. However, rear castor designs make the wheelchair less stable in lateral directions. Indoor wheelchairs typically have short wheelbases.

All wheelchairs are not propelled by the person sitting in the wheelchair. In many hospitals and long-term care facilities, wheelchairs are propelled by attendants. Attendant propelled wheelchair designs must consider the rider and the attendant as users. The rider must be transported safely and comfortably. The attendant must be able to operate and easily maneuver safely and with minimum physical strain.

Active users often prefer highly maneuverable and responsive wheelchairs which fit their physical and psychosocial character. The ultralight wheelchair evolved from the desire of wheelchair users to develop functional ergonomic designs for wheelchairs. Ultralight wheelchairs are made of such materials as aluminum, alloy steel, titanium, or composites. The design of ultralight wheelchairs allows a number of features to be customized by the user or be specified for manufacture. The most common features of all are the light weight, the high quality of materials used in their construction, and their functional design. Many people can benefit from ultralight wheelchair designs.

The desire to achieve better performance has led wheelchair users, inventors, and manufacturers to constantly develop specialized wheelchairs for sports. There is no real typical sports wheelchair as the design depends heavily on the sport. Basketball and tennis wheelchairs are often thought to typify sports wheelchair design. However, racing, field events or shooting wheelchairs have little in common with the former.

Some wheelchairs are made to change configuration from reclining to sitting, and from sitting to standing. Most stand-up wheelchairs cannot be driven in the stand-up setting in order to insure safe and stable operation. Standing gives the user the ability to reach cabinet and counter spaces otherwise inaccessible. Standing has the additional advantage of providing therapeutic benefits, i.e., hemodynamic improvements, and amelioration of osteoporosis.

Stairs and other obstacles persist despite the progress made in universal design. Stair-climbing wheelchairs are electrically powered wheelchairs designed to ascend and descend stairs safely under the occupant's control. Stair-climbing wheelchairs are quite complicated, and often reconfigure themselves while climbing stairs. The additional power required to climb stairs often reduces the range of the wheelchair when compared to standard power wheelchairs.

141.3 Wheelchair Structure and Component Design

Several factors must be considered when designing a wheelchair frame: what are the intended uses, what are the abilities of the user, what are the resources available, and what are the existing products available. These factors determine if and how the frame will be designed and built. Successful designs of wheelchairs can only be accomplished with continuous input from and interaction with wheelchair users. The durability, aesthetics, function, ride comfort, and cost of the frame are dependent on the materials for construction, the frame geometry, and fabrication methods. One of the issues that makes wheelchair design more complicated is the fact that many users are dependent upon wheeled mobility everyday, nearly all day.

Materials

Most wheelchairs are made of either aluminum or steel. Some chairs are made of titanium or advanced composite materials, primarily carbon fiber, and in the future composite frames will probably begin to become more available. All of these materials have their strengths and weaknesses.

Common aluminum wheelchairs are Tungsten Inert Gas (TIG) welded (i.e., electrically welded together in a cloud of inert gas). They are sometimes bolted together using lugs. Most aluminum wheelchair frames are constructed of round drawn 6061 aluminum tubing. This is one of the least expensive and most versatile of the heat treatable aluminum alloys. It has most of the desirable qualities of aluminum, has good mechanical properties and high corrosion resistance. It can be fabricated using most standard techniques.

Most steel wheelchairs are made of mild steel (1040 or 1060) or chromium-molybdenum alloy (4130 or 4140) seamless tubing commonly called chro-moly. Mild steel is very inexpensive and easy to work with. It is wildly available, and performs well for many applications. However, it has a low strength to weight ratio compared to other materials. Chro-moly is widely used because of its weldability, ease of fabrication, mild hardenability, and high fatigueability. Commonly wheelchairs are made of tubing 0.028-0.035 inches in wall thickness but diameters vary depending on the expected loads from between 0.25 to 1.25 in.

More and more of the high-end wheelchairs are made of titanium. Titanium is a lightweight, strong, nonferrous metal. Titanium wheelchair frames are TIG welded. Titanium is the most exotic of the metals used in production wheelchairs and the most expensive. Titanium requires special tooling and skill to be machined and welded. It has very good mechanical properties and high corrosion resistance. It is resilient to wear and abrasion. Titanium is used because of its availability, appearance, corrosion resistance, very good strength and light weight. A drawback of titanium, besides cost, is that titanium once worn or if flawed may break rapidly (i.e., it has a tendency towards brittle fractures).

Advanced composites have been in use in aerospace and industrial applications for a number of years. These materials include Kevlar, carbon fiber, and polyester limestone composite. These materials are now making the transition to wheelchair design. Kevlar is an organic fiber which is yellow in color and soft to the touch. It is extremely strong and tough. It is one of the lightest structural fabrics on today's market. Kevlar is highly resistant to impact, but its compression strength is poor. Carbon fibers are made by changing the molecular structure of Rayon fibers by extreme stretching and heating. Carbon fiber is very stiff (high modulus of elasticity), very strong (high tensile strength) and has very low density (weight for a given volume). Composites come as cloth or yarn. Composite cloth is woven into bidirectional or unidirectional cloth. Unidirectional weaves can add strength along a particular direction. Composites must be bound together by resin or epoxy. Generally, polyester resins or various specialty epoxies (e.g., Safe-T-Poxy) are used. To achieve greatest strength a minimum amount of epoxy must be used while wetting all of the fibers. This is often achieved through a process called bagging. To increase the strength and stiffness of structural components a foam (e.g., styrofoam, urethane or PVC) core is used. The strengthening occurs because of the separation of the cloth layers (it now becomes more like a tube than a flat sheet). Polyesther limestone composites have been used widely in industrial high voltage electrical component enclosures. A blend of polyester and limestone are used to form a mixture which can be molded under pressure and heated to form a stiff and durable finished product. Polyester limestone composites have high impact strength, and hold tolerances well, but have substantially lower strength to weight ratios than other composites. Their primary advantage is cost, polyester limestone composites are very inexpensive and readily available. Composites can be molded into elaborate shapes which opens a multitude of possibilities for wheelchair design.

Frame Design

Presently all common wheelchair frames center around tubular construction. The tubing can either be welded together, or bolted together using lugs. There are two basic common frame styles: box frame and the cantilever frame. The box frame is named such because of its rectangular shape, and that tubes outline the edges of the "box". Box frames can be very strong and very durable. A cantilever frame is named so because the front and rear wheels, when viewing the chair from the side, appear to be connected by only one tube; this is similar to having the front wheels attached to a cantilever beam fixed at the rear wheels. Both frame types require cross bracing to provide adequate strength and stiffness.

The box frame provides great strength and rigidity. If designed and constructed properly the frame only deflects minimally during normal loading, and most of the suspension is provided by the seat cushion, the wheels and the wheel mounting hardware. Many manufacturers do not triangulate their box frame designs to allow some flexibility. The cantilever frame is based upon a few basic principles: (1) the frame can act as a suspension, (2) there are fewer tubes and they are closer to the body which may make the chair less conspicuous, and (3) there are fewer parts and fewer welds which makes the frame easier to construct.

Wheels and Casters

Casters can be as small as 2 in. in diameter or as large as 12 in. in diameter for wheelchairs designed for daily use. Casters are either pneumatic, semi-pneumatic, or solid (polyurethane). Pneumatic casters offer a smoother ride at the cost of increased maintenance, whereas polyurethane casters are very durable. Semi-pneumatic tires offer a compromise. Most active users prefer 5 in. polyurethane casters or 8 inch pneumatic casters for daily use. An 8-in. caster offers a better ride comfort at the expense of foot clearance. Caster foot clearance is maximized with a 2-in. "Roller Blade" caster often used for court sports (e.g., basketball, tennis, and racquetball). Rear wheels come in three common sizes 22, 24, and 26 in. They come in two styles: spoked and MAG. MAG wheels are typically made of plastic and are die cast. MAG wheels require minimal maintenance, and wear well. However, spoked wheels are substantially lighter, more responsive, and are generally preferred by active manual wheelchair users. Rear tires can be two types; pneumatic or puncture proof. Pneumatic tires can either use a separate tube and tire or a combined tube and tire (sew-up). Commonly, a belted rubber tire with a Butyl tube (65 psi) is used. However, those desiring higher performance prefer sew-up tires or Kevlar belted tires with high pressure tubes (180 psi). Puncture proof tires are heavier, provide less suspension, and are less lively than pneumatic tires.

The chair must be designed to optimize the interaction of the wheels with the ground. Four critical performance factors need to be considered: (1) Castor Flutter, (2) Castor Float, (3) Tracking, and (4) Alignment. Castor flutter is the shimmy (rapid vibration of the front wheels) that may occur on some surfaces above certain speeds. When one of the castors does not touch the floor when on level ground, the wheelchair has castor float. Castor float decreases the stability and performance of the wheelchair. A manual wheelchair uses rear wheel steering via differential propulsion torque. Tracking is the tendency of the wheelchair/rider to maintain its course once control has been relinquished. Tracking is important, as the rider propels the handrims periodically (about every second) and if the chair does not track well it will drift from its course between pushes and force the rider to correct heading. This will waste valuable energy and reduce control over the chair. Alignment generally refers to the orientation of the rear wheels with respect to one another. Typically, it is desirable to have the rear wheels parallel to one another without any difference between the distance across the two rear wheels at the front and back. Misalignment on the order of 1/8 inch can cause a noticeable increase in the effort required to propel the wheelchair.

141.4 Ergonomics of Wheelchair Propulsion

The most important area of wheelchair design and prescription is determining the proper interaction between the wheelchair and the user. This can lead to reducing the risk of developing repetitive strain injury while maximizing mobility. Cardiovascular fitness can be improved through exercise which requires a properly fitted wheelchair.

Kinematics

Kinematic data by itself does not provide sufficient information for the clinician to implement appropriate rehabilitation intervention strategies or for the engineer to incorporate this information into wheelchair design changes. Kinematic data are commonly collected at 60 Hz, which is the maximum frequency of many videotape-based systems. Kinematic data analysis shows that experienced wheelchair users contact the pushrim behind top-dead-center and push to nearly 90° in front of top-dead-center. This is significantly longer than non-wheelchair users. Lengthening the stroke permits lowering the propulsion moment and may place less stress on the users' joints.

An important aspect of the evaluation, and possible retraining of wheelchair users is to determine the optimal stroke kinetics and kinematics [2]. However, there is typically some degree of variation from one stroke to another. Wheelchair propulsion kinematic data are typically cyclic (i.e., a person repeats or nearly repeats his/her arm motions over several strokes). Each marker of the kinematic model (e.g., shoulder, elbow, wrist, knuckle) of each subject generates an x and y set of data which is periodic. The frequencies of the x and y data are dependent upon the anthropometry of the individual, the construction of the wheelchair, and the speed of propulsion. The periodic nature of the kinematic data for wheelchair propulsion can be exploited to develop a characteristic stroke from a set of kinematic data (with the rear hub of the wheelchair chosen as the origin) including several strokes.

Kinetics

The SMARTwheel was developed to measure the pushrim forces required for evaluating net joint forces and moments to allow the clinician and researcher to study the level of stress experienced by the joint structures during wheelchair propulsion. The SMARTwheel uses a standard wheelchair wheel fitted with three beams 120° apart and each has two full strain gage bridges. The strain gage bridges are each interfaced through an instrumentation amplifier to a micro-controller which transmits the data through a mercury slip-ring to the serial port of a computer. Kinetics of wheelchair propulsion are affected by speed of propulsion, injury level, user experience, and wheelchair type and fit. Van Der Woude et al. have reported on an ergometer which detected torque by way of a force transducer located in the wheel center and attached to what is referred to as the "wheel/hand rim construction" [7]. The ergometer was adjusted for each subject's anthropometric measurements. Data were sampled at 100 Hz for 7.5 second periods with a digital filter cut-off frequency of 10 Hz. Mean and peak torque increased with mean velocity, a maximum mean peak torque of 31 N-m occurred at 1.27 m/s. Torque curves of inexperienced subjects showed an initial negative deflection and a dip in the rising portion of the curve. Torque curves were reported to be in agreement with results of previous investigations [8,9,10].

Brauer and Hertig [12] measured the static torque produced on push-rims which were rigidly restrained by springs and mounted independent of the tires and rims of the wheelchair. The spring system was adjustable for the subject's strength. The wheels were locked in a fixed position. Torque was measured using slide-wire resistors coupled to the differential movements between the push-rim and wheels and recorded using a strip chart recorder. Subjects were asked to grasp the push-rim at six different test positions (-10, 0, 10, 20, 30, and 40 degrees relative to vertical) and to use maximal effort to turn both wheels forward. Male subjects (combined ambulatory and wheelchair users) produced torques of 27.9 to 46.6 N-m and female subjects produced torques of 17.1 to 32.1 N-m [12]. Grip location, handedness, grip strength, and how well the test wheelchair fit the anthropometric measurements of the individual affected the torque generated. Problems encountered were slipperiness of the push-rims due to a polished finish and limited contact due to the small diameter of the push-rim tubing (12.7 mm or 1/2 in). The use of one wheelchair for all subjects presented the problem of variations due to inappropriate fit for some individuals.

Brubaker, Ross, and McLaurin [13] examined the effect of horizontal and vertical seat position (relative to the wheel position) on the generation of static push-rim force. Force was measured using a test platform with a movable seat and strain gauged beams to which the push-rims were mounted. Pushing and pulling forces were recorded using a strip chart recorder. Static force was measured for four grip positions (–30, 0, 30, and 60 degrees) with various seat positions. Push-rim force ranged from approximately 500 to 750 N and varied considerably with seat position and rim position [13].

Net Joint Forces and Moments

Net joint forces and moments acting at the wrist, elbow and shoulder during wheelchair propulsion provide scientists and clinicians with information related to the level of stress borne by the joint structure. Joint moments and forces are calculated using limb segment and joint models, anthropometric data, kinetic data, and kinematic data. Joint moments data shows that forces at each joint vary among subjects in terms of peak forces, where they occur during the propulsion phase, and how quickly they develop. Peak net joint moments occur at different joint angles for different subjects, and conditions (e.g., speed, resistance). Convention for joint angles is that 180 degrees at the elbow represents full extension; while at the wrist, this is the hand in the neutral position (flexion less than 180 degrees and extension greater than 180 degrees). Joint angles at the shoulder are determined between the arm and the trunk, with zero measured at the point where the trunk and arm are aligned. Wheelchair users show maximum net shoulder moment between 20° and 40° of extension. Some wheelchair users also show a rapid rise in the elbow extensor moment at the beginning of the stroke with the elbow at about 120°. This moment value begins to decrease between 150° and 170°. At the wrist, the peak moments occur between 190° and 220°. Net joint moments and force models need to account for hand center of pressure, inaccuracies in anthropometric data, and joint models related to clinical variables.

141.5 Power Wheelchair Electrical Systems

Some people are impaired to an extent that they would have no mobility without a power wheelchair. However, some people may have limited upper body mobility and may have the ability to propel a manual wheelchair for short distances. These people may liken using a power wheelchair to admitting defeat.

However, a power wheelchair may provide greater mobility. In such cases it may be best to suggest a power wheelchair for longer excursions, and a manual wheelchair for in home and recreational use.

User Interface

Power wheelchairs are often used in conjunction with a number of other adaptive devices. For people with severe mobility impairments, power wheelchairs may be used with communication devices, computer access devices, respirators, and reclining seating systems. The integration of the users' multiple needs must also be considered when designing or prescribing a power wheelchair.

The joystick is the most common control interface between the user and the wheelchair. Joysticks produce voltage signals proportional to displacement, force, or switch closures. Displacement joysticks are most popular. Displacement joysticks may use either potentiometers, variable inductors (coils) or optical sensors to convert displacement to voltage. Inductive joysticks are most common as they wear well and they can be made quite sensitive. Joysticks can be modified to be used for chin, foot, elbow, tongue or shoulder control. Typically, short throw joysticks are used for these applications. Force sensing joysticks use three basic transducers: simple springs and dampeners on a displacement joystick, cantilever beams with strain gages, and fluid with pressure sensors. Force sensing joysticks which rely on passive dampeners or fluid pressure generally require the user to have a range of motion within normal values for displacement joysticks users. Beam-based force sensing joysticks require negligible motion, and hence may be used for people with limited motion abilities.

People who exhibit intention or spastic tremor or with multiple sclerosis may require special control considerations. Signal processing techniques are often required to grant the user greater control over the wheelchair. Typically, signal averaging or a low pass filter with a cut-off frequency of below 5 Hertz is used. The signal processing is typically incorporated into the controller.

Some people lack the fine motor control to effectively use a joystick. An alternative for these people is to use a switch control or a head position control. A switch control simply uses either a set of switches or a single switch and a coded input, i.e., Morse code or some other simple switch code. The input of the user is latched by the controller and the wheelchair performs the task commanded by the user. The user may latch the chair to repeatedly perform a task a specified number of times, e.g., continue straight until commanded to do otherwise. Switch control is quite functional, but it is generally slower than joystick control. Switch inputs can be generated in many ways. Typically, low pressure switches are used. The input can come from a sip-and-puff mechanism which works off of a pressure transducer. A switch contact is detected when the pressure exceeds or drops below a threshold. The pressure sensor may be configured to react to pressure generated by the user blowing into or sipping from an input or by the user simply interrupting the flow in or out of a tube. Sip-and-puff may also be used as a combination of proportional and switch control. For example, the user can put the control in the "read speed" mode and then the proportional voltage output from the pressure transducer will be latched as the user-desired speed.

Simple switches of various sizes can be used to control the chair using many parts of the body. Switches may be mounted on the armrests or a lap tray for hand or arm activation, on the footrest(s) for foot activation, or on a head rest for head activation. The motion of the head can also be used for proportional control by using ultrasonic sensors. Ultrasonic sensors can be mounted in an array about the headrest. The signal produced by the ultrasonic sensors is related to the position of the head. Hence, motion of the head can be used to create a proportional control signal. Ultrasonic head control and switch control can be combined to give some users greater mastery over their power wheelchair. Switches can be used to select the controller mode, whereas the ultrasonic sensors give a proportional input signal.

A critical consideration when selecting or designing a user interface is that the ability of the user to accurately control the interface is heavily dependent upon the stability of the user within the wheelchair. Often custom seating and postural support systems are required for a user interface to be truly effective. The placement of the user interface is also critical to its efficacy as a functional control device.

Integrated Controls

People with severe physical impairments may only be able to effectively manipulate a single input device. Integrated controls are used to facilitate using a single input device (e.g., joystick, head switches, voice recognition system) to control multiple actuators (e.g., power wheelchair, environmental control unit, manipulator). This provides the user with greater control over the environment. The M3S-Multiple Master Multiple Slave bus is designed to provide simple, reliable access to a variety of assistive devices. Assistive devices include input devices, actuators, and end-effectors. M3S is based on the Computer Area Network (CAN) standard.

A wide range of organizations provide assistive devices which offer the opportunity of functioning in a more independent manner. However, many of these devices and systems are developed without coordination resulting in incompatible products. Clinicians and users often desire to combine products from various sources to achieve maximal independence. The result is to have several devices with their own input devices and overlapping functions. Integrated controls provide access to various end-effectors with a single input device. The M3S provides an electronic communication protocol so that the system operates properly.

M3S is an interface specification with a basic hardware architecture, a bus communication protocol, and a configuration method. The M3S standard incorporates CAN plus two additional lines for greater security (i.e., 7 wire bus, 2 power lines, 2 CAN lines, safety lines, 1 shield, and 1 harness line). The system can be configured to each individual's needs. An M3S system consists of a microcontroller in each device and a control and configuration module (CCM). The CCM insures proper signal processing, system configuration, and safety monitoring. The CCM is linked to a display (e.g., visual, auditory, tactile) which allows the user to select and operate each end-effector. Any M3S compatible device can communicate with another M3S compatible device. M3S is an International Organization of Standards (ISO) open system communication implementation.

Power System

To implement a motor controller, a servo-amplifier is required to convert signal level power (volts at milliamps) to motor power (volts at amps). Typically, a design requirement for series, shunt, and brushless motor drives is to control torque and speed, and hence, power. Voltage control can often be used to control speed for both shunt and series motors. Series motors require feedback to achieve accurate control.

Either a linear servo-amplifier or a chopper can be used. Linear servo amplifiers are not generally used with power wheelchairs primarily because of their lower efficiency than chopper circuits. A motor can be thought of as a filter to a chopper circuit, in this case, the switching unit can be used as part of a speed and current control loop. The torque ripple and noise associated with phase control drives can be avoided by the use of high switching frequencies. The response of the speed control loop is likewise improved with increasing switching frequency.

Motor torque is proportional to the armature current in shunt motors and to the square of the current in series motors. The conduction loss of the motor and servo-amplifier are both proportional to the current squared. Optimal efficiency is achieved by minimizing the form factor, (I_{rms}/I_{mean}). This can be done by increasing the switching frequency to reduce the amplitude of the ripple. Benefits of increased efficiency are increased brush life, gear life, and lower probability of field permanent magnet demagnetization.

Switching or chopper drives are classified as either unidirectional or bi-directional. They are further divided by whether they use dynamic braking. Typically, power wheelchairs use bi-directional drives without dynamic braking. However, scooters may use unidirectional drives. The average voltage delivered to the motor from a switching drive is controlled by varying the duty cycle of the input waveform. There are two common methods of achieving this goal: (1) fixed pulse width, variable repetition rate, and (2) pulse-width modulation (PWM). Power wheelchair servo-amplifiers typically employ PWM.

Pulse width modulation at a fixed frequency has no minimum on-time restriction. Therefore, current peaking and torque ripple can be minimized. For analysis, a d.c. motor can be modeled as a RL circuit, resistor and inductor, or in series with a voltage source. If the motor current is assumed continuous, then the minimum and maximum motor current can be represented by Eq. (141.1).

$$I_{min} = \frac{e^{-(R/L)t_{off}}\left(1-e^{-(R/L)t_{on}}\right)}{1-e^{-(R/L)(t_{on}-t_{off})}}\frac{V_s}{R} - \frac{V_{gen}}{R}$$

$$I_{max} = \frac{1-e^{-(R/L)t_{on}}}{1-e^{-(R/L)(t_{on}-t_{off})}}\frac{V_s}{R} - \frac{V_{gen}}{R}$$

Two basic design principles are used when designing switching servo-amplifiers: (1) I_{max} should be limited to five times the rated current of the motor to ensure that demagnetization does not occur, (2) the ripple, $(I_{max}-I_{min})/I_{avg}$, should be minimized to improve the form factor, and reduce the conduction loss in the switching devices. To achieve low ripple, either the inductance has to be large or the switching frequency has to be high. Permalloy powder cores can be used to reduce core loss at frequencies above a few kilohertz. However, this comes at the cost of the electrical time constant of the motor, degrading the motor response time. Hence, raising the switching frequency is most desirable. A power MOSFET has the ability to switch rapidly without the use of load-shaping components.

There are several motor types that may be suitable for use with power wheelchairs. Most current designs use permanent magnet direct current motors. These motors provide high torque, high starting torque and are simplest to control. Permanent magnet direct current motors can either be controlled in what are commonly called current mode or voltage mode. These modes developed out of designs based upon controlling torque and speed, respectively.

Alternating current motors can be designed to be highly efficient and can be controlled with modern power circuitry. Because of the development and wide spread dissemination of switching direct current converters, it is quite feasible to use alternating current motors with a battery supply. To date, alternating current motors have only been used in research on power wheelchairs. The output of the motor is controlled by varying the phase or the frequency.

The battery energy storage system is recognized as one of the most significant limiting factors in powered wheelchair performance. Battery life and capacity are important. If battery life can be improved, the powered wheelchair user will have longer, reliable performance from his/her battery. An increase in battery capacity will allow powered wheelchair users to travel greater distances with batteries that weigh and measure the same as existing wheelchair batteries. Most importantly, increases in battery capacity will enable the use of smaller and lighter batteries. Because batteries account for such a large proportion of both the weight and volume of current powered wheelchair systems, wheelchair manufacturers must base much of their design around the battery package.

Power wheelchairs typically incorporate 24 volt d.c. energy systems. The energy for the wheelchair is provided by two, deep cycle lead-acid batteries connected in series. Either wet cell or gel cell batteries are used. Wet cell batteries also cost about one-half as much as gel cell batteries. Gel cells may be required for transport by commercial air carriers.

Battery technology for wheelchair users remains unchanged despite the call for improvements by power wheelchair users. This may be in part due to the relatively low number of units purchased, about 500,000 per annum, when compared to automotive applications with about 6.6 million per annum by a single manufacturer. Wheelchair batteries are typically rated at 12 Volts and 30-90 ampere-hours capacity at room temperature. A power wheelchair draws about 10 amps during use. The range of the power wheelchair is directly proportional to the ampere-hour rating for the operating temperature.

Batteries are grouped by size. Group size is indicated by a standard number. The group size defines the dimensions of the battery as shown in Table 141.1. The ampere-hour rating defines the battery's capacity.

It is important that the appropriate charger be used with each battery set. Many battery chargers automatically reduce the amount of current delivered to the battery as the battery reaches full charge. This helps to prevent damage to the battery from boiling. The rate at which wet and gel cell batteries charge is significantly different. Some chargers are capable of operating with both types of batteries.

Table 141.1 Standard Power Wheelchair Battery Group Sizes

Group number	Length	Width	Height
U1	7-3/4	5-3/16	7-5/16
22NF	9-7/16	5-1/2	8-15/16
24	10-1/4	6-13/16	8-7/8
27	12-1/16	6-13/16	8-7/8

Note: Units are in inches.

Many require setting the charger for the appropriate battery type. Most wheelchair batteries connected in series are charged simultaneously with a 24 Volt battery charger.

Electromagnetic Compatibility

Powered wheelchairs have been reported to exhibit unintended movement. Wheelchair manufacturers and the US Food and Drug Administration Center for Devices and Radiological Health (FDA-CDRH) have examined the susceptibility of powered wheelchairs and scooters to interference from radio and microwave transmissions. These devices are tested at frequencies ranging from 26 MHz to 1 GHz, which is common for transmissions (e.g., radio, TV, microwave, telephones, mobile radios). Power wheelchairs incorporate complex electronics and microcontrollers which are sensitive to electromagnetic (EM) radiation, electrostatic discharge (ESD), and other energy sources.

Electric powered wheelchairs may be susceptible to electromagnetic interference (EMI) present in the ambient environment. Some level of EMI immunity is necessary to ensure the safety of power wheelchair users. Electromagnetic compatibility (EMC) is the term used to describe how devices and systems behave in an electromagnetic environment. Because of the complexity of power wheelchairs and scooters, and the interaction with an electromagnetic environment, susceptibility to interference cannot be calculated or estimated reliably. A significant number of people attach accessories (e.g., car stereos, computers, communication systems) to their power wheelchairs which also share the batteries. This may increase the susceptibility of other system components to EMI. A number of companies make electric powered devices designed to operate on power wheelchairs to provide postural support, pressure relief, environmental control, and motor vehicle operation. These devices may alter the EMI compatibility of power wheelchairs as provided by the original equipment manufacturer (OEM). Wheelchairs and accessories can be made to function properly within EM environments through testing.

Field strengths have been measured at 20 V/m from a 15 Watt hand held cellular telephone, and 8 V/m from a 1 Watt hand held cellular telephone [2]. The FDA-CDRH has tested power wheelchairs and scooters in a Gigahertz Transverse Electromagnetic (GTEM) cell, and in an anechoic chamber with exposure strengths from 3 V/m to 40 V/m. The US FDA requires that a warning sticker be placed on each power wheelchair or scooter indicating the risk due to EMI.

Two tests are commonly performed on chairs: brake release and variation in wheel speed. The device(s) used for measuring wheel speed and brake release must not significantly alter the field. The brakes shall not release or the wheels are not to move with a wheel torque equivalent to a 1:6 slope with a 100 kilogram rider when the wheelchair is exposed to EM radiation. Non-electrical contact methods (e.g., audio sensing, optical sensing) of measuring brake release or wheelchair movement are preferable. Nominal wheel speed may drift over the length of the test. This drift is primarily due to drop in battery charge over the test interval. Wheel speed must be recorded without EM interference between test intervals. The percentage change in wheel speed during exposure to EM interference shall be referenced to the nominal wheel speed for that test interval. The variation in absolute forward speed, $(v_{emR} + v_{emL})/2$, is to be within 30% of the nominal forward speed, $(v_{nomR} + v_{nomL})/2$. The differential speed between the two wheels should be within 30% of each other, $2 \cdot (v_{emR} - v_{emL})/(v_{nomR} + v_{nomL})$.The test frequency must be held long enough to accommodate the slowest time constant (time required to reach 63% of maximum or minimum) of parameters related to wheelchair driving behavior. Currently two seconds is used by the FDA-CDRH test laboratory.

141.6 Personal Transportation

Special adaptive equipment requirements increase with the degree of impairment and desired degree of independence in areas such as personal care, mobility, leisure, personal transportation, and employment. People are concerned that they receive the proper equipment for them to safely operate their vehicle. Access and egress equipment have the greatest maintenance requirements. Other devices such as hand controls, steering equipment, securement mechanisms, and interior controls require less maintenance. Most users of adaptive driving equipment are satisfied with the performance of such equipment. Most frequent equipment problems are minor and are repaired by consumers themselves.

Physical functional abilities such as range of motion, manual muscle strength, sensation, grip strength, pinch strength, fine motor dexterity, and hand-eye coordination all may be related to driving potential. Driving characteristics must also be evaluated when determining an individual's potential to safely operate a motor vehicle. Throttle force, brake force, steering force, brake reaction time, and steering reaction time are all factors which influence an individual's driving potential.

Vehicle Selection

While it is often a difficult task to find an automobile which meets the specific needs of a particular wheelchair user, currently, no automobile meets the needs of all wheelchair users. Automotive consumers with disabilities are also concerned about ease of entry, stowage space for the wheelchair, and seat positioning. Reduced size, increasingly sloping windshields, lower roofs, and higher sills of new cars make selecting a new vehicle difficult for wheelchair users. The ability to load the wheelchair into a vehicle is essential. Some individuals with sufficient strength and a suitable vehicle are able to stow their wheelchairs inside the vehicle without the use of assistive devices. Many people must rely on an external loading device. The selection of the appropriate vehicle should be based upon the client's physical abilities and social needs.

An approach some people have used to overcome the problems associated with a smaller car is to use a car-top wheelchair carrying device. These devices lift the wheelchair to the top of the car, fold, and stow it. They have been designed to work with four door sedans, light trucks, and compact automobiles.

There are several critical dimensions to an automobile when determining wheelchair accessibility. The wheelbase of the automobile is often used by auto manufacturers to determine vehicle size (e.g., full-size, mid-size, compact). Typical ranges for passenger vehicles are presented in Table 141.2.

Lift Mechanisms

Many wheelchair users who cannot transfer into a passenger vehicle seat or prefer a larger vehicle, drive vans equipped with wheelchair lifts. Platform lifts may use a lifting track, a parallelogram lifting linkage, or a rotary lift. Lift devices are either electromechanically or electrohydraulically powered. The platform often folds into the side doorway of the van. Crane lifts, also called swing-out lifts, have a platform which elevates and folds or rotates into the van. Lifts may either be semi-automatic or automatic. In many cases

TABLE 141.2 Typical Ranges of Accessibility Dimensions for Sedans

Wheelbase	93–108
Door height	33–47
Door width	41–47
Headroom	36–39
Max. Space behind seat	9–19
Min. Space behind seat	4–9
Seat-to-ground distance	18–22
Width of door opening	38–51

Note: Units in inches.

semi-automatic lifts require the user to initiate various stages (e.g., unlocking door, door opening, lowering lift) of the lifting process. Automatic lifts are designed to perform all lift functions. They usually have an outside key operated control box, or an interior radio controlled control box.

Some lifts use electrohydraulic actuators to lift and fold, with valves and gravity used to lower the lift. Crane lifts may swing out from a post at the front or rear of the side door. Interlocking mechanisms are available with some lifts to prevent the lift from being operated while the door is closed. The Society of Automotive Engineers (SAE) has developed guidelines for the testing of wheelchair lift devices for entry and exit from a personal vehicle. The standards are intended to set an acceptable level of reliability and performance for van lifts.

Wheelchair Restraint Mechanisms

Securement systems are used to temporarily attach wheelchairs to vehicles during transport. Many wheelchair users can operate a motor vehicle from their wheelchair, but are unable to transfer into a vehicle seat. Auto safety standards have reduced the number of US automobile accident fatalities despite an increase in the number of vehicles. The crash pulse determines the severity of the collision of the test sled, and hence, simulates real-world conditions. Securement systems are tested with a surrogate wheelchair at 30 miles per hour (48 +2/–0 kilometers per hour) with a 20g deceleration. Wheelchairs must be safely restrained when experiencing an impact of this magnitude and no part of the wheelchair shall protrude into the occupant space where it might cause injury.

Proper use of lap and shoulder belts is critical to protecting passengers in automobiles seats. A similar level of crash protection is required for individuals who remain in their wheelchairs during transportation. Wheelchairs are flexible, higher than a standard automobile seat, and not fixed to the vehicle. The passenger is restrained using a harness of at least one belt to provide pelvic restraint and two shoulder or torso belts that restrain both shoulders. A head support may also be used to prevent rearward motion of the head during impact or rebound. A three point restraint is the combination of a lap belt and a shoulder belt (e.g., pelvic torso restraint, lap-sash restraint, lap-shoulder restraint).

The relationship between injury criteria and the mechanics of restraint systems are important to insure the safety of wheelchair users in motor vehicles. Hip and head deflection are often used criteria for determining potential injury. The automotive industry has invested considerable effort for research and development to protect vehicle passengers. Research is not nearly extensive for the passenger who remains seated in a wheelchair while traveling. Many wheelchair and occupant restraint systems copy the designs used for standard automobile seats. However, this type of design may not be appropriate.

Crash tests have shown that for 10 or 20 g impacts of 100 millisecond duration, people may sustain injuries despite being restrained. When shoulder belts mounted 60 inches above the floor were used to restrain a 50th percentile male dummy, it was found that the torso was well controlled, and head and chest excursions were limited. When shoulder belts were anchored 36 inches above the floor they were ineffective in controlling torso movement. Kinematic results and head injury criteria (HIC) can be used to estimate the extent of injury sustained by a human passenger. A HIC of 1000 or greater indicates a serious or fatal head injury. Generally, a HIC value approaching or exceeding 1000 is indicative of head impact with some portion of the vehicle interior. The open space typically surrounding a wheelchair user in a public bus precludes impact with the buses' interior. High HIC values may occur when the torso is effectively restrained and there is a high degree of neck flexion. If the chin strikes the chest then there may be an impact great enough to cause head injury.

Hand Controls

Hand-controls are available for automatic and manual transmission vehicles. However, hand-controls for manual transmission automobiles must be custom made. There are also portable hand-controls, and long-term hand controls. Portable hand-controls are designed to easily attach to most common automobiles with a minimal number of tools. Hand-controls are available for either left or right hand control.

Many hand controls are attached to the steering column. Hand-controls either clamp to the steering column or are attached to a bracket which is bolted to the steering column or dash, typically where the steering column bolts to the dash. Installation of the hand-controls should not interfere with driver safety features (e.g., air bags, collapsible steering columns). The push rods of the hand-control either clamp directly to the pedals or the levers connected to them.

Most systems activate the brakes by having the driver push forward on a lever with a hand grip. This allows the driver to push against the back of the seat, creating substantial force, and braces the driver in the event of a collision. The throttle, or gas pedal, is operated in a number of ways. Some systems use a twist knob or motorcycle type throttle. Other systems actuate the throttle by pulling on the brake throttle lever. Another method is to rotate the throttle-brake lever downwards (i.e., pull the lever towards the thigh at a right angle to the brake application to operate the throttle). It is common to have the same vehicle driven by multiple people which may require the vehicle to be safely operated with hand-controls, and the OEM foot controls. Care must be taken to insure that the lever and brackets of the hand-controls do not restrict the driving motions of foot control drivers.

Many people have the motor control necessary to operate a motor vehicle, but they do not have the strength required to operate manual hand-controls. Automatic or Fly-By-Wire hand-controls use external actuators (e.g., air motors, servo mechanisms, hydraulic motors) to reduce the force required to operate various vehicle primary controls. Power steering, power brakes, six-way power seats, and power adjustable steering columns can be purchased as factory options on many vehicles.

Six-way power seats are used to provide greater postural support and positioning than standard automotive seats. They can be controlled by a few switches to move fore-aft, incline-recline, and superior-inferior. This allows the user to position the seat for easy entry and exit, and for optimal driving comfort. Power adjustable steering columns also make vehicles more accessible. By using a few buttons, the steering column can be tilted upwards or downwards allowing positioning for entry/exist into the vehicle, and for optimal driving control.

Custom devices are available for people who require more than the OEM options for power assistance. Microprocessor and electronic technology have dramatically changed how motor vehicles are designed. Many functions of an automobile are controlled electronically or with electromechanical-electrohydraulic controls. This change in vehicle design has made a wide variety of options available for people who require advanced vehicle controls. Many automobiles use electronic fuel injection. Electronic fuel injection systems convert the position of the accelerator pedal to a serial digital signal which is used by a microcontroller to inject the optimal fuel-air mixture into the automobile at the proper time during the piston stroke. The electronic signal for the accelerator position can be provided by another control device (e.g., joystick, slide-bar).

References

1. Adams T.C., Reger S.I. 1993. Factors affecting wheelchair occupant injury in crash simulation. *Proc. 16th Annual RESNA Conference,* Las Vegas, NV, pp. 80-82.
2. Adams T.C., Sauer B., Reger S.I. 1992. Kinematics of the wheelchair seated body in crash simulation. *Proc. RESNA Intl. '92,* Toronto, Ontario, Canada, pp. 360-362.;
3. Asato K.T., Cooper R.A., Robertson R.N., and Ster J.F. 1993. SMART^{Wheels}: Development and Testing of a System for Measuring Manual Wheelchair Propulsion Dynamics, *IEEE Trans. Biomed. Eng.* 40(12):1320-1324.
4. Aylor J.H., Thieme A., Johnson B.W. 1992. A battery state-of-charge indicator. *IEEE Trans. Ind. Electr.,* 39(5):398-409.
5. Boninger M.L., Cooper R.A., Robertson R.N., Shimada S.D. 1997, Three-dimensional pushrim forces during two speeds of wheelchair propulsion, *Am. J. Phys. Med. Rehab.*, 76:420-426.
6. Brauer R.L. and Hertig B.A. 1981. Torque generation on wheelchair handrims, *Proc. 1981 Biomechanics Symp., ASME/ASCE Mechanics Conference,* pp. 113-116.

7. Brienza D.M., Cooper R.A., Brubaker C.E. 1996, Wheelchairs and seating, *Curr. Opinion in Orthopedics*, 7(6):82-86.
8. Brubaker C.E.,Ross S., and McLaurin C.A. 1982. Effect of seat position on handrim force, *Proc. 5th Annual Conference on Rehabilitation Engineering*, p. 111.
9. Cooper R.A. 1998, *Wheelchair Selection and Configuration*, Demos Medical Publishers, New York, NY.
10. Gray D.B., Quatrano L.A., Lieberman M.L. 1998, *Designing and Using Assistive Technology*, Brookes Publishing Company, Baltimore, MD.
11. Cooper R.A., Trefler E., Hobson D.A. 1996, Wheelchairs and seating: issues and practice, *Technology and Disability*, 5:3-16.
12. Cooper R.A. 1995, Intelligent control of power wheelchairs, *IEEE Engineering in Medicine and Biology Magazine*, 15(4):423-431.
13. Cooper R.A. 1995, *Rehabilitation Engineering Applied to Mobility and Manipulation*, Institute of Physics Publishing, Bristol, United Kingdom.
14. Cooper R.A., Gonzalez J.P., Lawrence B.M., Rentschler A., Boninger M.L., VanSickle D.P. 1997, Performance of selected lightweight wheelchairs on ANSI/RESNA tests, *Arch. Phys. Med. Rehab.*, 78:1138-1144.
15. Cooper R.A. 1996, A perspective on the ultralight wheelchair revolution, *Technology and Disability*, 5:383-392.
16. Cooper R.A., Robertson R.N., Lawrence B., Heil T., Albright S.J., VanSickle D.P., Gonzalez J.P. 1996, Life-cycle analysis of depot versus rehabilitation manual wheelchairs, *J. Rehab. Res. Dev.*, 33(1):45-55.
17. Cooper R.A. 1993. Stability of a Wheelchair Controlled by a Human Pilot, *IEEE Trans. on Rehabilitation Engineering*. 1(4):193-206.
18. Cooper R.A., Baldini F.D., Langbein W.E., Robertson R.N., Bennett P., and Monical S. 1993. Prediction of Pulmonary Function in Wheelchair Users, *Paraplegia*. 31:560-570.
19. Cooper R.A., Horvath S.M., Bedi J.F., Drechsler-Parks D.M., and Williams R.E. 1992. Maximal Exercise Responses of Paraplegic Wheelchair Road Racers, *Paraplegia*. 30:573-581.
20. Cooper R.A. 1991. System Identification of Human Performance Models, *IEEE Trans. on Systems, Man, and Cybernetics*. 21(1):244-252.
21. Cooper R.A. 1991. High Tech Wheelchairs Gain the Competetive Edge, *IEEE Engineering in Medicine and Biology Magazine*. 10(4):49-55.
22. Kauzlarich J.J., Ulrich V., Bresler M., Bruning T. 1983. Wheelchair Batteries: Driving cycles and testing. *J. Rehab. Res. Dev.* 20(1):31-43.
23. MacLeish M.S., Cooper R.A., Harralson J., and Ster J.F. 1993. Design of a Composite Monocoque Frame Racing Wheelchair, *J. Rehab. Res. Dev.* 30(2):233-249.
24. Powell F.,Inigo R.M. 1992. Microprocessor based D.C. brushless motor controller for wheelchair propulsion. *Proc. RESNA Intl.* '92, Toronto, Canada, pp. 313-315.
25. Riley P.O., Rosen M.J. 1987. Evaluating manual control devices for those with tremor disability. *J. Rehab. Res. Dev.* 24(2):99-110.
26. Sprigle S.H., Morris B.O., Karg P.E. 1992. Assessment of transportation technology: Survey of driver evaluators. *Proc. RESNA Intl.* '92, Toronto, Ontario, Canada, pp. 351-353.
27. Sprigle S.H., Morris B.O., Karg P.E. 1992. Assessment of transportation technology: Survey of Equipment vendors. *Proc. RESNA Intl.* '92, Toronto, Ontario, Canada, pp. 354-356.
28. Schauer J., Kelso D.P., Vanderheiden G.C. 1990. Development of a serial auxiliary control interface for powered wheelchairs. *Proc. RESNA 13th Annual Conf.*, Washington, D.C., pp.191-192.

142

Externally Powered and Controlled Orthotics and Prosthetics

Dejan B. Popović
University of Belgrade, Belgrade

Rehabilitation of humans with disabilities requires effective usage of assistive systems for restoration of motor functions. Important features of an effective assistive system are: (1) reliability; (2) minimum increase of energy rate and cost with respect to able-bodied subjects performing the same task; (3) minimum disruption of normal activities when employing the assistive system; (4) cosmetics; and (5) practicality. The system should be easy to don and doff and available for daily use at home. These requirements and available technology have led to the development of externally powered orthoses and prostheses that interface directly or indirectly with the human neuromuscular system. These devices include battery-powered actuators or muscle stimulators, microprocessor-based controller and reliable biologic-like sensors.

Hand, arm, and lower extremities assistive systems for amputees and humans with paralysis are elaborated in this chapter. Two approaches for the restoration of movements in humans with paralysis are described: functional activation of paralyzed muscles called Functional Electrical Stimulation (FES) or Functional Neuromuscular Stimulation (FNS), and the combined usage of an FES and a mechanical orthosis. The latter is called Hybrid Assistive System (HAS). The section dealing with prosthesis relates to externally controlled and powered artificial limbs, with specific emphasis on so-called myoelectric controlled devices.

Available assistive systems meet many of the requirements listed above, but still have some drawbacks. The small number of externally powered system users on a daily basis is a significant indicator that these systems are not perfected. Recent neurophysiological findings on muscle properties and strengthening techniques, in addition to improved percutaneous or implantable stimulators, may increase the applicability of many assistive systems, specifically in cases of paralyzed limbs. Major limitations in daily application relate to insufficiently adaptive and robust control methods. The complexity of central nervous system (CNS) control and the interface between voluntary control and external artificial control are still challenging, unanswered questions. Hierarchical control methods, combining symbolic models at the highest level and analytic models at lower levels, give hope for further progress (see Chapter 165—Control of Movements).

0-8493-0462-8/00/$0.00+$.50

Sensory feedback is essential for effective control of FES systems. In addition to artificial sensors, hopes are directed towards the use of natural sensors and the development of an "intelligent" movement controller.

142.1 FES Systems

Functional Electrical Stimulation (FES) can help in regaining functional movements in numerous paralyzed humans. FES activates innervated but paralyzed muscles, by using an electronic stimulator to deliver trains of pulses to neuromuscular structures. The basic phenomenon of the stimulation is a contraction of muscle due to the controlled delivery of electric charge to neuromuscular structures.

FES systems can restore (1) goal-oriented (hand and arm) movements, and (2) cyclic (walking and standing) movements.

Restoration of Hand Functions

The objective of an upper extremity assistive system is directed toward establishing independence to the user. Efforts in developing upper extremity FES systems were targeted toward individuals with diminished, but preserved, shoulder and elbow functions, with lack of wrist control and grasping ability [1].

There have been several designs of FES systems. These systems can be divided among the source of control signals to trigger or regulate the stimulation pattern: shoulder control [2], voice control [3,4], respiratory control [5], joystick control [6], and position transducers [7, 8] or trigger [9, 6]. The division can be made based upon the method to which patterned electrical stimulation is delivered: one to three channel surface electrode systems [6-10], multichannel surface stimulation system [4], multichannel percutaneous systems with intramuscular electrodes [2,3,5], and fully implanted systems with epimysial electrodes [1]. Only a small number of FES grasping systems has been used outside the laboratory.

The first grasping system used to provide prehension and release [11] used a splint with a spring for closure and electrical stimulation of the thumb extensor for release. This attempt was unsuccessful mostly because of the state-of-the-art technology used, but also because of muscle fatigue and erratic contractile response. Rudel et al. [12], following the work of Vodovnik [8], suggested the use of a simple two-channel stimulation system and a position transducer (sliding potentiometer). The shift of the potentiometer forward from its neutral position causes opening by stimulating the dorsal side of the forearm; a shift backward causes closing of the hand by stimulating the volar side of the forearm.

The follow-up of initial FES system use was systematically continued in Japan. Japanese groups succeeded in developing the FES clinic in Sendai, Japan, where many subjects are implanted with up to 30 percutaneous intramuscular electrodes that are used for therapy, but not to assist grasping. The Japanese research approach in functional grasping relates to subjects lacking not only hand functions but also the elbow control (e.g., C4 complete CNS injuries), and the system uses either a voice or suck/puff control and preprogrammed EMG-based stimulation patterns. This preprogrammed EMG-based stimulation is developed by detailed studies of muscle activities with intramuscular electrodes in able-bodied subjects while reaching and grasping [3].

The approach taken at Ben Gurion University, Israel [4] used a voice controlled multichannel surface electrode system. As many as 12 bipolar stimulation channels and a splint are used to control elbow, wrist, and hand functions. There is very little practical experience with the system, and the system has to be tuned for the needs of every single user. Surface stimulation most probably does not allow control of small hand and forearm muscles necessary to provide dexterity while grasping. Daily mounting and fitting of the system is problematic. The work of Nathan resulted with a commercial device called Handmaster NMS1 [9,13]. This device is approved as a therapeutic device claiming that it is improving grasping functionality in humans after stroke. A detailed analysis of effects of FES systems after stroke is presented by Hines and colleagues [14].

The group at the Institut for Biokibernetik, Karlsruhe, Germany suggested the use of EMG recordings from the muscle, which is stimulated [15]. The aim of this device is to enhance the grasping by using weak muscles. Hence, in principle it could be possible to use retained recordings from the volar side of the forearm to trigger on and off the stimulation of the same muscles. In this case, it is essential to eliminate the stimulation artifact and the evoked potential caused by the stimulus in order to eliminate positive feedback effects which will generate a tetanic contraction that cannot be turned off using the method presented. This approach is further developed in Denmark [16].

The Case Western Reserve University (CWRU) fully implantable system has a switch to turn the system on and off and select the grasp and a joystick to proportionally control aperture of the hand for palmar and lateral grasp [17-19]. These two grasps are synthesized using a preprogrammed synergy. The joystick mounted on the contralateral shoulder voluntarily controls the preprogrammed sequence of stimulation. The palmar grasp starts from the extended fingers and thumb (one end position of the joystick), followed by the movement of the thumb to opposition and flexing of fingers (other end position of the joystick). The lateral grasp starts from the full extension of fingers and the thumb, followed by the flexion of fingers and adduction of the thumb. The system is applicable if the following muscles can be stimulated: extensor pollicis longus, flexor pollicis longus, adductor pollicis, opponens, flexor digitorum profundus and superficialis, and extensor digitorum communis. It is possible to surgically change the grasp permanently [1] (pining some joints, tendon transfer, etc.). An important feature of the grasping system is related to daily fitting of the joystick (zeroing the neutral position), and going to hold mode from the movement mode. The hold mode is the regime where the muscle nerves are stimulated at the level that the same force is maintained, and the user selects the level. At this time, the CWRU system uses the joystick with two degrees of freedom or velocity sensor software to switch from active control to hold mode. When the system is in the hold mode, joystick movements do not effect the grasp. The initial CWRU system [20] suggested the use of myoelectric signals obtained from a site with some regaining voluntary activity. The CWRU implantable system is the only system used for assistance in daily living function [1]. The functional evaluation of the system [21] showed that there is substantial improvement in simple grasping tasks, which is important to a person with tetraplegia, and an increasing number are in use around North America and Australia.

Prochazka [7] suggested the use of wrist position to control the stimulation of muscles to enhance the tenodesis grasping, and designed a device called the Bionic Glove (Fig. 142.1). A sensor is used to detect wrist movement, and trigger opening and closing of the hand. A microcomputer is built into the battery operated stimulation unit, which detects movements and controls three channels to stimulate thumb extension and flexion, and finger flexors. Clinical evaluation of the Bionic Glove [22] showed that the stimulation is beneficial for persons with tetraplegia both therapeutically and as an orthosis, but that the overall acceptance rate is still low.

Based on four-channel surface electrical stimulation for walking triggered from the lesion EMG recordings [23,24], a grasping system with up to three channels of surface stimulation provides similar function to the Prochazka's system [25]. The use of myoelectric signals from above lesion sites (forearm wrist extensors) to trigger the stimulation of thumb and finger flexors was presented. A threshold discrimination of the EMG is done, which activates the proper stimulus pattern to be applied to the surface electrodes. A new laboratory version of the myoelectric control of a grasping system has been developed and tried at the ETH Zurich [26]. An interesting approach of using myoelectric control to drive the CWRU system was evaluated by Scott and colleagues [27]. Using muscles that are not affected and easily controllable, a bilateral control of the CWRU system might be possible.

The idea of controlling the whole arm and assisting manipulation in humans lacking shoulder and elbow control has been getting more attention recently from the research team at the Case Western Reserve University in Cleveland, Ohio [28-31]. The system is designed to combine a fully implantable grasping system with some additional channels to control elbow extension, flexion, and shoulder movements. A different approach is taken by using surface stimulation to control elbow movements [32,33]. The clinical study at the Rehabilitation Institute in Belgrade with the so-called Belgrade grasping/reaching system showed that most persons with tetraplegia responding to stimulation of the biceps and triceps brachii muscles improve their function within six weeks to the stage at which they become reluctant to use FES assistance for manipulation [32,33].

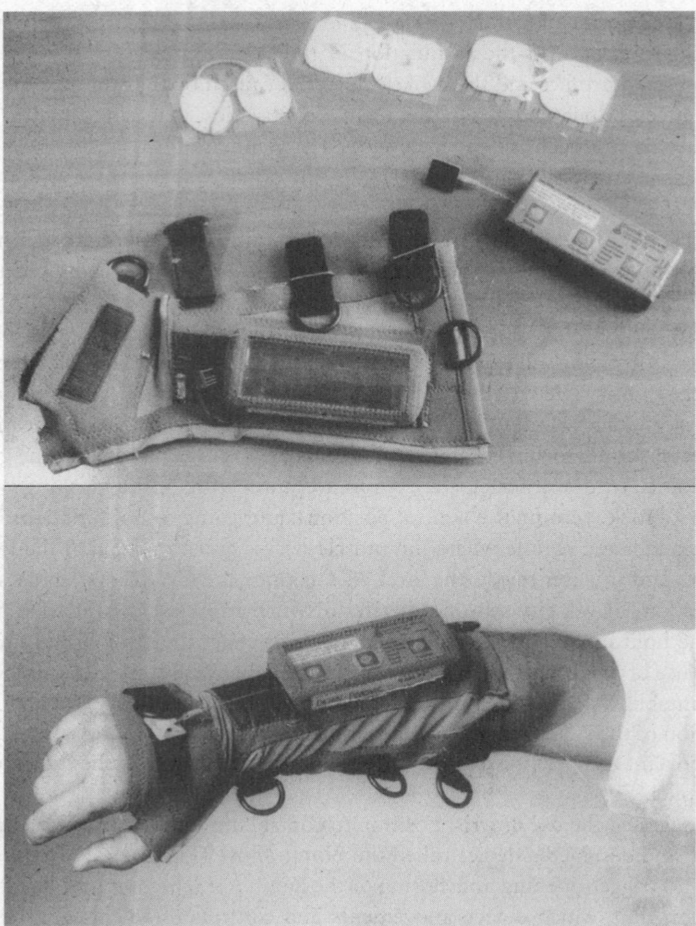

FIGURE 142.1 Bionic glove for restoration of grasping in persons with tetraplegia used for multicentre trials [7,22]. The upper panel shows electrodes, stimulator, and the glove, while the bottom panel shows the glove mounted and ready for use.

The development of implantable cuff electrodes to be used for sensing contact, slippage, and pressure [34-38] opens a new prospective in controlling grasping devices, and the Center for Sensory Motor Interactions in Aalborg, Denmark is pursuing a series of experiments combining their sensing technique with the fully implantable CWRU system.

Restoration of Standing and Walking

The application of FES to the restoration of gait was first investigated systematically in Ljubljana, Slovenia [39-44]. Currently, FES for gait rehabilitation is used in a clinical setting in several rehabilitation centers [45-57] and there is a growing trend for the design of devices for home use.

Current surface FES systems use various numbers of stimulation channels. The simplest one, from a technical point, is a single-channel stimulation system. This system is only suitable for stroke patients and a limited group of incomplete spinal cord injury patients. These individuals can perform limited ambulation with assistance of the upper extremities without an FES system, although this ambulation may be both modified and/or impaired. The FES in these humans is used to activate a single muscle group. The first demonstrated application of this technique was in stroke patients [40], even though the original patent came from the Liberson patent in 1961. The stimulation is applied to ankle dorsiflexors so the "foot-drop" can be eliminated. A commercial system has been designed by Stein and colleagues [58], which integrates a single-channel stimulator and a tilt sensor; thus, eliminating a foot switch which was proved for easily generating false triggering and malfunctioning. Single and dual channel correcting foot-drop is now a regular clinical treatment in some rehabilitation institutions [59].

A multichannel system with a minimum of four channels of FES is required for ambulation of a patient with a complete motor lesion of lower extremities and preserved balance and upper body motor control [49]. Appropriate bilateral stimulation of the quadriceps muscles locks the knees during standing. Stimulating the common peroneal nerve on the ipsilateral side, while switching off the quadriceps stimulation on that side, produces a flexion of the leg. This flexion, combined with adequate movement of the upper body and use of the upper extremities for support, allows ground clearance and is considered the swing phase of the gait cycle. Hand or foot switches can provide the flexion–extension alternation needed for a slow forward or backward progression. Sufficient arm strength must be available to provide balance in parallel bars (clinical application), and with a rolling walker or crutches (daily use of FES). These systems evolved into a commercial product called Parastep-1R (Sigmedics, Chicago, IL) which was approved for home usage in 1994 by the Food and Drug Administration.

Multichannel percutaneous systems for gait restoration were suggested [46,50,61,62]. The main advantage of these systems is the plausibility to activate many different muscle groups. A very similar preprogrammed stimulation pattern to the one in a human with no motor disorders is delivered to ankle, knee, and hip joints, as well as to paraspinal muscles. The experience of the Cleveland research team suggested that 48 channels are required for a complete SCI walking system to achieve a reasonable walking pattern, but they recently changed their stimulation strategy and included some external bracing. Fine-wire intramuscular electrodes are cathodes positioned close to the motor point within selected muscles. Knee extensors (rectus femoris, vastus medialis, vastus lateralis, vastus intermedius), hip flexors (sartorius, tensor fasciae latae, gracilis, iliopsoas), hip extensors (semimembranosus, gluteus maximus), hip abductors (gluteus medius), ankle dorsiflexors (tibialis anterior, peroneus longus), ankle plantar flexors (gastrocnemius lateralis and medialis, plantaris and soleus), and paraspinal muscles are selected for activation. A surface electrode is used as a common anode. Interleaved pulses are delivered with a multichannel, battery-operated, portable stimulator. The hand controller allows the selection of gait activity. These systems were limited to the clinical environment. The application was investigated in complete spinal cord lesions and in stroke patients [61,62]. The same strategy and selection criteria for implantation was used for both stroke and SCI patients. Recent developments use the CWRU system with eight channels per leg to be activated and improved control [61-63].

A multichannel totally implanted FES system [55] was proposed and tested in a few subjects. This system uses a 16-channel implantable stimulator attached to the epineurium electrodes. Femoral and gluteal nerves were stimulated for hip and knee extension. The so-called "round-about" stimulation was applied in which four electrodes were located around the nerve and stimulated intermittently. This stimulation method reduces muscle fatigue.

The development of the stimulation technology is providing new hope for patients and researchers alike. Two new techniques are especially important: (1) application of remotely controlled wireless microstimulators [64-66], and (2) so-called "stimulator for all seasons" [67]. There are several attempts to design effective wireless stimulators that are believed to be capable of selectively stimulating fascicles [68,69]. The technology used for cochlear implants is finding its way into applications for standing and walking restoration [70].

However, there are some unanswered questions which limit the effectiveness of FES systems that deal with muscle fatigue, reduced joint torques generated through FES in comparison to CNS activated torques in healthy subjects, modified reflex activities, spasticity, joint contractures, osteoporosis, and stress fractures. From the engineering point of view, the further development of FES systems has to address the following issues: the interface between an FES system and neuromuscular structures in the organism, biocompatibility of the FES system, and overall practicality. The least resolved questions in FES systems deal with control. Open-loop control, often called reference trajectory control, is not efficient because of the disturbances and model errors. Closed-loop control requires that we know in advance the desired trajectory and use multiple sensors within the system in which the actuators have a delay in the same time range as duration of some locomotion states. In principle, closed-loop systems offer substantial increases in input–output linearity and repeatability, along with a substantial decrease in the sensitivity of the system to parameter variations (internal disturbances) and load changes (external disturbances). Digital closed-loop methods using proportional (P) and proportional plus integral (PI) controllers were studied for both recruitment and fatigue of muscles [71].

A non-conventional method based on symbolic representation of motion in animals and humans, called artificial reflex control, was proposed [72-76]. Different solutions are discussed in great detail in Chapter 165 of this book. Many recent studies for improving control [77-82] aim to prolong standing with FES systems without arm supports. This is a very difficult task, considering that the human body acts as an inverted chain, and that FES systems generate non-physiological, fast fatiguing muscle contractions.

The research results indicate that the usual classification of CNS lesions is applicable as a guideline, but is not convenient as a selection criterion for FES candidates. FES users and an appropriate FES system must be selected through a functional diagnosis. The term "functional diagnosis" means that the functional status after the injury determines the type of treatment. General statements about FES systems indicate that it is suitable for subjects with preserved peripheral neuromuscular structures, moderate spasticity, without joint contractures and limited osteoporosis. Subjects should be able to control their balance and upper body posture using the upper extremities for assistance (parallel bars, walker, crutches, etc.). Subjects with pathologies that affect the heart, lungs, or circulation should be treated with extreme care, and often they should not be included in an FES walking program. A satisfactory mental and emotional condition is extremely important because an FES treatment requires a certain degree of intelligence and understanding of the technical side of the system. High motivation and good cooperation with medical staff is a significant aspect for the efficacy of FES. Surface stimulation may be applied only in subjects with limited sensation because it activates pain receptors. Subjects suitable for FES can include those with head and spinal cord injuries, cerebral paralysis, multiple sclerosis, and different types of myelitis and others.

The final current drawback to FES ambulation is the excessive energy cost while walking. FES walking should be described as a sequence of static states; long stance phase (several seconds) followed by brisk flexion movement and short step, followed by the same movement at the contralateral side. Dynamic walking is a necessary development which will introduce better use of inertial and gravity forces, and reduce energy rate and cost, increase the walking distance, and increase the speed of progression in parallel with decreased fatigue of muscles because the muscles will not be stimulated for long periods.

Table 142.1 summarizes differences in walking performances between able-bodied persons and persons with paralysis (energy cost, energy rate, amount of the use of upper extremities, cardiovascular stress measured through pulse rate, and blood pressure).

Hybrid Assistive Systems (HAS)

A specific approach to integrating two assistive systems (FES and an external mechanical orthosis) has been proposed (Fig. 142.2). These systems are called hybrid assistive systems (HAS) or hybrid orthotic systems (HOS) [83,84]. A few possibilities for HAS design have been suggested, which combine relatively simple rigid mechanical structures for passive stabilization of lower limbs during stance phase and FES systems. These systems combine the use of a reciprocating gait orthosis with multichannel stimulation [53,84], the use of an ankle–foot orthosis or an extended ankle–foot orthosis with a knee cage [85], or the use of a self-fitting modular orthosis [86]. A few more sophisticated laboratory systems were demonstrated [85,87-89]. Each trend in the design of HAS implies different applications as well as specific hardware and control problems. On the basis of accumulated experience, the following features can serve as criteria for a closer description of various HAS designs [88]: (1) partial mechanical support, (2) parallel operation of the biological and mechanical system, and (3) sequential operation of the biological and

TABLE 142.1 Best Performance with the Usage of an Assistive System Compared to Performance of Able-Bodied Subjects

	Oxygen Rate ml/kg/min	Oxygen Cost ml/kg/m	Speed vMAX m/min	Heart Rate b/min	Blood Pressure mmHg	Use of Upper Extremities [%]
Paralyzed subjects	13.6	0.6	60	118 (84 rest)	119/85	28
Able-bodied subjects	12.1	0.15	85	95 (81 rest)	120/80	0
Control method	Voluntary, hand-switch	Voluntary, sensory driven	Automatic, preprogrammed	Voluntary, hand-switch or sensory driven	Voluntary, hand-switch or sensory driven	Voluntary, hand-switch or sensory driven
Assistive system used	FES+SFMO FES+RGO	FES+SFMO	FES implant	FES+RGO FES+SFMO	FES+RGO	FES+RGO

The data is collected from literature [53, 83, 86, 88; 115-119]. Abbreviations used in the table are: RGO—reciprocating gait orthosis; SFMO—self-fitting modular orthosis; FES— functional electrical stimulation. Data for able-bodied subjects are from Waters et al. [1989].

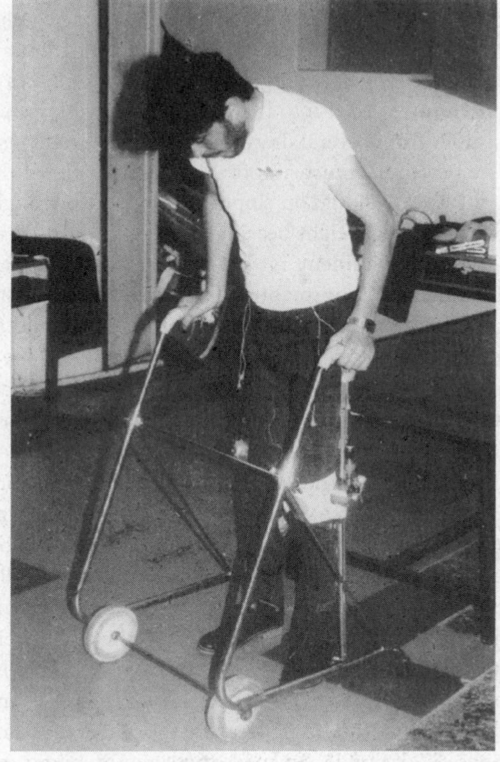

FIGURE 142.2 An incomplete tetraplegic subject walking with a hybrid assistive system. The particular system incorporates a six-channel surface stimulation system and a powered self-fitting modular orthosis [88].

the mechanical system. The partial mechanical support refers to the use of braces to assist FES only at specific events within a walking cycle [85]. The advanced version of powered orthoses to be used with FES is being developed by Durfee and colleagues [89]. Control of joints in mechanical orthosis is again becoming a target for research and development mainly because of new technological tools [90,91].

142.2 Active Prostheses

The role of active prosthesis is to extend the function provided by a "non-externally" powered and controlled artificial organ; hence, to improve the overall performance of motor function, ultimately providing a better quality of life.

Active Above-Knee Prostheses

Effective restoration of walking and standing of handicapped humans is an important element to improve the quality of life. Artificial legs of different kinds have been in use for a long time, but in many cases they are inadequate for the needs of amputees, specifically for high above-knee amputations (e.g., hip disarticulation), bilateral amputees, and patients who have demanding biomechanical requirements (e.g., subjects involved in sports).

Modern technology has led to improved designs of below-knee prostheses (BKP) in recent years. Below-knee amputees perform many normal locomotor activities and participate in many sports requiring running, jumping, and other jerky movements [92]. The biggest progress was made using readily available and easy-to-work-with plastic and graphite alloys for building the artificial skeletal portion of the shank and foot [93]. Below-knee prostheses are light, easy to assemble, and reliable. They provide good support and excellent energy absorption, which prevents impact, jerks, and allows the push-off phase in the gait cycle. Existing below-knee prostheses, even those without ankle joints, duplicate the behavior of the normal foot–ankle complex during swing and stance phases of the step cycle.

The same technology has been introduced into the design of above-knee prostheses (AKP). However, commercially available AKPs suffer from several drawbacks. The requirements for an AKP were stated by Wagner and Catranis [94]. The prosthesis must support the body weight of the amputee like the normal limb during the stance phase of level walking, on slopes and on soft or rough terrain. This implies that the prosthesis provides "stability" during weight bearing activity, i.e., it prevents sudden or uncontrolled flexion of the knee. The second requirement is that the body is supported such that undesirable socket/stump interface pressures and gait abnormalities due to painful socket/stump contact are prevented. The analysis of biomechanical factors that influence the shaping, fitting, and alignment of the socket is a problem in itself. If the fitting has been accomplished, allowing the amputee to manipulate and control the prosthesis in an active and comfortable manner, the socket and stump can be treated as one single body. The third requirement, which is somewhat controversial, is that the prosthesis should duplicate as nearly as possible the kinematics and dynamics of normal gait. The amputee should walk with a normal-looking gait over a useful range of speeds associated with typical activities for normal persons of similar age. The latter requirement has received attention in recent years and fully integrated systems, so called "self-contained" active AKPs are being incorporated into modern rehabilitation. The self-contained principle implies that the artificial leg contains the energy source, actuator, controller, and sensors (Fig. 142.3).

Devices such as the polycentric knee mechanism [95], the polycentric knee mechanism with a hydraulic valve, or the AKP with friction type brake [96] satisfy some of the above performance requirements. A logically controlled AKP with a hydraulic valve represents a further bridge between purely passive and fully controllable assistive devices [97,98]. Studies of the knee joint performance in the stance phase have been done [99-101]. From those studies it became clear that for several gait modes (ramp, stairs), active knee control in the swing phase is desirable. To meet these requirements, self-contained, multi-mode AKPs have been introduced [102-105]. The final prototype version of an active self-contained AKP allowing controlled flexion and extension throughout the gait cycle was developed by James [97] and is a product of Otto Bock Company, Germany. This leg uses an efficient knee controller with a custom-built hydraulic valve and a single chip microcontroller with a rule-based control scheme (see Chapter 165: Control of Movements).

The advantages of the externally powered and controlled leg (see, e.g., [104]) are an increased speed of locomotion, better gait symmetry, and lower energy cost and rate. These accomplishments are due to the fact that the powered leg allows controlled knee bounce after heel contact, limited push-off at the end of the stance phase, and effective flexion and controlled knee extension during the swing phase of gait. The external control allows the amputee to walk almost without any circumduction, which helps gait symmetry considerably. The knee joint of a standard endoskeletal prosthesis is fitted with an actuator having two independent braking units (friction drum type brake with an azbestum lining to allow control of the knee joint stiffness), and the extension-flexion driver with a ball-screw mechanism. The rechargeable battery power supply is designed for up to three hours of continuous level walking. The Motorola 68HC11-based microcontroller has been fitted into the interior of the prosthesis fulfilling self-containing principles. A hierarchical controller allows for intention recognition (volitional actions of the subject), adaptation to environmental changes, cyclical triggering throughout locomotion, and minimal jerk actuator control.

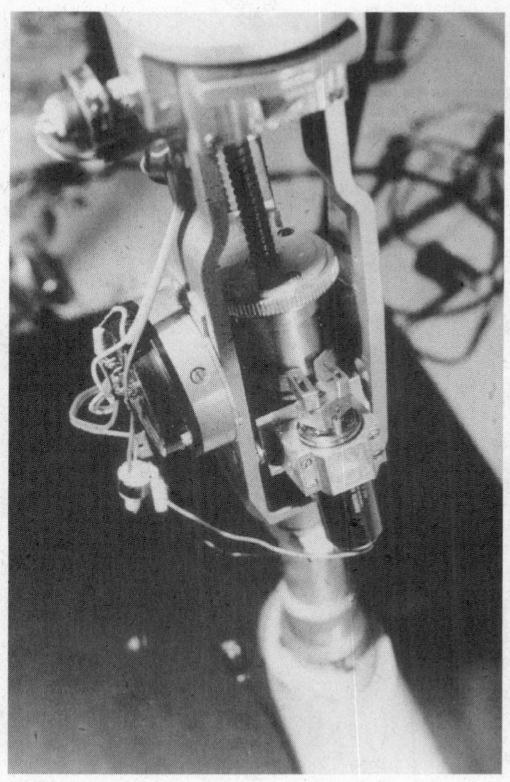

FIGURE 142.3 A self-contained, battery powered, microcomputer controlled above-knee prosthesis using artificial reflex control [104].

Myoelectric Hand and Arm Prostheses

The popularity of this control strategy comes from the fact that it is "very elegant" because the control signals going to the prosthesis may come from the myoelectric activity of the muscle which formerly controlled the lost limb [23,24,106,107]. Therefore, in some sense, the brain's own signals are used to control the motion of the prosthesis [109]. Although the validity of myoelectric control-based techniques seems reasonable, in practice the performances are somewhat disappointing. There are a number of reasons for this. In the cases of high level amputation, the electrical activity of the remaining muscles have only minor correlation with the normal arm movements and this makes coordinated control of several functions very difficult. A more important problem is that any EMG controlled prosthesis is inherently open-loop (preprogrammed) [110]. The absence of position proprioception and other related problems of reliability in executing movements [111] has contributed to the poor success rate of myoelectrically controlled prostheses.

Presently two systems based on myoelectric control are used with success (Fig. 142.4). The Otto Bock myoelectric hand [120] uses the technique of detection of the fixed threshold of the integrated rectified surface electromyography. Extension/flexion of the targeted muscle group serves as the on/off control, which causes opening and closing of the hand, respectively. Two sets of electrodes are used to detect this differential signal. The Utah arm [107,108] utilizes the myoelectric signals from two antagonists, as the proportional elbow control (Fig. 142.5). These signals are full wave rectified and differentiated to give the command signal. The Utah arm has a wrist rotator and it can be combined with different grasping systems. Most of the artificial hands used in everyday activities do not include the five finger dexterity, except in the devices that are mechanically preprogrammed. A new two-degree of freedom hand prosthesis with hiearachical grip was designed by Kyberd [112]. A new appealing version of a controller was designed by Kurtz [113].

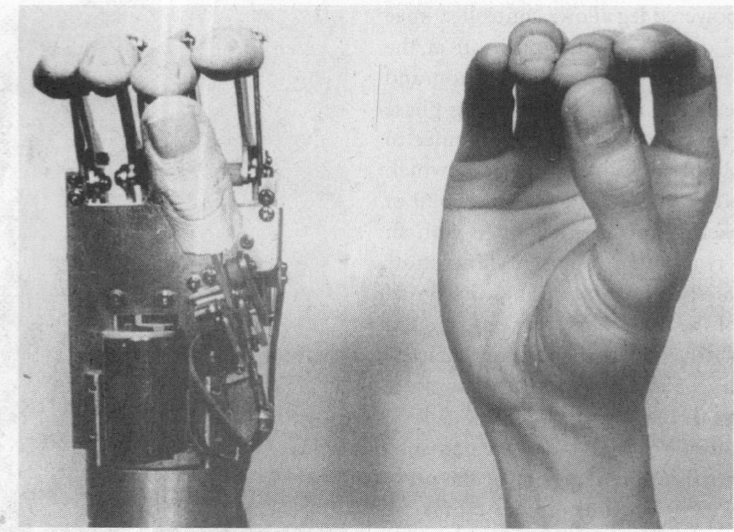

FIGURE 142.4 The Waseda University prototype of a myoelectric artificial hand.

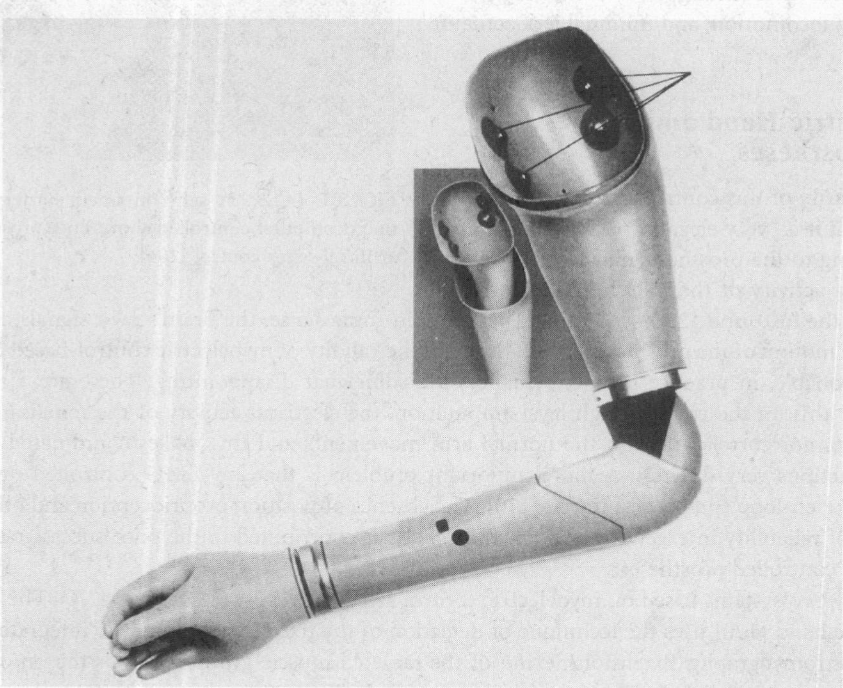

FIGURE 142.5 The Utah arm [107].

An interesting approach to overcoming the problems encountered with EMG control has been proposed by Simpson [110]. The control strategy is termed "extended physiological proprioception (EPP)", and uses the positions of intact joints as the controlling input signals to a prosthesis. This strategy is based on recognition of the fact that intact joints possess inherent position feedback, and that most goal oriented tasks are highly synergistic [32,33]. By establishing a one-to-one relationship between the position of the intact joint and the position in the space of the terminal device (prosthesis), the natural feedback of the intact joint can be "extended" to the prosthesis, and thereby provide it with proprioceptive information. Based on this idea, an externally controlled elbow–wrist–hand prosthesis was evaluated [110]. The shoulder joint position is measured with a potentiometer, mounted between the upper arm and trunk. An important feature of the system is that disabled humans can learn how to control the system within minutes using trial-and-error procedure. Bertos [114] presented a very elegant solution for position control based on EPP.

Defining Terms

Functional electrical stimulation or **Functional neuromuscular stimulation:** Patterned electrical stimulation of neuromuscular structures dedicated to restore motor functions.

Externally controlled assistive system: Assistive system for restoration of motor functions with automatic control.

Externally powered assistive system: Assistive system for restoration of motor functions which uses external power to either stimulate muscles or drive actuators.

Hybrid assistive systems: Combination of a functional electrical stimulation and a mechanical orthosis.

Myoelectric (EMG) control: Use of voluntary generated myoelectric activity as control signals for an externally controlled and powered assistive system.

Artificial reflex control: A sensory-driven control algorithm based on knowledge representation (production rule based system).

Reciprocating gait orthosis: A walking and standing assistive system with a reciprocating mechanism for hip joints which extends the contralateral hip when the ipsilateral hip is flexed.

Self-fitting modular orthosis: A modular, self-fitting, mechanical orthosis with a soft interface between the human body and the orthosis.

References

1. Peckham, P. H., and Keith, M. W., Motor prostheses for restoration of upper extremity function. In *Neural Prostheses: Replacing Motor Function After Disease or Disability,* In Stein, R. B., Peckham, P. H., and Popović, D., Oxford University Press, New York, 1992, 162-190.
2. Buckett, J. R., Peckham, H. P., et al., A flexible, portable system for neuro-muscular stimulation in the paralyzed upper extremities, *IEEE Trans. Biomed. Eng.*, BME-35:897-904, 1988.
3. Handa, Y., Handa, T., et al., Functional electrical stimulation (FES) systems for restoration of motor function of paralyzed muscles-versatile systems and a portable system, *Front. Med. Biol. Eng.*, 4:241-255, 1992.
4. Nathan, R.H. and Ohry, A., Upper limb functions regained in quadriplegia: A hybrid computerized neuromuscular stimulation system, *Arch. Phys. Med. Rehabil.*, 71:415-421, 1990.
5. Hoshimiya, N., Naito, N., Yajima, M., and Handa, Y., A multichannel FES system for the restoration of motor functions in high spinal cord injury patients: A respiration-controlled system for multi-joint upper extremity, *IEEE Trans. Biomed. Eng.*, BME-36:754-760 1989.
6. Popović, D., Popović, M., et al., Clinical evaluation of the Belgrade grasping system, *Proc. V Vienna Intern. Workshop on Functional Electrical Stimulation*, Vienna, 1998.
7. Prochazka, A., Gauthier, M., Wieler, M., and Kenwell, Z., The bionic glove: an electrical stimulator garment that provides controlled grasp and hand opening in quadriplegia, *Arch Phys. Med. Rehabil.*, 78:608-614, 1997.
8. Reberšek, S. and Vodovnik, L., Proportionally controlled functional electrical stimulation of hand, *Arch. Phys. Med. Rehabil.*, 54:378-382, 1973.
9. Nathan, R., Handmaster NMS—present technology and the next generation, In Popović, D., (Ed.) *Proc. II Intern. Symp. on FES*, Burnaby, pp. 139-140, 1997.
10. Vodovnik, L., Bajd, T., et al., Functional electrical stimulation for control of locomotor systems, *CRC Crit. Rev. Bioeng.*, 6:63-131, 1981.

11. Long, C. II and Masciarelli, C. V., An electrophysiologic splint for the hand, *Arch. Phys. Med. Rehabil.*, 44:499-503, 1963.
12. Rudel, D., Bajd, T., Reberšek, S., and Vodovnik, L., FES assisted manipulation in quadriplegic patients, In Popović, D. (Ed.), *Advances in External Control of Human Extremities VIII*, pp. 273-282, ETAN, Belgrade, 1984.
13. Nathan, R.H., An FNS-based system for generating upper limb function in the C4 quadriplegic, *Med. Biol. Eng. Comp.*, 27:549-556, 1989.
14. Hines, A. E., Crago, P. E., and Billian, C., Hand opening by electrical stimulation in patients with spastic hemiplegia, *IEEE Trans. Rehab. Eng.*, TRE-3:193-205, 1995.
15. Holländer, H. J., Huber, M., and Vossius, G., An EMG controlled multichannel stimulator, In Popović, D. (Ed.), *Advances in External Control of Human Extremities IX*, pp. 291-295, ETAN, Belgrade, 1987.
16. Sennels, S., Biering-Soerensen, F., Anderson, O. T., and Hansen, S. D., Functional neuromuscular stimulation control by surface electromyographic signals produced by volitional activation of the same muscle: adaptive removal of the muscle response from the recorded EMG-signal, *IEEE Trans. Rehab. Eng.*, TRE-5:195:206, 1997.
17. Lan, N., Feng, H. Q., and Crago, P. E., Neural network generation of muscle stimulation patterns for control of arm movements, *IEEE Trans. Rehab. Eng.*, TRE-4:213-224, 1994.
18. Hart, R. L., Kilgore, K. L., and Peckham, P. H., A comparison between control methods for implanted FES hand-grasp systems, *IEEE Trans. Rehab. Eng.*, TRE-6:208-218, 1998.
19. Smith, B. T., Tang, Z., Johnson, M. W., et al., An externally powered multichannel, implantable stimulator-telemeter for control of paralyzed muscles, *IEEE Trans. Biomed. Eng.*, BME-45:463-475, 1998.
20. Peckham, P. H., Mortimer, J. T., and Marsolais, E. B., Controlled prehension and release in the C5 quadriplegic elicited by functional electrical stimulation of the paralyzed forearm muscles, *Ann. Biomed. Eng.*, 8:369-388, 1980.
21. Wijman, A. C., Stroh, K. C., Van Doren, C. L., Thrope, C. L., Peckham C. H., and Keith, M. W., Functional Evaluation of Quadriplegic Patients Using a Hand Neuroprosthesis, *Arch. Phys. Med. Rehabil.*, 71:1053-1057, 1990.
22. Popović, D., Stojanović, A., et al., Clinical evaluation of the bionic glove, *Arch. Phys. Med. Rehabil.*, 80, 1999.
23. Graupe, D. and Kline, W.K., Functional separation of EMG signals via ARMA identification methods for prosthesis control purposes, *IEEE Trans. System, Man Cybern.*, SMC-5:252-259, 1975.
24. Graupe, D., EMG pattern analysis for patient-responsive control of FES in paraplegics for walker-supported walking, *IEEE Trans. Biomed. Eng.*, BME-36:711-719, 1989.
25. Saxena, S., Nikolic, S., and Popović, D., An EMG controlled FES system for grasping in tetraplegics, *J. Rehabil. Res. Develop.*, 32:17-23, 1995.
26. Keller, T., Curt, A., Dietz, V., and Morari, M., EMG controlled grasping neuroprosthesis for high lesioned tetraplegic patients, In Popović, D., (Ed.) *Proc. II Intern. Symp. on FES*, Burnaby, pp. 129-130, 1997.
27. Scott, T. R. D., Peckham, P. H., and Kilgore, K. L., Tri-state myoelectric control of bilateral upper extremity neuroprosthesis for tetraplegic individuals, *IEEE Trans. Rehab. Eng.*, TRE-4:251-263, 1996.
28. Grill, J. H. and Peckham, P. H., Functional neuromuscular stimulation for combined control of elbow extension and hand grasp in C5 and C6 quadriplegics, *IEEE Trans. Rehab. Eng.*, TRE-6:190-199, 1998.
29. Crago, P. E., Memberg, W. D., Usey, M. K. et al., An elbow extension neuroprosthesis for individuals with tetraplegia, *IEEE Trans. Rehab. Eng.*, TRE-6:1-6, 1998.
30. Lemay, M. A. and Crago, P. E., Closed loop stabilization in C4 and C4 tetraplegia, *IEEE Trans. Rehab. Eng.*, TRE-5:244-252, 1997.
31. Smith, B. T., Mulcahey, M. J., and Betz, R. R., Development of an upper extremity FES system for individuals with C4 tetraplegia, *IEEE Trans. Rehab. Eng.*, TRE-4:264-270, 1996.

32. Popović, M. and Popović, D., A new approach to reaching control for tetraplegic subjects, *J. Electromyog. Kinesiol.*, 4:242-253, 1994.
33. Popović, D. and Popović, M., Tuning of a nonanalytic hierarchical control system for reaching with FES, *IEEE Trans. Biomed. Eng.*, BME-45:203-212, 1998.
34. Haugland, M. and Sinkjaer, T., Cutaneous whole nerve recordings used for correction of foot-drop in hemiplegic man, *IEEE Trans. Rehab. Eng.*, TRE-3:307-317, 1995.
35. Haugland, M. K. and Hoffer, J. A., Slip information provided by nerve cuff signals: application in closed-loop control of functional electrical stimulation, *IEEE Trans. Rehab. Eng.*, TRE-2:29-37, 1994.
36. Haugland, M. K., Hoffer, J. A., and Sinkjaer, T., Skin contact force information in sensory nerve signals recorded by implanted cuff electrodes, *IEEE Trans. Rehab. Eng.*, TRE-2:18-27, 1994.
37. Sahin, M. and Durand, D. M., Improved nerve cuff electrode recordings with subthreshold anode currents, *IEEE Trans. Biomed. Eng.*, BME-45:1044-1050, 1998.
38. Upshaw, B. and Sinkjaer, T., Digital signal processing algorithms for the detection of afferent nerve recorded from cuff electrodes, *IEEE Trans. Rehab. Eng.*, TRE-6:172-181, 1998.
39. Bajd, T., Kralj, A., and Turk, R., Standing-up of a healthy subject and a paraplegic patients, *J. Biomech.*, 15:1-10, 1982.
40. Gračanin, F., Prevec, T., and Trontelj, J., Evaluation of use of functional electronic peroneal brace in hemiparetic patients, in *Advances in External Control of Human Extremities III*, ETAN, Belgrade, 1967, 198-210.
41. Kralj, A., Bajd, T., and Turk, R., Electrical stimulation providing functional use of paraplegic patient muscles, *Med. Prog. Technol.*, 7:3-9, 1980.
42. Kralj, A., Bajd, T., Turk, R., and Benko, H., Results of FES application to 71 SCI patients, in *Proc. RESNA 10th Ann. Conf. Rehabil. Tech.*, San Jose, 1987, 645- 647.
43. Vodovnik, L. and McLeod, W. D., Electronic detours of broken nerve paths, *Med. Electr.*, 20:110-116, 1965.
44. Vodovnik, L., Crochetiere, W. J., and Reswick, J. B., Control of a skeletal joint by electrical stimulation of antagonists, *Med. Biol. Eng.*, 5:97-109, 1967.
45. Andrews, B. J., Baxendale, R. H., et al., Hybrid FES orthosis incorporating closed loop control and sensory feedback, *J. Biomed. Eng.*, 10:189-195, 1988.
46. Brindley, G. S., Polkey, C. E., and Rushton, D. N., Electrical splinting of the knee in paraplegia, *Paraplegia*, 16:428-435, 1978.
47. Hermens, H. J., Mulder, A. K., et al., Research on electrical stimulation with surface electrodes, in *Proc. 2nd Vienna Intern. Workshop on Functional Electrostimulation*, Vienna, 1986, 321-324.
48. Jaeger, R., Yarkony, G. Y., and Smith, R., Standing the spinal cord injured patient by electrical stimulation: Refinement of a protocol for clinical use, *IEEE Trans. Biomed. Eng.*, BME-36:720-728, 1989.
49. Kralj, A. and Bajd, T., *Functional Electrical Stimulation, Standing and Walking After Spinal Cord Injury* CRC Press, Boca Raton, FL, 1989.
50. Marsolais, E. B. and Kobetic, R., Functional walking in paralyzed patients by means of electrical stimulation, *Clin. Orthop.*, 175:30-36, 1983.
51. Mizrahi, J., Braun, Z., Najenson, T., and Graupe, D., Quantitative weight bearing and gait evaluation of paraplegics using functional electrical stimulation, *Med. Biol. Eng. Comput.*, 23:101-107, 1985.
52. Petrofsky, J. S. and Phillips, C. A., Computer controlled walking in the paralyzed individual, *J. Neurol. Orthop. Surg.*, 4:153-164, 1983.
53. Solomonow, M., Biomechanics and physiology of a practical powered walking orthosis for paraplegics, in *Neural Prostheses: Replacing Motor Function After Disease or Disability*, (Eds.) Stein, R. B., Peckham, H. P. and Popović, D., (Eds.), Oxford University Press, New York, 1992, 202-230.
54. Stein, R. B., Prochazka, A., et al., Technology transfer and development for walking using functional electrical stimulation, in Popović, D. (Ed.), *Advances in External Control of Human Extremities X*, Nauka, Belgrade, 1990, 161-176.

55. Thoma, H., Frey, M., et al., Functional neurostimulation to substitute locomotion in paraplegia patients, in Andrade, D. et al. (Eds.), *Artificial Organs*, VCH Publishers, 1987, 515-529.
56. Vossius, G., Mueschen, U., and Hollander, H. J., Multichannel stimulation of the lower extremities with surface electrodes, in *Advances in External Control of Human Extremities IX*, Popović, D. (Ed.), ETAN, Belgrade, 1987, 193-203.
57. Waters, R. L., McNeal, D. R., Fallon W., and Clifford, B., Functional electrical stimulation of the peroneal nerve for hemiplegia, *J. Bone Joint Surg*, 67:792-793, 1985.
58. Wieler, M., Stein, R.B., et al., A Canadian multicentre trial using functional electrical stimulation for assisting gait, in Popović, D., (Ed.), *Proc. II Intern. Symp. on FES*, Burnaby, 1997, 160-161.
59. Aćimović, R., Stanič, U., et al., Routine clinical use of functional electrical stimulation in hemiplegia, in Popović, D. (Ed.), *Proc. II Intern. Symp. on FES*, Burnaby, 1997, 182-183.
60. Graupe, D. and Parastep I, Sigmedics, IL—the therapy for paraplegics or an orthosis for walking, personal communication, 1997.
61. Marsolais, E. B. and Kobetic, R., Implantation techniques and experience with percutaneous intramuscular electrode in the lower extremities, *J. Rehabil. Res*, 23:1-8, 1987.
62. Kobetic, R. and Marsolais, E. B., Synthesis of paraplegic gait with multichannel functional electrical stimulation, *IEEE Trans. Rehab. Eng.*, TRE-2:66-79, 1994.
63. Abbas, J. J. and Triolo, R. J., Experimental evaluation of an adaptive feedforward controller for use in functional neuromuscular stimulation systems, *IEEE Trans. Rehab. Eng.*, TRE-5:12-22, 1997.
64. Cameron, T., Loeb, G. E., Peck, R., et al., Micromodular implants to provide electrical stimulation of paralyzed muscles and limbs, *IEEE Trans. Biomed. Eng.*, BME-44:781-790, 1997.
65. Cameron, T., Liinama, T., Loeb, G. E., and Richmond, F. J. R., Long term biocompatibility of a miniature stimulator implanted in feline hind limb muscles, *IEEE Trans. Biomed. Eng.*, BME-45:1024-1035, 1998.
66. Cameron, T., Richmond, F. J. R., and Loeb, G. E., Effects of regional stimulation using a miniature stimulator implanted in feline posterior bicpes femoris, *IEEE Trans. Biomed. Eng.*, BME-45:1036-1045, 1998.
67. Strojnik, P., Whitmoyer, D., and Schulman, J., An implantable stimulator for all season, in Popović, D. (Ed.), *Advances in External Control of Human Extremities X*, Nauka, Belgrade, 1990, 335-344.
68. Ziaie, B., Nardin, M. D., Coghlan, A. R., and Najafi. K., A single channel implantable microstimulator for functional neuromuscular stimulation, *IEEE Trans. Biomed. Eng.*, BME-44:909-920, 1997.
69. Haugland, M. K., A miniature implantable nerve stimulator, in Popović, D. (Ed.), *Proc. II Intern. Symp. on FES*, Burnaby, 1997, 221-222.
70. Houdayer, T., Davis, R., et al., Prolonged closed-loop standing in paraplegia with implanted cochlear FES-22 stimulator and Andrews ankle-foot orthosis, in Popović, D. (Ed.), *Proc. II Intern. Symp. on FES*, Burnaby, 1997, 168-169.
71. Chizeck, H. J., Adaptive and nonlinear control methods for neuroprostheses, in Stein, R. B., Peckham, H. P., and Popović, D., (Eds.) *Neural Prostheses: Replacing Motor Function After Disease or Disability*, Oxford University Press, New York, 1992, 298-328.
72. Andrews, B. J., Baxendale, R. M., et al., A hybrid orthosis for paraplegics incorporating feedback control, in *Advances in External Control of Human Extremities IX*, ETAN, Belgrade, 1987, 297-310.
73. Tomović, R., Control of assistive systems by external reflex arcs, in Popović, D. (Ed.), *Advances in External Control of Human Extremities IX*, ETAN, Belgrade, 1984, 7-21.
74. Tomović, R., Popović, D., and Tepavac, D., Adaptive reflex control of assistive systems, in Popović, D. (Ed.), *Advances in External Control of Human Extremities IX*, ETAN, Belgrade, 1987, 207-214.
75. Veltink, P. H., Koopman, A. F. M., and Mulder, A. J., Control of cyclical lower leg movements generated by FES, in *Advances in External Control of Human Extremities X*, Popović, D. (Ed.), Nauka, Belgrade, 1990, 81-90.
76. Tomović, R., Popović, D., and Stein, R. B., *Nonanalytical Methods for Motor Control*, World Scientific Publishing, Singapore, 1995.

77. Riener, R. and Fuhr, T., Patient-driven control of FES supported standing up: A simulation study, *IEEE Trans. Rehab. Eng.*, TRE-6:113-124, 1998.
78. Veltink, P. H. and Donaldson, N., A perspective on the controlled FES-supported standing, *IEEE Trans. Rehab. Eng.*, TRE-6:109-112, 1998.
79. Matjačić, Z. and Bajd, T., Arm-free paraplegic standing — part I: Control model synthesis and simulation, part II: Experimental results, *IEEE Trans. Rehab. Eng.*, TRE-6: 125-150, 1998.
80. Davoodi, R. and Andrews, B. J., Computer simulation of FES standing up in paraplegia: A self-adaptive fuzzy controller with reinforcement learning, *IEEE Trans. Rehab. Eng.*, TRE-6:151-161, 1998.
81. Donaldson, N. and Yu, C., A strategy used by paraplegics to stand up using FES. *IEEE Trans. Rehab. Eng.*, TRE-6:162-166, 1998.
82. Hunt, K. J., Munih, M., and Donaldson, N., 1997. Feedback control of unsupported standing in paraplegia: Part I: Optimal control approach; Part II: Experimental results, *IEEE Trans. Rehab. Eng.*, TRE-5:331-352, 1997.
83. Popović, D., Tomović, R., and Schwirtlich, L., Hybrid assistive system—Neuroprosthesis for motion, *IEEE Trans. Biomed. Eng.*, BME-37:729-738, 1989.
84. Solomonow, M., Baratta, R., et al., Evaluation of 70 paraplegics fitted with the LSU RGO/FES, in Popović, D. (Ed.), *Proc. II Intern. Symp. on FES*, Burnaby, 1997, 159.
85. Andrews, B. J., Barnett, R. W., et al., Rule-based control of a hybrid FES orthosis for assisting paraplegic locomotion, *Automedica*, 11:175-199, 1989.
86. Schwirtlich, L. and Popović, D., Hybrid orthoses for deficient locomotion, in Popović, D. (Ed.), *Advances in External Control of Human Extremities VIII*, ETAN, Belgrade, 1984, 23-32.
87. Phillips, C. A., An interactive system of electronic stimulators and gait orthosis for walking in the spinal cord injured, *Automedica*, 11:247-261, 1989.
88. Popović, D., Schwirtlich, L., and Radosavljević, S., Powered hybrid assistive system, in Popović, D. (Ed.), *Advances in External Control of Human Extremities X*, Nauka, Belgrade, 1990, 191-200.
89. Goldfarb, M. and Durfee, W. K., Design of a controlled-brake orthosis for FES-aided gait, *IEEE Trans. Rehab. Eng.*, TRE-4:13-24, 1996.
90. Irby, S. E., Kaufman, K. R., and Sutherland, D.H, A digital logic controlled electromechanical long leg brace, *Proc. 15 Southern Biomed. Eng. Conf*, Dayton, OH, 1996, 28.
91. Kaufman, K. R., Irby, S. E., Mathewson, J. W., et al., Energy efficient knee-ankle-foot orthosis, *J. Prosthet. Orthot.*, 8:79-85, 1996.
92. Inman, V. T., Ralston, J. J., and Todd, F., *Human Walking*. Williams & Wilkins, Baltimore, MD, 1981.
93. Doane, N. E. and Holt, L. E., A comparison of the SACH foot and single axis foot in the gait of the unilateral below-knee amputee, *Prosth. Orthot. Intern*, 7:33-36, 1983.
94. Wagner, E. M. and Catranis, J. G., New developments in lower-extremity prostheses, in *Human Limbs and Their Substitutes*, Klopsteg, P. E., Wilson, P. D., et al. (Eds.), McGraw-Hill, New York, 1954, (reprinted 1968).
95. Cappozzo, A., Leo, T., and Cortesi, S., A polycentric knee-ankle mechanism for above-knee prostheses, *J. Biomech.*, 13:231-239, 1980.
96. Aoyama, F., Lapoc system leg, *Proc. Rehab. Eng. Seminar REIS '80*, Tokyo, Japan, 1980, 59-67.
97. James, K., Stein, R.B., Rolf, R., and Tepavac, D., Active suspension above-knee prosthesis, in Goh, D. and Nathan, A. (Eds.), *Proc VI Intern. Conf. Biomed. Eng.*, 1991, 317-320.
98. Bar, A., Ishai, P., Meretsky, P., and Koren, Y., Adaptive microcomputer control of an artificial knee in level walking, *J. Biomed. Eng*, 5:145-150, 1983.
99. Flowers, W. C. and Mann, R.W., An Electrohydraulic knee-torque controller for a prosthesis simulator, *J. Biomechan. Eng.*, 99:3-8, 1977.
100. Flowers, W. C., Rowell, D., Tanquary, A., and Cone, A., A microcomputer controlled knee mechanism for A/K prostheses, *Proc. 3rdCISM-IFToMM Int. Symp: Theory and Practice of Robots and Manipulators*, Udine, Italy, 1978, 28-42.

101. Stein, J. L. and Flowers, W. C., Above-knee prosthesis: A case study of the interdependency of effector and controller design, *ASME Winter Annual Meeting*, Chicago, IL, 1980, 275-277.
102. Koganezawa, E., Fujimoto, H., and Kato, I., Multifunctional above-knee prosthesis for stairs walking, *Pros. Orth. Intern.*, 11:139-145, 1987.
103. Kuzhekin, A. P., Jacobson, J. S., and Konovalov, V. V., Subsequent development of motorized above-knee prosthesis, in Popović, D. (Ed.), *Advances in External Control of Human Extremities VIII*, ETAN, Belgrade, 1984, 525-530.
104. Popović, D. and Schwirtlich, L., Belgrade active A/K prosthesis, in deVries, J. (Ed.), *Electrophysiological Kinesiology*, 804:337-343, Excerpta Medica, Amsterdam, Intern. Cong. Ser, 1988.
105. Tomović, R., Popović, D., Turajlic S., and McGhee, R.B., Bioengineering actuator with non-numerical control, *Proc. IFAC Conf. Orthotics & Prosthetics*, Columbus, OH, Pergamon Press, 1982, 145-151.
106. Lyman, J. H., Freedy, A., Prior, R., and Solomonow, M., Studies toward a practical computer-aided arm prosthesis system, *Bull. Prosth. Res.*, 213-224, 1974.
107. Jacobsen, S. C., Knutti, F. F., Johnson, R. T., and Sears, H. H., Development of the Utah artificial arm, *IEEE Trans. Biomed. Eng.*, BME-29:249-269, 1982.
108. Park, E. and Meek, S. G., 1995. Adaptive filtering of the electromyographic signal for prosthetic and force estimation, *IEEE Trans. Biomed. Eng.*, BME-42:1044-1052.
109. Sheridan, T. B. and Mann, R. W., Design of control devices for people with severe motor impairment, *Human Factors*, 20:321-3389, 1978.
110. Simpson, D. C., The choice of control system for the multimovement prosthesis: extended physiological proprioception (e.p.p.), in *The Control of Upper-Extremity Prostheses and Orthoses*, Herberts, P. (Ed.), C. Thomas. 1974, chap. 15.
111. Gibbons, D. T., O'Riain, M. D., and Philippe-Auguste, J. S., An above-elbow prosthesis employing programmed linkages, *IEEE Trans. Biomed. Eng.*, BME-34:251-258, 1987.
112. Kyberd, P. J., Holland, O. E., Chappel, P. H. et al., MARCUS: A two degree of freedom hand prosthesis with hierarchical grip control, *IEEE Trans. Rehab. Eng.*, TRE-3:70-76, 1995.
113. Kurtz, I., Programmable prosthetic controller, *Proc. MEC '97*, Frederciton, NB, 33, 1997.
114. Bertos, Y. A., Hechathorne, C. H., Weir, R. F., and Childress, D. S., Microprocessor-based EPP position controller for electric powered upper limb prostheses, *Proc. IEEE Intern. Conf. EMBS*, Chicago, IL, 1997.
115. Bowker, P., Messenger, N., Oglivie, C., and Rowley, D. I., Energetics of paraplegic walking, *J. Biomed. Eng.*, 14:344-350, 1992.
116. Nene, A. V. and Jennings, S. J., Physiological cost index of paraplegic locomotion using the ORLAU ParaWalker, *Paraplegia*, 30:246-252, 1992.
117. Petrofsky, J. S. and Smith, J. B., Physiologic cost of computer-controlled walking in persons with paraplegia using a reciprocating-gait orthosis, *Arch. Phys. Med. Rehabil.*, 72:890-896, 1991.
118. Waters, R. L. and Lunsford, B. R., Energy Cost of Paraplegic Locomotion, *J. Bone Joint Surg.*, 67-A:1245-1250, 1985.
119. Waters, R. L., Yakura, J. S., Adkins, R., and Barnes, G., Determinants of gait performance following spinal cord injury, *Arch. Phys. Med. Rehabil.*, 70:811-818, 1980.
120. Pike, personal communication.

Further Information

Stein, R. B., Peckham, H. P., and Popović, D., *Neural Prostheses: Replacing Motor Function After Disease or Disability*, Oxford University Press, New York, 1992.

Kralj, A. and Bajd, T., *Functional Electrical Simulation: Standing and Walking After Spinal Cord Injury*, CRC Press, Boca Raton, FL, 1989.

Agnew, W. V. and McCreery, D. B., *Neural Prostheses: Fundamental Studies*, Prentice-Hall, Englewood Cliffs, NJ, 1990.

Proceedings: Advances in External Control of Human Extremities I-IX, Yugoslav Committee for ETAN, Belgrade, Yugoslavia, 1962-1987 (9 books).

Popović, D. (Ed.), *Advances in External Control of Human Extremities X*, Nauka, Belgrade, 1990.

Tomović, R., Popović, D., and Stein, R. B., *Nonanalytical Methods for Motor Control*, World Scientific Publishers, Singapore, 1990.

143

Sensory Augmentation and Substitution

Kurt A. Kaczmarek
University of Wisconsin at Madison

This chapter will consider methods and devices used to present visual, auditory, and tactual (touch) information to persons with sensory deficits. *Sensory augmentation systems* such as eyeglasses and hearing aids enhance the existing capabilities of a functional human sensory system. *Sensory substitution* is the use of one human sense to receive information normally received by another sense. Braille and speech synthesizers are examples of systems that substitute touch and hearing, respectively, for information that is normally visual (printed or displayed text).

The following three sections will provide theory and examples for aiding the visual, auditory, and tactual systems. Because capitalizing on an *existing* sensory capability is usually superior to substitution, each section will consider first augmentation and then substitution, as shown below:

Human Sensory Systems

Visual	Auditory	Tactual
Visual augmentation	Auditory augmentation	Tactual augmentation
Tactual vision substitution	Visual auditory substitution	Tactual substitution
Auditory vision substitution	Tactual auditory substitution	

143.1 Visual System

With a large number of receptive channels, the human visual system processes information in a parallel fashion. A single glimpse acquires a wealth of information; the field of view for two eyes is 180 degrees horizontally and 120 degrees vertically [Mehr & Shindell, 1990]. The spatial resolution in the central (foveal) part of the visual field is approximately 0.5 to 1.0 minute of arc [Shlaer, 1937], although Vernier acuity, the specialized task of detecting a misalignment of two lines placed end to end, is much finer, approximately 2 seconds of arc [Stigmar, 1970]. Low-contrast presentations substantially reduce visual acuity.

0-8493-0462-8/00/$0.00+$.50

The former resolution figure is the basis for the standard method of testing visual acuity, the Snellen chart. Letters are considered to be "readable" if they subtend approximately 5 minutes of arc and have details one-fifth this size. Snellen's 1862 method of reporting visual performance is still used today. The ratio 20/40, for instance, indicates that a test was conducted at 20 ft and that the letters that were recognizable at that distance would subtend 5 minutes of arc of 40 ft (the distance at which a normally sighted, or "20/20," subject could read them). Although the standard testing distance is 20 ft, 10 ft and even 5 ft may be used, under certain conditions, for more severe visual impairments [Fonda, 1981].

Of the approximately 6 to 11.4 million people in the United States who have visual impairments, 90% have some useful vision [NIDRR, 1993]. In the United States, *severe visual impairment* is defined to be 20/70 vision in the better eye with best refractive correction (see below). *Legal blindness* means that the best corrected acuity is 20/200 or that the field of view is very narrow (<20 degrees). People over 65 years of age account for 46% of the legally blind and 68% of the severely visually impaired. For those with some useful vision, a number of useful techniques and devices for visual augmentation can allow performance of many everyday activities.

Visual Augmentation

People with certain eye disorders see better with higher- or lower-than-normal light levels; an illuminance from 100 to 4000 lux may promote comfortable reading [Fonda, 1981]. Ideal illumination is diffuse and directed from the side at a 45-degree angle to prevent glare. The surrounding room is preferably 20% to 50% darker than the object of interest.

Refractive errors cause difficulties in focusing on an object at a given distance from the eye [Mountcastle, 1980]. Myopia (near-sightedness), hyperopia (far-sightedness), astigmatism (focus depth that varies with radial orientation), and presbyopia (loss of ability to adjust focus, manifested as far-sightedness) are the most common vision defects. These normally can be corrected with appropriate eyeglasses or contact lenses and are rarely the cause of a disability.

Magnification is the most useful form of image processing for vision defects that do not respond to refractive correction. The simplest form of image magnification is getting closer; halving the distance to an object doubles its size. Magnifications up to 20 times are possible with minimal loss of field of view. At very close range, eyeglasses or a loupe may be required to maintain focus [Fonda, 1981]. Hand or stand magnifiers held 18 to 40 cm (not critical) from the eye create a virtual image that increases rapidly in size as the object-to-lens distance approaches the focal length of the lens. Lenses are rated in diopters ($D = 1/f$, where f is the focal length of the lens in centimeters). The useful range is approximately 4 to 20 D; more powerful lenses are generally held close to the eye as a loupe, as just mentioned, to enhance field of view. For distance viewing, magnification of 2 to 10 times can be achieved with hand-held telescopes at the expense of a reduced field of view.

Closed-circuit television (CCTV) systems magnify print and small objects up to 60 times, with higher effective magnifications possible by close viewing. Users with vision as poor as 1/400 (20/8000) may be able to read ordinary print with CCTV [Fonda, 1981]. Some recent units are portable and contain black/white image reversal and contrast enhancement features.

Electrical (or, more recently, magnetic) stimulation of the visual cortex produces perceived spots of light called *phosphenes*. Some attempts, summarized in Webster et al. [1985], have been made to map these sensations and display identifiable patterns, but the phosphenes often do not correspond spatially with the specific location on the visual cortex. Although the risk and cost of this technique do not yet justify the minimal "vision" obtained, future use cannot be ruled out.

Tactual Vision Substitution

With sufficient training, people without useful vision can acquire sufficient information via the tactile sense for many activities of daily living, such as walking independently and reading. The traditional long cane, for example, allows navigation by transmitting surface profile, roughness, and elasticity to the hand. Interestingly, these features are *perceived* to originate at the tip of the cane, not the hand where they are transduced; this is a simple example of *distal attribution* [Loomis, 1992]. Simple electronic aids such as the hand-held Mowat sonar sensor provide a tactile indication of range to the nearest object.

Braille reading material substitutes raised-dot patterns on 2.3-mm centers for visual letters, enabling reading rates up to 30 to 40 words per minute (wpm). Contracted Braille uses symbols for common words and affixes, enabling reading at up to 200 wpm (125 wpm is more typical).

More sophisticated instrumentation also capitalizes on the spatial capabilities of the tactile sense. The Optacon (*op*tical-to-*ta*ctile *con*verter) by TeleSensory, Inc. (Mountain View, Calif.) converts the outline of printed letters recorded by a small, hand-held camera to enlarged vibrotactile letter outlines on the user's fingerpad. The camera's field of view is divided into 100 or 144 pixels (depending on the model), and the reflected light intensity at each pixel determines whether a corresponding vibrating pin on the fingertip is active or not. Ordinary printed text can be read at 28 (typical) or 90 (exceptional) wpm.

Spatial orientation and recognition of objects beyond the reach of a hand or long cane are the objective of experimental systems that convert an image from a television-type camera to a matrix of electrotactile or vibrotactile stimulators on the abdomen, forehead, or fingertip. With training, the user can interpret the patterns of tingling or buzzing pints to identify simple, high-contrast objects in front of the camera, as well as experience visual phenomena such as looming, perspective, parallax, and distal attribution [Bach-y-Rita, 1972; Collins, 1985].

Access to graphic or spatial information that cannot be converted into text is virtually impossible for blind computer users. Several prototype devices have been built to display computer graphics to the fingers via vibrating or stationary pins. A fingertip-scanned display tablet with embedded electrodes, under development in our laboratory [Kaczmarek et al., 1997], eliminates all moving parts; ongoing tests will determine if the spatial performance and reliability are adequate.

Auditory Vision Substitution

Electronic speech synthesizers allow access to electronic forms of text storage and manipulation. Until the arrival of graphic user interfaces such as those in the Apple Macintosh® and Microsoft Windows® computer operating systems, information displayed on computer screens was largely text-based. A number of products appeared that converted the screen information to speech at rates of up to 500 wpm, thereby giving blind computer users rapid access to the information revolution. Fortunately, much of the displayed information in graphic operating systems is not essentially pictorial; the dozen or so common graphic features (e.g., icons, scroll bars, buttons) can be converted to a limited set of words, which can then be spoken. Because of the way information is stored in these systems, however, the screen-to-text conversion process is much more complex, and the use of essentially spatial control features such as the mouse await true spatial display methods [Boyd et al., 1990].

Automated optical character recognition (OCR) combined with speech synthesis grants access to the most common printed materials (letters, office memorandums, bills), which are seldom available in Braille or narrated-tape format. First popularized in the Kurzweil reading machine, this marriage of technologies is combined with a complex set of lexical, phonetic, and syntactic rules to produce understandable speech from a wide variety of, but not all, print styles.

Mobility of blind individuals is complicated, especially in unfamiliar territory, by hazards that cannot be easily sensed with a long cane, such as overhanging tree limbs. A few devices have appeared that convert the output of sonar-like ultrasonic ranging sensors to discriminable audio displays. For example, the Wormald Sonicguide uses interaural intensity differences to indicate the azimuth of an object and frequency to indicate distance [Cook, 1982]; subtle information such as texture can sometimes also be discriminated.

143.2 Auditory System

The human auditory system processes information primarily serially; having at best two receptive channels, spatial information must be built up by integration over time. This later capability, however, is profound. Out of a full orchestra, a seasoned conductor can pinpoint an errant violinist by sound alone.

Human hearing is sensitive to sound frequencies from approximately 16 to 20,000 Hz and is most sensitive at 1000 Hz. At this frequency, a threshold root-mean-square pressure of 20 Pa (200 μbar) can be perceived by normally hearing young adults under laboratory conditions. Sound pressure level (SPL) is measured in decibels relative to this threshold. Some approximate benchmarks for sound intensity are a whisper at 1 m (30 dB), normal conversion at 1 m (60 dB), and a subway train at 6 m (90 dB). Sounds increasing from 100 to 140 dB become uncomfortable and painful, and short exposures to a 160-dB level can cause permanent hearing impairment, while continuous exposure to sound levels over 90 dB can cause slow, cumulative damage [Sataloff et al., 1980].

Because hearing sensitivity falls off drastically at lower frequencies, clinical audiometric testing uses somewhat different scales. With the DIN/ANSI reference threshold of 6.5 dB SPL at 1 kHz, the threshold rises to 24.5 dB at 250 Hz and 45.5 dB at 125 Hz [Sataloff et al., 1980]. Hearing loss is then specified in decibels relative to the reference threshold, rather than the SPL directly, so that a normal audiogram would have a flat threshold curve at approximately 0 dB.

Auditory Augmentation

Loss of speech comprehension, and hence interpersonal communication, bears the greatest effect on daily life and is the main reason people seek medical attention for hearing impairment. Functional impairment begins with 21- to 35-dB loss in average sensitivity, causing difficulty in understanding faint speech [Smeltzer, 1993]. Losses of 36 to 50 dB and 51 to 70 dB cause problems with normal and loud speech. Losses greater than 90 dB are termed *profound* or *extreme* and cannot be remedied with any kind of hearing aid; these individuals require auditory substitution rather than augmentation.

Hearing loss can be caused by conduction defects in the middle ear (tympanic membrane and ossicles) or by sensorineural defects in the inner ear (cochlear transduction mechanisms and auditory nerve). Conduction problems often can be corrected medically or surgically. If not, hearing aids are often of benefit because the hearing threshold is elevated uniformly over all frequencies, causing little distortion of the signal. Sensorineural impairments differentially affect different frequencies and also cause other forms of distortion that cannot be helped by amplification or filtering. The dynamic range is also reduced, because while loud sounds (> 100 dB) are often still perceived as loud, slightly softer sounds are lost. Looked at from this perspective, it is easy to understand why the amplification and automatic gain control of conventional hearing aids do not succeed in presenting the 30-dB or so dynamic range of speech to persons with 70+ dB of sensorineural impairment.

Most hearing aids perform three basic functions. (1) Amplification compensates for the reduced sensitivity of the damaged ear. (2) Frequency-domain filtering compensates for hearing loss that is not spectrally uniform. For example, most sensorineural loss disproportionately affects frequencies over 1 kHz or so, so high-frequency preemphasis may be indicated. (3) Automatic gain control (ACG) compresses the amplitude range of desired sounds to the dynamic range of the damaged ear. Typical AGC systems respond to loud transients in 2 to 5 ms (attack time) and reduce their effect in 100 to 300 ms (recovery time).

Sophisticated multiband AGC systems have attempted to normalize the ear's amplitude/frequency response, with the goal of preserving intact the usual intensity relationships among speech elements. However, recent research has shown that only certain speech features are important for intelligibility [Moore, 1990]. The fundamental frequency (due to vocal cord vibration), the first and second formants (the spectral peaks of speech that characterize different vowels), and place of articulation are crucial to speech recognition. In contrast, the overall speech envelope (the contour connecting the individual peaks in the pressure wave) is not very important; articulation information is carried in second-formant and high-frequency spectral information and is not well-represented in the envelope [Van Tasell, 1993]. Therefore, the primary design goal for hearing aids should be to preserve and make audible the individual spectral components of speech (formants and high-frequency consonant information).

The cochlear implant could properly be termed an auditory augmentation device because it utilizes the higher neural centers normally used for audition. Simply stated, the implant replaces the function of the (damaged) inner ear by electrically stimulating the auditory nerve in response to sound collected by an external microphone. Although the auditory percepts produced are extremely distorted and noise-like due to the inadequate coding strategy, many users gain sufficient information to improve their lipreading and speech-production skills. The introductory chapter in this section provides a detailed discussion of this technology.

Visual Auditory Substitution

Lipreading is the most natural form of auditory substitution, requiring no instrumentation and no training on the part of the speaker. However, only about one-third to one-half of the 36 or so phonemes (primary sounds of human speech) can be reliably discriminated by this method. The result is that 30% to 50% of the words used in conversational English look just like, or very similar to, other words (homophenes) [Becker, 1972]. Therefore, word pairs such as buried/married must be discriminated by grammar, syntax, and context.

Lipreading does not provide information on voice fundamental frequency or formants. With an appropriate hearing aid, any residual hearing (less than 90-dB loss) often can supply some of this missing information, improving lipreading accuracy. For the profoundly deaf, technological devices are available to supply some or all of the information. For example, the Upton eyeglasses, an example of a cued-speech device, provide discrete visual signals for certain speech sounds such as fricatives (letters like *f* of *s*, containing primarily high-frequency information) that cannot be readily identified by sight.

Fingerspelling, a transliteration of English alphabet into hand symbols, can convey everyday words at up to 2 syllables per second, limited by the rate of manual symbol production [Reed et al., 1990]. American Sign Language uses a variety of upper body movements to convey words and concepts rather than just individual letters, at the same effective rate as ordinary speech, 4 to 5 syllables per second.

Closed captioning encodes the full text of spoken words on television shows and transmits the data in a nonvisible part of the video signal (the vertical blanking interval). Since July of 1993, all new television sets sold in the United States with screens larger than 33 cm diagonal have been required to have built-in decoders that can optionally display the encoded text on the screen. Over 1000 hours per week of programming is closed captioned [National Captioning Institute, 1994].

Automatic speech-recognition technology may soon be capable of translating ordinary spoken discourse accurately into visually displayed text, at least in quiet environments; this may eventually be a major boon for the profoundly hearing impaired. Presently, such systems must be carefully trained on individual speakers and/or must have a limited vocabulary [Ramesh et al., 1992]. Because there is much commercial interest in speech command of computers and vehicle subsystems, this field is advancing rapidly.

Tactual Auditory Substitution

Tadoma is a method of communication used by a few people in the deaf-blind community and is of theoretical importance for the development of tactual auditory substitution devices. While sign language requires training by both sender and receiver, in Tadoma, the sender speaks normally. The trained receiver places his or her hands on the face and neck of the sender to monitor lip and jaw movements, airflow at the lips, and vibration of the neck [Reed et al., 1992]. Experienced users achieve 80% keyword recognition of everyday speech at a rate of 3 syllables per second. Using no instrumentation, this the highest speech communication rate recorded for any tactual-only communication system.

Alternatively, tactile vocoders perform a frequency analysis of incoming sounds, similarly to the ear's cochlea [Békésy, 1955], and adjust the stimulation intensity of typically 8 to 32 tactile stimulators (vibrotactile or electrotactile) to present a linear spectral display to the user's abdominal or forehead skin. Several investigators [Blamey & Clark, 1985; Boothroyd & Hnath-Chisolm, 1988; Brooks & Frost, 1986; Saunders et al., 1981] have developed laboratory and commercial vocoders. Although vocoder users cannot recognize speech as well as Tadoma users, research has shown that vocoders can provide enough "auditory" feedback to improve the speech clarity of deaf children and to improve auditory discrimination and comprehension in some older patients [Szeto & Riso, 1990] and aid in discrimination of phonemes by lipreading [Hughes, 1989; Rakowski et al., 1989]. An excellent review of earlier vocoders appears in Reed et al. [1982]. The most useful information provided by vocoders appears to be the second-formant frequency (important for distinguishing vowels) and position of the high-frequency plosive and fricative sounds that often delineate syllables [Bernstein et al., 1991].

143.3 Tactual System

Humans receive and combine two types of perceptual information when touching and manipulating objects. *Kinesthetic* information describes the relative positions and movements of body parts as well as muscular effort. Muscle and skin receptors are primarily responsible for kinesthesis; joint receptors serve primarily as protective limit switches [Rabischong, 1981]. *Tactile* information describes spatial pressure patterns on the skin given a fixed body position. Everyday touch perception combines tactile and kinesthetic information; this combination is called *tactual* or *haptic* perception. Loomis and Lederman [1986] provide an excellent review of these perceptual mechanisms.

Geldard [1960] and Sherrick [1973] lamented that as a communication channel, the tactile sense is often considered inferior to sight and hearing. However, the tactile system possess some of the same spatial and temporal attributes as both of the "primary" senses [Bach-y-Rita, 1972]. With over 10,000 parallel channels (receptors) [Collins & Saunders, 1970], the tactile system is capable of processing a great deal of information if it is properly presented.

The human kinesthetic and tactile senses are very robust and, in the case of tactile, very redundant. This is fortunate, considering their necessity for the simplest of tasks. Control of movement depends on kinesthetic information; tremors and involuntary movements can result from disruption of this feedback control system. Surgically repaired fingers may not have tactile sensation for a long period or at all, depending on the severity of nerve injuries; it is known that insensate digits are rarely used by patients [Tubiana, 1988]. Insensate fingers and toes (due to advanced Hansen's disease or diabetes) are often injured inadvertently, sometimes requiring amputation. Anyone who has had a finger numbed by cold realizes that it can be next to useless, even if the range of motion is normal.

The normal sensitivity to touch varies markedly over the body surface. The threshold forces in dynes for men (women) are lips, 9 (5); fingertips, 62 (25); belly 62 (7); and sole of foot, 343 (79) [Weinstein, 1968]. The fingertip threshold corresponds to 10-μm indentation. Sensitivity to vibration is much higher and is frequency-and area-dependent [Verillo, 1985]. A 5-cm^2 patch of skin on the palm vibrating at 250 Hz can be felt at 0.16-μm amplitude; smaller areas and lower frequencies require more displacement. The minimal separation for two nonvibrating points to be distinguished is 2 to 3 mm on the fingertips, 17 mm on the forehead, and 30 to 50 mm on many other locations. However, size and localization judgments are considerably better than these standard figures might suggest [Vierck & Jones, 1969].

Tactual Augmentation

Although we do not often think about it, kinesthetic information is reflected to the user in many types of human-controlled tools and machines, and lack of this feedback can make control difficult. For example, an automobile with power steering always includes some degree of "road feel" to allow the driver to respond reflexively to minor bumps and irregularities without relying on vision. Remote-control robots (telerobots) used underwater or in chemical- or radiation-contaminated environments are slow and cumbersome to operate, partly because most do not provide force feedback to the operator; such feedback enhances task performance [Hannaford & Wood, 1992].

Tactile display of spatial patterns on the skin uses three main types of transducers [Kaczmarek & Bach-y-Rita, 1995; Kaczmarek et al., 1991]. *Static tactile* displays use solenoids, shape-memory alloy actuators, and scanned air or water jets to indent the skin. *Vibrotactile* displays encode stimulation intensity as the amplitude of a vibrating skin displacement (10 to 500 Hz); both solenoids and piezoelectric transducers have been used. *Electrotactile* stimulation uses 1- to 100-mm^2-area surface electrodes and careful waveform control to electrically stimulate the afferent nerves responsible for touch, producing a vibrating or tingling sensation.

Tactile rehabilitation has received minimal attention in the literature or medical community. One research device sensed pressure information normally received by the fingertips and displayed it on the forehead using electrotactile stimulation [Collins & Madey, 1974]. Subjects were able to estimate surface roughness and hardness and detect edges and corners with only one sensor per fingertip. Phillips [1988] reviews prototype tactile feedback systems that use the intact tactile sense to convey hand and foot pressure and elbow angle to users of powered prosthetic limbs, often with the result of more precise control of these devices.

Slightly more attention has been given to tactile augmentation in special environments. Astronauts, for example, wear pressurized gloves that greatly diminish tactile sensation, complicating extravehicular repair and maintenance tasks. Efforts to improve the situation range from mobile tactile pins in the fingertips to electrotactile stimulation on the abdomen of the information gathered from fingertip sensors [Bach-y-Rita et al., 1987].

Tactual Substitution

Because of a paucity of adequate tactual display technology, spatial pressure information from a robot or remote manipulator is usually displayed to the operator visually. A three-dimensional bar graph, for example, could show the two-dimensional pressure pattern on the gripper. While easy to implement, this method suffers from two disadvantages: (1) the visual channel is required to process more information (it is often already heavily burdened), and (2) reaction time is lengthened, because the normal human tactual reflex systems are inhibited. An advantage of visual display is that accurate measurements of force and pressure may be displayed numerically or graphically.

Auditory display of tactual information is largely limited to warning systems, such as excessive force on a machine. Sometimes such feedback is even inadvertent. The engine of a bulldozer will audibly slow down when a heavy load is lifted; by the auditory and vibratory feedback, the operator can literally "feel" the strain.

The ubiquity of such tactual feedback systems suggests that the human-machine interface on many devices could benefit from intentionally placed tactual feedback systems. Of much current interest is the *virtual environment*, a means by which someone can interact with a mathematic model of a place that may or may not physically exist. The user normally controls the environment by hand, head, and body movements; these are sensed by the system, which correspondingly adjusts the information presented on a wide-angle visual display and sometimes also on a spatially localized sound display. The user often describes the experience as "being there," a phenomenon known as *telepresence* [Loomis, 1992]. One can only imagine how much the experience could be enhanced by adding kinesthetic and tactile feedback [Shimoga, 1993], quite literally putting the user in touch with the virtual world.

Defining Terms

Distal attribution: The phenomenon whereby events are normally perceived as occurring external to our sense organs—but also see Loomis' [1992] engaging article on this topic. The environment or transduction mechanism need not be artificial; for example, we visually perceive objects as distant from our eyes.

Electrotactile: Stimulation that evokes tactile (touch) sensations within the skin at the location of the electrode by passing a pulsatile, localized electric current through the skin. Information is delivered by varying the amplitude, frequency, etc. of the stimulation waveform. Also called *electrocutaneous stimulation.*

Illuminance: The density of light falling on a surface, measured in lux. One lux is equivalent to 0.0929 foot-candles, an earlier measure. Illuminance is inversely proportional to the square of the distance form a point light source. A 100-W incandescent lamp provides approximately 1280 lux at a distance of 1 ft (30.5 cm). Brightness is a different measure, depending also on the reflectance of the surrounding area.

Kinesthetic perception: Information about the relative positions of and forces on body parts, possibly including efference copy (internal knowledge of muscular effort).

Sensory augmentation: The use of devices that assist a functional human sense; eyeglasses are one example.

Sensory substitution: The use of one human sense to receive information normally received by another sense. For example, Braille substitutes touch for vision.

Sound pressure level (SPL): The root-mean-square pressure difference from atmospheric pressure (≈100 kPa) that characterizes the intensity of sound. The conversion SPL = 20 log (P/P_0) expresses SPL in decibels, where P_0 is the threshold pressure of approximately 20 Pa at 1 kHz.

Static tactile: Stimulation that is a slow local mechanical deformation of the skin. It varies the deformation amplitude directly rather than the amplitude of vibration. This is "normal touch" for grasping objects, ect.

Tactile perception: Information about spatial pressure patterns on the skin with a fixed kinesthetic position.

Tactual (haptic) perception: The seamless, usually unconscious combination of tactile and kinesthetic information; this is "normal touch."

Vibrotactile: Stimulation that evokes tactile sensations using mechanical vibration of the skin, typically at frequencies of 10 to 500 Hz. Information is delivered by varying the amplitude, frequency, etc. of the vibration.

Virtual environment: A real-time interactive computer model that attempts to display visual, auditory, and tactual information to a human user as if he or she were present at the simulated location. The user controls the environment with head, hand, and body motions. A airplane cockpit simulator is one example.

References

Bach-y-Rita P. 1972. Brain Mechanisms in Sensory Substitution. New York, Academic Press.

Bach-y-Rita P., Kaczmarek, KA, Tyler, M. and Garcia-Lara, M. 1998. Form perception with a 49-point electrotactile stimulus array on the tongue. J. Rehab. Res. Dev. 35:427–430.

Bach-y-Rita P, Webster JG, Tompkins WJ, Crabb T. 1987. Sensory substitution for space gloves and for space robots. In Proceedings of the Workshop on Space Telerobotics, Jet Propulsion Laboratory, Publication 87–13, pp 51–57.

Barfield, W., Hendrix, C., Bjorneseth, O., Kaczmarek, KA and Lotens, W. 1996. Comparison of human sensory capabilities with technical specifications of virtual environment equipment. Presence 4:329–356.

Becker KW. 1972. Speechreading: Principles and Methods. Baltimore, National Educational Press.

Békésy GV. 1955. Human skin perception of traveling waves similar to those of the cochlea. J Acoust Soc AM 27:830.

Bernstein LE, Demorest ME, Coulter DC, O'Connell MP. 1991. Lipreading sentences with vibrotactile vocoders: Performance of normal-hearing and hearing-impaired subjects. J Acoust Soc Am 90:2971.

Blamey PJ, Clark GM. 1985. A wearable multiple-electrode electrotactile speech processor for the profoundly deaf. J Acoust Soc Am 77:1619.

Boothroyd A, Hnath-Chisolm T. 1988. Spatial, tactile presentation of voice fundamental frequency as a supplement to lipreading: Results of extended training with a single subject. J Rehabil Res Dev 25(3):51.

Boyd LH, Boyd WL, Vanderheiden GC. 1990. The Graphical User Interface Crisis: Danger and Opportunity. September, Trace R&D Center, University of Wisconsin-Madison.

Brooks PL, Frost BJ. 1986. The development and evaluation of a tactile vocoder for the profoundly deaf. Can J Public Health 77:108.

Collins CC. 1985. On mobility aids for the blind. In DH Warren, ER Strelow (eds), Electronic Spatial Sensing for the Blind, pp 35–64. Dordrecht, The Netherlands, Matinus Nijhoff.

Collins CC, Madey JMJ. 1974. Tactile sensory replacement. In Proceedings of the San Diego Biomedical Symposium, pp 15–26.

Collins CC, Saunders FA. 1970. Pictorial display by direct electrical stimulation of the skin. J Biomed Syst 1:3–16.

Cook AM. 1982. Sensory and communication aids. In AM Cook, JG Webster (eds), Therapeutic Medical Devices: Application and Design, pp 152–201. Englewood Cliffs, NJ, Prentice-Hall.

Fonda GE. 1981. Management of Low Vision. New York, Thieme-Stratton.

Geldard FA. 1960. Some neglected possibilities of communication. Science 131:1583.

Hannaford B, Wood L. 1992. Evaluation of performance of a telerobot. NASA Tech Briefs 16(2): item 62.

Hughes BG. 1989. A New Electrotactile System for the Hearing Impaired. National Science Foundation final project report, ISI-8860727, Sevrain-Tech, Inc.

Kaczmarek KA, Bach-y-Rita P. 1995. Tactile displays. In W Barfield, T Furness (eds), Virtual Environments and Advanced Interface Design. New York, Oxford University Press.

Kaczmarek, KA, Tyler, ME and Bach-y-Rita, P. 1997. Pattern identification on a fingertip-scanned electrotactile display. Proc. 19th Annu. Int. Conf. IEEE Eng. Med. Biol. Soc. pp. 1694–1697.

Kaczmarek KA, Webster JG, Bach-y-Rita, Tompkins WJ. 1991. Electrotactile and vibrotactile displays for sensory substitution systems. IEEE Trans Biomed Eng 38:1.

Loomis JM. 1992. Distal attribution and presence. Presence: Teleoperators Virtual Environ 1(1):113.

Loomis JM, Lederman SJ. 1986. Tactual perception. In KR Boff et al (eds), Handbook of Perception and Human Performance, vol II: Cognitive Processes and Performance, pp 31.1–31.41. New York, Wiley.

Mehr E, Shindell S. 1990. Advances in low vision and blind rehabilitation. In MG Eisenberg, RC Grzesiak (eds), Advances in Clinical Rehabilitation, vol 3, pp 121–147. New York, Springer.

Moore BCJ. 1990. How much do we gain by gain control in hearing aids? Acta Otolaryngol (Stockh) Suppl. 469:250.

Mountcastle VB (ed). 1980. Medical Physiology. St. Louis, Mosby.

National Captioning Institute, Falls Church, Va, 1994. Personal communication.

NIDRR. 1993. Protocols for choosing low vision devices. U.S. Department of Education. Consensus Statement 1(1–28).

Phillips CA. 1988. Sensory feedback control of upper- and lower-extremity motor prostheses. CRC Crit Rev Biomed Eng 16:105.

Rabischong P. 1981. Physiology of sensation. In R Tubiana (ed), The Hand, pp 441–467. Philadelphia, Saunders.

Rakowski K, Brenner C, Weisenberger JM. 1989. Evaluation of a 32-channel electrotactile vocoder (abstract). J Acoust Soc Am 86(suppl 1):S83.

Ramesh P, Wilpon JG, McGee MA, et al. 1992. Speaker independent recognition of spontaneously spoken connected digits. Speech Commun 11:229.

Reed CM, Delhorne LA, Durlach NI, Fischer SD. 1990. A study of the tactual and visual reception of fingerspelling. J Speech Hear Res 33:786.

Reed CM, Durlach NI, Bradia LD. 1982. Research on tactile communication of speech: A review. AHSA Monogr 20:1.

Reed CM, Rabinowitz WM, Durlach NI, et al. 1992. Analytic study of the Tadoma method: Improving performance through the use of supplementary tactile displays. J Speech Hear Res 35:450.

Sataloff J, Sataloff RT, Vassallo LA. 1980. Hearing Loss, 2d ed. Philadelphia, Lippincott.

Saunders FA, Hill WA, Franklin B. 1981. A wearable tactile sensory aid for profoundly deaf children. J Med Syst 5:265.

Sherrick CE. 1973. Current prospects for cutaneous communication. In Proceedings of the Conference on Cutaneous Communication System Development, pp 106–109.

Shimoga KB. 1993. A survey of perceptual feedback issues in dextrous telemanipulation: II. Finger touch feedback. In IEEE Virtual Reality Annual International Symposium, pp 271–279.

Shlaer S. 1937. The relation between visual acuity and illumination. J Gen Physiol 21:165.

Smeltzer CD. 1993. Primary care screening and evaluation of hearing loss. Nurse Pract 18:50.

Stigmar G. 1970. Observation on vernier and stereo acuity with special reference to their relationship. Acta Ophthalmol 48:979.

Szeto AYJ, Riso RR. 1990. Sensory feedback using electrical stimulation of the tactile sense. In RV Smith, JH Leslie Jr (eds), Rehabilitation Engineering, pp 29–78. Boca Raton, Fla, CRC Press.

Tubiana R. 1988. Fingertip injuries. In R Tubiana (ed), The Hand, pp 1034–1054. Philadelphia, Saunders.

Van Tasell DJ. 1993. Hearing loss, speech, and hearing aids, J Speech Hear Res 36:228.

Verrillo RT. 1985. Psychophysics of vibrotactile stimulation. J Acoust Soc Am 77:225.

Vierck CJ, Jones MB. 1969. Size discrimination on the skin. Science 163:488.

Webster JG, Cook AM, Tompkins WJ, Vanderheiden GC (eds). 1985. Electronic Devices for Rehabilitation. New York, Wiley.

Weinstein S. 1968. Intensive and extensive aspects of tactile sensitivity as a function of body part, sex and laterality. In DR Kenshalo (ed), The Skin Senses, pp 195–218. Springfield, Ill, Charles C Thomas.

Further Information

Presence: Teleoperators and Virtual Environments is a bimonthly journal focusing on advanced human-machine interface issues. In an effort to develop tactile displays without moving parts, our laboratory has demonstrated simple pattern recognition on the fingertip [Kaczmarek et al., 1997] and tongue [Bach-y-Rita et al., 1998] using electrotactile stimulation.

The Trace Research and Development Center, Madison, Wisc., publishes a comprehensive resource book on commercially available assistive devices, organizations, etc. for communication, control, and computer access for individuals with physical and sensory impairments.

Electronic Devices for Rehabilitation, edited by J. G. Webster (Wiley, 1985), summarizes the technologic principles of electronic assistive devices for people with physical and sensory impairments.

144 Augmentative and Alternative Communication

Barry Romich
Prentke Romich Company

Gregg Vanderheiden
University of Wisconsin

Katya Hill
Edinboro University of Pennsylvania and University of Pittsburgh

144.1 Introduction

The inability to express oneself through either speech or writing is perhaps the most limiting of physical disabilities. Meaningful participation in life requires the communication of basic information, desires, needs, feelings, and aspirations. The lack of full interpersonal communication substantially reduces an individual's potential for education, employment, and independence.

Today, through multidisciplinary contributions, individuals who cannot speak or write effectively have access to a wide variety of techniques, therapies and systems designed to ameliorate challenges to verbal communication. The field of augmentative and alternative communication (AAC) consists of many different professions, including speech-language pathology, regular and special education, occupational and physical therapy, engineering, linguistics, technology, and others. Experienced professionals now can become certified as Assistive Technology practitioners and/or suppliers (Minkel, 1996). Individuals who rely on AAC, as well as their families and friends, also contribute to the field (Slesaransky-Poe, 1998).

Engineering plays a significant role in the development of the field of AAC and related assistive technology. Engineering contributions range from relatively independent work on product definition and design to the collaborative development and evaluation of tools to support the contributions of other professions, such as classical and computational linguistics and speech-language pathology.

Augmentative communication can be classified in a variety of ways ranging from unaided communication techniques, such as gestures, signs, and eye pointing, to highly sophisticated electronic devices employing the latest technology. This chapter focuses on the review of technology-aided techniques and related issues.

Technology-based AAC systems have taken two forms: hardware designed specifically for this application and software that runs on mass market computer hardware. Three basic components comprise AAC systems. These are the language representation method (including acceleration techniques), the user interface, and the outputs. Generally, a multidisciplinary team evaluates the current and projected

0-8493-0462-8/00/$0.00+$.50

skills and needs of the individual, determines the most effective language representation method(s) and physical access technique(s), and then selects a system with characteristics that are a good match.

144.2 Language Representation Methods and Acceleration Techniques

The ultimate goal of AAC intervention is functional, interactive communication. The four purposes that communication fulfills are (1) communication of needs/wants, (2) information transfer, (3) social closeness, and (4) social etiquette (Light, 1988). From the perspective of the person who uses AAC, communicative competence involves the ability to transmit messages efficiently and effectively in all four of the interaction categories, based on individual interests, circumstances, and abilities (Buekelman & Miranda, 1998). To achieve communication competence, the person using an AAC system must have access to language representation methods capable of handling the various vocabulary and message construction demands of the environment. Professionals rely on the theoretical models of language development and linguistics to evaluate the effectiveness of an AAC language representation method.

Language is defined as an abstract system with rules governing the sequencing of basic units (sounds, morphemes, words, and sentences), and rules governing meaning and use (McCormick & Schiefelbusch, 1990). An individual knows a language when he or she understands and follows its basic units and rules. Knowledge of language requires both linguistic competence (understanding the rules), and linguistic performance (using these rules). Most language models identify the basic rules as phonology, semantics, morphology, syntax and pragmatics. The basic rules of language apply to AAC. For example, AAC research on vocabulary use has documented the phenomenon of core and fringe vocabulary (Yorkston et al., 1988) and the reliance on a relatively limited core vocabulary to express a majority of communication utterances (Vanderheiden & Kelso, 1987). Research on conversations has documented topic and small talk patterns (Stuart et al., 1993; King et al., 1995). An awareness of how a given AAC system handles these basic rules enables one to be critical of AAC language representation methods.

In addition to the language representation method, the acceleration technique(s) available in an AAC system contribute(s) to communication competence. Communication by users of AAC systems is far slower than that of the general population. Yet the speed of communication is a significant factor influencing the perceptions of the user's communication partner, and potential for personal achievement for the person relying on AAC. The development and application of techniques to accelerate communication rates is critical. Further, these techniques are most effective when developers pay attention to human factors design principles (Goodenough-Trepagnier, 1994).

Alphabet-based language representation methods involve the use of traditional orthography and acceleration techniques that require spelling and reading skills. AAC systems using orthography require the user to spell each word using a standard keyboard or alphabet overlay on a static or dynamic display. A standard or customized alphabet overlay provides use for all the rules and elements of a natural language; however, spelling letter-by-letter is a slow and inefficient AAC strategy without acceleration techniques.

Abbreviation systems represent language elements using a number of keystrokes typically smaller than that required by spelling. For example, words or sentences are abbreviated using principled approaches based on vowel elimination or the first letters of salient words. Abbreviation systems can be fast, but require not only spelling and reading skills, but memory of abbreviation codes. Typically, people with spelling and reading skills have large vocabulary needs and increased demands on production of text. Demasco (1994) offers additional background and proposes some interesting work in this area.

Word prediction is another acceleration technique available from many sources. Based on previously selected letters and words, the system presents the user with best guess choices for completing the spelling of a word. The user then chooses one of the predictions or continues spelling, resulting in yet another set of predictions. Prediction systems have demonstrated a reduction in the number of keystrokes, but recent research (Koester and Levine, 1994) reports that the actual communication rate does not represent a statistical improvement over spelling. The reason for word predication's failure to improve rate is that increased time is needed to read and select the word table choices. The cost of discontinuity and increased cognitive load in the task seems to match the benefits of reduced keystrokes.

Picture Symbol-based language representation methods involve the use of graphic or line drawn symbols to represent single word vocabulary or messages (phrases, sentences, and paragraphs). A variety of AAC symbol sets are available on devices or software depict the linguistics elements available through the system. Picture Communication Symbols (PCS), DynaSyms, and Blissymbols are popular symbol sets used in either dedicated or computer-aided systems. One taxonomy differentiates symbols according to several subordinate levels including static/dynamic, iconic/opaque, and set/system (Fuller, et.al., 1998)

Universal considerations regarding the selection of a symbol set include research on symbol characteristics such as size, transparency, complexity and iconicity (Fuller et al., 1997, Romski & Sevcik, 1988). Understanding symbol characteristics is necessary to make clinical decisions about vocabulary organization and system selection. The choice of picture communication language representation methods is facilitated when teams use a goal-driven graphic symbol selection process (Schlosser et al., 1996). For example, identification of vocabulary and language outcomes assists the selection of an AAC graphic symbol system.

Vocabulary size and representation of linguistic rules (grammar) are concerns for users relying on graphic symbol sets. Users of static display systems have a limited number of symbols available on any one overlay; however, they have ready access to that vocabulary. Dynamic display users have an almost unlimited number of symbols available as vocabulary for message generation; however, they must navigate through pages or displays to locate a word. Frequently, morphology and syntax are not graphically represented. Since research has strongly supported the need for users to construct spontaneous, novel utterances (Beukelman et al., 1984), neither method is efficient for interactive communication.

Semantic Compaction or Minspeak is perhaps the most commonly used AAC language representation method (Baker, 1994). With this method, language is represented by a relatively small set of multi-meaning icons. The specific meaning of each icon is a function of the context in which it is used. Semantic compaction makes use of a meaningful relationship between the icon and the information it represents; it does not require spelling and reading skills, and yet is powerful even for people with these skills. The performance of Minspeak stems from its ability to handle both vocabulary and linguistic structures as found in the Minspeak Application Programs (MAPS) of Words Strategy and Unity. Both MAPS support the concept of a core and fringe vocabulary. They provide the architecture for handling rules of grammar and morphology. The number of required keystrokes is reduced relative to spelling.

Predictive selection is an acceleration technique used with Minspeak. When this feature is enabled, only those choices that complete a meaningful sequence can be selected. With scanning, for example, the selection process can be significantly faster because the number of possible choices is automatically reduced.

144.3 User Interface

Most AAC systems employ a user interface based on the selection of items that will produce the desired output (Vanderheiden and Lloyd, 1986). Items being selected may be individual letters, as used in spelling, or whole words or phrases expressing thoughts, or symbols that represent vocabulary. The numerous techniques for making selections are based on either direct selection or scanning.

Direct selection refers to techniques by which a single action from a set of choices indicates the desired item. A common example of this method is the use of a computer keyboard. Each key is directly selected by finger. Expanded keyboards accommodate more gross motor actions, such as using the fist or foot. In some cases, pointing can be enhanced through the use of technology. Sticks are held in the mouth or attached to the head using a headband or helmet. Light pointers are used to direct a light beam at a target. Indirect pointing systems might include the common computer mouse, trackball, or joystick. Alternatives to these for people with disabilities are based on the movement of the head or other body part. Figure 144.1 depicts direct selection in that the desired location is pointed to directly.

Scanning refers to techniques in which the individual is presented with a time sequence of choices and indicates when the desired choice appears. A simple linear scanning system might be a clock face type display with a rotating pointer to indicate a letter, word, or picture. Additional dimensions of scanning can be added to reduce the selection time when the number of possible choices is larger. A common technique involves the arrangement of choices in a matrix of rows and columns. The selection process has two steps. First, rows are scanned to select the row containing the desired element. The selected row is then scanned to select the desired element. This method is called row-column scanning. See Fig. 144.2. Either by convention or by the grouping of the elements, the order might be reversed. For example, in the U.K., column-row scanning is preferred over row-column scanning. Other scanning techniques also exist and additional dimensions can be employed.

FIGURE 144.1 Direct selection.

Both direct selection and scanning are used to select elements that might not of themselves define an output. In these cases the output may be defined by a code of selected elements. A common example is Morse code by which dots and dashes are directly selected but must be combined to define letters and numbers. Another example, more common to AAC, has an output defined by a sequence of two or more pictures or icons.

Scanning, and to some degree direct selection, can be faster when the choices are arranged such that those most frequently used are easiest to access. For example, in a row-column scanning spelling system that scans top to bottom and left to right, the most frequently used letters are usually grouped toward the upper left corner.

Generally the selection technique of choice should be that which results in the fastest communication possible. Consideration of factors such as cognitive load (Cress and French, 1994), environmental changes, fatigue, aesthetics, and stability of physical skill often influence the choice of the best selection technique.

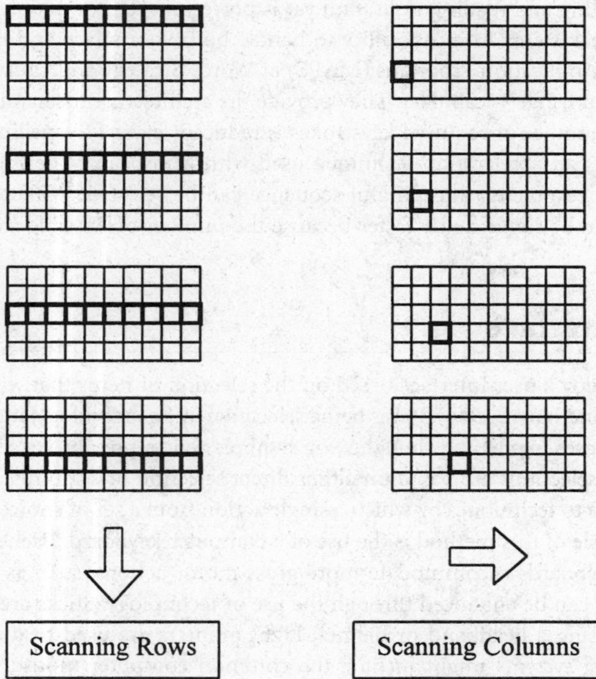

FIGURE 144.2 Row-column scanning.

144.4 Outputs

AAC system outputs in common use are speech, displays, printers, beeps, data, infrared, and other control formats. Outputs are used to facilitate the interactive nature of "real time" communication. Users may rely on auditory and visual feedback to enhance vocabulary selection and the construction of messages. The speech and display outputs may be directed toward the communication partner to support the exchange of information. Auditory feedback such as beeps and key clicks, while useful to the individual using the system, also provide the communication partner with the pragmatic information that a user is in the process of generating a message. Finally, outputs may be used to control other items or devices such as printers.

AAC speech output normally consists of two types: synthetic and digitized.

Synthetic speech usually is generated from text input following a set of rules. Synthetic speech is usually associated with AAC systems that are able to generate text. These systems have unlimited vocabulary and are capable of speaking any word or expression that can be spelled. Commonly used synthetic speech systems like DECtalk™ offer a variety of male, female, and child voices. Intelligibility has improved over recent years to the point that it is no longer a significant issue. With some systems it is actually possible to sing songs, again a feature that enhances social interaction. Most synthetic speech systems are limited to a single language, although bilingual systems are available. Further limitations relate to the expression of accent and emotion. Research and development in artificial speech technology is attacking these limitations.

Digitized speech is essentially speech that has been recorded into digital memory. Relatively simple AAC systems typically use digitized speech. The vocabulary is entered by speaking into the system through a microphone. People who use these systems can say only what someone else said in programming them. They are independent of language and can replicate the song, accent, and emotion of the original speaker.

RS-232c serial data output is used to achieve a variety of functional outcomes. Serial output permits the AAC system to replace the keyboard and mouse for computer access, a procedure known as emulation. The advantage of emulation is that the language representation method and physical access method used for speaking is used for writing and other computer tasks (Buning and Hill, 1998). Another application for the serial output is environmental control. Users are able to operate electrical and electronic items in their daily-living surroundings. Particular sequences of characters, symbols or icons are used to represent commands such as answering the telephone, turning off the stereo, and even setting a thermostat. In addition, the serial output also may be used to monitor the language activity for purposes of clinical intervention, progress reporting, and research.

Infrared communication now is available in many AAC devices. Infrared output supports the same functions as the RS-232c serial data output, but without requiring direct wiring or linking between operating devices. Infrared interfaces providing computer access improve the independence of AAC users because no physical connection needs to be manipulated to activate the system. Infrared interfaces can perform as environmental control units by learning the codes of entertainment and other electronic systems.

144.5 Outcomes, Intervention and Training

The choice of an AAC system should rely primarily on the best interests of the person who will be using the system. Since outcomes and personal achievement will be related directly to the ability to communicate, the choice of a system will have lifelong implications. The process of choosing the most appropriate system is not trivial and can be accomplished best with a multidisciplinary team focusing on outcomes. Team members should realize that the interests of the individual served professionally are not necessarily aligned with those of the providers and/or payers of AAC services. The temptation to select a system that is easy to apply or inexpensive to purchase frequently exists, especially with untrained teams. Teams that identify outcomes and have the goal of achieving interactive communication make informed decisions.

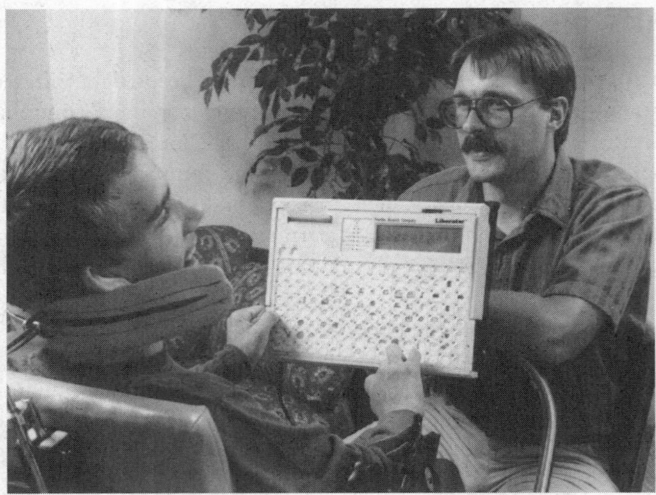

FIGURE 144.3 Individual using an electronic communication system. Photo courtesy of Prentke Romich Company, Wooster, OH.

Following the selection (and funding) of a system, the next step is the actual intervention program. For successful outcomes to be achieved, intervention must go beyond the technical use of the AAC system and include objectives for language development and communication pragmatics. Early in the history of AAC, Rodgers postulates (1984) that AAC systems are more like tools than appliances. To be effective, there is a need for much more than simply "plugging them in and turning them on." The individual who relies on AAC must develop the skills to become a communication craftsperson. In general, as well as specific to AAC, the fastest, most efficient use of a system occurs when the individual operates from knowledge that is in the head, rather than knowledge that must be gathered from the world (Norman, 1980). The implication for AAC is that an intervention program must include a significant training component to assure that the needed knowledge is put into the head (Romich, 1994). Further, a drill component of training develops automaticity (Treviranus, 1994).

Perhaps the single factor limiting the widespread use of assistive technology in general and AAC in particular is the lack of awareness of its availability and the impact it can have on the lives of the people who could benefit (Romich, 1993). This situation exists within not only the general public but also in many of the professions providing services to this population. Training opportunities continue to be limited for a majority of professionals working with persons who could benefit from AAC technology. University programs have lagged behind in their integration of this information into professional curricula.

Intervention and training should extend beyond the initial application of the system. People who rely on AAC frequently live unstable lives. Consequently they have a need for on-going services. For people with congenital (from birth) conditions, developmental delays may have occurred that result in educational and training needs past the normal age of public school services. For people with progressive neurological disorders, such as Lou Gehrig's disease, the physical skill level will change and additional accommodations will need to be evaluated and implemented.

144.6 Future

Technological development in and of itself will not solve the problems of people with disabilities. Technology advancements, however, will continue to provide more powerful tools allowing the exploration of new and innovative approaches to support users and professionals. As access to lower cost, more powerful computer systems increases, the availability of alternative sources of information, technical support and training is possible either through software or the internet. Users and professionals will have increased opportunities to learn and exchange ideas and outcomes.

A related development is the collection and analysis of data describing the actual long term use of AAC devices. Currently, clinicians and researchers have not utilized this information in the clinical intervention process, progress reporting, or research. Tools for monitoring, editing, and analyzing language activity are just now becoming available (Hill and Romich, 1998).

References

Baker BR. 1994. Semantic Compaction: An Approach to a Formal Definition. Proceedings of the Sixth Annual European Minspeak Conference, Swinstead, Lincs, UK: Prentke Romich Europe.

Beukelman DR, Yorkson K, Poblete M, Naranjo C. 1984. Frequency of word occurrence in communication samples produced by adult communication aid users, *J. Speech Hearing Dis.*, 49, 360-367.

Beukelman DR, Mirenda P. 1998. Augmentative and Alternative Communication: Management of Severe Communication Disorders in Children and Adults, Baltimore, Paul H. Brookes Publishing Co.

Buning ME, Hill K. 1999. An AAC device as a computer keyboard: More bang for the buck. AOTA Annual Conference, Indianapolis.

Cress CJ, French GJ. 1994. The Relationship between Cognitive Load Measurements and Estimates of Computer Input Control Skills, *Ass. Tech.* 6.1, 54-66.

Demasco P. 1994. Human Factors Considerations in the Design of Language Interfaces in AAC, *Ass. Tech.* 6.1, 10-25.

Fuller D, Lloyd L, Schlosser R. 1992. Further development of an augmentative and alterntive communication symbol taxonomy, *Aug. Alt. Comm.* 8:67-74.

Fuller D, Lloyd L, Stratton M. 1997. Aided AAC symbols. In L Lloyd, D Fuller, & H Arvidson (eds.), *Augmentative and alternative communication: Principles and practice*, pp 48-79, Needham Heights, MA, Allyn & Bacon.

Goodenough-Trepagnier D. 1994. Design Goals for Augmentative Communication, *Ass. Tech.* 6.1, 3-9.

Hill K, Romich B. 1998. Language Research Needs and Tools in AAC, *Ann. Biomed. Eng.* 26, 131.

Horstmann Koester H, Levine SP. 1994. Learning and Performance of Able-Bodied Individuals Using Scanning Systems with and without Word Prediction, *Ass. Tech.* 6.1, 42-53.

King J, Spoeneman T, Stuart S, Beukelman D. 1995. Small talk in adult conversations, *Aug. Alt. Comm.* 11, 244-248.

Light J. 1988. Interaction involving individuals using augmentative and alternative communication systems: State of the art and future directions, *Aug. Alt. Comm.* 4, 66-82.

McCormick L, Schiefelbusch RL. 1990. Early Language Intervention: An Introduction, 2nd ed., Columbus, Merrill Publishing Company.

Minkel J. 1996. Credentialing in Assistive Technology: Myths and Realities, RESNA News, 8(5), 1.

Norman DA. 1980. The Psychology of Everyday Things, New York, Basic Books, Inc..

Rodgers B. 1984. Presentation at Discovery '84 Conference, Chicago, IL.

Romich B. 1993. Assistive Technology and AAC: An Industry Perspective, *Ass. Tech.* 5.2, 74-77.

Romich B. 1994. Knowledge in the World vs. Knowledge in the Head: The Psychology of AAC Systems, *Comm. out.* 14(4).

Romski M, Sevcik R. 1988. Augmentative and alternative communication systems: Considerations for individuals with severe intellectual disabilities, *Aug. Alt. Comm.* 4, 83-93.

Schlosser RW, Lloyd LL, McNaughton S. 1996. Graphic symbol selection in research and practice: Making the case for a goal-driven process, Communication… Naturally: Theoretical and Methodological Issues in *Aug. Alt. Comm.* E. Bjorck-Akesson and P. Lindsay (eds.), *Proc. of the Forth ISAAC Research Symposium*, pp. 126-139.

Slesaransky-Poe G. 1998. ACOLUG: The Communication On-Line User's Group, *ISAAC Proc.*, 51-52.

Stuart S, Vanderhoof D, Beukelman D. 1993. Topic and vocabulary use patterns of elderly women, *Aug. Alt. Comm.* 9, 95-110.

Treviranus J. 1994. Mastering Alternative Computer Access: The Role of Understanding, Trust, and Automaticity, *Ass. Tech.* 6.1, 26-41.

Vanderheiden GC, Lloyd LL. 1986. In SW Blackstone (ed.), Augmentative Communication: An Introduction, Rockville, MD: American Speech-Language-Hearing Association.

Vanderheiden GC, Kelso D. 1987. Comparative analysis of fixed-vocabulary comunication acceleration techniques. *Aug. Alt. Comm.* 3:196-206.

Yorkston KM, Dowden PA, Honsinger MJ, Marriner N, Smith K. 1988. A Comparison of standard and user vocabulary list, *Aug. Alt. Comm.* 4:189-210.

Further Information

There are a number of organizations and publications that relate to AAC.

AAC (*Augmentative and Alternative Communication*) is the quarterly refereed journal of ISAAC. It is published by Decker Periodicals Inc., PO Box 620, L.C.D. 1, Hamilton, Ontario, L8N 3K7 CANADA, Tel. 905-522-7017, Fax. 905-522-7839. http://www.isaac-online.org

ACOLUG (Augmentative Communication On-Line User Group) is a listserve with primary participants being people who rely on AAC. Topics include a wide range of issues of importance to this population, their relatives, and friends, and those who provide AAC services. http://nimbus.ocis.temple.edu/~kcohen/listserv/homeacolug.html

American Speech-Language-Hearing Association (ASHA) is the professional organization of speech-language pathologists. ASHA has a Special Interest Division on augmentative communication. ASHA, 10801 Rockville Pike, Rockville, MD, 20852, Tel. 301-897-5700, Fax. 301-571-0457. http://www.asha.org.

Augmentative Communication News is published bi-monthly by Augmentative Communication, Inc., 1 Surf Way, Suite #215, Monterey, CA, 93940, Tel. 408-649-3050, Fax. 408-646-5428.

CAMA (Communication Aid Manufacturers Association) is an organization of manufacturers of AAC systems marketed in North America. CAMA, 518-526 Davis St., Suite 211-212, Evanston, IL, 60201, Tel. 800-441-2262, Fax. 708-869-5689. http://www.aacproducts.org.

Communicating Together is published quarterly by Sharing to Learn as a means of sharing the life experiences and communication systems of augmentative communicators with other augmentative communicators, their families, their communities, and those who work with them. Sharing to Learn, PO Box 986, Thornhill, Ontario, L3T 4A5, CANADA, Tel. 905-771-1491, Fax. 905-771-7153. http://www.isaac-online.org.

Communication Outlook is an international quarterly addressed to the community of individuals interested in the application of technology to the needs of persons who experience communication handicaps. It is published by the Artificial Language Laboratory, Michigan State University, 405 Computer Center, East Lansing, MI, 48824-1042, Tel. 517-353-0870, Fax. 517-353-4766. http://www.msu.edu/~artlang/CommOut.html.

Trace Research & Development Center, University of Wisconsin—Madison. http://www.trace.wisc.edu.

ISAAC is the International Society for Augmentative and Alternative Communication). USSAAC is the United States chapter. Both can be contacted at PO Box 1762 Station R, Toronto, Ontario, M4G 4A3, CANADA, Tel. 905-737-9308, Fax 905-737-0624. http://www.isaac-online.org.

RESNA is an interdisciplinary association for the advancement of rehabilitation and assistive technologies. RESNA has many Special Interest Groups including those on Augmentative and Alternative Communication and Computer Applications. RESNA, 1700 North Moore Street, Suite 1540, Arlington, VA, 22209-1903, Tel. 703-524-6686.

145

Measurement Tools and Processes in Rehabilitation Engineering

George V. Kondraske
University of Texas at Arlington

In every engineering discipline, measurement facilitates the use of structured procedures and decision-making processes. In rehabilitation engineering, the presence of "a human," the only or major component of *the system of interest*, has presented a number of unique challenges with regard to measurement. This is especially true with regard to the routine processes of rehabilitation that either do or could incorporate and rely on measurements. This, in part, is due to the complexity of the human system's architecture, the variety of ways in which it can be adversely affected by disease or injury, and the versatility in the way it can be used to accomplish various tasks of interest to an individual.

Measurement supports a wide variety of assistive device design and prescription activities undertaken within rehabilitation engineering (e.g., Leslie and Smith [1990], Webster et al. [1985], and other chapters within this section). In addition, rehabilitation engineers contribute to the specification and design of measurement instruments that are used primarily by other service providers (such as physical and occupational therapists). As measurements of human structure, performance, and behavior become more rigorous and instruments used have taken advantage of advanced technology, there is also a growing role for rehabilitation engineers to assist these other medical professionals with the proper application of measurement instruments (e.g., for determining areas that are most deficient in an individual's performance profile, objectively documenting progress during rehabilitation, etc.). This is in keeping with the team approach to rehabilitation that has become popular in clinical settings. In short, the role of measurement in rehabilitation engineering is dynamic and growing.

0-8493-0462-8/00/$0.00+$.50

In this chapter, a top-down overview of measurement tools and processes in rehabilitation engineering is presented. Many of the measurement concepts, processes, and devices of relevance are common to applications outside the rehabilitation engineering context. However, the nature of the human population with which rehabilitation engineers must deal is arguably different in that each individual must be assumed to be unique with respect to at least a subset of his or her performance capacities and/or structural parameters; i.e., population reference data cannot be assumed to be generally applicable. While there are some exceptions, population labels frequently used such as "head-injured" or "spinal cord-injured" represent only a gross classification that should not be taken to imply homogeneity with regard to parameters such as range of motion, strength, movement speed, information processing speed, and other performance capacities. This is merely a direct realization that many different ways exist in which the human system can be adversely affected by disease or injury and recognition of the continuum that exists with regard to the degree of any given effect. The result is that in rehabilitation engineering, compared with standard human factors design tasks aimed at the average healthy population, many measurement values must be acquired directly for the specific client.

Measurement in the present context encompasses actions that focus on (1) the human (e.g., structural aspects and performance capacities of subsystems at different hierarchical levels ranging from specific neuromuscular subsystems to the total person and his or her activities in daily living, including work), (2) assistive devices (e.g., structural aspects and demands placed on the human), (3) tasks (e.g., distances between critical points, masses of objects involved, etc.), and (4) overall systems (e.g., performance achieved by a human-assistive device-task combination, patterns of electrical signals representing the timing of muscle activity while performing a complex maneuver, behavior of an individual before and after being fitted with a new prosthetic devices, etc). Clearly, an exhaustive treatment is beyond the scope of this chapter. Measurements are embedded in every specialized subarea of rehabilitation engineering. However, there are also special roles served by measurement in a broader and more generic sense, as well as principles that are common across the many special applications. Emphasis here is placed on these.

There is no lack of other literature regarding the types of measurement outlined to be of interest here and their use. However, it is diffusely distributed, and gaps exist with regard to how such tools can be integrated to accomplish goals beyond simply the acquisition of numeric data for a given parameter. With rapidly changing developments over the last decade, there is currently no comprehensive source that describes the majority of instruments available, their latest implementations, procedures for their use, evaluation of effectiveness, etc. While topics other than measurement are discussed, Leslie and Smith [1990] produced what is perhaps the single most directly applicable source with respect to rehabilitation engineering specifically, although it too is not comprehensive with regard to measurement, nor does it attempt to be.

145.1 Fundamental Principles

Naturally, the fundamental principles of human physiology manifest themselves in the respective sensory, neuromuscular, information-processing, and life-sustaining systems and impact approaches to measurement. In addition, psychological considerations are vital. Familiarization with this material is essential to measurement in rehabilitation; however, treatment here is far beyond the scope of this chapter. The numerous reference works available may be most readily found by consulting relevant chapters in this *Handbook* and the works that they reference. In this section, key principles that are more specific to measurement and of general applicability are presented.

Structure, Function, Performance, and Behavior

It is necessary to distinguish between structure, function, performance, and behavior and measurements thereof for both human and artificial systems. In addition, hierarchical systems concepts are necessary both to help organize the complexity of the systems involved and to help understand the various needs that exist.

Structural measures include dimensions, masses (of objects, limb segments), moments of inertia, circumferences, contours, compliances, and any other aspects of the physical system. These may be considered hierarchically as being pertinent to the total human (e.g., height, weight, etc.), specific body segments (e.g., forearm, thigh, etc.), or components of basic systems such as tendons, ligaments and muscles.

Function is the *purpose* of the system of interest (e.g., to move a limb segment, to communicate, to feed and care for oneself). Within the human, there are many single-purpose systems (e.g., those that function to move specific limb segments, process specific types of information, etc.). As one proceeds to higher levels, such as the total human, systems that are increasingly more multifunctional emerge. These can be recognized as higher-level configurations of more basic systems that operate to feed oneself, to conduct personal hygiene, to carry out task of a job, etc. This multilevel view of just functions begins to help place into perspective the scope over which measurement can be applied.

In rehabilitation in general, a good deal of what constitutes measurement involves the application of structured subjective observation techniques (see also the next subsection) in the form of a wide range of rating scales [e.g., Fuhrer, 1987; Granger & Greshorn, 1984; Potvin et al., 1985]. These are often termed *functional assessment scales* and are typically aimed at obtaining a global index of an individual's ability to function independently in the world. The global index is typically based on a number of items within a given scale, each of which addresses selected, relatively high-level functions (e.g., personal hygiene, mobility, etc.). The focus of measurement for a given item is often in estimate of the *level* of independence or dependence that the subject exhibits or needs to carry out the respective function. In addition, inventories of functions that an individual is able or not able to carry out (with and without assistance) are often included. The large number of such scales that have been proposed and debated is a consequence of the many possible functions and combinations thereof that exist on which to base a given scale. Functional assessment scales are relatively quick and inexpensive to administer and have a demonstrated role in rehabilitation. However, the nature and levels of measurements obtained are not sufficient for many rehabilitation engineering purposes. This latter class of applications generally begins with a function at the level and of the type used as a constituent component of functional assessment scales, considers the level of performance at which that function is executed more quantitatively, and incorporates one or more lower levels in the hierarchy (i.e., the human subsystems involved in achieving the specific functions of daily life that are of interest and their capacities for performance).

Where functions can be described and inventoried, *performance* measures directly characterize *how well* a physical system of interest executes its intended function. performance is multidimensional (e.g., strength, range, speed, accuracy, steadiness, endurance, etc.). Of special interest are the concepts of *performance capacity* and *performance capacity measurement.* Performance capacity represents the *limits* of a given system's ability to operate in its corresponding multidimensional performance space. In this chapter, a resource-based model for both human and artificial system performance and measurement of their performance capacities is adopted [e.g., Kondraske, 1990, 1995]. Thus the *maximum* knee flexor strength available (i.e., the resource availability) under a stated set of conditions represents one unique performance capacity of the knee flexor system. In rehabilitation, the terms *impairment, disability,* and *handicap* [World Health Organization, 1980] have been prominently applied and are relevant to the concept of performance. While these terms place an emphasis on what is missing or what a person cannot do and imply not only a measurement but also the incorporation of an assessment or judgment based on one or more observations the resource-based performance perspective focuses on "what is present" or "what is right" (i.e., performance resource availability). From this perspective, an impairment can be determined to exist if a given performance capacity is found to be less than a specified level (e.g., less than 5th percentile value of a health reference population). A disability exists when performance resource insufficiency exists in a specified task.

While performance relates more to what a system can do (i.e., a challenge or maximal stress is implied), *behavior* measurements are used to characterize what a system does naturally. Thus a given variable such as movement speed can relate to both performance and behavior depending on whether the system (e.g., human subsystem) was maximally challenged to respond "as fast as possible" (performance) or simply observed in the course of operation (behavior). It is also possible to observe a system that it is behaving at one or more of its performance capacities (e.g., at the maximum speed possible, etc.) (see Table 145.1).

Subjective and Objective Measurement Methods

Subjective measurements are made by humans without the aid of instruments and objective measurements result from the use of instruments. However, it should be noted that the mere presence of an instrument does not guarantee complete objectivity. For example, the use of a ruler requires a human judgment in reading the scale and thus contains a subjective element. A length-measurement system with an integral data-acquisition system would be more objective. However, it is likely that that even this system would involve human intervention in its use, e.g., the alignment of the device and making the decision as to exactly what is to be measured with it by selection of reference points. Measures with more objectivity (less subjectivity) are preferred to minimize questions of bias. However, measurements that are intrinsically more objective are frequently more costly and time-consuming to obtain. Well-reasoned tradeoffs must be made to take advantage of the ability of a human (typically a skilled professional) to quickly "measure" many different items subjectively (and often without recording the results but using them internally to arrive at some decision).

It is important to observe that identification of the variable of interest is not influenced by whether it is measured subjectively or objectively. This concept extends to the choice of instrument used for objective measurements. This is an especially important concept in dealing with human performance and behavior, since variables of interest can be much more abstract than simple lengths and widths (e.g., coordination, postural stability, etc.) In fact, many measurement variables in rehabilitation historically have tended to be treated as if they were inextricably coupled with the measurement method, confounding debate regarding *what should be measured* with *what should be used to measure it* in a given context.

Measurements and Assessments

The basic representation of a measurement itself in terms of the actual units of measure is often referred to as the *raw form.* For measures of performance, the term *raw score* is frequently applied. Generally, some form of *assessment* (i.e., judgment or interpretation) is typically required. Assessments may be applied to (or, viewed from a different perspective, may require) either a single measure of groups of them. Subjective assessments are frequently made that are based on the practitioner's familiarity with values for a given parameter in a particular context. However, due to the large number of parameters and the amount of experience that would be required to gain a sufficient level of familiarity, a more formal and objective realization of the process that takes place in subjective assessments is often employed. This process combines the measured value with objectively determined reference values to obtain new metrics, or scores, that facilitate one or more steps in the assessment process.

For aspects of performance, *percent normal scores* are computed by expressing subject Y's availability of performance resource $k[R_{Ak}(Y)]$ as a fraction of the mean availability of that resource in a specified reference population $[R_{Ak}(\text{pop})]$. Ideally, the reference population is selected to match the characteristics of the individual as closely as possible (e.g., age range, gender, handedness, etc.).

$$\text{Percent normal} = \frac{R_{A_k}(\text{Y})}{R_{A_k}(\text{pop})} \times 100 \tag{145.1}$$

Aside from the benefit of placing all measurements on a common scale, a percent normal representation of a performance capacity score can be loosely interpreted as a probability. Consider grip strength as the

TABLE 145.1 The Scope of Measurement in Rehabilitation is Broad

Hierarchical Level	Structure	Function	Performance	Behavior
Global/Composite • Total human • Human with artificial systems	• Height • Weight • Postures • Subjective and instrumented methods	• Multifunction, reconfigurable system • High-level functions: tasks of daily life (working, grooming, recreation, etc.) • Functional assessment scales Single-number global index Level of indep. estimates	• No single-number direct measurement is possible •Possible models to integrate lower-level measures • Direct measurement (subjective and instrumented) of selected performance attribute for selected functions	• Subjective self- and family reports • Instrumented ambulatory activity monitors (selected attributes) • See notes under "function"
Complex Body Systems • Cognitive • Speech • Lifting, gait • Upper extremity • Cardiovascular/respiratory • Etc.	• Dimensions • Shape • Etc. • Instrumented methods	• Multifunction, reconfigurable systems • System-specific functions	• Function specific subjective rating scales Often based on impairment/ disability concepts Relative metrics • Some instrumented performance capacity measures Known also as "functional capacity" (misnomer)	• Subjective and automated (objective) videotape evaluation • Instrumented measures of physical quantities vs. time (e.g., forces, angles, motions) • Electromyography (e.g., muscle timing patterns, coordination)
Basic Systems • Visual information processors • Flexors, extensors • Visual sensors • Auditory sensors • Lungs • Etc.	• Dimensions • Shape • Masses • Moments of inertia • Instrumented methods	• Single function • System-specific functions	• Subjective estimates by clinician for diagnostic and routine monitoring purposes • Instrumented measures of performance capacities (e.g., strength, extremes/range of motion, speed, accuracy, endurance, etc.)	• Instrumented systems Measure and log electrophysiologic biomechanical, and other variables vs. time Post-hoc parameterization
Components of basic systems • Muscle • Tendon • Nerve • Etc.	• Mechanical properties • Instrumented methods/imaging	• Generally single-function • Component-specific functions	• Difficult to assess for individual subjects • Infer from measures at "basic system level" • Direct measurement methods with lab samples, research applications	• Difficult to assess for individual subject • Direct measurement methods with lab samples, research applications

Note: Structure, function, performance, and behavior are encompassed at multiple hierarchical levels. Both subjective and objective, instrumented methods of measurement are employed.

performance resource. Assume that there is a uniform distribution of demands of demands placed on grip strength across a representative sample of tasks of daily living, with requirements ranging from zero to the value representing mean grip strength availability in the reference population. Further assuming that grip strength was the only performance resource that was in question for subject Y (i.e., all other were available in nonlimiting amounts), the percent normal score would represent the probability that a task involving grip strength, randomly selected from those which average individuals in the reference population could execute (i.e., those for which available grip strength would be adequate), could be successfully executed by subject Y. While the assumptions stated here are unlikely to be perfectly true, this type of interpretation helps place measurements that are most commonly made in the laboratory into daily-life contexts.

In contrast to percent normal metrics, *z-scores* take into account variability within the selected reference population. Subject Y's performance is expressed in terms of the difference between it and the reference population mean, normalized by a value corresponding to one standard deviation unit (σ) of the reference population distribution:

$$z = \frac{R_{A_k}\left(\mathrm{Y}\right) - R_{A_k}\left(\mathrm{pop}\right)}{\sigma} \tag{145.2}$$

It is important to note that valid z-scores assume that the parameter in question exhibit a normal distribution in the reference population. Moreover, z-scores are useful in assessing measures of structure, performance, and behavior. With regard to performance (and assuming that measures are based on a resource construct, i.e., a larger numeric value represents better performance), a z-score of zero is produced when the subject's performance equals that of the mean performance in the reference population. Positive z-scores reflect performance that is better than the population mean. In a normal distribution, 68.3% of the samples fall between z-scores of –1.0 and +1.0, wile 95.4% of these samples fall between z-scores of –2.0 and +2.0. Due to variability of a given performance capacity within a healthy population (e.g., some individuals are stronger, faster, more mobile that others), a subject with a raw performance capacity score that produces a percent normal score of 70% could easily produce a z-score of –1.0. Whereas this percent normal score might raise concern regarding the variable of interest, the z-score of –1.0 indicates that a good fraction of healthy individuals exhibit lower level of performance capacity.

Both percent normal and z-scores require reference population data to compute. The best reference (i.e., most sensitive) is data for that specific individual (e.g., preinjury or predisease onset). In most cases, these data do not exist. However, practices such as preemployment screenings and regular checkups are beginning to provide individualized reference data in some rehabilitation contexts.

In yet another alternative, it is frequently desirable to use values representing demands imposed by tasks [R_{Dk}(task A)] as the reference for assessment of performance capacity measures. Demands on performance resources can be envisioned to vary over the time course of a task. In practice, an estimate of the worst-case value (i.e., highest demand) would be used in assessments that incorporate task demands as reference values. In one form, such assessments can produce binary results. For example, availability can be equal to or exceed demand (resource sufficiency),or it can be less than demand (resource insufficiency). These rule-based assessments are useful in identifying limiting factors, i.e., those performance resources that inhibit a specified type of task from being performed successfully or that prevent achievement of a higher level of performance in a given type of task.

$$\begin{aligned} &\textit{If } R_{A_k}\left(\text{subject Y}\right) \geq R_{D_k}\left(\text{task A}\right), \textit{ then} \\ &R_{A_k}\left(\text{subject Y}\right) \textit{ is sufficient, else} \\ &R_{A_k}\left(\text{subject Y}\right) \textit{ is insufficient} \end{aligned} \tag{145.3}$$

These rule-based assessments represent the basic process often applied (sometimes subliminally) by experienced clinicians in making routine decisions, as evidenced by statements such as "not enough strength," "not enough stability," etc. It is natural to extend and build on these strategies for use with objective measures. Extreme care must be employed. It is often possible, for example, for an individual to substitute another performance resource that is not insufficient for one that is. Clinicians take into account many such factors, and objective components should be combined with subjective assessments that provide the required breadth that enhances validity of objective components of a given assessment.

Using the same numeric values employed in rule-based binary assessments, a *preference capacity stress* metric can be computed:

$$\text{Performance capacity stress}\left(\%\right)=\frac{R_{D_k}\left(\text{task A}\right)}{R_{A_k}\left(\text{subject Y}\right)}\times 100 \qquad (145.4)$$

Binary assessments also can be made using this metric and a threshold of 100%. However, the stress value provides additional information regarding how far (or close) a given performance capacity value is from the sufficiency threshold.

145.2 Measurement Objectives and Approaches

Characterizing the Human System and Its Subsystems

Figure 145.1 illustrates various points at which measurements are made over the course of a disease or injury, as well as some of the purposes for which they are made. The majority of measurements made in rehabilitation are aimed at characterizing the human system.

Measurements of human structure [e.g., Pheasant, 1986] play a critical role in the design and prescription of components such as seating, wheelchairs, workstations, artificial limbs, etc. Just like clothing, these items must "fit" the specific individual. Basic tools such as measuring tapes and rulers are becoming supplemented with three-dimensional digitizers and devices found in computer-aided manufacturing.

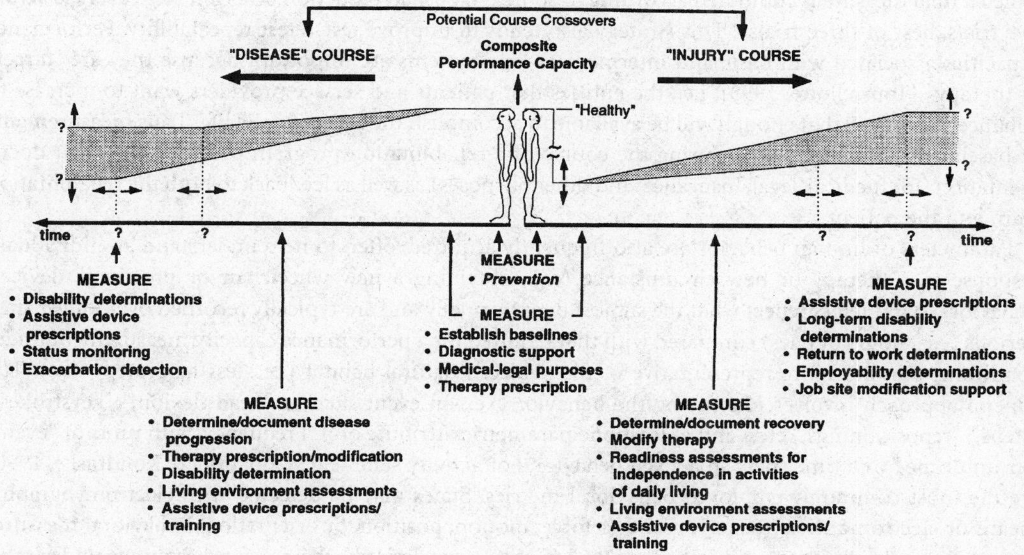

FIGURE 145.1 Measurements of structure, performance, and behavior serve many different purposes at different points over the course of a disease or injury that results in the need for rehabilitation services.

Measurements of structure (e.g., limb segment lengths, moments of inertia, etc.) are also used with computer models [Vasta & Knodraske, 1995] in the process of analyzing tasks to determine demands in terms of performance capacity variables associated with basic systems such s flexors and extensors.

After nearly 50 years, during which a plethora of mostly disease- and injury-specific functional assessment scales were developed, the *functional independence measure* (FIM) [Hamilton et al., 1987; Keith et al., 1987] is of particular note. It is the partial result of a task-force effort to produce a systematic methodology (Uniform Data System for Medical Rehabilitation) with the specific intent of achieving standardization throughout the clinical service-delivery system. This broad methodology uses subjective judgements exclusively, based on rigorous written guidelines, to categorize demographic, diagnostic, functional, and cost information for patients within rehabilitation settings. Its simplicity to use once learned and its relatively low cost of implementation have helped in gaining a rather widespread utilization for tracking progress of individuals from admission to discharge in rehabilitation programs and evaluating effectiveness of specific therapies within and across institutions.

In contrast, many objective measurement tools of varying degrees of technological sophistication exist [Jones, 1995; Kondraske, 1995, Smith & Leslie, 1990; Potvin et al, 1985; Smith, 1995] (also see Further Information below. A good fraction of these have been designed to accomplish the same purposes as corresponding subjective methods, but with increased resolution, sensitivity, and repeatability. The intent is not always to replace subjective methods completely but to make available alternatives with the advantages noted for situations that demand superior performance in the aspects noted. There are certain measurement needs, however, that cannot be accomplished via subjective means (e.g., measurement of a human's visual information-processing speed, which involves the measurement of times of less than 1 second with millisecond resolution). These needs draw on the latest technology in a wide variety of ways, as demonstrated in the cited material.

With regard to instrumented measurements that pertain to a specific individual, performance capacity measures at both complex body system and basic system levels (Fig. 145.1) constitute a major area of activity. A prime example is methodology associated with the implementation of industrial lifting standards [NIOSH, 1981]. Performance-capacity measures reflect the limits of availability of one or more selected resources and require test strategies in which the subject is commanded to perform at or near a maximum level under controlled conditions. Performance tests typically last only a short time (seconds or minutes). To improve estimates of capacities, multiple trials are usually included in a given "test" from which a final measure is computed according to some established reduction criterion (e.g., average across five trials, best of three trials). This strategy also tends to improve test-retest repeatability. Performance capacities associated with basic and intermediate-level systems are important because they are "targets of therapy" [Tourtellotte, 1993], i.e., the entities that patients and service providers want to increase to enhance the chance that enough will be available to accomplish the tasks of daily life. Thus measurements of baseline levels and changes during the course of a rehabilitation program provide important documentation (for medical, legal, insurance, and other purposes) as well as feedback to both the rehabilitation team and the patient.

Parameters of human behavior are also frequently acquired, often to help understand an individual's response to a therapy or new circumstance (e.g., obtaining a new wheelchair or prosthetic device). Behavioral parameters reflect what the subject does normally and are typically recorded over longer time periods (e.g., hours or days) compared with that required for a performance capacity measurement under conditions that are more representative of the subject's natural habitat (i.e., less laboratory-like). The general approach involves identifying the behavior (i.e., an event such as "head flexion," "keystrokes," "steps," "repositionings," etc.) and at least one parametric attribute of it. Frequency, with units of "events per unit time," and time spent in a given behavioral or activity state [e.g., Gonapthy & Kondraske, 1990] are the most commonly employed behavioral metrics. States may be detected with electromyographic means or electronic sensors that respond to force, motion, position, or orientation. Behavioral measures can be used as feedback to a subject as a means to encourage desired behaviors or discourage undesirable behaviors.

Characterizing Tasks

Task characterization or *task analysis*, like the organization of human system parameters, is facilitated with a hierarchical perspective. A highly objective, algorithmic approach could be delineated for task analysis in any given situation [Imrhan, 1995; Maxwell, 1995]. The basic objective is to obtain both descriptive and quantitative information for making decisions about the interface of a system (typically a human) to a given task. Specifically, function, procedures, and goals are of special interest. *Function* represents the purpose of a task (e.g., to flex the elbow, to lift an object, to communicate). In contrast, task goals relate to performance, or how well the function is to be excuted, and are quantifiable (e.g., the mass of an object to be lifted, the distance over which the lift must occur, the speed at which the lift must be performed, etc.). In situations with human and artificial systems, the term *overall task goals* is used to distinguish between goals of the combined human-artificial system and goals associated with the task of operating the artificial system. Procedures represent the process by which goals are achieved. Characterization of procedures can include descriptive and quantitative components (e.g., location of a person's hands at beginning and end points of a task, path in three-dimensional space between beginning and end points). Partial or completely unspecified procedures allow for variations in *style. Goals* and *procedures* are used to obtain numeric estimates of task demands in terms of the performance resources associated with the systems anticipated to be used to execute the task. Task demands are time dependent. Worst-case demands, which may occur only at specific instants in time, are of primary interest in task analysis.

Estimates of task demand can be obtained (1) direct measurement (i.e., of goals and procedures), (2) the use of physics-based models to map direct measurements into parameters that relate more readily to measurable performance capacities of human subsystems, or (3) inference. Examples of direct measurement include key dimensions and mass of objects, three-dimensional spatial locations between "beginning" and "end points" of objects in tasks involving the movement of objects, etc. Instrumentation supporting task analysis is available (e.g., load cells, video and other systems for measuring human position and orientation in real-time during dynamic activities), but it is not often integrated into systems for task analysis per se. Direct measurements of forces based on masses of objects and gravity often must be translated (to torques about a given body joint): this requires the use of static and dynamic models and analysis [e.g., Vista & Kondraske, 1995; Winter, 1990].

An example of an inferential task-analysis approach that is relatively new is nonlinear causal resource analysis (NCRA)[Kondraske, 1999; Kondraske et al., 1997; Kondraske, 1988]. This method was motivated by human performance analysis situations where direct analysis is not possible (e.g., determination of the amount of visual information-processing speed required to drive safely on a highway). Quantitative task demands, in terms of performance variables that characterize the involved subsystems, are inferred from a population data set that includes measures of subsystem performance, resource availabilities (e.g., speed, accuracy, etc.), and overall performance on the task in question. This method is based on the simple observation that the individual with the *least amount* of the given resource (i.e., the lowest performance capacity) who is still able to accomplish a given goal (i.e., achieve a given level of performance in the specified high-level task) provides the key clue. That amount of availability is used to infer the amount of demand imposed by the task.

The ultimate goal to which task characterization contributes is to identify limiting factors or unsafe conditions when a specific subject undertakes the task in question; this goal must not be lost while carrying out the basic objectives of task analysis. While rigorous algorithmic approaches are useful to make evident the true detail of the process, they are generally not performed in this manner in practice at present. Rather, the skill and experience of individuals performing the analysis are used to simplify the process, resulting in a judicious mixture of subjectives estimates and objective measurements. For example, some limiting factors (e.g., grip strength) may be immediately identified without measurement of the human or the task requirements because the margin between availability and demand is so great that quick subjective "measurements" followed by an equally quick "assessment" can be used to arrive at the proper conclusion (e.g., "grip strength is a limiting factor in this task").

Characterizing Assistive Devices

Assistive devices can be viewed as artificial systems that either completely or partially bridge a gap between a given human (with his or her unique profile of performance capacities, i.e., available performance resources) and a particular task or class of tasks (e.g., communication, mobility, etc.). It is thus possible to consider the aspects of the device that constitute the user-device interface and those aspects which constitute, more generally, the device-task interface. In general, measurements supporting assessment of the user-device interface can be viewed to consist of (1) those which characterize the human and (2) those which characterize tasks (i.e., "operating" the assistive device). Each of these was described earlier. Measurements that characterize the device-task interface are often carried out in the context of the complete system, i.e., the human-assistive device-task combination (see next subsection).

Characterizing Overall Systems in High-Level-Task Situations

This situation generally applies to a human-artificial system-task combination. Examples include an individual using a communication aid to communicate, an individual using a wheelchair to achieve mobility, etc. Here, concern is aimed at documenting how well the task (e.g., communication, mobility, etc.) is achieved by the composite or overall system. Specific aspects or *dimensions of performance* associated with the relevant function should first be identified. Examples include speech, accuracy, stability, efficiency, etc. The total system is then maximally challenged (tempered by safety considerations) to operate along one or more of these dimensions of performance (usually not more than two dimensions are maximally challenged at the same time). For example, a subject with a communication device may be challenged to generate a single selected symbol "as fast as possible" (stressing speed without concern for accuracy). Speed is measured (e.g., with units of symbols per second) over the course of short trial (so as not to be influenced by fatigue). Then the "total system" may be challenged to generate a subset of specific symbols (chosen at random from the set of those available with a given device) one at a time, "as accurately as possible" (stressing accuracy while minimizing stress on speed capacities). Accuracy is then measured after a representative number of such trials are administered (in terms of "percent correct," for example). To further delineate the speed-accuracy performance envelope, "the system" may be challenge to select symbols at a fixed rate while accuracy is measured. Additional dimensions can be evaluated similarly. For example, endurance (measure in units of time) can be determined by selecting an operating point (e.g., by reference to the speed-accuracy performance envelope) and challenging the total system "to communicate" for "as long as possible" under the selected speed-accuracy condition.

In general, it is more useful if these types of characterizations consider all relevant dimensions with some level of measurements (i.e., subjective or objective) than it would be to apply a high resolution, objective measurement in a process that considers only one aspect of performance.

145.3 Decision-Making Processes

Measurements that characterize the human, task, assistive device, or combination thereof are themselves only means to an end; the end is typically a decision. As noted previously, decisions are often the result of assessment processes involving one or more measurements. Although not exhaustive, many of the different types of assessments encountered are related to the following questions: (1) Is a particular aspect of performance normal (or impaired)? (2) Is a particular aspect of performance improving, stable, or getting worse? How should therapy be modified? (3) Can a given subject utilize (and benefit from) a particular assistive device? (4) Does a subject possess the required capacity to accomplish a given higher level task (e.g., driving, a particular job after a work-related injury, etc.)?

In Fig. 145.2, several of the basic concepts associated with measurement are used to illustrate how they enter into and facilitate systematic decision-making processes. The upper section shows raw score values as well as statistics for a healthy normal reference population in tabular form (left). It is difficult to reach any decision by simple inspection of just the raw performance capacity values. Tabular data are used to obtain percent normal (middle) and z-score (right) assessments. Both provide a more directly interpretable result regarding subject A's impairments. By examining the "right shoulder flexion extreme of motion" item in the figure, it can be seen that a raw score value corresponding to 51.2% normal yields a very large-magnitude, negative z-score (–10.4). This z-score indicates that virtually no one in the reference population would have a score this low. In contrast, consider similar scores for the "grip strength" item (56.2% normal, z-score = –1.99). On the basis of percent normal scores, it would appear that both of these resources are similarly affected, whereas the z-score basis provides a considerably different perspective due to the fact that grip strength is *much more variable* in healthy populations than the extreme angle obtained by a given limb segment about a joint, relatively speaking. As noted, z-scores account for this variability.

The lower section of Fig. 145.2 considers a situation in which the issue is a specific individual (subject A) considered in a specific task. Tabular data now include raw score values (which are the same as in upper section of the figure) and quantitative demands (typically worst case) imposed on the respective performance resources by task X. The lower-middle plot illustrates the process of individually assessing sufficiency of each performance resource in this task context using a rule-based assessment that incorporates the idea of a threshold (i.e., availability must exceed demand for sufficiency). The lower-right plot illustrates an analogous assessment process that is executed after computation of a stress metric for each of the performance capacities. Here, any demand that corresponds to more than a 100% stress level is obviously problematic. In addition to binary conclusions regarding whether a given capacity is or is not a limiting factor, it is possible to observe that of the two limiting resources (e.g., grip strength and right shoulder flexion extreme of motion), the former is more substantial. This might suggest, for example, that the task be modified so as to decrease the grip-strength demand (i.e., gains in performance capacity required would be substantial to achieve sufficiency) and the use of focused exercise therapy to increase shoulder flexion mobility (i.e., gains in mobility required are relatively small).

145.4 Current Limitations

Quality of Measurements

Key issues are measurement validity, reliability (or repeatability), accuracy, and discriminating power. At issue in terms of current limitations is not necessarily the quality of measurements but limitations with regard to methods employed to determine the quality of measurements and their interpretability.

A complete treatment of these complex topics is beyond the present scope. However, it can be said that standards are such [Potvin et al., 1985] that most published works regarding measurement instruments do address quality of measurements to some extent. Validity (i.e., how well does the measurement reflect the intended quantity) and reliability are most often addressed. However, one could easily be left with the impression that these are binary conditions (i.e., measurement is or is not reliable or valid), when in fact a continuum is required to represent these constructs. Of all attributes that relate to measurement quality, reliability is most commonly expressed in quantitative terms. This is perhaps because statistical methods have been defined and promulgated for the computation of so-called reliability coefficients [Winer, 1971]. Reliability coefficients range from 0.0 to 1.0, and the implication is that 1.0 indicates a perfectly reliable or repeatable measurement process. Current methods are adequate, at best, for making inferences regarding the relative quality of two or more methods of quantifying "the same thing." Even these comparisons require great care. For example, measurement instruments that have greater intrinsic resolving power have a great opportunity to yield smaller-reliability coefficients simply because they are capable of measuring the true variability (on repeated measurement) of the parameter in question within the system under test. While there has been widespread determination or reliability coefficients, there has been little or no effort directed toward determination of what value of a reliability coefficient is "good enough" for a particular application. In fact, reliability coefficients are relatively abstract to most practitioners.

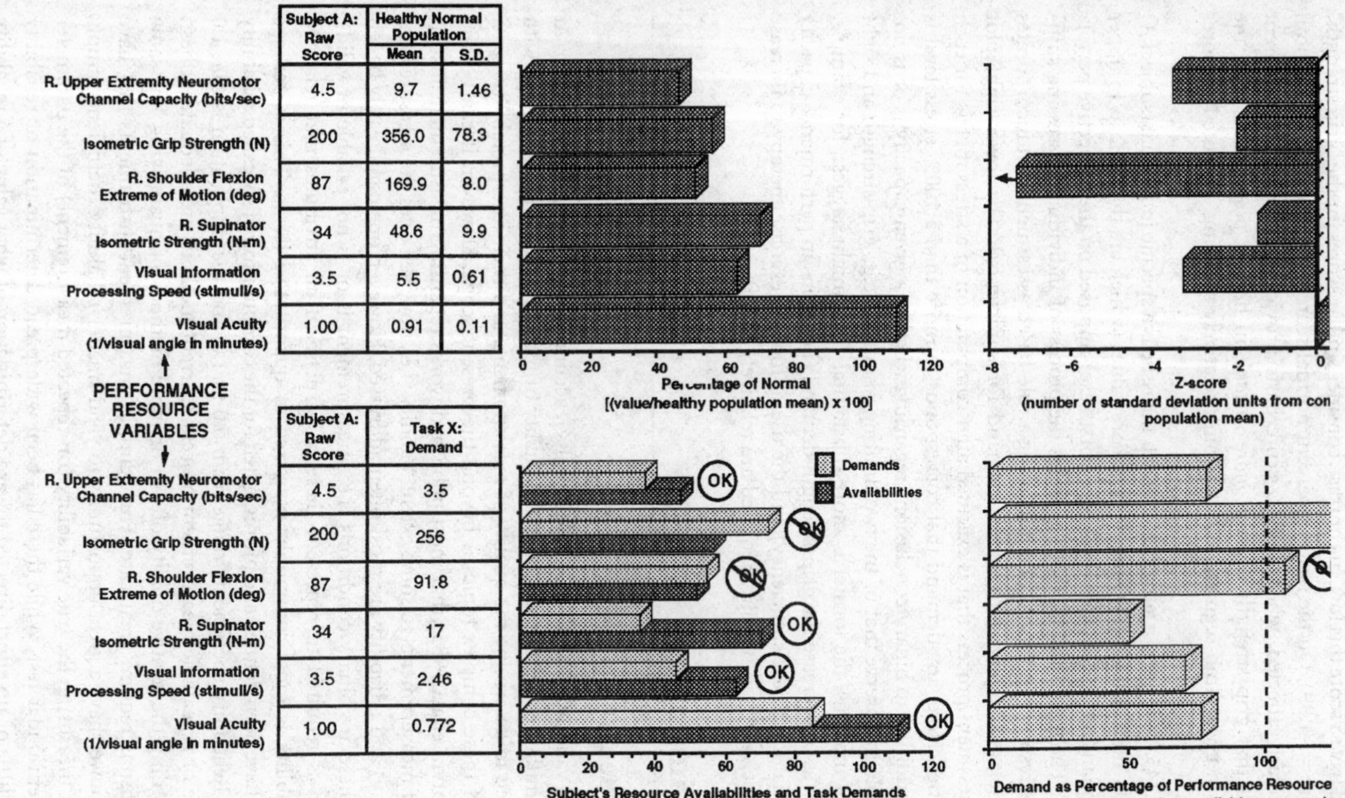

PERFORMANCE RESOURCE VARIABLES	Subject A: Raw Score	Healthy Normal Population Mean	Healthy Normal Population S.D.
R. Upper Extremity Neuromotor Channel Capacity (bits/sec)	4.5	9.7	1.46
Isometric Grip Strength (N)	200	356.0	78.3
R. Shoulder Flexion Extreme of Motion (deg)	87	169.9	8.0
R. Supinator Isometric Strength (N-m)	34	48.6	9.9
Visual Information Processing Speed (stimuli/s)	3.5	5.5	0.61
Visual Acuity (1/visual angle in minutes)	1.00	0.91	0.11

PERFORMANCE RESOURCE VARIABLES	Subject A: Raw Score	Task X: Demand
R. Upper Extremity Neuromotor Channel Capacity (bits/sec)	4.5	3.5
Isometric Grip Strength (N)	200	256
R. Shoulder Flexion Extreme of Motion (deg)	87	91.8
R. Supinator Isometric Strength (N-m)	34	17
Visual Information Processing Speed (stimuli/s)	3.5	2.46
Visual Acuity (1/visual angle in minutes)	1.00	0.772

FIGURE 145.2 Examples of different types of assessments that can be performed by combining performance capacity measures and reference values of different types. The upper section shows raw score values as well as statistics for a healthy normal reference population in tabular form (*left*). It is difficult to reach any decision by simple inspection of just the raw performance capacity values. Tabular data are used to obtain a percent normal assessment (*middle*) and a z-score assessment (*right*). Both of these provide a more directly interpretable result regarding subject A's impairments. The lower section shows raw score values (same as in upper section) and quantitative demands (typically worst case) imposed on the respective performance resources by task X. The lower-middle plot illustrates the process of individually assessing sufficiency of each performance resource in this task context using a threshold rule (i.e., availability must exceed demand for sufficiency). The lower-right plot illustrates a similar assessment process after computation of a stress metric for each of the performance capacities. Here, any demand that corresponds to more than a 100% stress level is obviously problematic.

Methods for determining the quality of a measurement process (including the instrument, procedures, examiner, and actual noise present in the variable of interest) that allow a practitioner to easily reach decisions regarding the use of a particular measurement instrument in a specific application and limitations thereof are currently lacking. At the use of different measurements increases and the number of options available for obtaining a given measurement grows, this topic will undoubtedly receive additional attention. Caution in interpreting literature, common sense, and the use of simple concepts such as "I need to measure range of motion to within 2 degrees in my application" are recommended in the meantime [Mayer et al., 1997].

Standards

Measurements, and concepts with which they are associated, can contribute to a shift from experience-based knowledge acquisition to rule-based, engineering-like methods. This requires (1) a widely accepted conceptual framework (i.e. known to assistive device manufacturers, rehabilitation engineers, and other professionals within the rehabilitation community), (2) a more complete set of measurement tools that are at least standardized with regard to the definition of the quantity measured, (3) special analysis and assessment software (that removes the resistance to the application of more rigorous methods by enhancing the quality of decisions as well as the speed with which they can be reached), and (4) properly trained practitioners. Each is a necessary *but not sufficient* component. Thus balanced progress is required in each of these areas.

Rehabilitation Service Delivery and Rehabilitation Engineering

In a broad sense, it has been argued that all engineers can be considered rehabilitation engineers who merely work at different levels along a comprehensive spectrum of human performance, which itself can represent a common denominator among all humans. Thus an automobile is a mobility aid, a telephone is a communication aid, and so on. Just as in other engineering disciplines, measurement must be recognized not only as an important end in itself (in appropriate instances) but also as an integral component or means within the overall scope of rehabilitation and rehabilitation engineering processes. The service-delivery infrastructure must provide for such means. At present, one should anticipate and be prepared to overcome potential limitations associated with factors such as third-party reimbursement for measurement procedures, recognition of equipment and maintenance costs associated with obtaining engineering-quality measurements, and education of administrative staff and practitioners with regard to the value and proper use of measurements.

Defining Terms

Behavior: A general term that relates to what a human or artificial system does while carrying out its function(s) under given conditions. Often, behavior is characterized by measurement of selected parameters or identification of unique system states over time.

Function: The purpose of a system. Some systems map to a single primary function (e.g., process visual information). Others (e.g., the human arm) map to multiple functions, although at any given time multifunction systems are likely to be executing a single function (e.g., polishing a car). Functions can be described and inventoried, whereas level of performance of a given function can be measured.

Functional assessment: The process of determining, from a relatively global perspective, an individual's ability to carry out tasks in daily life. Also, the result of such a process. Functional assessments typically cover a range of selected activity areas and include (at a minimum) a relatively gross indication (e.g., can or can't do; with or without assistance) of status in each area.

Goal: A desired endpoint (i.e., result) typically characterized by multiple parameters, at least one of which is specified. Examples include specified task goals (e.g., move an object of specified mass from point A to point B in 3 seconds) or estimated task performance (maximum mass, range, speed of movement obtainable given a specified elemental performance resource availability profile), depending on whether a reverse or forward analysis problem is undertaken. Whereas function describes the general process of task, the goal directly relates to performance and is quantitative.

Limiting resource: A performance resource at any hierarchical level (e.g., vertical lift strength, knee flexor speed) that is available in an amount that is less than the worst-case demand imposed by a task. Thus a given resource can be "limiting" only when considered in the context of a specific task.

Overall task goals: Goals associated with a task to be executed by a human-artificial system combination (to be distinguished from goals associated with the task of operating the artificial system).

Performance: Unique qualities of a human or artificial system (e.g., strength, speed, accuracy, endurance) that pertain to how well that system executes its function.

Performance capacity: A quantity in finite availability that is possessed by a system or subsystem, drawn on during tasks, and limits some aspect (e.g., speed, force, production, etc.) of a system's ability to execute tasks, or, the limit of that aspect itself.

Performance capacity measurement: A general class of measurements, performed at different hierarchical levels, intended to quantify one or more performance capacities.

Procedure: A set of constraints placed on a system in which flexibility exists regarding how a goal (or set of goals) associated with a given function can be achieved. Procedure specification requires specification of initial intermediate, and/or final states or conditions dictating how the goal is to be accomplished. Such specification can be thought of in terms of removing some degrees of freedom.

Structure: Physical manifestation and attributes of a human or artificial system and the object of one type of measurements at multiple hierarchical levels.

Style: Allowance for variation within a procedure, resulting in the intentional incomplete specification of a procedure or resulting from either international or unintentional incomplete specification of procedure.

Task: That which results from (1) the combination of specified functions, goals, and procedures or (2) the specification of function and goals and the observation of procedures utilized to achieve the goals.

References

Fuhrer MJ. 1987. Rehabilitation Outcomes: Analysis and Measurement. Baltimore, Brookes.

Ganapathy G, Kondraske GV. 1990. Microprocessor-based instrumentation for ambulatory behavior monitoring. J Clin Eng 15(6):459.

Granger CV, Greshorn GE. 1984. Functional Assessment in Rehabilitation Medicine. Baltimore, Williams & Wilkins.

Hamilton BB, Granger CV, Sherwin FS, et al. 1987. A uniform national data system for medical rehabilitation. In MJ Fuhrer (ed), Rehabilitation Outcomes: Analysis and Measurement, pp 137–147. Baltimore, Brookes.

Imrhan S. 2000. Task analysis and decomposition: Physical components. In JD Bronzino (ed), Handbook of Biomedical Engineering, 2nd ed., Boca Raton, Fla, CRC Press.

Jones RD. 1995. Measurement of neuromotor control performance capacities. In JD Bronzino (ed), Handbook of Biomedical Engineering. Boca Raton, Fla, CRC Press.

Keith RA, Granger CV, Hamilton BB, Sherwin FS. 1987. The functional independence measure: A new tool for rehabilitation. In MG Eisenberg, RC Grzesiak (eds), Advances in Clinical Rehabilitation, vol 1, pp 6–18. New York, Springer-Verlag.

Kondraske GV. 1988. Experimental evaluation of an elemental resource model for human performance. In Proceedings of the Tenth Annual IEEE Engineering in Medicine and Biology Society Conference, New Orleans, pp 1612–13.

Kondraske GV. 1990. Quantitative measurement and assessment of performance. In RV Smith, JH Leslie (eds), Rehabilitation Engineering, pp 101–125. Boca Raton, Fla, CRC Press.

Kondraske GV. 2000. A working model for human system-task interfaces. In JD Bronzino (ed), Handbook of Biomedical Engineering, 2nd ed., Boca Raton, Fla, CRC Press.

Kondraske GV, Johnston C, Pearson, A & Tarbox L. 1997. Performance Prediction and limiting resource identification with nonlinear causal resource analysis. *Proceedings, 19th Annual Engineering in Medicine and Biology Society Conference*, pp 1813–1816.

Kondraske GV, Vasta PJ. 2000. Measurement of information processing performance capacities. In JD Bronzino (ed), Handbook of Biomedical Engineering, 2nd ed., Boca Raton, Fla, CRC Press.

Maxwell KJ. 2000. High-level task analysis: Mental components. In JD Bronzino (ed), Handbook of Biomedical Engineering, 2nd ed., Boca Raton, Fla, CRC Press.

Mayer T, Kondraske GV, Brady Beals, S., & Gatchel RJ. 1997. Spinal range of motion: accuracy and sources of error with inclinometric measurement. *Spine*, 22(17), 1976–1984.

National Institute of Occupational Safety and Health (NIOSH). 1981. Work Practices Guide for Manual Lifting (DHHS Publication No. 81122). Washington, US Government Printing Office.

Pheasant ST. 1986. Bodyspace: Anthropometry, Ergonomics and Design. Philadelphia, Taylor & Francis.

Potvin AR, Tourtellotte WW, Potvin JH, et al. 1985. The Quantitative Examination of Neurologic Function. Boca Raton, Fla, CRC Press.

Smith RV, Leslie JH. 1990. Rehabilitation Engineering. Boca Raton, Fla, CRC Press.

Smith SS. 2000. Measurement of neuromuscular performance capacities. In JD Bronzino (ed), Handbook of Biomedical Engineering, 2nd ed., Boca Raton, Fla, CRC Press.

Tourtellotte WW. 1993. Personal communication.

Vasta PJ, Kondraske GV. 1994. Performance prediction of an upper extremity reciprocal task using nonlinear causal resource analysis. In Proceedings of the Sixteenth Annual IEEE Engineering in Medicine and Biology Society Conference, Baltimore.

Vasta PJ, Kondraske GV. 2000. Human performance engineering: Computer based design and analysis tools. In JD Bronzino (ed), Handbook of Biomedical Engineering, 2nd ed., Boca Raton, Fla, CRC Press.

Webster JG, Cook AM, Tompkin WJ, Vanderheiden GC. 1985. Electronic Devices for Rehabilitation. New York, Wiley.

Winer BJ. 1971. Statistical Principles in Experimental design, 2d ed. New York, McGraw-Hill.

Winter DA. 1990. Biomechanics and Motor Control of Human Movement, 2d ed. New York, Wiley.

World Health Organization. 1980. International Classification of Impairments, Disabilities, and Handicaps. Geneva, World Health Organization.

Further Information

The section of this *Handbook* entitled "Human Performance Engineering" contains chapters that address human performance modeling and measurement in considerably more detail.

Manufacturers of instruments used to characterize different aspects of human performance often provide technical literature and bibliographies with conceptual backgrounds, technical specifications and application examples. A partial list of such sources is included below. (No endorsement of products is implied.)

Baltimore Therapeutic Equipment Co.
7455-L New Ridge Road
Hanover, MD 21076-3105
http://www.bteco.com/

Chattanooga Group
4717 Adams Road
Hixson, TN 37343
http://www.chattanoogagroup.com/

Henley Healthcare
120 Industrial Blvd.
Sugarland, TX 77478
http://www.henleyhealth.com/

Human Performance Measurement, Inc.
P.O. Box 1996
Arlington, TX 76004–1996
http://www.flash.net/~hpm/

Lafayette Instrument
3700 Sagamore Parkway North
Lafayette, IN 47904–5729
http://www.lafayetteinstrument.com

The National Institute on Disability and Rehabilitation Research (NIDRR), part of the Department of Education, funds a set of Rehabilitation Engineering Research Centers (RERCs) and Research and Training Centers (RTCs). Each has a particular technical focus; most include measurements and measurements issues. Contact NIDRR for a current listing of these centers.

Measurement devices, issues, and application examples specific to rehabilitation are included in the following journals:

IEEE Transactions on Rehabilitation Engineering
IEEE Service Center
445 Hoes Lane
P.O. Box 1331
Piscataway, N.J. 08855–1331
http://www.ieee.org/index.html

Journal of Rehabilitation Research and Development
Scientific and Technical Publications Section
Rehabilitation Research and Development Service
103 South Gay St., 5th floor
Baltimore, MD 21202–4051
http://www.vard.org/jour/jourindx.htm

Archives of Physical Medicine and Rehabilitation
Suite 1310
78 East Adams Street
Chicago, IL 60603–6103

American Journal of Occupational Therapy
The American Occupational Therapy Association, Inc.
4720 Montgomery Ln.,
Bethesda, MD 20814–3425
http://www.aota.org/

Physical Therapy
American Physical Therapy Association
1111 North Fairfax St.
Alexandria, VA 22314
http://www.apta.org/

Journal of Occupational Rehabilitation
Subscription Department
Plenum Publishing Corporation
233 Spring St.
New York, NY 10013
http://www.plenum.com/

146

Rehabilitation Engineering Technologies: Principles of Application

Douglas Hobson
University of Pittsburgh

Elaine Trefler
University of Pittsburgh

Rehabilitation engineering is the branch of *biomedical engineering* that is concerned with the application of science and technology to improve the quality of life of individuals with disabilities. Areas addressed within rehabilitation engineering include wheelchairs and seating systems, access to computers, sensory aids, prosthetics and orthotics, alternative and augmentative communication, home and work-site modifications, and universal design. Because many products of rehabilitation engineering require careful selection to match individual needs and often require custom fitting, rehabilitation engineers have necessarily become involved in service delivery and application as well as research, design, and development. Therefore, as we expand on later, it is not only engineers that practice within the field of rehabilitation engineering.

As suggested above, and as in many other disciplines, there are really two career tracks in the field of rehabilitation engineering. There are those who acquire qualifications and experience to advance the state of knowledge through conducting research, education, and product development, and there are others who are engaged in the application of technology as members of service delivery teams. At one time it was possible for a person to work in both arenas. However, with the explosion of technology and the growth of the field over the past decade, one must now specialize not only within research or service delivery but often within a specific area of technology.

One can further differentiate between rehabilitation and assistive technology. *Rehabilitation technology* is a term most often used to refer to technologies associated with the acute-care rehabilitation process. Therapy evaluation and treatment tools, clinical dysfunction measurement and recording instrumentation,

0-8493-0462-8/00/$0.00+$.50

and prosthetic and orthotic appliances are such examples. *Assistive technologies* are those devices and services that are used in the daily lives of people in the community to enhance their ability to function independently, examples being specialized seating, wheelchairs, environmental control devices, workstation access technologies and services are now communication aids. Recognition and support of assistive technology devices and services are now embedded in all the major disability legislation that has been enacted over the last decade.

The primary focus of this chapter is on the role of the rehabilitation engineering practitioner as he or she carries out the responsibilities demanded by the application of assistive technology.

Before launching into the primary focus of this chapter, let us first set a conceptual framework for the *raison d'etre* for assistive technology and the role of the assistive technology professional.

146.1 The Conceptual Frameworks

The application of assistive technology can be conceptualized as minimizing the functional gap between the person and his or her environment. This reality is what technology does for all of us to varying degrees. For example, if you live in a suburban area that has been designed for access only by car and your car breaks down, you are handicapped. If your house has been designed to be cooled by air conditioning in the "dog days" of summer and you lose a compressor, your comfort is immediately compromised by your incompatibility with your environment. Similarly, if you live in a home that has only access by steps and you have an impairment requiring the use of a wheelchair, you are handicapped because you no longer have abilities that are compatible with your built environment. Because our environments, homes, workplaces, schools, and communities have been designed to be compatible with the abilities of the norm, young children, persons with disabilities, and many elderly people experience the consequences of their mismatch as a matter of course. The long-term utopian solution would be to design environments and their contents so that they can be used by all people of all ages, which is the essence of the universal design concept. However, given that today we do not have very many products and environments that have been universally designed, rehabilitation engineers attempt to minimize the effects of the mismatch by designing, developing, and providing technologies that will allow persons with disabilities to pursue their life goals in a manner similar to any other person. Of course, the rehabilitation engineer cannot accomplish this working in isolation but rather must function as a part of a consumer-responsive team that can best deal with the multiplicity of factors that usually impact on the successful application of assistive technology.

Let us now move to another conceptual framework, one that conceptualizes how people actually interact with technology.

The following conceptualization has been adapted from the model proposed by Roger Smith [Smith, 1992]. In Fig. 146.1, Smith suggests that there are three cyclic elements that come into play when humans interact with technology: the human and his or her innate sensory, cognitive, and functional abilities; the human factor's characteristics of the interface between the human and the technology; and the technical characteristics of the technology itself in terms of its output as a result of a specific input by the user. People with disabilities may have varying degrees of dysfunction in their sensory, cognitive, and functional abilities. The interface will have to be selected or adapted to these varying abilities in order to allow the person to effectively interact with the technology. The technology itself will need to possess specific electronic or mechanical capabilities in order to yield the desired outcome. The essence of assistive technology applications is to integrate all three of these element into a functional outcome that meets the specific needs of a user. This is usually done by selecting commercially available devices and technologies at a cost that can be met by either the individual or his or her third-party payment source. When technologies are not available, then they must be modified from existing devices or designed and fabricated as unique custom solutions. It is particularly in these latter activities that a rehabilitation engineer can make his or her unique contribution to team process.

In 1995, Cook and Hussey [Cook and Hussey, 1995], published an excellent text, *Assistive Technologies—Principles and Practice.* As well as comprehensively addressing many of the assistive technologies

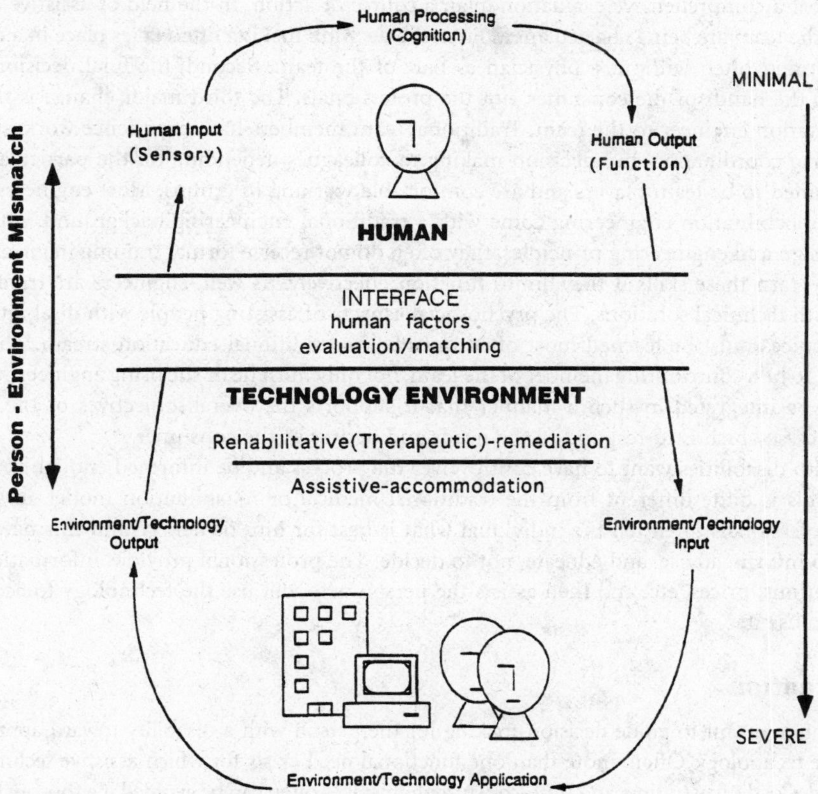

FIGURE 146.1 Conceptual framework of technology and disability. (Modified from Smith [1992].)

briefly covered in this chapter, they also present a conceptual framework which builds on the one developed by Smith above. They introduce the additional concepts of activity and context. That is, understanding of a person's activity desires and the context (social, setting, physical) in which they are to be carried out are essential components to successful assistive technology intervention.

It should be realized that there are several levels of assistive technology. The first level might be termed *fundamental technology* in contrast to advanced technology. Fundamental technologies, such as walkers, crutches, many wheelchairs, activities of daily living (ADL) equipment, etc., usually do not require the involvement of the rehabilitation engineer in their application. Others on the team can better assess the need, confirm the interface compatibility, and verify that the outcome is appropriate. The rehabilitation engineer is most often involved in the application of advanced technologies, such as powered wheelchairs, computerized workstation designs, etc., that require an understanding of the underlying technological principles in order to achieve the best match with the abilities and needs of the user, especially if custom modifications or integration of devices are required to the original equipment. The rehabilitation engineer is usually the key person if a unique solution is necessary.

Let's now discuss a few fundamental concepts related to the process by which assistive technology is typically provided in various service delivery programs.

146.2 The Provision Process

The Shifting Paradigm

In the traditional rehabilitation model of service delivery, a multidisciplinary team of professionals is already in place. Physicians, therapists, counselors, and social worker meet with the client and, based on the findings of a comprehensive evaluation, plan a course of action. In the field of assistive technology, the rules of the team are being charted anew. First, the decision making often takes place in a nonmedical environment and often without a physician as part of the team. Second, the final decision is rapidly moving into the hands of the consumer, not the professionals. The third major change is the addition of a rehabilitation engineer to the team. Traditional team members have experience working in groups and delegating coordination and decision making to colleagues depending on the particular situation. They are trained to be team players and are comfortable working in groups. Most engineers who enter the field of rehabilitation engineering come with a traditional engineering background. Although well versed in design and engineering principles, they often do not receive formal training in group dynamics and need to learn these skills if they are to function effectively. As well, engineers are trained to solve problems with technical solutions. The psychosocial aspects of assisting people with disabilities to make informed choices must be learned most often outside the traditional education stream. Therefore, for the engineer to be a contributing member of the team, not only must he or she bring engineering expertise, but it must be integrated in such a manner that it supports the overall objectives of the technology delivery process, which is to respond to the needs and desires of the consumer.

People with disabilities want to have control over the process and be informed enough to make good decisions. This is quite different from the traditional medical or rehabilitation model, in which well-meaning professionals often tell the individual what is best for him or her. Within this new paradigm, the role is to inform, advise, and educate, not to decide. The professional provides information as to the technical options, prices, etc. and then assists the person who will use the technology to acquire it and learn how to use it.

The Evaluation

An evaluation is meant to guide decision-making for the person with a disability toward appropriate and cost-effective technology. Often, more than one functional need exists for which assistive technology could be prescribed. Costly, frustrating, and time-consuming mistakes often can be avoided if a thorough evaluation based on a person's total functional needs is performed before any technology is recommended. Following the evaluation, a long-range plan for acquisition and training in the chosen technology can be started.

For example, suppose a person needs a seating system, both a powered and manual wheelchair, an augmentative communication device, a computer workstation, and an environmental control unit (ECU). Where does one begin? Once a person's goals and priorities have been established, the process can begin. First, a decision would likely be made about the seating system that will provide appropriate support in the selected manual chair. However, the specifications of seating system should be such that the components also can be interfaced into the powered chair. The controls for the computer and augmentative communication device must be located so that they do not interfere with the display of the communication device and must in some way be compatible wit the controls for the ECU. Only if all functional needs are addressed can the technology be acquired in a logical sequence and in such a manner that all components will be compatible. The more severely disabled the individual, the more technology he or she will need, and the more essential is the process of setting priorities and ensuring compatibility of technical components.

In summary, as suggested by the conceptual model, the process begins by gaining an understanding of the person's sensory, cognitive, and functional abilities, combined with clarification of his or her desires and needs. These are then filtered through the technology options, both in terms of how the interface will integrate with the abilities of the user and how the technology itself will be integrated to meet the defined needs. This information and the associated pros and cons are conveyed to the user, or in some cases their caregiver, who then has the means to participate in the ultimate selection decisions.

Service Delivery Models

People with disabilities can access technology through a variety of different service delivery models. A team of professionals might be available in a university setting where faculty not only teach but also deliver technical services to the community. More traditionally, the team of rehabilitation professionals, including a rehabilitation engineer, might be available at a hospital or rehabilitation facility. More recently, technology professionals might be in private practice either individually, as a team, or part of the university, hospital, or rehabilitation facility structure. Another option is the growing number of rehabilitation technology suppliers (RTSs) who offer commercial technology services within the community. They work in conjunction with an evaluation specialist and advise consumers as to the technical options available to meet their needs. They then sell and service the technology and train the consumer in its use. Local chapters of national disability organizations such as United Cerebral Palsy and Easter Seals also may have assistive technology services. In recent years, a growing number of centers for independent living (CILs) have been developed in each state with federal support. Some of these centers have opted to provide assistive technology services, in addition to their information and referral services, which are common to all CILs. And finally, there are volunteers, either in engineering schools or community colleges (student supervised projects) or in industry (high-technology industries often have staff interested in doing community service), such as the Telephone Pioneers. Each model has its pros and cons for the consumer, and only after thoroughly researching the options will the person needing the service make the best choice as to where to go with his or her need in the community. A word of caution. Only if there is timely provision and follow-up available is a service delivery system considered appropriate, even if the cost of the service is less.

A more extensive description of service delivery options may be reviewed in a report that resulted from a RESNA-organized conference on service delivery [ANSI/RESNA, 1987].

146.3 Education and Quality Assurance

Professionals on the assistive technology team have a primary degree and credential in their individual professions. For example, the occupational or physical therapist will have a degree and most often state licensure in occupational or physical therapy. The engineer will have recognized degrees in mechanical, electrical, biomedical, or some other school of engineering. However, in order to practice effectively in the field of assistive technology, almost all will need advanced training. A number of occupational therapy curriculums provide training in assistive technology, but not all. The same is true of several of the others. Consumers and payers of assistive technology need to know that professionals practicing in the field of assistive technology have a certain level of competency. For this reason, all professionals, including rehabilitation engineers, are pursuing the ATP (assistive technology practitioner) credential through RESNA.

RESNA

RESNA, an interdisciplinary association of persons dedicated to the advancement of assistive technology for people with disabilities, has a credentialing program that credentials individuals on the assistive technology team. As part of the process, the minimum skills and knowledge base for practitioners is tested. Ties with professional organizations are being sought so that preservice programs will include at least some of the knowledge and skills base necessary. Continuing education efforts by RESNA and others also will assist in building the level of expertise of practitioners and consumers. At this time RESNA has a voluntary credentialling process to determine if a person meets a predetermined minimal standard of practice in the field of Assistive Technology. Persons who meet the prerequisite requirements, pass a written exam, and agree to abide by the RESNA Standards of Practice can declare themselves as RESNA certified. They can add the ATP if they are practitioners or ATS if they are suppliers of assistive technology.

Payment for technology and the services required for its application is complex and changing rapidly as health care reform evolves. It is beyond the scope of this discussion to detail the very convoluted and individual process required to ensure that people with disabilities receive what they need. However, there are some basic concepts to be kept in mind. Professionals need to be competent. The documentation of need and the justification of selection must be comprehensive. Time for a person to do this must be allocated if there is to be success. Persistence, creativity, education of the payers, and documentation of need and outcomes are the key issues.

146.4 Specific Impairments and Related Technologies

Current information related to specific technologies is best found in brochures, trade magazines (*Report Rehab*), exhibit halls of technology-related conferences, and databases such as ABLEDATA. Many suppliers and manufacturers are now maintaining Websites, which provides a quick means to locate information on current products. What follows is only a brief introduction to specific disabilities areas to which assistive technology applications are commonly used.

Mobility

Mobility technologies include wheelchairs, walkers, canes, orthotic devices, FES (functional electrical stimulation), laser canes, and any other assistive device that would assist a person with a mobility impairment, be it motor or sensory, to move about in his or her environment. There are very few people who have a working knowledge of all the possible commercial options. Therefore, people usually acquire expertise in certain areas, such as wheelchairs. There are hundreds of varieties of wheelchairs, each offering a different array of characteristics that need to be understood as part of the selection process. Fortunately, there are now several published ways that the practitioner and the consumer can obtain useful information. A classification system has been developed that sets a conceptual framework for understanding the different types of wheelchairs that are produced commercially [Hobson, 1990]. *Paraplegic News* and *Sports and Spokes* annually publish the specifications on most of the manual and powered wheelchairs commonly found in the North American marketplace. These reviews are based on standardized testing that is carried out by manufacturers following the ANSI/RESNA wheelchair standards [ANSI/RESNA, 1990]. Since the testing and measurements of wheelchairs are now done and reported in a standard way, it is possible to make accurate comparisons between products, a tremendous recent advancement for the wheelchair specialist and the users they serve [Axelson et al., 1994].

Possibly the most significant advancement in wheelchairs is the development and application of industry, on an international scale, for testing the safety and durability of their products. These standards also mandate what and how the test information should be made available in the manufacturer's presale literature. The Rehabilitation Engineering Research Center and University of Pittsburgh [RERC, 1999] maintains a large Website, where among its many resources is a listing of wheelchair research publications and a general reference site, termed Wheelchairnet [Wheelchairnet, 1999]. The RERC site also tracks the current activities occurring in many of the wheelchair standards working groups. Finally, Cooper [1995, 1998] has published two excellent reference texts on rehabilitation engineering with emphasis on wheeled mobility.

Sitting

Many people cannot use the wheelchairs as they come from the manufacturer. Specialized seating is required to help persons to remain in a comfortable and functional seated posture for activities that enable them to access work and attend educational and recreational activities. Orthotic supports, seating systems in wheelchairs, chairs that promote dynamic posture in the workplace, and chairs for the elderly that fit properly, are safe, and encourage movement all fit into the broad category of sitting technology.

Sensation

People with no sensation are prone to skin injury. Special seating technology can assist in the preventions of tissue breakdown. Specially designed cushions and backs for wheelchairs and mattresses that have pressure-distributing characteristics fall into this category. Technology also has been developed to measure the interface pressure. These tools are now used routinely to measure and record an individual's pressure profile, making cushion selection and problem solving more of a science than an art.

Again, a classification system of specialized seating has been developed that provides a conceptual framework for understanding the features of the various technologies and their potential applications. The same reference also discusses the selection process, evaluation tools, biomechanics of supported sitting, and materials properties of weight-relieving materials [Hobson, 1990].

Access (Person-Machine Interface)

In order to use assistive technology, people with disabilities need to be able to operate the technology. With limitations in motor and/or sensory systems, often a specially designed or configured interface system must be assembled. It could be as simple as several switches or a miniaturized keyboard or as complex as an integrated control system that allows a person to drive a wheelchair and operate a computer and a communication device using only one switch.

Communication

Because of motor or sensory limitations, some individuals cannot communicate with spoken or written word. There are communication systems that enable people to communicate using synthesized voice or printed output. Systems for people who are deaf allow them to communicate over the phone or through computer interfaces. Laptop computers with appropriate software can enable persons to communicate faster and with less effort than previously possible. Some basic guidelines for selecting an augmentative communication system, including strategies for securing funding, have been proposed in an overview chapter by James Jones and Winifred Jones [Jones & Jones, 1990].

Transportation

Modified vans and cars enable persons with disabilities to independently drive a vehicle. Wheelchair tie-downs and occupant restraints in personal vehicles and in public transportation vehicles are allowing people to be safely transported to their chosen destination. Fortunately, voluntary performance standards for restraint and tie-down technologies have been developed by a task group within the Society for Automotive Engineers (SAE). Standards for car hand controls, van body modifications, and wheelchair lifts are also available from SAE. These standards provide the rehabilitation engineer with a set of tools that can be used to confirm safety compliance of modified transportation equipment. Currently in process and still requiring several more years of work are transport wheelchair and vehicle power control standards.

Activities of Daily Living (ADL)

ADL technology enables a person to live independently as much as possible. Such devices as environmental control units, bathroom aids, dressing assists, automatic door openers, and alarms are all considered aids to daily living. Many are inexpensive and can be purchased through careful selection in stores or through catalogues. Others are quite expensive and must be ordered through vendors who specialize in technology for independent living.

Ron Mace, now deceased and creator of the Center for Universal Design at the North Carolina State University, is widely acknowledged as the father of the Universal Design concept. The concept of universal design simply means that if our everyday built environments and their contained products could be designed to meet the needs of a wider range of people, both young and old, then the needs of more persons with disabilities would be met without the need for special adaptions [Center for Universal Design, 1999]. Others like Paul Grayson have also published extensively regarding the need to re-think how we design our living environments [Grayson, 1991]. Vanderheiden and Denno have prepared human factors guidelines that provide design information to allow improved access by the elderly and persons with disabilities [Denno et al., 1992; Vanderheiden & Vanderheiden, 1991; Trace Center, 1999].

School and Work

Technology that supports people in the workplace or in an educational environment can include such applications as computer workstations, modified restrooms, and transportation to and from work or school. Students need the ability to take notes and do assignments, and people working have a myriad of special tasks that may need to be analyzed and modified to enable the employee with the disability to be independent and productive. Weisman has presented an extensive overview of rehabilitation engineering in the workplace, which includes a review of different types of workplaces, the process of accommodation, and many case examples [Weisman, 1990].

Recreation

A component of living that is often overlooked by the professional community is the desire and, in fact, need of people with disabilities to participate in recreational activities. Many of the adaptive recreational technologies have been developed by persons with disabilities themselves in their effort to participate and be competitive in sports. Competitive wheelchair racing, archery, skiing, bicycles, and technology that enables people to bowl, play pool, and fly their own airplanes are just a few areas in which equipment has been adapted for specific recreational purposes.

Community and Workplace Access

There is probably no other single legislation that is having a more profound impact on the lives of people with disabilities then the Americans with Disabilities Act (ADA), signed into law by President Bush in August of 1990. This civil rights legislation mandates that all people with disabilities have access to public facilities and that reasonable accommodations must be made by employers to allow persons with disabilities to access employment opportunities. The impact of this legislation is now sweeping America and leading to monumental changes in the way people view the rights of persons with disabilities.

146.5 Future Developments

The field of rehabilitation engineering, both in research and in service delivery, is at an important crossroad in its young history. Shifting paradigms of services, reduction in research funding, consumerism, credentialing, health care reform and limited formal educational options all make speculating on what the future may bring rather hazy. Given all this, it is reasonable to say that one group of rehabilitation engineers will continue to advance the state of the art through research and development, while another group will be on the front lines as members of clinical teams working to ensure that individuals with disabilities receive devices and services that are most appropriate for their particular needs.

The demarcation between researchers and service providers will become clearer, since the latter will become credentialed. RESNA and its professional specialty group (PSG) on rehabilitation engineering are working out the final credentialing steps for the Rehabilitation Engineer RE and the Rehabilitation Engineering Technologist RET. Both must also be an ATP. They will be recognized as valued members of the clinical team by all members of the rehabilitation community, including third-party payers, who will reimburse them for the rehabilitation engineering services that they provide. They will spend as much or more time working in the community as they will in clinical settings. They will work closely with consumer-managed organizations who will be the gatekeepers of increasing amounts of government-mandated service dollars.

If these predictions come to pass, the need for rehabilitation engineering will continue to grow. As medicine and medical technology continue to improve, more people will survive traumatic injury, disease, and premature birth, and many will acquire functional impairments that impede their involvement in personal, community, educational, vocational, and recreational activities. People continue to live longer lives, thereby increasing the likelihood of acquiring one or more disabling conditions during their lifetime. This presents an immense challenge for the field of rehabilitation engineering. As opportunities grow, more engineers will be attracted to the field. More and more rehabilitation engineering education programs will develop that will support the training of qualified engineers, engineers who are looking for exciting challenges and opportunities to help people live more satisfying and productive lives.

References

ANSI/RESNA. 1990. Wheelchair Standards. RESNA Press, RESNA, 1700 Moore St., Arlington, VA 22209–1903.

Axelson P, Minkel J, Chesney D. 1994. A Guide to Wheelchair Selection: How to Use the ANSI/RESNA Wheelchair Standards to Buy a Wheelchair. Paralyzed Veterans of America (PVA).

Bain BK, Leger D. 1997. Assistive Technology. An Interdisciplinary Approach. Churchill Livingstone, New York.

Center for Universal Design, 1999. http://www.design.ncsu.edu/cud/

Cook AM, Hussey SM. 1995. Assistive Technologies: Principles and Practice. Mosby, St. Louis, MO.

Cooper RA. 1995. Rehabilitation Engineering Applied to Mobility and Manipulation. Institute of Physics Publishing, Bristol, U.K.

Cooper RA. 1998. Wheelchair Selection and Configuration. Demos Medical Publishing, New York.

Deno JH, et al. 1992. Human Factors Design Guidelines for the Elderly and People with Disabilities. Honeywell, Inc., Minneapolis, MN 55418 (Brian Isle, MN65–2300).

Galvin JC, Scherer MJ. 1996. Evaluating, Selecting, and Using Appropriate Assistive Technology, Aspen Publishers, Gaithersburg, MD.

Hobson DA. 1990. Seating and mobility for the severely disabled. In R Smith, J Leslie (eds), Rehabilitation Engineering, pp 193–252. CRC Press, Boca Raton, FL.

Jones D, Jones W. 1990. Criteria for selection of an augmentative communication system. In R Smith, J Leslie (eds), Rehabilitation Engineering, pp 181–189. CRC Press, Boca Raton, FL.

Medhat M, Hobson D. 1992. Standardization of Terminology and Descriptive Methods for Specialized Seating. RESNA Press, RESNA, 1700 Moore St., Arlington, VA 22209–1903.

Rehabilitation Technology Service Delivery—A Practical Guide. 1987. RESNA Press, RESNA, 1700 Moore St., Arlington, VA 22209–1903.

Smith RO. 1992. Technology and disability. AJOT 1(3):22.

Society for Automotive Engineers. 1994. Wheelchair Tie-Down and Occupant Restraint Standard (committee draft). SAE. Warrendale, PA.

Trace Center, 1999. http://trace.wisc.edu/

Vanderheiden G, Vanderheiden K. 19991. Accessibility Design Guidelines for the Design of Consumer Products to Increase their Accessibility to People with Disabilities or Who Are Aging. Trace R&D Center, University of Wisconsin, Madison, WI.

Weisman G. 1990. Rehabilitation engineering in the workplace. In R Smith, J Leslie (eds), Rehabilitation Engineering, pp 253–297. CRC Press, Boca Raton, FL.

WheelchairNet, 1999. http://www.wheelchairnet.org

Further Information

ABLEDATA, 8455 Colesville Rd., Suite 935, Silver Spring, Md. 20910–3319.

Historical Perspectives 4
Electromyography

Leslie A. Geddes
Purdue University

Early Investigations

The study of bioelectricity started with the Galvani-Volta controversy over the presence of electricity in frog muscle [see Geddes and Hoff, 1971]. Galvani likened the sciatic nerve-gastrocnemius muscle to a charged Leyden jar (capacitor) in which the nerve was the inner conductor and the surface of the muscle was the outer conductor. Therefore, Galvani thought that by joining the two with an arc of dissimilar metals, the biologic capacitor was discharged and the muscle twitched. Volta proved conclusively that it was the dissimilar metals in contact with tissue fluid that was the stimulus.

Interestingly, it was found that when the sciatic nerve of a nerve-muscle preparation was laid on the cut end of another frog muscle and the nerve was touched to the intact surface, the muscle of the nerve-muscle preparation twitched. Here was evidence of stimulation without metal conductors; this experiment was performed by Matteucci [1842].

With the first galvanometers, it was shown that current would be indicated when one electrode was placed on the cut end of a frog muscle and the other on the intact surface. This phenomenon became known as the *injury current* or *frog current*, the cut surface being negative to the intact surface.

Whereas the foregoing experiments showed that skeletal muscle possessed electricity, little was known about its relation to contraction. Matteucci [1842] conducted an ingenious experiment in which he placed the nerve of a second nerve-muscle preparation on the intact muscle of a first such preparation and stimulated the nerve of the first using an inductorium. He discovered that both muscles contracted. Here is the first evidence of the electric activity of contracting skeletal muscle.

Matteucci and DuBois-Reymond both found that the injury current disappeared when a muscle was contracted tetanically. This observation led directly to the concept of a resting membrane potential and its disappearance with activity [see Hoff and Geddes, 1957].

That human muscle, as well as frog muscle, produced electric activity was demonstrated by Du Bois-Reymond [1858] in the manner shown in Fig. HP4.1. With electrodes in saline cups connected to a galvanometer, Du Bois-Reymond stated that as soon as the fingers were placed in the cups, the galvanometer needle deflected, and it required some time for a position of equilibrium to be attained. Du Bois-Reymond [1858] stated:

> As soon as this state [equilibrium] is attained, the whole of the muscles of one of the arms must be so braced that an equilibrium may be established between the flexors and the extensors of all the articulations of the limb, pretty much as in a gymnastic school is usually done when one wants to let a person feel the development of one's muscles.
>
> As soon as this is done, the [galvanometer] needle is thrown into movement, its deflection being uniformly in such a sense as to indicate in the braced arm "an inverse current," according to Nobili's nomenclature; that is to say, a current passing from the hand to the shoulder. The braced arm then

0-8493-0462-8/00/$0.00+$.50

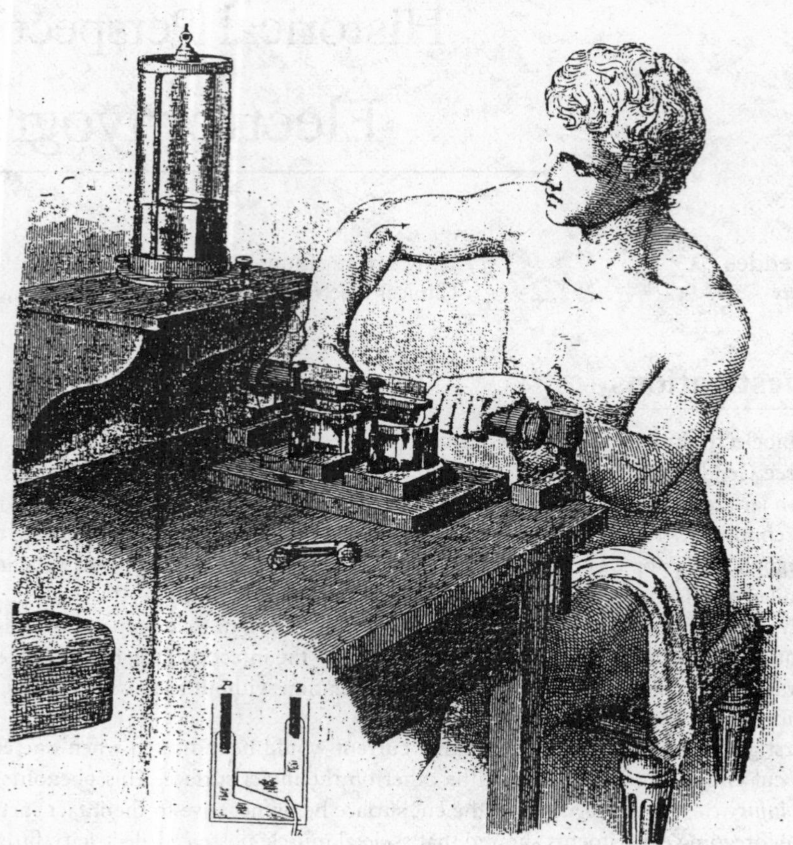

FIGURE HP4.1 The first evidence contracting skeletal muscle in man produces an electrical signal. (From Du Bois-Reymond [1858].)

acts the part of the copper in the compound arc of zinc and copper mentioned above. [Du Bois-Reymond was referring to the polarity of a voltaic cell in which zinc and copper are the positive and negative electrodes, respectively.]

The rheotome and slow-speed galvanometer were used to reconstruct the form of the muscle action potential. However, it was desired to know the time course of the electric change associated with a single muscle contraction (twitch), as well as its relationship to the electrical event (action potential). A second item of interest was the so-called latent period, that time between the stimulus and the onset of muscle contraction, which Helmholtz [1853] reported to be 10 ms for frog muscle.

Waller [1887] set himself the task of measuring the latent period and the relationship between the action potential and the force developed by frog gastrocnemius muscle in response to a single stimulus. He found that the onset of the twitch was later than the onset of the action potential, as shown in Fig. HP4.2. However, the true form of the muscle action potential and its relationship to the onset of the twitch had to await the development of the micropipet electrode, the vacuum-tube amplifier, and the cathode-ray oscilloscope. In 1957, Hodgkin and Horowicz [1957] recorded the twitch and action potential of a single muscle fiber of the frog. Figure HP4.3 is a copy of their record. Note that the onset of the action potential precedes the onset of muscle contraction by about 4 ms. We know that it is the action potential that triggers the release of mechanical energy.

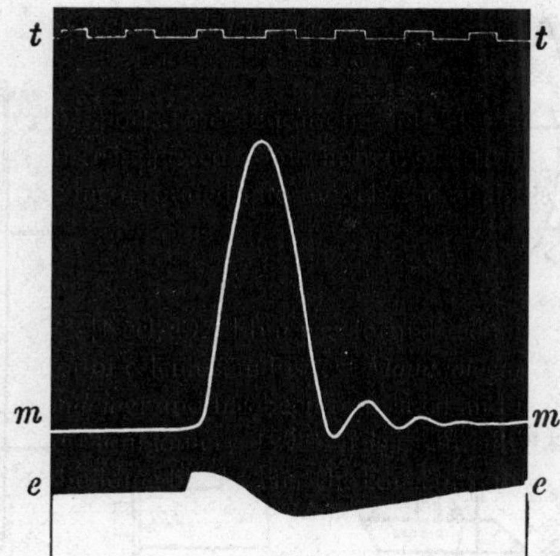

FIGURE HP4.2 Relationship between the twitch (recorded with a myograph, *m*) and the action potential (recorded with a capillary electrometer, *e*) of a frog gastrocnemius muscle. The time marks (*t*) are 1/20 s. (From Waller [1887].)

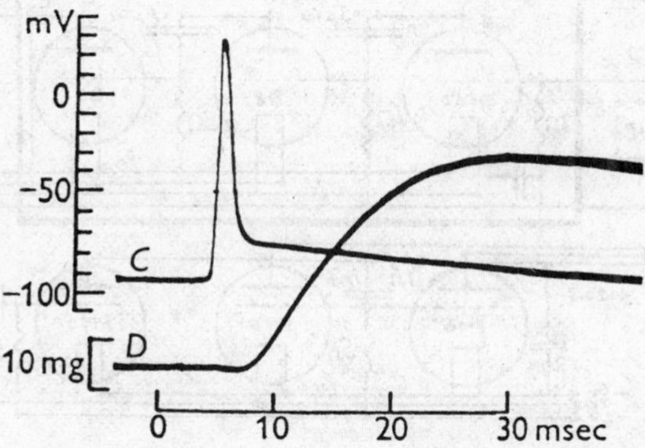

FIGURE HP4.3 The relationship between the muscle action potential and the force of concentration in a single skeletal muscle fiber in the frog. (From Hodgkin and Horowicz [1957].)

Clinical Electromyography

It was well known that when a nerve that innervates a skeletal muscle is cut, the muscle is paralyzed immediately; however, days to weeks later (depending on the species), on careful visual examination, the individual muscle fibers are seen to be contracting and relaxing randomly, i.e., fibrillating. The first to bring the facts together regarding normal muscle action potentials and denervation-fibrillation potentials were Denny-Brown and Pennybacker in the United Kingdom [1939]; the date and locale are highly significant. They distinguished between involuntary twitching of innervated muscle and fibrillation of denervated muscle by recording both the electric and mechanical activity of muscles. Two instrumental advances made their study possible: (1) the use of a hypodermic needle electrode inserted into the muscle and (2) the use of a rapidly responding, mirror-type photographic recorder, the Matthews [1928] oscillograph.

The cathode-ray tube was not generally available in the United Kingdom when Matthews [1928] constructed his moving-tongue mirror oscillograph. The device consisted of a piece of soft iron mounted on a steel leaf spring, as shown in Fig. HP4.4. A strong electromagnet attracted the soft iron, which bent

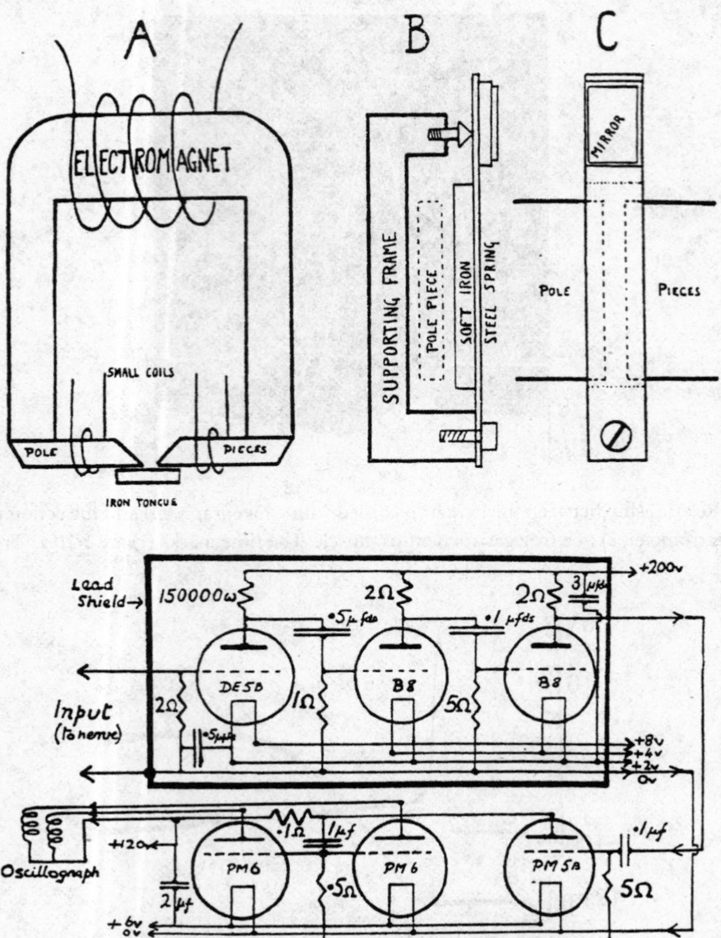

FIGURE HP4.4 The Matthews moving-tongue oscillograph and amplifier used with it. The electromagnetic coil (*A*) provided an attractive force on the tongue (soft iron and steel spring); the signal current applied to the small coils aided or opposed this force and caused the tongue to bend more or less and hence move the mirror which reflected a beam of light on to a recording surface. (From BHC Matthews, 1928, J. Physiol (Lond) 65:225, with permission.)

the steel (leaf-spring) support. Two coils mounted on the pole faces were connected to the output tubes of a five-stage, single-sided, resistance-capacitance coupled amplifier. The amplified potentials altered the current in the electromagnet coils, causing more or less bending of the leaf spring, thereby tilting the mirror mounted on the leaf spring and permitting photographic recording of action potentials.

With the Matthews oscillograph, Denny-Brown and Pennybacker [1939] laid the foundation for clinical electromyography when they reported as follows:

> Denervated muscle fibers contract periodically, and the confused medley of small twitches constitutes true fibrillation. The movement is so slight that it can seldom be seen in the clinic. The twitchings appear to be due to heightened excitability of the sarcolemma or rapidly conducting portion of the muscle fibre to traces of free acetylcholine in the tissues.
>
> Reinnervated muscle is free from such involuntary excitation, except for the curious "contracture" of facial muscle, which consists of periodic, intense repetitive discharges which suggest a central mechanism.

Earlier it was stated that 1939 was significant; this was the year when World War II broke out in Europe. Soon motor nerve injuries due to shrapnel wounds began to appear in large numbers, and the need for electromyography to identify denervation fibrillation potentials and their gradual disappearance with reinnervation was urgent. The first electromyograph in North America was developed by Herbert Jasper at McGill University (Montreal Neurological Institute). Starting in 1942, design concepts were developed, and by 1944 prototypes had been built and used clinically. In his report to the Committee on Army Medical Research, Capt. Jasper [1945] stated:

> The present equipment has been developed over a period of about 18 months experimentation with different designs of electromyograph for use on hospital wards. Numerous modifications of design have been incorporated in the present model in order to provide simplicity of operation, portability, freedom from electrical interference, and perfection of both the audible and visible analysis of the electrical activity of normal and diseased muscles. A portable clinical electromyograph has been developed which has proven to be practical in operation on hospital wards to aid in the diagnosis and prognosis of muscles paralyzed by traumatic injuries of their nerve supply. Four complete units have been constructed for use in special centers for the treatment of nerve injuries.

The Royal Canadian Army Medical Corps (RCAMC) electromyograph had many unique design features that were incorporated in all later commercially available electromyographs. It consisted of three units, a small battery-operated three-stage differential amplifier (Fig. HP4.5) and an oscilloscope (Fig. HP4.6), both placed on a loudspeaker cabinet on rubber-wheel casters (Fig. HP4.7). Thus simultaneous visual display and aural monitoring of normal motor units and fibrillation potentials was possible.

The preamplifier (Fig. HP4.5) was very carefully constructed, the input tubes being supported by rubber-mounted antimicrophonic sockets. The grid resistors (R9) and plate resistors (R8) were wire wound and carefully matched. A high common-mode rejection ratio was obtained by matching the input tubes and adjustment of the potentiometer (R7) in the screen-grid supply. A common-mode rejection ratio in excess of 10,000 was easily achieved. The overall frequency response extended from 3 to 10,000 Hz.

The cathode-ray oscilloscope (Fig. HP4.6) was of unusual design for that time because the sweep velocity was independent of the number of sweeps per second, a feature to appear much later in oscilloscopes. A linear sweep (time base) was obtained by charging a capacitor (0.1 μF) through a pentode (6U7G) which acted as a constant-current device. The sweep was initiated at a rate of about 7 times per second by the multivibrator (6N7), which also provided an output to enable stimulating a nerve, the stimulus occurring at the beginning of each sweep, thereby permitting nerve conduction-time measurements. The oscilloscope also contained the audio amplifier (6L6). The cathode-ray tube had a short-persistence, blue-white phosphor that produced brilliant blue-white images of remarkable clarity. A camera was used to obtain photographic records of the waveforms, which were optimized by listening to them via the loudspeaker (as advocated by Adrian) as the needle electrode was being inserted and adjusted. Fig. HP4.8 illustrates typical action potentials.

At the end of the war (1945), oscilloscopes became available, and Fig. HP4.7 shows the RCAMC electromyograph with a Cossor oscilloscope (right) and the high-gain differential amplifier (left), both on the loudspeaker cabinet, which was on casters. The recessed opening at the top of the loudspeaker cabinet face provided access to the on-off and volume controls.

In addition to the creation of a high-performance EMG unit, Jasper introduced the monopolar needle electrode system used in all subsequent EMGs. The needle electrode was insulated with varnish down to its tip and was paired with a skin-surface electrode of silver. The patient was grounded by another electrode taped to the same member that was being examined. Figure HP4.8 illustrates application of the electrodes and typical motor-unit and fibrillation potentials. The report by Jasper and Ballem [1949] summarized the experience with the RCAMC electromyograph and laid the foundation for diagnostic EMG.

World War II ended in 1945, after which electromyographs became available commercially. Their features were essentially the same as those embodied in the RCAMC electromyograph. From the beginning, these units were completely power-line-operated, the author's master's thesis describing the first of these units.

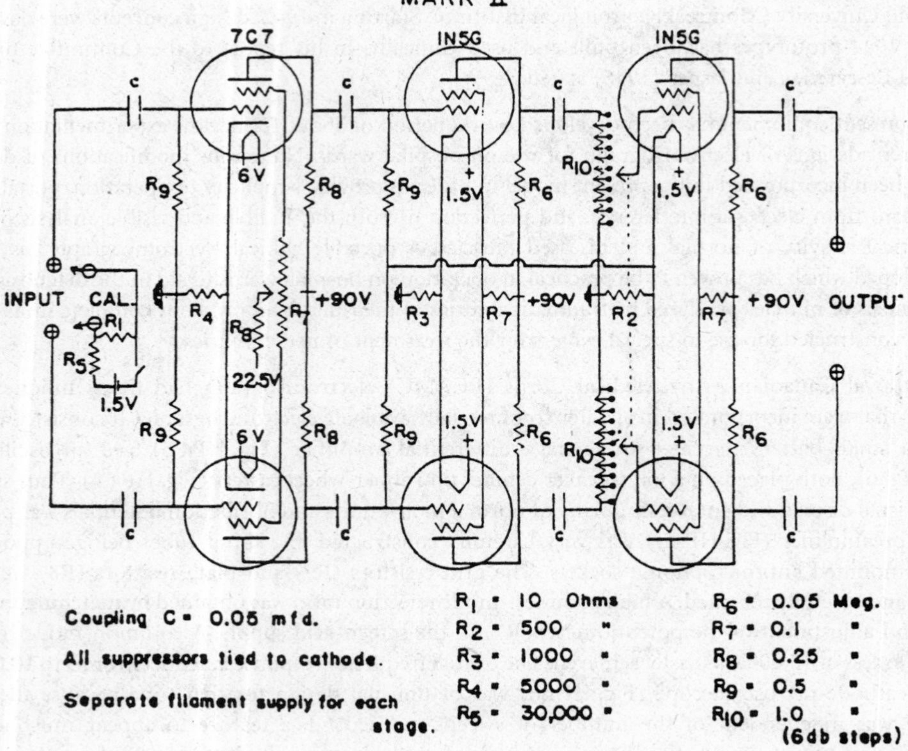

FIGURE HP4.5 The three-stage resistance-capacity coupled differential amplifier used is the RCAMC electromyography. (From Jasper et al. [1945].)

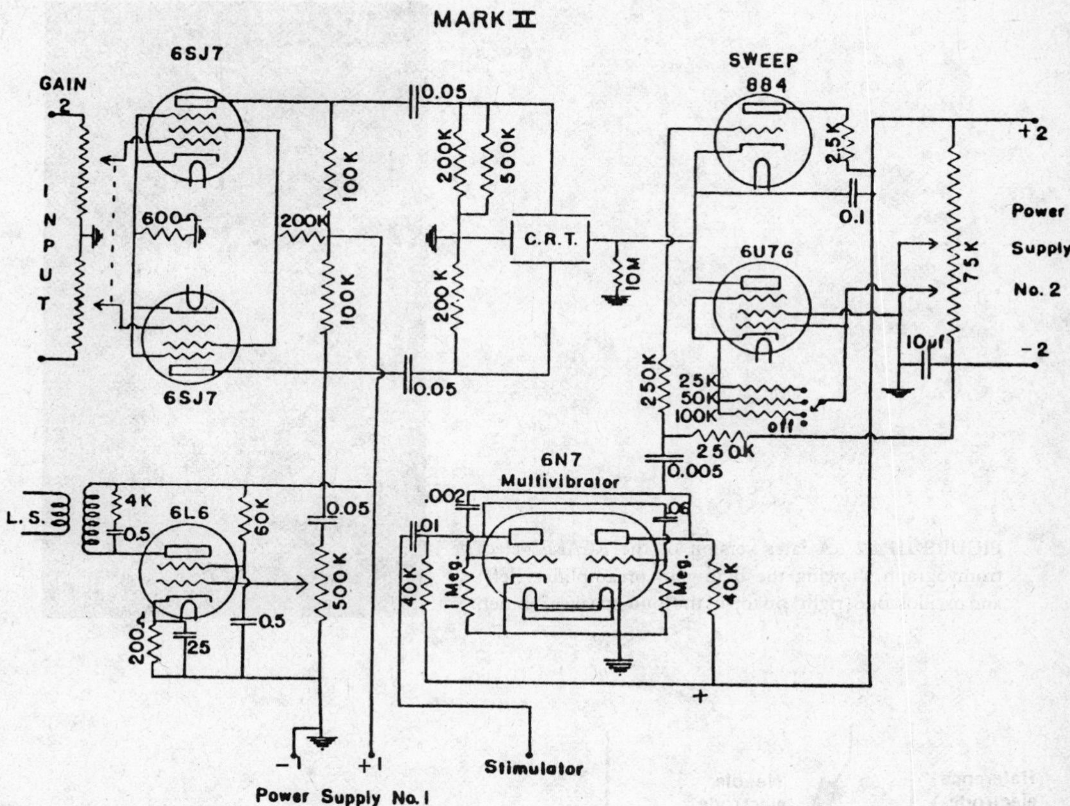

FIGURE HP4.6 The oscilloscope, loudspeaker amplifier, and stimulator unit of the RCAMC electromyograph. (From Jasper et al. [1945].)

FIGURE HP4.7 A later version of the RCAMC electromyograph showing the high-gain preamplifier (left) and oscilloscope (right) on top of the loud-speaker cabinet.

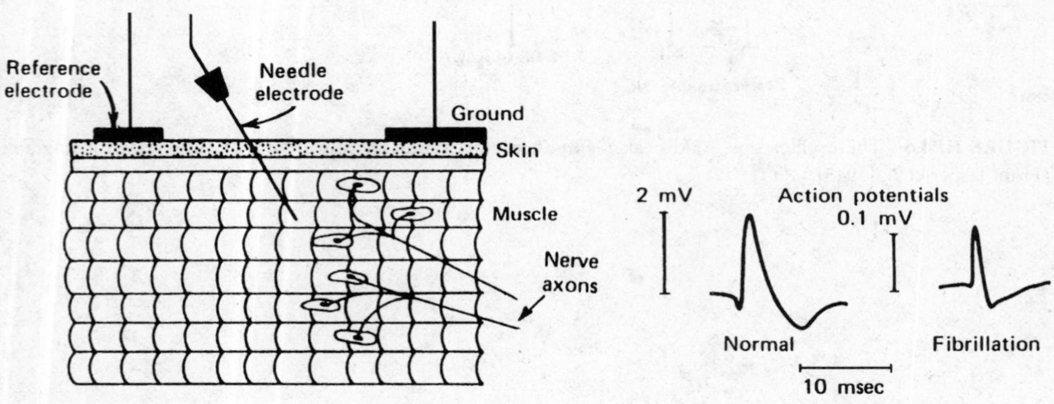

FIGURE HP4.8 Electrode arrangement and action potentials with the RCAMC electromyograph. (Redrawn from Jasper and Ballem [1949].)

References

Denny-Brown D, Pennybacker JB. 1939. Fibrillation and fasciculation in voluntary muscle. Brain 61:311.

Du Bois-Reymond E. 1858. Untersuchungen uber thierische Elekticitat. Moleschott's Untersuch. Z Natur Mensch 4:1.

Geddes LA, Hoff HE. 1971. The discovery of bioelectricity and current electricity (the Galvani-Volta controversy). IEEE Spect 8(12):38.

Helmholtz H. 1853. On the methods of measuring very small portions of time and their application to physiological purposes. Philos Mag J Sci 6:313.

Hodgkin AL, Horowicz P. 1957. The differential action of hypertonic solutions on the twitch and action potential of a muscle fiber. J Physiol (Lond) 136:17P.

Hoff HE, Geddes LA. 1957. The rheotome and its prehistory: A study in the historical interrelation of electrophysiology and electromechanics. Bull Hist Med 31(3):212.

Jasper HH, Ballem G. 1949. Unipolar electromyograms of normal and denervated human muscle. J Neurophysiol 12:231.

Jasper HH. 1945. The RCAMC electromyograph, Mark II. With the technical assistance of Lt. RH Johnston and LA Geddes. Report submitted to the Associate Committee on Army Medical Research, National Research Council of Canada, 27 April 1945.

Matteucci C. 1842. Duxieme memoire sur le courant electrique propre de la grenouille et sur celui des animaux a sang chaud. Ann Chim Phys 3S(6):301.

Matthews BHC. 1928. A new electrical recording system for physiological work. J Physiol (Lond) 65:225.

Waller AD. 1887. A demonstration on man of electromotive changes accompanying the heart's beat. J Physiol (Lond) 8:229.

References

[illegible]

[illegible]

[illegible]

[illegible] 1981 [illegible]

[illegible]

[illegible]

[illegible]

[illegible]

[illegible]

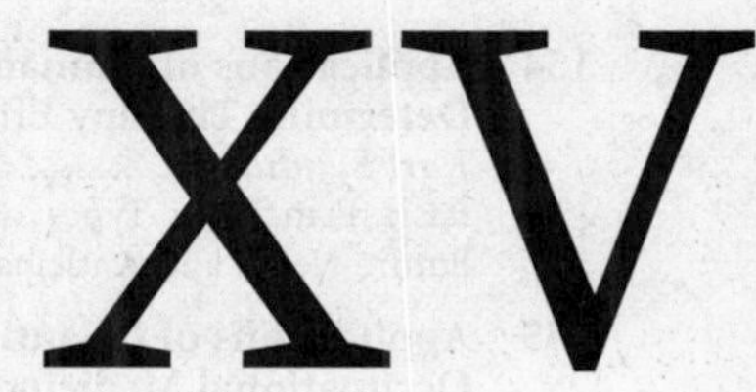

Human Performance Engineering

George V. Kondraske
University of Texas at Arlington,
Human Performance Institute

THE ULTIMATE GOAL OF HUMAN performance engineering is enhancement of the performance and safety of humans in the execution of tasks. The field (in a more formalized sense) was fueled initially by military applications but has become an important component in industrial settings as well. In a biomedical engineering context, the scope of definition applied to the term *human* not only encompasses individuals with capabilities that differ from those of a typical healthy individual in many possible different ways (e.g., individuals who are disabled, injured, unusually endowed, etc.) but also includes those who are "healthy" (e.g., health care professionals). Consequently , one finds a wide range of problems in which human performance engineering and associated methods are employed. Some examples include

- Evaluation of an individual's performance capacities for determining the efficacy of new therapeutic interventions or so-called level of disability for worker's compensation and other medical-legal purposes.
- Design of assistive devices and/or work sites in such a way that a person with some deficiency in his or her "performance resource profile" will be able to accomplish specified goals.
- Design of operator interfaces for medical instruments that promote efficient, safe, and error-free use.

In basic terms, each of these situations involves one or more of the following: (1) a human, (2) a task or tasks, and (3) the interface of a human and task(s). Human performance engineering emphasizes concepts, methods, and tools that strive toward treatment of each of these areas with the engineering rigor that is routinely applied to artificial systems (e.g., mechanical, electronic, etc.). Importance is thus placed on models (a combination of cause-and-effect and statistical), measurements (of varying degrees of sophistication that are selected to fit needs of a particular circumstance), and various types of analyses. Whereas many specialty areas within biomedical engineering begin with an emphasis on a specific subsystem and then proceed to deal with it at lower levels of detail (sometimes even at the molecular level) to determine how it functions and often why it malfunctions, human performance engineering emphasizes subsystems and their performance capacities (i.e., *how well* a system functions), the integration of these into a whole and their interactions, and their operation in the execution of tasks that are of ultimate concern to humans. These include tasks of daily living, work, and recreation. In recent years, there has been an increased concern within medical communities on issues such as quality of life, treatment outcome measures, and treatment cost-effectiveness. By linking human subsystems into the "whole" and discovering objective quantitative relationships between the human and tasks, human performance engineering can play an important role in addressing these and other related concerns.

Human performance engineering combines knowledge, concepts, and methods from across many disciplines (e.g., biomechanics, neuroscience, psychology, physiology, and many others) which, in their overlapping aspect, all deal with similar problems. Among current difficulties is that these wide-ranging efforts are not linked by a conceptual framework that is commonly employed across contributing disciplines. In fact, few candidate frameworks exist even within the relevant disciplines. One attempt to provide some unification and commonality is presented in Chapter 147 as basis for readers to integrate material in subsequent chapters and from other sources. In a further attempt to enhance continuity across this section, chapter authors have been requested to consider this perspective and to incorporate basic concepts and terms where applicable.

Chapters 148 through 150 look "toward the human" and focus on measurement of human performance capacities and related issues. Owing to a combination of the complexity of the human system (even when viewed as a collection of rather high-level subsystems) and limited space available, treatment is not comprehensive. For example, measurement of sensory performance capacities (e.g., tactile, visual, auditory) is not included in this edition. Both systems and tasks can be viewed at various hierarchical levels. Chapters 148 and 149 focus on a rather "low" systems level and discuss basic functional units such as actuator, processor, and memory systems. Chapter 150 moves to a more intermediate level, where speech, postural control gait, and hand-eye coordination systems could be considered. Measurement of structural parameters, which play important roles in many analyses, also is not allocated the separate chapter it deserves (as a minimum) due to space limitations. Chapter 151 and 152 then shift focus to consider the analysis of different types of tasks in a similar, representative fashion.

Chapters 153 through 155 are included to provide insight into a representative selection of application types. Space constraints, the complexity of human performance, and the great variety of tasks that can be considered limit the level of detail with which such material can reasonably be presented. Work in all application areas will begin to benefit from emerging computer-based tools, which is the theme of chapter 156. The section concludes with a look to the future (Chapter 157) that summarizes selected current limitations, identifies some specific research and development needs, and speculates regarding the nature of some anticipated developments.

Many have contributed their talents to this exciting field in terms of both research and applications, yet much remains to be done. I am indebted to the authors not only for their contributions and cooperation during the preparation of this section but also for their willingness to accept the burdens of communicating complex subject matter reasonably, selectively, and as accurately as possible within the imposed constraints.

147

A Working Model for Human System-Task Interfaces

George V. Kondraske
University of Texas at Arlington

Humans are complex systems. Our natural interest in ourselves and things that we do has given rise to the study of this complex system at every conceivable level ranging from genetic, through cellular and organ systems, to interactions of the total human with the environment in the conduct of purposeful activities. At each level, there are corresponding practitioners who attempt to discover and rectify or prevent problems at the respective level. Some practitioners are concerned with specific individuals, while others (e.g., biomedical scientists and product designers) address populations as a whole. Problems dealt with span medical and nonmedical contexts, often with interaction between the two. Models play a key role not only in providing conceptual understanding of key issues at each level but also in describing relationships between various levels and providing frameworks that allow practitioners to obtain reasonably predictable results in a systematic and efficient fashion. In this chapter, a working model for human system-task interfaces is presented. Any such model must, of course, not only consider the interface per se but also representations of the human system and tasks. The model presented here, the *elemental resource model* (ERM), represents the most recent effort in a relatively small family of models that attempt to address similar needs.

147.1 Background

The interface between a human and a task of daily living (e.g., work, recreation, or other) represents a level that is quite high in the hierarchy noted above. One way in which to summarize previous efforts directed at this level, across various application contexts, is to recognize two different lines along which study has evolved: (1) *bottom-up* and (2) *top-down.* Taken together, these relative terms imply a focus of interest at a particular level of convergence. Here, this special level is termed the human-task interface level. It is emphasized that bottom-up and top-down terms are used here to characterize the general

0-8493-0462-8/00/$0.00+$.50

course of development and not specific approaches applied at a particular instant in time. A broad view is necessary to grapple with the many previous efforts that either are, or could be, construed to be pertinent.

The biomedical community has approached the human-task interface largely along the bottom-up path. This is not surprising given the historical evolution of interest first in anatomy (human structure) and then physiology (function). The introduction of chemistry (giving rise to biochemistry) and the refinement of the microscope provided motivations to include even lower hierarchical levels of inquiry and of a substantially different character. Models in this broad bottom-up category begin with *anatomic components* and include muscles, nerves, tendons (or subcomponents thereof), or subsets of organs (e.g., heart, lungs, vasculature, etc.). They often focus on relationships between components and exhibit a scope that stays within the confines of the human system. Many cause-and-effect models have been developed at these lower levels for specific purposes (e.g., to understand lines of action of muscle forces and their changes during motion about a given joint).

As a natural consequence of linkages that occur between hierarchical levels and our tendency to utilize that which exists, consideration of an issue at any selected level (in this case, the human-task interface level) brings into consideration *all lower levels* and all models that have been put forth with the stated purpose of understanding problems or behaviors at the original level of focus. The amount of detail that is appropriate or required at these original, lower levels results in great complexity when applied to the total human at the human-task interface level. In addition, many lower-level modeling efforts (even those which are quantitative) are aimed primarily at obtaining a basic scientific understanding of human physiology or specific pathologies (i.e., pertaining to *populations* of humans). In such circumstances, highly specialized, invasive, and cumbersome laboratory procedures for obtaining the necessary data to populate models are justified. However, it is difficult and sometimes impossible to obtain data describing *a specific individual* to utilize in analyses when such models are extended to the human-task interface level. Another result of drawing lower-level models (and their approaches) into the human-task interface context is that the results have a specific and singular character (e.g., biomechanical versus neuromuscular control versus psychologic, etc.) [e.g., Card et al., 1986; Delp et al., 1990; Gottlieb et al., 1989; Hemami, 1988; Schoner & Kelso, 1988]. Models that incorporate most or all of the multiple aspects of the human system or frameworks for integrating multiple lower-level modeling approaches have been lacking. Lower-level models that serve meaningful purposes at the original level of focus have provided and will continue to provide insights into specific issues related to human performance at multiple levels of considerations. However, their direct extension to serve general needs at the human-task interface level has inherent problems; a different approach is suggested.

A top-down progression can be observed over the major history in human factors/ergonomic [Gilbreth & Gilbreth, 1917; Taylor, 1911] and vocational assessment [e.g., Botterbusch, 1987] fields (although the former has more recently emphasized a "human-centered" concept with regard to design applications). In contrast to the bottom-up path, in which anatomic components form the initial basis of modeling efforts, the focus along the top-down path begins with consideration of the *task or job* that is to be performed by the total human. The great variety in the full breadth of activities in which humans can be engaged gives rise to one aspect of complexity at this level that pertains to taxonomies for job and task classification [e.g., Fleishman & Quaintance, 1984; Meister, 1989; U.S. Department of Labor, 1992]. Another enigmatic aspect that quickly adds complexity with respect to modeling concerns the appropriate level to be used to dissect the items at the highest level (e.g., jobs) into lower-level components (e.g., tasks and subtasks). In fact, the choice of level is complicated by the fact that no clear definition has evolved for a set of levels from which to choose.

After progressing down through various levels at which all model elements represent tasks and are completely outside the confines of the human body, a level is eventually reached where one encounters the human. Attempts to go further have been motivated, for example, by desires to predict performance of a human in a given task (e.g., lifting an object, assembling a product, etc.) from a set of measures that characterizes the human. At the human-task interface level, difficulty is encountered with regard to the strategy for approaching a system as complex, multifaceted, and multipurpose as a human [Fleishman

& Quaintance, 1984; Wickens, 1984]. In essence, the full scope of options that have emerged from the bottom-up development path is now encountered from the opposite direction. Options range from relatively gross analyses (e.g., estimates of the "fraction" of a task that is physical or mental) to those which are much more detailed and quantitative. The daunting prospect of considering a "comprehensive quantitative model" has led to approaches and models, argued to be "more practical," in which sets of parameters are often selected in a somewhat mysterious fashion based on experience (including previous research) and intuition. The selected parameters are then used to develop predictive models, most of which have been based primarily on statistical methods (i.e., regression models) [Fleishman, 1967; Fleishman & Quaintance, 1984]. Although the basic modeling tools depend only on correlation, it is usually possible to envision a causal link between the independent variables selected (e.g., visual acuity) and the dependent variable to be predicted (e.g., piloting an aircraft). Models (one per task) are then tested in a given population and graded with regard to their prediction ability, the best of which have performed marginally [Kondraske and Beehler, 1994]. Another characteristic associated with many of the statistically based modeling efforts from the noted communities is the almost exclusive use of healthy, "normal" subjects for model development (i.e., humans with impairments were excluded). Homogeneity is a requirement of such statistical models, leading to the need for one model per task per population (at best). Also, working with a mindset that considers only normal subjects can be observed to skew estimates regarding which of the many parameters that one might choose for incorporation in a model are "most important." The relatively few exceptions that employ cause-and-effect models (e.g., based on physical laws) at some level of fidelity [e.g., Chaffin & Andersonn, 1984] often adopt methods that have emerged from the bottom-up path and are, as noted above, limited in character at the "total human" level (e.g., "biomechanical" in the example cited).

The issue is *not* that no useful models have emerged from previous efforts but rather that no clear comprehensive strategy has emerged for modeling at the human-task interface level. A National Research Council panel on human performance modeling [Baron et al., 1990] considered the fundamental issues discussed here and also underscored needs for models at the human-task interface level. While it was concluded that an all-inclusive model might be desirable (i.e., high fidelity, in the sense that biomechanical, information processing, sensory and perceptual aspects, etc. are represented), such a model was characterized as being highly unlikely to be achieved and perhaps ultimately not useful because it would be overly complex for many applications. The basic recommendation made by this panel was to pursue development of more limited scope submodels. The implication is that two or more submodels could be integrated to achieve a broader range of fidelity, with the combination selected to meet the needs of particular situations. The desire to "divide efforts" due to inherent complexity of the problem also surfaces within the histories of the bottom-up and top-down development paths discussed above. While a reasonable concept in theory, one component in the division of effort that has consistently been underrepresented is the part that ties together the so-called submodels. Without a conceptual framework for integration of relatively independent modeling efforts and a set of common modeling constructs, prospects for long-term progress are difficult to envision. This, along with the recognition that enough work had been undertaken in the submodel areas so that key issues and common denominators could be identified, motivated development of the ERM.

The broad objectives of the ERM [Kondraske, 1994] are most like those of Fleishman and colleagues [Fleishman, 1966, 1972,1982; Fleishman & Quaintance, 1984], whose efforts in human performance are generally well known in many disciplines. These are the only two efforts known that (1) focus on the total human in a task situation (i.e., directly address the human-task interface level); (2) consider tasks in general, and not only a specific task such as gait, lifting, reading, etc.; (3) incorporate all aspects of the total human system (e.g., sensory, biomechanical, information processing, etc.); and (4) aim at quantitative models. There are also some similarities with regard to the incorporation of the ideas of "abilities" (of humans) and "requirements" (of tasks). The work of Fleishman and colleagues has thus been influential in shaping the ERM either directly or indirectly through its influence of others. However, there are several substantive conceptual differences that have resulted in considerably different endpoints. Fleishman's work emerged from "the task" perspective and is rooted in psychology, whereas the ERM emerges from the perspective of "human system architecture" and is rooted in engineering methodology with regard to quantitative aspects of system performance but also incorporates psychology *and* physiology. Both approaches address humans *and* tasks, and both efforts contain aspects identifiable with psychology and engineering, as they ultimately must. These different perspectives, however, may explain in part some of the major differences. Aspects unique to the ERM include (1) the use of a resource construct for modeling and measurement of *all* aspects of a system's performance, (2) the use of cause-and-effect resource economic principles (i.e., the idea of threshold "costs" for achieving a given level of performance in any given high-level task), (3) the concept of monadology (i.e., the use of a finite set of "elements" to explain a complex phenomenon), and (4) a consistent strategy for identifying performance elements at different hierarchical levels.

The ERM attempts to provide a quantitative and relatively straightforward framework for characterizing the human system, tasks, and the interface of the human to tasks. It depends in large part on, and evolves directly from, a separate body of material referred to collectively as *general systems performance theory* (GSPT). GSPT was developed first and independently, i.e., removed from the human system context. It incorporates resource constructs exclusively for modeling of the abstract idea of *system performance*, including specific rules for measuring performance resource availability and resource economic principles to provide a cause-and-effect analysis of the interface of any system (e.g., humans) to tasks. The concept of a *performance model* is emphasized and distinguished from other model types.

147.2 Basic Principles

The history of the ERM and the context in which it was developed are described elsewhere [Kondraske, 1987*a*, 1990*b*, 2000]. It is important to note that the ERM is derived from the combination of GSPT with the philosophy of monadology and their application to the human system. As such, these two constituents are briefly reviewed before presenting and discussing the actual ERM.

General Systems Performance Theory

The concept of performance now pervades all aspects of life, especially decision-making processes that involve both human and artificial systems. Yet it has not been well understood theoretically, and systematic techniques for modeling and its measurement have been lacking. While a considerable body of material applicable to general systems theory exists, the concept of performance has not been incorporated in it, nor has performance been addressed in a general fashion elsewhere. Most of the knowledge that exists regarding performance and its quantitative treatment has evolved within individual application contexts, where generalizations can easily be elusive.

Performance is multifaceted, pertaining to how well a given system executes an intended function and the various factors that contribute to this. It differs from behavior of a system in that "the best of something" is implied. The broad objectives of GSPT are

1. To provide a common conceptual basis for defining and measuring all aspects of the performance of any system.
2. To provide a common conceptual basis for the analysis of any task in a manner that facilitates system-task interface assessments and decision making.
3. To identify cause-and-effect principles, or laws, that explain what occurs when any given system is used to accomplish any given task.

While GSPT was motivated by needs in situations where the human is "the system" of interest and it was first presented in this context [Kondraske, 1987*a*], application of it has been extended to the context of artificial systems. These experiences range from computer vision and sensor fusion [Yen & Kondraske, 1992] to robotics [Kondraske & Khoury, 1992; Kondraske & Standridge, 1988].

A succint statement of GSPT designed to emphasize key constructs is presented below in a steplike format. The order of steps is intended to suggest how one might approach any system or system-task interface situation to apply GSPT. While somewhat terse and "to-the-point," it is nonetheless an essential prerequisite for a reasonably complete understanding the ERM.

1. Within a domain of interest, select any level of abstraction and identify the system(s) of interest (i.e., the physical structure) and its function (i.e., purpose).
2. Consider "the system" and "the task" separately.
3. Use a resource construct to model the system's *performance.* First, consider the unique intangible qualities that characterize *how well a system executes its function.* Each of these is considered to represent a unique performance resource associated with a specific *dimension of performance* (e.g., speed, accuracy, stability, smoothness, "friendliness," etc.) of that system. Each performance resource is recognized as a *desirable* item (e.g., endurance versus fatigue, accuracy versus error, etc.) "possessed" by the system in a certain quantitative amount. Thus one can consider *quantifying* the amount of given *quality* available. As illustrated, an important consequence of using the resource construct at this stage is that confusion associated with duality of terms is eliminated.
4. Looking toward the system, identify all I dimensions of performance associated with it. In situations where the system does not yet exist (i.e., design contexts), it is helpful to note that dimensions of performance of the system are the same as those of the task.

5a. Keeping the resource construct in mind, define a parameterized metric for each dimension of performance (e.g., speed, accuracy, etc.). If the resource construct is followed, values will be produced with these metrics that are always nonnegative. Furthermore, a larger numeric value will consistently represent *more* of a given resource and therefore *more performance capacity.*

5b. Measure system performance with the system *removed from* the specific intended task. This is a reinforcement of Step 2. The general strategy is to *maximally stress* the system (within limits of comfort and/or safety, when appropriate) to define its *performance envelope*, or more specifically, the envelope that defines *performance resource availability,* $R_{A_S}(t)$. Note that $R_{A_S}(t)$ is a continuous surface in the system's nonnegative, multidimensional performance space. Also note that unless all dimensions of performance and parameterized metrics associated with each are defined using the resource construct, a performance envelope cannot be guaranteed. Addressing the issue of measurement more specifically , consider resource availability values $R_{A_i}|_{Q_{i,k}}(t)$ for $i = 1$ to I, associated with each of the I dimensions of performance. Here, each $\mathbf{Q}_{i,k}$ represents a unique condition, in terms of a set of values R_i along *other* identified dimensions of performance, under which a specific resource availability (R_{A_i}) is measured; i.e., $\mathbf{Q}_{i,k} = \{R_{1,k}, R_{2,k}, \ldots, R_{p,k}\}$ for all $p \neq i$ $(1 \geq p \geq I)$. The subscript k is used to distinguish several possible conditions under which a given resource availability (R_{A_1}, for example) can be measured. These values are measured using a set of "test tasks," each of which is designed to *maximally stress* the system (within limits of comfort and/or safety, when appropriate): (a) along each dimension of performance individually (where $\mathbf{Q}_{i,k} = \mathbf{Q}_{i,0} = \{0, 0, \ldots, 0\}$) or (b) along selected subsets of dimensions of performance simultaneously (i.e., $\mathbf{Q}_{i,k} = \mathbf{Q}_{i,n}$, where each possible $\mathbf{Q}_{i,n}$ has one or more nonzero elements). The points obtained $[R_{A_i}|_{Q_{i,k}}(t)]$ provide the basis to estimate the performance envelope $R_{A_S}(t)$. Note that if only on-axis points are obtained (e.g., maximally stress one specific performance resource availability with minimal or no stress on other performance resources, or the $\mathbf{Q}_{i,0}$ condition), a rectangular or idealized performance envelope is obtained. A more accurate representation, which would be contained within the idealized envelope, can be obtained at the expense of making additional measurements or the use of known mathematic functions based on previous studies that define the envelope shape in two or more dimensions.

5c. Define estimates of single-number *system figures-of-merit,* or *composite performance capacities,* as the mathematical product of all or any selected subset of $R_{A_i}|_{Q_{i,k}}(t)$. If more accuracy is desired and a sufficient number of data points is available from the measurement process described in Step 5b, composite performance capacities can be determined by integration over $R_{A_S}(t)$ to determine the volume enclosed by the envelope. The composite performance capacity is a measure of performance at a higher level of abstraction than any individual dimension of performance at the "system" level, representing the capacity of the system to perform tasks that place demands on *those performance resources availabilities included in the calculation.* Different composite performance capacities can be computed for a given system by selecting different combinations of dimensions of performance. Note that the definition of a composite performance capacity used here preserves dimensionality; e.g., if speed and accuracy dimensions are included in the calculation, the result has units of speed × accuracy. (This step is used only when needed, e.g., when two general-purpose systems of the same type are to be compared. However, if decision making that involves the interface of a specific system to a specific task is the issue at hand, a composite performance capacity is generally of any use only to *rule out* candidates).

6. Assess the "need for detail." This amounts to a determination of the number of hierarchical levels included in the analysis. If the need is to determine if the currently identified "system" can work in the given task or how well it can execute its function, go to Step 7 now. If the need is to determine the *contribution of one or more constituent subsystem* or why a desired level of performance is not achieved at the system level, repeat Steps 1 to 5 for all J functional units (subsystems), or a selected subset thereof based on need, that form the system that was originally identified in Step 1; i.e., go to the next-lowest hierarchical level.
7. At the "system" level, look toward the task(s) of interest. Measure, estimate, or calculate *demands* on system performance resources (e.g., the speed, accuracy, etc. required), $R_{D_i}|_{Q'_{i,k}}(t)$, where the notation here is analogous to that employed in Step 5b. This represents the quantitative definition and communication of *goals*, or the set of values P_{HLT} representing level of performance P desired in a specific high-level task (*HLT*). Use a worst-case or other less-conservative strategy (with due consideration of the impact of this choice) to summarize variations over time. This will result in a set of M points (R_{D_m}, for $m = 1$ to M) that lie in the multidimensional space defined by the set of I dimensions of performance. Typically, $M \geq I$.
8. Use *resource economic principles* (i.e., require $R_A \geq R_D$ for "success") at the system level *and* at all system-task interfaces at the subsystem level (if included) to evaluate success/failure at each interface. More specifically, for a given system-task interface, all task demand points (i.e., the set of R_{D_m} associated with a given task or subtask) must lie within the performance resource envelope $R_{A_S}(t)$ of the corresponding system. This is the key law that governs system-task interfaces. If a two-level model is used (i.e., "system" and "subsystem" levels are incorporated), map system-level demands to demands on constituent subsystems. That is, functional relationships between P_{HLT} and demands imposed on constituent subsystems [i.e., $R_{D_{i,j}}(t, P_{HLT})$] must be determined. The nature of these mappings depends on the type of systems in question (e.g., mechanical, information processing, etc.). The basic process includes application of Step 7 to the subtasks associated with each subsystem. If *resource utilization flexibility* (i.e., redundancy in subsystems of similar types) exists, select the "best" or optimal subsystem °configuration (handled in GSPT with the concept of *performance resource substitution*) and *procedure* (i.e., use of performance resources over time) as that which allows *accomplishment of goals* with *minimization of stress* on available performance resources across all subsystems and over the time of task execution. Thus redundancy is addressed in terms of a constrained performance resource optimization problem. Stress on individual performance resources is defined as $0 < R_{D_{i,j}}(t, P_{HLT})/ R_{A_{i,j}}(t) < 1$. It is also useful to define and take note of reserve capacity, i.e., the margin between available and utilized performance resources.

The preceding statement is intended to reflect the true complexity that exists in systems, tasks, and their interfaces when viewed primarily from the perspective of performance. This provides a basis for the judicious decision making required to realize "the best" *practical implementation* in a given situation where many engineering tradeoffs must be considered. While a two-level approach is described above, it should be apparent that it can be applied with any number of hierarchical levels by repeating the steps outlined in an iterative fashion starting at a different level each time. A striking feature of GSPT is the *threshold effect* associated with the resource economic principle. This nonlinearity has important implications in quantitative human performance modeling, as well as interesting ramifications in practical applications such as rehabilitation, sports, and education. Note also that no distinction is made as to whether a given performance resource is derived from a human or artificial system; both types of systems, or subcomponents thereof, can be incorporated into models and analyses.

Monadology

Monadology dates back to 384 B.C. [Neel, 1977] but was formalized and its importance emphasized by Gottfried Wilhelm Leibniz, inventor of calculus, in his text *Monadologia* in 1714. This text presents what is commonly called *Leibniz's Monadology* and has been translated and incorporated into contemporary philosophy texts (e.g., Leibniz and Montgomery, 1992). It is essentially the idea of "basic elements" *vis a vis* chemistry, alphabets, genetic building blocks, etc. The concept is thus already well accepted as being vital to the systematic description of human systems from certain perspectives (i.e., chemical, genetic). Success associated with previous applications of monadology, whether intentional or unwitting (i.e., discovered to be at play *after* a given taxonomy has emerged), compels its serious a priori consideration for other problems.

Insight into how monadology is applied to human performance modeling is perhaps more readily obtained with reference to a widely known example in which monadology is evident, such as chemistry. Prior to modern chemistry (i.e., prior to the introduction of the periodic table), alchemy existed. The world was viewed as being composed of an infinite variety of unique *substances.* The periodic table captured the notion that this infinite variety of substances could all be defined in terms of a finite set of basic elements. Substances have since been analyzed using the "language" of chemistry and organized into categories of various complexity, i.e., elements, simple compounds, complex compounds, etc. Despite the fact that this transition occurred approximately 200 years ago, compounds remain that have yet to be analyzed. Furthermore, the initial periodic table was incorrect and has undergone revision up to relatively recent times. Analogously, in the alchemy of human performance, the world is viewed as being composed of an infinite variety of unique tasks. A "chemistry" can be envisioned that first starts with the identification of the "basic elements" or, more specifically, *basic elements of performance.* Simple and complex tasks are thus analogous to simple and complex compounds, respectively. The analogy carries over to quantitative aspects of GSPT as well. Consider typical equations of chemical reactions with resources on the left and products (i.e., tasks) on the right. Simple compounds (tasks) are realized by drawing on basic elements in the proper combination and amounts. The amount of "product" (level of performance in a high-level task) obtained depends on availability of the *limiting resource.*

Another informative aspect of this analogy is the issue of how to deal with the treatment of hierarchical level. Clearly, the chemical elements are made up of smaller particles (e.g., protons, neutrons, and electrons). Physicists have identified even smaller, more elusive entities, such as bosons, quarks, etc. Do we need to consider items at this lowest level of abstraction each time a simple compound such as hydrochloric acid is made? Likewise, the term *basic* in basic elements of performance is clearly relative and requires the choice of a particular hierarchical level of abstractions for the identification of systems or basic functional units, a level that is considered to be both natural and useful for the purpose at hand. Just as it is possible but not always necessary or practical to map chemical elements down to the atomic particle level, it is possible to consider mapping a basic element of performance (see below) such as *elbow flexor torque production capacity* down to the level of muscle fibers, biochemical reactions at neuromuscular junctions, etc.

The Elemental Resource Model

The resource and resource economic constructs used in GSPT specifically to address performance have employed and have become well established in some segments of the human performance field, specifically with regard to attention and information processing [Navon & Gopher, 1979; Wickens, 1984]. However, in these cases, the term *resource* is used mostly conceptually (in contrast to quantitatively), somewhat softly defined, and applied to refer in various instances to systems (e.g., different processing

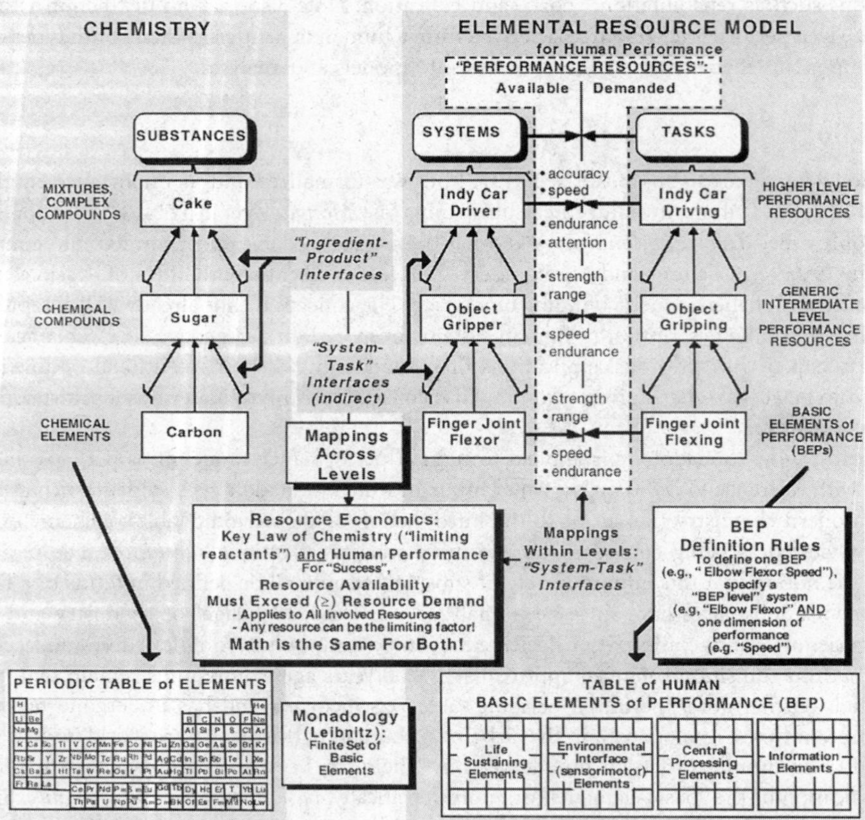

FIGURE 147.1 The elemental resource model contains multiple hierarchical levels. Performance resources (i.e., the basic elements) at the "basic element level" are finite in number, as dictated by the finite set of human subsystems and the finite set of their respective dimensions of performance. At higher levels, new "systems" can be readily created by configuration of systems at the basic element level. Consequently, there are an infinite number of performance resources (i.e., higher-level elements) at these levels. However, rules of general systems performance theory (refer to text) are applied at any level in the same way, resulting in the identification of the system, its function, dimensions of performance, performance resource availabilities (system attributes), and performance resource demands (task attributes).

centers), broad functions (e.g., memory versus processing), and sometimes to infer a particular aspect of performance(e.g., attentional resources). In the ERM, through the application of GSPT, these constructs are incorporated universally (i.e., applied to all human subsystems) and specifically to model "performance" at both conceptual and quantitative levels. In addition to the concept of monadology, the insights of others [Shoner & Kelso, 1988, Turvey et al., 1978] were valuable in reinforcing the basic "systems architecture" employed in the ERM and in refining description of more subtle, but important aspects.

As illustrated in Fig. 147.1, the ERM contains multiple hierarchical levels. Specifically, three levels are defined: (1) the basic element level, (2) the generic intermediate level, and (3) the high level. GSPT is to define performance measures at any hierarchical level. This implies that to measure performance, one must isolate the desired system and then stress it maximally along one dimension of performance (or more, if interaction effects are desired) to determine performance resource availability. For example, consider the human "posture stabilizing" system (at the generic intermediate level), which is stressed maximally along a stability dimension. As further illustrated below, the basic element level represents measurable stepping stones in the human system hierarchy between lower-level systems (i.e., ligaments, tendons, nerves, etc.) and higher-level tasks.

A summary representation emphasizing the basic element level of the ERM is depicted in Fig. 147.2. While this figure is intended to be more or less self-explanatory, a brief walk-through is warranted.

Looking Toward the Human. The entire human (lower portion of Fig. 147.2) is modeled as a pool of elemental *performance resources* that are grouped into one of four different domains: (1) life-sustaining, (2) environmental interface (containing purely sensory and sensorimotor components), (3) central processing, and (4) information. Within each of the first three domains, physical subsystems referred to as *functional units* are identified (see labels along horizontal aspect of grids) through application of fairly rigorous criteria [Kondraske, 1990*b*]. GSPT is applied to each functional unit, yielding first a set of dimensions of performance (defined using a resource construct) for each unit. A single *basic element of performance* (BEP) is defined by specifying two items: (1) the basic functional unit and (2) one of its dimensions of performance. Within a domain, not every dimension of performance indicated in Fig. 147.2 is applicable to every functional unit in that domain. However, there is an increasing degree of "likeness" among functional units (i.e., fewer fundamentally different types) in this regard as one moves from life-sustaining to environmental-interface to central-processing domains. The fourth domain, the information domain, is substantially different than the other three. Whereas the first three represent physical systems and their intangible performance resources, the information domain simply represents information. Thus, while memory functional units are located within the central-processing domain, the *contents of memory* (e.g., motor programs and associated reference information) are partitioned into the information domain. As illustrated, information is grouped, but within each group there are many specific skills. The set of available performance resources $[R_{A_{i,j}}(t)|_Q]$ consist of both BEPs (i = dimension of performance, j = functional unit) and information sets (e.g., type i within group j). Although intrinsically different, both fit the resource construct. This approach permits even the most abstract items such as motivation and friendliness to be considered with the same basic framework as strength and speed.

Note that resource availability in GSPT and thus in the ERM is potentially a function of time, allowing quantitative modeling of dynamic processes such as child development, aging, disuse atrophy, and rehabilitation. The notation further implies that availability of a given resource must be evaluated at a specific operating point, denoted as **Q.** At least conceptually, many parameters can be used to characterize this **Q** point. In general, the goal of measurement when looking "toward the human" is to isolate functional units and maximally stress (safely) individual performance resources to determine availability. Such measures reflect performance capacities. The simplest **Q** point is one in which there is stress along only one dimension of performance (i.e., that corresponding to the resource being stressed). Higher-fidelity representation is possible at the expense of additional measurements. The degree to which isolation can be achieved is of practical concern in humans. Nonetheless, it is felt that reasonable isolation can be achieved in most situations [Kondraske, 1990*a*, 1990*b*]. Moreover, this and similar issues of practical concern should be addressed as separate problems; i.e., they should not be permitted to obfuscate or thwart efforts to explain phenomenon at the human-task interface.

Looking Toward the Task. The midportion of Fig. 147.2 suggests the representation of any given task in terms of the unique set of demands $[R_{D_{i,j}}(t)]$ imposed on the pool of BEPs and information resources; i.e., this is the elemental level of task representation. Shading implies that demands can be represented quantitatively in terms of amount. The upper portion of the figure defines hierarchical mapping options, where mapping is the process of translating what happens in tasks typically executed by humans to the elemental level. Two such additional levels (for a total of three, including the elemental level) are included as part of the ERM: (1) generic intermediate-level tasks and (2) higher-level complex tasks. At all three levels of tasks (Figs. 147.1 and 147.2) are processes that occur over time and can be characterized by specific goals (e.g., in terms of speed, force, etc.) and related to systems at the same level that possess performance resources. Using established analytical techniques, even the most complex task can be divided into discrete task segments. Then mapping analyses, which take into account task procedures

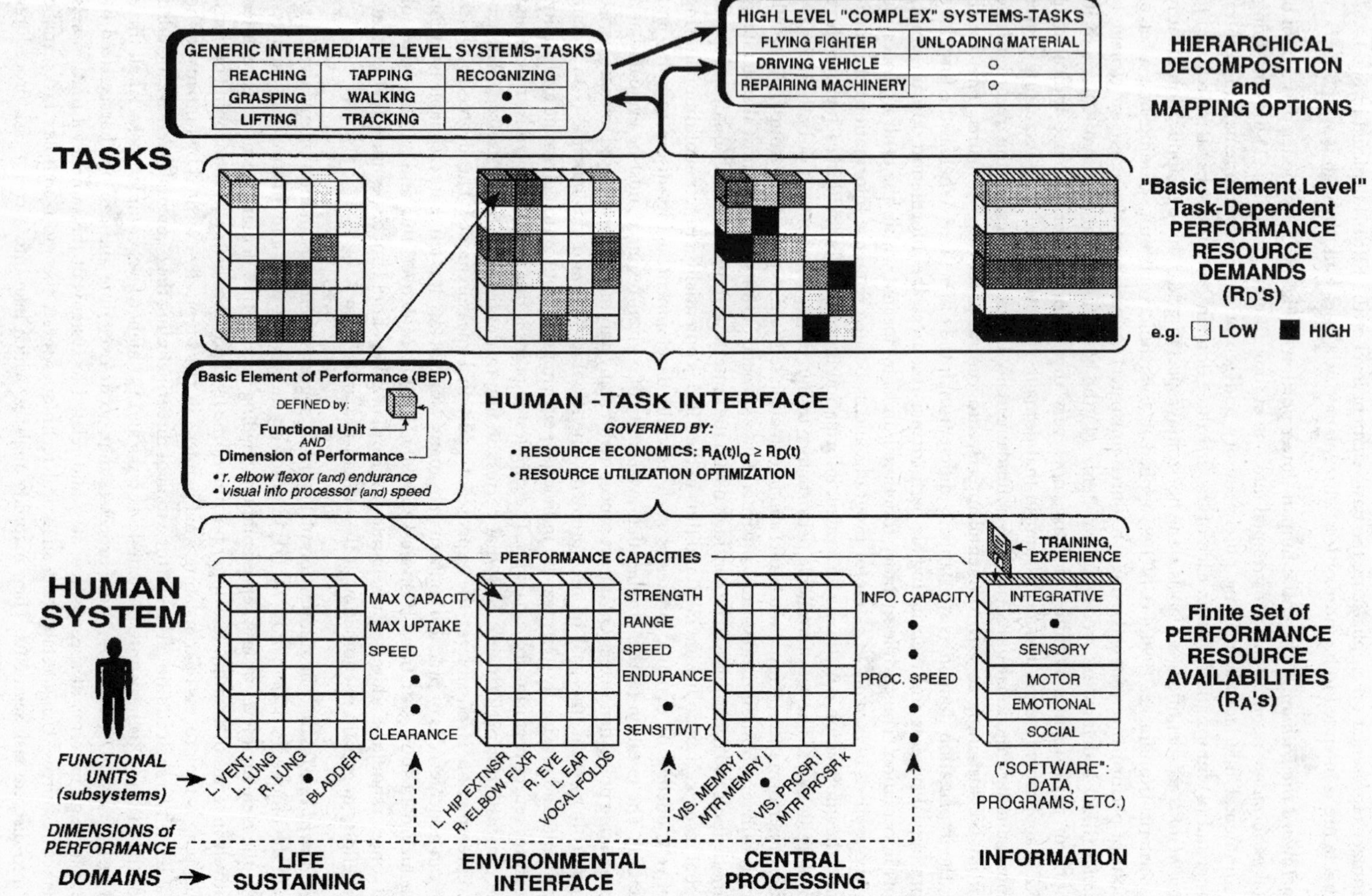

FIGURE 147.2 A summary of the major constructs of the elemental resource model, emphasizing the categorization of performance resources at the basic element level into four life-sustaining, environmental-interface, central-processing, and information domains.

(e.g., a squatting subject with two hands on the side of a box), are applied to each task segment to determine $R_{D_{i,j}}(t)$. Once this is found, the worst case or a selected percentile point in the resource demand distribution (over a time period corresponding to a selected task segment) can be used to obtain a single numeric value representation of demand for a given resource *and* the conditions under which the given demand occurs; i.e., the **Q′** point (e.g., at what speed and position angle does the worst-case demand on elbow flexor torque occur?). This reduction process requires parameterization algorithms that are similar to those used to process time-series data collected during tests designed to measure performance resource availability.

The Human-Task Interface. Using GSPT, success in achieving the goals of a given task segment is governed by resource economic principles requiring that $R_{A_{i,j}}(t)|_Q \geq R_{D_{i,j}}(t)$ for all i and j (i.e., $R_{A_{11}} \geq R_{D_{11}}$ AND $R_{A12} \geq R_{D12}$ AND $R_{A13} \geq R_{D13}$). In other words, all task demands, when translated to the individual subsystems involved, must fit within the envelopes that define performance resource availability. Adequacy associated with any one resource is a *necessary but not sufficient condition* for success. Concepts and observations in human performance referred to as *compensation* or *redundancy* are explained in terms of *resource utilization flexibility*, which includes the possibility of substituting one performance resource (of the same dimensionality) for another (i.e., *resource substitution*). It has been hypothesized [Kondraske, 1990*b*] that an optimal performance resource utilization is achieved through learning. Furthermore, the optimization rule suggested by GSPT is that the human system is driven to accomplish task goals and use procedures that minimize performance resource stress (i.e., the fraction of available performance resource utilized) over the duration of a given task segment and across all BEPs involved. Minimizing stress is equivalent to maximizing the margin between available and utilized performance resources. Thus optimization is highly dependent on the resource availability profile. It would be predicted, for instance, that two individuals with different resource availability profiles would not optimally accomplish the same task goals by using identical procedures.

147.3 Application Issues

Implications of the ERM and what it demonstrates regarding intrinsic demands imposed by nature and methods for creatively navigating these demands over both the short and long terms are considered. The ERM offers a number of flexibilities with regard to how it can be applied (e.g., "in whole" or "in part," "conceptually" or "rigorously," with "low tech" or sophisticated "high tech" tools, and to define performance measures or to develop predictive models). While immediate application is possible at a conceptual level of application, it also provides the motivation and potential to consider coordinated, collaborative development of sophisticated tools that allow rigorous and efficient solutions to complex problems by practitioners without extraordinary training.

The ERM provides a basis for obtaining insight into the nature of routine tasks that clinicians and other practitioners are expected to perform; there are both "troublesome" and "promising" insights in this regard. Perhaps the most obvious troublesome aspect is that the ERM makes it painfully evident that *many* BEPs are typically called into play in tasks of daily living, work, and/or recreation and that resource insufficiency associated with *any one* of this subset of BEPs can be the factor that limits performance in the higher-level task. The further implication is that for rigorous application, one must know (via measurement and analyses) the availability and demand associated with each and every one of these unique resources. An additional complexity with which practitioners must cope is the high degree of specificity and complexity of resources in the information domain (i.e., the "software"). There is no simple, rapid way to probe this domain to determine if the information required for a given task is correct; it requires methods analogous to those used to debug software source code.

Aspects that hold promise are associated with (1) the nature of hierarchical systems, (2) the threshold mathematics of resource economics, and (3) the fact that when n resources combine to address a single task, the mathematics of logical combination is employed to arrive at an overall assessment. That is, the individual "$R_A \geq R_D$?" questions result in a set of "OK" or "not OK" results that are combined with logical AND operations to obtain the final "OK" or "not OK" assessment (note that the OR operator is used when resource substitution is possible).

Conceptual, Low-Tech Practical Application

The ERM description alone can be used simply to provide a common conceptual basis for discussing the wide range of concepts, measurements, methods, and processes of relevance in human performance or a particular application area [Frisch, 1993; Mayer & Gatchel, 1988; Syndulko et al., 1988]. It also can be used at this level as a basis for structured assessment [Kondraske, 1988*b*, 2000] of individuals in situations, including therapy prescription, assistive device prescription, independent living decision making (e.g., self-feeding, driving, etc.), age or gender discrimination issues in work or recreational tasks, etc. At points in such processes, it is often more important to consider the full scope of different performance resources involved in a task using even a crude level of quantification than it is to consider just a select few in great depth. A "checklist" approach is recommended. The professional uses only his or her judgment and experience to consider both the specific individual and the specific task of interest to make quantitative but relatively gross assessments of resource adequacy using a triage-like categorization process (e.g., with "definitely limiting," "definitely not limiting," or "not sure" categories). This is feasible because of the threshold nature of the system-task interface; in cases where R_A and R_D are widely separated instrumented, high- resolution measurements are not required to determine if a given performance resource is limiting. Any resource(s) so identified as "definitely limiting" becomes an immediate focus of interest. If none is categorized as such, concern moves to those in the "not sure" category, in which case more sensitive measurements may be required. Purely subjective methods of measuring resource availability and/or demands can be augmented with selected, more objective and higher-resolution measurements in "hybrid applications."

Conceptual, Theoretical Application

The ERM can be used to reconsider previous work in human performance. For example, it can be employed to reason why Fleishman [Fleishman & Quaintance, 1984] (as well as others) achieved promising, but limited, success with statistically based predictions of performance in higher-level tasks using regression models with independent variables, which can now be viewed as representing lower-level performance resources (in most cases). Specifically, regression models rely heavily on an assumption that there exists some correlation between dependent and each independent variable, the latter of which typically represent scores from maximal performance tasks and therefore reflect resource availability (using GSPT and ERM logic). Brief reflection results in the realization that correlation is not to be anticipated between the level of performance attainable in a "higher-level task" and *availability* of one of the many performance resources essential to the task (e.g., if 4 cups of flour are needed for a given cake, having 40 cups available will not *alone* result in a larger cake of equal quality—availability of another ingredient may in fact be limiting). Rather, correlation *is* expected between high-level task performance and the amount of resource *utilization*. Unfortunately, as noted above, the independent variables used typically reflect resource availability. The incomplete labeling of performance variables in such studies reflects the general failure to distinguish between utilization and availability. Why not, then, just use measures of resource utilization in such statistical models? Resource availability measures are simple to obtain in the laboratory without requiring that the individual execute *the* high-level task of interest. Resource utilization measures can only be obtained experimentally by requiring the subject to execute the task in question, which is counterproductive with respect to the goal of using a set of laboratory measurements to extrapolate to performance in one or more higher-lever task situations. Furthermore, regression models based on linear combination of resource *availability* measures do not reflect the nonlinear threshold effect accounted for with resource economic, GSPT-based performance models. One potential alternative based on GSPT and termed *nonlinear causal resource analysis* has been proposed [Kondraske, 1988*a*; Vasta & Kondraske, 1994].

Application "in Part"

In this approach, whole domains (i.e., many BEPs) are assessed simultaneously, resulting in an estimate or well-founded assumption that states that "all performance resources in domain *X* are nonlimiting in task(s) *Y*." Such assumptions are often well justified. For example, it would be reasonable to assume that a young male with a sports-related knee injury would have only a reduction in performance resource availability in the environmental interface domain. More specifically, it is reasonable to assume that the scope of interest can be confined to a smaller subset of functional units, as in gait or speech. These can then be addressed with rigorous application. Examples of this level and manner of applying the ERM have been or are being developed for head/neck control in the context of assistive communication device prescription [Carr, 1989], workplace design [Kondraske, 1988*c*], evaluating work sites and individuals with disabilities for employment [Parnianpour & Marras, 1993], gait [Carollo & Kondraske, 1987], measurement of upper extremity motor control [Behbehani et al., 1988], speech production performance [Jafari, 1989; Jafari et al., 1989], and to illustrate changes in performance capacity associated with aging [Kondraske, 1989]. Additionally, in some applications only the generic intermediate level need be considered. For example, one may only need to know how well an individual can walk, lift, etc. While it is sometimes painfully clear just how complex the execution of even a relatively simple task is, it also can be recognized that relatively simple, justifiable, and efficient strategies can be developed to maintain a reasonable degree of utility in a given context.

Rigorous, High-Tech Application

While this path may offer the greatest potential for impact, it also presents the greatest challenge. The ultimate *goal* would be to capture the analytic and modeling capability (as implied by the preceding discussions) for a "total human" (single subject or populations) and "any task" in a desktop computer system (used along with synergistic "peripherals" that adopt the same framework, such as measurement tools [e.g., Kondraske, 1990*a*]). This suggests a long-term, collaborative effort. However, intermediate tools that provide significant utility are feasible and needed (e.g., [Allard, Stokes, and Blanchi, 1994; Vasta and Kondraske, 1995]). A promising example based directly on GSPT and the ERM is Nonlinear Causal Resource Analysis, or NCRA [Vasta and Kondraske, 1994; Kondraske et al, 1997]. This inferential method has recently been used [Kondraske et al, 1997] to develop models that relate performance resource demands on various human subsystems to level of performance attained in higher level mobility tasks (e.g., gait and stair climbing). In turn, these models have been used to predict level of performance in these higher level tasks with success that far exceeds that which has been obtained with regression models. Furthermore, the NCRA method inherently provides not only a prediction of high level task performance, but also identifies which performance resources are most likely to be preventing better performance (i.e., which ones are the "limiting performance resources"). While tools that are useful to practitioners are desirable, they are almost essential for the efficient conduct of in-depth experimental work with the ERM that must accompany the evolution of such tools.

The issue of biologic variability and its influence on numeric analyses can be raised in the context of rigorous numeric application. In this regard, the methods underlying GSPT and the ERM (or any similar cause-and-effect model) are noted to be analogous with those used to design artificial systems. In recent years, conceptual approaches and mathematic tools widely known as *Taguchi methods* [e.g., Bendell et al., 1988] have been shown to be effective for understanding and managing a very similar type of variability that surfaces in the manufacture of artificial systems (e.g., variability associated with performance of components of larger systems and the effect on aspects of performance of the final "product"). Such tools may prove useful in working through engineering problems such as those associated with variability.

147.4 Conclusion

The ERM is a step toward the goal of achieving an application-independent approach to modeling any human-task interface. It provides a systematic and generalizable (across all subsystem types) means of identifying performance measures that characterize human subsystems, as well as a consistent basis for performance measurement definition (and task analysis). It also has served to stimulate focus on a standardized, distinct set of variables that facilitates clear communication of an individual's status among professionals.

After the initial presentation, refinements in both GSPT and the ERM were made. However, the basic approaches, terminology, and constructs used in each have remained quite stable. More recent work has focused on development of various components required for application of the ERM in nontrivial situations. This entails using GSPT and basic ERM concepts to guide a full "fleshing out" of the details of measurement parameterizations and models for different types of human subsystems, definition of standard conventions and notations, and development of computer-based tools. In addition, experimental studies designed to evaluate key constructs of the ERM and to demonstrate the various ways in which it can be applied are being conducted. A good portion of the developmental work is aimed at building a capability to conduct more complex, nontrivial experimental studies. Collaborations with other research groups also have emerged and are being supported to the extent possible. Experiences with it in various contexts and it various levels of application have been productive and encouraging.

The ERM is one, relatively young attempt at organizing and dealing with the complexity of some major aspects of human performance. There is no known alternative that, in a specific sense, attempts to accomplish the same goals as the working model presented here. Is it good enough? For what purposes? Is a completely different approach or merely refinement required? The process of revision is central to the natural course of the history of ideas. Needs for generalizations in human performance persist.

Defining Terms

Basic element of performance (BEP): A modeling item at the basic element level in the ERM defined by identification of a specific system at this level and one of its dimensions of performance (e.g., functional unit = visual information processor, dimension of performance = speed, BEP = visual information processor speed).

Behavior: A general term that relates to what a human or artificial system does while carrying out its function(s) under given conditions. Often, behavior is characterized by measurement of selected parameters or identification of unique system states over time.

Composite performance capacity: A performance capacity at a higher level of abstraction, formed by combining two or more lower-level performance capacities (e.g., via integration to determine the area or volume within a performance envelope).

Dimension of performance: A unique quality that characterizes how well a system executes its function (e.g., speed, accuracy, torque production); one of axes or the label associated with one of the axes in a multidimensional performance space.

Function: The purpose of a system. Some systems map to a single primary function (e.g., process visual information). Others (e.g., the human arm) map to multiple functions, although at any given time multifunction systems are likely to be executing a single function (e.g., polishing a car). Functions can be described and inventoried, whereas level of performance of a given function can be measured.

Generic intermediate level: One of three major hierarchical levels for systems and tasks identified in the elemental resource model. The generic intermediate level represents new systems (e.g., postural maintenance system, object gripper, object lifter, etc.) formed by the combination of functional units at the basic element level (e.g., flexors, extensors, processors, etc.). The term *generic* is used to imply the high frequency of use of systems at this level in tasks of daily life (i.e., items at the "high level" in the ERM).

Goal: A desired endpoint (i.e., result) typically characterized by multiple parameters, at least one of which is specified. Examples include specific task goals (e.g., move an object of specified mass from point *A* to point *B* in 3 seconds) or estimated task performance (maximum mass, range, speed of movement obtainable given a specified elemental performance resource availability profile), depending on whether a reverse or forward analysis problem is undertaken. Whereas function describes the general process of a task, the goal directly relates to performance and is quantitative.

Limiting resource: A performance resource at any hierarchical level (e.g., vertical lift strength, knee flexor speed) that is available in an amount that is less than the worst-case demand imposed by a task. Thus a given resource can only be "limiting" when considered in the context of a specific task.

Performance: Unique qualities of a human or artifical system (e.g., strength, speed, accuracy, endurance) that pertain to how well that system executes its function.

Performance capacity: A quantity in finite availability that is possessed by a system or subsystem, drawn on during the execution of tasks, and limits some aspect (e.g., speed, force, production, etc.) of a system's ability to execute tasks; or, the limit of that aspect itself.

Performance envelope: The surface in a multidimensional performance space, formed with a selected subset of a system's dimensions of performance, that defines the limits of a system's performance. Tasks represented by points that fall within this envelope can be performed by the system in question.

Performance resource: A unique quality of a system's performance modeled and quantified using a resource construct.

Performance resource substitution: The term used in GSPT to describe the manner in which intelligent systems, such as humans, utilize redundancy or adapt to unusual circumstances (e.g., injuries) to obtain optimal procedures for executing a task.

Procedure: A set of constraints placed on a system in which flexibility exists regarding how a goal (or set of goals) associated with a given function can be achieved. Procedure specification requires specification of initial, intermediate, and/or final states or conditions dictating how the goal is to be accomplished. Such specification can be thought of in terms of removing some degrees of freedom.

Resource construct: The collective set of attributes that define and uniquely characterize a resource. Usually, the term is applied only to tangible items. A resource is desirable and measurable in terms of amount (from zero to some finite positive value) in such a manner that a larger numeric value indicates a greater amount of the resource.

Resource economic principle: The principle, observable in many contexts, that states that the amount of a given resource that is available (e.g., money) must exceed the demand placed on it (e.g., cost of an item) if a specified task (e.g., purchase of the item) is to be executed.

Resource utilization flexibility: A term used in GSPT to describe situations in which there are more than one possible source of a given performance resource type, i.e., redundant supplies exist.

Structure: Physical manifestation and attributes of a human or artificial system; the object of one type of measurements at multiple hierarchical levels.

Style: Allowance for variation within a procedure, resulting in the intentional incomplete specification of a procedure or resulting from either intentional or unintentional incomplete specification of procedure.

System: A physical structure, at any hierarchical level of abstraction, that executes one or more functions.

Task: That which results from (1) the combination of specified functions, goals, and procedures or (2) the specification of function and goals and the observation of procedures utilized to achieve the goals.

References

Allard P, Stokes IAF, Blanchi JP. 1994 Three-Dimensional Analysis of Human Movement. Champaign, Ill, Human Kinetics.

Baron S, Kruser DS, Huey BM (eds). 1990. Quantitative Modeling of Human Performance in Complex, Dynamic Systems. Washington, National Academy Press.

Behbehani K, Kondraske GV, Richmond JR. 1988. Investigation of upper extremity visuomotor control performance measures. IEEE Trans Biomed Eng 35(7):518.

Bendell A, Disney J. Pridmore WA. 1988. Taguchi Methods: Applications in World Industry. London, IFS Publishing.

Botterbusch KF. 1987. Vocational Assessment and Evaluation Systems: A Comparison. Menomonie, Wisc, Stout Vocational Rehabilitation Institute, University of Wisconsin.

Card SK, Moran TP, Newell A. 1986. The model human processor. In KR Boff, L Kaufman, and JP Thomas (eds), Handbook of Perception and Human Performance, vol II, pp 45.1–45.35. New York, Wiley.

Carollo JJ, Kondraske GV. 1987. The prerequisite resources for walking: Characterization using a task analysis strategy. In J Leinberger (ed), Proceedings of the Ninth Annual IEEE Engineering and Medical Biology Society Conference, p 357.

Carr B. 1989. Head/Neck Control Performance Measurement and Task Interface Model. M.S. thesis, University of Texas at Arlington, Arlington, Texas.

Chaffin DB, Andersson GBJ. 1984. Occupational Biomechanics. New York, Wiley.

Delp SL, Loan JP, Hoy MG, et al. 1990. An interactive graphics-based Model of the lower extremity to study orthopaedic surgical procedures. IEEE Trans Biomed Eng 37(8):757.

Fleishman EA. 1956. Psychomotor selection tests: Research and application in the United States Air Force. Personnel Psychology 9:449.

Fleishman EA. 1966. Human abilities and the acquisition of skill. In EA Bilodeau (ed), Acquisition of Skill, pp 147–167. New York, Academic Press.

Fleishman EA. 1967. Performance assessment based on an empirically derived task taxonomy. Human Factors 9:349.

Fleishman EA. 1972. Structure and measurement of psychomotor abilities. In RN Singer (ed), The Psychomotor Domain: Movement Behavior, pp 78–106. Philadelphia, Lea and Febiger.

Fleishman EA. 1982. Systems for describing human tasks. Am Psychol 37:821.

Fleishman EA, Quaintance MK., 1984. Taxonomies of Human Performance. New York, Academic Press.

Frisch HP. 1993. Man/machine interaction dynamics and performance analysis. In Proceedings of the NATO-Army-NASA Advanced Study Institute on Concurrent Engineering Tools and Technologies for Mechanical System Design. New York: Springer-Verlag.

Gilbreth FB, Gilbreth FM. 1917. Applied Motion Study. New York, Sturgis and Walton.

Gottlieb GL, Corcos DM, Agarwal GC. 1989. Strategies for the control of voluntary movements with one mechanical degree of freedom. Behav Brain Sci 12:189.

Hemami H. 1988. Modeling, control, and simulation of human movement. CRC Crit Rev Bioeng 13(1):1.

Jafari M. 1989. Modeling and Measurement of Human Speech performance Toward Pathology Pattern Recognition. Ph.D. dissertation, University of Texas at Arlington, Arlington, Texas.

Jafari M, Wong KH, Behbehani K, Kondraske GV. 1989. Performance characterization of human pitch control system: An acoustic approach. J Acous Soc Am 85(3):1322.

Kondraske GV. 1987*a*. Human performance: Science or art? In K Foster (ed), Proceedings of the Thirteenth Northeast Bioengineering Conference, pp 44–47.

Kondraske GV. 1987*b*. Looking at the study of human performance. SOMA: Eng Human Body (ASME) 2(2):50.

Kondraske GV. 1988*a*. Experimental evaluation of an elemental resource model for human performance. In G Harris, C Walker (eds), Proceedings of the Tenth Annual IEEE Engineering and Medical Biology Society Conference, pp 1612–1613. New York, IEEE.

Kondraske GV. 1988*b*. Human performance measurement and task analysis. In A Enders (ed), Technology for Independent Living Sourcebook, 2d ed. Washington, RESNA.

Kondraske GV. 1988*c*. Workplace design: An elemental resource approach to task analysis and human performance measurements. In Proceedings of the International Conference of the Association of Advanced Rehabilitation Technology, pp 608–611.

Kondraske GV. 1989. Neuromuscular performance: Resource economics and product-based composite indices. In Proceedings of the Eleventh Annual IEEE Engineering and Medical Biology Society Conference, pp 1045–1046. New York, IEEE.

Kondraske GV. 1990*a*. A PC-based performance measurement laboratory system. J Clin Eng 15(66):467.

Kondraske GV. 1990*b*. Quantitative measurement and assessment of performance. In RV Smith, JH Leslie (eds), Rehabilitation Engineering, pp 101–125. Boca Raton, Fla, CRC Press.

Kondraske GV. 1993. The HPI Shorthand Notation System for Human System Parameters. Technical Report 92-001R V1.5. Human Performance Institute, University of Texas at Arlington, Arlington, Texas.

Kondraske GV. 1994. An elemental resource model for the human-task interface. Int J Technol Assess Health Care, 11(2), 153–173.

Kondraske GV. 2000. Measurement tools and processes in rehabilitation engineering. In JD Bronzino (ed), Handbook of Biomedical Engineering, 2nd ed., Boca Raton, Fla, CRC Press.

Kondraske GV, Beehler PJH. 1994. Applying general systems performance theory and the elemental resource model to gender-related issues in physical education and sport. Women Sport Phys Act J 3(2)1–19.

Kondraske GV, Beehler PJH, Behbehani K, et al. 1988. Measuring human performance: Concepts, methods, and application examples. SOMA: Eng Human Body 2 (ASME), Jan 6.

Kondraske GV, Johnston, C., Pearson, A., and Tarbox, L. 1997. Performance Prediction and limiting resource identification with nonlinear causal resource analysis. *Proceedings, 19th Annual Engineering in Medicine and Biology Society Conference,*(pp. 1813–1816).

Kondraske GV, Khoury GJ. 1992. Telerobotic system performance measurement: Motivation and methods. In Cooperative Intelligent Robotics in Space III, PP 161–172. New York, SPIE.

Kondraske GV, Standridge R. 1988. Robot performance: conceptual strategies. In Conference of Digest IEEE Midcon/88 Technical Conference, pp 359–362. New York, IEEE.

Leibniz, G.W. and Montgomery, G.R. *Discourse on Metaphysics and the Monadology.* Prometheus Books, 1992.

Mayer TG, Gatchel RJ. 1991. Functional Restoration for Spinal Disorders: The Sports Medicine Approach, pp 66–77. Philadelphia, Lea & Febinger.

Meister D. 1989. Conceptual Aspects of Human Performance. Baltimore, Johns Hopkins University Press.

Navon D, Gopher D. 1979. On the economy of the human processing system. Psych Rev 96:214.

Neel A. 1977. Theories of Psychology: A Handbook. Cambridge, Mass, Schenkman.

Panianpour M, Marras WS. 1993. Development of clinical protocols based on ergonomics evaluation in response to American Disability Act (1992). In Rehabilitation Engineering Center Proposal to National Institute on Disability and Rehabilitation Research, Ohio State University, Columbus, Ohio.

Schoner G, Kelso JAS. 1988. Dynamic pattern generation in behavioral and neural systems. Science 239:1513.

Syndulko K, Tourtellotte WW, Richter E. 1988. Toward the objective measurement of central processing resources. IEEE Eng Med Biol Soc Magazine 7(1):17.

Taylor FW. 1911. The Principles of Scientific Management. New York, Harper and Brothers.

Turvey, MT, Shaw RE, Mace W. 1978. Issues in the theory of action: Degrees of freedom, coordinative structures, and coalitions, In J Requin (ed), Attention and Performance VII. Hillsdale, NJ, Lawrence Earlbaum.

US Department of Labor. 1992. Dictionary of Occupation Titles, 4th ed. Baton Rouge, La, Claitor's Publishing Division.

Vasta PJ, Kondraske GV. 1994. Performance prediction of an upper extremity reciprocal task using nonlinear causal resource analysis. In Proceedings of the Sixteenth Annual IEEE Engineering in Medicine and Biology Society Conference. New York, IEEE.

Wickens CD. 1984. Engineering Psychology and Human Performance. Columbus, Ohio, Charles E Merrill.

World Health Organization (WHO). 1980. International classification of impairments, disabilities, and handicaps. WHO Chron 34:376.

Yen SS, Kondraske GV. 1992. Machine shape perception: Object recognition based on need-driven resolution flexibility and convex-hull carving. In Proceedings of the Conference on Intelligent Robots and Computer Vision X: Algorithm Techniques, pp 176–187. New York, SPIE.

Further Information

General discussions of major issues associated with human performance modeling can be found in the following texts:

Fleishman EA, Quaintance MK. 1984. Taxonomies of Human Performance. New York, Academic Press.

Meister D. 1989. Conceptual Aspects of Human Performance. Baltimore, Johns Hopkins University Press.

Neel A. 1977. Theories of Psychology: A Handbook. Cambridge, Mass, Schenkman.

Requin J. 1978. Attention and Performance VII. Hillsdale, NJ, Lawrence Earlbaum.

Wickens CD. 1984. Engineering Psychology and Human Performance. Columbus, Ohio, Charles E Merrill.

More detailed information regarding general systems performance theory, the elemental resource model, and their application is available from the Human Performance Institute, P.O. Box 19180, University of Texas at Arlington, Arlington, TX 76019-0180 and at the HPI web site: http://www-ee.uta.edu/hpi.

148

Measurement of Neuromuscular Performance Capacities

Susan S. Smith
Texas Woman's University

Movements allow us to interact with our environment, express ourselves, and communicate with each other. Life is movement. Movement is constantly occurring at many hierarchical levels including cellular and subcellular levels. By using the adjective "human" to clarify the term "movement", we are not only defining the species of interest, but also limiting the study to observable performance and its more overt causes. Study of human performance is of interest to a broad range of professionals including rehabilitation engineers, orthopaedic surgeons, therapists, biomechanists, kinesiologists, psychologists, and so

0-8493-0462-8/00/$0.00+$.50

on. Because of the complexity of human performance and the variety of investigators, the study of human performance is conducted from several theoretical perspectives including, (1) anatomical, (2) purpose or character of the movement (such as locomotion), (3) physiological, (4) biomechanical, (5) psychological, (6) socio-cultural, and (7) integrative. The Elemental Resource Model (ERM), presented at the beginning of this section, is an integrative model which incorporates aspects of the other models into a singular system accounting for the human, the task, and the human-task interface [Kondraske, 1999].

The purposes of this chapter are to: (1) provide reasons for measuring four selected variables of human performance: extremes/range of motion, strength, speed of movement, and endurance; (2) briefly define and discuss these variables; (3) overview selected instruments and methods used to measure these variables; and (4) discuss interpretation of performance for a given neuromuscular subsystem.

148.1 Neuromuscular Functional Units

While the theoretical perspectives listed previously may be useful within specific contexts or within specific disciplines, the broader appreciation of human performance and its control can be gained from the perspective of an integrative model such as the ERM [Kondraske, 1999]. This model organizes performance resources into four different domains. Basic movements, such as elbow flexion, are executed by *neuromuscular functional units* in the environmental interface domain. Intermediate and complex tasks, such as walking and playing the piano, utilize multiple basic functional units. A person performing a movement operates the involved functional units along different dimensions of performance according to the demands of the task. Dimensions of performance are factors such as joint motion, strength, speed of movement, and endurance. Lifting a heavy box off the floor requires, among other things, a specific amount of strength associated with neuromuscular functional units of the back, legs, and arms according to the weight and size of the box. Reaching for a light weight box from the top shelf of a closet requires that the shoulder achieve certain extremes of motion according to the height of the shelf.

Whereas four dimensions of performance are considered individually in this chapter, they are *highly interdependent.* For an example, strength availability during a movement is partly dependent on joint angle. Despite interdependence, considering the variables as different dimensions is essential to studying human performance. The components limiting a person's ability to complete a task can only be identified and subsequently enhanced by determining, for example, that the reason a person cannot reach the box off the top shelf is not because of insufficient range of motion of the shoulder, but because of insufficient strength of the shoulder musculature required to lift the arm through the range of motion. Isolating the subsystems involved in a task and maximally stressing them along one or more "isolated" dimensions of performance is a key concept in the ERM. "Maximally stressing" the subsystems means that the maximum amount of the resource available is being determined. This differs from determining the amount of the resource which happened to be used while performing a particular task. Often the distinction between obtaining maximal performance vs. submaximal performance is in the instructions given to the subject. For example, in measuring speed, we say "move as fast as you can."

Purposes of Measuring Selected Neuromuscular Performance Capacities

Range of motion, strength, speed of movement, and/or endurance can be measured for one or more of the following purposes:

1. To determine the amount of the resource available and to compare it to the normal value for that individual. "Normal" is frequently determined by comparisons with the opposite extremity or with normative data when available. This information may be used to develop goals and a program to change the performance.
2. To assist in determining the possible effects of insufficient or imbalanced amounts of the variable on a person's performance of activities of daily living, work, sport, and leisure pursuits. In this

case, the amount of the variable is compared to the demands of the task, rather than to norms or to the opposite extremity.
3. To assist in diagnosis of medical conditions and the nature of movement dysfunctions.
4. To reassess status in order to determine the effectiveness of a program designed to change the amount of the variable.
5. To motivate persons to comply with treatment or training regimes.
6. To document status and the results of treatment or training and to communicate with other involved persons.
7. To assist in ergonomically designed furniture, equipment, techniques, and environments.
8. To provide information to combine with other measures of human performance to predict functional capabilities.

148.2 Range of Motion and Extremes of Motion

Range of motion (*ROM*) is the amount of movement that occurs at a joint. *Range of motion* is typically measured by noting the *extremes of motion* (*EOM*). The designated reference or zero position must be specified for measurements of the two extremes of motion. For example, to measure elbow (radiohumeral joint) flexion and extension, the preferred starting position is with the subject supine with the arm parallel to the lateral midline of the body with the palm facing upward [Norkin and White, 1995; Palmer and Epler, 1998]. Measurements are taken with the elbow in the fully flexed position and with the elbow in the fully extended position.

Movement Terminology

Joint movements are described using a coordinate system with the human body in an anatomical position. The anatomical position of the body is an erect position, face forward, arms at sides, palms facing forward, and fingers and thumbs in extension. The central coordinate system consists of three cardinal planes and axes with its origin located between the cornua of the sacrum [Panjabi, White, and Brand, 1974]. Figure 148.1 demonstrates the planes and axes of the central coordinate system. The same coordinate system can parallel the master system at any joint in the body by relocating the origin to any defined point.

The sagittal plane is the y, z plane; the frontal (or coronal) plane is the y, x plane; the horizontal (or transverse) plane is the x, z plane. Movements are described in relation to the origin of the coordinate system. The arrows indicate the positive direction of each axis. An anterior translation is $+z$; a posterior translation is $-z$. Clockwise rotations are $+\theta$, and counterclockwise rotations are $-\theta$.

Joints are described as having degrees of freedom (dof) of movement. Dof is the number of independent coordinates in a system that are necessary to accurately specify the position of an object in space. If a motion occurs in one plane and around one axis, the joint is defined as having one dof. Joints with movements in two planes occurring around two different axes, have two dof, and so on.

Angular movements refer to motions that cause an increase or decrease in the angle between the articulating bones. Angular movements are flexion, extension, abduction, adduction, and lateral flexion (See Table 148.1). Rotational movements generally occur around a longitudinal (or vertical) axis except for movements of the clavicle and scapula. The rotational movements occurring around the longitudinal axis (internal rotation, external rotation, opposition, horizontal abduction, and horizontal adduction) are described in Table 148.1. Rotation of the scapula is described in terms of the direction of the inferior angle. Movement of the inferior angle of the scapula toward the midline is a medial (or downward) rotation, and movement of the inferior angle away from the midline is lateral (or upward) rotation. In the extremities the anterior surface of the extremity is used as the reference area. Because the head, neck, trunk, and pelvis rotate about a midsagittal, longitudinal axis, rotation of these parts is designated as right or left. As can be determined from Fig. 148.1, axial rotation of the trunk toward the left can be described mathematically as $+\theta y$.

A communications problem often exists in describing motion using the terms defined in Table 148.1. A body segment can be in a position such as flexion, but can be moving toward extension. This confusion is partially remedied by using the form of the word with the suffix, *-ion*, to indicate a static position and using the suffix, *-ing*, to denote a movement. Thus, an elbow can be in a position of 90° flex*ion* and also extend*ing*.

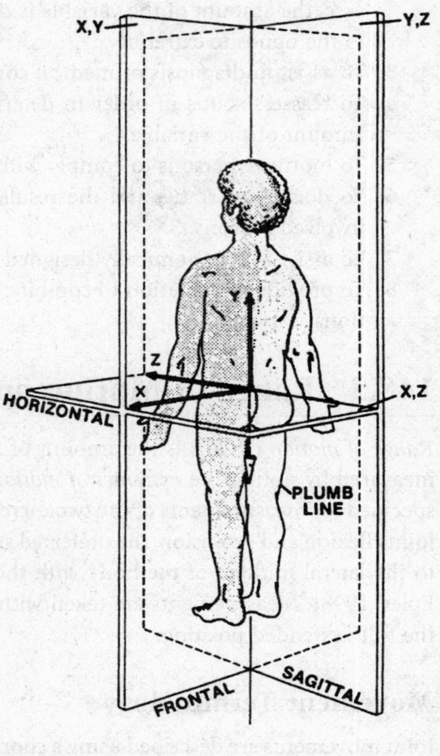

Factors Influencing ROM/EOM and ROM/EOM Measurements

The ROM and EOM available at a joint is determined by morphology and the soft tissues surrounding and crossing a joint, including the joint capsule, ligaments, tendons, and muscles. Other factors such as age, gender, swelling, muscle mass development, body fat, passive insufficiency (change in the ROM/EOM available at one joint in a two-joint muscle complex caused by the position of the other joint), and time of day (diurnal effect) also affect the amount of motion available. Some persons, because of posture, genetics, body type, or movement habits, normally demonstrate hypermobile or hypomobile joints. Dominance has not been found to significantly affect available ROM. See discussion in Miller [1985]. The shapes of joint surfaces, which are designed to allow movement in particular directions, can become altered by disease, trauma, and posture, thereby, increasing or decreasing the ROM/EOM. Additionally, the soft tissues crossing a joint can become tight (contracted) or overstretched altering the ROM/EOM.

FIGURE 148.1 Planes and axes are illustrated in anatomical position. The central coordinate system with its origin between the cornua of the sacrum is shown. *Source:* White III AA, Panjabi MM. 1990. *Clinical Biomechanics of the Spine*, 2nd ed., p 87. Philadelphia, JB Lippincott Company. With permission.

The type of movement, active or passive, also affects ROM/EOM. When measuring active ROM (AROM), the person voluntarily contracts muscles and moves the body part through the available motion. When measuring passive ROM (PROM), the examiner moves the body part through the ROM. PROM is usually slightly greater than AROM due to the extensibility of the tissues crossing and comprising the joint. AROM can be decreased because of restricted joint mobility, muscle weakness, pain, unwillingness to move, or inability to follow instructions. PROM is assessed to determine the integrity of the joint and the extent of structural limitation.

Instrumented Systems Used to Measure ROM/EOM

The most common instrument used to measure joint ROM/EOM is a goniometer. The universal goniometer, shown in Fig. 148.2a, is most widely used clinically. A variety of universal goniometers have been developed for specific applications. Two other types of goniometers are also shown in Fig. 148.2.

Table 148.2 lists and compares several goniometric instruments used to measure ROM/EOM. Choice of the instrument used to measure ROM/EOM depends upon the degree of accuracy required, time available to the examiner, the measurement environment, the body segment being measured, and the equipment available.

Non-goniometric methods of joint measurement are available. Tape measures, radiographs, photography, cinematography, videotape, and various optoelectric movement monitoring systems can also be

TABLE 148.1 Movement Terms, Planes, Axes, and Descriptions of Movements

Movement Term	Plane	Axis	Description of Movement
Flexion	Sagittal	Frontal	Bending of a part such that the anterior surfaces approximate each other. However, flexion of the knee, ankle, foot, and toes refers to movement in the posterior direction.
Extension	Sagittal	Frontal	Opposite of flexion; involves straightening a body part.
Abduction	Frontal	Sagittal	Movement away from the midline of the body or body part; abduction of the wrist is sometimes called *radial deviation.*
Adduction	Frontal	Sagittal	Movement towards the midline of the body or body part; adduction of the wrist is sometimes called *ulnar deviation.*
Lateral flexion	Frontal	Sagittal	Term used to denote lateral movements of the head, neck, and trunk.
Internal (medial) rotation	Horizontal	Longitudinal	Turning movement of the anterior surface of a part towards the midline of the body; internal rotation of the forearm is referred to as *pronation.*
External (lateral) rotation	Horizontal	Longitudinal	Turning movement of the anterior surface of a part away from the midline of the body; external rotation of the forearm is referred to as *supination.*
Opposition	Multiple	Multiple	Movement of the tips of the thumb and little finger toward each other.
Horizontal abduction	Horizontal	Longitudinal	Movement of the arm in a posterior direction away from the midline of the body with the shoulder joint in 90° of either flexion or abduction.
Horizontal adduction	Horizontal	Longitudinal	Movement of the arm in an anterior direction toward the midline of the body with the shoulder joint in 90° of either flexion or abduction.
Tilt	Depends on joint	Depends on joint	Term used to describe certain movements of the scapula and pelvis. In the scapula, an anterior tilt occurs when the coracoid process moves in an anterior and downward direction while the inferior angle moves in a posterior and upward direction. A posterior tilt of the scapula is the opposite of an anterior tilt. In the pelvis, an anterior tilt is rotation of the anterior superior spines (ASISs) of the pelvis in an anterior and downward direction; a posterior tilt is movement of the ASISs in a posterior and upward direction. A lateral tilt of the pelvis occurs when the pelvis is not level from side to side, but one ASIS is higher than the other one.
Gliding	Depends on joint	Depends on joint	Movements that occur when one articulating surface slides on the opposite surface.
Elevation	Frontal		A gliding movement of the scapula in an upward direction as in shrugging the shoulders.
Depression	Frontal		Movement of the scapula downward in a direction reverse of elevation.

used to measure or calculate the motion available at various joints. These methods are beyond the scope of this chapter.

Key Concepts in Goniometric Measurement

Numerous textbooks [Clarkson and Gilewich, 1989; Norkin and White, 1995; Palmer and Epler, 1998] are available that describe precise procedures for goniometric measurements of each joint. Unfortunately, there is a lack of standardization among these references.

In general, the anatomical position of zero degrees (*preferred starting position*) is the desired starting position for all ROM/EOM measurements except rotation at the hip, shoulder, and forearm. The arms of the goniometer are usually aligned parallel and lateral to the long axis of the moving and the fixed body segments in line with the appropriate landmarks. In the past, some authors contended that placement of the axis of the goniometer should be congruent with the joint axis for accurate measurement [West, 1945; Wiechec and Krusen, 1939]. However, the axis of rotation for joints changes as the body segment moves through its ROM; therefore, a goniometer cannot be placed in a position in line with the joint axis during movement. Robson [1966] described how variations in the placement of the goniometer's axis could affect the accuracy of ROM measurements. Miller [1985] suggested that the axis problem could be handled by ignoring the goniometer's axis and concentrating on the accurate alignment of the arms of the goniometer with the specified landmarks. Potentially some accuracy may be sacrificed,

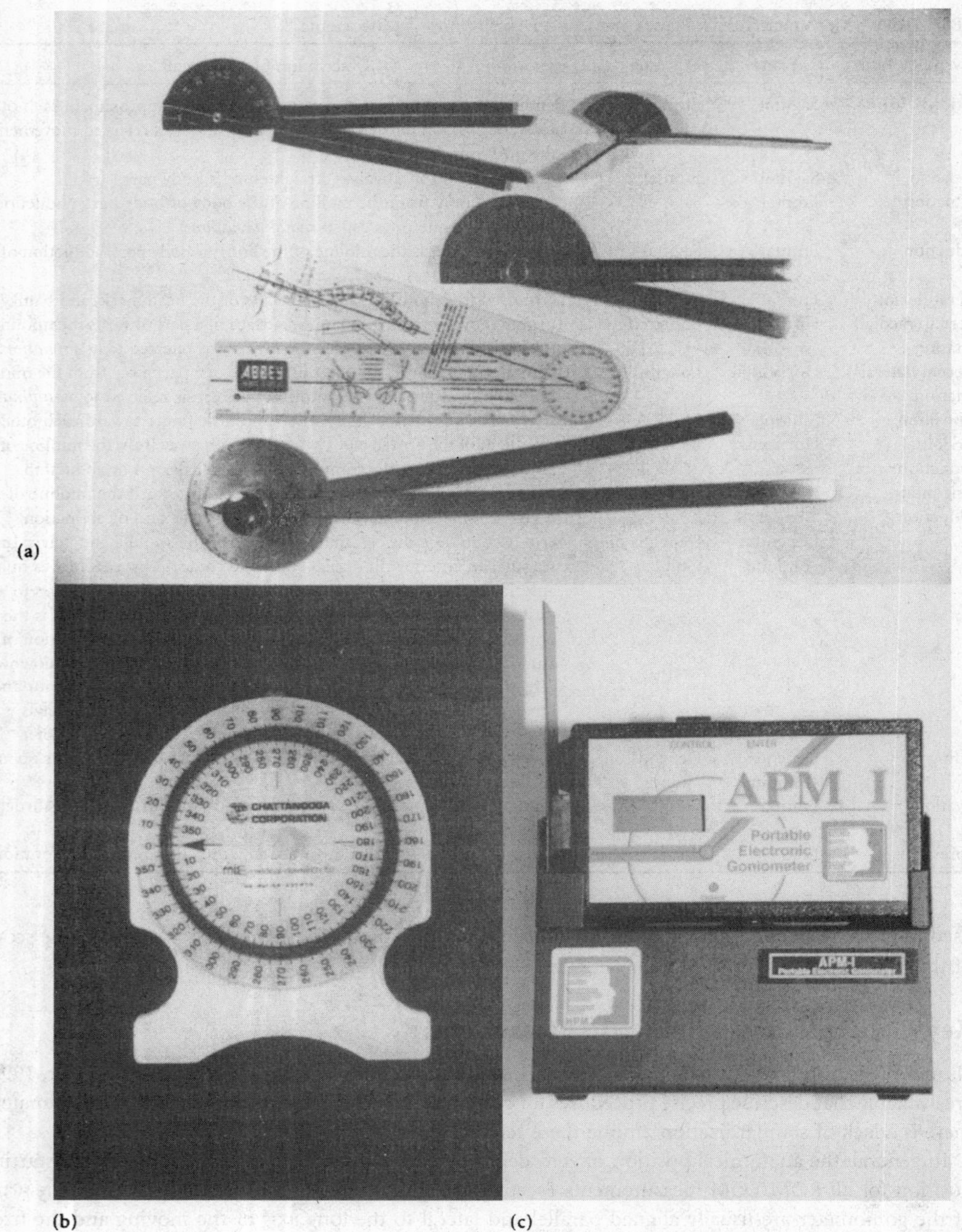

FIGURE 148.2 Three types of goniometric instruments used to measure range and extremes of motion are shown: (*a*) typical 180- and 360-degree universal goniometers of various sizes; (*b*) a fluid goniometer, which is activated by the effects of gravity; (*c*) an APM I digital electronic device that works similarly to a pendulum goniometer.

but the technique is simplified and theoretically more reproducible. The subject's movement is observed for unwanted motions that could result in inaccurate measurement. For example, a subject might attempt to increase forearm supination by laterally flexing the trunk.

TABLE 148.2 Comparison of Various Goniometers Commonly Used to Measure Joint Range and Extremes of Motion

Type of Goniometer	Advantages/Uses	Disadvantages/Limitations
Universal goniometer		
A protractor-like device with one arm considered movable and the other arm stationary; protractor can have a 180° or 360° scale and is usually numbered in both directions; available in a range of sizes and styles to accommodate different joints (See Fig. 148.2a).	Inexpensive; portable; familiar devices; size of the joint being measured determines size of the goniometer used; clear plastic goniometers have a line through the center of the arms to make alignment easier and more accurate; finger goniometers can be placed over the dorsal aspect of the joint being measured.	Several goniometers of different sizes may be required, especially if digits are measured; full-circle models may be difficult to align when the subject is recumbent and axis alignment is inhibited by the protractor bumping the surface; the increments on the protractors may vary from 1°, 2°, or 5°; placement of the arms is a potential source of error.
Fluid (or bubble) goniometer		
A device with a fluid-filled channel with a 360°–scale that relies on the effects of gravity (See Fig. 148.2b); dial turns allowing the goniometer to be "zeroed;" some models are strapped on and others must be held against the body part.	Quick and easy to use because it is not usually aligned with bony landmarks; does not have to conform to body segments; useful for measuring neck and spinal movements; using a pair of fluid goniometers permits distinguishing regional spinal motion.	More expensive than universal goniometers; using a pair of goniometers is awkward; useless for motions in the horizontal plane; error can be induced by slipping, skin movement, variations in amount of soft tissue owing to muscle contraction, swelling, or fat, and the examiner's hand pressure changing body segment contour; reliability may be sacrificed from lack of orientation to landmarks and difficulty with consistent realignment [Miller, 1985].
Pendulum goniometer		
A scaled, inclinometer-like device with a needle or pointer (usually weighted); some models are strapped on and others must be held against the body part (not shown).	Inexpensive; same advantages as for the fluid goniometer described above.	Some models cannot be "zeroed"; useless for motions in the horizontal plane; same soft tissue error concerns as described above for the fluid goniometer.
"Myrin" OB goniometer		
A fluid-filled, rotatable container consisting of compass needle that responds to the earth's magnetic field (to measure horizontal motion), a gravity-activated inclination needle (to measure frontal and sagittal motion), and a scale (not shown).	Can be strapped on the body part allowing the hands free to stabilize and move the body part; not necessary to align the goniometer with the joint axis; permits measurements in all three planes.	Expensive and bulky compared with universal goniometer; not useful for measuring small joints of hand and foot; susceptible to magnetic fields [Clarkson and Gilewich, 1989]; subject to same soft tissue error concerns described above under fluid goniometer.
Arthrodial protractor		
A large, flat, clear plastic protractor without arms that has a level on the straight edge (not shown).	Does not need to conform to body segments; most useful for measuring joint rotation and axioskeletal motion.	Not useful for measuring smaller joints, especially those with lesser ROMs; usually scaled in large increments only.
APM I		
Computerized goniometer with digital sensing and electronics; can perform either continuous monitoring or calculate individual ROM/EOM from a compound motion function (See Fig. 148.2c).	Easy to use; provides rapid digital read-out; measures angles in any plane of motion; one hand is free to stabilize and move body segments; particularly easy for measuring regional spinal movements.	Expensive compared to most other instruments described; device must be rotated perpendicular to the direction of segment motion only; unit must stabilize prior to measurement; excessive delays in recording must be avoided; subject to the same soft tissue error concerns as described above under fluid goniometer.
Electrogoniometer		
Arms of a goniometer are attached to a potentiometer and are strapped to the proximal and distal body parts; movement from the device causes resistance in the potentiometer which measures the ROM (not shown).	More useful for dynamic ROM, especially for determining kinematic variables during activities such as gait; provides immediate data; some electrogoniometers permit measurement in one, two, or three dimensions.	Aligning and attaching the device is time-consuming and not amenable to all body segments; device and equipment needed to use it are moderately expensive; essentially laboratory equipment; less accurate for measurement of absolute limb position; device itself is cumbersome and may alter the movement being studied.

Numerical Notation Systems

Three primary systems exist for expressing joint motion in terms of degrees. These are the *0–180 System*, the *180–0 System*, and the *360 System*. The 0–180 System is the most widely accepted system in medical applications and may be the easiest system to interpret. In the 0–180 System, the starting position for all movements is considered to be 0°, and movements proceed toward 180°. As the joint motion increases, the numbers on the goniometric scale increase. In the 180–0 System, movements toward flexion approach 0° and movements toward extension approach 180°. Different rules are used for the other planes of motion. The 360 System is similar to the 180–0 System. In the 360 System, movements are frequently performed from a starting position of 180°. Movements of extension or adduction which go beyond the neutral position approach 360°. Joint motion can be reported in tables, charts, graphs, or pictures. In the 0–180 System, the starting and ending ranges are recorded separately, as 0°–130°. If a joint cannot be started in the 0° position, the actual starting position is recorded, as 10°–130°.

148.3 Strength

Muscle strength implies the force or torque production capacity of muscles. However, to measure strength, the term must be operationally defined. One definition modified from Clarkson and Gilewich [1989] states that muscular strength is the maximal amount of torque or force that a muscle or muscle groups can voluntarily exert in one maximal effort, when type of muscle contraction, movement velocity, and joint angle(s) are specified.

Strength Testing and Muscle Terminology

Physiologically, skeletal muscle strength is the ability of muscle fibers to generate maximal tension for a brief time interval. A muscle's ability to generate maximal tension and to sustain tension for differing time intervals is dependent on the muscle's cross-sectional area (the larger the cross-sectional area, the greater the strength), geometry (including the muscle fiber arrangement, length, moment arm, and angle of pennation), and physiology. Characteristics of muscle fibers have been classified based on twitch tension and fatigability. Different fiber types have different metabolic traits. Different types of muscle fibers are differentially stressed depending on the intensity and duration of the contraction. Ideally, strength tests should measure the ability of the muscle to develop tension rapidly and to sustain the tension for brief time intervals. In order to truly measure muscle tension, a measurement device must be directly attached to the muscle or tendon. Whereas this direct procedure has been performed [Komi, 1990], it is hardly useful as a routine clinical measure. Indirect measures are used to estimate the strength of muscle groups performing a given function, such as elbow flexion.

Muscles work together in groups and may be classified according to the major role of the group in producing movement. The *prime mover*, or *agonist*, is a muscle or muscle group that makes the major contribution to movement at a joint. The *antagonist* is a muscle or muscle group that has an opposite action to the prime mover(s). The antagonist relaxes as the agonist moves the body part through the ROM. *Synergists* are accessory muscles that contract and work with the agonist to produce the desired movement. Synergists may work by stabilizing proximal joints, preventing unwanted movement, and joining with the prime mover to produce a movement that one muscle group acting alone could not produce.

A number of terms and concepts are important toward understanding the nature and scope of strength capacity testing. Several of these terms are defined below; however, there are no universally accepted definitions for these terms.

Dynamic contraction—the output of muscles moving body segments [Kroemer, 1991].

Isometric—tension develops in a muscle, but the muscle length does not change and no movement occurs.

Static—same as isometric.

Isotonic—a muscle develops constant tension against a load or resistance. Kroemer [1991] suggests the term, *isoforce*, more aptly describes this condition.

Concentric—a contraction in which a muscle develops internal force that exceeds the external force of resistance, the muscle shortens, and movement is produced [O'Connel and Gowitzke, 1972].

Eccentric—a contraction in which a muscle lengthens while continuing to maintain tension [O'Connel & Gowitzke, 1972].

Isokinetic—a condition where the angular velocity is held constant. Kroemer [1991] prefers the term, *isovelocity*, to describe this type of muscle exertion.

Isoinertial—a static or dynamic muscle contraction where the external load is held constant [Kroemer, 1983].

Factors Influencing Muscle Strength and Strength Measurement

In addition to the anatomical and physiological factors affecting strength, other factors must be considered when strength testing. The ability of a muscle to develop tension depends on the type of muscle contraction. Per unit of muscle, the greatest tension can be generated eccentrically, less can be developed isometrically, and the least can be generated concentrically. These differences in tension generating capacity are so great that the type of contraction being strength–tested requires specification.

Additionally, strength is partially determined by the ability of the nervous system to cause more motor units to fire synchronously. As one trains, practices an activity, or learns test expectations, strength can increase. Therefore, strength is affected by previous training and testing. This is an important consideration in standardizing testing and in retesting.

A muscle's attachments define the angle of pull of the tendon on the bone and thereby the mechanical leverage at the joint center. Each muscle has a moment arm length, which is the length of a line normal to the muscle passing through the joint center. This moment arm length changes with the joint angle which changes the muscle's tension output. Optimal tension is developed when a muscle is pulling at a 90° angle to the bony segment.

Changes in muscle length alter the force-generating capacity of muscle. This is called the *length-tension relationship. Active* tension decreases when a muscle is either lengthened or shortened relative to its resting length. However, after applying a precontraction stretch, or slightly lengthening a muscle and the series elastic component (connective tissue) prior to a contraction causes a greater amount of *total* tension to be developed [Soderberg, 1992]. Of course, excessive lengthening would reduce the tension-generating capacity.

A number of muscles cross over more than one joint. The length of these muscles may be inadequate to permit complete ROM of all joints involved. When a multijoint muscle simultaneously shortens at all joints it crosses, further effective tension development is prevented. This phenomena is called *active insufficiency.* For example, when the hamstrings are tested as knee flexors with the hip extended, less tension can be developed than when the hamstrings are tested with the hip flexed. Therefore, when testing the strength of multijoint muscles, the position of all involved joints must be considered.

The *load-velocity relationship* is also important in testing muscle strength. A load-velocity curve can be generated by plotting the velocity of motion of the muscle lever arm against the external load. With concentric muscle contractions, the least tension is developed at the highest velocity of movement and vice versa. When the external load equals the maximal force that the muscle can exert, the velocity of shortening reaches zero and the muscle contracts isometrically. When the load is increased further, the muscle lengthens eccentrically. During eccentric contractions, the highest tension can be achieved at the highest velocity of movement [Komi, 1973].

The force generated by a muscle is proportional to the contraction time. The longer the contraction time, the greater the force development up to the point of maximum tension. Slower contraction leads to greater force production because more time is allowed for the tension produced in contractile elements to be transferred through the noncontractile components to the tendon. This is the *force-time relationship.*

Tension in the tendon will reach the maximum tension developed by the contractile tissues only if the active contraction process is of adequate (even up to 300 msec) duration [Sukop and Nelson, 1974].

Subject effort or motivation, gender, age, fatigue, time of day, temperature, occupation, and dominance can also affect force or torque production capacity. Important additional considerations may be changes in muscle function as a result of pain, overstretching, immobilization, trauma, paralytic disorders, neurologic conditions, and muscle transfers.

Grading Systems and Parameters Measured

Clinically, the two most frequently used methods of strength testing are actually non-instrumented tests: the manual muscle test (MMT) and the functional muscle test (See Amundsen [1990] for more information on functional muscle tests). In each of these cases interval scaled grading criteria are operationally defined. However, a distinct advantage of using instruments to measure strength is that quantifiable units can be obtained, usually force or torque. Torque = force × the distance between the point of force application and the axis of rotation:

$$T = Fx. \tag{148.1}$$

An important issue in strength testing is deciding whether to measure force (a linear quantity) or torque (a rotational quantity). If the point of application of a force is closer to the axis of rotation, the muscle being assessed has a mechanical advantage as compared to when the point of contact is more distal. Therefore, when forces are measured, unless the measurement devices are applied at the same anatomical position for each test, force measurements can differ substantially even though actual muscle tension remains the same. If the strength testing device has an axis of rotation that can be aligned with the anatomical axis of rotation, then torque can be measured directly. When this is not the case, the moment arm can be measured and torque calculated. Force is more typically measured in whole–body exertions, such as lifting. Another issue is whether to measure and record peak or averaged values. However, if strength is defined as maximum torque production capacity, peak values are implied.

In addition to single numerical values, some strength measurement systems display and print force or torque (versus time) curves, angle-torque curves, and graphs. Computerized systems frequently compare the "involved" with the "uninvolved" extremity calculating "percent deficits." As strength is considered proportional to body weight (perhaps erroneously, see Delitto [1990]) force and torque measurements are frequently reported as a peak torque to body weight ratio. This is seemingly to facilitate use of normative data where present.

Methods and Instruments Used to Measure Muscle Strength

There are two broad categories of testing force or torque production capacity: one category consists of measuring the capacity of defined, local muscle groups (e.g., elbow flexors); the second category of tests consists of measuring several muscle groups on a whole–body basis performing a higher level task (e.g., lifting). The purpose of the test, required level of sensitivity, and expense are primary factors in selecting the method of strength testing. No single method has emerged as being clearly superior or more widely applicable. Like screwdrivers, different types and different sizes are needed depending upon job demands.

Many of the instrumented strength testing techniques, which are becoming more standardized clinically and which are almost exclusively used in engineering applications, are based on the concepts and methods of MMT. Although not used for performance capacity tests, because of the ease, practicality, and speed of manual testing, it is still considered a useful tool, especially diagnostically to localize lesions. Several MMT grading systems prevail. These differ in the actual test positions and premises upon which muscle grading is based. For example, the approach promoted by Kendall, McCreary, and Provance [1993] tests a specific muscle (e.g., brachioradialis) rather than a motion. The Daniels and Worthingham [Hislop and Montgomery, 1995] method tests motions (e.g., elbow flexion) which involve all the agonists and synergists used to perform the movement. The latter is considered more functional and less time consuming, but less specific. The reader is advised to consult these references directly for more information about MMT methods. Further discussion of non-instrumented tests is beyond the scope of this chapter.

An argument can be made for using isometric testing because the force or torque reflects actual muscle tension as the position of the body part is held constant and the muscle mechanics do not change. Additionally, good stabilization is easier to achieve, and muscle actions can be better isolated. However, some clinicians prefer dynamic tests, perceiving them as more reflective of function. An unfortunate fact is that neither static nor dynamic strength measurements alone can reveal whether strength is adequate for functional activities. However, strength measurements can be used with models and engineering analyses for such assessments.

Selected instrumented methods of measuring force or torque production capacity are listed and compared in Table 148.3. Table 148.3 is by no means comprehensive. More indepth review and comparisons of various methods can be found in Amundsen [1990] and Mayhew and Rothstein [1985]. Figure 148.3 illustrates three common instruments used to measure strength.

Key Concepts in Measuring Strength

Because of the number of factors influencing strength and strength testing (discussed in a previous section), one can become discouraged rather than challenged when faced with the need to measure strength. Optimally, strength testing would be based on the "worst case" functional performance demands required by an individual in his or her daily life. "Worst case" testing requires knowing the performance demands of tasks including the positions required, types of muscle contractions, and so on.

In the absence of such data, current strategy is to choose the instruments and techniques that maximally stress the system under a set of representative conditions that either (a) seem logical based on knowledge of the task, or (b) have been reported as appropriate and reliable for the population of interest. An attempt is made to standardize the testing in terms of contraction-type, test administration instructions, feedback, warm-up, number of trials, time of day, examiner, duration of contraction (usually 4 to 6 seconds), method and location of application of force, testing order, environmental distractions, subject posture and position of testing, degree of stabilization, and rest intervals between exertions (usually 30 seconds to 2 minutes) [Chaffin, 1975; Smidt and Rogers, 1982]. In addition, the subject must be observed for muscle group substitutions and "trick" movements.

148.4 Speed of Movement

Speed of movement refers to the rate of movement of the body or body segments. The maximum movement speed that can be achieved represents another unique performance capacity of an identified system that is responsible for producing motion. Everyday living, work, and sport tasks are commonly described in terms of the speed requirements (e.g., repetitions per minute or per hour). For physical tasks, such descriptions translate to translational motion speeds (e.g., as in lifting) as well as rotational motion speeds (i.e., movement about a dof of the joint systems involved). Thus, there is important motivation to characterize this capacity.

Speed of Movement Terminology

Speed of movement must be differentiated from *speed of contraction*. Speed of contraction refers to how fast a muscle generates tension. Two body parts may be moving through an arc with the same speed of movement; however, if one part has a greater mass, its muscles must develop more tension per unit of time to move the heavier body part at the same speed as the lighter body part.

Speed, velocity, and acceleration also can be distinguished. The terms velocity and speed are often used interchangeably; however, the two quantities are frequently not identical. Velocity means the rate of motion *in a particular direction*. *Acceleration* results from a change in velocity over time. General

TABLE 148.3 Comparison of Various Instrumented Methods Used to Measure Muscle Strength

Instrument/Method	Advantages/Uses	Disadvantages/Limitations
Repetition maximum		
Amount of weight a subject can lift a given number of times and no more; one determines either a one repetition maximum (1-RM) or a ten repetition maximum (10-RM). A 1-RM is the maximum amount of weight a subject can lift once; a 10-RM is the amount of weight a subject can lift 10 times; a particular protocol to determine RMs is defined [DeLorme and Watkins, 1948]; measures dynamic strength in terms of weight (pounds or kilograms) lifted.	Requires minimal equipment (weights); inexpensive and easy to administer; frequently used informally to assess progress in strength training.	Uses serial testing of adding weights which may invalidate subsequent testing; no control for speed of contraction or positioning; minimal information available on the reliability and validity of this method.
Hand-held dynamometer		
Device held in the examiner's hand used to test strength; devices use either hydraulics, strain gauges (load–cells), or spring systems (See Fig. 148.3a); used with a "break test" (the examiner exerts a force against the body segment to be tested until the part gives way) or a "make test" (the examiner applies a constant force while the subject exerts a maximum force against it); "make tests" are frequently preferred for use with hand-held dynamometers [Bohannon, 1990; Smidt, 1984]; measures force; unclear whether test measures isometric or eccentric force (this may depend on whether a "make test" or a "break test" is used).	Similar to manual muscle testing (MMT) in test positions and sites for load application; increased objectivity over MMT; portable; easy to administer; relatively inexpensive; commercially available from several suppliers; adaptable for a variety of test sites; provide immediate output; spring and hydraulic systems are non-electrical; load–cell based systems provide more precise digital measurements.	Stabilization of the device and body segment can be difficult; results can be affected by the examiner's strength; limited usefulness with large muscle groups; spring-based systems fatigue over time becoming inaccurate; range and sensitivity of the systems vary; shape of the unit grasped by the examiner and shape of the end-piece vary in comfort, and therefore the force a subject or examiner is willing to exert; more valuable for testing subjects with weakness than for less involved or healthy subjects due to range limits within the device (See discussion in Bohannon [1990]).
Cable tensiometer		
One end of a cable is attached to an immovable object and the other end is attached to a limb segment; the tensiometer is placed between the sites of fixation; as the cable is pulled, it presses on the tensiometer's riser which is connected to a gauge (See discussion in Mayhew and Rothstein [1985]); measures isometric force.	Mostly used in research settings; evidence presented on reliability when used with healthy subjects [Clarke, 1954; Clarke, Bailey and Shay, 1952]; relatively inexpensive.	Requires special equipment for testing; testing is time-consuming and some tests require two examiners; unfamiliar to most clinicians; not readily available; less sensitive at low force levels.
Strain gauge		
Electroconductive material applied to metal rings or rods; a load applied to the ring or bar deforms the metal and a gauge; deformation of the gauge changes the electrical resistance of the gauge causing a voltage variation; this change can be converted and displayed using a strip chart recorder or digital display; measures isometric force.	Mostly used in research settings; increased sensitivity for testing strong and weak muscles.	Strain gauges require frequent calibration and are sensitive to temperature variations; to be accurate the body part must pull or push against the gauge in the same line that the calibration weights were applied; unfamiliar to most clinicians; not commercially available; difficult to interface the device comfortably with the subject.
Isokinetic dynamometer		
Constant velocity loading device; several models marketed by a number of different companies; most consist of a movable lever arm controlled by an electronic servomotor that can be preset for selected angular velocities usually between 0° and 500° per second; when the subject attempts to	Permits dynamic testing of most major body segments; especially useful for stronger movements; most devices provide good stabilization; measures reciprocal muscle contractions; widespread	Devices are large and expensive; need calibration with external weights or are "self-calibrating;" signal damping and "windowing" may affect data obtained; angle-specific measurements may not be accurate if a damp is used because torque readings do not relate to the

TABLE 148.3 (continued) Comparison of Various Instrumented Methods Used to Measure Muscle Strength

Instrument/Method	Advantages/Uses	Disadvantages/Limitations
accelerate beyond the pre-set machine speed, the machine resists the movement; a load cell measures the torque needed to prevent body part acceleration beyond the selected speed; computers provide digital displays and printouts (See typical device in Fig. 148.3b); measures isokinetic-concentric (and in some cases, isokinetic-eccentric) and isometric strength; provides torque (or occasionally force) data; debate exists about whether data are ratio-scaled or not; accounting for the weight of the segment permits ratio-scaling [Winter, Wells, and Orr, 1981].	clinical acceptance; also records angular data, work, power, and endurance-related measures; provides a number of different reporting options; also used as exercise devices.	goniometric measurements; joints must be aligned with the mechanical axis of the machine; inferences about muscle function in daily activities from isokinetic test results have not been validated; data obtained between different brands are not interchangeable; adequate stabilization may be difficult to achieve for some movements; may not be usable with especially tall or short persons.
Hand dynamometer		
Instruments to measure gripping or pinching strength specifically for the hand; usually use a spring scale or strain-gauge system (See typical grip strength testing device in Fig. 148.3c); measures isometric force.	Readily available from several suppliers; easy to use; relatively inexpensive; widespread use; some normative data available.	Only useful for the hand; different brands not interchangeable; normative data only useful when reported for the same instrument and when measurements are taken with the same body position and instrument setting; must be recalibrated frequently.

velocity and acceleration measurements are beyond the intent of this chapter. Reaction speed and response speed are other related variables also not considered.

Factors Influencing Speed of Movement and Speed of Movement Measurements

Muscles with larger moment arms, longer muscle fibers, and less pennation tend to be capable of generating greater speed. Many of the same factors influencing strength, discussed previously, such as muscle length, fatigue, and temperature affect the muscle's contractile rate. The load-velocity relationship is especially important when testing speed of movement. In addition to these and other physiological factors, speed can be reduced by factors such as friction, air resistance, gravity, unnecessary movements, and inertia [Jensen and Fisher, 1979].

Parameters Measured

Speed of movement can be measured as a linear quantity or as an angular quantity. Typically, if the whole body is moving linearly in space as in walking or running, a point such as the center of gravity is picked, and translational motion is measured. Also, when an identified point on a body segment (e.g., the tip of the index finger) is moved in space, translational movement is observed, and motion is measured in translational terms. If the speed of a rotational motion system (e.g., elbow flexors) is being measured, then the angular quantity is determined. As the focus here is on measuring isolated neuromuscular performance capacities, the angular metric is emphasized. Angular speed of a body segment is obtained by: angular speed = change in angular position/change in time:

$$\sigma = \frac{\Delta\phi}{\Delta t}. \tag{148.2}$$

Thus, speed of movement may be expressed in revolutions, degrees, or radians per unit of time, such as degrees per second (deg/s).

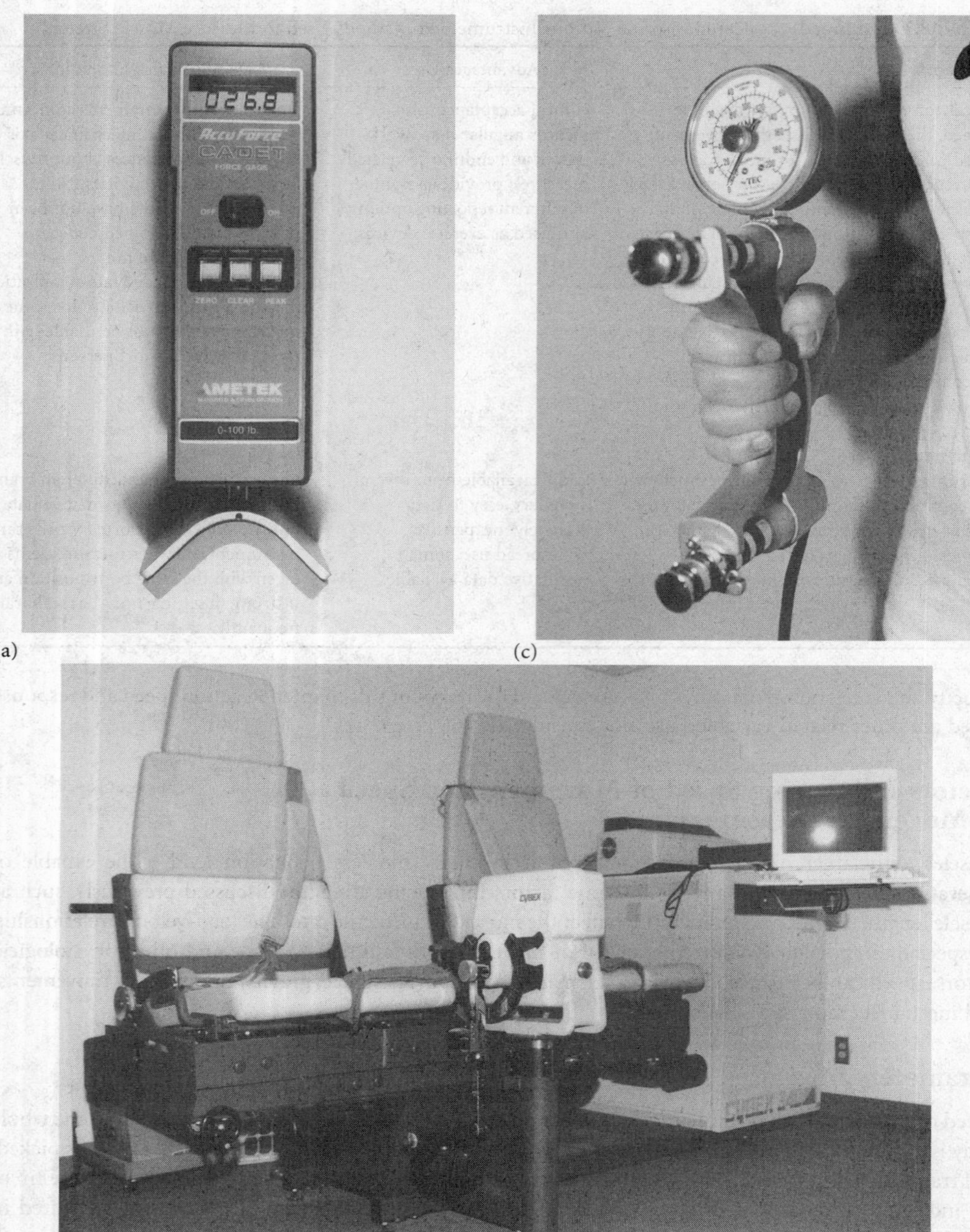

(a) (c) (b)

FIGURE 148.3 Three types of instrumented strength testing devices are shown: (*a*) a representative example of a typical hand-held dynamometer; (*b*) an example of an isokinetic strength testing device; (*c*) a hand dynamometer used to measure grip strength

Another type of speed measure applies to well-defined (over fixed angle or distance) cyclic motions. Here repetitions per unit time or cycles per unit time measures are sometimes used. However, in almost every one of these situations, speed can be expressed in degrees per second or meters per second. The latter units are preferred because they allow easier comparison of speeds across a variety of tasks. The only occasion when this is difficult is when translation motion is not in a simple straight line, such as when a person is performing a complex assembly task with multiple subtasks.

The issue of whether to express speed as maximum, averaged, or instantaneous values must also be decided based on which measure is a more useful indicator of the performance being measured. In addition to numerical reporting of speed data, time-history graphs of speed may be helpful in comparing some types of performance.

Instruments Used to Measure Speed of Movement

When movement time is greater than a few seconds and the distance is known, speed can be measured with a stop watch or with switch plates, such as the time elapsed in moving between two points or over a specified angle. With rapid angular joint movements, switch plates or electrogoniometers with electronic timing devices are required. Speeds can also be computed from the distance or angle and time data available from cinematography, optoelectric movement monitoring systems, and videotape systems. Some dynamic strength testing devices involve presetting a load and measuring the speed of movement.

In addition, accelerometers can be used to measure acceleration directly, and speed can be derived through integration. However, piezoelectric models have no steady-state response and may not be useful for slower movements. Single accelerometers are used to measure linear motion. Simple rotatory motions require two accelerometers. Triaxial accelerometers are commercially available that contain three pre-mounted accelerometers perpendicular to each other. Multiple accelerometer outputs require appropriate processing to resolve the vector component corresponding to the desired speed. Accelerometers are most appropriately used to measure acceleration when they are mounted on rigid materials. Accelerometers have the advantage of continuously and directly measuring acceleration in an immediately usable form. They can also be very accurate if well-mounted. Because they require soft tissue fixation and cabling or telemetry, they may alter performance and further error may be induced by relative motion of the device and tissues. The systems are moderately expensive (See discussion of accelerometers in Robertson and Sprigings [1987]).

Key Concepts in Speed of Movement Measurement

As discussed, maximum speed is determined when there is little stress on torque production resources. As resistance increases, speed will decrease. Therefore, the load must be considered and specified when testing speed. Because speed of movement data are calculated from displacement and temporal data, a key issue is minimizing error which might result from collecting this information. Error can result from inaccurate identification of anatomical landmarks, improper calibration, perspective error, instrument synchronization error, resolution, digitization error, or vibration. The sampling rate of some of the measurement systems may become an issue when faster movements are being analyzed. In addition, the dynamic characteristics of signal conditioning systems should be reported.

148.5 Endurance

Endurance is the ability of a system to sustain an activity for a prolonged time (static endurance) or to perform repeatedly (dynamic endurance). Endurance can apply to the body as a whole, a particular body system, or to specific neuromuscular functional units. High levels of endurance imply that a given level of performance can be continued for a long time period.

Endurance Terminology

General endurance of the body as a whole is traditionally considered cardiovascular endurance or aerobic capacity. Cardiovascular endurance is most frequently viewed in terms of $V0_{2_{max}}$. This chapter considers only endurance of neuromuscular systems. Although many central and peripheral anatomic sites and physiologic processes contribute to a loss of endurance, endurance of neuromuscular functional units is also referred to as *muscular endurance.*

Absolute muscle endurance is defined as the amount of time that a neuromuscular system can continue to accomplish a specified task against a constant resistance (load and rate) without relating the resistance to the muscle's strength. Absolute muscle endurance and strength are highly correlated. Conversely, strength and *relative muscle endurance* are inversely related. That is, when resistance is adjusted to the person's strength, a weaker person tends to demonstrate more endurance than a stronger person. Furthermore, the same relationships between absolute and relative endurance and strength are correlated by type of contraction; in other words, there is a strong positive correlation between isotonic strength and *absolute* isotonic endurance and vice versa for strength and *relative* isotonic endurance. The same types of relationships exist for isometric strength and isometric endurance [Jensen and Fisher, 1979].

Factors Influencing Neuromuscular Endurance and Measurement of Endurance

Specific muscle fiber types, namely fast-twitch fatigue-resistant fibers (FR), generate intermediate levels of tension and are resistant to short-term fatigue (a duration of about 2 minutes or intermittent stimulation). Slow-twitch fibers (S) generate low levels of tension slowly, and are highly resistant to fatigue. Muscle contractions longer than 10 seconds, but less than 2 minutes, will reflect local muscle endurance [Åstrand and Rodahl, 1986]. For durations longer than 2 minutes, the S fibers will be most stressed. A submaximal isometric contraction to the point of voluntary fatigue will primarily stress the FR and S fibers [Thorstensson and Karlsson, 1976]. Repetitive, submaximal, dynamic contractions continued for about 2 to 6 minutes will measure the capacity of FR and S fibers. Strength testing requires short duration and maximal contractions; therefore, to differentiate strength and endurance testing, the duration and intensity of the contractions must be considered.

Because strength affects endurance, all of the factors discussed previously as influencing strength, also influence endurance. In addition to muscle physiology and muscle strength, endurance is dependent upon the extensiveness of the muscle's capillary beds, the involved neuromuscular mechanisms, contraction force, load, and the rate at which the activity is performed.

Endurance time, or the time for muscles to reach fatigue, is a function of the contraction force or load [von Rohmert, 1960]. As the load (or torque required) increases, endurance time decreases. Also, as speed increases, particularly with activities involving concentric muscle contractions, endurance decreases.

Parameters Measured

Endurance is *how long* an activity can be performed at the required load and rate level. Thus, the basic unit of measure is time. Time is the only measure of how long it takes to complete a task. If we focus on a given variable (e.g., strength, speed, or endurance), it is necessary to either control or measure the others. When the focus is endurance, the other factors of force or torque, speed, and joint angle, can be described as conditions under which endurance is measured. Because of the interactions of endurance and load, or endurance and time, for examples, a number of *endurance-related* measures have evolved. These endurance-related measures have clouded endurance testing.

One endurance-related measure uses either the number of repetitions that can be performed at 20, 25, or 50 percent of maximum peak torque or force. The units used to reflect endurance in this case are number of repetitions at a specified torque or force level. One difficulty with this definition has been described previously, that is, the issue of relative versus absolute muscle endurance. Rothstein and Rose [1982] demonstrated that elderly subjects with selected muscle fiber type atrophy were able to maintain 50 percent of their peak torque longer than young subjects. However, if a high force level is required to perform the task, then the younger subject would have more endurance in that particular activity [Rothstein, 1982]. Another difficulty is that the "repetition method" can be used only for dynamic activities. If isometric activities are involved, then the time an activity can be sustained at a specified force or torque level is measured. Why have different units of endurance? Time could be used in both cases. Furthermore, the issue of absolute versus relative muscle endurance becomes irrelevant if the demands of the task are measured.

Yet another method used to reflect endurance is to calculate an endurance-related work ratio. Many isokinetic testing devices, such as the one shown in Fig. 148.3b, will calculate work (integrate force or torque over displacement). In this case, the total amount of work performed in the first five repetitions is compared with the total amount of work performed in the last five repetitions of a series of repetitions (usually 25 or more). Work degradation *reflects* endurance and is reported as a percentage. An additional limitation of using these endurance ratios is that work cannot be determined in isometric test protocols. Mechanically there is no movement, and no work is being performed.

Overall, the greatest limitation with most *endurance-related* approaches is that the measures obtained cannot be used to perform task-related assessments. In a workplace assessment, for example, one can determine how long a specific task (defined by the conditions of load, range, and speed) needs to be performed. Endurance-related metrics can be used to reflect changes over time in a subject's available endurance capacity; however, endurance-related metrics cannot be compared to the demands of the task. Task demands are measured in time or repetitions (e.g., 10) with a given rate (e.g., 1/0.5 h) from which total time (e.g., 5 h) can be calculated. A true endurance measure (versus an endurance-related measure) can serve both purposes. Time reflects changes in endurance as the result of disease, disuse, training, or rehabilitation and also can be linked to task demands.

Methods and Instruments Used to Measure Neuromuscular Endurance

Selection of the method or instrument used to measure endurance depends on the purpose of the measurement and whether endurance or endurance-related measures will be obtained. As in strength testing, endurance tests can involve simple, low level tasks or whole–body, higher level activities. The simplest method of measuring endurance is to define a task in terms of performance criteria and then time the performance with a stop watch. A subject is given a load and a posture and asked to hold it "as long as possible" or to move from one point to another point at a specific rate of movement for "as long as possible."

An example of a static endurance test is the Sorensen test used to measure endurance of the trunk extensors [Biering-Sorensen, 1984]. This test measures how long a person can sustain his or her torso in a suspended prone posture. The individual is not asked to perform a maximal voluntary contraction, but an indirect calculation of load is possible [Smidt and Blanpied, 1987].

An example of a dynamic endurance test is either a standardized or non-standardized, dynamic isoinertial (see previous description in the section on strength testing) repetition test. In other words, the subject is asked to lift a known load with a specified body part or parts until defined conditions can no longer be met. Conditions such as acceleration, distance, method of performance, or speed may or may not be controlled. The more standardized of these tests, particularly those which involve lifting capacity, are reported and projections about performance capacity over time are estimated [Snook, 1978]. Ergometers and some of the isokinetic dynamometers discussed previously measure work, and several can calculate endurance-related ratios. These devices could be adapted to measure endurance in time units.

Key Concepts in Measuring Muscle Endurance

Of the four variables of human performance discussed in this chapter, endurance testing is the least developed and standardized. Except for test duration and rest intervals, attention to the same guidelines as described for strength testing is currently recommended.

148.6 Reliability, Validity, and Limitations in Testing

Space does not permit a complete review of these important topics. However, a few key comments are in order. First, it is important to note that reliability and validity are not inherent qualities of instruments, but exist in the measurements obtained only within the context in which they are tested. Second, reliability and validity are not either present or absent but are present or absent along a continuum. Third, traditional quantitative measures of reliability might indicate how much reliability a given measurement method demonstrates, but not how much reliability is actually needed. Fourth, technology has advanced to the extent that it is generally possible to measure physical variables such as time, force, torque, angles, and speed accurately, repeatably, and with high resolution. Lastly, clinical generalizability of human performance capacity measures ultimately results from looking at the body of literature on reliability as a whole and not from single studies.

For these types of variables, results of reliability studies basically report that: (1) if the instrumentation is good, and (2) if established, optimal procedures are carefully followed, then results of repeat testing will usually be in the range of about 5 to 20% of each other. This range of repeatability depends on: (1) the particular variable being measured (i.e., repeated endurance measures will differ more than repeated measures of hinge joint EOM), and (2) the magnitude of the given performance capacity (i.e., errors are often in fixed amounts such as 3° for ROM; thus, 3° out of 180° is smaller percentage–wise than 3° out of 20°). One can usually determine an applicable working value (e.g., 5 or 20%) by careful review of the relevant reliability studies. Much of the difference obtained in test–retest results is because of limitations in how well one can reasonably control procedures and the actual variability of the parameter being measured, even in the most ideal test subjects. Measurements should be used with these thoughts in mind. If a specific application requires extreme repeatability, then a reliability study should be conducted under conditions that most closely match those in which the need arises. Reliability discussions specific to some of the focal measures of this chapter are presented in Amundsen [1990], Hellebrandt, Duvall, and Moore [1949], Mayhew and Rothstein [1985], and Miller [1985].

Measurements can be reliable but useless without validity. Most validity studies have compared the results of one instrument to another instrument or to known quantities. This is the classical type of validity testing which is an effort to determine whether the measurement reflects the variable being measured. In the absence of a "gold standard" this type of testing is of limited value. In addition to traditional studies of the validity of measurements, the issue of the *validity of the inferences* based on the measurements is becoming increasingly important [Rothstein and Echternach, 1993]. That is, can the measurements be used to make inferences about human performance in real life situations? Unfortunately, measurements which have not demonstrated more than content validity are frequently used as though they are predictive. The validity of the inferences made from human performance data needs to be rigorously addressed.

Specific measurement limitations were briefly addressed in Tables 148.2 and 148.3 and in the written descriptions of various measurement techniques and instruments. Other limitations have more to do with interpreting the data. A general limitation is that performance variables are not fixed human attributes. Another limitation is that population data are limited and available normative data are, unfortunately, frequently extrapolated to women, older persons, and so on [Chaffin and Andersson, 1991]. Some normative data suggest the amount of resources such as strength, ROM, speed of movement, and endurance *required* for given activities; other data suggest the amount *available*. As previously mentioned, these are two different issues. Performance measurements may yield information about the current status of performance, but testing rarely indicates the *cause* or the nature of dysfunction. More definitive, diagnostic studies are used to answer these questions. Whereas, considerable information exists with regard to measuring performance capacities of human systems, much less energy has been directed to understanding requirements of tasks. The link between functional performance in tasks and laboratory-acquired measurements is a critical question and a major limitation in interpreting test data. The ERM addresses several of these limitations by using a multidimensional, individualized, cause–and–effect model.

148.7 Performance Capacity Space Representations

In both the study and practice, performance of neuromuscular systems has been characterized along one or two dimensions of performance at a time. However, human subsystems function within a *multi*-dimensional performance space. ROM/EOM, strength, movement speed, and endurance capacities are not only interdependent, but may also vary uniquely within individuals. Multiple measurements are necessary to characterize a person's performance capacity space, and performance capacity is dependent on the task to be performed. Therefore, both the individual and the task must be considered when selecting measurement tools and procedures [Chaffin and Andersson, 1991].

In many of the disciplines in which human performance is of interest, traditional thinking has often focused on single number measures of ROM, strength, speed, etc. More recent systems engineering approaches [Kondraske, 1999] emphasize consideration of the performance envelope of a given system and suggest ways to integrate single measurement points that define the limits of performance of a given system [Vasta and Kondraske, 1997]. Figure 148.4 illustrates a three-dimensional performance envelope derived from torque, angle, and velocity data for the knee extensor system. The additional dimension of endurance can be represented by displaying this envelope after performing an activity for different lengths of time. A higher level, composite performance capacity, as is sometimes needed, could be derived by computing the volume enclosed by this envelope. Such representations also facilitate assessment of the given system in a specific task; that is, a task is defined as a point in this space that will either fall inside or outside the envelope.

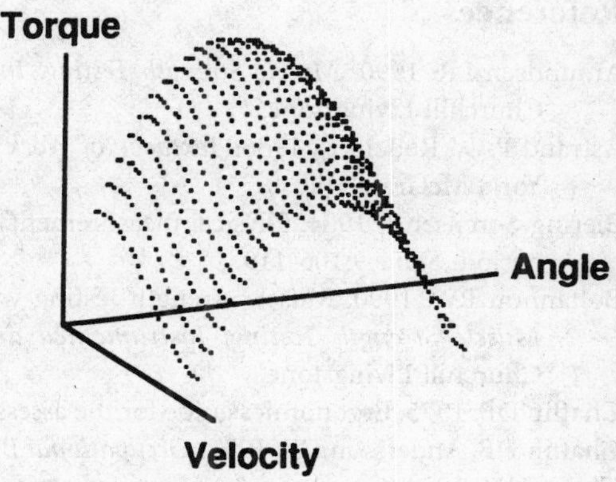

FIGURE 148.4 An example of a torque–angle–velocity performance envelope for the knee extensor system. *Source:* Vasta PJ, Kondraske GV. 1994. A multi-dimensional performance space model for the human knee extensor (Technical Report 94-001R). p 11. University of Texas at Arlington, Human Performance Institute, Arlington, Texas. With permission.

148.8 Conclusions

In conclusion, human movement is so essential that it demands interest and awe from the most casual observer to the most sophisticated scientists. The complexity of performance is truly inspiring. We are challenged to understand it! We want to reduce it to comprehensible units and then enhance it, reproduce it, restore it, and predict it. To do so, we must be able to define and quantify the variables. Hence, an array of instruments and methods have emerged to measure various aspects of human performance. To date, measurement of neuromuscular performance capacities along the dimensions of ROM/EOM, strength, speed of movement, and endurance represents a giant stride but only the "tip of the iceberg." Progress in developing reliable, accurate, and valid instruments and in understanding the factors influencing the measurements cannot be permitted to discourage us from the larger issues of applying the measurements toward a purpose. Yet, single measurements will not suffice; multiple measurements of different aspects of performance will be necessary to fully characterize human movement.

Defining Terms

Endurance: The amount of time a body or body segments can sustain a specified static or repetitive activity.

Extremes of motion (EOM): The end ranges of motion at a joint measured in degrees.

Muscle strength: The maximal amount of torque or force production capacity that a muscle or muscle groups can voluntarily exert in one maximal effort, when type of muscle contraction, movement velocity, and joint angle(s) are specified.

Neuromuscular functional units: Systems (that is, the combination of nerves, muscles, tendons, ligaments, and so on) responsible for producing basic movements.

Range of motion (ROM): The amount of movement that occurs at a joint, typically measured in degrees. ROM is usually measured by noting the extremes of motion, or as the difference between the extreme motion and the reference position.

Speed of movement: The rate of movement of the body or body segments.

References

Amundsen LR. 1990. *Muscle Strength Testing: Instrumented and Non-Instrumented Systems.* New York, Churchill Livingstone.

Åstrand P-O., Rodahl K. 1986. *Textbook of Work Physiology: Physiological Bases of Exercise,* 3rd ed. New York, McGraw-Hill.

Biering-Sorensen F. 1984. Physical measurements as risk indicators for low back trouble over a one year period. *Spine* 9:106-119.

Bohannon RW. 1990. Muscle strength testing with hand-held dynamometers. In LR Amundsen (ed), *Muscle Strength Testing: Instrumented and Non-Instrumented Systems.* pp 69-88. New York, Churchill Livingstone.

Chaffin DB. 1975. Ergonomics guide for the assessment of human strength. *Amer. Ind. Hyg. J.* 36:505-510.

Chaffin DB, Andersson GB. 1991. *Occupational Biomechanics,* 2nd ed. New York, John Wiley & Sons, Inc.

Clarke HH. 1954. Comparison of instruments for recording muscle strength. *Res. Q.* 25:398-411.

Clarke HH, Bailey TL, Shay CT. 1952. New objective strength tests of muscle groups by cable-tension methods. *Res. Q.* 23:136-148.

Clarkson HM, Gilewich GB. 1989. *Musculoskeletal Assessment: Joint Range of Motion and Manual Muscle Strength.* Baltimore, Williams & Wilkins.

Delitto, A. 1990. Trunk strength testing. In LR Amundsen (ed), *Muscle Strength Testing: Instrumented and Non-Instrumented Systems,* pp 151-162. New York, Churchill Livingstone.

DeLorme TL, Watkins AL. 1948. Technics of progressive resistive exercise. *Arch. Phys. Med. Rehabil.* 29:263-273.

Hellebrandt FA, Duvall EN, Moore ML. 1949. The measurement of joint motion: Part III—Reliability of goniometry. *Phys. Ther. Rev.* 29:302-307.

Hislop HJ, Montgomery J. 1995. *Daniel's and Worthingham's Muscle Testing: Techniques of Manual Examination,* 6th ed. Philadelphia, WB Saunders Company.

Jensen CR, Fisher AG. 1979. *Scientific Basis of Athletic Conditioning,* 2nd ed. Philadelphia, Lea & Febiger.

Kendall FP, McCreary EK, Provance PG. 1993. *Muscles: Testing and Function,* 4th ed. Baltimore, Williams & Wilkins.

Komi PV. 1973. Measurement of the force-velocity relationship in human muscle under concentric and eccentric contractions. In S Cerquiglini, A Venerando, J Wartenweiler (eds), *Biomechanics III,* pp 224-229. Baltimore, University Park Press.

Komi PV. 1990. Relevance of *in vivo* force measurements to human biomechanics. *J. Biomech.* 23 (suppl. 1): 23-34.

Kondraske GV. 1999. A working model for human system-task interfaces. In JD Bronzino (ed) *Biomedical Engineering Handbook,* 2nd ed. Boca Raton Fla, CRC Press, Inc.

Kroemer KHE. 1983. An isoinertial technique to assess individual lifting capability. *Human Factors* 25:493-506.

Kroemer KHE. 1991. A taxonomy of dynamic muscle exertions. *J. Hum. Muscle Perform.* 1:1-4.

Mayhew TP, Rothstein JM 1985. Measurement of muscle performance with instruments. In JM Rothstein (ed), *Measurement in Physical Therapy,* pp 57-102. New York, Churchill Livingstone.

Miller PJ. 1985. Assessment of joint motion. In JM Rothstein (ed), *Measurement in Physical Therapy,* pp 103-136. New York, Churchill Livingstone.

Norkin CC, White DJ. 1995. *Measurement of Joint Motion: A Guide to Goniometry,* 2nd ed. Philadelphia, FA Davis Company.

O'Connel AL, Gowitzke B. 1972. *Understanding the Scientific Bases of Human Movement.* Baltimore, Williams & Wilkins.

Palmer ML, Epler M. 1998. *Fundamentals of Musculoskeletal Assessment Techniques,* 2nd ed. Philadelphia, JB Lippincott Company.

Panjabi MM, White III AA, Brand RA. 1974. A note on defining body parts configurations. *J. Biomech.* 7:385-387.

Robertson G, Sprigings E. 1987. Kinematics. In DA Dainty, RW Norman (eds), *Standardizing Biomechanical Testing in Sport,* pp 9-20. Champaign, Ill, Human Kinetics Publishers, Inc.

Robson P. 1966. A method to reduce the variable error in joint range measurement. *Ann. Phys. Med.* 8:262-265.

Rothstein JM. 1982. Muscle biology: Clinical considerations. *Phys. Ther.* 62:1823-1830.

Rothstein JM, Echternach JL. 1993. *Primer on Measurement: An Introductory Guide to Measurement Issues.* Alexandria, Va, American Physical Therapy Association.

Rothstein JM, Rose SJ. 1982. Muscle mutability—Part II: Adaptation to drugs, metabolic factors, and aging. *Phys. Ther.* 62:1788-1798.

Soderberg GL. 1992. Skeletal Muscle Function. In DP Currier, RM Nelson (eds), *Dynamics of Human Biologic Tissues,* pp 74-96. Philadelphia, FA Davis Company.

Smidt GL. 1984. *Muscle Strength Testing: A System Based on Mechanics.* Iowa City, IA, SPARK Instruments and Academics, Inc.

Smidt GL, Blanpied PR. 1987. Analysis of strength tests and resistive exercises commonly used for low-back disorders. *Spine* 12:1025-1034.

Smidt GL, Rogers MR. 1982. Factors contributing to the regulation and clinical assessment of muscular strength. *Phys. Ther.* 62:1284-1290.

Snook SH. 1978. The design of manual handling tasks. *Ergonomics* 21:963-985.

Sukop J, Nelson RC. 1974. Effects of isometric training in the force-time characteristics of muscle contractions. In RC Nelson, CA Morehouse (eds), *Biomechanics IV,* pp 440-447. Baltimore, University Park Press.

Thorstensson A, Karlsson J. 1976. Fatiguability and fibre composition of human skeletal muscle. *Acta Physiol. Scand.* 98:318-322.

Vasta PJ, Kondraske GV. 1994. A multi-dimensional performance space model for the human knee extensor (Technical Report 94-001R). University of Texas at Arlington, Human Performance Institute, Arlington, TX.

Vasta PJ, Kondraske GV. 1997. An approach to estimating performance capacity envelopes: Knee extensor system example. *Proc. 19th Ann. Eng. Med. Biol. Soc. Conf.,* pp 1713-1716.

von Rohmert W. 1960. Ermittlung von erholungspausen fur statische arbeit des menschen, *Int. Z. Angew. Physiol.* 18:123-124.

West CC. 1945. Measurement of joint motion. *Arch. Phys. Med.* 26:414-425.

White III AA, Panjabi MM. 1990. *Clinical Biomechanics of the Spine,* 2nd ed. Philadelphia, JB Lippincott Company.

Wiechec FJ, Krusen FH. 1939. A new method of joint measurement and a review of the literature. *Am. J. Surg.* 43:659-668.

Winter DA, Wells RP, Orr GW. 1981. Errors in the use of isokinetic dynamometers. *Eur. J. Appl. Physiol.* 46:397-408.

Further Information

Human Performance Measurement, Inc. 1998. *APM I Portable Electronic Goniometer: User's Manual.* PO Box 1996, Arlington, TX 76004-1996.

Journals: *Clinical Biomechanics, Journal of Biomechanics, Medicine and Science in Sports and Exercise, Physical Therapy.*

Smith SS, Kondraske GV. 1987. Computerized system for quantitative measurement of sensorimotor aspects of human performance. *Phys. Ther.* 67:1860-1866.

Task Force on Standards for Measurement in Physical Therapy. 1991. Standards for tests and measurements in physical therapy practice. *Phys. Ther.* 71:589-622.

149

Measurement of Sensory-Motor Control Performance Capacities: Tracking Tasks

Richard D. Jones
Christchurch Hospital and University of Otago

149.1 Introduction

The human nervous system is capable of simultaneous and integrated control of 100-150 mechanical degrees of freedom of movement in the body via tensions generated by about 700 muscles. In its widest context, movement is carried out by a sensory-motor system comprising multiple sensors (visual, auditory, proprioceptive), multiple actuators (muscles and skeletal system), and an intermediary processor which can be summarized as a multiple-input multiple-output non-linear dynamic time-varying control system. This grand control system comprises a large number of interconnected processors and sub-controllers at various sites in the central nervous system of which the more important are the cerebral cortex, thalamus, basal ganglia, cerebellum, and spinal cord. It is capable of responding with remarkable accuracy, speed (when necessary), appropriateness, versatility, and adaptability to a wide spectrum of continuous and discrete stimuli and conditions. Certainly, by contrast, it is orders of magnitude more complex and sophisticated than the most advanced robotic systems currently available—although the latter also have what are often highly desirable attributes such as precision and repeatability and a much greater immunity from factors such as fatigue, distraction, and lack of motivation!

This chapter addresses the control function. First, it introduces several important concepts relating to sensory-motor control, accuracy of movement, and performance resources/capacities. Second, it provides an overview of apparatuses and methods for the *measurement* and *analysis* of complex sensory-motor

0-8493-0462-8/00/$0.00+$.50

performance. The overview focuses on measurement of sensory-motor control performance capacities of the *upper-limbs* and by means of **tracking tasks**.

149.3 Basic Principles

Sensory-motor Control and Accuracy of Movement

From the perspective of Kondraske's [Kondraske, 1995a; Kondraske, 1995b] *elemental resource model* of human performance, **sensory-motor control** is the function of the overall sensory-motor control system. This system can be considered as a hierarchy of multiple interconnected sensory-motor controllers cited in the *central processing and skills ("software") domains* (cf. environmental interface domain, comprising sensors and actuators, and life sustaining domains) of the elemental resource model. These controllers range from low-level elemental level controllers for control of movement around single joints, through intermediate-level controllers needed to generate integrated movements of an entire limb and involving multiple joints and degrees of freedom, and high-level controllers and processors to enable coordinated synergistic multi-limb movements and the carrying out of *central executive* functions concerned with allocation and switching of resources for execution of multiple tasks simultaneously.

Each of these controllers is considered to possess limited **performance resources (PRs)**—or **performance capacities**—necessary to carry out their control functions. PRs are characterized by **dimensions of performance**, which for controllers are *accuracy of movement* (including steadiness and stability) and *speed of movement.* Accuracy is the most important of these and can be divided into four major classes:

1. Spatial accuracy—Required by tasks which are *self-paced* and for which time taken is of secondary or minimal importance and includes tracing (e.g., map-tracking), walking, reaching, and, in fact, most activities of daily living. Limitation in speed PRs should have no influence on this class of accuracy.
2. Spatial accuracy with time constraints—Identical to "spatial accuracy" except that, in addition to accuracy, speed of execution of task is also of importance. Because maximal performance capacities for accuracy and speed of movement cannot, in general, be realized simultaneously, the carrying out of such tasks must necessarily involve speed-accuracy tradeoffs [Fitts, 1954; Fitts and Posner, 1967]. The extent to which accuracy is sacrificed for increased speed of execution, or vice versa, is dependent on the perceived relative importance of accuracy and speed.
3. Temporal accuracy—Required by tasks which place minimal demands on positional accuracy and includes single and multi-finger tapping and foot tapping.
4. **Spatiotemporal accuracy**—Required by tasks which place considerable demand on attainment of simultaneous spatial and temporal accuracy. This includes *paced* positional tasks such as tracking, driving a vehicle, ball games and sports, and video games. It should be stressed, however, that most of the above self-paced tasks also involve a considerable interrelationship between space and time.

Tracking tasks have become well established as being able to provide one of the most accurate and flexible means for laboratory-based measurement of spatiotemporal accuracy and, thus, of the performance capacity of sensory-motor control or sensory-motor coordination. In addition, they provide an unsurpassed framework for studies of the underlying control mechanisms of motor function (e.g., Cooper et al., 1989; Lynn et al., 1979; Neilson et al., 1993; Neilson et al., 1995; Neilson et al., 1998)—the potential for which was recognized as early as 1943 as seen in writings by Craik [1966]. They have achieved this status through their (a) ability to maximally stress the accuracy dimension of performance and, hence, the corresponding control PR, (b) the continuous nature and wide range of type and characteristics of input target signals they permit, (c) facility for a wide range of 1-D and 2-D **sensors** for measuring a subject's motor output, and (d) measure of continuous performance (cf. reaction time tests).

From this perspective, it will be of little surprise to find that tracking tasks are the primary thrust of this chapter.

The Influence of Lower-Level Performance Resources on Higher-Level Control Performance Resources

By their very nature, tasks which enable one to measure spatiotemporal accuracy are complex or higher-level sensory-motor tasks. These place demands on a large number of lower-level PRs such as visual acuity, dynamic visual perception, range of movement, strength, simple reaction times, acceleration/deacceleration, static steadiness, dynamic steadiness, prediction, memory, open-loop movements, concentration span, attention switching i.e., central executive function or supervisory attentional system (multitask abilities), utilization of preview, and learning.

It is, therefore, important to ask: If there are so many PRs involved in tracking, can tracking performance provide an accurate estimate of sensory-motor control performance capacities? Or, if differences are seen in tracking performance between subjects, do these necessarily indicate comparable differences in control performance capacities? Yes, they can, but only if the control resource is the *only* resource being maximally stressed during the tracking task. Confirmation that the other PRs are not also being maximally stressed for a particular subject can be ascertained by two means. First, by independently measuring the capacity of the other PRs and confirming that these are considerably greater than that determined as necessary for the tracking task in question. For example, if the speed range for a certain reference group on a non-target speed test is 650-1250 mm/s and the highest speed of a tracking target signal is 240 mm/s, then one can be reasonably confident that intra-group differences in performance on the tracking task are unrelated to intra-group differences in speed. Second, where this process is less straightforward or not possible, it may be possible to alter the demands imposed by the task on the PR in question. For example, one could see whether visual acuity was being maximally stressed in a tracking task (and hence, be a significant limiting factor to the performance obtained) by increasing or decreasing the eye-screen distance. Similarly, one could look at strength in this context by altering the friction, damping, or inertia of the sensor, or at range of movement by altering the gain of the sensor.

The conclusion that a task and a PR are unrelated for a particular group does not, of course, mean that this can necessarily be extrapolated to some other group. For example, strength may be completely uncorrelated with tracking performance in normal males yet be the primary factor responsible for poor tracking performance by the paretic arm of subjects who have suffered a stroke.

The foregoing discussion is based on the concept of a assumption that if a task requires less than the absolute maximum available of a particular PR then performance on that task will be independent of that PR. Not surprisingly, the situation is unlikely to be this simplistic or clear-cut! If, for example, a tracking task places moderate sub-maximal demands on several PRs, these PRs will be stressed to varying levels such that the subject may tend to optimize the utilization of those resources [Kondraske, 1995a; Kondraske, 1995b] so as to achieve an acceptable balance between accuracy, speed, stress/effort, and fatigue (physical and cognitive). Thus, although strength available is much greater than strength needed (i.e., Resource available $\gg$ Resource demand) for both males and females, could the differential in strength be responsible for males performing better on tracking tasks than females? [Jones et al., 1986]

149.3 Measurement of Sensory-Motor Control Performance

Techniques: An Overview

Tracking tasks are the primary methodological approach outlined in this chapter for measurement of sensory-motor control performance. There are, however, a large number of other approaches, each with their own set of apparatuses and methods, which can provide similar or different data on control performance. It is possible to give only a cursory mention of these other techniques in this chapter [see also Further Information].

Hand and foot test boards comprising multiple touch-plate sensors provide measures of accuracy and speed of lateral reaching-tapping abilities [Kondraske et al., 1984, 1988].

Measurement of steadiness and tremor in an upper-limb, lower-limb, or segment of either, particularly when sustended, can be made using variable size holes, accelerometers, or force transducers [Potvin et al., 1975]. A dual-axis capacitive transducer developed by Kondraske et al. [1984] provides an improved means of quantifying steadiness and tremor due to it requiring no mechanical connection to the subject (i.e., no added inertia) and by providing an output of limb position as opposed to less informative measures of acceleration or force. Interestingly, tests of steadiness can be appropriately considered as a category of tracking tasks in which the target is static. The same is also true for measurement of postural stability using force balance platforms, whether for standing [Kondraske et al., 1984; Milkowski et al., 1993] or sitting [Riedel et al., 1992][see also Further Information].

Tracking Tasks: An Overview

A tracking task is a laboratory-based test apparatus characterized by a continuous input signal—the target—which a subject must attempt to match as closely as possible by his/her output response by controlling the position of (or force applied to) some sensor. It provides unequalled opportunities for wide-ranging experimental control over sensors, displays, target signals, dimensionality (degrees of freedom), control modes, controlled system dynamics, and sensor-display compatibility, as well as the application of a vast armamentarium of linear and non-linear techniques for response signal analysis and systems identification. Because of this, the tracking task has proven to be *the* most powerful and versatile tool for assessing, studying, and modelling higher-level functioning of the human 'black-box' sensory-motor system.

There are three basic categories of tracking tasks differing primarily in their visual display and in the corresponding control system (Fig. 149.1). The pursuit task displays both the present input and output signals, whereas the compensatory task displays only the difference or error signal between these. The preview task [Poulton, 1964; Welford, 1968; Jones and Donaldson, 1986] (Fig. 149.2) is similar to the pursuit task except that the subject can see in advance where the input signal is going to be and plan

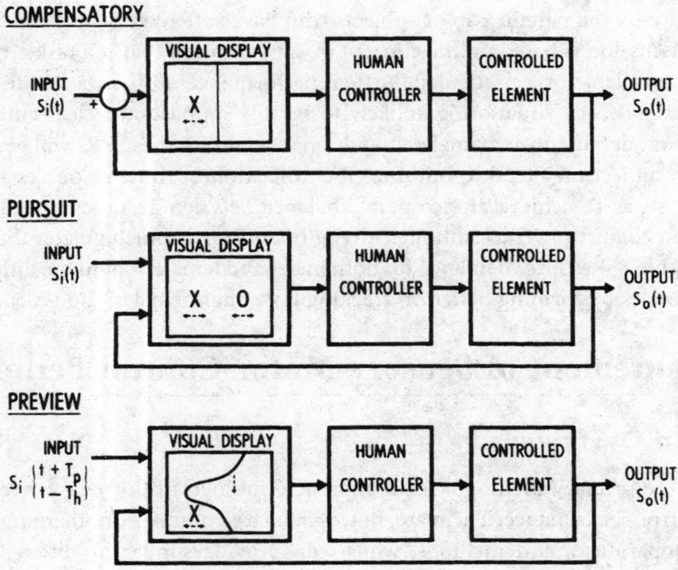

FIGURE 149.1 Modes of tracking (i) Compensatory: subject aims to keep resultant error signal **X** (= input signal—output signal = $s_i(t) - s_o(t)$) on stationary vertical line. (ii) Pursuit: subject aims to keep his output signal **X**, $s_o(t)$, on the target input signal **O**, $s_i(t)$. (iii) Preview: subject aims to keep his output signal **X**, $s_o(t)$, on the descending target input signal, $[s_i(t + T_p) - s_i(t - T_h)]$ (where t = present time, T_p = preview time, T_h = history or postview time).

accordingly to minimize the resultant error signal. Tracing tasks [Driscoll, 1975; Stern et al., 1983; Hocherman and Aharon-Peretz, 1994] are effectively self-paced 2-D preview tracking tasks.

The input-output nature of tracking tasks has made them most suitable for analysis using engineering control theories. This has led to the common view of pursuit tracking as a task involving continuous negative feedback [Notterman et al., 1982] but there is evidence that tracking viewed as a series of discrete events would be more appropriate [Bösser, 1984; Neilson et al., 1988]. The inclusion of preview of the input signal greatly complicates characterization of the human controller and Sheridan [1966] has suggested three models of preview control which employ the notions of constrained preview and nonuniform importance of input. Lynn et al. [1979] and Neilson et al. [1992] have also demonstrated how, by treating the neurologically impaired subject as a black-box, control analysis can lead to further information on underlying neurological control mechanisms.

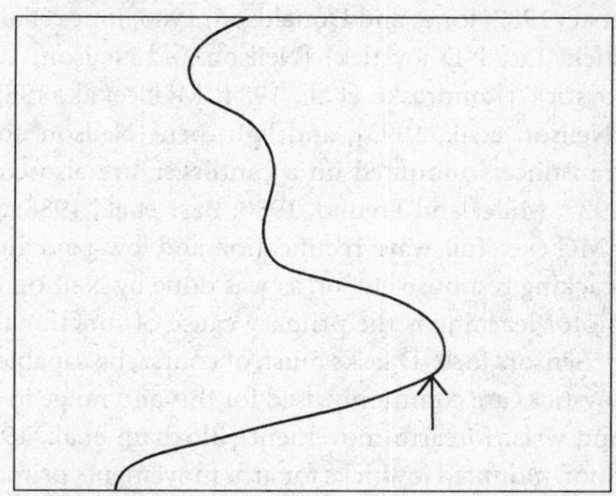

FIGURE 149.2 Visual display for random tracking with a preview of 8.0 s and postview of 1.1 s.

Despite the wide-spread utilization and acceptance of tracking tasks as a powerful and versatile means for quantifying and studying sensory-motor control capacities, there is little available on the market in this area. The most obvious exception to this is the photoelectric pursuit rotor which is ubiquitous in the motor behavior laboratories of university psychology departments and has been available since the 1950s [Welford, 1968; Schmidt, 1982; Siegel, 1985]. It is a paced 2-D task with a target with the periodic on each revolution. Although inexpensive, the pursuit rotor is a crude tracking task allowing limited control over target signals and possessing a very gross performance analysis in terms of time on target. Thus nearly all of the many and varied tracking tasks which have been used in countless experimental studies around the world have been developed by the users themselves for their specific objectives.

An improvement in this situation has occurred recently with the arrival on the market of a number of tracking devices from Human Performance Measurement, Inc. (Arlington, TX). These devices are a natural extension of those developed by Kondraske et al. [1984]. Off-the-shelf availability of computer-based tracking tests, including sensors for both upper- and lower-limbs, opens up the possibility of a much broader and widespread use of tracking tasks. In particular, one can look forward to a much greater utilization of tracking tasks outside of traditional research areas and in more routine assessment applications in clinical, rehabilitative, vocational, sports, and other environments.

Tracking Tasks: Options and Considerations

Whatever the reasons for needing quantification of sensory-motor control capacity via a tracking task, there are a number of options available and factors to be considered in choosing or designing a tracking task.

Sensors

Sensors for measuring a subject's motor output in 1-D tracking tasks can be categorized under (a) movements involving a single degree of freedom such as flexion-extension rotation around a single joint including elbow [Deuschl et al., 1996; Lynn et al., 1977; O'Dwyer et al., 1996; Soliveri et al., 1997], wrist [Johnson et al., 1996], or a finger, or pronation-supination of the wrist, and (b) movements involving two or more degrees of freedom of a body part (e.g., hand)—i.e., coordinated movement at multiple joints—which are either 1-D, such as some form of linear transducer [Patrick and Mutlusoy, 1982; Baroni et al., 1984; van den Berg et al., 1987] or 2-D, such as steering wheel [Buck, 1982; Ferslew et al., 1982; Jones and Donaldson, 1986; Jones et al., 1993], stirring wheel [De Souza et al., 1980], position stick (i.e., 1-D joystick) [Neilson and Neilson, 1980; O'Dwyer and Neilson, 1998; Potvin et al., 1977], joystick [Kondraske et al., 1984; Miall et al., 1985; Jones et al., 1993], finger-controlled rotating knob [Neilson et al., 1993], and light-pen [Neilson and Neilson, 1980]. Force sticks, utilizing strain-gauge transducers mounted on a cantilever, are also commonly used as sensors [Garvey, 1960; Potvin et al., 1977; Miller and Freund, 1980; Barr et al., 1988; Stelmach and Worrington, 1988]. Isometric integrated EMG (i.e., full-wave rectification and low-pass filtering of the raw EMG) can also be used to control the tracking response cursor, as was done by Neilson et al. [1990] to help show that impairment of sensory-motor learning is the primary cause of functional disability in cerebral palsy.

Sensors for 2-D tasks must, of course, be capable of moving with and recording two degrees of freedom. Joysticks are commonly used for this and range in size from small, for finger movement [Anderson, 1986] and wrist/forearm movement [Bloxham et al., 1984; Frith et al., 1986; Neilson et al., 1998], up to large floor-mounted joysticks for arm movements primarily involving shoulder and elbow function [Anderson, 1986; Behbehani et al., 1988; Dalrymple-Alford et al., 1994; Jones et al., 1993; Kondraske et al., 1984; Watson and Jones, 1998; Watson et al., 1997]. Other 2-D task sensors include a hand-held stylus for the photoelectric pursuit rotor [Schmidt, 1982; Siegel, 1985], plexiglass tracing [Driscoll, 1975], and tasks utilizing sonic digitizers [Hocherman and Aharon-Peretz, 1994; Stern et al., 1984; Viviani and Mounoud, 1990]. Abend et al. [1982] and Flash and Hogan [1985] used a two-joint mechanical arm to restrict hand movements to the horizontal plane in the investigation of CNS control of two-joint (shoulder and elbow) movements in trajectory formation. Stern et al. [1983] simply used the subject's finger as the sensor for a tracing task on a vertical plexiglass 'screen'; a video camera behind the screen recorded finger movements. Novel 'whole-body' 2-D tracking is also possible by having subjects alter their posture while standing on a dual-axis force platform [Kondraske et al., 1984].

1-D sensors can also be used in 2-D tasks by way of bimanual tracking. For example, O'Dwyer and Neilson [1995] used two 1-D joysticks to investigate dynamic synergies between the right and left arms.

Displays

Early tracking devices used mechanical-based displays such as a rotating smoked drum [Vince, 1948], the ubiquitous pursuit rotor, and a paper-strip preview task [Poulton, 1964; Welford, 1968]. An oscilloscope was, and still is to a much less degree, used in a large number of tracking tasks, initially driven by analog circuitry [Anderson, 1986; Flowers, 1976] but later by D/A outputs on digital computers [Kondraske et al., 1984; Miall et al., 1985; Sheridan, M.R. et al., 1987; Cooper et al., 1989]. Standard raster-based television screens have been used by some workers [Potvin et al., 1977; Beppu et al., 1984]. Non-raster vector graphics displays, such as Digital Equipment's VT11 dynamic graphics unit, proved valuable during the PDP-era as a means for generating more complex dynamic stimuli such as squares [Neilson and Neilson, 1980; Frith et al., 1986] and preview [Jones and Donaldson, 1986] (Fig. 149.2). More recently, raster-based color graphics boards have allowed impressive static displays and simple dynamic tracking displays to be generated on PCs. However, such boards are not, in general, immediately amenable for the generation of flawless dynamic displays involving more complex stimuli, such as required for preview tracking. Jones et al. [1993] have overcome this drawback by the use of specially-written high-speed assembly-language routines for driving their display. These generate a display of the target and the subject's response marker by considering the video memory (configured in EGA mode) as four overlapping planes, each switchable (via a mask), and each capable of displaying the background color and a single color from a palette. Two planes are used to display the target, with the remaining two being used to display the subject's pointer. The current target is displayed on one target plane, while the next view of the target, in its new position, is being drawn on the other undisplayed target plane. The role of the two planes is reversed when the computer receives a vertical synchronization interrupt from the graphics controller indicating the completion of a raster. Through a combination of a high update-rate of 60.34 Hz (i.e., the vertical interrupt frequency), assembly language, and dual display buffers, it has been possible to obtain an extremely smooth dynamic color display. Their system for tracking and other quantitative sensory-motor assessments is further enhanced through its facility to generate dynamic color graphics on two high-resolution monitors simultaneously: one for the tracking display, and one for use by the assessor for task control and analysis. The monitors are driven by a ZX1000 graphic controller (Artist Graphics Inc), at 800 × 600, and a standard VGA controller, respectively.

In contrast to the above CRT-based displays, Warabi et al. [1986] used a laser-beam spot to indicate a subject's hand position together with a row of LEDs for displaying a step target. Similarly, Gibson et al. [1987] used a galvonometer-controlled laser spot to display smooth and step stimuli on a curved screen together with a white-light spot controlled by subject. Leist et al. [1987], Viviani and Mounoud [1990], and Klockgether [1994] also used galvonometer-controlled spots but via back-projection onto a curved screen, transparent digitizing table, and plexiglass surface respectively. Van den Berg et al. [1987] used two rows of 240 LEDs each to display target and response. 2-D arrays of LEDs have also been used to indicate step targets in 2-D tracking tasks [Abend et al., 1982; Flash and Hogan, 1985].

Target Signals

Tracking targets cover a spectrum from smoothly-changing (low-bandwidth) targets, such as sinusoidal and random, through constant velocity ramp targets, to abrupt changing step targets.

Sinusoidal Targets

The periodicity, constancy of task complexity (over cycles), and spectral purity of sine targets make them valuable for measurement of within-run changes in performance (e.g., learning, lapses in concentration) [Jones and Donaldson, 1981], the study of ability to make use of the periodicity to improve tracking performance [Jones and Donaldson, 1989], and the study of the human frequency response [Leist et al., 1987]. Several other workers have also used sine targets in their tracking tasks [Ferslew et al., 1982; Johnson et al., 1996; Miller and Freund, 1980; Notterman et al., 1982; Potvin et al., 1977; Soliveri et al., 1997].

Bloxham et al. [1984] and Frith et al. [1986] extended the use of sinewaves into a 2-D domain by having subjects track a moving circle on the screen.

Random Targets

These are commonly generated via a sum of sines approach in which a number of harmonically or non-harmonically related sinusoids of random phase are superimposed [Baddeley et al., 1986; Barr et al., 1988; Cassell, 1973; Cooper et al., 1989; Frith et al., 1986; Hufschmidt and Lücking, 1995; Jones et al., 1993; Miall et al., 1985; Neilson and Neilson, 1980; van den Berg et al., 1987; Watson and Jones, 1998; Watson et al., 1997]. If harmonically related, this can effectively give a flat spectrum target out to whatever bandwidth is required. Thus, in Jones et al.'s [Jones et al., 1993; Watson and Jones, 1998; Watson et al., 1997] system the random signal generation program asks the user for the required signal bandwidth and then calculates the number of equal amplitude harmonics that must be summed together to give this bandwidth, each harmonic being assigned a randomly selected phase from a uniform phase distribution. Each target comprises 4096 (2^{12}) or more samples, a duration of at least 68 s (4096 samples/60.34 Hz), and a fundamental frequency of 0.0147 Hz (i.e., period of 68 s). By this means it is possible to have several different pseudo-random target signals that are non-periodic up to 68 s duration, have flat spectra within a user specified bandwidth, no components above this bandwidth, and whose spectra can be accurately computed by FFT from any 68 s block of target (or response). Another common approach to the generation of random targets is to digitally filter a sequence of pseudo-random numbers [Lynn et al., 1977; Potvin et al., 1977; Kondraske et al., 1984; van den Berg et al., 1987; Neilson et al., 1993] although this method gives less control over the spectral characteristics of the target. B'sser [1984] summed a number of these filtered sequences in such a way as to generate a target having an approximate 1/f spectrum. Another smooth pursuit target was generated by linking together short segments of sinewaves with randomly selected frequencies up to some maximum [Gibson et al., 1987] and was thus effectively a hybrid sinusoidal-random target.

Ramp Targets

These have been used in conjunction with sensory gaps of target or response to study predictive tracking and ability to execute smooth constant velocity movements in the absence of immediate visual cues in normal subjects [Flowers, 1978b] and subjects with cerebellar disorders [Beppu et al., 1987], stroke [Jones et al., 1989], and Parkinson's disease [Cooke et al., 1978; Flowers, 1978a].

Step Targets

These have been used in many applications and studies to measure and investigate subjects' abilities to predict, program, and execute ballistic (open-loop) movements. To enable this, spatial and temporal unpredictability have been incorporated into step tasks in various ways:

- Temporal predictability—The time of onset of steps has ranged from (a) explicitly predictable, with preview of the stimulus [Day et al., 1984; Jones et al., 1993], (b) implicitly predictable, with fixed interval between steps [Abend et al., 1982; Cooke et al., 1978; Deuschl et al., 1996; Flowers, 1978; Johnson et al., 1996; Potvin et al., 1977], to (c) unpredictable, with intervals between steps varied randomly over spans lying somewhere between 1.5 s and 7.0 s [Anderson, 1986; Angel et al., 1970; Baroni et al., 1984; Flowers, 1976; Gibson et al., 1987; Jones and Donaldson, 1986; Jones et al., 1993; Kondraske et al., 1984; Neilson et al., 1995; O'Dwyer and Neilson, 1998; Sheridan et al., 1987; Warabi et al., 1986; Watson et al., 1997].
- Amplitude predictability—The amplitude of steps has ranged from (a) explicitly predictable, where the endpoint of the step is shown explicitly before it occurs [Abend et al., 1982; Baroni et al., 1984; Deuschl et al., 1996; Jones et al., 1993; Sheridan et al., 1987; Watson et al., 1997], (b) implicitly predictable, where all steps have the same amplitude [Anderson, 1986; Angel et al., 1970; Cooke et al., 1978; Day et al., 1984; Johnson et al., 1996; Kondraske et al., 1984; O'Dwyer and Neilson, 1998; Potvin et al., 1977] or return-to-centre steps in variable-amplitude step tasks [Flowers, 1976; Jones and Donaldson, 1986; Jones et al., 1993], to (c) unpredictable, with between 2 and 8 randomly distributed amplitudes [Flowers, 1976; Jones and Donaldson, 1986; Warabi et al., 1986; Gibson et al., 1987; Sheridan, M.R. et al., 1987; Jones et al., 1993].
- Direction predictability—Previous step tasks have had steps whose direction of steps has ranged from (a) all steps explicitly predictable, alternating between right and left [Baroni et al., 1984; Cooke et al., 1978; Deuschl et al., 1996; Flowers, 1976; Johnson et al., 1996; Kondraske et al., 1984; O'Dwyer and Neilson, 1998; Potvin et al., 1977], or all in one direction (i.e., a series of discontinuous steps) [Sheridan et al., 1987], or between corners of an invisible square [Anderson, 1986], or having preview [Abend et al., 1982; Jones et al., 1993], (b) most steps predictable but with occasional "surprises" for studying anticipation [Flowers, 1978a], (c) a combination of unpredictable (outward) and predictable (back-to-center) steps [Angel et al., 1970; Jones and Donaldson, 1986; Jones et al., 1993; Watson et al., 1997], and (d) all steps unpredictable, with multiple endpoints [Warabi et al., 1986; Gibson et al., 1987] or resetting between single steps [Day et al., 1984].

The three elements of unpredictability can be combined in various ways to generate tasks ranging from completely predictable to completely unpredictable (Table 149.1). Several groups have implemented several variations of unpredictability both within and between step tracking tasks to investigate the possible loss of ability to use predictability to improve performance in, for example, Parkinson's disease [Flowers, 1978; Sheridan et al., 1987; Watson et al., 1997]. In addition to unpredictability, other characteristics can be built into step tasks including explicit target zones [Sheridan et al., 1987] and visual gaps in target [Flowers, 1976; Warabi et al., 1986].

An example of a 1-D step tracking task possessing full spatial and temporal unpredictability is that of Jones and colleagues [Jones and Donaldson, 1986; Jones et al., 1993] (Fig. 149.3a). The task comprises 32 abrupt steps alternating between displacement from and return to center screen. In the non-preview form, spatial unpredictability is present in the outward steps through four randomly distributed amplitude/direction movements (large and small steps requiring 90 and 22.5 deg. on a steering wheel respectively, and both to right and left of center) with temporal unpredictability achieved via four randomly

TABLE 149.1 Unpredictability in Step Tracking Tasks

		Temporal			Spatial—Amplitude			Spatial—Direction			Overall
		Full	Partial	None	Full	Partial	None	Full	Partial	None	Full
Angel et al. [1970]*	1-D	•				•		•		•	
Flowers [1976]*	1-D	•			•	•				•	
Potvin et al. [1977]	1-D		•			•				•	
Cooke et al. [1978]	1-D		•			•				•	
Flowers [1978a]	1-D		•			•			•		
Baroni et al. [1984]	1-D	•					•			•	
Day et al. [1984]	1-D			•		•		•			
Kondraske et al. [1984]	1-D	•				•				•	
Warabi et al. [1986]	1-D	•			•			•			•
Jones and Donaldson [1986]*	1-D	•			•	•		•		•	•
Gibson et al. [1987]	1-D	•			•			•			•
Sheridan et al. [1987]*	1-D	•			•		•			•	
Jones et al. [1993]*	1-D	•		•	•	•	•	•		•	•
Deuschl, et al. [1996]	1-D		•				•			•	
Johnson, et al. [1996]	1-D		•			•				•	
O'Dwyer and Neilson [1998]	1-D	•				•				•	
Abend et al. [1982]	2-D		•				•			•	
Anderson [1986]	2-D	•				•				•	
Watson et al. [1997]	2-D	•					•	•			

* Several authors have several variations of unpredictability within one task or between multiple tasks.

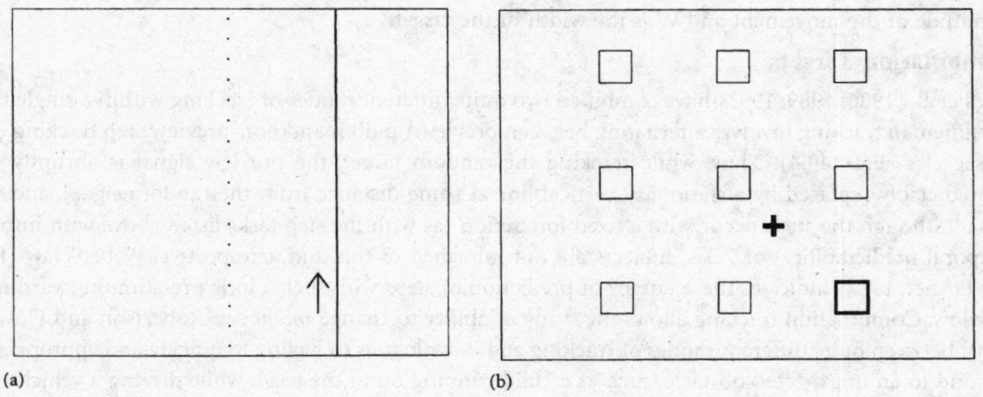

FIGURE 149.3 Visual displays for (a) 1-D step tracking task and (b) 2-D step tracking task (bottom-right square is current target).

distributed durations between steps (2.8, 3.4, 4.0, 4.6 s). This task has been used, together with preview random tracking, to demonstrate deficits in sensory-motor control in the asymptomatic arm of subjects who have had a unilateral stroke [Jones et al., 1989].

Watson and Jones [1997] also provide an example of a 2-D step tracking task with spatial and temporal unpredictability. In this task the subject must move a cross from within a central starting square to within one of eight 10 mm × 10 mm target squares that appear on the screen with temporal and spatial unpredictability (Fig. 149.3b). The centers of the eight surrounding targets are positioned at the vertices and midway along the perimeter of an imaginary 100 mm x 100 mm square centered on the central square. To initiate the task, the subject places the cross within the perimeter of the central target. After

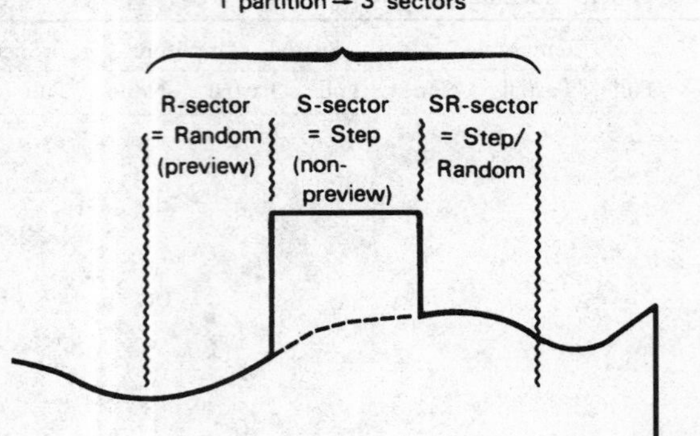

FIGURE 149.4 Section of input waveform in combination tracking in which the target alternates between preview-random and nonpreview-step.

a 2–5 s delay, one of the surrounding blue targets turns green and the subject moves the cross to within the green target square as quickly and as accurately as possible. After a further delay, the central target turns green indicating onset of the spatially predictive "back-to-center" target. The task, which comprises ten outward and ten return targets, was used to show that Parkinsonian subjects perform worse than matched controls on all measures of step tracking but are not impaired in their ability to benefit from spatial predictability to improve performance.

Step tasks with explicit target zones, in 1-D [Sheridan et al., 1987] or 2-D [Watson et al., 1997], provide the possibility of altering task difficulty by varying the size of the target. On the basis that subjects need only aim to get their marker somewhere within target zone (cf. close to center) then, according to Fitt's [1954] ratio rule, the difficulty of the primary movement is proportional to $\log_2(2A/W)$, where A is the amplitude of the movement and W is the width of the target.

Combination Targets

Jones et al. [1986, 1989, 1993] have combined two quite different modes of tracking within a single task. *Combination* tracking involves alternating between preview random and non-preview step tracking over 11 s cycles (Fig. 149.4). Thus, while tracking the random target, the preview signal is abruptly and unpredictably replaced by a stationary vertical line at some distance from the random signal, and vice versa. Although the steps occur with a fixed foreperiod (as with the step tasks listed above with implicit temporal predictability) of 7.3 s, subjects are not informed of this and, irrespective, Weber's law [Fitts and Posner, 1967] indicates the accuracy of prediction of steps with such a long pre-stimulus warning is very low. Combination tracking allows the study of ability to change *motor set* [Robertson and Flowers, 1990] between quite different modes of tracking and is analogous to having to quickly and appropriately respond to an unexpected obstacle, such as a child running on to the road, while driving a vehicle.

Dimensionality

The number of dimensions of a tracking task usually refers to the number of cartesian coordinates over which the *target* moves, rather than those of the response marker or sensor handle, or the number of degrees of freedom of the target or of the upper-limb. Some examples: (1) Most tasks with a 2-D joystick sensor or light-pen are 2-D but, if the target moves in the vertical direction only [Neilson and Neilson, 1980; Kondraske et al., 1984; Miall et al., 1985; Jones et al., 1993], the task is only considered 1-D, irrespective of whether the response marker is confined to vertical movements on the screen or not; (2) If the target trajectory is a circle [Bloxham et al., 1984], the task is 2-D despite the target having only one degree of freedom (i.e., radius r is constant); (3) A pursuit rotor is a 2-D task as it has a target which moves in two dimensions (whether cartesian or polar) as well as doing so with two degrees of freedom.

Watson and Jones [1998] compared random 2-D with 1-D performance but in doing so scaled their 2-D target down so as to have an average displacement and velocity equal to that of its 1-D horizontal and vertical components. By this means they were able to unequivocally demonstrate that there is poorer performance on 2-D tasks and that is due to both the increased dimensionality and increased position/speed demands of an unscaled 2-D task.

Dual-axis tracking is a variant of 2-D tracking in which the 2-D task comprises two simultaneous orthogonal 1-D tasks in which one or more of the target, input device, control dynamics, and on-line feedback are different between the two axes. It has been used to investigate mechanisms and characteristics of 2-D tracking, such as whether it is a single task or two separate orthogonal tasks [Fracker and Wickens, 1989; Navon et al., 1984].

Tracking Mode

The two primary modes of tracking—compensatory and pursuit—have been introduced above. The majority of tasks are of the pursuit type which is appropriate in that has a more direct parallel with real-world sensory-motor tasks than the more artificial compensatory task (e.g., [Vince, 1948; Garvey, 1960; Potvin et al., 1977; Miller and Freund, 1980; B'sser, 1984; Barr et al., 1988]) in which the subject is only shown the instantaneous value of the error signal. The compensatory mode may be preferentially chosen for control-theory modelling due to its simpler set of defining equations [Potvin et al., 1977]. The preview task [Welford, 1968; Jones and Donaldson, 1986] is an important variation of pursuit tracking in which a still greater correspondence with everyday tasks is achieved.

Controlled System Dynamics

It has been well established that subjects can deal satisfactorily with a variety of tracking systems incorporating different control characteristics [Neilson et al., 1995; Poulton, 1974]. That notwithstanding, the majority of tracking tasks have a zero-order controlled system in which the position of the response marker is proportional to the position of the sensor and the mechanical characteristics of friction, inertial mass, and velocity damping are simply those of the input device. Van den Berg et al. [1987] eliminated even these by feeding back a force signal from a strain gauge on the tracking handle to the power amplifier of a torque motor connected to their sensor. Conversely, Neilson et al. [1993] artificially introduced mechanical characteristics into the movement of their response marker by having a linear 2nd-order filter as the controlled system; by an appropriate transfer function ($H(z) = 0.4060/(1 - 1.0616z^{-1} + 0.4610z^{-2})$), they were able to introduce inertial lag and underdamping (resonant peak at 2.0 Hz). Miall et al. [1985] introduced an analog delay of 500 ms between their joystick and display so that they could study the effect of delayed visual feedback on performance. Soliveri et al. [1997] used both 1st-order (velocity) and zero-order (position) linear control dynamics to investigate differences in learning between parkinsonian and control subjects and between on- and off-medication. Navon et al. [1984] used a combination of velocity and acceleration control dynamics in their study on dual-axis tracking. Nonlinear transfer functions, such as 2nd-order Volterra (fading memory) nonlinearities, have also been used in tracking controlled systems, primarily as a means for investigating adaptive inverse modelling mechanisms in the brain relating to voluntary movement [Davidson et al., 1999; Sriharan, 1997].

Controlled system dynamics can also be changed during a task. In "critical tracking", a novel variation of pursuit tracking conceived by Jex [1966], the delay of the controlled system increases during the task. There is no external target but instead the subject's own instability acts as an input to an increasingly unstable controlled system, $Y(s) = K8/(s - 8)$, in which the level of instability, represented by the root 8 (= 1/T), is steadily increased during the task until a preset error is exceeded. The task has been described as analogous to driving a truck with no brakes down a hill on a winding road [Potvin et al., 1977]. The task has been applied clinically by Potvin et al. [1977] and Kondraske et al. [1984] and shown to be a reliable measure of small changes in neurological function [Potvin et al., 1977]. Alternatively, "gain-change step tracking" [Neilson et al., 1995], in which the gain of the control-display relation is increased or decreased without warning, has been used to investigate adaptive mechanisms in the brain [O'Dwyer and Neilson, 1998].

Having a torque motor as part of the sensor opens up several new possibilities. It can be operated as a "torque servo", in which applied torque is independent of position [Kondraske et al., 1984], or a "position servo", in which applied torque is proportional to position error (together with velocity damping if desired) [Thomas et al., 1976]. By adding external force perturbations, it is possible to measure and study neuromuscular reflexes and limb transfer function (i.e., stiffness, viscosity, and inertia), such as by applying constant velocity movements [Kondraske et al., 1984] or pulsatile [van den Berg et al., 1987], sinusoidal [Gottlieb et al., 1984], or random [Kearney and Hunter, 1983; van den Berg et al., 1987] force perturbations. Alternatively, the torque motor can be used to alter controlled system characteristics in tracking tasks for studies and/or improvement of voluntary movement. For example, van den Berg et al. [1987] cancelled unwanted controller characteristics. Chelette et al. [1995] used "force reflection" to improve tracking performance in both normal subjects and those with spasticity, and Johnson et al. [1996] used anti-viscous loading to investigate the cause of poor tracking in patients with Parkinson's disease.

Sensor-Display Compatibility

It is generally accepted that the level of compatibility between sensor and display in continuous tracking tasks influences the accuracy of performance [Neilson and Neilson, 1980]. The perfectly compatible sensor is the display marker itself [Poulton, 1974] where the subject holds and moves the response marker directly such as with a light-pen in tracking [Neilson and Neilson, 1980], rotary pursuit [Schmidt, 1982; Welford, 1968], handle on a two-joint mechanical arm [Abend et al., 1982], or in self-paced 2-D tracing tasks [Driscoll, 1975; Stern et al., 1984; Hocherman and Aharon-Peretz, 1994]. Similarly, van den Berg et al. [1987] achieve a high sensor-display compatibility by having the LED arrays for target and response displayed directly above a horizontally-moving handle. However, the majority of tracking tasks have sensors which are quite separate from the response marker displayed on an oscilloscope or computer screen. Sensor-display compatibility can be maximized in this case by having the sensor physically close to the display, moving in the same direction as the marker, and with a minimum of controlled system dynamics (e.g., zero-order). In the case of a joystick in a 2-D task, for example, direct compatibility (*Left-Right → Left-Right*) is easier than inverse compatibility (*Left-Right → Right-Left*), which is easier than non-compatibility (*Left-Right → Up-Down*). In contrast, fore-aft movements on a joystick appear to possess bidirectional compatibility in that *Fore-aft → Up-Down* seems as inherently natural as *Fore-aft → Down-Up* (i.e., no obvious inverse).

Sensor-display compatibility may not, however, be overly critical to performance. For example, Neilson and Neilson [1980] found no decrement in performance on random tracking of overall error scores, such as mean absolute error, between a light pen and a 1-D joystick; nevertheless, the latter did result in a decrease in gain, an increase in phase lag, and an increase in the non-coherent response component. Conversely, normal subjects find incompatible 2-D tracking very difficult to perform, taking up to 4 hours of practice to reach a level of performance equal to that seen on prepractice 2-D compatible tracking [Neilson et al., 1998].

Response Sampling Rates

Although some workers have manually analyzed tracking data from multichannel analog chart recordings [Flowers, 1976; Beppu et al., 1984] or analog processed results [Potvin et al., 1977], the majority have used computers, sometimes via a magnetic tape intermediary [Day et al., 1984; Miall et al., 1985], to digitize data for automated analyses. Sampling rates used have varied from 10 Hz [Neilson and Neilson, 1980], through 20 Hz [Cooper et al., 1989; Neilson et al., 1993; Neilson et al., 1998], 28.6 Hz [Jones and Donaldson, 1986], 30.2 Hz [Watson and Jones, 1998; Watson et al., 1997], 40 Hz [Frith et al., 1986], 60 Hz [Viviani and Mounoud, 1990], 60.3 Hz (= screen's vertical interrupt rate) [Jones et al., 1993], 66.7 Hz [O'Dwyer and Neilson, 1998], 100 Hz (all 2-D tasks) [Abend et al., 1982; Stern et al., 1984; Hocherman and Aharon-Peretz, 1994], to as high as 250 Hz [Day et al., 1984].

For the most part, a relatively low sampling rate is quite satisfactory for analysis of tracking performance as long as the Nyquist criterion is met and there is appropriate analog or digital low-pass filtering to prevent aliasing. Spectral analysis indicates that the fastest of voluntary arm movements have no power above about 8.7 Hz [Jones and Donaldson, 1986]. This is very similar to the maximal voluntary oscillations of the elbow of 4-6 Hz [Neilson, 1972; Leist et al., 1987] and to maximum finger tapping rates of

6-7 Hz [Muir et al., 1995]. The sampling rate can be reduced still further if the primary interest is only in *coherent* performance, whose bandwidth is only of the order of 2 Hz for both kinesthetic stimuli [Neilson, 1972] and visual stimuli [Leist et al., 1987; Neilson et al., 1993]; that is, performance above 2 Hz must be open-loop and, hence, learned and preprogrammed [Neilson, 1972]. Thus, from an information theory point of view, there is no need to sample tracking performance beyond, say, 20 Hz. However, a higher rate may well be justified on the grounds of needing better temporal resolution than 50 ms for transient or cross-correlation analysis, unless one is prepared to regenerate the signal between samples by some form of non-linear interpolation (e.g., sinc, spline, polynomial).

Other Measures

Several researchers have further extended the information which can be derived from upper-limb tracking performance by comparison with other simultaneously recorded biosignals. The most common of these is the EMG, particularly integrated EMG due to its close parallel to force of contraction [Neilson, 1972] and where the tracking movement is constrained to be around a single joint. The EMG has been used together with step tracking for fractionating reaction times into premotor and motor components [Anson, 1987; Sheridan et al., 1987] and confirmation of open-loop primary movements [Sittig et al., 1985; Sheridan et al., 1987]. In smooth tracking, correlation/cross-spectral analysis between the EMG and limb position has been used to study limb dynamics [Neilson, 1972; Barr et al., 1988].

In contrast, Cooper et al. [1989] measured the EEG at four sites during 2-D random tracking to show that slow changes in the EEG (equivalent to the Bereitschaftspotential preceding self-paced voluntary movement), particularly at the vertex, are correlated with the absolute velocity of the target.

Simultaneous measurement of hand and eye movements has been carried out by Warabi et al. [1986] and Leist et al. [1987] using EOG to measure horizontal eye movements and by Gibson et al. [1987] who used an infra-red limbus reflection technique. Interestingly, Leist et al. [1987] found that ocular pursuit and self-paced oscillations were limited to about 1 Hz and 2.2 Hz, respectively, whereas the equivalent values for arm movements are 2 Hz and 4-6 Hz, respectively.

Standard Assessment Procedures

Having designed and constructed a tracking task or set of tracking tasks with the characteristics necessary to allow measurement of the sensory-motor control performance capacities under investigation, it is essential that this process be complemented by a well formulated set of standard assessment procedures. These must include (a) standard physical setup, in which positioning of subject, sensor, and screen are tightly specified and controlled, as well as factors such as screen brightness, room lighting, etc. and (b) standardized instructions. The latter are particularly important in tasks where speed-accuracy tradeoff [Fitts, 1954; Welford, 1968; Agarwal and Logsdon, 1990] is possible. This applies particularly to step tracking in which leaving the tracking strategy completely up to subjects introduces the possibility of misinterpretation of differences in performance on certain measures, such as reaction time, risetime, and mean absolute error. For example, subjects need to know if it is more important to have the initial movement end up close to the target (i.e., emphasis on *accuracy* of primary movement) or to get within the vicinity of the target as soon as possible (i.e., emphasis on *speed* of primary movement); the latter results in greater under/overshooting but also tends to result in lower mean errors). The most common approach taken is to stress the importance of both speed and accuracy with an instruction to subjects of the form: "Follow the target as fast and as accurately as possible."

Test and Experimental Protocols

The design of appropriate test and experimental protocols is also a crucial component of the tracking task design process [Roscoe, 1975; Pitrella and Kruger, 1983]. When comparisons are made between different subjects, tasks, and/or conditions, careful consideration needs to be given to the paramount factors of *matching* and *balancing* to minimize the possibility of significant differences being due to some bias or confounding variable other than that under investigation. Matching can be achieved between experimental and control subjects in an inter-subject design by having average or one-to-one equivalence on age, gender, education, etc., or through an intra-subject design in which the subject acts as his/her own control in, say, a study of dominant versus non-dominant arm performance. Balancing is primarily needed to offset *order effects* due to learning which pervade much of sensory-motor performance [Welford, 1968; Poulton, 1974; Schmidt, 1982; Frith et al., 1986; Jones et al., 1990]. A study by Jones and Donaldson [1989] provides a good example of the application of these principles. Their study, aimed at investigating the effect of Parkinson's disease on predictive motor planning, involved 16 Parkinsonian subjects and 16 age and sex matched control subjects. These were then divided into 8 subgroups in a 3-way randomized cross-over design so as to eliminate between- and within-session order effects in determining the effect of target type, target preview, and medication on tracking performance.

149.4 Analysis of Sensory-Motor Control Performance

Accuracy

Analyses of raw tracking data can provide performance information which is objective and quantitative and which can be divided into two broad classes:

Measures of Global Accuracy of Performance

Measures of *global* (or overall or integrated) sensory-motor control capacities have proven invaluable for:

- *Vocational screening* of minimum levels of sensory-motor skills: Tracking tasks, in fact, have their origins in this area during World War II when they were used to help screen and train aircraft pilots [Poulton, 1974; Welford, 1968].
- *Clinical screening* for sensory-motor deficits (arising from one or more lesions in one or more sites in the sensory-motor system): An excellent example of the application of this in clinical practice is the provision of objective measures in off-road driving assessment programs [Jones et al., 1983; Croft and Jones, 1987].
- *Clinical and rehabilitation research* by measurement of longitudinal changes in sensory-motor function: There are many examples of subjects being assessed repeatedly on tracking tasks for periods up to 12 or more months. This has been done to quantify recovery following head injury [Jones and Donaldson, 1981] and stroke [Lynn et al., 1977; De Souza et al., 1980; Jones and Donaldson, 1981; Jones et al., 1990] as well as for studies of learning in tracking performance [Poulton, 1974; Jones and Donaldson, 1981; Schmidt, 1982; Frith et al., 1986; Jones et al., 1990]. They can also be used to quantify changes due to medication, such as in Parkinson's disease [Baroni et al., 1984; Johnson et al., 1996; Jones et al., 1996; Soliveri et al., 1997].

Measures of Characteristics of Performance

Measures of global accuracy of tracking performance can detect and quantify the presence of abnormal sensory-motor control performance capacities with considerable sensitivity [Potvin et al., 1977; Jones et al., 1989]. Conversely, they are unable to give any indication of which of the many subsystems or performance resources in the overall sensory-motor system are, or may be, responsible for the abnormal performance. Nor can they provide any particular insight into the underlying neuromuscular control mechanisms of normal or abnormal performance.

Four approaches can be taken to provide information necessary to help identify the sensory-motor subsystems and their properties responsible for the *characteristics* of observed normal and abnormal performance:

- Batteries of neurologic sensory-motor tests—These tests can be used to, at least ideally, isolate and quantify the various sensory, motor, cognitive, and integrative functions and subsystems involved in sensory-motor control performance as measured globally by, for example, tracking tasks.

- Functional decomposition—Fractionation of the various performance resources contributing to tracking performance.
- Traditional signal processing approaches—Time domain (ballistic and nonballistic) and frequency domain techniques.
- Graphical analysis—This has primarily been developed for measurement and investigation of changes in performance and underlying PRs over time.

Measures of Global Accuracy of Performance

The most commonly used measure of global or overall accuracy is the *mean absolute error* (MAE) [Jones and Donaldson, 1986] which indicates the average distance the subject was away from the target irrespective of side; it is also variously called average absolute error [Poulton, 1974], modulus mean error [Poulton, 1974], mean rectified error [Neilson and Neilson, 1980], or simply tracking error [Kondraske et al., 1984; Behbehani et al., 1988]. In contrast, the mean error, or constant position error, is of little value as it simply indicates only the extent to which the response is more on one side of the target than the other [Poulton, 1974]. Measures of overall performance which give greater weighting to larger errors include mean square error [Neilson et al., 1993], root mean square error [McRuer and Krendel, 1959; Navon et al., 1984; O'Dwyer and Neilson, 1995; Poulton, 1974], variance of error [Neilson and Neilson, 1980], and standard deviation of error [Poulton, 1974]. Relative or normalized error score equivalents of these can be calculated by expressing the raw error scores as a percentage of the respective scores obtained had subject simply held the response marker stationary at the mean target position [Poulton, 1974; Neilson and Neilson, 1980; Day et al., 1984]; i.e., no response = 100%. Alternatively, the relative root mean square error, defined as the square root of the ratio of the mean square value of the error signal to the mean square value of the target signal expressed as a percentage, allows tracking errors to be compared across tests using different target signals [Neilson et al., 1998].

Overall coherence is an important alternative to the above measures when it is wished to assess the similarity between target and response waveforms but there is a substantial delay between them. It provides an estimate of the proportion of the response that is correlated with the target over all frequencies [O'Dwyer et al., 1996; O'Dwyer and Neilson, 1995].

An issue met in viewing error scores from the perspective of Kondraske's [Kondraske, 1995a; Kondraske, 1995b] elemental resource model is the unifying requirement of its associated *general systems performance theory* (*GSPT*) that all dimensions of performance must be in a form for which a higher numerical value indicates a superior performance. Thus scores which state that a *smaller* score indicates a superior performance, including reaction times, movement times, and all error scores, need to be transformed into *performance scores* [Kondraske, 1988]. For example:

- *central response speed = 1/(reaction time)*
- *information processing speed = 1/(8-choice reaction time)*
- *movement speed = 1/(movement time)*
- *tracking accuracy = 1/(tracking error)*

As transformation via inversion is non-linear, the distributions of raw error scores and derived performance will be quite different. This has no effect on ordinal analyses, such as non-parametric statistics, but will have some effect on linear analyses, such as parametric statistics, linear regression/correlation, etc., and may include improvements due to a possible greater normality of the distributions of derived performances. An alternative transformation which would retain a linear relationship with the error scores is: *Tracking accuracy* = 100 – *Relative tracking error.* However, while this gives a dimension of performance with the desired 'bigger is better' characteristic, it also raises the possibility of negative values, implying an accuracy worse than zero!—the author can attest to the fact that some subjects do indeed end up with error scores worse than the "hands off" score. Irrespective of GSPT, there is no doubt that it is beneficial to deal conceptually and analytically with multiple performance measures when *all* measures are consistently defined in terms of "bigger is better".

Time on target is a much cruder measure of tracking performance than all of the above but it has been used reasonably widely due to it being the result obtained from the pursuit rotor. The crudeness generally reflects (a) a lack of spatiotemporal sampling during a task (i.e., simple integration of time on target only) preventing the possibility of further analysis of any form, and (b) a task's performance *ceiling* due to the target having a finite zone within which greater accuracy, relative to center of zone, is unrewarded. This latter factor can, however, be used to advantage for the case where the investigator wishes to have control over the *difficulty* of a task, to gain, for example, similar levels of task difficulty across subjects irrespective of individual ability. This attribute has been used very effectively with 2-D random tracking tasks to minimize the confounding effects of major differences in task load between experimental and control subjects in dual task studies of impairment of central executive function in subjects with Alzheimer's disease [Baddeley et al., 1986] and Parkinson's disease [Dalrymple-Alford et al., 1994].

Measures of Characteristics of Performance

Batteries of Neurologic Sensory-Motor Tests

Potvin and associates [Potvin and Tourtellotte, 1975; Potvin et al., 1985], now led by Kondraske et al. [Kondraske et al., 1984, 1988], have developed what is by far the most comprehensive battery of tests available for quantitative evaluation of neurologic function covering a number of sensory, motor, cognitive, and sensory-motor functions or performance resources. Similarly, Jones et al. [1989, 1993] have developed a battery of *component* function tests, most of which have been specifically designed to isolate and quantify the various performance resources involved in their tracking tasks. There is, therefore, a close resemblance between the component and tracking tests so as to maximize the validity of comparisons made between them.

Functional Decomposition of Tracking Performance

There are three main approaches whereby tracking performance can be fractionated or decomposed into its functional components: sensory, perceptual, cognitive, motor planning, and motor execution.

The first involves breaking the ballistic response in step tracking into reaction time, movement time, overshoot, and settling time [Behbehani et al., 1988; Evarts et al., 1981; Flowers, 1976; Jones and Donaldson, 1986; Kondraske et al., 1988] (see "Time domain (ballistic) analysis" below). This allows indirect deductions about cognitive, motor planning, and motor execution functions, although the distinction between cognitive and motor elements often remains imprecise.

The second involves calculation of differentials in tracking performance from inter-trial alterations in target and/or controlled system dynamics. This has been successfully used to study predictive motor planning [Bloxham et al., 1984; Day et al., 1984; Flowers, 1976; Flowers, 1978; Jones and Donaldson, 1989], acquisition/modification of motor sets [Frith et al., 1986], and reliance on visual feedback [Cooke et al., 1978; Flowers, 1976; Frith et al., 1986] in Parkinson's disease.

The third approach allows a more direct identification of the contribution of certain elemental resources to tracking performance during a specific tracking run. For example, by introducing the concept of a visuoperceptual buffer-zone, it is possible to estimate the contribution of visuoperceptual function to tracking performance [Jones et al., 1996]. This technique has been used to demonstrate that impaired visuoperceptual function in Parkinsonian subjects plays only a minor role in their poor tracking performance [Jones et al., 1996] but, conversely, that impaired tracking performance in stutterers is predominantly due to reduced dynamic visuospatial perception [A. J. White, R. D. Jones, K. H. C. Lawson, and T. J. Anderson, unpublished observations]. Furthermore, the visuoperceptual function can itself be fractionated into visual acuity, static perception, and dynamic perception [Jones and Donaldson, 1995].

In contrast to fractionation of performance on a high level task (e.g., tracking, driving), Kondraske et al. have developed techniques for the reverse process. They have shown how their hierarchical elemental resource model [Kondraske, 1995a; Kondraske, 1995b] can be used to *predict* performance on high level tasks from performance on a number of lower level tasks [Kondraske, 1987; Kondraske, 1995b; Kondraske et al., 1997; Vasta and Kondraske, 1994]. This approach has considerable potential in application areas such as rehabilitation. For example, it could be used in driving assessment programs to predict on-road driving ability by off-road measurement of performance on several key lower level tasks pertinent to driving, such as reaction time, visuospatial, cognitive, and tracking.

Time Domain (Non-Ballistic) Analysis of Tracking Performance

There are several run-averaged biases which can indicate the general *form* of errors being made, particularly when the tracking performance is subnormal. Positive *side of target* (%) and *direction of target* (%) biases reflect a greater proportion of errors occurring to the right of the target or while the target is moving to the right respectively [Jones and Donaldson, 1986] which, if substantial, may indicate the presence of some visuoperceptual deficit. Similarly, the *side of screen* bias (assuming mean target position is mid-screen) [Jones and Donaldson, 1986] is identical to the mean error or constant position error.

Perhaps the single most important measure of performance, other than mean absolute error, for non-transient targets is that of the average time delay, or *lag*, of a subject's response with respect to the target signal. The lag is most commonly defined as being the shift, τ, corresponding to the peak of the cross-correlation function, calculated directly in the time domain or indirectly via the inverse of the cross-spectrum in the frequency domain. Although simulation studies indicate that these techniques are at least as accurate and as robust to noise/remnants as the alternatives listed below [Watson, 1994], one needs to be aware of a bias leading to underestimation of the magnitude of the lag (or lead) due to distortion of the standard cross-correlation function, but specifically of the peak towards zero shift. The distortion arises due to the varying overlap of two truncated signals (i.e., the target and the response) resulting in the multiplication of the cross-correlation function by a triangle (maximum at $\tau = 0$ and zero at $\tau = NT_S$, assuming signals of equal length NT_S). This effect is minimal as long as both signals have a mean value of zero (i.e., zero d.c.). Temporal resolution is another factor deserving consideration. If desired, greater resolution than that of the sampling period can be obtained by interpolation of the points around the peak of the cross-correlation function by some form of curve fitting (e.g., parabola [Jones et al., 1993]).

An alternative estimate of the lag, which has proven accurate on simulated responses, can be gained from the *least squares time delay estimation* by finding the time shift between the response and target at which the mean square error is minimized [Fertner and Sjölund, 1986; Jones et al., 1993]. Another approach, *phase shift time delay* estimation, calculates lag from the gradient of the straight line providing a best least-squares to the phase points in the cross-spectrum [Watson et al., 1997]. This technique has, however, proven more sensitive to non-correlated remnants in the response than the other procedures [Watson, 1994].

Several measures used to help characterize within-run variability in performance include variance of error [Neilson and Neilson, 1980], standard deviation of error [Poulton, 1974], and inconsistency [Jones and Donaldson, 1986].

Time Domain (Ballistic) Analysis of Tracking Performance

Whether or not any of the above nonballistic analyses, evaluation of step tracking performance usually involves separate ballistic or transient analysis of each of the step responses. This generally takes the form of breaking up each response into three phases (Fig. 149.5): (1) reaction time phase, or the time between onset of step stimulus and initiation of movement defined by exit from a visible or invisible reaction zone, (2) primary movement phase, or the open-loop ballistic movement made by most normal subjects aiming to get within the vicinity of the target as quickly as possible, the end of which is defined as the first stationary point, and (3) secondary correction phase, comprising one or more adjustments and the remaining time needed to enter and stay within target zone. The step measures from individual steps can then be grouped into various step categories to allow evaluation of the effect of step size, spatial predictability, arm dominance, etc., on transient performance.

Accuracy of the primary aimed movement can also be characterized in terms of a constant error and a variable error (standard deviation of error), which are considered to be indices of accuracy of central motor programming and motor execution respectively [Guiard et al., 1983].

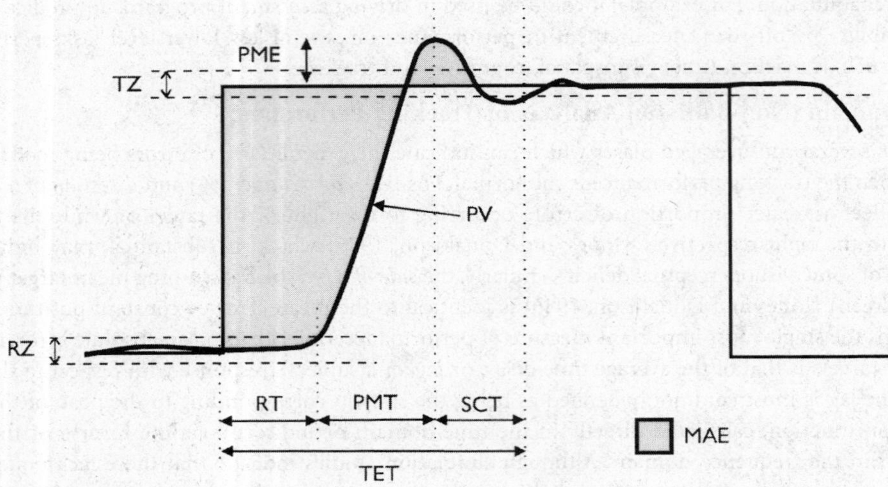

FIGURE 149.5 Transient response analysis.
Tolerance zones:
RZ = reaction zone
TZ = target zone
Performance parameters:
RT = reaction time
PMT = primary movement time
SCT = secondary correction time
TET = target entry time
PV = peak velocity
PME = primary movement error
MAE = mean absolute error over a fixed interval following stimulus

Phase-plane (velocity vs. position) plots provide an alternative means for displaying and examining the qualitative characteristics of step tracking responses. In particular, they have proven valuable for rapid detection of gross abnormalities [Potvin et al., 1985]. Behbehani et al. [1988] have introduced a novel quantitative element to phase-plane analysis by deriving an index of coordination: $I_C = V_m^2/A$, where V_m is the maximum velocity during an outward and return step and A is the area within the resultant loop on the phase-plane plot.

Frequency Domain Analysis of Tracking Performance

Cross-correlation and spectral analysis have proven invaluable tools for quantifying the frequency dependent characteristics of the human subject. The cross-spectral density function, or cross-spectrum $S_{xy}(f)$, can be obtained from the random target x(t) and random response y(t) by taking the Fourier transform of the cross-correlation function $r_{xy}(\tau)$, i.e., $S_{xy}(f) = F\{r_{xy}(\tau)\}$, or in the frequency domain via $S_{xy}(f) = X(f)Y(f)^*$, or by a nonparametric system identification approach (e.g., "spa.m" in MATLAB). The cross-spectrum provides estimates of the relative amplitude (i.e., gain) and phase-lag at each frequency. Gain, phase, and remnant frequency response curves provide objective measures of pursuit tracking behaviour, irrespective of linearity, and are considered a most appropriate "quasi-linear" tool for obtaining a quantitative assessment of pursuit tracking behaviour [Neilson and Neilson, 1980]. From the cross-spectrum one can also derive the *coherence function* which gives the proportion of the response signal linearly related to the target at each frequency: $(_{xy}^2 = {}^*S_{xy}(f)^{*2}/S_x(f)S_y(f)$. Lynn et al. [1977] emphasize, however, that one must be cognizant of the difficulty representing tracking performance by a quasi-linear time-invariant transfer function, especially if the run is of short duration or if the target waveform is of limited bandwidth, as the results can be so statistically unreliable as to make description by a 2nd- or

3rd-order transfer function quite unrealistic. Van den Berg et al. [1987] chose four parameters to characterize tracking performance: low-frequency performance via the mean gain of transfer function at the 3 lowest of 8 frequencies in target signal, high-frequency performance via the frequency at which the gain has dropped to less than 0.4, mean delay via shift of peak of cross-correlation function, and remnant via power in frequencies introduced by subject relative to total power. Spectral and coherence analysis have been used to demonstrate that the human bandwidth is about 2 Hz for both kinaesthetic tracking [Neilson, 1972] and visual tracking [Neilson et al., 1993], a much greater relative amplitude of second harmonic in the response of cerebellar subjects in sine tracking [Miller and Freund, 1980], a near constant lag except at low frequencies in normal subjects [Cassell, 1973], adaptation to time-varying signals [Bösser, 1984], 2-D asymmetry in postural steadiness [Milkowski et al., 1993], and that normal subjects can form non-dynamic and dynamic inter-limb synergies in a bimanual tracking task [O'Dwyer and Neilson, 1995].

Graphical Analysis of Tracking Performance

Most of the of the above analyses give quantitative estimates of some aspect of performance which is effectively assumed to be constant over time, other than for random fluctuations. This is frequently not the case, especially for more complex sensory-motor tasks such as tracking. Changes in performance over time can be divided into two major classes: class I: those for which the underlying PRs remain unchanged (these are due to factors such as practice, fatigue, lapses in concentration, lack of practice, and changes in task complexity) and class II: those for which one or more underlying PRs have changed (these are due to abrupt or gradual alterations at one or more sites in the sensory-motor system and include normal changes, such as due to age, and abnormal changes, due to trauma or pathology).

Studies of class I factors using tracking tasks are complicated most by the intra-run *difficulty* of a task not being constant. Changes in tracking accuracy *during* a run can be viewed via graphs of target, response, and errors [Jones et al., 1993]. The latter is particularly informative for sinusoidal targets for which the mean absolute errors can be calculated over consecutive epochs, corresponding to sine-wave cycles, and plotted both in a histogram form and as a smoothed version of this [Jones and Donaldson, 1986]. As complexity of task is constant over epochs (cf. random pursuit task), the error graph gives an accurate measure of a subject's time-dependent spatiotemporal accuracy that is not confounded by changes in task difficulty and, therefore, gives a true indication of changes in performance due to factors such as learning, fatigue, and lapses in concentration. Attempts by the author to derive an instantaneous or short epoch (up to several seconds) function or *index of task difficulty* which would allow equivalent graphs to be generated for random targets were unsuccessful.

Neilson et al. [1998] devised an alternative procedure for intra-run analysis, termed micromovement analysis. This involves segmentation of the X and Y deflections of the response cursor on the basis of discontinuities, flat regions, and changes in direction of the response. They used this to identify changes in visual-motor coupling during the first 4 min of tracking on a 2-D compatible task following 4 h of practice on a 2-D incompatible task. They propose that these changes are evidence of rapid switching between different sensory-motor models in the brain.

By comparison, as long as the task remains unchanged over successive runs, studies of class II factors using tracking tasks are complicated most by inter-run *learning*. Although most learning occurs over the first one or two runs or sessions, tracking performance can continue to improve over extended periods as evidenced by, for example, significant improvements still being made by normal subjects after nine weekly sessions [Jones and Donaldson, 1981]. Consequently, a major difficulty met in the interpretation of serial measures of performance following acute brain damage is differentiation of neurologic recovery from normal learning. Furthermore, it is not simply a matter of subtracting off the degree of improved performance due to learning seen in normal control subjects. Jones et al. [1990] have developed graphical analysis techniques which provide for the removal of the learning factor, as much as is possible, and which can be applied to generating recovery curves for individual subjects following acute brain damage such as stroke. They demonstrated that, for tracking, percentage improvement in performance (PIP) graphs give more reliable evidence of neurologic recovery than absolute improvement in performance

(PIA) graphs due to the former's greater independence from what are often considerably different absolute levels of performance.

Statistical Analysis

Parametric statistics (t-test, ANOVA) are by far the most commonly used in studies of sensory-motor/psychomotor performance due, in large part, to their availability and ability to draw out interactions between dependent variables. However, there is also a strong case for the use of non-parametric statistics. For example, the Wilcoxon matched-pairs statistic may be preferable for both between-group and within-subject comparisons due to its greater robustness over its parametric paired t-test equivalent, with only minimal loss of power. This is important due to many sensory-motor measures having very non-Gaussian skewed distributions as well as considerably different variances between normal and patient groups.

Defining Terms

Basic element of performance (BEP): Defined by a **functional unit** *and* a **dimension of performance**, e.g., right elbow flexor + speed.

Dimension of performance: A basic measure of performance such as speed, range of movement, strength, spatial perception, spatiotemporal accuracy.

Functional unit: A subsystem such as right elbow flexor, left eye, motor memory.

Performance capacity: The maximal level of performance possible on a particular dimension of performance.

Performance resource (PR): One of a pool of elemental resources, from which the entire human is modelled [Kondraske, 1995b], and which is available for performing tasks. These resources can be subdivided into life sustaining, environmental interface, central processing, and skills domains, and have a parallel with *basic elements of performances.*

Sensor: [in context of tracking tasks] A device for measuring/transducing a subject's motor output.

Sensory-motor control: The primary (but not only) performance resource responsible for accuracy of movement.

Spatiotemporal accuracy: The class of accuracy most required by tasks which place considerable demand on attainment of simultaneous spatial and temporal accuracy; this refers particularly to *paced* tasks such as tracking, driving, ball games, and video games.

Tracking task: A laboratory apparatus and associated procedures which have proven one of the most versatile means for assessing and studying the human "black-box" sensory-motor system by providing a continuous record of a subject's response, via some *sensor,* to any one of a large number of continuous and well-controlled stimulus or *target* signals.

References

Abend, W., Bizzi, E., and Morasso, P. 1982. Human arm trajectory formation. *Brain* 105:331-348.

Agarwal, G.C. and Logsdon, J.B. 1990. Optimal principles for skilled limb movements and speed accuracy tradeoff. *Proc. Ann. Int. Conf. IEEE Eng. Med. Biol. Soc.* 12:2318-2319.

Anderson, O.T. 1986. A system for quantitative assessment of dyscoordination and tremor. *Acta Neurol. Scand.* 73:291-294.

Angel, R.W., Alston, W., and Higgins, J.R. 1970. Control of movement in Parkinson's disease. *Brain* 93:1-14.

Anson J.C. 1987. Fractionated simple reaction time as a function of movement direction and level of pre-stimulus muscle tension. *Int. J. Neurosci.* 35:140.

Baddeley, A., Logie, R., Bressi, S., Della Salla, S., and Spinnler, H. 1986. Dementia and working memory. *Quart. J. Exp. Psychol.* 38A:603-618.

Baroni, A., Benvenuti, F., Fantini, L., Pantaleo, T., and Urbani, F. 1984. Human ballistic arm abduction movements: Effects of L-dopa treatment in Parkinson's disease. *Neurology* 34:868-876.

Barr, R.E., Hamlin, R.D., Abraham, L.D., and Greene, D.E. 1988. Electromyographic evaluation of operator performance in manual control tracking. *Proc. Ann. Int. Conf. IEEE Eng. Med. Biol. Soc.* 10:1608-1609.

Behbehani, K., Kondraske, G.V., and Richmond, J.R. 1988. Investigation of upper extremity visuomotor control performance measures. *IEEE Trans. Biomed. Eng.* 35:518-525.

Beppu, H., Suda, M., and Tanaka, R. 1984. Analysis of cerebellar motor disorders by visually-guided elbow tracking movement. *Brain* 107:787-809.

Beppu, H., Nagaoka, M., and Tanaka, R. 1987. Analysis of cerebellar motor disorders by visually-guided elbow tracking movement: 2. Contributions of the visual cues on slow ramp pursuit. *Brain* 110:1-18.

Bloxham, C.A., Mindel, T.A., and Frith, C.D. 1984. Initiation and execution of predictable and unpredictable movements in Parkinson's disease. *Brain* 107:371-384.

Bösser, T. 1984. Adaptation to time-varying signals and control-theory models of tracking behaviour. *Psychol. Res.* 46:155-167.

Buck, L. 1982. Location versus distance in determining movement accuracy. *J. Mot. Behav.* 14:287-300.

Cassell, K.J. 1973. The usefulness of a temporal correlation technique in the assessment of human motor performance on a tracking task. *Med. Biol. Eng.* 11:755-761.

Chelette, T.L., Repperger, D.W., Phillips, C.A. 1995. Enhanced metrics for identification of forearm rehabilitation. *IEEE Trans. Rehab. Eng.* 3:122-131.

Cooke, J.D., Brown, J.D., and Brooks, V.B. 1978. Increased dependence on visual information for movement control in patients with Parkinson's disease. *Canad. J. Neurol. Sci.* 5:413-415.

Cooper, R., McCallum, W.C., and Cornthwaite, S.P. 1989. Slow potential changes related to the velocity of target movement in a tracking task. *Electroenceph. Clin. Neurophysiol.* 72:232-239.

Craik, K.J.W. 1966. The mechanism of human action. In Sherwood SL (ed), *The Nature of Psychology—A selection of papers, essays, and other writings by the late Kenneth J.W. Craik*, Cambridge, Cambridge University Press.

Croft, D. and Jones, R.D. 1987. The value of off-road tests in the assessment of driving potential of unlicensed disabled people. *Brit. J. Occup. Ther.* 50:357-361.

Dalrymple-Alford, J.C., Kalders, A.S., Jones, R.D., and Watson, R.W. 1994. A central executive deficit in patients with Parkinson's disease. *J. Neurol. Neurosurg. Psychiatry* 57:360-367.

Davidson, P.R., Jones, R.D., Sirisena, H.R. et al., 1999. Evaluation of nonlinear generalizations of the Adaptive Model Theory. *Proc. Int. Conf. IEEE Eng. Med. Biol. Soc.* 21.

Day, B.L., Dick, J.P.R., and Marsden, C.D. 1984. Patient's with Parkinson's disease can employ a predictive motor strategy. *J. Neurol. Neurosurg. Psychiatry* 47:1299-1306.

De Souza, L.H., Langton Hewer, R., Lynn, P.A., Mller, S., and Reed, G.A.L. 1980. Assessment of recovery of arm control in hemiplegic stroke patients: 2. Comparison of arm function tests and pursuit tracking in relation to clinical recovery. *Int. Rehab. Med.* 2:10-16.

Deuschl, G., Toro, C., Zeffiro, T. et al., 1996. Adaptation motor learning of arm movements in patients with cerebellar disease. *J. Neurol. Neurosurg. Psychiatry* 60:515-519.

Driscoll, M.C. 1975. Creative technological aids for the learning-disabled child. *Amer. J. Occup. Ther.* 29:102-105.

Ferslew, K.E., Manno, J.E., Manno, B.R., Vekovius, W.A., Hubbard, J.M., and Bairnsfather, L.E. 1982. Pursuit meter II, a computer-based device for testing pursuit-tracking performance. *Percep. Mot. Skills* 54:779-784.

Fertner, A. and Sj'lund, S.J. 1986. Comparison of various time delay estimation methods by computer simulation. *IEEE Trans. Acoustics, Speech and Signal Proc.* 34:1329-1330.

Fitts, P.M. 1954. The information capacity of the human motor system in controlling the amplitude of movement. *J. Exp. Psychol.* 47:381-391.

Fitts, P.M. and Posner, M.I. 1967. *Human performance*, Brooks/Cole, California.

Flash, T. and Hogan, N. 1985. The coordination of arm movements: An experimentally confirmed mathematical model. *J. Neurosci.* 5:1688-1703.

Flowers, K.A. 1976. Visual 'closed-loop' and 'open-loop' characteristics of voluntary movement in patients with Parkinsonism and intention tremor. *Brain* 99:269-310.

Flowers, K.A. 1978a. Lack of prediction in the motor behaviour of Parkinsonism. *Brain* 101:35-52.

Flowers, K.A. 1978b. The predictive control of behaviour: Appropriate and inappropriate actions beyond the input in a tracking task. *Ergonomics* 21:109-122.

Frith, C.D., Bloxham, C.A., and Carpenter, K.N. 1986. Impairments in the learning and performance of a new manual skill in patients with Parkinson's disease. *J. Neurol. Neurosurg. Psychiatry* 49:661-668.

Garvey, W.D. 1960. A comparison of the effects of training and secondary tasks on tracking behavior. *J. App. Psychol.* 44:370-375.

Gibson, J.M., Pimlott, R., and Kennard, C. 1987. Ocular motor and manual tracking in Parkinson's disease and the effect of treatment. *Neurology* 50:853-860.

Gottlieb, G.L., Agarwal, G.C., and Penn, R. 1984. Sinusoidal oscillation of the ankle as a means of evaluating the spastic patient. *J. Neurol. Neurosurg. Psychiatry* 41:32-39.

Guiard, Y., Diaz, G., and Beaubaton, D. 1983. Left-hand advantage in right-handers for spatial constant error: Preliminary evidence in a unimanual ballistic aimed movement. *Neuropsychologica* 21:111-115.

Hocherman, S. and Aharon-Peretz, J. 1994. Two-dimensional tracing and tracking in patients with Parkinson's disease. *Neurology* 44:111-116.

Hufschmidt A, Lücking C. 1995. Abnormalities of tracking behavior in Parkinson's disease. *Mov. Disord.* 10:267-276.

Jex, H.R. 1966. A "critical" tracking task for manual control research. *IEEE Trans. Hum. Factors Electronics* 7:138-145.

Johnson, M.T.V., Kipnis, A.N., Coltz, J.D. et al., 1996. Effects of levodopa and viscosity on the velocity and accuracy of visually guided tracking in Parkinson's disease. Brain 119:801-813.

Jones, R.D. and Donaldson, I.M. 1981. Measurement of integrated sensory-motor function following brain damage by a preview tracking task. *Int. Rehab. Med.* 3:71-83.

Jones, R.D. and Donaldson, I.M. 1986. Measurement of sensory-motor integrated function in neurological disorders: three computerized tracking tasks. *Med. Biol. Eng. Comput.* 24:536-540.

Jones, R.D. and Donaldson, I.M. 1989. Tracking tasks and the study of predictive motor planning in Parkinson's disease. *Proc. Ann. Int. Conf. IEEE Eng. Med. Biol. Soc.* 11:1055-1056.

Jones, R.D. and Donaldson, I.M. 1995. Fractionation of visuoperceptual dysfunction in Parkinson's disease. *J. Neurol. Sci.* 131:43-50.

Jones, R.D., Donaldson, I.M., Sharman, N.B. 1996. A technique for the removal of the visuospatial component from tracking performance and its application to Parkinson's disease. *IEEE Trans. Biomed. Eng.* 43:1001-1010.

Jones, R., Giddens, H., and Croft, D. 1983. Assessment and training of brain-damaged drivers. *Amer. J. Occup. Ther.* 37:754-760.

Jones, R.D., Williams, L.R.T., and Wells, J.E. 1986. Effects of laterality, sex, and age on computerized sensory-motor tests. *J. Hum. Mot. Stud.* 12:163-182.

Jones, R.D., Donaldson, I.M., and Parkin, P.J. 1989. Impairment and recovery of ipsilateral sensory-motor function following unilateral cerebral infarction. *Brain* 112:113-132.

Jones, R.D., Donaldson, I.M., Parkin, P.J., and Coppage, S.A. 1990. Impairment and recovery profiles of sensory-motor function following stroke: single-case graphical analysis techniques. *Int. Disabil. Stud.* 12:141-148.

Jones, R.D., Sharman, N.B., Watson, R.W., and Muir, S.R. 1993. A PC-based battery of tests for quantitative assessment of upper-limb sensory-motor function in brain disorders. *Proc. Ann. Int. Conf. IEEE Eng. Med. Biol. Soc.* 15:1414-1415.

Kearney, R.E. and Hunter, I.W. 1983. System identification of human triceps surae stretch reflex dynamics. *Exp. Brain Res.* 51:117-127.

Kondraske, G.V. 1988. Experimental evaluation of an elemental resource model for human performance. *Proc. Ann. Int. Conf. IEEE Eng. Med. Biol. Soc.* 10:1612-1613.

Kondraske, G.V. 1995a. An elemental resource model for the human-task interface. *Int. J. Technol. Assess Health Care* 11:153-173.

Kondraske, G.V. 1995b. A working model for human system-task interfaces. In Bronzino JD (ed), *The Biomedical Engineering Handbook*, pp. 2157-2174. Boca Raton, FL, CRC Press.

Kondraske, G.V., Johnson, C., Pearson, A., Tarbox, L. 1997. Performance prediction and limited resource identification with nonlinear causal resource analysis. *Proc. Ann. Int. Conf. IEEE Eng. Med. Biol. Soc.* 19:1813-1816.

Kondraske, G.V., Potvin, A.R., Tourtellotte, W.W., and Syndulko, K. 1984. A computer-based system for automated quantitation of neurologic function. *IEEE Trans. Biomed. Eng.* 31:401-414.

Kondraske, G.V., Behbehani, K., Chwialkowski, M., Richmond, R., and van Maltzahn, W. 1988. A system for human performance measurement. *IEEE Eng. Med. Biol. Mag.* March:23-27.

Leist, A., Freund, H.J., and Cohen, B. 1987. Comparative characteristics of predictive eye-hand tracking. *Hum. Neurobiol.* 6:19-26.

Lynn, P.A., Reed, G.A.L., Parker, W.R., and Langton Hewer, R. 1977. Some applications of human-operator research to the assessment of disability in stroke. *Med. Biol. Eng.* 15:184-188.

Lynn, P.A., Parker, W.R., Reed, G.A.L., Baldwin, J.F., and Pilsworth, B.W. 1979. New approaches to modelling the disabled human operator. *Med. Biol. Eng. Comput.* 17:344-348.

McRuer, D.T. and Krendel, E.S. 1959. The human operator as a servo element. *J. Franklin Inst.* 267:381-403.

Miall, R.C., Weir, D.J., and Stein, J.F. 1985. Visuomotor tracking with delayed visual feedback. *Neurosci.* 16:511-520.

Milkowski, L.M., Prieto, T.E., Myklebust, J.B., lovett, and Klebust, B.M. 1993. Two-dimensional coherence: A measure of asymmetry in postural steadiness. *Proc. Ann. Int. Conf. IEEE Eng. Med. Biol. Soc.* 15:1181.

Miller, R.G. and Freund, H.J. 1980. Cerebellar dyssynergia in humans—a quantitative analysis. *Ann. Neurol.* 8:574-579.

Muir, S.R., Jones, R.D., Andreae, J.H., and Donaldson, I.M. 1995. Measurement and analysis of single and multiple finger tapping in normal and Parkinsonian subjects. *Parkinsonism Related Disord.* 1:89-96.

Navon, D., Gopher, D., Chillag, N., and Spitz, G. 1984. On separability of and interference between tracking dimensions in dual-axis tracking. *J. Mot. Behav.* 16:364-391.

Neilson, P.D. 1972. Speed of response or bandwidth of voluntary system controlling elbow position in intact man. *Med. Biol. Eng.* 10:450-459.

Neilson, P.D. and Neilson, M.D. 1980. Influence of control-display compatability on tracking behaviour. *Quart. J. Exp. Psychol.* 32:125-135.

Neilson, P.D., Neilson, M.D., and O'Dwyer, N.J. 1992. Adaptive model theory: Application to disorders of motor control. In: *Approaches to the Study of Motor Control and Learning*, J.J. Summers, (ed.), pp. 495-548. Amsterdam, Elsevier Science.

Neilson, P.D., Neilson, M.D., and O'Dwyer, N.J. 1993. What limits high speed tracking performance? *Hum. Mov. Sci.* 12:85-109.

Neilson, P.D., Neilson, M.D., and O'Dwyer, N.J. 1995. Adaptive optimal control of human tracking. In: *Motor Control and Sensory-Motor Integration: Issues and Directions*, D.J. Glencross and J.P. Piek (eds.), pp. 97-140. Amsterdam, Elsevier Science.

Neilson, P.D., Neilson, M.D., and O'Dwyer, N.J. 1998. Evidence for rapid switching of sensory-motor models. In *Motor Control and Human Skill: A Multidisciplinary Perspective,* J Piek (ed.), pp. 105-126. Champaign, IL, Human Kinetics.

Neilson, P.D., O'Dwyer, N.J., and Nash, J. 1990. Control of isometric muscle activity in cerebral palsy. *Devel. Med. Child Neurol.* 32:778-788.

Notterman, J.M., Tufano, D.R., and Hrapsky, J.S. 1982. Visuo-motor organization: Differences between and within individuals. *Percep. Mot. Skills* 54:723-750.

O'Dwyer, N.J., Ada, L., and Neilson, P.D. 1996. Spasticity and muscle contracture following stroke. *Brain* 119:1737-1749.

O'Dwyer, N.J. and Neilson, P.D. 1995. Learning a dynamic limb synergy. In *Motor Control and Sensory-Motor Integration: Issues and Directions,* D.J. Glencross and J.P. Piek (eds.), pp. 289-317. Amsterdam, Elsevier Science.

O'Dwyer, N.J. and Neilson, P.D. 1998. Adaptation to a changed sensory-motor relation: immediate and delayed parametric modification. In *Motor Behavior and Human Skill—A Multidisciplinary Approach*, J.P. Piek (ed.), pp. 75-104. Champaign, IL, Human Kinetics.

Sriharan, A. 1997. *Mathematical modelling of the human operator control system through tracking tasks.* Masters in Engineering Thesis. University of New South Wales, Sydney, Australia.

Patrick, J. and Mutlusoy, F. 1982. The relationship between types of feedback, gain of a display and feedback precision in acquisition of a simple motor task. *Quart. J. Exp. Psychol.* 34A:171-182.

Pitrella, F.D. and Kruger, W. 1983. Design and validation of matching tests to form equal groups for tracking experiments. *Ergonomics* 26:833-845.

Potvin, A.R. and Tourtellotte, W.W. 1975. The neurological examination: Advancement in its quantification. *Arch. Phys. Med. Rehabil.* 56:425-442.

Potvin, A.R., Albers, J.W., Stribley, R.F., Tourtellotte, W.W., and Pew, R.W. 1975. A battery of tests for evaluating steadiness in clinical trials. *Med. Biol. Eng.* 13:914-921.

Potvin, A.R., Doerr, J.A., Estes, J.T., and Tourtellotte, W.W. 1977. Portable clinical tracking-task instument. *Med. Biol. Eng. Comput.* 15:391-397.

Potvin, A.R., Tourtellotte, W.W., Potvin, J.H., Kondraske, G.V., and Syndulko, K. 1985. *The Quantitative Examination of Neurologic Function*, CRC Press, Boca Raton, FL.

Poulton, E.C. 1964. Postview and preview in tracking with complex and simple inputs. *Ergonomics* 7:257-266.

Poulton, E.C. 1974. *Tracking skill and manual control*, New York, Academic Press.

Riedel, S.A., Harris, G.F., and Jizzine, H.A. 1992. An investigation of seated postural stability. *IEEE Eng. Med. Biol. Mag.* 11:42-47.

Robertson, C. and Flowers, K.A. 1990. Motor set in Parkinson's disease. *J. Neurol. Neurosurg. Psych.* 53:583-592.

Roscoe, J.T. 1975. *Fundamental research statistics for the behavioral sciences*, 2nd ed., New York, Holt, Rinehart & Winston.

Schmidt, R.A. 1982. *Motor control and learning: A behavioral emphasis*, Champaign, IL, Human Kinetics.

Sheridan, M.R., Flowers, K.A., and Hurrell, J. 1987. Programming and execution of movement in Parkinson's disease. *Brain* 110:1247-1271.

Sheridan, T.B. 1966. Three models of preview control. *IEEE Trans. Hum. Factors Electron.* 7:91-102.

Siegel, D. 1985. Information processing abilities and performance on two perceptual-motor tasks. *Percep. Mot. Skills* 60:459-466.

Sittig, A.C., Denier van der Gon, J.J., Gielen, C.C., and van Wilk, A.J. 1985. The attainment of target position during step-tracking movements despite a shift of initial position. *Exp. Brain Res.* 60:407-410.

Stelmach, G.E. and Worrington, C.J. 1988. The preparation and production of isometric force in Parkinson's disease. *Neuropsychologica* 26:93-103.

Stern, Y., Mayeux, R., Rosen, J., and Ilson, J. 1983. Perceptual motor dysfunction in Parkinson's disease: a deficit in sequential and predictive voluntary movement. *J. Neurol. Neurosurg. Psychiatry* 46:145-151.

Stern, Y., Mayeux, R., and Rosen, J. 1984. Contribution of perceptual motor dysfunction to construction and tracing disturbances in Parkinson's disease. *J. Neurol. Neurosurg. Psychiatry* 47:983-989.

Thomas, J.S., Croft, D.A., and Brooks, V.B. 1976. A manipulandum for human motor studies. *IEEE Trans. Biomed. Eng.* 23:83-84.

van den Berg, R., Mooi, B., Denier van der Gon, J.J., and Gielen, C.C.A.M. 1987. Equipment for the quantification of motor performance for clinical purposes. *Med. Biol. Eng. Comput.* 25:311-316.

Vasta, P.J. and Kondraske, G.V. 1994. Performance prediction of an upper extremity reciprocal task using non-linear causal resource analysis. *Proc. Int. Conf. IEEE Eng. Med. Biol. Soc.* 16 (CD-ROM).

Vince, M.A. 1948. The intermittency of control movements and the psychological refractory period. *Brit. J. Psychol.* 38:149-157.

Viviani, P. and Mounoud, P. 1990. Perceptuomotor compatibility in pursuit tracking of two-dimensional movements. *J. Mot. Behav.* 22:407-443.

Warabi, T., Noda, H., Yanagisawa, N., Tashiro, K., and Shindo, R. 1986. Changes in sensorimotor function associated with the degree of bradykinesia of Parkinson's disease. *Brain* 109:1209-1224.

Watson, R.W. 1994. *Advances in zero-based consistent deconvolution and Evaluation of human sensory-motor function.* Ph.D. Dissertation, University of Canterbury, Christchurch, New Zealand.

Watson, R.W. and Jones, R.D. 1998. A comparison of two-dimensional and one-dimensional tracking performance in normal subjects. *J. Mot. Behav.* 30:359-366.

Watson, R.W., Jones, R.D., Sharman, N.B. 1997. Two-dimensional tracking tasks for quantification of sensory-motor dysfunction and their application to Parkinson's disease. *Med. Biol. Eng. Comput.* 35:141-145.

Welford, A.T. 1968. *Fundamentals of skill*, London, Methuen.

Further Information

The Quantitative Examination of Neurologic Function by Potvin et al. [1985], is a two volume book which provides a superb in-depth review of instrumentation and methods for measurement of both normal and abnormal neurologic function.

An excellent overview of "Control of Postural Stability" is contained in the theme section (edited by G. Harris) of the December 1992 issue of *IEEE Engineering in Medicine and Biology Magazine.*

150

Measurement of Information-Processing Performance Capacities

George V. Kondraske
University of Texas at Arlington

Paul J. Vasta
University of Texas at Arlington

The human brain has been the subject of much scientific research. While a tremendous amount of information exists and considerable progress has been made in unlocking its many mysteries, many gaps in understanding exist. However, it is not essential to understand in full detail how a given function of the brain is meditated in order to accept that the function exists or to understand how to maximally isolate it, stress it, and characterize at least selected attributes of its performance quantitatively. In this chapter, a systems view of major functional aspects of the brain is used as a basis of discussing methods employed to measure what can the termed *central processing performance capacities.* Central processing capacities are distinguished from the information that is processed. in humans, the latter can be viewed to represent the contents of memory (e.g. facts, "programs," etc.). Clearly, both the information itself and the characteristics of the systems that process it (i.e., the various capacities discussed below) combine to realize what are commonly observed as skills, whether perceptual, motor, cognitive, or other.

Investigations of how humans process information have been performed within various fields including psychology, cognitive science, and information theory with the primary motivation being to better understand how human information processing works and the factors that influence it. On the basis that it provided a rigorous definition for the measurement of information, it can be argued that Shannon's information theory [Shannon, 1948] has been and continues to be one of the more important developments to influence both the science and engineering associated with human information processing. Several early attempts to apply it to human information processing [Fitts, 1954; Hick 1952; Hyman, 1953] have stood the test of time and have provided the basis for subsequent efforts of both researchers and practitioners. These works are central to the material presented here. In addition, the work of Wiener [1955] is also noteworthy in that it began the process of viewing human and artificial information

0-8493-0462-8/00/$0.00+$.50

processing from a common perspective. Analogies between humans and computers have proven to be very useful up to certain limits.

While there is considerable overlap and frequent interchange, the roles of science and engineering are different. In the present context, the emphasis is on aspects of the latter. It has been necessary to engineer useful measurement tools and processes without complete science to serve a wide variety of purposes that have demanded attention in both clinical and nonclinical contexts. Whether purposeful or accidental, methods of systems engineering have been incorporated and have proven useful in dealing with the complexity of human brain structure and function. While scientific controversies continue to exist, much research has contributed to the now common view of separate (in both function and location) processing subsystems that make up the whole. Various versions of a general distributed, multiprocessor model of human brain function have been popularly sketched [Gazzaniga, 1985; Minsky, 1986; Ornstein, 1986]. For example, it is widely known that the occipital lobe of the brain is responsible for visual information processing, while other areas have been found to correspond to other functions. It is not possible to do justice here to the tremendous scope of work that has been put forth and therefore to the brain itself. Nonetheless, this compartmentalistic or systems approach, which stresses major functional systems that must exist based on overwhelming empirical evidence, has proven to be useful for explaining many normal and pathologic behavioral observations. This approach is essential to the development of meaningful and practical performance measurement strategies such as those described.

150.1 Basic Principles

Many of the past efforts in which performance-related measurements have played a significant role have been directed toward basic research. Furthermore, much of this research has been aimed at uncovering the general operational frameworks of normal human inforamtion processing and not the measurement of performance capacities and their use, either alone or in combination with other capacity metrics, to characterize humans of various types (e.g., normal, aged, handicapped, etc.). However, representative models and theories provide direction for, and are themselves shaped by, subsequent measurement efforts. While there are many principles and basic observations that have some relevance, the scope of material presented herein is limited to topics that more specifically support the understanding of human information-processing performance capacity measurement.

Functional Model of Central Information Processing

A simplified, although quite robust, model that is useful within the context of human information processing is illustrated in Fig. 150.1. With this figure, attention is called to systems, their functions, and major interconnectivities. At a functional level that is relatively high within the hierarchy of the human central nervous system, the central processing system can be considered to be composed of two types of subsystems: (1) information processors and (2) memories. As can be seen from the diagram, information from the environment is provided to the information-processing subsystems through human sensor subsystems. These not only include the obvious sensors (e.g., the eyes) that receive input from external sources but also those specifically designed to provide information regarding the internal environment, including proprioception and state of being. The capacity to process information input from multiple sources at a conscious level is finite. Constant overload is prevented by limiting the amount of information simultaneously received; specific sensory inforamtion of high priority is controlled by what may be termed an *attention processor.* Information associated with a given sensor system is processed by a corresponding processor. The model further suggests that there are memory subsystems for each sensory modality in bidirectional communications with the associated processor. Thus modality-specific (e.g., visual, auditory, etc.) information may be referenced and processed with new information. As information is received and processed at modality-specific levels, it is combined at a higher level by integrative processors to generate situation-specific responses. These may be in the form of musculoskeletal, cognitive, or attention-modifying events, as well as any combination of these.

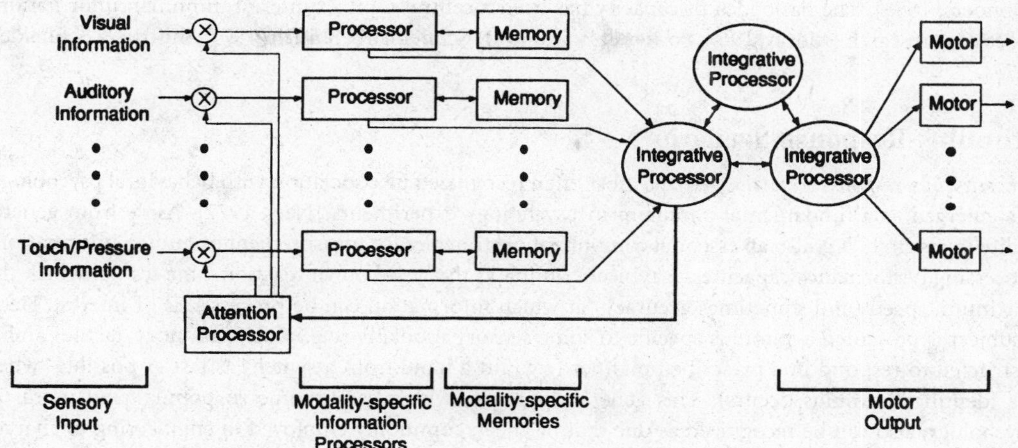

FIGURE 150.1 A functional systems-level block diagram of human information processing that facilitates description of the measurement of subsystem performance capacities.

General measurement issues of fidelity, validity, and reliability are natural concerns in measurement of information-processing performance capacities. Special issues emerge that increase the overall complexity of the problem when attempting to measure attributes that reflect *just* the information-processing subsystems. Unlike a computer, where one may remove a memory module, place it in a special test unit, and determine its capacity, the accessibility of the subsystems within the human brain is such that it is impossible to perfectly isolate and directly measure any of the individual components. When measuring characteristics of human information-processing subsystems, it is necessary to address measurement goals that are similar to those applied to analogous artificial systems within the constraint that human sensors and motoric subsystems also must be utilized in the measurement process.

Performance Capacities

In considering the performance of human information-processing systems, the resource-based perspective represented by the elemental resource model [Kondraske, 2000] is adopted here. This model for human performance encompasses all types of human subsystems and is the result of the application of a general theoretical framework for system performance to the human system and its subsystems. A central idea incorporated in this framework, universal to all types of systems, is that of *performance capacity.* This implies a finite availability of some quantity which thereby limits performance. A general two-part approach is used to identify unique performance capacities (e.g., visual information-processor speed): (1) identify the system (e.g., visual information processor) and (2) identify the dimension of performance (e.g., speed). In this framework, system performance capacities are characterized by availability of performance resources along each of the identified dimensions. These *performance resources* are to be distinguished from less rigorously defined general processing resources described by others [e.g., Kahneman, 1973, Wickens, 1984]. However, many of the important basic constructs associated with the idea of "a resource" are employed in a similar fashion in each of these contexts. For processors, key dimensions of performance are speed and accuracy. *Processor speed* and *processor accuracy* capacities are thus identified. For memory systems, key dimensions of performance are storage capacity, speed (e.g., retrieval), and accuracy (e.g., retrieval). Other important attributes of performance capacities are developed in other sections of this chapter.

Many aspects of information-processing performance have been investigated, resulting in discoveries that have provided insight into the capacities of subsystems as well as refinement to both system structure and function definitions [Lachman et al, 1979]. One of the oldest studies in which the basic concept of an information-processing performance capacity was recognized addressed "speed of mental processes"

[Donders, 1868]. The basic idea of capacity has been a central topic of interest in human information-processing research [Moray, 1967; Posner & Boies, 1971; Schneider & Shiffrin 1977; Shiffrin & Schneider, 1977].

Stimulus-Response Scenario

The *stimulus-response scenario,* perhaps most often recognized in association with behavioral psychology, has emerged as a fundamental paradigm in psychology experiments [Neel, 1977]. Aside from general utility in research, it is also an essential component of strategies for measurement of human information-processing performance capacities. A typical example is the well-known reaction-time test, in which the maximum speed (and sometimes accuracy) at which information can be processed is of interest. Here, a subject is presented a stimulus specific to some sensory modality (e.g., visual, auditory, tactile) and is instructed to respond in a prescribed manner (e.g., lift a hand from a switch "as fast as possible" when the identified stimulus occurs). This general approach, which has become so popular and useful in psychology, also can be recognized as one that has been commonly employed in engineering to characterize artificial systems (e.g., amplifiers, motors, etc.). A specific known signal (stimulus) is applied to the system's input, and the corresponding output (response) is observed. Specified, measurable attributes of the output, in combination with the known characteristics of the input, are used to infer various characteristics of the system under test. When the focus of interest is the performance limits of processing systems, these characteristics include processing speed, processing accuracy, memory storage capacity, etc.

In performance capacity tests, an important related component is the *prestimulus set,* or simply the way in which the system is "programmed to respond to the stimulus." This usually accomplished by one or more components of the instructions given to the subject under test (e.g., respond "as quickly as possible," etc.) just prior to the execution of an actual test.

Measurement of Information (Stimulus Characterization)

Within a given sensory modality, it is easy to understand that different stimuli place different demands or "loads" on information-processing systems. Thus, in order to properly interpret results of performance tests, it is necessary to describe the stimulus. While this remains a topic of ongoing research with inherent controversies, some useful working constructs are available. At issue is not simply a qualitative description, but the *measurement* of stimulus content (or complexity). Shannon's information theory [1948], which teaches how to measure the amount of information associated with a generalized information source, has been the primary tool used in these efforts. Thus a stimulus can be characterized in terms of the amount of information present in it. Simple stimuli (e.g., a light that is "on" or "off") possess less information than complex stimuli (e.g., a computer screen with menus, buttons, etc.). The best successes in attempts to quantitatively characterize stimuli have been achieved for simple, discrete stimuli (Hick, 1952; Hyman, 1953]. From Shannon, the amount of information associated with a given symbol i selected from a source with n such symbols is given by

$$I_i = \log_2 \left(1/p_i\right) \tag{150.1}$$

where p_i is the probability of occurrence of symbol i (within a finite symbol set), and the result has the units of "bits." Thus high-probability stimuli contain less information than low-probability stimuli. It is a relatively straightforward matter to control the probabilities associated with symbols that serve as stimuli in test situations. The application of basic information theory to the characterization of stimuli that are more complex (i.e., multiple components, continuous) is challenging both theoretically and practically (e.g., large symbol sets, different sets with different probability distributions that must be controlled, etc.)

If stimuli are not or cannot be characterized robustly in terms of a measure with units of "bits," operationally defined units are frequently used (e.g., "items," "chunks," "stimuli," etc.). In addition, as many stimulus attributes as possible are identified and quantified, and others are simply "described."

This at least maximizes the opportunity for obtaining repeatable measurements. However, the additional implication is that the number of "bits per item" or "bits per chunk" is waiting to be delineated and that perhaps a conversion could be substituted at a later time. While this state leaves much to be desired from a rigorous measurement perspective, it is nonetheless quite common in the evolution of measurement for many physical quantities and allows useful work to be conducted.

Speed-Accuracy Tradeoff

Fundamental to all human information-processing systems and tasks in the so-called *speed-accuracy tradeoff.* This basic limitation can be observed in relatively high-level everyday tasks such as reading, writing, typing, listening to a lecture, etc. Psychologists [see Wicklegren, 1977] have studied this tradeoff in many different context. Fitts [1966] demonstrated a relationship between measures reflecting actual performance (reaction time and errors) and incentive-based task goals (i.e., were subjects attempting to achieve high speed or high accuracy). As shown in Fig. 150.2, relationships that have been found suggest an upper limit to the combination of speed and accuracy available for information-processing tasks. In this figure, original reaction-time measures have been transformed by simple inversion to obtain true speed measures that also conform to the resource construct used in modeling performance capacities, as described earlier. The result then defines a two-dimensional (e.g., speed-accuracy) performance envelope for information-processing systems. The area within this envelope, given that each dimension represents a resource-based performance capacity, represents a higher-level or composite performance capacity metric (with units of speed × accuracy), analogous in some respects to a gain-bandwidth product or work (force × displacement) metric.

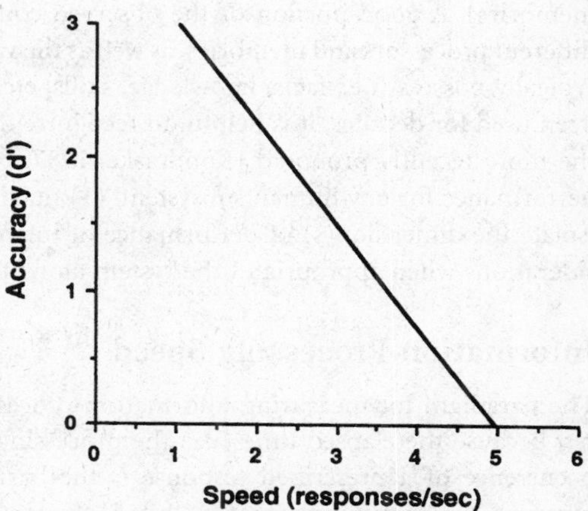

FIGURE 150.2 A typical information-processing performance envelope in two-dimensional speed-accuracy space. Data used were derived from Wicklegren [1977].

Divided Attention and Time Sharing

An approach sometimes used in a variety of measurement contexts incorporates a *dual-task or divided attention scenario* that is designed to require use of attention resources in two different simultaneously executed tasks [e.g., Wickens, 1984]. For example, visual tracking (primary task) accuracy and speed of response to an embedded visual stimulus (secondary task) can be measured. Details of the potential time-sharing possibilities at play are quite complex [Schneider & Shiffrin, 1977; Shiffrin & Schneider, 1977].

In comparison with single-task performance test situations, in which the attention processor may not be working at capacity, the additional demand is designed to change (compared with a single-task baseline) and sometimes maximizes the stress on attentional resources. Performance on both primary and secondary tasks, compared with levels attained when each task is independently performed, can provide an indirect measure of capacity associated with attention [Parasuraman & Davies, 1984]. This approach has been useful in determining relative differences in demand imposed by two different primary tasks by comparison of results from respective tests in which a fixed secondary task is used with different primary tasks. Of more direct relevance to the present context, an appropriate secondary task can be used to control in part the conditions under which a given performance capacity (defined in a standard way and measured in association with the primary task) is measured; for example, visual information-processing speed can be measured with no additional attentional load or with several different additional attentional load levels. While it may be possible to rank secondary tasks in terms of the additional load presented, there are no known methods to quantify attentional load in absolute terms.

150.2 General Measurement Paradigms

Despite the complexity of human information processing, a fairly small number of different measurement paradigms have emerged for quantification of the many unique human information-processing performance capacities. This is perhaps in part due to the limited number of system types (e.g., processors and memories). A good portion of the observed complexity can therefore be attributed to the number of different processors and memories, as well as the vastness and diversity of actual information that human's typically possess (i.e., facts, knowledge, skills, etc.). While most of the paradigms described below have been used for decades, it is helpful to recognize that they all conform to, or can be made to conform to, the more recently proposed [Kondraske, 1987, 1995] generalized strategy for measuring any aspect of performance for any human subsystem: (1) maximally isolate first the system of interest, (2) maximally isolate the dimension(s) of performance of interest, and (3) maximally stress (tempered by safety considerations when appropriate) the system along the dimension(s) (Table 150.1).

Information-Processing Speed

The paradigm for measuring information-processing speed is commonly referred to as a *reaction-time test* because the elapsed time (i.e., the processing time) between the onset of a given stimulus and the occurrence of a prescribed response is the basic measurable quantity. To obtain a processing-speed measure, the stimulus content in "bits," "chunks," or simply "stimuli" is divided by the processing time to yield measures with units of bits per second, chunks per second, or stimuli per second. Choice of stimulus modality isolates a specific sensory processor, whereas choice of a responder (e.g., index finger, upper extremity motion caused primarily by shoulder flexors, etc.) isolates a motoric processor associated with generation of the response. Much research has addressed the allocation of portions of the total processing time to various subsystem components [e.g., Sternberg, 1966]. The system of primary interest as presented here is the sensory processor. Processing times involved in the motoric processor involved in the response are substantially less in normal systems. However, it is recommended that given processor speed capacity be identified not only by the sensor subsystem stress but also by the responder (e.g., visual-shoulder flexor information processor speed). This identifies not only the test scenario employed but also the complete information path.

Information-Processing Accuracy and Speed-Accuracy Combinations

In contrast to the processor speed capacity, which describes limits on information rate, accuracy relates to the ability to resolve content. In general, paradigms for tests that include accuracy measures typically involve a finite set of symbols with a corresponding set of responses. Various stimulus presentation-response scenarios can be used. For example, a stimulus can be randomly selected from the set with the subject required to identify the stimulus presented. Alternately, a subset of stimuli can be presented and the subject asked to select the response corresponding to the symbol not in the subset (size of subsets should be small if it is desired to minimize stress on memory capacities). A key element is that, within a poststimulus response window, the subject is forced to select one of the available responses. This allows accuracy of the response to be measured in any of a number of ways ranging from relatively simple to complex [Green & Swets, 1966; Wicklegren, 1977]. The response window is either relatively open-ended, as short as possible, and variable (based on the speed of the subject's response) or fixed at a given length of time. When only accuracy is stressed, subjects are instructed to perform as accurately as possible, and the stress on speed is minimized (e.g., "take your time"). In a second type of speed-accuracy test, subjects are instructed to perform as fast and as accurately as possible, maximally stressing both dimensions, while both accuracy and speed of responses are measured. In a variation of this general paradigm, a fixed

TABLE 150.1 The Combination of the Human Sensor Subsystem (Determined by Stimulus Source) and Responder Isolates Unique Subsystems. Depending on objectives of the test task (as communicated to the subject under test) and metrics obtained, many different unique capacities can be measured.

Capacity Types	Stimulus Type	Responder*	Measurements
Attention	Visual Auditory	Upper extremity Vocal system	Length of time that task can be performed to specification and accuracy (if appropriate for task) when subject is instructed to perform for "as long as possible."
Processor speed	Visual Visual Visual Auditory Auditory Auditory Vibrotactile	Upper extremity Lower extremity — Upper extremity Lower extremity Vocal system Vocal system	Inverse of time required to react to stimulus (i.e. stimuli/second) when subject is instructed to respond "as quickly as possible."
Processor accuracy	Visual Visual Visual Auditory Auditory Auditory Vibrotactile	Upper extremity Lower extremity — Upper extremity Lowe extremity Vocal system Vocal system	Subject is asked to perform task (e.g., typically recognition of symbols from set with similar information content across symbols) "as accurately as possible" and without stress on speed of measures of accuracy. The percentage correct (out of a predetermined set size) is often used, although different accuracy measures have been proposed [Green & Swets, 1966; Wicklegren, 1977].
Processor speed-accuracy	Visual Visual Visual Auditory Auditory Auditory Vibrotactile	Upper extremity Lower extremeity — Upper extremity Lower extremity Vocal system Vocal system	Basically a combination of tests that stress speed and accuracy capacities individually; both speed and accuracy measures are obtained as defined above to get off-axis data points in two-dimensional speed accuracy space. Both speed and accuracy can be maximally stressed (e.g., by instructing subject to perform "as fast and accurately as possible"), or accuracy can be measured at different speeds by varying time available for responses and stressing accuracy within this constraint.
Memory storage capacity	Visual Visual Visual Auditory Auditory Auditory Vibrotactile	(Same options as above†	Maximum amount of information of type defined by stimulus which subject is able to recall. Stimuli usually consist of sets of symbols of varying complexity in different sets (e.g., spatially distributed lights, alphanumeric characters, words, motions, etc.) but similar amount of information per symbol in a given set. Units of "bits" are ideal but not often possible if number of bits per symbol is unknown. In such cases, units are often reported as "symbols," "chunks," or "items."

*General terms are used in table to illustrate combination options. Very specific definition that controls the motoric functional units involved as precisely as possible should be used for any given capacity test. For example, (1) upper extremity (shoulder flexor vs. elbow flexor vs. digit 2 flexor), (2) vocal system (lingua-dental "ta" vs. labial "pa" response), etc.

†Unless motor memory capacities are being tested, typically chosen to minimize stress on motoric processing and, compared with processor performance measures, not to isolate a unique memory system.

response window size is selected to challenge response speed. This provides a known point along the speed dimension of performance. In all three cases, measured response accuracy provides the second coordinate of a point in the two-dimensional speed-accuracy performance space. The combination of different paradigms can be used to determine the speed-accuracy performance envelope (see above).

Memory Capacity

As used here, *memory capacity* refers to the amount of information that can be stored and/or recalled. It is well accepted that separate memory exists for different sensory modalities. Also, short-, medium-, and long-term memory systems have been identified [Crook et al., 1986]. Thus separate capacities can be identified for each. Higher-level capacities also can be identified with processes that utilize memory, such as scanning [Sternberg, 1966]. A comprehensive assessment of memory would require tasks that challenge the various memory systems to determine resources available, singly and along different combinations of their various modalities, most directly relevant to the individual's development stage (infant, youth, adolescent, young adult, middle-aged adult, or older adult) and reflecting the diversity and depth of life experiences of that individual . No such comprehensive batteries are available. As noted by Syndulko et al. [1988], "Most typically, a variety of individual tests are utilized that evaluate selected aspects of memory systems under artificial conditions that relate best to college students." This general circumstance is in part due to the lack of generalized performance measurement strategies for memory. Despite such diversity, perhaps the most common variation of the many memory capacity tests involves providing a stimuli of the appropriate modality and requiring complete, accurate recall (i.e., a response) after some specified period of delay. The response window is selected to be relatively long so that processor speed is only minimally stressed. Typically, a test begins with a stimulus that has a low information content (i.e., one that should be within capacity limits of the lowest one expects to encounter). If the response is correct, the information content of the stimuli is increased, and another trial is administered. It is typically assumed that the amount of information added after each successful trial is fairly constant. Examples include adding another light to a spatially distributed light pattern, adding another randomly selected digit (0 to 9) to a sequence of such digits, etc. By continuing this process of progressively increasing the amount of stimulus information until an inaccurate response is obtained, the isolated memory capacity is assumed to be maximally stress. The result is simply the amount of information stored and recalled, in units of bits, chunks, or items.

Attention

Tests for basic attention capacity can be considered to be somewhat analogous to endurance capacity tests for neuromuscular systems. Simply put, the length of time over which a specified information-processing task can be executed provides a measure of attention capacity (or attention span). The test paradigm typically involves presentation at random time intervals of a randomly selected stimulus from a finite predefined symbol set for a relatively short, fixed time (e.g., 1 s). Within a predefined response window (with a maximum duration that is selected so that processing speed resources are minimally stressed), the subject must generate a response corresponding to the stimulus that occurred. Attention is maximally stressed by continuing this process until the subject either (1) makes no response within the allocated response window or (2) produces an incorrect response (i.e., a response associated with a stimulus that was not presented). These criteria thus essentially define the point at which the specified task (i.e., recognize stimuli and respond corretly within a generous allotted time) can no longer be completed. Once again, choice of stimulus modality isolates a specific sensory processor (e.g., visual, auditory, tactile, etc.). Clearly, stimulus complexity is also important. Stress on higher-level cognitive resources can be minimized by choice of simple stimuli (e.g. lights, tones, etc.), although motoric processors associated with response generation are also involved. However, these are minimally stressed, and as long as basic functionality is present, the choice of responder should have little influence on results compared with, for example, the influence occurring during processor speed capacity measurements.

150.3 Measurement Instruments and Procedures

Given the preceding functional systems model for human information processing (see Fig. 150.1) and review of measurement paradigms for information-processing subsystems, a general architecture for instruments capable of measuring information-processing performance capacities can be defined (Fig. 150.3). It is interesting to observe that this architecture parallels the human inforamtion-processing system (i.e., compare Fig. 150.1 and Fig. 150.3).

Most such instruments are computer-based, allowing for generation of stimuli with different contents, presentation of stimuli at precisely timed intervals, measurement of processing times (from which processing speeds are derived) with high accuracy (typically, to within 1 ms), and accurate recording of responses and determination of their correctness.

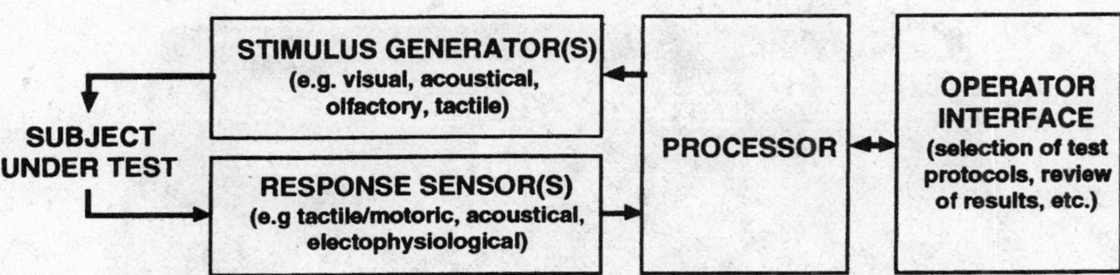

FIGURE 150.3 The general architecture for instruments used to measure human information-processing performance capacities.

A typical desktop computer system fits the general architecture presented in Fig. 150.3, and as such, it has been widely exploited "as is" physically with appropriate software to conduct information-processing tests. This is adequate for tests that specifically focus on higher-level, complex cognitive processing tasks due to the longer processing times involved. That is, screen refresh times and keyboard response times are smaller fractions of total times measured and contribute less error. However, for measurement of more basic performance capacities and to allow testing of subjects who do not possess normal physical performance capacities, custom test setups are essential. The general design goal for these stimuli generators and response sensors is to minimize the stress on human sensory, motor, and aspects of central processing resources that are not the target of the test. For example, the size of standard keyboard switches imposes demands on positioning accuracy when, for example, visual information-processing accuracy is being tested. According to Fitts' law [Fitts, 1954], the task component involved with hitting the desired response key (i.e., the component that occurs after the visual information processing component) requires longer processing times (associated with the motoric response task) for smaller targets. Thus, to minimize such influences, a relatively large target is suggested. Along the same lines of reasoning, these additional design guidelines are important: (1) key stimulus attributes that are not related to information content should be well above normal minimum human sensing thresholds (e.g., relatively bright lights that do not stress visual sensitivity, large characters or symbols that do not stress visual acuity), and (2) response speeds of stimulus generators and response sensors should be much faster than the fastest human response speed that is anticipated to be measured.

Even the many custom devices reported in the literature (as well as commercially available versions of some) have been devised primarily to support research into fundamental aspects of inforamtion processing and are used in studies that typically involve only healthy young college students. While such devices measure the stated or implied capacities well in this population, the often unstated assumption regarding other involved performance capacities (i.e., that they are available in "normal" amounts) is often severely challenged when the subject under test is a member of a population with impairments (e.g., head-injured, multiple sclerosis, Parkinson disease, etc.).

A representative instrument [Kondraske, 1990; Kondraske et al., 1984] is illustrated in Fig. 150.4. This device incorporates 8 LEDs and 15 high-speed touch sensors ("home", A1–A8, B1–B6) and a dedicated internal microprocessor to measure an array of different performance capacities associated with information processing. Measures at the most basic hierarchical level (visual inforamtion-processing speed, visual-spatial memory capacity, visual attention span, etc.) and intermediate level (coordination tasks involving different sets of more basic functional units) are included.

For example, visual-upper extremity information-processing speed is measured in a paradigm whereby the test subject first places his or her hand on the "home" plate. A random time after an audible warning cue, one out of eight LEDs (e.g., LED 4) stimulates the subject to respond "as quickly as possible" by lifting his or her hand off the home plate and touching the sensor (e.g., A4) associated with the lighted LED. Each LED represents a binary (ON of OFF) information source. Since an equiprobable selection scheme is employed, the $\log_2$(number of choices) represents the number of bits of information to be processed when determining which LED is lighted. In this example, the number of choices is eight; prior to the test, the subject was informed that any one of eight LEDs will be selected. Modes reflecting different

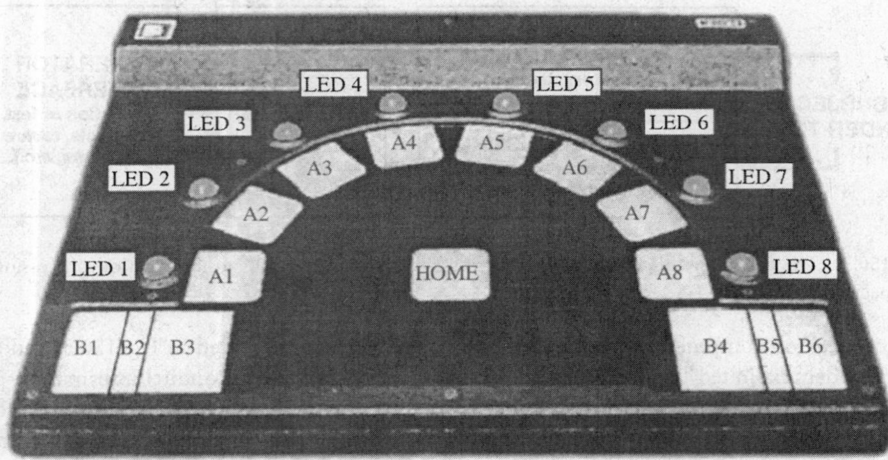

FIGURE 150.4 Major components of a specific microprocessor-based instrument for measurement of a selected subset of information-processing capacities incorporating high-intensity LEDs for visual stimuli and large-contact-area, fast-touch sensors to acquire responses (made by the upper extremity) from the subject under test. (Radius of semicircle is 15 cm.)

information loads (1, 2, 4, or 8 choices) are available. Processing speed (in stimuli per second) is obtained as a result from each trial by inverting the total processing time, measured as the time from presentation of the stimulus until the subject's hand is lifted off the home plate. By combining time measures from two or more tests with different information loads, a more direct measure of true processing speed in bits per second is obtained. For example, results from an 8-choice (3-bit load) test and a 1-choice test (0-bit load) are combined as follows: (3 bits – 0 bits)/$(t_3 - t_0)$, where t_n is the reaction time measured for an n-bit information load. This method removes transmission and other time delays not associated with the primary visual information-processing task. Multiple trials are performed (typically, 3 × number of choices, but no fewer than 5 trials for any mode) to improve test-retest reliability in the final measures, which are computed by averaging over a fraction (best 80%) of the trials. While even more trials might further improve repeatability of final capacity measures in some subjects, the overall test duration must be considered in light of the desire (from a test design perspective) to minimize stress on attention resources, the availability of which could vary widely over potential test subjects. Careful tradeoffs are necessary.

Visual-spatial memory capacity is measured using the same device. A random sequence of LEDs, beginning with a sequence length of "one," is presented to the subject. The subject responds by contracting (with the preferred hand) the subset of touch sensors (from A1 through A8) corresponding to the LEDs incorporated in a given stimulus presentation in the same order as the LEDs sequence was presented. A correct response causes a new trial to be initiated automatically with the sequence length increased by "one." The maximum length of this sequence (in items) that the subject can repeat without error is used as the measure of visual-spatial memory capacity. The best of two trials is used as a final score. The bright LEDs, relatively large contact area of the touch sensors (5 × 5cm), and a relatively long poststimulus response window (up to 10 s) ensure that other performance resources are minimally stressed.

A measure of visual attention span is obtained with a task in which random LEDs are lighted for 1 s at a random intervals ranging from 4 to 15 s. After each LED is lighted, a response window is entered (3 s maximum) during which the subject is required to indicate that a lighted LED was observed by contacting the corresponding touch sensor. The task continues until either the wrong touch sensor is contacted or no response is received within the defined response window. The time (in seconds) from the beginning of the test until an "end of test" criterion is met is used as the measure of visual attention performance capacity. The simple stimuli ensure that higher-level cognitive resources are minimally stressed. This allows a wide range of populations to be tested, including young children. Some normal subjects can achieve very long attention spans (e.g., hours) with this specific scenario, while some head-injured subjects (for example) produce times less than 60 s.

One example of a more comprehensive, computer-based memory performance test battery is that developed over the past 15 years by Crook and colleagues [1986]. This battery is directed at older adults, evaluates memory in the visual and auditory modalities at each of the three temporal process levels, and utilizes challenges that relate well to the daily activities of the target group. The challenges (i.e., stimulus sets) are computer generated and controlled, administered by means of a color graphics terminal, and in some cases, played for a videodisk player to provide a realistic a stimulus representation as possible outside of using actors in the everyday environment of the subject. The battery takes 60 to 90 minutes to administer and provides a profile of scores across the various measurement domains. Extensive age and gender norms are available to scale the scores in terms of percentiles. Despite its scope, it does not provide nor was it intended to provide, information about modalities other than visual and auditory (e.g., tactile, kinesthetic, or olfactory). It also does not address long-term memory storage beyond about 45 minutes (consider the test administration complexities in doing so). Nonetheless, it represents one of the most sophisticated memory performance measurement and assessment systems currently available.

150.4 Present Limitations

General methods and associated tools do no exist to quantify the information content of any arbitrary stimulus within a given modality (e.g., visual, auditory, etc.). Success has been achieved only for relatively simple stimuli (e.g., lights which are on or off, at fixed positions spatially, etc.) It can be said, however, that serious attempts to quantify complex stimuli even in one modality appear to be lacking. To the practitioner, this prevents the standardization that could occur across many information-processing performance capacity metrics. For example, information-processing speed capacities cannot yet be expressed universally in terms of bits per seconds. Perhaps more significant, this limitation not only impacts looking toward the human and the measurement of intrinsic performance capacities but also the analysis of information-processing tasks in quantitative terms. This, in turn, has limited the way in which information-processing performance capacity measures can be utilized. Whereas one can measure grip strength available in a subject (a neuromuscular capacity) and also the worst-case grip force required to support turning a door knob, it is not yet possible to quantify how much visual information one must process to reach out and find the door knob.

As was also noted, one cannot ideally isolate the systems under test. The required involvement of human sensory and motoric systems also requires the use of ancillary performance tests that establish at least some minimal levels of performance availabilities in these systems (e.g., visual acuity, etc.) in order to lend validity to measures of information-processing performance capacities. New imaging techniques and advanced electroencephalographic and magnetoencephalographic techniques such as brain mapping may offer the basis for new approaches that circumvent in part this limitation. While the cost and practicality of using these techniques routinely are likely to be prohibitive for the foreseeable future, the use of such techniques in research settings to optimize and establish validity of simpler methods can be anticipated.

Moray [1967] was one of many to use the analogy to data and program storage in a digital computer to facilitate understanding of a "limited capacity concept" with regard to human information processing. In the material presented, emphasis has been placed on the components analogous to those within the realm of computer hardware (i.e., processors and memories). These have been dealt with at their most basic operational levels for testing purposes, attempting to intentionally limit stresses on the test subject's acquired "skill" (i.e., stored information itself). It has been implied and is argued that knowledge of these hardware system capacities is in itself quite useful (e.g., it is helpful to know whether your desktop computer's basic information-processing rate is 100 or 400 MHz). However, many tasks that humans perform daily that involve information processing require not only processing and storage capacities but also the use of *unique information* acquired in the past (i.e., training or "programs"). These software components, which like processor and memory performance capacities also represent a necessary but not sufficient resource for completion of the specific task at hand, present even greater challenges to evaluation. If the analogy to computer programs is used (or even musical scores, plans, scripts, etc.), it can be observed that even the smallest "programming error" can lead to a failure of the total system in accomplishing its intended purpose with the intended degree of performance. Checking the integrity of human "programs" can be envisioned to be relatively easy if the desired functionality is in fact found to be present. However, if functionality is not present and diagnostic data points to a "software problem," a more difficult challenge is at hand. Methods that can be employed to reveal the source of the error in the basic information itself (i.e., the contents of memory) can be expected to be just as tedious as checking the integrity of code written for digital computers.

Defining Terms

Attention: A performance capacity representing the length of time that an information-processing system can carry out a prescribed task when maximally stressed to do so. This is analogous to endurance in neuromuscular subsystems.

Dual-task scenario: A test paradigm characteristic by a primary and secondary task and measurement of subject performance in both.

Memory: One of two major subsystem types of the human information-processing system that performs the function of storage of information for possible later retrieval and use.

Performance capacity: A quantity in finite availability that is possessed by a system or subsystem, drawn on during the execution of tasks, and limits some aspect (e.g., speed, accuracy, etc.) of a system's ability to execute tasks; or, the limit of that aspect itself.

Prestimulus set: The manner in which a human system is programmed to respond to an ensuing stimulus.

Processor: One of two major subsystem types within the human information-processing system that performs the general function of processing information. Multiple processors at different hierarchical levels can be identified (e.g., those specific to sensory modalities, integrative processors, etc.).

Processor accuracy: The aspect, quality, or dimension of performance of a processor that characterizes the ability to correctly process information.

Processor speed: The aspect, quality, or dimension of performance of an information processor that characterizes the rate at which information is processed.

Speed-accuracy tradeoff: A fundamental limit of human information-processing systems at any level of abstraction that is most likely due to a more basic limit in channel capacity; i.e., channel capacity can be used to achieve more accuracy at the expense of speed or to achieve more speed at the expense of accuracy.

References

Crook T, Salama M, Gobert J. 1986. A computerized test battery for detecting and assessing memory disorders. In Bés et al. (eds), Senile Dementia: Early Detection, pp 79–85. John Libbey Eurotext.

Donders FC. 1868–1869. Over de snelheid van psychische processen. Onderzoekingen gedaan in het Psyiologish Laboratorium der Utrechtsche Hoogescholl. In Attention and Performance II. Translated by WG Koster. Acta Psychol 30:412.

Fitts PM. 1954. The information capacity of the human motor system in controlling the amplitude of movement. J Exp Psychol 47:381.

Fitts PM. 1966. Cognitive aspects of information processing: III. Set for speed versus accuracy. J Exp Psychol 71:849.

Gazzaniga MS. 1985. The Social Brain. New York, Basic Books.

Green DM, Swets JR. 1966. Signal Detection Theory and Psychophysics. New York, Wiley.

Hick W. 1952. On the rate of gain of information. Q J Exp Psychol 4:11.

Hunt E. 1986. Experimental perspectives: Theoretical memory models. In LW Poon (ed), Handbook for Clinical Memory Assessment of Older Adults, pp 43–54. Washington, American Psychological Association.

Hyman R. 1953. Stimulus information as a determinant of reaction time. J Exp Psychol 45:423.

Kahneman D. 1973. Attention and Effort. Englewood Cliffs, NJ, Prentice-Hall.

Kondraske GV. 1987. Human performance: Science or art. In KR Foster (ed), Proceedings of the Thirteenth Northeast Bioengineering Conference, pp 44–47. New York, Institute of Electrical and Electronics Engineers.

Kondraske GV, Potvin AR, Tourtellotte WW, Syndulko K. 1984. A computer-based system for automated quantification of neurologic function. IEEE Trans Biomed Eng 31(5):401.

Kondraske GV. 1990. A PC-based performance measurement laboratory system. J Clin Eng 15(6):467.

Kondraske GV. 2000. A working model for human system-task interfaces. In JD Bronzino (ed), Handbook of Biomedical Engineering, 2nd ed., Boca Raton, Fla, CRC Press.

Lachman R, Lachman JL, Butterfield EC. 1979. Cognitive Psychology and Information Processing: An Introduction. Hillsdale, NJ, Lawrence Erlbaum.

Minsky M. 1986. The Society of Mind. New York, Simon & Schuster.

Moray N. 1967. Where is capacity limited? A survey and a model. Acta Psychol 27:84.

Neel A. 1977. Theories of Psychology: A Handbook. Cambridge, Mass, Schenkman Publishing.

Ornstein R. 1986. Multimind. Boston, Hougton Mifflin.

Parasuraman R, Davies DR (eds). 1984. Varieties of Attention. Orlando, Fla, Academic Press.

Posner MI, Boies SJ. 1971. Components of attention. Psychol Rev 78:391.

Sanders MS, McCormick EJ. 1987. Human Factors in Engineering and Design. New York, McGraw-Hill.

Schneider W, Shiffrin RM. 1977. Controlled and automatic human information processing: I. Detection, search, and attention. Psychol Rev 84:1.

Shannon CE. 1948. A mathematical theory of communication. The Bell System Technical Journal 27(3):379.

Shiffrin RM, Schneider W. 1977. Controlled and automatic human information processing: II. Perceptual learning, automatic attending, and a general theory. Psychol Rev 84:127.

Squire LR, Butters N. 1992. Neuropsychology of Memory. New York, Guilford Press.

Sternberg S. 1966. High-speed scanning in human memory. Science 153:652.

Syndulko K, Tourtellotte WW, Richter E. 1988. Toward the objective measurement of central processing resources. IEEE Eng Med Biol Soc Mag 7(1):17.

Wickens CD. 1984. Engineering Psychology and Human Performance. Columbus, Ohio, Charles E Merrill.

Wickelegren WA. 1977. Speed-accuracy tradeoff and information processing dynamics. Acta Psychol 41:67.

Wiener N. 1955. Cybernetics, or Control and Communications in the Animal and the Machine. New York, Wiley.

Further Information

A particularly informative yet concise historical review of the major perspectives in psychology leading up to much of the more recent work involving quantitative measurements of information processing and determination of information processing capacities is provided in *Theories of Psychology: A Handbook*, by A. Neel (Schenkman Publishing, Cambridge, Mass., 1977). Human information processing has long been and continues to be a major topic of interest of the U.S. military. Considerable in-depth information and experimental results are available in the form of technical reports, many of which are unclassified. *CSERIAC Gateway*, published by the Crew System Ergonomics Information Analysis Center (CSERIAC), is distributed free of charge and provides timely information on scientific, technical, and engineering knowledge. For subscription information, contact AL/CFH/CSERIAC, Bldg. 248, 2255 H St., Wright-Patterson Air Force Base, Ohio, 45433. Also, see the CSERIAC website: http://cseriac. flight. wpafb.af.ml/.

Also, the North Atlantic Treaty Organization AGAARD (Advisory Group for Aerospace Research and Development) Aerospace Medical Panel Working Group 12 has worked on standards for a human performance test battery for military environments (much of which addresses information processing) and has defined data exchange formats. Documents can be obtained through the National Technical Information Service (NTIS), 5285 Port Royal Road, Springfield, Va 22161.

Information regarding the design, construction, and evaluation of instrumentation (including quantitative attributes of stimuli, etc.) is often embedded in journal articles with titles that emphasize the major question of the study (i.e., not the instrumentation). Most appear in many journals associated with the fields of psychology, neurology, and neuroscience.

151

High-Level Task Analysis: Mental Components

Kenneth J. Maxwell
Departure Technology

As advanced information technologies have been increasingly integrated into communications, manufacturing, transportation, power, and medical systems, among others, the predominant requirements of many human-performed tasks have correspondingly shifted from manual to mental (i.e., information-processing) activities. Examples of tasks reflecting this shift include supervisory control of complex systems and operating advanced technology equipment. As a result, data, models, and metrics for describing and understanding mental tasks and the mental capabilities of persons performing them have become very important to engineering decisions making. *Task analysis* is a generic term that represents processes and methods for applying task data to improve engineering decisions. This chapter presents a process and associated methods for analyzing the mental components of tasks. It further describes the role of *mental task analysis* along with human performance and machine models (as applicable) in applied decision making and presents a quantitative, performance-based framework for performing a unified human-machine-task analysis. This chapter is not intended as a review. Selected techniques are presented to familiarize the reader with different major approaches that are currently applied to solving real-world engineering problems.

0-8493-0462-8/00/$0.00+$.50

151.1 Fundamentals

The goals of task analysis are to (1) decompose high-level tasks into their constituent, mutually exclusive, lower-level tasks; (2) describe intertask relationships and dependencies; and (3) define each task's goals, procedures, and performance and skill requirements. A major thesis of this chapter is that using task analysis for applied decisions will be improved to the degree that (1) the performance requirements are quantified and (2) they are quantified in a form that is consistent with quantitative models of human performance. A *task* is a goal-directed, procedural activity that is defined with regard to the resources (i.e., capabilities) that performing agents (i.e., humans or machines) must possess to accomplish the goal. *High-level tasks* refer to complex, composite activities that entail performing many simpler but interrelated tasks in meaningful ways. *Mental components* refer to the *perceptual and cognitive* information-processing activities and resources that are necessary or sufficient for accomplishing tasks.

Figure 151.1 depicts the mental task-analysis process in four major steps. Before any data are collected, required data should be defined on the basis of the characteristics of the system and the applied decision. Selection and use of data-collection methods, task models, and task metrics are driven by the system and applied decision.

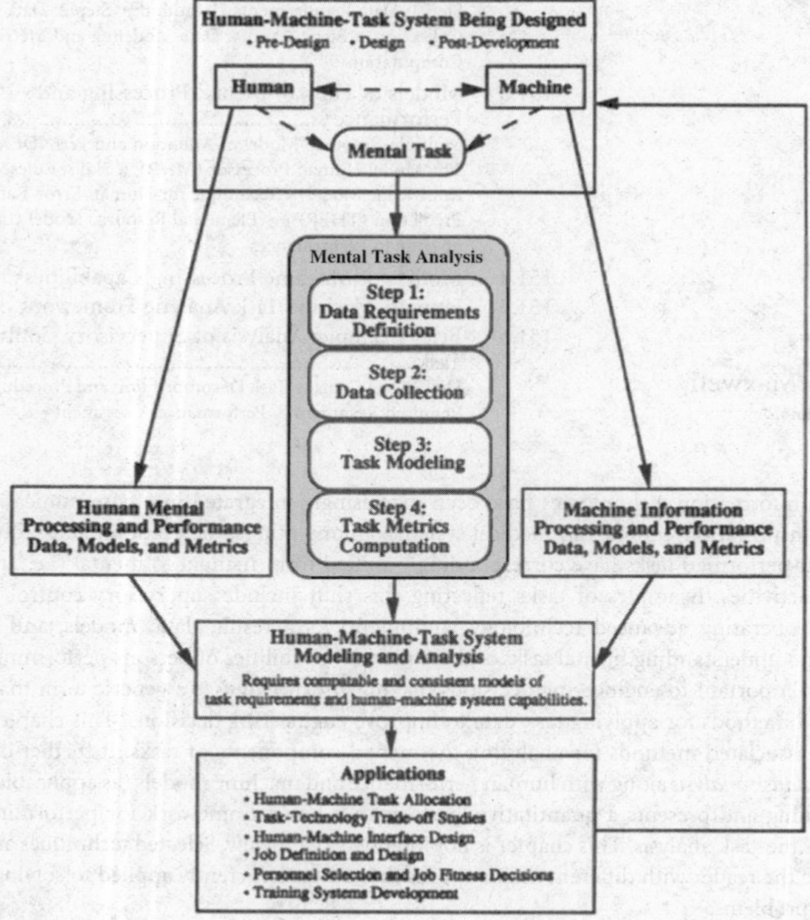

FIGURE 151.1 Mental task analysis is a four-step process that is used in combination with models of human and machine performance, as applicable, to make applied decisions.

Figure 151.1 illustrates how task analysis is used with human and machine models as a basis for solving applied problems. Task data by themselves are of limited use in solving applied problems because the task data address only one set of the total system variables that need to be considered. To be applied, task data must be analyzed with human performance and machine performance data (i.e., capacities), as applicable to the applied problem (i.e., *not all problems will involve machines.*). This combined human-machine-task system analysis requires that the analysis of each component is accomplished with consistent constructs. Further, the quality and precision of applied decisions are enhanced to the degree that these constructs provide for quantitative treatment. A framework for human-machine-task decision making (described below) combines the use of task decomposition and resource requirement modeling techniques with human performance and machine performance models.

151.2 Mental Task Analysis: Process and Methods

Step 1: Data Requirements Definition

In Step 1, the task data required for the analysis are identified and defined. There are many different dimensions and properties of tasks that can be analyzed. The functional scope of the human-machine-task system being analyzed, the point in the system design process at which the analysis is conducted, and the applied purpose of the analysis all contribute to determining what data are needed. In Fig. 151.1, the functional scope of the real-world, high-level mental task being performed by a human-machine system is represented generically. Identifying the task is necessary to bound the analysis. Task analysis can be applied to existing systems that perform well-defined tasks or to proposed systems for which tasks are in the process of being defined. At different points in the analysis, different decisions are applicable.

When task analysis is used for system design decisions, different types of task data can be collected as the fidelity of system prototypes progresses and operational system become available. These differences are apparent by examining three points in the system engineering process at which task analysis is of particular utility.

Predesign. After the functional and performance requirements for a system have been established, and analysis of major functions, together with data about technology availability and human performance, can be used to (1) define and allocate tasks, (2) develop procedures to accomplish the tasks, and (3) identify the needed human-machine interfaces. Note that the description of a task depends on the technology assumed. For example, a writing task would be accomplished differently with paper and pencil, a manual typewriter, or a computer with word processing software.

Design Phase. During the design phase, task scenarios, prototypes, and simulations can be used to detail and refine the task procedures, task requirements, allocation decisions, and human-machine interface design.

Postdevelopment. Analyzing well-defined tasks performed by an operational system provides a detailed description of task procedures and resource requirements that can be used to improve task performance and the system design.

The applied purpose of the task analysis has the greatest impact on what type of data are needed. Applications vary greatly and include design and nondesign decisions that may or may not involve machines. Applications include, but are not limited to, those described below.

Human-Machine Task Allocation. Allocation of tasks represents a major systems engineering decision. Task analysis is used to identify task load and performance requirements and contributes to assessing workload. Workload assessment provides a basis for assigning task responsibilities to humans and machines such that performance requirements can be satisfied and workload levels are not excessively low or high. These assessments augment qualitative allocation strategies (e.g., Fitts' lists) in which tasks are assigned on judgments of the relative performance capacities of humans and machines on generic tasks. Allocating tasks requires that the resource demands associated with each lower-level task are assessed and that the human and machine resources available can be matched against these at comparable levels of analysis.

Task-Technology Tradeoffs. Task analyses may be performed for systems that have yet to be developed but for which functional and performance requirements have been defined. In these cases, the analysis will be based on the defined requirements combined with data on human performance capacities and available machine technology. These analyses can be used to assess the workload, technology, and performance tradeoffs expected with different human-machine system configurations.

Human-Machine Interface Design. Mental task analysis provides not only a basis for allocating task responsibilities to humans and machines but also can be used to define, for example, the spatial aspects of display and control requirements for the interface between humans and machines.

Job Design. The roles humans satisfy in complex systems and organizations (e.g., manager, operator, maintainer) are generally designed to include many mental tasks. By identifying and describing the mental skills and resources required for task performance, task analysis provides a basis for optimizing job performance in new and existing systems.

Personnel Selection and Job Fitness. Mental task analysis yields a specification of information-processing resources (e.g., speed, accuracy) required for task performance. This specification can be used in combination with measurements of the available resources individuals possess. Assessing performance resource sufficiency provides a basis for selecting personnel and determining if injured, rehabilitated personnel are sufficiently recovered to safely return to work.

Training Systems Development. Task analysis identifies and characterizes the skills, knowledge, and mental capabilities that are necessary or sufficient for task performance. More specifically, it can be used to characterize differences in these abilities and in task performance between experts and novices. These differences serve as a basis for designing effective personnel training systems.

These applications reduce to answering the following types of questions:

- What is the best way to structure the task procedures?
- Does a specific individual have the needed mental resources to perform the task?
- What knowledge or skills does an individual need to acquire to perform the task with a given level of technology?
- Can the task be modified to accommodate the special needs of individuals?

If machines are involved:

- What human and machine combinations are most capable of performing the task?
- How should human and machine components interact in performing the task?

To answer these applied questions, the following data must be defined and described:

- The demands imposed on mental capacities by tasks (e.g., knowledge, required mental actions, speed and accuracy in performance actions).
- The mental capacities of humans in the system (e.g. knowledge, learning, processing speed, processing accuracy).

If machines are involved:

- The information-processing capacities of machine components (e.g., processing speed, algorithms, and data).
- The human-machine interaction demands imposed by the task.
- The human-machine interaction capacities of the humans in the system.
- The human-machine interaction capacities of applicable machines.

TABLE 151.1 Task Data-Acquisition Techniques

Technique	Data Collected	Requirements	Application Phase
Knowledge acquisition	Knowledge required to perform tasks	Task performers Interaction instruments, prototypes and simulations (dependent on techniques used)	All
Protocol analysis	Procedures Strategies Knowledge	Task performers Task scenarios	Design Postdevelopment
Instrumented human-Interface	Overt human action (e.g., keys pressed)	Task performers Applicable hardware and software	Design Postdevelopment
Observation	Overt machine behaviors Overt human behaviors Situation state and state changes	Task performers Prototypes or developed systems Monitoring Instruments	Design Postdevelopment

Step 2: Data Collection

Once the required task data are defined, they need to be collected. Table 151.1 lists four generic categories of techniques for acquiring data about mental tasks. The categories are not mutually exclusive. For example, all techniques use some form of observation, and instrumented human interface techniques and protocol analysis are used as knowledge-acquisition techniques. Each generic category is described in more detail below.

Knowledge-Acquisition Techniques. *Knowledge acquisition* refers to the process of eliciting from persons the knowledge they possess. This process was originally applied to elicit knowledge from experts in the development of knowledge-based systems. However, knowledge-acquisition techniques have been used more generally as a means of analyzing the cognitive components of tasks [Lehto et al., 1992]. Knowledge-acquisition methods include those listed below:

- Various interviewing techniques
- Automatic inductive methods that infer and specify knowledge
- Protocol analysis [Ericsson & Simon, 1984]
- Psychological scaling [Cooke & McDonald, 1988]
- Repertory grids from personal construct psychology theory [Boose, 1985]
- Observational techniques [Boy & Nuss, 1988]

Protocol Analysis. *Protocol analysis* refers to various techniques for collecting data in the form of verbal reports. In this methodology, persons generate thinking-aloud protocols of the unobservable (i.e., mental) activities they perform while accomplishing a task [Ericsson & Simon, 1984]. The verbal reports represent behavioral descriptions from which task strategies, requirements, procedures, and problem areas can be derived. Protocols provide data on the sequence in which tasks are performed and thus can be used to identify the strategies used by humans in accomplishing tasks that can be accomplished in multiple ways. There are several issues that need to be considered about the validity of using verbal reports as data and inferring thought processes from them. Ericsson and Simon [1984] address these concerns by demonstrating how verbal protocols are distinct from retrospective responses and classic introspection by trained observers and detailing the methods for collecting and analyzing protocols.

Instrumented Human-Interface Techniques. These techniques are used to passively and unobtrusively collect objective data on task performance and the procedures by instrumenting human-machine interfaces. These technique are used to record the timing, sequences, and frequencies of explicit actions. For a computer system, examples are keystrokes, button pushes, and mouse movements. In an aircraft, examples are steering and throttle commands, as well as any other actions controlling any onboard systems. By themselves, the data associated with actions can be used to infer workload and to study the frequency of actions, their temporal relationships, and human error. Although these data are objective, they are restricted to overt responses and, by themselves, do not provide a basis for analyzing mental components. Modeling the mental task components involves correlating these action data with data about the situations and system behavior that elicited the actions and the changes in the situations and system state caused by the actions, i.e., knowledge of context is vital.

Observation Techniques. *Observation* is a generic term referring to many techniques that can use a wide range of instruments to monitor situations and the behavior of systems and task performers. Two types of observation are distinguished. In *direct* observation, a human observer is with the task performer as he or she performs the task. In *indirect* observation, a human observer monitors the task performer remotely with audio and visual monitoring systems. Indirect observation can be done as the task is being performed or after the fact by using a recording. Observation techniques may be unobtrusive or invasive, record subjective or objective data, and may be used in controlled testing or real-world environments.

Steps 3 and 4: Data Modeling and Metric Computation

Steps 3 and 4 are the actual analysis steps. In Step 3, the task data are modeled in accordance with a selected modeling method. This task model is employed in Step 4 to compute various task metrics that are useful in solving the applied problem. Table 151.2 lists seven major approaches for analyzing the mental components of tasks. Each has been shown to be useful in applied decision making, and each is described below. For an extended review of techniques, see Linton et al. [1989].

Task-Timeline Analysis. Timeline analysis organizes and relates low-level tasks on the basis of their time of onset, duration, and concurrences in the context of performing a high-level task. Any task taxonomy or decomposition model can be used to define the lower-level tasks that will be mapped to the high-level task timeline. Generally, the task-timeline mapping is constructed by decomposing a high-level task in terms of specific scenarios because different scenarios can have very different task and time requirements.

This technique is useful for analyzing workload, allocating tasks, and determining human and system performance requirements. When the resource requirements for each lower-level task are mapped to the timeline, the result is a dynamic task-demand profile. Wickens [1984] discusses two limitations of timeline analysis. First, the technique often does not account for time-sharing capabilities in which the performance of multiple tasks can be accomplished efficiently when resource demands are low. Second, the technique is most effective when analyzing forced-pace activities. The more freedom the human has to schedule activities, the harder it is to construct a useful timeline.

TABLE 151.2 Analytic Techniques for Modeling Tasks

Technique	Modeled Dimensions/Characteristics	Analysis
Task time-line analysis	Temporal onset, duration, and concurrences	Taskload estimates Task complexity
GOMS, NGOMSL, and keystroke-level analysis	Task and system knowledge required by task performers Overt low-level actions Procedures Temporal relationships	Task complexity Knowledge requirements Performance estimates from production rule simulation
Task-action grammar	Task and system knowledge required by task performers Observable low-level actions Action sequences	Task complexity Task consistency Knowledge requirements
Fleishman and Quaintance's cognitive task taxonomy	Identifies twenty-three cognitive abilities underlying task performance. Tasks are modeled in terms of required abilities.	Cognitive ability requirements Performance requirements
Rasmussen's cognitive task analysis	Generic activities, knowledge states, and dependencies in cognitive task performance	Knowledge requirements Task complexity
Knowledge-representation techniques	Task and System knowledge required by task performers	Knowledge requirements
Elemental resource model	Mental resources needed to perform tasks Resources are defined in terms of function and performance requirements	Functional requirements Performance requirements

GOMS and NGOMSL Analysis. GOMS, an acronym for *g*oals, *o*perators, *m*ethods, and *s*election rules [Card et al., 1983, 1986], is an analytic technique for modeling (1) the knowledge about a task and a machine that a task performer must possess and (2) the operations that a task agent must execute in order to accomplish the task using the machine. It is based on work by Newell and Simon [1972] in human problem solving. The model was extended by Kieras [1988] into NGOMSL (for *n*atural GOMS *l*anguage), which affords a more detailed analysis and specification of tasks. In both GOMS and NGOMSL goals represent what a human is trying to accomplish. Operators are elementary perceptual, cognitive, or motor acts that may be observable or unobservable. Methods are sequences of operations that accomplish a goal. Selection rules are criteria used to select one method to apply when many methods are available.

The NGOMSL analysis uses this model to predict quantitative measures of the complexity of the knowledge required to perform a task with a system. These measures include learning time, amount of transfer, and task execution time. To accomplish this, methods (i.e., procedural knowledge) are represented in the form of production rules. These are IF-THEN rules that describe knowledge in approximately equal-sized units. The number of rules needed to represent a task provides a measure of the complexity of that task. Time predictions are obtained by modeling the production rules in a computer program that stimulates the tasks being analyzed.

The keystroke-level task model is an abbreviated application of the GOMS analysis [Card et al., 1983] that models only the overt actions (i.e., the observable operators and methods) taken by the task performer. In this way it does not require the inferences about mental processes required by the full GOMS approach. The model was developed with regard to a computer system and defines six operators: a keystroke or mouse button push, pointing using a mouse, moving the mouse, moving hands between the mouse and keyboard, mental preparation, and system response. Tasks are modeled by identifying the sequence of operators needed to perform the task.

Task-Action Grammar Analysis. *Task-action grammar* (TAG) [Payne & Green, 1986; Schiele & Green, 1990] is a formalism for modeling the knowledge required to perform simple tasks with a given system. Task knowledge is represented as sequences of simple acts. Similarities in the syntactic representation of simple tasks are used to derive higher-level schemas that apply across tasks possessing a family resemblance. The ability to derive schemas is used to assess the consistency of the tasks performed with a given user interface and has implications for ease of learning and overall usability. In similarity with NGOMSL, the number of grammatical rules and schemas provides a quantitative measure of task and interface complexity.

Rasmussen's Cognitive Task Analysis. Rasmussen [1986] and Rasmussen ad Goodstein [1988] define a framework for cognitive task analysis that was derived from an analysis of human-machine interaction in process-control tasks. One element of this framework is a schematic structure for describing cognitive tasks and information requirements. The schematic structure includes a sequence of (1) information-processing activities, (2) states of knowledge resulting from performing the information-processing activities, and (3) conditions (related to the experience of the human performing the task) that allow the processing sequence to be shortened. This structure constitutes a generic processing task model into which specific task data can be mapped. The information processing activities are

- Detecting a need for action
- Observing relevant data
- Identifying the present system state
- Interpreting the present system state
- Evaluating possible consequences of this state with reference to a goal
- Defining a desired target state
- Formulating procedures to achieve the desired state
- Executing the actions required by the procedures

Fleishman and Quaintance's Cognitive Task Taxomy. *Task taxonomies* classify tasks on the basis of various characteristics and properties. As such, they can serve a useful purpose in modeling tasks. Tasks of the same taxonomic classification will have similar characteristics and similar requirements. This inference provides a basis for efficiently creating models for analyzed tasks. Fleishman and Quaintance [1984] provide an extensive review of taxonomies of human performance and identify four classifications: (1) behavior descriptions, (2) behavior requirements, (3) ability requirements, and (4) task characteristics. They further detail a taxonomy of human abilities that identifies 23 cognitive factors underlying task performance. These factors were derived from an analysis of empirical performance data collected from a large number of diversified tasks and individuals. The factors constitute a structure for modeling cognitive tasks. That is, these factors can be mapped onto specific tasks and thus specify the cognitive abilities needed to perform the task.

Knowledge-Representation Techniques. *Knowledge representation* is a generic term that refers to several formalisms for modeling knowledge about tasks, systems, and the physical environment. Major representation schemes include production rules, frames, scripts, and cases. Production rules represent knowledge as a set of condition-consequence (i.e., IF-THEN) associations. Frames, scripts,, and cases are different forms of schemas that represent knowledge as typical stereotypical chunks. Rules and scripts generally represent procedural knowledge, while frames and cases represent declarative knowledge. By formally representing data acquired from knowledge-acquisition techniques, a model of the procedural and declarative knowledge required to perform a task is created.

Elemental Resource Analysis of Tasks. The *elemental resource model* (ERM) [Kondraske, 1995] defines human performance in terms of the performance and skill resources (capacities) that a human or system *possesses* and which can be brought to bear in performing tasks. Elemental resources are defined at a low-level such that many of them may be required to perform a high-level task. In addition, intermediate-level resources are defined to provide task-analysis targets at a less granular (i.e., higher hierarchical) level. In this model, a *resource* is defined as a paired construct consisting of a functional capability (e.g., visual word recognition) combined with a performance capability (e.g., recognition speed). Thus, even though the functional capability is the same, visual-word-recognition speed and visual-word-recognition accuracy are defined as two distinct resources because they model two distinct dimensions of performance. This model can be applied in task analysis by decomposing the mental components of a task in terms of the resources that a human or system performing them is *required* to process.

151.3 Models of Human Mental Processing and Performance

The analysis of task data by themselves is of limited use in solving applied problems because task data address only one set of the system variable that need to be considered. For example, a task-timeline analysis provides a model from which a measure of task demand can be computed. However, this analysis does not indicate whether the taskload is acceptable or how it could be optimally distributed among humans and machines. Making these judgments requires consideration of human and machine capabilities. Similarly, using the NGOMSL and TAG models, a measure of task complexity can be computed, indicating that one task or procedure is more complex than another. However, these judgments are based solely on the number of task steps and the amount of information needed to perform each step and do not consider how difficult each step is for a human or machine to perform. Using these task models without incorporating human and machine models forces a general assumption that task difficulty is a linear function of complexity (modeled as more task steps or information requirements). Even with this assumption, the task model by itself provides no indication of the acceptability of the task complexity or expected human or machine performance. Both these judgments require considerations of human and machine processing and performance models.

Because of these issues, task analysis is used in combination with human and machine models to make applied decisions. Using task, human, and machine models in concert requires that the mental components being modeled are described in consistent and compatible terms. Further, the quality and precision

TABLE 151.3 Models of Human Mental Processes and Performance

Model	Description	Modeled Components
Multiple resource model of attention and W/INDEX	Models attention as a collection of separable processing resources. W/INDEX is a computer model of workload that adopts the multiple resource construct and accounts for resource conflict from concurrent processing.	Identifies separable processing resources Time-sharing ability Automatic processing
Model human processor (HMP)	Models human information processing in terms of processing subsystems, memories, and performance parameters.	Performance is modeled by assigning time values to the model's parameters Perceptual, cognitive, motor processing subsystems Long-term and working memories
Keystroke-level performance model	Models the performance (i.e., time) estimated for executing each of six defined actions, from which overall estimated task completion times can be computed.	Performance is modeled by assigning time values to each of six modeled actions.
Skills, rules, and knowledge model	Models human information processing in terms of three levels of behavioral control: skill-based, rule-based, and knowledge-based.	Models the processes and requirements for each level of behavior. Can be used with quantitative models of human performance to estimate task time and errors.
Technique for human error-rate prediction (THERP)	Structure methodology for modeling human error and task completion in terms of probabilities.	Predicts human error and task completion probabilities using a human performance database and expert judgments
Elemental resource model of human and system performance (ERM)	Models human and system performance in terms of basic elements of performance and elemental resources	Performance is modeled in terms of elemental units that can be combined to provide estimates of high-level task performance and completability.

of the decision will be enhanced to the degree that the task model describes requirements quantitatively, while the human and machine models describe their capabilities quantitatively.

Five models of human mental processes and performance are listed in Table 151.3 and discussed below. These were selected because they (1) describe concepts that are generally applicable to a wide range of information-processing tasks and (2) are engineering-oriented models for use on applied problems. Connectionist (e.g., neural network) models [Rumelhart & McClelland, 1986; Schneider & Detweiller, 1987] and the optimal control model [Barron & Kleinman, 1969], which represent major but different approaches to modeling human processing, are intentionally not included due to space constraints.

Multiple-Resource Model of Attention and W/INDEX

In psychology, resource models of attention attribute the capability to perform the mental components of tasks to the capacity of applicable and available processing resources, time-sharing skills, and automatic processes. Attention is modeled as either a single processing resource or multiple, separable processing resources. In these models, the processing resource or resources afford a capacity to perform information-processing tasks (e.g., perception, cognition, and response selection) but have limited availability. The definition of resources here is different from that used in the ERM, which formally includes a specific performance dimension as part of the definition of a performance resource, which is distinguished from the system itself.

Norman and Bobrow [1975] adopted a single-resource model to examine the theoretical relationships between the application of the resource and performance on a single and multiple tasks. For a given task, performance may be limited by insufficient processing resources (resource-limited) or by insufficient data (data-limited). In multitask situations, a human performing the tasks is assumed to be capable of controlling the allocation of a portion of the processing resource to each task. Different allocation strategies result in different relative performance on each task. The relationship between allocation strategies and relative performance is plotted on a graph called the *performance operating characteristic.* Automated processing (e.g., skill-based) is assumed not to require processing resources.

The multiple-resource model of attention [Navon & Gopher, 1979; Wickens, 1980, 1984] extends and specifies the resource concept by identifying several separate processing resources that are exclusive to a subset of activities. Results of numerous dual-task studies were used to distinguish separable resources. The model identifies different resource structures for verbal and auditory modalities, spatial and verbal codes, manual and vocal responses, and between (1) selection and execution of responses and (2) perceptual and central processing stages.

W/INDEX, for *workload index* [North & Riley, 1989], is a computer model for predicting operator workload that uses the multiple-resource construct and accounts for time-sharing skills. A task-timeline combined with a design model that differentiates different interface channels (e.g., visual or auditory, manual or verbal) is used to compile an interface activity matrix. This matrix specifies, by subjective ratings, the amount of attention the task demands for each channel. A conflict matrix is generated with respect to each design channel that specifies a penalty resulting when the operator must attend simultaneously to multiple channels. The W/INDEX algorithm is then applied to these data, resulting in a workload profile.

The Model Human Processor (MHP)

MHP [Card et al., 1983, 1986] defines three interacting processing subsystems: perceptual, cognitive, and motor. The perceptual and cognitive subsystems include memories. Within each subsystem, processing parameters (e.g., perceptual processor, long-term memory, visual-image memory, and eye movements) and metrics (e.g., capacity, speed, and decay rate) are defined. Using data from numerous empirical studies, the MHP defines typical and range values for 19 parameters. To apply the model, processing parameters are associated with an analysis of the steps involved in accomplishing a task with a given system. Values for each parameter are assigned and summed, providing time estimates for task completion.

Skills, Rules, and Knowledge Model

Rasmussen [1983, 1986] developed a three-level model for describing qualitatively different modes of human information processing. *Skill-based processing* describes sensorimotor performance that is accomplished without conscious control. The behavior exhibited is smooth, automated, and integrated and only moderately based on feedback from the environment. Skill-based processes are generally simple automated behaviors, but they can be combined by higher-level conscious processes into long sequences to fit complex situations. *Rule-based processing* describes performance that is goal-directed and consciously controlled in a feed-forward manner from stored rules that have been developed from experience. Rule-based processing does not use feedback control. *Knowledge-based processing* is the highest level and applies to unfamiliar situations for which rules are available. Performance is goal-controlled and conscious. The goal and plans need to be explicitly developed, and reasoning is accomplished with knowledge (i.e., a mental model) of the system and environment.

This three-level model can be directly related to Rasmussen's framework for cognitive task analysis. Knowledge-based processing requires performing all the activities in the framework. Rule-based processing provides a means to bypass activities by applying rules that have been stored from familiarity. Skill-based processing provide a means to go directly from detection to execution. Thus the processing required is a function of the experience of the task performer and the nature of the task.

Technique for Human Error-Rate Prediction (THERP)

THERP [Swain & Guttmann, 1980] is a widely applied human reliability method [Meister, 1984] used to predict human error rates (i.e., probabilities) and the consequences of human errors. The method relies on conducting a task analysis. Estimates of the likelihood of human errors and the likelihood that errors will be undetected are assigned to tasks from available human performance databases and expert judgments. The consequences of uncorrected errors are estimated from models of the system. An event tree is used to track and assign conditional probabilities of error throughout a sequence of activities.

Elemental Resource Model (ERM) of Human Performance

ERM [Kondraske, 1995] was discussed briefly in the preceding section as a task-analysis technique. That discussion addressed how the ERM uses the same quantitative modeling construct (i.e., elemental performance resources) for task requirements and human capabilities. Major classes of resources include motor, environmental (i.e., sensing), central processing (i.e., perception and cognition), and skills (i.e., knowledge). Central processing and information (skill) resources are of interest in this chapter. When applied to humans or to machines (i.e., agents that perform tasks), the ERM describes the perceptual, cognitive, and knowledge resources available to the agent. If the agent has a deficiency in any of these elemental resources, a task requiring that resource cannot be completed by the agent. In this case, agents with sufficient resources need to be selected, agents with insufficient resources need to be trained, or the task goals or procedures need to be modified to accommodate the resources available to the agents targeted to perform the task.

151.4 Models of Machine Processing Capabilities

Table 151.4 describes models of machine and performance. These models vary greatly in scope and detail. They are presented to illustrate that quantitative analysis of performance resources using the ERM extends to machine capabilities. These models are not elaborated further.

151.5 A Human-Machine-Task Analytic Framework

Making applied decisions involves performing a combined human-machine-task analysis of available and required mental capacities. This analysis (1) requires that the mental components of tasks, the mental capabilities of humans, and the information processing capabilities of machines be modeled and analyzed in compatible and consistent terms and (2) is greatly enhanced if the terms are quantitative. From the

TABLE 151.4 Models of Machines Processes and Capabilities

Machine Model	Description	Components/Parameters
Process models	This refers to a category of models that describe the flow of information and the process performed on information through a system. It is useful for determining the consequences of failures in THERP.	Flow Transformations Dependencies Normal performance
Principles of operation models	In this category machines are modeled in terms of their operating principles. These are also useful for determining the consequences of failures needed to apply THERP.	Physical properties Mathematic formulas Mathematic relationships Logical relationships Normal range of performance
Human interaction models	In this category the characteristics of the machine side of the human-machine interface are modeled.	Physical properties Information content Information display format Performance metrics
Rasmussen's abstraction hierarchy	Developed for modeling process systems. Provides a task-independent description of systems at five levels.	Hierarchy levels that include process, operation, and interaction descriptions
ERM	Models machines in terms of the performance resources they possess. Resource definition is consistent across task, human, and machine domains.	Elemental information-processing performance resources available to the machine.

techniques and models described above for task analysis and human and machine models, a methodology for a unified, quantitative human-machine-task analysis of mental components is described.

The approach first quantifies task requirements by combining techniques that produce detailed task-decomposition analyses of goals, actions, and required procedures (e.g., GOMS, NGOMSL, TAG, keystroke-level task model, or Rasmussen's cognitive framework) with timeline analysis, where applicable, and the elementary resource analysis of tasks. The task-decomposition models are used to specify tasks at a level at which the elemental mental resources required to perform the tasks can be specified. The timeline provides a basis for specifying performance requirements. Other performance dimensions also should be used depending on the task.

The next step employs quantitative models of human performance and technology to reach an applied decision that best satisfies the required resources if sufficient resources are available. If sufficient resources are not available, additional performance and skill resources need to be obtained to complete the task, or the task requirements need to be reduced. Quantitative models of human performance include MHP, THERP, and ERM. Using the MHP, the human processing parameters needed for each task operation are identified. The time values assigned to each parameter are used to compute task time estimates. This approach is limited to estimating performance in terms of time. THERP maps operations to the tasks included in human reliability databases to estimate human error probabilities. The ERM provides a framework for specifying performance and functional capacities at the resource level. It is the only model that (1) incorporates all required dimensions of performance and skills and (2) uses consistent modeling constructs across tasks, humans, and machines.

Using the approach outlined above establishes a basis for making applied decisions objectively. The task-analysis portion of this approach is illustrated by the example below.

151.6 Brief Example: Analysis of Supervisory Control Task

Supervisory control refers to the role a human plays in operating a semiautomatic process or system. Examples include control of large systems such as a nuclear power plant and specific instrumentation such as a robotic or assistive device. Performing supervisory control is high-level task that predominantly consists of mental components. This task is used to generically illustrate the use of analytic techniques to model a task.

High-Level Goals

Two major modes of operation are distinguished [Wickens, 1984]. In one mode (normal), the system is behaving as expected. The primary role of the human controller is to monitor system activity and modify performance parameters in accordance with planned output goals. In the second mode (fault), the system is faulty. The goal of the human controller, in this case, is to detect and diagnose the fault and intervene to bring the process back into normal performance.

Task Decomposition and Procedures

Sheridan [1987] identifies 10 functions performed by a supervisory controller within these modes of operation. Rasmussen's cognitive task framework, which was developed with regard to process-control tasks, includes 8 functions applicable to the fault mode of operation. Figure 151.2 depicts these 8 generic functions as a top-level breakdown of the supervisory control task in fault mode. For this example, only selected tasks are decomposed to lower levels. At each level, the decomposition is not intended to be complete. Selected tasks at each level are intended to be illustrative and representative of increasingly specific and simple tasks. The arrow within each level provide procedural information. The goals of the lower-level tasks are defined by their titles (e.g., acquire data). Only the situation interpretation task is decomposed. In reference to Rasmussen's human information-processing model, the task is assumed to be knowledge-based. The lower-level tasks represent a hypothesis-testing strategy for knowledge-based

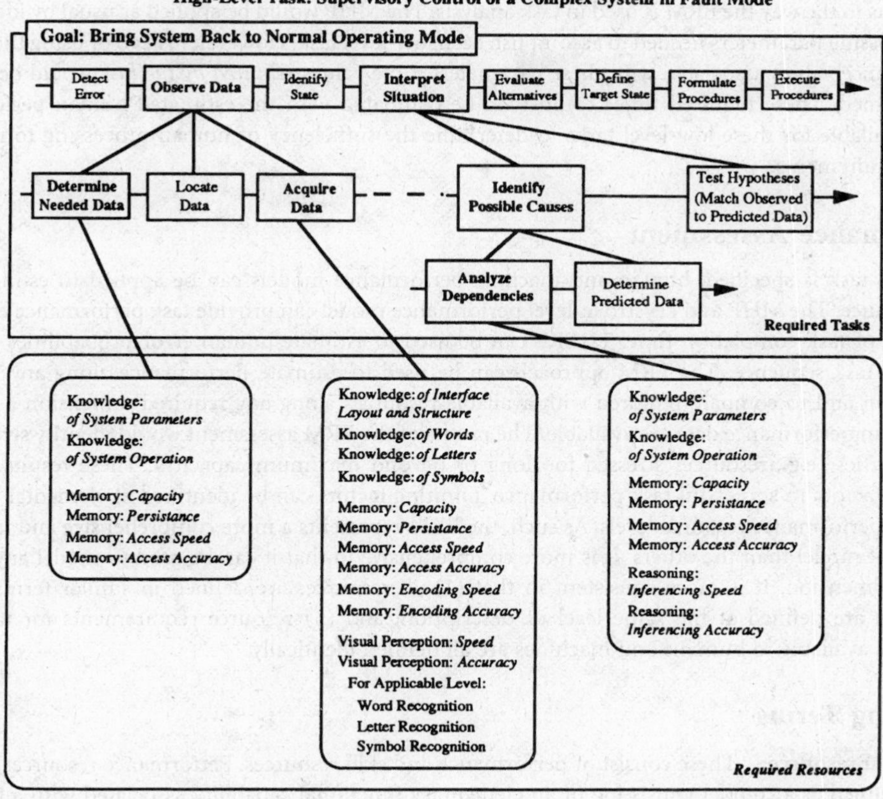

FIGURE 151.2 Selected task decomposition and resource requirements for a high-level supervisory control task.

diagnosis described by Rasmussen and Goodstein [1988]. However, other models of lower-level tasks could be used. Note that although a Rasmussen model is illustrated, the decomposition could be accomplished with other applicable models (e.g., GOMS).

Required Resources

Of the task-modeling techniques presented above, only Fleishman and Quaintance's cognitive abilities taxonomy and the ERM provide for explicit analysis of required resources. The approach defined by the ERM is illustrated in this example because it defines similar analyses for human and machine performance. As such, it is the only technique that models all three components of a human-machine-task system in the same way. For three of the lowest-level tasks, a set of elemental central processing resources is identified. Humans or machines responsible for performing each task are required to possess these resources to accomplish the task successfully. Four types of elemental mental resources are identified: knowledge, memory, perception, and reasoning. Resources are distinguished by their functional type and the performance dimension they specify. In the example, the performance dimensions are generically identified. In application, values should be assigned to these dimensions to the degree they are available or obtainable. Given time and cost constraints for the analysis, the ability to quantify all resources may not be practical. Task priorities can be used to selectively quantify resources, or the resource model can be used less rigorously as a guide to decision making if specific data are not available [Maxwell, 1995]. Once values are assigned, these required resources can be analyzed in concert with sets of available resources possessed by humans and machines. Available resources will have values that indicate a performance capability.

Although the MHP is a human processing and performance model, it can be adapted for use in a way analogous to the way the ERM is used in task analysis. The MHP would be applied as usual by identifying the processing parameters needed to accomplish each low-level task. However, instead of using the human performance values associated with these parameters, time values *required by the task* would be derived and assigned. These required times could then be compared with the estimated human performance times available for these low-level tasks to determine the sufficiency of human processing to meet the task's requirements.

Performance Assessment

After the task is specified, human and machine performance models can be applied to estimate task performance. The MHP and keystroke-level performance model can provide task performance estimates in terms of task completion time. THERP can be used to estimate human error probabilities for each task and task sequence. The ERM approach can be used to estimate performance along any required dimension and to compare required with available resources along any required dimension as long as the human performance data are available. The results of the ERM assessment would identify stress levels on capacities (e.g., resources stressed too long or beyond maximum capacity). These results indicate limiting factors to successful task performance. Limiting factors can be identified at elemental or intermediate performance resource levels. As such, the ERM represents a more comprehensive and internally consistent model than the others. It is more comprehensive in that it can be used to model any performance dimension. It is more consistent in that (1) all resources are defined in similar terms, (2) all resources are defined at the same level of description, and (3) resource requirements for tasks and resources available to humans and machines are all defined identically.

Defining Terms

Elemental resource: These consist of performance and skill resources. Performance resources are each defined as a couplet consisting of an elementary functional capability associated with a low-level undimensional quantitative capacity, either available or required, to perform the function. Skill resources are defined in terms of the knowledge, either available or required, to perform an elemental function.

Knowledge acquisition: Any of several techniques for eliciting knowledge about a task or system from a human.

Mental components: The perceptual, cognitive, and knowledge-processing components of a task.

Mental tasks: Tasks that consist predominantly of mental components and, in ERM terms, require mental performance and skill resources to perform.

Protocol analysis: Protocols are verbal reports generated contemporaneously by a person accomplishing a task that describe the unobservable (mental) processes being performed.

Task: A goal-directed, procedural activity that is defined with regard to the resources (i.e., capabilities) that performing agents (i.e., humans or machines) must possess to accomplish the goal.

Task analysis: The process of (1) decomposing high-level tasks into their constituent, mutually exclusive, lower-level tasks; (2) describing intertask relationships and dependencies; and (3) defining each task's goals, procedures, and performance and skill requirements.

Task taxonomy: A classification of tasks in accordance with a defined method, strategy, or set of criteria.

Task timeline: A time-based profile of a high-level task that maps its lower-level tasks by time-of-onset, duration, and concurrences onto a timeline.

References

Bailey RW. 1989. Human Performance Engineering, 2d ed. Englewood Cliffs, NJ, Prentice-Hall.

Barron S, Kleinman DL. 1969. The human as an optimal controller and information processor. IEEE Trans Man-Machine Syst MMS-10:9.

Boose JH. 1985. A knowledge acquisition program for expert systems based on personal construct psychology. Int J Man-Machine Stud 23:495.

Boy G, Nuss N. 1988. Knowledge acquisition by observation: Application to intelligent tutoring systems. In Proceedings of the Second European Knowledge Acquisition Workshop (EKAW-88), pp 11.1–14.

Card SK, Moran TP, Newell A. 1983. The Psychology of Human-Computer Interaction. Hillsdale, NJ, Erlbaum.

Card SK, Moran TP, Newell A. 1986. The model human processor. In KR Boff, L Kaufman, JP Thomas (eds), Handbook of Perception and Human Performance, vol. II, pp 45-1–45-35.

Cooke NM, McDonald JE. 1988. The application of psychological scaling techniques to knowledge elicitation for knowledge-based systems. In JH Boose, BR Gaines (eds), Knowledge-Based Systems, vol 1: Knowledge Acquisition for Knowledge-Based Systems, pp 65–82. New York, Academic Press.

Ericsson KA, Simon HA. 1984. Protocol Analysis. Cambridge, Mass, MIT Press.

Fleishman EA, Quaintance MK. 1984. Taxonomies of Human Performance: The Description of Human Tasks. Orlando, Fla, Academic Press.

Kieras DE. 1988. Towards a practical GOMS model methodology for user interface design. In M Helander (ed), Handbook of Human-Computer Interaction, pp 135–157. New York, Elsevier Science Publishers.

Kondraske GV. 1995. A working model of human-system-task interfaces. In JD Bronzino, (ed), The Biomedical Engineering Handbook. Boca Raton, Fla, CRC Press.

Lehto MR, Boose J, Sharit J, Salvendy G. 1992 Knowledge acquisition. In G Salvendy (ed), Handbook of Industrial Engineering, 2d ed. New York, Wiley.

Linton PM, Plamondon BD, Dick AO, et al. 1989. Operator workload for military system acquisition. In GR McMillan et al. (eds), Applications of Human Performance Models to System Design, pp 21–46. New York, Plenum Press.

Maxwell KM. 1995. Human-computer interface design issues. In JD Bronzino (ed), The Biomedical Engineering Handbook pp 2263–2277. Boca Raton, Fla, CRC Press.

Meister D. 1984. Human reliability. In FA Muckler (ed), Human Factors Review, pp 13–54. Santa Monica, Calif, Human Factors and Ergonomics Society.

Navon D, Gopher D, 1979. On the economy of the human processing system. Psychol Rev 86.

Newell A, Simon H. 1972. Human Problem Solving. Englewood Cliffs, NJ, Prentice-Hall.

Norman D, Bobrow D. 1975. On data-limited and resource-limited processing. J Cogn Psychol 7:44.

North RA, Riley VA. 1989. W/INDEX: A predictive model of operator workload. In GR McMillan et al. (eds), Applications of Human Performance Models to System Design, pp 81–90. New York, Plenum Press.

Payne SJ, Green T. 1986. Task-action grammars: A model of mental representation of task languages. Human-Comput Interact 2:93.

Rasmussen J. 1983. Skills, rules, and knowledge: Signals, signs, and symbols and other distinctions in human performance models. IEEE Trans Syst Man Cybernet SMC-13:257.

Rasmussen J. 1986. Information Processing and Human-Machine Interaction. New York, North-Holland.

Rasmussen J, Goodstein LP. 1988. Information technology and work. In M Helander (ed), Handbook of Human-Computer Interaction, pp 175–201, New York, Elsevier Science Publishers.

Rumelhart DE, McClelland JL. 1986. Parallel Distributed Processing, Explorations in the Microstructure of Cognition, vol I: Foundations. Cambridge, Mass, MIT Press.

Schiele F, Green T. 1990. HCI formalisms and cognitive psychology: The case of task-action grammar. In M Harrison, H Thimbleby (eds), Formal Methods in Human-Computer Interaction, pp 9–62. Cambridge, UK, Cambridge University Press.

Schneider W, Detweiller M. 1987. A connectionist/control architecture for working memory. In GH Bower (ed), The Psychology Learning and Motivation, vol 21, pp 53–119. Orlando, Fla, Academic Press.

Sheridan TB. 1987. Supervisory control. In G Salvendy (ed), Handbook of Human Factors, New York, Wiley.

Swain AD, Guttmann HE. 1980. Handbook of Human Reliability Analysis with Emphasis on Nuclear Power Plant Applications. Report no. NUREG/CR-1278, Nuclear Regulatory Commission, Washington.

Wickens CD. 1980. The structure of attentional resources. In R Nickerson, R Pew (eds), Attention and Performance VIII. Hillsdale, NJ, Erlbaum.

Wickens CD. 1984. Engineering Psychology and Human Performance. Columbus, Ohio, Charles E Merrill.

Further Information

Applications of Human Performance Models to System Design, edited by G.R. McMillan, D. Beevis, E. Salas, M.H. Strub, R. Sutton, and L. Van Breda, includes several chapters that describe task analysis and its relationship to human performance and workload analysis in the context of system design. *Behavioral Analysis and Measurement Methods*, by David Meister, provides a review and practical guide to applying task-analytic methods. The *Handbook of Perception and Human Performance*, Vols. I and II, edited by K.R. Boff, L. Kaufman, and J. P. Thomas, provides broad range of human factors and ergonomic data related to perceptual and cognitive task performance. The *Handbook of Human Factors*, edited by G. Salvendy, reviews task analytic methods, workload and resource models of mental processing, and human performance models in detail. The *Handbook of Industrial Engineering*, edited by G. Salvendy, reviews task analytic techniques and human performance models in the context of industrial design applications. The journal *Human Factors*, published by the Human Factors and Ergonomics Society, is a source for new task analytic techniques and application studies. *Cognitive Psychology and Information Processing*, by R. Lachman, J. L. Lachman, and E. C. Butterfield, presents an extensive discussion of the information-processing approach as applied to research in modern experimental and cognitive psychology.

152

Task Analysis and Decomposition: Physical Components

Sheik N. Imrhan
University of Texas at Arlington

A *task* can be viewed as a sequence of actions performed to accomplish one or more desired objectives. *Task analysis* and *decomposition* involve breaking down a task into identifiable elements or steps and analyzing them to determine the resources (human, equipment, and environmental) necessary for the accomplishment of the task. As indicated in earlier chapters, all human tasks require the interaction of mental and physical resources, but it is often convenient, for analytical purposes, to make a distinction between tasks that require predominantly physical resources of the person performing the tasks and tasks that require predominantly mental resources. These different types of tasks are often analyzed separately. Usually, in physical tasks, the mental requirements are described but are not analyzed as meticulously as the physical requirements. For example, a heavy-lifting task in industry requires musculoskeletal strength and endurance, decision making, and other resources, but the successful completion of such a task is limited by musculoskeletal strength. Decision-making and cognitive resources may be lightly tapped, and so the heavy-lifting task analysis will tend to focus only on those factors which modify the expression of musculoskeletal strength while keeping the other kinds of requirements at a descriptive level. The descriptions and analyses of manual materials handling tasks in Chaffin and Andersson [1991] and Ayoub and Mital [1991] exemplify this kind of focus. The degree with which a resource may be stressed depends on a number factors, among which are the resource availabilities (capacities) of the person performing the task compared with the task demands.

0-8493-0462-8/00/$0.00+$.50

This chapter deals with the analysis and decomposition of only the physical aspects of tasks. It is assumed that nonphysical resources are either available in nonlimiting quantities or are not crucial to task accomplishment. While the compartmentalization of task characteristics tends to blur the natural connections among the various human performance resources, it is still the most pragmatic method available for task analysis.

Physical task analysis first achieved scientific respectability in the early part of this century from the work of the industrial engineer Frederick Winslow Taylor [Taylor, 1911] and, shortly afterward, by Frank and Lillian Gilbreth [Gilbreth & Gilbreth, 1917]. Taylor and the Gilbreths showed how a task can be broken down into a number of identifiable, discrete steps that can be characterized by type of physical motions, energy expenditure, and time required to accomplish the task. By this method, they argued that many task steps, normally taken for granted, may contribute little or nothing to the accomplishment of the task and can therefore be eliminated. As a result of these kinds of analyses, Taylor and the Gilbreths were able to enhance productivity of individual workers on a scale that was considered unrealistic at that time. The basic approach to the highest level of physical task analysis today has evolved from the motion and time studies of Taylor and the Gilbreths (known as *taylorism*), both in the industrial and nonindustrial environments. The exact methods have become more sophisticated and refined, incorporating new technologies and knowledge accumulated about human-task interaction.

From its roots in time and motion studies, task analysis has become a very complex exercise. No longer is a single task considered in isolation, as shoveling was by Taylor and bricklaying by the Gilbreths. Today, different ideas in management and advanced data analytical methods have established the need for cohesion between physical tasks and (1) the more global processes, such as jobs and occupations encompassing them, and (2) the more detailed processes contained within them (the different types of subtasks and basic elements of performance). Campion and Medsker [1992] give examples of the first, and Kondraske [1995] gives examples of the second. This chapter describes physical task analysis, showing the approach for proceeding from higher-level tasks to lower-level ones. Figure 152.1 shows a summary of the overall approach.

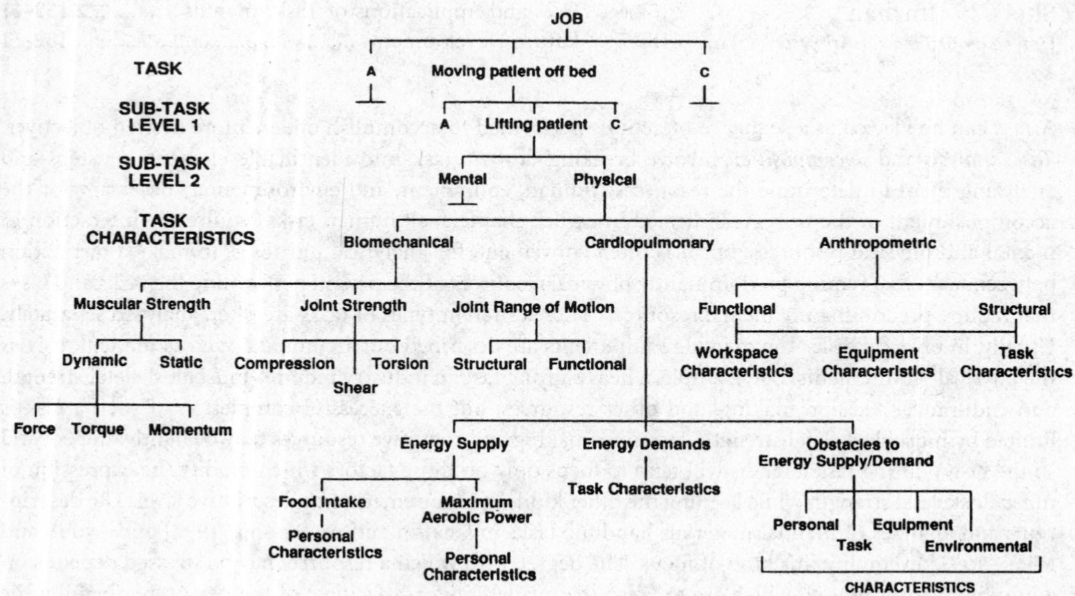

FIGURE 152.1 Diagram showing the concept of a hierarchical task analysis for physical tasks.

152.1 Fundamental Principles

The fundamental principle of physical task analysis is the establishment of the work relationships between the person(s) performing the task and the elements of the task under specified environmental and social conditions. The main driving forces are performance enhancement, protection from injuries and illnesses, and decision making. The order of importance depends on the particular context in which the task analysis is performed. Performance enhancement is a concern in all disciplines that deal with human performance, though the approaches and objectives may differ. In physical rehabilitation, for example, performance enhancement focuses on improving the performance of basic physical functions of the human body, such as handgrip, elbow flexion, etc., that have deteriorated from injury or illness. In athletics, it focuses on improving performance to the highest possible levels, such as the highest jump or the fastest 100-m run. In exercise and fitness, it aims at improving body functions to enhance health and quality of life. Finally, in the discipline of ergonomics, it aims at enhancing task performance as well as the health and safety of people at work.

Speed-Accuracy Principle

Performance enhancement and protection from injuries and illnesses are not necessarily mutually exclusive. In performance enhancement, the aim is to complete a task as accurately and quickly as possible. Both accuracy and speed are important in the work environment, especially in manufacturing, and are influenced by other factors. When lack of accuracy or precision leads to dire consequences, as in air traffic control jobs, then a decision to sacrifice speed is usually made. The basic concept of Fitt's law [1954] is relevant to these situations. It is difficult to find examples of cases where gross inaccuracy is tolerable, but in many cases more speed is required and less accuracy is tolerable, and a different point on the speed-accuracy continuum is selected. In the manufacturing environment, speed translates into productivity and accuracy into quality, and both determine profitability of the production enterprise. Speed and accuracy are rarely attainable at the highest levels simultaneously. In practice, a certain amount of one may be sacrificed for the other. Thus industrial engineers who set these tolerable limits talk of "optimum" speed instead of "fastest" speed and "optimum" quality instead of "perfect" quality. These facts must be borne in mind when analyzing physical tasks. A task analysis indicating that a person can increase his or her speed of performance should not necessarily imply that the task is being done inefficiently, because quality of performance may deteriorate with increased speed. This kind of flawed reasoning can occur when a variable is dealt with in isolation, without regard to its interaction with other variables.

Stress-Strain Concept

In task analysis, it is common practice to distinguish between *stress* and *strain*. However, these terms are sometimes used interchangeably and confusedly. In this chapter, *stress* refers to a condition that may lead to an adverse effect on the body, whereas *strain* refers to the effect of stress on the body. For example, working at a computer job in dim lighting often leads to headaches. The dim lighting is considered the stress, and the headache is the strain. The term *stressor* is also widely used as a synonym for stress. Strain has often been wrongly called stress. These terms must be clearly defined to determine which factors are causative ones (stresses) and which are consequences (strains). Stress is determined by task demands, while strain is determined by the amount of physical resources expended beyond some tolerable level, defined by the person's resource capacities. The stress-strain principle is pivotal to task analysis when one is concerned with errors, cumulative traumas, and injuries in the workplace and is applicable to task situations where task demands are likely to exceed human resource capacities.

Ayoub et al. [1987] developed a quantitative stress-strain index, called the *job severity index* (JSI), from empirical task analysis and epidemiologic data for manual materials handling tasks. This index computes the ratio of "job task physical demands to person physical capacities" from several interacting variables.

The application of the JSI is also detailed in Ayoub and Mital [1991]. Also, Kondraske [1995] defines a quantitative measure of stress that can be applied to individual performance resources. In this measure, defined as the "ratio of resource utilization to resource availability," an adverse effect may be noted when the stress level exceeds a threshold, which may be different for different types of performance resources (e.g., strength, range of motion, etc.).

Cumulative Strains

The elimination of motions and other physical actions that do not contribute directly to task performance (tayloristic practice) has led to a concentration of work at specific localities in the body and to consequent overwork at those localities, but this concentration often leads to rapid consumption of the limited resources of the working systems. For example, VDT data entry is a highly repetitive task with very little variation; work is concentrated heavily on the hands (rapid finger motions for operating the keyboard) and in the neck and trunk (prolonged static muscular contraction), and this often leads to cumulative trauma disorders of the hands, wrists, neck, and shoulders. By comparison, traditional typists performed a variety of tasks in addition to using the keyboard—changing paper on the typewriter, filing documents, using the phone, etc.—and this helped to stem the depletion of physiologic resources caused by rapid finger work or sustained static muscular contractions in the neck and shoulder. Also, it is now recognized that some functions of the body that have not been considered "productive" under taylorism are highly stressed and may even be the limiting factors in task performance. In data processing, the muscles of the neck and back maintain prolonged static contractions that often result in muscular strains in the neck and back and limit work time. Traditional task analysis does not always consider this static muscular work as productive, even though it is absolutely necessary for steadying the arms (for keyboard and mouse use) and the head (for viewing the screen and documents). Likewise, work pauses, which have been considered wasteful, are now valued for their recuperative effects on the physical (and mental) working systems of the body. The increasing awareness of the causes of work-related cumulative trauma disorders will continue to influence the interpretation of data derived from task analyses.

152.2 Early Task-Analysis Methods

The history of physical task analysis does not show a trend toward the development of a single generalized model. The direction of physical task analysis has been influenced by taylorism and by specific efforts in the military that were later adapted to the industrial and other nonmilitary environments. A list of influential methods, to which readers are referred, is given below:

1. Time studies of Fredrick Winslow Taylor [Taylor, 1911]
2. Time an motion studies of Frank and Lillian Gilbreth [Gilbreth & Gilbreth, 1917]
3. Job analysis developed by the U.S. Department of Labor in the 1930s [Drury et al., 1987]
4. The U.S. Air Force "method for man-machine task analysis" developed by R. Miller [1953]
5. Singleton's methods [Singleton, 1974]
6. Hierarchical task analysis (HTA) developed by Annette and Duncan [1974]
7. Checklist of worker activities developed by E.J. McCormick and others for the U.S. Office of Naval Research [Drury et al., 1987]
8. Position analysis questionnaire [McCormick et al., 1969]
9. The AET method (*Arbeitswissenchaftliches Erhenbungsverfahen zur Tatigkeitsanalyse*, or ergonomic job analysis) developed by Rohmert and Landau [Drury et al., 1987]

152.3 Methods of Physical Task Analysis

Methods of physical task analysis vary widely. Drury and colleagues [1987] point out that "the variation of task analysis format is the result of different requirements placed on the task analysis." While this is a pragmatic method for short-term results, a more general approach for long-term solutions is needed. Such an approach should aim to proceed from higher- to lower-level tasks, as discussed in other chapters [Kondraske, 1995]. The following general sequence of analysis is recommended:

1. Description of the system within which the tasks reside
2. Identification of jobs (if relevant) within the system
3. Identification of the tasks
4. Identification of subtasks that can stand independently enough to be analyzed as separate entities
5. Determination of the specific starting point of each task
6. Determination of the specific stopping point of each task
7. Characterization of the task as discrete, continuous, or branching [Drury et al., 1987]
8. Determination of task resources (human, equipment, and environmental) required to perform the task (this step is the essence of *task description*)
9. Determination of other support systems (e.g., in work situations, managerial and supervisory help may be needed)
10. Determination of possible areas of person-task conflicts
11. Determination of possible consequences of conflict, e.g. errors, slowing down of work rate, deterioration of quality of end product, decline in comfort, harm to health, and decline in safety

General Methods for Background Information

Many different techniques for gathering information in physical task analysis have evolved over the years, especially in the industrial and biomedical engineering environments. The quality and quantity of information that the techniques provide are limited by available equipment, and it seems that the quality and quantity of available equipment lag far behind the state of the art in technology. General data-gathering techniques, which are somewhat self-descriptive, include

- Direct observation (predominantly visual)
- Indirect observation (e.g., replay of videotape)
- Document review
- Questionnaire survey
- Personal interview

Specific Techniques

Specialized data-gathering techniques [Chaffin & Andersson, 1991; Niebel, 1993; Winter 1990] that have evolved into powerful tools in the industrial environment and which can be used for task analysis in other situations include

- Methods analysis—determines the overall relationships among various operations, the work-force, and equipment
- Operations analysis—determines the details of specific operations
- Worker-equipment relationship study—determines the interaction between people and their equipment (cases of person-machine mismatches are determined here)
- Motion study—details the motions made by the various segments of the body while performing the task
- Times study—details the time taken to perform each motion element
- Micromotion study—detailed description of motion using mainly videotapes
- Kinematic analysis—measures or estimates various motion parameters for specified body segments, e.g. displacement, velocity, acceleration. (It involves micromotion studies. The data are used for subsequent biomechanical analyses, e.g., to determine mechanical strain in the musculoskeletal system [Chaffin & Andersson, 1991]

- Kinetic analysis—measures or estimates forces produced by body segments and other biomechanical parameters (e.g., center of mass, moment of inertia, etc.) and physical parameters of objects (e.g., mass, dimension, etc.) (the data are used for subsequent biomechanical analyses)

Decomposition of Physical Tasks

Physical tasks can be viewed as exchanges of energy within a system consisting of the human, the task, the equipment, and the environment. In task analysis and decomposition, we look at the human on one side and the other system components on the other side and then determine the nature of the exchange. The energy is transmitted by muscular action. Tasks are hierarchical, however, and for systematic decomposition, it is useful to identify the various levels and proceed in a top-down sequence (see Fig. 152.1). The items in the various levels below are self-explanatory, but there is considerable content within each. Detailed descriptions of each is beyond the scope of this chapter, and readers should consult the references [Chaffin & Andersson, 1994; Corlett et al., 1979; Meister, 1985; Niebel, 1993; Singleton, 1974; Winter, 1990] for standard measurement procedures and analyses.

Level I. Identifying general types of tasks according to levels of energy requirements:

- Tasks requiring great muscular forces
- Tasks requiring medium to low levels of muscular forces

Tasks requiring very low levels of muscular forces are not predominantly physical. Their successful accomplishment is strongly dependent on cognitive resources. The application of strong and weak muscular forces is not necessarily mutually exclusive. Such forces my be exerted simultaneously by different muscular systems or may even alternate within the same system. For example, a typing task may require strong back extensor forces while sitting bent forward (if the workstation is poorly designed) and weak finger flexion forces for activating the keyboard keys, and a weak index finger flexion may be required to activate a trigger on a hand drill while strong handgrip forces stabilize the drill.

Level II. Identifying the crucial subsystems involved in the generation and transference of forces of varying intensities involved in tasks:

- Musculoskeletal—for great forces
- Cardiopulmonary—for medium to low forces
- General body posture
- Body motion
- Local segmental motion
- Local segmental configuration
- Anthropometry
- Workplace and equipment dimensions

The last three subsystems modify directly the expression of muscular force.

Level III. Identifying environmental requirements and constraints. Environmental conditions influence performance and functional capacities in humans. They should therefore be measured or estimated in task analyses. Table 152.1 describes common environmental constraints.

Level IV. Identifying body actions in relation to tasks. These actions can be decomposed into two major types:

1. Those which involve body motions. They are used for either changing body positions or moving objects.
2. Those which do not involve body motions. They are used for primarily for balancing and stabilizing the body for performing some task. Body posture is especially important here, especially when great muscular forces must be exerted.

TABLE 152.1 Environmental Factors in Task Analysis

Environmental factor	Constraint	Examples of effects
Illumination	Insufficient intensity	Errors
Heat	Excessive heat and humidity	Lowering of physiologic endurance
Cold	Excessive Cold	Numbness of hands—inability to grasp handtool properly
Vibration	Excessive amplitude	Loss of grip and control of vibrating tool
Noise	Excessive intensity	Inability to coordinate job with coworkers
Body supporting surface	Slipperiness	Difficulty in balancing body
Space	Insufficient workspace	Inability to position body properly

TABLE 152.2 Representative Actions with Motions

Class of Action	Subclass	Purpose of Action	Examples
Whole body		Positioning body	Walking, crawling, running, jumping, climbing
		Transporting object	Lifting, lowering, carrying horizontally, pushing, pulling, rotating
Gross segmental	Arm movement	Applying forces to object	Reaching for object, twisting, turning, lifting, lowering, transporting horizontally, positioning, pressing down, pushing, pulling, rotating
	Leg movement	Applying forces	Rotating (e.g., foot pedal), pushing
	Trunk movement	Generating momentum	Sitting and bending forward, backward, and sideways to retrieve items on a table
Fine segmental	Finger movement	Applying forces	Turning, pushing, pulling, lifting

TABLE 152.3 Representative Actions without Motions

Class of Action	Subclass	Purpose of Action	Examples
Whole body		Balancing body	Standing while preparing meal in kitchen
		Supporting or stabilizing object	Bracing ladder while coworker climbs on it
Segmental	Arm		Holding food item with one hand while cutting it with other hand
	Leg	Supporting body	Forcing foot on ground while sitting in constrained posture

These actions are not mutually exclusive during task performance. One part of the body may move, while another part, involved in the task, may be still. For example, the fingers move rapidly when using a keyboard, but the rest of the hand is kept motionless while the fingers are moving. Tables 152.2 through 152.5 depict practical decompositions of these actions as well as those requiring muscular forces.

Level V. Identifying anthropometric variables. Human anthropometry interacts with workplace and equipment dimensions. Tasks are performed better when the interaction defines a close "match" between human body size and task-related physical dimensions. Therefore, physical task analysis may be inadequate without measurements of the dimensions of people performing the task(s) and the corresponding dimensions of other aspects of the system of which the task is a part. This kind of matching is one of the basic thrusts in the discipline of ergonomics. Grandjean [1988] and Konz [1990] discuss these issues in many types of human-task environments, and Pheasant [1986] gives numerous body dimensions related to task design. Some important examples of human-system dimensions that must be considered during task analysis are shown in Table 152.6.

152.4 Factors Influencing the Conduct of Task Analysis

While task analysis traditionally focuses on what a person does and how he or she does it, other factors must be considered to complete the picture. In general, the characteristics of the task, the equipment

TABLE 152.4 Representative Actions Requiring Forces, for Different Types of Forces

Class of Action	Subclass	Sub-subclass	Body-Object Contact	Purpose of Action	Examples
Whole body	Whole body	Dynamic	One hand, two hands, or other contact	Generate great forces that must be transmitted over small localized area or large area. The effort produces motion of whole body.	Lift, push, pull
		Static	One hand, two hands, or other contact	Generate great forces that must be transmitted over small localized area or large area. The effort does not produce motion.	Lift, push, pull
Segmental		Static	One hand, two hands, one foot or two feet	Generate forces (over a wide range of magnitude) that must be transmitted over a relatively small localized area. The effort does not produce motion of the body segment.	Gripping with whole hand, pinching, pressing down, twisting, turning, pushing, pulling, or combinations
		Dynamic	One hand, two hands, one foot, or two feet	Generate great forces that must be transmitted over small localized area or large area. The effort produces motion of the body segment.	Pushing, pulling, pressing down

TABLE 152.5 Posture While Performing Tasks

Class	Subclass	Purpose	Examples
Whole body	Standing, sitting, kneeing, lying down, other	Optimize body leverage; position body for making proper contact with object; other	Squatting at beginning of a heavy lift to prevent excessive strain in the lower back
Segmental	Arms, legs, trunk, hands, other	Optimize segment leverage; position segment for making proper contact with object	Placing elbow at about right angles while using wrench to prevent excessive shoulder and muscle efforts

TABLE 152.6 Human Anthropometry and System Dimensions

Body Dimensions	System Dimensions
Height, arm length, shoulder width	Workspace—height of shelf, depth of space under work table, circumference of escape hatch
Leg length, knee height	Equipment size—distance from seat reference point to foot pedal of machine, height from floor to dashboard of vehicle
Hand length, grip circumference	Handtool size—trigger length, handle circumference
Crotch height, arm length	Apparel size—leg length of pants, arm length of shirt
Face height, hand width	Personal protective equipment—height of respirator, width of glove

used, and environmental conditions must all be accounted for. Human functional performance cannot be viewed in isolation because the person, the task, the equipment, and the environment form a system in which a change in one factor is likely to affect the way the person reacts to the others. For example, in dim lighting, a VDT operator may read from a document with marked flexion of the neck but with only slight flexion (which is optimal) when reading from the screen. Brighter ambient lighting may prevent sharp flexion when reading from documents but is likely to decrease contrast of characters on the screen. Screen character brightness must therefore be changed to maintain adequate reading performance to prevent sharp flexion of the neck again.

Knowing the Objectives of Task Analysis

Motion, speed, force, environment, and object (equipment) characteristics are common factors for consideration in task analysis. However, the reasons for performing a task are an equally important factor. Tasks that may involve the same equipment and which may seem identical may differ in their execution because of their objectives. Thus a person opening a jar in the battlefield in the midst of enemy fire to retrieve medicines will not perform that task in exactly the same way as a person at home opening a similar jar to retrieve sugar for sweetening coffee. In the former case, speed is more important, and both the type of grasp used and the manual forces exerted are likely to be markedly different from the latter case. Napier [1980] recognized this principle, in relation to gripping tasks, when he wrote that the "nature of the intended activity influences the pattern of grip used on an object." These kinds of considerations in task analysis can influence the design of objects and other aids and the nature of education and training programs.

Levels of Effort

The level of required effort when performing similar tasks in different environments is not necessarily the same, and the interpretation of task-analysis data is influenced by these differences. Tasks in the workplace are designed so that human physical effort is minimized, whereas tasks in the competitive athletics environment are designed to tap the limits of human performance resources. There are other differences. An athlete paces himself or herself according to his or her own progress. He or she has greater freedom to stop performing a task should physical efforts become painful or uncomfortable. In the workplace, however, there is often little freedom to stop. Many workers often push themselves to the limits of their endurance in order to maintain (flawed) performance goals set by their employers. This is why cumulative strains are more prevalent in the workplace than in many other environments. These differences in performance levels may influence the way task analyses are performed among the different disciplines and the application of the data derived therefrom.

Criticality of Tasks

An inventory of task elements is necessary for describing the requirements of the task. It can tell us about the person, equipment, and environmental requirements for performing a job, but also we must be able to identify specific elements or factors that can prevent the successful completion of the task or that an lead to accidents and injuries. These critical elements are usually measured in detail or estimated and used for developing quantitative models for predicting success or failure in performance. One such widely researched element is the compression force on the L5/S1 disk in the human spine while performing heavy-lifting tasks [Chaffin & Andersson, 1984]. This force is being used by the National Institute of Occupational Safety and Health as a criterion for determining how safe a lifting activity is. If analysis of task-analysis data indicates that the safety limit (3400 N of compression) is exceeded then work redesign, based on task-analysis information, must be implemented [National Technical Information Service, 1991; NIOSH, 1981].

152.5 Measurement of Task Variables

During task analysis, variables must be measured for subsequent data summary and analysis. It is important to know not only what the inventory of task-related variables is but also to what degree the variables are related to or affect overall performance. The measurement of task-related variables allows for quantification of the overall system. For example, the NIOSH lifting equation [National Technical Information Service, 1991] shows how task variables such as the horizontal distance of the load from the body, the initial height of the load, the vertical height of lift, the frequency of lifting, the angular displacement of the load from the saggital plane during lifting, and the type of hand-handle coupling affect the maximal weight of the load that can be lifted safely by most people in a manual materials-handling task. The number of task-related variables measured should be adequate for describing the task and for representing it quantitatively. However, there are often constraints. These include

1. The number of variables that can be identified
2. The number of variables that can be measured at an acceptable level of reliability and accuracy
3. The availability of measurement instruments

Merely summarizing individual measurements seldom yields the desired information. Task performance is essentially a multivariable operation, and variables often must be combined by some quantitative method that can yield models representative of the performance of the task. Sometimes a variable cannot be measure directly but can be estimated from other measured variables. The estimated variable may then be used in a subsequent modeling process. A good example of this is the intraabdominal pressure achieved during heavy lifting. Though it can be predicted by a cumbersome process of swallowing a pressure-sensitive pill, it also can be estimated from the weights of the upper body and load lifted and other variables related to lifting posture. Its estimate can then be used to estimate the compressive force in the L5/S1 disk and the tension in the erector spinae muscles during lifting. Imrhan and Ayoub [1988] also show how estimated velocity and acceleration of elbow flexion and shoulder extension can be used, with other variables, to predict linear pulling strength.

Task-Related Variables

Variables that are usually measured during physical task analysis include

1. Those related to the physical characteristics of task objects or equipment:
   - Weight of load lifted, pushed, carried, etc.
   - Dimensions of load
   - Location of center of mass of load
2. Those related to the nature of the task:
   - The frequency of performance of a cycle of the task
   - The range of heights over which a load must be lifted
   - The speed of performance
   - The level of accuracy of performance
3. Those related to the capacities of various physical resources of the person
   - Muscular strength
   - Joint range of motion
   - Joint motion (velocity and acceleration)
   - Maximal aerobic power
   - Anthropometry
4. Those related to the environment:
   - Temperature
   - Illumination
   - Vibration
5. Those related to workplace design:
   - Amount of space available for the task
   - Geometric and spatial relationships among equipment
   - Furniture dimensions
6. Those related to anthropometry:
   - Length, breadth, depth, or circumference of a body segment
   - Mass and mass distribution of a body segment
   - Range of motion of a skeletal joint

Instruments for Gathering Task-Analysis Data

A great number of instruments are available for measuring variables derived from task analyses. The choice of instruments depends on the type of variables to be measured and the particular circumstances. In general, there should be instruments for recording the sequence of actions during task, e.g., videotape with playback feature, and for measuring kinematic, kinetic, and anthropometric variables [Chaffin & Andersson, 1994; Winter, 1990]. The main kinematic variables include displacement (of a body part), velocity, and acceleration. Kinetic variables include force and torque. Anthropometric variables include body segment length, depth, width, girth, segment center of mass, segment radius of gyration, segment moment of inertia, joint axis of rotation, and joint angle. Some variables are measure directly, e.g., acceleration (with accelerometers), force applied at a point of contact between the body and an object (with load cells), and body lengths (with anthropometers); some may be measured indirectly, e.g., joint angle (from a videotape image) and intraabdominal pressure (using swallowed pressure pill); and others may be estimated by mathematic computations from other measured variables, e.g., compressive force on the lumbosacral (L5/S1) disk in the lower back. Posture targeting, a method for recording and analyzing stressful postures in work sampling [Corlett et al., 1979], is becoming popular.

The measurements of performance capacities associated with human functions should conform to certain criteria. Details can be found in Brand and Crownshield [1981] and Chaffin [1982]. A general set of psychometric criteria is also discussed by Sanders and McCormick [1993]. It includes measurement accuracy, reliability, validity, sensitivity, and freedom from contamination. Meeting these criteria depends not only on the instruments used but also on the methods of analysis and the expertise of the analyst. It is almost impossible to satisfy these criteria perfectly, especially when the task is being performed in its natural environment (as opposed to a laboratory simulation). However, the analyst must always be aware of them and must be pragmatic in measuring task variables. Meister [1985] lists the following practical requirements for measurements: (1) objective, (2) quantitative, (3) unobtrusive, (4) easy to collect, (5) requiring no special data-collection techniques or instrumentation, and (6) of relatively low cost in terms of money and effort by the experimenter. However, there are not necessarily mutually exclusive.

152.6 Uses and Applications of Task Analysis

The uses of task-analysis information depend on the objectives of performing the task, which, in turn, depend on the environment in which the task is performed. Table 152.7 gives the different situations.

152.7 Future Developments

To date, there is no reliable quantitative model that can combine mental and physical resources. A general-purpose task-analysis method that uses active links between the various compartments of human functions would be an ideal method, but our lack of knowledge about the way in which many of these "compartments" communicate precludes the development of such a method. The present approach to compartmentalization seems to be the most pragmatic approach. It offers the analyst access to paths from a gross task or job to the myriad of basic task elements of performance. Unfortunately, actual systems and subsystems used in practice seem to be loosely defined and inconsistent, often with confusing metrics and terminologies for important variables. Moreover, available task-analysis models deal only with specific classes of application [Drury, 1987]. A general-purpose model is badly needed for handling different mental and physical resources at the same time, dealing with wide ranges in a variable, and bridging the gaps across disciplines. Such a model should help us to perform task analyses and manipulate the resulting data from environments ranging from the work-place, the home (activities of daily living), athletics, exercise and fitness, and rehabilitation. Such a model also can help to eliminate the present trend of performing task analyses that are either too specific to apply to different situations or not specific enough to answer general questions. Fleishman [1982] deals with the issue. At least one recent effort is active [Kondraske, 1995].

TABLE 152.7 Uses and Applications of Task Analysis

Uses and Applications	Examples of Relevant Situations
Modeling human performance and determining decision-making strategies	A model showing the sequence of the various steps required to perform a task and the type of equipment needed at each step
Predicting human performance	Using a quantitative model to predict whether an elderly person has enough arm strength to lift a pot full of water from a cupboard onto a stove
Redesigning of the existing tasks or designing of new tasks	Eliminating an unnecessary step in the packaging of production items into cartons; designing a different package for a new item based on task analytic data gathered for a related item
Determining whether to use task performance aids	The analysis may show that most elderly persons do not possess enough strength to open many food jars and may, therefore, need a mechanical torqueing aid.
Personnel selection or placement	Matching personnel physical characteristics with task requirements can reveal which persons may be able to perform a task and, hence, be assigned to it.
Determining whether to use aids that enhance health or safety	Task analysis may show that too much dust always gets into the atmosphere when opening packages of a powered material and that workers should wear a respirator.
Determining educational and training procedures	Identification of difficult task steps or the need for using mechanical aids, together with knowledge of workers' skills, may indicate the level of education and training needed.
Allocating humans to machines and machines to humans	Matching people skills and capacities with the resource demands of machines
Determining emergency procedures	Task-analysis information may indicate which steps are likely to result in dangerous situations, and therefore, require contingency plans.

There is also a need for better quantification of performance. We need to know not only what kind of shifting in resources a person resorts to when one route to successful task accomplishment is blocked (e.g., insufficient strength for pressing down in a certain sitting posture) but also the quantity and direction of that shift. For example, we need to know not merely that a change in posture can increase manual strength for torqueing but also what the various postures can yield and what quantity of specific mental resources were involved in the change. The models today that answer the quantitative questions are mainly statistical regression ones, and they are limited by the number of variables and the number of levels of each variable that they can deal with.

Defining Terms

Anthropometry: The science that deals with the measure of body size, mass, shape, and inertial properties.

Cumulative trauma: The accumulation of repeated insults to body structures over a period of time (usually months or years) often leading to "cumulative trauma disorders."

Endurance: The maximum time for which a person can perform a task at a certain level under specified conditions without adverse effects on the body.

Fitts' law: The equation, derived by P. Fitts, showing the quantitative relationship between the time for a human (body segment) to move from one specific point to another (the target) as a function of distance of movement and width of the target.

Hierarchical task analysis: The analysis of a task by breaking it down to its basic components, starting from an overall or gross description (e.g., lifting a load manually) and moving down in a series of steps in sequence.

Job analysis: Any of a number of techniques for determining the characteristics of a job and the interactions among workers, equipment, and methods of performing the job.

Job Severity Index: An index indicating the injury potential of a lifting or lowering job. It is computed as the ratio of the physical demands of the job to the physical capacities of the worker.

Local segment: A specific body segment, usually the one in contact with an object required for performing a task or the one most active in the task.

Maximal aerobic power: The maximal rate at which a person's body can consume oxygen wile breathing air at sea level.

Motion study: The analysis of a task or job by studying the motions of humans and equipment related to its component activities.

Person-task conflict: The situation in which task demands are beyond a person's capacities and the task is unlikely to be performed according to specifications by that person.

Position analysis questionnaire: A checklist for job analysis, developed by the Office of Naval Research in 1969, requiring the analyst to rate or assess a job from a list of 187 job elements.

Safety limit: The maximum stress (e.g., load to be lifted) that most workers can sustain, under specified job conditions without adverse effects (injuries or cumulative strains) on their bodies.

Time study: The study of a task by timing its component activities.

Work pause: A short stoppage from work.

References

Ayoub MM, Bethea NJ, Deivanayagam S, et al. 1979. Determination and Modeling of Lifting Capacity. Final report, HEW (NIOSH), grant no. 5R01–OH–00545–02.

Ayoub MM, Mital A. 1991. Manual Materials Handling. London, Taylor and Francis.

Campion MA, Medsker GJ. 1992. Handbook of Industrial Engineering. New York, Wiley.

Chaffin DB, Andersson GD. 1991. Occupational Biomechanics. New York, Wiley.

Corlett EN, Madeley SJ, Manenica I. 1979. Postural targeting: A technique for recording work postures. Ergonomics 22(3):357.

Drury GD, Paramore B, Van Cott HP, et al. 1987. Task analysis. In J Salvendy (ed), Handbook of Human Factors, pp 371–399. New York, Wiley.

Fleishman EA. 1982. Systems for describing human tasks. Am Psychol 37(7):821.

Fitts P. 1954. The information capacity of the human motor system in controlling the amplitude of movement. J Exp Psychol 47:381.

Gilbreth FB, Gilbreth FM. 1917. Applied Motion Study. New York, Sturgis and Walton.

Grandjean E. 1988. Fitting the Task to the Man, 4th ed. New York, Taylor and Francis.

Imrhan SN, Ayoub MM. 1988. Predictive models of upper extremity rotary and linear pull strength. Hum Factors 30(1):83.

Kondraske GV. 1995. A working model for human system task interfaces. In JD Bronzino (ed), Handbook of Biomedical Engineering. Boca Raton, Fla, CRC Press.

Konz S. 1990. Work Design: Industrial Ergonomics. Worthington, Ohio, Publishing Horizon.

Meister D. 1985. Behavioral Analysis and Measurement Methods. New York, Wiley.

Napier JR. 1980. Hands. New York, Pantheon.

National Technical Information Service. 1991. Scientific Support Documentation for the Revised 1991 NIOSH Lifting Equation. PB91–226274. U.S. Department of Commerce, Springfield, Va.

Niebel BW. 1993. Motion and Time Study, 9th ed. Homewood, Ill, Richard D Irwin.

NIOSH, 1981. Work Practices Guide for Manual Lifting. NIOSH technical report no 81–122, US Department of Health and Human Services, National Institute of Occupational Safety and Health, Cincinnati, Ohio.

Pheasant S. 1986. Bodyspace. London, Taylor and Francis.

Singleton. 1974. Man-Machine Systems. London, Penguin.

Taylor FW. 1911. The Principles of Scientific Management. New York, Harper.

Winter DA. 1990. Biomechanics and Motor Control of Human Movement, 2nd ed. New York, Wiley.

Further Information

For the design of equipment in human work environments where task analysis is employed, see *Human Factors Design Handbook*, by W. E. Woodson (McGraw-Hill, New York, 1981). For human anthropometric data that are useful to task analysis, see *NASA Reference Publication 1024: Anthropometric Source Book*, vol 1: *Anthropometry for Designers* (Webb Associates, Yellow Springs, Ohio, 1978).

153

Human-Computer Interface Design Issues

Kenneth J. Maxwell
Departure Technology

Purpose of This Chapter

This chapter provides the reader with a structure for addressing human-computer interface (HCI) design issues and making design decisions on the basis of quantitative human performance assessments. HCI design issues are structured in accordance with a dual-component, dual-task HCI architecture and are analyzed using a performance-based usability model. This framework is general and not bound to specific interaction paradigms or limited to assumptions about specific technologies. As such, it is anticipated to serve the needs for making HCI design decisions for any biomedical application. It is beyond the scope of this chapter to identify, categorize, describe, and analyze all HCI design issues. Further, it is not the intent of this chapter to review and compare the specific characteristics of different user-interface systems—there are several commonly used—or to review in detail the voluminous research that has been generated in recent years on HCI design. References for this information are provided in the Further Information section.

HCI Design Challenges

Since the introduction of personal computers approximately 20 years ago, the manner in which humans and computers interact has dramatically change from computer-centered to user-centered. A major driver of change in HCI design has been a growing understanding of user needs and expectations and of the ways in which humans can use computers in accomplishing applied objectives. This increased user awareness in concert with emerging technology has dramatically expanded the uses and users of computers. This expansion is not simply limited to enabling users to do traditional tasks more effectively and

0-8493-0462-8/00/$0.00+$.50

efficiently. New technology can enable work to be done in ways that were previously impossible and with contructs that previously did not exist. For example, electronic documents can be merged, displayed, and stored in ways that do not have hardcopy analogues. New approaches to computer-based instructional media enable users to discover and experience concepts in ways that traditional multimedia education does not. The use of computer programs as aids for medical, legal, and business decision making has expanded individual's access to expertise.

That computers can be used to accomplish many different tasks and that both experienced and novice users need to use them present a complex challenge for HCI designers. The multitude of user needs creates a diverse set of functional and performance requirements that must be accommodated. Accommodating these diverse requirements is complicated by the advantages of designing user interfaces across applications and platforms using a standard set of generic interface devices and techniques. This approach reduces the need for task-specific devices and software and exploits user experience to reduce training requirements and improve interaction performance. However, it can cause the use of standard devices and techniques to be generalized and applied suboptimally. Another design goal, aimed at accommodating special user groups and individual user differences, pushes designs toward alternative devices and techniques that optimally meet specific requirements.

The emergence of new technologies combined with an awareness among software and hardware developers that users demand intuitive, direct, and easy-to-use interfaces promises to foster continued dramatic change in the near future. Reductions in size and integration of computers in specific instruments and personal devices have increased their portability and applied uses. The integration of digital information technology across information-gathering, communication, and information-delivery devices promises to dramatically change the model of a computer as a single device into a seamless assemblage of devices for assisting in all information-processing tasks. Not surprisingly, the design of human-computer interfaces has become prominent area of applied research in human factors and ergonomics.

153.1 Fundamentals

User Tasks

When humans and computers interact, two tasks are performed: an interaction task and an application task. The *interaction task* consists of the activities that the user performs to use the computer. These activities include moving a mouse device, typing keyboard inputs, viewing displays, searching for and managing files, managing displays, setting preferences, and executing programs. The *application task* refers to the user's functional objectives. These objectives vary widely and include writing a report, performing a statistical analysis, finding information in a database, controlling a process, making a medical diagnosis, and monitoring/testing the health of a system. The application task is a higher-level task than the interaction task, and consequently, the design issues associated with each task are different. The HCI design process should include a task analysis. Methods for analyzing tasks are presented by Maxwell [1995] and Imrhan [1995] in this *Handbook.*

A Dual-Component, Dual-Task-Level HCI Architecture

HCIs have many components. For the purposes of this chapter, two major types of components are of interest: mental and physical. These components are distinguishable in terms of the performance and skill resources [Kondraske, 1995] that a user needs to perform interaction and application tasks with the computer. Using the mental-interface component requires perceptual-cognitive performance and information (skill) resources. Using the physical-interface component requires sensory-motor performance and skill resources. The relationship of these HCI components with the user are illustrated in Fig. 153.1.

For each HCI component, a design specification associated with each user task is defined. The application task-level design specifies the structure and level of objects, actions, and information that the user

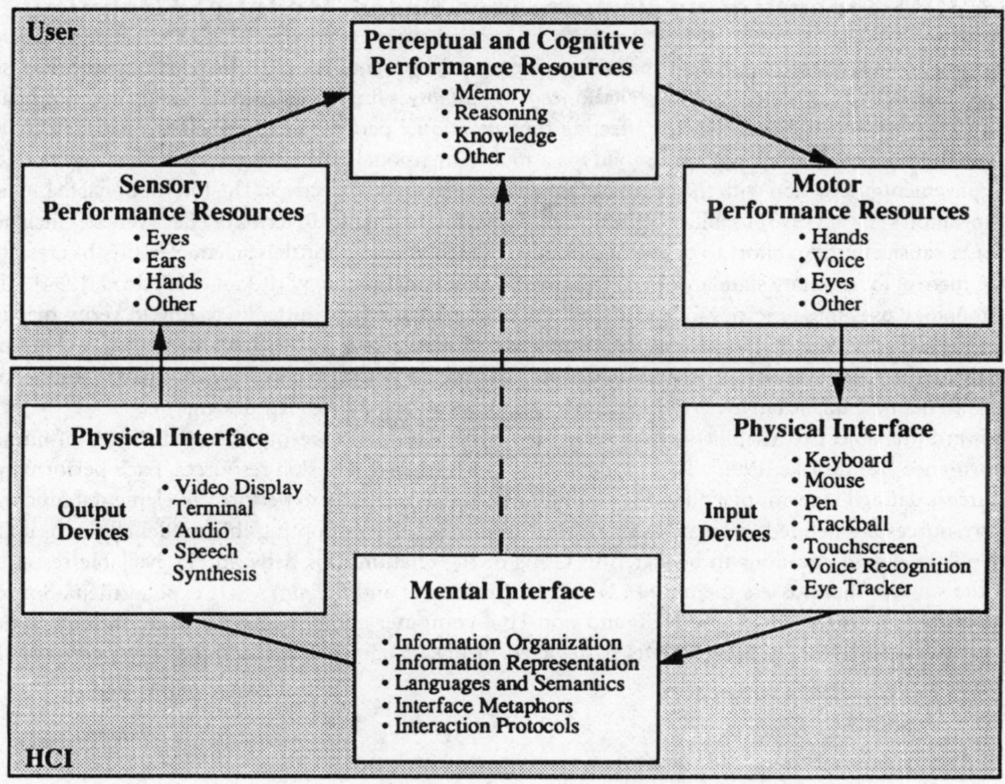

FIGURE 153.1 The mental and physical HCI components fit different user performance resources.

needs, and the procedures, both mental and physical, that a user must take to accomplish applied objectives. The interaction-task-level design specifies the representation and behavior of HCI elements associated with the mental and physical HCI components. This framework, illustrated in Fig. 153.2, reflects the need to enable two user tasks requiring both mental and physical performance and skill resources and is used to structure the analysis of HCI design issues.

Successful HCI design decisions must be made on the basis of (1) knowledge about the user's performance and skill resources, (2) the performance and skill resource requirements of the application task, and (3) the performance capabilities of available HCI technologies. The design of the physical-interface component requires knowledge of the user's sensory-motor performance and skill resources and of the resources required to use various physical HCI devices with regard to the application tasks to be performed. Design of the mental-interface component requires knowledge of the user's perceptual-cognitive performance and skill resources and of the resources required to process information in different organizations and representations with regard to the application tasks to be performed.

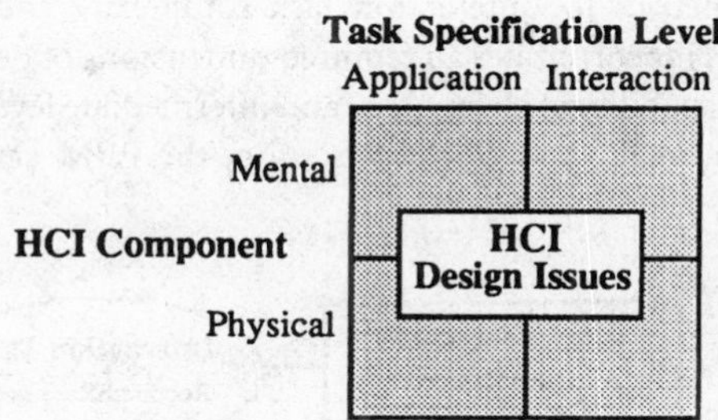

FIGURE 153.2 A dual-component, dual-task-level HCI architecture for design.

153.2 A Quantitative Performance-Based Model of Usability

Achieving a highly usable interface is the HCI designer's primary goal. All HCI design issues are addressed on the basis of their impact on usability. The concept of *usability* is fundamental to the design of any human-system interface and is a major factor affecting the operational performance, availability, maintainability, and reliability of the overall system. Usability is a multidimensional quality that affords the user practical and convenient interaction with the computer for achieving applied objectives. The International Standards Organization's (ISO) [1993] usability definition characterizes this quality in terms of effectiveness, efficiency, and user satisfaction. An effort to define quantitative usability metrics for these characteristics has resulted in the metrics for usability standards in computing (MUSiC) methodology [Bevan & Macleod, 1994]. This methodology uses these metrics for specifying usability goals and intermittently testing to verify meeting the goals as part of the design process. Generally, usability metrics focus on usability testing and are not directly useful early in the design process for defining and quantifying a highly usable interface. The objective here is to define a quantitative performance-based approach for guiding HCI design.

Toward this objective, usability is defined in terms of the elemental resource model (ERM) of human performance [Kondraske, 1995]. This model defines performance and skill resources. Each performance resource is defined in terms of a quantitative unidimensional capability to perform an elemental function. Skill resources are defined in terms of knowledge and experience. Adopting this model allows usability tradeoffs in design decisions to be stated in terms of the relationships between (1) available resources and the sources that possess them and (2) required resources and the sinks that expend them. Sources of resources include the user, the HCI, and non-HCI computer components. Sinks are the interaction and application tasks. The basic model is depicted in Fig. 153.3. The HCI designer has partial or full control over the following:

- HCI available resources
- Interaction-task required resources
- Application-task required resources
- Non-HCI computer components available resources

Why Use the ERM?

There are several methods for quantifying various aspects of HCI design. The GOMS (for goals, operators, methods, and selection rules) and keystroke-level models [Card et al., 1983, 1986], when used in combination with human performance models such as the model human processor, provide task performance estimates and knowledge requirements. The NGOMSL (for natural GOMS language)[Kieras, 1998] and TAG (task action grammer) [Payne & Green, 1986; Schiele & Green, 1990] models provide quantitative metrics for interaction task complexity and consistency. However, the ERM is the only model that (1) incorporates all required dimensions of performance and skills, (2) provides a performance-modeling capability at elementary and intermediate levels, and (3) uses consistent modeling constructs across tasks, humans, and computers. Also, the ERM can be used in a rigorous way when performance data are

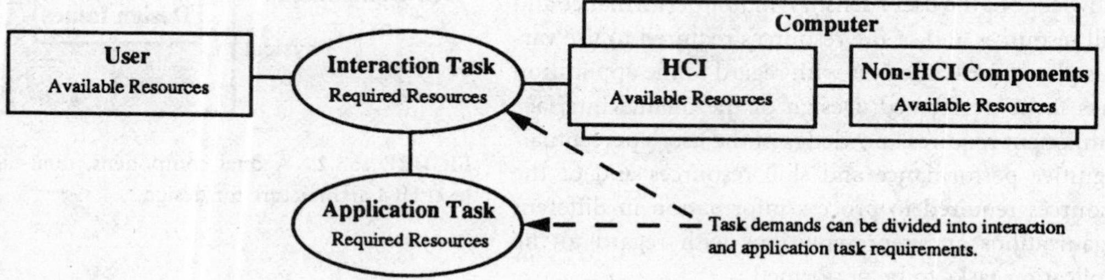

FIGURE 153.3 HCI design goal is to minimize interaction and application task demands.

obtainable or less rigorously as a set of principles to guide designing for enhanced usability. Finally, the ERM provides a framework for specifying performance and functional capacities that can be used in concert with other models and techniques at various design levels. This is illustrated in the example at the end of this chapter.

Minimal and Optimal Usability

In terms of the ERM, and HCI design is minimally usable when all the user's performance and skill resources needed to accomplish the interaction and application tasks are exhausted in achieving the desired level of application task performance. This means that the user can perform the task using the computer only with maximal effort and expenditure of resources. Note that this definition of usability does not address the design of the HCI specifically. Rather, it addresses the relationship between the performance that can be obtained with a given HCI design and the effort a user needs to expend to achieve it. This means that different users will have different points at which an HCI for a given task will be minimally usable. The concept of optimal usability is more complex. An optimized design will allow users maximal use of their performance and skill resources, allocated at their discretion to the application task, to maximize their performance on the application task. Thus an optimal design minimizes the resources required to perform the interaction task and enhances the utility of the user's resources in performing the application task. This means that the HCI will support the user in performing the portion of the application task allocated to the user. This is accomplished through application-level design issues associated with the organization and level of objects, actions, and information.

Performance-Based Usability Design Decisions

The design strategy for achieving usability follows from an analysis of available and required performance and skill resources and application of HCI design principles that act to reduce resource task demand on users (Fig. 153.4). For example:

- Usability will increase as the resources required of the user to perform the interaction and application tasks decrease.
- Usability will increase as the HCI available resources increase (provided the design principles for reducing resource demands are applied).

ERM-Based Approach to Usability Design Decisions

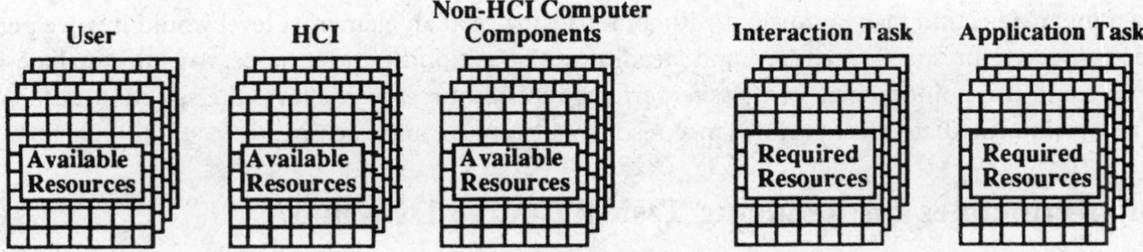

HCI Design Goal:
Use HCI performance resources to minimize performance and skill resource task demands on the user.

HCI Design Approach:

- **Apply design principles for reducing the user's performance and skill resource requirements**
  - Naturalness and Familiarity of Interaction
  - Simplicity of Interaction
  - Robustness of Design
  - Consistency of Design
  - Accommodate Different User Capabilities
- **Apply ERM quantitative analysis techniques**

FIGURE 153.4 A quantitative human-performance-based approach to HCI design.

A detailed discussion regarding performance resources is presented by Kondraske [1995]. Different resource sources and tasks impose different design objectives. One performance-based design objective is to apply HCI resources to minimize the number of performance resources the user needs to perform the interaction task. This approach permits the user to allocate more resources to performing the application task. Achieving this objective includes (1) incorporating appropriate input and display devices to meet the user's sensory-motor resources and (2) use of organizations, representations, and levels of object, actions, and information that fit within the limits of the user's knowledge and perceptual-cognitive resources.

A second performance-based design objective is to apply HCI resources to optimize the resources the user needs to perform the application task. Many of the objects, actions, and information with which the user interacts are application-task-specific. Organization and representation decisions should be based on these task-specific requirements. For example, if the application task involves drawing, the user needs general software and hardware drawing tools and devices. This consideration can be used to illustrate the relationship between application and interaction task-level design issues. A software drawing package can provide different levels of drawing elements (e.g., having a tool to directly draw *n*-sided polygons versus a line-drawing tool from which a polygon can be drawn segment by segment). Assume that the HCI provides a palette of drawing tools and that it includes primitive and complex tools. The user can draw an *n*-sided polygon using the line tool or using the polygon tool. Both tools are included in the palette. The issue here is how the decision to develop the polygon tool and include it in the product is made. The decision must be based on the user's application task need to draw polygons. This need is established by an analysis of the application task. The decision to include the polygon tool is an HCI decision at the application-task level. Once this applied need is established, the HCI design needs of how to represent, where to locate, and how to operate the tool are addressed. These design needs are HCI design issues at the interaction-task level.

The decisions at these two task levels can be further used to illustrate interactions in assessing performance-based usability. A decision to include the polygon tool made on the basis of the application-task requirements will increase usability at the application-task level in that the user will need to use fewer sensory-motor resources to create a polygon. However, the inclusion of the polygon tool complicates the interaction-task level in that more items are added to the tool palette and that procedures need to be defined for drawing it. As will be discussed, these complications are addressed by applying HCI design principles that reduce performance and skill resource demands.

Because the ERM defines performance resources at an elemental level, performance-based design can be conducted at the elemental level or at higher intermediate levels. Intermediate levels reduce the granularity of the analysis and provide for use of composite performance resources. For example, analyzing a menu-selection task performed with a mouse device at an elemental level would involve performance resources for speed, accuracy, and steadiness of positioning movements. An intermediate-level analysis could use pointing as a composite performance resource. At his level the selection task would involve two intermediate-level performance resources: pointing and clicking.

Design Principles for Reducing Task Resource Demands

Optimizing usability is accomplished through the application of resource-demand-reducing HCI design principles. As stated earlier, the ERM can be applied rigorously or as a set of guiding principles. Rigorous application requires an analysis of available and required performance and skill resources and performance data associated with each resource. These data are usually not obtainable for all resources. Less rigorous application uses the principles of resource economics, resource stress, and analyses of the performance and skill resources involved in tasks to guide the design toward lower resource demands on the user. Five major design principles that act to reduce task demands on performance and skill resources are described. Each principle can be applied to all HCI architectural design categories.

Naturalness and Familiarity of Interaction. Usability will be improved to the degree that the interaction is natural and familiar. Naturalness and familiarity reduce training requirements and increase available resources (e.g., available knowledge, sensory-motor performance). For physical components at a task level, this means that the user's sensory and effector modalities are properly matched to the required interactions. For example, keyboards used for typing, keypads for data entry, and mouse, pen, or touch-screen devices for pointing. For physical components at a feature level, this means that (1) user anthropometrics are employed in defining workstation and component configurations and dimensions and (2) that the physical characteristics of controls and displays (e.g., operating force, sensitivity, brightness, intensity, and size of displayed information) are congruous with the users and physical capabilities. For mental components at a task-level, natural interaction means that the HCI models the user's tasks in accordance with the user's knowledge and experience. For mental components at a feature level, natural interaction means that the user's perceptual and cognitive capabilities and limitations are employed in defining HCI features (e.g., icons, names, screen layouts, and the behavior of window controls such as scroll bars). *Direct manipulation* interaction techniques (e.g., drag and drop) are a form of natural interaction in that they create a software mechanism with a physical analogy.

Simplicity of Interaction. Simplicity applies across a wide variety of design issues and must be incorporated in every architectural category. Applying this principle reduces training requirements, memory demands, perceptual demands, and the cognitive demands for inferencing. For the physical interface component, applying this principle reduces the number of physical steps needed to select and invoke actions and reduces movement precision requirements.

Consistency of Design. Consistency reduces training requirements and memory requirements and promotes skill development. Consistency applies across a wide variety of design issues and must be incorporated in every architectural category. Examples include object, action, and information organizations, task flows, representations for objects and actions, and procedures. TAG [Payne & Green, 1986; Schiele & Green, 1990] provides a formal technique for analyzing consistency.

Robustness of Design. *Robustness* refers to the ability of the design to tolerate user error and to span the functional requirements of the user. Error tolerance reduces resources needed to recover from errors and reduces initial accuracy requirements, allowing the user to increase operating speed. Users may commit errors because of a lack of knowledge. In this case, the user may be exploring the system to learn it. These are not errors. But this type of behavior should be expected, and the design should allow the user to do this exploration and to recover from anything he or she may do. Users may commit errors because of a misunderstanding or negative transfer from other systems. Thus a user may infer that a menu item will elicit a given behavior because it does so on another system that he or she has used. Or the user may feel that he or she understands what a given menu item or icon means and be wrong. Users may commit errors through unintentional commands. For example, a user may inadvertently click a wrong button, hit a wrong key, or select a wrong item from a menu. Ensuring adequate functional coverage reduces the functional demands on the user.

User Capabilities Accommodation. In resource terms, applying this design principle increases the fit between an individual user's available mental and physical resources and those required by the interaction and application tasks. There are two major dimensions on which users vary: experience and capability.

Along the experience dimension users vary between experts and novices on either the interaction or application task. Thus a user could be very experienced with the applied task but not with doing that task on a computer. Similarly, a user may be very used to using computers but not skilled in the applied task. Novice users will (1) need more guidance and feedback, (2) require more structure to the task, (3) rely more on memory-aiding design features (e.g., menus), and (4) benefit more from on-line help facilities than expert users. Expert users (1) will want more flexibility in task control, (2) will want shortcut methods to do high-level actions, (3) will rely more on recall rather than memory aids, and

(4) will benefit more from the use of preferences and tailored configurations than novice users [Brown, 1988; Galitz, 1993].

Along the capacity dimension users vary between healthy and impaired. In the ERM, impairments are not defined differently from any other performance or skill differences among users. All differences are defined in terms of the amount of specific performance resources available to an individual. Impaired users will vary greatly in capacity and experience. Only through an analysis of resources demanded by the task and resources available to the user can an adequate design for an impaired user be accomplished successfully.

153.3 Selected HCI Design Issues, Goals, and Resource Requirements

Selected design issues in each architectural category are identified. Design goals for enhancing usability related to each issue are described. These goals are derived by applying the resource-demand-reducing design principles described earlier. The human performance resources associated with each issue are also identified. Thus for each issue the reader is presented with the areas of a human's performance resource profile that need to be assessed to achieve a design that enhances usability. For illustrative purposes, and example design process is presented using issues related to the design of a menu system.

Mental Interface: Application-Task-Level Specification

Design issues in this category are concerned with the organization and level of objects, actions, and information that users need to perform the interaction and applied tasks. Design decisions must consider the perceptual-cognitive performance and skill resources possessed by users. Design goals are to minimize the resources required to accomplish the applied task. Table 153.1 lists selected design issues of this type, along with the design goals and performance and skill resources involved in making the design decisions for each issue.

TABLE 153.1 Selected Mental-Interface/Application-Task-Level Design Issues

Design Issue	Selected Design Goals (Based on Design Principles)	Selected Performance and Skill Resources Required of the User
Application-task-level interaction structure	Match interaction structure to user's cognitive model of task Minimize the number of steps needed to complete tasks Consistent task syntax Consistent task semantics	User's knowledge of the application task structure
Design of dialogs and query languages	Structure syntax in accordance with natural language Define semantics in accordance with task usage Minimize number of dialogs and queries required Allow for flexible usage Consistency of usage	User's knowledge of the application task structure User's knowledge of application task terminology Dialog or query entry speed Dialog or query entry accuracy
Use of an application-task-level metaphor	Leverage user's prior experience Increase intuitiveness of application-task-level interaction	User's knowledge of the application task structure
Menu structure design	Optimize menu navigation Structure menus in accordance with user's knowledge of task Structure menus in accordance with user's assoiations among menu set.	User's knowledge of the application task structure

TABLE 153.2 Selected Mental-Interface/Interaction-Task-Level Design Issues

Design Issue	Selected Design Goals (Based on Design Principles)	Selected Performance and Skill Resources Required for the User
Interaction-level icon, symbol, and label selection	Select based on strength of association to represented objects and commands Establish consistent usage, i.e., identical icons, symbols, and labels should be used for the same objects and actions.	User's knowledge of computer objects and actions Perceptual accuracy Perceptual speed
Color selection	Consistency with cultural usage	User's existing color associations to application task and HCI objects and actions
Screen and window layout	Consistency of within-screen activity flow with other learned activities (e.g., left to right flow of activity for English speaking users)	Memory speed Memory accuracy Perceptual speed Perceptual accuracy
Number of controls and displays	Optimize Use of integrated displays where applicable	Memory speed Memory accuracy Perceptual speed Perceptual accuracy

Mental Interface: Interaction-Task-Level Specification

Design issues in this category are concerned with the representation and behavior of HCI elements. Design decisions must consider the perceptual-cognitive performance and skill resources possessed by users. Design goals are to minimize the resources required to accomplish the interaction task. Table 153.2 lists selected design issues of this type, along with the design goals and performance and skill resources involved in making design decisions for each issue.

Physical Interface: Application-Task-Level Specification

Design issues in this category are concerned with (1) the performance and structural characteristics of major physical elements of the workstation and (2) the physical actions required to perform tasks. Design decisions must consider the sensory-motor performance and skill resources possessed by users. Design goals are to minimize the resources required to accomplish the applied task. Table 153.3 lists selected design issues of this type, along with the design goals and performance and skill resources involved in making design decisions for each issue.

Physical Interface: Interaction-Task-Level Specification

Design issues at the feature level of the physical HCI are concerned with the physical performance parameters associated with using controls and displays. Design decisions must consider the sensory-motor performance and skill resources possessed by users. Design goals are to minimize the resources required to accomplish the interaction task. Table 153.4 lists selected design issues of this type, along with the design goals and performance and skill resources involved in making design decisions for each issue.

Brief Example Using Menu System Design

This example identifies the design issues associated with each HCI architectural category and traces the decision process for the design of a menu system. The issues and process are illustrated in Fig. 153.5. The design process begins with the results of an analysis of the application task, the performance and skill resources of the user, and available HCI technology. The analysis of the user will account for user knowledge of the application task, familiarity with computer interfaces, and user performance capabilities. The application-task analysis will provide information about the number of tasks, sequences of

TABLE 153.3 Selected Physical-Interface/Application-Task-Level Design Issues

Design Issue	Selected Design Goals (Based on Design Principles)	Selected Performance and Skill Resources Required for the User
Input device selection	Account for user familiarity Account for naturalness for task (e.g., keyboard for text entry, number pad for data entry, pen, mouse, or touch screen for pointing)	For manual and ocular controls: Movement speed Movement accuracy Movement steadiness Movement extension
Display selection	The greater the amount of information that has to be displayed simultaneously, the greater the size and resolution requirement. Greater size increases the number of required head and eye movements. Optimize size, resolution, and color needs.	Visual acuity Color Sensitivity Visual span angle
Action procedure specification	Minimize the required number of keystrokes, mouse movements, hand movements, head and eye movements Use of direct manipulation techniques (e.g., drag and drop interaction, scroll bars, mouse controls for moving and resizing windows)	For manual and ocular controls: Movement speed Movement accuracy Movement steadiness Movement extension
Use of multiple input and control methods (e.g., keyboard and mouse)	Provide the user with more than one action procedure for doing tasks	For manual and ocular controls: Movement speed Movement accuracy Movement steadiness Movement extension

TABLE 153.4 Selected Physical-Interface/Interaction-Task-Level Design Issues

Design Issue	Selected Design Goals (Based on Design Principles)	Selected Performance and Skill Resources Required for the User
Keyboard and button devices:		
Size of keys	Greater size will reduce the required accuracy and steadiness of movements Smaller size will reduce the required movement length and duration Optimize size across these considerations	For wrist, hand, and fingers: Movement accuracy Movement steadiness Movement speed Movement extension
Force required to press keys	Greater force requires more strength Too little force will not provide adequate feedback to the user. Optimize force across these considerations	Finger strength Proprioceptive feedback sensitivity
Use of audio/visual feedback when key is pressed	User requires feedback to know that the key press has been acknowledged.	Audio/visual acuity Audio/visual sensitivity
Visual display devices:		
Font	Low confusion rate	Visual acuity
Brightness	Optimal viewing	Visual sensitivity

required actions to perform tasks, the semantic associations among tasks and actions, and information to be manipulated. The HCI technology assessment will provide the designer with the set of resources available for application. Data from these analyses are used to make initial mental-interface component and physical-interface component decisions at an application-task level.

The decision to use a menu system to access actions and control the flow of the application task is based on the usability criteria discussed above with regard to performance and skill resources. Specifically, it is based on an assessment of the resources required to perform tasks using menu selection compared with other operating mechanisms. Alternative mechanisms would be command-line input from the user

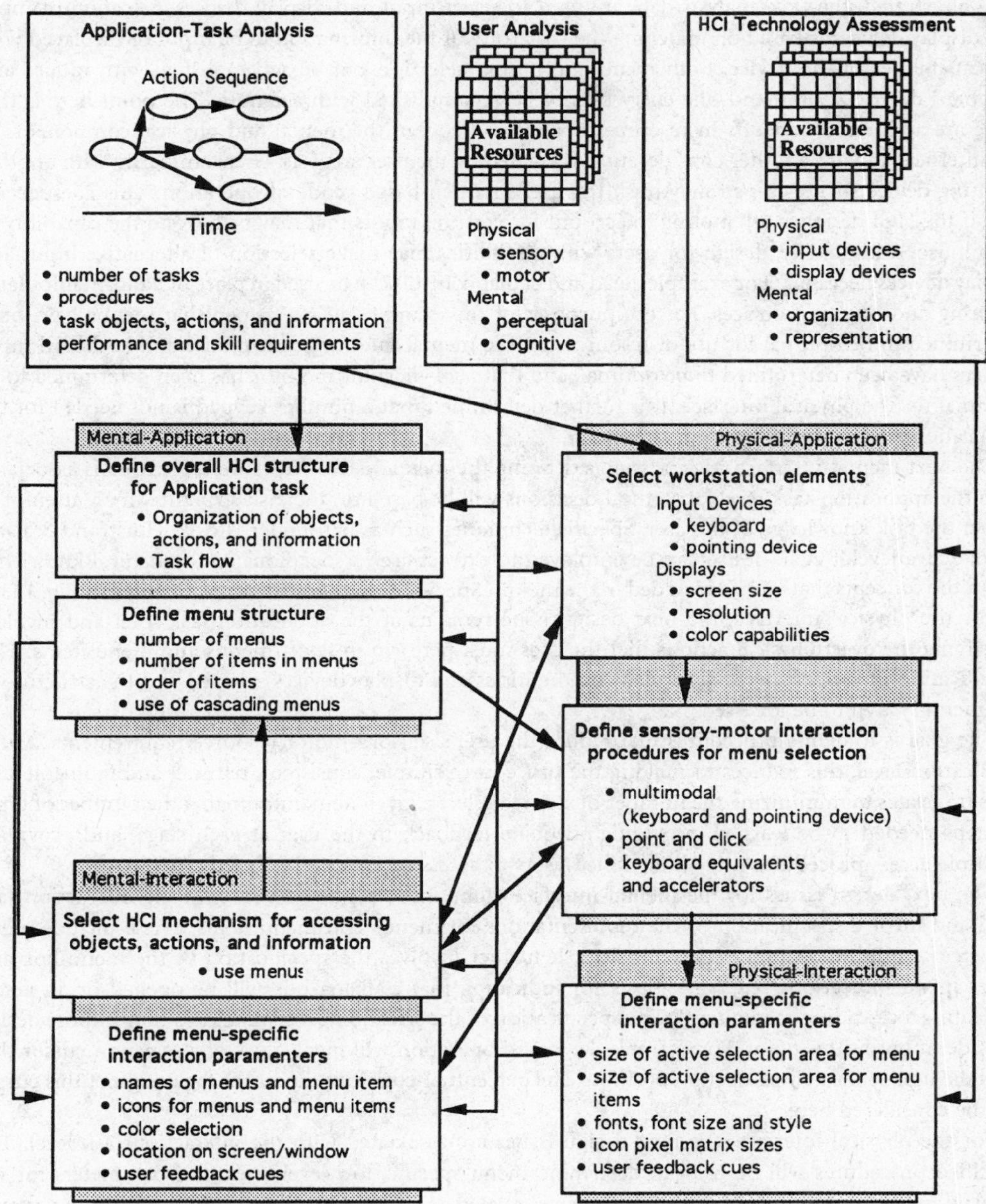

FIGURE 153.5 HCI design decisions for design of a menu system.

and visually presented buttons for each object and action. Menus and visual buttons both reduce the demands on the user's memory capacities because available actions are listed. In the case of menus, many actions are hidden but can be made explicit. If all actions were constantly visible, the user would have an easier access to individual actions (i.e., eliminating the step of calling up the menu) but at the cost of greater display size and clutter, requiring greater use of visual resources. If data specifically addressing each tradeoff in resources are not available, performance testing can be conducted to verify decisions. Before a final decision is made, however, the initial decisions made regarding the physical interface at the application task level need to be considered.

As illustrated, the task-analysis data are used to select input and display devices. Selection of input and display devices is based on matching the congruity of the information to be input or displayed with the capabilities of the device. Both menu and button selection can be accomplished with mouse and keyboard devices. Command-line entry cannot be accomplished with a mouse. The point here is that there are additional tradeoffs in resource capabilities between the mental and physical components of the interface. Finally, another consideration is created by the user analysis. User familiarity with another pointing device such as a pen may push the user to a pen-based mode of operation. This is especially true if the HCI design is attempting to account for gestural inputs that may be beyond the capability of the mouse. Additionally, design for users with disabilities may make selection of alternative input and display devices necessary. For example, head and ocular controls can be used in place of hand-manipulated pointing and selection devices. For the purposes of the example, it is assumed that menus have been determined to be optimal for use of resources on the mental interface, that mouse and keyboard input devices have been determined to be optimal, and that a 14-in. color monitor has been determined to be minimal for the physical interface. It is further determined that a number keypad is not needed for the application tasks of interest.

The next mental-interface design issue is defining the menu system structure. This issue is associated with the application-task level. Structural decisions will be based on the task structure in an attempt to reflect the task knowledge of the user. Specific techniques such as pathfinder [Roske-Hofstrand & Paap, 1986; Schvaneveldt et al., 1985] can be employed to extract user associations and structural knowledge about the concepts that will be included in the menus. Specific design decisions are included in Fig. 153.5.

For the physical interface, the next design issue remains at the application-task level and involves specifying the overt physical actions that the user must perform to select menus and menu items. This specification of procedures is dependent on the input and display devices selected and the structure of the menu system to be accessed.

The goal is to define procedures that reduce the user's sensory-motor resource requirements. At the application level, this reduces to making the procedures simple, consistent, natural, and fault-tolerant. This translates to minimizing the number of steps to select a given item, minimizing the number of types of steps needed to be learned, providing adequate feedback to the user at each stage, and providing multiple access procedures that allow skilled users to access actions more directly.

The next design issues for the mental-interface component are associated with the interaction-task level and involve specification of the representations for menus and menu items. These include menu and item names or graphic representations. It further involves the specification of the menu box and other menu indications (e.g., an ellipsis for indicating that a dialog box will be opened or an arrow indicating a cascading menu item). The specification of the selection procedures also may impose design considerations at this point. For example, keyboard operation will involve use of a selection cursor that cues the user to the active selection position. The perceptual-cognitive aspects of the design of this cursor will be considered here.

For the physical interface, the next design issues are associated with the interaction-task level. The specified procedures will be used to determine menu-specific, low-level operating parameters for the display and input devices. Menu-specific issues related to menu selection with a multibutton mouse device include the following:

- Size of the active screen selection area of each menu button. Greater size will translate to a reduced demand on pointing (positioning) accuracy resources.
- Size of the active screen selection area for each menu item. Greater size will require less precise pointing.
- Number of items in a selected menu. More items require more access time (or greater processing speed to keep access time fixed), more precise pointing, and create a greater probability of selection errors.
- Use of cascading menus. Selection requires more precise pointing.

Issues related to selection with a keyboard include the following:

- Key binding to selection action. Traditional, familiar, and consistent binding will reduce memory requirements and probability of user error.
- Distinguishability of the displayed selection cursor. Greater distinguishability requires less visual search and detection time.
- Size, sensitivity, required operating force of keyboard device.
- Response lag between keystroke and selection-cursor movement. Greater lag will create greater pointing error and increased pointing time.

Defining Terms

Application task: Refers to the applied objectives the user is employing the computer to accomplish (e.g., writing a report, performing a statistical analysis, finding information in a database, controlling a process, making a medical diagnosis, and monitoring/testing the health of a system).

Direct manipulation: A term coined by Schneiderman [1982, 1983]. Refers to an HCI or style of interaction with the following characteristics:

- Objects are continuously displayed.
- Physical action (e.g., point and click, drag and drop interaction) rather than command line interaction is employed.
- Impact of actions are rapid and directly observable.

Interaction task: Refers to the activities that the user performs to use the computer (e.g., moving a mouse device, typing keyboard inputs, viewing displays, searching for and managing files, managing displays, setting preferences, and executing programs).

Usability: Usability is a multidimensional quality that affords the user practical and convenient interaction with the computer for achieving applied objectives. The concept of usability is fundamental to the design of any human-system interface and is a major factor affecting the operational performance, availability, maintainability, and reliability of the overall system.

User-centered: User-centered design refers to a philosophy of human-machine system design that places the focus of the design process on the needs of the human who uses the system to accomplish a task. The following items characterize a user-centered design approach:

- The system is viewed as a tool that the user employs to accomplish objectives.
- The user is always performing a higher-level task than the system.
- Takes are allocated away from the user rather than to the user. This is not simply a matter of perspective but rather is fundamental to a human-centered approach. That is, the user is assumed to be performing a high-level task. Lower-level tasks that are better performed by the computer are allocated from the user to the computer. This approach is opposed to one that considers the computer as the high-level task performer and allocates to the user lower-level tasks that the machine cannot do well.
- Tasks allocated to the user are meaningful.
- The user's responsibilities provide job satisfaction and make use of the user's talents and skills.
- Users are involved in all phases of the design process.
- The user's skills and performance capabilities are explicitly considered in and accounted for in design decisions.

References

Bevan N, Macleod M. 1994. Usability measurement in context. Behav Inform Technol 13:132.

Brown CM. 1988. Human-Computer Interface Design Guidelines. Norwood, NJ, Ablex Publishing.

Card SK, Moran TP, Newell A. 1983. The Psychology of Human-Computer Interaction. Hillsdale, NJ, Erlbaum.

Card SK, Moran TP, Newell A. 1986. The model human processor. In KR Boff, L Kaufman, JP Thomas (eds), Handbook of Perception and Human Performance, vol II, pp 45-1–45-35. New York, Wiley.

Galitz WO. 1993. User-Interface Screen Design. Boston, QED Publishing Group.

Imrhan S. 1995. Analysis of high-level tasks: Physical components. In JD Bronzino (ed), The Biomedical Engineering Handbook. Boca Raton, Fla, CRC Press.

ISO. 1993. ISO CD 9241-11, Guidelines for Specifying and Measuring Usability.

Kieras DE. 1988. Towards a practical GOMS model methodology for user interface design. In M Helander (ed), Handbook of Human-Computer Interaction, pp 135–157. New York, Elsevier Science Publishers.

Kondraske GV. 1995. A working model for human-system-task interfaces. In JD Bronzino, (ed), The Biomedical Engineering Handbook, pp 2157–2174. Boca Raton, Fla, CRC Press.

Maxwell KJ. 1995. Analysis of high-level tasks: Mental components. In JD Bronzino (ed), The Biomedical Engineering Handbook, pp 2233–2248. Boca Raton, Fla, CRC Press.

Payne SJ, Green T. 1986. Task-action grammars: A model of mental representation of task languages. Human-Computer Interact 2:93.

Roske-Hofstrand RJ, Paap KR. 1986. Cognitive networks as a guide to menu organization: An application in the automated cockpit. Ergonomics 29:1301.

Schiele F, Green T. 1990. HCI formalisms and cognitive psychology: The case of task-action grammar. In M Harrison, H Thimbleby (eds), Formal Methods in Human-Computer Interaction, pp 9–62. Cambridge, UK, Cambridge University Press.

Schvaneveldt RW, Durso FT, Dearholt DW. 1985. PATHFINDER: Scaling with Network Structures. Report no. MCCS-85-9, Computing Research Laboratory, Las Cruses, NM.

Shneiderman B. 1982. The future of interactive systems and the emergence of direct manipulation. Behav Inform Technol 1:237.

Shneiderman B. 1983. Direct manipulation: A step beyond programming languages. IEEE Comput 16:57.

Further Information

The following publications provide specifications and style guidelines and standards associated with specific user-interface systems.

Apple Computer. 1992. Apple Human Interface Guidelines: The Apple Desktop Interface. Reading, Mass, Addison-Wesley.

IBM. 1989. Systems Application Architecture, Common User Access Advanced Interface Design Guide, SC26-4582-0.

IBM. 1991. Systems Application Architecture, Common User Access Advanced Interface Design Reference, SC26-4582-0.

IBM. 1991. Systems Application Architecture, Common User Access Guide to User Interface Design, SC34-4290-00.

Microsoft Programming Series. 1992. The Windows Interface: An Application Design Guide. Washington, Microsoft Press.

Open Systems Foundation. 1993. OSF/Motif Style Guide. Revision 1.2, Cambridge, Mass.

The following books provide general design standards, guidelines, and methods.

Card SK, Moran TP, Newell A. 1983. The Psychology of Human-Computer Interaction. Hillsdale, NJ, Erlbaum.

Helander M (ed). 1988. Handbook of Human Computer Interaction. New York, Elsevier Science Publishers.

Human Factors and Ergonomics Society (ed). 1988. ANSI Standard HFS 100. American National Standard for Human Factors Engineering of Visual Display Terminal Workstations. Santa Monica, Calif, Human Factors and Ergonomics Society.

ISO. 1993. ISO 9241, Ergonomic Requirements for Office Work and Visual Display Terminals.

ISO. 1993. ISO DIS 9241-10, Dialogue Principles.

ISO. 1993. ISO DIS 9241-14, Menu Dialogues.

Laurel B (ed). 1990. The Art of Human-Computer Interface Design. Reading, Mass, Addison-Wesley.

Mayhew DJ. 1991. Principles and Guidelines in Software User Interface Design. Englewood Cliffs, NJ, Prentice-Hall.

Norman DA, Draper SW (eds). 1986. User Centered System Design, Hillsdale, NJ, Erlbaum.

Peddie J. 1992. Graphical User Interfaces and Graphic Standards. New York, McGraw-Hill.

Shneiderman B. 1987. Designing the User Interface: Strategies for Effective Human-Computer Interaction. Reading, Mass, Addison-Wesley.

Smith SL, Moiser JN. 1986. Guidelines for Designing User Interface Software. Bedford, Mass, MITRE Corporation.

Tognazzini B. 1992. TOG on Interface. Reading, Mass, Addison-Wesley.

The *International Journal of Human-Computer Studies* and *Human Factors* are excellent sources for new research findings on HCI design. The Association for Computing Machinery (ACM) has a *Special Interest Group in Computer-Human Interaction* (SIGCHI). The Human Factors and Ergonomics Society has a technical group on computer systems. Both organizations are an excellent source of information on new research and development through publications, conferences, and professional contacts.

154

Applications of Human Performance Measurements to Clinical Trials to Determine Therapy Effectiveness and Safety

Pamela J. Hoyes Beehler
University of Texas at Arlington

Karl Syndulko
UCLA School of Medicine

A *clinical trial* is a research study involving human subjects and an intervention (i.e., device, drug, surgical procedure, or other procedure) that is ultimately intended to either enhance the professional capabilities of physicians (i.e., improve the service delivered), improve the quality of life of patients, or contribute to the field of knowledge in those sciences which are traditionally in the medical field setting—e.g., physiology, anatomy, pharmacology, epidemiology, neurology, cognitive psychology, etc. [Levin, 1986]. Clinical trials research is in the business of evaluating therapeutic interventions intended to benefit humans. Its value is directly related to the relevance of the questions "Do our treatments work?" "How well do our treatments work?" and "Are our treatments safe?" For example, drug *A* is designed and anticipated to relieve sinus congestion. Is there a drug interaction when drug *A* is taken with drug *B* and/or moderate levels of alcohol consumption such that while congestion is relieved (i.e., drug *A* is effective), human information processing capacities are reduced (i.e., is drug *A* safe?). And what is the time course of effects with regard to positive and negative (or adverse) effects? Thus it is clear that not only steady-state issues buy also dynamic questions are on interest in clinical trials research.

While clinical trials research incorporates many different components, the focus of this chapter is limited to study questions associated with human performance capacity variables and their measurement as they contribute to the determination of therapy effectiveness and safety. Such variables have been incorporated into trials since the use of controlled studies in the medical field began. However, the

0-8493-0462-8/00/$0.00+$.50

methodology employed to address human performance variables has been slowly but steadily shifting from mostly subjective to more objective instrumented methods [e.g., Tourtellotte et al., 1965] as both the understanding of the phenomena at play and the demand for improved quality of studies have increased. This chapter begins by briefly examining a classification of typical clinical trials study models, presents a summary of methods employed and key methodologic issues in both the design and conduct of studies (with special emphasis on issues related to the selection of measures and interpretation of results), and ends with a walk-through of a typical example that demonstrates the methods described. Brief discussion of the benefits that can be attributed to the use of objective, instrumented measures of human performance capacities as well as their current limitations in clinical trials research is also presented. While it is emphasized that most methods and issues addressed are applicable to any intervention, the use of human performance variables in pharmaceutical clinical trials has been most prevalent, and special attention has been given here to this application.

154.1 Basic Principles: Types of Studies

Depending on the set of primary and secondary questions to be addressed, considerable variety can exist with respect to the structure of a given trial and the analysis that is performed. Within the focus and scope of this chapter, clinical trials are classified into two categories for discussion: (1) safety-oriented and (2) efficacy-oriented. According to the Food and Drug Administration (FDA) [1977], four phases of research studies are required before a drug can be marketed in the United States. Phase I is known as *clinical pharmacology* and is intended to include the initial introduction of a drug into humans. These studies are safety-oriented; one issue often addressed is determination of the maximum tolerable dose. Phases II and III are known as *clinical investigation* and *clinical trials*, respectively, consisting of controlled and uncontrolled clinical trials research. Phase IV is the *postmarketing of clinical trials* to supplement premarketing data. Both safety and efficacy are addressed in phase II to IV investigations. As specified earlier, not all interventions studied are drugs. However, a similar phased approach is also characteristic of the investigation of therapeutic devices and treatments.

Safety-oriented trials usually involve vital sign measures (e.g., heart rate, blood pressure, respiration rate), clinical laboratory tests (e.g., blood chemistries, urinalysis, ECG), and adverse reaction tests (both mental and physical performance capacities) to evaluate the risk of the intervention. In this chapter we focus on the adverse reaction components, most of which historically have been addressed with subjective reporting methods. In the case of drug interventions, different doses (e.g., small to large) are administered within the trial, and the dose-related effects are examined. Thus the rate at which a drug is metabolized (*pharmacokinetics*), as evidenced by changes in drug concentrations in blood, cerebral spinal fluid, urine, etc., is usually addressed in safety-oriented drug studies, as well as the maximum dosage which subjects can tolerate [Baker et al,. 1985; Fleiss, 1986; Jennison & Turnbull, 1993; Tudiver et al., 1992].

Efficacy-oriented trials are usually conducted after initial safety-oriented trials have established that the intervention has met safety criteria, but they will always have safety questions and elements as well. In this type of study, the goal is to objectively determine the therapeutic effect. Whatever the type of intervention, these studies are designed around the bottom-line question "Is the intervention effective?" In many cases (e.g., drugs for neurologic disorders, exercise programs for musculoskeletal injuries, etc.), *effectiveness* implies improvement or retarding the rate of deterioration in one or more aspects of mental and/or physical performance. With regard to disease contexts, it is necessary to distinguish performance changes that merely reflect treatment of systems as opposed to those which reflect the slowing or reversal of the basic disease process. Studies typically include pre- and postintervention measurement points and, whenever feasible, a control group that is administered a placebo intervention [Fleiss, 1986; Tang & Geller, 1993; Weissman, 1991]. The situation is complicated because there are many performance capacities that could be affected. In response to the intervention, some capacities may improve, some may remain unchanged, and some may be adversely affected. Thus efficacy-oriented studies typically include a number of secondary questions that address specificity of effects. When the intervention is a drug, both the pharmacokinetics and *pharmacodynamics* are often addressed. Pharmacodynamic studies attempt to relate physiologic and/or metabolic changes in the concentration of the agent over time to corresponding changes in the therapeutic effect. Thus repeated measurement of selected performance capacities are required over relatively short periods of time as part of these protocols.

Clinical trials involving human performance metrics generally use the *randomized clinical trial* (RCT) design. The RCT is a way to compare the efficacy and safety of two or more therapies or regimens. It was originally designed to test new drugs [Hill, 1963]; however, over the last 25 years, it has been applied to the study of vaccinations, surgical interventions, and even social innovations such as multiphasic screening [Levin, 1986]. Levin [1986] describes four key elements of RCTs: (1) the trials are "controlled," i.e., part of the group of subjects receive a therapy that is tested while the other subjects receive either no therapy or another therapy; (2) the significance of its results must be established through statistical analysis; (3) a double-blind experimental design should be used whenever possible; and (4) the therapies being compared should be allocated among the subjects randomly. Levin[1986] further noted that "the RCT is the gold standard for evaluating therapeutic efficacy."

154.2 Methods

Studies are defined and guided by *protocols*, a detailed statement of all procedures and methods to be employed. Figure 154.1 summarizes the major steps in clinical trials in which human performance variables represent those of primary interest, although the general process is similar for most trial involving human subjects.

Selecting Study Variables

In many clinical trials, performance measurements must be focused on the specific disease or set of symptoms against which a medication or intervention is directed. The critical element is selection of an outcome assessment that will indicate directly whether the intervention can eliminate key signs or

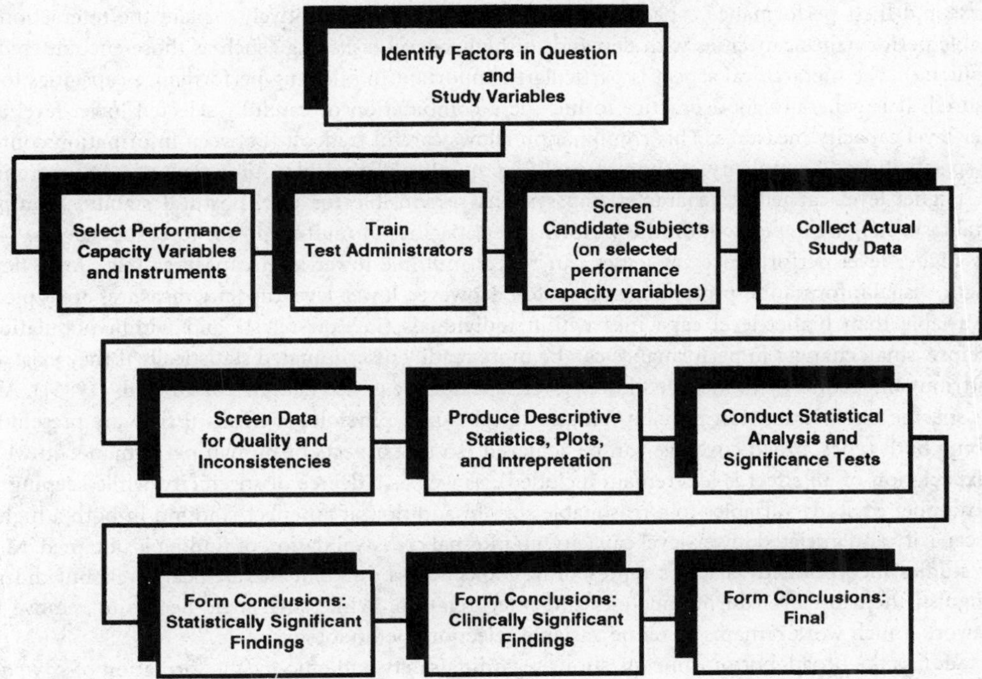

FIGURE 154.1 Summary of major steps in the conduct of a clinical trial involving human performance capacities.

symptoms of the disease, improve function affected by the disease, or even simply delay further disease progression and loss of function. Economic and statistical constraints of such trials often indicate that a minimal set of measures must be used to provide valid, reliable, and sensitive indicators of changes in underlying disease activity related to the treatments. In many instances, there is considerable controversy about what constitutes a valid and reliable assessment of the underlying disease process, or there may be de facto standards adopted by clinicians, pharmaceutical companies, and the FDA as to the choice of outcome assessments. In many diseases, these standards may be physician rating scales of disease severity. Thus objective studies may be required in the case of specific diseases and disorders to delineate the most sensitive set of performance measures for clinical trial evaluations [Mohr & Prouwers, 1991]. Further, it often must be demonstrated to clinicians, drug companies, and the FDA that the objective performance measures selected as outcome assessments actually do provide equal or more sensitive indicators of disease progression (and improvement) than do existing rating scale measures traditionally utilized in clinical trials. Experience with human performance testing in multiple sclerosis clinical trials, discussed below, illustrates these issues.

When objective measurement of performance capacities has been incorporated into many clinical trials, concepts and tools from human performance engineering can facilitate the selection of variables and shed some light on issues noted above. In either safety- or efficacy-oriented studies, study variable selection can be characterized as a two-step process: (1) identification of the factors in question (e.g., Table 154.1) and (2) selection of the relevant *performance capacities* to be measured and associated measurement instruments. This link between these two steps often represents a challenge to researchers for a number of reasons. First, duality in terminology must be overcome. Concerns about an intervention are typically initially identified with "negative" terms such as "dizziness" and not in terms of performance capacities such as "postural stability." Human performance models based on systems engineering concepts [e.g., Kondraske,1995] can be used to facilitate the translation of both formal and lay terms used to identify adverse effects to relevant performance capacities to be measured, as shown in Table 154.1.

In Kondraske's [1995] model, performance capacities are modeled as resources that a given system possesses and draws up to perform tasks. It also provides a basis for delineation of *hierarchical* human systems and their performance capacities as well as a basis to quantitatively explain the interaction of available performance capacities with demands of higher-level tasks (e.g., such as those encountered in daily living). The hierarchical aspect is particularly important in selecting performance capacities to be measured. It is generally good practice to include a combination of carefully selected lower-level and higher-level capacity measures. This combination allows careful tradeoffs between information content (i.e., specificity) and simplicity of the protocol (e.g. number of variables, time for test administration, etc.). Higher-level capacities (characterizing systems responsible for gait, postural stability, complex mental task, etc.) are dependent on the performance capacities of multiple lower-level subsystems. Thus a few higher-level performance capacities can reflect multiple lower-level capacities (e.g., knee flexor strength, visual information processing speed, etc.). However, lower-level capacity measures are typically less variable than higher-level capacities within individuals (i.e., on retest) and within populations. Therefore, small changes in performance can be more readily discriminated statistically if they exist and are attributable primarily to localized lower-level capacities (e.g., see data in Potvin et al. [1985]). Also, more specific information can provide valuable insights into physiologic effects if they are present. By combing both types, broad coverage can be achieved (so that aspects of human performance in which the expectation of an effect is less remain included), as well as a degree of specificity, while keeping the total number of study variables to a reasonable size. In addition, if an effect is found in both a higher-level capacity and a related lower-level capacity, an internal cross-validation of finding is obtained. Many early studies incorporated variables representing looks across different hierarchical levels but did not distinguish them by level during analysis and interpretation. While such logic begins to remove the guesswork, much work remains to refine variable-selection methodology.

To address the broad, bottom-line questions regarding safety and efficacy, the formation of *composite scores* (i.e., some combination of scores from two or more performance capacity challenges or "tests") is necessary. A resource model and the concept of hierarchical human systems also can be used to develop

TABLE 154.1 Representative Examples of Linking Factors in Question to Specific Measurable Performance Capacities

Factor in Question*	Selected Relevant Performance Capacities†
General central nervous system effects (of drug)	Selected activity of daily living execution speed
	Postural stability
	Upper extremity neuromotor channel capacity
	Manual manipulation speed
	Visual-upper extremity information processing speed
	Visual attention span
	Visual-spatial memory capacity
	Visual-numerical memory capacity
Alcohol interaction (of drug)	Selected activity of daily living execution speed
	Postural stability
	Visual information processing speed
	Visual-spatial memory capacity
Dizziness	Postural stability
Drowsiness	Selected activity of daily living execution speed
	Visual attention span
	Visual-upper extremity information procession speed
	Bond-Lader visual analog scale
Slowness of movement, psychomotor retardation	Visual-upper extremity information processing speed
	Upper extremity random reach movement speed
	Index finger (proximal intercarpal phalangeal joint), flexion-extension speed
Mood	Emotional stability (e.g., with Hamilton anxiety scale, Bond-Lader analog scale and other similar tools)
Speed	Reaction Time
Coordination	Finger tapping
Tremors, abnormal movements	Limb steadiness
	Postural stability
	Vocal amplitude and pitch steadiness
Joint stiffness or pain	Extremes and range of motion
	Isomeric strength (selected joins)
Weakness, strength	Postural stability (one leg, eyes open)
	Isometric grip strength
	Isometric strength (representative set of upper and lower extremity, proximal and distal muscle groups)
Sensation	Vibration sensitivity
	Thermal sensitivity

*Terms used include those often used by pharmaceutical companies to communicate adverse effects to general public.
†Lists are illustrative and not exhaustive. More or fewer items can be included depending on time available or willingness to expend for data collection, with commensurate tradeoffs between specificity and protocol simplicity.

composite variables for practical use in clinical trials. Component capacities can be selected that are theoretically or empirically most sensitive to the disease or condition under study. The composite variable is then created by calculating a more traditional weighted arithmetic sum [e.g., Potvin et al., 1985] or product [e.g., Kondraske, 1989] of the component scores. In both cases, the definition of component scores is important (i.e., whether a smaller or larger number represents better performance). Treatments are not approved if only a small subset of individuals is helped. Composite measures provide the only objective means of integrating multidimensional information about an intervention's effects. The primary advantages of the composite variable is the creation of a single, global, succinct measure of the disease, condition, or intervention, which is often essential as the primary outcome assessment for efficacy. The major disadvantage is loss of detailed information about the unique profile of performance changes that each subject of a group as a whole may show. Such tradeoffs are to be expected. Thus both composite and component measures play important but different roles in clinical trials.

It is imperative that the selection of the study measures also consider a test's objectivity (nonbiased), reliability (consistency), and validity (measures what is intends to measure) to add to the quality of the measurements. Many complex issues, which are beyond the present scope, are associated with measuring and interpreting the quality of a measurement. (See Baumgartner and Jackson [1993], Hastad and Lacy [1994], and Safrit [1990] for more information.)

Formation of the Subject Pool

Clinical research investigations usually involve a sample group of subjects from a defined population. Selection of the study population so that generalizations from that sample accurately reflect the defined population is a dilemma that must be adequately addressed. If *probability sampling* is chosen, each subject in the defined population theoretically has an equal chance of being included in the sample. The advantage of this kind of sampling is that differences between treatment groups can be detected and the probability that these differences actually exist may be estimated. If *nonprobability sampling* is chosen, there is no way to ensure that each subject had an equal chance of being included in the sample. Conclusions of nonprobability sampling therefore have less merit (not as generalizable) than those based on probability sampling.

Studies with healthy subjects are believed to be necessary before exposing sick persons to some interventions because persons with disease or injuries commonly have impaired function of various organs that metabolize drugs and may take medications that can alter the absorption, metabolism, and/or excretion rates of the intervention. Gender issues also should be a concern in clinical trials because of new FDA regulations [Cotton, 1993; Merkatz et al., 1993; Stone, 1993].

All subjects selected as study candidates should be informed of the procedures that will be utilized in the clinical trials investigation by signing an informed consent document, as required by the Department of Health and Human Services *Code of Federal Regulations* [1985] and other federal regulations applicable to research involving human subjects. Using all or a subset of study measurement variables (in addition to medical history and examinations as necessary), a screening procedure is recommended to determine if each subject meets the minimum performance criteria established for subject inclusion. When patients are part of the sample group, this performance screening also can be used to establish that the sample includes the desired balance of subjects with different amounts of "room for improvement" on relevant variables. An added benefit of this screening process is that subjects and test administrators obtain experience with test protocols, equipment, an procedures that will add to the validity of the study.

Unless available from a previous similar study, it is typical for pilot data to be collected to estimate the expected size of outcome effects. With this information, the power of the statistical analysis (i.e., the likelihood that a significant difference will be detected) can be estimated so that sample size can be determined [Cohchran & Cox, 1957; Fleiss, 1986; Kepple, 1982]. Another concern is that some of the original subject pool will not complete the study or will have incomplete data. High attrition rates may damage the credibility of the study, and every effort should be made to not lose subjects. DeAngelis [1990] estimates that attrition rates higher than 50% make the interpretation of clinical trial research very difficult. Some researchers believe that no attempt should be made to replace these subjects because even random selection of new subjects will not ensure bias caused by differential participation of all the subjects.

Data Collection

Investigators should seek to minimize sources of variability by careful control of the test conditions and procedures. Proper control can be attained by (1) using rooms that are reasonably soundproof or sound-deadened, well lighted, and of a comfortable temperature; (2) selecting chairs and other accessories carefully (e.g., no wheels for tests in which the subject is seated); (3) testing subjects one at a time without other subjects in the test room; (4) using standardized written instructions for each test to eliminate variability in what is stated; (5) allowing for familiarization with the test instruments and procedures; (6) not commenting about subject performance to avoid biased raising or lowering of expectations;

(7) arranging a test order to offset fatigue (mental and physical) or boredom and to include rest periods; and (8) training test administrators and evaluating their training using healthy subjects, especially in multicenter studies, so that consistent results can be obtained.

Data Integrity Screening

Despite all good efforts and features incorporated to ensure high-quality data, opportunities exist for error. It is therefore beneficial to subject data obtained during formal test sessions to quality screenings. This is a step toward forming the official study data set (i.e., the set that will be subjected to statistical and other analyses). Several independent analyses used to screen data, which are typically computer-automated, are described below.

Screening of Baseline Measures Against Inclusionary Criteria. If human performance study variables with specific performance criteria are used as part of the subject selection process as recommended above, each subject's baseline score can be compared with the score obtained during inclusionary testing. For most variables, baseline scores should not differ from inclusionary testing by more than 20% [Potvin et al., 1985]. Greater deviations point to examiner training, test procedure, or subject compliance problems.

Screening Against Established Norms. For variables that are not included in inclusionary screenings or in studies where performance variables are not used as part of the subjects inclusionary criteria, baseline data can be screened against established reference data (e.g., human performance means and standard deviations). Data are considered acceptable if they fall within an established range (e.g., two standard deviation units of the reference population mean). From the perspective of risk associated with not identifying a data point that could be a potential problem, this is a fairly liberal standard that can still identify problems in data collection and management. Standards that lessen risk can, or course, be employed at the discretion of the investigator, but at the expense of the possible identification of a larger number of data points that require follow-up.

Screening Against Anticipated Effects. It is more difficult to screen nonbaseline data for quality because of the possible influences of the intervention. However, criteria can be established based on (1) absolute level of variables (both maximums and minimums) and (2) rates and direction of change from one measurement period to the next. Even criteria that allow a rather wide range of data (or changes across repeated measures) can be useful in detecting gross anomalies and is recommended.

If anomalies are discovered using the screening methods described above, several outcomes are possible: (1) The anomaly can be traced and rectified (e.g., it may be attributable to a human or computer error with backup available); (2) the anomaly may be explainable as a procedural error, but it may not be possible to rectify; and (3) the anomaly may be unexplainable. Data anomalies that are explainable with supporting documentation could justify classification of the given data items as "missing data" or replacement of all data for the corresponding subject in the data set (i.e., the subject could be dropped from the study and replaced with another). There is no justification for eliminating or replacing data anomalies for which a documentable explanation does not exist.

Analysis and Interpretation of Results

Traditional inferential statistical analysis should be performed according to the statistical model and significance levels agreed on by the clinical research team when the study is designed. However, data also should be analyzed and interpreted from a clinical perspective as well.

In experimental research, tests of *statistical significance* are the most commonly used tools for assessing possible associations between independent and dependent variables as well as differences among treatment groups. The purpose of significance (i.e., statistical) tests is to evaluate the research hypothesis at a specific level of probability, or p value. For example, if a p value of 0.05 was chosen, the researcher is asking if the levels of treatment (for example) differ significantly so that these differences are not attributable to a chance occurrence more than 5 times out of 100. By convention, a p value of 0.01 (1%) or 0.05 (5%) is usually selected as the cutoff for statistical significance. Significance tests cannot accept a research hypothesis; all that significance tests can do is reject or fail to reject the research hypothesis [Thomas & Nelson, 1990]. Ultimately, significance tests can determine if treatment groups are different, but not why they are different. Good experimental design, appropriate theorizing, and sound reasoning are used to explain why treatment groups differ.

Exclusive reliance on tests of statistical significance in the present context can mislead the clinical researcher, and interpretation of *clinical significance* is required. The key issue here is the size of the observed effect; p values give little information on the actual magnitude of a finding (i.e., decrement or improvement in performance, etc.). Also, statistical findings are partially a function of sample size. Thus even an effect that is small in size can be detected with tests of statistical significance in a study with a large sample. For example, a mean difference in grip strength of 2 kg may be found in response to drug therapy. This difference is a quite small fraction of the variability in grip strength observed across normal individuals, however. Thus, although statistically significant, such a change would perhaps have a minimal impact on an individual's ability to function in daily activities which make demands on grip strength performance resources. Thus clinical significance addresses cause-and-effect relationships between study variables and performance in activities of daily life or other broad considerations such as the basic disease process. The danger of not defining clinical significance appropriately can be either that a good intervention is not used in practice (e.g., it may be rejected for a safety finding that is statistically significant but small) or that a poor intervention is allowed into practice (e.g., it may result in statistically significant improvements that are small).

Ideally, if empirical data existed that established what amount of a given performance capacity such as visual information processing speed (VIPS) was necessary to perform a task (e.g., driving safely on the highway), then it would be possible to interpret statistically significant findings (e.g., a decrement in VIPS of a known amount) to determine if the change would be a magnitude that would limit the individual from successfully accomplishing the given task. Unfortunately, while general cause-effect relationships between laboratory-measured capacities and performance in high-level tasks are evident, quantitative models do not yet exist, and completely objective interpretations of clinical significance are not yet possible. As such, the process by which clinical significance is determined is less well developed and structured. More recent concepts introduced by Kondraske [1988*a*, 1988*b*, 1989] based on the use of *resource economic* principles (i.e., the idea of a threshold "cost" associated with lower-level variables typically employed in clinical trials for achieving a given level of human performance in any given high-level task) may be helpful in defining objective criteria for clinical significance. This approach directly addresses cause-and-effect relationships between performance capacities and high-level tasks with an approach much like that which an engineer would employ to design a system capable of performing a specified task.

Despite known limitations, interpretation of clinical significance is always incorporated into clinical trials in some fashion. Any change in human performance for a given variable should be first documented to be statistically significant before it is considered to be a candidate for clinical significance (i.e., a statistically significant difference is a necessary but not a sufficient condition for clinical significance). Objective determination of clinical significance can be based in part on previous methods introduced by Potvin et al. [1985] in clinical trials involving neuromotor and central processing performance tests. They advocate the use of an objective criteria whereby a decrease or increase in a human performance capacity (with healthy test subjects, for example) should be greater than 20% to be classified as "clinically significant." Kondraske [1989] uses a similar approach that uses z-scores (i.e., number of standard deviation units form a population mean) as the basis for determining criteria. This accounts for population variability which is different for different performance capacity variables. A recent approach is to assess effect size, a statistical metric that is independent of sample size and which takes into account data variability. This method provides an objective basis for comparison of the magnitude of treatment effects among studies [Cohen, 1988; Ottenbacher & Barrett, 1991].

154.3 Representative Application Examples

Safety-Oriented Example: Drug-Alcohol Interaction

In this section, selected elements of an actual clinical trial are presented to further illustrate methods and issues noted above. To maintain confidentialities, the drug under test is simple denoted as drug *A*.

Identify Factors in Question. Upper respiratory infections are among the most frequent infections encountered in clinical practice and affect all segments of the general population. The pharmacologic agent of choice for upper respiratory infections are the antihistamines (reference drug), which possess a wide margin of safety and almost no lethality when taken alone in an overdose attempt. Although quite effective, antihistamines can produce several troublesome side effects (i.e., general CNS impairment, sedation, and drowsiness) and have been incriminated in automobile accidents as well as public transportation disasters. New drugs (e.g., drug *A*) are being developed to have similar benefits but fewer side effects. From prior animal studies, drug *A* has been shown to have minimal effects on muscle relaxation and muscular coordination as well as less sedative and alcohol-potentiating effects than those associated with classical antihistamines. Based on drug *A*'s history and concerns, the following factors in question were identified in this clinical trial: general CNS impairment, alcohol interaction, dizziness, drowsiness, mood, and slowness of movement/speed. The purpose of this investigation was to examine the effect of drug *A* on human performance capacity relative to a reference drug and placebo as well as the drug-alcohol potentiating interaction effects after multiple dose treatment.

Select Performance Capacity Variables and Instruments. With the factors in question identified, the following performance capacity variables and their testing instruments were selected for the clinical trial: visual information processing speed (VIPS), finger tapping speed (TS), visual arm lateral reach coordination (VALRC), visual spatial memory capacity (VSMC), postural performance (PP), digit symbol substitution task (DSST), and Bond-Lader Visual Analog Scale (BLS). These variables were collectively called the *human performance capacity test battery* (HPCTB) (Table 154.2). Due to space limitations, only one performance capacity variable (VMRS—visual information processing speed, i.e., VIPS) is discussed in this section.

Experimental Design. The first primary objective of this investigation was to examine after 8 days of oral dosing the relative effects of drug *A*, reference drug, and placebo on the human performance capacity of healthy male and female adult volunteers (see Table 154.2). This effect was determined by examining the greatest decrease in human performance capacity from day 1 of testing (i.e., baseline at –0.5 hour drug ingestion time) to day 8 of testing at drug ingestion times of 0.0, 1.0, 2.0, 4.0, and 6.0 hours. The

TABLE 154.2 Experimental Schedule for Human Performance Capacity Test Battery (HPCTB)

	Drug Ingestion Times (hours)					
Test Day	–0.5	0.0	1.0	2.0	4.0	6.0
1*	HPCTB					
8†		HPCTB	HPCTB	HPCTB	HPCTB	HPCTB
9†,‡		HPCTB	HPCTB	HPCTB	HPCTB	HPCTB

Note: The human performance capacity test battery (HPCTB) was administered in the following order: VIPS, TS, VALRC, VSMC, PP, DSST, and BLS.

*On day 1 of testing only, the –0.5 HPCTB was performed and utilized as the baseline value for test days 8 and 9.

†The peak effect was determined by comparing baseline values (day 1) against the greatest detriment in performance for the eighth and ninth testing days at drug ingestion times of 0.0, 1.0, 2.0, 4.0, and 6.0 hours.

‡On day 9 of testing, an alcohol drink was served immediately after drug administration and was ingested over a 15-minute period.

second primary objective was to examine after 9 days of oral dosing the relative effects of drug *A*, reference drug, and placebo in combination with a single dose of alcohol served immediately after drug administration (male alcohol dose 0.85 g/kg body weight, female alcohol dose 0.75 g/kg body weight) on the human performance capacity of healthy male and female adult volunteers (see Table 154.2). This effect was determined by comparing the greatest decrease in performance on the ninth day of study drug treatment (at drug ingestion times of 0.0, 1.0, 2.0, 4.0, and 6.0 hours) to baseline on day one of testing (–0.5 hour drug ingestion time) for the HPCTB (see Table 154.2). Independent variables were treatment group (1 = drug *A*, 2 = reference drug, and 3 = placebo) and gender (male/female). The dependent variable was maximum decrease in performance (peak effect) for each human performance capacity study variable and was determined by comparing the baseline value (day 1) against the greatest decrease in performance at drug ingestion times of 0.0, 1.0, 2.0, 4.0, and 6.0 hours for day 8 and day 9 of testing. Inferential statistical analysis of the data was performed using a 3 (treatment group: reference drug, drug *A*, and placebo) by 5 (hour; 0, 1, 2, 4, and 6) mixed factorial ANOVA with repeated measures on hour for each dependent variable. Statistical tests of significance of all ANOVAs were conducted at the 0.05 level.

Data Screening. For each dependent variable, baseline treatment data were compared against the criteria established for subject inclusion using each subject's performance score from the HPCTB to identify data anomalies. Then the expected increases/decreases in human performance capacity from baseline treatment data were compared against the drug therapy-influenced data to determine if these changes were within "reasonable" limits. Potential anomalies were detected in less than 5% of the data, with most occurring within records for only a few subjects. These cases were investigated, and in consultation with the principal investigator, decisions were made and documented to arrive at the official data set.

Descriptive Statistical Analysis. Figure 154.2 illustrates the visual information processing speed (VIPS) for treatment groups reference drug, drug *A*, and placebo. On day 1 of testing (baseline), all treatment groups had similar VIPSs, with means between 5.7 and 5.8 stimuli per second. On day 8 of testing and at drug ingestion time of 2 hours, VIPS decreased to 4.9 stimuli per second for drug *A*, to 5.1 stimuli per second for reference drug and to 5.65 stimuli per second for the placebo. On day 8, the greatest group impairment of VIPS occurred at drug ingestion time of 4 hours. Drug *A* appeared to have the greatest

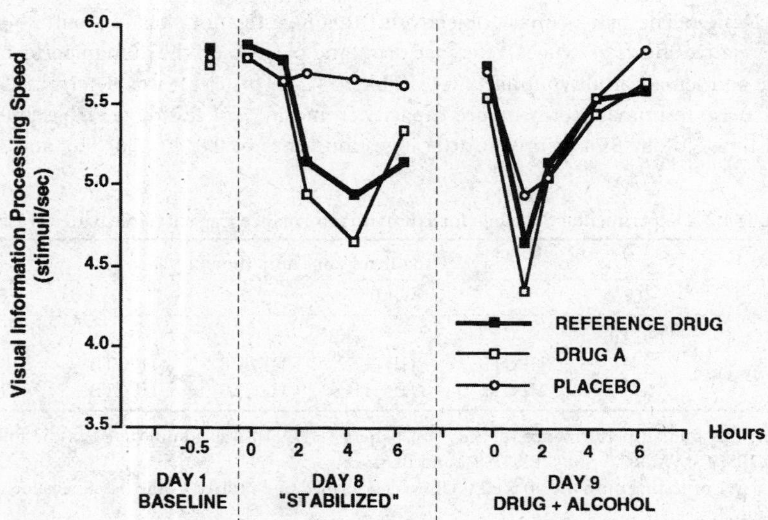

FIGURE 154.2 One performance capacity variable (visual information processing speed) for three treatment groups (drug *A*—the drug under test—reference drug, and placebo) taken from any actual efficacy-oriented study. Note changes from baseline to drug stabilization (day 8) and interaction with alcohol (day 9). See text for further explanation.

impairment of VIPS (4.6 stimuli per second), followed by the reference drug group (4.9 stimuli per second). This decrease in performance was not present with the placebo group (5.61 stimuli per second). By the sixth hour after drug ingestion on day 8, VIPS improved toward baseline for treatment groups reference drug (5.1 stimuli per second) and drug *A* (5.3 stimuli per second), while the placebo group remained unchanged (5.58 stimuli per second). One day 9 of testing, all treatment groups received their assigned drug and alcohol. By 1 hour after drug ingestion, all treatment groups' VIPSs were impaired from baseline, with drug *A* showing the greatest impairment (4.3 stimuli per second), followed by reference drug (4.6 stimuli per second) and placebo (4.9 stimuli per second). All treatment groups' VIPSs improved toward baseline values by the second hour after drug ingestion (5.0 to 5.1 stimuli per second) and the fourth hour after drug ingestion (5.4 to 5.5 stimuli per second), respectively. By the sixth hour after drug ingestion, reference drug and drug *A* plateaued between 5.55 and 5.6 stimuli per second, respectively, while the placebo group improved slightly above the baseline value (5.8 stimuli per second). The basic pattern observed (i.e., decrease from baseline, return to baseline, impairment of all groups, including placebo, with alcohol) lend validity to the overall study.

Statistical Analysis and Significance Test Results. On day 8 of testing, significant differences in VIPS for main effects treatment group and hour were observed. There was a significant interaction, however, between treatment group and hour of drug ingestion. Post hoc analysis demonstrated that treatment groups drug *A* and reference drug were significantly slower in VIPS than the placebo group at drug ingestion times of 2, 4, and 6 hours, but there was no difference in VIPS between treatment groups drug *A* and reference drug. (Simple main effects and simple, simple main effects showed that the placebo group had faster VIPSs than both reference drug and drug *A* treatment groups at drug ingestion times of 2, 4, and 6 hours.)

On day 9 of testing when the drug-alcohol potentiating interaction effect was of primary interest, significant differences in VIPS occurred for main effects treatment group and hour, but there were no interactions between treatment group and hour of drug ingestion. Post hoc analysis demonstrated that treatment groups drug *A* and reference group were significantly slower in VIPS than the placebo group at 1 hour after drug ingestion only, but there was no statistically significant difference in VIPS between treatment groups drug *A* and reference group at any hour after drug-alcohol ingestion. Also, the drug-alcohol interaction was significant 1 hour after drug-alcohol ingestion for all treatment groups.

Conclusions from Statistically Significant Findings. After 8 days of oral dosing, while a significant difference was found for both drug treatment groups compared with the placebo group, there was no significant difference in VIPS between drug *A* and reference drug treatment groups. The effect was apparent at approximately the same drug ingestion time of 4 hours. It was concluded that drug *A* had the same effect on safety as the reference drug in terms of CNS impairment as measured by VIPS.

After 9 days of oral dosing in combination with alcohol consumption, drug *A* and reference drug caused a greater decrease in VIPS compared with the placebo group. This effect was most apparent at 1 hour after drug ingestion. The placebo group also decreased its VIPS at 1 hour after drug ingestion at a statistically significant level. It was concluded after 9 days of oral dosing that drug *A* in combination with alcohol produced the same effect as the reference drug in combination with alcohol in terms of CNS impairment as measured by VIPS.

Conclusions from Clinically Significant Findings. Since there was no statistically significant difference between drug *A* and the reference drug on either day 8 or day 9, the only findings requiring an interpretation of clinical significance are the statistically significant findings for both drug *A* and reference drug relative to the placebo group. On both day 8 and day 9, these findings are clinically significant using either the Potvin et al. [1985] percentage change or Kondraske [1989] *z*-score approach. Using the resource economic model for human performance [Kondraske, 1988*a*, 1988*b*, 1989], it can be argued that, for example, during critical events during tasks such as driving, all of an individual's available VIPS resource is drawn on. Clearly, the substantial decrease in VIPS observed can compromise safety during such events.

Efficacy-Oriented Study: Experiences in Neurology

Performance testing in clinical trials in neurology has grown tremendously within the last decade [Mohr & Brouwers, 1991], but within specific disease areas there remains reluctance to accept the replacement of traditional rating-scale evaluations of disease presence and progression by objective performance testing [Syndulko et al., 1993]. The evolution of clinical evaluations in multiple sclerosis clinical trials illustrates this point.

Evaluation of disability change or deterioration in multiple sclerosis (MS) historically has been limited to physician rating scales that attempt to globally summarize some or most of the salient clinical features of the disease. General consensus in the use of rating scales for MS has been achieved in the specification of a minimal record of disability (MRD) for MS that incorporates several physician- and paramedical-administered rating scales for evaluating MS disease status and its effects on the patient's life [International Federation of Multiple Sclerosis Societies, 1984]. However, there has been general dissatisfaction with the use of rating scales in MS clinical trials because of issues of lack or sensitivity to disease change, lack of interrater reliability, and simply the inherent reliance on subjective ratings of signs, symptoms, and unmeasured performance changes [Paty et al., 1992].

As an alternative to MS rating scales, comprehensive, quantitative evaluation of human performance was originally proposed and developed over 25 years ago for use in MS clinical trials [Tourtellotte et al., 1965]. The first application of performance testing in a major MS clinical trial was the multicenter cooperative ACTH study [Rose et al., 1968]. The study proved that human performance testing could be incorporated into a multicenter MS clinical trial and that examiners at multiple centers could be trained to administer the tests in a standardized, repeatable fashion. The results generally supported fairly comparable levels of sensitivity among the outcome measures, including performance testing [Dixon & Kuzma, 1974]. Subsequent analyses showed that a priori composites that provided succinct summary measures of disease change in key functional areas could be formed from the data and that these composite measures were sensitive to treatment effects in relapsing/remitting MS patients [Henderson et al., 1978]. Despite the favorable results of the ACTH study, performance testing did not achieve general acceptance in MS clinical studies. Comprehensive performance testing as conducted in the ACTH trial was considered time-consuming, the instrumentation was not generally available, and despite the use of composites, the number of outcome measures remained too large. In a recent double-blind, placebo-controlled collaborative study in 12 medical centers, the study design included both rating-scale and performance measures as outcome assessments in the largest sample of chronic progressive MS patients studied to date [The Multiple Sclerosis Study Group, 1990]. Analyses of the change in performance composite scores from baseline over the 2-year course of the clinical trial showed that the drug-treated patients worsened significantly less than the placebo-treated patients. In contrast, the EDSS and other clinical rating scores did not show significant treatment effects for the subset of MS patients at the same center [Syndulko et al., 1993]. This indicated that the performance composites were more sensitive than the clinical disability measures to disease progression and treatment effects. A more recent comparative analysis of the full data set also supports the greater sensitivity of composites based on performance testing compared with rating scales to both MS disease progression and to a treatment effect [Syndulko et al., 1994*a*; Syndulko et al., 1944*b*]. Although the biomedical community remains divided regarding the best outcome assessment in MS clinical trials, the accumulating evidence favors performance testing over clinical measures.

Another type of efficacy-oriented study of particular note in the context of this chapter involves the combination of pharmacodynamic models with instrumented, objective, and sensitive performance capacity measurements to optimize therapeutic effectiveness by fine-tuning dose prescription for individual subjects. For example, the results from a Parkinson's disease study by Hacisalihzade et al. [1989] suggest that performance capacity measurements, which would play a key role in a strategy to determine pharmacodynamics of a drug for the individual (i.e., not a population), could become a component in the management of some patients receiving long-term drug therapy.

154.4 Future Needs and Anticipated Developments

The field of clinical trials in which human performance variables are of interest has expanded at such a rapid rate in recent years that it has been difficult for researchers to keep up with all the new technologies, instrumentations, and methodologies. The type of evaluation employed has steadily shifted from mostly subjective to more objective, which can be argued to provide improved measurement quality. Increased standardization of human performance measurements has led to greater cost-effectiveness due to computer-based test batteries [Kennedy et al., 1993; Woollacott, 1983]. Thus, even if the quality of measurement were the same with methodologies that used less sophisticated methods, the initial investment would frequently be saved with newer data management techniques. This sentiment has not yet been accepted by all segments of the relevant communities. Also, the speed at which data can now be collected (which is especially important when changes over short time periods are of interest, as in pharmacodynamic studies) has led to greater cost-effectiveness. Furthermore, since instrumented devices are becoming commercially available and more widely used, it is not necessary nor desirable for researchers whose primary interest is the intervention of disease process to "reinvent" measurements (possibly with subtle but significant changes that ultimately inhibit standardization) for use in studies. Also, new measurements that may be different—but not necessarily better—must undergo long and expensive studies prior to their use in clinical trials. Although certain human performance measures are becoming standard in selected situations, more experience is needed before a substantial degree of standardization in human performance measurement is achieved in such situations.

The perfect research strategy in clinical trials involving human performance capacities to determine therapy effectiveness and safety may not be possible due to financial and/or ethical considerations. Thus the challenge of clinical trials research in the future is to maintain scientific integrity while conforming to legal mandates and staying within economic feasibility; i.e., an optimization is almost always necessary. Multidisciplinary research efforts can be useful in these circumstances.

Defining Terms

Clinical significance: An additional level of interpretation of a statistically significant finding that addresses the size of the effect found.

Clinical trial: A research study involving human subjects and an intervention (i.e., device, drug, surgical procedure, or other procedure) that is ultimately intended to either enhance the professional capabilities of physicians, improve the quality of life of patients, or contribute to the field of knowledge in those sciences which are traditionally in the medical field setting.

Composite score: A score derived by combining two or more measured performance capacities.

Efficacy-oriented clinical trial: A study that is conducted to objectively determine an intervention's effectiveness along with any pharmacokinetic or pharmacodynamic interactions.

Nonprobability sampling: A method of sampling whereby no subject in a defined population has an equal chance of being included in the sample.

Performance capacities: Dimensional capabilities or resources (e.g., speed, accuracy, strength, etc.) that a given human subsystem (e.g., visual, central processing, gait production, posture maintenance, etc.) possesses to perform a given task.

Pharmacodynamics: The physiologic and/or metabolic changes in the concentration of a drug over time to corresponding measurable performance capacity changes in the therapeutic effect.

Pharmacokinetics: The rate at which a drug is metabolized, as evidenced by changes in drug concentrations in blood, cerebral spinal fluid, urine, etc.

Probability sampling: A method of sampling whereby each subject in a defined population theoretically has an equal chance of being included in the sample.

Protocol: A detailed statement of all procedures and methods to be employed in a research investigation.

Randomized clinical trial (RCT): A study that uses controlled trials whereby part of the group of subjects receives a therapy that is tested, while the other subjects receive either no therapy or another therapy.

Resource economics: A cause-effect type of relationship between available performance resources (i.e., capacities) and demands of higher-level tasks in which the relevant subsystems are used.

Safety-oriented clinical trial: A study that usually involves vital sign measures (e.g., heart rate, blood pressure, respiration rate), clinical laboratory tests (e.g., blood chemistries, urinalysis, ECG), and adverse reaction tests (both mental and physical performance capacities) to evaluate the risk of an intervention.

Sample: A proportion of the defined population that is studied. When used as a verb (sampling), it refers to the process of subject selection.

Significance (statistical) test: A tool used for assessing possible associations between independent and dependent variables as well as differences among treatment groups.

Statistical significance: An evaluation of the research hypothesis at a specific level of probability so that any differences observed are not attributable to a chance occurrence.

References

Baker SJ, Chrzan GJ, Park CN, Saunders JH. 1985. Validation of human behavioral tests using ethanol as a CNS depressant Model. Neurobehav Toxicol Teratol 7:257.

Baumgartner TA, Jackson AS. 1993. Measurement for Evaluation, 4th ed. Dubuque, Iowa, Wm. C. Brown.

Cochran WG, Cox GM. 1957. Experimental Designs, 2nd ed. New York, Wiley.

Cohen J. 1988. Statistical Power Analysis for the Behavioral Sciences. Hillsdale, NJ, Lawrence Erlbaum Associates.

Cotton P. 1993. FDA lifts ban on women in early drug tests, will require companies to look for gender differences. JAMA 269:2067.

DeAngelis C. 1990. An Introduction to Clinical Research. New York, Oxford University Press.

Department of Health and Human Services. 1985. Code of Federal Regulation of the Department of Health and Human Services, title 45, part 46.

Dixon WJ, Kuzma JW. 1974. Data reduction in large clinical trials. Comm Stat 3:301.

Food and Drug Administration. 1977. General Considerations for the Clinical Evaluation of Drugs. DHEW Publication No. (FDA) 77–3040. Washington.

Fleiss JL.1986. The Design and Analysis of Clinical Experiments. New York, Wiley.

Hacisalihzade SS, Mansour M, Albani C. 1989. Optimization of symptomatic therapy in Parkinson's disease. IEEE Trans Biomed Eng 36(3):363.

Hastad DN, Lacy AC. 1994. Measurement and Evaluation, 2nd ed. Scottsdale, Ariz, Gorsuch Scarisbrick.

Henderson WG, Tourtellotte WW, Potvin AR, Rose AS. 1978. Methodology for analyzing clinical neurological data: ACTH in multiple sclerosis. Clin Pharmacol Ther 24:146.

Hill AB. 1963. Medical ethics and controlled trials. Br Med J 1:1043.

Horne JA, Gibbons H. 1991. Effects of vigilance performance and sleepiness of alcohol given in the early afternoon (post lunch) vs early evening. Ergonomics 34:67.

International Federation of Multiple Sclerosis Societies. 1984. Symposium on a minimal record of disability for multiple sclerosis. Acta Neurol Scand 70(Suppl):101–217.

Jennison C, Turnbull BW. 1993. Group sequential tests for bivariate response: Interim analyses of clinical trials with both efficacy and safety endpoints. Biometrics 49:741.

Jones B, Kenward MG. 1989. Design and Analysis of Crossover Trials. New York, Chapman and Hall.

Kennedy RS, Turnage JJ, Wilkes RL. 1993. Effects of graded doses of alcohol on nine computerized repeated-measures tests. Ergonomics 36:1195.

Kepple G. 1982. Design and Analysis a Researcher's Handbook, 2nd ed. Englewood Cliffs, NJ, Prentice-Hall.

Kondraske GV. 1988*a*. Workplace design: An elemental resource approach to task analysis and human performance measurements. International Conference for the Advancement of Rehabilitation Technology, Montreal, Proceedings, pp 608–611.

Kondraske GV. 1988*b*. Experimental evaluation of an elemental resource model for human performance. Tenth Annual IEEE Engineering in Medicine and Biology Society Conference, New Orleans, Proceedings, pp 1612–1613.

Kondraske GV. 1989. Measurement science concepts and computerized methodology in the assessment of human performance. In T Munsat (ed), Quantification of Neurologic Deficit. Stoneham, Mass, Butterworth.

Kondraske GV. 1995. A working model for human system-task interfaces. In JD Bronzino (ed), Handbook of Biomedical Engineering. Boca Raton, Fla, CRC Press.

Levin RJ. 1986. Ethics and Regulation of Clinical Research, 2nd ed. Baltimore, Urban & Schwarzenberg.

Merkatz RB, Temple R, Sobel S. 1993. Women in clinical trials of new drugs: A change in Food and Drug Administration policy. N Engl J Med 329:292.

Mohr E, Prouwers P. 1991. Handbook of Clinical Trials: The Neurobehavioral Approach. Berwyn, Pa, Swets and Zeitlinger.

Ottenbacher KJ, Barrett KA. 1991. Measures of effect size in the reporting of rehabilitation research. Am J Phys Med Rehabil 70(suppl 1):S131.

Paty DE, Willoughby E, Whitakek J. 1992. Assessing the outcome of experimental therapies in multiple sclerosis. In Treatment of Multiple Sclerosis: Trial Design, Results, and Future Perspectives, pp 47–90. London, Springer-Verlag.

Potvin AR, Tourtellotte WWT, Potvin JH, et al. 1985. Quantitative Examination of Neurologic Function, vols I and II. Boca Raton, Fla, CRC Press.

Rose AS, Kuzma JW, Kurtzke JF, et al. 1968. Cooperative study in the evaluation of therapy in multiple sclerosis: ACTH vs placebo in acute exacerbations. Preliminary report. Neurology 18:1.

Safrit MJ. 1990. Evaluation in Exercise Science. Englewood Cliffs, NJ, Prentice-Hall.

Salame P. 1991. The effects of alcohol in learning as a function of drinking habits. Ergonomics 34:1231.

Stone R. 1993. FDA to ask for data on gender differences. Science 260:743.

Syndulko K, Ke D, Ellison GW, et al. 1994*a*. Neuroperformance assessment of treatment efficacy and MS disease progression in the Cyclosporine Multicenter Clinical Trial. Brain (submitted).

Syndulko K, Ke D, Ellison GW, et al. 1994*b*. A comparative analysis of assessments for disease progression in multiple sclerosis: I. Signal to noise ratios and relationship among performance measures and rating scales. Brain (submitted).

Syndulko K, Tourellotte WW, Baumhefner RW, et al. 1993. Neuroperformance evaluation of multiple sclerosis disease progression in a clinical trial. 7:69.

Tang D, Geller NL. 1993. On the design and analysis of randomized clinical trials with multiple endpoints. Biometrics 49:23.

The Multiple Sclerosis Study Group. 1990. Efficacy and toxicity of cyclosporine in chronic progressive multiple sclerosis: A randomized, double-blind, placebo-controlled clinical trial. Ann Neurol 27:591.

Thomas JR, Nelson JK. 1990. Research Methods in Physical Activity, 2nd ed. Champaign, Il, Human Kinetics Publishers.

Tourtellotte WW, Haerer AF, Simpson JF, et al. 1965. Quantitative clinical neurological testing: I. A study of a battery of tests designed to evaluate in part the neurologic function of patients with multiple sclerosis and its use in a therapeutic trial. Ann NY Acad Sci 122:480.

Tudiver F, Bass MJ, Dunn EV, et al. 1992. Assessing Interventions Traditional and Innovative Methods. London, Sage Publications.

Weissman A. 1991. On the concept "study day zero." Perspect Bio Med 34:579.

Willoughby PD, Whitaker JE. 1992. Assessing the outcome of experimental therapies in multiple sclerosis. In RA Rudick, DE Goodkin (eds), Treatment of Multiple Sclerosis: Trial Design, Results, and Future Perspectives, pp 47–90. London, Springer-Verlag.

Woollacott MH. 1983. Effects of ethanol on postural adjustments in humans. Exp Neurol 80:55.

Further Information

The original monograph, *Experimental and Quasi-Experimental Designs for Research*, by D.T. Campbell and J.C. Stanley (Boston: Houghton Mifflin, 1964), is a classic experimental design text. The book *Statistics in Medicine*, by T. Colton (Boston: Little, Brown, 1982), examines statistical principles in medical research and identifies common perils when drawing conclusions from medical research. The book *Clinical Epidemiology: The Essentials*, R.H. Fletcher, S.W. Fletcher, and E.H. Wagner (2nd ed., Baltimore: Williams & Wilkins, 1982), is a readable text with concise descriptions of various types of studies. *Clinical Trials: Design, Conduct, and Analysis*, By C.L. Meinert (New York: Oxford University Press, 1986), is a comprehensive text covering most details and provides many helpful suggestions about clinical trials research. R.J. Porter and B.S. Schoenberg, *Controlled Clinical Trials in Neurological Disease* (Boston: Kluwer Academic Publishers, 1990), is also a comprehensive text covering clinical trials with a special emphasis on neurologic diseases.

155

Applications of Quantitative Assessment of Human Performance in Occupational Medicine

Mohamad Parnianpour

Ohio State University

As early as 1700, Bernardino Ramazzini, one of the founders of occupational medicine, had associated certain physical activities with musculoskeletal disorders (MSD). He postulated that certain violent and irregular motions and unnatural postures of the body impair the internal structure [Snook et al., 1988]. Presently, much effort is directed toward a better understanding of work-related musculoskeletal disorders involving the back, cervical spine, and upper extremities. The World Health Organization (WHO) has defined occupational diseases as those work-related diseases where the relationship to specific causative factors at work has been fully established [WHO, 1985]. Other work-related diseases may have a weaker or unclear association to working conditions. They may be aggravated, accelerated, or exacerbated by workplace factors and lead to impairment of workers' performance. Hence obtaining the occupational history is crucial to proper diagnosis and appropriate treatment of work-related disorders. The occupational physician must consider the conditions of both the workplace and the worker in evaluation of injured workers. Biomechanical and ergonomic evaluators have developed a series of techniques for quantification of the task demands and evaluation of the stresses in the workplace. Functional capacity evaluation also has been advanced to quantify the maximum performance capability of workers. The motto of ergonomics is to avoid the mismatch between the task demand and functional capacity of individuals. A multidisciplinary group of physicians and engineers constitutes the rehabilitation team

0-8493-0462-8/00/$0.00+$.50

that will work together to implement the prevention measures. Through proper workplace design, workplace stressors could be minimized. It is expected that one-third of the compensatable low back pain in industry could be prevented by proper ergonomic workplace or task design. In addition to reducing the probability of both the initial and recurring episodes, proper ergonomic design allows earlier return to work of injured workers by keeping the task demands at a lower level. Unfortunately, ergonomists are often asked to redesign the task or the workplace after a high incidence of injuries has already been experienced. The next preventive measure that has been suggested is preplacement of workers based on the medical history, strength, and physical examinations [Snook et al., 1988]. Training and education have been the third prevention strategy in the reduction of musculoskeletal disorders. Some components of these educational packages such as "back schools" and the teaching of "proper body mechanics" have been used in the rehabilitation phase of injured workers as well.

Title I of the Americans with Disability Act [ADA, 1990] prohibits discrimination with regard to any aspect of the employment process. Thus the development of preplacement tests has been impeded by the possibility of discrimination against individuals based on gender, age, or medical condition. The ADA requires physical tests to simulate the "essential functions" of the task. In addition, one must be aware of "reasonable accommodations," such as lifting aids, that may make an otherwise infeasible task possible for a disabled applicant to perform. Healthcare providers who perform physical examinations and provide recommendations for job applicants must consider the rights of disabled applicants. It is extremely crucial to quantify the specific physical requirements of the job to be performed and to examine an applicant's capabilities to perform those specific tasks, taking into account any reasonable accommodations that may be provided. Hence task analysis and functional capacity assessment are truly intertwined.

Work-related disorders of the upper extremities, unlike low back disorders, can better be related to specific anatomic sites such as a tendon or compressed nerve. Examples of the growing number of cumulative trauma disorders of the upper extremities and the neck are carpal tunnel syndrome (CTS), DeQuervain's disease, trigger finger, lateral epicondylitis (tennis elbow), rotator cuff tendinitis, thoracic outlet syndrome, and tension neck syndrome. The prevalence of these disorders is higher among some specific jobs, such as meat cutters, welders, sewer workers, grinders, meat packers, and keyboard operators. Some of the common risk factors leading to pain, impairment, and physical damage in the neck and upper extremities are forceful motion, repetitive motion, vibration, prolonged awkward posture, and mechanical stress [Kroemer et al., 1994].

This chapter is intended to illustrate the application of some principles and practices of human performance engineering, especially quantification of human performance in the field of occupational medicine. I have selected the problem of low back pain to illustrate a series of concepts that are essential to evaluation of both the worker and the workplace, while realizing the importance of the disorders of the neck and upper extremities. By inference and generalization, most of these concepts can be extended to these situations.

155.1 Principles

Assessment of function across various dimensions of performance (i.e., strength, speed, endurance, and coordination) has provided the basis for a rational approach to clinical assessment, rehabilitation strategies, and determination of return-to-work potential for injured employees [Kondraske, 1990]. To understand the complex problem of trunk performance evaluation of low back pain (LBP) patients, the terminology of muscle exertion must first be defined. However, it should be noted that a number of excellent reviews of trunk muscle function have been performed [Andersson, 1991; Beimborn & Morrissey, 1988; Newton & Waddell, 1993; Pope, 1992]. I do not intend to reproduce this extensive literature here, since my motive is to provide a critical analysis that will lead the reader toward an understanding of the future of functional assessment techniques. A more extensive clinical application is provided elsewhere [Szpalski and Parnianpour, 1996; Parnianpour & Shirazi-adl, 1999].

Impairment, Disability, and Handicap

The tremendous human suffering and economic costs of disability present a formidable medical, social, and political challenge in the midst of growing healthcare costs and scarcity of resources. The WHO [1980] distinguished among impairment, disability, and handicap. *Impairment* is any loss or abnormality of psychological, physiologic, or anatomic structure or function—impairment reflects disturbances at the organ level. *Disability* is any restriction or lack of ability (resulting from impairment) to perform an activity in the manner or within the range considered normal for a human being—disability reflects disturbances at the level of person. *Handicap* is a disadvantage for a given individual, resulting from an impairment or a disability, that limits or prevents the fulfillment of a role that is normal (depending on age, sex, and social and cultural factors) for that individual. Since disability is the objectification of an impairment, handicap represents the socialization of an impairment or disability. Despite the immense improvement presented by the International Classification of Impairments, Disabilities, and Handicaps (ICIDH), the classification is limited from an industrial medicine or rehabilitation perspective. The hierarchical organization lacks the specificity required for evaluating the functional state of an individual with respect to task demands.

Kondraske [1990] has suggested an alternative approach using the principles of resource economics. The resource economics paradigm is reflective of the principal goal of ergonomics: fitting the demands of the task to the functional capability of the worker. The Elemental Resource Model (ERM) is based on the application of general performance theory that presents a unified theory for measurement, analysis, and modeling of human performance across all different aspects of performance, across all human subsystems, and at any hierarchical level. This approach uses the same bases to describe both the fundamental dimensions of performance capacity and task demand (available and utilized resources) of each functional unit involved in performance of the high-level tasks. The elegance of the ERM is due to its hierarchical organization, allowing causal models to be generated based on assessment of the task demands and performance capabilities across the same dimensions of performance [Kondraske, 1990].

Muscle Action and Performance Quantification

The details of the complex processes of muscle contraction in terms of the bioelectrical, biochemical, and biophysical interactions are under intense research. Muscle tensions is a function of muscle length and its rate of change and can be scaled by the level of neural excitation. These relationships are called the *length-tension* and *velocity-tension relationships*. From a physiologic point of view, the measured force or torque applied at the interface is a function of (1) the individual's motivation (magnitude of the neural drive for excitation and activation processes), (2) environmental conditions (muscle length, rate of change of muscle length, nature of the external load, metabolic conditions, pH level, temperature, and so forth), (3) prior history of activation (fatigue), (4) instructions and descriptions of the tasks given to the subject, (5) the control strategies and motor programs employed to satisfy the demands of the task, and (6) the biophysical state of the muscles and fitness (fiber composition, physiologic cross-sectional area of the muscle, cardiovascular capability). It cannot be overemphasized that these processes are complex and interrelated [Kroemer et al., 1994]. Other factors that may affect the performance of patients are misunderstanding of the degree of effort needed in maximal testing, test anxiety, depression, nociception, fear of pain and reinjury, as well as unconscious and conscious symptom magnification.

The following sections review some methods to quantify performance and lifting capability of isolated trunk muscles during a multilink coordinated manual materials handling task. Relevant factors that influence the static and dynamic strength and endurance measures of trunk muscles will be addressed, and the clinical applications of these assessment techniques will be illustrated.

The central nervous system (CNS) appropriately excites the muscle, and the generated tension is transferred to the skeletal system by the tendon to cause motion, stabilize the joint, and/or resist the effect of external forces on the body. Hence the functional evaluation of muscles cannot be performed without the characterization of the interfaced mechanical environment.

The four fundamental types of muscle exertion or action are isometric, isokinetic, isotonic, and isoinertial. In *isometric* exertion, the muscle length is kept constant, and there is no movement. Although mechanical work is not achieved, physiologic work, i.e., static work, is performed, and energy is consumed. When the internal force exerted by the muscle is greater than the external force offered by the resistance, then concentric, i.e., shortening, muscle action occurs, whereas if the muscle is already activated and the external force exceeds the internal force of the muscle, then eccentric, i.e., lengthening, muscle action occurs. When the muscle moves, either concentrically or eccentrically, dynamic work is performed. If the rate of shortening or lengthening of the muscle is constant, the exertion is called *isokinetic.* When the muscle acts on a constant inertial mass, the exertion is called *isoinertial. Isotonic* action occurs when the muscle tension is constant throughout the range of motion.

These definitions are very clear when dealing with isolated muscles during physiologic investigations. However, terminology employed in the literature of strength evaluation is imprecise. The terms are intended to refer to the state of muscles, but they actually refer to the state of the mechanical interface, i.e., the dynamometer. Isotonic exertion, as defined, is not as realizable physiologically because muscular tensions change as its lever arm changes despite the constancy of external loads. Special designs may vary the resistance level in order to account for changes in mechanical efficiency of the muscles. In addition, the rate of muscle length change may not remain constant even when the joint angular velocity is regulated by the dynamometer during isokinetic exertions. During isoinertial action, the net external resistance is not only a function of the mass (inertia) but also a function of the acceleration. The acceleration, however, is a function of the input energy to the mass. Hence, to fully characterize the net external resistance, we need to have both the acceleration and the inertial parameters (mass and moment of inertia) of the load and body parts. Future research should better quantify the inertial effects of the dynamometers, particularly during nonisometric and nonisokinetic exertions.

For any joint or joint complex, muscle performance can be quantified in terms of the basic dimensions of performance: strength, speed, endurance, steadiness, and coordination. Muscle *strength* is the capacity to produce torque or work by voluntary activation of the muscles, whereas muscle endurance is the ability to maintain a predetermined level of motor output—e.g., torque, velocity, range of motion, work, or energy—over a period of time. *Fatigue* is considered to be a process under which the capability of muscles diminish. However, neuromuscular adjustments take place to meet the task demands (i.e., increase in neural excitation) until there is final performance breakdown—endurance time. *Coordination,* in this context, is the temporal and spatial organizations of movement and the recruitment patterns of the muscle synergies.

Despite the proliferation of various technologies for measurement, basic questions such as "What needs to be measured and how can it best be measured?" are still being investigated. However, there is a consensus on the need to measure objectively the performance capability along the following dimensions: range of motion, strength, endurance, coordination, speed, acceleration, etc. Strength is one of the most fundamental dimensions of human performance and has been the focus of many investigations. Despite the general consensus about the abstract definition of strength, there is no direct method for measurement of muscle tension in vivo. Strength has often been measured at the interface of a joint (or joints) with the mechanical environment. A *dynamometer,* which is an external apparatus onto which the body exerts force, is used to measure strength indirectly.

Different modes of strength testing have evolved based on different levels of technologic sophistication. The practical implication of contextual dependencies on the provided mechanical environment of the strength measures must be considered during selection of the appropriate mode of measurement. In this regard, equipment that can measure strength in different modes is more efficient in terms of both initial capital investment, required floor space in the clinics or laboratories, and the amount of time it takes to get the person in and out of the dynamometer.

155.2 Low Back Pain and Trunk Performance

The problem of LBP is selected to present important models that could be used by the entire multidisciplinary rehabilitation team for the measurement, modeling, and analysis of human performance

[Kondraske, 1990]. The inability to relate LBP to anatomic findings and the difficulties in quantifying pain have directed much effort toward quantification of spinal performance. The problem is made even more complex by the increasing demand of the healthcare system to quantify the level of impairment of patients reporting back pain without objective findings.

There are three basic impairment evaluation systems, each having their merits and shortcomings: (1) anatomic, based on physical examination findings, (2) diagnostic, based on pathology, and (3) functional, based on performance or work capacity [Luck & Florence, 1988]. The earlier systems were anatomic, based on amputation and ankylosis. Although this approach may be more applicable to the hand, it is very inappropriate for the spine. The diagnostic-based systems suffer from lack of correspondence between the degree of impairment for a given diagnosis and the resulting disability and even more from the lack of a clear diagnosis. A large percentage of symptom-free individuals have anatomic findings detectable by the imaging technologies, while some LBP patients have no structural anomalies.

The function-based systems are more desirable from an occupational medicine perspective for the following reasons: They allow the rehabilitation team to rationally evaluate the prospect for return to light-duty work and the type of "reasonable accommodations" needed (such as assistive devices) that could reduce the task demand below the functional capability of the individual. By focusing on remaining ability and transferable skills rather than the disability or structural impairment of the injured worker, the set of feasible jobs can be identified. These points are extremely important, given the natural history of work disability after a single low back pain episode causing loss of work time: 40% to 50% of workers return to work by 2 weeks, 60% to 80% return by 4 weeks, and 85% to 90% return by 12 weeks. The small portion of disabled workers who become chronic are responsible for the majority of the economic cost of LBP. It is therefore the primary goal of the rehabilitation team to prevent the LBP, which is self-correcting in most cases, no matter what kind of therapy is used, from becoming a chronic disabling predicament. Injured workers should neither be returned to work too early nor too late, since both could complicate the prognosis. The results of functional capacity evaluation and task demand quantification should guide the timing for returning to work. It is clear that psychosocioeconomic factors become increasingly more important than physical factors as the disability progresses into "chronicity syndrome" and play a major role in defining the evolution of a low back disability claim. Future research should further establish the reliability and reproducibility of performance assessment tools to expedite their widespread use [Luck & Florence, 1988; Newton & Waddell, 1993].

Maximal and Submaximal Protocols

Biomechanical strength models of the trunk are usually based on static maximal strength measurement. In real-life work situations, individuals rarely exert lengthy or maximum static effort. In most clinical situations, submaximal protocols are recommended, especially in patients with pain or with cardiovascular problems. Also, submaximal testing is less susceptible to fatigue and injury. The activities of daily living also have a great deal of submaximal efforts at the self-selected pace. Hence it has been argued that testing at the preferred rate may be complementary to the maximal effort protocols. The preferred motion can be solicited by instructing the subject to perform repetitive movement at a pace and through the range of motion which he or she feels is the most comfortable. It has been shown that LBP patients and normal individuals have different resisted preferred flexion/extension motion characteristics. Having the subject perform against resistance is based on the hypothesis that, at higher resistance levels, the separation between the performance levels of patients and normal subjects becomes more evident. It has been shown, for example, that functional impairment of trunk extensors in LBP patients with respect to the normal population is larger at higher velocities during isokinetic trunk extension. However, the proponents of unconstrained testing have argued that separation of these groups can be performed based on the position, velocity, and acceleration profiles of the trunk during self-selected flexion/extension tasks. They have noted that pain and fear of reinjury may become the limiting factors. The sudden surge in acquiring performance measures of LBP patients during the initial rehabilitation process also underscores the validity of this concept.

Static and Dynamic Strength Measurements of Isolated Trunk Muscles

Weakness of the trunk extensor and abdominal muscles in patients with LBP was demonstrated using the cable tensiometer to measure isometric strength. The disadvantage of the cable tensiometer (which records applied force) is that it neglects to measure the lever arm distance from the center of trunk motion. It is also recommended that cable tensiometer be used to determine peak isometric torques rather than the stable average torque exerted over a 3-second period. Dynamometers used for testing dynamic muscle performances contain either hydraulic or servo motor systems to provide constant velocity, e.g., isokinetic devices, or constant resistance, e.g., isoinertial devices. The isokinetic devices can be further categorized into passive and active types. The robotics-based dynamometers can actively apply force on the body and hence allow eccentric muscle performance assessments, while only concentric exertions can be measured by the passive devices. Eccentric muscle action can stimulate the lowering phase of a manual materials handling task. Based on sports medicine literature, eccentric action has been implicated for its significant role in the muscle injury mechanism. Using isokinetic dynamometers, the isometric and isokinetic strengths of trunk extensor and abdominal muscles were shown to be weaker in LBP patients compared with healthy individuals. Dedicated trunk testing systems have become the cornerstone of objective functional evaluation and have been incorporated in the rehabilitation programs in many centers.

Two issues of importance for future research are the role of pelvic restraints and the significance of using newly developed triaxial dynamometers as opposed to more traditional uniaxial dynamometers. Studies on healthy volunteers have shown that trunk motions occur in more than one plane—lateral bending accompanies the primary motion of axial rotation. Numerous attempts have been made to measure the segmental range of motion three-dimensionally in the lumbar spine with the purpose of quantifying abnormal coupling and diagnosing instabilities.

The effect of posture on the maximum strength capability can be described based on the length-tension relationship of muscle action. Marras and Mirka [1989] studied the effect of trunk postural asymmetry, flexion angle, and trunk velocity (eccentric, isometric, and concentric) on maximal trunk torque production. It was shown that trunk torque decreased by about 8.5% of the maximum for every 15 degrees of asymmetric trunk angle. At higher trunk flexion angles, extensor strength increased. Complex, significant interaction effects of velocity, asymmetry, and sagittal posture were detected. The ranges of velocity studies were more limited (±30 degrees per second) than those used customarily in spinal evaluation. Tan et al. [1993] tested 31 healthy males for the effects of standing trunk-flexion positions (0, 15, and 35 degrees) on triaxial torques and did *electromyograms* (EMGs) of 10 trunk muscles during isometric trunk extension at 30%, 50%, 70%, and 100% of maximum voluntary exertions (MVE). Trunk muscle strength was significantly increased at a more flexed position. However, the accessory torques in the transverse and coronal planes were not affected by trunk postures. The recorded lateral bending and rotation accessory torques were less than 5% and 16% of the primary extension torque, respectively. The rectus abdominis muscles were inactive during all the tests. The EMGs of the erector spinae varied linearly with higher values of MVE, while the latissimus dorsi had a nonlinear behavior. The obliques were coactivated only during 100% MVE. The *neuromuscular efficiency ratio* (NMER) was constructed as the ratio of the extension torque over the processed (RMS) EMG of the extensor muscles. It was hoped that NMER could be used in clinical settings where generation of the maximum exertion are not indicated. However, the NMER proved to have a limited clinical utility because it was significantly affected by both exertion level and posture. The NMER of the extensor muscles increased at more flexed position. Studies that have combined the EMG activities and dynamometric evaluations have the potential of discovering the neuromuscular adaptation during different phases of injury and rehabilitation processes.

Static and Dynamic Trunk Muscle Endurance

The high percentage of type I fibers in the back muscles, in addition to the better vascularization of these muscle groups, contributes to their superior endurance. Physiologic studies indicate that at higher muscle utilization ratios (relative muscle loads), fatigue is detected earlier. Isometric endurance tests have been used to compute the median frequency (MF) of the myoelectrical activities of trunk muscles in both normal and LBP populations. The expected decline of the median frequency with fatigue is parameterized by the intercept (initial MF) and the slope of the fall. It has been shown that trunk range of motion (ROM) and isometric strength suffered from lower specificity and sensitivity than spectral parameters. Trunk muscle endurance does differ between healthy subjects and those reporting LBP. During isometric endurance testing, trunk flexors develop fatigue faster than extensors in symptom-free subjects. The flexor fatigability appeared significantly higher in patients with LBF as compared with controls. Chronicity also influences trunk muscle endurance. Chronic LBP patients showed reduced abdominal as well as back muscle endurance as compared with the healthy controls and lower back muscle endurance as compared with the intermittent LBP group. Individuals with a history of debilitating LBP demonstrated less isometric trunk extensor endurance than either normal individuals or patients with history of lesser LBP.

Soft tissues subjected to repetitive loading, due to their viscoelastic properties, demonstrate creep and load relaxation. The loss of precision, speed, and control of the neuromuscular system induced by fatigue reduces the ability of muscles to protect the weakened passive structure, which may explain many industrial, clinical, and recreational injury mechanisms. These results further indicate the necessity of relating clinical protocols to the job and show how short-duration maximal isometric testing alone cannot provide the complex functional interaction of strength, endurance, control, and coordination.

Parnianpour et al. [1988] studied the effect of isoinertial fatiguing of flexion and extension trunk movements on the movement pattern (angular position and velocity profile) and the motor output (torque) of the trunk. They showed that, with fatigue, there is a reduction of the functional capacity in the main sagittal plane. There is also a loss of motor control enabling a greater range of motion in the transverse and coronal planes while performing the primary sagittal task. Association of sagittal with coronal and transverse movements is considered more likely to induce back injuries; thus the effect of fatigue and reduction of motor control and coordination may be an important risk factor leading to injury-prone working postures. The endurance limit is a more useful predictor of incidence and recurrence of low back disorders than the absolute strength values. Although physiologic criteria used in the *National Institute for Occupational Safety and Health Lifting Guide* [NIOSH, 1981] considered cardiovascular demands of dynamic repetitive lifting tasks, the limits of muscular endurance were not explicitly addressed. Future research should fill this gap, since the maximum strength measures should not guide the design decisions. Maximum level of performance can only be maintained for short periods of time, and muscular fatigue should be avoided to prevent the development of MSD. This caveat should be applied to all dimensions of performance capability [Kondraske, 1990].

A prospective, randomized study among employees in a geriatric hospital showed that exercising during work hours to improve back muscle strength, endurance, and coordination proved cost-effective in preventing back symptoms and absence from work [Gundewall et al., 1993]. Every hour spent by the physiotherapist on the exercise group reduced the work absence by 1.3 days. In this study, both training and testing equipment were very modest. Endurance training is based on exercises with high repetition and low resistance, while strength training requires exercise with high resistance and low repetition.

Lifting Strength Testing

The National Institute for Occupational Safety and Health [NIOSH, 1981] recommended static, i.e., isometric, strength measurements as its standard as its standard for lifting tasks. This was based on the evidence that associated LBP with inadequate isometric strength. The incidence of an individual's sustaining an on-the-job back injury increases threefold when the task-lifting requirements approached or exceed the individual's strength capacity. However, lifting strength is not a true measure of trunk function but is a global measure taking into account arm, shoulder, and leg strength as well as the individual's lifting technique and overall fitness. It has been shown that strength tests were more valid and predictive of risk of low back disorders if they simulated the demands of the job. The clinicians must be aided with easy-to-use and validated instruments or questionnaires to gather information about the task demands in order to decide what testing protocol best simulates the applicant's spinal loading conditions.

Static strength measurements have been reported to underestimate significantly the loads on the spine during dynamic lifts. Comparing static and dynamic biomechanical models of the trunk, the predicted spinal loads under static conditions were 33% to 60% less than those under dynamic conditions, depending on the lifting technique. The recruitment patterns of trunk muscles (and thus the internal loading of the spine) are significantly different under isometric and dynamic conditions. General manual materials handling tasks require a coordinated multilink activity that can be simulated using classic psychophysical techniques or the robotics-based lift task simulators. Various lifting tests, including static, dynamic, maximal, and submaximal, are currently available. The experimental results of correlational studies have confirmed the theoretical prediction that strength will be dependent on the measurement technique. Since muscle action requires external resistance, the effect of muscle action will depend on the nature of the resistance. These results refute the implicit assumption that a generic strength test exists that can be used for preplacing workers (preemployment) and predicting the risk of injury or future occurrence of LBP. The psychometric properties of isokinetic and isoresistive modes of strength testing were recently addressed. The quantification of the surface response of strength as a function of joint angle and velocity was only possible for isokinetic testing, while isoresistive tests yielded a very sparse data set. Figure 155.1 and 155.2 illustrate these points graphically.

The wide conflicting results found in the literature regarding the relationship of an individual's strength to the risk of developing LBP may be due to inappropriate modes of strength measurements, i.e., lack of job specificity. Isometric strength testing of the trunk is still widely used, especially in large-scale industrial or epidemiologic studies, because it has been standardized and studied prospectively in industry. Compared with trunk dynamic strength testing protocols, the trunk isometric strength testing protocols are simpler and less expensive.

One outstanding issue during dynamic testing is the unresolved problem of how the wealth of information can be presented in a succinct and informative fashion. One approach has been to compare the statistical features of the data with the existing normal databases. This is particularly crucial because one does not have the option of comparing the results to the "contralateral asymptomatic joint," as one has

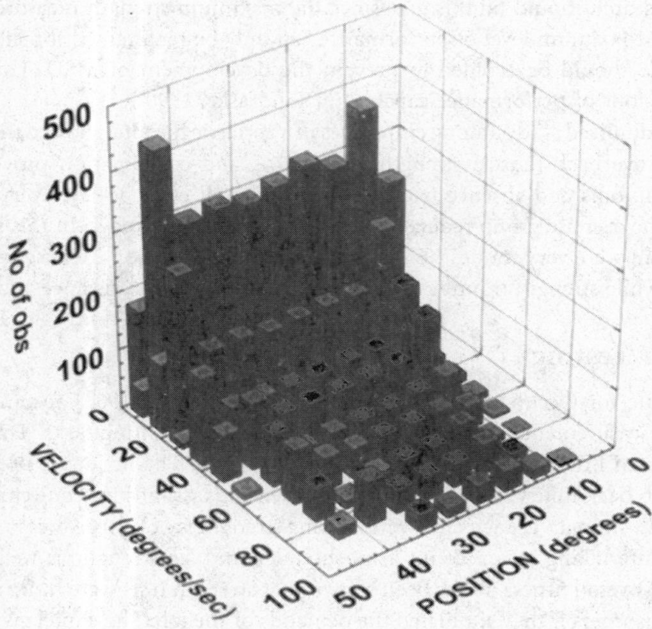

FIGURE 155.1 Bivariate distribution histogram of isokinetic trunk extension for 10 subjects.

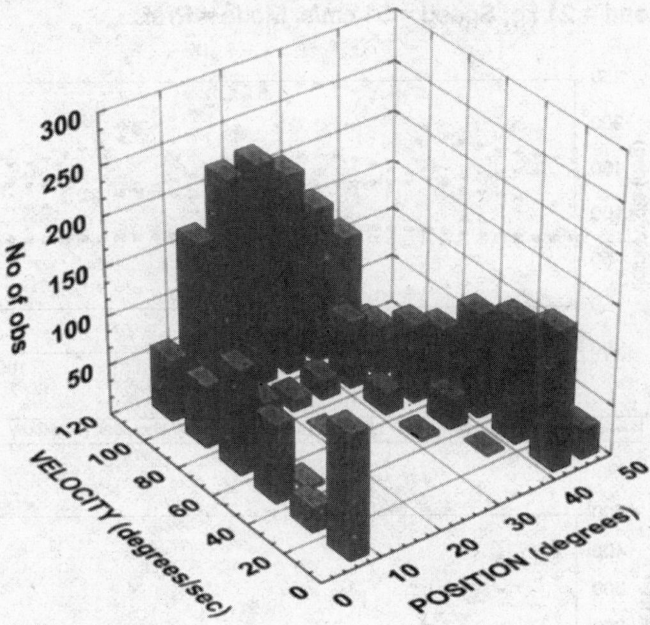

FIGURE 155.2 Bivariate distribution histogram of isotonic trunk extension for 10 subjects.

with lower or upper extremity joints. Given the large differences between individuals, I recommend comparison be made to job-specific databases. For example, it is more appropriate for the trunk strength of an injured construction worker to be compared with age- and gender-controlled healthy construction workers than with data from healthy college graduate students or office workers. However, given the scarcity of such data, I argue for comparison of performance capacity with job demand based on task analysis. The performance capacity evaluation is once again linked to task demand quantification.

Inverse and Direct Dynamics

A major task of biomechanics has been to estimate the internal loading of musculoskeletal structure and establish the physiologic loading during various daily activities. Kinematic studies deal with joint movement, with no emphasis on the forces involved. However, kinetic studies address the effect of forces that generate such movements. Using sophisticated experimental and theoretical stress/strain analyses, hazardous/failure levels of loads have been determined. The estimated forces and stresses are used to estimate the level of deformation in the tissues. This technique allows one to assess the risk of overexertion injury associated with any physical activity. Given repetitive motions and exertion levels much lower than the ultimate strength of the tissues, an alternative injury mechanism, the cumulative trauma model, has been used to describe much of the musculoskeletal disorders of the upper extremities.

The experimental data on the joint trajectories are differentiated to obtain the angular velocity and acceleration. Appropriate inertial properties of the limb segments are used to compute the net external moments about each joint. This mapping from joint kinematics to net moments is called *inverse dynamics. Direct dynamics* refers to studies that simulate the motion based on known actuator torques at each joint. The key issue in these investigations is understanding the control strategies underlying the trajectory planning and performance of purposeful motion. A highly multidisciplinary field has emerged to address these unsolved questions (see Berme and Cappozzo [1990] for a comprehensive treatment of these issues.)

It should be pointed out that determination of the external moments about different joints during manual materials-handling tasks is based on the well-established laws of physics (Fig. 155.3). However,

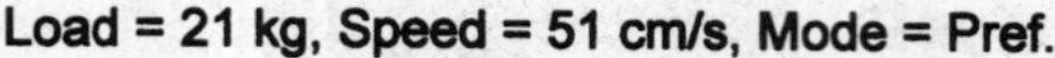

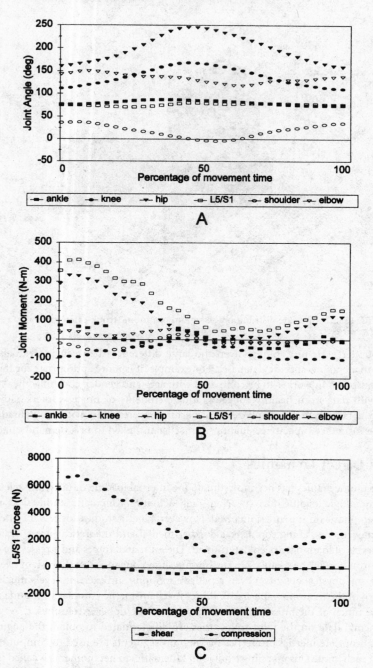

FIGURE 155.3 The dynamic analysis of a sagittal plane symmetrical isokinetic lift (load 21 kg, speed = 51 cm/s, mode = preferred method of lifting): (*a*) joint angle; (*b*) net muscular joint moment; and (*c*) the joint reaction forces at L5/S1.

the determination of human performance and assessment of functional capacity are based on other disciplines, e.g., psychophysics, that are not as exact or well developed. One can describe easily the job demand, in terms of the required moments about each joint, by analyzing the workers performing the tasks. However, one is unable to predict the ability to perform an arbitrary task based on the incomplete knowledge of functional capacities at the joint levels. A task is easily decomposed to its demands at the joint level; however, one cannot compose (construct) the set of feasible tasks based on one's functional capacity knowledge. The mapping from high-level task demands to the joint-level functional capacity for a given performance trial is unique. However, the mapping from joint-level functional capacity to the high-level task demand is one to many (not unique). The challenge to the human performance research community is to establish this missing link. Much of the integration of ergonomics and functional analysis depends on removal of this obstacle. The question of whether a subject can perform a task based on knowledge of his or her functional capacity at the joint level remains an area of open research. When ergonomists or occupational physicians evaluate the fitness of task demands and worker capability, the following clinical questions will be presented; (1) Which space should be explored for determining normalcy, fit, or equivalence? (2) Should we consider the performance of the multilink system in the joint space or end-effector (cartesian workspace)? These issues have profound effects on both the development of new technologies and the evaluation of trunk or lifting performance.

The enormous degrees of freedom existing in the neuromusculoskeletal system provide the control centers both the kinematic and actuator redundancies. The redundancies provide optimization possibility. Since one can lift an object from point *A* to point *B* with infinite postural possibilities, it can be suggested that certain physical parameters maybe optimized for the learned movements. The possible candidates for objective function to be optimized are movement time, energy, smoothness, muscular activities, etc. This approach, though still in its early stage, may be very important for spine functional assessment. One could compare the given performance with the optimal performance that is predicted by the model. This approach provides specific goals and gives biofeedback with respect to the individual's performance.

Comparison of Task Demands and Performance Capacity

The regression analysis was used to model the dynamic torque, velocity, and power output as a function of resistance level during flexion and extension using the B-200 Isostation [Parnianpour et al., 1990]. Results indicated that the measured torque was not a good discriminator of the tenth, fiftieth, and ninetieth percentile population. However, velocity and power were shown to effectively discriminate the three populations. Based on these data, it was suggested that during clinical testing, sagittal plane resistance should not be set at higher than about 80 N·m in order to minimize the internal loading of spine while taxing trunk functional capacity. This presentation of data may be useful to the physician or ergonomist in evaluating the functional capacity requirements of workplace manual materials-handling tasks. For example, a manual material-handling task that requires about 80 N·m (61 ft·lb) of trunk extensor strength could be performed by 90% of the population in the normal database if the required average trunk velocity does not exceed 40 degrees per second, while only 50% could perform the task if the velocity requirement exceeds 70 degrees per second. More important, only the top tenth percentile population could perform the task if the velocity requirement approaches 105 degrees per second (Fig. 155.4). A few versions of lumbar motion monitors that can record the triaxial motion in the workplace have been used to provide the trunk movement requirements. The preceding example also illustrates the importance of having the same bases for evaluation of both task and the functional capability of the worker.

155.3 Clinical Applications

Clinical studies have utilized quantitative human performance, i.e., strength and endurance measures, to predict the first incidence or recurrence of LBP and disability outcome and also as a prognosis measure

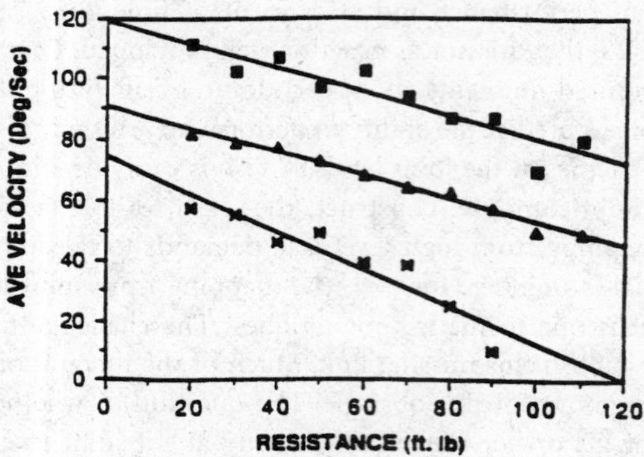

FIGURE 155.4 The average extension velocity measured during maximum trunk extension against a set resistance for 10th (cross), 50th (triangle), and 90th (square) percentile distribution.

during the rehabilitation process. Training programs to enhance the endurance and strength of workers have been implemented in some industries. More studies on the effectiveness of these programs are needed. It can be hypothesized that these programs complement the stress-management programs to enhance both worker satisfaction and coping strategies with regard to physical and nonphysical stressors at the workplace.

Functional-based impairment evaluation schemes traditionally have used spinal mobility. Given the poor reliability of range of motion (ROM), its large variability among individuals, and the static psychometric nature of ROM, the use of continuous dynamic profiles of motion with the higher-order derivatives has been suggested. Dynamic performances of 281 consecutive patients from the Impairment Evaluation Center at the Mayo Clinic were used. As part of the comprehensive physical and psychological evaluation, 281 consecutive LBP patients underwent isometric and dynamic trunk testing using the B200 Isostation. *Feature extraction* and *cluster analysis* techniques were used to find the main profiles in dynamic patient performances. The middle three cycles of movements were interpolated and averaged into 128 data points; thus the data were normalized with respect to cycle time. This allowed for comparison between individuals. Figure 155.5 presents the main profiles of sagittal trunk angular position. The number of patients in each group is also noted on the graph. Patients in the first ($n = 48$) and second ($n = 55$) groups had similar flexion mobility; however, those in the first group had more limited extension mobility. The time to peak sagittal position also varied among the five groups. Forty-seven patients in the fifth group showed extreme impairment in both flexion and extension. The third group (26 patients) showed differential impairments with respect to direction of motion. A marked improvement over the use of ROM has been achieved by preserving information in the continuous profiles. The LBP patients in this study are heterogeneous with respect to their movement profile. Uniform treatment of these patients is questionable, and rehabilitation programs should consider their specific impairments. Future research should incorporate the clinical profiles with these movement profiles to further delineate the heterogeneity of LBP patients. Marras et al. [1993] used similar feature-extraction techniques to characterize the movement profiles of 510 subjects belonging to normal ($n = 339$) and 10 LBP patient groups ($n = 171$). Subjects were asked to perform flexion/extension trunk movement at five levels of asymmetry, while the three-dimensional movement of the spine was monitored by the Lumbar Motion Monitor (an exoskeleton goniometer developed at the Biodynamics Lab of The Ohio State University). Trunk motions were performed against no resistance, and no pelvic stabilization was required. The quadratic discriminant analysis was able to correctly classify over 80% of the subjects. The same technology was used to develop logistic regression models to identify the high-risk jobs in industrial workplaces. Hence principles of human performance can be applied successfully to the worker and the task to avoid the mismatch between performance capability and task demand.

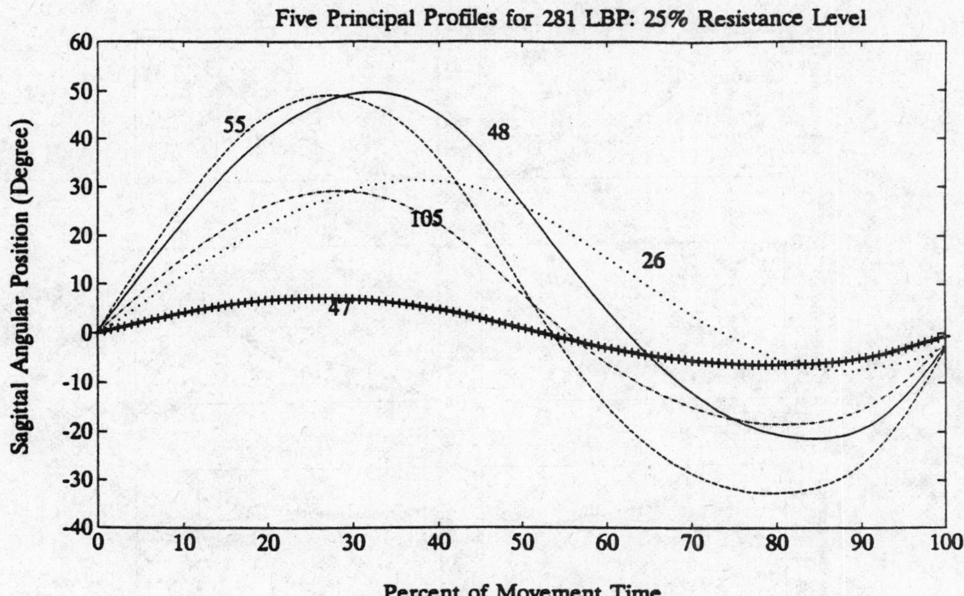

FIGURE 155.5 The five principal profiles of trunk sagittal movement for 281 low back pain patients.

The key limitations in the development of the discriminate functions for classification purposes, are the data-driven nature of the algorithms and the lack of theoretical orientation in the process of development and validation of these models. It is suggested that the mathematical simulation of flexion or extension trunk movement may identify an objective basis for the evaluation and assessment of trunk kinematic performance. A catalog of movement patterns that are optimal with respect to physical and biomechanical quantities may contribute to the emergence of a more theoretically based computational paradigm for the evaluation of kinematic performance of normal subjects and patients. It must be emphasized that in this paradigm one has no intention to claim that the central nervous system actually optimizes any single or composite cost function.

To provide clinical insight for interpretation of the distinctive features in the movement profiles, Parnianpour et al. [1999] have suggested an optimization-based approach for simulation of dynamic point-to-point sagittal trunk movement. The effect of strength impairment on movement patterns was simulated based on minimizing different physical cost functions: Energy, Jerk, Peak Torque, Impulse, and Work. During unconstrained simulations, the velocity patterns of all models are predicted, while time to peak velocity is distinct for each cost function (Figure 155.6). Imposing an 80% reduction in extensor muscle strength diminished the significant differences between unimpaired optimal movement profiles (Figure 155.7). The results indicate that the search for finding the objective function being used by central nervous system is an ill-posed problem since we are sure if we have included all the active constraints in the simulation. The four application areas of these results are: (1) providing optimized trajectories for biofeedback to patients during the rehabilitation process; (2) training workers to lift safely; (3) estimating the task demand based on the global description of the job; and (4) aiding the engineering evaluation to develop ergonomic and workplace interventions which are needed to accommodate individuals with prior disability.

155.4 Conclusions

The outcome of trunk performance is affected by the many neural, mechanical, and environmental factors that must be considered during quantitative assessment. The objective evaluation of the critical dimen-

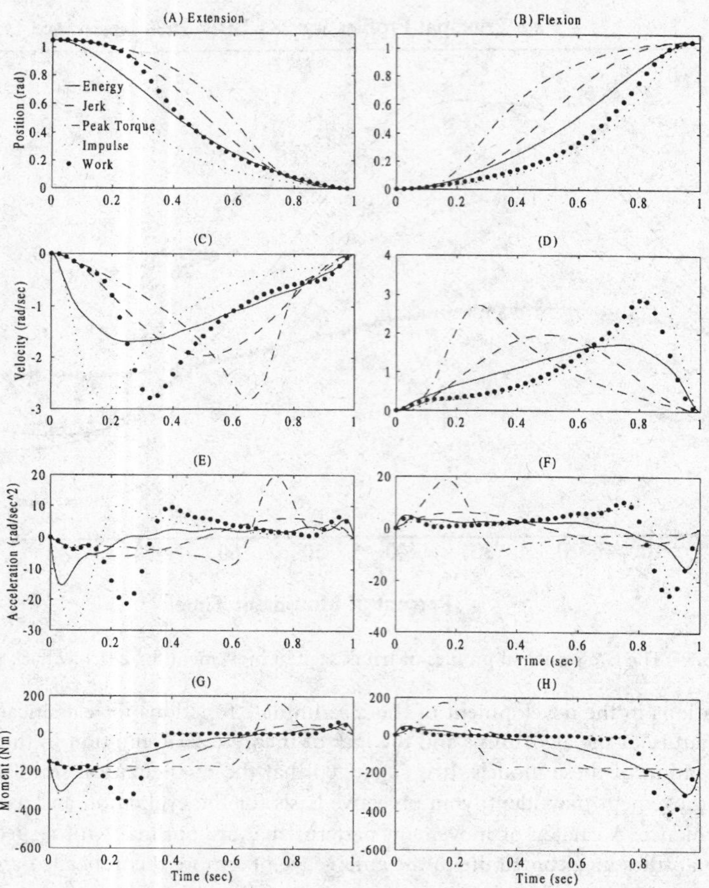

FIGURE 155.6 The optimized unconstrained trunk flexion and extension trajectories for the five cost functions: Energy, Jerk, Peak Torque, Impulse, and Work.

sions of functional capacity and its comparison with the task demands is crucial to the decision-making processes in the different stages of the ergonomic prevention and rehabilitation process. Knowing the tissue tolerance limits from biomechanical studies, task demands from ergonomic analysis, and function capacities from performance evaluation, the rehabilitation team will optimize the changes to the workplace or the task that will maximize the functional reserves (unutilized resources) to reduce the occurrence of fatigue and overexertion. This will enhance worker satisfaction and productivity while reducing the risk of the MSD. Based on ergonomic and motor control literature, the testing protocols that best simulate the loading conditions of the task will yield more valid results and better predictive ability. Ergonomic principles indicate that the ratio of the functional capacity to the task demand (utilization ratio) is critical to the development of muscular fatigue, which may lead to more injurious muscle recruitment patterns and movement profiles due to loss of motor control and coordination. However, large prospective studies are still needed to verify this. The most promising application for these quantitative measures is to be used as a benchmark for the safe return to work of injured workers, given the enormous variations within the normal population. With the advent of technologies to monitor trunk performance in the workplace, one can obtain estimates of the injurious levels of task demands (kinematic and kinetic parameters), which can be used to guide preplacement and rehabilitation strategies. The more functional the clinical tests become, the more clinicians need a complex interpretation scheme. An increasingly complex interpretation scheme opens the possibility of using mathematical modeling with intelligent computer interfaces.

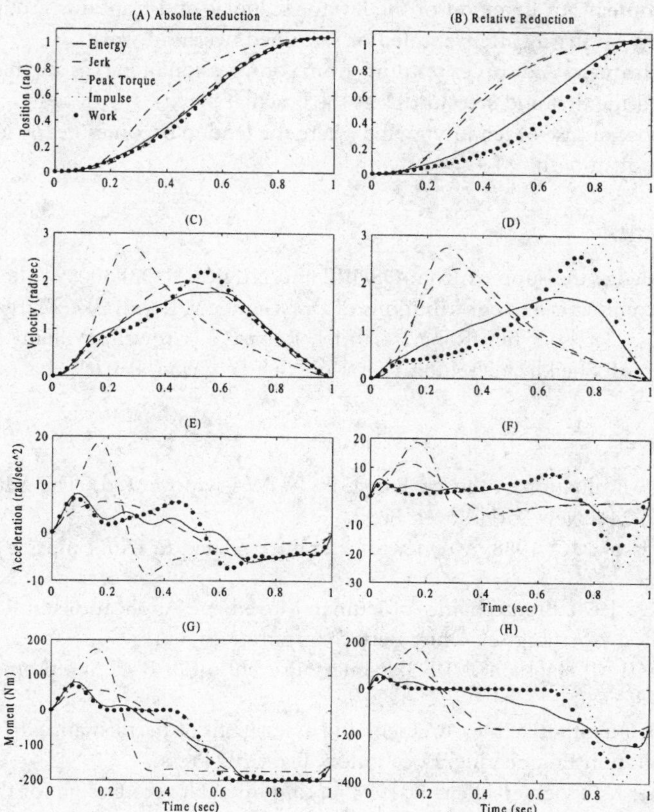

FIGURE 155.7 The optimized trunk flexion trajectories for the five cost functions with the relative and absolute peak extensor strength impairments. The relative strength reduction meant that the extensor strength was reduced to 80% of the peak extension moment required during unconstrained simulations. The absolute strength reduction meant that the extensor strength was set to 200 Nm.

The ability to identify subgroups of patients or high-risk individuals based on their functional performance will remain an open area of research to interested biomedical engineers within the multidisciplinary group of experts addressing neuromusculoskeletal occupational disorders.

Defining Terms

Carpal tunnel syndrome (CTS): The result of compression of the median nerve in the carpal tunnel of the wrist.

Cluster analysis: A statistical technique to identify natural groupings in data.

DeQuervain's disease: A special case of tendosynovitis (swelling and irritation of the tendon sheath) which occurs in the abductor and extensor tendons of the thumb, where they share a common sheath.

Electromyogram: Recordings of the electric potentials produced by muscle action.

Feature extraction: A statistical technique to allow efficient representation of variability in the original signals while reducing the dimensions of the data.

Lateral epicondylitis (tennis elbow): Tendons attaching to the epicondyle of the humerus bone become irritated.

Rotator cuff tendinitis: The irritation and swelling of the tendon or the bursae of the shoulder that is caused by continuous muscle or tendon effort to keep the arm elevated.

Tension neck syndrome: An irritation of the levator scapulae and trapezius group of muscles of the neck commonly occurring after repeated or sustained overhead work.

Thoracic outlet syndrome: A disorder resulting from compression of nerves and blood vessels between the clavicle and the first and second ribs at the brachial plexus.

Trigger finger: A special case of tendosynovitis where the tendon becomes nearly locked so that forced movement is not smooth.

Acknowledgments

The authors acknowledge the support from OSURF and NIDRR H133E30009. The authors would like to thank invaluable comments and contributions of Drs. George V. Kondraske, Margareta Nordin, Victor H. Frankel, Elen Ross, Jackson Tan, Robert Gabriel, Robert R. Crowell, William Marras, Sheldon R. Simon, Heinz Hoffer, Ali Sheikhzadeh, Jung Yong Kim, Sue Ferguson, Patrick Sparto, and Kinda Khalaf.

References

Andersson GBJ. 1991. Evaluation of Muscle Function. In JW Frymoyer (ed), The Adult Spine: Principles and Practice, p 241. New York, Raven Press.

Beimborn DS, Morrissey MC. 1988. A review of literature related to trunk muscle performance. Spine 13(6):655.

Berme N, Cappozzo A. 1990. Biomechanics of Human Movement: Applications in Rehabilitation, Sports and Ergonomics. Worthington, Ohio, Bertec Corporation.

Gundewall B, Liljeqvist M, Hansson T. 1993. Primary Prevention in Back Symptoms and Absence from Work. Spine 18(5):587.

Kondraske GV. 1990. Quantitative measurement and assessment of performance. In Smith RV, Leslie JH (eds), Rehabilitation Engineering, Boca Raton, Fla, CRC Press.

Kroemer KE, Kroemer H, Kroemer-Elbert K. 1994. Ergonomics: How to Design for the Ease & Efficiency. Englewood Cliffs, NJ, Prentice Hall.

Luck JV, Florence DW. 1988. A brief history and comparative analysis of disability systems and impairment evaluation guides. Office Practice 19:839.

Marras WS, Mirka GA. 1989. Trunk strength during asymmetric trunk motion. Human Factors 31(6):667.

Marras WS, Parnianpour M, Ferguson SA, et al. 1993. Quantification and classification of low back disorders based on trunk motion. European Journal of Medical Rehabilitation 3(6):218.

National Institute for Occupational Safety and Health (NIOSH). 1981. Work practice guide for manual lifting (DHHS Publication No. 81122). Washington, DC: U.S. Government Printing Office.

Newton M, Waddell G. 1993. Trunk strength testing with Iso-Machines, Part 1: Review of a decade of scientific evidence. Spine 18(7):801.

Parnianpour M, Nordin M, Kahanovitz N, et al. 1988. The triaxial coupling of torque generation of trunk muscles during isometric exertions and the effect of fatiguing isoinertial movements on the motor output and movement patterns. Spine 13:982.

Parnianpour M, Nordin M, Sheikhzadeh A. 1990. The relationship of torque, velocity and power with constant resistive load during sagittal trunk movement. Spine 15:639.

Parnianpour M, Shirazi-Adl A. 1999. Quantitative assessment of trunk performance. In Karwowski W, Marras WS (eds), Handbook of Occupational Ergonomics, pp 985–1006, Boca Raton, CRC Press.

Parnianpour M, Wang JL, Shirazi-Adl A, et al. 1999. A computational method for simulation of trunk motion: Towards a theoretical based quantitative assessment of trunk performance. Biomedical Engineering, Application, Basis and Communication 11:1-12.

Pope MH. 1992. A Critical Evaluation of Functional Muscle Testing. In Weinstein JN (ed), Clinical Efficacy and Outcome in the Diagnosis and Treatment of Low Back Pain, p 101, Ltd., New York, Raven Press.

Snook SH, Fine LJ, Silverstein BA. 1988. Musculoskeletal disorders. In Levy BS, Wegman DH (eds), Occupational Health: Recognizing and Preventing Work-related Disease, pp. 345–370, Little, Brown and Co., Boston/Toronto.

Szpalski M, Parnianpour M. 1996. Trunk performance, strength and endurance: measurement techniques and application. In Weisel S, Weinstein J (eds), The Lumbar Spine, second edition, pp. 1074–1105, Philadelphia, WB Saunders.

Tan JC, Parnianpour M, Nordin M, et al. 1993. Isometric maximal and submaximal trunk exertion at different flexed positions in standing: Triaxial torque output and EMG. Spine 18(16):2480.

World Health Organization. 1980. International Classification of Impairments, Disabilities, and Handicaps. Geneva. WHO.

World Health Organization. 1985. Identification and Control of Work-related Diseases. Technical report no. 174. Geneva, WHO.

156

Human Performance Engineering: Computer-Based Design and Analysis Tools

Paul J. Vasta
The University of Texas at Arlington Human Performance Institute

George V. Kondraske
The University of Texas at Arlington Human Performance Institute

156.1 Introduction

Computer software applications have been implemented in virtually every aspect of our lives and the field of human performance has not proven to be an exception. Due to the growing recognition of the role of human performance engineering in many different areas (e.g., clinical medicine, industrial design, etc.) and the resulting increase in requirements of the methods involved, having the "right tool for the job" is becoming of vital importance. Software developers bear the brunt of the responsibility of determining the qualities of a software application that define it as being "the right tool" for a specific job. In contrast, users of this class of tools must determine when their application extends a given tool beyond its intended scope. These abilities require a knowledge base spanning a number of different fundamental concepts and methods encompassing not only the obvious aspects of human performance and computer programing, but also many other less obvious issues related to database requirements, parameter standards, systems engineering principles, and software architecture. In addition, foresight of how specific components can best be integrated to fit the needs of a particular usage is also necessary.

This chapter addresses selected aspects of computer software tools specifically directed toward human performance design and analysis. The majority of tools currently available emphasize biomechanical models, and as such, this emphasis is reflected here. However, a much broader scope in terms of the

0-8493-0462-8/00/$0.00+$.50

body systems incorporated is anticipated, and an effort is made to consider the evolution of more versatile and integrated packages. Selected key functional components of tools are described and a representative sample of currently emerging state-of-the-art packages is used to illustrate not only a snapshot of the capabilities now available, but also those which are needed and options that exist in terms of the fundamental approach taken to address similar problems.

In general, the development of computer software applications, especially in maturing fields such as human performance engineering, serves multiple purposes. The most apparent is the relative speed and accuracy that can be achieved in computationally intensive tasks (e.g., dynamic analysis) compared to performing the processes by hand. In addition, computers can handle large amounts of data and help keep track of the multitude of parameters associated with the human architecture. This allows otherwise impossible procedures, such as the detailed analysis of a complex **human-task interface**, visualization of a human figure in a **virtual workspace**, or the computation of time-series multi-body joint torques, to be realized. Perhaps more important, though, are the indirect benefits provided. For example, relatively complex analytic methods utilized in research facilities can be directly implemented in the field by practitioners, with the need for only a complete knowledge of how to effectively *use* the capability available (which is substantially different than the knowledge required to *create* that ability). Also, because the software environment demands rigor (e.g., coding structure requirements, data handling and storage, etc.) and since the potential scope of end-user needs in terms of functional and parametric components encompassed is broad, decisions are forced and rules must be clearly delineated. These characteristics thus serve as a motivation to develop standards for (or at least decide upon the use of) many items including methods, parameters, parameter definition conventions, and units of measure. Such efforts can expose and correct inadequacies inherent in current standards while successful software tools (i.e., those which make their way into everyday use) facilitate dissemination of key knowledge and standards to both researchers and practitioners alike.

It can be anticipated that most analysis or decision-making tasks required of practitioners will be made available in computer-based form, including some that are not currently feasible to perform at all except, perhaps, by using intuitive methods that are inconsistently applied and produce results of questionable validity. Given the extent of the possible list of tools within a given package, and considering the substantial overlap of various support functions, classification of packages is difficult. This can already be seen in the relatively modest number of those currently available.

156.2 Selected Fundamentals

It is typical to think of a software capability in terms of a high-level process, for example "gait analysis," where the end result is of primary interest since the underlying processes are performed "invisibly". Yet these supportive processes, many of which have common mathematical methodologies and modeling approaches, determine the ultimate usability and applicability of the software to a given problem and, therefore, deserve closer examination. It makes sense to consider these methods generically not only because they are common to several existing packages, but also because they will undoubtedly be important in others.

Physics-based Models and Methods

Physics-based methods are implemented when direct analysis of the system and/or its resources is feasible. With regard to the human system, such methods are commonly applied in biomechanical analysis where unknown parameters, like joint torque, are derived from known characteristics, like segment lengths and reaction forces, based on established physical laws and relationships. Given that relatively simple biomechanical analyses typically include multiple segments and degrees of freedom, the feasibility of employing such methods depends not only on the availability of the required input data, but on the capability to process large amounts of information as well. As the analytical complexity of the task increases, e.g., with the inclusion of motion, the significance of implementing such processes within a computer software application becomes readily apparent.

Physical motion is common to most situations in which humans function and is therefore fundamental to the analysis of performance. Parameters such as segment position, orientation, velocity, and acceleration are derived using kinematic and/or dynamic analysis. This approach is equally appropriate for operations on a single joint system or **linked multi-body systems**, such as is typically required for human analysis. Depending on the desired output, forward (direct) or inverse analysis may be employed to obtain the parameters of interest. For example, inverse dynamic analysis can provide joint torque given motion and force data, while forward (direct) dynamic analysis uses joint torque to derive motion. Especially for three-dimensional analyses of multi-joint systems, the methods are quite complex and are presently a focal point for computer implementation [Allard, Stokes, and Blanchi, 1994].

Physics models used to address biomechanical aspects of human performance are relatively well established and utilized in both research and applied domains. In contrast, concepts and methods directed towards other aspects of human performance, such as information processing (in purely perceptual contexts as well as in neuromotor control contexts), are based on strong science and have been incorporated into one-of-a-kind computer models but, to our knowledge have not been engineered into any general purpose computer tools. It can be argued that Shannon's information theory provides the basis for a similar set of "physics-based" cause-and-effect models to support the incorporation of this aspect of human performance into computer-based modeling and analysis tools. Considerable work has gone into the application of information theory to human information processing [e.g., Lachman, Lachman, and Butterfield, 1979] within research domains. The definition of neuromotor **channel capacity** [Fitts, 1954] and metrics for the information content of a visual stimulus measured in bits [Hick, 1952; Hyman, 1953] are some examples of the influence of information theory that have become well established. Given the considerable science that now exists, useful computer-based analytic tools that use these ideas as basic modeling constructs are likely to emerge.

Inference-Based Models

Often, direct measurement of a **performance resource** or structural element is not feasible or practical and estimates must be inferred from models based on representative populations. Inferential based methods, therefore, utilize derived relationships between parameters to provide an estimation (or prediction) of an unknown quantity based on other available measures. One type of modeling approach in such a situation is through statistical regression. This process utilizes data measured from a population of subjects with characteristics similar to the subject or population of interest, to derive a function which represents a "typical" relationship between the independent (measured) variables and the dependent (desired) variable. Specifically, this is achieved through the determination of the function that minimizes the error between the actual and the predicted value across all observations of the independent variable. This process results in an approximation of the desired parameter with a quantified standard error [Remington and Schork, 1985]. Applications of this method require that the dependent variable is distributed normally and with constant variance, and correspondingly, that an estimation of its typical value (for a given population) is desired. Applied to the human system, regression is often implemented to develop **data models**, which are then used to estimate unknown structural parameter values from known parameters, such as the estimation of body segment moments of inertia from stature and weight [McConville, Churchill, Kaleps, Clauser, and Cuzzi, 1980]. Regression has also been used to predict task performance from a wide variety of other variables (e.g., height, weight, age, other performance variables, etc.). A specific example is the determination of the maximum acceptable load during lifting tasks as a function of variables such as body-weight and arm strength [Jiang and Ayoub, 1987].

A conceptually different and relatively new example of an inferential model, motivated by human performance problems specifically, is Non-linear Causal Resource Analysis (NCRA) [Vasta and Kondraske, 1994; Kondraske, 1988]. Quantitative task demands, in terms of performance variables that characterize the involved subsystems, are inferred from a population data set that includes measures of subsystem performance resource availabilities (e.g., speed, accuracy, etc.) and overall performance on the task in question. This method is based on the following simple concept. Consider a sample of 100 people, each with a known amount of cash (e.g., a fairly even distribution from $0 to $10,000). Each person is asked to try to purchase a specific computer, the "cost" of which is unknown. In the subgroup that was able to make the purchase (some won't have enough cash), the individual who had the least amount of cash provides the key clue. That "amount of cash availability" provides an estimate of the computer's cost (i.e., the unknown value). Thus, in human performance, demand is inferred from resource availabilities.

Fidelity

An important principle that has not often received the scrutiny it deserves is that of the level of fidelity desired, or perhaps more critically, necessary for a particular application. This includes anything that affects the range of applicability or quality of the results provided in a given application. Fidelity can be characterized in terms of three distinctly different components: (1) model scope, (2) computational quality (e.g., resolution limits, compounding of errors, convergence limits of numerical methods, etc.), and (3) data quality (e.g., noise, intrinsic resolution limits, etc.). Model scope considers the extent to which major systems are incorporated (e.g., neuromuscular, sensory, cognitive, etc.) and, as a separate issue, the extent to which these major subsystems are represented (e.g., peak torque and angle limits versus a three-dimensional torque-angle-speed envelope for neuromuscular systems of rotational joints).

Parameter Conventions

Often, the exact definition of parameters across (and even within) scientific disciplines varies, limiting communication of findings and inhibiting dissemination of knowledge. Careful selection and clear documentation of **parameter conventions** is an important principle in producing analytic software that can be understood, accepted, and used.

Within the broad scope of parameters that could possibly be incorporated in analytic and other software tools, there are many parameter convention challenges that arise. Due to the fact that of the reported analyses in which various parameters appear to have been of restricted scope and largely special purpose, more generalized situations where convention is important have escaped standardization in terms broad enough to support all application needs. As one example, consider the description of relative orientation between two object-attached coordinate systems in three dimensions (in the context of the human system, this specifies joint angle). There are two basic forms of angle set representations, each derived in terms of the method of rotation of one coordinate system (attached to a moving object) about a specific axis: (1) Fixed angle representation, which involves referencing each rotation of the moving system to some fixed reference frame and, (2) Euler angle representation, indicating that each consecutive rotation of the moving system is referenced to the coordinate axes of its present orientation. Given multiple degrees of freedom, multiple angles result representing the amount of rotation about a specific axis and at a defined position in the order of rotations. The specification of these parameters define the associated angle set convention (i.e., fixed or Euler). Utilization of the terms "roll, pitch, and yaw" [e.g., Chao, 1980] in communicating this convention, originally used to describe ship and aircraft orientation, can also lead to confusion. This is not only due to the lack of similarity between the defined reference frame of an aircraft and that of a human segment, but also due to its altered definition within other disciplines (e.g., fixed angle representation [Spong and Vidyasagar, 1989]). Thus, depending on the "type" of Euler angle used and the sequence of axes about which the rotations occur, two entirely different orientations are likely to result. This discussion does not even consider the clinical perspective on joint angles, where only angles measured in three orthogonal planes [Panjabi, White, and Brand, 1974] are considered. Despite the fact that the human architecture has remained constant (unlike many artificial systems), to our knowledge there is no standard convention that defines all angles in a total human link model for three-dimensional motion.

Data Formats

The utilization of data, especially within a software environment, involves both communication and manipulation not only within a single stand-alone application, but often across facilities, databases, and platforms. Data formats, in this light, can be considered among the most important of the components fundamental to analytic software. Problematic effects may result from aspects including inconsistent adoption of terminology (e.g., endurance vs. fatigue), units of measure specifications, file structures, parameter coding, and database structure. CAD environments in traditional engineering disciplines (e.g., drafting, mechanical design, etc.) have confronted such issues with standardization such as the .DXF file formats (used for two and three-dimensional geometric drawings; even accepted by some numerical machining systems) and the Gerber format (used to communicate a printed circuit board specification to board manufacturers). Within the realm of human performance engineering, standards for data formats remain at the forefront of developmental needs. Some weak standards are emerging, such as the relatively consistent use of ASCII text files with either labeled or position-dependent parameters. However, there are currently no known agreed upon or *de facto* standards for positions, labels, units of measure, etc.

156.3 Scope, Functionality, and Performance

As summarized in Fig. 156.1, CAD-like human performance software depends on conceptual issues as well as decisions regarding fidelity, functionality, and implementation. There is a also a great deal of interaction among these categories. Topics within each category were derived in part from a review of current packages (e.g., those described below) and an assessment of other issues and needs. These lists are intended to be illustrative and not exhaustive. While *every* major category applies to *any* given software tool, not all topics listed within a category will always be relevant.

Conceptual issues include taxonomies on which a given package is based (e.g., basic approaches to motor control, system performance, task categorizations and analysis, data estimation, etc.), parameter choices and codification (e.g., hierarchical level of representation, identification in structured input/output file formats) and compliance with or deviation from accepted standards with the varied communities

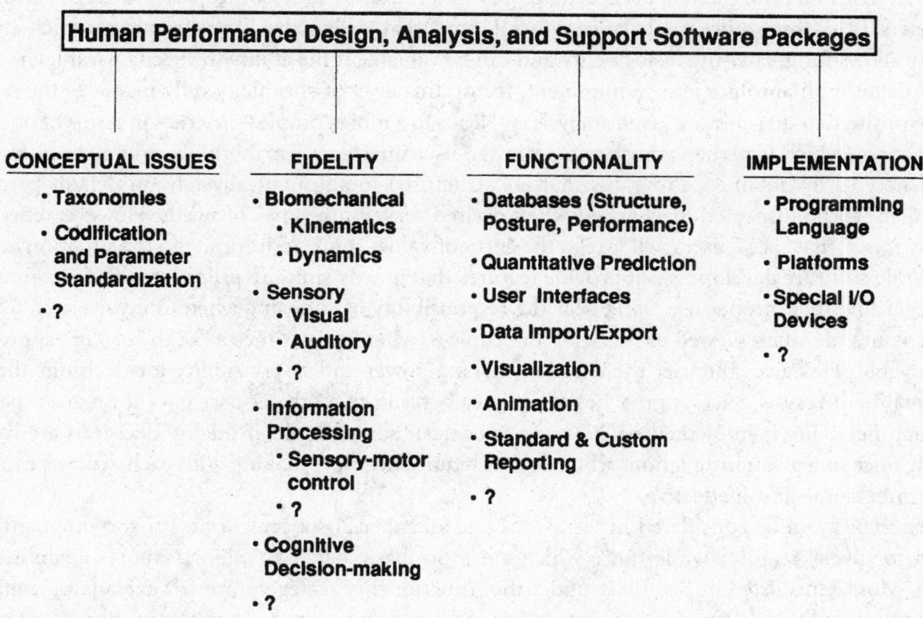

FIGURE 156.1 Related aspects of human design and performance software describe the diversity and scope of underlying issues.

that deal with human performance. They may well deserve special recognition given the impressionable developmental stage of the class of software tools addressed. Perhaps because developers are so familiar with a given perspective or body of knowledge, key conceptual issues are often overlooked or incorporated *de facto* from previous work (e.g., research studies) that is similar (but perhaps not identical) to the intended purpose of a given package, which is frequently broader and more general than the research efforts or projects that inspired it. Software packages impose on users the constructs on which they are based and this may result in conflicts within an already structured environment. Conceptual foundations and approaches are not yet as cleanly defined in human performance as they are in disciplines such as electrical and mechanical engineering. This is further complicated by the wide variety of disciplines (and therefore educational backgrounds) represented by those with interest in participating in software development. Clear and complete disclosure of the conceptual foundation used by developers is thus helpful to both users (potential and actual) and developers.

The concept and scope of fidelity in the present context has been delineated in the section titled "Fidelity" and is further represented by the subtopics in Fig. 156.1. We have chosen to apply it here in its broadest sense. In a fairly recent National Research Council panel on human performance modeling [Baron, Kruser, and Huey, 1990], recommendations were made toward problems regarding design and implementation issues. Though not specifically addressing computer software, the extension of the discussion is natural given present technology and implementation methods. While it was concluded that an all-inclusive model (i.e., high fidelity, in the sense that biomechanical, information processing, sensory and perceptual aspects, etc. are represented) might be desirable, it is highly unlikely that it could be achieved or would be useful, as the inherent complexity would impede effective usage. The basic recommendation made was to pursue more limited scope sub-models. The implication is that two or more computer-based versions of such sub-models could be integrated to achieve a wide range of fidelity, with the combination selected to meet the needs of particular situations. However, there is neither a general framework nor set of guidelines for developers of sub-models (i.e., an **open-architecture concept**) that would facilitate integration of relatively independent software development efforts. Thus, integration is left to the end-user and typically only cumbersome methods are available.

Fidelity should be considered in combination with productivity and both must be carefully assessed against needs. For example, a package that provides the user with control over a large number of parameters (e.g., joint stiffness, balance control, etc.) that define the human system under analysis, typically allows for greater intrinsic fidelity and can be valuable if the ability to specify parameters at that level of detail is an absolute user requirement. If not, this level of specificity only increases the number of prerequisite steps to achieve a given analysis and leads to a more complex program in terms of operation and function (which is perhaps most evident in the user interface). Furthermore, more accurate results are not necessarily obtained. Programs that are structured to automatically rely on default parameter values that can be inspected or changed when desired serve both types of needs. However, concern is typically raised that "lazy" users will forego the entry of values more appropriate to the situation at hand. Thus, while software developers can provide features that greatly simplify procedures and recommendations regarding their proper use, users bear the responsibility of choosing when to invoke such features. Fidelity issues are often viewed as "either"-"or" choices, when many times a "both" or "all" approach is quite feasible. This gives the user the decision-making power and responsibility for decisions that may affect quality of results. Such approaches also typically result in a greater user pool for a given package. How such flexibility is implemented, however, is critical. For example, if fidelity decisions are likely to be made once in a given installation, schemes that require decision-making with each analysis can prove to be cumbersome and ineffective.

Functionality can be considered in terms of basic and special subcategories. Far too often, attention is given to special features while those which are more basic (such as import/export capabilities) are ignored. Most items listed in Fig. 156.1 under the "functionality" category are self explanatory, although it should be noted that features such as importing and exporting are more complex in the present context than one might anticipate due in part to the lack of standards for parameters representation (see above).

Developers must carefully select functionality and potential users must carefully evaluate the impact of those choices against their specific needs. For example, the processing power required to display and animate contoured multi-segment human figures in real-time and in three dimensions is substantial. Programs with this ability typically require high-end platforms and also require that a large fraction of the programming effort and user interface be used for this functionality. The addition of processes such as a dynamic analysis places even a greater stress on processing power and complexity. Visualization of human figures in three dimensions along with some environmental components is essential for some applications. For many applications, such as those in which numerical results are required and conveyed to others with simple printed reports, such visualizations are not required, or contribute little, to achieving the desired end purpose (although they are almost always viewed as "attractive" by potential users). Functionality options are further developed and exemplified in the discussion of actual software packages below (Section 156.4).

The last of these four categories describes the support structure upon which the software packages are built. Software implementation in today's ever changing computer market requires careful thought. In addition to host platform and standard accessories, special input/output interfaces are often necessary (e.g., to allow direct access to laboratory acquired kinematic data). Support software for such special hardware is not always available on all platforms, preventing operation with analytic software in the same environment. Choice of programming language (supporting modules and libraries, platform diversity) and the operating system (Macintosh®, Windows™, Unix, etc.) are also critical. As exemplified in Section 156.4, no existing packages run on all platforms and under all operating systems, leaving users who wish to employ multiple packages with cumbersome and often costly solutions.

As computing power at the single user level (i.e., personal computer) increases, there is little doubt that a full set of state-of-the-art features covering a broad scope of human performance and data needs *could* be integrated into a single package. At present, though, limitations are real and necessary. These are the result of specific trade-offs that must be considered during development when determining target users. Operational costs include software acquisition, maintenance, basic platform acquisition and outfitting, and training costs. A large feature set and high performance are not always indicative of a software tool's value. Value is lost if a user's requirements are either unfulfilled or greatly exceed by the functionality and/or performance of a program. Unfortunately, general tradeoffs that are acceptable within different groups of users have not yet become evident.

156.4 Functional Overview of Representative Packages

In this section, a selected set of different types of software packages intended specifically for human design and performance analysis applications are described. At least one example of each type exists and is noted for reference and to illustrate the types of functionality developers are addressing in response to perceived needs. No endorsement of any example cited is implied. The packages included vary widely in function and performance and should not be considered to be directly comparable. Operational costs reflect this diversity, ranging from relatively inexpensive (e.g., approximately $100.00 plus the price of a moderately equipped DOS-based PC™) to the level defined by high-end workstations (e.g., in excess of $2,000 for software plus the cost of a Unix-based workstation). As with desktop computer applications, such as word processors and spreadsheets, the packages used to illustrate functionality are under constant development and changing feature-sets, performance, and cost are to be anticipated.

The types of packages included illustrate fundamental options with regard to general or dedicated analysis (e.g., user-defined tasks vs. gait), different levels of analysis (e.g., "muscle" vs. "joint" levels in biomechanical analyses), body-function specific or general purpose analysis (e.g., gait vs. user-defined function), fidelity in terms of scope (e.g., biomechanical, sensory, neuromotor control, etc.), and target user orientation (medical vs. non-medical design; although overlap is possible). It should be noted that while particular packages are highlighted under certain categories, underlying features and functions are common across many applications.

Human Parameter Databases

The need for data is evident in all aspects of human-related computer aided design and the analysis of human performance. In certain situations, requirements may be for a specific individual or a representative population. In the former case, there are certain parameters that are not readily attainable (e.g., location of center of mass for a segment, inertia, etc.) and therefore must be estimated, as in the latter case, from normative values derived from studies. For a number of years, these estimates have been available in book form primarily as lookup tables. This format, however, does not take advantage of current technology and is not sufficient for use with software-based analytic tools. Some currently available packages, in addressing this need, have included data tables that are utilized exclusively within the program. Programs such as *MannequinPro* and *SafeWork*, (detailed below) for example, each include a significant database of anthropometric measures for various populations. The user specifies design parameter values by choosing characteristics such as gender, ethnic origin, and/or morphology representative of the desired population. Various other stand-alone database software packages, for example *PeopleSize* (targeted for clothing design), allow data to be exported, thereby providing at least indirect support for external applications.

In addition to measures such as segment lengths, parameters including reach and range of motion for selected postures are frequently provided in databases (e.g., horizontal sitting reach, etc.). Because it is prohibitive (if not impossible) to measure and report all possible situations, this issue may be more adequately addressed through the use of parametric data models. Data derived from such models would provide estimates for all possible conditions as well as having the additional advantage of requiring lesser storage and management requirements. Caution is warranted with regard to issues such as model validity, statistical sample sizes and variations, combinatory effects of merging databases or data models, and traceability of data to original sources.

Biomechanical, Muscle-Level Modeling and Analysis

SIMM is a graphics-based tool for modeling and visualization of any human or animal musculoskeletal system and is directed at lower-level structures, i.e., individual muscles. The program allows users to model any type of musculoskeletal system using files which specify (1) the skeletal structure through polygonal surface objects, (2) the joint kinematics, and (3) the muscle architecture through specifications of a line of action for each muscle and isometric force data. *SIMM* contains various tools for observing and editing the model. The user can visualize and animate the model, display or hide the individual muscles (shown as line segments), manipulate joint angles, muscle forces, and moment arms. Joint torques (static) can be analyzed via plots and graphs and all of the muscle parameters may be edited including visual alteration of muscle lines of action. Finally, the user also has control over joint kinematics by altering the cubic splines used to control the animated joint movements. Developers claim that *SIMM* is aimed at biomechanics researchers, kinesiologists, workspace designers, and students who can benefit from visual feedback showing muscle locations and actions.

Musculographics, the developers of *SIMM*, also produce an add-on package specifically directed toward gait analysis that provides ground reaction force vectors and color-highlighted muscle activity through animation of the model. The software is designed to input data files written by movement analysis systems and can display plots of data such as muscle lengths and ground reaction forces. Another application, the *Dynamics Pipeline*, is a general-purpose package that provides forward and inverse dynamics to calculate motions resulting from forces or torques required to generate a given motion, respectively. A recent version of *SIMM* was made available for Microsoft Windows, however the original Unix-based version is still available.

Body-Function Specific Tools

Gait Lab is a public domain software package that allows fundamental parameters of human gait to be derived, analyzed, and inspected. The program is divided into two main sections (i.e., sub-programs that

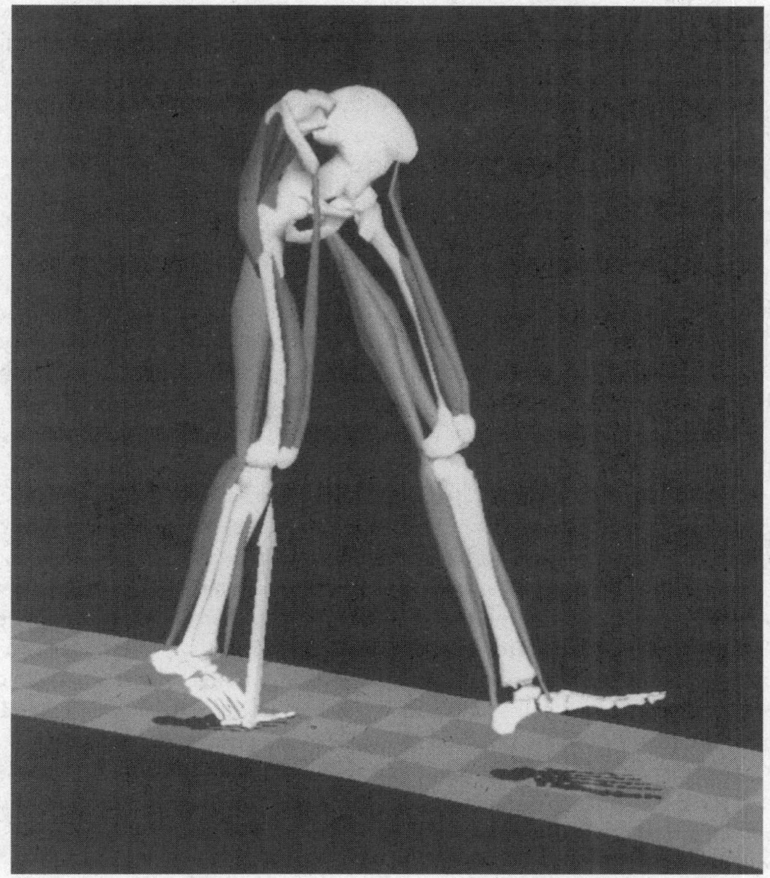

FIGURE 156.2 Muscle attachments and utilization (indicated by shading) for the lower extremities during gait, created and displayed using SIMM software. (Reprinted with permission from Musculo Graphics Inc.)

run under the "shell" of the main application): (1) a mathematics section which provides processes for generating text files representing various gait parameters throughout a gait cycle and (2) a graphics section that creates plots of the parameters as well as a representation of gait through an animated stick figure. A manual [Vaughan, Davis, and O'Connor, 1992], available separately, covers the gait parameters and equations associated with the software package as well as the underlying principles for their use in gait analysis. The software accepts four different, custom-formatted data files (ASCII text) based on direct measurements of a human subject. These include anthropometric, kinematic (i.e., three-dimensional positions of anatomic landmarks), force plate, and EMG measures. Sample files are included for a healthy adult male, female, and an adult male with cerebral palsy. From this data, text files are created for lower extremity body segment parameters (mass, center of gravity position, and moment of inertia), linear kinematics (joint center, heel, and toe locations, segment reference frames), center of gravity (position, velocity, and acceleration), angular kinematics (joint angles, angular velocity and acceleration), and dynamics (joint forces and torques). The package can also display up to three simultaneous plots of these parameters in any combination. Time, percent of gait cycle, and marker position scales are also available for plotting against any of the variables.

Gait Lab provides a useful function in its ability to generate the above mentioned parameters and is effective with its simplistic stick figure animation, allowing minimal processing power/speed requirements while providing the nuances of lower limb movement pertinent to gait analysis. The DOS-based software tested (no version number was indicated—copyright 1992), while certainly not published pre-Windows™, forces the user into the key-command mode (no mouse interface is provided) of using the enter, arrow, and function keys to operate menus and functions. Though not difficult to become acquainted with, it is not intuitive in layout or operation relative to today's user interface standards. Also, with the exception of the anthropometry file which can be created from within the software, no means is provided for editing data files. If new files are to be utilized, they must be in the required format (which must be inferred by inspection of the example files provided with the package).

Biomechanical, Joint-Level, Total Human Analysis

ADAMS is a software package designed for mechanical design and system simulation in general. It is an extensive package incorporating specific components for both modeling and kinematic/dynamic analysis. What sets *ADAMS* apart from other non-performance based simulation/dynamics packages is the availability of a module, *ADAMS/Android*, specifically designed to allow the user to create a human model. The human model may be edited as needed but is provided in a default configuration of a 15-joint body architecture with anthropometric parameters based on the 75th percentile male. Because the package focuses on mechanical systems in general, a great deal of control is provided to the user over the parameters of the model including contact forces, joint friction, damping, and non-linear stiffness as well as control parameters. Having this capability, the human model is able to affect control over objects in its environment as well as responding biomechanically to contact. Supplemental toolsets may be (custom) developed to provide proper control over model parameters such as motion resistance relationships, joint strength and range of motion limits, and delays in neuromuscular response. For a given model, kinematic or dynamic simulations may be run and values for any parameter can be plotted on screen. Both forward and inverse dynamic analysis are available and graphically based facilities such as the specification of trajectories via mouse input are available. Modules for animation and optimization are being developed at present but are expected to be released by the time this handbook is published.

Because *ADAMS*' capabilities are so extensive, it has become one of the most widely used mechanical system simulation packages. Outside of the realm of human performance analysis, the package is targeted at engineers and scientists who require complete observation and control of the model parameters across many hierarchical levels. While this format affords great fidelity and specificity in the analysis, there is also a great deal of complexity and detail that must be addressed in every new simulation run. Though **performance assessment** is not an inherent or recognized feature in the package, custom modules may be incorporated which would allow for direct comparison of system resource utilization and stress against the demands of the task.

Visualization for Low-End Computer Platforms

MannequinPro, while having inherent database functionality, is essentially a visualization tool that allows users to quickly create three-dimensional humanoid figures which may be represented in stick-figure form, as a wireframe "robot" with polygonal segments, or as its name implies, a mannequin with higher-resolution segments. These figures may be postured within preset (i.e., non-editable) range-of-motion limits, as well as viewed from many different perspectives. They are created automatically by the program according to user defined structural aspects chosen from a list of specifications for gender, population percentile, body type, age, and nationality. The package also provides a modest drawing (CAD-type) tool for creating various objects to be placed in the scene with the figure. Two additional features extend *MannequinPro* beyond the domain of a "simple" object rendering tool. The first is provided through animation effects, allowing the mannequin to walk via an internally generated gait sequence (a specific path may be defined through a virtual workspace), reach to a specific location, or move through any sequence of positions created on a frame-by-frame basis by the user. The second feature allows the user to specify forces acting on the mannequin's hands or feet from which static torques are calculated for both left and right wrists, elbows, shoulders, neck, back, hips, knees, and ankles.

As a visualization tool, *MannequinPro* provides a needed capability for a modest price and requires only a personal computer to run. Figures are easily produced to the anthropometric specifications available and posing, reaching, and walking functions, though limited, are simple to control. The package is primarily operated from mouse input with standard point, click, and drag operations that should be familiar to most users of current commercial software. Once the scene is created with humans and objects in their workspace, it is up to the user to extract any information regarding function and performance from the visual images presented by the package. These features, as well as others, are also contained within the software package *Jack* and *SafeWork* (described below), which provide greater functionality but are targeted toward workstation environments. In summation, *MannequinPro*, with a good price to performance ratio, is an effective tool when human visualization with moderate graphics is required at the single-user platform level.

Extended Visualization for the High-End Workstations

Jack, developed within the University of Pennsylvania Center for Human Modeling and Simulation [Badler, Phillips, and Webber, 1993], and *SafeWork* are high-end, CAD-oriented design and visualization packages. Similar in function to *MannequinPro* (above), but with considerably more flexibility and detail, they are designed to allow human factors engineers to place a human in a virtual workstation environment while still in the design phase. Both extend far beyond the capabilities of *MannequinPro*, trading simplicity for specificity in that they provide users with control over a large range of variables regarding many aspects of the human figure and its virtual environment. Human models are fully articulated and have preset or user-definable anthropometric parameters including segment dimensions, joint limits, moments of inertia, and strength allow highly-specific representation. The total functionality of theses package are far too great to list the individual aspects here, but examples include reach, hand gripping, balance (including reorientation behavior), collision detection, light and camera (view) manipulation, animation, and kinematic/dynamic analysis. *Jack* also allows for real-time posture tracking of a human operator through position and orientation sensors.

Similar to the functionality of *ADAMS*, these packages provide a great deal of control over the design and manipulation of the human figure. Because of the large number of controllable parameters and the difficulty involved in providing an interface allowing three-dimensional manipulation of high-end graphics in a two-dimensional environment, the majority of users will require training. *Jack* and *SafeWork* require a Silicon Graphics workstation to run.

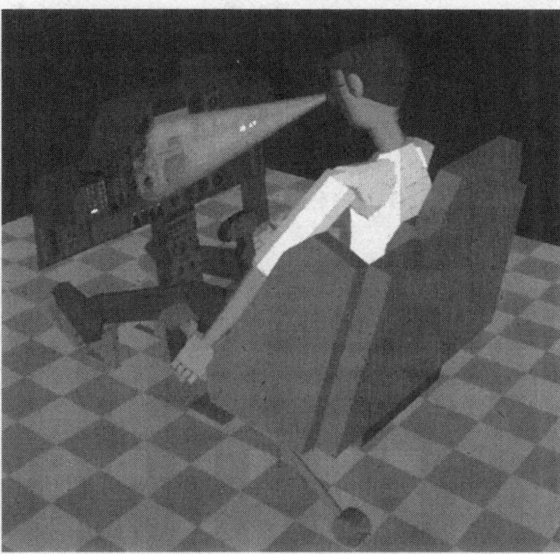

FIGURE 156.3 A human/workspace figure created and displayed using Jack software. (Reprinted with permission from the Center for Human Modeling and Simulation at the University of Pennsylvania.)

Total Human, Performance-based, Human-Task Interface Analysis

While now under development, *HMT-CAD* represents yet a different perspective in that it is based primarily on performance modeling constructs (e.g., Kondraske, 1995) and is aimed at producing bottom-line assessments regarding human-to-task, human-to-machine, and eventually machine-to-task interfaces using logic similar to that used by systems-level designers. In a joint effort between Photon Research Associates, Inc. and the University of Texas at Arlington Human Performance Institute, performance models and decision-making strategies are combined with advanced multi-body dynamic analysis.

HMT-CAD's initial development emphasized a top-down approach that set out with the goal of realizing a shell that would allow systematic incorporation of increasingly higher fidelity in a modular, step-wise fashion. The initial modules encompass biomechanical and neuromuscular aspects of performance that bases analyses on knowledge of range/extremes of motion, torque, and speed capacities. Gross total human (23 links and 41 degrees of freedom) and hand link models (16 links and 22 degrees of freedom each) are included. The Windows™ based tool provides a pre-defined framework for specifying human structure and performance parameters (e.g., anthropometry, strength, etc.) for individuals or populations. A task library is included based on an object-oriented approach to task analysis. HMT-CAD attempts to help communicate information about many parameters via a custom graphical-button interface. For example, one analysis result screen shows a human figure surrounded by buttons connected to body joints. Button color (e.g., red or green) communicates whether an analysis found limiting factors for the specified person in the specified task associated with that joint system. Using computer mouse input, clicking on red buttons produces a new window showing a list of performance capacities and stress levels associated with systems of that joint. Maximally stressed performance resources are tagged. *HMT-CAD* uses the same systems level modeling constructs for human and artificial systems. A novel technique that transparently draws upon a large set of data models as needed is used to provide "the best analysis possible" with the data provided by the user. This allows users to directly specify values for all parameters if desired, but does not require this for each analysis.

In the present prototype, analysis scope is very limited (e.g., upper extremity, object moving tasks). Databases are being populated to allow a similar fidelity of analysis for trunk and lower extremity analyses. Development of an optimization engine based on minimizing stress across performance resources [Kondraske and Khoury, 1992] to solve problems in which redundant approaches are possible is under way. No capability is currently provided for animation of human movement, as binary assessments (e.g., red and green buttons) and numerical results (stress levels on performance capacities) are emphasized. However, an interface to emerging general purpose animation tools is planned. Long-term development will proceed to increase fidelity in a stepwise fashion by the inclusion of performance capacities for other major systems (e.g., information processing associated with motor control) as well as additional performance capacities for major systems represented at any given time (e.g., the inclusion of neuromuscular endurance limits).

156.5 Anticipated Development

While it is relatively easy to outline the functionality and unique aspects of the software discussed above, it is a great deal harder to describe the level of difficulty required to achieve a desired end result and to characterize how well functions are performed. On-line "help" systems, multi-media components, associated software development tools, improved parameter measurement tools, and higher fidelity models will no doubt impact these aspects.

As mundane as it may seem, perhaps nothing is more important to the overall advancement of this class of software tools than the development of standards for parameters used, parameter conventions, and data formats. It is likely that *de facto* standards will emerge from individual commercial endeavors. Due to the diversity of professionals involved, it is difficult to envision the evolution of "standards by committee". Observation of success stories in analogous efforts also supports this opinion. This, in turn,

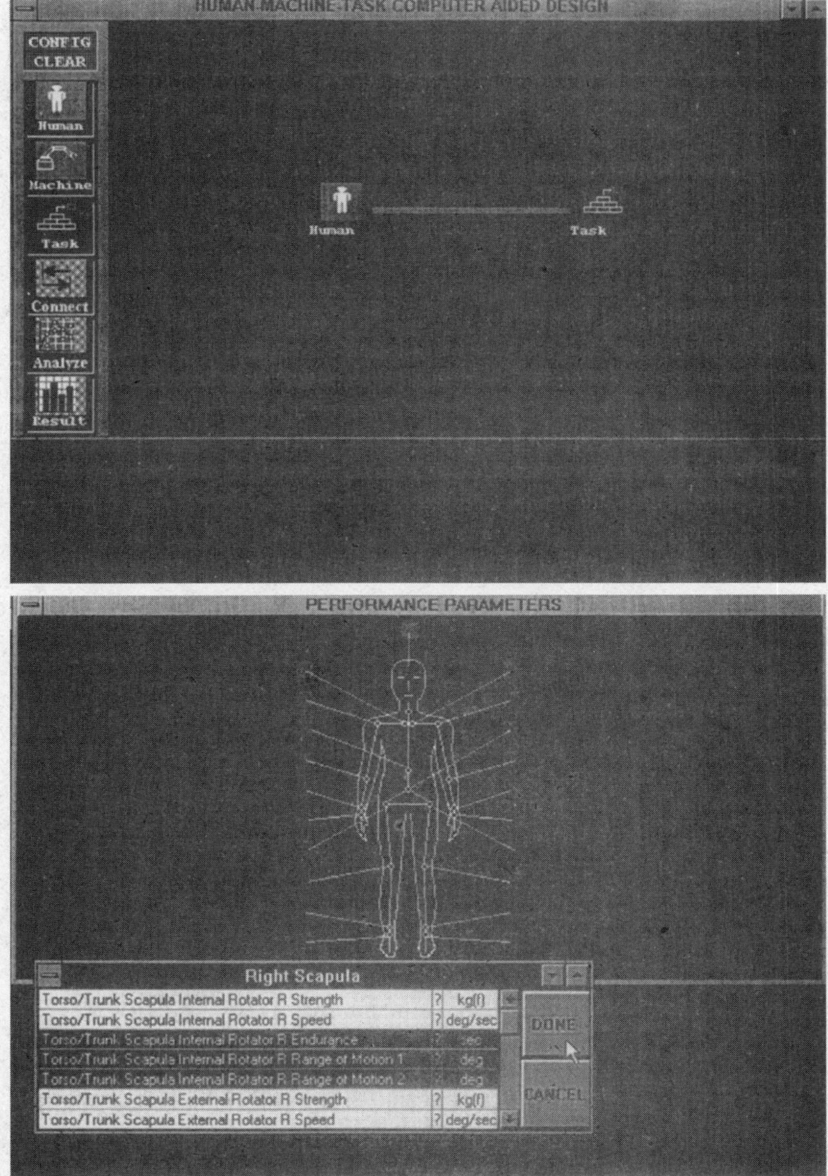

FIGURE 156.4 A desktop environment of the HMT-CAD performance assessment package including system model development and analysis result reporting.

will enable progress to be made in database and data model development, which is vital to many of the more routine functions such software will perform. Combined with modular object-oriented programming techniques, the emergence of true multi-tasking operating systems on desktop computers, and the support for inter-process communication (such as dynamic data exchange) will encourage developers. Many new tools, both similar to and completely different than those described here, can be anticipated.

Defining Terms

Channel capacity: The maximum rate of information flow through a specific pathway from source to receiver. In the context of the human, a sensory-motor pathway (e.g., afferent sensory nerves, processors, descending cerebro-spinal nerve and α-motoneuron) is an example of a channel through which motor control information flows from sensors to actuators (muscle).

Data model: A mathematical equation derived as a representation of the relationship between a dependent variable and one or more independent variables. Similar in function to a database, data models provide parameter values across conditions of other related parameters.

Human-task interface: The common boundary between the human and a task specified by the demands of a task on the respective resources of the human and the capacities of the human's resources involved in performing the task.

Linked multi-body system: A system of three or more individual segments joined (as an open or closed chain) in some manner with the degree of freedom between any two segments defined by the characteristics of the corresponding hinge.

Open-architecture concept: A general methodology for the development of functional modules that may be readily integrated to form a higher-level system through well-defined design and interface constructs.

Parameter conventions: Aspects and/or usage of a parameter that are specifically defined with respect to general implementation.

Performance assessment: The process of determining the level or degree in which a system can perform a specific task given the demands required and the availabilities of the resources of the system which are involved in performing the task.

Performance resource: Defined as a functional unit and an associated dimension of performance (e.g., knee extensor strength) that is available to a system.

Virtual workspace: A representation of a three-dimensional physical workspace generated by computer software and displayed on a video monitor or similar device. This enables, for example, the inclusion and manipulation of computer generated objects within the virtual workspace such that designs can be tested and modified prior to manufacturing.

References

Allard, P., Stokes, I.A.F., and Blanchi, J.P., eds. 1994. *Three-Dimensional Analysis of Human Movement*, Human Kinetics, Champaign, Illinois.

Badler, N.I., Phillips, C.B., and Webber, B.L. 1993. *Simulating Humans: Computer Graphics Animation and Control*, Oxford Univ. Press, New York.

Badler, N.I. 1989. Task-oriented animation of human figures. In *Applications of Human Performance Models to System Design*, G.R. McMillan, D. Beevis, and E. Salas, eds., Plenum Press, New York.

Baron, S., Kruser, D.S., and Huey, B.M., eds. 1990. *Quantitative Modeling of Human Performance in Complex, Dynamic Systems*, National Academy Press, Washington, D.C.

Chao, E.Y.S. 1980. Justification of triaxial goniometer for the measurement of joint rotation, *Biomechanics*, 13:989-1006.

Craig, J.J. 1989. *Introduction to Robotics Mechanics and Control*, 2nd ed., Addison-Wesley, New York.

Delp, S.L., Loan, J.P., Hoy, M.G., Zajac, F.E., Topp, E.L., and Rosen J.M. 1990. An interactive graphics-based model of the lower extremity to study orthopaedic surgical procedures. *IEEE Trans. Biomed. Eng.*, 37(8):757-67.

Fitts, P. 1954. The information capacity of the human motor system in controlling the amplitude of movement, *J. Exp. Psych.*, 47:381-91.

Hick, W. 1952. On the rate of gain of information, *Q. J. Exp. Psych.*, 4: 11-26.

Hyman, R. 1953. Stimulus information as a determinant of reaction time, *J. Exp. Psych.*, 45:423-32.

Jiang, B.C. and Ayoub, M.M. 1987. Modelling of maximum acceptable load of lifting by physical factors, *Ergonomics*, 30(3):529-38.

Khan, C. June 1991. Humanizing AutoCAD. *Cadence*.

Kondraske, G.V. 2000. A working model for human system-task interfaces. *Handbook of Biomedical Engineering*, J.D. Bronzino, ed., CRC Press Inc., Boca Raton, Florida.

Kondraske, G.V. 1988. Experimental evaluation of an elemental resource model for human performance, In: *Proc., Tenth Ann. IEEE Eng. Med. Biol. Soc. Conf.*, New Orleans, p. 1612-13.

Kondraske, G.V. and Khoury, G.J. 1992. Telerobotic system performance measurement: motivation and methods. In: *Cooperative Intelligent Robotics in Space III* (Proc. SPIE 1829), J.D. Erickson, ed. p. 161-172, SPIE.

Lachman, R., Lachman, J.L., and Butterfield, E.C. 1979. *Cognitive Psychology and Information Processing: An Introduction*, Lawrence Erlbaum Assoc., Hillsdale, New Jersey.

McConville, J.T., Churchill, T.D., Kaleps, I., Clauser, C.E., and Cuzzi, J. 1980. *Anthropometric Relationships of Body and Body Segment Moments of Inertia*, Air Force Aerospace Medical Research Laboratory, AFAMRL-TR-80-119, Wright-Patterson Air Force Base, Ohio.

Panjabi, M.M, White III, A.A., and Brand, R.A. 1974. A note on defining body parts configurations, *J. Biomechan.*, 7(4):385-87.

Pheasant, S.T. 1986. *Bodyspace: Anthropometry, Ergonomics and Design*, Taylor & Francis, Philadelphia.

Remington, R.D. and Schork, M.A. 1985. *Statistics with Applications to the Biological and Health Sciences*, Prentice-Hall, New Jersey.

Spong, M.W. and Vidyasagar, M. 1989. *Robot Dynamics and Control*, Wiley & Sons, New York.

Vaughan, C.L., Davis, B.L. and O'Connor, J.C. 1992. *Dynamics of Human Gait*, Human Kinetics, Champaign, IL.

Vasta, P.J. and Kondraske, G.V. 1994. Performance prediction of an upper extremity reciprocal task using non-linear causal resource analysis, In: *Proc., Sixteenth Ann. IEEE Eng. Med. Biol. Soc. Conf.*, Baltimore. 305–306.

Winter, D.A. 1990. *Biomechanics and Motor Control of Human Movement*, 2nd ed. Wiley, New York.

Further Information

For explicit information regarding the software packages included in this chapter, the developers may be directly contacted as follows:

ADAMS: Mechanical Dynamics, 2301 Commonwealth Blvd., Ann Arbor, MI 48105,USA, 800-88-ADAMS, http://www.adams.com

Gait Lab: originally published by Human Kinetics,1607 N. Market St., Box 5076, Champaign, IL 61825, (217) 351-5076, now available as public domain software on the International Society of Biomechanics web page at http://isb.ri.ccf.org/isb/data/gaitlab/.

HMT-CAD: Human Performance Institute, The University of Texas at Arlington, P.O. Box 19180, Arlington, TX 76019, (817) 272-2335

Jack: Center for Human Modeling and Simulation, The University of Pennsylvania, 200 South 33rd St., Philadelphia, PA 19104-6389, (215) 898-1488, http://www.cis.upenn.edu/~hms/jack.html

MannequinPro: HumanCAD, a division of Biomechanics Corporation of America, 1800 Walt Whitman Road, Melville, NY, 11747, (516) 752-3568, http://www.humancad.com

PeopleSize: Open Ergonomics Ltd., Loughborough Technology Centre, Epinal Way, Loughborough, LE11 3GE, UK, +44 (0) 1509 218 333, http://www.openerg.com/psz.htm

SafeWork: Genicom Consultants Inc., 3400 de Maisonneuve Blvd. West, 1 Place Alexis Nihon, Suite 1430, Montreal, Quebec, CANADA, H3Z 3B8, (514) 931-3000, http://www.safework.com

SIMM: Musculographics, Inc., 1840 Oak Ave., Evanston, IL 60201, (708) 866-1882, http://www.musculographics.com

A broad perspective of issues related to those presented here may be found in: Matilla, M. and Karwowski, W., eds. 1992. Computer applications in ergonomics, occupational safety, and health, In: *Proc. Int. Conf. Computer-Aided Ergonomics and Safety '92*, Tampere, Finland, North-Holland Publishers, Amsterdam.

Additionally, periodic discussions of related topics may be found in the following publications: *CSERIAC Gateway*, published by the Crew System Ergonomics Information Analysis Center. For subscription information contact AL/CFH/CSERIAC, Bldg. 248, 2255 H St., Wright-Patterson Air Force Base, Ohio, 45433; *IEEE Transactions on Biomedical Engineering* or *IEEE Engineering in Medicine & Biology Magazine.* For further information contact: IEEE Service Center, 445 Hoes Lane, P.O. Box 1331, Piscataway, NJ, 08855-1331.

Numerous other related sites can be found on the Internet. Among those of interest are:

Biomechanics World Wide: includes a very thorough listing of various topics related to biomechanics inlcuding links to educational, corporate, and related sites of interest. http://www.per.ualberta.ca/biomechanics/bwwframe.htm

Biomch-L: an internet list-server supporting discussions regarding topics relating to biomechanics. http://isb.ri.ccf.org/isb/biomch-l/.

Motion Lab Systems, Inc.: a collection of information and links to a variety of Internet sites focused on Electromyography, Biomechanics, Gait Analysis, and Motion Capture. Includes both commercial and non-commercial sites. http://www.emgsrus.com/links.htm

157

Human Performance Engineering: Challenges and Prospects for the Future

George V. Kondraske
University of Texas at Arlington

Human performance engineering is a dynamic area with rapid changes in many facets and constantly emerging developments that affect application possibilities. Material presented in other chapters of this section illustrates the breadth and depth of effort required as well as the variety of applications in this field. At the same time, the field is relatively young, and as is often the case in such instances, some "old" issues remain unresolved and new issues emerge with new developments. With the other material in this section as background, this chapter focuses on selected issues of a general nature (i.e., cutting across various human subsystem types and applications) in three major areas vital to the future of the field: (1) human performance modeling, (2) measurement, and (3) data-related topics. In addition to the identification of some key issues, some speculation is presented with regard to the lines along which future work may develop. Awareness of the specific issues raised is anticipated to be helpful to practitioners who must operate within the constraints of the current state of the art. Moreover, the issues selected represent scientific and engineering challenges for the future that will require not only awareness but also the collaboration of researchers and practitioners from the varied segments of the community to resolve.

157.1 Models

It is important to distinguish various types of models encountered in human performance. They include conceptual, statistically based predictive models, predictive models based on cause and effect (e.g., [Kondraske et al., 1997]), and data models [e.g., Vasta and Kondraske, 2000]. Issues related to data models are described later in this chapter, while those related to the remaining types are discussed in this section.

Traditionally, a tremendous amount of activity in various segments of the human performance community has been directed toward development of statistically based predictive models (e.g., regression

0-8493-0462-8/00/$0.00+$.50

TABLE 157.1 Representative Quotations Illustrating the Need for a Shift Toward the Development of Casual Models

Quote	Field	Reference
"... there has been little advancement of the science of ergonomics ... the degree of advancement is so low that, in my opinion, ergonomics does not yet qualify as a science ... There has been no progress in the accumulation of general knowledge ... because nearly all studies have not been general studies."	Human factors	Smith, 1987
"... experiments are performed year after year to answer the same questions, those questions—often fundamental ones—remain unanswered ... the methodology employed today is a hodgepodge of quick fixes that evolved over the years into a paradigm that is taught and employed as a sacrosanct, when in fact it is woefully inadequate and frequently incompetent ..."	Human factors	Simmon, 1987
"... authors and users of these instruments lacked a conceptual model ... lack of a well-developed conceptual model is unfortunately characteristic of the entire history of the field ..."	Rehabilitation	Frey, 1987

models). This remains the most popular approach. By their very nature, each of these is very application-specific. Even within the intended application, success of these models has ranged mostly from poor to marginal, with a few reports of models that look "very good" in the populations in which they were tested. However, the criteria for "good" have been skewed by the early performance achieved with these methods, and statistically significant *r* values of 0.6 have come to be considered "good." Even if predictive performance was excellent, the long-term merit of statistically based prediction methods can be questioned on the basis that this approach is intrinsically inefficient. A unique and time-consuming modeling effort, requiring new data collection with human subjects, is required for every situation. At the same time, powerful computer-based tools are commonly available that facilitate the generation of what is estimated to be hundreds and perhaps thousands of these regression models on an annual basis. It appears that many researchers and practitioners have forgotten that one purpose of such statistical methods is to provide insight into the cause-and-effect principles at play and into generalizations that might be more broadly applied. Statistically based regression models will continue to serve useful purposes within the specific application contexts for which they are developed. However, the future of predictive modeling in human performance engineering must be based on cause-and-effect principles. The sentiment for the need to shift from statistical to more physically based models is present and growing in several of the related application disciplines, as evidenced by a sampling of quotations from the literature presented in Table 157.1.

A National Research Council panel on human performance modeling [Baron et al., 1990] conducted one of the more broad-looking investigations regarding complex human performance models; i.e., those which consider one or more major attributes (e.g., biomechanical, sensory, cognitive, structural details, etc.) of a total or near-total human. The convening of this panel itself underscored the interest in and need for human performance models. As an interesting aside, while the panel's focus was on modeling, a large fraction of the efforts cited as background represented major software packages, most of which were developed to support defense industry needs. A major application for such models is in computer simulation of humans in various circumstances, which could be used to support prediction and other analysis needs. A series of useful recommendations was included in panel's report regarding the direction that should be taken in future work. However, recommendations were quite general in nature, and no specific plans were outlined regarding how some of these objectives might actually be achieved. Nonetheless, it was clear from the choice of previous work cited and from the recommendations that development of cause-and-effect that general-purpose models should receive priority. The working model presented for this section of this text [Kondraske, 2000] is one example of a causal, integrative model that reflects an initial attempt to incorporate many of this panel's recommendations.

Whether primarily statistical or causal in nature, frustration with what has been characterized or perceived as "limited success" achieved to date in efforts to unlock and formalize the underpinnings of complex human functions such as speech, gait, lifting, and memory have led to attacks in recent years on the basic reductionistic approach that has been pursued most aggressively. In fact, reductionism has been characterized as a "dead end" methodology—one that has been tried and has failed [Bunge, 1977; Gardner, 1985], as further evidenced by the following somewhat representative quote [Weismer & Liss, 1991]:

> This view held the scientific community spellbound until people started looking past the technological issues, and asking how well the reductionist observations were doing at explaining behavior, and the answer was, miserably. Many microscopic facts had been accumulated, and incredible technological advances had been made, but the sum of all of these reductionist observations could not make good sense of macroscopic levels of movement behavior.

Prior to total abandonment, it is perhaps wise not only to characterize the performance of reductionism but also to question why it might have failed *thus far.* For example, it may be prudent to entertain the proposition that reductionism may be "a correct approach" that has failed to date because of a problem with one or more components associated its implementation. Are we certain that this is not the case?

One potential explanation for an apparent failure of reductionism is simply the manner in which manpower has been organized (i.e., the research infrastructure) to attack problems of the magnitude of those considered, for example, by the National Research Council panel noted above. With some exceptions, the majority of funded research efforts are short term and involve very small teams. Is it possible that the amount of information "to reduce" is too much to produce results that would be characterized as "success" (i.e., does the equivalent of an undercapitalized business venture exist)? It is further observed that the value of *reduction* is often only demonstrable by the ability to *assemble.* As noted, the prime tool of assembly thus far has been statistical in nature, and relatively few causal models have been seriously entertained. With regard to human systems, it is clear that there are many such "items" to assemble and many details to consider if "assembly"—with high fidelity—is to result. Furthermore, it is clear that tools (i.e., special computer software) are needed to make such assembly efforts efficient enough to consider undertaking. Has the data management and analytical power of computers, which have only been readily available in convenient-to-use forms and with required capacities for less than a decade, been fully exploited in a "fair test" of reductionism?

Reductionism is clearly a methodology for understanding "that which is." Drawing by inference from engineering design in general, a close relationship can be observed with systems engineering synthesis methods used to define "that which will be." In artificial systems (as opposed to those which occur naturally, such as humans), reverse engineering (basically a reductionistic method) has been used quite successfully to create functionally equivalent replicas of products. The methodology employed is guided by knowledge of the synthesis process. The implementation of reductionism in human performance modeling can possibly benefit from the reverse mental exercise of "building a human," which may not be so abstract given current efforts and achievements in rehabilitation engineering. In looking toward the future, it is fair to ask, "To what extent have those who have attempted to apply reductionism, or who have dismissed it, attempted to bring to bear methods of synthesis?" The real support for reductionism is success obtained with methods based on this concept; new findings of an extremely encouraging nature are beginning to come forth (e.g., [Kondraske et al., 1997].

The issue of biologic variability is also frequently raised in the context of the desire to develop predictive models of human performance. It is perhaps noteworthy that variability is an issue with which one must deal in the manufacture of man-made products (i.e., particularly with regard to quality assurance in manufacturing), which are designed almost exclusively using causal principles. Taguchi methods [Bendell et al., 1988], a collection of mathematics and concepts, have been used in widespread fashion with remarkable success in modeling and controlling variability in the characteristics of a final product which is based on the combination of many components, each of which has multiple characteristics with values that also range within tolerance bands. Similarity of these circumstances suggests that Taguchi methods also may be valuable in human performance engineering. Investigations of this nature are likely to be a part of future work.

157.2 Measurements

Standardization of Variables and Conditions

Human performance literature is replete with different measures that characterize human performance. Despite the number and magnitude of efforts where human performance is of interest, a standard set of measurement variables has yet to emerge. Relatively loose, descriptive naming conventions are used to identify variables reported, leading to the perception of differences when, upon careful inspection, the variables used in two different studies or modeling efforts are the same. In other cases, the names of variables are the same, but conditions under which measures are acquired are different or not reported with enough detail to evaluate if numeric data are comparable (across the reports in question) or not.

The needs of science and engineering are considerably different. Although standardization is generally considered to be desirable in both arenas, the former stresses only that methods used be accurately reported so that they may be replicated. There has been, unfortunately, little motivation to achieve standardization in any broad sense in the human performance research world. In traditional areas of engineering, it is the marketplace that has forced the development of standards for naming of variables such as performance characteristics of various components (e.g., sensors, actuators, etc.) and for conditions under which these variables are measured. In more recent years, product databases (e.g., for electronic, mechanical, and electromechanical components) and systems-level modeling software (i.e., computer-aided design and computer-aided manufacturing), along with the desire to reduce concept-to-production cycle times, have further increased the levels of standardization in these areas.

Despite the somewhat gloomy circumstance at present, there are some signs of potential improvement. An increasing number of research groups are beginning to attack problems of larger scale, and interest is growing with regard to the ability to exchange data and models. Several small standards development efforts have surfaced recently in specific areas such as biomechanics to develop, for example, standards for kinematic representation of various body joints. The advent of an increasing number of commercially available measurement tools also has driven a trend to similarity in conventions used in some specific subareas. However, the quest for standardization in measurement that is often voiced is still a distant vision. A willingness to suffer the inconvenience of "standardizing" in the short run in exchange for a greater convenience and "power" over the long run must be recognized by a critical mass of researchers, instrument manufacturers, and practitioners before a significant breakthrough can occur. The large number of different professional bodies and societies that may potentially become involved in promulgating an official set of standards will most likely contribute to slow progress in this area.

New Instruments

Instruments to acquire measures of both human structure and performance have been improving at a rapid pace, commensurate with the improvement of base technologies (e.g., sensors, signal conditioning, microprocessors, and desktop computer systems, etc.). Compared with one decade ago, a practitioner or researcher can today assemble a relatively sophisticated and broad-based measurement laboratory with commercially available devices instead of facing the burden of fabricating his or her own instruments. However, a considerable imbalance still exists across the profile of human performance with regard to commercial availability of necessary tools. This imbalance extends to the profile of tools seen in use in contexts such as rehabilitation and sports training. In particular, devices that measure sensory performance capacities are not nearly as commonplace or well developed as those which measure, for example, strength and range of motion. Likewise, there are few commercially available instruments that measure aspects of neuromotor control in any general sense other than, for example, in a selected task such as maintaining stable posture. The complexity and diversity of higher-level human cognitive processes also have hindered development of measurement instruments in this domain that possess any true degree of commonality in content across products. A wide variety of computer-based test batteries has been proliferating that are more or less implementations of the great number of tests formerly administered in paper-pencil format.

From a practitioner's perspective, prospects for the future are both good and bad. There is some evidence that suppliers of instruments that cover areas of the performance where there has been vigorous competition (e.g., dynamic strength) are experiencing market saturation and uncertainty on the part of consumers regarding what is "the best way" to acquire necessary information. This may force some groups to drop product lines, while others are expressing subtle interests in expanding measurement instrument product lines to fill empty niches. This latter behavior is good in that it will likely increase the scope of measurement tools that is commercially available. However, problems associated with standardization often confuse potential users, and this, in general, has slowed progress in the adoption of more objective measurement instruments (compared with subjective methods) in some application areas. This, in turn, has inhibited product providers from taking the risks associated with introduction of new products.

Compared with the commercial availability of tools that characterize the performance of various human subsystems, instruments that quantify task attributes are much less prevelant. However, the perception that increased emphasis on this area will increase the utility of measures that characterize the human will likely motivate a substantial increase in the number and variety of products available for task characterization. In addition, factors such as the Americans with Disabilities Act (ADA), which encourages work-site evaluations and modifications to facilitate employment of individuals with disabilities, and the increase in work-site related injuries such as carpal tunnel syndrome have led to an increased demand for such tools.

There has been a subtle but noticeable developing sense of awareness in the various communities that there is room for more than one kind of instrument to measure a given variable. Thus debates regarding "the correct approach" which were commonplace are beginning to be replaced with debates concerned with determining "the best approach for a given situation." A prime example is in the area of strength measurement. Devices range from relatively inexpensive hand-held dynamometers that provide rapid measurements and meaningful results with neurologic patients to expensive devices with electric or hydraulic servomotor systems that are more suited, for example, for use in sports medicine contexts. This trend is quite healthy and is likely to spread.

Measuring the Measurements

Reliability and *validity* have been the key words associated with characterizing the quality of measures of human structure and performance [e.g., Potvin et al., 1985]. Often, these terms are used in a manner that implies that a given test "has it" or "does not have it" (i.e., reliability or validity) when in fact both reliability and validity should be recognized as continuums.

There are traditional, well-known methods used in academic circles to quantify reliability. The result produced implies an interpretation of "how much" reliability a given test has, but there is no corresponding way to determine how much is needed (in the same terms that reliability of a test is measured) so that one could determine if one has "enough." Other aspects of the academic treatment of human performance measurement quality are similarly troublesome with regard to the implications for widespread, general use of measurements. For example, there is an inherent desire (and need) to generalize the results of a given reliability study for a given instrument and, at the same time, a reluctance on the part of those who conduct or consume such studies to do so in writing. The strictly academic position is that one can *never* generalize a reliability study; i.e., it applies only to a situation which is identical to that reported (i.e., that instrument, those subjects, that examiner, that room, that time of day, etc.). Thus one may ask, "What is the value of reporting any such study?" The purely academic view may be the most correct, but strict interpretation is completely useless to a practitioner who needs to reach a conclusion regarding the applicability of a given instrument in given application. The mere fact that reliability studies are reported implies that there is an expectation of generalization to *some* degree. The question is, "How much generalization can one make?" This issue has not been adequately addressed from a general methodological standpoint. However, awareness of this issue and the need for improved methods are growing.

What is the general "quality" of measurement available today? The following is offered as a reasonably "healthy" perspective on this complex issue:

1. Technology has advanced to the point where it is now possible to measure many of the basic physical variables employing measurements such as time, force, torque, angles, linear distance, etc. very accurately, repeatably, and with high resolution without significant difficulty.
2. Most of the differences on test-retest are due to limitations on how well one can reasonably control procedures and actual variability of the parameter measured—even in the most ideal test subjects.
3. Many studies have been conducted on the reliability of a wide variety of human performance capacities. Some true generalization can be achieved by looking at this body of work as a whole and not from attempts to generalize from only single studies.
4. Across the many types of variables investigated, results of such studies are amazingly quite similar. Basically, they say that if your instrumentation is good, and if you carefully follow established, optimal test administration procedures, then it should be possible to obtain results on repeat testing that are within a range of 5% to 20% of each other. The exact location achieved in this range depends in large part on the particular variable in question (e.g., repeated measures associated with complex tasks with many degrees of freedom such as lifting or gait will differ more than repeated measures of hinge joint range of motions, for example). In addition, the magnitude of the given metric will influence such percentage characterizations since errors are often fixed amounts and independent of the size of the quantity measured. Thus, when measuring quantities with small magnitudes, a larger variability (i.e., more like 20%) should be anticipated. One can usually determine an applicable working value in these straightforward, usable terms (e.g., 5% or 20%) that allows direct comparison with an evaluation of needs by careful review of the relevant reliability studies.

A major point of this discussion is to illustrate that traditional methods used to "measure the measurements" do not adequately communicate the information that current and potential users of measurements often need to know. Traditional methods were valuable to some degree in providing relative indicators within the context of academic group studies. However, methods that are more similar to those used to characterize measurement system performance in physics and other traditional engineering areas (e.g., accuracy, signal-noise-ratio concepts, etc.) are needed. This will help practitioners who must make single measurements on individual subjects (i.e., not populations) to support clinical decision making [Mayer et al., 1997]. As the manner in which measurements are used continues to change to include more of the latter, new interest and methods for quantifying the quality of measurements can be anticipated.

157.3 Databases and Data Modeling

Much research has been conducted to define measurements of human structure and performance and to collect and characterize data as a basis to increase understanding and make inferences in different contexts. A large number of problems of major medical and other societal significance require the use of such data (1) for diagnostic purposes (i.e., when assessing measures obtained from a subject under test, such as to determine the efficacy of an intervention or treatment in a medical context), (2) for modeling and simulation (e.g., analyses for return-to-work decisions of injured employees, to support job-site modifications for individuals with disabilities, to design specifications for virtual reality systems), (3) for status/capacity evaluations (e.g., disability determinations for insurance settlements, worker's compensation claims, etc.), (4) to obtain a better understanding of disease and aging processes (e.g., associated with such problems as falls in elderly populations, etc.), (5) to gain insight into the impact of environmental and occupational factors (e.g., as in epidemiological studies, such as in lead toxicity or carpal tunnel syndrome), and (6) to support ergonomic design of consumer products and living environments for use by the public in general. Despite such a diverse range in needs for basically the same information, there has been little—if any—attempt to organize, integrate, and represent available data in a compact, accessible form that can serve the cited needs.

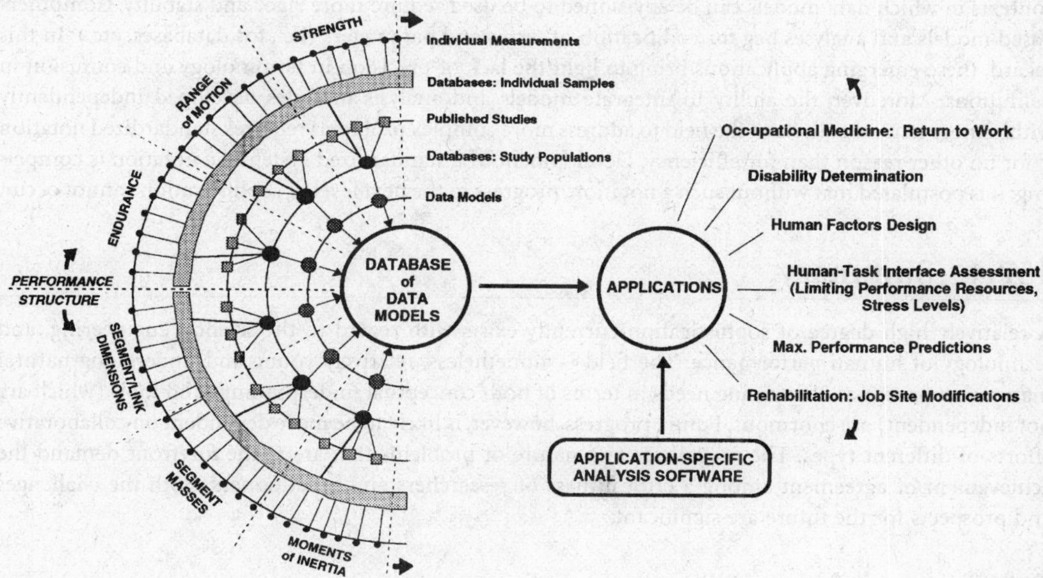

FIGURE 157.1 Summary of the complexity of the problem associated with integrating human structural and performance data to realize compact representations (i.e., parametric data models) to support analyses in multiple application areas.

Figure 157.1 illustrates the scope of the general problem and infers a potential approach to integration. At the most basic level, individual measurements for a given variable must be collected and databased. In addition, databases from published studies are required to identify specific gaps that limit either analytic functionality or fidelity in the types of tasks that data must support. Data from the literature are typically available only in summary form, i.e., in terms of means and standard deviations for defined populations. Future research is required to define and validate methods for integrating data (for example) from multiple studies to form a single, more robust data model for a given performance or structure parameter. This will involve testing data models developed from databases of individual measures and databases of studies against each other. Development of data warping methods (e.g., to adapt predictions of expected performance levels to subjects of different body sizes and/or ages) is required to provide models that would be transparent operationally and would allow analyses to proceed in data-limited situations with predictable tradeoffs in fidelity. In many areas, especially with regard to practitioners such as occupational and physical therapists, performance characterization has been viewed only unidimensionally. More multidimensional perspectives are needed for an adequate representation of fidelity [e.g., Kondraske, 2000]. Another major aspect of data modeling therefore involves the development of multivariate models of performance envelopes [Vasta and Kondraske, 1997] for individual human subsystems (e.g., knee flexors, visual information processors, etc.). Such models would support, for example, prediction of available torque production capacity under different specified conditions (e.g., joint angle, speed, etc.). In summary, a basic strategy is required for harvesting the vast amount of previously collected data to obtain compact, accessible representations. Means must be incorporated in this strategy to allow for integration of data from multiple sources and continuous update as new data as collected to enhance fidelity and range of applicability of composite models.

As previously noted, individuals within the human performance community typically adopt a variety of terms or identification schemes for structures and parameters employed in both the execution and communication of their work. The sheer number of parameters and variety of combinations in which they may be required to meet analysis needs may be sufficient to justify a more formal approach [e.g., Kondraske, 1993], thereby facilitating the development of common data structures. The engineering contexts in which data models can be envisioned to be used require more rigor and stability. Computer-based models and analyses beg for codification of terms and parameters (i.e., for databases, etc.). In this regard, these emerging applications bring to light the lack of precision in terminology and confusion in definitions. Moreover, the ability to integrate models and analysis modules developed independently within more narrow subsets of the field to address more complex problems requires standardized notation if for no other reason than for efficiency. Development of a standardized systematic notation is compelling; it is postulated that without such a notation, progress to the next level of sophistication cannot occur.

157.4 Summary

A relatively high degree of sophistication currently exists with regard to the science, engineering, and technology of human performance. The field is, nonetheless, relatively young and undergoing natural maturation processes. Remaining needs in terms of both conceptual underpinnings and tools (which are not independent) are enormous. Future progress, however, is likely to be more dependent on collaborative efforts of different types. The magnitude and nature of problems that are at the forefront demand the achievement of agreement among a critical mass of researchers and practitioners. Both the challenges and prospects for the future are significant.

References

Bendell A, Disney J, Pridmore WA. 1988. Taguchi Methods: Applications in World Industry. London, IFS Publishing.

Bunge M. 1977. Levels and reduction. Am J Physiol 233:R75.

Frey WD. 1987. Functional assessment in the '80's: A conceptual enigma, a technical challenge. In AS Halpern, MJ Fuhrer (eds), Functional Assessment in Rehabilitation. Baltimore, Paul H Brookes.

Gardner H. 1985. The Mind's New Science: A History of Cognitive Revolution. New York, Basic Books.

Kondraske GV. 1993. The HPI Shorthand Notation System for Human System Parameters (technical report 92-001R V1.5). Arlington, Texas, The University of Texas at Arlington, Human Performance Institute.

Kondraske GV, Johnston C, Pearson A, Tarbox L. 1997. Performance prediction and limiting resource identification with nonlinear causal resource analysis. Proceedings, 19th Annual Engineering in Medicine and Biology Society Conference, pp. 1813–1816.

Kondraske GV. 2000. A working model for human system-task interfaces. In JD Bronzino (ed), Handbook of Biomedical Engineering, 2nd ed., Boca Raton, Fla, CRC Press.

Mayer T, Kondraske GV, Brady Beals S, Gatchel RJ. 1997. Spinal range of motion: accuracy and sources of error with inclinometric measurement. Spine, 22(17):1976–1984.

Potvin AR, Tourtellotte WW, Potvin JH, et al. 1985. The Quantitative Examination of Neurologic Function. Boca Raton, Fla, CRC Press.

Simmon CW. 1987. Will egg-sucking ever become a science? Hum Fact Soc Bull 30(6):1.

Smith LL. 1987. Whyfore human factors. Hum Fact Soc Bull 30(2):1.

Vasta PJ, Kondraske GV. 1992. Standard Conventions for Kinematic and Structural Parameters for the "Gross Total Human" Link Model (technical report 92-003R V1.0). Arlington, Texas, The University of Texas at Arlington Human Performance Institute.

Vasta PJ, Kondraske GV. 1997. An approach to estimating performance capacity envelopes: knee extensor system example. Proceedings, 19th Annual Engineering in Medicine and Biology Society Conference, pp. 1713–1716.

Vasta PJ, Kondraske GV. 2000. Human performance engineering computer-based design and analysis tools. In JD Bronzino (ed), Handbook of Biomedical Engineering, 2nd ed., Boca Raton, Fla, CRC Press.

Weismer G, Liss JM. 1991. Reductionism is a dead-end in speech research. In CA Moore, KM Yorkston, DR Beukelman (eds), Dysarthria and Apraxia of Speech. Baltimore, Paul H Brookes.

Further Information

This section of this *Handbook* entitled "Human Performance Engineering" contains chapters that address human performance modeling and measurement in considerably more detail. Manufacturers of instruments used to characterize different aspects of human performance often provide technical literature and bibliographies of studies related to measurement quality, technical specifications of instruments, and application examples.

Future articles of relevance can be expected on a continuing basis in the following journals:

American Journal of Occupational Therapy
American Occupational Therapy Association, Inc.
4720 Montgomery Ln.
Bethesda, MD 20814-3425
http://www.aota.org/

Archives of Physical Medicine and Rehabilitation
W. B. Saunders Company, Periodicals Dept.
6277 Sea Harbor Dr, Orlando, FL 32887-4800
http://www.archives-pmr.org/apmr/index.html

Human Factors
Human Factors and Ergonomics Society
P.O. Box 1369
Santa Monica, CA 90406
http://www.hfes.vt.edu/HFES.html

IEEE Transactions on Rehabilitation Engineering and
IEEE Transactions on Biomedical Engineering
IEEE Service Center
445 Hoes Lane
P.O. Box 1331
Piscataway, NJ 08855-1331
http://www.ieee.org/index.html

Journal of Biomechanics
Pergamon Press
660 White Plains Road
Tarrytown, NY 10591

Journal of Occupational Rehabilitation
Plenum Publishing Corporation
233 Spring St.
New York, NY 10013
http://www.plenum.com/

Physical Therapy
American Physical Therapy Association
1111 North Fairfax St.
Alexandria, VA 22314
http://www.apta.org/

Rehab Management
4676 Admiralty Way
Suite 202
Marina Del Rey, CA 90292
http://www.cmrnet.com/cmrnet/CMR7386.htm

Journal of Rehabilitation Research and Development
Scientific and Technical Publications Section
Rehabilitation Research and Development Service
103 South Gay St., 5th Floor
Baltimore, MD 21202-4051
http://www.vard.org/jour/jourindx.htm

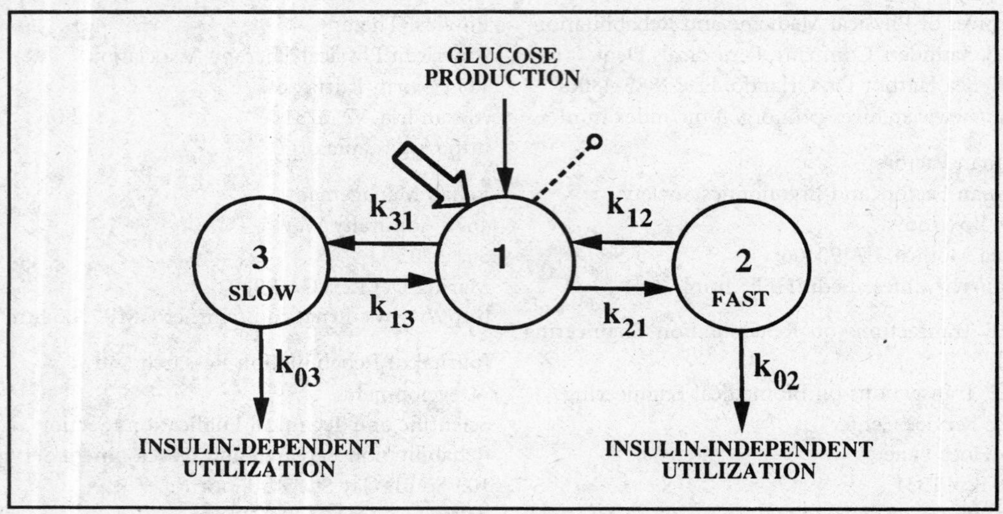

The linear compartmental model of glucose kinetics.

XVI

Physiological Modeling, Simulation, and Control

Joseph L. Palladino
Trinity College

The complexities of physiological systems often lead to the creation of models to aid in the understanding of measured experimental data and to predict new features of the system under study. Models may serve to compactly summarize known system properties, including biomechanical, chemical, and electrical phenomena. Experiments performed on the model yield new information that casts light on unknown mechanisms underlying these properties and stimulate the formulation of new questions and the design of new experiments. Further, such quantitative description often leads to a better understanding of the physical mechanisms underlying both normal and altered (diseased) system performance. Ultimately, modeling may facilitate clinical diagnosis of system failure at an early stage of disease development.

Models begin conceptually; for example, the concept that a blood vessel behaves as a fluid-filled pipe. Concepts may be developed into physical models, e.g., using a latex tube to describe a blood vessel, upon which experiments are performed. Often, concepts are realized as mathematical models, whereby the concept is described by physical laws, transformed into a set of mathematical equations, and solved via computer. Simulations, distinct from models, are descriptions that mimic the physiological system. The quantitative nature of physiological models allows them to be employed as components of systems for the study of physiological control, illustrated in several of this section's chapters.

Chapter 158 describes the iterative process of physiological modeling and presents three recent, widely different, modeling approaches in cardiovascular dynamics. The first, an arterial system model, combines features of two flawed classical methods into a hybrid model capable of interpreting a wide range of previously unexplained experimental observations. Next, a large-scale, distributed muscle model shows how mechanical description of muscle structure can predict a molecular mechanism responsible for heart muscle's complex mechanical function, i.e., structure predicts function. Finally, an approach for predicting ventricular hypertrophy based on cardiac function is presented, with function predicting structure. These latter two approaches serve to integrate molecular level and organ level function and structure.

Chapter 159 shows how mass balance is used to study material kinetics, e.g., pharmacokinetics. The production, distribution, transport, and interaction of exogenous materials, such as drugs or tracers, and endogenous materials, such as hormones, are also described. Examples of both linear and nonlinear compartmental models are presented, as well as parameter estimation, optimal experiment design, and model validation.

Chapter 160 demonstrates the importance of cardiovascular modeling in designing closed-loop drug delivery systems, such as intravenous infusion pumps for pharmacological agents. The chapter begins with an extensive review of segmental and transmission line models of the arterial system and time-varying elastance descriptions of the left ventricle. The author makes an argument for employing nonlinear cardiovascular-linked pharmacological models to aid in controller design, since data from these models can improve, and reduce the number of, animal and human experiments, and can be used to simulate human conditions not easily produced in animals.

Chapter 161 presents models used to study the continual interaction of the respiratory mechanical system and pulmonary gas exchange. Traditionally, the respiratory system is described as a chemostat—ventilation increases with increased chemical stimulation. Alternatively, the author proposes that quantitative description of the "respiratory central pattern generator", a network of neuronal clusters in the brain, is a much more sophisticated and realistic approach. Study of this dynamic, optimized controller of a nonlinear plant is interesting from both physiological and engineering perspectives, the latter due to applications of these techniques for new, intelligent control system design.

Chapter 162 introduces the use of neural network techniques and their application in modeling physiological control systems. As their name implies, neural networks are computational algorithms based upon the computational structure of the nervous system, and are characterized as distributed processing and adaptive. Neural networks have been used to describe the control of arm movements with electrical stimulation and the adaptive control of arterial blood pressure.

Chapter 163 surveys methods of system identification in physiology, the process of extracting models or model components from experimental data. Identification typically refers to model specification or model estimation, where unknown parameters are estimated within the specified model using experimental data and advanced computational techniques. Estimation may be either parametric, where algebraic or difference equations represent static or dynamic systems, or non-parametric, where analytical (convolution), computational (look-up tables), or graphical (phase-space) techniques characterize the system.

Chapter 164 shows how modeling can propose mechanisms to explain experimentally observed oscillations in the cardiovascular system. Low frequency flow oscillations are measured in individual vascular beds as well as in the entire systemic circulation. The authors modified a common windkessel description of the heart's systemic arterial load to include autoregulation, whereby peripheral resistance changes continuously to maintain cardiac output. This control system is characterized by a slow and delayed change in resistance due to smooth muscle activity. Experiments on this new model show oscillations in the input impedance frequency spectrum, and flow and pressure transient responses to step inputs consistent with experimental observations. This autoregulation model supports the theory that low frequency oscillations in heart rate and blood pressure variability spectra (Mayer waves) find their origin in the intrinsic delay of flow regulation.

Chapter 165 presents models to describe the control of human movement, primarily the dynamics of human extremities. Presented are the challenges associated with description of the extremities as rigid bodies using laws of mechanics such as D'Alembert's principle, the Newton-Euler method, and the LaGrange equations. Analytic limb models are presented first, described by sets of ordinary differential equations. Next, the complexities of limb movement control are described by nonanalytic models, for example an expert system denoted "artificial reflex control", learning algorithms and fuzzy logic. Finally, hybrid controller models, combining analytic and nonanalytic techniques, are discussed.

Chapter 166 describes physiological models of the fast eye movement, or saccade, that enables the eye to move quickly from one image to another. This common eye movement is encountered when scanning from the end of one line of printed text to the start of another. Early quantitative models helped define important characteristics of saccades. Increasingly complex models of the oculomotor plant were derived using improved models of the rectus eye muscle. Finally, application of systems control theory and simulation methods led to a saccade generator model in close agreement with experimental data.

158

Modeling Strategies in Physiology

J.L. Palladino
Trinity College

G.M. Drzewiecki
Rutgers University

A. Noordergraaf
University of Pennsylvania

158.1 Introduction

In physics and chemistry a multitude of experimental observations can be captured in a small number of concepts, often referred to as laws. A law is a statement of the relation between quantities which, as far as is known, is invariable within a range of conditions. For example, Ohm's law expresses the ratio between the voltage difference across and the current through an object as a constant. In physiology, experimental observations can likewise be ordered by relating them to natural laws. In such complex systems several laws may play their part, and the process is then referred to as development of a model. Model design, therefore, must meet stringent requirements and serves to rein in undisciplined speculation. Modeling attempts to identify the mechanism(s) responsible for experimental observations and is fundamentally distinct from simulation, in which anything that reproduces the experimental data is acceptable.

A successful model provides guidance to new experiments, which may modify or generalize the model, in turn suggesting new experiments. This iterative process is illustrated in Fig. 158.1. The quest for understanding of a body of experimental observations (block 1) leads to a tentative conceptual interpretation (block 2), which is transformed into quantitative statements by invoking established natural laws. The resulting mathematical model (block 3) consists of a set of equations that are solved generally with the aid of a computer. The equation solutions are quantitative statements about experimental observations, some of which were not part of the original body of experiments, extending knowledge about the

0-8493-0462-8/00/$0.00+$.50

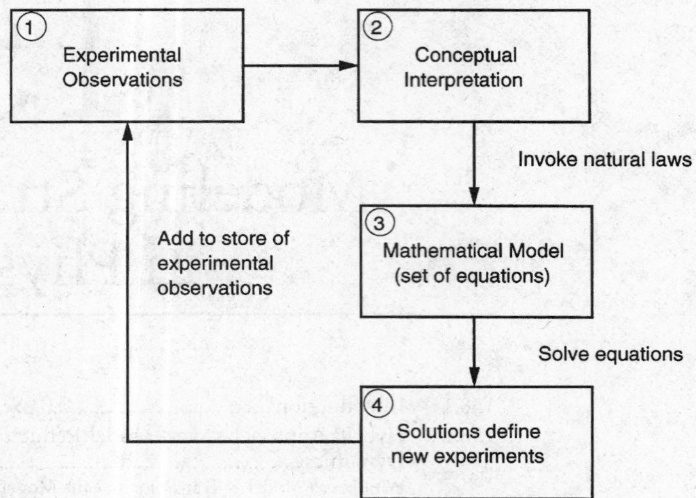

FIGURE 158.1 Iterative inter-relationships between models and experiments.

system under study. Consequently, new experiments become specified (block 4), adding to the store of experimental data. The enlarged body of experimental observations may cause modification in the investigator's conceptual interpretation, which in turn may give reason to modify the mathematical model. This iterative process should weed out flawed experiments and misconceptions in the mathematical model. If the entire effort is successful, a comprehensive, quantitative, and verified interpretation of the relevant experimental observations will result. This procedure recently led to the discovery that William Harvey's view of the circulation is unduly restrictive [32].

The first case study of this chapter outlines the development of an arterial system model using a hybrid approach. Features of two separate, flawed classical methods are combined to interpret a wide range of previously unexplained experimental observations. In the course of interaction between model and experimentation, inclusion of greater detail may expand model complexity and size, rendering it too unwieldy as a component to a larger physiological system. Reduction in size, distinct from simplification, may be necessary. The former is based on understanding, allowing intelligent separation of major and minor effects. The latter implies arbitrary action based on intuition. This case study closes by demonstrating systematic reduction of the arterial model.

A major challenge in the biosciences today is integration of a large body of molecular level information, with the ultimate goal of understanding organ function. Two different, novel modeling approaches to this goal are presented next. The first shows how a model based on muscle structure can predict a molecular mechanism responsible for a muscle's mechanical properties (function). The second presents an approach for predicting ventricular hypertrophy (structure) based on cardiac metabolic function. Due to space constraints, discussion is limited to the main points, however, details are available in the original research papers cited.

158.2 Hybrid Approaches and Model Reduction: Arterial Dynamics

Windkessel Model

This early focus was to develop a non-invasive method to measure cardiac output from noninvasively obtained arterial pressure information with the aid of a model of the systemic arterial circulation. Hales

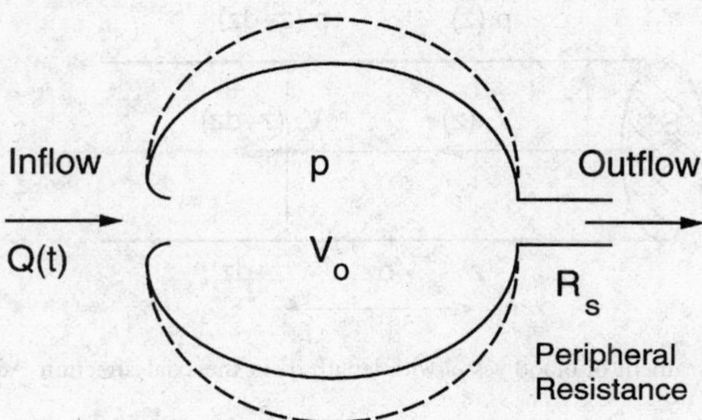

FIGURE 158.2 Windkessel description of arterial load. Adapted from [34].

(1733) believed that the aorta and major vessels act as a "compression chamber" which transforms pulsatile inflow from the heart into steady flow into the organs. Frank (1895) called this compression chamber a *windkessel*, after fire-fighting pumps that produce steady flow despite oscillating pumping.

Cardiac output is the product of stroke volume, the volume pumped with each beat, and heart rate. Heart rate is readily obtainable from the palpated pulse. Stroke volume was to be estimated using the following windkessel model of the arterial load. Fig. 158.2 shows left ventricular inflow, Q(t) into a compliant chamber (windkessel), arterial pressure *p*, and volume V_o coupled to a single resistor, R_s representing the peripheral resistance. Since all arterial properties are lumped, pulse wave velocity *c* is assumed infinite. The differential equation relating these variables is:

$$\frac{1}{E'}\frac{dp}{dt}+\frac{1}{R_s}p=Q(t) \tag{158.1}$$

where E′ represents the windkessel elastance dp/dV_o. During diastole Q(t) = 0, leading to the solution:

$$p(t)=p_o e^{-\frac{E'}{R_s}t} \tag{158.2}$$

By measuring the exponential decay of an experimental pressure curve and the (finite) pulse wave velocity *c*, R_s can be obtained, and thereby stroke volume V_s from:

$$V_s=\int Q(t)dt=\frac{1}{R_s}\int p(t)dt \tag{158.3}$$

The problem is that the pulse wave velocity is assumed to be both finite and infinite in the same model. This contradiction leads to errors in V_s of up to a few hundred percent. A large number of efforts to improve accuracy by modifying this model has achieved little.

Transmission Line Model

The interpretation of pulse wave speed has fascinated many researchers. Initially considering the aorta as a simple uniform, infinitely long transmission line, the Webers [43], [44] proposed the second version of a finite wave propagation theory. Consider a uniform vessel segment (Fig. 158.3) with axial direction *z*, cross-sectional area S, and radius r. Shown are two points: z and z + dz from left to right. Application of Newton's second law gives:

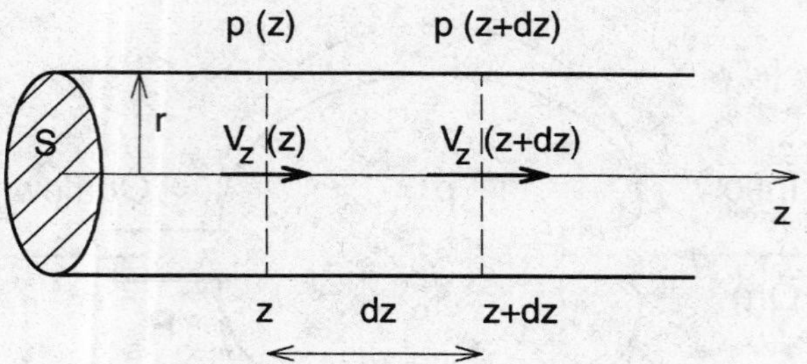

FIGURE 158.3 Segment of blood vessel with length dz in the axial direction. Adapted from [34].

$$S\left[p(z)-p(z+dz)\right]=S\,\rho\,dz\frac{\partial v_z}{\partial t} \tag{158.4}$$

where ρ is blood density, p is pressure, and v_z is axial velocity. This expression may be rewritten as the equation of motion:

$$-\frac{\partial p}{\partial z}=\rho\frac{\partial v_z}{\partial t} \tag{158.5}$$

The change in volume V with time is equal to the inflow minus the outflow:

$$S\left[v_z(z)-v_z(z+dz)\right]=\frac{dV}{dt} \tag{158.6}$$

which may be expressed as the equation of continuity:

$$-\frac{\partial v_z}{\partial z}=\frac{1}{S}\frac{\partial S}{\partial t} \tag{158.7}$$

Relating S to p via LaPlace's law and Hooke's law, and combining the result with Eq. 158.5 gives

$$\frac{\partial^2 p}{\partial t^2}=c^2\frac{\partial^2 p}{\partial z^2} \tag{158.8}$$

where:

$$c=\sqrt{\frac{hE}{2\rho r}} \tag{158.9}$$

with E denoting Young's modulus of elasticity. This last expression, the Moens-Korteweg equation [30], [26], predicts a frequency independent wave velocity, c. Experimental measurements show c to be highly frequency dependent. In spite of many attempts, the conflict between theory and experiment proved unresolvable, even if the vessel was given a finite length [25].

Combining the Windkessel and Transmission Line Theories

Broadening the view from pulse wave velocity to include other phenomena as well, experimental studies show that the arterial flow pulse decreases moving away from the heart, as expected for a passive transmission line. The arterial pressure pulse, however, increases, seemingly violating energy conservation. Further, input impedance $Z_{in} = P_{in}/Q_{in}$ is frequency dependent, similar to wave velocity. An arterial model capable of describing all of these features results by combining the windkessel and transmission line ideas [33].

The actual arterial system has bifurcations which should produce reflected pressure and flow waves. It is unfeasible to describe each branching vessel due to their very large number (10^9). A compromise is to take a finite number of interconnected uniform (lumped) segments. Each of 125 segments is represented by an electrical analog equivalent described by two partial differential equations in pressure and flow. The resulting transmission line, built of finite windkessel-like segments, was constructed as an electrical analog, well before the digital computer was sufficiently advanced, with voltage and current representing pressure and flow respectively. This model (Fig. 158.4) was the first to demonstrate c and Z_{in} correctly. The increasing pressure paradox was shown to be the result of constructive wave summation. Thus, the modification of the model to include branching proved to be the crucial step that allowed interpretation of a wide range of puzzling experimental observations [45].

Model Reduction

The level of insight with this large, distributed model permitted answering the question whether its size could be reduced, retaining the desired distributed system properties with a much reduced number of equations. A specific example, shown in Fig. 158.5, allows study of pulse wave transmission between the carotid and renal arteries [34]. This model reduces the number of equations from 250 to fewer than 40 by lumping large areas of the system outside the area under study.

If distributed system properties such as pulse wave reflection are not of interest, the arterial system load seen by the heart can be further reduced to a simple, three-element windkessel (Fig. 158.6). This concise model, denoted the Westkessel after its founder [46] is the most widely used arterial load representation today. It also allows accurate estimation of total arterial compliance, a quantity of clinical interest [40].

158.3 Predicting Function From Structure: Muscle Contraction

Description of muscle contraction has essentially evolved into two separate approaches—lumped whole muscle models and specialized crossbridge models of the sarcomere. The former seeks to interpret muscle's complex mechanical properties with a single set of model elements. Muscle experiments measure muscle force and length subjected to isometric (fixed length) conditions, isotonic (fixed load) conditions, and transient analysis where either length or load is rapidly changed.

Muscle force generation is believed to arise from the formation of crossbridge bonds between thick and thin myofilaments within the basic building block of muscle, the sarcomere. These structures, in the *nm-μm* range, must be viewed by electron microscopy or X-ray diffraction, limiting study to fixed, dead material. Consequently, muscle contraction at the sacromere level must be described by models that integrate metabolic and structural information.

Lumped Muscle Strip Models

Mayow (1674) proposed that muscle contraction arises from changes in its constituent elastic fibers as a byproduct of respiration, introducing the concept of muscle as an elastic material that changes due to metabolic processes. E. Weber (1846) viewed muscle as an elastic material, or spring, whose stiffness

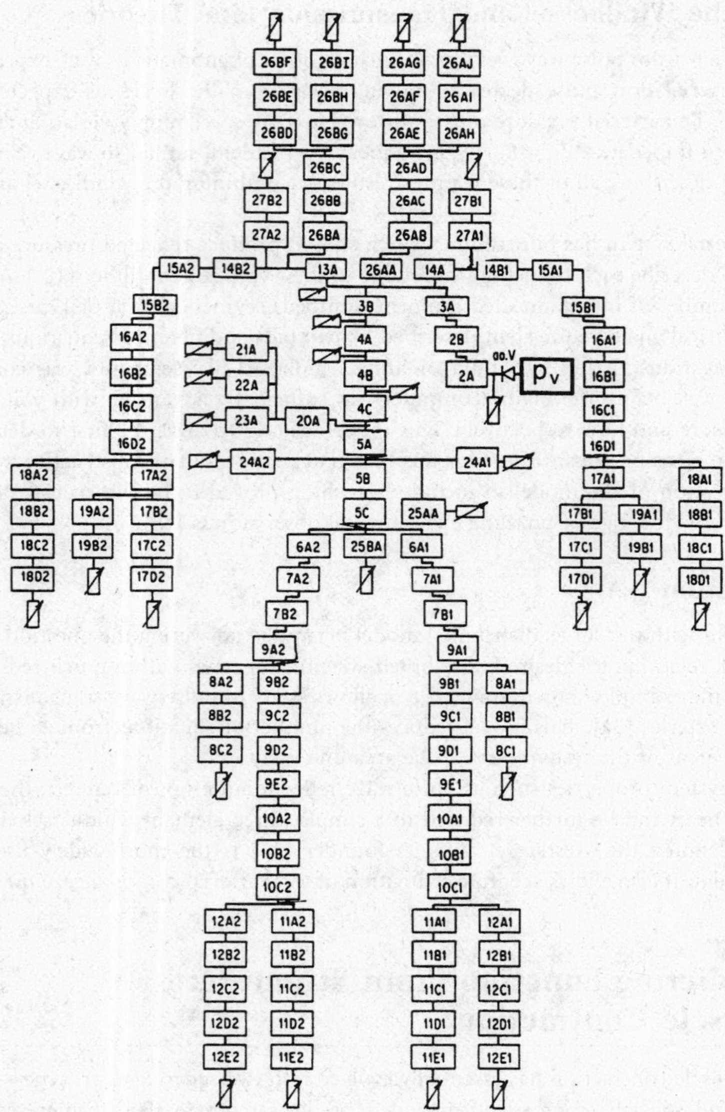

FIGURE 158.4 Electrical analog model of human systemic arterial circulation. Numbered boxes refer to windkessel-like segments, yielding 250 partial differential equations. This model combined features of the windkessel and transmission line schools of thought, producing a new approach capable of interpreting puzzling experimental observations. Adapted from [45].

depends on whether it is in a passive or active state. In 1890 Chauveau and Laulanié also considered muscle as an elastic spring with time dependent stiffness, and proposed that velocity of shortening plays a role in force production. Fick (1891) and Blix (1893) refuted a purely elastic description on thermodynamic grounds, finding that potential energy stored during stretching a strip of muscle was less than the sum of energy released during shortening as work and heat. Blix also noticed that at the same muscle length, muscle tension produced during stretch was greater than that during release. Tension would be the same during stretch and release for a simple spring. Further, Blix reported that he was unable to describe stress relaxation and creep in muscle with a visoelastic model.

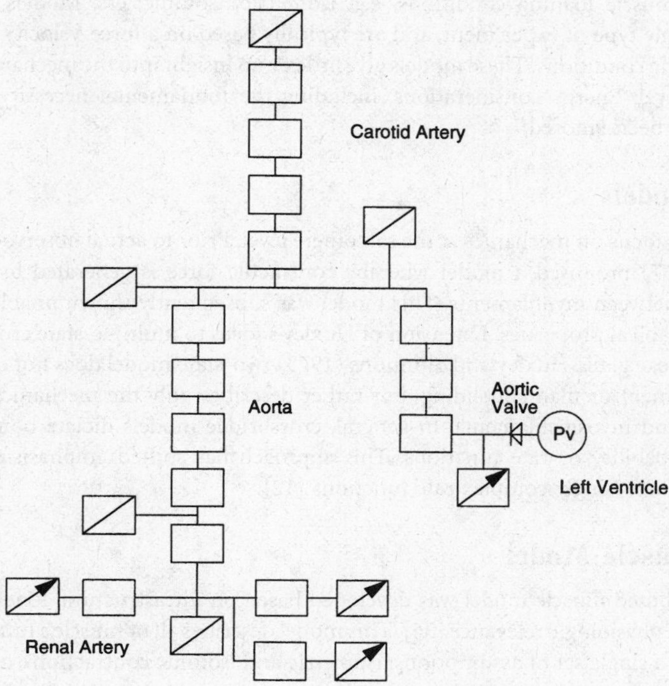

FIGURE 158.5 Reduced model of the systemic arterial tree of Fig. 158.4. This particular example retains distributed model elements between the area of interest, the carotid artery to the renal artery, and reduces the total number of system equations from 250 to fewer than 40 by lumping the remaining arterial segments. Adapted from [34].

Hill (1922) coupled the spring with a viscous medium, thereby reintroducing viscoelastic muscle models, after noticing that external work done accelerating an inertial load was inversely related to shortening velocity. A two-element viscoelastic model is objectionable on mechanical grounds: in series (Maxwell body), length is unbounded for a step change in force, in parallel (Voight body), force is infinite for a step change in length [37]. Further, Fenn (1923) reported that energy released by the muscle differs for isometric or shortening muscle. Subsequently, Hill (1939) developed the *contractile element*, embodied as an empirical hyperbolic relation between muscle force and shortening velocity. The contractile element was combined with two standard springs, denoted the parallel and series elastic elements. Current whole muscle models have, in general, incorporated Hill's contractile element, an empirically derived black-box force generator, in networks of traditional springs and dashpots (e.g. [31]). Since model elements are not directly correlated to muscle structure, there is no logical reason to stop at three elements—models with at least nine have been proposed [39].

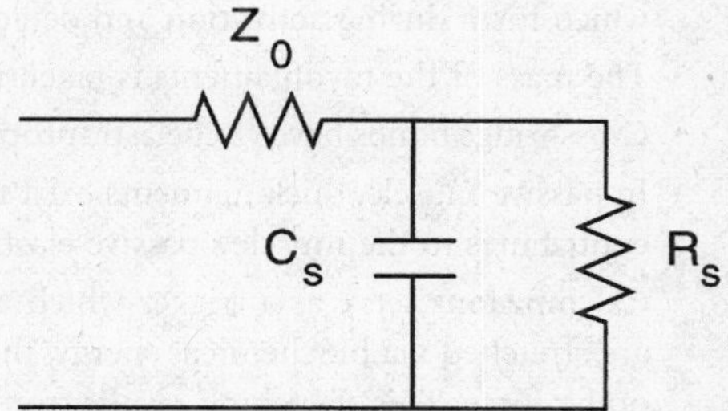

FIGURE 158.6 Equivalent systemic arterial load reduced to one differential equation. This modified windkessel, the Westkessel, after [46], enjoys widespread use as a concise description of the load seen by the left ventricle. Z_0 is characteristic impedance, C_s is arterial compliance and R_s is peripheral resistance.

This approach can be criticized on the following grounds. There exists little, if any, correlation between model elements and muscle anatomical structures. No single model has been shown to comprehensively describe different muscle loading conditions, e.g. isometric, isotonic, etc. Models tend to focus on description of a single type of experiment, and are typically based on a force-velocity relation measured under specific loading conditions. These models give little or no insight into the mechanism of contraction at the sarcomere level. Energy considerations, including the fundamental necessity of a biochemical energy source, have been ignored.

Crossbridge Models

Crossbridge models focus on mechanics at the sarcomere level. Prior to actual observation of crossbridge bonds, Huxley (1957) proposed a model whereby contractile force is generated by the formation of crossbridge bonds between myofilaments. This model was subsequently shown unable to describe muscle's transient mechanical properties. Extension of Huxley's ideas to multiple-state crossbridges has been very limited in scope, e.g. the Huxley and Simmons (1973) two-state model does not include crossbridge attachment, detachment or filament sliding, but rather describes only the mechanics of a single set of overlapping actin and myosin filaments. In general, crossbridge models dictate bond attachment and detachment by probability or rate functions. This approach has shifted emphasis away from muscle mechanics toward increasingly complex rate functions [12].

Distributed Muscle Model

A large-scale, distributed muscle model was developed based on ultrastructural kinetics and possessing direct anatomic and physiologic relevance [36]. This model describes all of muscle's fundamental mechanical properties with a single set of assumptions: isometric and isotonic contractions, and force transients resulting from rapid changes in muscle length during contraction. Variations in muscle properties with current loading conditions arise from the dynamic model structure, rather than adaptation of a particular force-velocity curve to describe the contractile element, or of a particular bond attachment or detachment function as for previous models. The following assumptions define the model:

- The sarcomere consists of overlapping thick and thin filaments connected by crossbridge bonds which form during activation and detach during relaxation.
- The mass of the myofilaments is taken into account.
- Crossbridge bonds have viscoelastic properties [22] and thus may exhibit stress relaxation and creep.
- In passive muscle, thick filaments exist in stable suspension at the center of the sarcomere, which contributes to the muscle's passive elastic properties [48].
- Calcium ions serve as a trigger which allows bond formation between filaments. Bonds formed are stretched via biochemical energy, thereby converting chemical to potential energy and developing force. This stretching results from crossbridge motion [22].
- The number of bonds formed depends upon the degree of thick-thin filament overlap [16].
- Asynchrony in bond formation and unequal numbers of bonds formed in each half sarcomere, as well as mechanical disturbances such as shortening and imposed length transients, cause movement of thick with respect to thin filaments. Since myofilament masses are taken into account these movements take the form of damped vibrations. Such vibrations occur with a spectrum of frequencies due to the distributed system properties.
- When the stress in a bond goes to zero the bond detaches [22].
- Myofilament motion and bond stress relaxation lead to bond detachment and therefore produce relaxation.

Description of the Sarcomere

Crossbridge bonds are described as linear viscoelastic material, each represented by a three-element spring-dashpot model (Fig. 158.7). During activation, the attached crossbridge head rotates, stretching

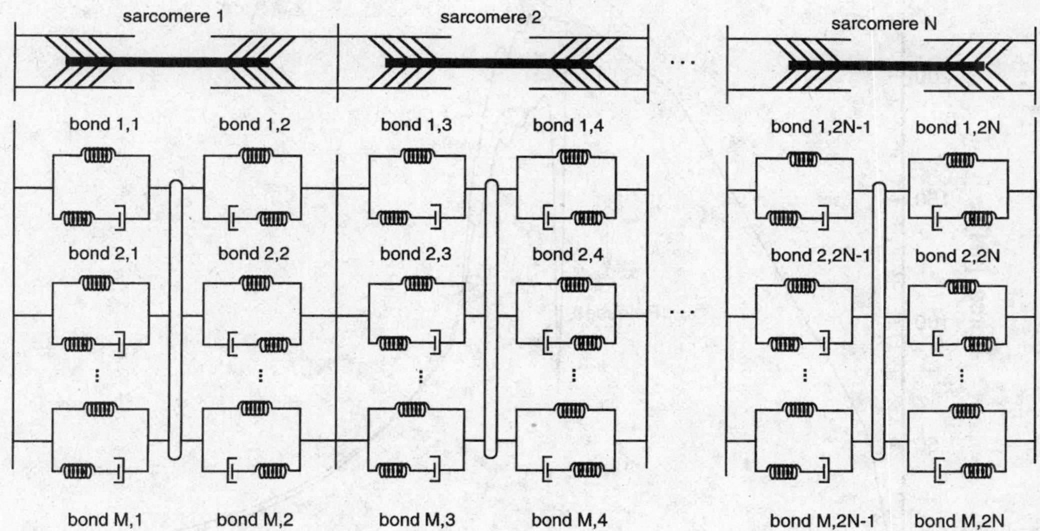

FIGURE 158.7 Schematic representation of a muscle fiber composed of N series sarcomeres each with M parallel pairs of active, viscoelastic bonds. Each bond is described by a three-element viscoelastic model. Adapted from [36].

the bond and generating force. A passive sarcomere is mechanically represented as myofilament masses interconnected by a matrix of permanently attached passive viscoelastic bonds. During activation, dynamically attached active bonds are formed as calcium ion and energy (ATP) allows.

The sarcomere model is extended to a muscle fiber by taking N sarcomeres in series, each with M parallel pairs of active bonds. The model currently consists of 50 series sarcomeres, each with 50 parallel pairs of crossbridges. Bond formation is taken to be distributed in two dimensions due to muscle's finite electrical propagation speed and the finite diffusion rate of calcium ions from the sarcoplasmic reticulum. The large set of resulting differential equations (5300) is solved by digital computer. Since the instantaneous number of crossbridge bonds, and therefore the model system equations, is continuously changing, the resulting model is strongly nonlinear and time-varying, despite its construction from linear, time-invariant components. Crossbridge bonds are added in a raster pattern due to the distributed system properties. After all bonds have been added, no further conditions are imposed. Crossbridge bonds subsequently detach simply from the small internal movements of myofilaments. Consequently, the model relaxes without separate assumptions.

This model was subjected to a wide range of loading conditions. A single set of model parameters is sufficient to describe the main features of isometric, isotonic and quick release and stretch experiments. The model is sensitive to mechanical disturbances, consistent with experimental evidence from muscle force curves [47], aequorin measurements of free calcium ion [1], and high speed X-ray diffraction studies [23]. The model is also consistent with sarcomere length feedback studies [22] where reduced internal motion delays relaxation.

Muscle Contraction

The outermost curve of Fig. 158.8 depicts an isometric contraction computed from the muscle fiber model. At each Δt available bonds are attached, adding their equations to the model, and force is generated via bond stretching. For the twitch contractions in Figs. 158.8 and 158.9 bonds are attached from $t = 0$ to 0.5 sec. Since bond formation is not perfectly symmetrical, due to the raster attachment pattern, small force imbalances within thick-thin units develop, leading to small myofilament motions that tend to detach bonds as their force goes to zero. Consequently, at each Δt of solution bonds are added during

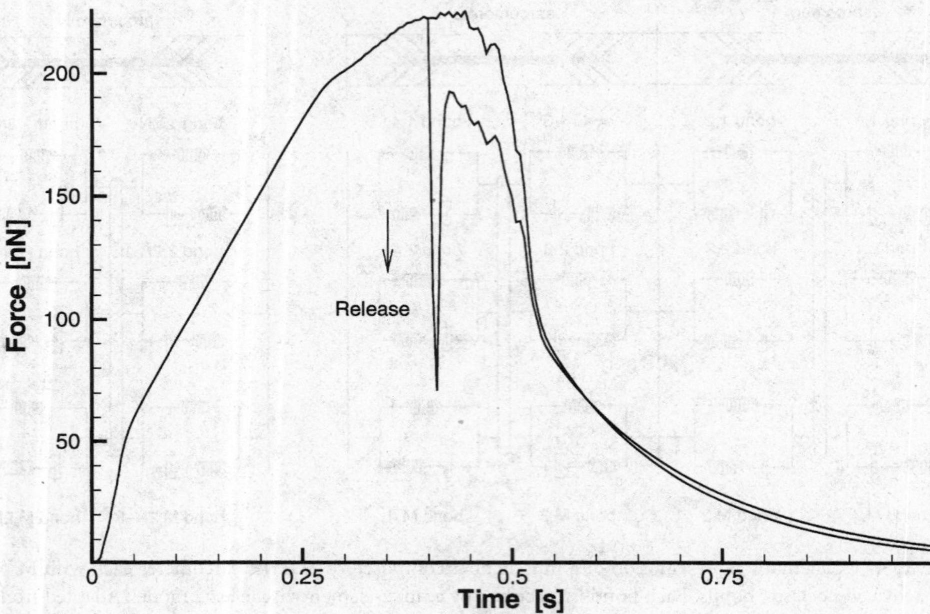

FIGURE 158.8 Outermost curve describes an isometric contraction computed from the distributed muscle fiber model. Also plotted is force computed during a quick release experiment, showing significant force deactivation. Adapted from [36].

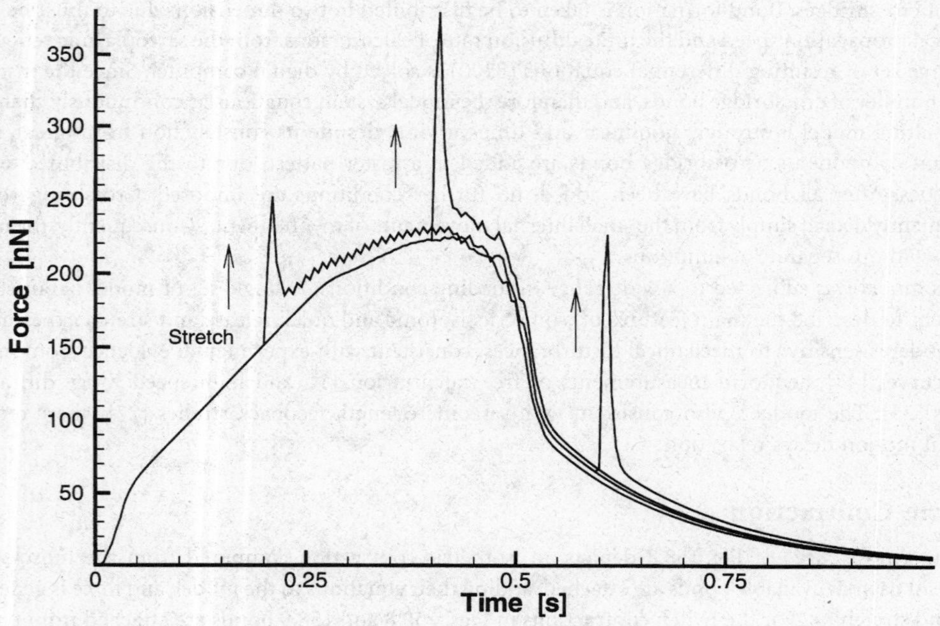

FIGURE 158.9 Quick stretches computed from the distributed muscle fiber model. Adapted from [36].

activation and detach continuously. Despite this very dynamic state, Fig. 158.8 shows that the net isometric fiber force generated smoothly rises during muscle activation. Peak isometric force of 250 nN is in close agreement with recent experimental measurements on isolated guinea pig myocytes [47].

The time of peak force corresponds to the end of bond attachment. Physiologically, this point is associated with a lack of available bond attachment sites. This may be due to lack of calcium ion for release necessary inhibition of actin-myosin interaction by troponin/ tropomyosin, lack of myosin heads in a state of steric for attachment, or some other mechanism. Force relaxation then results simply from bond detachment due to the same internal myofilament motion mechanism. Net force smoothly recovers to the resting force level, in contrast to the dynamic motions of individual bonds and myofilamets. Surprisingly, this smooth relaxation occurs without the added feature of cross-bridge recycling, i.e. bond reattachment after detachment.

Transient response to quick length change is a critical test of any muscle model. This distributed model predicts that rapid length changes enhance system vibration and thereby promote bond detachment and muscle relaxation, underlying muscle's force deactivation phenomenon. Figure 158.8 also shows force deactivation following quick release during an otherwise isometric twitch contraction. This particular release corresponds to 0.2% of sarcomere length over 10 msec. Quick release corresponding to 2% of sarcomere length applied during 10 msec was sufficient to bring fiber force to zero (not shown). Slower releases produce less force deactivation. These curves compare favorably with quick release transients measured on whole muscle strips and more recent experiments on isolated frog myocytes [4, 6], and for ferret papillary muscles [27].

Figure 158.9 shows quick stretches computed for the distributed fiber model. Stretches corresponding to 0.2% of sarcomere length were performed over 10 msec. Curves show rapid force overshoot, followed by rapid recovery and slow recovery back to the isometric force levels. Quick stretch response is in agreement with classic studies on muscle strip and recent studies on ferret papillary muscles [27].

The idea for this distributed muscle fiber model arose in 1990 [35]. At that time, muscle fibers were assumed to be functionally similar to muscle strips. Recently, experiments on isolated muscle fibers show this to be the case. Also, predictions from the model have recently been borne out, for example, peak isometric force [47]. Peak isometric stress measured on isolated rabbit myocytes (5.4 mN/mm^2) is very close to peak stress from rabbit papillary muscle strips (6.4 mN/mm^2) [3]. The distributed model generates peak isometric stress of 2.5 mN/mm^2. Other muscle phenomena measured on the isolated fiber include: a quadratic force-length relation [3], inotropic changes in contractile state [4], quick release and stretch [27, 4] and isotonic contractions [38].

Physiological modeling strives to create a framework that integrates experimental observations. A great strength of modeling is its capacity for predicting new experiments and physiological phenomena (Fig. 158.1). The distributed muscle fiber model predicts that mechanical external vibration of muscle should enhance its internal vibration and promote bond detachment and therefore relaxation. To test this hypothesis ultrasound was used to vibrate muscle. A single millisecond pulse of ultrasound delivered to the in vivo frog heart during systole was observed to reduce developed aortic pressure [7], [8]. Experiments were designed to eliminate heating effects. Ultrasound pulses applied late in the twitch had particularly strong effects. After peak muscle force is generated, bond attachment ends so detachment via vibration has no mechanism for recovery. This force (pressure) deactivation was reproducible and not permanently damaging as beats without irradiation produced normal pressure magnitudes and time courses. A mechanical effect is likely since moving the heart slightly out of the ultrasound beam produced no deactivation effect, despite having the same electrical environment. Ultrasound delivered during diastole was found to produce a premature ventricular contraction, as depicted in Fig. 158.10, an unexpected result. These experiments suggest that medical ultrasound may have strong effects on living tissue, and demonstrates how modeling can lead to the discovery of new physiological phenomena.

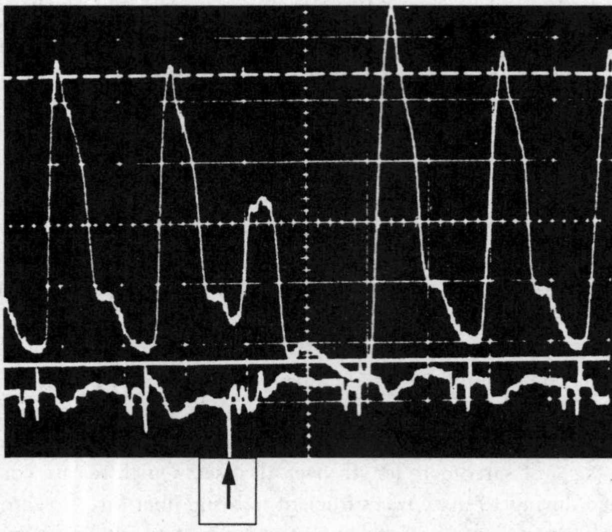

FIGURE 158.10 Premature ventricular contraction following ultrasound irradiation, during diastole, of the in vivo frog heart. The top trace is aortic pressure and below is the ECG. A lithotripter pulse of 20 MPa, 5 ms duration, is delivered at the vertical arrow. Adapted from [7].

158.4 Predicting Structure from Function: Cardiac Hypertrophy

Integration of Micro- and Macro-Dynamics

Modeling methodology often relies on the availability of structural parameters that have been determined from prior physical and quantitative experimentation. Alternatively, it is of vital interest to understand how a given structure arises, i.e., to understand nature's design. To accomplish this, it is necessary to allow the system to adapt to its external functional demands. Due to adaptive processes, a biological system may be undergoing synthesis and de-synthesis simultaneously. The balance of these opposing mechanisms results in a structure that meets the system's current functional demand. The advantage of this type of modeling approach is that it obviates measurement of every parameter. This enables the physiological modeler to provide theoretical insight outside the realm of conventional experimentation.

Often models become increasingly complex in the attempt to study smaller levels of structural detail. Ultimately, there may be a minimum level of structure at which the function is known and relatively constant, for example, a cell or a cardiac sarcomere. This structure is denoted the *micro-unit.* The assembly of such units comprises the biological system and is denoted the *macro-unit.* The macro-unit function may also be known, characterized, for example, by ventricular pressure and flow. It is the arrangement and interaction of multiple micro-units that introduces complexity in modeling. To resolve this problem, the micro-unit function is integrated while structure is varied to identify structural rules. The resulting rules are usually simple and thereby offer a special advantage to the integrative approach.

The concept of integrating micro-level function to solve for the macro-level of physiological function and overall geometry has been discussed earlier by Taylor and Weibel [41]. These researchers stated in qualitative terms that an organ system achieves "a state of structural design commensurate to functional needs resulting from regulated morphogenesis, whereby the formation of structural elements is regulated to satisfy but not exceed the requirements of the functional system." This process has been termed symorphosis.

Integrative Modeling

In its simplest form, this method matches function at two different levels of structure. A match is achieved by varying the macro-structure until energy utilization of the system is minimized. Structural variation can be accomplished analytically, or in the case of discontinuous or nonlinear functions by means of a computer. Energy is then minimized in terms of the micro-unit. Since the time frame for the adaptive process tends to be long, it may be difficult to accumulate experimental results. Alternatively, it would be difficult to design an "integrative experiment" whereby the macro-structure is actually altered. For example, it is not easy to add or subtract selected cardiac myocytes to or from an intact cardiac ventricle without trauma to the animal. Usually, the same heart is observed over long time periods for its response to various external loading conditions. A theoretical approach can overcome some of these difficulties.

Micro-unit

Integrative modeling begins identifying a minimal level of structure at which function is known. For an organ, this could be at the cellular level or, even lower, at the molecular level. The function of interest must be defined in terms of a physical quantity, such as electrical, mechanical, chemical, etc. The physical quantity of interest may also be time dependent, but the function of the micro-unit must be independent of change in the geometry of the macro- structure. For normal physiological conditions this rule is valid, but it may break down in the case of a disease process. Pathological conditions further require an analysis of how the micro-unit function alters with time, e.g. how myocyte function deteriorates over time in heart failure. The dimensions of the micro-unit must be defined. Since the micro-unit is chosen to be the smallest structural element, its mean dimensions are often treated as constants.

Macro-unit

The macro-unit, for example the structure at the organ level, consists of a group of organized micro-dynamic units. Macro-unit function must be defined in terms of the physical quantities of interest. The physical quantities at the two structural levels must be of the same type, for example, myocyte force and ventricular wall tension define force at different structural levels. Once the macrofunction is defined it must be independent of variation in the macro-structure, i.e., unchanged by the adaptive process over long time periods. Since the macro-function is important to the survival of the animal it may be independent of a disease process as well. This is generally true until structural changes cannot accommodate the deterioration of micro-function at a sufficiently rapid pace.

Organization and Equilibrium

The organization of the micro-units defines the macro-structure. If the structure is a simple geometry that models the anatomic shape, analytical expressions can be obtained. Use of the actual anatomical shape forces the modeling to be computational. Once the geometry is prescribed, the physical equilibrium equations can be derived that relate the micro-function to the macro-function. This is accomplished by mathematically integrating the micro-function over the entire geometry, hence the name integrative modeling. For example, the myocardial force is integrated over the ventricle in order to relate it to wall tension. The final equations contain the geometric dimensions of the model and are unknowns at the onset.

Energy Considerations

The micro-unit is the smallest building block employed to assemble the overall structure. The dimensions of the micro-unit therefore represent the limit of resolution for any physical quantity computed from the model. Of particular interest is the energy required by the micro-unit. When the micro-unit is fixed in dimensions, on average, it provides a gauge of energy consumption that is size independent. While the macro-unit is varied in size, the effect on the energy at the micro-unit level can then be observed. Energy considerations for biological systems are typically more complex than for standard engineering systems. For example, care must be taken to include the energy required for the basal tissue metabolism as well as the energy consumption necessary to accomplish functional tasks. This differs from a nonliving structure where only physical energy considerations suffice.

Macrostructure Dimensions

The final analysis solves for the dimensions of the macro-structure, accomplished by examining the micro-unit energy consumption, while the dimensions are varied. A solution is then obtained by solving for a zero derivative of energy with respect to each dimension and minimum in energy. Specific problems may require variance from the above procedure. For example, additional information may be available to assist in finding the solution or in placement of physiological constraints. This is particularly useful when a completely analytical approach is not possible. Lastly, it is not always necessary to obtain the solution relative to the micro-unit energy. Energy was chosen here because of its generality to all systems, but other physical quantities may be used. Whichever quantity is chosen it must be evaluated for the smallest functional unit. Ultimately, the resulting dimensions represent the optimal solution for the structure in terms of its smallest functional building block. These integrative modeling concepts are illustrated below in a study of cardiac hypertrophy.

Cardiac Hypertrophy

Current thinking dictates that the left ventricle attempts to normalize its peak wall stress values during systolic contractions and diastole by altering its dimensions [17]. There are several reasons to doubt the validity of this concept. First, anatomical locations of systolic and diastolic stress transducers have not been identified. Second, the value of wall stress cannot be determined from basic physical knowledge. Instead, wall stress is assumed to be regulated by an unknown mechanism. Lastly, this theory cannot explain ventricular size alterations due to metabolic disturbance, as opposed to a mechanical disturbance. Alternatively, integrative modeling is applied to the problem of cardiac hypertrophy.

It was hypothesized that cardiac size can be determined from energetics at the sarcomere level of structure, i.e., the micro-unit [10]. The cardiac sarcomere was redefined as a *sarcounit* (Fig. 158.11), a constant volume structure capable of axial stress generation. The function of the sarcounit was defined by its active and passive stress-length curves (Fig. 158.12), with the active function corresponding to maximal activation. Maximal stress was modulated by a periodic time function which was zero during diastole.

A cylindrical approximation to the geometry of the ventricle was chosen (Fig. 158.11). Thus, its structure was described by the length, diameter (D) and wall thickness (W) of the cylinder [9]. The

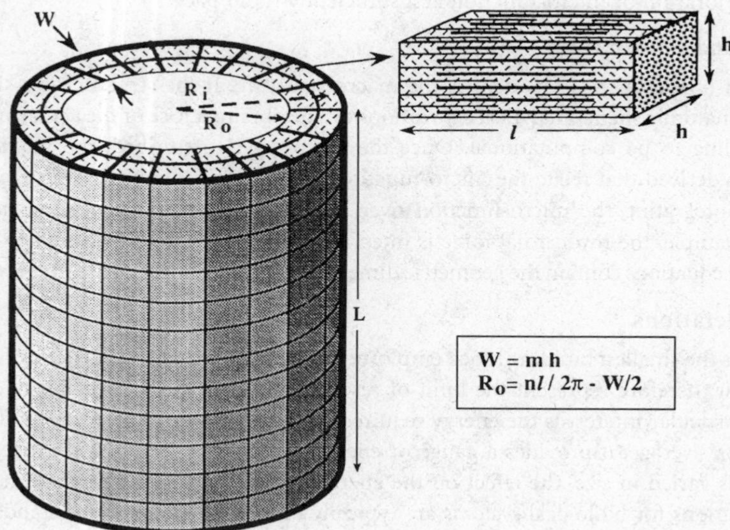

FIGURE 158.11 Cardiac sarcounit (upper right) and structural arrangement to form a cylindrical left ventricle. Adapted from [10].

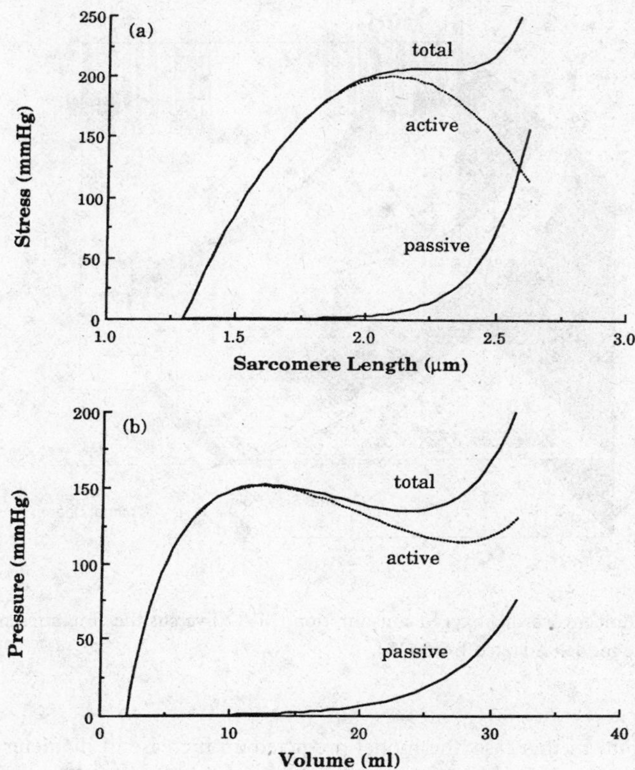

FIGURE 158.12 (top) Active and passive stress-length curves describing the sarcounit. (bottom) Active and passive pressure-volume curves for the initial cylindrical left ventricle model. Adapted from [9].

sarcounits were arranged in series and parallel so that ventricular diameter is determined by the number of series units and ventricular wall thickness and length by the number of parallel units.

Macro-function was defined by the mean aortic pressure and afterload impedance, using a modified windkessel arterial load. Preload was treated as a constant filling pressure in series with a constant flow resistance. Ventricular function was then equated with micro-function by integrating the stress and length of the sarcounits over the cylindrical structure. This yielded ventricular pressure and volume at any instant of time for given sarcounit stress and length at the corresponding time (Fig. 158.12). Heart rate was allowed to vary as necessary to maintain constant mean aortic pressure. The sarcounit energetics were evaluated from the rate of myocardial oxygen consumption (MVO_2). MVO_2 was computed by assuming a proportional relationship with the area of the stress—length loop of the sarcounit per unit time. All quantities were then computed numerically, allowing several cycles to achieve steady state. Each quantity was measured at the final steady-state beat.

To solve for the dimensions of the ventricle, ventricular diameter and wall thickness were varied while evaluating all quantities, including MVO_2. Results show that MVO_2 is minimized only at a specific shape of the cylinder, or W/D ratio (Fig. 158.13). Additionally, the level of MVO_2 varied with the overall size of the ventricle. When MVO_2 is also equated with the amount of oxygen supplied by the coronary circulation, a unique solution for left ventricular size and shape is obtained.

The above model was further employed to study observed patterns of cardiac hypertrophy. First, the initial size of the ventricle was consistent with that of a normal dog. Second, a pressure overload was created by raising mean aortic pressure. The model was then used to recompute ventricular size for the new afterload condition. Results show the classic wall thickening and increased wall thickness-to-diameter ratio (W/D) is consistent with observed *concentric hypertrophy.* Third, volume overload was imposed by

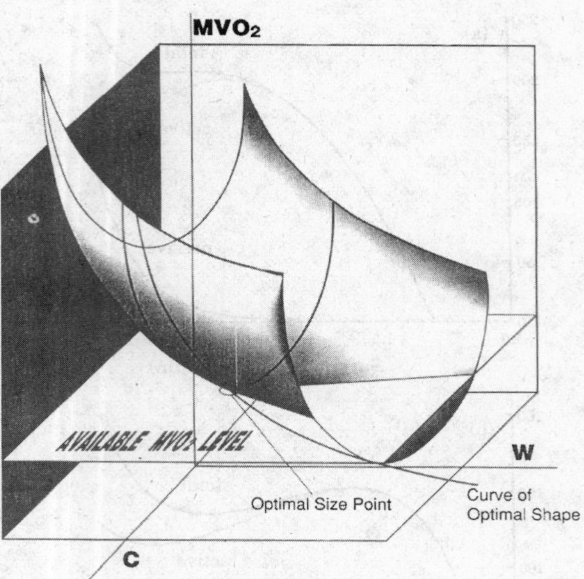

FIGURE 158.13 Sarcounit myocardial oxygen consumption (MVO_2) versus the circumference (C) and wall thickness (W) of the ventricle model. Adapted from [10].

elevating cardiac output. In this case, the model predicted an increase in diameter with relatively little increase in the W/D ratio, referred to as *eccentric hypertrophy.* Fourth, the sarcomere function was impaired and its effect on the ventricle evaluated. An eccentric form of hypertrophy resulted. The diameter to length ratio was found to increase along with the degree of myocardial dysfunction [11]. This was consistent with observations that show the ventricle assuming a progressively more spherical shape with further myocardial failure [24].

In conclusion, modeling by means of integrating micro- and macro-function was applied to the study of cardiac hypertrophy. This approach matches the functional information available at two different levels of structure. The ability to predict macro-structure dimensions is a result of achieving this match. The cardiac growth model suggests the conclusion that ventricular shape can be determined by the regulation of a single quantity, MVO_2, at the sarcomere level. This modeling approach is more general than the wall stress theory as it can predict growth response to metabolic anomaly as well as to mechanical overloads. As a primary benefit, this approach permits theoretical analysis that may be impractical to realize experimentally.

158.5 Summary

This chapter focuses on modeling as a tool to identify and quantify the mechanism(s) responsible for observed phenomena in physiology. The rather strict rules to which the modeling process is subject generally lead to the formulation of new, specific experiments, thereby broadening the pool of experimental data, and the data-base for modeling. On occasion, this process leads to areas of experimentation and the development of concepts originally judged to lie outside the scope of the original problem definition. Consequently, successful modeling aids in deepening a researcher's insight and breadth of view. Three examples taken from the cardiovascular field were summarized to illustrate procedures and techniques. A selection of benefits achieved was presented as well.

References

[1] Allen, D.G., Smith, G.L. and Nichols, C.G. 1988. Intracellular calcium concentration following length changes in mammalian cardiac muscle. in: *Starling's Law of the Heart Revisited*, H.E.D.J. ter Keurs and M.I.M. Noble, eds., Kluwer Academic Publishers, Dordrecht.

[2] Blix, M. 1893. Die Länge und die Spannung des Muskels, Skand. Arch. Physiol., 4:399.

[3] Bluhm, W.F., McCulloch, A.D., and Lew, W.Y.W. 1995. Technical Note: Active force in rabbit ventricular myocytes. J. Biomech. 28:1119–1122.

[4] Brandt, P.W., Colomo, F., Poggesi, C. and Tesi, C. 1993. Taking the first steps in contraction mechanics of single myocytes from frog heart. in: *Mechanism of Myofilament Sliding in Muscle Contraction*, H. Sugi ad G.H. Pollack, eds., Plenum Press, NY.

[5] Chauveau, A. 1890. L'élasticité active du muscle et l'énergie consacreé à sa création dans le cas de contraction statique. C.R. Acad. Sci. 111:19, 89.

[6] Colomo, F., Poggesi, C. and Tesi, C.1994. Force responses to rapid length changes in single intact cells from frog heart. J. Physiol. 475:347–350.

[7] Dalecki, D., Carstensen, E.L., Neel, D.S., Palladino, J.L. and Noordegraaf, A. 1991. Thresholds for premature ventricular contractions in frog hearts exposed to lithotripter fields. *Ultrasound Med. Biol.* 17:341–346.

[8] Dalecki, D., Raeman, C.H., Child, S.Z. and Carstensen, E.L. 1997. Effects of pulsed ultrasound on the frog heart: III. The radiation force mechanism. *Ultrasound Med. Biol.* 23:275–285.

[9] Drzewiecki, G.M., Karam, E. and Welkowitz, W. 1989. Physiological basis for mechanical time-variance in the heart: special consideration of non-linear function. J. Theor. Biol. 139:465–486.

[10] Drzewiecki, G.M., Karam, E., Li, J. K-J, and Noordegraaf, A. 1992. Cardiac adaptation of sarcomere dynamics to arterial load: a model of hypertrophy. Am. J. Physiol. 263:H1054–1063.

[11] Drzewiecki GM, Li J K-J, Knap B, Juznic S, Juznic G, Noordergraaf A. 1996. Cardiac hypertrophy and shape—A noninvasive index? J. Cardiovasc. Diag. Proced. 13:193–198.

[12] Eisenberg, E., Hill, T. L. and Chen, Y. 1980. Cross-bridge model of muscle contraction. *Biophys. J.* 29:195–227.

[13] Fenn, W. O. 1923. Quantitative comparison between energy liberated and work performed by isolated sartorius muscle of the frog. J. Physiol. 58:175–203.

[14] Fick, A. 1891. Neue Beitrage zur Kentniss von der Wärmeentwicklung im Muskel. Pflügers Arch., 51:541.

[15] Frank, O. 1895. Zur Dynamik des Herzmuskels. *Z. Biol.* (Munich) 32:370 [Eng. transl., see C.B. Chapman and E. Wasserman, Am. Heart J. 58:282–317, 476–478 1959.].

[16] Gordon, A., Huxley, A. and Julian, F. 1966. The variation in isometric tension with sarcomere length in vertebrate muscle fibers. *J. Physiol.* 184:170–192.

[17] Grossman, W. 1980. Cardiac hypertrophy: useful adaptation or pathologic process? *Am. J. Med.* 69:576–584.

[18] Hales, S. 1733, *Statistical Essays: Containing Haemostaticks, or an Account of some Hydraulick and Hydrostatical Experiments on the Blood and Blood-Vessels of Animals, etc.* vol. 2, Innys and Manby, London.

[19] Hill, A. V. 1922. The maximum work and mechanical efficiency of human muscles, and their most economical speed. *J. Physiol.*, 56:19.

[20] Hill, A. V. 1939. The heat of shortening and dynamic constants of muscle. *Proc. Roy. Soc. London (B)* 126:136.

[21] Huxley, A. F. 1957. Muscle structure and theories of contraction. *Prog. Biophys.* 7:255–318.

[22] Huxley, A. F. and Simmons, R. M. 1973. Mechanical transients and the origin of muscle force. Cold Spring Harbor Symp. *Quant. Biol.* 37:669–680.

[23] Huxley, H.E., Simmons, R. M., Faruqi, A. R., Kress, M., Bordas, J. and Koch, M.H.J. 1983. Changes in the X-ray reflections from contracting muscle during rapid mechanical transients and their structural implications. *J. Mol. Biol.* 169:469–506.

[24] Juznic S.C.J.E., Juznic G, Knap B. 1997. Ventricular shape: Spherical or cylindrical? in: *Analysis and Assessment of Cardiovascular Function*, G. Drzewiecki and J. Li, eds., Springer-Verlag, New York pp. 156–171.

[25] Kenner, T. and Wetterer, E. 1962. Experimentelle Untersuchungen über die Pulsformen und Eigenschwingungen zweiteiliger Schlauchmodelle. Pflügers Arch. Gesamte Physiol. Menschen Tiere 275:594.

[26] Korteweg, D.J. 1878. Über die Fortpflanzungsgeschwindigkeit des Schalles in elastischen Röhren. *Ann. Phys. Chem.* (Neue Folge) 5:525.

[27] Kurihara, S. and Komukai, K. 1995. Tension-dependent changes of the intracellular Ca^{++} transients in ferret ventricular muscles. *J. Physiol* 489:617–625.

[28] Laulanié, F. 1890. Principes et problèmes de la thermodynamique musculaire d'après les récents traveaux de M. Chauveau. *Rev. Vét.* 15:505.

[29] Mayow, J. 1907. On muscular motion and animal spirits, Fourth treatise in: *Medical-physical Works*, Alembic Club Reprint no. 17, published by the Alembic Club.

[30] Moens, A.I. 1878. *Die Pulskurve*, E.J. Brill, Leiden.

[31] Montevecchi, F.M. and Pietrabissa, R. 1987. A model of multicomponent cardiac fiber. *J. Biomech.* 20(4):365–370.

[32] Moser, M., Huang, J.W., Schwarz, G.S., Kenner, T. and Noordergraaf, A. 1998. Impedance defined flow. Generalization of William Harvey's concept of the circulation—370 years later. *Int. J. Cardiov. Med. Sci.*, 1:205–211.

[33] Noordergraaf, A. 1969. Hemodynamics, Ch 5 in: *Biological Engineering*, H.P. Schwan (ed), McGraw-Hill.

[34] Noordergraaf, A. 1978. *Circulatory System Dynamics.* Academic Press, New York.

[35] Palladino, J.L. 1990. *Models of cardiac muscle contraction and relaxation.* Ph.D. dissertation, University of Pennsylvania, Philadelphia. Univ. Microforms, Ann Arbor, MI.

[36] Palladino, J.L. and Noordergraaf, A. 1998. Muscle contraction mechanics from ultrastructural dynamics. Ch 3 in: *Analysis and Assessment of Cardiovascular Function*, G.M. Drzewiecki and J.K-J. Li, eds., Springer-Verlag, New York, pp. 33–57.

[37] Palladino, J.L. and Davis, R.B. 1999. Biomechanics, Ch 4 in: *Biomedical Engineering* J.D. Enderle, S.B. Blanchard and J.D. Bronzino, eds., Academic Press, San Diego, in press.

[38] Parikh, S.S., Zou, S-Z. and Tung, L. 1993. Contraction and relaxation of isolated cardiac myocytes of the frog under varying mechanical loads. *Circ. Res.* 72:297–311.

[39] Parmley, W.W., Brutsaert, D.L. and Sonnenblick, E.H. 1969. Effect of altered loading on contractile events in isolated cat papillary muscle. *Circ Res.* 24:521.

[40] Quick, C.M. 1998. *Reconciling windkessel and transmission descriptions of the arterial system and the mechanical stability of muscular arteries.* Ph.D. dissertation, Rutgers University, Piscataway. Univ. Microforms, Ann Arbor, MI.

[41] Taylor, C.R. and Weibel, E.R. 1981. Design of the mammalian respiratory system I: problem and strategy. Resp. Physiol. 44:1–10.

[42] Weber, E. 1846. *Handwörterbuch der Physiologie.* vol. 3B. R. Wagner, ed., Vieweg, Braunschweig.

[43] Weber, W. 1850. *Ueber die Anwendung der Wellenlehre auf die Lehre vom Kreislaufe des Blutes und ins besondere auf die Pulsehre*, Ber. Math. Physik. Cl. Königl. Sächs. Ges. Wiss. 18:353.

[45] Westerhof, N., Bosman, F., de Vries, C.J. and Noordergraaf, A. 1969. Analog studies of the human systemic arterial tree. J. Biomech. 2:21.

[46] Westerhof, N., Elzinga, G. and Sipkema, P. 1971. Artificial arterial system for pumping hearts. J. Appl. Physiol. 31:776.

[47] White, E., Boyett, M.R., and Orchard, C.H. 1995. The effects of mechanical loading and changes of length on single guinea-pig ventricular myocytes. *J. Physiol.* 482:93–107.

[48] Winegrad, S. 1968. Intracellular calcium movements of frog skeletal muscle during recovery from tetanus. *J. Gen. Physiol.* 51:65–83.

159

Compartmental Models of Physiologic Systems

Claudio Cobelli
University of Padova

Giovanni Sparacino
University of Padova

Andrea Caumo
San Raffaele Scientific Institute

Maria Pia Saccomani
University of Padova

Gianna Maria Toffolo
University of Padova

Compartmental models are a class of dynamic, i.e., differential equation, models derived from mass balance considerations which are widely used for quantitative study of the kinetics of materials in physiologic systems. Materials can be either exogenous, such as a drug or a tracer, or endogenous, such as a substrate or a hormone, and kinetics include processes such as production, distribution, transport, utilization, and substrate-hormone control interactions.

Compartmental modeling was first formalized in the context of isotopic tracer kinetics. Over the years it has evolved and grown as a formal body of theory [Carson et al., 1983; Godfrey, 1983; Jacquez, 1996].

Compartmental models have been widely employed for solving a broad spectrum of physiologic problems related to the distribution of materials in living systems in research, diagnosis, and therapy at the whole-body, organ, and cellular level. Examples and references can be found in books [Carson et al., 1983; Gibaldi and Perrier, 1982; Jacquez, 1996] and reviews [Carson and Jones, 1979; Cobelli, 1984; Cobelli and Caumo, 1998]. Purposes for which compartmental models have been developed include:

- identification of system structure, i.e., models to examine different hypotheses regarding the nature of specific physiologic mechanisms;
- estimation of unmeasurable quantities, i.e., estimating internal parameters and variables of physiologic interest;
- simulation of the intact system behavior where ethical or technical reasons do not allow direct experimentation on the system itself;
- prediction and control of physiologic variables by administration of therapeutic agents, i.e., models to predict an optimal administration of a drug in order to keep one or more physiologic variables within desirable limits;
- optimization of cost effectiveness of dynamic clinical tests, i.e., models to obtain maximum information from the minimum number of blood samples withdrawn from a patient;
- diagnosis, i.e., models to augment quantitative information from laboratory tests and clinical symptoms, thus improving the reliability of diagnosis;
- teaching, i.e., models to aid in the teaching of many aspects of physiology, clinical medicine, and pharmacokinetics.

0-8493-0462-8/00/$0.00+$.50

The use of compartmental models of physiologic systems has been greatly facilitated by the availability of specific software packages for the PC [Saam, 1998; Adapt, 1997], which can be used both for simulation and identification.

159.1 Definitions and Concepts

Let's start with some definitions. A *compartment* is an amount of material that acts as though it is well-mixed and kinetically homogeneous. A *compartmental model* consists of a finite number of compartments with specified interconnections among them. The interconnections represent a flux of material which physiologically represents transport from one location to another or a chemical transformation or both. An example of a compartmental model is illustrated in Fig. 159.1, where the compartments are represented by circles and the interconnections by arrows.

Given the introductory definitions, it is useful before explaining *well-mixed* and *kinetic homogeneity* to consider possible candidates for compartments. Consider the notion of a compartment as a physical space. Plasma is a candidate for a compartment; a substance such as plasma glucose could be a compartment. Zinc in bone could be a compartment as well as thyroxine in the thyroid. In some experiments, different substances could be followed in plasma: plasma glucose, lactate, and alanine provide examples. Thus, in the same experiment, there can be more than one plasma compartment, one for each of the substances being studied. This notion extends beyond plasma. Glucose and glucose-6-phosphate could be two different compartments inside a liver cell. Thus, a physical space may actually represent more than one compartment.

In addition, one must distinguish between compartments that are accessible and nonaccessible for measurement. Researchers often try to assign physical spaces to the nonaccessible compartments. This is a very difficult problem which is best addressed once one realizes that the definition of a compartment is actually a theoretical construct which may in fact lump material from several different physical spaces in a system; to equate a compartment with a physical space depends upon the system under study and assumptions about the model.

With these notions of what might constitute a compartment, it is easier to define the concepts of well-mixed and kinetic homogeneity. *Well-mixed* means that any two samples taken from the compartment at the same time would have the same concentration of the substance being studied and therefore be equally representative. Thus, the concept of well-mixed relates to uniformity of information contained in a single compartment.

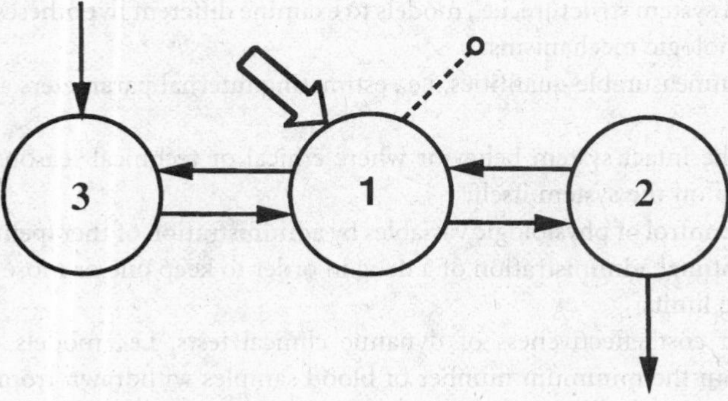

FIGURE 159.1 A three-compartment model. The substance enters *de novo* compartment 3 (arrow entering a compartment from "outside") and irreversibly leaves from compartment 2 (arrow leaving a compartment to the "outside"). Material exchanges occur between compartments 3 and 1 and compartments 1 and 2 and are represented by arrows. Compartment 1 is the accessible compartment: Test input and measurement (output) are denoted by a large arrow and a dashed line with a bullet, respectively.

Kinetic homogeneity means that every particle in a compartment has the same probability of taking the pathways leaving the compartment. Since when a particle leaves a compartment it does so because of metabolic events related to transport and utilization, it means that all particles in the compartment have the same probability of leaving due to one of these events.

The notion of a compartment, i.e., lumping material with similar characteristics into collections that are homogeneous and behave identically, is what allows one to reduce a complex physiologic system into a finite number of compartments and pathways. The number of compartments required depends both on the system being studied and on the richness of the experimental configuration. A compartmental model is clearly unique for each system studied, since it incorporates known and hypothesized physiology and biochemistry. It provides the investigator with insights into the system structure and is as good as the assumptions that are incorporated in the model structure.

159.2 The Compartmental Model

Theory

Let Fig. 159.2 represent the *i*th compartment of an *n*-compartment model; $q_i \geq 0$ denotes the mass of the compartment. The arrows represent fluxes into and out of the compartment: the input flux into the compartment from outside the system, e.g., *de novo* synthesis of material and/or exogenous test input, is represented by $u_i \geq 0$; the flux to the environment and therefore out of the system by F_{0i}; the flux from compartment *i* to *j* by F_{ji}, and the flux from compartment *j* to *i* by F_{ij}. All F_{hk} are ≥ 0. The general equations for the compartmental model are obtained by writing the mass balance equation for each compartment:

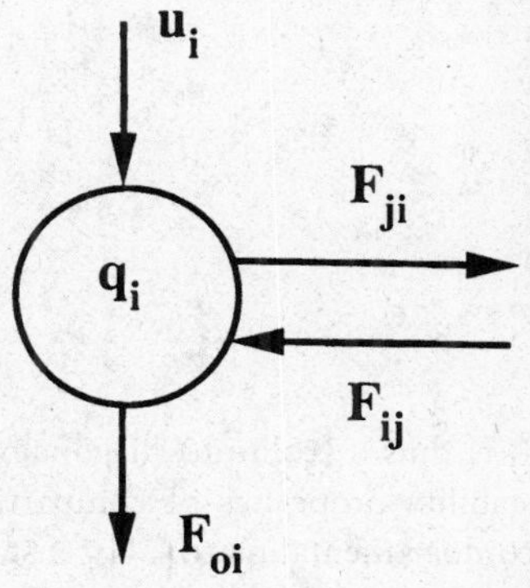

FIGURE 159.2 The *i*th compartment of an n-compartment model.

$$\dot{q}_i = \sum_{j \neq i} F_{ij} - \sum_{j \neq i} F_{ji} - F_{oi} + u_i \qquad q_i(0) = q_{io} \tag{159.1}$$

The fluxes u_i are generally constant or functions of only time. The fluxes F_{ji}, F_{ij}, and F_{0i} can be functions of $q_1, \ldots, q_n$ (sometimes also of time, but, for the sake of simplicity, we will ignore this dependence).

It is usually possible to write:

$$F_{ji}(\boldsymbol{q}) \equiv k_{ji}(\boldsymbol{q}) q_i \tag{159.2}$$

where $\mathbf{q} = [q_1, \ldots, q_n]^T$ is the vector of the compartmental masses. As a result, Eq. (159.1) can be written as:

$$\dot{q}_i = \sum_{j \neq i} k_{ij}(\boldsymbol{q}) q_j - \left(\sum_{j \neq i} k_{ji}(\boldsymbol{q}) + k_{oi}(\boldsymbol{q}) \right) q_i + u_i \qquad q_i(0) = q_{io} \tag{159.3}$$

The k_{ij}s are called *fractional transfer coefficients.* Equation (159.3) describes the *nonlinear* compartmental model. If the k_{ij}s do not depend on the masses q_is, the model becomes *linear.*

Defining $k_{ii} = -\left(\sum_{j \neq i} k_{ji} + k_{oi} \right)$ we can now write Eq. (159.3) as:

$$\dot{q}_i = \sum_{j} k_{ij}(\boldsymbol{q}) q_j + u_i \qquad q_i(0) = q_{io} \tag{159.4}$$

and the model of the whole system as

$$\dot{q} = K(q)q + u \qquad q(0) = q_o \tag{159.5}$$

where K is the *compartmental matrix* and $u = [u_1 \ldots u_n]^T$ is the vector of input fluxes into the compartments from outside the system.

For the linear case, K is constant and one has:

$$\dot{q} = Kq + u \qquad q(0) = q_o \tag{159.6}$$

The entries of the compartmental matrix K, for both the nonlinear (159.5) and the linear (159.6) model, satisfy

$$k_{ii} \leq 0 \text{ for all } i \tag{159.7}$$

$$k_{ij} \geq 0 \text{ for all } i \neq j \tag{159.8}$$

$$\sum_{i=1}^{n} k_{ij} = \sum_{i \neq j} k_{ij} + k_{jj} = -k_{oj} \leq 0 \text{ for all } j \tag{159.9}$$

K is thus a (column) diagonally dominant matrix. This is a very important property, and in fact the stability properties of compartmental models are closely related to the diagonal dominance of the compartmental matrix. For instance, for the linear model Eq. (159.6) one can show that no eigenvalues can have positive real parts and that there are no purely imaginary eigenvalues. This means that all solutions are bounded and if there are oscillations they must be damped. The qualitative theory of linear and nonlinear compartmental models have been reviewed in Jacquez and Simon [1993], where some new stability results on nonlinear compartmental models are presented as well.

The Linear Model

The linear model Eq. (159.6) has become very useful in applications due to an important result: the kinetics of a tracer in a constant steady-state system, linear or nonlinear, are linear with constant coefficients. An example is shown in Fig. 159.3, where the three-compartment model by Cobelli and co-workers [1984b] for studying tracer glucose kinetics in steady state at the whole-body level is depicted. Linear compartmental models in conjunction with tracer experiments have been extensively used in studying distribution of materials in living systems at the whole-body, organ, and cellular level. Examples and references can be found in Carson and co-workers [1983] and Jacquez [1996].

An interesting application of linear compartmental models at the organ level is in describing the exchange of materials between blood, interstitial fluid, and cell of a tissue from multiple tracer indicator dilution data. Compartmental models provide a finite difference approximation in the space dimension of a system described by partial differential equations which may be easier to resolve from the data. These models are discussed in [Jacquez, 1985] and an example of a model describing glucose transport and metabolism in the human skeletal muscle can be found in Saccomani and colleagues [1996].

The Nonlinear Model

The nonlinear model, Eq. (159.5), is frequently found in applications as well. For such models the entries of K are functions of q, most commonly k_{ij} is a function of only a few components of q, often q_i and q_j.

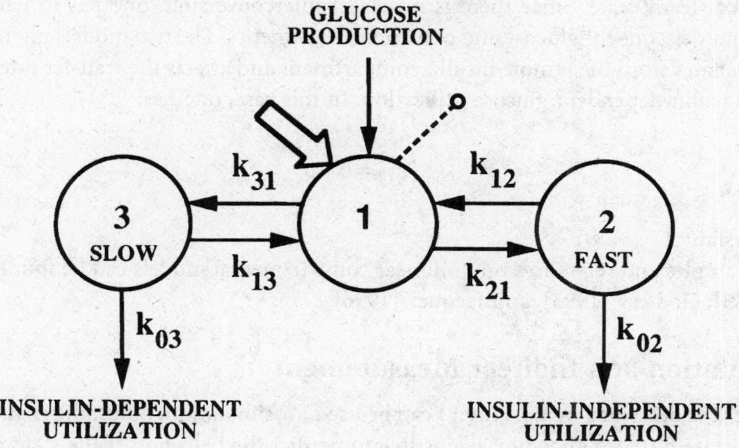

FIGURE 159.3 The linear compartmental model of glucose kinetics by Cobelli et al. [1984b].

Examples of k_{ij} function of q_i and q_j only are the Hill and Langmuir nonlinearities described, respectively, by:

$$k_{ij}\left(q_j\right) = \frac{\alpha q_j^{h-1}}{\beta + q_j^h} \tag{159.10}$$

$$k_{ij}\left(q_i\right) = 1 - \frac{q_i}{\gamma} \tag{159.11}$$

where α, β, γ are positive reals and h is a positive integer. For h = 1, Eq. (159.10) turns into the widely used Michaelis–Menten equation.

Other interesting examples arise in describing substrate-hormone control systems. For instance, the model of Fig. 159.4 has been proposed [Caumo and Cobelli, 1993; Vicini et al., 1997] to describe the control of insulin on glucose distribution and metabolism during a glucose perturbation, which brings

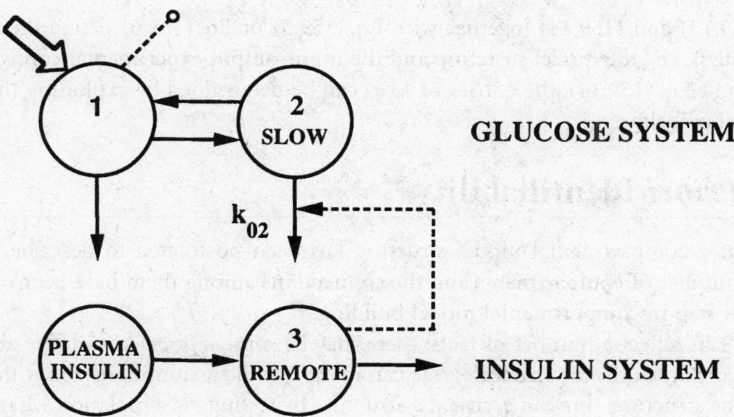

FIGURE 159.4 The nonlinear compartmental model of insulin control on glucose distribution and metabolism by Caumo and Cobelli [1993]. The dashed line denotes a control signal.

the system out of steady state. Since there is no direct interconversion, one has to have two separate compartmental models, one for glucose and one for insulin kinetics. The two models interact via a control signal which emanates from the remote insulin compartment and affects the transfer rate coefficient, k_{02}, responsible for insulin-dependent glucose utilization. In this case, one has:

$$k_{02}\left(q_3\right) = \delta + q_3 \tag{159.12}$$

where δ is a constant.

Additional examples and references on nonlinear compartmental models can be found in Carson and co-workers [1983], Godfrey [1993], and Jacquez [1996].

Use in Simulation and Indirect Measurement

Models such as those described by Eq. (159.5) can be used for simulation. Assuming that both the matrix $K(\boldsymbol{q})$ and initial state $\boldsymbol{q}(0)$ are known, it is possible to predict the behavior of the state vector $\boldsymbol{q}$ for any possible input function $\boldsymbol{u}$. A major use of compartmental models is in the indirect measurement of physiologic parameters and variables. For instance, given data obtained from compartments that are accessible to measurement, the model is used to estimate not directly measurable parameters and predict quantities related to nonaccessible compartments. In this case, $K(\boldsymbol{q})$ is unknown and input-output experiments are performed to generate data to determine $K(\boldsymbol{q})$. These data usually can be described by a linear algebraic equation

$$y = C\boldsymbol{q} \tag{159.13}$$

where $\boldsymbol{y} = [y_1 \ldots y_m]^T$ is the vector of m model outputs and C is a constant matrix. Since model outputs are often concentrations, the entries of C are inverse of compartment volumes. Note that usually only a small number of compartments is accessible to measurement.

Real data are noisy; thus, an error term $\boldsymbol{e}$ is usually added to $\boldsymbol{y}$ in Eq. (159.13) to describe the actual measurements

$$\boldsymbol{z} = \boldsymbol{y} + \boldsymbol{e} \tag{159.14}$$

$\boldsymbol{e}$ is usually given a probabilistic description, e.g., errors are assumed to be independent and often Gaussian. Data are usually collected at discrete time instants $t_1 \ldots t_N$ so that the measurements are not continuous functions of time $z(t)$ (assuming that only one output variable is observed), as in Eq. (159.14), but are denoted $z(t_1) \ldots z(t_N)$.

Equations (159.13) and (159.14) together with Eq. (159.5) or Eq. (159.6) define the compartmental model of the system, i.e., the model structure and the input-output experimental configuration.

From the data of Eq. (159.14), the entries of $K(\boldsymbol{q})$ can be determined by exploiting the identification techniques described below.

159.3 *A Priori* Identifiability

Let's assume that a compartmental model structure has been postulated to describe the physiologic system, i.e., the number of compartments and the connections among them have been specified. This is the most difficult step in compartmental model building.

The structure can reflect a number of facts: there may be some *a priori* knowledge about the system which can be incorporated in the structure; one can make specific assumptions about the system which are reflected in the structure; one can arrive at a structure by testing via simulation what is needed to fit the data. The result at this stage is a nonlinear, Eq. (159.5), or linear, Eq. (159.6), model which has as unknowns the parameters of the compartmental matrix K and also often those of the observation matrix C in Eq. (159.13).

Before performing the experiment to collect the data to be analyzed using the model or, if the experiment is already completed, before using the model to estimate the unknown parameters from the data, the following question arises: Do the data contain enough information to estimate all the unknown parameters of the postulated model structure? This question is usually referred to as the *identifiability* problem. It is set in the ideal context of error-free model structure and noise-free measurements, and is an obvious prerequisite to determine if parameter estimation from real data is well posed. In particular, if it turns out in such an ideal context that the postulated model structure is too complex for the particular set of ideal data, i.e., some model parameters are not identifiable from the data, there is no way in a real situation—where there is error in the model structure and noise in the data—that the parameters can be identified. To focus on the fact that the identifiability problem is worked out in the ideal context, one normally speaks of *a priori identifiability.*

A priori identifiability thus examines whether, given the ideal noise-free data *y*, Eq. (159.13), and the error-free compartmental model structure, Eqs. (159.5) or (159.6), it is possible to make unique estimates of all the unknown model parameters. A model can be *uniquely (globally) identifiable* or *nonuniquely (locally) identifiable*—that is, one or more of the parameters has more than one but a finite number of possible values- or *nonidentifiable*—that is, one or more of the parameters has an infinite number of solutions. For instance, the model of Fig. 159.3 is *nonidentifiable.*

The identifiability problem is, in general, a large nonlinear algebraic one and thus difficult to solve, since there is the need to solve a system of nonlinear algebraic equations which is increasing in the number of terms and nonlinearity degree with the model order, i.e., the number of compartments in the model. Various methods for testing identifiabilty of linear and nonlinear compartmental models are available [Carson et al., 1983; Cobelli and DiStefano, 1980; Godfrey, 1983; Jacquez, 1996]. For linear compartmental models, various specific methods based, e.g., on transfer function, similarity transformation, and graph topology, have been developed. Explicit identifiability results on catenary and mamillary compartmental models [Cobelli et al., 1979] and on the three-compartmental model [Norton, 1982] are also available. A parameter-bound strategy for dealing with nonidentifiable compartmental models has been also developed [DiStefano, 1983]. For nonlinear compartmental models, the problem is more difficult and for small models the output series expansion method [Pohjanpalo, 1978] is usually employed.

However, all the proposed methods apply to models of relatively low dimension; when applied to large models, the methods involve nonlinear algebraic equations too difficult to be solved even by resorting to symbolic algebraic manipulative languages (e.g., Reduce, Maple). These difficulties have stimulated new approaches to studying global identifiability based on computer algebra. In particular, for nonlinear models, approaches based on differential algebra have been investigated [Ljung and Glad, 1994; Saccomani et al., 1997], while for linear compartmental models a method based on the transfer function and the Groebner basis algorithm has been proposed and implemented in a software [Audoly et al., 1998].

If a model is *a priori* uniquely or nonuniquely identifiable, then identification techiques (e.g., least squares) can be used to estimate from the noisy data *z*, Eq. (159.14), the numerical values of the unknown parameters. If a model is *a priori* nonidentifiable, then not all its parameters can be estimated using identification techniques. Various strategies can be used, e.g., derivation of bounds for nonidentifiable parameters, model reparametrization (parameter aggregation), incorporation of additional knowledge, or design of a more informative experiment. An example of the use of some of these approaches for dealing with the nonidentifiable model of glucose kinetics of Fig. 159.3 is given in Cobelli and Toffolo [1987].

From the above considerations it follows that *a priori* unique identifiability is a prerequisite for well-posedness of parameter estimation and for reconstructability of state variables in nonaccessible compartments. It is a necessary step, but because of the ideal context where it is posed, it does not guarantee a successful estimation of model parameters from real input-output data.

159.4 Parameter Estimation

Given a uniquely identifiable parameterization of the compartmental model, one can proceed by estimating the values of the unknown parameters from the set of noisy real data. Usually the input-output experiment provides a limited set of discrete-time measurements, Eq. (159.14), for estimating the unknown parameters of the nonlinear, Eq. (159.5), and linear, Eq. (159.6), compartmental model. Problems arise in parameter estimation as a result of various sources of error, e.g., measurement error—$\boldsymbol{e}$ in Eq. (159.13)—and error in model structure—K in Eq. (159.5) or Eq. (159.6). It is normally not possible to consider explicitly errors in model structure: alternative model structures can be analyzed in order to minimize this error. The simplest situation is to consider the noisy real data as reflecting model output corrupted by an additive measurement error, like in Eq. (159.14).

Parameter estimates can be determined from Eq. (159.14) by several methods. These methods can be divided in two major categories. In the so-called Fisher estimation approach, only the data of Eq. (159.14) are supplied to the estimator in order to estimate the unknown model parameters. The second approach, known as the Bayes estimation approach, takes into account not only the data of Eq. (159.14), but also some statistical information which is *a priori* available on the unknown parameters of the model.

In the Fisher approach, parameter estimates can be obtained by nonlinear least squares or maximum likelihood together with their precision, i.e., a measure of *a posteriori* or numerical identifiability. Details and references on parameter estimation of physiologic system models can be found in Carson and coworkers [1983] and Landaw and DiStefano [1984].

Weighted nonlinear least squares is mostly used, and both direct and gradient-type search methods are implemented in estimation schemes. A correct knowledge of the error structure is needed in order to have a correct summary of the statistical properties of the estimates. This is a difficult task. Measurement errors are usually independent, and often a known distribution, e.g., Gaussian, is assumed. However, many properties of least squares hold approximately for a wide class of distributions if weights are chosen optimally, i.e., inverse of known variances or of their relative values if variances are known up to a proportionality constant. Under these circumstances, an asymptotically correct approximation of the covariance matrix of parameter estimates can be used to evaluate precision of parameter estimates. The approximation becomes exact for a sufficiently large sample size and/or decreasing variances of the measurement error. If measurement errors are Gaussian, then this approximation is the Cramer-Rao lower bound (inverse of the Fisher information matrix), i.e., the optimally weighted least squares estimator is also the maximum likelihood estimator. Care must be taken in not using lower bound variances as true parameter variances. Several factors corrupt these variances, e.g., inaccurate knowledge of error structure, limited data set. Monte Carlo studies are needed to assess robustness of Cramer-Rao lower bound in specific practical applications.

To examine the quality of model predictions to observed data, in addition to visual inspection, various statistical tests on residuals are available to check for the presence of systematic misfitting, nonrandomness of the errors, and accordance with assumed experimental noise. Model order estimation, i.e., number of compartments in the model, is also relevant here, and for linear compartmental models criteria such as the F-test, and those based on the parsimony principle like the Akaike and Schwarz criteria, can be used if measurement errors are Gaussian.

Bayes estimators, e.g., maximum *a posteriori* (MAP) and mean square (MS), assume the parameters vector to be a random vector and determine an estimate of its current realization according to a criterion connected to its *a posteriori* distribution (i.e., the probability distribution of the parameters vector conditioned on the knowledge of the data vector), which also reflects the distribution of the noise. The *a priori* information required on the unknown parameters vector can often be obtained from population studies. The concepts previously considered in the Fisher approach, i.e., assessment of the parameters estimates precision, choice of model order, can be addressed in the Bayes approach as well. Bayes estimation can be of relevant interest, since when statistical information on the unknown parameters of the model is *a priori* available and exploited, a possibly significant improvement in the parameters estimates precision with respect to Fisher estimation can be obtained.

159.5 Optimal Experiment Design

At this point, one has a compartmental model structure, a description of the measurement error, and a numerical value of the parameters together with the precision with which they can be estimated. It is now appropriate to address the optimal experiment design issue. The rationale of optimal experiment design is to act on design variables such as number of test input and outputs, form of test inputs, number of samples and sampling schedule, and measurement errors so as to maximize, according to some criterion, the precision with which the compartmental model parameters can be estimated [Carson et al., 1983; DiStefano, 1981; Landaw and DiStefano, 1984; Walter and Pronzato, 1990].

In the Fisher approach, the Fisher information matrix J, which is the inverse of the lower bound of the covariance matrix, is treated as a function of the design variables and usually the determinant of J (this is called D-optimal design) is maximized in order to maximize precision of parameter estimates, and thus numerical identifiability.

The optimal design of sampling schedules, i.e., the determination of the number and location of discrete-time points where samples are collected, has received much attention as it is the variable which is less constrained by the experimental situation. Theoretical and algorithmic aspects have been studied, and software is available, for both the single- and multi-output case [Cobelli et al., 1985; DiStefano, 1981; Landaw and DiStefano, 1984]. Optimal sampling schedules are usually obtained in an iterative manner: One starts with the model obtained from pilot experiments and the program computes optimal sampling schedules for subsequent experimentation. An important result for single-output linear compartmental models is that D-optimal design usually consists of independent replicates at P distinct time points, where P is the number of parameters to estimate.

Optimal sampling schedule design has been shown, e.g., to improve precision of the estimates as compared to schedules designed by intuition or other conventions and to optimize the cost effectiveness of a dynamic clinical test by reducing the number of blood samples withdrawn from a patient without significantly deteriorating their precision [Cobelli and Ruggeri, 1989; 1991].

The optimal input design problem has been relatively less studied, but some results are available on optimal equidose rectangular inputs (including the impulse) for parameter estimation in compartmental models [Cobelli and Thomaseth, 1987, 1988a,b].

In a Bayes estimation context, optimal experiment design is a more difficult task (see [Walter and Pronzato, 1990] for a survey). Applications have thus been less than those with a Fisher approach because the numerical implementation of the theoretical results is much more demanding.

159.6 Validation

Validation involves assessing whether the compartmental model is adequate for its purpose. This is a difficult and highly subjective task in modeling of physiologic systems, because intuition, understanding of the system, and so on play an important role. It is also difficult to formalize related issues such as model credibility, i.e., the use of the model outside its established validity range. Some efforts have been made, however, to provide some formal aids for assessing the value of models of physiologic systems [Carson et al., 1983; Cobelli et al., 1984]. A set of validity criteria have been explicitely defined, i.e., empirical, theoretical, pragmatic, and heuristic validity, and validation strategies have been outlined for two classes of models, broadly corresponding to simple and complex models. This operational classification is based on *a priori* identifiability and leads to clearly defined strategies as both complexity of model structure and extent of available experimental data are taken into account. For simple models, quantitative criteria based on identification, e.g., *a priori* identifiability, precision of parameter estimates, residual errors, can be used in addition to physiologic plausibility. In contrast with simple models, complex simulation models are essentially incomplete, as there will naturally be a high degree of uncertainty with respect to both structure and parameters. Therefore, validation will necessarily be based on less solid grounds. The following aids have been suggested: increasing model testability through model simplification; improved experimental design and model decomposition; adaptive fitting based on qualitative and/or quantitative feature comparison and time-course prediction; and model plausibility.

References

Adapt II User's Guide. 1997. Biomedical Simulation Resource, University of Southern California, Los Angeles.

Audoly, S., D'Angio', L., Saccomani, M.P., and Cobelli, C. 1998. Global identifiability of linear compartmental models. A computer algebra algorithm. IEEE Trans. Biom. Eng., 45: 36–41.

Carson, E.R. and Jones, E.A. 1979. The use of kinetic analysis and mathematical modeling in the quantitation of metabolic pathways *in vivo*: application to hepatic anion metabolism. New Engl. J. Med., 300:1016-1027 and 1078-1986.

Carson, E.R., Cobelli, C., and Finkelstein, L. 1983. The Mathematical Modelling of Metabolic and Endocrine Systems. John Wiley, New York.

Caumo, A. and Cobelli, C. 1993. Hepatic glucose production during the labelled IVGTT: estimation by deconvolution with a new minimal model. Am. J. Physiol. 264:E829-E841.

Cobelli, C. and Caumo, A. 1998. Using what is accessible to measure that which is not: necessity of model of system. Metabolism, 47: 1009–1035.

Cobelli, C., Lepschy, A., and Romanin Jacur G. 1979. Identifiability results on some constrained compartmental systems. Math. Biosci. 47:173-196.

Cobelli, C. and DiStefano III, J.J. 1980. Parameter and structural identifiability concepts and ambiguities: a critical review and analysis. Am. J. Physiol. 239: R7-R24.

Cobelli, C. 1984. Modeling and identification of endocrine-metabolic systems. Theoretical aspects and their importance in practice. Math. Biosci. 72:263-289.

Cobelli, C., Carson, E.R., Finkelstein, L., and Leaning M.S. 1984. Validation of simple and complex models in physiology and medicine. Am. J. Physiol. 246: R259-R266.

Cobelli, C., Toffolo, G., and Ferrannini, E. 1984. A model of glucose kinetics and their control by insulin. Compartmental and noncompartmental approaches. Math. Biosci. 72: 291-315.

Cobelli, C., Ruggeri, A., DiStefano III, J.J., and Landaw, E.M. 1985. Optimal design of multioutput sampling schedules: software and applications to endocrine-metabolic and pharmacokinetic models. IEEE Trans. Biomed. Eng. 32: 249-256.

Cobelli, C. and Thomaseth, K. 1987. The minimal model of glucose disappearance: optimal input studies. Math. Biosci. 83: 127-155.

Cobelli, C. and Toffolo, G. 1987. Theoretical aspects and practical strategies for the identification of unidentifiable compartmental systems. In Identifiability of Parametric Models, Walter, E., Ed., Pergamon, Oxford, 85-91.

Cobelli, C. and Thomaseth, K. 1988a. On optimality of the impulse input for linear system identification. Math. Biosci. 89: 127-133.

Cobelli, C. and Thomaseth, K. 1988b. Optimal equidose inputs and role of measurement error for estimating the parameters of a compartmental model of glucose kinetics from continuous-and discrete-time optimal samples. Math. Biosci. 89: 135-147.

Cobelli, C., Saccomani, M.P., Ferrannini, E., DeFronzo, R.A., Gelfand, R., and Bonadonna, R.C. 1989. A compartmental model to quantitate *in vivo* glucose transport in the human forearm. Am. J. Physiol. 257: E943-E958.

Cobelli, C. and Ruggeri, A. 1989. Optimal design of sampling schedules for studying glucose kinetics with tracers. Am. J. Physiol. 257: E444-E450.

Cobelli, C. and Ruggeri, A. 1991. A reduced sampling schedule for estimating the parameters of the glucose minimal model from a labelled IVGTT. IEEE Trans. Biomed. Engng. 38:1023-1029.

DiStefano III, J.J. 1981. Optimized blood sampling protocols and sequential design of kinetic experiments. Am. J. Physiol. 9:R259-R265.

DiStefano III, J.J. 1983. Complete parameter bounds and quasi identifiability conditions for a class of unidentifiable linear systems. Math. Biosci. 65:51-68.

Gibaldi, M. and Perrier, D. 1982. Pharmacokinetics. 2nd ed., Marcel Dekker, New York.

Godfrey, K. 1983. Compartmental Models and Their Application. Academic Press, London.

Jacquez, J.A. 1996. Compartmental Analysis in Biology and Medicine, 3rd ed., Biomedware, Ann Arbor, MI.

Jacquez, J.A. and Simon, C.P. 1993. Qualitative theory of compartmental systems. Siam Review 35:43-79.

Landaw, E.M. and DiStefano III, J.J. 1984. Multiexponential, multicompartmental, and noncompartmental modeling. II. Data analysis and statistical considerations. Am. J. Physiol. 246: R665-R677.

Ljung, L. and Glad, T. 1994. On global identifiability for arbitrary model parametrizations. Automatica 30:265-276.

Norton, J.P. 1982. An investigation of the sources of non-uniqueness in deterministic identifiability. Math. Biosci. 60:89-108.

Pohjanpalo, H. 1978. System identifiability based on the power series expansion of the solution. Math. Biosci. 41:21-33.

Saccomani, M.P., Audoly, S., Bellu, G., D'Angio', L., and Cobelli, C. 1997. Global identifiability of nonlinear model parameters. 1997. In Savaragi, Y. and Sagara, S., Eds., Proceedings of the SYSID '97, 11th IFAC Symposium on Systems Identification, pp.219-224, Kitakyushu (Japan), Nichigen Co. Ltd.

Saccomani, M.P., Bonadonna, R., Bier, D.M., De Fronzo, R.A., and Cobelli, C. 1996. A model to measure insulin effects on glucose transport and phosphorylation in muscle: a three-tracer study. Am J Physiol. 33: E170–E185.

SAAM II User Guide 1998. SAAM Institute FL-20, University of Washington, Seattle, WA.

Vicini, P., Sparacino, G., Caumo, A., and Cobelli, C. 1997. Estimation of hepatic glucose release after a glucose perturbation by nonparametric stochastic deconvolution. Comp. Meth. Progr. Biomed., 52: 147–156.

Walter, E. and Pronzato, L. 1990. Qualitative and quantitative experiment design for phenomenological models—a survey. Automatica 26: 195-213.

160
Cardiovascular Models and Control

William D. Timmons
University of Akron

This chapter reviews cardiovascular (CV) modeling for use in controller design, especially for closed-loop drug delivery systems (IV infusions of vasodilators, inotropes, anesthetics, neoplastic agents, and so on) and other pharmacologic applications. This first section describes the advantages and disadvantages of employing CV models for design and control and presents a brief history of CV modeling. The next section describes the basic principles and techniques for modeling the CV system, as well as the uptake, distribution, and action of cardio- and vasoactive pharmaceuticals. A short, annotated bibliography follows the reference section for those interested in further reading.

160.1 Advantages and Disadvantages of Desktop Patients

More and more industries are coming to rely on computer simulations to increase productivity and decrease costs as well as to increase reliability and safety of resulting products. The defense, automotive, and aerospace industries come most quickly to mind, but the medical field is rapidly gaining ground in this area, too. Recent ARPA[1] initiatives for defense technology conversion are probably most responsible for the recent surge in computer-aided health care. Today, computerized simulators teach medical students everything from basic anatomy to techniques in radial keratotomy, and skilled surgeons design, evaluate, and practice new and risky procedures without ever cutting into a single patient. Thanks to the ARPA initiative, telesurgery will soon be viable, so that in the near future, surgeons may operate on patients from half a world away. From the calculation of simple-dosage regimens to the design of sophisticated blood pressure controllers, cardiovascular medicine also benefits from this technology. CV models can be used to prove initial feasibility, reduce development time, decrease costs, and increase reliability and safety of cardiovascular devices and therapeutics.

[1]The Advanced Research Projects Agency of the U.S. Department of Defense, chair of the Defense Technology Conversion Council which administers the Technology Reinvestment Project.

0-8493-0462-8/00/$0.00+$.50

The increasing speed and decreasing costs of computers are having an enormous impact on controller design. There is a veritable explosion in the number of control designs as well as the theory to go with them. The ready availability of powerful computer workstations and sophisticated engineering design packages (e.g., MATLAB, MATRIXx, CC) allow for the interactive design of high-performance, robust controllers. Engineers can tweak a controller or perturb a system, then immediately assess the effect. When combined with modern visualization techniques, these tools allow one to readily see problems and solutions that might otherwise have taken years to find and solve. Furthermore, such capabilities allow for exhaustive testing and debugging, so that the transition from the workbench to the field becomes smooth. Many companies and agencies have reported flawless transitions from the simulator to the field, most notably NASA, with its many satellites and probes that have had to work right the first time. Various companies are using simulations to improve or correct equipment and systems already in the field. A recent example is BAE Automated Systems Inc. of Dallas, Texas, which is using simulations to help debug and correct the troubled baggage system at the new Denver International Airport [Geppert, 1994].

The risk of failure or the fear of malpractice often precludes the design and evaluation of controllers in live humans, so that devices and treatments are often tested on computer and animal models first. In the past, animals have been the preferred choice, but this approach is continually being reevaluated. Whereas most researchers readily agree that animal experimentation has been and will continue to be necessary for the advancement of medical science, most would welcome methods and procedures that might reduce or even eliminate the use of animals. Furthermore, animals are rarely perfect models of human diseases and conditions. Thus, there has been and will continue to be a demand for alternative models and procedures. Tissue cultures are beginning to fill some of this need, and it is likely that computer modeling and simulation may also fill this need.

Desktop patients, if used to complement animal studies, can actually increase the reliability and applicability of certain animal tests, while also reducing the number of required experiments. In certain instances, computer modeling may be the only way to test a device prior to human use, since some human conditions are extremely difficult if not impossible to reproduce in animals. Indeed, as computers become more powerful and models become better, it may be that some day computers will be used *in place of* certain animal and human trials. Meanwhile, computer models can be used to (1) demonstrate feasibility, (2) increase confidence in controller designs by complementing animal studies, (3) help design better animal and clinical experiments, and (4) reduce the number of required animal and human experiments.

The disadvantage of using computer simulation is that available models will almost certainly need to be tailored to the application at hand. Since there is generally no predefined method for modeling or for modifying an existing model, the accuracy and reliability of the resulting model will depend heavily on the skill of the modeler, the appropriateness of the modifications, and the model's intended scope. Furthermore, modeling typically requires a good understanding of the physics of the problem, and this in turn requires supporting data and experimentation. Thus, modifications to include new pharmaceuticals may be problematic, since their mechanisms may only be poorly understood. And, even if supporting data are available, if a basic model does not yet exist, a significant time commitment may be required to develop one. Despite this up-front cost, model development may save tremendous time and costs later in the development process.

Fortunately, there are a large number of CV models to pick from, and experienced modelers can determine which, if any, is best for a given application. Once determined, the model can usually be converted easily to a form compatible with the development platform. Once converted, model integrity and suitability must be checked. Modifications can then be made, after which the model integrity must again be checked to ensure that it has not been compromised. Once a model is obtained and suitably modified, adequate thought must be given to program flow. Ideally, program flow should be designed such that the controller code is the same for real time and for simulation. This may be facilitated by treating all facets of data acquisition as part of the controlled system, instead of as part of the controller. With this approach, all software for I/O and calibration would be bundled separately from the controller software. Windows-based systems facilitate this separation even further by allowing the controller and plant software to be compiled and run as two separate programs.

Other issues that must be considered involve the practical aspects of computer simulation. Even if a well-designed and validated model exists, many factors can affect the accuracy of the simulation. Stiffness, machine precision, program coding, integration method, step size, and even computer language can affect the results. Probably more insidious than all these factors is the operation of the model in regions outside its scope. This happens most often in large, complex programs when a variable takes on values outside its intended range. For example, a sudden increased cardiac preload may shift the end diastolic volume of the heart into a nonlinear region of the diastolic pressure-volume relationship. If the nonlinear behavior is not included in the model, misleading results might be produced. Finally, even in the best commercial code (let alone home-brewed code), bugs may affect simulation accuracy. For example, many commercial simulation packages are particularly poor at handling discrete events (such as a bolus infusion of a drug) when variable step-size algorithms are used to integrate continuous time models [Gustafsson, 1993].

160.2 History of Cardiovascular Modeling

William Harvey in the early 17th century was probably the first to clearly and convincingly demonstrate the role of the heart as a pump which caused the blood to flow in a unidirectional closed circuit through the systemic and pulmonary circulations. Prior to Harvey, the Galenic viewpoint from the 2d century AD predominated,[2] namely, that the blood ebbed to and fro in the veins to provide nutrition to the rest of the body, that the arteries (from the Greek *arteria*, meaning windpipe) carried air, that the lungs cooled the heart and removed sooty impurities from new blood as it entered from the intestines, and that the heart contained pores to allow spirit or gaseous exchange across the septum [Acierno, 1994].

It was Stephen Hales in his *Statistical Essays* nearly a century later who considered arterial elasticity and postulated its buffering effect on the pulsatile nature of blood flow [Hales, 1733]. He likened the depulsing effect to the fire engines of his day, in which a chamber with an air-filled dome ("inverted globe") acted to cushion the bolus from the inlet water pump so that "a more nearly equal spout" flowed out the nozzle. This analogy became the basis of the first modern cardiovascular models and antedated the ARPA movement by quite a few decades! In the translation from English to German, Hales's inverted globe became a *windkessel* (air kettle); his idea later became known as the Windkessel theory when it was more formally developed and propounded by the German physiologist Otto Frank (of the Frank-Starling law of the heart) near the beginning of our century [Noordergraaf, 1978].

Frank was initially a strong proponent of the Windkessel theory, and his work spawned the interest of many subsequent investigators and led to a proliferation of modified Windkessel-type models. A critical review of these early models can be found in Aperia [1940]. The development of the analog computer shortly after World War II lead to another burst in CV modeling, based on more sophisticated segmental and transmission line theories. Grodins [1959] was probably one of the first to use these new machines to simulate cardiovascular hemodynamics. Interestingly, another early analog computer model, PHYSBE [McLeod, 1966], became a popular benchmark that is still used today to evaluate computers and simulation languages. About the mid-1970s, the advent of the inexpensive digital computer led to another revolution in CV modeling, which has continued to the present with ever-increasingly detail and scope. As Rideout [1991] observed, model complexity seems to have been limited only by computer technology. If this trend continues, we can expect to see highly detailed models emerging within the next few years as parallel computers become economical.

The major criticism of the pure Windkessel theory is that it does not allow for finite wave propagation and reflection in the arteries. Whereas blood flow in the arterial vasculature is more properly formulated

[2]Galen, in turn, based many of his beliefs on those of the hippocratic era scholars—Hippocrates, Aristotle, Polybus, and Diocles (4th century BC).

in terms of space and time, the Windkessel theory ignores spacial considerations and instead lumps all the vasculature into a single point, or compartment, in space. Attempts to correct this shortcoming include adding inertial and damping factors as well as collapsible components. In addition, the great arteries can be partitioned into sections to produce segmental models, or, when electrical network theory is invoked, transmission line models. Frank himself realized that the Windkessel theory was flawed, and so he too added modifications to include traveling waves and reflections [Milnor, 1989; Noordergraaf, 1978; Rideout, 1991].

Despite its flaws, the pure Windkessel theory is still useful as a teaching tool because of its simplicity and clear, intuitive imagery. Furthermore, as mean pressure accounts for approximately 90% of the power in the arterial pressure waveform, it performs surprisingly well and displays many of the interesting phenomena seen in the vasculature. Add to this the fact that the equations are readily extended to include spacial effects (the segmental and transmission line models), and one can see why the theory has persisted into the late 20th century. Of course, one must keep in mind that any of these approaches is only an approximation; the segmental and transmission line approaches are in many ways only a mathematical convenience that allows one to fit an arbitrarily high-order model to observed data. These approaches do not necessarily lend themselves to extrapolations and predictions regarding the physiology.

Even before Frank, there was great interest in the pure mathematics of blood flow in distensible vessels. Many a famous mathematician and scientist explored wave propagation, reflection, and pressure-flow relations in blood vessels, including: Bernoulli, Euler, Young (of Young's modulus), Poiseuille, Navier and Stokes, the Weber brothers, Resal, Moens and Korteweg, Lamb, and Witzig; more recent investigators including Morgan and Kiely, and Womersley and McDonald [Fung, 1984; Milnor, 1989; Noordergraaf, 1978]. Womersley and McDonald's students continue to be influential in cardiovascular modeling today. Early studies of theoretical hemodynamics greatly influenced the field of fluid mechanics and paved the way for modern pulse-wave propagation and reflection models of the vasculature. These models are treated in the texts by Noordergraaf [1978], Fung [1985], Li [1987], and Milnor [1989].

The hydrodynamic pulse-wave CV models are not without criticisms either. For example, they are limited to short segments of the arterial or venous tree for which input and terminal impedances must be supplied. This restriction makes them currently impractical for the analysis of long-acting pharmaceutical agents and their feedback effects on the cardiovascular system. Pharmaceuticals with vaso- and cardio-active effects usually require the tracking of drug concentrations throughout the body, which, for these models, would require an exceedingly large number of components and would result in huge programs requiring large computers, with the simulations invariably being slow and cumbersome. Furthermore, numerical techniques such as the finite-element method must be used to solve many of the hydrodynamic equations, which over extended periods might incur significant round-off and integration errors. Nevertheless, this approach will certainly form the basis of future pharmacologically linked CV models as faster and more sophisticated computers and algorithms are developed.

Because of the current shortfall in computational technology, most pharmacologic simulations requiring a CV model employ simplified or reduced-order models mentioned earlier. Besides the segmental and transmission line approaches, purely black box models are also used. The reduced segmental and transmission line models are probably most appropriate for simulations of fast-acting CV pharmaceuticals, whereas compartmental models (see Chapter 157) are more appropriate for slower-acting drugs.

The distinction between segmental and transmission line models has become increasingly blurred, especially as elements of both are often combined into the same model. Transmission line models were used as early as 1905 by Frank, although Westerhof and Noordergraaf [1969] were probably most influential in promoting the transmission line approach. As a result, these models are often called Westkessel models. They employ black box, two-terminal circuits (Westkessel terminations) to reproduce the input impedances of lumped vascular trees. Segmental models, however, are generally three terminal circuits, with two terminals representing inertial flows and viscous resistances and the third representing the compliance of the segment with respect to an external reference (typically atmospheric pressure). The segmental approach is described in more detail below.

As mentioned above, reduction in model order is typically achieved by combining segmental and transmission line models. The idea here is to represent critical, nonlinear components by segments, while the less critical, more linear components are lumped together and represented by Westkessel terminations. Unfortunately, the black box Westkessel sections lose their physiologic meaning, which may present a problem in certain applications.

The highly simplified black box models are even worse in this regard. They are less concerned with the physiology than with the input-output relationships of the system and typically employ linear dynamics fit to observed patient responses. Though easy to implement, once defined, physiologically relevant perturbations may be difficult to produce. Furthermore, these models should not be used to infer properties or behaviors outside the range of the data sets. This is a potentially crippling drawback, especially when designing and evaluating CV controllers. However, if the goal is to generate clinically observed behaviors irrespective of the pathophysiology, then these models make an excellent choice.

Another potential drawback, which applies to both simplified and reduced-order models, is the loss of phenomena associated with pulse propagation and its effects on cardiac-arterial coupling and energetics. These effects are usually considered minor from a pharmacologic point of view, but their omission may lead to potentially serious design flaws when the resulting controller is used in certain patient populations. For example, isolated systolic hypertension (ISH) significantly contributes to strokes and heart disease in the elderly, and O'Rourke [1993] suggests that the therapy of choice, as well as the goal of treatment, should be different for ISH than for diastolic and mean arterial hypertension. It may be prudent, then, for control applications aimed at this population to be evaluated on more detailed models.

160.3 Cardiovascular Modeling

In this section, the concepts behind the segmental CV approach are described. Neurohumoral controls are also briefly discussed, as is the inclusion of pharmacokinetics and pharmacodynamics for drug uptake, distribution, and action.

An Idealized Segment of Artery

A short segment of artery can be modeled by an elastic, isobaric chamber attached to a rigid inlet and outlet tube (Fig. 160.1). The ability of the chamber to store fluid depends on its compliance, *C*, which is defined as

$$C = \frac{dV_c}{dP_c} \tag{160.1}$$

where V_c is total segment (chamber) volume and P_c is chamber pressure. Many modelers prefer to use the reciprocal of compliance (termed stiffness, *S*, or elastance, *E*). Here, viscoelastic (stress relaxation) effects are assumed negligible, so that compliance is not an explicit function of time. Thus, compliance becomes an instantaneous variable which can be obtained from the experimentally determined, steady-state pressure-volume (*P-V*) relationship of the segment. Furthermore, model order can now be reduced, as pressure can be mathematically represented as an empirical function of volume, either by a piecewise linear approximation, a polynomial quotient, or some other function fit to the *P-V* relationship. A typical *P-V* curve is illustrated in Fig. 160.2, along with its piecewise liner approximation.

In arteries, pressure is usually positive with small oscillations about a nominal operating point (point *A* in Fig. 160.2). Hence the piecewise linear approximation can be reduced to a single line with constant slope $1/C$:

$$P_c = \frac{\left(V_c - V_{cu}\right)}{C} \tag{160.2}$$

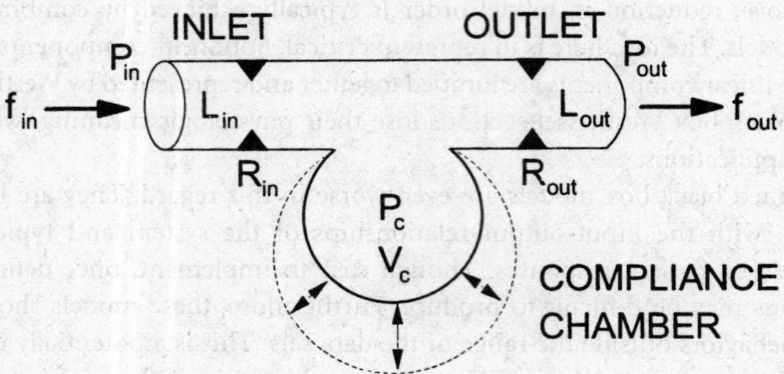

FIGURE 160.1 Idealized segment of artery approximated by an elastic, isobaric chamber attached to a rigid inlet and outlet tube. P_{in}, P_{out}, and P_c are *in*let, *out*let, and *c*hamber *p*ressures; f_{in} and f_{out} are *f*lows; L_{in} and L_{out} are inertances; R_{in} and R_{out} are viscous *r*esistances to flow; and V_c is total segment (*c*hamber) *v*olume. Compliance C is defined in the text.

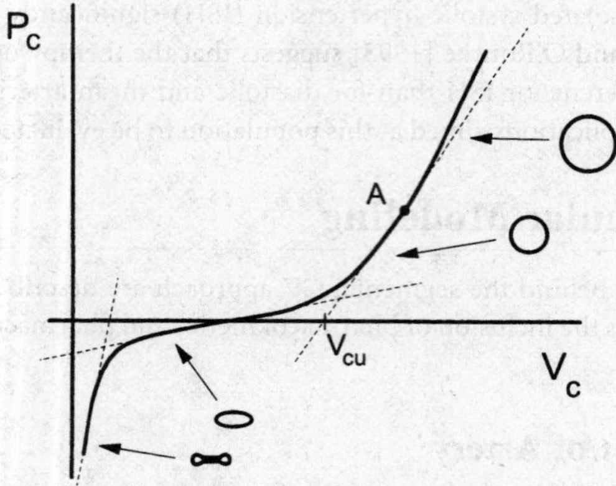

FIGURE 160.2 Static pressure-volume (P-V) relationship of a segment of artery.

where V_{cu} is the unstressed volume (the idealized zero-pressure volume intercept). Of course, if pressure fluctuates outside this region, then additional straight-line sections should be included.

Assuming that blood is incompressible and newtonian, that the flow profile is parabolic and unchanging with axial distance along a straight rigid tube, then flow (poisenillean) is linearly proportional to the pressure gradient across the ends of the tube. Hence, flow into and out of the chamber in Fig. 160.1 may be calculated given P_{in} and P_{out} (the inlet and outlet pressures); P_c; L_{in} and L_{out} (the inertances due to fluid mass); and R_{in} and R_{out} (the resistances due to viscous drag)

$$L_{in}\frac{df_{in}}{dt} + R_{in} f_{in} = P_{in} - P_c$$
$$L_{out}\frac{df_{out}}{dt} + R_{out} f_{out} = P_c - P_{out} \tag{160.3}$$

where f_{in} and f_{out} are the flows into and out of the segment. Volume can now be calculated from the difference between the inlet and outlet flows:

$$\frac{dV_c}{dt} = f_{in} - f_{out} \tag{160.4}$$

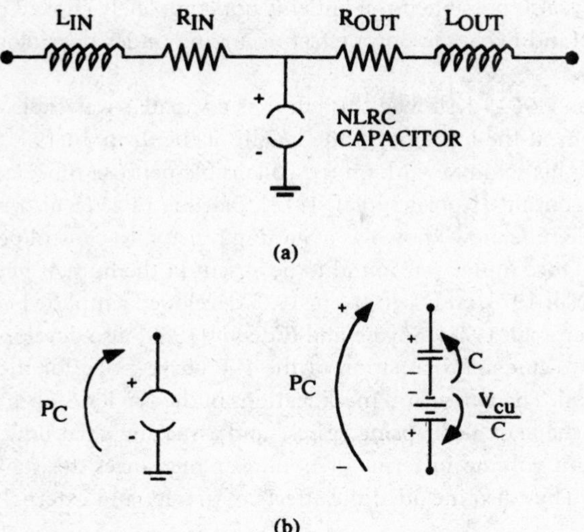

FIGURE 160.3 (*a*) The idealized segment of artery from Fig. 160.1 as an RLC circuit. (*b*) The nonlinear residual charge capacitor in (*a*). After Tsitlik et al. [1992].

P_c, V_c, f_{in}, and f_{out} can now be uniquely determined at any time given P_{in}, P_{out}, and the initial conditions of the segment. Determination of the parameters *L* and *R* are more problematic. They can be derived analytically, although probably because the modeling assumptions are not completely correct, an empirical fit generally produces better segmental properties.

The simplified segmental structure can be represented as an RLC circuit as shown in Fig. 160.3*a*. Tsitlik and coworkers [1992] remind us that each capacitor should contain a residual charge to simulate the respective segment's unstressed volume (V_{cu}). This can be achieved most easily by inserting a battery between a standard capacitor and ground (Fig. 160.3*b*) [Tsitlik et al., 1992]. Or, if the unstressed volume for the segment remains constant (say, because the pharmacologic drug of interest has little effect on it), then the unstressed volume (and the battery) can be eliminated by subtracting it from the total vascular blood volume. The segment will then contain the stressed volume only (V_c-V_{cu}), so that an ordinary capacitor will suffice (again assuming operation near point *A* on the *P-V* curve).

An Idealized Segment of Vein

In most of the venous circulation, the pressure inside the vessels is greater than the external pressure, so that the pressure-flow relationship is like that in the arteries. However, in the vena cava, in certain organs such as the lungs and during certain procedures such as resuscitation or measurement of blood pressure using a cuff, venous collapse may occur [Fung, 1984]. Thus, in contrast to an artery in which only one condition normally exists, two additional conditions must be considered for a vein.

The first condition occurs when the inlet and outlet transmural pressures are *positive.* [Transmural pressure is defined as the pressure gradient across the vessel wall (P_{inside}-$P_{outside}$).] In this instance, the vein can be treated as discussed above for an artery. A second condition occurs when the inlet and outlet transmural pressures are *negative.* In this instance, the vein will collapse, and flow will stop or be greatly reduced.

The third condition occurs when the *inlet* transmural pressure is *positive* and the *outlet* transmural pressure is *negative.* In this instance, blood flows into the chamber but not out because the outlet will be collapsed. As blood flows in, the pressure inside rises until it exceeds the external pressure, forcing the outlet open and allowing blood to flow out. But, as blood flows out, the internal pressure decreases, so that the outlet transmural pressure may become negative again, collapsing the outlet, and commencing the next cycle. This cycling is called flutter, and the collapsing at the outlet, choking. Besides flutter, a limited steady-state flow is also possible if the outlet is not completely choked off [Fung, 1984]. Limited flow is the predominant (and likely the only) effect occurring under physiologic conditions [Noordergraaf, 1978].

Purmutt and colleagues [1962] described this effect as a vascular waterfall. Water flowing over a falls depend only on conditions at the top, not on the length of the drop. In 1994 Holt used the term flow regulator for this effect in his seminal work on the collapsible penrose tube, and Conrad in 1969 called it a negative impedance conduit [Noordergraaf, 1978]. Starling in 1915 also made use of this concept when he used a collapsible tube, now known as a Starling resistor, to control peripheral resistance in his heart-lung preparations. Once flutter was found to be absent in the human physiology, Brower in 1970, Brower and Noordergraaf in 1973, and Griffiths in 1975 developed simplified equations of the pressure-flow relationship [Noordergraaf, 1978]. Snyder and Rideout [1969] also developed a model of collapsible vein. They employed a two-line approximation of the *P-V* curve—one for the uncollapsed region and one for the collapsed region. Then, based on modifications of the Navier-Stokes equations for an elliptical tube (the cross-sectional shape of a collapsing vessel), and assuming a flat flow profile, they were able to relate flow to pressure and volume in terms of nonlinear inertances, resistances, and compliances in Eqs. (160.2) and (160.3). They also included the effects of gravity and external pressures in the model.

Arterial and Venous Trees

By using the output of one segment as the input to a second, and the output of the second as the input to a third, and so on, differential equations for arbitrary lengths of vessel can be constructed (Fig. 160.4*a*). This method also allows bifurcations to be added. Another approach can be used to derive these same equations. First, partial differential equations are formulated to define flow along the desired length of vessel. The spacial differentials are then changed to finite differences, leaving only time derivatives [Noordergraaf, 1978].

Similar to an arterial section, the resulting lumped model of the vascular tree can be represented as an electrical circuit if desired (Fig. 160.4*b*), and nearby resistors and inductances can be combined to form Γ, T, or Π sections. A similar approach can be used for the venous circulation. To prevent ringing and other problems due to sudden impedance changes, it is often useful to add intermediate segments to taper the impedances between the heart and the capillaries [Rideout, 1991]. That is, the impedances

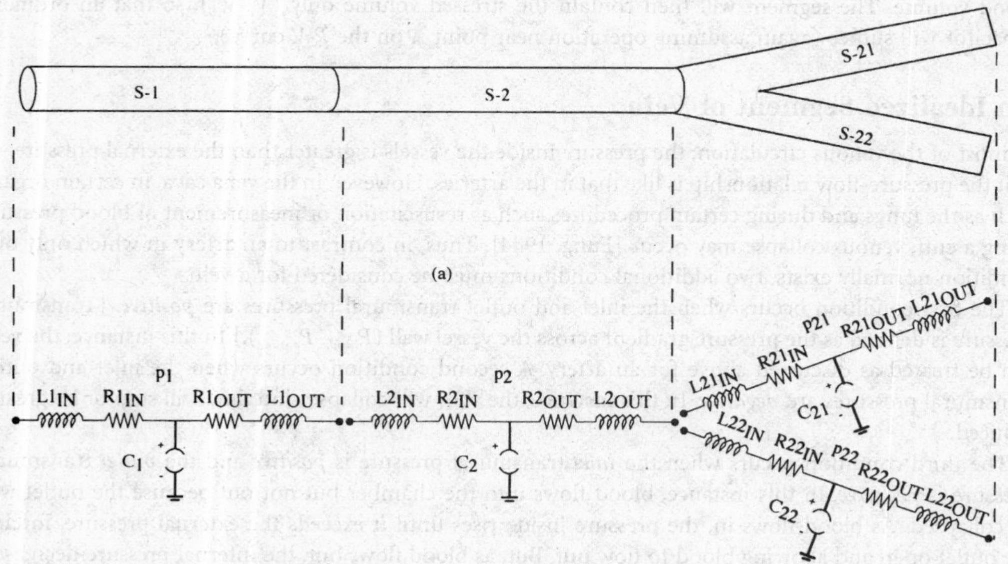

FIGURE 160.4 (*a*) A segmented, arbitrary length of vessel. (*b*) The equivalent RLC circuit.

of the large arteries should be tapered to match the higher resistive impedances of the arterioles and capillary beds and then reduced back again from the capillary beds to the large veins.

Arterial and venous branching can be roughly grouped by vessel size (large, medium, and small arteries, arterioles, capillaries, etc.) or by organ system (heart, head, legs, liver, fat, skin, muscle, etc.). Vessel grouping is useful when the model needs to track drug concentrations at whole-body effector sites, such as nitroprusside concentration in the systemic arteriolar and venous compartments. Organ grouping is useful when the model needs to track drug concentrations in certain organs, such as insulin in the liver, glucose in the pancreas, or antineoplastic agents in diseased tissues. Note that within each organ, the flows can be broken down further by vessel size or by regional or conceptual intraorgan blood flows. Once the desired organs and vessels are described, Westkessel-type impedances can be used for the rest of the vasculature.

Models of the Heart

The heart is a muscular pump with four chambers intimately linked into one organ (Fig. 160.5*a*). Nevertheless, in modeling the heart, each chamber is usually modeled independently of the others. Each is typically set up as an elastic compartment with inertance and resistance similar to an artery or vein (see above). The differences are (1) that valves are included to constrain flow to one direction, and (2) that elastance (the inverse of compliance) is treated as a time-varying parameter.

The valves are straightforward and are often implemented as ideal diodes (for electrical circuit models) or as IF-THEN-ELSE statements (for algorithmic models) to keep all flows nonnegative. Defects in the valves can be added to simulate heart defects (e.g., leaky diodes for regurgitation). Other types of heart defects are just as easily simulated. For example, Blackstone and coworkers [1976] placed an impedance between the atrial chambers to simulate a septal opening.

Mathematical representation of the time-varying elastance is more complicated and is generally based on the characterization of the cardiac work cycle in the pressure-volume plane. In the *P-V* diagram in Fig. 160.5*b*, the time course of the left ventricular cardiac cycle proceeds as follows.

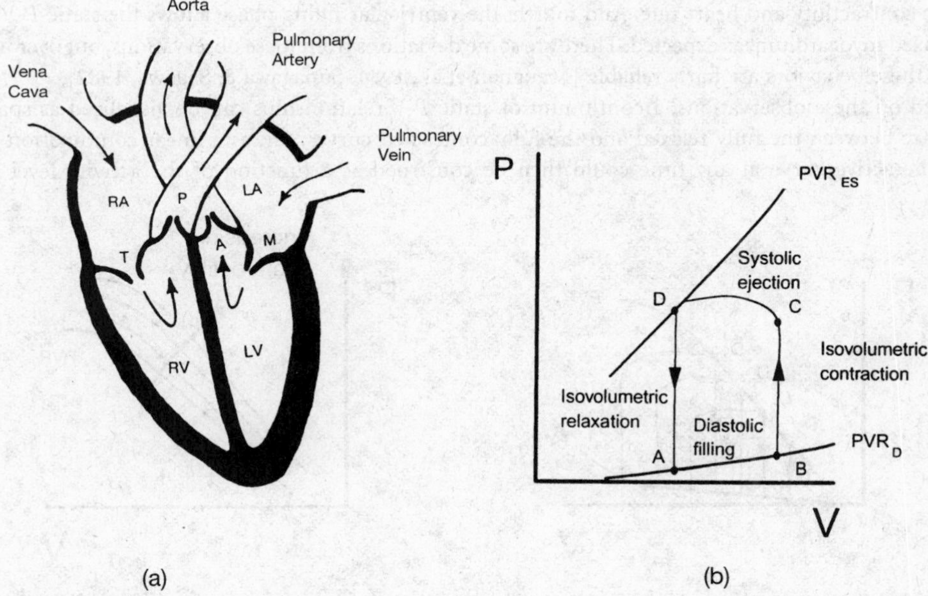

FIGURE 160.5 (*a*) The chambers and valves of the heart. LV, left ventricle; RV, right ventricle; LA, left atrium; RA, right atrium; M, mitral valve; A, aortic valve; T, tricuspid valve; P, pulmonary valve. (*b*) The left ventricular cardiac cycle in the *P-V* plane. See text.

During diastolic filling (A to B), ventricular pressure is below atrial and aortic pressures, so the mitral valve (between the left atrium and the left ventricle) is open, and the aortic valve (between the left ventricle and the aorta) is closed. Blood therefore flows from the atrium into the ventricle, based on the pressure gradient and the inertances and resistances of the inlet. Note that the pressure gradient will decrease as the ventricle fills due to the passive compliance of the chamber and the drop in pressure in the atrium.

After diastole, the ventricle is stimulated to contract so that pressure rises above the atrial pressure, closing the mitral valve. In this state, pressure is too low for the aortic valve to open, so that both valves are closed and no blood flow occurs. Meanwhile, the ventricle continues to contract, raising ventricular pressure. This period (B to C) is termed isovolumetric contraction. Here, the muscle contraction can be thought of as decreasing the compliance of the chamber.

Once ventricular pressure exceeds the pressure in the aorta, the aortic valve opens and systolic ejection commences (C to D). Blood flows out from the ventricle into the aorta based on the pressure gradient, the inertances, and the resistances of the outlet. The muscle continues to contract until the cardiac action potentials have run their course.

As the muscle begins to relax, pressure will drop until it falls below the aortic pressure. At this time, the aortic valve closes, and again blood flow ceases. Meanwhile, the ventricle continues to relax, decreasing the ventricular pressure. This period (D to A) is termed isovolumetric relaxation. Here, muscle relaxation can be thought of as *increasing the compliance* of the chamber. Finally, ventricular pressure falls below atrial pressure, the mitral valve opens, and the cycle begins anew.

Suga and Sagawa [1972, 1974] identified and characterized several key properties of the time-varying compliance in the pressure-volume plane. In their experiments, they varied preload, afterload, stroke volume, cardiac contractility, and heart rate for both ejecting beats (heart valves patent) and isovolumic beats (heart valves sewn shut). Four observations should be noted. First, the *P*-*V* point at the end of systole (Fig. 160.6*a*, points D_1–D_4) nearly always lies along the static *P*-*V* curve for activated myocardium. Second, the end systolic *P*-*V* curve is unaffected by changes in preload and afterload (Fig. 160.6*a*); however, it is affected by changes in cardiac contractility (Fig. 160.6*b*). Third, t_{max}, the time to end systole (point *D*), is also unaffected by changes in preload and afterload; however, it is affected by changes in cardiac contractility and heart rate. And fourth, the ventricular filling phase allows the static *P*-*V* curve for relaxed myocardium, as expected. There are some deviations from these observations, but over normal ranges these behaviors are fairly reliable [Maughan et al., 1984; Sunagawa & Sagawa, 1982].

Based on these observations, a continuum of static *P*-*V* relationships can be visualized as spanning the space between the fully relaxed and the fully contracted curves, say, as a linear combination of the two. The active curve at any time could then be construed as a function of the activity level of the

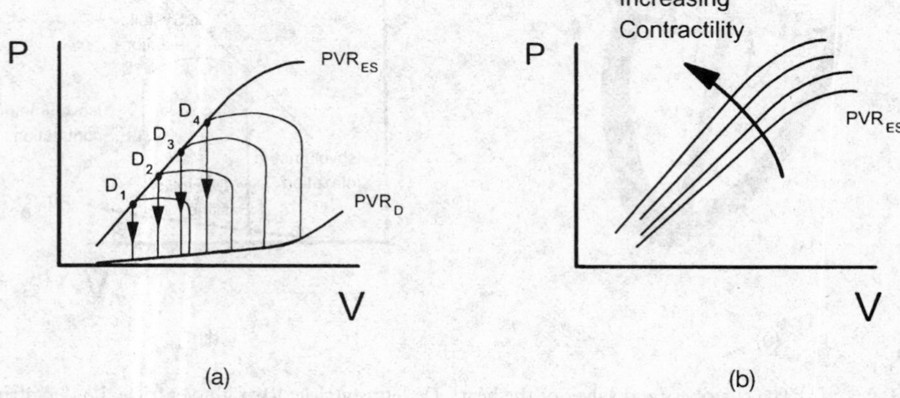

FIGURE 160.6 (*a*) Heart cycles with varying preload and afterload. (*b*) The end-systolic *P*-*V* curve with increasing contractility of the heart.

myocardium, α, which itself could be a function of time and t_{max}, and which would range from zero (inactive) to one (fully active):

$$PVR\left(\alpha\left(t,t_{max}\right)\right)=\alpha\left(t,t_{max}\right)\cdot PVR_{ES}+\left(1-\alpha\left(t,t_{max}\right)\right)\cdot PVR_{D} \tag{160.5}$$

Here, *PVR* (*p*ressure-*v*olume *r*elationship) is the active *P-V* curve, and PVR_{ES} and PVR_D are the end systolic and diastolic *P-V* curves, respectively. Using this approach, the PVR_{ES} curve would be parameterized in terms of the inotropic state of the myocardium, and t_{max} would be a function of inotropic state and heart rate.

Suga and Sagawa [1972] originally approximated the two curves, PVR_{ES} and PVR_D, with two straight lines that intersected in a common point on the volume axis (V_d, the unstressed, or dead, volume). This was a reasonable approximation, since, in their original experiments, the nonlinear portions of the *P-V* curves were not encountered. Based on this approach, the elastance *E* as a function of time becomes greatly simplified:

$$E\left(t\right)=\frac{P\left(t\right)}{V\left(t\right)-V_{d}} \tag{160.6}$$

With this formulation, $\alpha(t,t_{max})$ maps to $E(t)$ by the relation

$$E\left(t\right)=\alpha\left(t,t_{max}\right)\cdot E_{max}+\left(1-\alpha\left(t,t_{max}\right)\right)\cdot E_{min} \tag{160.7}$$

where E_{min} and E_{max} are the minimum and maximum elastances. E_{max} sets the inotropic state, and E_{min} sets the diastolic filling curve. A later modification allowed each curve to have its own dead volume [Sunagawa & Sagawa, 1982].

A common definition of $\alpha(t,t_{max})$ is a squared sine wave, time scaled and shifted to fit its first half-period into the systolic time interval, and zeroed elsewhere, e.g., see Martin and coworkers [1986]. Another common definition uses the first half-period of a sine wave, time-scaled and shifted to fit into the systolic interval. It is sometimes clipped and sometimes modified with a second harmonic to skew the waveform, e.g., see Rideout [1991]. Still others have approximated $\alpha(t,t_{max})$ as a square wave [Warner, 1959], a triangle wave [Katona et al., 1967; McLeod, 1966], a sum of charging and discharging exponentials [Sun & Chiaramida, 1992], and even as a sum of gaussian (bell-shaped) exponentials [Chung et al., 1994].

Pulsations can likewise be added to the other heart chambers, although the shape of the elastance curves are somewhat different [Sunagawa & Sagawa, 1982]. In the left ventricle, $E(t)$ appears more like a skewed sine wave while $V_d(t)$ is fairly constant during ejection; in the right ventricle, $E(t)$ appears more like a squared sine wave with a larger E_{max}, while $V_d(t)$ varies continuously throughout the systolic period; in the right atrium, $E(t)$ and $V_d(t)$ behave as in the right ventricle. A common simplification of this relationship was employed by Leaning and coworkers [1983]. They used the same function for each chamber's elastance (in this case a sine wave), but with a different E_{min}, E_{max}, and t_{max} for each, as well as a different time shift to contract them when appropriate, e.g., the atria prior to the ventricles. As is common, they also made V_d constant, though each chamber was provided with a different value.

Since the activation function is typically time-scaled to fit within the heart period, the effect of a changing heart rate, say, due to neural reflexes, can be explored. However, this requires that the heart period itself be partitioned into diastolic and systolic intervals. Various partitioning schemes exist. For the left ventricle, some fix the duration of systole [Grodins, 1959] or set it to a percentage of the heart period, with diastole taking up the rest of the period [Rideout, 1991]. A general clinic rule allocates one-third to systole and two-thirds to diastole. However, Beneken and DeWit [1967], after summarizing

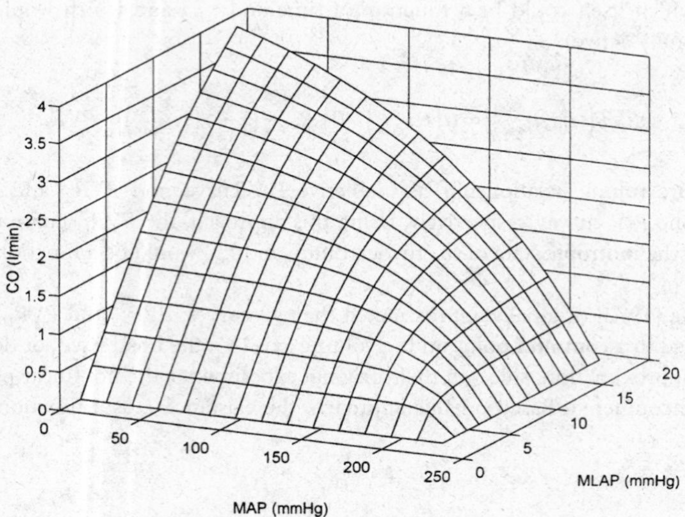

FIGURE 160.7 The Frank-Starling law of the heart; 3-D of Sagawa's [1967] left-ventricle equation for a 10-kg dog; CO, cardiac output; MAP, mean arterial pressure; MLAP, mean left arterial pressure.

several publications, suggest that ventricular systole is better approximated by one-fifth the heart period plus 0.16 s. Likewise, they determined the duration of atrial systole as 9/100 the heart period plus 0.1 s. These formulae have gained considerable popularity, e.g., see Leaning and coworkers [1983] and Rideout [1991].

In many drug-delivery applications, the time constants of interest are on the order of minutes to hours. In these situations, a pulsatile heart is not needed and can be traded advantageously for a mean flow model. Mean flow models are based either on the Frank-Starling law of the heart or on the pulsatile model above. In the Frank-Starling approach, a family of flow curves is constructed based on experimental observations of the heart under various conditions of preload and afterload (Fig. 160.7) [Sagawa, 1967]. These are then used to calculate stroke volume for each beat. In the other approach, the stroke volume is derived analytically from the equations of the pulsatile heart [Rideout, 1991]. When using the Frank-Starling law, attenuation factors are used to impose the effects of inotropic state and heart rate. Examples of nonpulsatile heat models include those by Sagawa [1967], Greenway [1982], Möller and coworkers [1983], and Tham and coworkers [1990]. The Guyton-Coleman model [Guyton et al., 1972, 1980] is also nonpulsatile, though it uses a very different approach (see below).

Some models of the uptake and distribution of slow-acting drugs do not explicitly use a heart, or, for that matter, a circulation. These fall under the category of purely compartmental models. In these models, the heart and circulation are subsumed into an assumption that each compartment in the model is uniformly mixed (See Chapter 159).

Models of Combined Heart and Circulation

Whole-body CV models couple a model of the heart to models of the vasculature. The complexity of the coupling depends on the particular need and can give rise to nonpulsatile left side (*left side* meaning the systemic circulation) only, nonpulsatile left and right side (*right side* meaning the pulmonary circulation), and pulsatile left and right side models [Rideout, 1991].

Left-side-only models are justified when the pharmaceutical agents of interest have little effect on the pulmonary circulation. In these cases, the right side is considered a follower circuit and hence can be eliminated to provide computational savings. Even if the agents of interest to affect the pulmonary circulation, an occamistic left-side-only model may still suffice. This approach may work for two reasons:

Either the model order is high enough that the right-side effects are inadvertently captured in the left-side model, or the drug effects are essentially the same on both sides. Keep in mind that these models have limited scope and hence should be used with care.

As an example of the modeling considerations for drug delivery, consider the following two drugs: propranolol, a beta blocker; and sodium nitroprusside (SNP), a vasodilator. Propranolol has a time constant on the order of 33 min, primarily affects the heart, and has only minor effects on the pulmonary circulation (although it may constrict the airways) [Ewy & Bressler, 1992; Gilman et al., 1993]. Hence, for propranolol, a nonpulsatile left-side-only model should be adequate for controller design. SNP, however, has a transport delay of 0.5 min and a time constant between 0.25 and 0.50 min. It affects the arterioles, the venous unstressed volume, and venous compliance. It also significantly decreases pulmonary wedge pressure [Ewy & Bressler, 1992; Gilman et al., 1993; Greenway, 1982]. Hence, for SNP, a nonpulsatile model is questionable, though it may be adequate. Furthermore, a two-sided model should be used, since SNP may have a significant effect on the pulmonary circulation. Greenway's [1982] two-sided nonpulsatile model seems to function adequately, although other investigators have switched to two-sided pulsatile models, e.g., see Yu et al., 1990.

Interestingly, Martin et al. [1986] may have finessed the field of nonpulsatile models by overcoming the speed limitations associated with pulsatile models. They did this by solving the pulsatile model equations semi-analytically, and were thus able to significantly reduce simulation time. Their model currently makes an attractive choice for studying and designing automated drug-delivery systems.

Despite the emphasis here on nonlinear CV-based modeling, probably the most used model for the design of blood pressure controllers is the model by Slate and Sheppard [1982]. It is based on a linear, first-order impulse response of the effects of nitroprusside on mean arterial blood pressure. The model includes a recirculation effect, as well as an occasionally observed nonlinear reflex (possibly due to the chemoreceptors). It has gained wide acceptance probably because of its simplicity. Nevertheless, it has severe limitations: It lumps the effects of the primary neural and humoral reflexes as well as the patient drug sensitivity into one gain; it does not account for the effect of changing cardiac output on drug transport; and it is not easily extended to include additional drugs or disease states.

Neural and Humoral Control

Once the heart and vasculature are linked, neural and hormonal controls need to be added. These include the baroreflex, the chemoreflex, the renin-angiotensin reflex, capillary fluid shift, autoregulation, stress relaxation, and renal–body-fluid balance (water intake and urine output).

Baroreceptors monitor the pressure in the carotid sinuses, the aortic arch, and other large systemic arteries and increase their firing rate when the pressure increases. Their response is nonlinear and depends on whether they are exposed to mean pressure only, pulsatile pressure only, or a combination of both. Katona and colleagues [1967] developed a model of baroreceptor feedback that has become the basis for many CV neural control models. The output of the baroreceptor model is often passed through a low pass filter representing the CNS and then mapped back to changes in heart rate, contractility, vascular resistances, and vascular unstressed volumes through the sympathetic and parasympathetic nervous systems (Fig. 160.8), e.g., see Yu and coworkers [1990].

In addition to baroreceptors, chemoreceptors may also mediate a strong CV response, especially when systemic arterial pressure falls below 80 mmHg [Guyton et al., 1972; see also Dampney, 1994]. This reflex, through the CNS, increases cardiac activity and peripheral resistance. Strangely, few models include this effect, although Slate and Sheppard [1982] may have inadvertently included it in their simplified black box model.

The renin-angiotensin reflex is mediated hormonally through the kidneys. When pressure falls below normal, renin is secreted into the blood stream by the kidneys, which then causes the release of free angiotensin. Angiotensin causes marked vasoconstriction and thus increases peripheral resistance and arterial pressure. This reflex is often included in models because it plays an important role in many forms of hypertension and heart failure.

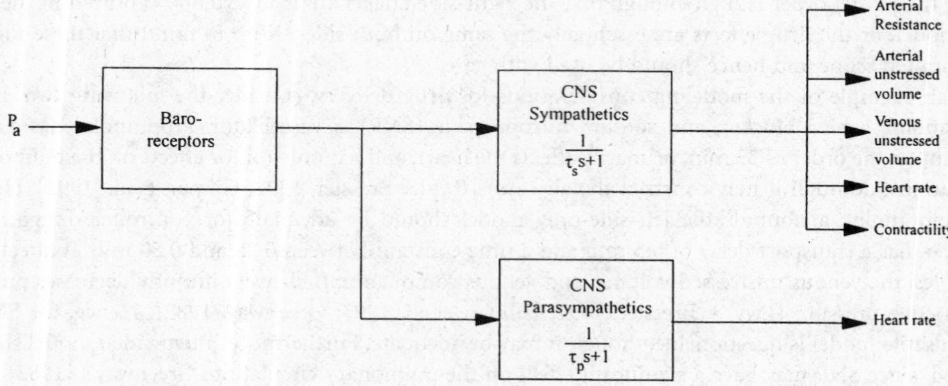

FIGURE 160.8 Baroreceptor firing rate is passed through a low-pass filter representing the CNS and then mapped back to changes in heart rate, contractility, vascular resistances, and vascular unstressed volumes.

The other CV control reflexes are only sometimes included, although these and others are elucidated and modeled by Guyton and his coworkers [1972, 1980]. In his elaborate model, there are five main empirically derived physiologic function blocks with many subcomponents. The model has nearly 400 parameters and is quite remarkable in its scope. Given current computer technology, however, it is somewhat cumbersome for drug delivery applications, although it promises to have great utility in the future.

Combined CV and Pharmacologic Models

For pharmacologic control studies, CV models readily lend themselves to the calculation of drug uptake and distribution to the various parts of the body (pharmacokinetics). Once the transport model has been built, models of drug action (pharmacodynamics) then need to be linked and related back to the CV model. Other models must also be linked if necessary. For example, models of inhalational anesthetic uptake, distribution, and effect on the respiratory and CV systems require not only a CV and pharmacologic model but also a lung model. Combined models such as these are known as multiple models, a term coined by Rideout's group [Beneken & Rideout, 1968; Rideout, 1991]. Again, the time course of the agents and their effects on the heart and circulation determine whether a pulsatile or nonpulsatile model is needed, as well as the requisite level of vascular detail.

To calculate drug uptake and transport, mass balance equations can be set up for each segment and compartment in the CV model. Additional compartments, such as a tissue or an effects compartment, can also be added (see Chapter 159). For example, given a particular compartment (or segment), the mass of drug in the compartment is calculated as follows:

$$\frac{dq_i}{dt} = f_{in,i}\, c_{i-1} - f_{out,i}\, c_i \tag{160.8}$$

where i identifies the compartment, q is the mass of drug, c is its concentration, f_{in} is the rate of blood flow into the compartment (normally equal to the flow out of the previous compartment), and f_{out} is the rate of blood flow out of the compartment. The instantaneous concentration in any compartment can then be calculated based on drug mass and total volume of the compartment (stressed plus unstressed volumes);

$$c_i = \frac{q_i}{V_i} \tag{160.9}$$

Once the concentration is known at the effector site, a pharmacodynamic model is used to describe its action. Saturation may need to be built into the effect, as well as a possible threshold concentration. Hill's sigmoidal equation can accommodate both:

$$EFF = \frac{c_e^n}{c_{50}^n + c_e^n} \tag{160.10}$$

where EFF is the normalized pharmacologic effect ($0 \leq \text{EFF} \leq 1$), c_e is the concentration at the effector site, c_{50} is the half-matrix concentration, and n parameterizes the steepness of the sigmoid. Hill's equation becomes the Michaelis-Menten equation when $n = 1$.

Once calculated, the drug effect can be used to modify the parameters in the CV model, much like the neural and humoral reflexes. Resistances, compliances (including the unstressed volumes) of each segment, neural and humoral feedback gains, heart rate, and contractility are all typically modified by drugs. For example, sodium nitroprusside causes vasodilation and blood pooling; therefore it would be made to primarily increase the compliance and unstressed volume of the veins, as well as to decrease arteriolar (peripheral) resistance [Greenway, 1982; Yu et al., 1990].

160.4 Conclusions

For designing controllers to automate cardiovascular therapeutics, the nonlinear, cardiovascular-linked pharmacologic models are well worth the added effort and time needed to construct and interface them to the design platform. Not only can they help prove initial feasibility, but they can also speed up the overall design process. For example, they can be used for interactive controller design; they can be used to design improved animal and human experiments, which can potentially reduce the total number of experiments; and they can be used to simulate human conditions that are not easily produced in animals.

The reduced compartmental and combined segmental and transmission line models are probably the most practical model form at this time given current computer technology. However, as computer technology grows, we can expect to see increasingly detailed pharmacologic models that include additional cardiovascular features such as pulse wave propagation and reflection as well as multiple short- and long-term neurohumoral feedback loops.

Acknowledgments

The author is indebted to Mr. F. Casas and Mr. S. Kumar for help in collecting the references; to Mr. F. Casas and Ms. O. Huynh for help in preparing the figures; and to Dr. S. E. Rittgers for his comments on the text.

References

Acierno LJ. 1994. The History of Cardiology, Pearl River, NY, Parthenon.

Aperia A. 1940. Hemodynamical studies. Scand Arch Physiol Supply 83:1.

Beneken JEW. 1963. Investigation on the regulatory system of the blood circulation. In A Noordergraaf, GN Jager, N Westerhof (eds), Circulatory Analog Computers, pp 16–28, Amsterdam, North-Holland.

Beneken JEW. 1972. Some computer models in cardiovascular research. In DH Bergel (ed), Cardiovascular Fluid Dynamics, vol 1, pp 173–223, New York, Academic Press.

Beneken JEW, DeWit B. 1967. A physical approach to hemodynamic aspects of the human cardiovascular system. In EB Reeve, AC Guyton (eds), Physical Bases of Circulatory Transport, pp 1–45, Philadelphia, W.B. Saunders.

Beneken JEW, Rideout VC. 1968. The use of multiple models in cardiovascular system studies: Transport and perturbation methods. IEEE Trans BME 15(4):281.

Blackstone EH, Gupta AK, Rideout VC. 1976. Cardiovascular simulation study of infants with transposition of the great arteries after surgical correction. In L Dekker (ed), Simulation of Systems, pp 599–608, Amsterdam, North-Holland.

Carson ER, Cobelli C, Finkelstein L. 1983. The Mathematical Modeling of Metabolic and Endocrine Systems, New York, John Wiley & Sons.

Chung DC, et al. 1994. A mathematical model of the canine circulation. In BW Patterson (ed), Modeling and Control in Biomedical Systems: Proceedings of the IFAC Symposium, pp 109–112, Galveston Tex.

Dampney RAL. 1994. Functional organization of central pathways regulating the cardiovascular system. Physiol Rev 74(2):323.

Dick DE, Rideout VC. 1965. Analog simulation of left heart and arterial dynamics. Proc 18th ACEMB, p 78, Philadelphia.

Ewy GA, Bressler R. 1992. Cardiovascular Drugs and the Management of Heart Disease, 2nd ed, New York, Raven Press.

Frank O. 1899. Die Grundform des arteriellen Pulses. Z Biol 37:483.

Fung YC. 1984. Biodynamics: Circulation, New York, Springer-Verlag.

Geppert L. 1994. Faults & failures. IEEE Spectrum 31(8):17.

Gilman AG, Rall TW, Nies AS, et al. 1993. Goodman and Gilman's The Pharmacological Basis of Therapeutics, 8th ed, New York, McGraw-Hill.

Greenway CV. 1982. Mechanisms and quantitative assessment of drug effects on cardiac output with a new model of the circulation. Pharm Rev 33(4):213.

Grodins FS. 1959. Integrative cardiovascular physiology: a mathematical synthesis of cardiac and blood vessel hemodynamics. Q Rev Biol 34:93.

Gustafsson K. 1993. Stepsize selection in implicit Runge-Kutta methods viewed as a control problem. Proc. 12th IFAC World Cong 5:137.

Guyton AC. 1980. Arterial Pressure and Hypertension, Philadelphia, W.B. Saunders.

Guyton AC, Coleman TG, Cowley AW, et al. 1972. Systems analysis of arterial pressure regulation and hypertension. Ann Biomed Eng 1:254.

Hales S. 1733. Statical Essays: Containing Haemastaticks, vol 2, London, Innys and Manby (Pfizer Laboratories, 1981).

Isaka S, Sebald AV. 1993. Control strategies for arterial blood pressure regulation. IEEE Trans BME. 40(4):353–363.

Katona PG. 1988. Closed loop control of physiological variables. Proc 1st IFAC Symp Modelling Control Biomed Sys, Venice.

Katona PG, Barnett GO, Jackson WD. 1967. Computer simulation of the blood pressure control of the heart period. In P. Kezdi (ed), Baroreceptors and Hypertension, pp 191–199, Oxford, Pergamon Press.

Kono A, Maughan WL, Sunagawa K, et al. 1984. The use of left ventricular end-ejection pressure and peak pressure in the estimation of the end-systolic pressure-volume relationship. Circulation 70(6):1057.

Leaning MS, Pullen HE, Carson ER, et al. 1983. Modelling a complex biological system: the human cardiovascular system—1. Methodology and model description. Trans Inst Meas Contr 5(2):71.

Li JK-J. 1987. Arterial System Dynamics, New York, New York University Press.

Linkens DA, Hacisalihzade SS. 1990. Computer control systems and pharmacological drug administration: A survey. J Med Eng Tech 14(2):41.

Martin JF, Schneider AM, Mandel JE, et al. 1986. A new cardiovascular model for real-time applications. Trans Soc Comp Sim 3(1):31.

Maughan WL, Sunagawa K, Burkhoff D, et al. 1984. Effect of arterial impedance changes on the end-systolic pressure-volume relation. Circ Res 54(5):595.

McDonald DA. 1974. Blood Flow in Arteries, 2nd ed, Baltimore, Williams & Wilkins.

McLeod J. 1966. PHYSBE … a physiological simulation benchmark experiment. Simulation 7(6):324.

Melchior FM, Srinivasan RS, Charles JB. 1992. Mathematical modeling of human cardiovascular system for simulation of orthostatic response. Am J Physiol 262(Heart Circ Physiol 31):H1920.

Middleman S. 1972. Transport Phenomena in the Cardiovascular System, New York, John Wiley & Sons.

Milnor WR. 1989. Hemodynamics, 2nd ed, Baltimore, Williams & Wilkins.

Möller D, Popović D, Thiele G. 1983. Modeling, Simulation and Parameter-Estimation of the Human Cardiovascular System, Friedr, Braunschweig, Vieweg & Sohn.

Noordergraaf A. 1978. Circulatory System Dynamics, New York, Academic Press.

O'Rourke MF. 1982. Arterial Function in Health and Disease. New York, Churchill Livingstone.

O'Rourke MF. 1993. Hypertension and the conduit and cushioning functions of the arterial tree. In ME Safar, MF O'Rourke (eds), The Arterial System in Hypertension, pp 27–37, the Netherlands, Kluwer Academic Pub.

Purmutt S, Bromberger-Barnea B, Bane HN. 1962. Alveolar pressure, pulmonary venous pressure, and the vascular waterfall. Med Thorac 19:239.

Rideout VC. 1991. Mathematical and Computer Modeling of Physiological Systems, Englewood Cliffs, NJ, Prentice-Hall (now distributed by Medical Physics Pub., Madison, WI).

Sagawa K. 1967. Analysis of the ventricular pumping capacity as a function of input and output pressure loads. In EB Reeve, AC Guyton (eds), Physical Bases of Circulatory Transport, pp 141–149, Philadelphia, W.B. Saunders.

Sagawa K. 1973. Comparative models of overall circulatory mechanics. In JHU Brown, JF Dickson III (eds), Advances in Biomedical Engineering, vol 3, pp 1–95, New York, Academic Press.

Slate JB, Sheppard LC. 1982. Automatic control of blood pressure by drug infusion. IEE Proc., Pt. A 129(9):639.

Snyder MF, Rideout VC. 1969. Computer simulation studies of the venous circulation. IEEE Trans BME 16(4):325.

Suga H, Sagawa K. 1972. Mathematical interrelationship between instantaneous ventricular pressure-volume ratio and myocardial force-velocity relation. Ann Biomed Eng 1:160.

Suga H, Sagawa K. 1974. Instantaneous pressure volume relationships and their ratio in the excised, supported canine left ventricle. Circ Res 35:117.

Sun Y, Chiaramida S. 1992. Simulation of hemodynamics and regulatory mechanisms in the cardiovascular system based on a nonlinear and time-varying model. Simulation 59(1):28.

Sunagawa K, Sagawa K. 1982. Models of ventricular contraction based on time-varying elastance. Crit Rev Biomed Eng 9:193.

Tham RQY, Sasse FJ, Rideout VC. 1990. Large-scale multiple model for the simulation of anesthesia. In DPF Moller (ed), Advanced Simulation in Biomedicine, pp 173–195, New York, Springer Verlag.

Tsitlik JE, Halperin HR, Popel AS, et al. 1992. Modeling the circulation with three-terminal electrical networks containing special nonlinear capacitors. Ann Biomed Eng 20:595.

Westenskow DR. 1986. Automating patient care with closed-loop control. MD Comput 3(2):14.

Westerhof N, Noordergraaf A. 1969. Reduced models of the systemic arteries. Proc 8th Int Conf Med Eng, Chicago.

Yu C, Roy RJ, Kaufman H. 1990. A circulatory model for combined nitroprusside-dopamine therapy in acute heart failure. Med Prog Tech 16:77.

Further Information

The text by Rideout [1991] has a particularly good tutorial on cardiovascular modeling and even includes ACSL source code. McDonald's [1974] and Noordergraaf's [1978] texts are classics, thoroughly reviewing cardiovascular physiology and the physical principles useful in modeling it. More recent texts on the cardiovascular system with a bioengineering slant include those by Milnor [1989], Li [1987], and Fung [1984]. Also, Guyton's 1980 text provides a detailed review of his model and its application for the analysis of hypertension and is well worth reading. Along this same line, Safar and O'Rourke's [1993] text, as well as O'Rourke's earlier work [1982], are also worth reading and provide many insights into hypertension and cardiovascular modeling. Melchior, Srinivasan, and Charles [1992] also present an overview of cardiovascular modeling, though with an emphasis on orthostatic response.

For historical interest, *The History of Cardiology* by Acierno [1994] provides fascinating reading, going back beyond even the early Greeks. In more recent history, a review of early CV models (early 1900s up to 1940) can be found in Aperia's monograph [1940], and a review of the post–World War II era models (late 1950s to early 1970s) that set the tone for almost all modern CV models can be found in Sagawa [1973].

For those interested in pharmacologic modeling, along with examples using the CV system, Carson, Cobelli, and Finkelstein's 1983 text is a good source. Middleman's 1972 text specifically targets in transport in the CV system, but it is somewhat dated. The current understanding of the anatomic and functional organization of the neurohumoral regulation pathways can be found in Dampney [1994].

For a survey of blood pressure controllers, the reader is referred to the recent article by Isaka and Sebald [1993]. Linkens and Hacisalihzade [1990], Katona [1988], and Westenskow [1986] also survey blood pressure controllers as well as other CV control applications.

161

Respiratory Models and Control

Chi-Sang Poon
Massachusetts Institute of Technology

The respiratory system is a complex neurodynamical system. It exhibits many interesting characteristics that are akin to other physiologic control systems. However, the respiratory system is much more amenable to modeling analysis than many other systems for two reasons. First, the respiratory system is dedicated to a highly specific physiologic function, namely, the exchange of O_2 and CO_2 through the motor act of breathing. This physiologic function is readily distinguishable from extraneous disturbances arising from behavioral and other functions of the respiratory muscles. Second, the respiratory control system is structurally well organized, with well-defined afferent and efferent neural pathways, peripheral controlled processes, and a central controller. The functional and structural specificity of the respiratory system—and the diverse neurodynamic behaviors it represents—make it an ideal model system to illustrate the basic principles of physiologic control systems in general.

Like any closed-loop system, the behavior of the respiratory control system is defined by the continual interaction of the controller and the peripheral processes being controlled. The latter include the respiratory mechanical system and the pulmonary gas exchange process. These peripheral processes have been extensively studied, and their quantitative relationships have been described in detail in previous reviews. Less well understood is the behavior of the respiratory controller and the way in which it processes afferent inputs. A confounding factor is that the controller may manifest itself in many different ways, depending on the modeling and experimental approaches being taken. Traditionally, the respiratory control system has been modeled as a closed-loop *feedback/feedforward regulator* whereby homeostasis of arterial blood gas and pH is maintained. Alternatively, the respiratory controller may be viewed as a *central pattern generator* in which rhythmic respiratory activity is produced in response to phasic afferent feedback. Finally, there is increasing evidence that the respiratory controller may function as an *adaptive self-tuning regulator* which optimizes breathing pattern and ventilation according to certain performance measure.

0-8493-0462-8/00/$0.00+$.50

Each modeling approach reveals a separate "law" of a multifaceted controller. However, these control laws must be somehow related to one another because they are governed by the same neural network that forms the respiratory controller. It is therefore instructive to examine not only how the various models work but also how they might be fit together to encompass the myriad of response behaviors of the respiratory control system.

161.1 Structure of the Respiratory Control System

The respiratory control system is a nonlinear, multioutput, delayed-feedback dynamic system which is constantly being perturbed by unknown physiologic and pathologic disturbances (Fig. 161.1). Rhythmic respiratory activity is produced by a respiratory central pattern generator (RCPG) which consists of a network of neuronal clusters in the medulla oblongata and pons areas of the brain stem. The RCPG forms the kernel of the respiratory controller. The control problem is defined by the characteristics of the controlled processes (plants) and the control objective.

Chemical Plant

Pulmonary Gas Exchange

The chemical plant is propelled by the ventilation of the lung, $\dot{V}_E$ which determines the alveolar PCO_2 and PO_2 according to the following mass-balance equations:

$$V_L CO_2 \cdot \frac{d}{dt} P_A CO_2 = \dot{V}_E \left(1 - \frac{V_D}{V_T}\right)\left(P_I CO_2 - P_A CO_2\right) + 863\ \dot{V}CO_2 \qquad (161.1)$$

$$V_L O_2 \cdot \frac{d}{dt} P_A O_2 = \dot{V}_E \left(1 - \frac{V_D}{V_T}\right)\left(P_I O_2 - P_A O_2\right) - 863\ \dot{V}O_2 \qquad (161.2)$$

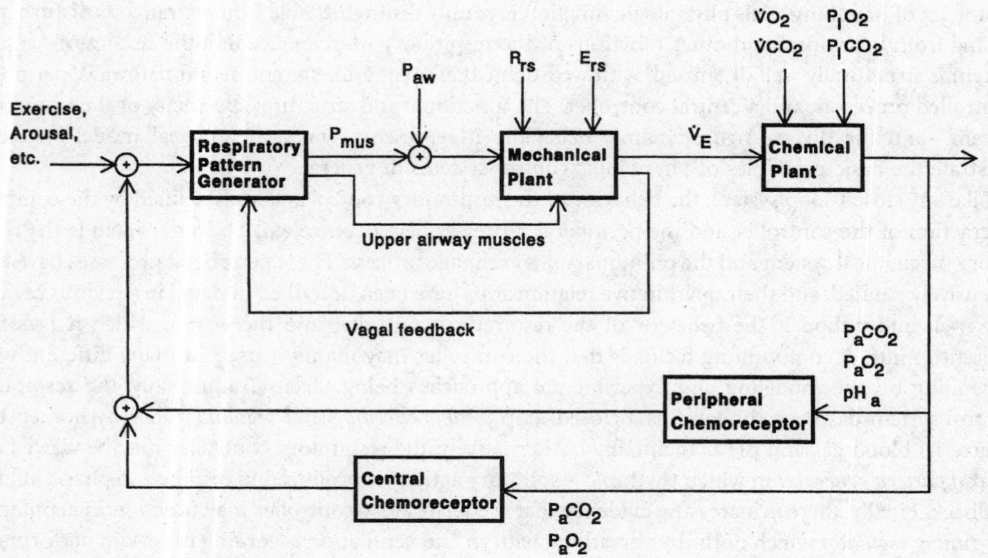

FIGURE 161.1 Block diagram of the respiratory control system.

The input-output relationships of the chemical plants are subject to several endogenous and exogenous disturbances. For example, increases in the metabolic production of CO_2 during muscular exercise, PCO_2 in the inspired air, and respiratory dead space during rebreathing—all contribute to an increase in CO_2 load to the lung. The added CO_2 is eliminated by an increase in pulmonary ventilation, but the effectiveness of the control action in restoring P_ACO_2 varies considerably with the type of disturbance. The input-output sensitivity $S_{pco2} \neq \delta P_A\ CO_2/\delta \dot{V}_E$ of the plant is highest with metabolic CO_2 load and lowest with inhaled CO_2 load, whereas the plant dynamics is slowest with dead space load.

In pulmonary disease the efficiency of the plant in removing CO_2 is further decreased because of pulmonary ventilation/perfusion maldistribution [Poon, 1987*a*]. Another source of disturbance is metabolic acidosis and alkalosis which alter arterial blood gas tensions and pH through acid-base buffering in blood. All these disturbances interact nonlinearly with the plant and are not directly sensed. How does the controller cope with such variety of disturbances effectively if its only means is to alter $\dot{V}_E$? Not surprisingly, the controller responds quite differently to the various forms of CO_2 disturbance and their combinations [Juratsch et al., 1982; Oren et al., 1981; Poon & Greene, 1985; Poon, 1989*a*, 1989*b*, 1992*b*; Sidney & Poon, 1995].

Chemosensory Feedback

The peripheral (arterial) chemoreceptors response to changes in P_aCO_2 may be modeled by linear first-order dynamics [Bellville et al., 1979]

$$\tau_p \cdot \frac{d}{dt} A_p = -A_p + G_p\left[P_aCO_2\left(t - T_p\right) - I_p\right] \tag{161.3}$$

where τ_p, G_p, I_p, are the time constant, gain, threshold of the peripheral chemoreceptor and T_p is the lung-to-chemoreceptor transit delay of blood flow. Similarly, the central (intracranial) chemoreceptor is responsive to brain tissue PCO_2

$$A_c = G_c\left\{P_bCO_2 - I_c\right\} \tag{161.4}$$

where the brain tissue PCO_2 is given by

$$V_bCO_2 \cdot \frac{d}{dt} P_bCO_2 = \dot{Q}_b\left\{P_aCO_2\left(t - T_c\right) - P_bCO_2\right\} + \frac{\dot{M}_bCO_2}{KCO_2} \tag{161.5}$$

and the cerebral blood flow [Vis & Folgering, 1980] is given by

$$\dot{Q}_b\left(t\right) = \dot{Q}_0 + \Delta\dot{Q}_b\left(t\right) \tag{161.6}$$

$$\tau_b \cdot \frac{d}{dt} \Delta\dot{Q}_b = -\Delta\dot{Q}_b + G_b \cdot P_aCO_2\left(t - T_c\right) \tag{161.7}$$

The central chemoreceptor normally has a greater sensitivity than the peripheral chemoreceptors (which may be silenced by hyperoxia). Their combined effects are presumably additive at low to moderate stimulation levels [Heeringa et al., 1979] but may become progressively saturated at higher levels [Eldridge et al., 1981]. The peripheral chemoreceptors have a shorter response time constant and delay than the central chemoreceptor, presumably due to their proximity to the lung and arterial blood gases. They are generally thought to be more responsive to rapid changes in blood chemistry; although the central chemoreceptor may also contribute substantially to the transient respiratory response to breath-to-breath fluctuations in chemical input during eupneic breathing in wakefulness [Bruce et al., 1992].

The dynamic response of the peripheral chemoreceptors to P_aO_2 may be modeled in a similar fashion but the steady-state response is hyperbolic [Cunningham et al., 1986]. Furthermore, hypoxia and hypercapnia have a multiplicative effect on carotid chemoreceptor discharge [Fitzgerald & Lahiri, 1986], although the effect of changes in pH is presumably additive [Cunningham et al., 1986]. Prolonged hypoxia has a depressant effect on central respiratory neurons [Neubauer et al., 1990] which may offset the stimulatory effect of peripheral chemosensitivity.

Inputs from the peripheral chemoreceptors are temporally gated throughout the respiratory cycle. Stimuli are effective only if they are delivered during the second half of the inspiratory or expiratory cycle [Eldridge & Millhorn, 1986; Hildebrandt, 1977; Teppema et al., 1985]. The effect of such stimuli on medullary respiratory neurons is excitatory during the inspiratory phase and inhibitory during the expiratory phase [Lipski & Voss, 1990]. The gating effect may be modeled by a gated time function

$$A_{epg} = gate\left\{A_p\left(t\right)\right\} \tag{161.8}$$

where gate $\{\cdot\}$ is a windowing function which reflects the phasic activity of peripheral chemoreceptor inputs.

The intrinsic transient delay and nonlinearity in the transduction of chemical signals are undesirable from an automatic control perspective. On the other hand, they may represent neural preprocessing in the sensory nervous system. Such transformations in the feedback path may have important bearing on the controller's ability to achieve the control objective.

Mechanical Plant

Respiratory Mechanics

Pulmonary ventilation results from tidal expansion and relaxation of the lung. The equation of motion is:

$$R_{rs}\cdot\frac{dV}{dt} = -E_{rs}\cdot V + P_{aw}\left(t\right) - P_{mus}\left(t\right) \tag{161.9}$$

Both R_{rs} and E_{rs} may be nonlinear. The mechanical load of the respiratory muscles is increased by airway constriction or lung restriction in pulmonary disease, or by the imposition of external resistance and elastance which cause a back pressure P_{aw} at the airway opening. Increases in ventilatory load may elicit a compensatory response in P_{mus} so that $\dot{V}_E$ is largely restored [Poon 1987*b*, 1989*a*, 1989*b*] except under severe ventilatory loading. Reverse compensatory responses are observed during ventilatory unloading [Poon et al., 1987*a*,*b*; Ward et al., 1982].

Respiratory Muscles

The mechanical plant is propelled by the respiratory muscles which serve as a mechanical pump. Normally, P_{mus} is sustained by the inspiratory muscles (principally the diaphragm and, to a lesser extent, the external intercostal and parasternal muscles). During hyperpnea or with increased expiratory load the expiratory muscles may be recruited [Marroquin, 1991], as are accessory muscles which may contribute significantly to the generation of P_{mus} in paraplegics.

Paradoxically, the respiratory muscles may also act to reduce ventilation. During quiet breathing some postinspiratory inspiratory activity may persist during the early expiratory phase, retarding lung emptying [Poon et al., 1987*b*]. In disease states inspiratory activity may also be recruited during the expiratory phase to prevent lung collapse [Stark et al., 1987] or overinflation. The upper airways abductor muscles may be activated by the controller or higher brain centers to increase R_{rs} in disease states [Stark et al., 1987] or in sleep [Phillipson & Bowes, 1986]. In awake states, respiration may be interrupted by behavioral, emotional, or defensive activities which compete for the respiratory apparatus under cortical or somatic sensory commands.

The mechanical efficiency of the respiratory muscles is limited by the force-velocity and force-length relationships which may be modeled as linear "active" resistance and elastance, respectively [Milic-Emili & Zin, 1986]. Neuromechanical efficiency may be impaired in respiratory muscle fatigue or muscle weakness, resulting in diminished P_{mus} for any given neural drive.

Thus the respiratory pump is equipped with various types of actuators to regulate V_E in the face of many exogenous and endogenous mechanical disturbances. Such disturbances are detected by mechanoafferents originating in the lungs, thorax, and airways.

Mechanosensory Feedback

Among the various types of mechanoafferents, vagal slowly adapting pulmonary stretch receptors (SAR) input exerts the greatest influence on respiration by inhibiting inspiratory activity and facilitating expiratory activity (Hering-Breuer reflex). Vagal SAR volume feedback is crucial for the compensatory response of the RCPG to mechanical disturbances.

The SAR characteristic is liner at low lung volumes but may exhibit saturation nonlinearity at elevated volumes. Also, at any given lung volume SAR discharge increases with increasing rate of lung inflation [Pack et al., 1986]. To a first approximation vagal volume feedback may be modeled by

$$vag(t) = sat\{a_0 + a_1\, V(t) + a_2\, \dot{V}(t)\} \tag{161.10}$$

where sat{·} is a saturation function; a_0 denotes tonic vagal activity [Phillipson, 1974]; and a_1, a_2 are sensitivity parameters.

Control Object

It is generally assumed that the prime function of respiration is to meet the metabolic demand by the exchange of O_2 and CO_2. The control problem is to accomplish this objective within some prescribed tolerance, subject to the physical as well as environmental constraints of the respiratory apparatus. Various models of the controller have been proposed as possible solutions to the control problem [Petersen, 1981], each representing a separate control strategy and degree of modeling abstraction. In what follows we provide a synopsis of the various modeling approaches and their physiologic significance.

161.2 Chemoreflex Models

The simplest controller model is a proportional controller [Cunningham et al., 1986; Grodins & Yamashiro, 1978]. In this approach attention is focused on the control of the chemical plant; the RCPG and the mechanical plant are lumped together to form a constant-gain controller. The control signal is taken to be $\dot{V}_E$, and the effects of the breathing pattern and vagal feedback are often neglected. Thus the controller is assumed to be a fixed relay station that regulates $\dot{V}_E$ via the chemoreflex loop. These simplification allow quantitative closed-loop analyses of the chemoreflex system.

Feedback Control

In the chemoreflex model the respiratory control system acts like a chemostat. This is consistent with the well-known experimental observation that ventilation increases with increasing chemical stimulation. The steady-state ventilatory response to CO_2 inhalation is given by:

$$\dot{V}_E = \alpha\left(PaCO_2 - \beta\right) \tag{161.11}$$

This empirical relationship is in agreement with the chemoreflex model which assumes that ventilatory output is proportional to the sum of chemosensory feedback:

$$\dot{V}_E = G_0\left(A_c + A_p\right) \tag{161.12}$$

Equation (161.11) follows from the steady-state solutions to the model Eqs. (161.3)–(161.7) and (161.12), provided that the effect of changes in cerebral blood flow may be neglected. The effects of hypoxia and hypoxic-hypercapnic interaction may be modeled in a similar fashion by a hyperbolic relation [Cunningham et al., 1986].

A comprehensive model of the dynamical chemoreflex system is due to the classic work of Grodins and coworkers [1967]. The dynamical response to CO_2 is given by the transient solutions of the system equations. Eqs. (161.3)–(161.7) and (161.12). Under closed-loop conditions system dynamics is also influenced by the nonlinear dynamical response of the chemical plant, Eq. (161.1), as well as CO_2 uptake in body tissues. Therefore, measurement of the CO_2 response curve, Eq. (161.8), is often a slow and tedious procedure. Open-loop conditions may be achieved by dynamic forcing of end-tidal PCO_2 with servo-control of P_tCO_2 which obviates the nonlinearity and slow equilibration of the lung and other body tissue stores. This technique has been used to experimentally estimate the model parameters in normal and peripheral-chemoreceptor denervated subjects by means of system identification techniques [Bellville et al., 1979]. Another open-loop technique is the CO_2 rebreathing procedure [Read, 1967] which causes a metabolically induced ramp increase in P_aCO_2 and, hence, a similar increase in $\dot{V}_F$ by way of chemoreflex. The rebreathing technique is much simpler to use than dynamic end-tidal PCO_2 forcing and is therefore suitable for clinical applications. However, this technique may not accurately determine the steady-state CO_2 response, presumably because of the variability in cerebral blood flow induced by changing P_aCO_2. More reliable techniques using pulse-step [Poon & Olson, 1985] or step-ramp [Dahan 1990] P_tCO_2 forcings have been proposed for rapid determination of the steady-state CO_2 response with minimal experimental requirements.

Stability of the chemoreflex model may be studied by linearization of the system about some nominal state. Instability may occur if the loop gain and phase shift exceed unity and 180°, respectively. This may explain the periodic breathing phenomena found in hypoxia or high altitude where peripheral chemoreceptor gain is increased and congestive heart failure in which transit delay is prolonged [Khoo et al., 1982]. However, chemical instability alone does not account for the periodic episodes in obstructive sleep apnea which may also be influenced by fluctuation in arousal state [Khoo et al., 1991]. A technique for estimation of chemoreflex loop gain using pseudorandom binary CO_2 forcing has been proposed [Ghazanshahi & Khoo, 1997].

Feedforward Control

The chemoreflex model provides a satisfactory explanation for the chemical regulation of ventilation as well as respiratory instability. However, it fails to explain a fundamental aspect of ventilatory control experienced by everyone in everyday life: the increase in ventilation during muscular exercise. Typically, $\dot{V}_E$ increases in direct proportion to the metabolic demand ($\dot{V}CO_2$, $\dot{V}O_2$) such that the outputs of the chemical plant, Eqs. (161.1) and (161.2), are well regulated at constant levels from rest to exercise. As a result, homeostasis of arterial blood chemistry is closely maintained over a wide range of work rates. The dilemma is: If increases in metabolic rate are not accompanied by corresponding increases in chemical feedback, then what causes exercise hyperpnea?

One possible explanation of exercise hyperpnea is the so-called set-point hypothesis [Defares, 1964; Oren et al., 1981] which stipulates that P_aCO_2 and P_aO_2 are regulated at constant levels by the chemoreflex controller. However, a set point is not evident during hypercapnic and hypoxic challenges where the homeostasis of P_aCO_2 and P_aO_2 is readily abolished. Furthermore, to establish a set point the controller gain must be exceedingly high, but this is not found experimentally in the hypercapnic and hypoxic ventilatory sensitivities. Finally, from a control systems perspective a high gain controller is undesirable because it could drive the system into saturation or instability.

Another possible explanation of exercise hyperpnea is that the proportional controller may be driven by two sets of inputs: a chemical feedback component via the chemoreflex loop and a feedforward component induced by some exercise stimulus [Grodins & Yamashiro, 1978]. The "feedforward-feedback control" hypothesis offers a simple remedy of the chemoreflex model, but its validity can be verified only if the postulated "exercise stimulus" is identified. Unfortunately, although many such signals have been proposed as possible candidates [Wasserman et al., 1986], none of them has so far been unequivocally shown to represent a sufficient stimulus for exercise.

Among the variously proposed mechanisms of exercise hyperpnea, the "PCO_2 oscillation" hypothesis of Yamamoto [1962] has received widespread attention. According to this hypothesis, the controller may be responsive not only to the mean value of chemical feedback but also to its oscillatory waveform which is induced by the tidal rhythm of respiration. This hypothesis is supported by the experimental finding that alterations of the temporal relationship of the P_ACO_2 waveform could profoundly modulate the exercise hyperpnea response [Poon, 1992*b*].

Regardless of the origin (or even existence) of the exercise stimulus, there is evidence that exercise and chemical signals are processed by the controller in a multiplicative fashion [Poon & Greene, 1985; Poon, 1989*a*, 1989*b*]. The input-output relationship of the controller may therefore be written as

$$\dot{V}_E = G_0\left(A_c + A_p\right)A_{ex} + G_{ex}A_{ex} \tag{161.13}$$

Note that $\dot{V}_E$ vanishes when A_{ex} is reduced to zero, in agreement with the finding of Phillipson and colleagues [1981].

161.3 Models of Respiratory Central Pattern Generator

Although the chemoreflex model is useful in studying the closed-loop control of ventilation and the chemical plant, it is lacking as to the control of breathing pattern and the mechanical plant. Models of the RCPG are often studied under open-loop conditions with chemical feedbacks being held constant while vagal volume feedback is experimentally manipulated. Modeling and analysis of the RCPG in closed-loop form are difficult because the input-output relationships of the RCPG and the mechanism of pattern generation are poorly understood. Several modeling approaches have been used to characterize its behavior under specific conditions.

Phase-Switching Models

The breathing cycle consists of two distinct phases: inspiration and expiration. It is therefore natural to model each phase separately and then combine them with a model of phase switching (for review, see Younes and Remmers [1981]). Each phase is controlled by a separate "center" which generates ramplike central inspiratory and expiratory activity, respectively. The resulting lung expansion and collapse are sensed by SAR, which causes the RCPG to switch from inspiration phase to expiration phase, or vice versa, when the volume-time relationship reaches some inspiratory or expiratory off-switch threshold. The off-switch mechanisms are dependent on both the threshold values and the corresponding volume-time histories. Also, inspiratory and expiratory off-switches may be mechanically linked, as suggested by the correlation between inspiratory and expiratory durations in consecutive breaths. Both the rate of rise of inspiratory activity and the inspiratory off-switch threshold increase with increasing chemical stimulation. Throughout the inspiratory phase, vagal volume feedback exerts graded and reversible inhibition on inspiratory activity until immediately before the inspiratory off-switch where inspiratory termination becomes irreversible [Younes et al., 1978]. Perturbations of the mechanical plant, such as increases in ventilatory load, alter respiratory rhythm by changing the volume-time profiles and off-switch thresholds

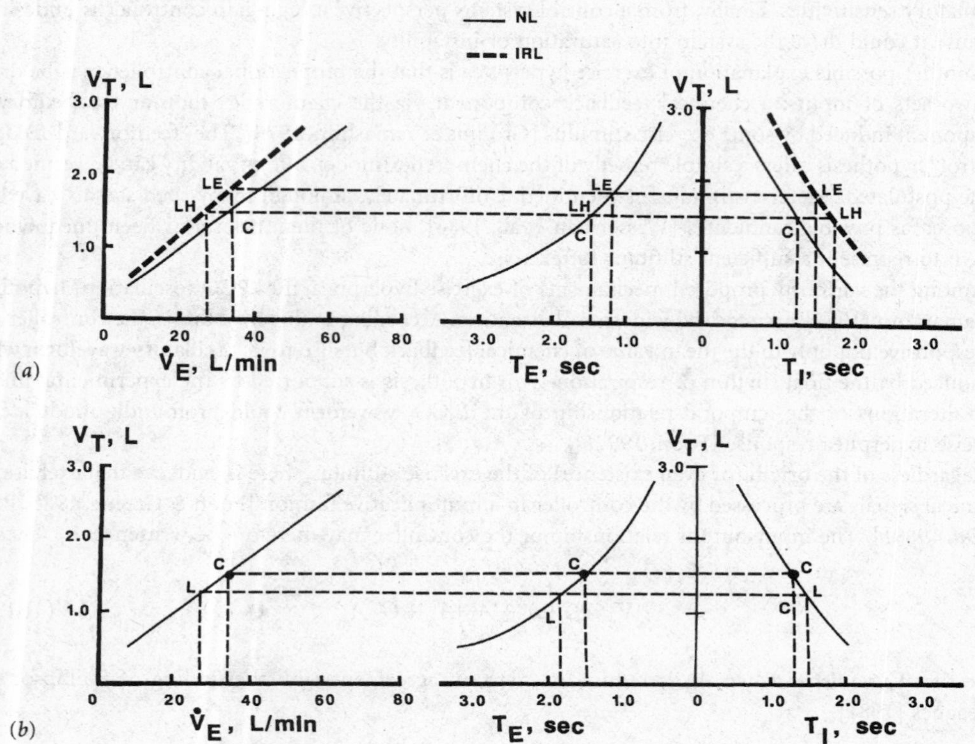

FIGURE 161.2 Stylized diagrams illustrating the Hey plots (*left*) and respiratory phase-switching curves (*right*) for subjects breathing normally (NL) or: (*a*) under an inspiratory resistive load (IRL): (*b*) under an inspiratory elastic load (IEL) at rest or during exercise. In (*a*), IRL causes the normal operating point (*C*) to move to a new point *LE* if the inspirate is free of CO_2, or to a point *LH* if CO_2 is added. In (*b*), IEL moves the operating point to the same point *L* whether CO_2 is present in the inspirate. Thus, the consequence of load compensation is highly dependent on the background CO_2 and the type of load. Relative scales only are shown on all axes. *Source:* Poon [1989*a*, 1989*b*], with permission.

(Fig. 161.2). These effects are abolished by vagotomy such that the RCPG always oscillates with a fixed rhythm. Various neuronal models of phase switching have been proposed [Cohen & Feldman, 1977], and putative phase-switching neurons in the medulla have been identified [Oku et al., 1992].

The two-phase respiratory cycle may be expanded into three phases if the expiratory phase is subdivided into an early postinspiratory inspiratory phase (stage I) and a late expiration phase (stage II). Vagal feedback has differential effects on neural activities during the two stages of expiration [St. John & Zhou, 1990].

An important link between RCPG and chemoreflex models is the Hey plot [Hey et al., 1966] which suggests that $\dot{V}_E$ and V_T are linearly related except when V_T approaches vital capacity. The slope of the Hey plot is altered by resistive loads but not elastic loads to ventilation (Fig. 161.2).

Neural Network Models

Respiratory rhythmicity is an emergent property of the RCPG resulting from mutual inhibition of inspiratory and expiratory related neurons. A minimal model due to Duffin [1991] postulated the early-burst inspiratory (*I*) neurons and Bötzinger complex expiratory (*E*) neurons to be the mutually inhibiting pair. Adaptation of the *I* neurons (e.g., by calcium-activated potassium conductance) results in sustained relaxation oscillation in the network under constant chemical excitation. Both neuron groups are assumed to have monosynaptic inhibitory projections to bulbospinal inspiratory (I_R) output neurons (Fig. 161.3). The model equations are:

$$T_t \frac{dx_t}{dt} - x_i = R_i - E_i + \sum C_n y_n$$
$$T_f \frac{dx_f}{dt} + x_f = R_f + C_f x_1 \quad (161.14)$$

where $y_i \equiv g(x_i - H_i)$; $g(X) \equiv \max(0, X)$; T_i, R_i, E_i, H_i are respectively the membrane time constant, resting activity, excitation, and threshold of the ith neuron; C_{ij} is the connection strength between the ith and jth neurons; and x_1, x_2, x_3, x_f correspond respectively to the activities of the *I, E,* I_R neurons and a fictive (*F*) neuron which represents the adaptation effect.

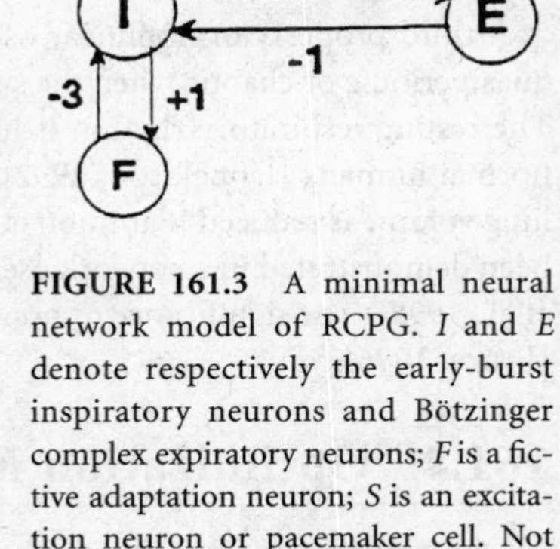

FIGURE 161.3 A minimal neural network model of RCPG. *I* and *E* denote respectively the early-burst inspiratory neurons and Bötzinger complex expiratory neurons; *F* is a fictive adaptation neuron; *S* is an excitation neuron or pacemaker cell. Not shown is the bulbospinal output neuron (I_g). Numbers denote connection strengths. *Source:* Matsugu and coworkers [1998], with permission.

The two-phase Duffin model may be extended to a three-phase pattern by incorporating other respiratory-related neurons. Richter and coworkers [Ogilvie et al., 1992; Richter et al., 1986] have proposed a three-phase RCPG model with mutual inhibition between *I* neurons and postinspiratory, late-inspiratory, and *E* neurons, all with adaptations. Botros and Bruce [1990] have described several variants of the Richter model with varying connectivity patterns, with and without adaptation.

Balis and coworkers [1994] proposed a RCPG model comprised of distributed populations of spiking neurons which are described by the Hodgkin-Huxley equations. Based on extensive spike-train analysis of neural recordings in vivo, the network connectivity differs from the Richter model in several essential ways. The model has been shown to mimick the normal and abnormal respiratory neural waveforms including the inspiratory ramp, inspiratory and expiratory off-switches, as well as apnea and apneusis. Rybak and colleagues [1997*a,b*] proposed a similar RCPG model that is modulated by both the membrane properties of the unit neuron and afferent feedback of the respiratory system.

Pacemaker and Hybrid Models

Respiratory rhythm may also originate from endogenous pacemaker cells. In neonatal rats, certain conditional bursting pacemaker neurons have recently been identified in a region of the ventrolateral medulla referred to as the pre-Botzinger complex [Smith et al., 1991]. Network oscillation by mutual inhibition is difficult in neonates because of immaturity of the GABAergic inhibitory system. During the postnatal developmental stage there may be progressive transformation from pacemaker to network oscillation with possibly some form of hybrid operation in the transition period [Smith, 1994]. A hybrid RCPG model composed of a neural network oscillator driven by a pacemaker cell (Fig. 161.3) has been proposed [Matsugu et al., 1998].

Nonlinear Dynamics

An empirical (black-box) approach to studying rhythmic behavior is to model the oscillation as limit cycles of a nonlinear oscillator. A classic example of limit-cycle oscillation is the Van der Pol oscillator:

$$\ddot{V} = \epsilon \left\{ 1 - \left(\frac{V}{a} \right)^2 \right\} V - V \quad (160.15)$$

where ε and a are shape and amplitude parameters, respectively. This model has been shown to mimick the phase resetting and phase singularity properties of the RCPG [Eldridge et al., 1989] which are typical of any oscillator with isolated limit cycles. Similar properties are exhibited by a network oscillator model of RCPG [Ogilvie et al., 1992] which produces similar limit cycles.

Another property of nonlinear oscillators is that they may be entrained by (or phase-locked to) other oscillators that are coupled to them. The respiratory rhythm has been shown to be entrained by a variety of oscillatory inputs including locomotion, mechanical ventilation, blood gas oscillation, and musical rhythm.

A third property of nonlinear oscillators is that the limit-cycle trajectory may bifurcate and become quasiperiodic or chaotic when the system is perturbed by nonlinear feedback or other oscillatory sources. The resting respiratory rhythm is highly rhythmic in vagotomized animals but may become chaotic in normal humans [Donaldson, 1992] and vagi-intact animals [Sammon & Bruce, 1991] especially when lung volume is reduced [Sammon et al., 1993]. Both entrainment quasiperiodic and chaotic regimes have been demonstrated in a network oscillator model of the RCPG that is driven by periodic inputs [Matsugu et al., 1998]. The significance of nonlinear variations in respiratory pattern is discussed in a recent review [Bruce, 1996].

161.4 Optimization Models

All control systems are meant to accomplish certain (implicit or explicit) control objectives, it is perhaps too simplistic to assume that the sole objective of respiration is to meet the metabolic demand; a complex physiologic system may subserve multiple objectives that are vital to animal survival. One approach to understanding respiratory control is therefore to discover the innate control objective that fits the observed behavior of the controller.

Optimization of Breathing Pattern

Any given level of ventilation may be produced by a variety of breathing patterns ranging from deep-and-slow to shallow-and-rapid breathing. How is the breathing pattern set for each ventilatory level? Rohrer [1925] was among the first to recognize that respiratory frequency at rest may be chosen by the controller to minimize the work rate of breathing, a notion that was subsequently advanced by Otis and colleagues [1950]. Mead [1960] showed that the resting frequency may be determined more closely on the basis of optimal inspiratory pressure-time integral, a measure of the energy cost of breathing. Neither the work nor energy measure, however, correctly predicts the inspiratory flow pattern [Bretschger, 1925; Ruttimann & Yamamoto, 1972]; such discrepancy has led to the suggestion that the optimization criterion may consist of a weighted sum of work and energy expenditures [Bates, 1986] or some higher-order terms [Hämäläinen and Sipilä, 1984; Yamashiro & Grodins, 1971].

Similarly, the optimization principle has also been applied to the prediction of airway caliber and dead space volume [Widdicombe & Nadel, 1963] as well as end-expiratory lung volume and respiratory duty cycle [Yamashiro et al., 1975]. In most cases, the cost functions are generally found to be relatively flat in the resting state but may become much steeper during CO_2 or exercise stimulations [Yamashiro et al., 1975]. This may explain the observed variability of breathing pattern which is generally more pronounced at rest than during hyperpnea or ventilatory loading.

Optimization of Ventilation

The classical optimization hypothesis of Rohrer [1925] and Bretschger [1925] suggests that conservation of energy may be an important factor in the genesis of breathing pattern. It thus appears that the controller is charged with two opposing objectives: to meet the metabolic demand by performing the work of breathing, and to conserve energy by minimizing the work. How does the controller reconcile such conflict? One possible solution is to establish priority. In a hierarchical model of respiratory control, metabolic needs take priority over energetic needs. At the higher hierarchy (outer/chemical feedback loop) ventilatory output is set by feedforward/feedback proportional control of the chemical plant to meet the metabolic demand, whereas at the lower hierarchy (inner/vagal feedback loop) breathing pattern is optimized by the RCPG for efficient energy utilization by the mechanical plant at a ventilation set by the higher hierarchy.

A potential drawback of such a hierarchical system is that it is nonrobust to perturbations. Changes in ventilatory load, for example, would disrupt the ventilatory command from the feedforward signal. This is at variance with the experimental observation of a load compensation response of the controller which protects ventilation against perturbations of the mechanical plant at rest and during exercise [Poon, 1989*a*, 1989*b*; Poon et al., 1987*a*, 1987*b*]. Furthermore, if the prime objective of the controller were indeed to meet the metabolic demand (i.e., to maintain chemical homeostasis), then the hierarchical control system seems to perform quite poorly; it is well-known that arterial chemical homeostasis is readily disrupted environmental changes.

Another form of conflict resolution is compromise. Poon (1983*a*, 1987] proposed that an optimal controller might counterbalance the metabolic needs versus energetic needs of the body, and the resulting compromise would determine the ventilatory response. The tug-of-war between the two conflicting control objectives may be represented by a compound optimization criterion which reflects the balance between the chemical and mechanical costs of breathing:

$$J\begin{cases} = J_c + J_m \\ = \left\{\alpha\left(P_eCO_2 - \beta\right)\right\}^2 + \ln \dot{W} \end{cases} \tag{161.16}$$

The power (quadratic) form and logarithmic form for J_c and J_m correspond, respectively, to the classical Steven's law and Fechner-Weber law of sensory perception [Milsum, 1966]. Assuming that the work rate of breathing $\dot{W} \sim \dot{V}_j^2$, Eq. (161.16) yields an optimal ventilatory response that conforms with the normal hypercapnic ventilatory response during CO_2 inhalation and the isocapnic ventilatory response during exercise [Poon, 1983*a*, 1987]. Furthermore, by generalizing Eq. (161.16) to include other chemical and work rate components, it is possible to predict the normal ventilatory responses to hypoxia and acidosis [Poon, 1983*a*], mechanical loading [Poon, 1987*b*], as well as breathing pattern responses to chemical and exercise stimulation [Poon, 1983*b*].

The ventilatory optimization model [Poon, 1983*a*, 1983*b*, 1987*b*] has several interesting implications. First, it provides a unified and coherent framework for describing the control of ventilation and control of breathing pattern with a common optimization criterion. Second, it offers a parsimonious explanation of exercise hyperpnea and ventilatory load compensation responses, without the need to invoke any putative exercise stimulus and load compensation stimulus. Third, it suggests that disruption of chemical homeostasis (e.g., during CO_2 inhalation) may represent an optimal response as much as maintenance of homeostasis during exercise.

Energetics of breathing is only one of many constraints that conflict with the metabolic cause of respiration. Another is the sensation of dyspnea which may be a limiting factor at high ventilatory levels [Oku et al., 1993]. A general optimization criterion may therefore include both energetic and dyspneic penalties as follows:

$$J = \left\{\alpha\left(P_aCO_2 - \beta\right)\right\}^2 + 2\ln \dot{V}_E + k\,\dot{V}_E^2 \tag{161.17}$$

where k is a weighting factor. At low to moderate ventilatory levels the energetic component (logarithmic term) dominates, and at high ventilatory levels the dyspneic component ($k \cdot \dot{V}_E^2$) dominates in counterbalancing the chemical component. In addition, the ventilatory apparatus may also be constrained by other factors such as behavioral and postural interference, which may further tip the balance of the optimization equation. It has been suggested that the periodic breathing pattern at extremely high altitudes may represent an optimal response for the conservation of chemical and mechanical costs of hypoxic ventilation [Ghazanshahi & Khoo, 1993].

Optimization of Neural Waveform

The neural output of the controller has been traditionally described in terms of measurable quantities such as breathing pattern and ventilation. A more accurate representation of the control signal is $P_{mus}(t)$ which drives the respiratory pump. The resulting continuous inspiratory and expiratory airflow, Eq. (161.9), then determine V_E and all other ventilatory patterns. The control problem the RCPG must solve is how to optimize $P_{mus}(t)$ in order to deliver adequate ventilation without incurring excessive energy (or other) losses. It has been shown by Poon and colleagues [1992] that many interesting characteristics of the RCPG conform to a general optimal control law that calls for dynamic optimization of $P_{mus}(t)$.

The model of Poon and coworkers [1992] assumes a compound optimization criterion, Eq. (161.16), with a mechanical penalty given by a weighted sum of the work rate of inspiration and expiration

$$\dot{W} = \dot{W}_t + \lambda \dot{W}_E \tag{161.18}$$

where

$$\begin{aligned} \dot{W}_t &= \frac{1}{T_T}\int_0^{T_I} \frac{P_{mus}(t)\dot{V}(t)}{\left[1 - P_{mus}(t)/P_{max}\right]^n \left[1 - \dot{P}_{mus}(t)/\dot{P}_{\max}\right]^n}\, dt \\ \dot{W}_E &= \frac{1}{T_T}\int_{T_I}^{T_T} P_{mus}(t)\dot{V}(t)\, dt \end{aligned} \tag{161.19}$$

The terms P_{max} and $\dot{P}_{max}$ denote the limiting capacities of the inspiratory muscles, and n is an efficiency index. The optimal $P_{mus}(t)$ output is found by minimization of J subjects to the constraints set by the chemical and mechanical plants, Eqs. (161.1) and (161.9). Because $P_{mus}(t)$ is generally a continuous time function with sharp phase transitions, this amounts to solving a difficult dynamic nonlinear optimal control problem with piecewise smooth trajectories. An alternative approach adopted by Poon and coworkers [1992] is to model $P_{mus}(t)$ as a biphasic function

$$P_{mus}(t)\begin{cases} = a_0 + a_1 t + a_2 t & 0 \le t \le T_1 \\ = P_{mus}(T_1)\exp\left[-\dfrac{(t - T_1)}{\tau}\right] & T_1 \le t \le T_T \end{cases} \tag{161.20}$$

where a_0, a_1, a_2 and τ are shape parameters; T_1 is the duration of the inspiratory phase of neural activity. The optimal P_{mus} waveform is given by the set of optimal parameters a_0*, a_1*, a_2*, τ*, T_1*, T_T* that minimize Eq. (161.16).

Poon and colleagues [1992] have shown that the dynamic optimization model predicts closely the $P_{mus}(t)$ trajectories under various conditions of ventilatory loading as well as respiratory muscle fatigue and weakness (Fig. 161.4). In addition, the model also accurately predicts the ventilatory and breathing pattern responses to combinations of chemical and exercise stimulation and ventilatory loading [Poon et al., 1992], all without invocation of any putative exercise stimulus or load compensation stimulus. The generality of model predictions strongly supports the hypothesis that neural processing in the RCPG is governed by an optimization law.

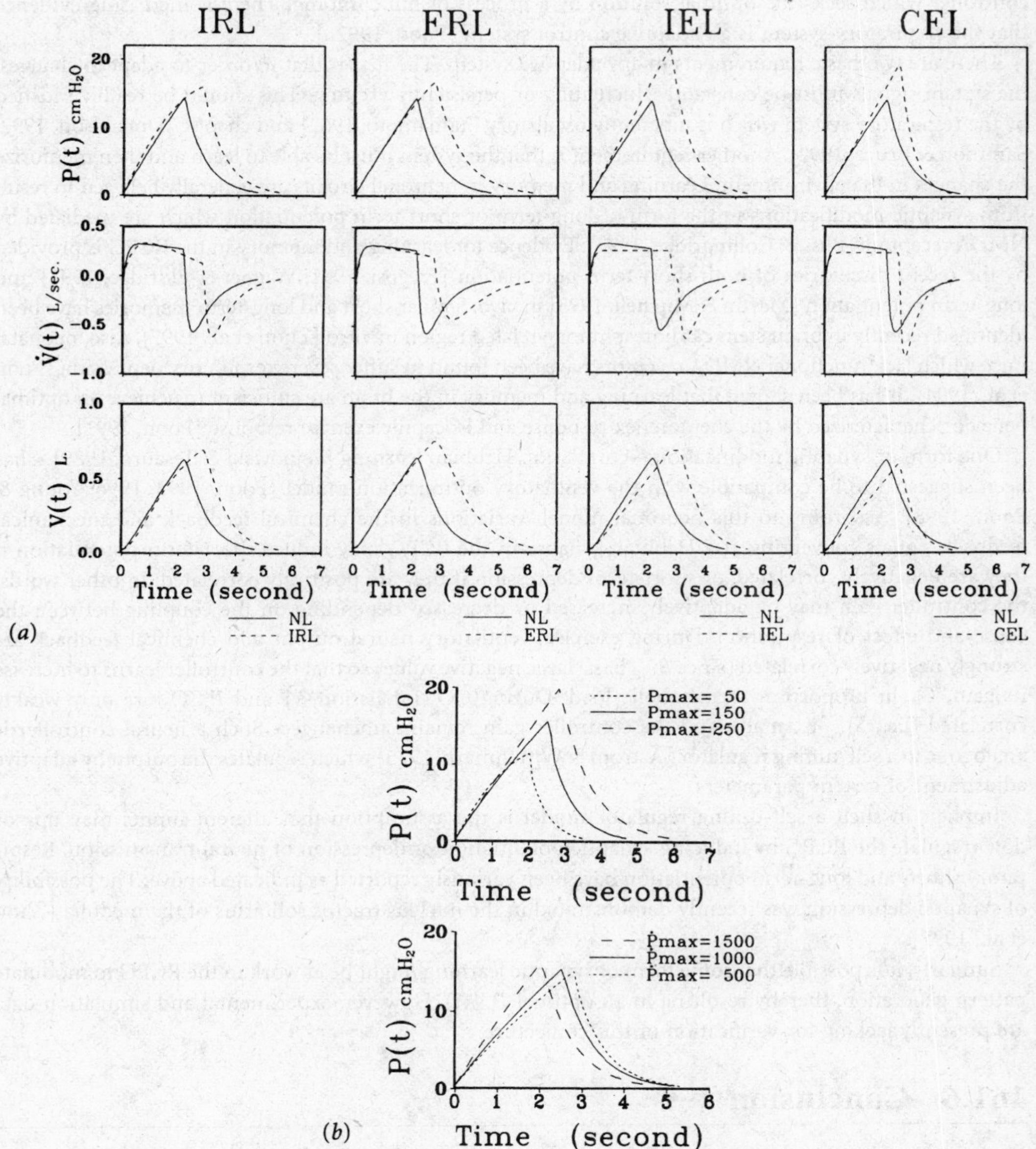

FIGURE 161.4 (*a*) Optimal waveforms for respiratory muscle driving pressure, *P*(*t*); respiratory airflow, *V̇*; and respired volume, *V*, during normal breathing (NL) or under various types of ventilatory loads; IRL and ERL, inspiratory and expiratory resistive load; IEL and CEL, inspiratory and continuous elastic load. (*b*) Optimal waveforms for *P*(*t*) under increasing respiratory muscle fatigue (amplitude limited; *upper* panel) and muscle weakness (rate limited; *lower* panel). *Source:* Poon and coworkers [1992], with permission.

161.5 Self-Tuning Regulator Models

Optimization models suggest an empirical objective function that might be obeyed by the controller. They do not, however, reveal the mechanism of such neural optimization. Priban and Fincham [1965] suggested that an optimal operating for arterial pH, PCO_2, and PO_2 might be achieved by a self-adaptive controller which seeks the optimal solution by a process of hill climbing. There is increasing evidence that the respiratory system is an adaptive control system [Poon, 1992*a*].

There are two basic requirements in any adaptive system. The first is that in order to adapt to changes, the system signals must be constantly fluctuating or persistently exciting. This should be readily satisfied by the respiratory system which is inherently oscillatory [Yamamoto, 1962] and chaotic [Donaldson, 1992; Sammon & Bruce, 1992]. Another requirement is that the system must be able to learn and then memorize the changes in the environment. Learning and memory in neuronal circuits are generally believed to result from synaptic modifications in the form of long-term or short-term potentiation which are mediated by NMDA receptors [Bliss & Collingridge, 1993]. Evidence for learning and memory in the RCPG is provided by the recent discoveries of both short-term potentiation [Fregosi, 1991; Wagner & Eldridge, 1991] and long-term potentiation [Martin & Mitchell, 1993] in vivo. Similar short and long-term memories have been identified recently in brain stem cardiorespiratory-related region in vitro [Zhou et al., 1997]. Also, neonatal mice which lack functional NMDA receptors have been found to suffer severe respiratory depression [Poon et al., 1994]. It has been shown that learning and memory in the brain are sufficient to achieve an optimal behavior characterized by the chemoreflex response and isocapnic exercise response [Poon, 1991].

One form of synaptic modification—correlation Hebbian learning [Sejnowski & Tesauro, 1989]—has been suggested to be compatible with the ventilatory optimization model [Poon, 1993, 1996; Young & Poon, 1998]. According to this neuronal model, variations in the chemical feedback and mechanical feedback signals converging at a Hebbian synapse in the RCPG may induce short-term potentiation if they are negatively correlated, or short-term depression if they are positively correlated. In other words, the controller gain may be adaptively increased or decreased depending on the coupling between the cause and effect of respiration. During exercise, ventilatory neural output and chemical feedback are strongly negatively correlated (since S_{pco2} has a large negative value) so that the controller learns to increase its gain, G_0, in proportion to metabolic load. During CO_2 inhalation, $\dot{V}_E$ and P_aCO_2 are only weakly correlated (i.e., S_{pco2} is small), and the controller gain remains unchanged. Such a neural controller is analogous to a self-tuning regulator [Åstrom & Wittenmark, 1975] which regulates the output by adaptive adjustment of system parameters.

Implicit in such a self-tuning regulator model is the assumption that afferent inputs may up- or downregulate the RCPG by inducing synaptic potentiation or depression of neural transmission. Respiratory short- and long-term potentiation have been variously reported as indicated above. The possibility of synaptic depression was recently demonstrated in the nucleus tractus solitarius of the medulla [Zhou et al., 1997].

Similarly, it is possible that some form of synaptic learning might be at work in the RCPG to modulate pattern generation, thereby resulting in an optimal $P_{mus}(t)$. However, experimental and simulation data are presently lacking for verification of this conjecture.

161.6 Conclusion

Many empirical and functional models have been proposed to describe various aspects of the respiratory control system. The classical chemostat model is useful in describing chemoreflex responses but may be too simplistic to explain the variety of system responses to exercise input and mechanical disturbances. The recent discovery of various complex behaviors of the controller such as nonlinear dynamics, optimization, and learning suggest that the RCPG may be endowed with highly sophisticated computational characteristics. Such computational capability of the RCPG may be important for the maintenance of optimal physiologic conditions under changing environments. The computational problem—which amounts to dynamic optimal control of a nonlinear plant—is a very difficult one even by the standard of modern digital computers but seems to be solved by the RCPG from instant to instant with relative ease. This remarkable ability of the respiratory neural network is interesting from both biologic and engineering standpoints. Understanding how it works may shed light on not only the wisdom of the body [Cannon, 1932] but also on the design of novel intelligent control systems with improved speed, accuracy and economy.

Nomenclature

A_c, A_p	Activity of central, peripheral chemoreceptor
α, β	Slope and intercept of CO_2 response curve; L/min/mmHg, mmHg
J_c, J_m, J	Chemical, mechanical, and total costs of breathing
KCO_2	Solubility constant of CO_2 in blood; $mmHg^{-1}$
M_bCO_2	Metabolic CO_2 production in brain tissues; L/min STPD
P_ACO_2, P_AO_2	Alveolar partial pressure of CO_2, O_2; mmHg
P_aCO_2, P_aO_2	Arterial partial pressure of CO_2, O_2; mmHg
P_bCO_2	Brain tissue PCO_2; mmHg
P_ICO_2, P_IO_2	Inspired partial pressure of CO_2, O_2; mmHg
P_{aw}	Airway pressure; cmH_2O
P_{mus}	Respiratory muscle driving pressure; cmH_2O
Q_b	Cerebral blood flow; L/min
$RCPG$	Respiratory central pattern generator
R_{rs}, E_{rs}	Total respiratory resistance, elastance; cmH_0/L/s, cmH_2O/L
S_{pco2}	Input-output sensitivity of chemical plant, $SP_ACO_2/S\dot{V}_F$ mmHg/L/min
T_I, T_E, T_T	Inspiratory, expiratory, and total respiration duration; s
V	Lung volume above relaxation volume; L
V_bCO_2	Brain tissue store of CO_2; L
$\dot{V}CO_2$, $\dot{V}O_2$	Whole-body metabolic CO_2 production, O_2 consumption; L/min STPD
V_D/V_T	Ratio of respiratory dead space to tidal volume
V_lCO_2; V_lO_2	Lung tissue store of CO_2, O_2; L
$\dot{V}_b$	Total ventilation of the lung; L/min BTPS
W	Work rate of breathing

References

Åstrom KJ, Wittenmark B. 1975. On self-tuning regulators. Automatica 9:185.

Bruce EN. 1996. Temporal variations in the pattern of breathing. J Appl Physiol 80:1079.

Balis UJ, Morris KF, Koleski J, et al. 1994. Simulations of a ventrolateral medullary neural network for respiratory rhythmogenesis inferred from spike train cross-correlation. Biol Cybern 70:311.

Bates JHT. 1986. The minimization of muscular energy expenditure during inspiration in linear models of the respiratory system. Biol Cybern 54:195.

Bellville JW, Whipp BJ, Kaufman RD, et al. 1979. Central and peripheral chemoreflex loop gain in normal and carotid body-resected subjects. J Appl Physiol 46:843.

Bliss TVP, Collingridge GL. 1993. A synaptic model of memory: Long-term potentiation in the hippocampus. Nature 361:31.

Botros SM, Bruce EN. 1990. Neural network implementation of the three-phase model of respiratory rhythm generation. Biol Cybern 63:143.

Bretschger HJ. 1925. Die geschwindigkeitskurve der menschlichen atemluft. Pflügers Arch für die Gesellsch Physiol 210:134.

Bruce EN, Modarreszadeh M, Kump K. 1992. Identification of closed-loop chemoreflex dynamics using pseudorandom stimuli. In Y Honda, Y Miyamoto, K Konno, et al. (eds), Control of Breathing and Its Modeling Perspective, pp 137–142, New York, Plenum Press.

Cannon WB. 1932. The Wisdom of the Body, New York, Norton.

Cohen MI, Feldman JL. 1977. Models of respiratory phase-switching. Federation Proc 36:2367.

Cunningham DJC, Robbins PA, Wolff CB. 1986. Integration of respiratory responses to changes in alveolar partial pressures of CO_2 and O_2 and in arterial pH. In NS Cherniack, JG Widdicombe (eds), Handbook of Physiology, sec 3, The Respiratory System, vol 2: Control of Breathing, part 2, pp 475–528, Washington, DC, American Physiological Society.

Dahan A, Berkenbosch A, DeGoede J, et al. 1990. On a pseudo-rebreathing technique to assess the ventilatory sensitivity to carbon dioxide in man. J Physiol 423:615.

Defares JG. 1964. Principles of feedback control and their application to the respiratory control system. In WO Fenn, H Rahn (eds), Handbook of Physiology, Respiration, sec 3, vol 1, pp 649–680, Washington, DC, American Physiological Society.

Donaldson GC. 1992. The chaotic behavior of resting human respiration. Respir Physiol 88:313.

Duffin J. 1991. A model of respiratory rhythm generation. NeuroReport 2:623.

Eldridge FL, Gill-Kumar P, Millhorn DE. 1981. Input-output relationships of central neural circuits involved in respiration in cats. J Physiol 311:81.

Eldridge FL, Millhorn DE. 1986. Oscillation, gating, and memory in the respiratory control system. In NS Cherniack, JG Widdicombe (eds), Handbook of Physiology, sec 3, The Respiratory System, vol 2, Control of Breathing, part 2, pp 93–134, Washington, DC, American Physiological Society.

Eldridge FL, Paydarfar D, Wagner P, et al. 1989. Phase resetting of respiratory rhythm: effect of changing respiratory "drive." Am J Physiol 257:R271.

Fitzgerald RS, Lahiri S. 1986. Reflex responses to chemoreceptor stimulation. In NS Cherniack, JG Widdicombe (eds), Handbook of Physiology, sec 3, The Respiratory System, vol 2, Control of Breathing, part 1, pp 313–362, Washington, DC, American Physiological Society.

Ghazanshahi SD, Khoo MCK. 1993. Optimal ventilatory patterns in periodic breathing. Ann Biomed Eng 21:517.

Ghazanshahi SD, Khoo MCK. 1997. Estimation of chemoreflex loop gain using pseudorandom binary CO_2 stimulation. IEEE Trans Biomed Eng 44:357.

Grodins FS, Buell SJ, Bart AJ. 1967. Mathematical analysis and digital simulation of the respiratory control system. J Appl Physiol 22:260.

Grodins FS, Yamashiro SM. 1978. Respiratory Function of the Lung and Its Control. New York, Macmillan.

Hämäläinen RP, Sipilä A. 1984. Optimal control of inspiratory airflow in breathing. Optimal Control Applications Methods 5:177.

Heeringa J, Berkenbosch A, de Goede J, et al. 1979. Relative contribution of central and peripheral chemoreceptors to the ventilatory response to CO_2 during hyperoxia. Respir Physiol 37:365.

Hey EN, Lloyd BB, Cunningham DJC, et al. 1966. Effects of various respiratory stimuli on the depth and frequency of breathing in man. Respir Physiol 1:193.

Hildebrandt JR. 1977. Gating: A mechanism for selective receptivity in the respiratory center. Federation Proc 36:2381.

Juratsch CE, Whipp BJ, Huntsman DJ, et al. 1982. Ventilatory control during experimental maldistribution of $\dot{V}_A/\dot{Q}$ in the dog. J Appl Physiol 52(1):245.

Khoo MCK, Kronauer RE, Strohl KP, et al. 1982. Factors including periodic breathing in humans: A general model. J Appl Physiol 53:644.

Khoo MCK, Gottschalk A, Pack AI. 1991. Sleep induced periodic breathing and apnea: A theoretical study. J Appl Physiol 70:2014.

Lipski J, Voss MD. 1990. Gating of peripheral chemoreceptor input to medullary inspiratory neurons: Role of Bötzinger complex neurons. In H Acker, A Trzebski, RG O'Regan, et al (eds), Chemoreceptors and Chemoreceptor Reflexes, pp 323–329, New York, Plenum Press.

Marroquin E Jr. 1991. Control of Respiration under Simulated Airway Compression. Thesis. Harvard-MIT Division of Health Sciences and Technology, Harvard Medical School, Boston.

Martin PA, Mitchell GS. 1993. Long-term modulation of the exercise ventilatory response in goats. J Physiol 470:601.

Masakazu M, Duffin J, Poon C-S. 1998. Entrainment, instability, quasi-periodicity, and chaos in a compound neural oscillator. J Computational Neurosci 5:35.

Mead J. 1960. Control of respiratory frequency. J Appl Physiol 15:325.

Milic-Emili J, Zin WA. 1986. Relationship between neuromuscular respiratory drive and ventilatory output. In PT Macklem, J Mead (eds), Handbook of Physiology, sec 3, The Respiratory System, vol 3, Mechanics of Breathing, part 2, pp 631–646, Washington, DC, American Physiological Society.

Milsum JH. 1966. Biological Control Systems Analysis, New York, McGraw-Hill.

Neubauer JA, Melton JE, Edelman NH. 1990. Modulation of respiration during brain hypoxia. J Appl Physiol 68(2):441.

Oku Y, Saidel GM, Altose MD, et al. 1993. Perceptual contributions to optimization of breathing. Ann Biomed Eng 21:509.

Ogilvie MD, Gottschalk A, Anders K, et al. 1992. A network model of respiratory rhythmogenesis. Am J Physiol 263:R962.

Oku Y, Tanaka I, Ezure K. 1992. Possible inspiratory off-switch neurones in the ventrolateral medulla of the cat. NeuroReport 3:933.

Oren A, Wasserman K, Davis JA, et al. 1981. Regulation of CO_2 set point on ventilatory response to exercise. J Appl Physiol 51:185.

Otis AB, Fenn WO, Rahn H. 1950. The mechanics of breathing in man. J Appl Physiol 2:592.

Pack AI, Ogilvie MD, Davies RO, et al. 1986. Responses of pulmonary stretch receptors during ramp inflations of the lung. J Appl Physiol 61(1):344.

Petersen ES. 1981. A survey of applications of modeling to respiration. In JG Widdicombe (ed), International Review of Physiology. Respiratory Physiology III, vol 23, ed. pp 261–326, Baltimore, University Park Press.

Phillipson EA. 1974. Vagal control of breathing pattern independent of lung inflation in conscious dogs. J Appl Physiol 37(2):183.

Phillipson EA, Bowes G. 1986. Control of breathing during sleep. In NS Cherniack, JG Widdicombe (eds), Handbook of Physiology, sec 3, The Respiratory System, vol 2, Control of Breathing, part 2, pp 649–689, Washington, DC, American Physiological Society.

Phillipson EA, Duffin J, Cooper JD. 1981. Critical dependence of respiratory rhythmicity on metabolic CO_2 load. J Appl Physiol 50(1):45.

Poon C-S. 1983*a*. Optimal control of ventilation in hypoxia, hypercapnia and exercise. In BJ Whipp, DM Wiberg (eds), Modelling and Control of Breathing, pp 189–196, New York, Elsevier.

Poon C-S. 1983*b*. Optimality principle in respiratory control. Proc Am Control Conf. 2nd, pp 36–40.

Poon C-S. 1987*a*. Estimation of pulmonary $\dot{V}/\dot{Q}$ distribution by inert gas elimination: State of the art. In C Cobelli, L Mariani (eds), Modelling and Control in Biomedical Systems, pp 443–453, New York, Pergamon Press.

Poon, C-S. 1987*b*. Ventilatory control in hypercapnia and exercise: Optimization hypothesis. J Appl Physiol 62(6):2447.

Poon C-S. 1989*a*. Effects of inspiratory elastic load on respiratory control in hypercapnia and exercise. J Appl Physiol 66(5):2400.

Poon C-S. 1989*b*. Effects of inspiratory resistive load on respiratory control in hypercapnia and exercise. J Appl Physiol 66(5):2391.

Poon C-S. 1991. Optimization behavior of brainstem respiratory neurons: A cerebral neural network model. Biol Cybern 66:9.

Poon C-S. 1992*a*. Introduction: Optimization hypothesis in the control of breathing. In Y Honda, Y Miyamoto, K Konno, et al. (eds), Control of Breathing and Its Modeling Perspective, pp 371–384, New York, Plenum Press.

Poon C-S. 1992*b*. Potentiation of exercise ventilatory response by CO_2 and dead space loading. J Appl Physiol 73(2):591.

Poon C-S. 1993. Adaptive neural network that subserves optimal homeostatic control of breathing. Ann Biomed Eng 21:501.

Poon C-S. 1996. Self-tuning optimal regulation of respiratory motor output by Hebbian covariance learning. Neural Networks 9:1367.

Poon C-S, Greene JG. 1985. Control of exercise hyperpnea during hypercapnia in humans. J Appl Physiol 59(3):792.

Poon C-S, Li Y, Li SX, et al. 1994. Respiratory rhythm is altered in neonatal mice with malfunctional NMDA receptors. FASEB J 8(4):A389.

Poon C-S, Lin S-L, Knudson OB. 1992. Optimization character of inspiratory neural drive. J Appl Physiol 72(5):2005.

Poon C-S, Olson RJ. 1985. A simple quasi-steady technique for accelerated determination of CO_2 response. Federation Proc 44:832.

Poon C-S, Ward SA, Whipp BJ. 1987*a.* Influence of inspiratory assistance on ventilatory control during moderate exercise. J Appl Physiol 62(2):551.

Poon C-S, Younes M, Gallagher CG. 1987*b.* Effects of expiratory resistive load on respiratory motor output in conscious humans. J Appl Physiol 63(5):1837.

Priban IP, Fincham WF. 1965. Self-adaptive control and the respiratory system. Nature Lond 208:339.

Read DJC. 1967. A clinical method for assessing the ventilatory response to carbon dioxide. Australasian Ann Med 16:20.

Richter DW, Ballantyne D, Remmers JE. 1986. How is the respiratory rhythm generated? A model. News Physiol Sci 1:109.

Rohrer F. 1926. Physilogie der Atembewegung. In ATJ Bethe, G von Bergmann, G Embden, et al. (eds), Handbuch der normalen und pathologischen Physiologie, vol 2, pp 70–127, Berlin, Springer-Verlag.

Ruttimann U, Yamamoto WS. 1972. Respiratory airflow patterns that satisfy power and force criteria of optimality. Ann Biomed Eng 1:146.

Rybak IA, Paton JFR, Schwaber JS. 1997*a.* Modeling neural mechanisms for genesis of respiratory rhythm and pattern: II. Network models of the central respiratory pattern generator. J Neurophysiol 77:2007.

Rybak IA, Paton JFR, Schwaber JS. 1997*b.* Modeling neural mechanisms for genesis of respiratory rhythm and pattern: III. Comparison of model performances during afferent nerve stimulation. J Neurophysiol 77:2027.

Sammon MP, Bruce EN. 1991. Vagal afferent activity increases dynamical dimension of respiration in rats. J Appl Physiol 70(4):1748.

Sammon M, Romaniuk JR, Bruce EN. 1993. Bifurcations of the respiratory pattern associated with reduced lung volume in the rat. J Appl Physiol 75(2):887.

Sejnowski TJ, Tesauro G. 1989. The Hebb rule for synaptic plasticity: algorithms and implementations. In JH Byrne, WO Berry (eds), Neural Models of Plasticity, pp 94–103, New York, Academic Press.

Sidney DA, Poon C-S. 1995. Ventilatory responses of dead space and CO_2 breathing under inspiratory resistive load. J Appl Physiol 78(2):555.

Smith JC. 1994. A model for developmental transformations of the respiratory oscillator in mammals. FASEB J 8(4):A394.

Smith JC, Ellenberger HH, Ballanyi K, et al. 1991. Pre-Bötzinger complex: A brainstem region that may generate respiratory rhythm in mammals. Science 254:726.

St. John WM, Zhou D. 1990. Discharge of vagal pulmonary receptors differentially alters neural activities during various stages of expiration in the cat. J Physiol 424:1.

Stark AR, Cohlan BA, Waggener TB, et al. 1987. Regulation of end-expiratory lung volume during sleep in premature infants. J Appl Physiol 62:1117.

Teppema LJ, Barts PWJA, Evers JAM. 1985. The effect of the phase relationship between the arterial blood gas oscillations and central neural respiratory activity on phrenic motoneurone output in cats. Respir Physiol 61:301.

Vis A, Folgering H. 1980. The dynamic effect of Pet_{CO2} on vertebral bloodflow in cats. Respir Physiol 42:131.

Wagner PG, Eldridge FL. 1991. Development of short-term potentiation. Respir Physiol 83:129.

Ward SA, Whipp BJ, Poon C-S. 1982. Density dependent airflow and ventilatory control during exercise. Respirat Physiol 49:267.

Wasserman K, Whipp B, Casaburi R. 1986. Respiratory control during exercise. In NS Cherniack, JG Widdicombe (eds), Handbook of Physiology, sec 3, The Respiratory System, vol 2, Control of Breathing, part 2, pp 595–619, Washington, DC, American Physiological Society.

Widdicombe JG, Nadel JA. 1963. Airway volume, airway resistance and work, and force of breathing—theory. J Appl Physiol 35:522.

Yamamoto WS. 1962. Transmission of information by the arterial blood stream with particular reference to carbon dioxide. Biophy J 2:143.

Yamashiro SM, Daubenspeck JA, Lauritsen TN, et al. 1975. Total work rate of breathing optimization in CO_2 inhalation and exercise. J Appl Physiol 38(4):702.

Yamashiro SM, Grodins FS. 1971. Optimal regulation of respiratory airflow. J Appl Physiol 18:863.

Younes M, Remmers JE. 1981. Control of tidal volume and respiratory frequency. In TF Hornbein (ed), Regulation of Breathing, part 1, pp 621–671, New York, Marcel Dekker.

Younes M, Remmers JE, Baker J. 1978. Characteristics of inspiratory inhibition by phasic volume feedback in cats. J Appl Physiol 45:80.

Young DL, Poon C-S. 1998. Hebbian covariance learning: A nexus for respiratory variability, memory, and optimization? In RL Hughson, DA Cunningham, J Duffin (eds), Advances in Modeling and Control of Ventilation, pp 73–83, New York, Plenum.

Zhou Z, Champagnat J, Poon C-S. 1997. Phasic and long-term depression in brainstem nucleus tractus solitarius neurons: differing roles of AMPA receptor desensitization. J Neurosci 17:5349.

162

Neural Networks for Physiological Control

James J. Abbas
The University of Kentucky

162.1 Introduction

This chapter is intended to provide a description of neural network control systems and of their potential for use in biomedical engineering control systems. Neural network techniques have been used by the engineering community in a variety of applications, with particular emphasis on solving pattern recognition and pattern classification problems [Carpenter 1989; Grossberg, 1988a,b; Hagan, 1996; Hecht-Nielsen, 1989; Nerrand, 1993; Pao, 1988; Sanchez-Sinencio 1992; Zurada, 1992]. More recently, there has been much research into the use of these techniques in control systems [Antsaklis, 1990; Miller, 1990b; White, 1992]. Much of this research has been directed at utilizing neural network techniques to solve problems that have been inadequately solved by other control systems techniques. For example, neural networks have been used in adaptive control of nonlinear systems for which good models do not exist. The success of neural network techniques on this class of problems suggest that they may be particularly well suited for use in a wide variety of biomedical engineering applications. It should be emphasized that the field of neural network control is a relatively new area of research and that it is not intended to replace traditional engineering control. Rather, the focus has been on the integration of neural network techniques into control systems for use when traditional control systems alone are insufficient.

There are several textbooks available on neural networks that provide good presentations of the implementation and applications of neural network techniques [Hagan, 1996; Sanchez-Sinencio, 1992; Simpson, 1990; Zurada, 1992]. Recently, a few books have been published that review the application of neural network techniques to control systems problems [Miller, 1990b; White, 1992]. This chapter is intended to provide an introduction to neural network techniques and a guide to their application to biomedical control systems problems. The reader is referred to recently published textbooks and numerous journal articles for the specific information required to implement a given neural network control system.

162.2 Neural Network Basics

The term *neural networks* is used to refer to a class of computational algorithms that are loosely based upon the computational structure of the nervous system. There is a wide variety among various neural

0-8493-0462-8/00/$0.00+$.50

network algorithms, but the key features are that they have a set of inputs and a set of outputs, they utilize distributed processing, and they are adaptive. The basic idea is that computation is collectively performed by a group of distinct units, or *neurons* (sometimes referred to as processing elements), each of which receives inputs and performs its own local calculation. These units interact by providing inputs to each other via synapses and some units interact with the environment via input/output signals. The connectivity of various units determines the structure, or *architecture*, of the network. Adjustments to the strengths of the synapses (i.e., adjustments to the synaptic weights) modify the overall input-output properties of the network and are adjusted via a *learning algorithm*. The design of a neural network includes the specification of the neuron, the architecture and the learning algorithm.

Most neural networks are based on a model of a neuron that captures the most basic features of real neurons: A neuron receives inputs from other neurons via synapses and the output from a neuron is a nonlinear function of a weighted summation of its inputs. There are many different sets of equations that capture these basic features; the following set of equations is commonly used:

$$y_j = \sum_{i=1}^{n} w_{ij} x_i$$

$$z_j = 1\Big/\left(1 + e^{-my_j}\right)$$

where y_j is the weighted summation of inputs to neuron j, x_i is the input from neuron i, n is the number of neurons providing synaptic input to neuron j, w_{ij} is the synaptic weight from neuron i to neuron j, z_j is the output from neuron j, and m is the slope of the sigmoidal output function. Note that in this set of equations, the output of a neuron is a static nonlinear function of the weighted summation of its inputs. Some commonly used variations on this model include: the use of different nonlinear output functions [Lippman, 1987], the use of recurrent inputs (i.e., past values of a neurons output would be included in the input summation) [Pineda, 1989; Williams and Zipser, 1989], and the use of dynamic neurons (i.e., Eq. (162.1) would be a differential equation rather than an algebraic one) [Carpenter, 1989; Grossberg, 1988a; Hopfield, 1982]. The use of recurrent inputs and/or dynamic neuron models may be particularly important for applications in the control of dynamic systems because they provide the individual neurons with memory and temporal dynamics [Pineda, 1989].

Most applications use networks of neurons that are arranged in layers. The most commonly used architecture is a three layer network which has an input layer, a hidden layer and an output layer. A feedforward neural network architecture is one in which each neuron receives input from neurons in the previous layer. The number of neurons in each layer must be specified by the designer. There are some heuristic rules, but no solid theory, to guide the designer in specifying the number of neurons [Zurada, 1992]. Some commonly used variations on this architecture include: the use of only a single layer of neurons [Kohonen, 1989; Widrow, 1962], the use of bidirectional connections between layers [Carpenter, 1989], the use of intra-layer connections [Kohonen, 1989], the use of pruning techniques which cut some of the connections for greater efficiency [Zurada, 1992], and the use of heterogeneous networks in which more than one neuron model is used to describe the various neurons in the network [Hecht-Nielson, 1989].

A general form of a neural network learning algorithm is given by:

$$\Delta w_{ij} = f\left(\eta, z_i, z_j, e_{ij}\right)$$

This equation states that the change in synaptic weight is a function of the learning rate (η), the activation of the pre-synaptic neuron (z_i), the activation of the post-synaptic neuron (z_j), and a training signal (e_{ij}). Several different learning algorithms have been developed [Fogel, 1990; Grossberg, 1988a; Hecht-Nielsen,

1990; Hinton, 1989; Rumelhart, 1988], most of which fit into this general form, but may not use all of the terms. An example is Hebbian learning [Brown, 1990; Hebb, 1949] in which the change in weight is proportional to the product of the pre-synaptic activation and the post-synaptic activation. Often, gradient descent techniques are used to adjust the synaptic weights. The most commonly used learning algorithm, error backpropogation, uses an error gradient descent technique and passes the output error backwards through the network to determine the training signal for a given neuron [Pineda, 1989; Rumelhart, 1988; Vogl 1988]. Most of the commonly used learning algorithms are classified as supervised learning techniques because they use a specification of the desired output of the network to determine the output error of the network. Other learning algorithms, such as reinforcement learning (discussed below) can be used in situations when it is not possible to directly specify the desired output of the network.

In most engineering applications, the neural network is used to perform a nonlinear mapping from the space of network inputs to the space of network outputs. In addition to the internal specifications of the network discussed above, the engineer using a neural network must select the signals to be used as input to the network and the signals to be used as output from the network. Although this decision appears to be trivial, in many applications it is not. For applications in control systems, the selection of input and output signals is dependent upon how the neural network fits into the overall structure of the system. Several different options are described below.

162.3 Neural Network Control Systems

Neural networks have often been applied in control systems where, at a block diagram level, the neural network replaces one component of the control system. Typically, the motivation for utilizing the neural network lies in the ability to perform nonlinear mappings, in the ability to handle a wide range of model structures, and/or in the ability of the network to adapt. While there has been activity in the development of a theoretical framework for neural network control systems, most applications to-date have been heuristically-based to one degree or another [Farotimi, 1991; Miller, 1990b; Nguyen, 1990; Quinn, 1993; Wang, 1992; White, 1992]. The field of neural network control has been described as bridging the gap between mathematically-based control systems engineering and heuristically-based artificial intelligence [Barto, 1990].

In designing a neural network control system, one must select the overall structure of the system and decide which components will utilize neural network algorithms. Several examples of control system structures are provided below, each of which utilizes one or more neural networks as described above. This section of the chapter provides a brief overview of some neural control systems that have potential for application in biomedical control systems. For excellent, thorough reviews of recent developments in neural network control systems, the reader is referred to Miller [1990b] and White [1992].

Supervised Control

In this structure (Fig. 162.1), the neural network is used to mimic, and eventually replace, an existing control system. Here, the neural network is used in a traditional feedback arrangement where it performs a mapping from the system outputs to the system inputs. The network would be trained using data collected from the actual system or a computer simulated model with the original controller active. The training signal for the network would be the difference between the output of the original controller and the output of the network. After the network learns to adequately mimic the original controller, the training is completed. Thus, the synaptic weights would no longer be adapted, and the original control system would be replaced by the neural network. This type of an arrangement could be useful in situations where the neural network could perform the task less expensively or more efficiently than the original one. One situation would be when the original control system requires such heavy computation that it is impractical for real-time use on an affordable computer system. In this case, the neural network might be able to learn to perform the same (or functionally equivalent) operation more efficiently such that implementation on an affordable general purpose computer or on specialized neural network hardware would be practical. A second situation would be when the original control system is a human operator that might be either expensive or prone to error. In this case the neural network is acting as an adaptive expert system that learns to mimic the human expert [Werbos, 1990]. A third situation would be when the neural network could learn to perform the control using a different, and more easily measured, set of output variables [Barto, 1990]. This type of application would be particularly well-suited for biomedical applications in which an invasive measurement could be replaced by a non-invasive one.

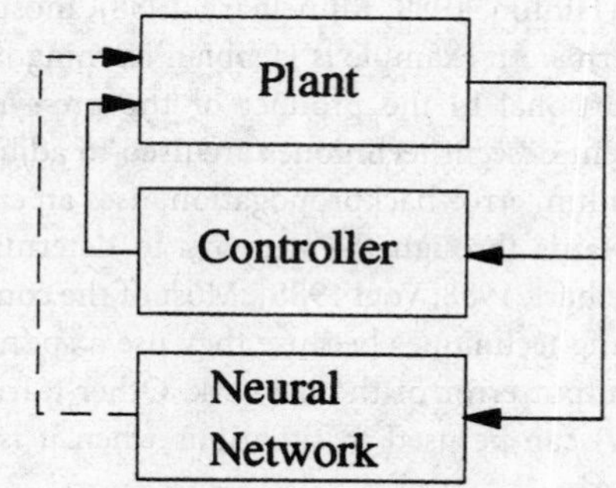

FIGURE 162.1 Supervised control system structure.

Direct Inverse Control

In direct inverse control, (Fig. 162.2), the neural network is used to compute an inverse model of the system to be controlled [Levin, 1991; Nordgren, 1993]. In classical linear control techniques, one would find a linear model of the system then analytically compute the inverse model. Using neural networks, the network is trained to perform the inverse model calculations, i.e., to map system outputs to system inputs. Biomedical applications of this type of approach include the control of arm movements using electrical stimulation [Lan, 1994] and the adaptive control of arterial blood pressure [Chen, 1997].

A variety of neural network architectures can be used in this direct inverse control system structure. A multi-layer neural network architecture with error backpropagation learning could be used off-line to learn an inverse model using a set of system input/output data. If on-line learning is desired, a feedforward model of the system can be trained while a feedback controller is active [Miller, 1990a; Miyamoto, 1988]. This technique, referred to as feedback error learning [Miyamoto, 1988], uses the control signal coming from the feedback controller as the training signal for the feedforward controller update. The idea used here is that if the feedforward control is inadequate, the feedback controller would be active and it would lead to updates in the feedforward controller. To achieve the rapid learning rates that would be desirable in on-line feedback error learning, typically single-layer networks have been used or the learning has been constrained to occur at one layer. This type of an approach is particularly useful in applications where on-line fine tuning of feedforward controller parameters is desired.

Model-based Approaches to Control

Neural network approaches have been used as an alternative to other nonlinear techniques for modeling physiological systems [Chon, 1998]. Several neural network control systems have utilized model-based approaches in which the neural network is used to identify a forward nonlinear system model. Often, these systems utilize neural network components in control system architectures that have been developed using linear or nonlinear components.

For many control applications, system output error can be readily measured but in order to adapt the controller parameters one needs to relate system output error to controller output (system input) error. One strategy would be to use the derivative of the system's output error with respect to its input to perform gradient descent adaptation of the controller parameters. However, there is no direct way of

FIGURE 162.2 Direct inverse control system structure.

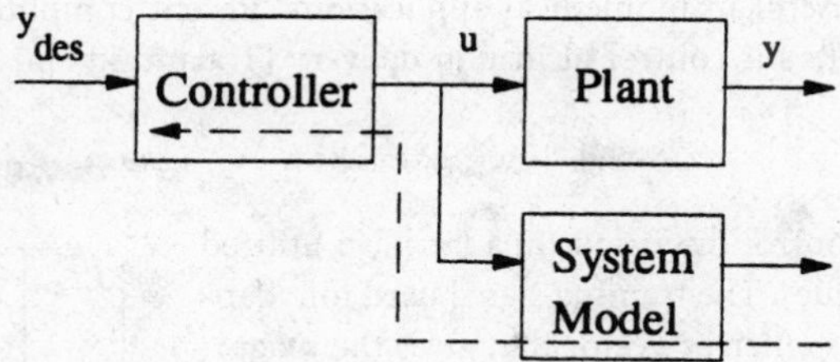

FIGURE 162.3 Backpropagating system output error through the system model.

calculating the derivative of the system's output error with respect to its input and therefore it must be approximated. To implement such an approximation, a pair of multi-layer neural networks (Fig. 162.3) have been used: one for identification of a forward model of the plant, and one for implementation of the controller [Barto, 1990]. The forward model is used to provide a means of computing the derivative of the model's output with respect to its input by backpropagating the error signal (the actual plant output tracking error) through the model of the plant to estimate the error in the control signal. The error in the control signal can then be backpropagated through the controller neural network. As described here, the backpropagation algorithm is used to minimize the output error of the system, which is a special case of maximization of utility. Although this is a very powerful technique with potential applications in a wide variety of nonlinear control problems, its primary limitation for many biomedical applications is the relatively slow training rate achieved by the backpropagation algorithm.

Neural networks have also been used in model reference adaptive control (MRAC) structures (Fig. 162.4) [Naranedra and Parthasarathy, 1990; Narendra, 1992]. This approach builds upon established techniques for adaptive linear control and incorporates neural networks to address the problem of controlling nonlinear systems. The MRAC approach is directed at adapting the controlled system such that it behaves like a reference model, which is specified by the control system designer. In the structure shown in Fig. 162.4, the identification model is a model of plant dynamics, the parameters of which are identified online using the identification error (the difference between the actual and the estimated plant outputs) as the training signal. The controller uses the estimates of the plant parameters, reference inputs, and system outputs to determine inputs to the system. In the neural network version of the control system structure, neural networks can be used for the identification of the system model, for the controller, or for both. Multi-layer neural networks using error backpropagation or a modified version of error backpropagation, termed *dynamic backpropagation,* have been used for adaptive control of nonlinear systems [Narendra and Parthasarathy, 1990; Narendra, 1992].

A third model-based approach is neural predictive control, which is a neural network version of nonlinear model predictive control [Trajanoski and Wach, 1998]. In this approach, the neural network is used for off-line identification of a system model, which is then used to design a nonlinear model predictive controller. This design may provide suitable control of nonlinear systems with time-delays and

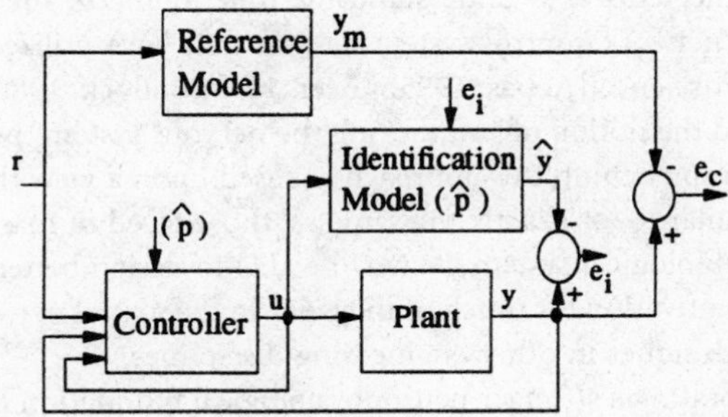

FIGURE 162.4 Indirect adaptive control system structure.

thus may be particularly useful in biomedical applications. Recent computer simulation studies have demonstrated positive results for control of insulin delivery [Trajanoski and Wach, 1998].

Adaptive Critics

All of the approaches to control discussed thus far have utilized supervised learning techniques. The training signal used for adaptation in these techniques is an error vector that gives the magnitude of the error in the system output signal and the direction in which it should change. For many complex, multi-variable control problems this type of system output error information is not readily available. The only measure available might be a scalar measure of system performance that is not directly related to the system outputs in a way that is understood by the control system designer. In such cases, a directed search of the control system parameter space could be performed in order to maximize (optimize) the scalar performance measure. Dynamic programming is a control systems engineering technique that has been used for optimal control of linear systems but these techniques are not well suited for large-scale systems or nonlinear systems. Neural networks have been used in an approach that is similar to dynamic programming in order to optimize system performance of large-scale nonlinear systems. These techniques, termed "adaptive critic methods", utilize a structure (Fig. 162.5) in which a critic module provides an evaluation signal to the controller module. The controller module utilizes a reinforcement learning algorithm in order to optimize performance. The details of this class of neural network controllers are beyond the scope of this chapter, but descriptions and example applications are given in [Barto, 1990; Barto, 1992; Werbos, 1992].

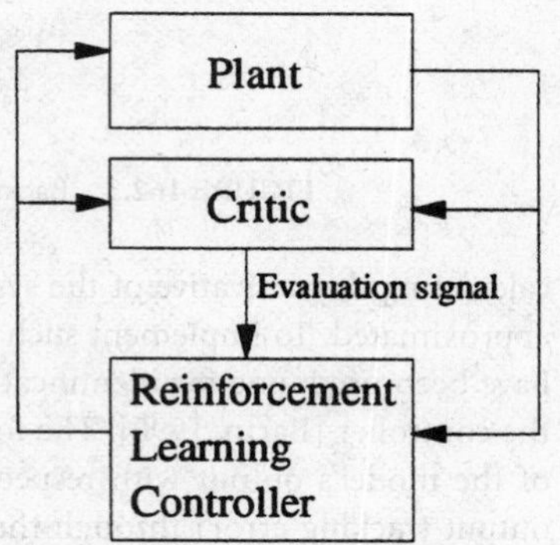

FIGURE 162.5 Adaptive critic control system structure.

Adaptive critic algorithms may be particularly useful in several biomedical engineering control systems where the relationship between system performance and measurable system outputs are ill-defined. While these methods that use reinforcement learning may be attractive because they are general and do not require detailed information about the system, this increased generality comes at a cost of increased training times [Barto, 1990]. Supervised learning methods, when they can be used, can be more efficient than reinforcement learning methods.

Neurophysiologically Based Approaches

All neural network systems are based upon neurophysiological models to some degree, but in most cases this biological basis is very superficial, as is evident in the description given above. Computational neuroscience is a term used to describe the development of computational models of the nervous system [Bower, 1998; Koch and Segev, 1989; Schwartz, 1990]. Such models are intended to be used by neuroscientists in their basic science efforts to understand the functioning of the nervous system. There are several examples of neural network control system designs that have utilized models developed in the field of computational neuroscience [Abbas, 1995b; Beer, 1993; Bullock, 1988; Houk, 1990; Taga, 1991]. These designs have extended the notion of "mimicking the nervous system" beyond what is used in most neural networks. The reasoning behind this approach is based upon a view that biological systems solve many problems that are similar to, or exactly the same as, those faced in many engineering applications and that by mimicking the biological system we may be able to design better engineering systems. This reasoning is similar to the motivation for much of the work in the neural network field and for the recent emphasis in biomimetic techniques in other engineering disciplines.

One approach to incorporating a stronger neurophysiological foundation has been to utilize an architecture for the overall control system that is based upon neurophysiological models, i.e., mimicking the neurophysiological system at a block diagram level. This type of approach has led to the development of hierarchical control systems that are based upon the hierarchical structure of the human motor control system [Kawato, 1987; Srinivasan, 1992].

A second approach to incorporating a stronger neurophysiological foundation has been to utilize more realistic models of neurons and of their interconnections in the design of the neural network [Beer, 1990; Houk, 1990; Missler and Kamangar, 1995]. An example of this approach is the design of a coupled oscillator neural network circuit that is based upon the locomotor control system of the cockroach [Beer, 1992; Quinn, 1993]. In this neural network, some of the neurons are capable of endogenously oscillating due to intrinsic membrane currents and the network is heterogeneous, meaning that each neuron is not described by the same model. Most of the parameters of the models are fixed. This network has been shown to robustly generate patterns for statically-stable gaits at various speeds.

A biomedical control system that utilizes a neurophysiologically-based approach has been developed for use in Functional Neuromuscular Stimulation (FNS) systems [Abbas, 1995; Abbas and Chizeck, 1995]. FNS is a rehabilitation engineering technique that uses computer-controlled electrical stimuli to activate paralyzed muscle. The task of a control system is to determine appropriate stimulation levels to generate a given movement or posture. The neural network control system utilizes a block diagram structure that is based on hierarchical models of the locomotor control system. It also utilizes a heterogenous network of neurons, some of which are capable of endogenous oscillation. This network has been shown to provide rapid adaptation of the control system parameters [Abbas and Chizeck, 1995; Abbas and Triolo, 1997] and has been shown to exhibit modulation of reflex responses [Abbas, 1995].

162.4 Summary

This chapter presents an overview of the relatively new field of neural network control systems. A variety of techniques are described and some of the advantages and disadvantages of the various techniques are discussed. The techniques described here show great promise for use in biomedical engineering applications in which other control systems techniques are inadequate. Currently, neural network control systems lack the type of theoretical foundation upon which linear control systems are based, but recently there have been some promising theoretical developments. In addition, there are numerous examples of successful engineering applications of neural networks to attest to the utility of these techniques.

Acknowledgments

The author gratefully acknowledges the support of the National Science Foundation (NSF-BCS-9216697), The Whitaker Foundation, the School of Engineering at The Catholic University of America and the Center for Biomedical Engineering at the University of Kentucky.

Defining Terms

Backpropagation: A technique used to determine the training signal used for adjusting the weights of a given neuron in a neural network.

Learning algorithm: An algorithm used to update the synaptic weights in a neural network.

Neural network: A term used to refer to a broad class of computational algorithms that are loosely based on models of the nervous system.

Reinforcement learning algorithms: Learning algorithms that utilize a system performance measure (that may or may not have a direct, known relationship to output error of the neural network) as a training signal for the neural network.

Supervised learning algorithms: Learning algorithms often used in neural networks that use the output error of the neural network as a training signal.

Synaptic weight: A scaling factor on the signal from one neuron in a network to another.

References

Abbas, J.J. 1995. Using Neural Models in the Design of a Movement Control System. In: *Computational Neuroscience.* J.M. Bower, ed. pp. 305-310. Academic Press, New York.

Abbas, J.J. and Chizeck, H.J. 1995. Neural network control of functional neuromuscular stimulation systems. *IEEE Trans. Biomed. Eng.*, 42(11):1117-1127.

Abbas, J.J. and Triolo, R.J. 1997. Experimental evaluation of an adaptive feedforward controller for use in functional neuromuscular stimulation systems. *IEEE Trans. Rehab. Eng.*, 5(1):12-22.

Antsaklis, P.J. 1990. Neural Networks in Control Systems. *IEEE Control Systems Magazine*, 10(3):3-5.

Barto, A.G. 1990. Connectionist Learning for Control: An Overview. In: *Neural Networks for Control.* W.T. Miller, R.S. Sutton, and P.J. Werbos, eds. pp. 5-58. MIT Press, Cambridge, MA.

Barto, A.G. 1992. Reinforcement Learning and Adaptive Critic Methods. In: *Handbook of Intelligent Control: Neural, Fuzzy and Adaptive Approaches.* D.A. White and D.A. Sofge, eds. pp. 469-492. Van Nostrand Reinhold, New York.

Beer, R.D. 1990. *Intelligence as adaptive behavior: an experiment in computational neuroethology,* Academic Press, Boston.

Beer, R.D., Chiel, H.J., Quinn, R.D., Espenschied, K.S., and Larsson, P. 1992. A Distributed Neural Network Architecture for Hexapod Robot Locomotion. *Neural Computation*, 4:356-365.

Beer, R.D., Ritzmann, R.E., and McKenna, T. 1993. *Biological Neural Networks in Invertebrate Neuroethology and Robotics,* Academic Press, New York.

Bower, J.M., 1998. Computational Neuroscience: Trends in Research, Plenum Press, New York.

Brown, T.H., Kairiss, E.W., and Keenan, C.L. 1990. Hebbian synapses: biophysical mechanisms and algorithms. *Annu. Rev. Neurosci.*, 13:475-511.

Bullock, D. and Grossberg, S. 1988. Neural Dynamics of Planned Arm Movements: Emergent Invariants and Speed-Accuracy Properties During Trajectory Formation. In: *Neural Networks and Natural Intelligence.* S. Grossberg, ed. pp. 553-622. MIT Press, Cambridge, MA.

Carpenter, G. 1989. Neural Network Models for Pattern Recognition and Associative Memory. *Neural Networks,* 2(4):243-258.

Chen, C.-T, Lin, W.-L., Kuo, T.-S, and Wang, C.-Y. 1997. Adaptive Control of Arterial Blood Pressure with a Learning Controller Based on Multilayer Neural Networks. *IEEE Trans. BME,* 44(7):601-609.

Chon, K.H., Holstein-Rathlou, N.-H., Marsh, D.J. and Marmarelis, V.Z. 1998. Comparative Nonlinear Modeling of Renal Autoregulation in Rats: Volterra Approach Versus Artificial Neural Networks. *IEEE Trans. Neural Networks*, 9(3):430-435.

Farotimi, O., Dembo, A., and Kailath, T. 1991. A General Weight Matrix Formulation Using Optimal Control. *IEEE Trans. Neural Networks*, 2(3):378-394.

Fogel, D. and Sebald, A.V. 1990. Use of Evolutionary Programming in the Design of Neural Networks for Artifact Detection. *Proc. IEEE/EMBS Conf.*, 12:1408-1409.

Grossberg, S. 1988a. *Neural Networks and Natural Intelligence*, MIT Press, Cambridge, MA.

Grossberg, S. 1988b. Nonlinear neural networks: principles, mechanisms and architectures. *Neural Networks,* 1:17-61.

Hagan, M.T., Demuth, H.B., and Beale, M. 1996. *Neural Network Design*, PWS Publishing Co., New York.

Hebb, D.O. 1949. *The Organization of Behavior, A Neuropsychological Theory,* John Wiley, New York.

Hecht-Nielson, R. 1989. *Neurocomputing,* Addison-Wesley, New York.

Hinton, G.E. 1989. Connectionist learning procedures. *Art. Intell.*, 40:185-234.

Hopfield. 1982. Neural networks and physical systems with emergent collective computational abilities. *Proc. Natl. Acad. Sci., USA*, 79:2554-2558.

Houk, J.C., Singh, S.P., Fisher, C., and Barto, A.G. 1990. An Adaptive Sensorimotor Network inspired by the Anatomy and Physiology of the Cerebellum. In: *Neural Networks for Control.* W.T. Miller, R.S. Sutton, and P.J. Werbos, eds. pp. 301-348. MIT Press, Cambridge, MA.

Kawato, M., Furukawa, K., and Suzuki, R. 1987. A Hierarchical Neural-Network Model for Control and Learning of Voluntary Movement. *Biol. Cybern.*, 57:169-185.

Koch, C. and Segev, I. 1989. *Methods in Neuronal Modeling: From Synapses to Networks,* MIT Press, Cambridge, MA.

Kohonen, T. 1989. *Self-Organization and Associative Memory,* Springer-Verlag, New York.

Lan, N., Feng, H.Q. and Crago, P.E. 1994. Neural Network Generation of Muscle Stimulation Patterns for Control of Arm Movements. *IEEE Trans. Rehabil. Eng.* 2(4):213-224.

Levin, E., Gewirtzman, R., and Inbar, G.E. 1991. Neural Network Architecture for Adaptive System Modeling and Control. *Neural Networks,* 4:185-191.

Lippman, R.P. 1987. An Introduction to Computing with Neural Nets. *IEEE Mag. on Acoustics, Signal and Speech Proc.,* April:4-22.

Miller, W.T., Hewes, R.P., Glanz, F.G., and Kraft, L.G.I.I.I. 1990a. Real-time Dynamic Control of an Industrial Manipulator Using a Neural-Network-Based Learning Controller. *IEEE Trans. Robotics & Automation,* 6(1):1-9.

Miller, W.T., Sutton, R.S., and Werbos, P.J. 1990b. *Neural Networks for Control,* MIT Press, Cambridge, MA.

Missler, J.M. and Kamangar, F.A. 1995. A Neural Network for Pursuit Tracking Inspired by the Fly Visual System. *Neural Networks,* 8(3):463-480.

Miyamoto, H., Kawato, M., Setoyama, T., and Suzuki, R. 1988. Feedback-error-learning neural network for trajectory control of a robotic manipulator. *Neural Networks,* 1:251-265.

Narendra, K.S. 1992. Adaptive Control of Dynamical Systems Using Neural Networks. In: *Handbook of Intelligent Control: Neural, Fuzzy and Adaptive Approaches.* D.A. White and D.A. Sofge, eds. pp. 141-184. Van Nostrand Reinhold, New York.

Narendra, K.S. and Parthasarathy, K. 1990. Identification and control of dynamical systems using neural networks. *IEEE Trans. Neural Networks,* 1(1):4-27.

Nerrand, O., Roussel-Ragot, P., Personnaz, L., Dreyfus, G., and Marcos, S. 1993. Neural networks and nonlinear adaptive filtering: unifying concepts and new algorithms. *Neural Computation,* 5:165-199.

Nguyen, D.H. and Widrow, B. 1990. Neural Networks for Self-learning Control Systems. *IEEE Control Systems Magazine,* 10(3):18-23.

Nordgren, R.E. and Meckl, P.H. 1993. An Analytical Comparison of a Neural Network and a Model-Based Adaptive Controller. *IEEE Trans. Neural Networks,* 4(4):685-694.

Pao, Y.H. 1989. *Adaptive Pattern Recognition and Neural Networks,* Addison-Wesley, Reading, MA.

Pined, F.J. 1989. Recurrent Backpropogation and the Dynamical Approach to Adaptive Neural Computation. *Neural Computation,* 1:161-172.

Quinn, R.D. and Espenschied, K.S. 1993. Control of a Hexapod Robot Using a Biologically Inspired Neural Network. In: *Biological Neural Networks in Invertebrate Neuroethology and Robotics.* R.D. Beer, R.E. Ritzmann, and T. McKenna, eds. pp. 365-382. Academic Press, New York.

Rumelhart, D.E. and McClelland, J.L. 1988. *Parallel Distributed Processing: Explorations in the Microstructure of Cognition,* MIT Press, Cambridge, MA.

Sanchez-Sinencio, E. and Lau, C. 1992. *Artificial Neural Networks: Paradigms, Applications and Hardware Implementations,* IEEE Press, New York.

Schwartz, E.L. 1990. *Computational Neuroscience,* MIT Press, Cambridge, MA.

Simpson. 1990. *Artificial Neural Systems,* Pergamon Press, New York.

Srinivasan, S., Gander, R.E., and Wood, H.C. 1992. A Movement Pattern Generator Model Using Artificial Neural Networks. *IEEE Trans. BME,* 39(7):716-722.

Taga, G., Yamaguchi, Y., and Shimizu, H. 1991. Self-organized control of bipedal locomotion by neural oscillators in unpredictable environment. *Biol. Cybern.,* 65:147-159.

Trajanoski, Z., and Wach, P. 1998. Neural Predictive Controller for Insulin Delivery Using the Subcutaneous Route. *IEEE Trans. BME,* 45(9):1122-1134.

Vogl, T.P., Mangis, J.K., Rigler, A.K., Zink, W.T., and Alkon, D.L. 1988. Accelerating the convergence of the back-propagation method. *Biol. Cybern.,* 59:257-263.

Wang, H., Lee, T.T., and Graver, W.A. 1992. A Neuromorphic Controller for a Three-Link Biped Robot. *IEEE Trans. Sys., Man, Cyber.,* 22(1):164-169.

Werbos, P.J. 1990. Overview of Designs and Capabilities. In: *Neural Networks for Control.* W.T. Miller, R.S. Sutton, and P.J. Werbos, eds. pp. 59-66. MIT Press, Cambridge, MA.

Werbos, P.J. 1992. Approximate Dynamic Programming for Real-Time Control and Neural Modeling. In: *Handbook of Intelligent Control: Neural, Fuzzy and Adaptive Approaches.* D.A. White and D.A. Sofge, eds. pp. 493-526. Van Nostrand Reinhold, New York.

White, D.A. and Sofge, D.A. 1992. *Handbook of Intelligent Control: Neural, Fuzzy, and Adaptive Approaches,* Van Nostrand Reinhold, New York.

Widrow, B. 1962. Generalization and Information Storage in Networks of Adaline 'Neurons'. In: *Self-Organizing Systems.* M.C. Jovitz, T. Jacobi, and Goldstein, eds. pp. 435-461. Spartan Books, Washington, D.C.

Williams, R.J. and Zipser, D. 1989. A learning algorithm for continually running fully recurrent neural networks. *Neural Computation,* 1:270-280.

Zurada, J., M. 1992. *Artificial Neural Systems,* West Publishing Co., New York.

Further Information

Very good introductions to the operation of neural networks are given in *Neural Network Design* by Hagan, Demuth & Beale and in *Artificial Neural Systems* by J. M. Zurada.

Detailed descriptions of the neural network control systems described in this chapter are provided in *Neural Network Control Systems* edited by Miller, Sutton & Werbos and *The Handbook of Intelligent Control* edited by White & Sofge.

For journal articles on neural network theory and applications, the reader is referred to *IEEE Transactions on Neural Networks, Neural Computation* (MIT Press), and *Neural Networks* (Pergamon Press).

For occasional journal articles on biomedical applications of neural networks, the reader is referred to *IEEE Transactions on Biomedical Engineering, IEEE Transactions on Rehabilitation Engineering* and the *Annals of Biomedical Engineering.*

163

Methods and Tools for Identification of Physiologic Systems

Vasilis Z. Marmarelis
University of Southern California

The problem of *system identification* in physiology derives its importance from the need to acquire *quantitative models* of physiologic function (from the subcellular to the organ system level) by use of experimental observations (data). Quantitative models can be viewed as summaries of experimental observations that allow scientific inference and organize our knowledge regarding the functional properties of physiologic systems. Selection of the proper (mathematical or computational) form of the model is based on existing knowledge of the system's functional organization. System identification is the process by which the system model is determined from data. This modeling and identification problem is rather challenging in the general use of a physiologic system, where insufficient knowledge about the internal workings of the system or its usually confounding complexity prevents the development of an explicit model. The models may assume diverse forms (requiring equally diverse approaches) depending on the specific characteristics of the physiologic system (e.g., static/dynamic, linear/nonlinear) and the prevailing experimental conditions (e.g., noise contamination of data, limitations on experimental duration). This chapter will not address the general modeling issue, but rather it will concentrate on specific methods and tools that can be employed in order to accomplish the system identification task in most cases encountered in practice. Because of space limitations, the treatment of these system identification methods will be consistent with the style of a review article providing overall perspective and guidance while deferring details to cited references.

Since the complexity of the physiologic system identification problem rivals its importance, we begin by demarcating those areas where effective methods and tools currently exist. The selection among candidate models is made on the basis of the following key functional characteristics: (1) static or dynamic; (2) linear or nonlinear; (3) stationary or nonstationary; (4) deterministic or stochastic; (5) single or multiple inputs and/or outputs; (6) lumped or distributed. These classification criteria do not constitute an exhaustive list but cover most cases of current interest. Furthermore, it is critical to remember that contaminating noise (be it systemic or measurement-related) is always present in an actual study, and experimental constraints often limit experimentation time and the type of data obtainable from the system. Finally, the computational requirements for a practicable identification method must not be extraordinary, and the obtained results (models) must be amenable to physiologic interpretation, in addition to their demonstrated predictive ability.

0-8493-0462-8/00/$0.00+$.50

A critical factor in determining our approach to the system modeling and identification task is the availability and quality of prior knowledge about the system under study with respect to the mechanisms that subserve its function. It can be said in general that, if sufficient knowledge about the internal mechanisms subserving the function of a system is available, then the development of an explicit model from first (physical or chemical) principles is possible, and the system modeling and identification task is immensely simplified by reducing to an estimation problem of the unknown parameters contained in the explicit model. Although the ease of this latter task depends on the manner in which the unknown parameters enter in the aforementioned explicit model, as well as on the quality of the available data, it is typically feasible. Furthermore, the obtained model is amenable to direct and meaningful physiologic interpretation. Unfortunately, it is rare that such prior knowledge (of adequate quality and quantity) is available in systems physiology. It is far more common that only limited prior knowledge is available, relative to the customary complexity of physiologic systems, that prevents the development of an explicit model from first principles and necessitates the search for a model that is compatible with the available input-output data.

Thus, the system identification problem is typically comprised of two tasks:

1. *Model Specification:* the selection or postulation of a model form suitable for the system at hand
2. *Model Estimation:* the estimation of the unknown parameters or functions contained within the specified model, using experimental data

All prior knowledge about the system under study is utilized in the model specification task. This includes results from specially designed preliminary experiments, which can be used to establish, for instance, whether the system is static or dynamic, linear or nonlinear, and so on. The model estimation task, however, relies on the quality of the available data (e.g., spectral characteristics, noise conditions, data length) and may set specific data-collection requirements.

This article will focus on practicable methods to perform both the model specification and model estimation tasks for systems/models that are *static or dynamic* and *linear or nonlinear.* Only the stationary case will be detailed here, although the potential use of nonstationary methods will be also discussed briefly when appropriate. In all cases, the models will take deterministic form, except for the presence of additive error terms (model residuals). Note that stochastic experimental inputs (and, consequently, outputs) may still be used in connection with deterministic models. The cases of multiple inputs and/or outputs (including multidimensional inputs/outputs, e.g., spatio-temporal) as well as lumped or distributed systems, will not be addressed in the interest of brevity. It will also be assumed that the data (single input and single output) are in the form of evenly sampled *time-series,* and the employed models are in discrete-time form (e.g., difference equations instead of differential equations, discrete summations instead of integrals).

In pursuing the model specification task, two general approaches have developed: *parametric* and *nonparametric.* In the parametric approach, algebraic or difference equation models are typically used to represent the input-output relation for static or dynamic systems, respectively. These models are accordingly linear or nonlinear, and contain a (typically small) number of unknown parameters. The latter may be constant or time-varying depending on whether the system or model is stationary or nonstationary. The precise form of these parametric models (e.g., degree/order of equation) must be determined in order to complete the model specification task. This precise form is either postulated a priori or guided by the data (in which case it is intertwined with the model parameter estimation task) as outlined in the following section.

In the nonparametric approach, the input-output relation is represented either analytically (in convolutional form through Volterra-Wiener expansions where the unknown quantities are kernel functions), computationally (i.e., by compiling all input-output mapping combinations to look-up tables) or graphically (in the form of operational surfaces/subspaces in phase-space). The graphical representation, of course, is subject to the three-dimensional limitation for the purpose of visual inspection. The model specification requirements for the Volterra-Wiener formulation consist of the order of system nonlinearity and system memory, and, for the computational or graphical approach, they consist of defining the appropriate phase-space mapping dimensions, as discussed below. Of the nonparametric approaches, the Volterra-Wiener (kernel) formulation has been used more extensively in a nonlinear context and will be the focus of this review, since nonlinearities are ubiquitous in physiology. Note that the nonparametric model estimation task places certain requirements on the experimental input (i.e., sufficient coverage of the frequency bandwidth and amplitude range of interest in each application) in order to secure adequate probing of the system functional characteristics.

An important hybrid approach has also developed in recent years that makes use of *block-structured* or *modular* models. These models are composed of parametric and/or nonparametric components properly connected to represent reliably the input-output relation. The model specification task for this class of models is more demanding and may utilize previous parametric and/or nonparametric modeling results. A promising variant of this approach, which derives from the general Volterra-Wiener formulation, employs *principal dynamic modes* as a canonical set of filters to represent a broad class of nonlinear dynamic systems. Another variant of the modular approach that has recently acquired considerable popularity but will not be covered in this review is the use of *artificial neural networks* to represent input-output nonlinear mappings in the form of *connectionist models*. These connectionist models are often fully parametrized, making this approach affine to parametric modeling, as well.

The relations among these approaches (parametric, nonparametric, and modular) are of critical practical importance and the subject of several recent studies [Marmarelis, 1994, 1997]. Considerable benefits may accrue from the combined use of these approaches in a cooperative manner that aims at securing the full gamut of distinct advantages specific to each approach.

In the following sections, an overview of these methodologies will be presented and the relative advantages and disadvantages in practical applications will be briefly outlined. The ultimate selection of a particular methodology hinges upon the specific characteristics of the application at hand and the prioritization of objectives by the individual investigator. Two general comments are in order:

1. No single methodology is globally superior to all others, i.e., excelling with regard to all criteria and under all possible circumstances. Judgment must be exercised in each individual case.
2. The general system identification problem is not solvable in all cases at present, and challenging problems remain for future research.

163.1 Parametric Approach

Consider the input and output time-series data, $x(n)$, respectively. If the system is *static and linear,* then we can employ the simplest (and most widely used) model of linear regression

$$y(n) = ax(n) + b + \epsilon(n) \tag{163.1}$$

to represent the input-output relation for every n, where $\varepsilon(n)$ represents the noise or error term (or model residual) at each n. The unknown parameters (a,b) can be easily estimated through least-squares fitting, using a well-developed set of linear regression methods (e.g., ordinary least-squares or generalized least-squares depending on whether $\varepsilon(n)$ is a white sequence).

This model can be extended to multiple inputs $\{x_1, x_2, \ldots, x_k\}$ and outputs $\{y_i\}$ as

$$y_i(n) = a_1x_1(n) + a_2x_2(n) + \ldots + a_kx_k(n) + b + \epsilon(n) \tag{163.2}$$

and well-developed multiple linear regression techniques exist that can be used for estimation of the unknown parameters $(a_1, a_2, \ldots, a_k, b)$ for each output y_i. Although these estimation methods are widely known and readily available in the literature—see, for instance, Eykhoff [1974] and Soderstrom and Stoica [1989]—they will be briefly reviewed at the end of this section.

In the event of nonstationarities, the regression coefficients (model parameters) will vary through time and can be estimated either in a piecewise stationary fashion over a sliding time window (batch processing) or in a recursive fashion using an adaptive estimation formula (recursive processing). The latter has been favored and extensively studied in recent years [Goodwin & Sin, 1984; Ljung, 1987; Ljung & Soderstrom, 1983], and it is briefly outlined at the end of this section.

If the system is *static and nonlinear,* then a nonlinear input-output relation

$$y(n) = \sum_{j=1}^{J} c_j P_j\left[x(n)\right] + \epsilon(n) \tag{163.3}$$

can be used as a parametric model, where the $\{P_j\}$ functions represent a set of selected nonlinear functions (e.g., powers, polynomials, sinusoids, sigmoids, exponentials or any other suitable of functions over the range of x), and $\{c_j\}$ are the unknown parameters that can be easily estimated through linear regression—provided that the $\{P_j\}$ functions do not contain other unknown parameters in a nonlinear fashion. In the latter case, nonlinear regression methods must be used (e.g., the gradient steepest-descent method) that are well developed and readily available [Eykhoff, 1974; Ljung, 1987; Soderstrom & Stoica, 1989], although their use is far from problem-free or devoid of risk of misleading results (e.g., local minima or noise effects). Naturally, the choice of the $\{P_j\}$ functions is critical in this regard and may depend on the characteristics of the system or the type of available data.

These cases of static systems have been extensively studied to date but have only limited interest or applicability to actual physiologic systems, since the latter are typically dynamic—i.e., the output value at time n depends also on input and/or output values at other previous times (lags). Note that the possible dependence of the present output value on previous output values can be also expressed as a dependence on previous input values. Thus, we now turn to the all-important case of dynamic systems.

For linear (stationary) dynamic systems, the discrete-time parametric model takes the form of an *auto-regressive moving average with exogenous variable* (ARMAX) equation:

$$\begin{aligned} y(n) = {} & \alpha_1 y(n-1) + \ldots + \alpha_k y(n-k) + \beta_0 x(n) + \beta_1 x(n-1) \\ & + \ldots + \beta_l x(n-l) + w(n) + \gamma_1 w(n-1) \\ & + \ldots + \gamma_m w(n-m) \end{aligned} \tag{163.4}$$

where $w(n)$ represents a white noise sequence. This ARMAX model is a difference equation that expresses the present value of the output, $y(n)$, as a linear combination of k previous values of the output (AR part), l previous (and the present) values of the input (X part), and m previous (and the present) values of the white noise disturbance sequence (MA part) that compose the model residual (error). When $\gamma_i = 0$ for all $i = 1, \ldots, m$, the residuals form a white sequence, and the coefficients ($\alpha_1, \ldots, \alpha_k, \beta_0, \beta_1, \ldots, \beta_l$) can be estimated through the ordinary least-squares procedure. However, if any γ_i is nonzero, then unbiased and consistent estimation requires generalized or extended least-squares procedures, similar to the one required for the multiple regression model of Eq. (163.2) when $\varepsilon(n)$ is a nonwhite error sequence (reviewed at the end of the section).

Although the estimation of the ARMAX model parameters can be straightforward through multiple linear regression, the model specification task remains a challenge. The latter consists of determining the maximum lag values (k, l, m) in the difference equation (163.4) from given input-output data, $x(n)$ and $y(n)$. A number of statistical procedures have been devised for this purpose (e.g., weighted residual variance, Akaike information criterion), all of them based on the prediction error for given model order (k, l, m) compensated properly for the remaining degrees of freedom (i.e., the total number of data minus the number of parameters). It is critical that the prediction error (typically, in mean-square sense) be evaluated on a segment of input-output data distinct from the one used for the estimation of the model parameters. Application of these criteria is repeated successively for ascending values of the model order (k, l, m), and the model specification process is completed when an extremum (minimum or maximum, depending on the criterion) is achieved [Soderstrom & Stoica, 1989].

For nonlinear (stationary) systems, the ARMAX model can be extended to the NARMAX model (nonlinear ARMAX) that includes nonlinear expressions of the variables on the right side of Eq. (163.4) [Billings & Voon, 1984]. For instances, a second-degree multinomial NARMAX model of order ($k = 2$, $l = 1$, $m = 0$) with additive white-noise residuals takes the form

$$\begin{aligned} y(n) = {} & \alpha_1 y(n-1) + \alpha_2 y(n-2) + \alpha_{1,1} y^2(n-1) + \alpha_{1,2} y(n-1) y(n-2) \\ & + \alpha_{2,2} y^2(n-2) + \beta_0 x(n) + \beta_1 x(n-1) + \beta_{0,0} x^2(n) \\ & + \beta_{0,1} x(n) x(n-1) + \beta_{1,1} x^2(n-1) + \gamma_{1,0} y(n-1) x(n) \\ & + \gamma_{2,0} y(n-2) x(n) + \gamma_{1,1} y(n-1) x(n-1) \\ & + \gamma_{2,1} y(n-2) x(n-1) + w(n) \end{aligned} \tag{163.5}$$

Clearly, the form of a NARMAX model may become rather unwieldy, and the model specification task (i.e., the form and degree of nonlinear terms, as well as the number of input, output, and noise lags involved in the model) is very challenging. Several approaches have been proposed for this purpose [Billings & Voon, 1984, 1986; Haber & Unbenhauen, 1990; Korenberg, 1988; Zhao & Marmarelis, 1994], and they are all rather involved computationally. However, if the structure of the NARMAX model is established, then the parameter estimation task is straightforward in a manner akin to multiple linear regression—since the unknown parameters enter linearly in the NARMAX model.

It is evident that all these multiple linear regression problems defined by Eqs. (163.2)–(163.5) can be written in a vector form:

$$y(n) = \phi^T(n)\theta + \epsilon(n) \tag{163.6}$$

where $\phi(n)$ represents the vector of all regression variables in each case, and θ denotes the unknown parameter vector. For a set of data points ($n = 1, \ldots, N$), Eq. (163.6) yields the matrix formulation

$$y = \Phi\boldsymbol{\theta} + \epsilon \tag{163.7}$$

The ordinary least-squares (OLS) estimate of θ

$$\hat{\theta}^{OLS} = \left[\Phi^T\Phi\right]^{-1} \Phi^T y \tag{163.8}$$

yields unbiased and consistent estimates of minimum variance, if $\epsilon(n)$ is a white sequence; otherwise, the generalized least-squares (GLS) estimator

$$\hat{\theta}^{GLS} = \left[\Phi^T\Sigma^{-1}\Phi\right]^{-1} \Phi^T\Sigma^{-1} y \tag{163.9}$$

ought to be used to achieve minimum estimation variance, where Σ denotes the covariance matrix of ϵ. Practical complications arise from the fact that Σ is not a priori known and, therefore, must be either postulated or estimated from the data. The latter case, which is more realistic in actual applications, leads to an iterative procedure that may not be convergent or yield satisfactory results (i.e., start with an OLS estimate of θ; obtain a first estimate of ϵ and Σ; evaluate a first GLS estimate of θ; obtain a second estimate of ϵ and Σ; and iterate until the process converges to a final GLS estimate of θ). As an alternative to this iterative procedure, an estimate of the *moving-average model* of the residual term:

$$\epsilon(n) = w(n) + \gamma_1 w(n-1) + \ldots + \gamma_m w(n-m) \tag{163.10}$$

may be obtained from initial OLS estimation, and the covariance matrix Σ may be estimated from Eq. (163.10). Although this alternative approach avoids problems of convergence, it does not necessarily yield satisfactory results due to the dependence on the noise model estimation of Eq. (163.10). Equivalent to this latter procedure is the *residual whitening* method, which amounts to prefiltering the data with the inverse of the transfer function corresponding to Eq. (163.10) (*prewhitening filter*), prior to OLS estimation. Finally, the parameter vector may be augmented to include the coefficient $\{\gamma_i\}$ of the moving-average model of the residuals in Eq. (163.10), leading to a *pseudo-linear* regression problem [since the estimates of $w(n)$ depend on $\hat{\theta}$] that is operationally implemented as an iterative *extended least-squares* (ELS) procedure [Billings & Voon, 1984, 1986; Ljung, 1987; Ljung & Soderstrom, 1983; Soderstrom & Stoica, 1989].

In the presence of nonstationarities, the parameter vector θ may vary through time, and its estimation may be performed either in a piecewise stationary manner (with segment-to-segment updates that may be obtained either recursively or through batch processing) or by introducing a specific parametrized time-varying structure for the parameters and augmenting the unknown parameter vector to include the additional parameters of the time-varying structure (e.g., a pth degree polynomial structure of time-varying model parameters will introduce p additional parameters for each time-varying term into the vector θ, that need be estimated from the data via batch processing).

Any batch processing approach folds back to the previously discussed least-squares estimation methods. However, the recursive approach requires a new methodological framework that updates continuously the parameter estimates on the basis of new data. This adaptive or recursive approach has gained increasing popularity in recent years, although certain important practical issues (e.g., speed of algorithmic convergence, effect of correlated noise) remain causes for concern in certain cases [Goodwin & Sin, 1984; Ljung, 1987; Ljung & Soderstrom, 1983]. The basic formulae for the recursive least-squares (RLS) algorithm are

$$\hat{\boldsymbol{\theta}}(n) = \hat{\boldsymbol{\theta}}(n-1) + \boldsymbol{\Psi}(n)\left[y(n) - \phi^{T}(n)\hat{\boldsymbol{\theta}}(n-1)\right] \tag{163.11}$$

$$\boldsymbol{\Psi}(n) = \gamma(n)P(n-1)\boldsymbol{\phi}(n) \tag{163.12}$$

$$P(n) = P(n-1) - \gamma(n)P(n-1)\boldsymbol{\phi}(n)\boldsymbol{\phi}^{T}(n)P(n-1) \tag{163.13}$$

$$\gamma(n) = \left[\boldsymbol{\phi}^{T}(n)P(n-1)\boldsymbol{\phi}(n) + \alpha(n)\right]^{-1} \tag{163.14}$$

where the matrix $P(n)$, the vector $\Psi(n)$ and the scalar $\gamma(n)$ are updating instruments of the algorithm computed at each step n. Note that $\{\alpha(n)\}$ denotes a selected sequence of weights for the squared prediction errors in the cost function and is often taken to be unity. This recursive algorithm can be used for on-line identification of stationary or nonstationary systems. A critical issue in the nonstationary case is the speed of algorithmic convergence relative to the time-variation of the system/model parameters.

Its initialization is commonly made as: $\hat{\theta}(0) = 0$ and $P(0) = c_0 I$, where c_0 is a large positive constant (to suppress the effect of initial conditions) and I is the identity matrix.

It is important to note that when output autoregressive terms exist in the model, the regression vector $\phi(n)$ is correlated with the residual $\epsilon(n)$, and, thus, none of the aforementioned least-squares estimates of the parameter vector will converge to the actual value. This undesirable correlation weakens when the predicted output values are used at each step for the autoregressive lagged terms (*closed-loop mode* or one-step predictive model) instead of the observed output values (*open-loop mode* or global predictive model). To remedy this problem, the *instrumental variable* (IV) method has been introduced that makes use of a selected IV that is uncorrelated with the residuals but strongly correlated with the regression vector $\phi(n)$ in order to evaluate the least-squares estimate [Soderstrom & Stoica, 1989]. The IV estimates can be computed in batch or recursive fashion.

In closing this section, we should note that a host of other estimation methods has been developed through the years and that these methods are operationally affine to the foregoing ones (e.g., prediction-error methods, stochastic approximation, gradient-based methods) or represent different statistical approaches to parameter estimation (e.g., maximum likelihood, bayesian approach) including the state-space formulation of the problem, that cannot be detailed here in the interest of space [for review, see Eykhoff, 1974; Goodwin & Payne, 1977; Goodwin & Sin, 1984; Ljung, 1987; Ljung & Soderstrom, 1983; Soderstrom & Stoica, 1989].

163.2 Nonparametric Approach

In the linear stationary case, the discrete-time nonparametric model takes the convolutional form:

$$y(n) = \sum_{m=0}^{M} h(m)\, x(n-m) + \epsilon(n) \tag{163.15}$$

where $\epsilon(n)$ is an output-additive error/noise term and $h(m)$ is the discrete impulse response of the linear time-invariant (stationary) system with memory extent M. For finite-memory systems (i.e., when $h(m)$ becomes negligible for $m > M$ and M is finite), the estimation of $h(m)$ from input-output data can be accomplished by multiple regression, since the model of Eq. (163.15) retains the form of a parametric model with no autoregressive terms. Note that any stable ARMAX model can be put in the form of Eq. (163.15), where $M \to \infty$ and $h(m)$ is absolute-summable. The estimation of $h(m)$ can be also accomplished in the frequency domain via discrete Fourier transforms (DFTs, implemented as FFTs), observing the fact that convolution turns into multiplication in the frequency domain. Particular attention must be paid to cases where the input DFT attains very small values to avoid numerical or estimation problems during the necessary division. In actual physiologic practice, many investigators have chosen to perform experiments with specific input waveforms to facilitate this identification task, i.e., the use of an impulsive input of unit strength directly yields $h(m)$, or the use of sinusoidal inputs of various frequencies yields directly the values of the DFT of $h(m)$ at the corresponding frequencies (covering the entire frequency range of interest).

In the nonlinear stationary case, the most widely used methodology for nonparametric modeling is based on the Volterra functional expansions and Wiener's theory that employs a gaussian white noise (GWN) test input in conjunction with a modified (orthogonalized) Volterra functional expansion. Wiener's critical contribution is in suggesting that GWN is an *effective test input* for identifying nonlinear systems of a very broad class and in proposing specific mathematical procedures for the estimation of the unknown system descriptors (kernels) from input-output data, as outlined below [for more details, see Marmarelis & Marmarelis, 1987; Rugh, 1981; Schetzen, 1980].

The input-output relation of a causal nonlinear stationary system in discrete time is seen as a mapping of a vector comprised of the input past (and present) values onto the (scalar) present value of the output:

$$y(n) = F\left[x(n'), n - M \leq n' \leq n\right] \tag{163.16}$$

were M is the system memory and F is a fixed multivariate function representing this mapping—note that, in continuous time, F is a functional. If the system is nonstationary, the function F varies through time. If the function F is analytic and the system is stable, then a discrete-time Volterra series expansion exists of the form:

$$y(n) = \sum_{i=0}^{\infty} \sum_{m_1=0}^{M} \cdots \sum_{m_i=0}^{M} k_i(m_1, \ldots, m_i) x(n - m_1) \ldots x(n - m_i) \tag{163.17}$$

This series converges for any stable system within the radius of convergence defined by the input ensemble. If the multivariate function F is not analytic but is continuous, then a finite-order Volterra model can be found that achieves any desirable degree of approximation (based on the Stone-Weierstrass theorem).

The multiple convolutions of the Volterra model involve kernel functions $\{k_i(m_1, \ldots, m_i)\}$ which constitute the descriptors of the system nonlinear dynamics. Consequently, the system identification task is to obtain estimates of these kernels from input-output data. These kernel functions are symmetric with respect to their arguments.

When a sinusoidal input is used, then the ith order Volterra functional (i.e., the i-tuple convolution in the Volterra model) gives rise to output harmonics of order i, $(i - 2)$, …, $(i - 2j)$ where j is the integer part $i/2$. When an impulse $A\delta(n)$ is used as input, then the ith order Volterra functional is contributing the term $A^i k_i(n, \ldots, n)$ to the output. This suggests that the Volterra kernels of a general system cannot be separated from each other and directly determined from input-output data, unless the Volterra expansion is of finite order. For a finite-order Volterra expansion, kernel estimation can be achieved through least-squares fitting procedures [Korenberg, 1988; Stark, 1968; Watanabe & Stark, 1975] or by use of specialized test inputs such as multiple impulses or step function [Schetzen, 1965; Shi & Sun, 1990; Stark, 1968] and sums of sinusoids of incommensurate frequencies [Victor et al., 1977], although the testing and analysis procedures can be rather laborious in those cases. The latter method yields estimates of the system kernels in the frequency domain at the selected input frequencies (and their harmonic/intermodulation frequencies) with increased accuracy but with limited frequency resolution.

The inability to estimate the Volterra kernels in the general case of an infinite series prompted Wiener to suggest the orthogonalization of the Volterra series when a GWN test input is used. The functional terms of the Wiener series are constructed on the basis of a Gram-Schmidt orthogonalization procedure requiring that the covariance between any two Wiener functionals be zero. The resulting Wiener series expansion takes the form:

$$y(n) = \sum_{i=0}^{\infty} G_i\left[h_i; x(n'), n' \leq n\right]$$

$$= \sum_{i=0}^{\infty} \sum_{j=0}^{[i/2]} \frac{(-1)^j i! P^j}{(i-2j)! j! 2^j} \sum_{m_1} \cdots \sum_{m_i} h_i(m_1, \ldots, m_{i-2j}, l_1, l_1, \ldots, l_j, l_j) \, x(n - m_1) \ldots x(n - m_{i-2j}) \tag{163.18}$$

where $[i/2]$ is the integer part of $i/2$ and P is the power level of the GWN input. The set of Wiener kernels $\{h_i\}$ is, in general, different from the set of Volterra kernels $\{k_i\}$, but specific relations exist between the two sets [Marmarelis & Marmarelis, 1978].

The Wiener kernels depend on the GWN input power level P (because they correspond to an orthogonal expansion), whereas the Volterra kernels are independent of any input characteristics. This situation can be likened to the coefficients of an orthogonal expansion of an analytic function being dependent on the interval of expansion. It is therefore imperative that Wiener kernel estimates be reported in the literature with reference to the GWN input power level that they were estimated with. When a complete set of Wiener kernels is obtained, then the complete set of Volterra kernels can be evaluated. Approximations of Volterra kernels can be obtained from Wiener kernels of the same order estimated with various input power levels. Complete Wiener or Volterra models can predict the system output to *any* given input. When the Wiener or Volterra model is incomplete, the accuracy of the predicted output will be, in general, different for the two models and will depend on each specific input.

The orthogonality of the Wiener series allows decoupling of the various Wiener functionals and the estimation of the respective Wiener kernels from input-output data through cross-correlation [Lee & Schetzen, 1965]

$$h_i\left(m_1, \ldots, m_i\right) = \frac{1}{i!\,P^i} E\left[y_i\left(n\right)x\left(n-m_1\right)\ldots x\left(n-m_i\right)\right] \tag{163.19}$$

where $y_i(n)$ is the ith response residual

$$y_i\left(n\right) = y\left(n\right) - \sum_{j=0}^{i-1} G_j\left(n\right) \tag{163.20}$$

The simplicity and elegance of the cross-correlation technique led to its adoption by many investigators in modeling studies of nonlinear physiologic systems [see McCann & Marmarelis, 1975; Marmarelis, 1987, 1989, 1994; Marmarelis & Marmarelis, 1978; Stark, 1968].

Since the ensemble average of Eq. (163.19) is implemented in practice by time-averaging over a finite data-record, and since GWN is not physically realizable, many importance practical issues had to be explored in actual applications of the cross-correlation technique. To name but a few: the generation of appropriate quasi-white test signals (that adequately approximate the ideal and not physically realizable GWN); the choice of input bandwidth relative to the system bandwidth; the accuracy of the obtained kernel estimates as a function of input bandwidth and record length; the effect of extraneous noise and experimental imperfections. An extensive study of some of these practical considerations can be found in Marmarelis and Marmarelis [1978] and in Marmarelis [1979].

A broad class of effective quasi-white test signals (CSRS) has been introduced that can be easily generated on the computer and may follow any zero-mean symmetric (discrete or continuous) amplitude probability density function [Marmarelis, 1977]. The use of this type of test signal allows through analysis of kernel estimation errors via cross-correlation and the optimization of the input bandwidth B (for given record length N) on the basis of a total mean-square error (TMSE) criterion for the ith-order kernel estimate given by:

$$\left(\text{TMSE}\right)_i = \frac{a_i}{B^4} + \frac{b_i B^{i-1}}{N} \tag{163.21}$$

where a_i and b_i are positive constants characteristic of the ith-order kernel [Marmarelis, 1979].

The cross-correlation technique results in considerable variance in the kernel estimates, unless very long data-records are available, because of the stochastic nature of the input. This has prompted the use of pseudo-random m-sequences [Billings & Fakhouri, 1981; Moller, 1983; Ream, 1970; Sutter, 1992]—which can reduce considerably the data-record requirements and yield improved estimation accuracy, provided that proper attention is given to certain problems in their high-order autocorrelation functions. To the same end, frequency-domain methods have been proposed for kernel estimation [Brillinger, 1970; French, 1976; Victor et al., 1977] yielding computational savings in some cases and offering a different perspective for the interpretation of results.

To reduce the requirements of long experimental data records and improve the kernel estimation accuracy, least-squares methods also can be used to solve the classical linear *inverse problem* described earlier in Eq. (163.6), where the parameter vector $\underline{\theta}$ includes all discrete kernel values of the finite Volterra model of Eq. (163.17), which is linear in these unknown parameters (i.e., kernel values). Least-squares methods also can be used in connection with orthogonal expansions of the kernels to reduce the number of unknown parameters, as outlined below. Note that solution of this inverse problem via OLS requires inversion of a large square matrix with dimensions $[(M + I + 1)!/((M + 1)!I!)]$, where M is the kernel memory and I is the (maximum) nonlinear order of the Volterra model. Since direct inversion of this matrix may require considerable computing effort and be subject to conditioning problems, it is often expedient to solve this inverse problem by use of QR decomposition that offers certain computational and numerical advantages. One such implementation is Korenberg's *exact orthogonalization method* [Korenberg, 1988], which estimates the discrete kernel values through least-squares fitting of a sequence of orthogonal vectors built from input values in accordance with the Volterra expansion (effectively forming the Q orthogonal matrix of the QR decomposition). Other numerical methods also can be used for matrix inversion, depending on the requirements of each application. Likewise, a variety of orthogonal bases can be used for kernel expansion in order to obtain a more concise representation of the kernels in each particular application. The use of the Laguerre orthogonal basis typically results in reduction of the number of unknown parameters for most physiologic system kernels (due to their built-in exponential structure), with consequent improvement in estimation accuracy [Marmarelis, 1993; Watanabe & Stark, 1975]. Thus, if $\{b_j(m)\}$ is a complete orthonormal basis defined over the system memory $[O, M]$, then the ith-order Volterra kernel can be expanded as

$$k_i\left(m_1, \ldots, m_i\right) = \sum_{j_1} \cdots \sum_{j_i} c_i\left(j_1, \ldots, j_i\right) b_{j_1}\left(m_1\right) \ldots b_{j_i}\left(m_i\right) \tag{163.22}$$

Then the Volterra model of Eq. (163.17) takes the form:

$$y\left(n\right) = \sum_{i} \left\{ \sum_{j_1} \cdots \sum_{j_i} c_i\left(j_1, \ldots, j_i\right) \mathrm{v}_{j_1}\left(n\right) \ldots \mathrm{v}_{j_i}\left(n\right) \right\} \tag{163.23}$$

where,

$$\mathrm{v}_j\left(n\right) = \sum_{m=0}^{M} b_j\left(m\right) x\left(n - m\right) \tag{163.24}$$

and the expansion coefficients $\{c_i(j_1, \ldots, j_i)\}$ of the unknown kernels can be estimated through multiple regression of $y(n)$ on the multinomial terms composed of the known functions $\{v_j(n)\}$. The kernel estimates are then reconstructed on the basis of Eq. (163.22). This kernel estimation method has been shown to be far superior to the conventional cross-correlation technique in terms of estimation accuracy and robustness to noise or experimental imperfections [Marmarelis, 1993].

Based on this observation, a general model for the Volterra class of systems can be proposed that is comprised of a set of parallel linear filters with impulse response functions $\{b_j(m)\}$ whose outputs are fed into a multi-input static nonlinearity, $y = f(v_1, v_2, \ldots)$, to produce the system output. If was Wiener who first proposed the use of Laguerre functions (albeit in continuous-time form) for the set $\{b_j(m)\}$, since they are defined over the interval $[0, \infty)$ compatible with causal systems and can be generated in analog form—a fashionable model at the time—by a simple RC ladder network. He also suggested that the multi-input static nonlinearity be expanded in terms of orthogonal hermite functions to yield a system characterization in terms of the resulting Hermite expansion coefficients that are estimated through covariance computations. For a review of this approach, see Schetzen [1980]. This approach has been viewed as rather unwieldy and has not found many applications to date. However, the use of Laguerre expansions of kernels (in discrete time) in conjunction with least-squares fitting has been shown to be rather promising for practical kernel estimation from short and noisy data records [Marmarelis, 1993] and consequently suitable for experimental studies of physiologic systems.

The greatest obstacle to the broader use of the Volterra-Wiener approach has been the practical limitations in estimating high-order kernels due to two reasons: (1) The amount of required computations increases geometrically with the order of estimated kernel; (2) kernel functions of more than three dimensions are difficult to inspect or interpret meaningfully. As a result, application of this approach has been limited to weakly nonlinear systems (second or third order).

A practical way of overcoming this limitation in the order of estimated kernels has been recently presented as a spin-off of studies on the relation between Volterra models and feedforward artificial neural networks [Marmarelis & Zhao, 1994]. According to this approach, a perceptronlike network with polynomial activation functions is trained with the experimental data using a modified version of the back-propagation algorithm. The parameters of the resulting network can be subsequently converted to kernel estimates of arbitrary high order (although the network itself is a nonlinear model of the system on its own right). This alternative to current kernel estimation algorithms seems to hold significant promise, since it relaxes many of the requirements of existing methods (e.g., input whiteness and low-order models) at the only apparent cost of greater computing effort.

In the event of nonstationarities, the kernel become dependent on time, and their estimation can be accomplished either in a piecewise stationary manner (when the nonstationary is slow relative to the system dynamic bandwidth) or through truly nonstationary estimation methods. In the former case, the piecewise stationary estimates can be obtained over adjacent segments or using sliding (overlapping) windows along the time record. They can be subsequently displayed in a time-ordered sequence (revealing the time-varying pattern), under the assumption of system quasi-stationarity over the data segment or sliding window. Kernel estimation through truly nonstationary methods may be accomplished by: (1) *adaptive methods,* employing the recursive least-squares formulae previously reviewed, which track changes in least-squares kernel estimates by continuous updating based on the output prediction error at each discrete step [Goodwin & Sin, 1984]; (2) *ensemble-averaging methods,* employing direct averaging of results obtained from many repetitions of identical experiments, which are rarely used because the latter is seldom feasible or practical in a physiologic experimental setting; (3) *temporal expansion methods,* employing explicit kernel expansions over time to avoid the experimental burden of the ensemble-averaging approach and the methodological constraints of adaptive methods, which utilize a single input-output data record to obtain complete nonstationary model representations under mild constraints [Marmarelis, 1981].

We note that the Volterra-Wiener approach has been extended to the case of nonlinear systems with multiple inputs and multiple outputs [Marmarelis & McCann, 1973; Marmarelis & Naka, 1974; Westwick & Kearney, 1992] where functional terms are introduced, involving cross-kernels which measure the nonlinear interactions of the inputs as reflected on the output. This extension has led to a generalization for nonlinear systems with spatio-temporal inputs that has found applications to the visual system [Citron et al., 1981; Yasui et al., 1979]. Extension of the Volterra-Wiener approach to systems with spike (action potential) inputs encountered in neurophysiology also has been made, where the GWN test input is replaced by a Poisson process of impulses [Krausz, 1975]; this approach has found many applications including the study of the hippocampal formation in the brain [Sclabassi et al., 1988]. Likewise, the case of neural systems with spike outputs has been explored in the context of the Volterra-Wiener approach, leading to efficient modeling and identification methods [Marmarelis et al., 1986; Marmarelis & Orme, 1993].

The mathematical relations between parametric (NARMAX) and nonparametric (Volterra) models have been explored, and significant benefits are shown to accrue from combined use [Zhao & Marmarelis, 1998]. These studies follow on previous efforts to relate certain classes of nonlinear differential equations with Volterra functional expansions in continuous time [Marmarelis, 1989; Rugh, 1981].

Only brief mention will be made of the computational and graphical nonparametric methods of phase-space mappings. The computational method requires prior specification of the number of input and output lags that are needed to express the present output value as

$$y(n) = G\left[y(n-1), \ldots, y(n-k), x(n), x(n-1), \ldots, x(n-l)\right] \tag{163.25}$$

where $G[\bullet]$ is an unknown function which maps the appropriate input-output vector $\mathbf{z}(n) = [y(n-1), \ldots, y(n-k), x(n), \ldots, x(n-l)]$ onto the present value of the output $y(n)$. Then, using experimental data, we compile on the computer a list of correspondences between $z(n)$ and $y(n)$ in the form of an empirical model that can predict the output for a given z by finding the closest correspondence within the compiled list (look-up table) or using a properly defined interpolation scheme.

The graphical method is based on the notion that the mathematical model of a discrete-time finite-order (stationary) dynamic system is, in general, a multivariate function $f(\bullet)$ of the appropriate lagged values of the input-output variables

$$f\left[y(n), y(n-1), \ldots, y(n-k), x(n-1), \ldots, x(n-l)\right] = 0 \tag{163.26}$$

This function (model) is a "constraint" among these input-output lagged (and present) values at each time instant n. The number of variables that partake in this relation and the particular form of $f(\bullet)$ are characteristic of the system under study. Therefore, in a geometric sense, the system (and its model) are represented by a subspace in a multidimensional space defined by the *coordinate system* of these variables. If one can define the appropriate space by choosing the relevant variables, then a model can be easily obtained by following the vector point corresponding to the data through time—provided that the input is such that the data cover the model subspace densely. If the input is ergodic, then sufficiently long observations will allow reliable estimation of the system model.

Although this is an old and conceptually straightforward idea, it has not been widely used (except in some recent studies of chaotic dynamics of autonomous systems, where no input variable exists) because several important practical issues must be addressed in its actual implementation, e.g., the selection of the appropriate coordinate variables (embedding space) and the impracticality of representation in high-dimensional spaces. If a low-dimensional embedding space can be found for the system under study, this approach can be very powerful in yielding models of strongly nonlinear systems. Secondary practical issues are the choice of an effective test input and the accuracy of the obtained results in the presence of extraneous noise.

163.3 Modular Approach

The practical limitations in the use of Volterra-Wiener models for strongly nonlinear systems have led some investigators to explore the use of block-structured or modular models in the form of cascaded or parallel configurations of linear subsystems (L) and static nonlinearities (N). This model representation often provides greater insight into the functional organization of the system under study and facilitates that identification task by allowing separate estimation of the various component subsystems (L and N), thereby avoiding the computational burden associated with the dimensionality of high-order kernel functions. The advantages afforded by this approach can be had only when prior specification of the structure of the modular model is possible.

Simple cascade models (e.g., L-N, N-L, L-N-L) have been studied extensively, yielding estimates of the component subsystems with moderate computational effort [Billings & Fakhouri, 1982; Korenberg & Hunter, 1986; Marmarelis & Marmarelis, 1978; Shi & Sun, 1990]. Distinctive kernel relationships (e.g., between first-order and second-order kernels) exist for each type of cascade model, which can be used for validation of the chosen modular model on the basis of the kernel estimates obtained from the data—a task that is often referred to as *structural identification.* Thus, the combined use of the modular and nonparametric approaches may yield considerable benefits. This idea can be extended to more complicated modular structures entailing multiple parallel and cascaded branches [Chen et al., 1990; Korenberg, 1991]. The case of nonlinear feedback in modular models attracts considerable interest, because of its frequent and critical role in physiologic control and autoregulation. A case of weak nonlinear feedback in sensory systems has received thorough treatment [Marmarelis, 1991]. Naturally, the analysis becomes more complicated as the complexity of the modular model increases. Particular mention should be made of a rather complex model of parallel L-N-L cascades (usually called the S_m model) that covers a broad class of nonlinear systems [Billings & Fakhouri, 1981; Rugh, 1981].

A method for the development of a general model for the Volterra-Wiener class of systems, which assumes a modular form, was originally proposed by Wiener and his associates and is reviewed in Schetzen [1980]. A general modular model, comprised of parallel L-N cascades, also has been proposed by Korenberg [1991], and another, employing *principal dynamic modes* as a filterbank whose outputs feed into a multi-input static nonlinearity, has been recently proposed by Marmarelis [1997].

This latter general model evolved from the original Wiener modular model. It was first adapted to studies of neural systems that generate spikes (action potentials), whereby a threshold-trigger is placed at the output of the general modular model that obviates the use of high-order kernels and yields a parsimonious complete model [Marmarelis & Orme, 1993]. The importance of this development is found in that, ever since the Volterra-Wiener approach was applied to the study of spike-output neural systems, it had been assumed that a large number of kernels would be necessary to produce a satisfactory model prediction of the timing of output spikes, based on the rationale that the presence of a spike-generating mechanism constitutes a *hard nonlinearity.* Although this rationale is correct if we seek to reproduce the numerical binary values of the system output using a conventional Volterra-Wiener model, the inclusion of a threshold trigger in our modular model yields a compact and complete model (of possibly low order) that predicts precisely the timing of the output spikes [Marmarelis et al., 1986]. This also led to a search for the principal pathways of dynamic transformations of neural signals using eigen-decomposition of a matrix composed of the Laguerre expansion coefficients of the first- and second-order kernels [Marmarelis & Orme, 1993]. The estimation of these principal dynamic modes is accomplished via the aforementioned eigen-decomposition [Marmarelis, 1994] or through the training of a specific artificial neural network (a modified perceptron with polynomial activation functions) using a modified back-propagation algorithm [Marmarelis & Zhao, 1997]. The latter method holds great promise in providing a practical solution to the ultimate problem of nonlinear system identification, because it is not limited to low-order nonlinearities and does not place stringent requirements on the necessary input-output data, but it does retain a remarkable degree of robustness in the presence of noise.

Acknowledgment

This work was supported in part by Grant No. RR-01861 awarded to the Biomedical Simulations Resource at the University of Southern California from the National Center for Research Resources of the National Institutes of Health.

References

Billings SA, Fakhouri SY. 1981. Identification of nonlinear systems using correlation analysis and pseudorandom inputs. Int J System Sci 11:261.

Billings SA, Fakhouri SY. 1982. Identification of systems containing linear dynamic and static nonlinear elements. Automatica 18:15.

Billings SA, Voon WSF. 1984. Least-squares parameter estimation algorithms for non-linear systems. Int J Systems Sci 15:601.

Billings SA, Voon WSF. 1986. A prediction-error and stepwise-regression estimation algorithm for non-linear systems. Int J Control 44:803.

Brillinger DR. 1970. The identification of polynomial systems by means of higher order spectra. J Sound Vib 12:301.

Chen H-W, Jacobson LD, Gaska JP. 1990. Structural classification of multi-input nonlinear systems. Biol Cybern 63:341.

Citron MC, Kroeker JP, McCann GD. 1981. Non-linear interactions in ganglion cell receptive fields. J Neurophysiol 46:1161.

Eykhoff P. 1974. System Identification: Parameter and State Estimation. New York, Wiley.

French AS. 1976. Practical nonlinear system analysis by Wiener kernel estimation in the frequency domain. Biol Cybern 24:111.

Goodwin GC, Payne RL. 1977. Dynamic System Identification: Experiment Design and Data Analysis. New York, Academic Press.

Goodwin GC, Sin KS. 1984. Adaptive Filtering, Prediction and Control. Englewood Cliffs, NJ, Prentice-Hall.

Haber R, Unbenhauen H. 1990. Structure identification of nonlinear dynamic systems—a survey of input/output approaches. Automatica 26:651.

Korenberg MJ. 1988. Identifying nonlinear difference equation and functional expansion representations: The fast orthogonal algorithm. Ann Biomed Eng 16:123.

Korenberg MJ. 1991. Parallel cascade identification and kernel estimation for nonlinear systems. Ann Biomed Eng 19:429.

Korenberg MJ, Hunter IW. 1986. The identification of nonlinear biological systems: LNL cascaded models. Biol Cybern 55:125.

Krausz HI. 1975. Identification of nonlinear systems using random impulse train inputs. Biol Cybern 19:217.

Lee YW, Schetzen M. 1965. Measurement of the Wiener kernels of a nonlinear system by cross-correlation. Int J Control 2:237.

Ljung L. 1987. System Identification: Theory for the User, Englewood Cliffs, NJ, Prentice-Hall.

Ljung L, Soderstrom T. 1983. Theory and Practice of Recursive Identification. Cambridge, Mass, MIT Press.

Marmarelis PZ, Marmarelis VZ. 1978. Analysis of Physiological Systems: The White-Noise Approach, New York, Plenum. Russian translation: Mir Press, Moscow, 1981. Chinese translation: Academy of Sciences Press, Beijing, 1990.

Marmarelis PZ, McCann GD. 1973. Development and application of white-noise modeling techniques for studies of insect visual nervous system. Kybernetik 12:74.

Marmarelis PZ, Naka K-I. 1974. Identification of multi-input biological systems. IEEE Trans Biomed Eng 21:88.

Marmarelis VZ (ed). 1987. Advanced methods of physiological system modeling: Volume I, Biomedical Simulations Resource, University of Southern California, Los Angeles.

Marmarelis VZ (ed). 1989. Advanced Methods of Physiological System Modeling: Volume II, New York, Plenum.

Marmarelis VZ (ed). 1994. Advanced Methods of Physiological System Modeling: Volume III, New York, Plenum.

Marmarelis VZ. 1977. A family of quasi-white random signals and its optimal use in biological system identification: I. Theory. Biol Cybern 27:49.

Marmarelis VZ. 1979. Error analysis and optimal estimation procedures in identification of nonlinear Volterra systems. Automatica 15:161.

Marmarelis VZ. 1981. Practicable identification of nonstationary nonlinear systems. Proc IEE, Part D 128:211.

Marmarelis VZ. 1989. Identification and modeling of a class of nonlinear systems. Math Comp Mod 12:991.

Marmarelis VZ. 1991. Wiener analysis of nonlinear feedback in sensory systems. Ann Biomed Eng 19:345.

Marmarelis VZ. 1993. Identification of nonlinear biological systems using Laguerre expansions of kernels. Ann Biomed Eng 21:573.

Marmarelis VZ. 1994. Nonlinear modeling of physiological systems using principal dynamic modes. In VZ Marmarelis (ed), Advanced Methods of Physiological Systems Modeling: Volume III, pp 1–28, New York, Plenum.

Marmarelis VZ. 1997. Modeling methodology for nonlinear physiological systems. Ann Biomed Eng 25:239.

Marmarelis VZ, Citron MC, Vivo CP. 1986. Minimum-order Wiener modeling of spike-output systems. Biol Cybern 54:115.

Marmarelis VZ, Orme ME. 1993. Modeling of neural systems by use of neuronal modes. IEEE Trans Biomed Eng 40:1149.

Marmarelis VZ, Zhao X. 1997. Volterra models and three-layer perceptions. IEEE Trans Neural Networks 8:1421.

McCann GD, Marmarelis PZ (eds). 1975. Proceedings of the First Symposium on Testing and Identification of Nonlinear Systems, Pasadena, Calif, California Institute of Technology.

Moller AR. 1983. Use of pseudorandom noise in studies of frequency selectivity: The periphery of the auditory system. Biol Cybern 47:95.

Ream N. 1970. Nonlinear identification using inverse repeat m-sequences. Proc IEEE 117:213.

Rugh WJ. 1981. Nonlinear System Theory: The Volterra/Wiener Approach. Baltimore, Johns Hopkins University Press.

Schetzen M. 1965. Measurement of the kernels of a nonlinear system of finite order. Int J Control 2:251.

Schetzen M. 1980. The Volterra and Wiener Theories of Nonlinear Systems, New York, Wiley.

Sclabassi RJ, Krieger DN, Berger TW. 1988. A systems theoretic approach to the study of CNS function. Ann Biomed Eng 16:17.

Shi J, Sun HH. 1990. Nonlinear system identification for cascade block model: An application to electrode polarization impedance. IEEE Trans Biomed Eng 37:574.

Soderstrom T, Stoica P. 1989. System Identification, London, Prentice-Hall International.

Stark L. 1968. Neurological Control Systems: Studies in Bioengineering, New York, Plenum.

Sutter E. 1992. A deterministic approach to nonlinear systems analysis. In RB Pinter, B Nabet (eds), Nonlinear Vision, pp 171–220, Boca Raton, Fla, CRC Press.

Victor JD, Shapley RM, Knight BW. 1977. Nonlinear analysis of retinal ganglion cells in the frequency domain. Proc Natl Acad Sci 74:3068.

Watanabe A, Stark L. 1975. Kernel method for nonlinear analysis: Identification of a biological control system. Math Biosci 27:99.

Westwick DT, Kearney RE. 1992. A new algorithm for identification of multiple input Wiener systems. Biol Cybern 68:75.

Yasui S, Davis W, Naka KI. 1979. Spatio-temporal receptive field measurement of retinal neurons by random pattern stimulation and cross-correlation. IEEE Trans Biomed Eng 26:263.

Zhao X, Marmarelis VZ. 1998. Nonlinear parametric models from Volterra kernels measurements. Math Comput Modelling 27:37.

164

Autoregulating Windkessel Dynamics May Cause Low Frequency Oscillations

Gary Drzewiecki
Rutgers, The State University of New Jersey

John K-J. Li
Rutgers, The State University of New Jersey

Abraham Noordergraaf
University of Pennsylvania

164.1 Introduction

The control of blood flow is a process involving neural and metabolic mechanisms. It is widely accepted that neural processes accomplish short term control of blood pre-capillary flow resistance such that blood pressure and flow are regulated. While metabolic processes dominate in the long term (Guyton, 1963) they ultimately determine the regulated blood flow levels in the various tissues in accordance with metabolic demands. This is achieved locally by the precapillary sphincters that adjust the duration and number of capillaries open and which, in turn, determine the value of the peripheral resistance, R_s. Another theory suggests that it is the mechanical effect of pressure and flow that stimulates vascular smooth muscle to react. However experiments have shown that tissue blood flow and systemic circulation is directly related to its metabolic demand (Berne and Levy, 1977). Experimental observations of low frequency flow oscillations can be found in studies of a single vascular bed as well as the entire systemic circulation. This differs from mechanical stimulation where researchers have employed a drug response. For example, Kenner and Ono (1972) have observed what they termed flow auto-oscillations in the carotid and femoral arteries of the dog, following intravenous administration of acetylcholine. Similar oscillations in flow have been observed following the infusion of adenosine into the coronary circulation

0-8493-0462-8/00/$0.00+$.50

(Wong and Klassen, 1991). These researchers attributed this response to opposing vasodilator and vasoconstrictor processes.

Other experimental evidence of low frequency resonance can be found in measurements of heart rate variability and blood pressure variability spectra (Chess et al., 1975; Akselrod et al., 1981; Pagani et al., 1986). Experiments in animals have shown that low frequency oscillations persist following denervation (Rimoldi et al., 1990), during external constant cardiac pacing (Wang et al., 1995), and in animals and humans with an artificial heart (Yambe et al., 1993; Cooley et al., 1998). To our knowledge, the origin of the low frequency resonance in the variability spectrum, the so-called Mayer waves (Mayer, 1876) is currently unexplained, although vasoregulatory dynamics has been proposed. The study by Yambe et al. (1993) revealed that the long waves were nearly 100% correlated with the periodicity of peripheral resistance.

Further autoregulatory instability has been observed in the pathological situation. For example, prominent low frequency oscillations have been observed during heart failure (Goldberger et al., 1984). While it has been proposed that these oscillations follow the respiratory rate, they have been observed to occur at even lower frequencies.

With innervation intact, the circulation at rest maintains constant ventricular stroke volume and blood pressure, on average. In addition, the regulation of cardiac output is accomplished primarily through the control of peripheral resistance. Then, the slow changes in heart rate (f_h) are directly related to cardiac output (CO) and inversely to peripheral resistance (R_s), so that, $f_h \alpha$ COα $1/R_s$ (Berne and Levy, 1977). This relationship is fundamental to the vascular theory of heart rate variation (Hering, 1924) and underscores the role of a time varying peripheral resistance. In this article, the dynamics of peripheral resistance control is examined analytically as an explanation of the very low frequency variation in heart rate.

164.2 Background

The control of blood flow resistance has been analyzed theoretically at both the organ and the microcirculatory level. Since metabolic demand of the tissue is related to oxygen uptake under aerobic conditions, Huntsman et al. (1978) developed a model for control of blood flow resistance for skeletal muscle based on this concept. In this study, it was found that the metabolic control considerably outweighed that of neural and myogenic. The time frame of observation was also important, where longer time periods emphasized the metabolic control factors. The model controlled flow by means of regulating the oxygen stored in the extra-vascular spaces. The control of peripheral resistance was affected by invoking the relationship between arterial diameter and oxygen tension. This model successfully predicted the large blood flow response following a brief occlusion.

The autoregulation response should ultimately arise from the action of multiple precapillary sphincters and resistance arterioles in the microcirculation. The control of flow in a microvascular model was analyzed by Mayrovitz et al. (1978). This model included muscular arterial and venous vasomotion, capillary filtration and reabsorption, and lymph flow. Tissue pressure was assumed to be regulated and was used to provide the control pathway for the activation of the precapillary sphincter. Local flow was found to vary considerably with periodic sphincter activity. This model demonstrated that autoregulation of flow is likely to find its genesis at the microcirculatory level.

Although our knowledge of pulse propagation in the arterial system is extensive, it has been useful to approximate the relationship between pressure and flow in a single artery or vascular bed. Such approximations have been referred to as reduced arterial system models of which the three element Windkessel is the most widely employed (Westerhof et al., 1971; Noordergraaf, 1978). This Windkessel model is modified here to study the effects of peripheral resistance.

The three element Windkessel model approximates the pulse propagation quality of the arterial system as a combination of an infinitely long tube, which is represented by its characteristic impedance, Z_o, in series with a parallel arrangement of a peripheral flow resistance, R_s, and a total arterial compliance, C_s. Within a single cardiac cycle, these three parameters are generally assumed to be constants (Toorop et al.,

1987). Recent studies have scrutinized this assumption. For example, the compliance has been shown to depend on the level of arterial pressure (Bergel, 1961). Other studies have called into question the accuracy of the various methods employed to experimentally determine their values (Stergiopulos et al., 1995a). This difficulty has been addressed by Quick et al. (1998) who showed that wave reflection imparts frequency dependence on the values of R_s and C_s.

The Windkessel model is often employed as a simplified load for the left ventricle. Under steady conditions and a fixed heart rate, this model is appropriate. But, given the case of varying ventricular function, the model responds inaccurately. The difficulty lies in the regulation of cardiac output, i.e., autoregulation is not provided. To correct this problem, our group has employed an iterative procedure that permits R_s to change such that cardiac output is maintained (Drzewiecki et al., 1992). This procedure, although providing autoregulation, does not account for the dynamics associated with this physiological mechanism.

In this chapter, the dynamics of autoregulation are incorporated into the modified Windkessel model. The frequency response of this autoregulating Windkessel is then predicted and compared with that of the standard three element model. The time response is also determined and discussed relative to the experimental observations in the literature. The stability of the model is also examined with respect to very low frequency oscillation in peripheral resistance.

The physical relationships that govern the Windkessel model are now reviewed. Time dependency is provided by the blood volume storage property of arterial compliance, C_s. The time derivative of peripheral arterial pressure, P_s, is proportional to the difference in blood flow into and out of the compliance, according to,

$$\frac{dP_s}{dt} = \left(Q - \frac{P_s - P_o}{R_s}\right)\frac{1}{C_s(P_s)} \tag{164.1}$$

where Q is the inflow. P_o accounts for the critical closing pressure (Alexander, 1977). The input pressure, P, is then determined from the sum of pressure due to flow through the characteristic impedance and the pressure, P_s,

$$P = QZ_o + P_s \tag{164.2}$$

The arterial compliance may be treated as a function of pressure also, accounting for large changes in blood pressure. The pressure dependence of compliance is then modeled by an exponential function (Liu et al., 1989; Li et al., 1990), as,

$$C_s = ae^{-bP_s} \tag{164.3}$$

where a and b are empirical constants evaluated from experimental data.

164.3 Modeling Methods

Steady Regulation

In this study, it will be assumed that the local tissue bed or whole animal is in a steady metabolic state. For this regulated condition, the level of blood flow is denoted Q_0, the peripheral resistance, R_{so}, and the pressure across it, P_{so}. The relation between these quantities reads,

$$P_{so} = R_{so}\, Q_o \tag{164.4}$$

When the flow is disturbed, the value of the peripheral resistance is adjusted in an attempt to restore flow to Q_o. Since the control of flow is not ideal, the new steady flow condition is described by

$$P_s = R_{reg}\, Q \tag{164.5}$$

where R_{reg} is the new value of peripheral resistance. Applying the simplest form of control, which is also imperfect, it is assumed that the change in peripheral resistance is proportional to the change in flow, i.e., linear proportional control. The change in peripheral resistance is then determined from the control equation,

$$\Delta R_s = R_{reg} - R_{so} = \frac{G(Q - Q_0)}{Q_0} \tag{164.6}$$

where G is defined as the autoregulatory gain.

Autoregulation Dynamics

There are two dynamic effects that must be accounted for to accurately model the physiological control mechanism. First, the change in peripheral resistance is delayed. This represents a delay from the time that the flow changes to when the vascular smooth muscle activity actually initiates the process of correcting the flow. This time delay in the change of peripheral resistance is denoted τ_Q. To account for this effect, the change in peripheral resistance determined from the steady value was written as $\Delta R_s(t - \tau_Q)$.

Physiologically, even when peripheral resistance adjusts, it cannot be changed instantaneously. To account for the slow change in resistance, a first-order rate equation was used to represent the slow action of vascular smooth muscle. Thus, the time dependent value of peripheral resistance was determined from the following,

$$\frac{dR_s}{dt} = \frac{1}{\tau_{sm}} \Delta R_s(t - \tau_Q) \tag{164.7}$$

where the time constant for smooth muscle action is τ_{sm}. This equation simply states that the rate of change of peripheral resistance with respect to time is proportional to desired change in resistance, ΔR_s.

The net effect of delay in autoregulation and smooth muscle response is to cause R_s to slowly approach its new control value, as defined by Eq. (164.6).

Regulatory Limitations

If the actual blood flow differs significantly from Q_0, the regulatory error can be large. Since a linear model is being employed here, the value of R_s predicted by Eqs. (164.4)–(164.7) may fall outside of the physiological range. Thus, the time dependent value of R_s was limited to maximum and minimum values of 1.6 R_{so} and 0.4 R_{so}, respectively. These values correspond with maximal vasoconstriction and vasodilation. When the computations reveal that the limit has been reached, the limited value of resistance was then inserted into Eq. (164.1) of the Windkessel model. Otherwise, the time dependent value obtained from Eq. (164.7) was employed.

The two differential equations (Eqs. 164.1 and 164.7) govern the time dependent characteristics of the autoregulating Windkessel model. They were integrated numerically by a computer. All parameters were evaluated for the canine systemic arterial system (Table 164.1).

164.4 Results

Autoregulation Curve

The standard experimental approach to measuring the autoregulation curve of a vascular bed has been to slowly vary input flow while recording pressure. Alternatively, one may alter pressure while measuring flow. In either case, the dependent variable stabilizes after several minutes. The set of final values results

TABLE 164.1 Parameters

Parameter	Value	Description
Z_o	0.25 mmHg/mL/sec	Characteristic impedance
C_s	0.50 mL/mmHg	Arterial compliance
R_{so}	3.10 mmHg/mL/sec	Basal peripheral resistance
a	3.0 mL/mmHg	Nonlinear compliance constant
b	0.02 mm Hg^{-1}	Nonlinear compliance constant
G	5.0	Autoregulation gain
Q_o	27.0 mL/sec	Regulated flow level
τ_Q	5.0 sec	Flow delay time
τ_{sm}	3.0 sec	Smooth muscle time constant

in the steady state autoregulation curve. This procedure was followed using the autoregulating Windkessel model (Fig. 164.1). To reduce computation time, the time constants of the model were shortened. This does not affect the result, but simply shortens the time to reach steady state.

The model generated autoregulation curve shows three distinct segments. The low flow segment was found to be a line of slope $0.4R_{so,}$ and corresponds with maximal vasodilation. There was no control of flow in this region either. The middle flow segment represents the autoregulatory range. The peripheral

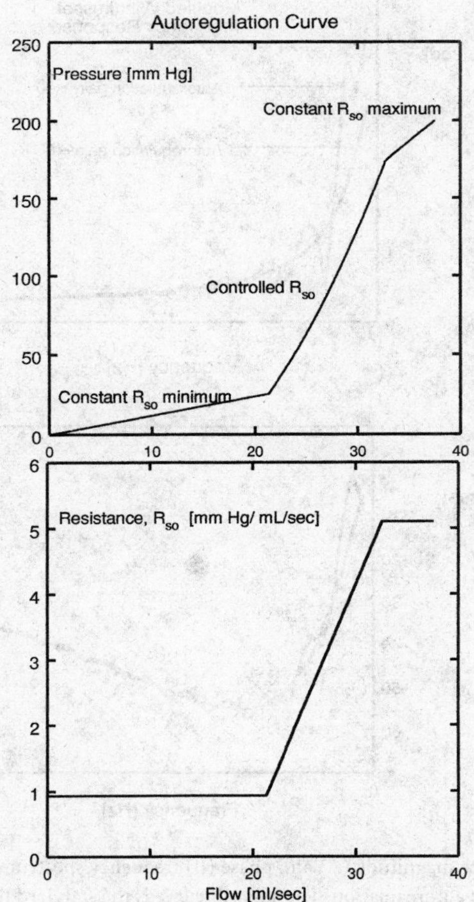

FIGURE 164.1 Pressure and flow relationship of the canine aorta obtained from the autoregulating Windkessel in the steady state. Flow regulation occurs in the range where the slope is steepest.

resistance was found to alter continuously in accordance with the feedback control relationship of Eq. (164.6) in this region. The slope of this segment was also proportionate to the autoregulatory gain factor, G, of the model. A greater slope reflects tighter control of flow, but, as will be shown later, can lead to autoregulatory instability.

Input Impedance

The frequency response of the autoregulating Windkessel was determined by applying a small constant amplitude flow sinusoid as input while the mean flow is held fixed. The resulting sinusoidal pressure response was computed from the model. The response was evaluated for as many cycles as were needed to obtain a sinusoidal steady state. For frequencies much below the heart rate, this required up to 30 seconds. The input impedance magnitude was then obtained by finding the ratio of the sinusoidal pressure and flow amplitudes over the range of frequencies. Phase was determined by finding the phase difference between the pressure and flow waves (Fig. 164.2a and b). In order to compare these results with the conventional modified Windkessel, the input impedance was re-evaluated when the autoregulatory gain is zero (G = 0) and is shown on the same graphs.

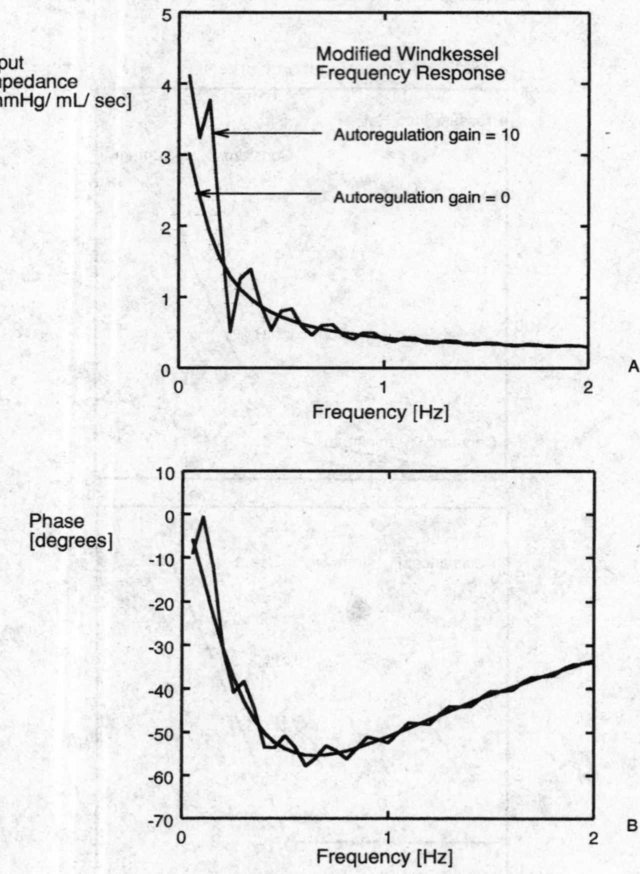

FIGURE 164.2 Input impedance magnitude (A) and phase (B) frequency spectrum obtained from the model. The oscillating curve represents active autoregulation. The smooth curve is the standard three – element Windkessel with constant peripheral resistance.

The high frequency impedance was found to be identical for both models. This might be expected, since autoregulation is slow compared with the heart rate. An unexpected result was that the impedance and phase were found to oscillate at about that of the modified Windkessel for frequencies near the heart rate. This oscillation was traced to the autoregulation delay time, τ_Q. The magnitude of these oscillations was found to increase further if the autoregulatory gain is increased. If the gain or the delay time was reduced to zero, the oscillations are eliminated. These oscillations in the frequency domain could have been predicted by referring to the Fourier transform of a delayed time function.

The very low frequency portion of the autoregulating impedance magnitude was higher in comparison with that of the unregulated Windkessel, where impedance approaches R_s at zero frequency. This can be explained by referring again to Fig. 164.1. Instead of converging to R_s, the autoregulating impedance approaches the slope of the pressure-flow curve. That is because the sinusoidal response measures the dynamic variation about the flow control point, Q_0. This is not contradictory, since the mean pressure divided by the mean flow is simultaneously $R_s = R_{so}$, at this point. And, the derivative of pressure, dP/dQ, for a given flow, must be the slope of the autoregulation curve in the steady state.

Flow Step Response

The pressure was computed from the model given a step increase in flow above the control level. A flow step was initiated 2.0 seconds from the beginning of the calculations (Fig. 164.3). An immediate increase in pressure occurs. This coincided with the time dependent increase in blood volume that shifts into the compliance, C_s. The peripheral resistance did not change during this time due to flow delay. It remains at the control value of R_{so}. Five seconds later, the resistance begins to increase towards the new value required by Eq. (164.4) and for autoregulation. Pressure increased nearly in proportion with resistance. The rate of change was determined by the time constant of smooth muscle response, τ_{sm}, and Eq. (164.5). An equilibrium value was reached about 20 seconds after the flow step occurred.

The above flow step response was repeated with a constant amplitude flow pulse added to the mean flow (Fig. 164.4). A flow step was initiated 10 seconds from the beginning of computations. The envelope of the pressure response follows that shown in Fig. 164.3, but additionally permits the observation of systolic and diastolic pressures. It was found that the pulse pressure amplitude increases following the step. This was associated with the decrease in pressure dependent compliance due to elevated pressure (Eq. 164.3).

Pressure Step Response

The flow was computed from the model following a step change in input pressure. The step change was initiated immediately by altering the initial condition for pressure from the steady state value of 83 to 90 mmHg (Fig. 164.5).

The graphs for flow and resistance were found to be 180° out of phase. This was expected, since they are inversely proportional for constant pressure. Part of the initial increase in flow was due to transient filling of the compliance, C_s. Flow was found to oscillate about its control value. This was reasonable since the control system consists of feedback with two time lags. Thus, the stability of control becomes a consideration. It was found that increasing τ_Q or the gain factor, G, led to unstable low frequency pressure oscillations. The smooth muscle time constant, τ_{sm}, was found to possess a damping effect and could be increased to eliminate oscillations. Thus, too rapid smooth muscle activation and accurate autoregulation were predicted to be a destabilizing influence on the circulation. Since the flow resistance was always limited to either maximal vasoconstriction or vasodilation, the amplitude of the oscillations was also constrained.

As another type of pressure step, an arterial occlusion was computed from the model. A two second occlusion was applied by rapidly reducing the input pressure to zero and then restoring it to the preocclusion level. The initial effect was to cause a rapid negative flow (Fig. 164.6). This represents blood

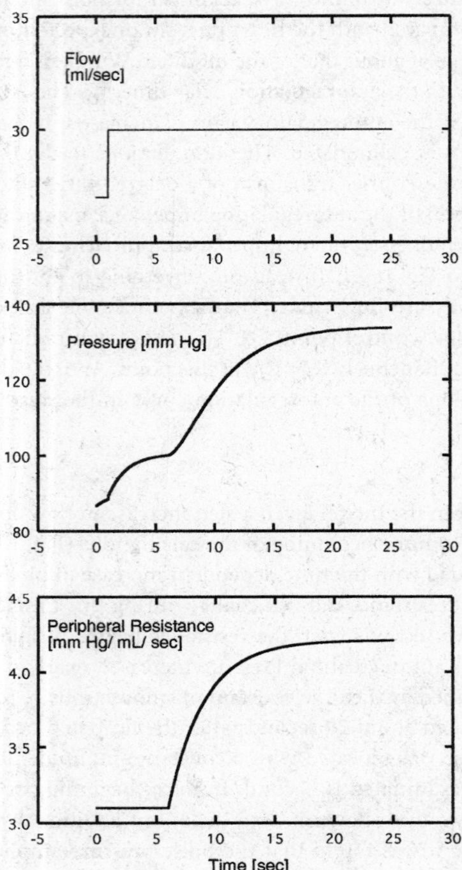

FIGURE 164.3 Flow step response of the model for steady levels of flow. The time responses of pressure and peripheral resistance are shown.

volume decrease of the arterial compliance. The effect was brief since the outflow time constant is less than one second. Peripheral resistance was unchanged until the end of the occlusion due to autoregulation delay. Afterwards, it rapidly dropped to the minimal resistance level of complete vasodilation. When the occlusion was released, the flow rapidly increased. Initially, the flow was briefly high at positive values as the blood volume of the compliance was restored. Flow remained elevated to relieve the ischemia. Following a delay time, flow was then restored to its initial value prior to occlusion. This was done in an oscillatory manner.

Minimal Circuit Model

As in much of hemodynamic research, the Windkessel model, as well as others, have been portrayed in electrical circuit form. The autoregulating Windkessel described here cannot be simplified completely in circuit form. But, a useful approximation was designed. It was assumed that the time constants of autoregulation and smooth muscle activation are minimal. Then, autoregulation is immediate. Additionally, it was assumed that vasoconstriction and vasodilation are not limited. Under these conditions, the circuit model of Fig. 164.7 can be used to approximate the autoregulating Windkessel. The constant current source represents the control value of autoregulated flow, Q_o. The resistance, R_{auto}, is the slope of the autoregulation line (Fig. 164.1). When the actual flow, $Q = Q_o$, the average blood pressure is equal

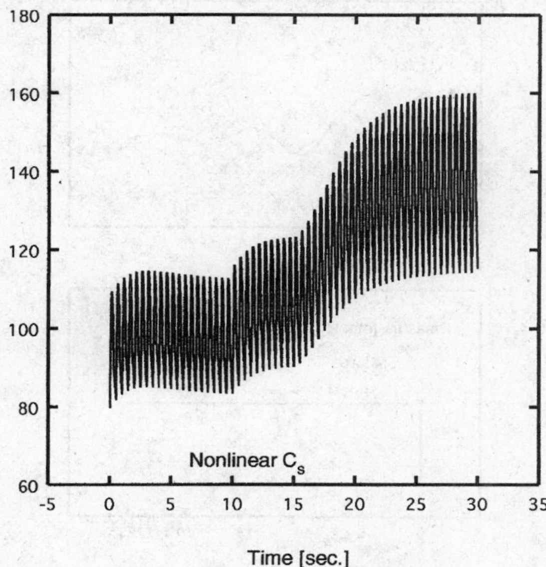

FIGURE 164.4 Flow step response as in Fig. 164.3, but with the addition of constant pulsatile flow. The effect of nonlinear compliance is evident as the pulse pressure increases following the increase in mean flow.

to the pressure source, MAP. Also, under this condition, $R_s = R_{so} = \frac{MAP}{Q_o}$. Note, that these expressions assume that the pressure drop across the characteristic impedance, Z_o, is negligible.

The impedance frequency spectrum was evaluated for the autoregulating circuit model. It was found to follow the result previously shown in Fig. 164.2, with the exception of impedance oscillations. The predicted impedance was in error by the amount of the oscillation amplitude. On average, over a range of frequencies, the error was negligible. Hence, for steady state conditions, the minimal circuit model offers a simple solution to the autoregulating Windkessel. The transient step responses, as predicted by the complete model, cannot be reproduced by this simplified circuit model.

164.5 Discussion

The autoregulation curve described in the current model differs from physiological data only in the sense that it is a segmented linear description. This was due to the fact that the model assumed linear feedback regulation. The actual flow regulation system possesses nonlinearities not accounted for in the current model. Most researchers have found that the steady state pressure flow curve is sigmoidal (Folkow, 1953; Granger and Guyton, 1969). Other researchers have expressed these nonlinearities as feedback gain that is flow dependent (Van Huis et al., 1985). Still others have found that a linear pressure-flow relationship is a good approximation within the normal physiological range of blood pressure (Sagawa and Eisner, 1975). In this case, the current model would be most accurate when flow is near Q_o and blood pressure is near MAP.

The autoregulation curve has been shown to be affected by the carotid baroreflex function (Shoukas et al., 1984). Increased carotid pressure leads to a decrease in peripheral resistance so that blood pressure can be maintained at a given cardiac output. In spite of the baroreceptor effect, these same researchers have shown that the steady state pressure-flow curve remains sigmoidal in shape. Thus, the concept of a steady state autoregulation curve applies just as well as when the baroreceptors are functional. The net effect of the baroreceptor response is to decrease the autoregulatory gain, as observed by a decrease in the slope of the steady state pressure-flow curve. This might be expected since pressure has the opposite effect on peripheral resistance to that of flow (Burattini et al., 1991). This would have the added benefit

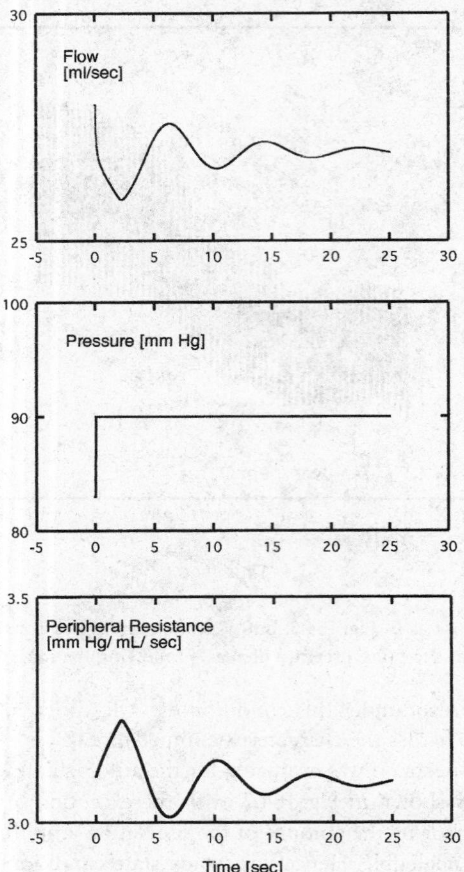

FIGURE 164.5 Pressure step response computed from the model ($\tau_Q = 2$, $\tau_{sm} = 5$, and $G = 10$). The step was produced by altering the initial condition for pressure at $t = 0$. Note that pressure and resistance are 180° out of phase. A damped resonant response is evident with a resonant frequency of ~0.1 Hz.

of stabilizing the autoregulation system by preventing its gain from becoming too high (see stability below). Thus, since the autoregulating Windkessel does not incorporate this response, its primary value is in the analysis of peripheral vascular beds. And, provided that the autoregulatory gain is reduced to account for the baroreceptor effect, it can be applied to the systemic circulation.

The autoregulating Windkessel incorporates the critical closing pressure phenomenon (Alexander, 1977). Critical closing causes the flow to cease at a slightly positive pressure, P_o. Debate over the mechanism that leads to P_o continues. Since its value is typically small compared with systemic blood pressure, its value was set to zero here.

The low frequency impedance of the autoregulating Windkessel was found to be greater than that predicted by the modified Windkessel (Fig. 164.2). The sinusoidal input impedance of a vascular bed has not been explored extensively below the frequency of the heart rate. Sipkema and Westerhof (1980) examined the low frequency range experimentally using an external sinusoidal pump. Under these conditions, it was found that the low frequency impedance approaches the slope of the autoregulation curve, as predicted by the autoregulating Windkessel. These researchers distinguish the low frequency impedance from peripheral resistance by defining it as a dynamic resistance.

Input impedance measurements have identified a variation in the frequency spectrum. While experimental studies have not provided an explanation for the origin of these variations, the current model ties them to autoregulation delay (τ_Q). The model predicts that the spectrum should follow an oscillatory

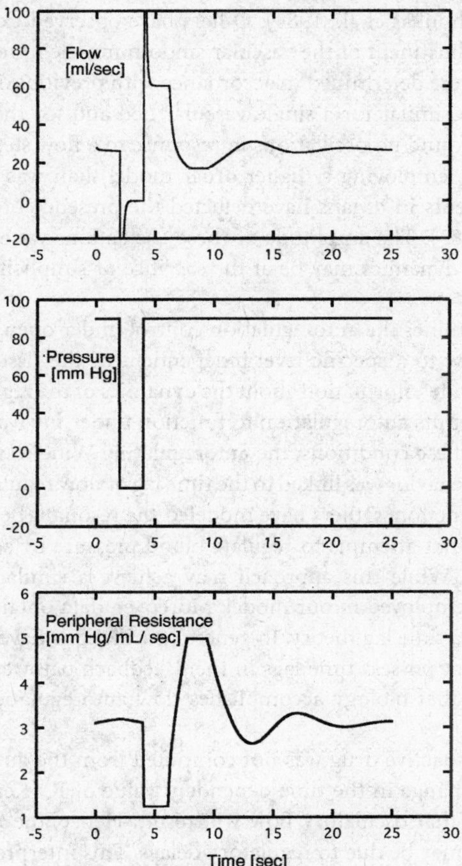

FIGURE 164.6 A vascular occlusion produced by reducing the input pressure to zero for two seconds. The flow response transients are due to compliance effects. Flow increases to maximum value corresponding with complete vasodilation. A damped oscillatory response returns the flow and resistance to control levels.

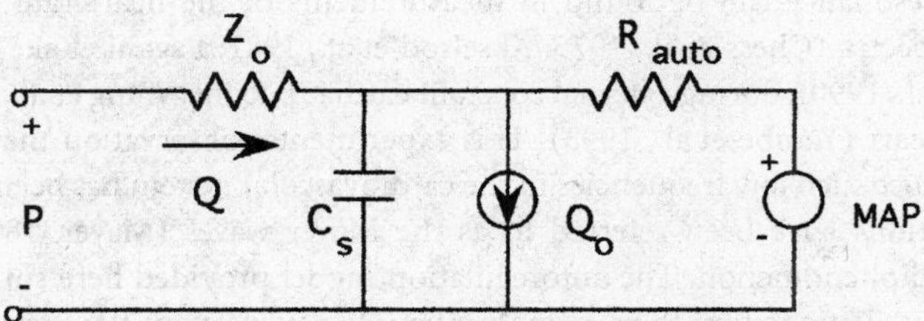

FIGURE 164.7 Simplified autoregulating Windkessel in electrical circuit form.

pattern, particularly for frequencies at and below the heart rate. Moreover, the sinusoidal steady state may not be easily achieved experimentally. The model required over one minute duration to obtain the steady state response. The model further predicts an erratic or random variation in the impedance spectrum if the steady state was not obtained. Another explanation could be that the different vascular beds may possess slightly different delay times. When measured at the aorta, this may appear as impedance scatter. Other researchers have attributed the source of impedance variations to nonlinear vascular properties (Stergiopulos et al., 1995b).

The flow step response of the model follows experimental studies quite precisely (Braakman et al., 1983; Van Huis et al., 1985; Shoukas et al., 1984). Others have observed experimentally an initial autoregulation delay and a slow adjustment of the vascular smooth muscle in response to a step in flow. The time constants of the model were determined in accordance with previous observations. The time course of the flow step response was similar for a single vascular bed and for the entire systemic circulation. Experimental measurements found no oscillations in response to a flow step for the systemic circulation (Shoukas et al., 1984). Hence, employing a higher order model than was used here would be of little value. But, *in vivo* measurements in organs have detected the presence of oscillations in the step flow response (Westerhof et al., 1983). The amplitude of these oscillations were small. Modifying the model to include these higher order dynamics may be at the sacrifice of simplicity and was not considered in this study.

The flow step response examines the autoregulation control under open loop conditions. This procedure essentially forces the flow to a specific level independent of the tissue's metabolic demands. The results of this experiment provide information about the dynamics of the regulatory process. Alternatively, the pressure step response permits autoregulation to function under *in vivo* conditions and provides the closed loop response. Under these conditions, the autoregulating Windkessel shows a damped resonant flow response. This resonant behavior was linked to the time lag of flow regulation. Damping was provided by the slow smooth muscle function. Others have modeled the resonant behavior of autoregulation as a second-order control system that attempts to regulate blood pressure by adjusting the peripheral resistance (Burattini et al., 1987). While this approach may achieve a similar response, the physiological mechanism differs from that employed in our model. Moreover, data obtained from the open loop flow response, supports the use of a time lag theory. In general, oscillations have been modeled and observed in other biological systems that possess time lags in their feedback pathways (An der Heiden, 1979.). It may be reasonable to expect that biology accomplishes flow autoregulation in a way similar to other types of homeostasis.

While the response to a vasoactive drug was not computed from the autoregulation model, it can be obtained by applying a step change in the time dependent value of R_s (Kenner and Ono, 1972). In this case, the model also predicts that oscillatory flow will result. Hence, the model suggests, alternatively, that the origin of oscillations may be due to regulatory delays. This interpretation, though, does not rely on conventional control system analysis and does not involve the parasympathetic and sympathetic reflex pathways (Wong and Klassen, 1991).

The model predicted that autoregulation can be unstable for some parameter values. When the parameters were calibrated to match flow step experiments, the model produced a stable result. But, the pressure step can best be characterized as a damped low frequency resonance. Experimental evidence of very low frequency resonance can be found in measurements of the heart rate variability and blood pressure variability spectra (Chess et al., 1975; Akselrod et al., 1981; Pagani et al., 1986), following denervation (Rimoldi et al., 1990), during external constant cardiac pacing (Wang et al., 1995), and in animals with total artificial heart (Yambe et al., 1993). This experimental observation that a nervous system is not necessary to produce such low frequencies in the cardiovascular system has been perplexing. So much so, that these oscillations have been referred to as the Mayer waves (Mayer, 1876), an observed but currently unexplained phenomenon. The autoregulation model provided here supports the theory that these low frequency oscillations find their origin in the intrinsic delay of flow regulation.

Model parameter values that lie outside the normal physiological range can further enhance autoregulatory instability. Abnormal values may occur in the pathological situation. Oscillations in pressure and flow were predicted by the model to lie well below heart rate, in the range below 0.1 Hz for the dog. For example, prominent low frequency oscillations have been observed during heart failure (Goldberger et al., 1984). While it has been proposed that these oscillations follow the respiratory rate, they have been observed to occur at even lower frequencies. The autoregulating Windkessel model offers a possible explanation of the source of these oscillations given pathological conditions. It also suggests the possibility that they may be used to noninvasively monitor the condition of the autoregulatory system.

164.6 Conclusions

The autoregulating Windkessel was found to accurately reproduce the experimental flow step response of both vascular beds and the systemic circulation. The model further provided the steady-state pressure-flow autoregulation curve, in linearized form. The impedance spectrum was predicted to differ from that of the three element Windkessel for frequencies below the heart rate. For frequencies near zero, the impedance approached the slope of the pressure-flow autoregulation curve, as opposed to peripheral resistance. But, simultaneously, the average pressure divided by the average flow of the model yielded the usual value of peripheral resistance.The low frequency impedance spectrum was also predicted to contain oscillations that were traced to autoregulatory delay. This result provides an explanation of the variability obtained experimentally during multi-beat examination of the impedance spectrum. The pressure step response of the autoregulating Windkessel revealed a damped low frequency resonant behavior. The model parameters that cause increased autoregulatory gain and decreased damping were found to lead to sustained flow oscillations at frequencies well below the heart rate. This resonance explains the origin of very low frequencies observed in the heart rate variability spectrum referred to as the Mayer waves.

References

Akselrod S, Gordon D, Ubel FA, Shannon DC, and Cohen RJ. 1981. Power spectrum analysis of heart rate fluctuation: A quantitative probe of beat-to-beat cardiovascular control. *Science* 213:220-222.

Alexander RS. Critical closure reexamined. 1977. *Circ. Res.* 40:531-535.

Bergel DH. The dynamic elastic properties of the arterial wall. 1961. *J. Physiol.* (London) 156:458-469.

Berne RM, Levy MN. 1977. *Cardiovascular Physiology* Mosby.

Braakman R, Sipkema P, Westerhof N. 1983. Steady state and instantaneous pressure-flow relationships: characterisation of the canine abdominal periphery. *Card. Res.* 17:577-588.

Burattini R, Borgdorff P, Gross DR, Baiocco B, Westerhof N. 1991. Systemic autoregulation counteracts the carotid baroreflex. *IEEE Trans. Biomed. Eng.* 38:48-56.

Burattini R, Reale P, Borgdorff P, Westerhof N. 1987. Dynamic model of the short-term regulation of arterial pressure in the cat. *Med. Biol. Eng. Comput.* 25:269-276.

Chess GF, Tam HK, Calarescu FR. 1975. Influence of cardiac neural inputs on rhythmic variations of heart period in the cat. *Amer. J. Physiol.* 228:775-780.

Cooley RL, Montano N, Cogliati C, Van de Borne P, Richenbacher W, Oren R, Somers VK. 1998. Evidence for a central origin of the low-frequency oscillation in the RR-interval variability. *Circulation* 98:556-561.

Drzewiecki GM, Karam E, Li J K-J, Noordergraaf A. 1992. Cardiac adaptation from sarcomere dynamics to arterial load: A model of physiologic hypertrophy. *Amer. J. Physiol.* 263: H1054-H1063.

Folkow B. 1953. A study of the factors influencing the tone of denervated blood vessels perfused at various pressures. *Acta. Physiol. Scand.* 27:99-117.

Goldberger AL, Findley LJ, Blackburn MR, Mandell AJ. 1984. Nonlinear dynamics in heart failure: Implications of long-wavelength cardiopulmonary oscillations. *Amer. Heart. J.* 107:612-615.

Granger HJ, Guyton AC. 1969. Autoregulation of the total systemic circulation following destruction of the nervous system in the dog. *Circ. Res.* 25:379-388.

Guyton AC. 1963. *Circulatory Physiology: Cardiac Output and Its Regulation.* Philadelphia, Saunders.

An der Heiden U. 1979. Delays in physiological systems. *J. Math Biol.* 8:345-364,.

Hering HE. 1924. Die Aenderung der Herzschlagzahl durch Aenderung des arteriellen Blutdruckes erfolgt auf reflectorischem Wege. *Pfluegers Arch. Gesamte Physiol. Menschen Tiere.* 206:721.

Huntsman LL, Attinger EO, Noordergraaf A. 1978. Metabolic autoregulation of blood flow in skeletal muscle. In: *Cardiovascular System Dynamics* (Baan J, Noordergraaf A, Raines J, Eds.), MIT Press, 400-414.

Kenner T, Ono K. 1972. Analysis of slow autooscillations of arterial flow. *Pflugers Arch.* 331:347-356.

Li J K-J, Cui T, Drzewiecki GM. 1990. A nonlinear model of the arterial system incorporating a pressure-dependent compliance. *IEEE Trans. Biomed. Eng. BME* 37:673-678.

Liu Z, Ting C-T, Zhu S, Yin FCP. 1989. Aortic compliance in human hypertension. *Hypertension* 14:129-136.

Mayer S. 1876. Studien zur physiologie des herzens und der blutgefasse. V. Uber spotane blutdruck-schwankungen. *Akad. Wiss. Wien. Math-Nat.* 74:281-307.

Mayrovitz HN, Wiedeman MP, Noordergraaf A. 1978. Interaction in the microcirculation. In: *Cardiovascular System Dynamics* (Baan J, Noordergraaf A, Raines J, Eds.), MIT Press, 194-204.

Noordergraaf A. 1978. *Circulatory System Dynamics.* New York, Academic Press.

Pagani M, Lombardi F, Guzzetti S, Rimoldi O, Furlan R, Pizzinelli P, Sandrone G, Malfatto G, Dell'Orto S, Piccaluga E, Turiel M, Baselli G, Cerutti S, Malliani A. 1986. Power spectral analysis of heart rate and arterial pressure variabilities as a marker of sympatho-vagal interaction in man and conscious dog. *Circ. Res.* 59:178-192.

Quick CM, Berger DS, Noordergraaf A. 1998. Apparent arterial compliance. *Amer. J. Physiol.* 274:H1393-H1403.

Rimoldi O, Pierimi S, Ferrari A, Cerutti S, Pagani M, Malliani A. 1990. Analysis of short-term oscillations of R-R and arterial pressure in conscious dogs. *Amer. J. Physiol.* 258:H967-H976.

Sagawa K Eisner A. 1975. Static pressure-flow relation in the total systemic vascular bed of the dog and its modification by the baroreceptor reflex. *Circ. Res.* 36:406-413.

Shoukas AA, Brunner MJ, Frankle AE, Greene AS, Kallman CH. 1984. Carotid sinus baroreceptor reflex control and the role of autoregulation in the systemic and pulmonary arterial pressure-flow relationships of the dog. *Circ. Res.* 54:674-682.

Sipkema P, Westerhof N. 1980. Peripheral resistance and low frequency impedance of the femoral bed. In *Cardiac Dynamics* (J. Baan, C. Alexander, A. Arntzenius, and E. L. Yellin, Eds.) M Nijhoff, Boston, p. 501-508.

Stergiopulos N, Meister J-J, Westerhof N. 1995a. Evaluation of methods for estimation of total arterial compliance. *Amer. J. Physiol.* 269:H1490-H1495.

Stergiopulos N, Meister J-J, Westerhof N. 1995b. Scatter in input impedance spectrum may result from the elastic nonlinearity of the arterial wall. *Amer. J. Physiol.* 269:H1490-H1495.

Toorop GP, Westerhof N, Elzinga G. 1987. Beat-to-beat estimation of peripheral resistance and arterial compliance during pressure transients. *Amer. J. Physiol.* 252:H1275-H1283.

Van Huis GA, Sipkema P, Westerhof N. 1985. Instantaneous and steady-state pressure-flow relations of the coronary system in the canine beating heart. *Card. Res.* 19:121-131.

Wang M, Evans J, Knapp C. 1995. Spectral patterns and frequency response characteristics of arterial pressure in heart paced dogs. *IEEE Trans. Biomed. Eng.* 42:708-717.

Westerhof N, Elzinga G, Sipkema P. 1971. An artificial arterial system for pumping hearts. *J. Appl. Physiol.* 31:776-781.

Wong AY, Klassen GA. 1991. Vasomotor coronary oscillations: a model to evaluate autoregulation. *Basic Res. Cardiol.* 86:461-475.

Yambe T, Nitta S-I, Sonobe T, Naganuma S, Kakinuma Y, Kobayashi S-I, Nanka S, Ohsawa N, Akiho H, Tanaka M, Fukuju T, Miura M, Uchida N, Sato N, Mohri H, Koide S, Abe K-I, Takeda H, Yoshizawa M. 1993. Origin of the rhythmical fluctuations in the animal without a natural heartbeat. *Artificial Organs* 17:1017-1021.

165

Control of Movements

Dejan B. Popović
University of Belgrade

The mechanical system being controlled is referred to as the *plant.* The configuration of the system at any instant in time comprises the plant *states.* The devices that power the system are called *actuators.* The signals driving the actuators are called *controls.* The *controller* is the process by which the controls are generated. The time histories of the plant states in response to the control signals are referred to as the system *trajectory.*

Controllers can be designed to function without sensors and therefore without knowledge of the actual plant trajectory (*open-loop control,* Fig. 165.1a). Reference-based open–loop controllers precompute and store control signals, and execute the desired motor task in real-time. To correct for disturbances and modeling errors, a feedback controller with ongoing knowledge of the effects of the disturbances can be used. *Closed-loop controllers* (Fig. 165.1b), or so-called *error driven controllers* use feedback information from sensors measuring the states. Both controllers described require full knowledge of the system parameters, desired trajectories of the system components, and allowed tracking errors. Operation of an open-loop or closed-loop controller can be simulated using analytic tools derived from mechanics and mathematics.

Nonanalytic control is the method developed following the biological control. The system operation, not the plant, is modeled as a set of mappings between inputs and outputs (Fig. 165.1c).

Control of movements in humans is a very complex task dealing with a large scale, nonlinear, time-variable, dynamically non-determined, redundant system [1]. A dynamical system may change its states in a continuous way. Without an optimization criterion, all trajectories of a system are equivalent. The term trajectory has a broad meaning; it does refer to the way the system coordinates vary in the course of transition from the current to the next state. An optimization criterion assigns to each transition trajectory or a subset of trajectories, a value so that they can be arranged in the order of preference. In some cases, the trajectories of the dynamical system may be well ordered so that a single solution stands in front. Such a trajectory is given the name, the optimal solution. Instances when optimal solutions of control tasks exist, or can be analytically determined, are relatively rare although this term is used rather loosely in everyday life.

165.1 Biological Control of Movements

Control of goal directed movements. Arm movements to a spatial target utilize sensory information (kinesthetic and visual) that is initially represented in different frames of reference, and the sensory signals

0-8493-0462-8/00/$0.00+$.50

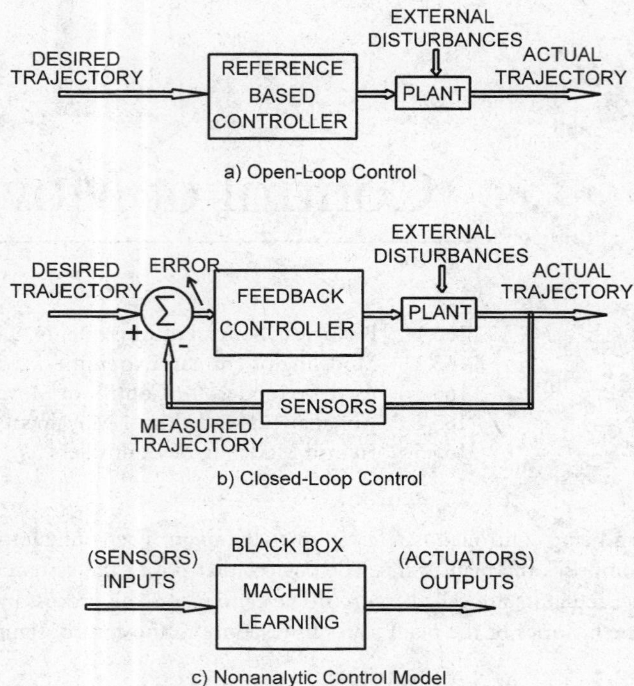

FIGURE 165.1 Open (a) and closed-loop (b) control systems. Nonanalytic model (c) for control of movements.

that specify target location need to be transformed into motor commands to arm muscles [2]. Arm movements illustrate an aspect of sensorimotor transformations. In arm movements the direction of force and movement do not coincide. The transformation between kinematics and kinetics is nontrivial in the case of arm motion. Not much is known about how this transformation might be implemented by neural circuits. Several investigators [3,4,5] have provided mathematical formulations of the problem. Other investigators have quantified biomechanical factors, such as muscle stiffness [6] and the changes in the muscle lever arms with posture, which also affect the relationship between force and movement. Arm muscle activation onset [7] and activation waveforms [8] have been empirically related to the direction of movement.

The automatic control of the arm reaching motion to accomplish grasping has been discussed in great detail [e.g., 9,10]. Imperfect models, wrong initial conditions, unmodeled perturbations and computational difficulties are serious problems in implementing analytical closed-loop control methods [11-14].

A simplified control method for reaching while grasping has been proposed [15]. This approach follows a so-called "extended physiological proprioception" control developed for bilateral amputees in Scotland [16]. The number of control variables can be reduced to a single input by establishing a synergy between joint angles.

Information about the target is provided by the visual system, whereas proprioceptors are adequate to signal initial arm posture [17,18]. Because proprioceptors sense muscle length and joint angles [19], the initial frames of reference for kinesthesis are fixed to the limb segments, i.e., elbow joint angles are initially sensed in the frame of reference fixed to the upper arm [20-22].

Control of cyclic movements. Three major requirements for cyclic movements (e.g., locomotion, cycling) are the following: 1) production of a basic cyclic rhythm which can control extremities and the trunk against gravity and propel it in the intended direction; 2) postural control of the moving body; and 3) adaptation of movements to meet environmental demands, tasks and preferred modalities of propulsion.

Because locomotion is achieved through the repetition of a well-defined movement, the emphasis in research to date has been on identifying principles and mechanisms that govern the generation of this basic rhythm. The idea of a central pattern generator for animal locomotion is generally attributed to G. Brown [23]. At that time, physiology was under the influence of strong individuals such as Sherrington [24] who had developed the concept of reflexes and favored the idea of locomotion being generated by a closed chain of such reflexes. The idea of autonomous central generators for a behavior such as locomotion was not fashionable and it took about 70 years for the concept to be generally accepted [25].

An important and relevant question is: Can the models and principles for control of locomotion developed from animal models be applied to humans? Animals afford far greater latitude in experimental paradigms than humans, and therefore are important for gathering physiological knowledge. The implicit assumption in using animal models is that knowledge gained can be applied to humans. Past experience has provided partial justification for this thinking. While walking, animals are mostly in a statically stable position, while in humans, the center of mass is outside the base of support for approximately 80% of a stride [26]. In quadrupeds, during locomotion, a tripodal support is ensured at all times, while in bipedal locomotion, a single support phase is interchanged with a double support phase. The contact surface of the foot to the ground is relatively large in humans compared to many animals. Traditionally human locomotor studies have concentrated mainly on a straight unobstructed gait. This has provided a large database for the characteristics of the basic locomotor patterns [e.g., 27,28].

The bipedal locomotion can be understood as a process whereby a pendulum like movement of the body about a point above the head (about 10 to 16%) is followed with an inverted pendulum–like movement about the leg in stance [29]. This behavior is repeated on the contralateral side of the body, and there is a temporal shift of 50% of the gait cycle. During the inverted pendulum phase of the locomotion the potential energy of the body is initially increased as the kinetic energy decreases and muscles push the body center of the mass against the gravitational force. This is followed by a decrease, when the center of the mass is falling down and moves faster. In the pendulum-like phase of the gait cycle the body swings with simultaneous shortening of the swinging leg to ensure ground clearance. Muscle activity varies with the speed of locomotion: for slow walking, muscle activity decelerates the body and diminishes the effects of gravity and inertial forces; for fast walking, muscles accelerate the body against gravity [30].

Locomotion is achieved by coordinated movements of the body segments employing an interplay of internal and external forces, e.g., muscular, inertial, gravitational and frictional. The muscular actions and their efficacy depend on the mastery of the usage of external forces. Analysis of pathological and normal gait patterns [31-33] suggests that it is impossible to define normal locomotion. However, there is a more or less well defined set of profiles for ground reaction forces and joint angles which can be recorded in most humans with no obvious musculo-skeletal and sensory injuries, and they highly correlate with patterns of muscle activities.

165.2 Modeling of Human Extremities

Modeling of a human body with the emphasis on the extremities is the prerequisite for the synthesis of control. Human extremities are unlike any other plant encountered in control engineering especially in terms of joints, actuators and sensors. This fact must be kept in mind when applying the general equations of mechanics to model the dynamics of functional motions. A simple extension of analytical tools used for the modeling of mechanical plants to the modeling of biomechanical systems may easily produce results in sharp discrepancy to reality.

A basic problem in motor control of human extremities is in the *planning* of motions to solve a previously specified task, and, then, *controlling* the extremity as it implements the commands. In terms of mechanics, extremities are rigid mechanical links with muscles, tendons, ligaments, and soft tissues that can set the joint angle to any value in its range. Muscles and tendons can also set the desired angular velocity, acceleration, and joint stiffness. The motions of extremities are called *trajectories* and consist of a sequence of positions, velocities, and accelerations of any part of the system (e.g., hand, fingertip, foot, etc.). It is anticipated that by using laws of mechanics one can determine necessary muscle forces to follow the given trajectory.

A body segment, when exposed to physiological loading, can be modeled as a rigid body, even though it contains bones and soft tissue (e.g., muscles, tendons, ligaments, skin etc.). Inertial properties of a rigid body are its mass and the inertia tensor. The inertia tensor for a rigid body provides information on inertial properties of the body when rotating. The mass for a rigid body describes its inertia for translation type of movements. In reality, the inertial properties change: the center of mass is shifting because the masses move relatively during contractions and passive stretching of muscles, relative movements of bones and other soft tissues modify the geometry and polycentric joints virtually change the length of a body segment.

The second issue in the modeling relates to joint structures. A joint is modeled as a *kinematic pair.* The theory of mechanisms defines connection of neighboring segments as kinematic pairs having up to five degrees of freedom (DOF). A typical human joint is a rotational kinematic pair having three DOF (ball joint) or one DOF (pin joint). However, many movements can not be simplified to a pure pin or ball joint rotation. Some human joints are actually double mechanical joints (e.g., ankle joint consists of one joint for internal–external rotation and inversion–eversion, and the other for movements in sagittal plane (flexion–extension). Sometimes the body segment rotation results from more complex structures such as two parallel bones connected between two neighboring joints (e.g., supination-pronation). Human joints are spatial and polycentric, and the displacement of their center of rotation results in the virtual change of the body segment.

Speaking of mechanics, one needs to distinguish the *forward kinematics* from the *inverse kinematics.* The direct kinematics is concerned with the determination of the position of body parts in the absolute reference frame, given joint angles, whereas the inverse kinematics computes joint angles for a given position of the body part. Of these two, the direct kinematic analysis is the simplest by far, with a straightforward solution for the unique end point of the body part corresponding to the given joint angles. On the other hand, the feasibility of an inverse kinematics solution depends on the redundancy of the structure as well as offsets and constraints.

The redundancy is an important feature in biological systems. The concept of *redundancy* requires a distinction between a degree of freedom and a permissible motion (e.g., rotation). The term DOF pertains exclusively to that permissible motion which is independent from other motions. A free rigid body has a maximum of six DOF, and a rigid body connected to another rigid body by a ball joint can have a maximum of three DOF; thus, any body segment has less than three DOF. Speaking of a structure as a whole, the total number of permissible rotations is termed system DOF. Some permissible rotations are not independent and they should not be called DOF (e.g., two parallel hinge joints). It is difficult to determine which permissible rotations should be controlled, and which will be constrained by themselves due to anatomical and physiological limitations.

Decomposing the linked structure into a set of single links is essential for numerical simulation. Internal forces within the body generated by muscle activities become external forces and torques when the dynamics of a single link are considered. The muscle is a linear actuator, with a very complex behavior depending on its length, velocity of shortening, firing rate, recruitment, and type. *Muscle models* [34] can be grouped to three classes: 1) Cross-bridge models including sarcomere–microscopic, conventional cross bridge, and unconventional cross-bridge models; 2) Fiber models; and 3) Whole muscle–macroscopic models (viscoelastic and black box models). The distinctions between these classes are not absolute (e.g., micro–macro).

The modeling and control of movements in this chapter relates to external control of muscles via so-called functional electrical stimulation. Macroscopic viscoelastic models started from the observation that the process of electrical stimulation transforms the viscoelastic material from a compliant, fluent state into the stiff, viscous state. Levin and Wyman [35] proposed a three-element model–damped and undamped elastic element in series. Hill's work [36] demonstrated that the heat transfer depends upon the type of contraction (isometric, slow contracting, etc.). The model includes the force generator, damping and elastic elements. Winters [37] generalized Hill's model in a simple enhancement of the original, which makes no claim to be based on the microstructure. Winters' model deals with joint moments:

$$M = M_0 - B\dot{\Theta},\ M_0 = AM_{0m}f(\Theta)$$

M is the muscle moment exerted about the joint, M_0 is the isometric moment (active state) with a maximum possible value M_{0m}, Θ is the joint angle corresponding to the current contractile element length, *A* is a normalized activation variable ($A = 0$ for relaxed muscle, $A = 1$ for a tetanized muscle), $f(\Theta)$ is the normalized tetanic length-tension relation, and *B* is a variable damping coefficient. For shortening muscle:

$$B = M_0(1+\alpha)/(\dot{\Theta}+\alpha\dot{\Theta}_0),\ \dot{\Theta}<0$$

whereas for lengthening muscle:

$$B = M_0(1-\beta)/(\dot{\Theta}+\gamma\dot{\Theta}_0),\ \dot{\Theta}>0$$

where a, β and γ are constants.

Zajac [38] proposed an appealing new version of the Hill model designed primarily for use in studies of mechanical interactions in multiple muscle systems. Muscles are connected to a skeleton with tendons, and they have spring like behavior when stretched. A variety of different non-rigid structures are included in the musculo-skeletal system (e.g., ligaments) adding different elastic and damping effects. Nonlinear models of electrically stimulated muscle have been reviewed by Durfee [39]. Block models are only one of several possible structures which can be used in biological system identification. Where there are only two blocks, one of which is nonlinear, the nonlinearity may be placed before or after the dynamic block, in so called Hammerstein or Wiener forms. While techniques for identifying the parameters for these models have been investigated by Kearney and Hunter [40], those working in the field of functional electrical stimulation have usually adopted the Hammerstein structure.

Dynamic analysis deals with two problems: determination of forces and torques when desired laws of motion are known (*forward dynamics*), and determination of trajectories and laws of motion when torques and forces are prescribed (*inverse dynamics*). Forward dynamics is a simple problem, and the only difficulty comes from errors generated through numerical derivation on noisy input data. Using laws of mechanics (D'Alembert principle [41,42], Newton-Euler Method [38, 43-47], Lagrange equations [48-51], Basic theorems of mechanics [52], etc.), in inverse dynamics, work only for open chain structures such as an arm when reaching, or a leg during the swinging, but it is not applicable for closed chain structures such as legs during the double support phase in walking, or the arm when contacting an immobile object, nor in redundant systems. Every closed and chained structure is dynamically undetermined and the only possibility to determine forces and torques is to use the theory of elasticity. The theory of elasticity assumes that the elements of the structure are solid (not rigid), and it is inappropriate for control design because of its extreme complexity. Approximate solution of the dynamics of a closed chain mechanism assumes how forces and torques are distributed in the structure. In the case of locomotion, it is necessary to introduce at least three equations in order to be able to determine the dynamics of the process. In most cases the chain is limited to a small number of admissible rotations. The reason for simplification is twofold: the problem can be solved, and the simulation results can be interpreted.

The complexity of a body segment dynamical model depends on a number and types of body segments, the joints connecting the segments, and the interaction among the segments and the environment. How complex a model is required for dynamic analysis of externally controlled walking, standing, manipulation, and grasping? Some decisions on how complex the model should be are simple while others are not. For

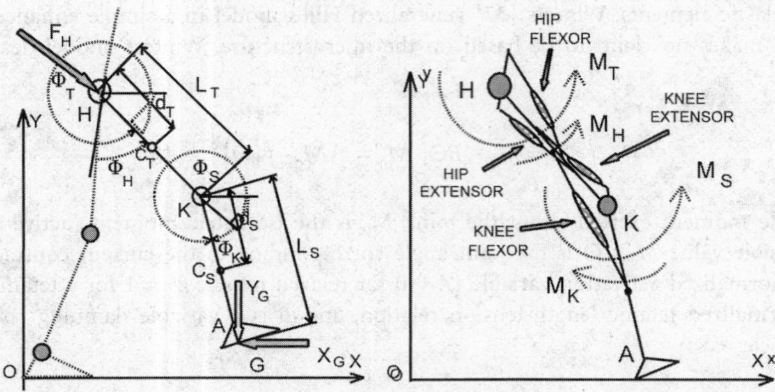

FIGURE 165.2 A sketch of a human body used for modeling of locomotion. Rigid body segments (left) are driven with two pairs of muscles (right). See the text for details.

instance, lower limbs can be assumed rigid during standing and walking. More controversial is the decision on how many segments should be assumed, and how many DOF should be assumed for each joint.

Several studies deal with strategies for prolonged standing [53-58] and functional electrical stimulation aided gait [59-64].

Recent studies are concerned with estimating parameters required for modeling and simulation [65-74]. While many of the results are very useful, they are incomplete and difficult to reproduce.

165.3 Analytic Models for Control of Movements

Attempts to analyze multi-joint structures are often characterized by reducing the number of DOF to a manageable level. Locomotion is often decomposed into a component in the sagittal plane, and smaller components in the frontal and horizontal planes. Using a smaller number of links is essential for simulation of gait.

A double, planar, pendulum with a moving hanging point, representing a human leg, is used to illustrate the modeling. The remaining part of the body is represented as a single force acting toward the moving hanging point (hip joint), called the hip joint torque (Fig. 165.2). Two pairs of muscles acting around the hip and knee joints actuate the leg. Both monoarticular and biarticular muscles add to the joint torques. Estimating the relative contribution of monoarticular and biarticular muscles is very difficult but possible [e.g., 75].

The model does not include ankle and phalangeal joints. The following system of differential equations describes the dynamics:

$$A_1\ddot{\phi}_S + A_2\ddot{\phi}_T\cos\left(\phi_T - \phi_S\right) + A_3\dot{\phi}_T^2\sin\left(\phi_T - \phi_S\right) - A_4\,\ddot{x}_H\sin\phi_S$$
$$-A_5\left(\ddot{y}_H + g\right)\cos\phi_S - X_G\,L_S\sin\phi_S + Y_G\,L_S\cos\phi_S = M_S \tag{165.1}$$

$$B_1\ddot{\phi}_T + B_2\ddot{\phi}_S\cos\left(\phi_T - \phi_S\right) + B_3\,\dot{\phi}_S^2\sin\left(\phi_T - \phi_S\right)$$
$$-B_4\ddot{x}_H\sin\phi_T - B_5\left(\ddot{y}_H + g\right)\cos\phi_T - x_G\,L_T\,\sin\phi_T + Y_G\,L_T\cos\phi_T = M_t$$

$$A_1 = J_{CS} + m_S\,d_S^2,\ B_1 = J_{CT} + m_S\,L_T^2 + m_T\,d_T^2 \tag{165.2}$$
$$A_2 = m_S\,d_S\,L_T,\ B_2 = A_2,\ A_3 = -A_2,\ B_3 = B_2$$
$$A_4 = m_S\,d_s,\ B_4 = m_S\,L_T + m_T\,d_T,\ A_5 = -A_4,\ B_5 = -B_4$$

There are several torques contributing to the total torques acting at the links:

$$\begin{aligned} &M_S = -M_K,\ M_T = M_K + M_H \\ &M_K = M_K^f - M_K^e - M_K^r,\ M_H = M_H^f - M_H^e - M_H^r \end{aligned} \tag{165.3}$$

The following notations are used: *S*—the shank segment (including the foot); *T*—the thigh segment; *CS, CT*—the center of the mass of the shank and thigh segments; *H*—the hip joint; *K*—the knee joint; *G*—the point of ground contact; d_s, d_T—distances of the proximal joint to the centers of the masses; L_s, L_T—lengths of the shank and thigh; m_s, m_T—masses of the shank and thigh; J_{CS}, J_{CT}—moments of inertia of the shank and thigh about the central axes perpendicular to xOy plane; *g*—gravitational acceleration; F_H—force acting at the hip joint; Y_G, Y_G—horizontal and vertical components of the ground reaction force; M_s, M_T—total torques acting at the shank and thigh segments; M_K, M_H—joint torques at the knee and hip joints; $\ddot{x}_H$, $\ddot{y}_H$—horizontal and vertical components of the hip acceleration; ϕ_S, ϕ_T—angles of the shank and thigh vs. the horizontal axis (Ox); ϕ_K, ϕ_H—angles of the knee and hip joint; ϕ_{TR}—angle of the trunk vs. the horizontal axis (Ox).

Flexion of a joint is defined to be the positive direction for angular changes; hence, the flexor torque is assumed to be positive. Index *f* is for the equivalent flexor muscle, and index *e* for the equivalent extensor muscle. The contribution of passive tissue crossing the joints is included by a 'resistive' torque (index *r*), which will be described below.

A three multiplicative factor model used is a modified version of the Hill model (Fig. 165.3). Active muscle force depends on the neural activation, muscle length and velocity of shortening or lengthening [41, 42]. The model is formulated as a function of joint angle and angular velocity, rather than muscle length and velocity. The three-factor model is given by:

$$M_a = A(u)\,f(\phi)\,g(\dot{\phi}) \tag{165.4}$$

where M_a is the active torque generated by a muscle contraction; $A(u)$ is the dependence of a torque on the level of muscle activity, *u*, depending on the stimulus amplitude, pulse width, and stimulus frequency; $f(\phi)$ is the dependence on the angle ϕ; and $g(\dot{\phi})$ is the dependence on the angular velocity $\dot{\phi}$. The parameter *u* is normalized to the range $0 \le u \le 1$. According to the literature [76], the muscle model described by (1) can predict the muscle torque with 85-90% accuracy during simultaneous, independent, pseudo-random variations of recruitment, angle and angular velocity.

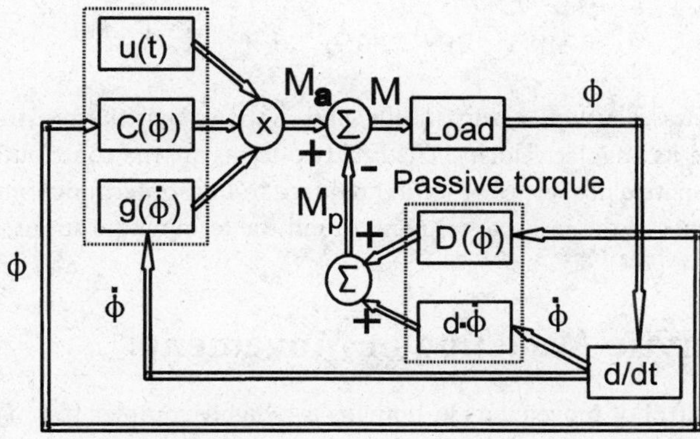

FIGURE 165.3 A three factor multiplicative model of a muscle (active torque). The model includes antagonist muscle (passive torque).

Activation A(u). In accordance with the literature [76-79] the muscle response to electrical stimulation was approximated by a second-order, critically damped [76,77], low-pass filter with a delay. Thus, the activation dynamics were assumed to be expressed by where $A(j\omega)$ is the Fourier transform of the muscle's contractile activity:

$$\frac{A(j\omega)}{U(j\omega)} = \frac{\omega_p^2}{\omega^2 + 2\, j\omega\, \omega_p + \omega_p^2}\, e^{-j\omega t_d} \tag{165.5}$$

$U(j\omega)$ is the Fourier transform of the muscle's electrical activity, ω_p is the muscle's natural (pole) frequency (typically in the range of 1-3 Hz, t_d is the excitation-contraction and other delays in the muscle.

Joint torque ($f(\phi)$) dependence on a joint angle. The nonlinear function relating torque and joint angle is simulated by a quadratic curve, $F = a_0 + a_1\,\phi + a_2\,\phi^2$. The torque generated can not be negative for any angle, so:

$$f(\phi) = \begin{cases} F \text{ if } F \geq 0 \\ 0 \text{ if } F < 0 \end{cases} \tag{165.6}$$

Coefficients a_i define the shape of the torque-angle curve and they are determined as the best fit through the experimental data recorded. The quadratic polynomial was selected as the simplest adequate fitting curve as described elsewhere [73,80]. Note that each of the quantities is non-negative.

Normalized joint torque ($g(\phi)$) dependence on the angular velocity. The normalized joint torques vs. joint angular velocities can be modeled piecewise linearly:

$$g(\dot{\phi}) = \begin{cases} c_{ij}, \dot{\phi} < (1 - c_{ij})/c_{ij} \\ 1 - c_{ij}\,\dot{\phi}, (1 - c_{ij})/c_{ij} \leq \dot{\phi} < 1/c_{ij} \\ 0, 1/c_{ij} \leq \dot{\phi}_K \end{cases} \tag{165.7}$$

The coefficients c_{ij} determine the slope and saturation level of the linearized torque vs. velocity of the muscle shortening. These coefficients can be determined using the method described in [81].

Resistive joint torque. Soft tissue, passive stretching of antagonistic muscles and ligaments introduce nonlinearities, which can be modeled as:

$$M_K^r = d_{11}(\phi_K - \phi_K 0) + d_{12}\,\dot{\phi}_K + d_{13}\,e^{d_{14}\phi_K} - d_{15}\,e^{d_{16}\phi_K} \tag{165.8}$$

The presented form was developed heuristically, and it shows that the resistive torques depend on both the joint angle and its angular velocity. The first two terms are the contributions of passive tissues crossing the joints (dissipative properties of joints) reduced to first-order functions. The other terms are the nonlinear components of the resistive torques around the terminal positions, and are modeled as a double exponential curve [73].

165.4 Nonanalytic Modeling of Movements

As stated earlier, the control of movements in humans is a highly complex task. The model of the body and extremities is complicated and not observable *per se*, the trajectories are not known in advance, the driving resources are user specific and time variable; and thus, real-time control using analytical tools is probably not feasible.

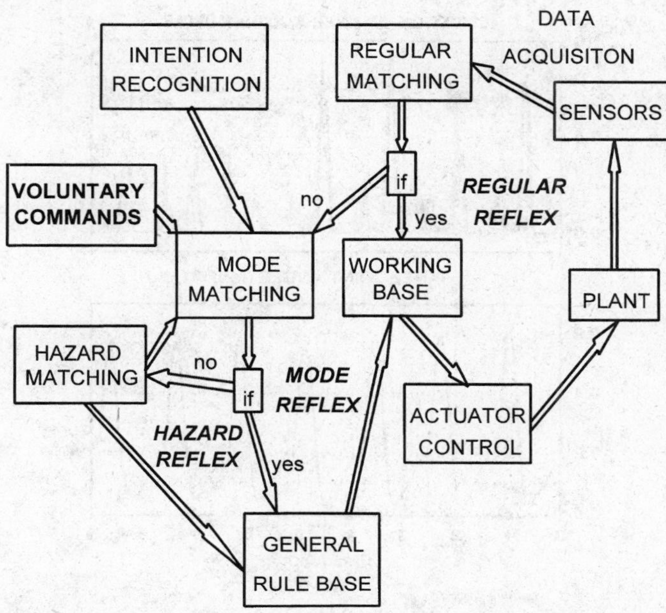

FIGURE 165.4 A scheme of the organization of a skill-based multilevel control for locomotion. Regular reflexes are sequences of system states connected with conditional If-Then expressions within a single mode of the gait, mode reflexes relate to the change of the mode of the gait (level, slope, stairs, etc.), and hazard reflexes are for situations where software or hardware cannot drive the system upon the preferred sequence of system states.

An alternative is to model the process of locomotion as a black box with multiple inputs and outputs. Tomović [82] proposed a non-numerical, control method, called Artificial Reflex Control (ARC). ARC (Fig. 165.4) refers to a skill-based expert using rules that have an *If-Then* structure [83]. The locomotion is presented as a sequence of discrete events. Each of these discrete events can be described in terms of a sensory pattern and a motor activity. A sensory pattern occurring during particular motor activity can be recognized with the use of human made (artificial or natural) sensors. The specific discrete event is called the state of the system by analogy to the state of finite-state automata [82,84]. This skill-based expert system is of the *On-Off* type and does not explicitly consider the system dynamics.

A rule-base control system has a hierarchical structure. The highest level is under volitional control of the user. Automatic adaptation to environmental changes and modes of gait is realized using artificial reflex control. The execution of the artificial reflex has to be tuned for smooth functional movements. Advantages of this hierarchical control method are the following: (1) adaptivity, (2) modularity, (3) ease of application and possibility of integration into a man-machine system. Rule-based control was tested in hybrid assistive systems [85,86] and artificial legs [83].

Nonanalytic control can be described as cloning the biological control [87,88]. The basic mechanisms for implementation of such algorithms are rule-based controllers (RBC) [89]. RBC were applied for assistive systems to restore locomotion with handcrafted [82,85,90,91], and lately an automatically determined knowledge base [92-97] with limited success. A RBC is a system implementing "If-Then" production rules, where the "If" part describes the sensory and motor state of the system, while the "Then" part of a rule defines the corresponding motor activity to follow.

It was hypothesized that machine learning (ML) can acquire the needed knowledge for RBC (see the Chapter of Abbas, this book). Learning in general can be described as capturing and memorizing of connectivisms between facts (Fig. 165.5). ML is a computerized capturing and memorizing process. In the most recent study, [93] used simulation results of a fully customized biomechanical model as inputs and outputs required for a ML. The following MLs were compared: 1) a multilayer perceptron (MLP)

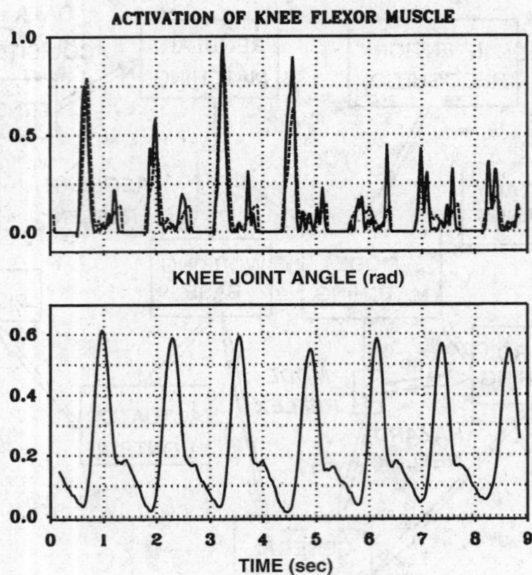

FIGURE 165.5 A mapping obtained as a result of the Radial basis function artificial neural network (RBF ANN) learning algorithm. The knee joint angle sensory information for first four consecutive strides (full line, bottom panel), the ground reactions and the hip accelerations were used as inputs for training. The activation of knee flexor muscles for the same four strides (full line, upper panel) was used for output during the training. The upper panel presents superimposed desired activation of muscles (full line) and a predicted activity of the muscle by a RBF ANN (dashed line) obtained from a trained network for seven consecutive steps. See text for details.

with the Levenberg-Marquardt improvement of a backpropagation (BP) algorithm [98]; 2) an adaptive-network-based fuzzy inference system (ANFIS) [99]; and 3) combination of an entropy minimization type of inductive learning (IL) technique [92,97,100] and a radial basis function (RBF) type of artificial neural network (ANN) [101,102] with an orthogonal least squares (OLS) learning algorithm [101].

A comparison of IL with adaptive logic networks (ALN) using restriction rules [95,96] shows that ALNs have some advantages over IL. Heller et al. [92] compared an IL method based on an algorithm called "hierarchical mutual information classifier" [103] with a MLP with a BP algorithm in reconstructing muscle activation from kinematic data during normal walking. The conclusion was that both techniques show comparable performance, although each technique has some advantages over the other one. A comparison of the IL method based on minimization of entropy and the RBF network in predicting muscle activation and sensory data from the history of sensory data for a human with spinal cord injury (SCI) [93], shows that the best generalization comes from combining both. The benefits of merging fuzzy logic and an ANN were explored extensively in the literature [99-104].

Investigations of multilayer perceptrons (MLPs) have been intensified since the formulation of the BP learning algorithm [98]. An MLP is a feed-forward network, typically consisting of several layers of nonlinear processing nodes called hidden layers with a linear output layer. Processing nodes take as input only the outputs of the previous layer, which are combined as a weighted sum and then passed through a nonlinear processing function known as the "activation function". This activation function is typically sigmoidal in shape. An MLP with three hidden layers can form arbitrarily complex decision regions and can separate classes that are meshed together. It can form regions as complex as those formed using mixture distributions and nearest neighbor classifiers [105].

The BP learning algorithm is a generalization of a gradient descent algorithm. The BP algorithm may lead to a local, rather than a global error minimum. If the local minimum found is not satisfactory, use of several different sets of initial conditions or a network with more neurons can be tried. A simple BP algorithm is very slow because it must use low learning rates for stable learning. There are ways to improve the speed and general performance of a BP algorithm. It can be improved in two different ways: by heuristics and by using more powerful methods of optimization. Speed and reliability of BP can be increased by techniques called momentum and adaptive learning rates (e.g., Levenberg-Marquardt modification of the BP algorithm).

Fuzzy sets are a generalization of conventional set theory. They were introduced by Zadeh [106] as a mathematical way to represent vagueness in everyday life. A formal definition of fuzzy sets that has been presented by many researchers is the following: a fuzzy set A is a subset of the universe of discourse X that admits partial membership. The fuzzy set A is defined as the ordered pair $A = \{x, m_A(x)\}$, where $x \in X$ and $0 \le m_A(x) \le 1$. The membership function $m_A(x)$ describes the degree to which the object x belongs to the set A, where $m_A(x) = 0$ represents no membership, and $m_A(x) = 1$ represents full membership.

One of the biggest differences between conventional (crisp) and fuzzy sets is that every crisp set always has a unique membership function, whereas every fuzzy set has an infinite number of membership functions that may represent it. This is at once both a drawback and advantage; uniqueness is sacrificed, but this gives a concomitant gain in terms of flexibility, enabling fuzzy models to be "adjusted" for maximum utility in a given situation.

The typical steps of "fuzzy reasoning" consist of: (1) Fuzzification, that is comparing the input variables with the membership functions of the premise (IF) parts in order to obtain the membership values between 0 and 1; (2) Weighing, that is applying specific fuzzy logic operators (e.g., AND operator, OR operator, etc.) on the premise parts membership values to get a single number between 0 and 1; (3) Generation, that is creating the consequent (THEN) part of the rule; and (4) Defuzzification, that is aggregating the consequent to produce the output.

There are several kinds of fuzzy rules used to construct fuzzy models. These fuzzy rules can be classified into the following three types according to their consequent form [107]:

1. crisply defined constant in the consequent:

$$\mathrm{R_i\text{: IF } x_1 \text{ is } A_{i1} \text{ and } \ldots\text{and } x_m \text{ is } A_{im} \text{ THEN } y \text{ is } c_i}$$

2. a linear combination of the system's input variables in the consequent:

$$\mathrm{R_i\text{: IF } x_1 \text{ is } A_{i1} \text{ and } \ldots\text{and } x_m \text{ is } A_{im} \text{ THEN } y \text{ is } g_i(x_1,\ldots,x_m) = b_0 + b_1x_1 + \ldots + b_mx_m}$$

3. fuzzy set in the consequent:

$$\mathrm{R_i\text{: IF } x_1 \text{ is } A_{i1} \text{ and } \ldots\text{and } x_m \text{ is } A_{im} \text{ THEN } y \text{ is } B_i}$$

where R_i is the i-th rule of the fuzzy system, x_j ($j = 1, 2,\ldots m$) are the inputs to the fuzzy system, and y is the output from the fuzzy system. The linguistic terms, A_{ij} and B_i, are fuzzy sets, c_i and b_j denote crisp constants.

The so-called zero-order Sugeno, or Takagi-Sugeno-Kang fuzzy model [108] has rules of the first type, whereas the first-order Sugeno fuzzy model has rules of the second type. The easiest way to visualize the first-order Sugeno fuzzy model is to think of each rule as defining the location of a "moving singleton" (a single spike from the consequent) depending on what the input is. Sugeno models are similar to the Mamdani model [109] which has rules of the third type, and which is more intuitive, but computationally less efficient. Fuzzification and weighing, are exactly the same, but generation and defuzzification are different [107].

For the type of fuzzy rules used in the Mamdani model various methods are available for defuzzification: the centroid of area, bisector of area, middle of maximum, largest of maximum etc. [110], but all of these methods are based on the calculation of the two-dimensional-shape surface, that is on the integration. The Sugeno-style enhances the efficiency of the defuzzification process because it greatly simplifies the computation; i.e., it has to find just the weighted average of a few data points. The implication method (generation) is simply multiplication, and the aggregation operator just includes all of the singletons.

Membership functions are subjective and context-dependent, so there is no general method to determine them. Currently, when fuzzy set theory is applied in control systems, the system designers are given enough freedom to choose membership functions and operators, usually in a trial and error way. After a hand-tuning process, the system can function effectively. However, the same methodology is hardly applicable when the system is a general purpose one, or when the context changes dynamically. This suggests an explanation why the most successful applications of fuzzy logic happen in control systems, rather than in natural language processing, knowledge base management, and general purposes reasoning.

The inductive learning (IL) is a symbolic technique, which uses supervised learning and generates a set of "if-then-else" decision rules. A method [97] is based on an algorithm called a "hierarchical mutual information classifier" [92, 103]. This algorithm produces a decision tree by maximizing the average mutual information at each partitioning step. It uses Shannon's entropy as a measure of information. Mutual information is a measure of the amount of information that one random variable contains about another random variable. It is a reduction in the uncertainty of one random variable due to the knowledge of the other. An effective method of integrating results of a mutual information algorithm into a production rule formalism, following the original work of Pitas [100,111] was done by Nikolić [112]. While generating the decision tree, the algorithm performs a hierarchical partitioning of the domain multidimensional space. Each new node of the decision tree contains a rule based on a threshold of one of the input signals. Each new rule further subdivides the example set. The training is finished when each terminal node contains members of only one class. An excellent feature of this algorithm is that it determines threshold automatically based on the minimum entropy. This minimum entropy method is equivalent to the determination of the maximum probability of recognizing a desired event (output) based on the information from the input.

Radial basis function (RBF) network is a feed-forward network. The RBF used here has a single output node and a single hidden layer which contains as many neurons as are required to fit the function within the specifications of error goal. The transformation from the input space to the hidden-unit space is nonlinear, whereas the transformation from the hidden-unit space to the output space is linear. A common learning algorithm for RBF networks is based on first randomly choosing some data points as radial basis function centers and then using singular value decomposition to solve for the weights of the network. An arbitrary selection of centers may not satisfy the requirement that centers should suitably sample the input domain. Furthermore, in order to achieve a given performance, an unnecessarily large RBF network may be required. Since a performance of an RBF network critically depends upon the chosen centers, we used an alternative learning procedure based on the OLS learning algorithm [101]. By providing a set of inputs and corresponding outputs, the values of weights and bias, and radial basis function centers (parameters for the RBF network) can be determined using the OLS algorithm in one pass of the learning data so that a network of an adequate size can be constructed.

When an input vector is presented to such a network, each neuron in the hidden layer will output a value according to how close the input vector is to the centers vector of each neuron. The result is that neurons with centers vector very different from the input vector will have outputs near zero. These small outputs will have a negligible effect on the linear output neurons. In contrast, any neuron whose centers vector is very close to the input vector will output a value near 1. If a neuron has an output of 1, its output weights in the second layer pass their values to the neuron in the second layer. The width of an area in the input space to which each radial basis neuron responds can be set by defining a spread constant for each neuron. This constant should be big enough to enable neurons to respond strongly to overlapping regions of the input space. The same spread constant is usually selected for each neuron.

165.5 Hybrid Modeling of Controllers

Feedback Error Learning (FEL) is a hybrid technique [113] using mapping to replace the estimation of parameters within the feedback loop in a closed-loop control scheme. FEL is a feed-forward neural

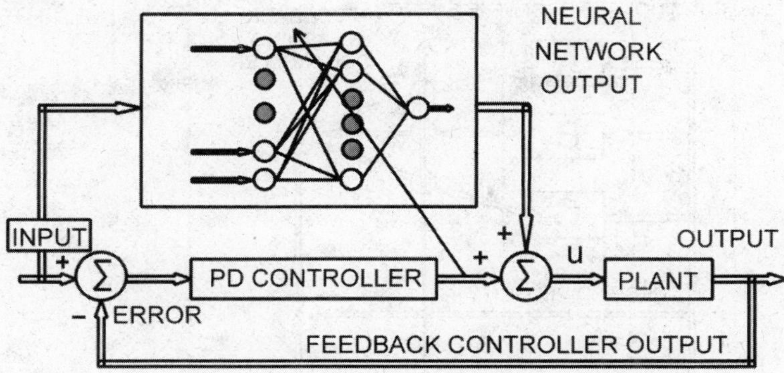

FIGURE 165.6 Model for the hybrid modeling of controllers using feedback error learning algorithm.

network structure, under training, learning the inverse dynamics of the controlled object. This method is based on contemporary physiological studies of the human cortex [114], and is shown in Fig. 165.6.

The total control effort, **u**, applied to the plant is the sum of the feedback control output and network control output. The ideal configuration of the neural network would correspond to the inverse mathematical model of the system's plant. The network is given information of the desired position and its derivatives, and it will calculate the control effort necessary to make the output of the system follow the desired trajectory. If there are no disturbances, the system error will be zero.

The configuration of the neural network should represent the inverse dynamics of the system when training is completed. It was prudent to use a total energy approach as the basis for the neural network, because only the commanded input and its first derivative is required for the FEL controller. By comparison, if the FEL were based on the mathematical model, the second derivative would be needed for the neural network to operate [115].

Figure 165.7 depicts a more in-depth explanation of the FEL strategy. The system input and output are labeled θ_d and θ, respectively. The proportional-plus-derivative (PD) feedback controller is included to provide stability during training of the neural network [114,117,118]. Enclosed in the dashed rectangle is an FEL controller, which outputs the necessary control signal, based on the desired inputs. The total energy of the system (nn) is calculated through parallel processing within the neural network, consisting of the functionals (A,B,C…) obtained from the total energy expression, and the synaptic weights w_i, w_j, and w_k. In addition to the above-mentioned synaptic weights, w_l is associated with damping losses. The training of the FEL controller is facilitated by changing the synaptic weights based on the output from a PD controller. The learning rule used, was proposed by Kawato, Furukawa, and Suzuki [119] and it is given as follows:

$$w_{i\text{new}} = w_{i\text{old}} + u_{PD} A\, \eta\, \Delta t$$

where $w_{i\,\text{new}}$ is the new value of the synaptic weight, $w_{i\,\text{old}}$ is the old value, u_{PD} is the output from the PD controller, A is the network functional associated with weight w_i, η is the learning rate and Δt is the integration step used in the computer simulation. A learning rate is included to control the rate of growth of the synaptic weights. The learning rule, as proposed by Kawato et al. [113], is based on the assumption of slow growth of the synaptic weights. The weights are initialized at zero, and the learning rates adjusted so the growths of the weights are uniform. This causes the weights to reach their final value at the same point in time, causing the error to approach zero. Subsequently, the learning as a function of error will level off, and the training of the neural network will be completed. However, if the growth of the weights is not homogeneous, it will result in an unbounded growth of the weights. The vertical line in Figure 165.7

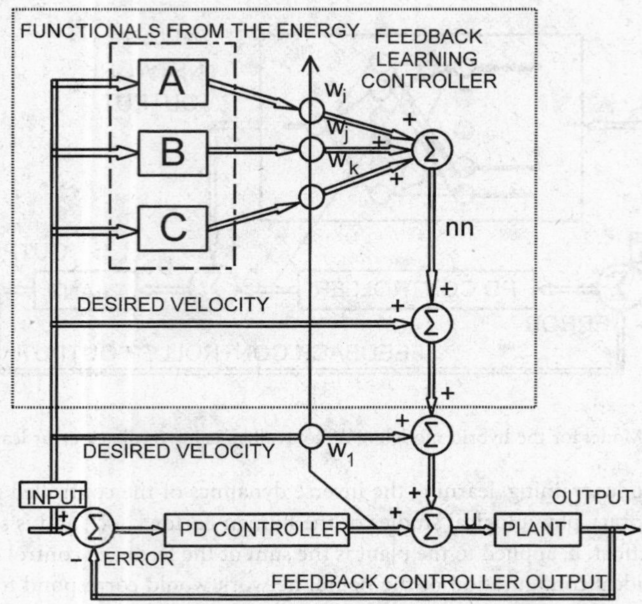

FIGURE 165.7 Scheme of the feedback error learning controller used for the one degree of freedom system. See text for details.

pointing upward through w_i, w_j, w_k and w_l symbolizes the learning. After the total energy is calculated, the time derivative is taken and divided by the desired velocity. The losses are calculated by multiplying the desired velocity by the weight w_l, and are then added to the control signal. Finally, the control effort from the FEL controller is added to the PD control effort.

In essence, the output of the feedback controller is an indication of the miss-match between the dynamics of the plant and the inverse-dynamics model obtained by the neural network. If the true inverse-dynamic model has been learned, the neural network alone will provide the necessary control signal to achieve the desired trajectory [118,120].

Defining Terms

Actuators: Device that powers a system.

Artificial reflex control: Expert system using rule-based control.

Closed-loop system: Control system which uses information about the output to correct the control parameters to minimize the error between the desired and actual trajectory.

Controller: Process by which the controls are generated.

Controls: The signals driving the actuators.

Degree of freedom: Independent variable defining the position. A free rigid body has six degrees of freedom, a ball joint three, and a hinge joint one.

Dynamic analysis: Analytic simulation of movements of the system considering forces, torques and kinematics. Forward dynamics uses the geometry and kinematics as input, and provides forces and torques as outputs; inverse dynamics starts from forces and torques and determines kinematics and geometry of the system.

Hierarchical control: Multilevel control allowing the vertical decomposition of the system.

Kinematic analysis: Analytic simulation of movements of the system considering positions, velocities and accelerations.

Kinematic pair: Connection of two neighboring segments.

Nonanalytic control: Mappings between inputs and outputs to be used for control.

Open-loop control: Control method that uses a prestored trajectory and the model of the system to control the plant.
Plant: The mechanical system being controlled.
Production rule: "If–Then" conditional expression used in expert systems.
Reference-based open loop control: Control which precomputes and stores control signals, and executes the desired motor task in real-time.
Rigid body: Set of material points with the distances between points being fixed.
State: The configuration of the system at any instant in time.
Trajectory: The time histories of the plant states in response to the control signals.

References

1. Latash, M. L., *Control of Human Movement*, Champaign, IL: Human Kinetics Publishers, 1993.
2. Soechting, J. F., Flanders, M., Moving in three dimensional space: frames of reference, vectors and coordinate systems, *Ann. Rev. Neurosci.* 5:167-191, 1992.
3. Hollerbach, J. M., Flash, T., Dynamic interactions between limb segments during planar arm movement, *Biol. Cybern.* 44:67-77, 1982.
4. Hoy, M. G., Zernicke, R. F., The role of intersegmental dynamics during rapid limb oscillations, *J. Biomech.* 19:867-877, 1986.
5. Zajac, F. E., Gordon, M. E., Determining muscles's force and action in multi-articular movement, In: *Exercise Sport Sci. Rev.* K. Pandoff, (Ed.) Baltimore: Williams and Wilkins, 17:187-230, 1989.
6. Mussa-Ivaldi, F. A., Hogan, N., Bizzi, E., Neural, mechanical and geometric factors subserving arm posture in humans, *J. Neurosci.* 5:2732-2743, 1985.
7. Hasan, Z., Karst, G. M., Muscle activity for initiation of planar, two-joint arm movements in different directions, *Exper. Brain Res.* 76:651-655, 1989.
8. Flanders, M., Temporal patterns of muscle activation for arm movements in three-dimensional space, *J. Neurosci.* 11:2680-2693, 1991.
9. Rugh, W. J., The Volterra/Wiener approach, In *Non-linear System Theory,* Baltimore: J. Hopkins University Press, 1981.
10. Hollerbach, J.M., Bennett, D. J., Feed-forward vs. feedback control of limb movements, In *Neural Prostheses, Replacing Motor function after disease or disability,* Stein, R. B. et al., (Eds.) Oxford University Press, New York, pp. 129-147, 1992.
11. Bennett, D. J., *The control of human arm movement: models and mechanical constraints,* Ph.D. Thesis, MIT, Cambridge, MA, 1990.
12. Eppinger, S. D., Seering, W. P., Understanding band-width limitations in robot force control, *Proc. IEEE Intern. Conf. Robotics and Automation,* pp. 904-909, 1987.
13. Jacobsen, S. C., Smith, C. C., Biggers, K. B., Iversen, E.K., Behavior based design for robot effectors, In: *Robotics Science,* Brady, M. (Ed.) Cambridge: MIT Press, pp. 505-539. 1989.
14. Stein, R.B., Feedback control of normal and electrically induced movements, In *Neural Prostheses, Replacing Motor function after disease or disability,* Stein, R. B. et al. (Eds.), New York: Oxford Univ. Press. pp. 281-297, 1992.
15. Popović, D., Popović, M., Tuning of a Nonanalytic Hierarchical Control System for Reaching with FES, *IEEE Trans. Biomed. Eng.* BME-45: 203-212, 1998.
16. Simpson, D.C., The choice of control system for the multimovement prosthesis: extended physiological proprioception, In *The control of upper-extremity prostheses and orthoses,* Herberts, P. (Ed.), C. Thomas. Chapter 15, 1974.
17. Bizzi, E., Accornero, N., Chapple, W., Hogan, N., Posture control and trajectory formation during arm movement, *J. Neurosci.* 4:2738-2744, 1984.
18. Hogan, N., The mechanics of multijoint posture and movement control, *Biol. Cybern.* 52:325-332, 1985.
19. McCloskey, D. I., Kinesthetic sensibility, *Physiol. Rev.* 58:762-820, 1978.

20. Soechting, J. F., Ross, B., Psychophysical determination of coordinate representation of human arm orientation, *Neurosci,* 13:595-604, 1984.
21. Morasso, P., Spatial control of arm movements, *Exp. Brain Res.* 42:223-227, 1981.
22. Morasso, P., Three dimensional arm trajectories, *Biol.Cybern,* 48:187-194, 1983.
23. Brown, T. G., The intrinsic factors in the act of progression in the mammal, *Proc. Roy. Soc.* B84:308-319, 1911.
24. Sherrington, C. S., *The Integrative Action of the Nervous System*, New Haven, CT: Yale University Press, 1906. (reprinted 1961).
25. Shik, M. L., Orlovsky, G. N., Neurophysiology of locomotor automatism, *Physiol. Rev.* 56: 465-501, 1976.
26. Patla, A., *Adaptability of human gait: Implication for the control of locomotion*, Elsevier Publ, 1991.
27. Winter, D. A., Foot trajectory in human gait: a precise and multifunctional motor control task, *Phys. Ther.* 72:45-56, 1992.
28. Inman, V. T., Ralston, H. J., Todd, F., *Human walking*, Baltimore, Williams & Wilkins, 1981.
29. Cappozzo, A., Figure, F., Leo, T., Marchetti, M., Movements and mechanical energy changes in upper part of the human body during walking, In: *Biomechanics VI-A*, Asmunssen, E. and Jorgensen, K. (Eds.) Baltimore: University Park Press, pp. 272-279, 1978.
30. Cavagna, G. A., Margaria, R., Mechanics of walking, *J. Appl. Phys.* 21:271-278, 1966.
31. Perry, J., *Gait Analysis: Normal and Pathological Function*, Thorofare, NJ: Slack Inc., 1992.
32. Winter, D. A., *Biomechanics and Motor Control of Human Movement*, 2nd ed. New York: Wiley-Interscience, 1990.
33. Winter, D. A., *The Biomechanics and Motor Control of Human Gait: Normal, Elderly and Pathological*, Waterloo, Ontario: University of Waterloo Press, 1991.
34. Zahalak, G. I., Modeling muscle mechanics (and energetics), In: *Multiple Muscle Systems: Biomechanics and Movement Organization,* J. M. Winters and S. L-Y., Woo (Eds.), New York: Springer Verlag, pp. 1-23, 1990.
35. Levin, A., Wyman, J., The viscous elastic properties of muscle, *Proc. R. Soc. London Biol.* 101:218-243, 1927.
36. Hill, T. L., The heat of shortening and the dynamic constants of muscle, *Proc. R. Soc. London Biol.* 126:135-195, 1938.
37. Winters, J. M., Hill-based muscle models: A systems engineering prospective. In: *Multiple Muscle Systems: Biomechanics and Movement Organization,* J. M. Winters and S. L-Y. Woo (Eds.), New York: Springer Verlag, pp. 66-93, 1990.
38. Zajac, F. E., Muscle and Tendon: Properties: Models, scaling, and application to biomechanics and motor control, *CRC Crit. Rev. Biomed. Eng.* 17:359-411, 1989.
39. Shue, G., Crago P. E., Chizeck H. J., Muscle-joint models incorporating activation dynamics, moment-angle and moment-velocity properties, *IEEE Trans Biomed Eng.* BME-42:212-223, 1995.
40. Kearney, R. E., Hunter, I. W., System identification of human joint dynamics, *CRC Crit. Rev. Biomed. Eng.* 18:55-87, 1990.
41. Huston, R. L., Passerello, C.E., Harlow, M.W., On human body dynamics, *Ann. Biomed. Eng.* 4:25-43, 1976.
42. Huston, R. L., Passerello, C. E., Harlow, M.W., Dynamics of multirigid-body systems, *J. Appl. Mech.* 45:889-894, 1978.
43. Stepanenko, Y., Vukobratović, M., Dynamics of articulated open-chain active mechanisms, *Math. Biosci.* 28:137-170, 1976.
44. Orin, D. E., McGhee, R. B., Vukobratović, M., Hartoch, G., Kinematic and kinetic analysis of open-chain linkages utilizing Newton-Euler methods, *Math. Biosci.* 43:107-130, 1979.
45. Marshall, R. N., Jensen, R. K., Wood, G. A., A general Newtonian simulation of an n-segment open chain model, *J. Biomech.* 18:359-367, 1985.

46. Khang, G., Zajac, F.E., Paraplegic standing controlled by functional electrical stimulation: part I—computer model and control-system design; part II—computer simulation studies, *IEEE Trans. Biomed. Eng.* BME-36:873-893, 1989.
47. Yamaguchi, G. T., Zajac, F. E., Restoring unassisted natural gait to paraplegics via functional neuromuscular stimulation: a computer simulation study, *IEEE Trans. Biomed. Eng.* BME-37:886-902, 1990.
48. Onyshko, S., Winter, D. A., A mathematical model for the dynamics of human locomotion, *J. Biomech.* 13:361-368, 1980.
49. Hatze, H., A complete set of control equations for the human musculo-skeletal system, *J. Biomech.* 10:799-805, 1977.
50. Hatze, H., Neuromusculoskeletal control systems modeling—a critical survey of recent developments, *IEEE Trans. Automat. Control*, AC-25:375-385, 1980.
51. Zheng, Y. F., Shen, J. S., Gait Synthesis for the SD-2 biped robot to climb sloping surfaces, *IEEE Trans. Robot. Automat.* RA-6:86-96, 1990.
52. Koozekanani, S. H., Barin, K., McGhee, R. B., Chang, H. T., A recursive free body approach to computer simulation of human postural dynamics, *IEEE Trans. Biomed. Eng.* BME-30:787-792, 1983.
53. Davoodi, R., Andrews B. J., Computer simulation of FES standing up in paraplegia: A self-adaptive fuzzy controller with reinforcement learning, *IEEE Trans. Rehab. Eng,* TRE-6:151-161, 1998.
54. Donaldson N., Yu, C., A strategy used by paraplegics to stand up using FES, *IEEE Trans. Rehab. Eng.* TRE-6:162-166, 1998.
55. Hunt K. J., Munih M., Donaldson, N., Feedback control of unsupported standing in paraplegia–Part I: Optimal control approach and Part II: Experimental Results, *IEEE Trans. Rehab. Eng.* TRE-5:331-352, 1997.
56. Matjačić, Z., Bajd, T., Arm-free paraplegic standing–part I: Control model synthesis and simulation and part II: Experimental results, *IEEE Trans. Rehab. Eng.* TRE-6: 125-150, 1998.
57. Riener, R., Fuhr, T., Patient-driven control of FES supported standing up: A simulation study, *IEEE Trans. Rehab. Eng.* TRE-6:113-124, 1998.
58. Veltink, P. H., Donaldson, N., A perspective on the controlled FES-supported standing, *IEEE Trans. Rehab. Eng.* TRE-6:109-112, 1996.
59. Abbas, J. J., Chizeck, H. J., Neural network control of functional neuromuscular stimulation systems: computer simulation studies, *IEEE Trans. Biomed. Eng.* BME-42:1117-1127, 1995.
60. Abbas, J. J., Triolo R. J., Experimental evaluation of an adaptive feedforward controller for use in functional neuromuscular stimulation systems, *IEEE Trans. Rehab. Eng.* TRE-5:12-22, 1997.
61. Gilchrist, L. A., Winter, D.A., A multisegment computer simulation of normal human gait, *IEEE Trans. Rehab. Eng.* TRE-5:290-299, 1997.
62. Graupe, D, Kordylewaki, H, Artificial neural network control of FES in paraplegics for patient responsive ambulation, *IEEE Trans. Biomed. Eng.* BME-42:699-707, 1995.
63. Kobetic, R., Marsolais, E. B., Synthesis of paraplegic gait with multichannel functional electrical stimulation, *IEEE Trans. Rehab. Eng.* TRE-2:66-79, 1994.
64. Kobetic, R., Triolo R. J., Marsolais, E. B., Muscle selection and walking performance of multichannel FES systems for ambulation in paraplegia, *IEEE Trans. Rehab. Eng.* TRE-5:23-29, 1997.
65. Crago, P.E., Muscle input-output model: The static dependence of force on length, recruitment and firing period, *IEEE Trans. Biomed. Eng.* BME-39:871-874, 1992.
66. Durfee, W. K., Mac Lean, K. E., Methods of estimating the isometric recruitment curve of electrically stimulated muscle, *IEEE Trans. Biomed. Eng.* BME-36:654-667, 1989.
67. Durfee, W. K., Model identification in neural prostheses system, In: *Neural Prostheses: Replacing Motor Function After Disease or Disability*, R. B. Stein, P. H. Pechkam, and D. Popović, (Eds.), New York: Oxford Univ. Press, pp. 58-87, 1992.
68. Durfee W. K., Palmer, K. I., Estimation of force activation, force-length, and force-velocity properties in isolated electrically stimulated muscle, *IEEE Trans. Biomed. Eng.* BME-41:205-216, 1994.

69. Franken, H. M., Veltink, P.H., et al., Identification of passive knee joint and shank dynamics in paraplegics using quadriceps stimulation, *IEEE Trans. Rehab. Eng.* TRE-1:154-164, 1993.
70. Franken H. M., Veltink, P. H., et al., Identification of quadriceps-shank dynamics using randomized interpulse interval stimulation, *IEEE Trans. Rehab. Eng.* TRE-4:182-192, 1995.
71. Hunter, I., Korenburg, M., The identification of nonlinear biological systems: Wiener and Hammersteiin cascade models, *Biol. Cybern.* 55:135-144, 1986.
72. Kearney, R. E., Stein, R. B., Parameswaran, L., Identification of intrinsic and reflex contribution of human ankle stiffness dynamics, *IEEE Trans. Biomed. Eng.* BME-44:493-504, 1997.
73. Stein, R. B., Zehr, E.P., et al., Estimating mechanical parameters of leg segments in individuals with and without physical disabilities, *IEEE Trans. Rehab. Eng.* TRE-4, pp. 201-212, 1996.
74. Xu, Y., Hollerbach, J.M., Identification of human joint mechanical properties from single trial data, *IEEE Trans. Biomed. Eng.* BME-45:1051-1060, 1998.
75. Herzog, W., derKeurs, H. E. D. J., Force-length relation of *in vivo* human rectus femoris muscles, *P. Flügens Arch*, 411:643-647, 1988.
76. Veltink, P. H., Chizeck, H. J., Crago, P. E., El-Bialy, A., Nonlinear joint angle control for artificially stimulated muscle, *IEEE Trans. Biomed. Eng.* BME-39:368-380, 1992.
77. Bajzek, T. J., Jaeger, R. J., Characterization and control of muscle response to electrical stimulation, *Ann. Biomed. Eng.* 15:485-501, 1987.
78. Baratta, R., Solomonow, M., The dynamic response model of nine different skeletal muscles, *IEEE Trans. Biomed. Eng.* BME-36:243-251, 1989.
79. Chizeck, H. J., Lan, N., Sreeter-Palmiere, L., Crago, P. E., Feedback control of electrically stimulated muscle using simultaneous pulse width and stimulus period variation, *IEEE Trans. Biomed. Eng.* BME-38: 1224-1234, 1991.
80. Scheiner, A., Stein, R B., Ferencz, D., Chizeck, H. J., Improved models for the lower leg in paraplegics, *Proc. IEEE Ann. Conf. EMBS*, San Diego, pp. 1151-1152, 1993.
81. Popović, D., Stein, R. B., et al., Optimal control of walking with functional electrical stimulation: A computer simulation study, *IEEE Trans. Rehab. Eng.* TRE-7(1), 1999.
82. Tomović, R., Control of Assistive Systems by External Reflex Arcs, In: *Advances in External Control of Human Extremities VIII,* D. Popović, (Ed.), Belgrade: Yugoslav Committee for ETAN. pp. 7-21, 1984.
83. Popović, D., Tomović, R., Schwirtlich, L., Tepavac, D., Control Aspects on Active A/K Prosthesis, *Int. J. Man—Machine Studies,* 35:750-767, 1991.
84. Tomović, R., Popović, D., Tepavac, D., Adaptive reflex control of assistive systems. In: *Advances in External Control of Human Extremities IX,* D. Popović (Ed.), Belgrade: Yugoslav Committee for ETAN, pp. 207-214, 1987.
85. Andrews, B. J., Barnett, R. W., et al., Rule-based control of a hybrid FES orthosis for assisting paraplegic locomotion, *Automedica,* 11:175-199, 1989.
86. Popović, D., Schwirtlich, L., Radosavljevic, S., Powered hybrid assistive system, In: *Advances in External Control of Human Extremities X,* D. Popović (Ed.), Belgrade: Nauka, pp. 177-187, 1990.
87. Popović, D., Finite state model of locomotion for functional electrical stimulation systems, *Progr. Brain Res.* 97:397-407, 1993.
88. Popović, D., Stein, R. B., et al., Sensory nerve recording for closed-loop control to restore motor functions, *IEEE Trans. Biomed. Eng.* BME-40:1024-1031, 1993.
89. Tomović, R., Popović, D., Stein R. B., *Nonanalutical Methods for motor control,* Singapore: World Sci. Publ., 1995.
90. Aeyels, B., Peeraer, L., Van der Sloten, J., Van der Perre, G., Development of an above-knee prosthesis equipped with a microprocessor-controlled knee joint: first test results, *J. Biomed. Eng.* 14:199-202, 1992.
91. Bar, A., Ishai, P., Meretsky, P., Koren, Y., Adaptive microcomputer control of an artificial knee in level walking, *J. Biomechan. Eng.* 5:145-150, 1983.

92. Heller, B., Veltink, P. H., et al., Reconstructing muscle activation during normal walking: a comparison of symbolic and connectionist machine learning techniques, *Biol. Cybern.* 69:327-335, 1993.
93. Jonić, S., Janković, T., Gajić, V., Popović, D., Three machine learning techniques for automatic determination of rules to control locomotion, *IEEE Trans. Biomed. Eng.* BME-46:300-311, 1999.
94. Kirkwood, C. A., Andrews, B. J., Mowforth, P., Automatic detection of gait events: A case study using inductive learning techniques, *J. Biomed. Eng.* 11:511-516, 1989.
95. Kostov, A., Stein, R. B., Popović, D. B., Armstrong, W. W., Improved Methods for Control of FES for Locomotion, *Proc. IFAC Symp. Biomed. Model*, Galveston, Texas, pp. 422-428, 1994.
96. Kostov, A., Andrews, B. J., Popović, D. et al., Machine learning in control of functional electrical stimulation (FES) for locomotion, *IEEE Trans. Biomed. Eng.* BME-42:541-552, 1995.
97. Nikolić, Z., Popović, D., Automatic rule determination for finite state model of locomotion, *IEEE Trans. Biomed. Eng.* BME-45:1081-1085, 1998.
98. Rumelhart, D. E., Hinton, G. E., Williams, R. J., Learning interval representation by error propagation, In: *Parallel Distributed Processing*, Ch. 8:318-361, Cambridge, MA: MIT Press, 1986.
99. Jang, J. S. R., ANFIS: Adaptive-Network-Based Fuzzy Inference Systems, *IEEE Trans. Sys. Man. Cybern.* vol. 23-SMC, pp. 665-685, 1993.
100. Watanabe, S., *Pattern Recognition*, New York: Wiley Interscience, 1985.
101. Chen, S., Cowan, C. F. N., Grant, P. M., Orthogonal least Squares learning Algorithm for Radial Basis Function Networks, *IEEE Trans. Neu. Net.* NN-2:302-309, 1991.
102. Nikolić, Z., *Automatic rule determination for finite state model of locomotion*, PhD Thesis, University of Miami, Miami, Florida, 1995.
103. Sethi, I. K., Sarvarayudu, G. P. R., Hierarchical classifier design using mutual information, *IEEE Trans. Pattern Anal. Mach. Intel.* PAM-4:441-445, 1992.
104. Lin, C. T., Lee, C. S. G., Neural-network-based fuzzy logic control and decision system, *IEEE Trans. Comput.* C-40:1320-1336, 1991.
105. Lippmann, R. P., An introduction to computing with neural nets, *IEEE ASSP Mag*, 3:4-22, 1987.
106. Zadeh, L. A., Fuzzy sets, *Information and Control*, 8:338-352, 1965.
107. Lee, K. M., Kwak, D. K., Lee-Kwang, H., Fuzzy Inference Neural Network for Fuzzy Model Tuning, *IEEE Trans. Sys. Man. Cybern.* SMC-26, pp. 637-645, 1996.
108. Sugeno, M., *Industrial applications of fuzzy control*, Elsevier Science, 1985.
109. Mamdani, E. H., Assilian, S., An experiment in linguistic synthesis with a fuzzy logic controller, *Int. J. Man-Machine Studies*, 7:1-13, 1975.
110. Hellendoorn, H., Thomas, C., Defuzzification in fuzzy controllers, *J. Intel. Fuzzy Syst.* 1:109-123, 1993.
111. Pitas, I., Milios, E., Venetsanopoulos, A. N., Minimum Entropy Approach to Rule Learning from Examples, *IEEE Trans. Sys. Man. Cybern.* SMC-22, pp. 621-635, 1992.
112. Nikolić, Z., Popović, D., Automatic detection of production rules for locomotion, *J. Autom. Control*, 6:81-94, 1996.
113. Kawato, M., Computational Schemes and Neural Network Models for Formation and Control of Multi-Joint Arm Trajectory, In: *Neural Network for Control*, W. T. Miller, R. T. Sutton, P. J. Werbos, (Eds.), MIT Press, 1990.
114. Kawato, M., Feedback-Error-Learning Neural Network for Supervised Motor Learning, In *Adv. Neural Computers*, pp. 365-372, 1990.
115. Kalanovic, V. D., Tseng, W. H. Back Propagation in Feedback Error Learning, In *Proc. Neural, Parallel and Sci. Comp*, vol.1:239-242, 1995.
116. Miyamoto, H., Kawato, M., Setoyama, T., Suzuki, R., Feedback-Error-Learning Neural Network for Trajectory Control of a Robotic Manipulator, *Neural Networks*, 1:251-265, 1988.
117. Nordgren, R. E., Meckl, P. M., An Analytical Comparison of a Neural Network and a Model-Based Adaptive Controller, *IEEE Trans. Neural Networks*, 4:595-601, 1993.

118. Roa, D. H., Bitner, D., Gupta, M. M., Feedback-error learning scheme using recurrent neural networks for nonlinear dynamic systems, *IEEE Proc.* 1994.
119. Kawato, M., Furukawa, K., Suzuki, R., A hierarchical neural-network model for control and learning of voluntary movement, *Biol. Cybern.* 57:169-185, 1987.
120. Szabo, P., Pandya, A. S., Neural network as robot arm manipulator controller, *IEEE Proc.* 1996.

Further Information

Latash, M., *Control of Human Movements*, Human Kinetics Publ., 1993.

Tomović, R., Popović, D., Stein, R.B., (Eds.) *Nonanalytical Methods for Motor Control*, Singapore: World Sci. Publ., 1995.

Winters, J. M., Woo, S. L-Y., (Eds.) *Multiple Muscle Systems: Biomechanics and Movement Organization*, New York: Springer Verlag, 1990.

Periodicals

IEEE Transactions on Robotics and Automation, IEEE Press.

IEEE Transactions on Biomedical Engineering, IEEE Press.

IEEE Transactions on Rehabilitation Engineering, IEEE Press.

IEEE Transactions on Neural Networks, IEEE Press.

IEEE Transactions on System, Man and Cybernetics, IEEE Press.

166

The Fast Eye Movement Control System

John Denis Enderle
University of Connecticut

This chapter presents a broad overview of the fast eye movement control system. A fast eye movement is usually referred to as a saccade, and involves quickly moving the eye from one image to another image. This type of eye movement is very common and is observed most easily while reading when the end of a line is reached, the eyes are moved quickly to the beginning of the next line. A qualitative description of the fast eye movement system is given first in the introduction and then followed by a brief description of saccade characteristics. Next, the earliest quantitative saccade model is presented, followed by more complex and physiologically accurate models. Finally, the saccade generator, or saccade controller, is then discussed on the basis of anatomic pathways and control theory. The purpose of this review is focused on mathematical models of the fast eye movement system and its control strategy, rather than on how visual information is processed. The literature on the fast eye movement system is vast, and thus this review is not exhaustive but rather a representative sample from the field.

The oculomotor system responds to visual, auditory, and vestibular stimuli, which results in one of five types of eye movements: fast eye movements, smooth pursuit eye movements, vestibular ocular movements, vergence eye movements, and optokinetic eye movements. Each of these movements is controlled by a different neuronal system, and all these controllers share the same final common pathway to the muscles of the eye. In addition to the five types of eye movements, these stimuli also cause head and body movements. Thus, the visual system is part of a multiple input–multiple output system.

Regardless of the input, the oculomotor system is responsible for movement of the eyes so that images are focused on the central 1/2° region of the retina, known as the *fovea.* Lining the retina are photoreceptive cells which translate images into neural impulses. These impulses are then transmitted along the optic nerve to the central nervous system via parallel pathways to the superior colliculus and the cerebral cortex. The fovea is more densely packed with photoreceptive cells than the retinal periphery; thus a higher resolution image (or higher visual acuity) is generated in the fovea than the retinal periphery. The purpose of the fovea is to allow us to *clearly* see an object, and the purpose of the retinal peripheral is

0-8493-0462-8/00/$0.00+$.50

to allow us to *detect* a new object of interest. Once a new object of interest is detected in the periphery, the saccade system redirects the eyes, as fast as possible, to the new object. This type of saccade is typically called a *goal-directed saccade.*

During a saccade, the oculomotor system operates in an open-loop mode. After the saccade, the system operates in a closed-loop mode to ensure that the eyes reach the correct destination. The reason that the saccade system operates without feedback during a fast eye movement is simple—information from the retina and muscle proprioceptors is not transmitted quickly enough during the eye movement for use in altering the control signal.

The oculomotor plant and saccade generator are the basic elements of the saccadic system. The oculomotor plant consists of three muscle pairs and the eyeball. These three muscle pairs contract and lengthen to move the eye in horizontal, vertical, and torsional directions. Each pair of muscles acts in antagonistic fashion due to reciprocal innervation by the saccade generator. For simplicity, the models described here involve only horizontal eye movements and one pair of muscles, the lateral and medial rectus muscle.

166.1 Saccade Characteristics

Saccadic eye movements, among the fastest voluntary muscle movements the human is capable of producing, are characterized by a rapid shift of gaze from one point of fixation to another. The usual experiment for recording saccades is for subjects to sit before a horizontal target display of small light emitting diodes (LEDs), with instructions to maintain their eyes on the lit LED by moving their eyes as fast as possible to avoid errors. A saccade is made by the subject when the active LED is switched off and another LED is switched on. Saccadic eye movements are conjugate and ballistic, with a typical duration of 20–100 ms and a latency of 150–300 ms. The latent period is thought to be the time interval during which the CNS determines whether to make a saccade, and if so, calculates the distance the eyeball is to be moved, transforming retinal error into transient muscle activity.

Generally, saccades are extremely variable, with wide variations in the latent period, time to peak velocity, peak velocity, and saccade duration. Furthermore, variability is well coordinated for saccades of the same size; saccades with lower peak velocity are matched with longer saccade durations, and saccades with higher peak velocity are matched with shorter saccade durations. Thus, saccades driven to the same destination usually have different trajectories.

To appreciate differences in saccade dynamics, it is often helpful to describe them with saccade main sequence diagrams [Bahill et al., 1975]. The main sequence diagrams plot saccade peak velocity–saccade magnitude, saccade duration–saccade magnitude, and saccade latent period–saccade magnitude. Shown in Fig. 166.1 are the main sequence characteristics for a subject executing 54 saccades. Notice that the peak velocity–saccade magnitude is basically a linear function until approximately 10°, after which it levels off to a constant for larger saccades. Many researchers have fit this relationship to an exponential function. The solid lines in Fig. 166.1*a* include an exponential fit to the data for positive and negative eye movements. The lines in the first graph are fitted to the equation

$$V=\alpha_i\left(1-e^{-\frac{x}{\beta_i}}\right) \tag{166.1}$$

where V is the maximum velocity, x the saccade size, and the constants α_i and β_i evaluated to minimize the summed error squared between the model and the data. Note that α_i is to represent the steady state of the peak velocity–saccade magnitude curve and β_i is to represent the time constant for the peak velocity–saccade magnitude curve. For this data set, α_i equals 825 and 637, and β_i equals 9.3 and 6.9, for positive and negative movements, respectively. The exponential shape of the peak velocity–saccade amplitude relationship might suggest that the system is nonlinear, if one assumes a step input to the

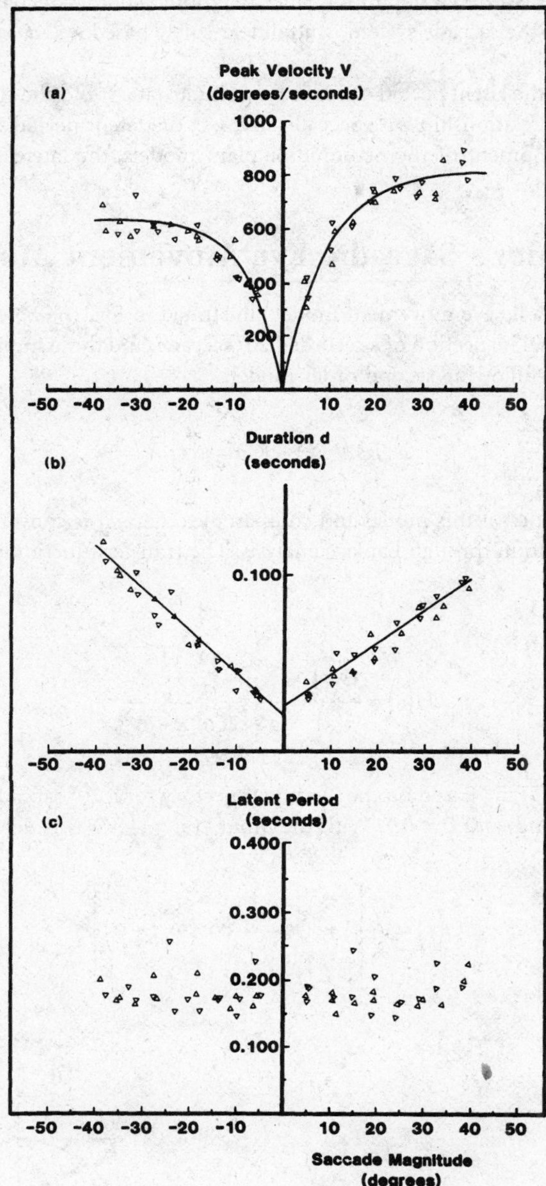

FIGURE 166.1 Main sequence diagrams. (*a*) peak velocity–saccade magnitude, (*b*) saccade duration–saccade magnitude, and (*c*) latent period–saccade magnitude for 54 saccadic eye movement by a single subject. *Source:* Enderle JD. 1988. Observations on pilot neurosensory control performance during saccadic eye movements. *Aviat Space Environ Med* 59:309. With permission.

system. A step input would provide a linear peak velocity–saccade amplitude relationship. In fact, the saccade system is not driven by a step input, but rather a more complex pulse-step waveform. Thus, one cannot conclude that the saccade system is nonlinear solely based on the peak velocity-saccade amplitude relationship.

Shown in Fig. 166.1*b* are data depicting a linear relationship between saccade duration–saccade magnitude. The dependent between saccade duration and saccade magnitude also might suggest that the system is nonlinear, if one assumes a step input. Since the input is not characterized by a step waveform, one cannot conclude that the saccade system is nonlinear solely based on the saccade duration–saccade magnitude relationship.

Shown in Fig. 166.1*c* is the latent period–saccade magnitude data. It is quite clear that the latent period does not show any linear relationship with saccade size, i.e., the latent period's value is independent of saccade size. In the development of the oculomotor plant models, the latent period will be implicitly assumed within the model.

166.2 Westheimer's Saccadic Eye Movement Model

The first quantitative saccadic eye movement model, illustrated in Fig. 166.2, was published by Westheimer [1954]. Based on visual inspection of a recorded 20° saccade, and the assumption of a step controller, Westheimer proposed the following second order model:

$$J\ddot{\Theta} + B\dot{\Theta} + K\Theta = \tau(t) \tag{166.2}$$

To analyze the characteristics of this model and compare it to data, it is convenient to solve Eq. (166.2) for peak velocity and duration through Laplace analysis. The transfer function of Eq. (166.2), written in standard form, is given by

$$H(s) = \frac{\Theta(s)}{\tau(s)} = \frac{\frac{\omega_n^2}{K}}{s^2 + 2\zeta\omega_n s + \omega_n^2} \tag{166.3}$$

where $\omega_n = \sqrt{\frac{K}{J}}$, and $\zeta = \frac{B}{2\sqrt{KJ}}$. Based on the saccade trajectory for a 20° saccade, Westheimer estimated $\omega_n = 120$ radians per second, and $\zeta = 0.7$. With the input $\tau(s) = \frac{\gamma}{K}$, $\Theta(t)$ is determined as

$$\Theta(t) = \frac{\gamma}{K}\left[1 + \frac{e^{-\zeta\omega_n t}}{\sqrt{1-\zeta^2}}\cos(\omega_d t + \phi)\right] \tag{166.4}$$

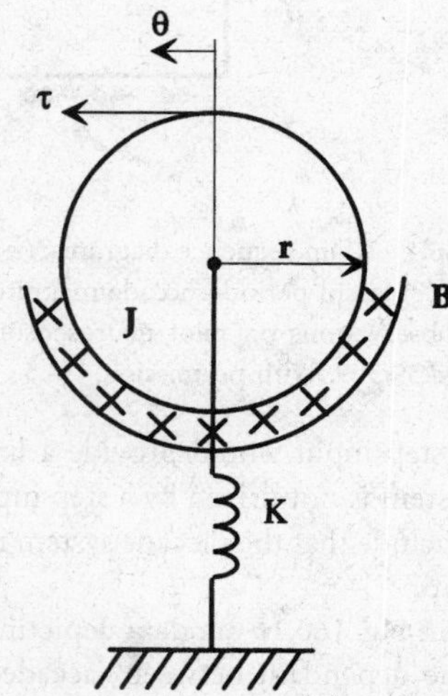

FIGURE 166.2 A diagram illustrating Westheimer's [1954] second-order model of the saccade system. The parameters J, B, and K are rotational elements for moment of inertia, friction, and stiffness, respectively, and represent the eyeball and its associated viscoelasticity. The torque applied to the eyeball by the lateral and medial rectus muscles is given by $\tau(t)$, and θ is the angular eye position.

where

$$\omega_d = \omega_n \sqrt{1-\zeta^2} \quad \text{and} \quad \phi = \pi + \tan^{-1}\frac{-\zeta}{\sqrt{1-\zeta^2}}.$$

Duration, T_p, is found by first calculating

$$\frac{\partial\Theta}{\partial t} = \frac{\gamma e^{-\zeta\omega_n t}}{K\sqrt{1-\zeta^2}}\left[-\zeta\omega_n \cos\left(\omega_d t+\phi\right) - \omega_d \sin\left(\omega_d t+\phi\right)\right] \tag{166.5}$$

then determining T_p from $\left.\frac{\partial\Theta}{\partial t}\right|_{t=T_p} = 0$, yielding

$$T_p = \frac{\pi}{\omega_n\sqrt{1-\zeta^2}} \tag{166.6}$$

With Westheimer's parameter values, T_p = 37 ms for saccades of all sizes, which is independent of saccade magnitude and not in agreement with the experimental data that have a duration which increases as a function of saccade magnitude.

Predicted saccade peak velocity, $\dot{\Theta}(t_{mv})$, is found by first calculating

$$\begin{aligned}\frac{\partial^2\Theta}{\partial t^2} = \frac{-\gamma e^{-\zeta\omega_n t}}{K\sqrt{1-\zeta^2}}&\left(-\zeta\omega_n\left(\zeta\omega_n \cos\left(\omega_d t+\phi\right)+\omega_d \sin\left(\omega_d t+\phi\right)\right)\right.\\ &\left.+\left(-\zeta\omega_n\omega_d \sin\left(\omega_d t+\phi\right)+\omega_d^2 \cos\left(\omega_d t+\phi\right)\right)\right)\end{aligned} \tag{166.7}$$

and the determining time at peak velocity, t_{mv}, from $\left.\frac{\partial^2\Theta}{\partial t^2}\right|_{t=t_{mv}} = 0$, yielding

$$t_{mv} = \frac{1}{\omega_d}\tan^{-1}\left(\frac{\sqrt{1-\zeta^2}}{\zeta}\right) \tag{166.8}$$

Substituting t_{mv} into Eq. (166.5) gives the peak velocity $\dot{\Theta}(t_{mv})$. Using Westheimer's parameter values, and with the saccade magnitude given by $\Delta\Theta = \frac{\gamma}{K}$ [based on the steady-state value from Eq. (166.3)], we have from Eq. (166.5)

$$\dot{\Theta}\left(t_{mv}\right) = 55.02\Delta\Theta \tag{166.9}$$

that is, peak velocity is directly proportional to saccade magnitude. As illustrated in the main sequence diagram shown in Fig. 166.1*a*, experimental peak velocity data have an exponential form and are not a linear function as predicted by the Westheimer model.

Westheimer noted the differences between saccade duration–saccade magnitude and peak velocity–saccade magnitude in the model and the experimental data and inferred that the saccade system was not linear because the peak velocity–saccade magnitude plot was nonlinear, and the input was not an abrupt step function. Overall, this model provided a satisfactory fit to the eye position data for a saccade of 20°, but not for saccades of other magnitudes. Interestingly, Westheimer's second-order model proves to be an adequate model for saccades of all sizes, if one assumes a different input function as described in the next section. Due to its simplicity, the Westheimer model of the oculomotor plant is still popular today.

166.3 Robinson's Model of the Saccade Controller

In 1964, Robinson performed an experiment to measure the input to the eyeballs during a saccade. To record the input, one eye was held fixed using a suction contact lens, while the other eye performed a saccade from target to target. Since the same innervation signal is sent to both eyes during a saccade, Robinson inferred that the input, recorded through the transducer attached to the fixed eyeball, was the same input driving the other eyeball. He estimated that the neural commands controlling the eyeballs during a saccade are a pulse plus a step, or simply, a pulse-step input.

It is important to distinguish between the tension or force generated by a muscle, called *muscle tension*, and the force generator within the muscle, called the active-state tension generator. The active-state tension generator creates a force within the muscle that is transformed through the internal elements of the muscle into the muscle tension. Muscle tension is external and measurable, and the active-state tension is internal and unmeasurable. Moreover, Robinson [1981] reported that the active-state tensions are not identical to the neural controllers, but described by low-pass filtered pulse-step waveforms. The neural control and the active-state tension signals are illustrated in Fig. 166.3. The agonist pulse input is required to get the eye to the target as soon as possible, and the step is required to keep the eye at that location.

Robinson [1964] also described a model for fast eye movements (constructed from empirical considerations) which simulated saccades over a range of 5–40° by changing the amplitude of the pulse-step input. Simulation results were adequate for the position-time relationship, but the velocity-time relationship was inconsistent with physiologic evidence. To correct this deficiency of the model, physiologic studies of the oculomotor plant were carried out during the 1960s through the 1970s that allowed the development of a more homeomorphic oculomotor plant. Essential to this work was the construction of oculomotor muscle models.

166.4 A Linear Homeomorphic Saccadic Eye Movement Model

In 1980, Bahill and coworkers presented a linear fourth-order model of the oculomotor plant, based on physiologic evidence, that provides an excellent match between model predictions and eye movement data. This model eliminates the differences seen between velocity predictions of the model and the data and the acceleration predictions of the model and the data. For ease in presentation, the modification of this model by Enderle and coworkers [1984] will be used.

Figure 166.4 illustrates the mechanical components of the oculomotor plant for horizontal eye movements, the lateral and medial rectus muscle, and the eyeball. The agonist muscle is modeled as a parallel combination of an active state tension generator F_{AG}, viscosity element B_{AG}, and elastic element K_{LT}, connected to a series elastic element K_{SE}. The antagonist muscle is similarly modeled as a parallel combination of an active-state tension generator F_{ANT}, viscosity element B_{ANT}, and elastic element K_{LT}, connected to a series elastic element K_{SE}. The eyeball is modeled as a sphere with moment of inertia J_p, connected to viscosity element B_p and elastic element K_p. The passive elasticity of each muscle is included in spring K_p for ease in analysis. Each of the elements defined in the oculomotor plant is ideal and linear.

Physiologic support for this model is based on the muscle model by Wilkie [1968], and estimates for the **extraocular** muscle elasticities and the passive tissues of the eyeball are based on experiments by Robinson [1981], Robinson and coworkers [1969], and Collins [1975], and studies of extraocular muscle viscosity by Bahill and coworkers [1980].

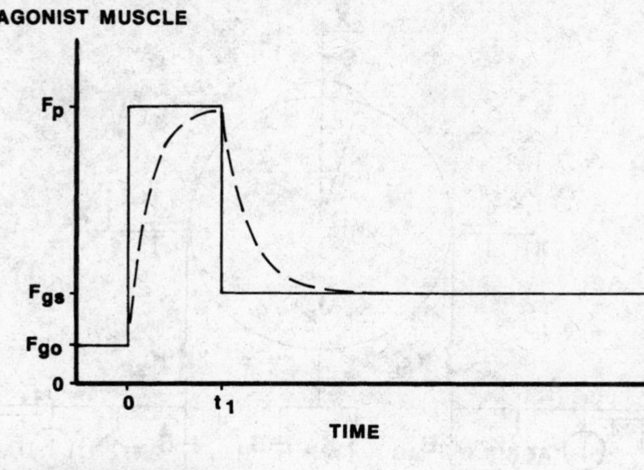

FIGURE 166.3 Agonist and antagonist neurologic control signals (solid lines) and the agonist and antagonist active-state tensions (dashed lines). Note that the time constant for activation is different from the time constant for deactivation. *Source:* Enderle JD, Wolfe JW. 1987. Time optimal control of saccadic eye movements. IEEE Trans Biomed Eng 34(1):43. With permission.

By summing the forces at junctions 2 and 3 (the equilibrium positions for x_2 and x_3) and the torques acting on the eyeball, using Laplace variable analysis about the operating point, the linear homeomorphic model, as shown in Fig. 166.4, is derived as

$$\begin{aligned} &\delta\Big(\mathrm{K}_{SE}\Big(\mathrm{K}_{ST}\big(\mathrm{F}_{AG}-\mathrm{F}_{ANT}\big)+\mathrm{B}_{ANT}\dot{F}_{AG}-\mathrm{B}_{AG}\dot{F}_{ANT}\Big)\Big) \\ &= \ddddot{\Theta}+P_3\dddot{\Theta}+P_2\ddot{\Theta}+P_1\dot{\Theta}+\mathrm{P}_0\Theta \end{aligned} \tag{166.10}$$

where

$$K_{ST}=K_{SE}+K_{LT},\ J=\frac{57.296\,J_P}{r^2},\ B=\frac{57.296\,B_P}{r^2},\ K=\frac{57.296\,K_P}{r^2},\ \delta=\frac{57.296}{rJB_{ANT}B_{AG}},$$

$$C_3=\frac{JK_{ST}\big(B_{AG}+B_{ANT}\big)+BB_{ANT}B_{AG}}{JB_{ANT}B_{AG}}$$

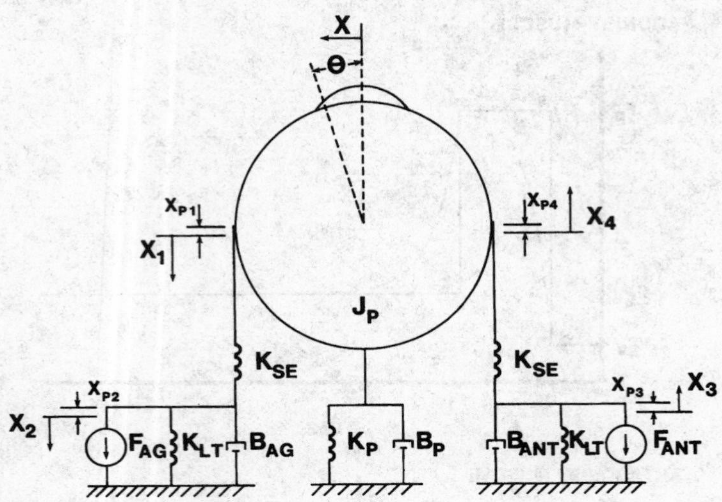

FIGURE 166.4 This diagram illustrates the mechanical components of the oculomotor plant. The muscles are shown to be extended from equilibrium, a position of rest, at the primary position (looking straight ahead), consistent with physiologic evidence. The average length of the rectus muscle at the primary position is approximately 40 mm and at the equilibrium position is approximately 37 mm. θ is the angle the eyeball is deviated from the primary position, and variable x is the length of arc traversed. When the eye is at the primary position, both θ and x are equal to zero. Variables x_1 through x_4 are the displacements from equilibrium for the stiffness elements in each muscle. Values x_{p1} through x_{p4} are the displacements from equilibrium for each of the variables x_1 through x_4 at the primary position. The total extension of the muscle from equilibrium at the primary position is x_{p1} plus x_{p2} or x_{p3} plus x_{p4}, which equals approximately 3 mm. It is assumed that the lateral and medial rectus muscles are identical, such that x_{p1} equals x_{p4} and x_{p3} and x_{p2}. The radius of the eyeball is r. *Source:* Enderle JD, Wolfe JW, Yates JT. 1984. The linear homeomorphic saccadic eye movement model—a modification. IEEE Trans Biomed Eng 31:717. With permission.

$$C_2 = \frac{JK_{ST}^2 + BK_{ST}\left(B_{AG} + B_{ANT}\right) + B_{ANT}B_{AG}\left(K + 2K_{SE}\right)}{JB_{ANT}B_{AG}}$$

$$C_1 = \frac{BK_{ST}^2 + \left(B_{AG} + B_{ANT}\right)\left(KK_{ST} + 2K_{SE}K_{ST} - K_{SE}^2\right)}{JB_{ANT}B_{AG}}$$

$$C_0 = \frac{KK_{ST}^2 + 2K_{SE}K_{ST}K_{LT}}{JB_{ANT}B_{AG}}$$

The agonist and antagonist active-state tensions are given by the following low-pass-filtered waveforms

$$\dot{F}_{AG} = \frac{N_{AG} - F_{AG}}{\tau_{AG}} \quad \text{and} \quad \dot{F}_{ANT} = \frac{N_{ANT} - F_{ANT}}{\tau_{ANT}} \tag{166.11}$$

where N_{AG} and N_{ANT} are the pulse-step waveforms shown in Fig. 166.3, and $\tau_{ag} = \tau_{ac}\,[u(t) - u(t - t_1)] + \tau_{de}u(t - t_1)$ and $\tau_{ant} = \tau_{de}[u(t) - u(t) - u(t - t_1)] + \tau_{ac}u(t - t_1)$ are the time-varying time constants [Bahill et al., 1980].

Based on an analysis of experimental evidence, Enderle and Wolfe [1988] determined parameter estimates for the oculomotor plant as: $K_{SE} = 125$ Nm^{-1}, $K_{LT} = 32$ Nm^{-1}, $K = 66.4$ Nm^{-1}, $B = 3.1$ Nsm^{-1},

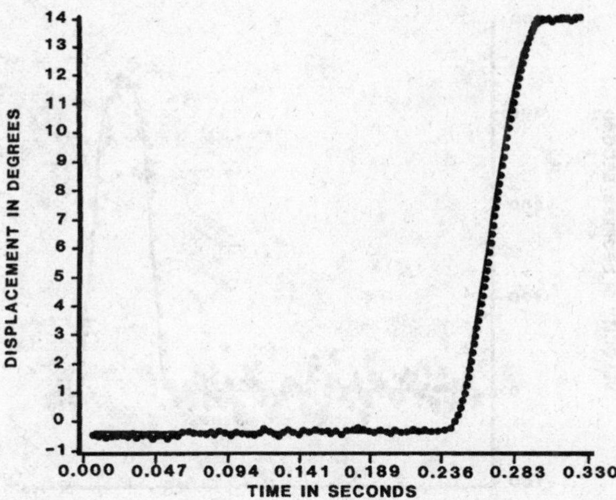

FIGURE 166.5 Saccadic eye movement in response to a 15° target movement. Solid line is the prediction of the saccadic eye movement model with the final parameter estimates computed using the system identification technique. Dots are the data. *Source:* Enderle JD, Wolfe JW. 1987. Time-optimal control of saccadic eye movements. IEEE Trans Biomed Eng 34(1):43. With permission.

$J = 2.2 \times 10^{-3}$ Ns²m⁻¹, $B_{AG} = 3.4$ *Nsm*⁻¹, $B_{ANT} = 1.2$ Nsm⁻¹, and $\delta = 72.536 \times 10^6$, and the steady-state active-state tensions as

$$F_{AG} = \begin{cases} 0.14 + 0.0185\Theta \; N \; for \; \Theta < 14.23° \\ 0.0283\Theta \; N \; for \; \Theta \geq 14.23° \end{cases} \tag{166.12}$$

$$F_{ANT} = \begin{cases} 0.14 - 0.00980\Theta \; N \; for \; \Theta < 14.23° \\ 0 \quad N \; for \; \Theta \geq 14.23° \end{cases} \tag{166.13}$$

Since saccades are highly variable estimates of the dynamic active-state tensions are carried out on a saccade-by-saccade basis. One method to estimate the active-state tensions is using the system identification technique, a conjugate gradient search program carried out in the frequency domain [Enderle & Wolfe, 1988]. Figures 166.5–166.7 show the system identification technique results for an eye movement response to a 15° target movement. A close fit between the data and model prediction is seen in Fig. 166.5. Figures 166.6 and 166.7 further illustrate the accuracy of the final parameter estimates for velocity and acceleration. Estimates for agonist pulse magnitude are highly variable for saccade to saccade, even for the same size—see Fig. 9 of Enderle and Wolfe [1988]. Agonist pulse duration is closely coupled with pulse amplitude; as the pulse amplitude increases, the pulse duration decreases for saccades of the same magnitude. Reasonable values for the pulse amplitude for this model range from about 0.6N–1.4N. The larger the magnitude of the pulse, the larger the peak velocity of the eye movement.

166.5 Another Linear Homeomorphic Saccadic Eye Movement Model

The previous linear model of the oculomotor plant is derived from a nonlinear oculomotor plant model by Hsu and coworkers [1976] and based on a linearization of the force-velocity curve [Bahill et al., 1980]. Muscle viscosity traditionally has been modeled with a hyperbolic force-velocity relationship. Using the linear model of muscle reported by Enderle and coworkers [1991], it is possible to avoid the linearization and derive an updated linear homeomorphic saccadic eye movement model.

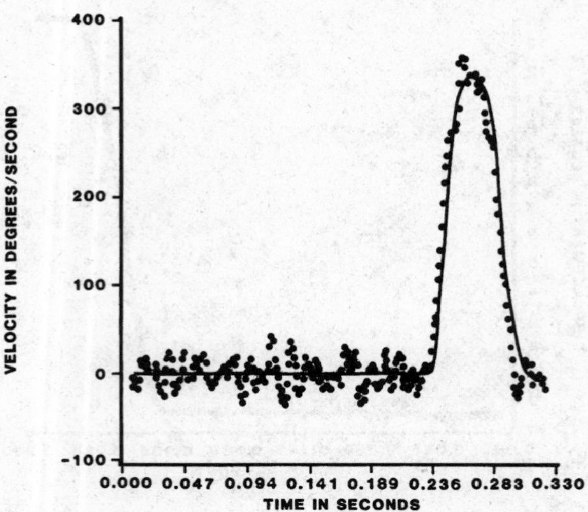

FIGURE 166.6 Velocity estimates for the saccadic eye movement illustrated in Fig. 166.5. Solid line is the saccadic eye movement model velocity prediction with the final parameter estimates computed using the system identification techniques. The dots are the two-point central difference estimates of velocity computed with a step size of 3 and a sampling interval of 1 ms. *Source:* Enderle JD, Wolfe JW. 1987. Time-optimal control of saccadic eye movements. IEEE Trans Biomed Eng 34(1):43. With permission.

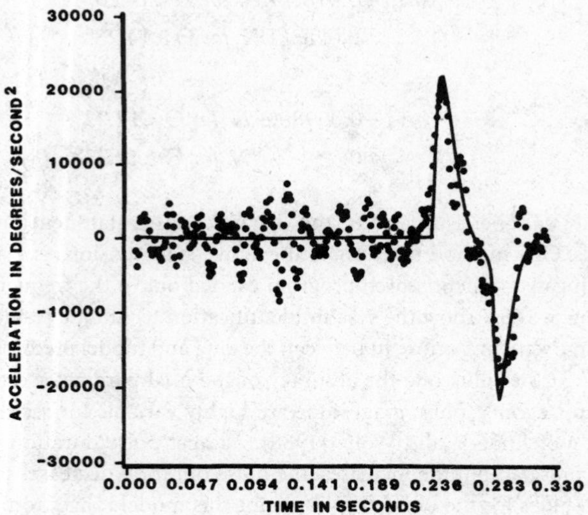

FIGURE 166.7 Acceleration estimates for the saccadic eye movement illustrated in Fig. 166.5. Solid line is the saccadic eye movement model acceleration prediction with the final parameter estimates computed using the system identification techniques. The dots are the 2-point central difference estimates of velocity computed with a step size of 4 and a sampling interval of 1 ms. *Source:* Enderle JD, Wolfe JW. 1987. Time-optimal control of saccadic eye movements. IEEE Trans Biomed Eng 34(1):43. With permission.

The linear muscle model has the static and dynamic properties of rectus eye muscle, a model without any nonlinear elements. The model has a nonlinear force-velocity relationship that matches muscle data using linear viscous elements, and the length tension characteristics are also in good agreement with muscle data within the operating range of the muscle. Some additional advantages of the linear muscle

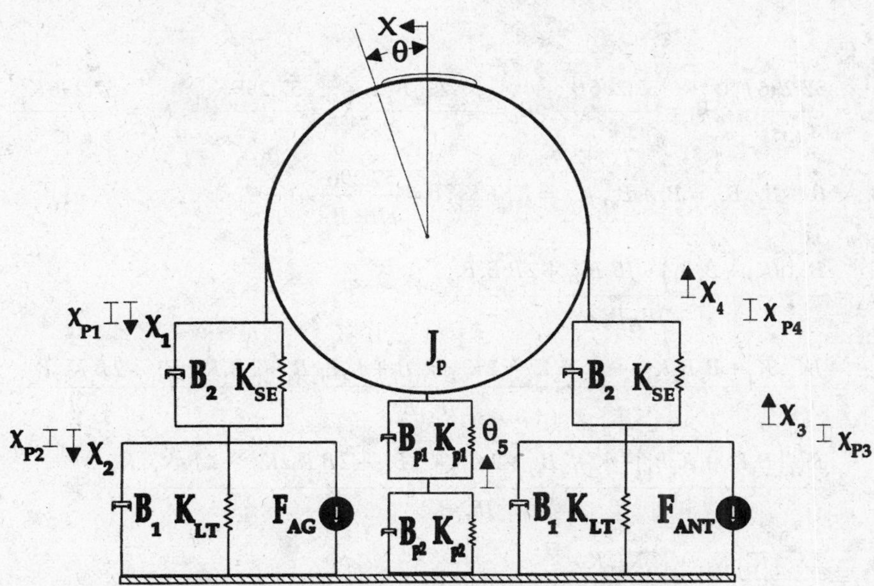

FIGURE 166.8 This diagram illustrates the mechanical components of the updated oculomotor plant. The muscles are shown to be extended from equilibrium, a position of rest, at the primary position (looking straight ahead), consistent with physiologic evidence. The average length of the rectus muscle at the primary position is approximately 40 mm and at the equilibrium position is approximately 37 mm. θ is the angle the eyeball is deviated from the primary position, and variable x is the length of arc traversed. When the eye is at the primary position, both θ and x are equal to zero. Variables x_1 through x_4 are the displacements from equilibrium for the stiffness elements in each muscle, and θ_5 is the rotational displacement for passive orbital tissues. Values x_{p1} through x_{p4} are the displacements from equilibrium for each of the variables x_1 through x_4 at the primary position. The total extension of the muscle from equilibrium at the primary position is x_{p1} plus x_{p2} or x_{p3} plus x_{p4}, which equals approximately 3 mm. It is assumed that the lateral and medial rectus muscles are identical, such that x_{p1} equals x_{p4} and x_{p3} equals x_{p2}. The radius of the eyeball is r.

model are that a passive elasticity is not necessary if the equilibrium point $x_e = 19.3°$, rather than 15°, and muscle viscosity is a constant that does not depend on the innervation stimulus level.

Figure 166.8 illustrates the mechanical components of the updated oculomotor plant for horizontal eye movements, the lateral and medial rectus muscle, and the eyeball. The agonist muscle is modeled as a parallel combination of viscosity B_2 and series elasticity K_{SE}, connected to the parallel combination of active state tension generator F_{AG}, viscosity element B_1 and length tension elastic element K_{LT}. Since viscosity does not change with innervation level, agonist viscosity is set equal to antagonist viscosity. The antagonist muscle is similarly modeled with a suitable change in active state tension to F_{ANT}. The eyeball is modeled as a sphere with moment of inertia J_p, connected to a pair of viscoelastic elements connected in series; the update of the eyeball model is based on observations by Robinson [1981]. Each of the elements defined in the oculomotor plant is ideal and linear.

The differential equation describing oculomotor plant model shown in Fig. 166.8 is derived by summing the forces acting at junctions 2 and 3 and the torques acting on the eyeball and junction 5, and using Laplace variable analysis about the operating point, and given by

$$\begin{aligned}\delta\Big(K_{SE}K_{12}\big(F_{AG}-F_{ANT}\big)+\big(K_{SE}B_{34}+B_2K_{12}\big)\big(\dot{F}_{AG}-\dot{F}_{ANT}\big)\\+B_2B_{34}\big(\ddot{F}_{AG}-\ddot{F}_{ANT}\big)\Big)=\ddddot{\Theta}+P_3\dddot{\Theta}+P_2\ddot{\Theta}+P_1\dot{\Theta}+P_0\Theta\end{aligned} \tag{166.15}$$

where

$$J = \frac{57.296 J_p}{r^2},\ B_3 = \frac{57.296 B_{p1}}{r^2},\ B_4 = \frac{57.296 B_{p2}}{r^2},\ K_1 = \frac{57.296 K_{p1}}{r^2},\ K_2 = \frac{57.296 K_{p2}}{r^2},$$

$$B_{12} = B_1 + B_2,\ B_{34} = B_3 + B_4,\ K_{12} = K_1 + K_2,\ \delta = \frac{57.296}{rJB_{12}B_4}$$

$$C_3 = \frac{B_{12}\left(JK_2 + B_3 B_4\right) + JB_4 K_{ST} + 2B_1 B_2 B_{34}}{JB_{12}B_4}$$

$$C_2 = \frac{JK_{ST}K_2 + B_3 B_4 K_{ST} + B_{12}B_3 K_2 + 2K_{SE}B_{34}B_1 + K_1 B_{12} B_4 + 2B_2 K_{LT} B_{34} + 2B_1 K_{12} B_2}{JB_{12}B_4}$$

$$C_1 = \frac{K_{ST}\left(B_3 K_2 + K_1 B_4\right) + K_1 K_2 B_{12} + 2K_{LT}K_{SE}B_{34} + 2B_1 K_{12} K_{SE} + 2B_2 K_{LT} K_{12}}{JB_{12}B_4}$$

$$C_0 = \frac{2K_{LT}K_{SE}K_{12} + K_1 K_{ST} K_2}{JB_{12}B_4}$$

Based on an analysis of experimental data, suitable parameter estimates for the oculomotor plant are: $K_{SE} = 125$ Nm^{-1}, $K_{LT} = 60.7$ Nm^{-1}, $B_1 = 4.6$ Nsm^{-1}, $B_2 = 0.5$ Nsm^{-1}, $J = 2.2 \times 10^{-3}$ Ns^{2}m^{-1}, $B_3 = 0.538$ Nsm^{-1}, $B_4 = 41.54$ Nsm^{-1}, $K_1 = 26.9$ Nm^{-1}, and $K_2 = 41.54$ Nm^{-1}. Based on the updated model of muscle and length tension data [Collins, 1975], steady-state active-state tensions are determined as described in Enderle and coworkers [1991] as

$$F = \begin{cases} 0.4 + 0.0175\Theta\ N\ for\ \Theta \geq 0° \\ 0.4 - 0.0125\Theta\ N\ for\ \Theta < 0° \end{cases} \tag{166.16}$$

Saccadic eye movements simulated with this model have characteristics which are in good agreement with the data, including position, velocity and acceleration, and the main sequence diagrams.

166.6 Saccade Pathways

Clinical evidence, lesion and stimulation studies all point toward the participation of vitally important neural sites in the control of saccades, including the cerebellum, superior colliculus, thalamus, cortex, and other nuclei in the brain stem; evidence shows also that saccades are driven by two parallel neural networks. From each eye, the axons of retinal ganglion cells exit and join other neurons to form the optic nerve. The optic nerve from each eye then join at the optic chiasm, where fibers from the nasal half of each retina cross to the opposite side. Axons in the optic tract synapse in the lateral geniculate nucleus (a thalamic relay) and continue to the visual cortex. This portion of the saccade neural network is concerned with the recognition of visual stimuli. Axons in the optic tract also synapse in the superior colliculus. This second portion of the saccade neural network is concerned with the location of visual targets and is primarily responsible for goal-directed saccades.

Saccadic neural activity of the superior colliculus and cerebellum, in particular, have been identified as the saccade initiator and terminator, respectively, for a goal-directed saccade. The impact of the frontal eye field and the thalamus, while very important, have less important roles in the generation of goal-directed saccades to visual stimuli. The frontal eye fields are primarily concerned with voluntary saccades, and the thalamus appears to be involved with corrective saccades. Shown in Fig. 166.9 is a diagram

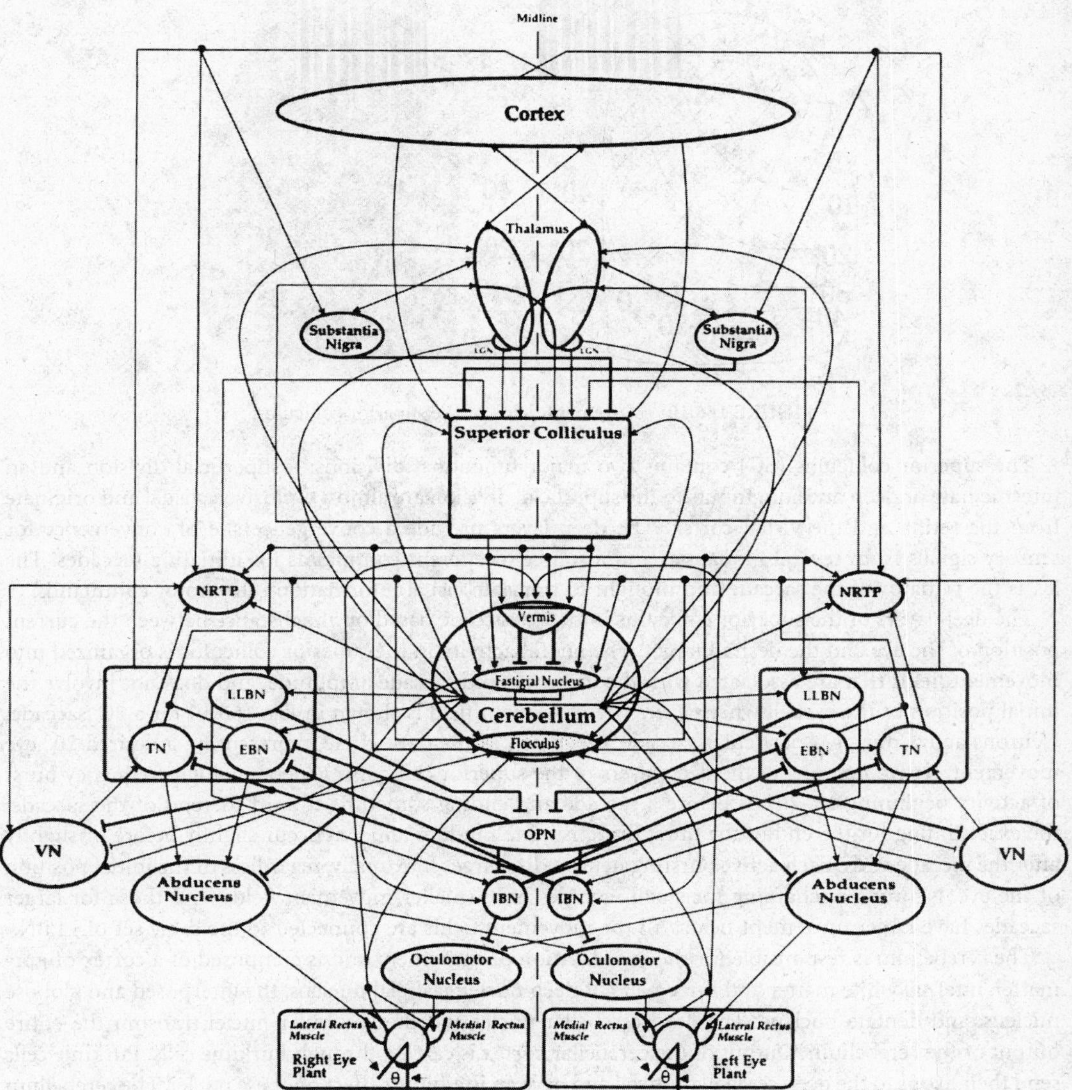

FIGURE 166.9 Shown is a diagram illustrating important sites for the generation of a conjugate horizontal saccade in both eyes. It consists of the familiar premotor excitatory burst neurons (EBN), inhibitory burst neurons (IBN), long lead burst neurons (LLBN), omnipause neurons (OPN), tonic neurons (TN), and the vestibular nucleus, abducens nucleus, oculomotor nucleus, cerebellum, substantia nigra, nucleus reticularis tegmenti pontis (NRTP), the thalamus, the deep layers of the superior colliculus (SC), and the oculomotor plant for each eye. Excitatory inputs are shown with an arrow, inhibitory inputs are shown with a ⊥. Consistent with current knowledge, the left and right structures of the neural circuit model are maintained. This circuit diagram was constructed after a careful review of the current literature. Each of the sites and connections is supported by firm physiologic evidence. Since interest is in goal-directed visual saccades, the cortex has not been partitioned into the frontal eye field and posterior eye field (striate, prestriate, and inferior parietal cortices).

illustrating important sites for the generation of a conjugate goal-directed horizontal saccade in both eyes [Enderle, 1994]. Each of the sites and connections detailed in Fig. 166.9 are fully supported by physiologic evidence. Some of these neural sites will be briefly described.

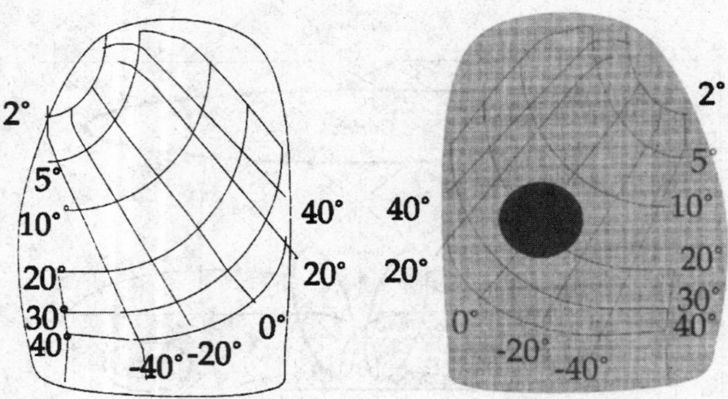

FIGURE 166.10 Movement fields of the superior colliculus.

The superior colliculus (SC) contains two major functional divisions: a superficial division and an intermediate or deep division. Inputs to the superficial division are almost exclusively visual and originate from the retina and the visual cortex. The deep layers provide a convergence site of convergence for sensory signals from several modalities and a source of efferent commands for initiating saccades. The SC is the initiator of the saccade and thought to translate visual information into motor commands.

The deep layers of the superior colliculus initiate a saccade based on the distance between the current position of the eye and the desired target. The neural activity in the superior colliculus is organized into movement fields that are associated with the direction and saccade amplitude, and does not involve the initial position of the eyeball whatsoever. The movement field is shown in Fig. 166.10 for a 20° saccade. Neurons active during a particular saccade are shown as the dark circle, representing a desired 20° eye movement. Active neurons in the deep layers of the superior colliculus generate a high-frequency burst of activity beginning 18–20 ms before a saccade and ending sometime toward the end of the saccade; the exact timing for the end of the burst firing is quite random and can occur slightly before or slightly after the saccade ends. Each active bursting neuron discharges maximally, regardless of the initial position of the eye. Neurons discharging for small saccades have smaller movement fields, and those for larger saccades have larger movement fields. All the movement fields are connected to the same set of LLBN.

The cerebellum is responsible for the coordination of movement and is composed of a cortex of gray matter, internal white matter and three pairs of deep nuclei: fastigial nucleus, the interposed and globose nucleus, and dentate nucleus. The deep cerebellar nuclei and the vestibular nuclei transmit the entire output of the cerebellum. Output of the cerebellar cortex is carried through Purkinje cells. Purkinje cells send their axons to the deep cerebellar nuclei and have an inhibitory effect on these nuclei. The cerebellum is involved with both eye and head movements, and both tonic and phasic activity are reported in the cerebellum. The cerebellum is not directly responsible for the initiation or execution of a saccade but contributes to saccade precision. Sites within the cerebellum important for the control of eye movements include the oculomotor vermis, fastigial nucleus, and the flocculus. Consistent with the operation of the cerebellum for other movement activities, the cerebellum is postulated here to act both as the coordinator for a saccade and as a precise gating mechanism.

The cerebellum is included in the saccade generator as a time-optimal gating element, using three active sites during a saccade: the vermis, fastigial nucleus, and flocculus. The vermis is concerned with the absolute starting position of a saccade in the movement field and corrects control signals for initial eye position. Using proprioceptors in the oculomotor muscles and an internal eye position reference, the vermis is aware of the current position of the eye. The vermis is also aware of the signals (dynamic motor error) used to generate the saccade via the connection with the NRTP and the superior colliculus.

With regard to the oculomotor system, the cerebellum has inputs from superior colliculus, lateral geniculate nucleus (LGN), oculomotor muscle proprioceptors, and striate cortex via NRTP. The cerebellum sends inputs to the NRTP, LLBN, EGN, VN, thalamus, and superior colliculus. The oculomotor vermis and fastigial nuclei are important in the control of saccade amplitude, and the flocculus, perihypoglossal nuclei of the rostral medulla, and possibly the pontine and mesencephalic reticular formation, are thought to form the integrator within the cerebellum. One important function of the flocculus may be to increase the time constant of the neural integrator for saccades starting at locations different from primary position.

The fastigial nucleus receives input from the superior colliculus, as well as other sites. The output of the fastigial nucleus is excitatory and projects ipsilaterally and contralaterally as shown in Fig. 166.9. During fixation, the fastigial nucleus fires tonically at low rates; 20 ms prior to a saccade, the contralateral fastigial nucleus bursts, and the ipsilateral fastigial nucleus pauses and then discharges with a burst. The pause in ipsilateral firing is due to Purkinje cell input to the fastigial nucleus. The sequential organization of Purkinje cells along beams of parallel fibers suggests that the cerebellar cortex might function as a delay, producing a set of timed pulses which could be used to program the duration of the saccade. If one considers nonprimary position saccades, different temporal and spatial schemes, via cerebellar control, are necessary to produce the same size saccade. It is postulated here that the cerebellum acts as a gating device which precisely terminates a saccade based on the initial position of the eye in the orbit.

The PPRF has neurons that burst at frequencies up to 1000 Hz during saccades and are silent during periods of fixation and neurons that fire tonically during periods of fixation. Neurons that fire at steady rates during fixation are called *tonic neurons* (TNs) and are responsible for holding the eye steady. The TN firing rate depends on the position of the eye (presumably through a local integrator type network). The TNs are thought to provide the step component to the motoneuron. There are two types of burst neurons in the PPRF called the *long-lead burst neuron* (LLBN) and a *medium-lead burst neuron* (MLBN); during periods of fixation, these neurons are silent. The LLBN burst at least 12 ms before a saccade and the MLBN burst less than 12 ms (typically 6–8 ms) before the saccade. The MLBN are connected monosynaptically with the abducens nucleus.

There are two types of neurons within the MLBN, the excitatory burst neurons (EBN) and the inhibitory burst neurons (IBN). The EBN and IBN labels describe the synaptic activity upon the motoneurons; the EBN excite and are responsible for the burst firing, and the IBN inhibit and are responsible for the pause. A mirror image of these neurons exists on both sides of the midline. The IBN inhibit the EBN on the contralateral side.

Also within the brain stem is another type of saccade neuron called the *omnipause neuron* (OPN). The OPN fires tonically at approximately 200 Hz during periods of fixation and is silent during saccades. The OPN stops firing approximately 10–12 ms before a saccade and resumes tonic firing approximately 10 ms before the end of the saccade. The OPN are known to inhibit the MLBN and are inhibited by the LLBN. The OPN activity is responsible for the precise timing between groups of neurons that causes a saccade.

Qualitatively, a saccade occurs according to the following sequence of events within the PPRF. First, the ipsilateral LLBN are stimulated by the SC, initiating the saccade. The LLBN then inhibits the tonic firing of the OPN. When the OPN cease firing, the MLBN is released from inhibition and begins firing (these neurons fire spontaneously and also stimulated by the fastigial nucleus). The ipsilateral IBN are stimulated by the ipsilateral LLBN and the contralateral fastigial nucleus of the cerebellum. The ipsilateral EBN are stimulated by the contralateral fastigial nucleus of the cerebellum and, when released from inhibition fire spontaneously. Except for the fastigial nucleus, there are no other accepted excitatory inputs to the EBN. The burst firing in the ipsilateral IBN inhibits the contralateral EBN and abducens nucleus and the ipsilateral oculomotor nucleus. The burst firing in the ipsilateral EBN causes the burst in the ipsilateral abducens nucleus, which stimulates the ipsilateral lateral rectus muscle and the contralateral oculomotor nucleus. With the stimulation of the ipsilateral lateral rectus muscle by the ipsilateral abducens nucleus and the inhibition of the ipsilateral medial rectus muscle via the oculomotor nucleus, a saccade occurs in the right eye.

Simultaneously, with the contralateral medial rectus muscle stimulated by the contralateral oculomotor nucleus, and with the inhibition of the contralateral lateral rectus muscle via the abducens nucleus, a saccade occurs in the left eye. A similar scenario is carried out in the right eye. Thus the eyes move conjugately under the control of a single drive center. The saccade is terminated with the resumption of tonic firing in the OPN via the fastigial nucleus.

166.7 Saccade Control Mechanism

Although the purpose for a saccadic eye movement is clear, that is, to quickly redirect the eyeball to the destination, the neural control mechanism is not. Until quite recently, saccade generator models involved a ballistic or preprogrammed control to the desired eye position based on retinal error alone. Today, an increasing number of investigators are putting forth the idea that visual goal-directed saccades are controlled by a local feedback loop that continuously drives the eye to the desired eye position. This hypothesis first gained acceptance in 1975 when Robinson suggested that saccades originate from neural commands that specify the desired position of the eye rather than the preprogrammed distance the eye must be moved. The value of the actual eye position is subtracted from the desired position to create an error signal that completes the local feedback loop that drives a high-gain burst element to generate the neural pulse. This neural pulse continuously drives the eye until the error signal is zero.

Subsequently, a number of other investigators have modified the local feedback mechanism proposed by Robinson [1975] to better describe the neural connections and firing patterns of brainstem neurons in the control of horizontal saccadic eye. In addition to the Robinson model, two other models describe a saccade generator [Enderle, 1994; Scudder, 1988]. All the models involve three types of premotor neurons; burst, tonic, and pause cells, as previously described, and involve a pulse-step change in firing rate at the motoneuron during a saccadic eye movement.

Although the general pattern of motoneuron activity is qualitatively accepted during a saccadic eye movement, there is little agreement on a quantitative discharge description. The saccade generator models by Robinson and Scudder are structured to provide a control signal that is proportionally weighted (or dependent) to the desired saccade size, as opposed to the saccade generator model structured to provide a control signal that is independent of saccade amplitude. Using time-optimal control theory and the system identification technique, Enderle and Wolfe [1988] investigated the control of saccades and reported that the system operates under a first-order time-optimal control. The concepts underlying this hypothesis are that each muscle's active-state tension is described by a low-pass filtered pulse-step waveform in which the magnitude of the agonist pulse is a maximum regardless of the amplitude of the saccade and that only the duration of the agonist pulse affects the size of the saccade. The antagonist muscle is completely inhibited during the period of maximum agonist stimulation. The saccade generator illustrated in Fig. 166.9 operates under these principles and provides simulations which match the data very well.

Saccadic eye movements were simulated using TUTSIM, a continuous time simulation program, for the saccade generator model presented in Fig. 166.9, and compared with experimental data. Neural sites (nucleus) are described via a *functional* block diagram description of the horizontal saccade generator model as shown in Fig. 166.11 and 166.12. Table 166.1 summarizes additional firing characteristics for the neural sites. The output of each block represents the firing pattern at each neural site observed during the saccade; time zero indicates the start of the saccade and T represents the end of the saccade. Naturally, the firing pattern observed for each block represents the firing pattern for a single neuron, as recorded in the literature, but the block represents the cumulative effect of all the neurons within that site. Consistent with a time optimal control theory, neural activity is represented within each of the blocks as pulses and/or steps to reflect their operation as timing gates. The superior colliculus fires maximally as long as the dynamic motor error is greater than zero, in agreement with the first-order time optimal controller and physiologic evidence. Notice that the LLBNs are driven by the superior colliculus as long as there is a feedback error maintained by the cerebellar vermis. In all likelihood, the maximal firing rate by the superior colliculus is stochastic, depending on a variety of physiologic factors such as the interest in tracking the target, anxiety, frustration, stress, and other factors. The actual firing patterns in the superior colliculus, the burst neurons in the PPRF (LLBN, EBN, and IBN) and abducens nucleus are

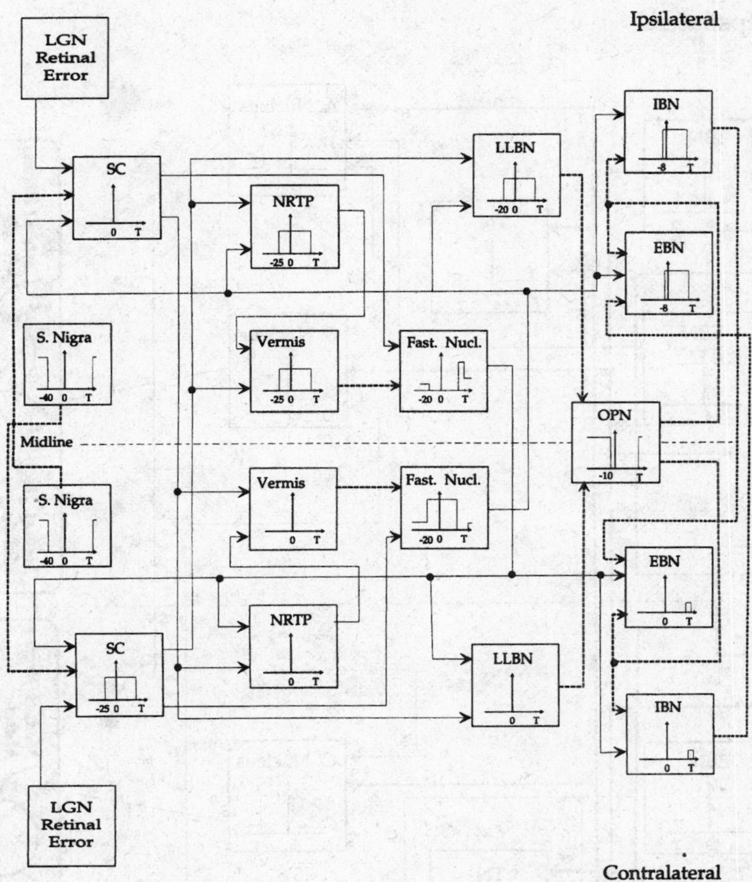

FIGURE 166.11 A functional block diagram of the saccade generator model. Solid lines are excitatory and dashed lines are inhibitory. This figure illustrates the first half of the network.

simulated with filtered pulse signals, consistent with the physical limitations of neurons. For the superior colliculus and the LLBN, this involves a single pulse, for the EBN and IBN, this involves two pulses with different filters (the first pulse describes the brief rise and subsequent fall within the first 10 ms during a saccade, and the second pulse describes the steady state pulse during the saccade) to match the electrophysiologic data.

Illustrated in Fig. 166.13 is an extracellular single unit recording for the EBN, eye position data, and a simulation for a 20° saccade. Details of the experimental are given in Enderle [1994]. A 20° saccade was simulated by using EBN data as input and the oculomotor plant in Fig. 166.8. Few differences between the data and the simulation results are observed for this movement, as well as other eye movements. The saccade generator model in Fig. 166.9 also provides an excellent description of the saccade system and matches the data very well for all naturally occurring saccades, including saccades with dynamic overshoot and glissadic behavior, without parametric changes [Enderle, 1994].

166.8 Conclusion

This chapter has focused on quantitative models and control of the fast eye movement system. Each of the oculomotor plant models described here are linear. Beginning with the most simple quantitative

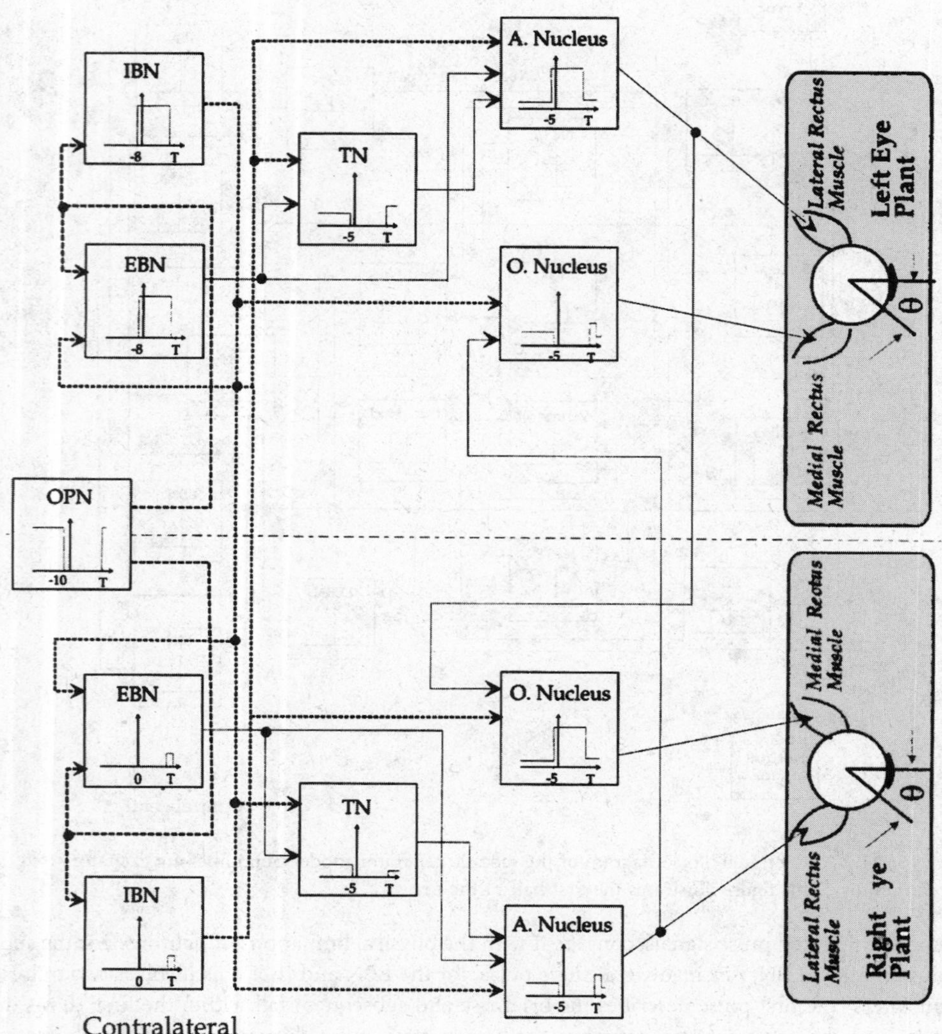

FIGURE 166.12 A functional block diagram of the saccade generator model. Solid lines are excitatory and dashed lines are inhibitory. This figure illustrates the second half of the network.

model of saccades by Westheimer [1954], important characteristics of saccades were determined as a means of evaluating the quality of saccade models. Next, models of increasing complexity were presented with the end goal of constructing a homeomorphic saccade model of the oculomotor plant. These plant models were driven by improved models of muscle that ultimately provided an excellent match of the static and dynamic properties of the rectus eye muscle. Finally, the control of saccades was considered from the basis of systems control theory and anatomic considerations. Many nonlinear models of the oculomotor plant exist, and readers interested in learning about them should consult Robinson [1981].

TABLE 166.1 Activity of Neural Sites During a Saccade

Neural Site	Onset Before Saccade	Peak Firing Rate	End Time
Abducens nucleus	5 ms	400–800 Hz	Ends approx. 5 ms before saccade ends
Contralateral fastigial nucleus	20 ms	200 Hz	Pulse ends with pause approx. 10 ms before saccade ends, resumes tonic firing approx. 10 ms after saccade ends
Contralateral superior colliculus	20–25 ms	800–1000 Hz	Ends approx. when saccade ends
Ipsilateral cerebellar vermis	20–25 ms	600–800 Hz	Ends approx. 25 ms before saccade ends
Ipsilateral EBN	6–8 ms	600–800 Hz	Ends approx. 10 ms before saccade ends
Ipsilateral fastigial nucleus	20 ms	Pause during saccade, and a burst of 200 Hz toward the end of the saccade	Pause ends with burst approx. 10 ms before saccade ends, resumes tonic firing approx. 10 ms after saccade ends
Ipsilateral FEF	>30 ms	600–800 Hz	Ends approx. when saccade ends
Ipsilateral IBN	6–8 ms	600–800 Hz	Ends approx. 10 ms before saccade ends
Ipsilateral LLBN	20 ms	800–1000 Hz	Ends approx. when saccade ends
Ipsilateral NRTP	20–25 ms	800–1000 Hz	Ends approx. when saccade ends
Ipsilateral substantia nigra	40 ms	40–100 Hz	Resumes firing approx. 40–150 ms after saccade ends
OPN	6–8 ms	150–200 Hz (before and after)	Ends approx. when saccade ends

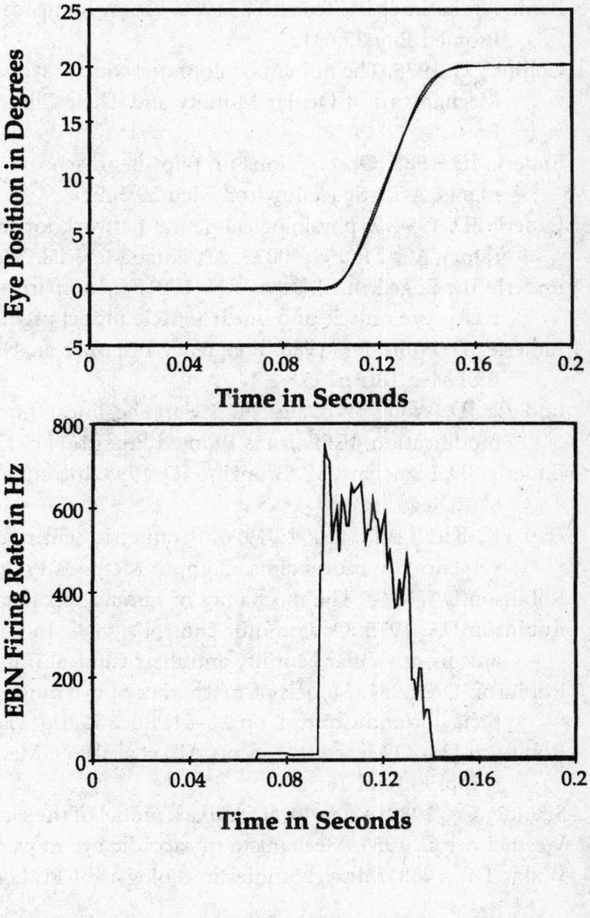

FIGURE 166.13 Simulated eye position in solid line shown in top graph, generated with EBN saccade data (shown in lower graph) and the oculomotor plant in Fig. 166.8. Actual eye movement data recorded with the EBN data during a saccadic eye movement, shown as dashed line in a top graph. Data provided by Dr. David Sparks from his laboratory at the University of Alabama.

Defining Terms

Active-state tension generator: The active-state tension generator describes the element within the muscle that creates a force. This force is different from muscle tension, which is the force due to the active-state tension generator and all the other elements within the muscle.

Extraocular muscles: The six muscles attached directly to the outside of the eyeball, consisting of the medial, lateral, superior, inferior recti, and the superior and inferior oblique muscles.

Homeomorphic: As close to reality as possible.

Latent period: The latent period is thought to be the time interval during which the CNS determines whether to make a saccade, and, if so, calculates the distance the eyeball is to be moved, transforming retinal error into transient muscle activity.

Main-sequence diagrams: Summary plots of the characteristics of saccades that allow one to compare inter- and intrasubject variations. Commonly used characteristics include: (1) peak velocity–saccade magnitude, (2) saccade duration–saccade magnitude, and (3) latent period–saccade magnitude.

Oculomotor system: The oculomotor system consists of the eyeball and extraocular muscles (also called the *oculomotor plant*) and the neural sites responsible for the eye movement.

Saccade: A fast eye movement.

References

Bahill AT, Clark MR, Stark L. 1975. The main sequence, a tool for studying human eye movements. Math Biosci 24:194.

Bahill AT, Latimer JR, Troost BT. 1980. Linear homeomorphic model for human movement. IEEE Trans Biomed Eng 27:631.

Collins CC. 1975. The human oculomotor control system. In G Lennerstrand, P Bach-y-Rita (eds), Basic Mechanisms of Ocular Motility and Their Clinical Implications, pp 145–180, Oxford, Pergamon Press.

Enderle JD. 1988. Observations on pilot neurosensory control performance during saccadic eye movements. Aviat Space Environ Med 59:309.

Enderle JD. 1994. A physiological neural network for saccadic eye movement control. Armstrong Laboratory/AO-TR-1994-0023. Air Force Material Command, Brooks Air Force Base, Texas.

Enderle JD, Engelken EJ, Stiles RN. 1991. A comparison of static and dynamic characteristics between rectus eye muscle and linear muscle model predictions. IEEE Trans Biomed Eng 38:1235.

Enderle JD, Wolfe JW. 1988. Frequency response analysis of human saccadic eye movements. Comput Biol Med 18(3):195.

Enderle JD, Wolfe JW, Yates JT. 1984. The linear homeomorphic saccadic eye movement model—a modification. IEEE Trans Biomed Eng 31(11):717.

Enderle, JD, Blanchard SB, Bronzino JD. 1999. Introduction to Biomedical Engineering, Academic Press, San Diego.

Hsu FK, Bahill AT, Stark L. 1976. Parametric sensitivity of a homeomorphic model for saccadic and vergence eye movements. Comput Methods Programs Biomed 6:108.

Robinson DA. 1964. The mechanics of human saccadic eye movement. J Physiol (London) 174:245.

Robinson DA. 1975. Oculomotor control signals. In G Lennerstrand, P Bach-y-Rita (eds), Basic Mechanisms of Ocular Motility and their Clinical Implication, pp 337–374, Oxford, Pergamon Press.

Robinson DA. 1981. Models of mechanics of eye movements. In BL Zuber (ed), Models of Oculomotor Behavior and Control, pp 21–41, Boca Raton, Fla, CRC Press.

Robinson DA, O'Meara DM, Scott AB, et al. 1969. Mechanical components of human eye movements. J Appl Physiol 26:548.

Scudder CA. 1988. A new local feedback model of the saccadic burst generator. J Neurophysiol 59(4):1454.

Westheimer G. 1954. Mechanism of saccadic eye movements. AMA Arch Ophthalmol 52:710.

Wilkie DR. 1968. Muscle: Studies in Biology, vol 11, London, Edward Arnold.

Further Information

Readers interested in additional information on the subject of fast eye movements should consult the following books. In addition, many journals publish articles on saccadic eye movements—for a sample of these journals, see the references listed within the following books as well.

Bahill AT. 1981. Bioengineering, Biomedical, Medical and Clinical Engineering, Englewood Cliffs, NJ, Prentice-Hall.

Carpenter RHS. 1988. Movements of the Eyes, 2d rev ed, London, Pion Ltd.

Wurtz RH, Goldberg ME. 1989. The Neurobiology of Saccadic Eye Movements, New York, Elsevier.

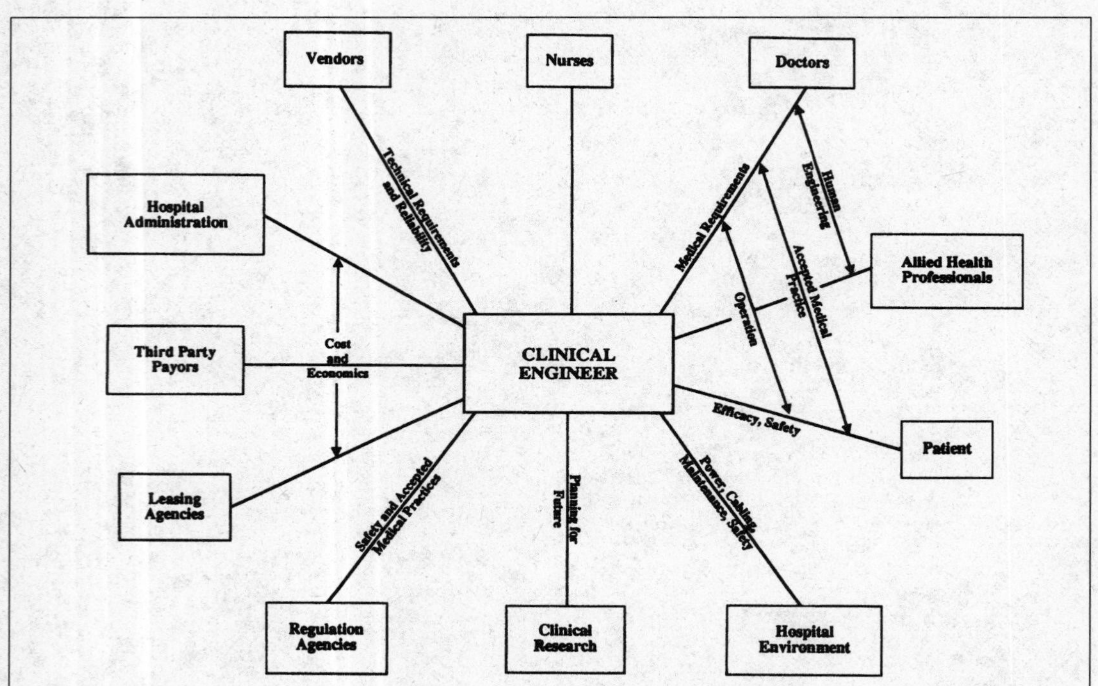

Human interactions of the clinical engineer.

XVII

Clinical Engineering

Yadin David
Texas Children's Hospital

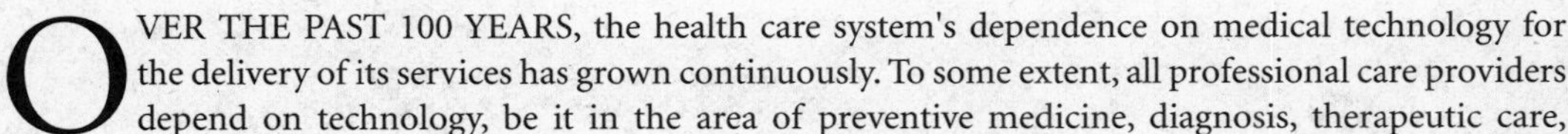

OVER THE PAST 100 YEARS, the health care system's dependence on medical technology for the delivery of its services has grown continuously. To some extent, all professional care providers depend on technology, be it in the area of preventive medicine, diagnosis, therapeutic care, rehabilitation, administration, or health-related education and training. Medical technology enables practitioners to intervene through integrated interactions with their patients in a cost-effective, efficient, and safe manner. As a result, the field of clinical engineering has emerged as the discipline of biomedical engineering that fulfills the need to manage the deployment of medical technology and to integrate it appropriately with desired clinical practices.

The health care delivery system presents a very complex environment where facilities, equipment, materials, and a full range of human interventions are involved. It is in this clinical environment that patients of various ages and conditions, trained staff, and the wide variety of medical technology converge. This complex mix of interactions may lead to unacceptable risk when programs for monitoring, controlling, improving, and educating all entities involved are not appropriately integrated by qualified professionals.

This section of clinical engineering focuses on the methodology for administering critical engineering services that vary from facilitation of innovation and technology transfer to the performance of technology assessment and operations support and on the management tools with which today's clinical engineer needs to be familiar. With increased awareness of the value obtained by these services, new career opportunities are created for clinical engineers.

In this section the authors have attempted to provide a description of the wide range of responsibilities clinical engineering professionals encounter. After presenting the evolution of the field of clinical engineering, Chapter 165 gives specific attention to the primary function of clinical engineers. Chapter 166 describes technology management and assessment in considerable detail, using examples and case studies in a large medical center. Chapter 167 focuses on a particular technology management tool for assessing the medical equipment risk factor that can help clinical engineers effectively manage/prioritize the services to be provided to each piece of equipment under their control. Chapter 168 takes an expansive view of career opportunities for clinical engineers, while Chapter 169 focuses on one particular role, namely, that of innovator and product developer. To further assist clinical engineers in managing their equipment, Chapters 170 and 171 examine an establishment of program indicators that can lead to quality improvement. The clinical engineering program can be based in a single community hospital, in a teaching medical center, within a chain of hospitals, as part of a government agency, or as a shared service organization. Chapter 172 describes the use of coordinated services as they are offered by the West of Scotland Health Board, and Chapters 173 and 174 present a review of the standards and regulatory agencies of interest to clinical engineers. Global programs present different challenges to practitioners in developing countries because the resources and logistics are different than they are in the United States. Chapter 175 looks at specific clinical engineering issues from the perspective of developing countries.

In addition to highlighting the important roles that clinical engineers serve in many areas, the section focuses on those areas of the clinical engineering field that enhance the understanding of the "bigger picture." With such an understanding, the participation in and contribution by clinical engineers to this enlarged scope can be fully realized. The adoption of the tools described here will enable clinical engineers to fulfill their new role in the evolving health care delivery system.

All the authors in this section recognize this opportunity and here recognized for volunteering their talent and time so that others can excel as well.

167

Clinical Engineering: Evolution of a Discipline

Joseph D. Bronzino
Trinity College/Biomedical Engineering Alliance for Connecticut (BEACON)

167.1 What Is a Clinical Engineer?

As discussed in the introduction to this *Handbook*, biomedical engineers apply the concepts, knowledge, and techniques of virtually all engineering disciplines to solve specific problems in the biosphere, i.e., the realm of biology and medicine. When biomedical engineers work within a hospital or clinic, they are more properly called *clinical engineers.* But what exactly is the definition of the term *clinical engineer*? In recent years, a number of organizations, e.g., the American Heart Association [1986], the American Association of Medical Instrumentation [Goodman, 1989], the American College of Clinical Engineers [Bauld, 1991], and the *Journal of Clinical Engineering* [Pacela, 1991], have attempted to provide an appropriate definition for the term, *clinical engineer.* For the purposes of this handbook, a *clinical engineer* is an engineer who has graduated from an accredited academic program in engineering or who is licensed as a professional engineer or engineer-in-training and is engaged in the application of scientific and technological knowledge developed through engineering education and subsequent professional experience within the health care environment in support of clinical activities. Furthermore, the clinical environment is defined as that portion of the health care system in which patient care is delivered, and clinical activities include direct patient care, research, teaching, and public service activities intended to enhance patient care.

167.2 Evolution of Clinical Engineering

Engineers were first encouraged to enter the clinical scene during the late 1960s in response to concerns about patient safety as well as the rapid proliferation of clinical equipment, especially in academic medical centers. In the process, a new engineering discipline—clinical engineering—evolved to provide the technological support necessary to meet these new needs. During the 1970s, a major expansion of clinical engineering occurred, primarily due to the following events:

- The Veterans' Administration (VA), convinced that clinical engineers were vital to the overall operation of the VA hospital system, divided the country into biomedical engineering districts, with a chief biomedical engineer overseeing all engineering activities in the hospitals in that district.

0-8493-0462-8/00/$0.00+$.50

- Throughout the United States, clinical engineering departments were established in most large medical centers and hospitals and in some smaller clinical facilities with at least 300 beds.
- Clinical engineers were hired in increasing numbers to help these facilities use existing technology and incorporate new technology.

Having entered the hospital environment, routine electrical safety inspections exposed the clinical engineer to all types of patient equipment that was not being maintained properly. It soon became obvious that electrical safety failures represented only a small part of the overall problem posed by the presence of medical equipment in the clinical environment. The equipment was neither totally understood nor properly maintained. Simple visual inspections often revealed broken knobs, frayed wires, and even evidence of liquid spills. Investigating further, it was found that many devices did not perform in accordance with manufacturers' specifications and were not maintained in accordance with manufacturers' recommendations. In short, electrical safety problems were only the tip of the iceberg. The entrance of clinical engineers into the hospital environment changed these conditions for the better. By the mid-1970s, complete performance inspections before and after use became the norm, and sensible inspection procedures were developed [Newhouse et al., 1989]. In the process, clinical engineering departments became the logical support center for all medical technologies and became responsible for all the biomedical instruments and systems used in hospitals, the training of medical personnel in equipment use and safety, and the design, selection, and use of technology to deliver safe and effective health care.

With increased involvement in many facets of hospital/ clinic activities, clinical engineers now play a multifaceted role (Fig. 167.1). They must interface successfully with many "clients," including clinical staff, hospital administrators, regulatory agencies, etc., to ensure that the medical equipment within the hospital is used safely and effectively.

Today, hospitals that have established centralized clinical engineering departments to meet these responsibilities use clinical engineers to provide the hospital administration with an objective option of equipment function, purchase, application, overall system analysis, and preventive maintenance policies.

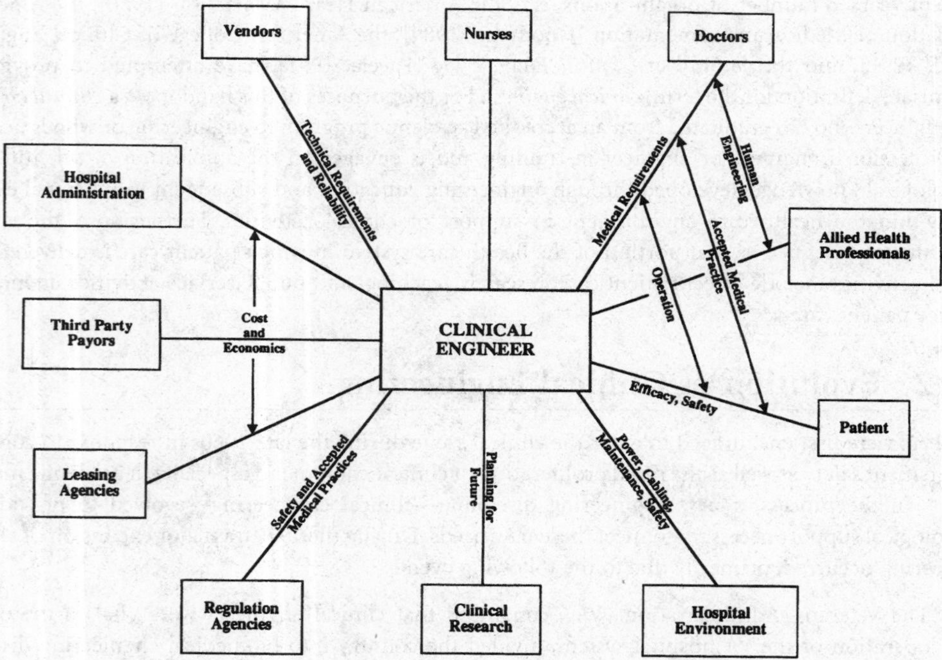

FIGURE 167.1 Diagram illustrating the range of interactions of a clinical engineer.

Some hospital administrators have learned that with the in-house availability of such talent and expertise, the hospital is in a far better position to make more effective use of its technological resources [Bronzino, 1986, 1992]. By providing health professionals with needed assurance of safety, reliability, and efficiency in using new and innovative equipment, clinical engineers can readily identify poor-quality and ineffective equipment, thereby resulting in faster, more appropriate utilization of new medical equipment.

Typical pursuits of clinical engineers, therefore, include

- Supervision of a hospital clinical engineering department that includes clinical engineers and biomedical equipment technicians (BMETs)
- Prepurchase evaluation and planning for new medical technology
- Design, modification , or repair of sophisticated medical instruments or systems
- Cost-effective management of a medical equipment calibration and repair service
- Supervision of the safety and performance testing of medical equipment performed by BMETs
- Inspection of all incoming equipment (i.e., both new and returning repairs)
- Establishment of performance benchmarks for all equipment
- Medical equipment inventory control
- Coordination of outside engineering and technical services performed by vendors
- Training of medical personnel in the safe and effective use of medical devices and systems
- Clinical applications engineering, such as custom modification of medical devices for clinical research, evaluation of new noninvasive monitoring systems, etc.
- Biomedical computer support
- Input to the design of clinical facilities where medical technology is used, e.g., operating rooms (ORs), intensive care units, etc.
- Development and implementation of documentation protocols required by external accreditation and licensing agencies.

Clinical engineers thus provide extensive engineering services for the clinical staff and, in recent years, have been increasingly accepted as valuable team members by physicians, nurses, and other clinical professionals. Furthermore, the acceptance of clinical engineers in the hospital setting has led to different types of engineering-medicine interactions, which in turn have improved health care delivery.

167.3 Hospital Organization and the Role of Clinical Engineering

In the hospital, management organization has evolved into a diffuse authority structure that is commonly referred to as the *triad model*. The three primary components are the governing board (trustees), hospital administration (CEO and administrative staff), and the medical staff organization [Bronzino and Hayes, 1988]. The role of the governing board and the chief executive officer are briefly discussed below to provide some insight regarding their individual responsibilities and their interrelationship.

Governing Board (Trustees)

The Joint Commission on the Accreditation of Healthcare Organizations (JCAHO) summarizes the major duties of the governing board as "adopting by-laws in accordance with its legal accountability and its responsibility to the patient." The governing body, therefore, requires both medical and paramedical departments to monitor and evaluate the quality of patient care, which is a critical success factor in hospitals today.

To meet this goal, the governing board essentially is responsible for establishing the mission statement and defining the specific goals and objectives that the institution must satisfy. Therefore, the trustees are involved in the following functions:

- Establishing the policies of the institution
- Providing equipment and facilities to conduct patient care
- Ensuring that proper professional standards are defined and maintained (i.e., providing quality assurance)
- Coordinating professional interests with administrative, financial, and community needs
- Providing adequate financing by securing sufficient income and managing the control of expenditures
- Providing a safe environment
- Selecting qualified administrators, medical staff, and other professionals to manage the hospital

In practice, the trustees select a hospital chief administrator who develops a plan of action that is in concert with the overall goals of the institution.

Hospital Administration

The hospital administrator, the chief executive officer of the medical enterprise, has a function similar to that of the chief executive officer of any corporation. The administrator represents the governing board in carrying out the day-to-day operations to reflect the broad policy formulated by the trustees. The duties of the administrator are summarized as follows:

- Preparing a plan for accomplishing the institutional objectives, as approved by the board
- Selecting medical chiefs and department directors to set standards in their respective fields
- Submitting for board approval an annual budget reflecting both expenditures and income projections
- Maintaining all physical properties (plant and equipment) in safe operating condition
- Representing the hospital in its relationships with the community and health agencies
- Submitting to the board annual reports that describe the nature and volume of the services delivered during the past year, including appropriate financial data and any special reports that may be requested by the board

In addition to these administrative responsibilities, the chief administrator is charged with controlling cost, complying with a multitude of governmental regulations, and ensuring that the hospital conforms to professional norms, which include guidelines for the care and safety of patients.

167.4 Clinical Engineering Programs

In many hospitals, administrators have established clinical engineering departments to manage effectively all the technological resources, especially those relating to medical equipment, that are necessary for providing patient care. The primary objective of these departments is to provide a broad-based engineering program that addresses all aspects of medical instrumentation and systems support.

Figure 167.2 illustrates the organizational chart of the medical support services division of a typical major medical facility. Note that within this organizational structure, the director of clinical engineering reports directly to the vice-president of medical support services. This administrative relationship is extremely important because it recognizes the important role clinical engineering departments play in delivering quality care. It should be noted, however, that in other common organizational structures, clinical engineering services may fall under the category of "facilities," "materials management," or even just "support services." Clinical engineers also can work directly with clinical departments, thereby bypassing much of the hospital hierarchy. In this situation, clinical departments can offer the clinical engineer both the chance for intense specialization and, at the same time, the opportunity to develop personal relationships with specific clinicians based on mutual concerns and interests [Wald, 1989].

Once the hospital administration appoints a qualified individual as director of clinical engineering, the person usually functions at the department-head level in the organizational structure of the institution

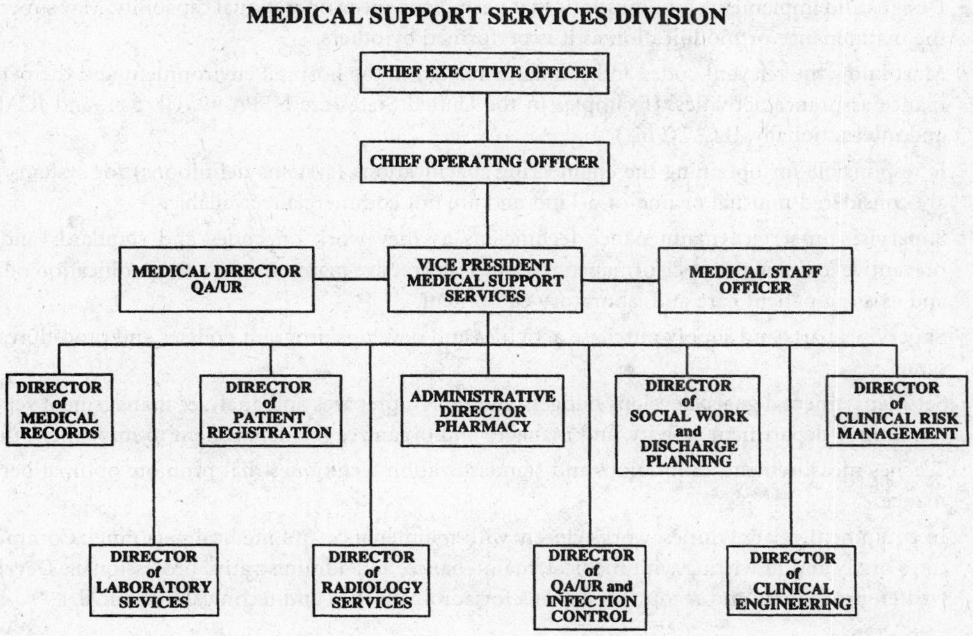

FIGURE 167.2 Organizational chart of medical support services division for a typical major medical facility. This organizational structure points out the critical interrelationship between the clinical engineering department and the other primary services provided by the medical facility.

and is provided with sufficient authority and resources to perform the duties efficiently and in accordance with professional norms. To understand the extent of these duties, consider the job title for "clinical engineering director" as defined by the World Health Organization [Issakov et al, 1990].

General Statement. The clinical engineering director, by his or her education and experience, acts as a manger and technical director of the clinical engineering department. The individual designs and directs the design of equipment modifications that may correct design deficiencies or enhance the clinical performance of medical equipment. The individual also may supervise the implementation of those design modifications. The education and experience that the director possesses enables him or her to analyze complex medical or laboratory equipment for purposes of defining corrective maintenance and developing appropriate preventing maintenance or performance assurance protocols. The clinical engineering director works with nursing and medical staff to analyze new medical equipment needs and participates in both the prepurchase planning process and the incoming testing process. The individual also participates in the equipment management process through involvement in the system development, implementation, maintenance, and modification processes.

Duties and Responsibilities. The director of clinical engineering has a wide range of duties and responsibilities. For example, this individual

- Works with medical and nursing staff in the development of technical and performance specifications for equipment requirements in the medical mission.
- Once equipment is specified and the purchase order developed, generates appropriate testing of the new equipment.
- Does complete performance analysis on complex medical or laboratory equipment and summarizes results in brief, concise, easy-to-understand terms for the purposes of recommending corrective action or for developing appropriate preventive maintenance and performance assurance protocols.

- Designs and implements modifications that permit enhanced operational capability. May supervise the maintenance or modification as it is performed by others.
- Must know the relevant codes and standards related to the hospital environment and the performance assurance activities. (Examples in the United States are NFPA 99, UL 544, and JCAHO, and internationally, IEC-TC 62.)
- Is responsible for obtaining the engineering specifications (systems definitions) for systems that are considered unusual or one-of-a-kind and are not commercially available.
- Supervises in-service maintenance technicians as they work on codes and standards and on preventive maintenance, performance assurance, corrective maintenance, and modification of new and existing patient care and laboratory equipment.
- Supervises parts and supply purchase activities and develops program policies and procedures for same.
- Sets departmental goals, develops budgets and policy, prepares and analyzes management reports to monitor department activity, and manages and organizes the department to implement them.
- Teaches measurement, calibration, and standardization techniques that promote optimal performance.
- In equipment-related duties, works closely with maintenance and medical personnel. Communicates orally and in writing with medical, maintenance, and administrative professionals. Develops written procedures and recommendations for administrative and technical personnel.

Minimum Qualifications. A bachelor's degree (4 years) in an electrical or electronics program or the equivalent is required (preferably with a clinical or biomedical adjunct). A master's degree is desirable. A minimum of 3 years' experience as a clinical engineer and 2 years in a progressively responsible supervisory capacity is needed. Additional qualifications are as follows:

- Must have some business knowledge and management skills that enable him or her to participate in budgeting, cost accounting, personnel management, behavioral counseling, job description development, and interviewing for hiring or firing purposes. Knowledge and experience in the use of microcomputers are desirable.
- Must be able to use conventional electronic trouble-shooting instruments such as multimeters, function generators, oscillators, and oscilloscopes. Should be able to use conventional machine shop equipment such as drill presses, grinders, belt sanders, brakes, and standard hand tools.
- Must possess or be able to acquire knowledge of the techniques, theories, and characteristics of materials, drafting, and fabrication techniques in conjunction with chemistry, anatomy, physiology, optics, mechanics, and hospital procedures.
- Clinical engineering certification or professional engineering registration is required.

Major Functions of a Clinical Engineering Department

It should be clear by the preceding job description that clinical engineers are first and foremost engineering professionals. However, as a result of the wide-ranging scope of interrelationships within the medical setting, the duties and responsibilities of clinical engineering directors are extremely diversified. Yet a common thread is provided by the very nature of the technology they manage. Directors of clinical engineering departments are usually involved in the following core functions:

Technology Management. Developing, implementing, and directing equipment management programs. Specific tasks include accepting and installing new equipment, establishing preventive maintenance and repair programs, and managing the inventory of medical instrumentation. Issues such as cost-effective use and quality assurance are integral parts of any technology management program. The director advises the administrator of the budgetary, personnel, space, and test equipment requirements necessary to support this equipment management program.

Risk Management. Evaluating and taking appropriate action on incidents attributed to equipment malfunctions or misuse. For example, the clinical engineering director is responsible for summarizing the technological significance of each incident and documenting the findings of the investigation. He or she then submits a report to the appropriate hospital authority and, according to the Safe Medical Devices Act of 1990, to the device manufacturer, the Food and Drug Administration (FDA), or both.

Technology Assessment. Evaluating and selecting new equipment. The director must be proactive in the evaluation of new requests for capital equipment expenditures, providing hospital administrators and clinical staff with an in depth appraisal of the benefits/ advantages of candidate equipment. Furthermore, the process of technology assessment for all equipment used in the hospital should be an ongoing activity.

Facilities Design and Project Management. Assisting in the design of new or renovated clinical facilities that house specific medical technologies. This includes operating rooms, imaging facilities, and radiology treatment centers.

Training. Establish and deliver instructional modules for clinical engineering staff as well as clinical staff on the operation of medical equipment.

In the future, it is anticipated that clinical engineering departments will provide assistance in the application and management of many other technologies that support patient care, including computer support, telecommunications, and facilities operations.

Defining Terms

JCAHO, Joint Commission on the Accreditation of Healthcare Organizations: Accrediting body responsible for checking hospital compliance with approved rules and regulations regarding the delivery of health care.

Technology assessment: Involves an evaluation of the safety, efficiency, and cost-effectiveness, as well as consideration of the social, legal, and ethical effects, of medical technology.

References

AHA, 1986. American Hospital Association Resource Center, Hospital Administration Terminology, 2nd ed. Washington, American Hospital Publishing.

Bauld TJ. 1991. The definition of a clinical engineer. J Clin Eng 16:403.

Bronzino JD. 1986. Biomedical Engineering and Instrumentation: Basic Concepts and Applications. Boston, PWS Publishing.

Bronzino JD. 1992. Management of Medical Technology: A Primer for Clinical Engineers. Boston, Butterworth.

Goodman G. 1989. The profession of clinical engineering. J Clin Eng 14:27.

ICC. 1991. International Certification Commission's Definition of a Clinical Engineer, International Certification Commission Fact Sheet. Arlington, Va, ICC.

Issakov A, Mallouppas A, McKie J. 1990. Manpower development for a healthcare technical service. Report of the World Health Organization, WHO/SHS/NHP/90.4.

Newhouse VL, Bell DS, Tackel IS, et al. 1989. The future of clinical engineering in the 1990s. J Clin Eng 14:417.

Pacela A. 1990. Bioengineering Education Directory. Brea, Calf, Quest Publishing.

Wald A. 1989. Clinical engineering in clinical departments: A different point of view. Biomed Instr Technol 23:58.

Further Information

Bronzino JD. 1992. Management of Medical Technology: A Primer for Clinical Engineers. Boston, Butterworth.

Journals: Journal of Clinical Engineering, Journal of Medical Engineering and Physics, Biomedical Instrumentation and Technology.

168

Management and Assessment of Medical Technology

Yadin David
Texas Children's Hospital

Thomas M. Judd
Kaiser Permanente

As medical technology continues to evolve, so does its impact on patient outcome, hospital operations, and financial efficiency. The ability to plan for this evolution and its subsequent implications has become a major challenge in most decisions of health care organizations and their related industries. Therefore, there is a need to adequately plan for and apply those management tools which optimize the deployment of medical technology and the facilities that house it. Successful management of the technology and facilities will ensure a good match between the needs and the capabilities of staff and technology, respectively. While different types and sizes of hospitals will consider various strategies of actions, they all share the need to manage efficient utilization of their limited resources and its monitoring. Technology is one of these resources, and while it is frequently cited as the culprit behind cost increases, the well-managed technology program contribute to a significant containment of the cost of providing quality patient care. Clinical engineer's skills and expertise are needed to facilitate the adoption of an objective methodology for implantation of a program that will match the hospital's needs and operational conditions.

Whereas both the knowledge and practice patterns of management in general are well organized in today's literature, the management of the health care delivery system and that of medical technology in the clinical environment has not yet reached that same high level. However, as we begin to understand the relationship between the methods and information that guide the decision-making processes regarding the management of medical technology that are being deployed in this highly complex environment, the role of the qualified clinical engineer becomes more valuable. This is achieved by reformulating the

0-8493-0462-8/00/$0.00+$.50

technology management process, which starts with the strategic planning process, continues with the technology assessment process, leads to the equipment planning and procurement processes, and finally ends with the assets management process. Definition of terms used in this chapter are provided at the end of the chapter.

168.1 The Health Care Delivery System

Societal demands on the health care delivery system revolve around cost, technology, and expectations. To respond effectively, the delivery system must identify its goals, select and define its priorities, and then wisely allocate its limited resources. For most organizations, this means that they must acquire only appropriate technologies and manage what they have already more effectively. To improve performance and reduce costs, the delivery system must recognize and respond to the key dynamics in which it operates, must shape and mold its planing efforts around several existing health care trends and directions, and must respond proactively and positively to the pressures of its environment. These issues and the technology manager's response are outlined here: (1) technology's positive impact on care quality and effectiveness, (2) an unacceptable rise in national spending for health care services, (3) a changing mix of how Americans are insured for health care, (4) increases in health insurance premiums for which appropriate technology application is a strong limiting factor, (5) a changing mix of health care services and settings in which care is delivered, and (6) growing pressures related to technology for hospital capital spending and budgets.

Major Health Care Trends and Directions

The major trends and directions in health care include (1) changing location and design of treatment areas, (2) evolving benefits, coverages, and choices, (3) extreme pressures to manage costs, (4) treating of more acutely ill older patients and the prematurely born, (5) changing job structures and demand for skilled labor, (6) the need to maintain a strong cash flow to support construction, equipment, and information system developments, (7) increased competition on all sides, (8) requirement for information systems that effectively integrate clinical and business issues, (9) changing reimbursement policies that reduce new purchases and lead to the expectation for extended equipment life cycles, (10) internal technology planning and management programs to guide decision making, (11) technology planning teams to coordinate adsorption of new and replacement technologies, as well as to suggest delivery system changes, and (12) equipment maintenance costs that are emerging as a significant expense item under great administrative scrutiny.

System Pressures

System pressures include (1) society's expectations—highest quality care at the lowest reasonable price, where quality is a function of personnel, facilities, technology, and clinical procedures offered; (2) economic conditions—driven often by reimbursement criteria; (3) legal-pressures—resulting primarily from malpractice issues and dealing with rule-intensive "government" clients; (4) regulatory—multistate delivery systems with increased management complexity, or heavily regulated medical device industries facing free-market competition, or hospitals facing the Safe Medical Devices Act reporting requirements and credentialling requirements; (5) ethics—deciding who gets care and when; and (6) technology pressures—organizations having enough capabilities to meet community needs and to compete successfully in their marketplaces.

The Technology Manager's Responsibility

Technology mangers should (1) become deeply involved and committed to technology planning and management programs in their system, often involving the need for greater personal responsibilities and expanded credentials; (2) understand how the factors above impact their organization and how technology can be used to improve outcomes, reduce costs, and improve quality of life for patients, (3) educate other health care professionals about how to demonstrate the value of individual technologies through involving financial, engineering, quality of care, and management perspective, and (4) assemble a team of caregivers with sufficient broad clinical expertise and administrators with planning and financial expertise to contribute their knowledge to the assessment process [16].

168.2 Strategic Technology Planning

Strategic Planning Process

Leading health care organizations have begun to combine strategic technology planning with other technology management activities in program that effectively integrate new technologies with their existingly technology base. This has resulted in high-quality care at a reasonable cost. Among those who have been its leading catalysts, ECRI (formerly the Emergency Care Research Institute) is known for articulating this program [4] and encouraging its proliferation initially among regional health care systems and now for single or multihospital systems as well [5]. Key components of the program include clinical strategic-planning, technology strategic planning, technology assessment, interaction with capital budgeting, acquisition and deployment, resource (or equipment assets) management, and monitoring and evaluation. A proper technology strategic plan is derived from and supports as well-defined clinical strategic plan [15].

Clinical and Technology Strategic Plan

Usually considered long-range and continually evolving, a clinical strategic plan is updated annually. For a given year, the program begins when key hospital participants, through the strategic planning process, assess what clinical services the hospital should be offering in its referral area. They take into account health care trends, demographic and market share data, and space and facilities plans. They analyze their facility's strengths and weaknesses, goals and objectives, competition, and existing technology base. The outcome of this process is a clinical strategic plan that establishes the organization's vision for the year and referral area needs and the hospital's objectives in meeting them.

It is not possible to adequately complete a clinical strategic plan without engaging in the process of strategic technology planning. A key role for technology managers is to assist their organizations throughout the combined clinical and technology strategic planning processes by matching available technical capabilities, both existing and new, with clinical requirements. To accomplish this, technology managers must understand why their institution's values and mission are set as they are, pursue their institution's strategic plans through that knowledge, and plan in a way that effectively allocates limited resources. Although a technology manager may not be assigned to develop an institution's overall strategic plan, he or she must understand and believe it in order to offer good input for hospital management. In providing this input, a technology manager should determine a plan for evaluating the present state of the hospital's technological deployment, assist in providing a review of emerging technological innovations and their possible impact on the hospital, articulate justifications and provisions for adoption of new technologies or enhancement of existing ones, visit research and laboratories and exhibit areas at major medical and scientific meetings to view new technologies, and be familiar with the institution and its equipment users' abilities to assimilate new technology.

The past decade has shown a trend toward increased legislation in support of more federal regulations in health care. These and other pressures will require that additional or replacement medical technology be well anticipated and justified. As a rationale for technology adoption, the Texas Children's Hospital focuses on the issues of clinical necessity, management support, and market preference, Addressing the issue of clinical necessity, the hospital considers the technology's comparison against medical standard of care, its impact on the level of care and quality of life, its improvement on intervention's accuracy and/or safety, its impact on the rate of recovery, the needs or desires of the community, and the change in service volume or focus. On the issue of management support, the hospital estimates if the technology will create a more effective care plan and decision-making process, improve operational efficiency in the current service programs, decrease liability exposure, increase compliance with regulations, reduce workload and dependence on user skill level ameliorate departmental support, or enhance clinical proficiency. Weighting the issue of market preference, the hospital contemplate if it will improve access to care, increase customer convenience and/or satisfaction, enhance the organization's image and market share, decrease the cost of adoption and ownership, or provide a return on its investment.

Technology Strategic Planning Process

When the annual clinical strategic planning process has started and hospital leaders have begun to analyze or reaffirm what clinical services they want to offer to the community, the hospital can then conduct efficient technology strategic planning. Key elements of this planning involve (1) performing an initial audit of existing technologies, (2) conducting a technology assessment for new and emerging technologies for fit with current or desired clinical services, (3) planning for replacement and selection of new technologies, (4) setting priorities for technology acquisition, and (5) developing processes to implement equipment acquisition and monitor ongoing utilization. "Increasingly, hospitals are designating a senior manager (e.g., an administrator, the director of planning, the director of clinical engineering) to take the responsibility for technology assessment and planning. That person should have the primary responsibility for developing the strategic technology plan with the help of key physicians, department managers, and senior executives" [4].

Hospitals can form a medical technology advisory committee (MTAC), overseen by the designated senior manager and consisting of the types of members mentioned above, to conduct the strategic technology planning process and to annually recommend technology priorities to the hospital strategic planning committee and capital budget committee. It is especially important to involve physicians and nurses in this process.

In the initial technology audit, each major clinical service or product line must be analyzed to determine how well the existing technology base supports it. The audit can be conducted along service lines (radiology, cardiology, surgery) or technology function (e.g., imaging, therapeutic, diagnostic) by a team of designated physicians, department heads, and technology managers. The team should begin by developing a complete hospital-wide assets inventory, including the quantity and quality of equipment. The team should compare the existing technology base against known and evolving standards-of-care information, patient outcome data, and known equipment problems. Next, the team should collect and examine information on technology utilization to assess its appropriate use, the opportunities for improvement, and the risk level. After reviewing the technology users' education needs as they relate to the application and servicing of medical equipment, the team should credential users for competence in the application of new technologies. Also, the auditing team should keep up with published clinical protocols and practice guidelines using available health care standards directories and utilize clinical outcome data for quality-assurance and risk-management program feedback [6].

While it is not expected that every hospital has all the required expertise in-house to conduct the initial technology audit or ongoing technology assessment, the execution of this planning process is sufficiently critical for a hospital's success that outside expertise should be obtained when necessary. The audit allows for the gathering of information about the status of the existing technology base and enhances the capability of the medical technology advisory committee to assess the impact of new and emerging technologies on their major clinical services.

All the information collected from the technology audit results and technology assessments is used in developing budget strategies. Budgeting is part of strategic technology planning in that a 2- to 5-year long-range capital spending plan should be created. This is in addition to the annual capital budget preparation that takes into account 1 year at a time. The MTAC, as able and appropriate, provides key information regarding capital budget requests and makes recommendations to the capital budget committee each year. The MTAC recommends priorities for replacement as well as new and emerging technologies that over a period of several years guides that acquisition that provides the desired service developments or enhancements. Priorities are recommended on the basis of need, risk, cost (acquisition, operational and maintenance), utilization, and fit with the clinical strategic plan.

168.3 Technology Assessment

As medical technology continues to evolve, so does its impact on patient outcome, hospital operations, and financial resources. The ability to manage this evolution and its subsequent implications has become a major challenge for all health care organizations. Successful management of technology will ensure a good match between needs and capabilities and between staff and technology. To be successful, an ongoing technology assessment process must b an integral part of an ongoing technology planning and management program at the hospital, addressing the needs of the patient, the user, and the support team. This facilitates better equipment planning and utilization of the hospital's resources. The manager who is knowledgeable about his or her organization's culture, equipment users' needs, the environment within which equipment will be applied, equipment engineering, and emerging technological capabilities will be successful in proficiently implementing and managing technological changes [7].

It is in the technology assessment process that the clinical engineering/technology manager professional needs to wear two hats: that of the manager and that of the engineer. This is a unique position, requiring expertise and detailed preparation, that allows one to be a key leader and contributor to the decision-making process of the medical technology advisory committee (MTAC).

The MTAC uses an ad hoc team approach to conduct technology assessment of selected services and technologies throughout the year. The ad hoc teams may incorporate representatives of equipment users, equipment service providers, physicians, purchasing agents, reimbursement mangers, representatives of administration, and other members from the institution as applicable.

Prerequisites for Technology Assessment

Medical technology is a major strategic factor in positioning and creating a positive community perception of the hospital. Exciting new biomedical devices and systems are continually being introduced. And they are introduced at a time when the pressure on hospitals to contain expenditures is mounting. Therefore, forecasting the deployment of medical technology and the capacity to continually evaluate its impact on the hospital require that the hospital be willing to provide the support for such a program. (*Note*: Many organizations are aware of the principle that an in-house "champion" is needed in order to provide for the leadership that continually and objectively plans ahead. The champion and the program being "championed" may use additional in-house or independent expertise as needed. To get focused attention on the technology assessment function and this program in larger, academically affiliated and government hospitals, the position of a chief technology officer is being created.) Traditionally, executives rely on their staff to produce objective analyses of the hospital's technological needs. Without such analyses, executives may approve purchasing decisions of sophisticated biomedical equipment only to discover later that some needs or expected features were not included with this installation, that those features are not yet approved for delivery, or that the installation has not been adequately planned.

Many hospitals perform technology assessment activities to project needs for new assets and to better manage existing assets. Because the task is complex, an interdisciplinary approach and a cooperative attitude among the assessment team leadership is required. The ability to integrate information from disciplines such as clinical, technical, financial, administrative, and facility in a timely and objective manner is critical to the success of the assessment. This chapter emphasizes how technology assessment fits within a technology planning and management program and recognizes the importance of corporate skills forecasting medical equipment changes and determining the impact of changes on the hospital's market position. Within the technology planning and management program, the focus on capital assets management of medical equipment should not lead to the exclusion of accessories, supplies, and the disposables also required.

Medical equipment has a life cycle that can be identified as (1) the innovation phase, which includes the concept, basic and applies research, and development, and (2) the adoption phase, which begins with the clinical studies, through diffusion, and then widespread use. These phases are different from each other in the scope of professional skills involved, their impact on patient care, compliance with regulatory requirements, and the extent of the required operational support. In evaluating the applicability of a device or a system for use in the hospital, it is important to note in which phase of its life cycle the equipment currently resides.

Technology Assessment Process

More and more hospitals are faced with the difficult phenomenon of a capital equipment requests list that is much larger than the capital budget allocation. The must difficult decision, then, is the one that matches clinical needs with the financial capability. In doing so, the following questions are often raised: How do we avoid costly technology mistakes? How do we wisely target capital dollars for technology? How do we avoid medical staff conflicts as they relate to technology? How do we control equipment-related risks? and How do we maximize the useful life of the equipment or systems while minimizing the cost ownership? A hospital's clinical engineering department can assist in providing the right answers to these questions.

Technology assessment is a component of technology planning that begins with the analysis of the hospital's existing technology base. It is easy to perceive then that technology assessment, rather than an equipment comparison, is a new major function for a clinical engineering department [8]. It is important that clinical engineers be well prepared for the challenge. They must have a full understanding of the mission of their particular hospitals, a familiarity with the health care delivery system, and the cooperation of hospital administrators and the medical staff. To aid in the technology assessment process, clinical engineers need to utilize the following tools: (1) access to national database services, directories, and libraries, (2) visits to scientific and clinical exhibits, (3) a network with key industry contacts, and (4) a relationship with peers throughout the country [9].

The need for clinical engineering involvement in the technology assessment process becomes evident when recently purchased equipment or its functions are underutilized, users have ongoing problems with equipment, equipment maintenance costs become excessive, the hospital is unable to comply with standards or guidelines (i.e., JCAHO requirements) for equipment management, a high percentage of equipment is awaiting repair, or training for equipment operators is inefficient due to shortage of allied health professionals. A deeper look at the symptoms behind these problems would likely reveal a lack of a central clearinghouse to collect, index, and monitor all technology-related information for future planning purposes, the absence of procedures for identifying emerging technologies for potential acquisition, the lack of a systematic plan for conducting technology assessment, resulting in an ability to maximize the benefits from deployment of available technology, the inability to benefit from the organization's own previous experience with a particular type of technology, the random replacement of medical technologies rather than a systematic plan based on a set of well-developed criteria, and/or the lack of integration of technology acquisition into the strategic and capital planning of the hospital.

To address these issues, efforts to develop a technology microassessment process were initiated at one leading private hospital with the following objectives: (1) accumulate information on medical equipment, (2) facilitate systematic planning, (3) create an administrative structure supporting the assessment process and its methodology, (4) monitor the replacement of outdated technology, and (5) improve the capital budget process by focusing on long-term needs relative to the acquisition of medical equipment [10].

The process, in general, and the collection of up-to-date pertinent information, in particular, require the expenditure of certain resources and the active participation of designated hospital staff in networks providing technology assessment information. For example, corporate membership in organizations and societies that provide such information needs to be considered, as well as subscriptions to certain computerized database and printed sources [11].

At the example hospital, and MTAC was formed to conduct technology assessment. It was chaired by the director of clinical engineering. Other managers from equipment user departments usually serve as the MTAC's designated technical coordinators for specific task forces. Once the committee accepted a request from an individual user, it identified other users that might have an interest in that equipment or system and authorized the technical coordinator to assemble a task force consisting of users identified by the MTAC. This task force then took responsibility for the establishment of performance criteria that would be used during this particular assessment. The task force also should answer the questions of effectiveness, safety, and cost-effectiveness as they relate to the particular assessment. During any specific period, there may be multiple task forces, each focusing on a specific equipment investigation.

The task force technical coordinator cooperates with the material management department in conducting a market survey, in obtaining the specified equipment for evaluation purposes, and in scheduling vendor-provided in-service training. The coordinator also confers with clinical staff to determine if they have experience with the equipment and the maturity level of the equipment under assessment. After establishment of a task force, the MTACs technical coordinator is responsible for analyzing the clinical experiences associated with the use of this equipment, for setting evaluation objectives, and for devising appropriate technical tests in accord with recommendations from the task force. Only equipment that successfully passes the technical tests will proceed to a clinical trial. During the clinical trial, a task force-appointed clinical coordinator collects and reports a summary of experiences gained. The technical coordinator then combines the results from both the technical tests and the clinical trial into a summary report for MTAC review and approval. In this role, the clinical engineer/technical coordinator serves as a multidisciplinary professional, bridging the gap between the clinical and technical needs of the hospital. To complete the process, financial staff representatives review the protocol.

The technology assessment process at this example hospital begins with a department or individual filling out two forms: (1) a request for review (RR) form and (2) a capital asset request (CAR) form. These forms are submitted to the hospital's product standards committee, which determines if an assessment process is to be initiated, and the priority for its completion. It also determines if a previously established standard for this equipment already exists (if the hospital is already using such a technology)—if so, an assessment is not needed.

On the RR, the originator delineates the rationale for acquiring the medical device. For example, the originator must tell how the item will improve quality of patient care, who will be its primary user, and how it will improve ease of use. On the CAR, the originator describes the item, estimates its cost, and provides purchase justification. The CAR is then routed to the capital budget office for review. during this process, the optimal financing method for acquisition is determined. If funding is secured, the CAR is routed to the material management department, where, together with the RR, it will be processed. The rationale for having the RR accompany the CAR is to ensure that financial information is included as part of the assessment process. The CAR is the tool by which the purchasing department initiates a market survey and later sends product requests for bid. Any request for evaluation that is received without a CAR or any CAR involving medical equipment that is received without a request for evaluation is returned to the originator without action. Both forms are then sent to the clinical engineering department, where a designated technical coordinator will analyze the requested technology maturity level and results of clinical experience with its use, review trends, and prioritize various manufactures' presentations for MTAC review.

Both forms must be sent to the MTAC if the item requested is not currently used by the hospital or if it does not conform to previously adopted hospital standards. The MTAC has the authority to recommend either acceptance or rejection of any request for review, based on a consensus of its members. A task force consisting of potential equipment users will determine the "must have" equipment functions, review the impact of the various equipment configurations, and plan technical and clinical evaluations.

If the request is approved by the MTAC, the requested technology or equipment will be evaluated using technical and performance standards. Upon completion of the review, a recommendation is returned to the hospital's products standard committee, which reviews the results of the technology assessment, determines whether the particular product is suitable as a hospital standard, and decides if its should be purchased. If approved, the request to purchase will be reviewed by the capital budget committee (CBC) to determine if the required expenditure meets with available financial resources and if or when it may be feasible to make the purchase. To ensure coordination of the technology assessment program, the chairman of the MTAC also serves as a permanent member of the hospital's CBC. In this way, there is a planned integration between technology assessment and budget decisions.

168.4 Equipment Assets Management

An accountable, systemic approach will ensure that cost-effective, efficacious, safe, and appropriate equipment is available to meet the demands of quality patient care. Such an approach requires that existing medical equipment resources be managed and that the resulting management strategies have measurable outputs that are monitored and evaluated. Technology managers/clinical engineers are well positioned to organize and lead this function. It is assumed that cost accounting is managed and monitored by the health care organization's financial group.

Equipment Management Process

Through traditional assets management strategies, medical equipment can be comprehensively managed by clinical engineering personnel. First, the management should consider a full range of strategies for equipment technical support. Plans may include use of a combination of equipment service providers such as manufacturers, third-party service groups, shared services, and hospital-based (in-house) engineers and biomedical equipment technicians (BMETs). All these service providers should be under the general responsibility of the technology manager to ensure optimal equipment performance through comprehensive and ongoing best-value equipment service. After obtaining a complete hospital medical equipment inventory (noting both original manufacturer and typical service provider), the management should conduct a thorough analysis of hospital accounts payable records for at least the past 2 years, compiling all service reports and preventative maintenance-related costs from all possible sources. The manager then should document in-house and external provider equipment service costs, extent of maintenance coverage for each inventory time, equipment-user operating schedule, quality of maintenance coverage for each item, appropriateness of the service provider, and reasonable maintenance costs. Next, he or she should establish an effective equipment technical support process. With an accurate inventory and best-value service providers identified, service agreements/contracts should be negotiated with external providers using prepared terms and conditions, including a log-in system. There should be an in-house clinical engineering staff ensuring ongoing external provider cost control utilizing several tools. By asking the right technical questions and establishing friendly relationships with staff, the manager will be able to handle service purchase orders (POs) by determining if equipment is worth repairing and obtaining exchange prices for parts. The staff should handle service reports to review them for accuracy and proper use of the log-in system. They also should match invoices with the service reports to verify opportunities and review service histories to look for symptoms such as need for user training, repeated problems, run-on calls billed months apart, or evidence of defective or worn-out equipment. The manager should take responsibility for emergency equipment rentals. Finally, the manager should develop, implement, and monitor all the service performance criteria.

To optimize technology management programs, clinical engineers should be willing to assume responsibilities for technology planning and management in all related areas. They should develop policies and procedures for their hospital's management program. With life-cycle costs determined for key high-risk or high-cost devices, they should evaluate methods to provide additional cost savings in equipment operation and maintenance. They should be involved with computer networking systems within the hospital. As computer technology applications increase, the requirements to review technology-related information in a number of hospital locations will increase. They should determine what environmental conditions and facility changes are required to accommodate new technologies or changes in standards and guidelines. Lastly, they should use documentation of equipment performance and maintenance costs along with their knowledge of current clinical practices to assist other hospital personnel in determining the best time and process for planning equipment replacement [12].

Technology Management Activities

A clinical engineering department, through outstanding performance in traditional equipment management, will win its hospital's support and will be asked to be involved in a full range of technology management activities. The department should start an equipment control program that encompasses routine performance testing, inspection, periodic and preventive maintenance, on-demand repair services, incidents investigation, and actions on recalls and hazards. The department should have multidisciplinary involvement in equipment acquisition and replacement decisions, development of new services, and planning of new construction and major renovations, including intensive participation by clinical engineering, materials management, and finance. The department also should initiate programs for training all users of patient care equipment, quality improvement (QI), as it relates to technology use, and technology-related risk management [13].

Case Study: A Focus on Medical Imaging

In the mide-1980s, a large private multihospital system contemplated the startup of a corporate clinical engineering program. The directors recognized that involvement in a diagnostic imaging equipment service would be key to the economic success of the program. They further recognized that maintenance cost reductions would have to be balanced with achieving equal or increased quality of care in the utilization of that equipment.

Programs startup was in the summer of 1987 in 3 hospitals that were geographically close. Within the first year, clinical engineering operations began in 11 hospitals in 3 regions over a two-state area. By the fall of 1990, the program included 7 regions and 21 hospitals in a five-state area. The regions were organized, typically, into teams including a regional manager and 10 service providers, serving 3 to 4 hospitals, whose average size was 225 beds. Although the staffs were stationed at the hospitals, some specialists traveled between sites in the region to provide equipment service. Service providers included individuals specializing in the areas of diagnostic imaging [x-ray and computed tomography (CT)], clinical laboratory, general biomedical instrumentation, and respiratory therapy.

At the end of the first 18 months, the program documented over $1 million in savings for the initial 11 hospitals, a 23% reduction from the previous annual service costs. Over 63% of these savings were attributable to "in-house" service x-ray and CT scanner equipment. The mix of equipment maintained by 11 imagining service providers—from a total staff of 30—included approximately 75% of the radiology systems of any kind found in the hospitals and 5 models of CT scanners from the three different manufacturers.

At the end of 3 years in 1990, program-wide savings had exceeded 30% of previous costs for participating hospitals. Within the imaging areas of the hospitals, savings approached and sometimes exceed 50% of initial service costs. The 30 imaging service providers—out of a total staff of 62—had increased their coverage of radiology equipment to over 95%, had increased involvement with CT to include nine models from five different manufacturers, and had begun in-house work in other key imaging modalities.

Tracking the financial performance of the initial 11 hospitals over the first 3 years of the program yields of the following composite example: A hospital of 225 beds was found to have equipment service costs of $540,000 prior to program startup. Sixty-three percent of these initial costs (or $340,000) was for the maintenance of the hospital's x-ray and CT scanner systems. Three years later, annual service costs for this equipment were cut in half, to approximately $170,000. That represents a 31% reduction in hospital-wide costs due to the imaging service alone.

This corporate clinical engineering operation is, in effect, a large in-house program serving many hospitals that all have common ownership. The multihospital corporation has significant purchasing power in the medical device marketplace and provides central oversight of the larger capital expenditures for its hospitals. The combination of the parent organization's leverage and the program's commitment to serve only hospitals in the corporation facilitated the development of positive relationships with medical device manufacturers. Most of the manufacturers did not see the program as competition but rather as a potentially helpful ally in the future marketing and sales of their equipment and systems. What staff provided these results? All service providers were either medical imaging industry or military trained. All were experienced at troubleshooting electronic subsystems to component level, as necessary. Typically, these individuals had prior experience on the manufacture's models of equipment under their coverage. Most regional managers had prior industry, third party, or in-house imaging service management experience. Each service provider had the test equipment necessary for day-to-day duties. Each individual could expect at least 2 weeks of annual service training to keep appropriate skills current. Desired service training could be acquired in a timely manner from manufactures and/or third-party organizations. Spare or replacement parts inventory was minimal because of the program's ability to get parts from manufacturers and other sources either locally or shipped in overnight.

As quality indicators for the program, the management measured user satisfaction, equipment downtime, documentation of technical staff service training, types of user equipment errors and their effect on patient outcomes, and regular attention to hospital technology problems. User satisfaction surveys indicated a high degree of confidence in the program service providers by imaging department mangers. Problems relating to technical, management, communication, and financial issues did occur regularly, but the regional manager ensured that they were resolved in a timely manner. Faster response to daily imaging equipment problems, typically by on-site service providers, coupled with regular preventive maintenance (PM) according to established procedures led to reduced equipment downtime. PM and repair service histories were captured in a computer documentation system that also tracked service times, costs, and user errors and their effects. Assisting the safety committee became easier with ability to draw a wide variety of information quickly from the program's documenting system.

Early success in imaging equipment led to the opportunity to do some additional value-added projects such as the moving and reinstallation of x-ray rooms that preserved exiting assets and opened up valuable space for installation of newer equipment and upgrades of CT scanner systems. The parent organization came to realize that these technology management activities could potentially have a greater financial and quality impact on the hospital's health care delivery than equipment management. In the example of one CT upgrade (which was completed over two weekends with no downtime), there was a positive financial impact in excess of $600,000 and improved quality of care by allowing faster off-line diagnosis of patient scans. However, opportunity for this kind of contribution would never have occurred without the strong base of a successful equipment management program staffed with qualified individuals who receive ongoing training.

168.5 Equipment Acquisition and Deployment

Process of Acquiring Technology

Typically, medical device systems will emerge from the strategic technology planning and technology assessment processes as required and budgeted needs. At acquisition time, a needs analysis should be conducted, reaffirming clinical needs and device intended applications. The "request for review" documentation from the assessment process or capital budget request and incremental financial analysis from the planning process may provide appropriate justification information, and a capital asset request (CAR) form should be completed [14]. Materials management and clinical engineering personnel should ensure that this item is a candidate for centralized and coordinated acquisition of similar equipment with other hospital departments. Typical hospital prepurchase evaluation guidelines include an analysis of needs and development of a specification list, formation of a vendor list and requesting proposals, analyzing proposals and site planning, evaluating samples, selecting finalists, making the award, delivery and installation, and acceptance testing. Formal request for proposals (RFPs) from potential equipment vendors are required for intended acquisitions whose initial or life-cycle cost exceeds a certain threshold, i.e., $100,000. Finally, the purchase takes place, wherein final equipment negotiations are conducted and purchase documents are prepared, including a purchase order.

Acquisition Process Strategies

The cost-of-ownership concept can be used when considering what factors to include in cost comparisons of competing medical devices. Cost of ownership encompasses all the direct and indirect expenses associated with medical equipment over its lifetime [15]. It expresses the cost factors of medical equipment for both the initial price of the equipment (which typically includes the equipment, its installation, and initial training cost) and over the long term. Long-term costs include ongoing training, equipment service, supplies, connectivity, upgrades, and other costs. Health care organizations are just beginning to account for a full range of cost-of-ownership factors in their technology assessment and acquisition processes, such as acquisition costs, operating costs, and maintenance costs (installation, supplies, downtime, training, spare parts, test equipment and tools, and depreciation). It is estimated that the purchase price represents only 20% of the life-cycle cost of ownership.

When conducting needs analysis, actual utilization information form the organization's existing same or similar devices can be very helpful. One leading private multihospital system has implemented the following approach to measuring and developing relevant management feedback concerning equipment utilization. It is conducting equipment utilization review for replacement planning, for ongoing accountability of equipment use, and to provide input before more equipment is purchased. This private system attempts to match product to its intended function and to measure daily (if necessary) the equipment's actual utilization. The tools they use include knowing their hospital's entire installed base of certain kinds of equipment, i.e., imaging systems. Utilization assumptions for each hospital and its clinical procedural mix are made. Equipment functional requirements to meet the demands of the clinical procedures are also taken into account.

Life-cycle cost analysis is a tool used during technology planning, assessment, or acquisition "either to compare high-cost, alternative means for providing a service or to determine whether a single project or technology has a positive or negative economic value. The strength of the life-cycle cost analysis is that it examines the cash flow impact of an alternative over its entire life, instead of focusing solely on initial capital investments" [15].

"Life-cycle cost analysis facilitates comparisons between projects or technologies with large initial cash outlays and those with level outlays and inflows over time. It is most applicable to complex, high-cost choices among alternative technologies, new service, and different means for providing a given service. Life-cycle cost analysis is particularly useful for decisions that are too complex and ambiguous for experience and subjective judgment alone. It also helps decision makers perceive and include costs that often are hidden or ignored, and that may otherwise invalidate results" [12].

"Perhaps the most powerful life-cycle cost technique is net present value (NPV) analysis, which explicitly accounts for inflation and foregone investment opportunities by expressing future cash flows in present dollars" [12].

Examples where LCC and NPV analysis prove very helpful are in deciding whether to replace/rebuild or buy/lease medical imaging equipment. The kinds of costs captured in life-cycle cost analysis, include decision-making costs, planning agency/certificate of need costs (if applicable), financing, initial capital investment costs including facility changes, life-cycle maintenance and repairs costs, personnel costs, and other (reimbursement consequences, resale, etc.).

One of the best strategies to ensure that a desired technology is truly of value to the hospital is to conduct a careful analysis in preparation for its assimilation into hospital operations. The process of equipment prepurchase evaluation provides information that can be used to screen unacceptable performance by either the vendor or the equipment before it becomes a hospital problem.

Once the vendor has responded to informal requests or formal RFPs, the clinical engineering department should be responsible for evaluating the technical response, while the materials management department should devaluate the financial responses.

In translating clinical needs into a specification list, key features or "must have" attributes of the desired device are identified. In practice, clinical engineering and materials management should develop a "must have" list and an extras list. The extras list contains features that may tip the decision in favor of one vendor, all other factors being even. These specification lists are sent to the vendor and are effective in a self-elimination process that results in a time savings for the hospital. Once the "must have" attributes have been satisfied, the remaining candidate devices are evaluated technically, and the extras are considered. This is accomplished by assigning a weighting factor (i.e., 0 to 5) to denote the relative importance of each of the desired attributes. The relative ability of each device to meet the defined requirements is then rated [15].

One strategy that strengthens the acquisition process is the conditions-of-sale document. This multi-faceted document integrates equipment specifications, performance, installation requirements, and follow-up services. The conditions-of-sale document ensures that negotiations are completed before a purchase order is delivered and each participant is in agreement about the product to be delivered. As a document of compliance, the conditions-of-sale document specifies the codes and standards having jurisdiction over that equipment. This may include provisions for future modification of the equipment, compliance with standards under development, compliance with national codes, and provision for software upgrades.

Standard purchase orders that include the conditions of sale for medical equipment are usually used to initiate the order. At the time the order is placed, clinical engineering is notified of the order. In addition to current facility conditions, the management must address installation and approval requirements, responsibilities, and timetable; payment, assignment, and cancellation; software requirements and updates; documentation; clinical and technical training; acceptance testing (hospital facility and vendor); warranty, spare parts, and service; and price protection.

All medical equipment must be inspected and tested before it is placed into service regardless of whether it is purchased, leased, rented, or borrowed by the hospital. In any hospital, clinical engineering should receive immediate notification if a very large device or system is delivered directly into another department (e.g., imaging or cardiology) for installation. Clinical engineering should be required to sign off on all purchase orders for devices after installation and validation of satisfactory operation. Ideally, the warranty period on new equipment should not begin until installation and acceptance testing are completed. It is not uncommon for a hospital to lose several months of free parts and service by the manufacturer when new equipment is, for some reason, not installed immediately after delivery.

Clinical Team Requirements

During the technology assessment and acquisition processes, clinical decision makers analyze the following criteria concerning proposed technology acquisitions, specifically as they relate to clinical team requirements: ability of staff to assimilate the technology, medical staff satisfaction (short term and long term), impact on staffing (numbers, functions), projected utilization, ongoing related supplies required, effect on delivery of care and outcomes (convenience, safety, or standard of care), result of what is written in the clinical practice guidelines, credentialling of staff required, clinical staff initial and ongoing training required, and the effect on existing technology in the department or on other services/departments.

Defining Terms

Appropriate technology [1]: A term used initially in developing countries, referring to selecting medical equipment that can "appropriately" satisfy the following constraints: funding shortages, insufficient numbers of trained personnel, lack of technical support, inadequate supplies of consumables/accessories, unreliable water an power utilities/supplies, and lack of operating and maintenance manuals. In the context of this chapter, appropriate technology selection must taken into consideration local health needs and disease prevalence, the need for local capability of equipment maintenance, and availability of resources for ongoing operational and technical support.

Clinical engineers/biomedical engineers: As we began describing the issues with the management of medical technology; it became obvious that some of the terms are being used interchangeably in the literature. For example, the terms *engineers, clinical engineers, biomedical equipment technicians,* *equipment managers,* and *health care engineers* are frequently used. For clarification, in this chapter we will refer to clinical engineers and the clinical engineering department as a representative group for all these terms.

Cost-effectiveness [1]: A mixture of quantitative and qualitative considerations. It includes the health priorities of the country or region at the macro assessment level and the community needs at the institution micro assessment level. Product life-cycle cost analysis (which, in turn, includes initial purchase price, shipping, renovations, installation, supplies, associated disposables, cost per use, and similar quantitative measures) is a critical analysis measure. Life-cycle cost also takes into account staff training, ease of use, service, and many other cost factors. But experience and judgement about the relative importance of features and the ability to fulfill the intended purpose also contribute critical information to the cost-effectiveness equation.

Equipment acquisition and deployment: Medical device systems and products typically emerge from the strategic technology planning process as "required and budgeted" needs. The process that follows, which ends with equipment acceptance testing and placement into general use, is known as the *equipment acquisition and deployment process.*

Health care technology: Health care technology includes the devices, equipment, systems, software, supplies, pharmaceuticals, biotechnologies, and medical and surgical procedures used in the prevention, diagnosis, and treatment of disease in humans, for their rehabilitation, and for assistive purposes. In short, technology is broadly defined as encompassing virtually all the human interventions intended to cope with disease and disabilities, short of spiritual alternatives. This chapter focuses on medical equipment products (devices, systems, and software) rather than pharmaceuticals, biotechnologies, or procedures [1]. The concept of technology also encompasses the facilities that house both patients and products. Facilities cover a wide spectrum—from the modern hospital on one end to the mobile imaging trailer on the other.

Quality of care (QA) and quality of improvement (QI): Quality assurance (QA) and Quality improvement (QI) are formal sets of activities to measure the quality of care provided; these usually include a process for selecting, monitoring, and applying corrective measures. The 1994 Joint Commission on the Accreditation of Healthcare Organizations (JCAHO) standards require hospital QA, programs to focus on patient outcomes as a primary reference. JCAHO standards for plant, technology, and safety management (PTSM), in turn, require certain equipment management practices and QA or QI activities. Identified QI deficiencies may influence equipment planning, and QI audits may increase awareness of technology overuse or under utilization.

Risk management: Risk management is a program that helps the hospital avoid the possibility of risks, minimize liability exposure, and stay compliant with regulatory reporting requirements. JCAHO PTSM standards require minimum technology-based risk-management activities. These include clinical engineering's determination of technology-related incidents with follow-up steps to prevent recurrences and evaluation and documentation of the effectiveness of these steps.

Safety: Safety is the condition of being safe from danger, injury, or damage. It is judgment about the acceptability of risk in a specified situation (e.g., for a given medical problem) by a provider with specified training at a specified type of facility equipment.

Standards [1]: A wide variety of formal standards and guidelines related to health care technology now exists. Some standards apply to design, development, and manufacturing practices for devices, software, and pharmaceuticals; some are related to the construction and operation of a health care facility; some are safety and performance requirements for certain classes of technologies, such as standards related to radiation or electrical safety; and others relate to performance, or even construction specifications, for specific types of technologies. Other standards and guidelines deal with administrative, medical, and surgical procedures and the training of clinical personnel. Standards and guidelines are produced and/or adopted by government agencies, international organizations, and professional and specialty organizations and societies. ECRI's *Healthcare Standards Directory* lists over 20,000 individual standards and guidelines produced by over 600 organizations and agencies from North America alone.

Strategic technology planning: Strategic technology planning encompasses both technologies new to the hospital and replacements for existing equipment that are to be acquired over several quarters. Acquisitions can be proposed for reasons related to safety, standard-of-care issues, and age or obsolescence of existing equipment. Acquisitions also can proposed to consolidate several service areas, expand a service areas to reduce cost of service, or add a new service area.

Strategic technology panning optimizes the way the hospital's capital resources contribute to its mission. It encourages choosing new technologies that are cost-effective, and it also allows the hospital to be competitive in offering state-of-the-art services. Strategic technology planning works for a single department, product line, or clinical service. It can be limited to one or several high-priority areas. It also can be used for an entire mulihospital system or geographic region [4].

Technology assessment: Assessment of medical technology is any process used for examining and reporting properties of medical technology used in health care, such as safety, efficacy, feasibility, and indications for use, cost, and cost-effectiveness, as well as social, economic, and ethical consequences, whether intended or unintended [2]. A primary technology assessment is one that seeks new, previously nonexistent data through research, typically employing long-term clinical studies of the type described below. A secondary technology assessment is usually based on published data, interviews, questionnaires, and other information-gathering methods rather than original research that creates new, basic data.

In technology assessment, there are six basic objectives that the clinical engineering department should have in mind. First, there should be ongoing monitoring of developments concerning new and emerging technologies. For new technologies, there should be an assessment of the clinical efficacy, safety, and cost/benefit ratio, including their effects on established technologies. There should be an evaluation of the short- and long-term costs and benefits of alternate approaches to managing specific clinical conditions. The appropriateness of existing technologies and their clinical uses should be estimated, while outmoded technologies should be identified and eliminated from their duplicative uses. The department should rate specific technology-based interventions in terms of improved overall value (quality and outcomes) to patients, providers, and payers. Finally, the department should facilitate a continuous uniformity between needs, offerings, and capabilities [3].

The locally based (hospital or hospital group) technology assessment described in this chapter is a process of secondary assessment that attempts to judge whether a certain medical equipment/product can be assimilated into the local operational environment.

Technology diffusion [1]: The process by which a technology is spread over time in a social system. The progression of technology diffusion can be described in four stages. The *emerging* or applied research stage occurs around the time of initial clinical testing. In the *new* stage, the technology has passed the phase of clinical trials but is not yet in widespread use. During the *established* stage, the technology is considered by providers to be a standard approach to a particular condition and diffuses into general use. Finally, in the *obsolete/outmoded* stage, the technology is superseded by another and/or is demonstrated to be ineffective or harmful.

Technology life cycle: Technology has a life cycle—a process by which technology is crated, tested, applied, and replaced or abandoned. Since the life cycle varies from basic research and innovation to obsolescence and abatement, it is critical to know the maturity of a technology prior to making decisions regarding its adoption. Technology forecast assessment of pending technological changes are the investigative tools that support systematic and rational decisions about the utilization of a given institution's technological capabilities.

Technology planning and management [3]: Technology planning and management are an accountable, systematic approach to ensuring that cost-effective, efficacious, appropriate, and safe equipment is available to meet the demands of quality patient care and allow an institution to remain competitive. Elements include in-house service management, management and analysis of equipment external service providers, involvement in the equipment acquisition process, involvement of appropriate hospital personnel in facility planning and design, involvement in reducing technology-related patient and staff incidents, training equipment users, reviewing equipment replacement needs, and ongoing assessment of emerging technologies [4].

References

1. ECRI. Healthcare Technology Assessment Curriculum. Philadelphia, August 1992.
2. Banata HD, Institute of Medicine. Assessing Medical Technologies. Washington, National Academy Press, 1985.
3. Lumsdon K. Beyond technology assessment: Balancing strategy needs, strategy. Hospitals 15:25, 1992.
4. ECRI. Capital, Competition, and Constraints: Managing Healthcare in the 1990s. A Guide for Hospital Executives. Philadelphia, 1992.
5. Berkowtiz DA, Solomon RP. Providers may be missing opportunities to improve patient outcomes. Costs, Outcomes Measurement and Management May-June: 7, 1991.
6. ECRI. Regional Healthcare Technology Planning and Management Program. Philadelphia, 1990.
7. Sprague GR. Managing technology assessment and acquisition. Health Exec 6:26, 1988.
8. David Y. Technology-related decision-making issues in hospitals. In IEEE Engineering in Medicine and Biology Society Proceedings of the 11th Annual International Conference, 1989.
9. Wagner M. Promoting hospitals high-tech equipment. Mod Healthcare 46, 1989.
10. David Y. Medical Technology 2001. CPA Healthcare Conference, 1992.
11. ECRI. Special Report on Technology Management, Health Technology. Philadelphia, 1989.
12. ECRI. Special Report on Devices and Dollars, Philadelphia, 1988.
13. Gullikson ML, David Y, Brady MH. An automated risk management tool. JCAHO, Plant, Technology and Safety Management Review, PTSM Series, no 2, 1993.
14. David Y, Judd T, ECRI. Special Report on Devices and Dollars, Philadelphia, 1988. Medical Technology Management. SpaceLabs Medical, Inc., Redmond, Wash 1993.
15. Bronzino JD (ed). Management of Medical Technology: A Primer for Clinical Engineers. Stoneham, Mass, Butterworth, 1992.
16. David Y. Risk Measurement For Managing Medical Technology. Conference Proceedings, PERM-IT 1997, Australia.

169

Risk Factors, Safety, and Management of Medical Equipment

Michael L. Gullikson
Texas Children's Hospital

169.1 Risk Management: A Definition

Inherent in the definition of *risk management* is the implication that the hospital environment cannot be made risk-free. In fact, the nature of medical equipment—to invasively or noninvasively perform diagnostic, therapeutic, corrective, or monitoring intervention on behalf of the patient—implies that risk is present. Therefore, a standard of acceptable risk must be established that defines *manageable risk* in a real-time economic environment.

Unfortunately, a preexistent, quantitative standard does not exist in terms of, for instance, mean time before failure (MTBF), number of repairs or repair redos per equipment item, or cost of maintenance that provides a universal yardstick for risk management of medical equipment. Sufficient clinical management of risk must be in place the can utilize safeguards, preventive maintenance, and failure analysis information to minimize the occurrence of injury or death to patient or employee or property damage. Therefore, a process must be put in place that will permit analysis of information and modification of the preceding factors to continuously move the medical equipment program to a more stable level of manageable risk.

Risk factors that require management can be illustrated by the example of the "double-edge" sword concept of technology (see Fig. 169.1). The front edge of the sword represents the cutting edge of technology and its beneficial characteristics: increased quality, greater availability of technology, timeliness of test results and treatment, and so on. The back edge of the sword represents those liabilities which must be addressed to effectively manage risk: the hidden costs discussed in the next paragraph, our dependence on technology, incompatibility of equipment, and so on [1].

For example, the purchase and installation of a major medical equipment item may only represent 20% of the lifetime cost of the equipment [2]. If the operational budget of a nursing floor does not include the other 80% of the equipment costs, the budget constraints may require cutbacks where they appear to minimally affect direct patient care. Preventive maintenance, software upgrades that address "glitches," or overhaul requirements may be seen as unaffordable luxuries. Gradual equipment deterioration without maintenance may bring the safety level below an acceptable level of manageable risk.

0-8493-0462-8/00/$0.00+$.50

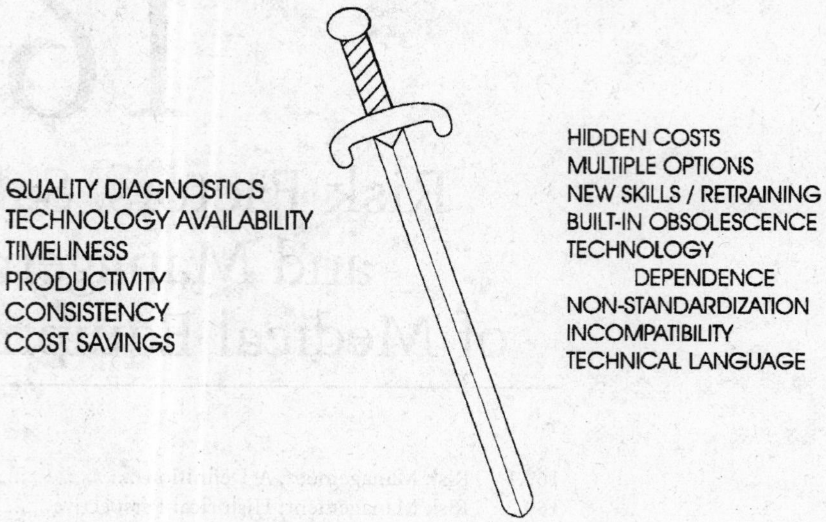

FIGURE 169.1 Double-edged sword concept of risk management.

Since economic factors as well as those of safety must be considered, a balanced approach to risk management that incorporates all aspects of the medical equipment lifecycle must be considered.

The operational flowchart in Fig. 169.2 describe the concept of medical equipment life-cycle management from the clinical engineering department viewpoint. The flowchart includes planning, evaluation, and initial purchase documentation requirements. The condition of sale, for example, ensures that technical manuals, training, replacement parts, etc. are received so that all medical equipment might be fully supported in-house after the warranty period. Introduction to the preventive maintenance program, unscheduled maintenance procedures, and retirement justification must be part of the process. Institutional-wide cooperation with the life-cycle concept requires education and patience to convince health care providers of the team approach to managing medical equipment technology.

This balanced approach requires communication and comprehensive planning by a health care team responsible for evaluation of new and shared technology within the organization. A medical technology evaluation committee (see Fig. 169.3), composed of representatives from administration, medical staff, nursing, safety department, biomedical engineering, and various services, can be an effective platform for the integration of technology and health care. Risk containment is practiced as the committee reviews not only the benefits of new technology but also the technical and clinical liabilities and provides a 6-month followup study to measure the effectiveness of the selection process. The history of risk management in medical equipment management provides helpful insight into its current status and future direction.

169.2 Risk Management: Historical Perspective

Historically, risk management of medical equipment was the responsibility of the clinical engineer (Fig. 169.4). The engineer selected medical equipment based on individual clinical department consultations and established preventive maintenance (PM) programs based on manufacturer's recommendation and clinical experience. The clinical engineer reviewed the documentation and "spot-checked" equipment used in the hospital. The clinical engineer met with biomedical supervisors and technicians to discuss PM completion and to resolve repair problems. The clinical engineer then attempted to analyze failure information to avoid repeat failure.

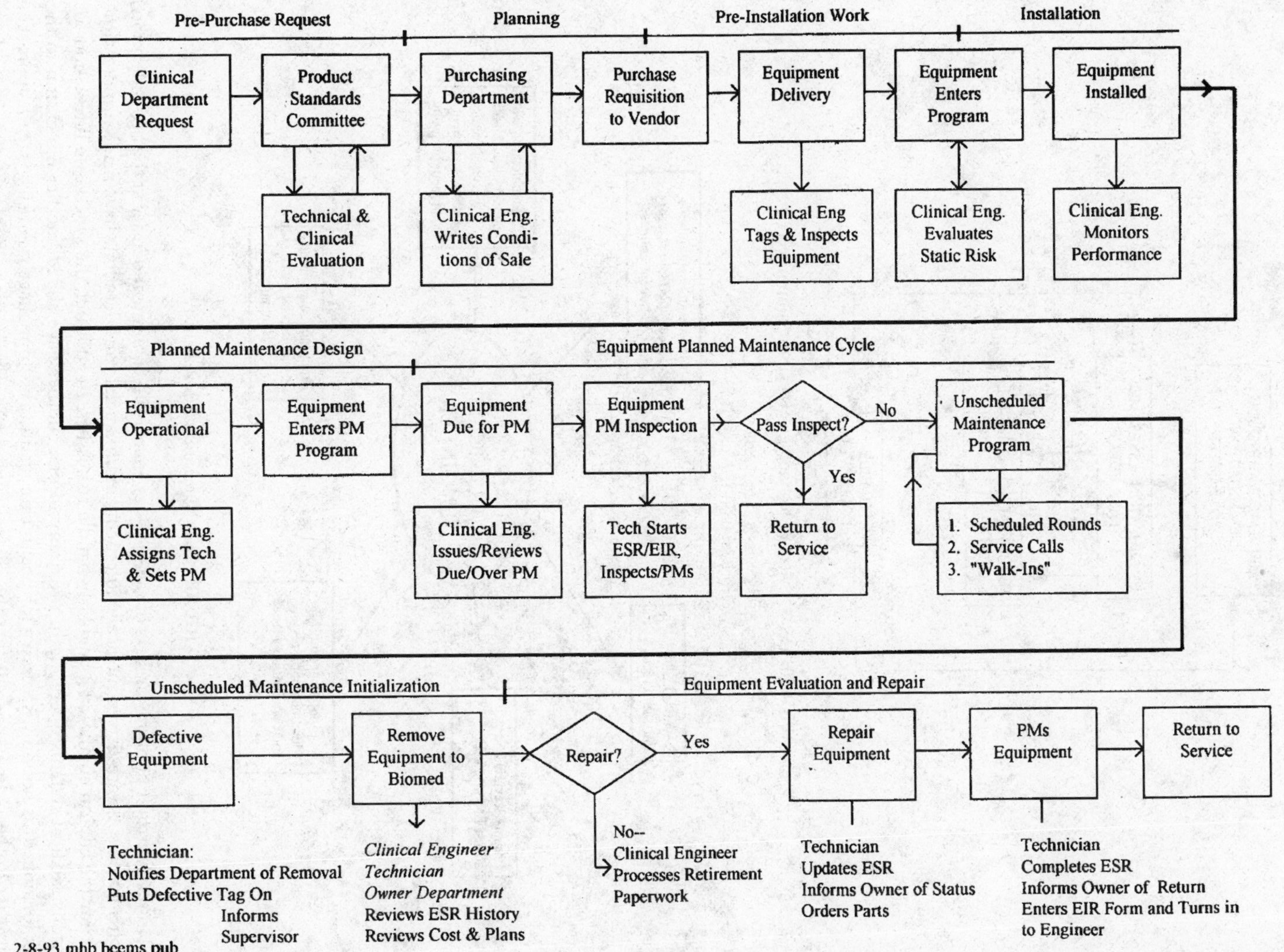

FIGURE 169.2 Biomedical engineering equipment management system (BEEMS).

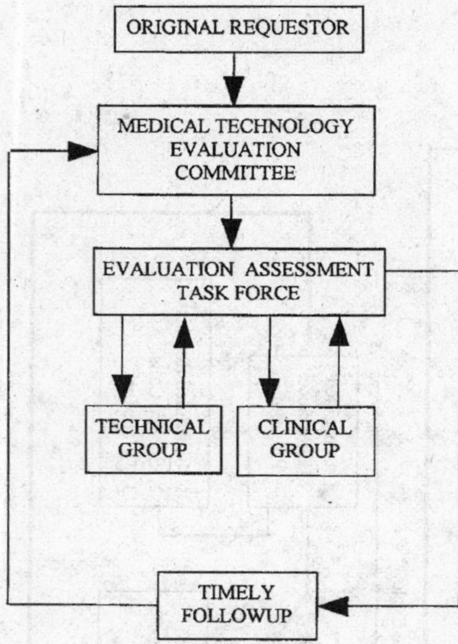

FIGURE 169.3 Medical technology evaluation committee.

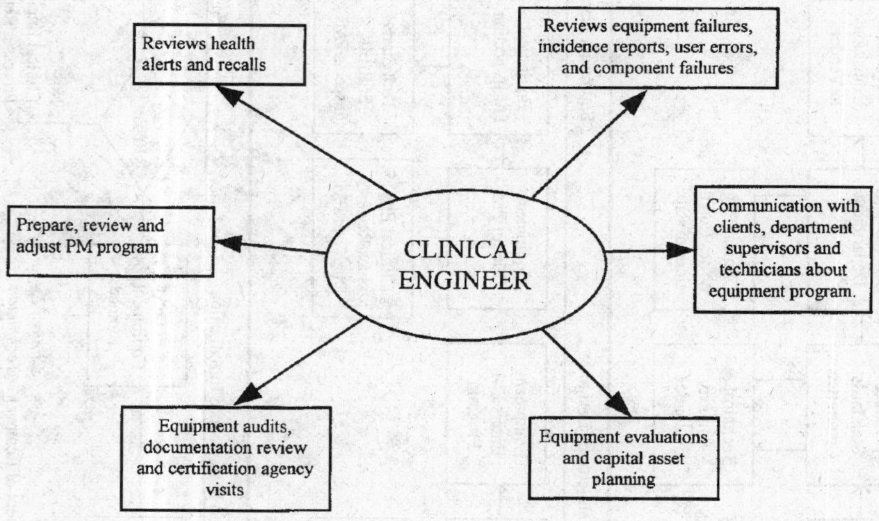

FIGURE 169.4 Operational flowchart.

However, greater public awareness of safety issues, increasing equipment density at the bed-side, more sophisticated software-driven medical equipment, and financial considerations have made it more difficult for the clinical engineer to singularly handle risk issues. In addition, the synergistic interactions of various medical systems operating in proximity to one another have added another dimension to the risk formula. It is not only necessary for health care institutions to manage risk using a team approach, but it is also becoming apparent that the clinical engineer requires more technology-intensive tools to effectively contribute to the team effort [3].

100	Medical Equipment Operator Error
101	Medical Equipment Failure
102	Medical Equipment Physical Damage
103	Reported Patient Injury
104	Reported Employee Injury
105	Medical Equipment Failed PM
108	Medical Equipment MBA

FIGURE 169.5 Failure codes.

169.3 Risk Management: Strategies

Reactive risk management is an outgrowth of the historical attitude in medical equipment management that risk is an anomaly that surfaces in the form of a failure. If the failure is analyzed and proper operational procedures, user in-services, and increased maintenance are supplied, the problem will disappear and personnel can return to their normal work. When the next failure occurs, the algorithm is repeated. If the same equipment fails, the algorithm is applied more intensely. This is a useful but not comprehensive component of risk management in the hospital. In fact, the traditional methods of predicting the reliability of electronic equipment from field failure data have not been very effective [4]. The health care environment, as previously mentioned, inherently contains risk that must be maintained at a manageable level. A reactive tool cannot provide direction to a risk-management program, but it can provide feedback as to its efficiency.

The engine of the reactive risk-management tool is a set of failure codes (see Fig. 169.5) that flag certain anomalous conditions in the medical equipment management program. If operator training needs are able to be identified, then codes 100, 102, 104, and 108 (MBA equipment returned within 9 days for a subsequent repair) may be useful. If technician difficulties in handling equipment problems are of concern, then 108 may be of interest. The key is to develop failure codes not in an attempt to define all possible anomaly modalities but for those which *can clearly be defined and provide unambiguous direction for the correction process.* Also, the failure codes should be linked to equipment type, manufacturer/model, technician service group, hospital, and clinical department. Again, when the data are analyzed, will the result be provided to an administrator, engineer, clinical departmental director, or safety department? This should determine the format in which the failure codes are presented.

A report intended for the clinical engineer might be formatted as in Fig. 169.6. It would consist of two part, sorted by equipment type and clinical department (not shown). The engineer's report shows the failure code activity for various types of equipment and the distribution of those failure codes in clinical departments.

Additionally, fast data-analysis techniques introduced by NASA permit the survey of large quantities of information in a three-dimensional display [5] (Fig. 169.7). This approach permits viewing time-variable changes from month to month and failure concentration in specific departments and equipment types.

The importance of the format for failure modality presentation is critical to its usefulness and acceptance by health care professionals. For instance, a safety director requests the clinical engineer to provide a list of equipment that, having failed, could have potentially harmed a patient or employee. The safety director is asking the clinical engineer for a clinical judgment based on clinical as well as technical factors. This is beyond the scope of responsibility and expertise of the clinical engineer. However, the request can be addressed indirectly. The safety director's request can be addressed in two steps: first, providing a list of high-risk equipment (assessed when the medical equipment is entered into the equipment inventory) and, second, a clinical judgment based on equipment failure mode, patient condition, and so

Source	Total PM	Fail PM Items	% Fail	Reported Fail-OK	Physical Damage	Patient Injury	Employee Injury	Back Again	Equip Fail	Equip Count	% Fail
10514 PULMONARY INSTR TEXAS CHILDREN'S HOSP											
1 NON-TAGGED EQUIPMENT	0		0.00	1	5				14		0.00
1280 THERMOMETER, ELECTRONIC	2		0.00						1	99	1.01
1292 RADIANT WARMER, INFANT	1		0.00						3	63	4.76
1306 INCUBATOR, NEONATAL	0	2	0.00		7				9	56	16.07
1307 INCUBATOR, TRANSPORT, NEONATAL	9	3	33.33		1			1 *	4	9	44.44
1320 PHOTOTHERAPY UNIT, NEONATAL	4		0.00						2	28	7.14
1321 INFUSION PUMP	35		0.00	15	9			3 *	34	514	6.61
1357 SUCTION,VAC POWERED,BODY FLUID	0		0.00						4	358	1.12
1384 EXAMINATION LIGHT, AC-POWERED	11		0.00						1	47	2.13
1447 CARDIAC MONITOR W/ RATE ALARM	0		0.00	1	1			1 *	1		0.00
1567 SURGICAL NERVE STIMULATOR /LOC	0		0.00						2	15	13.33
1624 OTOSCOPE	73		0.00						1	101	0.99
1675 OXYGEN GAS ANALYZER	0		0.00						3	44	6.82
1681 SPIROMETER DIAGNOSTIC	0		0.00						1	8	12.50
1703 AIRWAY PRESSURE MONITOR	1		0.00					1 *	7	9	77.78
1735 BREATHING GAS MIXER	13		0.00					1 *	4	38	10.53
1749 HYPO/HYPERTHERMIA DEVICE	1		0.00						1	3	33.33
1762 NEBULIZER	25		0.00						1	56	1.79
1787 VENTILATOR CONTINUOUS	96	7	7.29		1			1 *	11	99	11.11
1788 VENTILATOR NONCONTINUOUS	1		0.00		1				3	27	11.11
2014 HEMODIALYSIS SYSTEM ACCESSORIE	0		0.00						3	1	300.00
2051 PERITONEAL DIALYSIS SYS & ACC	4		0.00						1	5	20.00
2484 SPECTROPHOTOMETER, MASS	0		0.00						2	3	66.67
2695 POWERED SUCTION PUMP	16		0.00		1			1 *	1	32	3.13
5028 PH METER	0		0.00		1				1	4	25.00
5035 COMPUTER & PERIPHERALS	0		0.00						3	18	16.67
5081 OXYGEN MONITOR	0		0.00	3	1				15	102	14.71
5082 RESPIRATION ANALYZER	1		0.00						1	5	20.00
5097 EXAM TABLE	75		0.00						1	86	1.16
5113 PRINTER	2		0.00						1	12	8.33
5126 ADDRESSOGRAPH	2		0.00						4	1	400.00
9102 STADIOMETER	0		0.00						1	8	12.50
17211 ANESTHESIA MONITOR	23		0.00						2	23	8.70
90063 POWER SUPPLY, PORTABLE	20		0.00						1	25	4.00
Total for TEXAS CHILDREN'S HOSP	415	12		20	28			9	144	1899	

FIGURE 169.6 Engineer's failure analysis report.

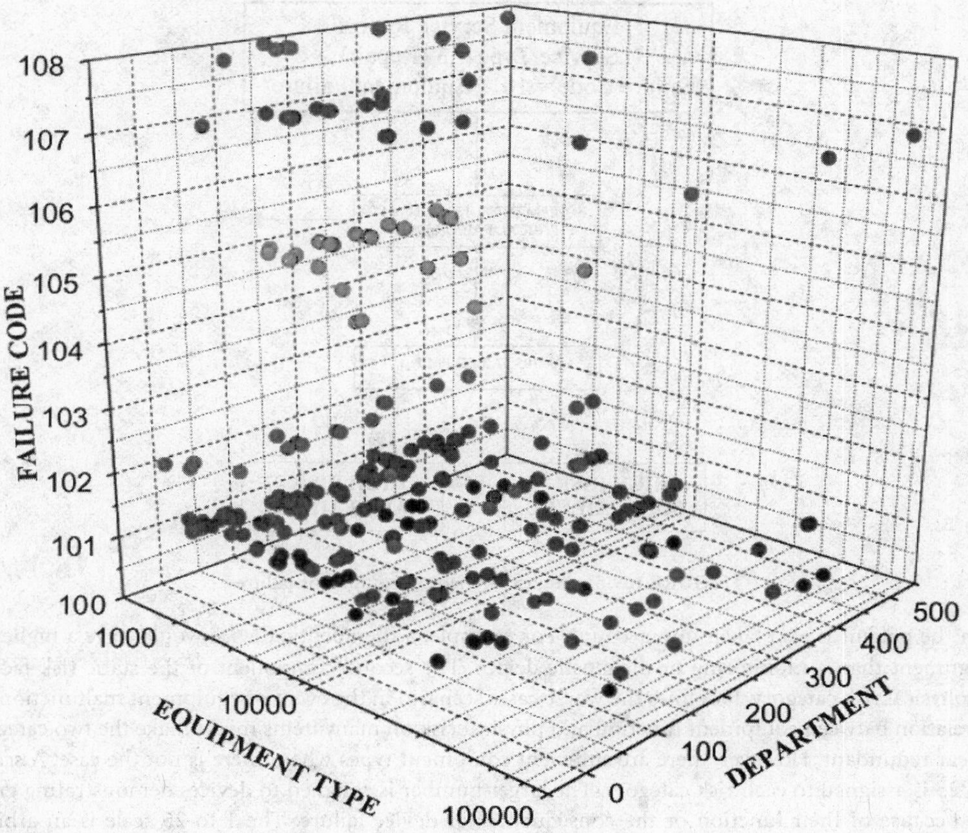

FIGURE 169.7 Failure code analysis using a 3-D display.

on. The flowchart in Fig. 169.8 provides the safety director with useful information but does not require the clinical engineer to make an unqualified clinical judgment. If the "failed PM" failure code were selected from the list of high-risk medical equipment requiring repair, the failure would be identified by the technician during routine preventive maintenance and most likely the clinician still would find the equipment clinically efficacious. This condition is a "high risk, soft failure" or a high-risk equipment item whose failure is *least* likely to cause injury. If the "failed PM" code were not used, the clinician would question the clinical efficacy of the medical equipment item, and the greater potential for injury would be identified by "high risk, hard failure." Monitoring the distribution of high-risk equipment in these two categories assists the safety director in managing risk.

Obviously, a more forward-looking tool is needed to take advantage of the failure codes and the plethora of equipment information available in a clinical engineering department. This proactive tool should use failure codes, historical information, the "expert" knowledge of the clinical engineer, and the baseline of an established "manageable risk" environment (perhaps not optimal but stable).

The overall components and process flow for a proactive risk-management tool [6] are presented in Fig. 169.9. It consists of a two-component static risk factor, a two-component dynamic risk factor, and two "shaping" or feedback loops.

The static risk factor classifies new equipment by a generic equipment type: defibrilator, electrocardiograph, pulse oximeter, etc. When equipment is introduced into the equipment database, it is assigned to two different static risk (Fig. 169.10) categories [7]. The first is the equipment function that defines the application and environment in which the equipment item will operate. The degree of interaction

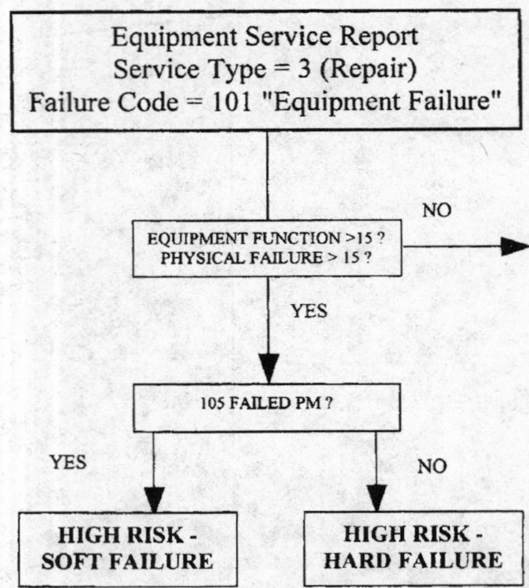

FIGURE 169.8 High-risk medical equipment failures.

with the patient is also taken into account. For example, a therapeutic device would have a higher risk assignment than a monitoring or diagnostic device. The second component of the static risk factor is the physical risk category. It defines the worst-cases scenario in the event of equipment malfunction. The correlation between equipment function and physical risk on many items might make the two categories appear redundant. However, there are sufficient equipment types where there is not the case. A scale of 1 to 25 is assigned to each risk category. The larger number is assigned to devices demonstrating greater risk because of their function or the consequences of device failure. The 1 to 25 scale is an arbitrary assignment, since a validated scale of risk factors for medical equipment, as previously described, is nonexistent. The risk points assigned to the equipment from these two categories are algebraically summed and designated the static risk factor. This value remains with the equipment type and the individual items within that equipment type permanently. Only if the equipment is used in a clinically variant way or relocated to a functionally different environment would this assignment be reviewed and changed.

The dynamic component (Fig. 169.11) of the risk-management tool consists or two parts. The first is a maintenance requirement category that is divided into 25 equally spaced divisions, ranked by least (1) to greatest (25) average manhours per device per year. These divisions are scaled by the maintenance hours for the equipment type requiring the greatest amount of maintenance attention. The amount of nonplanned (repair) manhours from the previous 12 months of service reports is totaled for each equipment type. Since this is maintenance work on failed equipment items, it correlates with the risk associated with that equipment type.

If the maintenance hours of an equipment type are observed to change to the point of placing it in a different maintenance category, a flag notifies the clinical engineer to review the equipment-type category. The engineer may increase the PM schedule to compensate for the higher unplanned maintenance hours. If the engineer believes the system "overacted," a "no" decision adjusts a scaling factor by a –5%. Progressively, the algorithm is "shaped" for the equipment maintenance program in that particular institution. However, to ensure that critical changes in the average manhours per device for each equipment type is not missed during the shaping period, the system is initialized. This is accomplished by increasing the average manhours per device for each equipment type to within 5% of the next higher maintenance requirement division. Thus the system is sensitized to variations in maintenance requirements.

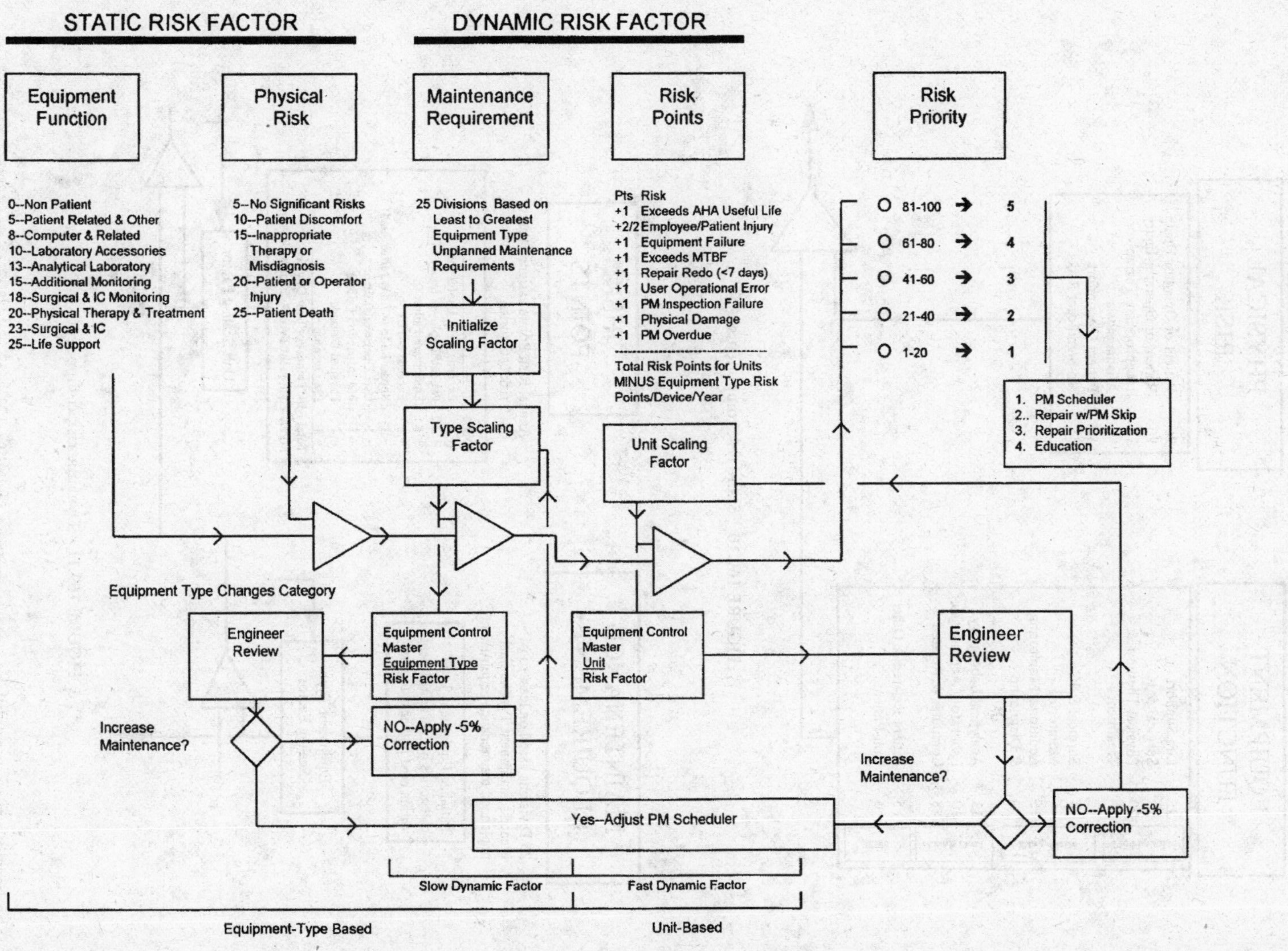

FIGURE 169.9 Biomedical engineering risk-management tool.

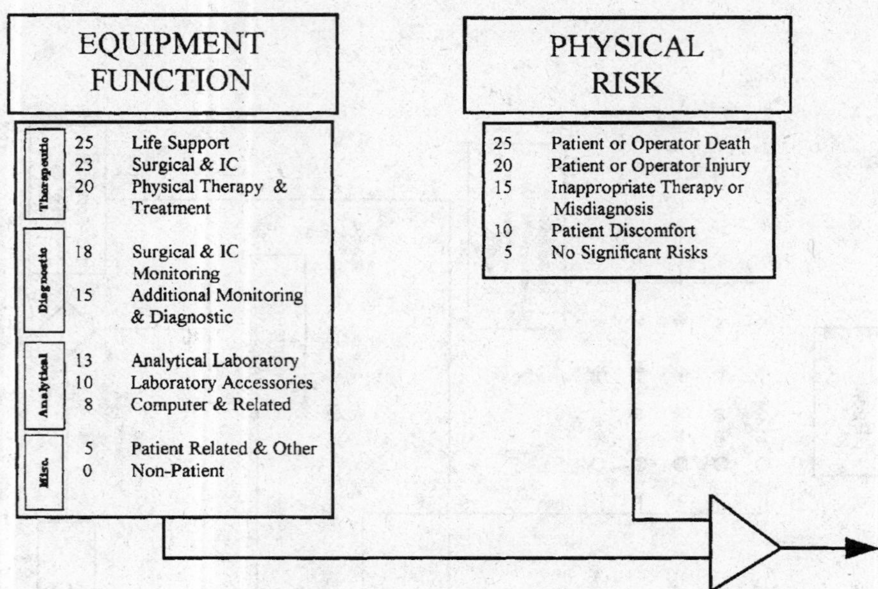

FIGURE 169.10 Static risk components.

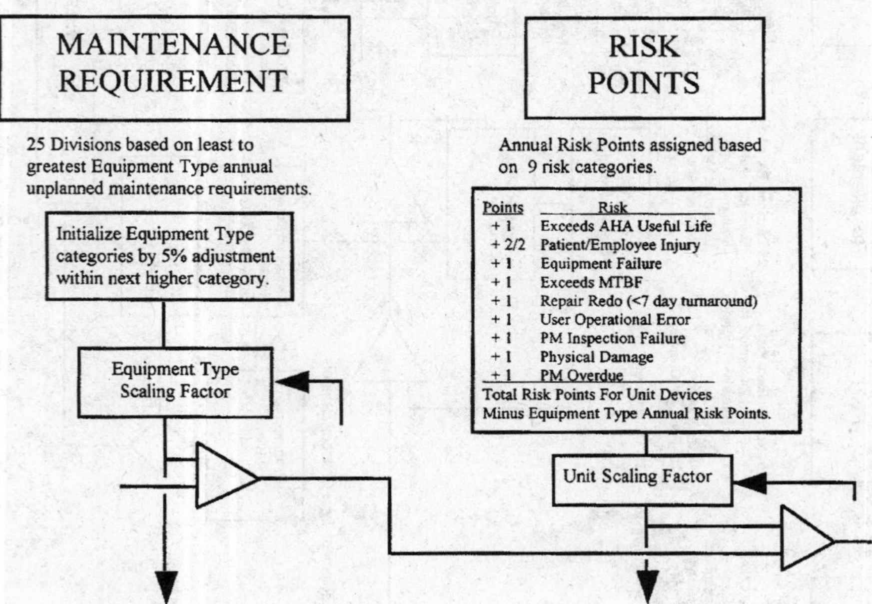

FIGURE 169.11 Dynamic risk components.

The baseline is now established for evaluating individual device risk. Variations in the maintenance requirement hours for any particular equipment type will, for the most part, only occur over a substantial period of time. For this reason, the maintenance requirement category is designated a "slow" dynamic risk element.

The second dynamic element assigns weighted risk points to *individual* equipment items for each unique risk occurrence. An *occurrence* is defined as when the device

- Exceeds the American Hospital Association Useful Life Table for Medical Equipment or exceeds the historical MTBF for that manufacturer and model
- Injures a patient or employee
- Functionally fails or fails to pass a PM inspection
- Is returned for repair or returned for rerepair within 9 days of a previous repair occurrence
- Misses a planned maintenance inspection
- Is subjected to physical damage
- Was reported to have failed but the problem was determined to be a user operational error

There risk occurrences include the failure codes previously described. Although many other risk occurrences could be defined, these nine occurrences have been historically effective in managing equipment risk. The risk points for each piece or equipment are algebraically summed over the previous year. Since the yearly total is a moving window, the risk points will not continue to accumulate but will reflect a recent historical average risk. The risk points for each equipment type are also calculated. This provides a baseline to measure the relative risk of devices within an equipment type. The average risk points for the equipment type are subtracted from those for each piece of equipment within the equipment type. If the device has a negative risk point value, the device's risk is less than the average device in the equipment type. If positive, then the device has higher risk than the average device. This positive or negative factor is algebraically summed to the risk values from the equipment function, physical risk, and maintenance requirements. The annual risk points for an individual piece of equipment might change quickly over several months. For this reason, this is the "fast" component of the dynamic risk factor.

The concept of risk has now been quantified in term of equipment function, physical risk, maintenance requirements, and risk occurrences. The total risk count for each device then places it in one of five risk priority groups that are based on the sum of risk points. These groups are then applied in various ways to determine repair triage, PM triage, educational and in-service requirements and test equipment/parts, etc. in the equipment management program.

Correlation between the placement of individual devices in each risk priority group and the levels of planned maintenance previously assigned by the clinical engineer have shown that the proactive risk-management tool calculates a similar maintenance schedule as manually planned by the clinical engineer. In other words, the proactive risk-management tool algorithm places equipment items in a risk priority group commensurate with the greater or lesser maintenance as currently applied in the equipment maintenance program.

As previously mentioned, the four categories and the 1 to 25 risk levels within each category are arbitrary because a "gold standard" for risk management is nonexistent. Therefore, the clinical engineer is given input into the dynamic components making up the risk factor to "shape the system" based on the equipment's maintenance history and the clinical engineer's experience. Since the idea of a safe medical equipment program involves "judgment about the acceptability of risk in a specified situation" [8], this experience is a necessary component of the risk-assessment tool for a specific health care setting.

In the same manner, the system tracks the unit device's assigned risk priority group. If the risk points for a device change sufficiently to place it in a different group, it is flagged for review. Again, the clinical engineer reviews the particular equipment item and decides if corrective action is prudent. Otherwise, the system reduces the scaling factor by 5%. Over a period of time, the system will be "formed" to what is acceptable risk and what deserves closer scrutiny.

169.4 Risk Management: Application

The information can be made available to the clinical engineer in the form of a risk assessment report (see Fig. 169.12). The report lists individual devices by property tag number (equipment control number), manufacturer, model, and equipment type. Assigned values for equipment function and physical risk are constant for each equipment type. The maintenance sensitizing factor enables the clinical engineer to control the algorithm's response to the maintenance level of an entire equipment type. These factors combine to produce the slow risk factor (equipment function + physical risk + maintenance requirements). The unit risk points are multiplied for the unit scaling factor, which allows the clinical engineer to control the algorithm's response to static and dynamic risk components on individual pieces of equipment. This number is then added to the slow risk factor to determine the risk factor for each item. The last two columns are the risk priority that the automated system has assigned and the PM level set by the clinical engineer. This report provides the clinical engineer with information about medical equipment that reflects a higher than normal risk factor for the equipment type to which it belongs.

The proactive risk management tool can be used to individually schedule medical equipment devices for PM based on risk assessment. For example, why should newer patient monitors be maintained at the same maintenance level as older units if the risk can be demonstrated to be less? The tool is used as well to prioritize the planned maintenance program. For instance, assume a PM cycle every 17 weeks is started on January 1 for a duration of 1 week. Equipment not currently available for PM can be inspected at a later time as a function of the risk priority group for that device. In other words, an equipment item with a risk priority of 2, which is moderately low, would not be overdue for 2/5 of the time between the current and the next PM start date or until the thirteenth week after the start of a PM cycle of 17 weeks. The technicians can complete more critical overdue equipment first and move on to less critical equipment later.

Additionally, since PM is performed with every equipment repair, is it always necessary to perform the following planned PM? Assume for a moment that unscheduled maintenance was performed 10 weeks into the 17 weeks between the two PM periods discussed above. IF the equipment has a higher risk priority of the three, four, or five, the equipment is PMed as scheduled in April. However, if a lower equipment risk priority of one or two is indicated, the planned maintenance is skipped in April and resumed in July. The intent of this application is to reduce maintenance costs, preserve departmental resources, and minimize the war and tear on equipment during testing.

Historically, equipment awaiting service has been placed in the equipment holding area and inspected on a first in, first out (FIFO) basis when a technician is available. A client's request to expedite the equipment repair was the singular reason for changing the work priority schedule. The proactive risk-management tool can prioritize the equipment awaiting repair, putting the critical equipment back into service more quickly, subject to the clinical engineer's review.

169.5 Case Studies

Several examples are presented of the proactive risk-assessment tool used to evaluate the performance of medical equipment within a program.

The ventilators in Fig. 169.13 show a decreasing unit risk factor for higher equipment tag numbers. Since devices are put into service with ascending tag numbers and these devices are known to have been purchased over a period of time, the X axis represents a chronological progression. The ventilator risk factor is decreasing for newer units and could be attributable to better maintenance technique or manufacturer design improvements. This device is said to have a *time-dependent risk factor.*

A final illustration uses two generations of infusion pumps from the same manufacturer. Figure 169.14 shows the older vintage pump as *Infusion Pump* 1 and the newer version as *Infusion Pump* 2. A linear regression line for the first pump establishes the average risk factor as 53 with a standard deviation of 2.02 for the 93 pumps in the analysis. The second pump, a newer version of the first, had an average risk factor of 50 with a standard deviation of 1.38 for 261 pumps. Both pumps have relatively time-independent risk factors. The proactive risk-management tool reveals that this particular brand of infusion pump in the present maintenance program is stable over time and the newer pump has reduced risk and variability of risk between individual units. Again, this could be attributable to tighter manufacturing control or improvements in the maintenance program.

Equip Control Number	Manuf	Model	Equipment Type	Equip Func	Phys Risk	Maint Requir	Avg Hours	Maint Sensitiz Factor	Slow Risk Factor	Unit Risk Points	Unit Scaling Factor	Risk Factor	Risk Priority	Equip Type Priority
Manager:	PHYSIOLOGICAL GROUP													
17407	322	4000A	NIBP SYSTEM	18	15	1	1.66	1.78	38	6.62	1.00	41	3	2
17412	322	4000A	NIBP SYSTEM	18	15	1	1.66	1.78	38	6.62	1.00	41	3	2
17424	322	4000A	NIBP SYSTEM	18	15	1	1.66	1.78	38	6.62	1.00	41	3	2
17431	322	4000A	NIBP SYSTEM	18	15	1	1.66	1.78	38	6.62	1.00	41	3	2
15609	65	BW5	BLOOD & PLMA WARMING DEVICE	5	5	1	2.51	1.17	14	10.64	1.00	22	2	1
3538	167	7370000	HR/RESP MONITOR	18	15	1	0.10	29.47	35	8.69	1.00	43	3	2
3543	167	7370000	HR/RESP MONITOR	18	15	1	0.10	29.47	35	7.69	1.00	42	3	2
15315	167	7370000	HR/RESP MONITOR	18	15	1	0.10	29.47	35	7.69	1.00	42	3	2
17761	167	7370000	HR/RESP MONITOR	18	15	1	0.10	29.47	35	6.69	1.00	41	3	2
18382	574	N100C	PULSE OXIMETER	18	15	1	0.70	4.21	35	7.54	1.00	42	3	2
180476	574	N100C	PULSE OXIMETER	18	15	1	0.70	4.21	35	7.54	1.00	42	3	2
16685	167	7275217	2 CHAN CHART REC	18	15	1	0.42	7.02	37	6.83	1.00	41	3	2

I have reviewed this risk analysis report and have investigated these equipment items for whcih the risk priority factor has exceeded the average risk for that equipment type. I have taken one of two actions:

1. investigated the equipment item and implemented changes to the maintenance program intended to reduce the risk priority value OR
2. indicated on the printout that the dynamic risk factor program has "overreacted" and the risk factor should be reduced by 5%.

ENGINEER: ____________________

DATE: __________

FIGURE 169.12 Engineer's risk-assessment report.

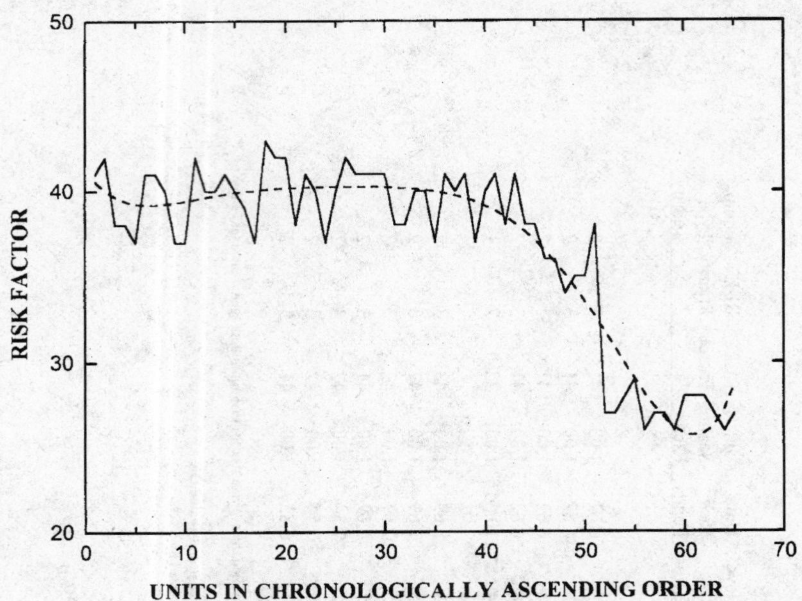

FIGURE 169.13 Ventilator with time-dependent risk characteristics.

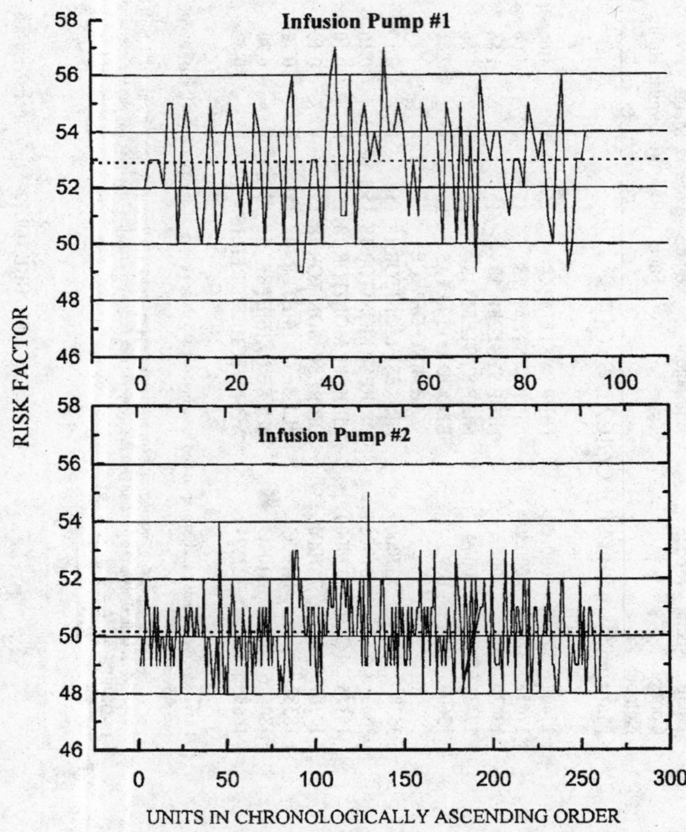

FIGURE 169.14 Time-independent risk characteristics infusion pump #1.

169.6 Conclusions

In summary, superior risk assessment within a medical equipment management program requires better communication, teamwork, and information analysis and distribution among all health care providers. Individually, the clinical engineer cannot provide all the necessary components for managing risk in the health care environment. Using historical information to only address equipment-related problems, after an incident, is not sufficient. The use of a proactive risk-management tool is necessary.

The clinical engineer can use this tool to deploy technical resources in a cost-effective manner. In addition to the direct economic benefits, safety is enhanced as problem equipment is identified and monitored more frequently. The integration of a proactive risk-assessment tool into the equipment management program can more accurately bring to focus technical resources in the health care environment.

References

1. Gullikson ML. 1994. Biotechnology Procurement and Maintenance II: Technology Risk Management. Third Annual International Pediatric Colloquium, Houston, Texas.
2. David Y. 1992. Medical Technology 2001. Health Care Conference, Texas Society of Certified Public Accountants, San Antonio.
3. Gullikson ML. 1993. An Automated Risk Management Tool. Plant, Technology, and Safety Management Series, Joint Commission on the Accreditation of Healthcare Facilities (JCAHO) Monograph 2.
4. Pecht ML, Nash FR. 1994. Predicting the reliability of electronic equipment. Proc IEEE 82(7):990.
5. Gullikson ML, David Y. 1993. Risk-Based Equipment Management Systems. The 9th National Conference on Medical Technology Management, American Society for Hospital Engineering of the American Hospital Association (AHA), New Orleans.
6. Gullikson ML. 1992. Biomedical Equipment Maintenance System. 27th Annual Meeting and Exposition, Hospital and Medical Industry Computerized Maintenance Systems, Association for the Advancement of Medical Instrumentation (AAMI), Anaheim, Calif.
7. Fennigkoh L. 1989. Clinical Equipment Management. Plant, Technology, and Safety Management Monograph 2.
8. David Y, Judd T. 1993. Medical Technology Management, Biophysical Measurement Series. Spacelabs Medical, Inc., Redmond, Wash.

170

Clinical Engineering Program Indicators

Dennis D. Autio
Dybonics, Inc.

Robert L. Morris
Dybonics, Inc.

The role, organization, and structure of clinical engineering departments in the modern health care environment continue to evolve. During the past 10 years, the rate of change has increased considerably faster than mere evolution due to fundamental changes in the management and organization of health care. Rapid, significant changes in the health care sector are occurring in the United States and in nearly every country. The underlying drive is primarily economic, the recognition that resources are finite.

Indicators are essential for survival of organizations and are absolutely necessary for effective management of change. Clinical engineering departments are not exceptions to this rule. In the past, most clinical engineering departments were task-driven and their existence justified by the tasks performed. Perhaps the most significant change occurring in clinical engineering practice today is the philosophical shift to a more business-oriented, cost-justified, bottom-line-focused approach than has been generally the case in the past.

Changes in the health care delivery system will dictate that clinical engineering departments justify their performance and existence on the same basis as any business, the performance of specific functions at a high quality level and at a competitive cost. Clinical engineering management philosophy must change from a purely task-driven methodology to one that includes the economics of department performance. Indicators need to be developed to measure this performance. Indicator data will need to be collected and analyzed. The data and indicators must be objective and defensible. If it cannot be measured, it cannot be managed effectively.

Indicators are used to measure performance and function in three major areas. Indicators should be used as internal measurements and monitors of the performance provided by individuals, teams, and the department. These essentially measure what was done and how it was done. Indicators are essential during quality improvement and are used to monitor and improve a process. A third important type of program indicator is the benchmark. It is common knowledge that successful businesses will continue to use benchmarks, even though differing terminology will be used. A business cannot improve its competitive position unless it knows where it stands compared with similar organizations and businesses.

0-8493-0462-8/00/$0.00+$.50

Different indicators may be necessary depending on the end purpose. Some indicators may be able to measure internal operations, quality improvement, and external benchmarks. Others will have a more restricted application.

It is important to realize that a single indicator is insufficient to provide the information on which to base significant decisions. Multiple indicators are necessary to provide cross-checks and verification. An example might be to look at the profit margin of a business. Even if the profit margin per sale is 100%, the business will not be successful if there are few sales. Looking at single indicators of gross or net profit will correct this deficiency but will not provide sufficient information to point the way to improvements in operations.

170.1 Department Philosophy

A successful clinical engineering department must define its mission, vision, and goals as related to the facility's mission. A mission statement should identify what the clinical engineering department does for the organization. A vision statement identifies the direction and future of the department and must incorporate the vision statement of the parent organization. Department goals are then identified and developed to meet the mission and vision statements for the department and organization. The goals must be specific and attainable. The identification of goals will be incomplete without at least implied indicators. Integrating the mission statement, vision statement, and goals together provides the clinical engineering department management with the direction and constraints necessary for effective planning.

Clinical engineering managers must carefully integrate mission, vision, and goal information to develop a strategic plan for the department. Since available means are always limited, the manager must carefully assess the needs of the organization and available resources, set appropriate priorities, and determine available options. The scope of specific clinical engineering services to be provided can include maintenance, equipment management, and technology management activities. Once the scope of services is defined, strategies can be developed for implementation. Appropriate program indicators must then be developed to document, monitor, and manage the services to be provided. Once effective indicators are implemented, they can be used to monitor internal operations and quality-improvement processes and complete comparisons with external organizations.

Monitoring Internal Operations

Indicators may be used to provide an objective, accurate measurement of the different services provided in the department. These can measure specific individual, team, and departmental performance parameters. Typical indicators might include simple tallies of the quantity or level of effort for each activity, productivity (quantify/effort), percentage of time spent performing each activity, percentage of scheduled IPMs (inspection and preventive maintenance procedures) completed within the scheduled period, mean time per job by activity, repair jobs not completed within 30 days, parts order for greater than 60 days, etc.

Process for Quality Improvement

When program indicators are used in a quality-improvement process, an additional step is required. Expectations must be quantified in terms of the indicators used. Quantified expectations result in the establishment of a threshold value for the indicator that will precipitate further analysis of the process. Indicators combined with expectations (threshold values of the indicators) identify the opportunities for program improvement. Periodic monitoring to determine if a program indicator is below (or above, depending on whether you are measuring successes or failures) the established threshold will provide a flag to whether the process or performance is within acceptable limits. If it is outside acceptable limits for the indicator, a problem has been identified. Further analysis may be required to better define the problem. Possible program indicators for quality improvement might include the number of repairs completed within 24 or 48 hours, the number of callbacks for repairs, the number of repair problems caused by user error, the percentage of hazard notifications reviewed and acted on within a given time frame, meeting time targets for generating specification, evaluation or acceptance of new equipment, etc.

An example might be a weekly status update of the percentage of scheduled IPMs completed. Assume that the department has implemented a process in which a group of scheduled IPMs must be completed within 8 weeks. The expectation is that 12% of the scheduled IPMs will be completed each week. The indicator is the percentage of IPMs completed. The threshold value of the indicator is 12% per week increase in the percentage of IPMs completed. To monitor this, the number of IPMs that were completed must be tallied, divided by the total number scheduled, and multiplied by 100 to determine the percentage completed. If the number of completed IPMs is less than projected, then further analysis would be required to identify the source of the problem and determine solutions to correct it. If the percentage of completed IPMs were equal to or greater than the threshold or target, then no action would be required.

External Comparisons

Much important and useful information can be obtained be carefully comparing one clinical engineering program with others. This type of comparison is highly valued by most hospital administrators. It can be helpful in determining performance relative to competitors. External indicators or benchmarks can identify specific areas of activity in need of improvement. They offer insights when consideration is being given to expanding into new areas of support. Great care must be taken when comparing services provided by clinical engineering departments located in different facilities. There are number of factors that must be included in making such comparisons; otherwise, the results can be misleading or misinterpreted. It is important that the definition of the specific indicators used be well understood, and great care must be taken to ensure that the comparison utilizes comparable information before interpreting the comparisons. Failure to understand the details and nature of the comparison and just using the numbers directly will likely result in inappropriate actions by managers and administrators. The process of analysis and explanation of differences in benchmark values between a clinical engineering department and a competitor (often referred to as *gap analysis*) can lead to increased insight into department operations and target areas for improvements.

Possible external indicators could be the labor cost per hour, the labor cost per repair, the total cost per repair, the cost per bed supported, the number of devices per bed supported, percentage of time devoted to repairs versus IPMs versus consultation, cost of support as a percentage of the acquisition value of capital inventory, etc.

170.2 Standard Database

In God we trust...all others bring data!

Florida Power and Light

Evaluation of indicators requires the collection, storage, and analysis of data from which the indicators can be derived. A standard set of data elements must be defined. Fortunately, one only has to look at commercially available equipment management systems to determine the most common data elements used. Indeed, most of the high-end software systems have more data elements than many clinical engineering departments are willing to collect. These standard data elements must be carefully defined and understood. This is especially important if the data will later be used for comparisons with other organizations. Different departments often have different definitions for the same data element. It is crucial that the data collected be accurate and complete. The members of the clinical engineering department must be trained to properly gather, document, and enter the data into the database. It makes no conceptual difference if the database is maintained on paper or using computers. Computers and their databases are ubiquitous and so much easier to use that usually more data elements are collected when computerized systems are used. The effort required for analysis is less and the level of sophistication of the analytical tools that can be used is higher with computerized systems.

The clinical engineering department must consistently gather and enter data into the database. The database becomes the practical definition of the services and work performed by the department. This standardized database allows rapid, retrospective analysis of the data to determine specific indicators identifying problems and assist in developing solutions for implementation. A minimum database should allow the gathering and storage of the following data:

In-House Labor. This consists of three elements: the number of hours spent providing a particular service, the associated labor rate, and the identity of the individual providing the service. The labor cost is not the hourly rate the technician is paid multiplied by the number of hours spent performing the service. It should include the associated indirect costs, such as benefits, space, utilities, test equipment, and tools, along with training, administrative overhead, and many other hidden costs. A simple, straightforward approach to determine an hourly labor rate for a department is to take the total budget of the department and subtract parts' costs, service contract costs, and amounts paid to outside vendors. Divide the resulting amount by the total hours spent providing services as determined from the database. This will provide an average hourly rate for the department.

Vendor Labor. This should include hours spent and rate, travel, and zone charges and any perdiem costs associated with the vendor supplied service.

Parts. Complete information on parts is important for any retrospective study of services provided. This information is similar for both in-house and vendor-provided service. It should include the part number, a description of the part, and its cost, including any shipping.

Timeless. It is important to include a number of time stamps in the data. These should include the date the request was received, data assigned, and date completed.

Problem Identification. Both a code for rapid computer searching and classification and a free text comment identifying the nature of the problem and description of service provided are important. The number of codes should be kept to as few as possible. Detailed classification schemes usually end up with significant inaccuracies due to differing interpretations of the fine gradations in classifications.

Equipment Identification. Developing an accurate equipment history depends on reliable means of identifying the equipment. This usually includes a department- and/or facility-assigned unique identification number as well as the manufacturer, vendor, model, and serial number. Identification numbers of provided by asset management are often inadequate to allow tracking of interchangeable modules or important items with a value less than a given amount. Acquisition cost is a useful data element.

Service Requester. The database should include elements allowing identification of the department, person, telephone number, cost center, and location of the service requester.

170.3 Measurement Indicators

Clinical engineering departments must gather objective, quantifiable data in order to assess ongoing performance, identify new quality-improvement opportunities, and monitor the effect of improvement action plans. Since resources are limited and everything cannot be measured, certain selection criteria must be implemented to identify the most significant opportunities for indicators. High-volume, high-risk, or problem-prone processes require frequent monitoring of indicators. A new indicator may be developed after analysis of ongoing measurements or feedback from other processes. Customer feedback and surveys often can provide information leading to the development of new indicators. Department management, in consultation with the quality-management department, typically determines what indicators will be monitored on an ongoing basis. The indicators and resulting analysis are fed back to individuals and work teams for review and improvement of their daily work activities. Teams may develop new indicators during their analysis and implementation of solutions to quality-improvement opportunities.

An *indicator* is an objective, quantitative measurement of an outcome or process that relates to performance quality. The event being assessed can be either desirable or undesirable. It is objective in that the same measurement can be obtained by different observers. This indicator represents quantitative, measured data that are gathered for further analysis. Indicators can assess many different aspects of quality, including accessibility, appropriateness, continuity, customer satisfaction, effectiveness, efficacy, efficiency, safety, and timeliness.

A program indicator has attributes that determine its utility as a performance measure. The reliability and variability of the indicator are distinct but related characteristics. An indicator is reliable if the same measurement can be obtained by different observers. A valid indicator is one that can identify opportunities for quality improvement. As indicators evolve, their reliability and validity should improve to the highest level possible.

An indicator can specify a part of a process to be measured or the outcome of that process. An outcome indicator assesses the results of a process. Examples include the percentage of uncompleted, scheduled IPMs, or the number of uncompleted equipment repairs not completed within 30 days. A process indicator assesses an important and discrete activity that is carried out during the process. An example would be the number of anesthesia machines in which the scheduled IPM failed or the number of equipment repairs awaiting parts that are uncompleted within 30 days.

Indicators also can be classified as sentinel event indicators and aggregate data indicators. A performance measurement of an individual event that triggers further analysis is called a *sentinel-event indicator.* These are often undesirable events that do not occur often. These are often related to safety issues and do not lend themselves easily to quality-improvement opportunities. An example may include equipment failures that result in a patient injury.

An aggregate data indicator is a performance measurement based on collecting data involving many events. These events occur frequently and can be presented as a continuous variable indicator or as rate-based indicators. A continuous variable indicator is a measurement where the value can fall anywhere along a continuous scale. Examples could be the number of IPMs scheduled during a particular month or the number of repair requests received during a week. A rate-based variable indicator is the value of a measurement that is expressed as a proportion or a ratio. Examples could be the percentage of IPMs completed each month or the percentage of repairs completed within one workday.

General indicators should be developed to provide a baseline monitoring of the department's performance. They also should provide a cross-check for other indicators. These indicators can be developed to respond to a perceived need within a department or to solve a specific problem.

170.4 Indicator Management Process

The process to develop, monitor, analyze and manage indicators is shown in Fig. 170.1. The different steps in this process include defining the indicator, establishing the threshold, monitoring the indicator, evaluating the indicator, identifying quality-improvement opportunities, and implementing action plans.

Define Indicator. The definition of the indicator to be monitored must be carefully developed. This process includes at least five steps. The event or outcome to be measured must be described. Define any specific terms that are used. Categorize the indicator (sentinel event or rate-based, process or outcome, desirable or undesirable). The purpose for this indicator must be defined, as well as how it is used in specifying and assessing the particular process or outcome.

Establish Threshold. A threshold is a specific data point that identifies the need for the department to respond to the indicator to determine why the threshold was reached. Sentinel-event indicator thresholds are set at zero. Rate indicator thresholds are more complex to define because they may require expert consensus or definition of the department's objectives. Thresholds must be identified, including the process used to set the specific level.

Monitor Indicator. Once the indicator is defined, the data-acquisition process identifies the data sources and data elements. As these data are gathered, they must be validated for accuracy and completeness. Multiple indicators can be used for data validation and cross-checking. The use of a computerized database allows rapid access to the data. A database management tool allows quick sorting and organization of the data. Once gathered, the data must be presented in a format suitable for evaluation. Graphic presentation of data allows rapid visual analysis for thresholds, trends, and patterns.

Evaluate Indicator. The evaluation process analyze and reports the information. This process includes comparing the information with established thresholds and analyzing for any trends or patterns. A *trend* is the general direction the indicator measurement takes over a period of time and may be desirable or undesirable. A *pattern* is a grouping or distribution of indicator measurements. A pattern analysis is often triggered when thresholds are crossed or trends identified. Additional indicator information is often required. If an indictor threshold has not been reached, no further action may be necessary, other than continuing to monitor this indicator. The department also may decide to improve its performance level by changing the threshold.

Factors may be present leading to variation of the indicator data. These factors may include failure of the technology to perform properly, failure of the operators to use the technology properly, and failure of the organization to provide the necessary resources to implement this technology properly. Further analysis of these factors may lead to quality-improvement activities later.

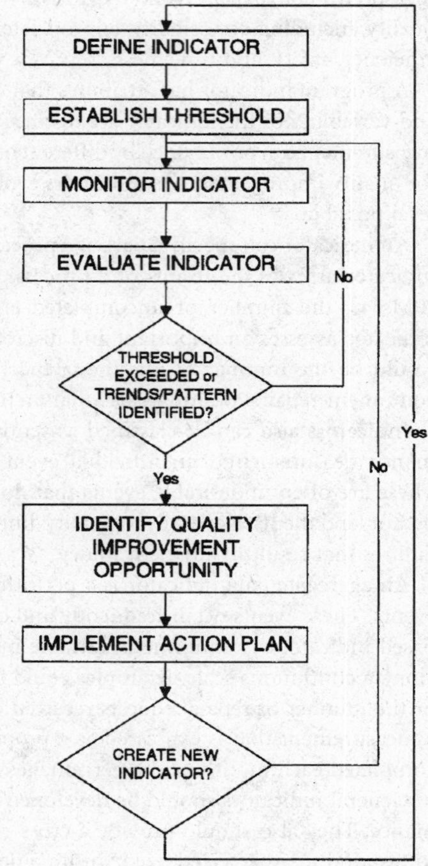

FIGURE 170.1 Indicator management process.

Identify Quality-Improvement Opportunity. A quality-improvement opportunity may present itself if an indicator threshold is reached, a trend is identified, or a pattern is recognized. Additional information is then needed to further define the process and improvement opportunities. The first step in the process is to identify a team. This team must be given the necessary resources to complete this project, a timetable to be followed, and an opportunity to periodically update management on the status of the project. The initial phase of the project will analyze the process and establish the scope and definition of the problem. Once the problem is defined, possible solutions can be identified and analyzed for potential implementation. A specific solution to the problem is then selected. The solution may include modifying existing indictors or thresholds to more appropriate values, modifying steps to improve existing processes, or establishing new goals for the department.

Implement Action Plan. An action plan is necessary to identify how the quality-improvement solution will be implemented. This includes defining the different tasks to be performed, the order in which they will be addressed, who will perform each task, and how this improvement will be monitored. Appropriate resources must again be identified and a timetable developed prior to implementation. Once the action plan is implemented, the indicators are monitored and evaluated to verify appropriate changes in the process. New indicators and thresholds may need to be developed to monitor the solution.

170.5 Indicator Example 1: Productivity Monitors

Defines Indicators. Monitor the productivity of technical personnel, teams, and the department. Productivity is defined as the total number of documented service support hours compared with the total number of hours available. This is a desirable rate-based outcome indicator. Provide feedback to technical staff and hospital administration regarding utilization of available time for department support activities.

Establish Thresholds. At least 50% of available technician time will be spent providing equipment maintenance support services (revolving equipment problems and scheduled IPMs). At least 25% of available technician time will be spent providing equipment management support services (installations, acceptance testing, incoming inspections, equipment inventory database management, hazard notification review).

Monitor Indicator. Data will be gathered every 4 weeks from the equipment work order history database. A trend analysis will be performed with data available from previously monitored 4–week intervals. These data will consist of hours worked on completed and uncompleted jobs during the past 4–week interval.

Technical staff available hours is calculated for the 4–week interval. The base time available is 160 hours (40 hours/week × 4 week) per individual. Add to this any overtime worked during the interval. Then subtract any holidays, sick days, and vacation days within the interval.

CJHOURS: Hours worked on completed jobs during the interval

UJHOURS: Hours worked on uncompleted jobs during the interval

AHOURS: Total hours available during the 4-week interval

$$\text{Productivity} = (\text{CJHOURS} + \text{UJHOURS})/\text{AHOURS}$$

Evaluate Indicator. The indicator will be compared with the threshold, and the information will be provided to the individual. The individual team member data can be summed for team review. The data from multiple teams can be summed and reviewed by the department. Historical indicator information will be utilized to determine trends and patterns.

Quality-Improvement Process. If the threshold is not met, a trend is identified, or a pattern is observed, a quality-improvement opportunity exists. A team could be formed to review the indicator, examine the process that the indicator measured, define the problem encountered, identify ways to solve the problem, and select a solution. An action plan will then be developed to implement this solution.

Implement Action Plan. During implementation of the action plan, appropriate indicators will be used to monitored the effectiveness of the action plan.

170.6 Indicator Example 2: Patient Monitors IPM Completion Time

Define Indicator. Compare the mean to complete an IPM for different models of patient monitors. Different manufacturers of patient monitors have different IPM requirements. Identify the most timely process to support this equipment.

Establish Threshold. The difference between the mean time to complete an IMP for different models of patient monitors will not be greater than 30% of the lessor time.

Monitor Indicator. Determine the mean time to complete an IPM for each model of patient monitor. Calculate the percentage difference between the mean time for each model and the model with the least mean time.

Evaluate Indicator. The mean time to complete IPMs was compared between the patient monitors, and the maximum difference noted was 46%. A pattern also was identified in which all IPMs for that one particular monitor averaged 15 minutes longer than those of other vendors.

Quality-Improvement Process. A team was formed to address this problem. Analysis of individual IPM procedures revealed that manufacturer X requires the case to be removed to access internal filters. Performing an IPM for each monitor required moving and replacing 15 screws for each of the 46 monitors.

The team evaluated this process and identified that 5 minutes could be saved from each IPM if an electric screwdriver was utilized.

Implement Action Plan. Electric screwdrivers were purchased and provided for use by the technician. The completion of one IPM cycle for the 46 monitors would pay for two electric screwdrivers and provide 4 hours of productive time for additional work. Actual savings were greater because this equipment could be used in the course of daily work.

170.7 Summary

In the ever-changing world of health care, clinical engineering departments are frequently being evaluated based on their contribution to the corporate bottom line. For many departments, this will require difficult and painful changes in management philosophy. Administrators are demanding quantitative measures of performance and value. To provide the appropriate quantitative documentation required by corporate managers, a clinical engineering a manager must collect available data that are reliable and accurate. Without such data, analysis is valueless. Indicators are the first step in reducing the data to meaningful information that can be easily monitored and analyzed. The indicators can then be used to determine department performance and identify opportunities for quality improvement.

Program indicators have been used for many years. What must change for clinical engineering departments is a conscious evaluation and systematic use of indicators. One traditional indicator of clinical engineering department success is whether the department's budget is approved or not. Unfortunately, approval of the budge as an indicator, while valuable, does not address the issue of predicting long-term survival, measuring program and quality improvements, or allowing frequent evaluation and changes.

There should be monitored indicators for every significant operational aspect of the department. Common areas where program indicators can be applied include monitoring interval department activities, quality-improvement processes, and benchmarking. Initially, simple indicators should be developed. The complexity and number of indicators should change as experience and needs demand.

The use of program indicators is absolutely essential if a clinical engineering departments is to survive. Program and survival are now determined by the contribution of the department to the bottom line of the parent organization. Indicators must be developed and utilized to determine the current contribution of the clinical engineering department to the organization. Effective utilization and management of program indicators will ensure future department contributions.

References

AAMI. 1993. Management Information Report MIR 1: Design of Clinical Engineering Quality Assurance Risk Management Programs. Arlington, Va, Association for the Advancement of Medical Instrumentation.

AAMI. 1993. Management Information Report MIR 2: Guideline for Establishing and Administering Medical Instrumentation Maintenance Programs. Arlington, Va, Association for the Advancement of Medical Instrumentation.

AAMI. 1994. Management Information Report MIR 3: Computerized Maintenance Management Systems for Clinical Engineering. Arlington, Va, Association for the Advancement of Medical Instrumentation.

Bauld TJ. 1987. Productivity: Standard terminology and definitions. J Clin Eng 12(2):139.

Betts WF. 1989. Using productivity measures in clinical engineering departments. Biomed Instrum Technol 23(2):120.

Bronzino JD. 1992. Management of Medical Technology: A Primer for Clinical Engineers. Stoneham, Mass, Butterworth-Heinemann.

Coopers and Lybrand International, AFSM. 1994. Benchmarking Impacting the Boston Line. Fort Myers, Fla, Association for Services Management International.

David Y, Judd TM. 1993. Risk management and quality improvement. In Medical Technology Management, pp 72–75. Redmond, Wash, SpaceLab Medical.

David Y, Rohe D. 1986. Clinical engineering program productivity and measurement. *J Clin Eng* 11(6):435.

Downs KJ, McKinney WD. 1991. Clinical engineering workload analysis: A proposal for standardization. Biomed Instrum Technol 25(2):101.

Fennigkoh L. 1986. ASHE Technical Document No 055880: Medical Equipment Maintenance Performance Measures. Chicago, American Society for Hospital Engineers.

Furst E. 1986. Productivity and cost-effectiveness of clinical engineering. J Clin Eng 11(2):105.

Gordon GJ. 1995. Break Through Management—A New Model For Hospital Technical Services. Arlington, Va, Association for the Advancement of Medical Instrumentation.

Hertz E. 1990. Developing quality indicators for a clinical engineering department. In Plant, Technology and Safety Management Series: Measuring Quality in PTSM. Chicago, Joint Commission on Accreditation of Healthcare Organizations.

JCAHO. 1990. Primer on Indicator Development and Application, Measuring Quality in Health Care. Oakbrook, Ill, Joint Commission on Accreditation of Healthcare Organizations.

JCAHO. 1994. Framework for Improving Performance. Oakbrook, Ill, Joint Commission on Accreditation of Healthcare Organizations.

Keil OR. 1989. The challenge of building quality into clinical engineering programs. Biomed Instrum Technol 23(5):354.

Lodge DA. 1991. Productivity, efficiency, and effectiveness in the management of healthcare technology: An incentive pay proposal. J Clin Eng 16(1):29.

Mahachek AR. 1987. Management and control of clinical engineering productivity: A case study. J Clin Eng 12(2):127.

Mahachek AR. 1989. Productivity measurement. Taking the first steps. Biomed Instrum Technol 23:16.

Selsky DB, Bell DS, Benson D, et al. 1991. Biomedical equipment information management for the next generation. Biomed Instrum Technol 25(1):24.

Sherwood MK. 1991. Quality assurance in Biomedical or clinical engineering. J Clin Eng 16(6):479.

Stiefel RH. 1991. Creating a quality measurement system for clinical engineering. Biomed Instrum Technol. 25(1):17.

171

Quality of Improvement and Team Building

Joseph P. McClain
Walter Reed Army Medical Center

In today's complex health care environment, quality improvement and team building must go hand in hand. This is especially true for Clinical Engineers and Biomedical Equipment Technicians as the diversity of the field increases and technology moves so rapidly that no one can know all that needs to be known without the help of others. Therefore, it is important that we work together to ensure quality improvement. Ken Blachard, the author of the *One Minute Manager* series, has made the statement that "all of us are smarter than any one of us"—a synergy that evolves from working together. Throughout this chapter we will look closely at defining quality and the methods for continuously improving quality, such as collecting data, interpreting indicators, and team building. All this will be put together, enabling us to make decisions based on scientific deciphering of indicators.

Quality is defined as conformance to customer or user requirements. If a product or service does what it is supposed to do, it is said to have high quality. If the product or service fails its mission, it is said to be low quality. Dr. W. Edward Demings, who is known to many as the "father of quality," defined it as surpassing customer needs and expectations throughout the life of the product or service.

Dr. Demings, a trained statistician by profession, formed his theories on quality during World War II while teaching industry how to use statistical methods to improve the quality of military production. After the war, he focused on meeting customer or consumer needs and acted as a consultant to Japanese organizations to change consumers' perceptions that "Made in Japan" meant junk. Dr. Demings predicted

0-8493-0462-8/00/$0.00+$.50

that people would be demanding Japanese products in just 5 years, if they used his methods. However, it only took 4, and the rest is history.

171.1 Deming's 14 Points

1. Create constancy of purpose toward improvement of product and service, with an aim to become competitive and to stay in business and provide jobs.
2. Adopt the new philosophy. We are in a new economic age. Western management must awaken and lead for change.
3. Cease dependence on inspection to achieve quality. Eliminate the needs for mass inspection by first building in quality.
4. Improve constantly and forever the system of production and service to improve quality and productivity and thus constantly decrease costs.
5. Institute training on the job.
6. Institute leadership: The goal is to help people, machines, and gadgets to do a better job.
7. Drive out fear so that everyone may work effectively for the organization.
8. Break down barriers between departments
9. Eliminate slogans, exhortations, and targets for the workforce.
10. Eliminate work standards (quota) on the factory floor.
11. Substitute leadership: Eliminate management by objective, by numbers, and numerical goals.
12. Remove barriers that rob the hourly worker of the right to pride of workmanship.
13. Institute a vigorous program of education and self-improvement.
14. Encourage everyone in the company to work toward accomplishing transformation. Transformation is everyone's job.

171.2 Zero Defects

Another well-known quality theory, called *zero defects* (*ZD*), was establishing by Philip Crosby. It got results for a variety of reasons. The main reasons are as follows:

1. A strict and specific management standard. Management, including the supervisory staff, do not use vague phrases to explain what it wants. It made the quality standard very clear: *Do it the right way from the start.* As Philip Crosby said, "What standard would you set on how many babies nurses are allowed to drop?"
2. Complete commitment everyone. Interestingly, Crosby denies that ZD was a motivational program. But ZD worked because everyone got deeply into the act. Everyone was encouraged to spot problems, detect errors, and prescribe ways and means for their removal. This commitment is best illustrated by the ZD pledge: "I freely pledge myself to make a constant, conscious effort to do my job right the first time, recognizing that my individual contribution is a vital part of the overall effort."
3. Removal of actions and conditions that cause errors. Philip Crosby claimed that at ITT, where he was vice-president for quality, 90% of all error causes could be acted on and fully removed by first-line supervision. In other words, top management must do its part to improve conditions, but supervisors and employees should handle problems directly. Errors, malfunctions, and/or variances can best be corrected where the rubber hits the road—at the source.

171.3 TQM (Total Quality Management)

The most recent quality theory that has found fame is called *TQM* (*Total Quality Management*). It is a strategic, integrated management system for achieving customer satisfaction which involves all managers and employees and uses quantitative methods to continuously improve an organization's processes. *Total*

Quality Management is a term coined in 1985 by the Naval Air Systems Command to describe its management approach to quality improvement. Simply put, Total Quality Management is a management approach to long-term success through customer satisfaction. Total Quality Management includes the following three principles: (1) achieving customer satisfaction, (2) making continuous improvement, and (3) giving everyone responsibility. TQM includes eight practices. These practices are (1) focus on the customer, (2) effective and renewed communications, (3) reliance on standards and measures, (4) commitment to training, (5) top management support and direction, (6) employee involvement, (7) rewards and recognition, and (8) long-term commitment.

171.4 CQI (Continuous Quality Improvement)

Step 8 of the total quality management practices leads us to the quality concept coined by the Joint Commission On Accreditation of Healthcare Organizations and widely used by most health care agencies. It is called *CQI* (*Continuous Quality Management*). The principles of CQI are as follows:

Unity of Purpose

- Unity is established throughout the organization with a clear and widely understood vision.
- Environment nurtures total commitment from all employees.
- Rewards go beyond benefits and salaries to the belief that "*We are family*" and "*We do excellent work.*"

Looking for Faults in the Systems

- Eighty percent of an organization's failures are the fault of management-controlled systems.
- Workers can control fewer than 20% of the problems.
- Focus on rigorous improvement of every system, and cease blaming individuals for problems (the 80/20 rule of J.M. Juran and the nineteenth-century economist Vilfredo Pareto).

Customer Focus

- Start with the customer.
- The goal is to meet or exceed customer needs and give lasting value to the customer.
- Positive returns will follow as customers boast of the company's quality and service.

Obsession with Quality

- Everyone's job.
- Quality is relentlessly pursued through products and services that delight the customer.
- Efficient and effective methods of execution.

Recognizing the Structure in Work

- All work has structure.
- Structure may be hidden behind workflow inefficiency.
- Structure can be studied, measure, analyzed, and improved.

Freedom Through Control

- There is control, yet freedom exists by eliminating micromanagement.
- Employees standardize processes and communicate the benefits of standardization.
- Employees reduce variation in the way work is done.
- Freedom comes as changes occur resulting in time to spend on developing improved processes, discovering new markets, and adding other methods to increase productivity.

Continued Education and Training

- Everyone is constantly learning.
- Educational opportunities are made available to employees.
- Greater job mastery is gained and capabilities are broadened.

Philosophical Issues on Training

- Training must stay tuned to current technology.
- Funding must be made available to ensure that proper training can be attained.
- Test, measurement, and diagnostic equipment germane to the mission must be procured and technicians trained on its proper use, calibration, and service.
- Creativity must be used to obtain training when funding is scarce.
  —Include training in equipment procurement process.
  —Contact manufacturer or education facility to bring training to the institution.
  —Use local facilities to acquire training, thus eliminating travel cost.
  —Allow employees to attend professional seminars where a multitude of training is available.

Teamwork

- Old rivalries and distrust are eliminated.
- Barriers are overcome.
- Teamwork, commitment to the team concept and partnerships are the focus.
- Employee empowerment is critical in the CQI philosophy and means that employees have the authority to make well-reasoned, data-based decisions. In essence, they are entrusted with the legal power to change processes through a rational, scientific approach.

Continuous quality improvement is a means for tapping knowledge and creativity, applying participative problem solving, finding and eliminating problems that prevent quality, eliminating waste, instilling pride, and increasing teamwork. Further it is a means for creating an atmosphere of innovation for continued and permanent quality improvement. Continuous quality improvement as outlined by the Joint Commission on Accreditation of Healthcare Organizations is designed to improve the work processes within and across organizations.

171.5 Tools Used for Quality Improvement

The tools listed on the following pages will assist in developing quality programs, collecting data, and assessing performance indicators within the organization. These tools include several of the most frequently used and most of the **seven tools of quality**. The seven tools of quality are tools that help health care organizations understand their processes in order to improve them. The tools are the cause-and-effect diagram, check sheet, control chart, flowchart, histogram, Pareto chart, and scatter diagram. Additional tools shown are the Shewhart cycle (PDCA process) and the bar chart. The Clinical Engineering Manager must access the situation and determine which tool will work best for his/her situational needs.

Two of the seven tools of quality discussed above are not illustrated. These are the scatter diagram and the check sheet. The scatter diagram is a graphic technique to analyze the relationship between two variations and the check sheet is simple data-recording device. The check sheet is custom designed by the user, which facilitates interpretation of the results. Most Biomedical Equipment Technicians use the check sheet on a daily basis when performing preventive maintenance, calibration, or electrical safety checks.

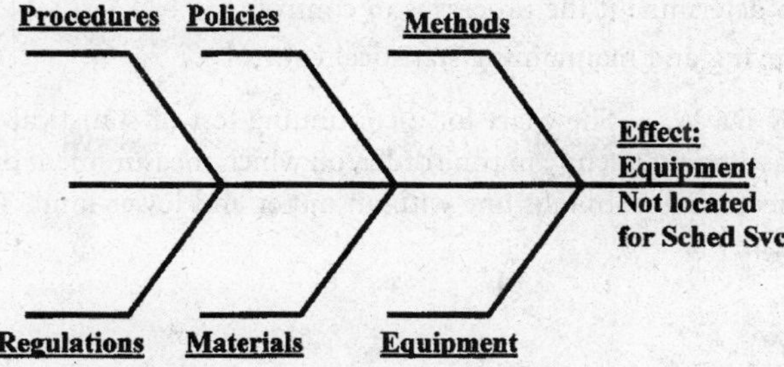

FIGURE 171.1 Cause-and-effect or Ishikawa chart.

Cause-and-Effect or Ishikawa Chart

This is a tool for analyzing process dispersion (Fig. 171.1). The process was developed by Dr. Karou Ishikawa and is also known as the *fishbone diagram* because the diagram resembles a fish skeleton. The diagram illustrates the main causes and subcauses leading to an effect. The cause-and-effect diagram is one of the seven tools of quality.

The following is an overview of the process:

- Used in group problem solving as a *brainstorming tool* to explore and display the possible causes of a particular problem.
- The *effect* (problem, concern, or opportunity) that is being investigated is stated on the right side, while the contributing *causes* are grouped in component categories through group brainstorming on the left side.
- This is an extremely effective tool for focusing a group brainstorming session.
- Basic components include environment, methods (measurement), people, money information, materials, supplies, capital equipment, and intangibles.

Control Chart

A control chart is a graphic representation of a characteristic of a process showing plotted values of some statistic gathered from that characteristic and one or two control limits (Fig. 171.2). It has two basic uses:

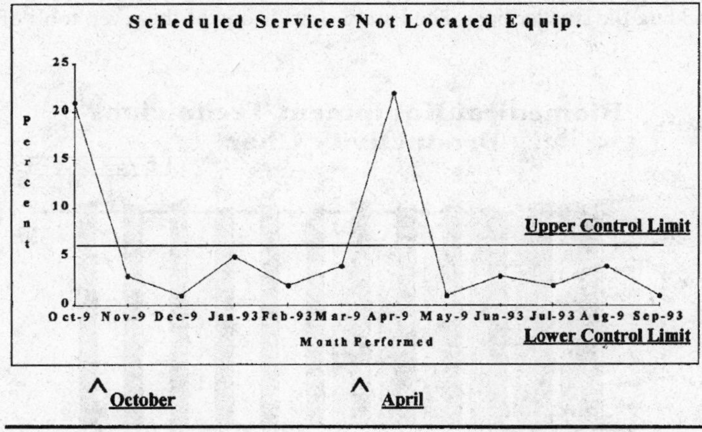

FIGURE 171.2 Control chart.

- as a judgment to determine if the process is in control
- as an aid in achieving and maintaining statistical control

(This chart was used by Dr. W. A. Shewhart for a continuing test of statistical significance.) A control chart is a chart with a baseline, frequently in time order, on which measurement or counts are represented by points that are connected by a straight line with an upper and lower limit. The control chart is one of the seven tools of quality.

Flowchart

A flowchart is a pictorial representation showing all the steps of a process (Fig. 171.3). Flowcharts provide excellent documentation of a program and can be a useful tool for examining how various steps in a process are related to each other. Flowcharting uses easily recognizable symbols to represent the type of processing performed. The flowchart is one of the seven tools of quality.

Start

Process Step

Decision

No

Fix

Yes

Stop

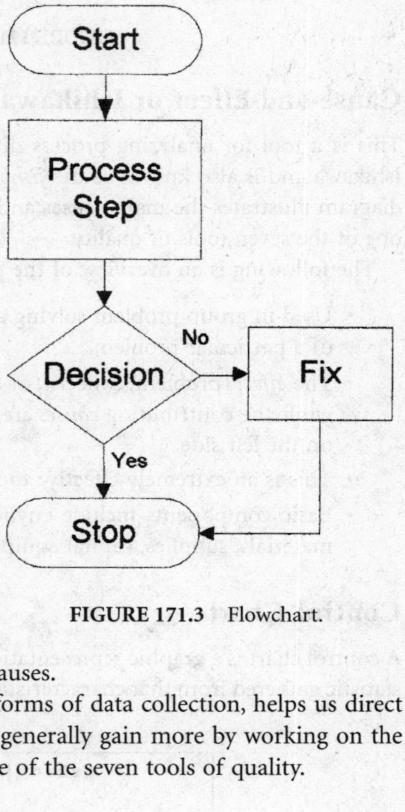

FIGURE 171.3 Flowchart.

Histogram

A graphic summary of variation in a set of data is a histogram (Fig. 171.4). The pictorial nature of the histogram lets people see patterns that are difficult to see in a simple table of numbers. The histogram is one of the seven tools of quality.

Pareto Chart

A Pareto chart is a special form of vertical bar graph that helps us to determine which problems to solve and in what order (Fig. 171.5). It is based on the Pareto principle, which was first developed by J. M. Juran in 1950. The principle, named after the nineteenth-century economist Vilfredo Pareto, suggests that most effects come from relatively few causes; i.e., 80% of the effects come from 20% of the possible causes.

Doing a Pareto chart, based on either check sheets or other forms of data collection, helps us direct our attention and efforts to truly important problems. We will generally gain more by working on the tallest bar than tackling the smaller bars. The Pareto chart is one of the seven tools of quality.

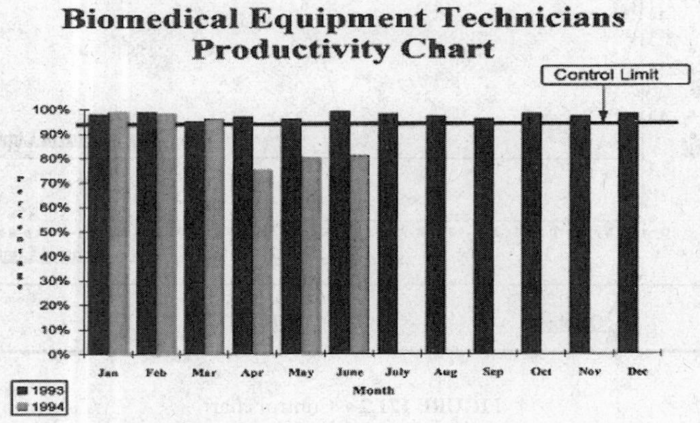

FIGURE 171.4 Histogram.

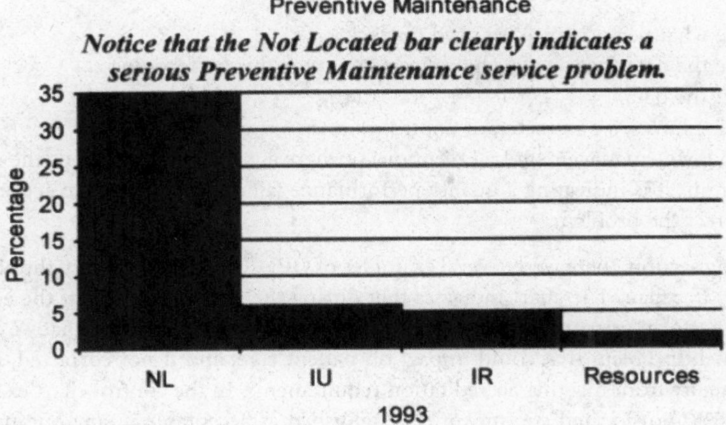

FIGURE 171.5 Pareto chart.

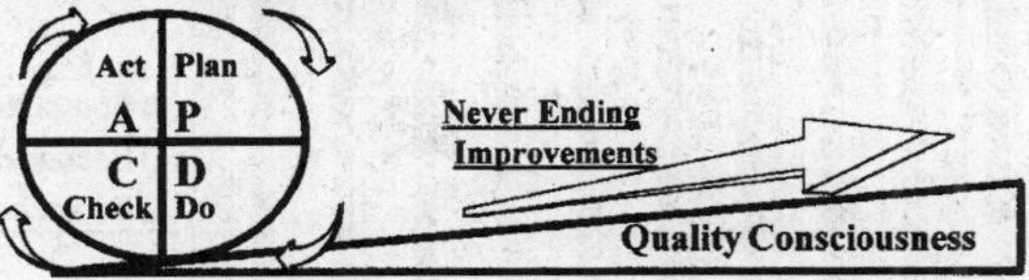

Step 1 - Plan (P): Collect data upon which a plan can be constructed.
Step 2 - Do (D): Take the necessary actions that further the developed plan.
Step 3 - Check (C): Check the results of our actions by collecting data to make sure that we have achieved what we planned.
Step 4 - Act (A): Act by making necessary changes to achieve Customer Satisfaction.
Act
Plan
A
P
C
D
Check
Do
Never Ending Improvements
Quality Consciousness

FIGURE 171.6 The Shewhart cycle.

The Plan-Do-Check-Act or Shewhart Cycle

This is a four-step process for quality improvement that is sometimes referred to as the *Deming cycle* (Fig. 171.6). One of the consistent requirements of the cycle is the long-term commitment required. The Shewart cycle or PDCA cycle is outlined here and has had overwhelming success when used properly. It is also a very handy tool to use in understanding the quality cycle process. The results of the cycle are studied to determine what was learned, what can be predicted, and appropriate changes to be implemented.

171.6 Quality Performance Indicators (QPI)

An *indicator* is something that suggests the existence of a fact, condition, or quality—an omen (a sign of future good or evil.) It can be considered as evidence of a manifestation or symptom of an incipient failure or problem. Therefore, quality performance indicators are measurements that can be used to ensure that quality performance is continuous and will allow us to know when incipient failures are starting so that we may take corrective and preventive actions.

QPI analysis is a five-step process:

Step 1: Decide what performance we need to track.
Step 2: Decide the data that need to be collected to track this performance.
Step 3: Collect the data.
Step 4: Establish limits, a parameter, or control points.
Step 5: Utilize BME (management by exception)—where a performance exceeds the established control limits, it is indicating a quality performance failure, and corrective action must be taken to correct the problem.

In the preceding section, there were several examples of QPIs. In the Pareto chart, the NL = not located, IU = in use, IR = in repair. The chart indicates that during the year 1994, 35% of the equipment could not be located to perform preventive maintenance services. This indicator tells us that we could eventually have a serious safety problem that could impact on patient care, and if not corrected, it could prevent the health care facility from meeting accreditation requirements. In the control chart example, an upper control limit of 6% "not located equipment" is established as acceptable in any one month. However, this upper control limit is exceeded during the months of April and October. This QPI could assist the clinical and/or Biomedical Equipment Manager in narrowing the problem down to a 2-month period. The histogram example established a lower control limit for productivity at 93%. However, productivity started to drop off in May, June, and July. This QPI tells the manager that something has happened that is jeopardizing the performance of his or her organization. Other performance indicators have been established graphically in Figs 171.7 and 171.8. See if you can determine what the indicators are and

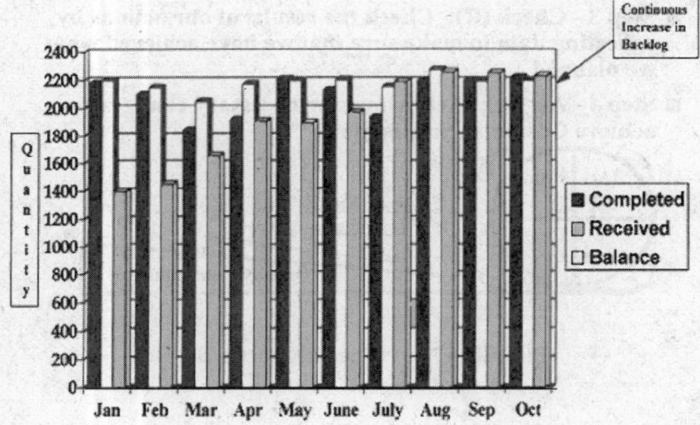

FIGURE 171.7 Sample repair service report.

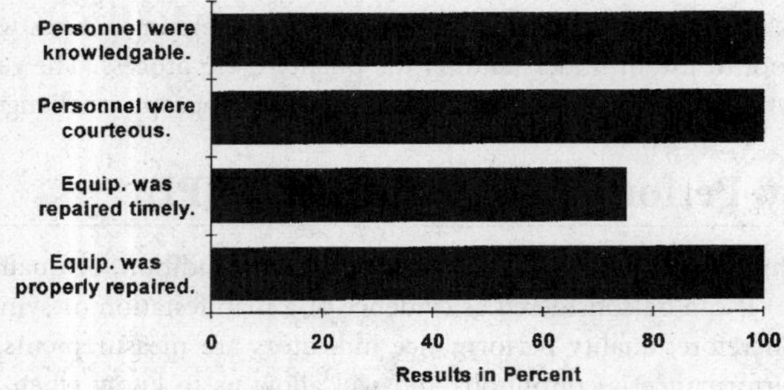

FIGURE 171.8 Customer satisfaction survey. August-September 1994.

what the possible cause might be. You may wish to use these tools to establish QPI tracking germane to your own organization.

171.7 Teams

A *team* is a formal group of persons organized by the company to work together to accomplish certain goals and objectives. Normally, when teams are used in quality improvement programs, they are designed to achieve the organization's vision. The organization's vision is a statement of the desired end state of the organization articulated and deployed by the executive leadership. Organizational visions are inspiring, clear, challenging, reasonable, and empowering. Effective visions honor the past, while they prepare for the future. The following are types of teams that are being used in health care facilities today. Some of the names may not be common, but their definitions are very similar if not commensurate.

Process Action Teams (PAT)

Process action teams are composed of those who are involved in the process being investigated. The members of a PAT are often chosen by their respective managers. The primary consideration for PAT membership is knowledge about the operations of the organization and consequently the process being studied. The main function of a PAT is the performance of an improvement project. Hence customers are often invited to participate on the team. PATs use basic statistical and other tools to analyze a process and identify potential areas for improvement. PATs report their findings to an Executive Steering Committee or some other type of quality management improving group ("*A problem well defined is half solved.*" John Dewey, American philosopher and educator; 1859–1952.)

Transition Management Team (TMT)

The transition management team (see *Harvard Business Review*, Nov–Dec 1993, pp 109–118) is normally used for a major organizational change such as restructuring or reengineering. The TMT can be initiated due to the findings of a PAT, where it has been indicated that the process is severely broken and unsalvageable. The TMT is not a new layer of bureaucracy or a job for fading executives. The TMT oversees the large-scale corporate change effort. It makes sure that all change initiatives fit together. It is made up of 8 to 12 highly talented leaders who commit *all* their time making the transition a reality. The team members and what they are trying to accomplish must be accepted by the power structure of the organization. For the duration of the change process, they are the CEO's version of the National Guard. The CEO should be able to say, "I can sleep well tonight, the TMT is managing this." In setting up a TMT, organizations should adopt a fail-safe approach: Create a position to oversee the emotional and behavioral issues unless you can prove with confidence that you do not need one.

Quality Improvement Project Team (QIPT)

A quality improvement project team can be initiated due to the findings of a PAT, where it has been indicated that the process is broken. The main agenda of the QIPT is to improve the work process that managers have identified as important to change. The team studies this process methodically to find permanent solutions to problems. To do this, members can use many of the tools described in this chapter and in many other publications on quality and quality improvement available from schools, bookstores, and private organizations.

Executive Steering Committee (ESC)

This is an executive-level team composed of the Chief Executive Officer (CEO) of the organization and the executive staff that reports directly to the CEO. Whereas an organization may have numerous QMBs, PATs, and QIPTs, it has only one ESC. The ESC identifies strategic goals for organizational quality improvement efforts. It obtains information from customers to identify major product and service requirements. It is through the identification of these major requirements that quality goals for the organization are defined. Using this information, the ESC lists, prioritizes, and determines how to measure the organization's goals for quality improvement. The ESC develops the organization's improvement plan and manages the execution of that plan to ensure that improvement goals are achieved.

Quality Management Board (QMB)

This is a permanent cross-functional team made up of top and midlevel managers who are jointly responsible for a specific product, service or process. The structure of the board intended to improve communication and cooperation by providing vertical and horizontal "links" throughout the organization.

171.8 Process Improvement Model

This process is following the Joint Commission on the Accreditation of Healthcare Organizations' Quality Cube, a method of assessing the quality of the organization.

PLAN:

a. Identify the process to be monitored.
b. Select important functions and dimensions of performance applicable to the process identified.
c. Design a tool for collection of data.

MEASURE (Under this heading, you will document how, when, and where data was collected):

a. Collect data
b. Select the appropriate tool to deliver your data (charts, graphs, tables, etc.)

ASSESS (Document findings under this heading):

a. Interpret data collected.
b. Design and implement change.
   (1) Redesign the process or tool if necessary.
   (2) If no changes are necessary, then you have successfully used the Process Improvement pathway.

IMPROVEMENT (Document details here):

a. Set in place the process to gain and continue the improvement.

OUTCOME (Document all changes here):

a. Positive changes made to improve quality of care based on Performance Improvement Activity.

Problem-Solving Model

The FOCUS-PDCA/PMAIO Process Improvement Model is a statistical-based quality control method for improving processes. This approach to problem-solving could be used by all Process Action Teams to ensure uniformity within an organization. FOCUS-PDCA/PMAIO is as follows:

F — Find a process to improve.
O — Organize a team that knows the process.
C — Clarify current knowledge of the process.
U — Understand the cause or variations.
S — Select the process to improve

P — Plan the improvement.	P — Plan
D — Do the improvement (pilot test).	M — Measure
C — Check the results of the improvement.	A — Assess
A — Act to hold the gain.	I — Improve
	O — Outcome

171.9 Summary

Although quality can be simply defined as conformity to customer or user requirements, it has many dimensions. Seven of them are described here: (1) performance, (2) aesthetics, (3) reliability (how dependably it performs), (4) availability (there when you need it), (5) durability (how long it lasts), (6) extras or features (supplementary items), and (7) serviceability (how easy it is to get serviced). The word *PARADES* can help you remember this:

PARADES Seven Dimensions of Quality

1. *Performance*: A product or service that performs its intended function well scores high on this dimension of quality.
2. *Aesthetics:* A product or service that has a favorable appearance, sound, taste, or smell is perceived to be of good quality.
3. *Reliability*: Reliability, or dependability, is such an important part of product quality that quality-control engineers are sometimes referred to as *reliability engineers.*
4. *Availability*: A product or service that is there when you need it.
5. *Durability*: Durability can be defined as the amount of use one gets from a product before it no longer functions properly and replacement seems more feasible than constant repair.
6. *Extras*: Feature or characteristics about a product or service that supplements its basic functioning (i.e., remote control dialing on a television).
7. *Serviceability*: Speed, courtesy, competence, and ease of repair are all important quality factors.

Quality Has a Monetary Value!

Good quality often pays for itself, while poor quality is expensive in both measurable costs and hidden costs. The hidden costs include loss of goodwill, including loss of repeat business and badmouthing of the firm. High quality goods and services often carry a higher selling price than do those of low quality. This information is evidenced by several reports in the *Wall Street Journal, Forbes Magazine, Money Magazine, Business Week Magazine,* etc. A good example is the turnaround of Japanese product sales using quality methodologies outlined in *The Deming Guide to Quality and Competitive Position,* by Howard S. and Shelly J. Gitlow. As Dr. Demings has stated *quality improvement* must be *continuous*!

Quality is never an accident; it has always the result of intelligent energy.

John Ruskins, 1819–1900
English art critic and historian
Seven Lamps of Architecture

References

Bittel LR. 1985. Every Supervisor Should Know, 5th ed, pp 455–456. New York, Gregg Division/McGraw-Hill.

DuBrin AJ. 1994. Essentials of Management, 3d ed. Cleveland, South-Western Publishing Co.

Duck JD. 1993. Managing change—The art of balancing. Harvard Business Review, November–December 1993, pp 109–118.

Gitlow HS, Gitlow SJ. 1987. The Deming Guide to Quality and Competitive Position. Englewood Cliffs, NJ, Prentice-Hall.

Goal/QPC. 1988. The Memory Jogger. A Pocket Guide of Tools for Continuous Improvement, Massachusetts, Goal/QPC.

Ishikawa K. 1991. Guide to Quality Control. New York, Quality Resources (Asian Productivity Organization, Tokyo, Japan).

Joint Commission on Accreditation of Healthcare Organizations—Comprehensive Accreditation Manual for Hospitals—Official Handbook, Library of Congress Number 96–076721 1998, Oakbrook Terrace, Illinois 60181.

Juran JM. 1979. Quality Control Handbook, 3d ed. New York, McGraw-Hill.

Katzenbach JR, Smith DK. 1994. The Wisdom of Teams. New York, Harper Business. A Division of Harper Collins Publishers.

Mizuno S. 1988. Management for Quality Improvement. The Seven New QC Tools. Boston, Productivity Press.

Sholters PR. 1993. The Team Handbook. How to Use Teams to Improve Quality. Madison, Wisc, Joiner Associates, Inc.

Walton M. 1986. The Deming Management Method. New York, Putnam Publishing Group.

172

A Standards Primer for Clinical Engineers

Alvin Wald
Columbia University

172.1 Introduction

The development, understanding, and use of standards is an important component of a clinical engineer's activities. Whether involved in industry, a health care facility, governmental affairs, or commercial enterprise, one way or another, the clinical engineer will find that standards are a significant aspect of professional activities. With the increasing emphasis on health care cost containment and efficiency, coupled with the continued emphasis on patient outcome, standards must be viewed both as a mechanism to reduce expenses and as another mechanism to provide quality patient care. In any case, standards must be addressed in their own right, in terms of technical, economic, and legal implications.

It is important for the clinical engineer to understand fully how standards are developed, how they are used, and most importantly, how they affect the entire spectrum of health related matters. Standards exist that address systems (protection of the electrical power distribution system from faults), individuals (means to reduce potential electric shock hazards), and protection of the environment (disposal of deleterious waste substances).

From a larger perspective, standards have existed since biblical times. In the Book of Genesis [Chap. 6, ver. 14], Noah is given a construction standard by God, "Make thee an ark of gopher wood; rooms shalt thou make in the ark, and shalt pitch it within and without with pitch." Standards for weights and measures have played an important role in bringing together human societies through trade and commerce. The earliest record of a standard for length comes from ancient Egypt, in Dynasty IV (circa 3000 B.C.). This length was the royal cubit, 20.620 inches (52.379 cm), as used in construction of the Great Pyramid.

The importance of standards to society is illustrated in the Magna Carta, presented by the English barons to King John in 1215 on the field at Runnymede. Article 35 states:

> "There shall be standard measures of wine, beer, and corn—the London quarter—throughout the whole of our kingdom, and a standard width of dyed, russet and halberject cloth—two ells within the selvedges; and there shall be standard weights also."

0-8493-0462-8/00/$0.00+$.50

The principles of this article appear in the English Tower system for weight and capacity, set in 1266 by the assize of Bread and Ale Act:

> "An English penny called a sterling, round and without any clipping, shall weigh thirty-two wheatcorns in the midst of the ear; and twenty ounces a pound: and eight pounds do make a gallon of wine, and eight gallons of wine do make a bushell, which is the eighth part of a quarter."

In the United States, a noteworthy use of standards occurred after the Boston fire of 1689. With the aim of rapid rebuilding of the city, the town fathers specified that all bricks used in construction were to be $9 \times 4 \times 4$ inches. An example of standardization to promote uniformity in manufacturing practices was the contract for 10,000 muskets awarded to Eli Whitney by President Thomas Jefferson in 1800. The apocryphal story is that Eli Whitney (better known to generations of grammar school children for his invention of the cotton gin) assembled a large number of each musket part, had one of each part randomly selected, and then assembled a complete working musket. This method of production, the complete interchangeability of assembly parts, came to be known as the "armory method," replacing hand crafting, which at that time had been the prevailing method of manufacturing throughout the world.

172.2 Definitions

A most general definition of a standard is given by Rowe [1983].

> "A standard is a multi-party agreement for establishing an arbitrary criterion for reference."

Each word used in the definition by Rowe corresponds to a specific characteristic that helps to define the concept of a standard.

Multi means more than one party, organization, group, government, agency, or individual.

Agreement means that the concerned parties have come to some mutually agreed upon understanding of the issues involved and of ways to resolve them. This understanding has been confirmed via some mechanism such as unanimity, consensus, ballot, or other means that has been specified.

Establishing defines the purpose of the agreement—to create the standard and carry forth its provisions.

Arbitrary emphasizes an understanding by the parties that there are no absolute criteria in creating the standard. Rather, the conditions and values chosen are based on the most appropriate knowledge and conditions available at the time the standard was established.

Criterions are those features and conditions that the parties to the agreement have chosen as the basis for the standard. Not all issues may be addressed, but only those deemed, for whatever reasons, suitable for inclusion.

A different type of definition of a standard is given in The United States Office of Management and Budget Circular A-119:

> "… a prescribed set of rules, conditions, or requirements concerned with the definition of terms; classification of components; delineation of procedures; specifications of materials, performance, design, or operations; or measurement of quality and quantity in describing materials, products, systems, services, or practices."

A code is a compilation of standards relating to a particular area of concern, i.e., a collection of standards. For example, local government health codes contain standards relating to providing of health care to members of the community. A regulation is an organization's way of specifying that some particular standard must be adhered to. Standards, codes, and regulations may or may not have legal implications, depending on whether the promulgating organization is governmental or private.

172.3 Standards for Clinical Engineering

There is a continually growing body of standards that affect health care facilities, and hence clinical engineering. The practitioner of health care technology must constantly search out, evaluate, and apply appropriate standards. The means to reconcile the conflicts of technology, cost considerations, the different jurisdictions involved, and the implementation of the various standards is not necessarily apparent. One technique that addresses these concerns and has proven to yield a consistent practical approach is a structured framework of the various levels of standards. This hierarchy of standards is a conceptual model that the clinical engineer can use to evaluate and apply to the various requirements that exist in the procurement and use of health care technology.

Standards have different purposes, depending on their particular applications. A hierarchy of standards can be used to delineate those conditions for which a particular standard applies. There are four basic categories, any one or all of which may be in simultaneous operation. (1) Local or proprietary standards (perhaps more properly called regulations) are developed to meet the internal needs of a particular organization. (2) Common interest standards serve to provide uniformity of product or service throughout an industry or profession. (3) Consensus standards are agreements amongst interested participants to address an area of mutual concern. (4) Regulatory standards are mandated by an authority having jurisdiction to define a particular aspect of concern. In addition, there are two categories of standards adherence: (1) voluntary standards, which carry no inherent power of enforcement, but provide a reference point of mutual understanding, and (2) mandatory standards, which are incumbent upon those to whom the standard is addressed, and enforceable by the authority having jurisdiction.

The hierarchy of standards model can aid the clinical engineer in the efficient and proper use of standards. More importantly, it can provide standards developers, users, and the authorities having jurisdiction in these matters with a structure by which standards can be effectively developed, recognized, and used to the mutual benefit of all.

172.4 A Hierarchy of Standards

Local, or proprietary standards, are developed for what might be called internal use. An organization that wishes to regulate and control certain of its own activities issues its own standards. Thus, the standard is local in the sense that it is applied in a specific venue, and it is proprietary in that it is the creation of a completely independent administration. For example, an organization may standardize on a single type of an electrocardiograph monitor. This standardization can refer to a specific brand or model, or to specific functional or operational features. In a more formal sense, a local standard may often be referred to as an institutional Policy and Procedure. The policy portion is the why of it; the procedure portion is the how. It must be kept in mind that standards of this type that are too restrictive will limit innovation and progress, in that they cannot readily adapt to novel conditions. On the other hand, good local standards contribute to lower costs, operational efficiency, and a sense of coherence within the organization.

Sometimes, local standards may originate from requirements of a higher level of regulation. For example, the Joint Commission for Accreditation of Healthcare Organizations [JCAHO] (formerly the Joint Commission for Hospital Accreditation (JCAH), a voluntary organization, (but an organization that hospitals belong to for various reasons, e.g., accreditation, reimbursement, approval of training programs) does not set standards for what or how equipment should be used. Rather, the JCAHO requires that each hospital set its own standards on how equipment is selected, used, and maintained. To monitor compliance with this requirement, the JCAHO inspects whether the hospital follows its own standards. In one sense, the most damaging evidence that can be adduced against an organization (or an individual) is that it (he) did not follow its (his) own standards.

Common interest standards are based on a need recognized by a group of interested parties, which will further their own interests, individually or collectively. Such standards are generally accepted by affected interests without being made mandatory by an authority; hence they are one type of voluntary standard. These standards are often developed by trade or professional organizations to promote uniformity in a product or process. This type of standard may have no inducement to adherence except for the benefits to the individual participants. For example, if you manufacture a kitchen cabinet that is not of standard size, it will not fit into the majority of kitchens and thus it will not sell. Uniformity of screw threads is another example of how a product can be manufactured and used by diverse parties, and yet be absolutely interchangeable. More recently, various information transfer standards allow the interchange of computer-based information amongst different types of instruments and computers.

Consensus standards are those that have been developed and accepted in accordance with certain well defined criteria so as to assure that all points of view have been considered. Sometimes, the adjective "consensus" is used as a modifier for a "voluntary standard." Used in this context, consensus implies that all interested parties have been consulted and have come to a general agreement on the provisions of the standard. The development of a consensus standard follows an almost ritualistic procedure to insure that fairness and due process are maintained. There are various independent voluntary and professional organizations that sponsor and develop standards on a consensus basis (see below). Each such organization has its own particular rules and procedures to make sure that there is a true consensus in developing a standard.

In the medical products field, standards are sometimes difficult to implement because of the independent nature of manufacturers and their high level of competition. A somewhat successful standards story is the adoption of the DIN configuration for ECG lead-cable connection by the Association for the Advancement of Medical Instrumentation [AAMI]. The impetus for this standard was the accidental electrocution of several children brought about by use of the previous industry standard lead connection (a bare metal pin, as opposed to the new recessed socket). Most (but not all) manufacturers of ECG leads and cables now adhere to this standard. Agreement on this matter is in sharp contrast to the inability of the health care manufacturing industry to implement a standard for ECG cable connectors. Even though a standard was written, the physical configuration of the connector is not necessarily used by manufacturers in production, nor is it demanded by medical users in purchasing. Each manufacturer uses a different connector, leading to numerous problems in supply and incompatibility for users. This is an example of a voluntary standard, which for whatever reasons, is effectively ignored by all interested parties.

However, even though there have been some failures in standardization of product features, there has also been significant progress in generating performance and test standards for medical devices. A number of independent organizations sponsor development of standards for medical devices. For example, the American Society for Testing and Materials [ASTM] has developed, "Standard Specification for Minimum Performance and Safety Requirements for Components and Systems of Anesthesia Gas Machines (F1161-88)." Even though there is no statutory law that requires it, manufacturers no longer produce, and thus hospitals can no longer purchase anesthesia machines without the built-in safety features specified in this standard. The Association for the Advancement of Medical Instrumentation has sponsored numerous standards that relate to performance of specific medical devices., such as defibrillators, electrosurgical instruments, and electronic sphygmomanometers. These standards are compiled in the AAMI publication, "Essential Standards for Biomedical Equipment Safety and Performance." The National Fire Protection Association [NFPA] publishes "Standard for Health Care Facilities (NFPA 99)," which covers a wide range of safety issues relating to facilities. Included are sections that deal with electricity and electrical systems, central gas and vacuum supplies, and environmental conditions. Special areas such as anesthetizing locations, laboratories, and hyperbaric facilities are addressed separately.

Mandatory standards have the force of law or other authority having jurisdiction. Mandatory standards imply that some authority has made them obligatory. Mandatory standards can be written by the authority having jurisdiction, or they can be adapted from documents prepared by others as proprietary or consensus standards. The authority having jurisdiction can be a local hospital or even a department within the hospital, a professional society, a municipal or state government, or an agency of the federal government that has regulatory powers.

In the United States, hospitals are generally regulated by a local city or county authority, and/or by the state. These authorities set standards in the form of health codes or regulations, which have the force of law. Often, these local bodies consider the requirements of a voluntary group, the Joint Commission for Accreditation of Healthcare Organizations, in their accreditation and regulatory processes.

American National Standards. The tradition in the United States is that of voluntary standards. However, once a standard is adopted by an organization, it can be taken one step further. The American National Standards Institute (ANSI) is a private, non-government, voluntary organization that acts as a coordinating body for standards development and recognition in the United States. If the development process for a standard meets the ANSI criteria of open deliberation of legitimate concerns, with all interested parties coming to a voluntary consensus, then the developers can apply (but are not required) to have their standard designated as an American National Standard. Such a designation does not make a standard any more legitimate, but it does offer some recognition as to the process by which it has been developed. ANSI also acts as a clearing house for standards development, so as to avoid duplication of effort by various groups that might be concerned with the same issues. ANSI is also involved as a U.S. coordinating body for many international standards activities.

An excellent source that lists existing standards and standards generating organizations, both nationally and internationally, along with some of the workings of the FDA (see below), is the "Medical Device Industry Fact Book" [Allen, 1996].

172.5 Medical Devices

On the national level, oversight is generally restricted to medical devices, and not on operational matters. Federal jurisdiction of medical devices falls under the purview of the Department of Health and Human Services, Public Health Service, Food and Drug Administration (FDA), Center for Devices and Radiological Health. Under federal law, medical devices are regulated under the "Medical Device Amendments of 1976" and the "Radiation Control for Health and Safety Act of 1968." Additional regulatory authorization is provided by the "Safe Medical Devices Act of 1990", the "Medical Device Amendments of 1992," the "FDA Reform and Enhancement Act of 1996," and the "Food and Drug Administration Modernization Act of 1997."

A medical device is defined by Section 201 of the Federal Food, Drug, and Cosmetic Act (as amended), as an:

> instrument, apparatus, implement, machine, contrivance, implant, *in vitro* reagent, or other similar or related article including any component, part, or accessory which is:
>
> recognized in the official National Formulary, or the United States Pharmacopeia, or any supplement to them;
>
> intended for use in the diagnosis of disease or other conditions, or in the care, mitigation, treatment, or prevention of disease, in man or other animals, or

intended to affect the structure of any function of the body of man or other animals; and which does not achieve its primary intended purposes through chemical action within or on the body of man…and which is not dependent upon being metabolized for the achievement of its primary intended purposes.

The major thrust of the FDA has been in the oversight of the manufacture of medical devices, with specific requirements based on categories of perceived risks. The 1976 Act (Section 513) establishes three classes of medical devices intended for human use:

> Class I. General controls regulate devices for which controls other than performance standards or premarket approvals are sufficient to assure safety and effectiveness. Such controls include regulations that (1) prohibit adulterated or misbranded devices; (2) require domestic device manufacturers and initial distributors to register their establishments and list their devices; (3) grant FDA authority to ban certain devices; (4) provide for notification of risks and of repair, replacement, or refund; (5) restrict the sale, distribution, or use of certain devices; and (6) govern Good Manufacturing Practices, records, and reports, and inspections. These minimum requirements apply also to Class II and Class III devices.
>
> Class II. Performance Standards apply to devices for which general controls alone do not provide reasonable assurance of safety and efficacy, and for which existing information is sufficient to establish a performance standard that provides this assurance. Class II devices must comply not only with general controls, but also with an applicable standard developed under Section 514 of the Act. Until performance standards are developed by regulation, only general controls apply.

Class III. Premarket Approval applies to devices for which general controls do not suffice or for which insufficient information is available to write a performance standard to provide reasonable assurance of safety and effectiveness. Also, devices which are used to support or sustain human life or to prevent impairment of human health, devices implanted in the body, and devices which present a potentially unreasonable risk of illness or injury. New Class III devices, those not "substantially equivalent" to a device on the market prior to enactment (May 28, 1976), must have approved Premarket Approval Applications (Section 510 k).

Exact specifications for General Controls and Good Manufacturing Practices (GMP) are defined in various FDA documents. Aspects of General Controls include yearly manufacturer registration, device listing, and premarket approval. General Controls are also used to regulate adulteration, misbranding and labeling, banned devices, and restricted devices. Good Manufacturing Practices include concerns of organization and personnel; buildings and equipment; controls for components, processes, packaging, and labeling; device holding, distribution, and installation; manufacturing records; product evaluation; complaint handling; and a quality assurance program. Design controls for GMP were introduced in 1996. They were motivated by the FDA's desire to harmonize its requirements with those of a proposed international standard (ISO 13485). Factors that need to be addressed include planning, input and output requirements, review, verification and validation, transfer to production, and change procedures, all contained in a history file for each device. Device tracking is typically required for Class III life-sustaining and implant devices, as well as postmarket surveillance for products introduced starting in 1991.

Other categories of medical devices include combination devices, in which a device may incorporate drugs or biologicals. Combination devices are controlled via intercenter arrangements implemented by the FDA.

Transitional devices refer to devices that were regulated as drugs, prior to the enactment of the Medical Device Amendments Act of 1976. These devices were automatically placed into Class III, but may be transferred to Class I or II.

A custom device may be ordered by a physician for his/her own use or for a specific patient. These devices are not generally available, and cannot be labeled or advertised for commercial distribution.

An investigational device is one that is undergoing clinical trials prior to premarket clearance. If the device presents a significant risk to the patient, an Investigational Device Exemption must be approved by the FDA. Information must be provided regarding the device description and intended use, the origins of the device, the investigational protocol, and proof of oversight by an Institutional Review Board to insure informed patient consent. Special compassionate or emergency use for a nonapproved device or for a nonapproved use can be obtained from the FDA under special circumstances, such as when there is no other hope for the patient.

Adverse Events. The Safe Medical Devices Act of 1990 included a provision by which both users and manufacturers (and distributors) of medical devices are required to report adverse patient events that may be related to a medical device. Manufacturers must report to the FDA if a device (a) may have caused or contributed to a death or serious injury, or (b) malfunctioned in such a way as would be likely to cause or contribute to a death or serious injury if the malfunction were to reoccur. Device users are required to notify the device manufacturer of reportable incidents, and must also notify the FDA in case of a device-related death. In addition, the FDA established a voluntary program for reporting device problems that may not have caused an untoward patient event, but which may have the potential for such an occurrence under altered circumstances.

New devices. As part of the General Controls requirements, the FDA must be notified prior to marketing any new (or modifying an existing) device for patient use. This premarket notification, called the 510(k) process after the relevant section in the Medical Device Amendments Act, allows the FDA to review the device for safety and efficacy.

There are two broad categories that a device can fall into. A device that was marketed prior to May 28, 1976 (the date that the Medical Device Amendments became effective) can continue to be sold. Also, a product that is "substantially equivalent" to a preamendment device can likewise be marketed. However, the FDA may require a premarket approval application for any Class III device (see below). Thus, these preamendment devices and their equivalents are approved by "grandfathering." (Premarket notification to the FDA is still required to assure safety and efficacy). Of course, the question of substantial equivalency is open to an infinite number of interpretations. From the manufacturer's perspective, such a designation allows marketing the device without a much more laborious and expensive premarket approval process.

A new device that the FDA finds is not substantially equivalent to a premarket device is automatically placed into Class III. This category includes devices that provide functions or work through principles not present in preamendment devices. Before marketing, this type of device requires a Premarket Approval Application by the manufacturer, followed by an extensive review by the FDA. (However, the FDA can reclassify such devices into Class I or II, obviating the need for premarket approval). The review includes scientific and clinical evaluation of the application by the FDA and by a Medical Advisory Committee (composed of outside consultants). In addition, the FDA looks at the manufacturing and control processes to assure that all appropriate regulatory requirements are being adhered to. Clinical (use of real patients) trials are often required for Class III devices in order to provide evidence of safety and efficacy. To carry out such trials, an Investigational Device Exemption must be issued by the FDA.

The Food and Drug Administration Modernization Act of 1997, which amends section 514 of the Food, Drug, and Cosmetic Act, has made significant changes in the above regulations. These changes greatly simplify and accelerate the entire regulatory process. For example, the law exempts from premarket notification Class I devices that are not intended for a use that is of substantial importance in preventing impairment of human health, or that do not present a potential unreasonable risk of illness or injury. Almost 600 Class I generic devices have been so classified by the agency. In addition, the FDA will specify those Class II devices for which a 510(k) submission will also not be required.

Several other regulatory changes have been introduced by the FDA to simplify and speed up the approval process. So-called "third party" experts will be allowed to conduct the initial review of all Class I and low-to-intermediate risk Class II devices. Previously, the FDA was authorized to create standards for medical devices. The new legislation allows the FDA to recognize and use all or parts of various appropriate domestic and internationally recognized consensus standards that address aspects of safety and/or effectiveness relevant to medical devices.

172.6 International Standards

Most sovereign nations have their own internal agencies to establish and enforce standards. However, in our present world of international cooperation and trade, standards are tending towards uniformity across national boundaries. This internationalization of standards is especially true since formation of the European Common Market. The aim here is to harmonize the standards of individual nations by promulgating directives for medical devices that address "Essential Requirements" [Freeman, 1993] (see below). Standards in other areas of the world (Asia, Eastern Europe) are much more fragmented, with each country specifying regulations for its own manufactured and imported medical devices.

There are two major international standards generating organizations, both based in Europe, the International Electrotechnical Commission (IEC) and the International Organization for Standardization (ISO). Nations throughout the world participate in the activities of these organizations.

The International Electrotechnical Commission [IEC], founded in 1906, oversees, on an international level, all matters relating to standards for electrical and electronic items. Membership in the IEC is held by a National Committee for each nation. The United States National Committee (USNC) for IEC was founded in 1907, and since 1931 has been affiliated with ANSI. USNC has as its members representatives from professional societies, trade associations, testing laboratories, government entities, other organizations, and individual experts. The USNC appoints a technical advisor and a technical advisory group for each IEC Committee and Subcommittee to help develop a unified United States position. These advisory groups are drawn from groups that are involved in the development of related U.S. national standards.

Standards are developed by Technical Committees (TC), Subcommittees (SC), and Working Groups (WG). IEC TC 62, "Electrical Equipment in Medical Practice," is of particular interest here. One of the basic standards of this Technical Committee is document 601-1, "Safety of Medical Electrical Equipment,

Part 1: General Requirements for Safety," 2nd Edition (1988) and its Amendment 1 (1991), along with Document 601-1-1, "Safety Requirements for Medical Electrical Systems" (1992).

The International Organization for Standardization [ISO]. oversees aspects of device standards other than those related to electrotechnology. This organization was formed in 1946 with a membership comprised of the national standards organizations of 26 countries. There are currently some 90 nations as members. The purpose of the ISO is to "facilitate international exchange of goods and services and to develop mutual cooperation in intellectual, scientific, technological, and economic ability." ISO addresses all aspects of standards except for electrical and electronic issues, which are the purview of the International Electrotechnical Commission. ANSI has been the official United States representative to ISO since its inception. For each Committee or Subcommittee of the ISO in which ANSI participates, a U.S. Technical Advisory Group (TAG) is formed. The administrator of the TAG is, typically, that same U.S. organization that is developing the parallel U.S. standard.

Technical Committees (TC) of the ISO concentrate on specific areas of interest. There are Technical Committees, Subcommittees, Working Groups and Study Groups. One of the member national standards organizations serves as the Secretariat for each of these technical bodies.

One standard of particular relevancy to manufacturers throughout the world is ISO 9000. This standard was specifically developed to assure a total quality management program that can be both universally recognized and applied to any manufacturing process. It does not address any particular product or process, but is concerned with structure and oversight of how processes are developed, implemented, monitored, and documented. An independent audit must be passed by any organization to obtain ISO 9000 registration. Many individual nations and manufacturers have adopted this standard and require that any product that they purchase be from a source that is ISO 9000 compliant.

The European Union was, in effect, created by the Single Europe Act, (EC-92) as a region "without internal frontiers in which the free movement of goods, persons, and capital is ensured." For various products and classes of products, the European Commission issues directives with regard to safety and other requirements, along with the means for assessing conformity to these directives. Products that comply with the appropriate directives can then carry the CE mark. EU member states ratify these directives into national law.

Two directives related to medical devices are the Medical Devices Directive (MDD), enacted in 1993 (mandatory as of June 15,1998), and the Active Implanted Medical Devices Directive (AIMDD), effective since 1995. Safety is the primary concern of this system, and as in the U.S., there are three classes of risk. These risks are based on what and for how long the device touches, and its effects. Safety issues include electrical, mechanical, thermal, radiation, and labeling. Voluntary standards that address these issues are formulated by the European Committee for Standardization (CEN) and the European Committee for Electrotechnical Standardization (CENELEC).

172.7 Compliance with Standards

Standards that were originally developed on a voluntary basis may take on mandatory aspects. Standards that were developed to meet one particular need may be used to satisfy other needs as well. Standards will be enforced and adhered to if they meet the needs of those who are affected by them. For example, consider a standard for safety and performance for a defibrillator. For the manufacturer, acceptance and sales are a major consideration in both the domestic and international markets. People responsible for specifying, selecting, and purchasing equipment may insist on adherence to the standard so as to guarantee safety and performance. The user, physician or other health care professional, will expect the instrument to have certain operational and performance characteristics to meet medical needs. Hospital personnel want a certain minimum degree of equipment uniformity for ease of training and maintenance. The hospital's insurance company and risk manager want equipment that meets or exceeds recognized safety standards. Third party payers, that is private insurance companies or government agencies, insist on equipment that is safe, efficacious, and cost effective. Accreditation agencies, such as local health agencies or professional societies, often require equipment to meet certain standards. More basically, patients, workers, and society as a whole have an inherent right to fundamental safety. Finally, in our litigatious society, there is always the threat of civil action in the case of an untoward event in which a "non-standard", albeit "safe", instrument was involved. Thus, even though no one has stated "this standard must be followed", it is highly unlikely that any person or organization will have the temerity to manufacture, specify, or buy an instrument that does not "meet the standard."

Another example of how standards become compulsory is via accreditation organizations. The Joint Commission for Accreditation of Healthcare Organizations has various standards (requirements). This organization is a private body that hospitals voluntarily accept as an accrediting agent. However, various health insurance organizations, governmental organizations, and physician specialty boards for resident education use accreditation by the JCAHO as a touchstone for quality of activities. Thus, an insurance company might not pay for care in a hospital that is not accredited, or a specialty board might not recognize resident training in such an institution. Thus, the requirements of the JCAHO, in effect, become mandatory standards for health care organizations.

A third means by which voluntary standards can become mandatory is by incorporation. Existing standards can be incorporated into a higher level of standards or codes. For example, various state and local governments incorporate standards developed by voluntary organizations, such as the National Fire Protection Association, into their own building and health codes. These standards then become, in effect, mandatory government regulations, and have the force of (civil) law. In addition, as discussed above, the FDA will now recognize voluntary standards developed by recognized organizations.

172.8 Limitations of Standards

Standards are generated to meet the expectations of society. They are developed by organizations and individuals to meet a variety of specific needs, with the general goals of promoting safety and efficiency. However, as with all human activities, problems with the interpretation and use of standards do occur. Engineering judgment is often required to help provide answers. Thus, the clinical engineer must consider the limits of standards, a boundary that is not clear and is constantly shifting. Yet clinical engineers must always employ the highest levels of engineering principles and practices. Some of the limitations and questions of standards and their use will be discussed below.

Non-Compliance with a Standard. Sooner or later, it is likely that a clinical engineer will either be directly involved with or become aware of deviation from an accepted standard. The violation may be trivial, with no noticeable effect, or there may be serious consequences. In the former case, either the whole incident may be ignored, or nothing more may be necessary than a report that is filed away, or the incident can trigger some sort of corrective action. In the latter case, there may be major repercussions involving investigation, censure, tort issues, or legal actions. In any event, lack of knowledge about the standard is not a convincing defense. Anyone who is in a position that requires knowledge about a standard should be fully cognizant of all aspects of that standard. In particular, one should know the provisions of the standard, how they are to be enforced, and the potential risks of non-compliance. Nonetheless, non-compliance with a standard, in whole or in part, may be necessary to prevent a greater risk or to increase a potential benefit to the patient. For example, when no other recourse is available, it would be defensible to use an electromagnet condemned for irreparable excessive leakage current to locate a foreign body in the eye of an injured person, and thus save the patient's vision. Even if the use of this device resulted in a physical injury or equipment damage, the potential benefit to the patient is a compelling argument for use of the non-compliant device. In such a case, one should be aware of and prepared to act on the possible hazard (excessive electrical current, here). A general disclaimer making allowance for emergency situations is often included in policy statements relating to use of a standard. Drastic conditions require drastic methods.

Standards and the Law. Standards mandated by a government body are not what is called "black letter law", that is a law actually entered into a criminal or civil code. Standards are typically not adopted in the same manner as laws, i.e., they are not approved by a legislative body, ratified by an elected executive, and sanctioned by the courts. The usual course for a mandated standard is via a legislative body enacting a law that establishes or assigns to an executive agency the authority to regulate the concerned activities. This agency, under the control of the executive branch of government, then issues standards that follow the mandate of its enabling legislation. If conflicts arise, in addition to purely legal considerations, the judiciary must interpret the intent of the legislation in comparison with its execution. This type of law falls under civil rather than criminal application.

The penalty for non-compliance with a standard may not be criminal or even civil prosecution. Instead, there are administrative methods of enforcement, as well as more subtle yet powerful methods of coercion. The state has the power (and the duty) to regulate matters of public interest. Thus, the state can withhold or withdraw permits for construction, occupancy, or use. Possibly more effective, the state can withhold means of finance or payments to violators of its regulations. Individuals injured by failure to abide by a standard may sue for damages in civil proceedings. However, it must be recognized that criminal prosecution is possible when the violations are most egregious, leading to human injury or large financial losses.

Incorporation and Revision. Because of advances in technology and increases in societal expectations, standards are typically revised periodically. For example, the National Fire Protection Association revises and reissues its "Standard for Health Care Facilities" (NFPA 99) every three years. Other organizations follow a five year cycle of review, revision, and reissue of standards. These voluntary standards, developed in good faith, may be adapted by governmental agencies and made mandatory, as discussed above. When a standard is incorporated into a legislative code, it is generally referenced as to a particular version and date. It is not always the case that a newer version of the standard is more restrictive. For example, ever since 1984, the National Fire Protection Association "Standard for Health Care Facilities" (NFPA 99) does not require the installation of isolated power systems (isolation transformers and line isolation monitors) in anesthetizing locations that do not use flammable anesthetic agents, or in areas that are not classified as wet locations. A previous version of this standard, "Standard for the Use of Inhalation Anesthetics, (Flammable and Nonflammable)" (NFPA 56A-1978) did require isolated power. However, many State Hospital Codes have incorporated, by name and date, the provisions of the older standard, NFPA 56A. Thus, isolated power may still be required, by code, in new construction of all anesthetizing locations, despite the absence of this requirement in the latest version of the standard that addresses this issue. In such a case, the organization having jurisdiction in the matter must be petitioned to remedy this conflict between new and old versions of the standard.

Safety. The primary purpose of standards in clinical practice is to assure the safety of patient, operator and bystanders. However, it must be fully appreciated that there is no such thing as absolute safety. The more safety features and regulations attached to a device, the less useful and the more cumbersome and costly may be its actual use. In the development, interpretation, and use of a standard, there are questions that must be asked: What is possible? What is acceptable? What is reasonable? Who will benefit? What is the cost? Who will pay?

No one can deny that medical devices should be made as safe as possible, but some risk will always remain. In our practical world, absolute safety is a myth. Many medical procedures involve risk to the patient. The prudent physician or medical technologist will recognize the possible dangers of the equipment and take appropriate measures to reduce the risk to a minimum. Some instruments and procedures are inherently more dangerous than others. The physician must make a judgment, based on his/her own professional knowledge and experience, as well as on the expectations of society, whether using a particular device is less of a risk than using an alternative device or doing nothing. Standards will help—but they do not guarantee complete safety, a cure, or legal and societal approval.

Liability. Individuals who serve on committees that develop standards, as well as organizations involved in such activities, are justifiably concerned with their legal position in the event that a lawsuit is instituted as a result of a standard that they helped to bring forth. Issues involved in such a suit may include restraint of trade, in case of commercial matters, or to liability for injury due to acts of commission or of omission. Organizations that sponsor standards or that appoint representatives to standards developing groups often have insurance for such activities. Independent standards committees and individual members of any standards committees may or may not be covered by insurance for participation in these activities. Although in recent times only one organization and no individual has been found liable for damages caused by improper use of standards (see following paragraph), even the possibility of being named in a lawsuit can intimidate even the most self-confident "expert". Thus, it is not at all unusual for an individual who is asked to serve on a standards development committee first to inquire as to liability insurance coverage. Organizations that develop standards or appoint representatives also take pains to insure that all of their procedures are carefully followed and documented so as to demonstrate fairness and prudence.

The dark side of standards is the implication that individuals or groups may unduly influence a standard to meet a personal objective, e.g., to dominate sales in a particular market. If standards are developed or interpreted unfairly, or if they give an unfair advantage to one segment, then restraint of trade charges can be made. This is why standards to be deemed consensus must be developed in a completely open and fair manner. Organizations that sponsor standards that violate this precept can be held responsible. In 1982, the United States Supreme Court, in the Hydrolevel Case [Perry, 1982], ruled that the American Society of Mechanical Engineers was guilty of antitrust activities because of the way some of its members, acting as a committee to interpret one of its standards, issued an opinion that limited competition in sales so as to unfairly benefit their own employers. This case remains a singular reminder that standards development and use must be inherently fair.

Inhibition. Another charge against standards is that they inhibit innovation and limit progress [Flink, 1984]. Ideally, standards should be written to satisfy minimum, yet sufficient, requirements for safety, performance, and efficacy. Improvements or innovations would still be permitted so long as the basic standard is followed. From a device users point of view, a standard that is excessively restrictive may limit the scope of permissible professional activities. If it is necessary to abrogate a standard in order to accommodate a new idea or to extend an existing situation, then the choice is to try to have the standard changed, which may be very time consuming, or to act in violation of the standard and accept the accompanying risks and censure.

Ex Post Facto. A question continually arises as to what to do about old equipment (procedures, policies, facilities, etc.) when a new standard is issued or an old standard is revised so that existing items become obsolete. One approach, perhaps the simplest, is to do nothing. The philosophy here being that the old equipment was acquired in good faith and conformed to the then existing standards. As long as that equipment is usable and safe, there is no necessity to replace it. Another approach is to upgrade the existing equipment to meet the new standard. However, such modification may be technically impractical or financially prohibitive. Finally, one can simply throw out all of the existing equipment (or sell it to a second-hand dealer, or use the parts for maintenance) and buy everything new. This approach would bring a smile of delight from the manufacturer and a scream of outrage from the hospital administrator. Usually what is done is a compromise, incorporating various aspects of these different approaches.

Costs. Standards cost both time and money to propose, develop, promulgate, and maintain. Perhaps the greatest hindrance to more participation in standards activities by interested individuals is the lack of funds to attend meetings where the issues are discussed and decisions are made. Unfortunately, but nonetheless true, organizations that can afford to sponsor individuals to attend such meetings have considerable influence in the development of that standard. On the other hand, those organizations that do have a vital interest in a standard should have an appropriate say in its development. A consensus of all interested parties tempers the undue influence of any single participant.

From another viewpoint, standards increase the costs of manufacturing devices, carrying out procedures, and administering policies. This incremental cost is, in turn, passed on to the purchaser of the goods or services. Whether or not the increased cost justifies the benefits of the standard is not always apparent. It is impossible to realistically quantify the costs of accidents that did not happen or the confusion that was avoided by adhering to a particular standard. However, it cannot be denied that standards have made a valuable contribution to progress, in the broadest sense of that word.

172.9 Conclusions

Standards are just like any other human activity, they can be well used or a burden. The danger of standards is that they will take on a life of their own; and rather than serve a genuine need will exist only as a justification of their own importance. This view is expressed in the provocative and iconoclastic book by Bruner and Leonard [1989], and in particular in their Chapter 9, 'Codes and Standards: Who Makes the Rules?' However, the raison d'être of standards is to do good. It is incumbent upon clinical engineers, not only to understand how to apply standards properly, but also how to introduce, modify, and retire standards as conditions change. Furthermore, the limitations of standards must be recognized in order to realize their maximum benefit. No standard can replace diligence, knowledge, and a genuine concern for doing the right thing.

References

Allen, A. (Editor) 1996. *Medical Device Industry Fact Book*, 3rd Edition. Canon Communications, Santa Monica, CA.

American Society for Testing and Materials [ASTM] 1916 Race Street, Philadelphia, PA 19103.

Association for the Advancement of Medical Instrumentation [AAMI] 3330 Washington Boulevard, Suite 400, Arlington, VA 22201.

Bruner, J. M. R. and Leonard, P. F. 1989. *Electricity, Safety and the Patient*, Year Book Medical Publishers, Chicago.

Flink, R. 1984. Standards: Resource or Constraint?, *IEEE Eng. Med. Biol. Mag.*, 3(1):14-16.

Food and Drug Administration, Center for Devices and Radiological Health, 5600 Fishers Lane, Rockville, MD 20857. URL: http://www.fda.gov/

Freeman, M. 1993. The EC Medical Devices Directives, *IEEE Eng. Med. Biol. Mag.*, 12(2):79-80.

International Organization for Standardization [ISO], Central Secretariat, 1 rue de Varembe, Case postale 56, CH 1211, Geneva 20, Switzerland. URL: http://www.iso.ch/index.html

International Electrotechnical Commission [IEC], Central Office, 3 rue de Varembé, P. O. Box 131, CH-1211, Geneva 20, Switzerland. URL: http://www.iec.ch/

Joint Commission on Accreditation of Health Care Facilities [JCAHO] 1 Renaissance Boulevard, Oakbrook, IL 60181.

National Fire Protection Association [NFPA] Batterymarch Park, Quincy, MA 02269.

Perry, T. S. 1982. Antirust Ruling Chills Standards Setting, *IEEE Spectrum*, 19(8):52-54.

Rowe, W. D. 1983. Design and Performance Standards. In: *Medical Devices: Measurements, Quality Assurance, and Standards*, C. A. Caceres, H. T. Yolken, R. J. Jones, and H. R. Piehler, (Eds.) p. 29-40, American Society for Testing and Materials, Philadelphia, PA.

173

Regulatory and Assessment Agencies

Mark E. Bruley
ECRI

Vivian H. Coates
ECRI

Effective management and development of clinical and biomedical engineering departments (hereafter called clinical engineering departments) in hospitals requires a basic knowledge of relevant regulatory and technology assessment agencies. Regulatory agencies set standards of performance and record keeping for the departments and the technology for which they are responsible. Technology assessment agencies are information resources for what should be an ever expanding role of the clinical engineer in the technology decision-making processes of the hospital's administration.

This chapter presents an overview of regulatory and technology assessment agencies in the U.S., Canada, Europe, and Australia that are germane to clinical engineering. Due to the extremely large number of such agencies and information resources, we have chosen to focus on those of greatest relevance and/or informational value. The reader is directed to the references and sources of further information presented at the end of the chapter.

173.1 Regulatory Agencies

Within the healthcare field, there are over 38,000 applicable standards, clinical practice guidelines, laws, and regulations [ECRI, 1999]. Voluntary standards are promulgated by more than 800 organizations; mandatory standards by more than 300 state and federal agencies. Many of these organizations and agencies issue guidelines that are relevant to the vast range of healthcare technologies within the responsibility of clinical engineering departments. Although many of these agencies also regulate the manufacture and clinical use of healthcare technology, such regulations are not directly germane to the management of a clinical department and are not presented.

For the clinical engineer, many agencies promulgate regulations and standards in the areas of, for example, electrical safety, fire safety, technology management, occupational safety, radiology and nuclear medicine, clinical laboratories, infection control, anesthesia and respiratory equipment, power distribution, and medical gas systems. In the U.S., medical device problem reporting is also regulated by many state agencies and by the U.S. Food and Drug Administration (FDA) via its MEDWATCH program. It is important to note that, at present, the only direct regulatory authority that the FDA has over U.S. hospitals is in the reporting of medical device related accidents that result in serious injury or death. Chapter 172 discusses in detail many of the specific agency citations. Presented below are the names and addresses of the primary agencies whose codes, standards, and regulations have the most direct bearing on clinical engineering and technology management.

0-8493-0462-8/00/$0.00+$.50

American Hospital Association
One North Franklin
Chicago, IL 60606
(312) 422-3000
Website: www.aha.org

American College of Radiology
1891 Preston White Drive
Reston, VA 22091
(703) 648-8900
Website: www.acr.org

American National Standards Institute
11 West 42nd Street
13th Floor
New York, NY 10036
(212) 642-4900
Website: www.ansi.org

American Society for Hospital Engineering
840 North Lake Shore Drive
Chicago, IL 60611
(312) 280 5223
Website: www.ashe.org

American Society for Testing and Materials
1916 Race Street
Philadelphia, PA 19103
(215) 299-5400
Website: www.astm.org

Association for the Advancement of Medical Instrumentation
3330 Washington Boulevard
Suite 400
Arlington, VA 22201
(703) 525-4890
Website: www.aami.org

Australian Institute of Health and Welfare
GPO Box 570
Canberra, ACT 2601
Australia
(61) 06-243-5092
Website: www.aihw.gov.au

British Standards Institution
2 Park Street
London, W1A 2BS
United Kingdom
(44) 071-629-9000
Website: www.bsi.org.uk

Canadian Healthcare Association
17 York Street
Ottawa, ON K1N 9J6
Canada
(613) 241-8005
Website: www.canadian-healthcare.org

CSA International
178 Rexdale Boulevard
Etobicoke, ON M9W 1R3
Canada
(416) 747-4000
Website: www.csa-international.org

Center for Devices and Radiological Health
Food and Drug Administration
9200 Corporate Boulevard
Rockville, MD 20850
(301) 443-4690
Website: www.fda.gov/cdrh

Compressed Gas Association, Inc.
1725 Jefferson Davis Highway
Suite 1004
Arlington, VA 22202
(703) 412-0900

ECRI
5200 Butler Pike
Plymouth Meeting, PA 19462
(610) 825-6000
(610) 834-1275 fax
Websites: www.ecri.org
www.ecriy2k.org
www.mdsr.ecri.org

Environmental Health Directorate
Health Protection Branch
Health Canada
Environmental Health Centre
19th Floor, Jeanne Mance Building
Tunney's Pasture
Ottawa, ON K1A 0L2 Canada
(613) 957-3143
Website: www.hc-sc.gc.ca/hpb/index_e.html

Therapeutic Products Programme
Health Canada
Holland Cross, Tower B
2nd Floor
1600 Scott Street

Address Locator #3102D1
Ottawa, ON K1A 1B6
(613) 954-0288
Website: www.hc-sc.gc.ca/hpb-dgps/therapeut

Food and Drug Administration
MEDWATCH, FDA Medical Products Reporting Program
5600 Fishers Lane
Rockville, MD 20857-9787
(800) 332-1088
Website: www.fda.gov/cdrh/mdr.html

Institute of Electrical and Electronics Engineers
445 Hoes Lane
P.O. Box 1331
Piscataway, NJ 08850-1331
(732) 562-3800
Website: www.standards.ieee.org

International Electrotechnical Commission
Box 131
3 rue de Varembe, CH 1211
Geneva 20
Switzerland
(41) 022-919-0211
Website: www.iec.ch

International Organization for Standardization
1 rue de Varembe
Case postale 56, CH 1211
Geneva 20
Switzerland
(41) 022-749-0111
Website: www.iso.ch

Joint Commission on Accreditation of Healthcare Organizations
One Renaissance Boulevard
Oakbrook Terrace, IL 60181
(630) 792-5600
Website: www.jcaho.org

Medical Devices Agency
Department of Health
Room 1209
Hannibal House
Elephant and Castle
London, SE1 6TQ
United Kingdom
(44) 171-972-8143
Website: www.medical-devices.gov.uk

National Council on Radiation Protection and Measurements
7910 Woodmont Avenue, Suite 800
Bethesda, MD 20814
(310) 657-2652
Website: www.ncrp.com

National Fire Protection Association
1 Batterymarch Park
PO Box 9101
Quincy, MA 02269-9101
(617) 770-3000
Website: www.nfpa.org

Nuclear Regulatory Commission
11555 Rockville Pike
Rockville, MD 20852
(301) 492-7000
Website: www.nrc.gov

Occupational Safety and Health Administration
US Department of Labor
Office of Information and Consumer Affairs
200 Constitution Avenue, NW
Room N3647
Washington, DC 20210
(202) 219-8151
Website: www.osha.gov

ORKI
National Institute for Hospital and Medical Engineering
Budapest dios arok 3, H-1125
Hungary
(33) 1-156-1522

Radiation Protection Branch
Environmental Health Directorate
Health Canada
775 Brookfield Road
Ottawa, ON K1A 1C1
Website: www.hc-sc.gc.ca/ehp/ehd/rpb

Russian Scientific and Research Institute
Russian Public Health Ministry
EKRAN, 3 Kasatkina Street
Moscow
Russia 129301
(44) 071-405-3474

Society of Nuclear Medicine, Inc.
1850 Samuel Morse Drive
Reston, VA 20190-5316
(703) 708-9000
Website: www.snm.org

Standards Association of Australia
PO Box 1055
Strathfield, NSW 2135
Australia
(61) 02-9746-4700
Website: www.standards.org.au

Therapeutic Goods Administration
PO Box 100
Wooden, ACT 2606
Australia
(61) 2-6232-8610
Website: www.health.gov.au/tga

Underwriters Laboratories, Inc.
333 Pfingsten Road
Northbrook, IL 60062-2096
(847) 272-8800
Website: www.ul.com

VTT
Technical Research Center of Finland
Postbox 316
SF-33101 Tampere 10
Finland
(358) 31-163300
Website: www.vti.fi

173.2 Technology Assessment Agencies

Technology assessment is the practical process of determining the value of a new or emerging technology in and of itself or against existing or competing technologies using safety, efficacy, effectiveness, outcome, risk management, strategic, financial, and competitive criteria. Technology assessment also considers ethics and law as well as health priorities and cost-effectiveness compared to competing technologies. A "technology" is defined as devices, equipment, related software, drugs, biotechnologies, procedures, and therapies; and systems used to diagnose or treat patients. The processes of technology assessment are discussed in detail in Chapter 168.

Technology assessment is not the same as technology acquisition/procurement or technology planning. The latter two are processes for determining equipment vendors, soliciting bids, and systematically determining a hospital's technology related needs based on strategic, financial, risk management, and clinical criteria. The informational needs differ greatly between technology assessment and the acquisition/procurement or planning processes. This section focuses on the resources applicable to technology assessment.

Worldwide, there are nearly 400 organizations (private, academic, and governmental), providing technology assessment information, databases, or consulting services. Some are strictly information clearing houses, some perform technology assessment, and some do both. For those that perform assessments, the quality of the information generated varies greatly from superficial studies to in-depth, well referenced analytical reports. In 1997, the U.S. Agency for Health Care Policy and Research (AHCPR) designated 12 "Evidence-Based Practice Centers" (EPC) to undertake major technology assessment studies on a contract basis. Each of these EPCs are noted in the list below and general descriptions of each center may be viewed on the internet at the AHCPR Website http://www.ahcpr.gov/clinic/epc/.

Language limitations are a significant issue. In the ultimate analysis, the ability to undertake technology assessment requires assimilating vast amounts of information, most of which exists only in the English language. Technology assessment studies published by the International Society for Technology Assessment in Health Care (ISTAHC), by the World Health Organization, and other umbrella organizations are generally in English. The new International Health Technology Assessment database being developed by ECRI in conjunction with the U.S. National Library of Medicine contains more than 30,000 citations to technology assessments and related documents.

Below are the names, mailing addresses, and Internet Website addresses of some of the most prominent organizations undertaking technology assessment studies:

Agence Nationale pour le Developpement de l'Evaluation Medicale
159 Rue Nationale
Paris 75013
France
(33) 42-16-7272
Website: www.upml.fr/andem/andem.htm

Agencia de Evaluacion de Technologias Sanitarias
Ministerio de Sanidad y Consumo
Instituto de Salud Carlos III, AETS
Sinesio Delgado 6, 28029 Madrid
Spain
(34) 1-323-4359
Website: www.isciii.es/aets

Alberta Heritage Foundation for Medical Research
125 Manulife Place
10180-101 Street
Edmonton, AB T5J 345
(403) 423-5727
Website: www.ahfmr.ab.ca

American Association of Preferred Provider Organizations
601 13th Street, NW
Suite 370 South
Washington, DC 20005
(202) 347-7600

American Academy of Neurology
1080 Montreal Avenue
St. Paul, MN 55116-2791
(612) 695-2716
Website: www.aan.com

American College of Obstetricians and Gynecologists
409 12th Street, SW
Washington, DC 20024
(202) 863-2518
Website: www.acog.org

Australian Institute of Health and Welfare
GPO Box 570
Canberra, ACT 2601
Australia
(61) 06-243-5092
Website: www.aihw.gov.au

Battelle Medical Technology Assessment and Policy Research Center (MEDTAP)
901 D Street, SW
Washington, DC 20024
(202) 479-0500
Website: www.battelle.org

Blue Cross and Blue Shield Association
Technology Evaluation Center
225 N Michigan Avenue
Chicago, IL 60601-7680
(312) 297-5530
(312) 297-6080 (publications)
Website: www.bluecares.com/new/clinical
(*An EPC of AHCPR*)

British Columbia Office of Health Technology Assessment
Centre for Health Services & Policy Research, University of British Columbia
429-2194 Health Sciences Mall
Vancouver, BC V6T 1Z3
CANADA
(604) 822-7049
Website: www.chspr.ubc.ca

British Institute of Radiology
36 Portland Place
London, W1N 4AT
United Kingdom
(44) 171-580-4085
Website: www.bir.org.uk

Canadian Coordinating Office for Health Technology Assessment
110-955 Green Valley Crescent
Ottawa ON K2C 3V4
CANADA
(613) 226-2553
Website: www.ccohta.ca

Canadian Healthcare Association
17 York Street
Ottawa, ON K1N 9J6
Canada
(613) 241-8005
Website: www.canadian-healthcare.org

Catalan Agency for Health
Technology Assessment
Travessera de les Corts 131-159
Pavello Avenue
Maria, 08028 Barcelona
Spain
(34) 93-227-29-00
Website: www.aatm.es

Centre for Health Economics
University of York
York Y01 5DD
United Kingdom
(44) 01904-433718
Website: www.york.ac.uk

Center for Medical Technology Assessment
Linköping University
5183 Linköping, Box 1026 (551-11)
Sweden
(46) 13-281-000

Center for Practice and Technology Assessment
Agency for Health Care Policy
and Research (AHCPR)
6010 Executive Boulevard, Suite 300
Rockville, MD 20852
(301) 594-4015
Website: www.ahcpr.gov

Committee for Evaluation and Diffusion
of Innovative Technologies
3 Avenue Victoria
Paris 75004
France
(33) 1-40-273-109

Conseil d'evaluation des technologies
de la sante du Quebec
201 Cremazie Boulevard East
Bur 1.01
Montreal, PQ H2M 1L2
Canada
(514) 873-2563
Website: www.msss.gouv.qc.ca

Danish Hospital Institute
Landermaerket 10
Copenhagen K
Denmark DK1119
(45) 33-11-5777

Danish Medical Research Council
Bredgade 43
1260 Copenhagen
Denmark
(45) 33-92-9700

Danish National Board of Health
Amaliegade 13
PO Box 2020
Copenhagen K
Denmark DK1012
(45) 35-26-5400

Duke Center for Clinical Health Policy Research
Duke University Medical Center
2200 West Main Street, Suite 320
Durham, NC 27705
(919) 286-3399
Website: www.clinipol.mc.duke.edu
(*An EPC of AHCPR*)

ECRI
5200 Butler Pike
Plymouth Meeting, PA 19462
(610) 825-6000
(610) 834-1275 fax
Websites:www.ecri.org
www.ecriy2k.org
www.mdsr.ecri.org
(*An EPC of AHCPR*)

Finnish Office for Health Care
Technology Assessment
PO Box 220
FIN-00531 Helsinki
Finland
(35) 89-3967-2296
Website: www.stakes.fi/finohta

Frost and Sullivan, Inc.
106 Fulton Street
New York, NY 10038-2786
(212) 233-1080
Website: www.frost.com

Health Council of the Netherlands
PO Box 1236
2280 CE
Rijswijk
The Netherlands
(31) 70-340-7520

Health Services Directorate Strategies
and Systems for Health
Health Promotion
Health Promotion and Programs Branch
Health Canada
1915B Tunney's Pasture
Ottawa, ON K1A 1B4
Canada
(613) 954-8629
Website: www.hc-sc.gc.ca/hppb/hpol

Health Technology Advisory Committee
121 East 7th Place, Suite 400
PO Box 64975
St. Paul, MN 55164-6358
(612) 282-6358

Hong Kong Institute of Engineers
9/F Island Centre
No. 1 Great George Street
Causeway Bay
Hong Kong

Institute for Clinical PET
7100-A Manchester Boulevard
Suite 300
Alexandria, VA 22310
(703) 924-6650
Website: www.icpet.org

Institute for Clinical Systems Integration
8009 34th Avenue South
Minneapolis, MN 55425
(612) 883-7999
Website: www.icsi.org

Institute for Health Policy Analysis
8401 Colesville Road, Suite 500
Silver Spring, MD 20910
(301) 565-4216

Institute of Medicine (U.S.)
National Academy of Sciences
2101 Constitution Avenue, NW
Washington, DC 20418
(202) 334-2352
Website: www.nas.edu/iom

International Network of Agencies for
Health Technology Assessment
c/o SBU, Box 16158
S-103 24 Stockholm
Sweden
(46) 08-611-1913
Website: www.sbu.se/sbu-site/links/inahta

Johns Hopkins Evidence-based Practice Center
The Johns Hopkins Medical Institutions
2020 E Monument Street, Suite 2-600
Baltimore, MD 21205-2223
(410) 955-6953
Website: www.jhsph.edu/Departments/Epi/
(*An EPC of AHCPR*)

McMaster University Evidence-based
Practice Center
1200 Main Street West, Room 3H7
Hamilton, ON L8N 3Z5
Canada
(905) 525-9140 ext. 22520
Website: http://hiru.mcmaster.ca.epc/
(*An EPC of AHCPR*)

Medical Alley
1550 Utica Avenue, South
Suite 725
Minneapolis, MN 55416
(612) 542-3077
Website: www.medicalalley.org

Medical Devices Agency
Department of Health
Room 1209
Hannibal House
Elephant and Castle
London, SE1 6TQ
United Kingdom
(44) 171-972-8143
Website: www.medical-devices.gov.uk

Medical Technology Practice Patterns Institute
4733 Bethesda Avenue, Suite 510
Bethesda, MD 20814
(301) 652-4005
Website: www.mtppi.org

MEDTAP International
7101 Wisconsin Avenue, Suite 600
Bethesda MD 20814
(301) 654-9729
Website: www.medtap.com

MetaWorks, Inc.
470 Atlantic Avenue
Boston, MA 02210
(617) 368-3573 ext. 206
Website: www.metawork.com
(*An EPC of AHCPR*)

National Institute of Nursing Research, NIH
31 Center Drive, Room 5B10. MSC 2178
Bethesda, MD 20892-2178
(301) 496-0207
Website: www.nih.gov/ninr

National Commission on Quality Assurance
2000 L Street NW, Suite 500
Washington. DC 20036
(202) 955-3500
Website: www.ncqa.org

National Committee of Clinical Laboratory Standards (NCCLS)
940 West Valley Road, Suite 1400
Wayne, PA 19087-1898
(610) 688-0100
Website: www.nccls.org

National Coordinating Center for Health Technology Assessment
Boldrewood (Mailpoint 728)
Univ of Southampton SO16 7PX
United Kingdom
(44) 170-359-5642
Website: www.soton.ac.uk/~hta/address.htm

National Health and Medical Research Council
GPO Box 9848
Canberra, ACT
Australia
(61) 06-289-7019

New England Medical Center
Center for Clinical Evidence Synthesis
Division of Clinical Research
750 Washington Street, Box 63
Boston, MA 02111
(617) 636-5133
Website: www.nemc.org/medicine/ccr/cces.htm
(*An EPC of AHCPR*)

New York State Department of Health
Tower Building
Empire State Plaza
Albany, NY 12237
(518) 474-7354
Website: www.health.state.ny.us

NHS Centre for Reviews and Dissemination
University of York
York Y01 5DD
United Kingdom
(44) 01-904-433634
Website: www.york.ac.uk

Office of Medical Applications of Research
NIH Consensus Program Information Service
PO Box 2577
Kensington, MD 20891
(301) 231-8083
Website: odp.od.nih.gov/consensus

Ontario Ministry of Health
Hepburn Block
80 Grosvenor Street
10th Floor
Toronto, ON M7A 2C4
(416) 327-4377

Oregon Health Sciences University
Division of Medical Informatics and Outcomes Research
3181 SW Sam Jackson Park Road
Portland, OR 97201-3098
(503) 494-4277
Website: www.ohsu.edu/epc
(*An EPC of AHCPR*)

Pan American Health Organization
525 23rd Street NW
Washington, DC 20037-2895
(202) 974-3222
Website: www.paho.org

Physician Payment Review Commission (PPRC)
2120 L Street NW
Suite 510
Washington, DC 20037
(202) 653-7220

Prudential Insurance Company of America
Health Care Operations and Research Division
56 N Livingston Avenue
Roseland, NJ 07068
(201) 716-3870

Research Triangle Institute
3040 Cornwallis Road
PO Box 12194
Research Triangle Park, NC 27709-2194
(919) 541-6512
(919) 541-7480
Website: www.rti.org/epc/
(*An EPC of AHCPR*)

San Antonio Evidence-based Practice Center
University of Texas Health Sciences Center
Department of Medicine
7703 Floyd Curl Drive
San Antonio, TX 78284-7879
(210) 617-5190
Website: www.uthscsa.edu/
(*An EPC of AHCPR*)

Saskatchewan Health
Acute and Emergency Services Branch
3475 Albert Street
Regina, SK S4S 6X6
(306) 787-3656

Scottish Health Purchasing Information Centre
Summerfield House
2 Eday Road
Aberdeen AB15 6RE
Scotland
United Kingdom
(44) 0-1224-663-456, ext. 75246
Website: www.nahat.net/shpic

Servicio de Evaluacion de Technologias Sanitarias
Duque de Wellington 2
E01010 Vitoria-Gasteiz
Spain
(94) 518-9250
E-mail: osteba-san@ej-gv.es

Society of Critical Care Medicine
8101 E Kaiser Boulevard
Suite 300
Anaheim, CA 92808-2259
(714) 282-6000
Website: www.sccm.org

Swedish Council on Technology Assessment in Health Care
Box 16158
S-103 24 Stockholm
Sweden
(46) 08-611-1913
Website: www.sbu.se

Southern California EPC-RAND
1700 Main Street
Santa Monica, CA 90401
(310) 393-0411 ext. 6669
Website: www.rand.org/organization/health/epc/
(*An EPC of AHCPR*)

Swiss Institute for Public Health
Health Technology Programme
Pfrundweg 14
CH-5001 Aarau
Switzerland
(41) 064-247-161

TNO Prevention and Health
PO Box 2215
2301 CE Leiden
The Netherlands
(31) 71-518-1818
Website: www.tno.n1/instit/pg/index.html

University HealthSystem Consortium
2001 Spring Road, Suite 700
Oak Brook, IL 60523
(630) 954-1700
Website: www.uhc.edu

University of Leeds
School of Public Health
30 Hyde Terrace
Leeds L52 9LN
United Kingdom
Website: www.leeds.ac.uk

USCF-Stanford University EPC
University of California, San Francisco
505 Parnassus Avenue, Room M-1490
Box 0132
San Francisco, CA 94143-0132
(415) 476-2564
Website: www.stanford.edu/group/epc/
(*An EPC of AHCPR*)

U.S. Office of Technology Assessment
(former address) 600 Pennsylvania Avenue SE
Washington, DC 20003
Note: OTA closed on 29 Sep, 1995. However, documents can be accessed via the internet at www.wws.princeton.edu/§ota/html2/cong.html. Also a complete set of OTA publications is available on CD-ROM; contact the U.S. Government Printing Office (www.gpo.gov) for more information.

Veterans Administration
Technology Assessment Program
VA Medical Center (152M)
150 S Huntington Avenue
Building 4
Boston, MA 02130
(617) 278-4469
Website: www.va.gov/resdev

Voluntary Hospitals of America, Inc.
220 East Boulevard
Irving, TX 75014
(214) 830-0000

Wessex Institute of Health Research and Development
Boldrewood Medical School
Bassett Crescent East
Highfield, Southampton SO16 7PX
United Kingdom
(44) 01-703-595-661
Website: www.soton.ac.uk/~wi/index.html

World Health Organization
Distribution Sales
CH 1211
Geneva 27
Switzerland 2476
(41) 22-791-2111
Website: www.who.ch
Note: Publications are also available from the WHO Publications Center, USA, at (518) 436-9686.

References

ECRI. Healthcare Standards Official Directory. ECRI, Plymouth Meeting, PA, 1999.

Eddy DM. A manual for assessing health practices & designing practice policies: The explicit approach. American College of Physicians, Philadelphia, PA, 1992.

Goodman C, Ed. Medical technology assessment directory. National Academy Press, Washington, D.C., 1988.

van Nimwegen Chr, Ed. International list of reports on comparative evaluations of medical devices. TNO Centre for Medical Technology, Leiden, the Netherlands, 1993.

Marcaccio KY, Ed. Gale Directory of databases. Volume 1: Online databases. Gale Research International, London, 1993.

Further Information

A comprehensive listing of healthcare standards and the issuing organizations is presented in the *Healthcare Standards Directory* published by ECRI. The Directory is well organized by keywords, organizations and their standards, federal and state laws, legislation and regulations, and contains a complete index of names and addresses.

The *International Health Technology Assessment* database is produced by ECRI. A portion of the database is also available in the U.S. National Library of Medicine's new database called *HealthSTAR*. Internet access to *HealthSTAR* is through Website address http://igm.nlm.nih.gov. A description of the database may be found at http://www.nlm.nih.gov/pubs/factsheets/healthstar.html.

174

Applications of Virtual Instruments in Health Care

Eric Rosow
Hartford Hospital

Joseph Adam
Premise Development Corporation

174.1 Applications of Virtual Instruments in Health Care

Virtual Instrumentation (which was previously defined in Chapter 88, "Virtual Instrumentation: Applications in Biomedical Engineering") allows organizations to effectively harness the power of the PC to access, analyze, and share information throughout the organization. With vast amounts of data available from increasingly sophisticated enterprise-level data sources, potentially useful information is often left hidden due to a lack of useful tools. Virtual instruments can employ a wide array of technologies such as multidimensional analyses and Statistical Process Control (SPC) tools to detect patterns, trends, causalities, and discontinuities to derive knowledge and make informed decisions.

Today's enterprises create vast amounts of raw data and recent advances in storage technology, coupled with the desire to use this data competitively, has caused a data glut in many organizations. The healthcare industry in particular is one that generates a tremendous amount of data. Tools such as databases and spreadsheets certainly help manage and analyze this data; however databases, while ideal for extracting data are generally not suited for graphing and analysis. Spreadsheets, on the other hand, are ideal for analyzing and graphing data, but this can often be a cumbersome process when working with multiple data files. Virtual instruments empower the user to leverage the best of both worlds by creating a suite of user-defined applications which allow the end-user to convert vast amounts of data into information which is ultimately transformed into knowledge to enable better decision making.

This chapter will discuss several virtual instrument applications and tools that have been developed to meet the specific needs of healthcare organizations. Particular attention will be placed on the use of quality control and "performance indicators" which provide the ability to trend and forecast various metrics. The use of SPC within virtual instruments will also be demonstrated. Finally, a non-traditional application of virtual instrumentation will be presented in which a "peer review" application has been developed to allow members of an organization to actively participate in the Employee Performance Review process.

0-8493-0462-8/00/$0.00+$.50

Example Application #1: The EndoTester™—A Virtual Instrument-Based Quality Control and Technology Assessment System for Surgical Video Systems

The use of endoscopic surgery is growing, in large part because it is generally safer and less expensive than conventional surgery, and patients tend to require less time in a hospital after endoscopic surgery. Industry experts conservatively estimate that about 4 million minimally invasive procedures were performed in 1996. As endoscopic surgery becomes more common, there is an increasing need to accurately evaluate the performance characteristics of endoscopes and their peripheral components.

The assessment of the optical performance of laparoscopes and video systems is often difficult in the clinical setting. The surgeon depends on a high quality image to perform minimally invasive surgery, yet assurance of proper function of the equipment by biomedical engineering staff is not always straightforward. Many variables in both patient and equipment may result in a poor image. Equipment variables, which may degrade image quality, include problems with the endoscope, either with optics or light transmission. The light cable is another source of uncertainty as a result of optical loss from damaged fibers. Malfunctions of the charge coupled device (CCD) video camera are yet another source of poor image quality. Cleanliness of the equipment, especially lens surfaces on the endoscope (both proximal and distal ends) are particularly common problems. Patient variables make the objective assessment of image quality more difficult. Large operative fields and bleeding at the operative site are just two examples of patient factors that may affect image quality.

The evaluation of new video endoscopic equipment is also difficult because of the lack of objective standards for performance. Purchasers of equipment are forced to make an essentially subjective decision about image quality. By employing virtual instrumentation, a collaborative team of biomedical engineers, software engineers, physicians, nurses, and technicians at Hartford Hospital (Hartford, CT) and Premise Development Corporation (Avon, CT) have developed an instrument, the EndoTester™, with integrated software to quantify the optical properties of both rigid and flexible fiberoptic endoscopes. This easy-to-use optical evaluation system allows objective measurement of endoscopic performance prior to equipment purchase and in routine clinical use as part of a program of prospective maintenance.

The EndoTester™ was designed and fabricated to perform a wide array of quantitative tests and measurements. Some of these tests include: (1) Relative Light Loss, (2) Reflective Symmetry, (3) Lighted (Good) Fibers, (4) Geometric Distortion, and (5) Modulation Transfer Function (MTF). Each series of tests is associated with a specific endoscope to allow for trending and easy comparison of successive measurements.

Specific information about each endoscope (i.e., manufacturer, diameter, length, tip angle, department/unit, control number, and operator), the reason for the test (i.e., quality control, pre/post repair, etc.), and any problems associated with the scope are also documented through the electronic record. In addition, all the quantitative measurements from each test are automatically appended to the electronic record for life-cycle performance analysis.

Figures 174.1 and 174.2 illustrate how information about the fiberoptic bundle of an endoscope can be displayed and measured. This provides a record of the pattern of lighted optical fibers for the endoscope under test. The number of lighted pixels will depend on the endoscope's dimensions, the distal end geometry, and the number of failed optical fibers. New fiber damage to an endoscope will be apparent by comparison of the lighted fiber pictures (and histogram profiles) from successive tests. Statistical data is also available to calculate the percentage of working fibers in a given endoscope.

In addition to the two-dimensional profile of lighted fibers, this pattern (and all other image patterns) can also be displayed in the form of a three-dimensional contour plot. This interactive graph may be viewed from a variety of viewpoints in that the user can vary the elevation, rotation, size, and perspective controls.

Figure 174.2 illustrates how test images for a specific scope can be profiled over time (i.e., days, months, years) to identify degrading performance. This profile is also useful to validate repair procedures by comparing test images before and after the repair.

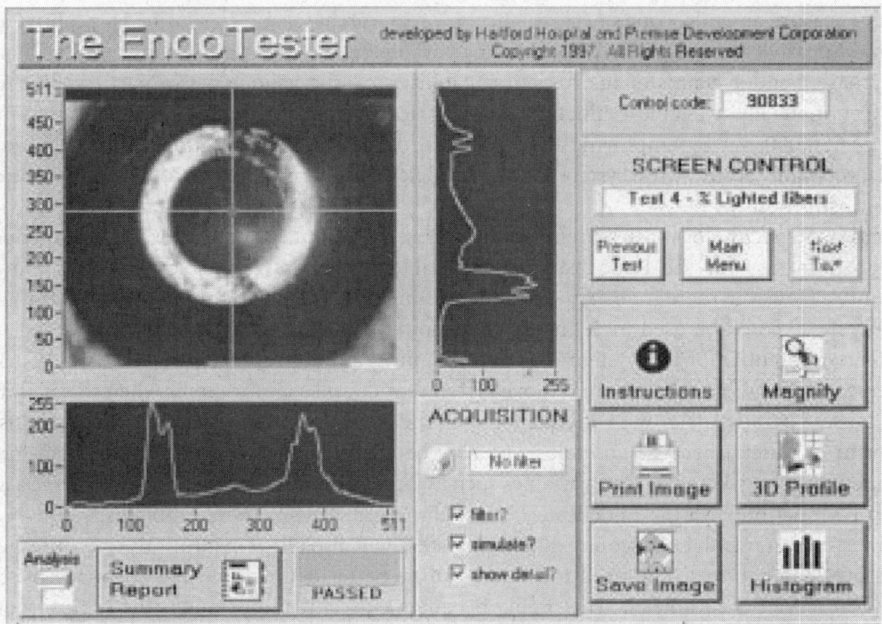

FIGURE 174.1 Endoscope tip reflection.

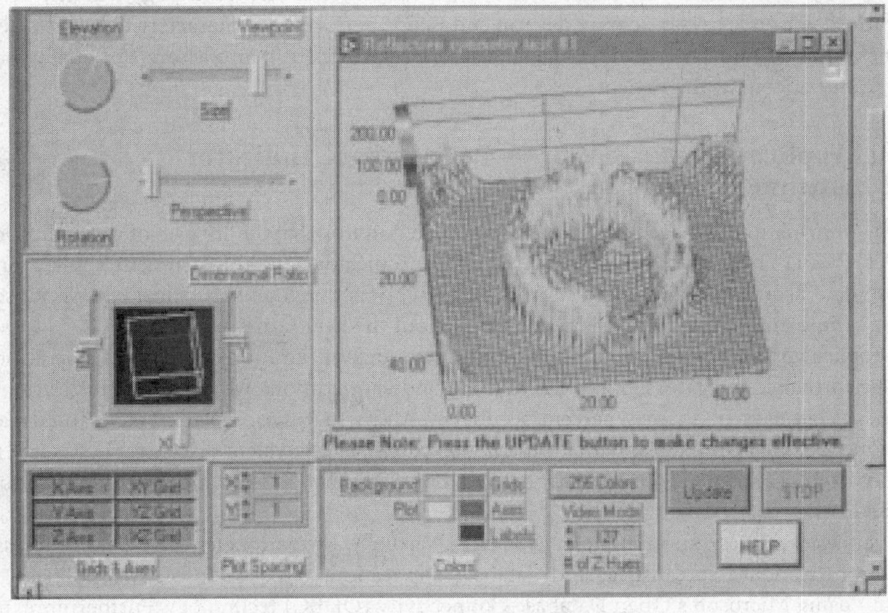

FIGURE 174.2 Endoscope profiling module.

The EndoTester™ has many applications. In general, the most useful application is the ability to objectively measure an endoscope's performance prior to purchase, and in routine clinical use as part of a program of prospective maintenance. Measuring parameters of scope performance can facilitate equipment purchase. Vendor claims of instrument capabilities can be validated as a part of the negotiation process. Commercially available evaluation systems (for original equipment manufacturers) can cost upwards of $50,000, yet by employing the benefits of virtual instrumentation and a standard PC, an affordable, yet highly accurate test system for rigid and flexible fiberoptic endoscopes can now be obtained by clinical institutions.

In addition to technology assessment applications, the adoption of disposable endoscopes raises another potential use for the EndoTester™. Disposable scopes are estimated to have a life of 20 to 30 procedures. However, there is no easy way to determine exactly when a scope should be "thrown away." The EndoTester™ could be used to define this end-point.

The greatest potential for this system is as part of a program of preventive maintenance. Currently, in most operating rooms, endoscopes are removed from service and sent for repair when they fail in clinical use. This causes operative delay with attendant risk to the patient and an increase in cost to the institution. The problem is difficult because an endoscope may be adequate in one procedure but fail in the next which is more exacting due to clinical variables such as large patient size or bleeding. Objective assessment of endoscope function with the EndoTester™ may eliminate some of these problems.

Equally as important, an endoscope evaluation system will also allow institutions to ensure value from providers of repair services. The need for repair can be better defined and the adequacy of the repair verified when service is completed. This ability becomes especially important as the explosive growth of minimally invasive surgery has resulted in the creation of a significant market for endoscope repairs and service. Endoscope repair costs vary widely throughout the industry with costs ranging from $500 to $1500 or more per repair. Inappropriate or incomplete repairs can result in extending surgical time by requiring the surgeon to "switch scopes" (in some cases several times) during a surgical procedure.

Given these applications, we believe that the EndoTester™ can play an important role in reducing unnecessary costs, while at the same time improving the quality of the endoscopic equipment and the outcome of its utilization. It is the sincere hope of the authors that this technology will help to provide accurate, affordable and easy-to-acquire data on endoscope performance characteristics which clearly are to the benefit of the healthcare provider, the ethical service providers, manufacturers of quality products, the payers, and, of course, the patient.

Example Application #2: PIVIT™—Performance Indicator Virtual Instrument Toolkit

Most of the information management examples presented in this chapter are part of an application suite called PIVIT™. PIVIT is an acronym for "Performance Indicator Virtual Instrument Toolkit" and is an easy-to-use data acquisition and analysis product. PIVIT was developed specifically in response to the wide array of information and analysis needs throughout the healthcare setting.

PIVIT applies virtual instrument technology to assess, analyze, and forecast clinical, operational, and financial performance indicators. Some examples include applications which profile institutional indicators (i.e., patient days, discharges, percent occupancy, ALOS, revenues, expenses, etc.), and departmental indicators (i.e., salary, non-salary, total expenses, expense per equivalent discharge, DRGs, etc.). Other applications of PIVIT include 360° Peer Review, Customer Satisfaction Profiling, and Medical Equipment Risk Assessment.

PIVIT can access data from multiple data sources. Virtually any parameter can be easily accessed and displayed from standard spreadsheet and database applications (i.e., Microsoft Access, Excel, Sybase, Oracle, etc.) using Microsoft's Open Database Connectivity (ODBC) technology. Furthermore, multiple parameters can be profiled and compared in real-time with any other parameter via interactive polar plots and three-dimensional displays. In addition to real-time profiling, other analyses such as SPC can

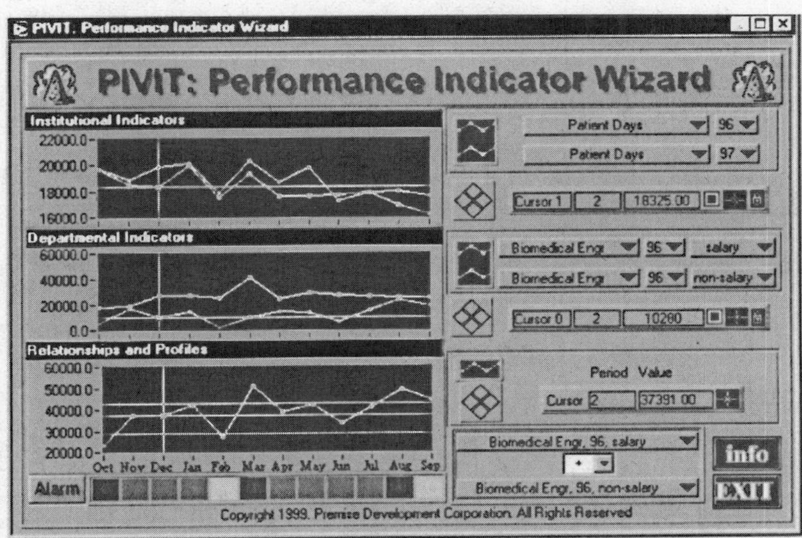

FIGURE 174.3 PIVIT™—Performance Indicator Wizard displays institutional and departmental indicators.

be employed to view large data sets in a graphical format. SPC has been applied successfully for decades to help companies reduce variability in manufacturing processes. These SPC tools range from Pareto graphs to Run and Control charts. Although it will not be possible to describe all of these applications, several examples are provided below to illustrate the power of PIVIT.

Trending, Relationships, and Interactive Alarms

Figure 174.3 illustrates a virtual instrument that interactively accesses institutional and department specific indicators and profiles them for comparison. Data sets can be acquired directly from standard spreadsheet and database applications (i.e., Microsoft Access®, Excel®, Sybase®, Oracle®, etc.). This capability has proven to be quite valuable with respect to quickly accessing and viewing large sets of data. Typically, multiple data sets contained within a spreadsheet or database had to be selected and then a new chart of this data had to be created. Using PIVIT, the user simply selects the desired parameter from any one of the pull-down menus and this data set is instantly graphed and compared to any other data set.

Interactive "threshold cursors" dynamically highlight when a parameter is over and/or under a specific target. Displayed parameters can also be ratios of any measured value, for example, "Expense per Equivalent Discharge" or "Revenue to Expense Ratio". The indicator color will change based on how far the data value exceeds the threshold value (i.e., from green to yellow to red). If multiple thresholds are exceeded, then the entire background of the screen (normally gray) will change to red to alert the user of an extreme condition.

Finally, multimedia has been employed by PIVIT to alert designated personnel with an audio message from the personal computer or by sending an automated message via e-mail, fax, pager, or mobile phone.

PIVIT also has the ability to profile historical trends and project future values. Forecasts can be based on user-defined history (i.e., "Months for Regression"), the type of regression (i.e., linear, exponential, or polynomial), the number of days, months, or years to forecast, and if any offset should be applied to the forecast. These features allow the user to create an unlimited number of "what if" scenarios and allow only the desired range of data to be applied to a forecast. In addition to the graphical display of data values, historical and projected tables are also provided. These embedded tables look and function very much like a standard spreadsheet.

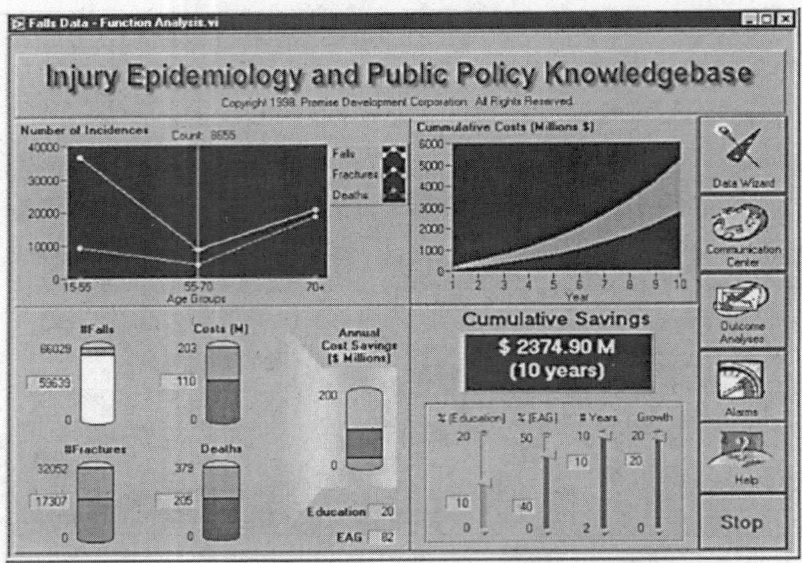

FIGURE 174.4 Injury epidemiology and public policy knowledgebase.

Data Modeling

Figure 174.4 illustrates another example of how virtual instrumentation can be applied to financial modeling and forecasting. This example graphically profiles the annual morbidity, mortality, and cost associated with falls within the state of Connecticut. Such an instrument has proved to be an extremely effective modeling tool due to its ability to interactively highlight relationships and assumptions, and to project the cost and/or savings of employing educational and other interventional programs.

Virtual instruments such as these are not only useful with respect to modeling and forecasting, but perhaps more importantly, they become a "knowledgebase" in which interventions and the efficacy of these interventions can be statistically proven. In addition, virtual instruments can employ standard technologies such as Dynamic Data Exchange (DDE), ActiveX, or TCP/IP to transfer data to commonly used software applications such as Microsoft Access® or Microsoft Excel®. In this way, virtual instruments can measure and graph multiple signals while at the same time send this data to another application which could reside on the network or across the Internet.

Another module of the PIVIT application is called the "Communications Center." This module can be used to simply create and print a report or it can be used to send e-mail, faxes, messages to a pager, or even leave voice-mail messages. This is a powerful feature in that information can be easily and efficiently distributed to both individuals and groups in real-time.

Additionally, Microsoft Agent® technology can be used to pop-up an animated help tool to communicate a message, indicate an alarm condition, or can be used to help the user solve a problem or point out a discrepancy that may have otherwise gone unnoticed. Agents employ a "text-to-speech" algorithm to actually "speak" an analysis or alarm directly to the user or recipient of the message. In this way, on-line help and user support can also be provided in multiple languages.

In addition to real-time profiling of various parameters, more advanced analyses such as SPC can be employed to view large data sets in a graphical format. SPC has been applied successfully for decades to help companies reduce variability in manufacturing processes. It is the opinion of this author that SPC has enormous applications throughout healthcare. For example, Fig. 174.5 shows how Pareto analysis can be applied to a sample trauma database of over 12,000 records. The Pareto chart may be frequency or percentage depending on front panel selection and the user can select from a variety of different parameters by clicking on the "pull-down" menu. This menu can be configured to automatically display

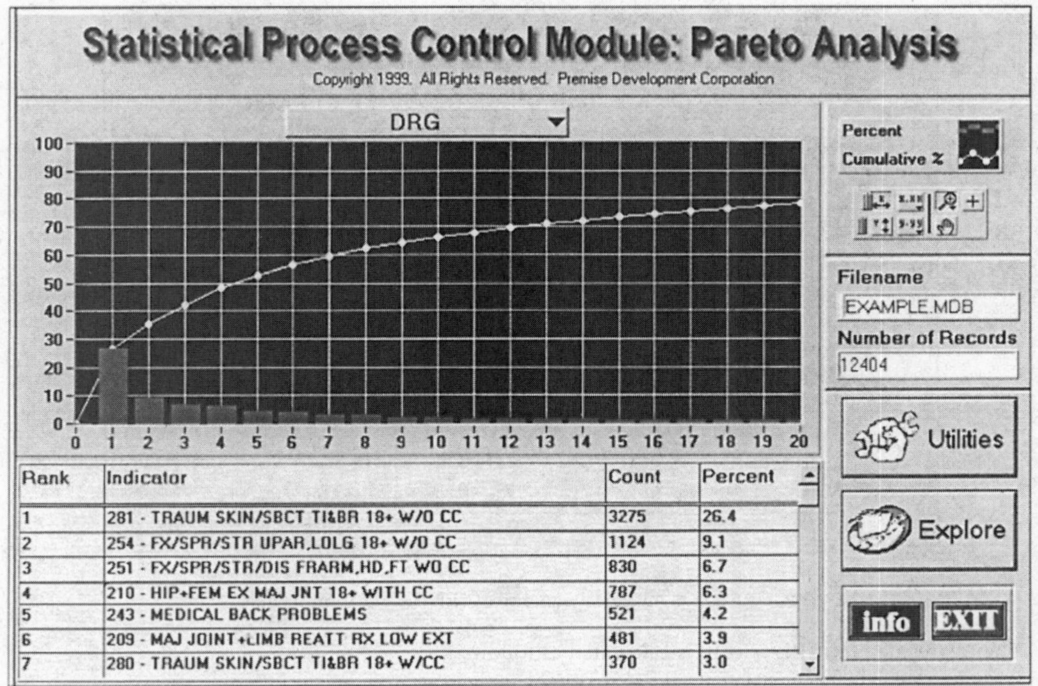

FIGURE 174.5 Statistical process control—Pareto analysis of a sample trauma registry.

each database field directly from the database. In this example, various database fields (i.e., DRG, Principal Diagnosis, Town, Payer, etc.) can be selected for Pareto analysis. Other SPC tools include run charts, control charts, and process capability distributions.

Medical Equipment Risk Criteria

Figure 174.6 illustrates a virtual instrument application which demonstrates how four "static" risk categories (and their corresponding values) are used to determine the inclusion of clinical equipment in the Medical Equipment Management Program at Hartford Hospital. Each risk category includes specific sub-categories that are assigned points, which when added together according to the formula listed below, yield a total score which ranges from 4 to 25.

Considering these scores, the equipment is categorized into five priority levels (High, Medium, Low, Grey List, and Non-Inclusion into the Medical Equipment Management Program). The four static risk categories are:

Equipment Function (EF): Stratifies the various functional categories (i.e., therapeutic, diagnostic, analytical, and miscellaneous) of equipment. This category has "point scores" which range from 1 (miscellaneous, non-patient related devices) to 10 (therapeutic, life support devices).

Physical Risk (PR): Lists the "worst case scenario" of physical risk potential to either the patient or the operator of the equipment. This category has "point scores" which range from 1 (no significant identified risk) to 5 (potential for patient and/or operator death).

Environmental Use Classification (EC): Lists the primary equipment area in which the equipment is used and has "point scores" which range from 1 (non-patient care areas) to 5 (anesthetizing locations).

Preventive Maintenance Requirements (MR): Describes the level and frequency of required maintenance and has "point scores" which range from 1 (not required) to 5 (monthly maintenance).

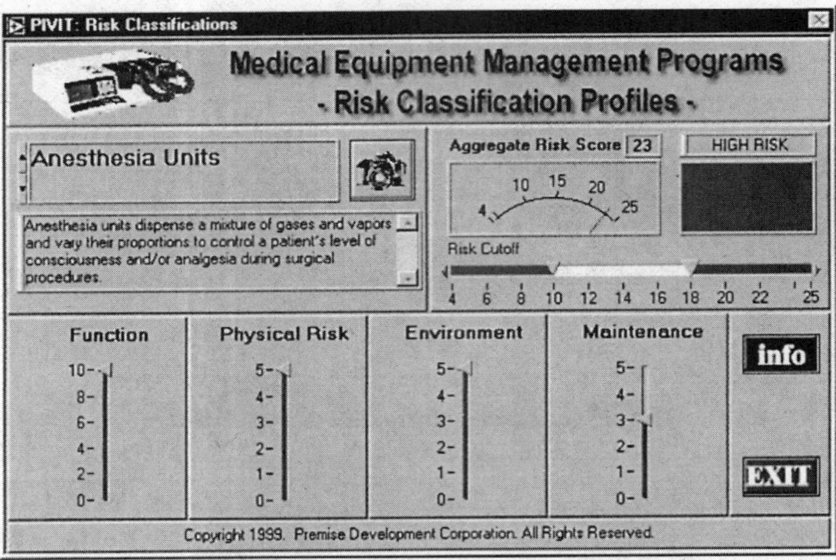

FIGURE 174.6 Medical equipment risk classification profiler.

The Aggregate Static Risk Score is calculated as follows:

$$\text{Aggregate Risk Score} = \text{EF} + \text{PR} + \text{EC} + \text{MR}$$

Using the criteria's system described above, clinical equipment is categorized according to the following priority of testing and degree of risk:

High Risk: Equipment that scores between and including 18 to 25 points on the criteria's evaluation system. This equipment is assigned the highest risk for testing, calibration, and repair.

Medium Risk: Equipment that scores between and including 15 to 17 points on the criteria's evaluation system.

Low Risk: Equipment that scores between and including 12 to 14 points on the criteria's evaluation system.

Hazard Surveillance (Gray): Equipment that scores between and including 6 and 11 points on the criteria's evaluation system is visually inspected on an annual basis during the hospital hazard surveillance rounds.

Medical Equipment Management Program Deletion: Medical equipment and devices that pose little risk and scores less than 6 points may be deleted from the management program as well as the clinical equipment inventory.

Future versions of this application will also consider "dynamic" risk factors such as: user error, mean-time-between failure (MTBF), device failure within 30 days of a preventive maintenance or repair, and the number of years beyond the American Hospital Association's recommended useful life.

Peer Performance Reviews

The virtual instrument shown in Fig. 174.7 has been designed to easily acquire and compile performance information with respect to institution-wide competencies. It has been created to allow every member of a team or department to participate in the evaluation of a co-worker (360° peer review). Upon running the application, the user is presented with a "Sign-In" screen where he or she enters their username and

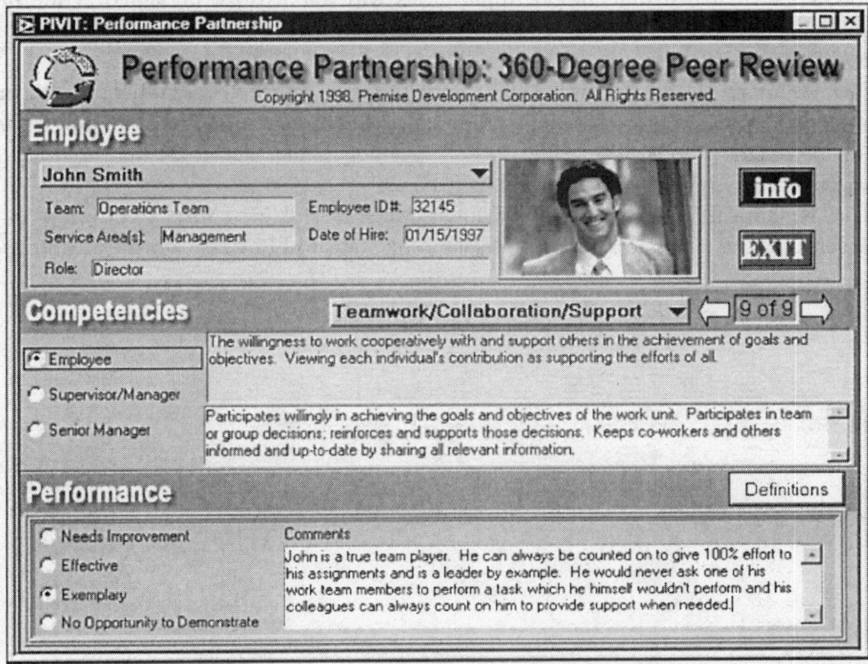

FIGURE 174.7 Performance reviews using virtual instrumentation.

password. The application is divided into three components. The first (top section) profiles the employee and relevant service information. The second (middle section) indicates each competency as defined for employees, managers, and senior managers. The last (bottom) section allows the reviewer to evaluate performance by selecting one of four "radio buttons" and also provide specific comments related to each competency. This information is then compiled (with other reviewers) as real-time feedback.

References

1. American Society for Quality Control. American National Standard. Definitions, Symbols, Forumulas, and Tables for Control Charts, 1987.
2. Breyfogle, F. W., Statistical Methods for Testing, Development and Manufacturing, John Wiley & Sons, New York, 1982.
3. Carey, R. G. and Lloyd, R. C. Measuring Quality Improvement in Healthcare: A Guide to Statistical Process Control Applications, 1995.
4. Fennigkow, L. and Lagerman, B., Medical Equipment Management. 1997 EC/PTSM Series/No. 1; Joint Commission on Accreditation of Hospital Organizations, 1997; 47-54.
5. Frost & Sullivan Market Intelligence, file 765, The Dialog Corporation, Worldwide Headquarters, 2440 W. El Camino Real, Mountain View, CA 94040.
6. Inglis, A., Video Engineering, McGraw Hill, New York, 1993.
7. Kutzner J., Hightower L., and Pruitt C. Measurement and Testing of CCD Sensors and Cameras, SMPTE Journal, pp. 325-327, 1992.
8. Measurement of Resolution of Camera Systems, IEEE Standard 208, 1995.
9. Medical Device Register 1997, Volume 2, Montvale NJ, Medical Economics Data Production Company, 1997.
10. Montgomery, D. C., Introduction to Statistical Quality Control, 2nd ed., John Wiley & Sons, 1992.

11. Rosow, E., Virtual Instrumentation Applications with BioSensors, presented at the Biomedical Engineering Consortium for Connecticut (BEACON) Biosensor Symposium, Trinity College, Hartford, CT, October 2, 1998.
12. Rosow, E., Adam, J., and Beatrice, F., The EndoTester™: A Virtual Instrument Endoscope Evaluation System for Fiberoptic Endoscopes, Biomedical Instrumentation and Technology, pp. 480-487, September/October 1998.
13. Surgical Video Systems, Health Devices, 24(11), 428-457, 1995.
14. Walker, B., Optical Engineering Fundamentals, McGraw Hill, New York, 1995.
15. Wheeler, D. J. and Chambers, D. S., Understanding Statistical Process Control, SPC Press, 2nd ed., 1992.

XVIII

Medical Informatics

Luis G. Kun
CIMIC/Rutgers University

IN THE LAST 20 YEARS the field of medical informatics has grown tremendously both in its complexity and in content. As a result, two sections will be written in this *Handbook*. The first one, represented in the chapters, will be devoted to areas that form a key "core" of computer technologies. These include: hospital information systems (HIS), computer-based patient records (CBPR or CPR), imaging, communications, standards, and other related areas. The second section includes the following topics: artificial intelligence, expert systems, knowledge-based systems neural networks, and robotics. Most of the techniques describe in the second section will require the implementation of systems explained in this first section. We could call most of these chapters the *information infrastructure* required to apply medical informatics' techniques to medical data. These topics are crucial because they not only lay the foundation required to treat a patient within the walls of an institution but they provide the roadmap required to deal with the patient's lifetime record while allowing selected groups of researchers and clinicians to analyze the information and generate outcomes research and practice guidelines information.

As an example a network of associated hospitals in the East Coast (a health care provider network) may want to utilize an expert system that was created and maintained at Stanford University. This group of hospitals, HMOs, clinics, physician's offices, and the like would need a "standard" computer-based patient record (CPR) that can be used by the clinicians from any of the physical locations. In addition in order to access the information all these institutions require telecommunications and networks that will allow for the electronic "dialogue." The different forms of the data, particularly clinical images, will require devices for displaying purposes, and the information stored in the different HIS, Clinical Information Systems (CIS), and departmental systems needs to be integrated. The multimedia type of record would become the data input for the expert system which could be accessed remotely (or locally) from any of the enterprise's locations. On the application side, the expert system could provide these institutions which techniques that can help in areas such as diagnosis and patient treatment. However, several new trends such as: total quality management (TQM), outcome research and practice guidelines could be followed. It should be obvious to the reader that to have the ability to compare information obtained in different parts of the world by dissimilar and heterogeneous systems, certain standards need to be followed (or created) so that when analyzing the data the information obtained will make sense.

Many information systems issues described in this introduction will be addressed in this section. The artificial intelligence chapters which follow should be synergistic with these concepts. A good understanding of the issues in this section is *required prior* to the utilization of the actual expert system. These issues are part of this section of medical informatics, other ones, however, e.g., systems integration and process reengineering, will not be addressed here in detail but will be mentioned by the different authors. I encourage the reader to follow up on the referenced material at the end of each chapter, since the citations contain very valuable information.

Several current perspectives in information technologies need to be taken in consideration when reading this section. One of them is described very accurately in the book entitled *Globalization, Technology and Competition* (*The Fusion of Computers and Telecommunications in the 1990s*) by Bradley, Hausman and Nolan (1993). The first chapter of this book talks about new services being demanded by end users which include the integration of computers and telecommunications. From their stages theory point of view, the authors describe that we are currently nearing the end of the micro era and the beginning of the network era. From an economy point of view, the industrial economy (1960s and 1970s) and the transitional economy (1970s and 1980s) is moving into an information economy (1990s and beyond). Also a leadership survey done in 1994 by the Healthcare Information and Management System Society (HIMSS) on trends in health care computing to mainly chief information officers, directors, and the likes of health care providers showed the following results:

1. In a market driven by cost containment, the most important forces driving increased computerization in health care were: (*a*) movement to manage care (25%), (*b*) outcomes data requests (24%), and (*c*) movement to health care networks (17%).
2. The most important systems priority for the next 2 years: (*a*) integrating across separate facilities (31%), (*b*) implementing a computer-based patient record (CPR) (19%), and (*c*) integrating departmental systems (13%) and reengineering to patient focused care (13%).
3. 56% felt that the information superhighway was essential for health care.
4. In the next 3 years the most significant health-care related computer development affecting the average consumer would be (*a*) more streamlined health care encounters (49%), (*b*) access to health information/services from home (20%), and (*c*) health care "smartcards" (17%).

5. Although 49% claimed to use the Internet, their health care facilities are using it for: (*a*) point-to-point E-mail (81%), (*b*) clinicians querying research databases (69%), (*c*) consumer-provider exchange (31%), and (*d*) two-way medical consultations (22%).
6. Clinicians will share computerized patient information in a nationwide system: (*a*) by the year 2000 (39%), (*b*) not happen for at least ten years (38%), and (*c*) within 1 to 3 years (14%). Many other questions and answers reflected some of the current technological barriers and users needs. Because of these trends it was essential to include in this *Handbook* technologies that today may be considered state of the art but when read about 10 years from now will appear to be transitional only. Information technologies are moving into a multimedia environment which will require special techniques for acquiring, displaying, storing, retrieving, and communicating the information. We are in the process of defining some of these mechanisms.

In some instances such as imaging, this *Handbook* contains a full section dedicated to the subject. That section contains the principles, the associated math algorithms, and the physics related to all medical imaging modalities. The intention in this section is to address issues related to imagining as a form of medical information. These concepts include issues related then to acquisition, storage/retrieval, display and communications of document and clinical images, e.g., picture archival and communications systems (PACS). From a CPR point of view, clinical and document images will become part of this *electronic chart,* therefore many of the associated issues will be discussed in this section more extensively.

The state of the telecommunications has been described as a revolution; data and voice communications as well s full-motion video have come together as a new dynamic field. Much of what is happening today is a result of technology evolution and need. The connecting thread between evolutionary needs and revolutionary ideas is an integrated perspective of both sides of multiple industries. This topic will also be described in more detail in this section.

In the first chapter Allan Pryor provides us with a tutorial on hospital information systems (HIS). He describes not only the evolution of HIS and departmental systems and clinical information systems (CIS), but also their differences. Within the evolution he follows these concepts with the need for the longitudinal patient record and the integration of patient data. This chapter includes patient database strategies for the HIS, data acquisition, patient admission, transfer and discharge functions. Also discussed are patient evaluation and patient management issues. From an end-user point of view, a terrific description on the evolution of *data-driven* and *time-driven* systems is included, culminating with some critical concepts on HIS requirements for decision support and knowledge base functionality. His conclusions are good indication of his vision.

Michael Fitzmaurice follows with "Computer-Based Patient Records" (CBPR or CPR). In the introduction, it is explained what is the CPR and way it is a necessary tool for supporting clinical decision making and how it is enhanced when it interacts with medial knowledge sources. This is followed by clinical decision support systems (CDSS): knowledge server, knowledge sources, medical logic modules (MLM), and nomenclature. This last issue in particular is one which needs to be well understood. The nomenclature used by physicians and by the CPRs differ among institutions. Applying *logic* to the wrong concepts can produce misinterpretations. The scientific evidence in this chapter includes patient care process, CDSS hurdles, CPR implementation, research data bases, telemedicine, hospital and ambulatory care systems. A table of hospital and ambulatory care computer-based patient records systems concludes this chapter.

Today it is impossible to separate computers and telecommunications (communications and networks). Both are part of information systems. Soumitra Sengupta provides us n this chapter with a tutorial-like presentation which includes an introduction and history, impact of clinical data, information types, and platforms. The importance of this section is reflected both in the contents reviewed under current technologies—LANs, WANs, middleware, medical domain middleware; integrated patient data base, and medical vocabulary—as well as in the *directions and challenges* section which includes improved bandwidth, telemedicine, and security management. In the conclusions the clear vision is that *networks* will become the de facto fourth utility after electricity, water, and heat.

"Non-AI Decision Making" is covered by Ron Summers and Ewart Carson. This chapter includes an introduction which explains the techniques of procedural or declarative knowledge. The topics covered in this section include: *analytical models*, and *decision theoretic models*, including *clinical algorithms* and *decision trees*. The section that follows cover a number of key topics which appear while querying large clinical databases to yield evidence of either diagnostic/treatment or research value; statistical models, database search, regression analysis, statistical pattern analysis, bayesian analysis, Depster-Shafer theory, syntactic pattern analysis, causal modeling, artificial neural networks. In the summary the authors clearly advise the reader to read this section in conjunction with the *expert systems* chapters that follow.

The standards section is closely associated with the CPR chapter of this section. Jeff Blair does a terrific job with his overview of standards related to the emerging health care information infrastructure. This chapter should give the reader not only an overview of the major existing and emerging health care information *standards* but an understanding of all current efforts, national and international, to coordinate, harmonize, and accelerate these activities. The introduction summarizes how this section is organized. It includes *identifier* standards (patient's, site of care, product, and supply labeling), *communications* (message format) standards, *content and structure* standards. This section is followed by a summary of *clinical data representations*, guidelines for *confidentiality, data, security, and authentication.* After that *quality indicators* and *data sets* are described along with *international* standards. Coordinating and promotion organizations are listed at the end of this chapter including points of contact which will prove very beneficial for those who need to follow up.

Design issues in developing clinical decision support and monitoring systems by John Goethe and Joseph Bronzino provide insight for the development of *clinical decision support systems.* In their introduction and throughout this chapter the authors provide a step-by-step tutorial with practical advice and make recommendations on design of the systems to achieve end-user acceptance. After that a description of a clinical monitoring system, developed and implemented by them for a psychiatric practice, is presented in detail. In their conclusions the *human engineering* issue is discussed.

The authors of this section represent industry, academia and government. Their expertise in many instances is multiple from developing to actual implementing these technical ideas. I am very grateful for all our discussions and their contributions.

175

Hospital Information Systems: Their Function and State

T. Allan Pryor
University of Utah

The definition of a hospital information system (HIS) is unfortunately not unique. The literature of both the informatics community and health care data processing world is filled with descriptions of many differing computer systems defined as an HIS. In this literature, the systems are sometimes characterized into varying level of HISs according to the functionally present within the system. With this confusion from the literature, it is necessary to begin this chapter with a definition of an HIS. To begin this definition, I must first describe what it is not. The HIS will incorporate information from the several departments within the hospital, but an HIS is not a departmental system. Departmental systems such as a pharmacy or a radiology system are limited in their scope. They are designed to manage only the department that they serve and rarely contain patient data captured from other departments. Their function should be to interface with the HIS and provide portions of the patient medical/administrative record that the HIS uses to manage the global needs of the hospital and patient.

A clinical information system is likewise not an HIS. Again, although the HIS needs clinical information to meets its complete functionality, it is not exclusively restricted to the clinical information supported by the clinical information systems. Examples of clinical information systems are ICU systems, respiratory care systems, nursing systems. Similar to the departmental systems, these clinical systems tend to be one-dimensional with a total focus on one aspect of the clinical needs of the patient. They provide little support for the administrative requirements of the hospital.

If we look at the functional capabilities of both the clinical and departmental systems, we see many common features of the HIS. They all require a database for recording patient information. Both types of systems must be able to support data acquisition and reporting of patient data. Communication of information to other clinical or administrative departments is required. Some form of management support can be found in all the systems. Thus, again looking at the basic functions of the system one cannot differentiate the clinical/departmental systems from the HIS. It is this confusion that makes defining the HIS difficult and explains why the literature is ambiguous in this matter.

0-8493-0462-8/00/$0.00+$.50

The concept of the HIS appears to be, therefore, one of integration and breadth across the patient or hospital information needs. That is, to be called an HIS the system must meet the global needs of those it is to serve. In the context, if we look at the hospital as the customer of the HIS, then the HIS must be able to provide global and departmental information on the state of the hospital. For example, if we consider the capturing of charges within the hospital to be an HIS function, then the system must capture all patient charges no matter which departmental originated those charges. Likewise all clinical information about the patient must reside within the database of the HIS and make possible the reporting and management of patient data across all clinical departments and data sources. It is totality of function that differentiates the HIS from the departmental or restricted clinical system, not the functions provided to a department or clinical support incorporated within the system.

The development of an HIS can take many architectural forms. It can be accomplished through interfacing of a central system to multiple departmental or clinical information systems. A second approach which has been developed is to have, in addition to a set of global applications, departmental or clinical system applications. Because of the limitation of all existing systems, any existing comprehensive HIS will in fact be a combination of interfaces to departmental/clinical systems and the applications/database of the HIS purchased by the hospital.

The remainder of this chapter will describe key features that must be included in today's HIS. The features discussed below are patient databases, patient data acquisition, patient admission/bed control, patient management and evaluation applications, and computer-assisted decision support. This chapter will not discuss the financial/administrative applications of an HIS, since those applications for the purposes of this chapter are seen as applications existing on a financial system that may not be integral application of the HIS.

175.1 Patient Database Strategies for the HIS

The first HISs were considered only an extension of the financial and administrative systems in place in the hospital. With this simplistic view many early systems developed database strategies that were limited in their growth potential. Their databases mimicked closely the design of the financial systems that presented a structure that was basically a "flat file" with well-defined fields. Although those fields were adequate for capturing the financial information used by administration to track the patient's charges, they were unable to adapt easily to the requirement to capture the clinical information being requested by health care providers. Today's HIS database should be designed to support a longitudinal patient record (the entire clinical record of the patient spanning multiple inpatient, outpatient encounters), integration of all clinical and financial data, and support of decision support functions.

The creation of a longitudinal patient record is now a requirement of the HIS. Traditionally the databases of the HISs were encounter-based. That is, they were designed to manage a single patient visit to the hospital to create a financial record of the visit and make available to the care provider data recorded during the visit. Unfortunately, with those systems the care providers were unable to view the progress of the patient across encounters, even to the point that in some HISs critical information such as patient allergies needed to be entered with each new encounter. From the clinical perspective, the management of a patient must at least be considered in the context of a single episode of care. This episode might include one or more visits to the hospital's outpatient clinics, the emergency department, and multiple inpatient stays. The care provider to manage properly the patient, must have access to all the information recorded from those multiple encounters. The need for a longitudinal view dictates that the HIS database structure must both allow for access to the patient's data independent of an encounter and still provide for encounter-based access to adapt to the financial and billing requirements of the hospital.

The need for integration of the patient data is as important as the longitudinal requirement. Traditionally the clinical information tended to be stored in separate departmental files. With this structure it was easy to report from each department, but the creation of reports combining data from the different proved difficult if not impossible. In particular in those systems where access to the departmental data was provided only though interfaces with no central database, it was impossible to create an integrated patient evaluation report. Using those systems the care providers would view data from different screens at their terminal and extract with pencil onto paper the results from each departmental (clinical laboratory, radiology, pharmacy, and so on) the information they needed to properly evaluate the patient. With the integrated clinical database the care provider can view directly on a single screen the information from all departments formatted in ways that facilitate the evaluation of the patient.

Today's HIS is no longer merely a database and communication system but is an assistant in the management of the patient. That is, clinical knowledge bases are an integral part of the HIS. These knowledge bases contain rules and/or statistics with which the system can provide alerts or reminders or implement clinical protocols. The execution of the knowledge is highly dependent on the structure of the clinical database. For example, a rule might be present in the knowledge base to evaluate the use of narcotics by the patient. Depending on the structure of the database, this may require a complex set of rules looking at every possible narcotic available in the hospital's formulary or a single rule that checks the presence of the class narcotics in the patient's medical record. If the search requires multiple rules, it is probably because the medical vocabulary has been coded without any structure. With this lack of structure there needs to be a specific rule to evaluate every possible narcotic code in the hospital's formulary against the patient's computer medication record. With a more structured data model a single rule could suffice. With this model the drug codes have been assigned to include a hierarchical structure where all narcotics would fall into the same hierarchical class. Thus, a single rule specific only to the class "narcotics" is all that is needed to compare against the patient's record.

These enhanced features of the HIS database are necessary if the HIS is going to serve the needs of today's modern hospital. Beyond these inpatient needs, the database of the HIS will become part of an enterprise clinical database that will include not only the clinical information for the inpatient encounters but also the clinical information recorded in the physician's office or the patient's home during outpatient encounters. Subsets of these records will become part of state and national health care databases. In selecting, therefore, and HIS, the most critical factor is understanding the structure and functionality of its database.

175.2 Data Acquisition

The acquisition of clinical data is key to the other functions of the HIS. If the HIS is to support an integrated patient record, then its ability to acquire clinical data from a variety of sources directly affect its ability to support the patient evaluation and management functions described below. All HIS systems provide for direct terminal entry of data. Depending on the system this entry may use only the keyboard or other "point and click" devices together with the keyboard.

Interfaces to other systems will be necessary to compute a complete patient record. The physical interface to those systems is straightforward with today's technology. The difficulty comes in understanding the data that are being transmitted between systems. It is easy to communicate and understand ASCII textual information, but coded information from different systems is generally difficult for sharing between systems. This difficulty results because there are no medical standards for either medical vocabulary or the coding systems. Thus, each system may have chose an entirely different terminology or coding system to describe similar medical concepts. In building the interface, therefore, it may be necessary to build unique translation tables to store the information from one system into the databases of the HIS. This requirement has limited the building of truly integrated patient records.

Acquisition of data from patient monitors used in the hospital can either be directly interfaced to the HIS or captured through an interface to an ICU system. Without these interfaces the acquisition of the monitoring data must be entered manually by the nursing personnel. It should be noted that whenever possible automated acquisition of data is preferable to manual entry. The automated acquisition is more accurate and reliable and less resource intensive. With those HISs which do not have interfaces to patient monitors, the frequency of data entry into the system is much less. The frequency of data acquisition affects the ability of the HIS to implement real-time medical decision logic to monitor the status of the patient. That is, in the ICU where decisions need to be made on a very timely manner, the information on which the decision is based must be entered as the critical event is taking place. If there is no automatic entry of the data, then the critical data needed for decision making may not be present, thus preventing the computer from assisting in the management of the patient.

175.3 Patient Admission, Transfer, and Discharge Functions

The admission application has three primary functions. The first is to capture for the patient's computer record pertinent demographic and financial/insurance information. A second function is to communicate that information to all systems existing on the hospital network. The third is to link the patient to previous encounters to ensure that the patient's longitudinal record is not compromised. This linkage also assists in capturing the demographic and financial data needed for the current encounter, since that information captured during a previous encounter may need not to be reentered as part of this admission. Unfortunately in many HISs the linkage process is not as accurate as needed. Several reasons explain this inaccuracy. The first is the motivation of the admitting personnel. In some hospitals they perceive their task as a business function responsible only for ensuring that the patient will be properly billed for his or her hospital stay. Therefore, since the admission program always allows them to create a new record and enter the necessary insurance/billing information, their effort to link the patient to his previous record may not be as exhaustive as needed.

Although the admitting program may interact with many financial and insurance files, there normally exists two key patient files that allow the HIS to meet its critical clinical functions. One is a master patient index (MPI) and the second is the longitudinal clinical file. The MPI contains the unique identifier for the patient. The other fields of this file are those necessary for the admitting clerk to identify the patient. During the admitting process the admitting process the admitting clerk will enter identifying information such as name, sex, birth date, social security number. This information will be used by the program to select potential patient matches in the MPI from which the admitting clerk can link to the current admission. If no matches are detected by the program, the clerk creates a new record in the MPI. It is this process that all too frequently fails. That is, the clerk either enters erroneous data and finds no match or for some reason does not select as a match one of the records displayed. Occasionally the clerk selects the wrong match causing the data from this admission to be posted to the wrong patient. In the earlier HISs where no longitudinal record existed, this problem was not critical, but in today's system, errors in matching can have serious clinical consequences. Many techniques are being implemented to eliminate this problem including probabilistic matching, auditing processes, postadmission consolidation.

The longitudinal record may contain either a complete clinical record of the patient or only those variables that are most critical in subsequent admissions. Among the data that have been determined as most critical are key demographic data, allergies, surgical procedures, discharge diagnoses, and radiology reports. Beyond these key data elements more systems are beginning to store the complete clinical record. In those systems the structure of the records of the longitudinal file contain information regarding the encounter, admitting physician, and any other information that may be necessary to view the record from an encounter view or as a complete clinical history of the patient.

175.4 Patient Evaluation

The second major focus of application development for the HIS is creation of patient evaluation applications. The purpose of these evaluation programs is to provide to the care giver information about the patient which assists in evaluating the medical status of the patient. Depending on the level of data integration in the HIS, the evaluation applications will be either quite rudimentary or highly complex. In the simplest form these applications are departmentally oriented. With this departmental orientation the care giver can access through terminals in the hospital departmental reports. Thus, laboratory reports, radiology reports, pharmacy reports, nursing records, and the like can be displayed or printed at the hospital terminals. This form of evaluation functionality is commonly called *results review*, since it only allows the results of tests from the departments to be displayed with no attempt to integrate the data from those departments into an integrated patient evaluation report.

The more clinical HISs as mentioned above include a central integrated patient database. With those systems patient reports can be much more sophisticated. A simple example of an integrated patient evaluation report is a diabetic flowsheet. In this flowsheet the caregiver can view the time and amount of insulin given, which may have been recorded by the pharmacy or nursing application, the patient's blood glucose level recorded in the clinical laboratory or again by the nursing application. In this form the caregiver has within single report, correlated by the computer, the clinical information necessary to evaluate the patient's diabetic status rather than looking for data on reports from the laboratory system, the pharmacy system, and the nursing application. As the amount and type of data captured by the HIS increases, the system can produce ever-more-useful patient evaluation reports. There exist HISs which provide complete rounds reports the summarize on one to two screens all the patient's clinical record captured by the system. These reports not only shorten the time need by the caregiver to locate the information, but because of the format of the report, can present the data in a more intuitive and clinically useful form.

175.5 Patient Management

Once the caregiver has properly evaluated the state of the patient, the next task is to initiate therapy that ensures an optimal outcome for the patient. The sophistication of the management applications is again a key differentiation of HISs. At the simplest level management applications consist of order-entry applications. The order-entry application is normally executed by a paramedical personnel. That is, the physician writes the order in the patient's chart, and another person reviews from the chart the written order and enters it into the computer. For example, if the order is for a medication, then it will probably be a pharmacist who actually enters the order into the computer. For most of the other orders a nurse or ward clerk is normally assigned this task. The HIS records the order in the patient's computerized medical record and transmits the order to the appropriate department for execution. In those hospitals where the departmental systems are interfaced to the HIS, the electronic transmission of the order to the departmental system is a natural part of the order entry system. In many systems the transmission of the order is merely a printout of the order in the appropriate department.

The goal of most HISs is to have the physician responsible for management of the patient enter the orders into the computer. The problem that has troubled most of the HISs in achieving this goal has been the inefficiency of the current order-entry programs. For these programs to be successful they have to complete favorably with the traditional manner in which the physician writes the order. Unfortunately, most of the current order-entry applications are too cumbersome to be readily accepted by the physician. Generally they have been written to assist the paramedic in entering the order resulting with far too many screens or fields that need to be reviewed by the physician to complete the order. One approach that has been tried with limited success is the use of order sets. The order sets have been designed to allow the physician to easily from a single screen enter multiple orders. The use of order sets has improved the acceptability of the order-entry application to the physician, but several problems remain preventing universal acceptance by the physicians. One problem is that the order set will never be sufficiently complete to contain all orders that the physician would want to order. Therefore, there is some subset of patients orders that will have to be entered using the general ordering mechanisms of the program. Depending on the frequency of those orders, the acceptability of the program changes. Maintenance issues also arise with order sets, since it may be necessary to formulate order sets for each of the active physicians. Maintaining of the physician-specific order sets soon becomes a major problem for the data processing department. It becomes more problematic if the HIS to increase the frequency of a given order being present on an order set allows the order sets to be not only physician-defined but problem-oriented as well. Here it is necessary to again increase the number of order sets or have the physicians all agree on those orders to be included in an order set for a given problem.

Another problem, which makes use of order entry by the physician difficult, is the lack of integration of the application into the intellectual tasks of the physician. That is, in most of the systems the physicians are asked to do all the intellectual work in evaluating and managing the care of the patient in the traditional manner and then, as an added task, enter the results of that intellectual effort into the computer. It is at this last step that is perceived by the physician as a clerical task at which the physician rebels. Newer systems are beginning to incorporate more efficiently the ordering task into other applications. These applications assist the physical throughout the entire intellectual effort of patient evaluation and management of the patient. An example of such integration would be the building of evaluation and order sets in the problem list management application. Here when the care provider looks at the patient problem list he or she accesses problem-specific evaluation and ordering screens built into the application, perhaps shortening the time necessary for the physician to make rounds on the patient.

Beyond simple test ordering, many newer HISs are implementing decision support packages. With these packages the system can incorporate medical knowledge usually as rule sets to assist the care provider in the management of patients. Execution of the rule sets can be performed in the foreground through direct calls from an executing application or in the background with the storing of clinical data in the patient's computerized medical record. This latter mode is called *data-driven execution* and provides an extremely powerful method of knowledge execution and alerting. that is, after execution of the rule sets, the HIS will "alert" the care provider of any outstanding information that may be important regarding the status of the patient or suggestions on the management of the patient. Several mechanisms have been implemented to direct the alerts to the care provider. In the simplest form notification is merely a process of storing the alert in the patient's medical record to be reviewed the next time the care provider accesses that patient's record. More sophisticated notification methods have included directed printouts to individuals whose job it is to monitor the alerts, electronic messages sent directly to terminals notifying the users that there are alerts which need to be viewed, and interfacing to the paging system of the hospital to direct alert pages to the appropriate personnel.

Execution of the rule sets are sometimes, *time-driven.* This mode results in sets of rules being executed at a particular point in time. The typical scenario for time-driven execution is to set a time of day for selected rule set execution. At that time each day the system executes the given set of rules for a selected population in the hospital. Time drive has proven to be a particularly useful mechanism of decision support for those applications that require hospitalwide patient monitoring.

The use of decision support has ranged from simple laboratory alerts to complex patient protocols. The responsibility of the HIS is to provide the tools for creation and execution of the knowledge base. The hospitals and their designated "experts" are responsible for the actual logic that is entered into the rule sets. Many studies are appearing in the literature suggesting that the addition of knowledge base execution tot he HIS is the next major advancement to e delivered with the HIS. This addition will become a tool to better manage the hospital in the world of managed care.

The inclusion of decision support functionality in the HIS requires that the HIS be designed to support a set of knowledge tools. In general a knowledge bases system will consist of a knowledge base and an inference engine. The knowledge base will contain the rules, frames, and statistics that are used by the inference applications to substantiate a decision. We have found that in the health care area the knowledge base should be sufficiently flexible to support multiple forms of knowledge. That is, no single knowledge representation sufficiently powerful to provide a method to cover all decisions necessary in the hospital setting. For example, some diagnostic decisions may well be best suited for bayesian methods, whereas other management decisions may follow simple rules. In the context of the HIS, I prefer the term *application manager* to *inference engine.* The former is intended to imply that different applications may require different knowledge representations as well as different inferencing strategies to traverse the knowledge base. Thus, when the user selects the application, he or she is selecting a particular inference engine that may be unique to that application. The tasks, therefore, of the application manager are to provide the "look and feel" of the application, control the functional capabilities of the application, and invoke the appropriate inference engine for support of any "artificial intelligence" functionality.

175.6 Conclusion

Today's HIS is no longer the financial/administrative system that first appeared in the hospital. It has extended beyond that role to become an adjunct to the care of the patient. With this extension into clinical care the HIS has not only added new functionality to its design but has enhanced its ability to serve the traditional administrative and financial needs of the hospital as well. The creation of these global applications which go well beyond those of the departmental/clinical systems is now making the HIS the *patient -focused* system. With this global information the administrators and clinical staff together can accurately access where there are inefficiencies in the operation of the hospital from the delivery of both the administrative and medical care. This knowledge allows changes in the operation of the hospital that will ensure that optimal care continues to be provided to the patient at the least cost to the hospital. These studies and operation changes will continue to grow as the use of an integrated database and implementation of medical knowledge bases become increasingly routine in the functionality of the HIS.

References

1. Pryor TA, Gardner RM, Clayton PD, et al. 1983. The HELP system. J Med syst 7:213.
2. Pryor TA, Clayton PD, Haug PJ, et al. 1987. Design of a knowledge driven HIS. Proc. 11th SCAMC, 60.
3. Bakker AR. 1984. The development of an integrated and co-operative hospital information system. Med Inf 9:135.
4. Barnett GO. 1984. The application of computer-based medical record systems in ambulatory practice. N Engl J Med 310:1643.
5. Bleich HL, Beckley RF, Horowitz GL, et al. 1985. Clinical computing in a teaching hospital. N Engl J Med 312:756.
6. Whiting-O'Keefe QE, Whiting A, Henke J. 1988. The STOR clinical information system. MD Comput 5:8.
7. Hendrickson G, Anderson RK, Clayton PD, et al.1992. The integrated academic information system at Columbia-Presbyterian Medical Center. MD Comput 9:35.
8. Safran C, Slack WV, Bleich HL. 1989. Role of computing in patient care in two hospitals. MD Comput 6:141.
9. Bleich HL, Safran C, Slack WV. 1989. Departmental and laboratory computing in two hospitals. MD Comput 6:149.
10. ASTM E1238–91. 1992. Specifications for transferring clinical observations between independent computer systems. Philadelphia, American Society for Testing and Materials.
11. Tierney WM, Miller ME, Donald CJ. 1990. The effect on test ordering of informing physicians of the charges for outpatient diagnostic tests. N Engl J Med 322:1499.
12. Stead WW, Hammond WE. 1983. Functions required to allow TMR to support the information requirements of a hospital. Proc 7th SCAMC 106.
13. Safran C, Herrmann F, Rind D, et al. 1990. Computer-based support for clinical decision making. MD Comput 7:319.
14. Tate KE, Gardner RM, Pryor TA. 1989. Development of a computerized laboratory alerting system. Comp Biomed Res 22:575.
15. Orthner HF, Blum BI (eds). 1989. Implementing Health Care Information Systems, Springer-Verlag.
16. Dick RS, Steen EB (eds). 1991. The Computer-Based Patient Record, National Academy Press.

176
Computer-Based Patient Records

J. Michael Fitzmaurice
Department of Health and Human Services

176.1 Introduction

The objective of this section is to present the computer-based patient record (CPR) as a powerful tool for organizing patient care data to improve patient care and strengthen communication of patient care data among health care providers. The CPR is even more powerful when used in a system that retrieves applicable medical knowledge to support clinical decision making. Evidence exists that the use of CPR systems (CPRS) can change both physician behavior and patient outcomes of care. As the speed and cost efficiency of computers rise, the cost of information storage and retrieval falls, and the breadth of ubiquitous networks becomes broader, it is essential that CPRs and systems that use them be evaluated for the improvements in health care that they can bring.

The primary role of the CPR is to support the delivery of medical care to a particular patient. Serving this purpose, ideally the CPR brings past and current information about a particular patient to the physician, promotes communication among health care givers about that patient's care, and documents the process of care and the reasoning behind the choices that are made. Thus, the data in a CPR should be acquired as part of the normal process of health care delivery, by the providers of care and their institutions to improve data accuracy and timeliness of decision support.

The CPR can also be an instrument for building a clinical data repository that is useful for collecting information about which medical treatments are effective in the practice of medicine in the community and for improving population-based health care. Additional applications of CPR data beyond direct patient care can improve population-based care. These applications bring personal and public benefits, but also raise issues that must be addressed by health care policy makers. Clinical data standards and communication networks, critical for using CPRs effectively, are treated separately in other sections in this chapter.

176.2 Computer-Based Patient Record

A CPR is a collection of data about a patient's health care in electronic form. The CPR is part of a system (a CPRS, usually maintained in a hospital or physician's office) that encompasses data entry and presentation,

0-8493-0462-8/00/$0.00+$.50

storage, and access to the clinical decision maker—usually a physician or nurse. The data is entered by keyboard, dictation and transcription, voice recognition and interpretation, light pen, touch screen, hand-held computerized notepad (perhaps wireless) with gesture and character recognition and grouping capabilities, and other means. Entry may also be by direct instrumentation from electronic patient monitors and bedside terminals, nursing stations, analysis by other computer systems such as laboratory autoanalyzers and magnetic resonance imagers, or another provider's CPRS.

Patient care data collected by a CPRS may be stored centrally or it may be stored in many places (distributed) for retrieval at the request of an authorized user through a database management system. The CPR may present data to the physician as text, tables, graphs, sound, images, full-motion video, and signals. The CPR may also point to the location of additional patient data that cannot be easily incorporated into the CPR.

In many current clinical settings (hospitals, physicians' offices, and ambulatory care centers), data pertaining to a patient's medical care are recorded and stored in a paper medical record. If the paper record is out of its normal location, or accompanying the patient during an off-site study or procedure, it is not available to the nurse, the attending physician, or the expert consultant. In paper form, data entries are often illegible and not easily retrieved and read by multiple users. On the other hand, an electronic form provides legible, clinical information which can be available to all users simultaneously, thus improving timely access to patient care data and communication among care providers.

Individual hospital departments (laboratory or pharmacy, for example) often lose the advantages of automated data when their own computer systems print the computerized results onto *paper.* The pages are then sent to the patient's hospital floor and assembled into a paper record. The lack of standards for the electronic exchange of this data hinders the integration of computerized departmental systems. Searching electronic files is often, but not always, easier than searching through paper. Weaknesses of paper medical record systems for supporting patient care and health care providers have been detailed [Korpman, 1990 and 1991].

Many of the functions of a CPR and how it operates within a health-care information system to satisfy user demands are explained in the Institute of Medicine's report, *The Computer-Based Patient Record: An Essential Technology for Health Care* [1991].

176.3 Clinical Decision Support Systems

Another role of the CPR is to enable a clinical decision support system (CDSS)—computer software designed to aid clinical decision making—to provide the physician with medical knowledge that is pertinent to the care of the patient. Diagnostic suggestions, testing prompts, therapeutic protocols, practice guidelines, alerts of potential drug-drug and drug-food reactions, treatment suggestions, and other decision support services can be obtained through the interaction of the CPR with a CDSS.

Knowledge Server. Existing knowledge about potential diagnoses and treatments, practice guidelines, and complicating factors pertinent to the patient's diagnosis and care is needed at the time treatment decisions are made. The go-between that makes this link is a "knowledge server," which acquires the necessary information for the decision maker from the knowledge server's information sources. The CPR can provide the knowledge server with proper context, i.e., specific data and information about the patient's identification and condition(s) [Tuttle, et al., 1994].

Knowledge Sources. Knowledge sources include a range of options, from internal development and approval by a hospital's staff, for example, to sources outside the hospital, such as the National Guidelines Clearinghouse [www.guideline.gov] sponsored by the Agency for Health Care Policy and Research, the American Medical Association, and the Association of American Health Plans; the Physicians Data Query program at the National Cancer Institute; other consensus panel guidelines sponsored by the National Institutes of Health; and guidelines developed by medical and other specialty societies. Additional sources of knowledge include the medical literature, which can be searched for high-quality, comprehensive review articles and for particular subjects using the "Grateful Med" program to explore the MEDLINE literature database available through the National Library of Medicine. Evidence-based Practice Center reports summarizing scientific evidence on specific topics of medical interest are available to support guideline development [www.guideline.gov].

Medical Logic Modules. If medical knowledge needs are anticipated, acquired beforehand, and put into a medical logic module (MLM), software can provide rule-based alerts, reminders and suggestions for the care provider at the point (time and place) of health service delivery. One format for MLMs is the Arden Syntax, which standardizes the logical statements [ASTM, 1992]. For example, an MLM might be interpreted as, "If sex = female, and age is greater than 50 years, and no Pap smear test result appears in the CPR, then recommend a Pap smear test to the patient." If MLMs are to have a positive impact on physician behavior and the patient care process, then physicians using MLMs must agree on the rules in the logical statements or conditions and recommended actions that are based on interactions with patient care data in the CPR.

Nomenclature. Because MLMs are independent, the presence or absence of one MLM does not affect the operation of other MLMs in the system. If done carefully and well, MLMs developed in one health-care organization can be incorporated in the CPRSs of other health-care organizations. However, this requires much more than using accepted medical content and logical structure. If the medical concept terminology (the nomenclature used by physicians and by the CPR) differs among organizations, the knowledge server may misinterpret what is in the CPR, apply logic to the wrong concept, or select the wrong MLM. Further, the physician receiving its message may misinterpret the MLM. [Pryor and Hripcsak, 1994]

For widespread use of CDSSs, a uniform medical nomenclature, consistent with the scientific literature is necessary. Medical knowledge is information that has been evaluated by experts and converted into useful medical concepts and options. For CDSSs to search through a patient's CPR, identify the medical concepts, retrieve appropriate patient data and information, and provide a link to the relevant knowledge, the CDSS has to recognize the names used in the CPR for the concepts [Cimino, 1993]. Providing direction for coupling terms and codes found in patient records to medical knowledge is the goal of the Unified Medical Language System (UMLS) project of the National Library of Medicine.

176.4 Scientific Evidence

Patient Care Processes. Controlled trials have shown the effectiveness of CDSS for modifying physician behavior using preventive care reminders. In a review of the scientific literature up to February 1992, Johnston, et al., [1994] reported that controlled trials of CDSSs have shown significant, favorable effects on care processes from (1) providing preventive care information to physicians and patients [McDonald, et al., 1984; Tierney, et al., 1986], (2) supporting diagnosis of high-risk patients [Chase, et al., 1983], (3) determining the toxic drug dose for obtaining the desired therapeutic levels [White, et al., 1987], and (4) aiding active medical care decisions [Tierney, 1988]. Johnston found clinician performance was generally improved when a CDSS was used and, in a small number of cases (3 of 10 trials), significant improvements in patient outcomes.

In a randomized, controlled clinical trial, one which randomly assigned some teams of physicians to computer workstations with screens designed to promote cost-effective ordering (for example, of drugs and laboratory tests), Tierney, et al., [1993] reported patient lengths of stay were .89 days shorter, and charges generated by the intervention teams were $887 lower, than for the control teams of physicians. These gains were not without an offset. Time and motion studies showed that intervention physician teams spent 5.5 minutes longer per patient during 10-hour observation periods. This study is a rare controlled trial that sheds light on the resource impact of a CDSS.

In this setting, physician behavior was changed and resources were reduced by the application of logical algorithms to computer-based patient record information. Nevertheless, a different hospital striving to attain the same results would have to factor in the cost of the equipment, installation, maintenance, and software development plus the need to provide staff training in the use of a CDSS.

Malpractice. In some emergency rooms, CDSS are used to prompt physicians to document the patient's record and to suggest the possibility of missed diagnoses—ones that could lead to serious patient harm and malpractice liability—such as myocardial infarctions. In 1990, all emergency room physicians in Massachusetts were offered lower malpractice premiums if they regularly used a particular CDSS [Institute of Medicine, 1991]. An additional benefit of a CDSS in ambulatory settings can be faster decisions and reduced time for the patient in the emergency room, (for example, for triage of patients with symptoms of acute cardiac ischemia) [Sarasin, et al., 1994].

Evaluation. CDSSs should be evaluated according to how well they enhance the user's performance in the user's environment [Nykanen, et al., 1992]. Extending this concept, society should judge CDSSs not only on enhanced physician performance, but also on whether patient outcomes are improved and system-wide health care costs are contained.

CDSS Hurdles. In a review of medical diagnostic decision support systems, Miller [1994] examines the development of CDSSs over the past 40 years, and identifies several hurdles to be overcome before large-scale, generic CDSSs grow to widespread use. These hurdles include determining: (1) how to support medical knowledge base construction and maintenance over time, (2) the amount of reasoning power and detailed representation of medical knowledge required (e.g., how strong a match of medical terms is needed to join medical concepts with appropriate information), (3) how to integrate CDSSs into the clinical environment to reduce the costs of patient data capture, and (4) how to provide flexible user interfaces and environments that adjust to the abilities and desires of the user (e.g., with regard to typing expertise and pointing devices).

CPR Implementation. The real and perceived barriers to implementing a CPR include physicians' reluctance to enter data. Yet direct physician interaction with the system is important because it reduces transcription errors and allows the system to respond to the physician with preventive medicine prompts, drug contraindication alerts, and other reminders during routine patient care. During Beth Israel Hospital's transition to a CPR, data entry may not have been as large a barrier as the burden of recording clinical data on *both* paper and a CPR. Moreover, there is reason to believe that physicians have a greater concern about the confidentiality and privacy of their text notes, than about other clinical data in the CPR [Rind and Safran, 1994]. Older physicians may make significantly fewer CPR system inquiries per case than younger physicians, but Clayton, et al., [1994] reported that physician age explained only a small amount of variation (3 percent) in CPR use. In contrast, he found the average use by physicians categorized by hospital department (medical specialty) to vary by a factor of 10 between high- and low-utilization departments.

Although Massaro [1993a, 1993b] and Dambro, et al., [1988] report on the problems faced and lessons learned during the installation period of their information technology systems, there is generally insufficient documentation in the medical literature about identifying and overcoming the barriers to successfully installing comprehensive, integrated clinical information systems in hospitals' and physicians' offices.

Research Databases. CPRs can have great value for developing research databases, medical knowledge, and quality assurance information that would otherwise require an inordinate amount of manual resources to obtain in their absence. An example of CPR use in research is found in a study undertaken at Latter Day Saints (LDS) Hospital. Using the HELP CPR system to gather data on 2,847 surgical patients, this study found that administering antibiotics prophylactically during the 2-hour window before surgery (as opposed to earlier or later within a 48-hour window) minimized the chance of surgical-wound infection. It also reduced the surgical infection rate for this time category to 0.59 percent, compared to the 1.5 percent overall infection rate for all the surgical patients under study [Classen, et al., 1992].

The same system was used at LDS Hospital to link the clinical information system data (including a measure of nursing acuity) with the financial systems' data. Using clinical data to adjust for the severity of patient illness, Evans, et al., [1994] measured the effect of adverse drug events due to hospital drug administration on hospital length of stay and cost. The attributable difference due to adverse drug events among similar patients was estimated to be an extra 1.94 patient days and $1,939 in costs.

Telemedicine. A CPR may hold and exchange radiological and pathological images of the patient taken or scanned in digital form. The advantage is that digital images may be transferred long distances without a reduction in quality of appearance. This allows patients to receive proficient medical advice even when they and their local family practitioner are far from their consulting physicians.

However, interpreting digital representations (images that are scanned, not originally digital) of radiographs, or X-ray images, while feasible, is inferior in some applications to reading conventional radiographs. Radiologists prefer customary X-ray films and view boxes to their digitized images, and would require significant training and specialized experience to achieve optimal performance with digitized images. Digital images, such as magnetic resonance images (MRIs), should appear no different whether viewed at the patient's site of care or by a radiologist hundreds of miles away from the patient—if the algorithms for compressing (to reduce the costs of data transmission) and decompressing the digital data do not result in a discernable loss of quality. This feature lends itself to the development of expert systems for reading MRIs of difficult cases, for example, particularly when a patient's medical facility is isolated from academic medical centers. For the most part, only radiology and pathology applications of telemedicine have been extensively evaluated and found medically feasible. The cost effectiveness of these applications has not been rigorously evaluated [Grigsby and Kaehny, 1993].

Telemedicine is, of course, more than teleradiology and telepathology. Much can be accomplished with telephone, facsimile machines, and teleconferencing that may not require substantial technical support or training. The financial concern is that payment for telemedicine services may increase the cost of care without improving patient outcomes. Research and project evaluation findings are needed to show where telemedicine applications improve the quality of care and are cost effective.

Additionally, when personally identifiable health care data are transported electronically across state borders for telemedicine uses, the applicability of state laws regarding confidentiality and privacy of this data is not often obvious. This raises legal questions for organizations that wish to move this data over national networks for patient management, business, or analytical reasons. A Federal privacy protection law that applies generally to medical data about patients does not currently exist [Waller, 1991], nor is there a Federal law or licensing system to protect physicians and their patients against the consequences of telemedicine-failure malpractice.

176.5 Hospital and Ambulatory Care Systems

A review of the literature to identify CPRS studies (with a focus on ambulatory care and on the research use of their data) revealed the systems shown in Table 176.1 [Agency for Health Care Policy and Research, 1993]. Table 176.1 presents the system acronym, its meaning, its location, and a brief description. Some systems have been in use for over 25 years and, for the most part, were developed in academic medical centers. In some cases, they are actually data registry systems. Six of these well-studied systems—COSTAR, TMR, HELP, STOR, BIHS, and RMRS—have built vast patient databases derived from the regular delivery of patient care [Tierney and McDonald, 1991]. Two of these systems, COSTAR and HELP, are also commercial systems.

There are numerous commercial computerized hospital information systems (HISs). Most of them focus primarily on administrative functions (admission, discharge, transfer, scheduling, and billing) and secondarily on order entry and results reporting. Although no one commercial HIS is operating today that integrates all the patient's clinical data from all sources and makes the data easily accessible to clinicians, nine commercial HIS vendors offering systems were identified in a study by Abt Associates, Inc., that

> "are designed to be used by clinicians (especially physicians), are comprehensive in functionality and well integrated or take a unique approach to integration, and that meet moderately rigorous standards for integration of clinical, financial, and analytic information are: 3M (HELP), Bell Atlantic (StatLAN Oacis), Cerner (CareNet), Health Data Sciences (Ulticare), IDX (OCM), Meditech (HIS), Phamis (LastWord), SMS (Invision), and TDS (7000)" [1993, 2-2].

Systems of the future may rely on intranet networks within the health enterprise and on web-based exchanges of patient information using web browsers, object-oriented technology, and document formatting languages, e.g., hypertext markup language (http) and extensible markup language (XML).

TABLE 176.1 Hospital and Ambulatory Care Computer-Based Patient Record Systems

Private Sector Systems	
ARAMIS	Arthritis, Rheumatism, and Aging Medical Information System [TOR (Time Oriented Record) is the collection instrument]; Stanford University Medical Center (MA); National chronic disease data bank.
ATHOS	AIDS Time-Oriented Health Outcome Study; Stanford University Medical Center (CA); National chronic disease data bank.
BICS	Brigham and Women's Hospital Integrated Computing System; Brigham and Women's Hospital (MA); Hospital and ambulatory care patient record.
BIHS	Beth Israel Hospital System; Beth Israel Hospital (MA); Hospital and ambulatory care patient record.
CIS	Clinical Information System; Columbia-Presbyterian Medical Center (NY); Hospital and ambulatory care patient record.
COSTAR	Computer-Stored Ambulatory Record; Massachusetts General Hospital (MA); Ambulatory care patient record (also commercial).
DIOGENE	University Hospital of Geneva (Switzerland); Hospital and ambulatory care patient record [non-U.S. but identified in AHCPR, 1993].
HCHP/AMRS	Harvard Community Health Plan Ambulatory Medical Record; HCHP (MA); Ambulatory care patient record.
HELP	Health Evaluation Through Logical Processing; LDS Hospital (UT); Hospital and ambulatory care patient record (also commercial).
MARS	Medical Archival System; University of Pittsburgh (PA); Hospital and ambulatory care patient record (under development).
PIMS/PIPS	Patient Information Management System/Patient Information Protocol System; Loyola University Medical Center (IL); Hospital and ambulatory care patient record.
RMRS	Regenstrief Medical Records System; Indiana University Medical Center (IN); Hospital and ambulatory care patient record.
RPMS	Resource and Patient Management System; Indian Health Service (AZ); Hospital and ambulatory care patient record.
STOR	Summary Time-Oriented Record; University of California at San Francisco (CA); Ambulatory care patient record.
THERESA	Grady Memorial Hospital (GA); Hospital and ambulatory care patient record.
TMR	The Medical Record; Duke University Medical Center (NC); Hospital and ambulatory care patient record.
VIPOR	Vermont Integrated Problem-Oriented Record; University of Vermont Health Center (VT); Ambulatory care patient record.
Federal CPR systems	
DCHP	Decentralized Hospital Computer Program; Department of Veterans Affairs; Hospital patient record.
CHCS	Composite Health Care System; Department of Defense; Hospital patient record (6 modules).
IHS	Indian Health Service, Department of Health and Human Services; Hospital and ambulatory care patient record.

176.6 Extended Uses of CPR Data

Data produced by such systems have additional value beyond supporting the care of specific patients. For example, subsets of individual patient care data from CPRs can be used for research purposes, quality assurance purposes, developing and assessing patient care treatment paths (planned sequences of medical services to be implemented after the diagnoses and treatment choices have been made), assessments of treatment strategies across a range of choices, and assessments of medical technologies in use in the community after their approval by the Food and Drug Administration. When linked with data measuring patient outcomes, CPR data may be used to help model the results achieved by different treatments, providers, sites of care, and organizations of care.

If these data were uniformly defined, accurately linked, and collected into databases pertaining to particular geographical areas, they would be useful for research into the patient outcomes of alternative treatments for specific conditions treated in the general practice of medicine and for developing information to assist consumers, health care providers, health plans, payers, public health officials, and others in making choices about treatments, technologies, sites and providers of care, health plans, and community health needs. This is currently an ambitious vision considering the presently limited use of CPRs and insufficient incentives for validating and storing electronic patient record data. Many such decisions are now based on data of inferior quality; but the importance of these decisions to the health care market is driving higher the demand for uniform, accurate clinical data.

Federal Programs

Uniform, electronic clinical patient data could be useful to many Federal programs that have responsibility for improving, safeguarding, and financing America's health. For example, the Agency for Health Care Policy and Research (AHCPR) is charged with undertaking research to "identify the most effective methods for preventing, diagnosing, treating, and managing various health conditions" in community settings. [PL 101-239, 1989]. The research should improve patient outcomes of care, quality measurement, address cost, and access problems. To examine the influence on patient outcomes of alternative treatments for specific conditions, this research needs to account for the simultaneous effects of many patient risk factors, such as diabetes and hypertension. Health insurance claims data do not have sufficient clinical detail for many research, quality assurance, and evaluation purposes. Therefore, claims data must be supplemented with data abstracted form the patients' medical records. In many cases, the data must be identified and collected prospectively from patients (with their permission) to ensure availability and uniformity. The use of a CPR could reduce the burden of this data collection, support practice guideline development in the private sector, and support the development, testing, and use of quality performance measures.

Medical effectiveness research could benefit from access to uniform, accurate, nationwide data from CPRs—for example, on the last 10,000 cases of a given condition such as benign prostatic hyperplasia, low back pain, or cataracts. Because CPRs do not typically contain information on patient outcomes after discharge from a hospital (or a physician's office), CPR data would have to be supplemented with patient outcome measures.

Other Federal, State, and local health agencies also could benefit from CPR-based data collections. For example, the Food and Drug Administration, which conducts postmarketing monitoring to learn the incidence of unwanted effects of drugs after they are approved, could benefit from analyzing the next 20,000 cases in which a particular pharmaceutical is prescribed, using data collected in a CPR. Greater confidence in postmarket surveillance could speed approval of new drug applications. The Health Care Financing Administration could provide guidance and information to its Professional Review Organizations (PROs) about local and nationwide medical practice patterns founded on analysis of national and regional CPR data about Medicare beneficiaries; the state-based, Medicare PROs could analyze data from CPRs in their own states to provide constructive, quality-enhancing feedback to hospitals and physicians. As a further example, the Centers for Disease Control and Prevention could more quickly and completely monitor the incidence and prevalence of communicable diseases with access to locally available CPR data on patient care. State and local public health departments could allocate resources more quickly to address changing health needs with early recognition of a community health problem. Many of these uses require linked data networks and data repositories (that the communities trust with their health data).

176.7 Selected Issues

While there are personal and public benefits to be gained from extended use of CPR data beyond direct patient care, the use of personal medical information for these uses, particularly if it contains personal identification, brings with it some requirements and issues that must be faced. Some of the issues that must be addressed by health care policy makers, as well as by private markets, are as follows:

Standards are needed for the nomenclature, coding, and structure of clinical patient care data; the content of data sets for specific purposes; and the electronic transmission of such data to integrate data efficiently across departmental systems within a hospital and data from the systems of other hospitals and health care providers. If benefits are to be realized from rapidly accessing and transmitting patient care data for managing patient care, consulting with experts across long distances, linking physician offices and hospitals, undertaking research, and other applications, data standards are essential [Fitzmaurice, 1994a]. The Computer-based Patient Record Institute has taken an active interest in the development of such standards, and the Healthcare Informatics Standards Board of the American National Standards Institute has coordinated the standard developing organizations that work on such standards in the United States and provided an official link to standards organizations in other countries, such as the European Standards Committee (CEN). The Organization of International Standards (ISO) Technical Committee 215, Health Informatics, was formed in 1998 to provide a forum for international coordination of health informatics standards. The Board of Directors of the American Medical Informatics Association [1994] has recommended specific approaches for many patient data standards' areas—adopt the best of what exists and move quickly toward continued refinement.

Administrative health data standards are mandated in the Health Insurance Portability and Accountability Act (HIPAA) of 1996 (Public Law 104-191). In this law, the Secretary of Health and Human Services (HHS) is directed to adopt standards for nine common health transactions (enrollment, claims, payment, and others) that must be used if those transactions are conducted electronically. Penalties are capped at $25,000 per year for each transaction violated. Digital signatures, when adopted by the Secretary, may be deemed to satisfy federal and state statutory requirements for written signatures for HIPAA transactions. The provisions of HIPAA are expected to take effect in 2001.

Confidentiality and privacy of individually identifiable patient care and provider data are very likely the most important issues. For most purposes, including much research, the user does not need to know patient identities if the data have been accurately linked by a trusted party. The benefits from extended uses of this data must be balanced against the opportunities for personal harm from unwanted disclosure (which must be avoided). A Federal privacy law that spells out the allowable uses of personally identifiable health care data and the conditions under which it can be used, with civil and criminal punishments for violating disclosure or redisclosure prohibitions, might appropriately address this issue. It should establish safeguards for privacy based on uniform and fair information practices [Gostin, et al., 1993; Privacy Protection Commission, 1977; Secretary's Advisory Committee on Automated Personal Data Systems, 1973]. HIPAA requires the Secretary of HHS to promulgate final regulations containing standards with respect to the privacy of individually identifiable health information transmitted in HIPAA transactions in 2000, if privacy legislation is not enacted in 1999.

The *quality of stored clinical data* may be questioned in the absence of organized programs and criteria to assess the reliability, validity, and sufficiency of this data for undertaking research and for providing useful information to consumers, medical care organizers, and payers. For proper analysis, the data should be sufficient to measure and assess the relevant risk factors influencing patient outcomes.

Electronically stored records in one state may be considered to be legally the same as paper records, but not in another state. In law, regulation, and practice, many states require pen and ink recording and signatures, apparently ruling out electronic records and signatures. This inconsistency, and uncertainty, creates a barrier for the electronic communication and legal acceptance of patient care data, images, medical advice, and claims across state boundaries for medical and other publicly acceptable purposes [Waller, 1991].

Standard unique identifiers for patients, health care providers, institutions and payers are needed to obtain economies and accuracy when linking patient care data at different locations, and patient care data with other relevant data. Under HIPAA, the Secretary of HHS must adopt standards for uniquely identifying providers, health plans, employers, and individuals. Because of concerns about the confidentiality of personal health information that may be linked using the unique individual health identifier, final implementation of that identifier will await explicit approval by the administration and Congress [Public Law 105-277].

Malpractice concerns arise as telemedicine technology allows physician specialists to give medical advice across state borders electronically to other physicians, other healthcare providers, and patients. Physicians are licensed by a state to practice within its own state borders. Does a physician who is active in telemedicine need to obtain a license from each state in which he or she practices medicine from outside the state? If the expert physician outside the patient's state gives bad advice, which state's legal system has jurisdiction for malpractice considerations?

System security and integrity become improtant as more and more data and information for patient treatment and other uses are exchanged through national networks. Not only does this issue relate to purposeful violations of privacy, but also to the accuracy of medical knowledge for patient benefit. If the system fails to transmit accurately what was sent to a physician—for example, an MRI, a practice guideline, or a clinical research finding—and if a physician's judgment and recommendation is based on a flawed image or other misreported medical knowledge, who bears the legal responsibility for a resulting inappropriate patient outcome? National standards for the security of health information are mandated by HIPAA.

Benefit-cost analysis methodologies must be developed and applied to inform investment decision makers about the most productive applications of CPR systems. There is a need for a common approach to measuring the benefits and the costs to be able to compare alternative applications. Accurate research findings applied to business risk and benefit assessments can advance commercial progress for developing and implementing CPR applications.

Regional health data repositories for the benefit of employers, hospital groups, consumers, and state health and service delivery programs raise issues about the ownership of patient care data, the use of identifiable patient care data, and the governance of the data repository. A study by the Institute of Medicine [1994] examines the power of regional health data repositories for improving public health, supporting better private health decisions, recognizing medically and cost-effective health care providers and health plans, and generally providing the information necessary to improve the quality of health care delivery in all settings. Some of the data in these repositories may be based on CPR data. This data may include personally identifiable data and move outside the environment in which it was created.

Patient medical record information is the subject of study by the National Committee on Vital and Health Statistics (NCVHS). The function of this expert public advisory is to advise the Secretary of HHS on issues relating to health data, statistics, privacy, and national health information policy [http://aspe.dhhs.gov/ncvhs/]. Mandated by HIPAA, NCVHS must "study the issues related to the adoption of uniform data standards for patient medical record information and the electronic exchange of such information" and "shall report to the Secretary not later than [in 2000] … recommendations and legislative proposals for such standards and electronic exchange." Current issues dealing with computer-based patient record data–formatting, definitions, codes, confidentiality, and technology–are frequently the subject of public NCVHS hearings.

Acknowledgments

The author is Senior Science Advisor for Information Technology, Agency for Health Care Policy and Research, in the U.S. Department of Health and Human Services. He appreciates the helpful comments given by Robert Esterhay, M.D., and Kathleen McCormick, Ph.D., R.N., on an early draft of this section.

References

Abt Associates Inc. 1993. *Overcoming Barriers to Integration and Implementation of Clinical Information Management Systems: Feasibility Study*, AHCPR Contract Final Report 282-92-0064. NTIS Accession No. PB 94 159886.

Administrative Simplification (HIPAA), [*http://aspe.os.dhhs.gov/admnsimp/index.htm*], accessed January 26, 1999.

Agency for Health Care Policy and Research. 1993. *Automated Data Sources for Ambulatory Care Effectiveness Research,* M.L. Grady and H.A. Schwartz, (Eds.) Agency for Health Care Policy and Research, Washington, D.C.

American National Standards Institute, Healthcare Informatics Standards Board. 1997. *HISB Inventory of Health Care Information Standards Pertaining to the Health Insurance Portability and Accountability Act of 1996, P.L. 104-191.* New York.

American National Standards Institute, Healthcare Informatics Standards Board. 1998. Inventory of Clinical Information Standards, New York. [http://web.ansi.org/rooms/room_41/public/docs.html]

ASTM. 1992. *E1460-92: Standard Specifications for Defining and Sharing Modular Health Knowledge Bases (Arden Syntax for Medical Logic Modules).* ASTM, Philadelphia, PA.

Board of Directors of the American Medical Informatics Association. 1994. Standards for medical identifiers, codes, and messages needed to create an efficient computer-stored medical record. *J. Am. Med. Inform. Assoc.* 1:1-7.

Brannigan, V.M. 1994. Protection of patient data in multi-institutional medical computer networks: Regulatory effectiveness analysis. *AMIA Proceedings. Seventeenth Annual Symposium on Computer Applications in Medical Care,* McGraw-Hill, Inc., New York, p. 59-63.

Chase, C.R., Vacek, P.M., Shinozaki, T., Giard, A.M., and Ashikaga, T. 1983. Medical information management: Improving the transfer of research results to presurgical evaluation. *Med. Care.* 21:410-424.

Cimino, J.J. 1993. Saying what you mean and meaning what you say: Coupling biomedical terminology and knowledge. *Acad. Med.* 68(4):257-260.

Classen, D.C., Evans, R.S., Pestotnik, S.L., Horn, S.D., Menlove, R. L., and Burke, J.P. 1992. The timing of prophylactic administration of antibiotics and the risk of surgical wound infection. *NEJM.* 326(5):281-285.

Clayton, P.D., Pulver, G.E., and Hill, C.L. 1994. Physician use of computers: Is age or value the dominant factor? *AMIA Proceedings. Seventeenth Annual Symposium on Computer Applications in Medical Care,* McGraw-Hill, Inc., New York, p. 301-305.

Clinton, W.J. and Gore, A. 1993. *Technology for America's Economic Growth, A New Direction to Build Economic Strength.* Executive Office of the President, Washington, D.C.

Committee on Information and Communication, National Science and Technology Council. 1994. *High Performance Computing and Communications: Technology for the National Information Infrastructure,* Executive Office of the President, Washington, D.C.

Dambro, M.R., Weiss, B.D., McClure, C.L., and Vuturo, A.F. 1988. An unsuccessful experience with computerized medical records in an academic medical center. *J. of Med. Ed.* 63:617-623.

Donaldson, M.S. and Lohr, K.N., (Eds.) 1994. *Health Data in the Information Age: Use, Disclosure, and Privacy.* National Academy Press, Washington, D.C.

Evans, R.S., Pestotnik, S.L., Classen, D.C., and Burke, J.R. 1993. Development of an automated antibiotic consultant. *M.D. Computing.* 10(1):17-22.

Evans, R.S., Classen, D.C., Stevens, M.S., Pestotnik, S.L, Gardner, R.M., Lloyd, J.F., and Burke, J.P. 1994. Using a health information system to assess the effects of adverse drug events. *AMIA Proc. Seventeenth Ann. Symp. Comp. Appl. Med. Care,* McGraw-Hill, Inc., New York, p 161-165.

Fitzmaurice, J.M. 1994b. Health Care and the NII. In: *Putting the Information Infrastructure to Work: Report of the Information Infrastructure Task Force Committee on Applications and Technology,* p 41-56. National Institute of Standards and Technology, Gaithersburg, MD.

Fitzmaurice, JM. 1998. A new twist in US health care data standards development: Adoption of electronic health care transactions standards for administrative simplification. *Int. J. Medical Informatics.* 48:19-28.

Gostin, L.O, Turek-Brezina, J., Powers, M., Kozloff, R., Faden, R.,and Steinauer, D. 1993. Privacy and security of personal information in a new health care system. *JAMA.* 270(20):2487-2493.

Grigsby, J and Kaehny, M.M. 1993. Analysis of expansion of access to care through use of telemedicine and mobile health services. In: *Report 1: Literature review and analytic framework.* Center for Health Policy Research, Denver, (Report to the Health Care Financing Administration under Contract No. 500-92-0046 in September 1993).

Information Infrastructure Task Force. 1993. *The National Information Infrastructure: Agenda for Action.* Executive Office of the President, Washington, D.C.

Institute of Medicine. 1991., Revised edition 1997. *The Computer-Based Patient Record: An Essential Technology for Health Care,* D.E. Detmer, R.S. Dick and E.B. Steen, (Eds.) National Academy Press, Washington, D.C.

Institute of Medicine. 1994. *Health Data In the Information Age: Use, Disclorure, and Privacy,* M.S. Donaldson and K.N. Lohr (Eds.), National Academy Press, Washington, D.C.

Johnston, M.E., Langton, K.B., Haynes, R.B., and Mathieu, A. 1994. Effects of computer-based clinical decision support systems on clinician performance and patient outcome. *Ann. Int. Med.* 120:135-142.

Korpman, R.A. 1990. Patient care automation; the future is now. part 2. The current paper system—can it be made to work? *Nurs. Econ.* 8(4):263-267.

Korpman, R.A. 1991. Patient care automation; the future is now, part 8. Does reality live up to the promise? *Nurs. Econ.* 9(3): 175-179.

Massaro, T.A. 1993a. Introducing physician order entry at a major academic medical center: I. Impact on organizational culture and behavior. *Acad. Med.* 68(1):25-30.

Massaro, T.A. 1993b. Introducing physician order entry at a major academic medical center: II. Impact on medical education. *Acad. Med.* 68(1):20-25.

McDonald, C.J., Hui, S.J., Smith, D.M., Tierney, W.M., Cohen, S.J., Weinberger, M., et al. 1984. Reminders to physicians from an introspective computer medical record. A two-year randomized trial. *Ann. Int. Med.* 100:130-138.

Miller, R.A. 1994. Medical diagnostic decision support systems—past, present, and future. *J. Am. Med. Informatics Assoc.* 1(1):8-27.

National Coordination Office for High Performance Computing and Communication. 1994. *HPCC FY 1995 Implementation Plan.* Executive Office of The President, Washington, D.C.

Nykanen, P., Chowdhury, S., and Wiegertz, O. 1992. Evaluation of decision support systems in medicine. In: *Yearbook of Medical Informatics 1992.* p. 301-309.

Privacy Protection Study Commission. 1977. *Personal Privacy in an Information Society.* U.S. Government Printing Office, Washington, D.C.

Pryor, T. Allan and Hripcsak, George. 1994. "Sharing MLMs: An Experiment Between Columbia-Presbyterian and LDS Hospital." *AMIA Proc. Seventeenth Ann. Symp. Comp. Appl. Med. Care,* McGraw-Hill, Inc., New York, p. 399-403.

Public Law 101-239. the Omnibus Budget Reconciliation Act of 1989.

Public Law 104-191, the Health Insurance Portability and Accountability Act of 1996, August 21, 1996.

Public Law 105-277, Department of Transportation and Related Agencies Appropriations Act, 1999, October 21, 1998.

Rind, D.M. and Safran, C. 1994. Real and imagined barriers to an electronic medical record. *AMIA Proc. Seventeenth Ann. Symp. Comp. Appl. Med. Care,* McGraw-Hill, Inc., New York, p. 74-78.

Sarasin, F.P., Reymond, J., Griffith, J.L., Beshansky, J.R., Schifferli, J.A., Unger, P., Sherrer, J., and Selker, H.P. 1994. Impact of the acute cardiac ischemia time-insensitive predictive instrument (ACI-TIPI) on the speed of triage decision making for emergency departent patients presenting with chest pain: a controlled clinical trial. *J. Gen. Int. Med.* 9:187-194.

Secretary's Advisory Committee on Automated Personal Data Systems. 1973. *Records Computers and the Rights of Citizens.* U.S. Department of Health, Education & Welfare, Washington, D.C.

Silva, J.S. and Ball, M.J. The professional workstation as enabler: Conference recommendations. *Int. J. Biomed. Comput.* 34(1-4):3-10.

Tierney, W.N. and McDonald, C.M. 1991. Practice Databases and their uses in clinical research. *Statistics in Medicine.* 10:541-557.

Tierney, W.M., McDonald, C.J., Hui, S.J. and Martin, D.K. 1986. Computer predictions of abnormal test results. Effects on outpatient testing, *JAMA.* 259:1194-1198.

Tierney, W.M., Miller, M.E., Overhage, J.M., and McDonald, C.J. 1993. Physician inpatient order writing on mcirocomputer workstations. *JAMA.* 269(3):379-383.

Tuttle, M.S., Sherertz, D.D., Fagan, L.M., Carlson, R.W., Cole, W.G., Shipma, P.B., and Nelson, S.J. 1994. Toward an interim standard for patient-centered knowledge-access. *AMIA Proc. Seventeenth Ann. Symp. Comp. Appl. Med. Care,* McGraw-Hill, Inc., New York, p. 564-568.

Waller, A.A. 1991. Legal aspects of computer-based patient records and record systems. In: R.S. Dick and E.B. Steen, (Eds.), *The Computer-Based Patient Record: An Essential Technology for Health Care,* National Academy Press, Washington, D.C., p. 156-179.

White, K.S., Lindsay, A., Pryor, T.A., Brown, W.F., and Walsh, K., 1984. Application of a computerized medical decision-making process to the problem of digoxin intoxication. *J. Am. Coll. Cardiol.* 4:571-576.

White, R.H., Hong, R., Venook, A.P., Dashbach, M.M., Murray, W., and Mungall, D.R. 1987. Initiation of warfarin therapy: comparisons of physician dosing with computer-predicted dosing. *J. Gen. Int. Med.* 2:141-148.

177

Computer Networks in Health Care

Soumitra Sengupta
Columbia University

Computer technology plays a prominent role today in health care delivery centers throughout the world. Information technology is gradually transforming the basic practice of medicine by improving the quality and quantity of information used by clinicians and administrators. The fundamental compute technology that accomplishes transparent and efficient flow of information from the creators of the information to the ultimate end users is *Computer Networking* [Tanenbaum 1988].

In a medical center, and to a lesser degree, in a health care practice center, computers are used for many purposes: in ancillary, departmental settings such as laboratory, pharmacy, radiology, pathology; in administrative contexts such as billing, patient management, transportation, payroll; in clinical and scholarly purposes such as electronic medical records, imaging, searchings for medical references; and in basic research functions such as molecular modeling, genetics, robotics surgery [Rennels 1989; Duisterhout et al., 1991]. The concept of distributed computing is fast becoming a reality in medical domains, and it deals with the collection, integration and presentation of data distributed over several computers. Seamless information exchange requires the ubiquitous presence of a network, which, comprised of hardware and software components or layers, interconnects these computers. Well-defined methods and rules to communicate between computers are called *communication protocols*, and a collection of specific layers and protocols with its implementation is called a *communication standard*. Several communication standards, proprietary and open, exist today.

177.1 History

During the 1970s and the early 1980s, computing at hospitals was targeted to administrative needs. Central mainframe computers were employed for financial data processing. Even then, laboratory functions were starting to be automated due to their sheer volume and importance to the clinical community. Advent of personal computers (PCs) truly opened up automation opportunities for the ancillary departments as it became cost-effective to collect both departmental data for billing purposes and clinical data for clinical purposes. During the late 1980s, these stand-alone systems became permanent fixtures.

0-8493-0462-8/00/$0.00+$.50

The networking technology also matured in the latter part of the 1980s. Networks' initial deployment within health care practice, however, was haphazard at best because of the large capital cost of laying cables and little understanding of the benefits at that time. Stand-alone systems started to need data from other systems: A single bill has to be sent to a patient for all services even though pharmacy and laboratory are different departments; ancillaries want to use the same demographic information that was entered in the hospital patient management system, and so on. Individual departments experimented with networking and succeeded in creating local solutions. This eventually became a problem when networks came to be appreciated as institutional resources and the local islands of information had to be rearranged for the larger institutional goals o networking efficiency and reliability.

177.2 Impact of Clinical Data

Clinical computing addresses the need for data related to patient care. These data must be delivered to the providers (physicians, nurses, technicians, therapists, social workers) in a timely fashion. The data must be accurate and be delivered in patient care settings such as bedside or nursing stations, intensive care units (ICUs), emergency rooms, operating theaters, clinics, and physicians' offices with varying degree of delivery efficiency. The needs for data in ICUs, for example, are more stringent than in a clinic. The primary data of importance in an inpatient setting are laboratory results. Other important sources of data include pharmacy, radiology, pathology, cardiology, vital signs, neurology, nursing documentation, obstetrics/gynecology, operative reports, discharge summaries, and visit notes. Whereas the paper-based patient chart has traditionally served the need for such information, it is clear that computers are far more efficient and accurate in organizing, maintaining, and disseminating these records. The ability to use clinical information to support automated decision-making applications does not exist in paper-based systems. Hence, the concept of an all electronic medical record has become popular in the past few years.

The delivery of clinical data from many different ancillary systems to diverse end-user settings requires a reliable networking infrastructure to transport the data. Clinical data may be collected from different sources and stored in a (conceptually) central clinical database. Here, a network is necessary to interconnect ancillary systems to a back-end central repository. In another scenario, a front-end workstation may query individual ancillaries and may dynamically construct an integrated view for a provider. Here, a network is necessary to connect ancillary systems to a front-end workstation. In reality, a comprehensive network that connects all computing entities within an institution enables all scenarios.

177.3 Information Types

Health care data have tremendous variety, all of which are important in patient care and have different implications for the performance required of a network. Most of the clinical data available today are in textual form as narrative reports. Radiology, pathology, and so on generate reports about specific tests; discharge summaries explain the entire episode of the patient's stay in the hospital; operative reports explain the procedure followed during operations. These reports range anywhere from a few sentences (in a specific type of pathology report) to several pages (autopsy reports).

Laboratory data tend to be in coded form, where each data item is represented as a code (alphanumeric, integers). This is very useful as providers need to look at trends in data over time (examples include temperature readings and anion count in blood). Furthermore, for computer-based clinical decision support, more data need to be in coded form so that a computer-based expert system can understand the values and act upon them to send alerts to providers, if necessary. Coded data, in general, are small in size, but in some cases such as an EKG tracing, each point may be a code, and thus the report may be of substantial size.

Hospitals also use a large amount of multimedia data. All radiologic tests are performed as visual interpretation of images and video sequences (X-ray images, CAT scans, PET scans, SPECT scans). To a lesser degree, ultrasound sonograms, pathology slides, gated blood pool studies and neurology are other sources of these data. A network that can support these kinds of data requires special consideration and design and is also more expensive.

In academic institutions, a large amount of computer-based educational information is graphical and video-based. Many stand-alone educational products are being actively used today. Looking toward the future, these data (depicting topics such as how to conduct a "physical and history" examination on a patient and how to conduct an arthroscopic surgery of the knee) will be available over the same network and on demand to all students, faculty, and practitioners.

177.4 Platforms

Individuals departments have typically purchased turnkey systems (or have purchased hardware and then developed the necessary software themselves) that are *best-of-breed* variety for their applications. Once the benefits of data exchange were understood, networking has become one of the key considerations for these systems. Since there are several standards of networking, and not all are supported by all systems, heterogeneity of computing platforms has become an issue. The solutions include *application gateways*, computers that translate from one protocol to another while passing the data between them transparently. Gateways are acceptable as an exception, but too many gateways cause severe performance and reliability problems in the network. The current norm is to choose a restricted set of protocol standards that are fully supported within an institution. A strict control on resources helps in addressing the network needs of the majority.

177.5 Current Technologies

Networks are usually defined as a set of *layers*, each layer providing service to the layer above it. The networking infrastructure suitable in a campus environment today is called a *local area network* (LAN). An organization as full control over its LANs. In contrast, when communication is required over long distances (e.g., and organization which has offices all over the world), services from a long distance carrier company are required to establish what is called a *wide area network* (WAN). The protocols at higher layers work transparently over LANs and WANs, thus providing transparency to applications. Currently, software solutions are appearing that use distributed programs in order to achieve total network transparency and data integration from the end-user viewpoint. These are popularly termed *middleware.* Finally, there are specific solutions and standards pertinent to the health care environment which address medical domain problems which may be called *medical domain middleware.* In the following, we elaborate further on each of these concepts.

Local Area Networks

The lowermost layer in networking is the *physical* medium, which can be copper cables (with different degrees of insulation, twists, and impedance), fiber cables (with different diameter and refraction indices), or even the atmosphere in the case of wireless, communication based on radiofrequency transmissions. Examples of physical medium standards include 10BaseT, 10Base5, and STP. The next higher layer, the *link* layer consists of electronics or optics to drive signals on the physical layer. The most popular LAN standards in this layer today are Ethernet (or IEEE 802.3) at a rated 10Mbits/s speed (or bandwidth) and Token Ring (IEEE 802.5) at 4 or 16Mbits/speed. Above these layers, several proprietary protocols and layer stacks may coreside in typical medical practice center LANs. These include IBM SNA and NetBIOS, Novell SPX/IPX, DEC DECNet and LAT, and Apple AppleTalk. Among open protocols, the TCP/IP stack is the most popular standard.

The *topology* of a LAN is the layout of networking segments (e.g., Ethernet, Token Ring) and their interconnections by devices such as repeaters, bridges, and routers. Some of the factors that influence topology design are traffic estimates and patterns, redundancy, isolation and security, physical plant including building and cabling layout, and organizational structure. A practical topology is that of redundant campus backbone segments that then connect to building segments, thus forming a hierarchy of segments. A medical center LAN needs to accommodate heterogeneous sets of computer and consequently needs to support several higher-layer protocols. The choice of higher-layer protocols also influences the choice of bridges and router devices. Although initially driven by the integration of existing systems, a medical center LAN settles on a small subset of protocols for most of its mainstream communication which is practically motivated by the support and maintenance issues.

In addition to medical domain applications, users require the usual file and print services for day-to-day applications such as spreadsheets, word processors, presentation programs, and electronic mail. As networked PCs have become popular as user workstations, Novell SPX/IPX in the IBM PC world and Apple AppleTalk in the Apple Macintosh world protocols are increasingly used to provide these services. A widely used application on the PCs today is terminal emulation by which a PC behaves as a terminal to an ancillary or central computer on the network, and thus subsequently has access to all medical applications on that application server computer. LANs, instead of the traditional point-to-point connections, have become the primary vehicle for computer communication at health care practice centers to the extent that all new computers are expected to be LAN-compatible. A special group of LAN specialists is required to support the LANs. Since LAN technology is constantly evolving toward better and faster communications, it is important to maintain up-to-date knowledge in this field.

Wide Area Networks

Health care centers employ WANs mainly for three purposes: to connect networks at physically distant buildings, offices, or clinics, to connect networks to different organizations (affiliated hospitals, insurance agencies, state or federal government regulatory bodies) or services (Internet, disaster recovery backup locations); and to provide dial-in-and dial-out capabilities for employees. The difference between the first two purposes is that of control and security: access to medical information requires better control when connected to an external organization. The concept of a *firewall* as part of networking topology has become popular when connecting to the Internet in order to deter unwarranted access. A physician's office may connect to insurance networks for billing, as well as to on-line services to connect to the large Internet. In all cases, an external telecommunications carrier's service is used to provide underlying long-distance, point-to-point communications capability.

The aggregate bandwidth in a WAN connection is typically much smaller than the LANs bandwidth because of the higher costs of WAN connections. Popular leased line connections have speeds ranging from 56 Kbits/s to 1.56 Mbits/s (T-1 circuits) to 45 Mbits/s (T-3 circuits), and fractions thereof. Other modes of such connections are accomplished by dedicated microwave or laser communications. In all cases, end devices such as modems are required to drive these connections. Additionally, multiplexers are employed to effectively use the available bandwidth. It is common to extend the LAN protocols to run over the WAN connections, so that applications run transparently regardless of location, but perhaps at a slower speed due to the restricted bandwidth.

Dial-in and dial-out access is provided over asynchronous phone lines with speeds ranging from 1200 bits to 28.8 Kbits/s. Integrated Services Digital Network (ISDN) solutions (at 64 Kbits/s) are becoming available and cost-effective. With implementations of the point-to-point protocol (PPP) standard, the dialing-in PC will become a full peer node on the organizational network, instead of becoming a terminal to some computer or going through a surrogate PC on the network.

Middleware

At higher layers of networking, the client-server computing paradigm is becoming an alternative to central host-based computing. By distributing the work over several nodes on the network, more efficient and flexible solutions are created. These solutions integrate data for the end-users at the front end while hiding the multiplicity of origins of the data at the back end. A reliable, distributed file system, distributed printing services, consolidated directory services, network-based security service, and transparent invocation of applications residing elsewhere are examples of middleware. In the medical center context, it is extremely important to establish standards at this level to effectively tame the heterogeneity. Specific examples of middleware include Open Software Foundation's Distributed Computing Environment (DCE), SUN's Open Network Computing (ONC), OSI X.500 directory services, MIT kerberos security service, and OMG Common Object Request Broker Architecture (CORBA).

Medical Domain Middleware

With the availability of standard networks, applications, and enabling middleware technologies, clinical computing concentrates on the collection, integration, and dissemination of clinical data. In addition, the major focus is to convert data into knowledge in order to help the provider make better clinical decisions. In order to accomplish these transformations, several medical domain middleware tools and concepts need to be constructed and understood to operate in a distributed computing environment.

Integrated Patient Database

A comprehensive clinical information system requires construction of a logically integrated repository of all clinical information for each patient. As stated previously, clinical data originate at many ancillary systems. All data may be collected at a single, physically central repository, using the network as and when data are produced at the ancillaries. In this case, a data review application needs only to query (over the network) the central repository for all data, which reduces the complexity at the front end. Alternatively, data review applications may query several ancillaries dynamically at the time specific patient's data are being reviewed, thus creating a virtual patient repository. In either case, data integration is transparent to the end-use.

Networks permit very efficient distributions of the patient database. Since, from a clinical viewpoint one patient's data are completely independent of another patient's data, each patient's data constitutes a separate database. Therefore, it is possible to isolate a specific patient database (or a set of patient databases) on one host. One logical distribution strategy is that of dedicating a fully functional database for each nursing station within a hospital with a well-defined strategy of migrating patients from one database to another. Once again, however, the operations must be transparent to the end-user, and the explicit reference to hosts and patient information must be hidden within the database functions.

The need for clinical and hospital data exchange among heterogeneous systems has required the creation of common standards for the definition of medical events, the collection of data associated with these events, and the recognition of events that may be useful to participating systems [McDonald & Hammond, 1989]. The only practical standard today is *Health Level* 7 (HL-7), commonly followed by many vendors today. Examples of events defined in HL-7 include admit, discharge, transfer, order, and order results. The use of the standard avoids the need to make unnecessary translations.

The success of these early databases has raised the interesting question: How can a patient's information be collected from many institutions (doctor's office, clinics, hospitals) to create a comprehensive record that is readily available over the network regardless of the patient's geographic location? Many efforts in electronic medical records at the national level include discussion of the pros and cons of merging of information for clinical care of the patient as well as for epidemiological research for the society at large.

Medical Vocabulary

A common medical vocabulary is fast becoming a necessity within clinical information systems. The vocabulary addresses the problem of understanding the semantics of medical concepts, answering questions about whether two concepts are the same or different, and if the latter, how so [Cimino et al., 1994]. For example, cardiomegaly and enlargement of heart are the same concepts, a fact that cannot be derived syntactically. A medical vocabulary defines medical concepts as entities and supports links between entities that collectively define that concept. In the exchange of information between systems, the vocabulary serves as the semantic standard of the meaning of the data items exchanged, much as HL-7 serves as the syntactic standard. Without the vocabulary, the end-user must know how each medical concept is described in each of the ancillaries to formulate his queries correctly. With the vocabulary, the concepts are semantically translated to either a canonical code or to the native ancillary code for that concept. The vocabulary makes the heterogeneity of terminology introduced by several ancillary systems transparent to the user.

Directions and Challenges

Medical computing will compute to take advantage of the ever-increasing expansion of the computing technology. Options not viable in the past are quickly becoming mainstream in the areas of improved networking bandwidth. In addition, the current changes in the health care business dictate more efficient forms of car delivery, which may be facilitated only by adopting information technologies. In the following sections, we discuss a few directions and challenges in this field.

Improved Bandwidth

The currently prevalent networks with 4, 10, and 16 Mbits. a bandwidth are inadequate for pervasive imaging and multimedia applications. Due to the richness of medical information, these applications are extremely important and have a tremendous impact on the required network bandwidth. Existing high-speed networks at 100 Mbits/s, primarily Fiber Distributed Data Interface (FDDI), have been used as the campus backbones but have failed to capture the marketplace due to the high cost of fiber cable deployment and adapter cards. Instead, new standards of100 Mbits/s communication (100Base-T or Fast Ethernet, 100VG-AnyLAN) are fast becoming reality. These technologies use the same cables (UTP grade 5, STP) as for the slower networks, but at a higher cost of equipment. If fiber optic cables are used, then communication can reach Gigabits/s (1000 Mbits/s) range, although most implementations today support 155Mbits/s. This is being accomplished by the new standard of *asynchronous transfer mode* (ATM) communication and corresponding devices called ATM switches. The long-distance carriers are standardizing on interfaces called synchronous optical network (SONET) or synchronous digital hierarchy (SDH) that support variable speeds ranging from 1.56 Mbits/s (T-1 circuits) to 155 Mbits/s (OC-3c) to 655 Mbits/s (OC-7c) over fiber optic cable. Since ATM switches can use SONET connections, ATM may be the enabling technology that allows LANs and WANs to work together seamlessly at high speeds without any performance penalties.

ATM technology is available in various forms. It can be used as a campus LAN backbone switch that connects routers, servers, and other switches. The second form is that of LAN extenders, whereby an Ethernet or Token Ring is extended to another location over high-speed ATM connections; in this form, ATM switches work as virtual LAN repeaters. ATM adapters are available for many workstations such as SUN SparcStations, IBM RS/6000, HP 9000, and DEC AlphaStations. The technology itself is flexible in that it can be supported over any type of cable (copper and fiber). The communication occurs over fixed length *cells* of 53 bytes (as opposed to variable length *packet* in X.25), which makes it possible to guarantee service to time-dependent (or isochronous) applications and allows very fast hardware-based cell switching to switching route the cells through the switches. ATM networking also capitalizes on today's high speed, reliable communication links by forgoing error/detection/correction at each intermediate node within the network.

The ATM standard, set by the International Telegraph and Telephone Union Telecommunications Sector (ITU-T), is promoted by an ATM Forum consisting of over 500 members. The Forum has also set standards for interfaces required to communicate between various ATM nodes, and between ATM layers and other protocols. Residing within the traditional link layer, several parallel stacks called *ATM adaptation layers* (AAL) define protocols and quality of services for voice circuit emulation (AAL 1) timing-sensitive video/audio transfer (AAL 2), connection-oriented and connectionless data transfer for bursty data (AAL 3/4), and simple and efficient adaptation layer for local, high speed LAN installations

(AAL 5). Although the ATM technology has great potential, the ATM solutions must integrate with the legacy networks in an orderly fashion to achieve a smooth transition. The accelerated integration of ATM technology into the desktop, hubs, bridges, and routers indicates that ATM has a high probability of dominating the marketplace.

At many institutions today, campus backbones are composed of fiber optic cables, and some institutions even have fiber optic cables at the desktops. At a few well-known radiology centers, *picture archiving communications systems* (PACS), a technology for creating, storing, and disseminating radiologic images, have been installed using networks exclusively dedicated to PACS information exchange. The PACS systems may be considered one of the ancillary systems serving one department's needs. A faster networking technologies become universally available and as workstations become powerful enough to provide on-the-fly decompression, a true multimedia view of patient information may be constructed for the first time at the provider's desktop.

Another example of applications requiring higher bandwidth is *telemedicine.* It is a concept whereby health care may be delivered remotely over computer networks by providers who are physically far away from the patient [Halpern et al., 1992]. This technology is being actively pursued by the federal government and by several state governments through the National Information Infrastructure initiative and through statewide information highway initiatives. Several tertiary-care institutions are arranging affiliation with primary and secondary care institutions to provide expert referral services. The concept is also popular in rural settings, where more sophisticated care may be available at a limited number of locations.

One of the fundamental requirements for telemedicine is a networking bandwidth that is capable of supporting simultaneous video, audio, image and data transfer in real time. Technologies such as ATM and SONet promise to be the solutions to the problem. Today, telemedicine is restricted to asynchronous referral services for medical images, remote review of clinical information, and occasional video conferences. In prototype situations, however, complicated scenarios involving telesurgery have been shown to be possible using remote robotic hands and head-mounted displays in an augmented reality setting.

Security

Networks have made flexible data access possible. Recently data interconnections have exploded among care facilities, private physicians' offices, nursing homes, insurance agencies, health maintenance organizations, research institutions, and state and federal regulatory agencies. Internet is the precursor to the eventual information super-highway. The problem of security has become ever more important in this new world of accessible yet confidential patient information—including issues of who owns the data, how to protect the data, how to convert it securely to meet societal needs for statistical information, and so on. Smart cards, encryption digital signatures, guaranteed authentication and authorization, and constant accounting and auditing are several techniques being considered in intense discussions addressing the security aspects of medical information networking [US Congress 1993].

Management

Managing networks and distributed applications is orders-of-magnitude harder than managing a single, monolithic system. It is, however, extremely important to invest in and learn how to manage the hundreds of networking and computing components that bring value to the networks. Although there are some good tools and standards available to monitor the lower layers of networks and devices such as bridges and routers, there is a total lack of management capabilities for middleware and nascent medical domain middleware. Monitoring of applications helps determine future demands for resources, performance guarantees, and opportunities to make a system more reliable. The primary management communication standard is simple network management protocol (SNMP), which is not yet at the advanced stage of managing higher-layer protocols and applications.

177.6 Conclusions

The use of networks at health care practice centers has provided value by increasing data availability, flexibility, and efficiency of operations. The graphical-user-interface-based applications, along with client-server paradigm of computing have begun to appear on these networks. The next generation of multimedia applications will require higher bandwidth networks, which have already begun to appear. In addition, the national efforts to link health care industries and providers in order to improve efficiency will now be possible due to the networks that have become the fourth utility—after electricity, water, and heat—in many medical centers. Individual doctor's offices, clinics, and practice settings are connecting to computer networks that accomplish the communication functions of the larger health care business networks.

Defining Terms

Communication protocols: A set of rules and methods by which computers communicate with each other.

Ethernet, Token Ring: Examples of specific kind of communication protocols suitable for local area networks.

Fiber distributed data interface (FDDI), asynchronous transfer mode (ATM): High-speed communication protocols suitable for both local and wide area networks.

Local area network: Computer networking infrastructure suitable for a campus (or smaller) environment where the owning organization controls connection and protocols.

Medical domain middleware: Middleware addressed specifically to transparently access health care information on the network.

Middleware: Software components that help create a transparent access to resources regardless of their physical location in the network.

Network topology: A detailed map of physical and conceptual connections among networking devices.

Unshielded or shielded twisted pair (UTP, STP): Types of cables with different electrical characteristics and prices.

Wide area network: Computer networking infrastructure that connects several local area networks distributed over great geographical distances using external services.

References

Cimino JJ, Clayton PD, Hripcsak G, et al. 1994. Knowledge-based approaches to the maintenance of a large controlled medical terminology. J Am Med Informatics Assoc 1:35.

Duisterhout JS, Hasman A, Salamon R (eds). 1991. Telematics in Medicine, Amsterdam, Elsevier Science Publishers.

Halpern EJ, Newhouse JH, Amis ESJ, et al. 1992. Evaluation of teleradiology for interpretation of intravenous urograms. J Digital Imaging 5(2):101.

McDonald CT, Hammond WE. 1989. Standard formats for electronic transfer of clinical data. Ann Intern Med 110:333.

Rennels GD, Shortliffe EH. 1987. Advanced computing for medicine. Sci Am 257(4):154.

Tanenbaum AS. 1988. Computer Networks, 2d ed, Englewood Cliffs, NJ, Prentice-Hall.

US Congress, Office of Technology Assessment. 1993. Protecting Privacy in Computerized Medical Information, OTA-TCT-576. Washington DC, US Government Printing Office.

178

Overview of Standards Related to the Emerging Health Care Information Infrastructure

Jeffrey S. Blair

IBM Health Care Solutions

As the cost of health care has become a larger percentage of the gross domestic product of many developed notions, the focus on methods to improve health care productivity and quality has increased. To address this need, the concept of a health care information infrastructure has emerged. Major elements of this concept include patient-centered care facilitated by computer-based patient record systems, continuity of care enabled by the sharing of patient information across information networks, and outcomes measurement aided by the greater availability and specificity of health care information.

The creation of this health care information infrastructure will require the integration of existing and new architectures, products and services. To make these diverse components work together, health care information standards (classifications, guides, practices, terminology) will be required [ASTM, 1994]. This chapter will give you an overview of the major existing and emerging health care information standards, and the efforts to coordinate, harmonize, and accelerate these activities. It is organized into the major topic areas of:

- Identifier standards
- Communications (message format) standards
- Content and structure standards

0-8493-0462-8/00/$0.00+$.50

- Clinical data representations (codes)
- Confidentiality, data security, and authentication
- Quality indicators and data sets
- International Standards
- Coordinating and promotion organizations
- Summary

178.1 Identifier Standards

There is a universal need for health care identifiers to uniquely specify each patient, provider, site of care and product; however, there is not universal acceptance and/or satisfaction with these systems.

Patient Identifiers

The social security number (SSN) is widely used as a patient identifier in the United States today. However, critics point out that it is not an ideal identifier. They say that not everyone has an SSN; several individuals may use the same SSN; and the SSN is so widely used for other purposes that it presents an exposure to violations of confidentiality. These criticisms raise issues that are not unique to the SSN. A draft document has been developed by the American Society for Testing and Materials (ASTM) E31.12. Subcommittee to address these issues. It is called the "Guide for the Properties of a Universal Health Care Identifier" (UHID). It presents a set of requirements outlining the properties of a national system creating a UHIC, includes critiques of the SSN, and creates a sample UHD (ASTM E31.12, 1994). Despite the advantages of a modified/new patient identifier, there is not yet a consensus as to who would bear the cost of adopting a new patient identifier system.

Provider Identifiers

The Health Care Financing Administration (HCFA) has created a widely used provider identifier known as the Universal Physician Identifier Number (UPIN) [Terrell et al., 1991]. The UPIN is assigned to physicians who handle Medicare patients, but it does not include nonphysician care givers. The National Council of Prescription Drug Programs (NCPDP) has developed the standard prescriber identification number (SPIN) to be used by pharmacists in retail settings. A proposal to develop a new national provider identifier number has been set forth by HCFA [HCFA, 1994]. If this proposal is accepted, then HCFA would develop a national provider identifier number which would cover all caregivers an sites of care, including Medicare, Medicaid, and private care. This proposal is being reviewed by various state and federal agencies. It has also been sent to the American National Standards, Institute's Health Care Informatics Standards Planning Panel (ANSI HISPP) Task Force on Provider Identifiers for review.

Site-of-Care Identifiers

Two site-of-care identifier systems are widely used. One is the health industry number (HIN) issued by the Health Industry Business Communications Council (HIBCC). The HIN is an identifier for health care facilities, practitioners, and retail pharmacies. HCFA has also defined provider of service identifiers for Medicare usage.

Product and Supply Labeling Identifiers

Three identifiers are widely accepted. The labeler identification code (LIC) identifies the manufacturer or distributor and is issued by HIBCC [HIBCC, 1994]. The LIC is used both with and without bar codes for products and supplies distributed within a health care facility. The universal product code (UPC) is maintained by the Uniform Code Council and is typically used to label products that are sold in retail settings. The national drug code is maintained by the Food and Drug Administration and is required for reimbursement by Medicare, Medicaid, and insurance companies. It is sometimes included within the UPC format.

178.2 Communications (Message Format) Standards

Although the standards in this topic area are still in various stages of development, they are generally more mature than those in most of the other topic areas. They are typically developed by committees within standards organizations and have generally been accepted by users and vendors. The overviews of these standards below were derived from many sources, but considerable content came from the Computer-based Patient Record Institute's (CPRI) "Position Paper on Computer-based Patient Record Standards" [CPRI, 1994] and the Agency for Health Care Policy and Research's (AHCPR) "Current Activities of Selected Health Care Informatics Standards Organizations" [Moshman Associates, 1994].

ASC X12N

This committee is developing message format standards for transactions between payors and providers. It is rapidly being accepted by both users and vendors. It defines the message formats for the following transaction types [Moshman Associates, 1994]:

- 834—enrollment
- 270—eligibility request
- 271—eligibility response
- 837—health care claim submission
- 835—health care claim payment remittance
- 276—claims status request
- 277—claims status response
- 148—report of injury or illness

ASC X12N is also working on the following standards to be published in the near future:

- 257, 258—Interactive eligibility response and request. These transactions are an abbreviated form of the 270/271.
- 274, 275—patient record data response and request. These transactions will be used to request and send patient data (tests, procedures, surgeries, allergies, etc.) between a requesting party and the party maintaining the database.
- 278, 279—health care services (utilization review) response and request. These transactions will be used to initiate and respond to a utilization review request.

ASC X12N is recognized as an accredited standards committee (ASC) by the American National Standards Institute (ANSI).

American Society for Testing and Materials (ASTM) Message Format Standards

The following standards were developed within ASTM Committee E31. This committee has applied for recognition as an ASC by ANSI.

- ASTM E1238 standard specification for transferring clinical observations between independent systems. E1238 was developed by ASTM Subcommittee E31.11. This standard is being used by most of the largest commercial laboratory vendors in the United States to transmit laboratory results. It has also been adopted by a consortium of 25 French laboratory system vendors. Health Level Seven (HL7), which is described later in this topic area, has incorporated E1238 as a subset within its laboratory results message format [CPRI, 1994].
- ASTM E1394 standard specification for transferring information between clinical instruments. E1394 was developed by ASTM Subcommittee E31.14. This standard is being used for communication of information from laboratory instruments to computer systems. This standard has been

developed by a consortium consisting of most U.S. manufacturers of clinical laboratory instruments [CPRI, 1994].

- ASTM 1460 specification for defining and sharing modular health knowledge bases (Arden Syntax). E1460 was developed by ASTM Subcommittee E31.15. The Arden Syntax provides a standard format and syntax for representing medical logic and for writing rules and guidelines that can be automatically executed by computer systems. Medical logic modules produced in one site of care system can be sent to a different system within another site of care and then customized to reflect local usage [CPRI, 1994].
- ASTM E1467 specification for transferring digital neurophysical data between independent computer systems. E1467 was developed by ASTM Subcommittee E31.17. This standard defines codes and structures needed to transmit electrophysiologic signals and results produced by electroencephalograms and electromyograms. The standard is similar in structure to ASTM E1238 and HL7; and it is being adopted by all the EEG systems manufacturers [CPRI, 1994].

Digital Imaging and Communications [DICOM]

This standard is developed by the American College of Radiology—National Electronic Manufacturers' Association (ACR–NEMA). It defines the message formats and communications standards for radiologic images. DICOM is supported by most radiology Picture Archiving and Communications Systems (PACS) vendors and has been incorporated into the Japanese Image Store and Carry (ISAC) optical disk system as well as Kodak's PhotoCD. ACR-NEMA is applying to be recognized as an accredited organizations by ANSI [CPRI, 1994].

Health Level Seven [HL7]

HL7 is used for intrainstitution transmission of orders; clinical observations and clinical data, including test results; admission, transfer, and discharge records; and charge and billing information. HL7 is being used in more than 300 U.S. health care institutions including most leading university hospitals and has been adopted by Australia and New Zealand as their national standard. HL7 is recognized as an accredited organization by ANSI [Hammond, 1993; CPRI, 1994].

Institute of Electrical and Electronic Engineers, Inc. (IEEE) P1157 Medical Data Interchange Standard (MEDIX)

IEEE Engineering in Medicine and Biology Society (EMB) is developing the MEDIX standards for the exchange of data between hospital computer systems [Harrington, 1993; CPRI, 1994]. Based on the International Standards Organization (ISO) standards for all seven layers of the OSI reference model, MEDIX is working on a framework model to guide the development and evolution of a compatible set of standards. This activity is being carried forward as a joint working group under ANSI HISPP's Message Standards Developers Subcommittee (MSDS). IEEE is recognized as an accredited organization by ANSI.

IEEE P1073 Medical Information Bus (MIB)

This standard defines the linkages of medical instrumentation (e.g., critical care instruments) to point-of-care information systems [CPRI, 1994].

National Council for Prescription Drug Programs (NCPDP)

These standards developed by NCPDP are used for communication of billing and eligibility information between community pharmacies and third-party payers. They have been is use since 1985 and now serve almost 60% of the nation's community pharmacies. NCPDP has applied for recognition as an accredited organization by ANSI [CPRI, 1994].

178.3 Content and Structure Standards

Guidelines and standards for the content and structure of computer-based patient record (CPR) systems are being developed within ASTM Subcommittees E31.12 and E31.19. They have been recognized by other standards organizations (e.g., HL7); however, they have not matured to the point where they are generally accepted or implemented by users and vendors.

A major revision to E1384, now called a *standard description for content and structure of the computer-based patient record,* has been made within Subcommittee E31.19 [ASTM, 1994]. This revision includes work from HISPP on data modeling and an expanded framework that includes master tables and data views by user.

Companion standards have been developed within E31.19. They are E1633, A Standard Specification for the Coded Values Used in the Automated Primary Record of Care (ASTM, 1994), and E1239-94, A Standard Guide for Description of Reservation/Registration-A/D/T Systems for Automated Patient Care Information Systems (ASTM, 1994). A draft standard is also being developed for object-oriented models for R-A/D/T functions in CPR systems. Within the E31.12 Subcommittee, domain specific guidelines for nursing, anesthesiology, and emergency room data within the CPR are being developed [Moshman Associates, 1994; Waegemann, 1994].

178.4 Clinical Data Representations (Codes)

Clinical data representations have been widely used to document diagnoses and procedures. There are over 150 known code systems. The codes with the widest acceptance in the United States include:

- International Classification of Diseases (ICD) codes, now in the ninth edition (ICD-9), are maintained by the World Health Organization (WHO) and are accepted worldwide. In the United States, HCFA and the National Center for Health Statistics (NCHS) have supported the development of a clinical modification of the ICD codes (ICD-9-CM). WHO has been developing ICD-10; however, HCFA projects that it will not be available for use within the United States for several years. Payers require the use of ICD-9-CM codes for reimbursement purposes, but they have limited valued for clinical and research purposes due to their lack of clinical specificity [Chute, 1991].
- Current Procedural Terminology (CPT) codes are maintained by the American Medical Association (AMA) and are widely used in the United States for reimbursement and utilization review purposes. The codes are derived from medical specialty nomenclatures and are updated annually [Chute 1991].
- The Systematized Nomenclature of Medicine (SNOMED) is maintained by the College of American Pathologists and is widely accepted for describing pathologic test results. It has a multiaxial (11 fields) coding structure that gives it greater clinical specificity than the ICD and CPT codes, and it has considerable value for clinical purposes. SNOMED has been proposed as a candidate to become the standardized vocabulary for computer-based patient record systems (Rothwell et al., 1993).
- Digital Imaging and Communications (DICOM) is maintained by the American College of Radiology—National Electronic Manufacturers' Association (ACR-NEMA). It sets forth standards for indicies of radiologic diagnoses as well as for image storage and communications [Cannavo, 1993].
- Diagnostic and Statistical Manual of Mental Disorders (DSM), now in its fourth edition (DSM-IV), is maintained by the American Psychiatric Association. It sets forth a standard set of codes and descriptions for use in diagnoses, prescriptions, research, education, and administration [Chute, 1991].
- Diagnostic Related Groups (DRGs) are maintained by HCFA. They are derivatives of ICD-9-CM codes and are used to facilitate reimbursement and case-mix analysis. They lack the clinical specificity to be of value in direct patient care or clinical research [Chute, 1991].

- Unified Medical Language System (UMLS) is maintained by the National Library of Medicine (NLM). It contains a metathesaurus that links clinical terminology, semantics, and formats of the major clinical coding and reference systems. It links medical terms (e.g., ICD, CPT, SNOMED, DSM, CO-STAR, and D-XPLAIN) to the NLM's medical index subject headings (MeSH codes) and to each other [Humphreys, 1991; Cimino et al., 1993].
- The Canon Group has not developed a clinical data representation, but it is addressing two important problems: clinical data representations typically lack clinical specificity and/or are incapable of being generalized or extended beyond a specific application. "The Group proposes to focus on the design of a general schema for medical-language representation including the specification of the resources and associated procedures required to map language (including standard terminologies) into representations that make all implicit relations 'visible,' reveal 'hidden attributes', and generally resolve…ambiguous references" [Evans et al., 1994].

178.5 Confidentiality, Data Security, and Authentication

The development of compute-based patient record systems and health care information networks have created the opportunity to address the need for more definitive confidentiality, data security, and authentication guidelines and standards. The following activities address this need:

- During 1994, several bills were drafted in Congress to address health care privacy and confidentiality. They included the Fair Health Information Practices Act of 1994 (H.R. 4077), the Health Care Privacy Protection Act (S. 2129), and others. Although these bills were not passed as drafted, their essential content is expected to be included as part of subsequent health care reform legislation. They address the need for uniform comprehensive federal rules governing the use and disclosure of identifiable health, information about individuals. They specify the responsibilities of those who collect, use, and maintain health information about patients. They also define the rights of patients and provide a variety of mechanisms that will allow patients to enforce their rights.
- ASTM Subcommittee E31.12 on Computer-based Patient Records is developing Guidelines for Minimal Data Security Measures for the Protection of Computer-based patient Records [Moshman Associates, 1994].
- ASTM Subcommittee E31.17 on Access, Privacy, and Confidentiality of Medical Records is working on standards to address these issues [Moshman Associates, 1994].
- ASTM Subcommittee E31.20 is developing standard specifications for authentication of health information [Moshman Associates, 1994].
- The Committee on Regional Health Data Networks convened by the Institute of Medicine (IOM) has completed a definitive study and published its findings in a book entitled *Health Data in the Information Age: Use, Disclosure, and Privacy* [Donaldson & Lohr, 1994].
- The Computer-based Patient Record Institute's (CPRI) Work Group on Confidentiality, Privacy and Legislation has completed white papers on "Access to Patient Data" and on "Authentication," and a publication entitled "Guidelines for Establishing Information Security: Policies at Organizations using Computer-based Patient Records" [CPRI, 1994].
- The Office of Technology Assessment has completed a two-year study resulting in a document entitled "Protecting Privacy in Computerized Medical Information." It includes a comprehensive review of system/data security issues, privacy information, current laws, technologies used for protection, and models.
- The U.S. Food and Drug Administration (FDA) has created a task force on Electronic/Identification Signatures to study authentication issues as they relate to the pharmaceutical industry.

178.6 Quality Indicators and Data Sets

The Joint Commission on Accreditation of Health Care Organizations (JCAHO) has been developing and testing obstetrics, oncology, trauma, and cardiovascular clinical indicators. These indicators are intended to facilitate provider performance measurement. Several vendors are planning to include JCAHO clinical indicators in their performance measurement systems [JCAHO, 1994].

The Health Employers Data and Information Set (HEDIS) version 2.0 has been developed with the support of the National Committee for Quality Assurance (NCQA). It identifies data to support performance measurement in the areas of quality (e.g., preventive medicine, prenatal care, acute and chronic disease, and mental health), access and patient satisfaction, membership and utilization, and finance. The development of HEDIS has been supported by several large employers and managed care organizations [NCQA, 1993].

178.7 International Standards

The International Organization for Standardization (ISO) is a worldwide federation of national standards organizations. It has 90 member countries. The purpose of ISO is to promote the development of standardization and related activities in the world. ANSI was one of the founding members of ISO and is representative for the United States [Waegemann, 1994].

ISO has established a communications model for Open Systems Interconnection (OSI) IEEE/MEDIX and HL7 have recognized and built upon the ISO/OSI framework. Further, ANSI HISPP has a stated objective of encouraging compatibility of U.S. health care standards with ISO/OSI. The ISO activities related to information technology take place within the Joint Technical Committee (JTC) 1.

The Comite Europeen de Noramalisation (CEN) is a European standards organization with 16 technical committees (TCs). Two TCs are specifically involved in health care: TC 251 (Medical Informatics) and TC 224 WG12 (Patient Data Cards) [Waegemann, 1994].

The CEN TC 251 on Medical Informatics includes work groups on: Modeling of Medical Records; Terminology, Coding, Semantics, and Knowledge Bases; Communications and Messages; Imaging and Multimedia; Medical Devices; and Security, Privacy, Quality and Safety. The CEN TC 251has established coordination with health care standards development in the United States through ANSI/HISPP.

In addition to standards developed by ISO and CEN, there are two other standards of importance. United Nations (U.N.) EDIFACT is a generic messaging-based communications standard with health-specific subsets. It parallels X12 and HL7, which are transaction-based standards. It is widely used in Europe and in several Latin America countries. The READ Classification System (RCS) is a multiaxial medical nomenclature used in the United Kingdom. It is sponsored by the National Health Service and has been integrated into computer-based ambulatory patient record systems in the United Kingdom [CAMS, 1994].

178.8 Standards Coordination and Promotion Organizations

In the United States, two organizations have emerged to assume responsibility for the coordination and promotion of health care standards development: the ANSI Health Care Informatics Standards Planning Panel (HISPP) and the Computer-based Patient Record Institute (CPRI). The major missions of an ANSI HISPP are:

1. To coordinate the work of the standards groups for health care data interchange and health care informatics (e.g., ACR/NEMA, ASTM, HL7, IEEE/MEDIX) and other relevant standards groups (e.g., X3, X12) toward achieving the evolution of a unified set of nonredundant, non-conflicting standards.
2. To interact with and provide input to CEN TC 251 (Medical Informatics) in a coordinated fashion and explore avenues of international standards development.

The first mission of coordinating standards is performed by the Message Standards Developers Subcommittee (MSDS). The second mission is performed by the International and Regional Standards Subcommittee. HISPP also has four task groups: (1) Codes and Vocabulary, (2) Privacy, Security and Confidentiality, (3) Provider Identification Numbering Systems, and (4) Operations. Its principle membership is comprised of representatives of the major health care standards development organizations (SDOs), government agencies, vendors, and other interested parties. ANSI HISPP is by definition a planning panel, not an SDO (Hammond, 1994; ANSI HISPP, 1994).

The CPRI's mission is to promote acceptance of the vision set forth in the Institute of Medicine Study report "The Computer-based Patient Record: An Essential Technology for Health Care." CRPI is a nonprofit organization committed to initiating and coordinating activities to facilitate and promote the routine use of computer-based patient records. The CPRI takes initiatives to promote the development of CPR standards, but it is not an SDO itself. CPRI members represent the entire range of stakeholders in the health care delivery system. Its major work groups are the: (1) Codes and Structures Work Group; (2) CPR Description Work Group; (3) CPR Systems Evaluation Work Group; (4) Confidentiality, Privacy, and Legislation Work Group; and (5) Professional and Public Education Work Group [CPRI, 1994].

Two work efforts have been initiated to establish models for principle components of the emerging health care information infrastructure. The CPR Description Work Group of the CPRI is defining a consensus-based model of the computer-based patient record system. A joint working group to create a common data description has been formed by the MSDS Subcommittee of ANSI HISPP and IEEE/MEDIX. The joint working group is an open standards effort to support the development of a common data model tat can be shared by developers of health care informatics standards [IEEE, 1994].

The CPRI has introduced a proposal defining a public/private effort to accelerate standards development for computer-based patient record systems [CPRI, 1994]. If funding becomes available, the project will focus on obtaining consensus for a conceptual description of a computer-based patient record system; addressing the need for universal patient identifiers; developing standard provider and sites-of-care identifiers; developing confidentiality and security standards; establishing a structure for and developing key vocabulary and code standards; completing health data interchange standards; developing implementation tools; and demonstrating adoptability of standards in actual settings. This project proposes that the CPRI and ANSI HISPP work together to lead, promote, coordinate and accelerate the work of SDOs to develop health care information standards.

The Workgroup on Electronic Data Interchange (WEDI) is a voluntary, public/private task force which was formed in 1991 as a result of the call for health care administrative simplification by the director of the Department of Health and Human Services, Dr. Louis Sullivan. They have developed an action plan to promote health care EDI which includes: promotion of EDI standards, architectures, confidentiality, identifiers, health cards, legislation, and publicity [WEDI, 1993].

178.9 Summary

This chapter has presented an overview of major existing and emerging health care information infrastructure standards and the efforts to coordinate, harmonize, and accelerate these activities. Health care informatics is a dynamic area characterized by changing business and clinical processes, functions, and technologies. The effort to create health care informatics standards is therefore also dynamic. For the most current information on standards, refer to the "For More Information" section at the back of this chapter.

References

American National Standards Institute's Health Care Informatics Standards Planning Panel, 1994. Charter statement. New York.

American Society for Testing and Materials (ASTM). 1994. Guide for the properties of a universal health care identifier. ASTM Subcommittee E31.12, Philadelphia.

American Society for Testing and Materials (ASTM). 1994. Membership information packet: ASTM Committee E31 on computerized systems, Philadelphia.

American Society for Testing and Materials (ASTM). 1994. A standard description for content and structure of the computer-based patient record, E1384–91/1994 revision. ASTM Subcommittee E31.19, Philadelphia.

American Society for Testing and Materials (ASTM). 1994. Standard guide for description of reservation registration-admission, discharge, transfer (R-ADT) systems for automated patient care information systems, E1239–94. ASTM Subcommittee E31.19, Philadelphia.

American Society for Testing and Materials (ASTM). 1994. A standard specification for the coded values used in the automated primary record of care, E1633. ASTM Subcommittee E31.19, Philadelphia.

Cannavo MJ. 1993. The last word regarding DEFF & DICOM. Healthcare Informatics 32.

Chute CG. 1991. Tutorial 19: Clinical data representations. Washington, DC, Symposium on Computer Applications in Medical Care.

Cimino JJ, Johnson SB, Peng P, et al. 1993. From ICD9-CM to MeSH Using the UMLS: A How-to-Guide, SCAMC, Washington DC.

Computer Aided Medical Systems Limited (CAMS). 1994. CAMS News 4:1.

Computer-based Patient Record Institute (CPRI). 1994. CPRI-Mail 3:1.

Computer-based Patient Record Institute (CPRI). 1994. Position paper computer-based patient record standards. Chicago.

Computer-based Patient Record Institute (CPRI). 1994. Proposal to accelerate standards development for computer-based patient record systems. Version 3.0, Chicago.

Donaldson MS, Lohr KN (eds). 1994. Health Data in the Information Age: Use, Disclosure, and Privacy. Washington, DC, Institute of Medicine, National Academy Press.

Evans DA, Cimino JJ, Hersh WR, et al. 1994. Toward a medical-concept representation language. J Am Med Informatics Assoc 1(3):207.

Hammond WE. 1993. Overview of health care standards and understanding what they all accomplish. HIMSS Proceedings, Chicago, American Hospital Association.

Hammond WE, McDonld C, Beeler G, et al. 1994. Computer standards: Their future within health care reform. HIMSS Proceedings, Chicago, Health Care Information and Management Systems Society.

Harrington JJ. 1993. IEEE P1157 MEDIX: A standard for open systems medical data interchange. New York, Institute of Electrical and Electronic Engineers.

Health Care Financing Administration (HCFA). 1994. Draft issue papers developed by HCFA's national provider identifier/national provider file workgroups. Baltimore.

Health Industry Business Communications Council (HIBCC). 1994. Description of present program standards activity. Phoenix.

Humphreys B. 1991. Tutorial 20: Using and assessing the UMLS knowledge sources. Symposium on Computer Applications in Medical Care, Washington, DC.

Institute of Electrical and Electronics Engineers (IEEE). 1994. Trial-use standard for health care data interchange—Information model methods: Data model framework. IEEE Standards Department, New York.

Joint Commission on Accreditation of Health Organizations (JCAHO). 1994. The Joint Commission Journal on Quality Improvement. Oakbrook Terrace, Ill.

Moshman Associates, Inc. 1994. Current activities of selected health care informatics standards organizations. Bethesda, MD. Office of Science and Data Development, Agency for Health Care Policy and Research.

National Committee for Quality Assurance. 1993. Hedis 2.0: Executive summary. Washington, DC.

Rothwell DJ, Cote RA, Cordeau JP, et al. 1993. Developing a standard data structure for medical language—The SNOMED proposal. Washington, DC, SCAMC.

Terell SA, Dutton BL, Porter L., et al. 1991. In search of the denominator: Medicare physicians—How many are there? Baltimore, Health Care Financing Administration.

Waegemann CP. 1994. Draft—1994 resource guide: Organizations involved in standards and development work for electronic health record systems. Newton, Mass, Medical Records Institute.

Workgroup for Electronic Data Interchange (WEDI). 1993. WEDI report: October 1993. Convened by the Department of Health and Human Services, Washington DC.

Further Information

For copies of standards accredited by ANSI, you can contact the American National Standards Institute, 11 West 42d St., NY, NY 10036, (212) 642–4900. For information on ANSI Health Care Informatics Standards Planning Panel (HISPP), contact Steven Cornish, (212) 642–4900.

For copies of individual ASTM standards, you can contact the American Society for Testing and Materials, 1916 Race Street, Philadelphia, PA 19103–1187, (215) 299–5400.

For copies of the "Proposal to Accelerate Standards Development for Computer-based Patient Record Systems," contact the Computer-based Patient Record Institute (CPRI), Margaret Amatayakul, 1000 E. Woodfield Road, Suite 102, Schaumburg, IL 60173, (708) 706–6746.

For information on provider identifier standards and proposals, contact the Health Care Financing Administration (HCFA), Bureau Program Operations, 6325 Security Blvd., Baltimore, MD 21207, (410) 966–5798. For information on ICD-9-CM codes, contact HCFA, Medical Coding, 401 East Highrise Bldg. 6325 Security Blvd., Baltimore, MD 21207, (410) 966–5318.

For information on site-of-care and supplier labeling identifiers, contact the Health Industry Business Communications Council (HIBCC), 5110 N. 40th Street, Suite 250, Phoenix, AZ 85018, (602) 381–1091.

For copies of standards developed by Health Level 7, you can contact HL7, 3300 Washtenaw Avenue, Suite 227, Ann Arbor, MI 48104, (313) 665–0007.

For copies of standards developed by the Institute of Electrical and Electronic Engineers/ Engineering in Medicine and Biology Society, in New York City, call (212) 705–7900. For information on IEEE/MEDIX meetings, contact Jack Harrington, Hewlett-Packard, 3000 Minuteman Rd., Andover, MA 01810, (508) 681–3517.

For more information on clinical indicators, contact the Joint Commission on Accreditation of Health Care Organizations (JCAHO), Department of Indicator Measurement, One Renaissance Blvd., Oakbrook Terrace, IL 60181, (708) 916–5600.

For information on pharmaceutical billing transactions, contact the National Council for Prescription Drug Programs (NCPDP), 2401 N. 24th Street, Suite 365, Phoenix, AZ 85016, (602) 957–9105.

For information on HEDIS, contact the National Committee for Quality Assurance (NCQA), Planning and Development, 1350 New York Avenue, Suite 700, Washington, D.C. 20005, (202) 628–5788.

For copies of ACR/NEMA DICOM standards, contact David Snavely, National Equipment Manufacturers Association (NEMA), 2101 L. Street N.W., Suite 300, Washington, D.C. 20037, (202) 457–8400.

For information on standards development in the areas of computer-based patient record concept models, confidentiality, data security, authentication, and patient cards, and for information on standards activities in Europe, contact Peter Waegemann, Medical Records Institute (MRI), 567 Walnut, P.O. Box 289, Newton, MA 02160. (617) 964–3923.

179

Non-AI Decision Making

Ron Summers
Loughborough University

Derek G. Cramp
City University, London

Ewart R. Carson
City University, London

Non-AI decision making can be defined as those methods and tools used to increase information content in the context of some specific clinical situation without having cause to refer to knowledge embodied in a computer program. Theoretical advances in the 1950s added rigor to this domain when Meehl argued that many clinical decisions could be made by statistical rather than intuitive means [1]. Evidence of this view was supported by Savage [2], whose theory of choice under uncertainty is still the classical and most elegant formulation of subjective Bayesian decision theory, and was much responsible for reintroducing Bayesian decision analysis to clinical medicine. Ledley and Lusted provided further evidence that medical reasoning could be made explicit and represented in decision theoretic ways [3]. Decision theory also provided the means for Nash to develop a "Logoscope" which might be considered as the first mechanical diagnostic aid [4].

An information system developed using non-AI decision-making techniques may comprise procedural or declarative knowledge. Procedural knowledge maps the decision-making process into the methods by which the clinical problems are solved or clinical decisions made. Examples of techniques which form a procedural knowledge base are those which are based on algorithmic analytical models, clinical algorithms, or decision trees. Information systems based on declarative knowledge comprise what can essentially be termed a database of facts about different aspects of a clinical problem; the causal relationships between these facts form a rich network from which explicit (say) cause-effect pathways can be determined. Semantic networks and causal probabilistic networks are perhaps the best examples of information systems based on declarative knowledge. There are other types of clinical decision aids, based purely on statistical methods applied to patient data, for example, classification analyses based on 1ogistic regression, relative frequencies of occurrence, pattern-matching algorithms, or neural networks.

The structure of this chapter mirrors to some extent the different methods and techniques of non-AI decision making mentioned above. It is important to distinguish between analytical models based on quantitative or qualitative mathematical representations and decision theoretic methods typified by the

0-8493-0462-8/00/$0.00+$.50

use of clinical algorithms, decision trees, and set theory. Most of the latter techniques add to an information base by way of procedural knowledge. It is then, that advantage can be taken of the many techniques which have statistical decision theoretic principles as their underpinning.

This section begins with a discussion of simple linear regression models and pattern recognition, but then more complex statistical techniques are introduced, for example, the use of Bayesian decision analysis which leads to the introduction of causal probabilistic networks. The majority of these techniques add information by use of declarative knowledge. Particular applications are used throughout to illustrate the extent to which non-AI decision making is used in clinical practice.

179.1 Analytical Models

In the context of this chapter, the analytical models considered are qualitative and quantitative mathematical models that are used to predict future patient state based on present state and a historical representation of what has passed. Such models could be representations of system behavior that allow test signals to be used so that response of the system to various disturbances can be studied, thus making predictions of future patient state.

For example, Leaning and co-workers [5,6] produced a 19-segment quantitative mathematical model of the blood circulation to study the short-term effects of drugs on the cardiovascular system of normal, resting patients. The model represented entities such as the compliance, flow, and volume of model segments in what was considered a closed system. In total, the quantitative mathematical model comprised 61 differential equations and 159 algebraic equations. Evaluation of the model revealed that it was fit for its purpose in the sense of heuristic validity, that is it could be used as a tool for developing explanations for cardiovascular control, particularly in relation to the CNS.

Qualitative models investigate time-dependent behavior by representing patient state trajectory in the form of a set of connected nodes, the links between the nodes reflecting transitional constraints placed on the system [7]. The types of decision making supported by this type of model are assessment and therapy planning. In diagnostic assessment, the precursor nodes and the pathway to the node (decision) of interest define the causal mechanisms of the disease process. Similarly, for therapy planning, the optimal plan can be set by investigation of the utility values associated with each link in the disease-therapy relationship. These utility values refer to a cost function, where cost can be defined as the monetary cost of providing the treatment and cost benefit to the patient in terms of efficiency, efficacy, and effectiveness of alternative treatment options. Both quantitative [8] and qualitative [9] analytical models can be realized in other ways to form the basis of rule-based systems; however that excludes their analysis in this chapter.

179.2 Decision Theoretic Models

Clinical Algorithms

The clinical algorithm is a procedural device that mimics clinical decision making by structuring the diagnostic or therapeutic decision processes in the form of a classification tree. The root of the tree represents some initial state, and the branches yield the different options available. For the operation of the clinical algorithm, the choice points are assumed to follow branching logic with the decision function being a yes/no (or similar) binary choice. Thus, the clinical algorithm comprises a set of questions that must be collectively exhaustive for the chosen domain, and the responses available to the clinician at each branch point must be mutually exclusive. These decision criteria pose rigid constraints on the type of medical problem that can be represented by this method, as the lack of flexibility is appropriate only for a certain set of well-defined clinical domains. Nevertheless, there is a rich literature available; examples include the use of the clinical algorithm for acid-base disorders [10] and diagnosis of mental disorders [11].

Decision Trees

A more rigorous use of classification tree representations than the clinical algorithm can be found in decision tree analysis. Although from a structural perspective, decision trees and clinical algorithms are similar in appearance, for decision tree analysis the likelihood and cost-benefit for each choice are also calculated in order to provide a quantitative measure for each option available. This allows the use of optimization procedures to gauge the probability of success for the correct diagnosis being made or for a beneficial outcome from therapeutic action being taken. A further difference between the clinical algorithm and decision tree analysis is that the latter has more than one type of decision node (branch point): at decision nodes the clinician must decide on which choice (branch) is appropriate for the given clinical scenario; at chance nodes the responses available have no clinician control, for example, the response may be due to patient specific data; and outcome nodes define the chance nodes at the "leaves" of the decision tree. That is, they summarize a set of all possible clinical outcomes for the chosen domain.

The possible outcomes from each chance node must obey the rules of probability and sum to unity; the probability assigned to each branch reflects the frequency of that event occurring in a general patient population. It follows that these probabilities are dynamic, with accuracy increasing, as more evidence becomes available. A utility value can be added to each of the outcome scenarios. These utility measures reflect a trade-off between competing concerns, for example, survivability and quality of life, and may be assigned heuristically.

When the first edition of this chapter was written in 1995 it was noted that although a rich literature describing potential applications existed [12], the number of practical applications described was limited. The situation has changed and there has been an explosion of interest in applying decision analysis to clinical problems. Not only is decision analysis methodology well described [13-16] but there are numerous articles appearing in mainstream medical journals, particularly *Medical Decision Making*. An important driver for this acceleration of interest has been the desire to contain costs of medical care, while maintaining clinical effectiveness and quality of care. Cost-effectiveness analysis is an extension of decision analysis and compares the outcome of decision options in terms of the monetary cost per unit of effectiveness. Thus, it can be used to set priorities for the allocation of resources and to decide between one or more treatment or intervention options. It is most useful when comparing treatments for the same clinical condition. Cost-effectiveness analysis and its implications are described very well elsewhere [17,18].

Influence Diagrams

In the 1960s researchers at Stanford Research Institute (SRI) proposed the use of influence diagrams as representational models when developing computer programs to solve decision problems. However, it was recognized somewhat later by decision analysts at SRI [19] that such diagrams could be used to facilitate communication with domain experts when eliciting information about complex decision problems. Influence diagrams are a powerful mode of graphic representation for decision modeling. They do not replace but complement decision trees, and it should be noted that both are different graphical representations of the same mathematical model and operations. Recently, two exciting papers have been published that make use of influence diagrams accessible to those interested in medical decision making [20,21].

179.3 Statistical Models

Database Search

Interrogation of large clinical databases yields statistical evidence of diagnostic value and in some representations form the basis of rule induction used to build expert systems [22]. These systems will not be discussed here. However the most direct approach for clinical decision making is to determine the relative frequency of occurrence of an entity, or more likely a group of entities, in the database of past cases. This enables a prior probability measure to be estimated [23]. A drawback of this simple, direct approach to problem solving is the apparent tautology of the more evidence available leading to fewer matches in the database being found; this runs against common wisdom that more evidence leads to an increase in probability of a diagnosis being found. Further, the method does not provide a weight for each item of evidence to gauge those that are more significant for patient outcome.

Regression Analysis

Logistic regression analysis is used to model the relationship between a response variable of interest and a set of explanatory variables. This is achieved by adjusting the regression coefficients, the parameters of the model, until a 'best fit' to the data set is achieved. This type of model improves upon the use of relative frequencies, as logistic regression explicitly represents the extent to which elements of evidence are important in the value of the regression coefficients. An example of clinical use can be found in the domain of gastroenterology [24].

Statistical Pattern Analysis

The recognition of patterns in data can be formulated as a statistical problem of classifying the results of clinical findings into mutually exclusive but collectively exhaustive decision regions. In this way, not only can physiologic data be classified but also the pathology that they give rise to and the therapy options available to treat the disease. Titterington [25] describes an application in which patterns in a complex data set are recognized to enhance the care of patients with head injuries. Pattern recognition is also the cornerstone of computerized methods for cardiac rhythm analysis [26]. The methods used to distinguish patterns in data rely on discriminant analysis. In simple terms, this refers to a measure of separability between class populations.

In general, pattern recognition is a two stage process as shown in Fig. 179.1. The pattern vector, P, is an n-dimensional vector derived from the data set used. Let Ω_p be the pattern space, which is the set of all possible values P may assume, then the pattern recognition problem is formulated as finding a way of dividing Ω_p into mutually exclusive and collectively exhaustive regions. For example, in the analysis of the electrocardiogram the complete waveform may be used to perform classifications of diagnostic value. A complex decision function would probably be required in such cases. Alternatively (and if appropriate), the pattern vector can be simplified to investigation of sub features within a pattern. For cardiac arrhythmia analysis, only the R-R interval of the electrocardiogram is required which allows a much simpler decision function to be used. This may be a linear or non-linear transformation process:

$$X = \tau P$$

where X is termed the feature vector and τ is the transformation process.

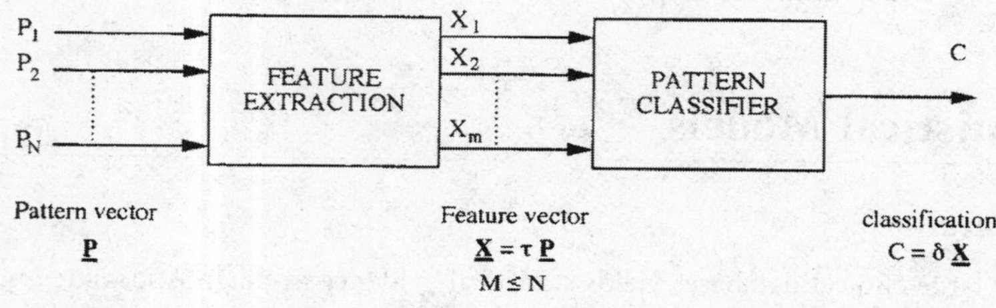

FIGURE 179.1

Just as the pattern vector P belongs to a pattern space Ω_p, so the feature vector X belongs to a feature space Ω_X. As the function of feature extraction is to reduce the dimensionality of the input vector to the classifier, some information is lost.

Classification of Ω_X can be achieved using numerous statistical methods including: discriminant functions (linear and polynomial); kernel estimation; k-nearest neighbor; cluster analysis; and Bayesian analysis.

Bayesian Analysis

Ever since their reinvestigation by Savage in 1954 [2], Bayesian methods of classification have provided one of the most popular approaches used to assist in clinical decision making. Bayesian classification is an example of a parametric method of estimating class conditional probability density functions. Clinical knowledge is represented as a set of prior probabilities of diseases to be matched with conditional probabilities of clinical findings in a patient population with each disease. The classification problem becomes one of a choice of decision levels, which minimizes the average rate of misclassification or to minimise the maximum of the conditional average loss function (the so-called minimax criterion) when information about prior probabilities is not available. Formally, the optimal decision rule which minimizes the average rate of misclassification is called the Bayes rule; this serves as the inference mechanism that allows the probabilities of competing diagnoses to be calculated when patient specific clinical findings become available.

The great advantage of Bayesian classification is that a large clinical database of past cases is not required, thus allowing the time taken to reach a decision to be faster compared with other database search techniques; furthermore, classification errors due to the use of inappropriate clinical inferences are quantifiable. However, a drawback of this approach to clinical decision making is that the disease states are considered as complete and mutually exclusive, whereas in real life neither assumption may be true.

Nevertheless, Bayesian decision functions as a basis for differential diagnosis have been used successfully, for example, in the diagnosis of acute abdominal pain [27]. De Dombal first described this system in 1972, but it took another 20 years or so for it to be accepted via a multi-center multinational trial. The approach has been exploited in ILIAD; this is a commercially available [28] computerized diagnostic decision support system with some 850–990 frames in its knowledge base. As it is a Bayesian system, each frame has the prevalence of a disease for its prior probability. There is the possibility however, that the prevalence rates may not have general applicability. This highlights a very real problem, namely the validity of relating causal pathways in clinical thinking and connecting such pathways to a body of validated (true) evidence. Ideally, such evidence will come from randomized controlled clinical or epidemiological trials. However, such studies may be subject to bias. To overcome this, Eddy and co-workers devised the Confidence Profile Method [29]. This is a set of quantitative techniques for interpreting and displaying the results of individual studies (trials); exploring the effects of any biases that might affect the internal validity of the study; adjusting for external validity; and, finally, combining evidence from several sources. This meta-analytical approach can formally incorporate experimental evidence and, in a Bayesian fashion, also the results of previous analytical studies or subjective judgments about specific factors that might arise when interpreting evidence. Influence diagram representations play an important role in linking results in published studies and estimates of probabilities and statements about causality.

Currently, much interest is being generated as to how, what is perceived as the Bayesian action-oriented approach can be used in determining health policy, where the problem is perceived to be a decision problem rather than a statistical problem, (see for instance Lilford and Braunholz [30]).

Dempster-Shafer Theory

One way to overcome the problem of mutually exclusive disease states is to use an extension to Bayesian classification put forward by Dempster [31] and Shafer [32]. Here, instead of focusing on a single disorder, the method can deal with combinations of several diseases. The key concept used is that the set of all possible diseases is partitioned into n-tuples of possible disease state combinations. A simple example will illustrate this concept. Suppose there is a clinical scenario in which four disease states describe the whole hypothesis space. Each new item of evidence will impact on all the possible subsets of the hypothesis space and is represented by a function, the basic probability assignment. This measure is a belief function that must obey the laws of probability and sum to unity across the subsets impacted upon. In the example, all possible subsets comprise: one which has all four disease states in it; four which have three of the four diseases as members; six which have two diseases as members; and finally, four subsets which have a single disease as a member. Thus, when new evidence becomes available in the form of a clinical finding, only certain hypotheses, represented by individual subsets, may be favored.

Syntactic Pattern Analysis

As demonstrated above, a large class of clinical problem solving using statistical methods involves classification or diagnosis of disease states, selection of optimal therapy regimes, and prediction of patient outcome. However, in some cases the purpose of modelling is to reconstruct the input signal from the data available. This cannot be done by methods discussed thus far. The syntactic approach to pattern recognition uses a hierarchical decomposition of information and draws upon an analogy to the syntax of language. Each input pattern is described in terms of more simple subpatterns which themselves are decomposed into simpler subunits, until the most elementary subpatterns, termed the pattern primitives, are reached. The pattern primitives should be selected so that they are easy to recognize with respect to the input signal. Rules which govern the transformation of pattern primitives back (ultimately) to the input signal are termed the grammar.

In this way a string grammar, G, which is easily representable in computer-based applications, can be defined:

$$G = \{V_T, V_N, S, P\}$$

where, V_T are the terminal variables (pattern primitives); V_N are the nonterminal variables; S is the start symbol; and P is the set of production rules which specify the transformation between each level of the hierarchy. It is an important assumption that in set theoretic terms, the union of V_T and V_N is the total vocabulary of G, and the intersection of V_T and V_N is the null (empty) set.

A syntactic pattern recognition system therefore comprises three functional subunits (Fig. 179.2): a pre-processor—this manipulates the input signal, P, into a form that can be presented to the pattern descriptor; the pattern descriptor which assigns a vocabulary to the signal; and the syntax analyzer which classifies the signal accordingly. This type of system has been used successfully to represent the electrocardiogram [33, 34] and the electroencephalogram [35] and for representation of the carotid pulse wave [36].

Causal Modeling

A causal probabilistic network (CPN) is an acyclic multiply-connected graph which at a qualitative level comprises nodes and arcs [37]. Nodes are the domain objects and may represent, for example, clinical

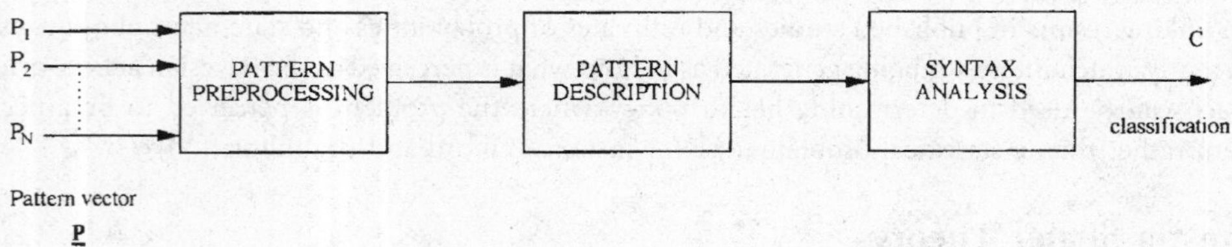

FIGURE 179.2

findings, pathophysiologic states, diseases, or therapies. Arcs are the causal relationships between successive nodes and are directed links. In this way the node and arc structure represent a model of the domain. Quantification is expressed in the model by a conditional probability table being associated with each arc, allowing the state of each node to be represented as a binary value, or more frequently as a continuous probability distribution. In root nodes the conditional probability table reduces to a probability distribution of all its possible states.

A key concept of CPNs is that computation is reduced to a series of local calculations, using only one node and those that are linked to it in the network. Any node can be instantiated with an observed value; this evidence is then propagated through the CPN via a series of local computations. Thus, CPNs can be used in two ways: to instantiate the leaf nodes of the network with known patterns for given disorders to investigate expected causal pathways; or to instantiate the root nodes or nodes in the graphical hierarchy with, for example, test results to obtain a differential diagnosis. The former method has been used to investigate respiratory pathology [38], and the latter method has been used to obtain pathologic information for electromyography [39].

Artificial Neural Networks

Artificial neural networks (ANNs) mimic their biologic counterparts, although at the present time on a much smaller scale. The fundamental unit in the biological system is the neuron. This is a specialized cell which, when activated, transmits a signal to its connected neighbors. Both activation and transmission involve chemical transmitters which cross the synaptic gap between neurons. Activation of the neuron takes place only when a certain threshold is reached. This biologic system is modelled in the representation of an artificial neural network. It is possible to identify three basic elements of the neuron model: a set of weighted connecting links that form the input to the neuron (analogous to neurotransmission across the synaptic gap); an adder for summing the input signals; and an activation function which limits the amplitude of the output of the neuron to the range (typically) –1 to +1. This activation function also has a threshold term that can be applied externally and forms one of the parameters of the neuron model. Many books are available which provide a comprehensive introduction to this class of model [e.g., 40].

ANNs can be applied to two categories of problems: prediction and classification. It is the latter which has caught the imagination of biomedical engineers for its similarity to diagnostic problem solving. For instance, the conventional management of patients with septicemia requires a diagnostic strategy that takes up to 18 to 24 h before initial identification of the causal microorganism. This can be compared to a method in which an ANN is applied to a large clinical database of past cases; the quest becomes one of seeking an optimal match between present clinical findings and patterns present in the recorded data. In this application pattern matching is a non-trivial problem as each of the 5000 past cases has 51 data fields. It has been shown that for this problem the ANN method outperforms other statistical methods such as k-nearest neighbor [41].

179.4 Summary

This chapter has reviewed what are normally considered to be the major categories of approach available to support clinical decision making which do not rely on what is classically termed artificial intelligence (AI). They have been considered under the headings of analytical, decision theoretic, and statistical models, together with their corresponding subdivisions. It should be noted, however, that the division into non-AI approaches and AI approaches that is adopted in this volume (see Chapter 184) is not totally clear-cut. In essence, the range of approaches can in many ways be regarded as a continuum. There is no unanimity as to where the division should be placed and the separation adopted; here is but one of a number that are feasible. It is therefore desirable that the reader should consider these two chapters together and choose an approach that is relevant to their particular clinical context.

References

1. Meehl R 1954. *Clinical versus Statistical Prediction,* Minnesota, University of Minnesota Press.
2. Savage LI. 1954. *The Foundations of Statistics,* New York, Wiley.
3. Ledley RS, Ludsted LB. 1959. Reasoning foundations of medical diagnosis. *Science,* 130:9.
4. Nash FA. 1954. Differential diagnosis: An apparatus to assist the logical faculties. *Lancet* 4:874.
5. Leaning MS, Pullen HE, Carson ER, et al. 1983. Modelling a complex biological system: The human cardiovascular system: 1. Methodology and model description. *Trans. Inst. Meas. Contr.* 5:71.
6. Leaning MS, Pullen HE, Carson ER, et al. 1983. Modelling a complex biological system: The human cardiovascular system: 2. Model validation, reduction and development. *Trans. Inst. Meas. Contr.* 5:87.
7. Kuipers BJ. 1986. Qualitative simulation. *Artif. Intell.* 29:289.
8. Furukawa T, Tanaka H, Hara S. 1987. FLUIDEX: A microcomputer-based expert system for fluid therapy consultations. In: MK Chytil, R Engelbrecht (Eds.), *Medical Expert Systems,* pp. 59-74, Wilmslow, Sigma Press.
9. Bratko I, Mozetic J, Lavrac N. 1988. In: D Michie, I Bratko (Eds.), *Expert Systems: Automatic Knowledge Acquisition,* pp. 61-83, Reading, Mass, Addison-Wesley.
10. Bleich HL. 1972. Computer-based consultations: Electrolyte and acid-base disorders. *Amer. J. Med.* 53:285.
11. McKenzie DP, McGary PD, Wallac et al.1993. Constructing a minimal diagnostic decision tree. *Meth. Inform. Med.* 32:161.
12. Pauker SG, Kassirer JP. 1987. Decision analysis. *N. Engl. J. Med.* 316:250.
13. Weinstein MC, Fineberg HV. 1980. *Clinical Decision Analysis.* London, Saunders.
14. Watson SR, Buede DM. 1994. *Decision Synthesis.* Cambridge, Cambridge University Press.
15. Sox HC, Blatt MA, Higgins MC, Marton KI. 1988. *Medical Decision Making.* Boston, Butterworth Heinemann.
16. Llewelyn H, Hopkins A. 1993. *Analysing How We Reach Clinical Decisions.* London, Royal College of Physicians.
17. Gold MR, Siegel JE, Russell LB, Weinstein MC. (Eds.) 1996. *Cost-effectiveness in Health and Medicine.* New York, Oxford University Press.
18. Sloan FA. (Ed.). 1996. *Valuing Health Care.* Cambridge, Cambridge University Press.
19. Owen DL. 1984. The use of influence diagrams in structuring complex decision problems. In: RA Howard and JE Matheson. (Eds.). *Readings on the Principles and Applications of Decision Analysis.* vol. 2. Menlo Park, CA, Strategic Decisions Group, pp. 763-72.
20. Owens DK, Shachter RD, Nease RF. 1997. Representation and analysis of medical decision problems with influence diagrams. *Med. Dec. Mak.* 17:241.
21. Nease RF, Owens DK. 1997. Use of influence diagrams to structure medical decisions. *Med. Dec. Mak.* 17:263.
22. Quinlan JR. 1979. Rules by induction from large collections of examples. In: D Michie (Ed.), *Expert Systems in the Microelectronic Age,* Edinburgh, Edinburgh University Press.
23. Gammerman A, Thatcher AR. 1990. Bayesian inference in an expert system without assuming independence. In: MC Golumbic (Ed.), *Advances in Artificial Intelligence,* pp. 182-218, New York, Springer-Verlag.
24. Spiegelhalter DJ, Knill-Jones RP. 1984. Statistical and knowledge-based approaches to clinical decision-support systems with an application in gastroenterology. *J. Roy. Stat. Soc. A* 147:35.
25. Titterington DM, Murray GD, Murray LS, et al. 1981. Comparison of discriminant techniques applied to a complex set of head injured patients. *J. Roy. Stat. Soc. A* 144:145.
26. Morganroth J. 1984. Computer recognition of cardiac arrhythmias and statistical approaches to arrhythmia analysis. *Ann. NY Acad. Sci.* 432:117.
27. De Dombal FT, Leaper DJ, Staniland JR, et al. 1972. Computer-aided diagnosis of acute abdominal pain. *Br. Med. J.* 2:9.

28. *Applied Medical Informatics,* Salt Lake City, UT.
29. Eddy DM, Hasselblad V, Shachter R. 1992. *Meta-Analysis by the Confidence Profile.* London, Academic Press.
30. Lilford RJ, Braunholz D. 1996.The statistical basis of public policy: a paradigm shift is overdue. *Br. Med. J.* 313: 603.
31. Dempster A. 1967. Upper and lower probabilities induced by multi-valued mapping. *Ann. Math Stat.* 38:325.
32. Shafer G. 1976. *A Mathematical Theory of Evidence,* Princeton, NJ, Princeton University Press.
33. Belforte G, De Mori R, Ferraris E 1979. A contribution to the automatic processing of electrocardiograms using syntactic methods. *IEEE Trans. Biomed. Eng.* BME-26 (3):125.
34. Birman KP. 1982. Rule-based learning for more accurate ECG analysis. *IEEE Trans. Pat. Anal. Mach. Intell.* PAMI-4 (4):369.
35. Ferber G. 1985. Syntactic pattern recognition of intermittant EEG activity. *Meth. Inf. Med.* 24 (2):79.
36. Stockman GC, Kanal LN. 1983. Problem reduction in representation for the linguistic analysis of waveforms. *IEEE Trans. Pat. Anal. Mach. Intell.* PAMI-5 (3):287.
37. Andersen SK, Jensen FV, Olesen KG. 1987. *The HUGIN Core-Preliminary Considerations on Inductive Reasoning: Managing Empirical Information in AI Systems.* Riso, Denmark.
38. Summers R, Andreassen S, Carson ER, et al. 1993. A causal probabilistic model of the respiratory system. In: *Proc. IEEE 15th Ann. Conf. Eng. Med. Biol. Soc.,* New York, IEEE, pp. 534-535.
39. Jensen FV, Andersen SK, Kjaerulff U, et al. 1987. MUNIN: On the case for probabilities in medical expert systems—a practical exercise. In: J Fox, M Fieschi, R Engelbrecht (Eds.), *Proc. First Conf. Europ. Soc. AI Med.,* pp. 149-160, Heidelberg, Springer-Verlag.
40. Haykin S. 1994. *Neural Networks: A Comprehensive Foundation,* New York, Macmillan.
41. Worthy PJ, Dybowski R, Gransden WR, et al. 1993. Comparison of learning vector quantisation and nearest neighbour for prediction of microorganisms associated with septicaemia. In: *Proc. IEEE 15th Annu. Conf. Eng. Med. Biol. Soc.,* New York, IEEE, pp. 273-274.

180

Design Issues in Developing Clinical Decision Support and Monitoring Systems

John W. Goethe
The Institute of Living

Joseph D. Bronzino
Trinity College/Biomedical Engineering Alliance for Connecticut (BEACON)

As discussed in previous chapters, health care facilities presently use computers to support a wide variety of administrative, laboratory, and pharmacy activities. However, few institutions have installed computer systems that provide ongoing monitoring of care and assist in clinical decision making. Despite some promising examples of the use of artificial intelligence to enhance patient care, very few products are routinely used in clinical settings [Morelli et al., 1987; Shortliffe & Duda, 1983]. One of the important factors that that has limited the acceptance of decision support tools for clinicians is the lack of medical staff input in system development. Thus, a major task for designers of computer systems that are to be incorporated into the daily clinical routine is to involve practitioners in the design process. This chapter presents a set of design recommendations, the goal of which is to ensure end-user acceptance, and describes a comprehensive clinical system that was designed using this approach and is now in routine use in a large psychiatric hospital.

180.1 Design Recommendations

The design phase of any initiative requires attention to the social and organizational context in which the new product or program will exist. Although a full discussion of the issue is beyond the scope of this chapter, it has been extensively covered in a number of texts. (See especially Peters and Tseng [1983] for a theoretical overview of the theory and illustrative case studies.) In addition to these general principles of organizational change, several specific design steps are critical to end-user acceptance.

Establish a Project Team. The composition of the project team should reflect the scope of the activities within the facility and the organizational structure (hierarchy). It should include at least one outside consultant and at least one member of the medical/professional staff who will serve as the liaison to a clinician task force representing all practitioners.

0-8493-0462-8-6/00/$0.00+$.50

Establish a Clinician Task Force. In addition to the project team, which represents the entire institution, it is important to form a clinician task force of five to seven members of the medical/professional staff. In some facilities all care is provided by physicians while in other settings psychologists, social workers, nurses, and other professionals function as primary clinicians. The task force must represent *all* these professionals.

Meetings of the task force must begin well before the hardware and software decisions have been made so that there will be adequate time to incorporate suggestions from the clinical staff. The implementation schedule must allow time for resolution of the issues raised by the end-users. (*Resolution* here does not imply that *all* parties will be pleased with *all* decisions, but the design team should ensure adequate time for discussion of each issue raised by the clinical staff and for reaching closure.)

Know the Limits within Which the Project Team Must Work. Part of the design task is to determine the limitations present and the level of expectation for system use. For example, is the number of terminals and their location sufficient for all clinicians to have easy access? Does the administration expect clinicians to type rather than dictate patient summaries? If an administrator is not on the project team, interviews with the key executives should be held as soon as possible, with follow-up meetings regularly scheduled.

Identify All Institutional Initiatives and Needs. It is important for system designers to be aware of all institutional plans. For example, the facility may have recently modified its clinical documentation procedures in response to a change in reimbursement regulations or as part of a new managed care contact. Many of these changes will have an impact on the system design, and some may create opportunities for cost or time savings through automation.

Identify the Types of Patient Care Activities Typical for the Institution. Using the language/action approach of Winograd [1987] one can classify all clinical care activities within an institution. As discussed by Morelli, and coworkers [1993] all actions in a psychiatric hospital can be grouped into six categories: assessment notation, medication order, nonmedication order, notification of a critical event, request by the clinician for additional information (consultation), and request of the clinician for additional information (e.g., to justify a treatment or hospital admission). This list, while not necessarily applicable to all clinical settings, is useful as a template. (See Lyytinen [1987] for a review of other approaches to understanding the component activities within an organization.) Once all possible clinical actions are categorized, list those that the system is intended to support. Such detailed assessment of the structure and nature of clinician actions allows designers to determine prior to implementation how the system will affect existing practices. For example, will clinicians be expected to use the computer to enter orders, eliminating the need for nurse transcription of handwritten orders? Will clinicians have to respond to computer notices and document why a treatment standard was not followed?

Illustrate How the New System Will Be Integrated into the Existing Environment. Computerization of one aspect of clinical care may have an unexpected impact on other areas of practice. Thus, a detailed map of the interactions involved in formulating and executing medical orders is a useful step [Morelli et al., 1993]. Even desirable changes in the clinical routine, such as reducing handwritten progress notes, must be examined since the existing nonautomated methods may have served some additional (e.g., educational) purpose not evident without careful mapping of the interaction. Both automated and manual procedures should appear logical and relevant to the staff. Screen formats and input/retrieval features, for example, must follow practices acceptable to the facility. Interactions with the computer must be viewed as nonintrusive and must not require extensive data entry by clinicians.

Clinician Interaction with the Terminal Should Be Part of Daily Clinical Activity. A computer system should not be a stand-alone tool; rather, it should be fully integrated into the clinical routine. For example, clinicians should have direct and easy access to on-line displays of current medications and results of laboratory tests. Furthermore, clinicians should be able to query the system for additional information when laboratory or other data indicate a potential problem.

The System Should Provide Real-time Feedback to Clinicians. The timeliness of an information system is a critical factor in health care applications. Although all functions do not have to operate literally in real time, they do have to operate in a manner consistent with clinician practice. For example, the system should notify clinicians immediately of events such as drug-drug interactions, but in other situations (e.g., a reminder that a follow-up test is due) a notification within 2 days may be sufficient. Laboratory results and treatment data must reflect the orders entered as of the day the system is being queried.

Conduct Site Visits to Assess Computer Use in the Delivery of Clinical Service. Important aspects of system design may be overlooked if unique characteristics of a given site are not taken into account. For example, the system designers may plan to install a computer terminal in the area from which medications are dispensed in each nursing station. If on one unit, because of size or location, this area cannot accommodate a standard terminal, an alternative device may have to be purchased or architectural changes made.

Include Support for Key Administrative as well as Clinical Care Tasks. Administrative activities directly related to clinician practice are quality assurance, peer review, continuing medical education (including information dissemination and needs assessment), drug utilization evaluations, patient outcomes assessment, and continuous quality improvement programs. Since both administrative and clinical tasks are intended to serve a common purpose, the delivery of the best available services to the patient, they should be integrated whenever possible. A detailed listing of all such tasks, specifying the structure of each and the individuals involved, may identify important areas in which the clinical information system can be usefully applied.

Evaluate the System Prior to Implementation. Once the design is complete, an evaluation step is necessary to test the utility of the system in an actual clinical practice.

The System Must Be Adaptable to Individual Clinician and Program Needs. Since rapid changes are commonplace in health care, the system must be able to be modified and updated without extensive additional programming.

180.2 Description of a Clinical Monitoring System

The Institute of Living's Clinical Evaluation and Monitoring System (CEMS) was developed to provide comprehensive support for clinical services in a psychiatric hospital. It represents an extension of earlier work that resulted in a prototype psychopharmacology monitoring system described elsewhere [Bronzino et al., 1989] The current system provides decision support and/or automated monitoring for each key component of care (assessment, diagnosis, and treatment) and consists of four modules: treatment standards ("pharmacotherapy guideline"), diagnostic checklists (DCLs), information alerts, and outcome assessment. Table 180.1 shows the components of care that are supported by each of the four modules of the CEMS.

The treatment standards are accessed from the menu available to all clinical staff. This document summarizes key information about the use of selected psychiatric medications (dosages, therapeutic serum levels, indications, and side effects) and presents decision trees to guide drug selection. The manual is continuously updated and serves as both a reference and a vehicle for dissemination of new information.

TABLE 180.1 Clinical Evaluation and Monitoring System (CEMS)

CEMS	Components of Care				
	Assessment				
Modules	Pt. History/Symptoms	Labs	Outcome	Diagnosis	Treatment
Treatment Standards		X		X	X
DCLs*	X		X	X	
Information alerts	X	X		X	X
Outcome assessment			X		X

*DCLs = diagnostic checklists

It is not linked electronically to the other modules, but the content is the database for the decision rules in the information alerts (described below).

The diagnostic checklists (DCLs) provide an automated method for assuring documentation of the key symptoms and behavioral issues that support the assigned diagnosis and for noting at subsequent evaluation points (e.g., at discharge) the degree of change in each symptom/behavior. In contrast to diagnostic assessments in other specialties, there are few procedures or laboratory tests that definitively establish a psychiatric diagnosis. The Diagnostic and Statistical Manual (DSM) of the American Psychiatric Association [1987] and the International Classification of Diseases (ICD) published by the World Health Organization [1978] provide standard criteria on which to base diagnoses. In both DSM and ICD there is a specific algorithm that includes the type, number, and intensity of symptoms required for each diagnosis. The algorithm for all DSM diagnoses, modified to provide additional data, are used in the CEMS, and there is a separate 1-page checklist for each disorder. The clinician's admission diagnosis determines which form is presented. The form can be completed by the clinician at the terminal or a paper version can be used with subsequent input by clerical staff. The clinician indicates if each symptom/behavior is present or absent, thereby ensuring complete documentation of the findings relevant to that diagnosis. If the symptoms specified fail to meet DSM criteria for the disorder selected, a message is generated along with an explanation. The DCL is also completed at the time of discharge, at which time the system generates a new version of the form that contains only those items that the clinician indicated were present on admission. (For implementation in settings with longer periods of treatment, the DCLs could also be used for serial assessments prior to discharge.)

The information alerts are computer-generated notifications that assist in ensuring compliance with practice guidelines and medication protocols. Clinicians may respond to these notices by changing the treatment plan or diagnosis or by following up on missing or abnormal laboratory orders as directed by the alert (i.e., the clinician is then in compliance with the treatment standard). The system also allows clinicians to document the reason for deviating from the standard. This step is initiated via a function key (see Screen 1); an item is then selected from a list of reasons for "nonstandard" practice or a free-text

```
TM0197-01I. A.            ALERT REVIEW BY CLINICIAN              DIAL700B
GOETHE, JOHN DR.                                   Alert Details as of 7/12/94

          Alert Nbr:  76
        Description:  SUBSTANCE USE DX; ON BENZODIAZEPINE

           Notified:  DR. GOETHE

          Issued on:  07/11/94   00:50
   Last Notified on:  07/12/94   00:41
   Notification Nrb:  2

       Patient Name:  Doe, John Q.
              MRNR:   123456
            Account:  123456789

        Patient DOB:  1/1/01
                Sex:  M

              F3: LAB    F5: RESTART   F11: SUSPEND   F13: HELP
   F2: MEDS   F4: DIAG   F6: HISTORY   F10: RETURN    F14: GUIDE
```

Screen 1

```
TM0197-01I. A.                 ALERT REVIEW BY CLINICIAN                         DIAL710A
GOETHE, JOHN DR.                                                  Alert Details as of 7/12/94

     Patient:  Doe, John Q.
       MRNR:  123456
     Account:  123456789
    Alert #12  MAJOR DEPRESSION NO ANTIDEPRESSANT

               Suspension Reasons
          01:  Medical work-up in progress to rule out organic cause of depression
          02:  Unable to tolerate antidepressants
          03:  Refusing antidepressants
          04:  Washout period
          05:  Drug-free trial
          06:  Patient's affective episode characterized by prominent thought and/or perceptual
               disturbance, and is receiving a trial of antipsychotic medication
          07:  Patient is receiving an antidepressant from an outside pharmacy
    Selection:  Type:

  F1: SELECT                   F5: RESTART                     F7: NXPG
                               F10: RETURN                     F14: GUIDE
```

Screen 2

explanation is entered (see Screen 2). The alert statement with the attached explanation is printed on the next *treatment plan review*, a medical records form summarizing treatments and patient progress that the clinician must sign. Thus, the information alerts provide ongoing monitoring and critical event notification and ensure documentation of the rationale for any nonstandard practice. (All notifications and the clinician responses are reviewed by the medical director or designee daily.) The direct feedback provided by the alerts also serves an educational and information dissemination function. New information about medications, for example, can be incorporated into one or more alerts as well as added to the treatment guidelines described above.

The system currently has 130 alerts and can be expanded to search as many as 2000 records for up to 200 critical events each day. Additional alerts can be added without additional programming, and changes in the specifications for an existing alert (e.g., dosage parameters) can be made by accessing the appropriate table from the *maintenance* screen. Alerts can include data elements on any medication prescribed, any laboratory value, any psychiatric diagnosis or a variety of historical/demographic items (e.g., gender, age). The limitation in the last of these four categories is the amount of information from the medical record that is currently on-line. With a completely electronic medical chart, this system could generate an alert for almost any clinical event.

180.3 Outcome Assessment

At admission, discharge, and regular intervals postdischarge nursing staff complete a functional assessment survey on all patients. These data are entered into the system using an optical scanner. Regular feedback is provided to hospital staff and used to evaluate clinical services and practice.

In a typical scenario a clinician, after logging on, would select "Show My Alerts" from the menu (Screen 3). Alerts for all of the patients assigned to that clinician are displayed as shown on Screen 4. (The medical director or other supervising clinician can select all active alerts by type, by practitioner, or by patient.) Patient name and alert message are given, but no further detail can be obtained nor can

TM0197-01	GUIDE	PSEC030A
GOETHE, JOHN DR.	MEDICAL STAFF	12 JUL 94 09:19

01: CHART REQUEST	11: OUTPATIENT ACTIVITY
02: CLINICAL DATA REVIEW	12: PATIENT EDUCATION MODULE
03: CLINICIAN PATIENT LIST	13: SHOW MY CHART DEFICIENCIES
04: DAY HOSPITAL ATTENDANCE REVIEW	14: SHOW MY PATIENTS
05: I.A. SHOW MY STANDING ALERTS	
06: INPATIENT ACTIVITY	
07: INPATIENT LOCATOR	
08: LIST MY PATIENTS (CLINICIAN)	
09: O.E. LAB ORDER REVIEW	
10: O.E. PATIENT DATA R/O	

SELECTION: 0
**END OF LIST

F1: SELECT

Screen 3

TM0197-01I. A.	ALERT REVIEW BY CLINICIAN	DIAL700A
GOETHE, JOHN DR.		Alert Details as of 7/12/94

Date	Count	Patient	Alert	Description
01:07/11/94	2	Doe, John	10	NO LITHIUM LEVEL > 10 DAYS
02: 06/24/94	10	Doe, Mary	5	ANTIPSYCHOT+ANTIDEPRESSANT+ANTIPARK
03: 07/12/94	1	Smith, John	12	MAJOR DEPRESSION NO ANTIDEPRESSANT
04: 07/05/94	6	Jones, J.P.	13	ANTIPSYCHOTIC NO PSYCHOTIC DIAG
05: 07/08/94	3	Brown, Jane	69	THYROID FUNCTION TEST ABNORMAL

Selection:

F1: SELECT	F3: LAB	F5: RESTART	F11: SUSPEND	F13: HELP
F2: MEDS	F4: DIAG	F6: HISTORY	F10: RETURN	F14: GUIDE

Screen 4

a direct response to the alert be made without selecting a specific patient and a specific alert for that patient. When the clinician goes to a specific alert, there is the option (Screen 1) to suspend the alert as described above or to query the system about the patient's diagnosis, current medication, laboratory values, or the "history" of that alert (i.e., if the alert has previously been issued, a log of all activity on that alert is given). From the current medication and laboratory screens there is an additional option that allows access to all *historical* lab values/medications. This historical feature is especially valuable for tracking patients who require frequent drug serum level determinations (e.g., with lithium carbonate) or who have medical disorders that necessitate periodic laboratory tests (e.g., chronic kidney or liver disease).

180.4 Conclusion

Computers are increasingly *user* friendly, but the clinical environment may often not be computer friendly. In addition to the obvious computer engineering tasks, there is a sizable human engineering issue that must be addressed if automated tools are to be accepted and fully utilized in clinical settings. Thus, designers must take into account the social and organizational structure of the health care delivery network the system is intended to serve, involve clinician staff in its development, and adhere to a design strategy that will ensure end-user acceptance. Although the system outlined in this chapter was developed for psychiatric practice, the components of care described are common to all branches of medicine, and the design steps recommended are applicable for a wide range of clinical settings.

Acknowledgements

Many elements of the system now in place at the Institute of Living represent extensions of the earlier efforts of a number of individual. Peter Ericson, former head of the Department of Information Services at the Institute, Bernard C. Glueck, Jr., M.D., former director of research, were among the pioneers in applying information systems to clinical care. Major contributors to one or more components described in this chapter include Pawel Zmarlicki, David Cole, David Warchol, Russell Dzialo-Evans, Bonnie Szarek, R.N.

Defining Terms

Decision trees: Decision trees or flow charts are a common way to present medical information that is intended to support the decision-making process. Many decisions are hierarchical in nature, but there may be more than one acceptable option at each nodal point on the decision tree.

Language/action model: The language/action model of human-computer communication is a term taken from the work of Winograd, who introduced the term "conversation for action" to describe how human behavior is coordinated within an organization. To quote Winograd "We work together by making commitments so that we can successfully anticipate the actions of others and coordinate them with our own." According to this model, computer systems are part of the conversational structure of the organization and, along with the human employees, engage in a variety of well-defined interactions (communications). Winograd's approach is based on earlier work by J. L. Austin and J. R. Searle.

Map: A map of the interactions is a detailed assessment, often diagrammatically expressed, of how the work of the organization is accomplished. In a medical facility, the steps involved include the physical assessment of the patient by the clinician, various procedures performed on the patient, and the recording, storage, retrieval, and analysis of data. Such maps can be informed by but do not necessarily have to follow Winograd's language/action approach (described above) or any other formal model of analysis.

References

American Psychiatric Association. 1987. Diagnostic and Statistical Manual of Mental Disorders, 3d ed rev. Washington, DC, American Psychiatric Association.

Bronzino JD, Morelli RA, Goethe JW. 1989. Overseer: A prototype expert system for monitoring drug treatment in the psychiatric clinic. IEEE Trans Biomed Eng 36:533.

Lyytinen K. 1987. Different perspectives on information systems: Problems and solutions. ACM Computer Survey 19:5.

Morelli RA, Bronzino JD, Goethe JW. 1987. Expert Systems in psychiatry: A review. Proceedings 20th Hawaii International Conference System Science, 3:84.

Morelli RA, Bronzino JD, Goethe JW. 1993. Conversations for action: A speech act model of human-computer communications in a psychiatric hospital. J Intelli Sys 3:87.

Peters JP, Tseng S. 1983. Managing Strategic Change in Hospitals, Chicago, American Hospital Association.

Shortliffe EH, Duda R. 1983. Expert systems research. Science 220:261.

Winograd T. 1987. A language/action perspective on the design of cooperative work. Report No. STAN-CS-87-1158, Stanford University.

World Health Organization. 1978. Mental Disorders: Glossary and Guide to Their Classification in Accordance with the Ninth Revision of the International Classification of Disease, Geneva, World Health Organization.

Artificial Intelligence

Stanley M. Finkelstein
University of Minnesota

THE FOCUS OF THIS SECTION is on artificial intelligence and its use in the development of medical decision systems with clinical application. The methodologic basis for expert systems and artificial neural networks in the medical/clinical domain will be discussed. This is one aspect of the growing field of medical or health informatics, the application of engineering, computer, and information sciences to problems in health and life sciences. Other informatics applications are addressed in Section XVII. This section contains eight original contributions that cover the history, methods, and future directions in the field. It intentionally omits specific computer hardware and software details related to currently available systems because they continue to change as technology changes and are likely to be outdated even as this *Handbook* goes to press. Furthermore, chapters representing complete examples of specific decision systems are not included in this section. Numerous example can be found in the current literature, and may have been cited in the chapters of this section to illustrate particular aspects of decision system structure, knowledge acquisition and representation, the user interface, and system evaluation and testing.

In Chapter 181, Dr. Kulikowski presents the history of the development of artificial intelligence (AI) methodology for medical decision making, from the early statistical and pattern-recognition methods of the 1960s to the continuing development of knowledge-based systems of the 1990s. Dr. Kulikowski has described the evolution of these AI applications. The early application explored various approaches to handling knowledge within the specific domains of interest, utilizing either causal networks, modular rule-based reasoning, or frame/template representation of the knowledge describing the clinical domain. These varied approaches pointed out the difficulties involved in the actual acquisition of the domain expertise needed for each application and generated research investigations into acquisition strategies ranging from literature review, case study evaluation, and detailed interviews with the experts. Alternative approaches to knowledge representation, the development and assessment of decision system applications, efforts to test and evaluate decision system performance, and the problems associated with reasoning in uncertain environments have become important areas of investigation. Currently, there is active interest in qualitative reasoning representation, the importance of the temporal framework of the decision process, and the effort to move toward more practical systems that embody decision support for diagnostic or treatment protocols rather than the fully automated decision system. Dr. Kulikowski points out that expert clinical systems have not become the indispensable clinical tool that many had predicted in the early days of AI in medicine, describes reasons for the shortfall, and looks forward to advanced software environments and medical devices that are beginning to routinely incorporate these ideas and methods in their development.

In Chapter 182, Drs. Micheli-Tzanakou and Zahner focus on artificial neural network (ANN) methodologies and their applications in the medical and health care arena. ANNs consist of a large number of interconnected neurons or nodes that can process information in a highly parallel manner. They are specified by their processing-element characteristics, their network topology, and the training rules they use to learn how to achieve correct pattern classification from an array of multiple inputs. Drs. Micheli-Tzanakou and Zahner provide the mathematical details for the back-propagation and ALOPEX training algorithms and discuss the benefits and deficiencies of these approaches in teaching the ANN to correctly classify input patterns presented to the system. Finally, the performance of several ANN approaches in mammography and chromosome and genetic sequence classification applications are reviewed and compared.

The contribution by Nykänen and Saranummi (Chapter 183) also provides a history of AI applications in medical decision systems but focuses on the clinical reasoning process, systems development within the clinical environment, and the critical issues for system acceptance. The early introduction of expert system shells provided a strong impetus for system development and indeed resulted in many published applications. The availability of inexpensive and powerful microcomputers and dedicated workstations for knowledge engineering also has contributed to the development of medical expert systems. The authors point out that data and knowledge are often less quantitative and consistent in clinical medicine than in the physical sciences and that moving from the knowledge level to the knowledge use level posed significant problems. Appropriate, objective, and standardized evaluation testing, and validation are often lacking as a part of the system development process, as discussed in this chapter, and this contributes to the lack of widespread acceptance of these systems. The authors conclude with a list of challenges for future clinical decision systems posed by the need for integration with new information technologies and the developing health care information infrastructure.

The chapter by Summers, Carson, and Cramp (Chapter 184) discusses specific methodologies used in expert system development and provides some general examples of rule-based and semantic network approaches to knowledge representation in medical applications. The authors use MYCIN and CASNET, two of the earliest medical expert system applications to achieve widespread recognition and serve as models for subsequent developments, to demonstrate and compare these two approaches for knowledge representation. MYCIN uses production rules to represent causal relationships between knowledge items to diagnose and treat microbial infections. CASNET uses causal or semantic networks to represent knowledge relationships in the diagnosis and treatment of glaucoma.

Dr. Garbay's contribution (Chapter 187) continues along these directions by providing a somewhat more conceptual approach to the question of knowledge domains, acquisition, and representation. Knowledge-acquisition techniques are reviewed, and several knowledge-acquisition tools designed to assist in the process are described. Dr. Garbay also presents fundamental ideas related to rule-based, case-based, and causal reasoning. She discusses models for the uncertainty and imprecision that is central to the clinical environment. The use of temporal knowledge and its incorporation into clinical decision systems is also discussed. The ideas of shallow and deep knowledge systems are introduced from the perspective of the system's explanation facilities. Shallow knowledge systems rely primarily on reasoning based on experience and experimental results, while deep knowledge systems are based on detailed knowledge regarding the structure and function of the underlying system such as those employing physiologic models within the knowledge base. Dr. Garbay, as have many contributors to this section, also introduces the new challenges for clinical decision systems with regard to the rapid developments in networking, communications, and information sciences.

A new and expanding application area involving intelligent patient monitoring and management in critical care environments is discussed in Chapter 186, by Drs. Dawant and Norris. Such systems involve the context-dependent acquisition, processing, analysis, and interpretation of large amounts of possibly noisy and incomplete data. These systems can be viewed within four distinct functional levels. At the signal level, the system acquires raw data from patient monitors, which include analog-to-digital conversion and some low-level signal processing. Data validity checks and artifact removal occur at the validation level. The transformation from numerical features that characterize the signals to a symbolic representation, such as normal or abnormal, is performed at the signal-to-symbol level. Finally, the inference level consists of the reasoning elements described in previous chapters to arrive at diagnoses, explanations, prediction, or initiation of control actions. Examples of both shallow and deep knowledge systems developed for patient monitoring are presented

Systems development and widespread dissemination often have been stymied by the lack of standardization of knowledge representation, as described in all the preceding contributions. The last two chapters look at this question from the perspective of medical terminology and the status of natural language processing in biomedicine.

In discussing applications of medical knowledge bases, Dr. Kerkhof (Chapter 187) states that while technological advances in computer processors and storage media permit virtually unlimited storage and fast retrieval capability, the interpretation of natural language constitutes a major obstacle in the advancement of knowledge-based systems in medicine. In this chapter, classification and coding systems are identified, and their content is described. The richness and depth of natural language are often the focus of difficulties when attempting to develop such systems, encountering differences in such elementary concerns as definitions, spelling, usage, and precision. Dr. Kerkhof offers several approaches to handling linguistic problems associated with such systems, but all have their own inherent limitations.

In Chapter 188, Dr. Johnson describes the techniques of natural language processing as a means to bridge the gap between textual and structured data so that users can interact with the system using familiar natural language, while computer applications can effectively process the resulting data. Applications are classified according to the levels of language competence embodied in their design. Speech-recognition and -synthesis systems deal with basic data representation, wile lexical, syntactic, and discourse systems function at increasing levels of complexity from single words to entire discourses.

While the specific applications described in this section relate to biomedical concerns, the definitions and methods overview relate to the development and utilization of expert systems and artificial neural networks for a wide variety of decision systems. Detailed implementation protocols for such systems are beyond the scope of this section but can be found in the references cited at the conclusion of each chapter.

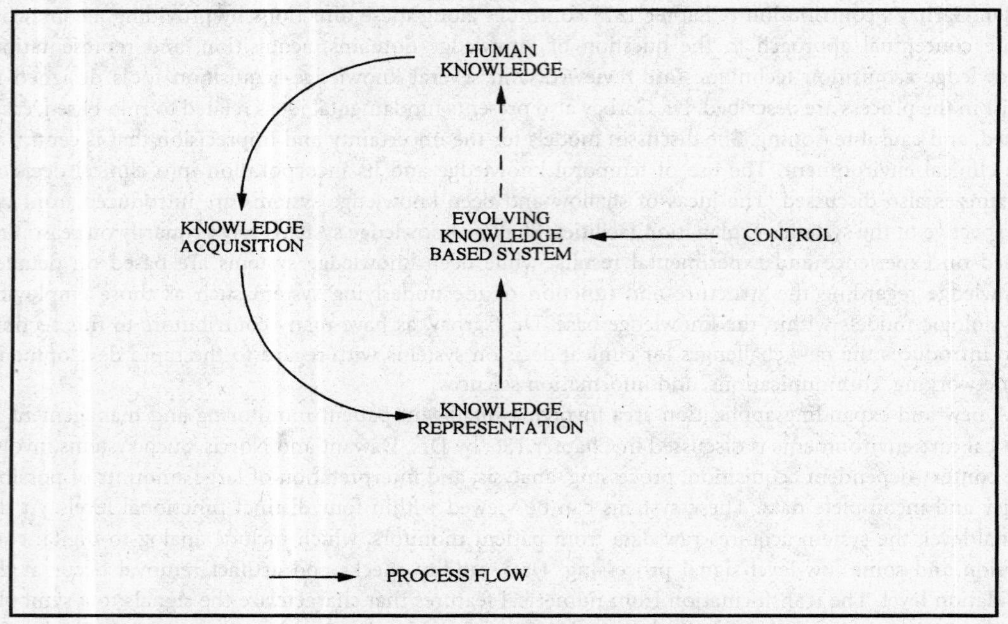

Basic steps in the development of an expert system.

181

Artificial Intelligence in Medical Decision Making: History, Evolution, and Prospects

Casimir Kulikowski
Rutgers University

Artificial intelligence (AI) for medical decision making and decision support had its origin in the knowledge-intensive expert consultation systems that were introduced in the early 1970s. These differed considerably from the data-intensive statistical and pattern-recognition methods that had been applied to medical reasoning problems since the 1960s and which saw a resurgence in the 1980s as new types of computationally more powerful artificial neural networks (ANNs) were developed, and sophisticated models of Bayesian and other belief networks were designed to capture the nuances of clinical reasoning. For the past decade it has become incresingly clear that a variety of approaches is needed to make computer-based medical decision systems useful and effective, preferably embedding them within systems that are based on an electronic medical record (EMR). Meanwhile, advanced software environments increasingly and routinely incorporate AI ideas and methods as they seek to facilitate the tasks of building, validating, and testing medical knowledge bases.

0-8493-0462-8/00/$0.00+$.50

181.1 Early Models of Medical Decision Making

The earliest efforts to formalize medical decision making involved the application of statistical decision methods (ROC curves) in radiographic interpretation. Precursor attempts at automating the logic of diagnostic reasoning by sorting symptoms and selecting diagnoses that matched a particular combination involved multiple slide-rule and early card and computer sorting techniques.

As computers became more powerful and easily programmable, they were the natural tool for representing diagnostic and treatment decision making. A watershed article by Ledley and Lusted [1959], in *Science*, described the reasoning bases for medical decision making in both logical and statistical terms. The main paradigm for representing medical decisions in the 1960s was statistical, whether Bayesian, hypothesis testing, or discriminant function analysis. The growth of standardized laboratory test panels and multiphasic screening methods had the effect of popularizing statistical techniques as an objective approach to the selection of decision thresholds. Bayesian methods were predominant and allowed incorporation of subjective estimates of prior probability into the calculation of diagnostic probabilities. Likelihood-ratio (hypothesis-testing) methods did not, but instead required a determination of sensitivity/specificity trade-offs with threshold selection dependent on a choice of clinically tolerable levels for them. There also were heuristic pattern-recognition methods being developed if more flexibility was required, but all these approaches at that time suffered from a common drawback. While they might perform very well for circumscribed problems with well-defined statistics or adequate samples from which to estimate the statistics, they were rarely acceptable to practicing physicians beyond their site of origin. A reason frequently given for this was the difficulty of explaining decisions based on strictly computational theories of probability in terms of the qualitative language and arguments familiar to physicians.

During the 1960s, the alternative to formal statistical methods was the coding of the sequence of expert decisions in a branching logic diagram or flowchart, often described as a clinical decision algorithm. This approach had the advantages of clarity and ease of explanation but was usually too rigid to capture the nuances of context without becoming very large, complex, and computationally expensive. During this period, most mainstream AI research appeared to have little to offer medical decision making, since it was concerned primarily with the computer representation and reasoning for general problem solving without considering explicit representations of uncertainty.

181.2 Emergence of the Knowledge-Based AI Methods for Medical Consultation

In the early 1970s, several groups involving both AI and biomedical researchers decided to explore different, more knowledge-intensive approaches to interpretive problem solving, including medical decision making. Several common themes emerged from the work of the four groups that initiated these efforts at Rutgers University, Stanford University, the University of Pittsburgh, and the MIT/Tufts collaborative group. All of them, at about the same time, recognized some of the shortcomings of existing methods and proposed a set of alternative representation and inferencing approaches that involved:

- Representing uncertainty more flexibly and qualitatively than appeared possible with probabilities.
- Representing more of the medical knowledge that motivated and justified a diagnostic, prognostic, or therapeutic decision.
- Developing a descriptive component of medical knowledge to which some general problem solving or inferencing strategy could be applied.

Early prototypes of AI consultation systems that incorporated distinct approaches to medical decision making were CASNET [Kulikowski and Weiss, 1972], MYCIN [Shortliffe et al., 1974], DIALOG/INTERNIST-1 (Pople et al., 1975], the Present Illness Program (PIP) [Pauker et al., 1976], and the Digitalis Advisor Program [Gorry et al., 1978].

That a domain-specific model could be used as the basis for interpretive decision making was first demonstrated in the DENDRAL system [Buchanan et al., 1970]. DENDRAL interpreted mass spectra by using rules that were constrained by a model of the possible chemical structures that could have given rise to the observed spectral data. This was quite different from the domain-independent, pattern-recognition approaches previously applied to the problem.

The first AI approach to decision making in medicine itself evolved within the Rutgers Research Resource on Computers in Biomedicine, which was established in 1972. One of the goals of our research group was to find general methods of representing medical reasoning that would take advantage of the knowledge of specific diseases in ways similar to those used by medical specialists. In a collaboration with Dr. Aran Safir, an ophthalmologist at the Mt. Sinai School of Medicine (New York), we explored methods for describing and applying medical knowledge for computer-based diagnosis and management. With Sholom Weiss, who worked on his doctoral dissertation on this project, we found that a very natural way for characterizing the underlying mechanisms of disease was through cause-and-effect relations. A network of causal links among pathophysiologic conditions (or abnormal physiologic states) could then be used to describe the many pathways through which a disease might manifest itself. Each pathophysiologic state could be inferred independently from the patient's condition and asserted with some degree of confidence depending on its pattern of supportive evidence. A prototype system for testing these ideas was developed in the area of glaucoma diagnosis and management. This proved to be a good problem, because glaucoma is a major cause of loss of vision and blindness with mechanisms that are sufficiently well understood to determine the course of management; yet at the same time sufficiently complex that advice on difficult cases (particularly those resistant to conventional treatment) must frequently be sought from specialist consultants.

A variety of reasoning strategies seemed to apply to different stages of the consultation. For example, while starting to gather data on a patient, the major problem for a consultant is to elucidate the patient's problem and develop diagnostic leads by asking the right questions and acquiring the relevant data. When enough data have accumulated, a different mode of reasoning is characteristically needed. The findings must be attributed to specific causes, whether single or multiple. Finally, all these explanations must be combined into a differential diagnosis. This serves to predict what will happen if the patient remains untreated (prognosis) and suggests possible ways of managing the disease (treatment or therapeutic planning). As we gained experience in extracting the knowledge from the expert practitioners, it became clear that different management strategies relied on different types of medical knowledge, and these strategies had to be sufficiently general and independent of the particular disease being modeled. Weiss built a prototype system called CASNET (for casual associational network) based on these ideas [Weiss, 1974; Weiss et al., 1978]. We were fortunate in being able to draw on the expertise of leading glaucoma specialists, who provided the in-depth knowledge needed to describe the disease and tested the system independently in their own laboratories. After two years of improving the initial prototype, the system was given a blind test by presenting it with new, previously unseen cases during a panel discussion at the major national meeting of the Academy of Ophthalmology in 1976. CASNET made no errors in its recommendations, though it was unable to answer some questions for which the knowledge was absent in the computer model [Lichter and Anderson, 1977].

During the early 1970s, another kind of biomedical AI resource was established at Stanford as the result of the DENDRAL project collaboration that had developed between AI researchers (E. Feigenbaum and B. Buchanan) and biomedical researchers (J. Lederberg and S. Cohen). This SUMEX-AIM Resource, established in 1974, provided advanced time sharing on a PDP-10 computer to a national network of investigators of AI in medicine, with appropriate systems and AI language support for what was then a very novel computational mode, serving an initial nucleus of investigators, including those at Rutgers and the University of Pittsburgh [Freiherr, 1979].

The medical component of the SUMEX-AIM research centered around the consultation system for infectious disease treatment developed by Shortliffe [Shortliffe et al., 1974; Shortliffe, 1976]. Its representation of medical knowledge was in the form of heuristic rules with a new certainty factor formalism, which decoupled the way in which positive and negative evidence was credited toward a hypothesis, thus providing more heuristic flexibility than previous probabilistic frameworks. The system's strategy was goal driven for gathering the data needed to reach the diagnostic conclusions necessary to choose the appropriate covering therapies. The modularity of MYCIN's rules turned out to be a critical design choice, since it was soon found that it was easy to modify the rules independently of the reasoning strategy and its inference engine, giving rise to a very flexible and updatable knowledge base. The certainty-factor model turned out to be less long-lived, since it was soon found to implicitly involve assumptions and constraints that did not adequately represent the complex dependencies that often exist among data and hypotheses. However, MYCIN, by providing an easy means of directly and modularly encoding the rules of medical reasoning (without an underlying disease model as in CASNET), gave rise to the most powerful representation for the first generation of knowledge-based AI systems—the rule-based paradigm [Buchanan and Shortliffe, 1984].

Another group to work intensely on AI approaches to medical decision making was that of Gorry at MIT and Pauker and Schwartz at Tufts, who led the development of the Digitalis Advisor Program (Gorry et al., 1978]. This system introduced two interesting new concepts: a patient-specific model (PSM) and a mathematical model for describing the mechanism being regulated and interpreted (digitalis uptake). The PSM grouped together all the information known about a patient during a consultation session (the findings, history, test results, etc.), as well as the hypotheses about the patient (diagnostic, prognostic, and therapeutic) produced by the computer diagnostic/therapeutic model. The mathematical model was a compartmental model of digitalis uptake, which was able to relate dosages to their expected therapeutic and toxic effects.

In a related AIM project, Pauker and Kassirer joined Gorry and Schwartz in formulating the Present Illness Program (PIP) [Pauker et al., 1976], which was designed to focus in on a medical problem based on a brief description of the patient's complaints (the present illness). Here, Minsky's frame formalism [Minsky, 1975], which had been developed to describe a template of expected objects in computer vision recognition tasks, was used as a template for grouping typical or characteristic information about a disease. Each frame contained slots that had to be filled with descriptors of the disease, with some of these slots containing rules for reasoning about the disease. In this way, diseases were related through rules, but these rules were grouped by their respective frames. An interesting observation that came from this work was that the probabilistic scoring of hypotheses, while important in diagnostic reasoning, is often overshadowed by the use of categorical or deterministic reasoning, which has to be flexible and suitably richly represented to be useful [Szolovitz and Pauker, 1978]. Subsequently, the ABEL system for modeling acid-base balance problems [Patil et al., 1981] took advantage of many of the ideas from PIP and combined them with a causal description of disease at multiple levels of abstraction and aggregation, which helped provide a more detailed description of pathophysiologic processes.

Contemporaneously, Dr. Harry Pople at the University of Pittsburgh had been investigating biomedical applications of AI—specifically abductive reasoning with causal graphs for scientific theory formation [Pople, 1973]. In a collaboration with Dr. Jack Myers, an eminent internist, Pople began a very broad and ambitious project: to capture all of Dr. Myers' diagnostic knowledge of internal medicine into a system that could reason automatically from the facts of a case. Over the next few years, they developed a taxonomic and causal representation for characterizing diseases and their manifestations and proceeded to develop a set of alternatives for describing the various strategies of diagnostic reasoning: the preliminary assessment of findings, the attribution of causes and explanations for each of the findings, the computation of confidences and resulting ranking of diagnostic hypotheses, the development of a differential diagnosis, and the confirmation of a diagnosis once it is already strongly indicated (through specialized tests). The DIALOG system was developed to test these ideas and codify Dr. Myers' knowledge [Pople et al., 1975]. Its later version, called INTERNIST-1 [Pople, 1977; Miller, 1984], demonstrated how the very large domain-specific knowledge base of internal medicine could be assembled and validated with complex cases of disease, both from the clinic and journal CPC reports.

Research in AI in medicine (AIM) was initially presented in a series of AIM workshops, starting at Rutgers University [Ciesielski, 1978], subsequently rotating among the AIM community. Later workshops were organized in conjuction with the American Association for Artificial Intelligence (AAAI) as part of its spring symposia, while sessions on medical expert systems were to be increasingly found at most relevant scientific, engineering, medical, and informatics conferences (AAAS, IEEE, ACM, AAAI, SCAMC, MEDINFO, and AMIA).

181.3 The Transition to Expert Systems and the Ascendancy of Rule-Based Systems—1976–1982

The experience in developing, testing, and disseminating the prototypes of the first-generation medical consultation systems, combined with similar experiences of AI researchers outside medicine, led to a shift in AI research away from general problem solving to more domain-specific knowledge-intensive problems. Feigenbaum (1978) defined generalizations of the MYCIN system for dealing with any problem where advice-giving knowledge could be captured in the form of rules. He emphasized the separation of rule-based systems from the inference engine that reasoned with them and coined the term knowledge engineering to describe the art of building a knowledge-based system. This process was centered around the interviewing of domain specialists by knowledge engineers (typically computer scientists or engineers), who would attempt to understand the problem being modeled, learn about the expertise that the specialist applied in solving it, and finally formalize all this into a computer representation of the problem-solving or consultative knowledge. The early stages of the process, usually called knowledge acquisition, frequently became the major bottleneck in developing an expert system, particularly if the knowledge engineer had difficulties in understanding the field of expertise or came with preconceived notions about it that did not match the expert's own. Another difficulty with first-generation expert system shells was that the knowledge had to be fitted to a predetermined computer representation, which might or might not match that of the domain problem. Most shells worked well with advice-giving reasoning that involved classification-type problems that could be reduced to the selection of one or several alternatives from a large set of candidate hypotheses. In this sense, their overall problem-solving paradigm did not differ much from the statistical and pattern-recognition methods that preceded them. Their major departure was in the architectural modularity of the rule base, flexible choice of reasoning strategies, and the representation of intermediate hypotheses and reasoning constructs with which to support and assemble a final conclusion. Researchers interested in the cognitive processes of human expert diagnosticians analyzed the problem-solving behavior of experts confronted with real and simulated cases to discern their reasoning strategies [Elstein et al., 1978; Kassirer and Gorry, 1978]. These were usually found to be sufficiently complex that they did not have an effect on the design of the rapidly spreading rule-based systems of the time. Rather, they influenced the development of more sophisticated systems based on deeper medical knowledge in the following decade.

The years 1976–1978 saw the testing, critiquing, and elaboration of the prototype systems. This resulted in changes and generalizations of the initial designs, leading to the first general system building frameworks (EXPERT [Weiss and Kulikowski, 1979], EMYCIN [van Melle, 1979], AGE [Nii and Aiello, 1979]). These came about because in the course of the research it became clear that incorporating expert knowledge about advice-giving consultation could be carried out with various kinds of reasoning: either or both hypothesis- and data-driven strategies, with corresponding backward or forward chaining of rules. Furthermore, inferences incorporated into the rules could be either interpreted (as in EMYCIN) or compiled (as in EXPERT), and knowledge could be chunked or clustered by subtasks of the problem solving with the aid of various other representational devices (context trees in EMYCIN, knowledge sources in AGE, rule clusters in EXPERT, and frames in PIP).

Building various expert systems capitalized on the increasing experience with alternative ways of representing knowledge. For example, the IRIS system [Trigoboff, 1976] applied a semantic network model (as was being developed in PROSPECTOR [Hart and Duda, 1977]) to represent glaucoma consultation knowledge with multiple strategies of inference; the CRYSALIS system [Englemore and Terry, 1979] which explicated the structure of chemical substances from X-ray diffraction data, used the blackboard model from speech understanding; the VM system developed a real-time variant of a rule-based system for ventilation management [Fagan et al., 1979]; and the MDX system pursued a distributed conceptual model for diagnosis [Chandrasekaran et al., 1979]. The CENTAUR system [Aikins, 1979] showed that both rules and frames could be usefully combined to represent expert reasoning, and AI/RHEUM [Lindberg, 1980] demonstrated how rule-based systems with specialized medical semantics could be adapted to represent already formalized knowledge found in rheumatologic diagnostic criteria tables. Meanwhile, many other systems were developed showing that rule-based methods could be widely and systematically applied [Reggia et al., 1980; Horn et al., 1980; Speedie et al., 1981; Buchanan and Shortliffe, 1984; Weiss and Kulikowski, 1984].

The evaluation of the first-generation systems was pursued for CASNET [Lichter and Anderson, 1977], MYCIN [Yu et al., 1979], the Digitalis Advisor [Swartout, 1981], and INTERNIST-1 [Pople, 1977]. Critical examination of knowledge structures in INTERNIST-1 led to their being rerepresented in INTERNIST-2 [Pople, 1977]. General methodologies for evaluating clinical predictions were also being developed [Shapiro, 1977]. This testing and evaluation led to the first efforts at automating knowledge acquisition and maintenance of the rule bases. For instance, the TEIRESIAS system was designed to improve the performance of MYCIN through heuristic analysis of individual cases leading to changes in rules that were incorrectly invoked or designed [Davis, 1979].

Likewise, Politakis and Weiss [1980] developed methods for testing and improving EXPERT rule bases by providing logical and statistical performance evaluation and rule-modification tools in an interactive environment. In 1980, a first workshop on general expert systems was held to compare the different techniques for knowledge representation and the systems that implemented them. A common problem was selected in advance without the knowledge of the different research groups and presented to them at the workshop for implementation in their system. The results of this experiment are summarized in Hayes-Roth et al. [1983]. While all the existing shells were able to represent a complex advice-giving problem (the diagnosis and tracking of a chemical spill), those with fixed inference engines proved the easiest to use for rapid prototyping. As might have been expected, the more general AI languages designed to capture expertise did permit more flexible, expressive, and detailed descriptions of the problem but at the cost of a much greater investment of time and effort in building both knowledge bases and specialized inference procedures.

The development and assessment of rule-based and frame-based systems demonstrated that while they could effectively capture large quantities of domain-specific expertise, there were still many unanswered questions about how the expertise should be best applied to specific problems while at the same time abstracting out general representations of knowledge. The complexity of human problem-solving processes revealed by cognitive analyses and simulations of clinical reasoning [Elstein et al., 1978] pointed to the difficulties of reconciling these goals.

181.4 Exploration of Alternative Representations for Medical AI and the Search for Performance—1983–1987

Two almost opposite trends in medical AI emerged from the generalization of expert systems, which continue to the present. On the one hand, driven by the goal of practical clinical applications, researchers tried to adapt existing representations and obtain high performance from them. This led to various systems with very specific goals, such as the interpretation of laboratory tests, which even reached commercial implementation and dissemination [Weiss et al., 1983]. At the opposite end of the spectrum, dissatisfaction with the oversimplified cognitive style of rule-based expert systems and the inadequate coverage and performance of large knowledge bases led to research into alternative representations of deeper medical knowledge [Chandrasekaran et al., 1979].

Explanation and Early Knowledge Level Work in AI Systems

In an attempt to reuse the MYCIN knowledge base for tutorial purposes in the GUIDON system [Clancey, 1989] it was found that generating explanations from the original rule base was not only difficult but also revealed that all kinds of implementational details for consultative inferencing were mixed in with the more abstract, descriptive medical knowledge that supported the inferences. Besides adding metarules for guiding the discourse involved in tutoring, the experience in building GUIDON led Clancey to reconsider the generality of the MYCIN rule base and its representation. The NEOMYCIN system [Clancey and Letsinger, 1981] completely separated descriptive medical knowledge from details of reasoning task implementation so that it would better match the problem-solving tasks of human experts. This coincided with Newell's more general observation across AI systems: that they needed to describe problem solving at the knowledge level, free of the encumberances of specific programming details [Newell, 1982]. In medicine, a similar experience had been reported in a follow-up to the digitalis project by Swartout who designed the XPLAIN system [Swartout, 1981] as an automatic programming approach for generating a consultation model from the specification of abstract system goals. Explanations and justifications would then be automatically built into the performance system.

The role of explanatory reasoning became a major focus of attention for many medical AI groups during this period. One of the earliest and major motivations for the AI approaches to reasoning was that they were able to explain their logic in terms that were much closer to the arguments of physicians. However, by the end of the first decade, most AIM researchers found that the explanatory capabilities of their systems, while useful for tracing logical connections through their knowledge, left much to be desired in terms of naturalness and flexibility. Traditional statistical decision systems could produce an explanation by accounting for each manifestation's contribution to the total probability (or weight) of a diagnosis. In CASNET, an explanation could be generated by tracing the pathway of confirmed causal states that led to the final diagnosis. The confidence in the confirmation of each state could, in turn, be explained by the pattern of observed evidence that supported it. In MYCIN and other derived rule-based systems, explanations could be easily generated by tracing the rules involved in producing a decision. In the Digitalis Advisor, an explanation could be generated in terms of the causal underpinnings of the patient-specific model, while in INTERNIST-1, the diagnostic strategy could be traced in terms of the sequence of manifestations activated or covered by a particular set of hypotheses. All these approaches, while initially satisfying to the developers, upon further consideration proved to be inadequate. Tracing every detail of logic by which a system arrived at a diagnostic or therapeutic conclusion could be very useful for debugging the knowledge base during system developments but it rapidly became tedious to the expert reviewer or the practitioner using the system. What was missing was a clear summary of the underlying justification of the reasoning. A major reason for this was that the first-generation representations, in their emphasis on symbolic and qualitative descriptions of knowledge, as opposed to the earlier numerical representations, failed to separate or abstract out information at the true knowledge level from the finer-grained symbolic details of the implementation.

NEOMYCIN [Clancey and Letsinger, 1981] demonstrated the need to reorganize the knowledge in a rule base in order to be able to produce "intelligently studied reasons" for a decision. The ABEL system showed the critical importance of hierarchical causal descriptions using composition and decomposition operators to explain interactions among subprocesses contributing to an overall pathophysiologic disease process [Patil, 1983]. In a restructuring of the INTERNIST-1 knowledge base, Pople [1982] showed how combinations of very elementary operators (causal, hierarchical) could be used to construct and explain alternative complexes of hypotheses with different interpretations for the same set of confirmed states. MDX stressed the importance of describing problem-solving tasks in diagnosis and using procedural knowledge to capture the knowledge of expert practitioners [Chandrasekaran et al., 1979]. This research, applied to several liver diagnostic problems and a red cell identification problem [Smith et al., 1985], among others, led Chandrasekaran [1986] to develop a theory of generic problem-solving tasks in order to advance our understanding of diagnostic and other expert reasoning processes.

A completely different form of reasoning was implemented in the ATTENDING system [Miller, 1983]. Rather than modeling consultative reasoning directly, Miller analyzed and critiqued the plans that physicians had developed for anesthesia administration, using an augmented transition network (ATN) of states. This novel departure helped define a new type of critiquing system that was very useful for instructional purposes and therefore more likely to be used by physicians who either felt threatened by automated consultative systems or else found them lacking and not essential to their practice.

Performance of Expert Medical Systems

In trying to deal with practical clinical problems, many AIM researchers found that the first general expert systems shells did not satisfy their requirements. The shells frequently had many options that would never be used in a particular application while being difficult to adapt to problem-specific reasoning methods and domain-specific knowledge structures. Furthermore, for an expert system to be really useful, they found that design of easy-to-use human interfaces, while not scientifically rewarding at first glance, was actually quite essential. In this regard, the next major medical project of the Heuristic Programming Project at Stanford, the ONCOCIN system [Shortliffe et al., 1981], demonstrated that well-engineered interfaces mimicking on a screen, the input of charts, images, and other data for cancer treatment protocols could directly improve data acquisition and were essential in guiding the application of the protocols. Knowledge acquisition for ONCOCIN was provided by a very flexible graphics-oriented system, called OPAL [Musen et al., 1986]. An important representational issue also was settled in the ONCOCIN experiments: the need for event-driven reasoning as well as goal-driven reasoning in rule-based systems, as had been earlier advocated in the EXPERT scheme.

Exploiting other, natural knowledge representations became a theme in an application of the EXPERT system for building a rheumatology knowledge base [Lindberg et al., 1983]. Here it was found that domain-specific decision rules had already been defined by the American Rheumatological Association in the form of diagnostic criteria tables. These, however, needed more detailed elaboration in terms of observational criteria for findings, as well as customization of the diagnostic criteria within a working expert system—AI/RHEUM [Kingsland et al., 1983].

The goal of producing high-performance systems led researchers during this period to experiment with technology transfer, disseminating the knowledge of experts within the medical community. The first expert system on a chip was pioneered at Rutgers with the SPE/EXPERT system for serum protein electrophoresis analysis [Weiss et al., 1983]. In this system, a knowledge base was developed by a leading specialist in the field, using the high-level EXPERT language, and then automatically translated into algorithmic form and compiled into the assembly language of the ROM that existed to process signals from the scanning densitometer of the Cliniscan instrument (TM-Helena Laboratories, Beaumont, Texas). Similar technology was used to develop an advice-giving system for primary eye care on one of the earliest hand-held computers [Kastner et al., 1984].

In a very large undertaking, the INTERNIST-1 knowledge base was recast into a form that would make it more easily accessible by flexible querying in the QMR system [Miller, 1984], rather than being restricted to the consultative mode it had been before. The intelligent retrieval facilities of QMR have made it one of the most widely disseminated results of expert system research [Miller, 1994]. In a similar trend, other earlier non-AI systems were reengineered to have more modular architectures, as with the HELP system, which provided advice as part of the medical information system at the Latter Day Saints Hospital. Its successor, the ILLIAD system, has subsequently become an integral part of quality assurance functions in the hospital network [Lau and Warner, 1992]. Likewise, the DXplain general decision support system was made widely available to physicians through the AMA/NET in the 1980s [Barnett et al., 1987].

With the increasing dissemination of AIM systems came parallel efforts to test and evaluate their performance while exploring new ways of dealing with the difficult issues of reasoning in uncertain environments.

An empirical approach to performance evaluation was taken with the AI/RHEUM system testing and resulted in the first system for refining a rule base using the accumulated experience of expertly solved cases [Politakis and Weiss, 1984]. This system, called SEEK, employed performance heuristics to suggest potential improvements in the rules (generalizations and specializations), which could be tested by incorporation into the model and evaluation over the entire database of cases. In SEEK 2, this work was extended to generate automatically and test the potential improvements, selecting the most successful improvements for inclusion in an updated knowledge base [Ginsberg, 1986]. Related to these approaches is the basic problem of uncovering underlying causal relations through the analysis of time-oriented data, such as was carried out in the RX system [Blum, 1982]. During this period, causality became a major concern of researchers seeking strong theoretical underpinnings that would explain reasoning in complex diagnostic consultation models [Patil, 1986].

Formal Theories of Reasoning Under Uncertainty in Medicine

Many different formal methods of inference were explored and applied in knowledge-based medical decision making. These included the application of multihypothesis Bayesian methods (Ben-Bassat et al., 1980], of fuzzy logic as in the CADIAG family of systems [Addlassnig and Kolarz, 1982] and the SPHINX system [Fieschi et al., 1983], or of the Dempster-Shaffer theory of evidence to structured medical problems [Gordon and Shortliffe, 1985]. Reggia et al. [1983] devised set covering strategies to parsimoniously explain diagnostic hypotheses within a probabilistic framework [Peng and Reggia, 1987].

Two major interrelated problems afflicting the handling of uncertainty in clinical decision making were how to combine evidence from multiple related sources in a non-mutually exclusive multihypothesis inference problem and how to represent the flow of uncertainty among intermediate causal and hierarchically related hypotheses. These problems had existed from the very begining of medical AI, and their formal treatment was avoided through the use of empirical, heuristic methods in the first generation of systems. Now researchers embarked on a more principled search for solutions. A probabilistic approach to propagating the effect of local changes in probabilistic networks was developed by Lauritzen and Spiegelhalter [1987]. The Dempster-Shaffer theory provided an alternative formalism for incorporating partial belief information into a complex network of hypotheses. However, in practice, it often proved computationally expensive or might produce very broad and uninformative confidence bounds on hypotheses at the end of long reasoning chains. In an attempt to provide a link between the semantics of hypotheses related at different levels of causation and abstraction, Pearl [1987] developed methods for propagating uncertainty within a structured Bayesian framework of statistical decision making. The more systematic application of these techniques and their implications for specific medical reasoning problems continues [Pearl, 1995].

The revival of interest in connectionist, neural network methods for learning and decision making was strongly stimulated around this time by the introduction of backpropagation methods [Rumelhard and McClelland, 1986]. These used differentiable nonlinear threshold functions that enabled the learning of much more complex decision rules from data than had been possible with the simpler perceptron methods of the 1960s. As researchers and practitioners came to realize that the expert systems and knowledge-engineering approaches required large amounts of investment by the most costly and indispensable consultants in the specialty being modeled, efforts again began to emphasize methods that learned directly from data through artificial neural networks (ANNs), as well as simpler but more explainable structures, such as decision trees [Breiman et al., 1984].

181.5 The Past Decade—Structure, Formalism, and Empiricism

Deep Medical Knowledge for Representation and Problem Solving

The trends described above have continued since 1987 and throughout the present decade as researchers continued to work on representations for the deeper medical knowledge that could help explain and justify decision making. One approach has continued to emphasize the centrality of causal reasoning and simulation in providing knowledge-based explanations for the physiologic underpinnings of medical reasoning [Kuipers, 1985; Long, 1987, 1991; Hunter and Kirby, 1991; Widman, 1992].

More generally, Newell's knowledge-level approach [Newell, 1982] became the inspiration for the second-generation of expert systems, which emphasized the multilevel nature of modeling problem-solving and related interacting knowledge structures [Chandrasekaran and Johnson, 1993].

Applications of Newell's SOAR architecture [Smith and Johnson, 1993] enabled a very general type of problem-solving mechanism (chunking) to roam over flexible spaces of goals, tasks, and specific knowledge structures, thereby providing a common framework for reasoning about decisions in individual cases or with groups of cases, as for learning rules or other changes to the knowledge base. In this, as in the KADS approach [Wielinga et al., 1992], expertise is viewed as a dynamic process that cannot be captured in strictly static knowledge structures but rather evolves as part of an ongoing interactive modeling of systems, agents, problems, and environments [Clancey, 1989]. A discussion of how this problem-solving cognitive approach affects medical decision making can be found in Kleinmuntz [1991]. Meanwhile, many of the lessons learned from earlier prototypes were applied to more sophisticated knowledge-based explanation and system-building schemes [Bylander et al., 1993; Koton, 1993; Swartout and Moore, 1993; Lanzola et al., 1995]. Relationships between abductive reasoning and temporal reasoning were investigated [Console and Torasso et al., 1991], and an epistemologic framework for structuring medical problem solving stated [Ramoni et al., 1992]. It is interesting to note that a related cognitive approach to eliciting and structuring medical knowledge grew up independently in the former Soviet Union through the work of Gelfand and collaborators [Gelfand et al., 1987; Gelfand, 1989].

Formal Decision-Making Methods and Empirical Systems

Formal decision-making approaches to medical reasoning took off in the 1980s with the founding of the journal, *Medical Decision Making*, which involved physicians as well as decision scientists in the process of understanding how formal statistical and logical methods can apply to the modeling and interpretation of medical problems. This renewed interest in the modeling of reasoning under risk and uncertainty was motivated by the twin needs of evaluating decision-support systems and of modeling more general epidemiologic effects of medical interventions, with and without computer-based systems being involved. While not directly related to knowledge-based AI approaches, this trend nevertheless interacted with developments in operations research (OR) and AI through the confluence of the influence diagram representation of abstract states for describing processes [Shachter, 1986] and belief network representations for reasoning with causal structured hypothesis networks [Pearl, 1987]. Together, they helped make Bayesian belief networks a practical tool for modeling many of the dynamic aspects of clinical cognition [Cowell et al., 1991] and provided a combined statistical-structural representation for decision-making preferences in computer-based systems [Lehman and Shortliffe, 1990; Farr and Shachter, 1992; Heckerman and Shortliffe, 1992; Jimison et al., 1992; Neapolitan, 1993]. The related use of the decision theoretic approach to integrate knowledge-based and judgmental elements has also been an important development in assisting the evaluation of systems [Cooper, 1993] and knowledge-based tools for specifying large expert system models [Olesen and Andreassen, 1993, Lazola et al., 1995] and building decision models are also becoming more prevalent [Sonneberg et al., 1994]. In addition, tools from meta-analysis have begun to be applied to assess conflicting results from related though different clinical studies [Littenberg and Moses, 1993] and merge diagnostic accuracy results [Midgette et al., 1993]. The temporal nature of medical processes and data has also led to some initial models for decision making that make temporal dependencies explicit in their representation [Kahn, 1991; VanBeek, 1991; Hazen, 1992; Cousins et al., 1993, Shuval and Musen, 1993]. Practical systems in the ICU or other monitoring situations typically have to use a temporal representation in the context of online expert systems [Schecke et al., 1991; Mora et al., 1993; Rutledge et al., 1993; Lehmann et al., 1994, Becker et al., 1997]. Generalized representations of temporal knowledge that enable reasoning with abstractions of event sequences linked to ontologies of problem solving have been developed for a wide variety of medical problems where the time course of an illness is critical [Haimowitz and Kohane, 1995, Long, 1996, Keravnou, 1996, Shahar and Mussen, 1996].

With more sophisticated methods for learning in artificial neural networks (ANNs), the last decade has seen a concomitant increase in their application to the modeling and learning of medical decisions [Barreto and DeAzevedo, 1993; Cherkassy and Lari-Najafi, 1992; Cho and Reggia, 1993; Forsstrom et al., 1991a; Hudson et al., 1991; Sittig and Orr, 1992, Dorfnerr and Porenta, 1994, Stevens et al., 1996]. Logic-based methods have also seen some popularity in structuring and building knowledge bases in medicine, particularly in Europe and Japan [Fox et al., 1990; Krause et al., 1993; Lucas, 1993]. Fuzzy logic methods have also shown their usefulness in a number of monitoring and image interpretation tasks [Becker et al., 1997, Park et al., 1998].

The evaluation of expert systems and decision support systems remains an important ongoing research and practical problem [Miller, 1986; Potoff et al., 1988; Willems et al., 1991; Bankowitz et al., 1992; Nohr, 1994] to which a variety of machine learning techniques can be applied for the performance assessment [Weiss and Kulikowski, 1991], knowledge base refinement [Widmer et al., 1993], and the retrospective analysis of results [Forsstrom et al., 1991b]

181.6 Prospects for AI in Medicine—Problems and Challenges

Usefulness of AI Decision Support Systems

As we approach the end of the 1990s, methodological developments in both the representation of knowledge and inferencing methods in medicine, however impressive in terms of their improvements over those of 20 or 30 years ago, still have not resulted in systems that are routinely used in the clinic. Reasons given for this usually include limitations in scope or ability to customize advice, lack of trust in automated advising systems, inadequate timeliness in critical situations, perception of the computer as a competitor, and most importantly, that decision support systems have yet to become indispensable adjuncts in medical practice situations. Problems and opportunities for medical AI in the 1990s have been identified and discussed with varying degrees of optimism or pessimism, depending on the author [Shortliffe, 1993; Uckun, 1993; Coeira, 1996; Kulikowski, 1996]. It has been suggested that the emphasis on decision support is misguided, and should be redirected towards medical education which would empower health care workers [Lillehaug and Lajoie, 1997]. This would tap into the developing theories of clinical cognition and the use of clinical simulations to train and enhance learning at various levels of expertise [Evans and Patel, 1989; Patel et al., 1995]. Some support for such a position comes from a comparative study of four of the most widely disseminated diagnostic decision support systems [Berner et al., 1994], which resulted in an editorial comment to the effect that their overall rating left much to be desired—a grade of C [Kassirer, 1994]. However, the systems reviewed were general diagnostic aids, which most physicians find less than indispensible given their own training and expertise. Another comparative study [Johnston et al., 1994] reported positive contributions to health care in three out of ten decision support systems reviewed. Most of the shortcomings came in systems that did not have to perform complex modeling or numerical computations, where computer-based assistance is often indispensable. Fields such as specialized laboratory test interpretation, treatment dosage planning and review, instrumentation monitoring and control, and multimodality imaging all suggest more essential domains for the application of AI methods [Adlassnig and Horak, 1995; Barahona, 1994; DeMelo et al., 1992; Kaye et al., 1995; Kulikowski et al., 1995; Menzies and Compton, 1997, Mora et al., 1993; Siregar and Sinteff, 1996, Taylor et al., 1997].

Opportunities: Knowledge-Bases, Systems Integration and the Web

The great computing success story of the 1990s has been the World Wide Web, and the proliferation of web-based information services. The ease of designing good graphical user interfaces (GUIs) has resulted in a concomitant ease of interaction with our networked machines. E-mail has become ubiquitous, and various standard data interchange formats across platforms have been developed in medicine as in other fields. This presents AI in medical decision-making with a number of challenges and opportunities. The first is to characterize and build the knowledge-bases that can help improve medical decision-making; the second is to integrate them into useful, working, indispensable clinical systems; the third is to validate, standardize, and share medical knowledge as widely and economically as is possible. Development of standard medical vocabularies and nomenclatures; coding schemes, and the Unified Medical Language System (UMLS) [Humphreys et al., 1995], are all essential to the standarization of medical knowledge representation. The availability of large digital image datasets has resulted in the creation of systems and knowledge bases for handling spatial information in anatomy [Brinkley, 1995; Hohne et al., 1995; Rosse et al., 1998], and has been accelerated by the widely disseminated multimodality image data from the Visible Human Project [Ackerman, 1995].

The re-usability of software is greatly facililtated by the development of ontologies for different types of knowledge and problem solving—a classical type of AI research which has received considerable new impetus from the needs of rapid knowledge transfer and scalability of web-based systems. In medicine, the PROTEGE system has become a model for such software re-use experiments [Tu et al., 1995], while other systems based on ontologies have also been developed [Smith et al., 1995, Muller et al., 1996].

Software integration for multiple uses is increasingly spurred by initiatives such as IAIMS [Stead, 1997], and inclusion of useful decision support modules may evolve naturally as productive computing environments become more commonplace. One method for connecting clinical databases to knowledge-bases used in decision support that is gaining acceptance is the Arden syntax [Hripcsak et al., 1990]. Automatic extraction of clinical data from narrative reports is also progressing due to a combination of statistical and linguistic methods [Chute and Yang, 1995; Hripcsak et al., 1996]. Meanwhile, AI systems are being designed to provide patients with intelligent access to medical information [Buchanan et al., 1995].

The present situation, then, can be characterized as one of consolidation in integrating a variety of methods and techniques for different types of medical decision tasks while at the same time involving considerable reassessment and change in terms of expectations about the imminent application of decision-making systems within health care environments. On the one hand, the forthcoming increase in networking of various knowledge sources, digital libraries, and multimedia information opens up a whole new set of opportunities for filtering knowledge to the overwhelmed practitioner, whether in medicine, law, engineering, or other professions. On the other hand, the hard issues of understanding the role of human judgment and responsibility for clinical decisions and actions when using increasingly complex computerized instrumentation and systems presents an ongoing challenge to researchers in medicine, biomedical engineering and computer science, medical informatics, and all other fields essential to the solution of these problems.

References

Ackerman MJ. 1991. The Visible Human Project. *J. Biocommun.* 18(2):14.

Adlassnig KP, Kolarz G. 1982. *CADIAG-2: Computer-Assisted Medical Diagnosis Using Fuzzy Subsets. Approximate Reasoning in Decision Analysis,* pp. 219–247.

Adlassnig KP, Horak W. 1995. Development and retrospective evaluation of HEPAXPERT-1. A routinely used expert system for interpretive analysis of hepatitis A and B serologic findings. *Artif. Intell. Med.* 7(1):1.

Aikins J. 1979. Prototype and production rules: An approach to knowledge representation for hypothesis information. In: *Proc. 6th Int. Joint Conf. Artificial Intell.,* pp. 1–3. Tokyo, Japan.

Bankowitz RA, Lave JR, McNeil MA. 1992. A method for assessing the impact of a computer-based decision support system on health care outcomes. *Meth. Inform. Med.* 31:3.

Barahona P. 1994. A causal and temporal reasoning model and its use in drug therapy applications. *Artif. Intell. Med.* 6(1):1.

Barnet GO, Cimino JJ, Hupp JA, Hoffer EP. 1987. DXplain: An evolving diagnostic decision-support system. *JAMA* 258:67.

Barreto JM, DeAzevedo FM. 1993. Connectionist expert systems as medical decision aid. *Artif. Intell. Med.* 5(6):515.

Becker K., Thull B, Kasmacher-Leidinger H, et al., 1997. Design and validation of an intelligent patient monitoring and alarm system based on a fuzzy logic process model. *Artif. Intell. Med.* 11(1):33.

Ben-Bassat M, Carlson RW, Pun VK, et al., 1980. Pattern-based interactive diagnosis of multiple disorders: The MEDAS system. *IEEE Trans. Pat. Anal. Mach. Intel.* (PAMI) 2(2):148.

Berner ES, Webster GD, Shugerman AA, et al., 1994. Performance of four computer-based diagnostic systems. *N. Engl. J. Med.* 330(25):1792.

Blum R. 1982. Discovery confirmation and incorporation of casual relationships from a large time-oriented data base: The RX project. *Comp. Biomed. Res.* 15:164.

Breiman L, Friedman J, Olshen R, Stone C. 1984. *Classification and Regression Trees.* Monterey, CA, Wadsworth.

Brinkley JF, Bradley SW, Sundsten JW, Rosse C. 1997. The Digital Anatomist information system and its use in the generation and delivery of web-based anatomy atlases. *Comp. Biomed. Res.* 30:472.

Buchanan BG, Sutherland G, Feigenbaum EA. 1970. Rediscovering some problems of artificial intelligence in the context of organic chemistry. In: B Meltzer, D Michie (Eds.), *Machine Intelligence,* pp. 209-254. Edinburgh, Edinburgh University Press.

Buchanan BG, Shortliffe EH. 1984. *Rule-Based Expert Systems: The MYCIN Experiments in the Stanford Heuristic Programming Project.* Reading, MA, Addison-Wesley.

Buchanan BG, Moore JD, Forsythe DE, et al., 1995. An intelligent interactive system for delivering individualized information to patients. *Artif. Intell. Med.* 7(2) 117.

Bylander T, Weintraub M, Simon SR. 1993. QUAWDS: Diagnosis using different models for different subtasks. In: J-M David et al., (Eds.), *Second Generation Expert Systems,* pp. 110–130. Berlin, Springer-Verlag.

Chandrasekaran B, Gomez F, Mittal 5, Smith J. 1979. An approach to medical diagnosis based on conceptual schemes. In: *Proc. 6th Int. Joint Conf. Artif. Intell.*, pp. 134–142. Tokyo, Japan.

Chandrasekaran B. 1986. Generic tasks in knowledge-based reasoning. High-level building blocks for expert system design. *IEEE Expert Intelligent Systems and Their Applications* 1(3):23.

Chandrasekaran B, Johnson TR. 1993. Generic tasks and task structures: History, critique and new directions. In: JM David, et al., (Eds.), *Second Generation Expert Systems,* pp. 232–272. Berlin, Springer-Verlag.

Cherkassky V, Lari-Najafi H. 1992. Data representation for diagnostic neural networks. *IEEE Expert* 7(5):43.

Cho S, Reggia JA. 1993. Multiple disorder diagnosis with adaptive competitive neural networks. *Artif. Intell. Med.* 5(6):469.

Chute CG, and Yang Y. 1995. An overview of statistical methods for the classificatrion and retrieval of patient events. *Methods Med. Inf.* 34(1): 104.

Ciesielski V (Ed.). 1978. *Proceedings of the Fourth Annual AIM Workshop,* Rutgers University Technical Report.

Clancey WJ. 1979. Tutoring rules for guiding a case method dialogue. *Int. J. Man-Machine Stud.* 11:25.

Clancey WJ, Letsinger R. 1981. NEOMYCIN: Reconfiguring a rule-based expert system for application to teaching. In: *Proc. 7th Int. Joint Conf. Artif. Intell.*, pp. 829–836.

Clancey WJ. 1989. The knowledge level reinterpreted: Modeling how systems interact. *Mach. Learn.* 4:285.

Clarke K, O'Moore R, Smeets R, et al., 1994. A methodology for evaluation of knowledge-based systems in medicine. *Artif. Intell. Med.* 6(2):107.

Console L, Torasso P. 1991. On the co-operation between abductive and temporal reasoning in medical diagnosis. *Artif. Intell. Med.* 3(6):291.

Cooper G 1993. Probabilistic and decision-theoretic systems in medicine. *Artif. Intell. Med.* 5(4):289.

Cousins SB, Chen W, Frisse ME. 1993. A tutorial introduction to stochastic simulation algorithms for belief networks. *Artif. Intell. Med.* 5(4):315.

Cowell RG, Dawid AP, Hutchinson T, Spiegelhalter DJ. 1991. A Bayesian expert system for the analysis of an adverse drug reaction. *Artif. Intell. Med.* 3(5):257.

Das AK and Musen MA. 1994. A temporal query system for protocol-directed decision support. *Meth. Inf. Med.* 33:358.

Davis R. 1979. Interactive transfer of expertise: Acquisition of new inference rules. *Artif. Intell.* 12:121.

DeMelo AS, Grarner J, Bronzino JD. 1992. SPINEX: An expert system to recommend the safe dosage of spinal anesthesia. In: *Proc. 14th Ann. Int. Conf. IEEE Eng. in Med. and Biol. Soc.*, pp. 902–903. Paris, France.

Duda RO, Hart PE, Nilsson NJ. 1976. Subjective Bayesian methods for rule-based inference systems. In: *Proc. Natl. Comput. Conf.*, New York.

Elstein AS, Shulman LS, Spratka SA. 1978. *Medical Problem Solving: An Analysis of Clinical Reasoning.* Cambridge, MA, Harvard University Press.

Englemore RS, Terry A. 1979. Structure and function of the Crysalis system. In: *Proc. Sixth Int. Joint Conf. Artif. Intell.*, pp. 250-256.

Evans DA and Patel VL. (Eds.). 1989. *Cognitive Science in Medicine.* Cambridge, MA, MIT Press.

Fagan LM, Kunz JC, Feigenbaum EA, Osborn JJ. 1979. Representation of dynamic clinical knowledge: Measurement interpretation in the intensive care unit. In: *Proc. 6th Int. Joint Conf. Artif. Intell.*, pp. 250–256.

Farr BR, Shachter RD. 1992. Representation of preferences in decision-support systems. *Comp. Biomed. Res.* 25(4):324.

Feigenbaum EA. 1978. The art of artificial intelligence: Themes and case studies of knowledge engineering. In: *Proc. Nail Comput. Conf.*, p 221. New York.

Fieschi M, Joubert M, Fieschi D, et al., 1983. A production rule expert system for medical consultations. *Medinfo* 83:503.

Forsatrom J, Eldund P, Virtanen H, et al., 1991a. DIAGAID: A connectionist approach to determine the diagnostic value of clinical data. *Artif. Intell. Med.* 3(4): 193.

Forsstrom J, Nuutila P, lijala K. 1991b. Using the ID3 algorithm to find discrepant diagnoses from laboratory databases of thyroid patients. *Med. Decis. Mak.* 11(3): 171.

Fox J, Glowinski A, Gordon C, et al., 1990. Logic engineering for knowledge engineering: Design and implementation of the Oxford system of medicine. *Artif. Intell. Med.* 2(6):323.

Freiher G. 1979. The seeds of artificial intelligence: SUMEX-AIM. In: *Div. Res. Res.*, pp. 80-2071. NIH.

Gelfand IM, Rosenfeld BI, Shifrin A. 1987. Data Structuring in Medical Problems. Moscow, USSR Academy of Sciences.

Gelfand IM. 1989. *Two archetypes in the psychology of man.* Kyoto Prize Lecture.

Ginsberg A, Weiss SM, Polikakis P. 1985. Seek-2: A generalized approach to automatic knowledge base refinement. In: *Proc. 9th Int. Joint Conf. Artif. Intell.*

Ginsberg A. 1986. A metalinguistic approach to the construction of knowledge base refinement systems. *Proc. AAAI* 86:436.

Gordon J, Shortliffe EH. 1985. The Dempster-Shaffer theory of evidence in rule based expert systems. In: BG Buchanan, EH Shortliffe (Eds.), *Rule-Based Expert Systems,* pp. 272–292. Reading, MA, Addison Wesley.

Gorry GA. 1970. Modeling the diagnostic process. *J. Med. Ed.* 45:293.

Gorry GA, Silverman H, Pauker SG. 1978. Capturing clinical expertise: A computer program that considers clinical responses to digitalis. *Am. J. Med.* 64:452.

Haimowitz IJ, Le PP, and Kohane IS. 1995. Clinical monitoring using regression-based trend templates. *Artif. Intell. Med.* 7(6):473.

Hart PE, Duda RE. 1977. PROSPECTOR—A computer-based consultation system for mineral exploration. *SRI Tech. Rep.* 155.

Hayes-Roth F, Waterman D, Lenat D. 1983. *Building Expert Systems.* Reading, MA, Addison Wesley.

Hazen GB. 1992. Stochastic trees: A new technique for temporal medical decision modeling. *Med. Decision Making* 12(3):163.

Heckerman DE, Shortliffe EH. 1992. From certainty factors to belief networks. *Artif. Intell. Med.* 4(1):35.

Heckerman DE, Horvitz EJ, and Nathwani BN. 1992. Toward normative expert systems. Part I—The Pathfinder system. *Methods Inf. Med.* 31 (2): 90.

Hohne KH, Pflesser B, Pommert A, et al., 1995. A new representation of knowledge concerning human anatomy and function. *Nat. Med.* 1:506.

Horn W, Buchstaller W, Trappl R. 1980. Knowledge structure definition for an expert system in primary medical care. In: *Proc. 7th Intl. Joint Conf. Art. Intell.*, pp. 850–852.

Hripcsak G, Johnson SB, Clayton PD. 1993. Desperately seeking data: Knowledge base-database links. In: *Proc. 17th Ann. Symp. Comp. App. Med. Care*, 639.

Hripcsak G, Friedman C, Alderson PO, et al., 1995. Unlocking clinical data from narrative reports: A study of natural language processing. *Ann. Intern. Med.* 122:681.

Hudson DL, Cohen ME, Anderson ME 1991. Use of neural network techniques in a medical expert system. *Int. J. Intell. Syst.* 6:213.

Humphreys B, Lindberg DAB, Schoolman HM, and Barnett O. 1998. The Unified Medical Language System. An informatics research collaboration. *J. Amer. Med. Inform. Assoc.* 5(1): 1.

Hunter J, Kirby I. 1991. Using quantitative and qualitative constraints in models of cardiac electrophysiology. *Artif. Intell. Med.* 3:41.

Johnston ME, Langton KB, Haynes RB, Mathieu A. 1994. Effects of computer-based clinical decision-support systems on clinician performance and patient outcome. A critical appraisal of research. *Ann. Intern. Med.* 120: 135.

Jimison HB, Fagan LM, Shachter RD, Shortliffe EH. 1992. Patient-specific explanation in models of chronic disease. *Artif. Intell. Med.* 4(3): 191.

Kahn MG. 1991. Modeling time in medical decision-support programs. *Med. Decision Making* 11(4):249.

Kahn MG, Fagan LM, Sheiner LB. 1989. Model-based interpretation of time-varying medical data. In: *Proc. 13th Ann. Symp. Comp. Appl. Med. Care*, pp. 28–32. Washington, IEEE Computer Society Press.

Kassirer IP. 1994. A report card on computer-assisted diagnosis—The grade: C. *N. Engl. J. Med.* 330(25): 1824.

Kassirer JP, Gorry GA. 1978. Clinical problem solving: A behavioral analysis. *Ann. Intern. Med.* 89:245.

Kastner JK, Dawson CR, Weiss SM. et al., 1984. An expert consultation system for frontline health workers in primary eye care. *J. Med. Syst.* 8:389.

Kaye J, Primiano FP and Metaxas D. 1995. Anatomical and physiological simulation for respiratory mechanics. *J. Imig. Guid. Surg.* 1(3): 164.

Keravnou ET. 1996. Temporal diagnostic reasoning based on time-objects. *Artif. Intell. Med.* 8(3):235.

Kim JH, Pearl J. 1987. CONVINCE: A conversational inference consolidation engine. *IEEE Trans. Syst. Man. Cybernet.* 17:120.

Kingsland L, Sharp G, Capps R, et al., 1983. Testing of a criteria-based consultant system in rheumatology. *Medinfo.* p. 514.

Kleinmuntz B, McLean RS. 1968. Diagnostic interviewing by digital computer. *Behav. Sci.* 13:75.

Kleinmuntz B. 1991. Computers as clinicians: An update. *Comput. Biol. Med.* 22(4):227.

Koton PA. 1993. Combining causal models and case-based reasoning. In: J-M David et al., (Eds.), *Second Generation Expert Systems*, pp. 69-78. Berlin, Springer-Verlag.

Krause P, Fox J, O'Neil M, Glowinski A. 1993. Can we formally specify a medical decision support system? *IEEE Expert* 8(3):56.

Kuipers BJ. 1985. The limits of qualitative simulation. In: M Kaufmann (Ed.), *Proc. 9th Int. Joint Conf. Artif. Intell.*, pp. 128-136. Los Altos, Calif.

Kuipers BJ. 1987. Qualitative simulation as causal explanation. *IEEE Trans. Syst. Man. Cybernet.* 17:432.

Kulikowski CA, Weiss 5. 1972. Strategies of database utilization in sequential pattern recognition. *Proc. IEEE Conf. Decision Control*, 103.

Kulikowski CA. 1980. Artificial intelligence methods and systems for medical consultation. *IEEE Trans. Pat. Anal. Mach. Intel.* (PAMI) 2(5):464.

Kulikowski CA, Ostroff J. 1980. Constructing an expert knowledge base for thyroid disease using generalized AI techniques. In: *Proc. 4th Ann. Symp. Comp. Health Care*, pp. 175–180.

Kulikowski CA, Gong L, and Mezrich RS. 1995. Knowledge-based medical image analysis for integrating context definition with the radiological report. *Methods Med. Inf.* 34(1): 96.

Kulikowski CA. 1996. AIM. Quo Vadis? *J. Amer. Med. Inform. Assoc.* 3(6):432.

Lanzola G, Quaglini S, Stefanelli M. 1995. Knowledge-acquisition tools for medical knowledge-based systems. *Methods Inf. Med.* 34(1): 25.

Lau LM, Warner HR. 1992. Performance of a diagnostic system (Iliad) as a tool for quality assurance. *Comp. Biomed. Res.* 25(4):314.

Lauritzen SL, and Spiegenhalter DJ. 1987. Local computations with probabilities and their applications to expert systems. *J R Stat. Soc.* (Series B) 50: 157.

Lehmann ED, Deutsch T, Carson ER, Sonksen PH. 1994. Combining rule-based reasoning and mathematical modelling in diabetes care. *Artif. Intell. Med.* 6(2):137.

Lehmann HP, Shortliffe EH. 1990. Thomas: Building Bayesian statistical expert systems to aid in clinical decision making. In: *Proc. 14th Symp. Applic. Med. Care*, pp. 58–64. New York, IEEE Press.

Lichter P, Anderson D. 1977. *Discussions on Glaucoma.* New York, Grune and Stratton.

Lillehaug SI, and Lajoie SP. 1998. AI in medical education—another grand challenge for medical informatics. *Artif. Intell. Med.* 12:197.

Lindberg DA. 1980. Computer based rheumatology consultant. *Medinfo.* 80:1311.

Lindberg DA, Sharp GC, Kay DR, et al., 1983. The expert consultant as teacher. *Moebius* 3:30.

Littenberg B, Moses LE. 1993. Estimating diagnostic accuracy from multiple conflicting reports. *Med. Decision Making.* 13(4):313.

Long WJ. 1987. The development and use of a causal model for reasoning about heart failure. In: *Proc. 11th Symp. Comp. Applic. Med. Care*, pp. 30-36.

Long WJ. 1991. Flexible reasoning about patient management using multiple methods. *Artif. Intell. Med.* 3:3.

Long WJ, Fraser H, Naimi S. 1997. Reasoning requirements for diagnosis of heart disease. *Artif. Intell. Med.* 10(1):5.

Lucas PIF. 1993. The representation of medical reasoning models in resolution-based theorem provers. *Artif. Intell. Med.* 5(5):395.

Menzies T, and Compton P. 1997. Applications of abduction. Hypothesis testing of neuroendocrinological qualitative compartmental models. *Artif. Intell. Med.* 10:145.

Mora FA, Passariello G, Carrault G, LePichon JP. 1993. Intelligent patient monitoring and management systems: A review. *IEEE Eng. Med. Biol.* 12:23.

Midgette AS, Stuiel TA, Littenberg B. 1993. A meta-analytic method for summarizing diagnostic test performances. *Med. Decis. Mak.* 13(3):253.

Miller PL. 1983. Attending: Critiquing a physician's management plan. *IEEE Trans. Pat. Anal. Mach. Intel.* (PAMI) 5:449.

Miller PL. 1986. The evaluation of artificial intelligence systems in medicine. *Comput. Meth. Prog. Biomed.* 22:5.

Miller PL. 1998. Tools for immunization guideline knowledge maintenance I. Automated generation of the logic "kernel" for immunization forecasting. *Comp. Biomed. Res.* 31(3): 172.

Miller RA, Pople HE, Myers JD. 1982. An experimental computer based diagnostic consultant for general internal medicine. *N. Engl. J. Med.* 307:468.

Miller RA. 1984. Internist-1/Caduceus: Problems facing expert consultant programs. *Meth. Inform. Med.* 23:9.

Miller RA. 1994. Medical diagnostic decision support systems: Past, present, and future: A threaded bibliography and commentary. *J. Am. Med. Info. Assoc.* 1(1):8.

Minsky N. 1975. A framework for representing knowledge. In: P Winston (Ed.), *The Psychology of Computer Vision.* New York, McGraw-Hill.

Muller R, Thews O, Rohrbach C, et al., 1996. A graph-grammar approach to represent causal, temporal, and other contexts in an oncological patient record. *Methods Inf. Med.* 35:137.

Musen MA, Fagen LM, Combs DM, Shortliffe EH. 1986. Facilitating knowledge entry for an oncology therapy advisor using a model of the application area. *Medinfo* 86:46.

Neapolitan RE. 1993. Computing the confidence in a medical decision obtained from an influence diagram. *Artif. Intell. Med.* 5(4):341.

Newell A. 1982. The knowledge level. *Artif. Intell.* 18:87.

Nii HP, Aiello N. 1979. AGE: A knowledge-based program for building knowledge-based programs. In: *Proc. 6th Int. Joint Conf. Artif. Intell.*, pp. 645–655. Tokyo, Japan.

Nohr C. 1994. The evaluation of expert diagnostic systems—How to assess outcomes and quality parameters? *Artif. Intell. Med.* 6(2):123.

Olesen KG, Andreassen S. 1993. Specification of models in large expert systems based on causal probabilistic networks. *Artif. Intell. Med.* 5(3):269.

Park W, Hoffman EA, Sonka M. Segmentation of intrathoracic airway trees: A fuzzy logic approach. *IEEE Trans. Med. Imag.* 17(4):489.

Patel VL, Kaufman DR, and Arocha JF 1995. Steering through the murky waters of a scientific conflict. Situated and symbolic models of clinical cognition. *Artif. Intell. Med.* 7(5):413.

Patil, RS, Szolovitz, P, Schwartz WB. 1981. Causal understanding of patient illness in medical diagnosis. In: *Proc. 7th Int. Joint Conf. Artif. Intell.*, pp. 893–899.

Patil R. 1983. Role of causal relations in formulation and evaluation of composite hypotheses. *IEEE Med. Comp.*

Patil R. 1986. Review of causal reasoning in medical diagnosis. In: *Proc. 10th SCAMC*, pp. 11–16.

Pauker SG, Gorry GA, Kassirer JP, Schwartz WB. 1976. Towards the simulation of clinical cognition: Taking a present illness by computer. *Am. J. Med.* 60:981.

Pearl J. 1987. *Probabilistic Reasoning in Intelligent Systems: Networks or Plausible Inference.* San Mateo, CA, Morgan Kaufmann.

Pedrycz W, Bortolan G, Degani R. 1991. Classification of electrocardiographic signals: A fuzzy pattern matching approach. *Artif. Intell. Med.* 3(4):21 1.

Peng Y, Reggia JA. 1987. A probabilistic causal model for diagnostic problem solving: Integrating symbolic causal inference with numeric probabilistic inference. *IEEE Trans. Syst. Man. Cybernet.* 17:146.

Politakis P, Weiss SM. 1980. A system for empirical experimentation with expert knowledge. In: *Proc. 15th HICCS*, pp. 675–683.

Politakis P, Weiss SM. 1984. Using empirical analysis to refine expert system knowledge bases. *Artif. Intell.* 22:23.

Poon AD, Fagan LM, Shortliffe EH. 1996. The PEN-Ivory project. Exploring user-interface design for selection of items from large controlled vocabularies of medicine. *J. Am. Med. Inform. Assoc.* 3:168.

Pople HE. 1973. On the mechanization of abductive logic. In: *Proc. Int. Joint Conf. Artif. Intell.*, pp. 147–182.

Pople H, Myers J, Miller R. 1975. DIALOG: A model of diagnostic logic for internal medicine. In: *Proc. 4th Int. Joint Conf. Artif. Intell.*, pp. 848–855.

Pople H. 1977. The formation of composite hypotheses in diagnostic problem solving. In: *Proc. 5th Int. Joint Conf. Artif. Intell.*, pp. 1030-1037.

Pople H. 1982. Heuristic methods for imposing structure on ill-structured problems: The structuring of medical diagnoses. In: P Szolovits (Ed.), *Artificial Intelligence in Medicine.* Boulder, CO, Westview Press.

Potthoff P, Rothemund M, Schwefel D, et al., 1988. *Expert Systems in Medicine.* Cambridge, England, Cambridge University Press.

Ramoni M, Stefanelli M, Magnani L, Barosi G. 1992. An epistemological framework for medical knowledge-based systems. *IEEE Tran. Sys. Man. Cybernet.* 22(6):1361.

Reggia JA, Pula TP, Price TR, Perricone BT. 1980. Towards an intelligent textbook of neurology. In: *Proc. 4th Ann. Symp. Comp. Appl. Med. Care*, pp. 190–199. Washington.

Reggia JA, Nau DS, Wang PY. 1983. Diagnostic expert systems based on a set covering model. *Int. J. Man-Machine Stud.* 19:437.

Rosse C, Mehino, JL, Modayur BR et al., 1998. Motivation and organizational principles for anatomical knowledge representaiton: The Digital Anatomist symbolic knowledge base. *J. Amer. Med. Inf. Assoc.* 5(1):17.

Rumelhart DE, McClelland JL. 1986. *Parallel Distributed Processing: Explorations in the Microstructure of Cognition.* Cambridge, MA, MIT Press.

Rutledge GW, Thomsen GE, Farr BR, et al., 1993. The design and implementation of a ventilator management advisor. *Artif. Intell. Med.* 5(1):67.

Schecke TH, Rau G, Popp HI, et al., 1991. A knowledge-based approach to intelligent alarms in anesthesia. *IEEE Eng. Med. Biol.* 10:38.

Schwartz WB. 1970. Medicine and the computer: The promise and problems of change. *N. Engl. Med.* 283:1257.

Shachter RD. 1986. DAVID: Influence diagram processing system for the Macintosh. In: *Proc. Workshop on Uncertainty in Artificial Intelligence*, AAAI, Philadelphia, PA, pp. 311–318.

Shahar Y, and Mussen MA. 1993. RESUME. A temporal-abstraction system for patient monitoring. *Comp. Biomed. Res.* 26:255.

Shapiro AR. 1977. The evaluation of clinical predictions: A method and initial application. *N. Engl. Med.* 296:1509.

Shortliffe EH, Axline SG, Buchanan BG, Cohen SN. 1974. An artificial intelligence program to advise physicians regarding antimicrobial therapy. *Comput. Biomed. Res.* 6:544.

Shortliffe EH. 1976. *Computer-Based Medical Consultation: MYCIN.* New York, Elsevier.

Shortliffe EH, Scott AC, Bitchoff MB, et al., 1981. An expert system for oncology protocol management. In: *Proc. 7th Int. Joint Conf. Artif. Intell.*, pp. 876–881.

Shortliffe EH. 1993. The adolescence of AI in medicine: will the field come of age in the 1990s? *Artif. Intell. Med.* 5:93.

Sirengar P, and Sinteff J-P. 1996. Introducing spatio-temporal reasoning into the inverse problem in electroencephalography. *Artif. Intell. Med.* 8(2):97.

Sittig DF, Orr IA. 1992, A parallel implementation of the backward error propagation neural network training algorithm: Experiments in event identification. *Comput. Biomed. Ret.* 25(6):547.

Smith JW, Svirbely J, Evans C, et al., 1985. RED: A red-cell antibody identification expert module. *J. Med. Syst.* 9(3):121.

Smith JW, Johnson TR. 1993. A stratified approach to specifying, designing, and building knowledge systems. *IEEE Expert* 8(3).

Smith JW, Bayazitoglu A, Johnson TR, et al., 1995. One framework, two systems. Flexible abductive methods in the problem-space paradigm applied to antibody identification and biopsy interpretation. *Artif. Intell. Med.* 7(3):201.

Sonnenberg FA, Hagerty CG, Kulikowski CA. 1994. An architecture for knowledge based construction of decision models. *Med. Decis. Mak.* 13(30):27.

Speedie SM, Palumbo FB, Knapp DA, Beardsley R. 1981. Rule-based drug prescribing review: An operational system. In: *Proc. 5th Ann. Symp. Comp. Appl. Med. Care*, pp. 598–602. Washington.

Stead WW. 1997. The evolution of the IAIMS: Lessons from the next decade. *J. Amer. Med. Inform. Assoc.* 4(2):S4

Stevens RH, Lopo AC, Wang P. 1996. Artificial neural networks can distinguish novice and expert strategies during complex problem solving. *J. Am. Med. Inf. Assoc.* 3(2): 131.

Summers R, Carson ER, Cramp DG. 1993. Ventilator management. *IEEE Eng. Med. Biol.* 12:50.

Swartout WR. 1981. Explaining and justifying expert consulting programs. In: *Proc. 7th Int. Joint Conf. Artif. lntell.*, pp. 815–822.

Swartout WR, Moore ID. 1993. Explanation in Second Generation Expert Systems. In: J-M David et al. (Eds.), *Second Generation Expert Systems,* pp. 543–615. Berlin, Springer-Verlag.

Szolovits P, Pauker SG. 1978. Categorical and probabilistic reasoning in medical diagnosis. *Artif. Intell.* 11:115.

Taylor P, Fox J, and Todd-Pokropek A. 1997. A model for integrating image processing into decision aids for diagnostic radiology. *Artif. Intell. Med.* 9(3):205.

Tu SW, Eriksson H, Gennari JH et al., 1995. Ontology-based configuration of problem-solving methods and generation of knowledge-acquisition tools. Application of PROTEGE II to protocol-based decision support. *Artif. Intell. Med.* 7(3): 257.

Uckun S, Dawant BM, Lindstrom DP. 1993. Model-based diagnosis in intensive care monitoring: The YAQ approach. *Artif. Intell. Med.* 5(1)31.

Uckun S. 1993. Artificial intelligence in medicine: state of the art and future prospects. *Artif. Intell. Med.* 5:89.

VanBeck P. 1991. Temporal query processing with indefinite information. *Artif. Intell. Med.* 3(6)325.

van Melle W. 1979. A domain independent production-rule system for consultation programs. In: *Proc. 6th Int. Joint Conf. Artif. Intell.*, pp. 923–925, Tokyo, Japan.

Wagner MM, Pankaskie M, Hogan W, et al., 1997. Clinical event monitoring at the University of Pittsburgh. *Proc. AMIA Fall Symp.* 188.

Weiss SM. 1974. *A system for model-based computer-aided diagnosis and therapy.* Thesis, Rutgers University.

Weiss S, Kulikowski C, Amarel S, Safir A. 1978. A model-based method for computer-aided medical decision-making. *Artif. Intell.* 11:145.

Weiss S, Kulikowski C. 1979. EXPERT: A system for developing consultation models. In: *Proc. 6th Int. Joint Conf. Artif. Intell.*, pp. 942–950. Tokyo, Japan.

Weiss SM, Kulikowski CA, Galen RS. 1983. Representing expertise in a computer program: The serum protein diagnostic program. *J. Clin. Lab. Automation* 3:383.

Weiss SM, Kulikowski CA. 1984. *A Practical Guide to Designing Expert Systems.* Totowa, Rowman and Allenheld.

Weiss SM, Kulikowaki CA. 1991. *Computer Systems That Learn.* San Mateo, Calif., Morgan Kaufniano.

Widman LE. 1992. A model-based approach to the diagnosis of the cardiac arrhythmias. *Artif. Intell. Med.* 4(l):1.

Widmer G, Horn W, Nagele B. 1993. Automatic knowledge base refinement: Learning from examples and deep knowledge in rheumatology. *Artif. Intell. Med.* 5(3):225.

Wielinga BJ, Schreiber AT, Breuker JA. 1992. KADS: A modelling approach to knowledge engineering. *Knowledge Acquisition* 4:5.

Willems JL, Abreu-Lima C, VanBemmel AP, et al., 1991. The diagnostic performance of computer programs for the interpretation of electrocardiograms. *N. Engl. J. Med.* 325:1767.

Yu VL, Buchanan BG, Shortliffe EH, et al., 1979. An evaluation of the performance of a computer-based consultant. *Comput. Progr. Biomed.* 9:95.

182

Artificial Neural Networks: Definitions, Methods, Applications

Daniel A. Zahner
Rutgers University

Evangelia Micheli-Tzanakou
Rutgers University

The potential of achieving a great deal of processing power by wiring together a large number of very simple and somewhat primitive devices has captured the imagination of scientists and engineers for many years. In recent years, the possibility of implementing such systems by means of electro-optical devices and in very large scale integrations has resulted in increased research activities.

Artificial neural networks (ANNs) or simply neural networks (NNs) are made of interconnected devices called *neurons* (also called *neurodes, nodes, neural units,* or simply *units*). Loosely inspired by the makeup of the nervous system, these interconnected devices look at patterns of data and learn to classify them. NNs have been used in a wide variety of signal-processing and pattern-recognition applications and have been applied successfully in such diverse fields as speech processing [1–4], handwritten character recognition [5–7], time-series prediction [8–9], data compression [10], feature extraction [11], and pattern recognition in general [12]. Their attractiveness lies in the relative simplicity with which the networks can be designed for a specific problem along with their ability to perform nonlinear data processing.

As the neuron is the building block of a brain, a neural unit is the building block of a neural network. Although the two are far from being the same, or from performing the same functions, they still possess similarities that are remarkably important. NNs consist of a large number of interconnected units that give them the ability to process information in a highly parallel way. The brain as well is a massively parallel machine, as it has been long recognized. As each of the 10^{11} neurons of the human brain integrates incoming information from all other neurons directly or indirectly connected to it, an artificial neuron sums all inputs to it an creates an output that is carrying information to other neurons. The connection from one neuron's dendrites or cell body to another neuron's processes is called a *synapse.* The strength by which two neurons are influencing each other is called a *synaptic weight.* In an NN, all neurons are connected to all other neurons by synaptic weights that can have seemingly arbitrary values, but in reality, these weights show the effect of a stimulus on the neural network and the ability or lack of it to recognize that stimulus.

0-8493-0462-8/00/$0.00+$.50

In the biological brain, two types of processes exist: static and dynamic. *Static* brain conditions are those which do not involve any memory processing, while *dynamic* processes involve memory processing and changes through time. Similarly, NNs can be distinguished into static and dynamic: the former being such that do not involve any previous memory and only depend on current inputs and the latter having memory and being able to be described by differential equations that express changes in the dynamics of the system through time.

All NNs have certain architectures, and all consist of several layers of neuronal arrangements. The most widely used architecture is that of the perceptron first described in 1958 by Rosenblatt [13]. In the sections that follow we will build on this architecture, but not necessarily on the original assumptions of Rosenblatt.

Since there are many names in the literature that express the same thing and usually create a lot of confusion for the reader, we will define the terms to be used and use them throughout the chapter. Terminology is a big concern for those involved in the field and for organizations such as IEEE. A standards committee has been formed to address issues such as nomenclature and paradigms. In this chapter, whenever possible, we will try to conform to the terms and definitions already in existence.

Some methods for training and testing of NNs will be described in detail, although many others will be left out due to lack of space, but references will be provided for the interested reader. A small number of applications will be given as examples, since many more are discussed in another chapter of this *Handbook* and it would be redundant to repeat them here.

182.1 Definitions

Neural nets (NNs) go by many other names, such as *connectionist models, neuromorphic systems,* and *parallel distributed systems,* as well as *artificial NNs* to distinguish them from the biologic ones. They contain many densely interconnected elements called *neurons* or *nodes,* which are nothing more than computational elements nonlinear in nature. A single node acts like an integrator of its weighted inputs. Once the result is found, it is passed to other nodes via connections that are called *synapses.* Each node is characterized by a parameter that is call *threshold* or *offset* and by the kind of nonlinearity through which the sum of all the inputs is passed. Typical nonlinearities are the hardlimiter, the ramp (threshold logic element), and the widely used sigmoid.

The simplest NN is the single-layer perceptron [13, 14], and it is a simple net that can decide whether an input belongs to one of two possible classes. Figure 182.1 is a schematic representation of a perceptron the output of which is passed through a nonlinearity called an *activation function.* This activation function is of different types, the most popular being a sigmoidal logistic function.

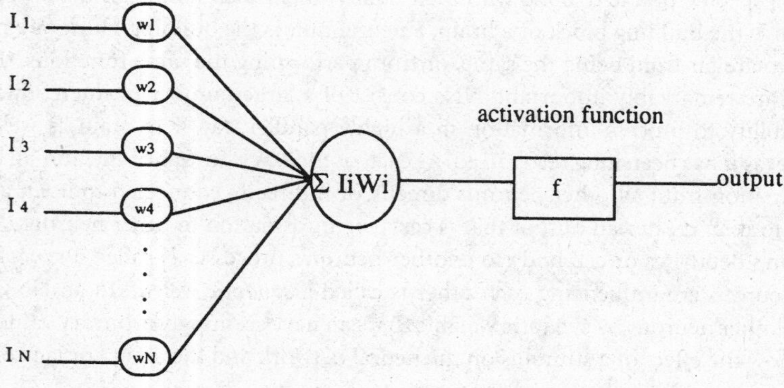

FIGURE 182.1 Artificial neuron.

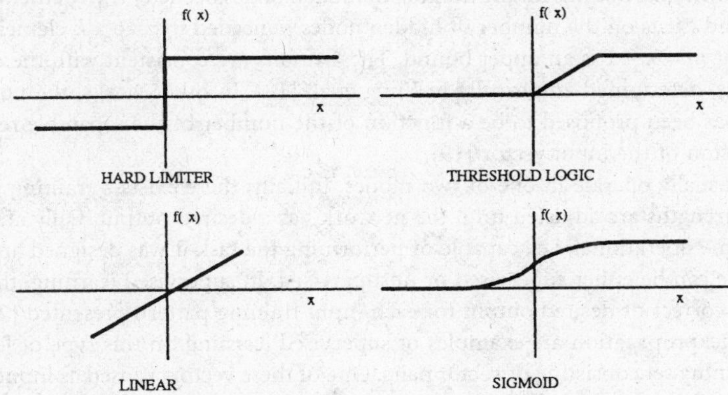

FIGURE 182.2 Typical activation functions.

Figure 182.2 is a schematic representation of some activation functions such as the handlimiter (or step), the threshold logic (or ramp), a linear, and a sigmoid. The neuron of Fig. 182.1 receives many inputs I_i each weighted by a weight W_i ($i = 1,2,\ldots,$ N). These inputs are then summed. The sum is then passed through the activation function *f*, and an output *y* is calculated only if a certain threshold is exceeded. Complex artificial neurons may include temporal dependencies and more complex mathematical operations than summation [15]. While each node has a simple function, their combined behavior becomes remarkably complex when organized in a highly parallel manner.

NNs are specified by their processing element characteristics, the network topology, and the training or learning rules they follow in order to adapt the weights W_i. Network topology falls into two broad classes: feedforward (nonrecursive) and feedback (recursive) [16]. Nonrecursive NNs offer the advantage of simplicity of implementation and analysis. For static mappings, a nonrecursive network is all one needs to specify any static condition. Adding feedback expands the network's range of behavior, since now its output depends on both the current input and network states. But one has to pay a price: longer times for teaching the NN to recognize its inputs.

Obviously, the NN of Fig. 182.1 is quite simple and inadequate is solving real problems. A multilayer perceptron (MLP) is the next choice. A number of inputs are now connected to a number of nodes at a second layer called the *hidden layer.* The outputs of the second layer may connect to a third layer and so on, until they connect to the output layer. In this representation, every input is connected to every node in the next layer, and the outputs of one hidden layer are connected to the nodes of the next hidden layer, and so on. Figure 182.3 depicts a three-layer feedforward network, the simplest MLP.

The optimal number of hidden neurons needed to perform arbitrary mapping is yet to be determined. Methods used in practice are mainly intuitive determination or are found by trial and error. Recent work

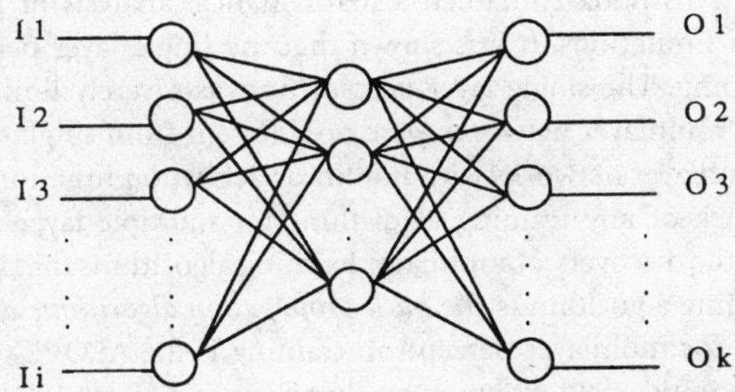

FIGURE 182.3 Three-layer feedforward network.

has attempted to formulate bounds for the number of hidden nodes needed [17]. Mathematical derivation proves that a bound exists on the number of hidden nodes m needed to map a k-element input set. The formulation is that $m = k - 1$ is an upper bound. These results are consistent with the optimal number of hidden neurons, determined empirically in Kung et al. [18]. In other works, the number of hidden nodes necessary has been proposed to be a function of the number of the separable regions needed as well as the dimension of the input vector [19].

Artificial NNs usually operate in one of two modes. Initially, there exists a training phase where the interconnection strengths are adjusted until the network has a desired output. Only after training does the network become operational, i.e., capable of performing the task it was designed and trained to do. The training phase can be either supervised or unsupervised. In supervised learning, there exists information about the correct or desired output for each input training pattern presented [20]. The original perceptron and backpropagation are examples of supervised learning. In this type of learning, the NN is trained on a training set consisting of vector pairs. One of these vectors is used as input to the network; the other is used as the desired or target output. During training, the weights of the NN are adjusted in such a way as to minimize the error between the target and the computed output of the network. This process might take a large number of iterations to converge, especially because some training algorithms (such as backpropagation) might converge to local minima instead of the global one. If the training process is successful, the network is capable of performing the desired mapping.

In unsupervised learning, no a priori information exists, and training is based only on the properties of the patterns. Sometimes it is also called *self-organization* [20]. Training depends on statistical regularities that the network extracts from the training set and represents them as weight values. Applications of unsupervised learning have been limited. However, hybrid systems of unsupervised learning combined with other techniques produce useful results [21–23]. Unsupervised learning is highly dependent on the training data, and information about the proper classification is often lacking [21]. For this reason, most NN training is supervised.

182.2 Training Algorithms

After McCulloch and Pitts [24] demosntrated, in 1943, the computational power of neuron-like networks, much effort was given to developing networks that could learn. In 1949, Donald Hebb proposed the strengthening of connections between presynaptic and postsynaptic units when both were active simultaneously [25]. This idea of modifying the connection weights, as a method of learning, is present in most learning models used today. The next major advancement in neural networks was by Frank Rosenblatt [13, 14]. In 1960, Windrow and Hoff proposed a model, called the *adaptive linear element* (ADALINE), which learns by modifying variable connection strengths, minimizing the square of the error in successive iterations [26]. This error-connection scheme is now known as the *least mean square (LMS) algorithm*, and it has found widespread use in digital signal processing.

There was great interest in NN computation until Minsky and Papert published a book in 1969 criticizing the perceptron. This book contained a mathematical analysis of perceptron-like networks, pointing out many of their limitations. It was shown that the single-layer perceptron was incapable of performing the XOR mapping. The single-layer perceptron was severely limited in its capabilities. For linear activation functions, multilayer networks were no different from single-layer models. Minsky and Papert pointed out that multilayer networks with nonlinear activation functions could perform complex mappings. However, the lack of any training algorithms for multiple-layer networks made their use impossible. It was not until the discovery of multilayer learning algorithms that interest in NNs resurfaced. The most widely used training algorithm is the *back-propagation algorithm*, as already mentioned.

Another algorithm used for multilayer perceptron training is the ALOPEX algorithm. ALOPEX was originally used for visual receptive field mapping by Tzanakou and Harth in 1973 [28–30] and has since been applied to a wide variety of optimization problems. These two algorithms are described in detail in the next subsections.

Backpropagation Algorithm

The backpropagation algorithm is a learning scheme where the error is backpropagated layer by layer and used to update the weights. The algorithm is a gradient-descent method that minimizes the error between the desired outputs and the actual outputs calculated by the MLP. Let

$$E_p = \frac{1}{2}\sum_{i=1}^{N}\left(T_i - Y_i\right)^2 \tag{182.1}$$

be the error associated with template p. N is the number of output neurons in the MLP, T_i is the target or desired output for neuron i, and Y_i is the output of neuron i calculated by the MLP. Let $E = \Sigma E_p$ be the total measure of error. The gradient-descent method updates an arbitrary weight w in the network by the following rule:

$$w(n+1) = w(n) + \Delta w(n) \tag{182.2}$$

where

$$\Delta w(n) \propto -\eta \frac{\partial E}{\partial w(n)} \tag{182.3}$$

where n denotes the iteration number, and η is a scaling constant. Thus the gradient-descent method requires the calculation of the derivatives $\partial E/[\partial w(n)]$ for each weight w in the network. For an arbitrary hidden-layer neuron, its output H_j is a nonlinear function f of the weighted sum of all its inputs (net_j):

$$H_j = f(net_j) \tag{182.4}$$

where f is the activation function. The most commonly used activation function is the sigmoid function given by

$$f(x) = \frac{1}{1+e^{-x}}\,. \tag{182.5}$$

Using the chain rule, we can write

$$\frac{\partial E}{\partial w_{ij}} = \frac{\partial E}{\partial net_j}\frac{\partial net_j}{\partial w_{ij}} \tag{182.6}$$

and since

$$net_j = \sum_{j=1}^{n} w_{ij} I_i \tag{182.7}$$

we have

$$\frac{\partial net_j}{\partial w_{ij}} = I_i \tag{182.8}$$

and thus Eq. (182.6) becomes

$$\frac{\partial E}{\partial w_{ij}} = \frac{\partial E}{\partial net_j} I_i \tag{182.9}$$

$$\frac{\partial E}{\partial net_j} = \sum_{k=1}^{m} \frac{\partial E}{\partial net_k} \frac{\partial net_k}{\partial H_j} \frac{\partial H_j}{\partial net_j}. \tag{182.10}$$

Recalling that

$$net_k = \sum_{j=1}^{n} w_{jk} H_j \tag{182.11}$$

it follows the

$$\frac{\partial net_k}{\partial H_j} = w_{jk}. \tag{182.12}$$

Also,

$$\frac{\partial H_j}{\partial net_j} = f'\left(net_j\right). \tag{182.13}$$

Therefore,

$$\frac{\partial E}{\partial net_j} = f'\left(net_j\right) \sum_{k=1}^{n} \frac{\partial E}{\partial net_k} w_{jk}. \tag{182.14}$$

Assuming f to be the sigmoid function of Eq. (182.5), then

$$f'\left(net_j\right) = Y_i\left(1 - Y_i\right). \tag{182.15}$$

Equation (182.14) gives the unique relation that allows the backpropagation of the error to all hidden layers. For the output layer,

$$\frac{\partial E}{\partial net_j} = \frac{\partial E}{\partial H_j} f'\left(net_j\right) \tag{182.16}$$

$$\frac{\partial E}{\partial H_j} = -\left(T_i - Y_i\right). \tag{182.17}$$

In summary, then, first, the output Y_i for all the neurons in the network is calculated. The error derivative needed for the gradient-descent update rule of Eq. (182.2) is calculated from

$$\frac{\partial E}{\partial w} = \frac{\partial E}{\partial net} \frac{\partial net}{\partial w}. \tag{182.18}$$

If j is an output neuron, then

$$\frac{\partial E}{\partial net_j} = -\left(T_i - Y_i\right)Y_i\left(1 - Y_i\right). \tag{182.19}$$

If j is a hidden neuron, then the error derivative is backpropagated by using Eqs. (182.14) and (182.15). Substituting, we get

$$\frac{\partial E}{\partial net_j} = Y_i\left(1 - Y_i\right)\sum_{k=1}^{m}\frac{\partial E}{\partial net_k} w_{jk}. \tag{182.20}$$

Finally, the weights are updated as in Eq. (182.2).

There are many modifications to the basic algorithm that have been proposed to speed the convergence of the system. *Convergence* is defined as a reduction in the overall error between a minimum threshold. It is the point at which the network is said to be fully trained. One method [31] used is the inclusion of a momentum term in the updated equation such that

$$w\left(n+1\right) = w\left(n\right) - \eta\frac{\partial E}{\partial w\left(n\right)} + \alpha\Delta w\left(n\right) \tag{182.21}$$

η is the learning rate and is taken to be 0.25. α is a constant momentum term that determines the effect of past weight changes on the direction of current weight movements.

Another approach used to speed the convergence of backpropagation is the introduction of random noise [32]. It has been shown that while inaccuracies resulting from digital quantization are detrimental to the algorithm's convergence, analog perturbations actually help improve convergence time.

One of these variations is the modification by Fahlman [33], called the *quickprop*, that uses second-derivative information without calculating the hessian needed in the straight backpropagation algorithm. It requires saving a copy of the previous gradient vector, as well as the previous weight change. Computation of the weight changes use only information associated with the weight being updated:

$$\Delta w = \frac{\dfrac{\nabla w_{ij}\left(n\right)}{\nabla w_{ij}\left(n-1\right) - \nabla w_{ij}\left(n\right)}}{\Delta w_{ij}\left(n-1\right)} \tag{182.22}$$

where $\nabla w_{ij}(n)$ is the gradient vector component associated with the weight w_{ij} at iteration n. This algorithm assumes that the error surface is parabolic, concave upward around the minimum, and that the slope change of the weight $\nabla w_{ij}(n)$ is independent of all other changes in weights. There are obviously problems with these assumptions, but Fahlman suggests a "maximum growth factor" μ in order to limit the rate of increase of the step size, namely, that if $\Delta w_{ij}(n) > \mu\Delta w_{ij}(n-1)$, then $\Delta w_{ij}(n) = \mu\Delta w_{ij}(n-1)$. Fahlman also used a hyperbolic arctangent function to the output error associated with each neuron in the output layer. This function is almost linear for small errors, but it blows up for large positive or large negative errors. Quickprop is an attempt to reduce the number of iterations needed by straight back-propagation, and it succeeded doing so by a factor of 5, but this factor is problem-dependent. This method also requires several trials before the parameters are set to acceptable values.

Backpropagation has achieved widespread use as a training algorithm for NNs. Its ability to train multilayer networks has led to resurgence of interest in the field. Backpropagation has been used successfully in applications as adaptive control of dynamic systems and in many general NN applications. Dynamic systems require monitoring of time in ways that monitor the past. In fact, the biological brain performs in an admirable way just because it has access to and uses values of different variables from previous instances. Backpropagation through time is another extension of the original algorithm proposed by Werbos in 1990 [34] and has been applied previously in the "truck backer-upper" by Nguyen and Widrow [35]. In this problem, a sequence of decisions must be made without an immediate indication of how effective these steps are. No indication of performance exists until the truck hits the wall. Backpropagation through time solves the problem, but it has its own inadequacies and performance difficulties. Despite its tremendous effect on NNs, the algorithm is not without its problems. Some of the problems have been discussed above. In addition, the complexity of the algorithm makes hardware implementations of it very difficult.

The ALOPEX Algorithm

The ALOPEX process is an optimization procedure that has been demonstrated successfully in a wide variety of applications. Originally developed for receptive field mapping in the visual pathway of frogs, ALOPEX's usefulness and flexible form have increased the scope of its applications to a wide range of optimization problems. Since its development by Tzanakou and Harth in 1973 [28], ALOPEX has been applied to real-time noise reduction [36], pattern recognition [37], adaptive control systems [38], and multilayer NN training, to name a few.

Optimization procedures, in general, attempt to maximize or minimize a function $F()$. The function $F()$ is called the *cost function*, and its value depends on many parameters or variables. When the number of parameters is large, finding the set (x_1 x_2 … x_N) that corresponds to the optimal (maximal or minimal) solution is exceedingly difficult. If N were small, then one could perform an exhaustive search of the entire parameter space in order to find the "best" solution. As N increases, intelligent algorithms are needed to quickly locate the solution. Only an exhaustive search can guarantee that a global optimum will be found; however, near-optimal solutions are acceptable because of the tremendous speed improvement over exhaustive search methods. ALOPEX iteratively updates all parameters simultaneously based on the cross-correlation of local changes ΔX_i and on the global response change ΔR, plus an additive noise. The cross-correlation term $\Delta X_i \Delta R$ helps the process move in a direction that improves the response. Table 182.1 shows how this can be used to find a global maximum of R.

TABLE 182.1

	ΔX		ΔR	$\Delta X \Delta R$
$X\uparrow$	+	$R\uparrow$	+	+
$X\uparrow$	+	$R\downarrow$	–	–
$X\downarrow$	–	$R\uparrow$	+	–
$X\downarrow$	–	$R\downarrow$	–	+

All parameters X_i are changed simultaneously at each iteration according to

$$X_i(n) = X_i(n-1) + \gamma \Delta X_i(n) \Delta R(n) + r_i(n) \qquad (182.23)$$

The basic concept is that this cross-correlation provides a direction of movement for the next iteration. For example, take the case where $X_i \downarrow$ and $R\uparrow$. This means that the parameter X_i decreased in the previous iteration, and the response increased for that iteration. The product $\Delta X_i \Delta R$ is a negative number, and thus X_i would be decreased again in the next iteration. This makes perfect sense because a decrease in X_i produced a higher response; if you are looking for the global maximum, then X_i should be decreased again. Once X_i is decreased and R also decreases, then $\Delta X_i \Delta R$ is now positive, and X_i increases.

These movements are only tendencies, since the process includes a random component that will act to move the weights unpredictably, avoiding local extrema of the response. The stochastic element of the algorithm helps it to avoid local extrema at the expense of a slightly longer convergence or learning period.

The general ALOPEX updating equation (Eq. 182.23) is explained as follows: $X_i(n)$ are the parameters to be updated, n is the iteration number, and $R()$ is the cost function, of which the "best" solution in terms of X_i is sought. Gamma (γ) is a scaling constant, and $r_i(n)$ is a random number from a Gaussian distribution whose mean and standard deviation are to find the "best" solution. As N increases, intelligent algorithms are needed to quickly locate the solution. Only an exhaustive search can guarantee that a global optimum is found; however, near-optimal solutions are acceptable because of the tremendous speed improvement over exhaustive search methods.

Backpropagation described earlier, being a gradient-descent method, often gets stuck in local extrema of the cost function. The local stopping points often represent unsatisfactory convergence points. Techniques have been developed to avoid the problem of local extrema, with simulated annealing [39] being the most common. Simulated annealing incorporates random noise, which acts to dislodge the process from local extrema. Crucial to the convergence of the process is that the random noise be reduced as the system approaches the global optimum. If the noise is too large, the system will never converge and can be mistakenly dislodged from the global solution. ALOPEX is another process that incorporates a stochastic element to avoid local extremes in search of the global optimum of the cost function. The cost function or response is problem-dependent and is generally a function of a large number of parameters. ALOPEX iteratively updates all parameters simultaneously based varied, and $\Delta X_i(n)$ and $\Delta R(n)$ are found by

$$\Delta X_i(n) = X_i(n-1) - X_i(n-2) \tag{182.24}$$

$$\Delta R(n) = R(n-1) - R(n-2) \tag{182.25}$$

The calculation of $R()$ is problem-dependent and can be easily modified to fit many applications. This flexibility was demonstrated in the early studies of Harth and Tzanakou [29]. In mapping receptive fields, no a priori knowledge or assumptions were made about the calculation of the cost function; instead, a "response" was measured. By using action potentials as a measure of the response [28, 29, 40, 41], receptive fields could be determined by using the ALOPEX process to iteratively modify the stimulus pattern until it produced the largest response.

It should be stated that due to its stochastic nature, efficient convergence depends on the proper control of both the additive noise and the gain factor γ. Initially, all parameters X_i are random, and the additive noise has a gaussian distribution with mean 0 and standard deviation σ initially large. The standard deviation σ decreases as the process converges to ensure a stable stopping point. Conversely, γ increases with iterations. As the process converges ΔR becomes smaller and smaller, and an increase in γ is needed to compensate for this.

Additional constraints include a maximal change permitted for X_i for one iteration. This bounded step size prevents the algorithm from drastic changes from one iteration tot he next. These drastic changes often lead to long periods of oscillation, during which the algorithm fails to converge.

Two-Dimensional Template Matching with ALOPEX

In this application, ALOPEX is used in template matching. A pattern is compared with a list of j templates, and for each template j, a response R_j is calculated. ALOPEX then iteratively modifies the initial pattern according to the global response calculation, local changes in the pattern, and an additive noise element. The updating equations are as described in the preceding subsection (Eqs. 182.23 to 182.25). An explanation of the response calculation used in template matching follows $X_i(n)$ are the parameters to be optimized, and φ_{ij} are the stored templates; j is the number of the templates.

The response, in this case, is given by

$$R_j(n) = \sum_{i=1}^{N^2} X_i(n)\varphi_{ij} \tag{182.26}$$

$$R(n) = CR'(n) \tag{182.27}$$

where

$$R'(n) = \text{Max}\left[R_j(n)\right]. \tag{182.28}$$

Other methods [40] of calculating $R'(n)$ include making it equal to a weighted sum of all the R_j:

$$R'(n) = \sum_{j=1}^{M} R_j(n) W_j(n) \tag{182.29}$$

where

$$W_j(n) = \frac{R_j(n)}{\sum_{j=1}^{M} R_j(n)}. \tag{182.30}$$

The weighting of R_j by W_j and summing tends to slow the convergence of the algorithm, and hence Eq. (182.28) is used to select $R(n)$ from all $R_j(n)$.

Using the simple response calculation in Eq. (182.26) for a pixel-by-pixel multiplication between the pattern and template *j*, good results were obtained for binary (two-valued) images. The algorithm has certain constraints that dictate how it finds the maximum response. However, when multivalued templates were used, there was no longer proper convergence.

Using a response calculation based on pixel-by-pixel correlation is not suited for intermediate-valued optimization. In binary-valued optimization, the tendency to lump the maximum allowed luminance into a single pixel does not exist; therefore, the problem can go undetected. The basic problem is that in supervised learning the ideal image or solution should have maximal response, and for the example shown above, this was not the case.

To solve this problem, a new response calculation had to be implemented. Understanding the previous problem and goal of any optimal response calculation, a new function was used. The new response function given by

$$R_j(n) = \sum_{i=1}^{N^2} -\left(X_i - \varphi_{ij}\right)^2 + \varphi_{ij} \tag{182.31}$$

defines a downward-facing parabola with its apex (or maximum) occurring only when $X = \varphi$. This is basically a least square error calculation that has been inverted so as to produce a maximum instead of a minimum. Also, the maximum contribution for each pixel is exactly φ; this is important because now the maximal response of $R_j(n)$ is simply equal to the total luminance. $R(n) = CR'(n)$ and $R'(n) = \text{Max}[R_j(n)]$ are the same as before. Using the new response calculation, much better convergence was achieved.

Multilayer Perceptron Network Training by ALOPEX

An MLP also can be trained for pattern recognition using ALOPEX. A response is calculated for the *j*th input pattern based on the observed and desired output

$$R_j(n) = \mathbf{O}_k^{des} - \left[\mathbf{O}_k^{obs}(n) - \mathbf{O}_k^{des}\right]^2 \tag{182.32}$$

where $\mathbf{O}_k^{obs}$ and $\mathbf{O}_k^{des}$ are vectors corresponding to O_k for all k. The total response for iteration n is the sum of all the individual template responses $R_j(n)$:

$$R(n)=\sum_{j=1}^{m} R_j(n) \tag{182.33}$$

In Eq. (182.33), m is the number of templates used as inputs. ALOPEX iteratively updates the weights using both the global response information and local weight histories, according to the following:

$$W_{ij}(n)=r_i(n)+\gamma\Delta W_{ij}(n)\Delta R(n)+W_{ij}(n-1) \tag{182.34}$$

$$W_{jk}(n)=r_i(n)+\gamma\Delta W_{jk}(n)\Delta R(n)+W_{ik}(n-1) \tag{182.35}$$

where γ is an arbitrary scaling factor, $r_i(n)$ is an additive gaussian noise, ΔW represents the local weight change, and ΔR represents the global response information. These values are calculated by

$$\Delta W_{ij}(n)=W_{ij}(n-1)-W_{ij}(n-2) \tag{182.36}$$

$$\Delta W_{jk}(n)=W_{jk}(n-1)-W_{jk}(n-2) \tag{182.37}$$

and

$$\Delta R(n)=R(n-1)-R(n-2). \tag{182.38}$$

Besides its universality to a wide variety of optimization procedures, the nature of the ALOPEX algorithm makes it suitable for VLSI implementation. ALOPEX is a biologically influenced optimization procedure that uses a single-value global response feedback to guide weight movements toward their optimum. This single-value feedback, as opposed to the extensive error-propagation schemes of other NN training algorithms, makes ALOPEX suitable for fast VLSI implementation.

182.3 VLSI Applications of Neural Networks

Much debate exists as to whether digital or analog VLSI design is better suited for NN applications. In general, digital designs are an easier to implement and better-understood methodology. In digital designs, computational accuracy is only limited by the chosen word length. While analog VLSI circuits are less accurate, they are smaller, faster, and consume less power than digital circuits [42]. For these reasons, applications that do not require great computational accuracy are dominated by analog designs.

Learning algorithms, especially backpropagation, require high precision and accuracy in modifying the weights of the network. This has led some to believe that analog circuits are not well suited for implementing learning algorithms [43]. Analog circuits can achieve high precision, at the cost of increasing the circuit size. Analog circuits with high precision (8 bits) tend to be equally large as their digital counterpart [44]. Thus high-precision analog circuits lose their size advantage over digital circuits. Analog circuits are of greater interest in applications requiring only moderate precision.

Early studies show that analog circuits can realize learning algorithms, provided that the algorithm is tolerant to hardware imperfections such as low precision and inherent noise. In a paper by Macq et al. [45], a full analog implementation of a Kohonen map [20], one type of neural network, with on-chip learning is presented. With analog circuits having been shown capable of the computational accuracy necessary for weight modification, they should continue to be the choice of NN research.

Size, speed, and power consumption are areas where analog circuits are far superior to digital circuits, and it is these areas that constrain most NN applications. To achieve greater network performance, the size of the network must be increased. The ability to implement larger, faster networks is the major motivation for hardware implementation, and analog circuits are superior in these areas. Power consumption is also of major concern as networks become larger [46]. As the number of transistors per chip increases, power consumption becomes a major limitation. Analog circuits dissipate less power than digital circuits, thus permitting larger implementations.

One of the leaders in the field of analog VLSI neural systems is Carver Mead, who has presented a methodology for implementing biologically inspired networks in analog VLSI [47]. A silicon retina modeled after the biological retina is constructed from simple analog circuits. These simple analog circuits function as building blocks and have been used in many other designs. Some of Mead's other work includes the design of a tracker [48], a sensorimotor integration system, capable of tracking a bright spot of light. An artificial cochlea for use in auditory localization also has been designed in analog VLSI [49].

Other methodologies of VLSI design of NNs include an analog/digital hybrid approach using pulse-stream networks. Pulse-stream encoding has been used in communication systems for many years and was first reported in the context of NNs by Murray and Smith in 1987 [50]. NN pulse-stream techniques attempt to combine the size and speed efficiency of analog circuits with the accuracy of digital circuits [51].

Recently, a digital VLSI approach to implementing the ALOPEX algorithm was undertaken by Pandya et al. [52]. Results of their study indicated that ALOPEX could be implemented using a single-instruction multiple-data (SIMD) architecture. A simulation of the design was carried out, in software, and good convergence for a 4×4 processor array was demonstrated.

In our laboratory, an analog VLSI chip has been designed to implement the ALOPEX algorithm by Zanher et al. [53]. By making full use of the algorithm's tolerance to noise, an analog design was chosen. As discussed earlier, analog designs offer larger and faster implementations than do digital designs.

182.4 Applications in Biomedical Engineering

Expert Systems and Neural Networks

Computer-based diagnosis is an increasingly used method that tries to improve the quality of health care. Systems that depend on artificial intelligence (AI), such as knowledge-based systems or expert systems, as well as hybrid systems, such as the above combined with other techniques, like NNs, are coming into play. Systems of this sort have been developed extensively in the last 10 years with the hope that medical diagnosis and therefore medical care would improve dramatically. Hatziglygeroudis et al. [54] are developing such a system with three main components: a user interface, a database management system, and an expert system for the diagnosis of bone diseases. Each rule of the knowledge representation part is an ADALINE unit that has as inputs the conditions of the rule. Each condition is assigned a significance factor corresponding to the weight of the input to the Adaline unit, and each rule is assigned a number, called a *bias factor*, that corresponds to the weight of the bias input of the unit. The output is calculated as the weighted sum of the inputs filtered by a threshold function.

Hudson et al. [55] developed an NN for symbolic processing. The network has four layers. A separate decision function is used for layer three and a threshold for each node in the same layer. If the value of the decision function exceeds the corresponding threshold value, a certain symbol is produced. If the value of the decision function does not exceed the threshold, then a different symbol is produced. The so generated symbols of adjacent nodes are combined at layer four according to a well-structured grammar. A grammar provides the rules by which these symbols are combined [56]. The addition of a symbolic processing layer enhances the NN in a number of ways. It is, for instance, possible to supplement a network that is purely diagnostic with a level that recommends further actions or to add additional connections or nodes in order to more closely simulate the nervous system.

With increasing network complexity, parameter variance increases, and the network prediction becomes less reliable. This difficulty can be overcome if some prior knowledge can be incorporated into the NN to bias it [57]. In medical applications in particular, rules can either be given by experts or can be extracted from existing solutions to the problems. In many cases the network is required to make reasonable predictions before it has gone through any sufficient training data, relying only on a priori knowledge. The better this knowledge is initially, the better is the performance and the shorter is the training [58, 59].

Applications in Mammography

One of the leading causes of death of women in the United States is breast cancer. Mammography has been proved to be an effective diagnostic procedure for early detection of breast cancer. An important sign in its detection is the identification of microcalcifications of mammograms, especially when they form clusters. Chan et al. [60] have developed a computer-aided diagnosis (CAD) scheme based on filtering and feature-extracting methods. In order to improve on the false-positive results, Zhang et al. [61] applied an artificial NN that is shift-invariant. They evaluated the performance of the NN by the "jack-knife" method [62] and receiver operating characteristic analysis [63, 64]. A shift-invariant NN is a feedforward NN with local, spatially invariant interconnections similar to those of the neocognition [65] but without the lateral interconnections. Backpropagation also was used for training for individual microcalcifications, and a cross-validation technique also was used in order to avoid overtraining. In this technique, the data set is divided into two sets, one used for training and the other for validating the predetermined intervals. The training of the network is terminated just before the performance of the network for the validating set decreases. The shift-invariant NN was proven to be much better in dropping the false-positive classifications by almost 55% over previously used NNs.

In another study, Zheng et al. [66] used a multistage NN for detection of microcalcification clusters with almost 100% success and only one false-positive result per image. The multistate NN consists of more than one NN connected in series. The first stage is called the *detail network*, with inputs the pixel values of the original image, while the second network, the *feature network*, gets as inputs the output from the first stage and a set of features extracted from the original image. This approach has higher sensitivity of classifications and a lower false-positive detection than the previous reports.

Another approach was used by Floyd et al. [67], where radiologists read the mammograms and come up with a list of eight findings, which were used as features for an NN. The results from biopsies were taken as the truth of diagnosis. For indeterminate cases, as classified by radiologists, the NN had a performance index of 0.86, which is quite high.

Downes [68] used similar techniques to identify stellate lesions. He used texture quantification via fractal-analysis methods instead of using the raw data. In mammograms, specific textures are usually indicative of malignancy. The method used for calculating the fractal dimension of digitized images was based on the relationship between the fractal dimension and the power spectral density.

Giger et al. [69] aligned the mammograms of left and right breasts and used a subtraction technique to find initial candidate masses. Various features were then extracted and used in conjunction with NNs in order to reduce false-positive assessments resulting from bilateral subtraction. Receiver operating characteristic (ROC) analysis was applied to evaluate the output of the NN. The methods used were evaluated using pathologically confirmed cases. This scheme yielded a sensitivity of 95% at an average of 2.5 false-positive detections per image.

Chromosome and Genetic Sequences Classification

Several clinical disorders are related to chromosome abnormalities that are difficult to identify accurately and also classify the individual chromosome. Automated systems can greatly help the human capabilities in dealing with some of the problems involved. One way to deal with this problem is the use of NNs. Several studies have already been done toward enhancing the ability of an automated computerized system to analyze chromosome identification [70]. One such study by Sweeney and Musavi [71] analyzes the metaphase of chromosome spreads employing probablistic NNs (PNNs), which have been used as alternatives to various classification problems. First introduced by Specht [72, 73], PNNs are combinations of a kernel-based estimator for estimation of probability densities and the Bayes rule for classification decision. The estimation with the highest value specifies the correct class. Thus training of PNNs means to find appropriate kernel functions, usually taken to be gaussian densities, and therefore, the problem is reduced to the selection of a scalar parameter, namely, the standard deviation, of the gaussian. A way to improve the accuracy of a PNN for chromosome classification is to use the knowledge that there can be a maximum of only two chromosomes assigned to each class. This knowledge can be easily incorporated into the NN. Similar or better results were obtained to the classic backpropagation-trained NN.

A hybrid symbolic/NN machine learning algorithm was introduced by Noordewier et al. [74] for the recognition of genetic sequences. The system uses a knowledge base of hierarchically structured rules to form an artificial NN in order to improve the knowledge base. They used this system in recognizing genes in DNA sequences. The learning curve of this system was compared with that of a randomly initialized, fully connected two-layer NN. The knowledge-based NN learned much faster than the other one, but the error of the randomly initialized NN was slightly lower (5.5% versus 6.4%). Methods also have been devised to investigate what the NN has learned by an automatic translation into symbolic rules of trained NN initialized by the knowledge-based method [75].

Medial axis transform (MAT)-based features as inputs to an NN have been used in studying human chromosome classification [76]. Prenatal analysis, genetic syndrome diagnosis, and others make this research very important. Human chromosome classification based on NN requires no a priori knowledge or even assumptions on the data. MAT is a widely used method for transformations of elongated objects and requires less storage and time while preserving the topologic properties of the object. MAT also allows for a transformation from a 2D image to a 1D representation of it. The so obtained features are then fed as inputs to a two-layer feedforward NN trained by back-propagation, with almost perfect results in classifying chromosomes. An optimization on an MLP also was done [77].

References

1. Mueller P, Lazzaro J. 1986. Real time speech recognition. In J Dember (ed), Neural Networks for Computing. New York, American Institute of Physics.
2. Bourland H, Morgan N. 1990. A continuous speech recognition system embedding a Multi-Layer Perceptron into HMM. In D Touretzky (ed), Advances in Neural Information Processing Systems 2, pp 186–193. San Mateo, Calif, Morgan Kauffman.
3. Bridle JS, Cox SJ. 1991. RecNorm: Simultaneous normalization and classification applied to speech recognition. In RP Lippman et al (eds), Advances in Neural Information Processing Systems 3, pp 234–240.
4. Lee S, Lippman RP. 1990. Practical characteristics of neural networks and conventional classifiers on artificial speech problems. In DS Touretzky (ed), Advance in Neural Information Processing Systems 2, pp 168–177, San Mateo, Calif, Morgan Kauffman.
5. Fukushima K. 1983. Neocognition: A neural network model for a mechanism of visual pattern recognition. IEEE Trans Syst Man Cybernet SMC-13(5):826.
6. Dasey TJ, Micheli-Tzanakou E. 1994. An unsupervised system for the classification of hand-written digits, comparison with backpropagation training. IEEE Trans Neural Networks.
7. LeCun Y, Boser B, Denker JS, et al. 1989. Backpropagation applied to handwritten ZIP code recognition. Neural Comp 1(4):541.
8. Hakim N, Kaufman JJ, Cerf G, Medows HE. 1990. A discrete time neural network model for system identification. Proc of IJCNN 90(V3):593.
9. Hesh D, Abdallah C, Horne B. 1991. Recursive neural networks for signal processing and control. In Proceedings of the First IEEE-SP Workshop on Neural Networks for Signal Processing, Princeton, NJ, pp 523–532.

10. Cottrell GW, Munro PN, Zipser D. 1989. Image compression by backpropagation: A demonstration of extensional programming. In Advances in Cognitive Science, vol. 2. Norwood, NY, Ablex.
11. Oja E, Lampinen J. 1994. Unsupervised learning for feature extraction. In JM Zurada et al (eds), Computational Intelligence Imitating Life, New York, IEEE Press.
12. Fogelman Soulie F. 1994. Integrating neural networks for real world applications. In JM Zurada et al (eds), Computational Intelligence Imitating Life. New York , IEEE Press.
13. Rosenblatt F. 1958. The perceptron: A probabilistic model for information storage and organization in the brain. Psychol Rev 65:386.
14. Rosenblatt F. 1962. Principles of Neurodynamics. New York, Spartan Books.
15. Lippman, RP. 1987. An introduction to computing with neural nets. IEEE ASSP Mag 4–22.
16. Moore K. 1992. Artificial neural networks: Weighing the different ways to systemize thinking. IEEE Potentials 23.
17. Huang S, Huang Y. 1991. Bounds on the number of hidden neurons in multilayer perceptrons. IEEE Trans Neural Networks 2(1):47.
18. Kung SY, Hwange J, Sun S. 1988. Efficient modeling for multilayer feedforward neural nets. Proc IEEE Conf Acoustics, Speech Signal Processing, New York, pp 2160–2163.
19. Mirchandani G. 1989. On hidden nodes for neural nets. IEEE Trans Circuits Syst 36(5):661.
20. Kohonen T. 1988. Self-Organization and Associative Memory. New York, Springer-Verlag.
21. Hecht-Nielsen R. 1987. Counterbackpropagation networks. In Proc of the IEEE First International Conf on Neural Networks, vol 2, pp 19–32.
22. Dasey TJ, Micheli-Tzanakou E. 1992. The unsupervised alternative to pattern recognition: I. Classification of handwritten digits. In Proc 3rd Workshop on Neural Networks, Auburn, Ala, pp 228–233.
23. Dasey TJ, and Micheli-Tzanakou E. 1992. The unsupervised alternative to pattern recognition: II. Detection of multiple sclerosis with the visual evoked potential. In Proc 3rd Workshop on Neural Networks, Auburn, Ala, pp 234–239.
24. McCulloch WC, Pitts W. 1943. A logical calculus of the ideas immanent in nervous activity. Bull Math Biophys 5:115.
25. Hebb D. 1949. The Organization of Behavior. New York, Wiley.
26. Widrow B, Lehr MA. 1990. 30 years of adaptive neural networks: Perceptron, Madaline, and backpropagation. Proc IEEE 78(9):1415.
27. Minsky M, Papert S. 1969. Perceptrons: An Introduction to Computational Geometry. Cambridge, Mass, MIT Press.
28. Tzanakou E, Harth E. 1973. Determination of visual receptive fields by stochastic methods. Biophys J 15:(42a).
29. Harth E, Tzanakou E. 1974. Alopex: A stochastic method for determining visual receptive fields. Vis Res 14:1475.
30. Tzanakou E. Michalak R, Harth E. 1979. The ALOPEX process: Visual receptive fields by response feedback. Biol Cybernet 35:161.
31. Rumelhart DE, McClelland JL (eds). 1986. Parallel Distributed Processing Cambridge, mass, MIT Press.
32. Holstrom L, Koistinen P. 1992. Using additive noise in backpropagation training. IEEE Trans Neural Networks 3(1):24.
33. Fahlmann SE. 1988. Faster learning variations of backpropagation: An empirical study. In D Touretsky et al (eds), Proc of the Connectionist Models Summer School. San Mateo, Calif, Morgan Kaufmann.
34. Werbos PJ. 1990. Backpropagation through time: What it does and how to do it. Proc IEEE 78(30):1550.
35. Nguyen D, Widrow B. 1989. The truck backer-upper: An example of self-learning in neural networks. In Proc of the Int Joint Conf Neural Networks, vol II, pp 357–361. New York, IEEE Press.

36. Ciaccio E, Tzanakou E. 1990. The ALOPEX process: Application to real-time reduction of motion artifact. Annu Int Conf IEEE EMBS 12(3):1417.
37. Dasey TJ, Micheli-Tzanakou E. 1990. A pattern recognition application of the Alopex process with hexagonal arrays. In Int Joint Conf Neural Networks, vol II, pp 119–125.
38. Venugopal K, Pandya A, Sudhakar R. 1992. ALOPEX algorithm for adaptive control of dynamical systems. Proc IJCNN 2:875.
39. Kirkpatrick S, Gelatt CD, Vecchi MP. 1983. Optimization by simulated annealing. Science 220:671.
40. Micheli-Tzanakou E. 1984. Non-linear characteristics in the frog's visual system. Biol Cybernet 51:53.
41. Micheli-Tzanakou E. 1983. Visual receptive fields and clustering. Behav Res Methods Instrum 15(6):553.
42. Mead C, Ismail (eds). 1991. Analog VLSI Implementation of Neural Systems. Boston, Kluwer Academic Publishers.
43. Ramacher and Ruckert (eds). 1991. VLSI Design of Neural Networks. Boston, Kluwer Academic Publishers.
44. Graf HP, Jackel LD. 1989. Analog electronic neural network circuits. IEEE Circuits Devices Mag 44.
45. Macq D, Verlcysen M, Jespers P, Legat J. 1993. Analog implementation of a Kohonen map with on-chip learning. IEEE Trans Neural Networks 4(3):456.
46. Andreou A, et al. 1991. VLSI neural systems. IEEE Trans Neural Networks 2(2):205.
47. Mead C. 1989. Analog VLSI and Neural Systems. Reading, Mass, Addison-Wesley.
48. Maher M, Deweerth S, Mahowald M, Mead C. 1989. Implementing neural architectures using analog VLSI circuits and systems. 36:(5):643.
49. Lazzaro J, Mead C. 1989. Silicone models of auditory localization. Neural Comput 1(1):1.
50. Murray AF, Smith A. 1987. Asynchronous arithmetic for VLSI neural systems. Electron Lett 23(12):642.
51. Murray AF, Corso D. Tarassenko L. 1991. Pulse-stream VLSI neural networks mixing analog and digital techniques. IEEE Trans Neural Networks 2(2):193.
52. Pandya S, Shandar R, Freytag L. 1990. An SIMD architecture for the Alopex neural networks. SPIE Parallel Architectures for Image Processing 1246:275.
53. Zahner AD. 1994. Design and implementation of an analog VLSI, adaptive algorithm with applications in neural networks, Master's thesis, Rutgers University.
54. Hatziglygeroudis I, Vassilakos PJ, Tsakalidis A. 1994. An intelligent medical system for diagnosis of bone diseases. In Proc of the Int Conf on Med Physics and Biomed Eng., Cyprus, vol 1, pp 148–152.
55. Hudson DL, Cohen ME, Deedwania PC. 1993. A neural network for symbolic processing. In Proc of the 15th Annual Int Conf of the IEEE/EMBS, vol 1, pp 248–249.
56. Hoperoft JE, Ullman JD. 1969. Formal Languages and Their Relation to Automata. Reading Mass, Addison-Wesley.
57. Roscheisen M, Hofmann R, Tresp V. 1992. Neural control for running mills: Incorporating domain theories to overcome data deficiency. In Advances in Neural Information Processing Systems 4.
58. Towell GG, Shavlik JW, Noordemier MO. 1990. Refinement of approximately correct domain theories by knowledge-based neural networks. In Proc of the 8th National Conf Artif Intell, pp 861–866.
59. Tresp V, Hollatz J, Ahmad S. 1994. Network structuring and training using rule-based knowledge. In Advances in Neural Information Processing Systems 5, pp 871–878.
60. Chan H-P, Doi K, Vyborny CJ, et al. 1990. Improvement in radiologists' detection of clustered microcalcifications on mammograms: The potential of computer aided diagnosis. Invest Radiol 25:1102.
61. Zhang W, Giger ML, Nishihara RM, Doi K. 1994. Application of a shift-invariant artificial neural network for detection of breast carcinoma in digital mammograms. In Proc of the World Congress on Neural Networks, vol 1, pp 45–52.

62. Fukunaga K. 1990. Introduction to Statistical Pattern Recognition, 2nd ed. New York, Academic Press.
63. Metz CE. 1988. Current problems in ROC analysis. In Proc Chest Imaging Conf Madison Wisc, pp 315–336.
64. Metz CE. 1989. Some practical issues of experimental design and data analysis in radiological ROC studies. Invest Radiol 24:234.
65. Fukushima K, Miyake S, Ito T. 1983. Neocognitron: A neural network model for a mechanism of visual pattern recognition. IEEE Trans Syst Man Cybernet SMC-13:826.
66. Zheng B, Qian W, Clarke LP. 1994. Artificial neural network for pattern recognition in mammography. In Proc of the World Congress on Neural Networks, vol. I, pp 57–62.
67. Floyd CE Jr, Yun AJ, Lo JY, et al. 1994. Prediction of breast cancer malignancy for difficult cases using an artificial neural network. In Proc of the World Congress on Neural Networks, vol I, pp 127–132.
68. Downes P. 1994. Neural network recognition of multiple mammographic lesions. In Proc of the World Congress on Neural Networks, vol I, pp 133–136.
69. Giger ML, Lu P, Huo Z, Zhang W. 1994. Application of artificial neural networks to the task of merging feature data in computer-aided diagnosis schemes. In Proc of the World Congress on Neural Networks , vol I, pp 43–46.
70. Piper J, Granunn E, Rutovitz D, Ruttledge H. 1990. Automation of chromosome analysis. Sig Proc 2(3):109.
71. Sweeney WP Jr, Musavi MT. 1994. Application of neural networks for chromosome classification. In Proc of the 15th Annual Int Conf of the IEEE/EMBS, vol 1, pp 239–240.
72. Specht DF. 1988. Probablistic neural networks for classification mapping or associative memory. In Proc of the IEEE Int Conf on Neural Networks, vol 1, pp 525–532. New York, IEEE Press.
73. Specht DF. 1990. Probablistic neural networks. 3(1):109.
74. Noordewier MO, Towell GG, Shavlik JW. 1993. Training knowledge-based neural networks to recognize genes in DNA sequences. Adv Neural Info Proc Sys 3:530.
75. Towell GG, Graven M, Shavlik JW. 1991. Automated interpretation of knowledge-based neural networks. Tech rep Univ of Wisc Computer Sci Dept, Madison, Wisc.
76. Lerner B, Rosenberg B, Levistein M, et al. 1994. Medial axis transform based features and a neural network of human chromosome classification. In Proc of the World Congress on Neural Networks, vol 3, pp 173–178.
77. Lerner B, Guterman H, Dinstein J, Romem Y. 1994. Learning curves and optimization of a multilayer perceptron neural network for chromosome classification. In Proc of the World Congress on Neural Networks, vol 3, pp 248–253.

183
Clinical Decision Systems

Pirkko Nykänen
VTT Information Technology

Niilo Saranummi
VTT Information Technology

183.1 Introduction

Clinical decision systems are information systems that support and assist health care professionals in clinical decision making tasks like diagnosis, therapy planning and monitoring. Their purpose is not to replace or substitute for human abilities or skills. The value of clinical decision systems depends on their role and contribution to the patient care process. Their role may be to act as a reference book of relevant medical knowledge or similar patient cases, or as a watchdog to screen routines or detect deviations, or they may remind the user of extreme possibilities or forecast outcomes of different alternatives. Additionally, these systems may produce information from data by interpretation, work as tutoring or guidance systems, and represent a more experienced consultant for a less experienced user.

183.2 Clinical Decision Making

The objective of health care is to understand, prevent and treat human diseases. In medical practice a clinician is involved in a complex cognitive task where he applies different kinds of knowledge on several cognitive and epistemic levels. A clinician develops, through education and experience, an understanding of fundamental anatomic, physiologic and pathologic processes, an understanding of how to apply diagnostic methods and procedures, and an understanding of the effects of drugs and treatments on patients. In a decision making situation a clinician has to understand the clinical state of the patient, the relevant processes going on in the patient, and he must have an understanding of the purpose of his intervention and of the possible actions he can initiate.

The central goal of medical research is to improve clinical practice. This can be achieved in many different ways. The focus may be on improving the theoretical understanding of the physiologic and pathologic processes on a high level of abstraction, or on concentrating in the study of diagnostic methods, or on evaluating new therapies, or on developing preventive measures [Pedersen et al. 1993]. In some cases medical research is close to basic natural science, in other situations it is close to applied engineering research.

0-8493-0462-8/00/$0.00+$.50

A medical decision-maker acquires and manages several different forms of knowledge. Basically two types of knowledge are involved in decision-making [van Bemmel and Musen 1997, Pedersen et al. 1993]: scientific or formal knowledge and experiential knowledge. Scientific knowledge deals with understanding the scientific principles and relationships between pathophysiological conditions and disease symptoms. This type of knowledge is found in scientific literature and articles. Scientific knowledge helps to explain and justify empirical phenomena and it tells to what extent simplifications and approximations make sense in real life situations. Experiential knowledge originates from well-documented patient cases and validated guidelines. Experiential knowledge helps the clinician to recognize a disease based on his experience. Experiential knowledge in decision making needs to be justified and validated within the framework of scientific knowledge, and when this is done, the most efficient computational models can be achieved. Scientific and experiential knowledge can be characterized as deep and shallow, or surface knowledge, respectively. The terms tacit and explicit knowledge provide another viewpoint on knowledge. Tacit knowledge in a sense is a descriptor for the skills of a clinician. Often during the course of his professional life, he may have operationalized his experiential and scientific knowledge to a degree that he can no longer explain "what he knows" explicitly.

Medicine is a complex area for clinical decision systems. The complexity arises from many sources. Medicine is not a formal science; it has an empirical basis. We have shallow or limited understanding of many of the disease processes and also of the medical decision making. The challenge of a physician is to be up-to-date on current "best practices" in medicine. Unfortunately a large part of medical knowledge is based on experience rather than on hard facts. "Evidence based medicine" as a large-scale initiative aims to improve that ratio [Cochrane].

In medical decision making, scientific and experiential knowledge are interwoven. A normal situation is that a clinician knows enough and has enough reliable data available to make a decision in that situation. Medical expertise, or medical competence, is situational and it is used in a data-intensive environment. The kind of knowledge which is relevant for decision making is huge, even in restricted subspecialities of medicine [Miller 1988]. Diagnosis is often said to be the main task of the physician and a lot of attempts have been made to model the diagnostic process. However, diagnosis is not the only decision making task of a medical professional. He is also involved in decisions that concern selection of treatments, monitoring and follow up, interpretation of signs and symptoms, prediction of consequences and effects, selection of drugs and procedures, etc. Each clinician applies his own pathway to decision making depending on the case at hand, taking into account the patient and external factors like the urgency of the situation, difficulty of the case, role that the physician assumes and possible choices and actions that can be taken.

In clinical decision making, medical knowledge is used to interpret patient data. This raises the issue of what patient data is available. The problem with patient data is that it is highly context sensitive. A major part of patient data collected during clinical episodes is unstructured. Referrals and discharge letters are mainly unstructured summaries accompanied by structured elements like laboratory findings. Each clinician has his own way of embedding the context into these; and other clinicians need to know this before they can truly interpret each other's notes. Several ways have been developed to summarize this information. These include classifiers like ICD-10, SNOMED, Read codes and DRGs. Much more ambitious are the two major projects which were launched on both sides of the Atlantic some years ago to develop "translators" for communication between clinicians, the Unified Medical Language System (UMLS) and Generalized Architecture for Languages, Encyclopedias and Nomenclatures in Medicine (GALEN).

183.3 Clinical Decision Systems History

The first approaches to providing computer support for medical decision making were based on Bayesian statistics. Programs for diagnostic problems showed that impressive diagnostic accuracy could be achieved if the computer programs were supported by reliable data [Shortliffe, 1988]. However, the demand for independence between variables in these approaches was found to be too restrictive in medicine and a lot of problems were related to the accessibility of reliable data. It was also noticed that to be able to tackle the information problem, simultaneous perspectives were needed from other disciplines such as, decision theory, linguistics, mathematical modeling, and cognitive psychology [Gremy, 1988].

Research with artificial intelligence in medicine started in the 1960s based on the understanding that when computer systems are used to emulate medical decision making steps, they need relevant medical knowledge and this knowledge has to be represented in a context-sensitive way. Much of this knowledge must be extracted from the experts in the field; the statistical data approach is not enough. Consequently, symbolic reasoning methods are needed. The early work resulted in knowledge-based consultations systems like MYCIN [Shortliffe, 1976] and INTERNIST [Miller et al., 1982] among others.

The focus of this early work was in the development of methodologies and tools to support modeling of a domain, knowledge acquisition including machine learning approaches, modeling of reasoning strategies, and knowledge representations. Interests in reasoning strategies were focused on causal, temporal, heuristic, strategic, and anatomic reasoning and reasoning from multiple sources of knowledge. Issues of validation of knowledge and evaluation of prototype systems were also raised. The term expert system was introduced and much of the research was focused on how to represent expert knowledge, how to acquire it and how to use it in problem solving. Certainty factors and other heuristic scoring measures were developed to manage probabilities and weights of evidence. Also issues related to users and their environments were tackled, like human-computer interaction and interfaces to databases and database management systems, integration of natural language processing with knowledge-based systems, and integration of medical decision analysis with artificial intelligence techniques [Buchanan and Shortliffe, 1984, Miller, 1988, Shortliffe et al., 1990].

In the late 1980s the rapid evolution both in AI hardware and software caused a big change. The expert system shells (i.e., tools to develop expert systems where the inference engine is separated from the knowledge base) were introduced and widely used. Cheap and effective microcomputers as well as dedicated professional workstations for knowledge engineering offered integrated and interactive environments for expert systems development. Progress in computer graphics, multimedia, and imaging provided further support for this.

The focus shifted to attempts to find general solutions to problems through modeling approaches. The final result of the development of a knowledge-based system was seen to be a model of expertise, which in turn is implemented in the knowledge base. These models of expertise incorporated assumptions about knowledge representations, knowledge base contents, problem-solving methods and application tasks [Musen, 1989]. Knowledge level modeling was widely recognized to be the major way to generalize and structure domain and task knowledge. The knowledge level model describes the problem-solving process in terms of general abstract conceptualizations without reference to the implementations [Newell, 1982].

Moving from the knowledge level to the knowledge-use level, however, posed a problem [Newell, 1982, Steels 1990]. Several frameworks were suggested for this: Heuristic classification with the focus on the inference structures [Clancey, 1985], distinctions between deep and shallow knowledge at the knowledge level [Chandrasekaran and Mittal, 1982], a problem-solving method approach focusing on how the problem might be solved at the knowledge-use level [Bennet, 1985] and finally the generic task approach where the analysis of expertise is performed in terms of the tasks and ordering of the tasks [Chandrasekaran, 1986].

From the process point of view, medical reasoning may be modeled by fundamental medical tasks, e.g., diagnosis, therapy planning and monitoring. An AI-based clinical decision system requires formulation of an inference model (i.e., the reasoning process) which exploits an appropriate ontology (i.e., the medical process). An architecture then provides a conceptual description of these components at the knowledge and computational level [Stefanelli, 1993].

In spite of that progress, AI based systems did not gain widespread use. What are the reasons for this? Most AI systems were developed within a narrow framework for a few dedicated users and for well-defined tasks and restricted domains. Also, development projects were mostly very technology- or methodology-oriented. Knowledge acquisition was also a major barrier. It has been realized that expertise cannot be elicited by interviews and cannot be completely expressed as formal representations as experts cannot explicate their tacit knowledge.

Today, decision support systems are no longer knowledge-based systems, but rather, knowledge-based approaches have been integrated with traditional information technologies resulting in decision support systems. Also the connectionist approaches of classical AI, e.g., artificial neural networks, have gained more use and there exist many well-functioning neural network applications in restricted, data-rich problem areas. The Internet is a potential platform for the future of AI-based approaches because the flexibility for representations and programs which was typical for AI approaches can be found on the Internet. The Internet also allows new possibilities for distribution and access to clinical decision support.

183.4 Clinical Application Areas and Types of Systems

The first application areas were diagnosis and treatment planning. The scope broadened soon to systems addressing tasks of monitoring and follow-up of patients, diseases, and treatments, especially in the domains of chronic diseases and intensive care. Applications have tackled medical, managerial, administrative, and financial decisions in health care. Most clinical decision systems were planned for highly specialized hospital environments, and very rarely were problems of community care, primary care, or occupational health examined. Most successful clinical decision systems exist in the following situations (van Bemmel and Musen 1997):

- When systems offer support for data validation like supporting systems, which are interfaced with laboratory information systems. These kinds of systems help to increase data reliability, to administer data processing and data flow activities, to transmit data, and to manage quality control.
- When systems support data reduction, i.e., they reduce, transform, and present data in such a way that it is easier for the health care professional to interpret data and screen the massive data flow. Typical areas for this type of system are intensive care and operating rooms.
- When systems support data acquisition like in medical imaging where data is processed to give the medical professional the best possible view.

Most stand-alone systems work as consulting systems in such a way that they require direct interaction from the user to start the dialogue. Interfaced systems are connected with information systems like laboratory systems or electronic patient records and they retrieve data from that system for their input. Integrated systems are directly connected or embedded into medical devices like diagnostic ECG recorders.

183.5 Requirements and Critical Issues for Clinical Decision Systems Development

Clinical decision systems can be successful if they are planned to support human professionals in tasks where support is needed and where support can be offered with information technology. For instance, information enhancement is needed in many decision making situations as a precursor for the next decision step, data reduction is necessary to interpret or identify significant findings, and data validation is needed to have reliable data.

Representational issues, both semantic and syntactic, have to be solved in such a way that no misunderstandings or misinterpretations are met. We have to be conscious also of the system's brittleness, because the model is an abstraction and approximation of the reality, and transferring a model from one environment to another may result in an invalid model and implementation.

In developing models and specifying the system, users need to be involved. On the other hand, they need tools which enable them to analyze, structure, and visualize domain knowledge without strong support by technology personnel, i.e., tools that support modeling and reasoning strategies and system architectures for different tasks, domains, and environments [Brender et al., 1993].

Clinical decision systems must be acceptable to the users, serve real users' needs, be cost-effective and their effects and impacts must be known. In the early expert systems, explanation facilities were important for acceptability by users. These explanation facilities served both educational purposes and through them the system's output was made clear for the user. Critiquing was more acceptable to the users than the consulting mode [Miller 1988], because the critiquing system does not give advice to the user, but takes the user's proposal and discusses its pros and cons in relation to other possibilities.

Medical textbooks, journals, and other reference materials are the stores of experiential and scientific knowledge. Accessing the right source at the right time is the question. Care guidelines and protocols are an approach to compiling medical knowledge into an operational form leading to protocol-based care. The acts needed in handling the problem of a patient can be seen as an instantiation of a template containing that "care package". Clinical protocols are used routinely in clinical research to determine the efficacy and effectiveness of new drugs and medical procedures. They are also commonly used in cancer therapy and in other medical specialities.

However, while medical knowledge is universal, clinical practice is local. Therefore guidelines and protocols need to be locally adapted. Furthermore, as each patient is an individual, protocols cannot be applied in every detail. Instead the user must have the freedom to apply a template according to the needs of that particular case. Consequently what started as protocols and decision trees in the knowledge-based system era of the 1980s has gradually changed so that today they are called clinical guidelines and clinical pathways [Prestige]. Consequently, decision support is gradually becoming embedded into clinical information systems compared to the stand alone decision systems of the past.

183.6 Evaluation of Clinical Decision Systems

Evaluation and validation are necessary tasks during development and before accepting the systems into use [Clarke et al., 1994]. They are means to control the development and guarantee the quality of the results; thus it supports development of good clinical practice. The users need guidance and support to perform evaluations and the developer or the vendor of the system must provide the user with the necessary material needed for evaluation [Nykänen, 1990]. The evaluations of clinical decision systems performed and reported in the literature show that a global, generally accepted methodology for evaluation does not exist. The reported results are subjective and evaluations have been performed without generally accepted objectives and standards.

In a study [Wyatt, 1987] 14 medical decision support systems were evaluated according to a three-stage procedure where in stage one, the main consideration was the system's clinical role, stage two considered if the system had been tested with a proper sample with predefined goals for accuracy and utility, and stage three considered if the performance of the doctor was tested when he was aided or not aided by the system. The result was that only three systems had their clinical role defined, seven systems had paper tests, and only one system had been subjected to a field test. Medical experts were used as the "Gold Standard," in most cases only one expert had been used. From these 14 systems, only two have ever been used in practice. Another study [Lundsgaarde, 1987] concluded that only about 10% of the medical decision support systems had reached laboratory testing. A further study of 64 medical decision support systems [Pothoff et al., 1988] concluded that most systems were still in their developmental stage and only eight systems were developed to the stage to enter routine clinical use.

Most of the decision support systems are planned to work as supporting devices for the clinicians and thus they have to be integrated into the user's environment. Evaluation of their fitness for that has been neglected in reported evaluations with some exceptions. General practitioners used the SPHINX system [Botti et al., 1989] for six months. Though this system was used by the practitioners via the French MINITEL-network, the legal issues were not considered in the studies. It should, however, be of particular importance for a system operating on a network [Clarke et al., 1994].

In a study by [Nykänen et al. 1998], some health telematics projects were studied from the validation methodology perspective. Results show that importance on evaluation and validation has been understood only very recently. Too little resources are usually reserved for evaluation and usable evaluation methodologies still do not exist as well as experience and skills on how to apply them. Validation and evaluation are not yet fully integrated with the development process so that evaluation results could be used as feedback for the systems' development.

In summary, organizational and human aspects in a health care environment require more attention. In the user environment, evaluation should at least consider if the system helps the user to improve the outcome of his work in the organizational setting and if it is usable and cost-effective [Friedman and Wyatt 1997, Brender 1997].

It is difficult to foresee all the legal implications of decision support systems on clinical practice. Health professionals may even think that using computer applications is less harmful than not using them [Hafner et al. 1989]. Scientists have tried to stimulate a discussion of legal responsibilities of clinicians that have not been giving state-of-the-art treatment because they are not able to find relevant information. This is becoming one of the leading arguments in favor of decision support systems in health care in the future [Forström and Rigby 1998]. A general interpretation of the liability principle has also been that decisions suggested or supported by a computer system are ultimately the responsibility of the doctor who puts them into effect.

A project in the Telematics Applications Programme of the European Union (VATAM, Validation of Telematic Applications in Medicine) develops guidelines for validation. These guidelines will include a repository of existing methodologies, a glossary of terms, and a library of tools and checklists [Talmon et al. 1998]. The purpose of the guidelines is to report on experiences, to inform on possible approaches, and to assist in selecting an experimental study design, methodology, methods, and metrics for evaluation. The guidelines are targeted for the various stakeholders, the different types of health telematic systems and they cover the system life cycle. Information and descriptions of various evaluation approaches and methodologies can also be found in [Friedman and Wyatt 1997, Brender 1997].

183.7 Summary

First, clinical decision systems applied statistical methods and these systems were developed for restricted data reduction problems in data rich domains. Most problems at this phase were related to acquisition of required valid data. Next, a generation of clinical support systems were developed using artificial intelligence approaches and methods. These systems aimed at emulating human intelligent behavior. However, modeling and representation of human intelligence and medical reasoning proved to be more difficult than expected and the resulting systems were restricted in performance and brittle for their knowledge and domain. The biggest problems with artificial intelligence-based systems were related to knowledge acquisition because medical expertise could not be elicited through interviews as experts cannot explicate their tacit knowledge. Current clinical decision systems have been integrated with traditional information technologies.

Integration of decision support systems with information technologies and with the developing health care information infrastructure offers real challenges for clinical decision systems in the future. Systems can be integrated, interfaced or embedded with other information technology systems and products and data acquisition can be performed via networks and support with the best modality offered via networks, where and when needed. The future users of decision support systems include not only health care professionals and physicians, but also patients and citizens at home and in need of support.

References

Bennet J S. 1985. ROGET: A knowledge-based system for acquiring the conceptual structure of a diagnostic expert system. *J. Automatic Reasoning*, 1:49-74.

Botti G, Joubert M, Fieschi D, Proudon H, and Fieschi M. 1989. Experimental use of the medical expert system SPHINX by general practitioners: Results and analysis. In: Barber B, Cao D, Qin D and Wagner G (Eds.), *Proc. Sixth Conf. Med. Inform. (MEDINFO 1989)*, p. 67-71, North-Holland, Amsterdam.

Brender J, Talmon J, Nykänen P, O'Moore R, Drosos P and McNair P. 1993. KAVAS-2: Knowledge Acquisition, Visualisation and Assessment System. In: *European Conference on Artificial Intelligence in Medicine,* Andreassen S, Engelbrecht R and Wyatt J, (Eds.) Munich, 1993, pp. 417-420. IOS Press, Amsterdam.

Brender J. 1997. Methodology for assessment of medical IT-based systems in an organisational context. *Technology and Informatics* 42, IOS Press, Amsterdam.

Buchanan B G and Shortliffe E H (Eds.). 1984. *Rule-based expert systems: The MYCIN experiments of the Stanford Heuristic Programming Project.* Addison-Wesley, Reading, MA.

Chandrasekaran B and Mittal S. 1982. Deep versus compiled knowledge approaches to diagnostic problem solving. In *Proc. Amer. Assoc. Artif. Intell. '82,* pp. 349-354.

Chandrasekaran B. 1986. Generic tasks in knowledge-based reasoning: High building blocks for expert systems. *IEEE Expert,* Fall: 23-30.

Clancey W J. 1985. Heuristic classification. *Art. Intell.* 27: 289-350.

Clarke K, O'Moore R, Smeets R, Talmon J, Brender J, McNair P, Nykänen P, Grimson J and Barber B. 1994. A methodology for evaluation of knowledge based systems in medicine. *Int. J. Art. Intell. Med.,* 6(2):107-122.

Cochrane: *The Cochrane Collaboration.* http://hiru.mcmaster.ca/cochrane/default.htm

Friedman CP and Wyatt J, *Evaluation methods in medical informatics.* Springer-Verlag, New York, 1997

Forsström JF and Rigby M, Addressing the Quality of the IT Tool – Assessing the Quality of Medical Software and Information Services. *Int. J. Med. Inform.,* 1998 (accepted for publication).

Galen: *Generalised Architecture for Languages, Encyclopaedias and Nomenclatures in medicine, a European Union supported project in the Telematics Applications Program, sector Health.* http://www.cs.man.ac. uk/mig/giu

Gremy F. 1988. The role of informatics in medical research. In: *Data, information and knowledge in medicine, developments in medical informatics in historical perspective. Meth. Inf. Med. Special Issue,* pp. 305-309.

Hafner AW, Filipowicz AB, Whitely WP. 1989. Computers in medicine: Liability issues for physicians. *Int. J. Clin. Monit. Comput.,* 6:185-94.

Lundsgaarde H P. 1987. Evaluating medical expert systems. *Social Science and Medicine,* 24 (10): 805-819.

Miller R A, Pople H E and Myers J D. 1982. Internist-1: An experimental computer-based diagnostic consultant for general internal medicine. *N. Eng. J. Med.* 307: 468-476.

Miller P L (Ed.). 1988. *Selected Topics in Medical Artificial Intelligence.* Springer-Verlag, New York.

Musen M A. 1989. *Generation of model-based knowledge acquisition tools for clinical trial advice systems.* PhD Dissertation, Stanford University, USA.

Newell A. 1982. The knowledge level. *Art. Intell.,* 18:87-127.

Nykänen P (Ed.). 1990. *Issues in evaluation of computer-based support to clinical decision making. Report of the SYDPOL-5 Working Group,* Oslo University Press, Reserach Report 127, Oslo.

Nykänen P, Enning J, Talmon J, Hoyer D, Sanz F, Thayer C, Roine R, Vissers M and Eurling F. 1998. Inventory of validation approaches in selected health telematics projects. *Int. J. Med. Inform.,* (accepted for publication).

Pedersen S A, Jensen P F and Nykänen P. 1993. *Epistemological analysis of deep medical knowledge in selected medical domain.* Public Report of the KAVAS-2 (A2019) Project. Commission of the European Communities, Telematic systems in health care.

Pothoff P, Rothemund M, Schwebel D, Engelbrecht R and van Eimeren W. 1988. Expert systems in medicine. Possible future effects. *Int. J. Technology Assessment in Health Care,* 4:121-133.

Prestige: *Patient record supporting telematics and guidelines, a European Union supported project in the Telematics Applications Program, sector Health* http://www.rbh.thames.nhs.uk/rbh/itdept/r&d/projects/prestige.htm

Shortliffe E H. 1976. *Computer-based consultations: MYCIN.* Elsevier, New York.

Shortliffe E H. 1988. Editorial: Medical knowledge and medical decision making. In: *Data, information and knowledge in medicine, developments in medical informatics in historical perspective. Meth. Inf. Med. Special Issue 1988,* 209–218.

Shortliffe E H, Perreault L E, Wiederhold G and Fagan L (Ed.). 1990. *Medical Informatics: Computer applications in health care.* Addison-Wesley, Reading, MA.

Steels L. 1990. Components of expertise. *AI Magazine, Summer 12:* 28-49.

Stefanelli M. 1993. European research efforts in medical knowledge-based systems. *Int. J. Art. Intell. Med.* 5: 107-124.

Talmon J, Enning J, Castadena G, Eurlings F, Hoyer D, Nykänen P, Sanz F, Thayer C, Vissers M. 1998. The VATAM guidelines. *Int. J. Med. Inform.,* (accepted for publication) Guidelines accessible via web-server: http://www-vatam.unimaas.nl

UMLS: *Unified Medical Language System,* http://nlm.nih.gov/research/umls/UMLSDOC.html

Van Bemmel JH and Musen M (Eds.) 1997. *Handbook of Medical Informatics.* Bohn Stafleu Van Loghum, Houten, the Netherlands.

Wyatt J. 1987. The evaluation of clinical decision support systems: A discussion of the methodology used in the ACORN project. In: *Proc. Art. Intell. Med.,* Fox J, Fieschi M and Engelbrecht R, (Eds.) Springer-Verlag, Berlin Heidelberg, pp. 15-24.

184

Expert Systems: Methods and Tools

Ron Summers
Loughborough University

Ewart R. Carson
City University, London

Derek Cramp
City University, London

Biomedical engineering is a fertile domain for the application of computer-based techniques that have the potential to enhance patient care. These techniques come in many forms and can be classified according to the level of processing required. Low-level processing includes techniques such as signal interpretation, where, for example, a threshold value has to be reached before an appropriate action is taken. This type of system uses a shallow data model to transform data into a more useful form. However, in this chapter, the systems of interest are computer programs which provide a further transformation, that of information into knowledge, employing symbolic logic rather than numerical calculations alone to provide appropriate explanation and justification of concepts used and conclusions reached. These so-called expert systems have various synonyms, including *intelligent systems, knowledge-based systems,* and *decision-support systems.*

Expert systems have an interesting historical development. The 1970s saw much of the work concentrated on the problem of knowledge representation. As the introduction of the ubiquitous personal computer aided dissemination of the technology on smaller hardware platforms, the work in the 1980s looked much more deeply into the issues involved in knowledge acquisition. By the 1990s, even with commercial exploitation, the number of expert systems in everyday use is disappointingly small. Research into their evaluation may be one way in which their use will eventually be accepted.

There are no methods and tools that cover every phase of expert system development (Fig. 184.1), so to aid the reader this chapter is subdivided into sections for knowledge acquisition and knowledge representation, respectively; a final catch-all section is used for those methods and tools which did not fit into either section.

184.1 Expert System Process Model

Before embarking on an investigation of the tools and methods available for expert system design and implementation, it is important to identify the distinct phases in their development (see Fig. 184.1). The problem is one in which knowledge from a human expert (or a set of experts) must be transferred to a computer program for dissemination. Domain-specific knowledge is elicited from the human expert by

0-8493-0462-8/00/$0.00+$.50

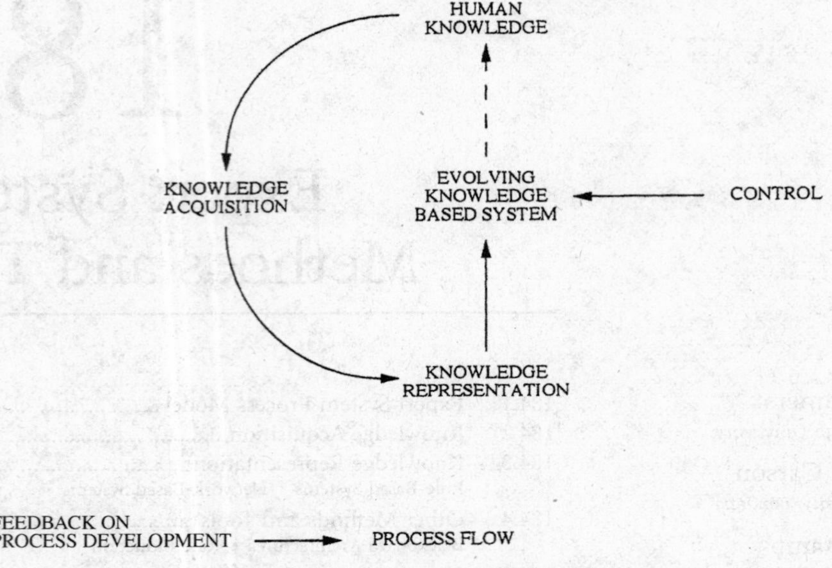

FIGURE 184.1 Knowledge engineering design.

knowledge-acquisition methods. The output of this phase of expert system design is a "paper-based" system. To be of any use, this system must be represented in a form that can be implemented as a computer program. A number of methods and tools are available to perform such tasks; these are detailed below.

A different type of knowledge is required for the control strategy, i.e., decisions on when to use or "fire," specific rules in the knowledge base. In practice, these procedures are equated with different types of search strategies, such as depth-first, breadth-first, or various implementations of optimized search strategies.

What is not captured in Fig. 184.1 are issues dealing with evaluation of the system of human-computer interactions. These are interesting topics worthy of investigation but are outside the scope of this chapter.

184.2 Knowledge Acquisition

The classic method of knowledge acquisition is the interview. Both experts and knowledge engineers are very familiar with the method because it is a part of everyday routine. However, it is not very efficient in capturing and formalizing the knowledge required for an expert system. To try and preserve the interview situation, a more formal method termed a *structured interview* is employed. In this method, the expert is given a list of variables used to describe the domain (e.g., clinical findings) and a second list that contains all possible outcomes of the system (e.g., disease states). The expert is asked to link the two lists, from which a set of production rules can be generated (see "Rule-Based Systems" below for a description of a production rule). It is crucial that all rules are formalized; although a rule of thumb suggests that 90% of the rules are captured in 10% of the development time, it is important that the remaining 10% of the knowledge base is captured. To aid this process, the second stage of the structured interview involves the expert looking at the paper-based rule set generated thus far. The knowledge engineer goes through each rule to test for conditions that affect their appropriateness. From the use of this procedure new rules may be generated.

Advantages of this method are that the experts enjoy the process, which is not too demanding of their time (experts tend to be busy), and it can be used for any type of domain. Disadvantages of the technique, as of any interview, are that it is sometimes difficult to get experts to obey instructions; experts are prone to give anecdotes, and although these may be useful, efficiency of the method deteriorates. The knowledge engineer also requires a certain amount of prior knowledge about the domain to be able to remain astute and alert to changes in theme during the interview.

Protocol analysis is a third method of knowledge acquisition. The knowledge engineer records routine activity by "shadowing" the expert during daily activity. This is a time-consuming method that generally elicits knowledge by the appropriate use of "How?" and "When?" type queries.

Graphic-based methods of knowledge acquisition also have been tried, such as the laddered grid technique, which investigates domain knowledge as a hierarchy. This is also a favored method for automatic elicitation techniques, such as inferring rules from training examples using the ID3 algorithm (see "Rule-Induction").

184.3 Knowledge Representation

Knowledge representation has two main formats: rule-based systems, which take advantage of production rule technology, and semantic networks, which provide a graphic representation of the knowledge base. Below are case studies in both areas, and although dated, it can be argued that they still convey the best exemplar material of their genre. The DENDRAL system was certainly one of the first functioning expert systems, which provided a landmark for further research. MYCIN is one of the most cited expert systems in the literature and one can infer the most well known. CASNET, with its semantic net representation, provides a link with Chapter 181. Further development of this type of model provided the basis for causal probablistic networks and other statistical network approaches to problem solving.

Rule-Based Systems

DENDRAL interpreted mass spectrographic data and worked at a high intellectual level. Although the scope of the system was very narrow, it assumed expert knowledge in what was a difficult task domain [1]. At the time of its first implementation, DENDRAL was a precursor to expert systems and was responsible for the foundation of many of the techniques used in knowledge engineering. These include both knowledge elicitation and knowledge representation. The former was identified to be a "bottleneck" in the design and implementation of expert systems, because it needed prior knowledge of the expertise to be embodied. This expertise could only be elicited in discrete stages due to the time demands made on the human experts involved.

MYCIN was developed by Shortliffe and others at Stanford University [2]. The domain of interest was aiding the identification and recognizing the significance of organisms causing microbial infection and then recommending an optimal treatment protocol.

MYCIN is an example of an expert system that uses "production rules" to represent the causal relationships between individual items of factual knowledge held on the knowledge base. The form of a production rule is as follows:

IF premise assertions are true
THEN consequent assertions are true
with confidence weight *X*.

Notice that each production rule has associated with it a measure of certainty. This aids in the reasoning strategy of the system, where a value of –1 represents complete disbelief and a value of +1 represents complete belief in the consequent assertion(s). The assertions can be boolean combinations of clauses, each of which consists of a predicate statement triple: (attribute, object, value). For example, (Gram stain, *E. coli*, gramneg), indicates that the Gram stain of the *E. coli* organism is gram-negative.

The uniformity of representation for both domain-specific inferences and reasoning goals makes it possible for MYCIN to use a very general and simple control strategy, i.e., a goal-directed backward-chaining of the production rules. This approach can be described in the following way: the first rule to be evaluated is the one that contains the highest-level goal, which for MYCIN is "To determine if there are any organisms, or classes of organisms, that require therapy." To deduce the need for therapy requires knowledge of the infections, which is usually unknown in the first instance. Therefore, the system tries to satisfy subgoals that originate in the premise of the top goal that will allow the infections to be inferred. *Rule chaining* is the name given to the process in which the production rule hierarchy is linked together; the premise portion of each subgoal rule fires a new set of subgoals. This process is repeated until the most fundamental level of the hierarchy is reached, where the rules become assertions that can only be confirmed or denied by directly questioning the user for the appropriate information.

After MYCIN determines the significant infections, found by assessing the overall certainty factor that combines the individual degrees of confidence associated with each production rule, the organisms that account for the infections are found deterministically. Then, if appropriate, the system proceeds to recommend an antimicrobial therapy. To reach its decision, the MYCIN therapy selector uses a description of the infection(s) present, the causal organisms together with a ranking of drugs by their sensitivity, and a set of drug-preference categories. The algorithm used within the therapy selector also calculates the drug dose required and contains knowledge to modify the value if, for example, the patient is in renal failure. An advantage of this therapy selection system is that it can accept and critique a treatment protocol proposed by the user. The therapy selector has an appropriate facility to generate explanation and justification for its choice of action. "Why?" queries are dealt with by displaying the rule it is trying to imply, and if "Why?" is asked again, the query is answered by ascending the goal-tree hierarchy. "How?" queries are interpreted as the chain of rules that are fired to get to that particular conclusion, and if "How?" is repeated, the query is answer by descending the goal-tree hierarchy.

An advantage of using a production rule system is that each rule is a small "packet" of knowledge, each one being independent of all the others. This has two consequences: (1) changing or adding knowledge to the MYCIN knowledge base is relatively easy, and (2) addition of new rules is facilitated by having a modular data structure.

The MYCIN system can deal with both inexact and incomplete information. Inexact information is inherent in this domain because many test results are qualitative in nature, and relatively few statements can be made with absolute certainty. Incomplete information may arise from the time constraints involved in the identification of an organism, i.e., the time taken for an identifying laboratory test to be completed and the result of it made known to the clinician, which in some cases can be the order of 24 to 48 hours.

A disadvantage of the system is that disease states cannot always be adequately described by a rule. Also, it may not always be possible to map a series of desired actions into a set of production rules. Another disadvantage is that although new knowledge can be added by inserting a new rule, this may not interact with the existing rules in the anticipated way.

The MYCIN system is encoded in LISP, and a normal consultation takes on average 20 minutes, which includes time allowed for the optional use of the explanation facility. MYCIN has been evaluated as having a diagnostic success rate of 72%. This relatively poor performance (for an *expert* system) combined with clinician suspicion and resistance to new technology contributed to the fact that the system was not used in a clinical setting. However, as a research tool it was the subject of extensive interest in the AI fraternity, being responsible for a number of successful descendants. One of these, the PUFF system, developed to interpret pulmonary function test results, is one of the few medical expert systems in daily clinical use [3].

Network-Based Systems

CASNET is an expert system for consultation in the diagnosis and therapy of glaucoma [4]. A feature of this system is that the medical knowledge used in the patient-specific reasoning process is encompassed in a causal-associational network model of the specific disease process. Such a network, termed a *semantic net*, allows the structure of the medical knowledge covered by the system to be more coherent. A semantic network comprises nodes connected by links, where nodes correspond to either the condition or action part of the rules and the links are the inferences between the two. The model of disease is separate from the decision-making strategy, which allows the updating of both data structures to be facilitated more easily (compare with "Causal Probablistic Modeling" in Chapter 181).

The CASNET model has a descriptive component that comprises four sets of elements as follows: observations (these consist of symptoms and laboratory test results, etc. and form the direct evidence that a disease is present), pathophysiologic states (these describe internal abnormal conditions or mechanisms that can directly cause the observed findings; the causal relations between states are of the form

$$n_i \xrightarrow{a_{ij}} n_j$$

where n_i and n_j are pathophysiologic states, and a_{ij} is the causal frequency with which state n_i, when present in a patient, leads to state n_j, disease categories (each category consists of a pattern of states and observations and is therefore conceptually at the highest level of abstraction), and treatment plans and therapies (composed of sets of related treatments or treatment plans).

Other components of the CASNET model are the decision rules, which state the degree of confidence with which an inference of a pathophysiologic state can be made from an observed pattern of findings. In rule form, this translates to

$$t_i \xrightarrow{Q_{ij}} n_j$$

where t_i is a finding (or observation) or boolean combination of findings, n_j is pathophysiologic state as before, and Q_{ij} is a number in the range –1 to +1 representing the confidence with which t_i is believed to be associated with n_j. The value of Q is a certainty factor that indicates the belief that the patient is in state n_j. A threshold function is then used to determine whether or not the certainty factor confirms, denies, or leaves undetermined a particular state. Rules for associating disease categories and pathophysiologic states to treatment protocols are in the form of a classification table that consists of ordered triples,

$$\left(n_1, D_1, T_1\right), \left(n_2, D_2, T_2\right), \ldots, \left(n_i, D_i, T_i\right)$$

where n_i is the pathophysiologic state, D_i is the disease process arising from it, and T_i are the preferred treatment regimes for disease D_i.

The pathogenesis and mechanisms of a disease process are described in terms of cause-and-defect relationships between pathophysiologic states. In this way, complete or partial disease processes can be characterized by pathways through the network. When a set of a cause-and-effect relationships is specified, the resulting network can be described as an acyclic graph of states. This state network is defined by a four-tuple (S, F, N, X), where S is the set of starting states (i.e., those states which have no antecedent causes), F is the set of final states, N are the number of states visited between S and F, and X are the causal relationships between the states visited (in the form of a list).

An important test for any expert is how it copes with contradictory information. This can occur in the CASNET system, e.g., when a particular state is confirmed even though all its potential causes in the network are denied. When such a situation occurs, the CASNET system advises the user that this has happened. There are two explanations for the origin of this contradictory information: the model of the disease process may be incomplete (i.e., one or more nodes together with their respective causal relationships are missing from the knowledge base), or it may be that the threshold function associated with one of the existing causal links has been set at an inappropriate level. The modular structure of the knowledge base enables easy access to the underlying reason for the contradiction, so updating of the knowledge base can be facilitated.

An advantage of the CASNET system is that it can present alternative expertise derived from different consultants. It is this, coupled with the fact that glaucoma is a well-defined clinical problem, that results in an accuracy of diagnosis, which contributed to its relative success. The diagnostic accuracy of CASNET has been evaluated at greater than 75% for particularly difficult cases and greater than 90% for cases that constitute a broad clinical spectrum. Unfortunately, the clinical utility of the system was not assessed as high, which resulted in CASNET never being used in a routine clinical environment. However, the system was a very useful tool for research into AI in medicine and spawned further interest in other approaches to causal modeling.

184.4 Other Methods and Tools

Blackboard Architectures

The HEARSAY-II speech understanding system is a research program of Carnegie-Mellon University [5]. A feature of this system is its general problem-solving framework, termed the *blackboard architecture*, which has become the basic control model for real-time intelligent signal understanding systems [6]. The blackboard architecture contains four elements: *entries* (which is the name given to intermediate by-products of the problem-solving strategy), *knowledge sources* (these are diverse and independent subprograms that are capable of solving a specific problem efficiently; they can be though of as a condition-action duplex, where the condition premise describes the situations to which each knowledge source can contribute, and the action component specifies the interaction of that particular knowledge source in the overall problem-solving strategy; it is the interaction of that particular knowledge source in the overall problem-solving strategy; it is the activated knowledge sources that produce the entries), the *blackboard* itself (which can be considered as a structured database; it serves two roles; first, it represents intermediate states of problem-solving activity, and second, since all communication between the individual knowledge sources is carried out via the blackboard, it can accommodate hypotheses from one particular knowledge source that may activate a number of others), and an *intelligent control mechanism* (which has sufficient knowledge embodied within it to decide if and when any particular knowledge source should be activated, thus generating entries that are recorded on the blackboard).

The blackboard control structure supports a hierarchical arrangement of knowledge sources. For instance, in the HEARSAY-II system, the top level of the hierarchy is the spoken sentence, followed by phrase, word sequence, word, syllable segment, and phoneme, respectively. The control postulate is that of "establish-and-refine," i.e., each activated knowledge source in the hierarchy tries to confirm or reject itself. If it is confirmed, then it refines itself by calling on the next immediate layer in the hierarchy to enter the same cycle. This procedure continues until (if successful) the bottom layer of the hierarchy is reached. Once the phonemes have established themselves, the spoken sentence has been modeled completely. The spoken sentence is then recognized as syntactically valid, which is one of the definitions used for speed "understanding." Further signal-processing techniques can be applied to obtain a more comprehensive understanding, such as translation into another language or paraphrasing an argument.

The blackboard architecture as been used successfully in providing interpretation of the results of blood-gas analysis [7].

Rule Induction

An expert system shell based on the rule-induction method of knowledge acquisition and representation is available and is based on the ID3 induction algorithm [8]. The original goal of the ID3 algorithm was to seek out the valuable information from large volumes of data by means of a classification rule. ID3 uses known results described in terms of a fixed set of attributes and produces a decision tree that is able to classify all possible results correctly. To optimize the decision function, the concept of "information entropy" is used.

These features are faithfully reproduced in the 1st Class Expert System shell (registered trademark of 1st Class Expert Systems, Inc.). A hypertext facility is added to provide explanation or justification as well as further information for any attribute or result obtained. Using optimization of the decision function ensures that the minimum set of questions are asked of the user to obtain the required result.

As a consequence, this method is often more efficient and thorough than the human expert. The system based on the 1st Class Expert System shell is robust to missing and conflicting data. When classification is not possible, explanations can be provided by the hypertext facility. This system has been used in biomedical engineering applications such as machine malfunction, where alarms generated by the expert system are quicker and more informative than those set by the manufacturer.

184.5 Summary

The process of knowledge engineering design has been discussed in some detail. The phases of knowledge acquisition and knowledge representation have been presented, and the methods used in each phase have been discussed. The advantages of the structured interview method of knowledge acquisition over other available methods can be deduced. Knowledge representation is presented as a choice between rule-based systems and semantic networks.

MYCIN is an example of a procedural expert system, to be compared with CASNET, which is an example of a declarative expert system. The distinction between the two types hinges on the relationship between the encoded knowledge representation and the control algorithm. In systems where the knowledge representation and control algorithm are combined, then a procedural definition applies. This is exemplified by the production rule methodology, where causality is implicit in the IF...THEN rules. The ability to provide explanation of action, one of the prerequisites of an expert system, cannot naturally be accommodated. However, to overcome this problem, context sensitive explanations can be used. These are a set of fixed files, each one specific to a particular production rule, which can be fired whenever the production rule they represent is used. A disadvantage of this approach is that the explanation can appear "stale" to an experienced user of the system. In systems where the knowledge representation and control algorithm are separate, then a declarative definition applies. The knowledge representation is explicitly defined, allowing a set of algorithms to act on it. This can include the test and hypothesis diagnostic cycle, an explanation-generating algorithm, and an algorithm that produces text which constitutes computer-based advice.

References

1. Lindsay RK, Buchanan BG, Feigenbaum EA, Lederberg J. 1980. Applications of Artificial Intelligence for Organic Chemistry: The DENDRAL Project. New York, McGraw-Hill.
2. Shortliffe EH. 1976. Computer-Based Medical Consultations: MYCIN. New York, Elsevier.
3. Aikins JS, Kunz JC, Shortliffe EH, Fallat RJ. 1983. PUFF: An expert system for interpretation of pulmonary function data. Comput Biomed Res 16:199.
4. Weiss S, Kulikowski CA, Amarel S, Safir A. 1978. A model-based method for computer-aided medical decision making. Artif Intell 11:145.
5. Erman LD, Hayes-Roth F, Lesser VR, Reddy DR. 1980. The Hearsay-II speech understanding system: Integrating knowledge to resolve uncertainty. Comput Surv 12:213.
6. Hayes-Roth F, Waterman DA, Lenat DB. 1983. Building Expert Systems. Reading, Mass, Addison-Wesley.
7. Chelsom JJL. 1990. The interpretation of data in intensive care medicine: An application of knowledge-based techniques. Ph.D. thesis, City University, London.
8. Quinlan JR. 1979. Rules by induction from large collections of examples. In D Michie (ed), Expert Systems in the Microelectronic Age. Edinburgh, Edinburgh University Press.

185

Knowledge Acquisition and Representation

Catherine Garbay
Laboratoire TIMC/IMAG

185.1 Introduction

The scope of designing knowledge-based systems in medicine has evolved during these last few years, from pure decision making automata to open environments able to acquire, formalize, validate, as well as handle and disseminate elements of the human expertise. The application fields have expanded in a correlated manner, from the sole field of symbolic processing to data analysis, information management, simulation, and planning.

Therefore, the scope of knowledge acquisition and representation has been shifted from the mere transcription of expert knowledge to the modeling of pathophysiological phenomena of increasing complexity [Stefanelli, 1993]. To reflect such evolution, the notion of models is made central in this chapter and the nature and specificity of medical expertise is addressed in the first section. Knowledge acquisition is considered afterwards, by successively considering human-centered and data-driven approaches to knowledge extraction. Knowledge representation issues are finally discussed in the framework of their evolution from basic representation schemes to second generation systems.

185.2 Medical Expertise: Domain and Control Knowledge

Expert knowledge is usually described as involving domain knowledge and control knowledge [Schreiber, 1993]. Domain knowledge is meant to represent the various elements, e.g., facts, features, relations, and concepts, that are used to represent a case, while control knowledge is meant to represent the expert know-how, that is, the way to reason about a case.

Domain Knowledge

Domain knowledge may be seen as a kind of formal knowledge tying names, properties and relations to the whole empirium of data, facts, observations, hypotheses or results that are extracted from a case.

0-8493-0462-8/00/$0.00+$.50

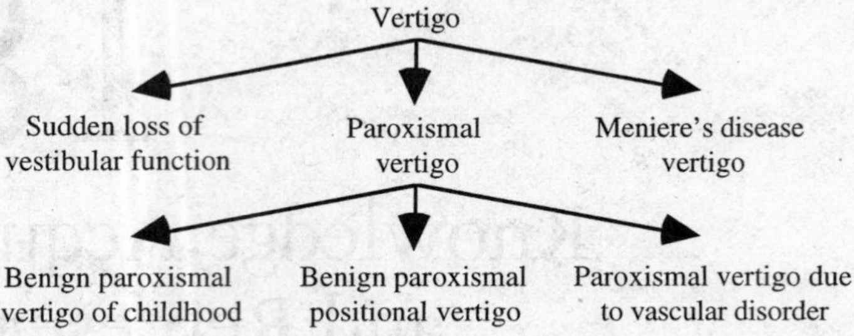

FIGURE 185.1 Vertigo taxonomy (partial view). (*Source*: Schmid [1987].)

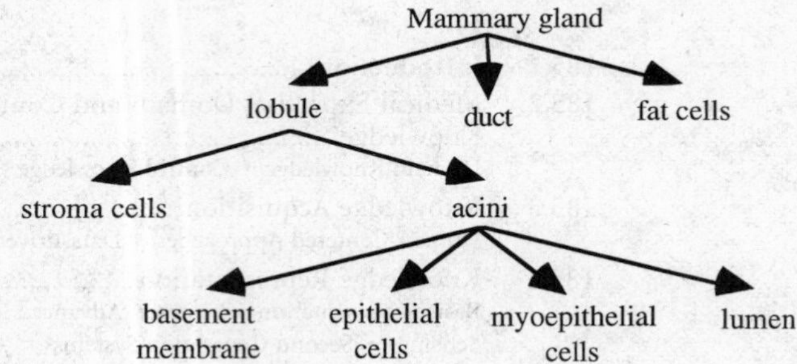

FIGURE 185.2 A "part of" hierarchy example (partial view). (*Source*: Garbay and Pesty [1988].)

Domain ontologies are used to structure and organize domain knowledge according to well-identified knowledge types or relations.

Taxonomic ontologies have been widely used in medicine because of their capacity to classify and organize diseases into hierarchical structures. As may be seen in Fig. 185.1, paroxismal vertigo is classified as a particular kind of vertigo, and may be further characterized as benign paroxismal vertigo of childhood, benign paroxismal positional vertigo, or paroxismal vertigo due to vascular disorder [Schmid, 1987].

"Part of" hierarchies make it possible to describe the compositional characteristics of concepts, and are particularly useful for describing anatomical properties. As may be seen in Fig. 185.2, a mammary gland lobule may be described as composed of stroma cells and acini, which in turn are composed of a basement membrane, epithelial and myoepithelial cells, and a lumen [Garbay and Pesty, 1988].

Knowledge may also be organized according to its role in the reasoning process. According to Patel et al. [1989], this process is decomposed into several levels, from the empirium (the whole universe of medical problem solving data) to the global complex level (complete understanding of the patient case). Several intermediate levels are distinguished in terms of observations, findings, facets (complex gathering of findings understood as the consequence of a pathophysiologic process) and diagnostic, as summarized in Fig. 185.3.

Current research emphasizes the notion of causal ontologies (see Fig. 185.11), where a deep functional understanding of the process leading to a disease is sought [Jensen et al., 1987; Stefanelli, 1993].

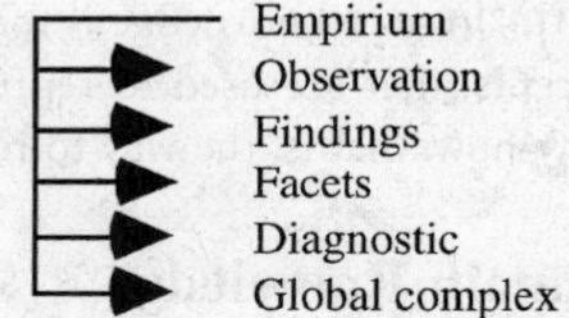

FIGURE 185.3 Organizing knowledge according to its role in the reasoning process. (*Source*: Patel et al. [1989].)

Control Knowledge

Reasoning may be seen as the process of handling facts, observations and hypotheses, to reach a final understanding of a situation. Various mental abstraction levels are traversed in this process, which

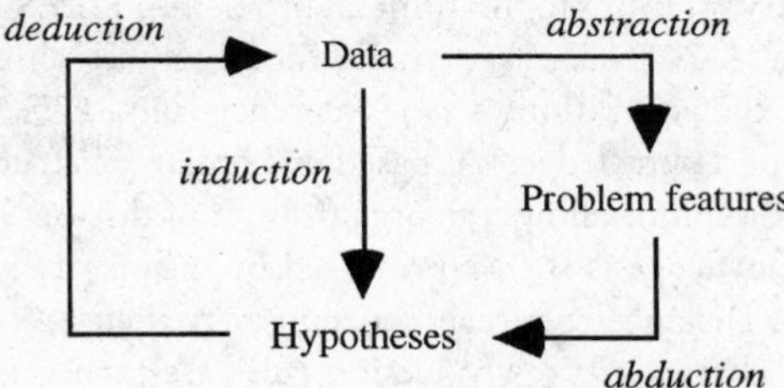

FIGURE 185.4 An epistemological model of diagnostic reasoning. (*Source*: Lanzola and Stefanelli [1992].)

may involve different inference schemes, depending on the type of data to be processed (e.g., uncertain, geometric or temporal), and on the type of inferencing to be applied (e.g., deductive, inductive, or non-monotonic).

An epistemological model of diagnostic reasoning is proposed by Lanzola and Stefanelli [1992] in terms of abstraction, abduction, deduction, and eliminative induction (Fig. 185.4). The abstraction corresponds to identification of the clinical features that diagnosis should be able to explain. The abduction corresponds to formulation of a set of hypotheses from which expected manifestations are deduced. Eliminative induction is finally used to reduce uncertainty of inferred hypotheses by matching their consequences against observed data.

The global approach to a case nowadays is often formalized in terms of tasks [Bylander and Chandrasekaran, 1987; Steels, 1990; Schreiber et al., 1993], which makes it possible to structure the overall reasoning process in terms of goals and sub-goals, and thus to organize knowledge and know-how according to a problem-solving rationale [Newell, 1982].

185.3 Knowledge Acquisition

Knowledge acquisition has been traditionally considered as the process of extracting and transcribing the expert knowledge into a computer-usable form. This complex process has been shown to critically depend on the knowledge engineering tool at hand, since most representation formalisms convey restrictive and rigid semantics, which do not refer to a clear ontology [Schreiber, 1993]. Current trends are to consider knowledge acquisition as an integral part of the designing process, centered on the notion of a model.

Knowledge discovery, data mining, and machine learning techniques have recently attracted considerable attention, due to the growing amount of data available, and to the growing necessity to base the reasoning on evidence taken from physical measurements. Data-driven approaches to knowledge extraction have been developed as a consequence, complementing the more traditional human-centered approaches, by enabling systems to create new knowledge, update existing knowledge, and improve their performance without intervention and reprogramming.

Human-Centered Approaches

Interviewing Techniques

Knowledge is usually acquired through direct interaction with a human expert, but may also be obtained from many other sources, such as textbooks, reports, databases, or case studies [Buchanan, 1983]. A systematic and structured approach is necessary to ensure the completion and consistency of the interviewing process, which is usually driven by a knowledge engineer. Other techniques are often used to complement the mere interviewing approach, such as observing the expert in daily working situations, writing scenarios and reports, or filling out questionnaires.

Knowledge Acquisition Tools

Several computerized tools have been designed in the past to assist the knowledge acquisition/transcription process. The reference work done in TEIRESIAS [Davis and Lenat, 1982] and ETS [Boose, 1985] appears to bring complementary solutions to this difficult problem. TEIRESIAS is rather meant to support the testing and refinement of a "MYCIN like" system. It proceeds by searching for missing or inconsistent information and may suggest the generation of new inference rules. ETS, on the contrary, is meant to support direct elicitation of expert knowledge. It is based on bipolar constructs structuring the perception of the world (e.g., the grade of a tumor allows the organization of diseases according to their prognostic significance). The acquired knowledge is then represented by means of so-called repertory grids tying concepts and their properties. Distances between concepts may then be computed, as well as the discriminatory power of a given characteristic. Production rules are then generated and submitted to the expert to be modified or tested.

Knowledge Acquisition Environments

Knowledge acquisition environments have been designed to integrate several acquisition techniques. The expert is usually assisted by a cognitive engineer to use the software environment. AQUINAS [Boose and Bradshaw, 1987] and MORE [Kahn et al., 1985] are particularly representative of this approach. AQUINAS is based on the cooperation between several interviewing methods and has been designed as an extension of ETS. MORE may interview the expert according to eight different strategies and integrate several facilities such as a rule generator, an inference engine to test the rules, and online assistance to the user.

Knowledge Acquisition Methodologies

The term "knowledge acquisition methodologies" refers to approaches aiming at a close integration between knowledge acquisition and system design, in which knowledge acquisition is considered as a modeling process rather than as an extraction/transcription problem. The modeling approach permits the investigator to work at the "right" abstraction level, often called the knowledge level [Newell, 1982], and to refer a level at which knowledge may be expressed without referring to any implementation constraint [Schreiber et al., 1993]. It is formulated in opposition to symbol level approaches, which support implementation level modeling.

KADS (KBS Analysis and Design Structured Methodology) is among the best known methodologies [Breuker and Wielinga, 1987], and has been conceived to guide the whole process of expert system design. It permits the progressive evolution from real world to implementation by the building of four successive models that organize the domain ontology into four knowledge levels: domain level (description of concepts and the relationship between them), inference level (reasoning primitives), task level (reasoning steps, goals and ways to reach them, structured task) and strategy level (control and goal scheduling).

Data-Driven Approaches

Knowledge Discovery and Data Mining

Knowledge discovery may be defined as "the non-trivial extraction of implicit, previously unknown, and potentially useful information from given data" [W.J. Frawley et al., 1991]. Such a process makes use of various techniques, involving segmentation, filtering, statistical as well as model-based analysis, or supervised induction.

It is most often considered as a pre-processing step in the design of decision-support systems, by providing computerized support to the identification of salient attributes, to the computation of representative values for these attributes, or to the structurization of the database into semantically meaningful clusters, thus improving performance. Factor analysis may, for example, be used in this respect. Such an approach has been used in a system devoted to the development of *in vitro* fertilization treatment plans, where a case-based reasoning module has been extended by a knowledge mining component [Juristica et al., 1998].

Learning

Three main approaches to machine learning are usually distinguished, i.e., inductive learning, analogical reasoning and explanation-based learning. Inductive learning techniques have been widely used in medicine, in the framework of decision-trees or version space/conceptual clustering methods, and more recently by the introduction of the evolutionary and connectionist paradigms.

A typical example of inductive learning is ID3, a decision-tree learning technique developed by Quinlan [1986]. ID3 constructs a classification tree in a recursive, top-down, divide and conquer fashion, using the concept of entropy as a quantitative measure of the "information" conveyed by each attribute. Conversely to ID3 which proceeds by division, the approach of version space/conceptual clustering proceeds in a bottom-up way, by grouping sets of domain concepts to form more general ones.

Genetic-based learning has been conceived as an extension of the evolutionary programming paradigm: it has been designed as a generate-and-test process that maintains a pool of competing problem-solving procedures (most often decision rules), tests them to a limited extent, creates new ones from those that performed well in the past, and prunes poor ones from the pool. Connectionist approaches are revealed to be particularly useful in domains where it is impossible to structure and represent the domain knowledge in logic form.

Machine learning is now introduced in an increasing number of systems and application domains in medicine, especially in domains where the knowledge is difficult to acquire by hand. Current approaches make use of real data and domain models to drive the knowledge acquisition process.

Hau and Coiera [1997] for example, describe a learning system that takes real-time patient data obtained during cardiac bypass surgery to create models of normal and abnormal cardiac physiology. Conversely, qualitative models of heart disease were used early in Kardio to learn diagnostic rules [Bratko et al., 1989]. Examples of "faulty" and "non-faulty" behaviors are generated by the model and transmitted to a learning program, which can then learn diagnostic rules for each particular fault.

Machine learning may also be used to refine existing knowledge bases in cases where there is a disagreement between the expert-provided domain knowledge and actual cases. An application to acute abdominal pain has been experimented with for effective performance improvement [Dzeroski et al., 1997], using the theory revision system NEITHER. As a consequence of this process, rule R1, for example, has been specialized into rules R6 and R7:

rule R1: OPERATE ← (RIGIDITY YES)
rules R6: OPERATE ← (RIGIDITY YES) (TYPE STEADY) (SITE-RUQ NO)
rules R7: OPERATE ← (RIGIDITY YES) (TENDER-RLQ SURFACE)

185.4 Knowledge Representation

The representation of medical knowledge is a very active research field, characterized by a wide range of tools, models, and languages, which, together with the presence of increasing computer capabilities, allows one to specify and emulate systems of a growing complexity. Knowledge representation schemes indeed have known an important evolution, from basic schemes supporting a rather heuristic approach, to advanced schemes involving a deeper consideration of the various dependencies between knowledge elements. Second generation systems have finally been introduced to cope with complex integration and modeling constraints.

Basic Representation Schemes

Basic schemes for the representation of domain and control knowledge are considered in this section. The representation of concepts, their structure and properties is considered first. How to model the expert way of reasoning is examined afterwards. It is examined successively as the ability to apply well-experimented inference schemes (usually represented by rules) and as the capacity to compare and retrieve learned situations (so-called case-based reasoning). One main feature of human reasoning, e.g., its capacity to handle uncertain and imprecise information, is finally considered.

Frame Representations and Object-Oriented Languages

Frame-based representations have been introduced as a way to organize prototypical knowledge about the world, and are currently among the most widely used tools for the construction of AI systems. A

Frame Cytomegalovirus
- sort-of — intestitial pneumonia
- epidemiological data
  - other kind of pneumonia — yes
  - CD4 — less than 200/mmc
- clinical picture
  - symptoms — fever, non-productive cough, dyspnea
  - signs — cyanosis tachypnea
- laboratory tests
  - PO2 — lowering
  - LDH — growing

FIGURE 185.5 Using frames to represent diagnostic knowledge. (*Source*: Fiore et al. [1993]).

frame defines the prototypical description of concepts sharing similar properties and behavior. It is defined as a set of slots describing the concept attributes and their values. A variety of other slots may be introduced, including procedures, default and current values, or type/value constraints. The frames are organized in hierarchies, tying classes and instances. The basic inference mechanism is *instanciation*, in which the attribute values of the new instance are obtained either by inheritance, by computation (using the procedural attachments), or by default.

In a knowledge-based system to diagnose HIV-pneumonias [Fiore et al., 1993], frames are used to describe the physiopathological states of the disease for each particular pneumonia, in terms of several slots describing the epidemiological data, clinical picture, laboratory tests, and diagnosis (Fig. 185.5).

The object-oriented paradigm is nowadays increasingly considered as a way to cope simultaneously with knowledge modeling and software engineering issues, as in HELIOS [Coignard et al., 1992]. HELIOS has been conceived as a software engineering environment to facilitate the development of medical applications. It provides a core set of medical classes, from which to build a new application. Several services are also provided, which include the acquisition and management of medical information, the processing of natural language queries, as well as data driven decision support.

Semantic Networks and Conceptual Graphs

Semantic Networks [Quillian, 1968] have been proposed as a flexible formalism to represent the semantics of concepts (the nodes in the network) and their relations (the arcs) in the framework of a graph-like structure. The reasoning proceeds by unification between an unknown fact and a known concept or subgraph in the network.

The approach has rapidly been extended to represent causal relationhips as well as hierarchical taxonomies and to include frames. It has been applied successfully to numerous applications in medicine, as in CASNET [Weiss et al., 1978], a consultation system for glaucoma, INTERNIST [Pople, 1977], a diagnostic consultant in internal medicine, and PIP [Pauker et al., 1976] a system devoted to renal disease. The reasoning strategy in these systems is of the event-driven type: initial data triggers a number of hypotheses, which are then to be confirmed [Kulikowski, 1984]. The subgraph of confirmed or undetermined hypotheses constitute a patient-specific model. Various weights and scores are usually introduced to render the reasoning strategy more flexible.

Conceptual graphs [Sowa, 1984] have been designed as an extension of the previous formalism in which an explicit representation of the links between concepts is sought. A conceptual graph is a directed graph comprising two kinds of nodes: concept nodes and conceptual relationship nodes. The representation is grounded on first-order logic and is currently being applied in a number of projects for the modeling and representation of medical terminology, as in GALEN [Alpay et al., 1993]. One major goal of GALEN (Generalized Architecture for Languages Encyclopedias and Nomenclatures in Medicine) is to form a general Terminology Server able to conciliate a number of classifications and nomenclatures. It is based on a Semantical Encyclopedia of Terminology connected to a Multilingual Information Module.

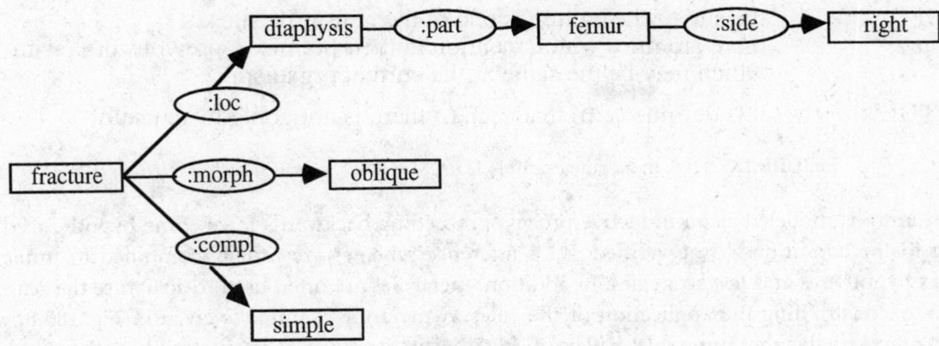

FIGURE 185.6 Conceptual graph representing the concept "simple oblique diaphyseal fracture of the right femur." (*Source*: Bernauer and Goldberg [1993]).

Conceptual graphs are used in this project to generate natural language expressions from information coded according to the GALEN Master Notation.

Conceptual graphs may also be used as a formalism that supports both expressiveness and classification purposes, i.e., as a knowledge representation formalism able to integrate both terminological and domain knowledge, as shown in Fig. 185.6 [Bernauer and Goldberg, 1993].

In the example in Fig. 185.6, the composite concept "simple oblique diaphyseal fracture of the right femur" is described as a nested set of 2-tuples each tying a relation (:loc, for example) and a composite concept (diaphyis, for example). Taxonomic as well as compositional relations are used to support classification. The equivalent nested notation for the graph in Fig. 185.6 is the following:

```
(fracture
        (:loc       (diaphysis    (:part   (femur     (:side    (right))))))
        (:morph     (oblique))
        (:compl     (simple)))
```

Rule-Based Reasoning

The use of if-then rules (also called *production rules* or *condition-action pairs*) is a straightforward and popular way to represent the expert know-how. A rule example is given in Fig. 185.7, in which a deduction is made about the presence of bacteroids based on a composite observation.

This type of knowledge is often said to be shallow or heuristic, since it is a largely empirical and nonformalized representation mechanism. Given a set of facts, the reasoning process is then modeled as the successive activation of rules, which in turn produce new facts to be considered, until no applicable rule can be found.

The rules are handled by means of an inference engine, which may work under two inferencing modes, the data driven or goal driven mode. In the data-driven mode, the reasoning is modeled as a deductive process proceeding from a given premise to some conclusion. In the goal-driven mode, on the contrary,

IF	the infection is primary-bacteremia
and	the site of the culture is one of the sterile sites
and	the suspected portal of entry of the organism is the gastro-intestinal tract
THEN	there is a suggestive evidence (0.7) that the identity of the organism is bacteroids

FIGURE 185.7 A production rule example from MYCIN. (*Source*: Shortlife [1976].)

IF	the site of the culture is one of the nonsterile sites
and	there are rules which mention in their premise a previous organism which may be the same as the current organism
THEN	it is definite (1.0) that each of them is not going to be useful

FIGURE 185.8 A meta-rule example from MYCIN. (*Source*: Shortlife [1976].)

the reasoning is modeled as an inductive process proceeding backwards from some hypothesized conclusion to the conditions to be verified. Both inference schemes are usually combined to implement complex hypotheses and test strategies. In addition, metarules are often used to structure the reasoning process by constraining the application of the rules. A metarule example is given in Fig. 185.8, which permits one to predict that some rules will be poorly informative, given the non-sterility of the culture site.

MYCIN, a system for diagnosing and treating infectious blood diseases, is one of the earliest and most widely known expert system in medicine [Shortliffe, 1976].

Uncertain and Imprecise Reasoning

How to model the uncertainty and imprecision that is central to the human reasoning style has raised a number of research activities during the last decades [van Ginneken and Smeulders, 1988]. The Bayesian model has been for a long time the primary numerical approach to uncertain reasoning. It is based on assigning a probability distribution to each of the variables representing the problem at hand. These probabilities express the uncertainty and likelihood of occurrence of symptoms as well as diseases. Given *P(Di)*, the *a priori* probability of disease *Di*, given *P*(*S*|*Di*), the probability that symptom *S* may occur in the context of disease *Di*, Bayes' rule (1) allows one to compute *P*(*Di*|*S*), the probability that disease *Di* occurs when *S* is observed, according to the following formula:

$$P\left(Di \middle| S\right) = \frac{P\left(S \middle| Di\right) * P\left(Di\right)}{\Sigma_j \; P\left(S \middle| Dj\right) * P\left(Dj\right)} \quad (185.1)$$

The most important assumption underlying the use of Bayes' rule is that all disease hypotheses must be mutually exclusive and exhaustive. It means that only one disease is assumed to be present. If this condition is met, Bayes has a completely predictable performance and guarantees a conclusion with minimum overall error [van Ginneken and Smeulders, 1988].

Some intuitive methods for the combination and propagation of these numbers have been suggested and used, as in MYCIN (see the rules in Figs. 185.7 and 185.8, where coefficients are used to weight their conclusions). The empirical character of the approach has often been criticized. However, it is consistent with the heuristic style that is inherent to rule-based modeling.

The theory of possibility [Zadeh, 1978] has been used in medicine to represent the vagueness of clinical predicates. These vague predicates are represented by means of a fuzzy set and a possibility distribution (or membership function). Fuzzy reasoning may then be implemented by means of rules handling fuzzy facts.

Examples of membership function modeling criteria for the classification of rheumatoid arthritis are given in Fig. 185.9 [Leitich et al., 1996]. Seven criteria are defined in total, from which at least four must be present in conjunction to establish a diagnosis.

Case-Based Reasoning

CBR (Case-based reasoning) emerged in recent years as a powerful problem-solving technique applicable to a wide range of tasks in Artificial Intelligence. It is based on the hypothesis that new situations are analyzed by reference to past experience, i.e., by considering their similarity to well-experimented situations, a notion that is made central in this kind of reasoning [Campbell and Wolstencroft, 1990]. New problem-solution pairs may then be learned as a consequence: this inherent combination of problem

Criterion s_i	Membership function $\mu_{si}(x)$
s_1: morning stiffness lasting at least one hour	$\mu_{s1}(x) = \begin{cases} 0 & \text{if } t \leq 15 \\ (t-15)/45 & \text{if } 15 < t < 60 \\ 1 & \text{if } t \geq 60 \\ \nu & \text{if } t \text{ is unknown} \end{cases}$ where t is the duration of morning stiffness in minutes
s_2: arthritis of 3 or more joint areas	$\mu_{s2}(x) = \begin{cases} 0 & \text{if } n \leq 1 \\ 0.5 & \text{if } n = 2 \\ 1 & \text{if } n \geq 3 \\ \nu & \text{if } n \text{ is unknown} \end{cases}$ where n is the number of involved joint areas

FIGURE 185.9 Membership functions examples in the field of rheumatoid arthritis. (*Source*: [Leitich et al., 1996].)

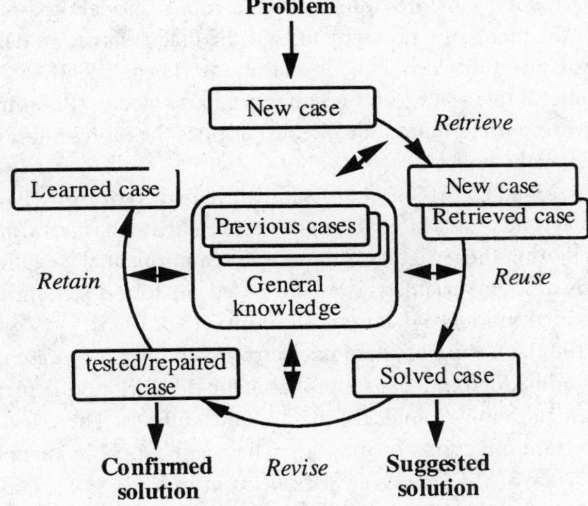

FIGURE 185.10 The Case-based reasoning cycle, a general view. (*Source*: [Aamodt and Plaza, 1994].)

solving with learning through problem solving experience gives a particular strength to CBR over most other methods [Aamodt and Plaza, 1994].

Case-based reasoning finally appears as a cycle comprising four stages, as illustrated in Fig. 185.10: retrieve the most similar case, reuse the retrieved information to solve the problem by analogical reasoning, revise the proposed solution, and finally retain the part of the experience likely to be useful for future problem solving.

The advantage of the approach is two-fold: first of all, it is based on the description of experiences, rather than on the modeling of generic or abstract knowledge, second, such a system may evolve in an incremental way, as cases grow [Goos and Schewe, 1993]. However, the choice of the similarity measure turns out to be critical with respect to the reasoning accuracy. The structurization of the case base also has to be carefully considered in order to avoid combinatorial search [Gierl et al., 1993].

Current trends [Reateguiet al., 1997] are to combine CBR with complementary forms of reasoning such as rule-based, model-based, or neural networks (NN) to solve diagnosis problems. The NN here are used during the consultation process to make hypotheses of possible diagnosis solutions and to guide the search of similar cases, restricting the research to cases with similar interpretations (e.g., diagnoses).

Advanced Representation Schemes

Recent approaches to knowledge representation consider as central the capacity to better situate the reasoning with respect to contextual, causal, and temporal modeling of knowledge elements. A number of knowledge domains are concerned, involving therapy and monitoring, pathophysiology, or planning.

Context-Based Reasoning

It is now well recognized that advanced medical decision systems should situate their reasoning with respect to the context in which a problem is considered, including the patient clinical context, but also the chronology of events together with their causal relationships.

In connection with the development of the knowledge-based system TheMPO (Therapy Management in Pediatric Oncology), which supports therapy and monitoring in pediatric oncology, a graph-grammar approach has been used to design and implement a graph-oriented patient model which allows the representation of non-trivial (causal, temporal etc.) clinical contexts [Müller et al., 1996]. A graphical interface has been designed to facilitate the specification and retrieval of contexts by the physician.

Causal Reasoning

The last decade has known the development of solid foundations for diagnostic systems employing logic, probability theory, and set theory, thus providing rich and robust modeling tools to the developer, and allowing him to cope with the increasing necessity to base the diagnosis on models of the disease process and models of the structure and function of the human body [Lucas, 1997].

There are at least two alternative ways to represent causal knowledge [Ramoni et al., 1992]. The first method is based on a network representation, the second implies the representation of both the structure and behavior of a physiological system.

Network representations, called CPN (causal probabilistic netwoks) or BBN (Bayesian belief networks), are in the form of a direct acyclic graph in which nodes represent stochastic variables and arcs represent conditional dependencies among the variables. These representations make explicit the dependency and independency assumptions among variables. They are based on sound semantics and easily extend to compact representations called influence diagrams [Ramoni et al., 1995].

MUNIN [Jensen et al., 1987], an expert system in electromyography, is representative of this approach. The domain knowledge is embedded in a causal probabilistic network and is further divided into three levels, representing diseases, pathophysiological features, and findings. These levels are linked by causal relations: diseases cause certain affections in muscles. These affections, in turn, cause expectations for certain findings. A partial view of the causal network is given in Fig. 185.11. Diseases are grouped into a single node (on the left), and characterized by their grades. The disease node is connected to several pathophysiologic nodes describing the pathophysiologic status of a muscle in terms of discrete properties (the muscle structure is rather normal, there is no myotonia, and no muscle loss). Some expectations regarding, for example, the muscle force and atrophy or the presence of spontaneous myotonic distrophy are finally obtained. Probabilities are used to characterize each state in the network (the horizontal bars in the figure) and propagated through the causal links.

Despite its effective representation power, the approach still bears strong limitations, due to the large amount of information required: current propagation algorithms require that all the conditional probabilities defining a conditional dependency be known, together with all the *a priori* probabilities attached to the root variables. MUNIN, for example, consists of about 1100 discrete random variables linked together, thus rendering its managing hardly tractable by current computer technology. Several methods have been investigated by the author of MUNIN to reduce the memory space and calculation time [Suojanen et al., 1997]. A methodology for automatically inducing Bayesian network has been introduced by Larranaga et al. [1997], based on genetic algorithms, and applied to the prediction of survival in malignant skin melanoma.

An alternative way is to represent both the structure and behavior of a physiological system [Ironi et al., 1993]. The system structure is simply given in terms of variables and relations. The system behavior is then modeled by a set of mathematical equations tying related variables, which represent potential

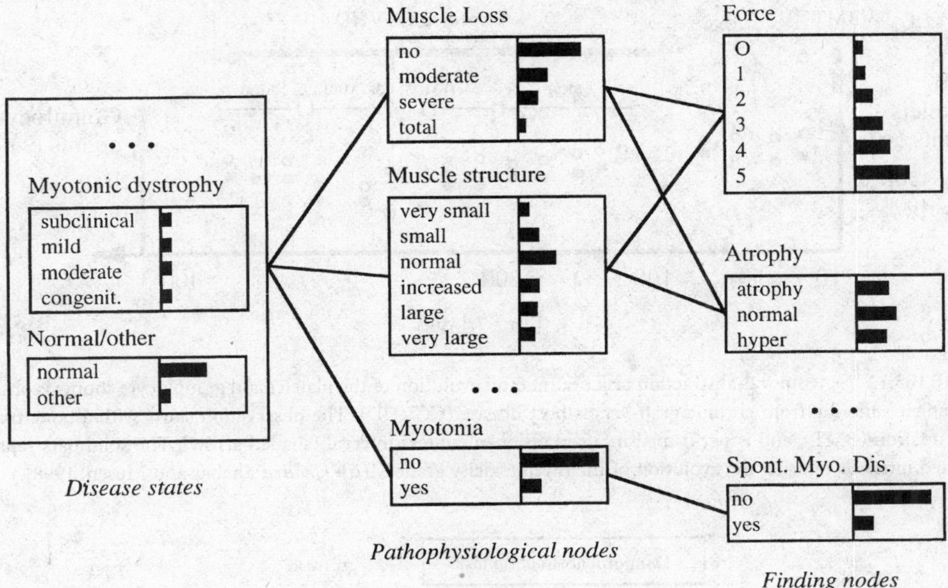

FIGURE 185.11 The domain knowledge in MUNIN (partial view). (*Source*: [Jensen et al., 1987].)

perturbations of the system state. Compartmental system theory is often used in this respect because it provides a robust modeling framework based on differential calculus. It implies, however, that a precise quantitative modeling of the pathophysiologic phenomena is possible. Qualitative models, on the contrary, allow one to cope with the fuzziness and incompleteness of pathophysiological knowledge. Among these, QSIM is the most widely applied formalism in medicine [Stefanelli, 1993]. This formalism provides a descriptive language to represent the structure of a physiologic system and a simulation algorithm to infer its qualitative behavior. The language consists in qualitative constraints that abstract the relationships in a differential equation. This approach has been applied to a variety of fields in medicine [Ironi et al., 1990]

It should be noted that representing pathophysiologic knowledge requires a special emphasis on the notion of time [Stefanelli, 1993], a problem early addressed by Fagan [1979]. In addition, specific knowledge acquisition issues are raised that have been only recently addressed [Ironi et al., 1993].

Temporal Knowledge

Time constitutes an integral and important aspect of medical concepts, and its explicit modeling is increasingly considered as central to the design of advanced medical systems. However, such modeling still remains challenging, due to the necessity to consider compound objects (e.g., disorders, treatments, or patient states) exhibiting different temporal existences and complex interactions, through mechanisms that are not completely understood. The definition of adapted temporal ontologies conveying a clear semantic is currently an important subject of debate among the community [Keravnou, 1995; Shahar and Musen, 1998].

The ability to reason about time is known to depend on the ability to provide short, informative, and context-sensitive summaries of time-oriented clinical data, in the form of temporal abstractions able to summarize clinical features over a period of time (Fig. 185.12). Such an approach has been considered in RESUME [Shahar and Musen, 1998], a system for forming high-level concepts from raw time-oriented clinical data which has been applied in three domains: monitoring of children's growth, care of diabetes patients, and protocol-based care.

In RESUME, a temporal abstraction is characterized as a quadruplet: {⟨parameter, value, context⟩, interval}, meaning that a given parameter has exhibited a certain value in a certain context for a certain

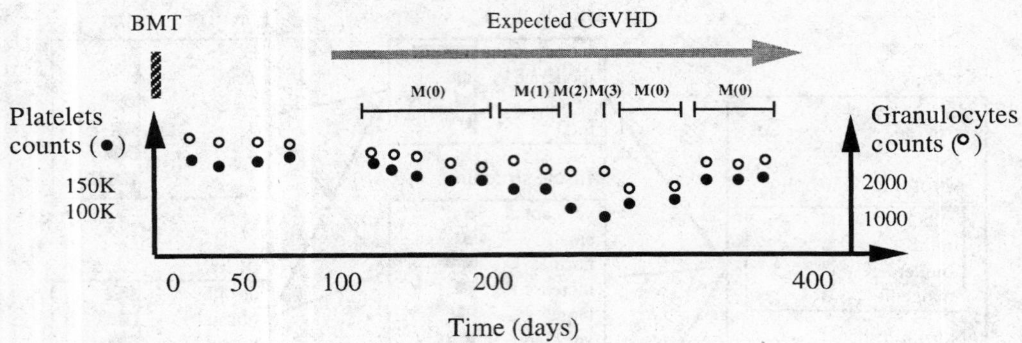

FIGURE 185.12 A temporal-abstraction task example: an evolution of the platelet and granulocyte counts is observed for a patient suffering from chronic graft-versus-host disease (CGVHD). The observation starts with a bone marrow transplantation (BMT), and is performed in a certain open context interval (shaded arrow). The solid bars represent abstracted intervals figuring the evolution of the myelotoxicity grade M(n). (*Source*: Shahar and Musen, 1998].)

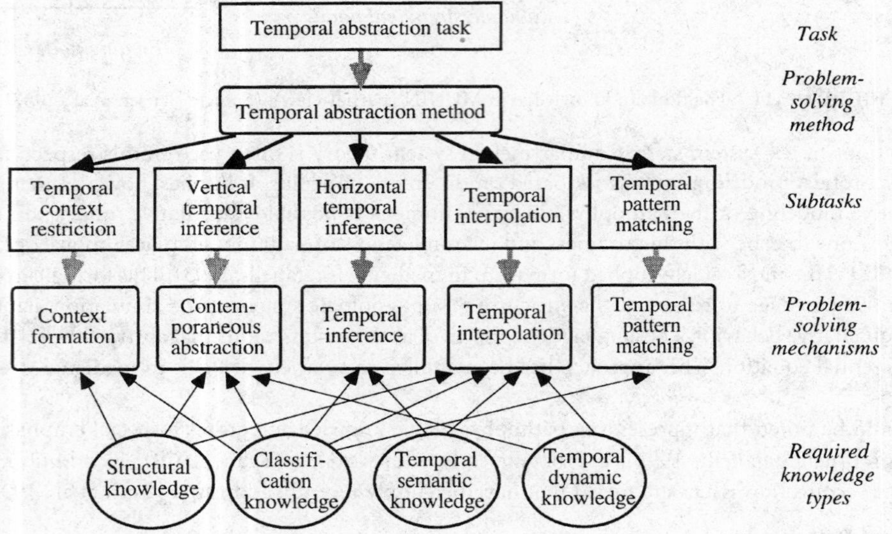

FIGURE 185.13 The temporal abstraction task. Grey arrow: decomposed into, thin arrow: solved by, bold arrow: used-by (*Source*: Shahar and Musen, 1998].)

period of time. The temporal abstraction task is modeled as a complex task decomposed into specific subtasks and problem-solving mechanisms, themselves depending on various knowledge types (Fig. 185.13). Such decomposition defines a general problem solving method (as defined in the section titled "Task Oriented Modeling") for interpreting data in time-oriented domains.

Current efforts concern the integration of temporal and causal modeling, thus allowing the modeling of multiple interacting mechanisms operating over a variety of time periods. Such models have been applied to the domain of heart disease [Long, 1996] and to the interpretation of time-series of blood glucose measurements coming from home monitoring [Riva and Bellazzi, 1996]. However, severe problems come from the computational burden induced by the learning and the managing of these models.

Protocols & Guidelines

Various attempts have been made to design knowledge representation systems which are capable of capturing and handling clinical procedures, from the early work on ONCOCYN [Tu et al., 1989]. The

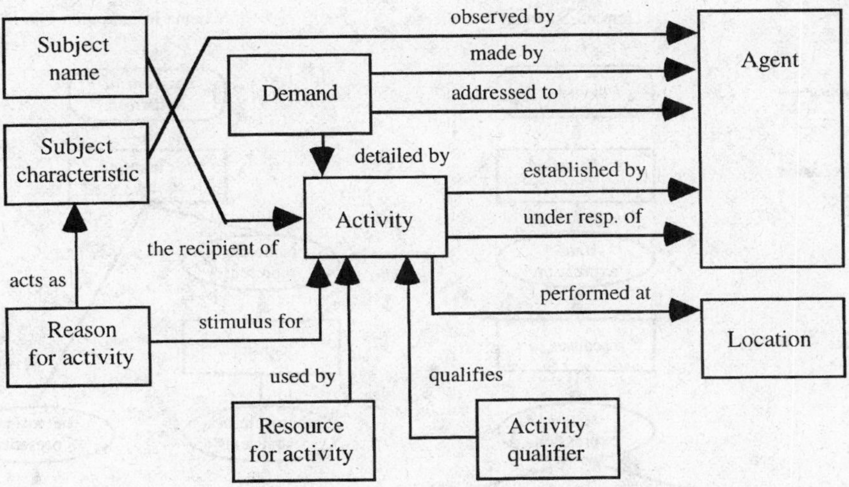

FIGURE 185.14 Model of a prescription activity in primary care (partial view). (*Source*: Gordon et al. [1993].)

objectives are to guide the user in the accurate and correct application of protocols but also to disseminate shared consensus guidelines among the community.

A generic modeling of clinical protocol has been developed in DILEMMA [Gordon et al., 1993], a system delivering guideline-based decision support for shared care in oncology, cardiology, and primary care. A *protocol* is a formal description of a way to achieve an objective or goal, given a class of problems. It is used as a template from which to derive the specific activity to be performed for a given clinical context. Figure 185.14 depicts a data model of a prescription activity, in which precise information is given about the patient, the agent responsible for the drug prescription and administration, together with administration guidelines, and a description of potential side effects and expected consequences.

The development of standard languages for guideline modeling is witnessing a growing interest, from mere procedural approaches like the Arden syntax [Hripsack et al., 1994] to high-level representation languages based on task ontologies like in the Proforma approach [Fox et al., 1997].

Current developments concern the integration of time-dependent patient information like in EON (Musen et al., 1995), as well as the development of web-based environments, the aim being to produce clinical guidelines which can be widely shared between humans from different institutions, thus dealing with patient and organization preferences [Quaglini et al., 1997]. GLIF (GuideLine Interchange Format), for example, has been introduced to allow the sharing of representations among different sites [Patel et al., 1997].

Second Generation Systems

Second generation systems organize their knowledge into modules and multilevel structures to solve problems of increasing complexity. The objectives are (i) to better structure the knowledge, facilitate its management and allow its reuse, and (ii) to allow the design of hybrid systems combining heterogeneous knowledge and reasoning schemes.

Task-Oriented Modeling

Task-oriented modeling has been proposed to model general problem solving methods, based on two classes of reusable components [Bylander and Chandrasekaran, 1987]: (i) domain-independent problem-solving methods, e.g., standard methods to perform prototypical tasks (see Fig. 185.13) and (ii) domain ontologies, e.g., description of the main concepts and their relations (see Fig. 185.15). These generic components are then instantiated to cope with the specificity of given expertise domains. This notion may be used as a model to drive knowledge acquisition and system design, since knowledge and

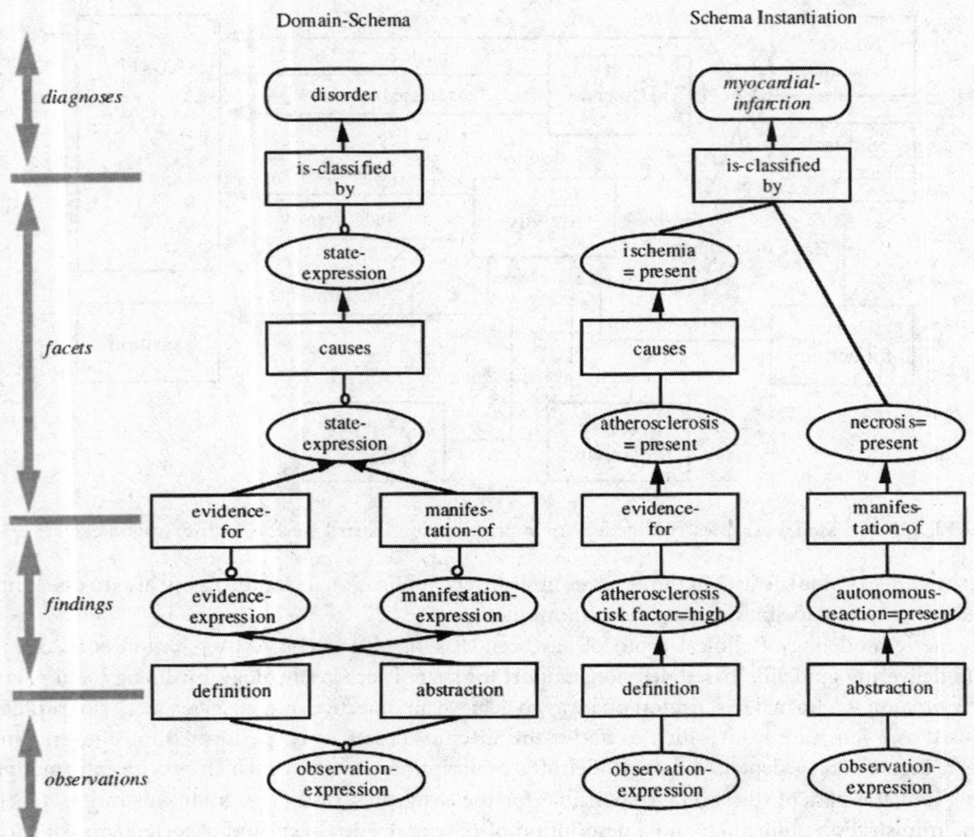

FIGURE 185.15 A domain ontology example, from FreeCall, a system for emergency-call-handling support. (*Source*: Post et al. [1996].) Note that the knowledge is organized according to the classification proposed by Patel et al. [1989] (see Fig. 185.3).

know-how specific to a task are specified at the right abstraction level [Steels, 1990]. Several knowledge acquisition tools specific to different task types may indeed be used to guide the expert through a structured elicitation protocol, for that specific problem type. Second generation expert systems also allow the reuse of knowledge, since some knowledge is generic and can be applied across several domains.

Hybrid Systems

There is a growing need for the integration of multimodal knowledge representation and reasoning schemes. Such integration is needed for at least two complementary reasons, i.e., the need to analyze data from different modalities and abstraction levels and the need to dynamically adapt the knowledge representation formalism and inference scheme to the situation at hand and to the current problem solving state.

The first viewpoint is adopted in AMNESIA [O'Hare et al., 1992], a system devoted to the diagnosis of memory related disorder illnesses. In such systems, the knowledge and reasoning skills are distributed among several specialists that cooperate towards the solving of the problem, each specialist being devoted to handling a precise knowledge domain or task.

The second viewpoint has been emphasized in NEOANEMIA, a system able to recognize disorders causing anemia [Ramoni et al., 1992]. An abductive inference scheme is first of all applied, to generate an initial set of hypotheses. A deductive inference scheme is then applied to compute the expected manifestations, which is based on a separate and explicit representation of taxonomic and causal ontology.

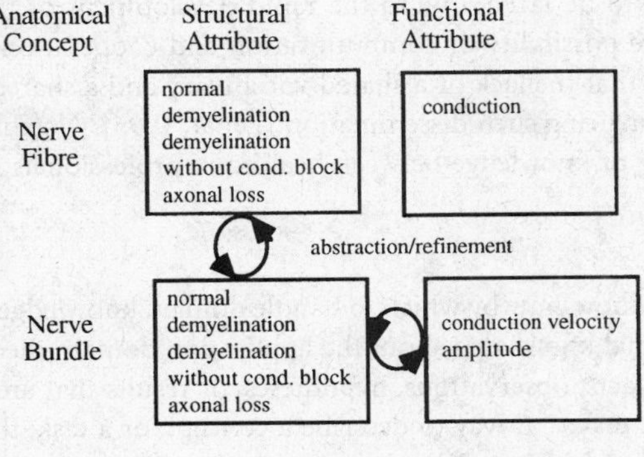

FIGURE 185.16 A hybrid knowledge representation scheme in the field of EMG diagnosis. (*Source*: Cruz et al. [1997].)

Inductive reasoning is finally used to prune the set of hypotheses, by matching their expected manifestations against the available data.

Structural, functional, and causal knowledge have been combined at various abstraction levels in a system dedicated to EMG diagnosis [Cruz et al., 1997]. Various anatomical structures like the nervous fiber and the nervous bundle are described in terms of their state and function, as illustrated in Fig. 185.16. Abstraction and refinement operations are associated with these attributes and defined in a probabilistic way to cope with the non-deterministic character of this knowledge.

Changes caused by normal or abnormal functioning of structures, as well as those caused by external agents, are modeled by means of processes expressed at the appropriate level of abstraction. A local neuropathy for example is a lesion that causes deep pathologic processes of local demyelination and axon loss, as modeled below:

process: local neuropathy
causes: neuropathy
(point$_i$ of nerve$_N$, severity)
constraints: transversal dispersion
longitudinal dispersion
effects: segment$_{kl}$ of fiber bundle$_{FB}$
normal = N
demyelination = D
demyelination without cond. block = DB
axonal loss = A

To characterize a neuropathy depends on the proportion of damaged fibers, on one hand, and on the extension of the damage along the nerve fibers, on the other hand. Transversal and longitudinal dispersion functions are moreover used to compute the lesion severity. The diagnosis finally proceeds through successive hypothesis refinement steps, progressively exploiting the various knowledge elements at hand.

185.5 Conclusion

The acquisition and representation of knowledge has been presented as an actively evolving research field characterized by modeling and software engineering issues of increasing complexity. Representing domain ontologies has been shown to raise a variety of issues, such as using conceptual graphs to represent the semantics of medical terms, using frames to describe domain taxonomies or modeling pathophysiological processes through causal networks. Modeling the expert control knowledge has been considered in terms of reasoning as well as planning abilities, which in turn leads to specific modeling issues.

In addition, the scope of designing knowledge-based systems in medicine currently evolves from the mere diagnostic task to the broader issue of patient management, which implies a better integration in existing hospital information systems.

New challenges are now to be faced, due to the rapid development of the internet communication facilities, which increase the possibility of communication and cooperation among health care professionals. It is clear, however, that the lack of a shared vocabulary and a shared understanding of medical terminology is currently impairing such dissemination [Lenat, 1995]. A stable ontological foundation is needed to allow the sharing of knowledge between healthcare professionals [Rossi Mori et al., 1997].

Defining Terms

Control knowledge: The know-how by which to handle domain knowledge and to reason about a case.

Domain knowledge: Formal knowledge about the application domain theory as well as actual knowledge about the data, facts, observations, hypotheses, or results that are attached to a given case.

Generic modeling, Generic task: A way to describe a concept, or a task, that is domain dependent in the sense that it is related to a medical domain, but that is generic in the sense that it may apply to several application areas in medicine.

Heuristic knowledge: Knowledge that reflects an empirical and non-formalized way to reason about a situation or solve a problem.

Knowledge level: A notion introduced by Newell, to express the rationale behind the exploitation of knowledge, in terms that are free from any implementation constraint.

Model, Modelling approach: A formal representation of the concepts, relationships, and reasoning schemes that constitue the domain and control knowledge of the application domain at hand; a designing approach that is driven by a formal representation of the domain and control knowledge, in which a close integration of the knowledge acquisition process is performed. This approach allows one to work at a level that is free from any implementation constraints, the knowledge level.

Ontology (domain, taxonomic, causal): The domain ontology characterizes the structure of the domain knowledge, in terms of concepts and relationships. Taxonomic and causal ontologies are two major methods for organizing domain knowledge; taxonomic ontologies emphasize the hierarchical relations between concepts, while causal ontologies emphasize their deep causal relations.

Shallow (vs deep) knowledge: Shallow knowledge is often used in opposition to deep knowledge, a knowledge based on a formal theory of the domain of interest (knowledge of a pathophysiological process for example). The term shallow refers to shortened reasoning schemes, that are often gained by experience.

References

Aamodt A, Plaza E. 1994. Case-based reasoning: foundational issues, methodological variation, and system approaches, *AI Commun.*, 7(1): 39-59.

Alpay L, Baud R, Rassinoux AM, et al. 1993. Interfacing Conceptual Graphs and the Galen Master Notation for Medical Knowledge Representation and Modelling. In: S Andreassen et al., (Eds.), *Artificial Intelligence in Medicine*, pp. 337-347, Amsterdam, IOS Press.

Bernauer J, Goldberg H. 1993. Compositional Clasification based on Conceptual Graphs. In: *Artificial Intelligence in Medicine*, S. Andreassen, R. Engelbrecht and J. Wyatt, (Eds.) pp. 348-359. Amsterdam, IOS Press.

Boose JH, Bradshaw JM. 1987. Expertise Transfer and Complex Problems: Using AQUINAS as a Knowledge Acquisition Workbench for Knowledge-Based System, *Int. J. Man-Mach. Stud.*, 26:3-28.

Boose JH. 1985. A Knowledge Acquisition Program for Expert Systems Based on Personal Construct Psychology. *Int. J. Man-Mach. Stud.*, 23: 495-525.

Bratko I, Mozetic I, Lavrac N. 1989. *KARDIO: A Study in Deep and Qualitative Knowledge for Expert Systems.* Cambridge, MA: MIT Press.

Breuker J, Wielinga B. 1987. Use of Models in Interpreting Verbal Data. In: *Knowledge Elicitation for Expert Systems: a Practical Handbook.* A. Kidd. (Ed.) pp. 17-44. Plenum Press.

Buchanan BG, Barstow D, Bechtel et al. 1983. Construction of an Expert System. In: *Building Expert Systems.* F. Hayes-Roth, D.A. Waterman and D.B. Lenat. (Eds.) pp. 127-167. Reading, MA, Addison-Wesley Publishing Company, Inc.

Bylander T, Chandrasekaran B. 1987. Generic Tasks for Knowledge-Based Reasoning: the "Right" Level of Abstraction for Knowledge Acquisition. *Int. J. Man-Mach. Stud.*, 26:231-243.

Campbell JA, Wolstencroft J. 1990. Structure and Significance of Analogical Reasoning. *Art. Intell. Med..* 2:103-118.

Coignard J, Jean F, Degoulet P, et al. 1992. HELIOS: Object-Oriented Software Engineering in Medicine, In: *Health Technology and Informatics.* vol. 2. pp. 141-149. Amsterdam, IOS Press.

Cruz J, Barahona P. 1997. A causal functional model applied to EMG diagnosis. In: *AIME 97: Proc. Euro. Conf. Art. Intell. Med., Lecture Notes in Artificial Intelligence.* E Keravnou et al. (Eds.), pp. 249-260. Berlin, Springer-Verlag.

Davis R, Lenat D. 1982. *Knowledge-Based System in Artificial Intelligence.* New York, MacGraw-Hill.

Dzeroski S, Potamias G, Moustakis V, et al. 1997. Automated revision of expert rules for treating acute abdominal pain in children. In: E Keravnou et al. (Eds.), *AIME 97: Proc. Euro. Conf. Art. Intell. Art. Intell. Med., Lecture Notes in Artificial Intelligence.* pp. 98-109. Berlin, Springer-Verlag.

Fagan LM. 1979. Representation of Dynamic Clinical Knowledge: Measurement Interpretation in the Intensive Care Unit. In: *Proc. of the 6th IJCAI.* pp. 260-262. Stanford, CA, Stanford University, Department of Computer Science.

Fiore M, Sicurello F, Vigano M et al. 1993. A Knowledge-Based System to Classify and Diagnose HIV-Pneumonias. In: *Artificial Intelligence in Medicine,* S. Andreassen, R. Engelbrecht and J. Wyatt, (Eds.), pp. 185-190. Amsterdam, IOS Press, Inc.

Fox J, Johns N, Rahmanzadeh A. 1997. Protocols for medical procedures and therapies: a provisional description of the Pro*forma* language and tools, In: *AIME 97: Proc. Euro. Conf. Art. Intell. Med., Lecture Notes in Artificial Intelligence.* E Keravnou et al., (Eds.) pp. 21-38. Berlin, Springer-Verlag.

Frawley W, Piatesky-Shapiro G, Matheus C. 1991. Knowledge discovery in databases: an overview, In: *Knowledge Discovery in Databases,* G Piatessky-Shapiro, W Frawley (Eds.), pp. 1-27, MIT Press.

Garbay C, Pesty S. 1988. Expert Systems for Biomedical Image Interpretation. In: *Artificial Intelligence and Cognitive Sciences,* J. Demongeot, T. Hervé, V. Rialle, and C. Roche (Eds.) pp. 323-345. Manchester University Press.

Gierl L, Schmidt R, Pollwein B. 1993. ICONS: Cognitive Basic Functions in a Case-Based Consultation System for Health Care. In: *Artificial Intelligence in Medicine,* S. Andreassen, R. Engelbrecht and J. Wyatt, (Eds.) pp. 230-236. Amsterdam, IOS Press, Inc.

Goos K, Schewe S. 1993. Case Based Reasoning in Clinical Evaluation. In: *Artificial Intelligence in Medicine,* S. Andreassen, R. Engelbrecht and J. Wyatt (Eds.), pp. 445-448. Amsterdam, IOS Press.

Gordon C, Herbert SI, Jackson-Smale A et al. 1993. Care Protocols and Health Informatics. In: *Art. Intell. Med.,* S. Andreassen, R. Engelbrecht and J. Wyatt (Eds.), pp. 289-309. Amsterdam, IOS Press, Inc.

Hau DT, Coiera EW. 1997. Learning Qualitative Models of Dynamic Systems, *Mach. Learn.,* 26:177-212.

Hripsack G, Ludeman P, Pryor TA et al. 1994. Rationale for the Arden syntax, *Comp. Biomed. Res.,* 27:291-324.

Ironi L, Cattaneo A, Stefanelli M. 1993. A Tool for Pathophysiological Knowledge Acquisition. In: *Artificial Intelligence in Medicine,* S. Andreassen, R. Engelbrecht and J. Wyatt (Eds.), pp. 13-31. Amsterdam, IOS Press, Inc.

Ironi L, Stefanelli M, Lanzola G. 1990. Qualitative Models in Medical Diagnosis. *Art. Intell. Med.* 2: 85-101.

Jensen FV, Andersen SK, Kjaerulff U et al. 1987. MUNIN – On the Case for Probabilities in Medical Expert Systems – a Practical Excercise. In: *AIME 87: Proc. Euro. Conf. Art. Intell. Med., Lecture Notes in Medical Informatics.* J. Fox, M. Fieschi and R. Engelbrecht (Eds.). pp. 149-160. Berlin Heidelberg, G. Springer-Verlag.

Juristica I, Mylopoulos J, Glasgow J et al. 1998. Case-based reasoning in IVF: prediction and knowledge mining, *Art. Intell.,* 12: 1-24.

Kahn G, Nowlan S, McDermott J. 1985. MORE: an Intelligent Knowledge Acquisition Tool *Proc. AAAI '85.* pp. 581-584.

Keravnou ET. 1995. Modelling medical concepts as time-objects. In: *AIME 95: Proc. Euro. Conf. Art. Intell. Med.* P. Barahona et al. (Eds.) pp. 67-78. Berlin, Springer-Verlag.

Kulikowski CA. 1984. Artificial Intelligence Methods and Systems for Medical Consultation. In: *Readings in Medical Artificial Intelligence.* W.J. Clancey and E.H. Shortliffe (Eds.). p.72-97. Reading, MA, Addison-Wesley Publishing Company.

Lanzola G, Stefanelli M. 1992. A specialized Framework for Medical Diagnosis Knowledge-Based Systems. *Comput. Biomed. Res.* 25:351-365.

Larranaga P, Sierra B, Gallego MJ, et al. 1997. Learning Bayesian networks by genetic algorithm: a case study in the prediction of survival in malignant skin melanoma. In: *AIME 97: Proc. Euro. Conf. Art. Intell. Med.* E Keravnou et al. (Eds.) pp. 261-272. Berlin, Springer-Verlag.

Leitich H, Adlassning K-P, Kolarz G. 1996. Development and evaluation of fuzzy criteria for the diagnosis of rheumatoid arthritis, *Meth. Inform. Med.*, 35: 334-342.

Lenat DB. 1995. Steps to sharing knowledge, In: *Towards Very Large Knowledge Bases.* NJI Mars (Ed.) pp. 3-6. IOS Press.

Long W. 1996. Temporal reasoning for diagnosis in a causal probabilistic knowledge base, *Art. Intell. Med.* 8: 193-215.

Lucas PJF. 1997. Model-based diagnosis in medicine, *Art. Intell. Med.*, 10:201-208.

Mc Graw KL, Harbison-Briggs K. 1989. *Knowledge Acquisition: Principles and Guidelines.* New Jersey, Prentice-Hall.

Müller R, Thews O, Rohrbach C, et al. 1996. A Graph-Grammar Approach to Represent Causal, Temporal and Other Contexts in an Oncological Patient Record. *Meth. Inform. Med.*, vol. 35 (2): 127-141.

Musen MA, Tu SW, Das AK et al. 1995. A component-based architecture for automation of protocol-directed therapy. In: *AIME 95: Proc. Euro. Conf. Art. Intell. Med.* P. Barahona et al. (Eds.) pp. 3-13. Berlin, Springer-Verlag.

Newell A. 1982. The knowledge level. *Art. Intell.*, 18:87-127.

O'Hare et al. 1992. AMNESIA—Implementing a Distributed KBS using RAPIDO. In: *Proc. 12th. Ann. Inter. Conf. of the British Computer Society Specialist Group on ES.* Bramer and Milne (Eds.). pp. 1-29, Cambridge University Press.

Patel VL, Allen VG, Arocha JF et al. 1997. *Representing Clinical Guidelines in GLIF: Individual and Collaborative Expertise.* SMI Report Number: SMI-97-0694.

Patel VL, Evans DA, Kaufman, D.R. 1989. A Cognitive Framework for Doctor-Patient Interaction. In: *Cognitive Science in Medicine.* D.A. Evans and V.L. Patel (Eds.). pp. 257-312. MIT Press.

Pauker SG, Gorry GA, Kassirer JP et al. 1976. Towards the Simulation of Clinical Cognition – Taking a Present Illness by Computer. *Am. J. Med.* 60:981-996.

Pople H. 1977. The Formation of Composite Hypothesis in Diagnostic Problem Solving: an Exercise in Synthetic Reasoning. In: *Proc. of the 5th IJCAI.* pp. 1030-1037. Pittsburgh, PA, Carnegie-Mellon University, Department of Computer Science.

Post WM, Koster W, Sramek M et al. 1996. FreeCall, a system for emergency-call-handling support, *Meth. Inform. Med.*, 35(3):242-255.

Quaglini S, Saracco R, Stefanelli M et al. 1997. Supporting tools for guideline development and dissemination. In: *AIME 97: Proc. Euro. Conf. Art. Intell. Med.* E Keravnou et al. (Eds.) pp. 39-50. Berlin, Springer-Verlag.

Quillian MR. 1968. Semantic Memory. In: *Semantic Information Processing.* M. Minsky (Ed.). p.216-270, Cambridge, MA, MIT Press.

Quinlan JR. 1986. Induction of Decision Trees, *Mach. Learn.* 1 (1):81-106.

Ramoni M, Riva A, Stefanelli M, et al. 1995. Medical decision making using ignorant influence diagrams. In: *AIME 95: Proc. Euro. Conf. Art. Intell. Med.* P Barahona et al. (Eds.) pp. 139-150. Berlin, Springer-Verlag.

Ramoni M, Stefanelli M, Barosi G et al. 1992. An Epistemological Framework for Medical Knowledge-based Systems. *IEEE Trans. on Sys., Man and Cyb.* 22:1361-1375.

Reategui EB, Campbell JA, Leao BF. 1997. Combining a neural network with case-based reasoning in a diagnostic system, *Art. Intell. Med.* 9:5-27.

Riva A, Bellazzi R. 1996. Learning temporal probabilistic causal models form longitudinal data, *Art. Intell. Med.* 8: 217-234.

Rossi Mori A, Gangemi A, Steve G, et al. 1997. An ontological analysis of surgical deeds. In: *AIME 97: Proc. Euro. Conf. Art. Intell. Med.* E Keravnou et al. (Eds.) pp. 361-372. Berlin, Springer-Verlag.

Schmid R. 1987. An Expert System for the Classification of Dizziness and Vertigo. In: *AIME 87: Proc. Euro. Conf. Art. Intell. Med., Lecture Notes in Medical Informatics.* J. Fox, M. Fieschi and R. Engelbrecht (Eds.). pp. 45-53. Berlin, Heidelberg, Germany, Springer-Verlag.

Schreiber ATh, van Heijst G, Lanzola G et al. 1993. Knowledge Organisation in Medical KBS Construction. In: *Artificial Intelligence Medicine,* S. Andreassen, R. Engelbrecht and J. Wyatt (Eds.), pp. 394-405. Amsterdam, IOS Press, Inc.

Shahar Y, Musen MA. 1996. Knowledge-based temporal abstraction in clinical domains, *Art. Intell.* 8: 267-298.

Shortliffe EH. 1976. *Computer-Based Medical Consultations: MYCIN.* New York, Elsevier.

Sowa JF. 1984. *Conceptual Structures: Information Processing in Mind and Machine.* Reading, MA, Addison-Wesley.

Steels L. 1990. Components of Expertise. *AI Magazine.* Summer 1990.

Stefanelli M. 1993. European Research Efforts in Medical Knowledge-Based Systems. *Art. Intell. Med.* 5:107-124.

Suojanen M, Olesen KG, Andreassen S. 1997. A method for diagnosing in large medical expert systems based on causal probabilistic networks. In: *AIME 97: Proc. Euro. Conf. Art. Intell. Med.* E Keravnou et al. (Eds.) pp. 285-295. Berlin, Springer-Verlag.

Tu SW, Kahn MG, Musen MA, et al. 1989. Episodic skeletal plan-refinement on temporal data, *Communication of the ACM,* 32:1439-1455.

van Ginneken AM, Smeulders AWM. 1988. An Analysis of Five Strategies for Reasoning in Uncertainties and their Suitability in Pathology. In: *Pattern Recognition and Artificial Intelligence.* E.S. Gelsema and L.N. Kanal LN (Eds.). pp. 367-379. North Holland, Elsevier Science Publisher B.V.

Weiss SM, Kulikowski C, Safir, A. 1978. Glaucoma Consultation by Computer. *Comp. Biol. Med.* 8(1):25.

Wyatt J. 1997. Computer-assisted decision support. In: *Internet, Telematics and Health,* M Sosa-Iudicissa et al. (Eds.) pp. 229-237. IOS Press.

Further Information

Artificial Intelligence in Medicine is a bimonthly journal published by Elsevier Science Publishers B.V. This international journal publishes articles concerning the theory and practice of medical artificial intelligence.

Studies in Health Technology and Informatics is a series published by IOS Press, in which two volumes are of particular interest to the reader interested in the new trends in health telematics:

- Volume 16, edited by C. Gordon and J.P. Christensen, entitled *Health Telematics for Clinical Guidelines and Protocols* (1995);
- Volume 36, edited by M. Sosa-Iudicissa et al., entitled *Internet, Telematics and Health* (1997).

Lecture Notes in Artificial Intelligence is a sub-series of *Lecture Notes in Computer Science* published by Springer-Verlag, in which the proceedings of the 5th and the 6th Conference on Artificial Intelligence in Medicine Europe, edited by P. Barahona et al. in 1995, and E. Keravnou et al. in 1997, gives an excellent outlook of the most recent advances of the field throughout Europe.

186

Knowledge-Based Systems for Intelligent Patient Monitoring and Management in Critical Care Environments

Benoit M. Dawant
Vanderbilt University

Patrick R. Norris
Vanderbilt University

186.1 Introduction

Patient monitoring and management in critical care environments such as intensive care units (ICUs) and operating rooms (ORs) involves estimating the status of the patient, reacting to events that may be life-threatening, and taking actions to bring the patient to a desired state. This complex process includes the interaction of physicians and nurses with diverse data (ranging from clinical observations to laboratory results to online data) provided by bedside medical equipment. New monitoring devices provide health care professionals with unsurpassed amounts of information to support decision making. Ironically, rather than helping these professionals, the amount of information generated and the way the data is presented may overload their cognitive skills and lead to erroneous conclusions and inadequate actions. New solutions are needed to manage and process the continuous flow of information and provide efficient and reliable decision support tools.

Patient monitoring can be conceptually organized in four layers [Coiera, 1993]: (1) the *signal level,* which acquires and performs low-level processing of raw data; (2) the *validation level,* which removes data artifacts; (3) the *signal-to-symbol transform level,* which maps detected features to symbols such as normal, low, or high; and (4) the *inference level,* which relies on a computer representation of medical knowledge to derive possible diagnoses, explanations of events, predictions about future physiologic states, or to control actions. In addition to these four layers, medical decision support systems need data interfaces to other clinical information systems as well as carefully designed user interfaces to facilitate rapid and accurate situation assessment by care providers.

0-8493-0462-8/00/$0.00+$.50

A number of systems have been developed to address problems faced by clinicians in critical care environments. These range from low-level signal analysis applications for detecting specific features in monitored signals to complete architectures for signal acquisition, processing, interpretation, and decision support. As always, specificity and generality are conflicting requirements. Systems developed for specific applications are usually successful in their limited domain of expertise. However, the lessons and problem-solving strategies learned in one domain are often difficult to generalize. Conversely, generic architectures aim at providing support for modeling and developing a wide range of applications. They also strive for flexibility, modularity, and ease of expansion. This generality is often at the expense of expertise and performance in specific domains [Uckun et al., 1993]. This chapter reviews some of the problems and proposed solutions associated with the four layers of intelligent monitoring systems.

186.2 Intelligent Patient Monitoring and Management

Signal Acquisition and Low-Level Processing

Typical signals found in critical care environments include vital signs (ECG, EEG, arterial pressure, intracranial pressure, etc.) and information provided by therapeutic devices, infusion pumps, and drainage devices. Modern monitoring devices are also capable of providing derived and computed information in addition to raw data. Heavily instrumented patients frequently have up to 20 medical devices monitoring them, producing up to 100 pieces of clinically relevant information. These instruments are often stand-alone, and interconnection requires developing dedicated software in-house, usually with substantial effort. To address this problem, the IEEE standards committee, IEEE 1073, has been working since 1984 on a standard for medical device data communication in critical care environments called the Medical Information Bus (MIB). Although much of the standard has been adopted by ANSI, and despite long-term efforts by individual institutions [Gardner et al., 1992], a standard adhered to by major medical device manufacturers is still lacking. Today, this absence of interconnectivity remains a major obstacle to the development and implementation of intelligent monitoring systems.

Current bedside monitors typically provide instantaneous values for the monitored variables. To complement this information, numerous algorithms have been proposed to detect features in the signals. In particular, detection of significant trends has received much attention [Avent and Charlton, 1990, Haimowitz and Kohane, 1996]. Methods based on median filters [Makivirta et al., 1991] and fuzzy logic [Sittig et al., 1992; Steinman and Adlassnig, 1994] have also been proposed. More recently, techniques such as sub-band adaptive filtering, chaos analysis [Karrakchou et al. 1996], and wavelet transform [Unser and Aldroubi, 1996] have been investigated. Multivariate and data fusion methods have been used to reveal interactions between signals and applied to problems such as ventricular rhythm tracking [Thoraval and Carrault, 1997]. Traditional waveform analysis also continues to be applied to signals such as intracranial pressure [Czosnyka et al., 1996] and arterial blood pressure [Karamanoglu, 1997]. Interpreting the output of these algorithms and assessing their impact on patient care is an ongoing task.

Data Validation

The risk of noise contamination, inadequate wiring, or instrument failure is significant, especially as the amount of sampled data increases. Unreliable information can drastically reduce the practitioner's ability to rapidly assess and act on monitor data. In addition, false alarms due to erroneous or incomplete data reduce clinicians' confidence in the instrumentation resulting in alarms being disabled or ignored. Tsien and Fackler [1997] report that after 298 monitored hours in a pediatric ICU, 86% of the 2,942 alarms were found to be false-positive alarms while an additional 6% were deemed irrelevant true alarms.

Methods relying on data redundancy and correlation (thus assuming some level of interconnectivity between bedside instruments) as well as contextual information to eliminate false alarms, validate data, and diagnose malfunctions in the monitoring equipment have been proposed. Rule-based systems [Jiang, 1991; van der Aa, 1990] and neural networks [Orr, 1991] have been developed for detecting faults and malfunctions in the breathing circuit during anesthesia. Other intelligent alarming systems focusing on data validation have been implemented in the operating room, and include rule-based systems for respiratory-circulatory management [Garfinkel et al., 1988] and for patient state-dependent data collection and processing [Arcay-Varela and Hernandez-Sande, 1993]. Rule-based systems have been combined with fuzzy logic [Schecke et al., 1992] and neural networks for false alarm reduction [Navabi et al., 1991]. Laursen [1994] investigated causal probabilistic networks to reduce false alarms in the ICU. In addition to systems specifically designed for data validation and false-alarm rejection, diagnostic and advice-giving systems often include a data-validation step in their overall strategy.

Signal-to-Symbol Transformation and Giving Advice

The next layer in information processing involves transforming numerical values into symbolic information and using these symbols for diagnosis and giving advice. A representative sample of these systems is discussed below. For a more complete compilation the reader is referred to Uckun [1994]. Although a strict taxonomy of diagnostic and advice-giving systems is difficult to create, systems with so-called shallow and deep knowledge will be distinguished. *Shallow* knowledge expresses empirically based, experiential knowledge [Mora et al., 1993] generally called *heuristics*. Reasoning in these systems is mostly associative, and knowledge is usually represented in terms of an IF (antecedent) THEN (action) form. Shallow knowledge systems exhibit a high degree of competence in their limited domain of application, but performance usually degrades quickly near the edge of the domain of expertise. At the other end of the spectrum, *deep* knowledge refers to knowledge about the structure, behavior, or function of a system. Model-based reasoning involves using this knowledge to diagnose, predict, or explain the behavior of the system over time.

Shallow Knowledge Systems

Typically, these systems perform a specific task and have similar architectures: (1) a module responsible for transforming quantitative data into qualitative or symbolic information, (2) a module capable of translating symbolic information into patient state, and (3) a module designed to provide therapeutic advice based on the estimated patient state.

Ventilator management has received much attention in intelligent monitoring research, following the pioneering efforts of Fagan et al. [1984] on the VM project. VM was developed in the late 1970s as an extension of the rule based MYCIN formalism to interpret online quantitative data in the ICU setting and to provide assistance for postoperative patients undergoing mechanical ventilation. VM addressed the importance of interpreting data differently depending on patient status.

COMPAS [Sittig et al., 1989] was designed to assist in management of patients with adult respiratory distress syndrome (ARDS). COMPAS is a rule-based system incorporating a blackboard architecture with three main types of knowledge sources (KSs): (1) the data-processing KSs compute derived information from the primary data; (2) the data-classification KSs transform numerical information into symbolic information; and (3) the protocol-instruction KSs suggest therapeutic actions. These knowledge sources roughly correspond to the upper three layers of intelligent monitoring systems (see above). The raw data level consists of information provided by clinical or blood gas laboratories, x-ray departments, and respiratory therapists. Although COMPAS is no longer used the knowledge it contains has served as a basis for a system used clinically [Henderson et al., 1991].

The AIRS system provides three levels of decision support to match three phases of ventilation [Summers et al., 1993]: (1) the startup phase matches the patient state to one of 14 predetermined states and suggests initial ventilator settings; (2) the maintaining phase provides context-dependent values for set point and trend alarms; and (3) the weaning phase contains knowledge represented as PROLOG premise-actions rules to provide advice in the weaning process.

Vie-Vent [Miksch et al., 1993] is an advice-giving system for the monitoring and management of mechanically ventilated neonates. It consists of three modules: (1) the data-acquisition and validation module performs simple range checking and measurement correlation; (2) the data-abstraction module translates numerical values into normal, low, and high ranges based on predetermined rules; and (3) the rule-based therapy-planning module proposes ventilator settings based on the qualitative values for pH, pCO_2, and PO_2 provided by the data abstraction module. These rules are adapted to reflect the site of measurement and the mode of ventilation. Vie-Vent is designed to operate in real-time and it includes a data validation component using time-point, time-interval, and trend-based methods [Horn et al., 1997].

PATRICIA [Moret-Bonillo et al., 1993] is a system designed to assist in monitoring and managing mechanically ventilated patients in ICUs. It consists of two main components: the deterministic module, which transforms incoming information into symbolic ranges, and the heuristic module, which interprets these symbolic ranges using rules to infer patient status and prescribe therapeutic advice. It uses so-called natural contexts, which include demographic information, patient history, and diagnosis information, to provide patient-dependent ranges for the symbolic variables. Based on these symbolic values, the rule-based system performs data validation both in the deterministic and the heuristic modules. It is also capable of detecting "prealarm" situations by predicting a temporal evolution of monitored parameters. PATRICIA is currently working off-line and underwent an extensive retrospective validation [Moret-Bonillo et al., 1997]

Other systems developed for ventilator management include ESTER [Hernandez-Sande et al., 1989], KUSIVAR (developed for the management of patients with ARDS) [Rudowski et al., 1989], and WEAN-PRO (designed for assisting physicians in weaning postoperative cardiovascular patients) [Tong, 1991]. Expert-system methodology also has been applied to the monitoring and management of patients in the cardiac care unit (CCU) [Mora et al., 1993, Sukuvaara et al., 1993b].

As opposed to the systems discussed before, GANESH [Dojat et al., 1992] has been designed for use as a closed-loop controller. Its domain of application is the weaning of patients under pressure-supported ventilation (PSV). Based on the respiratory rate, tidal volume, and end-tidal partial pressure of carbon dioxide, a rule-based controller determines the patient status and action to be taken and then acts on the ventilator. Preliminary clinical evaluations of the system have shown its ability to maintain patients in the region of comfort and to progressively reduce the pressure of weanable patients. Recently, the theory of Event Calculus has been proposed to capture and use temporal knowledge for this application [Chittaro and Dojat, 1997].

Blom [1991] also proposes an expert-system approach for closed-loop control of arterial pressure during surgery. In this approach, heuristic rules supervise a PID controller and adjust its parameters. The result is a robust controller capable of coping with artifacts in the signal, low signal-to-noise ratio, and patient sensitivity to nitroprusside, a drug used to achieve controlled hypotension during surgery. Huang and Rozear [1998] also describe a fuzzy logic-based system to control administration of multiple IV drugs for hemodynamic regulation. It consists of: (1) a fuzzy decision analysis module to assign the patient to one of six possible states using information derived from monitor data; (2) a hemodynamic management module to determine the dosage changes of each drug based on the current state; and (3) a therapeutic assessment module for scheduling drug administration based on drug properties such as pharmacological delay.

Deep Knowledge Systems

Despite its advantage over associative reasoning for explanation, robustness, and prediction, model-based reasoning has been used in few medical systems. Uckun [1992] suggests this may be attributed to the relatively shallow level of knowledge regarding most disease processes and the complexity and inherent variability of human anatomy, physiology, and pathophysiology. These difficulties still challenge our ability to develop accurate quantitative, qualitative, or hybrid models. Despite the difficulties involved, several model-based reasoning systems have been proposed for patient monitoring.

Rutledge and colleagues' VentPlan [1993] recommends ventilator settings for ICU patients. It has three main components: a belief network, a mathematical modeling module, and a plan evaluator. The belief network computes probability distributions for the physiologic model parameters from quantitative and qualitative information. The model predicts the distribution of future values for the dependent variables based on this data and current ventilator settings, and can simulate the evolution of important variables under various ventilator settings to evaluate proposed therapy plans. The plan evaluator predicts p_aCO_2 and O_2 delivery to rank therapy plans using a multiattribute scoring system and to propose ventilator settings.

The patient model component of SIMON [Uckun et al., 1993; Dawant et al., 1993] was built on an extension of the qualitative process theory (QPT) [Forbus, 1984] to diagnose neonates suffering from respiratory distress syndrome (RDS), to predict the patient's evolution, and to dynamically adjust alarm parameters based on patient state. As opposed to QPT, this approach does not exhaustively predict all distinct states the system may reach, since this is of little use when external variables (i.e., ventilator settings) are not constant. It determines patient status based on the temporal evolution of clinical events, and can modify the model based on discrepancies between observed and predicted values. SIMON underwent a limited retrospective evaluation using neonatal ICU data.

KARDIO [Lavrac et al., 1985] and CARDIOLAB [Siregar et al., 1993] are examples of models developed for model-based ECG simulation, prediction, and interpretation. KARDIO generates qualitative descriptions of the ECG signals corresponding to various arrhythmia combinations, based on the relation between the ECG and 30 elementary arrhythmias. Observed ECG can then be compared with the generated descriptions, and the associated arrhythmias can be retrieved from the database. Without a model permitting its automatic generation, the creation of an exhaustive database associating arrhythmias with ECG description would be a daunting, if not impossible, task.

186.3 Moving Toward Computer Architectures for Real-Time Intelligent Monitoring

Previously highlighted is the fact that patient monitoring is a complex, multifaceted task, and that much research has been done to facilitate the monitoring tasks of clinicians. But despite such efforts, few of the systems described earlier are in clinical use. None of them have been fielded or tested in other institutions. This may be due to the specificity of these systems and to the difficulty of modifying their structure for slightly different applications. Rather than focusing on specific tasks, other researchers are attempting to develop generic architectures that can support the development and implementation of a wider range of applications. These architectures are designed to support diverse reasoning schemes, guarantee real-time response, and permit the integration of the entire spectrum of tasks required for intelligent patient monitoring. This section describes a few such systems.

GUARDIAN [Hayes-Roth et al., 1992] is a blackboard system. Based on a general architecture for intelligent agents, it has been applied in respiratory and cardiovascular monitoring domains. GUARDIAN emerges from a long history of research on reasoning and problem solving, and much emphasis is put on the generic problem of coordinating and selecting one of possible perceptual, cognitive, or action operations under real-time constraints. It integrates heuristic, structural, or functional knowledge of (1) common respiratory problems, (2) respiratory, circulatory, metabolic, and mechanical ventilator systems, and (3) generic flow, diffusion, and metabolic systems. A system capable of instantiating treatment protocols based on (1) current patient context, (2) executing plans and close-loop control actions, (3) monitoring the execution of these plans and actions, and (4) modifying plan execution as necessitated by patient response has also been designed to work with GUARDIAN.

The Intelligent Cardiovascular Monitor (ICM) [Factor et al., 1991] is an application of the multitrellis software architecture, which supports modeling of real-time tasks in terms of an acyclic hierarchical network of decision processes. In ICM, the lowest level corresponds to raw data, and processes at this level implement signal processing algorithms to extract features from the signal. Information is abstracted along the hierarchy in a way reminiscent of information flow in a blackboard architecture, until reaching the highest level corresponding to physiologic processes or events. As opposed to GUARDIAN, the process trellis architecture assumes sufficient resources to execute the entire program and meet real-time constraints. This architecture has also been used for the development of DYNASCENE [Cohn et al., 1990], which models hemodynamic abnormalities in terms of "scenes" and temporal relationships between scenes. Each scene corresponds to one physiologic process (e.g., increased pericardial pressure) and is related to modules for computing specific information from the incoming data. DYNASCENE and ICM do not work in conjunction with patient models and hence are not capable of predicting patient behavior.

SIMON [Dawant et al., 1993] is a framework designed to support the development of real-time intelligent patient monitoring systems. It is organized around three main modules: the patient model, the data-acquisition and abstraction module, and a flexible user interface. The patient model is responsible for (1) estimating the physiologic state of the patient, (2) predicting the temporal evolution of the monitored variables, and (3) defining a dynamic monitoring *context*. The context includes normal and abnormal ranges for monitored variables, expected temporal evolutions of these variables, significant values and patterns to be detected in the incoming signals (events), and display configuration information (i.e., critical information to be displayed). The data abstraction module adjusts alarm limits and thresholds based on the current context and implements strategies to detect complex events. These may involve multiple variables and temporal relationships between events detected on each variable. SIMON has been applied off-line to the problem of monitoring neonates in the intensive care unit.

Following the recommendations of the INFORM project [Hunter et al., 1991], Sukuvaara et al. [1994] proposed an object-oriented framework to implement systems based on the blackboard approach to problem solving. Data on the blackboard are separated into four main categories: (1) source data (raw data acquired from monitors, hospital systems, or users), (2) preprocessed variables, (3) physiologically interpreted variables, and (4) patient status. Preprocessing knowledge sources (KSs) span the first two levels and extract features from the incoming data. Transformation KSs span the second and third levels in the blackboard hierarchy, and map values of preprocessed variables onto (low, high) intervals. Patient status assessment KSs get input from physiologically interpreted data and determine the patient state. The system also assumes sufficient resources to execute all its knowledge sources and meet real-time constraints.

186.4 Discussion and Conclusions

Intelligent patient monitoring and management are complex tasks involving all aspects of information processing. They require context-dependent acquisition, processing, analysis, and interpretation of large amounts of possibly noisy and incomplete data. These systems also need to interface to hospital-wide patient data management systems (PDMS) to gather data from the laboratories, the pharmacy, or radiology. Finally, these systems should include efficient mechanisms to present relevant and timely information to clinical users.

The systems described here illustrate that it is not easy to choose between specificity and generality. Systems that undergo clinical evaluation are typically developed for a very narrow, specific application which lends itself to rule-based approaches. However, these systems are often of limited clinical utility due to their narrow scope. Research aimed at developing deeper models to infer a wider range of behaviors, on the other hand, has failed to produce overwhelming results. Models for predicting patient behavior or response to therapeutic actions are clearly desirable, but currently are only available in very limited or constrained situations. Original modeling methods and approaches have been proposed, but require assumptions that either drastically reduce the scope of the model or make it unrealistic for clinical use.

Complete architectures aim at identifying the essence of patient monitoring tasks and at providing support for the development of a class of systems. These systems should permit experimentation with a number of problem-solving strategies, the integration of diverse reasoning mechanisms, and should ultimately span the entire spectrum of tasks required for efficient and reliable patient monitoring. However, to move from the research laboratory to the clinical arena, these systems will have to be fitted with rich and flexible modeling tools capable of overcoming the complexity of generic architectures. In addition, they will have to interface with a variety of devices and systems, efficiently manage vast amounts of diverse data, and successfully interact with clinical users.

To address the issue of information presentation, web-based technology is being integrated with clinical information systems, and prototypes have been developed to provide integrated clinical user interfaces to ICU patient data [Norris 1997, Nenov 1995] at or close to real time, via a web browser.

However, the fact remains that despite years of research and development and despite the potential of techniques being developed, intelligent monitoring systems have failed to make a significant impact on the quality of patient care. Saranummi [1997] contends that the main reason for this phenomenon is a lack of interaction between clinicians, industry, and academic researchers. Clinicians and academic researchers must interact closely to define the problem, develop solutions, and evaluate these solutions. When this is done, industry needs to be involved to assess the commercial viability of the project. To foster interaction Saranummi advocates the use of new technical platforms that support "plug-in" modules to existing monitoring and PDMS systems. These modules extend the functionality of existing monitoring systems without interfering with routine operation, permitting the *in situ* evaluation of new technical solutions. Efforts such as the IMPROVE project sponsored by the European Union, aimed at gathering large annotated data sets of ICU patients, could also facilitate the evaluation of novel information processing methods in close to real world situations [Nieminen et al., 1997].

Coiera [1994] also suggests that current systems are not adapted to their task. He notes that a major effort has been put into developing systems to assist the physician in making diagnoses and taking therapeutic actions. He suggests, however, that the main problem facing health care professionals is not choosing a diagnosis or therapeutic action, but rather establishing a clear picture of the world leading to these choices. Following this observation, Coiera advocates the creation of a "task layer" designed to provide a good match between displayed information and the cognitive processes of physicians and nurses. More research on the way clinicians interact with intelligent monitoring devices is necessary, though, to support the design of these interfaces.

The design of intelligent monitoring systems is thus an ongoing task. Context-based acquisition, processing, analysis, and display of the information are an essential concept in these systems. Elements defining this context are, however, complex and diverse. On the one hand, patient status and history define an interpretation and monitoring context that can be used to determine normal and abnormal limit values, trends, normal relations between observed variables, and significant events to be detected in the signals. Models could automatically establish these contexts, which, in turn, could be used to configure the monitoring system. On the other hand, cognitive tasks faced by the clinician, such as organ system function, treatment course, or resource utilization define the type and form of information presented. This information-processing context may be more difficult to automate and will require development of better interaction mechanisms between the system and its users.

References

Arcay-Varela B, Hernandez-Sande C. 1993. Adaptive monitoring in the ICU: A dynamic and contextual study. *J. Clin. Eng.* 18:67.

Avent RK and Charlton JD. 1990. A critical review of trend-detection methodologies for biomedical monitoring systems. *Crit. Rev. Biomed. Eng.* 17:621.

Blom JA. 1991. Expert control of the arterial blood pressure during surgery. *Int. J. Clin. Monit. Comput.* 8:25.

Chittaro L and Dojat M. 1997. Using a general theory of time and change in patient monitoring: experiment and evaluation. *Compu. Biol. Med.* 27:435

Cohn AI, Rosenbaum S, Factor M, and Miller PL. 1990. DYNASCENE: An approach to computer-based intelligent cardiovascular monitoring using sequential clinical "scenes". *Meth. Inform. Med.* 29:122.

Coiera E. 1993. Intelligent monitoring and control of dynamic physiological systems. *Artif. Intell. Med.* 5:1.

Coiera E. 1994. Design for decision support in a clinical monitoring environment. In: *Proceedings of the International Conference on Medical Physics and Biomedical Engineering,* Nicosa, Cyprus 130–142.

Czosnyka M, Guazzo E, Whitehouse M, et al. 1996. Significance of intracranial pressure waveform analysis after head injury. *Acta Neurochirurgica* 138:531.

Dawant BM, Uckun S, Manders EJ, and Lindstrom DP. 1993. The SIMON project: Model-based signal acquisition, analysis, and interpretation in intelligent patient monitoring. *IEEE Eng. Med. Biol. Mag.* 12:92.

Dojat M, Brochard L, Lemaire F, and Harf A. 1992. A knowledge-based system for assisted ventilation of patients in intensive care units. *Int. J. Clin. Monit. Comput.* 9:239.

Dojat M, Pachet F, Guessoum Z, et al. 1997. NeoGanesh: a working system for the automated control of assisted ventilation in ICUs. *Artif. Intell. Med.* 11:97.

Factor M, Gelertner DH, Kolb CE, et al. 1991. Real-time data fusion in the intensive care unit. *IEEE Computer* 24:45.

Fagan LM, Shortliffe EH, Buchanan BG. 1984. Computer-based medical decision making: From MYCIN to VM. In: *Readings in Medical Artificial Intelligence: The First Decade,* WJ Clancey and EH Shortliffe, (Eds.), pp. 241-255. Reading, MA, Addison-Wesley.

Forbus KD. 1984. Qualitative Process Theory. Ph.D. Dissertation, Massachusetts Institute of Technology, Cambridge, MA.

Gardner RM, Hawley WL, East TD, et al. 1991. Real time data acquisition: Recommendations for the Medical Information Bus (MIB). *Int. J. Clin. Monit. Comput.* 8:251.

Garfinkel D, Matsiras P, Lecky JH, et al. 1988. PONI: An intelligent alarm system for respiratory and circulation management in the operating rooms. In: *Proceedings of the 12th Symposium on Computer Applications in Medical Care,* pp. 13-17.

Haimowitz IJ and Kohane IS. 1996. Managing temporal worlds for medical trend diagnosis. *Artif. Intell. Med.* 8:229.

Hayes-Roth B, Washington R, Ash D, et al. 1992. Guardian: A prototype intelligent agent for intensive-care monitoring. *Artif. Intell. Med.* 4:165.

Henderson S, Crapo RO, Wallace CJ, et al. 1991. Performance of computerized protocols for the management of arterial oxygenation in an intensive care unit. *Int. J. Clin. Monit. Comput.* 8:271.

Hernandez-Sande C, Moret-Bonillo V, Alonso-Betanzos A. 1989. ESTER: An expert system for management of respiratory weaning therapy. *IEEE Trans. Biomed. Eng.* 36:559.

Horn W, Miksch S, Egghart G, et al. 1997. Effective data validation of high-frequency data: Time-point-, time-interval-, and trend-based methods. *Comput. Biol. Med.* 27:389.

Huang JW and Rozear MP. 1998. Multiple drug hemodynamic control using fuzzy decision theory. *IEEE Trans. Biomed. Eng.* 45:213.

Hunter J, Chambrin MC, Collinson P, et al. 1991. INFORM: Integrated support for decisions and activities in intensive care. *Int. J. Clin. Monit. Comput.* 8:189.

Jiang A. 1991. The design and development of a knowledge-based ventilatory and respiratory monitoring system. Ph.D. Dissertation, Vanderbilt University, Nashville, TN.

Karamanoglu M. 1997. A system for analysis of arterial blood pressure waveforms in humans. *Comput. Biomed. Res.* 30:244.

Karrakchou M, Vibe-Rheymer K, Vesin J-M, et al. 1996. Improving cardiovascular monitoring through modern techniques. *IEEE Eng. Med. Biol. Mag.* 1996:68.

Laursen P. 1994. Event detection on patient monitoring data using causal probabilistic networks. *Meth. Inform. Med.* 33:111.

Lavrac N, Bratko I, Mozetic I, et al. 1985. KARDIO-E: An expert system for electrocardiographic diagnosis of cardiac arrhythmias. *Expert Syst.* 2:46.

Makivirta A, Koski E, Kari A, and Sukuvaara T. 1991. The median filter as a preprocessor for a patient monitor limit alarm system in intensive care. *Comput. Methods. Progr. Biomed.* 34:139.

Miksch S, Horn W, Popow C, Paky F. 1993. VIE-VENT: Knowledge-based monitoring and therapy planning of the artificial ventilator of newborn infants. In: *Artificial Intelligence in Medicine: Proceedings of the European Conference.* (AIME), S Andreassen, et al. (Eds.) pp. 218–229. Amsterdam, IOS Press.

Mora A, Passariello G, Carrault G, Le Pichon J-P. 1993. Intelligent patient monitoring and management systems: A review. *IEEE Eng. Med. Biol. Mag.* 12(4):23.

Moret-Bonillo V, Alonso-Betanzos A, Martin EG, et al. 1993. The PATRICIA project: A semantic-based methodology for intelligent monitoring in the ICU. *IEEE Eng. Med. Biol. Mag.* 12:59.

Moret-Bonillo V, Mosqueira-Rey E, Alonso-Betanzos A. 1997. Information analysis and validation of intelligent monitoring systems in intensive care units. *IEEE Trans. Inf. Tech. Biomed.* 1:89.

Navabi MJ, Watt RC, Hameroff SR, and Mylrea KC. 1991. Integrated monitoring can detect critical events and improve alarm accuracy. *J. Clin. Eng.* 16:295.

Nenov VI, Klopp J. 1995. Remote analysis of physiologic data from neurosurgical ICU patients. *J. Amer. Med. Informatics. Assoc.* 2:273.

Nieminen K, Langford RM, Morgan CJ, et al. 1997. A clinical description of the IMPROVE data library. *IEEE Eng. Med. Biol. Mag.* 16(6):25.

Norris PR, Dawant BM, Geissbuhler A. 1997. Web-Based Data Integration and Annotation in the Intensive Care Unit. In: *Proceedings of the American Medical Informatics Association Annual Fall Symposium*, pp. 794–798.

Orr JA. 1991. An anesthesia alarm system based on neural networks. Ph.D. dissertation, University of Utah, Salt Lake City, UT.

Rudowski R, Frostell C, and Gill H. 1989. A knowledge-based support system for mechanical ventilation of the lungs: The KUSIVAR concept and prototype. *Comput. Methods. Progr. Biomed.* 30:59.

Rutledge GW, Thomsen GE, Farr BR, et al. 1993. The design and implementation of a ventilator-management advisor. *Artif. Intell. Med.* 5:67.

Saranummi N, Korhonen I, van Gils M, and Kari A. 1997. Framework for biosignal interpretation in intensive care and anesthesia. *Meth. Inform. Med.* 36:340.

Schecke T, Langen M, Popp HJ, et al. 1992. Knowledge-based decision support for patient monitoring in cardioanesthesia. *Int. J. Clin. Monit. Comput.* 9:1.

Siregar P, Coatrieux JL, and Mabo P. 1993. How can deep knowledge can be used in CCU monitoring. *IEEE Eng. Med. Biol. Mag.* 12(4):92.

Sittig DF, Pace NL, Gardner RM, et al. 1989. Implementation of a computerized patient advice system using the HELP clinical information system. *Comput. Biomed. Res.* 22:474.

Sittig DF, Cheung KH, Berman L. 1992. Fuzzy classification of hemodynamic trends and artifacts: Experiments with the heart rate. *Int. J. Clin. Monit. Comput.* 9:251.

Steinman F and Adlassnig K-P. 1994. Clinical monitoring with fuzzy automata. *Fuzzy Sets Syst.* 61:37.

Sukuvaara T, Sydanma M, Nieminen H, et al. 1993. Object-oriented implementation of an architecture for patient monitoring. *IEEE Eng. Med. Biol. Mag.* 12:69.

Sukuvaara T, Koski E, and Mäkivirta A. 1993. A knowledge-based alarm system for monitoring cardiac operated patients: Technical construction and evaluation. *Int. J. Clin. Monit. Comput.* 10:117.

Summers R, Carson ER, and Cramp DG. 1993. Ventilator management: The role of knowledge-based technology. *IEEE Eng. Med. Biol. Mag.* 12(4):50.

Thoraval L, Carrault G, Schleich, JM, et al. 1997. Data fusion of electrophysiological and hemodynamic signals for ventricular rhythm tracking. *IEEE Eng. Med. Biol. Mag.* 16(6):48.

Tong DA. 1991. Weaning patients from mechanical ventilation: A knowledge-based system approach. *Comput. Meth. Progr. Biomed.* 35:267.

Tsien, CL and Fackler, JC. 1997. Poor prognosis for existing monitors in the intensive care unit. *Crit. Care. Med.* 25:614.

Uckun S. 1992. Model-based reasoning in biomedicine. *Crit. Rev. Biomed. Eng.* 19(4):261.

Uckun S, Dawant BM, Lindstrom DP. 1993. Model-based diagnosis in intensive care monitoring: The YAQ approach. *Artif. Intell. Med.* 5:31.

Uckun S. 1994. Intelligent systems in patient monitoring and therapy management: A survey of research projects. *Int. J. Clin. Monit. Comput.* 11:241.

Uckun S. 1996. Instantiating and monitoring skeletal treatment plans. *Meth. Inform. Med.* 35:324.

Unser M and Aldroubi A. 1996. A review of wavelets in biomedical applications. *Proc. IEEE* 84(4):626.

van der Aa JJ. 1990. Intelligent alarms in anesthesia: A real time expert system application. Ph.D. dissertation, Eindhoven University of Technology.

187

Medical Terminology and Diagnosis Using Knowledge Bases

Peter L.M. Kerkhof
Medwise Working Group

"... this man with two blue hands and abnormal blood gases and a negative chest X-ray. And high blood pressure. ...(one physician) said he didn't know what the hell was going on. Would I take a look at him? Just a curbside consultation—that's all. Just an opinion. I said O.K.—be glad to. It was entirely appropriate. I was a physician of record."

From *Two Blue Hands*, by Berton Roueché

187.1 Introduction

Computers in medicine assist the process of communication and support the integration of information. In the above example given by Roueché, the consulting physician could be replaced by a computer system that stores and handles information that was previously obtained from a number of experts in various medical specialties. Ideally, the victim described here carries a patient data card which, in an emergency case, presents valuable information to the physician. The Medical Records Institute [1981] is actively engaged in the development of an electronic patient record.

Computer-mediated reasoning systems have been described by Williams [1982]: Knowledge Base (KB) systems are used for interpretation of data about a specific problem, in the light of knowledge represented in the KB, to develop a problem specific model and then to construct plans for problem solution. The KB contains the descriptive or factual knowledge pertaining to the domain of interest. Candidate hypotheses are then derived through some pattern matching system. Next, a reasoning "engine" (also termed *inference machine*) carries out the manipulation specified to reach a decision. Knowledge obviously can be represented in several ways: it can be characterized by symbols, plain words, and definitions along with their interrelations. Knowledge may be expressed by spoken or written words, flow charts, (mathematical) equations, tables, figures and so forth, depending on the level of abstraction.

0-8493-0462-8/00/$0.00+$.50

Aspects of language and text interpretation are central issues in artificial intelligence. Language is regarded as a metaphor of thought. When it provides the primary abstraction of knowledge about reality, it seems reasonable to expect that a powerful abstraction of language should also provide a powerful representation of knowledge. Various strategies have been explored: semantic networks offer a versatile tool for representing knowledge of virtually any type that can be captured in words by employing nodes (representing things) and links (referring to meaningful relationships between things), thus expressing causal, temporal, taxonomic and associational connections. Other approaches (such as frame-systems and production systems) have also been investigated. *Conceptual Graphs* (ANSI X3H4) [Sowa, 1984] are an emerging standard for knowledge representation, and the method is particularly suited to the representation of natural language semantics. Weed [1991] has advanced the idea of "knowledge coupling" i.e., a combination of a computerized medical record and decision-support software.

An ideal medical KB must be comprehensive (while integrating text, graphics, video, and sound), accurate and verifiable, easily accessible where doctors see patients, and the system should be adaptable to their own preferred terms or abbreviations [Wyatt, 1991]. Future developments will certainly include the use of artificial neural networks [Baxt, 1991], and examples realized thus far include myocardial infarction, diabetes mellitus, epilepsy, bone fracture healing, appendicitis, dermatology diagnosis, and EEG topography recognition.

Currently, most internationally available textbooks of (internal) medicine and medical dictionaries are retrievable on CD-ROM. Lexi-Comp's *Clinical Reference Library* is an example, featuring user-defined hyperlinks. A variety of titles are available as digital books on data cards suitable for palm computers, such as the *Pocket PDR*. In addition, the Internet offers numerous sites with relevant information [Smith and Edwards, 1997]. The September 1997 issue of the journal, *MD Computing*, focuses on medical resources on the Web.

This chapter addresses several approaches employed to represent medical definitions and knowledge. First, a survey will be presented on available coding systems, and next an overview is given of current KB systems which primarily refer to the process of establishing a diagnosis. Finally, an anthology concerning problems related to medical terminology is presented, along with directions for obtaining a solution.

187.2 Classification and Coding Systems

With the exception of one British system, all classification or coding systems have been developed in the USA. This survey lists all projects along with some of their characteristics.

The International Classification of Diseases (ICD) [International Classification of Diseases, 1994] system entered its tenth version, succeeding the ninth edition released in 1978. With more than 8000 codes, it is applied worldwide for classifying diagnoses and also permits diagnosis related group (DRG) assignment employed for billing and reimbursement purposes.

Systematized Nomenclature of Medicine (SNOMED) [Systematized Nomenclature of Medicine, 1993] offers a structured nomenclature and classification for use in human as well as in veterinary medicine. It consists of eleven modules with multiple hierarchies covering about 132,600 records, with a printed as well as a CD-ROM version available.

Physicians' Current Procedural Terminology (CPT) [Physicians' Current Procedural Terminology, 1992] provides uniform language for diagnostic as well as surgical and other interventional services. This system is currently in its fourth edition which is distributed by the American Medical Association (AMA), and has been incorporated in the Medicare program.

Medical Subject Headings (MeSH) [1993] is a systematic terminology hierarchy which is used to index the MEDLINE medical publications system. It forms the standard for coding keywords related to the contents of articles in the field of medicine. The system is updated yearly.

The National Library of Medicine (NLM) in 1986 started a project called Unified Medical Language System (UMLS) [Lindberg et al., 1993]. The project aims to address the fundamental information access problem caused by the variety of independently constructed vocabularies and classifications used in different sources of machine-readable biomedical information. The UMLS approach will be to compensate for differences in the terminologies or coding schemes used in different systems, as well as for differences in the language employed by system users, rather than to impose a single standard vocabulary on the biomedical community. The Metathesaurus contains 311,000 terms and maintains cross reference to CPT [Physicians' CPT, 1992], ICD, SNOMED [SNOMED, 1993] and other indexes. NLM encourages broad experimentation with the fourth edition, and the tools are (with certain provisions) available on the basis of a one-year agreement. The complete files are distributed on four CD-ROMs, while the application programs are limited to browsing tools for the metathesaurus.

Gabrieli [1992] constructed a computer-oriented medical nomenclature based on taxonomic principles. His system covers 150,000 preferred terms and a similar number of synonyms. The partitioning method employed for medical classification readily permits replacement of English names with terms of any other language, thus creating the perspective of a world-wide standard.

Read from the U.K. designed a classification for various computer applications [Chisholm, 1990]. It is designed in accordance with six key criteria: to be comprehensive, hierarchical, coded, computerized, cross-referenced, and dynamic. The system is closely connected with the British National Health Service (NHS), and includes 100,000 preferred terms, 250,000 codes, and 150,000 synonyms.

187.3 An Electronic Medical Encyclopedia at Your Fingertips: Knowledge Bases and Diagnosis

In this era, most information (as stored in written documents or computer records) consists of *natural language* text. Traditional retrieval systems use sets of keywords (index terms). It is called a *full text retrieval system*, when all (nontrivial) words of that text are indexed for future identification during search procedures. Information requests can then be formulated with the aid of *Boolean operators* (namely: **and, or, not**). Synonymous terms can be specified by employing the **or**-operator. The *fuzzy-set retrieval model* still uses Boolean logic but refines the query process by assigning different weight factors to the individual index terms. As an attractive alternative, the *vector space processing system* evaluates the similarity where both the index terms and the stored texts are represented by weighted term vectors [Salton, 1991]. The assumption that index terms are independent (i.e., orthogonal vector space) obviously implies a shortcoming of this retrieval method.

Within the domain of medicine, various KB systems have been developed and current systems will be briefly characterized here. Again, all projects in this area originate in the USA, with the exception of two (namely OSM and Medwise). One approach (namely CONSULTANT) addresses the field of veterinary medicine. A tabulated survey of the situation created in 1987 has been published [Kerkhof, 1987]. An occasional system is in the public domain, whereas others sell at a single-user price anywhere between \$125 and \$2000 depending on the configuration, discount, and other factors.

CMIT developed by the AMA [Finkel, 1981] forms a reference for the selection of preferred medical terms including certain synonyms, and generic terms with built-in arrangements to provide maximum convenience in usage, currency, and timely publication. Thus, CMIT is both a system of reference and a "distillate of a vast amount of medical knowledge." The system is available in electronic form.

Blois was the first physician to apply CMIT as a diagnostic tool in his *RECONSIDER* project [Blois, 1984]. The application was released in 1981 and covered 3,262 disease entities, while 21,415 search terms were listed in a directory along with their frequency of occurrence (serving as an indicator of impact for each search term).

DXplain [Barnett et al., 1987] is also based on CMIT [Finkel, 1981] and was released in 1986. The project has close connections with AMA, and information is distributed using the World Wide Web. The KB contains information on 2,000 diseases and understands over 4,700 terms, with 65,000 disease-term relationships.

QMR patient diagnostic software [Middleton, 1991] covers 660 disease profiles and over 5,000 findings. It is the personal computer version of the INTERNIST-I prototype, sold by First DataBank in San Bruno at $495. The program is versatile, easy to use, and offers an attractive user-interface, but the size of its KB remained remarkably constant over the years.

The program *MEDITEL* [Waxman and Worley, 1990] addresses the issue of diagnosis in adults, and was marketed by Elsevier Publishers. Over the last few years, not much news has been reported in the literature.

ILIAD version 4.0 [Bergeron, 1991] is a software package designed to aid students and residents in their clinical decision logic. The ILIAD project is an outgrowth of the Health Evaluation through Logical Processes (HELP) system developed at LDS hospital in Salt Lake City, UT. The KB includes 1,300 diseases and 5,600 manifestations, and is built upon the experience of experts in nine subspecialties of internal medicine. The KB is updated semi-annually for revision and expansion. Future versions will support access to video laser discs, thereby providing an actual view of X-rays, echocardiograms, skin lesions, and other physical findings.

The *Oxford System of Medicine (OSM)* project was initiated by the Imperial Cancer Research Fund for use in primary care, and to help general practitioners during routine clinical work to support decision-making tasks. Such tasks include (1) diagnosing illnesses, (2) planning investigations, and patient treatment schedules, (3) prescribing drugs, (4) screening for disease, (5) assessing the risk of a particular disease, and (6) determining a referral to a specialist [Krause et al., 1993].

Medwise is a medical KB founded in 1983 and now covers some 3,900 disease entities, with 29,000 different keywords [Kerkhof et al., 1993]. It includes a separate KB with almost 500 equivalent terms, each referring to an average of three related terms e.g., extremity, leg, and limb. Equivalents are automatically generated to assist the user during the process of data entry. The matrix structure of the Medwise KB permits semantic differentiation of particular observations, e.g., lymphadenopathy as detected during palpation versus lymphadenopathy as interpreted from an X-ray picture. In the Medwise system both refer to different elements of the matrix, and their corresponding weight factors are individually incorporated when calculating the score for disease profile matching. For obvious reasons the language employed is English, but a complete dictionary that offers translations into Dutch is available [Kerkhof, 1996].

The *Framemed* system [Bishop and Ewing, 1993] divides medical information into 26 domains (comparable to the matrix in Medwise) [Kerkhof et al., 1993], and arranges the items in a hierarchical sequence; formation of the hierarchies yields a logical framework for a standardized terminology that is in the public domain. The objective is to achieve a standard coded terminology to which all existing systems can relate, with obvious use as an electronic encyclopedia and for differential diagnosis. Framemed offers synonyms, and claims as an advantage a simple overall structure that facilitates updating.

STAT!-Ref [STAT, 1994] offers the contents of a first choice medical library (including several standard textbooks, e.g., on Primary Care) as well as Medline (either primary care, cardiology, or oncology) on CD-ROM.

MD-Challenger [MD-Challenger, 1994] offers a clinical reference and educational software for acute care and emergency medicine (everything from abdominal pain to zygoapophyseal joint arthritis), with nearly 4,000 annotated questions and literature references. The system also includes Continuing Medical Education (CME) credits.

Labsearch/286 is a differential diagnosis program allowing input of up to two abnormal laboratory findings plus information on symptoms and signs. Laboratory data concentrate on body fluids (blood, urine, cerebrospinal, ascitic, synovial, and pleural fluid) entered as high or low [Labsearch, 1994]. The system includes 6,500 diseases and 9,800 different findings. No details are available on its diagnostic performance, although the restriction of only two laboratory data certainly will limit its potential.

CONSULTANT is a KB for veterinary medicine developed by White [White and Lewkowicz, 1987]. This database for computer-assisted diagnosis and information management is available on a fee-for-service basis and is actually used in hundreds of private practices and institutions in North America.

Dambro [1998] compiled a book (approx. $50) that is updated annually, *Griffith's 5 Minute Clinical Consult.* The first edition appeared in 1993 and indeed contains a realm of chart-like presented practical information on a thousand topics with reference to their ICD-code. A CD-ROM version is also on the market. Similar to the design of Framemed [Bishop and Ewing, 1993], the information is compiled by a group of contributing authors whose names are listed in conjunction with each disease profile. Tilly and Smith [1997] presented a comparable book for the field of veterinary medicine along with a CD-ROM version: *The 5 Minute Veterinary Consult.*

The *Birth Defects Encyclopedia* edited by Buijse [1992] has the significant subtitle "The comprehensive, systematic, illustrative reference source for the diagnosis, delineation, etiology, biodynamics, occurrence, prevention and treatment of human anomalies of clinical relevance." A unique feature of this printed encyclopedia is the BDFax-service which allows anyone in the world to call the Center for Birth Defects Information Services [Buyse, 1992] and request a current, daily updated version of any article in the KB. The information will then be faxed. Furthermore, BRS Information Technologies makes the full text available for online search and retrieval. Finally, the related Birth Defects Information System (BDIS) is a sophisticated computer-based profile matching system that reduces the research and diagnostic assistance tasks associated with complex syndromes.

187.4 Problems Related to Medical Terminology

kiotomy (ki-ot'o-me) excision of the vulva (in Dorland's pocket *Medical Dictionary,* 21st ed., p. 342; of vulvectomy on p. 684).

Inspection of other sources suggests that in the word vulva the first letter v should be replaced by the letter u, followed by rearrangement of letters to obtain an anagram, namely **uvula** *(being a portion of the soft palate).*

Medical language forms one of the greatest obstacles for the practical use of any type of KB designed for application in the field of medicine [Kerkhof, 1992]. Just imagine the confusion introduced when a visiting surgeon from an exotic country is requested to perform a kiotomy, and consults the above mentioned popular dictionary to look up what is supposed to be excised.

Natural language often has remote roots e.g., adrenaline and epinephrine are the same chemical substances [Kerkhof, 1996]. Various words can be used in two senses: as noun and as adjective, e.g., antipyretic and adrenal. The Latin word "os" means both "mouth" and "bone". And what's in a phrase? "*... a neurologist found a nonfluent dysphasia, without circumlocution or paraphasias*" (Is there a single term for computer entry?). Or: "*The CT scan showed accentuation of the peripheral margins of the bilateral parieto-occipital forceps major, and splenium low-absorptive abnormalities.*" What does it mean, anyway? These examples illustrate the problem of translating medical phrases into concise "computer-storable language".

Besides problems inherent in the understanding of natural language, additional difficulties pertaining to medical terminology can be indicated:

American vs. British spelling. Two standard differences are evident, namely the use of the digraph in British spelling (e.g., anaemia vs. anemia) and preference for using *c* (e.g., in leucocyte) rather than *k* (as in the American word leukocyte). However, combination of both rules does not apply in the British equivalent of the American spelling of the word leukemia, where the anglicized version is spelled as "leukaemia" [Kerkhof, 1996].

Synonyms. For some reason "epistaxis" is identical to "nosebleed", and "pruritus" equals "itching"; it is not all that difficult, but the major problem is that you have to recall this every time you use either of them. Thrombocytosis and thrombocythemia are two words to indicate that the number of platelets in the peripheral circulation is in excess of 350,000 per microliter.

Eponyms. Many disease names refer to the first author (e.g., Boeck's disease for sarcoidosis) who described the particular disorder, to the first patient analyzed in detail (e.g., Mortimer's disease, again for sarcoidosis), or to the area (e.g., Lyme disease) where the illness was first detected. An epidemiologist ever attempted to replace the term Bornholm disease just to honor a friend (Sylvest). Geographical variations also occur: Grave's disease (or Parry's disease) as it is known in the U.S., U.K., and Australia (referring to goiter), is termed Basedow's disease on the European continent, but Flajani's disease in Italy.

Preferred terminology. In radiology "air" means gas within the body, regardless of its composition or site, but the term should be reserved for inspired atmospheric gas. With reference to pneumothorax, subcutaneous emphysema, or the contents of the gastrointestinal tract, the preferred term is "gas". Sometimes the preferred terminology refers to simplicity; there is no virtue in talking about "male siblings" if we mean "brothers". On other occasions the preferred terminology pertains to technical vocabulary which permits high precision and resolution descriptions if the available information is extremely exact. In these circumstances a valuable tool is blunted, if carelessness creeps in. For example the word "clumsiness" describes defective coordination of movement, whereas the term "dysdiadokokinesis" refers to the well-defined phenomenon of a defect in the ability to perform rapid movements of both hands in unison [Murphy, 1976].

Different expressions and their exact meaning. A straightforward example is: tympanism, tympanites, tympanitis and tympany, particularly in relation to: tympanal, tympanic, tympanous, and tympanitic [Kerkhof, 1996]. One annotated example is presented: (1) heterogenous (not originating in the body); (2) heterogeneous (not of uniform composition, quality or structure); (3) heterogenetic (pertaining to asexual generation); and (4) heterogenic (the same as heterogeneous).

Implicit information. A particular statement may imply a multitude of information components, e.g., if urinalysis is found to be normal, then the preceding examination implies (at least) the absence of the quadruplet proteinuria, hematuria, glucosuria, and casts. Also *mirror-terms* may apply: "dry cough" implies "no productive cough" (which can be entered as a negative finding). Likewise, leukopenia in particular implies "no leukocytosis". This mutual exclusion principle applies to all antonyms, especially all terms beginning with hypo- or hyper-.

Imprecise terminology [Yu, 1983]. Some terms may carry a vague meaning, e.g., tumor, swelling, mass, and lump. To a large extent, however, the use of such terms reflects the uncertainty around an observation. In that respect it refers to a justifiable "law of preservation of uncertainty". In other words, it would be incorrect to specify an observation in greater detail than the facts permit. This notion has consequences for the selection of equivalents.

Certainty vs. uncertainty. Decision analysis itself does not reduce the uncertainty about the true state of nature, but as long as some choice needs to be made it does enable one to make rational decisions in the light of uncertainty [Weinstein and Fineberg, 1980]. Yet another aspect of certainty vs. uncertainty deserves mentioning, namely where percent-wise figures about prognosis or outlook are subjectively interpreted [Eraker and Politser, 1982]. Reduction of the probabilities by a factor of one-tenth completely reversed the preferences of a set of respondents. This "certainty effect" shows that outcomes perceived with certainty are overweighted relative to uncertain outcomes. Thus, the formulation of information affects its interpretation by humans.

Knowledge engineering implies various levels of translation [Adeli, 1990]. The expert formulates as precisely as possible his thoughts, the engineer provides feedback using his/her own phrases to ensure an exact match between both minds, and subsequently the resulting expression is translated to a format usable for the computer program. These three steps involve transformations of language, while assuming that the occasional user of the program appreciates the full scope of the original thoughts of the expert.

Information source vs. actual patient [Kerkhof, 1987]. Current medical information sources tend to adhere to preference terminology to promote the use of uniform medical language. However, such standard vocabulary is not used by the average patient to describe their individual health problems [Hurst, 1971]. Then it is left to the clinician to transpose, e.g., "puffy face" and "moon face" if appropriate. Indeed, better health care can be realized by educating the patient about the value of structured communication with the physician [Verby and Verby, 1977].

Subspecialty interpretation and jargon. In chest radiology "type I bulla" means the same as "bleb" or "air cyst", while pneumatocele is not regarded as a proper synonym [Glossary, 1989]. When naming a "hollow space" you may choose anything out of the following set: (1) cavity, (2) crypt, (3) pouch, (4) gap, (5) indentation, (6) dell, (7) burrow, (8) crater, (9) concavity, (10) excavation, (11) gorge,

(12) pocket, (13) cave, (14) cavern, (15) cistern, or (16) lacuna. However, every expression may exhibit a nuance within a certain context. Then there is a jargon: the terms (1) show, (2) engagement, (3) lightening, and (4) station, for example, have a particular meaning within the field of obstetrics [Kerkhof, 1996]. The term "streaking" has a meaning which differs for the microbiologist and the radiologist.

Multilingual approaches. The relation between a concept and the various corresponding terms in different languages is, in general, not unique. This implies that a multitude of different words from different syntactical categories may represent a single concept. Particularly the European countries are confronted with additional natural language problems. The Commission of the European Communities supports research activities in this area such as EPILEX (a multilingual lexicon of epidemiological terms containing Catalan, Dutch, English, French, German, Italian, Portuguese, and Spanish) [Du V Florey, 1993] and the development of a multilingual natural language system [Baud et al., 1994].

Frequency of occurrence. The meaning of semi-quantitative indicators like "always, often, etc." is not transparent when screening a medical text. The intuitive interpretation of some quasi-numerical determinants is summarized elsewhere [Kong et al., 1986].

Noise terms. A study [Kerkhof, 1993] revealed that input information consisted on average of 75 terms per patient case; required for establishing the primary diagnosis were only 15 terms. This implies that 80% of the input data consisted of "noise terms", which do not directly contribute to the confirmation of the correct diagnosis. Rather such overwhelming portions of information may blur the process of hypothesis formation, and in particular when the diagnostic task is solely left to the human being who is prone to confusion if the number of contributing elements is more than about ten.

187.5 Solution for Discrepancies

"Representation of knowledge is an art, not a science" [D.B. Lenat, 1989].

Various scientists have described techniques to help and solve language-related problems in the communication area of medicine. An available online medical dictionary can be an important tool for research and application in natural language processing. Dorland's *Illustrated Medical Dictionary* has been converted to an online interactive computer-based version. To assure exact description of definitions, a coding system has been advocated because of: (1) ease of handling, (2) redundancy to avoid errors, (3) specificity, (4) indication of relationships (e.g., by using common code groups for similar terms), (5) equating equivalent terms and linking them in different languages, and (6) the feasibility to cross-link terms on the basis of common portions of their codes [Bishop and Dombrowski, 1990].

Attempts have also been made to automatically translate existing medical terminology systems [Cimino and Barnett, 1990]. For example, using a semantic network for mapping, the closest match between the MeSH term "portography" was "portal contrast phlebogram" in the ICD-procedures directory.

Two independent routes are available to attack the majority of linguistic problems related to a medical KB: (1) Use uniform input data, i.e., force the user of the KB to enter only controlled terms as they already exist in the KB (terms allowed by the editorial committee responsible for the KB). This route implies tedious work for the user, and the job of selecting recognizable entries has to be carried out every time the KB is consulted. (2) Apply an additional KB with equivalent expressions. Once the user enters a particular term, the KB automatically generates a list of equivalent expressions to choose from. The construction of such an auxiliary KB has to be realized only once, apart from the obvious updating process desired as the system evolves. Furthermore, a matrix structure as applied in the Medwise KB proves to be helpful in organizing information, and incorporating a substantial portion of semantic interpretation when dealing with medical knowledge. Inclusion of the auxiliary KB also enhances user-friendliness, and accelerates the process of data entry [Kerkhof et al., 1993].

Discussion

While advances in computer technology permit virtually unlimited storage and fast retrieval capabilities, interpretation of natural language still constitutes a major obstacle for the application of medical KBs. Even common words often have highly ambiguous meanings [Murphy, 1976], e.g., "weakness" can mean lethargy or paralysis. Therefore, the user of any KB should be aware of the fact that justifiable adoption of a particular equivalent expression may depend on the precise interpretation of the complete case report.

On the other hand, the use of uniform input data on the basis of a thesaurus would imply a substantial loss of refinement with respect to terminology. Ideally, there is a maximum degree of freedom in the selection of words and their nuances, both for the KB and for the description of the individual case study.

An auxiliary KB with synonyms [SNOMED, 1993; Gabrieli, 1992; Chisholm, 1990; Kerkhof et al., 1993; Lindberg et al., 1993] indeed may offer considerable help during the process of matching input terms with information sources [Kerkhof, 1993]. An adequate solution to (at least a part of the) linguistic problems, will substantially enhance the performance of diagnostic computer programs.

Summary

A multitude of projects address the issue of how to handle an ever increasing amount of medical information. Coding schemes have been designed and are regularly refined, while other approaches aim to collect, structure, and disclose this gigantic corpus of information that features an unconfined potential to thrive.

The lack of standardized medical language limits the optimal use of computers in medicine. Major obstacles concern: (1) imprecise terminology, (2) the limited scope of a thesaurus, (3) multilingual conversion problems, (4) discrepancies between information sources vs. actual clinical cases, as well as (5) the ever expanding range of synonyms, acronyms, antonyms and eponyms. Incorporation of a separate knowledge base containing equivalent expressions appears to be indicated for the practical use of medical information systems.

Defining Terms

AMA: American Medical Association, with headquarters in Chicago. This organization is actively involved in new developments concerning medical informatics. CMIT [Finkel, 1981] and CPT [Physicians' Current Procedural Terminology, 1992] as well as the cooperation with DXplain [Barnett et al., 1987] are prototypes of their activities.

ICD: International Classification of Diseases, a widely accepted system to organize all possible medical diagnoses. The tenth version is translated for worldwide application.

IMIA: International Medical Informatics Association, consisting of the national societies from about 40 countries.

NLM: National Library of Medicine, in Bethesda (Maryland), USA.

UMLS: Unified Medical Language System, a project initiated by the NLM and distributed on CD-ROM. Cooperation with parties to implement the system within their own environment is encouraged, but requires a contract [Lindberg et al., 1993].

References

Adeli, H. 1990. *Knowledge Engineering*, Vol. I and II. New York, McGraw-Hill.

Barnett, G.O., Cimino, J.J., Hupp, J.A., and Hoffer, E.P. 1987. DXplain, an evolving diagnostic decision-support system. *JAMA* 258:67–74.

Baud, R., Lovis, C., Alpay, L., Rassinoux, A.M., Scherrer, J-R., Nolan, A., and Rector, A. 1994. Modelling for natural language understanding. *SCAMC* 93:289–293.

Baxt, W.G. 1991. Use of an artificial neural network for the diagnosis of myocardial infarction. *Ann. Intern. Med.* 115:483–848.

Bergeron, B. 1991. ILIAD, a diagnostic consultant and patient simulator. *M.D. Comput.* 8:46– 53.

Bishop, C.W. and Dombrowski, T. 1990. Coding, why and how. *M.D. Comput.* 7:210–215.

Bishop, C.W. and Ewing, P.D. 1993. Framed, a prototypical medical knowledge base of unusual design. *M.D. Comput.* 10:184–192.

Blois, M.S. 1984. *Information and Medicine.* Berkeley, Univ. California Press.

Buyse, M.L. (Ed.), *Birth Defects Encyclopedia.* Center for Birth Defects Information Services, Inc., Dover, MA; BDFax service (508) 785-2347.

Chisholm, J. 1990. Read Clinical Classification. BMJ 300:1092.

Cimino, J.J. and Barnett, G.O. 1990. Automated translation between medical terminologies using semantic definitions., *M.D. Comput.* 7:104–109.

Dambro, M.R. 1998. *Griffith's 5 Minute Clinical Consult.* Baltimore, Williams & Wilkins.

Dorland's pocket MEDICAL DICTIONARY, 21st edition p.342; cf. vulvectomy on p.684.

Du V Florey, C. 1993. EPILEX, a multilingual lexicon of epidemiological terms, Office for official publications of the European Communities (ISBN 92-826-6211-X), Luxembourg.

Eraker, S.A. and Politser, P.E. 1982. How decisions are reached; physician and patient. *Ann. Intern. Med.* 97:262–268.

Finkel, A.J., Ed. 1981. *Current Medical Information and Terminology, 5th ed.* American Medical Association, Chicago.

Gabrieli, E.R. 1992. A New American Electronic Medical Nomenclature, *Automedica* 14:23–30.

Glossary of "chest" radiological terms. 1989. Nomenclature Committee of the Fleischner Society.

Hurst, J.W. 1971. The art and science of presenting a patient's problems. *Arch. Intern. Med.* 128:463.

International Classification of Diseases, tenth Clinical Modification. 1994. Commission on Professional and Hospital Acitivities, Ann Arbor, Michigan.

Kerkhof, P.L.M. 1987. Dreams and realities of computer-assisted diagnosis systems in medicine (Editorial). *Automedica* 8:123–134.

Kerkhof, P.L.M. 1992. Knowledge base systems and medical terminology. *Automedica* 14:47– 54.

Kerkhof, P.L.M. 1996. *Woordenboek der geneeskunde N/E and E/N.* (*Medical Dictionary, 2nd ed.*), Houten, The Netherlands, Bohn Stafleu Van Loghum (ISBN 90-6016-858-5).

Kerkhof, P.L.M. 1998. *Veterinair Woordenboek E/N and N/E.* (*Veterinary Dictionary*), Maarssen, The Netherlands, Elsevier Publ. (ISBN 90-6348-240-X).

Kerkhof, P.L.M., van Dieijen-Visser, M.P., Koenen, R., Schreuder, J.J.A., de Bruin H.G., and Gill, K. 1993. The Medwise Diagnostic Module as a Consultant. *Expert Systems with Applications* 6:433–440.

Kong, A., Barnett, G.O., Mosteller, F., and Youtz, C. 1986. How medical professionals evaluate expressions of probability. *New Engl. J. Med.* 315:740–744.

Krause, P., Fox, J., O'Neil, M., and Glowinski, A. 1993. Can we formally specify a medical decision support system? *IEEE Expert* 8:56–61.

Labsearch/286, software available from REMIND/1 Corp., P.O. Box 752, Nashua, NH 03061, USA.

Lenat, D.B. and Guha, R.V. 1989. *Building Large Knowledge Base Systems.* Addison-Wesley Publ. Co.

Lindberg, D.A., Humphreys, B.L., and McCray, A.T. 1993. The Unified Medical Language System. *Methods Inf. Med.* 32:281–291.

MD-Challenger, 70 Belle Meade Cove, Eads, TN 38028, USA.

Medical Records Institute, P.O. Box 289, Newton, MA 02160, USA.

MeSH, *Medical Subject Headings.* 1993. National Library of Medicine, Bethesda.

Middleton, B., Shwe M.A., Heckerman D.E., Henrion M., Horvitz E.J., Lehmann H.P., and Cooper G.F. 1991. Probabilistic diagnosis using a reformulation of the INTERNIST-1/QMR knowledge base; part II evaluation of diagnostic performance. *Meth. Inform. Med.* 30:15–22.

Murphy, E.A. 1976. *The Logic of Medicine.* Johns Hopkins Univ. Press, Baltimore, MD.

Physicians' Current Procedural Terminology. 4th ed., 1992. American Medical Association, Chicago.

Roueché, B. 1982. *The Medical Detectives*, Washington Square Press, New York, p. 317.

Salton, G. 1991. Developments in automatic text retrieval. *Science* 253:974–980.

Smith, R.P. and Edwards, M.J.A. *The Internet for Physicians.* New York, Springer (ISBN 0- 387-94936-4).

SNOMED International. 1993. College of American Pathologists, Northfield, IL.

Sowa, J.F. 1984. *Conceptual Graphs, Information Processing in Mind and Machine.* Reading, MA, Addison-Wesley.

STAT!-Ref. 1994. Available from Teton Data Systems, P.O. Box 3082, Jackson, WY.

Tilly L.P., and Smith W.K. 1997. *The 5 Minute Veterinary Consult.* Baltimore, Williams, and Wilkins.

Verby, J. and Verby, J. 1977. *How to Talk to Doctors.* Arco Publ. Co., Inc., New York.

Waxman, H.S. and Worley, W.E. 1990. Computer-assisted adult medical diagnosis, subject review and evaluation of a new microcomputer-based system. *Medicine (Baltimore)* 69:125–136.

Weed, L.L. 1991. *Knowledge Coupling.* New York, Springer-Verlag.

Weinstein, M.C. and Fineberg, H.V. 1980. *Clinical Decision Analysis.* W.B. Saunders, Philadelphia p.61.

White, M.E. and Lewkowicz, J. 1987. The CONSULTANT database for computer-assisted diagnosis and information management in veterinary medicine. *Automedica* 8:135–140.

Williams, B.T. 1982. *Computer Aids to Clinical Decisions.* CRC Press Inc., Boca Raton, Florida Vol. II pp. 81–99.

Wyatt, J. 1991. Computer-Based Knowledge Systems. *Lancet* 338:1431–1436.

Yu, V.L. 1983. Conceptual Obstacles in Computerized Medical Diagnosis. *J. Med. Philos.* 8:67–83.

Further Information

The fields of medical knowledge bases and medical terminology are rapidly developing. There is no single comprehensive source of information available, but the reader is advised to scrutinize the following journals for new information:

M.D. Computing, published by Springer Verlag (New York and Berlin), reports on research in the field of medical informatics. Of special interest to the clinician.

IEEE Expert appears four times a year and presents the latest on artificial intelligence and expert systems. Contact P.O. Box 3014, Los Alamitos, CA 90720. For those interested in technical details on intelligent systems and their applications. IEEE (P.O. Box 1331, Piscataway NJ 08855-1331, USA) publishes a number of related journals e.g., on knowledge engineering and on multimedia.

The National Library of Medicine (NLM, 8600 Rockville Pike, Bethesda MD 20894, USA) releases news bulletins, and provides information on UMLS and contracts for cooperation.

Ongoing projects on medical nomenclature are surveyed elsewhere in this chapter.

Pertinent Internet Web Sites:

British Medical Journal: http://www.bmj.com
JAMA: http://www.ama-assn.org/public/journals
The Lancet: http://www.thelancet.com
The New England Journal of Medicine: http://www.nejm.org
Grateful Med: http://igm.nlm.nih.gov/
Physician's Desk Reference: http://www.pdrnet.com/
About ICD-9-CM: http:\www.icd-9-cm.org/
About ICD-10: http:\www.cdc.gov.nchswww/about/major/dvs/icd10des.htm
Medical guide: http://www.medmatrix.org
Search engine for Health on the Net: http://www.hon.ch/
Alternative Medicine: http://www.yahoo.com/health/alternative_medicine
Dermatology images: http://tray.dermatology.uiowa.edu/dermimag.htm
Focus on molecular medicine: http://www.aston.it/icamm

Obviously, annual meetings form the forum for presentation of the latest developments: IMIA conferences are well known besides the world congress organized every four years. IMIA publishes a Yearbook of Medical Informatics; contact Schattauer Publishers, Subscriptions Dept., P.O. Box 104545, W-7000 Stuttgart 10, Germany.

188

Natural Language Processing in Biomedicine

Stephen B. Johnson
Columbia University

188.1 Introduction

Natural language is the primary means of communication in all complex social interactions. In biomedical areas, knowledge and data are disseminated in written form, through articles in the scientific literature, technical and administrative reports, and hospital charts used in patient care. Much vital information is exchanged verbally, in interactions among scientists, clinical consultations, lectures, and in conference presentations. Increasingly, computers are being employed to facilitate the process of collecting, storing, and distributing biomedical information. Textual data is now widely available in an electronic format, through the use of transcription services, and word processing. Important examples include articles published in the medical literature and reports dictated during the process of patient care (e.g., radiology reports and discharge summaries).

While the ability to access and review narrative data is highly beneficial to researchers, clinicians, and administrators, the information is not in a form amenable to further computer processing, for example, storage in a database to enable subsequent retrievals. At present, the most significant impact of the computer in medicine is seen in processing *structured data*, information represented in a regular, predictable form. This information is often numeric in nature, e.g., measurements recorded in a scientific study, or made up of discrete data elements, e.g., elements selected from a predefined list of diseases.

The techniques of natural language processing provide a means to bridge the gap between textual and structured data, allowing humans to interact using familiar natural language, while enabling computer applications to process data effectively.

188.2 Linguistic Principles

Natural language processing (or computational linguistics) is a branch of computer science concerned with the relationship between information as expressed in natural language (sound or text) and represented in formalisms that facilitate computer processing. Natural language *analysis* studies how to convert natural language input into a structured form, while natural language *generation* investigates how to

0-8493-0462-8/00/$0.00+$.50

produce natural language output from structured representations. Natural language processing investigates and applies scientific principles of linguistics through development of computer systems. One of the most important principles is that natural language is built up from several layers of structure, each layer defined as a set of restrictions on the previous layer [Harris 1991]:

TABLE 188.1 Layers of Linguistic Structure

Linguistic Layer	Description
Phonological	Mapping of sounds to phonemes (letters)
Morphological	Grouping of phonemes into morphemes (roots and inflections)
Lexical	Combining of morphemes into words
Syntactic	Combining of words into sentence structure
Semantic	Mapping of sentence structure into literal meanings
Pragmatic	Combining of sentence meanings into discourse meaning

These layers correspond roughly to the subfields of linguistics, and to the areas of study in natural language processing, which seeks to develop them into computer models. Knowledge about the rules of language structure at any or all of its levels is called *competence.* Most natural language processing applications in biomedicine do not even begin to approach the language competence of humans. For example, a transcription system may have knowledge about the sounds of a language but know nothing about the syntax of sentences. Similarly, a program that indexes scientific articles may know which terms to look for in the text, but have no ability to express how these terms are related to one another. However, it is important to emphasize that such applications may perform extremely useful tasks, even with very limited competence.

Many natural language processing applications exploit the fact that biomedical fields are *restricted semantic domains,* which means that natural language information associated with that field is focused on a narrow range of topics. For example, an article about cell biology is limited to discussions of cells and tissues, and is unlikely to mention political or literary issues. The natural language of a restricted semantic domain is called a *sublanguage.* Sublanguages tend to have specialized vocabularies and specialized ways of structuring sentences (e.g., the "telegraphic style" of notes written about patients in hospital charts) and ways of organizing larger units of discourse (e.g., the format of a technical report) [Grishman and Kittredge 1986, Kittredge and Lehrberger 1982].

These properties of sublanguages allow the use of methods of analysis and processing that would not be possible when processing the language of newspaper articles or novels. For example, a program that indexes medical articles can select index terms from a list of terminology known to be of interest to researchers; a speech recognition system can exploit the fact that only certain words can be uttered by a user in response to a given prompt; a system that analyzes clinical reports can look for predictable semantic patterns that are characteristic of the given domain.

188.3 Applications in Biomedicine

There are four forms of natural language exchange possible between human and the computer: (1) the human can supply information to the computer using natural language; (2) the human can retrieve information from the computer through natural language; (3) the computer can supply information to the human in the form of natural language output; and (4) the computer can request information from the human using natural language questions. Applications of natural language processing may support one or several of these modes of exchange. For example, an automated questionnaire for patients may generate multiple choice questions but be able to receive only numbered answers, while a database interface may be able to accept natural language questions submitted by a researcher and return natural language replies. Applications can be classified according to the levels of language competence embodied in their design: (1) speech systems, (2) lexical systems, (3) syntactic and semantic systems, and (4) discourse systems.

Speech Systems

Speech recognition systems [Grasso 1995] process voice data into phonemic representations (such as text), while speech synthesis systems generate spoken output from such representations. Speech processing applications may work only at this level of language competence, e.g., software that transcribes spoken input into text, or software that generates speech when visual information cannot be employed, such as on the telephone. Applications in this category include systems that capture structured endoscopic reports [Johannes and Carr-Locke 1992], and emergency room notes [Linn et al., 1992]. Applications may also employ speech technology as part of a larger system. Examples include interfaces to diagnostic systems [Landau et al., 1989], [Shiffman et al., 1992], and systems for taking patient histories [Johnson et al., 1992].

Lexical Systems

Lexical systems work with language at the level of words and terms (short sequences of words). A *lexicon* is a special type of database that provides information about words and terms which may include pronunciation, morphology (roots and affixes), and syntactic function (noun, verb, etc.). The Specialist Lexicon provides information about medical terms and general English words [National Library of Medicine 1993]. A thesaurus groups synonymous terms into semantic classes, which are frequently organized into a hierarchical classification scheme. The Systematized Nomenclature of Medicine (SNOMED) classifies medical terms in several hierarchies, such as (1) topography, (2) morphology, (3) etiology, (4) function, (5) disease, (6) occupation, and (7) procedure [Rothwell et al., 1993]. Medical Subject Headings (MeSH) classifies terms used in medical literature [National Library of Medicine 1990]. The Metathesaurus of the Unified Medical Language System (UMLS) combines SNOMED, MeSH, and other thesauri [Lindberg 1993].

Semantic classes are used to index natural language documents to facilitate their retrieval from a database. For example, MeSH is used to index the medical literature in the MEDLINE database [Bachrach 1978]. Semantic classes are also used to define the set of data elements (*controlled vocabulary*) that can be processed by a computer application. Controlled vocabularies are used by diagnostic systems and clinical information systems—by the programs that collect data, store it in databases, and display it [Linarrson and Wigertz 1989].

Syntactic and Semantic Systems

Lexical techniques can only approximate the meaning of an article or some text entered by the user of a computer application. One approach to obtaining a deeper understanding is to produce a representation of the syntactic structure of a sentence (e.g., determining the subject, the verb, and the object), and then map this structure into a representation of its meaning (such as predicate calculus). A computer program that analyzes the structure of sentences is called a *parser*, and uses a lexicon and a *grammar* (a formal representation of the syntactic rules of a language).

Representations of semantic content (or knowledge representation) vary, but most representations are variations of a structure called a *frame*, in which predefined slots are filled with information from natural language sentences. A number of systems process clinical reports into their own specialized frame representations, for example the Linguistic String Project [Sager et al., 1986], MedLEE [Friedman 1994], and SymText [Haug 1997]. The representation of "conceptual graphs" [Sowa 1984] has emerged as a potential standard for semantics and is used for medical language processing by a number of systems which include METEXA [Schroder 1992], MENELAS [Bouaud et al., 1997], and RECIT [Baud et al., 1992].

Discourse Systems

Information from the sentences of a discourse (e.g., an article or clinical report) combine in complex ways; for example, cross-references are made (e.g., using pronouns), and events are related temporally or causally. Understanding a discourse fully may require information about the context in which the natural language exchange occurs, or the background knowledge of a technical field. Systems based on syntax and semantics usually carry out an analysis of the discourse and context information as a subsequent stage. Other systems attempt to approximate the semantic analysis of sentences, using medical knowledge to guide the overall processing of a text by defining the possible sequences of topics and subtopics characteristic of the domain. Domains in which these methods have been attempted include: (1) the history and physical sections of a patient chart [Archbold and Evans 1989]; (2) echocardiography reports [Canfield et al., 1989]; (3) discharge summaries [Gabrieli and Speth 1987]; and (4) radiology reports [Ranum 1988].

Natural language is also a convenient mode for the output of applications, for example in producing clinical reports [Kuzmak and Miller 1983]. Some expert systems have used natural language to express their recommendations [Rennels et al., 1989], and to provide explanations of their reasoning process [Lewin 1991].

188.4 Challenges

The design of controlled vocabularies is an important area of research [Cimino et al., 1989]. As yet, no comprehensive vocabulary for clinical medicine exists, and the effort to create one is beyond the means of any single institution. Indexing of articles and reports is currently a labor-intensive process, and quality control is a significant problem. Index terms are a poor approximation of the meaning of a text. Similarly, systems that capture clinical information through the entry of individual findings fall short of capturing the patient history. Natural language processing systems with greater understanding of syntax and semantics may help by providing richer information representations.

No natural language processing system currently covers the complete range of language interaction with significant competence at all levels—from speech to discourse. In the near future, the systems with the greatest practical success will specialize in performing selected language processing tasks in well-defined domains.

Defining Terms

Controlled vocabulary: A set of well-defined data elements intended for use in a computer application.

Frame representation: A structure for representing complex information, consisting of fixed, named "slots" which may contain data values or frames.

Grammar: A formal representation of language structure, usually syntactic structure.

Language competence: Knowledge about the rules of a language.

Lexicon: A formal compilation of the words of a language, with information about phonetics, morphology, syntax, and semantics (depending on requirements of computer applications).

Morphology: Field studying how sounds combine into morphemes, and how morphemes combine to form words.

Parser: A computer program that uses a grammar and lexicon to analyze sentences of a language, producing a syntactic or semantic representation.

Phonology: Field studying the rules that govern the sounds used in languages.

Pragmatics: Field studying how sentences are combined to form larger units of discourse, and how to represent meaning in the context of a situation.

Semantics: Field studying the relation between natural language sentences and formal representations of meaning.

Structured data: Data that has a explicit, unambiguous, and regular structure, making it amenable to computer processing.

Sublanguage: The natural language of a restricted semantic domain that is focused on a specific set of tasks or purposes, e.g., interpreting X-ray images.

Syntax: Field studying how words combine to form sentences.

References

Allen, J. 1995. *Natural Language Understanding*; Second Edition. Redwood City, CA: The Benjamin Cummings Publishing Company, Inc.

Archbold, A. and Evans, D. 1989. On the Topical Structure of Medical Charts. In: *Proceedings of the 13th Annual SCAMC*, p. 543–547. IEEE Computer Society Press, Washington, D.C.

Bachrach, C.A. and Chaen, T. 1978. Selection of MEDLINE contents, the development of its thesaurus, and the indexing process, *Medical Informatics.* 3(3): 237–254.

Baud, R.H., Rassinoux, A.M., and Scherrer, J.R. 1992. Natural Language Processing and semantical representation of medical texts. *Meth. Inform. Med.* 31(2):117–125.

Bouaud, J., Zweigenbaum, P., et al., 1997. *A Semantic Composition Method Driven by Domain Knowledge Models.* Twenty-First Annual Symposium of the American Medical Informatics Association.

Canfield, K., Bray, B., Huff, S., and Warner, H. 1989. Database capture of natural language echocardiography reports: A Unified Medical Language System approach. In: *Proceedings of the 13th Annual SCAMC*, p. 559–563. IEEE Computer Society Press, Washington, D.C.

Cimino, J.J., Hripcsak, G., Johnson, S.B., Friedman, C., Fink, D.J., and Clayton, P.D. 1989. *Designing an Introspective, Multi-Purpose Controlled Medical Vocabulary.* Proceedings of the 13th Annual SCAMC, p. 513–518. IEEE Computer Society Press, Washington, D.C.

Friedman, C., Alderson, P.O., Austin, H.M., Cimino, J.J., and Johnson, S.B. 1994. A general natural language text processor for clinical radiology. *JAMIA* 1(2): 161–174.

Gabrieli, E. and Speth, D. 1987. *Computer Processing of Discharge Summaries. In: Proceedings of the 11th Annual SCAMC*, p. 137–140. IEEE Computer Society Press, Washington, D.C.

Haug, P.J. et al., 1994. *A Natural Language Understanding System Combining Syntactic and Semantic Techniques.* Eighteenth Annual Symposium of the American Medical Informatics Association.

Grasso, M.A. Automated speech recognition in medical applications. *MD Computing*, 1995;12(1):16–23.

Grishman, R. and Kittredge, R., Eds. 1986. *Analyzing Language in Restricted Domains: Sublanguage Description and Processing.* Erlbaum Associates, Hillsdale, New Jersey.

Harris, Z. 1991. *A Theory of Language and Information—A Mathematic Approach.* Clarendon Press, Oxford.

Johannes, R.S. and Carr-Locke, D.L. 1992. *The Role of Automated Speech Recognition in Endoscopic Data Collection. Endoscopy.* 24(Suppl 2): 493–498.

Johnson, K., Poon, A., Shiffman, S., Lin, R., and Fagan, L. 1992. A history taking system that uses continuous speech recognition. In: *Proceedings of the 16th Annual SCAMC*, p. 757–761. IEEE Computer Society Press, Washington, D.C.

Kittredge, R. and Lehrberger, J., Eds. 1982. *Sublanguage—Studies of Language in Restricted Semantic Domains.* De Gruyter, New York.

Kuzmak, P.M. and Miller, R.A. 1983. Computer-aided generation of result text for clinical laboratory tests. In: *Proceedings of the 7th Annual SCAMC*, p. 275–278. IEEE Computer Society Press, Washington, D.C.

Landau, J.A., Norwich, K.N., and Evans, S.J. 1989. *Automatic Speech Recognition—Can it Improve the Man-Machine Interface in Medical Expert Systems? Int. J. Biomed. Comp.*. 24(2): 111–117.

Lewin, H.C. 1991. HF-Explain: a natural language generation system for explaining a medical expert system. In: *Proceedings of The 15th Annual SCAMC*, p. 644–648. IEEE Computer Society Press, Washington, D.C.

Lindberg, D.A.B., Humphreys, B.L., and McCray, A.T. 1993. *The Unified Medical Language System.* In: *Yearbook of Medical Informatics,* van Bemmel, J.H., McCray, A.T., Eds., p. 41–51. International Medical Informatics Association, Amsterdam.

Linarrson, R. and Wigertz, O. 1989. The data dictionary—a controlled vocabulary for integrating clinical databases and medical knowledge bases. *Meth. Inform. Med.* 28(2): 78–85.

Linn, N.A., Rubenstein, R.M., Bowler A.E., and Dixon, J.L. 1992. *Improving the Quality of Emergency Room Documentation Using the Voice-Activated Word Processor: Interim Results.* In: *Proceedings of the 16th annual SCAMC,* p. 772–776. McGraw Hill, New York.

National Library of Medicine. 1990. Medical Subject Headings (NTIS NLM-MED-90-01). National Library of Medicine. Bethesda, Maryland.

National Library of Medicine. 1993. The Specialist Lexicon. Natural language systems group, National Library of Medicine, Bethesda, MD.

Ranum, D. 1988. *Knowledge Based Understanding of Radiology Text.* In: *Proceedings of the 12th Annual SCAMC,* p. 141–145. IEEE Computer Society Press, Washington, D.C.

Rennels, G., Shortliffe, E., Stockdale, F., and Miller, P. 1989. A computational model of reasoning from the clinical literature. *AI Magazine.* 10(1): 49–57.

Rothwell, D.J., Palotay, J.L., Beckett, R.S., and Brochu, L., Eds. 1993. *The Systematized Nomenclature of Medicine.* SNOMED International. College of American Pathologists, Northfield, Illinois.

Sager, N., Friedman, C., and Lyman, M. 1987. *Medical Language Processing—Computer Management of Narrative Data.* Addison-Wesley, Reading, Mass.

Scherrer, J.R., Cote, R.A., and Mandil, S.H. 1989. *Computerized natural language medical processing for knowledge representation.* North Holland, Amsterdam.

Schroder, M. 1992. *Knowledge-Based Processing of Medical Language: A Language Engineering Approach.* In: *Advances in Artificial Intelligence,* 16th German conference on AI, p. 221–234. Springer Verlag, Berlin.

Shiffman, S., Lane, C.D., Johnson, K.B., and Fagan, L.M. 1992. The integration of a continuous speech recognition system with the QMR diagnostic program. In: *Proceedings of the 16th Annual SCAMC,* p. 767–771. IEEE Computer Society Press, Washington, D.C.

Spyns, P. 1996. Natural Language Processing in Medicine. *Meth. Inform. Med.* 35:285–301.

Sowa, J.F. 1984. *Conceptual Graphs: Information Processing in Mind and Machine.* Addison-Wesley, Reading MA.

Van Bemmel, J.H. (Ed.). *Meth. Inform. Med.,* 1998:4(5).

Further Information

For general information about the field of natural language processing, see [Allen 1987], and [Covington 1994]. Surveys of research in sublanguage can be found in [Kittredge and Lehrberger 1982] and [Grishman and Kittredge 1986]. A variety of papers on medical language processing are collected in [Scherrer et al., 1989 and Van Bemmel 1998].

Historical Perspectives 5
Electroencephalography

Leslie A. Geddes
Purdue University

Historical Background

Hans Berger (1929) was the first to record electroencephalograms from human subjects. However, before then, it was well known that the brain produced electrical signals. In fact, in Berger's first paper there is a short history of prior studies in animals. Interestingly, the first person to demonstrate the electrical activity of the brain did not make recordings. In 1875, Richard Caton in the United Kingdom used the Thomson (Kelvin) sensitive and rapidly responding reflecting telegraphic galvanometer to display the electrical activity of exposed rabbit and monkey brains. His report [Caton, 1875], which appeared in the *British Medical Association Journal*, occupied only 21 lines of a half-page column. In part, the report stated:

> In every brain hitherto examined, the galvanometer has indicated the existence of electric currents. The external surface of the grey matter is usually positive in relation to the surface of a section through it. Feeble currents of varying direction pass through the multiplier [galvanometer] when the electrodes are placed on two points of the external surface, or one electrode on the grey matter, and one on the surface of the skull. The electric currents of the grey matter appear to have a relation to its function. When any part of the grey matter is in a state of functional activity, its electric current usually exhibits negative variation. For example, on the areas shown by Dr. Ferrier to be related to rotation of the head and to mastication, negative variation of the current was observed to occur whenever those two acts respectively were performed. Impressions through the senses were found to influence the currents of certain areas; e.g., the currents of that part of the rabbit's brain which Dr. Ferrier has shown to be related to movements of the eyelids, were found to be markedly influenced by stimulation of the opposite retina by light.

No recordings of the movement of the spot of light on the scale of the Kelvin galvanometer have been found, perhaps because, at that time, telegraphic operators used to read the dots and dashes of the Morse code by watching the movements of the spot of light on the galvanometer scale. Nonetheless, Caton's description clearly shows than he witnessed the fluctuating potentials that we now know exist. Also important is the fact that Caton was the first to report visual-evoked potentials.

Berger, a psychiatrist in Jena, Germany, was aware of the several prior electroencephalographic animal studies and had conducted experiments using dogs. The only recording devices available to him were the string galvanometer, developed by Einthoven [1903] for electrocardiography and the capillary electrometer developed by Marey [1876]. Although there were a few mirror-type oscillographs available for recording waveforms from alternating current (50 to 60 Hz) generators and transformers, the sensitivity of such devices was very low, and they could not be used for bioelectric recording without a vacuum-tube amplifier.

The voltage appearing on the scalp produced by the brain is only about one-tenth that of the ECG detected with limb leads. To enable recording brain activity with the string galvanometer, the tension in the string was reduced, which increased the sensitivity but reduced the speed of response. This was the method used by Berger when he found that the capillary electrometer was unsatisfactory.

0-8493-0462-8/00/$0.00+$.50

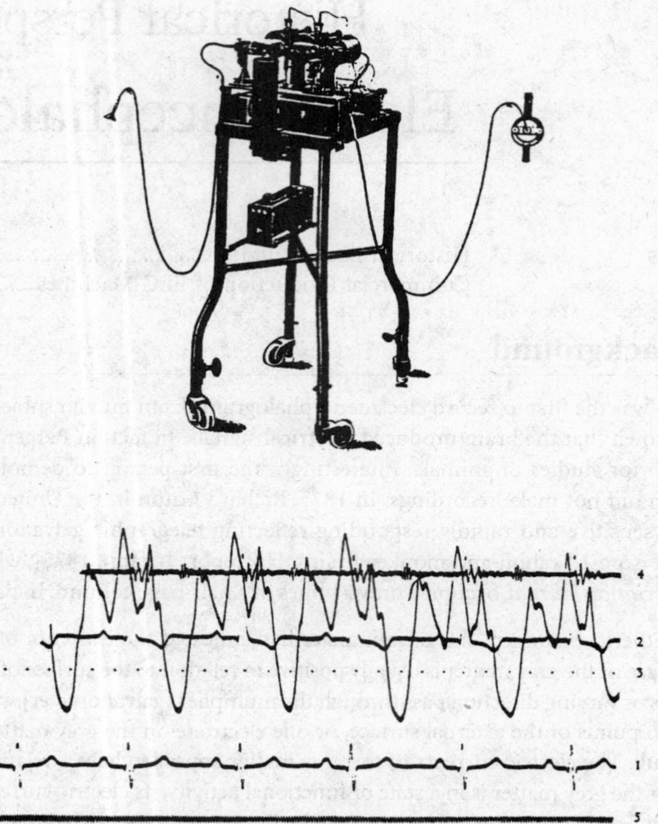

FIGURE HP5.1 (*a*) The Edelmann double-string galvanometer of the type used by Berger. On the left it is a funnel for detecting the arterial or venous pulse and on the right is a device for detecting the heart sounds, all of which can be recorded along with the two channels of ECG, and a time signal on 12-cm-wide photographic paper. (*b*) The phonocardiagram, venous-pulse record, and two recordings of the ECG. (From Zusatz apparate zum Elektrokardiographen, Siemens-Reiniger-Veifa Berlin ca. 1926.)

Apart from problems of the lack of sensitivity and speed of response in the recording apparatus, Berger faced severe electrode problems. Because of the low amplitude of the cortical signals, the electrodes had to be very stable, producing no voltages that could be seen on the electrocortical recordings. In other words, the electrode noise and stability had to be in the low-microvolt range. Zinc electrodes were popular at that time, and Berger used them in his first dog studies just after the turn of this century. For human use, he used zinc-plated needles (insulated down to the tip) and inserted them through existing trephine holes of skull defects so that the tip was epidural. The electrodes were sterilized with 10% formalin solution. Berger first used the single-string Edelmann galvanometer designed for electrocardiography. Later he used a two-string unit of the type shown in Fig. HP5.1 so that he could record the ECG along with the EEG. Such units also were equipped with devices to record the venous or arterial pulses and heart sounds. The photographic recording paper was 12 cm wide and 50 m in length.

The string galvanometer is a current-drawing device; therefore, a low electrode-subject resistance was necessary for adequate sensitivity. Berger stated that it was difficult to obtain a low resistance and reported a value of 1600 Ω for his needle electrodes when placed 5 to 6 cm apart with the tips in the epidural space. A high-resistance electrode pair would reduce the amplitude of the recorded activity. Later Berger used chlorided silver needle electrodes to help solve this problem.

Difficulties with the zinc-plated needle electrodes led Berger to develop very thin lead-foil electrodes, wrapped in flannel, and soaked in 20% NaCl solution. The combination of saline-soaked flannel and the use of a rubber bandage to hold the electrodes on the scalp permitted recording the EEG for hours without the saline evaporating. The resistance measured between a pair of such electrodes ranged from 500 to 7600 Ω depending on the size of the electrodes.

Berger's final improvement to his recording equipment consisted of placing a capacitor in series with the electrodes and string galvanometer to block the steady potential difference due to slight electrochemical differences in the two lead electrodes. Recall that the amplitudes being recorded were in the range of tens of microvolts, and any steady difference in electrode potential would cause a steady deflection of the baseline of the recording. With the capacitor, this steady offset potential did not deflect the galvanometer baseline.

Describing his first studies with the zinc-plated needle electrodes, Berger stated [translation by Gloor, 1969]:

> As a general result of these recordings with epidural needle electrodes I would consequently like to state that it is possible to record continuous current oscillations, among which two kinds of waves can be distinguished, one with an average duration of 90 σ, the other with one of 35 σ. The longer waves of 90 σ are the ones of larger amplitude, the shorter, 35 σ waves are of smaller amplitude. According to my observations there are 10 to 11 of the larger waves in one second, of the smaller ones, 20 to 30. The magnitude of the deflections of the larger 90 σ waves can be calculated to be about 0.0007 to 0.00015 V, that of the smaller 35 σ waves, 0.00002 to 0.00003 V. [The symbol σ was used for milliseconds.]

In his first paper, Berger called the dominant low-frequency first-order and the higher-frequency waves second order. In his second paper he stated:

For the sake of brevity I shall subsequently designate the waves of first order as alpha waves = α – *w*, the waves of second order as beta waves = β – w, just as I shall use "E. E. G." as the abbreviation for the electroencephalogram and "E. C. G." for the electrocardiogram.

The average duration of the waves reproduced in [his] Figure 1 is for the α – w = 120 σ and for the β –w = 30 to 40 σ.

Berger objected to the term *electrocerebrogram* to designate a record of the electrical activity of the brain. He stated:

> Because for linguistic reasons, I hold the word "electrocerebrogram" to be a barbarism, compounded as it is of Greek and Latin components, I would like to propose, in analogy to the name "electrocardiogram" the name "electroencephalogram" for which here for the first time was demonstrated by me in man.

Berger's papers contain many EEGs from patients, but those from his son Klaus are discussed frequently. In fact, Klaus was used as a subject for electrode testing. For example, Berger wrote:

> Klaus' records were taken with every other possible type of electrodes: silver, platinum, lead electrodes, etc; also, different arrangements of these on the skin surface of the head were used. However, time and again it was found that the best arrangement was that with electrodes placed on the forehead and occiput. Of Klaus' many records, I only want to show another small segment of a curve obtained in this manner [Fig. HP5.2]. In this instance head-band electrodes were applied to the forehead and

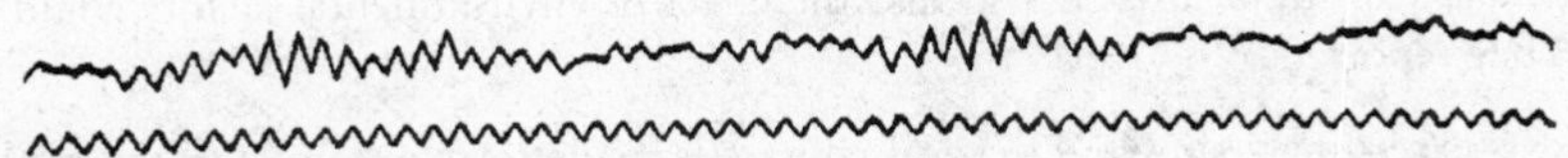

FIGURE HP5.2 Klaus at the age of 15. Double-coil galvanometer. Condenser inserted. Recording from forehead and occiput with head-band electrodes. (*Top*) The record obtained from the scalp; (*bottom*) time in 1/10 second.

occiput and were fixed with rubber bandages. From these head-band electrodes, records were taken with galvanometer 1 of the double-coil galvanometer; galvanometer 2 was set at its maximum sensitivity and was used as a control to make sure that no outside currents were entering the galvanometer circuit to disturb the examination. At that time I was still very distrustful of the findings I obtained and time and again I applied such precautionary measures. The record of galvanometer 2 ran as completely straight line, without my oscillation.

From the lengthy discussions in Berger's papers, it is easy to see that few believed that his recordings originated in the brain. To dispel some of the uncertainty, Berger usually recorded the ECG along with the EEG. Occasionally, he recorded heart sounds and the arterial pulse. In addition in the electrical activity of the heart, various critics proposed that Berger's recorded activity was due to friction of blood in the cerebral arteries, pulsations of the brain and/or scalp, respiration, contraction of piloerector or skeletal muscle, and glandular activity. Berger dealt with all the potential artifacts, pointing out that their time course and frequency were different from those of the EEG and showed that the EEG continued with transient slowing of the heart rate. Finally, he stated (perhaps in exasperation):

> I therefore believe I have discussed all the principal arguments against the cerebral origin of the curves reported here which in all their details have time and again preoccupied me, and in doing so I have laid to rest my own numerous misgivings.

Among the first to publish an English-language verification of Berger's observations were Jasper and Carmichael [1935], then at Brown University in the United States. Silver electrodes (1 to 2 cm in diameter), covered with flannel and soaked in saline, were connected to an amplifier/mirror-oscillograph system. They were able to confirm Berger's findings and extend them, showing that with a two-channel system used to record the EEG of a girl with a convulsive disorder, the alpha wave frequency was 10 per second on the left side of the head and 6 to 8 per second on the right side, one of the early indications that the EEG was altered by brain pathology. Figure HP5.3*a* shows one of the records obtained by Jasper and Carmichael.

Jasper later came to the Montreal Neurological Institute (McGill University) and created the Electropysiological Laboratory, which was formally opened with a celebration meeting held on February 24–26, 1939. In attendance were the world leaders in electrophysiology.

Clinical EEG at the Montreal Neurological Institute was inaugurated using a machine built by Andrew Cipriani. The inkwriter was of unique design and featured a strong magnetic field produced by an electromagnet. In this field was a circular coil coupled to an inkwriting pen. It is interesting to observe that this principle had been used by d'Arsonval [1891] with pneumatic coupling to a tambour that caused a pen to write on a rotating drum, as shown in Fig. HP5.4*a*. Later in the 1930s, this coil design was coupled to a conical diaphragm and became the first dynamic loudspeaker. The pen motor devised by Cipriani is sketched in Fig. HP5.4*b*. A small vane, affixed to the writing stylus, dipped into an oil chamber (dashpot) to provided damping. This four-channel instrument was in routine use when the author first came in contact with Jasper in the early 1940s; it was replaced in 1946 by a six channel model III Grass EEG.

Meanwhile, at Harvard University (Boston, Mass), Gibbs, Davis, and Lennox [1935], were pursuing their interest in epilepsy. Recognizing the potential of the EEG for the diagnosis of epilepsy, they initiated a series of studies that would occupy the next many decades. Their recorder consisted of an ink-writing telegraphic recorder called the Undulator. In December of 1935, they published their first paper on EEG which carried a footnote that read: "This paper is no, XVII in a series entitled Studies in Epilepsy." Citing the Berger papers and that by Jasper and Carmichael, Gibbs et al. stressed the importance of direct-inking pens for immediate viewing of the EEG so that the effect of environmental factors could be identified immediately. They reported:

> The method is exceedingly simple. Electrical contact is made to two points on the subject's head. Except for the study of grand mal epileptic seizures we regularly employ as electrodes two hypodermic

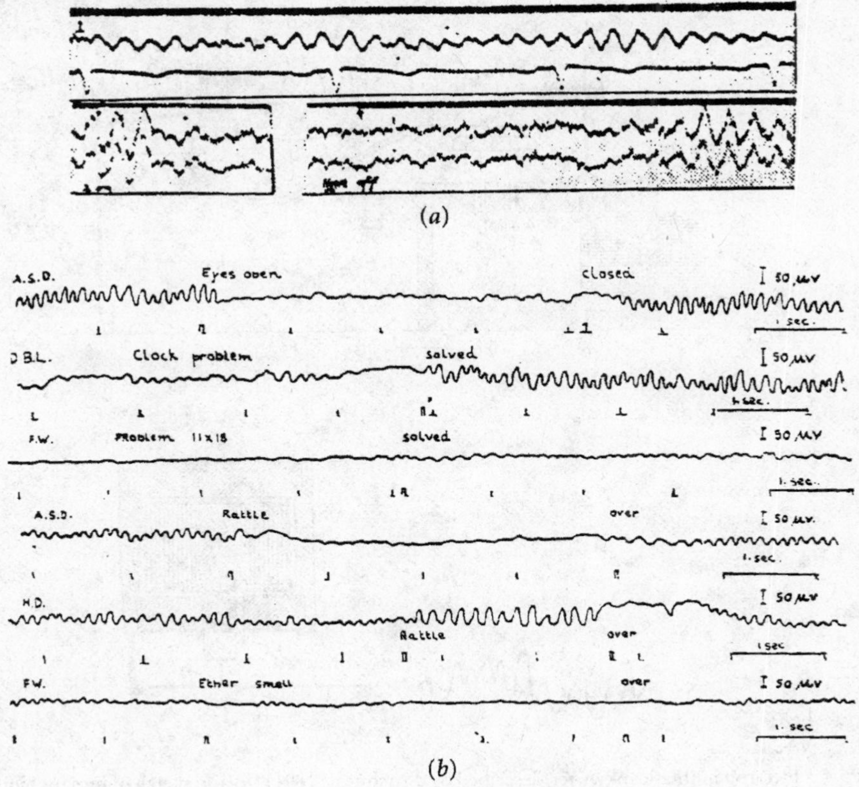

FIGURE HP5.3 The first U.S. records of the EEG to confirm Berger's report. (*a*) The records obtained by Jasper and Carmichael (1935). The first channel shows the alpha waves, and the second shows the electrical activity detected by electrodes on the leg above the knee. The second record shows alpha inhibition by illumination of the retina, and the third record shows return of the alpha waves when the light was extinguished. (*b*) The first records published by Gibbs et al. (1935) showing alterations in normal subjects by various types of sensory stimulation eyes open and closed, problem solving, noise (rattle), and smelling ether. (Both by permission.)

needles inserted one into the scalp at the vertex of the skull and the other into the lobe of the left ear. Enough procaine hydrochloride is injected previously to insure the continued comfort of the subject.

Figure 3*b* is a reproduction of the first EEG obtained by Gibbs et al. Soon Gibbs produced his well-known *Atlas of Electroencephalography*, which first appeared in 1941 and became the "bible" for training electroencephalographers.

The first recording equipment used by Gibbs et al. [1935] was built by Lovett Garceau. It consisted of a four-stage, singled-sided, resistance-capacity-coupled amplifier made with high-grain, screen-grid tubes driving a distinct-linking telegraphic recorder called the Undulator (*U* in Fig. 5) obtained from the Western Union Telegraph Company, Fig. 5 shows the circuit diagram. the overall high-frequency response extended to 25 Hz [Grass, 1984].

Slightly earlier, in Germany, Toennies [1932] had developed a direct-linking recorder that he called the Neurograph. Fig. HP5.6 shows a picture of the instrument and a record of the human electrocardiogram and the response of canine eyes to light. Note that the chart is running in the opposite direction to conventional recordings. The time marks are 1/5 second.

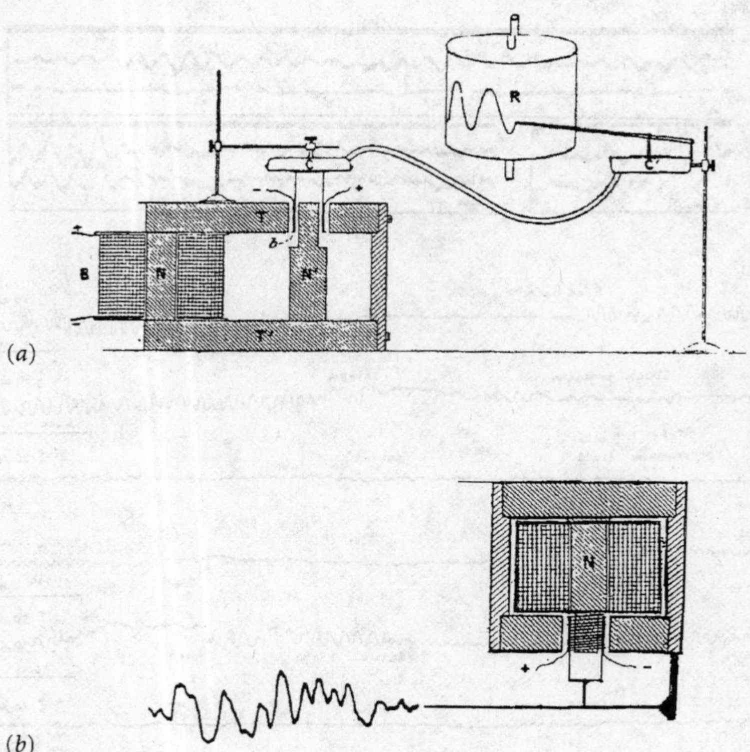

FIGURE HP5.4 Electropenumatic inkwriter, described by d'Arsonval (1891) (*a*) and sketch of moving-coil inkwriter devised by Cipriani in the late 1930s for EEG at the Montreal Neurological Institute (*b*).

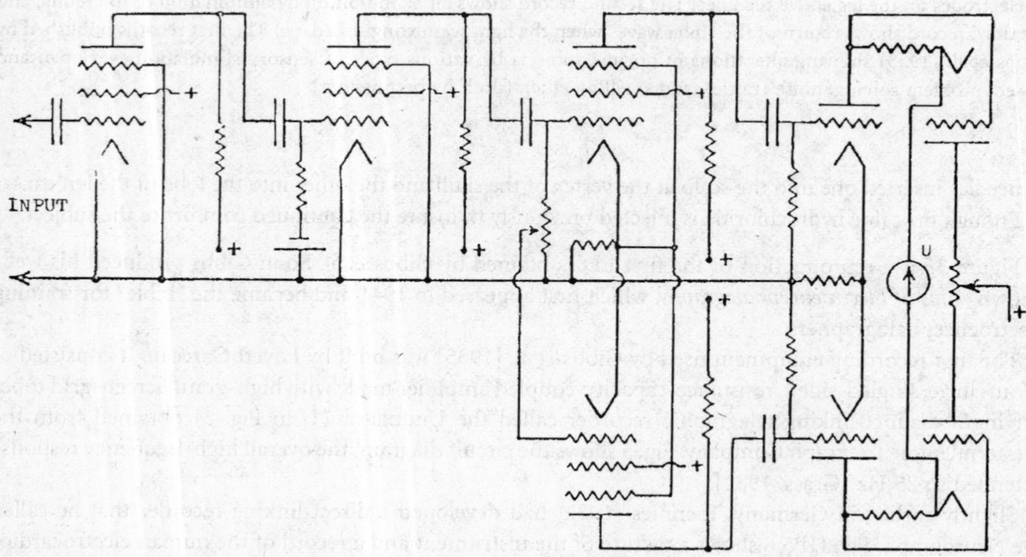

FIGURE HP5.5 The single-sided, five-stage, resistance-capacitor-coupled amplifier developed by Gareau to drive the Western Union inkwriting telegraphic recorder (Undulator) used by Gibbs et al. to record EEGs (From Garceau et al. 1935. Arch Neurol Psychiatry. With permission.)

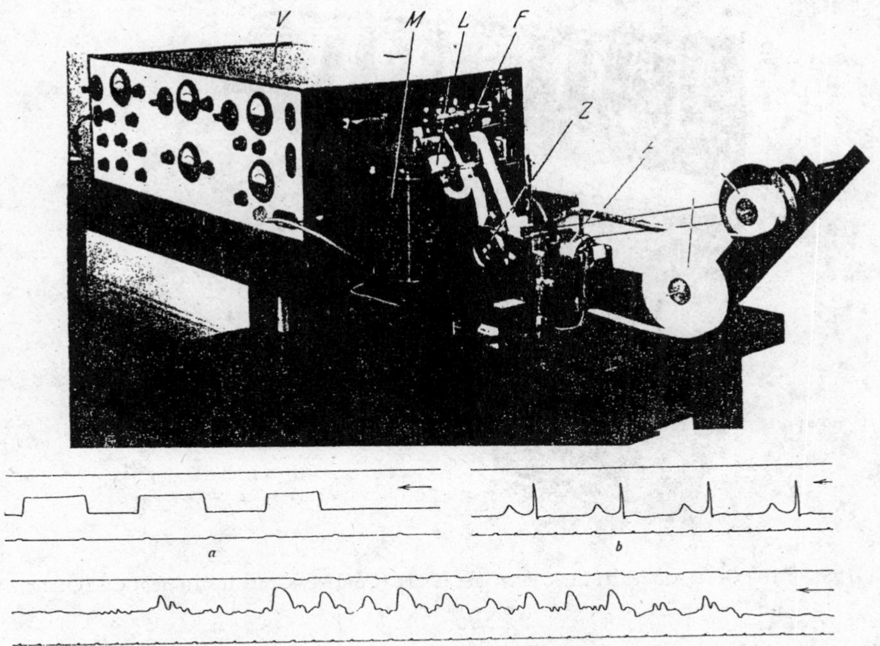

FIGURE HP5.6 Toennies Neurograph and recordings showing a 1-mV calibration, the human ECG, and electrical activity from the canine eye, time marks 1/5 s. (From Toennies [1932].)

Commercial Production of EEG Machines

The excellent collaboration between Gibbs and Grass in 1935 resulted in replacement of the Undulator telegraphic inkwriter with a robust, d'Arsonval-type inkwriter in early 1936. Meanwhile, the Grass Instrument Company had been founded (1935) to produce electrophysiologic equipment at the well-known address, 101 Old Colony Avenue, Quincy, Mass; the address in the same today.

In 1937, Grass adopted folding chart paper and by 1939 was providing three-, four-, and six-channel EEGs. The Model III shown in Fig. HP5.7 is the machine recognized by all and founded many EEG laboratories. It featured a knee-hole console with the pens at the right and ample viewing space for the record as it evolved. The chart speed was 30 mm/s, which ultimately became the standard.

At about the same time Grass was building EEG machines in Quincy, Mass., Franklin Offner, a research associate of Ralph Gerard at the University of Chicago, started his own company at 5320 North Kedzie Avenue, Chicago, to produce an EEG using the Crystograph, a high-speed, piezoelectric inkwriter that was described by Offner and Gerard [1936]. It consisted of two slabs of rochelle salt crystals, three corners of each were clamped, and the fourth was free to move when a voltage was applied to electrodes on the crystals. The moving corners were mechanically linked by a slender brass belt that caused motion of the rod that carried the inkwriting stylus. The sinusoidal frequency response extended uniformly to 100 Hz; the chart speed 25 mm/s, the same as for ECG. However, a gear shift provided chart speeds above and below 25 mm/s.

The Crystograph was used with the first Offner EEG machines, and it was ideally suited for high efficiency energy transfer from vacuum-tube amplifiers because of its high impedance. The differential amplifiers were housed in two 19-in relay tracks, as shown in Fig. HP5.8*a*; the Crystograph rested on an adjacent table; a six-channel Crystograph is illustrated. Damping was adjusted by a series variable resistor to achieve an excellent response to a step function. The record shown in Fig. HP5.8*a* was produced by a 3-μV (peak-peak) square wave, showing the excellent transient response and the remarkably low internal noise level of the amplifiers.

FIGURE HP5.7 The Grass Model III EEG, *circa* late 1940s. (Courtesy Grass Instrument Co., Quincy, Mass.)

Although the Cyrstograph EEG produced elegant records, temperature and humidity played havoc with the rochelle salt crystals, and Offner replaced it with the Dynagraph, shown in Fig. HP5.8*b*, the recorder being a low-impedance, robust d'Arsonval-type moving-coil inkwriter.

Efficient coupling between the output amplifier stage and the moving-coil inkwriter was always a problem that was solved by Offner in his Dynagraph. A synchronous vibrator sampled the signal from the preamplifier, amplified the pulses, and recombined them after passing through a step-down transformer, thereby providing an efficient impedance match between the amplifier output stage and penwriter. Fig. HP5.8*b* is an illustration of the Offner Dynagraph that featured a sit-down console for viewing the record. The antiblocking feature of this design is shown at the bottom of Fig. HP5.8*b* by a sine wave record in which a transient, 100 times larger than the recording, was presented and the recording was restored a fraction of a second after the transient. Offner later replaced the vacuum tubes with transistors, bringing out the first transistorized EEG.

In the late 1940s, Warren Gilson, who was a pioneer in devising physiologic instruments, started production of EEGs in Madison, Wisc.; Fig. HP5.9 is a photograph of one of his later instruments (*circa* 1959).

It is noteworthy that following the Grass Model III EEG, all subsequent instruments were of the console type with the penwriter at the right end of the desktop of the console. Not easily identified in any of the foregoing figures are the two rotary switches for each channel. Each switch could connect each side of the differential amplifier input to any of the 21 electrodes of the 10–20 system. Switching also was provided to apply a step-function calibrating signal to all channels simultaneously, the step function typically being produced by depressing a pushbutton or rotating a knob. In addition, a low-current ohmmeter was provided to permit measurement of the resistance of any pair of electrodes. Some EEGs included a high-frequency filter for each channel to exclude muscle artifacts often seen when patients clenched their jaws.

Electroencephalographs have changed little since their development in the post-World War II days. Transistors have replaced the vacuum tubes, but the chart speed and frequency response are the same as those established by the first manufacturers of EEG machines, although many new recording techniques have been introduced.

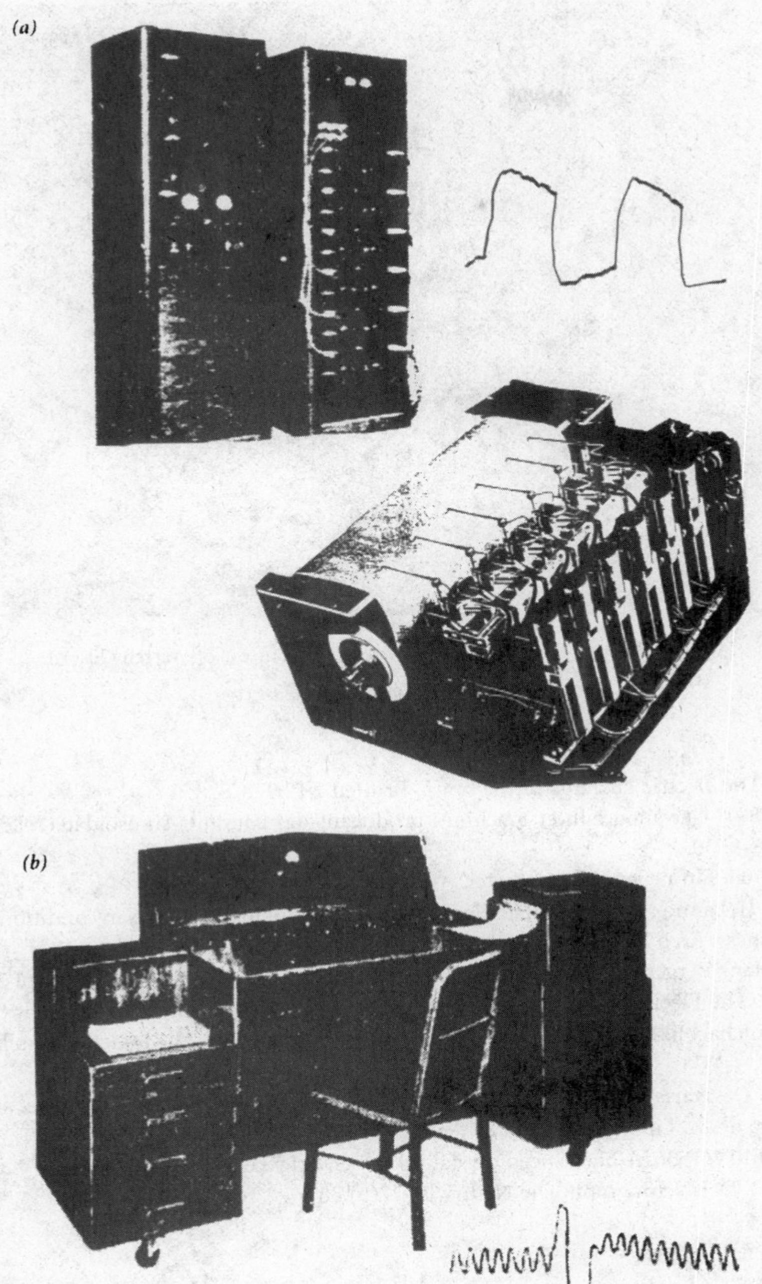

FIGURE HP5.8 The Offner EEG which used the Crystograph recorder (*a*) and the Offner Dynagraph EEG (*b*), *circa* late 1940s (Courtesy of Franklin Offner.)

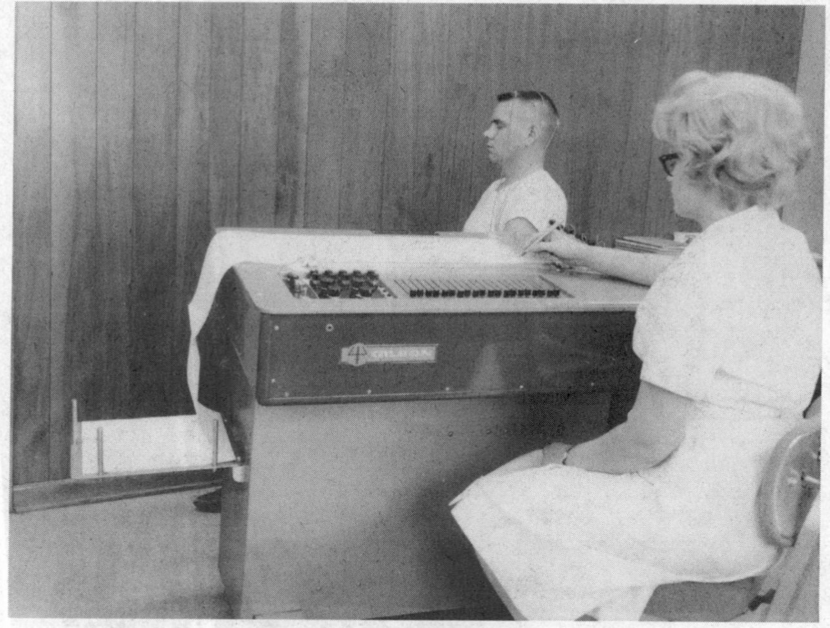

FIGURE HP5.9 The Gilson EEG, *circa* 1950 (Courtesy of Warren Gilson).

References

Caton R. 1875. The electric currents of the brain. Br Med J 12:278.

d'Arsonval A. 1891. Galvanograph et machines producant des courants sinusoidaux. Mem Soc Biol 3 (new series):530.

Einthoven W. 1903. Ein neues Galvanometer. Ann Phys 12 (suppl 4):1059.

Gibbs FA, Davis H, Lennox WG. 1935. The electroencephalogram in epilepsy as in conditions of impaired consciousness. Arch Neurol Psychiatry 34(6):1135.

Gloor P. 1969. Hans Berger on the electroencephalogram. EEG Clin Neurophysiol (suppl 28):350.

Grass AM. 1984. The Electroencephalographic Heritage. Quincy, Mass, Grass Instrument Co.

Jasper HH, Carmichael L. 1935. Special article: Electrical potentials from the intact human brain. Science 81:51.

Marey EJ. 1876. Des variations electriques des muscles du coeur en particulier etudiee au moyen de l'ectometre de M. Lippman, C R Acad Sci 82:975.

Offner F, Gerard RW. 1936. A high-speed crystal inkwriter. Science 84:209.

Toennies JF. 1932. Der Neurograph. Die Naturwiss 27(5):381.

XX

Ethical Issues Associated with the Use of Medical Technology

Subrata Saha
Clemson University

Joseph D. Bronzino
Trinity College/Biomedical Engineering Alliance for Connecticut (BEACON)

Biomedical engineering is responsible for many of the recent advances in modern medicine. These developments have led to new treatment modalities that have significantly improved not only medical care, but the quality of life for many patients in our society. However, along with such positive outcomes new ethical dilemmas and challenges have also emerged. These include: (1) involvement of humans in clinical research, (2) definition of death and the issue of euthanasia, (3) animal experimentation and human trials for new medical devices, (4) patient access to sophisticated and high cost medical technology, (5) regulation of new biomaterials and devices. With these issues in mind, this section discusses some of these topics. The first chapter focuses on the concept of professional ethics and its importance to the practicing biomedical engineer. The second chapter deals with the role medical technology has played in the definition of death and the dilemmas posed by advocates of euthanasia. The third chapter focuses on the use of animals and humans in research and clinical experimentation. The final chapter addresses the issue of regulating the use of devices, materials, etc. in the care of patients.

Since the space allocated in this Handbook is limited, a complete discussion of the many ethical dilemmas encountered by practicing biomedical engineers is beyond the scope of this section. Therefore, it is our sincere hope that the readers of this Handbook will further explore these ideas from other texts and articles, some of which are referenced at the end of the chapter. Clearly, a course on biomedical ethics should be an essential component of any bioengineering curriculum.

With new developments in biotechnology and genetic engineering, we need to ask ourselves not only if we can do it, but also "should it be done?" As professional engineers we also have an obligation to educate the public and other regulatory agencies regarding the social implications of such new developments. It is our hope that the topics covered in this section can provide an impetus for further discussion of the ethical issues and challenges faced by the bioengineer during the course of his/her professional life.

189

Professional Ethics in Biomedical Engineering

Daniel E. Wueste
Clemson University

189.1 A Variety of Norms Govern Human Conduct

Various norms or principles govern our activities. They guide us and provide standards for the evaluation of conduct. For example, our conduct is governed by legal and moral norms. These two sets of norms overlap, but it is clear that the overlap is not complete. Morality requires some acts that are not legally required of us and vice versa. Not surprisingly, then, we speak of legality and morality, of legal obligations and moral obligations. This is appropriate primarily because (1) law and morality are distinct sources of obligation and (2) it is possible for legal and moral obligations to conflict. Recognizing that moral and legal obligations have distinct sources, it is easier to appreciate the nature of a conflict between them and work toward its resolution. The same thing can be said *within* the sphere of morality. Moral rules and principles can and do give rise to conflicting obligations. Here too, the way people speak is revealing. People speak of professional obligations and social responsibilities, as well as the duties of "ordinary morality." Such talk is increasingly heard in the high-tech fields of biomedicine and biomedical engineering, for here the development and use of sophisticated technology intersect with the rights and interests of human beings in an especially profound way.

189.2 Professional Ethics and Ethics Plain and Simple

Talk of professional ethics presupposes a distinction between the constraints that arise "from what it means to be a decent human being" (Camenisch, 1981) and those that come with one's role or attach to the enterprise in which one is engaged. Paul Camenisch calls the former "ethics plain and simple." The latter are the elements of an occupational or role morality. They only apply to persons who occupy a specific role. The idea here is simple enough. A father, for example, has responsibilities that a man who is not a father does not have. So too, a college teacher, a cleric, or a police officer has responsibilities that persons not occupying these roles do not have.

0-8493-0462-8/00/$0.00+$.50

A professional ethic is an occupational or role morality. It is like the law in several ways. For example, it is not a mere restatement of the norms of ordinary morality. Another point of similarity is that like the law, its scope is limited; it has a jurisdiction in the sense that what it requires or permits is role specific. Thus, for example, the norms of the lawyer's professional ethic do not impose obligations on people who are not lawyers. Still another similarity is that a professional ethic may allow (or even require) acts that ordinary morality disallows or condemns. An explanation of this fact begins with the observation that, in general, as professionals do their work they are allowed to put to one side considerations that would be relevant and perhaps decisive in the ethical deliberations of nonprofessionals. So, for instance, an attorney is free to plead the statute of limitations as a bar to a just claim against his/her client or to block the introduction of illegally seized evidence in a criminal trial even though his/her client committed the alleged offense. Now one thing that might be said here is that the attorney is simply doing his/her job. The point would be well taken. However, it leaves something rather important unsaid, namely, that the conduct in question is obligatory for the attorney.

Thinking along lines such as these is not confined to the paradigm professions of law and medicine. For example, scientists have been known to claim that as scientists they are free to (indeed must) put to one side social, political, and moral concerns about the uses to which their discoveries may be put. Indeed, some have claimed that scientists are morally obligated not to forgo inquiry even if what emerges from it can be put to immoral or horrendous uses. Plainly, such an appeal to a professional ethic can excite controversy. And this reveals another respect in which a professional ethic (role morality) is like law: it is subject to moral critique. Indeed, it is not only subject to moral criticism, its validity depends on its being morally justified (i.e., justified in terms of "ethics plain and simple").

One important way in which a professional ethic differs from law is that it is widely held that its standards are in some sense higher than those of ordinary morality. More is expected of professionals than nonprofessionals. They are expected to act on the basis of the knowledge that sets them apart and in doing so they are expected to put the interests of clients or patients *ahead* of their own interests. Clearly, if professionals are governed by special, higher standards, if the rights and duties of professionals differ from the rights and duties of nonprofessionals, then it makes a difference whether one's occupation counts as a profession.

189.3 Professions

As it happens, there is no generally accepted definition of the term "profession". However, several writers have suggested that some characteristics common to recognized professions are necessary or essential. The idea that these writers share is that with these characteristics in mind one can mark a serviceable distinction between professional and nonprofessional occupations.

Bayles [1989] maintains that three features of a profession are necessary: (1) extensive training, (2) that involves a significant intellectual component, and (3) puts one in a position to provide an important service to society. To be sure, other features are common. For example: (1) the existence of a process of certification or licensing, (2) the existence of a professional organization, (3) monopoly control of tasks, (4) self-regulation, and (5) autonomy in work. But, according to Bayles, they are not essential. The crucial point in the argument that they are not essential is that a large (and growing) number of professionals work in organizations (e.g., HMOs or institutions such as hospitals) where tasks are shared and activities are directed and controlled by superiors. Two things are noteworthy here. First, Bayles's analysis does not include normative features among those that distinguish professions from non professions. Second, in his analysis an occupation may count as a profession though it lacks certain features common to most professions. While Bayles's decision not to include normative features among the distinguishing features of professions has been criticized by several writers who insist that such features are indeed essential, the second point has met with widespread agreement.

It will be well to avoid the controversy respecting what is and what is not essential to a profession. Happily, a survey of the substantial literature on professions reveals agreement on several points that, taken together, provide a helpful picture of professions and professional activity. This picture is rather like a sketch made by a police artist who works with descriptions provided by various witnesses. It has five elements. The first is the centrality of abstract, generalized, and systematic knowledge in the performance of occupational tasks. The second element is the social significance of the tasks the professional performs; professional activity promotes basic social values. The third element is the claim to be better situated/qualified than others to pronounce and act on certain matters. This claim reaches beyond the interests and affairs of clients. Professionals (experts) believe that they should define various aspects of society, life, and nature and we generally agree. For example, we defer to them (or at least our elected representatives do) in matters of public policy and national defense. Moreover, in certain settings, a hospital for example, it is simply impossible not to defer to the judgment of experts/professionals. The crucial premise can scarcely be doubted: in the contemporary world there is more to know (much of it having immediate practical application) than any one person is capable of knowing. The fourth point is that, on the basis of their expertise and the importance of the work that requires it, professionals claim that as practitioners they are governed by role-specific norms—a professional ethic—rather than the norms that govern human conduct in general. Now, it cannot be denied that particular applications of the relevant norms have been a source of controversy. However, since controversy of this sort generally presupposes the applicability and validity of the role-specific norms, it only serves to confirm the general point. The final element of the composite picture is that now most professionals work in bureaucratic organizations/institutions. The romantic appeal of the model of the professional as solo-practitioner may incline one to balk at this. But romance must not be allowed to prevail over evidence. And in this case the evidence is substantial. In fact, some of it is arresting. For example, it was recently reported that eighty percent of recent medical school graduates are salaried employees of HMOs, clinics, or hospitals (*US News and World Report,* 1999).

189.4 The Profession of Biomedical Engineering

As we have seen, it is not an idle question whether biomedical engineering is a profession. If professionals are governed by special, higher standards, if the rights and duties of professionals differ from the rights and duties of nonprofessionals, then it makes a difference whether one's occupation counts as a profession.

Since there is no standard definition, no set of necessary and sufficient conditions the satisfaction of which would be decisive, in answering the question we will have to proceed in a different way. It is suggested that we think in terms of characteristics shared by recognized professions that, taken together, constitute a composite picture akin to what a police sketch artist might draw. What is seen when one looks at this picture? In particular, does the picture match the reality of biomedical engineering practice? Looking at this composite picture of a profession it is seen that (1) abstract, generalized, and systematic knowledge is crucial to the performance of occupational tasks (2) these tasks promote basic social values, (3) practitioners claim to be better situated/qualified than others to pronounce and act on certain matters, (4) the conduct of practitioners is governed by role-specific norms, and (5) most of the work done by practitioners is done within bureaucratic institutions. The fit between the reality of biomedical engineering practice and this composite picture of professions is tight. Indeed, looking to this picture to answer the question whether biomedical engineering is a profession, it can scarcely be doubted that the answer is yes.

One possible objection to this answer might be that there is no code of ethics for bioengineers, and thus bioengineering does not count as a profession because the fourth requirement is not satisfied. The objection could be met by pointing to a code for bioengineers, if there was one. But a better response is that the objection itself is misplaced. It is based on the mistaken assumption that there is a set of conditions the satisfaction of which is necessary and sufficient for the ascription of the term "profession" to an occupation. But the existence of a code of ethics is neither necessary nor sufficient for an occupation to count as a profession; what we are working with is a composite picture, not a definition. The objection is misplaced for a second reason as well. Even if there is no code of ethics for bioengineers, codes of ethics are well known in engineering. For example, the National Society of Professional Engineers and the Institute of Electrical and Electronic Engineers have codes of ethics. In addition, the National

Committee on Biomedical Engineering (Australia) has produced a set of professional standards. Moreover, a professional ethic may develop by other than quasi-legislative means; in this respect it is like law which has both legislative and customary forms. Thus, there is nothing here to impugn the claim that bioengineering is a profession.

189.5 Two Sources of Professional Ethics

It is important to be clear about the fact that a professional ethic may develop in more than one way. A professional ethic is a role morality. The norms of a profession's role morality need not be expressly "legislated" by, for example, a professional organization, because most of them are implicitly legislated in practice. Thus, one way to identify the norms of a profession's role morality is to reflect on the expectations one has respecting the conduct of one's peers. The stable interactional expectancies of practice can constitute what amounts to a customary morality of a profession, an uncodified professional ethic. To be sure, these customary norms may be codified (and often are). However, just as in the case of customary law, codification is not necessary for their validity. Indeed, it is entirely appropriate to say that codification of some such norms (like some laws) is the result of the codifier's recognition of their independent validity.

It should be noted here that the norms of a customary morality are valid only if they are accepted in practice. It is important, however, that while acceptance is necessary, it is not sufficient. It cannot be sufficient, because if it were, the idea that something is right simply because someone or some group believes it is right would have to be granted. And that, of course, is patently false. What, then, are the additional conditions for the validity of such norms? This is surely a fair question. However, answering it completely would lead us far afield. Consequently, a short answer will have to suffice. The case for the validity of a norm of customary morality (in other words, for its status as a *moral* norm) turns on whether, in addition to being accepted in practice, compliance with it has good consequences and does not infringe upon the rights of other persons. It will be noticed that these are precisely the sorts of considerations that persons charged with the task of rule making do, or at any rate should, regard as decisive in doing their work. In any case, whether the norms of a professional ethic arise in practice or are expressly "legislated" by a group, they are "dual aspect norms" (Wellman, 1985). They are in play for role agents who are trying to decide what action to take; they are also in play for others within and outside of the profession who observe or by other means become aware of deviation from (or conformity to) them and react accordingly.

189.6 Professional Ethics in Biomedical Engineering

It is clear that biomedical engineering has an ethical dimension. After all, human well being is at stake in much if not most of what a biomedical engineer does. Indeed, error or negligence on the part of a biomedical engineer can result in unnecessary suffering or death. Of course, much the same can be said of other engineering fields. Yet, there is something distinctive here. The National Committee on Biomedical Engineering has identified three ways in which biomedical engineering differs from other branches of engineering. First, biomedical engineers work with biological materials that behave differently from and have different properties than the materials that most engineers work with. Second, preparation for a career as a biomedical engineer involves study of both engineering principles and the life sciences. Third, and most important for present purposes, is "the indirect and very often direct responsibility of biomedical engineers for their work with patients." Such responsibility for the well-being of others is a clear indicator that a role has an ethical dimension.

Many things in biomedical engineering that fall under the rubric of professional ethics have to do with policies or procedures. For example, the development of

1. methods for obtaining informed consent and criteria for justified departure from these methods,
2. means for identifying subjects for clinical trials,

3. criteria of thorough testing,
4. standards to obviate or mitigate conflicts of interest, as well as mechanisms for their application and enforcement, and
5. criteria for just distribution of scarce biomedical resources (expertise and technology).

All of these things (in this far from exhaustive list) fall under the rubric of professional ethics. However, they are institutional in the sense that they call for decisions about policies or procedures rather than individual action. Decisions in these areas are not decisions to be made by individual practitioners nor are they decisions to be taken case by case. They are decisions about structures of practice that require quasi-legislative activity. Relying on others with requisite expertise, and soliciting input from persons whose interests are at stake in biomedical engineering practices, biomedical engineers should work to develop structures of practice that satisfy legal requirements and ensure, as far as possible, that their professional practice manifests a commitment to safety and the promotion of human well-being. These carefully designed structures of practice should be part of the explicitly quasi-legislated portion of the professional ethic of biomedical engineering.

The question that arises naturally here is how to proceed in this undertaking. Precisely which of the dominant approaches in ethics—utilitarian, deontological, or aretaic—is best is a subject of vigorous debate among philosophers. But this issue will not be debated here. Instead, a sketch will be presented of an approach that can be employed in designing ethical structures of practice, and then, with one difference to be explained, used in making individual ethical decisions in one's capacity as a professional.

189.7 Tools for Design and Decision in Professional Ethics

Multiple analyses or several independent judges are often relied on in making decisions. In general, this is done when something significant turns on the final decision. For example, physicians frequently call for a consultation and patients are encouraged to seek a second opinion before an invasive procedure is performed. Similarly, hospitals and universities rely on panels or commissions—an Institutional Review Board, for example—to make decisions about proposed research or other pressing issues. In such cases it is assumed, rightly, that relying on multiple modes of analysis or several judges is wise even though (1) they produce the same judgment in many—hopefully most—cases and, thus, appear to involve redundancy; and (2) in some cases they produce conflict that could have been avoided by relying on a single mode of analysis or single judge. Why is this a wise course? One part of the answer is that our confidence is bolstered when the same conclusion is reached by different trustworthy means or judges. Here redundancy is a value. The second part of the answer is that being open to conflicting opinions and analyses can help us to avoid errors that would occur otherwise. This will happen when the conflict prompts reexamination of the question that reveals facts previously overlooked or undervalued or mistaken analyses. It should be noted that the approach recommended here is not political; logical and evidentiary considerations rather than simple consensus justify the judgments it produces. Randy Barnett sums up the case for such an approach:

> The virtue of adopting multiple or redundant modes of analysis is…twofold: (a) convergence (or agreement) among them supports greater confidence in our conclusions; and (b) divergence (or conflict) signals the need to critically reexamine the issue in a search for reconciliation. In sum, convergence begets confidence, divergence stimulates discovery. [Barnett, 1990]

In the context of professional ethics an approach of this sort would involve reliance on three modes of analysis: (1) utilitarian, (2) rights-based deontological, and (3) role-based institutional. A brief description of each mode of analysis is presented in the following paragraphs.

A utilitarian analysis begins with the assumption that rightness is a function of value and tells us that what is morally required of us is the production of the greatest amount of good possible in a situation for all of the affected parties. Utilitarian thinking leads quite directly to an embrace of the familiar principles of nonmaleficence and beneficence. Deontological analysis denies the essential connection between rightness and goodness asserted by utilitarianism. Unlike utilitarians, deontologists hold that some actions are intrinsically right and some actions are intrinsically wrong. More particularly, they insist that the fact that the consequences of an action are the best possible in a given situation does not show that the action is right. The most famous of deontological theories, that of Immanuel Kant, teaches that what is morally required of us is that wherever found, in ourselves or others, humanity is always treated as an end and never as a mere means to an end. In other words, persons have intrinsic worth (as Kant said, they have a dignity rather than a price) and must never be treated as if their value were merely instrumental (as if they were things). That is our duty; the other side of that coin is the right others have to receive that sort of treatment from us. What matters on this approach, then, is whether one is treating people as they deserve to be treated. Thus, deontological thinking leads directly to an embrace of the familiar principles of autonomy and justice. These two modes of analysis are alike in this: both proclaim independence from what the commonly accepted ideas regarding right and wrong are in a community; on both views morality is not something that is instituted (made); rather, it is discovered (not by empirical investigation, but by ratiocination). Thus, these modes of analysis contrast sharply with the third mode of analysis to be discussed here, namely, the role-based institutional mode.

Thinking in terms of role morality rightness is a function of conformity with the stable interactional expectations—accepted norms of conduct—associated with a role. Responsibilities and rights are tied to the function of a role agent within an institution. A role morality is institutional in the quite basic sense that it is instituted, that is, brought into existence by human beings. Sometimes, of course, this is accomplished by quasi-legislative means. But such activity is neither necessary nor the most common means of creation. A role morality is implicit in practice, it is established through mutually beneficial interaction over time. It is a customary or conventional morality. (There is an analogy here to the law in its customary and legislative forms.) One final point, noted earlier, is that considerations of non-institutional morality play a critical role in the validation of the norms of a role morality. The familiar maxim in medicine, *primum non nocere*, as well as the implicit rule of lawmaking that lawmakers must promulgate the laws they make (no secret laws), are examples of principles that would be readily embraced by those thinking along these lines and validated by considerations of non-institutional morality.

We can summarize this brief description of the three modes of analysis in the following way. The key question for the utilitarian is one of maximal value; for the rights-based deontological thinker it is one of deserved or rightful treatment; with role morality it is one of conformity to established custom or practice.

The recommended approach for the design of ethical structures of practice is to use all three modes of analysis. The hope is that they will converge on the same result. When they do we can be confident that implementation of the principle or policy is justifiable. That is the case, for example, with the rules of professional practice requiring fidelity, confidentiality, privacy, and veracity.

Using this approach one hopes for convergence on the same result. That is to be expected in easy cases. But not all cases are easy cases. Our analyses may diverge rather than converge. What then? Barnett suggests that divergence "stimulates discovery." The idea is that achieving convergence may be difficult, but at least sometimes it may be achieved on a second pass by retracing one's steps—going through the analyses again, paying special attention to the input—or rethinking the analyses themselves, paying special attention to previously identified sources of difficulty. For example, utilitarian and deontological analyses may diverge because they deal with individual rights in different ways. Deontologists treat rights as trumps; utilitarians simply include them among the considerations that count in their calculations respecting likely consequences. It may be that the two modes of analysis diverge because of an erroneous assignment of weight to the rights that are in play. If so, convergence can be achieved by rethinking the weight assigned to the rights in the utilitarian calculations. It must be admitted, however, that divergence may not be eliminated by such means. What then? In anticipation of such cases, a presumption should be made in favor of one of the modes of analysis.

When one's project is the design of ethical structures of practice, the presumption should be made in favor of rights-based deontological analysis, it being understood that utilitarian considerations may rebut the presumption favoring rights in some circumstances. When one's project is not the design of decision devices (structures of practice), but deciding what one ought to do as a professional, a presumption should be made in favor of institutional responsibilities, i.e., professional ethics, it being understood that utilitarian or deontological considerations may rebut the presumption in favor of role responsibilities under some circumstances. The point of this presumption is that the burden of justification is properly placed on those who would depart from valid norms. Two things argue in favor of this: (1) a presumption in favor of institutional responsibilities (professional ethics) presupposes a justification of the sort provided by the convergence of the three modes of analysis and (2) making this presumption guards against the dangers of failing to take the professional ethic seriously and robbing the earlier work—constructing ethical structures of practice—of its point.

189.8 Professional Integrity, Responsibility, and Codes

Professional ethics involves more than merely complying with the norms of a code of ethics. This is true for several reasons, not the least of which is that there may not be a code. But even if there is a code, for example, the NSPE code, or the code of the AMA or ABA, there is still much more to professional ethics than compliance with the norms of that code. The rules and principles of a code set out the criteria for distinguishing malpractice from minimally acceptable practice. They do not reach to, nor do they define responsible practice. Indeed, they cannot do this, for responsible practice is more than doing one's duty (thus we speak of responsibilities rather than duties); responsible practice involves discretion and judgment in an essential way. Moreover, it involves the integration of professional judgment (expertise) and moral judgment [Whitbeck, 1998]. Here the boundaries between fact and value are fluid or at any rate they vary much as boundaries marked by a river, the course of which changes over time.

It is perhaps best to conceive professional ethics as a call for responsible conduct on the part of practitioners. The call is justified because the integrity of individual practitioners is required for the integrity of a profession, which, in turn, is necessary to justify the trust of others essential to the success of professional practice in any area. There is much work to be done in making clear what the demands of responsible practice are and in maintaining integrity in practice (which is produced by adherence to the standards of responsible practice). There is no ethical algorithm; responsible judgment and action are essential in the development, interpretation, and application of the normative principles governing the profession of biomedical engineering.

References

Abbott, A. *The System of Professions*, Chicago: University of Chicago Press, 1988.

Barber, B. Some problems in the sociology of the professions. In: Lynn, K.S. and the editors of *Deadulus*, Ed. *The Professions in America* Boston: Houghton Mifflin Company, 1965.

Barker, R. Recent work in the history of medical ethics and its relevance to bioethics. *American Philosophical Association Newsletter on Philosophy and Medicine*, 1996; 96 Fall (1): 90–96.

Barnett, R. Foreword: of chickens and eggs—the compatibility of moral rights and consequentialist analysis. *Harv. J. L. & Pub. Pol'y* 1989, 12:611.

Barnett, R. The virtues of redundancy in legal thought. *Cleveland State Law Review* 1990, 38:153.

Baum, R. Engineers and the public: sharing responsibilities. In: *Professional Ethics and Social Responsibility*, Wueste, D.E. Ed. Lanham, MD: Rowman and Littlefield, 1994.

Bayles, M. *Professional Ethics*, 2nd ed., Belmont CA.: Wadsworth, 1989.

Beauchamp, T.L. and Childress, J.F. *Principles of Biomedical Ethics* 2nd ed., New York: Oxford University Press, 1983.

Beer, J.J. and Lewis, W.D. Aspects of the professionalization of science. In: Lynn, K.S. and the editors of *Deadulus*, Ed. *The Professions in America* Boston: Houghton Mifflin Company, 1965.

Behrman, J.N., *Essays on Ethics in Business and the Professions*, Englewood Cliffs, N.J.: Prentice Hall, 1988.

Callahan, J. Professions, institutions, and moral risk. In: *Professional Ethics and Social Responsibility*, Wueste, D.E. Ed. Lanham, MD: Rowman and Littlefield, 1994.

Camenisch, P. Business ethics: On getting to the heart of the matter, *Business and Professional Ethics Journal* 1981; 1:59.

Emmet, D. *Rules, Roles and Relations*, Boston: Beacon Press, 1975.

Hardwig, J. Toward an ethics of expertise. In: *Professional Ethics and Social Responsibility*, Wueste, D.E. Ed. Lanham, MD: Rowman and Littlefield, 1994.

Hughes, E.C. Professions. In: *The Professions in America* Lynn, K.S. and the editors of *Deadulus*, Ed. Boston: Houghton Mifflin Company, 1965.

National Committee on Biomedical Engineering, *Professional Standards in Biomedical Engineering*, The Institution of Engineers, Australia, 1983.

Saha, P. and Saha, S. Ethical responsibilities of the clinical engineer, *J. Clin, Eng.*, 1986; 11(1):17.

Saha, P. and Saha, S. Clinical trials of medical devices and implants: Ethical concerns, *IEEE Eng. in Med. and Bio. Magazine* 1988 (June):85.

U.S. News and World Report, March 15, 1999, 12.

Wellman, C. *A Theory of Rights*, Totowa, NJ: Rowman and Allanheld, 1985.

Whitbeck, C. *Ethics in Engineering Practice and Research*, New York: Cambridge University Press, 1998.

Wueste, D.E. Professions, professional ethics and bioengineering, *Critical Reviews in Biomedical Engineering*, 1997 25(2):127.

Wueste, D.E. Role moralities and the problem of conflicting obligations. In: *Professional Ethics and Social Responsibility*, Wueste, D.E. Ed. Lanham, MD: Rowman and Littlefield, 1994.

190

Beneficence, Nonmaleficence, and Technological Progress

Joseph D. Bronzino
Trinity College/Biomedical Engineering Alliance for Connecticut (BEACON)

190.1 Introduction

Two moral norms have remained relatively *constant across* the various moral codes and oaths that have been formulated for health-care deliverers since the beginnings of Western medicine in classical Greek civilization, namely **beneficence**—the provision of benefits—and **nonmaleficence**—the avoidance of doing harm. These norms are traced back to a body of writings from classical antiquity known as the *Hippocratic Corpus.* Although these writings are associated with the name of Hippocrates, the acknowledged founder of Western medicine, medical historians remain uncertain whether any, including the *Hippocratic Oath,* were actually his work. Although portions of the Corpus are believed to have been authored during the sixth century BC, other portions are believed to have been written as late as the beginning of the Christian Era. Medical historians agree, though, that many of the specific moral directives of the *Corpus* represent neither the actual practices nor the moral ideals of the majority of physicians of ancient Greece and Rome.

Nonetheless, the general injunction, *"As to disease, make a habit of two things—to help or, at least, to do no harm,"* was accepted as a fundamental medical ethical norm by at least some ancient physicians. With the decline of Hellenistic civilization and the rise of Christianity, beneficence and nonmaleficence became increasingly accepted as the fundamental principles of morally sound medical practice. Although beneficence and nomaleficence were regarded merely as concomitant to the craft of medicine in classical Greece and Rome, the emphasis upon compassion and the brotherhood of humankind, central to Christianity, *increasingly made* these norms the only acceptable motives for medical practice. Even today the provision of benefits and the avoidance of doing harm are stressed just as much in virtually all contemporary Western codes of conduct for health professionals as they were in the oaths and codes that guided the health-care providers of past centuries.

Traditionally, the ethics of medical care have given greater prominence to nomaleficence than to beneficence. This priority was grounded in the fact that, historically, medicine's capacity to do harm far exceeded its capacity to protect and restore health. Providers of health care possessed many treatments

0-8493-0462-8/00/$0.00+$.50

that posed clear and genuine risks to patients but that offered little prospect of benefit. Truly effective therapies were all too rare. In this context, it is surely rational to give substantially higher priority to avoiding harm than to providing benefits.

The advent of modern science changed matters dramatically. Knowledge acquired in laboratories, tested in clinics, and verified by statistical methods has increasingly dictated the practices of medicine. This *ongoing alliance* between medicine and science became a critical source of the plethora of technologies that now pervades medical care. The impressive increases in therapeutic, preventive, and rehabilitative capabilities that these technologies have provided have pushed beneficence to the forefront of medical morality. Some have even gone so far as to hold that the old medical ethic of *"Above all, do no harm"* should be superseded by the new ethic that *"The patient deserves the best."* However, the rapid advances in medical technology capabilities have also produced great uncertainty as to what is most beneficial or least harmful for the patient. In other words, along with increases in ability to be beneficent, medicine's technology has generated much debate about what actually counts as beneficent or nonmaleficent treatment. To illustrate this point, let us turn to several specific moral issues posed by the use of medical technology (Bronzino, 1992; 1999).

190.2 Defining Death: A Moral Dilemma Posed by Medical Technology

Supportive and resuscitative devices, such as the respirator, found in the typical modern intensive care unit provide a useful starting point for illustrating how technology has rendered medical morality more complex and problematic. Devices of this kind allow clinicians to sustain respiration and circulation in patients who have suffered massive brain damage and total permanent loss of brain *function.* These technologies force us to ask: precisely when does a human life end? When is a human being indeed dead? This is not the straightforward factual matter it may appear to be. All of the relevant facts may show that the patient's brain has suffered injury grave enough to destroy its functioning forever. The facts may show that such an individual's circulation and respiration would permanently cease without artificial support. Yet these facts do not determine whether treating such an individual as a corpse is morally appropriate. To know this, it is necessary to know or perhaps to decide on those features of living persons that are essential to their status as "living persons." It is necessary to know or decide which human qualities, if irreparably lost, make an individual identical in all morally relevant respects to a corpse. Once those qualities have been specified, deciding whether total and irreparable loss of brain function constitutes death becomes a straightforward factual matter. Then, it would simply have to be determined if such loss itself deprives the individual of those qualities. If it does, the individual is morally identical to a corpse. If not, then the individual must be regarded and treated as a living person.

The traditional criterion of death has been irreparable cessation of heart beat, respiration, and blood pressure. This criterion would have been quickly met by anyone suffering massive trauma to the brain prior to the development of modem supportive technology. Such technology allows indefinite artificial maintenance of circulation and respiration and, thus, forestalls what once was an inevitable consequence of severe brain injury. The existence and use of such technology therefore challenges the traditional criterion of death and forces us to consider whether continued respiration and circulation are in themselves sufficient to distinguish a living individual from a corpse. Indeed, total and irreparable loss of brain function, referred to as "brainstem death," "whole brain death," and, simply, "brain death," has been widely accepted as the legal standard for death. By this standard, an individual in a state of brain death is legally indistinguishable from a corpse and may be legally treated as one even though respiratory and circulatory functions may be sustained through the intervention of technology. Many take this legal standard to be the morally appropriate one, noting that once destruction of the brain stem has occurred, the brain *cannot function* at all, and the body's regulatory mechanisms will fail unless artificially sustained. Thus mechanical sustenance of an individual in a state of brain death is merely postponement of the inevitable and sustains nothing of the personality, character, or consciousness of the individual. It is merely the mechanical intervention that differentiates such an individual from a corpse and a mechanically ventilated corpse is a corpse nonetheless.

Even with a consensus that brainstem death is death and thus that an individual in such a state is indeed a corpse, hard cases remain. Consider the case of an individual in a persistent vegetative state, the condition known as "neocortical death." Although severe brain injury has been suffered, enough brain function remains to make mechanical *sustenance of* respiration and circulation unnecessary. In a persistent vegetative state, an individual exhibits no purposeful response to external stimuli and no evidence of self-awareness. The eyes may open periodically and the individual may exhibit sleep-wake cycles. Some patients even yawn, make chewing motions, or swallow spontaneously. Unlike the complete unresponsiveness of individuals in a state of brainstem death, a variety of simple and complex responses can be elicited from an individual in a persistent vegetative state. Nonetheless, the chances that such an individual will regain consciousness virtually do not exist. Artificial feeding, kidney dialysis, and the like make it possible to sustain an individual in a state of neocortical death for decades. This sort of condition and the issues it raises were exemplified by the famous case of Karen Ann Quinlan. James Rachels (1986) provided the following description of the situation created by Quinlan's condition:

> In April 1975, this young woman ceased breathing for at least two 15-minute periods, for reasons that were never made clear. As a result, she suffered severe brain damage, and, in the words of the attending physicians, was reduced to a "chronic vegetative state" in which she "no longer had any cognitive function." Accepting the doctors' judgment that there was no hope of recovery, her parents sought permission from the courts to disconnect the respirator that was keeping her alive in the intensive care unit of a New Jersey hospital.
>
> The trial court, and then the Supreme Court of New Jersey, agreed that Karen's respirator could be removed. So it was disconnected. However, the nurse in charge of her care in the Catholic hospital opposed this decision and, anticipating it, had begun to wean her from the respirator so that by the time it was disconnected she could remain alive without it. So Karen did not die. Karen remained alive for ten additional years. In June 1985, she finally died of acute pneumonia. Antibiotics, which would have fought the pneumonia, were not given.

If brainstem death is death, is neocortical death also death? Again, the issue is not a straightforward factual matter. For, it too, is a matter of specifying which features of living individuals distinguish them from corpses and so make treatment of them as corpses morally impermissible. Irreparable cessation of respiration and circulation, the classical criterion for death, would entail that an individual in a persistent vegetative state is not a corpse and so, morally speaking, must not be treated as one. The brainstern death criterion for death would also entail that a person in a state of neocortical death is not yet a corpse. On this criterion, what is crucial is that brain damage be severe enough to cause failure of the body's regulatory mechanisms.

Is an individual in a state of neocortical death any less in possession of the characteristics that distinguish the living from cadavers than one whose respiration and circulation are mechanically maintained? Of course, it is a matter of what the relevant characteristics are, and it is a matter that society must decide. It is not one that can be settled by greater medical information or more powerful medical devices. Until society decides, it will not be clear what would count as beneficent or nonmaleficent treatment of an individual in a state of neocortical death.

190.3 Euthanasia

A long-standing issue in medical ethics, which has been made more pressing by medical technology, is euthanasia, the deliberate termination of an individual's life for the individual's own good. Is such an act ever a permissible use of medical resources? Consider an individual in a persistent vegetative state. On the assumption that such a state is not death, withdrawing life support would be a deliberate termination of a human life. Here a critical issue is whether the quality of a human life can be so low or so great a liability to the individual that deliberately taking action to hasten death or at least not to postpone death is morally defensible. Can the quality of a human life be so low that the value of extending its quantity is totally negated? If so, then Western medicine's traditional commitment to providing benefits and avoiding harm would seem to make cessation of life support a moral requirement in such a case.

Consider the following hypothetical version of the kind of case that actually confronts contemporary patients, their families, health-care workers, and society as a whole. Suppose a middle-aged man suffers a brain hemorrhage and loses consciousness as a result of a ruptured aneurysm. Suppose that he never regains consciousness and is hospitalized in a state of neocortical death, a chronic vegetative state. He is maintained by a surgically implanted gastronomy tube that drips liquid nourishment from a plastic bag directly into his stomach. The care of this individual takes seven and one-half hours of nursing time daily and includes (1) shaving, (2) oral hygiene, (3) grooming, (4) attending to his bowels and bladder, and so forth.

Suppose further that his wife undertakes legal action to force his care givers to end all medical treatment, including nutrition and hydration, so that complete bodily death of her husband will occur. She presents a preponderance of evidence to the court to show that her husband would have wanted this result in these circumstances.

The central moral issue raised by this sort of case is whether the quality of the individual's life is sufficiently compromised by neocortical death to make intentioned termination of that life morally permissible. While alive, he made it clear to both family and friends that he would prefer to be allowed to die rather than be mechanically maintained in a condition of irretrievable loss of consciousness. Deciding whether the judgment in such a case should be allowed requires deciding which capacities and qualities make life worth living, which qualities are sufficient to endow it with value worth sustaining, and whether their absence justifies deliberate termination of a life, at least when this would be the wish of the individual in question. Without this decision, the traditional norms of medical ethics, beneficence and nonmaleficence, provide no guidance. Without this decision, it cannot be determined whether termination of life support is a benefit or a harm to the patient.

An even more difficult type of case was provided by the case of Elizabeth Bouvia. Bouvia, who had been a lifelong quadriplegic sufferer of cerebral palsy, was often in pain, completely dependent upon others, and spent all of her time bedridden. Bouvia, after deciding that she did not wish to continue such a life, entered Riverside General Hospital in California. She desired to be kept comfortable while starving to death. Although she remained adamant during her hospitalization, Bouvia's requests were denied by hospital officials with the legal sanction of the courts.

Many who might believe that neocortical death renders the quality of life sufficiently low to justify termination of life support, especially when this agrees with the individual's desires, would not arrive at this conclusion in a case like Bouvia's. Whereas neocortical death completely destroys consciousness and makes purposive interaction with the individual's environment impossible, Bouvia was fully aware and mentally alert. She had previously been married and had even acquired a college education. Televised interviews with her portrayed a very intelligent person who had great skill in presenting persuasive arguments to support her wish not to have her life continued by artificial means of nutrition. Nonetheless, she judged her life to be of such low quality that she should be allowed to choose to deliberately starve to death. Before the existence of life support technology, maintenance of her life against her will might not have been possible at all and at least would have been far more difficult.

Should Elizabeth Bouvia's judgment have been accepted? Her case is more difficult than the care of a patient in a chronic vegetative state because, unlike such an *individual, she* was able to engage in meaningful interaction with her environment. Regarding an individual who cannot speak or otherwise meaningfully interact with others as nothing more than living matter, as a "human vegetable," is not especially difficult. Seeing Bouvia this way is not easy. Her awareness, intelligence, mental acuity, and ability to interact with others means that although her life is one of discomfort, indignity, and complete dependence, she is not a mere "human vegetable."

Despite the differences between Bouvia's situation and that of someone in a state of neocortical death, the same issue is posed. Can the quality of an individual's life be so low that deliberate termination is morally justifiable? How that question is answered is a matter of what level of quality of life, if any, is taken to be sufficiently low to justify deliberately acting to end it or deliberately failing to extend it. If there is such a level, the conclusion that it is not always beneficent or even nonmaleficent to use life-support technology must be accepted.

Another important issue here is respect for individual autonomy. For the cases of Bouvia and the hypothetical instance of neocortical death discussed above, both concern voluntary euthanasia, that is, euthanasia voluntarily requested by the patient. A long-standing commitment, vigorously defended by various schools of thought in Western moral philosophy, is the notion that competent adults should be free to conduct their lives as they please as long as they do not impose undeserved harm on others. Does this commitment entail a right to die? Some clearly believe that it does. If one owns anything at all, surely one owns one's life. In the two cases discussed above, neither individual sought to impose undeserved harm on anyone else, nor would satisfaction of their wish to die do so. What justification can there be then for not allowing their desires to be fulfilled?

One plausible answer is based upon the very respect of individual autonomy at issue here. A necessary condition, in some views, of respect for autonomy is the willingness to take whatever measures are necessary to protect it, including measures that restrict autonomy. An autonomy-respecting reason offered against laws that prevent even competent adults from voluntarily entering lifelong slavery is that such an exercise of autonomy is self-defeating and has the consequence of undermining autonomy altogether. By the same token, an individual who acts to end his own life thereby exercises his autonomy in a manner that places it in jeopardy of permanent loss. Many would regard this as justification for using the coercive force of the law to prevent suicide. This line of thought does not fit the case of an individual in a persistent vegetative state because his/her autonomy has been destroyed by the circumstances that rendered him/her neocortically dead. It does fit Bouvia's case though. Her actions indicate that she is fully competent and her efforts to use medical care to prevent the otherwise inevitable pain of starvation is itself an exercise of her autonomy. Yet, if allowed to succeed, those very efforts would destroy her autonomy as they destroy her. On this reasoning, her case is a perfect instance of limitation of autonomy being justified by respect for autonomy and of one where, even against the wishes of a competent patient, the life-saving power of medical technology should be used.

Active Versus Passive Euthanasia

Discussions of the morality of euthanasia often distinguish active from passive euthanasia in light of the distinction made between killing a person and letting a person die, a distinction that rests upon the difference between an act of commission and an act of omission. When failure to take steps that could effectively forestall death results in an individual's demise, the resultant death is an act of omission and a case of letting a person die. When a death is the result of doing something to hasten the end of a person's life (giving a lethal injection, for example), that death is caused by an act of commission and is a case of killing a person. When a person is allowed to die, death is a result of an act of omission, and the motive is the person's own good, the omission is an instance of passive euthanasia. When a person is killed, death is the result of an act of commission, and the motive is the person's own good, the commission is an instance of active euthanasia.

Does the difference between passive and active euthanasia, which reduces to a difference in how death comes about, make any moral difference? It does in the view of the American Medical Association. In a statement adopted on December 4, 1973, the House of Delegates of the American Medical Association asserted the following (Rachels, 1978):

> The intentional termination of the life of one human being by another—mercy killing—is contrary to that for which the medical profession stands and is contrary to the policy of the American Medical Association (AMA).
>
> The cessation of extraordinary means to prolong the life of the body where there is irrefutable evidence that biological death is imminent is the decision of the patient and immediate family. The advice of the physician would be freely available to the patient and immediate family.

In response to this position, Rachels (1978) answered with the following:

> The AMA policy statement isolates the crucial issue very well, the crucial issue is "intentional termination of the life of one human being by another." But after identifying this issue and forbidding "mercy killing," the statement goes on to deny that the cessation of treatment is the intentional termination of a life. This is where the mistake comes in, for what is the cessation of treatment in those circumstances (where the intention is to release the patient from continued suffering), if it is not "the intentional termination of the life of one human being by another?"

As Rachels correctly argues, when steps that could keep an individual alive are omitted for the person's own good, this omission is as much the intentional termination of life as taking active measures to cause death. Not placing a patient on a respirator due to a desire not to prolong suffering is an act intended to end life as much as the administration of a lethal injection. In many instances the main difference between the two cases is that the latter would release the individual from his pain and suffering more quickly than the former. Dying can take time and involve considerable pain even if nothing is done to prolong life. Active killing can be done in a manner that causes death painlessly and instantly. This difference certainly does not render killing, in this context, morally worse than letting a person die. Insofar as the motivation is merciful (as it must be if the case is to be a genuine instance of euthanasia) because the individual is released more quickly from a life that is disvalued than otherwise, the difference between killing and letting one die may provide support for active euthanasia. According to Rachels (1978), the common rejoinder to this argument is the following:

> The important difference between active and passive euthanasia is that in passive euthanasia the doctor does not do anything to bring about the patient's death. The doctor does nothing and the patient dies of whatever ills already afflict him. In active euthanasia, however, the doctor does something to bring about the patient's death: he kills the person. The doctor who gives the patient with cancer a lethal injection has himself caused his patient's death; whereas if he merely ceases treatment, the cancer is the cause of death.

According to this rejoinder, in active euthanasia someone must do something to bring about the patient's death, and in passive euthanasia the patient's death is caused by illness rather than by anyone's conduct. Surely this is mistaken. Suppose a physician deliberately decides not to treat a patient who has a routinely curable ailment and the patient dies. Suppose further that the physician were to attempt to exonerate himself by saying, "I did nothing. The patient's death was the result of illness. I was not the cause of death." Under current legal and moral norms, such a response would have no credibility. As Rachels (1978) notes, *"it would be no defense at all for him to insist that he didn't do anything. He would have done something very serious indeed, for he let his patient die."*

The physician would be blameworthy for the patient's death as surely as if he had actively killed him. If causing death is justifiable under a given set of circumstances, whether it is done by allowing death to occur or by actively causing death is morally irrelevant. If causing someone to die is not justifiable under a given set of circumstances, whether it is done by allowing death to occur or by actively causing death is also morally irrelevant. Accordingly, if voluntary passive euthanasia is morally justifiable in the light of the duty of beneficence, so is voluntary active euthanasia. Indeed, given that the benefit to be achieved is more quickly realized by means of active euthanasia, it may be preferable to passive euthanasia in some cases.

Involuntary and Non-Voluntary Euthanasia

An act of euthanasia is involuntary if it hastens the individual's death for his own good but against his wishes. To take such a course would be to destroy a life that is valued by its possessor. Therefore, it is no different in any morally relevant way from unjustifiable homicide. There are only two legitimate reasons for hastening an innocent person's death against his will: self-defense and saving the lives of a larger number of other innocent persons. Involuntary euthanasia does not fit either of these justifications. By definition, it is done for the good of the person who is euthanized and for self-defense or saving innocent others. No act that qualifies as involuntary euthanasia can be morally justifiable.

Hastening a person's death for his own good is an instance of non-voluntary euthanasia when the individual is incapable of agreeing or disagreeing. Suppose it is clear that a particular person is sufficiently self-conscious to be regarded a person but cannot make his wishes known. Suppose also that he is suffering from the kind of ailment that, in the eyes of many persons, makes one's life unendurable. Would hastening his death be permissible? It would be if there were substantial evidence that he has given prior consent. This person may have told friends and relatives that under certain circumstances efforts to prolong his life should not be undertaken or continued. He might have recorded his wishes in the form of a Living Will (below) or on audio- or videotape. Where this kind of substantial evidence of prior consent exists, the decision to hasten death would be morally justified. A case of this scenario would be virtually a case of voluntary euthanasia.

But what about an instance in which such evidence is not available? Suppose the person at issue has never had the capacity for competent consent or dissent from decisions concerning his life. It simply cannot be known what value the individual would place on his life in his present condition of illness. What should be done is a matter of what is taken to be the greater evil—mistakenly ending the life of an innocent person for whom that life has value or mistakenly forcing him to endure a life that he radically disvalues.

> To My Family, My Physician, My Clergyman, and My Lawyer:
>
> If the time comes when I can no longer take part in decisions about my own future, let this statement stand as testament of my wishes: If there is no reasonable expectation of my recovery from physical or mental disability, I, ______________________________, request that I be allowed to die and not be kept alive by artificial means or heroic measures. Death is as much a reality as birth, growth, maturity, and old age—it is the one certainty. I do not fear death as much as I fear the indignity of deterioration, dependence, and hopeless pain. I ask that drugs be mercifully administered to me for the terminal suffering even if they hasten the moment of death.
>
> This request is made after careful consideration. Although this document is not legally binding, you who care for me will, I hope, feel morally bound to follow its mandate. I recognize that it places a heavy burden of responsibility upon you, and it is with the intention of sharing that responsibility and of mitigating any feelings of guilt that this statement is made.
>
> Signed: ______________________________
>
> Date: ______________________________
>
> Witnessed by:
>
> ______________________________
>
> ______________________________

Living Will statutes have been passed in at least 35 states and the District of Columbia. For a Living Will to be a legally binding document, the person signing it must be of sound mind at the time the will is made and shown not to have altered his opinion in the interim between the signing and his illness. The witnesses must not be able to benefit from the individual's death.

Should Voluntary Euthanasia be Legalized?

The recent actions of Dr. Kavorkian have raised the question: "Should voluntary euthanasia be legalized?" Some argue that even if voluntary euthanasia is morally justifiable, it should be prohibited by social policy nonetheless. According to this position, the problem with voluntary euthanasia is its impact on society as a whole. In other words, the overall disutility of allowing voluntary euthanasia outweighs the good it could do for its beneficiaries. The central moral concern is that legalized euthanasia would eventually erode respect for human life and ultimately become a policy under which "socially undesirable" persons would have their deaths hastened (by omission or commission). The experience of Nazi Germany is often cited in support of this fear. What began there as a policy of euthanasia soon became one of eliminating individuals deemed racially inferior or otherwise undesirable. The worry, of course, is that what happened there can happen here as well. If social policy encompasses efforts to hasten the deaths of people, respect for human life in general is eroded and all sorts of abuses become socially acceptable, or so the argument goes.

No one can provide an absolute guarantee that the experience of Nazi Germany would not be repeated, but there is reason to believe that its likelihood is negligible. The medical moral duty of beneficence justifies only voluntary euthanasia. It justifies hastening an individual's death only for the individual's benefit and only with the individual's consent. To kill or refuse to save people judged socially undesirable is not to engage in euthanasia at all and violates the medical moral duty of nomaleficence. As long as only voluntary euthanasia is legalized, and it is clear that involuntary euthanasia is not and should never be, no degeneration of the policy need occur. Furthermore, such degeneration is not likely to occur if the beneficent nature of voluntary euthanasia is clearly distinguished from the maleficent nature of involuntary euthanasia and any policy of exterminating the socially undesirable. Euthanasia decisions must be scrutinized carefully and regulated strictly to ensure that only voluntary cases occur, and severe penalties must be established to deter abuse.

References

Bronzino, J. D. Chapter 10 Medical and Ethical Issues in Clinical Engineering Practice. In: *Management of Medical Technology.* Butterworth, 1992.

Bronzino, J. D. Chapter 20 Moral and Ethical Issues Associated with Medical Technology. In: *Introduction to Biomedical Engineering.* Academic Press, 1999.

Rachels, J. "Active and Passive Euthanasia," In: *Moral Problems,* 3rd ed., Rachels, J., (Ed.), Harper and Row, New York, 1978.

Rachels, J. *Ethics at the End of Life: Euthanasia and Morality,* Oxford University Press, Oxford, 1986.

Further Information

Daniels, N. *Just Health Care.* Cambridge University Press, Cambridge, 1987.

Dubler, N. N. and Nimmons, D. *Ethics on Call.* Harmony Books, New York, 1992.

Jonsen, A. R. *The New Medicine and the Old Ethics.* Harvard University Press, Cambridge, MA, 1990.

Murphy, J. and Coleman, J. *The Philosophy of Law,* Rowman and Allenheld, 1984.

191
Ethical Issues of Animal and Human Experimentation in the Development of Medical Devices

Subrata Saha
Clemson University

Pamela S. Saha
Clemson University

191.1 Introduction

The number and sophistication of new medical devices is transforming modern medicine at a rate never experienced before. These developments have saved the lives of many more patients and improved their quality of life drastically. Artificial joint replacements alone have transformed the field of orthopaedic surgery, and over 250,000 total hip and total knees are implanted annually in the United States. Ventricular assistive devices (VASs) have extended lives by years, lives that before would have certainly ended at the point the device would have been needed. An estimated two to three million artificial or prosthetic parts, manufactured by hundreds of different companies, are implanted in America each year. The massive production of such devices is not only big science but big business.

New medical devices, however, require thorough testing for safety and efficacy as well as submission for approval by the federal Food and Drug Administration (FDA) before being put on the market for public use. Testing of a new product takes considerable time and expense and is not without problems beyond those of a technical nature. This paper considers the ethical questions that arise when the demands of science, economics, and progress are not entirely compatible with issues raised about the rights and obligations toward human beings and animals.

One of the first levels of testing where ethical debate is more prominent is at the stage when the biomedical scientist is faced with the need to use animal subjects.[1] Over the last 30 years there has been a growing debate over whether or not the use of experimental animals is even appropriate. Some animal rights activists have made violent protests and vandalized research facilities where animal experimentation takes place. Increased public sensitivity did help promote an effort by the government, as well as the

0-8493-0462-8/00/$0.00+$.50

scientific community, to regulate the use of animal subjects and to educate the public as to the importance of such research to the health and well being of both animals and humans. As a result efforts to promote the humane use of experimental animals that are used to advance knowledge of biomedical sciences are to be supported.

Inevitably human beings are involved during the final stages of testing medical devices or systems. While this imposes numerous ethical concerns and has stirred much discussion, clinical trials are necessary as the alternative would mean an end of learning anything new for the betterment of medical science and continued use of unsupported practices based on conjecture.[2]

Consequently, engineers involved with the development and design of medical technology need to become familiar with various aspects of clinical trials and animal research as well as the ethical issues that they raise. Normally, conduction of human experimentation is not a part of the training of an engineer,[3] nor are complications presented by ethical concerns a traditional part of engineering education.[4] In this chapter various ethical concerns are examined that are essential elements in the animal testing and the clinical trial of any new treatment modality.

191.2 Clinical Trials

The Reason for Clinical Trials

Clinical trials are designed to ascertain the effectiveness and safety of a new medical device as compared with established medical practice. This form of rigorous scientific investigation in a controlled environment for the assessment of new treatment modalities is superior to the is forum of private opinion and individual chance taking. Holding the practice of medicine at the status quo and discontinuing all innovative work would be the only way to prevent exposing patients to some unforeseen risks in new treatments that come with the promise of improved care. Yet, even this does not protect patients from uncontrolled experimentation as conditions change even if the practice of medicine stands still. The changing effectiveness of antibiotics as well as the increase in antibiotic resistant strains through uninformed overuse of such drugs is a perfect example of how even standard care held stationary can lead to a decline rather than just a status quo in level of medical care. If we wish to continue to seek out new ways to conquer disease or even maintain a current level of care, then we are essentially forced to decide on a manner in which human beings are to shoulder the risks involved.

Total Mastectomies for Stage I and II breast tumors and extracranial-intracranial anastomosis for internal carotid atherosclerosis were once routinely practiced. However, randomized trials have brought the necessity of these surgical procedures under closer scrutiny. Certain procedures which are continually used, such as the now twenty-year-old use of obstetric ultrasonography, are losing support in the battle to cut medical costs as these do not have the benefit of randomized clinical trials to prove their value. Answers are needed to questions concerning health and economic risks and benefits, and the burden of false positive and negative results, especially in this time of multiple options in a profit-driven atmosphere, along with dwindling resources and government aid to the sick, old, and poor.[5]

A clinical trial is the most reasonable means to test a device as well as control risks and prevent abuse of human subjects because of the following six major factors: (1) A limited number of closely monitored subjects in a controlled environment are in a safer situation than are the same subjects unobserved in relation to one another within the larger population, where the same kinds of risks are imposed by uncertainty about the efficacy of new devices. (2) Clinical trials give conclusive answers to important medical inquiries that otherwise could only be answered by guessing. Medical decisions based on proven results are certainly superior to those dependent on untested clinical opinion. One author has stated that "trials were introduced because personal opinion was so notoriously fragile, biased, and unreliable."[6] Risks would indeed be greater and potential for harm magnified if doctors are made to make decisions in an environment of general uncertainty and when medical products available are unsupervised by review boards. In an uncontrolled situation, the individual practitioner can be influenced by motives of economic profit, the need to appear knowledgeable and abreast of the new, or by high-pressure salesmanship. In the zeal to argue for individual rights in clinical trials, worry should be placed on how those same rights are threatened in an environment without objective controls. (3) A society that restricts and oversees the advancement of medicine through regulated human experimentation will prevent the subjection of people to needless risks brought on by a plurality of devices that may cause harm and offer only a maintenance of the status quo as a benefit. There are a plentitude of types of redundant consumer goods on the market today, e.g., soft drinks that offer multiple ways to relieve one condition—thirst. However, in medicine the concerns are different. The onslaught of numerous types of drugs to relieve nausea during pregnancy should be limited to the testing of a few medicines that offer the greatest benefit. Control of the market place in medicine through government or private testing of the few most promising forms of treatment is a much safer environment than an extensive supermarket full of many possibilities and many potential risks. (4) Clinical trials advance expedient corroboration of medical theories so that research can be channeled in directions that show meaningful results. Promising research is rapidly pinpointed, and harmful products are removed from the hiding place of private opinion and promotional tactics. (5) Clinical trials are the answer to the moral imperative to thoroughly test all new medical devices. Such testing is deficient without a controlled study on human subjects. No researcher can state confidently that a product is safe and effective for human use without such a test. (6) All clinical trials are evaluated by specialized committees formed for the objective of supervising the ethical conduct of investigators using human beings as subjects. In this manner, individual rights are protected and ethical guidelines effectively imposed in a manner superior to what could be expected in the isolated world of private clinical practice, with its competition for patients and pressures from manufacturers.

Dilemmas Presented by Clinical Trials

The Problem of Informed Consent

One of the most controversial issues generated by clinical trials is that of informed consent.[7] Informed consent protects certain human rights, such as the patient's freedom to decide what risks to take with his/her own body, the right to the truth from the doctor in the doctor-patient relationship, and a just distribution of goods in accordance with a standard of equity and access to redress for undeserved harm. These values cannot be sacrificed for any sort of anticipated benefit from research. "The loss of such values is so harmful that benefits become meaningless."[8]

Clinical Trials and the Doctor-Patient Relationship

Another issue currently debated is the conflict of clinical trials with the therapeutic obligation. Some authors argue that we must face the fact that if we are to expand the knowledge needed for obtaining high-quality treatment, we must sacrifice our therapeutic obligation. There are others who take a more apprehensive view of clinical trials, stating that trials on healthy subjects are condemned by the Nuremburg Code, the Tokyo Declaration, and the Helsinki Declaration of the World Medical Association.[9]

However, outright condemnation of experimentation on healthy human subjects ignores the vital need for progress in preventive as well as remedial medical treatment. For example, the development and use of vaccines carry minimal but real risks to their recipients. Yet few dispute that vaccine research using human subjects is morally justified and may even be compulsory, despite the reality that persons can and do die from experimental as well as FDA approved vaccines. The control of crippling and deadly diseases, and their eventual elimination (i.e., small pox), is due to the study and implementation of vaccines. Those who support clinical trials say that a validated medical practice is a far better alternative for both the individual and society as a whole than subjection to treatments whose effectiveness are not validated by controlled trials.

One article offers several ideas for the elicitation of informed consent and assignment to randomized groups. The best model suggested for general use begins with selection of eligible patients, who are prerandomized and given the entire protocol with an explanation of benefits and risks of all the options; consent is then sought, the patient knowing his or her group assignment.[10]

Proposed deviations from the above standard could be defended before review boards; for example, investigations that would be impeded by the patient knowing his or her group's assignment but that promise the patient major benefits. However, the patient should be informed of any such stipulations and of the relative risks and benefits.

Standards of research require further consideration as a consensus needs to be achieved. For example, in order to test the effectiveness of transplanted embryonic pig cells into the brain for the treatment of Parkinson's disease, a control group is required in which surgical boring of a hole into the skull occurs without the addition of any cells. This is to remove any placebo effect or any other unforeseeable effect of a mock surgical procedure on Parkinson's disease. However, the question of simply boring a hole does subject a human being to a procedure with significant risk with little bases for suspecting a beneficial outcome. The design of scientific studies needs to weigh the risks to the human subjects against the need for scientific purity. Such opinions have been voiced by Arthur Caplan, director of the University of Pennsylvania's Center for Biomedical Ethics who recently told the Boston Herald that, "striving for scientific accuracy is a commendable goal, but asking someone to have a hole drilled in their head for no purpose is putting science ahead of the subject's interest."

There has been much effort particularly in recent years toward the protection of individual rights. This is not surprising in this age of rising autonomy and declining paternalism. This concern has led to the consideration of whether or not treatment should be changed mid-course in an experiment due to the "appearance" of a tendency toward one result as opposed to another. At some point in collecting data, the experimenter may begin to surmise a particular outcome or even come to expect it (although he/she may not be certain). In these circumstances, the question arises as to whether the individual subject should be granted the hypothetical benefit of the "good guess," or should the guess be treated as merely a guess and, hence, a gamble. To do so would mean that the best treatment for the individual might not be one that has been proven valid. In fact, such a choice implies that the best treatment is a perpetual gamble. Nonetheless, treatment methods that maximize the good guess have been presented.[11]

Another issue that has surfaced from the debate on informed consent is the idea of the "therapeutic misconception," which refers to an unyielding expectation of personalized care on the part of patients during a clinical trial. This finding shows a need for better communication between the patient and the doctor. However, such "false expectations" could be due to a fundamental trust on the patient's part believing that through cooperation with the physician in research, personal health needs will eventually be met according to the best available data at any given time. Perhaps, this notion should be addressed during a clinical trial.[12]

A local newspaper reported that some companies are enticing private physicians to register patients for their studies that have significant fees. What used to be the domain of academic researchers motivated by the drive for new discoveries, fame, and career advancement is now a multibillion dollar industry with numerous companies working with thousands of doctors in private practice having a profound impact on the doctor-patient relationship. Often the patient is unaware that significant amounts of money are involved in their recruitment for a study.[13] In addition to the expected problems of human research that have been debated in the past, the influence of a growing industry invites the need for continued inspection and control not only for the prevention of abuse of human subjects but for even the possible effect such methods of recruiting could have on scientific integrity.

The Need for Double-Blind Trials

The double-blind study has magnified the issues concerning clinical trials already mentioned. Obviously, this type of trial means that both the physicians and the patients involved will not be given full information about the experiment. The double-blind study is the best safeguard against biased results. Indeed, the purpose of the double-blind trial is to bring about a treatment plan based on objective fact rather than on biased personal belief or guess work. Even well intentioned and honest researchers can fall victim to seeing their data too subjectively. For instance, an investigator may have a vested interest in the study because of personal prestige, financial gain, or merely normal enthusiasm and faith in his own work.[14] These represent obstacles to objective results in research that are not blind. The originator and/or sponsor of a project may suffer from a conflict of interest when either becomes directly involved in a study. The double-blind approach protects against such conflict. Although some investigators have argued that it is unethical for them to deprive control subjects of their product because they believe so strongly in its efficacy, personal conviction of the usefulness of one's own work is not a good ethical reason for failing to thoroughly test a new device for safety and effectiveness. The moral imperative to thoroughly test is a basic concept to be followed to achieve a standing of good biomedical engineering.[15]

The biomedical engineer must face the current debate over the use of medical technology in accordance with the ethical, professional, and scientific imperative to thoroughly test his or her innovation. It is now time for biomedical engineers to stand up and support the sound moral use of clinical trials for the purposes of (1) scientific credibility of the biomedical engineering field, (2) protection of the individual from improper medical care due to unsupervised, unethical market-place forms of human research, and (3) the promotion of medical care based on sound reasoning and scientific fact. These principles are at the heart of the professionalism required of biomedical engineering research.[16]

191.3 Animal Experimentation

Animal Testing

Animal experimentation is an important step in the development of new implants and devices to determine their safety and effectiveness prior to their use in humans. There is no current alternative to the use of animal models to evaluate the biocompatibility of new materials and the response by a host.[17] However, animal rights groups have raised important ethical considerations and have challenged the scientific community to justify the use of animals in research. The scientific community should respond and debate these issues. They should help to educate the public and increase understanding of the need for such research. A more proactive approach in this matter is imperative. It is also important that efforts be made by scientists to demonstrate to the public that animal research is done humanely, sparingly, and only when alternatives do not exist.[18]

The Need for Animal Research

Biomedical engineering has brought remarkable advancement to the field of medicine in a very short time frame. Just in the last quarter of this century we have seen achievements such as (1) pacemakers, (2) total joint replacement, (3) artificial hearts, (4) CAT scan machines, and (5) improved surgical techniques made possible through the use of fiber optics, lasers, and ultrasonic devices. Along with these remarkable advances came the need for closer monitoring of the safety and effectiveness of inventions before their release for public use. More regulations, standards, and testing protocols were devised to ensure that each new product is subjected to a uniform system of scrutiny and procedure for approval. As part of that process, the testing of a product in animals is a likely necessary step. For example, *NIH Guidelines for the Physicochemical Characterization of Biomaterials* outlines a stepwise process for testing blood-contacting devices progressively in animals. A possible hierarchy of testing starting from *in vitro* to *in vivo* systems may be as follows:[19]

Cell culture cytotoxicity (mouse L929 cell line)
Hemolysis (rabbit or human blood)
Mutagenicity [human or other mammalian cells or Ames test (bacterial)]
Systemic injection acute toxicity (mouse)
Sensitization (guinea pig)
Pyrogenicity (limulus amebocyte lysate [LAL] or rabbit)
Intracutaneous irritation (rat, rabbit)
Intramuscular implant (rat, rabbit)
Blood compatibility (rat, dog)
Long-term implant (rat, rabbit, dog, primate)

Animal testing is a necessary means of evaluating a device in the internal environment of the living system. The suggestion that this can be replaced by computer modeling can only be made by one unaware of the lack of knowledge about *in vivo* conditions.[20] The nearly inscrutable chemical pathways, the complex milieu of high chemical, mechanical and electrical activity, and the incalculable number of interactions inside the living organism cannot possibly be modeled theoretically today. Add to this the fact that normal conditions are different from disease states such as thrombosis or inflammation, and we find ourselves still further removed from the prospects of using simulation methods. Yet to thoroughly rule out both long and short-term failure of these products, they must be tested sufficiently to rule out risks of toxicity and even possible carcinogenic effects over many years. In response to questions raised about animal research, the American Association for the Advancement of Science issued a resolution in 1990 that supports the use of animals in research while emphasizing that experimental animals be treated humanely.

Regulations/Guidelines Related to Animal Research

A biomedical engineer must become familiar with regulations with regard to animal research. The National Research Council published a *Guide for the Care and Use of Laboratory Animals* initially as early as 1963 and most recently in 1996.

In addition to becoming familiar with this Guide, it is important to be aware of all applicable federal, state, and local laws, regulations, and policies such as the Animal Welfare Regulations and the Public Health Service Policy on Humane Care and Use of Laboratory Animals.[22]

The general objective is to have a cogent reason for conducting a particular experiment. It must be important to the health and well-being of humans or animals.[23] It must be demonstrated that alternative means would not achieve the necessary goal. The choice of species must be shown to be essential and that species lower on the evolutionary ladder would not be suitable. An effort must be made to minimize the number of animals required and to prevent unnecessary duplication of tests. Pain and discomfort must be minimized unless important to the conduction of the experiment. Standards must be followed concerning (1) space allotment, (2) type of confinement, (3) availability of food and water, (4) care in transit, and (5) periods of exercise. The experiment must have clear limits and be performed under the close supervision of individuals appropriately trained. Provisions for veterinary care, environmental conditions, and euthanasia must also be made. If palliation of discomfort or pain must be withheld, a detailed explanation must be made available.

In 1985 congress passed The Food Security Act which amended the Animal Welfare Act of 1966. This required that research institutions have a committee of no less than three persons who are qualified to monitor animal care and practices in experimentation.[17] These persons are to represent society's interest in the proper treatment of animal subjects. The committee requires one veterinarian and one other person not connected with the institution. This committee must inspect all animal research and care facilities at the institution twice a year. The institution should in turn keep a report of each inspection on file for 3 years and report any information that notes violation of legal standards on the treatment of animals to federal officials. The committee must review specific areas of treatment of experimental animals, including pain management, veterinary care, and pre- and post-surgical care. A single animal must not be subjected to major survival surgery more than once unless justified scientifically. Dogs must be exercised and environmental conditions must be conducive to psychological well-being of primates. Note that policies by the National Science Foundation now extend to all vertebrate animals which is broader than the Animal Welfare Act that did not include (1) rats, (2) mice, (3) farm animals, and (4) birds.

The Public Debate

The debate about animal experimentation is marred by extremes on both sides of the issue.[24] Scientists have been physically threatened and assaulted and their laboratories vandalized by extremist animal rights groups.[25] Unwarranted charges that scientists only conduct animal research for personal gain and that these experiments are cruel and unnecessary demonstrate the extent of misinformation that exists for the public.

The actions of irresponsible activists must not obscure the serious philosophical issues raised by animal rights advocates. Some of them argue that there exists no justification for the claim that animals can be exploited for human benefit. This line of reasoning is that animals have rights equal to that of humans and as animals cannot give informed consent for an experiment they should be precluded from use. This represents a radical change that would have consequences beyond the use of animals in science. When animals are used as beasts of burden for example, should this be considered slavery? What of the use of animals for food, clothing, or even simply in sport. One author noted that a 1990 study showed that although 63% of literature advocating animal rights addresses only their use in research, such use annually is only 0.003% of the number of animals used as food.[26] What about protection against pests in the home as well as in agriculture? Even the ownership of pets may be called into question, as a case of imprisonment. Of all possible targets for basing a case against human exploitation of animals, the use of them in medical experiments is the least worthy. Scientists not only make the least use of animals but do so under strict regulation and with all attempts being made to minimize their use. If we are seriously going to elevate animals to a moral standing equal to human beings, then the research laboratory is not the appropriate place to start. The burden of explaining how all future interaction should take place with the rest of the animal kingdom lies with those who make such a claim.

The other extreme however, is the claim by some scientists that animals have no moral standing or intrinsic value. A report by the National Research Council in 1988 on the Use of Laboratory Animals in Biomedical and Behavioral Research says "Our society does, however, acknowledge that living things have inherent value."[27] This is consistent with society's tendency to treat animals differently than inanimate objects. Even a mouse merits higher consideration than a stone. That animals have value, moral standing, or even rights is not equivalent to the suggestion that they are equal to human beings in their value, standing, and rights.

Humans are not the only beneficiaries of research on animals. Veterinary medicine would suffer without such research and the care of animals has significantly improved as a result of responsible animal experimentation. Today, human psychoactive drugs are being used for pets with behavioral problems.[28]

Conclusion

While there is continued consensus within the scientific community on the need for animal research, there is also increased sensitivity and awareness of the need for humane treatment of animals and the intrinsic value and moral standing of non-human animals. This has lead to improved guidelines and regulation on such use as well as much needed discussion on the purpose and ethics of animal research. While the debate will certainly continue, the use of violence, slander, and media sensationalism should be discouraged. This will only lead to further polarization of extreme views. Constructive and responsible discussion will help bring about a consensus on this very important matter that concerns the health and welfare of animals and humans alike.

References

1. An, Y.H. and Friedman, R.J. *Animal Models in Orthopaedic Research*, CRC Press, Boca Raton, FL, 1998.
2. Saha, P. and Saha, S. "Clinical trials of medical devices and implants: Ethical concerns," *IEEE Eng. Med. Y Biol. Mag.*, 7, 85–87, 1988.
3. Saha, P. and Saha, S. Ethical responsibilities of the clinical engineer, *J. Clin. Eng.*, 11, 17–25, 1986.
4. Saha, P. and Saha, S. The need of biomedical ethics training in bioengineering, In *Biomedical Engineering I: Recent Developments*, S. Saha, Ed., Pergamon Press, New York, 369–373, 1982.
5. Bracken, M.B. Clinical trials and the acceptance of uncertainty, *British Med. J.*, 294, 1111–1112, London, 1987.

6. Vere, D. W. Problems in controlled trials: A critical response. *J. Med. Ethics*, 9(2), 85–89, 1983.
7. Kaufmann, C.L. Informed consent and patient decision making: Two decades of research, *Soc. Sci. Med.*, 21, 1657–1664, 1983.
8. Dyck, A.J. and Richardson, H.W. The moral justification for research using human subjects, In *Biomedical Ethics and the Law*, J.M. Humber and R. F. Almeder, Eds., Plenum Press, New York, 243–259.
9. Arpzilange, P., Dion, S., Mathe, G. Proposal for ethical standards in therapeutic trials, *British Med. J.*, 291, 887–889, 1985.
10. Kopelman, L. Randomized clinical trials consent and the therapeutic relationship, *Clin. Res.*, 31(1), 1–11, 1983.
11. Meler P. Terminating a trial: The ethical problem, *Clin. Pharmacol. Therap.*, 25, 637–640, 1979.
12. Appelbaum, P., Lidz, C.W., Benson, P., et al. False Hopes and Best Data: Consent to Research and the Therapeutic Misconception, *Hastings Center Report*, 17(2), 20–24, 1987.
13. Eichenwald, K. and Kolata, G. Drug trials threaten doctors' credibility, *Anderson Independent Mall*, Sunday, May 16, 1999, 13A.
14. Saha, S. and Saha, P. Bioethics and applied biomaterials, *J. Biomed Mat. Res., App. Biomat.*, 21(A–2), 181–190, 1987.
15. Saha, S. and Saha, P. Biomedical ethics and the biomedical engineer: A review, *Crit. Rev. Biomed. Eng.*, 25(2), 163–201, 1997.
16. Mappes, T.A. and Zembaty, J.S. *Biomedical Ethics*, 2nd ed., McGraw-Hill, New York, 1986.
17. Saha, P. and Saha, S. Ethical Issues on the Use of Animals in the Testing of Medical Implants, *J. Long-Term Effects of Med. Impl.*, 1(2), 127–134, 1991.
18. Lukas, V. and Podolsky, M.L. *The Care and Feeding of an IACVC*, CRC Press, Boca Raton, FL, 1999.
19. Vale, B.H., Wilson, J.E., and Niemi, S. M. Animal models, In *Biomaterials Science*, Academic Press, San Diego, CA, 240, 1996.
20. Malakoff, D. Alternatives to animals urged for producing antibodies, *Science*, 284, 230, 1999.
21. National Research Council, *Guide for the Care and Use of Laboratory Animals*, 2nd ed., 1996.
22. Rollin, B.E. and Kesel, M.D. *The Experimental Animal in Biomedical Research*, CRC Press, Boca Raton, FL, 1, 1990.
23. Saha, S. and Saha, P. Biomedical ethics and the biomedical engineer: A review, *Crit. Rev. Biomed. Eng.*, 25(2), 163–201, 1997.
24. Saha, S. and Saha, P. Biomedical engineering and animal research, *BMES Bulletin*, 16(2), 22, 1992.
25. Kower, J. Activists ransack Minnesota labs, *Science*, 284:410–411, 1997.
26. Conn, P.M and Parker, J. Animal Rights: Reaching the Public, *Science*, 282:1417, 1998.
27. Herzog, H. A. Jr. Informed Opinions on Animal Use Must Be Pursued, *ILAR News*, Institute of Laboratory Animal Resources, 31(2), Spring 1989.
28. Bunk, S. Market Emerges for Use of Human Drugs on Pets, *The Scientist*, 1 & 10, April 12, 1999.

192

Regulation of Medical Device Innovation

Joseph D. Bronzino
Trinity College/Biomedical Engineering Alliance for Connecticut (BEACON)

192.1 Introduction

Responsibility for regulating medical devices falls to the Food and Drug Administration (FDA) under the Medical Device Amendment of 1976. This statute requires approval from the FDA before new devices are marketed and imposes requirements for the clinical investigation of new medical devices on human subjects. Although the statute makes interstate commerce of an unapproved new medical device generally unlawful, it provides an exception to allow interstate distribution of unapproved devices in order to conduct clinical research on human subjects. This investigational device exemption (IDE) can be obtained by submitting to the FDA *"a protocol for the proposed clinical testing of the device, reports of prior investigations of the device, certification that the study has been approved by a local institutional review board, and an assurance that informed consent will be obtained from each human subject"* (Bronzino et al., 1990a, b).

With respect to clinical research on humans, the FDA distinguishes devices into two categories: devices that pose significant risk and those that involve insignificant risk. Examples of the former included orthopedic implants, artificial hearts, and infusion pumps. Examples of the latter include various dental devices and daily-wear contact lenses. Clinical research involving a significant risk device cannot begin until an institutional review board (IRB) has approved both the protocol and the informed consent form and the FDA itself has given permission. This requirement to submit an IDE application to the FDA is waived in the case of clinical research where the risk posed is insignificant. In this case, the FDA requires only that approval from an IRB be obtained certifying that the device in question poses only insignificant risk. In deciding whether to approve a proposed clinical investigation of a new device, the IRB and the FDA must determine the following (Bronzino et al., 1990a, b):

1. Risks to subjects are minimized.
2. Risks to subjects are reasonable in relation to the anticipated benefit and knowledge to be gained.
3. Subject selection is equitable.
4. Informed consent materials and procedures are adequate.
5. Provisions for monitoring the study and protecting patient information are acceptable.

The FDA allows unapproved medical devices to be used without an IDE in three types of situations: emergency use, treatment use, and feasibility studies.

0-8493-0462-8/00/$0.00+$.50

192.2 Ethical Issues in Feasibility Studies

Manufacturers seeking more flexibility in conducting investigations in the early developmental stages of a device have submitted a petition to the FDA, requesting that certain limited investigations of significant risk devices be subject to abbreviated IDE requirements (Bronzino et al., 1990a, b). In a feasibility study, or "limited investigation," human research on a new device would take place at a single institution and involve no more than ten human subjects. *The sponsor of a limited investigation would be required to submit to the FDA a "Notice of Limited Investigation," which would include a description of the device, a summary of the purpose of the investigation, the protocol, a sample of the informed consent form, and a certification of approval by the responsible IRB. In certain circumstances, the FDA could require additional information, or require the submission of a full IDE application, or suspend the investigation* (Bronzino et al., 1990a, b).

Investigations of this kind would be limited to certain circumstances: *(1) investigations of new uses of existing devices, (2) investigations involving temporary or permanent implants during the early developmental stages, and (3) investigations involving modification of an existing device* (Bronzino et al., 1990a).

To comprehend adequately the ethical issues posed by clinical use of unapproved medical devices outside the context of an IDE, it is necessary to utilize the distinctions between practice, nonvalidated practice, and research elaborated in the previous pages. How do those definitions apply to feasibility studies?

Clearly, the goal of this sort of study, i.e., generalizable knowledge, makes it an issue of research rather than practice. Manufacturers seek to determine the performance of a device with respect to a particular patient population in an effort to gain information about its efficacy and safety. Such information would be important in determining whether further studies (animal or human) need to be conducted, whether the device needs modification before further use, and the like. The main difference between use of an unapproved device in a feasibility study and use under the terms of an IDE is that the former would be subject to significantly less intensive FDA review than the latter. This, in turn, means that the responsibility for ensuring that use of the device is ethically sound would fall primarily to the IRB of the institution conducting the study.

The ethical concerns posed here are best comprehended with a clear understanding of what justifies research. Ultimately, no matter how much basic research and animal experimentation has been conducted on a given device, the risks and benefits it poses for humans cannot be adequately determined until it is actually used on humans.

The benefits of research on humans lie primarily in the knowledge that is yielded and the generalizable information that is provided. This information is crucial to medical science's ability to generate new modes and instrumentalities of medical treatment that are both efficacious and safe. Accordingly, for necessary but insufficient condition for experimentation to be ethically sound, it must be scientifically sound (Capron, 1978; 1986).

Although scientific soundness is a necessary condition of ethically acceptable research on humans, it is not of and by itself sufficient. Indeed, it is widely recognized that the primary ethical concern posed by such investigation is the use of one person by another to gather knowledge or other benefits where these benefits may only partly or not at all accrue to the first person. In other words, the human subjects of such research are at risk of being mere research resources, as having value only for the ends of the research. Research upon human beings runs the risk of failing to respect them as people. The notion that human beings are not mere things but entities whose value is inherent rather than wholly instrumental is one of the most widely held norms of contemporary Western society. That is, human beings are not valuable wholly or solely for the uses to which they can be put. They are valuable simply by being the kinds of entities they are. To treat them as such is to respect them as people.

Respecting individuals as people is generally agreed to entail two requirements in the context of biomedical experimentation. First, since what is most generally taken to make human beings people is their autonomy—their ability to make rational choices for themselves—treating individuals as people means respecting that autonomy. This requirement is met by ensuring that no competent person is subjected to any clinical intervention without first giving voluntary and informed consent. Second, respect for people means that the physician will not subject a human to unnecessary risks and will minimize the risks to patients in required procedures.

Much of the ethical importance of the scrutiny that the FDA imposes upon use of unapproved medical devices in the context of an IDE derives from these two conditions of ethically sound research. The central ethical concern posed by use of medical devices in a feasibility study is that the decreased degree of FDA scrutiny will increase the likelihood that either or both of these conditions will not be met. This possibility may be especially great because many manufacturers of medical devices are, after all, commercial enterprises, companies that are motivated to generate profit and thus to get their devices to market as soon as possible with as little delay and cost as possible. These self-interested motives are likely, at times, to conflict with the requirements of ethically sound research and thus to induce manufacturers to fail (often unwittingly) to meet these requirements. Note that profit is not the only motive that might induce manufacturers to contravene the requirements of ethically sound research on humans. A manufacturer may sincerely believe that its product offers great benefit to many people or to a population of especially needy people and so from this utterly altruistic motive may be prompted to take shortcuts that compromise the quality of the research. Whether the consequences being sought by the research are desired for reasons of self-interest, altruism, or both, the ethical issue is the same. Research subjects may be placed at risk of being treated as mere objects rather than as people.

What about the circumstances under which feasibility studies would take place? Are these not sufficiently different from the "normal" circumstances of research to warrant reduced FDA scrutiny? As noted above, manufacturers seek to be allowed to engage in feasibility studies in order to investigate new uses of existing devices, to investigate temporary or permanent implants during the early developmental stages, and to investigate modifications to an existing device. As also noted above, a feasibility study would take place at only one institution and would involve no more than ten human subjects. Given these circumstances, is the sort of research that is likely to occur in a feasibility study less likely to be scientifically unsound or to fail to respect people in the way that normal research upon humans does in "normal" circumstances?

Such research would be done on a very small subject pool, and the harm of any ethical lapses would likely affect fewer people than if such lapses occurred under more usual research circumstances. Yet even if the harm done is limited to a failure to respect the ten or fewer subjects in a single feasibility study, the harm would still be ethically wrong. To wrong ten or fewer people is not as bad as to wrong in the same way more than ten people but it is to engage in wrongdoing nonetheless. In either case, individuals are reduced to the status of mere research resources and their dignity as people is not properly respected.

Are ethical lapses more likely to occur in feasibility studies than in studies that take place within the requirements of an IDE? Although nothing in the preceding discussion provides a definitive answer to this question, it is a question to which the FDA should give high priority in deciding whether to allow this type of exception to IDE use of unapproved medical devices. The answer to this question might be quite different when the device at issue is a temporary or permanent implant than when it is an already approved device being put to new uses or modified in some way. Whatever the contemplated use under the feasibility studies mechanism, the FDA would be ethically advised not to allow this kind of exception to IDE use of an unapproved device without a reasonably high level of certainty that research subjects would not be placed in greater jeopardy than in "normal" research circumstances.

192.3 Ethical Issues in Emergency Use

What about the mechanism for avoiding the rigors of an IDE for emergency use?

> "*The FDA has authorized emergency use where an unapproved device offers the only alternative for saving the life of a dying patient, but an IDE has not yet been approved for the device or its use, or an IDE has been approved but the physician who wishes to use the device is not an investigator under the IDE* (Bronzino et al., 1990a, b).

Because the purpose of emergency use of an unapproved device is to attempt to save a dying patient's life under circumstances where no other alternative is at hand, this sort of use constitutes practice rather than research. Its aim is primarily benefit to the patient rather than provision of new and generalizable information. Because this sort of use occurs prior to the completion of clinical investigation of the device, it constitutes a nonvalidated practice. What does this mean?

First, it means that while the aim of the use is to save the life of the patient, the nature and likelihood of the potential benefits and risks engendered by use of the device are far more speculative than in the sort of clinical intervention that constitutes validated practice. In validated practice, thorough investigation, including preclinical studies, animals studies, and studies on human subjects of a device has established its efficacy and safety. The clinician thus has a well-founded basis upon which to judge the benefits and risks such an intervention poses for his patients.

It is precisely this basis that is lacking in the case of a nonvalidated practice. Does this mean that emergency use of an unapproved device should be regarded as immoral? This conclusion would follow only if there were no basis upon which to make an assessment of the risks and benefits of the use of the device. The FDA requires that a physician who engages in emergency use of an unapproved device must "*have substantial reason to believe that benefits will exist. This means that there should be a body of preclinical and animal tests allowing a prediction of the benefit to a human patient*" (Bronzino et al., 1990a, b).

Thus, although the benefits and risks posed by use of the device are highly speculative, they are not entirely speculative. Although the only way to validate a new technology is to engage in research on humans at some point, not all nonvalidated technologies are equal. Some will be largely uninvestigated, and assessment of their risks and benefits will be wholly or almost wholly speculative. Others will at least have the support of preclinical and animal tests. Although this is not sufficient support for incorporating use of a device into regular clinical practice, it may however represent sufficient support to justify use in the desperate circumstances at issue in emergency situations. Desperate circumstances can justify desperate actions, but desperate actions are not the same as reckless actions, hence the ethical soundness of the FDA's requirement that emergency use be supported by solid results from preclinical and animal tests of the unapproved device.

A second requirement that the FDA imposes on emergency use of unapproved devices is the expectation that physicians "*exercise reasonable foresight with respect to potential emergencies and make appropriate arrangements under the IDE procedures. Thus, a physician should not "create" an emergency in order to circumvent IRB review and avoid requesting the sponsor's authorization of the unapproved use of a device*" (Bronzino et al., 1990a, b).

From a Kantian point of view, which is concerned with protecting the dignity of people, it is a particularly important requirement to create an emergency in order to avoid FDA regulations which prevent the patient being treated as a mere resource whose value is reducible to a service of the clinician's goals. Hence, the FDA is quite correct to insist that emergencies are circumstances that reasonable foresight would not anticipate.

Also especially important here is the nature of the patient's consent. Individuals facing death are especially vulnerable to exploitation and deserve greater measures for their protection than might otherwise be necessary. One such measure would be to ensure that the patient, or his legitimate proxy, knows the highly speculative nature of the intervention being offered. That is, to ensure that it is clearly understood that the clinician's estimation of the intervention's risks and benefits is far less solidly grounded than in the case of validated practices. The patient's consent must be based upon an awareness that the particular device has not undergone complete and rigorous testing on humans and that estimations of its potential are based wholly upon preclinical and animal studies. Above all the patient must not be lead to believe that there is complete understanding of the risks and benefits of the intervention. Another important point here is to ensure that the patient is aware that the options he is facing are not simply life or death but may include life of a severely impaired quality, and therefore that even if his life is saved, it may be a life of significant impairment. Although desperate circumstance may legitimize desperate actions, the decision to take such actions must rest upon the informed and voluntary consent of the patient, especially when he/she is an especially vulnerable patient.

It is important here for a clinician involved in emergency use of an unapproved device to recognize that these activities constitute a form of nonvalidated practice and not research. Hence, the primary obligation is to the well-being of the patient. The patient enters into the relationship with the clinician with the same trust that accompanies any normal clinical situation. To treat this sort of intervention as if it were an instance of research and hence justified by its benefits to science and society would be to abuse this trust.

192.4 Ethical Issues in Treatment Use

The FDA has adopted regulations authorizing the use of investigational new drugs in certain circumstances where a patient has not responded to approved therapies. This "treatment use" of unapproved new drugs is not limited to life-threatening emergency situations, but rather is also available to treat "serious" diseases or conditions (Bronzino et al., 1990a, b).

The FDA has not approved treatment use of unapproved medical devices, but it is possible that a manufacturer could obtain such approval by establishing a specific protocol for this kind of use within the context of an IDE.

The criteria for treatment use of unapproved medical devices would be similar to criteria for treatment use of investigational drugs: (1) the device is intended to treat a serious or life-threatening disease or condition, (2) there is no comparable or satisfactory alternative product available to treat that condition, (3) the device is under an IDE, or has received an IDE exemption, or all clinical trials have been completed and the device is awaiting approval, and (4) the sponsor is actively pursuing marketing approval of the investigational device. The treatment use protocol would be submitted as part of the IDE, and would describe the intended use of the device, the rationale for use of the device, the available alternatives and why the investigational product is preferred, the criteria for patient selection, the measures to monitor the use of the device and to minimize risk, and technical information that is relevant to the safety and effectiveness of the device for the intended treatment purpose (Bronzino et al., 1990a, b).

Were the FDA to approve treatment use of unapproved medical devices, what ethical issues would be posed? First, because such use is premised on the failure of validated interventions to improve the patient's condition adequately, it is a form of practice rather than research. Second, since the device involved in an instance of treatment use is unapproved, such use would constitute nonvalidated practice. As such, like emergency use, it should be subject to the FDA's requirement that prior preclinical tests and animal studies have been conducted that provide substantial reason to believe that patient benefit will result. As with emergency use, although this does not prevent assessment of the intervention's benefits and risks from being highly speculative, it does prevent assessment from being totally speculative. Here too, although desperate circumstances can justify desperate action, they do not justify reckless action. Unlike emergency use, the circumstances of treatment use involve serious impairment of health rather than the threat of premature death. Hence, an issue that must be considered is how serious such impairment must be to justify resorting to an intervention whose risks and benefits have not been solidly established.

In cases of emergency use, the FDA requires that physicians not use this exception to an IDE to avoid requirements that would otherwise be in place. This particular requirement would be obviated in instances of treatment use by the requirement that a protocol for such use be previously addressed within an IDE.

As with emergency use of unapproved devices, the patients involved in treatment use would be particularly vulnerable patients. Although they are not dying, they are facing serious medical conditions and are thereby likely to be less able to avoid exploitation than patients under less desperate circumstances. Consequently, it is especially important that patients be informed of the speculative nature of the intervention and of the possibility that treatment may result in little or no benefit to them.

192.5 The Safe Medical Devices Act

On November 28, 1991, the Safe Medical Devices Act of 1990 (Public Law 101-629) went into effect. This regulation requires a wide range of healthcare institutions, including hospitals, ambulatory-surgical facilities, nursing homes, and outpatient treatment facilities, to report information that "reasonably suggests" the likelihood that the death, serious injury, or serious illness of a patient at that facility has been caused or contributed to by a medical device. When a death is device-related, a report must be made directly to the FDA *and* to the manufacturer of the device. When a serious illness or injury is device-related, a report must be made to the manufacturer *or* to the FDA in cases where the manufacturer is not known. In addition, summaries of previously submitted reports must be submitted to the FDA on a semiannual basis. Prior to this regulation, such reporting was voluntary. This new regulation was designed to enhance the FDA's ability to quickly learn about problems related to medical devices. It also supplements the medical device reporting (MDR) regulations promulgated in 1984. MDR regulations require that reports of device-related deaths and serious injuries be submitted to the FDA by manufacturers and importers. The new law extends this requirement to users of medical devices along with manufacturers and importers. This act represents a significant step forward in protecting patients exposed to medical devices.

References

Bronzino, J. D., Flannery, E. J., and Wade, M. L. "Legal and Ethical Issues in the Regulation and Development of Engineering Achievements in Medical Technology," Part I *IEEE Engineering in Medicine and Biology,* 1990a.

Bronzino, J. D., Flannery, E. J., and Wade, M. L. "Legal and Ethical Issues in the Regulation and Development of Engineering Achievements in Medical Technology," Part II *IEEE Engineering in Medicine and Biology,* 1990b.

Bronzino, J. D., Chapter 10 "Medical and Ethical Issues in Clinical Engineering Practice" In: *Management of Medical Technology.* Butterworth, 1992.

Bronzino, J. D., Chapter 20 "Moral and Ethical Issues Associated with Medical Technology" in: *Instruction to Biomedical Engineering.* Academic Press, 1999.

Capron, A. "Human Experimentation: Basic Issues." In: *The Encyclopedia of Bioethics* vol. II. The Free Press, Glencoe, IL, 1978.

Capron, A. "Human Experimentation." In: (J. F. Childress, et al., eds.) University Publications of America, 1986.

Further Information

Daniels, N. *Just Health Care.* Cambridge University Press, Cambridge, 1987.

Dubler, N. N. and Nimmons, D. *Ethics on Call.* Harmony Books, New York, 1992.

Jonsen, A. R. *The New Medicine and the Old Ethics.* Harvard University Press, Cambridge, MA, 1990.

Murphy, J. and Coleman, J. *The Philosophy of Law.* Rowman and Allenheld, 1984.

Appendix A

The Role of Professional Societies in Biomedical Engineering

Swamy Laxminarayan
New Jersey Institute of Technology

Joseph D. Bronzino
Trinity College/Biomedical Engineering Alliance for Connecticut (BEACON)

Jan E. W. Beneken
Eindhoven University of Technology

Shiro Usai
Toyohashi University of Technology

Richard D. Jones
Christchurch Hospital

Professionals have been defined as an aggregate of people finding identity in sharing values and skills absorbed during a common course of intensive training. Parsons [1954] stated that one determines whether or not individuals are professionals by examining whether or not they have internalized certain given professional values. Friedson [1971] redefined Parson's definition by noting that a professional is someone who has internalized professional values and is to be recruited and licensed on the basis of his or her technical competence. Furthermore, he pointed out that professionals generally accept scientific standards in their work, restrict their work activities to areas in which they are technically competent, avoid emotional involvement, cultivate objectivity in their work, and put their clients' interests before their own.

The concept of a profession that manages technology encompasses three occupational models: science, business, and profession. Of particular interest in the contrast between science and profession. Science is seen as the pursuit of knowledge, its value hinging on providing evidence and communicating with colleagues. Profession, on the other hand, is viewed as providing a service to clients who have problems they cannot handle themselves. Science and profession have in common the exercise of some knowledge, skill, or expertise. However, while scientists practice their skills and report their results to knowledgeable colleagues, professionals—such as lawyers, physicians, and engineers—serve lay clients. To protect both the professional and the client from the consequences of the layperson's lack of knowledge, the practice of the profession is regulated through such formal institutions as state licensing. Both professionals and scientists must persuade their clients to accept their findings. Professionals endorse and follow a specific

0-8493-0462-8/00/$0.00+$.50

code of ethics to serve society. On the other hand, scientists move their colleagues to accept their findings through persuasion [Goodman, 1989].

Consider, for example, the medical profession. Its members are trained in caring for the sick, with the primary goal of healing them. These professionals not only have a responsibility of the creation, development, and implementation of that tradition, they also are expected to provide a service to the public, within limits, without regard of self-interest. To ensure proper service, the profession itself closely monitors licensing and certification. Thus medical professionals themselves may be regarded as a mechanism of social control. However, this does not mean that other facets of society are not involved in exercising oversight and control over physicians in their practice of medicine.

Professional Development. One can determine the status of professionalization by noting the occurrence of six crucial events: (1) the first training school, (2) the first university school, (3) the first local professional association, (4) the first national professional association, (5) the first state license law, and (6) the first formal code of ethics [Wilensky, 1964; Goodman, 1989; Bronzino, 1992].

The early appearances of training school and the university affiliation underscore the importance of the cultivation of a knowledge base. The strategic innovative role of the universities and early teachers lies in linking knowledge to practice and creating a rational for exclusive jurisdiction. Those practitioners pushing for prescribed training then form a professional association. The association defines the task of the profession: raising the quality of recruits, redefining their function to permit the use of less technically skilled people to perform the more routine, less involved tasks, and managing internal and external conflicts. In the process, internal conflict may arise between those committed to established procedures and newcomers committed to change and innovation. At this stage, some form of professional regulation, such as licensing or certification, surfaces because of a belief that it will ensure minimum standards for the profession, enhance status, and protect the layperson in the process.

The latest area of professional development is the establishment of a formal code of ethics, which usually includes rules to exclude the unqualified and unscrupulous practitioners, rules to reduce internal competition, and rules to protect clients and emphasize the ideal service to society. A code of ethics usually comes at the end of the professionalization process.

In biomedical engineering, all six critical steps mentioned above have been clearly taken. Therefore, biomedical engineering is definitely a profession. It is important here to note the professional associations across the globe that represent the interest of professionals in the field.

A.1 Biomedical Engineering Societies in the World

Globalization of biomedical engineering (BME) activities is underscored by the fact that there are several major professional BME societies currently operational throughout the world. The various countries and continents to have provided concerted "action" groups in biomedical engineering are Europe, the Americas, Canada, and the Far East, including Japan and Australia. while all these organizations share in the common pursuit of promoting biomedical engineering, all national societies are geared to serving the needs of their "local" memberships. The activities of some of the major professional organizations are described below.

American Institute for Medical and Biological Engineering (AIMBE)

The United States has the largest biomedical engineering community in the world. Major professional organizations that address various cross sections of the field and serve over 20,000 biomedical engineers include (1) the American College of Clinical Engineering, (2) the American Institute of Chemical Engineers, (3) The American Medical Informatics Association, (4) the American Society of Agricultural Engineers, (5) the American Society for Artificial Internal Organs, (6) the American Society of Mechanical Engineers, (7) the Association for the Advancement of Medical Instrumentation, (8) the Biomedical Engineering Society, (9) the IEEE Engineering in Medicine and Biology Society, (10) an interdisciplinary

Association for the Advancement of Rehabilitation and Assistive Technologies, (11) the Society for Biomaterials, (12) Orthopedic Research Society, (13) American Society of Biomechanics, and (14) American Association of Physicist in Medicine. In an effort to unify all the disparate components of the biomedical engineering community in the United States as represented by these various societies, the American Institute for Medical and Biological Engineers (AIMBE) was created in 1992. The AIMBE is the result of a 3-year effort funded by the National Science Foundation and led by a joint steering committee established by the Alliance of Engineering in Medicine and Biology and the U.S. National Committee on Biomechanics. The primary goal of AIMBE is to serve as an umbrella organization "for the purpose of unifying the bioengineering community, addressing public policy issues, identifying common themes of reflection and proposals for action, and promoting the engineering approach in society's effort to enhance health and quality of life through the judicious use of technology" [Galletti, 1994].

AIMBE serves its role through four working divisions: (1) the Council of Societies, consisting of the 11 constituent organizations mentioned above, (2) the Academic Programs Council, currently consisting of 46 institutional charter members, (3) the Industry Council, and (4) the College Fellows. In addition to these councils, there are four commissions, Education, Public Awareness, Public Policy, and Liaisons. With its inception in 1992, AIMBE is a relatively young institution trying to establish its identity as an umbrella organization for medical and biologic engineering in the United States. As summarized by two of the founding officials of the AIMBE, Profs Nerem and Galletti:

> What we are all doing, collectively, is defining a focus for biological and medical engineering. In a society often confused by technophobic tendencies, we will try to assert what engineering can do for biology, for medicine, for health care and for industrial development, We should be neither shy, nor arrogant, nor self-centered. The public has great expectations from engineering and technology in terms of their own health and welfare. They are also concerned about side effects, unpredictable consequences and the economic costs. Many object to science for the sake of science, resent exaggerated or empty promises of benefit to society, and are shocked by sluggish or misdirected flow from basic research to useful applications. These issues must be addressed by the engineering and medical communities. For more information, contact the Executive Office, AIMBE, 1901 Pennsylvania Avenue, N.W., Suite 401, Washington DC 20006–3405 (Tel: 202–496–9660; fax: 202–466–8489; email: AIMBE@aol.com).

IEEE Engineering in Medicine and Biology Society (EMBS)

The Institute of Electrical and Electronic Engineers (IEEE) is the largest international professional organization in the world and accommodates 37 different societies under its umbrella structure. Of these 37, the Engineering in Medicine and Biology Society represents the foremost international organization serving the needs of nearly 8000 biomedical engineering members around the world. The field of interest of the EMB Society is application of the concepts and methods of the physical and engineering sciences in biology and medicine. Each year, the society sponsors a major international conference while cosponsoring a number of theme-oriented regional conferences throughout the world. A growing number of EMBS chapters and student clubs across the major cities of the world have provided the forum for enhancing local activities through special seminars, symposia, and summer schools on biomedical engineering topics. These are supplemented by EMBS's special initiatives that provide faculty and financial subsidies to such programs through the society's distinguished lecturer program as well as the society's Regional Conference Committee. Other feature achievements of the society include its premier publications in the form of three monthly journals (*Transactions on Biomedical Engineering, Transactions on Rehabilitation Engineering, and Transactions on Information Technology in Biomedicine*) and a bi-monthly *EMB Magazine* (the *IEEE Engineering in Medicine and Biology Magazine*). EMBS is a transnational voting member society of the International Federation for Medical and Biological Engineering. For more information, contact the Secretariat, IEEE EMBS, National Research Council of Canada, Room 393, Building M-55, Ottawa, Ontario K1A OR8, Canada. (Tel: 613–993–4005; fax: 613–954–2216; email: soc.emb@ieee.org).

Canadian Medical and Biological Engineering Society

The Canadian Medical and Biological Engineering Society (CMBES) is an association covering the fields of biomedical engineering, clinical engineering, rehabilitation engineering, and biomechanics and biomaterials applications. CMBES is affiliated with the International Federation for Medical and Biological Engineering and currently has 272 full members. The society organizes national medical and biological engineering conferences annually in various cities across Canada. In addition, CMBES has sponsored seminars and symposia on specialized topics such as communication aids, computers, and the handicapped, as well as instructional courses on topics of interest to the membership. To promote the professional development of its members, the society as drafted guidelines on education and certification for clinical engineers and biomedical engineering technologists and technicians. CMBES is committed to bringing together all individuals in Canada who are engaged in interdisciplinary work involving engineering, the life sciences, and medicine. The society communicates to its membership through the publication of a newsletter as well as recently launched academic series to help nonengineering hospital personnel to gain better understanding of biomedical technology. For more information, contact the Secretariat, The Canadian Medical and Biological Engineering Society, National Research Council of Canada, Room 393, Building M-55, Ottawa, Ontario K1A OR8, Canada (Tel: 613–993–1686; fax: 613–954–2216).

European Society for Engineering in Medicine (ESEM)

Most European countries are affiliated organizations of the International Federation for Medical and Biological Engineering (IFMBE). The IFMBE activities are described in another section of this chapter. In 1992, a separate organization called the European Society for Engineering in Medicine (ESEM) was created with the objective of providing opportunities for academic centers, research institutes, industry, hospitals and other health care organizations, and various national and international societies to interact and jointly explore BME issues of European significance. These include (1) research and development, (2) education and training, (3) communication between and among industry, health care providers, and policymakers, (4) European policy on technology and health care, and (5) collaboration between eastern European countries in transition and the western European countries on health care technology, delivery, and management. To reflect this goal the ESEM membership constitutes representation of all relevant disciplines from all European countries while maintaining active relations with the Commission of the European Community and other supranational bodies and organizations.

The major promotional strategies of the ESEM's scientific contributions include its quarterly journal *Technology and Health Care, ESEM News,* the Society's Newsletter, a biennial European Conference on Engineering and Medicine, and various topic-oriented workshops and courses. ESEM offers two classes of membership: the regular individual (active or student) membership and an associate grade. The latter is granted to those scientific and industrial organizations which satisfy the society guidelines and subject to approval by the Membership and Industrial Committees. The society is administered by an Administrative Council consisting of 13 members elected by the general membership. For more information, contact the Secretary General, European Society for Engineering in Medicine, Institut für Biomedizinische Technik, Seidenstrasse 36, D-70174 Stuttgart, Germany. (Fax: 711–121–2371

French Groups for Medical and Biological Engineering

The French National Federation of Bioengineering (*Genie Biologique et Medical, GMB*) is a multidisciplinary body aimed at developing methods and processes and new biomedical materials in various fields covering prognosis, diagnosis, therapeutics, and rehabilitation. These goals are achieved through the creation of 10 regional centers of bioengineering, called the *poles.* The poles are directly involved at all levels, from applied research through the industrialization to the marketing of the product. Some of the actions pursued by these poles include providing financial seed support for innovative biomedical engineering projects, providing technological help, advice, and assistance, developing partnerships among universities and industries, and organizing special seminars and conferences. The information dissemination of all scientific progress is done through the *Journal of Innovation and Technology in Biology and Medicine.* For more information, contact the French National Federation of Bioengineering, Coordinateur de la Federation Francaise des Poles GBM, Pole GBM Aquitaine-Site Bordeaux-Montesquieu, Centre de Resources, 33651 Martillac Cedex, France.

International Federation for Medical and Biological Engineering (IFMBE)

Established in 1959, the International Federation for Medical and Biological Engineering (IFMBE) is an organization made up from an affiliation of national societies including membership of transnational organizations. The current national affiliates are Argentina, Australia, Austria, Belgium, Brazil, Bulgaria, Canada, China, Cuba, Cyprus, Slovakia, Denmark, Finland, France, Germany, Greece, Hungary, Israel, Italy, Japan, Mexico, Netherlands, Norway, Poland, South Africa, South Korea, Spain, Sweden, Thailand, United Kingdom, and the United States. The first transnational organization to become a member of the federation is the IEEE Engineering in Medicine and Biology Society. At the present time, the federation has an estimated 25,000 members from all of its constituent societies.

The primary goal of the IFMBE is to recognize the interests and initiatives of its affiliated member organizations and to provide an international forum for the exchange of ideas and dissemination of information. The major IFMBE activities include the publication of the federation's bimonthly journal, the *Journal of Medical and Biological Engineering and Computing*, the *MBEC News*, establishment of close liaisons with developing countries to encourage and promote BME activities, and the organization of a major world conference every 3 years in collaboration with the International Organization for Medical Physics and the International Union for Physical and Engineering Sciences in Medicine. The IFMBE also serves as a consultant to the United Nations Industrial Development Organization and has nongovernmental organization status with the World Health Organization, the United Nations, and the Economic Commission for Europe. For more information, contact the Secretary General, International Federation for Medical and Biological Engineering, AMC, University of Amsterdam, Meibergdreef 15, 1105 AZ Amsterdam, the Netherlands. (Tel: 20–566–5200, ext. 5179; fax 20–691–7233; email: ifmbe@amc.uva.nl)

International Union for Physics and Engineering Sciences in Medicine (IUPESM)

The IUPESM resulted from the IFMBE's collaboration with the International Organization of Medical Physics (IOMP), culminating into the joint organization of the triennial World Congress on Medical Physics and Biomedical Engineering. Traditionally, these two organizations held their conferences back to back from each other for a number of years. Since both organizations were involved in the research, development, and utilization of medical devices, they were combined to form IUPESM. Consequently, all members of the IFMBE's national and transnational societies are also automatically members of the IUPESM. The statutes of the IUPESM have been recently changed to allow other organizations to become members in addition to the founding members, the IOMP and the IFMBE.

International Council of Scientific Unions (ICSU)

The International Council of Scientific Unions is nongovernmental organization created to promote international scientific activity in the various scientific branches and their applications for the benefit of humanity. ICSU has two categories of membership: scientific academies or research councils, which are national, multidisciplinary bodies, and scientific unions, which are international disciplinary organizations. Currently, there are 92 members in the first category and 23 in the second. ICSU maintains close working relations with a number of intergovernmental and nongovernmental organizations, in particular with UNESCO. In the past, a number of international programs have been launched and are being run in cooperation with UNESCO. ICSU is particularly involved in serving the interests of developing countries.

Membership in the ICSU implies recognition of the particular field of activity as a field of science. Although ICSU is heralded as a body of pure scientific unions to the exclusion of cross and multidisciplinary organizations and those of an engineering nature, IUPESM, attained its associate membership in the ICSU in the mid-1980s. The various other international scientific unions that are members of the ICSU include the International Union of Biochemistry and Molecular Biology (IUBMB), the International Union of Biological Sciences (IUBS), the International Brain Research Organization (IBRO), and the International Union of Pure and Applied Biophysics (IUPAB). The IEEE is an affiliated commission of the IUPAB and is represented through the Engineering in Medicine and Biology Society [ICSU Year Book, 1994]. For more information, contact the Secretariat, International Council of Scientific Unions, 51 Boulevard de Montmorency, 75016 Paris, France. (Tel: 1–4525–0329; fax: 1–4288–9431; email: icsu@paris7.jussieu.fr)

Biomedical Engineering Societies in Japan

The biomedical engineering activities in Japan are promoted through several major organizations: (1) the Japan Society of Medical Electronics and Biological Engineering (JSMEBE), (2) the Institute of Electronics, Information and Communication Engineering (IEICE), (3) the Institute of Electrical Engineers of Japan (IEEJ), (4) the Society of Instrument and Control Engineers (SICE), (5) the Society of Biomechanisms of Japan (SBJ), (6) the Japanese Neural network Society (JNNS), (7) Japan Ergonomics Research Society (JERS), and (8) the Japan Society of Ultrasonics in Medicine (JSUM). The various special sister societies that are affiliated under the auspicies of these organizations mainly focus on medical electronics, biocybernetics, neurocomputing, medical and biologic engineering, color media and vision system, and biologic and physiologic engineering. The JSMEBE has the most BME concentration, with two conferences held each year in biomedic engineering and three international journals that publish original peer-reviewed papers. The IEICE, which is now 77 years old, is one of the largest international societies, constituting about 40,000 members. The aim of the society is to provide a forum for the exchange of knowledge on the science and technology of electronics, information, and communications and the development of appropriate industry in these fields.

BME Activities in Australia and New Zealand

The BME activities in Australia and New Zealand are served by one transnational organization called the Australasian College of Physical Scientists and Engineers in Medicine (ACPSEM) covering Australia and New Zealand and a national organization called the College of Biomedical Engineers of the Institute of Engineers, (CBEIE) serving the Australian member segment.

The Australasian College of Physical Scientists and Engineers in Medicine was founded in 1977 and comprises 6 branches and 339 members. The membership is made up of 76% from Australia, 17% from New Zealand, and the rest from overseas. A majority of members are employed in pubic hospitals, with most of these in departments of medical physics and clinical engineering. The primary objectives of the college are (1) to promote and further the development of the physical sciences and engineering in medicine and to facilitate the exchange of information and ideas among members of the college and others concerned with medicine and related subjects and (2) to promote and encourage education and training in the physical sciences and engineering in medicine. Entry to ordinary membership of the college requires applicants to have an appropriate 4-year bachelor's degree and at least 5 years of experience as a physical scientist or engineer in a hospital or other approved institution.

Bioengineering in Latin America

Latin American countries have demonstrated in the past decade significant growth in their bioengineering activities. In terms of IEEE statistics, Latin America has the fastest growing membership rate. Currently, there are over 10,000 IEEE members alone in this region, of which about 300 are members of the Engineering in Medicine and Biology Society. In an effort to stimulate this growth and promote active international interactions, the presidents of the IEEE, EMBS, and the IFMBE met with representatives of biomedical engineering societies from Argentina, Brazil, Chile, Columbia, and Mexico in 1991 [Robinson, 1991]. This meeting resulted in the formation of an independent Latin American Regional Council of Biomedical Engineering, known by its spanish and portugese acronym as CORAL (*Consejo Regional de Ingenieria Biomedica para Americana Latina*). Both the EMBS and the IFMBE are the founding sponsoring members of the CORAL. The main objectives of CORAL are (1) to foster, promote and encourage the development of research, student programs, publications, professional activities, and joint efforts and (2) to act as a communication channel for national societies within Latin American region and to improve communication between societies, laboratories, hospitals, industries, universities, and other groups in Latin America and the Caribbean. Since its inception, CORAL has already provided the centerpiece for bioengineering activities in Latin America through special concerted scientific meetings and closer society interactions both in a national and international sense. For more information, contact the Secretary General, CORAL, Centro Investigacion y de Estudios, Avanzados Duel Ipn, Departamento Ingenieria Electrica, Seccion Bioelectronica, Av. Instituto Politecnico Nacional 2508, Esg. Av. Ticoman 07000, Mexico Apartado Postal 14–740, Mexico.

A.2 Summary

The field of biomedical engineering, which originated as a professional group on medical electronics in the late fifties, has grown from a few scattered individuals to very well-established organization. There are approximately 50 national societies throughout the world serving an increasingly growing community of biomedical engineers. The scope of biomedical engineering today is enormously diverse. Over the years, many new disciplines such as molecular biology, genetic engineering, computer-aided drug design, nanotechnology, and so on, which were once considered alien to the field, are now new challenges a biomedical engineer faces. Professional societies play a major role in bringing together members of this diverse community in pursuit of technology applications for improving the health and quality of life of human beings. Intersocietal cooperations and collaborations, both at national and international levels, are more actively fostered today through professional organizations such as the IFMBE, AIMBE, CORAL, and the IEEE. These developments are strategic to the advancement of the professional status of biomedical engineers. Some of the self-imposed mandates the professional societies should continue to pursue include promoting public awareness, addressing public policy issues that impact research and development of biologic and medical products, establishing close liaisons with developing countries, encouraging educational programs for developing scientific and technical expertise in medical and biologic engineering, providing a management paradigm that ensures efficiency and economy of health care technology [Wald, 1993], and participating in the development of new job opportunities for biomedical engineers.

References

Fard TB. 1994. International Council of Scientific Unions Year Book, Paris, ICSU.

Friedson E. 1971. Profession of Medicine. New York, Dodd, Mead.

Galletti PM, Nerem RM. 1994. The Role of Bioengineering in Biotechnology. AIMBE Third Annual Event.

Goodman G. 1989. The profession of clinical engineering. J Clin Eng 14:27.

Parsons T. 1954. Essays in Sociological Theories. Glencoe, Ill, Free Press.

Robinson CR. 1991. Presidents column. IEEE Eng Med Bio Mag.

Wald A. 1993. Health care: Reform and technology (editors note). IEEE Eng Med Bio Mag 12:3.

Index

0-8493-0462-8/00/$0.00+$.50

C

D

E

F

M

N

O

Q

R

T

U

V